Proceedings of the
Twentieth International Cryogenic Engineering Conference

(ICEC 20)

Proceedings of the
Twentieth International Cryogenic Engineering Conference
(ICEC 20)

Beijing, China, 11-14 May 2004

Edited by

Liang ZHANG, Liangzhen LIN and Guobang CHEN
Technical Institute of Physics and Chemistry
Chinese Academy of Sciences, Beijing, CHINA

2005

ELSEVIER

Amsterdam Boston Heidelberg London New York Oxford Paris
San Diego San Francisco Singapore Sydney Tokyo

ELSEVIER B.V.
Radarweg 29
P.O. Box 211, 1000 AE Amsterdam
The Netherlands

ELSEVIER Inc.
525 B Street, Suite 1900
San Diego, CA 92101-4495
USA

**ELSEVIER Ltd
The Boulevard, Langford Lane
Kidlington, Oxford OX5 1GB
UK**

ELSEVIER Ltd
84 Theobalds Road
London WC1X 8RR
UK

First edition 2005

ISBN: 0-080-44559-4

Printed in Great Britain.

PREFACE

The Twentieth International Cryogenic Engineering Conference (ICEC 20) is the Twentieth in the International series of biennial conference on cryogenic engineering held by the International Cryogenic Engineering Committee. It is the second such conference held in Beijing (ICEC 13, 1990).

ICEC 20 which was held on May 11-14, 2004 in Beijing, brought together around 400 delegates from 17 counties, including 322 registered participants and dozens exhibitors from 31 companies.

The technical program of ICEC-20 consisted of 345 contributed papers and 9 plenary invited papers, including the Mendelssohn Award speech. There are 19 parallel oral sessions with 114 papers, including a special oral session dedicated to the space cryogenics and 231 papers presented in 7 poster sessions. Besides, the program also consisted of a half-day technical visits to the Institute of No.101, China Academy of Launch Vehicle Technology; Technical Institute of Physics and Chemistry, Institute of Electric Engineering, Chinese Academy of Sciences.

The Proceedings of ICEC-20 documents 9 plenary papers and 5 leading papers, and 238 contributed papers, which were selected from 312 papers after peer reviewing and some of them were revised according to the review results.

The International Cryogenic Engineering Committee presented the Mendelssohn Award during the Conference to Prof. Gustav Klipping and Dr. Ingrid Klipping in recognition of their outstanding contributions to the development and promotion of Cryogenics world-wide and also to the promotion of ICEC over its whole time since the beginning.

We would like to thank all of our friends and colleagues who devoted their efforts for the successful holding of the Conference. Our appreciation is also to the numerous reviewers for their excellent contributions.

We are looking forward to seeing you at the ICEC 21, which will be held in July 2006, Prague, Czech.

Liang ZHANG	Chairman, Organizing Committee
Liangzhen LIN	Chairman, Program Committee
	Vice Chairman, Organizing Committee
Guobang CHEN	Vice Chairman, Program Committee

International Cryogenic Engineering Committee

Guy GISTAU-BAGUER, Chairman
Kazuo FUNAKI
Tomiyoshi HARUYAMA, Secretary
Peter KITTEL
Peter KOMAREK
Philippe LEBRUN, Vice Chairman
Liangzhen LIN
Kishan NARAYANKHEDKAR
Carlo RIZZUTO
Horst ROGALLA
Stanislas SAFRATA
Ralph SCURLOCK
Alistair STEEL
Steve Van SCIVER
Konstantin YUSHCHENKO
Liang ZHANG, Vice Chairman
Bruno ZIEGLER, Treasurer

Honorary Members

Gustav KLIPPING
Martin WOOD

Local Organizing Committees

Organizing Committee

Liang ZHANG , Chairman
Liangzhen LIN, Vice Chairman
Ming LI, Vice Chairman
Chunzheng CHEN
Guangming CHEN
Anzhong GU
Hongwei GU
Fangzhong GUO
Jianguo HAN
Zhenghe HAN
Zezhao HUA
Bencheng HUANG
Li KONG
Yanzhong LI
Jingtao LIANG
Zhongyue QIU
Yuejin TANG
Ruzhu WANG
Weiyang WANG
Yan WANG
Peide WENG
Peiheng WU
Liye XIAO
Yanru YANG
Pingxiang ZHANG
Jiwen ZHENG
Yuan ZHOU

Programme Committee

Liangzhen LIN, Chairman
Guobang CHEN, Vice Chairman
Liezhao CAO
Jun JI
Xiaohua JIANG
Laifeng LI
Qing LI
Jing LIU
Tao LIU
Limin QIU
Jiasu WANG
Yinshun WANG
Qiansheng YANG
Daole YIN
Guansen YUAN

Advisory Board

Chaosheng HONG, Chairman
Wenshan ZHAN, Vice Chairman
Chingwu CHU
Zizhao GAN
Yiyi LI
Luguang YAN
Guozhen YANG
Zhongxian ZHAO
Lian ZHOU

ICEC 20 SPONSOR

International Cryogenic Engineering Committee

ICEC 20 CO-SPONSORS

International Institute of Refrigeration
Chinese Academy of Sciences
National Natural Science Foundation of China
Chinese Association of Refrigeration
Technical Institute of Physics and Chemistry, Chinese Academy of Sciences
Institute of Electrical Engineering, Chinese Academy of Sciences
China Refrigeration and Air Conditioning Industry Association

ICEC 20 EXHIBITORS

Air Liquide
Alstom Magnets and Superconductors
AMAC International Inc.
Barber Nichils Inc.
Beijing Saisiai M&E Equipment Co . , Ltd
Beijing Bloom-Top Co.Ltd.
Chengdu Golden Phoenix Liquid Nitrogen Container Co., Ltd.
Cryofab Inc.
Cryogenmash
Cryomagnetics Inc.
Cryomech Inc.
DeMaCo
DY Refrigeration Co., Ltd.
East changing Technologies Inc.
Huanghe Oxygen Plant Factory
Kaifeng Air Separation Group Company, Ltd.
Lanzhou vacuum Equipment co., Ltd.
Linde Kryotechnik AG
Messer Cryotherm GmbH
Nippon Sanso Corporation
Oxford Instruments Superconductivity
PBS
Ricor-Cryogenic and Vacuum System
Scientific Instruments Inc.
Shandong Dongyue Chemical Co., Ltd.
Shanghai Hanbell Precise Machinery Co., Ltd.
Stars (Guangzhou) Refrigerating Equipment Manufacturing Co., Ltd .
Sumitomo Electric Industries, Ltd.
Taylor-Wharton Beijing Cryogenic Equipment Co., Ltd.
Tianjin Huanhu Hospital
Tristan
Vacree Technologies Co., Ltd.
Velan SA
Weka AG
Wessington Cryogenics

CONTENTS

Preface   *v*

International Cryogenic Engineering Committee and Honorary Members   *vi*

Local Organizing Committees   *vii*

ICEC 20 Sponsor, ICEC 20 Co-Sponsors and ICEC 20 Exhibitors   *viii*

Mendelssohn Award Lecture: Our life with Cryogenics   1
I. Klipping, G. Klipping

LARGE SCALE REFRIGERATION

Cryogenics in Russia: Development of Basic Cryogenic Cycles (***Plenary Invited Paper***)   11
V.M. Brodianski, A. Alexeev

Cryogenics in Russia (***Plenary Invited Paper***)   19
A.Alexeev, G. Klipping

Conceptual Design of a 2kW Cryogenic System for the Neutrino Superconducting Beam
Line Magnet System in the J-PARC 50GeV Proton Accelerator   27
Y. Makida, Y. Doi, T. Haruyama, N. Kimura, T. Kobayashi, T. Nakamoto, T. Ogitsu,
H. Ohhata, A. Yamamoto

The Neutral Beam Test Facility Cryopumping Operation: Preliminary Analysis and Design of
the Cryogenic System   31
B. Gravil, D. Henry, M. Dremel, JJ. Cordier, R. Hemsworth, D. Van Houtte

Performance of a Cryogenic System for the Superconducting RF Cavity at NSRRC   35
F.Z. Hsiao, H.C. Li, W.S. Chiou, S.H. Chang, J.R. Chen

The Cryogenic System of BESIII in Preliminary Design   39
L. Liu, Z. Zhu, S. Han, Y. Makida, H. Yamaoka, K. Tsuchiya, W. Liang, L. Gong, L. Zhang

Design Issues of the Distribution Valve Box of the KSTAR Helium Refrigerator   43
H.-S. Chang, G.M. Gistau Baguer, K.W. Cho, Y.S. Kim, G.S. Lee, P. Briend, P. Fuenzalida,
A. Fossen

A Retrospect on the Successful Commissioning of the Linde 18kW Refrigerators for the LHC
Particle Accelerator at CERN   47
A. Kuendig, B. Chromec, U. Fleck, Th. Voigt

Helium Refrigeration System for Wendelstein 7-X   51
A. Kuendig, F. Schauer, Y. Bozhko, C.P. Dhard

CRYOGENIC SYSTEMS AND TEST FACILITIES

Cryogen Free Refrigeration to 270mK by Adiabatic Expansion of He^3 Gas   55
(Plenary Invited Paper)
V. Mikheev, P. Noonan, A. Adams

Development of the Cryogenic Ground Support Equipment (CGSE) for the Superconducting
Magnet of the Alpha Magnetic Spectrometer (AMS-02)   63
A.Gu, A. Grechko, S. Harrison, H. Hofer, W. Lin, Y. Shi, S.C.C. Ting, S. Vostrikov, R. Wang

Unique Method for Liquid Nitrogen Precooling of a Plate Fin Heat Exchanger in a Helium
Refrigeration Cycle   67
T.B. Weber, A. Candia, G. Howell, M. Racine, J.G. Weisend II

Test Results and Analysis of Cryogenic System for Cooling HTS Cables and the Next
Stage Plan   71
Y.F. Fan, L.H. Gong, X.D. Xu, L.F. Li, L. Zhang, L.Y. Xiao

Cryogenic Facility Design in BEPCII Superconducting Upgrade   75
L.X. Jia, L. Wang, G.D,. Yang, X.K. Liu, X.B. Zhang, H.M. Tang, H.P. Du, Z.H. Xing,
C.S. Liu, G. Han, F.Y. Xu, T.H. Wang, C.Z.Yang, H. Xiao, D. Liu, L.H. Shi, M.X. Huang,
X.Y. Gai, F.Y. Xu, Z.M. Ji, W. Jiang, Z.L. Yao, A.B. Chen

First Cold Test of the ISAC-II Medium Beta Cryomodule   79
I.Sekachev, R.E. Laxdal, C. Marshall, T. Ries, G. Stanford

Experimental Study of Self-pressurizing Transfer System for Cryogenic Liquid   83
Q.F. Li, G.B. Chen, X.M. Xie, R. Bao

Design of a Cooling System for the Cold Test of the ITER TF Coils before Installation   87
G.R. Zahn, J.L. Duchateau, W.H. Fietz, B. Gravil, R. Heller, F. Millet, S. Nicollet,
P. Chesny

Study on Key-tech Problems in Development of a Cold Neutron Source in China Advanced
Research Reactor   91
F. Quanke, Y. Qingfeng, Z. Cun, S. Feng, Y. Luzheng

Experiences from the Operation Tests with the ITER Model Pump   95
H. Haas, A. Antipenkov, Chr. Day, M. Dremel, A. Mack, D.K. Murdoch

Design of Cryogenic System for the KSTAR Device   99
Y.S. Kim, H.S. Chang, K.W. Cho, G.M. Gistau Baguer, J.S. Bak, G.S. Lee, P. Briend,
P. Fuenzalida, A. Fossen

Air Liquid Cryogenic System for the KSTAR Device   103
P. Briend, P. Dauget, F. Delcayre, I. Abe, E. Fauve, A. Fossen, P. Fuenzalida, L.A. Lee,
G. Gistau, Y.S. Kim, K.W. Cho, H.S. Chang

Cryogenic System for BEPCII Cavity IR Quadrupole and Detector Solenoid Magnets   107
L.X. Jia, L. Wang

Status of the ISAC-II Cryogenic System   111
I.Sekachev, W. Andersson, R.E. Laxdal, C. Marshal, G. Stanford

Overview of Different Standard Helium Liquefiers/Refrigerators for Various Applications   115
A.Caillaud, M. Bonneton, F. Delcayre, V. Grabié, D. Grillot, B. Hilbert, A. Praud, E. Walter

Standard Liquefier-Test Results with Improved Turbines   119
H. Schoenfeld, D. Cretegny, K. Loehlein

Thermal Fluid Modeling of BEPCII IR Quadrupole Magnet Cryostat   123
L. Wang, H.M. Tang, X.B. Zhang, G.D. Yang, L.X. Jia

COOLERS

Recent Developments on Cryocoolers in Europe (***Invited Paper***)   127
A.Ravex, T. Trollier

Experimental Study on a Split Free-Piston Stirling Cooler for Space Applications   137
W. Zhang, Y. Wu, G.H. Lu

The Effects of the Clearance in the Cold Finger on the Performance of the Stirling
Refrigerator   141
Y-J. Hong, D-Y. Koh, S-J. Park, H-B. Kim, Y-D. Choi

Development of a Stirling Cryocooler with Linear Guidway Supports   145
G. Hong, Q. Li, Y.M. Xi, Z.S. Nie

Novel Regenerator Design and its Experimental Characteristics   149
K. Nam, S. Jeong

Investigation of a Single-Stage G-M Refrigerator with High Cooling Capacity for
HTS Devices   153
Z-C. Fang, X-T. Su, L-Q. Liu, L-H. Gong, L. Zhang

Preliminary Study on GM Refrigerator with Heat Exchanger   157
W.Q. Liang, L.H. Gong, L. Zhang

Dimensional Analysis on Multi-Component Mixed-Refrigerants Flow Characteristics in
Adiabatic Capillary Tube   161
Y.F. Qi, M.Q. Gong, Z.H. Sun, E.C. Luo, J.F. Wu

An Experimental Research on a Small Flow Vortex Tube at Low Temperature Ranges   165
J.Y. Liu, M.Q. Gong, J.F. Wu, Y. Cao, E.C. Luo

Performance of Different Throttle Refrigeration Cycles   169
M.Q. Gong, J.F. Wu

A 70 K J-T Cryocooler Based on the Fractional Mixed-Refrigerant Cycle   173
M.Q. Gong, J.F. Wu

Development of a Cryogenic Water-Trap Based on the Recuperative Mixture Refrigeration
Cycle   177
M.Q. Gong, J.F. Wu, Q.G. Hu

Room Temperature Magnetic Refrigerator using Both Metal Gd or/and Gd-Si-Ge Alloys
and a Permanent Magnetic Field Source   181
D.W. Lu, H.B. Wu, G.Q. Yuan, Y.S. Han, X.N. Xu, X. Jin

Properties of Thermoelectric Device with $(Bi_xSb_{1-x})_2(Te_ySe_{1-y})_3$ Materials 185
H-J. Liu, G. Li, L-F. Li

PULSE TUBE COOLERS

Future Trend of Pulse Tube Cryocooler Research (***Plenary Invited Paper***) 189
Y. Matsubara

Performance of a Pulse Tube Cryocooler with Two Separate Stages using ^{3}He in the
Second Stage 197
N. Jiang, U. Lindemann, G.B. Chen, G. Thummes

Experimental Analysis of the Oscillating Flow Characteristics for High Frequency
Regenerators 201
X-L. Wang, M-G. Zhao, J-H. Cai, J-T. Liang, Y. Zhou

Regenerative Recuperator for Tandem 4 K Pulse Tube Refrigerator 205
J. Jung, S. Jeong, Y. Choi, K. Nam

Experimental Investigation on DC Flow Suppression and Refrigeration Characteristics of a
High-Performance Single-Stage GM-Type PTC 209
Y. Jiang, G. Chen, G. Thummes

Use of Ceramic Part in G-M Refrigerator 213
X-T. Su, Z-C. Fang, L-H. Gong, L-F. Li, L. Zhang

A Study on the In-Line Type Inertance Tube Pulse Tube Cryocooler 217
S-J. Park, D-Y. Koh, Y-J. Hong, H-B. Kim, S-Y. Kim, W-S. Jung

Experiment on a Single-Stage GM Type Pulse Tube Cryocooler near 20K 221
Z.H. Gan, H.Z. Liu, Z.Z. Cheng, L.M. Qiu, G.B. Chen

Design and Test of a 80 K Miniature Pulse Tube Cryocooler 225
M-G. Zhao, X-L. Wang, X-F. Hou, J-T. Liang, J-H. Cai, Y. Zhou

Development of the 3.5 W at 80 K High-Frequency Pulse Tube Cryocooler 229
W. Jing, G.P. Wang, J.T. Liang, J.H. Cai

The Effect of Timing Ratio of Injecting to Rejecting on G-M Type PTCs Cooling
Temperature 233
Y-Y. Wang, J-H. Cai, J-T. Liang

Comparison of Oscillating Flow Characteristics of a Metallic and a Nonmetallic Regenerator
at High Frequencies 237
X-L. Wang, H-Z. Dang, J-H. Cai, J-T. Liang, Y. Zhou

Analysis of Inertance Tube in the Pulse Tube Refrigerator 241
N. Chen, C.G. Yang, L. Xu, R.P. Xu

Single Stage Double-Inlet Pulse Tube Refrigerator with Integral Thermo-Electric Cooler 245
R. Karunanithi, S. Jacob, S. Kasthurirengan, N. Durgesh, B. Upendra, K. Tamizhanban

Experimental Study of Wall Temperature Distribution of Stirling Type Pulse Tube
Refrigerator (SPTR) 249
Y.W. Liu, Y.L. He, J. Huang, X.Z. Li, C.Z. Chen

Studies on a Single Stage GM Type Pulse Tube Cryocooler using an Indigenous Helium Compressor   253
S. Kasthurirengan, S. Jacob, R. Karunanithi, B. Upendra, D.S. Nadig, B. Anand

Experimental Investigation on the Performances of Gas Flow in Oscillating Tube   257
Z.C. Li, L. Xu

Optimal Vector Analysis of the Phase Shifter in the Pulse Tube Refrigerator   261
N. Chen, C.G. Yang, L. Xu

Characteristics of 4 K Pulse Tube Cryocoolers in Applications   265
C. Wang

Experimental Investigation on Heat Transfer Performance of a Cryogenic Thermosyphon   269
W.Z. Li, L.M. Qiu, X.J. Zhang, P. Chen, Y.L. He

Reduction of Convective Heat Losses in Pulse Tube Refrigerators by Additional DC Flow   273
M. Shiraishi, Y. Fujisawa, M. Murakami, A. Nakano

Interference Characterization of Stirling-Type Nonmagnetic and Nonmetallic Pulse Tube Cryocoolers for High-Tc SQUIDs Operation   277
H.Z. Dang, J.T. Liang, Y.L. Ju, Y. Zhou

Performance of Vibration-Free Pulse Tube Cryocooler System for a Gravitational Wave Detector   281
T. Haruyama, T. Tomaru, T. Suzuki, T. Shintomi, N. Sato, A. Yamamoto, Y. Ikushima, R. Li, T. Akutsu, T. Uchiyama, S. Miyoki

Influence of He-H_2 Mixture Proportion on the Performance of Pulse Tube Refrigerator   285
Y.H. Huang, G.B. Chen, Z.H. Gan, K. Tang, R. Bao

Three Dimensional Numerical Computation of the Flow Field in a Pulse Tube   289
W.J. Ding, Y.L. He, W.Q. Tao

Comparison of the Performance of Regenerators to Counterflow Heat Exchangers   293
M.E. Will, A.T.A.M. de Waele

Temperature Hysteresis in a GM-Type Orifice Pulse Tube Refrigerator   297
Z.H. Gan, G. Thummes

THERMO-ACOUSTIC REFRIGERATION SYSTEM

Theoretical Prediction on the Coupling of Thermoacoustic Prime Mover and RC Load   301
K. Tang, G.B. Chen, Z.Z. Jia, N. Jiang, R. Bao

Lattice Gas Modeling of Thermoacoustic Oscillations   305
X.Q. Zhang, Y. Chen, X. Liu

Characteristics of a Miniature Pulse Tube Cooler Driven by Thermoacoustic Standing Wave Engine   309
W. Dai, E.C. Luo, Y. Zhou, W.X. Zhu

Experimental Investigation on the Dynamic Character of Thermoacoustic Engine and Parameters of Regenerator   313
X. Chen, Q. Li, Z.Y. Li, Z.B. Yu, X.J. Xie

Basic Operation Principle of a Novel Cascade Thermoacoustic Prime Mover   317
E. Luo, E. Dai, H. Ling

Study on a Linear Compressor with Metal-Bellow Replacing Conventional Cylinder-Piston Unit   321
Z. Wu, E. Luo, W. Dai, J. Hu

Analytical Model for Onset Temperature Difference of Thermoacoustic Stirling Prime Movers   325
J. Hu, E. Luo, H. Ling

Two-Dimensional Numerical Simulation of the Inertance Tube   329
Y. Zhang, W. Dai, E. Luo, R. Radebaugh, M. Lewis

Numerical Simulation on a Novel Cascade Thermoacoustic Prime Mover   333
H. Ling, E. Luo, W. Dai, X. Li

New Perspective on Thermodynamic Cycles of Oscillating Flow Regenerators   337
E. Luo, W. Dai, R. Radebaugh

Experimental Investigation on a Room-Temperature Traveling-Wave Thermoacoustic Refrigerator   341
Y. Huang, E. Luo, W. Dai, Z. Wu

Study on Thermoacoustic Resonance Pipe Driven by Loudspeaker using Two-Microphone Method   345
L. Fan, B.R. Wang, T. Jin, Y. Shen, S.Y. Zhang

Stack and Frequency's Coupling Characteristic Analysis in Thermoacoustic Systems   349
F.Z. Zhang, R.Q. Jiang, Q. Li, X. Chen

Numerical Simulation of Loudspeaker-Driven Thermoacoustic Refrigerator   353
Q. Tu, Z.J. Chen, J.X. Liu, X.Q. Zhang, Q. Li, F.Z. Guo

Study on the Optimal Characteristic Dimension of Regenerator in a Thermoacoustic Engine   357
Z.B. Yu, Q. Li, X. Chen, F.Z. Guo and X.J. Xie

Investigation on Acoustic Characteristics of Regenerator in the Thermoacoustic Apparatus   361
X.J. Xie, Q. Li, X. Chen, Z.B. Yu

Discussion on Resonant Frequency in Thermoacoustic Systems   365
T. Jin, L. Fan, B.S. Zhang, B.R. Wang, G.B. Chen

Figure of Merit for Regenerator Working in Acoustic Field   369
F.Z. Guo

Investigation of Acoustic Streaming in a Thermoacoustic-Stirling Heat Engine   375
P. Duthil, E. Bretagne, J. Wu, M.-X. François, Q. Li

Investigation on a Thermoacoustically Driven Pulse Tube Cooler Working at 82.5 K   379
L. Qiu, D. Sun, W. Yan, P. Chen, Z. Gan, G. Chen

Thermoacoustic Turbulent-Flow Model for Inertance Tubes used for Pulse Tube Refrigerators   383
E. Luo, R. Radebaugh, W. Dai, M. Lewis, Z. Wu, Y. Zhang

Thermoacoustically Driven Pulse Tube Refrigeration below 90 K   387
K. Tang, G.B. Chen, T. Jin, R. Bao

SPACE CRYOGENICS

New Generation of Cooled Infrared Space Telescopes (*Plenary Invited Paper*)   391
D. Lemke

Study on Electricity Generation Performance of Thermoelectric Device in Extreme
Environmental Conditions   401
Z.-L. Wu, Y-X. Zhou, T. Li, J. Liu

Experimental Investigation on He II Porous Plug Liquid-Vapor Phase Separator   405
X. Yu, Q. Li, Z. Li, Qiang Li

Novel Cryogenic Motion Control for Aerospace   409
Z.T. Cao, H.P. Chen, G.G. He, J.J. Cai

Spitzer Space Telescope Thermal/Cryogenic System Flight Performance   413
P.T. Finley, R.A. Hopkins, R.B. Schweickart

Thermal Validation of the Design of the CFRP Support Members to be used in the Spatial
Framework of the Herschel Space Observatory   419
P.C. McDonald, E. Jaramillo

Development and Testing of a Novel Thermal Switch   423
J.G. You, D.P. Dong, W.Y. Wang, Z.W. Li

Cryocoolers Development and Integration for Space Applications at Air Liquide   427
A.Ravex, T. Trollier, L. Sentis, F. Durand, P. Crespi

Overview of the Air Liquide Cryogenic Ground Space Activities for Launchers and Satellites   437
S. Duval, P. Bravais, JC. Courty, S. Crispel, G. Marot

Thermal Analysis of Cryogenic Rocket Engine with One, Two and Three Dimensional
Approaches   441
B.T. Kuzhiveli, S.C. Ghosh, G.K. Kuruvila, V.G. Gandhi

Liquid Hydrogen Droplet Generation from a Vibrating Orifice   445
J. Xu, D. Celik, M. Y. Hussaini, S.W. Van Sciver

The Working Performance of Radiant Cooler in Thermal Vacuum Test   449
H.Y. Xu, D.P. Dong, W.Y. Wang, Z. Li

A Study on the Conic Support in the Radiant Cooler   453
L. Y.Fu, F.H. Tu, W.Y. Wang

Analysis of the Influences of View Angle on Paraboloid W Style Radiant Cooler's Working
Performance   457
X.J. Wang, Y.P. Pan, Y.L. He, C.Z. Chen

CRYOGENIC COMPONENTS AND MACHINERY

Design and Manufacturing of Cryosorption Pumps for Testbeds of ITER Relevant
Neutral Beam Injectors ... 461
M. Dremel, C. Day, A. Mack, H. Haas, V. Hauer, E. Speth, H.D. Falter, R. Riedl, B. Gravil

Simple Method for Calculation of Cryogenic Transfer Line Cool Down ... 465
A.Alexeev, Th. Pelle

Study of New Aerodynamic Foil Thrust Gas Bearing with Elastic Support for Cryogenic
Turbo-Expander ... 469
C.Z. Chen, S.N. Lin, Y. Hou, Z.H. Zhu

Design of Small Turbo Braton Cycle Air Refrigerator Test Rig ... 473
Y. Hou, Y. Sun, J. Wang, C.Z. Chen

Analysis of Centrifugal Stresses in the Rotor of a Cryogenic Turboexpander ... 477
P. Ghosh, S.K. Sarangi

Study of Temperature Distribution of Flat Steel Ribbon Wound Cryogenic High-pressure
Vessel ... 483
X.L. Cui, G.M. Chen

The Design and Study of a Reciprocating Helium Compressor ... 487
G. Lei, L-H. Gong, W-H. Lu, L. Zhang

SUPERCONDUCTOR BEHAVIOUR

Strain Effects on the Critical Current of Bi-2223/Ag Tape ... 491
Y.L. Wu, Z.D. Wang, Y.F. Xiong

Thermal, Electrical Properties and Microstructure of High Purity Niobium for High Gradient
Superconducting Cavities ... 495
H.M. Wen, X. Singer, W. Singer, D. Proch, E. Vogel, D. Hui

AC Loss Characteristics of Bi-2212 Multifilamentary Round Wires and their Cables ... 499
K. Funaki, M. Iwakuma, K. Kajikawa, N. Hirano, S. Nagaya, N. Ohtani, T. Hasegawa

Property of Joint Resistance of Bi2223 Multi-filamentary Tape by Using Sn-Pb Solder ... 503
C. Gu, C. Zhuang, J.Y. Zhang, T.M. Qu, Z. Han

Quench Characteristics of High-T_c Superconducting Tape for Alternating Over-Current
using DFT ... 507
S-H. Lim, S-W. Yim, J-H. Lee, S.C. Ko, S-D. Hwang, B-S. Han

Dip-coated $Yba_2Cu_3O_{7-x}$ Films on Ag Substrate by MOD Method ... 511
J. Dong, M. Liu, D.M. Liu, Y. Zhao, J.X. Liang, J.L. Sheng, H.L. Suo, M.L. Zhou

Influence of Ag Substrate Surface Condition on Surface Morphology of YBCO Film ... 517
M. Liu, D.M. Liu, J. Dong, Y. Zhao, J. Sheng, J.X. Liang, H.L. Suo, M.L. Zhou

Boiling Heat Transfer to Liquid Nitrogen Pool from Ag Sheathed PbBi2223 Tapes
Carrying Over-Current ... 521
W. Bailey, E.A. Young, Y. Yang, C. Beduz

Equivalent Circuit Model for High-T_c Superconducting Flux Flow Transistor with a
Dual-Gate Structure   525
S. Ko, H-G. Kang, S-H. Lim, J-H. Lee, B-S. Han

Direct Observation of Percolation and Phase Separation in $Pr_{5/8}Ca_{3/8}MnO_3$ System   529
G.X. Cao, J.C. Zhang, P.L. Li, J. Yu, S.P. Wang, C. Jing, S.X. Cao, X.C. Shen

The Reentrant Behaviour of Spin-Glass Phase in $(La,Pr)_{0.5}Ca_{0.5}MnO_3$ Manganites   533
J.C. Zhang, G.X. Cao, Y.N. Sha, S.P. Wang, J. Yu, G.Q. Jia, C. Jing, S.X. Cao, X.C. Shen

Anisotropic Transport and Magnetic Properties of $Pr_{1/2}Sr_{1/2}MnO_3$ Single Crystal Sample   537
S.X. Cao, X.Y. Wang, Y.F. Zhang, L.M. Yu, C. Jing, J.C. Zhang, X.C. Shen

Dependence of Structure and Magnetic Properties on Magnetic Field and Temperature
for Bi-Mn Alloy   541
Y.S. Liu, J.C. Zhang, G.Q. Jia, X.Y. Zhang, Z.M. Ren, C. Jing, X. Li, S.X. Cao, K. Deng

Physical Properties in Layered Transition-Metal Oxide Crystals and Anisotropic
Transport Measurement   545
K.Q. Ruan, Y. Yu, S.L. Huang, H.L. Li, S. Qian, L.Z. Cao

Design of a Cryogenic Giant Magnetostrictive Actuator using HTS   549
J.J. Cai, Z.T. Cao, H.P. Chen, G.G. He

A New Method of Zooming in the Cool Power of the Room Temperature Magnetic Refrigeration   553
D.W. Lu, H.B. Wu, G.Q. Yuan, Y.S. Han, X.N. Xu, X. Jin

Extended Power Law of Nonlinear Transport Properties of Superconducting Materials   557
Z.H. Ning, X. Hu, D.L. Yin, Z. Qi, F.R. Wang, J.D. Guo, C.Y. Li

AC Loss Characteristics of HTS Tapes Subject to Bending Strains   561
O. Tsukamoto, J. Ogawa, H. Suzuki

LTS SUPERCONDUCTING DEVICES

EAST Superconducting Tokamak Device (***Plenary Invited Paper***)   565
P.D. Weng and EAST team

Challenges of Large Scale Superconducting Accelerators
(***Plenary Invited Paper – Abstract***)   571
D. Proch

Cryogen-free Superconducting Magnet Fabricated by a React-and-Wind Method
Employing Nb_3Sn Wires with CuNbTi Reinforcement   573
K. Watanabe, G. Nishijima, S. Awaji, K. Takahashi, K. Miyoshi, S. Megro, M. Ishizuka,
T. Hasebe

Design and Manufacture of a NbTi Insert Module Relevant for the ITER PF Magnets Conductor   577
A.della Corte, L. Affinito, S. Chiarelli, P. Gislon, G. Messina, L. Muzzi, G. Pasotti,
S. Turtù, M. Mariani, A. Matrone, S. Rossi

Multi-Tube Power Leads Tower for BEPCII IR Magnets   581
L.X. Jia, X.B. Zhang, L. Wang T.H. Wang, Z.L. Yao

The MICE Focusing Solenoids and their Cooling System 585
M.A. Green, G. Barr, W. Lau, R.S. Senanayake, S.Q. Yang

Evaluation of the ITER Cable in Conduit Conductor Heat Transfer 589
S. Nicollet, D. Ciazynski, J.L. Duchateau, B. Lacroix, B. Renard

The BESIII Detector Magnet 593
*Z. Zhu, L. Zhao, L. Wang, Z. Hou, S. Huang, H. Yang, J. Hu, J. Zhou, S. Han, C. Yi,
H. Chen, Q. Xu, L. Liu, Y. Makida, H. Yamaoka, K. Tsuchiya, B. Wang, B. Wahrer,
C. Taylor, C. Chen*

Stability Analysis for Cryocooled Aluminum-Stabilized Superconducting Magnet 597
H. Liu, Q. Wang, Y. Yu, Y. Bai

Thermal Design of BESIII Magnet 601
Q. Xu, Z. Zhu, L. Liu, S. Wang, C. Yi, H. Chen, L. Zhang

A Numerical Code to Simulate Quenching in Conduction Cooled Superconducting
Magnet 605
Y. Dai, B. Blau, H. Hofer, Y. Yu, Q. Wang

Numerical Simulation on Thermal Stress and Magnetic Forces of Current Leads in
BEPCII 609
X.B. Zhang, L.X. Jia

Design and Manufacture of a Low-Current ADR Magnet for a Space Application 613
C. Brockley-Blatt, S. Harrison, I. Hepburn, R. McMahon, S. Milward, R. Stafford Allen

Study of Stress, Strain in Super-Conducting Magnet by Fiber-Bragg Grating 617
Z.A. Feng, Q.L. Wang, F. Dai, G.J. Huang

HTS SUPERCONDUCTING DEVICES

Design Study of Conduction-cooled High Temperature Superconducting Magnet 621
Q.L. Wang, H. Huang, S.S. Song, Y.M. Dai, B.Z. Zhao, Y.J. Yun, L.G. Yan

Testing the Coils of the Superconducting Magnet for the Alpha Magnetic
Spectrometer 625
*R. McMahon, S. Harrison, S. Milward, R. Stafford Allen, H. Hofer, J. Ulbricht,
G. Viertel, S.C.C. Ting*

Characterization of the Quenching Process and the Protection of an ac High
Temperature Superconductor Coil 629
X.H. Li, Y.S. Wang, L.Y. Xiao

Technical Analyses of HTS Bulk used in a Rotating Machine 633
M. Qiu, H.K. Huo, Z. Xu, Z.H. Yao, D. Xia, L.Z Lin

Investigation of the Subcooled LN2 Characteristics in the Prototype Cryostat for the
Resistive Superconducting Fault Current Limiter 637
S. Cho, H.S. Yang, D.L. Kim, W.M. Jung, D.H. Kim, H.R. Kim, O.B. Hyun

LN2 Forced Flow Cooling on HTS Power Cables 641
D.-Y. Koh, H.-K. Yeom, K.-S. Lee

Shape Optimization of Field Windings in a High Temperature Superconducting
Synchronous Generator with Finite Element Method   645
F.C. Song, G.Q. Zhang, B. Chen, S.Z. Yu, G.B. Gu

Development and Testing of 10.5kV/1.5kA HTS Power Cable   649
Y.B. Lin, Z.Y. Gao, Z. Ning, S.T. Dai, Y.P. Teng, L. Xu, L.Z. Lin, L.Y. Xiao

Normal Transition in YBCO Thin Film and its Effect on Current Limiting
Characteristics   653
T. Onishi, K. Sasaki, T. Kayoda

Study on a Hybrid Superconducting Magnetic Bearing System   657
J.R. Fang, L.Z. Lin, L.G. Yan

Electromagnetic Design and Analysis of a PM Motor with HTS Bulk   661
H.K. Huo M. Qiu, D. Xia, Z. Xu, L.Z. Lin

Finite-Element Simulation of HTS Bulk Reluctance Motor   665
Z. Xu, M. Qiu, Z.H. Yao, D. Xia

A Current Compensation Type Superconducting Fault Current Limiter   669
D.F. Chen, C.H. Zhao, L.Y. Xiao

Evaluation on Power System Transient Stability with Resistor Type SFCL and
Parameter Design   673
Y. Ye, L.Y. Xiao

Effects of Bias Power Supply on Bridge Type Superconducting FCL   677
Z. F.Zhang, L.Y. Xiao

20kA HTS Current Leads for EAST Tokamak Project   681
Y.F. Bi, X.Q. Chen, D.K. Ma, X.L. Liu, S.T. Wu, J.G. Li

Design of Cooling System and Temperature Characteristics Analysis of High
Temperature Superconducting Demonstrative Synchronous Generator   685
B. Chen, G.Q. Zhang, F.C. Song, S.Z. Yu, G.B. Gu

A Prototype 4m, 2kA, AC HTS Power Cable System   689
H.X. Xi, B. Hou, Y.F. Bi, X.C. Yang, H.K Ding, Y. Xin

Development of a Three-Phase HTS Power Transformer   693
*Y.S. Wang, X. Zhao, H. Li, J. Han, L.Y. Xiao, L.Z. Lin, Y. Guan, Q. Bao, G. Lu, Z. Zhu,
X. Xu, Y. Lu, S. Dai, D. Hui*

Thermal Stability and Quench Propagation in a Cryocooler-Cooled Pancake Coil of
BI-2223 Tape Operating Between 35K and 65K   697
A.P. Johnstone, Y. Yang, C. Beduz

A New Concept of Bridge Fault Current Limiter-SMES for Interline Application to
Improve Power Quality   701
C.H. Zhao, L.Y. Xiao, L.Z. Lin, Y.J. Yu

Design and Primary Experiment for a Single Phase 220V/100A/6kW Bridge Fault
Current Limiter-SMES 705
C.H. Zhao, Y. Zhang, S. Li, X.H. Huang, D.F. Chen, L.Y. Xiao, L.Z. Lin, Y.J. Yu

Fabrication and Characterization of Bi-2223 Coils for Generator Applications 709
B. Xu, M.K. Al-Mosawi, Y. Yang, C. Beduz

Shape Optimization of HTS Magnets using Hybrid Genetic Algorithms 713
C. Wang, Q.L. Wang

An Adaptive Neuro-Fuzzy Control Strategy for the Bridge Type Superconducting Fault Current
Controller 717
W.Y. Guo, C.H. Zhao, L.Y. Xiao

APPLICATIONS

Cryoelectronics – An Experimental Examination of the Behaviour of Power
Electronic Devices at Low Temperatures 721
C.J. Hawley, S.A. Gower

Operational Amplifiers in the Temperature Range from 300 to 4 K 725
W. Nawrocki, J. Pajҳkowski

Development and Test of a Cryopulverizer for Reprocessing Plastic Wastes 729
S. Jacob, S. Kasthurirengan, R. Karunanithi, Upendra Behera, D.S. Nadig

A Cryostat with Optics Windows for Study of ICF Cryogenics Target 733
Y. Liu, Y.Q. Zhu, G. Zhou, C. Jiang

Temperature Effects on Formation of a Uniform Layer of Isotopes Inside a
Cryogenic ICF Target 737
G. Zhou, Y.Q. Zhu, Y. Liu, C. Jiang, Q. Li

GAS LIQUEFACTION AND SEPARATION

Optimization Analysis of Peakshaving Cycle to Liquefy the Natural Gas 741
Y. Shi, A. Gu, R. Wang, G. Zhu

The Experimental Study on the Performance of a Small-Scale Oxygen Concentrator
by PSA 745
Y.W. Zhang, Y.Y. Wu, J.Y. Gong, J.L. Zhang

The Thermodynamic Analysis and Adjusting Methods for Reliquefaction Plant on
Board LEG Carriers 749
Z.L. Zhao, A.Z. Jiang, L.L. Xiong

A Study of the Optimum Temperature for Hydrogen Storage on Carbon
Nanostructures 753
Q.R. Zheng, A.Z. Gu, X.S. Lu, W.S. Lin

Design of Vortex Tubes and Experimental Program on LOX Separation using
Cryogenic Vortex Tubes 757
*S. Jacob, Upendra Behera, P.J. Paul, S. Kasthurirengan, R. Karunanithi, S.N. Ram,
K. Dinesh*

HEAT TRANSFER AND THERMAL INSULATION

Heat Transfer Enhancement of He II Co-Current Two-Phase Flow in the Presence of Atomisation 761
B. Rousset, P. Thibault, S. Perraud, L. Puech, P.E. Wolf, R. van Weelderen

Design Consideration Validation of Cryo-Components for Current Feeder System of
SST-1 using Flexibility Analysis 765
N.C. Gupta, B. Sarkar, G. Gupta, Y.C. Saxena

Contractible Thermosyphon for Conduction Cooled Superconducting Magnets 769
C. Noh, S. Kim, S. Jeong, H. Jin

Experimental Characterization of the Hydraulic Behavior of Cooling for Toroidal Field Magnet of
SST-1 773
*B. Sarkar, N.C. Gupta, A.K. Sahu, G. Bansal, S. Pradhan, C.P. Dhard, Misra Ruchi,
Panchal Pradip, Tank Jignesh, G. Gupta, Y.C. Saxena*

Performance Analysis of Counter Flow Plate-Fin Heat Exchangers Considering the
Effect of Longitudinal Conduction 777
T.C. Mi, Y.Z. Li, J. Wang, Y.H. Zhu

Investigation of Header Configuration and its Effect on Flow Maldistribution in Plate-Fin Heat
Exchanger 781
J. Wen, Y.Z. Li, K. Zhang

Experimental Investigation on High Performance Cryogenic Heat Transfer Based upon Natural
Circulation Cooling 785
L.M. Qiu, W.Z. Li, Z.H. Gan, X.J. Zhang, B. Jiao, G.B. Chen, C.J. Yang

Film Boiling Modes in Weakly Subcooled He II around Lambda Pressure 789
M. Nozawa, N. Kimura, M. Murakami, P. Zhang

Modeling of Multilayer Vacuum Insulation – Complexity Versus Accuracy 793
M. Chorowski, J. Polinski

Experimental Study of the Narrow Channel Heat Transfer in Liquid Nitrogen 797
P. Zhang, G.C. You, X. Ren, R.Z. Wang

Observation of Various Boiling States in Nearly Saturated He II 801
P. Zhang, M. Murakami, M. Nozawa

Heat Transfer During Natural Convective Boiling in a Vertical Uniformly Heated
Closed Tube Submerged in Saturated Helium 805
B. Baudouy

Several Problems in the Thermal Contact Resistance Research of Solid Interfaces at
Low Temperature 809
L. Xu, Z.C. Li

CFD Approach to Thermal Pulse Propagation and Heat Transfer in Forced Flow
Field of He II 813
M. Yanase, M. Murakami

Investigation of Thermal Insulation for HTS Cable Systems   817
D.H. Kim, D.L. Kim, H.S. Yang, W.M. Jung, S.D. Hwang

Experimental Investigation on Pool Boiling Heat Transfer of Pure Refrigerants and
Binary Mixtures   821
Z.H. Sun, M.Q. Gong, Y.F. Qi, E.C. Luo, J.F. Wu

Modeling the Thermal Mechanical Behavior of a 300 K Vacuum Vessel that is
Cooled by Liquid Hydrogen in Film Boiling   825
S.Q. Yang, M.A. Green, W. Lau

Fractal Description of Thermal Contact Resistance between Rough Surfaces at Low Temperature   829
R.P. Xu, H.D. Feng, T.X. Jin, L. Xu, L.P. Zhao

Study of Two Phase Flow Distribution and its Influence in a Plate-Fin Heat Exchanger   833
Q. Xu, Y.Z. Li, Z. Zhang

A Thermodynamic Optimization of Counterflow Recuperative-Type Heat Exchangers   837
Q. Wang, H.F. Zhang, G.M. Chen

A New Type of Condenser-Evaporator Safely Operated in Large Air Separation Plant   841
Y.Y. Wu, L.F. Chen, T.H. Wu

Heat Transfer in Gas Filled Pipes with Closed Warm End under Different Orientations   845
M. Hieke, Ch. Haberstroh, H. Quack

Normalized Representation for Steady State Heat Transport in a Channel Containing
He II Covering Pressure Range up to 1.5 Mpa   849
A.Sato, M. Maeda, M. Yuyama, Y. Kamioka

Thermal Insulation of the Wendelstein 7-X Superconducting Magnet System   853
*F. Schauer, M. Nagel, Y. Bozhko, M. Pietsch, S.Y. Shim, F. Kufner, F.Leher, A. Binni,
H. Posselt*

Modeling Free Convection Flow of Liquid Hydrogen Within a Cylindrical Heat
Exchanger Cooled to 14 K   857
S.Q. Yang, M.A. Green, W. Lau

Experiment on Natural Convection of Subcooled Liquid Nitrogen   861
Y.S. Choi, H-M. Chang, S.W. Van Sciver

CRYOGENS

Numerical Simulation of Flow Boiling of Cryogenic Liquids in Tubes   865
X.D. Li, R.S. Wang, A.Z. Gu

A Density Equation for Saturated Helium-3   869
X.Y. Li, Y.H. Huang, G.B. Chen, V. Arp

The Characterization of Carbon Nanofibres Based on N_2 Adsorption Isotherms at 77 K   873
C. Zhang, X. Lu, A. Gu

Experimental Investigation into Storage of Confined Cryogenic Liquids without Evaporation Venting   877
Z. Huang, R. Wang, Y. Shi, A. Gu

Thermal Stress Analysis of Cryogenic Adhesive Joints for Non-Magnetic Dewar   881
H.M. Zhu, T.X. Jin, L. Xu, H. Sun, Y.M. Xiao

PROPERTIES OF MATERIALS AT CRYOGENIC TEMPERATURES

Development of Cryogenic Structural Materials (*Plenary Invited Paper*)   885
L.F. Li, Y.T. Zhang, Y.Y. Li

Thermal Conductivity of Materials and Measuring System at Cryogenic Temperature   895
B. Li, B.M. Wu, D.S. Yang, W.H. Zheng, L.Z. Cao

Cryogenic Properties of Polymer Composite Materials – A Review   899
S.Y. Fu, Y. Li, Y.H. Zhang, Q.Y. Pan, C.J. Huang

CRYOFLUIDS – A Software for Physical Property Data of Cryogenic Fluids   903
P. Lalkoshti, R. Banerjee, A. Tarafder, S.K Sarangi

Discussion on Gibbons' Equation of State for Helium-3   911
Y.H. Huang, G.B. Chen, X.Y. Li

An Internet-Based Working Fluid Properties Database   915
J. Huang, Y.L. He, Y.W. Zhang

A New Process for the Preparation of Polyimide/Silica Hybrid Films   919
Y. Li, S.Y. Fu, Q.Y. Pan, Y.H. Zhang, D.J. Lin

CRYOGENIC INSTRUMENTATION AND PROCESS CONTROL

Design of a Fibre Reinforced Plastic Anticryostat for Magnetorelaxometric Measurements   923
T. Koettig, P. Weber, S. Prass, P. Seidel

Measurement of Heat Load on the Current Lead Using Boil-Off Calorimetry   927
H.S. Yang, D.L. Kim, D.H. Kim, W.M. Jung, S. Cho

Alignment of the ISAC-II Medium Beta Cryomodule with a Wire Monitoring System   931
W.R. Rawnsley, Y. Bylinski, K. Fong, R.E. Laxdal, G. Dutto, D. Giove

Digital Temperature Sensors in the Range of 77-300 K   935
W. Nawrocki, A. Jurkowski, J. Pająkowski

Measurement of Characteristic Feature of Cavitation Flows of He I and He II   939
K. Harada, R. Tsukahara, M. Murakami

Application of PIV Technique to Cavitating Flows of Liquid Helium   943
R. Tsukahara, M. Murakami, K. Harada

Measurement of the Cryogenic Tensile Properties of Polymer Composites   947
Q.Y. Pan, S.Y. Fu, Y.H. Zhang, Y. Li

High Precise Thermopower Measurement System and its Applying to Organic
Conductors (TMTSF)$_2$ClO$_4$ in a Direction
951
Y.S. Chai, H.S. Yang, J. Liu, L. Zhu, L.Z. Cao

Multichannel Temperature Monitor Compatible with PC
955
Yu.P. Filippov, S.V. Alpatov, B.N. Sveshnikov

Interchangeability of Diode Temperature Sensors for Various Non-Standard Excitation Currents
959
S. Courts, C. Yeager

The Laser Megajoule CryoTarget Thermal Regulation
963
*V. Lamaison, D. Brisset, B. Cathala, G. Paquignon, P. Bonnay, D. Chatain,
D. Communal, J-P. Périn*

Discussion of the Protection of Pressure Vessels by using Safety Valves-Rupture
Disc-Combinations
967
M. Süßer

New Temperature and Magnetic Field Sensors for Cryogenic Applications Developed
Under a European Project
971
*V.F. Mitin, P.C. McDonald, F. Pavese, N.S. Boltovets, V.V. Kholevchuk, I.Yu. Nemish,
V.V. Basanets, V.K. Dugaev, P.V. Sorokin, E.F. Venger, E.V. Mitin*

CRYOGENICS IN MEDICINE AND BIOLOGY

Characterization Studies of Sputter Deposited Cermet Cryogenic Heaters
975
C. Yeager, S. Courts, L. Chapin

Study of Skin Burns due to Contact with Cold Mediums at Extremely Low
Temperature
979
Z-S. Deng, J. Liu

A New Method for Uniformly Cooling Engineered Tissues
983
L-N. Yu, Z-S. Deng, J. Liu, Y-X. Zhou

Comparative Analysis of the Cryogens used in Cryomedical Applications
987
M. Chorowski, A. Piotrowska

LHC

Measured Performance of Four New 18 kW @4.5 K Helium Refrigerators for the LHC Cryogenic
System
991
H. Gruehagen, U. Wagner

Overview of the Air Liquide Cryogenic Systems Designed for CERN LHC
995
*P. Dauguet, P. Briend, F. Delcayre, B. Hilbert, C. Mantileri, G. Marot, E. Monneret,
E. Walter*

1.8 K Refrigeration Units for the LHC: Performance Assessment of Pre-Series Units
999
S. Claudet, G. Ferlin, F. Millet, L. Tavian

The Applicability of Ultrasonic Oxygen Deficiency Hazard Detectors in the LHC
Accelerator Tunnel
1003
K. Brodzinski, M. Chorowski, W. Gizicki, W. Ostropolski, G. Riddone

Numerical Study of the Final Cooldown from 4.5 K to 1.9 K of the Large Hadron Collider   1007
G. Riddone, L. Liu, L. Tavian

Flow and Thermo-Mechanical Analysis of the LHC Sector Helium Relief System   1011
M. Chorowski, J. Fydrych, G. Riddone

Cryogenic Infrastructure for Testing of LHC Series Superconducting Magnets   1015
J. Axensalva, V. Benda, L. Herblin, JP. Lamboy, A. Tovar-Gonzalez, B. Vullierme

Final Testing of the ATLAS Central Solenoid before Installation   1019
Y. Doi, T. Haruyama, M. Kawai, T. Kondo, Y. Kondo, Y. Makida, A. Yamamoto, F. Haug, J. Metselaar, G. Passardi, O. Pavlov, M. Pezzetti, O. Pirotte, R. Ruber, E. Sbrissa, H. Ten Kate, H. Tyrvainen

Control System and Operation of the Cryogenic Test Facilities for LHC Series Superconducting Magnets   1023
J. Axensalva, L. Herblin, JP. Lamboy, A. Tovar-Gonzalez, B. Vullierme

Studies on Cooling of the TOTEM Particle Detector at the LHC   1027
F. Haug

Cryogenics Safety Review of the ATLAS Experiment at CERN   1031
F. Haug

Series Production of 13 kA Current Leads with Dry and Compact Warm Terminals   1035
T.P. Andersen, V. Benda, B. Vullierme

Measurement of the Thermal Properties of Prototype Lambda Plates for the LHC   1039
R. Marie, L. Metral, A. Perin, J.-M. Rieubland

The Cold-Compressor Systems for the LHC Cryoplants at CERN   1043
A.Kuendig, H. Schoenfeld, Th. Voigt, K. Kurtcuoglu, S. Yoshinaga, N. Saji, T. Shimba, T. Honda

Two 100 m Invar® Transfer Lines at CERN: Design Principles and Operating Experience for Helium Refrigeration   1047
S. Claudet, G. Ferlin, F. Millet, E. Roussel, JP. Sengelin

Experience with the String2 Cryogenic Instrumentation and Control System   1051
P. Gomes, Ch. Balle, E. Blanco, J. Casas, S. Pelletier, M.A. Rodriguez, L. Serio, A. Suraci, N. Vauthier

The Management of Cryogens at CERN   1055
D. Delikaris, K. Barth, G. Passardi, L. Serio. L. Tavian

Conceptual Design of the Cryogenic Electrical Feedboxes and the Superconducting Links of LHC   1059
T. Goiffon, J. Lyngaa, L. Metral, A. Perin, P. Trilhe, R. van Weelderen

Author Index   1063

Mendelssohn Award Lecture

Our life with cryogenics

Ingrid and Gustav Klipping

Limastrasse 38, D-14163 Berlin, Germany
(formerly: Free University Berlin, Department of Physics, Low Temperature Laboratory)

We are very grateful and feel truly honoured by receiving the Mendelssohn Award. And that it happens just here, at the 20th International Cryogenic Engineering Conference, makes it even more special. The information on this decision of the ICE Committee came as a big surprise – and of course it made us look back on our 47 years of joint activities before and with cryogenics.

Don't worry, neither I nor my husband, Gustav, will speak about new developments in physics or engineering – that was much better done during the scientific sessions of this conference. We will just tell you a bit about our personal experiences with cryogenics.

We met a very long time ago, in 1946, one year after Word War II, when we both enrolled at the Technical University Berlin, Gustav for studying Engineering and Physics and I for studying Chemistry. At that time studying was a kind of adventure: Berlin had been severely destroyed by bombs and fighting. From TV reports on this topic you will know what this means. At our university only one larger lecture hall had survived the general destruction, laboratories were rare and overcrowded, and laboratory work meant a lot of improvisation. During the first term the university had only 1200 students, and the number of professors with a clean political background was quite limited. Living conditions were complicated. But soon friendships among the students were born from the need to help each other, and in spite of all hardships we had a strong feeling that things could only get better, and life was full of promises. For the two of us these promises came true in a way which we could not even dream of at that time.

At the university and during our professional life we had the good fortune to meet a number of great personalities who became our guides, mentors and friends and also gave us the chance of new, lasting experiences. This is a good opportunity to remember them with gratitude.

The first was Iwan N. Stranski (Fig. 1), Professor of Physical Chemistry at the Technical University and a well-known specialist in crystal growth. Students of physics as well as chemistry had to take his courses, and both of us were soon fascinated by the subject as well as the man and decided for him as mentor of our Diploma and Doctoral theses. Due to the lack of laboratories at the university, we got the chance to do our thesis work at a research institute, the former "Kaiser-Wilhelm-Institut für Physikalische und Elektrochemie", today known as "Fritz Haber Institute of the Max Planck Society", where Prof. Stranski headed the Department of Crystal Growth. Gustav joined this institute in 1950, I followed in 1952, and we worked both with Prof. Stranski for a number of years and on different topics, first as students and then as assistants. Among other subjects we learned a lot about crystal growth during this period.

Figure 1 Iwan N. Stranski
(1897-1979)

After having spent the larger part of our youth in Hitler's Germany with its restrictions and uniform, aberrant ideas, Prof. Stranski opened new horizons to us. He was an internationally experienced, open-minded, critical teacher as well as an intellectually stimulating, quick-witted speaker, and in his seminars and as his assistants we got to know many well-known scientists from around the world. He had no

2

illusions about the world and the behaviour of human beings and could be quite cynical, but nevertheless he taught us above all to be generous, whatever the question might be. Over the years he became a fatherly friend whom we owe thanks for very much.

Here a short digression may be allowed: our student friendship had by and by developed into a closer relation, we married in 1957 (Fig. 2), had two daughters in the early sixties, and this average family of four has by now become a rather large clan with two sons in law and the astonishing number of seven grandchildren. – Although the photo may look like it, our life consisted not entirely of science or cryogenics!

In 1951, around the time when we joined the Fritz-Haber-Institute, Max von Laue (Fig. 3), world-famous theoretical physicist and Nobel Prize Winner of 1914 for the detection of X-ray diffraction in crystal lattices, became its Director. Under Hitler he had stood up for Albert Einstein and what was then called "Jewish Physics" and had therefore been forced to retire early from his professorship in Berlin. After the war he actively engaged himself in the rebuilding of German scientific institutions and, in spite of his age of 72, agreed immediately when he was asked to overtake the burden of heading our institute. He had worked with Max Planck in Berlin for several years before 1910 and had been one of the group of seven Nobel Prize winners in Berlin who

Figure 2 The authors in 1957

Figure 3 Max von Laue
(1879 – 1960)

in the 1920s met weekly in a famous seminar for discussing the latest developments in physics and chemistry. For us, he and the numerous famous physicists he knew well and invited to the institute somehow bridged the gap between our time and the glorious heyday of science in Berlin at the beginning of our century before Hitler came into power.

Professor von Laue became another fatherly mentor, and he gave our life the deciding turn towards cryogenics. Although he was a theoretical physicist, he was fascinated by modern technology in all its varieties. In the 1930es he had seen the attempts of the Kaiser-Wilhelm-Gesellschaft to set up a Low Temperature Laboratory in Berlin, which failed due to the beginning of the war. When in the mid-1950es the first Helium liquefiers came into the market, he was immediately taken by the idea to realise this earlier plan by installing such a plant at our institute, where many researchers were interested in Helium temperatures for their projects.

Before the mid-1950s production of low temperatures and research in this temperature range was the sole domain of only a handful of specialised institutes around the world, like the well-known Kamerlingh-Onnes Laboratory at Leiden, The Netherlands, or the Clarendon Laboratory at Oxford, England. At the average university, low temperature physics was not a subject of teaching, and 'cryogenic engineering' was yet unknown. As a consequence specialists were scarcely available when cryogenics began to emerge as a field of engineering

In January 1957 Prof. von Laue therefore asked Gustav if he would be willing to set up a low temperature laboratory. He knew that before becoming a physicist Gustav had been a navy engineer during the war and was trained in the operation of complicated machinery. A 4-l-Meissner Helium liquefier, the very first one manufactured by the Linde Company, was installed (Fig. 4), and the Fritz-Haber-Institute became thus the first non-low-temperature institute in Germany with low temperature facilities. Well aware of this unique position,

Figure 4 The first 4-l-Linde liquefier at the Low Temperature Laboratory, Berlin, in 1957

Max von Laue generously invited researchers from other institutes in Berlin or elsewhere to come as guests and use these facilities for their experiments.

The tasks of the new laboratory were defined as shown in Table 1. These tasks remained valid throughout the next 35 years. Max von Laue took a lively interest in the work going on, but he did not give any directions – the head of this laboratory was free in his decisions regarding handling of the plant as well as the work of his group. Fortunately, even this condition was never questioned in later years.

At the beginning we could not imagine what it meant to enter a new, developing field of research and development with this amount of freedom in one's decisions. The laboratory was well equipped with technical and academic staff, and even funding was never really a problem because fortunately we had always a sufficient number of research and development projects going on.

Although I continued to work with Prof. Stranski for a couple of years until his retirement, I slowly 'diffused' into cryogenics by reading and starting a literature card file – such helpful tools as computers were not yet available! Later I was lucky enough to get a position in the Low Temperature Laboratory, and from then on we worked together until our retirement. Over the years, at work we became a kind of 'Cooper pair' – not directly linked but efficiently interacting over a certain distance, which in the average was less than 50 meters. Gustav has often been asked by colleagues, especially from the industry, how he could stand it to have his wife around even at work! Well, they obviously didn't know much about Cooper pairs! For us this close co-operation was fulfilling and rewarding, and we never got bored of each other due to a lack of topics of mutual interest.

Being an engineer, Gustav attempted to run the Helium plant under the engineering aspects valid for any machine: as continuously as possible and up to its limits. In the beginning this was a rather unusual attitude in the case of Helium liquefaction. Maintenance and repair time of Helium liquefiers still exceeded or was at best equal to the running time, and for many years most liquefiers in German institutes were operated only occasionally. Under these circumstances our total output of meagre 600 l in the first year, 1958, was considered a record. However, conditions improved, by and by bigger plants were installed (Fig. 5) (Fig. 6), and thus the liquefaction capacity was stepwise increased in accord with the growing demand (Fig. 7). 40 years after its start the laboratory supplied about 200.000 l liquid Helium per year to users in Berlin. Throughout this long time, we could always deliver any required amount of liquid exactly in time.

When more and more low temperature

Table 1 Tasks of the Low Temperature Laboratory, Berlin

- Liquefaction of Helium and Nitrogen
- Development of cooling methods and cryogenic equipment for any required application
- Know how transfer to the users
- Low temperature research

Figure 5 The second Helium plant at the Low Temperature Laboratory Berlin: capacity 10 l/h, CTi, USA

Figure 6 The actual Helium Plants at the Low Temperature Laboratory Berlin: Capacity 2 x 50 l/h; CTi, USA

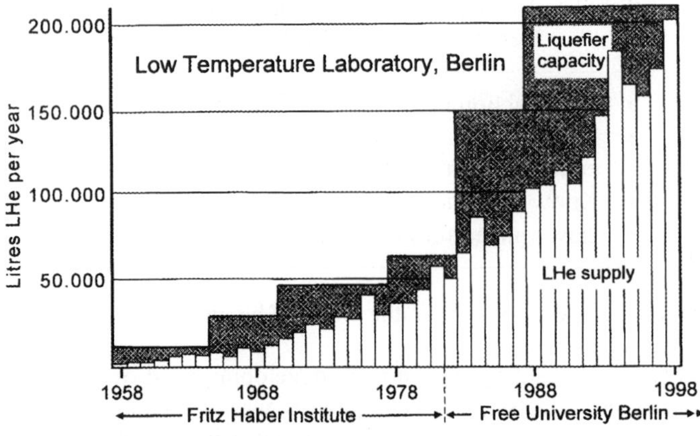

Figure 7 LHe production capacity and amount of liquid supplied

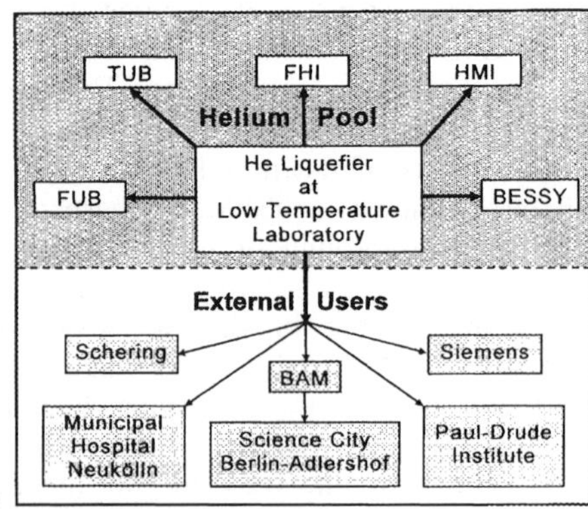

Figure 8 LHe supply to users in Berlin by the central Low Temperature Laboratory

experiments were set up in different institutes in Berlin, a number of them decided to economize Helium supply by setting up a non-profit "Helium Pool" (Fig. 8). The Low Temperature Laboratory acted as central facility which delivered liquid Helium to the members at cost price and provided know how regarding the economic use of the coolant as well as the development of cooling methods or design of cryostats for specific applications It also served a number of "external" users, institutes and also industrial laboratories, who did not belong to the Pool. In 1982 the expanding laboratory moved with plants, experiments and staff into the newly built Department of Physics of the Free University Berlin. Over the years this university had become one of main users of liquid Helium and could now provide ample space for the laboratory in its new building.

Regarding research and development at the Low Temperature Laboratory I will just give a few examples (Table 2). We began with the focus on *cryopumping* – which is nothing but crystal growth on cold surfaces, something for which we were well prepared from our work with Prof. Stranski. In the 1950es the development of efficient cryopumps was still in its infancy. Many basic questions regarding growth and evaporation of the condensed gas layers needed answers. Another main subject was the development of *continuous cooling methods* with accurate adjustability of any desired temperature in the range from 1.5 to 300 K. The invention of the continuous flow cryostat was the first step, and this problem never lost actuality throughout the years. In this context we also took up research on *Helium II properties and its flow behaviour* in narrow channels, with the aim of

Table 2 Main research and development areas of the Low Temperature Laboratory, Berlin

- Cryopumping
- Continuous cooling methods for temperatures T = 1.5 – 300 K
- Heluim II properties and its flow behaviour in narrow channels
- Space cryogenics
- MRI cooling

developing reliable automatic He II refilling devices. This came in useful when in the early 1970es we became involved with *space cryogenics*. He II cooling of infrared space telescopes and the development of qualified cryocomponents for space application became subjects which occupied us for about 20 years. Finally, I should mention *Magnet Resonance Imaging*. When in 1983 the first MRI system was installed at a university clinic in Berlin, we were asked for support on the cryogenic side. The early MRI systems needed refilling of about 500 l LHe every 4 weeks. We did not only deliver the liquid Helium and refill the MRI cryostat, but soon saw possibilities for improving the refilling process. By stepwise changes in the refilling and control system we could achieve a reduction of the refilling time from 7 hours to only 1 hour – a fact which was much appreciated by the medical doctors in view of the high treatment costs: at that time about 1700 DM per hour.

Gustav's election as Member of the International Cryogenic Engineering Committee in 1969 as well as our engagement in the scientific and technical cooperation between Germany and India were essential steps towards activities on the international level – something which Gustav will talk about in more detail. Our world widened immensely. We worked in joint projects with colleagues in the USA, India, Japan, China and the USSR, spent time in these countries and had many guests from abroad in our laboratory as well as in our home. This gave us new insights, brought memorable experiences and gave us the chance to make friends around the world – something for which we are deeply grateful.

Our relation to cryogenics can best be summarised in a short Limerick verse:

> Although we are now grey and old,
> we are still in love with the cold.
> It was always demanding,
> surprising, rewarding,
> and a challenge which still does hold.

At first let me once more express our thanks for being honoured by the Mendelssohn Award. We were lucky in many aspects to get involved with cryogenics at an early stage and are grateful for the many chances we had by working in this field. As Ingrid already has indicated, being a Member of the International Cryogenic Engineering Committee certainly belongs to them. I will give here a short summary of some experiences related to the Committee's work.

Up to the 1950s low temperature physicists could not foresee that temperatures $T < 77$ K might be used for large-scale technical application already in the near future. The change came with the growing (military) interest in powerful long-range rockets, which needed large amounts of liquid Oxygen and liquid Hydrogen for fuel. Cryogenic engineering began to emerge as a recognised field, especially in the Soviet Union and in the USA. The beginning of the space age, which is marked by the start of the very first satellite "Sputnik" by the Soviet Union in October 1957, gave a further strong impetus to its development.

The US National Bureau of Standards' Cryogenic Laboratory – a government funded central laboratory – was founded at that time and officially opened in 1954. In conjunction with this event a successful Cryogenic Engineering Conference (CEC) was held, which turned out to be the beginning of the long and successful series of CECs. Although international attendance at these conferences increased over the years, they did not really cover the needs of the developing cryogenic communities in Europe and Asia. This problem was early recognised by Kurt Mendelssohn (Fig. 9), who headed a research group at the Clarendon Laboratory in Oxford, England. He was born and grew up in Berlin and studied physics at Berlin's Humboldt University in the 1920s. Here and later at the University of Breslau he worked in the group of Franz Simon (later: Sir Francis Simon), where in the context of research on Nernst's third law of

Figure 9 Kurt Mendelssohn
(1906 – 1980)

thermodynamics he became involved in low temperature production and measurements. In 1933 Franz Simon, Nicholas Kurti and also Kurt Mendelssohn emigrated to England, where F.A. Lindemann (later: Lord Cherwell), former student of Walter Nernst in Berlin and since then profoundly interested in low temperature research, found places for them at the Clarendon Laboratory, which in turn became one of the leading laboratories in this field.

Kurt Mendelssohn discussed the imbalance of information on cryogenic technology in Europe and Asia compared to the USA foremost with Keichi Oshima (Fig. 10), Professor of Technical Physics

Figure 10 Keichi Oshima
(1921 – 1988)

and Director of the Low Temperature Laboratory at the University of Tokyo. The latter served his country in many important positions at national as well as at international level (OECD) and was a strong promoter of new technologies in Japan. Often he expressed his opinion that it would be one of the essential requirements for a positive development of world politics to overcome the poverty in Asia. And new technologies he considered to be a suitable means for reaching this goal. Being aware that conferences focussing on the aims and possibilities of upcoming new technologies are essential for activating support for these fields, he shared Kurt Mendelssohn's view that conferences similar to CEC would be required in Europe and Asia. Correspondingly, he proposed to organise an International Cryogenic Engineering Conference at Kyoto, Japan, in 1967 as a first try. Its success led them to think about an International Cryogenic Engineering Committee which could act as organiser of a new series of international conferences to be held more or less alternately in Europe and Asia. Kurt Mendelssohn became active and soon convinced a number of cryogenic specialists of this idea. Already in March 1969 the International Cryogenic Engineering Committee was set up at a Meeting in Paris (Table 3). As legal structure, a *"Verein" (society) according to Swiss law, domiciled in Zürich and without commercial interests* was chosen for making the Committee independent of governmental or administrative influence and ensure the freedom to handle its affairs in the interest of cryogenic engineering. One of its main aims was to promote the cooperation between research institutes and industry in cryogenics – and the Members were chosen correspondingly.

Table 3   International Cryogenic Engineering Committee

FOUNDED:	18. März 1969 by *Kurt Mendelssohn*, FRS, Clarendon Laboratory, Oxford, and *Keichi Oshima*, Professor, Technichal Physcis, Tokio University
STATUTES:	"Verein" (Society)according to the Swiss Civil Code, domiciled in Zürich, Switzerland, without commercial interests and independent of government institutions
MEMBERS:	maximal 20 specialists elected by the Committee
OBJECT:	To promote the development of cryogenics, above all by holding *International Conferences* at varying places worldwide

The CEC organisers were not outright pleased with this new, internationally aimed competitor in the field of cryogenic engineering conferences. However, ICEC expressed the wish for co-operation and elected two American colleagues as Committee Members. After two years of partly controversial discussion about the most sensible number and schedule of conferences in the field, CEC finally agreed to change from the so far annual to a biennial sequence of their conferences and alternation with the ICEC events. From then on the co-operation was peaceful and satisfying. Members of both committees attended the mutual conferences, and the CEC conferences became a convenient opportunity for the ICE Committee Meetings between ICECs. It is regrettable that the actual situation in world politics led to certain restrictions regarding scientific exchange which in turn have negative effects on the relations between scientists in the USA and in Europe and Asia. One can but hope for fast improvement of the general conditions.

The concept of the International Cryogenic Engineering Conferences proved to be a good one – after all we have now ICEC 20! For about twenty years the conferences (Table 4) were evenly distributed between Japan and European countries, with a ratio of 1:3. Keichi Oshima's strong engagement should be emphasized here: he organised and chaired three of the first nine conferences, two in Kyoto and one in Kobe, all of them most successful. Meanwhile China and India have become stronger players in the cryogenic community, and other countries are getting nearer to the breakthrough stage. Asia's weight is increasing, and the distribution of conferences between Europe and Asia is changing correspondingly. A look into the Journal CRYOGENICS confirms the healthy growth of cryogenics in Asia: in the first four issues of this year two thirds of the contributions came from there.

Each of the ICEC conferences had its specific highlights and boundary conditions. I will mention here only three examples – ICEC 3 in Berlin, ICEC 13 in Beijing and ICEC 18 in Mumbai – because they mark basic concepts of the ICE Committee's work.

The second ICEC conference in 1968 was rather provisionally organised in Brighton, England, during the preparation phase for the International Committee. When the Committee had been formally

Table 4 International Cryogenic Engineering Conferences

		Year	Place	Local Chairman
ICEC 1	–	1967	Kyoto, JAPAN	K. Oshima
ICEC 2	–	1968	Brighton, ENGLAND	A.A. Smailes
• ICEC 3	–	1970	Berlin, GERMANY	G. Klipping
ICEC 4	–	1972	Eindhoven, NETHERLANDS	J.W.L. Köhler
ICEC 5	–	1974	Kyoto, JAPAN	K. Oshima
ICEC 6	–	1976	Grenoble, FRANCE	A. Lacaze
ICEC 7	–	1978	London, ENGLAND	J.B. Gardner
ICEC 8	–	1980	Genova, ITALY	C. Rizzuto
ICEC 9	–	1982	Kobe, JAPAN	K. Oshima
ICEC 10	–	1984	Otaniemi, FINLAND	O.V. Lounasmaa
ICEC 11	–	1986	Berlin, GERMANY	G. Klipping
ICEC 12	–	1988	Southampton, ENGLAND	R.G. Scurlock
• ICEC 13	–	1990	Beijing, CHINA	C.S. Hong
ICEC 14	–	1992	Kiev, UKRAINE	K.A. Yushchenko
ICEC 15	–	1994	Genova, ITALY	C. Rizzuto
ICEC 16	–	1996	Kitakyushu, JAPAN	K. Yamafuji
ICEC 17	–	1998	Bournemouth, ENGLAND	R.G. Scurlock
• ICEC 18	–	2000	Mumbai, INDIA	K.G. Narayankhedkar
ICEC 19	–	2002	Grenoble, FRANCE	G. Gistau-Baguer
ICEC 20	–	2004	Beijing, CHINA	L. Zhang

founded at the Paris Meeting in 1969, one was looking for a suitable place for ICEC 3 in another European country, but none of the Members really volunteered. Rather spontaneously I proposed Berlin as a good place for ICEC 3 and my readiness to organize it. I had assisted Prof. Stranski 1952 with the organisation of the first international symposium in Berlin after the war and had attended two CECs in

Figure 11 Committee Meeting at Prague in September 1969 (1: G. Klipping, 2: K. Mendelssohn, 3: S. Safrata, 4: J.W.L. Köhler, 5: Mrs. Safrata, 6: J. Olsen, 7: I. Klipping, 8: J.B. Gardner, and Czech colleagues

America, with a certain interest for the organisational aspects. As a born optimist I considered this to be sufficient experience for organising an international conference. However, my suggestion was met with some reservation, especially on Kurt Mendelssohn's side. German activities in international bodies were not yet readily accepted at that time. Moreover, Berlin seemed to be a rather risky place due to its abnormal political situation as a wall-surrounded island in the middle of USSR-inclined East-Germany. And the recent student revolts clouded the picture additionally. However, Stan Safrata, the Czechoslovakian Committee Member, argued that politics and science are different shoes. In science one should not allow the existence of non-accessible "white spots" on the world map, which he emphasized by inviting the Committee for its next Meeting six months later to Prague, just because of the politically uncertain situation in Czechoslovakia at that time, only half a year after the forced end of the Dubček era. His arguments convinced the majority: not only the invitation was accepted (Fig. 11) but also Berlin as place for holding ICEC 3 in May 1970.

8

Experience from the previous two conferences was scarcely available or applicable, and professional conference organisers were still unknown. Locally we had to rely on our own and our collaborators' and students' potential and imagination for the organisational tasks, and all of them did their job with much enthusiasm. The scientific concept was developed in close co-operation with the Members of the ICE Committee. Regarding the practical concept of the event we asked ourselves what we would expect from a well organised and enjoyable conference and did our best to realise this picture. Fortunately, the drawbacks of Berlin's insular situation were politically compensated by ample resources of the city government for attracting international events as a contribution to normality. Thus we obtained considerable financial support for ICEC 3, including an invitation of all 400 participants to a high-class opera performance, Orff's *Carmina Burana*, which became a highlight of these days. This helped a lot to make the conference attractive and a success. At the end we were richer by a host of experiences, and Kurt Mendelssohn, Keichi Oshima and the other Committee Members left Berlin as friends.

In 1980 Chao-sheng Hong (Fig. 12) , Professor and Head of the Academia Sinica's Cryogenic Laboratory in Beijing, the Senior of cryogenics in China, had been elected as Committee Member. Cryogenics in China developed rapidly during the following years, and his proposal to hold ICEC 13 in Beijing in 1990 had been gladly accepted. Quite unexpectedly, some aspects of this conference came to resemble those given in Berlin 20 years before. While the conference preparations were in full swing, the Tiananmen incident occurred, and suddenly economic and scientific interaction with China became a political issue, with negative tendencies in many countries. Their Ministries of Foreign Affairs partly went as far as suggesting to call off all scientific exchange with China. Understandably enough, even in our Committee doubts were voiced if Beijing could and should be kept as conference site. However, we had not forgotten the deciding arguments in the case of ICEC 3 and remained true to our position not to allow politics to interfere with scientific exchange. Beijing remained the place of choice, and ICEC 13 took place at the Fragrant Hill Hotel in April 1990 as planned. It was a well organised, successful and enjoyable event in a beautiful surrounding and a wonderful experience of Chinese hospitality and culture. Even higher powers supported it by surprising the foreign delegates with the spectacle of a sand storm.

During the conference Chinese students and younger scientists were eager to discuss with the foreign attendants living and working conditions in western countries, and many of them expressed their interest in working abroad for some time. We had visited China twice before, and it was a pleasure for us that over the years a number of successful co-operations between the Low Temperature Laboratory in Berlin, the Academia Sinica and Chinese universities materialized. Frequently we had guest scientists from China in our laboratory and several Chinese students worked for their doctoral theses in Berlin.

Figure 12 Chao-sheng Hong

The last of the three conferences to be mentioned is ICEC 18, which was held in Mumbai in February 2000 and had a quite different background. Cryogenics in India, a developing country, had a past history of about 3 decades, which was essentially influenced by a series of long-term Agreements between India and Germany on co-operation in science and technology. Germany became partner country of the Indian Institute of Technology, Madras, one of five high-standard technical universities founded in India in the early 1950es for filling the gap in higher technical education in the country. For many years the development of the IIT Madras was supported with large funds for equipment, personnel exchange and joint research projects. In 1969 it was decided to install a low temperature laboratory at the Department of Physics. I was asked to assist with planning and later also with practical advice and lectures at Madras. This turned out to be the beginning of 30 years of close co-operation with regular mutual visits, training of personnel in Berlin, and joint projects. With growing success the co-operation was extended to other German universities and the Max-Planck-Institute for Solid State Physics.

My counterpart at Madras was Prof. R. Srinivasan (Fig. 13), a theoretical physicist with technical leanings, an outstanding talent for organisation and the gift of conveying enthusiasm to everybody working with him. He made not only the low temperature laboratory at Madras flourish, but also became the motor of the development of cryogenics in India. In due course further cryogenic laboratories came

Figure 13   R. Srinivasan

into being. Among these the Centre for Cryogenic Technology at the Indian Institute of Science, Bangalore, headed by Prof. S. Jacob, and the School of Cryogenics at the Indian Institute of Technology, Mumbai, headed by Prof. K. G. Narayankhedkar, developed particularly well. Another interesting laboratory to be mentioned is the Inter-University Consortium for DAE Facilities IUC-DAEF at Indore, which was founded in the 1980s and was headed during its building phase from nearly zero to a fully equipped, efficiently working low temperature institute by Prof. R. Srinivasan after his retirement from the IIT Madras. As the name indicates, the IUC is a facility open to researchers from all universities in India which otherwise do not have access to low temperatures and the corresponding experimental equipment.

However, in spite of all progress a back-lag in interaction and exchange at international level remained a serious problem for the Indian cryogenic community. For improving the basis for interaction the ICEC Committee elected Prof. Srinivasan in 1985 as Member. When he retired, Prof. Narayankhedkar succeeded him in 1996, and eventually it was decided to hold ICEC 18 in Mumbai at the campus of the IIT Bombay. The conference certainly was a success. Exemplary organisation and the beautiful, tropical surroundings already made it special. In addition, it was highlighted by an impressive social program which showed the participants a variety of facets of the Indian culture. For us this event was a kind of culmination of our long co-operation with Indian colleagues, and we do hope that it has far-reaching effects with regard to the recognition and development of cryogenics inside the country as well as to the improvement of interaction between India and the world-wide cryogenic community. At the present conference a number of Indian colleagues are participating which can be taken as a good sign.

Ingrid and I are both grateful for the chances which my membership in the Committee gave us for actively contributing to the development of cryogenics and international relations within the cryogenic community. Getting to know other cultures and mentalities by working together in scientific or technical projects or in the preparation of conferences is most stimulating and rewarding. It deepens the understanding of each other and friendships develop – far beyond the mere sphere of work.

If one is working in the one and the same field over a long time, it is quite usual that profession and hobby tend to melt into each other. When a couple is working together and sharing not only their private but also their professional interests, this probably holds even more: we two certainly have gone "cryogenic" – and we are quite happy with it!

Cryogenics in Russia: Development of Basic Cryogenic Cycles

V.M. Brodianski*, A.Alexeev**

*Moscow Power Engineering Institute, Moscow, Russia, kryobrod@kryos.mpei.ac.ru
** Messer Cryotherm GmbH, Euteneuen 4, 57548 Kirchen (Sieg) Germany, Alexander.Alexeev@nexgo.de

Three important and very successful ideas were born in the former Soviet Union, moved out and introduced worlwide: (i) The Low Pressure Expander (turbine) and low pressure cycles for air separation, developed and introduced by P.L. Kapitza, (ii) The One Flow Cascade with Multicomponent Refrigerant for liquefaction and treatment of natural gas, developed and introduced by A.P. Klimenko (Klimenko Cycle), and (iii) "Mixed Gas Refrigeration", a refrigeration concept developed since the 1960ies at the Moscow Power Engineering Institute. The history and thermodynamics of these cycles will be discussed.

KAPITZA CYCLE

It is well known that the air liquefaction system developed by Carl Linde in 1895 was the first commercially successful cryogenic system. Most of the modern industrial cryogenics was later developed based on inventions of Carl Linde.

The works of Linde were highly appreciated by D. Mendeleew and N. Umov in Russia. Moscow University purchased the air liquefier made by Linde in 1898, thus the first cryogenic laboratory for education purposes was organized in Russia.

Though the process developed by Linde (Figure 1a) was relatively simple and reliable, it had some disadvantages, the major one of which was poor efficiency and thus high power consumption. Today we know, that the thermodynamic efficiency of the cold part of the Linde´s system η_e was lower than 10 % due to the huge temperature difference between warm and cold streams in the heat exchanger.

Of course, C. Linde, as well as other engineers and inventors, tried to improve the efficiency of the process (Figure 1b). For example, C. Linde introduced an additional precooling system based on a classic ammonia refrigerator. That way he managed to reduce a temperature difference between the warm and cold streams in heat exchanger, consequently reducing the temperature of the stream before throttling valve, and increasing efficiency of the process.

George Claude (Figure 1c) developed another solution in 1902; he introduced in to the cycle an additional cryogenic piston-expander. This idea can be interpreted as introducing a precooling circuit, which allows reducing the temperature difference in heat exchangers. The other advantage of the Claude's process is that the energy of the high-pressure air was better utilized in the cycle, due to the fact that the expansion in the engine is more efficient compare to expansion through a throttle valve. Utilizing his cycle G. Claude received almost the same result as C. Linde, but at a lower operation pressure (~60 bar compared to 200—220 bar in the Linde´s system). That was a great innovation.

Peter Kapitza developed further the idea of G. Claude; he developed a highly efficient turbine-expander (isentropic efficiency higher than 80 %) and reduced the operation pressure further to 6—8 bar. The temperature difference in the heat exchanger was reduced, the expansion losses in the turbine were small, so the cold part of this process was practically loss-free compare to the original Linde´s system; the efficiency of the cycle was increased essentially.

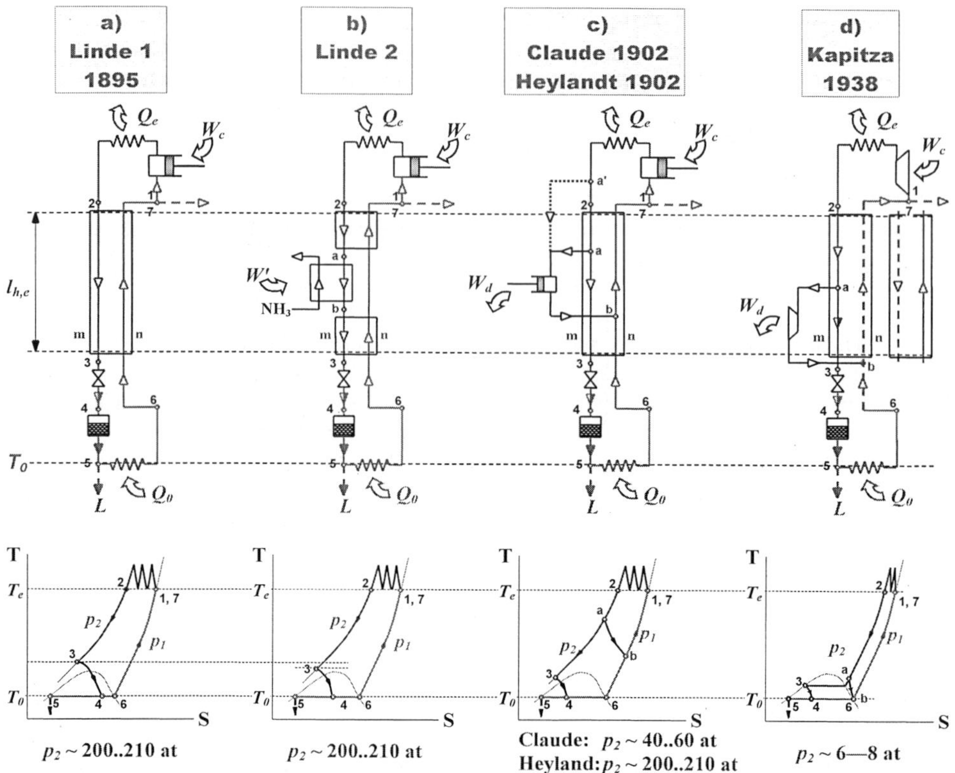

Figure 1 Basic Cryogenic cycles for air liquefaction

The idea to reduce operating pressures from customary 200-220 bar to much lower 6-8 bar was so new, that the Peter Kapitza had some problems with cryogenic community accepting his idea, especially in Germany, where the leading cryogenic engineer Helmuth Hausen (Linde, Höllriegelskreuth b. München) was very cautious about Kapitza´s invention. In a publication [11] Mr. Hausen wrote "the air liquefaction process of Kapitza does not bring any advantages conc. power consumption compared to the conventional high pressure processes even in spite of low operation pressure, probably [it does not bring any advantages] conc. mass, required space and production costs…The highest value of the efficiency 83 %, which is given by Kapitza, cannot be taken as proper value for the cooling capacity[1]".

Incidentally H.Hausen conceded that the low pressure process could be valuable for air separation systems, if the predicted efficiency of the Kapitza turbine would be confirmed. A similar position was voiced by P.Grassmann [10]. But the caution (discretion) of H. Hausen was interpreted by many other specialists as a kind of pessimism acc. Kapitza cycle. Finally it decelerated introducing of the process proposed by Kapitza in the air separation industry though could not stop it.

In this connection it is useful to remember the discussion between Kapitza on one side and several Russian professors on the other side. The position of Kapitza´s opponents was based on two following claims:
1. It is impossible to have an expander with adiabatic efficiency $\eta_{ad} > 0.83$ (the best turbine-expander of Linde-company had $\eta_{ad} = 0.6..0.65$ at that time).
2. The cycle of low pressure air separation must inevitable have efficiency lower than the cycle of high pressure.

---

[1] original text „…das Luftverflüssigungsverfahren von Kapitza trotz der Anwendung sehr niedriger Drücke gegenüber den bekannten Hochdruckverfahren keine Ersparnis an Energieaufwand, vermutlich auch nicht an Gewicht, Platzbedarf und Herstellkosten zu bringen vermag…Der höchste Wert des Wirkungsgrades von 83%, den Kapitza für seine Turbine angibt, kann aber … nicht als maßgebend für die Kälteleistung angesehen werden".

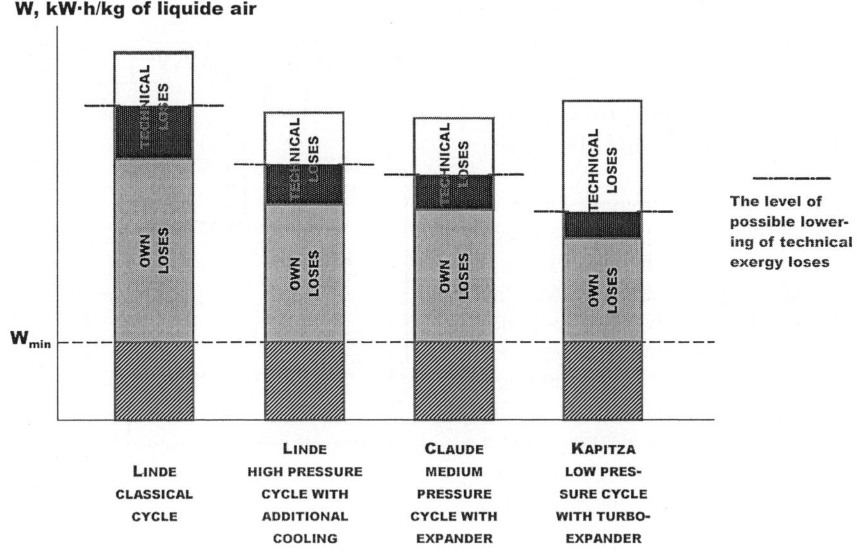

Figure 2   Distribution of losses in basic cryogenic cycles for air liquefaction

Both these statements were based on experience of that time, which later collapsed, proven by time. To understand it, it is necessary to analyze the problem with the help of thermodynamics. Today we know that all the thermodynamic losses can be divided into two groups — own losses Σd_0 and technical losses Σd_t (Figure 2). The own losses are inherent to given object and truly can't be eliminated. In opposite the technical losses can be reduced by proper operations in principle to „zero". P. Kapitza realized it in full scale. Introducing his radial inflow low temperature turbine-expander Kapitza eliminated most technical losses Σd_t by radical transformation of turbine-design, taking into consideration the high density of the cold compressed air even in the region of wet vapor).

The situation acc. the second claim was more complicated. The Kapitza's opponents were absolutely correct in evaluation of the Kapitza cycle-performance. But they did not take into consideration the possible significant reduction of technical lossed Σd_t in Kapitza's process by further development, especially concerning application of Kapitza-cycle for air separation. Of course, the own losses Σd_0 are increasing by lowering of working pressure by changing from Claude-process to Kapitza-process (the opponents of Kapitza were correct). But the technical losses d_t can be lowered still further at the same times by low pressure p_m. In the long run the total efficiency significantly grew, especially by large systems.

What we can learn from it. At the first, for analysis of given process it is necessary to divide the losses of exergy on d_t and d_0. The final decisions must be made based on the analysis of both Σd_t and Σd_0. In the case of Kapitza's cycle the own losses in turbine-expander are absent at all; in this system there are only technical losses of exergy, which can be lowered in principle to zero.

Moreover, we can learn from it, that it is sometimes necessary not to restrict the attention on the present situation, but to analyze other possibilities and future trends, which could be connected not with thermodynamic aspects only, but also with economical aspects too.

Kapitza's opponents were brilliant specialists, but they think on the level of preconceived ideas based on the better models of low temperature systems in that time. In opposite Kapitza was a physicist, he wasn't tied with cryogenics engineering traditions. He was a master of scientific original decisions, and used methods of scientific investigation. Sometimes it is of advantage to look on the problem from outside to achieve an essential improvement of an established technology.

Today the low-pressure air separation plants use different variations of the processes proposed by Peter Kapitza and run successfully all over the world.

14

KLIMENKO CYCLE

A.P.Klimenko worked in 1950's on the liquefaction of the natural gas. The main challenge was that the classic liquefaction processes known by 1950's (from C. Linde to P. Kapitza) were developed for pure substances[2], mainly for a gaseous cryogen (the liquid fraction develops, collected and withdrawn from the process just after the throttle valve) and could not maintain liquid neither in the heat exchangers, nor in the expanders. In contrast, the natural gas begins to liquefy at temperatures close to the ambient temperature, because the natural gas is a mixture, which consists of many components (most of them are hydrocarbons, like methane, ethane, propane, butane etc; other components are of non-organic nature like nitrogen, helium, carbon dioxide, H_2S, etc). Therefore it is very difficult to use the classical methods to liquefy natural gas.

The conventional method for liquefaction of natural gas used in 1950's employed a cascade system consisted of at least three separate refrigerators with different refrigerants. Propane

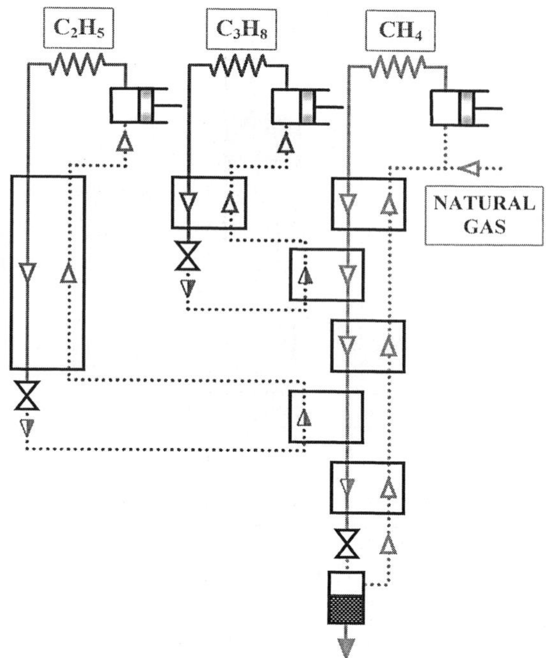

Figure 3   Classical cascade method for liquefaction of the natural gas (Keesom).

was conventionally used for the first stage with cooling temperatures of ca. 240 K, ethylene for the second stage with cooling temperatures of ca. 200 K, and methane for the third stage (Figure 3). The cascade system is a very efficient system. But it has some disadvantages, the main of them being the complexity and high number of hardware-components (for example at least three compressors).

A.P.Klimenko realised that it was necessary to develop new liquefaction methods, taking in consideration special properties of natural gas. His idea was very simple; he decided to take the refrigerants "required" for a cascade system (methane, ethane/ethylene and propane) from natural gas directly during the process of liquefaction. This way, Klimenko transformed the complex composition of the natural gas from disadvantage into advantage.

The simplified flowsheet of the Klimenko process is shown in Figure 4. The refrigeration required for the cooling of the natural gas is provided by a circulating mixed refrigerant stream containing components such as butane, propane, ethane, methane and nitrogen. The heat of compression is removed in the aftercooler with water. In the first stage the mixed refrigerant stream, as a result of being compressed and cooled, is partially condensed, and the two phases are separated in the first separator. The vapor proceed as separate stream to the next heat exchanger, where heat is transferred to the returning refrigerant stream. The vapor is thereby partially condensed and proceeds from the first heat exchanger to the second phase separator, where the phases again separated. The liquid phase from the first separator is expanded in the first throttle valve before being separated from the cycle. These steps – partial condensation, separation, subcooling, and expansion repeated in the second and the third stage of the separation process.

Consequently, some hardware components (for example several compressors) are not required any longer; it makes the system simpler and more reliable. The capital expenditure can be reduced over the conventional cascade system. Additionally, the Klimenko process is very versatile; it is possible to produce not only LNG, but other gas products, like gaseous or liquid ethane, propane, butane and other natural gas components.

---

[2] Although the air is a mixture of nitrogen, oxygen, argon and other gases, the thermodynamic of the air is very similar to the thermodynamic of an single component refrigerant like nitrogen or oxygen.

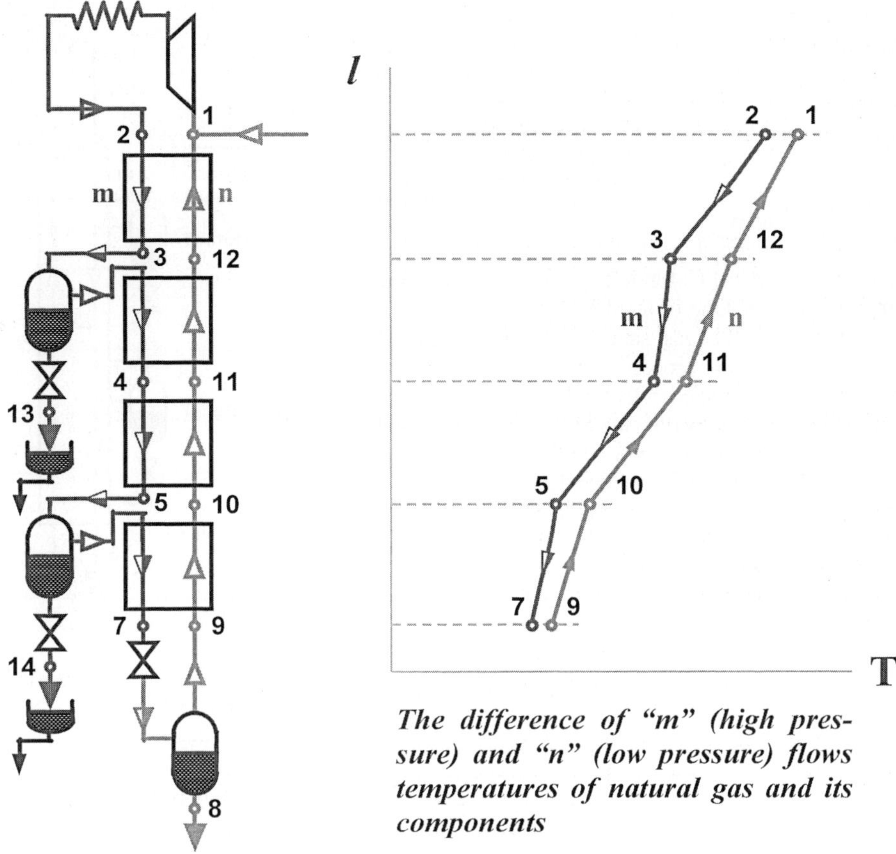

The difference of "m" (high pressure) and "n" (low pressure) flows temperatures of natural gas and its components

Figure 4   Flowsheet of the Kleemenko cycle

The thermodynamic analysis of the Klimenko cycle shown, that the temperature differences in the heat exchanger are very small, consequently the thermodynamic losses in the heat exchanger are small. Surprisingly, the losses in the throttle device are also very small. That is because of throttling of liquefied refrigerant.

One can say, that Klimenko follows the same principle like C. Linde, G. Claude and P. Kapitza. But (and that is absolutely new) the losses in the cycle were reduced neither by introducing an external refrigerator (like C. Linde), nor by introducing an expander (G. Claude, P. Kapitza). That was (using of mixed refrigerants) absolutely new in the low temperature refrigeration.

The Idea of Klimenko was developed further intensively in 60'ies and 70'ies. Now each industrial gas company develops systems for liquefaction of natural gas based on the mixed refrigerants. These processes are based on the Klimenko cycle.

MIXED REFRIGERANT CYCLE

In the early seventies, the author and Gresin worked on a refrigeration system for electronic component cooling with the goal of developing a high efficiency one stage cooling system without any additional precooling or machinery based on an a Klimenko idea to improve the classical refrigeration process developed by C. Linde by using of mixed refrigerants.

As a first step the Klimenko process was realized in small scale as a closed cycle (Figure 5) [1]. The system had an efficiency comparable to that of classical refrigeration cycles with expander. That system turned to be impractical due to the high number of hardware components, such as separators and throttle valves.

In a further development investigations showed that it was possible to reduce the number of separators in the system, and finally the separators were eliminated due to the use of a special refrigerant mixture. A mixed refrigerant Joule Thomson system for liquid nitrogen temperatures was invented. It combined the simplicity of the classical C. Linde process and high efficiency. The report about our mixed refrigerant system was presented to the International Congress of Refrigeration.

The mixed refrigerant Joule Thomson system looks very similar to the classical C. Linde´s system, but with some fundamental differences. One of them is that Linde´s process takes place mainly in the gas state, where as the mixed refrigerant Joule Thomson process happens mainly in the two-phase region – the mixed refrigerant condenses in the high pressure flow m at the warm side of heat exchanger and evaporates in the low pressure stream n at the cold side of the heat exchanger. Therefore the apparent heat capacity of a mixed refrigerant is a composition of the real heat capacity c_p and the heat of vaporization r:

at the high pressure side: $C_m = r_m + c_{p,m}$,

at the low pressure side: $C_n = r_n + c_{p,n}$.

Because the heat of vaporization r is higher than the real heat capacity c_p ($r >> c_p$) and the heat of vaporization r is only slightly pressure dependent ($r_n \approx r_m$), the heat capacity of the high pressure stream and the heat capacity of the low pressure streams are very close. As a result the temperature difference in the heat exchanger and the losses are very small (the mean temperature difference is less than 15 K for optimized mixtures). This is an essential feature of a mixed refrigerant Joule Thomson process.

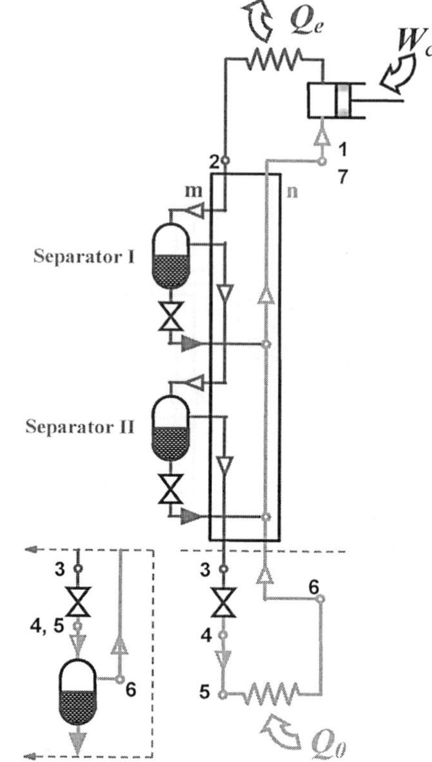

Figure 5 The first small-scale mixed refrigerant system

In the classical processes developed by Linde (JT with precooling), Claude and Kapitza the relative high losses in heat exchangers are compensated by additional precooling or machinery. In the mixed refrigerant process the losses in the heat exchanger are so small, that precooling and/or expanding is not necessary.

The next steps in optimization of the mixture composition were related to different requirements as refrigeration below 77 K, shortening cool down time, etc. The further development shows the other interesting effects:

- the splitting of some liquid mixtures (like nitrogen-hydrocarbons) into two separate unmixible liquids: low-boiling liquid (mainly nitrogen) and high-boiling liquid (mainly hydrocarbons). Because the low- boiling liquid (mainly nitrogen) boils at the constant temperature, it is possible to build the mixed refrigerant systems with very constant cooling temperature.

- The very short cool down time, essentially shorter than for the conventional nitrogen JT-systems, may be achieved.

- The cooling temperatures below 77 K (practically down to solid temperature of nitrogen - 63 K) can be achieved with a single stage mixed refrigerant JT-system, however the efficiency of the system is less in this case.

Based on the mixed refrigerant JT technics the refrigerators as well as liquefiers and gas cooling systems can be developed.

All the methods developed lead to the invention of a new class of small scale industrial systems based on a mixed refrigerant Joule Thomson system.

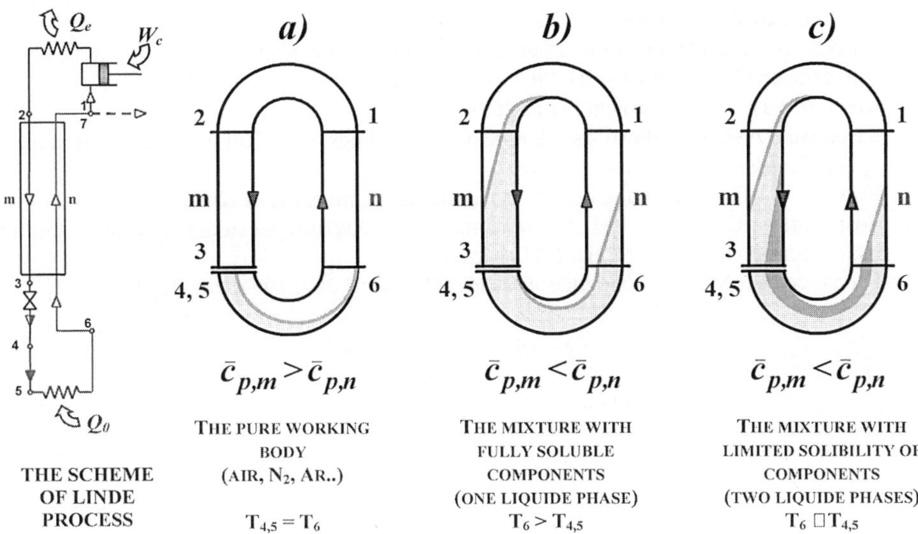

$$\bar{c}_{p,m} > \bar{c}_{p,n}$$

$$\bar{c}_{p,m} < \bar{c}_{p,n}$$

$$\bar{c}_{p,m} < \bar{c}_{p,n}$$

$$T_{4,5} = T_6$$

$$T_6 > T_{4,5}$$

$$T_6 \square T_{4,5}$$

Figure 6   Mixed refrigerant Joule Thomson system

CONCLUSION

Three important and very successful ideas were born in the former Soviet Union, moved out and introduced worlwide: (i) The Low Pressure Expander (turbine) and low pressure cycles for air separation, developed and introduced by P.L. Kapitza, (ii) The One Flow Cascade with Multicomponent Refrigerant for liquefaction and treatment of natural gas, developed and introduced by A.P. Klimenko (Klimenko Cycle), and (iii) "Mixed Gas Refrigeration", a refrigeration concept developed since the 1960ies at the Moscow Power Engineering Institute.

REFERENCES

1. Alfeev V. Brodyansky V. and others. "Refrigerant for a cryogenic throttling unit". Pat. specification №1336892, London 1971.
2. Боярский М.Ю. и др., "Автономные криорефрижераторы малой мощности", Энергоатомиздат, Москва.
3. Boiarski M.J., Brodianski V.M. Loungsworth R.C. Retrospective of mixed refrigerant technology and modern status of cryocoolers. Adv. of Cryog. Eng. Vol. 43 (1998).
4. Brodianski V., Boiarski M., Lunin A. The exergy analysis of the throttle refrigerating systems on pure and mixed refrigerants. Proc. of the ESDA, Pd vol 64–3 Eng Sys. Des Anl, Vol.3, ASME 1994.
5. Бродянский В.М., Семёнов "Термодинамические основы криогенной техники", "Энергия", 1980, Москва (In Russian, there are the Chinese translation).
6. Brodianski V.M., Sorin M.V. and P. Le Goff "The Efficiency of Industrial Processes: Exergy Analysis and optimization", "Elsevier" 1994.
7. Бродянский В.М. "От твёрдой вода до жидкого гелия. История холода", – М.: Энергоатомиздат, 1995.
8. Brodyansky V.M., Fratsher V., Mikhalek K. Exergy method and its application. – M: Energoatomizdat, 1988. (German variant – "Exergy", Leipzig, 1986)
9. Brodyansky V.M., Gresin A.K., Gromov E.M. and other. The use of mixtures as the working body in throttle (Joule–Thompson) Cryogen Refrigerators. Paper 1.105. XIII Int. Congress of Cold. May 12, 1971. V. 1, p. 43..45.

10. Grassmann P., Neuere Verfahren zur Gewinnung flüssiger Luft and flüssiges Sauerstoffs, "VDI Zeitschr.", Bd.85, N 23, 1941, p. 26..27.

11. Hausen H. Aussichten der Luftterflussigungsverfahrens von Kapitza, "Zeitsch. für die ges. Kälteindustrie", H. 2, 1941 p. 24..28.

12. Keesom W.H. Comm. Leiden, Suppl. №76, 1933.

13. Kleemenko A.P. "One flow Cascade Cycle". Proceedings of the Xth Int. Cong. Refr. Copenhagen 1, 34..39, 1959

14. Клименко А.П. Разделение природных углеводородных газов. Киев, "Техника", 1964

15. Боярский М.Ю., Лунин А.И., Могорычный В.И. "Характеристики криогенных систем при работе на смесях". Изд. МЭИ, 1990, Москва.

16. Бродянский В.М. "Кислородная эпопея". "Природа" №4, 1994, стр. 32..41.

Cryogenics in Russia

Alexeev, A. and Klipping, G.*

Messer Cryotherm, Euteneuen 4, D-57548 Kirchen (Sieg), Germany; Alexander.Alexeev@nexgo.de
(formerly: Moscow Power Engineering Institute (Technical University), Low Temperature Dept.)
*Limastr. 38, D-14163 Berlin, Germany; KLIPPING@aol.com

The actual state of Russian cryogenics is described, including academy and government institutes, universities as well as the industry. After the catastrophic break-down of funding following Perestroika and the ensuing brain drain, during the last few years things began to change for the better. The Russian cryogenic industry developed rapidly, above all the air separation industry. In the production of high-purity rare gases Russia belongs to the big players. Scientific institutes successfully use their excellent design and production capacity for by-order manufacturing of high-tech equipment for large research projects around the world. Moreover, new funding programs support their recovery

INTRODUCTION

The low temperature research and cryogenic engineering have a long tradition in Russia. Every physicist or engineer working in this field certainly knows the name of Peter (Pjotr) L. Kapitza, who built one of the first helium liquefiers with expansion engine in 1934, and who in 1978 received the Nobel Prize. He actively promoted the development of low temperature R&D in Russia by founding a number of research institutes and was during his later years head of the large Russian Institute in Moscow devoted to all aspects of low temperature R&D which now bears his name. Another large Low Temperature Institute in Kharkov, Ukraine, was since 1960 headed by Boris Verkin (and is now named after him), whose interests ranged from solid state physics to many aspects of low temperature application. For example, he initiated, founding of perhaps the first independent Institute of Cryobiology and Cryomedicine. Other Russian physicists well-known in low temperature physics are Lev Landau (Nobel Prize 1962), Vitaly Ginsburg (Nobel Prize 2003) and Alexey Abrikosov (Nobel Prize 2003), who all essentially contributed to the theoretical understanding of superfluidity and superconductivity.

Although many reports appeared in the media during the last fifteen years and much information is accessible via the internet today, the knowledge of conditions in the Russian Federation (in the following shortened to Russia) is still limited. This holds especially for the organisation of research and development (R&D). Quite a number of Russian scientific institutes are famous and have international contacts, but nevertheless it seems to be still difficult for scientists abroad to understand the structures in the different fields of science and technology, and to find out who should be approached for making contacts, for getting information on national conferences or for distributing information on international conferences. Therefore, with cryogenics taken as example, the special organisation as well as the actual situation of R&D in Russia are described in this paper.

SCIENCE AND TECHNOLOGY IN THE FORMER SOVIET UNION

For understanding the actual situation it is necessary to have a look at the past. The structures of scientific research and technical development in the former Soviet Union (USSR) developed differently from those in most Western countries. In the West numerous university laboratories, research institutes of different size – independent as well as national or international – and industrial laboratories are all engaged in research on a varying scale, and many of them are striving for application of their results in technical developments. The different research groups have ample possibilities for exchange of ideas at conferences or in direct interaction, and co-operation of different parties in specific projects has always been quite common, is encouraged by the government and to a large part promoted by special funding programs.

Other than in the West, in the USSR the universities' task was in the first line education while R&D was concentrated in very large governmental institutes, often with a staff size of up to several thousand and a wide variety of research areas. They usually started at a small scale with a specific field of interest and then grew and diversified more and more. This process is well described in a book published in 1998 on occasion of the Ioffe Physico-Technical Institute's 80th anniversary [1]. It originated from the Physico-Technical Department of the State Institute of X-Rays and Radiology founded at St. Petersburg in 1918, became an independent institute devoted to fundamental and applied research with "a staff numbering in tens" in 1923, and then grew bigger and bigger taking up more areas of research, partly due to new fields of science opening up, but also to government demands regarding development of the country's industry and later military issues.

The development of the other governmental research institutes with activities in cryogenics like the Kapitza Institute in Moscow and the Verkin Institute in Kharkov, which were mentioned before, the Budker Institute of Nuclear Physics at Novosibirsk, the Ioffe Institute at St. Petersburg or the Kurchatov Institute in Moscow followed much the same pattern.

These large institutes were a kind of rather independent universes, where almost everything was made in-house, ranging from screws over laboratory equipment to the manufacture of complete products like satellites or Tokamaks and the likes. Consequently, in addition to their high-standard research departments all of them had large design departments and industry-size workshops or better production sites – a feature which clearly distinguishes them from comparable large research institutes in the West where production as far as possible is given to the industry. The Russian research institutes did high-level research, educated graduate students, developed and built any size of experimental or technical equipment, all in parallel, and partly also served the surrounding industry. The Verkin Institute supplied, for example, tons of cryogenic liquids to industrial users in Kharkov and its surrounding. And these institutes worked mostly independent of each other, although occasionally specific services, e.g. radiation treatment of materials, were mutually provided.

This specific organisational structure has various reasons:
• The USSR government put huge investments into its science and technology base, at a level perhaps unparalleled in other countries. That meant, however, also a large amount of control by the government and specific demands regarding support of industry and military.
• The lack of the "infrastructure" which is usual in Western countries: in a centrally steered economy with its tendency of seeing mainly the general needs and its wish for overall control, there is no room for the small- and medium-size highly specialised private companies which in the West provide special services or deliver special technical components required by research institutes or larger industries.
• A significant part of the former Soviet science base was engaged in defence-related R&D and received its funding from military sources. This resulted in the demand for strict secrecy: communication with other institutions or industry inside the country was kept at a minimum, scientists working on different projects in the same institute did not discuss their work with each other [2,3]. As quintessence of this process, which is typical for military-oriented states, so-called "Science Towns", secret closed towns, were built, which could not be found on any map (in a recent BBC report their number was given as about 60 with a total population of about 3 million [4]). This obsessive secrecy undermined the traditional control per review. Combined with guaranteed, non-competitive public funding and the lack of public accountability it had the unfortunate consequence that highest quality R&D

Table 1: Comparison of country and population size of the Russian Federation, the USA, China and Europe (rounded figures) [8]

	Russian Federation	USA	China	Europe
Size of country {Million km^2]	17	9.6	9.6	2.3
Population [Million]	144	290	1 287	361
Population density [Persons/km^2]	8.4	30	134	157

existed beside undisturbed activities at a much lower level.

• The special geographical conditions (long distances and low population density, see Table 1) made communication between research teams difficult. Realisation of complex projects with several participants was therefore time-consuming and expensive.

PERESTROYKA AND THE SITUATION AFTER 1989

The end of the Cold War was one of the most remarkable events in recent human history. In Russia it has been followed more or less immediately by a move away from the old communist centrally planned economy to a market economy. As with almost all rapid transitions, this process has been a painful one with a withdrawal of public funding from many sectors that were once relying and dependent on it, and it is and probably will remain for yet a rather long time difficult to get funding from other sources. The demand for defence-related R&D services also was reduced, and many of the Research Institutes concerned face severe economic hardship. A detailed description of the situation in the mid-90es is given in the 80 years anniversary report of the Ioffe Institute, St. Petersburg: miserable salaries which were not paid regularly, problems with the payment for electricity, heating and water, the equipment becoming out of date and needing repair [1]. The continuing financial stringencies in Russian government funding means that quite a number of research institutes may not survive.

Not only the research institutes but also the industry was hit by the change. The initial shock of Russia's move to creating a democratic market economy, with privatisation of the former Soviet big industrial companies, their sudden exposure to competition and, in addition, the breaking away of former markets in the Eastern block, led to a collapse of many factories.

This had the negative effect of increasing unemployment and the related ills of social and economic deprivation. On the other hand, it also had the positive effect of a visibly reduced tendency towards secrecy. Today the large research institutes and also universities provide their own websites with detailed information, Russian researchers are free to communicate with colleagues abroad and can discuss at conferences or with visitors without restrictions. Although secret closed Science Towns still exist, some of them – like Sarov (formerly Arsamas-16) Novouralsk, Snezhinsk, Angarsk, Zelenogorsk, Zheleznogorsk and others – have meanwhile appeared on ordinary maps and are open to visitors, though these still need appropriate authorisation. Sarov and Los Alamos, nuclear centers of Russia and the USA, are twin towns today, something which could only be dreamed of until recently.

<u>Brain drain</u>
In view of the difficult economic situation, the abolition of travel restrictions was followed by emigration at a rather large scale, either for good or for a limited period of time. Depending on the source, the figures differ slightly, but an estimated loss of about 2.5 Million people (including 400 000 Jews and 800 000 repatriated people of German origin) after Perestroyka should be rather correct. Emigration unavoidably means a brain drain, since many of those who decide to go are among the most talented and active ones. According to Russian Government statistics, before the break-up of the Soviet Union there were about 800 000 scientific researchers in Russia as a whole, while today their number is only 426 000, the estimated number of scientists who emigrated (those who do not expect to return) during the period 1991-2001 was 20 600, and approximately 80 000 to 100 000 left for temporary contract work during the same period [5]. The emigration of Russian scientists reached a maximum between 1995 and 1996 but has stabilised thereafter. The Director of the Moscow State Lomonosov University, Victor Sadovnichi, was quoted as commenting the brain drain as „one of the most serious Russian problems" [6].

The consequences for a specific institute are described in the Ioffe Institute's report of 1998. They had to give "leave" to more than 100 experienced and active researchers with academic degrees who went to Europe or USA for scientific work. "Earlier, just these scientists had played the most active role in the education of future scientific generations, implementation of the ideas, and laying foundations for new scientific ideas" [1]. A serious consequence of the loss of young talented scientists is the disproportionate number of rather old and lesser talented ones among those who stay. According to S. A. Tsyplyayev, representative of the President of the Russian Federation in St. Petersburg, half of all science workers in St. Petersburg are 40 years or older, and 50% of those holding the doctorate are already receiving pensions [7]. The National Science Foundation, a funding agency for R&D, stated that most of the Russian scientists best known abroad are reaching the end of their research careers, and that these are extraordinarily difficult times for young scientists to take their place [7]. A German daily newspaper gave the average age of the Members of the Russian Academy of Sciences as about 72 years and that of most teachers and researchers at universities and research institutes as near to retirement with an average of about 57 years [6].

In the long run, the brain drain also may have positive aspects, however. The scientists abroad, even those who plan to stay there, keep contacts to their former institutes in Russia and can thus help to improve international contacts and exchange. Some of them are working in joint projects with participation of both sides and the exchange of ideas and results. And those who return after some time abroad will come home with international experience and new impressions which will certainly be helpful for the future development of R&D in Russia. Similar experiences have been made in Germany after World War II and in China during the last 20 years.

As one answer to the brain drain Russia's efforts regarding education have been intensified. Traditionally the Russian education system is achievement-oriented: talented children attend better schools, and the best of them later study at the best universities. As a consequence the range of differentiation among schools and universities is rather wide, and Perestroika even intensified competition between them. The highly valued "Special Schools" of Soviet times, schools focusing on special fields, have developed into so-called Lyceums, Colleges and Gymnasiums, most of them with entrance examinations for keeping a high standard. Competition-based selection combined with specific and systematic training of the students depending on their individual talents makes education in Russia very efficient. The number of dropouts is rather low. An important feature of Russian education is the

Table 2: Government Research institutes with cryogenic activities in Russia and the Ukraine

Institute	Location Website	R+D and Production in Cryogenics
Russian Research Centre Kurchatov Institute	Moscow www.kiae.ru	• Cryogenics for Tocamacs • Superconducting magnets • Applied cryogenics
P.L.Kapitza Institute for Physical Problems	Moscow kapitza.ras.ru,	• Applied cryogenics
P.N.Lebedev Physical Institute Russian Academy of Sciences	Moscow www.lpi.msk.su	• Applied cryogenics
Bochvar Institute – All-Russian Scientific Research Institute of Inorganic Materials	Moscow www.bochvar.ru	• High temperature superconductors
Joint Institute for Nuclear Research	Dubna (Moscow area) www.jinr.ru	• Cryogenics for high energy accelerators • Superconducting magnets • Applied cryogenics
Institute for High Energy Physics	Protvino (Moscow area) www.ihep.su	• Cryogenics for high energy accelerators • Superconducting magnets • Applied cryogenics
State Research Center of Russian Federation – Troitsk Institute for Innovation and Fusion Research	Troitsk (Moscow area) www.triniti.troitsk.ru	• Cryogenics for fusion • Superconducting magnets • Applied cryogenics
Ioffe Institute	Saint-Petersburg www.ioffe.rssi.ru	• Applied cryogenics
The Budker Institute of Nuclear Physics	Novosibirsk www.inp.nsk.su	• Cryogenics for accelerators • Superconducting magnets • Applied cryogenics
B.Verkin Institute for Low Temperature Physics and Engineering	Kharkov, Ukraine www.ilt.kharkov.ua	• Space cryogenics • Applied cryogenics

cooperation between institutions: integrated projects of schools/universities and scientific centres/industrial companies, in which schoolchildren and students attend lectures of leading scientists and work actively in the institutes' laboratories, are aimed at keeping a high educational level [1].

RUSSIAN CRYOGENICS TODAY

The following overview reflects the momentary situation, and the picture may be different in a couple of years since changes in Russia's scientific and industrial landscape will probably continue. However, since R&D institutes and universities as well as industrial companies are meanwhile present in the Internet, new developments can more easily be followed in the future.

<u>Research institutes</u>
Several large research institutes are active in cryogenics (Table 2, see former page; the Verkin Institute, Ukraine, is listed here for its national importance in the former USSR). All of them have a wide range of interests in research, and thus a good basis for various applications in cryogenic R&D. Differences as well as overlapping of focus areas in cryogenics reflect the allocation of definite tasks during Soviet times.

These research institutes include huge well-equipped high-standard design departments, workshops and production facilities, which have a first-class scientific back-up by their research departments. They turned out to be a valuable asset after Perestroyka. Even during Soviet times, in spite of all difficulties regarding external contacts, many of these institutes had international connections not only with countries in the Eastern block but, depending on their area of work, also in Western Europe and the USA. These could now be intensified, and soon a number of them – for example the Kurchatov Institute and the Budker Institute – became contractors for large research installations at national and international institutions around the world. They deliver all kinds of equipment, much of it involving ultra-high vacuum and cryogenics, and are known and estimated for high quality, reliability, in-time delivery and, last but not least, economic prices. Institutes or institute departments which are taking part in this new line of business are better off than in old times, they can pay higher salaries and buy modern equipment, which in turn leads to less problems regarding brain drain.

This also demonstrates the potential for the development of new, highly specialised industries which can successfully compete in the world market. In the West many successful companies in high-tech areas started as offspring from academic institutions, and similarly new cryogenic companies may soon be founded in Russia.

<u>Universities</u>
The total number of universities and colleges in Russia is 1100, of which about 600 are state universities and state colleges. The number of technical universities is more than 100. The actual total number of students is about 4.1 Mio, which corresponds to about 280 students per 10 000 residents. Compared to 180 students/10 000 (1991) before Perestroyka this is a remarkable increase. It reflects the fact that the number of colleges increased essentially in this time – several hundreds of private universities and colleges were founded, which are now in hard competition with the state system of higher education.

Table 3 (see next page) lists Russian Universities with teaching as well as R&D in cryogenics. It can be seen that the interest in refrigeration and cryogenics is wide spread, and students can specialise in these areas at different schools and with different focus. The Odessa State Academy of Refrigeration (OSAR), Ukraine, which also is listed in the table, is even completely devoted to this field. After Perestroyka the demand for cryogenic engineers declined and with it also the number of students in this field. As a consequence, many of the of the former Departments of Cryogenics have meanwhile changed into Departments of Refrigeration and offer education not only in cryogenic engineering but also in conventional refrigeration, air conditioning and other related areas.

According to the annual rating of Russian state universities, five of those listed in Table 3 (see next page) entered the first ten in the year 2002 (exceptions: Moscow State University of Ecological Engineering, Odessa State Academy of Refrigeration and Saint-Petersburg State University of Refrigeration and Food Technologies).

Table 3: Universities with cryogenic activities in Russia and the Ukraine

University	Location Website	R+D and Teaching in Cryogenics
Moscow Power Engineering Institute – Technical University – (Russian: МЭИ)	Moscow www.mpei.ac.ru	• Cryogenic Engineering
State Baumann Technical University – (Russian: МВТУ)	Moscow www.bmstu.ru	• Cryogenic Engineering
Moscow State University of Ecological Engineering – (Russian: МГУИЭ бывш. МИХМ)	Moscow www.mguie.ru	• Cryogenic Engineering
Moscow Institute of Physics and Technology – Technical University – (Russian: МФТИ)	Moscow www.mipt.ru	• Applied cryogenics • Cryobiology • Cryoelectronics
Moscow State Lomonosov University – (Russian: МГУ)	Moscow www.msu.ru	• Applied cryogenics • Cryochemistry • Cryoelectronics
Moscow Engineering Physics Institute – State University – (Russian: МИФИ)	Moscow www.mephi.ru	• Applied cryogenics • Cryoelectronics
Saint-Petersburg State University of Refrigeration and Food Technologies – (Russian: Санкт-Петербургский государственный университет низкотемпературных и пищевых технологий)	Saint Petersburg www.sarft.spb.ru	• Cryogenic Engineering
Odessa Stase academy of refrigeration – OSAR – (Russian: Одесская академия холода)	Odessa, Ukraine www.osar.odessa.ua	• Cryogenic Engineering

Table 4: Cryogenic industry in Russia and the Ukraine

Company	Location Website (Language)	Products
Cryogenmash	Balashika/Moscow www.cryogenmash.ru (Russian, English)	**Large scale cryogenic equipment:** • Air separation plants (up to 70 000 m3/h) • LHe, LH2, LNG plants • Cryogenics for launch sites (Baikonur etc.) • Cryogenics for tocamac/accelerator/MHD etc. • Turbines, heat exchangers, vessels etc. • Engineering
Heliummash	Moscow www.geliymash.aha.ru (Russian, English)	**Large and medium scale cryogenic equipment:** • LHe, LH2 plants • Turbines, heat exchangers, vessels, etc • Space cryogenics • Engineering
Sybkryotechnika	Omsk www.sibcryo.com (Russian, English)	**Small scale cryogenics equipment:** • Cryocoolers for military and space applications • GM-coolers and cryovacuum pumps • Mobile air separation plants • Vessels, etc. • Engineering
Uralkryotechnika	Ekaterinburg www.cryotech.narod.ru (Russian)	**Small and medium scale cryogenic equipment:** • Vessels • Transfer lines etc.
Iceblick	Moscow www.iceblick.com (Russian, English)	**High purity rare gases:** • Neon • Krypton • Xenon • Helium
Kislorodmash	Odessa, Ukraine www.kislorodmash.com (Russian, English)	**Large and medium scale cryogenic equipment:** • Air separation plants • Heat exchangers, vessels, etc. • Engineering

<u>Industry</u>
With the exception of Iceblick (founded in 1990 "as a result of creative and business collaboration of companies from Russia, Ukraine and the USA" according to the website), all companies listed in Table 4 (see former page) existed long before Perestroyka (here again, an Ukrainian company, Kislorodmash, is included as an essential part of former USSR cryogenics). The different areas of cryogenic engineering were distributed among them, with the claims well staked out all over Eastern Europe. This distribution caused a lack of competition and had unavoidably some negative effects. But it had also advantages like stable order quantity, balanced workload, high turnover and extremely accelerated growth. With Perestroyka the situation changed almost overnight. With privatisation the companies faced quite suddenly the rules of a free market, and competition among each other and with Western companies as well as the search for new fields of activity became a must.

Cryogenmash is the largest of the Russian cryogenic companies. Its founding and development after World War II were actively promoted by P.L. Kapitza. At Soviet times Cryogenmash was among the biggest of its kind worldwide. In the USSR and Eastern Europe it had the monopoly for large cryogenic plants and built air separation plants as well as all kinds of large-scale cryogenic facilities for space launch sites like Baikonur (today: Kasachstan) or Plesetsk and for research projects like accelerators or fusion. The bad economic situation in the 1990es, especially in the metal-producing and chemical industry, resulted in an abrupt decrease of the demand for cryogenic, and the drastic cut in public funding for large research projects did the rest. Orders dropped alarmingly.

Thus Cryogenmash had no choice but to look for new markets outside Russia. In addition to a clever marketing policy this required above all technical improvements so that more modern and technically better solutions could be offered to the customers. These efforts led, for example, to the development of a new generation of air separation plants which can rather successfully compete with Western products. As a result new customers abroad were found and the total order value recovered. Cryogenmash delivered, for example, hydrogen liquefaction plants and the complete oxygen, hydrogen and nitrogen supply system for the Indian space launch site SHAR and the Sea Launch Site (Russia-Ukraine-USA-Norway). A 300 l/h helium liquefier was delivered to India, and in Europe Cryogenmash built a 1.5 km helium transfer line for LEP at CERN in Switzerland. Until 2003 the total turnover could be increased by the factor of 6 to about 1.5 billion Ruble, which means ca. 50 million $.

The development of the other large companies followed similar lines after Perestroyka. However, a new feature in the scenery of Russian cryogenics are a number of smaller, more specialised cryogenic companies which aim at specific demands and niche-applications like cryoequipment for medical purposes (dermatology, cosmetology, gynaecology, proctology), electronic components for high-temperature superconductor devices, special cryostats and accessories, high resolution temperature sensors and the necessary electronics, and equipment for freezing and long-term preservation of bio-products. Their success can not yet be estimated, but their existence is an indication that by and by in Russia an infrastructure in the Western sense may develop.

CRYOGENIC SOCIETIES, CONFERENCES, EXHIBITIONS

Russia is Member of the IIR, the International Institute of Refrigeration, a scientific and technical intergovernmental organisation "to promote knowledge of refrigeration technology and all its applications". IIR covers the whole range of refrigeration and cryogenics, that is, room temperature down to helium temperature. A number of Commissions represent the various aspects of low temperature generation and application ranging from air conditioning over food processing down to ultra low temperatures. The IIR and its activities are well known among the Russian specialists, and Commission Conferences as well as a General Conference, the XIVth International Congress on Refrigration (Moscow 1975), have been held in the country.

In other countries it was felt in the 1950s that the emerging field of cryogenics – technical applications in the liquid helium temperature range – with its importance and potential for new technologies needed specific representation in addition to IIR. National platforms – like the Cryogenic Engineering Conference (CEC) and the Cryogenic Society of America, the British Cryogenic Council or the Cryogenic Association of Japan – were founded, with simple administrative rules and high flexibility in reaction to new developments. However, in Russia no such national platform for furthering contacts and exchange between science and industry in this field exists until now, in spite of the long-standing, extended and successful activities in cryogenics.

A first step in this direction is a series of exhibitions beginning in 2002 and organised by the exhibition company "Mir-Expo", which is supported by the Russian Ministry of Industry, Science and Technology. The *Cryogen-Expo 2003 – 2nd Specialised Trade Fair* was held at the All-Russian Exhibition Center in Moscow in November 2003. This exhibition is aimed at establishing and strengthening co-operation in the field of cryogenics, to exchange experience, to develop and implement applied cryogenic technologies. A two-days cryogenic engineering workshop with more than 20 presentations was part of the by-program. At the moment it is mainly a national event, but international participation is welcome: in 2003 a company from Czech Republic was among the exhibitors (more than 20, total exhibition area about 1500 m^2), and there were also some visitors from other countries. Information on the *Cryogen-Expo 2004* (Nov. 16-19) can be obtained from websites [9].

OUTLOOK

From its founding in 1969 the International Cryogenic Engineering Committee (ICEC) always had one or two Members from the USSR. However, after Perestroyka repeated attempts at finding a new Russian Member remained unsuccessful. In spring 2002 both authors discussed in which way it might be possible to make use of the meanwhile free exchange of ideas on national and international level for improving the communication between cryogenic specialists in Russia and the worldwide cryogenic community. These discussions were continued on a broader basis between Members of the ICEC and Russian conference participants during ICEC 19 at Grenoble, and CEC-ICEM at Anchorage.

As a consequence, this paper was prepared in close cooperation with Russian colleagues in the country and all around the world. It is the result of many detailed and interesting discussions. We got to know each other better during this process, and a group of engaged persons has developed who can answer almost any question regarding cryogenics in Russia (questions can be addressed to author A.A. who will forward them for competent answers). We don't consider our efforts to be finished with this presentation but are ready to continue and hope for lively resonance from Russia too, perhaps even proposals for an ICE Commission Member. It is planned to install an open information platform in the internet, which we hope will support the communication between Russian and foreign members of the worldwide cryogenic community, in which a stronger Russian presence is welcome and looked forward to.

REFERENCES

1. Grigor'yants, V. (Ed.): Ioffe Institute: 1918 – 1998 Publishing Dept. Ioffe Physical-Technical Institute, St. Petersburg (1998) 1; 9; 40; 46
2. Golovin, I. N.: Beeilt Euch Genossen – Stalins Atombombenprogramm, Wissenschaft und Frieden (1995)
3. Kubbig, B.W.: Es war eine Räubertat der Amerikaner – Interview mit Dr. Igor N. Golovin, Wissenschaft und Frieden (1995)
4. BBC Monitoring: Over 50 per cent of Russia's scientific, industrial potential is in closed towns, Ren TV, Moscow, in Russian 1730 gmt 10 Jun 02 BBC-Report, (www.cdi.org/russia/johnson/6303-3.cfm)
5. Sher G. S., President and Executive Director, U.S. Civilian Research and Development Foundation (CRDF) for the Independent States of the Former Soviet Union, (sher@crdf.org), E-mail message 2002, June 6, (www.cdi.org/russia/johnson/6324-9.cfm)
6. Avenarius, T.: Im Armenland der Geistreichen, Süddeutsche Zeitung (2002)No. 187, 14./15 August, 11
7 A View of Russian Science After Soros, Report No. 88, National Science Foundation (NSF/Europe Office), (1997) April (www.nsf.gov/home/int/europe/reports/88.htm)
8 The CIA World fact book, (www.cia.gov/cia/publications/factbook/)
9 Exhibition: (www.mirexpo.ru/eng/exhibitions/cryogen04.shtml); Accompanying Conference: (www.mirexpo.ru/eng/exhibitions/cryogen_conf..shtml)

Conceptual design of a 2kW cryogenic system for the neutrino superconducting beam line magnet system in the J-PARC 50GeV proton accelerator.

Makida Y., Doi Y., Haruyama T., Kimura N., Kobayashi T., Nakamoto T., Ogitsu T., Ohhata H., Yamamoto A..

High Energy Accelerator Research Organization, KEK, Oho 1-1, Tsukuba, Ibaraki, Japan

An intense neutrino beam generated by the J-PARC 50 GeV proton accelerator is planned as a second generation long-baseline neutrino-oscillation experiment to study neutrino oscillation. A superconducting magnet system is required for the arc section of the beam line to bend the beam within the available space. One string of 14 cryostats containing two magnets is cooled by supercritical helium (SHE) flow re-cooled by two-phase helium (2PHe) counter flow. The string with a length of 150 m is set in the neutrino beam line tunnel 12m bellow the ground. Cooling flow is supplied by a helium cryogenic system with a capacity of 2 kW at 4.5K. The conceptual design of the cryogenic system is described.

INTRODUCTION

A second generation long-baseline neutrino-oscillation experiment has been proposed as one of the main research projects by using the J-PARC 50 GeV proton accelerator that is a JAERI-KEK joint facility constructed in the Tokai-campus by 2007 [1], [2]. The neutrino-oscillation experiment needs artificial neutrino beam being injected into the Super-Kamiokande detector located 295km west of J-PARC. To produce such neutrino beam, the primary proton beam is transported from the main synchrotron to an aluminum target (see Figure 1). This beam line has a bending section with a radius of 105 m, where a superconducting magnet system with a dipole field of 2.6 T and quadrapole filed of 19T/m is required.

A helium refrigerator is constructed to keep these magnets in the superconducting stage. Not only thermal load estimation but also cooling characteristics depending on flow condition should be considered to make specifications of the refrigerator. This paper describes a conceptual consideration about the magnet cooling scheme and the refrigerator system.

THERMAL LOAD AT MAGNETS AND TRANSFER LINES

Superconducting combined function magnets & cryostats.

The magnets system consists of 14 pairs of single layer left/right asymmetric coils with same dipole function in combination with focus/defocus quadrupole function [3, 4]. The main parameters of the magnet are summarized in Table 1.

Each pair of coils is set in one cryostat with a unit length of 10.5m including an inter-connections (see Figure 2), and 14 cryostats are strung along the 150 m arc region (see Figure 1). The cryostat design is based on the LHC arc dipole magnets, so that some parts, support posts and shield trays etc., are purchased from common mass products to reduce the cost. The radiation trays are set onto the suitable anchor point of the posts and are kept at 50-80 K by cold He gas flow. The estimated heat leakage into the cold mass are listed in Table 2. Several beam monitors and collector coils will be inserted into some suitable interconnections, where the additional

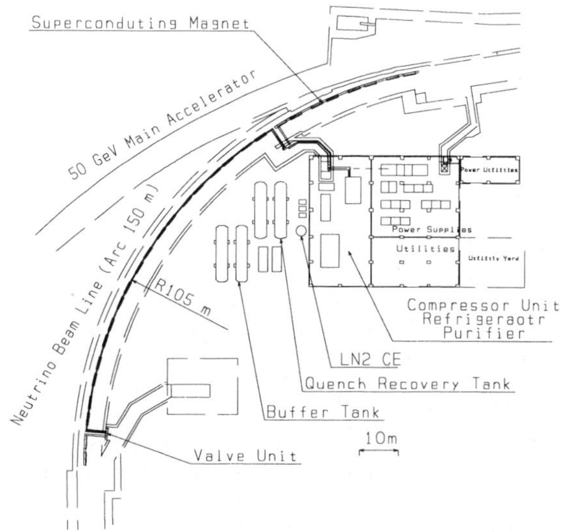

Figure 1 Arc section of the neutrino beam line and layout of the magnets and cryogenic components.

Table 1 Main design parameters of the superconducting combined function magnet

Physical length	3609 mm	Cold vessel diameter	570 mm
Magnetic length	3300 mm	Cold vessel length	4150 mm $\times 2$
Coil inner & outer diameter	173.4 & 204.0 mm	Vacuum vessel diameter	914.4 mm
Yoke inner & outer diameter	244 & 550 mm	Vacuum vessel length	8438 mm
Shell (Vessel) outer diameter	570 mm	Interconnect length	2062 mm
Weight	6.2 ton	Fluid capacity (in cold vessel)	70 ℓ
Operating current	7345 A		
Dipole field	2.59 T	Cooling type	SHE forced flow
Quadrupole filed	18.7 T/m		w/ 2PHE recooler
Stored energy	285 kJ		
Inductance	12.9 mH		

Table 2 Thermal load at the magnet cryostat

		300 K to 50-80 K (radiation shield)	50-80 K(radiation shield) to 4.5 K
Cryostat	Radiation	24.8 W	0.26 W
	Conduction	28.1 W	3.0 W
	Sub total	52.9 W	3.26 W
Interconnect	Radiation	20.0 W	0.024 W
	Conduction		0.80 W
	Sub total	20.0 W	0.824 W
Beam loss			150 W (1.0 W/m (maximum))
Total (14 cryo.+ 13 Inter. + beam loss)		1001 W	206 W

In future design, heat leakage through beam monitors and collector magnets (3-5 W/equipment) will be added.

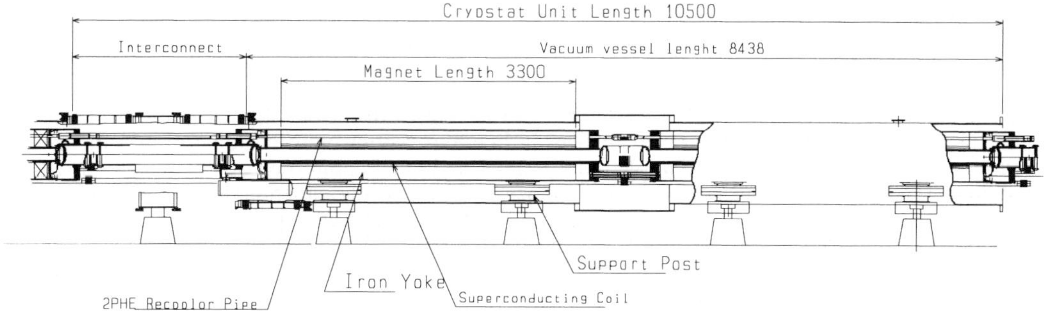

Figure 2　　Magnet cryostat

heat-leakage of 3-5 W/component is estimated.

Dynamic heat dissipation due to the high intensity beam is predicted as beam halos of 1W/m in maximum along the beam line continuously during the accelerator running.

The total heat load at the magnet string is estimated up to about 206 W at this moment, and predicted additional loads due to beam monitors, collector magnets and beam pipes would not exceed 100 W in total.

Magnet cooling circuit

A superconductive critical temperature of the coil at the operational current of 7345 A is 6.4 K, consequently the cryogenic system must keep the coils below 5.0 K with stability margin. The cold mass is cooled by supercritical helium (SHE) flow directly. Saturated helium gas flow (2 phase helium, 2PHE) is not chosen as coil coolant because of its low dielectric strength and low heat transfer in the vapor stagnation. But utilizing the latent heat in 2PHE is essential to improve thermodynamic efficiency of the cryogenic system. Otherwise several times larger flow rate is required in case of single SHE flow. Figure 3 shows the conceptual flow diagram. Supplied SHE flow enters one end of the string, goes through the magnets in sequence, and then reaches a valve unit at the opposite end of the string. SHE expands into 2PHE at the JT valve there. 2PHE counter flow re-cools SHE in the cryostat, finally returns to the refrigerator. φ 34 mm copper tube for 2PHE flow is inserted in a φ 50 mm bore through iron yoke and a part of SHE flow can be re-cooled with 2PHE (see Figure 2, 3). If whole SHE flow is heat-exchanged with 2PHE, lower SHE temperature is realized. But the string doesn't have a space for installation of usual heat exchangers.

Transfer line

Main components of the cryogenic system, a compressor, a cold box and buffer tanks etc., are located on the surface area. 40 m sub-tunnel is constructed to make a way from this surface site to the main beam line at -12 m under the ground. The 4.5K cold helium (SHE supply, 2PHe return) and 50-80K shield cooling gas are transferred through pipes arranged within a common vacuum jacket laid in the sub-tunnel (see Figure 1). Between a main-sub tunnel junction and one end of the magnet string, SHE supply, 2PHE return and shield supply pipes in the common jacket are lined along the magnets. Between opposite string end and the JT valve box another multi-pipe transfer line is

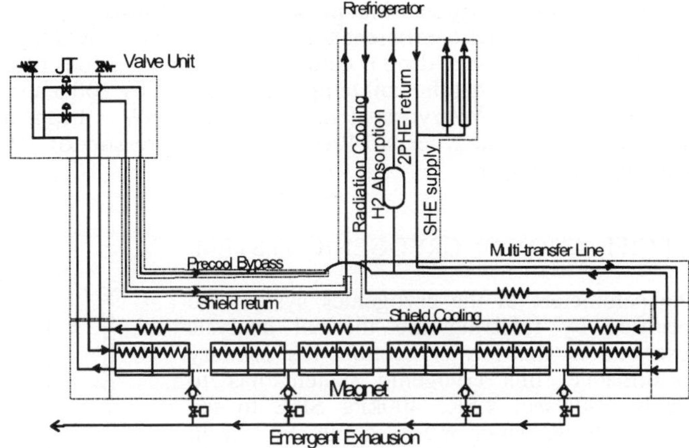

Figure 3   Conceptual flow diagram of the cryogenic system

trained. One bypass line for pre-cooling magnets and one shield return line are laid along the magnets between the valve unit and the main-sub junction individually. The average heat leaks are specified as follows [5]:

4.5 K supply and return surround by 80K radiation shield	<0.5 W/m (0.1W/m possible)
50 K supply – 80 K return	<1.3 W/m
Bypass without radiation shield	<1.0 W/m

By using these parameters, heat loads at transfer lines are estimated as shown in Table 3.

Total thermal load

Other thermal loads in the valve unit of 9W, current leads of 1g/sec/pair have to be taken into account. The total thermal loads are 410 W on the cold mass and 1339 W on the shield (see Table 4).

Table 3 Thermal load at the transfer lines

Location	Transfer	Length	Fluid	Load
Sub tunnel	Main-sub tunnel junction ↔ Cold box	40 m	SHE supply & 2PHE return	40 W
			Shield gas supply & return	104 W
Main tunnel (1)	Upper end of the string ↔ Main-sub junction	50 m	SHE supply & 2PHE return	50W
			Shield gas supply	65 W
Main tunnel (2)	Down end of the string ↔ Main-sub junction	100 m	Pre-cool bypass	50 W
			Shield return	130 W
Main tunnel (3)	Down end of the string ↔ Valve unit	15 m	SHE supply & 2PHE return	15 W
			Shield gas supply	39 W
Total thermal load (Pre-cool bypass is not taken into account)			Cold mass (4.5K)	105 W
			Shield	338 W

MAGNET COOLING CONDITON

It is important for the refrigerate specification to estimate magnet cooling characteristics depending on cold helium flow behavior. Obviously the larger heat transfer into 2PHE from SHE flow removing thermal loads results in the lower SHE and coil temperature. But,

Table 4 Summary of thermal load

	Cold mass	Shield
Magnet	206 W	1001 W
Transfer line	105 W	338 W
Current leads	90 W (1g/s)	
Valve unit	9 W	
Total	410 W	1339 W

according to the present cryostat design, a part of SHE beside the 2PHE copper pipe is re-cooled. So, the rest SHE, that is, the coil rise up to higher temperature expected by whole SHE/PHE flow. The cooling characteristics depending on SHE/2PHE flowing pattern is analyzed to find a proper flow condition. The analysis boundaries are as follows:

- Supplied SHE pressure and temperature are fixed at 400 kPa and 4.5 K respectively.
- The maximum coil temperature must be lower than 5.0K for superconductive margin.
- The return 2PHE pressure must be larger than 120 kPa for return to the compressor.

Some features of the calculating method are as follows:
- The heat transfer coefficient at 2PHE is calculated by Klimenko's functions.
- The 2PHE pressure drop is calculated on homogeneous model.

Figure 4 shows that the coil temperature cooled by 100 g/sec flow rises up to 4.8 K under the predicted thermal load of 15 W/cryostat (heat leak of 3.26W + beam halo 150 W/14 cryostats). The expansion pressure at the JT valve needs 140 kPa for return pressure of 120 kPa. Optimum flow ratio χ of 0.5 is also found by this analysis.

SPECIFICATION OF CRYOGENIC SYSTEM

The parameters of the cryogenics system are specified by referring the thermal analysis described in the former section. As primary performance, this cryogenic system has to supply 100g/sec, 4.5K, 400kPa SHE to the magnet string, which means about 2.0 kW at 4.5 K. For radiation shield cooling of 1500 W, the cryogenic system also supply 100 g/s, 50 K helium gas to the magnet string. Corresponding to the refrigeration power, a compressor should have capacities of 600 – 700 kW, 1800 kPa and 200 g/sec.

Two considerable phenomena induced by irradiation are predicted in the magnet cryostats. One is that hydrogen is emitted from the organic material, for example, insulation sheets and plastic spacers. According to recent experimental study about molecular emission from irradiated polymers, hydrogen is major component and may be stored up to 0.2 g during a half year beam run. This value is low enough to be adsorbed in the system.

The other phenomenon is that helium is transformed into tritium because of nuclear transmutation. The transmutation into tritium has an average of 54 Bq/cc in helium volume at 1 W/m beam loss. And tritium also is stocked up to 95 μg in the string after half year beam run. Though the tritium quantity is negligible from view point of refrigerator operation, the value of 54 Bq/cc is beyond a HTO criterion of < 5mBq/cc in the tritium control regulation. This means that the whole cryogenic system must be inside of a restrained radioactive area once tritium diffuses into the cold box or the compressor. In the revised design work, the tritium treatment must be solved.

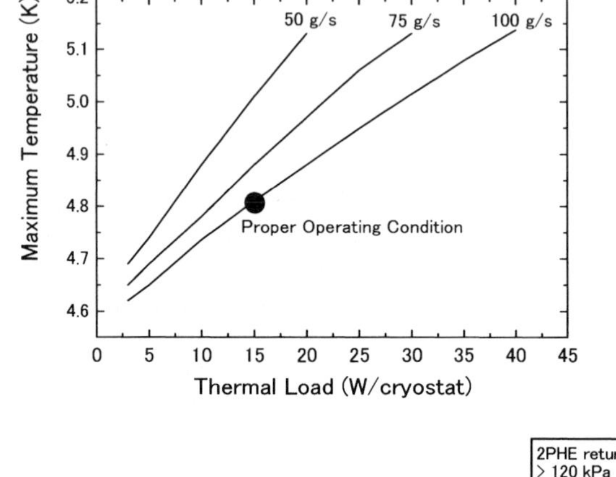

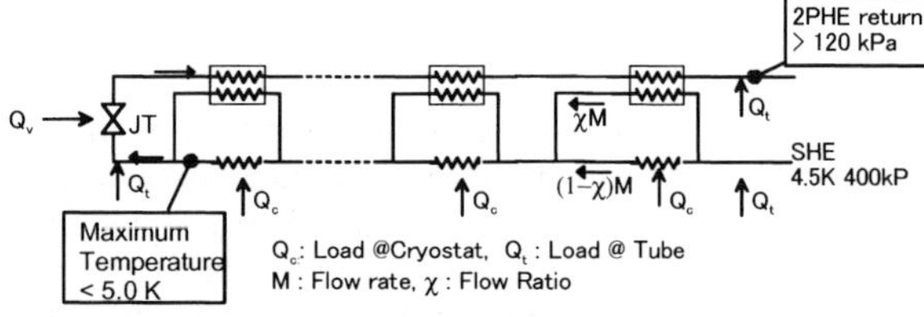

Figure 4   Cooling characteristic depending on flow condition.

SUMMARY

A 2 kW helium cryogenic system at 4.5 K is to be built to cool the superconducting magnet system for the neutrino beam experiment in J-PARC. 100 g/s SHE/2PHE circulation is required to keep the magnets below 5.0 K.

REFERENCES

1. Itow, Y., The JHF-Kamioka neutrino project, hep-ex/0106019
2. Furusawa, M., Hino, R., Ikeda, Y. et al., The joint project for high intensity proton accelerators, KEK Report 99-4; JAERI-Tech 99-056; JHF-99-3 (1999)
3. Ogitsu, T., Makida, Y., Kobayashi, T. et al., Superconducting magnet system at the 50 GeV proton beam line for the J-PARC neutrino experiment, to be published in IEEE Transactions on Applied Superconductivity
4. Nakamoto, T., Higashi, N., Ogitsu, T. et al., Design of superconducting combined function magnets at the 50 GeV proton beam line for the J-PARC neutrino experiment, to be published in IEEE Transactions on Applied Superconductivity
5. Barton, H.R., Clausen, M., Horlitz, G. et al., The refrigeration system for the superconducting proton ring of the electron proton collider HERA, Advances in Cryogenic Engineering (1986) 31 635-645

THE NEUTRAL BEAM TEST FACILITY CRYOPUMPING OPERATION : PRELIMINARY ANALYSIS AND DESIGN OF THE CRYOGENIC SYSTEM .

B. Gravil[1], D. Henry[1], M. Dremel[2], JJ. Cordier[1], R. Hemsworth[1], D. Van Houtte[1]

[1]Association Euratom-CEA, CEA/DSM/Département de Recherches sur la Fusion Contrôlée, C.E. Cadarache, F- 13108 Saint-Paul lès Durance Cedex FRANCE
[2] FZK, Institut für Technische Physik, Karlsruhe 76021, Germany

ABSTRACT

The ITER neutral beam heating and current drive system is to be equipped with a cryosorption cryopump made up of 12 panels connected in parallel, refrigerated by 4.5 K 0.4 MPa supercritical helium. The pump is submitted to a non homogeneous flux of H_2 or D_2 molecules, and the absorbed flux varies from 3 $Pa.m^{-3}.s^{-1}$ to 35 $Pa.m^{-3}.s^{-1}$. In the frame of the "ITER first injector and test facility CSU-EFDA task" (TW3-THHN-IITF1), the ITER reference cryosystem and cryoplant designs have been assessed and compared to optimised designs devoted to the Neutral Beam Test Facility (NBTF). The 4.5 K cryopanel, which has a mass of about 1000 kg, must be periodically regenerated up to 90 K and occasionally to 470 K. The cool-down time after regeneration depends strongly on the refrigeration capacity. Fast regeneration and cool-down of the cryopanels are not considered a priority for the test facility operation, and an analysis of the consequences of a limited cold power refrigerator on the cooling down time has been carried out and will be discussed. This paper presents a preliminary evaluation of the NBTF cryoplant and the associated process flow diagram.

INTRODUCTION

In the frame of the "ITER first injector and test facility CSU-EFDA task" (TW3-THHN-IITF1), the ITER reference cryosystem and cryoplant designs have been assessed and then compared to optimised designs devoted to the neutral beam test facility (NBTF). The NBTF includes a large cryopump which must operate under ITER representative operating conditions.

The cryogenic system will have to be able to refrigerate the NB Test Facility Cryopump in "acceptable" and "representative" operating conditions during short (20 s) and long pulses (3600 s). It is important to consider the system reliability, cost optimisation and the procurement and installation schedule.

CRYOPUMP DESIGN OVERVIEW

The cryopump [1] designed by FZK is made up of three parts which are: the 4.5 K cryopanels, the 80 K shields and the 80 K chevrons baffles. The cryopanels located in between the external shields and the chevrons baffles each consist of fourteen panels connected in series, which are coated on one face with activated charcoal.

OPERATING REQUIREMENTS

During standby and pumping operating mode, the cryopanels are maintained at a temperature of 4.5 K, by supercritical helium circulation, whilst the shields and chevrons baffles are cooled to ≈80 K by gaseous He. The activated charcoal on the cryopanels must be warmed up to 90 K to release all the H_2, or D_2, trapped during beam operation. In this phase, the shields and the chevrons baffles have to be maintained at a temperature close to 90 K. Regeneration at 470 K is foreseen for the ITER injector cryopumps in order to remove other (impurity) gases from the charcoal. This is not essential on the NBTF but its feasibility could be demonstrated. In this operation, only the cryopanels have to be heated up to 470 K. The shields and baffles could be maintained at a temperature close to 300 K if the cryopump design can accept the consequent differential thermal expansion. If this is not the case, the cryopanels, the shields and the chevrons baffles can be fed in series in order to limit the temperature differences between the three parts. The disadvantage of such a process is that a greater mass has to be heated which increases the time needed to cool the pump back down to the operating temperatures.

CRYOGENIC LOSSES

In long pulse pumping mode: The refrigeration power required by the cryopanels at 4.5 K results from the various heat loads such as: radiation from shields and baffles; radiation transmitted through the baffles; conduction from shields and baffles through the supports and the residual gas; and thermalisation of the incoming gas. The refrigeration power needed to maintain the external shields and the chevron baffles at 80 K results also from the various heat loads such as: radiation from the external wall; radiation from the beamline components; pre-cooling of the pumped gas rom 350 K to 80 K, conduction through the residual gas from the external wall to the shields, and conduction through the supports. The total heat loads are: 150 W on the 4.5 K cryopanels, and 18.5 kW on the 80 K shields and chevrons baffles (FZK and CEA estimations).

During regeneration at 90 K, the released H_2 or D_2 will result in a fast rise in pressure and consequently an increase of the heat load by conduction through the gas from the 300 K external wall to the 80 K shields and from the internal beamline components to the chevrons baffles. The load due to conduction through the residual gas, which is proportional to the distance between warm and cold surfaces, can reach 30 kW in the ITER reference design where there is only 60 mm between vessel wall and the external 80 K shields. The heat transfer by conduction between shields, baffles and cryopanels will tend to reduce the temperatures differences between the various parts of the cryopump.

ASSESSMENT OF CRYOGENIC ITER COMPONENTS

The total cold power needed for ITER at 4.5 K is $43.2 \text{ kW} + 0.17 \text{ kg.s}^{-1}$ of LHe, which is to be supplied by 4 identical LHe process modules each of 18 kW at 4.5 K, whereas the cold power required at 80 K, 987 kW, is supplied by two 80 K helium loops and an LN2 subsystem each of 500 kW. Taking into account the very big difference between the needs of ITER and of the NBTF (150 W at 4.5 K), the ITER reference solution appears not to be well adapted to the NBTF, (see Figure 1).

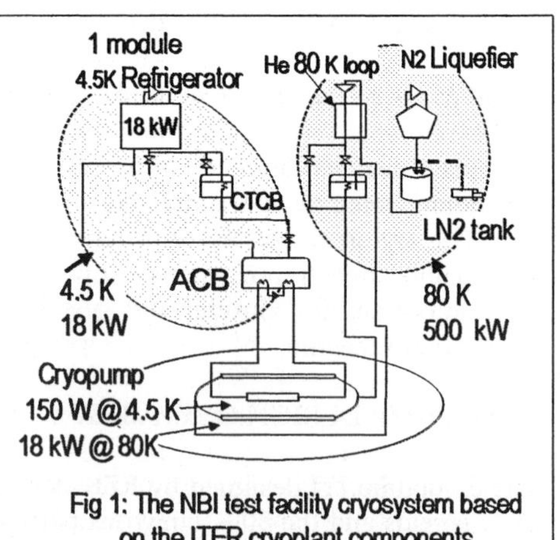

Fig 1: The NBI test facility cryosystem based on the ITER cryoplant components

ALTERNATIVE CRYOSYSTEM PROPOSED FOR THE NBTF

Two alternative solutions are proposed for the NBTF cryoplant, that are optimised in terms of reliability, 4.5 K cooling power, 80 K cooling power, cost and development. A standard, industrial,

4.5 K refrigerator cryosystem is proposed, (see Figure 2). This has an available maximum cold power of 500 W at 4.5 K in pure refrigerator mode and 150 l/h of LHe in pure liquefier mode. The cold power at 80 K has to be produced by an extra 80 K refrigeration module, for which two possible alternative options are proposed, (see Figure 3). The first system uses a cold circulator to generate a helium flow which is thermalised by liquid nitrogen at atmospheric pressure. The second system operates with a helium Bryton cycle that includes mainly a compressor, a counter flow

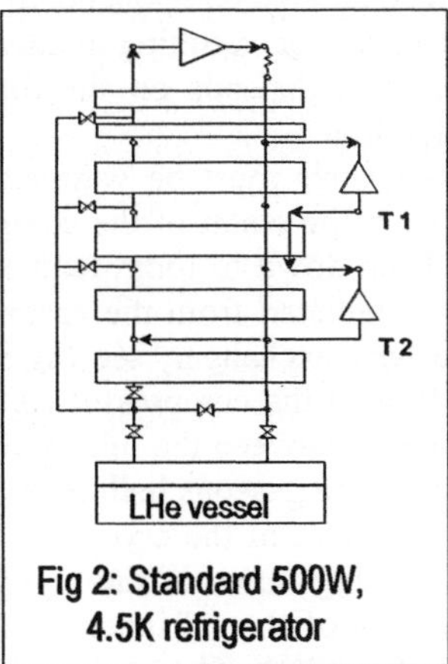

Fig 2: Standard 500W, 4.5K refrigerator

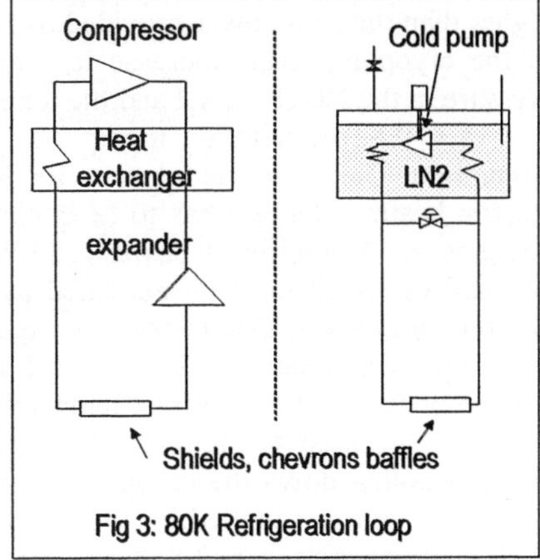

Fig 3: 80K Refrigeration loop

heat exchanger and an expander. In order to compare the two systems, one can say that the Bryton system has a lower operating cost and allows a progressive cooling down of the shields and baffles, and that using an 80 K cold circulator and LN2 bath results in a very high LN2 consumption ($\approx$10000 l per day) during operation. However, for optimised efficiency a Bryton cycle system must be operated with a high delta T to minimise the required helium flow rate and consequently the electric power consumption of the compressor. The CEA prefers to promote the Bryton cycle solution. The process diagram is represented on (Figure 4).

OPERATING CONSTRAINTS

In long pulse operating mode: The cryopanels are fed by the 4.5 K Joule-Thomson supercritical helium flow with a mass flow rate of 0.03 kg.s^{-1} inducing 1.5 K of delta T with a pressure in the cryopanels higher than 0.4 MPa. Liquid helium contained in a storage tank imposes a constant temperature at the inlet of the cryopanels and allows an added cold power by evaporating liquid helium if necessary. This Joule-Thomson flow is then expanded throw the J-T valve producing liquid helium that is stored in the liquid helium tank.

Through the second 80 K refrigerator option, the shields and chevrons baffles are refrigerated from the Bryton cycle by gaseous helium with an inlet temperature of 66 K and an outlet temperature of 90 K corresponding to 0.150 kg.s^{-1} of He flow. The cold power supplied by the expander is about 22 kW and the electric consumption is $\approx$250 kW. In 90 K regeneration mode: the Bryton cycle supplies the cryopanels, shields and baffles which are connected in series, the outlet of the shields being coupled with the inlet of the cryopanels in order to fix the temperature at 90 K at the inlet of the cryopanels. An outgassing process will start progressively

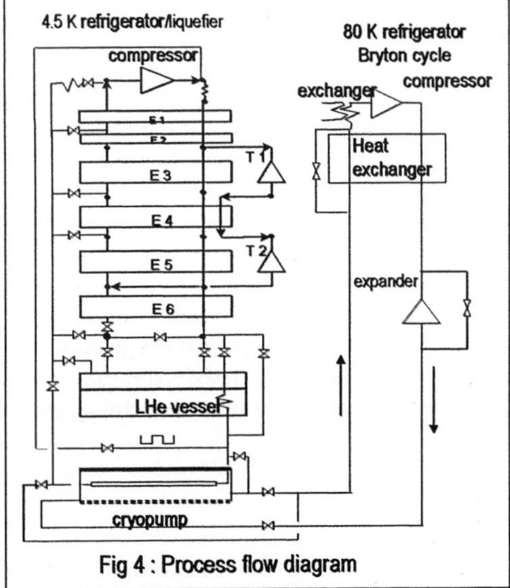

Fig 4 : Process flow diagram

with the rise in temperature of the first cryopanels. The released gas will be mainly pumped by the other cold cryopanels and partly by the forepumping system. The outgassing will be completed when the last cryopanel is warmed-up to 90 K. If the quantity pumped by the cryopanels is far from the saturation capacity, a maximum of outgassing is foreseen when the last cryopanel is heated. During this phase the additional heat load by conduction through the residual gas from the vessel wall to the external shields and from internal beamline components to the baffles will rise becoming higher than the cold power provided by the expander, resulting in a temperature increase of all parts of the cryopump .This increase in temperature depends on the forepumping efficiency via the pressure in the NBTF vessel and the time duration.

In 470 K regeneration mode: The cryopanels must be warmed up to 470 K by hot He gas circulation. The He gas is taken at 300 K from the outlet of the compressor and then heated by an electric heater. This gas has to be cooled back down to room temperature before returning to the compressor. A significant power (15 kW) is radiated from the cryopanels at 470 K to the shields and baffles which must be extracted by a heat exchanger, see Fig 5, before flowing back to the Bryton compressor. Due to the low mass flow at the compressor (0.080 kg.s^{-1}) that is available to warm-up the cryopanels, a high delta T appears between the inlet and outlet of the cryopanels. The consequence is that a high inlet temperature of the gaseous helium is required in order to be able to reach the specified 470 K temperature at the outlet of the cryopanel. The required power is about 40 kW.Cooling down the cryopump: The cooling down of the cryopanels is carried out with the 1 kW cold power supplied by the two turbines of the refrigerator. The time duration to cool the 1000 kg cryopanels from 300 K to 4.5 K is about 26 h. The cooling down of the shields and baffles is progressively achieved by the Bryton cycle with the cold power supplied by the expander. The time duration to cool down the 5700 kg of shields and baffles from 300 K to 80 K is about 10 h.

Cooling down after the 90 K regeneration: The cryopanels have to be cooled down using the flow from the refrigerator, thermalised at 4.5 K in the liquid helium bath. The available mass flow rate of 0.030 kg.s^{-1} leads to a cool down time from 90 K to 4.5 K of 550 s. During this phase a part of the cold power is delivered by the evaporation of LHe contained in the 1000 l liquid helium storage tank.

Cooling down after the 470 K regeneration: The cool down of the cryopanels from 470 K to 4.5 K is divided in two phases, first the cooling down from 470 K to 300 K room temperature, then the cooling down from room temperature to 4.5 K. During the first phase the helium flow available of 0.080 kg.s^{-1} is directly taken from the outlet of the compressor at room temperature. The time duration of this phase is about 1 h. The second phase is identical to the cooling down describe above.

CONCLUSION

The presented solution which include a standard 4.5 K refrigerator/liquefier and an 80 K helium Bryton cycle appears well adapted to supply the neutral beam test facility cryopump. Nevertheless some points still remain to be clarified, specially during transient of operation. The 90 K regeneration constitute one of the phases that are rather difficult to predict:

-firstly from the point of view of the pressure evolution during outgassing of hydrogen or deuterium due to a progressive warm-up of the cryopanels that are connected in series.

-secondly from the point of view of the additional heat loads due to the molecular conduction from the warm vessel wall to the shields and from the internal beamline components to the chevron baffles. This present work has to be considered as the first stage in defining a cryosystem which could satisfy the expressed operating specifications of the NBTF cryopump. It will have to be subjected to more detailed studies by the suppliers of the cryogenic refrigerators.

REFERENCES

1.Riedl,R et al. Design and integration of the FZK cryopump, CCNB Meeting Padua June 5-2003

Performance of a cryogenic system for the superconducting RF cavity at NSRRC

Hsiao F. Z., Li H. C., Chiou W. S., Chang S. H., and Chen J. R.

National Synchrotron Radiation Research Center, 101 Hsin-Ann Road, Hsinchu 30077, Taiwan

A helium cryogenic system that is designed based on the modified Claude cycle provides cooling power to the superconducting RF cavity and the superconducting magnet in the synchrotron light source. This paper presents the performance and the features achieved by the cryogenic system. A maximum cooling capacity 469 W and a maximum liquefaction rate 139 L/h is obtained at the storage dewar. Commission with the magnet is successful and shows small impact on the cryogenic system as coil quenches. Finally, the failure events which interrupted the system operation is summarized.

INTRODUCTION

A helium cryogenic system that is designed based on the modified Claude cycle provides a maximum cooling power 450 W at 4.5 K or a maximum liquefaction rate 110 l/hr using liquid nitrogen for precooling. Construction of this system started from Feb. 2001 and the installation began in Oct. 2002. The commission phase started from Jan. and ended in Oct. 2003. This helium system is dedicated for the cooling of a 500 MHz superconducting radio frequency (SRF) cavity which is the key item of an upgrade project to enhance the beam current and stability of NSRRC electron storage ring. Besides, this helium system keeps a superconducting magnet at 4.3 K (1.085 bar), where the magnet is cooled at the liquefaction mode. The cryogenic system is a turnkey system and provided by the Air Liquide Company. The system includes one 315 kW compressor, one 45 kW recovery compressor, one 10 kW refrigerator, one 2000 liter dewar, and two 100 m^3 gas helium buffer tanks. Inside the cold box, two expansion turbines connected in series provide the major cooling power for helium gas stream. Two 80 K adsorbers and one 20 K adsorber are installed in the cold box. The switch and regeneration of the two 80 K adsorbers are fully automatic and performed without interrupting the cold box. There are two analyzers monitoring the impurity level of the helium flow. One analyzer monitors the oxygen and humidity level at the cold box side, the other one monitors the oil aerosol and nitrogen level at the compressor side. The designed thermal dynamic state and preliminary test result of this helium system can be found in references [1,2].

PERFORMANCE

During the commission phase the target of the cryogenic system focused on the maximum refrigeration power and the maximum liquefaction rate. Status of the system is archived and trends of the data are analyzed. Table 1 summarizes the commission result obtained at the 2000 l main dewar. The liquefaction rate is obtained by measuring the dewar rate-of-rise with built-in level sensor and under the condition cold gas returned to the cold box. The refrigeration capacity is measured by the dewar built-in DC heater. The system provides a maximum liquefaction rate 139 l/hr with liquid nitrogen for precooling and a liquefaction rate 59 l/hr without liquid nitrogen. For the refrigeration capacity we obtained a cooling power 258 W with 11 l/hr liquefaction rate without liquid nitrogen and a cooling power 452 W with 5 l/hr liquefaction rate using liquid nitrogen. By linear extrapolating the data we have a cooling power 327.2 W (469.5 W) at refrigeration mode without (with) liquid nitrogen for precooling. During measuring the liquefaction rate and refrigeration capacity without liquid nitrogen precooling, we observed the mass flow rates are 24 g/sec and 44 g/sec respectively; the values are 39

36

g/sec and 82 g/sec with liquid nitrogen precooling.

Table 1 Feature of the cryogenic system during capacity measurement at main dewar

Item	Units	Design value				Measurement value			
		Refrigeration		Liquefaction		Refrigeration		Liquefaction	
		w/o LN2	w/ LN2	w/o LN2	w/ LN2	w/o LN2	w/ LN2	w/o LN2	w/ LN2
Cooling capacity	W	255	450	-	-	327.2	469.5	-	-
Liquefaction rate	l/hr	-	-	51	115	-	-	52	134
Mass flow rate	g/s	51.2	73.9	42.3	59.6	44	82	23.5	39
Frequency driver	Hz	35	52	30	42	42	57.5	33.2	43.5
Discharge pressure	bar	12.2	15	15	15	15	15	15	15
Suction pressure	bar	1.05	1.05	1.05	1.05	1.05	1.05	1.05	1.05
Suction-pressure fluctuation	mbar	$+/-3$	$+/-3$	$+/-3$	$+/-3$	$+/-2$	$+/-1$	$+/-2$	$+1/-2$
Main dewar pressure	bar	1.25	1.3	1.25	1.3	1.25	1.272	1.25	1.25
Pressure fluctuation in main dewar	mbar	$+/-3$	$+/-3$	$+/-3$	$+/-3$	$+/-1$	$+3/-2$	$+/-1$	$+/-1$
Warm turbine speed	Hz	2724	2730	3340	2658	2912	2751	3200	2754
Cold turbine speed	Hz	-	2069	-	-	1837	1969	1846	1988

A further capacity verification after the 5 m long, nitrogen thermal shielding, multi-channel line was performed. The liquefaction and refrigeration capacity was measured in a 500 l test dewar, which was connected to the multi-channel line via an adopter and 1.6 m long flexible vacuum jacket lines. Table 2 summarizes the measured result obtained at the test dewar. The measurement for the liquefaction rate with liquid nitrogen precooling showed a rising rate 114 l/hr in the test dewar and a rate 25 l/hr in the main dewar. A liquefaction rate 59 l/hr in the test dewar was obtained without liquid nitrogen. Thus a maximum refrigeration capacity 327.1 W (447.9 W) without (with) liquid nitrogen precooling is obtained in the test dewar. For all the measurement in the test dewar, a higher flow rate into the cold box is required to achieve a comparable capacity as the case measured in the main dewar.

Table 2 Feature of the cryogenic system during capacity measurement at test dewar

Item	Units	Measurement value			
		Refrigeration		Liquefaction	
		w/o LN2	w/ LN2	w/o LN2	w/ LN2
Cooling capacity	W	327.1	447.9	-	-
Liquefaction rate	l/hr	-	-	59	139
Mass flow rate	g/s	50.5	89	29.5	48.2
Frequency driver	Hz	56.1	59	47.9	50.7
Discharge pressure	bar	15	14.699	15	15
Suction pressure	bar	1.05	1.05	1.05	1.05
Main dewar pressure	bar	1.38	1.4	1.38	1.38
Test dewar pressure	bar	1.3	1.3	1.18	1.19
Warm turbine speed	Hz	2993	2770	3200	2750
Cold turbine speed	Hz	1950	2042	1931	2050

Stand-alone operation of the cryogenic system has the following features. The suction line pressure has a variation smaller than +/-2 mbar and the variation is smaller than +3/-3 mbar for the dewar pressure. A frequency driver that implements an energy-saving feature varies the frequency in the range 30 Hz to 60 Hz such that the mass flow of helium changes from 44 g/s to 89 g/s. In the refrigerator, two 80 K cryogenic adsorbers automatically switch at every 100 h and the off-duty one automatically proceeds to on-line regeneration for the next operation. It takes 4.5 hr to complete the regeneration process. The interlock logic keeps the cryogenic system safe from utility failure and a small-capacity compressor recovers the helium to the storage tanks and maintains a suction line pressure at 1.1 bar. The emergency power maintains continuous operation of the control system and the recovery compressor to minimize helium loss in case of an electricity failure. The time required to warm up the cold box to room temperature is 18 hr; it takes 38 days to naturally warm up the main dewar at zero level to room temperature. Initial cooling down of the cold box and the main dewar from room temperature to 4.5 K takes 36 hr, and it takes 10 hr to store the liquid helium to 80 % main dewar level with liquid nitrogen precooling. During the liquid helium accumulation period, value of the dewar level sometimes stagnates at 40% and 82%. As dewar level approaching the maximum setting value, the turbines will turn down their speeds and the heater will turn on to avoid over-level event.

COMMISSION WITH MAGNET

A superconducting magnet SW6 with helium vessel of volume 200 l was installed in the storage ring in Jan. 2004. The cold box did not provide liquid helium to the magnet until the severe damaged warm turbine was replaced with a new one. Figure 1 shows connection of the cold box, the main dewar, the test dewar, and the SW6 magnet. Location of the SRF cavity to be installed in Nov. 2004 is also indicated. A control valve box is connected to the downstream of the multi-channel transfer line. The SW6 and the test dewar are then connected to the valve box via flexible transfer lines. The cold gas from the test dewar returns to the cold box, while the gas from SW6 returns to the compressor suction line. Operation pressure for the SW6, the test dewar, and the main dewar is 1.08 bar, 1.24 bar, and 1.4 bar respectively. The SW6 is designed with small margin to its critical field and the cryostat is operated close to 4.3 K. Instead of high Tc material the current lead of the coil is made of copper and cooled by the cold helium gas. Thus the measured temperature of the return helium gas is around 60-70 K, which is lower than the estimated temperature 100 K. Because of the low operation pressure and the high return gas temperature, the returned gas from SW6 is warmed up and sent back to the compressor.

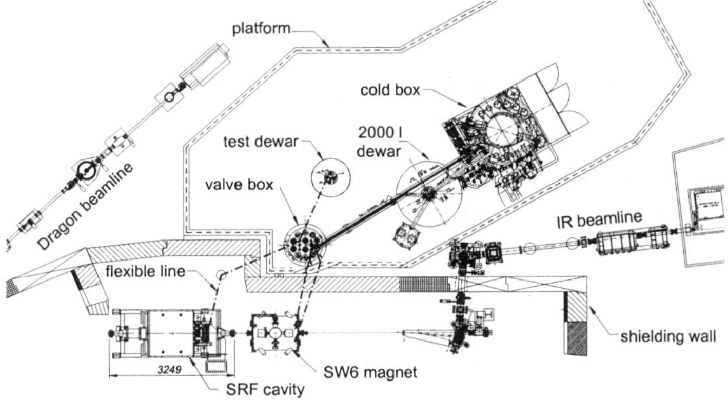

Figure 1   Layout of the cold box, main dewar, valve box, test dewar and SW6

Commission of the SW6 is quite successful. Figure 2 shows the training records of the SW6 coil. After ten times of training, the coil was charged to the design current 285 A and a magnet field 3.2 T was maintained. Three hours after the successful training of SW6, the beam current of the storage ring was stored at 200 mA without difficulty. Each coil training caused a quench and then vaporized a maximum 10 liter liquid helium. The stored energy of the coil is 50 kJ and part of the stored energy is dumped to the flywheel diode. The diode is cooled partly by liquid helium and partly by cold helium gas. During the training period the cold box and the test dewar was isolated from SW6 and the compressor kept running to recover the large amount of helium gas from SW6 cryostat. Consumption more than 60 l/hr was required to continuous fill liquid helium to the SW6 cryostat and kept a constant level. Another measurement indicates that the heat load of SW6 cryostat consumes liquid helium up to 3 l/hr as the coil current is 285 A. We concluded most of the helium is vaporized to accommodate the heat loss during the helium transfer phase. A total heat loss 30 W is estimated for the control valve box, the multi-channel line, the flexible transfer line, and the connectors.

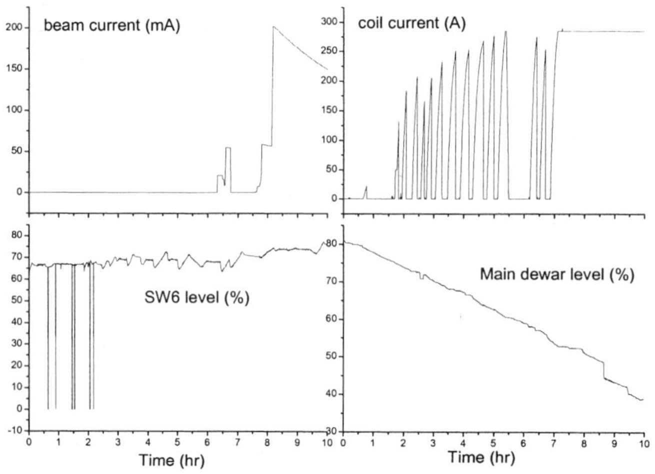

Figure 2   Training records of SW6

Without liquid nitrogen precooling the cold box can not maintain the main dewar level if no cold gas returns to the cold box.   Two strategies are used to solve this problem.   We open the by-pass valve in between the helium supply line and cold return line in the valve box and let the cold gas come back to the cold box.   In the meanwhile the helium level for SW6 cryostat is regulated at On-Off control filling mode with inlet valve opened at 5% and triggered from +/-1 % cryostat level fluctuation.   Small impacts on the cryogenic system are observed as the SW6 quenches: a pressure fluctuation +/-2 mbar in the suction line, a pressure fluctuation +/-3 mbar in main dewar; a temperature increase 0.3 K in cold return gas from the valve box; and a speed fluctuation +/-10 Hz at cold turbine.   The On-Off filling mode for SW6 causes a pressure fluctuation +/-2 mbar at the test dewar.

TRIP RECORDS

Several events interrupted the operation of the cryogenic system.   Table 3 summarizes the number of trips caused from different parts.   The oxygen sensor failed due to some unknown reason.   Failure of the compressor motor is believed caused from the poor lubrication of motor bearing.   Failure of the output control card of the frequency driver may be owing to the overheating of its cabinet inside temperature.   Damage of the warm turbine may cause from impurity or dust.   A poor controlled cooling water temperature for the compressor leaded to an oscillating helium temperature in the discharge line. This unstable temperature disturbed the whole system and finally the compressor was tripped due to too big temperature oscillation as compressor speed was high.   Cooling water with well-controlled temperature will avoid this kind of event.

Table 3   Trip records

Event	Count	Cause
Warm turbine failure	3	Damaged by impurity or dust
Oxygen sensor abnormal reading	3	Sensor cell damage
Compressor motor damage	1	Motor overheated due to poor bearing lubrication
Frequency driver failure	1	Output control card failure
Electricity shortage	3	Electricity trips
Cooling water failure	2	Small flow rate; unstable water temperature

REFERENCES

1.   Hsiao, F. Z., et al., The Liquid Helium Cryogenic System for the Superconducting Cavity in SRRC, Proceedings of the 2001 Particle Accelerator Conference (2001), 1604-1606
2.   Hsiao, F. Z., et al., The Pilot-Runs of the Helium Crygoneic System for the TLS Superconducting Cavity, Proceedings of the 2003 Particle Accelerator Conference, Oregon, Portland (2003).

The cryogenic system of BESIII in preliminary design

Liu L.[1], Zhu Z.[2], Han S.[2], Makida Y.[3], Yamaoka H.[3], Tsuchiya K.[3], Liang W.[1], Gong L.[1] and Zhang L.[1]

[1] Technical Institute of Physics and Chemistry, Chinese Academy of Sciences, Beijing 100080, China
[2] Institute of High Energy Physics, Chinese Academy of Sciences, Beijing 100039, China
[3] High Energy Accelerator Research Organization, KEK, Tsukuba, Ibaraki, 305-0801, Japan

The cryogenic system of the superconducting (SC) solenoid magnet consists of a refrigerator, compressor, buffer tank, recovery tank, liquid nitrogen storage tank, control dewar and SC solenoid magnet, where the refrigerator, compressor, buffer tank and recovery tank are shared with another superconduting user of Beijing Electron Positron Collider II (BEPC II), a pair of superconducting interaction region quadrupole magnets (SCQ). The solenoid magnet is indirectly cooled by forced, two-phase, helium flow during steady state operation and switched to a thermo-syphon mode in case of emergency stop of the refrigerator. This paper describes the preliminary design of this cryogenic system, including the heat load estimation, the cooling flow scheme, the operation modes, the structural design of the pipes, control dewar and cryostat of this SC solenoid magnet.

INTRODUCTION

The cryogenic system of the SC solenoid magnet for the Beijing Spectrometer III (BESIII) detector system will be constructed at the Institute of High Energy Physics (IHEP) in Beijing. In order to realize the goal of increasing the luminosity of BEPC by two orders of magnitude, three superconducting hardware systems, which are a pair of SCQ, a pair of superconducting radio frequency cavities, and the superconducting detector (SCD) solenoid magnet, need to be developed in BEPCII. The SCD solenoid magnet which is inside the BESIII detector is designed to produce axial steady magnetic field of 1.0 Tesla over the tracking volume and to meet the requirement of particle momentum resolution to particle detectors.

This paper describes the preliminary design of the BESIII cryogenic system including the heat load estimation, the cooling flow scheme, thermodynamic design, the configuration of the control dewar and cryostat, and the operation modes of this SC solenoid magnet.

BASIC PARAMETERS OF THE BESIII MAGNET

We adopt a design featuring single layer of coil, indirect cooling by forced flow of LHe, pure aluminium based stabilizer and NbTi/Cu superconductor for the magnet. A serpentine tube with a bore diameter of 25 mm is welded on the outer surface of the support cylinder. One of main reasons that the pool boiling method is unfavorable is possibility of an explosive boiling off resulting in the rapid loss of cryogen in a quench. The force cooling method also has the advantage that a smaller quantity of cryogen is required.

The overall dimension of the magnet is 3.89 m in length, 2.65 m in inner diameter, and 3.4 m in outer diameter. The coil itself is 3.5 m in length and 2.9 m in effective diameter. The general parameters of the magnetic solenoid of the BESIII detector are summarized in TABLE 1.

Table 1 Basic parameters of the BESIII magnet

Items		Parameter
Cryostat		
	Inner radius	1.325 m
	Outer radius	1.700 m
	Length	3.890 m
Coil		
	Effective radius	1.450 m
	Length	3.500 m
	Conductor dimension	3.70 mm × 20.0 mm
Electrical parameters		
	Central field	1.0 T
	Nominal current	3150 A
	Inductance	2.0 H
	Stored energy	9.2 MJ
Cold mass		3.5 ton

CRYOGENIC SYSTEM OVERVIEW

The system consists of a helium compressor, a cold box, transfer lines, a control dewar, a magnet cryostat, a buffer tank, a recovery tank, and a liquid nitrogen storage tank, where the helium compressor, cold box, buffer tank, recovery tank, and liquid nitrogen storage tank are shared with a pair of SCQs. Therefore, a flow diagram of the cryogenic system only including the control dewar, magnet cryostat, and transfer lines is shown in Figure 1.

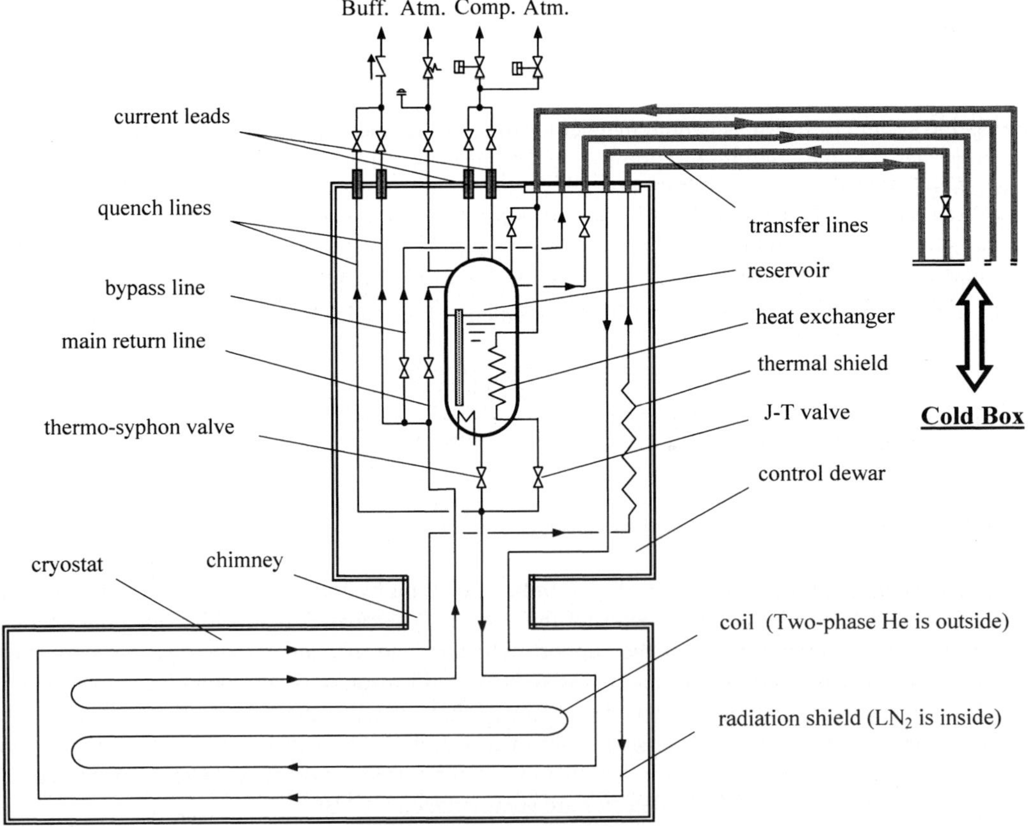

Figure 1 Flow diagram of the cryogenic system

The compressed helium gas at 1.65 MPa from the compressor flows into the cold box and comes out at 5.4 K and 0.3 MPa after the J-T valve in the cold box. Passing through the transfer line and the heat exchanger inside the control dewar at the top of the magnet, the helium gas is sub-cooled to 4.52 K at a pressure of 0.285 MPa. At this point, the helium condition is supercritical. After the J-T valve inside the control dewar, the helium is in the two-phase state at 4.46 K and 0.125 MPa and flows into the magnet through the chimney port. In the steady state, the magnet is indirectly cooled by forced, two-phase flow at a mass flow rate of 10 g/s. After cooling the magnet, two-phase helium returns to the control dewar. The boil-off helium gas from the control dewar flows into the cold box via transfer line. The returning cold gas comes back to the compressor, cooling the input helium line. Part of the helium returning from the magnet is used to cool the current leads. The returning route of the helium can take one of two possible paths. One path is the usual route, to compressor suction, and the other is to the atmosphere.

OPERATION MODES

There are eight operation modes designed for BESIII cryogenic system, which are gas charging mode, cooldown mode, normal operation mode, quench mode, quench recover mode, refrigerator failure mode, warmup mode and shut down mode.

In the case of a quench, the magnet is separated automatically from the control dewar and the boil-off helium gas from the magnet flows to the buffer tank directly through the quench lines.

In the case of the refrigerator suspension, the thermo-syphon valve will be opened to make the LHe in the reservoir circulate through the coil pipe gravitationally. This thermo-syphon cooling mode may be possible for the coil to be kept at a sufficiently low temperature until the discharge is completed.

In the case of cooldown, warmup and quench recovery, if the temperature of the return helium gas is higher than 20 K, the return gas will flow back through the bypass line instead of the main return line to meet the demand of the cold box.

HEAT LOAD

The upper thermodynamic parameters are decided on the basis of the estimation on the heat load shown in Table 2. What should be mentioned here is the dynamic heat load listed in this table. Except the steady head load, during the excitation or discharge, the heat caused by the electromagnetic induction in support cylinder should also be taken into accounted. If we set the discharge ramping rate dI/dt is $3300 / 1800 = 1.833$ A/s, we can obtain the dynamic heat load $Q_{dy} = E^2 / R = 3.39$ W, where E is the induced voltage and R is the electricity resistance of the support cylinder.

Table 2 Heat load estimation

Items of Heat Load	77 K	4.5 K
Support rods in cryostat	26.527 W	1.038 W
Radiation in cryostat	73.801 W	3.236 W
Current leads	——	7.920 W + 0.421 g/s
Radiation in chimney & control dewar	10.025 W	0.407 W
Support rods in chimney & control dewar	3.900 W	0.022 W
Bayonet and valves in control dewar	46.000 W	13.000 W
Measuring wires	5.311 W	0.831 W
Dynamic heat load	——	3.390 W
Total	165.5 W	29.85 W + 0.421 g/s
Heat load used （× safety factor of 1.5）	**248 W**	**44.8 W + 0.63 g/s**

HELIUM AND NITROGEN PIPES

According to the estimation on heat load and using the safety factor 1.5, the mass flow rates of nitrogen and helium have been determined, which is 1.89 g/s (that is 8.42 L/h) for nitrogen and 10g/s for helium whose inlet pressure is 0.285MPa and temperature is lower than 5.5K.

Based on the mass flow, the inner diameters of nitrogen and helium pipes have also been decided, which are 14 mm for nitrogen and 16mm for helium. Considering the quench mode, the inner diameter of helium pipe in the coil cryostat has been increased to 25 mm. Under this design, the maximal pressure drop in nitrogen pipe is expected to 0.113 MPa and the pressure drop will be 3000 Pa during the normal operation. Concerning the helium pipe, the maximal pressure drop is 0.216 MPa, the normal pressure drop is 2500 Pa, and the quench pressure drop is 4100 Pa.

CONTROL DEWAR AND CHIMNEY

The following principle is used to determine the volume of the control dewar: the system should operate still normally for four times the discharge time (the discharge time of 0.5 hour is assumed) in the case of the refrigerator suspension. The volume of the dewar we decided to use is 200 L, whose main dimension parameters are: diameter and height of the reservoir: $\phi_o 722 \times 794$ mm, thickness of the reservoir: 9 mm, diameter and height of the vaccum vessel: $\phi_o 1500 \times 2100$ mm, thickness of the vaccum vessel cylinder: 8 mm, thickness of the end plates: 40 mm. The superconducting bus lines from the coil enter the reservoir through insulated feedthroughs and are connected with helium-gas-cooled Cu current leads.

With respect to the heat exchanger inside the reservoir, where the super-critical helium flow is cooled inside the pipe and the saturated liquid helium boils outside the pipe. Through the thermodynamic analysis, the general heat resistance R = 0.0301 m•K/W was obtained. Hence the calculated length of the pipe L is 8.14 m.

The chimney containing all the cryogenic and electrical lines from the coil crosses the barrel yoke, where the superconducting current leads are indirectly cooled through thermal conduction from the helium return line.

CRYOSTAT

The BESIII superconducting solenoid consists of the coils in a cryostat. The outer high-strength aluminum hoop restrains the conductor and the coolant is supplied through the aluminum tubes axially attached to the outer surface of this support cylinder. Two coaxial aluminum cylinders, with close end plates cooled by LN_2, act as the radiation heat shield.

Having set up a simplified thermodynamic model of the cooling tube and support cylinder, through the thermal analysis, we found if the heat load of 45 W is distributed evenly on the support cylinder and the number of the axial turns of the helium cooling pipe is 24, the maximal temperature difference on the outer surface is about 0.002 K. Considering the unevenly distributed head load can not be avoided, the safety factor 10 has been introduced. In this case, the maximal temperature difference will be 0.02 K, which is still safe enough. Therefore, we finally adopt the following helium cooling pipe: the diameter is $\phi 31/25$ mm, the length of per turn is 3.5 m, the number of turns is 24.

CONCLUSIONS

The design of the cryogenic system of the superconducting solenoid magnet of BESIII has been done. In this design, the heat load of the whole system has been estimated, the cooling flow scheme and the operation modes determined, and the structural design of the pipes, control dewar and cryostat accomplished.

Design issues of the distribution valve box of the KSTAR helium refrigerator

Chang H.-S.[a], Gistau Baguer G. M. [a], Cho K. W. [a], Kim Y. S. [a], and Lee G. S. [a]
Briend P.[b], Fuenzalida P. [b], and Fossen A.[b]

[a]Korea Basic Science Institute, Yusung-Gu, Daejeon, 305-333, Korea
[b]Air Liquide DTA, BP 15- 38360 Sassenage, France

The operation mode of the KSTAR (Korea Advanced Superconducting Tokamak Research) helium refrigerator is based on the day average thermal load of the KSTAR cold components. However, since the thermal load on the magnets and current-feeding devices of KSTAR are strongly time-dependent, a system which smoothens the heat load peaks and distributes the cryogenic helium at the proper instant is required. In this proceeding, we report the design issues of such a system, the distribution valve box of the KSTAR helium refrigerator, which are essential to cope with various operation modes and probable abnormal events of the KSTAR device.

INTRODUCTION

KSTAR (Korea Superconducting Tokamak Advanced Research) is a tokamak device with 30 fully superconducting (SC) magnet coils. The main duty of the KSTAR helium refrigerator is to keep all cold components of KSTAR (SC coils, magnet structures, SC bus-lines, current lead system, and thermal shields) at suitable temperatures in order to operate the SC magnet coils consistent with the operation scenario of KSTAR.

The normal operation scheme of the refrigerator is based on the day average thermal load of the KSTAR cold components [1]. However, the actual thermal load on the magnet system (coil and structure) [2] and current-feeding devices (current lead system and SC bus-lines) [1] are strongly time dependent and exhibit large peak values during the normal operation of KSTAR. In order to smoothen the time variation of the thermal loads, a distribution valve (DV) box, equipped with cryogenic control valves (CV's), cryogenic circulators, and heat exchangers (HX's) submerged in a huge liquid helium (LHe) bath (thermal damper), intervenes the cryogenic helium via cryogenic transfer lines (CTL's) between the refrigerator cold box and the KSTAR cold components [3]. Besides "damping" the temporal thermal load variation of KSTAR, the DV box also has to distribute and arrange the helium in order to cope with the following tasks [1]:

- Cool-down (warm-up) of the KSTAR cold components from any temperature down to their operating cryogenic temperature (up to room temperature) within the constraints of time and temperature difference between the components;
- Protection of the KSTAR cold components and refrigerator from damaging in case of probable abnormal events;
- Simulation of the thermal load and pressure variation in the KSTAR cold components to pre-commission the refrigerator;
- SC coil/bus-line CICC (cable-in-conduit conductor) cleaning.

The thermal loads on the thermal shields (TS's), which are cooled by high-pressure (HP) gaseous helium (GHe) produced in the cold box and distributed via the DV box, are estimated to be almost static [4]. On the other hand, the current lead (CL) system has its own LHe distribution system. Therefore, in this proceeding the cryogen distribution scheme of the TS's and CL's will not be considered and only the supercritical helium (SHe) circuits of the DV box will be presented.

NORMAL OPERATION

The normal operation of KSTAR consists of idle, toroidal-field (TF) ramp/de-ramp, stand-by, and plasma-shot mode [1]. During the idle mode the refrigerator switches to a partial liquefaction state and LHe is stored in the storage tank [4]. This LHe is consumed during the other modes by transferring the LHe from the storage tank to the thermal damper (TD) via valve L1 and vaporizing it through valve G1, as shown in Figure 1 (from now on "capital letter(s)+number" is a cryogenic valve of Figure 1, unless otherwise stated).

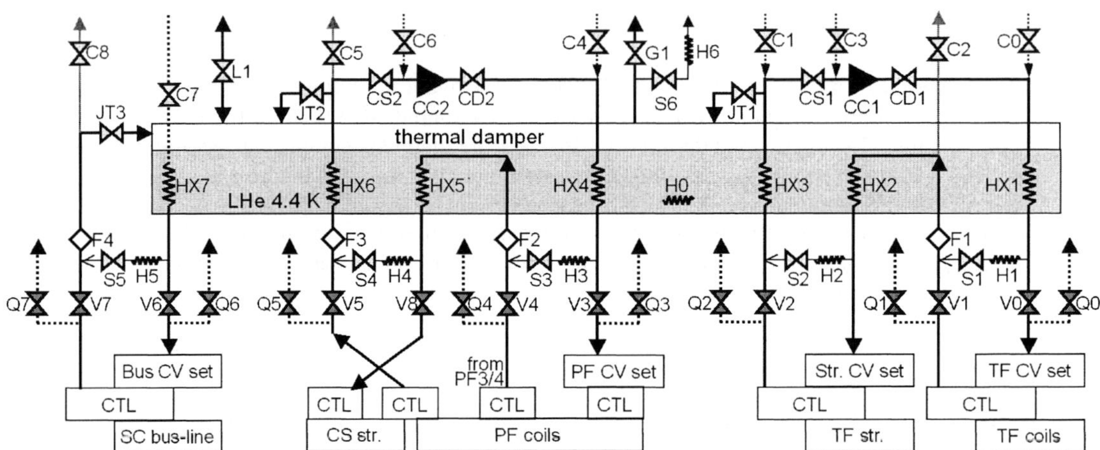

Figure 1: A schematic illustration of the SHe circuits in the DV box. All valves are cryogenic valves and all HX's (HX1~7) are of brazed aluminum plate fin exchanger type.

TF (PF) magnet cooling

Except during the idle mode, a constant SHe mass flow rate of 300 g/s is forced by the cryogenic circulator CC1 (CC2) to the TF (PF) coils. Before entering the coils, the SHe is cooled in HX1 (HX4) down to 4.5 K and distributed via the CTL's to each particular coil by manipulating the TF (PF) CV set. The return SHe from the TF (PF3/4) coils is then filtered by filter F1 (F2), mainly to protect CC1 (CC2) from damaging due to particles which may have been separated from the coil CICC. The thermal load of the TF (PF3/4) coils is deposited on the TD by exchanging heat between SHe of the circuit and LHe of the TD via HX2 (HX5). The 4.5 K SHe is then supplied to the TF (PF1~4) magnet structure and the return SHe from the TF magnet structure (PF1~4 magnet structure and PF1/2/5~7 coils) is cooled down to 4.5 K in HX3 (HX6) before entering the suction side of CC1 (CC2). F3 filters the particles of PF1/2/5~7 coils.

The temporal evolution of the magnet thermal load [2] which will be deposited on the LHe of the TD via HX1~6 is shown in Figure 2(a). The first 70 seconds represent the plasma-shot mode where all coils are charged and during the remaining time interval, which is the stand-by mode, only the TF coils are charged. When a plasma disruption (VDE) occurs inside the KSTAR plasma vacuum-vessel, most of the energy is deposited on the TF magnet structures. The extra thermal load resulting from this VDE is deposited through HX3 on the LHe of the TD. However, the curves in Figure 2(b) show that a TD with 5000 liter of 4.4 K LHe and 10% ullage is sufficient to cope with such excess thermal load [4].

During the idle mode where no coils are charged, the thermal load on the magnets is much less than those during the other modes. Therefore, instead of operating CC1 (CC2), which heat loss is about 0.9 kW during nominal operation, CS1 (CS2) and CD1 (CD2) are closed and about 60 (30) g/s of SHe is directly issued from the cold box through C0 (C4) to the TF (PF) coils and finally expanded thorough JT1

(JT2) into the TD [1,4]. This idle-mode cooling scheme saves about 1.8 kW of the refrigerator cooling power for 14 hours a day.

<u>SC bus-line cooling</u>
SHe coming from the cold box and cooled down to 4.5 K in HX7 is distributed via the CTL's to each particular bus-line group by manipulating the bus CV set. The return SHe is filtered by F4 and expanded into the TD by JT3 [4]. The value of the SHe mass flow rate depends on the specific operation mode.

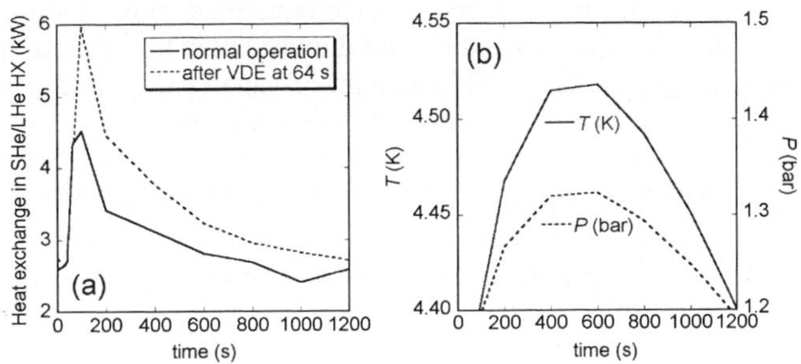

Figure 2: (a) Temporal evolution of the magnet thermal load which will be deposited on the TD via HX1~6 of Figure 1. (b) Pressure and temperature time evolution of a TD with 5000 liter 4.4 K LHe and 10% ullage in case of a VDE occurrence during the normal operation of KSTAR.

COOL-DOWN, WARM-UP, AND CICC CLEANING [1]

One constraint imposed on the TF magnet cool-down and warm-up procedure is that the temperature of the magnet structures always has to be lower than that of the coils. To achieve this, CS1 and CD1 are closed; helium is fed from the HP line of the cold box into the coils and structures via C0 and C1, respectively; and released through C2. To cool-down or warm-up the PF coils (SC bus-lines), helium is issued from the HP line of the cold box through C4 (C7) and returned through C5 (C8). When the temperature of the TF (PF) magnet is low enough, helium is fed via C3 (C6) to cool down CC1 (CC2).

To clean the CICC of the TF coils (PF coils; SC bus-lines), V0 and V1 (V3~5 and V8; V6 and V7) are closed and pressurized GHe is issued from the refrigerator external purifier by way of Q0 (Q3; Q6) to the TF coils (PF coils; SC bus-lines). The returned GHe is gathered via Q1 (Q4 and Q5; Q7), its impurity analyzed, and finally sent to the refrigerator recovery compressor.

ABNORMAL EVENTS [1]

When the refrigerator process control system receives a quench signal of the TF coils (PF coils; SC bus-lines) from the KSTAR main control system, V0~2 and the TF structure CV set (V3~5 and V8; V6 and V7) are quickly closed and the HP hot helium is evacuated through Q0~2 (Q3~5; Q6 and Q7) to the quench collector.

In the case of a fast discharge of the TF coils, V2 and the TF structure CV set are quickly closed and the HP hot helium is evacuated through Q2 to the quench collector. Quench in the TF coils and/or in the SC bus-lines of the TF coils are always accompanied by a TF coil fast discharge event.

A replacement of Q0~7 and their corresponding safety valves by CERN's fast opening "Quench valves" [5] is under consideration.

THERMAL LOAD SIMULATION [1]

To simulate the time variation of the thermal load and pressure in the TF coils (TF magnet structures; PF3/4 coils; PF1~4 magnet structure and PF1/2/5~7 coils; SC bus-lines), V0 and V1 (TF structure CV set and V2; V3 and V4; V8 and V5; V6 and V7) have to be closed first. Then the heater H1 (H2; H3; H4; H5) simulates the CTL heat loss and the thermal load which is estimated to be deposited on the LHe of the TD via HX2 (HX3; HX5; HX6; JT3). On the other hand, the estimated pressure variation of the SHe is simulated by S1 (S2; S3; S4; S5).

The heater H0 is used to simulate the time evolution of the total thermal load of the KSTAR cold components, including the heat loss of the refrigerator itself. In this way, without using the other heaters, it is possible to do a simple test of the refrigerator cold box cooling power and the LHe production/consumption balance. However, the main duty of H0 is to vaporize the amount of LHe which is estimated to be consumed by the CL's. This amount of vaporized helium is extracted through S6 and heated up to 300 K by H6 to simulate the thermal load of the CL system.

The simulation of the thermal load and pressure variation in the TS's can be accomplished by installing a by-pass line, which is equipped with a heater and CV, between the main supply and main return helium lines of the TS's.

CONCLUSION

The major task of the DV box of the KSTAR helium refrigerator is the distribution of helium during various normal operation modes and the cool-down/warm-up of the cold components of KSTAR. However, by installing proper valves, heaters, and by-pass lines to the helium circuit of the DV box, it is possible to perform other tasks, such as CICC cleaning, protection of KSTAR and the refrigerator in case of abnormal events, and the pre-commissioning of the refrigerator by simulating the time variation of the thermal load and pressure in each cold component of KSTAR before being connected to the refrigerator.

ACKNOWLEDGMENT

This work was supported by Korea Ministry of Science and Technology under the KSTAR Project Contract.

REFERENCES

1. Kim Y. S. *et al.*, Technical Requirement for the KSTAR Helium Refrigeration System (2004).
2. Alekseev A. *et al.*, Thermo-hydraulic analysis of the KSTAR magnet system (2002).
3. Fuenzalida P. *et al.*, KSTAR Helium Refrigeration System TECHNICAL PROPOSAL (2004).
4. Briend P. *et al.*, KSTAR Helium Refrigeration System PROCESS JUSTIFICATION FILE (2004).
5. Velan S.A.S., Helium Cryogenic Safety Relief Valves for LHC Project.

A Retrospect on the Successful Commissioning of the Linde 18kW Refrigerators for the LHC Particle Accelerator at CERN

A. Kuendig, B. Chromec, U. Fleck, Th. Voigt,
Linde Kryotechnik AG, Dättlikonerstrasse 5, CH-8422 Pfungen, Switzerland

Six years ago, a call for tender for four new identical refrigerators was published by CERN, the European laboratory for particle physics. Within the tender, an exergetic equivalent capacity of 18 kW at 4.5 K per unit was specified. Finally CERN split the scope between the two bidders and LINDE KRYOTECHNIK AG got the order for two units. With completion of the 18kW cold boxes end of 2001, beginning of 2002 the installation at CERN took place. Acceptance test was started and accomplished in 2003. The paper highlights their characteristic features, like the low power input, which was honored by CERN with a bonus for the performance, the excellent process stability, as well as the remarkable compact design.

INTRODUCTION

CERN, the European Laboratory for Particle Physics, is presently engaged in the construction of the Large Hadron Collider (LHC). This LHC will make use of high field super-conducting magnets operating in super-fluid helium and thus will require a huge refrigeration capacity at several temperature levels. The refrigeration capacity will be generated and distributed by eight refrigerator stations located along the colliders tunnel.

Four of these stations will be equipped with the LEP-12KW refrigerators which will be upgraded for this purpose. The Linde Kryotechnik AG got the order for two of the new refrigerators. Table 1 gives a short overview on the capacity requirements.

Table 1: The specified heat loads to the LHC 18 kW refrigerators in the main operating modes

Operating Mode	4.5 K isothermal [W]	4.5 K – 20 K cold compressors [W]	20 K – 280 K current leads [W]	50 K – 75 K radiation shield [W]
Installed	4'400	20'700	55'400	33'000
Normal	2'600	12'400	36'500	22'000
Low-Intensity	1'600	7'700	36'500	22'000
Injection Standby	1'200	5'700	14'900	22'000
75 K Standby	0	0	0	22'000

Table 2: A useful comparison: the 10 years old LEP12 KW refrigerators

	4.5 K isothermal [W]	4.5 K – 280 K current leads [W]	50 K – 75 K radiation shield [W]
Design-Values	10'000	20'500	6'700

THE OUTLINED PROCESS

The refrigeration process which the Linde Kryotechnik AG selected for the new refrigerators is outlined as follows:
- A Brayton cycle with three pressure levels and five stages of expansion with a total of 10 turbines as indicated in the schematic temperature-entropy-diagram (Fig 1) and in the cold-box flow sheet (Fig3).
- Two compressor stages with three screw compressor units in the first stage and two screw compressor units in the second stage.

- The high pressure level is automatically adjusted to the load. The medium pressure is floating, it is reached by equalizing the flow from the coldbox with the compressors capacity.
- Liquid nitrogen pre-cooling is performed for cool-down purposes and for the liquid filling of the magnets of LHC. The liquefaction rate with liquid nitrogen support is 5800 l/h.
- Two switchable full flow adsorbers at 80 K and one full flow adsorber with a bypass at 20 K. The regeneration of the adsorbers is performed automatically by one common blower system.
- The coldbox is horizontally designed. Only the heat-exchangers in the temperature range below 20 K are in vertical position. The coldbox length is 12m, the coldbox diameter 3.5 m.

Fig.1: Schematic Temperature-Entropy diagram of the Linde LHC 18kW refrigerator.

Fig.2: Schematic Temperature-Entropy diagram of the 10 years old Linde LEP 12 kW refrigerator.

There is a remarkable difference in the process design between the LHC 18 kW refrigerators and the LEP 12 kW refrigerators which now are ten years old. The figures 1 and 2 show schematically the two process versions. Here are some motives which made Linde to change their design of large refrigerators and liquefiers.

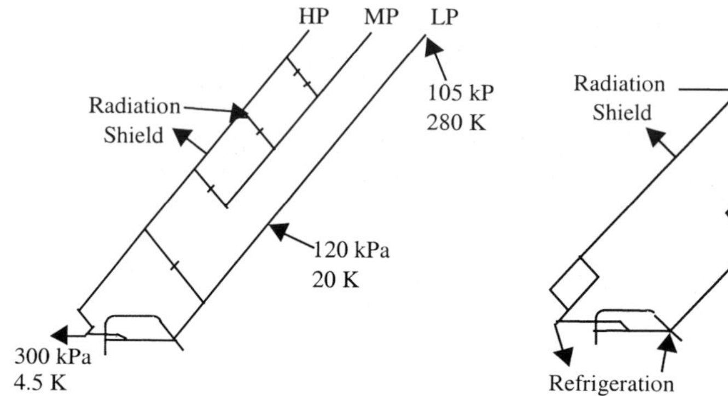

Fig. 3 Process Flow Diagram (PFD) of the coldbox

- More simple piping between turbines and heat-exchangers
- More simple and consequently more efficient heat-exchangers. The old process required alternately three or four heat-exchanger streams. The redistribution of these flows was costly and it required a considerable part of cold-box space.
- More efficient heat-exchangers. In the new process the helium flow is cooled down as far as needed. In the old process the total flow was cooled to 20 K.
- More pressure ratio on the single turbines. The new process design enables the same performance with smaller turbines.
- The slightly lower efficiencies of the turbines are compensated by the increased efficiency of the heat-exchangers.
- The new process is easy to control. The different independent expansion

stages enable to maintain independently optimal temperatures for the adsorbers beds and the radiation shield.

- The new turbine arrangement makes possible to replace a turbine while most of the other turbines are running.

EXERGY ANALYSIS

Saving power input was a declared design criteria in the Technical Specification from CERN and the electrical power consumption in three different operating modes had to be guaranteed by the bidders in their offer. Positive deviation from the guaranteed values during the acceptance tests were sadly penalized, negative deviations were honored. The Linde Kryotechnik refrigerators were low in the guarantee and during the acceptance tests even a lower power input was realized. CERN paid a remarkable bonus.

In spite of this excellent reputation, the overall adiabatic efficiency of the Linde - LHC 18 kW refrigerator box in the installed mode is barely 30.0 %. This efficiency value is the product of the following values.

Isothermal efficiency of screw compressor and motor efficiency:	54.0%
Efficiency factor for pressure drops in after-coolers, oil removal system and warm piping	96.5%
Adiabatic coldbox efficiency	57.6%

Compared to other refrigerators, the coldbox efficiency value seems to be rather moderate. Even the ten years old LEP 12 kW refrigerator is slightly better and performes an adiabatic efficiency value of 58.2 %. Is there a design failure in the new refrigerators or is the adiabatic efficiency an inappropriate reference number in the comparison of cryogenic refrigerators for different applications?

The second is true. All the helium expansion turbines today are wasting their shaft power. The use of gas bearing turbines makes it difficult to recuperate this power. Thus, refrigeration at 4.5 K can be produced with a higher adiabatic efficiency than refrigeration at higher temperatures. By the same reason liquefiers perform a lower adiabatic efficiency than 4.5 K refrigerators. Of course, the LEP12 refrigerators, after being modified and upgraded for LHC will perform a lower efficiency too.

Table 4: Efficiency of Linde-LHC 18 kW refrigerators in different modes

Installed	Normal	Low Intensity	Injection Standby	75K standby	Liquefaction
57.6%	54.7%	48.3%	48.1%	45.4%	48%

DEVIATION BETWEEN DESIGNED AND MEASERED VALUES

The main difference between the designed process values and the measurements on site was caused by the compressor system. The total volume flow in the three first stage compressors was 4% lower than expected, the total volume flow of the second stage was 7% more than expected. The power input differed proportional to the flows, thus the compressor efficiency was as expected.

The reduced first stage flow limited the maximal heat load at 4.5 K and from 4.5 K to 20 K. Fortunately there was a margin of 5% calculated and the guaranteed capacity was still fulfilled. The higher second stage flow resulted in a lowered medium pressure and consequently a higher pressure ratio on the turbines. It finally was possible to adjust a reduced discharge pressure.

CONCLUSIONS

The acceptance of the two Linde LHC 18kW refrigerators by CERN in 2003 was a milestone in the huge project of the cryogenic system for LHC. A previous milestone was fulfilled a year before with the tests of the pre-series cold compressing system[4)5)]. This year, in the queue of orders for LHC there is the commissioning of a refrigerator for the Atlas experiment provided and the next step then will be the commissioning of the four cold compressing systems in the tunnel. The Linde Kryotechnik AG further has a order for four pre-cooling and purifying systems which are provided for the upgraded LEP 12kW refrigerators. Finally, to complete the series, an order for the upgrade of two Linde-LEP 12kW refrigerators is expected for 2005. There will be stuff for further contributions in future conferences.

Picture 1: The compressor system

Picture 2: The two Linde LHC 18kW refrigerators ready for shipping

Picture 3: The coldbox top plate

Picture 4: The purge panel (left side) and the electrical cabinets (right side)

REFERENCES:

[1] A high efficient 12 kW helium refrigerator for the LEP 200 project at CERN,
     B. Chromec et al.,   "Supercollider 5," P. Hale, ed., Plenum Press, New York (1994)
[2] Specification of four new large 4.5 K helium refrigerators for the LHC, to be published
     S. Claudet et al, in: "Advances in Cryogenic Engineering Vol. 45," Plenum Press, New York (2000).
[3] Two large 18 kW (equivalent power at 4.5 K) refrigerators for CERN's LHC project supplied by Linde Kryotechnik
     AG, J. Bösel et al, CEC 1999
[4] Control Considerations of Multi Stage Cold Compression Systems in Large Helium Refrigeration Plants
     A. Kündig et al. ICEC 19, 2002
[5] The cold-compressor systems for the LHC cryoplants at CERN
     A. Kuendig et al. ICEC 20, 2004

Helium Refrigeration System for Wendelstein 7-X

Kuendig, A., Schauer, F.*, Bozhko, Y.*, Dhard C.P.*

Linde Kryotechnik AG, Daettlikonerstrasse 5, CH-8422 Pfungen, Switzerland
*Max-Planck-Institut für Plasmaphysik, Greifswald Branch, Euratom Association, Wendelsteinstrasse
1, D-17491 Germany

The Greifswald Branch of the Max-Planck-Institute for Plasma Physics (IPP) is currently constructing WENDELSTEIN 7-X, a new fusion experiment of the stellarator type. This paper gives an overview on the cooling requirements as well as on the refrigeration system chosen for the superconducting magnet system and cryo vacuum pumps of the experiment.

A flexible refrigerator was developed which is able to run in quite different working conditions. Special features are single phase forced flow cooling with adjustable supply temperatures down to 3.4 K using cold circulators and cold compressors. The various operating modes, combined with the requirement to limit the power input, day-night load leveling, and the spatial boundary conditions for the system layout were a challenge for process and construction designers.

INTRODUCTION:

With a complex arrangement of 70 superconducting coils the Wendelstein 7-X (W7-X) stellarator type fusion experiment will achieve a confinement field up to 3 T on the plasma axis. The cooling requirements for this magnet system, and principal refrigeration solutions were described earlier [1-3]. This paper presents an overview of the finally chosen refrigeration system and coolant distribution.

The total refrigeration capacity exergetic equivalent to 4.5 K will be about 7 kW without LHe support. This is a modest value as compared to the LHC refrigerators at CERN which perform a 4.5 K equivalent capacity of 18 kW. Nevertheless, the W7-X refrigerator will hold another record: It will be equipped in the final stage with thirteen turbo-machines, namely seven turbo-expanders, two cold compressors and four supercritical helium pumps. In the first stage all turbines but only one cold compressor and two pumps will be installed which allow W7-X operation up to the standard field of 2.5 T without cryo-vacuum pumps (CVPs). Suitable interfaces will be provided for the remaining equipment.

REFRIGERATION REQUIREMENTS

Six different consumers will be supplied with cryogenic refrigeration. These are the coils, the coil housings and coil support structures, the cryo-vacuum pumps, the current leads, the W7-X cryostat heat radiation shield, and the shield of the CVPs. For the CVP-shield, a two phase forced flow of liquid nitrogen is foreseen as refrigerant; for all the other consumers forced flow of single phase helium is used.

A particularity in the technical specification of this refrigerator is the considerable load variation between standby operation and the operation during experiments. During experiments not only the heat load is increased, the refrigerator has to provide more supercritical flow to the different consumers and this flow is required at a lower temperature. More flow increases the pressure drop in the coils which in return results in a disproportionately increased heat load from the circulators.

The refrigeration requirements are mainly characterized by a long time base load at standby conditions, and short-lived peaks during the W7-X experiments (s. Table 1). A cold buffer in form of a 10000 l helium dewar is used to cover refrigeration peaks during limited periods.

Table 1 Specified refrigeration capacities and operation times.
(Exergetic powers represent reversible values.)

Peak Power Mode (3.0 T – operation) 50 hours per year		Coils	Housings Structure	CVP	Current Leads	Radiation Shield	CVP-Shield
Fluid		Helium	Helium	Helium	Helium	Helium	Nitrogen
Heat load	[W]	1100	1800	450	39000	14000	6000
Mass-flow	[g/s]	450	800	250	25	135	300
Supply temperature	[K]	3.4	3.4	3.4	4.5*	≈50	80
Return temperature	[K]	4.4	4.1	4.2	300	≈70	80
Supply pressure	[kPa]	530	340	360	120	<1700	200-800
Pressure drop	[kPa]	230	40	60	15	200	80
Exergetic power input	[kW]	138.3	159.1	44.2	134.7	68.4	16.5

Standard Mode (2.5 T – operation) 700 hours per year		Coils	Housings Structure	CVP	Current Leads	Radiation Shield	CVP-Shield
Fluid		Helium	Helium	Helium	Helium	Helium	Nitrogen
Heat load	[W]	800	1800	450	23400	14000	6000
Mass-flow	[g/s]	200	300	250	15	135	300
Supply temperature	[K]	3.9	3.9	3.9	4.5*	≈50	80
Return temperature	[K]	4.9	5.1	4.4	300	≈70	80
Supply pressure	[kPa]	370	310	360	120	<1700	200-800
Pressure drop	[kPa]	70	10	60	15	200	80
Exergetic power input	[kW]	60.8	118.1	39.8	82.8	68.4	16.5

Standby 8000 hours per year		Coils	Housings Structure	CVP	Current Leads	Radiation Shield	CVP-Shield
Fluid		Helium	Helium	Helium	Helium	Helium	Nitrogen
Heat load	[W]	250	1800	450	7800	14000**	6000
Mass-flow	[g/s]	12	86	22	5	135	300
Supply temperature	[K]	7	7	7	4.5*	≈50.	80
Return temperature	[K]	10	10	10	300	≈70	80
Exergetic power input	[kW]	8.3	60	15.1	27.6	68.4	16.5

*) saturated vapor of 4.5K   **) 28000 W during plasma vessel baking

THE COLD BUFFER

During 90% of the scheduled operating time the refrigerator will be in standby mode. Only 10% of the time is planned for W7-X experiments, and the refrigeration for most of these experiments is performed in the standard (2.5 T) mode. Only for a few tests, which currently are scheduled for less than 1% of total operating time, the peak power mode will be required. Thus, it was decided to design a refrigerator, which performs the specified capacities in the

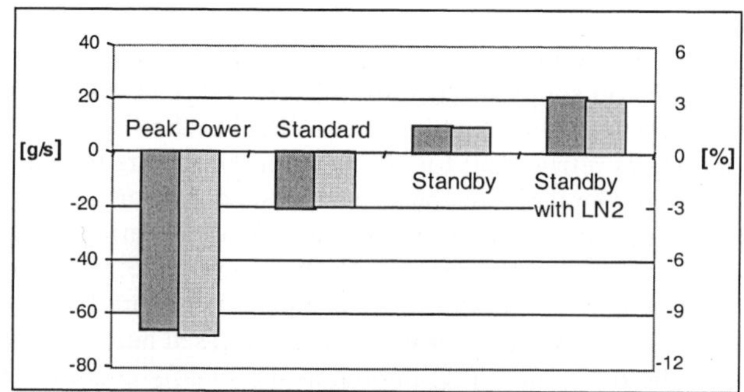

Figure 1 Liquefaction (+) and evaporation of liquid (-) in different modes in g/s and as percentage of the high pressure flow.

standard mode and in the peak power mode only with the support of liquid helium from a dewar. A single test at peak power conditions will take not longer than four hours according to the LHe tank which is available. During the standby periods the refrigerator shall re-liquefy the helium. In the warm state the helium inventory is stored in a 1000 m^3 tank battery, in the cold state in a dewar of 10 m^3. Figure 1 shows the quantities of helium which swaps between the two buffers. The refrigerator is tuned to get the best rate of yield from the stored coldness. It maintains the dewar permanently at 120 kPa and redraws the liquid helium into one of the sub-atmospheric sub-coolers. There, by the heat of evaporation, this helium contributes to the refrigeration. The vapor enthalpy compensates for the refrigeration of the current leads; it perfectly covers the exergetic losses of the heat exchangers, and it contributes to the shield cooling. The turbines provide the refrigeration for the Joule Thompson stage, for the cold compressors, and for most of the radiation shield cooling. During the peak power mode, the turbines T1 and T2 (Fig.2) are not operated.

During standby the circulators for supercritical liquid helium are switched off, and the temperatures of all the turbines are slightly increased. The high pressure level is reduced, the medium pressure level increased, and all the turbines keep running, by this way providing a high liquefaction rate (Fig. 2).

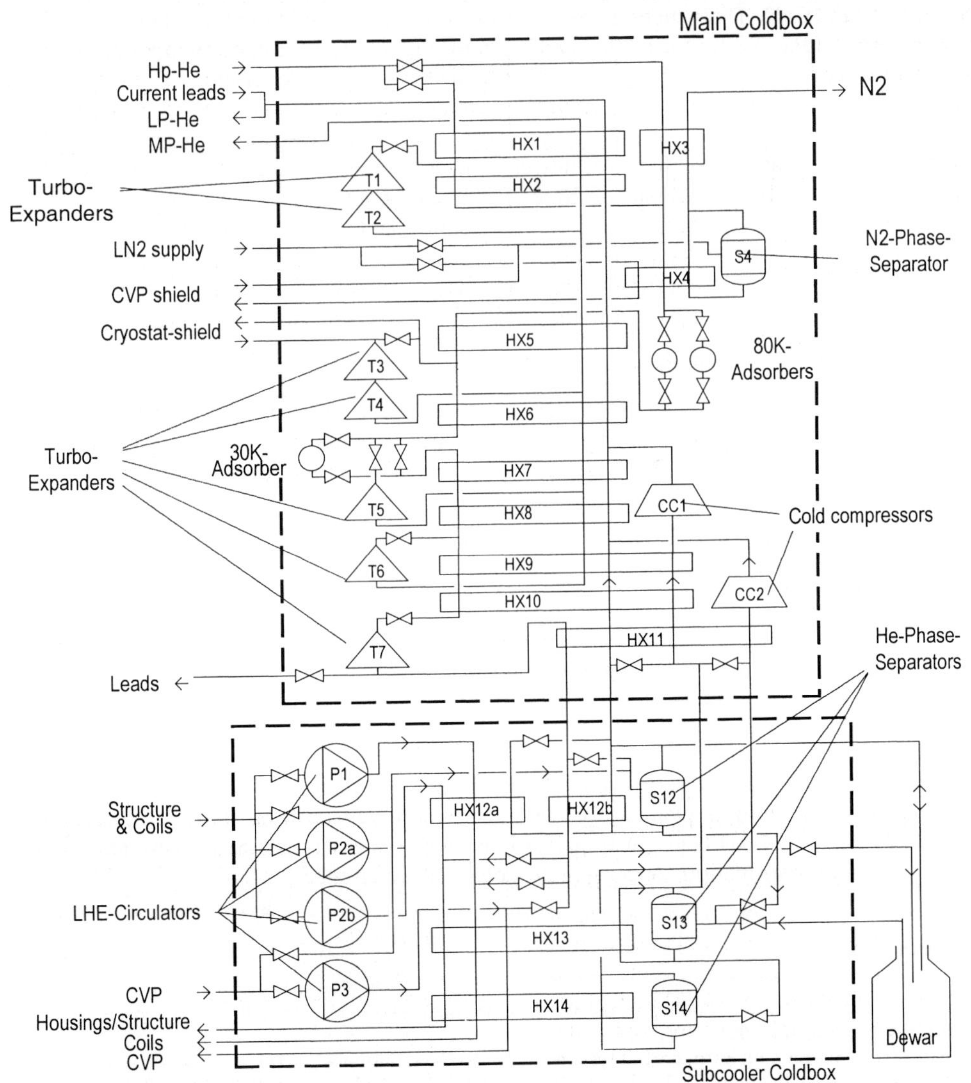

Figure 2: Simplified flow sheet of the refrigerator. Only main flow lines are indicated. Special lines for standby and transient modes are not indicated (P2b, P3, and CC2 will be installed at a later stage).

If necessary, the liquefaction rate of helium may be increased with liquid nitrogen pre-cooling. A two phase nitrogen flow is also used to provide 6 kW of refrigeration to the shields of the cryo-vacuum pumps. This nitrogen flow returns to the refrigerator mainly as vapor which is further used to support the helium cooling in the heat-exchangers HX1 and HX2. An interface is provided to boost this nitrogen flow at a later stage with an additional pump, if necessary.

THE SUB-COOLING - AND DISTRIBUTION SYSTEM

Special attention is paid to the sub-cooling and distribution system (Fig.2). Each of the supercritical streams to the refrigeration consumers can be circulated either by a liquid helium pump, or for the purpose of redundancy, by the redirected Joule Thompson stream of the refrigerator.

Three helium evaporation stages at three different temperature levels provide conditioning of these streams. The vapor phase of the first stage is connected to the refrigerators low pressure line at 120 kPa and holds the saturation temperature of 4.4 K. The two other stages are operating at sub-atmospheric pressures. The vapor returning from them is recompressed to low pressure level in the two cold turb compressors CC1 and CC2. Their suction pressures are adjustable by varying the rotation speed. The provided pressures are 66 kPa (3.8 K) for CC1 and 37 kPa (3.3 K) for CC2. Supply temperatures to the refrigeration consumers of more than 3.8 K are provided with only CC1 in operation. This arrangement provides the following advantages:

- The three evaporators do not only save energy, they save investment costs too. Applying a system with only two evaporators, the total power input into the system during the peak power mode would be 20% more. Only a small part of these 20% could be compensated by drawing more liquid helium from the dewar. Inescapably a bigger compressor size and more turbine capacity would be required.
- At the specified heat loads, in the peak power mode as well as in the standard mode, the cold compressor CC1 is running at optimal conditions. The cold compressor CC2 is running at design conditions in the peak power mode and in the standard mode it is simply turned off.
- Due to the different pressure ratios the two cold compressors can be placed at different temperature levels in the process. This is helpful for the layout of turbines and heat exchangers.
- By using cold turbo compressors instead of a warm process vacuum system, none of the heat exchangers above 15 K requires more than three streams, and the used streams flow in all the operating modes. This enables a heat exchanger design with a comparably small cross section and with a good heat transfer rate.

CONCLUSIONS

With the W7-X refrigerator, Linde Kryotechnik AG and the Max Planck Institute for Plasma Physics broke new grounds. Thanks to the well directed use of the cold buffer as well as the extremely optimized sub-cooling system the refrigerator will save costs and energy. Due to the application of cold compressors and time distributed loads the warm compressor system will be comparably small. Apart from saving costs, space within the buildings is saved as well.

The W7-X refrigerator was ordered a few months ago, and currently we are in the phase of basic engineering. Commissioning is planned for 2008 and experiments will be started in 2010. The Linde Kryotechnik AG and the Max-Planck-Institute for Plasma Physics will use the opportunities of future conferences to give more information concerning progress and experiences.

REFERENCES

1. Nagel, M. and Schauer, F., Cooling of the W7-X Superconducting Coils, Proc. 19th Int. Cryog. Eng. Conf. (ICEC 19), Grenoble, France (2002) 677-680
2. Bozhko, Y. and Schauer, F., Refrigeration system for W7-X, Nucl. Fusion 43 (2003) 835-841
3. Schauer, F., Bau, H., Bozhko, Y., Brockmann, R., Nagel, M., Pietsch, M., Raatz, S., Cryotechnology for Wendelstein 7-X, Fusion Eng. and Design 66-68 (2003) 1045-1048

Cryogen Free Refrigeration to 270mK by Adiabatic Expansion of He^3 Gas

Mikheev V., Noonan P., Adams A.

Oxford Instruments Superconductivity, Tubney Woods, Abingdon, Oxon OX13 5QX, UK

Using the well-know phenomena of adiabatic expansion of gas and adsorption pumping, two new types of He^3 refrigerator have been built at OIS. The first type operates in a liquid helium environment at 4.2K (so called "wet"). The second design is cryogen free and uses a Pulse Tube Cryocooler ("dry"). This ULT system is embodied in a commercial He^3 Cryofree refrigerator Heliox AC-V, developed by Oxford Instruments. This system uses single-phase electrical power, does not require a water cooler and can be turn key in operation. Cooling performance and other features and instrumentation services will be discussed.

A lot of fascinating and intriguing phenomena in Physics are unravelled at ultra low temperatures (ULT). That is why ULT refrigerators are still in high demand. The first stage of ULT is temperatures below 1K, which can be achieved by the so-called He^3 refrigerator. For many decades typical He^3 refrigerators have comprised two refrigerators in one system. The first is a He^4 refrigerator (so called 1K pot) for condensing He^3 gas into liquid. The second is the He^3 refrigerator itself, which produces cooling to a temperature below 300mK by pumping vapour above liquid He^3 typically using a cryogenic adsorption pump [1].

An essential attribute for any adsorption He^3 system is the 1K pot. This device produces the low temperature required for condensation of He^3 gas. Usually this refrigerator reaches ~1.5K and the He^3 liquefaction fraction is greater than 90% of the initial gas content. Starting He^3 vapour pumping at such a low temperature means that very little He^3 liquid needs to be evaporated to achieve base temperature. This leads to 2-3 days hold time with a residual heat leak of order 10 microWatt to the He^3 pot – the coldest part of refrigerator. This type of the He^3 refrigerator is reasonably simple to construct and operate, provides good temperature stability and is easy to control. Commercial He^3 systems are usually automated [1]. The only complication is the necessity for a second refrigerator with liquid He^4 with its needle valve or impedance and all pumping accessories, including a rotary pump, pumping line, valves etc. This complication makes the system expensive, where the main cost contribution is the additional He^4 refrigerator.

However it has been shown that the He^4 refrigerator is not required and can be avoided by using the procedure of adiabatic expansion of He^3 gas. A suitable "expander" can be made using an adsorption pump at variable temperature [2]. This new method of He^3 gas liquefaction reduces the cost of the whole system.

ADIABATIC (ISENTROPIC) EXPANSION OF HE^3 GAS AS A SOLUTION FOR THE REPLACEMENT OF THE 1K POT

The adiabatic expansion of gases is a well known phenomenon and method of cooling gases. The majority of industrial liquefiers utilise adiabatic expansion for cooling and liquefaction of air, hydrogen and noble gases including He^4. Current commonly used cryogenic expansion machines are reciprocating piston or turbine expansion devices.

Only He^3 has been excluded from this process. This has happened for obvious reasons. Expansion machines are expensive and complicated mechanical devices with a significant volume of operating gas. The 1K pot is simpler, smaller and cheaper than standard expansion devices and the quantity of He^3 gas is very limited so the 1K pot arrangement is commonly used.

From the end of 1995 we were developing novel He^3 systems at Oxford Instruments and an adiabatic method was under consideration from the beginning. To avoid a complicated expansion machine an attempt was made to cool He^3 by expanding it into a reservoir. The cooling effect was seen straight away and some liquefaction of He^3 was observed. Later we found an article by Troitskii and Fradkov, who observed this effect and used it for operating a He^3 refrigerator [2]. Their result was very modest when compared with commercial He^3 systems. The lowest temperature obtained was ~0.36K and could be maintained for only 9 hours. This may explain why the method of adiabatic expansion of He^3 gas was not recognised until now.

The challenge was to improve the adiabatic expansion method to achieve the specification required for a commercial product. The results of this work are reported below.

Let us to remind ourselves briefly of the main, simple thermodynamic equations for gas adiabatic expansion and its implementation for He^3.

As it is well known for an ideal gas one can write

$$PV = mRT \qquad\qquad (1)$$

Where P is the pressure, V is the volume, R is the gas constant and m is the mass of the gas at the temperature T. The reversible adiabatic (isentropic) process means that entropy of the system is constant, hence

$$\Delta S = c_v \, ln(\frac{T_2}{T_1}) + Rln(\frac{V_2}{V_1}) = 0 \qquad\qquad (2)$$

The relationships between thermodynamic parameters of the system are defined as

$$(\frac{T_2}{T_1}) = (\frac{P_2}{P_1})^{(\gamma-1)/\gamma}; \qquad\qquad (\frac{T_2}{T_1})(\frac{V_2}{V_1})^{R/c_v} = 1 \qquad\qquad (3)$$

He^3 is a monoatomic gas, hence $\gamma = 5/3$

When the pressure of He^3 gas drops from P_1 to P_2 the final temperature T_2 is defined by the expression

$$T_2 = T_1(\frac{P_2}{P_1})^{0.4} \qquad\qquad (4)$$

For quantative evaluation of (4) let's consider a very modest set of conditions:

- initial temperature of He^3 gas T_1= 4.2K
- the pressure of the gas drops adiabatically from 3 Bar to 1 Bar.

According to (4) T_2 should decrease from 4.2 K to 2.7K , which means it has dropped below the boiling point of 3.2K for 1 bar. This means that liquefaction should take place under these conditions. This effect is easy to demonstrate on the Entropy -Temperature diagram, obtained by Daunt [3] and shown on Fig.1

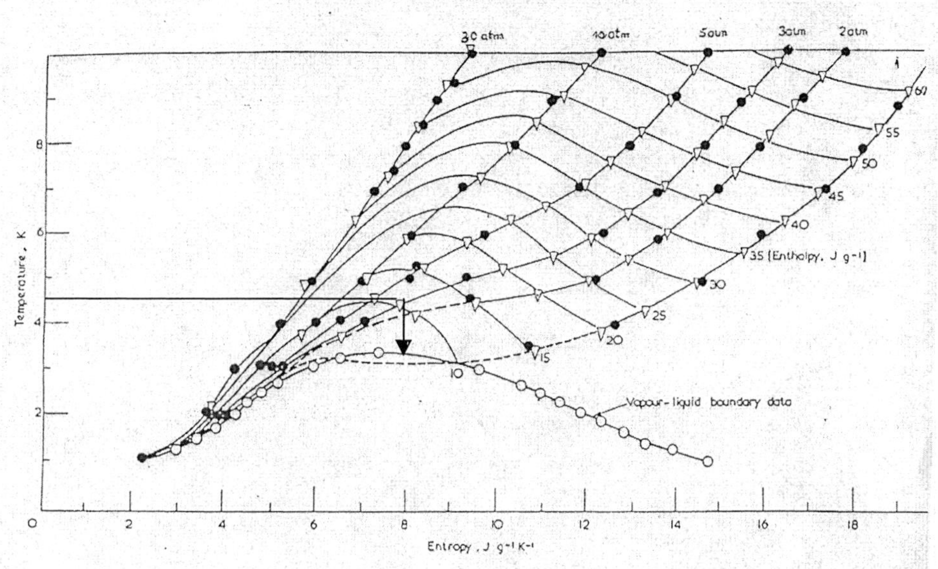

Fig 1. Isobars and isenthalps on the temperature, T, versus entropy, S, chart for He^3

The intersection of the horizontal line at the temperature 4.2K with the 3 bar Isobar (3 atm on the picture) corresponds to the initial state of the He^3 gas. The vertical line down from 3 bar isobar to 1 isobar describes the adiabatic expansion process and the intersection of this line with the vapour–liquid curve corresponds to He^3 liquefaction. In fact this process will start slightly earlier at the critical point for He^3 which is 3.3K. The liquefaction process continues until the pressure of the He^3 gas is above the saturated vapour pressure of the liquid.

The liquid fraction achieved is a key parameter and strongly depends on the initial gas conditions of pressure and temperature. It is important to note that to have a significant high pressure of He^3 at 4.2K in the sealed volume one would need a very high gas pressure at room temperature.

There are advantages to expanding into an additional reservoir that is being cooled. A cooled reservoir with a given mass of gas will have a reduced pressure according to (1) as T_{hot}/T_{cold} =P_{hot} /P_{cold}. If this reservoir is connected to our He^3 pot at high pressure it will cause expansion of He^3 from the He^3 pot to the cooled vessel. Neglecting friction and any heat loses due to thermal conductivity, the expansion process is adiabatic and the He^3 gas produces work during its flow from the He^3 pot to the additional

58

reservoir. This process is essentially single shot, but in the case of He^3 gas even single shot expansion is very effective because the operating temperature range is wide. For example cooling of the reservoir from 80K down to 4K corresponds to a pressure drop of 20 times. This effect could be described in different language as an increase of its effective volume by factor 20 according to $T_{hot}/T_{cold} = V_{cold}/V_{hot}$ at constant pressure (analogy with the piston expander).

At low temperatures the effect of pressure drop will be increased because this reservoir is filled with an adsorbent and becomes an adsorption pump, which at a temperature below 20K starts to adsorb He^3 gas and decrease the pressure still further. After liquefaction of part of the He^3 gas the process automatically goes to the standard operation mode of pumping vapour above the He^3 liquid.

Analysis shows that to make such a single expansion refrigerator effective for operation from 4.2K one needs to create a very high initial pressure and optimise the volumes in the whole system and its temperature profile. That is because this system has as its essential element an adsorption pump with limited sorption capacity for He^3 gas, at a pressure below 10^{-3} mbar, which corresponds T~300mK. The adsorption pump should operate without saturation the whole time until all liquid He^3 in the He^3 pot has been evaporated .

The simple calculations of the expansion conditions are based on the simplified picture Fig2.
The maximum pressure for this system will be defined by the quantity of He^3 gas, adsorbed at T=4.2K and P= 10^{-3} mbar. For simplicity we neglect the volume of the connecting tube. Then

$$P_{max} = C_0 \rho V_0 \, / \, [V_0(T_0) + V_3(T_3)] \qquad\qquad (5)$$

where ρ is density of adsorbent, and C_0 –sorption capacity at 10^{-3} mbar , V_3-volume of He^3 Pot, T_0 ,T_3- initial temperatures of adsorption pump and He^3 pot before expansion respectively.

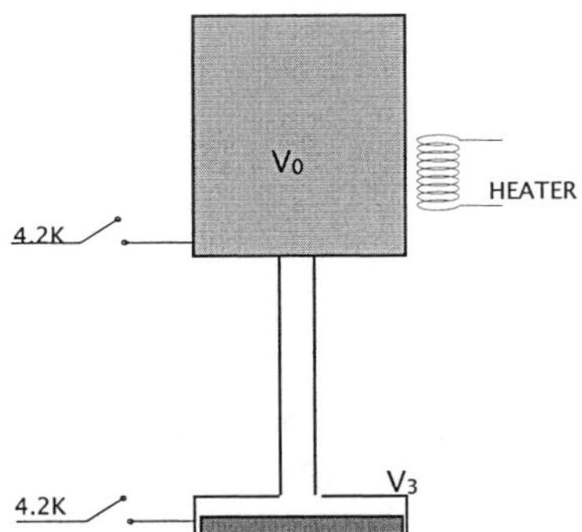

Fig.2 Schematic of a single expansion He^3 refrigerator vessel with a reservoir V_0 with variable temperature. The heater and heat switches provide for heating and cooling of components to 4.2K.

The maximum pre-expansion pressure will be defined by the amount of adsorbed He^3 gas, the volume of the He^3 pot and the initial temperature of the adsorption pump. The limited amount of He^3 in the system will lead to a small liquefied fraction and restrict the refrigerator to a short hold time at the base temperature. In practice to get a reasonable liquid fraction after expansion the He^3 gas initial pressure should be more then 100 bar at the room temperature. Naturally, such a high pressure in the He^3 system impacts the system design and safety requirements.

In this work we describe another method of significantly increasing the liquid fraction without needing a very high initial pressure. We called it the multi-expansion process [4].
The main feature of this method is to deliberately saturate the adsorption pump with He^3 at the initial stage with the aim of being able to increase the gas pressure. For this we utilised the pressure dependence of the sorption capacity. For example at a pressure of about 100 mbar the adsorption pump can accept a double quantity of He^3 gas compared with 10^{-3} mbar. According to (5) P_{max} will rise by the same factor, which is good, but we should remember that after expansion with such a large amount of gas the adsorbent will be saturated. It will not then be able to operate as a pump at the pressure of 10^{-3} mbar to obtain the base temperature.

To solve this problem we used a multi-expansion process, which needs an additional volume sufficient to accommodate all the extra He^3 gas during the first stage of expansion. For simple expansion the extra gas can be transferred from the He^3 pot to the room temperature reservoir [5]. To accommodate more He^3 during the first stage of expansion we made a pre-expansion reservoir with small volume, but cooled by attachment to the first stage of the pulse tube refrigerator (PTR). As a result the room temperature pressure in our system does not exceed 10 Bar (good safety feature).
The multi-expansion layout is shown in Fig3. The procedure is as follows:

1. The adsorption pump is cold and has adsorbed the maximum amount of He^3 gas from the cold dump (valve is open) until it saturated at approximately 100 mbar.
2. The-adsorption pump is warmed up to 60-80K and the valve is closed to create maximum pressure in the He^3 pot.
3. Wait for about 20-30min until the He^3 Pot has pre-cooled to ~4.2K.
4. The first stage of expansion, with the adsorption pump warm, happens by opening the valve and discharging high pressure He^3 gas into the cold dump
5. The second expansion stage occurs by cooling the adsorption pump with the valve closed.
After cooling of the adsorption pump below 20K it works like a standard He^3 evaporating refrigerator down to the base temperature of <300mK. The typical fraction of He^3 liquid obtained with this procedure is ~50%.

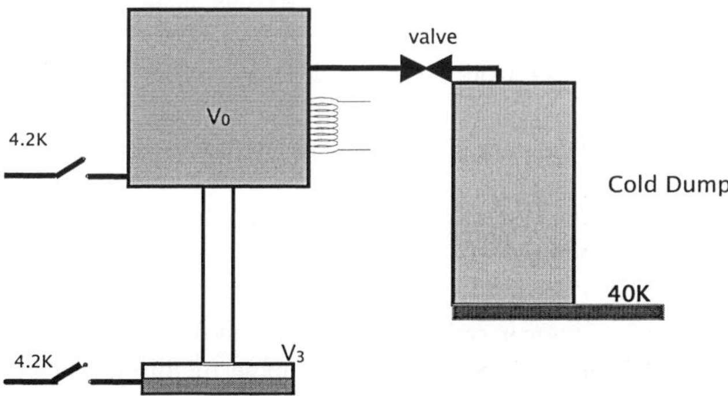

Fig 3. Schematic view of multi-expansion layout

NOTE: It is important to note that the valve operation does not have a significant effect on the refrigerator's vibration performance because the valve's operation takes place only during the regeneration mode and not at base temperature.

Utilising the multi-expansion method we have built two types of He^3 refrigerator, i.e. "wet" and "dry", at Oxford Instruments Superconductivity.

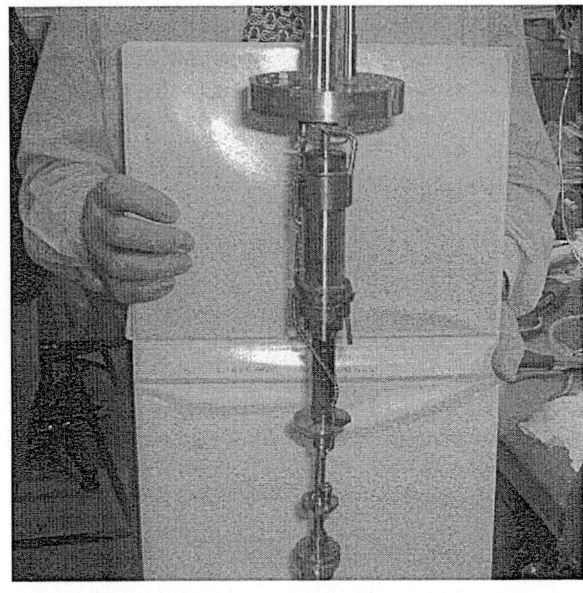

Fig.4 Experimental "wet" prototype of expansion He^3 refrigerator for operating in liquid He^4

The "wet" system means that the He^3 refrigerator is operating in a dewar with liquid He^4 at 4.2K. The refrigerator itself has a vacuum jacket with a capillary inside for liquid He^4 flow through heat exchangers for pre-cooling the He^3 gas and the adsorption pump. The development prototype was tested in January 2003 and an its external view shown in Fig.4.

The "wet" system shown in Fig.4 (vacuum jacket has been removed) has achieved the following specification:

Base temperature <270 mK, Hold time >70hours, Cooling Power 100 microWatt at T<350mK Total He^3 amount ~7 L STP

This system is genuinely vibration less and could satisfy the most serious demands for vibration sensitive measurement.

The second type of refrigerator is based on the "dry" version. It is a so-called cryogen free system based on a pulse tube cryocooler (Cryomech PT405).

The base temperature and hold time of this Cryofree system is very similar to the "wet" system. It was also established that the adiabatic He^3 refrigerator is rather economical and does not require the high cooling power of the PT 405, which is 0.5 Watt at 4.2K. The disadvantage of this type PTR's is the existence of a water-cooled compressor, which makes the cryocooler very bulky and has high electrical Power consumption.

Recently Cryomech Ltd has developed a special pulse tube PT403 with an air-cooled 3KW compressor and the cooling power 300 mW at 4.2K.

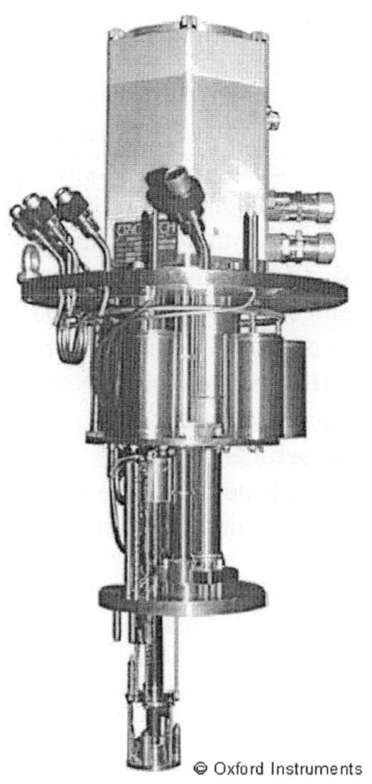

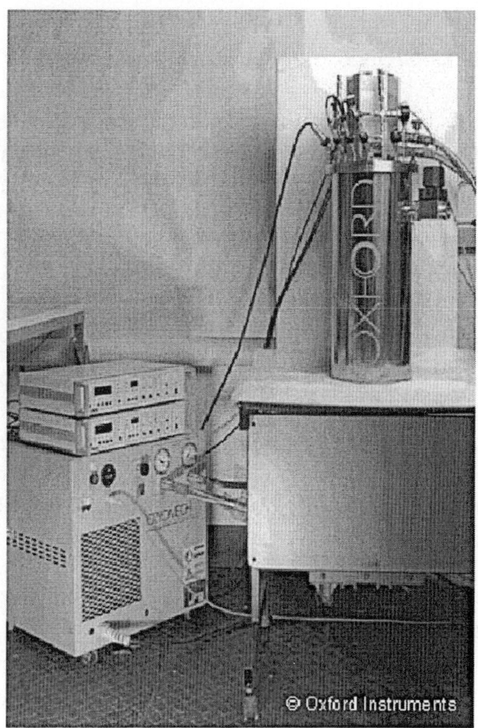

Fig 5 The Cold Head of PT403 with adiabatic expansion He^3 refrigerator and the general view of Cryogen free system Heliox AC-V

Fig 6. Cooling Power curve for Heliox AC-V

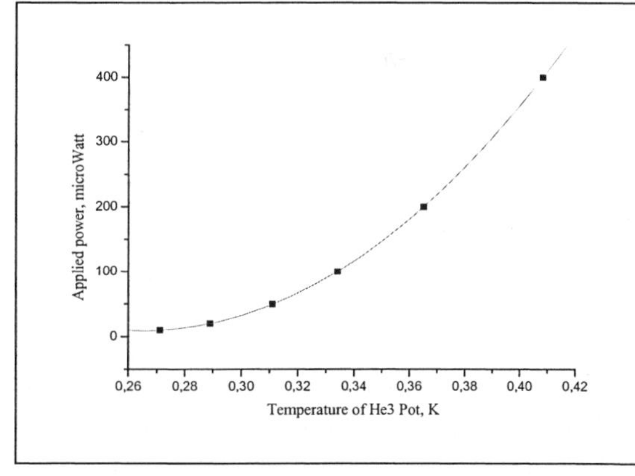

That is sufficient cooling power for operation of our adiabatic system and OIS has developed a product Heliox AC-V based on the air-cooled PT403.

Additional specification parameters:
-       Temperature stability -   better than ±2 mK at < 5K
-       Regeneration time 1 hour typically
-       Cooldown <10 hours typically

CONCLUSIONS

Oxford Instruments Superconductivity has developed and built two types of He^3 refrigerator with liquefaction of He^3 gas by the adiabatic expansion process. Both systems have a similar specification. The "wet" system is a genuine vibrationless refrigerator for the most sensitive measurements. The "dry" system is operated from single-phase electrical power, does not need a water cooler and hence became the first ultra low temperature " kitchen" refrigerator. It is important to mention the reliability of the expansion refrigerator. Because current cryocoolers produce cooling down to <3K, this causes an extra liquid fraction by direct condensation into the He^3 pot. However this temperature is close to the He^3 critical point so any failure in the pulse tube and subsequent deterioration of the minimum temperature achieved by the PTR will completely stop the operation of a traditional He^3 refrigerator. For our He^3 system this limit is much higher at 4.2K or even 5K, where adiabatic expansion of He^3 still produces liquefaction of He^3 and hence ultra low temperatures. This fact was proven experimentally.

We believe that the described here method will progress in the future. This point is supported by a recent publication [6] by a group from Roma University of a "wet" refrigerator based on adiabatic expansion of He^3.

REFERENCES:

[1] Oxford Instruments brochure with product guide.
[2] V.F.Troitskii, A.B.Fradkov, Instruments and Experimental Techniques 27, N2, 483-486, 1984
[3] J.G. Daunt, Cryogenics, N12, 473-479, 1970
[4] OIS patent applications US20040089017, EP1387133 and JP2003-310723"
[5] F. Simon expansion described in the book "Advanced Cryogenics", edited C A Bailey, Plenum Press 1971, p 40
[6] A. Graziani, G Dall'Oglio, L Martinis, L Pizzo, L Sabbatini, Cryogenics, 43, 659-662, 2003

Development of the Cryogenic Ground Support Equipment (CGSE) for the Superconducting Magnet of the Alpha Magnetic Spectrometer (AMS-02)

Gu A.[1], Grechko A.[2], Harrison S.[4], Hofer H.[3], Lin W.[1], Shi Y.[1], Ting S.C.C.[5], Vostrikov S.[2], Wang R.[1]

[1] Shanghai Jiao Tong University, 1954 Huashan Rd., Shanghai, 200030, China
[2] Kurchatov Institute, Kurchatov Square, 1, Moscow, 123182, Russia
[3] Swiss Federal Institute of Technology, Laboratory for High Energy Physics, Building HPK, ETH-Honggerberg, CH-8093, Zurich, Switzerland
[4] Space Cryomagnetics Ltd., E1 Culham Science Centre, Abingdon, England
[5] Massachusetts Institute of Technology, Cambridge, MA 02139, USA

AMS-02 (http://ams.cern.ch/AMS/ams_homepage.html) is a particle detector based on the International Space Station (ISS). Its mission is to search for antimatter, dark matter and missing matter in space. At the heart of the detector is a large, powerful superconducting magnet. The CGSE is required to cool down the 2000 kg magnet to its operating temperature of 1.8 K, and to fill the magnet helium vessel with 2500 l of superfluid helium on the ground and before launch.

INTRODUCTION

The Alpha Magnetic Spectrometer (AMS-02) is a particle detector designed to search for anti-matter, dark matter and the origin of cosmic rays in space. The detector will be assembled in CERN, Geneva and installed on the International Space Station (ISS). The planned duration of the experiment is between 3 and 5 years. One of the major components of the detector is a superconducting magnet, currently under construction by Space Cryomagnetics Ltd of Culham, England. The magnetic dipole field is achieved by an arrangement of 14 superconducting coils. The magnet system [1, 2] consists of a pair of large Helmholtz-like coils together with two series each of six racetrack coils, circumferentially distributed between them. This arrangement was mainly chosen to minimize the stray field outside of the magnet and so reduce the magnetic dipole moment. The magnet generates a field of 0.9 T in the centre of the 1.1 m diameter clear bore. All the superconducting coils were wound from a high-purity, aluminium-stabilised mono-strand niobium-titanium (NbTi) conductor. The coils are located inside a toroidal vacuum vessel, and are indirectly cooled by superfluid helium at 1.8 K. The cooling circuit is thermally connected to a 2500 l superfluid helium vessel that serves as a cold reservoir. This cooling system is designed to ensure that any loss of superfluid helium, following a quench on orbit, is minimised.

The Cryogenic Ground Support Equipment (CGSE) for AMS-02 will be used to cool down the magnet and to fill the vessel of the magnet with superfluid helium. It is currently under development by the magnet group of the AMS collaboration and will be manufactured in China. The main requirements of the system are as follows.

- Compactness and mobility for testing the detector in different parts of the world (China, Europe, USA). Ultimately it has to be located in the confined space of one of the launch pads at the Kennedy Space Center, and it must be able to be disconnected quickly and safely only hours before the shuttle launch.
- Reliability and safety are especially important during preparation of the detector for operations at the launch pad. The last top-up with superfluid helium is planned to be just 88 hours before launch to ensure as much helium as possible is available to keep the magnet cold on orbit. All electrical equipment used on the launch pad has to be explosion-proof because of the hazards peculiar to this environment.

- The helium used for filling the magnet has to be particularly pure. Any impurities - even small solid particles of frozen gases or water - can prevent sealing of a cryogenic valve, leading to high heat loads and rapid consumption of the superfluid helium. This would limit the useful lifetime of the experiment.
- Controlled, gradual and uniform cooling down of the magnet, with temperature gradient not more than 50 K across the coils. This eliminates the risk of damage to the magnet due to thermal stresses at temperatures above 90 K.

CHOICE OF THE METHOD FOR GENERATING SUPERFLUID HELIUM

During development of the system, it is particularly important to consider how to cool down to superfluid helium temperature. One notable feature of the AMS-02 magnet system is that – for various reasons - its cryogenic valves are rather small. This makes it relatively difficult to pump helium vapour from the helium vessel owing to the substantial hydraulic resistance, which limits the flow rate during pumping.

Usually, superfluid helium can be produced by pumping vapour from the vessel. In this case approximately half of the initial volume of liquid helium is lost. Then the vessel has to be topped up. For this purpose helium is supplied to the vessel through a Joule-Thomson (J-T) valve to provide the required pressure drop between the supply Dewar and the target vessel. The temperature and pressure of the liquid helium before the J-T valve are of great importance in determining how much vapour will be produced by the expansion across the valve. Fig.1 shows that, if normal liquid helium at 4.2 K is expanded through the J-T valve, the downstream vapour fraction exceeds 40%. To pump such a large quantity of vapour from the flight vessel would be a very difficult and time-consuming operation. It is therefore better to pre-cool the liquid helium before throttling. One possibility is to use a filling line with heat exchange between the supply flow of liquid helium and the return flow of low temperature pumped helium vapour. In this manner it is possible to reduce the temperature of the liquid helium to around 3 K before the valve: this will reduce the vapour mass fraction to 21%. This method was used for the launch of the Infrared Space Observatory (ISO), a European Space Agency (ESA) mission. Experience with this system showed that this is a large amount of vapour and it is difficult to pump through narrow flight valves. It is best to reduce the amount of vapour pumped from the vessel as much as possible.

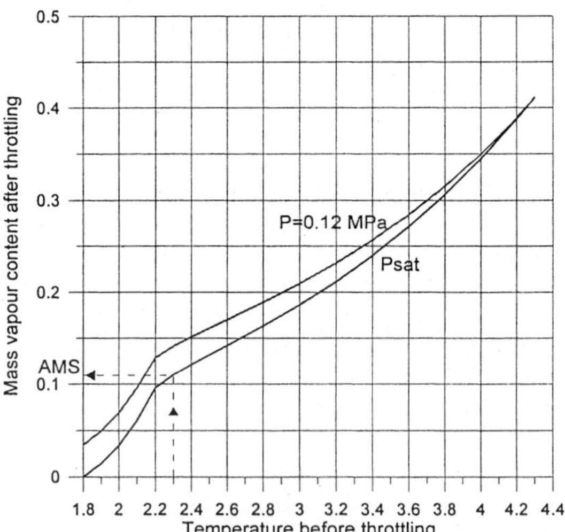

Figure 1 Mass vapour content after throttling versus temperature before throttling at pressure 0.12 MPa and at saturated pressure for helium

The following procedure has been developed for filling AMS magnet with superfluid helium. First, the AMS magnet is pumped to the operating conditions of 1.8 K at the saturation pressure of 16 mbar. Then, during the top-up operation, the master liquid helium Dewar is pumped to a state close to superfluidity (2.3 K, 53 mbar): this pre-cooled helium flows through the J-T valve and enters the AMS magnet. In this case the mass vapour content will be only 11%, which is not too difficult to pump. This method for topping up has been experimentally tested in the Kurchatov Institute, Russia and showed good results.

SHORT DESCRIPTION OF THE SYSTEM

The main components of the CGSE are (Fig. 2): a system for controlled cool down and warm up in the temperature range between 300 and 80 K; liquid nitrogen tank; main (master) 1000 l Dewar for liquid helium permanently connected to the system; a few interchangeable 1000 l Dewars for supply with liquid helium; cold valve box; a vacuum pump system consisting of two Leybold vacuum pumps RUTA WS2001FU/SV630F/A with total capacity 2x2000 m^3/h at 8 mbar; a gaseous helium supply for the Superfluid Cooling Loop (SCL) of the magnet; set of cryogenic lines; and a cryostat to simulate the AMS magnet.

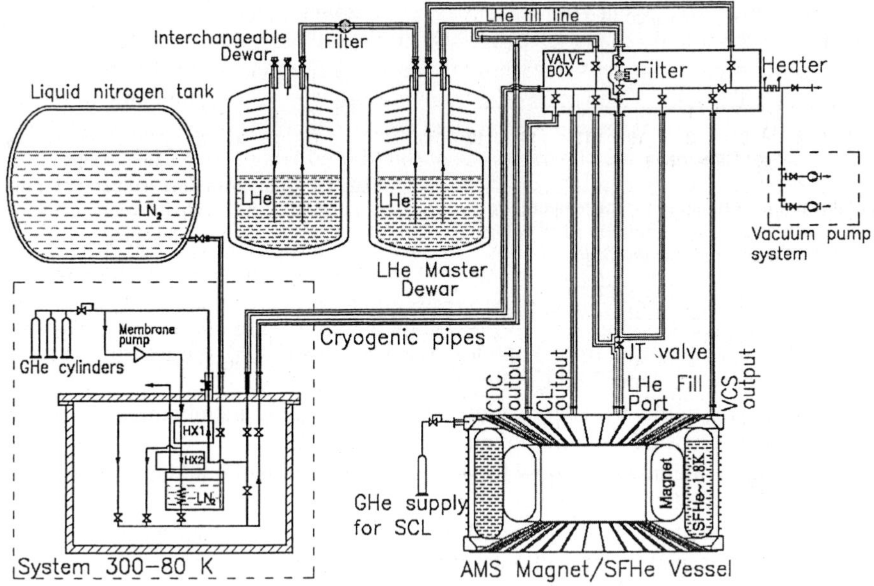

Figure 2 Simplified layout of the CGSE

The principle for cooling between 300 and 80 K is to circulate the helium gas through heat exchangers HX1, HX2 and a liquid nitrogen bath, with the return gas flowing through heat exchangers. The temperature of the cooling flow is controlled by mixing warm or intermediate and cold flows to provide smooth cooling of the magnet without large thermal stresses. At temperatures between 80 and 4.2 K the magnet is cooled down by liquid helium from Dewars. From 4.2 to 1.8 K the vessel is pumped down to 16 mbar. Liquid helium is added from Dewars through the J-T valve.

Table 1 Key CGSE parameters

Parameter	Value
Cooled mass (magnet, helium vessel)	2000 kg
Temperature range of cooling	300~1.8 K
Maximal temperature gradient during at the magnet at the range 300 − 90 K	50K
Volume of the magnet helium vessel	2500 l
Maximal pressure in the helium vessel	1.6 bar
Consumption of cryogens for one cool down and filling cycle	5 m^3 LHe +2m^3 LN$_2$

The minimum time required to cool down from 300 to 4.2 K is 14 days (Fig.3). The time taken to pump from 4.2 to 1.8 K is not more than 22 hours (Fig.4). The flow rate of superfluid helium during the top-up operation is between 40 and 100 l/h.

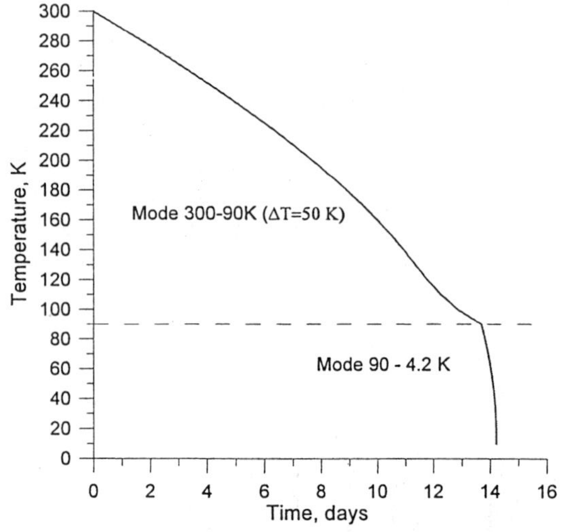

Figure 3 Cool down time estimation for the temperature range 300-4.2 K. Calculated for two 4 mm ID AMS valves in series and maximum flowrate 2 g/s

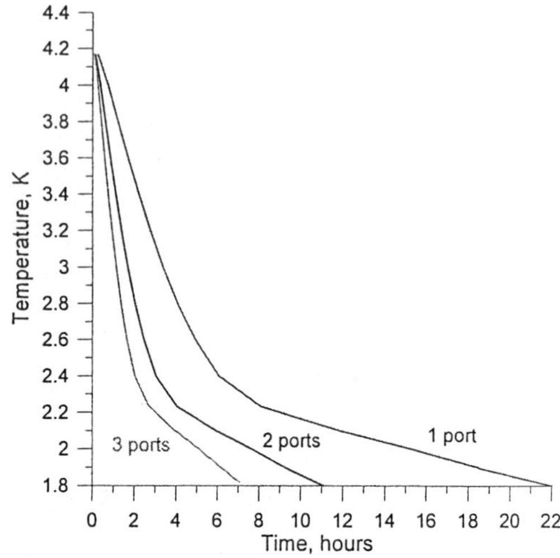

Figure 4 Cool down time estimation for the temperature range 4.2-1.8 K with different numbers of ports for pumping

Some special features of the system.

- 5 ports for helium connections to AMS-02: liquid helium fill port, output of helium from the Cool Down Circuit (CDC), output from Vapour Cooled Shields (VCS), output from current leads (CL), supply of helium to the superfluid Cooling Loop (SCL).
- Clean compressors without oil lubrication and with a double membrane system are used in the system 300-80 K to ensure the purity of circulating helium.
- Utilization of cold helium return flow for cooling shields in the liquid helium filling line to ensure low heat inleak to the helium supply
- Fine filters installed at the liquid helium filling line to remove any solid particles. The filter in the valve box can be cleaned in situ without opening the system
- The main (master) Dewar can work with internal vacuum. It is equipped with a cold coupling with specially-designed helium lines to guarantee leak tightness and eliminate any possibility of air leakage into the system. For the same reason burst disks are used throughout the system instead of relief valves which cannot guarantee tightness when the internal pressure is sub-atmospheric.
- A cryostat simulating the AMS magnet allows the system to be tested before commissioning with the flight magnet.

SUMMARY

The procedure for the production of superfluid helium has been chosen and experimentally demonstrated. The design of the AMS-02 CGSE is in progress. The test of the system is scheduled for the end of 2005.

REFERENCES

1. Blau, B., Harrison, S.M., et al., The superconducting magnet system of AMS 02 - a particle physics detector to be operated on the International Space Station, IEEE Transactions on Applied Superconductivity (2002) 12 349-352
2. Harrison, S.M., Ettlinger, E., et al., Cryogenic system for a large superconducting magnet in space, IEEE Transactions on Applied Superconductivity (2003) 13 1381-1384

UNIQUE METHOD FOR LIQUID NITROGEN PRECOOLING OF A PLATE FIN HEAT EXCHANGER IN A HELIUM REFRIGERATION CYCLE.

WEBER, T. B.; CANDIA, A.; HOWELL, G.; RACINE, M.; WEISEND II, J.G.

Stanford Linear Accelerator Center, 2575 Sand Hill Road, Menlo Park, CA 94025, USA

Precooling of helium by means of liquid nitrogen is one the oldest and most common process features used in helium refrigerators. The principal tasks are to permit a rapid cool down to 80 K of the plant, to increase the cooling power of the plant in low temperature operation and to increase the rate of pure liquid production. The advent of aluminum plate fin heat exchangers in the design of helium refrigerators has made this task more complicated because of the potential damage to these heat exchangers.

INTRODUCTION

SLAC was just one of the places where damage to the main Helium/Helium/Nitrogen heat exchanger has occurred. The most common reason has been the failure to maintain the barrier between the helium and the nitrogen gas flows. After suffering such a failure at SLAC, a new method was sought when the heat exchanger was replaced. The solution is unique as it uses helium in all parts of the heat exchanger, thus avoiding the problems associated with nitrogen. Helium is circulated in a semi-closed loop through a liquid nitrogen bath and is then sent via a short transfer line to the liquid nitrogen inlet on the main plant. The inlet to the system is tied to the helium return line from the main plant, providing a constant source of low-pressure helium gas to the precooler system. This also prevents the intrusion of air into the system because the low-pressure side is always maintained above atmospheric pressure by the main system. The advantages are that only helium is used in the main plant and no high-pressure helium is used in the precooler, making the full output of the main compressor available to the plant for liquefaction.

THE NATURE OF THE PROBLEM

When the time came to replace the Aluminum plate fin heat exchanger in the Research Yard Refrigerator [1], a study was undertaken to try and understand the failure mode and come up with a solution. In private conversations with Jim Kilmer of Fermi National Accelerator Laboratory and engineers from Altec, the replacement manufacturer of our new heat exchanger, two theories were put forward for the failure modes in the heat exchangers.
1. A plant upset where the nitrogen is frozen by the lower temperature return Helium
2. The failure was the result of 2-phase flow in the nitrogen path with the resulting failure coming from the rapid localized cooling when the liquid vaporized. This excess cooling would then exceed the thermal parameters of the heat exchanger.
Either way, it appears that the solution was to find a way to not use nitrogen in the heat exchanger. An earlier work-around was used on the SLC final focus when some part of the high-pressure stream was diverted, cooled and sent through the leaking He/N2 path. This fix required a reduction in flow to the turbines with the resulting reduction in cooling power.

THE SOLUTION

A calculation was done to see how much flow would be needed to use Helium in the Nitrogen path in the heat exchanger. The entropy values were compared for each gas at 77.4 K and 300 K (Table 1)[2].

Table 1

Entropy for Helium and Nitrogen

Temperature K	Nitrogen h(J/g)	Helium h(J/g)
77.4	14.1	430.00
260	420.5	1365.00
300	462.1	1572.70
Delta h @ 288 K	449.62	1510.39

The T-S Diagram [3] from the manufacturer of the plant requires a Nitrogen precool at 20 g/s for a flow of 140 g/s in the main flow path. However, our capacity is only 100 g/sec so a scaling was done giving 14.25 g/sec of Nitrogen. This scaling reduced the required cooling capacity of the nitrogen from 8414 watts to 6407 watts at 288 K. Helium flow rates (Table2)[2] relates the Nitrogen flow to a thermally equivalent Helium flow.

Table 2

Helium flow rates for a given Nitrogen requirement

	Nitrogen	Helium
g/s	14.25	4.24
J/s	6407	6407
Density g/l		6.15
Flow m^3/h		93.97

These numbers indicated that it would be possible to replace the nitrogen precooler with a helium loop with no loss of performance. A search was started around the lab to identify possible parts for the system. A nitrogen precooler heat exchanger was located from an older experiment (Figure 1). Also located was an unused Corken DA 690 dry compressor (Figure 2).

The Corken was rated at 102.28 m^3/h at a maximum rpm of 825. With all the required items needed, the system was designed to take suction gas from the CTI-4000 return line at 0.11 MPa and send the helium to the precooler at 0.239 MPa. Flow from the precooler was then sent via a vacuum jacketed transfer line to the liquid nitrogen inlet on the CTI-4000 (Figure 3). To prevent over cooling the top plate and compressor, a control loop was added to the LabView control system to stop the compressor should there be an upset in the CTI-4000 operations.

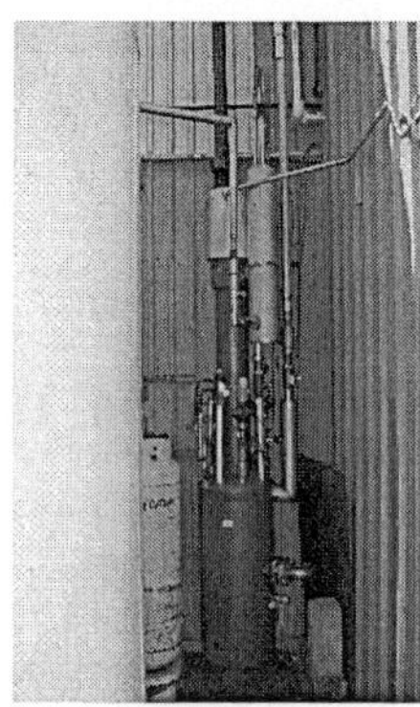

Figure 1

LN$_2$ Precooler

Figure 2

Corken Compressor Installation

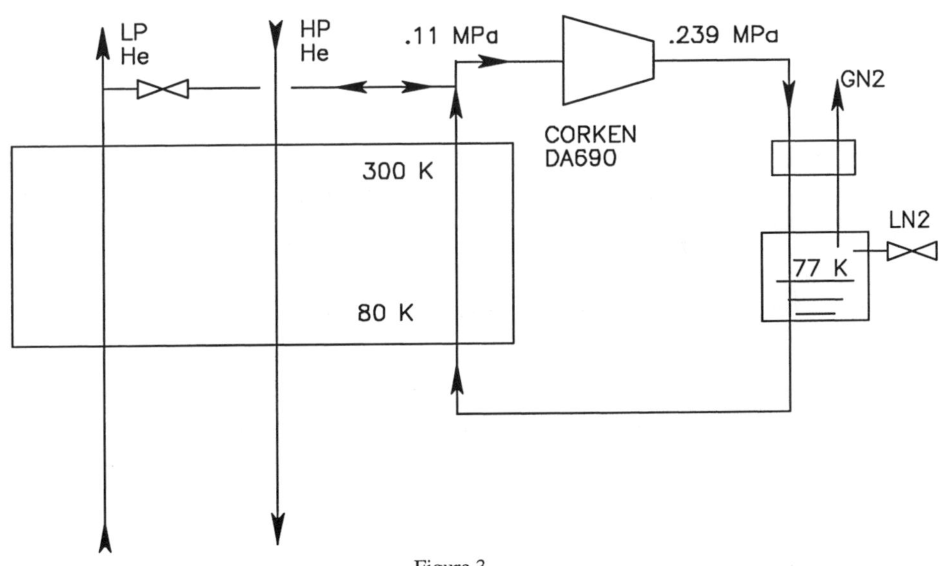

Figure 3

Flow Path of Helium Precooler

OPERATION

The precooler was operated for the last two high power runs of SLAC Experiment E-158. The experiment placed a heat load of 1 kW on the cooling system at 17.5 K. This required the plant be operated at or near maximum conditions (Table 3.)

Table 3

Operational parameters

Flow High Pressure Grams/second	Return pressure MPa	Precooler flow m^3/h	Temperature Kelvin
102	0.110	101.94	89

FUTURE IMPROVEMENTS

The present precooler does not have enough capacity to permit pumping the LN_2 to a lower temperature. If the need occurs, the old heat exchanger has three flow paths of which two are now connected. This will permit the operation in two flow paths. We can then use this heat exchanger in the new precooler with larger flow paths in a pumped regime and operate the precooler with a 67 K outlet temperature.

CONCLUSIONS

A different and unique method for nitrogen precooling a plate fin heat exchange was developed. It provided the needed cooling power for SLAC E-158 and has the ability to be expanded to give more cooling power should the need arise in the future.

ACKNOWLEDGEMENTS

This work would not have been possible with out the dedicated support of the SLAC Experimental Facilities Department, in particular the cryogenic technicians led by A. Candia. Work supported by Department of Energy contract DE-AC03-76SF00515.

REFERENCES

1.   Weisend II, J. G., et al, The Cryogenic System for the SLAC E-158 Experiment
Advances in Cryogenic Engineering (2001) 47A 170-179
2.   Jensen, J.E., Stewart, R.B., Tuttle, W.A., Bubble Chamber Group, Selected Cryogenic Data Notebook, (1972) II-F-1 & VI-F-1.1
3.   CTI-Cryogenics Operator's Manual for Model 4000 Helium Refrigerator Oct. (1976)

Test results and analysis of cryogenic system for cooling HTS cables and the next stage plan

Fan Y-F.[1,2], Gong L-H.[1], Xu X-D.[1], Li L-F.[1], Zhang L.[1], Xiao L-Y.[3]

[1] Technical Institute of Physics and Chemistry, Chinese Academy of Sciences, Beijing 100080, China
[2] Graduate School of Chinese Academy of Sciences, Beijing 100039, China
[3] Institute of Electrical Engineering, Chinese Academy of Sciences, Beijing 100080, China

A 10-m long 10.5 kV / 1.5 kA three-phase alternating current high-temperature superconducting power cables prototype was developed and tested in China in August 2003. The sub-cooled liquid nitrogen was used to cool HTS cables and their terminations in this prototype. The cryogenic system comprised two parts: the sub-cooled liquid nitrogen cycle and the refrigeration station. Test results will be presented and analyzed in this paper. In addition, the next stage plan, which is the development of 75-m three-phase alternating current high-temperature superconducting power cables, will be installed in the power grid this year and tested the next year.

INTRODUCTION

In the past decade or so, the high temperature superconducting (HTS) power cable started to be developed and became the focus of research as one of the application of high temperature superconductor in the world [1,2]. In 2002, Pirelli Cables and Systems and Detroit Edison installed and operated the world's first HTS power cable to deliver electricity in a utility network. Its successful integration was a milestone in the process of HTS application and realized the transition of HTS cable technology from the laboratory to the real-world field [1]. In China, the research and development (R&D) project on HTS power cable was initiated at the end of 1997. In 1998, a 1-m long 1 kV / 1 kA HTS cable model was developed and successfully tested by the Institute of Electrical Engineering (IEE), CAS. Later, they studied a 6-m long 2 kV / 2 kA HTS power cable prototype to increase the critical current and to adopt the improved fabrication technique of HTS cable in 2000 [3]. Last year, collaborating with Technical Institute of Physics and Chemistry (TIPC), CAS, they successfully developed and tested a 10-m long 10.5 kV / 1.5 kA three-phase alternating current (AC) HTS power cables prototype.

With the development of HTS power cable, the cryogenic system for cooling HTS cables also won the chance of development. The cryogenic system of LN2 immersion was adopted in the 1-m and 6-m HTS power cable models. It was unfit for the longer and larger HTS power cable in the power grid, so the cryogenic system with high-efficiency and high-stability was needed for the real-world application of HTS power cable. We developed a cryogenic system for the 10-m HTS power cables prototype [4].

TEST RESULTS AND ANALYSIS

As of February 2004, the HTS power cables had been operated for 10 times. Duration of the operation at 1.5 kA was more than 5 hours each time. Meanwhile, the cryogenic system had been cumulatively

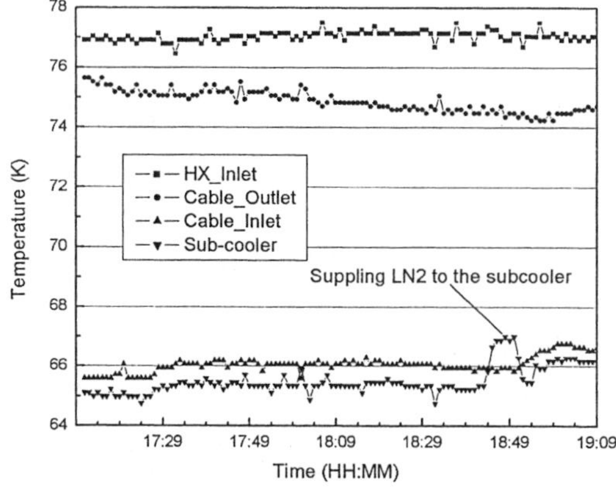

Figure 1   Temperature of LN2 during the running

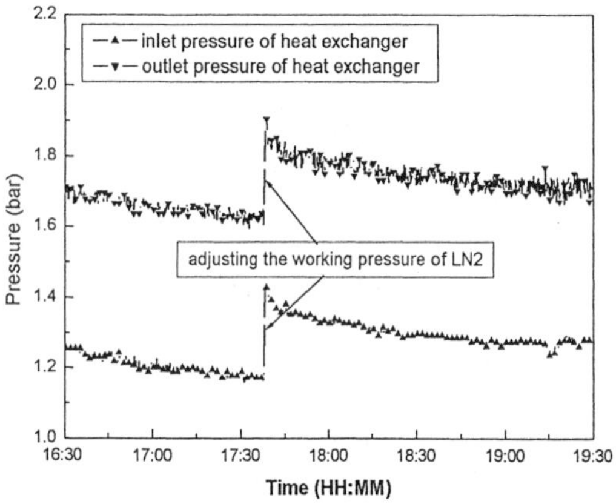

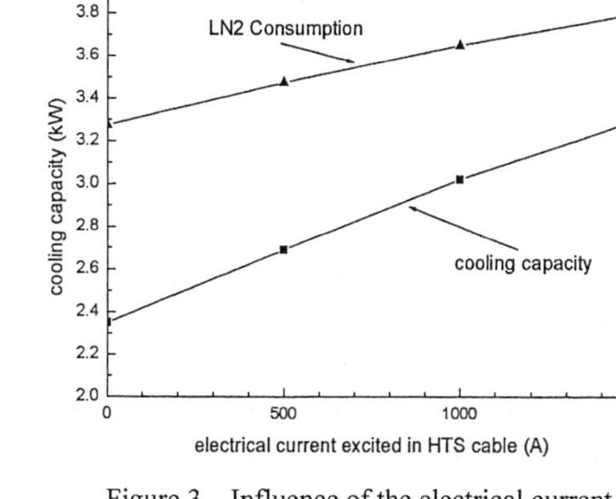

Figure 2   Pressure of LN2 during the running

worked for 80 hours. The test result, which was gotten on August 15, 2003, proved that the LN2 inlet and outlet temperatures of HTS cables were 65.5K and 75.5K respectively, when the LN2 flow rate was 800 L/h, and the fluctuation of the temperatures was about 1.0 K, as shown in Figure 1. Figure 2 showed the LN2 pressure distribution for the inlet and outlet pressure of heat exchanger inside the sub-cooler. In order to research the influence of electrical current on the heat load (corresponding to the cooling power) and LN2 consumption, we did a series of HTS cables tests under different electrical currents. The results were shown in Figure 3.

Figure 3   Influence of the electrical current

In January 2004, we carried out some experiments on cryogenic system to research the influence of LN2 flow rate and the inlet temperature of HTS power cables on the background heat load of whole HTS cable installation, including the cryogenic system. The background heat load means the heat load without electrical current passing the HTS cables. Figure 4 and Figure 5 respectively show the influences of LN2 flow rate on the temperature difference and pressure difference of the heat exchanger inside sub-cooler in the condition of different inlet temperatures. Seen from Figure 4, the temperature difference (ΔT) decreases as the flow rate increasing and the ratio of temperature difference to flow rate approximately keeps a negative constant. From the first principle of thermodynamics, the background heat load could be calculated by Equation 1, which follows:

$$Q = \dot{m} \cdot C_p \cdot \Delta T \tag{1}$$

with Q: heat load, $\dot{m}$: LN2 mass flow rate, C_p: LN2 heat capacity of constant pressure, ΔT: LN2 temperature difference of heat exchanger inside sub-cooler. The calculated value of heat load is about 2.35 kW. Also, Seen from Figure 5, the pressure difference (Δp) increases as the flow rate increasing and the ratio of them approximately keeps a positive constant.

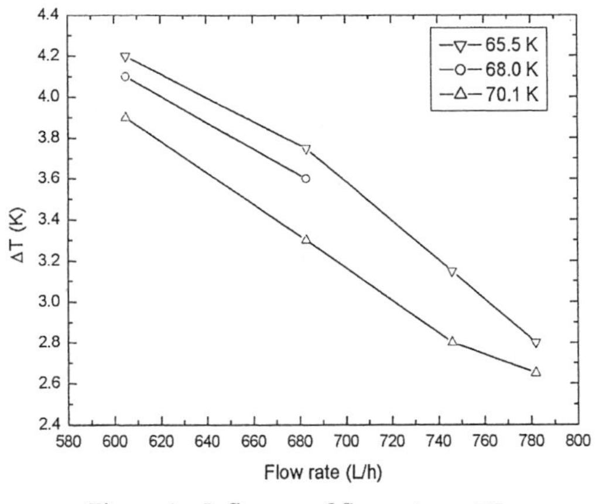

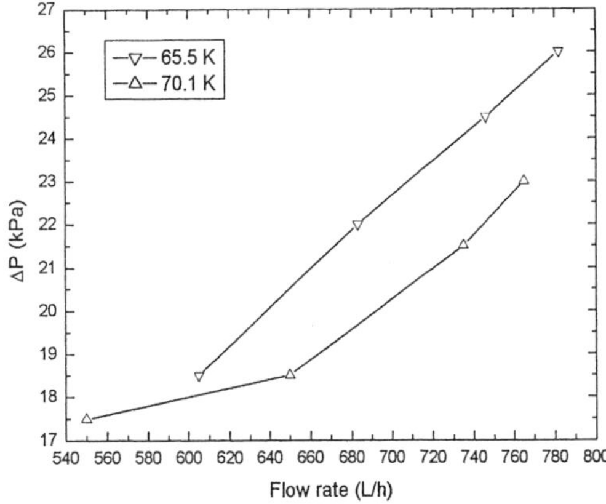

Figure 4   Influence of flow rate on ΔT Figure 5   Influence of flow rate on Δp

As mentioned in above paragraph, test results have indicated that cryogenic system could have operated stably and reliably for cooling HTS power cables and have verified that both design and installation are correct and successful. Cryogenic system of using the sub-cooled LN2 cycle to cool the HTS power cables could satisfy the demand of the high-efficiency and high-stability for the real-world application of HTS power cable. Because of the special property of low boiling point to LN2, the trouble-free startup of LN2 pump using in a LN2 cycle provides the valuable experience for us. Additionally, a data acquisition and control system has been built, and heat load, LN2 flow rate, temperature, and pressure have also been measured correctly. All results provide confidence in the design approach of cryogenic system for 75-m HTS cables installation. On the other hand, the real background heat load is 2~3 times larger than that estimated theoretically. It may result from two main parts: conducting heat load of the LN2 feeding line and HTS cables terminations, and conducting heat load of the six current leads from room temperature. Perhaps use of the Super Insulated Vacuum (SIV) lines and optimization of the design and structure of current leads [5] could effectively reduce the heat load.

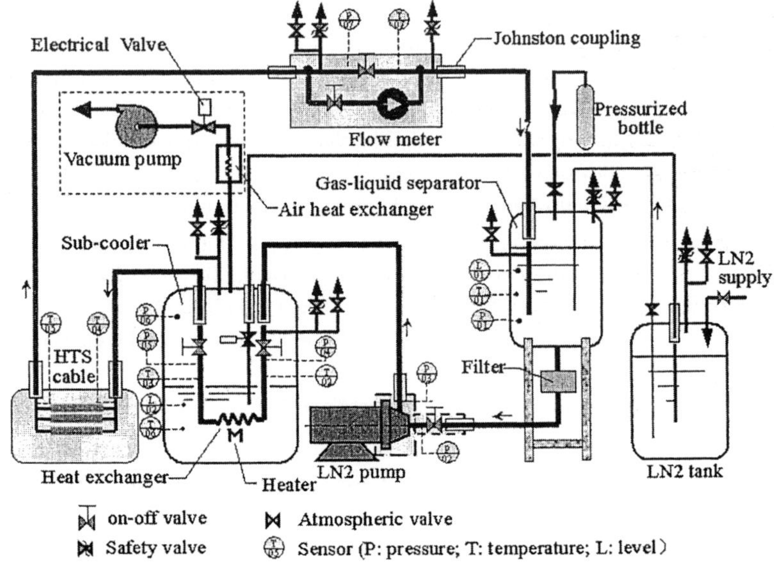

Figure 6   Simplified flow diagram of the cryogenic system for 75-m HTS cables installation

NEXT STAGE PLAN

Based on the 10-m HTS power cables prototype, we will develop the 75-m three-phase AC HTS power cables installation in the next two years. The 75-m HTS power cables installation will be developed together by IEE of CAS, TIPC of CAS and Gansu Changtong Cable Group. The cables will be installed and operated in the power grid in Baiyin, Gansu province. The project is being built now and will be finished at the year-end and tested the next spring. Similarly to the 10-m HTS power cables prototype, the cryogenic system for the 75-m HTS power cables adopts the sub-cooled LN2 cycle to cool the HTS cables and their terminations. It consists of two parts: the sub-cooled LN2 cycle and the refrigeration station. The simplified flow diagram is shown in Figure 6. The cooling capacity is produced by the method of decompressing LN2 in the sub-cooler to sub-atmosphere.

CONCLUSIONS

In this paper, we have presented the test results of the cryogenic system for cooling the 10-m HTS cables prototype. The results include the distribution for the temperature and pressure of LN2 during the HTS cables running, the influence of electrical currents on the heat load and LN2 consumption, and the influence of LN2 flow rate on the temperature difference and the pressure difference. Besides, some helpful experience has been summarized through results analysis and is the basis of design and manufacture of the cryogenic system for the 75-m HTS power cables. Finally, the next stage plan: development of the cryogenic system for the 75-m HTS power cables installation has been introduced.

ACKNOWLEDGEMENTS

This project is derived from The National High Technology Research and Development Program (Project No. 2002AA306161) and together supported by the Ministry of Science and Technology and Gansu Changtong Cable Group.

REFERENCES

1. Pietro Corsaro, Massimo Bechis, Paola Caracino, et al., Manufacturing and commissioning of 24 kV superconducting cable in Detroit. <u>Physica C 378-381</u> (2002) 1168-1173

2. Takato Masuda, Takeshi Kato, Hiroyasu Yumura, et al., Experimental results of a 30 m, 3-core HTSC cable. <u>Physica C 372-376</u> (2002) 1555-1559.

3. Y.B. Lin, L.Z. Lin, Z.Y. Gao, H.M. Wen, L. Xu, L. Shu, J. Li, L.Y. Xiao, L. Zhou, and G.S. Yuan. Development of HTS transmission power cable. <u>IEEE Trans. on Applied Superconductivity</u> (2001) <u>11</u> 2371-2374

4. Y.F. Fan, L.H. Gong, X.D. Xu, L.F. Li, L. Zhang, L.Y. Xiao, Cryogenic system with the sub-cooled liquid nitrogen for cooling HTS power cable. <u>Cryogenics</u> (accepted)

5. J.M. Lock, Optimization of current leads into a cryostat. <u>Cryogenics</u> (1969) <u>9</u> 438-442

Cryogenic Facility Design in BEPCII Superconducting Upgrade

Jia L.X.*,Wang L., Yang G.D., Liu X.K., Zhang X.B., Tang H.M., Du H.P., Xing Z.H., Liu C.S., Han G., Xu F.Y., Wang T.H., Yang C.Z., Xiao H., Liu D., Shi L.H., Huang M.X., Gai X.Y., Xu F.Y., Ji Z.M, Jiang W., Yao Z.L., Chen A.B.,

*Brookhaven National Laboratory, Upton, New York 11973, USA
Institute of Cryogenics and Superconductivity Technology, Harbin Institute of Technology, Harbin 150001, China

Three kinds of superconducting device are to be constructed at interaction regions in the upgrade of Beijing Electron-Positron Collider (BEPCII). Two sets of refrigerators with each capacity of 500W at 4.5K are adopted to provide the refrigeration for them. The cryogenic systems to support the operation of the superconducting facilities are under design by Harbin Institute of Technology in China. This paper presents the current design of main cryogenic facilities.

INTRODUCTION

In order to support the operation of three superconducting devices in BEPCII upgrade, a cryo-plant with the cooling capacity of 1.0 kW at 4.5K is under design [1, 2]. It is composed of two refrigerator systems with each capacity of 500W at 4.5K. One is to cool a pair of superconducting quadrupole magnets (SCQ) together with a superconducting solenoid magnet (SSM) located at the first interaction region, and the other is to cool a pair of SRF cavities at the second interaction region.

The main tasks of the cryogenic system involve production of refrigeration power, preparation and distribution of cryogens, cooling of cryomodules, handling of different modes of operation, recovery, purification and storage of cryogens as liquid or gas, process control and monitoring, and cryogenic safety. Correspondingly, the cryogenic system is made of various cryogenic facilities such as refrigerators, control dewar, control valve boxes, subcoolers, transfer lines, storage tanks and purifiers. The engineering design of the BEPCII cryogenic system has been carrying out by the Institute of Cryogenics and Superconductivity Technology at Harbin Institute of Technology since March, 2003, and almost completed. This paper presents the current design of the main cryogenic components in detail.

CRYOGENIC FACILITIES DESIGN FOR SUPERCONDUCTING MAGNETS

The refrigerator systems can be divided into three parts, which are the helium compressor and refrigerator system, LN2 system, and cryogen production and distribution system. The three parts are connected by low temperature supply and return transfer lines and 300K supply and return lines. The cooling system for the magnets shares the helium purification, recovery and storage system with that for the SRF cavities [4]. Two LN2 tank with each volume of 30m3 serves to supply the liquid nitrogen for precooling of the refrigerator and cooling of heat shields inside the cryomodule, which are respectively located close to the first and second colliding hall.

The SCQ magnets are to be cooled by single-phase helium in order to eliminate the flow instability in constraint cooling channel [3], and the SSM magnet is to be cooled by forced two-phase flow. The cryogen production and distribution system for the magnets mainly includes a control dewar with a built-in heat exchanger acting as a thermal subcooler, a distribution valve box, service cryostats for magnets, and cryogenic transfer lines.

Control dewar and subcooler

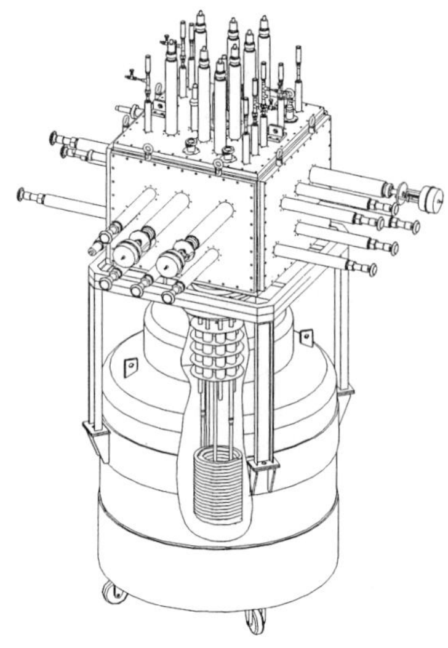

Figure 1 Control dewar, subcooler and distribution valve box

The subcooler is a coil-shaped heat exchanger residing in the control dewar with the volume of 1000 liter (see Figure 1). The neck of the control dewar is modified to allow the installation of the heat exchanger and connected with the distribution valve box. The heat loads through the neck to the dewar have to be minimized. The subcooler is used to further cool the supercritical helium or two-phase helium flow from the J-T circuit of the refrigerator before it goes into the magnet cooling circuit. The liquid pressure in the control dewar is lower than that in the heat exchanger and while the outgoing helium temperature in the heat exchanger can reach to the saturation temperature at control dewar pressure. It produces low temperature helium with high pressure. The control dewar can server as an oscillation dumper for helium supply to the magnets and a phase separator for the helium returning from the cryostats. It can also act as a buffer vessel that can provide additional cooling by using the liquid stored in the dewar during times when the thermal load exceeds the capacity of the refrigerator, or withdraw the LHe when needed.

The supercritical helium at 2.7bar and around 5.2K with the mass flow rate of 35.25g/s from the refrigerator cold box flows through the subcooler and can be cooled to about 4.6K. The heat exchanger is made of copper tubing with the inner diameter of 17mm and the outer diameter of 19mm. The copper tubing with 40m in length is currently wound as three layers with the diameters of 200mm, 245mm, and 290mm, respectively. The height of the heat exchanger is about 350mm. The size and structure of the subcooler is largely restricted by the neck size and inner space of the liquid helium dewar. The numerical optimization for the subcooler is being carried out, and the design will be further updated later.

Distribution valve boxes

The distribution valve box is one of substantial facilities in the BEPCII cryogenic system through which the supercritical or two-phase helium from the refrigerator is sub-cooled and distributed to the two SCQ magnets and SSM magnet. It is deigned as a vacuum-insulated cuboid vessel in order to reduce the radiation heat load from room temperature (Fig.1). The vessel is composed of a support frame and six demountable stainless steel plates. The bottom plate is located on the top of the 1000L dewar, and the other five plates are easy to disassemble. The valve box is 1m wide, 1m long and 0.8m high. There are totally thirteen cryogenic valves for control of helium flow, two valves for LN$_2$ delivery respectively to the SCQ and SSM, nine safety valves, one vacuum relief valve, several pressure transducers and temperature sensors as well as relevant electric feedthroughs and service ports arranged tightly and compactly in or on the valve box. Thirteen cryogenic bayonets are applied to connect the low temperature transfer lines among the refrigerator cold box, the valve box and the magnets' cryostats, which allow the disconnection of the transfer lines without destroying the vacuums inside the valve box and transfer lines themselves. To decrease the conductive heat leak induced by the parts fixed or supported on the chamber walls, the valves and bayonets with long stems are employed. The piping for the safety valves and pressure transducers is made as long as possible.

Besides the heat leak, thermal stress is another concern for the design of the valve box. The displacement caused by the thermal stress for the piping must be considered when the temperature changes during cool down and warm up process in order to avoid the mechanical failure. For the six vessel plates, incase of cracking of the piping in the box, the low temperature helium or liquid nitrogen would spill into the vacuum vessel, and then the plates would deform or even to be broken due to the thermal stress. The numerical simulations and analyses on the thermal stress and strain for the piping and the plates at the above scenario were performed by ICST. The thickness of the plates is finally chosen as 30mm, and five flexible lines are used to compensate the piping contraction during the cool-down process.

<u>Service Cryostats for magnets</u>

The service cryostat is designed to provide a connection between the warm electrical power cable and the superconductor wires from the coils. Gas cooled current leads for SC coils are installed inside the cryostat. The current leads are made of copper tubes with the insulators at each end. The electrical power cables are attached to the bus bars of the current lead at the top and run to the power supply box.

The cryostat also serves as a cryogenic distribution box that controls the cooling flows for the superconducting magnets, thermal shields, and current leads by cryogenic valves at various operating modes including magnet quench. It is connected to the magnet cryostat through a cryogenic chimney. The heat leak and thermal stress are also two of major concerns during their design similar to the distribution box. There are totally three service cryostats in the first colliding hall, of which two are for SCQ magnets (see Figure 2), and one is for SSM magnet (see Figure 3). The cylindrical vacuum-insulated vessel made of stainless steel is 800mm in diameter and 1040mm in height for SCQ, and 1000mm in diameter and 1400mm in height for SSM. The SCQ service container can be easily assembled or disassembled from the flange at its lower portion, while the SSM service cryostat can be disassembled from the bottom flange for easy maintenance. The vacuum ports are set at the lower portion of the vacuum chamber. On the top flanges are tightly arranged the cryogenic valves and bayonets to control and connect the supply and return of helium and nitrogen, and helium back to quench line, as well as safety valves, pressure transducers and electric feedthroughs. The SC cables and various transfer lines can be disconnected at the bottom of the cryostat.

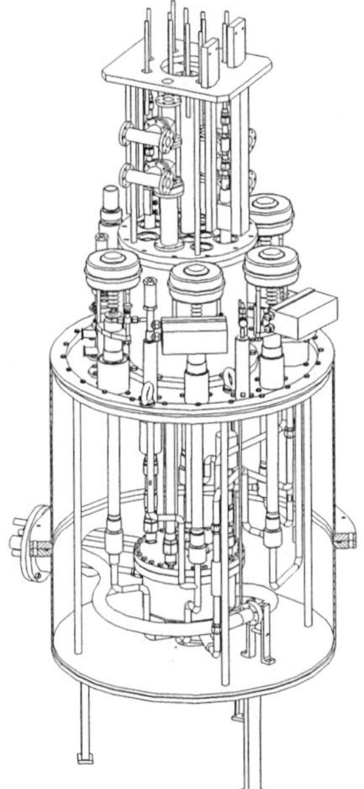

Figure 2 SCQ service cryostat

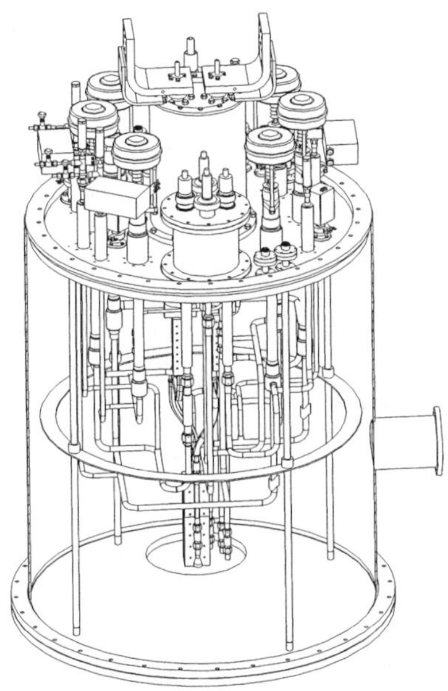

Figure 3 SSM service cryostat

CRYOGENIC FACILITIES DESIGN FOR SUPERCONDUCTING CAVITIES

The cryogen production and distribution system for the magnets mainly includes a control dewar of 2000 liter, a distribution valve box, and cryogenic transfer lines, which all are placed at the second interaction region. The control dewar performed the same function as that of the 1000L control dewar at the first interaction region.

<u>2000L control dewar</u>

The modified control dewar can server as both an oscillation dumper and a phase separator for the helium supply and return to and from the cavities. It can also act as a buffer vessel that can provide additional cooling by using the liquid stored in the dewar during times when the thermal load exceeds the capacity of the refrigerator, or withdraw the LHe when needed.

SRF distribution valve box

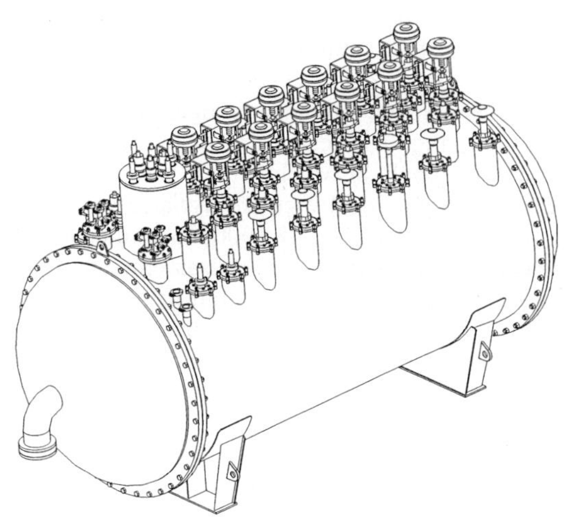

Figure 4 SRF valve box

The SRF valve box is to distribute and control the helium and nitrogen flows to the helium vessels and thermal shields of two SFR cavities at on-line or test spots (see Figure 4). It consists of a horizontal cylindrical vacuum vessel with two ellipse end caps and supports, twenty-four cryogenic valves, twenty-two bayonets, twelve safety relief valves and various ports for instrumentation. All the components including the piping inside are vertically placed at its upper portion, which will be pre-assembled outside by customized tooling, and then inserted into the vessel. The valves are connected with the container by flanges to allow easy disassembly. The vacuum port is set at one end cap. The vessel is 1400mm in inner diameter, 10mm in thickness and 2794mm in length. The overall length of the facility is 3952mm. It is one of the most complicated designed facilities in BEPCII cryogenic system.

CONCLUSIONS

This paper presents the current design of main cryogenic components in BEPCII. They are designed based on the operation requirement and space limitation. The heat leaks induced from room temperature and thermal stress during cool-down, warm-up and failure modes are considered.

REFERENCES

1. Institute of Cryogenics and Superconductivity Technology, Conceptual design report of BEPCII cryogenic system, Harbin Institute of Technology, (2002).
2. Jia, L.X. and Wang, L., Cryogenics in BEPCII upgrade, Proceedings of ICEC19 (2002).
3. He, Z.H. and Wang, L., CFD analyses on LHe cooling for SCQ magnets in BEPCII upgrade, to be published in Advances in Cryogenic Engineering (2004), 49.
4. Wang, L. and Jia, L.X., Computational simulation of refrigeration process for BEPCII superconducting facilities, to be published in Advances in Cryogenic Engineering (2004), 49.

First Cold Test of the ISAC-II Medium Beta Cryomodule

I. Sekachev, R.E. Laxdal, C. Marshall, T. Ries, G. Stanford,

TRIUMF, 4004 Wesbrook Mall, Vancouver, BC, Canada, V6T 2A3

ISAC-II is an upgrade of the ISAC radioactive beam facility that includes the addition of 43 MV of heavy ion superconducting accelerator. A medium-beta cryomodule comprises four superconducting quarter-wave cavities and one superconducting solenoid magnet. An initial cryomodule has been assembled and cooled with LHe from a supply dewar. This paper will summarize the design of the medium-beta cryomodule and will present the results of the first cold test and alignment.

INTRODUCTION

TRIUMF is now constructing an extension to the ISAC facility, ISAC-II [1], to permit acceleration of radioactive ion beams up to energies of at least 6.5 MeV/u for masses up to 150. Central to the upgrade is the installation of 43 MV of heavy ion superconducting linac. The superconducting linac is composed of two-gap, bulk niobium, quarter wave RF cavities, for acceleration, and superconducting solenoids, for periodic transverse focusing, housed in several cryomodules and grouped into low, medium and high-beta sections. Each cryomodule has a single vacuum system for thermo-isolation and beam. This demands extreme cleanliness of internal components to avoid superconducting surface contamination. An initial stage to be completed in 2005 includes the installation of the medium-beta section consisting of five cryomodules. An initial cryomodule has been designed and assembled at TRIUMF.

CRYOMODULE ENGINEERING

The medium-beta cryomodule, described elsewhere[1], is a stainless steel vacuum tank containing four superconducting quarter wave RF cavities and one superconducting 9-Tesla focusing solenoid. The superconducting elements are supported on a beam that is suspended from the lid by struts (Fig. 1). The struts are slung from three adjustable points, two upstream and one downstream, that are mounted on a plate that is laterally and for/aft adjustable. There is an independently mounted liquid helium reservoir (120 L inventory) suspended from the lid attached to the superconducting elements by soft bellows. All the superconducting elements are aligned with respect to each other on the support beam and then aligned relative to the vacuum tank beam ports with a measured cold-offset to allow for thermal contraction. The efficiency of cooldown is improved by a manifold and distribution system 'spider', connected to the LHe transfer line, that delivers cold gas and liquid to the bottom of each element through 5mm Cu tubing. Once the liquid begins to collect in the reservoir a pre-cool valve on the distribution manifold is opened and

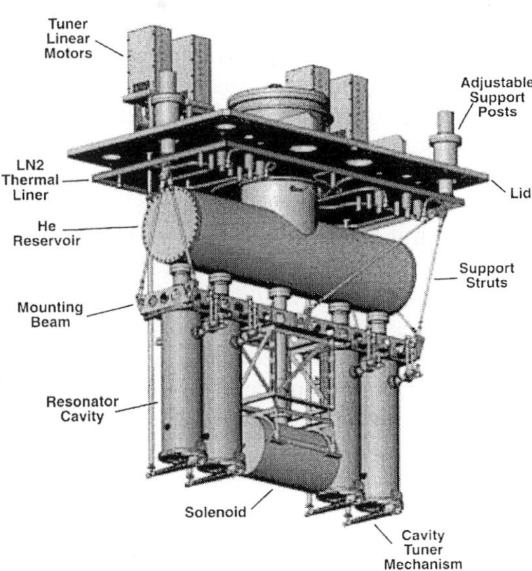

Figure 1: Cryomodule top plate assembly.

LHe flows directly into the reservoir. Except for the niobium cavities the cold mass is predominantly made from 316L stainless steel.

The entire cold mass is surrounded by a forced flow, liquid nitrogen cooled, thermal shield. The shield consists of several Cu panels riveted together to form a box with 10mm ID Cu tubing soldered to the panels to form a serial LN2circuit. After soldering the panels are nickel plated. A μ-metal magnetic shield is attached to the inside of the vacuum tank outside the LN2 shield. A single LN2 panel and μ-metal shield suspended from the lid make up the top thermal and magnetic enclosure respectively.

Assembly and Alignment

A wire position monitor (WPM) system[2] is used to provide an off-axis measure of the cavity and solenoid positions during cooldown. WPMs, each consisting of four strip lines, are attached to the cavities and solenoid by an `L'-bracket, positioning them at the horizontal beam plane 0.31-m transverse to the beam direction. A wire running parallel to the beam axis and through the monitors carries an RF signal that is converted to an x-y position. The lid is accurately located to the vacuum tank via a pair of dowels. The beam ports in the vacuum tank are accurately located to these dowels via a transfer fixture. An assembly stand (Fig. 2), which mimics the vacuum tank, is used to assemble all the tank internals to the lid. The top flange is machined flat and dowel pins are used to precisely position the top plate on the stand. The transfer jig is used to transfer the lid reference to a line of sight replicating the reference beam line and the WPM wire line. An outrigger bracket at either end of the stand allows the mounting of telescopes, and end plates are equipped with dummy' beam flanges and

Figure 2: Top plate assembly in assembly stand.

WPM flanges. Both cryomass elements and WPM striplines are prealigned at room temperature to the theoretical line of sights with telescopes and alignment targets fitted in the beam tubes of the cavities, solenoid and striplines.

The cryomodule tank is also outfitted with a pair of optical windows and alignment targets to set up and monitor an external optical reference line with a telescope. As well a pair of optical targets are installed in the upstream and downstream cavities. After the warm alignment in the assembly frame the top assembly is lowered into the vacuum tank, the wire is attached to the end flanges and tensioning device, and the cryomodule is prepared for pumpdown. Optical measurements, taken periodically, serve to check for unexpected differences between the WPM position and the position of the cold mass. During cooldown they provide a calibration of the thermal contraction of the WPM brackets. The cold tests not only establish the integrity of the cryostat and the static load but also determines the required warm offset of each element such that they are aligned at cold temperature. After the initial cooldown the cryomodule is warmed and opened, the elements adjusted to their warm offset position. The tank is then reassembled, recooled and the alignment checked with both WPM and optical targets. The required alignment tolerance is ±400 μm and ±200 μm for the cavities and solenoid respectivly. Once the internal alignment is complete the cryomodule is installed in the accelerator vault and the beam ports are aligned to the theoretical beam centerline. Three target posts on the lid can be characterized in space relative to building benchmarks such that, in future, when the accelerator is operational, a cryomodule can be removed and reinstalled to the same position.

Expected Cooldown Performance

A flow of 6 liquid liters/hour of 100% liquid nitrogen is expected to cooldown the medium-beta cryomodule LN_2 shield in ~24 hours. The bulk of the cold mass will cool by radiation to ~250K. With this flow and assuming 36-m of 10-mm pipe inside the cryomodule, the pressure drop falls from 13.1-kPa while warm to 0.69-kPa while cold. During design it was estimated that the cryomodule would have a static load of 13~W and this number was used to estimate the required refrigerator power. It is estimated that a flow of 100ltr/hr for seven hours would be required to collect liquid.

CRYOMODULE COMMISSIONING

The cryomodule assembly and commissioning tests are conducted in the clean laboratory area in the new ISAC-II building. LN2 for pre-cooling and shield cooling is fed from a transfer line coming from an external tank. Helium is presently fed from local dewars; plans are underway to add a transfer line from the soon to be installed LHe refrigerator[3]. Exhaust gases from the cryomodule are passed through vaporizers and the flow is monitored by gas meters. A local computer records WPM information. Control and data acquisition are done with an EPICS control system. All pertinent signals are logged for future analysis.

Warm measurements

At room temperature and atmosphere the lid is installed, WPM assembled, and bolts tightened to 45 ft-lbs. The bolts are then loosened and the lid lifted off the O-ring by a few mm, so as not to disturb the WPM assembly, then reinstalled. The lifting and tightening are repeated five times and the wpm position coordinates are taken for each. The results are shown in Fig. 3. In all cases the vertical measurements repeat within ±20 μm. The horizontal position changes after the first cycle, `A', as the cavity support frame skews but stays within ±75 μm after this first cycle. The horizontal `walk' may be due to the strut geometry at the downstream end. In this case the two main struts are slung from the same support tower while in the upstream end the two main struts are slung from separate towers.

The vacuum pump down is repeated twice. The final position of the WPM's in both horizontal and vertical planes is repeatable to within 20μm.

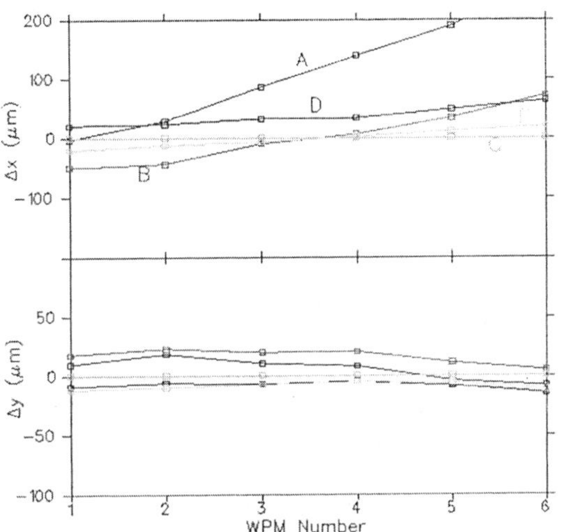

Figure 3: Repeatability of WPM position after repeated (`A' to `E') lifting of lid and torquing down bolts.

Cold Measurements

The cold tests can be grouped into alignment tests and cryogenic tests. The tests include three temperature cycles from room temperature to LN2 temperature and one cooldown to helium temperature.

Alignment: The main goal is to determine the repeatability of the cooldown process and to establish offset values for each cavity and the solenoid to enable warm positioning compatible with alignment at cold temperatures. Cold vertical and horizontal wpm positions are given in Fig. 4 for the four thermal cycles. Due to the different materials involved the solenoid experienced more vertical contraction, with ~4.4mm at LN2 and ~5mm at LHe temperatures while the cavities contracted ~3.3mm at LN2 and 3.8mm at LHe temperatures. The contraction of the cavities was somewhat position dependent with WPM1 and WPM2 shifting vertically about 0.2mm more than WPM5 and WPM6. The vertical contraction at cold tempatures varies somewhat since the temperature of the side shield hence the average strut temperature is not constant. In the horizontal direction things are far less clear. It is expected that the WPM mould move upon cooling since the off-axis stainless steel brackets would contract. However there is a 0.45mm difference between the movement of WPM1 and WPM6, evidence

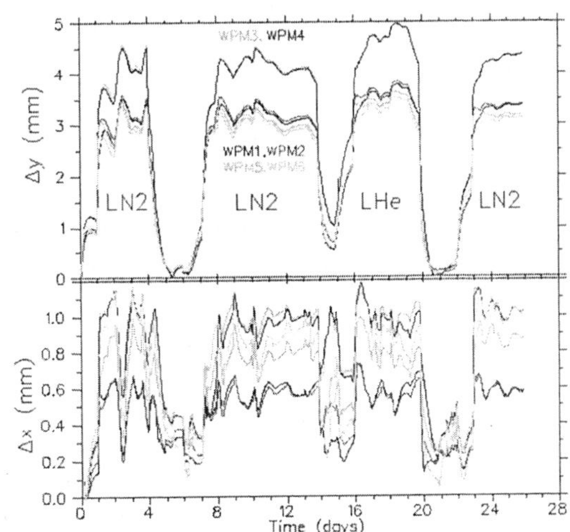

Figure 4: X and Y WPM position history during the four cooldown cycles.

that the cavity support beam is yawing laterally during cooldown. The position of the cold mass is analyzed at three cold LN2 temperatures where the sensors on the cold mass are coincident. The differences in position between cycle 2 and cycle 1 and 3 are summarized in Fig. 5. The positions are repeatable to within ±50 μm vertically and ±100μm horizontally. Again the horizontal position of the downstream end is found to be less stable than the upstream end. The optical target psotions are recorded periodically during the cold cycling. A comparison between the optical targets and the visual targets gives the calibration for WPM position relative to the beam ports. The differences are due to the different contraction of the bracket material and the niobium cavity. A change of pressure in the helium space inflates the bellows resulting in a net force on the cavity support frame yielding a measured deflection of 30μm/100Torr pressure variation. The same pressure variation also caused a side load on the support frame skewing the frame by 10μm/100Torr.

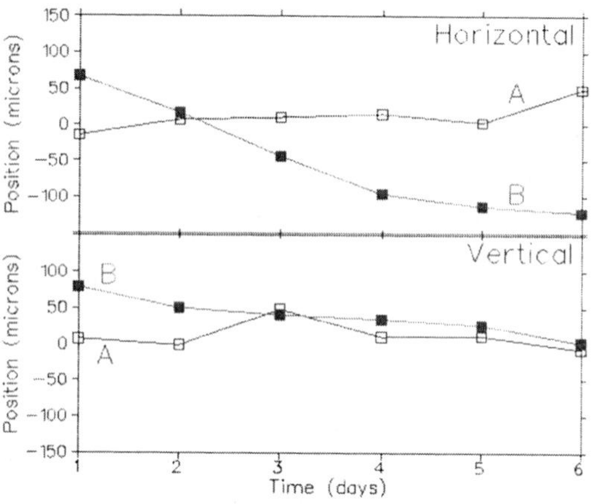

Figure 5: Comparison of WPM positions during three LN2 cold periods with second cycle used as reference.

Cryogenic Tests

The LHe fill was preceded by an LN2 fill. The LN2 is boiled off by spoiling the vacuum with 1Torr helium plus internal cold mass heaters. The helium space was pumped to determine that no LN2 remained then the vacuum was restored before the fill. The cold-mass began the fill at an average temperature of 210K. Initially three 100L dewars were used to start liquid collection in the cavities after five hours. At this point the average cavity temperature reached 20K. The fill shows clearly that the distribution `spider' effectively distributes the helium to the bottom of all cold masses. A further 250LHe is required to fill the magnet. In all 1400LHe is consumed over a period of three days to fully thermalize the cryomodule. The final static load on the helium, after thermalization, as measured from the gas boil-off is 10W (Fig. 6). The LN2 flow required to keep the side shield less than 100K is determined to be ~5ltr/hr matching design estimates.

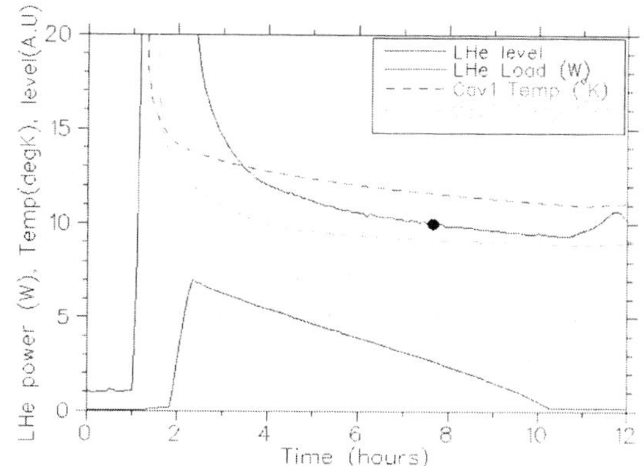

Figure 6: LHe level (A.U.)and static load (W) measured from gas boil off. Also shown are temperature of top flange of cavity 1 and 2 showing the thermalization.

CONCLUSIONS

The first medium beta cryomodule cold test has been a total success. The distribution `spider' distributes the cold gas evenly to the bottom of the cold mass. The static load and required LN2 flow are 10W and 5ltr/hr respectively, within design estimates. A WPM six monitor system developed at TRIUMF is now operational. The system monitors continuously the position of the structure during cold cycling. The three point mount system produces a slight temperature dependent skew in the cavity support frame.

REFERENCES

[1] G. Stanford, et al, Design of the Medium Beta Cryomodule for the ISAC-II Superconducting Heavy Ion Linac, Alaska Cryogenic Conference
[2] W. Rawnsley, et al, Alignment of the ISAC-II Medium Beta Cryomodule with a Wire Monitoring System, this conference.
[3] I. Sekachev, et al, Status of the ISAC-II Cryogenic System, this conference.

Experimental study of self-pressurizing transfer system for cryogenic liquid

Li Q.F., Chen G.B., Xie X.M., Bao R.

Cryogenics Laboratory, Zhejiang University, Hangzhou, China, 310027

A series of experiments have been made to investigate cryogenic liquid transfer performance by self-pressurizing method for liquid oxygen tank. The simulation propellant is liquid nitrogen and the pressurizing gas is N_2 gas containing CO_2. By varying the content of CO_2 in pressurizing gas, both "liquid oxygen gasified pressurization mode" and "abundant in oxygen pressurization mode" are simulated. The experiments show that the simulation arrangement is feasible.

INTRODUCTION

To choose an appropriate pressurizing method for achieving the pressurizing transfer process for liquid propellant is a very important problem for cryogenic rocket design. The way of self-pressurizing transportation is an essential factor, by which the deadweight of rocket can be reduced, and the delivery capability and reliability may also be improved.

In order to carry out the simulation experiment of self-pressurizing transfer of cryogenic propellant rocket, a special down-scale test-bed has been designed. It follows the principle that the content of the pressurize gas in the simulation experiment is the same as the self-pressurize gas in the liquid propellant rocket, and the mass flux of impurity (carbon dioxide and water vapor) passing the engine filter in the experiment is also the same as in flight state.

In the self-pressurizing transfer system, the pressurizing gas is oxygen gas containing some carbon dioxide and water vapor. Based on calculation and data analysis, it has been recognized that the contents of carbon dioxide [1] and water vapor [2] in the system are far higher than their saturation solubility in the oxygen, and solid particles may form at low temperatures. In order to simulate the floating state of the solid particles in liquid, liquid nitrogen is used as simulation propellant instead of liquid oxygen, and the content of water vapor is converted into a proper content of carbon dioxide. Thus nitrogen gas containing carbon dioxide is used as pressurizing gas, instead of the combustion product used in the liquid oxygen-kerosene rocket.

EXPERIMENTAL APPARATUS

The experimental apparatus shown in Figure 1 mainly consists of a gas distribution system, a liquid transfer system and the data acquisition and control system. According to the requirement of the simulation, the pressurized gases of 0.4MPa[*] are prepared in advance in the gas tank(left side in Figure 1), which will be regulated to 0.15MPa when passing the pressure regulator to transfer the cryogenic liquid in LN_2 tank right side in Figure 1. The tank filter, liquid flow meter and engine filter are installed in the pipe of the liquid transfer system.

---

[*] The pressure readings in the paper all use gauge pressure.

In the experiment, the pressure difference between the two sides of the tank filter and the engine filter, the system temperatures, the flow rates of pressurized gas and liquid are measured by the measuring system. Whether the engine filter is jammed by solid particles formed in the transfer process can be observed accordingly.

The compositions of pressurized gases, which are validated by a GC112A gas chromatograph before experiments, are listed in Table 1.

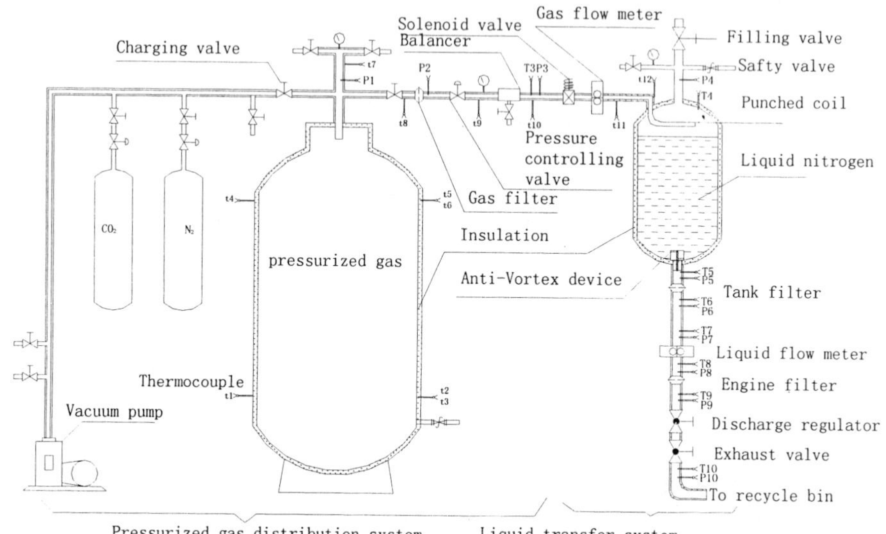

Figure 1 Schematic of experimental apparatus

Table 1 Composition of pressurized gases (Volume percent)

Mode	Required composition			Composition used in experiment	
	H_2O	CO_2	O_2	CO_2	N_2
Abundant in oxygen pressurization	7.08	3.10	89.82	7.346	92.654
Liquid oxygen gasified pressurization	0.142	0.063	99.795	0.1429	99.8571

EXPERIMENTAL RESULT AND DISSCUSSION

Liquid oxygen gasified pressurization mode

The pressurized gas is N_2 gas containing 0.143% CO_2. The engine filter installed is a stainless steel screen (150 μm, 960 mm²). The opening of the discharge regulator is about 1/3 of the total flow area. Detailed experimental results are shown in Fig.2.

The experimental system must be well precooled in advance. At 29s after data acquisition and control system start, the experiment begins with both the exhaust and solenoid valves open. T5~T9 retain at LN_2 temperature and T10 drops down to LN_2 temperature dramatically (see Fig. 2(a)). At the same time, P3 to P7 drop down at the beginning, and then go back quickly (see Fig. 2(b)). 47s later, P3~P7 begin to be stable, that means the liquid transfers stably. In Figure 2(c), the pressure difference between the two sides of the engine filter (P8-P9) is about 2kPa in the transfer period. After 263s, all the pressure readings begin to drop down to zero as the liquid was emptied in the pipe.

In Fig. 2(d), the liquid flow rate is quite stable in the transfer period. The maximum value is 1.5L/s, and the integral value is 339.21L. At the beginning of the experiment, the pressurized gas flow reaches a peak value, and then drops down quickly. Then it gradually increases, and has a reading about 0.0172 kg/s at 262s. The integral gas flow is 3.94kg.

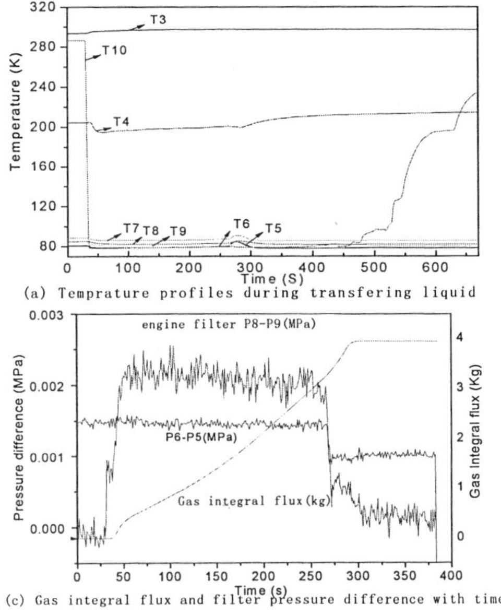

(a) Temprature profiles during transfering liquid

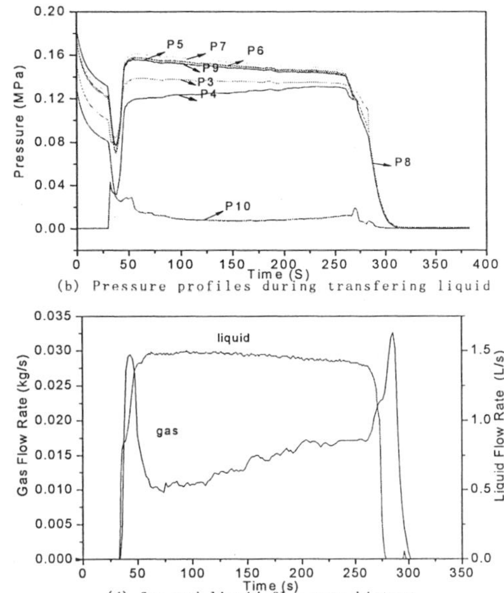

(b) Pressure profiles during transfering liquid

(c) Gas integral flux and filter pressure difference with time

(d) Gas and liuqid flow rate history

Figure 2 Temperature, pressure and flux profiles of gasified pressurization

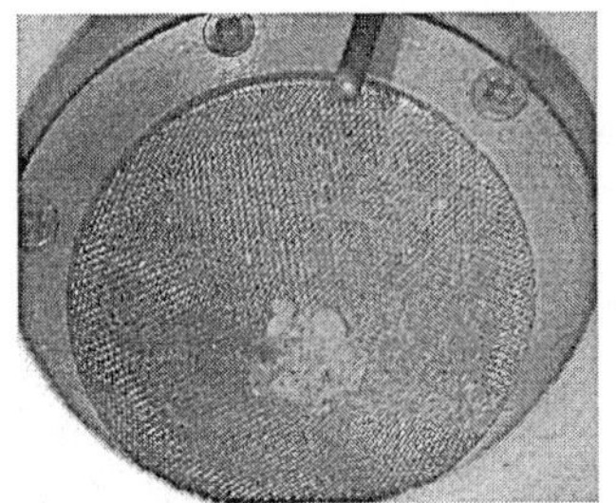

Figure 3 Surface status of engine filter

The engine filter is dismantled as soon as the experiment stopped. As shown in Fig. 3, at the bottom of the tapered filter there are a few white crystal particles adhered to other impurity. The GC112A gas chromatograph says the particles are CO_2.

In summary, the liquid flux in pressurized transfer process is stable, and the engine filter has not been jammed by CO_2 crystal particles piled on the bottom of the filter.

Abundant in oxygen pressurization mode

The pressurized gas used in this experiment is N_2 gas containing 11% CO_2. Both engine filter(150 μm, 960 mm^2) and tank filter(150 μm, 5000 mm^2) installed are stainless steel screens. The opening of discharge regulator holds about 1/3 of the total flow area. The experimental results are shown in Fig.4.

From Fig. 4(a) we can see that T5~T9 remain at LN_2 temperature and T10 drops down to LN_2 temperature dramatically. At the same time, P3~P7 drop down at the beginning, and then go back quickly, see Fig. 4(b). After about 34s, P3~P7 begin to be stable. It means that the liquid transfers stably. In liquid transfer period, the pressure difference between two sides of the engine filter (P8-P9) is about 1.5kPa, and that of the tank filter (P6-P5) is also 1.5kPa, and the pressure reading drops down to zero in 243s (Fig. 4(c)).

As shown in Figure 4(d), the liquid flow is quite stable in the transfer period. The maximum liquid flux is1.5L/s and the integral is 321.76 L. The integral of gas flow is 3.29kg.

The engine filter is dismantled after experiment and a few crystal particles under a layer of frost are found on the screen, see Fig. 5(b). It is certified by a GC112A gas chromatograph that the particles are CO_2. The frost on the screen came from air (see Fig. 5(b)).

The result shows that the solid particles may be kept back by the tank filter in the transfer pipe. It means that the particles have not jammed the engine filter and this additional tank filter has no obvious effect on the liquid transfer process.

Crystallization experiment

A simple crystallizing experiment is carried out to observe the real condensation state of CO_2 in liquid nitrogen.

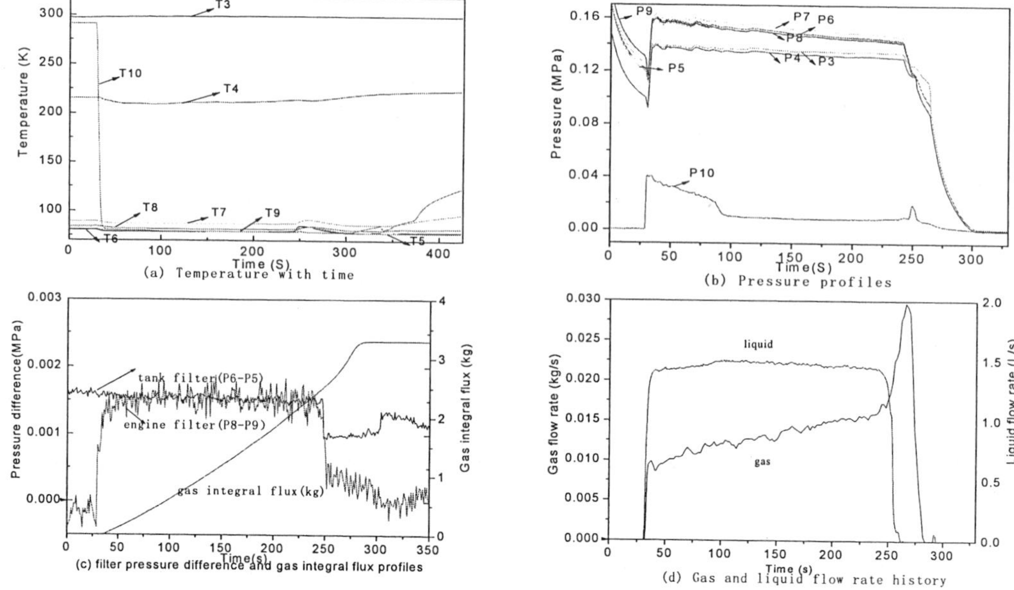

Figure 4 Temperature, pressure and flux profiles of abundant in oxygen pressurization

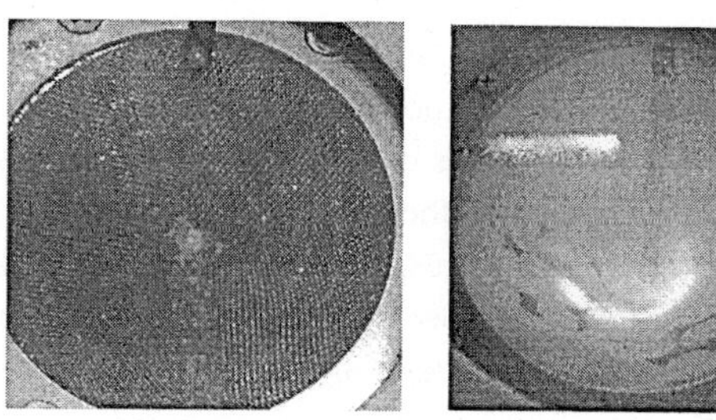

(a) Dismantled engine filter   (b) Dismantled tank filter

Figure 5 Filter surface statuses

The experiment shows that liquid nitrogen filled with CO_2 becomes an emulsion. It is a highly dispersive and unstable system, in which CO_2 solid particles about 1~10 μm diameter [3] are dispersed in liquid nitrogen. The emulsion may possibly be delaminated, distorted, and demulsified along with the time.

CONCLUSION

The experiments show that the simulation test-bed works effectively. In the transfer pipe, the pressure difference between the two sides of the filter shows no obvious change along with the time and the variation of the flow flux. That means the filter has not been obviously jammed, and the pressurized gas containing carbon dioxide can be used to transfer the cryogenic liquid.

The simple crystallizing experiment illuminates that the feasibility of two modes of the self-pressurizing experiment depends on the solution character of carbon dioxide and water vapor in the liquid nitrogen or oxygen, especially on their crystallizing rate.

REFERENCES

1. Li QF, Chen GB, Xie XM, Solubility of Solid Carbon Dioxide in Liquid Nitrogen and Oxygen, <u>Cryogenics</u> (2003) <u>134</u> 8-15 (in Chinese)

2. R.Reliai, R.G. Scurlock et al., High Solubility of Water in Liquid Nitrogen and Other Cryogenic, <u>Advances in Cryogenic Engineering</u>, Plenum Press, New York, NY, USA <u>29</u> 1005-1012

3. Chen ZQ, Dai MG, <u>Colloid Chemistry</u>, Higher Education Press, Beijing, China (1984) 316-317 (in Chinese)

Proceedings of the Twentieth International Cryogenic Engineering Conference
(ICEC 20), Beijing, China. © 2005. Published by Elsevier Ltd

DESIGN OF A COOLING SYSTEM FOR THE COLD TEST OF THE ITER TF COILS BEFORE INSTALLATION

Zahn G.R.[1], Duchateau J.L[2], Fietz W.H.[1], Gravil B.[2], Heller R.[1], Millet F.[3], Nicollet S.[2], Chesny P.[4]

[1] Forschungszentrum Karlsruhe (FZK), EURATOM Association, Institut fuer Technische Physik, Postfach 3640, D-76021 Karlsruhe, Germany
[2] Association Euratom-CEA, Département de Recherches sur la Fusion Contrôlée, CEA-Cadarache, F-13108 Saint-Paul-lez-Durance, France
[3] CEA-Grenoble, CEA/DSM/DRFMC/Service Des Basses Températures, F-38054 Grenoble, France
[4] CEA-Saclay, CEA/DSM/DAPNIA/SACM, F-91191 Gif sur Yvette, France

ABSTRACT

The ITER superconducting magnet system consists, besides six poloidal field coils and the central solenoid, of 18 toroidal field coils with the size of 16.6 m x 9 m and a weight of 400 t each. For each of these TF coils an intensive cold test is recommended prior to the installation into the torus in order to minimize the risk of malfunction which would cause tremendous costs and massive time delay. This paper presents the testing possibilities and a preliminary design of such a cold test facility including the process engineering of a cooling system.

INTRODUCTION

The magnet system of ITER consists of 18 toroidal field coils, the central solenoid, six poloidal field coils and several smaller correction coils. The construction principle of the large coils has been proven in a series of model coil tests, e.g. [1], where it was shown that such coils can be built with confidence from industry. However, from these experiments as well as from the construction of other actual front end superconducting coils it has been learned that tests with respect to high voltage, leak test, mass flow and current operation are indispensable for such coils. Therefore it is well accepted that superconducting coils of that challenge have to be tested before installation to ensure a proper functioning.

For a cold test of the ITER TF coils a large variety of test options exist, starting simplest with the test of TF coil parts e.g. a double pancake and ending with a possible test of a set of complete coils up to the rated current. Because for a test with rated current a massive reinforcement would be necessary [2] the most promising test is the test of one complete coil up to such a reduced coil current which does not require a reinforcement structure. This test allows checking high voltage toughness, leak tightness at cryogenic temperature, joint and forced flow cooling behavior. In this paper an outline for a cold test facility is presented based on the testing of a single TF coil at 4.5 K.

TESTING POSSIBILITIES

A cold test of one complete coil with reduced current offers the following test possibilities:
- Temperature-controlled cool-down and warm-up to check the temperature distribution
- Leak testing of the winding including the manifold region and the coil case
- High voltage insulation test at room and helium (He) temperature as well as in the "Paschen minimum" condition
- Investigation of the mass flow distribution and the pressure drop in the pancakes and the casing up to the reduced current to validate the operation parameters and to get the input for the simulation codes

- Measurement of the joint resistances
- Test of the sensors installed in the coil (temperature, pressure, pressure drop, balance of quench detectors, temperature compensation of strain gauges)
- Fast discharge from the maximum allowable current for a free standing coil

Parallel to the test of the ITER TF coils, components of ITER can be easily checked in the same cold test facility. In particular this could include:

- Test of ITER current leads with respect to He consumption and heat load at the cold end
- Test of components as cold He circulation pumps that still have to be developed for the required mass flow rate of the ITER cooling system
- Test of ITER current feeder system
- Test of the proposed control strategy of the ITER cooling system [3] as bypass system in the cooling circuit including heat load cycling tests

TEST FACILITY

The preliminary design of a vacuum vessel for vertical installation of a single TF coil is shown in Fig. 1a (left) and the arrangement of such a test facility in the required building in Fig. 1b (right).

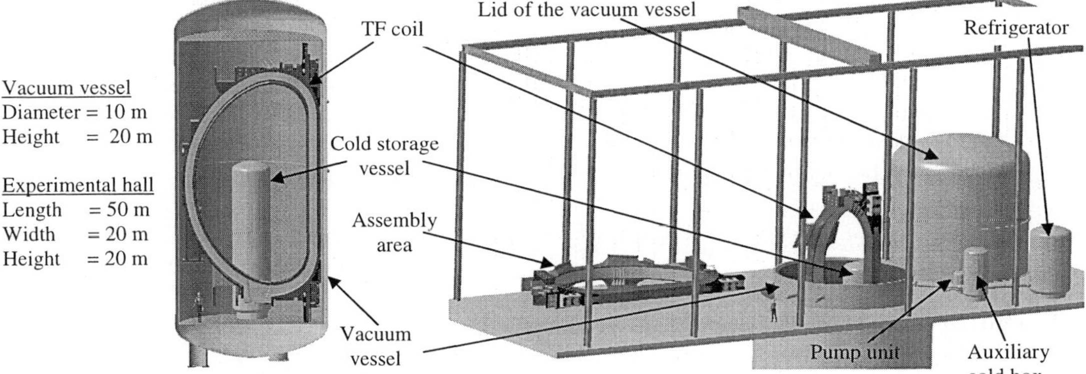

Figure 1 Preliminary design of a facility for the cold test of an ITER TF coil

The main components of the cold test facility are the vacuum vessel including liquid nitrogen shield for hosting the TF coil (Fig. 1a) and a cold storage vessel for the collection of the expelled He after a safety discharge, a He refrigerator/liquefier and an auxiliary cold box (ACB) for the distribution of the cryogenic supply from the refrigerator and for the heat load removal from the coil. Of course compressors, purifiers, warm gas storage and a control room are necessary.

Location of the test facility

Two possible locations are considered for the test of the complete ITER TF coils.

- The first possibility is to perform the test at the manufacturer site. The main advantage of this option is the possible repair in case of any defect on the coil without additional shipment.
- The second possibility is to perform the test at the ITER site. This option is the most promising solution because the whole infrastructure, buildings and maybe one of the 4 refrigerators required for ITER can be used. Furthermore the ITER staff can be trained in this facility for the ITER operation and the coil test facility would be available after the coil tests for examination of other components.

CRYOGENIC COOLING SYSTEM OF THE COLD TEST FACILITY

The cooling system should be designed similar to the TFMC test facility at the Forschungszentrum Karlsruhe, including the gained experiences [4] and also to that of ITER [3, 5]. A supercritical He cooling circuit will be used for circulating the high mass flow rate of the TF coil with cold circulation pumps. The pumps can be included in the ACB or in a separate pump unit. A second pump planned for redundancy reasons may not be required if an exchange is possible without interruption of the coil cooling, but this depends on the design of the pumps. Fig. 2 shows a tentative flow diagram of the He cooling system required for the test of the TF coil.

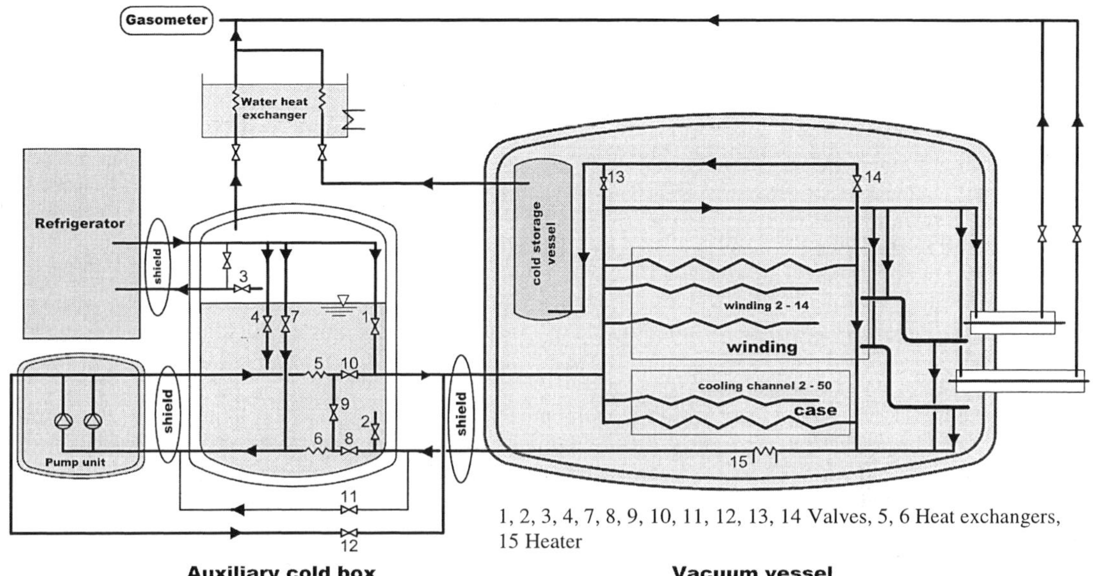

Figure 2 Preliminary design of a cooling system for the cold test of a single ITER TF coil

All operation modes mentioned hereafter are standard modes in the TOSKA facility and their safe and reliable operation was demonstrated during the coil tests (LCT, W7-X, ITER TFMC) over 20 years [4].

Cool down

During the cool down the He will be supplied from the refrigerator to all components in parallel through the ACB and valve 1. The return flow is expanded into the ACB through valve 2 and fed back to the refrigerator over valve 3. The ACB is cooled in parallel and after the cool down is completed, the refrigerator is liquefying into the ACB via the Joule-Thomson valve 4.

Steady state operation

For the steady state cooling of winding, case and current leads, a He mass flow rate up to 350 g/s will be circulated in the secondary cooling loop by means of circulation pumps. The He compression heat load will be transferred after the pumps to the liquid He in the ACB via the immersed heat exchanger (HX) 5 in order to provide He for the coil cooling at a temperature level of 4.5 K. In the immersed HX 6, the heat load of winding, case and transfer lines is also transferred to the liquid He in the ACB. This facilitates to obtain a stable inlet temperature to the pump unit. Valve 7 is foreseen for the stabilization of the pressure in the secondary cooling loop at the low pressure side to 3.5 bar, but the pressure can be increased to the design values of ITER. The stabilization at the outlet of the coil has the advantage that the pressure can be chosen at a low value independently of the pressure drop with no danger to run into the two phase region. For cooling the current leads the required He mass flow rate is subtracted from the secondary cooling loop inside the vacuum vessel with the advantage that no separate transfer line is necessary and the He is supplied at the same low temperature as for the coil. The main preliminary cooling requirements for reduced current operation of 30 kA are listed in the following Table 1. For comparison the corresponding numbers for rated current operation are also included.

Standby operation

In standby mode only the mass flow rate of the refrigerator will be used without running the pumps for cooling the complete test arrangement. The He is supplied via valve 7 into the secondary loop and is flowing through the pumps, the HX 5, then cooling the coil, returning to the ACB and expanded in valve 2 (valve 8 closed and 10 open) and finally liquefied into the ACB. Also a small portion (~1 g/s) of He is subtracted from the secondary loop for the standby cooling of the current leads.

Table 1 Preliminary cooling requirements for the cold test of a ITER TF coil

Operation	T	P	Mass flow rate						Heat load						
			W	C	FS	RS	CL	**Total**	W	C	RS	FS	PP	Facility	**Total**
Unit	K	bar	g/s	g/s	g/s	g/s	g/s	**g/s**	W	W	W	W	W	W	**W**
Rated (68 kA)	4.5	3.5	112	200	24	100	10	**446**	150	400	200	70	590	300	**1710**
Reduced (30 kA)	4.5	3.5	112	200	24	0	4	**340**	50	600	0	50	450	300	**1450**

T = temperature, P = Outlet pressure of the TF coil and inlet pressure of the pump, $\dot{m}$ = mass flow rate, $\dot{Q}$ = heat load
W = Winding, C = Case, FS = Feeder system, RS = Reinforcement structure, CL = Current leads, PP = Pumping power

<u>Collecting and storing the expelled He after a safety discharge</u>
The expelled He after a safety discharge or a quench, will be collected in a cold storage vessel and stored at low temperature by opening the valves 13 and 14 automatically with the quench or safety discharge trigger. This storage vessel will be installed inside the vacuum vessel and LN_2 shield as shown in Fig. 1.

<u>Test of the cooling system as proposed for ITER</u>
According to the ITER Design Description Document (DDD) [5] and also as explained in [3] the cooling system of the ITER TF coil is slightly different than proposed here for the cold test facility. In the secondary cooling loop no HX is foreseen in front of the pump. This operation can be simulated by closing the valve 8, opening valve 11 and bypassing HX 6. Furthermore, the ITER cooling system will be operated with a constant mass flow rate in the secondary loop also after ITER plasma pulsing with nuclear heat load fluctuations. This can be simulated to certain limits by introducing heat pulses using the heater 15 and bypassing partly HX 5 by opening the valve 12 in order to avoid an overloading of the ACB and the refrigerator.

<u>Refrigerator/liquefier cryogenic plant</u>
As listed in Tab. 1, the required cooling capacity is in the range between 1.5 kW and 2 kW. In the case of a test at the manufacturer site a suitable refrigerator has to be ordered. For a cold test facility at the ITER site the use of one of the four 18 kW, 4.5 K refrigerator cryogenic plants foreseen for ITER [3] can be considered. However, an adapted separate refrigerator/liquefier plant offers the advantage of low power consumption, the availability during ITER operation for pre-testing of components and the use of four identical plants for the ITER machine.

CONCLUSION

The cold test of a complete ITER TF coil at 4.5 K at a reduced coil current of about 30 kA is the most favourite option. This test scenario allows to detect high voltage, leak, joint, sensor and cooling failure before installation into the ITER machine and therefore to avoid enormous disassembly and re-installation work and as consequence large additional costs and delay. In addition the proposed cold test facility allows to prove the cooling circuit planed for ITER and will allow to train the ITER staff when the cold test facility is built on ITER site.

REFERENCES

1. Maix R.K., Completion of the ITER toroidal field model coil, <u>Fusion Engineering und Design,</u> (2001) <u>58-59</u>, 159-164.
2. Raff A. et al., The Test the EURATOM LCT COIL at the Outermost Limits at 1.8 K as Example of Single Coil Testing of D Shaped Toroidal Field Coils, <u>MT-15,</u> Science Press, (1997) <u>20-24</u>, 373-376,
3. Millet F. et al., Design and Performance Analysis of the ITER Cryoplant and Cryo-Distribution Subsystem, <u>Proceedings of ICEC19,</u> 105-108, Grenoble, France (2002)
4. Zahn G. R. et al., , Cryogenic test results of the ITER TF model coil test in Toska, <u>Proc. CEC,</u> (2003), Anchorage, USA, accepted for publication in Adv. Cryo. Eng. (2004)
5. Kalinin V., Detailed Design Document for ITER cryogenic system, <u>DDD 3.4</u> (2001)

Study on key-tech problems in development of a cold neutron source in China advanced research reactor

Quanke F., Qingfeng Y., Cun Z., Feng S[*]., Luzheng Y[*].

School of Energy and Power Engineering, Xi'an Jiaotong University, 710049, China
[*]Dept of Reactor Engineering, China Institute of Atomic Energy, Beijing, China

This paper introduces main design schemes of the cold neutron source in CARR. Because of large reactor power, high nuclear heating and high heat flux in moderator cell, special solutions were proposed. Moderator cell is directly cooled by cryogenic helium. Higher saturation vapor pressure of moderator, hydrogen, is adopted. A composite condenser and a single-tube thermal-siphon loop for moderator are used. All these designs will provide the CNS with excellent operation performances.

INTRODUCTION

Cold neutrons have very wide-ranging applications in modern science and technology researches, and in performance improving of industrial and agricultural products. Specially, it has unique advantage in the detection of molecular structure of organic substances such as protein, nucleic acid etc. The development and application level of cold neutrons characterizes the level of modern science and technique to a country.

In the past 30 years, about 20 sets of cold neutron sources have been established in many research reactors in the world. Most of them belong to mini-type cold neutron sources and their heat loads generated by nuclear heating are mainly from 100 to 1000W. To get strong flux and high quality (high proportion of cold neutrons with long wavelength) of cold neutrons requires larger reactor power, bigger cell size, the more reasonable shape of liquid moderator zone and the lower void fraction in liquid moderator.

One of important difficulties in design of a large cold neutron source is how to remove the heat load of moderator cell out. Power of CARR is 60MW. It has been decided to use Liquid hydrogen as moderator and two-phase thermal-siphon loop as cycling apparatus to transport nuclear heat from the cell to the hydrogen condenser. All nuclear heating generated from cell material and liquid hydrogen should be transported by vaporized hydrogen in the case of conventional cold cell structure (Fig 1). As a result, high heat flux and small cross section area in the moderator cell will cause the void fraction in the moderator liquid hydrogen to exceed the limited amount (20~25%). Therefore, a special structure of moderator cell and thermal-siphon loop should be developed.

MAIN DEMANDS ON CNS DESIGN OF CARR

1) The cold neutron source designed for CARR should supply high flux and high profit factor of cold neutrons.

2) Material of the moderator cell should be as little as possible, and its construct should be as simple as possible.

3) The designed moderator cell should have excellent ability of moderating neutrons and bearing-pressure

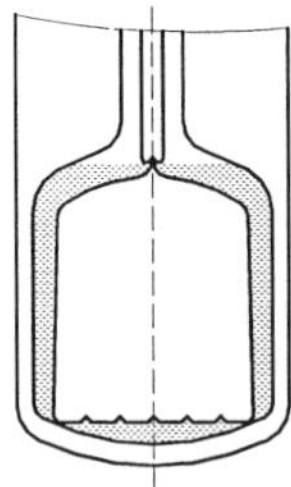

Figure 1 Conventional moderator cell structure

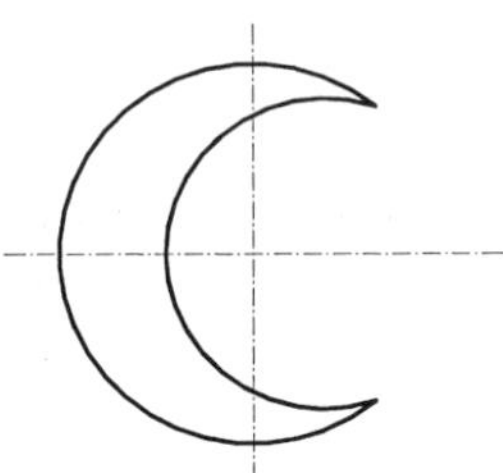

Figure 2 Moderator cell with crescent moon-shape

(inner and outer pressure). It should work well in different work conditions, specially referring to quite different temperature and pressure.

4) Void fraction of liquid hydrogen in the moderator cell should be controlled bellow 20%, had better bellow 15%.

5) Another common work conditions of the cold neutron source may occur like this: the reactor is running without operation of helium refrigerator. All nuclear heating load of cell material should be carried away easily by coolant fluid under room-temperature condition. This means that circulating helium should easily reach all the walls in the moderator cell.

6) Level of liquid hydrogen in the moderator cell should be maintained at constant height to assure the steady operation of the CNS.

7) Hydrogen facilities, specially, inner the reactor should have multi-layer safety barriers.

STRUCTURAL FEATURES OF THE CNS IN CARR

<u>Design of the moderator cell</u>

The Monte Carlo MCNP code of neutronics is used to calculate and analyze the structure of the moderator cell designed for CARR. The results show that, ideal moderator cell should have 30~35mm thickness of moderator in the square window(200×200mm) of neutron scattering and have thicker layer of moderator around the window's brim. It is well known that an annular-cylinder type of moderator cells used in the CNS of FRANCE ORPHEE is successful example of moderator cell structure, for it has annulus-ring of moderator zone and has advantages described above.

Furthermore, according to the principle of neutron moderation, it had better no moderator zone existing in the exit of cold neutron, shown as in a crescent moon-shape moderator cell (in Fig.2). For this shape of moderator cells has a function of gathering moderated neutrons, it is beneficial to increase cold neutron flux in scattering channel. So it has more remarkable advantages over annular-cylinder type of moderator cells in moderating neutrons. But it has a fatal disadvantage in structural stability when high pressure in moderator cavity acts on the thin inner-wall of the cell. This may be the main reason why crescent moon-shape moderator cells have been rarely used in different CNS in the world.

According to the analysis described above, a special structure of moderator cells is designed for CARR by authors. As shown in Fig.3, the inner cup of the moderator cell in Fig 1 is moved near to one side of the hydrogen cell, so that a crescent moon-shape moderator cell is constructed. Furthermore, an interlayer is constructed at outside of the moderator cell, in which cold helium is introduced to help cooling moderator and its cell material.

This kind of moderator cells has crescent moon-shape moderator zone and good inner-pressure-bearing character like annulus cylindrical-type moderator cells. Therefore, it may be an idea kind of moderator cells.

The function of the helium assistant-cooling system outside the moderator cell is so powerful that more than 2kW nuclear heating power in the moderator cell can be directly transported out by it. At first, all helium with temperature 17.5K from refrigerator is guided into the interlayer

Figure 3 New designed moderator cell for CARR

of the moderator cell. Then it flows into hydrogen-helium heat exchanger to condense hydrogen steam into liquid hydrogen. Nuclear heating power carried by vaporized hydrogen from the cell to the condenser is only 800~900W. In this way, void fraction of liquid hydrogen in the moderator cell will be reduced to a large extent.

Flow rate of hydrogen stream in moderator cell is predicted as 0.70l/s according to hydrogen latent heat and the nuclear heating power for vaporized hydrogen to carry off. In order to exactly predict void fraction in liquid moderator, mock-up tests were conducted. Different kinds of Working fluids are used to simulate working cycles of two-phase hydrogen in thermal-siphon loop. They are water, Freon 113, alcohol and liquid nitrogen. The facility for mock-up test is shown in Fig 4.

From all the results of mock-up tests, quantitative relations were determined between void fraction in moderator and their main influencing factors such as bubble-rising velocity in liquid hydrogen, liquid hydrogen density, viscosity and surface tension. From all information, we can predict that the void

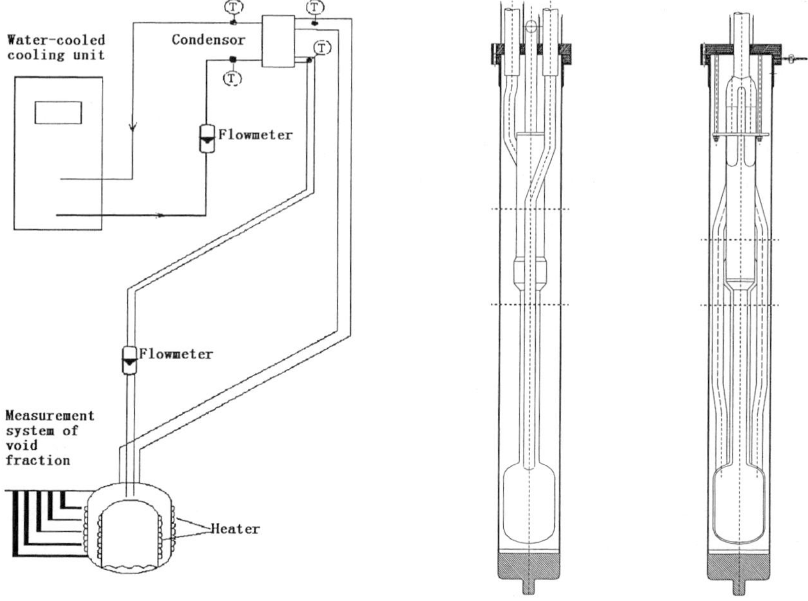

Figure 4 Facility for mock-up tests   Fig 5 Structure of CNS facility in reactor pool of CARR

fraction in liquid hydrogen is about 10~15% when steam-generating rate is 0.70l/s for the moderator cell described above. This result can entirely satisfy the requirement of neutron moderation.

Considering complexity of the cell structure designed for CARR and performances of 6061 aluminum alloys, feasibility of fabrication process must be carefully considered, especially welding procedure.

<u>Design of condenser and loop</u>

In the CNS of CARR, all thermal-siphon loop including moderator cell, hydrogen-helium heat exchanger and their connecting pipe are placed in a six-meter long vertical channel, which is made of 6061 aluminum alloys and inserted in heavy water pool of the reactor. Inner-diameter of the vertical channel is 260mm. high vacuum is kept in channel to ensure excellent thermal-isolation between the CNS and wall of the channel.

After carefully analyzing various affecting factors and comparing different schemes, Layout of all in-pile components of the CNS in CARR is proposed as shown in Fig 5. Hydrogen steam going up to the condenser and liquid hydrogen falling down to the cell are completed in a single tube.

The cold helium from refrigerator is directly guided into the moderator cell from the upper end of the vertical channel. After surrounding-flowing outside of the moderator cell, helium goes up through a annular pipe to condenser. Then it flows out of the condenser and returns to refrigerator.

This arrangement shown in Fig.5 has another important advantage, that is, when reactor keeps operating and refrigerator stops working, the helium cycling system can still send room-temperature helium into the moderator cell by a fan to carry out nuclear heating of cell material and to keep temperature of all cell wall bellow 110℃.

CONCLUSIONS

Based on the research results described above, follow conclusions can be obtained:

1) The power of nuclear heating of the moderator cell in CARR is quite large and heat flux in moderator is very high. Conventional structures of moderator cell and thermal-siphon loop can't supply normal operation of the CNS for CARR.

2) There are many requests and limits in the design of CNS, so a reasonable design should harmonize the conflicts among these requests and limits.

3) In the CNS of CARR, the moderator cell is directly cooled by cold helium, higher working pressure is chosen for the hydrogen loop, and a moderator cell with crescent moon-shape moderator zone is used. All of these will effectively contribute to reduce void fraction of liquid hydrogen in moderator cell and to increase the cold neutron gain.

REFERENCES

1.   Klause G, Present and future cold neutron sources in Garching, Germany, <u>International Workshop on Cold Neutron Utilization</u>, Korea (1997) 43-52

2.   D.L.Selby, Scientific upgrades at the high flux isotope reactor at oak national laboratory, <u>International Workshop on Cold Neutron Utilization</u>, Korea (1997) 83-87

3.   Shen Feng, Yuan LZ, Conceptual study of cold-neutron source in china advanced research reactor, <u>Physica B311</u>(1-2): Jan 2002, 152-157

4.   B.Farnous, New cold neutron sources of the Orphee reactor [A]. <u>Proceeding the 5th Asian Symposium on Research Reactor</u>[C]. Taejon: Korea (1996) 599-609

Proceedings of the Twentieth International Cryogenic Engineering Conference
(ICEC 20), Beijing, China. © 2005 Elsevier Ltd. All rights reserved.

Experiences from the operation tests with the ITER model pump

Haas, H.[*], Antipenkov, A.[*], Day, Chr.[*], Dremel, M.[*], Mack, A.[*], Murdoch, D.K.[**]

[*]Forschungszentrum Karlsruhe GmbH, ITP, PO Box 3640, 76021 Karlsruhe, Germany
[**]EFDA CSU, c/o Max-Planck-Institut für Plasmaphysik, Boltzmannstr. 2, 85748 Garching, Germany

Within the development work for ITER (International Thermonuclear Experimental Reactor), the test bed TIMO (Test facility for the ITER model pump) was built at Forschungszentrum Karlsruhe. The aim of this test facility is to check the suitability of the 1:2 ITER torus model cryopump under all aspects of operation. This model pump is the result of extensive preliminary examinations focusing on different aspects of cryosorption vacuum pump operation in the ITER fuel cycle. To pump out the required plasma exhaust gas flow of 120 (Pa·m³)/s, which essentially consists of all mix hydrogen isotopes as well as helium and various impurities, the pump surfaces cooled to 5 K were coated with activated charcoal embedded in an inorganic cement. Apart from all the gases which are condensed at the operation temperature of 5 K, helium as well as hydrogen isotopes are fixed on the activated charcoal-coated pump area of 4 m² by means of sorption.

TIMO TEST BED FOR TESTING THE ITER MODEL PUMP

TIMO provides all the necessary tools to examine all operation modes foreseen for the ITER vacuum pumps [1].

Process gas supply in TIMO is realised with seven supply lines between a gas storage and the metering system. For the different tests with the model pump, pure gases, such as protium, deuterium, helium, neon, argon, nitrogen, and an ITER-relevant gas mixture, are available. The necessary gas throughputs are controlled with four flow meters in the range between $1.7\ 10^{-2}$ and 169 (mbar·l/s)

For standard operation, coolant supply of the two cooling circuits in the model pump, the 80 K shielding and sorption panels, gaseous helium at 80 K, and supercritical helium at 4.5 K are required. The coolant flows are controlled in a valve box which is installed between the model pump, a 4.5 K control cryostat, and a 80 K facility.

The 4.5 K helium flow at a max. pressure of 6 bar is supplied to a control cryostat by the 2 kW LINDE refrigerator [2]. From this control cryostat which consists of a 3500 l liquid helium tank, the 5 K panel circuit of the model pump is provided with a Joule Thomson flow of 60 g/s as well as with a helium mass flow of 250 g/s by a centrifugal pump. The 80 K circuits of the model pump are supplied by the 80 K facility. This 80 K facility includes a 600 l liquid nitrogen tank, a heat exchanger, and a centrifugal pump to transfer the maximal mass flow of 200 g/s at 15 bar.
For the regeneration and reactivation process after a pump test performed with the sorption panel, a helium mass flow of max. 50 g/s is available at a temperature level of 300 K as well as 450 K.

To simulate the ITER duct conditions during pumping, the TIMO test vessel can be warmed up to a maximal temperature of 480 K.

With the process control system (Simatic S5 135U), an ITER-relevant cycle operation can be simulated with the model pump.

To analyse the actual gas compositions during the different steps of the pump test, an analytical device of the type of quadrupole gas mass spectrometer "Balzers GAM 400" was installed in the TIMO test vessel. With this mass spectrometer, a standard mass range between 1-128 AMU can be measured. In addition to this normal analysis mode, high-resolution operation is possible over the mass range of 1-22 AMU. In this high-resolution operation, the separation of helium (4.0026 AMU) and deuterium (4.0282 AMU) can be achieved.

OPERATIONAL TESTS

To check the operational behaviour of a cryosorption pump, TIMO was built at the Forschungszentrum Karlsruhe. During several test campaigns with the ITER model pump, a lot of data were collected, which improved the understanding of the mechanism of cryosorption pumps on the prototype scale [3,4,5]. Within a new test campaign at TIMO, further tests were performed to check the feasibility of ITER neutral beam heating and current drive (H&CD) system requirements and to study the behaviour of this type of cryopump in view of heat losses as well as in safety aspects.

NBI simulation tests

The ITER Neutral Beam Injectors (NBI) consists of two operational and one diagnostic injector line, including different pump sections [6]. One NBI produces a deuterium beam of 16.7 MW at 1 MeV during maximal operation pulses of up to 3600 s. The aim of the simulation tests in the TIMO test bed was to check with the ITER model pump the required operation conditions of the NBI cryopumps. A very essential point of these tests was to study the behaviour of the charcoal-coated cryopanels under typical conditions of the NBI pumps. To fix comparable parameters for the simulation tests, the size of the pump areas in the different NBI pump sections, the required gas loads, the operation pressures, and the type of panel coating were taken into account.

The tests were performed at the temperature levels of 15 K, 20 K, and 25 K. During the tests at these temperature levels, the operation was simulated for the two types of neutralisers, the standard NBI line and the diagnostic NBI line as well as for the NBI calorimeter. For each of these NBI pump sections, the different panel configurations, coated on one side and coated on both sides, were studied. To specify the necessary gas load for the two different cases, the distribution of the incoming gas between the panel

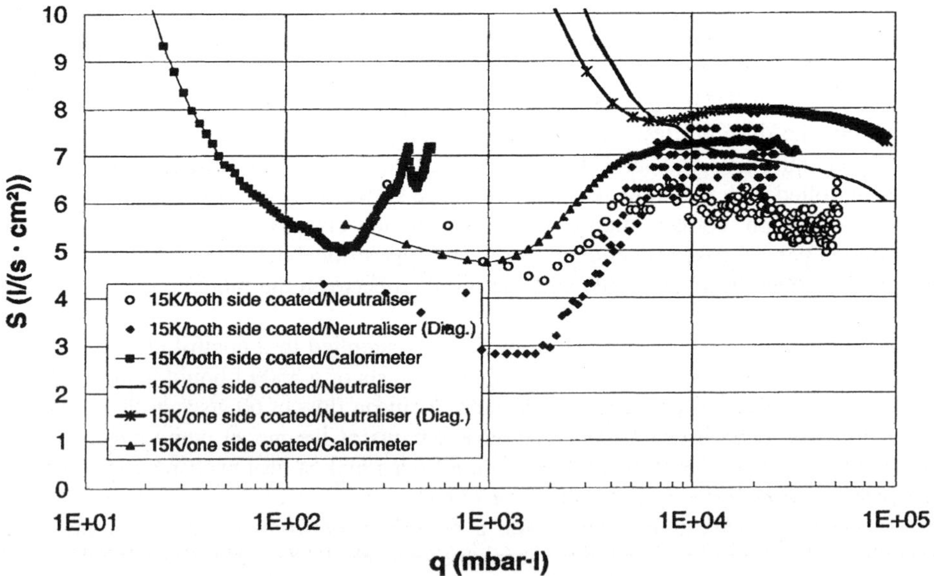

Figure 1: Results of the NBI simulation tests at a temperature level of 15 K.

sides was checked using a Monte Carlo code. As a result of the preparatory work, it was found that for the one-side-coated panel tests, the simulated metering rate must be higher by a factor 6.

The test results demonstrate that using a charcoal-coated panel, the required pumping speed of the NBI cryopumps (~1 l/s·cm²) can be fulfilled. Concerning the high gas loads in the long pulse, NBI pump operation could not be examined due to the limited deuterium inventory allowed for TIMO. Further tests in a future NBI test bed are necessary to clarify open questions regarding the feasible capacities of the charcoal-coated pump surface during long-pulse operations.

<u>Heat load tests</u>

To check the heat loads of a cryopump of ITER-comparable design, tests were performed in TIMO with the model pump. The data collected served to support the working group responsible for the design of the ITER cryoplants.

During these tests, both cooling circuits of the ITER model pump were operated in a steady-state mode at constant temperature. This means that after adjusting the cooling flows in the 80 K shields and 5 K panels to a constant value, helium was metered into the volume of the model pump. The heat loads of both cooling circuits were measured at pressure levels between 10^{-4} mbar and 10 mbar.

The measured values show that depending on the pressure inside the model pump and the coolant flows, the heat loads in the 80 K circuit of the ITER model pump rise up to max. values of 4500 W. In the 5 K panel circuit of the model pump at 10^{-1} mbar, a heat load of ~480 W was measured. At higher pressures of up to 10 mbar, the results rose up to 950 W. Nevertheless, such high heat loads are only expected during unusual disturbances of the regular pump process.

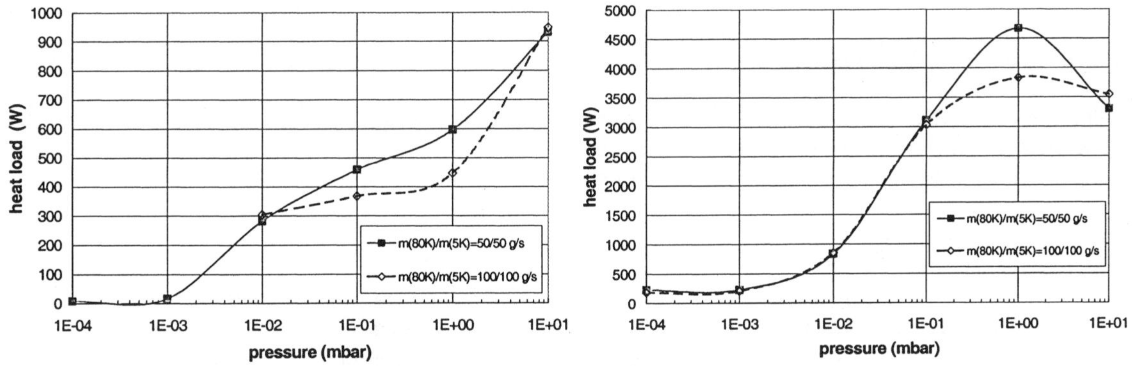

Figure 2: Measured heat loads of the 5 K panel circuit and the 80 K shielding, respectively.

<u>Safety tests</u>

While preparing the design of the different ITER components, also safety studies concerning the risk associated with the high hydrogen inventory were executed parallel to the development activities. As a result of these studies, assumptions were defined for an air inbreak accident into the ITER vacuum vessel. So far, the pressure increase has been assumed to be ~2.7 mbar/s. To support the modelling work for the safety scenarios, safety tests were carried out with the ITER model pump.

These safety tests in TIMO were aimed at collecting experimental data for a fast release of the pumped gas inventory inside the cryopump, followed by high transfer inside the pump volume. To simulate an air inbreak accident with pressure increase rates of 0.5 to 20 mbar/s, a new metering line was integrated in the TIMO test bed. For reasons of safety, nitrogen was used for these tests. During the tests, the different pressure gradients were examined. In addition, the behaviour of the cryopump was studied for operation with and without active cooling, respectively.

The first test results demonstrate that the hydrogen released will be fixed on the pumping surfaces of the connected cryopumps. In further studies, the operation sequences of the different cryopumps of the ITER vacuum system will have to be taken into account to better assess the real risk potential of the hydrogen inventory inside the fusion reactor.

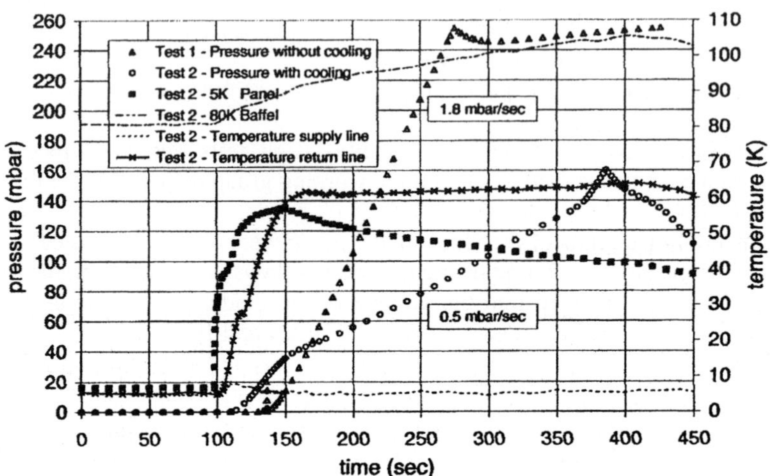

Figure 3: Pressure and temperature measured during the safety tests performed.

CONCLUSIONS AND OUTLOOK

The results of the NBI simulation tests demonstrate that the panel coating technology developed by Forschungszentrum Karlsruhe can also be used for the ITER NBI pumps. The results of the heat load tests and the safety tests supported the further development steps of the ITER cryopump system.

ACKNOWLEDGEMENTS

The authors wish to thank the operators of the TIMO facility, J. Weinhold, D. Zimmerlin, and H. Jensen, for their help as well as Mrs. Edinger for her support during the evaluation of the collected data. This work has been performed within the framework of the Nuclear Fusion Project of Forschungszentrum Karlsruhe, supported by the European Communities under the European Fusion Technology Programme.

REFERENCES

1. Haas, H. et al., Test facility TIMO for testing the ITER model cryopump, Proc. 17th IAEA Fusion Energy Conference, Yokohama, Japan, October 1998, Vol. 3, pp. 1077-1080.
2. Spath, K. et al., Performance tests of a 2 kW He refrigerator for SC magnets tests down to 3.3 K, Advances in Cryogenic Engineering, Vol. 39 (1994), 563-570.
3. Mack, A. et al.; First operation experience with ITER-FEAT model pump, Fusion Engineering and Design 58-59 (2001) 365-369.
4. Mack, A. et al., Design of the torus cryopump, Proc. ISFNT-6, San Diego, US, April 2002.
5. Haas, H. et al., Performance tests of the ITER model pump, Fusion Engineering and Design 69 (2003) 91-95 Helsinki.
6. Dremel, M. et al., Design and manufacturing of cryosorption pumps for test beds of ITER relevant neutral beam injectors, this Conference.

Proceedings of the Twentieth International Cryogenic Engineering Conference
(ICEC 20), Beijing, China. © 2005 Elsevier Ltd. All rights reserved.

Design of cryogenic system for the KSTAR device

Kim Y. S. [a], Chang H. S. [a], Cho K. W. [a], Gistau Baguer G. M. [a], Bak J. S. [a] and Lee G. S. [b]
Briend P. [b], Fuenzalida P. [b], and Fossen A. [b]

[a]Korea Basic Science Institute, Yusung-Gu, Daejeon, 305-333, Korea
[b]Air Liquide DTA, BP 15- 38360 Sassenage, France

The KSTAR tokamak device for advanced research in the field of fusion experiment requires a large amount of refrigeration power at several temperature level. The design of the KSTAR helium refrigeration system is based on the refrigerator and distribution system. The helium refrigerator is designed to operate stably to cope with the pulse heat loads of the superconducting magnet operation in tokamak device. The helium distribution system is designed to distribute refrigeration to KSTAR cooling components such as 30 superconducting coils, magnet supporting structures, thermal shields and current leads with a minimal temperature variation. In this paper, the development of the cryogenic system for the KSTAR tokamak device is described.

INTRODUCTION

The purpose of the KSTAR helium (He) refrigerator is to keep the superconducting (SC) components of the KSTAR device below their current sharing temperature and maintain the thermal shields (TS) and current leads (CL) at a suitable cryogenic temperature. The SC coils, their corresponding structures, and the SC buslines are cooled with forced flow of supercritical He (SHe) whereas the CL and TS are kept at the proper operating temperature by liquid He (LHe) and pressurized gaseous He (GHe), respectively [1]. For the KSTAR helium refrigeration system, a safe and stable operation and fully automatic control is required because of the long-term operation time and a various cooling modes such as cool-down and warm-up, normal operation and abnormal events such as coil quench and electrical power failure.

KSTAR COOLING COMPONENTS

The duty of the refrigerator concerning the SC coils is to remove the static thermal load due to radiation and conduction and the dynamic load due to current charging in order to maintain the coils far below a certain critical temperature. Table 1 shows the thermal-hydraulic parameters of the SC coils. The required mass flow rate of the SHe to cool the SC coils is about 600 g/s during the normal operation. All magnet structures are cooled with forced flow supercritical helium. The magnet supporting posts are installed below the TF coil structures. These posts absorb the thermal shrinkage of the magnets and support the magnet weight and plasma disruptions load. The supporting ring and posts are cooled with 4.5 K SHe and 55 K GHe, respectively [1].

The power transmission system for the KSTAR device requires current leads to link the SC bus-lines at 4.5 K to the normal bus-lines at 300 K. The current lead system consists of 11 pairs of brass overloaded vapor-cooled leads for the CS and PF coils, one pair of copper vapor-cooled leads for the TF coils, and two current lead boxes which consist of cryostat, thermal radiation shield, and liquid helium reservoir. The KSTAR SC bus-lines have 2 ducts and 12 pairs of CICCs. Each bus duct consists of NbTi CICCs, thermal shields, electrical insulations, and support structures. The average length of one bus duct is about

10 m. The thermal shield reduces the thermal radiation from the room temperature side to the coil temperature (4.5 K) region. The cryopanel has been designed to maintain a maximum temperature of 80 K during normal operation and 100 K during vacuum vessel baking. In table 2, the cryogenic characteristics of KSTAR cooling components are summarized.

Table 1. Thermal-hydraulic parameters of the KSTAR SC coils

	Unit	TF coils	CS coils				PF coils		
			CS1U&L	CS2U&L	CS3U&L	CS4U&L	PF5U&L	PF6U&L	PF7U&L
Strand			Nb3Sn				NbTi		
# of coils		16	2	2	2	2	2	2	2
Total coil mass	ton	46	13				37		
Coil connection			serial			parallel			serial
Cryogen			SHe						
He inlet	ea	32	10	8	2+2	3+3	4+4	4+4	6
He outlet	ea	33	11	9	3+3	4+4	5+5	5+5	7
cooling channel/coil	ea	4	10	8	4	6	8	8	6
total cooling channel	ea	64	20	16	4+4	6+6	8+8	8+8	12
L coil	m	~640	~660	~540	~285	~410	~1550	~2510	~1670
L cooling channel	m	~160	~66	~68	~71	~68	~194	~314	~278
DH	mm	~0.5							
AHe	mm2	138	107						
Tin	K	4.5							
$Tout$	K	5.0	6.6				5.0		
Pin	bar	5.5-5.7	5.5-7.8						
$Pout$	bar	3.0							
m dot	g/s	300	183				117		

Table 2. Summary of KSTAR cooling components

Device		No	Cold mass (ton)	Operating temp. (K)	Cryogen	Amount of cryogen
SC coils	TF coil	16	100	4.5	SHe	3000 liter
	CS coil	8 (4 pairs)				
	PF coil	6 (3 pairs)				
Magnet structu	TF/CS/PF	16/1/80	140			500 liter
Supporting structures	1 ring	10				
	8 posts			55	GHe	
Current lead	Lead	12 pairs		4.5	LHe	2000 liter
	Lead box	2		55	GHe	
SC bus-line	Bus-lines	12 pairs	20	4.5	SHe	
	Bus duct	2		55	GHe	
Thermal shield	CRTS, VVTS	20				500 liter

OPERATION SCENARIO

The annual operation mode of KSTAR device consists of clean-up, cool-down, warm-up and operation period. The operation period is about 8 month. Before cool-down process, each part of the cryogenic passages (pipes, magnet passages, heat exchangers) must to be cleaned by a stream of purified helium. Helium flow for the cleanup process is supplied and controlled by the refrigerator. A special procedure will be applied to the coil and bus-line CICC: pure helium provided by the purifier will supplied to the CICC until moisture and impurity content at the discharge side is corrected. All cold mass should be cooled down by the refrigerator system from 300 K to the operation temperature. Cool-down time from 300 K to 4.5 K is estimated about 1 month. SC magnet system requires a limitation of the temperature difference between any components of KSTAR should be less than 50 K and the temperature of the TF coils should be always higher than that of the TF coil structures.

In order to estimate the refrigeration power for the KSTAR device, the baseline operation mode was considered. During the baseline mode, a daily operation time is 8 hr/day (18 shot/day), the plasma shot time is 70 sec and pulse repetition time is 1200 sec. During the shot mode, TF coils are fully charged to 35 kA, PF coils pulsed operated to its maximum current and plasma current charged up to 2 MA. The daily operation scenario is listed in table.

THERMAL LOADS DURING NORMAL OPERATION

Table 3 shows the various thermal loads of the KSTAR cryogenic system components at each operating modes [2]. Considered thermal loads during each operating modes are;

- Idle mode: conduction loss, radiation loss
- TF ramp/de-ramp, stand-by mode: conduction loss, radiation loss, joint DC loss, eddy current loss, AC loss.
- Shot mode: conduction loss, radiation loss, joint DC loss, joint AC loss, Eddy current loss, AC loss, nuclear heat loss, friction loss.

Table 3. Thermal loads at each operating modes of KSTAR cooling components

Mode	Period (hr)	SC Coil (kW@4.5K)	SC structure (kW@4.5K)	SC bus (kW@4.5K)	Lead (g/s @ Lhe)	TS (kW@70K)
Idle	14	0.5	0.8	0.06	10	13
TF ramp	1	0.6	0.8	0.1	16	13
Standby	7.65	0.6	0.8	0.1	16	13
Shot	0.35	11	4.0	1.0	52	13
TF deramp	1	0.6	0.8	0.1	16	13

The design power of the He refrigerator is based on the "day average" thermal load of the KSTAR device and the constant thermal losses of the refrigerator and the cryogenic He transfer lines [3]. Since the temporal variation of the KSTAR thermal load is very large (the thermal load during the shot mode is about 10 times that of the other modes), the additional cooling power required during the various operation modes is obtained by vaporizing LHe which has been liquefied and stored in the LHe storage tank during the idle mode (day average cooling scheme). The day averaged thermal load of the KSTAR cold components is about 3.8 kW @ 4.5 K. This value does not incorporate the heat losses of the liquid helium storage tank, cryogenic circulators, and the distribution losses of the cryogenic transfer lines and thermal dampers, etc.

DESIGN OF THE REFRIGERATOR

The present design specifications of the KSTAR He refrigerator is summarized in Table 4 [4]. In the compressor station GHe is isothermally compressed. By expanding the high pressure GHe via turbines and valves, He at cryogenic temperatures is produced in the cold box and transferred to the distribution valve (DV) box. The function of the DV box is to supply to each cooling objects with the proper type and amount of cryogenic He, in accordance with the day average cooling scheme. Especially, the amount of SHe to cool the SC coils and structures is far beyond the capacity of the cold box SHe production. To overcome this problem a circulator, which is a cryogenic compressor, is used to circulate the SHe in a closed circuit. The thermal load of the coils and structures is then absorbed via SHe/LHe heat exchangers (HX's) immersed in the so called by thermal damper which is a temporary LHe storage located in the DV box. The present LHe capacity of the thermal damper has been determined on the basis of a continuous operation of the TF magnets even in the case of a plasma disruption.

All the cold components of the KSTAR system have to be linked to the distribution box through the vacuum insulated transfer lines. At the limit of the distribution box, all circuits must be isolated by valve. The helium distribution system should be with vacuum insulated housing, thermal damper, cryogenic circulator and cryogenic valves, etc. to supply various kind of helium to the KSTAR cold components. The energy that is released by the various components is dumped into the liquid helium that is stored in the thermal damper capacity. The quantity of liquid that is vaporized depends on the cooling power. The liquid helium in the thermal damper shall be in 4.3 K, 1.1 bar. The required SHe flow rate for the all SC coils is estimated as a 600 g/s. This flow rate is generated by cryogenic circulators because the JT mass flow is not sufficient to circulate all loads. The cryogenic circulator is operating at almost constant pressure.

Table 4. The design specification of KSTAR helium refrigerator system

Helium Compressors	**an oil flooded screw compressors** **Total mass flow rate : ~1 kg/s at 18 bara** **Oil removal system : under 10 ppbw**
Helium purifier	**Mass flow capacity : 32 g/s at 17 bar** **Air reduction : inlet : 500 ppm, outlet : < 1 ppm**
Cold Box	**Total Capacity : 7 kW at 4.5 K (day average)** **Operate without Liquid Nitrogen pre-cooling** **6 oil-free, static gas bearing turbo-expanders** **12 aluminum plate fin heat exchangers**
Storages	**Liquid Helium Tank : 20,000 liter** **Warm Helium Storage : 900 m³ at 20 bara** **Recovery Gas Bag : 100 m³** **Liquid Nitrogen Tank : 20,000 liter**
Distribution System	**Thermal damper : LHe at 4.3 K** **SHe circulation capacity : 600 g/s (300 g/s X2 EA)** **Circulation SHe pressure : 3 bar ~ 6 bar** **No. of cryogenic valves : ~ 100 ea**

CONCLUSION

Basically, the KSTAR helium refrigeration system was designed by considering a various requirements of the KSTAR cold components and operation modes. The static and pulse heat loss of the KSTAR cold components was estimated. The main design feature for the KSTAR helium refrigerator is peak power save system by using the thermal damper and LHe storage tank because the KSTAR tokamak will be pulse operated device. The typical day average heat losses of each device are 1.7 kW at 4.5 K of supercritical helium, 13 kW at 70 K of gaseous helium and 13 g/s of liquid helium.

ACKNOWLEDGMENT

This work was supported by Korea Ministry of Science and Technology under the KSTAR Project Contract.

REFERENCES

1. Kim Y.S. *et al*. "Cryogenic System for the KSTAR device", International Cryogenic Engineering Conference, Grenoble, France (2002) 109-112
2. Kim Y.S. *et al*. "ANALYSIS OF THE THERMAL LOADS ON THE KSTAR CRYOGENIC SYSTEM", CEC/ICMC 2003, Alaska, USA (2003)
3. Kim Y. S. *et al*., Technical Requirement for the KSTAR Helium Refrigeration System (2004).
4. Fuenzalida P. *et al*., KSTAR Helium Refrigeration System TECHNICAL PROPOSAL (2004).

Air liquide cryogenic system for the KSTAR device

Briend P., Dauget P., Delcayre F., Abe I., Fauve E., Fossen A., Fuenzalida P., Lee L.A., Gistau G.[*], Kim Y.S.[**], Cho K.W.[**], Chang H.S.[**]

Air Liquide, Advanced Technology Division, B.P. 5, 38360 Sassenage, France
[*] Consultant, 44 chemin de la Buisse, 38330 Biviers, France
[**] Korean Basic Science Institute, Yeoeun-Dong, Yusung-Gu, Taejeon, 305-333, Korea

The cold components of the KSTAR tokamak require forced flow of supercritical helium for magnets/structure, boiling liquid helium for current leads, and gaseous helium for thermal shields. The cryogenic system should provide stable operation and full automatic control. A three-pressure helium cycle composed of six turbines has been simulated. A design operating mode defined by an exergy approach results with a system composed of a 7 kW refrigerator, using gas and liquid storages for mass balancing. During Shot/Standby mode, the heat loads are highly time-dependent. A thermal damper is used to smooth these variations and will allow stable operation.

INTRODUCTION

The tokamak developed in the KSTAR (Korean Superconducting Tokamak Advanced Research) project makes intensive use of superconducting magnets operated at 4.5K. The purpose of this discussion is to give a brief overview of the proposed cryogenic system for the KSTAR tokamak. This paper begins by quantifying the thermal loads of the refrigerator in the normal modes of operation and by describing the thermo-dynamical and process approach leading to the refrigerator design power.   Also included is a discussion of the main results of the process calculations in the two main modes (Idle and Shot/Standby). An explanation of the main process control principle of the refrigerator follows.

DUTIES OF THE REFRIGERATOR AND DESIGN MODE

The cryogenic system will provide safe and stable operation and full automatic control to accommodate both normal operating modes and abnormal events. Normal operating modes occur in 24-hour cycles and involve the annual cooling/warming of 250 tons coils and structures. Ramp, Shot, Standby, De-Ramp, and Idle are examples of operating modes. Abnormal events include quenching, plasma disruption, and vacuum disclosure.

The dedicated VINCENTA code [1], [2] details the heat loads and liquid helium consumption related to the various tokamak operations. Table 1 summarizes these results. The Isothermal Refrigeration power @ 4.4K includes the heat loads due to both the supercritical helium circulation and the components of the DVB (distribution valves box). These heat loads are estimated from the preliminary design of the system. The equivalent refrigeration power varies between a maximum value of 9 kW in Shot/Standby mode and a minimum value of 5 kW in Idle mode. Designing the refrigerator regarding only the maximum case would lead to an oversized system. Therefore, a mean value of refrigeration power, approximately 7 kW, is selected as the design operating mode.

Table 1 Cryogenic System Duties in Normal Operation

Heat Loads	Operating Modes		
	Shot / Standby 8hrs/day	Idle 14 hrs/day	Ramp / De-Ramp 2 hrs/day
Thermal Shields @ 55K-75K	18 kW	18 kW	18 kW
Isothermal Refrigeration @ 4.4K	5.4 kW	2.6 kW	4.8 kW
LHe to current leads	23 g.s^{-1}	19 g.s^{-1}	20 g.s^{-1}
Total Refrigeration Power equivalent @ 4.5K	**9 kW**	**5.7 kW**	**8 kW**

The excess refrigeration power in Idle mode is used to produce liquid helium (LHe). LHe production is routed to a rising-level storage. In Shot/Standby mode, the lack of refrigeration power is balanced by liquid helium withdrawal. At the same time, the equivalent amount of gaseous helium fills gas cylinders at ambient temperature.

Table 2 gives the helium mass balance of the two main modes: Shot/Standby and Idle. In Shot/Standby mode, the process is close to that of a pure refrigerator whereas in Idle mode, the process is 50% liquefaction. Therefore, a careful study of the refrigeration system design is necessary to determine the optimal component that will be capable of satisfying a wide range of operations.

Table 2 Helium Mass Balance

Balance Components	Operating Modes	
	Shot/Standby	Idle
Total Refrigeration Power	7 kW	7 kW
LHe to current leads	23 g.s^{-1}	19 g.s^{-1}
Liquid Helium production	2 g.s^{-1}	32 g.s^{-1}
% Refrigeration	97 %	54%
Liquid Helium storage	-21 g.s^{-1}	+13g. s^{-1}

HELIUM CYCLE DESCRIPTION

The following description is related to the Shot/Standby mode and is illustrated by Figure 1. To simplify this discussion, the letters HX and T refer to a heat exchanger and a turbine, respectively. The convention used for referring to Figure 1 during this description employs [1] for corresponding to (1) on Figure 1 and so on. The HP (high pressure) helium (995 g/s) is cooled down to 188K in the counter-current heat exchanger HX1. At the outlet of HX1, the flow rate is divided into two flow: one to the turbines T1/T2 (148 g/s) and the other dedicated to the colder turbines [1]. At the cold end of HX6, the helium HP flow is at 55K, part of which (220g/s) is dedicated to the thermal shields[2]. Flow from the thermal shields is expanded in T3. Through HX7 / HX8 / T4, the helium gas is cooled down to 21.7K[3]. The next exchanger section (HX9 / HX10 / T5) cools HP helium gas from 21.7K to 12.8K[4].

At about 20K, neon and hydrogen impurities are trapped in an adsorber. The expanded flow rate (634 g/s) at medium pressure in T2, T3, T4 and T5 is warmed in MP (medium pressure) circuits (HX1 to HX10) and sent to the MP compressor suction. Most of the remaining flow (291g/s) is expanded to 2.6 bar through T6, cooled down to 4.55K through HX12, and poured in LHe storage at 1.3bar[5].

The remaining HP flow is sent to the SC (superconducting) bus lines[6]. LHe storage feeds the current leads circuit and the thermal damper in the DVB (distribution valves box)[7]. The TD (thermal damper) is dedicated to absorb 4.4K refrigeration loads (5346W). The low pressure helium vapor at 1.19 bar is warmed and sent to the low pressure suction of compression station[8]. In Idle mode, the circulators in the DVB are stopped. The required tokamak supercritical helium flow rate is taken at 3 bars at the T6 outlet.

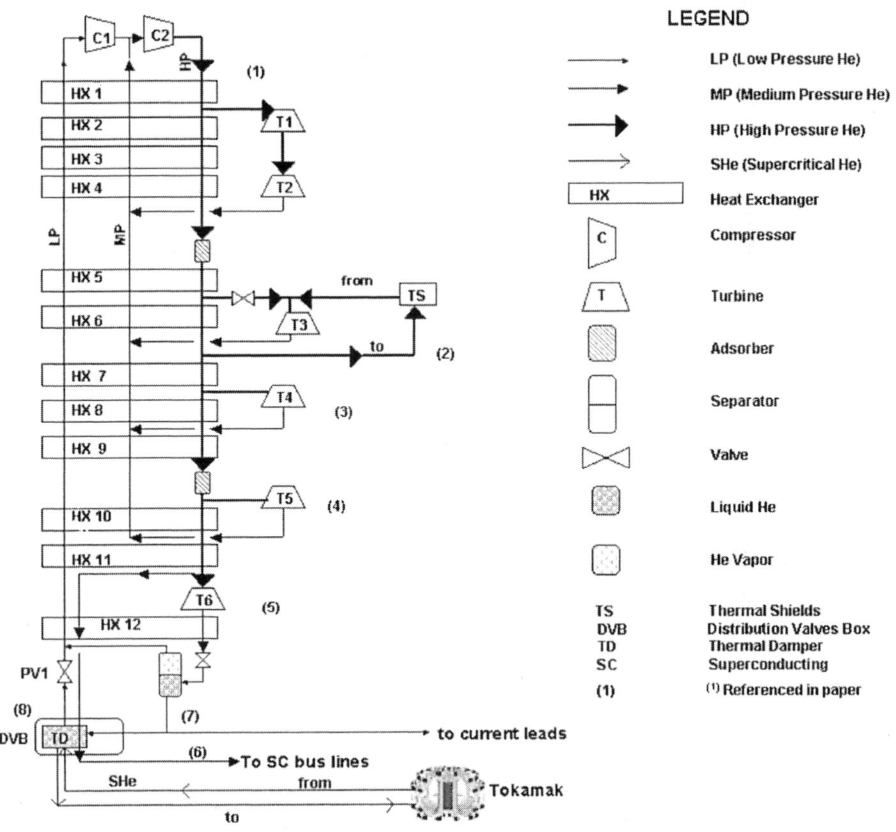

Figure 1 KSTAR Helium Cycle Process Flow Diagram

PROCESS CONTROL

One of the key control points of the refrigeration unit is the stable operation of the refrigerator coupled with the time-dependent heat load. The time-dependent heat load shown in Figure 2 is a result of the periodic and unsteady Shot/Standby operation mode of the tokamak. The main target of the control procedure is to keep the cold vapor flow rate constant when returning to the cold box. Using the PV1 control valve to maintain a given set point of the LP pressure will provide a constant mass flow rate to the volumetric LP compressor. The heat balance in the damper is the sum of the energy deposit from thermal loads minus the exhausted energy from both the vaporization and evacuation towards the cold box. The flow control will allow either the storage or the release of energy in the thermal damper and will thus ensure the stability of the refrigeration system.

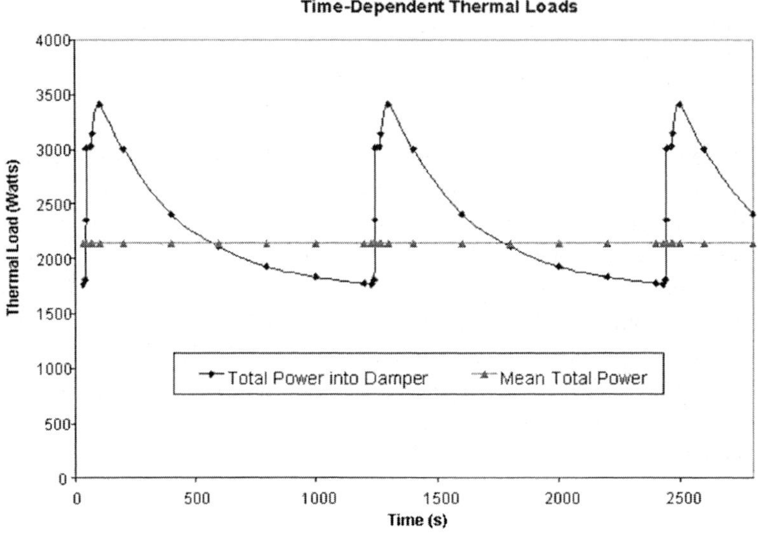

Figure 2   Shot/Stand-by Mode: Total Transferred Power into the Damper vs. Time

CONCLUSION

A helium cycle consisting of three pressure settings and six turbines has been simulated for the KSTAR cooling requirements. A design operating mode has been selected using an exergy approach. The result is a 7 kW constant power refrigerator and the use of gas and liquid storages for mass balance. The time-dependent heat load deposit from the tokamak during Shot/Standby mode is buffered into a thermal damper. A flow control method is proposed and will allow a very stable operation of the refrigerator.

REFERENCES

1. Kim Y.S. et al, Design of the Cryogenic System for the KSTAR Device, Proceedings of the Twentieth International Cryogenic Engineering Conference Beijing, China (2004) 20
2. Chang H. S. et al., Design Issues in the Distribution Valve Box of the KSTAR Helium Refrigerator, Proceedings of the Twentieth International Cryogenic Engineering Conference Beijing, China (2004) 20
3. Alekseev A., SINTEZ Final Report Review on Structural and Thermal Stability of The KSTAR Magnet System , Saint Petersbourg , Russia (2002)

Cryogenic System for BEPCII SRF Cavity IR Quadrupole and Detector Solenoid Magnets

Jia L.X.[1] and Wang L.[2]

[1]Brookhaven National Laboratory, Upton, New York 11973, USA
[2]Institute of Cryogenics and Superconductivity Technology, Harbin Institute of Technology,
Harbin 150001, CHINA

Beijing Electron-Positron Collider Upgrade (BEPCII) requires three types of superconducting facilities, including one pair of superconducting radio frequency (SRF) cavities, one pair of interaction region quadrupole magnets, and one detector solenoid magnet. The cryo-plant for BEPCII has a total cooling capacity of 1kW at 4.5K, which is composed of two separate helium refrigerators of 500 W each. The two refrigerators comprise the same gas storage and recovery system. The engineering design for the cryogenic systems, power leads, control dewars, subcooler, cryogenic valve boxes, cryogenic transfer-lines and cryogenic controls, is completed. The production of their subsystems is under way. This paper summarizes the progress in cryogenics of the BEPCII project.

INTRODUCTION

The cryogenic facilities of the BEPCII cryoplant are distributed over four areas; the first colliding hall, the second colliding hall, the compressor hall, and the gas tank farm (see Figure 1). The BESIII detector solenoid magnet and a pair of interaction region quadrupole magnets are located in the first colliding hall. The two SRF cavities are located in the second colliding hall. The two colliding halls are in a distance of 100 meters. To reduce the length of cryogenic transfer line and to meet the different loading requirements in two operation modes, colliding mode and the synchrotron radiation mode, two separate helium refrigerators of 500W each are used for the two colliding hall [1,2].

THE FIRST COLLIDING HALL

The layout of the cryogenic subsystem in the first colliding hall is shown in Figure 2. There are three superconducting facilities: the detector solenoid magnet with its valve box and two quadrupole magnets with a separated valve box. To provide cooling capacity to these facilities, a 500 W helium refrigerator (Linde TCF50) and a 1000 litter liquid helium control dewar are installed in the adjacent refrigerator building. The liquid helium produced by the refrigerator cold box is delivered to the subcooler that serves the control dewar. On the top of the subcooler is a cryogenic valve box that serves as a distributor to deliver subcooled helium to the valve boxes of the detector magnet and the two quadrupole magnets. The valve box for each magnet contains the power leads, cryogenic valves, relief valves, and pressure and temperature instrument ports. The connecting vacuum jacketed cryogenic transfer lines are of multiple types with tubes for helium and nitrogen. The multiple transfer line between quadrupole magnet and its valve box also contains the superconducting cables. The quench gas return lines and lead cooling gas

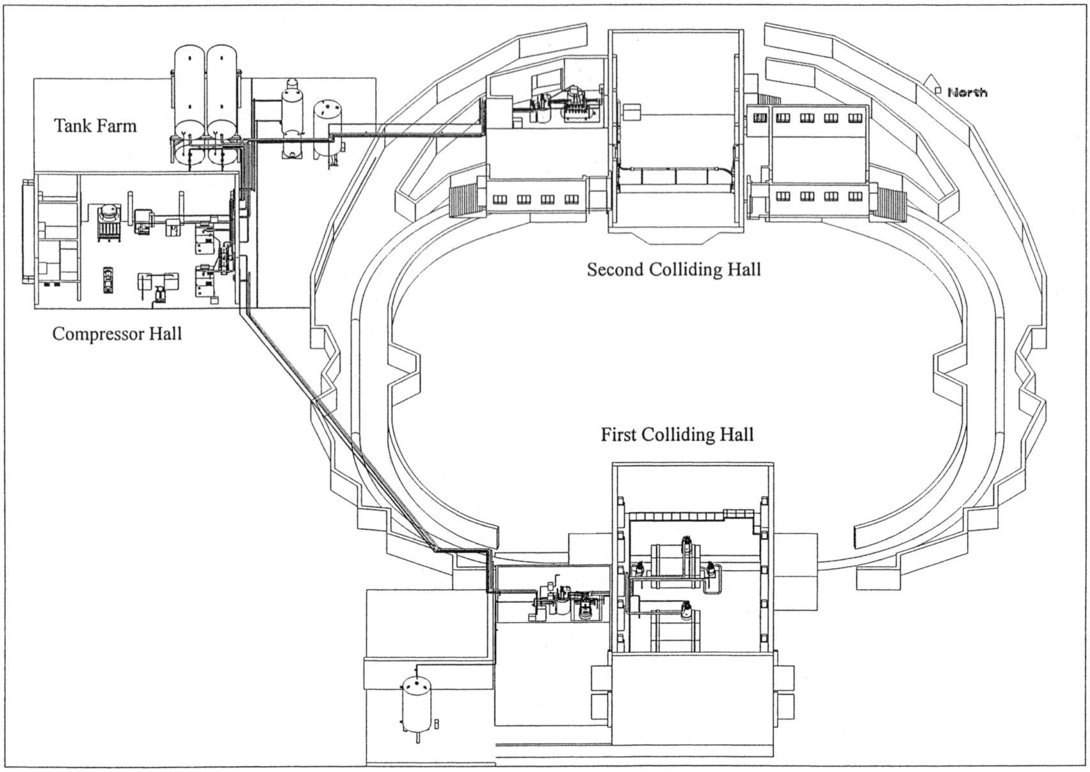

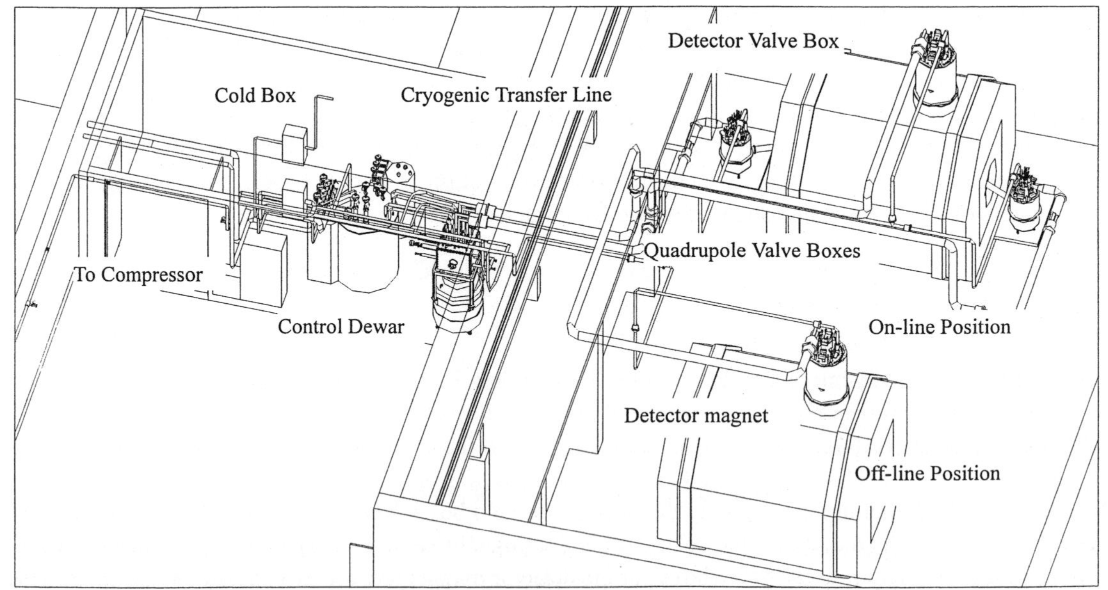

Figure 1   BEPCII ring and its cryogenic facilities

Figure 2   The first colliding hall and its cryogenic system

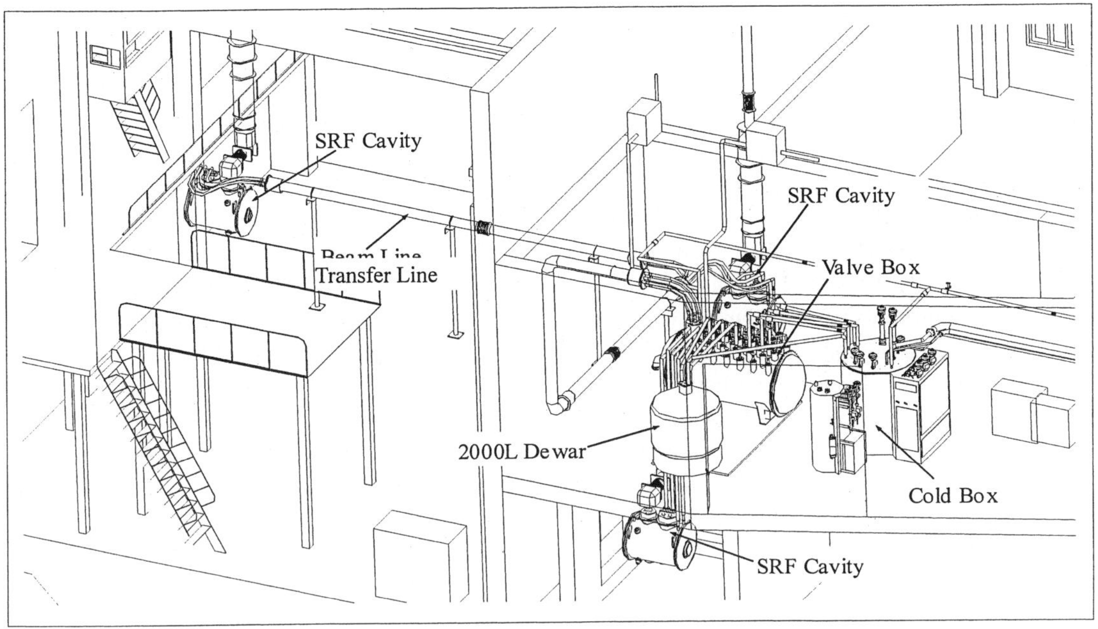

Figure 3   The second colliding hall and its cryogenic system

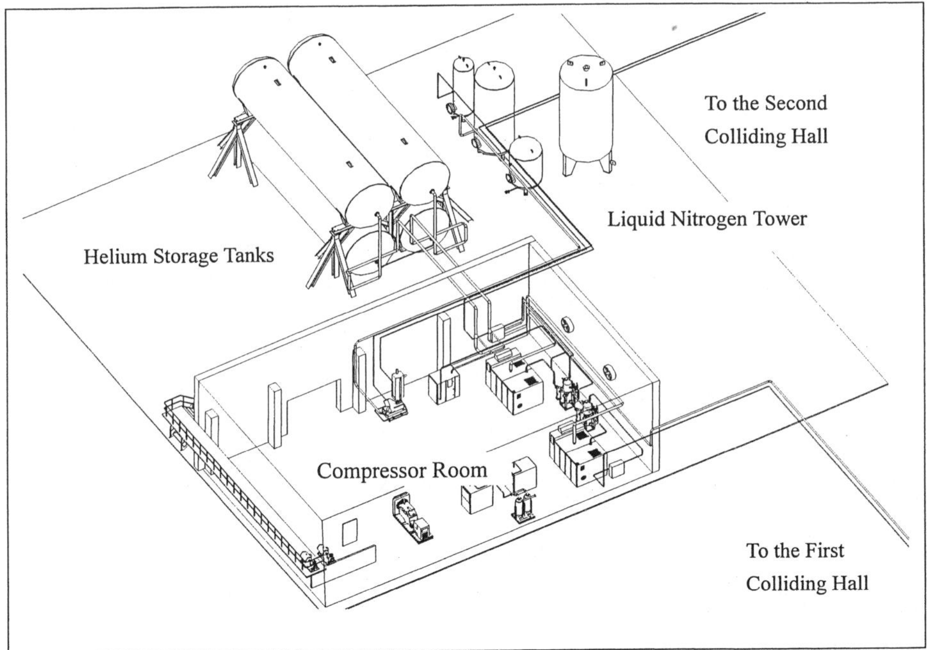

Figure 4   Compressor room and gas tank farm

return lines are directly connected to the compressor suction. The high pressure warm gas lines and low pressure warm gas line run from the cold box to remotely located compressor building. A liquid nitrogen tower for the refrigerator cold box and magnet heat shield is located just outside of the first colliding hall. The vacuum jacketed transfer lines for the detector solenoid magnet is designed to accommodate its off-line test positions as well as the in-line operation position. The vacuum separation barriers are arranged in various portions of the transfer lines that allows the quick reassemble and relocation of each cry modules.

THE SECOND COLLIDING HALL

The layout of the cryogenic subsystem in the second colliding hall is shown in Figure 3. Three superconducting RF cavities: two in-lined cavities and one off-line test cavity. In the refrigerator building adjacent to the second colliding hall, there are four major cryo equipments: the refrigerator cold box with the cooling capacity of 500W (Linde TCF50), the external cryogenic helium purifier, the 2000 litter liquid helium dewar, and the SRF distribution valve box. The 2000 litter dewar is used to maintain available liquid helium for each cavity for cool down and operation. Each cavity has a liquid volume of 300 litters. The main cryogenic valve box distributes liquid to three SRF cavities. The valve box is so designed that three cavities can be cold at the same time or anyone of three cavities can be cold when others are warm. This horizontal valve box contains all necessary cryogenic valves, relief valves, and pressure sensor ports. The multiple cryogenic transfer lines in the second colliding hall are designed with several vacuum barriers for operating and testing of these cavities.

THE COMPRESSOR HALL AND TANK FARM

The layout of the helium compressor hall and the helium gas tank farm is shown in Figure 4. The compressor hall is remotely located in a new building that is about 100 meters away from the first colliding hall and 70 meters away from the second colliding hall. Two compressors with separated oil removal, cooling water, and VFD systems are located in the compressor hall. Each compressor supports each cold box in the first and the second colliding halls. The piping for the two compressors is designed as to support individual cold box at different colliding hall and also to be the backup of each other. The compressor hall also contains the instrument air pumping system and the emergency electrical power generator. The cryogenic control room for the entire cryoplant is also located in the same building. Adjacent to the compressor hall is the tank farm in which there are four high pressure gas tanks, 130 m^3 each, one impurity helium gas tank, one instrument air tank, and one liquid nitrogen tower of 30 m^3 in capacity.

CONCLUSION

The engineering design work for the BEPCII cryogenic system is carried out in the framework of collaboration among three institutes: the Institute of High Energy Physics, the Institute of Cryogenics and Superconductivity Technology, and the Brookhaven National Laboratory. The design work started in March 2003 and basically completed in May 2004. The construction will be soon started as this report is presented. The mechanical fabrication of the cryogenic system is to be conducted in China with some imported components such as the refrigerators and cryogenic valves.

REFERENCES

1. Jia, L.X. and Wang, L., 1kW Cryoplant for BEPCII Superconducting Facilities, <u>Proceedings of ICCR'2003</u>, Hangzhou, China (2003) 47-50
2. Jia, L.X. and Wang, L., Cryogenics in BEPCII Upgrade, <u>Proceedings of ICEC-19</u> Grenoble France (2002) 39-42

Status of the ISAC-II Cryogenic System

Sekachev I., Andersson W., Laxdal R.E., Marshal C., Stanford G.
TRIUMF, 4004 Wesbrook Mall, Vancouver, BC, Canada, V6T 2A3

ISAC-II is an upgrade of the ISAC radioactive beam facility that includes the addition of 43 MV of heavy ion superconducting accelerator. A first stage comprising a 500W class LHe refrigerator and distribution to five cryomodules is scheduled for installation by early 2005. The paper will present the design and status of the Phase I installation including a description of the distribution piping, details of the refrigeration system, and a summary of the helium storage and gas recovery.

INTRODUCTION

The ISAC-II superconducting accelerator is now under construction at TRIUMF[1] (Fig. 1). A first stage includes the installation of 20MV of accelerating structure is expected to be completed before the end of 2005 with a second equal stage due for completion ~3 years later. The new building is to accomodate the ISAC extension is now complete.

The refrigerator system will be installed in two equal stages. The first stage includes a 500 W helium refrigerator with associated compressors, oil removal system (ORS), helium buffer tank, helium dewar, room temperature piping, helium distribution transfer lines, and nitrogen distribution network.

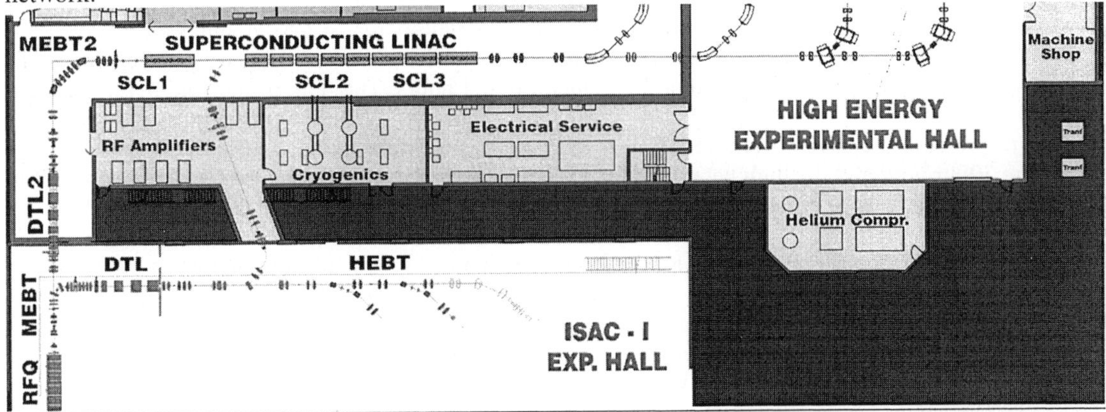

Figure 1: ISAC-II addition showing new superconducting linac, cryogenics (refrigerator room) and helium compressor building.

OVERVIEW OF THE ISAC-II CRYOGENIC SYSTEM

The ISAC-II phase I cryogenic system supplies 4.5K helium to five medium beta cryomodules[1] and to two high beta cryomodules. The refrigerator cold box delivers LHe to a main supply dewar located in the cryogenics (refrigerator room). The main supply dewar supplies LHe to the main distribution trunk line with a moderate push pressure. The ISAC-II cryomodules (CMs) are fed in parallel (Fig. 2) from the main distribution line with U-tube transfer lines. Each cryomodule has two associated female bayonet cans one for the delivery bayonet and one for the cold return bayonet. There is also a main warm return line that takes warmer gas (>20K) from the CM back to the suction side of the main compressor during cooldown. There is also gas from the solenoid power leads and for cooling of the stack that is valved into the warm return line through throttle valves.

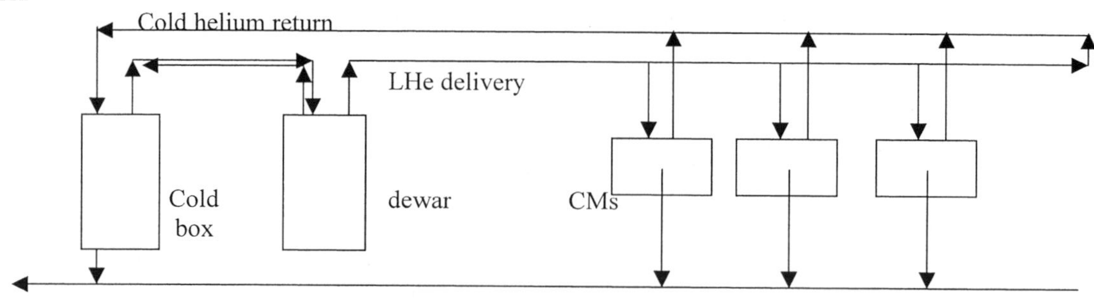

Figure2 : Schematic of the parallel flow concept to the cryomodule's.

The cryomodules contain a LN2 cooled thermal shield consisting of a ½" copper tube soldered to a copper box. Liquid nitrogen is supplied to the cryomodules via a parallel distribution scheme from a local nitrogen dewar/phase seperator. The dewar is fed LN2 from an external 9000 USG nitrogen tank via existing nitrogen vacuum isolated piping. A separate LN2 circuit is used to cool the vacuum jacketed cold distribution system.

During normal operation the linac will be cold and the rf will be on to accelerate the beam. The total expected cryogenic load for phase I is 390W. This can be broken down into 192W of static losses and 198W of active rf loss. The static loss budget is split half and half with 96W for the cold distribution system and 96W for the cryomodules. This assumes static losses of 13W and 14W for the medium and high beta CM's respectively. The liquification loss from the cryomodule power leads is expected to be 0.7gm/sec. There will be times where a cryomodule or cryomodules will have to be cooled down from room temperature. A flow of 6 liquid liters/hour of 100% liquid nitrogen will cooldown the medium-beta cryomodule LN_2 shield in ~24 hours. The bulk of the cold mass will cool by radiation to ~250-K. Cold tests indicate that a flow of 20 liquid liters/hour of helium is a sufficient flow to pre-cool the cavity from 250-K to 4-K in about four hours. The medium beta cryomodules require 100ltr/hour and the high beta modules 150ltr/hr for efficient cooling. The CM's are cooled one at a time and each filled cryomodule is topped up periodically as the cold mass thermalizes.

REFRIGERATION SYSTEM

The phase I refrigeration system has been purchased from Linde Kryotechnik AG, Switzerland. Refrigerator. The refrigerator-liquefier on order is the Linde TCF50 with liquefaction rate of about 5.2 g/sec and refrigeration rate of of 530 Watts at 4.5 K with simultaneous liquification rate of 0.71 g/sec.

The vendor supplied system consists of a cold box/liquifier, a main and recovery compressor, as well as oil removal and gas management systems. TRIUMF will supply helium gas storage tanks (170 m^3 per phase), the main LHe dewar and warm and cold piping. In addition TRIUMF will supervise the installation to Linde's specifications and Linde will commission the system to the supply dewar. In the near future a gas analyzer and gas purifier system will be added.

The compressor room will house the main compressors and recovery compressors, oil removal system and gas management system. A cryogenic room will house the cold box/liquefier and liquid helium dewar. This room is adjacent to the accelerator vault.

<u>Helium Gas Storage Tanks</u>
The total inventory of helium must be able to be stored as a gas before the system can be operated and also at any time the refrigeration system is shut down. Two 114 m^3 horizontal tanks will be provided to achieve this for phase 1; an additional tank will be added for phase 2. These will be located outside the experimental hall close to the compressor room. They will be stacked 3 high on a steel support frame. They will operate when full at 14 bara (203 psia, a safety pressure relief valve is set at 220 psig). It has been found that standard 30,000 USG (114 m^3) propane storage tanks can be utilized for

this application and modified for additional penetrations plus thoroughly cleaned inside to avoid helium contamination.

Compressor Room

The compressor room will house 2 compressors for phase 1 and an additional 2 when phase 2 is added. The main compressor is a Kaeser ESD441SFC direct drive screw compressor of 268 KW producing a compressed helium flow of 79 g/s at 14 bara pressure to be delivered to the cold box/liquefier in the cryogenic room via the warm piping. The main compressor has a variable frequency drive option allowing part load performance during normal operating modes depending on the active load from the accelerator. The second compressor is a recovery compressor allowing the helium gas inventory to be recovered and compressed into the storage tanks. It is a Kaeser BSD62 screw compressor of 37.5 KW producing a compressed helium flow of 12 g/s at 14 bara pressure. The compressor room ventilation system controls the compressors cooling air inlet temperature by directing all the air out of the room in hot weather and partially in colder weather. The main compressor also utilizes water as well as air for cooling. The main compressor and the recovery compressor do not operate at the same time.

The gas management system manages helium flow to keep the main compressor inlet pressure at the correct level and to redirect flow, when the main compressor is shut off, to the recovery compressor. It also provides an oil absorber for oil removal from the helium gas. There is provision for the installation hook up of a helium gas purifier and gas analyzer in the near future.

Refrigeration System Piping And Installation

Installation will be done by a local refrigeration contractor up to completion, upon which a First Certification is issued allowing operation of the system. As part of the specification to do this work TRIUMF is providing Process and Instrumentation Drawing as well as complete 3D Installation Drawings and Bills of Material. Linde will be providing special installation instructions for their equipment. The refrigeration contractor will provide labour and the required piping to the requirements of Safety and Refrigeration Codes that are applicable.

Cryogenic Room

The cryogenic room will house the cold box/liquefier, control system rack and the liquid helium dewar. A duplicate system will be installed for phase 2 at a later date. The room lies adjacent to the accelerator vault and the transfer line runs from the dewar to the vault distribution system through an opening in the wall. The helium vapour return and cool down warm gas return also run through this opening.

The TCF50 cold box is capable of delivering a continuous stable pure liquefaction rate of 5.2 g/s and a continuous stable refrigeration rate of at least 530 watts at 4.5 K with a simultaneous liquefaction rate of 0.71 g/s. It will be capable of continuous stable operation with loads varying from 270 W to 530 W depending on the accelerator mode of operation. Linde will also be supplying a cryogenic transfer line to transfer the liquid helium to the adjacent dewar. Liquid nitrogen will be supplied to the cold box through an existing LN_2 distribution system.

Liquid helium is supplied to a 1000 litre dewar adjacent to the cold box (phase 2 will require another) through a Linde supplied transfer line; dewar helium vapour also returns through this transfer line. The liquid helium will exit the dewar through a TRIUMF supplied distribution system transfer line. A spare port is provided for possible hook up to an adjacent dewar. Two level probes will be provided in the dewar – one active, one spare, to monitor level information for the control system. Two replaceable heaters will also be provided – one active, one spare, to control liquid level when required and also it will be used during refrigeration system commissioning prior to the linac accelerator installation.

Gas Purifier and Analyzer

During commissioning Linde will provide this equipment. Connections on the gas management package allow for their installation. Maintaining gas purity to <5 ppm of contaminants is essential to the proper operation of the system and TRIUMF will be purchasing their own purifier and analyzer in the near future.

<u>Control System</u>

The control of the refrigeration system is done by a local Siemans controller supplied by Linde. The system will be monitored by the standard ISAC EPICS control system via Prophi-bus communication. Initially alarms and status will be passed to EPICS while in the future it is planned that some process control will be available. The control of the distribution system and cryomodule interface will be implimented by TRIUMF via PLC with the ISAC EPICS interface.

LIQUID HELIUM DISTRIBUTION SYSTEM

The liquid helium distribution and vapour return system will be a parallel flow system from common trunk lines running parallel to the linac. Current thinking embraces the idea of a transfer 'U' tube between the cryomodule and a bayonet box associated with a proportional valve on the supply and standard valve on vapour return. These transfer 'U' tubes would not be liquid nitrogen cooled but would employ vacuum jacket with super insulation. The rest of the distribution system would be vacuum jacketed and cooled with liquid nitrogen. A separate line would also run from each cryomodule in a non-vacuum jacketed but insulated line for the return of warm helium during cool down which gets sent back to the compressor.

The entire distribution and return piping system will be contracted to a supplier of this type of specialized equipment.

Distribution and return piping will run parallel to the linac at an elevation to accommodate the 'U' tube bayonet box and valve. It will be supported by a steel framework that will also serve as an access walkway for cryomodule maintenance and 'U' tube removal in the event that a cryomodule would require removal from the linac for internal maintenance. In this case it can be valved off and removed and replaced with a beam tube allowing continued operation of the linac.

STATUS

The compressors are expected at TRIUMF towards the end of this month. Installation will commence in June. The cold box is expected by the end of July. Specifications for dewar procurement will be issued this month with an order expected to be placed in June for delivery in August 2004. The order for the phase 1 buffer tanks should be placed this month. Commissioning of the refrigerator system to the main dewar is expected before year end. The distribution system has been delayed for cash flow reasons. The plan is to complete the specifications by Oct. 2004 with delivery and installation scheduled for early 2005.

An initial medium beta cryomodule has just been undergone the first cryogenic tests[3]. Both static load and LN2 consumption are within specifications. The first cold rf test is expected in June 2004. First beams from the linac are scheduled for Nov. 2005.

REFERENCES

1. W. Andersson, R.E. Laxdal, I. Sekachev, G. Stanford, Overview of the Cryogenic System for the ISAC-II Superconducting Linac at TRIUMF, Proceedings of the Nineteenth International Cryogenic Engineering conference (ICEC 19), Grenoble, France, 2002
2. G. Stanford, R.E. Laxdal, C. Marshall, T. Ries and I. Sekachev, Design of the Medium-Beta Cryomodule for the ISAC-II Superconducting Heavy Ion Accelerator, <u>CEC-ICMC conference</u> (2003), Anchorage, Alaska.
3. G. Stanford, R.E. Laxdal, C. Marshall, T. Ries and I. Sekachev, First Cold Test of the ISAC-II Medium Beta Cryomodule, this conference.

Overview of different standard helium liquefiers / refrigerators for various applications

Caillaud A., Bonneton M., Delcayre F., Grabié V., Grillot D., Hilbert B., Praud A., Walter E.

Division des Techniques Avancées, Air Liquide, Rue de Clémencière, B.P.15, 38360 Sassenage, France

The design of Air Liquide DTA standard helium liquefaction / refrigeration systems was upgraded three years ago. Several of these systems were built and started up over the last years. Since then, these standard machines have been supplying different applications with either LHe or refrigeration power at 4.5 K. The standard design of this range of machines will be presented in relation with typical applications such as SRF cavities or SMES magnets operated at 4.5 K. This standard design can be modified to accommodate cryogenic power needs at temperatures ranging from 2 K to several tens of Kelvin. An overview of some design adaptations will be presented in relation with different applications such as superconducting magnet operated at 3 K or space simulation chamber.

EXAMPLES OF APPLICATIONS FOR HELIAL LIQUEFIERS/REFRIGERATORS

HELIAL machines are fully automatically controlled systems which can be used either as liquefiers, either as refrigerators, or both, depending on the final use of liquid helium.

Synchrotron light sources [1], with ranges from IR to hard X-rays, are produced by a beam of electrons circulating at high energies in a storage ring. The radio-frequency (RF) power and voltage required for storing the electron beam are provided by means of cryomodules containing superconducting RF cavities. The cavities made of Niobium deposited on Copper are bath-cooled with saturated liquid helium near the atmospheric pressure. Standard HELIAL are then used to supply with liquid helium the cryostats in which the cavities are installed.

HELIAL can also be used for SMES (Superconducting Magnetic Energy Storage) applications [2]. SMES devices store energy in magnetic fields. The devices consist of closed coils of superconducting wires. Superconductors are the only appropriate materials for SMES devices because they have no electrical resistance and thus can be operated in a persistent current mode without being connected to a power supply. A SMES device is the only method of storing electrical energy without first converting it into mechanical or chemical energy. SMES scale can be large (from a few Wh up to more than 1 GWh).

From materials to complete satellites, some equipment are sent into space and will have to face extreme conditions. Therefore, they are tested in space simulation chambers which are cooled in order to recreate space conditions. These chambers are maintained at a cold temperature thanks to cryopanels supplied with cold helium gas which can be produced by HELIAL adapted to temperatures above 10K.

In a NBI (Neutral Beam Injector) [3], the neutral beam is produced by charge exchange of ions extracted from plasma box and accelerated by an extractor. The pressure over the beam propagation has to be kept low so as to reduce the reionization losses. Cryocondensation pumps enable to maintain this low pressure, by freezing hydrogen or deuterium on Helium panels. The refrigerator HELIAL is then used to supply continuously these panels with LHe at 3.8K.

Basically, HELIAL are small-to-medium liquefiers. Some laboratories in physics or biotechnologies need permanently small quantities of liquid helium to perform experiments. They can recover helium gas from their users and liquefy it again to have their own production adapted to their needs.

STANDARD HELIAL RANGE

Air Liquide DTA have updated its range of the so-called HELIAL standard helium liquefiers /

116

refrigerators. Three machines have been defined : HELIAL 1000, HELIAL 2000 and HELIAL 3000. The table 1 presents the performances guaranteed for each machine.

Table 1 HELIAL performances

	HELIAL 1000	HELIAL 2000	HELIAL 3000
Liquefaction capacity without LN2	30 l/h	85 l/h	180 l/h
Liquefaction capacity with LN2	65 l/h	175 l/h	300 l/h
Refrigeration capacity without LN2	130 W	415 W	560 W
Refrigeration capacity with LN2	160 W	500W	875 W

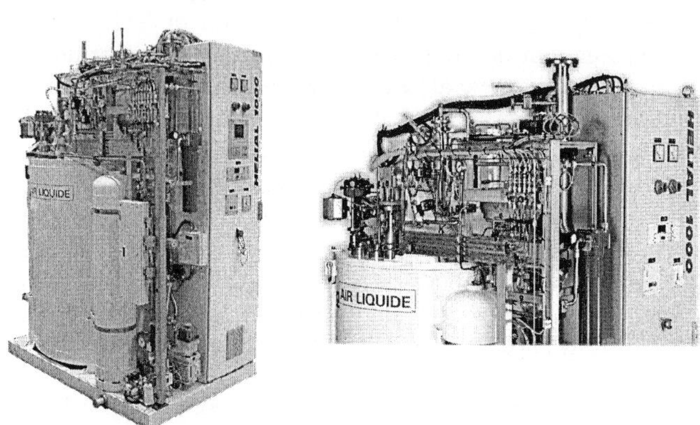

Figure 1 HELIAL 1000 cold box and warm panel

A standard HELIAL is basically composed of three modules :
- The screw compressor with a bulk oil separator, and coolers;
- The oil removal system;
- The cold box.

Additional equipment can be proposed in option such as variable frequency driver for compressor capacity adaptation, liquid helium dewar, HP or MP gas storage capacities, warm or cryogenic piping, internal or external helium purifier, recovery compressor, supervision control system...

CYCLE DESIGN

The HELIAL cycle is based on a Claude cycle which is the combination of a Brayton and a Joule-Thomson cycles. Two turbines associated in series compose the Brayton, and the Joule-Thomson function is ensured by the JT valve. The figure 1 presents a schematic view of the cycle associated to the corresponding T-S diagram.

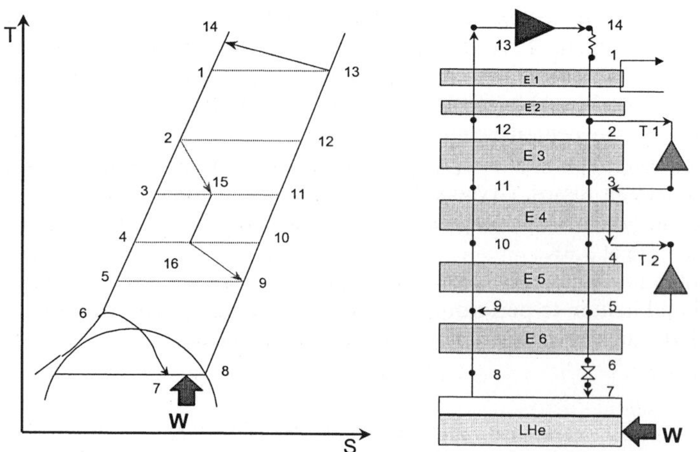

Figure 2 HELIAL T-S diagram and cycle design

Helium gas at ambient temperature and atmospheric pressure is compressed into the cycle compressor up to maximum 15 bar abs (see Figure 2, 13 to 14). After the compression, a first oil separation is then performed in the bulk oil separator integrated in the compressor casing, followed by a cooling of the helium gas (14 to 1). The helium stream is sent to the final oil removal module where the small quantity of remaining oil (either aerosols or vapours) is removed in two coalescing cartridges and a charcoal adsorber.

The high pressure gas enters the cold box (1). The flow is fed to the heat exchangers to be cooled by the low pressure returning stream. Liquid nitrogen can be used in the first heat exchanger E1 to help cooling the high pressure stream down to 80K. At 80K, the flow is then supplied to the 80K adsorber filled with charcoal in order to remove air impurities. The stream exits the adsorber (2) and is split into two parts :
- the gas which is expanded in the turbines in the so-called Brayton cycle;
- the gas which leads to the production of liquid helium in the so-called Joule-Thomson cycle.

The Brayton cycle flow
The power extracted by the turbines compensates for :
- the power which is necessary to compensate the liquefaction/refrigeration load;
- the power dissipated by the temperature difference at the warm end of the heat exchanger (warm end heat exchanger pinch);
- the power corresponding to the heat-in leaks into the cold box and eventual associated equipment (dewar, cryogenic interconnecting pipes…).

In between the turbines 1 and 2, the medium pressure stream is cooled in a heat exchanger to about 20K (3 to 4), and finally expanded in the cold turbine from which it exits at about 12K (4 to 9).

The Joule-Thomson cycle
The stream exiting the 80K adsorber and not sent to the turbines is cooled against the low pressure cold stream in the heat exchangers E3 and E4 (2 to 4). It is then purified in the 20K adsorber to remove low levels of air contamination. It is fed to the heat exchangers E5 and E6 (4 to 6), the so-called Joule-Thomson exchanger, and finally expanded from the cycle high pressure to 1.2 bar abs in the JT valve (6 to 7). The expansion produces a mixture of liquid and gas at 4.5K. The liquid is stored in the dewar whereas the cold gas is returned to the cold box low pressure stream (8).

The cold box low pressure stream
The low pressure gas returning from the dewar is mixed to the gas which was expanded through the turbines. This stream flows through LP channels of the heat exchangers where it cools down the HP counter flow. It finally exits the cold box at ambient temperature (13) and then returns back to the compressor suction side.

ADAPTATIONS OF STANDARD HELIAL FOR SPECIFIC APPLICATIONS

The standard HELIAL can be adapted for specific applications at temperature below or above 4.5 K..

Temperatures below 4.5K
HELIAL 1000, for NBI-IPR project, has been used to provide cold power at 3.8K. To achieve such a low temperature, the liquid helium bath must be pumped below atmospheric pressure. This is performed thanks to a warm vacuum screw compressor which enables to pump liquid helium down to 660 mbar abs. The technical solution presented here is specific to this project. Other technical solutions can be proposed for refrigeration at temperature below 4.5K.

The cycle is kept identical to a standard liquefier. Liquid helium is withdrawn from the main dewar towards a phase separator, connected to the suction of the vacuum screw compressor. The vapour phase is pumped and flows through a recuperator heat exchanger in order to warm the cold gas up to 300K before entering the vacuum screw compressor. The cold enthalpy of the gas is recovered inside this heat exchanger by HP gas coming from the inlet of the cold box. This HP flow is cooled down and expanded in a second JT valve at the outlet of the main JT valve, producing additional gas/liquid mixture at 4.5K.

The liquid at 3.8K is withdrawn to the customer's cryopumps and evaporated inside, producing cold gas which goes back to the phase separator. The circulation from the phase separator towards the cryopumps is performed thanks to thermosiphon effect.

The vacuum screw compressor enables to compress helium from 600 mbar abs up to 1.2 bar abs. Oil is removed from helium in an additional oil removal system. Finally helium gas is sent back to the main compressor suction.

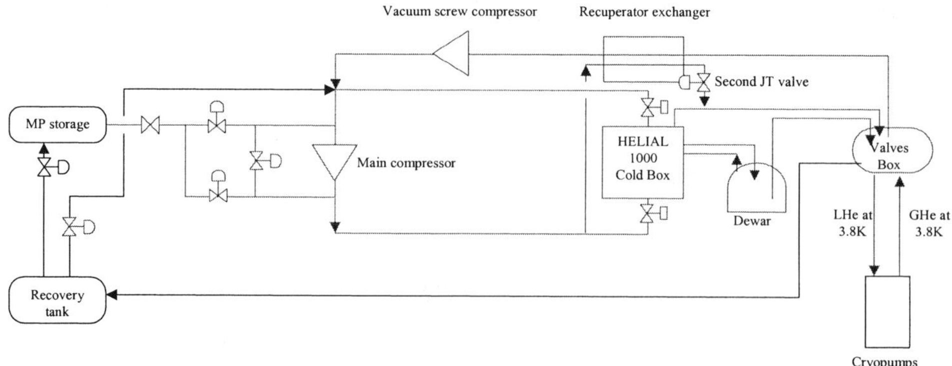

Figure 3 Standard HELIAL refrigerator adapted for 3.8K refrigeration system

Temperatures above 10K

For a space simulation chamber in China, the design of HELIAL 1000 has been adapted in order to provide cold helium gas at 13.5 K to cool cryopanels.

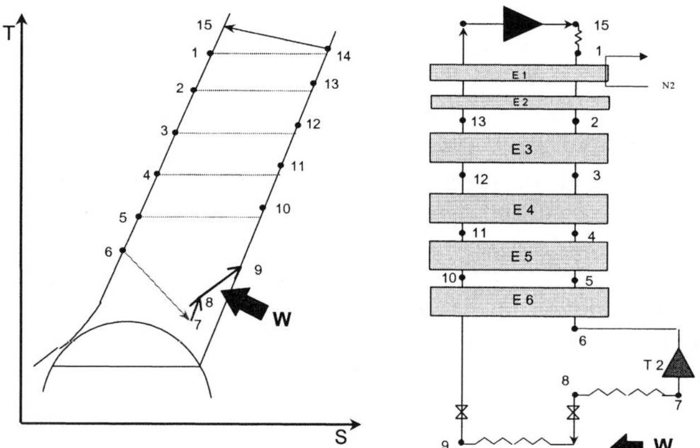

The Joule-Thomson circuit has been removed (see figure 4), keeping only the Brayton cycle composed of a single turbine. The warm helium gas is compressed in the compressor station up to 15 bar abs and sent to the oil removal system before entering the cold box. It is cooled down inside the heat exchangers E1 to E6 thanks to the cold LP counter flow. The turbine is connected at the end of the heat exchanger E6 and expands at 1.6 bar abs the gas which is directly sent to the customer's cryopanels.

Figure 4 HELIAL adaptation for refrigeration power @13.5K

The delivered temperature can be controlled thanks to an electrical heater which has been added inside the cold box. The gas is warmed in the cryopanels and sent back to the cold box heat exchangers to cool down the HP flow. The heat exchangers are kept identical compared to the standard ones. This HELIAL 1000 is able to provide 600 W between 13.5K and 18K.

ACKNOWLEDGMENT

The authors would like to thank their colleagues P. Cadeau, R. Fabre, G. Gaillard, and F. Floreancig, as well as their customers teams, for their contributions to the design study of HELIAL standard machines.

REFERENCES

1. Technical Specifications for the Engineering, Design, Manufacturing, Testing and Delivery of the cryogenic plant of the SOLEIL superconducting RF system, Same volume, Gif-Sur-Yvette, France (2003) 5.
2. Committee on the Impact of Selling the Federal Helium Reserve, Board on Physics and Astronomy, Commission on Physical Sciences, Mathematics, and Applications, National Materials Advisory Board, Commission on Engineering and Technical Systems, National Research Council, The Impact of Selling the Federal Helium Reserve, National Academy Press, Washington D.C., U.S.A. (2000) 26-39.
3. Design, Fabrication, Testing and Supply, Installation and Commissioning of 110 W at 3.8K Helium Refrigerator System at IPR, SST/ENQ/F/001, Gandhinagar, India (2001).

Standard Liquefier-test results with improved turbines

Schoenfeld, H., Cretegny, D., Loehlein, K.

Linde Kryotechnik AG, Daettlikonerstrasse 5, CH-8422 Pfungen, Switzerland

Gas bearing turbine efficiency has been increased. These turbines are used in helium plants for refrigeration and liquefaction. Turbine design modifications have been presented in [1]. The impact of higher turbine efficiency on liquefaction capacity was predicted for small scale Helium liquefiers. Measurements were performed to confirm the predictions. The performance results on capacity and operation are presented in this paper.

INTRODUCTION

A large number of standard helium liquefiers is used in scientific institutions and in industry. Standard technology guarantees high reliability. Cost effectiveness in both investment and operation has become more and more important. However, low operational costs often demand higher investment costs. Linde Kryotechnik AG has demonstrated how to increase the efficiency of a standard liquefier by using improved turbines. In future customers will benefit from a more cost effective liquefaction process with same or lower investment costs.

TCF20 HELIUM LIQUEFIER

For the field test a Linde TCF20 standard helium liquefier was used. It covers a range of 20 - 40 l/h. Linde Kryotechnik AG has taken over this kind of plant after more than 80 units were built by Linde CryoPlants Ltd. (UK). The process flow diagram is shown in figure (1).

Gaseous helium compressed at ambient temperature enters the coldbox through the high pressure (HP) pipe. In a first step the compressed helium is cooled in heat exchanger HX1. Heat exchanger HX1 is a counter flow plate fin heat exchanger with an integrated liquid nitrogen (LN2) evaporator. The cooling capacity of the heat exchanger is supplied from the returning low pressure (LP) stream of helium and the liquid nitrogen (LN2). Once the helium gas passed HX1 it will enter the heat exchanger HX2, which is separated into two sections. Between the two sections the HP stream is split into the turbine and the Joule-Thomson (JT) stream. A large fraction of the helium gas will flow through the turbine string, passing turbine inlet valve and filter before expanding in turbine TU1. After being further cooled in heat exchanger HX3 the helium gas flows to turbine TU2, where it is expanded to an even lower pressure before joining the returning JT-stream. The rest of the helium gas from HX2 continuous as the JT-stream to be cooled down in heat exchangers HX3 - HX5. After the helium exits heat exchanger HX5 it is transferred to the dewar. A liquid fraction of the two phase flow is kept in the dewar. The gaseous fraction is returning as LP stream to the coldbox. It is warmed in the heat exchangers before returning to the intake of the compressor.

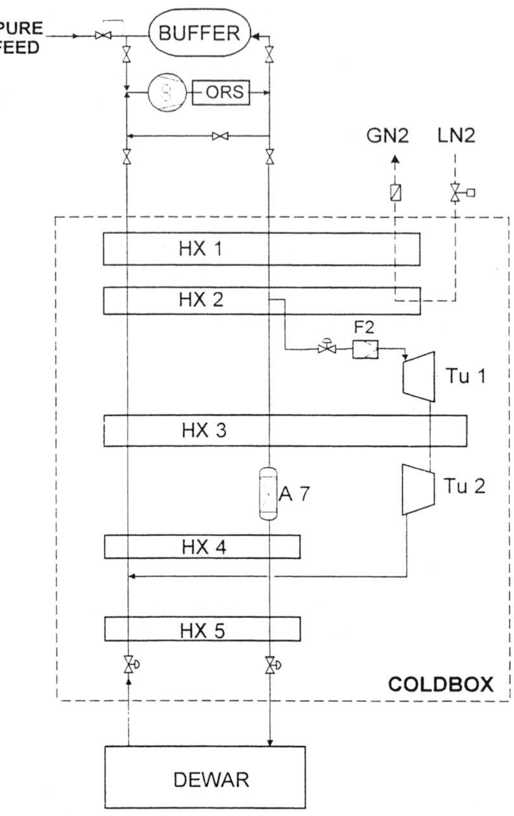

Figure 1 Process flow diagram of a TCF20 standard liquefier

IMPROVED TURBINE DESIGN

Linde Kryotechnik AG has been using well proven turbine technology. Experience of several hundred applications, reliability and low manufacturing costs were convincing arguments for the former design. Presently, Linde Kryotechnik AG has been achieving significant improvements in the design of dynamic gas bearing turbines. Most outstanding improvement is the increase in turbine efficiency.

Expansion turbines are used to extract work from helium process gas. HP helium gas is expanded over a rotating turbine wheel. Process gas temperature is decreased. Turbine efficiency is expressed as actual turbine work relatively to isentropic (ideal) expansion work.

$$\eta_s = \frac{w_a}{w_s} = \frac{h_0 - h_{3a}}{h_0 - h_{3s}}$$

Improved blade design, reduced heat inleaks and increased rotational speed are the main reasons for reaching higher turbine efficiencies. Enthalpies at inlet and outlet of the turbine are calculated from measured pressure and temperature data. Experimental results from the test bench were presented in [1]. The same instrumentation was now brought into a standard TCF20 liquefier. For the turbine chain it consists of pressure transducers and Rhodium- iron temperature sensors at the inlet and outlet of the turbines.

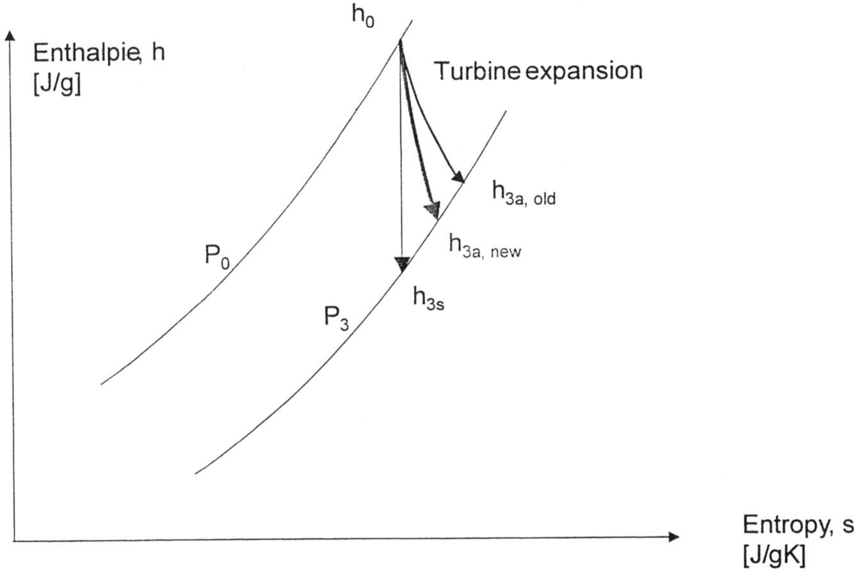

Figure 2 isentropic and real expansion in a turbine

The following table shows turbine efficiencies before (0) and after (1) the development work.

Table 1 Turbine efficiency comparison

Design case	0	0	1	1
Turbine position	Tu 1	Tu 2	Tu 1	Tu 2
Isentropic efficiency	61%	67%	80.9%	79.3%

Improved design of the axial bearing widens the operating field of the turbine. A further improvement, which was achieved with the new turbine generation, is larger robustness of the turbine. The rotor is kept by a spiral - grooved dynamic gas bearing. The use of carbon material allows that touching of the bearing does not lead to a turbine damage. It only results in an tolerable, increased wear of the bearing.

IMPACT ON LIQUEFIER CAPACITY

The test was carried out with a constant level of liquid helium in the dewar and constant pressure in the buffer tank. Liquid helium was constantly taken out of the dewar and warmed up to ambient. Mass flow was taken by a thermal flow meter.

The liquefaction rate was 1.36 g/s. In comparison to a TCF 20 under the same test conditions with conventional gas bearing turbines this is an improvement of 52 %.

The accuracy of the mass flow measurement was 4%. The accuracy of the turbine efficiency measurement was 1% for turbine 1 and 3.5 % for turbine 2.

FURTHER ADVANTAGES FOR THE PROCESS

The enlarged operating field of the turbine will allow improved cool down performance of liquefiers and refrigerators. Especially for refrigerators with different operation points optimal turbine design can be found less dependent from restricting axial force considerations.

REFERENCES

1. Cretegny D. et al. *Efficiency improvements of small gas bearing turbines - impact on standard helium liquefier performance,* CEC/ICMC 2003

Thermal Fluid Modeling of BEPCII IR Quadrupole Magnet Cryostat

Wang L, Tang H.M., Zhang X.B., Yang G. D., Jia L.X. [*]

Institute of Cryogenics and Superconductivity Technology, Harbin Institute of Technology, Harbin 150001, China
[*]Brookhaven National Laboratory, Upton, New York 11973, USA

A pair of superconducting interaction region quadrupole magnets for BEPCII was designed and fabricated at Brookhaven National Laboratory, USA. The cryogenic system for the IR magnets was designed at Harbin Institute of Technology, China. This paper provides the results of thermal fluid modeling for the magnet cryostat. The numerical analyses were carried out for two types of cooling methods, the subcooled liquid helium and the supercritical helium flow. The pressure and temperature changes in the cooling circuits are given.

INTRODUCTION

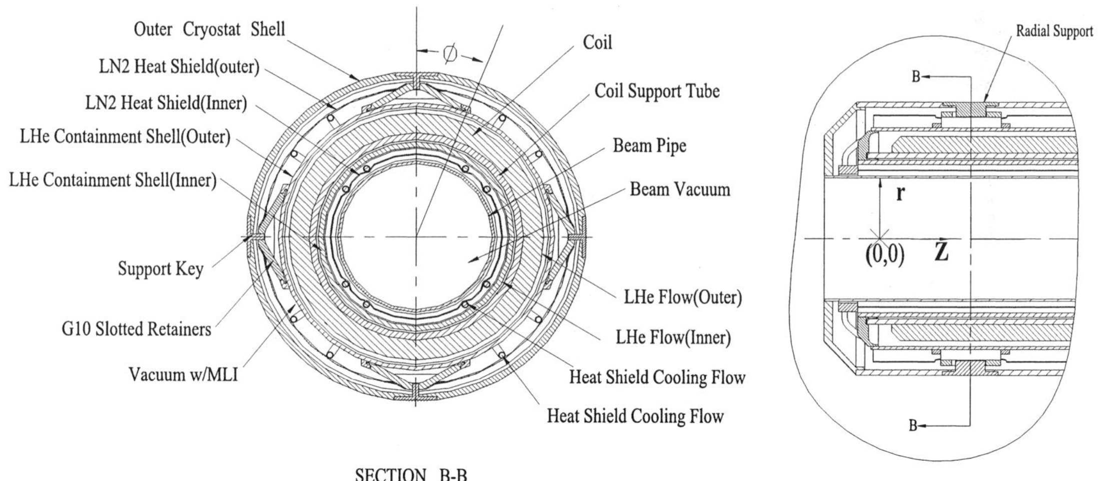

Figure 1   Cryostat of SCQ magnet in BEPCII

Two identical iron-free and non-collared magnets with active shielding are proposed for the BEPCII interaction region upgrade. Each magnet consists of seven coils wound in a common stainless steel cylinder [1, 2]. Due to the requirement of small size and low field leakage as well as the limitation of available space, the magnet cryostat was designed compactly. Figure 1 shows the cross section of the cryostat. The main components include, from inner to outer, the cryostat inner vacuum vessel, the inner heat shield with a series of liquid nitrogen cooling tubes, the inner helium vessel, coil support tube, outer helium vessel, outer heat shield, and outer vacuum vessel (see Figure 1). Each magnet is fixed by eight

124

radial supports and one axial support. The heat loads at 4K for the SCQ magnet are mainly due to the thermal conductions of the radial key supports and the axial support, the thermal conduction and radiation through multi-layer insulation (MLI). The stainless steel support keys, aligned with and welded to the outer vacuum vessel, engage slots in G-10 retainers, arranged in a 90° pattern around the circumference of the outer helium channel, in two axial locations. The G-10 retainers are also mechanically and thermally stationed to the 80K heat shield. The inner diameter of the outer cryostat wall is 290mm and the outer diameter of the inner cryostat wall is 138mm. The gaps of outer and inner helium cooling passages are respectively 6.0mm and 2.0mm. The cooling tubes of heat shields are 5.0mm in diameter. In order to avoid the flow instabilities in the constrained cooling channels, the single-phase helium fluid including the subcooled liquid or supercritical flow will be adopted to cool the magnets and were studied in this paper by numerical simulation.

PHYSICAL MODEL AND BOUNDARY CONDITIONS

The single-phase helium flow goes through the outer and the inner cooling channel of the magnet in turn. The numerical simulation for the cooling channel was performed by a commercial computational code of FLUENT. The simplified physical model is given in Figure 2. Without considering the axial heat load, one quarter of the channel is modeled due to symmetry of structure and radial thermal loads. The helium fluid adsorbs the heat along the passage induced by the heat conduction through the radial supports and radiation from 80K heat shields. In the model, for the outer shell of the outer cooling channel, the heat conduction fluxes are added to the contact surfaces with the support keys, and the radiation heat is evenly guided into the whole area of the outer shell. For the inner cooling passage, only the radiation heat is taken into consideration. The lengths of the outer and inner channel are about 900mm and 1000mm, respectively. The material of the channel is stainless steel, and the properties are used as the function of the temperature.

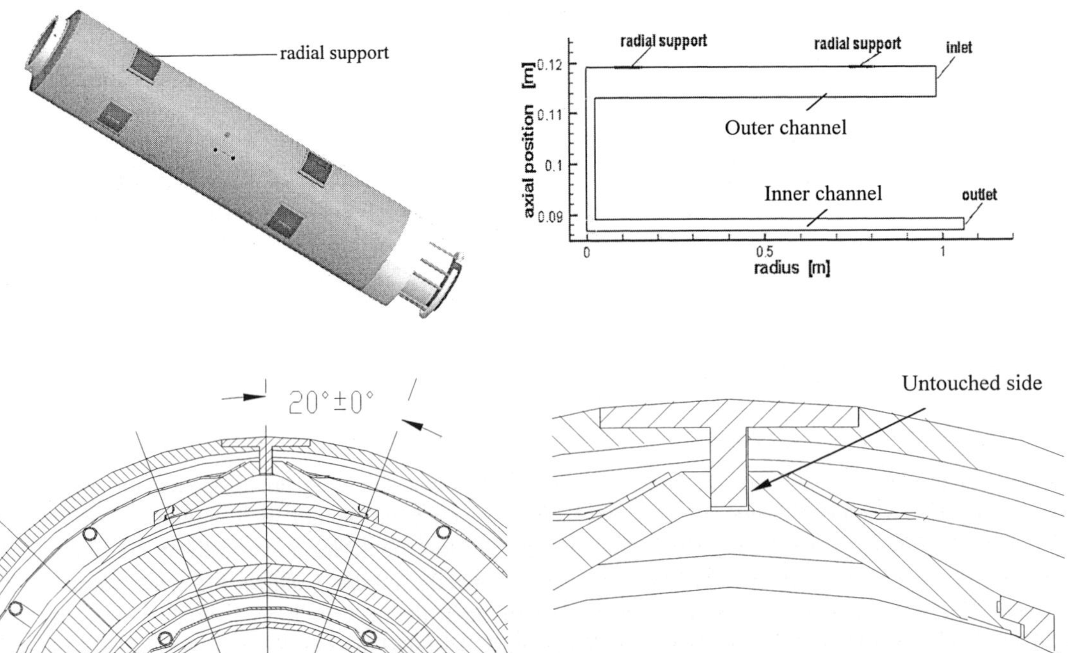

Figure 2   Physical model

SUPERCRITICAL HELIUM COOLING MODELING

The supercritical helium used to cool the magnet is produced by the subcooler. The helium at 2.7bar and 5.2K from the J-T circuit of the refrigerator flows through the heat exchanger immersed in the LHe control dewar and cooled to 4.46K by the saturated liquid helium at 1.2bar in the dewar. And then the helium goes into the magnet cryostat. The steady state, three-dimensional turbulent flow was modeled. The standard $k - \varepsilon$ momentum equation and standard wall function were adopted for the simulation. The coordinates of the circumferential ϕ, radial r and axial z are defined as shown in Figure 1. The origins are set at the end of the outer channel. The mass flow rate is 12g/s.

The temperature change at different radius at the same circumferential ϕ is show in Figure 3. At ϕ=20° and ϕ=-20°, along the radial direction, the conduction heat goes into the cold mass at liquid helium temperature region through the local small contact surface between the G-10 retainers and the outer wall of the outer cooling channel. For the worse case, the stainless steel key only touches the left surface and the surface along the positive Z direction of the G-10 retainers (Figure 2). It will cause the different heat flux to the cooling flow at different spots along the passage. Based on the above structure and the calculated curves, the temperature changes are different for the spots at ϕ=20° (Figure 3a) and ϕ=-20° (Figure 3b). The highest temperature of the supercritical helium channel is about 6.0K at ϕ=-20°, but 5.5K at ϕ=20°, which happens nearby the outer wall of the channel. The closer to the cold mass, the lower is the temperature of the fluid. The helium fluid close to the magnet keeps lower than 4.8K even at the hot spots. The pressure drop is about 10Pa to be neglected. In the inner helium channel, due to only radiation heat flux, the temperature is almost uniform. In order to lower down the highest temperature, the mass flow rate should be increased.

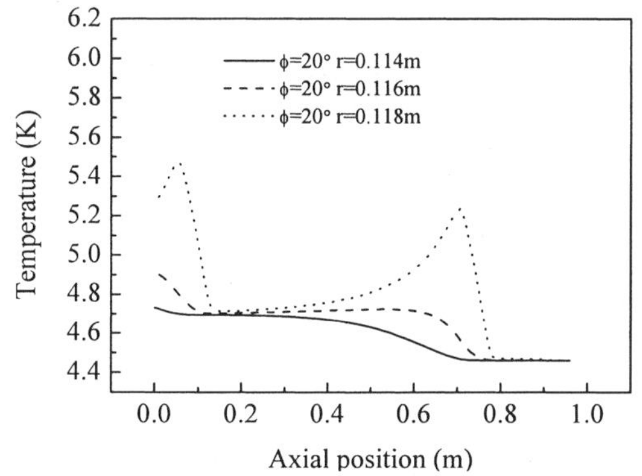

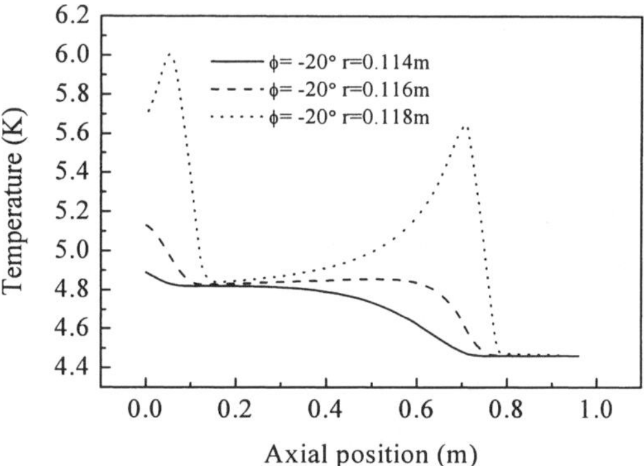

Figure 3a   Temperature vs. axial position at ϕ=20°   Figure 3b   Temperature vs. axial position at ϕ=-20°

SUBCOOLED HELIUM COOLING MODELING

For the subcooled helium cooling modeling, the standard $k - \varepsilon$ momentum equation and standard wall function were also adopted for the simulation. The inlet pressure and temperature of subcooled liquid helium are defined at 1.4bar and 4.46K, and the mass flow rate is also of 12g/s. Figure 4 presents the simulation results. By comparison, at the same spot and mass flow rate, the temperature of the subcooled helium inside the outer cooling channel is a little lower than that of the supercritical helium. The highest temperature at ϕ=20° close to the outer wall is about 5.0K. The difference is due to the difference of specific heat for the subcooled liquid helium and the supercritical helium. The former is around 6000J/kg-K, and the latter is less than 5000J/kg-K. The temperature difference of the fluid close to the cold mass is less than 0.1K between the inlet and outlet along the outer passage at the ϕ=10° and ϕ=30°

for both cooling methods. The temperature of the magnet keeps around 4.5K.

Even though less subcooled flow can be used for the magnets cooling than the supercritical flow, one major concern for the subooled flow is when the temperature of the subcooled helium rises up to the saturated temperature of 4.6K at 1.4bar, it will turn to the two-phase and probably cause the flow instability. Therefore, the supercritical flow is preferred when the cooling capacity is plentiful.

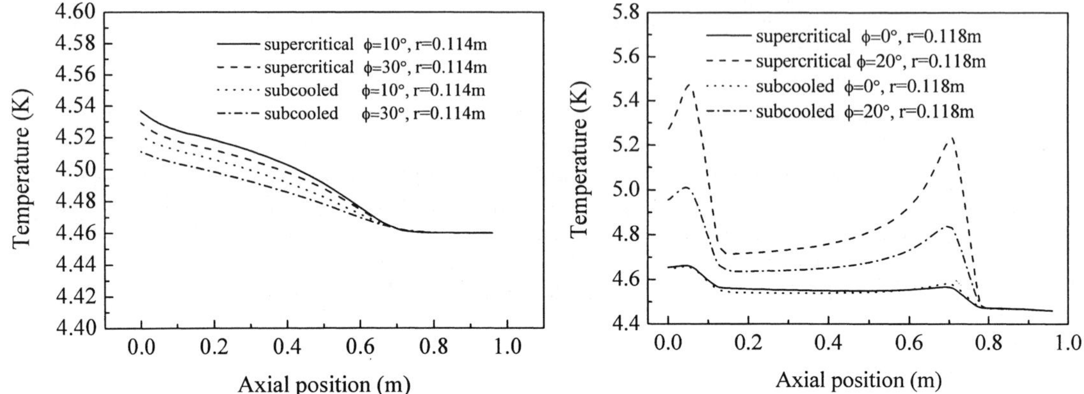

Figure 4   Comparison of the temperature vs. axial position for the supercritical and subcooled helium

CONCLUSION

The numerical simulations were performed for the magnets cooling in BEPCII upgrade. The temperature changes in the cooling circuits are analyzed in detail. By comparison, at the same spot and mass flow rate, the temperature of the subcooled helium inside the cooling channel is a little lower than that of the supercritical helium. The selection of cooling schemes is up to various factors. The supercritical flow is preferred when the cooling capacity is plentiful.

REFERENCES

1. Brookhaven National Laboratory, Design report for the BEPC-II superconducting IR magnets, Magnet Division Note: AM-NO-328 (2003).
2. Institute of Cryogenics and Superconductivity Technology, Conceptual design report for the BEPCII cryogenic system, Harbin Institute of Technology (2002).
3. Jia, L. X. and Wang, L., Cryogenics in BEPCII upgrade, Proceedings of ICEC19, Narosa Publishing House, New Delhi (2002) 39-42.

RECENT DEVELOPMENTS ON CRYOCOOLERS IN EUROPE

A. Ravex and T.Trollier

Air Liquide, Division Techniques Avancées
2,rue de Clémencière – B.P. 15
38360 Sassenage, France

ABSTRACT

New markets such as high temperature superconductors applications and space applications are requiring new cooling performances, efficiency and high reliability for future cryocoolers . New emerging technologies such as flexure bearings and Pulse Tube have demonstrated potentially interesting capabilities for future generations of cryocoolers. In this paper we present the recent developments performed in Europe in this field.

INTRODUCTION

Since several decades, cryocoolers are commercially available, based essentially on low frequency Gifford Mac Mahon (GM), high frequency Stirling and Joule Thomson (JT) expansion cycles in the 4K to 200K temperature range with cooling powers ranging from a few hundred milliwatts up to a few hundred watts. Few years ago, a 4K low frequency GM type Pulse Tube cooler (PTC) has been commercially introduced in the US by Cryomech [1]. Also, 4He-3He dilution refrigerators are commercially available for ultra low temperatures down to a few milliKelvin. Obviously these commercial products meet well identified applications cooling and operation specifications. GM cryocoolers are essentially integrated in vacuum cryopumps and MRI cryostats. Compact Stirling cryocoolers are widely used in infrared detection modules mostly for military night vision optical systems. JT expansion cryocoolers in open cycles, fed with high pressure gas storage bottles, are mainly integrated in single shoot missile guidance infrared systems. 4K GM cryocoolers, GM type PTC and dilution fridges are common cooling solutions in experimental devices for physicists in Research Laboratories.

Recently emerging applications such as high temperature superconductors based electronic (filters for Telecommunication, AD converters, software defined radio) or electro technical (transformers, fault current limiters, motors and generators rotors, power cables) devices are requiring new performances in terms of operation temperature (25K – 65K), cooling power, reliability, integration and maintenance. Space borne applications (biological samples freezing and preservation on board the International Space Station, earth observation missions, astrophysics instruments) are also requiring high efficiency and reliability cryocoolers able to survive launching conditions even in the ultra low temperature domain.

To meet this new requirements, several technologies have emerged (flexure bearings, PTC, sorption coolers) and are actively developed to provide a new generation of cryocoolers. Ongoing developments performed in this field in Europe are presented here after.

GM CRYOCOOLERS AND GM TYPE PTC DEVELOPMENTS

Large cooling capacity single stage coolers

The emergence of electro technical devices such as fault current limiters, transformers or generators/motors rotors implementing high temperature superconductors technology has induced a need for large cooling capacity cryocoolers in the temperature range from 25K up to 65K. GM cryocoolers and GM type PTC are a potential solution mainly for cooling powers from a few ten watts up to a few hundred watts. Few years ago Cryomech in the US has introduced a new series of large cooling power GM cryocoolers (AL 230 and AL 330) with optimized performances below 50K (introducing lead as regenerative material in the lower part of the regenerator): typically 45W and 75W at 25K and 135W and 260W at 65K for 5.5kW and 7kW electrical input and 60Hz operation.

Leybold [2] has also commercialized such a GM cryocooler (model 120T, 25W at 20K with 6kW electrical input).

The main inconvenients of GM coolers are the level of induced vibrations by the cold head at its operation frequency (from 1Hz to 2.4Hz, depending on the cold head motor and electrical supply frequency 50/60Hz) and the need for periodical maintenance (friction seals) requiring generally to stop and heat up the system (unless complicated or thermally inefficient coupling of the cold head with the thermal load are implemented). The PTC technology, with no moving part in the cold finger, is a promising option to keep clear of these drawbacks.

Giessen University is developing such in partnership with Leybold [3] and Siemens a low temperature, large capacity single stage PTC in its TransMIT-Center of Adaptative Cryotechnology and Sensors. A prototype has demonstrated 20W to 45W (depending on the type - RW4000, 5kW or RW6000, 7kW - and number of Leybold compressors operated) cooling power at 26K, corresponding to a 0,43% COP in the best configuration, comparable with the above mentioned commercial GM cryocoolers. Ultimate temperature as low as 14K has been achieved. This PTC has been used as a liquid Neon condenser to operate a thermo siphon cooling loop for a HTc motor rotor prototype in partnership with Siemens.

At Jena University [4], work is also performed on large capacity GM type PTC. Implementing lead coated stainless steel or bronze wire meshes in the regenerator of a Four Valve PTC, an ultimate temperature of 17K has been demonstrated as well as a cooling power of 40W at 34K (which was the ultimate temperature of this PTR before implementing lead wire mesh) with Leybold RW600 (7kW) compressor.

Multi stage coolers

Multistage GM and GM type PTC in the 4K to 10K temperature range have been commercially introduced by Cryomech and Sumitomo few years ago. A potential large scale use of these coolers is for the thermal shielding and Helium boil-off condensation (or even direct conductive cooling of dry magnets) in medical MRI or research NMR cryostats. Prototypes of such zero boil-off MRI units are under development and tests.

Achieving temperature even below 4K with such mechanical cryocoolers is a challenge of interest for some scientific applications.

Giessen University in Germany is still developing 4K PTCs based on further development of the earlier 4K Giessen PTC. These coolers are built and sold in small quantities. A first prototype has demonstrated a cooling capacity of 0.6W at 4.2K and simultaneously 20W at 65K with a 6kW compressor: it has been used for a 5.5T magnet cooling and is foreseen for precooling 3He sorption coolers. A second prototype has demonstrated a cooling capacity of 0.2W at 4.2K and simultaneously 1W at 68K with a 1.75kW compressor: it has been used for a Josephson Voltage Standard prototype. A new design with independent gas circuits of the two stages allows to operate only the smaller 2nd stage with ^{3}He: a low temperature $T_{min} = 1.27$ K has been demonstrated.

Eindhoven University in the Netherlands has designed a three-stage small in size and weight GM type PTC (total volume of the tubes and the regenerator is only 0.28 liter, meaning small amount of working gas and regenerator material). This cooler operates at a frequency of 1.83 Hz and is driven by a 4 kW compressor. The lowest temperature achieved with ^{4}He is 2.13 K, with ^{3}He the temperature of 1.73 K has been reached. The cooling power at 4.2K is 80 mW with ^{4}He and 124 mW with ^{3}He.

STIRLING CRYOCOOLERS AND STIRLING TYPE PTC DEVELOPMENTS

High reliability: flexure bearing technology

To achieve high reliability for demanding applications, it is now commonly agreed that the flexure bearing technology introduced by Oxford University more than a decade ago is a unique feature to ensure the required frictionless operation and clearance sealing in linear Stirling coolers. Several companies in the US (Ball Aerospace, Lockheed Martin, NGST formerly TRW) have developed for space applications specific coolers generally associating a two back to back pistons compressor with an actively driven cold finger, implementing flexure bearings, linear motors and phase shift control electronics.

Thales [5] has undertaken the development and commercialization of a new Stirling cryocoolers series (cooling power ranging from 600 mW up to 2W at 80K) implementing the flexure bearings twin pistons linear compressor with moving magnet motor (reducing risks of cycle gas pollution by motor coil out-gassing and of flying wires and feed-through failures) technology and a pneumatically driven cold finger (reduced complexity: no motor, no active phase shift control electronics). A picture of two cooler types extracted form the LSF coolers family is shown in the Figure 1 and the performances are presented in Table 1.

LSF9180
600 mW @ 80 K / 23°C
40 W input power
$\varnothing$ 5 mm coldfinger
2.4 kg
LSF9188
1600 mW @ 80 K / 23°C
60 W input power
$\varnothing$ 10 mm coldfinger
2.4 kg

Table 1: cooler characteristics.

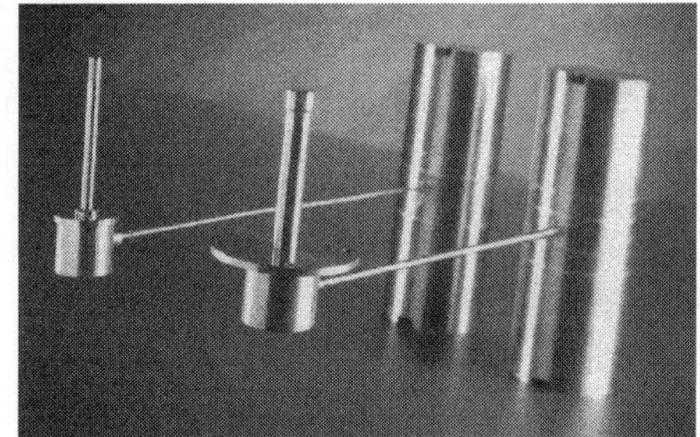

Figure 1: LSF9180 and LSF9188 Thales coolers.

These coolers are intended to offer a high reliability (ongoing endurance tests have already demonstrated more than 28.000 hours of operation without any performance degradation) with selling prices close to those of conventional sliding piston tactical coolers.

In partnership with Thales, based on their flexure bearing technology, Air Liquide has developed a specific Stirling cooler for space applications. The prototype, developed for Cryosystem - a freezer to be flown on board the International Space Station – under ESA funding, implements also flexure bearings on the cold finger and material and design changes in compressor to take into account specific operation conditions in space. Performances (9.3W at 80K with 150W electrical input) are presented elsewhere [6] in this conference.

AIM and Leybold have also recently developed new Stirling coolers implementing flexure spring (not flexure bearing) for the compressor pistons central position and resonant operation control, allowing to reduce radial wear with cylinder as compared to conventional helicoidal spring technology.

Stirling type Pulse Tube Coolers

The main drawback of the Stirling coolers in terms of reliability and integration easiness is their cold finger: the moving displacer technology and the associated linear motor are a potential source of failure and exported vibrations, the reduced clearance between the moving regenerator/displacer and its shelve induces strict mechanical load restrictions on the cold finger cold tip to avoid any internal friction which generally results in very complicated thermal link design, the active phase shift control loop complicates the drive and control electronics.

The PTC technology with no moving part in the cold finger and passive pneumatic phase shift control is recognized as an attractive alternative to Stirling coolers. Recent developments, such as phase shift control by inertance, have greatly improved the PTC temperature stability and overall efficiency which is now comparable with Stirling coolers.

Air Liquide, in partnership with CEA/SBT and Thales has undertaken since several years the development of PTCs. Under European Space Agency funding, a Miniature Pulse Tube Cooler (MPTC) has been recently successfully designed and tested [7, 8]. Typical performances of this MPTC are reported in Table 2 and a picture is shown in Figure 2. The integration of this MPTC in future space missions is under consideration.

Performance	1 W @ 80 K
Input Power	35 W
Frequency	50 Hz
Mass	2.8 kg
Compressor OD	63 mm
Compressor length	157 mm
Cold finger OD	100 mm
Cold finger length	179 mm

Table 2: MPTC characteristics.

Figure 2: MPTC EM cooler.

The need for larger cooling capacity PTCs has been clearly identified, first for ground based commercial applications (i.e. superconducting filters for telecom) but also for future earth observation or other space applications. Air Liquide and its partners have developed such a larger heat lift capacity PTC demonstrator. The performances achieved are 7.7W at 80K for 200W input power in in-line configuration. U-shape design has been prototyped and a coaxial version is under industrialization at Thales for future commercialization (Figure 3).

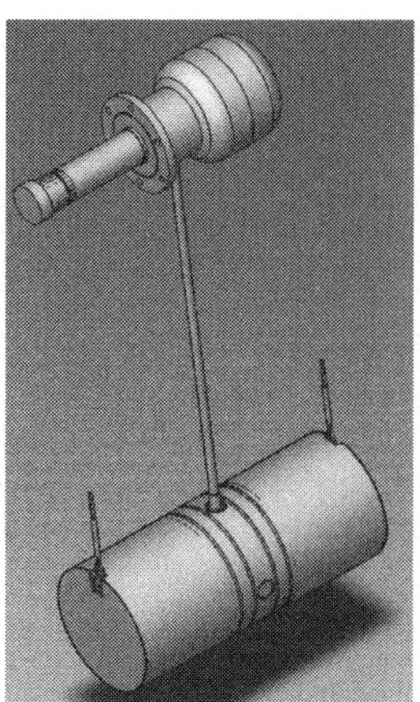

Figure 3: U-shape PTC prototype and coaxial shape PTC CAD design.

In the framework of the European SuperADC program, a two staged PTC is also under development by the same partners with the aim to produce simultaneous cooling power at 80K and 20K-30K. Preliminary results have been presented recently elsewhere [9].

Giessen University, in partnership with AIM and Leybold, in a project on "HTS for future telecommunication" supported by the German Government has also developed Stirling type PTCs with medium cooling capacities. U-shaped, inline and coaxial cold finger configurations have been developed and driven either by AIM SL200/400 linear compressor (100W nominal input power) or by Leybold Polar SC7 linear compressor (200W nominal input power). Experimental results are summarized in Table 3 [10].

PTC/Compressor/ Input power	Geometry	No-load temperature (K)	Cooling power @ 80 K (W)	COP @ 80 K
PT03/SL400/100 W	U-shaped	41.3*	3.3*	3.3 %*
PT07/SL400/100 W	Inline	41.9*	3.55*	3.55 %*
PT08/SL400/100 W	Coaxial	36.4*	3.6*	3.6 %*
	Coaxial	49.7	3.3	3.3 %
PT05/Polar SC7/200 W	U-shaped	44.5	7.4	3.7 %
PT04/Polar SC7200 W	Inline	45.0	8.7	4.4 %
PT09/Polar SC7/200 W	Coaxial	47.0	7.4	3.7 %

*) Data marked by * are obtained with 2nd inlet; all other results are with inertance only.
Table 3: Summary of results (water cooling of warm end at 20 °C)

ILK Dresden has also developed an inertance U-shape PTC (1W:80K/100W electrical) driven by an AIM linear pressure oscillator. A prototype based on an active reservoir principle driven by a rotary machinery implementing a pressure wave generator and an expander is presently under development with a 15W at 80K performance goal.

Large cooling capacity Stirling and Stirling type Pulse Tube Coolers

Large cooling capacity Stirling cryocoolers (1 to 4 kW at 80K for single stage units and several hundred watts at 20K for two staged units) have been developed several decades ago by Philips mainly for Air, Nitrogen, Neon and Hydrogen liquefaction. These cryocoolers with rotating motor for compression piston drive via crankshaft and mechanical expander phase shift control are still commercialized by Cryogenics & Refrigeration B.V. Their technology generates high level of vibration and short interval maintenance.

In view of potential future HTS applications in motors/generators or power transmission cables, the interest for such large cooling capacity Stirling type PTC has been increasing. In the US, Praxair and American Superconductor Company have initiated development programs and demonstrated 200W cooling power at 80K with 4kW electrical input in linear flexure bearing compressors (provide by Clever Fellow Inc. or Stirling Technologies).

In Europe, Air Liquide and Giessen University have recently initiated similar development programs.

JT EXPANSION CRYOCOOLERS

The demand for miniature JT expansion coolers operated around liquid Nitrogen temperature in open loop for infrared modules cooling has been stable during the past years. Few improvements or innovations have been published. A common requirement for these products is the capability to deliver in a first time a large flow for rapid initial cooling down from room temperature and later on to provide the minimal flow required for the detector temperature stabilization. Most of the existing systems are implementing flow control valves with a needle moving in a calibrated orifice and controlled either by gas actuated bellows or axial differential thermal extension based devices. These flow control systems are complicated and expensive to integrate. Air Liquide has developed and patented a new simple and compact system based on the radial differential thermal expansion in an annular geometry.

The use of mixture (freons or hydrocarbons with nitrogen, neon, hydrogen) as cycle fluid in JT expansion close cycle coolers has been an active subject of development in the recent years. The main advantages of using such mixtures (the physical properties of mixture and their potential interest have been extensively described in an old patent from (Refrigerant for a cryogenic throttling unit. V. N. Alfeev. Patent specification N°1.336.892, 1971) is a higher specific cooling power with lower expansion ratio compared to pure nitrogen, allowing for close cycle operation using GM coolers type compressors. A commercial product (CryoTiger) has been introduced by APD few years ago with a few tens of watt of cooling power around 100K. Developments are also progressing at Dresden Technical University in this field [11].

Twente University is performing an innovative work related to JT expansion coolers aimed to develop miniature systems driven by thermal compressors (using sorption/desorption properties of activated charcoal or other types of adsorbent under thermal cycling) for electronic components or detectors cooling. Etching technology is used for counter flow heat exchangers and JT expansion capillary manufacturing. A prototype using Xenon as fluid and charcoal as adsorber has been developed to demonstrate the technology with cooling at 165K, using a Peltier thermo element for precooling [12]. A major interest of this technology is its potential for low vibration (no moving mechanical parts, except check valves operated at very low frequency in the thermal compressor). A development [13] is supported by ESA in the perspective of the future Astrophysics mission Darwin. The goal is to develop a two-stage sorption JT cooler. The first stage (H2/Charcoal), precooled by the first stage, should reject 3W at 80K (radiator) and provide 25mW cooling at 14,5K with radiative precooling at 50K. The second stage (He/Charcoal) should reject 5W at 50K (radiator) and provide 10mW cooling at 4,5K.

SUBKELVIN CRYOCOOLERS

SubKelvin cooling is mainly required on ground by scientists for laboratory experiences or in space for detectors cooling in Astrophysics missions.

Commercial systems for laboratory use

A double stage (^{4}He/^{3}He) sorption cooler developed by CEA/SBT [14] has been coupled by Air Liquide with a commercial Cryomech 4K PTC for precooling. This system [15] is the first cryogen free system able to provide 3 orders of magnitude in temperature. It provides ultimate temperature down to 265mK and typical cooling power of 20μW at 290mK. Hold time in excess of 2 days have been measured at temperature below 300mK. Drive electronic is under development and in a near future the complete system will be fully automated.

A similar approach has been used to associate an Air Liquide standard dilution fridge with a 4K PTC. This cryogen free system [15] provides more than 4 orders of magnitude in temperature with a continuous operation. An ultimate temperature of 8mK has been achieved with a cooling power of about 250 μW at 100 mK.

These SubKelvin commercial mechanical coolers are presented on Figure 4.

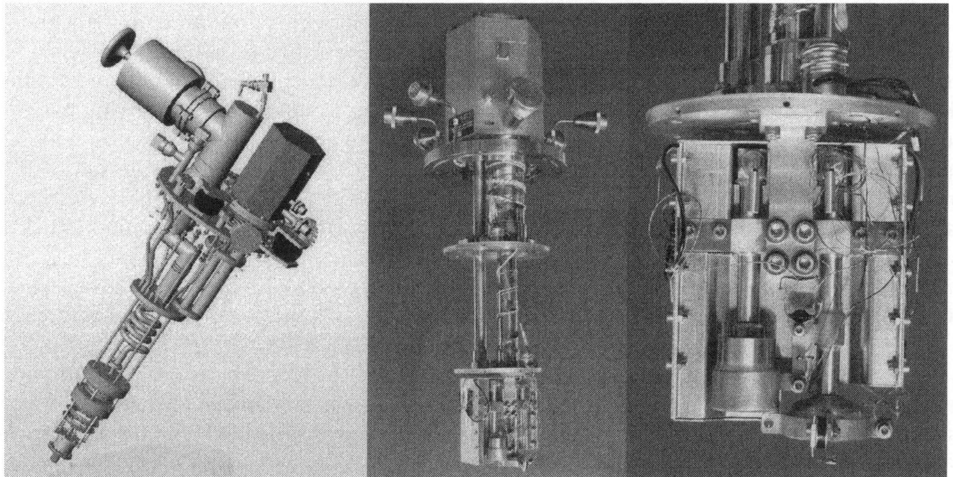

Figure 4: SubKelvin cryocoolers (4K PTC coupled with ^{4}He/^{3}He sorption or dilution stages)

Developments for space applications

Astrophysics missions in space requiring subKelvin cooling are already planed (Planck, Herschel) or foreseen in the future (Darwin).

Air Liquide has developed in partnership with CNRS/CRTBT and space qualified a dilution refrigerator for the cooling from 4 K down to 0,1 K in microgravity environment of the bolometers array of HFI (one of the Planck mission instruments). The CNRS/CRTBT has patented a unique design using 3He and 4He isotopes stored in high pressure vessels to feed a continuous mixing/separation process in a capillary pipe and rejecting the mixture (after a Joule Thomson expansion providing extra cooling at 1.6K) in space. This original process has been already demonstrated on ground against gravity (up side down cryostat) and in the Archeops stratospheric balloon mission. The qualification model has demonstrated the required cooling power of 100nW at 100mK [16] and has successfully survived the launching vibration tests.

CEA/SBT has developed and qualified a double stage (^{4}He/^{3}He) sorption cooler for two instruments (SPIRE and PACS) of Herschel mission. The technology of this cooler, including the gas heat switches required for cooler operation, has been extensively demonstrated on ground (operation against gravity) and in rocket missions. The qualification model has successfully passed the vibrations tests and demonstrated the specified performances: 10μW at 290mK, 2h recycling time and 46h hold time [17]. Flight models are under manufacturing.

Adiabatic Demagnetisation Refrigerators (ADR) are also candidates for subKelvin cooling in space or even in refrigerators for laboratory use. Few years ago, Cryogenic Spectrometers GmbH (CSP) has developed and coupled an ADR with a 4K PTC developed at Giessen University. A temperature of 60mK has been reached and 8 hours holding time at 100mK was provided. Under ESA contract, Oxford Instruments and Mullard Space are presently developing an ADR stage [18]. CEA/SBT has also undertaken such a development. Their specificity is to associate their ADR stage with an 3He sorption cooler: starting with precooling from 300mK will greatly decrease the magnetic field necessary to obtain an efficient demagnetization cycle [19].

An innovative superfluid vortex cooler (SVC) has been developed at Eindhoven TU [20]. SVC is a combination of a fountain pump and a vortex cooler. The working fluid in the SVC is superfluid ^{4}He. It is a small closed-cycle cooler with no moving parts. A certain amount of heat, supplied to the fountain part of the cooler, causes a circulation of helium in it. The total volume of the SVC is 1.25 cm^3. The cooler is gravity independent and requires little additional infrastructure. In principle a compact cryogen-free cooler with no moving parts in the cold area going down to temperatures below 1 K can be obtained. The PTR that achieves the required precooling temperature has been developed at the University of Giessen. It is a two-stage GM type PTR providing 8 mW of cooling power at 1.45 K. In the preliminary experiments with the integrated system a lowest temperature of 1.19 K has been reached.

CONCLUSION

New performances and reliability requirements for future HTS devices or space missions have simulated the development of new technologies. European academic institutions and industrial partners have initiated developments of different cryocooler technologies covering a large range of cooling power and cooling temperatures. Commercial products are already proposed by the industry and several demonstrators have demonstrated their performances and potential capability to be shortly industrialized. Challenging developments are also performed for future space missions.

REFERENCES

1   specs@cryomech.com (Air Liquide is distributing Cryomech coolers in Europe).
2   www.leyboldvac.de
3   H.U. Haefner, "Powerful 25K single stage Pulse Tube cryocooler". Challenges of Applied Cryoelectrics II, 26-27 Mai 2003, 308. WE-Heraeus-Seminar

4   A. Waldauf, T. Köttig, S. Moldenhauer, M. Thürk, and P. Seidel "Improved Cooling Power by means of a Regenerator made from Lead Wire Mesh". Proceedings of the thirteenth International Cryocooler Conference, ICC13, to be published.

5   info.cryo@nl.thalesgroup.com

6   A. Ravex, T. Trollier, L. Sentis, F. Durand, P.Crespi "Cryocoolers development and integration for space applications at Air Liquide". Proceedings of the International Cryogenic Engineering Conference, ICEC20, to be published.

7   T. Trollier, A. Ravex, L. Duband, I. Charles, T. Benschop, J. Mullié, P. Bruins and M. Linder, "Miniature 50 to 80 K Pulse Tube Cooler for Space Applications". Cryocoolers 12, Monterey, Kluwer Academic Publisher/ Plenum Publishers, 2003, p 165-171.

8   T. Trollier, A. Ravex, L. Duband, I. Charles, T. Benschop, J. Mullié, P. Bruins and M. Linder, "Performance Test Results of a Miniature 50 to 80K Pulse Tube Cooler". Proceedings of the thirteenth International Cryocooler Conference, ICC13, to be published.

9   I. Charles, J. M. Duval, L. Duband, T. Trollier, A. Ravex and J.Y. Martin, "High Frequency large cooling power pulse tube". Proceedings of the nineteenth International Cryogenic Engineering Conference, ICEC19, p395-398.

10   L.W.Yang and G.Thummes, "Development of Coaxial Stirling-type Pulse Tube Cryocooler". Proceedings of the thirteenth International Cryocooler Conference, ICC13, to be published

11   A.Alexeev and E.Mantwill, "Auto-refrigeration Cascade and Mixed Gas Joule Thomson Refrigerator". Proceedings of the nineteenth International Cryogenic Engineering Conference, ICEC19, p335-338.

12   J.F. Burger, H.J.M. ter Brake, H.J. Holland, M. Elwenspoek, H. Rogalla, "Construction and Operation of a 165K Microcooler with a sorption compressor and micromachined cold stage". Cryocoolers 12, Monterey, Kluwer Academic Publisher/ Plenum Publishers, 2003, p 643-649.

13   J.F. Burger, H.J.M. ter Brake, H.J. Holland, G. Venhorst, E. Hondebrick, R.J. Meijer, T.T. Veensra, H. Rogalla, "Development of a 4K Sorption cooler foe ESA Darwin Mission: System-Level design considerations". Proceedings of the thirteenth International Cryocooler Conference, ICC13, to be published

14   L. Duband, *Cryocoolers 11*, Kluwer Academic/Plenum Publishers, New York p. 561 (2001)

15   S. Triqueneaux, A.Ravex and P.Hernandez, "SubKelvin mechanical coolers". Proceedings of the Cryogenic Engineering – CEC, Anchorage 2003, to be published.

16   L. Sentis, J. Butterworth, P. Camus, Y. Blanc, G. Guyot, " Cryogenic test of a 0.1K Dilution Freezer for Planck HFI". Proceedings of the thirteenth International Cryocooler Conference, ICC13, to be published.

17   L. Duband, L. Clerc, L. Guillemet, R. Vallcorba, "Herschel Sorption Coolers – Qualification Models". Proceedings of the thirteenth International Cryocooler Conference, ICC13, to be published.

18   I.D. Hepburn, "Refrigeration in space to temperatures in the region 10 to 100 mK". Proceedings of. Space Cryogenics Workshop, 20-21 July 1998, ESTEC, Noordwijk, ESA WPP-157, March 1999.

19   N. Luchier, "Small Adiabatic Demagnetization Refrigerator for Space Missions". Proceedings of the thirteenth International Cryocooler Conference, ICC13, to be published.

20   I.A. Tanaeva and A.T.A.M. de Waele,"Helium-3 Pulse Tube Cryocooler". Cryocoolers 12, Monterey, Kluwer Academic Publisher/ Plenum Publishers, 2003, p 283-292.

Experimental Study on a Split Free-Piston Stirling Cooler for Space Applications

Wu, Zhang, Yinong Wu, Guohua, Lu

Shanghai Institute of Technical Physics (SITP), Chinese Academy of Sciences, Shanghai, 200083,
P.R,China

With the purpose of lightening Stirling cooler used for space, a split Free-Piston
Stirling cooler is designed and assembled based on a previous double-driven split
Stirling cooler operated successfully. Compared with double-driven Stirling
cooler, the stroke and phase of pneumatically driven displacer in the Free-Piston
Stirling cooler are hard to control. In this paper the way of adjusting the stroke
and phase, which will give a preferable stroke and phase, is studied
experimentally in detail. And some methods of adjusting correlative parameters
are achieved. Driven by the same compressor, the cooling performance of the
Free-Piston split Stirling cooler under optimal parameters setting is close to that
of double-driven Stirling cooler.

INTRODUCTION

Since clearance seal with flexure spring technology was introduced in "Oxford 80K cooler" [1], long life
stirling cooler has been used in space widely as one of the important cooling mode. Many companies
have participated in the development of this kind of cooler and so does our company. A split Oxford-type
Stirling cooler double-driven by linear motors had been developed by SITP and successfully applied in
the Shenzhou No.3 space ship mission. It could produce 500mW cooling power at 80K in consuming less
than 30w input power. However, the size and the weight of this kind of stirling cooler are more and more
difficult to meet some missions with the development of small satellites. It is necessary to make the
long-life stirling cooler with smaller size as well as lighter weight.

DESIGN FEATURES AND EXPERIMANTAL APPARATUS

The displacer in stirling cryocooler can be seen as a single-degree-of-freedom spring-mass-damped system (see Figure 1). The movement of displacer is decided together by the mass of displacer m, the spring constant K and the damping coefficient C.

The natural frequency f_0 of this system is:

$$f_0 = \left(K/m \right)^{0.5} / 2\pi \tag{1}$$

And the damping ratio ζ of this system is:

$$\zeta = C/(2\sqrt{K \bullet m}) \tag{2}$$

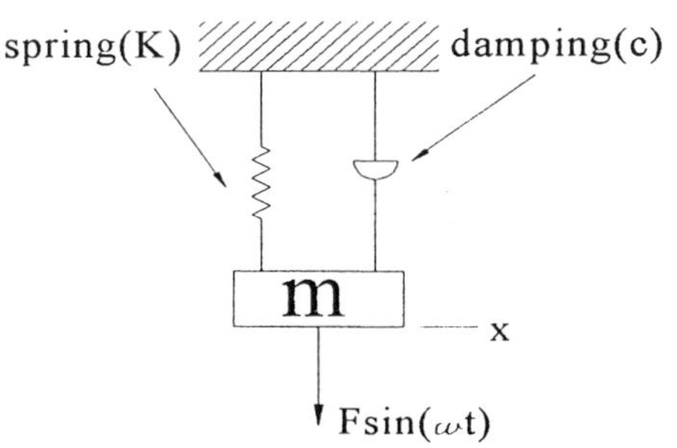

Figure 1 spring-mass-damped system

According to classical analysis of single-degree-of spring-mass-damped system [2], amplitude gain will increase with the decrease of the damping ratio and maximum amplitude gain will occur at a frequency that is somewhat less than the natural frequency in the case of modest degree of damping. The corresponding frequency is referred as resonant frequency and decided by the natural frequency and the damping ratio. It's relatively easy to achieve the natural frequency we prefer through adjusting the spring stiffness, but difficult to get appropriate damping ratio.

In many stirling coolers the seal between piston and cylinder as well as the damping was fulfilled by dry-friction. These coolers are small but at the cost of short lifetimes. Oxford 80K Stirling cooler solved the problem of contact seal by using clearance seal, but in order to produce appropriate damping ratio a linear motor had to be added in the expander. As mentioned above, the motor increases the size and the weight of the Stirling cooler.

In order to meet the long-life requirement and have lighter weight, based on our previous split Oxford-type stirling cooler double-driven by linear motor, an improved free piston expander is designed. Different with previous expander, the motor in it has been abandoned and a new damping modulating structure has been adopted. The schematic of damping modulating structure as well as the measurement system is shown in the figure 2.

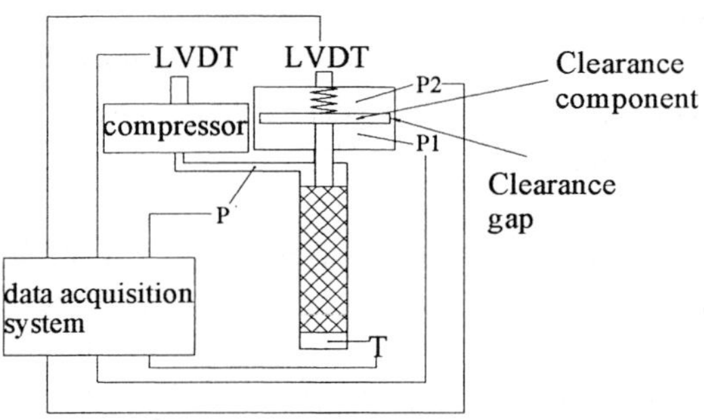

Figure 2 Structure and measurement system

The clearance component divides the pneumatic chamber into two parts. Gas will flow from one chamber into the other chamber through the gap when the clearance component moves reciprocatingly. The gap width being very small, it will hold back the flow of the gas and synchronously the pressure difference between the two chambers will be established. Namely the system has produced a force. So long as the force is in opposite phase with the movement, the force can be seen as a damping force. For better understanding the influence of clearance component on the system, calculation on the movement of displacer is carried out based on the following assumptions:

- ✧ For the purpose of simulating the effect of the clearance component on the system, the resistance of the regenerator is neglected;
- ✧ Ideal gas and constant properties;
- ✧ One-dimensional flow

In order to compare the cooling performance with the previous cooler used in Shenzhou No.3 space ship, the compressor remains the same. Two pressure sensors are placed in the two chambers respectively and a pressure sensor is installed in the pipe to measure the system pressure. Two linear variable differential transformers (LVDT) are used to measure the displacement signal of compressor piston and expander piston respectively. A PT100 is used to measure the temperature of cold finger. Those six signals are all transferred to the data acquisition system, which it's easy to know the phase difference of any two fluctuating signals.

THEORETICAL CALCULATION AND EXPERIMENTAL RESULTS

In order to research the adjusting method on the stroke and phase as well as the optimum matching condition between the compressor and expander, lots of experiments are done based on the various dynamic parameters such as the natural frequency of displacer, the width of clearance gap, charging pressure and the driven frequency. Besides, some theoretical calculation results are utilized to validate some experimental results.

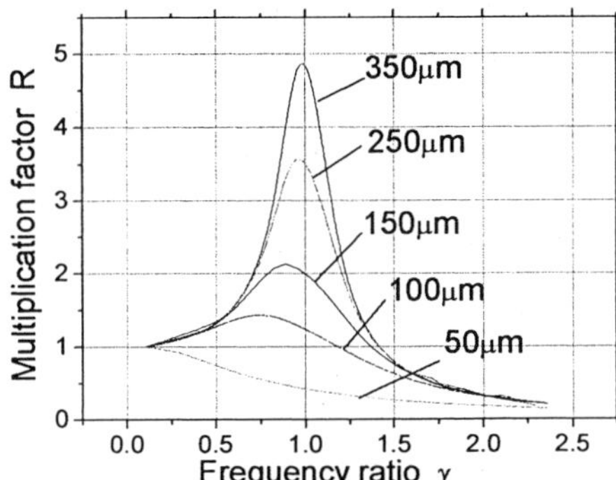

Figure 3 Theoretical calculation on frequency response for different gap width

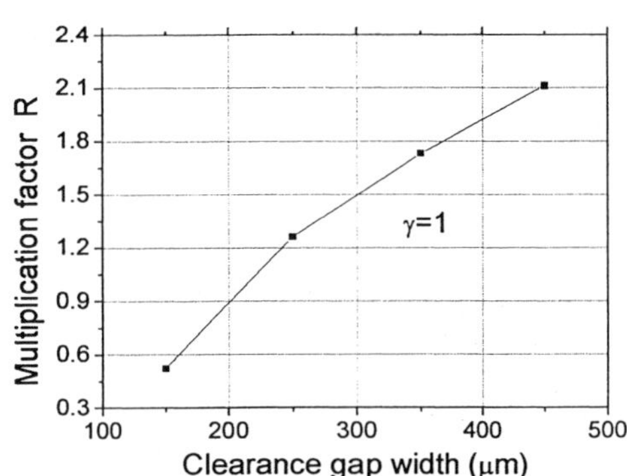

Figure 4 Experimental result on multiplication factor for different gap width

Seen from the curves in Figure 3, similar with the classical frequency response curves in single-degree-of-freedom system, with the increase of the clearance gap width the displacer amplitude gain will increase and the resonant frequency will be closer to the natural frequency. That is smaller the clearance gap width is, bigger the damping ratio is. The experimental results in figure 4 also illustrate the above conclusion. Because of not considering the resistance of the regenerator the actual multiplication factor is less than that in calculation. In order to avoid too big or too small damping ratio, selecting appropriate width of the clearance gap is very important.

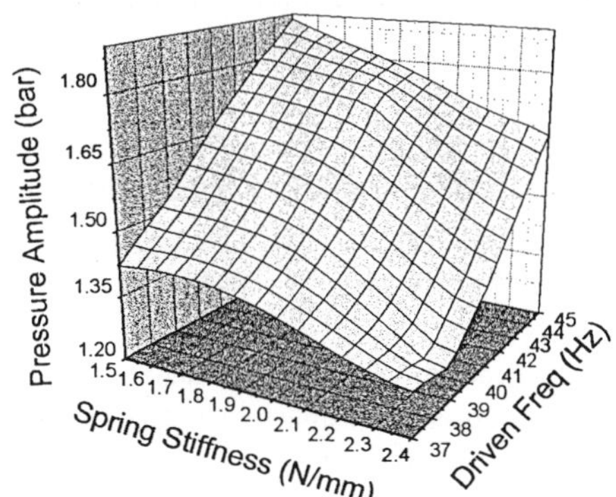

Figure 5 Pressure amplitude curve with different spring stiffness and driven frequency

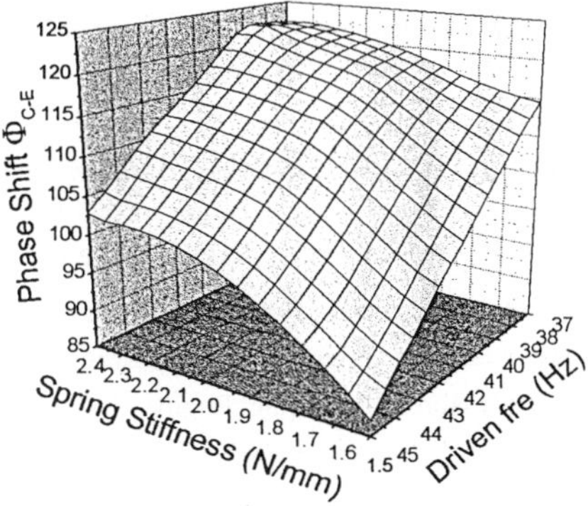

Figure 6 Phase shift curve with different spring stiffness and driven frequency

Seen from the above experimental results in figure 5 and 6, with the decrease of the driven frequency and the increase of the spring stiffness the phase shift between compressor and expander will increase, while the pipe pressure amplitude will decrease. According to [2] the phase shift between driven force and the displacement of displacer will increase with the rise of the driven frequency. Moreover, the phase of the pressure at low temperature is lag to that at room temperature and the phase shift will increase with the rise of the driven frequency according to [3]. This trend is disadvantageous to our stirling cooler system. The optimum cooling performance will occur when maximum driven force and appropriate phase shift are all satisfied.

The main optimal dynamic parameters of the cooler are as follows: the charging pressure is 15bar, the resonance frequency of the displacer is 46Hz, the breadth of the clearance is $300\,\mu$ m and the operating frequency is 43Hz. The optimal performances of the model cooler are as follows: the lowest cooling temperature is 51K, the maximal cooling power at 80K cooling temperature is 632mw, the input power is 27.3W and the corresponding COP is 2.31%(see figure 7). The results are close to the level that our previous split Oxford-type double-driven Stirling cooler had reached.

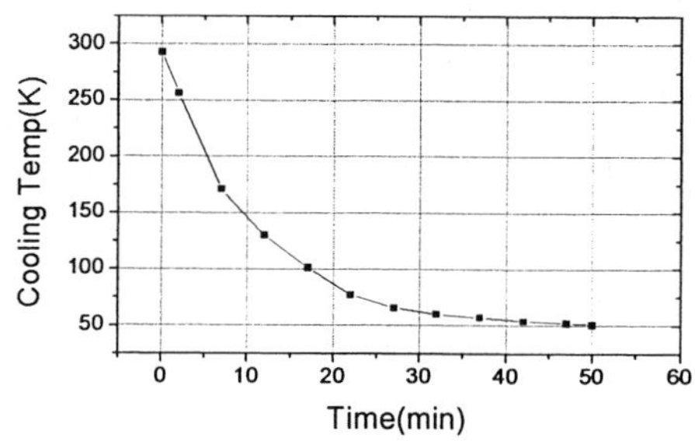

Figure 7 Optimal cool-down curve for our improved cooler

CONCLUSION

An improved expander with clearance component is introduced, which will meet the long life and lighter weight requirement.

The movement of the clearance component can be seen as a kind of damping and it can adjust the stroke and phase of the displacer effectively.

The new expander can obtain 632mW cooling power at 80K in consuming 27.3W input power, which is proved to be a promising method to use in space.

REFERENCES

1. S.T.Werrett, G.D.Peskett, G.Davey, T.W.Bradshaw, J.Delderfield, "Development of A Small Stirling Cycle Cooler for Space-flight Applications," Oxford University, et al, <u>Advances in Cryogenic Engineering</u> (1986) <u>31</u> 791-799

2. Wenjun Shen, Jinghui Zhang, <u>Mechanical vibration—theory and application,</u> The publishing house of national defense industry, Beijing, (1984) (in Chinese)

3. Kwanwoo Nam, Sangkwon Jeong, Measurement of cryogenic regenerator characteristics under oscillating flow and pulsating pressure, Cryogenic Engineering Laboratory, Korea Advanced Institute of Science and Technology, Taejon, Korea, <u>Cryogenics</u> (2003) <u>43</u> 575-581

Proceedings of the Twentieth International Cryogenic Engineering Conference
(ICEC 20), Beijing, China. © 2005 Elsevier Ltd. All rights reserved.

The effects of the clearance in the cold finger on the performance of the Stirling refrigerator

Yong-Ju Hong, Deuk-Yong Koh, Seong-Je Park, Hyo-Bong Kim, Young-Don Choi*

HVAC & Cryogenic Engineering Group, Korea Institute of Machinery & Materials, 305-600, P. O. Box 101, Yu-Sung, Taejeon, Korea
* Department of Mechanical Engineering, Korea University, 136-701, Seoul, Korea

The Stirling refrigerator has been widely used for the cooling of the infrared and cryo-sensor. The Stirling refrigerator has the clearance in the compressor and cold finger. The cooling capacity of the Stirling refrigerator comes from the relation of the pressure in the expansion space and motion of displacer. In the radial clearance of the cold finger, the parasitic dynamic losses from the leakage flow and shuttle heat losses are inevitably occurred. Therefore, to get the maximum performance of the refrigerator, the radial clearance should be optimized. In this study, the influences of the radial clearance on the performance were investigated by numerical simulation. The results show the effects of the radial clearance, operating frequency and the stroke of the displacer on the cooling performance of the Stirling refrigerator (the enthalpy flow through the regenerator and clearance, the shuttle loss).

INTRODUCTION

Over the past decade and a half there has been rapid development of the Stirling refrigerators, mainly for military and space application. The refrigerators working on the Stirling cycle are characterized by high efficiency, fast cool down, small size, light weight, low power consumption and high reliability. The Stirling refrigerators have been widely used for the cooling of the infrared sensors and high temperature superconducting filters to the liquid nitrogen temperature.

In the Stirling refrigerator, the clearances between the cylinder and the moving object (the piston or the displacer) are inevitably occurred. The clearance between the cylinder and piston in the compressor has the effects on the performance and reliability of the compressor, and the clearance between the displacer and the cylinder in the cold finger has the significant effects on the thermal performance of the refrigerator.

The gross refrigeration of the Stirling refrigerator comes from the relation of the pressure in the expansion space and motion of displacer. The reduction in the amplitude of pressure would result from the excessive leakage flow through the clearance in the cold finger. It is well known that a conventional refrigerator with displacers usually uses seal rings to reduce the leakage flow. The cooling capacity can be calculated by subtracting all the loss terms from gross refrigeration. The clearance would lead to the parasitic dynamic losses. The axial temperature gradient along the expander causes the shuttle heat loss[1,2]. The repetitive filling and emptying of the clearance cause the appendix loss[3].

In this study, the effects of the clearance in the cold finger on the cooling performance of the Stirling refrigerator were investigated by analysis.

DESCRIPTION OF ANALYSIS

A Stirling refrigerator comprises of a compressor and a cold finger connected by a tube. The cold finger contains a displacer with regenerator and an expansion space as shown Figure 1.

The displacer with regenerator is actuated by the mechanical driving mechanism. For the purpose of providing a refrigerator with regenerator with displacer having a good cooling performance, the seal rings

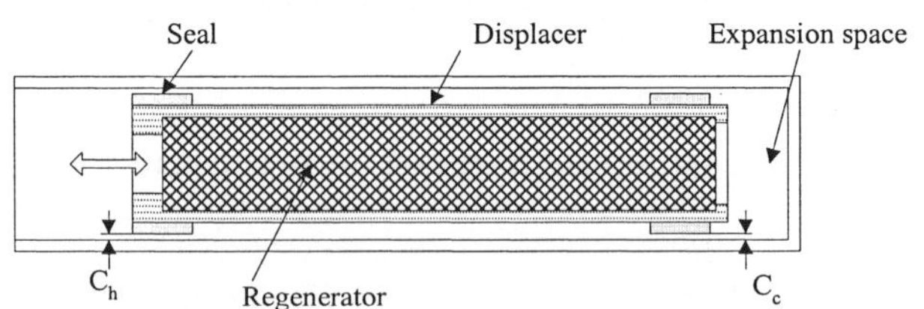

Figure 1   The schematic diagram of a cold finger of the Stirling refrigerator

are employed at the hot and cold end of the displacer.

Table 1 shows the specifications of the Stirling refrigerator in the calculation. The SAGE[4] was used to investigate the effects of the operating frequency, the stroke of the displacer and the radial clearance between the seal rings and the expander on the cooling performance.

Table 1   The specifications of the refrigerator

Working fluid	Helium (real gas)	Piston/Displacer phase	90 deg
Charging pressure	2.2 MPa	Regenerator	#250 (Porosity 0.69)
Piston diameter	10 mm	Regenerator diameter	7.1 mm
Stroke of piston	4 mm	Regenerator length	70 mm
Expander diameter	8 / 7.7 mm	Hot temperature	300 K
Displacer diameter	7.4 / 7.1 mm	Cold temperature	80 K

Table 2 shows the calculation results for the operating frequency of 50 Hz, the stroke of the displacer of 3 mm and the radial clearance between the seal rings and the expander of 0.015 mm. The result shows the enthalpy flow through the refrigerator is a dominant heat loss. The losses by the clearance amount to the 33% of the total loss. The conduction loss through the wall of the expander and displacer is relative small. Figure 2 shows the PV diagram of the compression space and the expansion space.

Figure 3 shows the mass flow rate flowing to the expansion space. The flow through the clearance has a phase lag of 80 degree to the flow through the regenerator and small amplitude.

Table 2   The calculation results

PV work in compression space	21.33 W
Gross refrigeration	3.557 W
Enthalpy flow through the regenerator	1.162 W
Enthalpy flow through the clearance	0.1744 W
Shuttle heat loss	0.5314 W
Conduction heat loss	0.2732 W
Cooling capacity	1.416 W

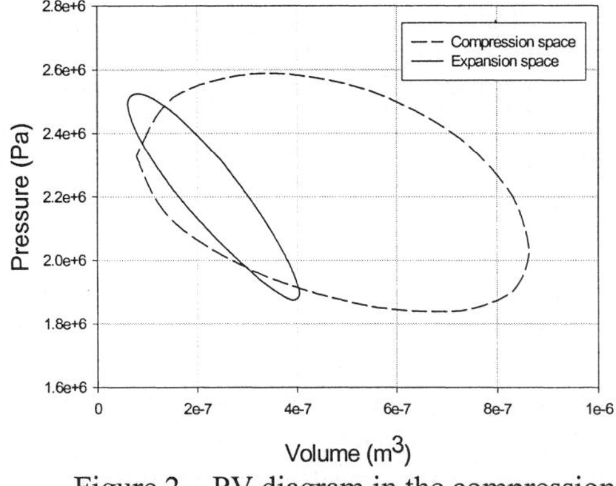

Figure 2   PV diagram in the compression and expansion space

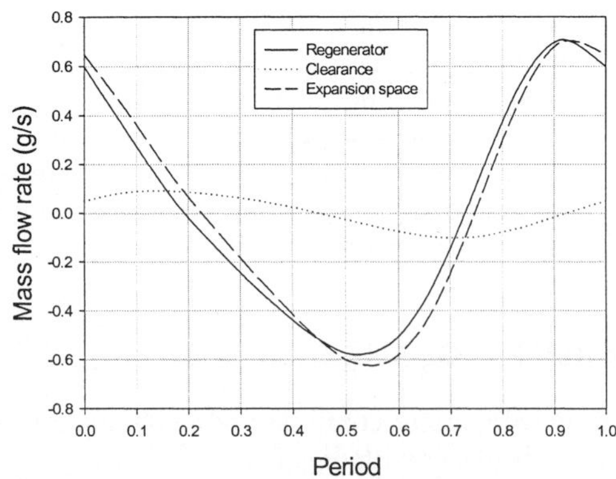

Figure 3   mass flow rate flowing to the expansion space

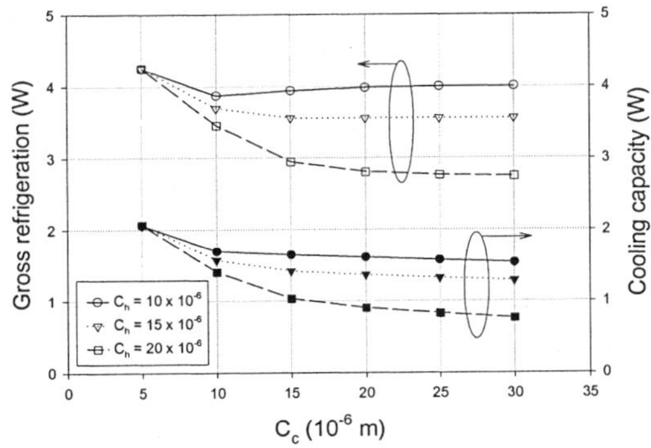

Figure 4   Gross refrigeration and cooling capacity vs. radial clearance

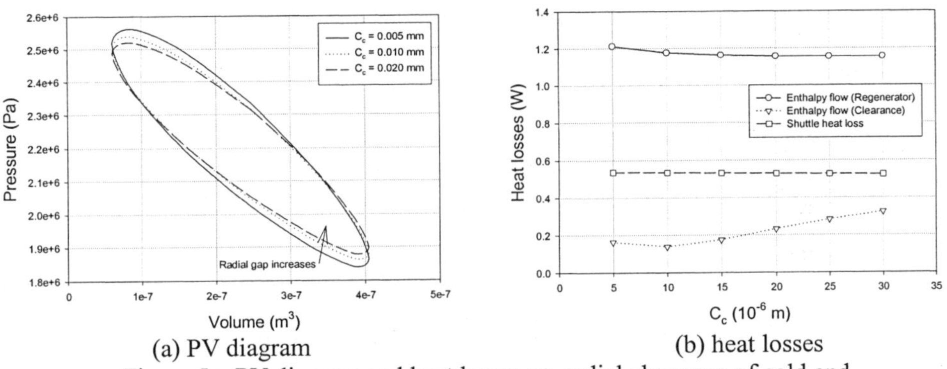

(a) PV diagram   (b) heat losses

Figure 5   PV diagram and heat losses vs. radial clearance of cold end

Figure 4 show the calculation results of gross refrigeration and cooling capacity when the radial clearances of hot end (C_h) are 0.01, 0.015, 0.02 mm as function of different radial clearance of cold end (C_c). The results show the cooling capacity is maximized when the C_c is small. The cooling capacity decreases as the C_c increases. At the given C_c, the gross refrigeration and cooling capacity increase as the C_h decreases. The influence of the C_h on the cooling capacity is small when the C_c has a small value.

Figure 5 show the PV diagram of expansion space, the enthalpy flow through the regenerator and the losses from clearance when the the C_h is 0.015 mm as a function of the C_c. It is obvious that the larger C_c result in the smaller amplitude of pressure in the expansion space due to the leakage flow through the clearance. The enthalpy flow through the regenerator decreases as the C_c increases, whereas the enthalpy flow through the clearance increases. So the increase of the enthalpy flow through the clearance leads to the decrease of the cooling capacity when the C_c increases.

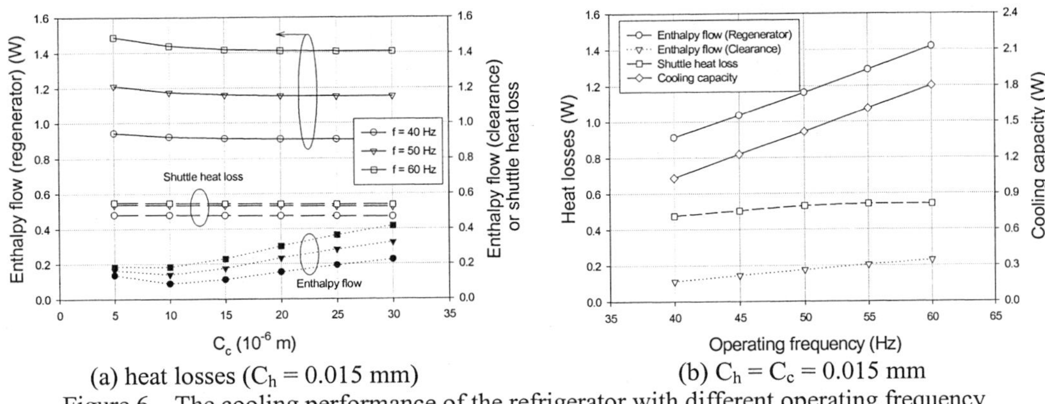

(a) heat losses (C_h = 0.015 mm)   (b) C_h = C_c = 0.015 mm

Figure 6   The cooling performance of the refrigerator with different operating frequency

144

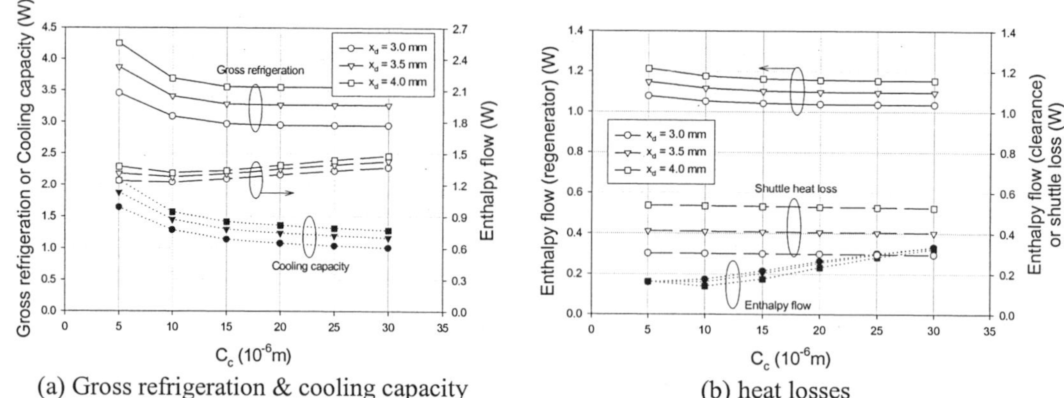

(a) Gross refrigeration & cooling capacity   (b) heat losses

Figure 7   The cooling performance of the refrigerator with different stroke

Figure 6 show the cooling performance of the refrigerator with different operating frequency(f) and C_c when the stroke of the displacer is 4 mm and the C_h is 0.015 mm. As shown as Figure 6, the enthalpy flow through the regenerator is proportional to the operating frequency. The enthalpy flow through the clearance doe not have a great role in the losses when the operating frequency is low and C_c is small.

Figure 7 show the cooling performance of the refrigerator with different stroke of displacer(x_d) and C_c when the operating frequency is 50 Hz and the C_h is 0.015 mm. At the high x_d, the increases of the cooling capacity are relative small than the increases of the gross refrigeration due to the larger shuttle heat loss. As shown as Figure 7, when x_d is small and C_c is large, the amount of the enthalpy flow through the clearance exceed the shuttle loss.

It is clear that the enthalpy flow through the clearance depends on the operating frequency and the radial clearance of the cold end, whereas the shuttle loss depends on the stroke of the displacer. The leakage flow through the clearance has the effects on the cooling capacity through the gross refrigeration and the enthalpy flow.

SUMMARY

In this study, the effects of the clearance in the cold finger on the cooling performance of the Stirling refrigerator were investigated by analysis. The SAGE was used to investigate the effects of the operating frequency, the stroke of the displacer and the radial clearance between the seal rings and the expander on the cooling performance. The calculation results show that the enthalpy flow through the refrigerator is a dominant heat loss. The leakage flow through the clearance has the effects on the cooling capacity through the gross refrigeration, the enthalpy flow and the shuttle heat loss. The enthalpy flow through the clearance depends on the operating frequency and the radial clearance of the cold end, whereas the shuttle loss depends on the stroke of the displacer.

ACKNOWLEDGEMENT

This work was carried out under "Dual use technology program" with Nexen as industrial partners.

REFERENCE

1. Rios PA, An approximate solution to the shuttle heat transfer losses in a reciprocating machine, <u>ASME J Engrg Power</u> (1971) <u>4</u> 177-182
2. Ho-Myung Chang, Dae-Jong Park, Sangkwon Jeong, Effect of gap flow on shuttle heat transfer, <u>Cryogenics</u> (2000) <u>40</u> 159-166
3. G. Walker, <u>Miniature refrigerators for cryogenic sensors and cold enectronics</u>, Clarendon Press(1989)
4. David Gedeon, <u>Sage User's Guide</u>, Gedeon Associates (1999)

Development of a Stirling cryocooler with linear guidway supports

Hong G., Li Q., Li Q., Xi Y., Nie Z.

Technical Institute of Physics and Chemistry, CAS, P.O.Box 2711, Beijing 100080

The MTTF of a rotary Stirling cryocooler is usually 2500 hours and hardly achieve longer beyond that due to the wear characters of the piston ring materials. In this paper, a rotary cryocooler is introduced which utilizes inexpensive linear guidways to support the moving parts of the cooler. This transfers the sliding wears between the piston and cylinder or between the displacer and cold cylinder to rolling friction on the linear guidways, and the piston and displacer get less wear. Thus, the cooler can operate with higher reliability and longer lifetime.

INTRODUCTION

It seems that rotary Stirling cryocoolers have some disadvantages in the applications and are being buried in oblivion. But don't forget their advantages any other Stirling coolers could not be comparable. For instance, the displacements and phase angle of the piston and displacer of rotary Stirling coolers are fixed due to the crankcase mechanism, so its kinetics is not influenced by the thermodynamics (there is no feedback from the thermal features in rotary cooler), and the cooler has much more steady thermal performances at this point. In addition, the rotary motor for the cooler, which is mature merchant goods, can be easily gotten in market. The main shortage of the rotary cooler is that it has larger lateral force of the piston (or displacer) against their cylinders, and the serious wear between them restricts the lifetime of the cooler. The MTTF of a rotary cooler is usually 2500 hours and hardly achieve longer beyond that due to the wear characters of the piston ring materials. In order to take their advantages, avoid their shortages to the maximum limitation and keep their low cost, we suggested to utilize inexpensive linear guidways to support the moving parts of the rotary cooler. This method transfers the sliding wear between piston and cylinder, which mainly limits the cooler lifetime, to rolling friction on the linear guidways, and reduction or elimination of the wear between piston and cylinder can be achieved. Thus, the cooler can operate with higher reliability and longer lifetime.

DESIGN PHILOSOPHY

As the piston and displacer of a rotary Stirling cooler are driven by the crankcase mechanism, the displacements of the piston and displacer and the phase angle between them are invariable while the cooler operates. The kinetics of the cooler is not influenced by the thermal performance (there is no feedback from the thermal features in rotary cooler), and the cooler has much more steady thermal performances at this point. Moreover, the rotary motor used in the cooler is commercial product that makes the cooler easy for industrialization. Therefore, the rotary Stirling cryocooler was selected. But the piston in this type Stirling cooler has much larger vertical force against the cylinder wall while the piston moves reciprocally inside the cylinder. So does the displacer in the cold cylinder. This leads to more wear between piston and cylinder or between displacer and cold cylinder, and shorter lifetime of the cooler. In order to take the advantages and avoid the shortages of the rotary Stirling cooler, linear guidways are used to support the piston and displacer, keeping the clearance uniform between piston and cylinder or between displacer and cold cylinder. In this way, the sliding wear between piston and cylinder or between displacer and cold cylinder is transferred to rolling wear on the linear guidway bearings and is nearly eliminated. So the cooler can get higher reliability and longer operating time.

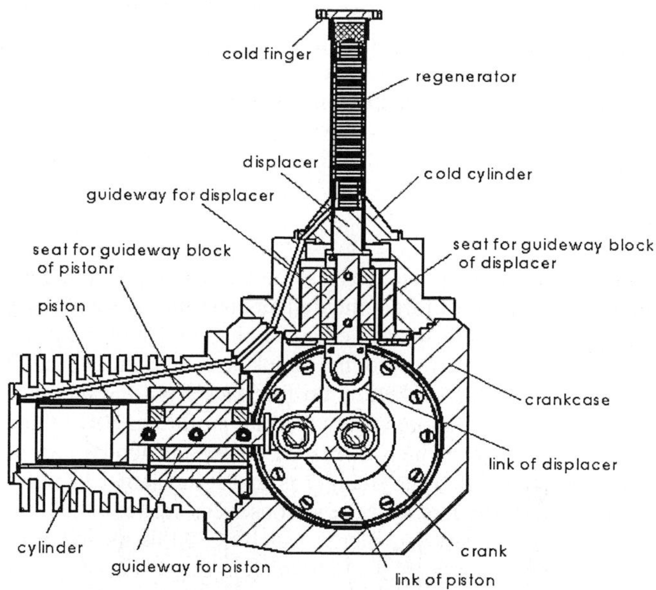

Figure 1   Stirling cryocooler with linear guideways supports

As shown in Figure 1, the cooler consists of a compressor, a cold head, a rotary motor and a crankcase mechanism with a certain pressure helium gas inside. The compressor, cold head and rotary motor are set in perpendicular to each other. The piston of the compressor is driven by the rotary motor through the cranks and get reciprocating movement inside the cylinder, while the displacer of the cold head is also driven by the same rotary motor through the same cranks and get reciprocating movement inside the cold cylinder. The movement phase angle between the piston and displacer is 90° with the displacer leading the piston.

A linear guidway (Figure 2, Figure 3) makes up of a guide-way, a guide-block and ball bearings. It is installed between the piston and cylinder with the guide-way fixed to the piston and the guid-block to a seat which is mounted on the cylinder flange. Due to the linear guidway, the radial position of the piston is kept un-deviational while it moves axially in the cylinder, and the wear between the surfaces of the piston and cylinder is reduced or eliminated.

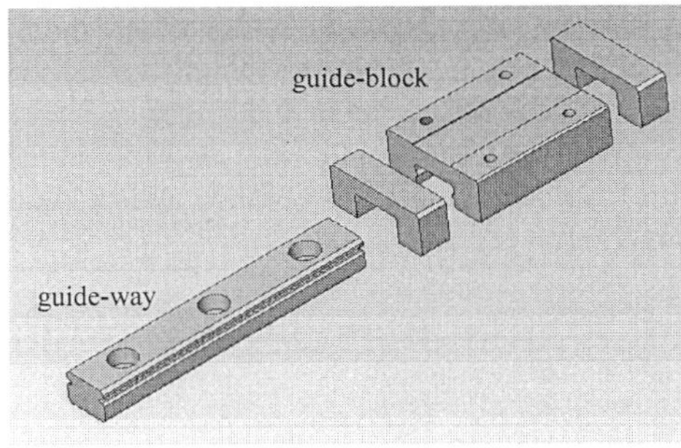

Figure 2   Linear guideway used in the cooler

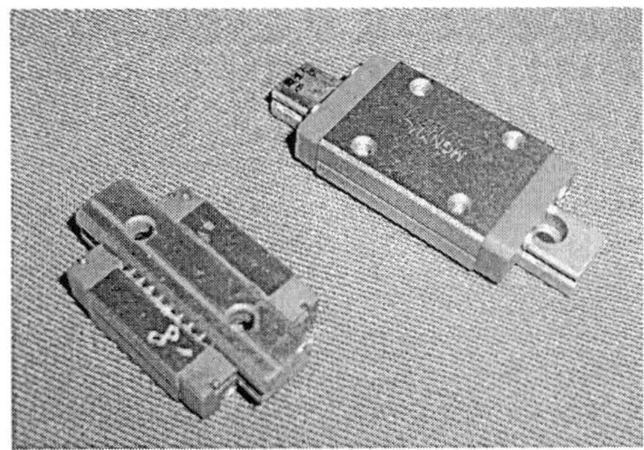

Figure 3   Photograph of the linear guideways

There is also a linear guidway installed between the displacer and cold cylinder. The guide-way is fixed to the displacer and the guide-block to a seat which is mounted on the cold cylinder flange. Due to the linear guidway, the radial position of the displacer is kept un-deviational while it moves axially in the cylinder, and the wear between the surfaces of the displacer and cold cylinder is reduced or eliminated.

In this cooler, clearance seals are applied. The clearance is about 10 μm between the piston and cylinder and between the displacer and cold cylinder. And sorbent are also used for keeping the high pure helium gas inside the cooler chamber. The total weight of the cooler is 2.5 kg. Figure 4 is a photograph of the cooler.

Figure 4   Photograph of the cooler

TEST RESULTS

148

A preliminary test on the cooler has been taken. Multiple tests on the cooler have been performed with different displacements of piston and displacer (actually different cranks and links). But due to the problem of manufacturing of the cooler, the tests showed that the cooler got a bad performance. For instance, the lowest temperature of the cold finger was 10 K higher than last test while all the test conditions keep same. That is likely the results of reduction of displacements of the piston and displacer due to the loose (clearance) fittings between the cranks (which was made for easy change of different cranks and links).

Figure 5 shows one of the test results. The lowest cooling temperature is 60 K and the total input power is 53 W with 7 mm and 4 mm displacements of the piston and displacer respectively. The rotating speed of the motor is 2100 turns per minutes.

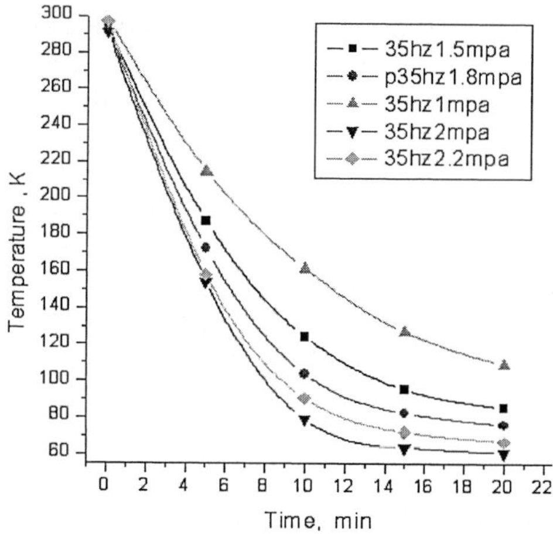

Figure 5   Cooling performance of the cooler

CONCLUSIONS

The purpose of applying linear guidways as the supports to moving piston and displacer of a rotary Stirling cryocooler is to keep the radial position of the piston and displacer while they reciprocate in the axial direction. Thus the rub or wear between piston and cylinder or between displacer and cold cylinder is reduced and eliminated, and the lifetime and reliability of the cooler are improved. In this method, a low cost Stirling cryocooler with more than 5000 hours (expected) lifetime was developed and some preliminary test data were gained. The cooling performance of the cooler is not satisfying due to the loose fitting of the cranks. The final test on the re-manufactured cooler will be taken later. Also a test equipment for confirming the wear lifetime of the linear guideways will also be established and tested. A future paper will describe the test results.

Novel regenerator design and its experimental characteristics

Kwanwoo Nam and Sangkwon Jeong

Cryogenic Engineering Laboratory, Korea Advanced Institute of Science and Technology
Guseong-dong, Yuseong-gu, Daejeon, Republic of Korea

This paper discusses a new regenerator with parallel wires and its experimental characteristics. Pressure drop characteristics of the parallel wire regenerator were compared to that of the screen mesh regenerator. It was found that the friction factor of the parallel wire type was three to five times smaller than that of the screen mesh type. Ineffectiveness was determined by measuring the instantaneous pressure, the flow rate and the gas temperature at the warm and cold sides of the regenerator. Parallel wire regenerator made of stainless steel showed poor thermal performance due to its excessive axial conduction loss. To reduce the axial conduction loss, segmentation of the regenerator was suggested and the ineffectiveness result was presented for the segmented regenerator.

INTRODUCTION

A regenerator with parallel heat transfer components to the oscillating flow theoretically has a better performance than screen mesh or random wire type regenerators that are commonly used in cryocoolers. Advantages of the parallel geometry are the small friction factor and the low void fraction. In reality, however, parallel geometry regenerators have been easily suffered from flow maldistribution and axial conduction. Several previous researchers have attempted to realize the parallel geometry type regenerators [1], [2], [3], [4]. Some of them failed to obtain a desired performance and most of the previous studies showed the refrigerator performance, which is an indirect method to measure the regenerator characteristics. In this paper, parallel wire geometry which is similar to the parallel plate or tube was suggested for a regenerator and series of experiments were carried out to characterize the regenerator directly. Figure 1 shows a conceptual diagram of the parallel wire regenerator. Bundle of fine wires are tightly fixed in a housing and gas flows through the void space formed by wires. This paper describes the detailed fabrication process and the direct measurement results of the parallel wire regenerator.

FABRICATION PROCEDURE

Figure 2 shows the fabrication procedure of the regenerator with parallel wire bundle. The detailed description for each fabrication step is as follows.
(a) : Fine stainless steel wire was wound around in a house-made reel. Wire

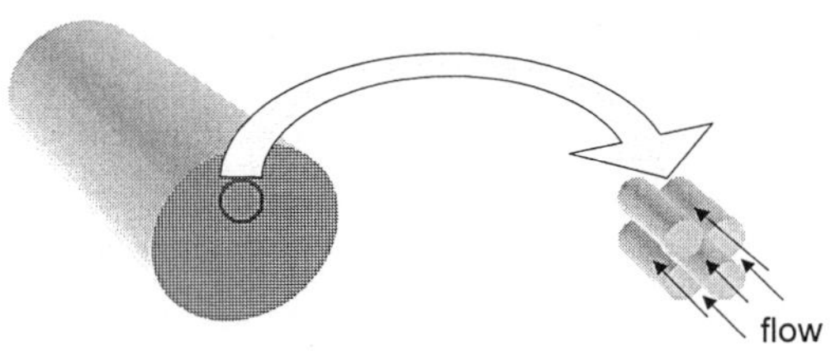

Figure 1 Schematic diagram of parallel wire regenerator

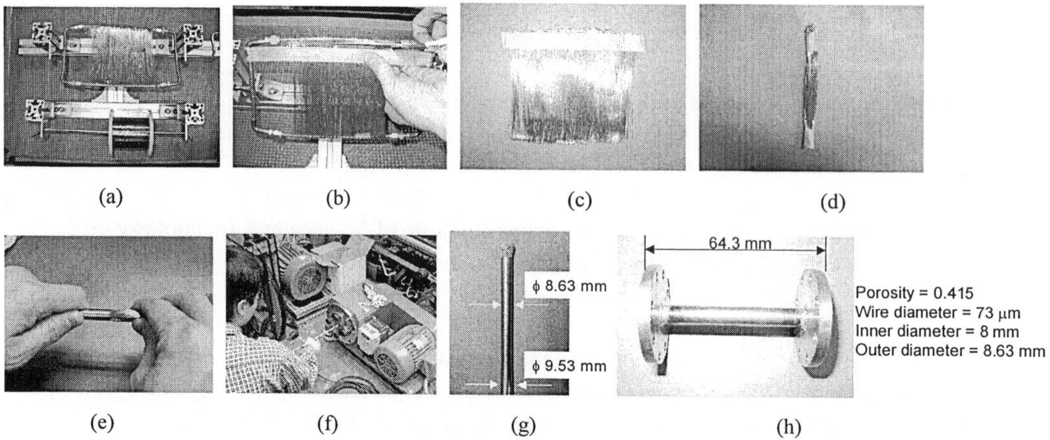

Figure 2 Fabrication procedure for the parallel wire regenerator

diameter was equal to 73 μm. It was not necessary to align wires each other but it was better to wind all wires tightly.

(b) : After the wire bundle was fixed with a sticky tape, the wire bundle was cut off with scissors.

(c) and (d) : Flat bundle was rolled round and the end of the bundle was fixed with the tape.

(e) : Round bundle was inserted into the tube.

(f) and (g) : The tube with the wire bundle was put into a swaging machine in order to reduce the diameter of the tube. This process made the porosity become small and the wire bundle be stuck in the tube.

(h) : The ends of the swaged tube were cut away with a wire-EDM(Electro Discharge Machine). Then, the wire bundle inside the tube was cleaned up carefully. Finally, flanges were attached at the ends of the regenerator to connect the sample with the experimental setup.

EXPERIMENTAL RESULTS

Friction factor

After a prototype regenerator was fabricated, the friction factor was measured at steady flow condition. The Fanning friction factor is defined as follows.

$$f_F \triangleq \frac{\Delta P d_h}{2 \rho u^2 L_r}, \quad u = \frac{\dot{m}}{\rho A_g}, \quad d_h = \frac{e_v d_w}{1 - e_v} \qquad (1)$$

where A_g : free flow cross sectional area of the regenerator, e_v : porosity, d_w : wire diameter, L_r : regenerator length. Pressure drop and flow rate were measured by a variable reluctance type sensor (Validyne model DP-10) and a mass flow controller (Bronkhorst model F-113AC), respectively. Screen regenerators were also tested to be compared with the parallel wire regenerator. Tested samples were made of #200, #250 and #400 mesh. Weave style was a twill type and the porosity was about 0.68. The friction factor of the

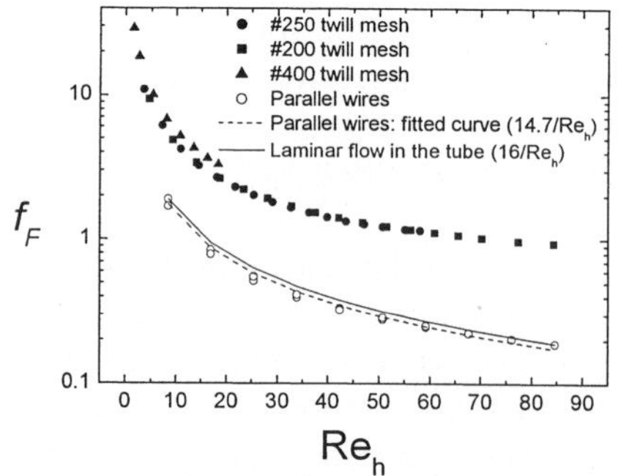

Figure 3 Steady flow friction factor

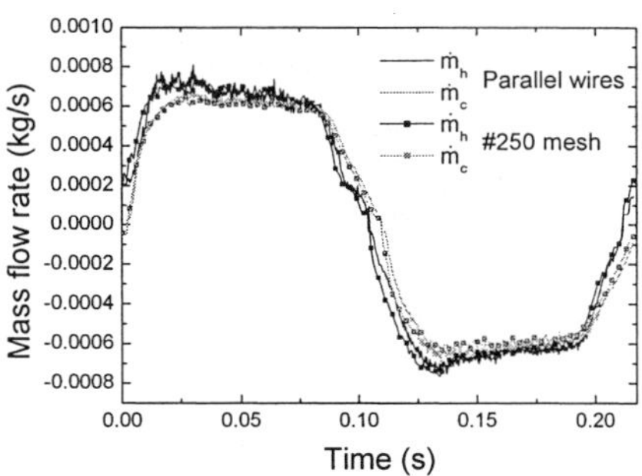
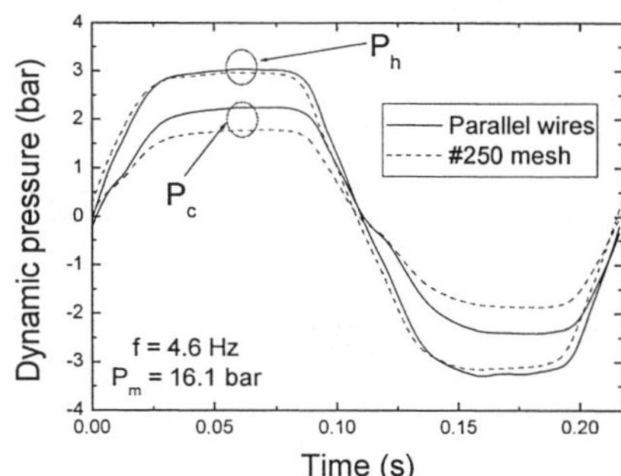

Figure 4 Operating conditions for ineffectiveness measurement (T_h = 277.1 K, T_c = 106.9 K for the parallel wire regenerator and T_h = 287.9 K, T_c = 93.3 K for the #250 mesh regenerator)

parallel wire regenerator was three to five times smaller than those of the screen regenerators as shown in Figure 3. Furthermore, the friction factor of the parallel wire geometry was very similar to the simple analytical result calculated for fully developed laminar flow in a tube [5].

Ineffectiveness
Ineffectiveness is a quantitative performance factor for thermal efficiency of cryocooler regenerator. It is defined as the following equation [6].

$$\lambda \triangleq \left(\int_0^\tau \dot{m}hdt \right)_{cold-end} \Bigg/ \left(\left(\int_0^{\tau_1} \dot{m}hdt \right)_{warm-end} + \left(\int_{\tau_1}^\tau \dot{m}hdt \right)_{cold-end} \right) \tag{2}$$

where h : enthalpy of gas, the time interval from 0 to τ_1 : the duration of warm gas flow period, the time interval from τ_1 to τ : the duration of cold gas flow period. The instantaneous mass flow and the gas temperature were measured to calculate the ineffectiveness under actual operating conditions of cryocooler. Full explanation of the experimental setup was presented elsewhere [7]. Testing conditions of regenerators are shown in Figure 4. The #250 mesh regenerator was also tested and its total heat transfer area was the same as that of the parallel wire regenerator. The geometric properties and the experimental results are summarized in Table 1. It was revealed that the ineffectiveness of the parallel wire regenerator was larger than that of the screen regenerator in the operating condition of the experiment. This phenomenon was mainly due to excessive axial conduction in the parallel wire regenerator. Estimated conduction loss was approximately 0.74 W for the parallel wire type and about 0.04 W for the #250 screen. It was assumed that the conduction degradation factor for the packed screens was 0.1. From the ineffectiveness measurement result, we concluded that it was absolutely necessary to reduce the axial conduction loss of the parallel wire regenerator. One of the simplest methods is to segment the regenerator as described in the next section.

SEGMENTATION

Figure 5 shows the schematic of the segmented regenerator. The parallel wire regenerator was

Table 1 Geometric properties and experimental results for the screen and the parallel wire regenerator

	Parallel wires	#250 twill
Inner diameter (mm)	8	8
Length (mm)	64.3	64.3
Wire diameter (mm)	0.073	0.038
Porosity	0.415	0.68
Heat transfer area (m²)	0.108	0.108
Max. pressure drop (bar)	0.99	1.41
λ (%)	3.1	1.4

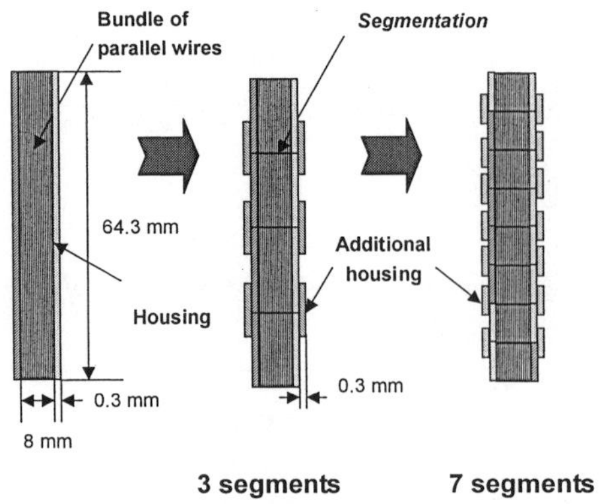

Figure 5 Schematic diagram of the segmented parallel wire regenerator

configured into three and seven-segmented regenerator using a wire-EDM cut. The segmentation of the wire bundle results in thermal contact resistance in axial direction, which presumably reduces the axial conduction thermal loss of regenerator. Ineffectiveness was measured for two types of the segmented regenerators. Operating conditions were similar to the case of the non-segmented regenerator (Figure 4). Ineffectiveness of the three-segmented and the seven-segmented regenerator are 2.7% and 2.2%, respectively. This experimental result indicates that the segmentation clearly improves the thermal performance of the parallel wire regenerator.

SUMMARY

A parallel wire regenerator was conceived and its tested performance data was presented. The friction factor of the parallel wire regenerator was only 20 to 30 % of that of screen mesh regenerator, but the ineffectiveness was more than doubled in a typical operating condition of cryocooler. The poor thermal characteristic of the parallel wire regenerator was alleviated by segmentation to decrease axial conduction loss.

ACKNOWLEDGEMENTS

This research was supported by a grant from Center for Applied Superconductivity Technology of the 21st Century Frontier R&D program funded by the Ministry of Science and Technology and Brain Korea 21 project, Republic of Korea.

REFERENCES

1. Yaron, R., Shokralla, S., Yuan, J., Bradley, P.E. and Radebaugh, R., Etched foil regenerator, Advances in Cryogenic Engineering (1996), 41 1339-1346
2. Rawlins, W.C. and Timmerhaus, K.D., Measurement of the performance of a spiral wound polyimide regenerator in a pulse tube refrigerator, Advances in Cryogenic Engineering (1992), 37B 947-953
3. Hofmann, A., Wild, S. and Oellrich, L.R., Parallel flow regenerator for pulse tube cooler application, Advances in Cryogenic Engineering (1998), 43 1627-1633
4. Chafe, J.N., Green, G.F. and Hendricks, J.B., A Neodymium plate regenerator for low-temperature Gifford-McMahon refrigerators, Cryocoolers 9 (1997), 653-662
5. White F. M., Fluid Mechanics, McGraw Hill, New York, USA (1994) 307-328
6. Radebaugh, R., Linenberger, D. and Voth, R.O., Methods for the Measurement of Regenerator Ineffectiveness, NBS Special Publication (1981), 607 70-81
7. Nam, K. and Jeong, S., Measurement of cryogenic regenerator characteristics under oscillating flow and pulsating pressure, Cryogenics (2003) 43 575-581

Investigation of a single-stage G-M refrigerator with high cooling capacity for HTS devices

Fang Z-C.[1,2], Su X-T.[1,2], Liu L-Q.[1], Gong L-H.[1], Zhang L.[1]

[1]Technical Institute of Physics and Chemistry, Chinese Academy of Sciences, Beijing, 100080, China
[2]Graduate School of Chinese Academy of Sciences, Beijing, 100039,China

Because the cooling of HTS device needs the cryogenic refrigerator working at 30 K, and the investigation of HTS devices are more and more considerable, the requirement for the G-M refrigerator with the cooling capacity of 50 W/30 K increases gradually. But there is no any product of the G-M refrigerator that can meet the request of cooling HTS device. In the paper we summarized the experimental work of single-stage G-M refrigerators. At present, we can get the lowest temperature of 17 K operating at 1 Hz. The cooling capacity of 27 W has been obtained at 30 K.

INTRODUCTION

In order to operate high temperature superconducting (HTS) applications under high magnetic field, liquid nitrogen is not sufficient since the temperature between 20~30 K has to be maintained continuously in order to achieve the goal of super-conduction [1,2]. Since conventional single stage G-M refrigerators cannot reach the temperature below 20 K, they cannot provide the high cooling capacity at 30 K. At the same time, the conventional double stage G-M refrigerators have a lowest temperature of the second stage, about 7 K. However, they are not efficient in the temperature range of 30~50 K. Therefore the conventional single and double stage G-M refrigerators cannot achieve the goal of cooling HTS devices. Now we are developing the single-stage G-M refrigerators that can supply the high capacity, high efficiency, and 50 W/30 K cooling requirement.

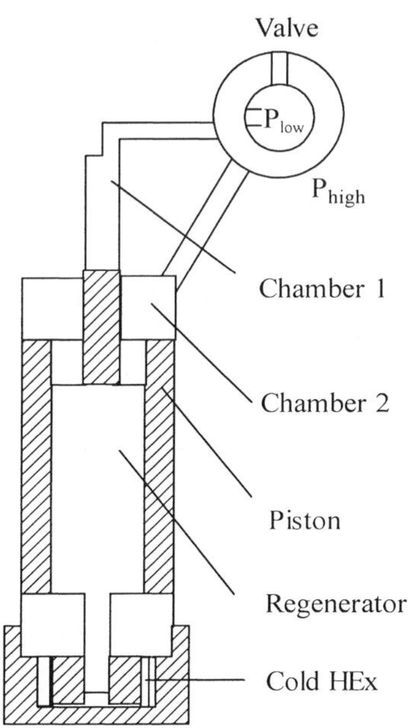

Figure 1   Schematic diagram of the G-M refrigerator

DESIGN PRINCIPLE OF THE SINGLE STAGE G-M REFRIGERATOR

The movement of the displacer is achieved by an unbalance of pressures in the variable chambers, resulting from the delivery and exhausting of gas through the rotary valve. The timing of the movement is achieved by the rotation of the valve disc on the stationary valve plate. Depending on the location of the valve disc, the supply pressure to chambers 1 and chamber 2 may be high or low [3]. Depending on the angle difference of the valve disc, cooling capacity and the bottom temperature of the cold head will change with the timing of the exhausting and delivery. Figure 1 shows the schematic diagram of the G-M refrigerator.

The theoretical calculation of the heat-loss is listed below, as Table 1 shows.

Table 1   Analysis of heat-loss of single-stage G-M cycle

List	Parameter	Calculation value （W）	Percentage of supplied
1	Theoretical Cooling power	175.8	
2	Actual Cooling power	—	
3	Loss		
3.1	P-V Loss	—	~40%
3.2	Conduction Loss	9.68	5.5%
	(a) Cylinder	(6.99)	—
	(b) Piston	(0.34)	—
	(c)Regenerative material	(2.35)	—
3.3	Motional heat loss	38.5	21.9%
	(a) Shuttle loss	(20.9)	—
	(b) Pump loss	(17.6)	—
3.4	Regenerative loss	13.6	7.7%
3.5	Radiation loss	0.209	1.2%

DEVELOPING THE SINGLE STAGE G-M REFRIGERATOR FOR HTS DEVICES

With the help of numerical simulation of single stage G-M refrigerator, we have assembled a single stage G-M refrigerator for HTS devices. The results of experiments and analysis of the results will be presented below, which will give us a way to develop a more efficient G-M refrigerator.

<u>Influence of stroke on the capability of refrigerator</u>
The ideal theoretical cooling capacities of each cycle were associated with the stroke, which is the main factor of cold chamber volume.

Both the bottom temperature of cold head and the cooling capacities depend on the displacer stroke. The lower temperature, 26.6 K, was obtained with 25 mm stroke, while 32.6 K of bottom temperature with 20 mm stroke. The lowest temperature may be limited by the regenerator efficiency and shuttle loss. The E1, E2 represents the corresponding experimental results of 20 mm and 25 mm displacer stroke. Table 2 gives us the results of the experiments E1, E2.

Table 2   The bottom temperature and drop speed of E1、E2

Experiment	Drop speed （K/min）	Bottom Temperature (K)
E1	8	32.6
E2	9	26.6

<u>Influence of heat exchanger on the capability of refrigerator</u>
The development of heat exchanger or cold head is the most important part. How to transfer the heat efficiently is the key point of the refrigerator. At first, we utilize the bronze screen to increase the heat

transfer area. Figure 1 may present us some schematic diagram of cold head exchanger. Later, we use the conventional heat exchanger. F1, F2 indicate the results of different kinds of cold heat exchanger.

Table 3   Results of experiments (F1、F2)

Experiment	Drop speed (K/min)	Bottom temperature (K)	P_h/P_1 (MPa/MPa)
F1	8	26.6	2.4/0.62=3.9
F2	9	17.7	2.4/0.45=5.3

We analyze the results and list the reasons as follows:

The former design has the character of large heat transfer area and high heat exchange efficiency. At same time the pressure loss was very big.

The conventional design makes good use of the narrow channels to pass through the He medium. The flow resistance and the empty volume loss are very small. But both the heat transfer area and coefficient cannot supply the efficient heat transfer.

Influence of different rotation frequency on the capability of refrigerator

The rotation frequency is associated with the moving way of displacer and intake and exhaust of high pressure and lower pressure gas. Meanwhile much conventional loss is also laid on the moving way, such as the shuttle loss, pump loss. Figure 2 and Figure 3 presented us the bottom temperature and cooling power of cold head at different frequency. Adjusting rotation frequency actually changed the timing of intake and exhaust.

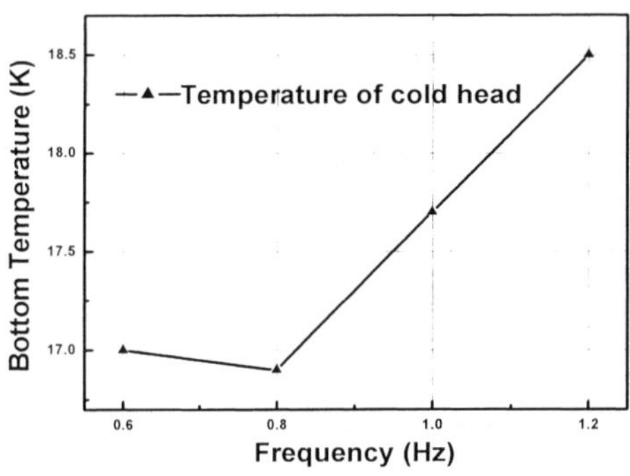

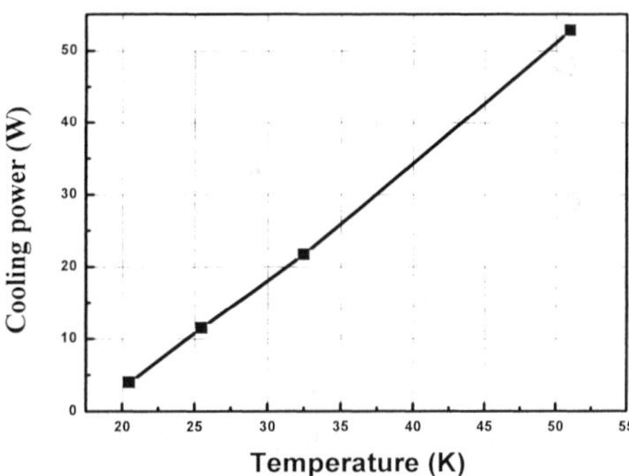

Figure 2   Bottom temperature at different rotation frequency   Figure 3   The cooling power of cold head

Influence of ratio of regeneration to cold chamber volume on the capability of refrigerator

The results of numerical simulation show us the relation between the loss of regenerator and the ratio of Vre/Vo. See Figure 4 below.

At different ratio of Vre/Vo, we can almost get the same bottom temperature of cold head. When Vre/Vo equates 5.1, the bottom temperature is 17.7 K while 18.0 K at 6.2. At fact, we can make the conclusion that the volume of regeneration is sufficient.

Influence of ratio of regenerative material on the capability of refrigerator

For the different specific heat of regenerative material, the ratio of regenerative material may influence the capability of regenerator. Figure 5 gives the curve of the loss of regenerator at different ratio of Lcu/Lregeneraor. Lcu and Lregeneraor represent the length of Cu regenerative material and the length of the whole regenerator, respectively. Table 4 shows us the results at different ratio of L_{Cu}/L_{Re}.

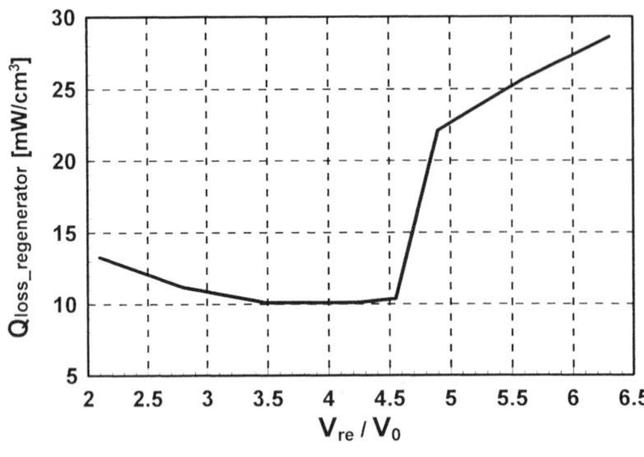

Figure 4   Loss of regenerator with the ratio of Vre/Vo

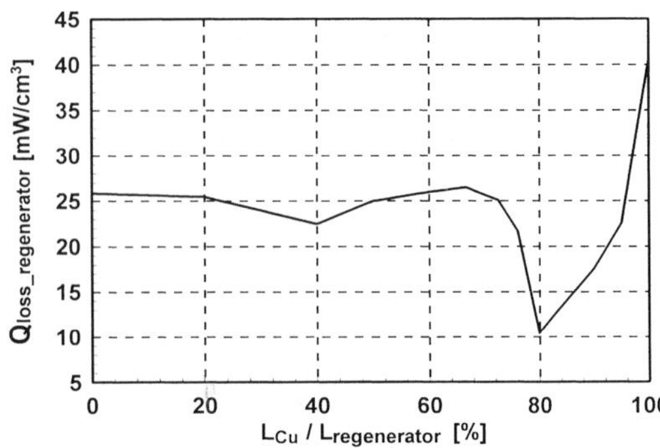

Figure 5   Loss of regenerator with the ratio of Lcu/Lre

We can find that the capacity of refrigerator varied with the ratios of Lcu/Lre. The temperature field of the regeneration was related with these two regenerative materials for the different specific heat. 70% may be the optimum value of Lcu/Lre through experiments.

Table 4   Results of experiment at different ratio value of L_{Cu}/L_{Re}

Experiment	L_{Cu}/L_{Re}	Drop speed (K/min)	Bottom temperature (K)	Ph/Pl (MPa/MPa)
H1	0.785	9	26.6	2.3/0.6=3.9(ZC40)
H2	0.70	8	17.7	2.4/0.45=5.3(ZC60)
H3	0.73	8	18.1	2.35/0.44=5.3(ZC60)

Note: ZC40/ZC60 represent the type of compressor

CONCLUSIONS

The single stage G-M refrigerator with high reliability, high capacity and high efficiency can meet 30 K cooling requirements of many of the new HTS devices. We can get the cooling power of 27 W at the 30 K and 17 K of the bottom temperature. The heat exchanger is the key part of the refrigerator in the future. We will show new results of single stage G-M refrigerator for HTS devices.

REFERENCES

1.   Fiedler, A., Gerban, J. and Haefner, H.U., "Efficient Single Stage Gifford-MacMahon Refrigerator Operating at 20K", Advances in Cryogenic Engineering, Plenum Press, New York (1997) 43B, 1823-1830

2.   R.C.Riedy. "Low temperature, high performance G-M refrigerator", Cryogenics (1993) 33,653-658

3.   Wang C. "Advanced Cryocooler Developments at Cryomech and Their Application", In: Proceedings of Beijing International Conferences on Cryogenics. Beijing, China (2000) 105-112

Preliminary study on GM refrigerator with heat exchanger

Liang W., Gong L. and Zhang L.

Technical Institute of Physics and Chemistry, Chinese Academy of Sciences, Beijing 100080, China

A new type of GM refrigerator with a heat exchanger has been in preliminary study in which two same GM refrigerators work in opposite phase. A heat exchanger will be used to replace the regenerators and a special rotary valve working at low temperature will be employed to generate pressure wave. The motion of the piston is analyzed to study the relations of the theoretical cooling power (TCP) and other parameters. All losses are ignored in the calculation.

INTRODUCTION

In this research two identical GM refrigerators work in the opposite phases when one is in the intake period while another is in the discharging period. The gases from different GM refrigerators exchange cooling power in the heat exchanger in which there is no AC-Flow and void volume. This design can avoid the losses caused by the AC-Flow and the void volume and hope to improve the thermal efficiency of GM refrigerator. The thermal efficiency of the heat exchanger is a key to this type of GM refrigerator.

Also a rotary valve is employed to generate the pressure wave. The friction heat of the valve can cause the loss of the cooling power at low temperature. It is a disadvantage to use such a rotary valve contrasted with the traditional GM refrigerator.

STRUCTURE OF THE NEW TYPE OF REFRIGERATOR

As shown in Fig.1, the new type of refrigerator has one heat exchanger, one rotary valve and two identical GM cylinders. The rotary valve works at the low temperature so that the gas flows in the heat exchanger are direct. This design is used to avoid an AC-Flow heat exchanger.

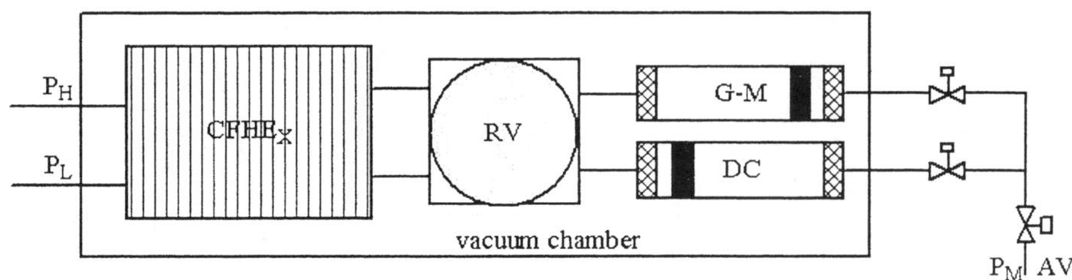

Figure 1 Brief structure of the new type of the refrigerator

THE ANALYSIS OF TCP THE AND MOTION OF THE PISTON

The motion of piston has great influence on the mass wave and TCP. The forces added on the piston are analyzed on the base of the following assumptions.

158

1. The pressure in the cylinder is:

$$P_{\mathrm{H}} = 7.0 + 7.0 \times (1 + \sin\theta) \quad (\text{bar}) \tag{1}$$

2. The gases in the cylinder and the damping chamber (DC) are ideal.

3. P_{M} is constant.

4. The flowing mass between the damping chamber is determined by using equations (2) and (3):

$$m_0 = \mu A_0 \sqrt{2\frac{\kappa}{\kappa-1}\frac{P}{\upsilon}\left[\left(\frac{P_r}{P}\right)^{2/\kappa} - \left(\frac{P_r}{P}\right)^{(\kappa+1)/\kappa}\right]} \quad (P \geq P_r) \tag{2}$$

$$m_0 = -\mu A_0 \sqrt{2\frac{\kappa}{\kappa-1}\frac{P_r}{\upsilon_r}\left[\left(\frac{P}{P_r}\right)^{2/\kappa} - \left(\frac{P}{P_r}\right)^{(\kappa+1)/\kappa}\right]} \quad (P \leq P_r) \tag{3}$$

5. The mass and the pressure of the gas in the damping chamber are calculated using the equation (4):

$$\frac{PV^\kappa}{m^\kappa} = const \tag{4}$$

6. The mass of the piston is defined as: $M = 0.5$ （Kg）
7. The diameter is defined at: $D = 26$ (mm)
8. The stroke of piston is: $S = 30.0$ (mm)
9. The motion of the piston is determined by the equation (5):

$$M\frac{\partial L^2}{\partial^2 t} = \sum_{i=1}^{n} F_i \quad \left.\frac{\partial L}{\partial t}\right|_{L=0,S} = 0 \tag{5}$$

The motion of the piston will be calculated by solving the equations (1)(2)(3)(4)(5). The motion is mainly influenced by P_{M} and D_0 (the opening of the adjusted valve (AV)). Figures 2 and 3 show the pressure wave and the mass wave in the damping chamber (DC). Figure 5 shows the motion of the piston. The influence of the friction on the TCP is shown in the Figure 6. In the Figure 7 the relations between the P_{M}, D_0 and the TCP are shown and the optimal results are got when P_{M} is 14.5 bar and D_0 is 0.6 mm. Figures 2,3,4,5 are got when P_{M} is 14.5 bar and D_0 is 0.6 mm.

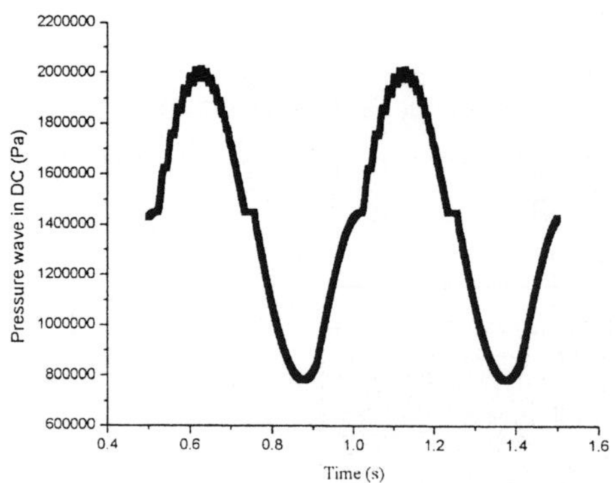

Figure 2 The pressure wave in DC

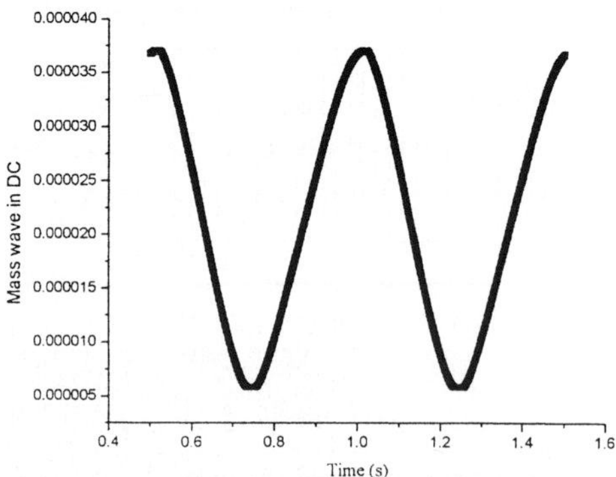

Figure 3 The mass wave in DC

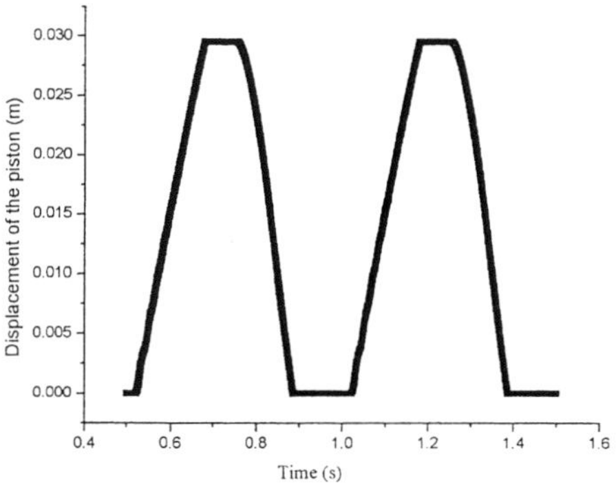

Figure 5 The motion of the piston

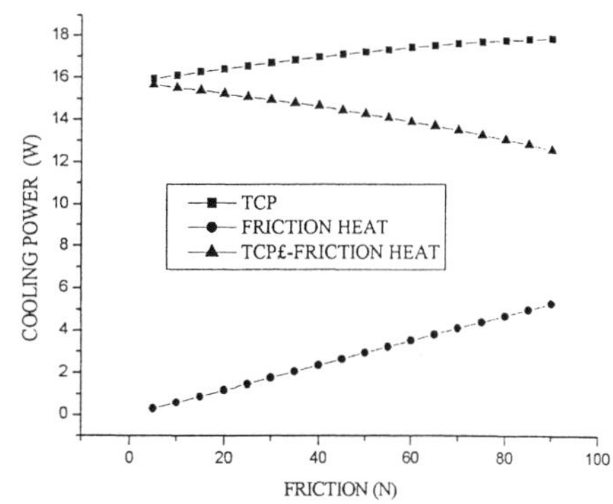

Figure 6 The influence of the friction on TCP

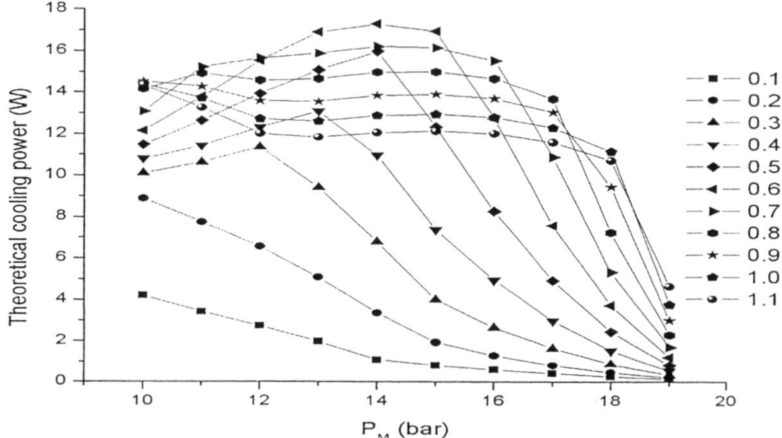

Figure 7 The relations between the P_M, D_0 and the TCP (D_0 in mm)

THE CALCULATION OF THE HEAT EXCHANGER

The structure of the heat exchanger is shown in the Figure 7. According to the calculation its theoretical heat exchanging efficiency is about 98.3%. Fifty pieces of the copper net are sintered to a block. About twenty blocks are made and there is an adiabatic mat between two blocks. The following Table 1 will show the detail parameters.

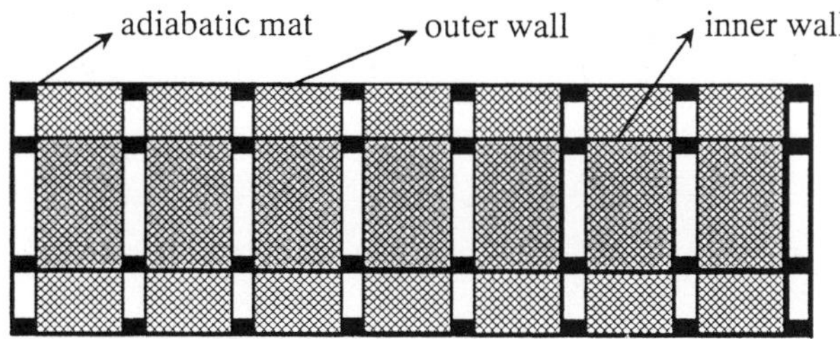

Figure 8. Schematic representation of the heat exchanger

Table 1 The parameters of the heat exchanger

Outer diameter (mm)	Inner diameter (mm)	Length (mm)	Efficiency (%)	Thickness of the mat (mm)
64	43	168	98.3	0.5

CONCLUSION

We have analyzed the relations between the motion of the piston and TCP and get the optimal datum. Two programs are developed to analyze the motion of the piston and the heat exchanger. Most parts of the results are shown in the above figures and table.

ACKNOWLEDGMENT

Thanks for Prof.dr. A.T.A.M. de Waele of TU/e instructing and helpful discussion of this research.

REFERENCES

1.   Will,M.E., Zeegers,J.C.H., De Waele,A.T.A.M., Counter-flow pusle-tube refrigerator, <u>Proceedings of ICEC-19</u>, Narosa Publishing House, New Delhi(2002) 407-410

Dimensional analysis on multi-component mixed-refrigerants flow characteristics in adiabatic capillary tube

Qi Y.F., Gong M.Q., Sun Z.H., Luo E.C., Wu J.F.

Technical Institute of Physics and Chemistry, Chinese Academy of Sciences, Beijing, 100080，China

In multi-component mixed-refrigerants Joule-Thomson cryocooler, the refrigerants flow in the capillary tube is very complicated. Most design methods in the literatures were developed on a theoretical simulation of the two-phase flow through the tube with some simple assumptions such as homogenous two-phase flow and constant friction factor along the tube. Some correlations or methods were valid only for limited refrigerants. In this paper, instead of the complicated mathematical model, a dimensionless correlation is developed to predict the mixed-refrigerants flow characteristics through adiabatic capillary tube. This ease-to-use and fairly accurate method, suitable for hand calculations, can calculate the refrigerants flow rate directly.

INTRODUCTION

Capillary tube is usually used as an expansion and refrigerant flow controlling device in small industry refrigeration systems, air-conditioning systems and household refrigerators because of its simplicity and low cost. It is a long hollow copper with an inside diameter between 0.33 to 4.0 mm and a length from 1 to 6 m. Though a capillary tube itself is very simple, the flow within the tube is quite complicated. The process of refrigerant flow through the capillary tube is a flash process, in which the refrigerant flows at a high speed and changes from liquid to vapor-liquid blend. In practical consideration, main concern is the refrigerants flowing characteristics for matching the system to get the highest performance.

The selection of the proper diameter and length of a capillary tube is important for a given system to run properly. In the most works found in public literatures [1-3], the refrigerants flowing characteristics were analyzed on numerical simulation with an assumption of homogenous two-phase flow. Some other correlations [4-7] were developed based on the Buckingham π-theorem to calculate the flow characteristics, and this ease-to-use and fairly accurate method, suitable for hand calculations, can calculate the refrigerants flow rate directly.

In recent years, there has been a remarkable development of the mixed-refrigerant J-T cryocooler, which cover a large temperature range from below liquid nitrogen temperature (77 K) to 230 K with corresponding mixture refrigerants. There are usually 5 to 7 components in the cryocooler. The mixed-refrigerants flow in the capillary tube is more complicated than pure refrigerants in the refrigeration system. The mixed-refrigerants flow characteristics are very important for the system design. However, the research on mixed-refrigerants flow characteristics is few in the open literatures. In this paper, a dimensionless correlation is developed to predict the mixed-refrigerants flow characteristics through the adiabatic capillary tube based on the experimental data. The correlation can be used directly to give a guide for system design.

DEVELOPMENT OF DIMENSIONLESS CORRELATION

Dimensionless correlation has been developed in order to predict refrigerant mass flow rate in several capillary tubes operating at different conditions. Dimensionless parameters are selected considering the different parameters' effect on flow characteristics based on the Buckingham π-theorem [8].

Dimensionless parameters

The first step of π-theorem was to select parameters that have influence on the mass flow rate through capillary tube.

162

These parameters are the geometric parameters of the capillary tube (inner diameter d and length L), inlet conditions (inlet pressure P_{in}, degree of subcooling T_{sub}, and quality x), outlet condition (outlet pressure P_{out}), and refrigerant properties (specific volume v, viscosity μ of liquid and gas phase, liquid specific heat C_{Pl}, surface tension σ, and latent heat of vaporization h_{lg}). Therefore, mass flow rate can be represented as function of several parameters shown in Equation (1).

$$m = f_1(d, L, P_{in}, P_{out}, T_{sub}, x, v_l, v_g, C_{Pl}, \mu_l, \mu_g, \sigma, h_{lg}) \tag{1}$$

$$\pi_0 = f_2(\pi_1, \pi_2, \pi_3, \pi_4, \pi_5, \pi_6, \pi_7, \pi_8) \tag{2}$$

$$\pi_0 = C \cdot \pi_1^{e1} \cdot \pi_2^{e2} \cdot \pi_3^{e3} \cdot \pi_4^{e4} \cdot \pi_5^{e5} \cdot \pi_6^{e6} \cdot \pi_7^{e7} \cdot \pi_8^{e8} \tag{3}$$

Then the parameters d, v_l, μ_l and C_{Pl} are defined as the repeating variables and nine dimensionless π terms presented in Table 1 are obtained. Buckingham π-theorem guarantees that the functional relationship must be the equivalent form shown in Equation (2).

To obtain the function f_2, Equation (2) is rewritten in the form with unknown constants and exponents as shown in Equation (3). Constant C and exponents of the eight π parameters are obtained by applying the least squares method to minimize the errors between mass flow rate calculated from Equation (3) and experimental values.

Table 1 Definitions of dimensionless π parameters

π-terms	Significances	Definitions
π_0	Flow rate	$m/(d \cdot \mu_l)$
π_1	Geometry effect	L/d
π_2	Inlet pressure effect	$(d^2 P_{in})/(v_l \mu_l^2)$
π_3	Outlet pressure effect	$(d^2 P_{out})/(v_l \mu_l^2)$
π_4	Liquid inlet effect	$(d^2 C_{Pf} T_{sub})/(v_l^2 \mu_l^2)$
	Two-phase inlet effect	x
π_5	Density effect	v_g / v_l
π_6	Viscosity effect	$(\mu_l - \mu_g)/\mu_g$
π_7	Vaporization effect	$(d^2 h_{lg})/(v_l^2 \mu_l^2)$
π_8	Metastable effect	$(d\sigma)/(v_l \mu_l^2)$

Correlations for pure refrigerants

In Table 2, some dimensionless correlations for pure refrigerants in public literatures are listed. The outlet pressure (π_3 term) was not included in all correlations because the choked flow condition was easily reached for typical steady-state applications. Under this condition, the mass flow rate is dependent on the inlet pressure of the capillary tube only, and the pressure at the outlet has no relation with the mass flow rate. Also the metastable effect (π_8 term) was neglected because the influence is little.

In addition, it should be noted that the physical properties were evaluated at the corresponding bubble point temperature at the inlet pressure under the condition that the inlet is in sub-cooled liquid state. Only the Bittle's [4] correlation shown as Equation (5) was developed for two-phase flow inlet condition, in which the physical properties were evaluated at the state of inlet pressure and temperature.

It is important to select proper dimensionless π term to reflect the different parameters' contribution to the flow characteristics in a capillary tube. For sub-cooled liquid inlet condition, only the geometry effect (π_1 term), the inlet pressure effect (π_2 term) and the liquid inlet effect (π_4 term) were included in Equations (6), (7), (8) and (9) for one kind of refrigerant; the density effect (π_5 term), the viscosity effect (π_6 term) and the vaporization effect (π_7 term) were considered in Equations (4) and (10) because the correlations were applicable to three or four kinds of refrigerant. For the two-phase inlet condition, the density effect (π_5 term) was added in Equation (5).

Table 2 Dimensionless correlations for pure refrigerants

Authors	Refrigerants	Inlet state	Correlations	
Bittle [4]	R22, R134a, R152a, R410A	Liquid	$\pi_0 = 1.893 \cdot \pi_1^{-0.484} \cdot \pi_2^{1.369} \cdot \pi_4^{0.019} \cdot \pi_5^{0.773} \cdot \pi_6^{0.265} \cdot \pi_7^{-0.824}$	(4)
		Two-phase	$\pi_0 = 836.9 \cdot \pi_1^{-0.740} \cdot \pi_2^{0.417} \cdot \pi_4^{0.981} \cdot \pi_5^{-0.646}$	(5)
Melo [5]	R12	Liquid	$\pi_0 = 0.06135 \cdot \pi_1^{-0.509} \cdot \pi_2^{0.495} \cdot \pi_4^{0.160}$	(6)
	R134a	Liquid	$\pi_0 = 0.125 \cdot \pi_1^{-0.552} \cdot \pi_2^{0.460} \cdot \pi_4^{0.178}$	(7)
	R600a	Liquid	$\pi_0 = 0.719 \cdot \pi_1^{-0.590} \cdot \pi_2^{0.403} \cdot \pi_4^{0.157}$	(8)
Wolf [6]	R134a	Liquid	$\pi_0 = 0.0129 \cdot \pi_1^{-0..387} \cdot \pi_2^{0.492} \cdot \pi_4^{0.187}$	(9)
	R22, R134a, R410A	Liquid	$\pi_0 = 1.892 \cdot \pi_1^{-0..484} \cdot \pi_2^{1.369} \cdot \pi_4^{0.0187} \cdot \pi_5^{0.773} \cdot \pi_6^{0.265} \cdot \pi_7^{-0.824}$	(10)
Wei [7]	R407C	Liquid	$\pi_0 = 108.54 \cdot \pi_1^{-0.4125} \cdot \pi_2^{0.2892} \cdot \pi_4^{0.2377} \cdot \pi_5^{0.2446}$	(11)

<u>Correlation for mixed-refrigerants</u>

In order to obtain the dimensionless correlation for multi-component mixed-refrigerants, a set-up based on the real J-T cryocooler cycle was established [9], and the experimental conditions are listed in Table 3.

Table 3 Experimental conditions in this study

Mixed-refrigerants	$N_2+CH_4+C_2H_6+C_3H_8+iC_4H_{10}$
Inner diameter of capillary tube (mm)	1.2, 1.7
Length of capillary tube (m)	1, 1.5, 2.0, 3.0, 4.0
Inlet Pressure (bar)	14.9~21.9
Outlet pressure (bar)	2.7~4.4
Temperature range (K)	100~200
Inlet quality	0.1~0.43

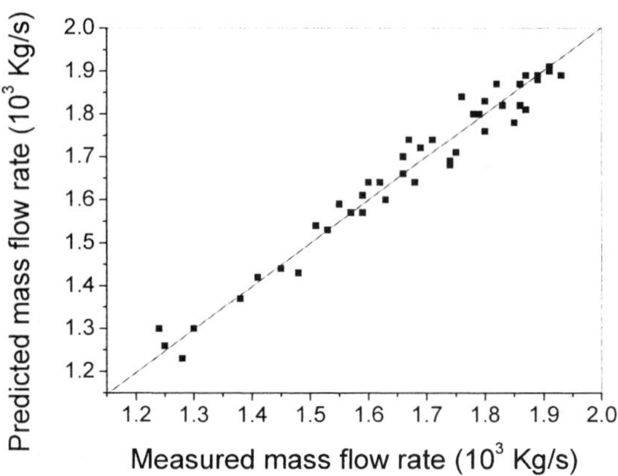

Figure 1 Comparisons of predicted mass flow rate with experimental data

In the multi-component mixed-refrigerants flow characteristics experiments, the state at the inlet of the capillary tube was in two-phase state and the choked flow was not reached at the outlet of the capillary tube. Therefore, the metastable (π_8 term) should be excluded and the outlet pressure effect (π_3 term) should be considered in the correlation. Based on the experimental data, the correlation for mixed-refrigerants flow characteristics shown in Equation (12) was obtained.

It should be noted that the physical properties were evaluated as an average value of the inlet and outlet state in π_0, π_5, π_6 and π_7 terms because the outlet pressure effect was considered in the correlation. The physical properties in π_2 and π_3 terms were evaluated at the corresponding inlet and outlet state.

Figure 1 shows comparisons of the predicted data with measured mass flow rate. Approximately 95.3 % of the data are correlated within a relative deviation of ± 5 %, and the average deviation is 1.80 %.

$$\pi_0 = 2679940.9 \cdot \pi_1^{-0.25744} \cdot \pi_2^{-0.02996} \cdot \pi_3^{0.01504} \cdot \pi_4^{-0.11064} \cdot \pi_5^{-0.44198} \cdot \pi_6^{-0.91566} \cdot \pi_7^{-0.02619} \tag{12}$$

$$m = f_3 \left[d^{0.9777}, \left(\frac{\mu_l}{\mu_g} - 1 \right)^{-0.91566}, \left(\frac{v_g}{v_l} \right)^{-0.44198}, L^{-0.25744}, \dots \dots \right] \tag{13}$$

Expanding the dimensionless π terms in Equation (12), four parameters that have more great influence on the mass flow rate are obtained as shown in Equation (13). Therefore, for a given multi-component refrigerants mixture, the primary factors that influence the flow characteristics in the adiabatic capillary tube are inner diameter of capillary tube, viscosity ratio and specific volume ratio between the liquid phase and gas phase, and the length of capillary tube.

SUMMARY

1. The dimensional analysis method was developed to predict multi-component mixed-refrigerants flow characteristics in the capillary tube. Some dimensionless parameters that influence mass flow rate were obtained on the basis of the Buckingham π-theorem.
2. A dimensionless correlation for predicting mixed-refrigerants flow characteristics was obtained based on the experimental data. The average deviation of the correlation is 1.80 %. The correlation can be used to the calculation of mass flow rate for a real system design.
3. Different from the correlations in the room-temperature refrigeration system, the pressure at the outlet of the capillary tube should be considered in correlation for mixed-refrigerants because the critical condition was not reached.
4. According to the correlation, the primary factors that influence mixed-refrigerants flow characteristics in adiabatic capillary tube were obtained. The result can give a guide for system design.

ACKNOWLEDGMENT

The authors acknowledge with gratitude the continued support received from the National Natural Sciences Foundation of China under the contract number of 50206024.

REFERENCES

1. Bittle, R.R, Pate M.B., A theoretical model for predicting adiabatic capillary tube performance with alternative refrigerants, ASHRAE Transactions (1996) 2 52-64
2. Bansal, P.K., Rupasinghe, A.S., A homogeneous model for adiabatic capillary tubes, Applied Thermal Engineering (1998) 3-4 207-219
3. Liang, S.M., Wong, T.N., Numerical modeling of refrigerant flow through adiabatic capillary tubes, Applied Thermal Engineering (2001) 21 1035-1048.
4. Bittle, R.R., Wolf, D.A. Pate, M.B., A generalized performance prediction method for adiabatic capillary tubes, International Journal of HVAC&R Research (1998) 4 27-43
5. Melo, C., Ferreira, R.T.S., Neto, C.B., Goncalves, J.M., Mezavilam, M.M., An experimental analysis of adiabatic capillary tubes, Applied Thermal Engineering (1999) 19 669-684
6. Wolf, D.A., Bittle, R.R., Pate, M.B., Adiabatic capillary tube performance with alternative refrigerants. ASHRAE research project PR-762, (1995)
7. Wei, C.Z., Lin, Y.T., Wang, C.C., Leu, J.S., An experimental study of performance of capillary tubes for R-407C refrigerant. ASHRAE Transactions (1999) 105 pp.634-638
8. Buckingham, E., On physically similar systems: illustrations of the use of dimensional equations, Physical Review (1914) 4 345–376.
9. Qi, Y.F., Cao, Y., Gong, M.Q., Luo, E.C., Wu, J.F., Experimental study on the adiabatic capillary tube expansion device in mixed-refrigerant J-T cryocooler. Proceedings of ICEC19, Grenoble, France (2002) 339-342

An experimental research on a small flow vortex tube at low temperature ranges

Liu J.Y.[1,2], Gong M.Q.[1], Wu J. F.[1], Cao Y.[1], Luo E.C.[1]

[1] Technical Institute of Physics and Chemistry, Chinese Academy of Sciences, Beijing, 100080, China
[2] Graduate School of the Chinese Academy of Sciences, Beijing, 100039, China

A small flow vortex tube operating in low temperature below 80 K was developed. Experimental results show that there is still significant separation effect of vortex tube below 80 K for some inert gases. With LN_2 pre-cooling, the lowest temperature of 60.2 K and 3.3 W refrigeration at 67.5 K has been first achieved using a closed-cycle vortex tube refrigeration system. And an interesting inverse temperature effect was found when the cold mass fraction is less than 0.38. These achievements can help thoroughly understand the mechanism of energy separation in the vortex tube and use the vortex tube in some potential situations.

INTRODUCTION

The separation of a gas stream entering into a tube through one or several tangential nozzles into two streams with different stagnation temperatures (the Ranque-Hilsch effect) is a phenomenon which has been investigated long time. Because it has no moving parts and consists mainly of a simple tube, the vortex tube has been used in some special fields [1]. Many works found in open literatures have focused on experimental and analytical investigations of describing the characteristics of the vortex tube. Hilsch [2] and Bruun [3] conducted important experimental investigations on vortex tube. Theoretical and analytical descriptions of the energy separation and the temperature and velocity profiles in a vortex tube were given by Deissler and Perlmutter [4] and Ahlborn [5]. Stephan [6] performed the mathematical investigation of energy separation in the vortex tube. Lewins and Bejan [7] developed the heat-exchanger analogy modeling. Recently, Neills and his colleagues [8] attempted to use vortex tube to replace the traditional expansion instruments in refrigeration system. Using a vortex tube instead of the traditional throttle valve, Luo etc. [9] developed a new mix-refrigerant auto-cascade refrigeration cycle. However, most of them paid attention on the performance of the vortex tube only in normal ambient temperature range, and there are very few articles concerning the characteristic of the vortex tube below 80 K.

The performance of the vortex tube depends on many geometrical and physical parameters. In order to understand the influence of the absolute gas temperature and to apply the vortex tube in low temperature conditions, it is necessary to carry out some experiments at low temperature range. The goal of this work is to investigate the characteristics of a vortex tube at low temperature.

EXPERIMENTAL RESEARCH

A closed-cycle vortex tube refrigeration system has been built in our laboratory (see Figure 1). The compressed gas without any solid particles or lubricating oil is supplied by a diaphragm compressor. The after cooler, LN_2 pre-cooler and several counter-flow heat exchangers cool the high pressure gas to required temperature before entering the vortex tube. The high pressure gas with a temperature less than 80 K expands in the vortex tube and then is divided into cold and hot streams. The cold gas leaves from the central orifice near the entrance nozzle, while the hot gas discharges from the periphery at the far end of the tube. The flow ratio of the cold gas to the hot gas is controlled by a control valve. A copper electric heating coil installed on the cold end of the vortex tube is used to test the refrigeration of the vortex tube. Two returning gases are measured by two rotameters respectively before flowing back to the inlet of the compressor. For the system operating in low temperature, the thermal insulating performance is very important. So the cryostat is placed into a vacuum tank and a high-performance vacuum pump is employed to guarantee a good thermal insulation.

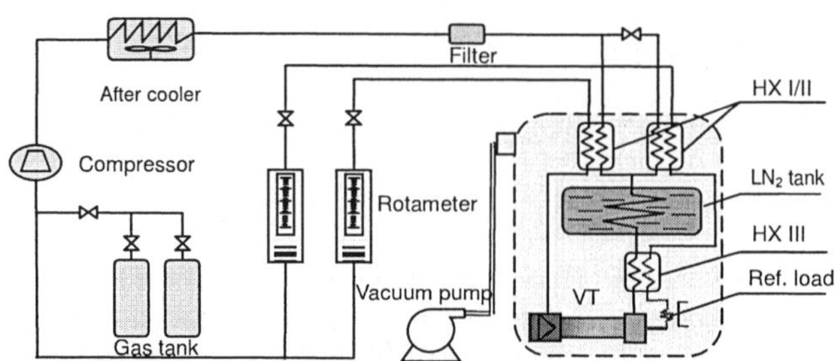

Figure 1   The schematic diagram of the experimental system

The vortex tube investigated in this work (see Figure 2) is built based on our preliminary experiments. It consists of a long conical tube, a vortex chamber, several nozzles, a control valve and some fixing and sealing parts. All parts are made by stainless steel. The essential dimensions are presented in Table 1.

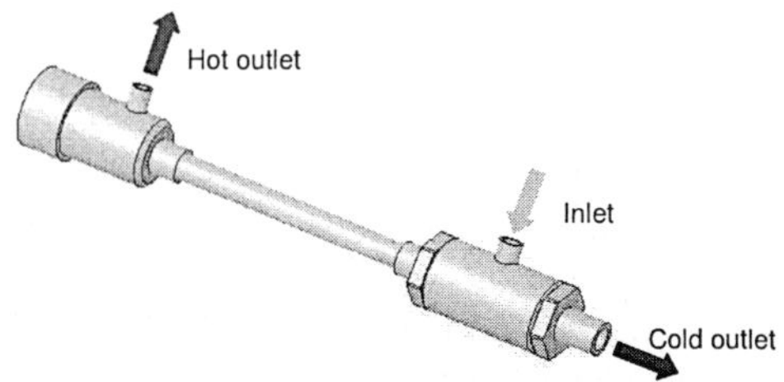

Figure 2   3D appearance of the vortex tube

Table 1   Specifications of the vortex tube

Tube Characteristics	
Length of tube, L(mm)	120
Cone angle of tube	2°32'
Diameter of tube, D(mm)	5
Number of inlet nozzle	2 or 4
Diameter of nozzle, d_n(mm)	0.3
Diameter of orifice, d_c(mm)	3

Two rotameters with an accuracy of ±2.5 % are used to measure the cold and hot flow stream rate, respectively. All temperatures are measured by Pt100 thermometers, which are calibrated in a large temperature range from 52 K to 323 K with an accuracy of ±0.1 K. All temperatures are recorded by a high precision data acquisition system [Keithley Model 2700]. Three manometers with an accuracy of ±0.4 % are used to measure the pressures of the inlet and two outlets.

EXPERIMENTAL RESULTS AND DISSCUSSION

The performance of the vortex tube can be evaluated with following parameters: cold mass flow fraction μ, cold gas temperature difference ΔT_c, hot gas temperature difference ΔT_h, isentropic temperature efficiency η_s and isentropic temperature difference ΔT_s, which can be expressed as follows:

$$\mu = \frac{\dot{m}_c}{\dot{m}_i} \tag{1}$$

$$\Delta T_c = T_c - T_i \tag{2}$$

$$\Delta T_h = T_h - T_i \tag{3}$$

$$\eta_s = \Delta T_c / \Delta T_s \tag{4}$$

$$\Delta T_s = T_i - T_s = T_i \left[1 - \left(\frac{p_c}{p_i} \right)^{(k-1)/k} \right] \tag{5}$$

With $\dot{m}_c$: cold flow mass, $\dot{m}_i$: inlet flow mass, T_i: temperature of inlet gas, T_c: temperature of cold flow, T_h: temperature of hot flow, p_i: pressure of inlet, p_c: pressure of cold outlet.

A series of experiments were carried out at different conditions. Figure 3 shows a typical cool-down curve with LN_2 pre-cooling. It shows that the temperatures of inlet, cold and hot outlets sharply decreased within first 35 minutes after start-up. After about 70 minutes, the temperature of cold end of the vortex tube reached 62 K. And the temperatures of hot and cold ends are less than that of inlet of the vortex tube. The variation of temperatures of inlet, cold and hot gas versus the cold mass flow fraction are shown in Figure 4, which shows a significant temperature separation effect as same as that of the common vortex tube operating in normal ambient temperature. And a surprising and seldom reported abnormal phenomenon was found that when the cold mass flow fraction decreased to about 0.38, the temperature of cold end was higher than that of hot end. We called this "inverse temperature effect".

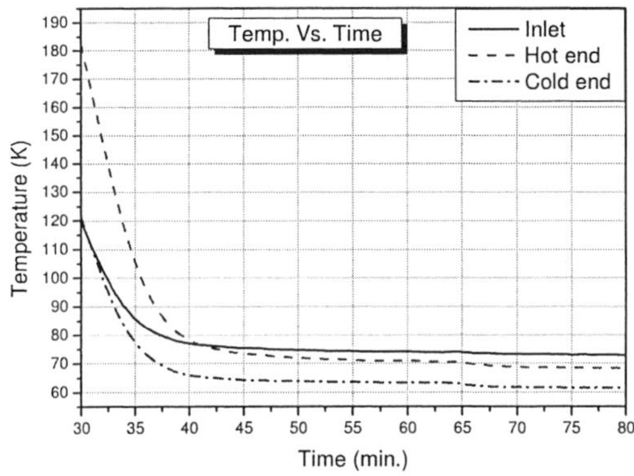

Figure 3   Operating curve

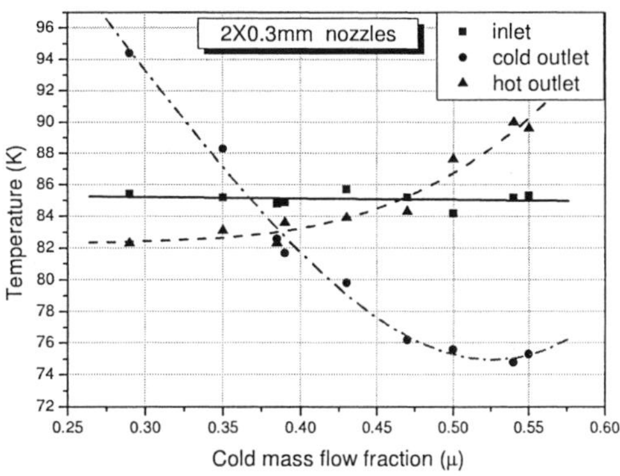

Figure 4   Typical performance at P_i=2.60MPa

The influence of the inlet pressure on the cold and hot gas temperature differences is also investigated by changing the inlet pressure from 1.70 MPa to 2.60 MPa. Figure 5 shows the variations of the cold and hot gas temperature differences versus the cold mass flow fraction at different inlet pressures. As shown in result curves, a higher inlet pressure increases the absolute value of the cold gas temperature difference and decreases the hot gas temperature difference.

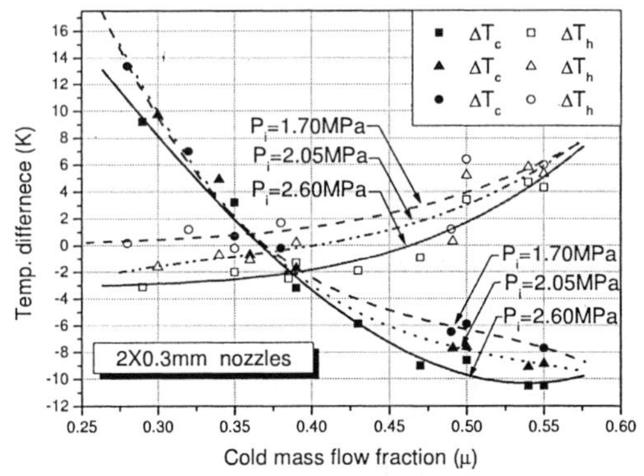

Figure 5   Performance at different inlet pressures

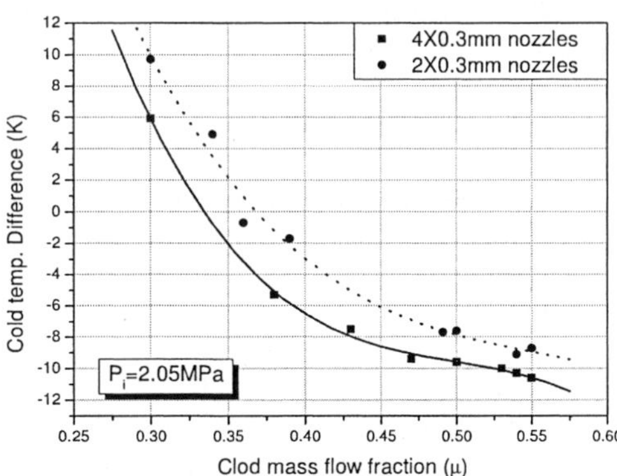

Figure 6   Performance with different inlet nozzles

To study the effect of the inlet nozzle on the performance, two different nozzle types were designed and fabricated. One has two nozzles with inner diameter of 0.3 mm uniformly distributed along periphery, and the other has four equal ones. The variation of the cold gas temperature difference versus the cold mass flow fraction is shown in Figure 6. The result shows that the vortex tube with four 0.3 mm nozzles gives a better performance than the other from the point of the cold gas temperature difference. In order to gain the lowest refrigerating temperature, the entrance with four 0.3 mm nozzles is used.

To know the influence of the substances on the performances of the vortex tube, neon and helium were used in our experiments, respectively. The maximum cold gas temperature difference was measured at inlet pressure of 2.05 MPa. In experiment using neon, the maximum cold gas temperature difference is 9.1 K, which is higher than the 4.9 K attained using helium as working gas. Therefore, neon shows a higher performance for gaining the lowest refrigerating temperature in this experimental system.

Table 2 presents the lowest refrigerating temperatures at differential inlet pressures under optimum conditions. With LN_2 pre-cooling, the lowest refrigerating temperature of 60.2 K and 3.3 W refrigeration at 67.5 K has been successfully achieved by a closed-cycle vortex tube refrigeration system.

Table 2   Lowest refrigerating temperature at different inlet pressures

Inlet pressure P_i (MPa)	Total flow rate (Nm^3/h)	Inlet temp. T_i (K)	Cold temp. T_c (K)	Hot temp. T_h (K)	Efficiency η_s (%)
2.03	8.6	74.0	63.3	70.8	20.0
2.30	9.2	73.0	61.7	68.5	21.8
2.37	9.4	72.5	60.8	67.5	22.8
2.40	9.5	71.6	60.2	66.9	22.7

CONCLUSIONS

A new fabricated vortex tube and closed cycle refrigeration system operating at low temperature were developed. Systematic experiments at different geometrical and physical conditions were carried out to test the performance of the vortex tube at temperature less than 80 K. Such conclusions can be summarized as follows:

1. At the temperature below 80 K, there is still evident temperature separation effect in the vortex tube, this make it possible to use the vortex tube in low temperature range.
2. The pressure of the inlet has a great influence on the temperature separation. Increasing inlet pressure simultaneously decreases the hot and cold temperatures of the vortex tube. And for the cold effect, neon is found better than helium. The four 0.3 mm nozzles configuration gives a better performance.
3. The cold mass flow fraction is a key parameter for the performance of the vortex tube. When μ decreases to less than 0.38, an "inverse temperature effect" will emerge, this provides some new evidences for the thorough understanding of the energy separation mechanism in vortex tube.
4. With LN_2 pre-cooling, the lowest temperature of 60.2 K and 3.3 W refrigeration at 67.5 K has been first time achieved.

ACKNOWLEDGEMENT

This work is financially supported by the National Natural Sciences Foundation of China (№50076044).

REFERENCES

1. Bruno T.T, Laboratory application of the vortex tube, Journal of Chemical Education (1987) 64 987-988
2. Hilsch R., The use of the expansion of gases in a centrifugal field as a cooling process, Review Sciences Instruments (1947) 18 108-113
3. Bruun H.H., Experimental investigation of the energy separation in vortex tube, Journal Mechanical Engineering science (1969) 11 567-582
4. Deissler R.G., Perlmutter. M, Analysis of the flow and energy separation in a turbulent vortex, International Journal of Heat & Mass Transfer (1960) 1 173-191
5. Boye Ahlborn, Stuart Groves, Secondary flow in a vortex tube, Fluid Dynamics Research (1997) 21 73-86
6. Stephan K., Lin S., An investigation of energy separation in a vortex tube, International Journal of Heat & Mass Transfer (1983) 26 341-348.
7. Lewins Jeffery, Bejan Adrian, Vortex tube optimization theory, Energy (1999) 24 931-943
8. Gregory F.Nellis, The application of vortex tube to refrigeration cycles, The 9th International Refrigeration and Air Conditioning Conference, Purdue, USA (2002) 85-92
9. Luo Er.C, Gong M.Q, etc., A new type cascade refrigeration cycle using mix-refrigerant fraction and vortex tube, Chinese patent, CN00316710.2, (2002)

Proceedings of the Twentieth International Cryogenic Engineering Conference
(ICEC 20), Beijing, China. © 2005 Elsevier Ltd. All rights reserved.

Performance of different throttle refrigeration cycles[*]

Gong M.Q., Wu J.F.

Technical Institute of Physics and Chemistry, Chinese Academy of Sciences
P.O. Box 2711, Beijing, 100080, China

Numerous cycle configurations were presented based on the Joule-Thomson effect of real gases for different refrigeration applications from cryogenic temperature to room temperature. Using throttle device to produce refrigeration is the general feature among those refrigeration cycles, such as the vapor-compression cycle, the cascade cycle, the single-stage recuperative cycle with mixed-refrigerants, and the dual mixed-refrigerant cascade recuperative cycle. In this paper, thermodynamic models were presented to analyze the thermodynamic performances of those throttle refrigeration cycles, especially the recuperative mixed-refrigerant cycles for low temperature refrigeration applications. The results show that with optimal mixture and cycle configuration, the low-temperature recuperative refrigeration cycle can achieve a high performance as good as the commercial vapor-compression cycle.

INTRODUCTION

The vapor-compression refrigeration cycle is widely used in commercial refrigeration applications near room temperature range from 230 K to 320 K, such as industry and domestic refrigerators, air -conditioners, heat pumps, etc. The effective lowest refrigeration temperature for the single-stage vapor-compression cycle is about 230 K, limited by the refrigerant properties and the compressor characteristics. When a lower refrigeration temperature of 210 K is required, traditional measurement is using a two-stage compression cycle to provide such low temperature refrigeration. However, when the refrigeration temperature is even lower e.g. 190 K, usually a two-stage cascade cycle is used, which uses a warm refrigeration loop to precool a cold refrigeration loop. For much lower temperature applications, multistage cascade cycle is used, for example, in the former natural gas liquefaction system, three stage cascade cycle using three refrigeration loops of pure propane, ethylene, and methane to reach a cryogenic temperature of 110 K. The theoretical number of stages is unrestricted, but in practice the number is determined by economic and complexity considerations. In recent years, the number of stage is usually no more than two.

In recent years, there has been a remarkable development of the mixed-refrigerant recuperative refrigeration cycle [1-3]. Driven by an oil-lubricated single-stage compressor, the mixed-gases recuperative refrigerator has many merits over other type coolers. Those obvious merits are high reliability because of no moving parts at the coldest end, high efficiency with optimal mixtures, low cost, and easy to be built in industry scale, etc. Without any modification of the hardware, and only with different mixed-refrigerants charged, the mixed-gases Joule-Thomson refrigerator can produce cryogenic

[*] This work is financially supported by the National Natural Sciences Foundation of China under the contract number of 50206024.

170

and ultra-low cooling covering a large temperature range from below liquid nitrogen temperature (77K) to warmer than 230K. Therefore, this refrigeration method is quite reliable and flexible in many applications, such as infrared detectors, cryosurgical devices, HTS devices, material studies, biomedical storage, gas chiller or liquefaction, etc. Numerous mixed-gases refrigeration cycle configurations were developed in the past few years. Among those cycle configurations, the single-stage recuperative cycle has the simplest cycle configuration. Former study shows that this cycle with optimal mixture has the highest thermodynamic efficiency [1, 4].

It may be wonder what about the thermodynamic performances of those throttle refrigeration cycles with different configurations and refrigerants at ideal conditions. In this paper, the thermodynamic performances of those throttle cycles were studied. The Carnot efficiencies of four typical throttle refrigeration cycles at different cooling temperature levels were calculated, including a vapor-compression cycle, a two-stage cascade cycle, a single-stage mixed-refrigerant recuperative cycle, and a dual mixed-refrigerant recuperative cycle.

THEORETICAL EFFICIENCIES OF THE CYCLES

The expressions of Carnot efficiencies of these four throttle cycles can be listed as the following.

Vapor-compression cycle (VCC)

Figure 1 shows a typical vapor-compression refrigeration cycle. The efficiency of this cycle can be listed as [6]:

$$COP_{VCC} = \frac{h_1 - h_5}{h_{2r} - h_1} \tag{1}$$

$$\eta_{VCC} = COP_{VCC} \times \frac{T_0 - T_c}{T_c} = \frac{h_1 - h_5}{h_{2r} - h_1} \times \frac{T_0 - T_c}{T_c} \tag{2}$$

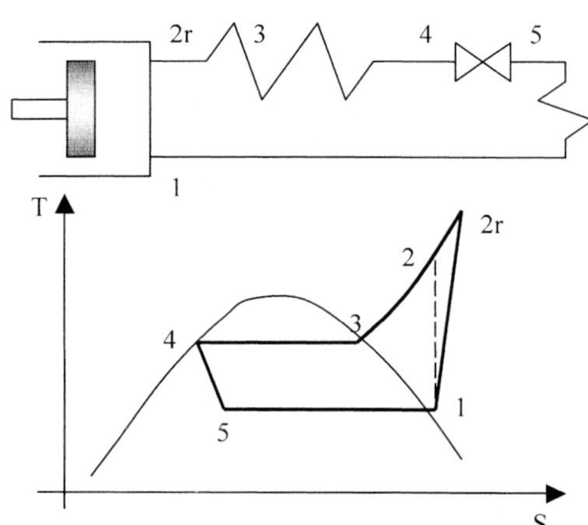

Figure 1 A vapor-compression cycle

where T_0 and T_c are ambient temperature and refrigeration temperature, and h is specific enthalpy, η is the Carnot efficiency.

Cascade refrigeration cycle (CRC)

Figure 2 shows a typical two-stage cascade refrigeration cycle. The efficiency of this cycle can be expressed as [6]:

$$COP_{CRC} = \frac{h_{1s} - h_{5s}}{(h_2 - h_1)\beta + (h_{2s} - h_{1s})} \tag{3}$$

$$\eta_{CRC} = \frac{h_{1s} - h_{5s}}{(h_2 - h_1)\beta + (h_{2s} - h_{1s})} \times \frac{T_0 - T_c}{T_c} \tag{4}$$

where $\beta = m_w/m_c$ is the flow rate ratio of the warm and cold refrigeration loops; m_w and m_c are flow rates of the warm and low refrigeration loops; all subscript expressions denote the steady states in the cycle.

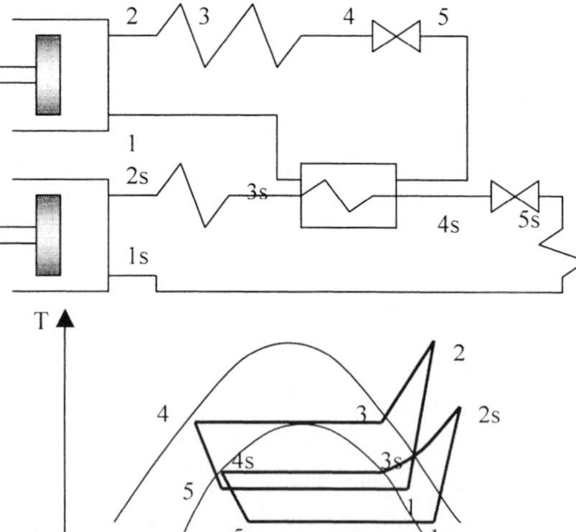

Figure 2 Schematic diagram of a two-stage cascade refrigeration cycle

<u>Single-stage mixed-refrigerant recuperative refrigeration cycle (MRC)</u>
Figure 3 shows the schematic diagram of a typical single-stage mixed-refrigerant recuperative refrigeration cycle. For mixed-refrigerant recuperative cycle, the thermodynamic performance can also be presented as [1, 5]:

$$q_c = h_6 - h_5 = h_6 - h_4 = h_1 - h_3 = \min(h(T)_{pl} - h(T)_{ph}) - (h_{1s} - h_1) = \Delta h_T - q_{loss} \tag{5}$$

$$q_{loss} = h_{1s} - h_1, \quad \Delta h_T = \min(h(T)_{pl} - h(T)_{ph}) \tag{6}$$

$$COP_{MRC} = \frac{q_c}{h_2 - h_{1s}} \tag{7}$$

$$\eta_{MRC} = COP_{MRC} \times \frac{T_0 - T_c}{T_c} = \frac{q_c}{h_2 - h_{1s}} \times \frac{T_0 - T_c}{T_c} \tag{8}$$

where q_{loss} is the loss of recuperator.

<u>Dual mixed-refrigerant recuperative cycle (DMRC)</u>
A schematic diagram of a typical dual mixed-refrigerant recuperative cycle is shown in Figure 4. The efficiency of the cycle can also be expressed as [1, 5]:

$$q_c = \min(h(T)_{pl} - h(T)_{ph}) - (h_{1s} - h_1) = \Delta h_T - q_{loss} \tag{9}$$

$$\Delta h_T = \min(h(T)_{pl} - h(T)_{ph}), \quad q_{loss} = h_{1s} - h_1 \tag{10}$$

$$COP_{DMRC} = \frac{q_c}{(h_2 - h_{1s}) + (h_b - h_a)\beta} \tag{11}$$

$$\eta_{DMRC} = COP_{DMRC} \times \frac{T_0 - T_c}{T_c}$$
$$= \frac{q_c}{(h_2 - h_{1s}) + (h_b - h_a)\beta} \times \frac{T_0 - T_c}{T_c} \tag{12}$$

where β is the flow rate ratio of the warm and cold refrigeration loops, can be calculated with Eq. (4).

SIMULATION RESULTS

Above equations describe the thermodynamic performance of those throttle refrigeration cycles. The Carnot efficiencies of those cycles can be calculated with Equations (2), (4), (8) and (12). It should be mentioned here that there should be no temperature profile cross in recuperative heat exchangers in MRC and DMRC cycles [3-5]. Using an appropriate optimization method, the performance of the cycles can be optimized. Typical optimization results of those cycles were presented in Table 1.

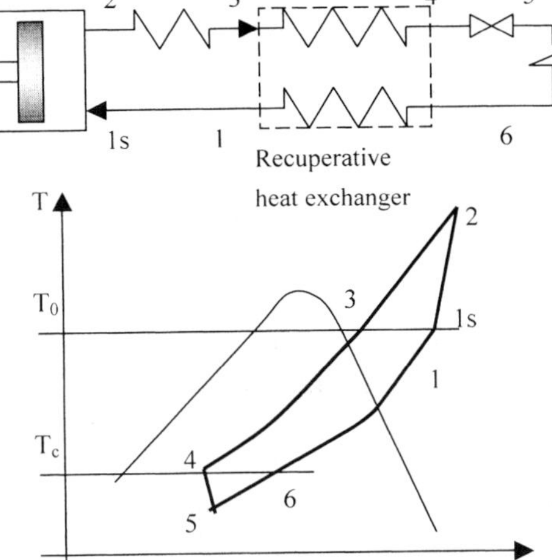

Figure 3 A mixed-refrigerant recuperative cycle

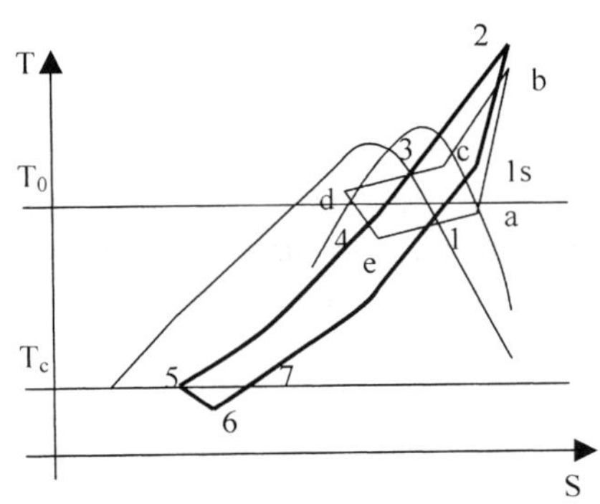

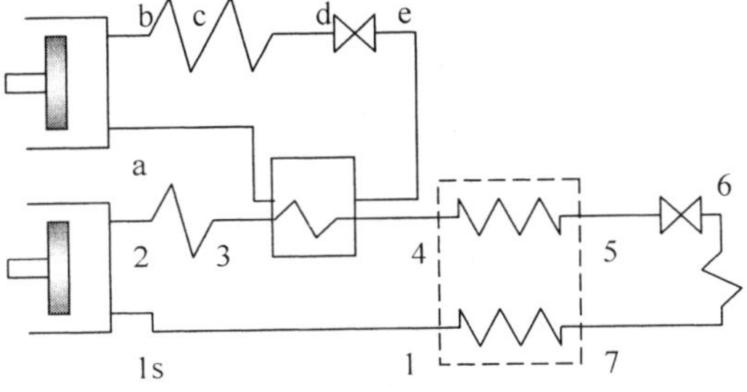

Figure 4 Schematic diagram of a dual mixed-refrigerant recuperative cycle

Table 1 Thermodynamic performance comparisons of the four throttle cycles

Items	VCC	CRC		MRC	DMRC	
Refrigerants	R407C	Warm loop: R407C	Cold loop: R23	Multicomponent mixtures consisting of N_2, CH_4, C_2H_6, C_3H_8, iC_4H_{10}, iC_5H_{10}	Warm loop: R407C	Cold Loop: multicomponent mixtures of N_2, CH_4, C_2H_6, C_3H_8, iC_4H_{10}
T_0, K	300	300	300	300	300	300
T_c, K	245.1	241.9	198	120	251.8	120
p_h, MPa	1.3	1.2	1.2	1.8	1.3	1.8
p_l, MPa	0.15	0.15	0.15	0.3	0.15	0.3
COP	2.28	1.67 / 1.00	/	0.341	2.36 / 0.338	/
η, %	51.2	51.3		51.2	50.7	

From Table 1, it can be found that all the throttle cycles can provide almost equivalent values of the Carnot efficiencies for different refrigeration temperatures from cryogenic to near room temperature ranges. The calculation results were obtained at ideal compression conditions that there is an ideal adiabatic compressing process without entropy generation in the compressor. The results shown in Table 1 indicate that the mixed-refrigerant recuperative throttle cycle can achieve quite good performance as high as the commercial vapor-compression refrigeration cycle. However, because of the additional loss in the recuperators and heat isolation at low temperatures, the practical performance of the low temperature recuperative cycle is hard to achieve a high performance as good as the vapor-compression cycle.

SUMMARY

Thermodynamic performances of throttle cycles including the widely used commercial vapor-compression cycle, the cascade cycle, the single-stage mixed-refrigerant recuperative cycle, and the dual mixed-refrigerant recuperative cycle were illustrated and analyzed in this paper. The Carnot efficiency was used as a comparison criterion in the comparisons of all those throttle cycles at different refrigeration temperature ranges. The optimized results at ideal compressing-process conditions show that all the throttle cycles can provide near equivalent values even providing refrigeration from cryogenic temperature to near room temperature ranges.

REFERENCES

1. Gong, M.Q. Thermodynamic analysis and experimental investigation on multicomponent mixture recuperative throttle refrigerators operating in 80-230K temperature ranges, Ph.D. dissertation, Beijing, China, (2002)
2. Luo, E.C. Gong, M.Q., Wu J.F., and Zhou Y., Properties of gas mixtures and their use in mixed-refrigerant Joule-Thomson refrigerators, in Adv. in Cryo. Eng. (2004) 49, (in press)
3. Boiarski, M. et al., Modern trends in designing small-scale throttle-cycle coolers operating with mixed refrigerants, in Cryocoolers (2001) 11 513-522
4. Gong, M.Q. et al., Thermodynamic design principle of mixed-gases Kleemenko refrigeration cycles, in Adv. Cryogenic. Eng. (2002) 47 873-880
5. Gong, M.Q. et al., Optimum composition calculation for multicomponent cryogenic mixture used in Joule-Thomson refrigerators, in Adv. Cryogenic. Eng. (2000) 45 283-290
6. Li S.S, et al, in Refrigeration Mechanics and Theory (Chinese version), China Machine Press, (1988)

A 70 K J-T cryocooler based on the fractional mixed-refrigerant cycle[*]

Gong M.Q., Wu J.F.

Technical Institute of Physics and Chemistry, Chinese Academy of Sciences
P.O. Box 2711, Beijing, 100080, China

An effort was made in this paper to reach about 65 to 70 K temperature range by using a mixed-gases Joule-Thomson system. A special cycle configuration was disclosed utilizing a new developed fractional phase-separator with inner heat and mass transfer. This separator avoids cold clogging essentially. An optimal mixture was used to improve the system performance. A stable coldest temperature of 65 K and a cooling capacity of 1.06 W at 70 K were achieved driven by a commercial single-stage hermetic compressor with a nominal input power of 500 W. The detailed analysis of mixture properties and cycle configuration are also presented in this paper.

INTRODUCTION

The last two to three decades has seen a remarkable development of the mixed-gases Joule-Thomson refrigerator (MJTR). Driven by an oil-lubricated commercial compressor makes the MJTR more competitive with other type coolers. This new revival refrigeration method with multi-component mixture has demonstrated high performance in the cooling temperature range from 80 K to 230 K in many applications, such as cooling infrared sensors, gas chiller or liquefaction, cryosurgery, cryo-preservation, water vapor cryo-trapping, etc. The merits of mixed-gases refrigeration is obvious such as low cost, high reliability, high efficiency, and easily to be produced in industry scale, etc. However, for applications of cooling high temperature superconductivity devices, the appropriate temperature should be lower than 80 K. That is, providing cooling capacity at 70 K is accepted for commercialized applications. Difficulties are brought by using high-boiling point components and oil lubricant in the system in the traditional MJTR system because of the coldest clogging and low efficiency. The detailed information of the thermodynamic consideration of mixed-refrigerant and design of cycle configuration for reaching 70 K are presented in this paper.

THERMODYNAMIC DESIGN

For properly operating at low temperature range, a careful thermodynamic design was made from two aspects of gas properties and cycle configuration, which were discussed in the following.

Properties of the cryogenic gases for low temperature refrigeration

For operating at 70 K temperature range, two kind of properties of cryogenic gases are most important.

---

[*] This work is financially supported by the National Natural Sciences Foundation of China under the contract number of 50206024.

174

One is the liquid-solid (L-S) equilibrium characteristics which essentially influence the reliability of the refrigeration system, the other is the Joule-Thomson effect which represents the thermodynamic performance of the system. For most substances, low boiling temperature component has low freezing point. A list of the solubility limits for different substance is shown in Table 1 [1].

Table 1 Solubility limit for some hydrocarbon substances at low temperature range [1]

Temperature /K	Solubility limits (mole fraction)				
	CH_4	C_2H_6	C_3H_8	iC_4H_{10}	iC_5H_{12}
65	0.62	0.23	0.21	0.03	0.01
70	0.70	0.33	0.33	0.05	0.02
75	0.78	0.47	0.50	0.08	0.04
80	0.85	0.62	0.71	0.13	0.07
85	0.92	0.80	0.97	0.20	0.10
90	0.99	1	1	0.29	0.16

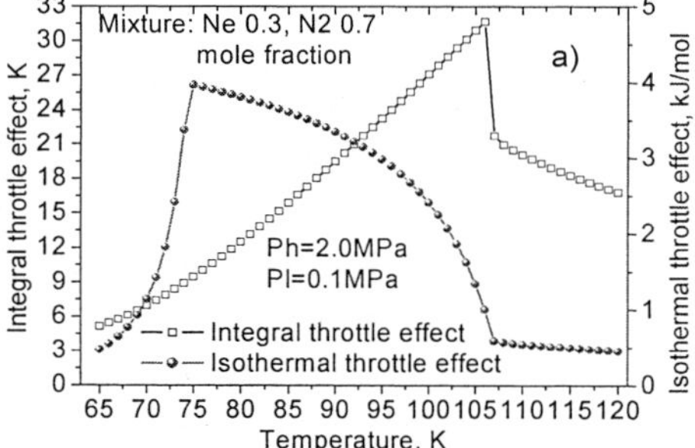

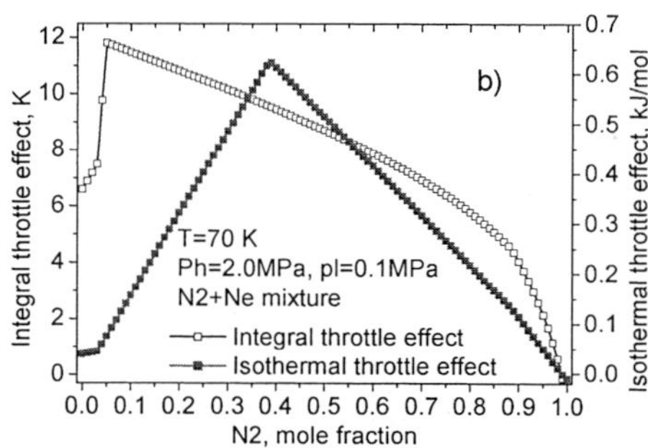

Figure 1 Throttle effects of a mixture at low temperature, a) influence of temperature, b) influence of mole concentration

From Table 1, it can be found that there is an acceptable solubility limit for the most used substance in the low temperature. For most mixtures used in this temperature range, the mole fractions of those substances are far less than this solubility limit because of the phase separation in the cycle configuration which will be discussed later. So a high reliability of the refrigeration system can be reached.

Joule-Thomson effects, especially the isothermal Joule-Thomson are important parameters in the design of the throttle refrigeration system. In the 70 K temperature range, the candidate gases are still nitrogen and neon. The isothermal effect Δh_T and integral throttling effect ΔT_h can be obtained with the following equations [2]:

$$\Delta T_h = \int \mu dp = \int \left(\frac{\partial T}{\partial P} \right)_h dp = T_0 - T_1 \tag{1}$$

$$\Delta h_T = h_0 - h_1 \tag{2}$$

where T_0, T_1 are the temperatures of the gas before and after the throttling, h_0 is the enthalpy of the gas before being compressed, and h_1 is the enthalpy of the gas after being isothermally compressed, μ is the differential throttle effect, and p is pressure. Figure 1 shows the calculation results of the throttle effects of nitrogen and neon mixture at different conditions, in which p_h and p_l denote the high and low pressures, respectively. Figure 1 a) gives an suggestion that the separation temperature in the refrigeration cycle should be better in the range from 100 K to 120 K, and Figure 1 b) shows that the mixture composition after separation should be better that the mole fraction of nitrogen is about 0.4 in the mixture, and the rest component is neon.

The optimization of the multicomponent mixture is another very important thing in the design of the 70 K refrigeration system. The normal boiling point and freezing point should be considered while optimizing the mixtures. From the discussion of the properties of mixture in the above section, an optimal mixture was used in this refrigeration system. Indeed, for different refrigeration temperature ranges the mixture component candidates are different. For the 70 K temperature range of this cryocooler, seven component mixtures are adopted, which consists of hydrocarbons, fluorinated hydrocarbons, nitrogen and neon. The optimized mixture is designed to minimize the irreversibility losses of the heat exchanger over the operating temperature range, and to be compatible with the compressor lubricating oil and the materials of construction of the refrigeration system.

Design of the refrigeration cycle configuration

The schematic diagram of the refrigeration system for operating at 70 K temperature range is shown in Figure 2 [3]. The refrigeration system consists of two separate modules: the compressor unit and the cryostat, which are connected by gas lines. The compressor unit consists of a compressor, an air-cooled condenser or so-called after cooler, and an oil pre-removal unit. The compressor is a hermetic single stage oil-lubricated air-conditioning compressor with a nominal input power of 0.5 kW. The disadvantage is the oil contamination in the refrigeration stream. If the oil contamination in the mixture exceeds compatible quantities, this would result in clogging in the cold end at low temperature. Having a good oil removal unit is much more important in the low temperature refrigeration system. However, it is not enough just using an oil removal unit in the system. So special phase separators are used in the design of such cycle configuration for low temperature refrigeration. Special structures have been incorporated in the phase separator, which is known as the reflux heat exchanger, also called the dephlegmation separator, instead of the traditional liquid vapor separator to ensure the liquid and vapor separation. Two dephlegmation separators are used in this cycle to ensure that there will be no solid generated in the coldest section of the system. One separator is set at ambient temperature, and the other is near 110 K temperature. The separation temperature is determined by the recuperative heat exchanger size.

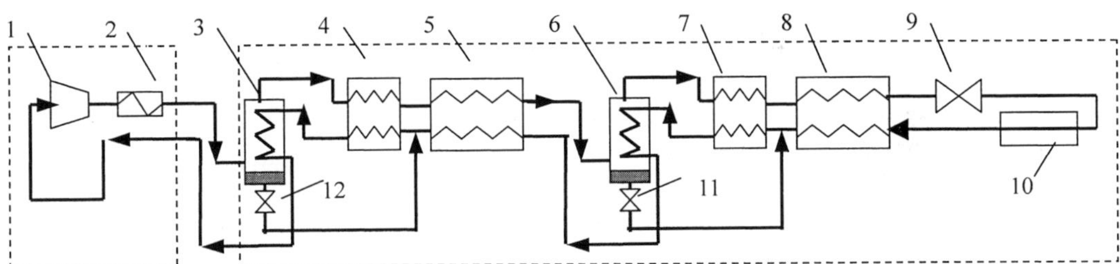

Figure 2 Schematic diagram of the refrigeration cycle configuration with a dephlegmation separator, 1: compressor 2: after cooler 3,6: simplified dephlegmation separator 4,5,7,8: heat exchanger 9,11,12: throttle valve 10: evaporator

EXPERIMENTAL RESULTS

Experimental rig

Based on above considerations, an experimental setup was built and tested. The features of experimental setup were listed in Table 2.

Experimental results

Extensive experimental test were carried out with the setup described above. Figure 3 shows the typical cooling down curves of the temperatures and pressures of the refrigerator. Figure 4 shows the flow rate variation in the cool-down process. The diagram of cooling capacity vs. refrigeration temperature is shown in Figure 5. From those figures, it can be found that a lowest stable refrigeration temperature of 66K and 1.06 W cooling power at 70 K were achieved. From Figure 3, the cool-down process is a little

176

bit long because of the large heat capacity of the cryostat and poor vacuum isolation. This can be improved by a future design of the cryostat especially the recuperative heat exchanger and improvement of the vacuum system.

Table 2 Specifications of the experimental rig

Components	Number	Features
Compressor	1	Single-stage; oil-lubricated; hermetic; 0.5 kW input power
Heat exchanger	4	7 inner tubes: inner/outer diameter: 1.8/3 mm; outer tube inner/outer diameter: 10/12 mm
JT valve	3	Capillary tube
Mixed-refrigerant	/	Seven components of light hydrocarbons form C1 to C5; nitrogen; neon

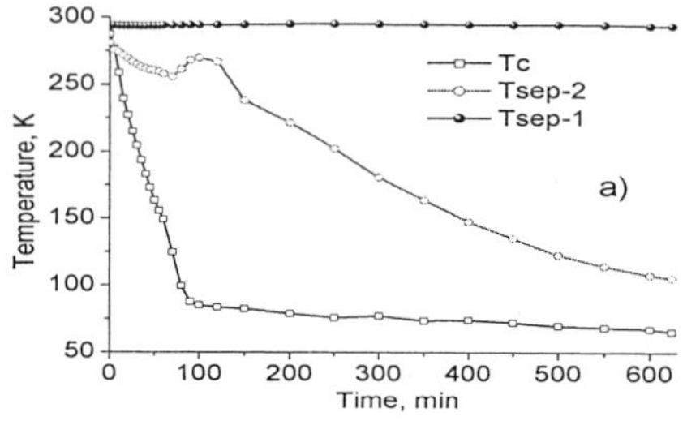
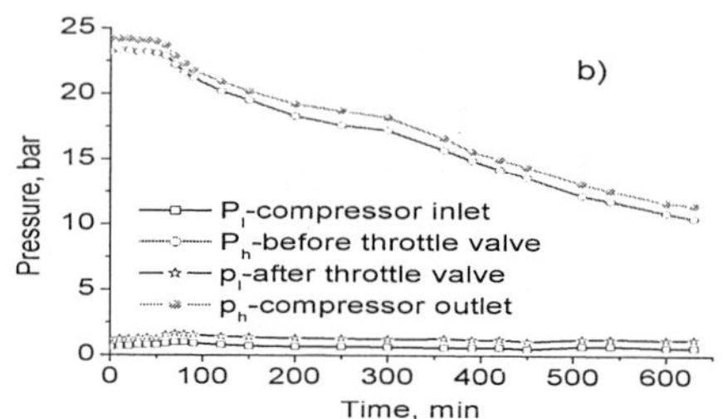

Figure 3 cooling down curve, a) temperature, b) pressure

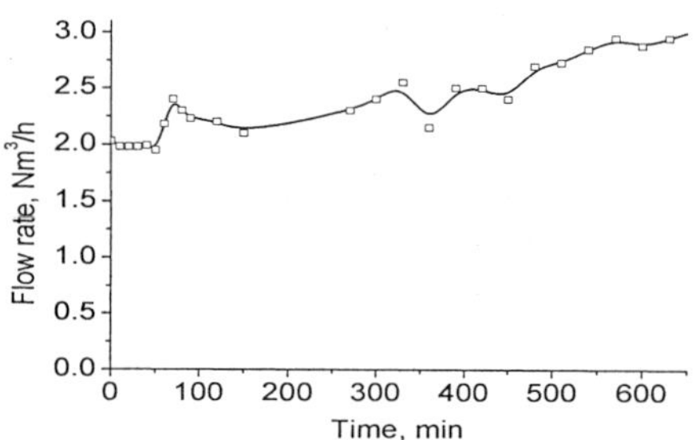
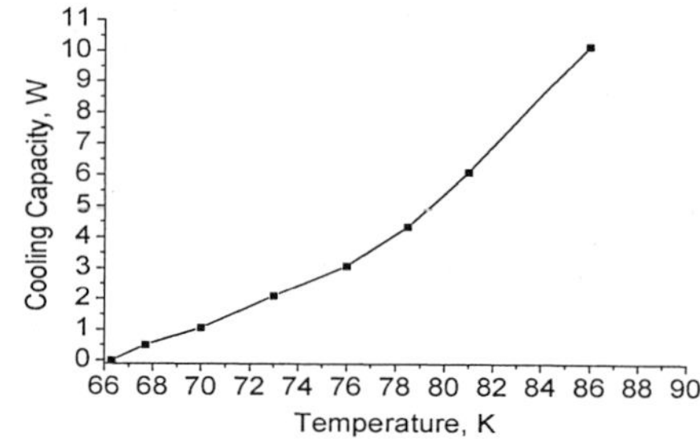

Figure 4 Flow rate variation curve

Figure 5 Cooling capacity vs. temperature

SUMMARY

A mixed-refrigerant Joule-Thomson refrigeration system is developed, built, and tested in our experiment for operating at 70 K temperature range. A stable lowest temperature at 66 K and 1.06 W cooling capacities at 70 K were achieved with an input power of 0.5 kW. The results obtained in this investigation shows that with optimal mixed-refrigerant and cycle configuration, the mixed-refrigerant J-T cryocooler can provide refrigeration at 70 K temperature properly. This provides a low cost, high reliability refrigerator for 70 K temperature range application, especially for cooling high temperature superconductivity devices.

REFERENCES

1. Luo E.C. Study on mixed-refrigerant Joule-Thomson cryocoolers in liquid nitrogen temperature range, in Ph.D. thesis, Beijing, (1997)

2. Cravalho E. G., Smit J. L., in Engineering thermodynamics, Pitman, Boston, (1981)

3. Wu, J.F., Gong, M.Q., et al, A new type mixture auto-cascade refrigeration cycle with a partial condensation and separation reflux exchanger and its preliminary experimental test, in Adv. in Cryo. Eng. (2002) 47 887-892

Proceedings of the Twentieth International Cryogenic Engineering Conference
(ICEC 20), Beijing, China. © 2005 Elsevier Ltd. All rights reserved.

Development of a cryogenic water-trap based on the recuperative mixture refrigeration cycle[*]

Gong M.Q., Wu J.F., Hu Q.G.

Technical Institute of Physics and Chemistry, Chinese Academy of Sciences
P.O. Box 2711, Beijing 100080, P.R. China

In this paper, the details of a cryogenic water-trap based on the recuperative mixed-refrigerant refrigeration are disclosed. The cryogenic water-vapor trap is driven by an oil-lubricated single-stage compressor with nominal input power of 500 W. The lowest temperature is 110 K, at which the saturated vapor pressure of water is about 1e-10 Pa. A comparison was made between a vacuum system with and without the cryogenic water-trap. The result is that the pumping down time with the cryogenic water-trap can be reduced greatly as ten times. This will be important for most vacuum systems, e.g. vacuum deposition system, semiconductor industry, etc. Compared with the traditional cryogenic pump, there is no requirement of periodic maintenance. Moreover, this new developed cryogenic water-trap has high reliability, low cost, and easy to be built in large scale based on numerous merits of mixed-refrigerant refrigeration system.

INTRODUCTION

Water vapor is a major problem in most vacuum systems. It increases the pump-down time to reach the required base pressure. When the vacuum achieves 1e-4 Pa, more than 60 to 90 percent in residual gases is water vapor [1]. Traditional mechanical vacuum pumps including the molecular turbo-pump are not efficient in pumping water vapor. However, a cold interface at a low temperature range has a high water vapor pumping efficiency. This is the so-called cryogenic water-vapor trap. Indeed, the cryogenic water-vapor trap can not work independently. The work mechanism of the cryogenic water vapor pump is just using a cold interface to condensate the water from vapor phase to solid phase. So it is a passive working method. Usually, the water trap is used accompanied with an active vacuum pump such as the molecular turbo pump, oil diffusion pump, etc.

The temperature of the cold-trap interface varies from 110 K to 150 K based on the required vacuum associated with different base pressure required. The base pressure of the vacuum system is determined by the partial pressure of water vapor in the vacuum. Figure 1 shows relationship of between temperature and the partial pressure [2]. The equilibrium pressure of water is determined by the temperature. Therefore, the partial pressure of the water vapor varies from 1.0e-10 to 1.0e-7 Pa associated to the temperature range from 110 to 150 K. This required temperature range is just suitable for the low temperature mixed-refrigerant refrigerator. Therefore, a low temperature water-vapor trap was developed based on a mixed-refrigerant Joule-Thomson refrigerator. The details of the design and test performance of a cryogenic water-vapor trap using a mixed-refrigerant refrigerator will be presented in this paper.

---

[*] This work is financially supported by the National Natural Sciences Foundation of China under the contract number of 50206024.

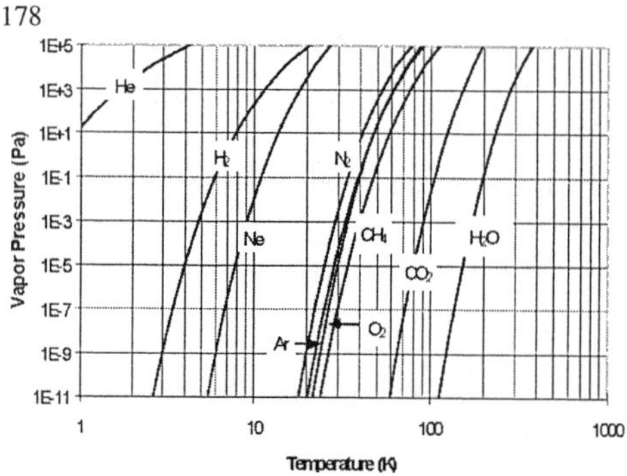

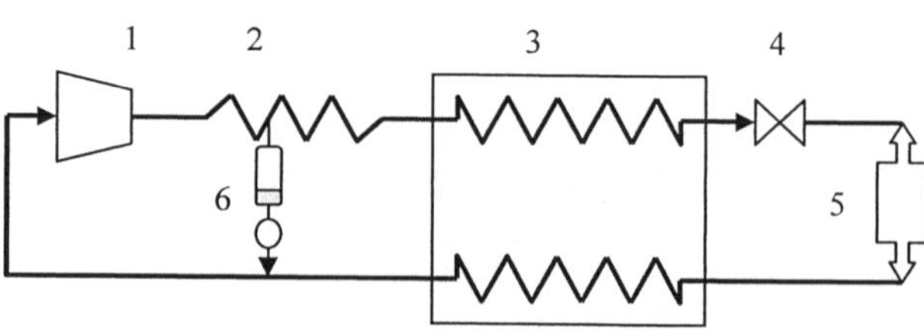

Figure 1 Equilibrium vapor pressure curves for common gases   Figure 2 Schematic diagram of the refrigerator

DESIGN OF THE CRYOGENIC WATER TRAP

Design of the mixed-refrigeration subsystem

The refrigeration power required at the temperature range described above is different for different vacuum vessel size, which determines the vacuum pump used. In this experiment, a standard vacuum test setup with a nominal inner diameter of 160 mm is used to study the performance of the new developed cryogenic water vapor trap. Therefore, a suitable mixed-gases Joule-Thomson refrigerator was used to improve the evacuation performance by removing the extra water vapor in the vacuum system. Figure 2 shows the schematic diagram of the refrigerator. The compressor used in this refrigerator has a nominal input power of 500 W. A tubes-in-tube heat exchanger was used as the recuperator. A capillary tube is used as the throttle valve. The evaporator is designed as the cold trap interface, which was fitted in the inlet of the turbo vacuum pump. An optimal mixture with six components was used as the refrigerant. The details of the refrigerator can be found in the reference [3].

Configuration of the cryogenic water trap system

A new configuration was developed in this cryogenic water trap system. Figure 3 shows the schematic diagram of the system. In this system configuration, the cold trap interface which is also acting as the evaporator in the refrigeration system, was put in the vacuum passage between the inlet of the vacuum pump and the outlet of vacuum vessel. The cold stage of the refrigeration system mainly consisting of the recuperator is put in a vacuum Dewar, which is evacuated by the first stage pump. On the other hand, the cold stage of the recuperator, in which the coldest temperature is almost as low as the cold trap interface, provides another cold trap for the first stage vacuum pump*. This obviously decreases the pressure after the turbo pump, and improves the performance of the vacuum system greatly.

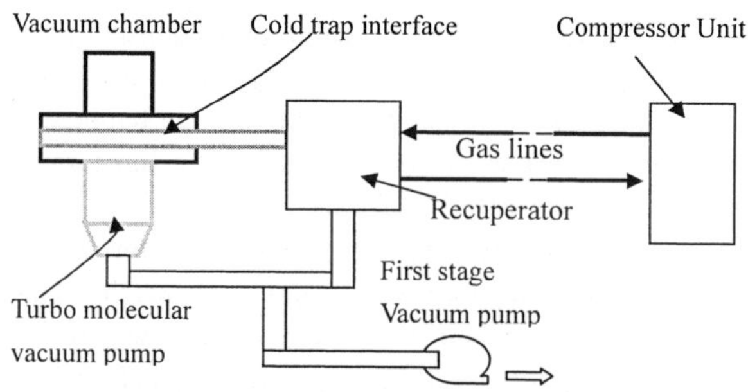

Figure 3 Schematic diagram of the system

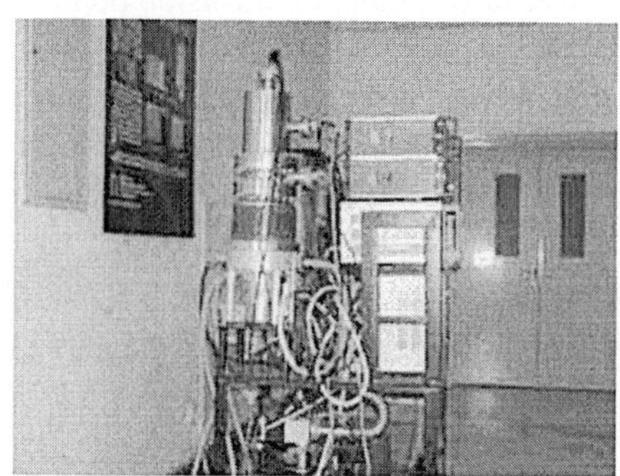

Figure 4 Photo of the system

---

* Patent pending

Figure 5 Photo of the cold trap interface

Figure 6 Configuration of the trap interface

EXPERIMENTAL TEST RESULTS

A prototype of the cryogenic water vapor trap, based on the design consideration described above, is developed and tested. Figures 4, 5 and 6 show the photos of the prototype and the cold trap interface. This prototype was tested comprehensively to know its performance. Comparisons of the pumping speed and the ultimate pressure of the system were made with and without the cryogenic water vapor trap. Figure 7 shows the results of the comparison. From Figure 7 a), it is obvious that the pump speed with the cryogenic trap is almost about ten times faster than that without the cryogenic trap. Moreover, the ultimate vacuum pressure is also ten times lower than that without the cryogenic trap. The improvement of the performance is same for a similar cryogenic water-vapor trap – Aquatrap from Polycold Company [4, 5], which is shown in Figure 7 b). The temperature cool-down curve of the cold interface is shown in Figure 8 as well as the variation of the first stage pressures at two situations that with or without the cold trap.

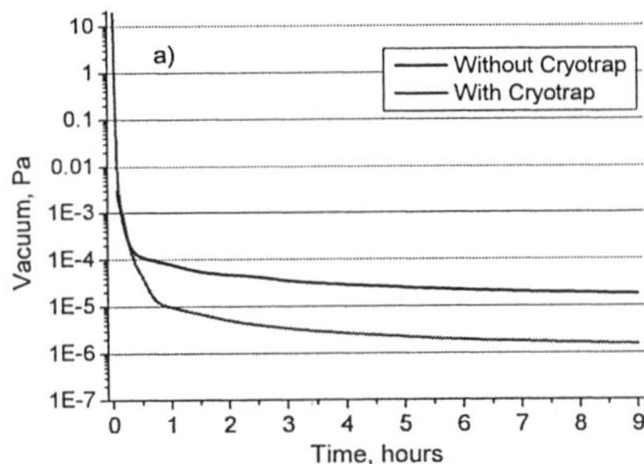

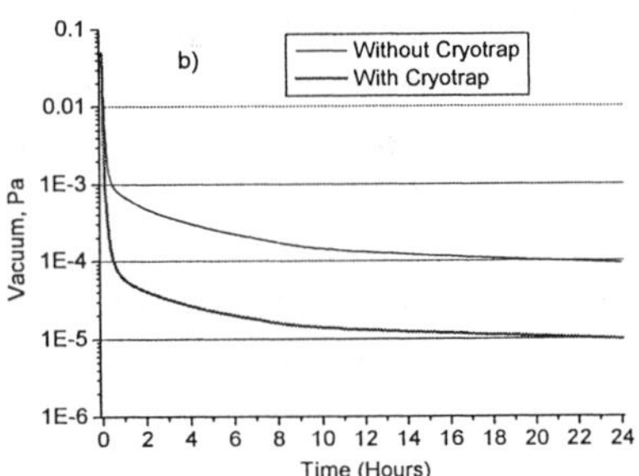

Figure 7 Vacuum performance of the cryo-trap, a) prototype built in this work, b) Aquatrap [5]

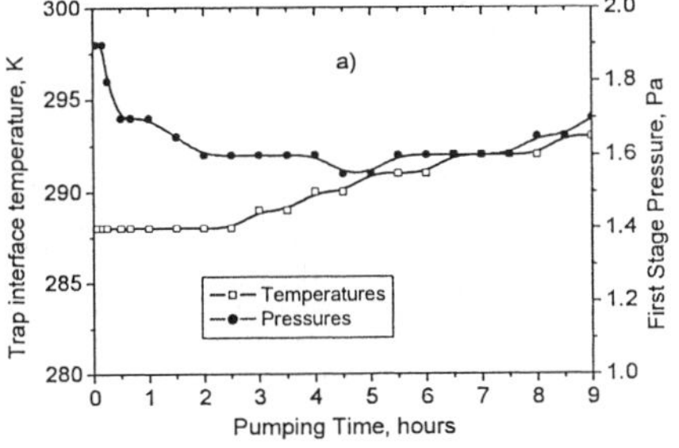

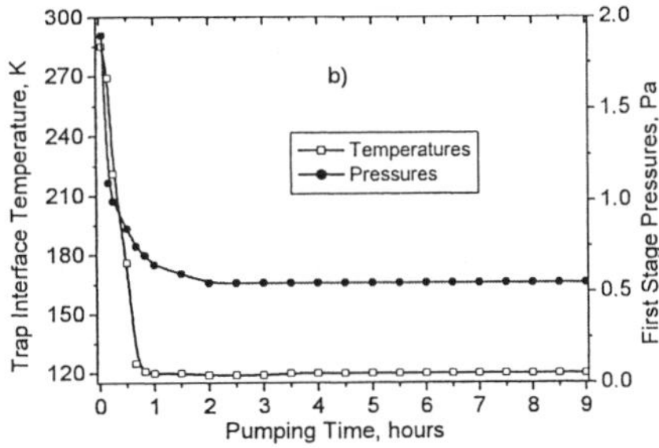

Figure 8 Variations of interface temperature and the first stage pressures, a) without b) with cold trap

In the prototype built in this work, when the cryogenic trap does not operate, the first stage steady operation pressure is about 1.6 Pa, and the vacuum ultimate pressure is 2.0e-5 Pa after 9 hours operation. However, when the cryogenic trap works, the first stage pressure decreases down to about 0.55 Pa in the final steady state operation, and the ultimate pressure reaches up to 1.6e-6 Pa. The decease of the first stage pressure can improve the performance of the turbo-molecular pump. The final temperature of the trap interface is about 120 K in the steady state operation.

The vacuum system pumped by a turbo-molecular pump combined with a low cost mixed-refrigerant Joule-Thomson refrigerator can provide the similar evacuating performance as a cryogenic pump. The traditional cryogenic pump evacuates the vessel by capturing gases with a colder interface which is cooled by a cryogenic refrigerator e.g. GM cryocooler at a cryogenic temperature as low as 20 K. Therefore, the built cost of a cryogenic pump is much larger than the vacuum system consists of a turbo pump and a cryogenic water trap. Moreover, the cryogenic pump needs regular maintenance, e.g. replacing the oil removing unit in the cryocooler compressor. However, the built cost cryogenic water-vapor trap cooled by a mixed-refrigerant Joule-Thomson Refrigerator is very low because of the usage of large-scale commercial compressor widely used in room temperature refrigeration field. Seldom maintenance is needed in a long operation period.

SUMMARY AND DISCUSSION

A cryogenic water-vapor trap based on the mixed-gases Joule-Thomson refrigerator was designed, built, and tested in this paper. The overall pumping speed of the system that combined with a turbo molecular pump is higher ten times than that without the cold trap. The cryogenic water-vapor trap presented in this work offers the highest pump speed for water vapor, and the lowest cost. When used with other high vacuum pumps, it offers the best overall performance available today for large vacuum systems operating in the range from atmosphere to 1.0e-6 Pa.

Indeed, all vacuum pumps including mechanical, oil diffusion, cryopump, ion, titanium sublimation, etc., have limitations. These limitations include cost, contamination, operational difficulty, selective gas pumping, pumping speed, pressure range and throughput. When attempting to select the high vacuum pump for any given application, it becomes obvious that there is no universal or ideal pump. The practice is to make serious compromises usually based on cost trade-offs. The proper approach is to utilize the advantages of the most applicable pumps and use a combination of pumps. The cryogenic water-vapor presented in this work combined with a turbo pump has many advantages over other kind pump include cost, availability, ease of operation, and faster response time in cool-down. They are easy to install on existing vacuum systems; it only requires locating a cryogenic interface on the chamber.

REFERENCES

1. http://www.igc.com/polycold/images/pdf_images/high_speed_water_vapor_cryopumping.pdf

2. Wagner L.C., Cryopumps applied to high temperature vacuum processes, 41st Technical Conference Proceedings (1998) 14-19

3. Gong M.Q., Wu J.F., Liu J.L., Hu Q.G., and Zhou Y., A mixed-refrigerant auto-cascade cryocooler operating in 120 K-150 K temperature range. Proceedings of 6th Sino-Japan seminar on cryogenics (2000) 258-263

4. Khatri, A. Boiarski, M. and Nesterov, S., Water trap refrigerated by a throttle-cycle cooler using mixed gas refrigerant, Adv. Cryo. Eng (1998) 43 1693-1702

5. Former IGC-APD Aquatrap advertising booklet

Room temperature magnetic refrigerator using both metal Gd or/and Gd-Si-Ge alloys and a permanent magnetic field source

Lu D.W., Wu H.B., Yuan G.Q., Han Y.S., Xu X.N., Jin X.
National Laboratory of Solid Microstructures, Department of Physics, Nanjing University, Nanjing 210093, P.R.China

Wu W.
Dept. Materials Science and Engineering of Sichuan Institute of Technology, P R China

A room temperature magnetic refrigerator using both metal Gd or/and Gd-Si-Ge alloys and permanent magnetic field source was introduced in the present paper. The permanent magnet was assembled in a shape of cylinder in which there are 1.4 Tesla uniform magnetic fields. A temperature span of 25 K was reached for the alloys when environment temperature was about 290 K. But if the alloys are short of Ga, the less temperature range could be reached at the same temperature. In addition, the magnetic refrigerator was also filled with metal Gd as the refrigerant; a temperature span of 26 K was reached.

INTRODUCTION

Since the discovery of magnetic-caloric effect of some materials, the great effort has been made to make use of it in refrigeration. To utilize it cooling in low temperature of liquid helium has been quite successful. Compared with this cooling refulgence of low temperature range, the use in near room temperature is not gratifying. But there was still much progress in room temperature magnetic refrigerator. Brown [1], for example, designed a demonstration experiment of magnetic refrigeration around room temperature in 1976 which shown us the possibility of room temperature magnetic refrigeration (RTMR) and realization clue. Zimm group[2] in 1997 made the 1st magnetic refrigeration unit around room temperature, which is of high heat efficiency. But there is a common character that they all used the superconductor-coils to generate the magnetic field. Since room temperature superconductors has not been found so far, say nothing of using them in practice, maintaining a strong magnetic field by traditional superconductors-coils is quite expensive and inconvenient. Maybe it is a feasible method to substitute a permanent magnet for superconductor-coils, for it seems that enhancing the magnetic field intensity of permanent magnet by using new high-powered permanent magnetic materials and new assembling techniques is more hopeful than finding the room temperature superconductors these days. As a try, we used the assembled permanent magnet arrays to apply the working field and run the room temperature magnetic refrigerator in a reciprocating way.

As regards the magnetic refrigerant, we think that metal Gd may be suitable for a demonstrator for its large magnetocaloric effect. But using only Gd seems hard to gain an enough cooling power. We tried Ga-Si-Ge alloys [3] commended by Ames Lab to compare its performance with pure metal Gd in the demonstration device of magnetic refrigeration, the result was goodish although the alloys did not increase 30% cooling power as expected beforehand.

PRINCIPLE AND AN INTRODUCTION TO THE DEMONSTRATION DEVICE

For a general magnetic system, during the external magnetic field changes, the entropy of the system may be calculated according to the following expression (the Maxwell relation):

$$\Delta S = \mu_0 \int_{H_i}^{H_F} \left(\frac{\partial M}{\partial T} \right)_H dH \tag{1}$$

where T is the absolute temperature, S is the entropy of the system, μ_0 is vacuum magnetic conductivity, H is the external magnetic field, M is magnetic moment, H_i is the initial field, and H_f is the last field.

Expression (1) is a universal equation (e.g. for Gd) unless the first order magnetic phase transition occurs during the process. For the 1st order phase transition, we have the following expression:

$$\Delta S = \mu_0 \Delta M \frac{\Delta H}{\Delta \Theta} \tag{2}$$

where $\Delta \Theta$ and ΔH are the changes of temperature and applied field ,respectively, along the phase equilibrium curve, and the value ΔS and ΔM are both the differences between the two phases.
It is of dissension [4] if expression (1) still keeps correct during the process of magnetic phase transition or does not. We think that in a common situation, when applied field, the system does not only undergo the first order transition, but also undergoes the changes within the respective certain phase. We suggest a new mixed processing method to deal with this sort of problem of Gd$_5$Si$_2$Ge$_2$ alloy and other materials of the first order magnetic transition. We will introduce the processing method in other paper. But the final result still keeps true for Ga-Si-Ge alloys as behaves by using Maxwell relations. Under such a consideration we chose the materials as the representative of giant magneto-caloric effect materials to circulate in the demonstration device.

Because we use the permanent magnet arrays, the magnetocaloric temperature effect is about 2 K/Tesla, if we want to accumulate a large temperature varying range. We took it for granted that the AMR (active magnetic regenerator)[5] cycle should be employed.

The above is a brief summary of the magnetic refrigeration principle, the following is a description about the demonstration device and magnetic refrigerant.

Working materials
At the beginning, we chose metal Gd to fill in the magnetic refrigerant beds as a test. The Gd is of commercial quality in purity. Because the heat transfer time for material Gd is a bit too long, when its size is not small enough, we hope the material to be smaller than 0.2mm in diameter. To meet the need, we cut the clump Gd into thin sheets about 0.2mm in thickness, and then get them to be grinded in a tin with a spheroidizing mill. At last the Gd articles are basically shaped small spheres after such a process.

The typical heat transfer time for such a size of Gd spheres is about 2 seconds. The period of the AMR cycle is 6 seconds or so when the Gd spheres are used in the magnetic refrigerator. The total weight of Gd spheres filled in the two beds is about 1 kilogram.

Gd$_5$Si$_2$Ge$_2$ alloy and Gd$_5$Si$_{1.985}$Ge$_{1.985}$Ga$_{0.03}$ alloy were also filled in the beds of the AMR to circulate. The sphere shape of Gd-Si-Ge alloys is more regular than that of Gd spheres. The total weight of the alloys is also about 1 kilogram. By the magnetizing curves of the alloys, the alloys quality was equivalent with that prepared by Ames Laboratory.

The permanent magnet arrays

The permanent magnet system is made of many small cubic magnet units (see Figure 1). Every magnet unit is shaped as a small sector in which the magnetic field orients to a certain direction regularly. The angle between the neighbor directions is 45 degrees. The integral magnet arrays are shaped like a cylinder whose diameter is 140 centimeters. There is a hole at the center position of the magnet arrays cylinder, the diameter of the hole is 3 cm. All the magnet arrays co-generate a quite uniform magnetic field about 1.4 Tesla in the hole. There is a well-ranged magnetic field distribution (see Figure 3). The length of the cylinder is 0.2 m and the magnetic working materials will be magnetized in the hole of working space and demagnetized outside of the hole.

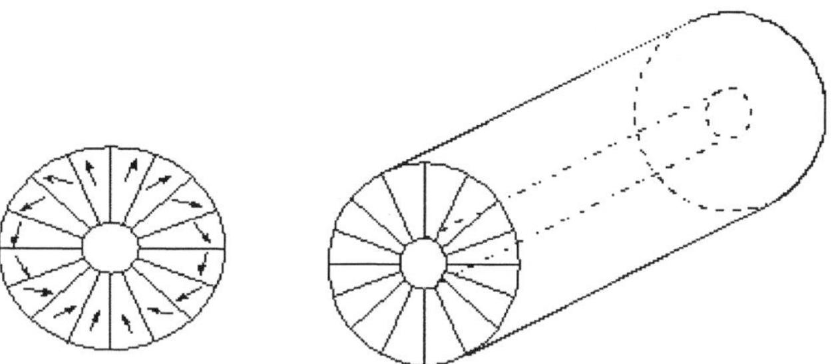

Figure 1   The field of O-shaped magnet arrays cylinder

Working style of the device

The room temperature magnetic refrigerator device works in a reciprocating way. There are two beds (see figure 2) in which the magnetic refrigerants such as small Gd spheres et. al.. We devised the magnetic refrigeration demonstration device using the above permanent magnet arrays, and the pneumatic drives are employed both for the movement of magnetic refrigerant beds and the flow of cycle water. The cycle water is sealed from the air and is confined between two chambers of a cylinder which is driven by another cylinder as the role of a pneumatic driver. The piston of the later drives that of the former to move to and fro, the air within the later is so compressed that the cycle water will be impelled to flow in the AMR. The connection tube between the cold sink and the cylinder is plastic. Because the heat transfer of the plastic tube and the air is much less than that of water and metal materials, the main heat loss of the system is from AMR. As for the heat transfer of the AMR, we did a simple estimation with the size of the AMR system. The AMR consists of a stainless tube and working particles in it. If the small particles are Gd, the calculating result shows that the size of the AMR is quite reasonable for the minimal heat loss.

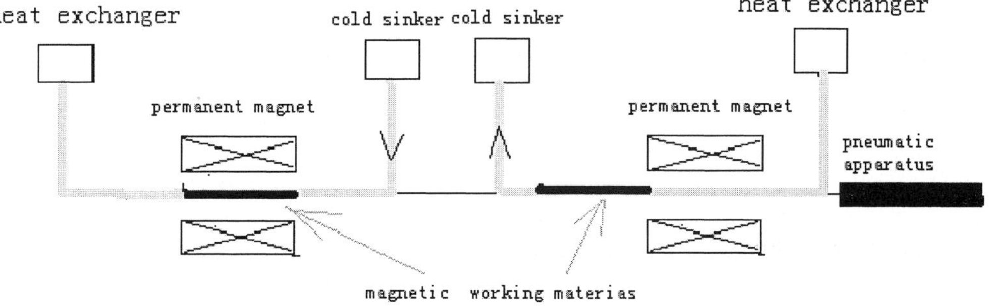

Figure 2   The sketch of a reciprocating magnetic refrigerator

RESULTS AND CONCLUSIONS

In real circulation, the system reached a temperature span more than 26 K when Gd spheres were used, and a temperature span of 25 K when $Gd_5Si_{1.985}Ge_{1.985}Ga_{0.03}$ spheres were used. When we substituted these two kinds of working materials as $Gd_5Si_2Ge_2$ alloy, the temperature span was less than 20 K. We refer the result about $Gd_5Si_2Ge_2$ to an ill-suited working temperature region.

We tried to couple using the working materials in the real cycle, the experiment did not go far, for there was something wrong with the equilibrium of forces applied on two beds. But the alloy $Gd_5Si_{1.985}Ge_{1.985}Ga_{0.03}$ worked well in the real circulating at any case. For there were some difficulties in shaping control of working materials, we did not succeed in preparing the three working materials exactly in the same shape of sphere. Although we cannot conclude that Gd-Si-Ge alloys have a better cooling effect than metal Gd, but the alloys are indeed valuable because of its optimal chemical stableness.

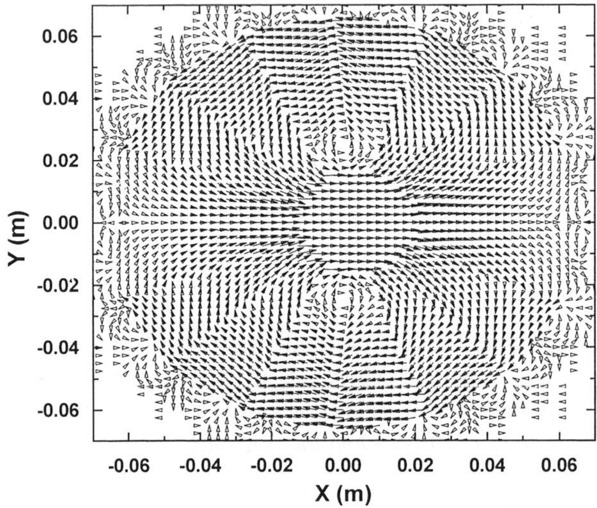

Figure 3   The magnetic field distribution in the hole of the cylinder

ACKNOWLEDGEMENT

This work was supported financially partly by the project of the Key Technologies of Research and Development (99-A30-01-03), and partly by the HTRDP(863) of P.R. China government, the serial number of the project is 2002AA324010.

REFERENCES

1.Brown, G. V., J. Appl. Phys. (1976), 47, 3673

2.Zimm C., Jastrab A., Sternberg A., Pecharsky V., Gschneidner K. Jr., Osborne M., and Anderson I., Adv. Cryog. Eng. (1998), 43, 1759

3.Pecharsky V. K. and Gschneidner K. Jr., Phys. Rev. Lett. (1997), 78, 4494

4.Giguere A., Foldeaki M. Ravi B., Chahine R., Bose T., Frydman A. and Barclay J., Phys. Rev.Lett.(1999), 83 2262

5.Barclay, J. A., and Steyert W. A. Cryogenics (1982b), 22, 73

Properties of thermoelectric device with $(Bi_xSb_{1-x})_2(Te_ySe_{1-y})_3$ materials

Liu H-J.[1,2], Li G.[1], Li L-F.[1]

[1] Technical Institute of Physics and Chemistry, Chinese Academy of Sciences, Beijing 100080, China
[2] Graduate School of the Chinese Academy of Sciences, Beijing 100039, China

Thermoelectric materials are of interest for applications as heat pumps and power generators. The COP of thermoelectric devices is quantified by the figure of merit, ZT, where Z is a measure of thermoelectric devices and T is the absolute temperature. The $(Bi_xSb_{1-x})_2(Te_ySe_{1-y})_3$ alloys are state-of-the-art room temperature materials. This paper focuses on the properties of refrigeration of these kinds of materials. The electrical properties of thermoelectric materials were measured from 80-300 K. Power factor of materials were given, respectively. The properties of thermoelectric devices are studied at different temperature, the cooling power and cooling temperature of thermoelectric devices are given.

INTRODUCTION

Traditional approaches to cooling are based on thermodynamic cycles involving compression and expansion of refrigerant gases. Thermoelectric refrigeration devices, in contrast, do not rely on dynamic cycles of gases. It relies on physical a phenomena called the Peltier effect. When an electric current passes through a thermoelectric material, the heat transported by charge carriers (electron or hole) leads to a temperature gradient [1]. Heat is absorbed on the cold side and rejected at the sink, thus providing a cooling temperature.

Thermoelectric devices have several distinct advantages, including non-moving parts, non-vibration, quiet performance and spot cooling. More importance, it is a "green" refrigeration with no freon refrigerants and a negligible global warming potential. They will be applied in the cooling of CCDs (charge-coupled devices), infrared detectors, low-noise amplifiers, and computer chips. Thermoelectric refrigeration is also being considered in the automobile industry for use in the "next-generation vehicle". The most common application of these materials is the thermoelectric cooler and warmer, which are sold at many local stores. The essence of defining a good thermoelectric material lies in material's figure of merit, $Z = S^2\sigma/K$, where S is the Seebeck coefficient (defined as $\Delta V/\Delta T$, V is voltage and T is absolute temperature), σ is the electrical conductivity, and K is the total thermal conductivity ($K=K_e+K_L$, the lattice and electronic contributions, respectively). Alloys based on the $(Bi_xSb_{1-x})_2(Te_ySe_{1-y})_3$ system for room temperature applications have higher performance for thermoelectric refrigeration. For all the current state-of-the-art materials, the dimensionless figure of merit, ZT, is about 1. There is no theoretical or thermodynamic reason why it cannot be larger. Researchers have done lots of work to improve thermoelectric properties of materials. Up till now, how to reduce the lattice thermal conductivity becomes more important in improving materials' figure of merit. This concept is related to Slack's earlier assertion of a "phonon-glass/electron-crytal" model, which suggests that a good thermoelectric material should have the electronic properties of a crystalline material and the thermal properties of a glass. In this paper, we focused on the properties of thermoelectric refrigeration devices at different temperature. The electrical conductivity and Seebeck coefficient of n-type and p-type materials were measured, and the cooling power and cooling temperature (Δ T) of the thermoelectric device are also given.

PROPERTIES OF MATERIALS

Three kinds of materials were employed in our experiments. Alloy A (p-type) and alloy B (n-type) are commercial products from Northern Refrigeration Company (Harbin, China). Alloy C is prepared by mechanical alloying method in our laboratory with the milling time for 80hrs, and its chemical composition is $(Bi_2Te_3)_{0.25}(Sb_2Te_3)_{0.75}+3wt\%Te$.

XRD patterns
The XRD patterns of the alloy A and alloy B are shown in Figure 1. There are not Bi, Te, Sb and Se traces in the diffraction patterns of alloy A and alloy B. XRD patterns of alloy A and alloy B materials are almost same because they have same crystal structure.

Electrical properties
The electrical conductivity of the studied materials was measured using the four-probe method. The results of the measurements are shown in Figure 2. For impurity semiconductor, concentration of current carriers almost doesn't change with the

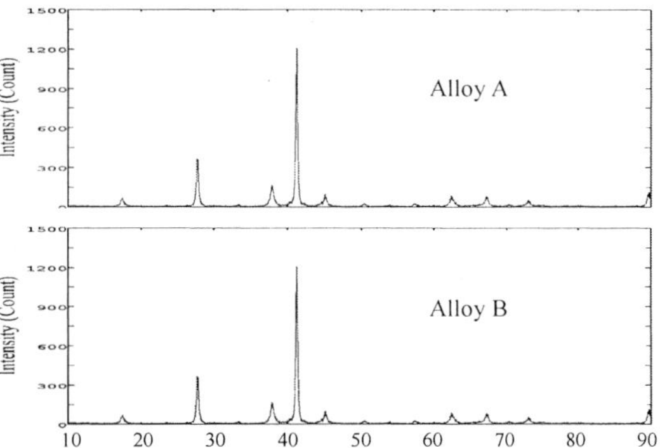

Figure 1 X-ray diffraction patterns (Cu K$_\alpha$ radiation) for A and B materials

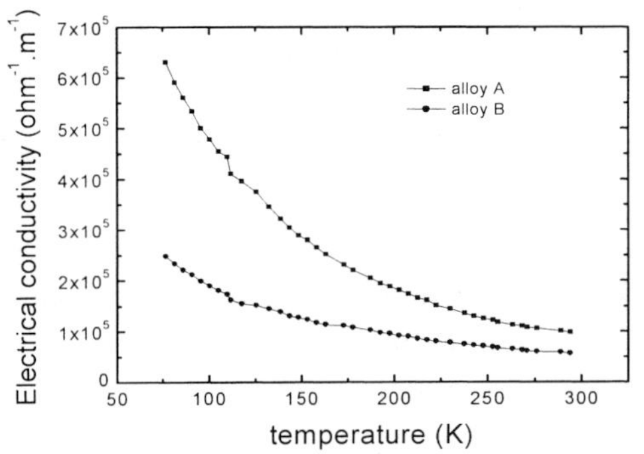

Figure 2   Materials' Electrical conductivity changes with temperature 80-300 K

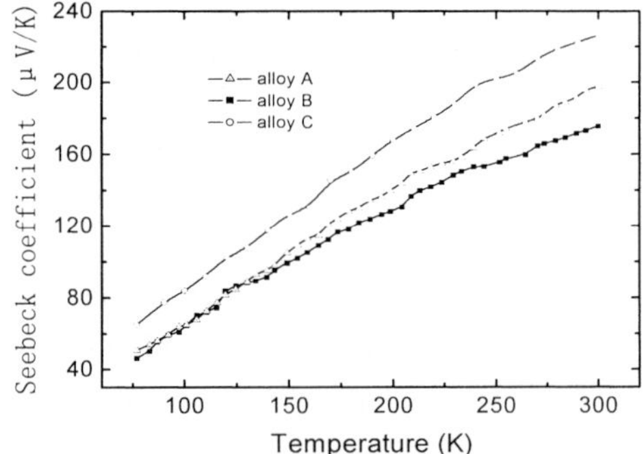

Figure 3   Seebeck coefficients of materials as a function of temperature 80-300 K

change of temperature when intrinsic excitation is not very strong. Lattice vibration has a strong impact on carrier mobility. With the increase of temperature, lattice vibration becomes stronger, so as to lead the decrease of the carrier mobility. The experimental results show that the electrical conductivity of alloy A (p-type) and alloy B (n-type) are $0.99\times10^5\ \Omega^{-1}.m^{-1}$ and $0.68\times10^5\ \Omega^{-1}.m^{-1}$ at 300K, respectively.

Figure 3 is the measurement results of Seebeck coefficient. It indicates that the Seebeck coefficient as a function of temperature, increases with increase of temperature. It is noted that, to aid comparison, the absolute value of Seebeck coefficient of p-type materials was used in Figure 3. The alloy C (p-type) prepared by mechanical alloying method in our laboratory, is also studied, and the result (curve C) was also shown in Figure 3. The Seebeck coefficients of alloy B and alloy C are 197.6 and 226.3 μ V/K at 300 K, respectively. Seebeck coefficient of alloy C is higher than those of the commercial materials. The thermoelectric properties of alloy C, will be studied in the future, including thermal conductivity and electrical conductivity.

The power factor of alloy A and alloy B is shown in Figure 4. It shows that power factor increases with increase of temperature below 200K and almost invariable above 200K. The data of power factor of alloy A and alloy B are 3.8×10^{-3} and 2.2×10^{-3} W/m • K^2 at 300 K, respectively.

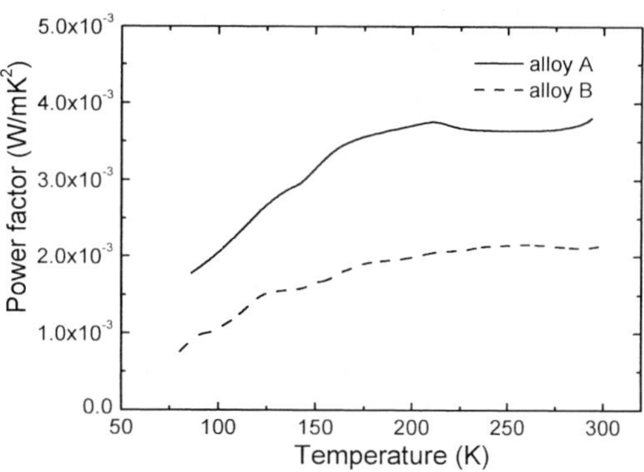

Figure 4　Power factor as a function of temperature from 80 K to 300 K

PERFORMANCE OF THERMOELECTRIC DEVICE

It is apparent that the figure of merit Z, as defined by $Z = S^2\sigma/K$, is a characteristic not of a pair of materials but, rather, of a particular couple. For a given pair of materials, we can write Z as following equation [6]:

$$Z = (S_p - S_n)^2 / [(\kappa_p \rho_p)^{1/2} + (\kappa_n \rho_n)^{1/2}]^2. \tag{1}$$

Actually, above equation is rather cumbersome when attempting to find a good thermoelectric material, since it involves the properties of both thermoelements.

A Peltier module is composed of thermoelectric couples （alloy A and alloy B）, heat conducting ceramic plates and some electrically conducting wire. Ceramic plates form the cold and hot surfaces of the module and provide good heat transfer and low electrical conductivity. The ceramic face size of the tested device was $3.5 \times 4.0 cm^2$, the test was carried out at 1atm. The hot side plate was cooled using water

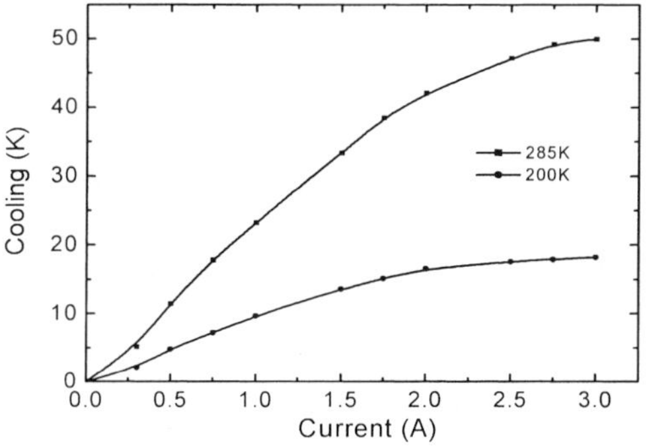

Figure 5　Observed cooling temperature as a function of current in a 3. 5×4 cm^2 rectangle

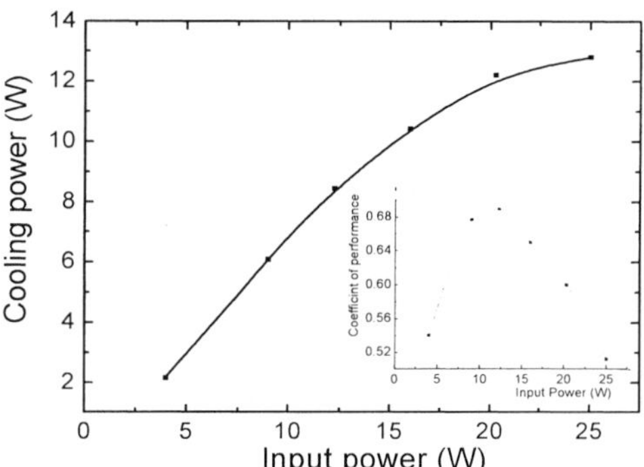

Figure 6　Cooling power as a function of input power (hot side temperature was 285 K)

188

at 285 K and LN₂ as a cold background at 200 K. Other researchers have done a lot of work in thermoelectric device at room and low temperature [2-5].

The observed cooling temperatures and cooling power (285 K and 200 K hot side plate temperature) as functions of current or input power are shown in Figure 5 and Figure 6 respectively. There are 49.5 K and 17.5 K cooling temperatures under hot side temperatures of 285 K and 200 K, respectively. In this bulk device, 13 W output power at 25 W input power is obtained when the hot side temperature is 285 K and cooling temperature is 20 K.

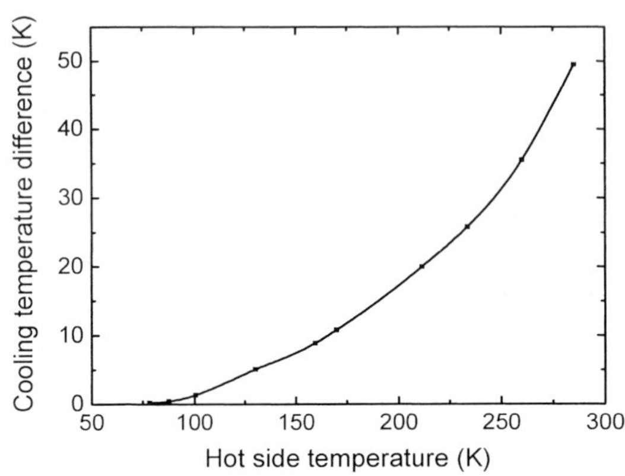

Figure 7   Cooling temperature as a function of hot side temperature

Figure 7 shows the change of cooling temperature differences at different hot side temperatures. The temperature difference of 1.4 K was obtained when the hot side temperature was 100 K. The temperature difference was only 0.45 K when the hot side temperature was 80 K.

SUMMARY

In summary, the electrical properties of materials have been measured as a function of temperature. The properties of bulk thermoelectric device have been measured. When the hot side temperatures are 285 K and 200 K, the cooling temperature differences are 49.5 K and 17.5 K, respectively. The maximum cooling power obtained is 13 W when the input power is 25 W.

REFERENCES

1.Goldsmid, H. J., Electronic Refrigeration , Pion Limited, London, (1986)

2.Vedernikov, M.V., Kuznetsov, V. L., "Cooling Thermoelements with Superconducting Leg", CRC Handbook of Thermoelectrics, Edited by Rowe, D.M., CRC Press, Boca Raton, (1994), 609-616

3.Xuan, X. C., Ng, K.C., et al., "A two-stage cuboid-styled thermoelectric cooler with switched polarity", Twentieth international conference on thermoelectrics, Beijing, China, (2001) 444-447

4.Venkatasubramanian, R., et al, Nature, (2001), vol. 413, 597

5.Venkatasubramanian, R., "Thin film superlattice and quantum-well structure-a new approach to high-performance thermoelectric materials", Naval Res. Rev, (1996), 58, 31-40

6.Goldsmid, H.J., CRC Handbook of Thermoelectrics, Edited by Rowe, D.M., CRC Press, Boca Raton, (1994), 19-25

Future trend of pulse tube cryocooler research

Matsubara, Y.

2-3-3 Oanakita Funabashi Chiba 274-0068, Japan

The flexibility of a pulse tube allows integrating the cryocooler with its application. However, the drawback will be a deterioration of thermodynamic efficiency because of the application oriented design philosophy. To minimize the additional thermodynamic losses, the extending cycle analysis, which can simulates the requested configuration, will be required. Therefore, one of the important subjects for the future progress of such a flexible pulse tube coolers is a substantial improvement of the software as a design tool of the cryocooler, which could be used to the actual application. Several proposals based on a basic cycle analysis are presented.

INTRODUCTION

Earlier stage of pulse tube studies was mainly focused on the improvement of thermodynamic efficiency, which results in the development of various phase-shift mechanisms. Many phase-shift mechanisms have been developed in this stage [1-3]. As the thermodynamic efficiency of the pulse tube approaches to that of Gifford-McMahon (GM) or Stirling cycle cryocoolers, the research target was set to achieve the possibility of lower cooling temperature by utilization of multistage pulse tube [4]. Efficient multistage method was also developed [5,6]. From the viewpoint of cryocooler application, the importance of the object oriented cooler development has been well recognized and some of the cryocoolers are already specialized from such a viewpoint of cost reduction, reduced mechanical vibration or extended operating lifetime.

In the case of pulse tube cryocoolers, however, non-obstructive or invisible cooler development will be possible by step into the application field furthermore. For example, the cold part of the pulse tube cooler could be utilized as a part of the cooling object such as, the supporting rod of the cold mass. The design of such an integrated pulse tube cold part without restraint of optimized cooler configuration leads to the deterioration of thermodynamic efficiency of the cooler. To minimize the additional thermodynamic losses, the extensive cycle analysis will be required to simulate the requested configuration. Therefore, one of the important issues for the future progress of such a flexible pulse tube cooler is a substantial improvement of the software as a design tool of the cryocooler, which could be applied to the actual application. This paper describes several examples of the valved and the valve-less pulse tube coolers.

BACKGROUND OF PULSE TUBE COOLER DEVELOPMENT

Cryocoolers are classified into two different types from the viewpoint of the operating gas flow patterns as shown in Figure 1. The circulating flow type composed of a turbo-expander or a low temperature valved reciprocating expander with a counter flow heat exchanger, while the oscillating flow type consists of the valve-less expander and a regenerator. Thus, the compact system can be fabricated, using the latter type cryocooler. In fact, many small-scale cryocoolers based on the oscillating flow type are applied to the wide range of application fields. It has been considered that the pulse tube cooler, as one of the oscillating flow type, has a potential to replaces the other type of coolers such as, Stirling, GM, Solvay and Vuillemier cycle. However, one of the difficulties to develop the pulse tube cooler is its systematic design; the interactions of composed components are complicated and it is hard to apply thermodynamic analyses.

190

In the case of circulating flow type, the function of each component, such as a compressor, a counter flow heat exchanger, an expander or a JT valve, are rather independent of each other. However, the function of regenerator for oscillating flow type cryocoolers has a strong dependency of other components.

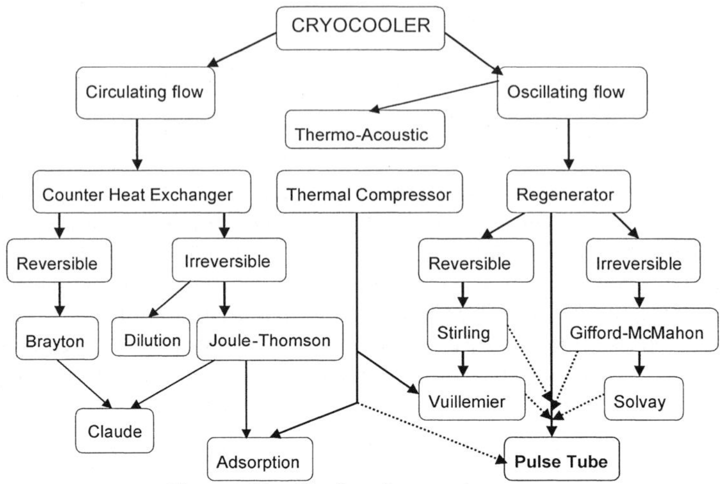

Figure 1 Classification of cryocoolers

Therefore, the design of oscillating flow type cryocooler is much more complicated than the design of circulating flow type. The pulse tube cryocooler, which is classified as one of the oscillating gas flow type, is further complicated to design the components independently each other, because it has no solid displacer or expander. The concept of equivalent PV work makes the simulation easier. The minimum requirement of each component will be predicted by means of these simplified simulation models.

GM-TYPE (VALVED) PULSE TUBE COOLER

Figure 2(a) shows generalized schematics of a valved type regenerative cryocooler. The pressure wave is generated by the combination of a compressor and switching valves consist of V1 for the high pressure intake and V2 for the low pressure exhaust.

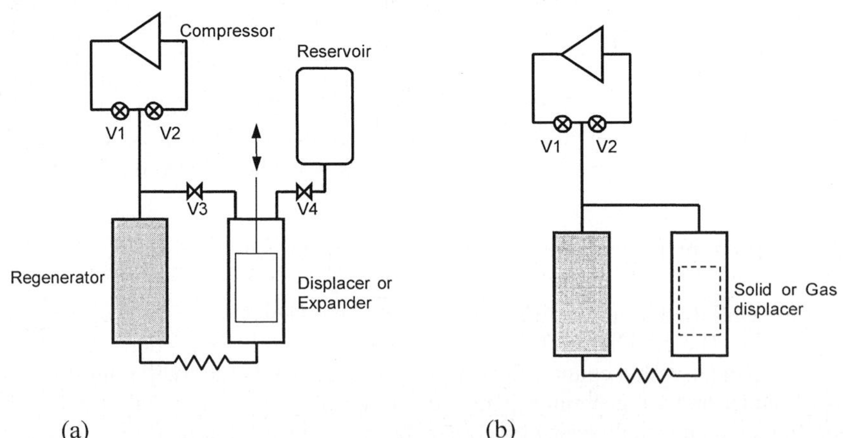

Figure 2   Generalized schematics of valved type pulse tube cooler

If V4 is fully closed and V3 is opened, it becomes a GM cycle cooler, which requires external mechanical force to drive the displacer: controlled opening rate of V3 gives the minimum driving force of the displacer. If V3 is fully closed and V4 is opened, it becomes Solvay cycle cooler: controlled opening rate of V4 gives pneumatically controlled expander. If the solid displacer is removed, it becomes the orifice (V3 close, V4 open), the double inlet (V3 and V4 open) and the by-pass (V3 open, V4 close) pulse tube coolers, respectively. Here the third case named as the by-pass pulse tube is not used commonly, however, it would be used when the total volume of the cooler must be minimized, because it does not requires a large reservoir volume.

The valved type cryocooler is commonly used because of its flexibilities to the applications; however, it has a lower thermodynamic efficiency than that of Stirling type cryocooler, in general. This lower thermodynamic efficiency is primarily caused by the work loss at the switching valve system. Therefore, to find out the way to reduce this loss by the appropriate analytical method would be an important issue for the future development of cooler application. The loss at the valve will be classified into two groups. One is a limited size of the valve port, which gives a large pressure drop and is mainly depending on the valve design. The other is a loss due to the large pressure difference just before the valve opening and is mainly depending on the phase shifter process of the cycle. Therefore, the later one could be minimized by selecting the optimum relation of the expansion volume with the pressure ratio.

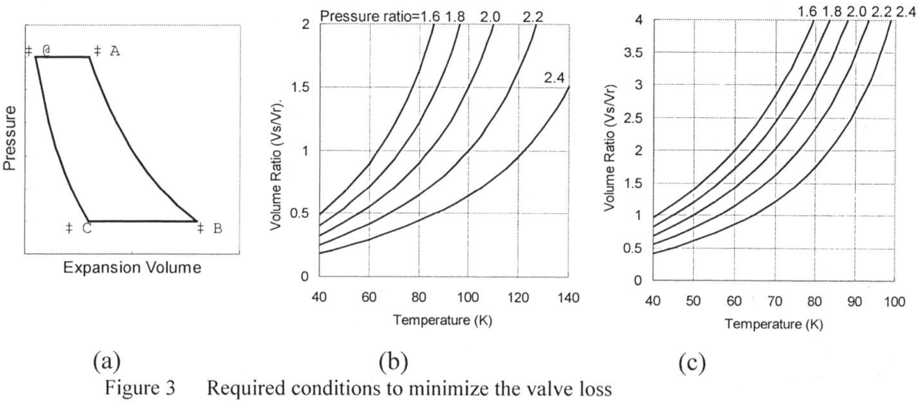

Figure 3   Required conditions to minimize the valve loss

In the case of GM cycle, this optimization is limited because the total volume is fixed. Figure 2(b) represents a simplified model of GM cooler having the regenerator void volume of Vr and the piston displacement volume of Vs. To minimize the work loss at V1 and V2, a PV diagram at the cold end of the piston as shown in Figure 3(a) is desirable. When the piston is located at the cold end, V1 is opened at the point ①. The piston is moved up, then the gas flow to the cold end volume with the pressure of Ph. As V1 is closed at the point ② and keep moving the piston, the pressure decreases. The pressure becomes Pl at the end of the piston stroke, the point ③. V2 is opened and the piston moves downward. V2 is closed at the point ④ and keep moving the piston, then the pressure increases. The pressure becomes Ph at the end of the stroke and the cycle is completed. These conditions are realized when the ratio of the swept volume and the regenerator void volume is selected with particular sets of pressure ratio and the temperature ratio as shown in Figure 3(b). For example, if the pressure ratio is 2 and the cooling temperature is 80 K, then the volume ratio of 0.9 should be selected to minimize the valve loss.

Modification of GM-type pulse tube cooler
Extending this consideration, optimization of GM-type pulse tube cooler will be obtained. Similar calculation has been done for the pulse tube cooler simply replacing the solid piston of Figure 2(b) to a compressible gas piston. The result is shown in Figure 3(c). Similarity in the operating condition of GM cooler can be seen, although the available pressure ratio at the given temperature ratio is more critical than the case of GM cooler.

In the case of Solvay cycle, however, the total volume can be changed by changing the expander stroke volume without any restriction discussed above. If the required cooling temperature is lowered, the regenerator performance becomes more important. In such a case, it is recommended that the Solvay cycle type phase shifter, which released from the fixed total volume specialized in the GM cycle, should be selected to realize the efficient valved pulse tube cooler.

Multi-staged pulse tube cooler

Figure 4   Displacers with no pressure seal for 4K GM

cooler (Photo is offered by SHI)

In the case of multi-staged pulse tube, especially 2-staged 4K pulse tube cooler, is always compared with a 4K GM cooler. Thermodynamic performance of 4 K GM cooler is better than that of 4K pulse tube cooler, in general. Demerits of 4K GM cooler are frequently regarded as the large mechanical vibration, a short maintenance interval due to the existence of the solid displacer and its pressure seal at the low temperature part. The recent progress of 4K GM cooler, however, does not require any low temperature seal as shown in Figure 4. Therefore, if the thermodynamic efficiency and the cooler setting orientation are the most important issue, the 4K GM cooler still has superiority over the 4K pulse tube cooler. The merit of 4K pulse tube will be appeared when the single stage pulse tube using He3 is precooled by another single stage pulse tube cooler. This method could give an efficient cooling performance at the temperature below 4K, using the minimum amount of He3 [7]. It is noted that the configuration of 4K GM displacer without pressure seal resembles well with the double inlet line of the pulse tube cooler as shown in figure 2(b).

STIRLING TYPE (VALVE LESS) PULSE TUBE COOLER

Stirling type pulse tube cooler has been developed as an alternative of the Stirling cooler. The flexure spring supported a linear compressor, first developed at Oxford University [8], has been successfully used for Stirling cooler and now it applied to the pulse tube coolers. With this compressor, which is now frequently called as a pressure wave generator, a single stage small-scale pulse tube cooler is already replaceable with Stirling cooler even from the viewpoint of thermodynamic efficiency. Recently, a medium or a large scale Stirling type pulse tube cooler becomes an important issue for application to the superconducting system of electric power field.

To realize such a cooler with a higher thermodynamic efficiency, however, the extensive analyses of each component and its interrelation is essential. This requires user-friendly software as a design tool of the cooler. Analysis of Stirling type pulse tube cooler always requires momentum equation because of its higher driving frequency up to a few tenth hertz. This induces more complicated simulation program to have optimum sizes of critical components.

The Stirling type pulse tube cooler utilizes a linear driven pressure wave generator as a work source and an inertance tube phase shift mechanism as a work receiver, which is shown schematically in Figure 5. To predict the inertance effect as a phase shifter, the numerical analysis has been executed [3]. An example calculated result in comparison with experimental result is shown in Figure 6(a) and (b). The boundary condition is selected as the piston movement of the pressure wave generator because it can also be obtained from the experiment. The measured data is based on the inertance tube of 10.7mm inner diameter and 1.9 meters length. Driving frequency was 49 Hz and the PV work of 1.5 kW, dashed line in Figure 6(b), was required when the cold end temperature is 80 K with heat load of 75 watts. Calculated PV work corresponding this operating condition is shown by doted line. Work flow <W> and enthalpy flow <H> through the regenerator, a pulse tube and an inertance tube are calculated as shown in Figure 6(a). Work flow is consumed within the inertance tube.

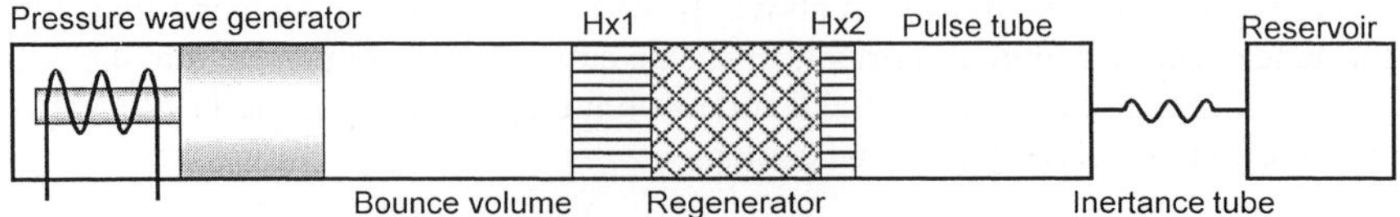

Figure 5   Valve-less pulse tube cooler

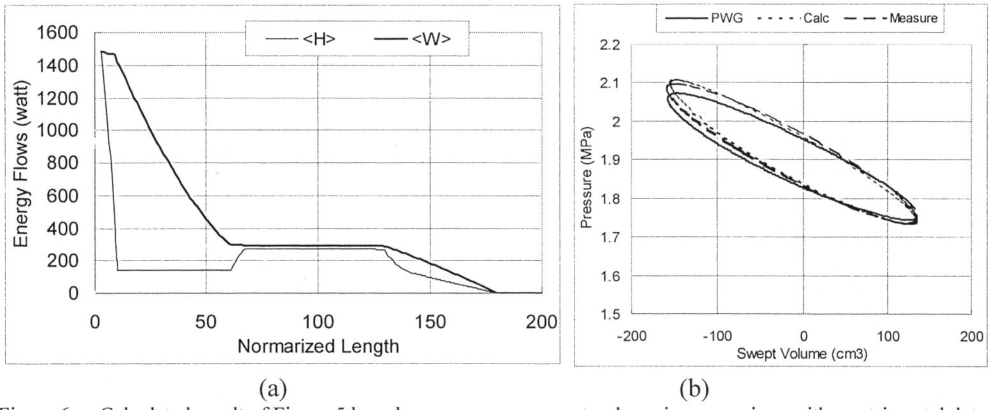

(a)   (b)

Figure 6   Calculated result of Figure 5 based on compressor swept volume in comparison with experimental data

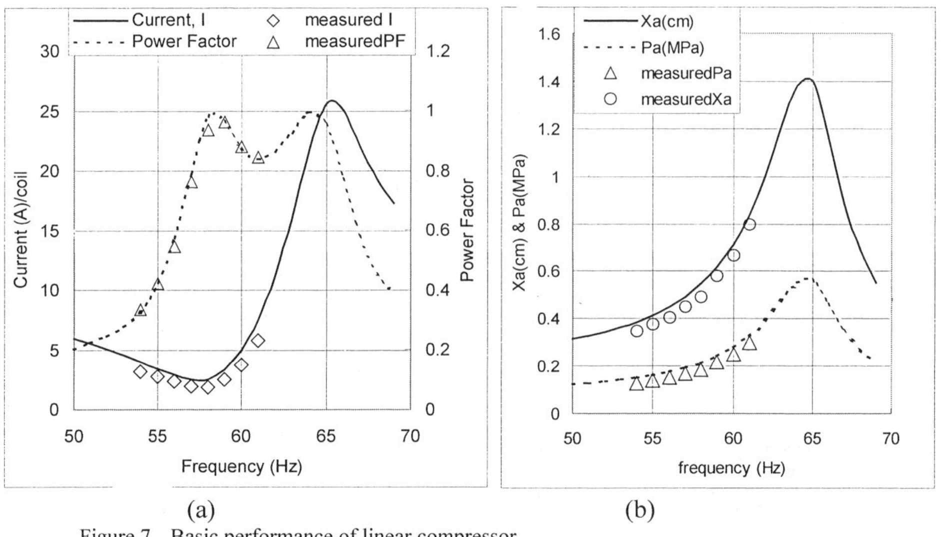

(a)   (b)

Figure 7   Basic performance of linear compressor

Performance of the linear motor for the pressure wave generator is separately calculated as shown in Figure 7(a) and (b). Frequency dependency of the performance is well described with the experimental data. It is noted that a bounce volume inserted between the compressor piston and the regenerator gives an important effect to the performance. Figure 8 shows the effect of the bounce volume at the fixed frequency. It indicates the similar effect of frequency dependency. This implies that an optimum volume for the driving frequency is existed.

If the inertance part is replaced by a set of solid piston and mechanical spring as shown in Figure 9 schematically, required momentum equations reduces to only two sets. Figure 10 shows the result of example calculation of this method. Figure 10(a) and (b) well represent Figure 7(a) and (b) qualitatively. Distinctive feature of this method is that work flows at each important section are able to evaluate as shown in Figure 10(c).

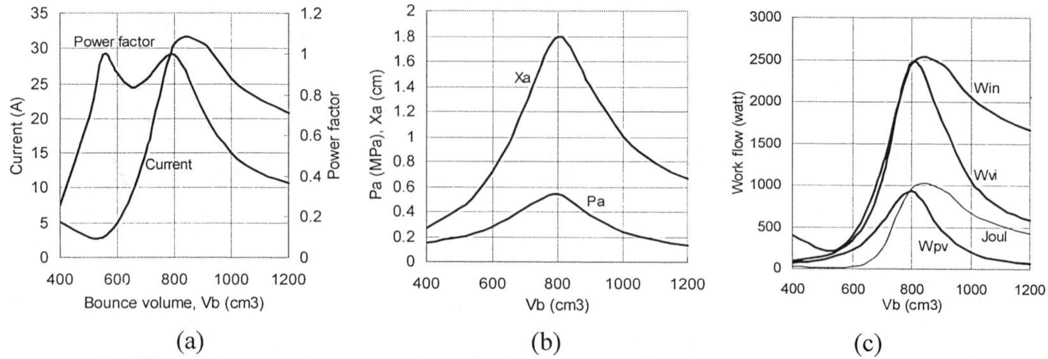

Figure 8   Effect of bounce volume at the fixed driving frequency (Xa: piston stroke, Pa: pressure amplitude, Wpv: PV-work at the piston head, Win: effective input power, Joul: joul loss of the winding, Wvi: total input power)

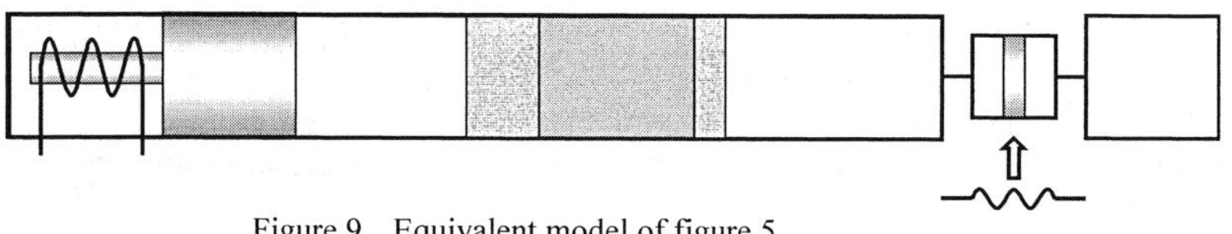

Figure 9   Equivalent model of figure 5

It is noted that the relation of the total input work (Wvi), the effective input work (Win), PV work at the compressor head and the work flow within the pulse tube (Wex) are well explained. The best operating frequency of this particular case is predicted as about 47 Hz, which gives the maximum of Wex/Wvi. This kind of simplified simulation method will be effective to optimize the each component of pulse tube cooler driven by the linear motor system.

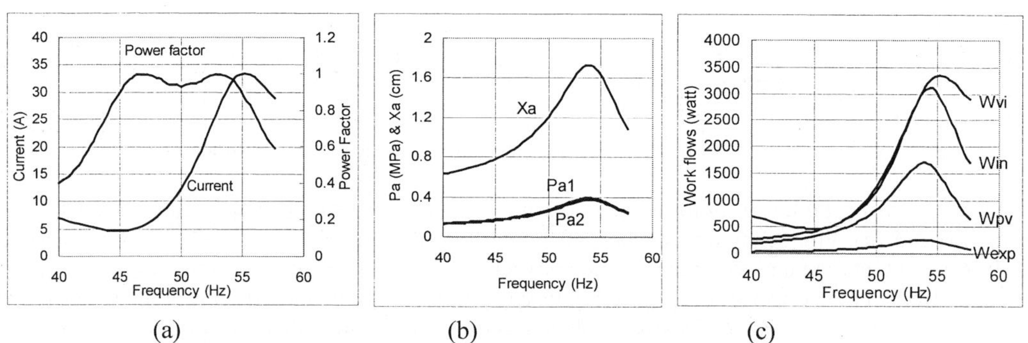

Figure 10   Performance of resonating warm expander with a linear compressor

Modification of warm expander method
Simulation program used for Figure 9 is applicable to other type of phase shift mechanisms. Figure 11(a) shows one of the examples. The warm expander of Figure 9 becomes a resonant warm displacer and the calculated result is given in Figure 11(b). In this example, the same temperature at the both end of the regenerator has been used. This result indicates the possibility of a phase shift mechanism without using the buffer volume. It is found that the work flows though the pulse tube (W1) is recovered as (W2) through the warm displacer, where (WL1) and (WL2) are work loss at the regenerator and the displacer respectively. Therefore, the required work input to the pressure wave generator is smaller than the work input to the regenerator. This means the theoretical limit of the thermodynamic efficiency of this method approaches to Carnot efficiency.

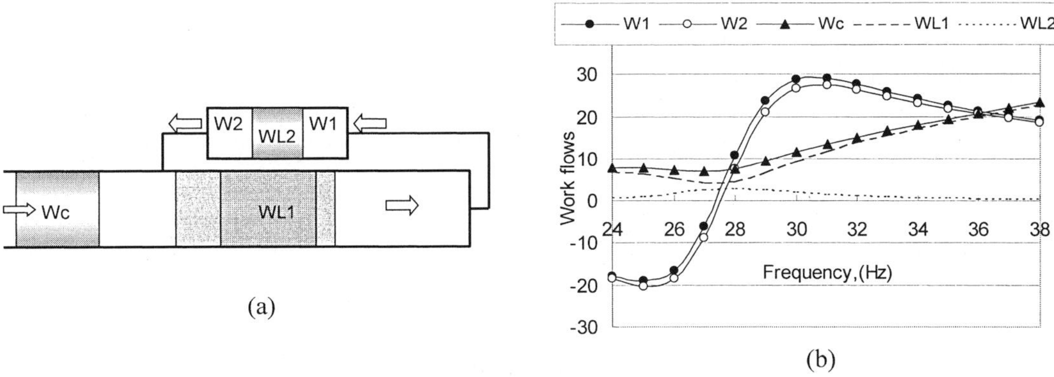

Figure 11   Frequency dependency of feedback type resonating displacer

Modification of split-Stirling type pulse tube cooler

One of the disadvantages of the Stirling type pulse tube cooler is the mechanical vibration caused by the integrated compressor unit to the cold head. The split-Stirling type, as shown in figure 12(a), has been developed to compensate this particular disadvantage. Separation of the cold head gives user convenience, however, the degradation of the cooler efficiency is proportional to increase the split tube length due to the pressure drop within the tube.

The possibility to extend the split tube length by introducing the effect of acoustic resonation is considered. The split-tube is replaced by an inertance tube with the minimum acoustic resistance. Based on the modified computer program used for an inertance tube phase shifter [3], an example calculation has been conducted. The tube length is 1 meter and its inner diameter is 16.5 mm. When the frequency is 50 Hz, the work transfer ratio, defined as a work output divided by a work input, of 95 % has been predicted for the case of 1.5 kW work-flow. To understand this phenomenon more clearly, a simplified equivalent model was introduced as shown in figure 12(b).

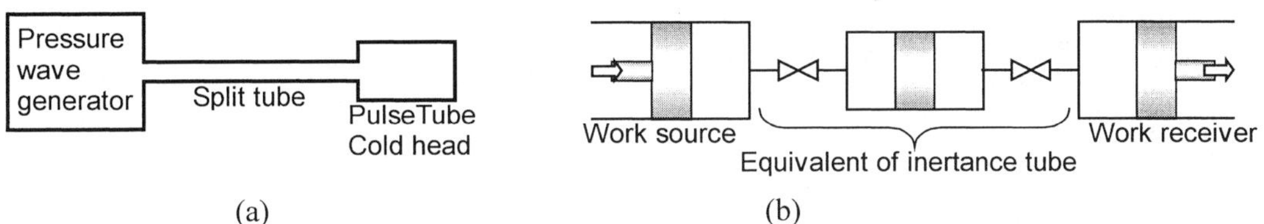

Figure 12   (a) Schematics of split-Stirling type pulse tube cooler, (b) Modified split method.

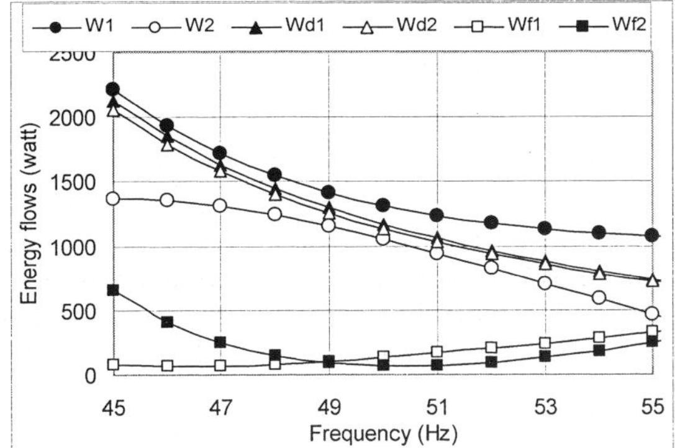

Figure 13   Performance of modified split-Stirling type pulse tube cooler, W's indicate the workflow, 1 is for the work source and 2 is for the work reciever While, d1 and d2 are for the left and the right side of displacer, respectively.

Figure 13 shows an example calculation of the fixed diameter and the length of the transfer tube. The workflows are plotted as a function of driving frequency. Wf1 and Wf2 are the work loss at the valve mark in the figure 12.(b), which represent the loss due to the flow friction at the tube wall. It is noted that the pressure ratio at the regenerator side is much larger than the compressor side. It shows the optimum frequency is a function of insert volume at the inlet of the transfer tube. The maximum work transfer efficiency was 81 % at 50 Hz. This result indicates the possibility of an efficient split-Stirling type pulse tube cooler by use of a resonant transfer tube.

CONCLUSIONS

There are many analytical methods of pulse tube coolers. Some of them are precise analysis of a particular component; such as a regenerator, a pulse tube phase shifter and others are the simplified analysis of total system. Most of them, however, are not easy to use it extensively for the new users. In other words, the developed numerical codes are generally not easy to understand. In these situations, there are still some rooms to develop the user-friendly software, which is able to design the pulse tube cooler to meet the requirement for the particular application. However, the additional thermodynamic losses could not be avoided in most cases.

To keep the thermodynamic efficiency high enough for the applications, several approaches have been described based on the simplified computer analysis. For valved type pulse tube coolers, the operating condition to minimize the valve loss was given.

The use of resonant displacer as a phase shifter for valve-less pulse tube cooler gives Carnot efficiency as its maximum efficiency. Possibility of such a phase shift mechanism without use of an orifice and a reservoir was introduced by a simplified calculation. An efficient split-Stirling type pulse tube cooler is proposed by the use of resonating tube and a bounce volume.

Further analytical study to overcome the various configurations requested from the application fields should be performed, which makes the best use of the basic advantages of the pulse tube cooler.

ACKNOWLEDGMENT

Part of this paper was supported by Central Research Institute of Electric Power Industry. The author appreciates to R. Maekawa (NIFS) for his assistance.

REFERENCES

1. Matsubara, Y. and Miyake, A., Alternative Methods of the Orifice Pulse Tube Refrigerator, Proc. of 5th International Cryocooler Conference, Naval Postgraduate School, (1989) 127-135.

2. Zhu, S., Wu, P. and Chen, Z., Double inlet pulse tube refrigerators: an important improvement, Cryogenics 30, (1990) 514-520.

3. Zhu, S., Zhou, S. L., Yoshimura, N. and Matsubara, Y.; Phase Shift Effect of the Long Neck Tube for the Pulse Tube Refrigerator: Cryocooler 9, Plenum Press, (1997) 269-278.

4. Matsubara, Y. and Gao J. L., Novel configuration of three-stage pulse tube refrigerator for temperature below 4 K, Cryogenics 34, 259 (1994).

5. Wang, C., Thummes, G. and Heiden, C., A two-stage pulse tube cooler operating below 4 K, Cryogenics 37, 159 (1997).

6. Gao, J. L. and Matsubara, Y., Development and application of multi-stage pulse tube refrigerators, Proceedings of ICCR 2003, (2003) 86-92.

7. Matsubara,Y., Kobayashi, H. and Zhou, S.L., Thermally Actuated He3 Pulse Tube Cooler, Cryocoolers 11, KA/Plenum Press (2001) 273-280.

Performance of a Pulse Tube Cryocooler with Two Separate Stages using ^{3}He in the Second Stage

Jiang N. [a, b], Lindemann U. [a], Chen G.B. [b], Thummes G. [a]

[a] Institute of Applied Physics, University of Giessen, D-35392 Giessen, Germany
[b] Cryogenics Lab, Zhejiang University, 310027 Hangzhou, P. R. China

In this paper, the performance of a two-stage pulse tube cryocooler with ^{3}He as the 2nd stage working fluid is presented. A minimum no-load temperature of 1.27 K at the 2nd stage with the 1st stage at 23 K has been achieved. At an input power of 1.49 kW to the 2nd stage, the cooler provides a maximum cooling power of 735 mW at 4.2 K with ^{3}He.

INTRODUCTION

The development in phase shifting methods and the utilization of magnetic regenerative materials have led to great progress in pulse tube refrigeration. So far, the lowest no-load temperature of 2.07 K was obtained with ^{4}He by means of a liquid nitrogen precooled two-stage pulse tube cryocooler (PTC) [1].

For regenerative cryocoolers using ^{4}He as the working fluid, a refrigeration temperature limit of 2K or so is induced by the λ superfluid phase transition of ^{4}He. The sharp increase of the fluid specific heat accompanying the superfluid phase transition makes regenerators disabled. Besides, the thermal expansion coefficient of ^{4}He also equals zero near 2 K, which means that adiabatic compression or expansion occurs without change in temperature, and therefore no refrigeration will be produced by the pulse tube cooler. Different research groups [2-4] succeeded to break the temperature barrier of ^{4}He by using ^{3}He, whose superfluid phase transition occurs in the Millikelvin temperature region and whose thermal expansion coefficient approaches zero close to 1 K (see [2, 4] for more details).

By a three-stage pulse tube cryocooler, a minimum no-load temperature of 1.78 K with ^{3}He was achieved by the Eindhoven group [2]. A minimum no-load temperature of 1.47 K with ^{3}He was obtained by a two-stage GM cooler employing a novel rare earth regenerator material [3].

Recently, in our research group a lowest temperature of 1.27 K has been achieved by means of a newly designed PTC [4]. This cooler consists of two pulse tube stages with separate gas supply. To save the amount of ^{3}He, only the smaller second stage is charged with ^{3}He, while the first stage still operates with ^{4}He as working fluid. Here we present new results on the cooling performance of this PTC including cooling load maps with ^{3}He as working fluid.

EXPERIMENTAL SET-UP

A schematic of the two-stage PTC is shown in Fig. 1. The cooler consists of two parallel separate pulse tube stages. The incoming gas to the second stage is precooled by a heat exchanger thermally connected to the cold end of the first stage. Compared to that of coolers with common gas supply, such design presents higher flexibility in distribution of cooling capacity of the stages and allows an easier adjustment of dc-flow. In addition, this arrangement makes possible to operate the two stages with different compressors and at different operating frequencies.

The matrix of the coldest part of the second stage regenerator consists of layers of Pb, ErNi and $HoCu_2$ spheres, as indicated in Fig. 1. Two commercial helium compressors, Leybold CP4000 (or CP6000) and RW2, are employed for driving the first and second stage, respectively. A more detailed description of the cooler has been given previously [4].

The temperatures of the 1st and 2nd stage cold platforms are measured by means of calibrated platinum and Cernox resistance thermometers, respectively. Net cooling powers of the 1st and 2nd stage are measured using resistive heaters attached to the cold platforms. Dynamic pressures are recorded by piezoelectric pressure sensors connected to the inlet of each stage regenerator.

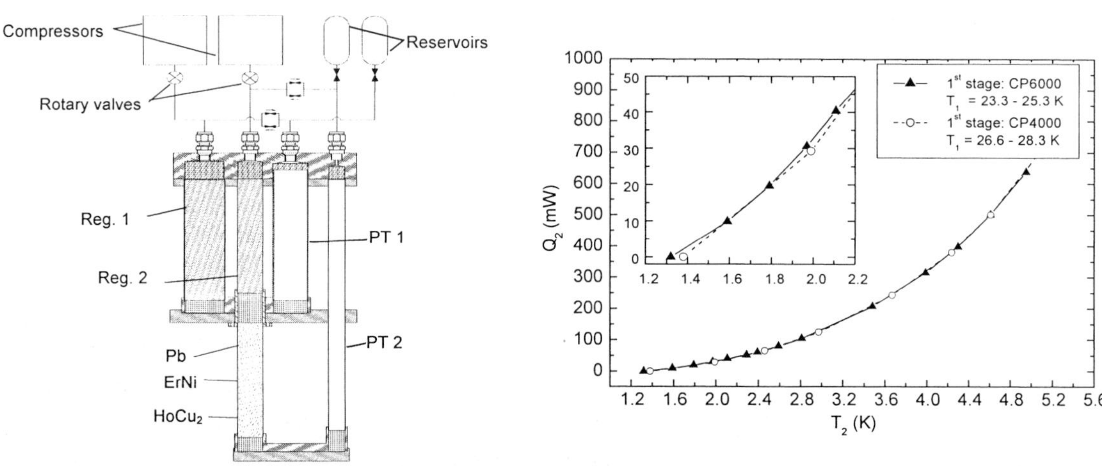

Figure 1   Schematic of the two-stage pulse tube cooler with separate stages.

Figure 2 2nd stage cooling power versus temperature with the 1st stage driven by CP6000 or CP4000.

EXPERIMENTAL RESULTS

All of the cooler tests are performed with the 1st stage charged with ^{4}He to an (absolute) pressure of 17.5 bar at ambient temperature. The 2nd stage is operated with ^{3}He at a charging pressure of 17.0 bar, if not stated otherwise. The rotary valve timings (expansion to compression interval in a cycle) for both stages are fixed. In the following Q_1, T_1 and Q_2, T_2 denote the net cooling power and temperature of the 1st and 2nd stage, respectively.

Fig. 2 displays the net cooling power of the 2nd stage as a function of temperature either with the 1st stage driven by the CP4000 or CP6000 compressor. The operating parameters for both cases are optimised without additional heat load on both stages ($Q_1 = Q_2 = 0$). The 1st stage temperature is around 24 K with CP6000 (input power: 6.1 kW) and around 27 K with CP4000 (input power: 4.3 kW). The load curves in Fig. 2 are almost superposed, which indicates that the performance of the 2nd stage is not sensitive to 1st stage temperature except at temperatures well below 2 K (see inset to Fig. 2). With CP6000 operation of the 1st stage, the 2nd stage minimum no-load temperature is 1.32 K, which is 0.06 K lower than that with CP4000 ($T_2 = 1.38$ K).

With CP6000 on the 1st stage, the cooler provides 33 mW, 134 mW and 371 mW at 2.0 K, 3.0 K and 4.2 K, respectively. With CP4000, the cooling power at 2.0 K is slightly reduced to 30 mW at 2.0 K, while at higher temperatures it is essentially the same as with CP6000. The electrical power consumption by the 2nd stage (RW2 compressor) is about 1.3 kW.

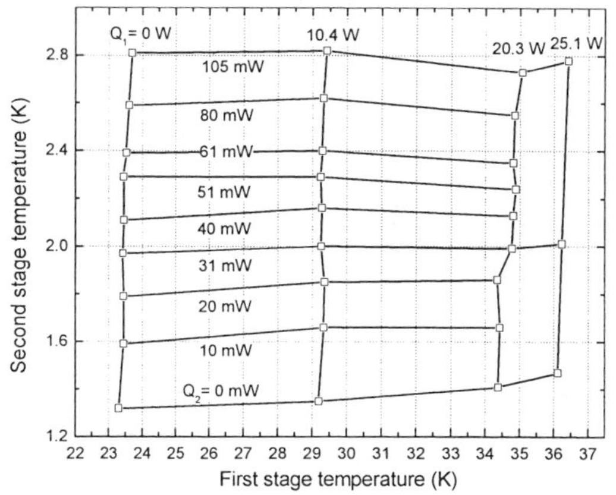

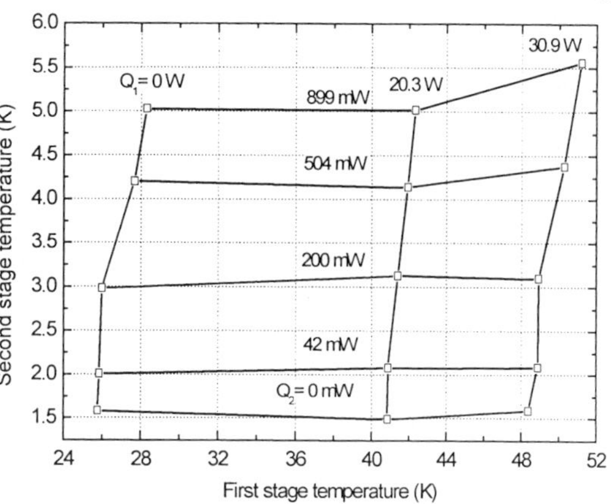

Figure 3   Cooling load map below 3 K; CP6000 compressor on 1st stage; parameters optimised for minimum temperature at $Q_1 = Q_2 = 0$.

Figure 4   Cooling load map with the 1st stage operated on the CP4000 compressor; parameters optimised at $Q_1 = 0$ and $Q_2 = 50$ mW.

A more detailed behaviour of the cooling performance with load on both stages and with the 2nd stage operating below 3 K can be seen from the cooling load map in Fig. 3. The data were obtained with CP6000 on the 1st stage after optimising the cooler for minimum temperature without load on both stages. It follows from Fig. 3 that the 2nd stage temperature at fixed cooling power depends only slightly on the 1st stage temperature and thus on Q_1. For example, at $Q_2 = 31$ mW the 2nd stage temperature is 1.97 K at $T_1 = 23.4$ K ($Q_1 = 0$) and rises to 2.01 K at $T_1 = 36.2$ K ($Q_1 = 25.1$ W). The measured input powers to the CP6000 and RW2 compressor in Fig. 3 are 6.11 - 6.36 kW and 1.26 - 1.36 kW, respectively, depending on the heat load on each stage.

Higher 2nd stage cooling powers above ≈ 1.8 K are achieved by optimising the cooler under an applied heat load ($Q_2 > 0$) at the cost of an increased no-load temperature. Fig. 4 shows a cooling load map that was measured after optimisation at $Q_1 = 0$ and $Q_2 = 50$ mW. In these tests the 1st stage was operated on the CP4000 compressor. The minimum no-load temperature ($Q_2 = 0$) is now 1.58 K, 1.49 K, and 1.59 K for 1st stage cooling powers of $Q_1 = 0$, 20 W, and 31 W, respectively. The 2nd stage provides cooling powers of 42 mW, 38 mW and 34 mW at 2.0 K, as well as 518 mW, 528 mW and 464 mW at 4.2 K with $Q_1 = 0$, 20 W and 31 W, respectively.

The thermal expansion coefficient of ^{3}He approaches zero at lower temperatures when the pressure decreases [2, 4]. So does the temperature limit of a pulse tube cooler with ^{3}He working fluid. Therefore the cooling performance of the 2nd stage has also been tested under different ^{3}He working pressures. Fig. 5 displays the cooling power of the 2nd stage for three average working pressures of <p> = 9.2 bar, 10.2 bar, and 12.4 bar, where <p> = 10.2 bar corresponds to the standard ^{3}He charging pressure of 17.0 bar. In the tests in Fig. 5 the 1st stage was driven by the CP6000 compressor. The cooling performances under the three average working pressures are very close together. The solid curve in Fig. 5 for <p> = 10.2 bar, which gives a minimum temperature of 1.32 K, corresponds to the load curve with CP6000 in Fig. 2. A lower no-load temperature of 1.27 K is obtained by decreasing the ^{3}He average working pressure to 9.2 bar. At this low working pressure, the electric power consumption of the 2nd stage is only 1.17 kW.

Fig. 6 gives the cooling load map with increased ^{3}He average working pressure of 12.4 bar and with the cooler optimised under a high heat load to the 2nd stage ($Q_1 = 0$, $Q_2 = 545$ mW). The minimum temperature of the 2nd stage is now 1.47 K with $T_1 = 35.5$ K at $Q_1 = 20.3$ W. A cooling power of 735 mW at 4.2 K is obtained with $T_1 = 24.6$ K at $Q_1 = 0$.

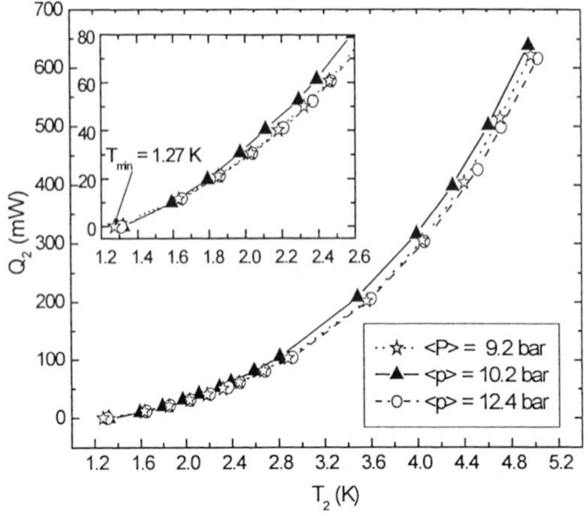

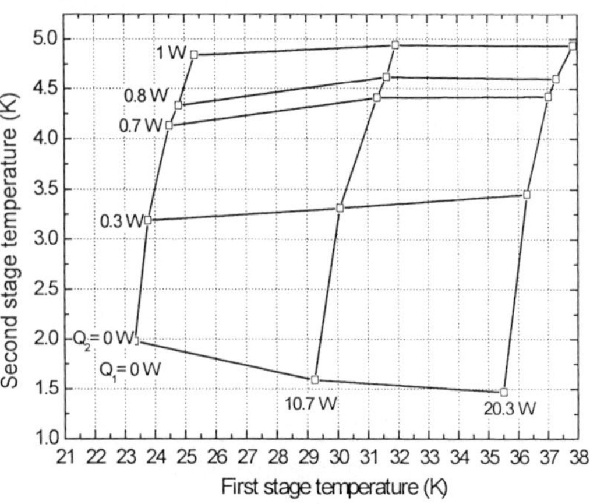

Figure 5   Cooling power of the 2nd stage operated at different ^{3}He working pressures; 1st stage driven by CP6000; $Q_1 = 0$, $T_1 = 23 - 26$ K; parameters optimized for minimum temperature at $Q_1 = Q_2 = 0$.

Figure 6   Cooling load map at an increased ^{3}He average pressure of $<p> = 12.4$ bar; CP6000 compressor on 1st stage; parameters optimized for minimum temperature at $Q_1 = 0$ and $Q_2 = 545$ mW.

SUMMARY

A minimum no-load temperature of 1.27 K with ^{3}He in the 2nd stage is achieved by the present two-stage PTC, which is the lowest temperature obtained by regenerative cryocoolers up to now. The cooler can provide 42 mW at 2.0 K with 1.3 kW input power to the 2nd stage and 4.3 kW to the 1st stage. After optimization at $Q_1 = 0$ and $Q_2 = 545$ mW, the maximum cooling power at 4.2 K is 735 mW with an input power of 1.49 kW to the 2nd stage at a 1st stage temperature of 24.6 K and a 1st stage input power of 6.2 kW.

ACKNOWLEDGEMENTS

The authors thank Y. Kuecuekkaplan (Giessen) for his help with the experimental setup. Jiang Ning thanks the German Academic Exchange Service (DAAD) for a scholarship.

REFERENCES

1.   Thummes G, Bender S, Heiden C. Approaching the ^{4}He lambda line with a liquid nitrogen precooled two-stage pulse tube refrigerator. Cryogenics (1996) 36(9) 709-11

2.   Xu MY, De Waele ATAM, and Ju YL. A pulse tube refrigerator below 2K. Cryogenics (1999) 39 865-869

3.   Satoh T, Numazawa T. Cooling performance of a small GM cryocooler with a new ceramic magnetic regenerator material. Cryocoolers 12 (2002) 397-402

4.   Jiang N, Lindemann U, Giebeler F, and Thummes G. A ^{3}He pulse tube cooler operating down to 1.3 K. Cryogenics, submitted for publication.

Proceedings of the Twentieth International Cryogenic Engineering Conference
(ICEC 20), Beijing, China. © 2005 Elsevier Ltd. All rights reserved.

Experimental analysis of the oscillating flow characteristics for high frequency regenerators

Wang X-L.[1,2], Zhao M-G.[1,2], Cai J-H.[1], Liang J-T.[1], Zhou Y.[1]

[1]Technical Institute of Physics and Chemistry, CAS, P.O.Box 2711, Beijing 100080
[2]Graduate School of the Chinese Academy of Sciences, Beijing 100039

In this paper, effects of different operating conditions on amplitudes of the dynamic pressure and mass flow rate and phase shift between them at the cold-end of the regenerator under high frequency oscillating flow are studied in details. The instantaneous velocity at the outlet-end of the regenerator is obtained by using a hot wire anemometer. The results show that there are optimal values for orifice settings, operating frequencies, mean pressures, and pressure ratios to improve the refrigeration performance of high frequency pulse tube refrigerators. The reasons of the optimal value existed are discussed.

INTRODUCTION

For optimization of the regenerative machines, it is important to predict accurately friction and heat transfer losses of regenerators. Recently, some experimental investigations [1,2,3,4] on flow characteristics of regenerators under reciprocating flow have been made, in which have shown that flow friction factors of regenerators under the oscillating flow are larger than those of under steady flow. These results are useful for the understanding of the mechanisms and the prediction of performance and design of cryogenic regenerators.

In this paper, pressure and velocity performance of the regenerator under oscillating flow are measured in a carefully designed apparatus to simulate actual operation of cryocoolers. The effects of operating frequencies, orifice settings, mean pressures and pressure ratios at the warm-end of regenerators on the amplitude and phase difference of dynamic pressures and velocities are discussed. It is very useful to understand operating mechanisms in a pulse tube refrigerator.

EXPERIMENTAL APPARATUS AND METHODS

Figure 1 shows the schematic diagram of the experimental apparatus in which the dynamic velocities and pressures at both sides of the regenerator are measured. Details of experimental arrangement and measurement approaches can be found in the authors' other research papers [2, 5].

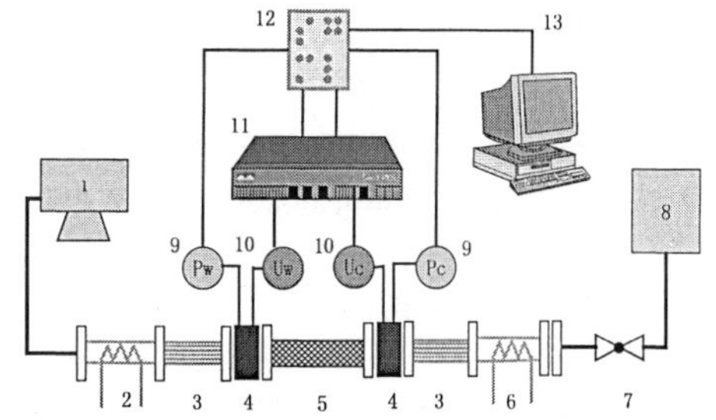

Figure 1   Schematic diagram of experimental apparatus

RESULTS AND DISCUSSIONS

Properties of tested stainless steel regenerators are summarized in Table 1. The wire diameter and pitch are measured using a profile projector. The porosity and hydraulic diameter are determined from the equations in the reference [2].

Table 1   Properties of the tested regenerators

Length (mm)	Diameter (mm)	Number of packed screens	Mesh No.	Wire diameter (mm)	Pitch (mm)	Porosity	Hydraulic Diameter (mm)
104	7.9	1380	400	0.03	0.0635	0.63	0.051
74	11	852	400	0.03	0.0635	0.63	0.051

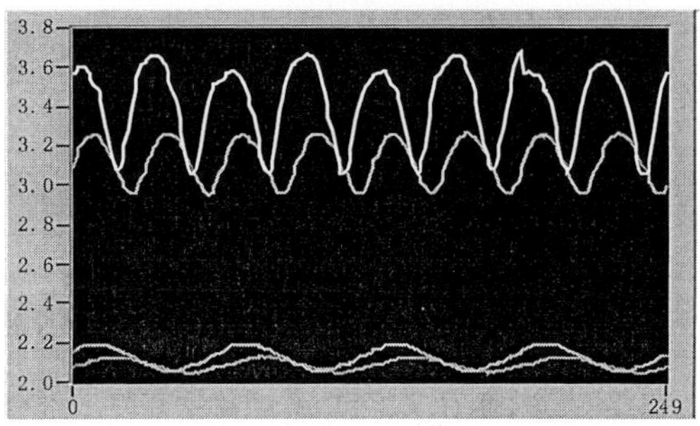

Figure 2   Typical raw signals of velocity and pressure waves

Figure 3   Processed experimental data

The typical raw experimental data of oscillating flow is shown in Figure 2. The raw experimental data measured from the hot wire anemometer and the pressure transducers require data processing which include velocity transformation and correction. Figure 3 shows the real pressure and velocity wave at the inlet (U_w, P_w) and the outlet (U_c, P_c) of the test regenerator. Pressure and velocity for high frequency clearly show phase shift and amplitude attenuation.

According to Redebaugh [6], an enthalpy flow in a pulse tube can be expressed as:

$$(H) = \langle P_d \dot{V} \rangle = (1/2) P_1 \dot{V_1} \cos\theta = (1/2) R T_0 \dot{m_1} (P_1 / P_0) \cos\theta \tag{1}$$

where P_1 is the amplitude of pressure variation, P_0 is the mean pressure, $\dot{m_1}$ is the amplitude of the sinusoidal mass flow rate, θ is the phase angle between the pressure and the flow. It can be seen from this equation that, P_1 , $\dot{m_1}$, θ are three crucial parameters to improve the refrigeration performance of a pulse tube refrigerator. In this paper, effects of the orifice settings, the mean pressure, the operating frequency and the pressure ratio at the warm-end of the regenerator on three parameters above at the cold-end of the regenerator are studied in details. The experimental results are presented as follows.

Effect of orifice settings

Figure 4   shows the effects of the orifice settings on the phase angle ($\Delta\theta_{mc,Pc}$) and amplitude of dynamic pressure (Pc_{max})and mass flow rate (mc_{max}) at the cold-end of the regenerator. The measurements are made at mean pressure of 2.0 MPa, with the operating frequency of 30 Hz. The orifice opening setting is from the closed position (0) to the 60 grids that it is fully opened. it is clearly found that the phase angle $\Delta\theta_{mc,\,Pc}$ and the pressure amplitude (Pc_{max}) at the outlet end of the regenerator decreases rapidly, while the amplitude of the mass flow rate (mc_{max}) increases greatly with the opening value of the orifice valve increasing. The optimum position of the orifice is determined from integrating the three factors. This shows that the opening value of the orifice valve is crucially important to the refrigeration performance of a pulse tube refrigerator.

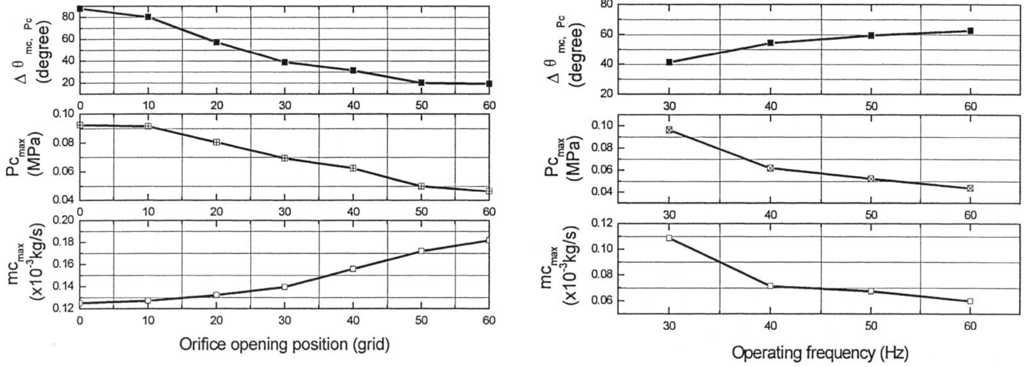

Figure 4   Effect of the orifice setting   Figure 5   Effect of the operating frequency

Effect of operating frequencies

The experimental data are carefully chosen to be sure that only the frequency is varying and that other parameters are kept constant. The effects of difference operating frequencies on the flow characteristics at the outlet-end of the regenerator are shown in Figure 5. The phase shift $\Delta\theta_{mc,Pc}$ goes up, while the amplitude of mass flow rate (mc_{max}) and pressure (Pc_{max}) at the outlet-end of the regenerator goes down when operating frequency varied from 30 Hz to 60 Hz in our experimental conditions. The results show that the theoretical refrigerator power is inverse proportional to the operating frequency under high frequency conditions in a pulse tube refrigerator. However, there is also a limit imposed by the optimum frequency of the compressor. The efficiency of the compressor is advantageous under its optimum frequency. So the optimum frequency in practice is determined from the two factors as mentioned above. Each of them should be considered carefully in our design and experimental work.

Effect of mean pressures

The objective of the third experiment is to verify the dependency of refrigeration performance on mean pressures. The frequency is kept constant at 30 Hz, and the orifice value is set at 20 grids. The measurements are made at mean pressures of 1.0, 1.5, 2.0, 2.5, 3.0 MPa. To show more clearly the effect of mean pressures, the variation of the pressure ratio (Pc_{max}/P_0) of the Pc_{max} to the mean pressure is plotted in Figure 6. we can see from the figure that mc_{max}, Pc_{max}/P_0 increases rapidly, and $\Delta\theta_{mc,Pc}$ is reduced with the mean pressure varied from 1.0-2.0 MPa, as a result, the refrigeration capacity is improved. Though $\Delta\theta_{mc,Pc}$ is decreased, the pressure ratio is also lessoned rapidly, and mc_{max} has little change when the mean pressure is from 2.0 MPa to 3.0 MPa, So we can predict there is an optimum mean pressure in a pulse tube refrigerator.

Effect of pressure ratios

Figure 7 illustrates the influences of the pressure ratio at the inlet-end of the regenerator on the outlet parameters of the cold end with the frequency of 30 Hz, the mean pressure of 2.0 MPa. The different pressure ratios are achieved by adjusting input electric power of the compressor. mc_{max}, Pc_{max}, $\Delta\theta_{mc,P2}$ have a steep increase when the pressure ratio increased. The results show that the refrigerator capability is enhanced rapidly with the elevation of the pressure ratio. On the other hand, the increase of the mass flow rate with the pressure ratio becoming high leads to large regenerator losses. So an optimum pressure ratio is required to improve refrigeration performance in pulse tube refrigerators.

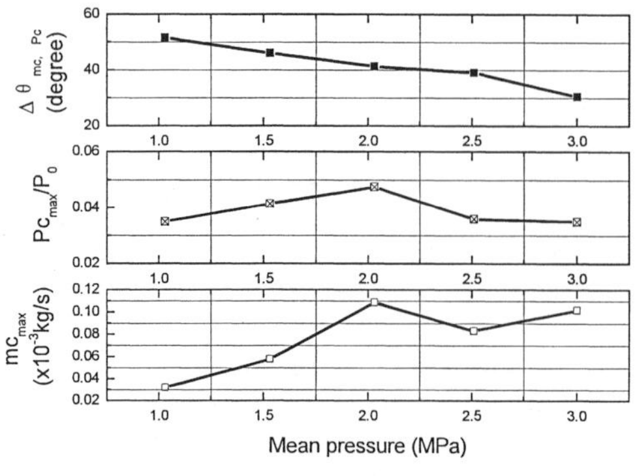

Figure 6   Effect of the mean pressure

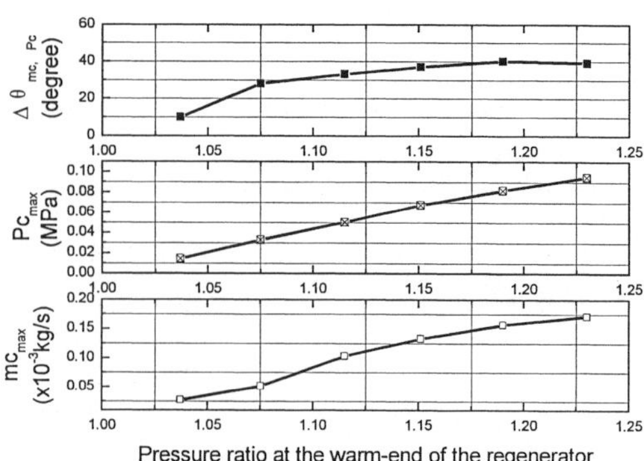

Figure 7   Effect of the pressure ratio

CONCLUSIONS

Detailed experimental data on the oscillating flow characteristics of high frequency regenerators in a pulse tube cryocooler are presented in this paper. The influences of factors, including the orifice setting, operating frequency, system mean pressure, and the pressure ratio at the warm-end of the regenerator on the flow characteristics of the regenerator are achieved. For a given pulse tube refrigerator under high frequency conditions, the orifice position, operating frequencies, the mean pressure, and the pressure ratio have an optimal value existed, and the existent reasons are presented. The results avail the desirable design and experimental studies of pulse tube refrigerators.

ACKNOWLEDGEMENTS

The National Natural Science Foundation of China funds this work. (Grant No. 50206025)

REFERENCES

1. Helvensteijn, B.P.M., Kashani, A. Spivak,A.L., et al., Pressure drop over regenerators in oscillating flow, CEC/ICMC 97, Portland, Oregon, USA (1997)

2. Yonglin Ju, Yan jiang and Yuan Zhou, Experimental study of the oscillating flow characteristics for a regenerator in a pulse tube cryocooler, Cryogenics (1998) 38 649-656

3.Kwanwoo Nam and Sankwon Joeng, Measurement of cryogenic regenerator characteristics under oscillating flow and pulsating pressure, Cryogenics (2003) 43 575-581

4.Sungryel Choi, Kwanwoo Nam, Sangkwon Jeong, Investigation on the pressure drop characteristics of cryocooler regenerators under oscillating flow and pulsating pressure conditions, Cryogenics (2004) 44 203-210

5 X.L.wang, M.G.Zhao, J.H.Cai, W.Dai, J.T.Liang, Experimental Flow Characteristics Study of a High Frequency Pulse Tube Regenerator, Cryocoolers 13, (2004)   (in press)

6.Ray Radebaugh, Develepment of the pulse tube refrigerator as an efficient and reliable cryocooler. Proc. Institute of Refrigeration (1999-2000), London

Regenerative recuperator for tandem 4 K pulse tube refrigerator

Jung J., Jeong S., Choi Y.*, Nam K.

Department of Mechanical Engineering, KAIST, 373-1, Guseong-dong, Yuseong-gu, Daejeon, Korea
*SPEC Corporation, 1695-5, Sinil-dong, Daedeok-gu, Daejeon, Korea

A recuperator for a tandem recuperative 4 K pulse tube refrigerator is under development. It is constructed with micro structure to have low thermal ineffectiveness. The structure is formed on a stainless steel plate by chemical etching and the etched plates are stacked and bonded through a vacuum brazing process to compose a plate-type heat exchanger. The pressure drop test at room temperature and the thermal ineffectiveness test between room temperature and liquid nitrogen temperature are carried out to estimate low-temperature performance. With the experimental data, the pressure drop of 30 kPa and the thermal ineffectiveness of 5 % are estimated at the helium flow rate of 0.41 g/s between 4 K and 40 K.

INTRODUCTION

As the temperature goes down especially near 4 K, all the regenerative cryocoolers suffer from poor regenerator efficiency. The thermal ineffectiveness of the 2nd stage regenerator may reach above 5 % depending on the temperature range. Exotic solid materials such as Er_3Ni or $HoCu_2$ have been utilized as the regenerator matrix with their high magnetic entropy change. However, these magnetic materials are usually very expensive and limited with narrow working temperature range. Naturally, the idea of utilizing helium's heat capacity as regenerative medium has been conceived. Typical specific heat value of high-pressure helium above 1 MPa exceeds that of common regenerator material. The regenerator in this case may behave like a recuperator in exchanging heat between two helium flows all the time.

Although the recuperative 4 K pulse tube refrigerator (see Figure 1) is not a brand-new concept, it is not successfully realized [1], [2]. In the realization of the recuperator concept, the unavailability of the heat exchanger to be substituted for the 2nd stage regenerator is a big obstacle. The requirements of the so-called 2nd stage recuperator as shown in Fig. 1 are low pressure drop, low thermal ineffectiveness and small void volume, which are the same as those of the regenerator. Physically, they are fulfilled when the flow path of the heat exchanger has a very small hydraulic diameter with very large heat transfer area [3]. In other words, the recuperator should be constructed with micro structure. However, the technology of such a heat exchanger is not well established yet. The development of the recuperator with micro structure is demanded first in order to proceed to the recuperative 4 K pulse tube refrigerator. This paper describes the design, fabrication and performance test of such a recuperator.

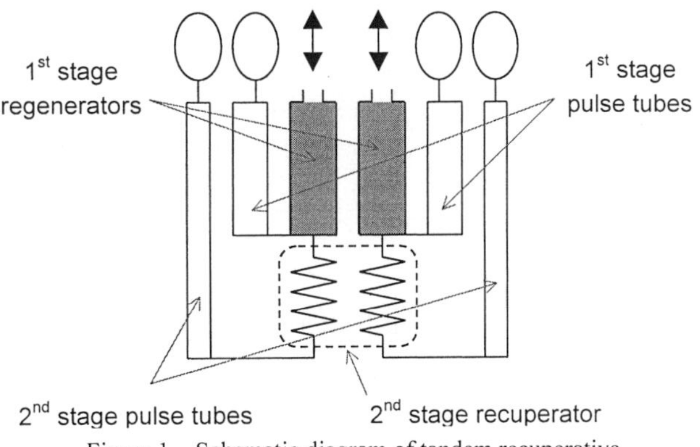

Figure 1   Schematic diagram of tandem recuperative 4 K pulse tube refrigerator

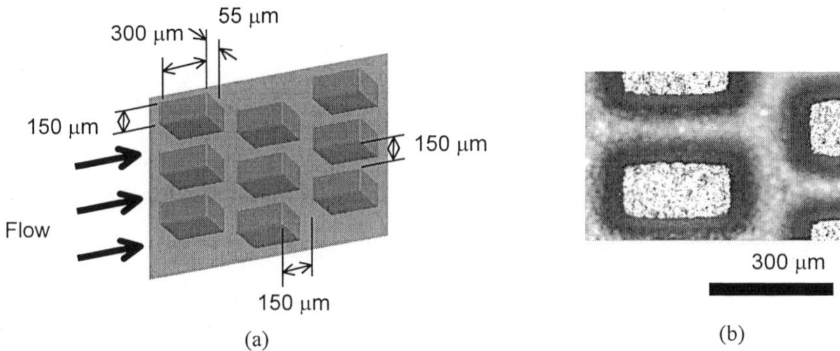

Figure 2   Designed micro structure (a) and etched pattern (b)

RECUPERATOR DESIGN

Micro structure in recuperator

The studies on micro structure and their fabrication are active in chemical engineering as the concept of micro reactors. The formation of micro structure is accomplished through etching process of thin stainless steel plates. Then, the etched plates are stacked and bonded through a vacuum brazing or a diffusion bonding to compose a complete system such as a micro reactor or a micro heat exchanger. This fabrication procedure is adopted for the recuperator in this study.

Since the fabrication procedure inherently implies stacked layers, the recuperator fabricated in such a way belongs to a micro plate-type heat exchanger with small void volume. The etched structure in this study is a micro channel with the staggered rectangular fins (see Figure 2). For the enhanced heat transfer and the structural rigidity, the structure is etched as narrowly as possible. However, it is not an optimized shape and may need further shape optimization.

Hydrodynamic and thermal design of recuperator

With the designed micro structure, the macro shape of the recuperator is determined. Three design parameters for the shape determination are the width, the aspect ratio and the number of stackings (see Figure 3). Since the macro shape is directly related to the performance of the recuperator, the performance criteria for the 2nd stage application should be provided before determining the design parameters. The criteria will be the same as that is applied to the 2nd stage regenerator. However, because of insufficient data on it, we could not find appropriate values and, therefore, built up performance criteria (Table 1) in a subjective way.

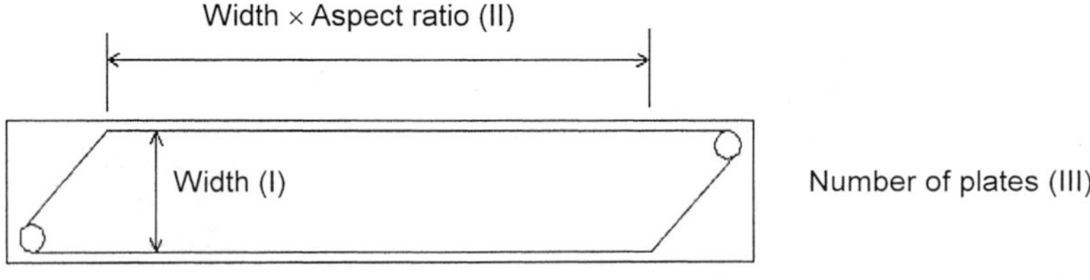

Figure 3   Parameters of recuperator shape

Table 1   Criteria on recuperator performance for 2nd stage application

Mass flow rate	1 g/s
Operating temperature	4 ~ 40 K
Allowable pressure drop	below 100 kPa
Allowable thermal ineffectiveness	below 5 %
Allowable void volume	below 10 cc per flow path

Table 2 Design values and expected performance of recuperator at steady operation

Width of recuperator	35 mm
Aspect ratio of recuperator	5:1
Number of stackings in recuperator	25 per flow path
Mass flow rate	1 g/s
Operating temperature	4 ~ 40 K
Pressure drop	below 10 kPa
Thermal ineffectiveness	3.5 %
Void volume	6.7 cc per flow path

The design parameters are determined to satisfy the criteria. Among various combinations of the design values that satisfy the criteria, those values in Table 2 are decided. The decision is based on the easiness of practical fabrication. The performance is evaluated at the steady operation (Table 2). The pressure drop is evaluated with the friction factor for flow between flat plates without any fins. Thus, the evaluation must be under-estimated. The thermal ineffectiveness is evaluated numerically [3] with the heat transfer coefficient for flow between flat plates without any fins. Thus, the evaluation may be over-estimated.

RECUPERATOR PERFORMANCE

The actual recuperator is fabricated according to the design values. However, the vacuum brazing, which is in the fabrication procedure, is not a strictly controlled process. It is not precisely known how much of the brazing filler metal fills the micro channel. The micro channels after the vacuum brazing might be slightly deformed from the etched shape. This conjecture is supported by direct observation of the cross-sectional view of the recuperator.

<u>Hydrodynamic performance</u>
The pressure drop at steady flow operation is measured in the recuperator. Since the recuperator has two flow paths, the measurement of the pressure drop is carried out for each path. While the inlet pressure is fixed at 1550 kPa, the mass flow rate is measured, changing the outlet pressure. The working fluid is nitrogen (N_2) gas and the experiment is carried out at room temperature. The pressure drop data is plotted in Figure 4. The pressure drops for two flow paths differ by approximately 10 %. The void volume is indirectly measured with an external vessel of known volume and its pressure change after connection to the evacuated channels. The measured void volume of the channels is 1.97 cc (channel 1) and 2.07 cc (channel 2), respectively. They differ by approximately 5 % and are only one third of the design value.

<u>Thermal performance</u>
The thermal ineffectiveness at steady flow operation is measured with helium gas in the recuperator. The measurement is carried out between room temperature and liquid nitrogen temperature. The result is listed in Table 3. The inlet and the outlet temperatures for each of the flow paths are measured. When the recuperator is perfectly insulated from environment, ΔT at the hot end and ΔT at the cold end should be equal for helium at the temperature range of the experiment. However, the experimental data indicates 2.1 W of heat inflow to the recuperator. The heat inflow may be from conduction through the connection lines or from radiation of the vacuum chamber. Since the estimated conduction is smaller than 0.1

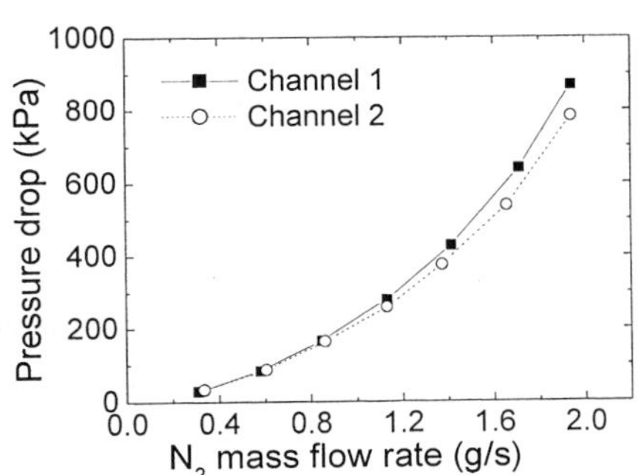

Figure 4 Pressure drop vs. mass flow rate in recuperator

Table 3   Thermal ineffectiveness measurement

Helium mass flow rate	0.41 g/s
Hot stream inlet temperature	300 K
ΔT at hot end	2.5 K
Cold stream inlet temperature	83 K
ΔT at cold end	3.5 K
Thermal ineffectiveness	1.6 % (calculated from ΔT at cold end)

W, most of the heat inflow is speculated to be due to radiation. In the experiment, because of the high pressure drop in the recuperator and the limited pressure range of the helium compressor, the helium flow rate is restrained to 0.41 g/s.

Performance estimation at low temperature
The pressure drop is related with the fluid properties and the mass flow rate through friction factor. The friction factor is a function of Reynolds number. For low-temperature helium flow of 0.41 g/s between 4 K and 40 K, the pressure drop is estimated to 30 kPa. In the same manner, the thermal ineffectiveness is related with the fluid properties and the mass flow rate through the dimensionless parameter of $hW^2/\dot{m}c_\mathrm{p}$ (h : heat transfer coefficient, W : width of recuperator, m : mass flow rate, c_p : heat capacity)[3]. Since h is proportional to k (thermal conductivity of helium), the variation of h with respect to the temperature is taken into account in the estimation of the thermal ineffectiveness through the variation of k. At the helium flow of 0.41 g/s, the thermal ineffectiveness is estimated to 5 % between 4 K and 40 K.

CONCLUSION

The recuperator for a tandem recuperative 4 K pulse tube refrigerator is designed and fabricated. With the pressure drop test at room temperature and the thermal ineffectiveness test between room temperature and liquid nitrogen temperature, the low-temperature performance of the recuperator is estimated to 30 kPa of the pressure drop and 5 % of the thermal ineffectiveness at the helium flow rate of 0.41 g/s between 4 K and 40 K. This result is not satisfactory. The recuperator is supposed to be just on the criteria or slightly below the criteria. However, the recuperator has a large room for improvement. In the recuperator design, both of the micro structure and the macro shape designs can be optimized for 4 K application.

ACKNOWLEDGEMENTS

This work is supported by a grant from the Center for Applied Superconductivity Technology of the 21st Century Frontier R&D Program funded by the Ministry of Science and Technology, Republic of Korea, and supported by the Brain Korea 21 Project.

REFERENCES

[1] de Waele, A. T. A. M., Zeegers, J. C. H., Counter-flow pulse tube refrigerators, Advances in Cryogenic Engineering (2002) 47 617-624
[2] Liang, J., Zhang, C., Cai, J., Luo, E., Zhou, Y. and Xu, L., Pulse tube refrigerator with low temperature switching valve: concept and experiments, Cryogenics (1997) 37 497-503
[3] Jung, J., Jeong, S., Development of recuperator for 4 K pulse tube refrigerators operating at opposite phases, Advances in Cryogenic Engineering (2004) 49 to be published

Experimental Investigation on DC Flow Suppression and Refrigeration Characteristics of a High-Performance Single-Stage GM-Type PTC

Jiang Yanlong [1, 2, 3], Chen Guobang[2], Thummes Guenter [3]

[1]Department of Man-Machine and Environment Engineering, Nanjing University of Aeronautics and Astronautics, Nanjing, 210016, P. R. China
[2]Cryogenics laboratory, Zhejiang University, Hangzhou, 310027, P. R. China
[3]Institute of Applied Physics, University of Giessen, 35392, Germany

The influence of DC flow induced by the introduction of double-inlet on the refrigeration performance of the pulse tube cooler is experimentally examined. A double-valved double-inlet configuration instead of conventional single-valved one is successfully used in experiments to reduce the DC flow. Besides, the performance of the cooler with various regenerator matrices and supply power of compressors has been extensively investigated under different operation modes. Operating under double-inlet mode, 18.4K and 14.7K minimum temperature has been reached driven by RW2 and CP4000 compressor respectively.

INTRODUCTION

The introduction of the double inlet usually leads to an increased performance of the pulse tube cooler. However, the asymmetric flow impedance of the double-inlet valve, regenerator and pulse tube will cause a DC flow around the loop of them. The DC flow from the warm end to the cold end adds an additional thermal load to the cold end of the cooler, and greatly deteriorates the refrigeration performance due to the large temperature difference between two ends, although its value is quite small compared with AC flow [1]. In addition, the DC flow is also a key point to induce the temperature instability of the cooler in some cases [2]. For the better performance of the cooler, the DC flow should be well suppressed to a certain small extent by several ways [3-4]. Based on the detailed analysis and comparison for all kinds of the suppression ways, two parallel-placed needle valves with opposite flow direction referred as two-valved configuration instead of traditional single-valved one as double-inlet are introduced in the paper to eliminate the DC flow and are proved to be a successful way. Besides, the performance of the cooler with various regenerator matrices and supply power of compressors has been extensively investigated under different operation modes. Operating under double-inlet mode, 18.4K and 14.7K minimum temperature has been reached driven by RW2 and CP4000 compressor respectively. Further optimization of the regenerator matrix and operation parameters, better performance of the cooler could be possible. This development will lead the single-stage pulse tube cooler to be used in 20K temperature range instead of the multi-stage pulse tube cooler using at present.

EXPRIMENTAL DETAILS

Fig.1 shows the sketch of the test cooler. This is a single double-inlet pulse tube cooler. The sizes of the pulse tube and regenerator are $\Phi28\times0.5\times155$ and $\Phi32.35\times0.5\times129$ respectively. Two different regenerator matrices are tested. matrix1: all space of the regenerator filled with 247 screens of stainless steel mesh and matrix2: 1/3 space of the stainless steel mesh at the cold part of the regenerator was replaced by lead balls of 0.25mm diameter. A calibrated carbon-glass resistance thermometer is used to monitor the temperature at the cold end of the pulse tube. The cooling power is measured by applying a heat load via a resistive heater.

For the flow asymmetry of a needle valve, the DC flow direction will be opposite when the valve placed in opposite flow direction. So the structure shown in Fig.1 with two parallel-placed needle valves with opposite flow direction called as double-valved configuration, instead of conventional single-valved configuration as the double-inlet is used in this experiment to eliminate the DC flow. In Fig.1 DIV1 expresses the main flow direction of the valve from the regenerator to the pulse tube as indicated with arrow by the manufacturer, DIV2 means the reverse direction.

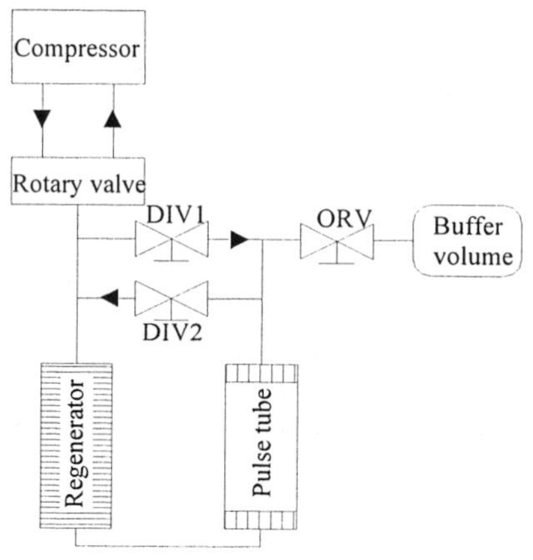

Fig.1   sketch of the pulse tube cooler

Fig.2   Cooling power under different operation mode

RESULTS AND ANALYSIS

Fig.2 shows the cooling power of the cooler with double-valved configuration compared to single-valve one with matrix1 driven by RW2 compressor. As it is shown in Fig.2, with the single-valve configuration, the minimum temperature obtained is 31.5K, which is only 1.0K lower than that of the orifice mode, while the minimum temperature of about 27K could be reached with the double-valved configuration. Besides, a 5W more cooling power in the measured temperature range has been obtained for the double-valved arrangement. These results show that the double-valved configuration is an effective way to suppress the DC flow.

When the temperature is below 70K, the volumetric specific heat of the stainless steel mesh is lower than the helium. Matrix1's regeneration capacity will greatly degrade, which hamper the temperature further lowering. Replacing 1/3 space of the stainless steel mesh by lead balls (matrix2), which provide larger volumetric specific heat at the cold part of regenerator, maybe leads to a lower temperature and larger cooling power. Due to the difference of the orifice and double-inlet pulse cooler in operating principles, the effect of the matrix arrangements on cooling performance will also be some different.

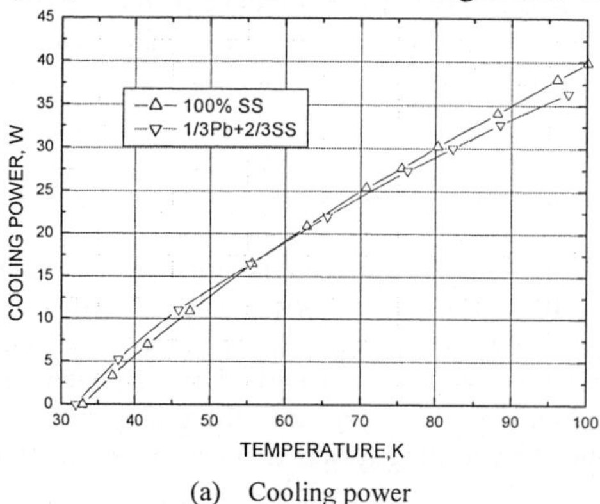

(a)   Cooling power

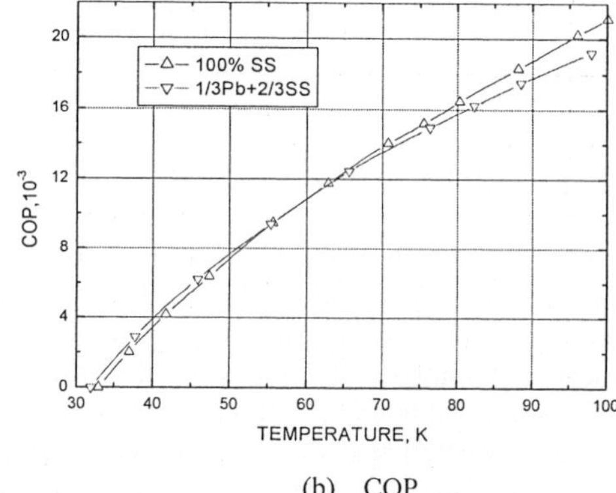

(b)   COP

Fig.3   Cooling power and COP vs. temperature with various regenerator matrices (Orifice mode,RW2)

Fig.3 gives the cooling power and COP of the cooler operating under orifice mode with various regenerator matrices. As it is shown in figures, the minimum temperature with matrix2 is only little lower than that with matrix1. And the cooling power and COP is little higher than that with matrix1 when the temperature below 54K, while temperature above 54K, it is lower than that with matrix1, which could be well explained with conventional regeneration theory.

From the phase shift and fluid network theories, the introduction of the double-inlet, not only increases the phase shift capacity of the mass flow and pressure, but also the flow impedance could be reduced. Both of them lead to a better performance of the cooler. As we see in Fig.4, when the cooler operating under double-inlet mode, the temperature drops from 27K to 18.4K with matrix2 and 11.5W cooling-power at 30K could be obtained. At 40K and 50K, the COP with matrix2 could be 56% and 31% more than that with matrix1. These results show that with matrix2 the positive effect of the lead balls on regeneration capacity will take dominant position than the passive effect in consequence of increased flow resistance. We also could see in Fig.4b, when the temperature is higher than 40K, the slope of the COP decreases quite sharp and at 50K it will be lower than that with matrix1. This change tendency is the same as the orifice case, which depends on the thermo-property of the lead balls. While for the measured temperature range, with matrix2 the COP is always higher than the case with matrix1. The possible reason of the refrigeration characteristics difference is: operating under double-inlet mode, the DC flow plays a great role on the performance of the cooler, with matrix2 the DC flow could be better suppressed which leads to a better performance of the cooler [5].

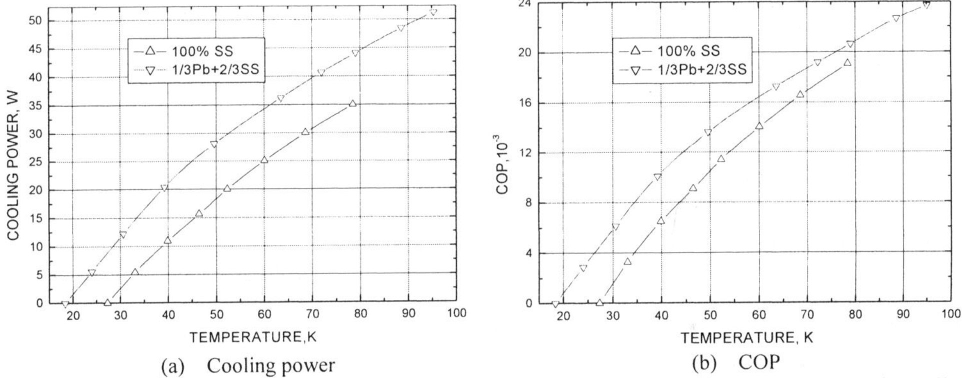

(a) Cooling power   (b) COP

Fig.4 Cooling power and COP vs. temperature with various regenerator matrices (double-inlet mode,RW2)

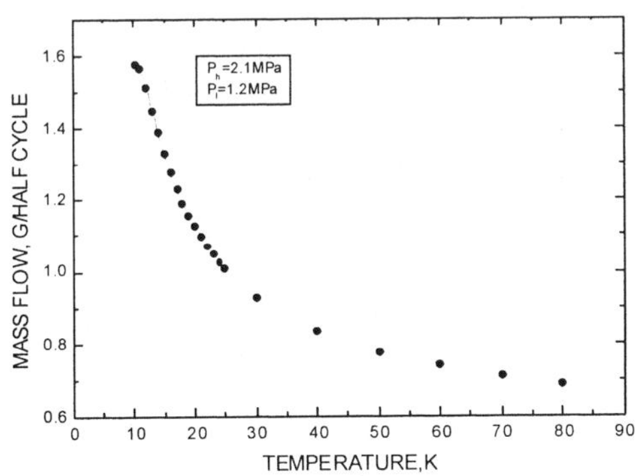

Fig.5 Calculated gas flow per half cycle vs. temperature

As we know, when refrigeration temperature below 20K, the real gas property of helium will impose a great influence on the performance of the cooler, which needs more mass flow to further increase its performance. Fig.5 shows the required mass flow to maintain the high- and low-pressure 2.1Mpa and 1.2Mpa respectively for a half cycle of the cooler. The calculation was carried out based on the temperature profiles of the regenerator and pulse tube are linear. In 300K-80K temperature range, we calculate the gas properties with mean temperature and then the mass flow; while for the range from 80K to minimum temperature the gas properties of 20 differential temperature segments are used to calculate the mass flow. From Fig.5, we can see that with temperature decreases, especially for a temperature below 20K, the required gas of the cooler increases very fast.

Fig.6 shows the cooling power and COP curves vs. temperatures driven by different compressors. It is seen in Fig.6a that the minimum temperature of the cooler is as low as 14.7K, and 29.5W cooling power has been obtained at 30K driven by CP4000. This is one of the best results for a single stage pulse tube cooler could reach so far. However, the slope of the COP is not as steep as the case driven by RW2 shown in Fig.6b. Especially when the temperature higher than 31K, with RW2 could get higher COP than that driven by CP4000. This means the cooler which uses a compressor with low power supply could get higher efficiency.

We also tested the performance of the cooler with CP6000 compressor. As shown in Fig.6a, no more cooling power than that of CP4000 has been obtained driven by CP6000, while the COP are much lower. One of the possible reasons causing this result is that the larger gas flow provided by CP6000, leads to too large pressure ratio in pulse tube to expand sufficiently. So a proper compressor selection is very important for a certain cooler to obtain good performance when the required cooling power is fulfilled. The other possible reason is that, when the temperature below 20K, the volumetric specific heat of the helium is higher than that of lead balls, the regenerative efficiency of the regenerator is very low. More gas flow will further deteriorate regenerator performance, and as a results the cooling performance are limited. In this case, through further optimization of the regenerator filling matrix (for example, using magnetic regenerative material), lower temperature and higher COP may be obtained.

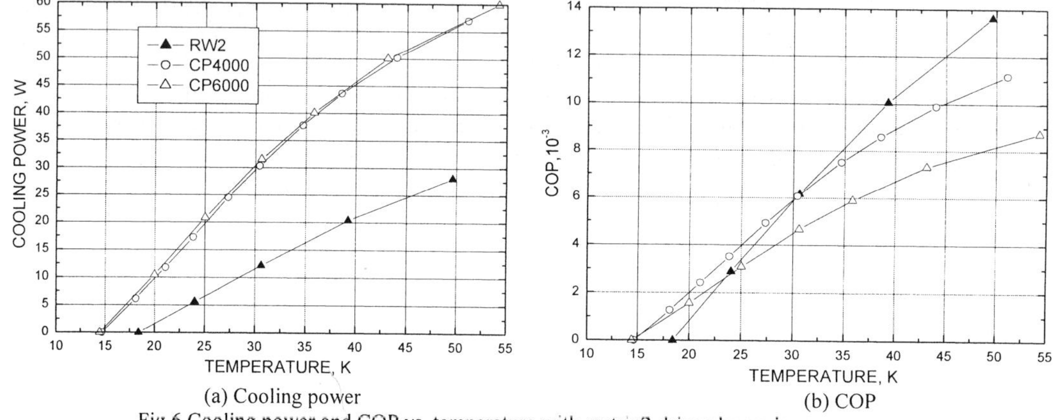

(a) Cooling power   (b) COP

Fig.6 Cooling power and COP vs. temperature with matrix2 driven by various compressors

CONCLUSIONS

The existence of DC flow deteriorates refrigeration performance of a double-inlet pulse tube cooler greatly. A double-valved configuration instead of conventional single-valved one is successfully used in experiments to suppress the DC flow to a small extent. The effect of the regenerator matrices on performance of the cooler not only depends on the material thermodynamic property, but also the operation mode of the cooler. Operating under double-inlet mode, 18.4K and 14.7K minimum temperature has been reached driven by RW2 and CP4000 compressor respectively.

ACKNOWLEDGEMENT

The financial support from the German academic exchange service (DAAD) to one of the authors (Yanlong Jiang) should be greatly appreciated.

REFERENCES

1. Gedeon, D., DC gas flows in Stirling and pulse tube cryocoolers, Cryocoolers 9, Plenum Press, New York, USA(1997) 385-392
2. Siegel, A., Haefner, H.U., Investigation to the long-term operational behavior of GM-Pulse tube cryocooler, Adv. Cryog. Eng. 47, Melville, New York, USA(2002) 903-907
3. Gardner, D.L. and Swift, G.W., Use of inertance in orifice pulse tube refrigerators, Cryogenics (1997)24 117-121
4. Wang, C., Thummes, G. and Heiden, C., Control of DC gas flow in a single-stage double-inlet pulse tube cryocooler, Cryogenics(1998)38 843-847
5. Jiang, Y. L., Chen, G. B., Thummes, G., Influence of regenerate characteristic on the performance of pulse tube cooler, Proceeding of ICCR'2003, International Academic Publishers, Beijing, P. R. China (2003) 93-96

Use of ceramic part in G-M refrigerator

Su X-T.[1,2], Fang Z-C.[1,2], Gong L-H.[1], Li L-F.[1], Zhang L.[1]

[1]Technical Institute of Physics and Chemistry, CAS, P.O.Box2711, Beijing 100080,China
[2]Graduate School of Chinese Academy of Sciences, Beijing 100039,China

Engineering ceramics has good property of wear resistance. In this paper, the application of toughening ZrO_2 ceramics for wear part in G-M refrigerator is presented. In order to increase life span and reliability of refrigerator, ZrO_2 ceramics has been applied to fabricate rotary unit, which is a wear part of G-M refrigerator. We also have studied the wear particles formed in Teflon-metal sliding and ZrO_2-ZrO_2 sliding respectively. The thickness of wear particles is estimated from analyses based on a model. The results show ZrO_2 ceramics has the highest application potential for wear part in G-M refrigerator.

INTRODUCTION

Engineering ceramics

Various industrial processes require the use of wear-resistance materials to prevent or decrease wear loss and to reduce downtime of the equipment running in contact with abrasive environment, and also to increase the performance and quality of the processes. Traditionally used hard irons and steels and some polymers are quickly destroyed. Engineering ceramics with their high hardness and high resistance offer substantial advantages over metallic or polymeric materials. They have a growing application potential for the wear parts. Ceramics used most for wear-protection are dense or low-porous ZrO_2 ceramics, and some other oxide-based ceramics. Among the ceramics used in industry at the present time, ZrO_2 ceramics, and silicon carbide-based ceramics have the highest application potential. They demonstrate excellent wear-resistance and high mechanical properties. ZrO_2-based materials are well documented for low thermal-conductivity, high fracture toughness, as well as offering excellent wear resistance in adverse environments. Toughening ZrO_2 ceramics with its enhanced toughness appears to be ideal wear-resistance materials in a variety of engineering applications.

Rotary unit of G-M refrigerator

Rotary unit is a key part of G-M refrigerator. Its running reliability directly affects the performance and life of refrigerator. As Figure1 shown, the rotary valve fits onto the square shaft on the motor and rides on the face of the valve plate. The rotary

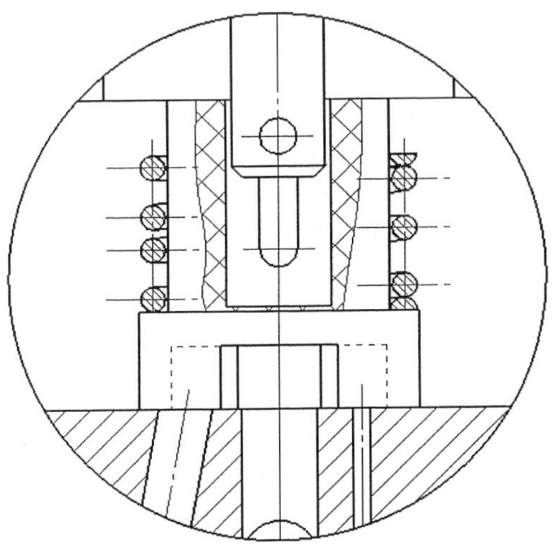

Figure 1   Schematic diagram of the rotary unit

valve controls the cycling of high and low pressure gas for the expansion process as well as for driving the displacer. For every half revolution of the rotary valve, the cold head cycles once. Originally rotary valve and rotary plate are made of Teflon and metal material respectively. After running for a long time, serious abrasion will happen between two parts. The performance and life of the refrigerator will be affected seriously. The application of toughening ZrO_2 ceramics for wear part in G-M refrigerator is presented in this paper.

THE STUDY OF TOUGHENING ZrO_2 CERAMICS FOR THE USE IN VALVE UNIT

Properties of ceramics

We compared properties of ZrO_2 ceramics with other materials commonly used in cryogenic engineering. Hardness of ZrO_2 ceramics is far larger than that of other materials, also ZrO_2 ceramics with a lubricant action is appreciated. ZrO_2 ceramics with their high mechanical properties and wear resistance offer notable advantages over metallic or polymeric materials[1,2,3].

Considerations of design

Designing reliable engineering component with ceramics is considerably more difficult and unquestionably different than designing with metal materials. In order to avoid failures, ceramics component must be stressed very low and ceramics of good quality and strength must be used[4]. We have attempted to design rotary unit because ZrO_2 ceramics provide necessary wear resistance and strength properties. Ceramic materials are not able to dissipate stress concentrations by plastic deformation due to their brittle behavior. So the engineering design differs from that of metal material. To avoid stress concentration on the ceramic component, the ceramic part is embedded into metal containment (see Figure2).

Figure 2   Hybrid rotary unit with metal and ZrO_2 ceramics

Estimation of wear particle thickness

We estimate the wear particles thickness formed in Teflon-metal sliding and ZrO_2-ZrO_2 sliding respectively by an analytical approach, which used a model for the formation of such a particle. The model considers an adhesive junction formed as a result of contact between asperities on the sliding surface and subjected to a compressive stress due to normal load and a sheer stress because of the relative motion. It produces reversal of stresses in the substrate and after a large number of such occurrences may result in the formation of a wear particle resembling a flattened ellipsoid. In order that a wear particle can be formed the elastic strain energy due to recovery in the loaded condition must be equal to or greater than the surface energy of the particle.

$$Ee \geq Es \tag{1}$$

Where Ee and Es denote the elastic strain energy of the junction and the surface energy of the particle material respectively. The elastic strain energy for an ellipsoid is given by

$$Ee = P_0^2 V_p / 2E \tag{2}$$

Where P_0 is the normal stress acting on the junction, E is Young's modulus of elasticity of the particle material and $V_p = \pi ABC/6$ is the volume of the ellipsoidal wear particle. A and B are the major and minor axes of the elliptical particle surface and C is the thickness of the wear particle. The normal stress and shear stress acting on the junction raise to the junction growth phenomenon so that

$$P_0^2 + aS^2 = \sigma_y^2 \qquad\qquad S = \mu P_0$$

Where σ_y is the yield strength of particle material, S is the shear stress acting at the junction and a a constant. Combining above two equations the normal stress can be expressed as $P_0 = \sigma_y \big/ \left(1 + a\mu^2\right)^{1/2}$.

$$Ee = \frac{\pi}{12} \cdot \frac{\sigma_y^2}{(1 + a\mu^2)E} \cdot ABC \tag{3}$$

The fracture surface area As which results in the formation of a flattened ellipsoidal wear particle can be expressed as $As = \pi \left(\dfrac{AB}{4} + \left(\dfrac{A^2 + B^2}{2} \right)^{\frac{1}{2}} \dfrac{1}{2} \times \dfrac{2}{3} C \right)$. Where the perimeter of an ellipse is $\pi \left(\dfrac{A^2 + B^2}{2} \right)^{\frac{1}{2}}$.

Also, the thickness of a flattened ellipsoid is about two-thirds the thickness of an ellipsoid because the volume of the ellipsoids should be equal, i.e.

$$\left(\frac{\pi}{4}\right)ABC' = \left(\frac{\pi}{6}\right)ABC \qquad\qquad C' = \frac{2}{3}C$$

The total surface energy required is therefore given by

$$Es = 2\gamma\pi \left(\frac{AB}{4} + \left(\frac{A^2 + B^2}{2} \right)^{\frac{1}{2}} \frac{1}{3}C \right) \tag{4}$$

Thus the thickness of wear particle can be expressed as

$$C \geq \frac{\gamma AB}{\left\{ \dfrac{\sigma_y^2}{6(1 + a\mu^2)E} \right\} AB - \dfrac{4}{3}\gamma \left\{ \dfrac{\left(A^2 + B^2\right)}{2} \right\}^{\frac{1}{2}}} \tag{5}$$

Where γ is the surface energy of the particle material. The constant a, a value of 3 is assumed.

The wear particles formed in the manner described above may either escape from the interface or be trapped between the high points of the sliding surfaces. What actually happens to the particle depends on the location of the particle in the contact zone at the instant it is formed. If the wear particle is trapped another strong bond is likely to be formed between the steel surface and the loose wear particle because of the much higher surface energy of the steel material compared with that of the polymers. This results in the formation of a transfer film on the steel surface. Afterwards the bond will therefore be established between the newly formed polymer wear particle and the polymer-sliding surface or the polymer layer deposited on the steel surface. This particle will be liberated from the surface only if the elastic strain energy stored in it becomes greater than or equal to the adhesion energy acting on the interface, i.e.

216

$$Ee \geq Ea \tag{6}$$

Where Ea is the adhesion energy. When this particle is relieved from the normal and shear stresses acting at the interface due to sliding motion the adhesion at the interface will prevent the particle from contracting and there will be a residual stress of magnitude $\varepsilon\sigma_y/\left(1+a\mu^2\right)^{1/2}$ where ε is Poisson's ratio for the particle material. Therefore

$$Ee = \frac{1}{2}\frac{\varepsilon^2\sigma_y^{\,2}}{(1+a\mu^2)E}\cdot\frac{\pi}{6}ABC_1 \tag{7}$$

$$Ea = W_{ab}\cdot A_{adh} = W_{ab}\cdot\pi\left(\frac{AB}{4}+\left(\frac{A^2+B^2}{2}\right)^{1/2}\frac{1}{3}C_1\right) \tag{8}$$

Where A_{adh} is the area over which adhesion occurs and W_{ab} is the specific energy of adhesion, i.e. the energy required to separate $1cm^2$ of the interface between materials a and b involved in adhesion, which is equal to 2γ for identical materials and equal to the arithmetic average of the surface energies for incompatible materials. Thus from eqns. (6), (7) and (8) we obtain

$$C_1 \geq \frac{W_{ab}AB}{\left\{\dfrac{\sigma_y^{\,2}\varepsilon^2}{3(1+a\mu^2)E}\right\}AB - \dfrac{4}{3}W_{ab}\left\{\dfrac{(A^2+B^2)}{2}\right\}^{\frac{1}{2}}} \tag{9}$$

The two kinds of material have different wear properties, so the thickness of a wear particle of Teflon was calculated from eqns. (5) and equaled to $9.04\,\mu\,m$, the thickness of a wear particle for ZrO_2 ceramics was also calculated using eqns. (9) and equaled to $1.62\,\mu\,m$. The results show the thickness of wear particle from ZrO_2 material is far less than that of the traditional material. Wear rate of materials increases with the size of wear particle. So toughening ZrO_2 ceramics has distinct advantage of wear resistance for wear part in G-M refrigerator.

CONCLUSIONS

Engineering ceramics with their high hardness and wear resistance offer substantial advantages over metallic or polymeric materials. Toughening ZrO_2 ceramics with its enhanced toughness appears to be ideal wear-resistance materials for wear part in G-M refrigerator. It can improve the lifetime and reliability of the refrigerator.

REFERENCES

1.John B. Wachtman, <u>Mechanical properties of ceramics</u>, Wiley, New York, (1996)

2.Dietrich Munz and Theo Fett, <u>Ceramics mechanical properties, failure behavior, materials selection</u>, Springer, New York, USA (1999)

3. Ernest Robinowicz, <u>Friction and wear of materials</u>, Wiley, New York (1999)

4.W.E.C. Creyke and R. Morrell, <u>Design with non-ductile materials</u>, Applied Science Publishers, New York, USA (1995)

A study on the in-line type inertance tube pulse tube cryocooler

Seong-Je Park, Deuk-Yong Koh, Yong-Ju Hong, Hyo-Bong Kim, Seon-Young Kim*, Woo-Seok Jung*

HVAC & Cryogenic Engineering Group, Korea Institute of Machinery & Materials, 305-600,
P. O. Box 101, Yu-Sung, Taejeon, Korea
*DA Lab., LG Electronics Inc., 327-23, Gasan-Dong, KeumChun-Gu, Seoul 153-023, Korea

The Experimental results of the in-line type inertance tube pulse tube cryocooler
for cooling superconductor RF filter are presented in this paper. The purpose of
this study is to analyze the characteristics of in-line type inertance tube pulse tube
refrigerator (IPTR), and to get main factor to improve the performance of the in-
line type IPTR. Firstly, design parameters of the in-line IPTR are discussed by
ARCOPTR program, and then to find optimal conditions of in-line type IPTR,
cool down characteristics according to the variations of the charging pressure,
inertance tube volume, regenerator volume and pulse tube volume are measured
by the experiment.

INTRODUCTION

The pulse tube cryocooler was first described by W.E.Gifford and R.C. Longsworth in 1964. This type of
the pulse tube cryocooler is now called as the basic pulse tube cryocooler. The performance of this pulse
tube refrigerator has been greatly improved by introducing an orifice and a buffer volume added to the
hot end of the pulse tube. This type of the pulse tube refrigerator, which is called as the orifice pulse tube
refrigerator, was modified by R.Radebaugh et al. in 1986.

In 1990, the double inlet pulse tube refrigerator, in which a bypass tube is connected between a
pressure wave generator and the hot end of the pulse tube, was suggested by S.Zhu et al. [1]. The
refrigeration power per unit mass flow rate through the regenerator was greatly increased in the double
inlet pulse tube refrigerator. In the case of the high frequency operation, there is another phase.

Commonly used means to achieve the optimum performance of the Stirling type pulse tube
refrigerator is the inertance tube [2,3,4,5]. A detailed analysis of the IPTR was reported by Zhu et al. [6].
They carried out analysis providing the performance as a function of the diameter and length of the long
neck tube (inertance tube). The analysis was verified by an experiment in which a long tube was
connected directly between the reservoir and compressor volume.

In this paper, discussions about the net cooling with volume of the regenerator and the volume ratio
of the regenerator to the pulse tube are discussed first by ARCOPTR (Ames Research Center Orifice
Pulse Tube Refrigerator) program of NASA Ames Research Center and then the effects of the charging
pressure, the volume of the regenerator and the diameter and length of the inertance tube on the cool-
down characteristics of the pulse tube cryocooler are investigated by experiment.

DESIGN OF THE IN-LINE IPTR BY ARCOPTR AND EXPERIMENTAL DESCRIPTION

Schematic diagram of the in-line IPTR is shown in Figure 1. The inertance tube pulse tube cryocooler
consists of a linear compressor, transfer line, heat exchanger, regenerator, pulse tube, inertance tube,
buffer and vacuum chamber. Linear compressor consists of linear motor, inner and outer yoke,
permanent magnet, coil, cylinder, piston and flexure bearing. The pressure oscillation is generated by
using a single acting helium compressor (Sunpower compressor revised by LG) for Stirling cryocooler.

Figure 2, Figure 3 and Figure 4 show the net cooling capacity, P-V work and COP with volume of
the regenerator and the volume ratio of the regenerator to the pulse tube, respectively. As the diameter of
the regenerator decreases, maximum cooling capacity and COP increase and volume ratio at the

maximum cooling capacity and COP decrease, but the range of the volume ratio at the maximum cooling capacity and COP is very narrow. Therefore, although cooling capacity and COP are low, the IPTR is operated stably in the large volume of the regenerator. Operating frequency of the IPTR is 60 Hz.

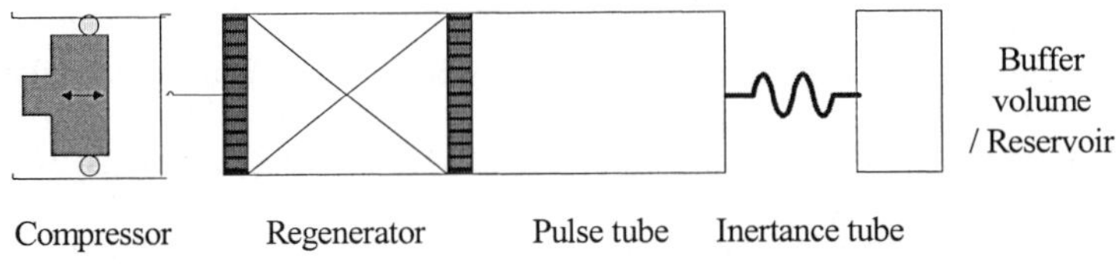

Figure 1 Schematic diagram of the in-line type IPTR

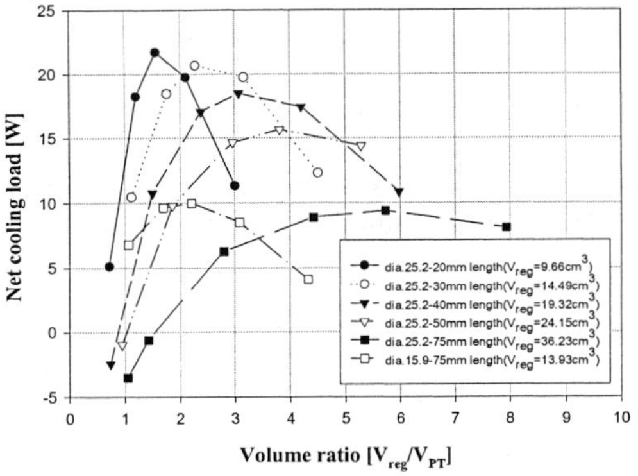

Figure 2 Net cooling capacity of IPTR

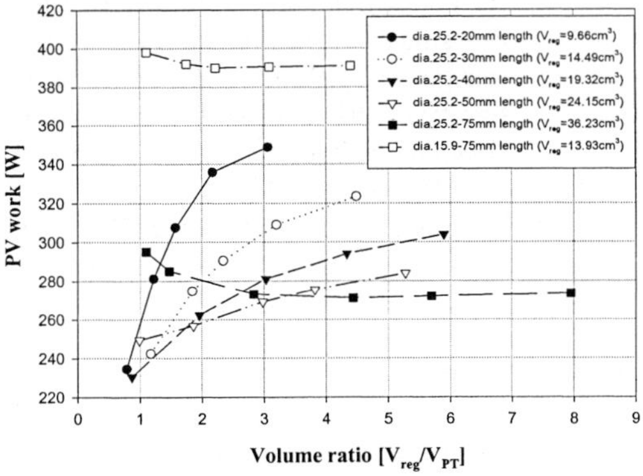

Figure 3 P-V work of IPTR

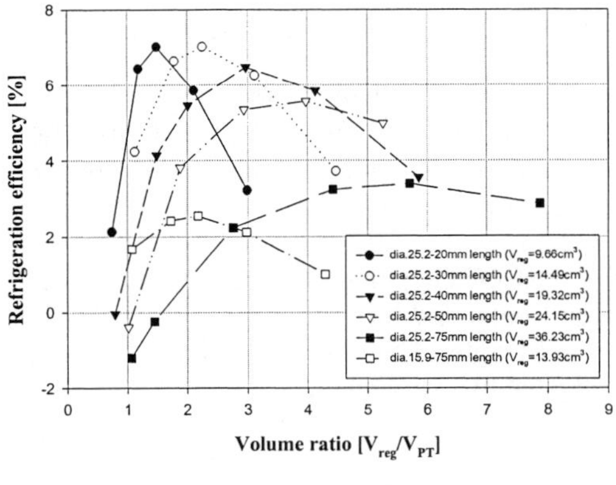

Figure 4 COP of IPTR

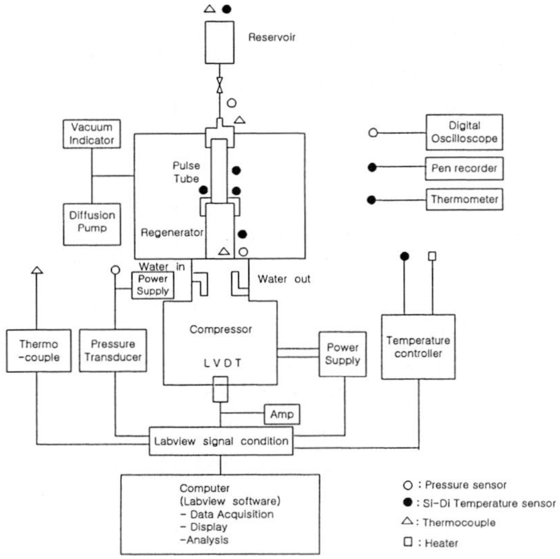

Figure 5 Experimental apparatus of IPTR

Figure 5 shows experimental apparatus of the in-line type IPTR. An AC power supply is used to supply and control the operating frequency and input voltage of the linear compressor. Piezoelectric pressure sensors are installed on the exit of the compressor, hot end of the pulse tube and buffer to measure pressure waves. The silicon diode thermometer is attached to measure the temperature at cold end, and the heater is provided at the cold end of the pulse tube to measure the cooling capacity. The cold end is installed to vacuum chamber, and the pressure of the vacuum chamber is maintained below 10^{-5} Torr to reduce the thermal loss during measurements. After the regenerator of the pulse tube cryocooler is cleaned by evacuating and purging with clean high-pressure helium gas, the pulse tube cryocooler is connected to the compressor.

EXPERIMENTAL RESULTS

Figure 6 shows the variations of the lowest temperature with length and diameter of the inertance tube and the charging pressure at the regenerator length of 40 mm. The lowest temperature was measured at the charging pressure of 21 atm and inertance tube of 1.6 m in length and 3 mm in diameter.

Figure 7 and Figure 8 show the variations of the lowest temperature with length and diameter of the inertance tube and the charging pressure at the regenerator length of 50 mm, 60mm, respectively. The lowest temperature in the regenerator length of 60 mm was measured at the charging pressure of 27 atm and inertance tube of 1.6 m in length and 3 mm in diameter. That is to say, as the charging pressure increased, the lowest temperature decreased.

Figure 9 shows cooldown characteristics of the in-line type IPTR at the cycle frequency of 60Hz with the charging pressure of 27 atm, inertance tube of 1.6 m in length and 3 mm in diameter and regenerator length of 75mm. The lowest temperature of the cold end was about 50 K.

Figure 10 and Figure 11 show cooling capacity and COP of the in-line type IPTR with the charging pressure, respectively. As shown in Figure 10, cooling capacity was the highest in the charging pressure of 32 atm and 5W at 72K. On the other hand, COP of the in-line IPTR was highest in the charging pressure of 21 atm and 0.018 at 77K.

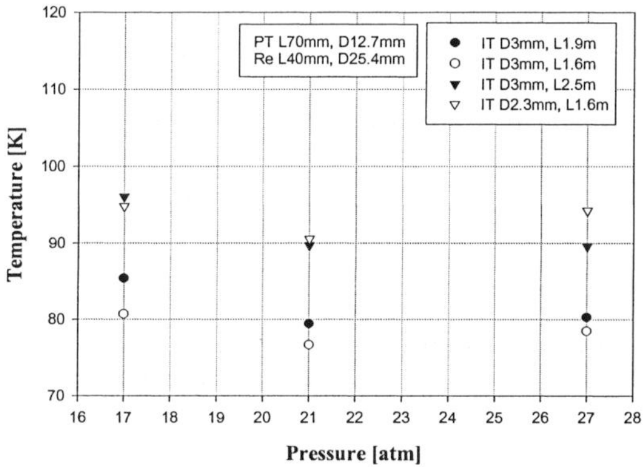

Figure 6 The lowest temperatures with charging pressure and inertance tube (regenerator 40mm long)

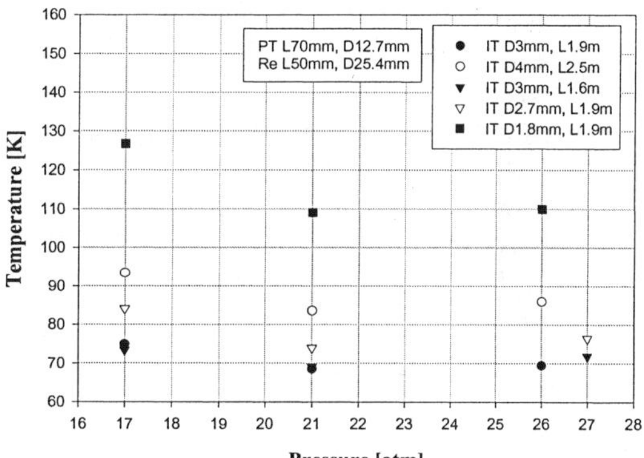

Figure 7 The lowest temperatures with charging pressure and inertance tube (regenerator 50mm long)

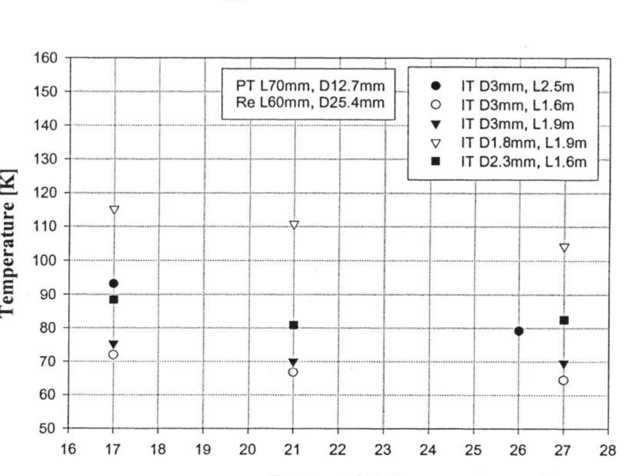

Figure 8 The lowest temperatures with charging pressure and inertance tube (regenerator 40mm long)

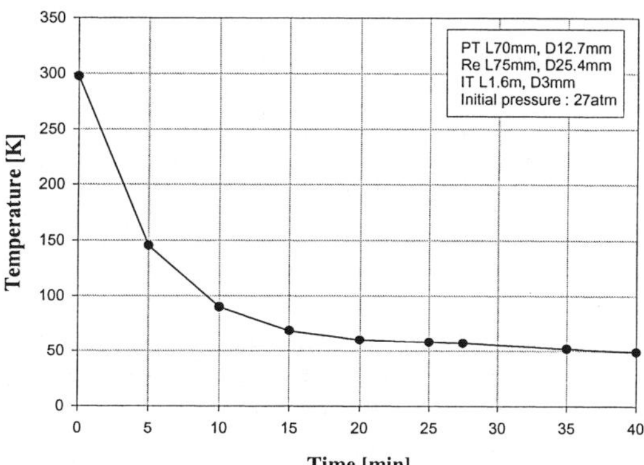

Figure 9 Cooldown characteristics of IPTR (regenerator 75mm long)

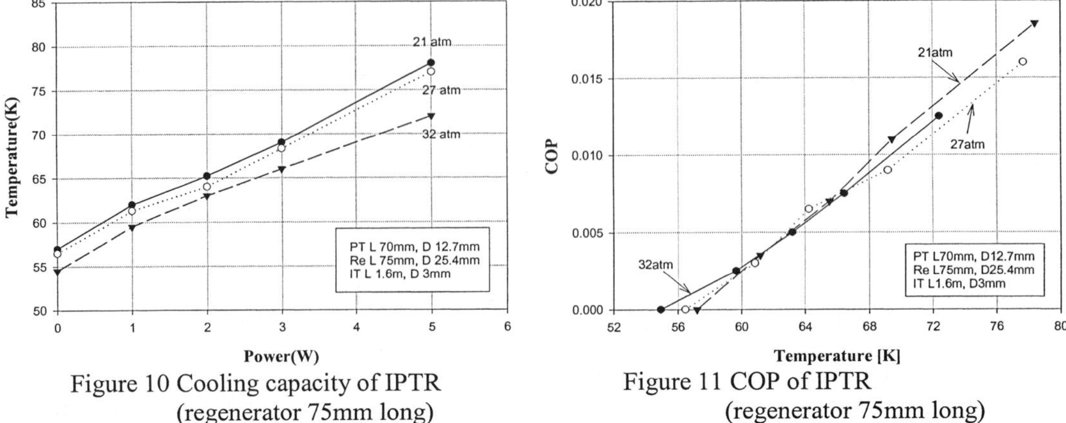

Figure 10 Cooling capacity of IPTR
(regenerator 75mm long)

Figure 11 COP of IPTR
(regenerator 75mm long)

CONCLUSION

We have undertaken analysis about the net cooling with volume of the regenerator and the volume ratio of the regenerator to the pulse tube by ARCOPTR program. And then, the effects of the charging pressure, the volume of the regenerator and the diameter and length of the inertance tube on the cool-down characteristics of the pulse tube cryocooler were investigated by experiment. The following conclusions are drawn from the experimental results.

(1) In the analysis by ARCOPTR program, as the diameter of the regenerator decreased, maximum cooling capacity and COP increased and volume ratio at the maximum cooling capacity and COP decreased, but the range of the volume ratio at the maximum cooling capacity and COP was very narrow, therefore the IPTR was operated stably in the large volume of the regenerator.

(2) As volume of the regenerator increased, the performances of the lowest temperature, cooling capacity and COP were improved.

(3) Cooling capacity was the highest in the charging pressure of 32 atm and 5W at 72K. On the other hand, COP of the in-line type IPTR was the highest in the charging pressure of 21 atm and 0.018 at 77K.

ACKNOWLEDGEMENT

This work was supported by the "Dual use technology program"

REFERENCE
1. Zhu, S. et al., (1990), Double inlet Pulse Tube Refrigerators: An Important improvement, <u>Cryogenics</u>, 30, 514-520.
2. Kanao, K., Watanabe, N. and Kanazawa, Y., (1994), A miniature Pulse Tube Refrigerator for Temperature below 100K, <u>ICEC Supplement, Cryogenics</u>, Vol. 34, pp. 167-170.
3. Pat R.Roach and Ali Kashani, (1990), Pulse Tube Coolers with an Inertance Tube: Theory, Modeling and Practice, <u>Advances in Cryogenic Engineering</u>, <u>43</u>, 1895-1902.
4. D.L.Gardner and G.W.Swift, (1997), Use of Inertance in Orifice Pulse Tube Refrigerators, <u>Cryogenics</u>, 1997, <u>37</u>, 117-121.
5. K.V.Ravikumar and Y.Matsubara, (1998), Pulse tube refrigerator based on fluid inertia, <u>Advances in Cryogenic Engineering</u>, Vol.43, 1911-1918.
6. Zhu, S. W., Zhou, S. L., Yoshimura, N. and Matsubara, Y., (1997), Phase Shift Effect of the Long Neck Tube for the Pulse Tube Refrigerator", <u>Cryocoolers 9</u>, pp. 269-278.

Proceedings of the Twentieth International Cryogenic Engineering Conference
(ICEC 20), Beijing, China. © 2005 Elsevier Ltd. All rights reserved.

Experiment on a Single-stage GM Type Pulse Tube Cryocooler near 20K

Z.H. Gan, H.Z. Liu, Z.Z. Cheng, L.M. Qiu, G.B. Chen

Cryogenics lab., Zhejiang University, Hangzhou, P.R.China   310027

An experiment on a single-stage GM type pulse tube cryocooler was presented in this paper. The minimum temperature 22.4K and cooling power 5.65W at 80K were obtained with 2kW input power, when the double-inlet measure consists of two parallel needle valves in the inverse directions, and the regenerator matrix consists of phosphor bronze screens and stainless steel screens of 247 mesh alternation. During 24-hour uninterrupted operation, the cold head temperature fluctuation is less than 0.3K.

INTRODUCTION

With the development of superconductor, the need for stable and reliable cryocoolers near 20K is increasing. Many cryocoolers can meet this requirement, including Stirling, GM and pulse tube coolers[1]. Without moving parts at the cold end, pulse tube cryocooler is better than Stirling and G-M cooler in the aspects of operating time and reliability. So it is significant to make practicable researches on pulse tube cryocoolers.

During the past years, much progress on pulse tube cryocooler has been made. With a four-valve structure, a minimum temperature 20.5K was obtained when 2.4kW power was inputted [2]. With sintering matrix in the cold end, a minimum temperature 24K was obtained by W.J. Sun [3]. When the double-inlet mode consists of two parallel needle valves in the inverse directions (named double-valved configuration), and the matrix was filled with lead sphere at 1/3 space of regenerator near cold head, a minimum temperature 14.7K was obtained with 4kW power input [4]. Furthermore, the minimum temperature of a single-stage pulse tube cryocooler with the double-valved configuration was below 13K, driven by a 13kW compressor [5]. It indicated that DC flow in the pulse tube refrigerators will be controlled well with double-valved configuration. This new structure is verified again in our experiment on a single-stage GM type pulse tube cryocooler and presented in this paper.

EXPERIMENTAL APPARATUS

A single-stage pulse tube cryocooler used in this experiment is shown in Figure 1. The experimental apparatus consists of a helium compressor (1), rotary valve (2), regenerator (3), cold end heat exchanger(4), pulse tube(5), hot end heat exchanger(6), double-inlet valves (7,8), orifice valve (9), reservoir (10). The regenerator and pulse tube are made of stainless steel tubes with outer diameter, wall thickness and length of $\phi20\times0.3\times210$mm and $\phi14\times0.5\times220$mm, respectively. The reservoir volume is 0.5l. The regenerator matrix consists of phosphor bronze screens and stainless steel screens of 247 mesh alternation. The double-inlet mode is double-valved configuration, which consists of two parallel needle valves (SWAGELOK, SS-6MG-MM) in the inverse directions.

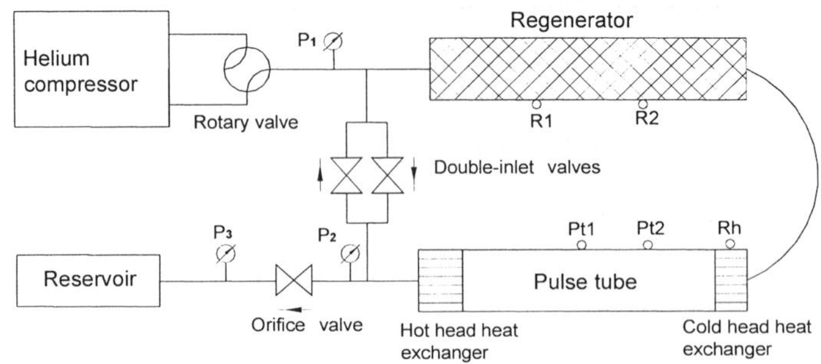

Figure 1 Schematic diagram of a single-stage GM type pulse tube cryocooler

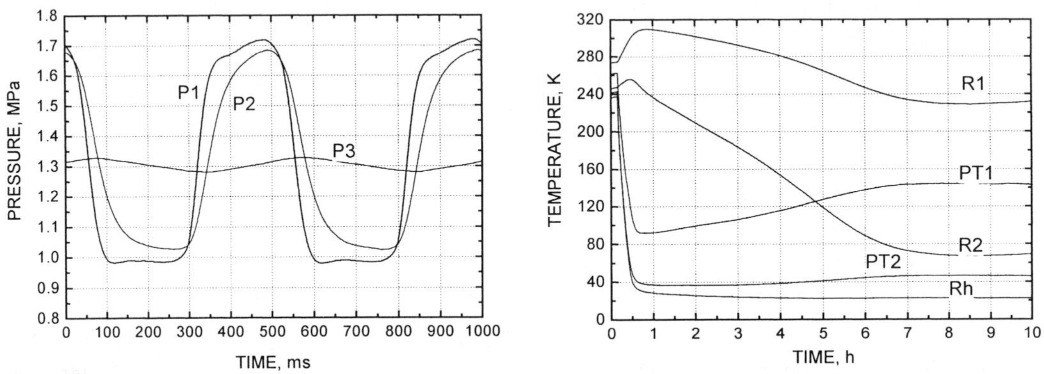

Figure 2 Pressure wave at the minimum temperature   Figure 3 Cooling temperature curve

A rhodium-iron thermometers and four Pt100 thermometers (R1, R2, PT1, PT2) are used to measure the temperatures of the cryocooler. Three piezoelectric pressure sensors (type KPY 46R, Siemens) are used for monitoring the dynamic pressure at the hot end of the regenerator and pulse tube, and reservoir, as shown in Figure 1. Based on the heat balance method, manganin wires wrapped around the cold head are used to measure the cooling power at the cold head. Temperatures and cooling power recordings are accomplished by means of a PREMA multimeter (type 5017SC) through one NI GPIB card into a computer together with pressure recordings through one NI Lab-PC-1200 data acquisition card. All of them can be displayed easily by LabVIEW software in the computer.

RESULTS AND DISCUSSION

The experiments operate with a charge pressure of 1.40 MPa (absolute pressure) and a frequency of 2 Hz. Pressure waves of various measuring points at the minimum temperature are shown in Figure 2. The mean value of P1, the pressure located at the hot end of the regenerator is 1.35MPa, and its pressure ratio is 1.75. The amplitude of P2, the pressure located at the hot end of the pulse tube is a little smaller than that of P1, and its pressure ratio is 1.64. The amplitude of P3, the pressure of the reservoir, is very small, and its pressure ratio is 1.04.

After the optimization of the valves opening, the minimum temperature 22.4K with no load was obtained, with the input power of the compressor 2kW. As shown in Figure 3, after the cryocooler operated for 3 hours, the temperature at the cold end of the pulse tube reached about 24K, another 6 hours later, the minimum temperature 22.4K was obtained. R1 and R2, the temperatures located on the regenerator, increased in first half an hour, then began to decrease, which may be due to the influence of

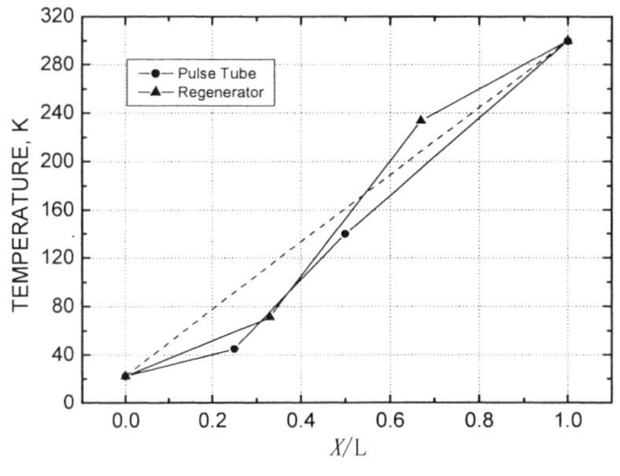

Figure 4 Temperature distribution

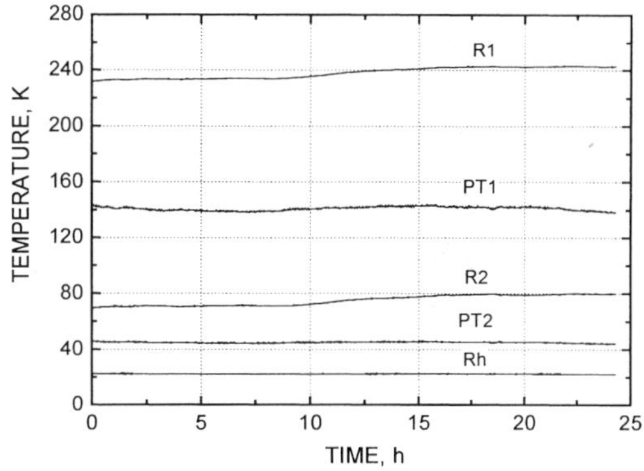

Figure 5 Temperature fluctuation

the length of the regenerator. By shortening its length, better performance of the cryocooler is expected.

Considering the temperature gradient at the measuring points during the stable period of the whole system, the temperatures of hot end of both pulse tube and regenerator are assumed to be 300K(ambient temperature). The temperature gradient of the whole system was analogous with linear distribution, as shown in the Figure 4. However, the temperature R1 was a little higher than expected, maybe due to the excessive length of the regenerator.

Temperature stability is quite important to decide the refrigeration performance of pulse tube cryocoolers. In the pulse tube cryocoolers, the refrigeration process is controlled by the mass flow rate through each component in the system. The mass flow rate through each component has a different amplitude and different phase with each other. The stable mass flow rate through each component (such as regenerator, pulse tube, orifice, double-inlet valves and heat exchangers) is such an important factor to keep the refrigeration temperature stable. Temperature fluctuations at various measuring points are shown in Figure 5 and 6. The temperature fluctuation amplitude of Rh, the minimum temperature, is lower than 0.3K. The rest: $PT1 \leq 2K$, $PT2 \leq 0.5K$, $R1 \leq 5K$, $R2 \leq 5K$, which is analogous with the experiment results of a single-stage pulse tube cryocooler by Y.L.Jiang and a two-stage pulse tube cryocooler by J.L.Gao [6,7].

Considering the maximum cooling power as the goal, opening of the valves was optimized. And a cooling power of 5.65W at 80K was obtained, shown in Figure 7. When the valves were set to the optimum opening at which the minimum temperature was obtained, a cooling power of only 4.30W could be obtained at 80K. On the contrary, at the optimum opening at which maximum cooling power at 80K was obtained, the temperature with no load was only 31K. In sum, the opening of the valves in different temperature ranges is different with one another, which is important to the future gas mixture refrigeration experiment.

CONCLUSION

With the double-inlet mode consisting of two parallel needle valves in the inverse directions, the minimum temperature of 22.4K and a cooling power of 5.65W at 80K were obtained on a single-stage GM type pulse tube cryocooler. During 24-hour uninterrupted operation, the minimum temperature fluctuation is less than 0.3K, which could meet the requirement of the application in the realm of superconductivity. At about 20K, the volume specific heat capacity of helium is much higher than those of the stainless steel screens and phosphor bronze screens, so the performance of the regenerator was deteriorated. Through optimization of the regenerator matrix (filled with lead spheres or magnetic materials), and shortening the length of the regenerator, the minimum temperature with no load can be expected to be lower, and the refrigeration performance could still be improved, all of which are what we

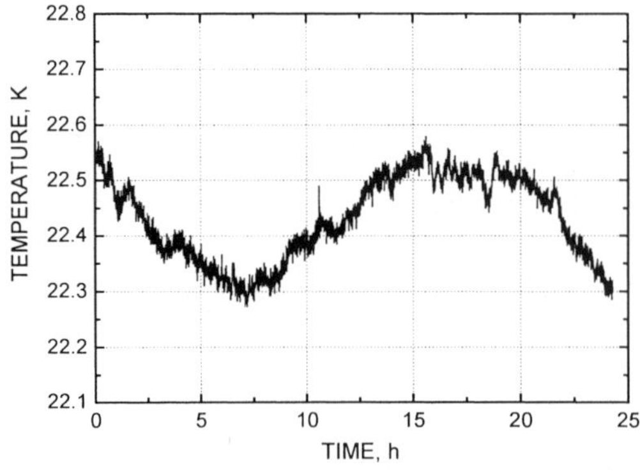

Figure 6. Cold head temperature
fluctuation for 24 hours

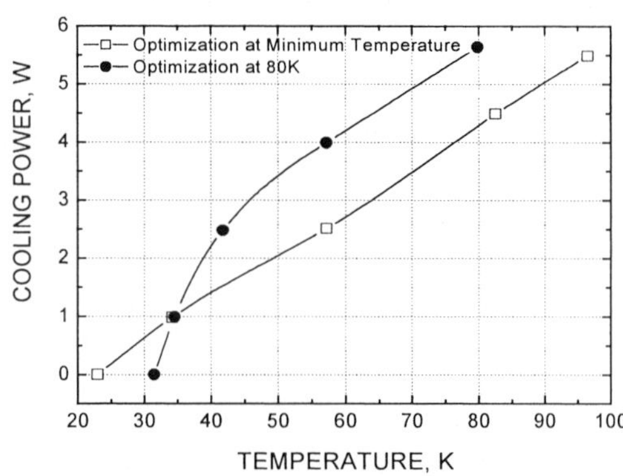

Figure 7. Cooling power with different
valve setting

are planning to do.

ACKNOWLEDGEMENT

This work is financially supported by National Natural Science Foundation of China (50106013) and Natural Sciences Foundation of Zhejiang Province (501134). And thanks to the equipments donated by Deutscher Akademischer Austauschdienst (DAAD) (335.104401.019).

REFERENCES

1. J. Yuan, J. Maguire, etc. 30-50K Single Stage Pulse Tube Refrigerator for HTS Applications. Cryocoolers (2001) 11 235-241

2. R.Li, K.Kanao, N.Watanabe, etc. A Four-valve Pulse Tube Cryocooler with a Cooling Power Over 30W at 80K. Adv. Cryo. Eng (1998) 43 (B): 1991

3. W.J. Sun, L.Y. Sun, etc. High Power Pulse Tube Refrigeration with Improved Cold End. Vacuum & Cryogenics (2001) 7(1) 27-32

4. Y.L. Jiang, G.B. Chen, etc. Experimental Investigation on DC Flow Control in a Single-stage Pulse Tube with Temperature below 20K. Cryogenics (2002) 30(6) 11-15

5. F. Giebeler, Private message or http://www.cryo.transmit.de/

6. Y.L. Jiang, G.B. Chen, G. Thummes. Influence of Regenerator Flow Resistance on Stability of Pulse Tube Cooler, CRYOGENICS AND REFRIGERATION - PROCEEDINGS OF ICCR'2003. Editors: Chen GB, Hebral B, Chen GM (2003) 93-96

7. J. L. Gao. IGC-APD Advanced Two-stage Pulse Tube Cryocoolers. Adv. Cryo. Eng (2001) 47 683-690

Design and test of a 80 K miniature pulse tube cryocooler

Zhao M-G. [1,2], Wang X-L.[1,2], Hou X-F.[1,2], Liang J-T.[1], Cai J-H.[1] and Zhou Y.[1]

[1]Technical Institute of Physics and Chemistry, CAS, P.O.BOX 2711, Beijing, 100080, China
[2]Graduate School of the Chinese Academy of Sciences, Beijing, 100039, China

A 80 K miniature pulse tube cryocooler has been designed, fabricated and tested. The pulse tube cryocooler is designed to provide 0.5 W of cooling power at 80 K with input electrical power less than 40 W. The cryocooler system incorporates a compressor with U-shaped pulse tube configuration. A computer simulation program based on the theory of thermodynamic non-symmetry is developed for performance predictions, optimizations, and as a guide for system design. Dimensions of the pulse tube and the regenerator are optimized further by experimental methods. The double inlet valve and the inertance tube are both used for phase shifting. The lowest temperature of 59.3 K is achieved.

INTRODUCTION

The pulse tube cryocooler (PTC) was originally described in 1964, and is now called the basic pulse tube cryocooler. The performance of the PTC has been greatly improved by introducing an orifice and a buffer volume added to the hot end of the pulse tube in 1984, and this is now called the orifice pulse tube cryocooler. The double inlet pulse tube cryocooler was introduced in 1990, which greatly increased the performance of the PTC. And the inertance pulse tube cryocooler was investigated in 1994.

Because of the absence of moving parts at low temperature, the PTC driven by a linear compressor has the potential to achieve high reliability and very long lifetime. With the technical development of the PTC, the performance of the PTC in high frequency operation can now compare with the Stirling cryocooler in some cases.

Here we report on the state of development of a 80 K miniature pulse tube cryocooler. Compactness, low cost, lightweight, high performance and high reliability are required. Based on the demands, U-shaped miniature PTCs were designed, fabricated and tested. The design and the test results of the PTCs are presented.

DESIGN OF PULSE TUBE CRYOCOOLER

A computer simulation program based on the theory of thermodynamic non-symmetry is developed for performance prediction, optimization, and as a guide for system design. Three samples of miniature PTCs were designed by the simulation program, and then fabricated and tested. The PTCs use U-shaped configuration. In principle the U-shaped pulse tube configuration has better performance than that of the co-axial configuration, and is more convenient to get cryogenic cooling from the pulse tube cold head than the in-line configuration.

A schematic diagram of the PTC is shown in Figure 1. The cryocooler system incorporates a compressor with a U-shaped pulse tube configuration and a reservoir. The compressor is made in our laboratory with a linear motor supported by flexure springs. Maximum swept volume of the compressor is 2 cm^3, which leads to a lightweight cryocooler. Double inlet valve and inertance tube are used for phase shifting.

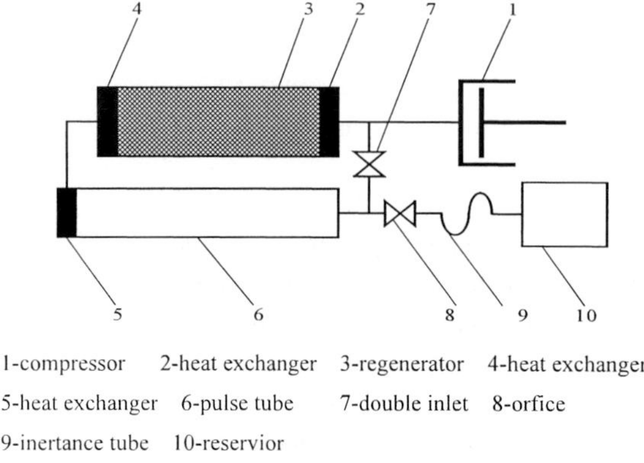

1-compressor   2-heat exchanger   3-regenerator   4-heat exchanger

5-heat exchanger   6-pulse tube   7-double inlet   8-orfice

9-inertance tube   10-reservior

Figure 1   Schematic diagram of the U-shaped pulse tube cryocooler

Figure 2   Cool down profile with no heating load

The phase difference between of the mass flow and pressure wave of the regenerator is measured with our dynamic experimental apparatus. By the measurement of phase, we can optimize dimensions of the pulse tube, regenerator and inertance tube more effectively.

ANALYSIS

Enthalpy Flow Theory shows that if the flow and pressure are sinusoidal functions of time, the time–average enthalpy flow rate can be written as:

$$< \dot{H} > = < P_d \dot{V} > = (1/2)\,|P\,\|\,\dot{V}\,|\cos\theta = (1/2)RT_0\,|\dot{m}\,|\,(|P|/P_0)\cos\theta$$

where P_d is dynamic pressure, $\dot{V}$ is volume flow rate. P is sinusoidal pressure oscillation, θ is phase angle between flow and the pressure, R is gas constant per unit mass, $\dot{m}$ is sinusoidal mass flow rate and P_0 is average pressure. We can use it for qualitative analysis.

Effect of operational frequency on cooling capacity

To predict the cooling capacity, the mass flow rate, the ratio between the pressure and average pressure, and the phase between the pressure and the mass flow rate are needed. The pressure reduces with the rising of operational frequency, which will lead the pressure ratio reduction. The mass flow rate and the phase between the pressure and the mass flow rate will be reduced too. Therefore, the cooling capacity of the cooler will increase with lowing of operational frequency.

Effect of average pressure and input electrical power

According to an analysis as before, the mass flow rate and the phase between pressure and mass flow rate increase with an increase of average pressure. And the pressure ratio first increases then decreases with the pressure rising. The cooling capacity increases then decreases by the average pressure increasing, it exists an optimum average pressure.

With the input electrical power increasing, the pressure and the mass flow rate increase. While the phase between the pressure and the mass flow rate increases with the input electrical power increasing. So the cooling capacity increases then decreases with the input electrical power increasing. It also exists an optimum input electrical power.

TEST OF PULSE TUBE CRYOCOOLER

Figure 2 is a typical cool-down profile of the PTC. The cooler can obtain the lowest temperature of 61.5 K within 40 minutes with no load. The experiments are operated at an average pressure of 4.0 MPa and an operational frequency of 35 Hz with 45.5 W of input electrical power. In the experiments, double inlet and inertance tube are both used to shift the phase. The length of the inertance tube is 2.4 m and the inner diameter of the inertance tube is 1mm.

Figure 3 shows the effect of different average pressures on the lowest temperature of the PTC. The lower temperature goes down with an increasing of average pressure. The temperature of the cold head changes little when the average pressure is higher than 3.5 MPa.

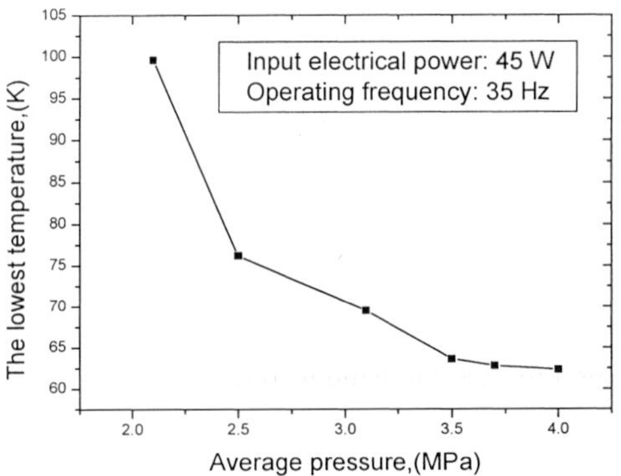

Figure 3   Effect of average pressure on the lowest temperature

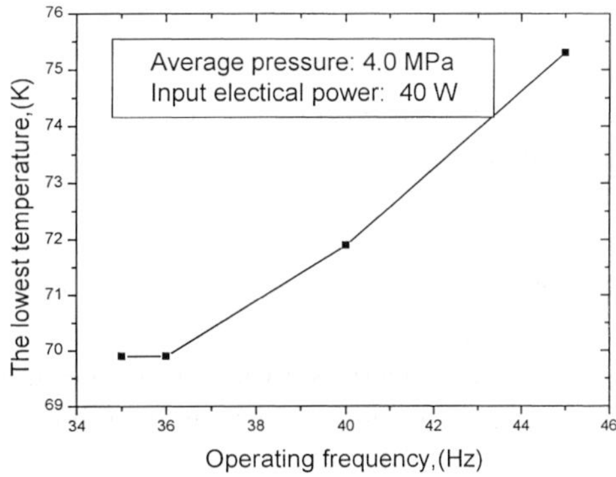

Figure 4   Relationships between the lowest temperature and operating frequency

Figure 4 shows the effect of different frequencies on the lowest temperature. The lowest temperature varies with frequency, and 35 Hz is the optimal frequency for the lowest temperature of the PTC.

The cooling capacity of the PTC with different input electrical power is shown in Figure 5. The average pressure is 4.0 MPa and operating frequency is 35 Hz. The lowest temperature is 59.3 K when the input electrical power is 54.3 W. The lowest temperature of 66.3 K was got with 40 W of electrical power with no load.

Figure 6 shows the cooling capacity of different input electrical power. The PTC can provide 0.5 W of cooling capacity at 80 K with 54.3 W of input electrical power. But only 0.25 W of cooling capacity is got at 80 K with 40 W of input electrical power. The ratio between cooling capacity and input electrical power increases with an increase of input electrical power. The measurements shows, there is a large pressure difference between the warm end and the cold tip of the regenerator, which causes the mass flow rate to decrease. On the other hand, the pressure difference is smaller in the cold tip. These two effects cause the cooling capacity of the PTC to reduce, and we can conclude the flow resistance in the regenerator is too large. It shows that the performance of the PTC is not in its optimum situation.

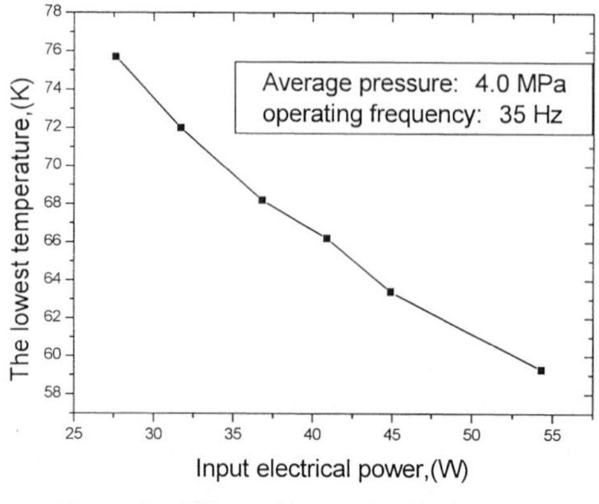

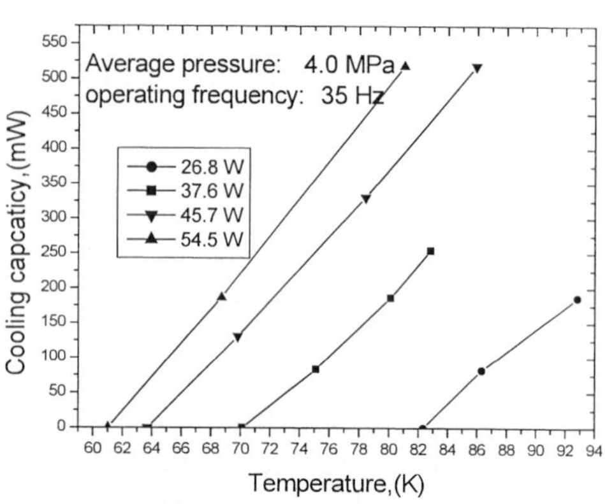

Figure 5   Effect of input electrical power
on the lowest temperature.

Figure 6   Cooling capacities as a function of cold end
temperature with different input electrical power

By analyzing the above test results of the PTC we have designed, the dimensions of the pulse tube and the regenerator geometry can be optimized in further.

CONCLUSION

The lowest temperature of 59.3 K has been achieved and the PTC can provide a cooling capacity of 0.5 W at 80 K with an input electric power of 54.3 W. But the cooling capacity at 80 K with 40 W of input electrical power is not sufficient. Because this PTC was adjusted for lower temperature, it is not in the best performance at a temperature of 80 K. The next step is to optimize the dimensions of the PTC and experimental apparatus of the PTC system.

ACKNOWLEDGEMENT

Research supported by National Natural Science Foundation of China. (Grant No.50206025)

REFERENCES

1. Liang J.T., Cai J.H., Zhou Y., et al, "Development of 40-80 K Linear-Compressor Driven Pulse Tube Cryocoolers," Cryocooler 12 Kluwer Academic/Plenum Publishers (2003), pp. 115-121.

2. Radebaugh R., "Development of the Pulse Tube Refrigerator as an efficient and Reliable Cryocooler," Proc. Institute of Refrigeration (London) 1999-2000.

3. Liang J.T., "Development and Experimental Verification of a Theoretical Model for Pulse Tube Refrigeration," PHD thesis 1991-1993.

4. Yang L.W. and Thummes G., "High-frequency Pulse Tube Coolers for HTS Applications," Cryogenics and refrigeration /Proceedings of ICCR' 2003, pp. 69-72.

Development of the 3.5 W at 80 K high-frequency pulse tube cryocooler

Jing W., Wang G.P., Liang J.T., Cai J.H.

Technical Institute of Physics and Chemistry, Chinese Academy of Sciences,
P.O.Box 2711, Beijing, 100080, China

This paper reports a 3.5 W at 80 K high-frequency pulse tube cryocooler (PTC), which is developed in the Cryogenic Laboratory of the Chinese Academy of Sciences (CL/CAS). The U-shaped configuration is adopted for the pulse tube refrigerator. Up to now, the high-frequency PTC has reached the lowest temperature of 38.8 K and the cooling capacity is about 3.58 W at 80 K, with the input power of 200 W of the linear motor, and the optimal frequency is about 40 Hz. And also, this paper introduces the necessary analysis on the performance and discusses the approaches to improve the performance. The research is still on the way.

INTRODUCTION

In recent years, the pulse tube cryocoolers (PTCs) have experienced a rapid development with the goal to eventually replace Stirling- and GM-coolers in various applications of cryoelectronics [1]. An emerging market for highly reliable cryocoolers, either Stirling-coolers or Stirling-type PTCs, can be expected in wireless communication, where the use of high-temperature superconductor devices (e.g. high-quality microwave receiver filters) is moving forwards to commercial applications [2, 3, 4].

In order to meet the different requirements in space and military applications, a Stirling-type high-frequency PTC has been developed in the CL/CAS. The PTC has the advantage of operating with no moving displacer at the cold end and nearly no mechanical vibration. The PTC can be used to cool electronic devices and high-temperature superconductor devices in mobile communications and to keep a low temperature for far infrared devices in space applications.

This paper reports three aspects of the PTC mentioned above: the description of the PTC, the experimental data and the result analysis, so as to improve the performance of the PTC.

DESCRIPTION OF THE PTC

The PTC is driven by a commercial linear compressor, the Leybold Polar SC7, with a maximum swept volume of 10 cm^3.The U-shaped configuration, which allows easy access to the cold tip, is adopted to design the pulse tube cold tip, as shown in Figure 1. And also, the cold tip configuration of this PTC is designed to be readily adaptable to meet different requirements, such as temperature and cooling capacity. The regenerator and the pulse tube are made of thin-wall stainless steel tubes. A stack of stainless steel screens of mesh No. 400 server as the regenerator matrix. Phase shifting is adjusted by use of an orifice valve at the warm end of the pulse tube and a double-inlet valve between the warm end of the pulse tube and the inlet of regenerator. Moreover, the inertance tube has been used as flow impedance. The input power of the compressor can be adjusted from 0 W to 250 W, and its frequency can be changed between 30 Hz and 60 Hz.

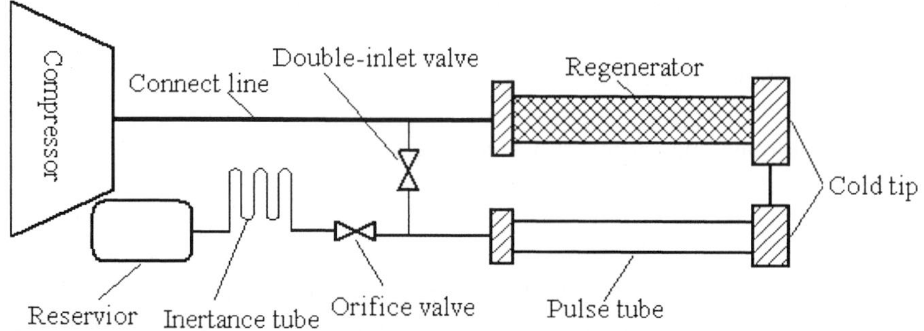

Figure 1 Schematic diagram of the Stirling–type PTC with U-shaped configuration

EXPERIMENTAL RESULT

The experimental data are shown in Figures 2-5 to describe the PTC performances clearly. All the test results are obtained at the frequency between 33 Hz and 40 Hz, the helium mean pressure is between 2.0 MPa and 3.0 MPa, as the optimum operating parameters are found in the ranges mentioned above.

Figure 2 shows the curve of the cold tip temperature in a typical process, while the mean pressure is 2.8 MPa and the operating frequency is 33 Hz. From step"1" to step "4", the input power of the compressor has changed from 100 W to 200 W, the water cooling and the double-inlet valve are taken to operate in the process. Finally, a lowest temperature of 38.8 K is available, as seen in Figure 2 (Point A).

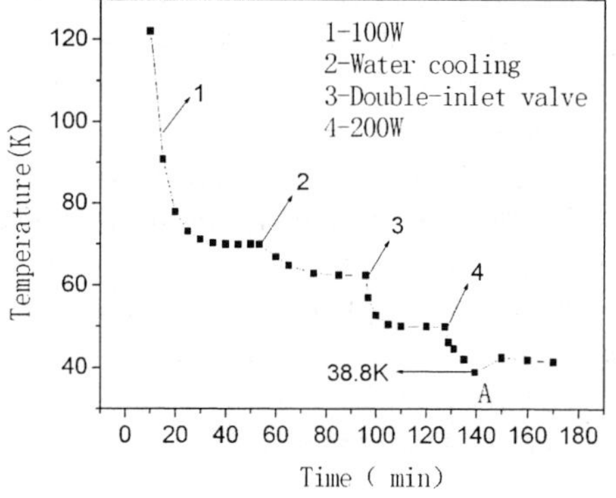

Figure 2 Curve of the cold tip temperature

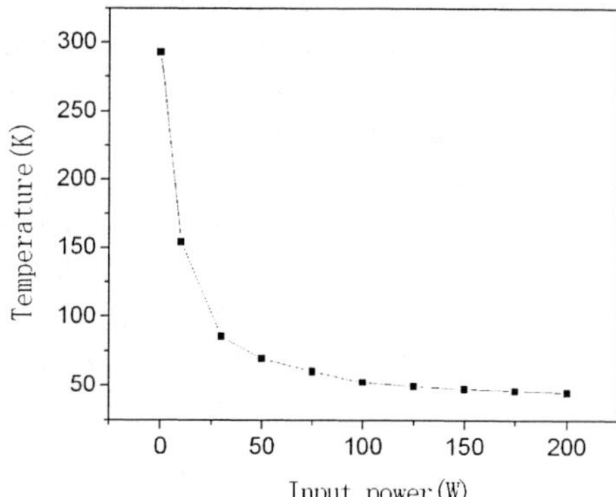

Figure 3 Lowest temperatures of the PTC at different input powers

Figure 3 shows the lowest temperature of the PTC at different input powers to the compressor, while the mean pressure is 2.8 MPa and the operating frequency is 33 Hz. At 100 W to the compressor, the lowest temperature is 53 K, the lowest temperature is 47 K with input power of 150 W, the lowest temperature is 45 K with input power of 200 W. However, as the input power increases from 150 W to 200 W, the decrease of the lowest temperature is just 2 K, this is because that the efficiency of the compressor is not in the optimal case.

Figure 4 shows the cooling capacity versus the temperature of the cold tip for the U-shaped PTC at different input power of the compressor, while the mean pressure is 2.8 MPa and the operating frequency is 40 Hz. After optimization position of value, the PTC provides a cooling capacity of 3.58 W at 80 K with the input power of 200 W, and the cooling capacity of 4.25 W at 80 K with the input power of 250 W. The cooling capacity changes linearly with the cold tip temperature.

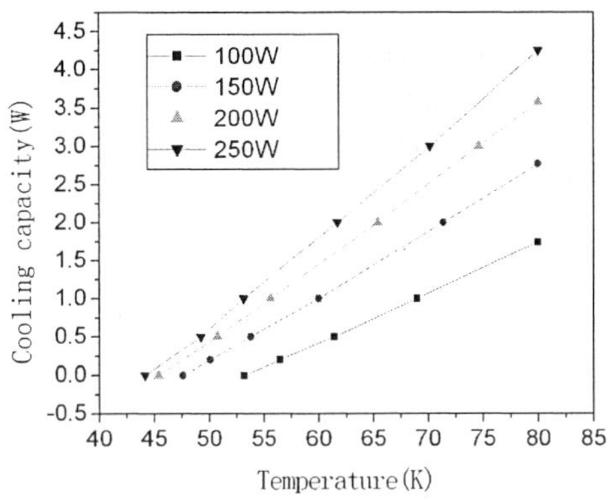

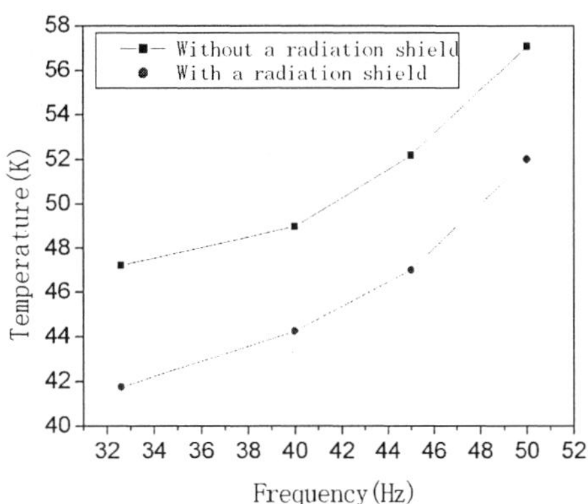

Figure 4 Cooling capacity versus the cold tip temperature at different input powers to the compressor

Figure 5 Temperature of the cold tip at different frequencies, Operation with and without a radiation shield

For improving the performance of the PTC, the radiation shield has been taken to decrease the heat loss. While the mean pressure is 2.8 MPa and the input power is 200 W, the temperature of the cold tip with the radiation shield is lower than that without the shield, the temperature difference is nearly 5 K, as shown in Figure 5.

RESULT ANALYSIS

In this paper, the compressor efficiency η_1 is defined as the ratio of the output PV power W_1 to the input power W_2 (W_2 includes W_1, Joule heat loss and friction loss), that is: $\eta_1 = W_1/W_2$; the cold tip efficiency η_2 is defined as the ratio of the cooling capacity P to the output PV power W_1, that is: $\eta_2 = P/W_1$; the PTC efficiency is denoted as the COP, which includes the compressor efficiency η_1 and the cold tip efficiency η_2, that is: $COP = \eta_1 * \eta_2$ [5]. While the mean pressure is 2.8 MPa, the COPs of the PTC at different frequencies are shown in Table 1.

Table 1 The COPs of the PTC at different frequencies

Frequency (Hz)	Input power W_2 (W)	Cooling capacity P at 80 K (W)	η_1	η_2	COP
33	200	3.45	53.3%	3.24%	1.73%
40	200	3.58	66.5%	2.7%	1.79%
46	200	3.00	63.2%	2.37%	1.5%

As shown in Table 1, the optimal COP of the PTC is 1.79% with 40 Hz of the operating frequency. The optimal η_1 is 63.2% with 46 Hz of the operating frequency and the optimal η_2 is 3.24% with 33 Hz of the operating frequency. That is, when the operating frequency decreasing from 40 Hz, the compressor efficiency and the cold tip efficiency will be improved respectively.

CONCLUSIONS

The present work demonstrates a Stirling-type high-frequency PTC with a U-shaped cold tip. The lowest temperature of 38.8 K and the cooling capacity of 3.58 W at 80 K have been achieved with the optimal frequency of 40 Hz and the input power of 200 W.

Based on the analysis, the better matching between the compressor and the cold tip, and the reduction of heat loss of the PTC and friction loss of the compressor can lead to an improvement of the PTC efficiency. The further work on the optimization of the PTC performance can be expected.

ACKNOWLEDGEMENTS

This work is supported by the National Natural Science Foundation of China (Grant No. 50206025).

REFERENCES

1.   Radebaugh, R., Development of the pulse tube refrigerator as an efficient and reliable cryocooler, Proceedings of the Institute of Refrigeration, vol.96, London (2001), pp.11-31.

2.   Häfner, H.-U., Fiedler,A., and Rolff, N., Long-life Stirling cooler for HTS-electronics qualification and application, 8th International Superconductive Electronics Conference,-Extended Abstracts-, Osaka(2001), paper P1-E9.

3.   Kotsubo,V., Olson,J.R., Champagne,P., Williams,B., Clappier, B., and Nast T.C., Development of pulse tube cryocoolers for HTS satellite communications, Cryocoolers10, Kluwer Academic/Plenum Publishers, New York(1999), pp.171-179.

4.   L.W.Yang, N.Rolff, G.Thummes, and H.U. Häfner, Development of a 5 W at 80 K Stirling-Type Pulse Tube Cryocooler, Cryocoolers12, Kluwer Academic/Plenum Publishers (2003), pp.149-155.

5.   Jing W., Hou Y.K., Ju Y.L., Liang J.T. and Cai J.H., The Design of Linear Compressor Driven Pulse Tube Cryocooler, ICCR2003, Oct. 28-Nov. 1, 2003, HangZhou, China, pp.109-112.

The effect of timing ratio of injecting to rejecting on G-M type PTC's cooling temperature

Wang, Y-Y.[1,2], Cai, J-H.[1], Liang, J-T.[1]

[1]Technical Institute of Physics and Chemistry, CAS, Beijing 100080, PR China
[2]Graduate School of the Chinese Academy of Sciences, Beijing 100039, PR China

The compression to expansion timing ratio of gas in PTC has effect on the cooling temperatures. The timing ratio can be changed through changing the dimensions of rotary valve's heads. In this paper we will study the effects under different dimensions, and get different cooling temperatures. The lowest temperature has reached 35K by using the single-stage G-M type PTC. Finding the optimum dimensions of rotary valve's heads is helpful to get lower cooling temperatures of PTC and to improve the performance of the rotary valve, although these results are not universal.

INTRODUCTION

It is well known that there is a growing requirement of the cryocoolers, especially in the fields of superconducting devices, astronomy and military etc. G-M type pulse tube coolers (PTC) are one kind of cryocoolers that can get the temperature below the temperatures of liquid nitrogen (LN_2) and liquid helium (LHe). Nowadays the single-stage GM-type PTC has achieved the lowest temperature 14.7K and the multi-stage GM type PTC has achieved the lowest temperature 1.78K. The two-stage PTC produced by Cryomech has been in application.

Rotary valve is an important component in the G-M type pulse tube coolers as well as G-M cryocoolers. It can be used to adjust the compression-expansion time of working gas and the frequency in PTC, although this can be substituted by the solenoid valve which can be controlled more easily by computer. The advantage of rotary valve is that its frequency can be adjusted continuously, and has longer life than that of the solenoid valve.

As well known that the ratio of compression to expansion time of gas in the PTC, that is to say the timing ratio has effect on the PTC's cooling temperature. In this paper the effect is analyzed. Firstly, the times of injecting and ejecting gas are computed respectively according to the dimensions and structure of the rotary valve's heads during a cycle. Simply to suppose that injecting time equals to the compression time of gas in PTC, and ejecting time equals to the expansion time. Then through doing the thermodynamic analysis about the compression and expansion processes, the different ratios of injecting to ejecting time of gas are got, and the ratios are analyzed to get the optimum. According to the thermodynamic analysis, the longer expansion time has higher COP and lower cooling temperature because this makes the gas expansion completely. While too long expansion time will make the compression time too short and result in less gas supply, which will affect the COP and cooling temperature of PTC. Therefore there exists an optimum timing ratio for the PTC. Secondly, we use a single stage G-M type PTC to test the above theoretical analysis. In experiment the DC flow wave of working gas is got, through which we can get the time ratios. Using different rotary valve's heads with different dimensions and structure we can get different DC flow waves with different time ratios and cooling temperatures. The experimental results indicate that the timing effect of the ratio exists under the nearly same other conditions, and the cooling temperatures have distinctive difference due to the effect of timing ratio. According to this we can analyze the optimal dimensions and structure of rotary valve's heads, and in the next the optional dimensions and structure of rotary valve to different PTC can be analyzed, so as to improve the performance of GM-type PTC.

There are two methods that can control the timing ratio in G-M type PTC. One method is controlled by the solenoid valves. The main advantage of this method is it's flexibility. It is also helpful to study the effects of wave form on the refrigerating performance, and can be used to find the optimal wave form of the PTC. This method is thought to be the most effective method to the application of the PTC. Of course there are some disadvantages using this method, such as short-life, noise, irreversible loss.

The other method is using rotary valve, which is modified through the rotary valve of GM cooler. It consist of an electric motor, rotary components (rotary heads) and valve body. Here there are also two methods to control the waveform. One is that the speed of electric motor is fixed but only the configuration of the rotary components may be changed, as depicted in this paper. The other is that the configuration of the rotary components is fixed while the speed of the electric motor may be changed. If we want to prolong the time of compressing gas, only the speed of electric motor need be slowed, and it is convenient to control this using a computer. The researchers of Giessen University have done some contribution in this aspect.

ANALYSIS

In the rotary valve the valve heads, which are the rotary parts include two components as depicted in Figure 1. In this figure, H11 and H12 are the high pressure inlets (where the dimensions of H11 and H12 are equal to each other.), and H2 is the low pressure outlet. In Figure 1 we can change the values of H11, H12 and H2, but the dimensions in Figure 1(b) are not changed. Therefore, the timing ratio is changed through the changes of the dimensions in Figure 1(a).

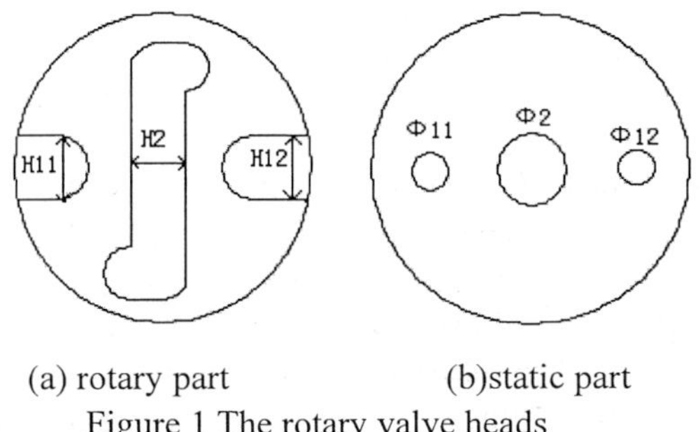

(a) rotary part   (b)static part

Figure 1 The rotary valve heads

When H11 and H12 in Figure 1(a) fit in with Φ11 and Φ12 in Figure 1(b), the high pressure gas is injected into the regenerator. When H2 in Figure 1(a) fit in with Φ11 and Φ12 in Figure 1(b), the low pressure gas is rejected from the regenerator. This is a recycle. As well known, it includes two processes, as depicted in Figure 2. The injecting process and the rejecting process are depicted in Figure 2(a) and (b) respectively. In Figure 2(a), the high pressure gas is injected into the regenerator from point A to point B, and this process is synchronous with the gas compressing process in PTC. In Figure 2(b), the low pressure gas is rejected from the regenerator from point C to point D, and this process is synchronous with the gas expending process in PTC.

If T_1 denotes the time interval from point A to point B and T_2 denotes the time interval from point C to point D. We can simply suppose that T_1/T_2 is the ratio of injecting to ejecting time of gas in PTC.

Supposing rotary velocity to be n (circles/min), then

$$T_1 = 60\frac{\theta 1}{2\pi n} \tag{1}$$

$$T_2 = 60\frac{\theta 2}{2\pi n} \tag{2}$$

where $\theta 1$ and $\theta 2$ are angles of in the interval of injecting and ejecting gas. The change of dimensions of H11 and H2 leads to the change of $\theta 1$ and $\theta 2$. So we can get the different ratios T_1/T_2, which is connected with the ratio of H11 to H2.

In the interval of T_1 the gas is compressed and the heat released by gas is dissipated. In the interval of T_2 the gas in PTC expands and absorbs the heat. Suppose that

$$Q_{dissipated} = W_{compressing} = \int_{r_1} d(pv) \tag{3}$$

$$Q_{absorbed} = W_{expanding} = \int_{r_2} d(pv) \tag{4}$$

where $Q_{dissipated}$ is the quantity of dissipated heat in the hot end and $Q_{absorbed}$ is the quantity of absorbed heat in the cold end of PTC respectively.

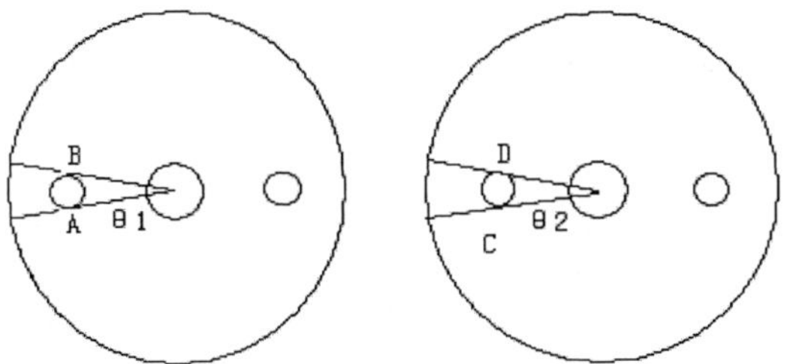

(a) Injecting gas Process   (b) Ejecting gas process

Figure 2 The time intervals of injecting and rejecting of gas

According to the thermal equilibrium of Q_{diss} and Q_{abs}, the more is Q_{diss}, the more is Q_{abs}. These two values relate to integrating times T_1 and T_2. If letting T_1 be longer , that is to say, letting the injecting time be longer, we can make the value of Q_{abs} be larger. But the longer compressing time means the shorter expanding time. If the expanding time is too short, the gas will expand not enough, which make the cooling temperature and the cooling power not optimal. So there is an optimum of timing ratio. Through changing the dimensions of H11, H12 and H2, we can approximate the optimum of T_1/T_2.

EXPERIMENTAL RESULTS

Firstly, the dimension of H11 is changed. Secondly the dimension of H2 is changed too. Under this we can get different ratios of H11/H2, which reflect the ratio of T_1/T_2.

In the experiments, the orifice and double inlets are adjusted in order to get the lower cooling temperature. The experiment results are depicted in Figure 3. The temperature is the average of several experimental results under the same conditions.

In Figure 1 we can see that with the rising of the ratio H11/H2 (that is to say with the rising of ratio of T_1/T_2) the cooling temperature begins lower. But cooling temperature begins rising when this ratio reaches approximately 1.25 with the rise of the ratio H11/H2. Therefore, we concluded that there is an optimum of ratio approximately equaling to 1.25. This also means that there is an optimum of the timing ratio. The ratio has effect on the cooling temperature and power of the PTC.

The reason is that when the injecting time is longer, the gas in PTC does more work which relates to the cooling power. In the more we can achieve the lower cooling temperature. While it will be opposed to this when the ratio is too large as we have observed in the experiment because of the insufficient expanding of gas.

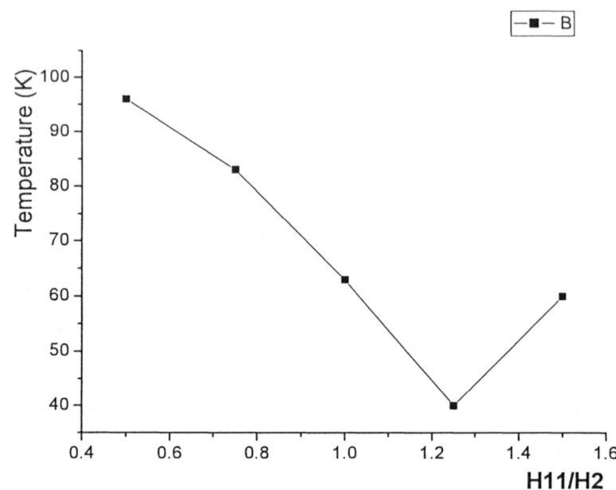

Figure 3 The relationship between the ratio H1/H2 and the cooling temperature

CONCLUSION

The effects of the timing ratio on the cooling temperature have been analyzed through analyzing the angles swept by the rotary valve heads. According to the experimental results the ratios of the H11/H2 have been obtained, which is connected with the timing ratio of injecting to ejecting time of gas and have effects on the cooling temperature of the PTC. And according to the experimental results there exists an optimum of this ratio. In this paper it is approximately 1.25. And it is helpful to find the exact optimal timing ratio and lower the cooling temperature of cold end of PTC. I think that the next step is to make the optimum is universal.

ACKNOWLEDGMENTS

The work is founded by the National Natural Science Foundation of China (grant no.50176052) and Technical Institute of Physics & Chemistry, Chinese Academy of Sciences.

REFERENCES

1. Jiang Y. L., Chen G. B.,Thummes G., Experimental investigation on DC flow control in a single-stage pulse tube operating below 20K, ICCR2003, Hangzhou, April 22-25, 2003.

2. Xu M. Y., de Waele, A. T. M., Ju Y. L. , A pulse tube refrigerator below 2K,Cryoigenics(1999), 39 965-869.

3. Ju, Y. L., Thermodynamic analysis of GM-type pulse tube coolers, Cryogenics (2001), 41 513-520.

Comparison of oscillating flow characteristics of a metallic and a nonmetallic regenerators at high frequencies

Wang X-L.[1,2], Dang H-Z.[1,2], Cai J-H.[1], Liang J-T.[1], Zhou Y.[1]

[1]Technical Institute of Physics and chemistry, CAS, P.O.Box 2711, Beijing 100080
[2]Graduate School of the Chinese Academy of Sciences, Beijing 100039

The oscillating flow characteristics of two regenerators are studied carefully under constant temperature condition in this paper. They are identical in geometrical dimension, but packed with stainless steel wire screens and nylon screens, respectively. The results show that cycle-averaged pressure drop in oscillating flow is 1.2-2 times higher than that in steady flow for a stainless steel regenerator, which is close to experimental results in previous work, while the oscillating flow pressure drop in a nylon wire screen regenerator is larger than that in a stainless steel regenerator at the same mass flow rate, which is three to five times higher than that in steady flow.

INTRODUCTION

An accurate prediction of the pressure drop across a regenerator subjected to an oscillating flow is crucially important for the design of a regenerative refrigerator. In recent years, some works [1,2,3,4] have been performed on the oscillating flow pressure-drop in the regenerator packed with stainless steel wire-screens under different experimental conditions. Some useful correlation equations between the friction factor f and Reynolds number Re were obtained. These experimental results are beneficial to the analysis and optimum design of an oscillating flow regenerator.

Recently, high frequency Stirling-type nonmagnetic and nonmetallic pulse tube cryocoolers for cooling Superconducting Quantum Interference Devices (SQUIDs) have been developed in our laboratory. The main components of the refrigerator are made of nonmagnetic and nonmetallic materials. The refrigeration performances of the pulse tube cryocooler have been studied [5,6]. The oscillating flow characteristics of the nonmetallic regenerator packed with nonmetallic materials have not yet been studied at present. The present work is an attempt at closing this gap in the database.

EXPERIMENTS

Regenerator samples

Two samples of regenerators are tested to measure oscillating flow characteristics. The regenerator 1# is packed with the stainless steel wire screens, which are widely used as popular regenerator materials for cryocoolers. The regenerator tube is also stainless steel tube. For the regenerator 2#, the regenerator tube is made of glassfilled epoxy, and the regenerator matrix consists of a stack of nylon screens, which are the same as the materials in the reference [5]. To be compared, the geometric dimension of the two regenerators is same. The regenerator has an inner diameter of 11 mm, a wall thickness of 1.25 mm and a length of 70 mm. Table 1 shows the specification of the two regenerators as mentioned above.

Table 1 Properties of sample regenerators

Regenerator NO.	Wire screen materials	Number of packed screens	Mesh No.	Wire diameter (mm)	Pitch (mm)	Porosity	Hydraulic Diameter (mm)
1#	Stainless steel	853	400	0.03	0.0635	0.63	0.051
2#	Nylon	672	350	0.0426	0.0726	0.54	0.050

Experimental configurations and methods

An experimental apparatus is designed for the measurement of the dynamic pressures and velocities of the oscillating flow gas at the two ends of the regenerator. The schematic diagram of the experimental apparatus can be found in the author's other research paper [7] in this conference. The instantaneous velocities of the oscillating gas flow are obtained by using a hot wire anemometer (DANTEC, 55P11). Two differential pressure transducers are used for the pressure measurements at two sides of the test section. The two heat exchangers are provided at the two ends to maintain a constant temperature through the whole test section. A linear compressor is used to generate oscillating pressure and flow at various frequencies (30-60 Hz).

RESULTS AND DISCUSSIONS

Pressure drop characteristics

Figure 1 shows amplitudes of pressure drop across the two regenerators for different amplitudes of mass flow rate ($m1_{max}$) at the inlet-end of the test section. Figure 2 presents the phase difference between the pressure drop and the mass flow rate. 1# (st.st.) in these figures denotes the regenerator packed with the stainless steel wire screens. The different amplitudes of mass flow rate are achieved by adjusting the input electric power of the compressor. The experimental data is obtained in operating frequencies of 30 Hz and 40 Hz and the mean pressure of 2.0 MPa. It is clear from Figures 1,2 that the pressure drop amplitude and the phase difference is proportional to the amplitude of mass flow rate, and the frequency has little effect on the pressure drop amplitude and the phase difference, and pressure drop always lags the mass flow rate at the inlet end. The pressure drop and phase difference across the stainless steel regenerator and the nonmetallic regenerator have obvious difference. The pressure drop and phase difference across the regenerator 2# with Nylon wire screens are larger than that across the regenerator 1# with stainless steel

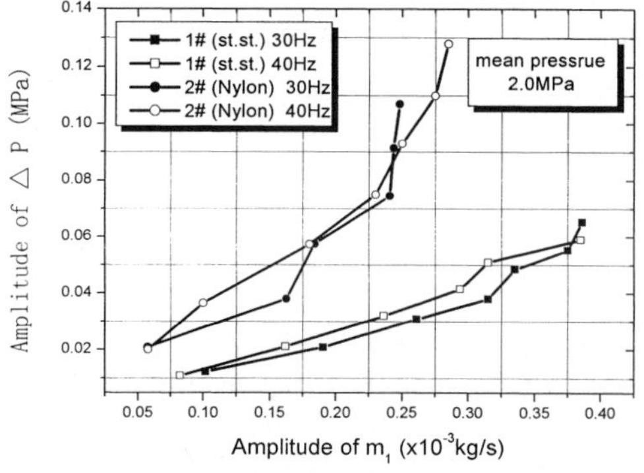

Figure 1 Pressure drop characteristics

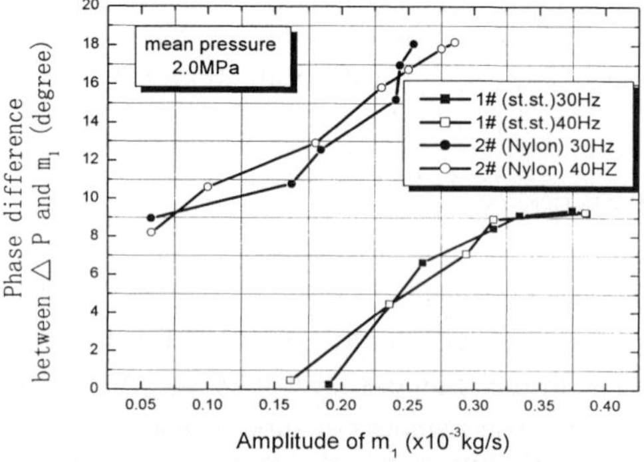

Figure 2 Phase shift characteristics

239

wire screens under the same mass flow rate. The results show that the oscillating flow pressure drop correlation equations and models about the stainless steel regenerator obtained in other works [1,2,3], are not applicable to the nonmetallic regenerator. The pressure drop characteristics in nonmetallic regenerator should be considered carefully.

The comparison of the pressure drop under oscillating flow and steady flow

We compared the oscillating flow pressure drop obtained from the experimental data with that from steady flow. The following correlation equation in steady flow is used for predicting the friction factor of a steady flow through a stack of woven screens [8]:

$$f_{st} = 33.6/\operatorname{Re}_l + 0.337 \qquad f_{st} = \Delta P_{st}/0.5\rho u^2 n \qquad u = u_0/\beta \tag{1}$$

where, ΔP_{st}: the steady flow pressure drop, u: the cross sectional mean flow velocity in the packed column, n: the number of screens, Re_l : Reynolds number defined as $\operatorname{Re}_f = l \cdot u / v, l$:mesh distance. Based on the experimental data, the cycle-averaged pressure drop of oscillating flow ($\Delta \overline{P}$) may be computed by the following equation:

$$\Delta \overline{P} = \frac{1}{N_i} \sum_{j=1}^{N_i} \sqrt{\Delta \widetilde{P}_j^{\,2}} \tag{2}$$

with N_i being the total number of sampling intervals in one cycle and $\Delta \widetilde{P}_j^{\,2}$ being the ensemble-averaged data at j^{th} interval. The compared results of pressure drop of oscillating flow and steady flow are shown in Figure 3. The ratio of cycle-averaged pressure drop to the steady flow pressure drop ($\Delta \overline{P}/\Delta P_{st}$) is also described in Figure 4. The results show that the oscillating flow pressure drop in a stainless steel regenerator is 1.2-2 times higher than that of the steady pressure drop at the same cross-sectional mean velocity. The consistent results have been obtained in other works [1,3], which ensured the reliability and creditability of our experimental results. It is found clearly that the oscillating flow pressure drop for a nonmetallic regenerator is three to five times higher than that of a steady flow at the same Reynolds number. The difference of the pressure drop and the phase shift between stainless steel regenerators and non-metallic regenerators subjected to oscillating flows is mainly determined from the

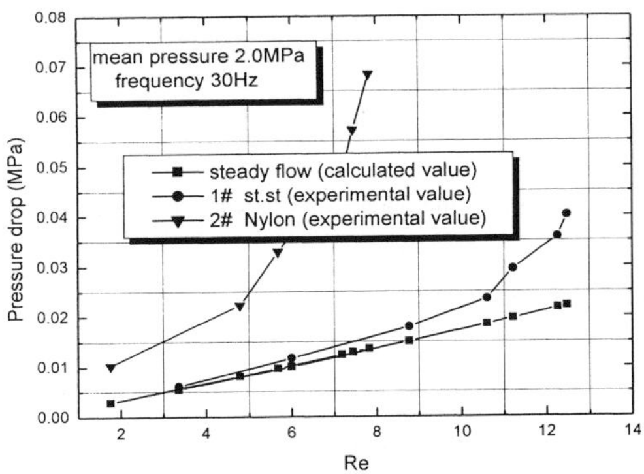

Figure 3   Comparison between the amplitude of the experimental pressure drop and the calculated value from the steady flow

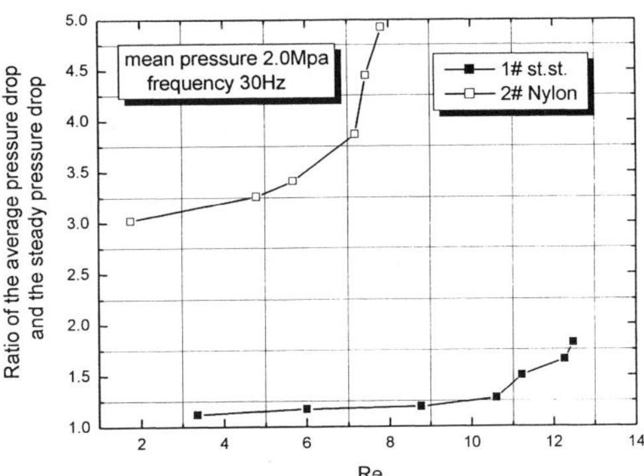

Figure 4   Ratio of the oscillating flow pressure drop to the steady flow pressure drop

characteristics of the wire screens. The effect mechanism of non-metallic wire screens on the oscillating flow characteristics is probably distinct from that of the stainless steel wire screens, which required to be considered carefully. The results in this paper should be useful for the optimum design of a non-metallic and non-magnetic pulse tube cooler.

CONCLUSIONS

An extensive set of experimental data has been collected to compare pressure drops for a metallic regenerator and a non-metallic regenerator at the same Reynolds number under high frequency oscillating flow. The experimental results of the metallic regenerator are identical in magnitude with conclusions obtained in previous works. However, averaged pressure drop in a nonmetallic regenerator packed with nylon wire screens is larger than that in a metallic regenerator packed with stainless steel wire screens at the same mass flow rate. The oscillating flow pressure drop in a nonmetallic regenerator is three to five times larger than that in a steady flow at the same Reynolds number.

ACKNOWLEDGEMENTS

The National Natural Science Foundation of China funds this work. (Grant No. 50206025)

REFERENCES

1. Helvenstcijn, B.P.M., Kashani, A. Spivak,A.L., et al., Pressure drop over regenerators in oscillating flow, CEC/ICMC 97, Portland, Oregon, USA (1997)

2. Zhao T S, Cheng, P. Oscillating pressure drops through a woven-screen packed column subjected to a cycle flow. Cryogenics (1996) 36 333-341

3. Yonglin Ju, Yan jiang and Yuan Zhou, Experimental study of the oscillating flow characteristics for a regenerator in a pulse tube cryocooler, Cryogenics (1998) 38 649-656

4. Sungryel Choi, Kwanwoo Nam, Sangkwon Jeong, Investigation on the pressure drop characteristics of cryocooler regenerators under oscillating flow and pulsating pressure conditions, Cryogenics (2004) 44 203-210

5. Dang, H.Z., Ju, Y.L., Liang, J.T., and Zhou Y., On the development of a non-metallic non-magnetic miniature pulse tube cooler, Cryocoolers 12 (2002) 799-804

6. Dang, H.Z., Ju, Y.L. and Zhou, Y, System design and measurement of a non-metallic non-magnetic miniature pulse tube cooler, Proceedings of ICEC 19 (2002) 423-426

7. Wang X-L., Zhao M-G., Cai J-H., Liang J-T., Zhou Y., Experimental analysis of the oscillating flow characteristics for high frequency regenerators, ICEC 20 (2004) (in press)

8. Miyabe, H, Takahashi S, Hamaguchi K. An approach to the design of stirling engine regenerator matrix using packs of wire gauzes, Proc.17th IECEC (1982) 1 1839-1844

Analysis of inertance tube in the pulse tube refrigerator

Chen N, Yang C.G, Xu L, Xu R.P.

Institute of refrigeration and cryogenics, Shanghai Jiao Tong University, Shanghai 200030,P R China

Refrigeration effect in pulse tube coolers can be explained in terms of phase shift between oscillations of pressure and mass flow rate at the cold end of the pulse tube. In this paper a linear ordinary differential equation of the second order with variable coefficient has been developed to describe the time-dependent phase shifting effect due to variable frictional coefficient. The comparison between integral average value of the transient parameter on a cycle and the experimental and published data has been conducted. It reveals that different inertance tubes have their own optimal working conditions, and the inertance tube with small internal diameter is more suitable to work under high frequency conditions.

INTRODUCTION

Aiming to improve the performance of pulse tube refrigerator (PTR) further, especially for high frequency operation or large refrigerating capacity cryocooler, the intertance tube has been introduced as a phaser in recent years. Until now there are many different theoretic methods which all focus on the derivation of the impedance of intertance tube based on the thermo electrical analogy. The difference between them is the way to obtain the inductive, capacitive and resistive reactance of interance tube. In general, there are two different ways to derivate the impedance. The first one is that based on the navier-stokes equations of working gas, the nonlinear inertia term has firstly been neglected [1], as a result the gas velocity of the one-dimensional flow in the tube can be derived easily; The other is that based on the analysis of the volume element of gas in the tube, the governing equations have been set up in which the pressure gradient in the inertance tube is taken to be the sum of a resistance term and an inertance term [2]. Then this one dimension equation can be solved. In this paper, a new theoretical model has firstly been developed which reveals the transient oscillatory phase shifting characteristic of inertance tube as well as reservoir. It has been simplified to a linear ordinary differential equation of the second order with variable coefficient. Actually the distributed parameter model has obtained all of relations of physical parameters. Then integrating these equations has got lumped coefficient. In this model the attenuation of pressure amplitude along the longitudinal direction has been involved. Furthermore the fraction coefficient is expressed the function of position, time and geometry size. Finally Comparison has been done to verify this model and some useful characteristics have also been found.

MATHEMATIC MODELS

Mathematical model of intertance tube

Considering a hydrodynamically fully developed reciprocating flow in a pipe, the governing conservation equations of mass and momentum for an incompressible fully developed flow is

$$\frac{\partial u}{\partial x} = 0$$

$$\frac{\partial u}{\partial t} = -\frac{1}{\rho}\frac{\partial p}{\partial x} + v\left(\frac{\partial^2 u}{\partial r^2} + \frac{1}{r}\frac{\partial u}{\partial r}\right)$$

(1)

with x and r: axial and radial coordinates, u : axial velocity, p: pressure, ρ : density of gas, v : kinematic viscosity of gas.

It is assumed that the reciprocating flow is driven by a sinusoidal varying pressure gradient in the hot end of the pulse tube. From the published experimental data [3], it is clear that the frequency and phase angle of the pressure wave do not vary,however its amplitude attenuates along the tube. So we assume that the attenuation of pressure amplitude is linear along the tube, the boundary conditions are given by

$$p(x) = a + b \cdot x , \quad x = 0, p(0) = p_{hotend} \sin \omega t \; , x = l, p(l) = p_{reservior} \sin \omega t \tag{2}$$

with p : amplitude of pressure, l :length of inertance tube, ω : frequency of pressure. The coefficients a and b can be got by substituting the boundary conditions into the expression p (x).

An exact solution for the axial velocity profile of a fully developed reciprocating flow in a round tube has been obtained by T.S.Zhao [4]. The velocity distribution (u) is given as

$$u = \frac{kD^2}{4\alpha^2 v}[B \cos \omega t + (1 - A) \sin \omega t] \tag{3}$$

Then integrating the equation (3), the average velocity (u_m) on the cross section of the tube can be obtained. According to the definition of shearing stress at the wall we get the friction force (τ_w)between the gas and the wall. By integrating u_m and τ_w along the tube, the average friction force is

$$F_1 = \int_0^l \tau_w \cdot 2\pi R dx = \pi R \frac{32 F_w}{A_0} \sin (\phi + \phi_l) \frac{D^4 \sigma^2}{(32 v)^2} \int_0^l p^2(x) dx \tag{4}$$

$$\bar{u}_m = \frac{1}{l} \int_0^l \frac{kD^2 \sigma}{32 v} \sin \phi dx = \frac{1}{l} \cdot \frac{D^2 \sigma}{32 v} \sin \phi \int_0^l p(x) dx \tag{5}$$

The relation between F_1 and $\bar{u}_m$ is given as

$$F_1 = \pi R \frac{32 F_w}{A_0} \sin (\phi + \phi_l) \frac{D^2 \sigma}{32 v} \cdot \frac{1}{\sin \phi} \cdot \frac{\int_0^l p^2(x) dx}{\int_0^l p(x) dx} \cdot \bar{u}_m \tag{6}$$

with A , B, F_w and σ : expressions of Bessel function, Φ : phase difference, D internal diameter of pipe, A_0: internal cross-sectional area of pipe.

Oscillatory flow is a very complex phenomenon. Until now the theory on laminar oscillatory flows has almost been established, with less unknowns in contrast with the theory of transitional and turbulent oscillatory flows [5]. In fact the flow pattern in the inertance tube is turbulent in most cases, because the velocity of the gas in the tube reaches 20-25m/s [3], in this case Re number of working gas is above 10000. So some equations [6] have been adopted to correct the coefficient of friction.

Mathematical model of gas reservoir

The real physical process happening in the gas reservoir is that gas is to be charged and exhausted periodically. From the viewpoint of thermodynamics this process is complex, because it can be influenced by many different thermodynamic and geometry parameters. It is found that the geometry dimension of the gas reservoir is smaller than that of the wavelength of the gas wave, as a result the parameters of the gas in the reservoir can be regarded as a lump parameter. In some degree it acts as the gas spring, the spring force can be imposed on the outlet section of reservoir.

The first law of thermodynamics, perfect gas and mass conservation equation are used to describe the process of gas in the reservoir. Solving those equations, the expression of the pressure imposed on the terminal of the inertance tube by the gas reservoir is given by

$$dF = Sdp = S(\frac{dQ}{C_v V} + k \frac{T_{in}}{V} dm_{in}) \tag{7}$$

The phase shifter including inertance tube and gas reservoir works together in the vacuum chamber, so it is reasonable to assume that the gases in the reservoir undergo adiabatic process. Meanwhile, due to the frictional heat, the gases in the inertance tube undergoes polytropic process, so the expression of pressure (F_2) is finally given

$$F_2 = \frac{k}{V} T_{hotend} \left(\frac{a + bP_{hotend}}{P_{hotend}} \right)^{\frac{n-1}{n}} \rho A_0 \cdot x \tag{8}$$

with V : volume of reservoir, k: adiabatic coefficient, n: polytropic coefficient.

<u>Mathematical model of phase shifter</u>

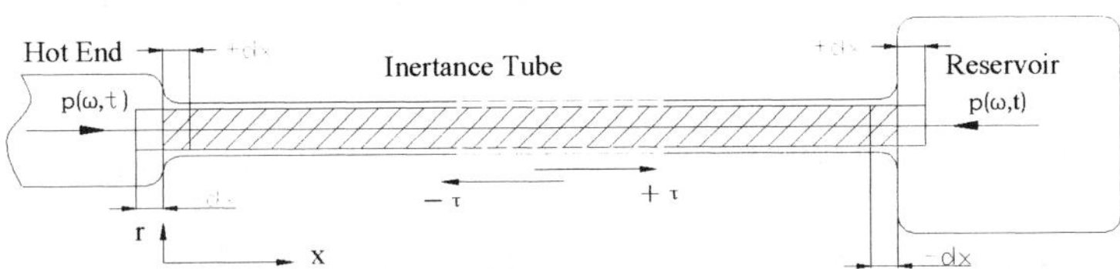

Figure 1 Phase shifter including inertance tube and gas reservior

From published experimental data [3], it is found that the length of the inertance tube has the same dimension as the wavelength of gas parameters. So according to the acoustic theory, it is reasonable to regard the mass in the tube as a lumped mass. A linear ordinary differential equation of the second order with variable coefficient has been derived to describe the physical process in the phase shifter (see Figure 1).

$$\rho A_0 l \frac{d^2 x}{dt^2} + f_1(t, \omega) \frac{dx}{dt} + f_2(t_r, V_r) x = pA_0 \sin \omega t \tag{9}$$

The three terms of at the left-hand of the equation represent the inertia force, friction resistance and the spring force of the gas reservoir respectively. The right-hand term describes the sinusoidal exciting force. We can find that both the friction resistance and the spring force vary with the time and frequency, which is different from the previous theoretical model in which the friction force is regarded as independent of the time.

VERIFICATION

The method of electrical analogy has been proved as an effective way to evaluate the characteristic of a phase shifter. From the viewpoint of it, the resistor of resistance, R, the inductor of inductance, L, and the capacitor of capacitance, C, deduced from the equation (9), are $f_1(t, \omega)/A_0$, ρl and $1/f_2(t_r, V_r)A_0$ respectively. Those expressions reveal the time-dependent phase shifter effect, which has been ignored previously because more attention has been paid to the average effect in a cycle. In order to verify this model, we firstly calculate the phase shifting characteristic of the IPTR [2], the calculating results agree with the experimental data. Then the further analyses have been conducted to try to get the performance of inertance tubes with different geometry size and under different working conditions. Their dimensions and working condition are shown in Table 1 respectively.

Table 1 Calculation conditions of inertance tube

NO	Dimensions	Working conditions				
	Length×Radius(mm)	Frequency (Hz)	Temperature in hot end (K)	Charge pressure (MPa)	Reservoir (cm³)	Press ratio
1	1000×0.5	30~70	303	2.5	59	1.17
2	2000×1.0	30~70	303	2.5	59	1.15
3	3000×2.0	30~70	303	2.5	59	1.10

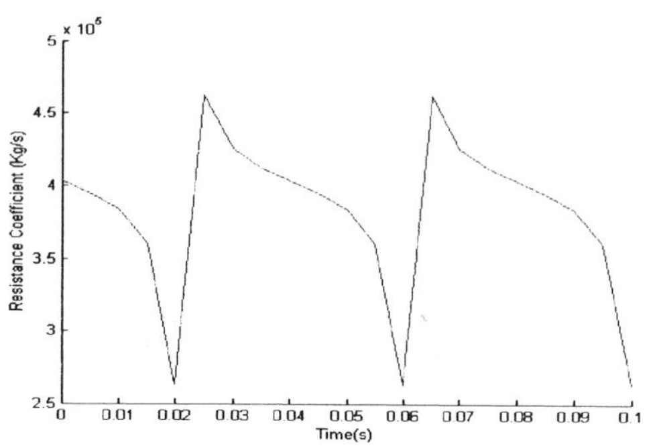

Figure 4 Resistance coefficient as a function of time

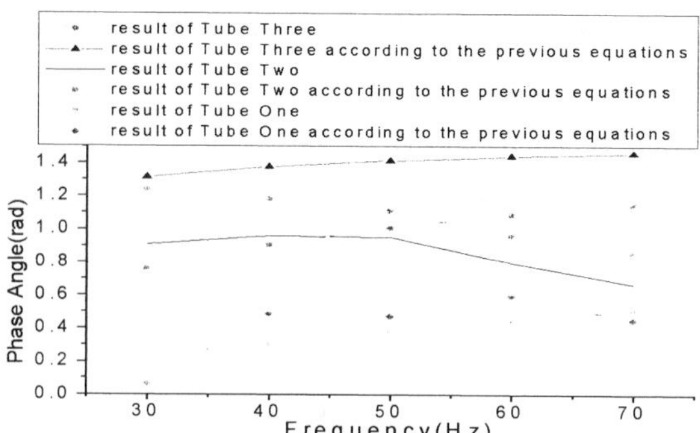

Figure 5 Shifting angle vs frequency for inertance tube with different geometry size

Figure 4 shows the relationship between the coefficient of resistance and the time for the case of tube one (see Table 1) working under 50Hz. We find that the period of resistance coefficient is twice as large as the pressure wave. Meanwhile this coefficient is a strong nonlinear function of the time. It also reveals that there exists an asymmetry in the behavior of the inertance tube in spite of their symmetric structure. In practice, it is undeniable that this effect is very difficult to be observed directly by means of experimental measures because the response time of the testing instruments is always longer than that of the periodical parameters. So by integrating the periodical parameter on a cycle, the average effect of the phase shifting is obtained. In contrast to the common idea about the phase shifting effect [7], according to which the phase shifting angle is directly proportional to the working frequency, there exists an optimal working frequency about a corresponding geometry size (see Figure 5), which is in agreement with some experimental data [2]. In general, it is found that different inertance tubes have their own optimal working conditions, and the inertance with small internal diameter is more suitable to work under high frequency conditions.

CONCLUSION

In this paper a model has been developed to describe the time-dependent phase shifting characteristic of an inertance tube. This model has been simplified to a linear ordinary differential equation of the second order with variable coefficient. In this model the resistance coefficient of the inertance tube is expressed as a function of time, frequency and some other parameters, which are deduced from momentum and mass differential equations of the working gas. Consequently, the phase shifting angle provided by the interance tube also varies with the time. By integrating, the average phase angle on a cycle agrees generally with the published result. It is also found that different inertance tube have their own optimal working conditions, and the inertance with small internal diameter is more suitable to work under high frequency conditions.

REFERENCES

1.   Y.L.Ju, G.Q.He,Y.K.Hou,J.T.Liang,Y.Zhou，Experimental measurements of the flow resistance and inductance of intertance tubes at high acoustic amplitudes, Cryogenics(2003) 43 1-7
2.   P.C.T.de Boer, Performance of the inertance pulse tube, Cryogencis (2002) 42 209-221
3.   S.W.Zhu, S.L.Zhou, N.Yoshimura and Y.Matsubara, The long neck tube for the pulse tube refrigerator, Cryocoolers 9, Plenum Press,New York（1997）269-278
4.   T.S.Zhao, P.Cheng, The friction coefficient of a fully developed laminar reciprocating flow in a circular pipe, Int.j.heat and fluid flow (1996) 17 167-172
5.   Munekazu OHMI ，Manabu IGUCHI，Phase shift effect of Flow pattern and frictional losses in pulsating piple flow part 7 wall shear stress in a turbulent flow, bulletin of the JSME （1981）24 1764-1771
6.   Pat R. Roach, Ali Kashani, Pulse tube coolers with an inertance tube: theory,modeling,and practice, Advances in cryogenic engineering(1998) 43 1895-1902
7.   Y.K.Hou,Y.L.Ju,L.W.Yan,J.T.Liang,Y.Zhou, Experimental study on a high frequency miniature pulse tube refrigerator with inertance tube, Advances in Cryogenic Engineering(2000) 47 730-738

Single stage double-inlet pulse tube refrigerator with integral thermo-electric cooler

Karunanithi R., Jacob S, Kasthurirengan S, Durgesh Nadig, Upendra Behera and Tamizhanban K.

Centre for Cryogenic Technology, Indian Institute of Science, Bangalore 560012, India

The use of Thermo-electric Cooler (TEC) at the warm end heat exchanger of a pulse tube refrigerator is explored in this paper. An attempt is made to keep the warm end of regenerator at a temperature lower than room temperature using TEC. Additionally, a cam operated pressure wave generator used in this system enhances the durability compared to that of both solenoid and rotary valve based systems. When operated with an air-cooled helium compressor of 1.6 kW capacity in double-inlet mode, it gives about 7 W refrigeration at 77 K and a no load temperature of about 40 K.

INTRODUCTION

For over two decades, in the area of Pulse Tube Refrigerator (discovered by Gifford and Longsworth [1] - Basic Pulse Tube) there have been significant advancements. Due to the absence of mechanical moving component at the cold side of the system, pulse tube refrigerators find applications in semiconductor fabrication, Charge Coupled Devices for astronomical telescopes, SQUID detection for non-destructive tests and medical applications where vibration and noise levels should be very low and the system should have high reliability.

DESCRIPTION

The schematic of a Double Inlet Pulse Tube Refrigerator (DIPTR) with integral TEC is shown in figure 1.

The system without the bypass valve, orifice valve and the reservoir forms the Basic Pulse Tube Refrigerator [1] in which the hot end of the pulse tube is closed. The addition of reservoir and the orifice valve converts it to Orifice Pulse Tube Refrigerator [2]. In the Double Inlet Pulse Tube Refrigerator [3], there is a bypass valve between the regenerator and the pulse tube at the room temperature end. A medium pressure (17-20 bar) helium compressor and the pressure wave generator produce the required pressure oscillations. The regenerator contains a porous matrix, which absorbs heat from the gas when it goes towards the pulse tube and gives it off when the gas flows in the opposite direction. The pulse tube is an empty tube with heat exchangers (hot end HE and cold end HE) at both the ends. In the Basic Pulse tube mode, when the system is subjected to oscillating gas flow, due to thermo-acoustic effect (acoustically driven temperature effect), open end of the pulse tube is cooled and the heat is transported to the hot end where it is rejected in the HE. In the Orifice Pulse Tube, introduction of a needle valve

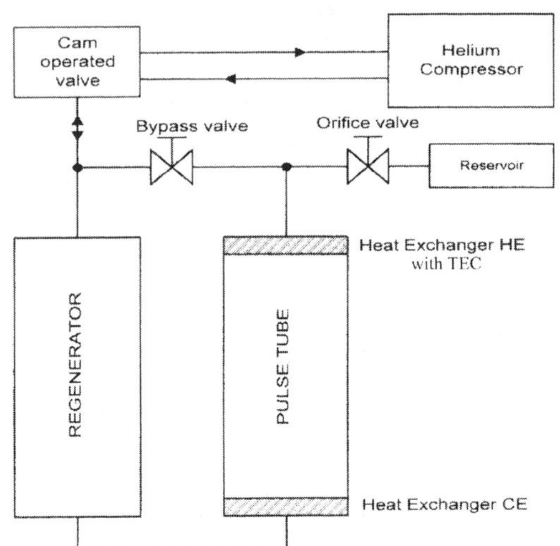

Figure 1. Schematic of DIPTR with TEC

(orifice) and a buffer volume (reservoir) enhance the performance. This arrangement creates the required phase shift between the pressure wave and the mass flow at the warm end, so that they are in phase at the cold end. Introduction of the bypass valve (Double Inlet Pulse Tube) improves the performance further by sending a part of the gas directly to the warm end of the pulse tube reducing the mass flow of the gas through the regenerator. Commercial needle valves by Hoke (1315G4B) are used for the above. The problem of non-symmetry in the flow coefficients in the forward and reverse directions is avoided by connecting a set of identical valves back to back in series. This way, the flow coefficient in the reverse direction becomes the limiting factor and the opening has to be accordingly increased.

The warm end heat exchanger consists of four numbers of Melcor make thermoelectric coolers (PT8-12-40) with suitable finned heat exchangers to dissipate the heat from the hot end of the TEC. With this arrangement, the hot end heat exchanger could be cooled to 288 K (15°C) when the room temperature is 310 K while the pulse tube is being operated. This arrangement eliminates the water circulation system with or without mechanical refrigerator needed in the conventional systems. Additionally, another thermoelectric cooler kept inside the vacuum jacket is used to keep the warm end of the regenerator at 275 K. A switch mode power supply (SMPS) of 48 V and 6 A capacity is used for driving all the TECs. Each TEC module consists of 127 couples connected thermally parallel and electrically serial way. When operated with a 12 D.C. power supply, it draws 6 A current and is capable of removing around 60 W heat for zero ΔT between hot and cold faces. The maximum ΔT that could be achieved with the above power supply for Q = 0 (amount of heat that is removed from the cold face) is around 50°C. The size of each module is 40 mm x 40 mm square and 3.3 mm width. Using CPU cooling fan of Pentium 4 computer, the heat from the warm end is dissipated. As 12 V D.C. is needed for operating this fan also, it is connected parallel to the TEC module. Figure 2 shows the design of the TEC setup coupled to the hot end heat exchanger of the pulse tube replacing the water circulating system.

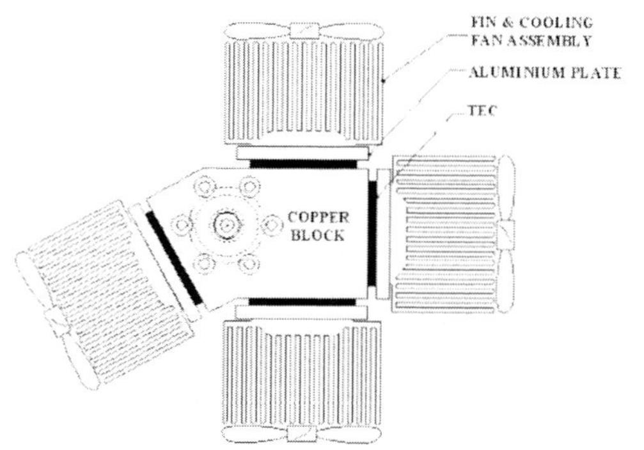

Figure 2. TEC arrangement at warm end heat exchanger

One of the critical components of the pulse tube system is the pressure wave generator. Usually a solenoid operated or a rotary valve operated pressure wave generator is used in the pulse tube systems [4]. While the solenoid operated pressure wave generator is prone to failure beyond a limited number of operations, the rotary valve system is likely to develop leaks in the sealing between the high/low pressure inlets and the outlet. For commercializing the pulse tube system, and for reliable and continuous operation for a very long period without failure, addressing the above problem and the use of TECs at warm end heat exchangers are important steps forward.

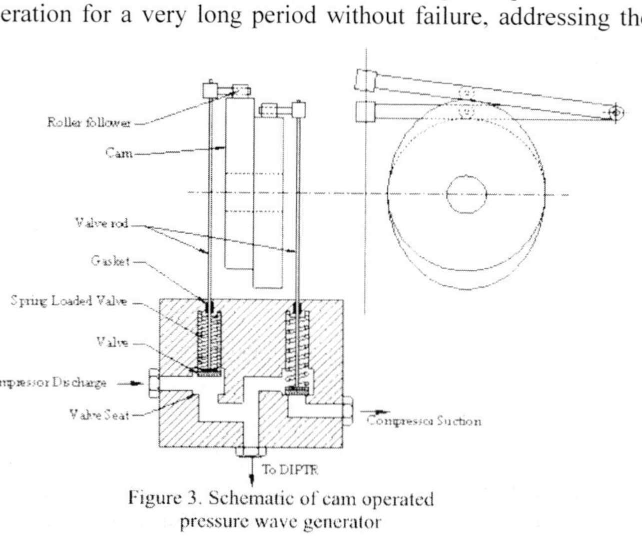

The schematic of the concept of a cam operated spring-loaded valve based pressure wave generator is shown in Figure 3. It consists of a set of spring-loaded valves coupled to the suction and discharge sides of the compressor. The outlets of both the valves are coupled together to form the common inlet to the DIPTR. A set of valve rods couples the valves to the cam mechanism. The rods are always kept under tension to avoid the leak at the valve seat. The cams can be rotated at varying speeds using a motor coupled to a variable frequency drive. Similar cam operated systems are

Figure 3. Schematic of cam operated
pressure wave generator

employed in the expansion engine of Collin's helium liquefier for connecting high and low pressure gas streams to the expansion engine. But to our knowledge, such a system has not been employed as a pressure wave generator for pulse tube refrigerators elsewhere. Instead of developing such a system, if one has access to the cold head drive of a commercial G-M cryocooler, it can be suitably modified and used as a pressure wave generator of pulse tube refrigerator [5]. The cold head drive performs two functions in the G.M. Cryocooler.

The cold head drive converts the circular motion of the motor to reciprocal motion to move the stem up and down connecting integral regenerator-displacers of two stages. This is done in synchronization with the opening and closing of the high and low-pressure sides of the compressor by way of operating the corresponding cam operated spring-loaded valves. For the pulse tube application, as there is no mechanical displacer in the system, it is sufficient if only the pressure wave is generated by using the cam operated spring-loaded valves. Hence, the components needed for moving the

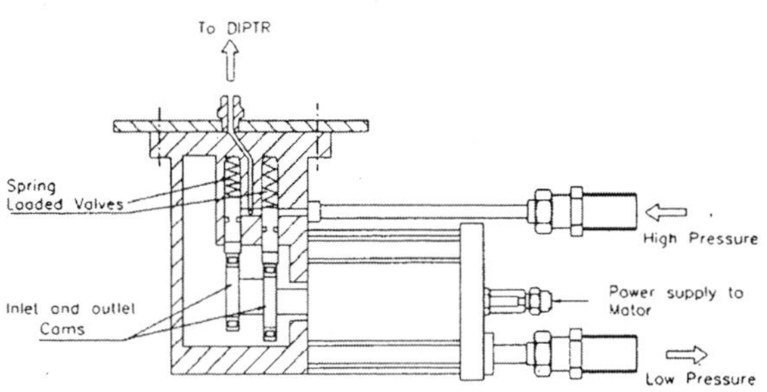

Figure 4. Schematic of modified pressure wave generator

integral regenerator-displacer up and down are discarded from the cold head drive in the present system and only the parts needed for generating the pressure wave are retained. The schematic of modified pressure wave generator is shown in Figure 4. The compressor discharge and suction sides are coupled to the device with the help of quick disconnect coupling (Aeroquip coupling). The common output from the device is connected to the warm end of the regenerator. As it is necessary to change the frequency of operation to maintain the pressure wave in-phase with the mass flow at the cold end of the pulse tube to get maximum refrigeration power, a variable frequency drive of approximately 750 W is used to operate the cold head drive motor. By varying the frequency of the drive, it is possible to change the operating range of pressure wave generated from 1 Hz to 10 Hz. It is also possible to vary the duty cycle of high and low-pressure profile by adjusting the cam setting.

PERFORMANCE OF THE SYSTEM

A typical cooldown characteristics of the pulse tube system using the above pressure wave generator and TECs as warm end heat exchangers is shown in figure 5. The performance is almost the same as that of the conventional water circulating

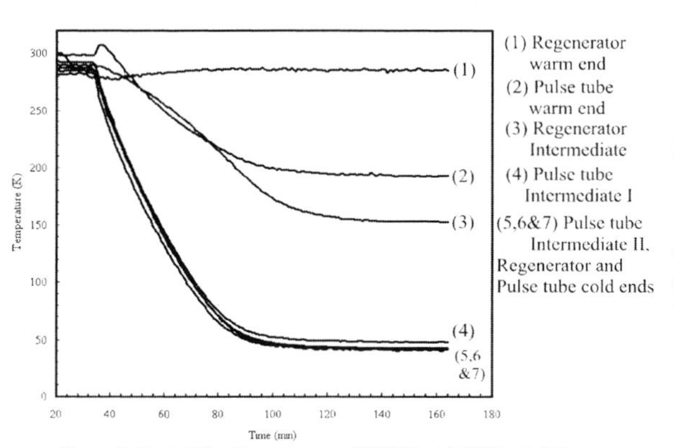

Figure 5. Typical Cooldown pattern of DIPTR with TEC at 1.5 Hz

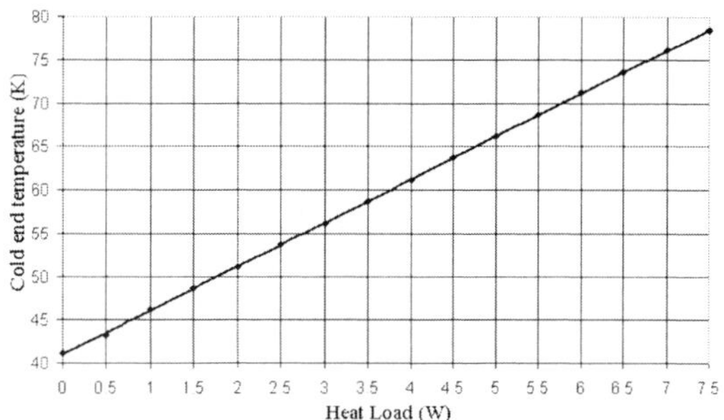

Figure 6. Heat load curve of the DIPTR wih TEC

system as warm end heat exchanger [6]. The temperature profile of warm end of regenerator is maintained at around 280 K with the help of TEC. The temperature of the warm end of the pulse tube is reduced in this case. The heat load curve shown in figure 6 shows that marginally higher refrigeration power could be obtained with this system compared to that of the water-cooling system showing that the

TECs can be used to eliminate the water circulating system. Further optimization could result in better performance. The results of the cam-operated system developed by us are compared to those of a rotary valve based system [4] in Table 1.

Table 1.Comparision of the pressure data of cam operated system and the rotary valve system

Parameters	Cam operated pressure wave generator	Rotary valve based pressure wave generator [4]
P_{Max}	17 bar	14.5 bar
P_{Min}	9.75 bar	10 bar
ΔP	7.25 bar	4.5 bar
P_{Max}/P_{Min}	1.75	1.45
$P_{Average}$	13.5 bar	12 bar

From the data, it can be concluded that the performance of this system is comparable to that of a rotary valve based system. Further, from the waveform pattern shown in figure 7, we can conclude that the system presented in this paper performs as efficiently as the rotary valve based system with better durability

CONCLUSIONS

The performances of a single stage double inlet pulse tube refrigerator operated with a cam operated pressure wave generator and thermo-electric coolers are compared with those of the

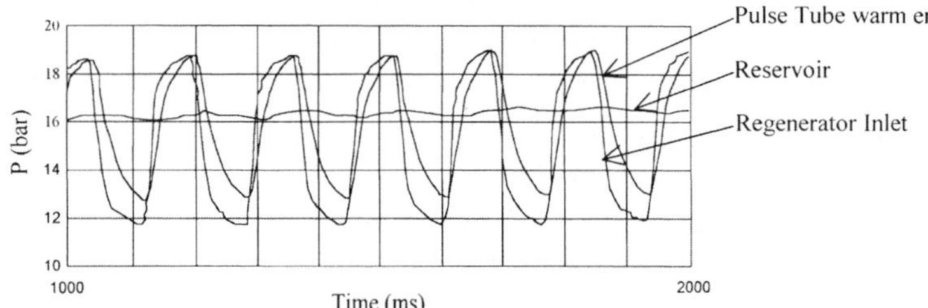

Figure 7. Pressure wave patterns at regenerator inlet, pulse tube warm end and at reservoir

conventional systems. From the data, we can conclude that the cam operated pressure wave generator performs as efficiently as the other types of the system with better reliability. The water circulation system at the hot end heat exchanger can be eliminated by the TECs without compromising on the performance. The D.C. flow due to non-symmetry of the commercial needle valves used is corrected by using a set of two identical valves back to back in series. Further optimizations in terms of cooling the regenerator warm end using the TEC are in progress. It is expected that this can result in lower no load temperature and better refrigeration capacity.

REFERENCES

1. Gifford, W.E and Longsworth, R.C, Pulse Tube Refrigeration, Trans. ASME, J. Eng. Ind. 86: (1964) 264
2. Mikulin, E.I, Tarasov, A.A and Shkrebyonock, M.P, Low-Temperature expansion pulse tubes, Advances in Cryogenic Engineering, Vol. 29, Plenum press, New York (1984) 629
3. Zhu, S. P, Wu and Chen, Z, Double inlet Pulse tube refrigerators: an important improvement, Cryogenics, Vol. 30 (1990) 514
4. Liang, J, Development & Experimental verification of a theoretical model for pulse tube refrigerator, Ph.D. Thesis, Institute of Engineering thermophysics, Chinese academy of sciences (1993) 145
5. Karunanithi, R, Jacob, S, Kasthurirenga, S, Upendra Behera and Nadig, D.S, Design and Development of a Single stage double inlet pulse tube refrigerator Proceedings of the Eighteenth Internation Cryogenic Engineering Conference (ICEC 18) Mumbai, India (2000) 539 – 542
6. Karunanithi, R, Jacob, S, Kasthurirenga, S, Upendra Behera and Nadig, D.S, Development of a reliable and simple pressure wave generator for pulse tube refrigerators To be published in Review of Scientific Instruments

Experimental study of wall temperature distribution of stirling type pulse tube refrigerator (SPTR)

Liu Y.W., He Y.L., Huang J., Li X.Z.[*], Chen C.Z.

State Key Laboratory of Multiphase Flow in Power Engineering, School of Energy & Power Engineering, Xi'an Jiaotong University, 28 Xianning West Road, Xi'an 710049, P.R.China
* Air Conditioning and Refrigeration Center, Department of Mechanical and Industrial Engineering, University of Illinois at Urbana-Champaign, 1206 West Green Street, Urbana, IL 61801,USA

In this paper, the experimental study on wall temperature distribution of a Stirling type pulse tube refrigerator has been carried out. The influences of structural and working parameters on performance of SPTR have been studied, especially on wall temperature distributions of pulse tube and regenerator. Through a number of experiments, the rule of wall temperature distributions and the relationship of it's dependence on structural parameters and working parameters has been found, giving some hints to further improve the performance of a given SPTR, and DC flow has also been validated from temperature distribution curve.

INTRODUCTION

Gifford and Longsworth [1] developed the original pulse tube refrigerator. Mikulin et al. [2] and Zhu et al. [3] proposed new concepts known respectively as orifice type and double inlet type. It is well known that pulse tube refrigerator is a good alternative to Stirling and G-M refrigerators, because there is no moving part in the cold head. In order to further improve the performance of Stirling type pulse tube refrigerator, it is important to study the wall temperature distribution curve of pulse tube and regenerator of SPTR. The wall temperature distribution curve of the regenerator can be regarded as an experimental criterion if the design and cooling capacity of SPTR is the best.

EXPERIMENTAL EQUIPMENT

High-performance pulse tube refrigerator usually has an inline arrangement. However, in inline configuration the cold head is located in the middle, which makes the cooling of thermal interfacing with the device rather difficult. Therefore, for some practical reasons and experimental demand we have chosen a U-shaped configuration. A schematic view of the setup is shown in Figure 1.

In our experiments, SPTR was driven by oil-less-lubricated and valve-less helium compressor, whose frequency was adjusted from 5 to 30Hz by a transducer (FRN1.5G9S—4CE type). The regenerator was composed of a stainless steel tube with a wall thickness of 0.2 mm, an inner diameter of 14 mm and a length of 88 mm packed with 250-meshes stainless steel screen. The pulse tube was made of stainless steel tube with a wall thickness of 0.2 mm, an inner diameter of 10 mm and a length of 118 mm. At the end of the regenerator and the pulse tube, two copper heat exchangers were connected with them and cooled by water. The connecting tube in experiment was made of a standard copper tube with outside diameter of 6 mm and a wall thickness of 1 mm. The orifice valve and the double-inlet valve were adjustable needle valve for space application. The reservoir with a volume of about 500 cm^3 was connected to the hot end of pulse tube through the orifice valve. The regenerator and the pulse tube were placed in a vacuum vessel with a pressure of 1 mmPa and were maintained for thermal insulation from the environment at the room temperature.

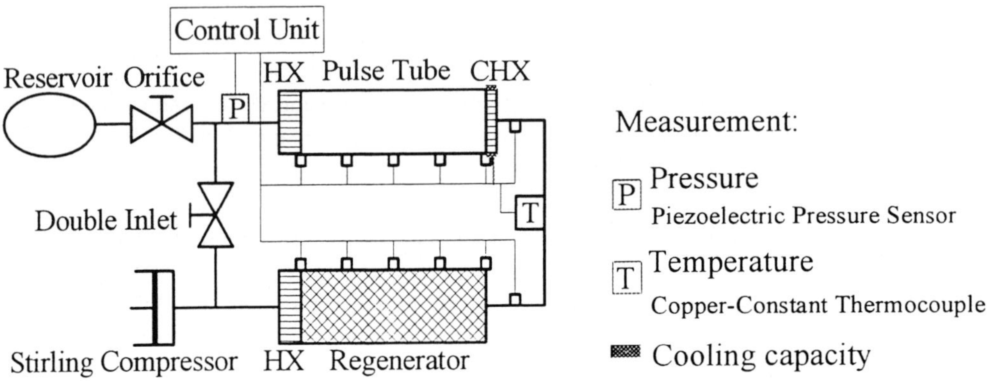

Figure 1 Schematic of Stirling-type PTR with U-shaped configuration

Measurement probes for pressure and temperature were installed as shown in the figure 1. A piezoelectric pressure sensor (CYG-1105 type) was installed at the hot end of the pulse tube to monitor the dynamic pressure wave. Based on the heat balance method, manganin wires wrapped around the cold head, which are supplied by a 0-24V steady voltage power source, are used to measure the cooling capacity. Thermocouples of T-type (Copper-Constant) were soldered at five positions on the pulse tube wall and the regenerator wall. All data were recorded by a data acquisition system controlled by a personal computer.

RESULTS AND DISCUSSION

In our experiment, we have obtained wall temperature distribution curves of SPTR depended on structural parameters (orifice valve and double-inlet valve), working parameters (working frequency and charging pressure), and cooling capacity of the cold head.

Effect of structural parameters on wall temperature distribution

Figure 2 shows the experimental results of the dependence of wall temperature profiles of the regenerator and the pulse tube on the openings of orifice valve (a) and double-inlet valve (b). In condition of charging pressure of 1.24 MPa, working frequency of 15.7 Hz, we adjusted the orifice valve and double-inlet valve to find the optimum opening, and measured the temperature profiles on steady conditions at the same time. The optimum opening of the orifice valve is 130^0 and the one of the double-inlet valve is 160^0. From figure 2, it can be found that adjusting orifice valve and double inlet valve can all improve the cooling performance of the refrigerator. From figure 2a, temperature distribution of the regenerator approximated to ideal linear distribution. The effect of orifice valve on the temperature distribution of the regenerator and the pulse tube was regular.

However, the effect of double-inlet valve was not. From figure 2b, Temperature distribution was mostly deviated from linearity and become very irregular and complex. The temperature distribution of the pulse tube and the regenerator could become worse. The reason is that the double-inlet valve can arouse the airflow short-circuit, and then lead the more complex airflow in the refrigerator, such as DC flow and second flow. Temperatures along the regenerator and along the pulse tube are very sensitive indicators for change of DC flow. DC flow result in irregular change of temperature distribution. Therefore, we can validate DC flow in SPTR from temperature distribution curve.

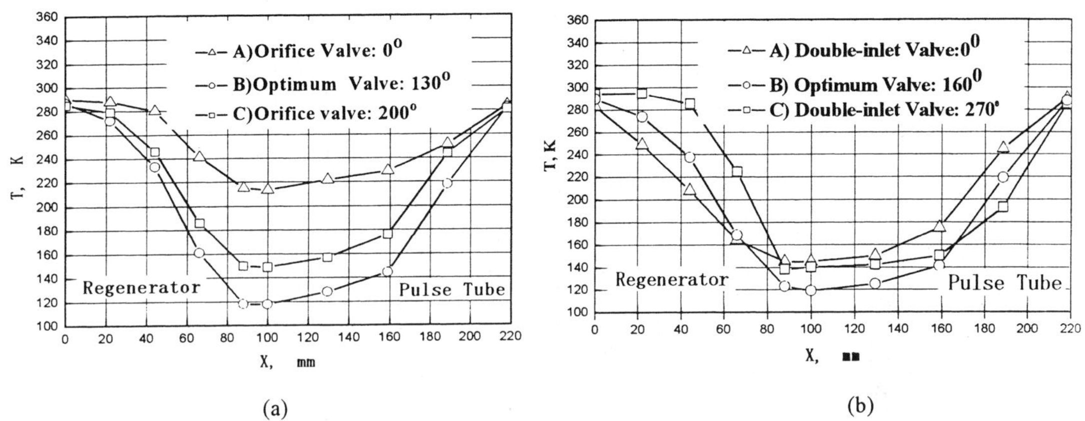

Figure 2 Temperature in regenerator and in pulse tube versus orifice valve (a) and double-inlet valve (b)

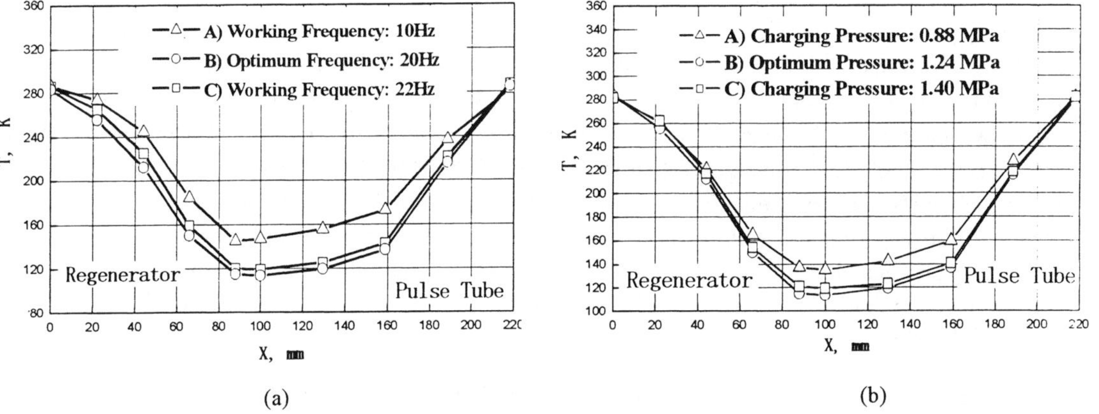

Figure 3 Temperature in regenerator and in pulse tube versus working frequency (a) and charging pressure (b)

<u>Effect of working parameters on wall temperature distribution</u>
Figure 3 shows the experimental results of the dependence of the wall temperature profiles of the regenerator and the pulse tube on working frequency (a) and charging pressure (b). In condition of orifice valve of 130^0 and double-inlet valve of 160^0, we changed working frequency and charging pressure to find the optimum value and measured the temperature profiles on steady conditions. The optimum working frequency is 20Hz and the optimum charging pressure is 1.24 MPa. From figure 3, it can be found that working frequency and charging pressure can further adjust the temperature distribution. The experimental results shows that when the working frequency and charging pressure are in optimum condition, the wall temperature distribution of the regenerator and the pulse tube are better. Especially in the regenerator, temperature distribution is closed to ideal linear.

<u>Effect of cooling capacity on wall temperature distribution</u>
Figure 4 shows the experimental results of the dependence of the wall temperature profiles of the

regenerator and the pulse tube on cooling capacity. In condition of charging pressure of 1.24 MPa, working frequency of 20 Hz, orifice valve of 130^0 and double-inlet valve of 160^0, we changed cooling capacity and measured the temperature profiles on steady conditions. From figure 4, we found that the cooling capacity only affect the slope of temperature distribution of the regenerator and the pulse tube when the structural parameters and working parameters were on the optimum valve.

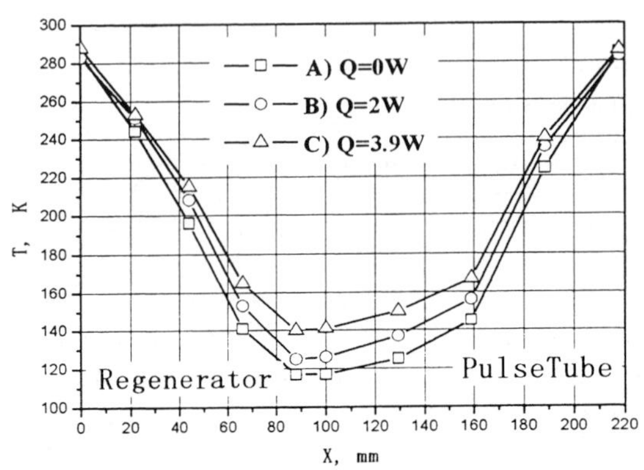

Figure 4 Temperature in regenerator and in pulse tube versus cooling capacity

CONCLUSION

Detailed experimental research on wall temperature distribution of the pulse tube and the regenerator of a Stirling type pulse tube refrigerator has been carried in this paper. Through analyzing our experimental result, we can not only find some general characteristic of SPTR, but also can we obtain the rule of wall temperature distribution and relationship of its dependence on structural parameters and working parameters. The experimental results show that wall temperature distribution curve of SPTR can be regarded as experimental criterion if the design and cooling capacity of SPTR is the best, and be helpful to further improvement and design of SPTR. At the same time, we can also validate DC flow from temperature distribution curve.

ACKNOWLEDGMENT

The project was financially supported by the National Key of Fundamental R&D of China (No.G2000026303) and National Natural Science Foundation of China (No. 50276046).

REFERENCES

1. Gifford,W.E., Longsworth,R.C., Pulse tube Refrigeration, Trans.ASME. Series B (1964), 86 264-268

2.Mikulin,E.I.,Tarasow,A.A.,Shkrebyonock,M.P., Low temperature expansion pulse tubes, Advances In Cryogenic Engineering (1984), 29 629-637

3.Zhu Shaowei, Wu Peiyi, Chen Zhongqi, Double inlet pulse tube refrigerators: An important improvement, Cryogenics (1990), 30 514-520

Studies on a Single Stage GM type Pulse Tube Cryocooler Using an Indigenous Helium Compressor

Kasthurirengan S., Jacob S., Karunanithi R., Upendra Behera, Nadig D.S., Anand Bahuguni

Centre for Cryogenic Technology, Indian Institute of Science, Bangalore 560 012, India

A GM type single stage Pulse Tube Cryocooler has been designed, fabricated and operated with an indigenous Helium Compressor developed using a 2 kW Freon Compressor suitably modified with cooling arrangements for oil and gas. Using a regenerator of 19 mm od and 250 mm length, housing 1850 wire screens of 250 mesh size, a lowest temperature of 37.5 K and a refrigeration power of ~10W at 77 K have been obtained for a Pulse Tube of 14 mm od and 300 mm length, with an indigenous rotary valve. The performances of this configuration have been compared with those when operated using a 3 kW imported helium compressor. Studies show that due to the increased pressure ratio (of high to low pressure) of the indigenous compressor, ~ 50% more cooling power is obtained in this case. A comparative study of results relating to, cool-down, pressure waveform, cooling power, and the angular variation of cold end temperature, when the system is operated with these compressors, is presented here.

INTRODUCTION

Cryocoolers or Cryorefrigerators produce the required refrigeration power at specific low temperature. Several types of cryocoolers based on different cycles such as GM, Stirling and Pulse Tube have been used for a variety of applications such as sensor cooling, radiation shield cooling, cooling of superconducting magnets, cryopumping etc. Of the above, Pulse Tube Cryocoolers [1-2] are preferred due to the absence of moving parts at cryogenic temperatures, which leads to increased reliability and long-term performance. In a Pulse Tube cooler, the high and low pressures from a helium compressor are alternately applied to the empty Pulse Tube through a regenerator, using a rotary valve.

We have developed a single stage Pulse Tube experimental system that uses an imported water-cooled helium compressor of 3 kW along with an indigenous rotary valve. Also, an indigenous helium compressor (using a 2kW reciprocating type Freon compressor) has been developed and the experimental studies have been repeated using the latter. In this paper, the performances of the Pulse Tube cooler with these compressors are compared.

EXPERIMENTAL SETUP

The schematic of experimental set up is shown in Figure 1. The Pulse Tube and the Regenerator housings are made up of AISI 304 stainless steel. The dimensions of Pulse Tube are 14 mm od., 250 mm length, while those of the Regenerator are 19mm od. and 210 mm length. The heat exchangers and the flow straighteners are made of copper of electrolytic grade. The regenerator matrix is made of stainless steel wire meshes of size 250 and contains ~ 1850 meshes [3-4]. The warm end of the Pulse Tube is connected to a heat exchanger through which cold water is circulated to maintain it at ambient temperature. Both the Pulse Tube and the regenerator are mounted to the top flange of the vacuum jacket by o-ring seals. The cold ends of the Pulse Tube and regenerator ends in a copper heat exchanger with indium seals.

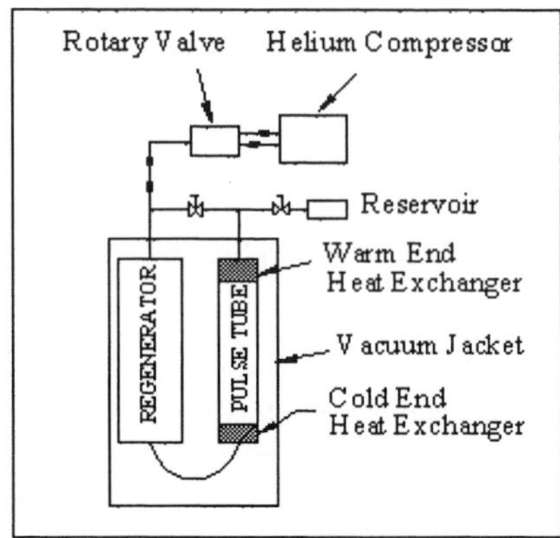

Figure 1 Schematic of Single Stage Pulse Tube Cryocooler

The Helium compressor serves as the source for high and low pressure gas supply to the Pulse Tube. The design of the setup was such that the Pulse Tube can be operated either in basic, or orifice or double inlet mode with the help of needle valves (Swagelok) along with a stainless steel buffer volume of 0.5 litres.

Platinum sensors (PT100) have been used as temperature sensors while piezoelectric transducers (KPY47R, Siemens) are used as pressure sensors. The vacuum jacket housing the Pulse Tube and the regenerator is fixed to a rotatable horizontal axle, by which the orientation of the Pulse Tube with respect to gravity can be varied from 0^0 to 180^0. At the cold end of the Pulse Tube, a manganin heater of $62.5\ \Omega$ is fixed and energized by using a power supply at constant current mode.

Experimental studies have been conducted with two different helium compressors (a) An imported 3kW rotary type water-cooled helium compressor (RICOR) and (b) An indigenously developed 2kW reciprocating type helium compressor (using a Tecumseh make refrigeration compressor). Before the discussion of the experimental results, some details of this development are presented below.

INDIGENOUS COMPRESSOR DEVELOPMENT

The main purpose of this development is the absence of any local suppliers to provide Helium gas compressors. Also, most of the imported Helium compressors are based on conventional refrigeration compressors with suitable oil / gas cooling arrangements for use with helium gas. In the Pulse Tube system, the major cost is that of the Helium compressor. Hence indigenisation of this component will considerably reduce the total cost of the system. Further, such a compressor may also be used for other gases.

Adaptation of refrigeration compressor for helium gas
When a conventional Freon compressor is used for refrigeration, the cooling of the compressor occurs by the cold refrigerant itself. However, when Freon is replaced by helium gas in the above, the heat of compression considerably increases due to increase in the ratio of specific heats for helium ($\gamma_{He} \sim 1.67$) This causes (a) considerable heating of the compressor and (b) cracking of the oil in the compressor. To take care of these, methods for cooling either the compressor body as a whole or the lubricant oil in the compressor should be adopted. In the present development, both the above methods have been used.

Cooling of compressor body by cold water
The compressor body is wound tightly with 3/8''o.d. copper pipe and bonded with metal paste. Cold water is circulated through the copper pipe to reduce the compressor body temperature.

<u>Cooling the lubricant oil of the compressor</u>
The cooling of the oil can be carried out either externally or internally. We have used the external cooling arrangement as shown in Figure 2 and is described below.

A copper pipe welded at the bottom of the compressor and the oil from this pipe collects in a buffer vessel and is cooled to ambient temperature. This oil is fed back to the compressor by high-pressure helium with the help of two solenoid valves A and B operated by an electronic timer. Due to the cyclic opening and closing of the solenoid valves the oil is taken from the bottom of the compressor and is circulated back to the compressor.

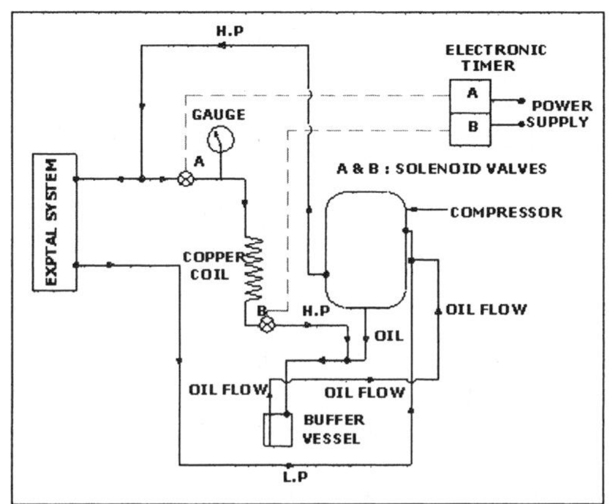

Figure 2: Schematic of modified Freon compressor for helium gas Indicating oil and gas flow circuits

RESULTS AND DISCUSSION

<u>Cool down behaviour</u>
The cool down behaviour of the Pulse Tube refrigerator is similar when operated with either of the compressors. The typical cool down is compared for the 14 mm Pulse Tube in the double inlet mode at 2.3 Hz in Figure 3. The lowest temperature reached is ~ 43.8 K, for the imported compressor and ~ 42.5K for the indigenous compressor. The typical cool down time is marginally higher for the imported compressor by about 10 minutes.

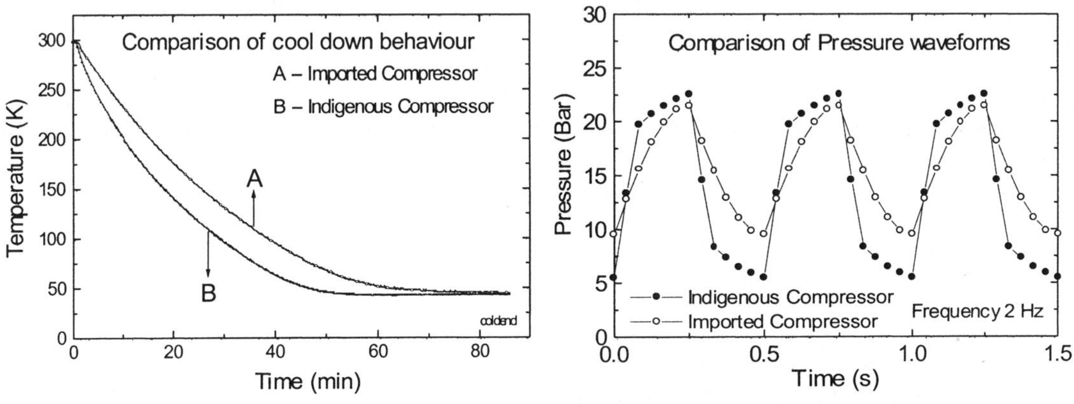

Figure 3 Cool down behaviour of Pulse Tube Refrigerator   Figure 4 Comparison of pressure waveforms of compressors

<u>Pressure ratio</u>
The pressure ratio is the ratio of high to low pressure at the outlet of the rotary valve and is shown in Figure 4. The pressure ratio is 22.5 / 5.5 for the indigenous compressor and 21.5 / 9.5 for the imported

256

compressor. The pressure waveform is more to trapezoidal for the indigenous compressor, where as it is nearly sinusoidal for the imported compressor. It is observed that the increased pressure ratio and the modified waveform have a significant effect on the performance of the Pulse Tube refrigerator.

<u>Refrigeration Power of the Pulse Tube</u>
The cooling power is measured by applying a given heat load and monitoring the steady temperature reached at the cold end of the Pulse Tube. The heat load is raised in known steps up to 10W. The typical experimental results with indigenous and imported helium compressors for the 14mm Pulse Tube are shown in Figure 5. It is observed that at 77 K, refrigeration powers of about 7W and 10 W are achieved for the imported and indigenous compressors respectively, indicating clearly the effect of increased pressure ratio.

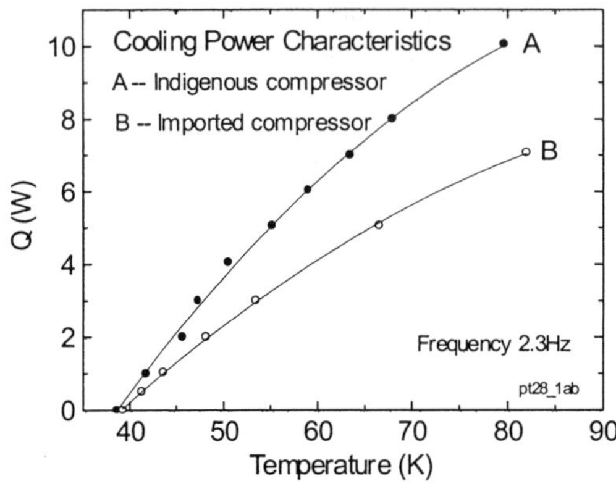

Figure 5 Cooling power characteristics of PTR

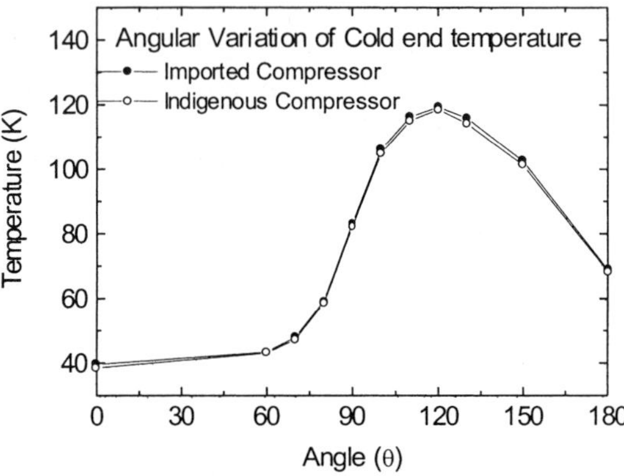

Figure 6 Angular variations of cold end temperature of PTR

<u>Angular variation of cold end temperature of PTR</u>
Due to the convection effects occurring in the Pulse Tube, the cold end temperature depends on the orientation of the Pulse Tube with respect to the gravity. The typical angular variation behaviour of 14 mm Pulse Tube is compared for both the compressors in Figure 6. It is observed that the behaviour is nearly the same, except that the maximum temperature at 120° is somewhat lower in the case of indigenous compressor.

CONCLUSION

This work discusses the performances of the single stage Pulse Tube refrigerator, when operated with imported and indigenous helium compressors. The results show that higher the pressure ratio and more trapezoidal the pressure waveform, higher is the cooling power of the Pulse Tube refrigerator, as has been observed in our studies with the indigenous helium compressor.

REFERENCES

1. Thummes, G., Landgraf, R., Giebeler, F., Muck, M. and Heiden, C., Pulse Tube Refrigerator for High Tc SQUID operation, <u>Advances in Cryogenic Engineering</u> (1996) <u>41B</u> 1463-1470

2. Ravex, A., Poncet, J.M., Charles, I. and Bleuze, P., Development of Low frequency Pulse Tube Refrigerators, <u>Advances in Cryogenic Engineering</u> (1998) <u>43B</u> 1957-1964

3. Kasthurirengan, S., Thummes, G. and Heiden, C., Reduction of Convection Heat Losses in Low Frequency Pulse Tube Coolers with Mesh Inserts, <u>Advances in Cryogenic Engineering</u> (2000) <u>45</u> 135-142

4. Kasthurirengan, S., Thummes, G. and Heiden, C., Elliptical and Circular Pulse Tubes: A comparative Experimental Study, Proceedings of ICEC18, (2000) 547-550

Experimental investigation on the performances of gas flow in oscillating tube

Li Z.C., Xu L

Institute of Refrigeration & Cryogenics Engineering, Shanghai Jiao Tong University, Shanghai 200030, P. R. China

The gas oscillating tubes are the key part of the gas wave refrigerator for heat change with environment. The flow progress of the gas in the oscillating tubes is very complex. A set of experimental apparatus has been designed and manufactured. The work medium is compressed air supplied by the oiless air compressor. The oscillating tube with shock wave damping tank and circumfluence tube have been performed with this apparatus. The experimental results show that the reflecting shock wave in the gas oscillating tubes were weaken and the temperature drops and isentropic refrigerating efficiency of the oscillating tube can be increased by coupling with shock wave damping tank or circumfluence tube.

INTRODUCTION

Gas wave refrigerator, also called pressure wave refrigerator or thermal separator, is a kind of refrigerator utilizes pressure energy of gas to achieve low temperature. The gas oscillating tubes are the key parts of the gas wave refrigerator for heat change with environment. The flow progress of the gas in the oscillating tubes is very complex.

Shock and expansion waves are two kinds of basic moving waves in the oscillating tube. Their movement and interaction between them produce the refrigerating effect. When supersonic flow from nozzle meet the immobile gas in oscillating tube, the shock wave forms in front of the interface and move to the hot end of the tube. The moving speed is supersonic contrast to the flow before shock wave, but is subsonic contrast to the flow after shock wave. The entropy increases when the gas compressed by shock wave. The shock wave stronger is, and the entropy increases larger is.

The research shows that the reflected shock wave is the main factor which influences the refrigerating effect [1]. Shock wave is reflected on rigid wall by shock wave, and expansion wave is reflected on free surface. A stronger shock wave is formed when two shock waves interaction, but the intensity weakens when it acted by expansion wave. The shock waves in oscillating tube can be eliminate or weaken by increasing friction of the wall of the tube, increasing damp in hot end, reducing the energy of the gas or acting by expansion wave.

EXPERIMENTAL APPARATUS

Considering similar to the industrial equipment and operation expediently, a set of experimental apparatus for has been designed and manufactured [2], as shown in figure 1.

The work medium is compressed air supplied by the oiless air compressor. The temperature drops and refrigerating efficiency are calculated by measuring the temperatures of the gas flow into and out the oscillating tube. The thermocouples and Agilengt 34970A data acquisition/switch unit are applied to measure the temperature. The Cy-yd-205 type pressure sensors are used to measure the pressure of the gas in oscillating tube, the signal from them are amplified by electric charge amplifier and received by an A/D card fixed in computer.

It can be know by analyzing the refrigeration mechanisms and the factors effect isentropic efficiency of gas wave refrigerator, that there are two ways to weaken the reflected shock wave and increase the isentropic efficiency: 1). To decrease the enthalpy of the gas oscillating in the tube as far as possible by enhancing heat transfer outside the oscillating tubes [3]; 2). To weaken the reflecting shock waves by changing the gas flow in the tube.

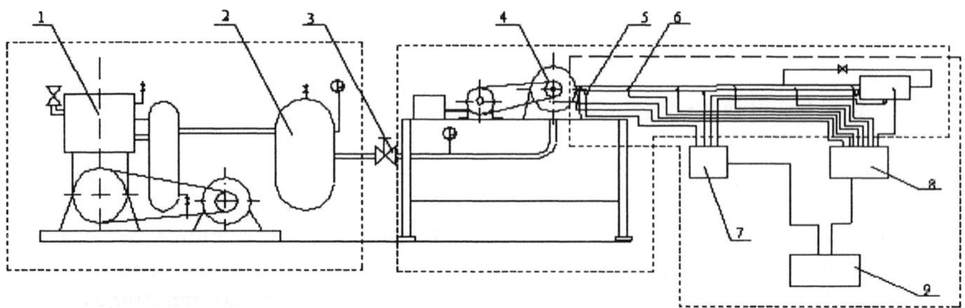

1.ZW-1.0/8 oilless air compressor   2.gas storage tank   3.valve   4. Pulse tube-Gas wave refrigerator 5.pressure sensor   6.thermocouple   7.electric charge amplifier   8.Agilengt 34970A data acquisition/switch unit   9.computer

Figure 1 Experimental apparatus for gas wave refrigerator

The gas flow in the oscillating tubes can be changed by changing the structure of the oscillating tubes, especially the structure of the hot end of the tubes. The reflected shock waves can be weakened by designing the structure of the oscillating tubes properly or adjusting the phases of shock waves and expansion waves. Shock wave is reflected on rigid wall by shock wave, and expansion wave is reflected on free surface. So a tank was introduced on the hot end of the oscillating. When the shock waves move to the end of the oscillating tube and come into contact with the section expanding suddenly, the shock waves will weaken and expansion waves produced and move towards the cool end in the oscillating tube. A valve and a circumfluence tube were fixed to adjust the gas flow. Then, three projects for experiment were put forward, as shown in figure 2.

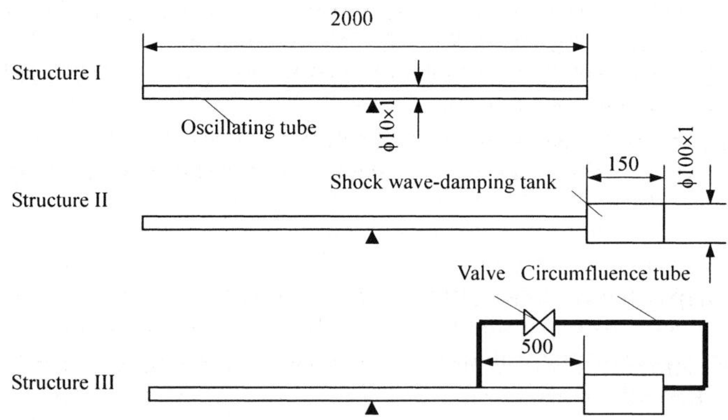

Structure I: straight tube   Structure II: tube with shock wave damping tank
Structure III: tube with circumfluence tube   ▲ location of pressure measurement
Figure 2 Schematic of different structures of oscillating tubes

FLOW IN THE OSCILLATING TUBE

In order to research the gas flow in the oscillating tubes with different structures, the pressure waves in three oscillating tubes above-mentioned were measured. At the same time, the temperatures and pressures of the entering and discharged gas were measured. The pressure of the gas entering into the tube is 0.5MPa, and the pressure of the discharged gas is 0.1MPa ($\varepsilon = 5.0$). The pressure waves in the tubes were shown in figure 3.

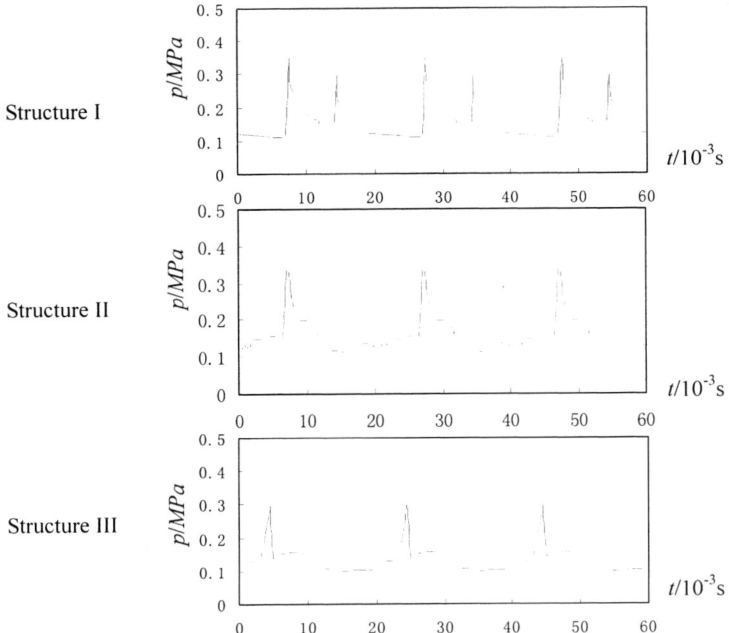

Figure 3 Pressure waves in different oscillating tubes

In the pressure waves in structure I, two peak pressures appear. The higher one is the main shock wave moving from the cool end to hot end, and the lower one is the shock wave reflected on the rigid wall of the tube. Though the peak value of the reflected shock wave is lower than the main shock wave, it is more high compared with it in structure II and structure III.

In the structure II, the reflected wave is weaker than it in structure I. It indicates that the tank on the hot end of the oscillating tube can weaken the shock wave effectively.

In the structure III, the reflected wave is weaker tan it in structure I and II. The pressure wave becomes smoother after the main shock wave. Also, the time of the shock wave acting in the tube become shot.

TEMPERATURE DISTRIBUTION ALONG THE OSCILLATING TUBE

Figure 4 shows the temperature distribution on the outer surface of the oscillating tube. In structure I, the temperature is most high in the three structures. The temperature climbs from the open end to the hot end. The highest temperature appears on the end of the oscillating tube. A small peak appears near the end of tube as a result of the reflected shock wave influencing in the tube. In structure III, the highest temperature appears near the middle of the oscillating tube. The temperature becomes stable after the peak point. It indicates that the gas flow stably in the tail part of the oscillating tube.

REFRIGERATING EFFECT IN DIFFERENT OSCILLATING TUBES

Figure 5 and 6 show the temperature drops and the isentropic refrigerating efficiency with different structures of the oscillating tubes (under different pressures of the entering gas).

The experimental results show temperature drops and isentropic refrigerating efficiency of the oscillating tube can be increased by coupling with shock wave damping tank or circumfluence tube. Especially by coupling with the circumfluence tube, the temperature drops and isentropic refrigerating efficiency increase by 7.6K and 5.0% ($\varepsilon = 8.0$) respectively contrast to the shock damping tank structure.

260

Therefore, the max temperature drop of the experimental oscillating tube is 48.2K and the max refrigerating efficiency is 49.3%.

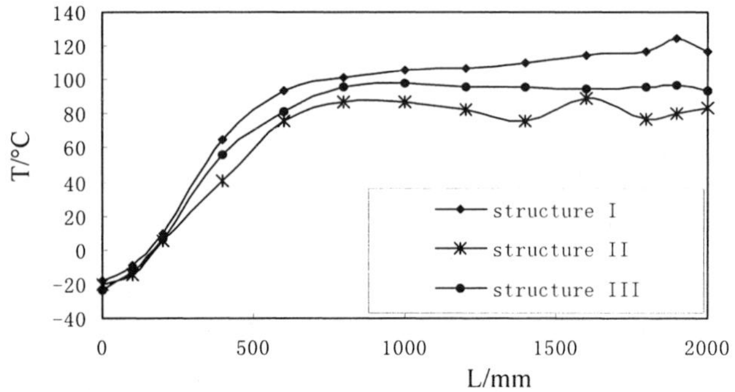

Figure 4 Temperature distributions along oscillating tube with different structures

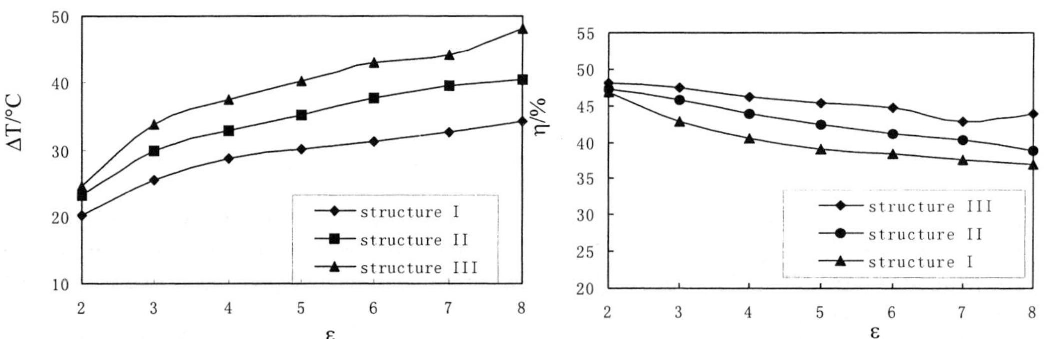

Figure 5 Temperature drops with different structures   Figure 6 Refrigerating efficiency with different structures

CONCLUSIONS

Shock wave damping tank and circumfluence tube can weaken the influence of reflecting shock wave and adjust the gas flow. In the oscillating tube. The temperature drops and isentropic refrigerating efficiency of the oscillating tube can be increased by coupling with shock wave damping tank or circumfluence tube.

REFERENCES

[1] Li X.L etc. Peak-Oscillating Effect of Pressure Wave Refrigerator. Cryogenics and refrigeration-Proceedings of ICCR'98, Hangzhou, China, (1998) 534~537
[2] Li Zhaoci. Study on performance in coupling of pulse tube-gas wave refrigerator [Dissertation]. Shanghai Jiaotong University, China, (2002)
[3] Li Xuelai etc.. Experimental research on the enhancing heat transfer outside the oscillating tube in pressure wave refrigerator. Refrigeration (China). (1996) 57 7-9

Optimal vector analysis of the phase shifter in the pulse tube refrigerator

Chen N, Yang C.G, Xu L

Institute of refrigeration and cryogenics, Shanghai Jiao Tong University, Shanghai 200030,P R China

Based on vector analysis, the effect of the performance of the phase shifter in a pulse tube refrigerator on the refrigeration capacity and compressor work has been studied. Refrigeration capacity, compression work and efficiency have been expressed as the function of phase angle and amplitude of mass rate flowed through the phase shifter. Then this complex function has been illustrated in a phasor diagram with the simple geometry relation. By analyzing this geometry relation, some interesting results have been obtained.

INTRODUCTION

Pulse tube refrigeration is based on a cyclic process such that a gas column with cold and warm end temperature is compressed, displaced towards the warm end, expanded and redisplaced towards the cold end. This process is realized by feeding adequate gas flow to both ends of the tube. The phase angle between pressure and mass flow rate at the cold end of pulse tube refrigerators (PTR) is the key parameter for the performance of PTR. The various types of PTR differ mainly by the performance of the phase shifter, which can by realized by use of passive or active elements [1]. Various authors have developed their own theory to explain the mechanism of the PTR. Among them, the phasor analysis is an effective way [2]. In this analysis thermodynamic quantities are expanded in a Fourier series. Only the direct component and fundamental oscillating terms are considered. All higher frequency terms are ignored. Due to different parameters shown in the same phasor diagram, it can be used to predict the performance of PTR directly and clearly.

In this paper, the effect of the performance of the phase shifter in PTR on the refrigeration capacity and compression work has been studied based on vector analysis. In order to get the exact amount of the optimal complex impedance of phase shifter, the complex functions between refrigeration capacity, compression work and the parameters of components have firstly been simplified to the geometrical relationship shown clearly in the phasor diagram. Then through analyzing this geometry relation, many interesting results have been obtained which will be helpful to the design of PTR or the choice of suitable phase shifter.

PHASE ANALYSIS

As a starting point for analyzing the pulse tube, the following idealizations will be used:
1) In an ideal regenerator there is no dead volume, entropy production and pressure drop is zero;
2) In an ideal pulse tube, entropy production is zero;
3) The compressor and pulse tube are adiabatic, i.e. there is no heat transfer to the walls;

262

4)   All oscillating quantities are of small amplitude;

5)   The reservoir is large enough for its pressure and temperature to be considered constant.

Then all quantities can be written in the form

$$x(z,t) - x_0(z) = x_1(z)\cos(\omega t + \theta_x)$$

(1)

which can be represented by the real part of a phaser in phase space. Phasors behave as vector. But it is only the real part that is of interest. A simple model of a pulse tube is developed below. This model emphasizes the different components of mass flows and pressure and the relationships between them. The pressure and temperature can firstly be expressed as the form of equation (1). Then the ideal gas law can be applied to the adiabatic process and expanding the pressure term into a Taylor series and keeping only the lowest order term, the mass that must flow into the pulse tube through the regenerator to allow the change in m_p is

$$m_p = -(3\omega/5R)P_1\sin(\omega t)\int_0^L T_0^{-1}dz$$

(2)

where R is the gas constant(per unit mass), P_1 is the amplitude of pressure wave, T_0 is the temperature of charged working gas.

In the same way, the mass rate due to the phase shifter (m_0) is:

$$m_0(z,t) = (R + Qi)p_1 T(L)\cos(\omega t)/T(z)$$

(3)

The mass flow in the hot end of PTR is the sum of the two mass flows. The refrigeration power is just the enthalpy flow in the pulse tube，that is

$$\langle H \rangle = \int_0^\tau (m_p + m_0)dt + \frac{2}{5}\frac{C_p T_0}{\tau p_0}\int_0^\tau p_1 \cos(\omega t)(m_p + m_0)dt$$

(4)

For the basic pulse tube (BPT), $m_0=0$ since there is no phase shifter, the first term of right-hand of equation is unequal to zero, but the integral average value on a cycle is equal to zero.

With H: enthalpy, m: mass flow, p: pressure, t: time, ω: frequency, C_p: specific heat. τ: period.

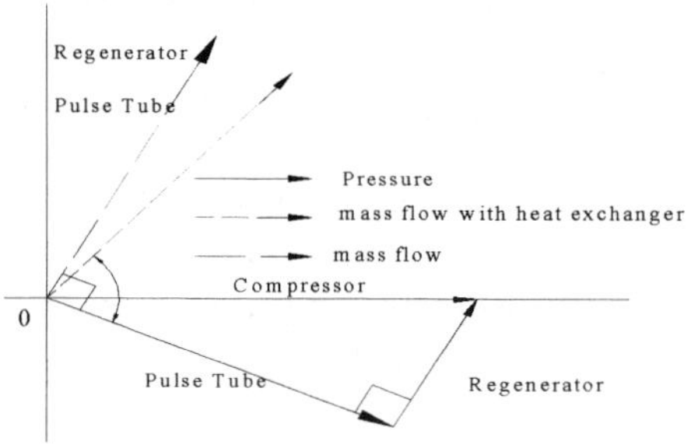

Figure 1 Pressure and mass flow phasor diagram

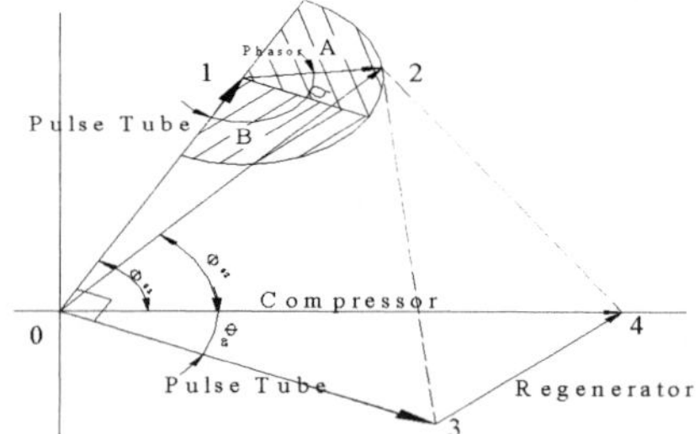

Figure 2 Phasor diagram with phase shifter

Figure 1 shows a phasor diagram for a basic pulse tube. For a basic pulse tube, $m_0=0$. The mass flow (m_p) is proportional to and shifted $90°$ from the pressure in the pulse tube. The mass flow is also parallel and proportional to the pressure drop in the regenerator. The latter relationship is a result of the regenerator pressure drop driving the mass flow. The pressure in the pulse tube plus the pressure drop in the regenerator equals the pressure in the compressor. Since the pressure in the pulse tube is $90°$ from the mass flow there is no refrigeration. In practice, heat transfer between the gas and the pulse tube wall causes a small phase shift resulting in small refrigeration. The compressor pressure times mass flow is proportional to the work done by the compressor. The compressor is doing work, yet there is no refrigeration. all the work is going into the regenerator loss which is proportional to m Δ p.

A phasor diagram for a pulse tube refrigerator with phase shifter is shown in Figure 2. The relationship between the pulse tube pressure and m_p is the same as in the basic pulse tube. In terms of the angle (α), the phase shifter can be divided into two types. When α is less than 90° (Area B in Figure 2), the first type of phase shifter can just provide resistance and capacitive impedance, such as orifice, throttle; when α is bigger than 90° (Area A in Figure 2), the inductive reactance can be provided by the second phase shifer i.e. inertance tube.

VECTOR ANALYSIS

From the phasor diagram (see Figure 2), according to the physical parameters they represent, there exists the following geometrical relationship

$$l_{01} + l_{12} = l_{02}, \; l_{03} + l_{34} = l_{04}, \; l_{34} \, /\!/ \, l_{02} \tag{5}$$

Enthalpy flow through the cold end of the PTR and power consumption of the pulse tube are the directly proportional to $H_{\Delta 203}$, $H_{\Delta 204}$ respectively, which is the area of some corresponding region.

$$H_{\Delta 203} = l_{02} \times l_{03} \times \cos(\phi_{\angle 203}) \;\; , \;\; H_{\Delta 204} = l_{02} \times l_{04} \times \cos(\phi_{\angle 204}) \tag{6}$$

There exists geometrical relation as follows

$$H_{\Delta 203}^{\,2} + (2S_{\Delta 203})^2 = (l_{02} \times l_{03})^2, \;\; H_{\Delta 204}^{\,2} + (2S_{\Delta 204})^2 = (l_{02} \times l_{04})^2 \tag{7}$$

With l_{01}, Φ_{01}: magnitude and phase angle of mass rate flowing from refrigerator into the pulse tube, l_{12}, Φ_{12}: magnitude and phase angle of mass rate passing through the phase shifter, l_{02}, Φ_{02}: magnitude and phase angle of mass rate passing through the cold end of PTR, l_{03}, Φ_{03}: magnitude and phase angle of pressure in pulse tube, l_{34}: magnitude of pressure in refrigerator, l_{04}: magnitude of pressure provided by compressor. $\Phi_{\angle 012}$: angle formed between vector l_{01} and l_{12}. The others are named in the same manner.

In order to analyze the effect of phase shifter on the performance of PTR, $H_{\Delta 203}$ and $H_{\Delta 204}$ have been expressed as the function of l_{12} and Φ_{12} as follows.

$$H_{\Delta 203} = l_{03} l_{12} \sin \phi_{\angle 012}, \;\; H_{\Delta 204} = l_{03}(l_{01} - l_{12})(\tan\phi_{\angle 102} + \tan\phi_{\angle 304})/(1 - \tan\phi_{\angle 102} \tan\phi_{\angle 304}) \tag{8}$$

with $\quad \phi_{\angle 102} = \arctan(l_{12} \sin(\phi_{\angle 012})/(l_{01} - l_{12} \cos(\phi_{\angle 012})))$

According to the definition of the coefficient of performance $\eta = W/H$, it results

$$\eta = l_{12} \sin(\phi_{\angle 102})/((l_{01} - l_{12}) \tan(\phi_{\angle 102} + \phi_{\angle 304})) \tag{9}$$

From the Equation (8), (9), it is found that the characteristic of the phase shifter can influence the performance of PTR directly. Unfortunately the globally optimal working condition (l_{12}, Φ_{12}) can not be obtained by differentiating the method of multi-variable function. However, Equation (8) still provides much important information on the pulse tube. When l_{12} is equal to zero, it becomes the basic pulse tube refrigeration, which is no refrigeration at ideal conditions. If the angle ($\Phi_{\angle 204}$) keeps constant, the longer the vector (l_{02}) is, the bigger the refrigeration capacity will get. That is the reason why the volume of the reservoir is always adopted as big as possible and the orifice or valve must provide enough resistance. However, we can also find out that with the increase of length of l_{02} the compression work will be increased correspondingly. So it is not the optimal choice to obtain good performance of PTR. When the length of l_{02} is invariable, the refrigeration capacity varies with the angle of $\Phi_{\angle 012}$. If the angle of $\Phi_{\angle 012}$ equals to 90°, there exists a maximum of refrigeration capacity. According to this conclusion, it is impossible for the first type of phase shifter to get its maximum refrigeration power because the capacitance caused by the reservoir with finite volume cannot be neglected in practical applications. Whereas it is possible for the other kind of phase shifter to shift by 90° from the mass flow rate of gas in the pulse tube because the inductance and capacitance can counteract each other. This is the reason why many pulse tube refrigerators using the inertance tube as phase shifter achieve better performance than those using orifice or throttle. It is also found that while the inductance and capacitance are counteracting with each other, in other words while working under the resonant condition in term of electrical analogy, the biggest refrigeration capacity can be achieved. When $\cos(\Phi_{\angle 012})$ equal to l_{12}/l_{01}, the compression work get its minimum value, in which l_{02} will contact the circle with the radius of l_{12}. So we can adjust the phase shifter to this condition for minimal work input.

What is mentioned above can be applied as guidance to the design or adjustment of the phase shifter. If the resistive characteristic as well as the geometry size is known, the optimal design could be obtained. For example [3], under the resonant working condition optimal parameters of the inertance tube can be obtained by iteration from the following two expressions

$$\frac{4\omega^2 \rho l}{\pi d^2} = \frac{RT}{V} \quad , \quad \frac{128\,\mu l}{\pi d^4} = M \tag{10}$$

with V: volume of gas reservoir, d: diameter of tube, l: length of tube, M is the constant.

The first expression of Equation (10) represents the phase shifter working under the resonant condition. In other words, $\Phi_{\angle 012}$ equals to 90° in Figure 2. When this expression has been satisfied, the constant, M, in the second expression should be as large as possible to get enough refrigeration capacity. Because the length of l_{12} corresponds to the magnitude of M, equation (8) explains this reason.

CONCLUSION

Effect of the phase shifter on the refrigeration capacity, compression work and efficiency has been studied by means of phase analysis. In the phasor diagram, the geometrical relation between the vectors, which make the analysis clear and simple, express all of these complex functions. By analyzing these relations, it is found that under the resonant condition the refrigeration capacity could get the maximum value. Meanwhile the condition under which the compression work gets its minimum value is also deduced. According to this result, a method to design the phase shifter has subsequently been introduced. In addition, many other interesting results have also been obtained by this easy and direct way.

REFERENCES

1.   A.Hofmann, H.Pan, Phase shifting in pulse tube refrigerators, Cryogenics(1999) 39 529-537

2.   P.Kittel, A.Kashani,J.M.Lee,P.R.Roach,General pulse tube theory, Cryogenics(1996) 36 849-857

3.   Y.K.Hou,Y.L.Ju,L.W.Yan,J.T.Liang,Y.Zhou, Experimental study on a high frequency miniature pulse tube refrigeration with inertance tube, Advances in cryogenics engineering(2002) 47 731-738

Characteristics of 4 K pulse tube cryocoolers in applications

Chao Wang

Cryomech, Inc., 113 Falso Drive, Syracuse, NY 13211, USA

Cryomech has developed and commercialized 4 K pulse tube cryocoolers, Models PT403, PT405, PT407 and PT410, which provide cooling capacities from 0.25 W to 1.0 W at 4.2K. The latest developments at Cryomech enabled the pulse tube cryocoolers to have almost the same capacity and efficiency as GM cryocoolers. The pulse tube cryocoolers have opened many applications and demonstrated their advanced features with respect to long meantime between maintenance, very low vibration and small magnetic field distortion from rare earth materials.

INTRODUCTION

The 4 K pulse tube cryocooler is a new generation of cryo-refrigeration system that can provide cooling capacities below 4 K. It has no moving parts at cryogenic temperatures and leads to advanced features over the 4 K GM cryocooler.

Cryomech, Inc. commercialized the world's first 4 K pulse tube cryocooler, Model PT405 in 1999[1]. In recent years, we have continually developed and commercialized a series of 4 K pulse tube cryocoolers, Models PT403, PT407 and PT410 that provide cooling capacity from 0.25 W to 1.0 W at 4.2 K[2]. These 4 K pulse tube cryocoolers have opened many challenging applications in cooling NMR and MRI magnets, precooling dilution refrigerator, ADR and sorption cooler, cooling sensitive devices like SQUID magnetometer, etc. These applications demonstrate great advantages of pulse tube cryocoolers over GM cryocoolers in the field.

This paper introduces the Cryomech 4 K pulse tube cryocoolers and their performances. The characteristics of the cryocoolers in applications are presented and compared with 4 K GM cryocoolers.

4 K PULSE TUBE CRYOCOOLERS

Figure 1 shows photographs of the 4 K pulse tube cryocoolers, Models PT403, PT405, PT407 and PT410. The configurations of them have been described in reference 1. The PT405 has the same layout geometry as the PT407. Their specifications are given in Table 1.

It has been confirmed that the vibrations in the pulse tube cold heads mainly come from the stretching of the tubes generated by gas compression and expansion[2]. The rotary valve and motor have been integrated in the warm end for the standard cold heads. A special version with a remote rotary valve has also been developed for all of our two-stage pulse tube cryocoolers. In this version (see Figure 2), the rotary valve and motor is separated from the pulse tube expander by 3 feet through a S.S. flexible line. An electrical isolator made of non-metal material is mounted between the rotary valve and the S.S. flexible line to isolate the EMI and RF noise from the driving motor for the rotary valve. The performance of this split version is approximately 5% less than that of the standard integrated version. These split 4 K pulse tube cryocoolers are used for cooling sensitive devices, such as SQUIDs magnetometer, etc.

The 10 K pulse tube cryocoolers developed at Cryomech have achieved the same cooling capacity and efficiency as the 10 K GM cryocoolers[3]. The latest improvements on a laboratory PT410 increases its performance to 1.2W@4.2K and 45W@40K simultaneously for 7.8 kW power input. This unit provides almost the same capacity and efficiency as the 4 K GM cryocooler. This performance will enable the pulse tube cryocooler to replace the GM cryocoolers in many applications in the near future.

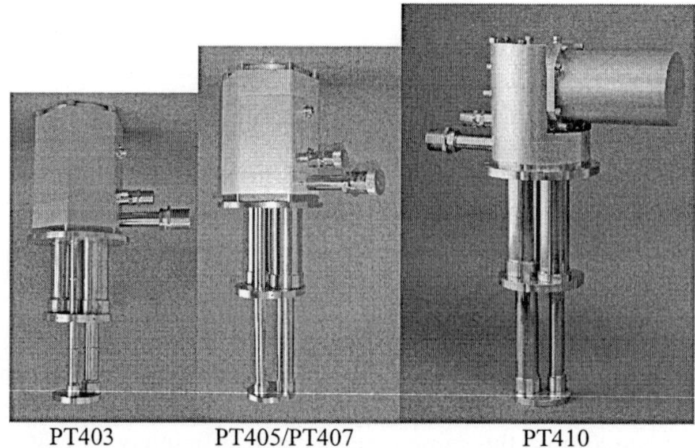

Figure 1. Photographs of the 4 K pulse tube cryocoolers

Table 1. Specifications of the 4 K pulse tube cryocoolers

	PT403	PT405	PT407	PT410
Specification	0.25W@4.2K & 10W@65K	0.5W@4.2K & 30W@65K	0.7W@4.2K & 30W@55K	1.0W@4.2K & 40W@45K
Power input	1 phase, 3 kW	3 phase, 4.6 kW	3 phase, 7 kW	3 phase, 8 kW

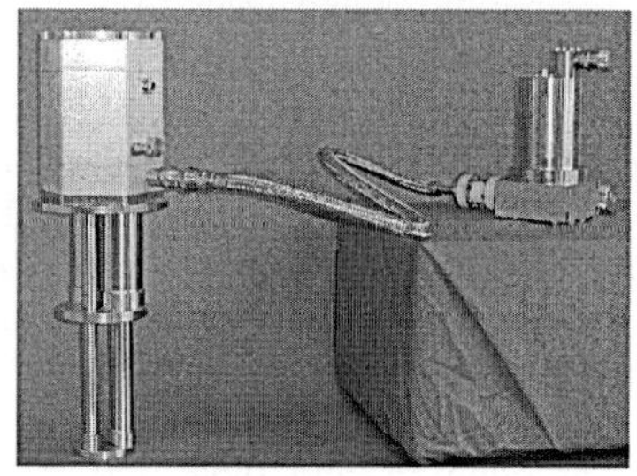

Figure 2. PT405/PT407 with remote rotary valve/motor

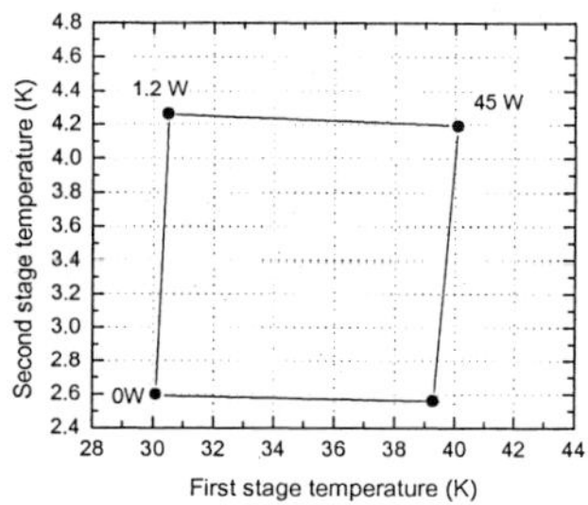

Figure 3. Latest performance of PT410

CHARACTERISTICS OF THE 4 K PULSE TUBE CRYOCOOLER

The 4 K pulse tube cryocoolers demonstrated their advanced features in the applications when compared with the 4 K GM cryocoolers. These features are presented below.

Magnetic field distortion

Small magnetic field distortion from the 4 K pulse tube cryocooler was found in NMR, MRI and SQUIDs systems. The amplitude of the magnetic distortion is approximately 20 nT compared to that of 200 nT from a SHI-SRDK408 4 K GM cryocooler. This magnetic field distortion is caused by the rare earth regenerative materials in the 2nd stage regenerator. Figure 4 shows the variations of magnetization of the rare earth materials of $HoCu_2$ and Er_3Ni at different temperatures. The rare earth materials were put into a very sensitive solenoid to measure their magnetization at external magnetic field of 0.1 Oe and 1 Oe. There are temperature oscillations of 2nd stage regenerative materials at the same frequency of refrigeration. The temperature oscillation could be a few Kelvins[4] and generate a magnetic field fluctuation. This is schematically shown in Figure 5 (a). For a 4 K GM cryocooler (Figure 5 (b)), the magnetic field distortions are generated not only by the temperature swing, but also the motion of the rare earth materials with the displacer. The magnetic field distortion from a 4 K GM cryocooler is normally ten times higher than that from a 4 K pulse tube cryocooler.

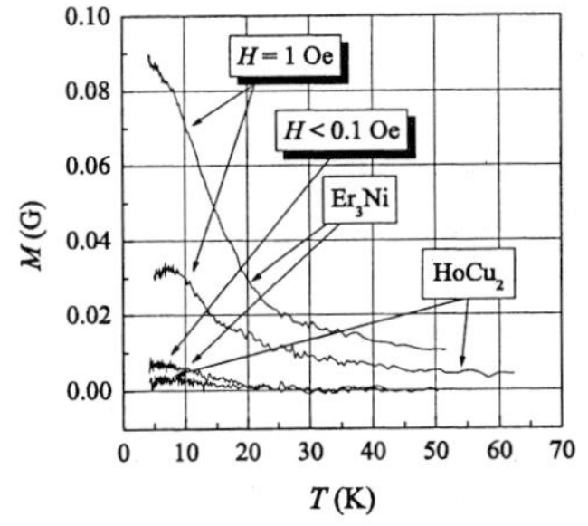
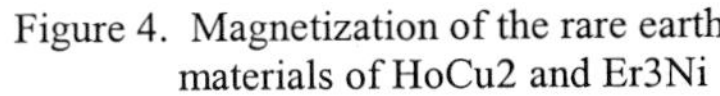

Figure 4. Magnetization of the rare earth materials of HoCu2 and Er3Ni

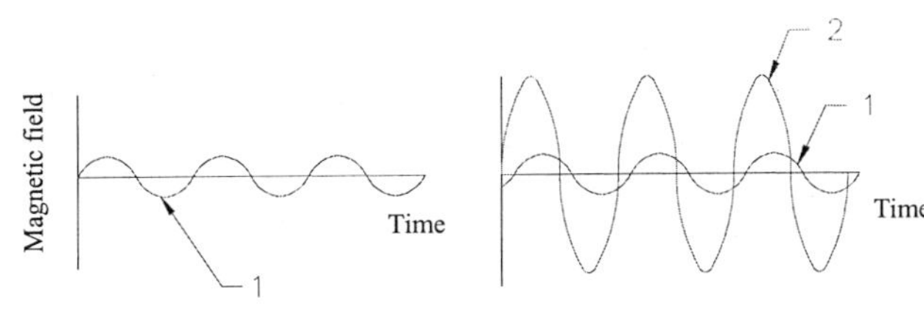

(a) Pulse tube   (b) GM

Figure 5. Magnetic field fluctuation generated by the cryocoolers. 1. generated by temperature oscillation; 2. generated by motion of the GM displacer.

Vibration

Figure 6 shows the installation of the pulse tube and GM cryocooler on the cryostat. Vibration of 4 K pulse tube cryocooler is so small that in most applications it can be directly mounted on the cryostat.

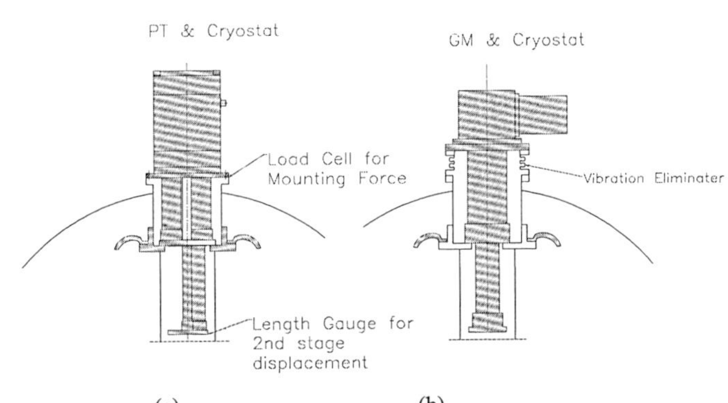

Figure 6 Installation of PT and GM cryocoolers on cryostat

Table 2. Vibration from PT and GM cryocooler

	Mounting force (amplitude)	2nd stage displacement (amplitude)
4 K GM cryocooler (SRDK408)	38 Lb	42 μm
4 K PT cryocooler (PT405)	1.0 Lb	11 μm

For example, a PT405 pulse tube cryocooler was installed in a MRI cryostat. Spin echo testing was performed on it to check the various methods of vibration isolation used for the GM cryocooler before (see Figure 6 (b)). No vibration isolation methods were necessary for the Pulse Tube cooled MRI magnet. No magnetic fluctuations that come from vibration have been observed when directly mounting the pulse tube cryocooler on the MRI cryostat.

The vibration of PT and GM cryocoolers are compared and given in Table 2. A load cell, mounted under the room temperature flanges of the cryocoolers (see figure 6(a)), is used to measure the mounting force. A length gauge which contacts the bottom of the 2nd stage heat exchanger measures the displacement. The mounting force from the 4 K GM cryocooler is 38 times that of the 4 K pulse tube cryocooler, and the displacement is 4 times greater.

Meantime between maintenance (MTBM)

Currently, the maintenance interval of the 4 K GM cryocooler is ~10,000 hours. Cryomech's goal is to provide the 4 K pulse tube cryocooler with MTBM > 5 years (43,800 hours). Three possible service requirements for the 4 K Pulse Tubes in 5 years were investigated and given below.

1. Adsorber in the compressor package. The lifetime of the adsorber is mainly determined by the oil carryover which passes through the oil separator and reaches the adsorber. Figure 7 shows the oil carryover in the CP900 series compressors used for pulse tube cryocoolers. The CP900s are controlled to

have oil carryover of less than 80 mg/day (29 g/year). The adsorbers for the CP900s have been tested and have an ability to adsorb > 300 g oil. It ensures system operation for more than 5 years without service.

2. Lifetime of rotary valve and valve plate in the cold head. The rotary valve and valve plate in the pulse tube cryocooler have less wear since there are no wear particles generated from displacer seals in the GM cryocooler. The valve and valve plate material have been studied and selected. It was found that there was only 0.03 mm wearing away for the valve and no significant wear on the valve plate after 12,000 hours running. We predict that the valve and valve plate will last more than 5 years.

3. Contamination in the cold head. Impact of air contamination in a PT405 pulse tube cryocooler has been investigated and shown in Figure 8. The pulse tube cryocooler has less sensitivity than GM cryocoolers to air contamination (78% N_2, 20% O_2). After adding 600 Torr·Liter air in the system, the first stage lost 2W at the temperature of 65 K and the 2nd stage temperature increased by 0.1 K. This feature enables the pulse tube cryocooler to operate for a long time without needing cold head service.

Since the first PT405 pulse tube cryocooler was delivered to a user in July 1999, a few hundred pulse tube cryocoolers are working in the field. Many of them have operated for 15,000~25,000 hours. So far, there has been no report on the performance degradation of our pulse tube cryocoolers. All of this information supports us toward our goal of providing the Pulse Tubes with 5 years MTBM.

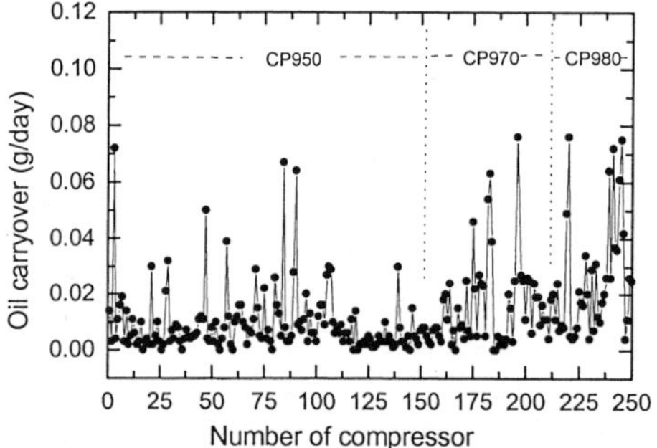

Figure 7 Oil carryover in CP900 series compressor

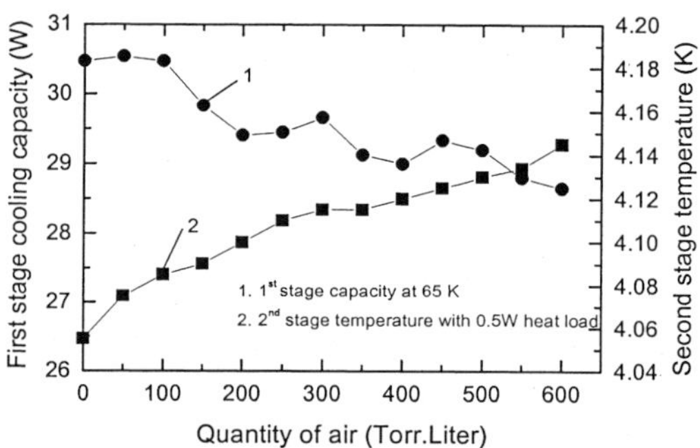

Figure 8. Impact of air contamination in the PT405

CONCLUSION

Cryomech has developed and commercialized 4 K pulse tube cryocoolers to provide cooling capacities from 0.25 W to 1.0W at 4.2K. The 4 K pulse tube cryocoolers have opened many challenging applications and demonstrated their advanced features with respect to very low vibration, low magnetic field distortion and long MTBM.

ACKNOWLEDGMENT

The author would like to thank Dr. V. Ankudinov at Moscow Power Engineering Institute for providing magnetization of the rare earth materials.

REFERENCE

1. Wang, C. and Gifford, P.E., 0.5W Class Two-Stage 4 K Pulse Tube Cryorefrigerator, in: *Advances in Cryogenic Engineering* (2000), 45A, pp.1-8.
2. Wang, C. and Gifford, P.E., "Development of 4 K Pulse Tube Cryocoolers at Cryomech", in: *Advances in Cryogenic Engineering* (2002), 47B, pp. 641-648.
3. Wang, C., Efficient 10 K Pulse Tube Cryocoolers, to be published in *Cryocooler 13.*
4. Wang, C., "Numerical analysis of 4 K pulse tube coolers: Part II. Performances and Internal Processes ", *Cryogenics,* (1997), vol.37, pp.215-220

Proceedings of the Twentieth International Cryogenic Engineering Conference
(ICEC 20), Beijing, China. © 2005 Elsevier Ltd. All rights reserved.

Experimental Investigation on Heat Transfer Performance of A Cryogenic Thermosyphon

Weizheng Li, Limin Qiu, Xuejun Zhang, Ping Chen, Yonglin He

Cryogenics Lab，Zhejiang University，Hangzhou，310027, China

As a high performance heat transfer device, the thermosyphon has many promising applications in cryogenic technique and superconductive magnet cooling. In this paper, the thermosyphon made of stainless steel has an operating length of 6.0×10^{-1} m and an inner diameter of 0.9×10^{-2} m. The effects of the filling ratio and inclination angle on its heat transfer rate are experimentally investigated. For this self-made thermosyphon, the optimum filling ratio is about 18.5% and the optimum inclination angle to the horizon is about 50°.

INTRODUCTION

With advancements in cryocooler-based applications such as cryocooler-cooled superconducting magnets, it will be required to search for effective means of heat transfer between components to be cooled and cryocoolers. At present, the most common means of heat transfer in these applications is heat conduction by copper bar. But in dealing with many applications, where the heat transfer distance would be large, this method will impose a limit on the heat transfer by its cross-sectional area available for heat conduction. To alleviate this constraint on the heat transfer by the conduction bar cross-section, one would need to look for heat transfer means that can transfer much more heat for the same temperature difference.

As an attractive alternative, heat pipes are known to be effective in such heat transfer. The cryogenic heat pipe (working temperature is from 0 to 150 K) reported in the literature can be categorized into four groups[1]: thermosyphon, wick-based heat pipe, cryogenic capillary-pumped loop and cryogenic loop heat pipe. The thermosyphon is a gravity assisted wickless heat pipe which has been utilized for cooling electrical devices, extracting a thermal energy from solar heat or geothermal heat sources. It is a practical heat transfer device due to its simple structure and low production cost. Here, we focused on a cryogenic two-phase nitrogen thermosyphon. It is important to grasp the effects of filling ratio and inclination angle on the heat transfer rate of thermosyphon. In this study, the effects of both filling ratio and inclination angle on heat transfer characteristics are experimentally investigated.

EXPERIMENTAL SET-UP

The schematic illustration of the experimental apparatus is shown in Figure 1. It consists of a thermosyphon, a cryostat, a liquid nitrogen container, an evacuation device for evacuating the inside of the thermosyphon and the cryostat, and a reservoir tank for measuring the amount of charged gas[2]. The thermosyphon made of stainless steel (1Cr18Ni9Ti) has an operating length of 6×10^{-1} m, an inner diameter of 0.9×10^{-2}m and a tube wall thickness of 0.1×10^{-2} m. The evaporator section is electrically heated. The condenser section is cooled by liquid nitrogen in a liquid nitrogen container.

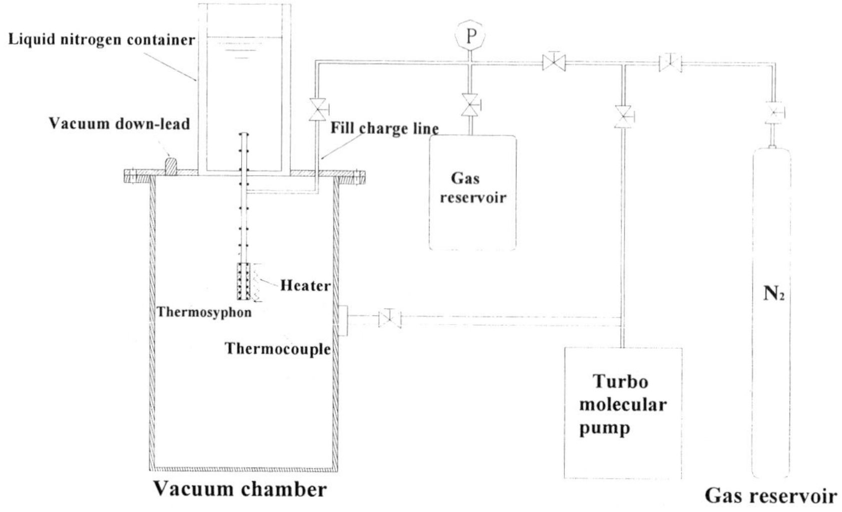

Figure 1 Schematic illustration of the experimental set-up

The thermosyphon is divided into the evaporator section, the adiabatic section and the condenser section.   Through changing the length of evaporator section, two cases are experimentally investigated. The lengths of the three parts are shown in Table 1, respectively.

Table 1: Length of thermosyphon

	Case A	Case B
L_e (cm)	5	10
L_a (cm)	45	40
L_c (cm)	10	10

After sufficiently evacuating the inside of the thermosyphon, the gaseous nitrogen is charged into the thermosyphon through the fill charge line from the gas reservoir.   The gaseous nitrogen is condensed into liquid on the cooled condenser wall.   Then, the liquid flows down to the evaporator section and forms a pool.   The amount of the liquid nitrogen fill charge is calculated from measuring the pressure difference of pre and post liquefaction in the reservoir tank.   The filling ratio ε is defined as the volume ratio of charged liquid nitrogen to the whole volume of the thermosyphon.

In this experiment, temperatures of the outer surface of evaporator section are measured by the copper-constantan thermocouples (see Figure 1).   The temperatures are acquired by the Keithley2700 type multimeter, which starts as soon as the thermosyphon works.

RESULTS AND DISCUSSION

The temperature measurements are carried out in two cases as mentioned above.   And the latter analysis, the heating power Q is fixed.   In each case, the temperature in evaporator section changes drastically along with the position, and the change trend of temperature is the same despite of the filling ratio ε as shown in Fig. 2 and Fig.3. The temperatures of the condenser section are fixed during a series of measurements. The temperatures at the evaporator section are the average values of the output from the two thermocouples on both sides, as shown in Fig. 1.

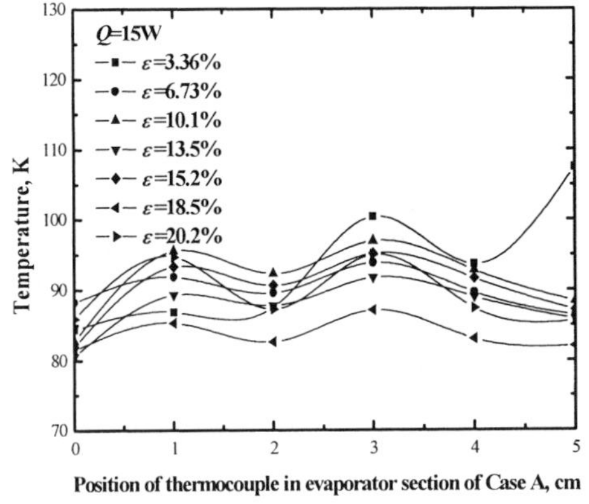

Figure 2   Optimum filling ratio of Case A

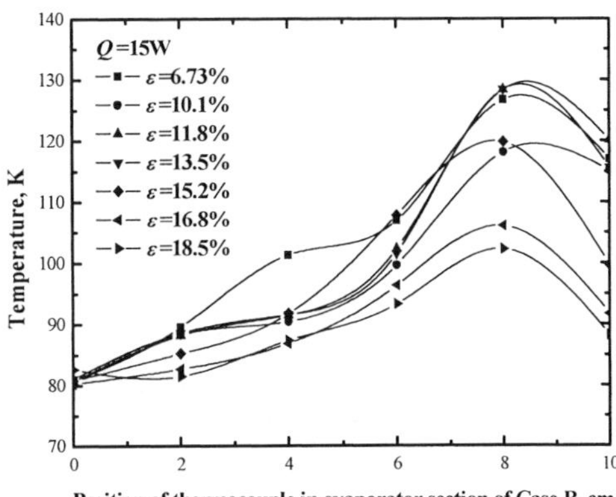

Figure 3   Optimum filling ratio of Case B

Fig. 2 shows the typical results of the steady temperature distribution along the axis of the evaporator with various values of filling ratio in the case A.   For the given thermosyphon in this paper, when ε =18.5%, the mean temperature in the evaporator section is minimum as shown in Fig. 2.   The lower this mean temperature, the better the cooling effect.   So an optimum filling ratio of this thermosyphon exists, and should be in the range of 15% to 20%.   This is because, less filling mass will result in the liquid working fluid drying off, whereas much filling mass will result in the liquid working fluid boiling when the thermal resistance of vapor film in liquid pool of evaporator section should be considered.   It should be noted that the working fluid of thermosyphon is in the supercritical state at ambient temperature, excessive working fluid will be dangerous too.

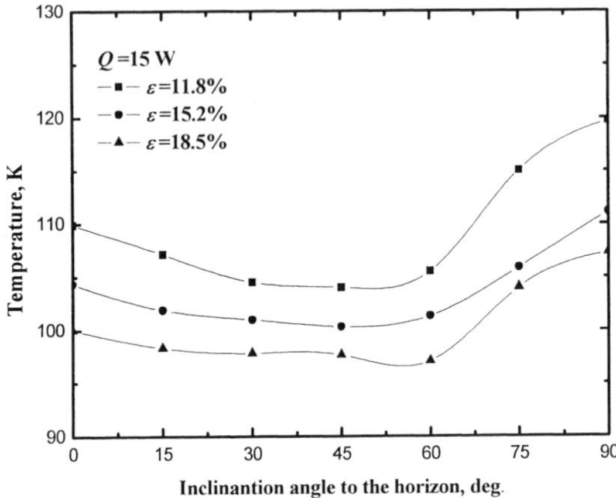

Figure 4   Effects of inclination angle in Case B

Fig. 3 shows the typical results of the steady temperature distribution along the axis of the evaporator for various values of filling ratio in the case B.   The mean temperature with a certain filling ratio in this case is higher than that in case A correspondingly.   This is because, the length of evaporator section in case A is shorter than that in case B, where the working fluid condensed is not enough to give a rapid supply to the evaporator section, which will result in the temperature of the evaporator section rising.

In the ground applications, the condensate on the inner surface of the condenser section flows down to the evaporator section at a certain speed, which results from the gravity.   The speed is varied with the inclination angle of thermosyphon.   To study the effect of inclination angle to the horizon on the heat transfer rate, the inclination angle is varied from 0 to 90° as shown in Fig. 4.   The temperatures in this figure are the average values of the whole evaporator section.   Similar to our earlier shown results, for any given inclination angle, the optimum filling ratio is always 18.5%, because the mean temperature of the whole evaporator section is the lowest.   And for a chosen filling ratio, with an initial increase in the inclination angle, the mean temperature decreases, which effects high heat transfer rate; but beyond a certain value of about 50°, the mean temperature increases, which results in decreasing heat transfer rate. So there exists an optimum inclination angle too.

CONCLUSIONS

In this study, with regard to the filling ratio and inclination angle, the heat transfer rate of a self-made cryogenic thermosyphon is experimental investigated in terms of mean temperature in the evaporator section.   The following conclusions are obtained:
1. There exists an optimum filling ratio for any thermosyphon.   Too much or too little filling mass will result in bad effects on the heat transfer rate of a thermosyphon.   For this self-made thermosyphon, the optimum filling ratio is about 18.5%.
2. The inclination angle also has strong effect on the heat transfer rate of a thermosyphon in the ground application.   For this self-made thermosyphon, the optimum inclination angle is about 50°.

ACKNOWLEDGEMENTS

The project is financially supported by the Foundation for the Author of National Excellent Doctoral Dissertation of China (200033), Excellent Young Teacher Project of Department of Education.

REFERENCES

1.   Chandratilleke, R., Hatakeyama, H., Nakagome, H., Development of cryogenic loop heat pipe, <u>Cryogeinics</u> (1998) <u>38</u> 263-269
2.   Nakano, A., Shiraishi, M., Nishio, M., Murakami, M., An experimental study of heat transfer characteristics of a two-phase nitrogen thermosyphon over a large dynamic range operation, <u>Cryogeinics (1998)</u> <u>38</u> 1259-1266.

Reduction of convective heat losses in pulse tube refrigerators by additional DC flow

Shiraishi M., Fujisawa Y.*, Murakami M.* and Nakano A.

National Institute of AIST, Tsukuba-East, Namiki 1-2-1, Tsukuba, Ibaraki, 305-8564, Japan
*University of Tsukuba, Tennoudai 1-1-1,Tsukuba, Ibaraki, 305-8573, Japan

The effectiveness of using an additional DC flow to improve cooling performance by reducing convective heat losses caused by gravity-driven secondary flow in inclined orifice pulse tube refrigerators was studied by visualization of flow and by measurement of gas temperatures at the cold and hot ends of the pulse tube. Results revealed that the gas temperature decreased at the cold end and increased at the hot end by decreasing the secondary flow to an optimal level in the core region. An additional DC flow thereby improved the cooling performance (temperature difference between hot and cold ends) by over 10%.

INTRODUCTION

The cooling performance of a pulse tube refrigerator is extremely sensitive to convective heat losses caused by convective secondary flows. Two typical convective secondary flows induced in a pulse tube are acoustic streaming induced by pressure oscillating [1] and gravity-driven secondary flow induced by gravity in inclined pulse tubes [2]. If these secondary flows are induced, convective heat losses occur because the flow from the hot end to the cold end in the core region transports fluid that has a relatively higher temperature compared to the cold-end temperature, and thus the fluid releases heat to the cold end. Decreasing the flow to the cold end should therefore reduce convective heat loss. Reduction of convective heat losses by controlling secondary flow is an effective method for improving the cooling performance of pulse tube refrigerators.

For practical applications, pulse tube refrigerators are used not only in vertical positions but also in inclined positions. When a refrigerator is used in an inclined position, a gravity-driven secondary flow is induced, which significantly influences the cooling performance. The cooling performance changes drastically when the refrigerator is increasingly inclined from 0° to 180° from vertical [2]. Here, we defined the inclination angle θ as shown in the inset of Fig. 1. In a previous study, we found that an effective approach was to reduce the convective heat loss in inclined pulse tube refrigerators was the use of an additional DC flow generated by a second orifice valve [3].

Here, the objective was to clarify in further detail the effectiveness of using an additional DC flow to reduce convective heat losses in inclined pulse tube refrigerators. First, the oscillating flow in an orifice pulse tube refrigerator was visualized. Then, the gas temperature distribution at the hot and cold ends of the pulse tube was measured as an indicator of cooling performance.

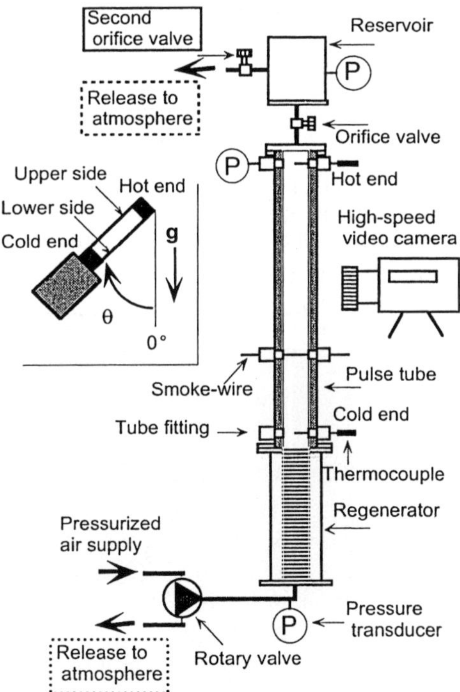

Figure 1 Schematic of the experimental apparatus used to visualize flow in an inclined orifice pulse tube refrigerator with a second orifice valve. The second orifice valve released air to the atmosphere, thus inducing additional DC flow toward the hot end through the pulse tube. Inset shows the definition of inclination angle θ.

EXPERIMENT

Figure 1 shows the experimental apparatus used to visualize oscillating flow in an orifice pulse tube refrigerator at various θ and to measure the gas temperature at the cold and hot ends. Details of the apparatus are described elsewhere [3]. In brief, the pulse tube was made of transparent plastic and was 16 mm in diameter and 320 mm long. The regenerator was made of #100 stainless-steel screen and was 18 mm in diameter and 170 mm long. The volume of reservoir was about 10 times larger than that of the pulse tube. The first orifice valve connected the hot end of the pulse tube to the reservoir, and the second orifice valve was fitted to the reservoir and had its outlet to the atmosphere. Thus, a pressure oscillation of air as the working gas was generated between the high pressure of the pressurized air of about 0.2 MPa and the low pressure of the atmosphere by introducing pressurized air into the rotary valve during the compression phase and releasing it into the atmosphere during the expansion phase. The second orifice valve released air from the reservoir into the atmosphere, generating an additional DC flow toward the hot end through the pulse tube. Each valve was fully open after 6 turns of the valve, and each turn was graduated into 25 divisions, thus totaling 150 "turn-divisions" at the fully open position. The valve opening was expressed in these divisions as arbitrary units; Vo and Vso are openings of the first and second orifice valves, respectively. Thermocouples were installed at the cold and hot ends to measure the gas temperature. Three pressure transducers were respectively installed near the warm end of the regenerator, at the hot end of the pulse tube, and in the reservoir. The smoke-wire was a 0.1-mm-diameter tungsten wire. Both ends of the wire were soldered to copper supports acting both as electrodes and as supports to keep the wire taut, and the wire was tightened by using the tube fittings located about one-third of the pulse-tube length from the cold end.

Suitable experimental conditions were determined based on the results of preliminary experiments; frequency was 6 Hz, amplitude of the pressure wave (defined as the compression ratio between the high and low pressures) was 1.2, and Vo was set at 10 turn-divisions to maximize the cooling performance at $\theta = 0°$ with a closed second orifice valve [4]. Visualization was done by increasing Vso in intervals of 10 turn-divisions up to a maximum 30 turn-divisions. Then, the visualization and the evaluation of cooling performance (i.e., gas temperature measurement) were done under these conditions at five different θ (0°, 90°, 120°, 150°, and 180°) typical for inclined pulse tube refrigerators. The movement of the smoke-line was recorded for more than five cycles by using a high-speed video camera with a frame rate of 400 frames/sec.

RESULTS AND DISCUSSION

Here, only the results for $\theta = 120°$ are presented because this angle is one of the best angles to evaluate the effectiveness of this approach in the inclination region where the effect of θ on cooling performance is strongest. Figure 2 shows typical visualization results for Vso = 0 turn-divisions (i.e., second orifice valve was closed) for the oscillation of the smoke-line during the first three cycles of oscillation as representative results. The smoke-line was emitted at the moment when the flow direction changed (due to a pressure oscillation) from expansion to compression in the cycle. Frames 1-5 show smoke-lines at three representative points in a cycle (explained in the figure caption). The smoke-line near the wall was elongated toward the hot end, while the smoke-line in the core region lagged toward the cold end. The smoke-line profiles gradually became asymmetric in which the leading edge of the smoke-line in the core region drifted toward the upper side wall of the pulse tube (left side of each frame) due to the gravity-driven secondary flow [3].

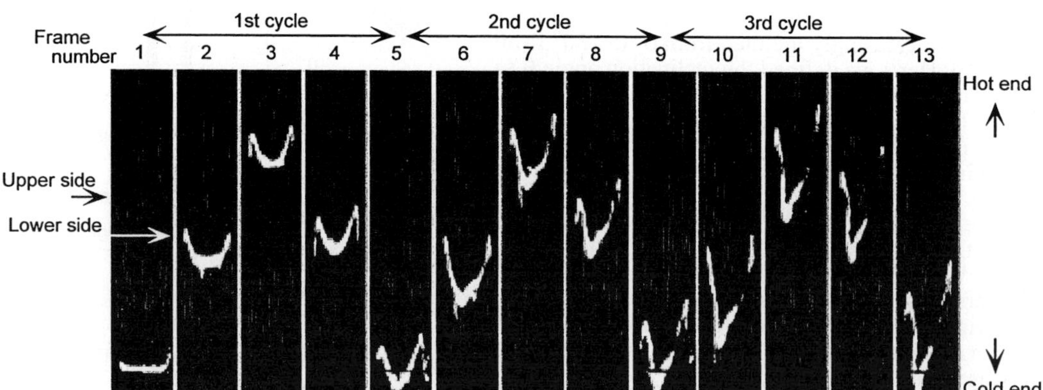

Figure 2 Typical visualization results for Vso = 0 turn-divisions (closed second orifice valve) during the first three cycles at $\theta = 120°$. Left and right sides of each frame correspond to the upper and lower side-walls of the pulse tube, respectively (see Fig. 1). The top and bottom of each frame are towards the hot and cold ends of the pulse tube, respectively. The smoke-line was emitted at the moment when the flow direction changed (by a pressure oscillation) from expansion to compression in a cycle: just after the smoke-line was emitted (Frame 1), at the turning point of the hot end where the smoke-line changed its direction of motion (Frames 3, 7, and 11), at the turning point of the cold end (Frames 5, 9, and 13), and halfway between these turning points (Frames 2, 4, 6, 8, 10, and 12).

275

The effect of Vso on the secondary flow was evaluated based on the smoke-line at the hot-end turning point (see caption of Fig. 2) of every cycle, assuming that the change in the smoke-line shape before and after a cycle was due only to secondary flow. Figure 3 shows the smoke-lines at the hot-end turning point for Cycles 1, 3, and 5 for Vso = 0, 20 and 30 turn-divisions. In the figure, three representative points on the smoke-line are marked A, B and C (in the frame for Vso = 0 turn-divisions and cycle 5), corresponding to the leading edges in the core region, near the upper wall, and near the lower walls, respectively. When the second orifice valve was closed (Vso = 0 turn-divisions), the smoke-line was elongated toward both sides of the cold end (downward) and hot end (upward), while its center remained at the initial position seen for Cycle 1. In contrast, when the second orifice valve was open (Vso > 0 turn-divisions), the smoke-lines for Vso = 20 and 30 turn-divisions were less elongated in the axial direction than that when the valve was closed (Vso = 0 turn-divisions), as evidenced by the distance between the leading edges of the smoke-line in the core region (A) and near the walls (B and C). This less elongation indicates that the secondary flow for Vso = 20 and 30 turn-divisions was slower than that for Vso = 0 turn-divisions. Moreover, for Vso = 20 turn-divisions, the leading edge in the core region (A) roughly remained as it was at the position of cycle 1, and for Vso = 30 turn-divisions, the leading edge in the core region (A) gradually shifted toward the hot end with time as seen from the change in the position of the leading edge.

From the smoke-lines in Fig. 3, we evaluated the velocities of the representative points A, B, and C on a smoke-line as V-core, V-upper, and V-lower, respectively. Here, negative and positive velocities in the figures correspond to flows going towards the cold end and hot end, respectively. Figure 4 shows the result. With increasing Vso, V-core monotonously changed from negative to positive velocity when Vso was between 20 and 30 turn-divisions, and therefore the absolute value of V-core was minimum when Vso was between 20 and 30 turn-divisions. In contrast, V-upper and V-lower had only positive values, indicating that the flow was only toward the hot end.

To clarify the effect of secondary flow on gas temperatures at the cold end (Tg-cold) and hot end (Tg-hot), the measured Tg-cold and Tg-hot were plotted together with V-core in Figure 5. Also plotted in the figure is the difference in these two temperatures (i.e., Tg-diff = Tg-hot - Tg-cold) by which the cooling performance was evaluated. Each temperature was normalized by the difference from its respective temperature for Vso = 0 turn-divisions. With increasing Vso up to about 10 turn-divisions, Tg-cold decreased and then increased, while Tg-hot increased slightly (compared to Tg-cold) and then decreased. Thus, Tg-diff (the cooling performance) increased with increasing Vso, peaking when Vso approached 10 turn-divisions, and then decreased, indicating that the cooling performance was improved when Vso was around 10 turn-divisions.

This result can be explained as follows. When the second orifice valve (Vso) was opened to about 10 turn-divisions, an additional DC flow appeared in the pulse tube. In the core region, this induced flow counteracted the convective secondary flow, causing a decrease in V-core, thus decreasing the flow to the cold end. Because this decrease in flow to the cold end also decreased the heat transfer, the gas temperature decreased at the cold end and increased at the hot end. Thus, the cooling performance improved, as revealed by an increase in Tg-diff. In contrast, when Vso was further increased to more than 10 turn-divisions, although the absolute value of V-core remained lower than that near Vso = 0 or 30 turn-divisions, the gas temperature increased at the cold end and decreased at the hot end. As a result of this change in gas temperatures, the cooling performance degraded, as evidenced by a decrease in Tg-diff. The additional DC flow

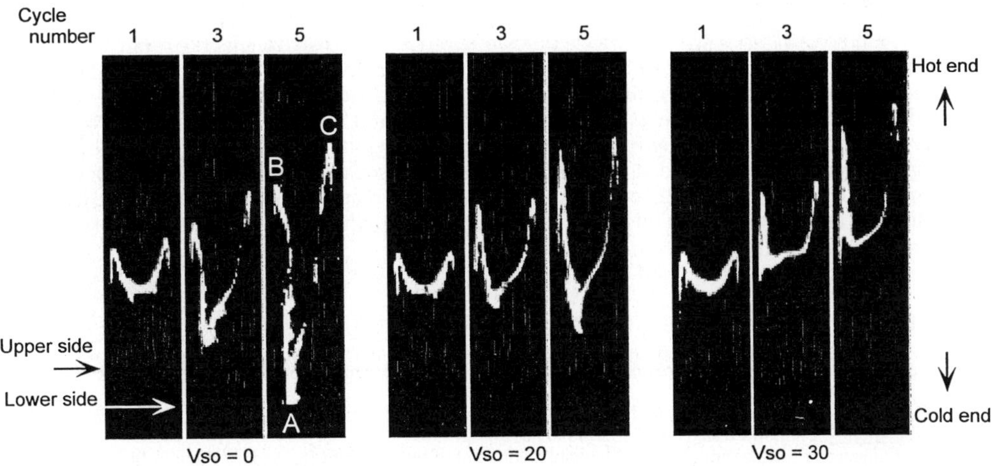

Figure 3 Smoke-lines at the turning point of the hot end for three representative cycles (Cycles 1, 3, and 5) for Vso = 0, 20, and 30 turn-divisions and $\theta = 120°$. Marks A, B, and C in the figure indicate representative points on the smoke-line.

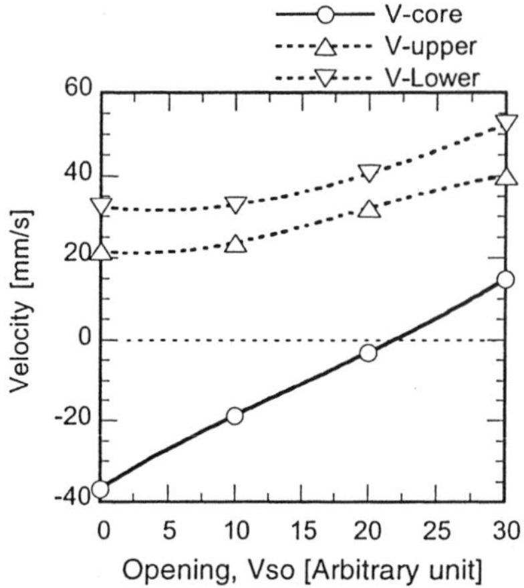

Figure 4 Effect of orifice opening (Vso) on flow velocities (V-core, V-upper, and V-lower) in a pulse tube refrigerator at θ = 120°. Velocities were obtained from the change in points A, B, and C, respectively, in Fig.3. Positive and negative values of velocity correspond to flow to the hot and cold ends, respectively.

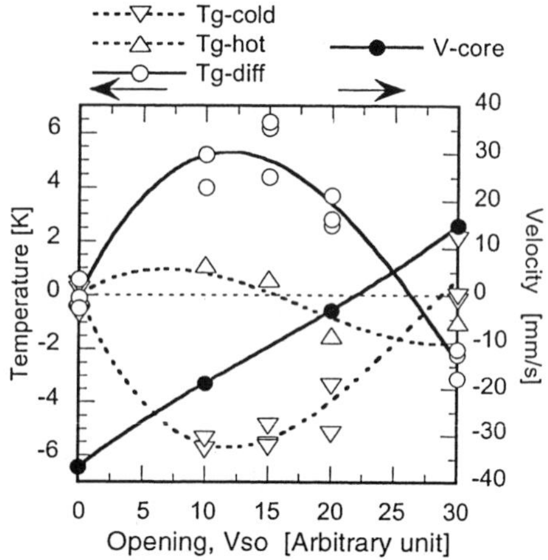

Figure 5 Gas temperatures at the cold and hot ends (Tg-cold and Tg-hot, respectively), gas temperature difference (Tg-diff) between the hot and cold ends (as an indicator of cooling performance), and V-core versus Vso at θ = 120°. Each temperature was normalized by difference from respective temperatures for Vso = 0 turn-divisions.

caused V-core to decrease, and caused the flow to the hot end to increase, as revealed by the increase in V-upper and V-lower (Fig. 4). The flow to the hot end transported fluid that had a relatively lower temperature compared to the hot-end temperature, and thus the increase in this flow caused a decrease in the hot-end temperature. Moreover, because the additional DC flow passed through the refrigerator, it introduced additional heat loss to the cold end where heat was transferred through the regenerator by the additional DC flow. Therefore, too large an increase in the DC flow, such as when Vso > 20, causes an increase in the gas temperature at the cold end and a decrease at the hot end, thus degrading the cooling performance.

Vso for maximum Tg-diff (i.e., Vso ~12 turn-divisions estimated from Fig. 5), and thus maximum cooling performance, did not coincide with that for V-core = 0 mm/s (Vso ~22 turn-divisions estimated from Fig. 5). V-core corresponds to the maximum velocity of a velocity profile of the secondary flow in the core region, whereas heat transfer from the hot end to the cold end by the flow in the core region depends on its mass flow rate. This mass flow rate roughly varies as (V-core)^3, if the velocity profile is assumed a conical profile. Consequently, with increasing Vso, heat loss by the heat transfer abruptly decreases compared to a decrease in V-core. Therefore, the optimum second-valve opening for cooling performance, namely, an optimum Vso ~12 turn-divisions (estimated from Fig. 5) was much lower than Vso for V-core = 0 mm/s.

SUMMARY

The effectiveness of using an additional DC flow to reduce convective heat losses in inclined orifice pulse tube refrigerators was studied by clarifying how an additional DC flow affects convective heat losses. Results revealed that the gas temperature decreased at the cold end and increased at the hot end by decreasing the convective secondary flow to an optimum level in the core region generated by the additional DC flow, and that the gas temperature increased at the cold end and decreased at the hot end when the additional DC flow exceeded this level. These results confirmed that convective heat loss can be effectively reduced by using an additional DC flow, if the level of the flow is optimized, and that the cooling performance of the refrigerator can thereby be improved.

REFERENCES

1. Olson, J. R. and Swift, G. W., Acoustic streaming in pulse tube refrigerators: tapered pulse tubes, Cryogenics (1997) 37, 769-776.
2. Thummes, G., Schreiber, M., Landgraf, R.and Heiden, C., Convective heat losses in pulse tube coolers: Effect of pulse tube inclination, Cryocoolers 9, Plenum Press, New York, (1997) 393-402.
3. Shiraishi, M., Fujisawa, Y., Murakami, M. and Nakano, Reduction of the Effect of Secondary Flow in Inclined Orifice Pulse Tube Refrigerators by Additional DC Flow Generated by a Second Orifice Valve, Advances in Cryogenic Engineering 49, American Institute of physics, New York, (2004), (to be published).
4. Shiraishi, M., Takamatsu, K., Murakami, M. and Nakano, Visualization study of secondary flow in an inclined pulse tube refrigerator, Advances in Cryogenic Engineering 45, Kluwer Academic/Plenum Publishers, New York, (2000) 119-125.

Proceedings of the Twentieth International Cryogenic Engineering Conference
(ICEC 20), Beijing, China. © 2005 Elsevier Ltd. All rights reserved.

Interference characterization of Stirling-type nonmagnetic and nonmetallic pulse tube cryocoolers for high-Tc SQUIDs operation

Dang H.Z., Liang J.T., Ju Y.L., Zhou Y.

Cryogenics Laboratory, P.O. Box 2711, Technical Institute of Physics and Chemistry, Chinese Academy of Sciences, Beijing 100080, P.R. China

In order to achieve portable low-noise cryogen-free cooling systems for continuous high-Tc SQUIDs operation and realize their direct coupling with the sensors, Stirling-type non-magnetic and non-metallic pulse tube cryocoolers (NNPTCs) have been developed and a series of experiments performed. The optimizations of the cooler geometry and working parameters have been carried out. The design and optimization of the system are described, and the analysis and evaluation of the typical cooling performance and interference characterization of the new-type pulse tube coolers are presented.

INTRODUCTION

Superconducting Quantum Interference Devices (SQUIDs) are the most sensitive magnetic flux-to-voltage sensors known so far and have developed a wide variety of important applications [1]. In 1970's, the concept of supplying a cryogen-free mechanical cooling system for low-Tc SQUIDs were proposed and reduced to practice. In the late 1980's, the advent of high-Tc SQUIDs relaxed the cooling requirements, and many attempts have been made to cool high-Tc SQUIDs by various types of cryocoolers from then on.

GM and Stirling cryocoolers were ever the workhorse for cooling high Tc SQUIDs. But the moving displacers introduce severe mechanical vibrations and electro-magnetic interference (EMI) signals, which are often fatal for the SQUID operation, so that the coolers have to be separated from the sensors in time or space, which is named as *time-separation* or *space-separation* method accordingly. *Time-separation* method results in the non-continuous operation of SQUIDs, while the *space-separation* method usually significantly adds to the incompactness and complexity of the system. Compared with the two conventional regenerative cryocoolers, pulse tube cryocoolers (PTCs) are more attractive candidates for high Tc SQUIDs cooling because of the absence of moving parts at low temperatures, which introduces much less mechanical vibrations and EMI. Therefore, there exists the possibility of continuous operation of SQUID sensors when they are near or even attached directly to the coolers. According to the drivers, PTCs could be divided into "Stirling-type" and "G-M type". Stirling-type PTCs have advantages over their G-M type counterparts in terms of compactness, flexibility, portability and low vibration, due to the much smaller volume and much lighter weight (a reduction in volume or weight by a factor of above 5-10) and the absence of rotary valves, which adds the possibility of the direct coupling. Therefore, we attempt to develop a better cooling system for high Tc SQUIDs based on Stirling-type PTCs other than GM type ones.

NNPTCS, THE WAY TO REALIZE THE DIRECT COUPLE

An intractable problem has to be solved in advance to realize the direct coupling of the SQUIDs with the cooler. Nowadays metallic materials or materials that exhibit marked remanent magnetization are widely used to fabricate the cooler components. The magnetic components introduce direct magnetic interferences, and the metallic ones in the vicinity of the SQUID pick-up loops could generate Johnson noise and then cause distortion of environmental fields. To solve the intractable problem in a simple and advisable way, we fabricate all key components of the PTCs by non-magnetic, non-metallic and electrically insulating materials. Based on a series of systematical experiments, we chose a special machinable ceramic for the cold heads, and stacked Nylon screens for the regenerator matrix. We fabricate the regenerator tube by a special glassfilled epoxy resin and the pulse tube by a kind of Nylon plastics. The vacuum chambers and connecting flanges are made of acrylic glass, and all straighteners of polytetrafluoroethylene plastic. The types of PTCs are named as NNPTCs for short, and they will eliminate the interference introduced by the materials and could realize the direct coupling with the SQUIDs.

NEGATIVE EFFECT OF NONMETALLIC MATERIALS AND EXPERIMENTAL OPTIMIZATION

It is interesting that the use of nonmetallic materials for tubes and regenerator matrix results in a beneficial byproduct, that is, the axial thermal conduction losses are reduced greatly due to the much lower thermal conductivities [2]. But the selected nonmetallic matrix has a negative effect on the performance of the regenerator. For a given regenerator housing, the volumetric heat capacity of matrix materials, c_m, and the thermal penetration depths in solid matrix are two important factors that influence the performance of regenerators. The c_m indicates how much heat per volume can be stored. We have gotten the variations of c_m of stainless steel and three nonmetallic materials, Nylon, Kapton, and Teflon, with the temperature, as shown in **Figure 1**. The c_m of Nylon is the highest one in the three nonmetallic materials, but it is only about 0.38~0.55 times that of stainless steel at temperature ranges of 60~300 K, which indicates a poorer regenerator performance compared with those of the metallic counterparts.

To avoid no contributing to the effective heat capacity, the thermal penetration depth in the solid should not be smaller than the characteristic matrix dimension. For high-frequency miniature PTCs working at liquid nitrogen temperature ranges, we usually needn't consider the effect of the thermal penetration depth [3]. But for Nylon screens, when the frequencies are between 30~80 Hz and the temperatures between 60~300 K, we could not make sure that all of the wires will contribute toward the effective heat capacity. In order to weaken the negative effect, either lower operating frequency or screens with larger mesh number should be adopted. It is difficult to produce the nonmetallic screens with too large mesh number (such as, more than 500). As for lower frequency, according to calculation [2], when the temperature is between 60~300 K, if we guarantee the same thermal penetration depths, the frequency of the NNPTC should be between 1.575~5.85 Hz, which is impossible for Stirling-type PTCs. So the poor performance of the NNPTCs can be expected, and the optimization of the coolers becomes important.

The optimizations focus on the investigations of the effect of the cooler geometry and working parameters on the cooling performance. For the optimization of cooler geometry, a theoretical model based on the analyses of thermodynamic and hydrodynamic behaviors of gas parcels in the oscillating flow regenerators has been developed, and the preliminary rough scope of optimum dimensions obtained. Then a series of experiments have been performed for the practical design and optimization. In our experiments, the optimum aspect ratio of the regenerator and the pulse tube is 5~6 and 13~16, respectively. The experiments have also been conducted for the optimization of the ratio of the volume of the pulse tube to the void volume of the regenerator, and the optimum value is between 0.5 and 0.6 [2].

Based on the geometry optimizations, a series of experiments have been performed for the optimization of the working parameters, which include the operating frequency, charge pressure, input power, temperature of the hot end, opening of the valves, direction of asymmetric double-inlet valve, orientation of the cold head, etc. For example, we got the frequency dependence of the no-load temperature of the cold head for the coolers, and their optimum operating frequencies are between 36~40 Hz. With the same parameters, the optimum frequencies of the metallic PTCs developed in the same laboratory are around 50 Hz. The lower optimum frequencies may result from smaller heat penetration depth discussed previously.

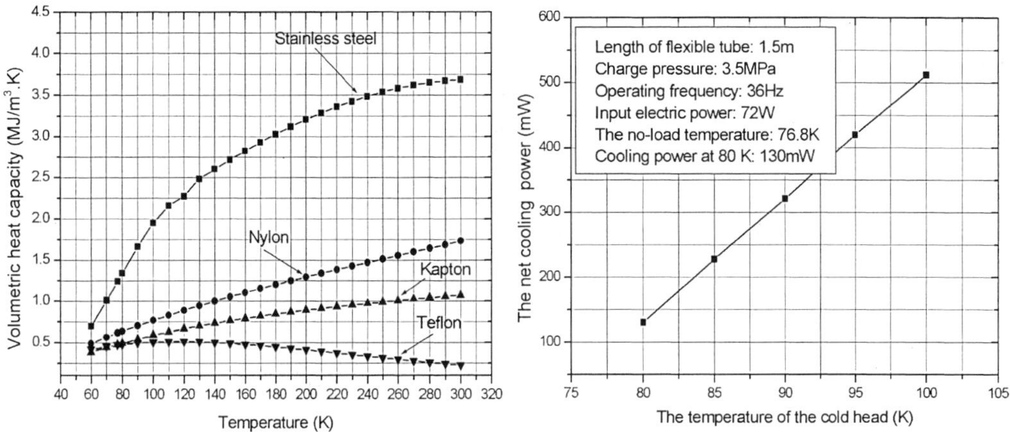

Figure 1. Variations of volumetric heat capacity of the various materials with temperature

Figure 3. Temperature dependence of the net cooling power of the NNPTC

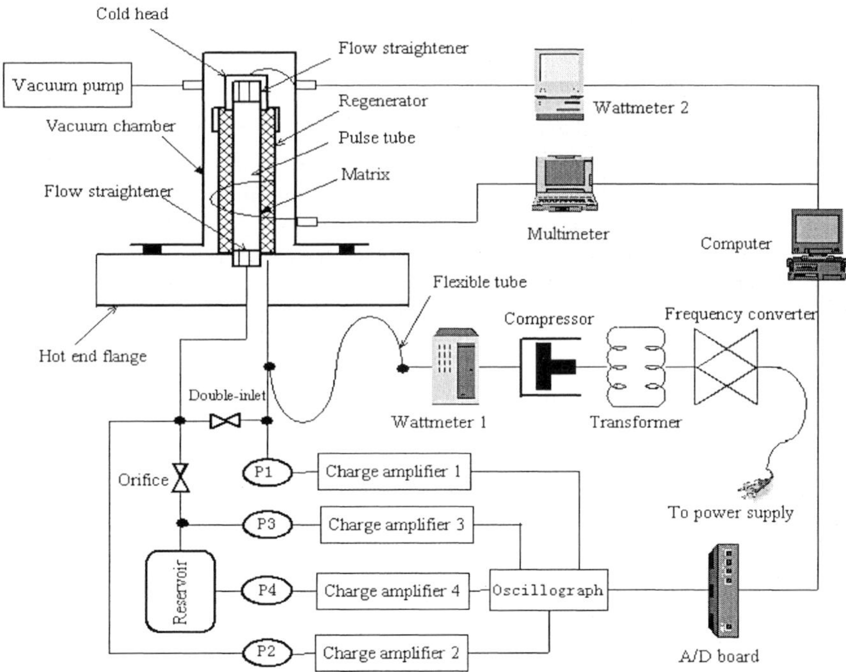

Figure 2. The schematic diagram of the system designed for high Tc SQUIDs.

SYSTEM DESCRIPTION

The schematic diagram of the system specially designed for cooling the high Tc SQUIDs is shown in **Figure 2**. A special synthetic epoxy resin adhesive is used to realize the connection of tubes to the cold head and flanges. The compressor system consists of a frequency converter, a transformer, and a linear compressor. The transformer protects the compressor, and the frequency converter adjusts the operating frequency. To reduce the vibrations generated by the compressor, the connection of it and the NNPTC is realized by a flexible small-diameter tube. The NNPTC is of single stage and coaxial type.

TYPICAL COOLING PERFORMANCE AND INTERFERENCE CHARACTERIZATION

A typical cooling power of over 100mW at 80K with 70 W input electric power for the Stirling-type co-axial NNPTCs developed has been achieved. **Figure 3** shows the temperature dependence of the cooling power of one of the cooler between 80 and 100 K. The efficiency of the NNPTCs at 80 K is about 0.6% of Carnot efficiency, which is much smaller than 4~11%, the corresponding values of their metallic counterparts [3]. The much poorer performance confirms the previous analysis.

The remanent magnetic field and Johnson noise in the NNPTC can be negligible because all the components in the vicinity of the SQUID pick-up loops are made of nonmagnetic and nonmetallic materials. The interferences introduced by the mechanical vibrations include the interference induced by the translation and rotation of the cold head. According to the measurements, the sum is between $50\sim90 \, fT/\sqrt{Hz}$, which is less than the maximum acceptable value of most high Tc SQUIDs (usually $100 \, fT/\sqrt{Hz}$). Another troublesome interference is introduced by thermal fluctuations, which affect the London penetration depth, and then change the effective sensing area of the SQUIDs. In our experiments, the maximal temperature fluctuation of the cold head at 80K with 130mW applied heat load in 800 seconds is about 40mK, which is well below the upper limit of the requirement (about 100mK).

CONCLUSIONS

A portable low-noise cryogen-free cooling system based on Stirling-type co-axial NNPTCs for high-Tc SQUIDs operation has been designed and constructed.

A typical cooling power of 130mW at 80K for 70 W of input power has been achieved, which can meet the basic cooling requirements of the high Tc SQUIDs. Preliminary results indicate the interference characteristics of the system can meet the basic noise requirements of the sensors.

A disadvantage is that the performances of the Stirling-type NNPTCs are much poorer than those of their conventional metallic counterparts, and in practical applications, larger cooling power and lower working temperatures are usually desired. So the performance optimization of the coolers is underway.

ACKNOWLEDGMENT

This work was partially supported by the National Natural Science Foundation of China under grant No.50176052 and 50206025.

REFERENCES

1. Cantor, R., DC SQUIDs: Design, optimization and practical applications, <u>SQUID Sensors: Fundamentals, Fabrication and Applications,</u> Kluwer Academic Publishers, Dordrecht (1996) 179-233.

2. Dang, H. Z. Ju, Y. L. Liang, J. T. Cai, J. H. and Zhou Y. Performance investigation of Stirling-type nonmagnetic and nonmetallic pulse tube cryocoolers for high-Tc SQUIDs Operation, <u>The 13th International Cryocooler Conference</u>, New Orleans, U.S.A., (2004)

3. Radebaugh, R., Development of the pulse tube as an efficient and reliable cryocooler, <u>Proc. Institute of Refrigeration(London)</u>, (2000), <u>96</u> 11-29.

Performance of vibration-free pulse tube cryocooler system for a gravitational wave detector

T. Haruyama, T. Tomaru, T. Suzuki, T. Shintomi, N. Sato, A. Yamamoto, Y. Ikushima*, R. Li*,
T. Akutsu**, T. Uchiyama** and S. Miyoki**

KEK, High Energy Accelerator Research Organization, 1-1, Oho, Tsukuba 305-0801, Japan,
*Sumitomo Heavy Industries Ltd., 2-1-1 Yato, Nishitokyo, Tokyo, 188-8585, Japan
**Institute for Cosmic Ray Research (ICRR), The University of Tokyo, 5-1-5, Kashiwanoha, Kashiwa,
Chiba, 277-8582, Japan

A pulse tube cryocooler system with a quite low vibration level has been developed for a gravitational wave detector. The origin of two types of vibration in a cryocooler system was identified, and a vibration-reduction method has been developed. A thermal link by using bundles of thin pure aluminum wires was applied for the vibration reduction at the cold stages, and a bellows was used for vibration reduction at the cold head. A cryocooler system with vibration of 50 nm in the Z-direction at the vibration reduction stage, and with a cooling power of 0.4 W at 4.2K and 15 W at 45 K has been developed. An overall vibration level was also confirmed as being identical with that of the seismic level at Kamioka mine. Results of development and the final performance of a vibration-free cryocooler system are presented.

INTRODUCTION

A feasibility study for the Large-scale Cryogenic Gravitational wave Telescope (LCGT) project has been promoted in Japan [1]. At present, a proto-type system, the Cryogenic Laser Interferometer Observatory (CLIO), is under construction in Kamioka mine, Japan [2]. To enhance the sensitivity of the gravitational wave detector, the sapphire mirrors should be cooled down around 20 K by using a cryocooler. Since the mirrors are extremely sensitive to vibrations, it is required that the vibration of the cryocooler must be reduced to the same level as that of the seismic vibration in Kamioka mine where the LCGT is expected to be installed. Then, one of the key technologies for realizing the LCGT project is the development of a cryocooler cooling system with a quite low vibration level.

   We have investigated the origin of vibration in a pulse tube cryocooler system with the measuring method described in Ref.3, and found that there are two different origins of vibration in the cryocooler [3]. Based on the commercial two-stage pulse tube cryocooler, SRP-052A, by Sumitomo Heavy Industries, ltd., a vibration-free cryocooler system, which can reduce these two types of vibration, has been developed. In this paper, the origin of the vibration in a pulse tube cryocooler system, the vibration reduction method, the achieved performance of cooling power and vibration level, and the progressing test overview is reported.

ORIGIN OF VIBRATIONS IN A CRYOCOOLER

A measurement of the vibration in a pulse tube cryocooler has been carried out by using optical sensors; it was found that there exist two different vibrations in the cryocooler [3]. The first vibration is an intrinsic one at the cold stages of the cryocooler, which originates in a pressure wave inside of the pulse tube and the regenerator. The second one is vibration at the cold head of the cryocooler that mainly originates in the rotary valve and compressor. Regarding the first intrinsic vibration, the level at the cold stage is almost comparable between a pulse tube cryocooler and a GM cryocooler. The typical vibration level at the cold stage is about ~10 μm at 1 Hz. On the other hand, the second one, the acceleration level at the cold head of a pulse tube cryocooler, is about two orders of magnitude smaller than that of the GM cryocooler. However, even for a pulse tube cryocooler, both the vibration level at the cold stage and the acceleration level at the cold head are still far from the requirement of the CLIO.

VIBRATION REDUCTION SYSTEM

In order to meet the low-vibration requirement, innovative methods were applied to a commercially available basic 4 K pulse tube cryocooler, SRP-052A, a product of Sumitomo Heavy industries, Ltd. The original cooling power of this cryocooler is 0.5 W at 4.2 K and 20 W at 45 K by using a 7 kW GM type compressor [4]. The cryocooler consists of a two-stage cold head, a rotary valve unit, flexible hoses of 20 m long for discharge and return gas and a helium compressor. A schematic diagram of the developed vibration-free pulse tube cryocooler system is shown in Figure 1.The main components of the system are a cold head mounted on a rigid frame, a split rotary valve unit on the table and a cryostat with an isolation bellows and the vibration-reduction (VR) stages with heat link.

<u>Reduction of intrinsic displacement due to pressure oscillation</u>
A couple of reports on the reduction of the cold stage by using flexible thermal links have already been published for specific applications with low-vibration requirements [5,6]. It has been reported that this method can reduce the vibration level down to ~1/10 of the original cold stage. We used braided wires of oxygen-free copper for the high-temperature 1st stage, and braided aluminum wires with high purity for the 2nd stage. The diameter of each thin wire is 0.1 mm. The Young's modulus of the aluminum wire is ~1/3 of that of the copper wire. Moreover, the spring constant becomes much smaller with keeping the same heat conduction by using the thinner wires. Two VR stages are supported by the alumina fiber FRP (AFRP) pipes from the cryostat lower flange.

<u>Cold head supporting stage and rotary valve table</u>
As shown in Figure 1, the support stage of four I-beam poles sustains the weight of the cold head and any kinds of vibration coming into the cold head will be short-cut to the ground. The VR stages are isolated from the cold head by a welding bellows. Since a driving motor for the rotary valve and the compressor generate a mechanical vibration, the rotary valve table is introduced to cut off the vibration. Flexible tubes from the compressor and rigid tubes to the rotary valve are all fixed by cramps. One of the features of the system is a connecting rigid tube of 40 cm long to separate the rotary valve from the cold head. Figure 2 shows pictures of the system and the details of the vibration-reduction stages.

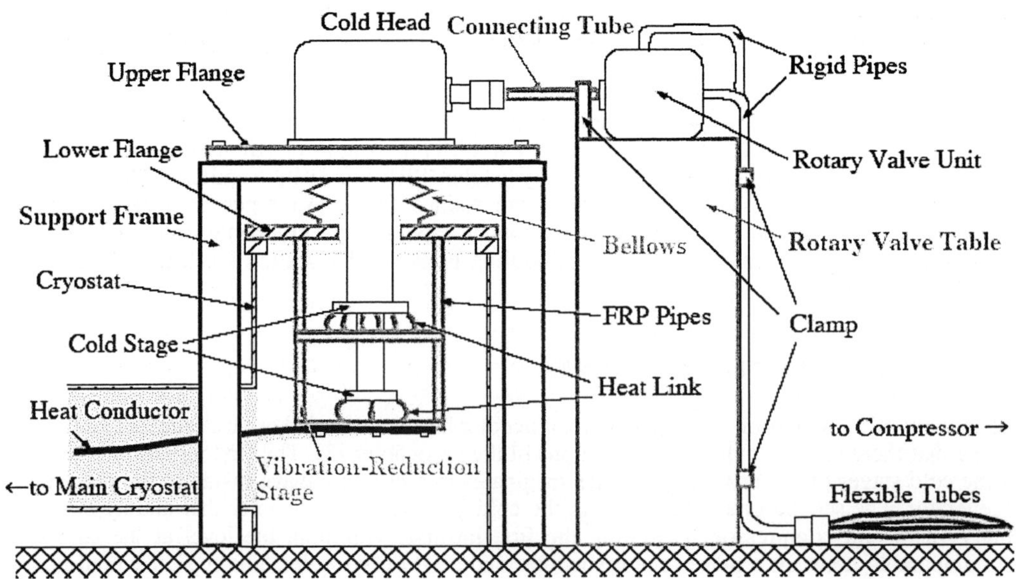

Figure 1. Schematic diagram of the vibration-reduction system for a 4K PT cryocooler. The compressor is not shown.

(a)   (b)

Figure 2. Pictures of the vibration-reduction pulse tube cryocooler system. (a) Support frame and rotary valve unit on the table, (b) Heat link and the alumina FRP support rods.

PERFORMANCE OF THE SYSTEM

Cooling power with VR stages

In the CLIO system, a nominal cooling power of 15 W at 45 K and 0.4 W at 4.2 K is required. Table 1 summarizes the measured cooling power in various configurations. Due to a kind of thermal resistance at the contact surface of the heat link, the temperatures at each VR stage are 2-3 K for the 1st VR stage and 0.2-0.3 K for the 2nd VR stage higher than that of each cold stage [7].

Table 1. Comparison of the cooling power achieved in various configurations (by a 7 kW compressor).

Configuration	1st stage temperature	2nd stage temperature
(1) SRP-052A original	40.8 K @20 W	4.08 K @0.5 W
(2) 40 cm connecting tube	42.2 K @20 W	4.17 K @0.5 W
(3) With VR stage -Cold stage direct -Reduction stage	41.2 K **43.7 K @15W**	4.15 K **4.43 K @0.5 W**

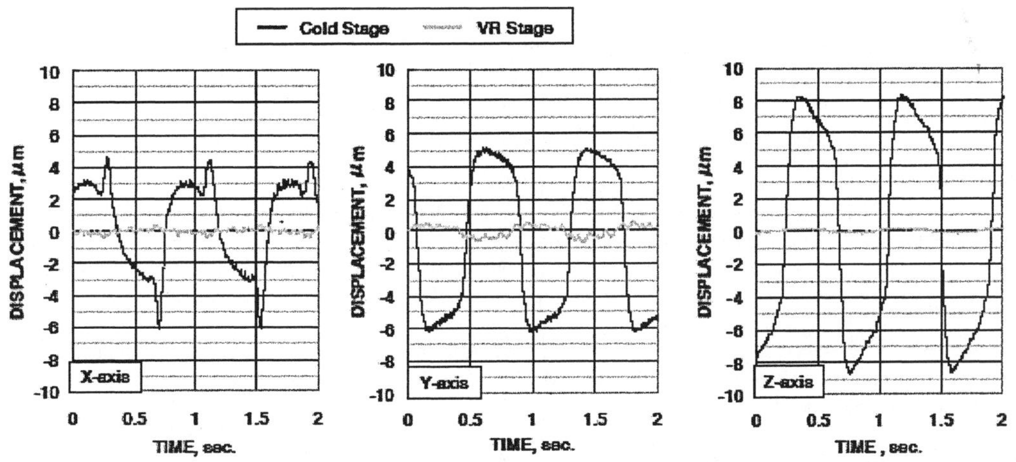

Figure 3. Measured typical vibration of the second VR stage and the cold stage.

284

<u>Vibration at the VR stage</u>
Figure 3 shows the measured vibration at the VR stage and the cold stage, which were measured by a laser displacement sensor (LC-2420, Keyence, Co., with a resolution of 10 nm). The vertical direction parallel to the pulse tube was defined as the Z-axis, and the direction of 15 degrees rotated from the central axis of the connecting tube was defined as the X-axis. The detailed vibrations in the X, Y and Z directions at the 2nd cold stage and the VR stage are as follows: X-axis, +/-5.3 to +/-0.42 μm; Y-axis, +/-5.7 to +/-0.65 μm; Z-axis, +/-8.5 to +/- 0.05 μm. The effective vibration reduction is verified by using this method, especially in the Z-direction. The vibration at the 1st stage is also evaluated the same in way as those of the 2nd stage [7].

<u>Overall vibration measurement in Kamioka mine</u>
All of the system shown in Figure 1 was temporally installed in Kamioka mine for vibration measurements. It was confirmed that the vibration level of the lower flange of the system was identical to the seismic vibration level in Kamioka mine [8].

ON-GOING AND FURTHER PERFORMANCE TESTS

Before the final set of cryocoolers in the CLIO system, a set of 4 K and 80 K vibration-free cryocooler units was installed in the prototype cryostat for the final evaluation. The evaluation will be completed by this summer, and the following installation for the CLIO system will be started in Kamioka mine.

SUMMARY

A vibration-free 4 K pulse tube cryocooler system has been developed for the gravitational wave detector. The origin of two types of vibration in the cryocooler system was identified, and a vibration-reduction method has been developed. A thermal link by using bundles of thin pure aluminum wires was applied for vibration reduction at the cold stages, together with a bellows for vibration reduction at the cold head. The performance of vibration of 50 nm in the Z-direction at the second VR stage, and with a cooling power of 0.4 W at 4.2K and 15 W at 45 K has been achieved by using a 7 kW GM-type compressor. The overall vibration level was also confirmed as being identical with that of the seismic level at Kamioka mine. A vibration-free cryocooler system is now available for the gravitational wave detectors in the CLIO.

ACKNOWLEDGEMENTS

This research was supported by a Grant-in-Aid for Scientific Research on Priority Area (14047217) prepared by Ministry of Education, Culture, Sports, Science and Technology in Japan.

REFERENCES

1.   Kuroda, K. et al., "Large-Scale Cryogenic Gravitational Wave Telescope", Int. J. Mod. Phys. D, Vol. 8 (1999) p.557.
2.   Ohashi, M. et al., "Design and Construction Status of CLIO", *Class. Quantum Grav.*, Vol. 20 (2003) pp. S599-S607
3.   Tomaru, T., Suzuki, T., Haruyama, T., Shintomi, T., Yamamoto, A., Koyama, T., Li, R. and Matsubara, Y., "Vibration Analysis of Cryocoolers", *Cryogenics*, Vol. 44, (2004), pp.309-317.
4.   Xu, M.Y., Yan, P.D., Koyama, T., Ogura, T. and Li, R., "Development of a 4 K Two-stage Pulse Tube Cryocooler", *Cryocooler 12*, Kluwer Academic/Plenum Publishers, New York (2003), pp. 301-307.
5.   Wang, C. and P.E. Gifford, "Performance Characteristics of a 4 K Pulse Tube in Current Applications", *Cryocooler 11*, Kluwer Academic/Plenum Publishers, New York (2001), pp. 205-212.
6.   C. Lienerth, G. Thummers and C. Heiden, "Low-Noise Cooling of HT-SQUIDs by Means of a Pulse Tube Cooler with Additional Vibration Compensation", *Proceedings of the 18th International Cryogenic Engineering Conference*, Narosa Publishing House, New Delhi (2000), pp. 555-558.
7.   Li, R., Ikushima, Y., Koyama, T., Tomaru, T., Suzuki, T., Haruyama, T., Shintomi, and Yamamoto, "Vibration-free pulse tube Cryocooler system for Gravitational Wave Detectors II – Cooling Performance and Vibration –", to be published in *Cryocooler 13*, Kluwer Academic/Plenum Publishers, New York (2005).
8.   Tomaru, T., Suzuki, T., Haruyama, T., Shintomi, T., Sato, N., Yamamoto, A., Ikushima, Y., Li, R., Akutsu, T., Uchiyama, T. and Miyoki, S., "Vibration-free pulse tube Cryocooler system for Gravitational Wave Detectors I – Vibration-Reduction Method and Measurement –", to be published in *Cryocooler 13*, Kluwer Academic/Plenum Publishers, New York (2005).

Influence of He-H$_2$ Mixture Proportion on the Performance of Pulse Tube Refrigerator [*]

Huang Y.H., Chen G.B., Gan Z.H., Tang K., Bao R.

Cryogenics Laboratory, Zhejiang University, Hangzhou 310027, P.R.China

The performance of pulse tube refrigerator (PTR) can be improved by using mixtures as its working fluid. Based on an experimental work on a two-stage PTR with He-H$_2$ mixture whose hydrogen percentage rising from 0% to 100%, an optimal proportion of H$_2$ in the He-H$_2$ mixture for the cooling temperature around 30K was found. At this temperature region, the cooling power and COP of the PTR reach their maximum both with an increment about 30-40% against pure helium when the hydrogen percent is about 60-70%.

INTRODUCTION

Unlike the Stirling or Gifford-McMahon refrigerators, the pulse tube refrigerator has no moving parts at the cold end, which enables itself to be a promising long-life and reliable cryocooler. Improvement in its efficiency have occurred rapidly since the orifice type pulse tube was recommended in 1984 by Mikulin[1]. We embarked on the study of the influence of mixture working fluids on the pulse tube refrigeration performance[2,3] in 1995. Taking into account the thermodynamics, heat transfer and fluid flow characteristics of mixture working fluids, performance prediction[4] of pulse tube refrigeration cycle with mixtures was developed. Experimental work was repeated times to confirm the results in the last two years. Both the theoretical and experimental investigations indicate that higher cooling power and coefficient of performance (COP) could be achieved with He-H$_2$ mixtures as working fluids than those with pure He for PTRs working in the 30K and 80K cooling temperature regions. But the molar fraction of hydrogen in the He-H$_2$ mixture was limited in 60% in our previous work by the experimental setup. Recently, high hydrogen percentage He-H$_2$ mixtures were charged in the same PTR. A better result of the cooling power and COP has been obtained.

EXPERIMENTAL SETUP

The outline of the experimental two-stage PTR is presented in Figure 1. The main dimensions and regenerative materials are tabulated in Table 1.The pressure wave needed by the PTR is generated by a C100W homemade helium compressor and a rotatory valve. A pressure sensor is located at the inlet of the regenerator, as shown in Figure 1, to measure the input pressure wave. The temperatures at cold ends of the first and the second stages are measured by two Rh-Fe resistance thermometers. The electric power consumed by the compressor is measured by a powermeter. Heat balance method is used to measure the cooling power. Eight Cu-Constantan thermocouples are fixed along the first and second stage pulse tubes

[*] Funded by the National Natural Science Foundation (Grant No. 50376055 and 50106013) and the Special Research Fund for Doctoral Training in Universities by the National ministry of Education of China (Grant No. 20010335010).

and regenerators to show the temperature distributions. This is helpful for the adjustment of the two-stage PTR in the operation.

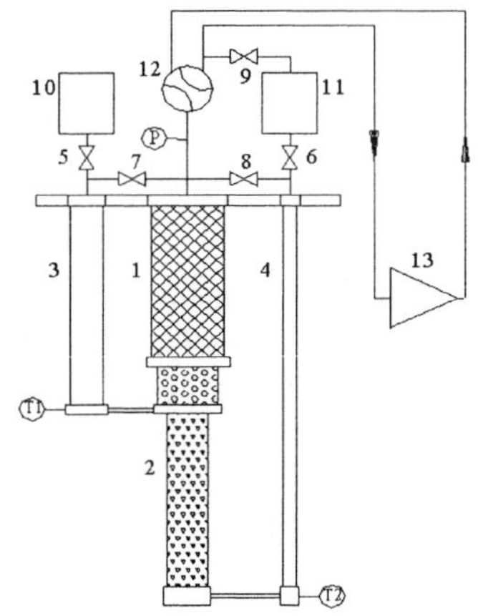

Table1　Dimensions and regenerative packing of two-stage PTR

Stage	Pulse tube (mm)	Regenerator (mm)	Regenerative packing
1st	$\phi26\times280$	$\phi59\times130$ $+\phi45\times60$	1200 pieces of 250 mesh phosphor bronze screen $+\phi0.3mm$ lead 744g
2nd	$\phi12\times370$	$\phi19\times190$	Er3Ni 285g

Figure1　Outline of two-stage PTR

1.2. 1st and 2nd stage regenerators, 3.4. 1st and 2nd stage pulse tubes, 5.6. orifices, 7.8. double inlets, 9. 2nd orifice, 10.11. reservoirs, 12. rotary valve, 13. compressor, $\textcircled{P}$-pressure sensor, $\textcircled{T}$-temperature sensors

RESULTS AND ANALYSIS

Different He-H_2 mixtures with 1.4MPa gauge pressure were charged into the two-stage PTR. Meanwhile, the settings of all valves and regenerator materials were kept the same. The relationship between the cooling power or COP of the PTR and the second stage cooling temperature are shown in Figure 2. It can be seen that higher cooling power and COP are achieved with He-H_2 mixture of 7~70% H_2 than those with pure helium in the 15~50K temperature region. The characteristics of pure hydrogen curve seem to be special. The performance of the PTR with pure hydrogen is much lower below 34K than that with mixtures whose H_2 percentage is less than 70%.

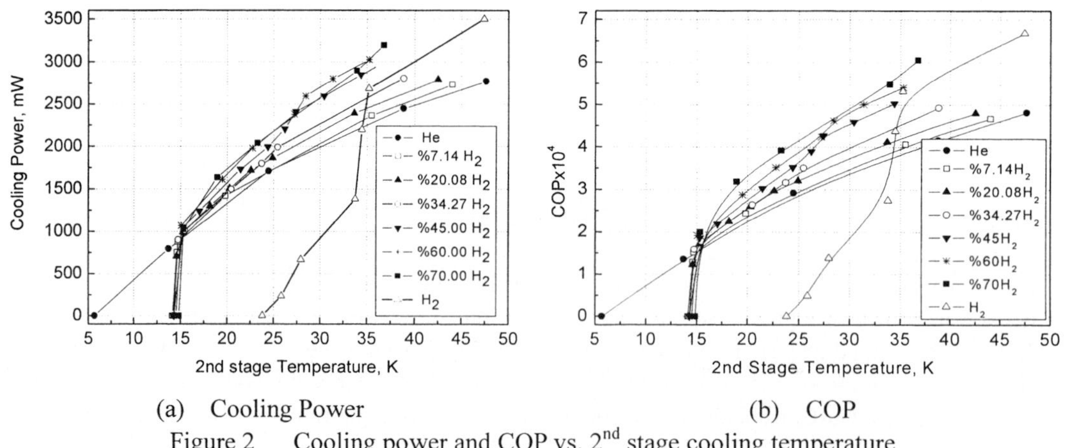

(a)　Cooling Power　　　　　　　　　　　(b)　COP

Figure 2　Cooling power and COP vs. 2nd stage cooling temperature

Figure 3 presents the relation between cooling power or COP and molar fraction of H_2 at the 2nd

stage cooling temperatures 25, 30 and 35K. This figure says that both the cooling power and COP increase with the accretion of H_2 molar fraction and attain their maximum values with He-H_2 mixtures of 60-70% H_2.

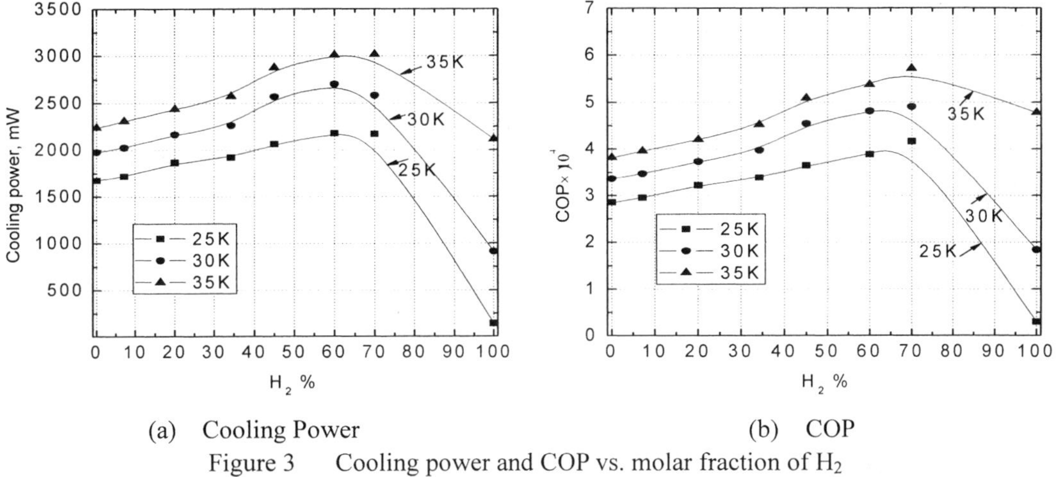

(a)   Cooling Power   (b)   COP

Figure 3   Cooling power and COP vs. molar fraction of H_2

Figure 4 gives the tendency of the specific cooling power and specific COP against molar fraction of H_2 in He-H_2 mixture corresponding to Figure 3. With 60-70% hydrogen, both the specific cooling power and specific COP reach their maximum and have 30-45% increment compared to pure helium.

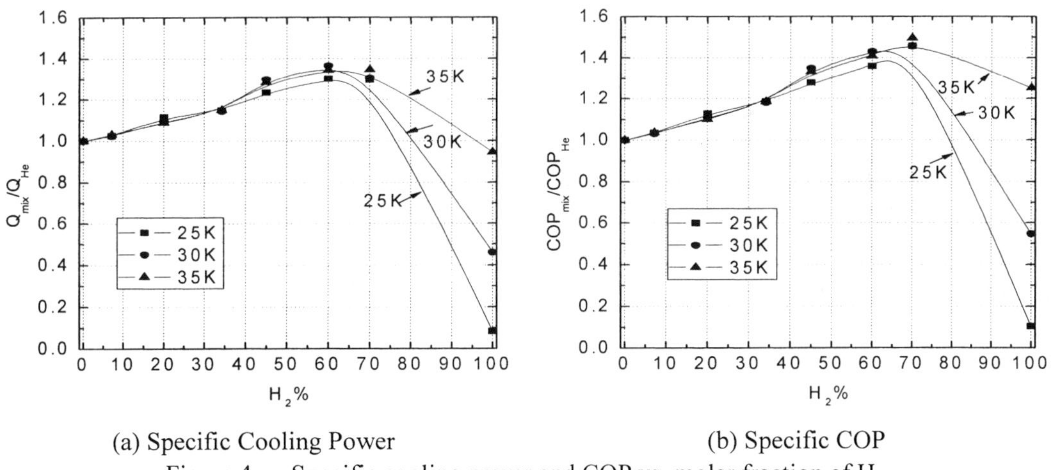

(a) Specific Cooling Power   (b) Specific COP

Figure 4   Specific cooling power and COP vs. molar fraction of H_2

In Figures 2-4, the experimental results indicate that the refrigeration performance of the PTR in the 30K temperature region is obviously enhanced by using proper He-H2 mixture as working fluid. Excellent thermodynamic performance and the reasonable heat transfer and flow properties of He-H_2 mixture contribute to the performance improvement of PTR. But we find these results are quite beyond the values predicted by the modified Brayton Cycle[4]. So far, it is found that the magnetic regenerative material Er_3Ni is able to absorb (or adsorb) H_2 and produces Er_3NiH_x. The volume specific heat of this metallic hydride is much higher than that of Er_3Ni and Pb in the temperature region of 15~50K[5], that is, Er_3NiH_x has better regenerative performance. This is an important factor of the enhancement of the PTR performance.

CONCLUSIONS

Mixtures as working fluid can improve the performance of pulse tube refrigerator. Experimental research was carried out on a two-stage PTR with He-H$_2$ mixture. The pulse tube refrigeration performance with He-H$_2$ mixtures is better than that with pure helium in 30K cooling temperature region when the hydrogen fraction is less than about 80%. A 30-45% increment of both the cooling power and COP has been obtained with 60-70% H$_2$ in He-H$_2$ mixtures. The great performance improvement of the PTR may be attributed to not only the excellent cycle thermodynamic performance and the reasonable heat transfer and flow properties of He-H$_2$ mixtures, but also the high volume specific heat of Er$_3$NiH$_x$ regenerative materials.

ACKNOWLEDGEMENT

The project is financially supported by the National Natural Science Foundation of China (Grant No. 50106013 and 50376055) and the Special Research Fund for Doctoral Training in Universities by the National ministry of Education of China (Grant No. 20010335010).

REFERENCES

1.   Mikulin, E.I., Tarasov, A.A., et al., Low temperature expansion pulse tubes, Adv. Cryo. Eng. (1984) 29 629-637

2.   Yu, J.P., Chen G.B., Gan Z.H., Investigation on the regenerator performance using gas mixtures, International Conference of Cryogenics and Refrigeration Hangzhou, P.R. China (1998) 409-412

3.   Chen, G.B., Yu, J.P., Gan, Z.H., et al, Experimental investigation on pulse tube refrigerator with binary mixtures. Adv. Cryo. Eng. (2000) 45 183-187

4.   Chen, G.B., Gan, Z.H., Thummes, G., et al., Thermodynamic performance prediction of pulse tube refrigeration with mixture fluids, Cryogenics (2000) 40(4-5) 261-267

5.   Chen, G.B., Tang, K., Huang, Y.H., et al., He-H$_2$ mixture and Er$_3$NiH$_x$ packing for the refrigeration enhancement of pulse tube refrigerator, Chinese Science Bulletin (2004) 49(5) 527-530

Three dimensional numerical computation of the flow field in a pulse tube

Ding W.J., He Y.L., Tao W.Q.

State Key Laboratory of Multiphase Flow in Power Engineering,
School of Energy & Power Engineering, Xi'an Jiaotong University, Xi'an 710049, China

A physical and numerical model of the pulse tube in the pulse tube refrigerator has been set up in this paper. Three-dimensional computation of a compressible and oscillating flow field in pulse tube was numerically investigated using a self-developed code. The distributions of the velocity and pressure waves and the velocity vector in the pulse tube were provided in this paper. The velocity wave near the hot end of the pulse tube lags behind that near cold end, and at the some cross section of middle part of the pulse tube the direction of the axial velocity reverses.

INTRODUCTION

Pulse tube refrigerator (PTR) is an attracting device of small cooling capacity widely used in aerospace engineering and for military purpose because of its inherent advantages such as no moving parts in the cold stage, low manufacturing cost, reduced mechanical vibration, etc [1]. The advantages and developments achieved in recent years enable PTR to have brought more and more interest and many laboratories around the world are presently working on the subject.

The internal working process in the PTR is very complex due to the unsteady, oscillating compressible gas flow, the porous media in the regenerator, etc. In recent years, it has been recognized that in the pulse tube of PTR secondary flows exist [2,3]. Such secondary flows are resulted from the combined effect of forced convection of the pulsating streaming and the natural convection caused by the large temperature difference between the hot and cold ends. The secondary flow and the mass streaming in a pulse tube can be a major heat loss mechanism. It carries heat from the hot heat exchanger(i.e. hot end) to the cold heat exchanger(cold end), thereby reducing the cooling power of a PTR. In reference [4], the characteristics of DC-Flow phenomenon in PTR are analyzed theoretically. The analytical results demonstrated that the gas velocity and density fluctuations in phase and amplitudes owing to the pressure fluctuations are the root to produce the DC-Flow in the PTR. Obviously, one-dimensional numerical simulations for PTR can not describe the complex secondary flows and DC-Flow phenomena in the PTR. In order to have a better understanding of the working processes in the PTR, improve their refrigeration performance and eliminate their drawbacks, we must set up a multidimensional numerical model for the PTR.

In recent years the NASA Ames research center declared that they have set up the two-dimensional physical and numerical model of predigestion PTR and whose numerical results agree well with the experimental data. Except America many other countries have started to research the multidimensional physical and numerical model of PTR, such as Japan, South Korea, Germany, etc. A multidimensional model of PTR are being investigated in our research center. Three-dimensional

290

numerical model will be used in the pulse tube, and the heat exchanger will be simulated by a two-dimensional model, while the connection tube will be calculated using one-dimensional mode. In a simplified way, the method of our research is called 3-2-1 model of the PTR.

As he first step for 3-2-1 model, a three-dimensional physical and numerical model of the pulse tube in the PTR has been set up in this paper. The compressible and oscillating flow field in pulse tube was numerically investigated using a self-developed code. The distributions of the velocity and pressure waves in the pulse tube were revealed, and the velocity vector in the longitudinal section of the pulse tube was provided in this paper.

NUMERICAL METHODS AND GOVERNING EQUATIONS

The program for solving viscous flows problem at all Mach numbers was developed based on the SIMPLEC algorithm, using the compressible form [5]. On the basis of the compressible program the influence of the oscillating was considered in this paper. Only the pulse tube was taken as a model for the calculation. The velocity and pressure waves coming from the compressor was directly input to the cold end of the pulse tube as the calculation boundary condition. The calculation model was presented in Figure 1.

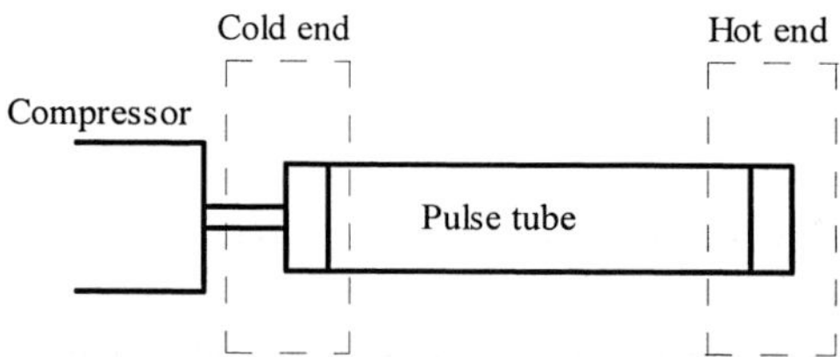

Figure 1   The calculation model

The three dimensional governing equations for compressible and oscillating fluid flow in the pulse tube take following form

$$\frac{\partial}{\partial t}(\rho\phi)+\frac{\partial}{\partial z}(\rho u\phi)+\frac{1}{r}\frac{\partial}{\partial r}(r\rho v\phi)+\frac{1}{r}\frac{\partial}{\partial\varphi}(\rho w\phi)=$$
$$\frac{\partial}{\partial z}(\Gamma\frac{\partial\phi}{\partial z})+\frac{1}{r}\frac{\partial}{\partial r}(\Gamma r\frac{\partial\phi}{\partial r})+\frac{1}{r}\frac{\partial}{\partial\varphi}(\frac{\Gamma}{r}\frac{\partial\phi}{\partial\varphi})+S \tag{1}$$

where ϕ is the general variable, representing u, v and w, Γ is the general diffusion coefficient, and S is

the general source term. The calculation frequency is 13.3Hz, the charge pressure is 12bar, the working fluid is helium, the inner diameter of the pulse tube is 27.8mm and its length/diameter ratio is 9.

RESULTS AND DISCUSSION

<u>The distributions of the velocity and the pressure waves in the pulse tube</u>
The distribution of the velocity waves is shown in Figure 2, and Figure 3 shows pressure distribution in the pulse tube. From Figure 2 we can see that the velocity wave near the hot end of the pulse tube lags

behind that near cold end, and the value of the velocity is gradually become less and less from the cold end to the hot end of the pulse tube. The lag phenomena of the velocity wave was caused by the compressible character of the working fluid.

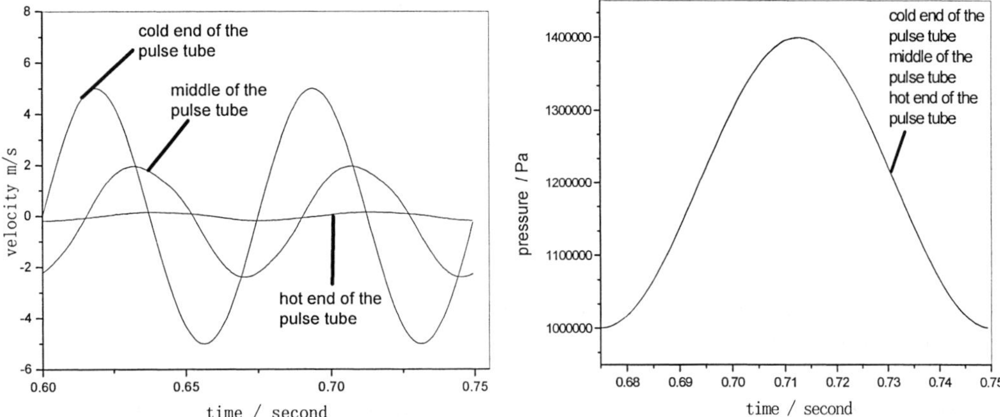

Figure 2 The distributions of velocity waves   Figure 3 The distributions of pressure waves

Velocity vector in longitudinal section

A general view of the velocity distribution in the longitudinal cross section across the axis of the pulse tube is shown in Figure 4. For the purpose of clarity, the picture is not drawn in scale. To see the flow direction more clearly, the local velocity vectors near the hot and cold ends are magnified and presented in Figure 5. In both Figures 4 and 5, the crank angle is 90 degree.

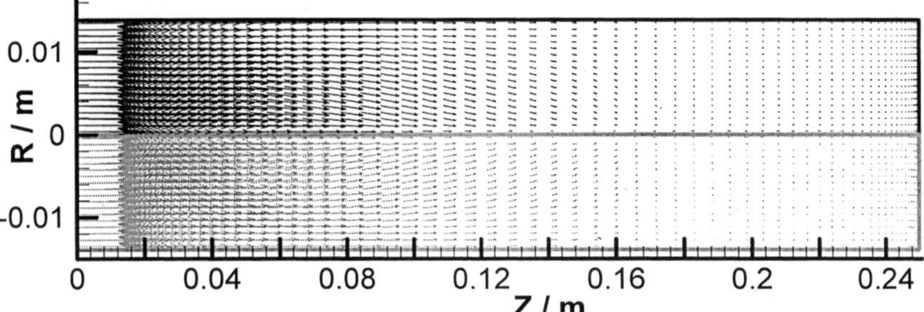

Figure 4 Velocity vector in longitudinal section for the crank angle of 90 degree

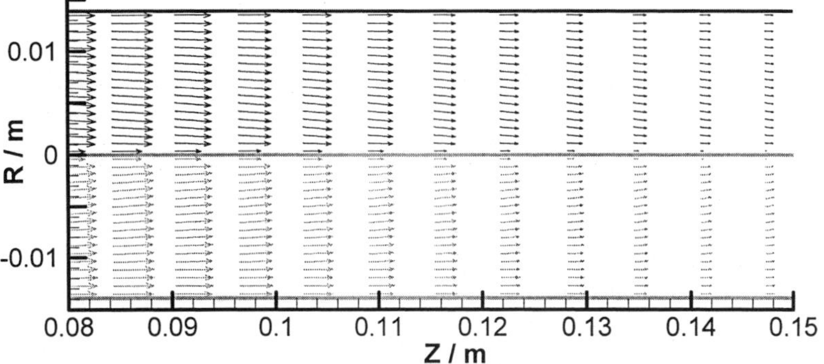

(a) Local velocity field near cold end

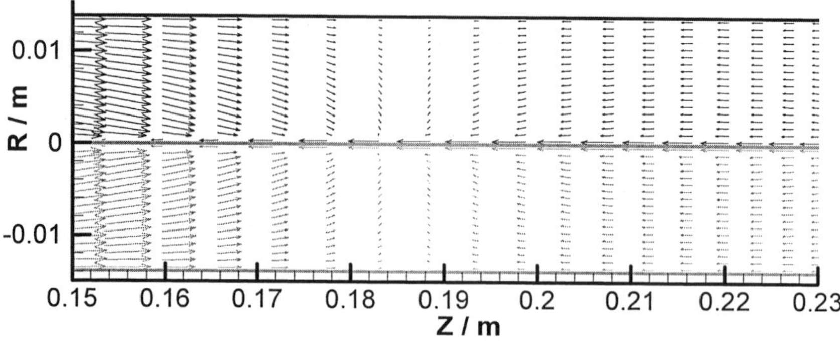

(b) Local velocity field near hot end

Figure 5 Details of flow pattern in longitudinal section for the crank angle of 90 degree

It can be clearly seen from Figures 4、5 that the direction of the axial velocity reverses at the some cross section of the middle part of the pulse tube. The changing of the axial velocity direction starts from the center part of the pulse tube, that is the direction changing of the axial velocity of the center part of the pulse tube makes the whole cross section to reverse the axial velocity direction. So there is some radial velocity in the cross section where the axial velocity reverses. From Figure5 (a)、(b) following feature may be noted: The axial velocity of the center part of the tube becomes smaller and smaller from the cold end to the hot end of the pulse tube, at the some cross section the axial velocity of the center part of the tube disappears, then there is reverse axial velocity at the center part of the tube, and the axial velocity of the other part of the tube is gradually reverses caused by the working fluid in the center part of the tube.

CONCLUSIONS

The velocity wave near the hot end of the pulse tube lags behind that near cold end. At the some cross section of the middle part of the pulse tube the direction of the axial velocity reverses, the changing of the axial velocity direction starts from the center part of the pulse tube.

ACKNOWLEDGMENT

The work reported here was supported by the National Key Project of R & D in China(G2000026303) and the National Natural Science Foundation of China (Grant No. 50276046).

REFERENCES

1.   He, Y.L., Theoretical and experimental investigations on the performance improvements of split-Stirling refrigerator and pulse tube cryocooler, Ph D thesis, School of Energy & Power Engineering, Xi'an Jaiotong University, Xi'an, China(2002)

2.   Jeong, E.S., Secondary flow in basic pulse tube refrigerators, Cryogenics(1996) 36 317-323

3.   Lee, J.M., Kittel, P., Timmergaus, K.D., Radebaugh, R., Flow patterns intrinsic to the pulse tube refrigerator, In:Proceedings of the Seventh International Crycooler Conference, Kirtland AFB, NM 87117-5776, Phllips Laboratory(1993) 125-139

4.   Ju, Y.L., Dynamic Experimental Study and Numerical Simulation of the Oscillations Flow in the Pulse tube Refrigerator, Ph D thesis, Institute of Mechanics Chinese Academy of Sciences(1998)

5.   Karki, K.C., Patankar, S. V., Pressure based calculation procedure for viscous flowsat all speeds in arbitrary configurations, AIAA J(1989) 27 1167-1174

Proceedings of the Twentieth International Cryogenic Engineering Conference
(ICEC 20), Beijing, China. © 2005 Elsevier Ltd. All rights reserved.

Comparison of the performance of regenerators to counterflow heat exchangers

Will, M.E., de Waele, A.T.A.M.

Department of Applied Physics, Eindhoven University of Technology, P.O. Box 513, NL-5600 MB
Eindhoven, The Netherlands

Irreversible processes in regenerators and heat exchangers limit the performance of cryocoolers. In our research we study the possibility to avoid regenerators in pulse-tube refrigerators (PTR's) by using two identical PTR's operating in opposite phase. The two regenerators are replaced by one counterflow heat exchanger. In this contribution we treat the performances of regenerators and heat exchangers from a fundamental point of view. The losses in the two systems are calculated from the entropy production due to the various irreversible processes. The expressions are brought in special forms which make comparison relatively easy.

INTRODUCTION

Irreversible processes in regenerators and heat exchangers limit the performance of cryocoolers. In our research we study the possibility to avoid regenerators in pulse-tube refrigerators (PTR's) by using two identical PTR's operating in opposite phase. The two regenerators are replaced by one counterflow heat exchanger [1]. The performances of heat exchangers and regenerators can be compared by calculating the entropy-production rates by the four different irreversible processes: axial thermal conduction in the gas, axial thermal conduction in the material, flow resistance and heat exchange between the gas and the material.

The regenerator is supposed to be filled with spherical particles with diameter d_h. The free flow area of the regenerator is $A_\mathrm{g} = (1 - f) A$ where f is the filling factor and A is the area of the cross section. The heat exchanger is supposed to consist of N parallel tubes with diameter d_1 for the high- and N tubes for the low-pressure side. The total cross section for the gas flow is $A_\mathrm{g} = 2N\pi d_1^2/4$. The wall thickness of one tube at pressure p is $\delta_\mathrm{w} = d_1 p/p_\mathrm{c}$ with p_c the breaking stress of the material. As $\delta_\mathrm{w} \ll d_1$ we will disregard the difference between the total area A and the area of the gas flow A_g. So in the case of the heat exchanger $A \approx A_\mathrm{g}$. It will turn out that, in the optimum situation, the tube diameter and the corresponding Reynolds numbers, will be very small so we only consider laminar flow. We will further assume ideal-gas conditions. In order to avoid unnecessary complications we will treat the various contributions to lowest order in the temperature differences. The basic expressions for the entropy-production rates are derived from a paper by De Waele et al. [2] and the thesis from Steijaert [3].

LOSSES

Heat conduction

The entropy-production rate per unit length due to heat conduction in the gas in the axial direction can be written as

$$\frac{\mathrm{d}\dot{S}_{\mathrm{cg}}}{\mathrm{d}l} = A_\mathrm{g}\frac{N_\mathrm{u}\kappa_\mathrm{g}}{T^2}\left(\frac{\mathrm{d}T}{\mathrm{d}l}\right)^2.\tag{1}$$

Here κ_g is the coefficient of thermal conductivity of the gas, T is the temperature, and l the length co-ordinate. The parameter N_u represents the Nusselt number which is different for the regenerator and the heat exchanger.

The entropy production due to axial heat conduction through the material can be given as

$$\frac{\mathrm{d}\dot{S}_{cm}}{\mathrm{d}l} = A_s \frac{\kappa_s}{T^2} \left(\frac{\mathrm{d}T}{\mathrm{d}l}\right)^2 \tag{2}$$

with κ_s the thermal conductivity of the solid material. For the regenerator $A_s = fA$ and $\kappa_s = C_k \kappa_m$ with C_k a factor taking into account the bad thermal contact between the grains and κ_m the thermal conductivity of the material. For the heat exchanger $A_s = 2N\pi\delta_w d_1 = 4Ap/p_c$ and $\kappa_s = \kappa_m$.

Flow resistance
The entropy production due to flow resistance can be written as

$$\frac{\mathrm{d}\dot{S}_f}{\mathrm{d}l} = \eta \frac{z}{AT} \overset{*}{n}^2 V_m^2 \tag{3}$$

with η the viscosity of the gas, V_m the molar volume, $\overset{*}{n}^2$ the mean square molar flow, and z a geometrical factor. For the regenerator $z = 1600/d_h^2$ and for the heat exchanger $z = 128/d_1^2$.

Heat exchange
The entropy production due to the heat exchange between gas and material can be written as

$$\frac{\mathrm{d}\dot{S}_e}{\mathrm{d}l} = g_e \frac{C_p^2 \overset{*}{n}^2}{A\kappa_g N_u} \frac{1}{T^2} \left(\frac{\mathrm{d}T}{\mathrm{d}l}\right)^2 \tag{4}$$

with C_p the molar heat capacity of the gas and g_e a geometrical factor which is $g_e = d_h^2/12f$ for the regenerator and $g_e = d_1^2$ for the heat exchanger.

OPTIMIZATION

The optimal regenerator and heat exchanger can be found by minimizing the entropy productions. If we are looking for the optimum grain size d_h we can write the total entropy production for the regenerator per unit volume by adding up the four contributions and dividing by A. The result is

$$\sigma_r = (1-f) N_{ur} \frac{\kappa_g}{T^2} \left(\frac{\mathrm{d}T}{\mathrm{d}l}\right)^2 + 0.16 f \frac{\kappa_m}{T^2} \left(\frac{\mathrm{d}T}{\mathrm{d}l}\right)^2 + \eta \frac{1600}{Td_h^2} j^2 V_m^2 + \frac{d_h^2}{12 N_{ur} f} \frac{C_p^2 j^2}{\kappa_g} \frac{1}{T^2} \left(\frac{\mathrm{d}T}{\mathrm{d}l}\right)^2 \tag{5}$$

with $j = \overset{*}{n}/A$ the molar flux. For the heat exchanger

$$\sigma_e = N_{ue} \frac{\kappa_g}{T^2} \left(\frac{\mathrm{d}T}{\mathrm{d}l}\right)^2 + \frac{4p}{p_c} \frac{\kappa_m}{T^2} \left(\frac{\mathrm{d}T}{\mathrm{d}l}\right)^2 + \eta \frac{128}{Td_1^2} j^2 V_m^2 + \frac{d_1^2}{N_{ue}} \frac{C_p^2 j^2}{\kappa_g} \frac{1}{T^2} \left(\frac{\mathrm{d}T}{\mathrm{d}l}\right)^2 . \tag{6}$$

The last two terms in Eqs.(5) and (6) depend on the grain size d_h and the tube diameter d_1 respectively. The coefficient of thermal conductivity and the viscosity of the gas are approximated by the relations $\kappa_g = \kappa_0 \sqrt{T}$ and $\eta = \eta_0 \sqrt{T}$ with κ_0 and η_0 contants. The optimum values for the diameters are given by

$$d_{h0}^4 = 3072 \frac{f N_{ur} \eta_0 \kappa_0 T^4}{p^2 \left(\frac{\mathrm{d}T}{\mathrm{d}l}\right)^2} \tag{7}$$

$$d_{10}^4 = \frac{512}{25} \frac{N_{ue} \eta_0 \kappa_0 T^4}{p^2 \left(\frac{\mathrm{d}T}{\mathrm{d}l}\right)^2} \tag{8}$$

respectively. They result in minimum contributions of the flow and the heat exchange to the entropy-production rates of

$$\sigma^2_{\text{rfe0}} = \frac{10000}{3}\frac{1}{N_{\text{ur}}f}\frac{\eta_0}{\kappa_0}\frac{R^4 j^4}{Tp^2}\left(\frac{\mathrm{d}T}{\mathrm{d}l}\right)^2 \tag{9}$$

$$\sigma^2_{\text{efe0}} = 3200\frac{1}{N_{\text{ue}}}\frac{\eta_0}{\kappa_0}\frac{R^4 j^4}{Tp^2}\left(\frac{\mathrm{d}T}{\mathrm{d}l}\right)^2. \tag{10}$$

Note that these terms vary as T^{-1} while the heat conduction terms vary as T^{-2}, so the relative importance of the heat conduction terms increases with lower temperatures. The ratio of the optimum diameters is given by

$$\frac{d^4_{\text{h0}}}{d^4_{\text{10}}} = 150\frac{fN_{\text{ur}}}{N_{\text{ue}}} \tag{11}$$

and the ratio of the minimum dissipation rates is

$$\frac{\sigma^2_{\text{rfe0}}}{\sigma^2_{\text{efe0}}} = \frac{25}{24}\frac{N_{\text{ue}}}{N_{\text{ur}}f} \tag{12}$$

so, within the limitations of the validity of our calculations, the ratios are only determined by the filling factor and the ratio between the Nusselt numbers and independent of the many possible other parameters.

NUMERICAL VALUES

The gas properties are detremined by the values for helium $\kappa_0 = 0.008$ W/K$^{3/2}$m, $\eta_0 = 1.05$ μPas/K$^{1/2}$, and $C_{\text{p}} = 20.8$ J/molK. For stainless steel $\kappa_{\text{m}} = 15$ W/Km and $p_{\text{c}} = 250$ MPa. The Nusselt number is for the regenerator $N_{\text{ur}} = 10$ and for the heat exchanger $N_{\text{ue}} = 4.36$. In our calculations we use somewhat arbitrarily $p = 1.75$ MPa, $j = 100$ mol/sm^2, $f = 0.5$, $C_{\text{k}} = 0.16$, $T = 200$ K, $\mathrm{d}T/\mathrm{d}l = 200$ K/m. With these values we can calculate the various contributions to the entropy production. For the regenerator

$$\sigma_{\text{r}} = \left[0.56 + 1.2 + 1.07\frac{\text{mm}^2}{d_{\text{h}}^2} + 0.64\frac{d_{\text{h}}^2}{\text{mm}^2}\right]\frac{\text{W}}{\text{Km}^3}. \tag{13}$$

The diameter dependence of the two last terms is similar to the one given in Fig.1. In the optimum situation $d_{\text{h0}} = 1.14$ mm and $\sigma_{\text{rf0}} = \sigma_{\text{re0}} = 0.83$ W/Km^3. The total entropy-production rate density in the optimum situation is $\sigma_{\text{r0}} = 3.43$ W/Km^3.

For the heat exchanger we get the relation

$$\sigma_{\text{e}} = \left[0.49 + 0.42 + 0.086\frac{\text{mm}^2}{d_1^2} + 8.7\frac{d_1^2}{\text{mm}^2}\right]\frac{\text{W}}{\text{Km}^3}. \tag{14}$$

The most interesting are the diameter dependent terms. The entropy-production rates of the two last terms together with the total is given in Fig.1 as a function of the tube diameter. In the optimum situation the tube diameter $d_{10} = 0.314$ mm and $\sigma_{\text{ef0}} = \sigma_{\text{ee0}} = 0.87$ W/Km^3. The total is $\sigma_{\text{e0}} = 2.65$ W/km^3.

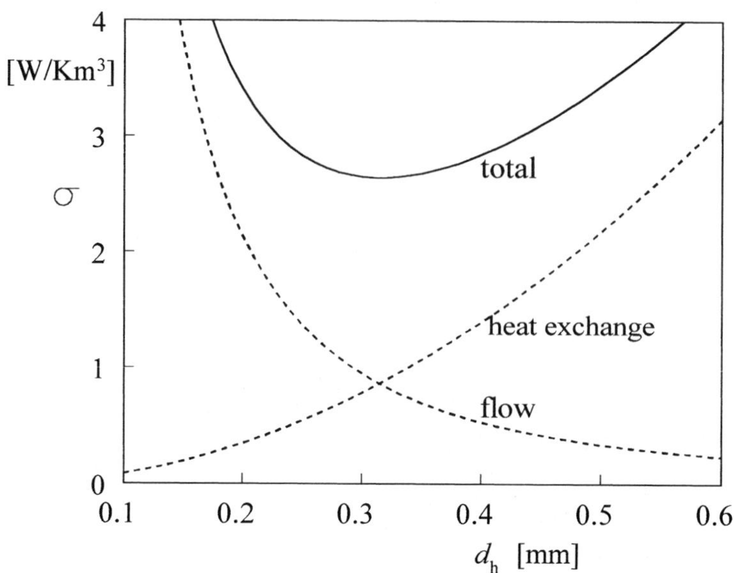

Figure 1: Entropy-production rate densities for a heat exchanger due to the flow resistance and the heat exchange between the gas and the matrix. The total shows a minimum which corresponds to the optimum performance of the regenerator.

DISCUSSION AND CONCLUSIONS

The entropy production in an optimized heat exchanger is less than in an optimized regenerator, so a PTR with a heat exchanger should work better. At temperatures much lower than 200 K the contributions from the flow resistance and the heat exchange become significantly less than the contributions due to the heat conduction in the axial directions. Therefore, one has some freedom to choose diameters which are more convenient than the small values found in the optimization procedure.

ACKNOWLEDGMENT

This project is supported by the Dutch Technology Foundation (STW).

REFERENCES

1.   Will, M.E., Zeegers, J.C.H., De Waele, A.T.A.M., "Counter-Flow Pulse-Tube Refrigerators", Proceedings of ICEC-19, Narosa Publishing House, New Delhi (2002) 407.
2.   De Waele, A.T.A.M., Steijaert, P.P., Gijzen, J., "Thermodynamical Aspects of Pulse Tubes", Cryogenics, (1997), 37 313.
3.   Steijaert, P.P., Thermodynamical Aspects of Pulse-Tube Refrigerators, PhD Thesis, Eindhoven University of Technology, (1999).

Temperature Hysteresis in a GM-Type Orifice Pulse Tube Refrigerator[1]

Gan Z.H. [*,**], Thummes G. [**]

* Cryogenics Lab. Zhejiang University, Hangzhou, 310027, P.R.China
**Institute of Applied Physics, University of Giessen, D-35392, Germany

Compared to the instability in some double-inlet pulse tube refrigerators due to the DC-flow in the cooling system, the single orifice pulse tube refrigerator (OPTR) is usually considered to operate stably. However, in recent experiments we found a new phenomenon in form of a temperature hysteresis in a single stage GM-type OPTR that depends on the heat load and on the adjustment of the needle valve connecting the pulse tube warm end with the buffer volume. The lowest temperature of this cooler, when operated on a 6 kW compressor, was 18.8 K with double inlet and 30.4 K with single orifice.

INTRODUCTION

The double-inlet mode introduced by Zhu et al. in 1990 [1] is one of the most efficient configurations in pulse tube refrigerators. Nowadays the no-load temperature can be as low as 13 K in a high power single-stage pulse tube refrigerator [2]. However, The double-inlet bypass opens up the possibility of DC-flow in the cold head, which can give rise to temperature instabilities in some pulse tube refrigerators [3, 4]. Although some effective DC flow control methods have been introduced [5, 6], the temperature instabilities are still existing, especially at high heat load.

Normally the single orifice pulse tube refrigerator (OPTR) is considered to operate stably since the configuration prevents the occurrence DC-flow. But in recent experiments we found new temperature hysteresis effects in a single stage GM-type OPTR, which are described in this paper.

EXPERIMENT SET-UP

Figure 1 shows the schematic of the GM-type single-stage pulse tube refrigerator system, which consists of a helium compressor (Leybold, model RW6000, nominal input power: 6 kW), a rotary valve and a single stage pulse tube cold head. The design of the cooler is similar to that in Ref. [2]. The regenerator and pulse tube are made of stainless steel tubes each with a length of about 200 mm, and with outer diameters of 48 mm and 41 mm, respectively. The regenerator matrix consists of stainless steel screens in the warm part and of lead spheres in the cold part. The pulse tube refrigerator is equipped with a 2.5 litre reservoir and three needle valves (Swagelok, SS-ORS3MM), one needle valve for the orifice (ORV), and two anti-parallel needle valves (DIV1, DIV2) that allow adjustment of the flow symmetry of the second inlet.

Temperatures are measured by means of Pt100 resistance thermometers (T1, T2, T3, and T5 in Figure 1) and a Cernox thermometer (T4). Piezoelectric pressure sensors (Siemens, KPY46R) are used for monitoring the dynamic pressures at the hot end of regenerator (P1) as well as in the reservoir (P0).

The cooler operates with a charging pressure of 17.5 bar (absolute pressure) and a frequency of 1.4 Hz.

---

[1] The work was carried out at the Institute of Applied Physics, University of Giessen.

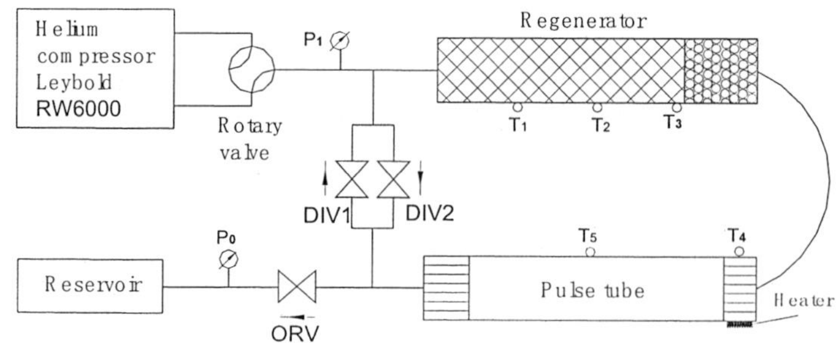

Figure 1 Schematic of the GM-type single stage pulse tube refrigerator

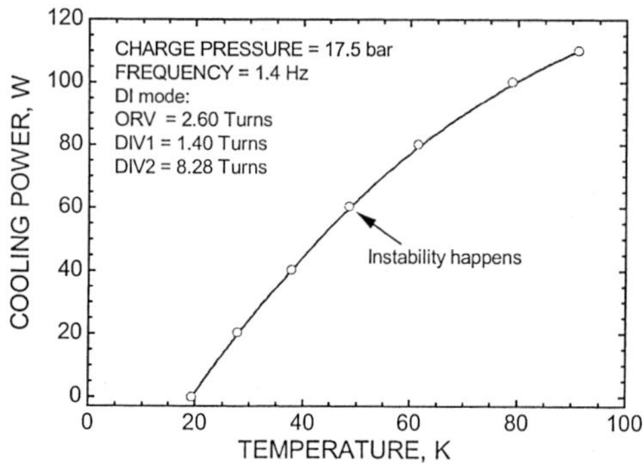

Figure 2   Cooling power vs. temperature in double-inlet mode.

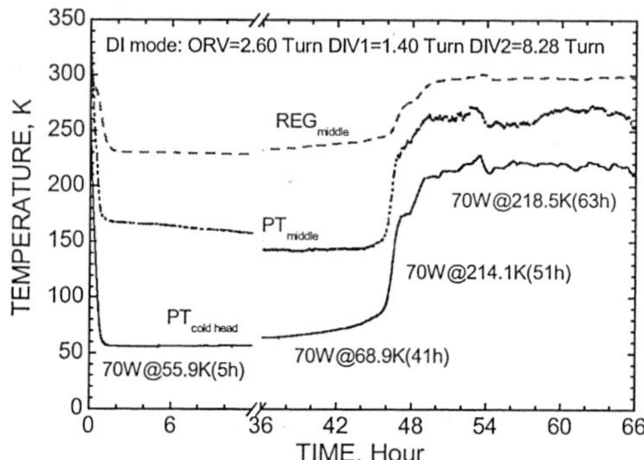

Figure 3   Instability at 70 W heat load

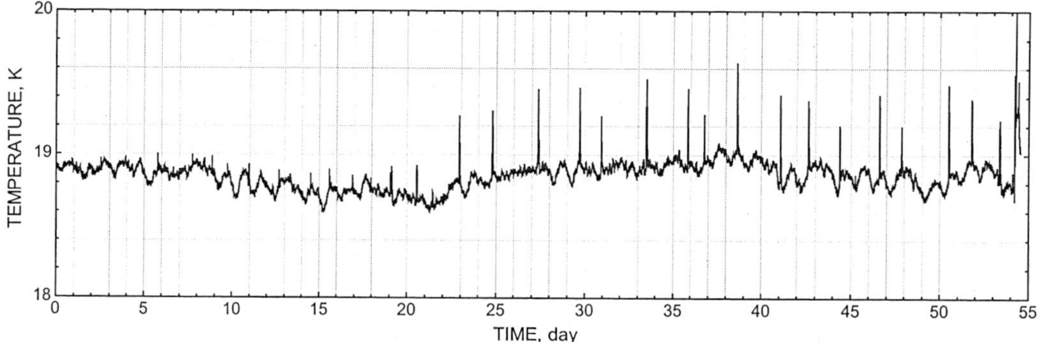

Figure 4   Temperature stability without heat load

EXPERIMENTAL RESULTS AND DISCUSSION

Temperature instability in double-inlet mode

Figure 2 shows the cooling power of the cooler with double-valved configuration. The minimum temperature obtained is 19.3 K and a cooling power of 100 W is available at 79K.

Temperature instabilities where observed after several hours of operation, when the heat load was above 60 W. As an example, Figure 3 shows the temperature instability when the cooling power is 70 W. It is seen from Figure 3 that the temperature increases slowly from initially 56 K to 69 K after 41 hours of operation, then rises sharply to about 220 K, and then fluctuates near 220 K.

When the cooling power is below 40 W (38 K), the temperature is stable for a long time running. Figure 4 shows the temperature stability for nearly 55 days without heat load as an example. The temperature oscillation is less than 0.5 K around the average temperature of 18.8 K.

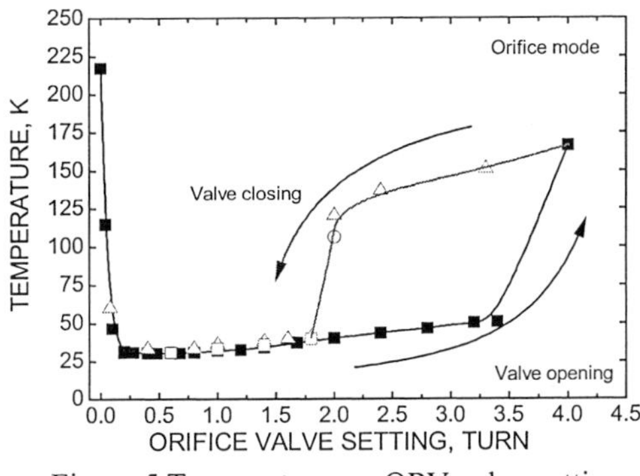

Figure 5 Temperature vs. ORV valve settings

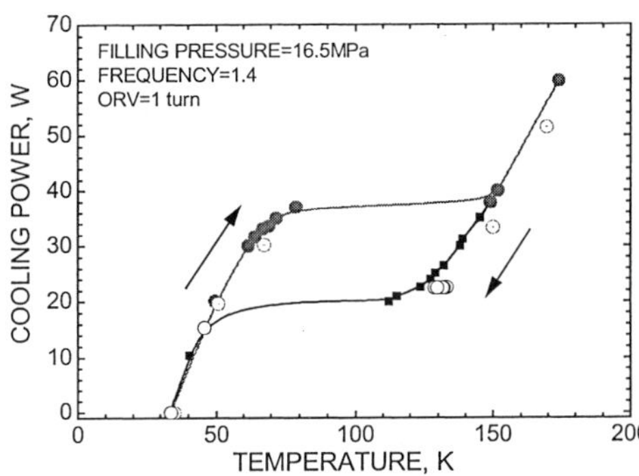

Figure 6 Cooling power vs. temperature
in orifice mode

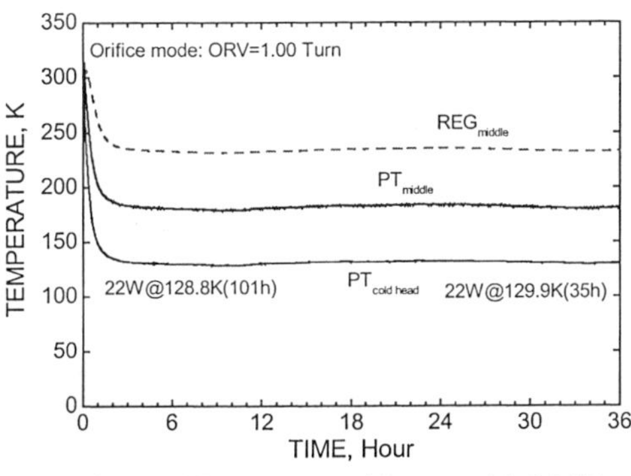

Figure 7 Temperature history with 22 W
heat load at the beginning

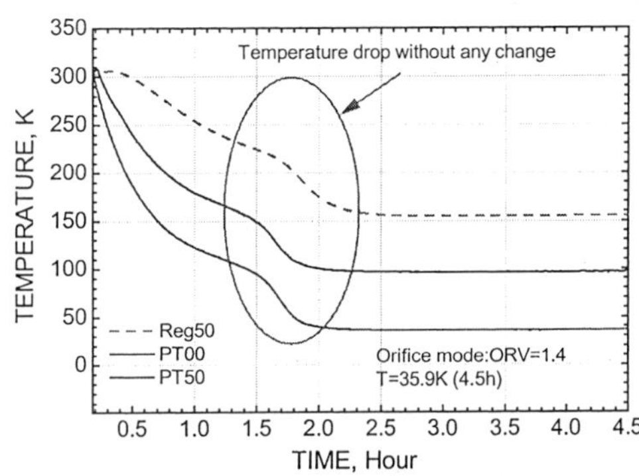

Figure 8 Cooling down process in orifice mode

This indicates that the double-valve configuration between the hot end of pulse tube and regenerator is an effective method to control the DC-flow and to improve the cooler performance, in particular below 40 W heat load but cannot prevent the occurrence of instabilities at heat loads above 60W in the tested cooler system.

A new phenomenon: temperature hysteresis in orifice mode

When the two valves DIV1 and DIV2 in Figure 1 are closed, the cooler system operates in single orifice mode. A new phenomenon of temperature hysteresis, which depends on the orifice needle valve adjustment and on the heat load, and up to now has not been reported for other OPTRs, is observed in the tested cooler in single orifice mode of operation.

Figure 5 shows the cold tip temperature as function of the orifice valve setting. Starting with the cooler in basic mode that yields about the 217 K, the temperature of the cold tip of the pulse tube drops to a stable temperature of 30.4 K by opening the orifice valve by one turn. Upon further opening of the valve the temperature increase gradually to 51 K, corresponding to about 3.5 turns. Then, at 4 turns the temperature rises sharply to 167 K, which is also accompanied by a hysteresis effect, i.e. in order to achieve the initial low temperatures the needle valve had to be closed again to about 1.8 turns.

The hysteresis is also seen upon cool down of the cooler when the orifice valve is set to more than 2 turns. The cooler then only cools to a higher stationary temperature level (above 120 K), and further valve closing below 1.8 turns is necessary to reach the initial low temperatures. That is, the temperature hysteresis occurs when the orifice valve setting is between 1.8 and 4 turns.

The expected variation of cooling power with cold tip temperature is like that in Figure 2, independent of the heating process. However, for the present cooler a hysteresis is also found in the cooling power as function of temperature as shown in Figure 6. When the heat load is gradually applied from zero to 37 W, the cold tip temperature rises from 33.6 K to 79 K with the orifice valve set to 1 turn. With a slightly higher heat load the temperature increases rapidly to another stable state above 149 K. When the heating process in the opposite order, then the heat load has to be decreased to below about 20 W to switch back to the initial load line. This temperature hysteresis occurs when the cooling power is between 20 and 38 W.

As shown in Figure 7, when cooling down the system with a heat load of 22 W the temperature can only reach a stationary value of 129 K, which corresponds to the lower branch of the hysteresis curve in Figure 6. This means that in this state at high stationary temperature the cooler does not provide enough cooling power so that the heat load prevents the transition to the state with low stationary temperature.

Temperature drops are also seen in the cool down curves in orifice mode in the indicated area in Figure 8. In this case the temperature drop is probably related to the heat capacity of the regenerator matrix which consists of stainless steel screens and lead spheres. The heat capacity of lead is larger than that of stainless steel when the temperature is below about 70 K.

CONCLUSION

A temperature of 19 K and 100 W of cooling power at 79 K are obtained in this single-stage pulse tube refrigerator in double-inlet mode of operation that employs two anti-parallel valves for DC-flow control. Temperature instabilities occur in the cooler when the cooling power is above 60 W, while stable operation for more than 50 days is observed when the cooling power is below 40 W.

A new temperature hysteresis effect is observed in orifice mode. The hysteresis depends on the heat load and on the adjustment of the orifice valve setting. This phenomenon might be useful also to understand the instability in the double-inlet mode of pulse tube refrigerators.

ACKNOWLEDGEMENT

Gan Z. H. gratefully acknowledges a re-invitation scholarship of the German Academic Exchange Service (DAAD) and support by the National Natural Science Foundation of China (50106013). Thanks are also due to Dr. Yang L.W. (Giessen) for helpful discussion.

REFERENCES

1. Zhu, S.W., Wu, P.Y., and Chen, Z.Q., Double inlet pulse tube refrigerator: an important improvement, Cryogenics(1990) 30 514
2. Haefner, H.U., Giebeler, F., and Thummes G. Einstufiger 25 K Pulsrohrkuehler fuer HTS-Energie-Applikationen, DKV-Tagungsbericht 2003, vol. 1, Deutscher Kaelte- und Klimatechnischer Verein, Stuttgart (2003), pp. 173-183 (ISBN: 3-932 715-35-7), in German
3. Nobuaki Seki, Shuichi Yamasaki, Junpei Yuyama, Masahiko Kasuya, Kenji Arasawa, Shinji Furuya, and Hidetoshi Morimoto. Temperature stability of pulse tube refrigerators. Proc. of ICEC 16/ICMC. (1996) 267-270
4. Toyoichiro Shigi, Yoshiaki Fujii, Masahiro Yamamoto, Masaki Nakamura, Minoru Yamaguchi, Yoshiko Fujii, Tomio Nishitani, Tetsuya Araki, Etsuji Kawaguchi, and Masayoshi Yanai. Anomaly of one-stage double-inlet pulse tube refrigerator. Proc. of ICEC 16/ICMC. (1996) 263-266
5. Wang, C., Thummes, G., and Heiden, C., Control of DC gas flow in a single-stage double-inlet pulse tube cooler, Cryogenics (1998) 38 843-847
6. Jiang, Y. L., Chen, G.B., and Thummes, G., Experimental investigation on DC flow control in a single-stage pulse tube refrigerator operating below 20 K. Cryogenics and Refrigeration – Proc. of ICCR'2003. Editors: Chen GB, Hebral B, Chen GM (2003) 77-80

Theoretical prediction on the coupling of thermoacoustic prime mover and RC load

Tang K., Chen G.B., Jia Z.Z., Jiang N., Bao R.

Cryogenics Laboratory, Zhejiang University, Hangzhou 310027, China

The coupling of thermoacoustic prime mover and its load is of great importance for the performance of thermoacoustic system. A standing wave thermoacoustic prime mover with RC (resistance and capacitance) load is simulated, and the influence of RC load and dimensions of the resonance tube on the behavior of the thermoacoustic system is discussed according to the computed results.

INTRODUCTION

The ultimate purpose of the investigations on the thermoacoustic prime mover is to use it to drive a load, such as a pulse tube refrigerator or a thermoacoustic refrigerator etc. When the load is connected to the prime mover, the coupling between them is of great importance for the performance of the thermoacoustic system. In order to investigate the coupling relation, a standing wave thermoacoustic prime mover connected with an RC (resistance and capacitance) load is simulated with linear thermoacoustics [1]. The influence of some structure parameters (e.g. resistance and volume of RC load, length and inner diameter of resonance tube) on the performance of the system (e.g. frequency, acoustic power output, pressure amplitude, hot end temperature of stack) is discussed based on the computed results.

NUMERICAL SIMULATION

According to linear thermoacoustics [1], momentum, continuity and energy equations for a short channel dx can be written as:

$$\frac{dp_1}{dx} = -\frac{i\omega\rho_m}{1-f_v}\frac{U_1}{A} \tag{1}$$

$$\frac{dU_1}{dx} = -\frac{i\omega A}{\gamma p_m}\left[1+(\gamma-1)f_\kappa\right]p_1 + \frac{f_\kappa-f_v}{(1-f_v)(1-\mathrm{Pr})}\frac{1}{T_m}\frac{dT_m}{dx}U_1 \tag{2}$$

$$\frac{dT_m}{dx} = \frac{\dot{H}_2 - \frac{1}{2}\mathrm{Re}\left[p_1\tilde{U}_1\left(1-\frac{f_\kappa-\tilde{f}_v}{(1+\mathrm{Pr})(1-\tilde{f}_v)}\right)\right]}{\frac{\rho_m c_p|U_1|^2}{2A\omega(1-\mathrm{Pr}^2)|1-f_v|^2}\mathrm{Im}(f_\kappa+\mathrm{Pr}\,\tilde{f}_v)-(Ak+A_{solid}k_{solid})} \tag{3}$$

where p_1 and U_1 are pressure and velocity amplitudes; ω is angular frequency; ρ_m, T_m, c_p, γ, k and Pr are mean density, temperature, isobaric specific heat, specific heat ratio, thermal conductivity and Prandtl number of working fluid, respectively; f_v and f_κ are viscous and thermal functions; A is flow area of the channel; A_{solid} and k_{solid} are cross section area and thermal conductivity of the solid forming the channel;

$\dot{H}_2$ is total power; i is imaginary unit. Superscript $\sim$ means the conjugation of a complex quantity.

The simulation for sanding wave thermoacoustic prime mover with an RC load, as shown in Figure 1, is carried out with Eqs (1)-(3). The main dimensions of the thermoacoustic prime mover and the RC are tabulated in Table 1. Working gas, heating power and working pressure are considered as He, 2000W and 2.0MPa, respectively, in the computation. Although R (resistance) of RC load is decided by the opening of the valve in practice, the value of R is directly given in the computation for simplifying the simulation. C (capacitance) of RC load can be calculated with $C = V/\rho_m a^2$, where V is the volume of the capacitance, ρ_m and a are mean density and acoustic velocity, respectively.

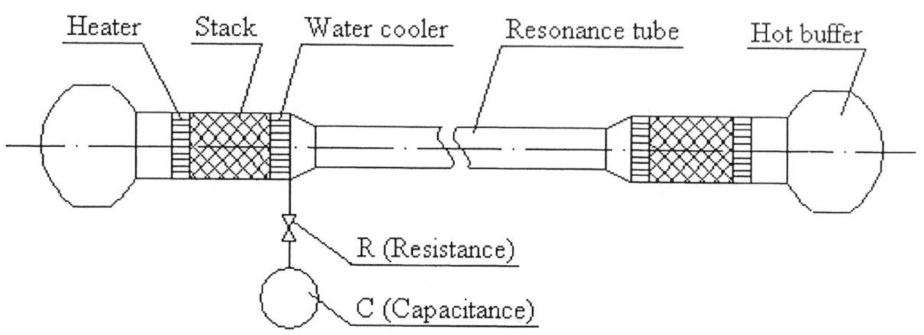

Figure 1   Schematic of thermoacoustic engine with RC load

Table 1   Dimensions of thermoacoustic engine and RC load

	Heater	Stack	Water cooler	Resonance tube	Hot buffer	Volume of C
Diameter (mm)	54	56	56	36-100 optional	/	/
Length (mm)	65	285	64	2000-10000 optional	(1.2L)	(0.25-1L optional)

COMPUTED RESULTS AND ANALYSIS

The influence of RC load on the performance of the thermoacoustic system is discussed with length and inner diameter of the resonance tube given as 4m and 36mm, respectively. Computed frequency with different volumes of C is shown in Figure 2. It can be seen that frequency increases slowly with accretion of R, and larger volume of C results in lower frequency, when R is smaller than 10^6 N.s/m^5. However, when R is in the range of 10^6 -10^8 N.s/m^5, frequency increases rapidly with accretion of R, and the increasing speed of larger volume of C is higher than that of smaller volume of C. When value of R is beyond 10^8 N.s/m^5, the increase of R leads to the decrease of frequency. Computed acoustic power output to RC load is presented in Figure 3. There is a peak along each of the curve with different volumes of C. Larger volume of C leads to a higher peak and a peak-shift in the direction of lower R. When R approaches to zero or infinity, the acoustic power output is almost zero. Figure 4 shows computed pressure amplitude at the inlet of RC load. In contrast with Figure 3, there is a valley along each of the curve with different volumes of C in Figure 4. Larger volume of C leads to a deeper valley and a valley-shift in the direction of lower R. The computed hot end temperature of the stack is presented in Figure 5. We can see that the forms of the curves in Figure 5 are similar to those in Figure 3.

Based on the comparison of all the forementioned figures, it is found that there is always a marked variation in the range of 10^6 -10^8 N.s/m^5 for R along all the curves. Further analysis indicates that the peaks in Figure 3 occur when R is almost equal to $1/\omega C$, that is, the equality of R and $1/\omega C$ results in maximum acoustic power output. These peaks lead to the valleys in Figure 4 and the peaks in Figure 5. It is because considerable part of the acoustic power generated by the stack is consumed by RC load in the

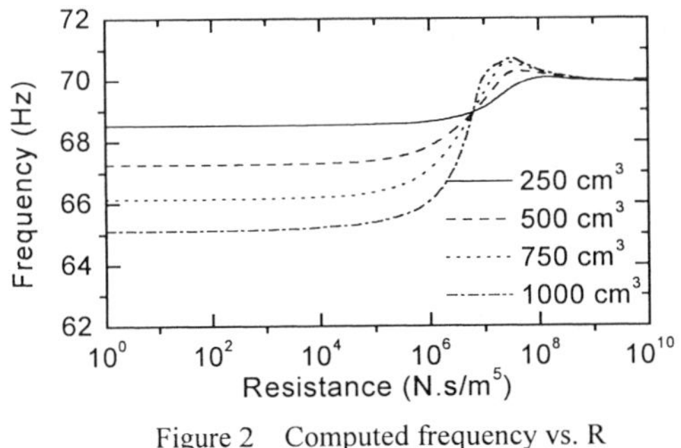

Figure 2　Computed frequency vs. R

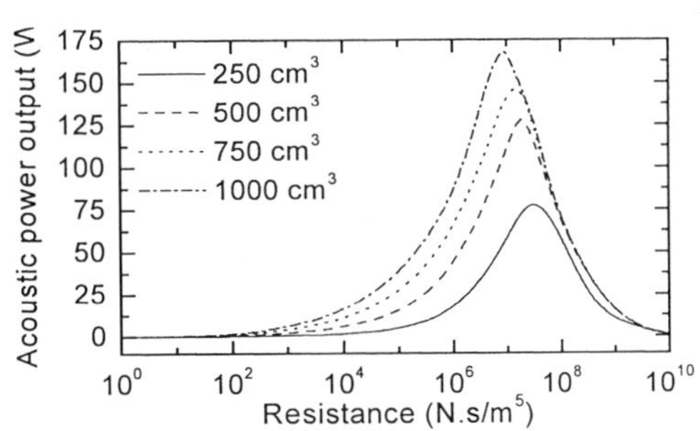

Figure 3　Computed acoustic power output vs. R

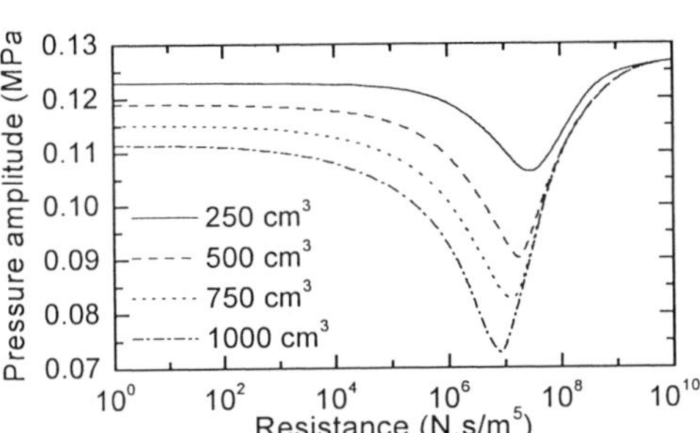

Figure 4　Computed pressure amplitude vs. R

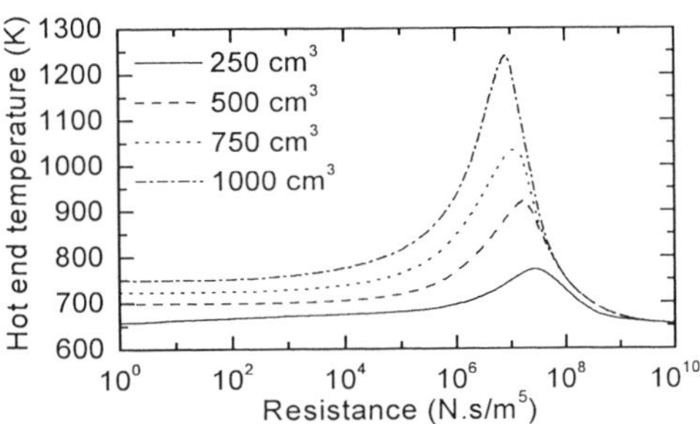

Figure 5　Computed hot end temperature of stack vs. R

peak region of Figure 3. This weakens the oscillating flow in the thermoacoustic system, and then, decreases the heat transfer in the stack, and induces an increase of stack's hot end temperature. Meanwhile, it is also found that the volume of C is a decisive factor for the performance of the thermoacoustic system when $1/\omega C$ is larger than R. In this case, an increase of volume of C may result in decreases of both frequency and pressure amplitude, as shown in Figure 2 and Figure 4, and may lead to increases of acoustic power output and stack's hot end temperature, as presented in Figure 3 and Figure 5. However, when R is larger than $1/\omega C$, the performance of thermoacoustic system is mainly dominated by R. Larger R induces lower frequency, less acoustic power output and lower hot end temperature of the stack, but higher pressure amplitude.

It is worthy of attention that the equalities of R and $1/\omega C$ are not always the best. Although acoustic power output is the maximum in this case, pressure amplitude is the minimum and hot end temperature of the stack is the highest. For a thermoacoustically driven pulse tube refrigerator, not only the input acoustic power but also the pressure amplitude is of great importance for the refrigeration performance. In addition, the hot end temperature of the stack can not be too high to the safety.

R and volume of RC load are fixed as 10^8 N.s/m^5 and 500 cm^3 for the computation of the resonance tube. Figure 6 presents the relation of frequency and L (length of the resonance tube) with different inner diameters. We can see that a longer and thinner resonance tube may realize a lower frequency. Computed pressure amplitude at the inlet of RC load is shown in Figure 7. It can be seen that the pressure amplitude increases with an accretion of resonance tube length initially, then decreases. The curves of computed acoustic power output to RC load, as shown in Figure 8, are similar to those of pressure amplitude in Figure 7. In order to explain the trend of the curves in Figure7 and Figure 8, acoustic power losses of components are computed with 56mm inner diameter resonance tube as an example and presented in Figure 9. Figure 9 indicates that the acoustic power loss of thermoacoustic core (composed of stack, heater and water cooler) is much high than that of any other component and dominates the behavior of thermoacoustic system when the resonance tube is short. In this case, a decrease of the acoustic power loss of thermoacoustic core with an increase of the resonance tube length enhances the oscillation and increases both the pressure amplitude and the acoustic power output. However, the acoustic power loss of resonance tube increases and catches up to that of the thermoacoustic core when the resonance tube is

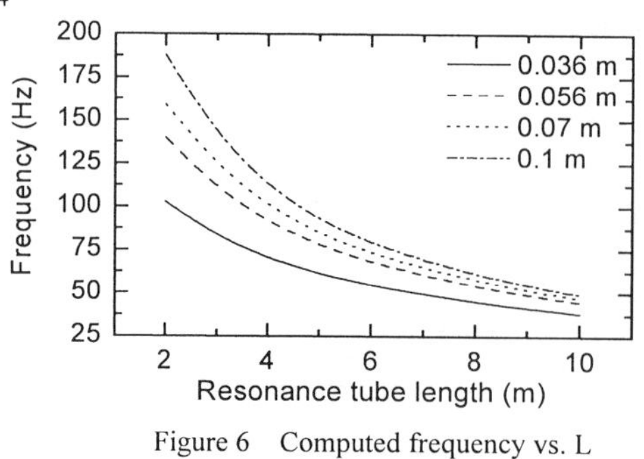

Figure 6   Computed frequency vs. L

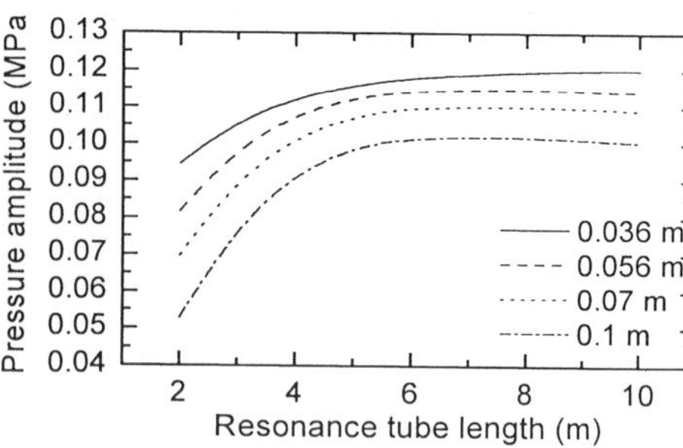

Figure 7   Computed pressure amplitude vs. L

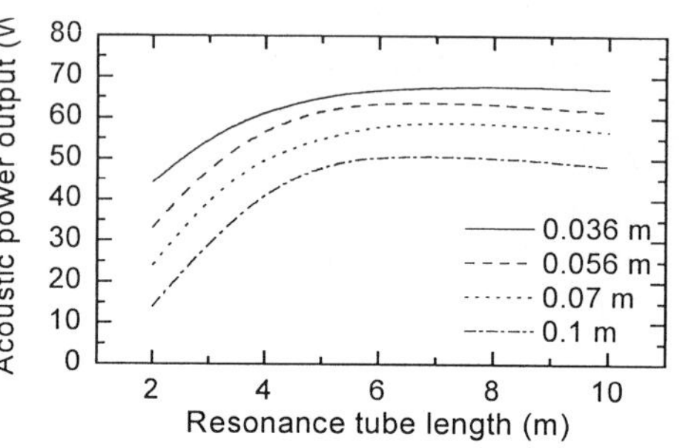

Figure 8   Computed acoustic power output vs. L

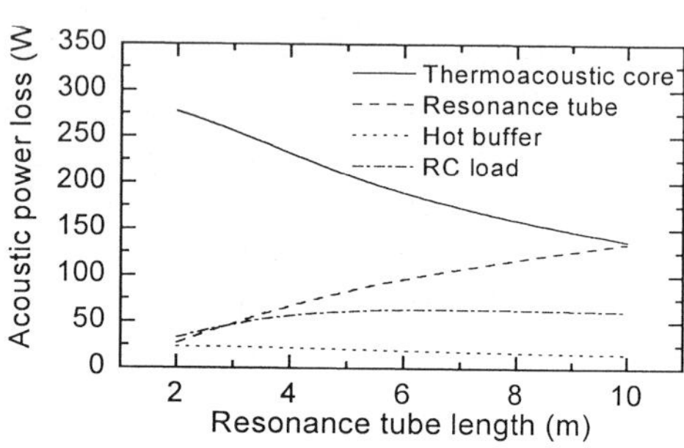

Figure 9   Computed acoustic power loss vs. L

prolonged further. Both the pressure amplitude and the acoustic power output decrease with an increase of resonance tube length when the acoustic power loss of the resonance tube is comparable or even beyond that of the thermoacoustic core.

CONCLUSIONS

The computed results indicate that the behavior of the thermoacoustic system is greatly influenced by the impedance of the load and the dimensions of the resonance tube. C of RC load is a key influencing factor when $1/\omega C$ is larger than R, while R is of great importance, when $1/\omega C$ is less than R. The maximum acoustic power output can be achieved when $1/\omega C$ equals to R, while pressure amplitude is the minimum and hot end temperature of the stack is the highest. Besides, there is an optimal value of resonance tube length for the maximum acoustic power output. A thinner resonance tube may lead to lower frequency, higher pressure amplitude and larger acoustic power output in the computation range.

ACKNOWLEDGMENT

The authors hope to express their appreciation for the National Natural Sciences Foundation of China (50376055) and the University Doctoral Subject Special Foundation of China (20010335010).

REFERENCES

1.   Swift, G.W., Thermoacoustic engine, J.Acoust.Soc.Am. (1988) 84 1145-1180

Lattice gas modeling of thermoacoustic oscillations

Zhang X. Q., Chen Y., Liu X.

Department of Physics, Tsinghua University, Beijing 100084, P. R. China

A two-dimensional 9-bit thermal lattice gas model was applied to simulate self-sustained oscillations in a thermoacoustic resonant tube. The time-evolution of pressure oscillation and the change in temperature with time and space were successfully simulated using the present model. The simulated results from our work are qualitatively in good agreement with that measured or simulated from other researchers in earlier literature. The application of a lattice-gas method to thermoacoustics in the present paper has demonstrated that it would be a useful approach for further thermoacoustic research and applications.

INTRODUCTION

The study of thermoacoustic devices is presently attracting considerable interest. Numerical simulations have played important roles in the advance of modern thermoacoustics and in the developments of thermoacoustic devices for a variety of commercial, military and industrial applications. In the present paper, a totally different numerical method — a lattice gas method (LGM), was applied to thermoacoustic research. Compared to other conventional computational methods, the LGM is simple and unconditionally stable in the algorithm. More important feature is that the LGM can offer access, theoretically or numerically, to regimes that cannot be easily reached by other methods. These features of the LGM are attractive to thermoacoustic research.

The LGM has been developed since the1980s. It was originally proposed for the investigation of hydrodynamic problems [1-2] and was later applied thermodynamics problems [3-4] and acoustics [5-6]. At present, lattice-gas automata methods have been applied to a number of classes of problems. These application fields involved to date have hydrodynamic instabilities, flow in porous media, phase transitions, reactive systems, etc. Thermoacoustic oscillations studied in the present paper involve the inter-disciplinary fields of thermodynamics and acoustics. The application of a lattice-gas method to thermoacoustics and the results from our work have demonstrated that it would be a useful approach for further thermoacoustic research and applications.

A TWO-DIMENSIONAL 9-BIT LATTICE GAS MODEL FOR THERMOACOUSTIC ENGINES

A lattice gas model can be viewed as a simple, fully discrete microscopic model, in which the time, space and fluid are all discrete. Therefore, the fluid consists of discrete particles that reside on a finite region of a regular lattice. At each lattice node one particle at most is permitted according with the exclusion principle. These particles move at regular time intervals from one lattice node to another along the lattice links, and collide at a lattice node obeying the mass, the momentum and the energy conservation laws for a thermal lattice gas model. The theoretical and numerical researches [2] on the lattice gas method have manifested that the statistics of large quantity microscopic particle motions can exhibit the physically realistic macroscopic behavior of fluid motion.

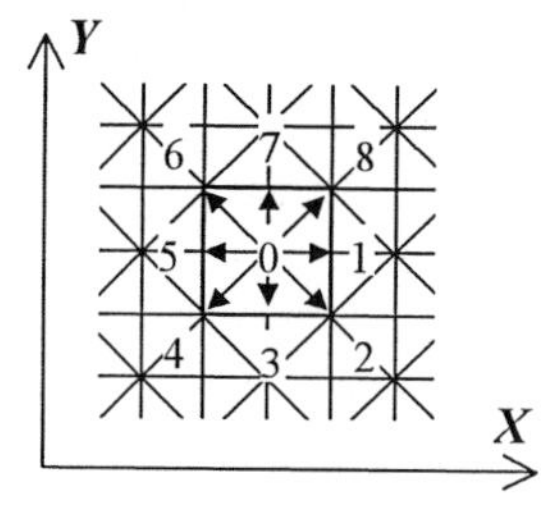

Figure 1 Two-dimensional 9-bit lattice.

We used a 9-velocity thermal model with three speeds $\left(0,1,\sqrt{2}\right)$ on a two-dimensional square lattice in the present study. The two-dimensional square lattice with corresponding velocity vectors is depicted in Fig. 1. All particles have unit mass. The gas particles moving along the axes have speed 1, while particles along the diagonals have speed $\sqrt{2}$, and the particles resting on the lattice have speed 0. The exclusion principle enables a natural Boolean description and encoding of the states of lattice gases. Any possible internal state of an individual automaton can thus be represented by 9 Boolean variables in the present model, which is called 9-bit lattice gas.

In the present model, we used similar two-body collision rules to that stated in Ref. [3]. In order to increase the collision rate and accelerate the rate of approach to equilibrium, thus shorten the computation time, we also considered multi-body collision rules in our model besides these two-body collision rules. This consideration is important for practical computations. These multi-body rules are obtained based on the two-body rules, satisfying the exclusion principle in pre-collision and post-collision. In addition, in order to simulate the thermal boundary condition imposed on the stack and heat exchangers in a thermoacoustic engine, some special collision rules and different probabilities of heat exchange were applied to the stack and heat exchangers in our model [7]. All of these rules used in the present model are reversible and determinate, namely, each input state corresponds to only one output state, or vice versa. All of two-body rules and part of multi-body rules applied in our model are depicted in Fig. 2.

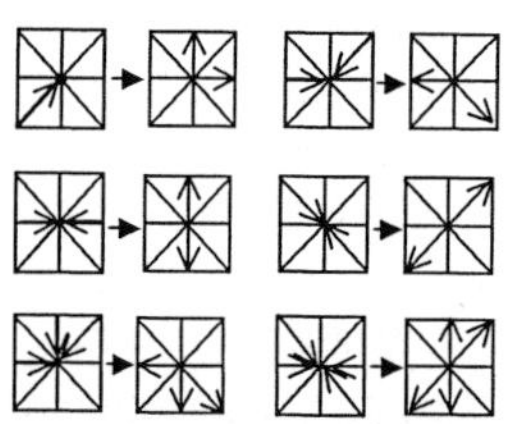

Figure 2 Collision rules

Studies [3] show that a 9-bit lattice-gas model satisfies some macroscopic equations such as the Navier-Stokes equation subject to some particle speed distribution condition. In the low-density approximation, this condition is $d_1/d_2 = 4$, $d_0/d_1 = 4$, where d_0, d_1, and d_2 are particle densities corresponding respectively to three speeds 0, 1 and $\sqrt{2}$ in a 9-bit model. This condition for particle speed distribution above was applied in our simulation.

The noise of lattice gas method is its intrinsic nature and can be reduced by a statistic method. The larger the statistical cell, the smaller the residual noise as well as the poorer the effective space resolution. In order to be able to both describe wave characteristic and obtain a macroscopic variable, the size of region for the statistic average should be chosen properly so that it comes from the result of a compromise between the accepted noise level and the desired physical resolution.

SIMULATION RESULTS AND ANALYSIS

The physical model under consideration for lattice gas simulation is a thermoacoustic prime mover as shown in Fig.3. This model engine consists of the working gas air, a medium for thermal rectification called the stack, heat exchangers at each end of the stack, and a 1/2 wavelength of resonator. The thermoacoustic stack in addition to two heat exchangers is generally called thermoacoustic components, which are positioned near the left end of the tube. The temperature distribution along the boundary of tube was prescribed and a temperature gradient was applied to the region of stack, as shown in Fig.3. Thus the stack and the two heat exchangers were imposed a thermal boundary condition under which the heat exchange between local gas particle and the solid boundary would occur. For the regions of the resonant tube outside the thermoacoustic components both thermal and adiabatic boundary conditions were applied with a probability of 0.01 for the thermal boundary condition.

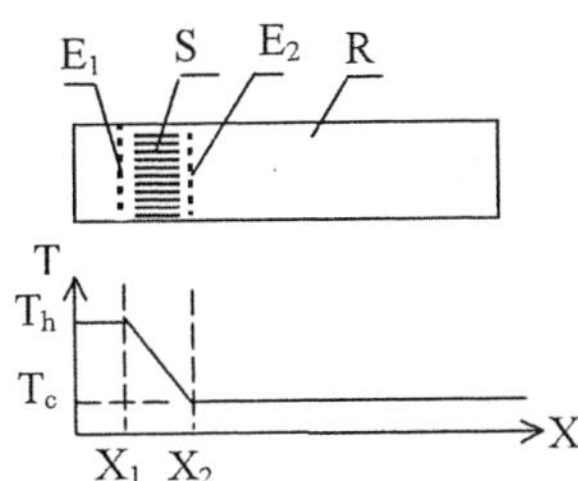

Figure 3 The physical model.
E1—Hot heat exchanger
E2—Cold heat exchanger
S—Stack, R—Resonant tube.

The lattice gas simulation was performed on a rectangular cross-section of a model engine of 2000×100 lattice nodes. The distance of the left end of stack from the left end of tube was 450 in lattice units. The length of stack was 100 in lattice units, and the lengths of two heat exchangers are all 10 in lattice units. The thickness and spacing of plates of stack and heat exchangers are all taken to be 1 lattice unit and 3 lattice units, respectively. The temperature layer in which heat exchanges between the fluid particles and walls would be expected to occur was set at 1 lattice unit. More meaningful conclusions from this model would require more lattice gas units, which will be considered in our next work. The hot temperature of the stack end was fixed at 2.0, while the cold temperature at 1.33, with temperature difference across the stack 0.67. In the initial calculations, the particle density of lattice gas was set at 2.5. The cell size of statistical region was set at 10×10 in lattice gas unit.

The onset behavior of oscillation in a thermoacoustic engine was successfully simulated in our present work. The buildup process of sound pressure at the cold end of tube is demonstrated clearly in Fig. 4. Thermoacoustic oscillation is excited in the resonant tube when the temperature gradient across the stack is greater than some value, and then the amplitude of acoustic pressure increases and finally reaches saturation. The self-excited oscillation will remain stable with the temperature condition along with other parameters constant. These onset oscillation features in a thermoacoustic prime mover were also observed by the experiment in early publication [8] and also simulated in conventional numerical method [9]. By comparing the results from our model with those of the experimental measurement and calculated [8-9], it can be found that the onset behaviors of self-excited thermoacoustic oscillation are qualitatively in good agreement. The pressure wave shape in Fig.4(a) shows the noise nature of the lattice gas method, and the pressure wave curve in Fig.4(b) shows somewhat non-linear feature. The asymmetry of profile curve in the top figure is also evidence of the non-linearity. It could be the consequence of the large temperature gradient across the stack and large probability of heat exchange.

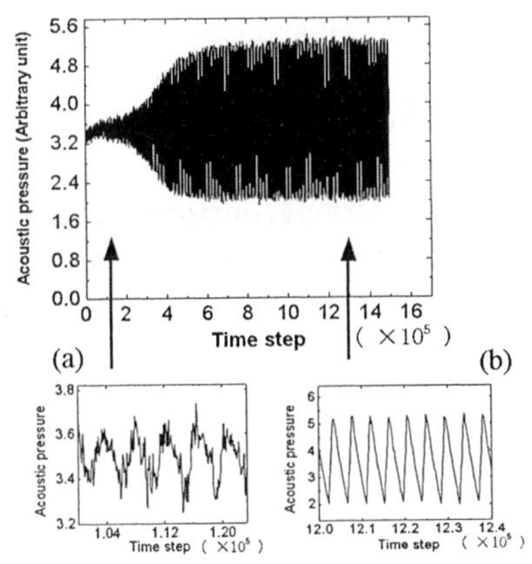

Figure 4 Acoustic oscillations at $\Delta T = 0.67$ applied across stack

We studied numerically as well the distribution of temperature field in the thermoacoustic engine. The contours of temperature in engine are shown in Fig.5, in which the isothermal lines can be observed. The time evolution process of temperature and its propagation process from the hot region towards two cold regions of tube ends can be clearly observed from Fig. 5(a). It can also be seen from Fig. 5 that at the unstable stage of onset oscillation the temperature distribution is inhomogeneous in the hot region, while at the stable-state of oscillation the temperature distribution becomes homogeneous in the hot tube. But, in the stack region the temperature shows the oscillation feature, and the temperature gradient is established and stable. Such a temperature re-distribution is just the consequence of thermoacoustic effect in stack. It can be observed as well from Fig. 5(b) that in the hot region of the tube the temperature is highest in the engine, while at the cold region of the tube (the right hand end) is a slightly higher temperature region. That could be because the dissipation heat of acoustic power in the overall engine is accumulated in the tube end, and/or because the heat is pumped to the right end region of tube by thermal acoustics generated at the left end of tube.

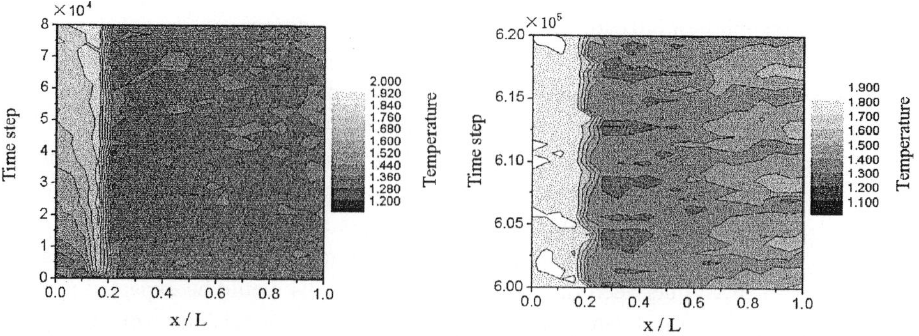

Figure 5 Distribution of temperature field in a thermoacoustic engine. (a) At onset phase of oscillation. (b) At stable-state of oscillation. x denotes position in engine, and L length of the engine.

CONCLUSIONS

In this paper we used a two-dimensional 9-bit thermal lattice gas model to simulate thermoacoustic oscillations. The building process of self-excited oscillation and the onset behavior of an acoustic wave in a thermoacoustic prime mover were successfully simulated. The temperature distribution and temperature time evolution in a thermoacoustic engine were obtained. The results from our model showed qualitatively good agreement with those from other researchers in earlier literature. The present paper work demonstrated that the lattice gas method would be a new powerful approach to investigate some complicated thermoacoustic problems. To reduce model noise and to extend its applications to thermoacoustic refrigerators and other thermoacoustic systems with complex geometric boundaries, an improved lattice gas model and more experimental work would be desirable. Efforts in all these directions are currently underway.

REFERENCES

1.   M. J. Biggs and S. J. Humby, Lattice-gas automata methods for engines, Trans Ichem E (1998), 76 162-174

2.   J.-P. Rivet, and J. P. Boon, Lattice Gas Hydrodynamics, Cambridge University Press (2001)

3.   S. Chen, M. Lee, K. H. Zhao and G. D. Doolean, A lattice gas model with temperature, Physica D (1989), 37 42-59

4.   S. P. Das and M. H. Ernst, Thermal transport properties in a square lattice gas, Physica A (1992) 187 191-209

5.   Y. Sudo and V. W. Sparrow, Sound Propagation Simulation Using Lattice Gas Methods, AIAA Journal (1995), 33(9) 1582-1589

6.   P. Lavallee, Attenuation of sound waves in lattice gases, Physics Letters A (1992), 163 392-396

7.   Yu Chen, Xiaoqing Zhang, Xu Liu, A lattice gas model of thermoacoustic engines, J. Acoust. Soc. Am. (2004), (Accepted)

8.   Wheatley J., Hofler T., Swift G. W., and A.Migliori, Understanding some simple phenomena in thermoacoustics with application to heat engines, Am. J. Phys. (1985), 53(2) 147-161

9.   S. Karpov and A. Prosperetti, A nonlinear model of thermoacoustic devices, J. Acoust. Soc. Am. (2002), 112(4) 1431-1444

Characteristics of a miniature pulse tube cooler driven by thermoacoustic standing wave engine

Wei Dai, Ercang Luo, Yuan Zhou, Wenxiu Zhu

Refrigeration and Cryogenic Division, Technical Institute of Physics and Chemistry,
Chinese Academy of Sciences, P.O.Box 2711, Beijing 100080, CHINA

Thermoacoustically driven pulse tube cooler is very promising due to no moving components. As a first step toward realizing a compact-sized pulse tube cooler, we built an ordinary 1/4 wavelength standing wave engine which has performed quite well. Then a miniature co-axial pulse tube cooler is coupled with the engine. By modifying hot reservoir, resonance tube, stacks and other parameters, we have achieved a no-load temperature of 84.4K at pulse tube cold end with pressure ratio 1.12 and frequency 53.9Hz. Some experimental details are introduced here.

INTRODUCTION

Thermoacoustic engines can serve as thermal compressors with no moving pistons. They have attracted lots of attentions recently. When combined with pulse tube coolers, cryocoolers with no moving components can be built. This is very attractive in many application fields. There are some reports on developing thermoacoustically-driven pulse tube cooler[1,2]. The published lowest temperature is about 90 Kelvin. However, the necessity of long resonance tube leads to a size incompatibleness which is an obstacle to putting it into real applications. Recently, the success of replacing resonance tube with dual-opposed supported pistons[3] ignites our interest in building a compact-sized thermoacoustically-driven pulse tube cooler to reach liquid nitrogen temperature. As a first step, we build an ordinary 1/4 wavelength standing wave engine with improved heat exchanger design and use it to drive a miniature pulse tube cooler. The details are introduced below.

EXPERIMENTAL DETAILS

Standing wave engine configuration and performance

The illustration of the standing wave engine configuration is shown in figure 1. The hot reservoir, hot end heat exchanger, stack and water-cooled heat exchanger have an inner diameter of about 80mm. Through a transitional part, the resonance tube of 50mm inner diameter is connected. The length is often changed to find a suitable working condition. Finally, an ambient reservoir of a volume of about 6 liters is connected to the other end of the resonance tube. The heating is realized by cartridge heaters from Watlow company. Hot parts of the system are wrapped by material made from ceramic fibre.

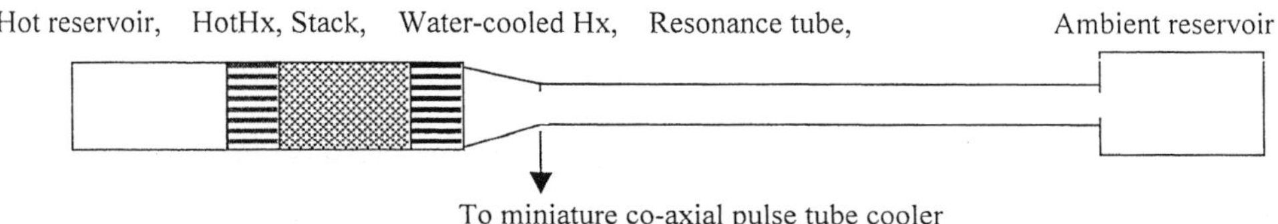

Figure 1. Illustration of integral system

Specially for the two heat exchangers which are important parts in the engine, we use EDM (electric discharge machining) methods to manufacture. Figure 2 shows a typical configuration of the water-cooled heat exchanger. The shown channels on the bulk cooper block is for the gas to flow through. Circular channels and holes under the uncutted cooper area, which are hidden by the housing, are for the water to flow. Determination of the fin length and thickness is based on the consideration of heat transfer efficiency and flow resistance. The widths of gas passage for hot heat exchanger and water-cooled heat exchanger is 1mm and 0.5mm, respectively.

Figure 2. Photo of the water-cooled heat exchanger

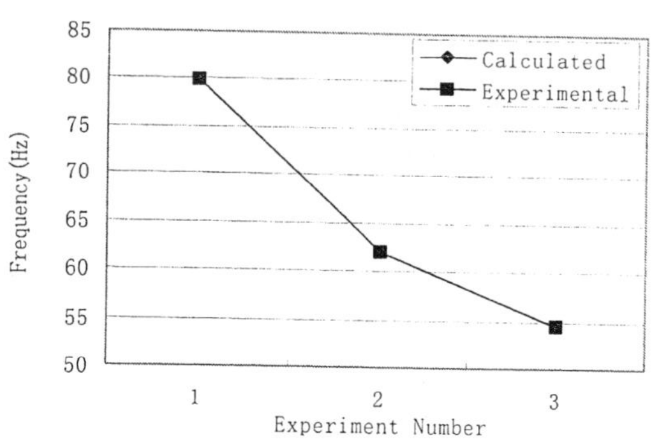

Figure 3. Frequency comparison between experiments and calculations

With the above mentioned configurations we have done experiments to optimize performance. Table 1 lists some typical results. Highest pressure ratio with helium is 1.15, while with nitrogen is 1.22. These results are obtained with no load connected.

Besides experiments, we also use our own program to simulate the performance of the engine. The calculation shows good agreement on frequency with the experimental data. Based on this, we found the engine actually works at a mode somewhere between 1/4 wavelength and 1/2 wavelength. This indicates the ambient reservoir is not large enough. However, calculation of the pressure amplitude and temperature distribution gives rather big deviation, which needs to be further improved.

Table 1. Typical performance of the thermoacoustic engine

Average pressure (bar)	Resonance tube length (m)	Stack mesh number	Heating power (Kw)	Pressure ratio	Frequency (Hz)
35	3.5	15	2.0	1.15	82.7
33.4	5.0	15	2.0	1.12	64
35.7	6	8	2.5	1.12 (with PTR)	53.9

Pulse tube cooler performance

The pulse tube cooler is provided by PTR group in our institute. It is co-axial and works with double-inlet mode. The outside diameter of the regenerator is about 10mm. The regenerator material is stainless steel mesh with mesh number being 400. Due to the convenience of driving the pulse tube cooler with linear compressor, we test the basic performance of the cooler with similar working conditions which may be provided by the thermoacoustic engine. Table 2 gives some typical experimental results. After these preliminary tests, we have a rough feel of how well the pulse tube cooler performs.

Table 2. Typical experimental results of the pulse tube cooler driven by a linear compressor

Frequency (Hz)	Average pressure(bar)	Pressure amplitude(bar)	Cold end temperature (K)
50	30	2.0	86.0
50	30	2.25	83.7
60	30	2.0	92.6
60	30	2.25	88.5

<u>Performance of the pulse tube cooler driven thermoacoustically</u>
Hot end of the pulse tube cooler introduced above is originally cooled by a fan. The heat transfer capability is quite limited as shown through experiments. So when coupling with the thermoacoustic engine, we insert a water-cooled heat exchanger before the inlet of the pulse tube cooler's regenerator, which helps to improve the performance. Figure 4 gives the best result obtained up to now. With pressure ratio 1.12 and frequency 53.9Hz, the lowest no-load temperature of pulse tube cold end is 84.4K. The heating power is 2.5Kw. Unfortunately, when the cold end was still slowly cooling down, some trouble happened, which forced us to stop the experiment for a while.

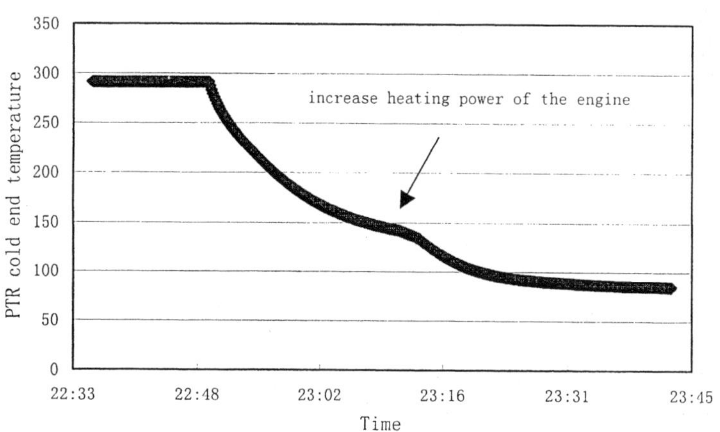

Figure 4. Cool-down curve of the pulse tube cooler cold end

CONCLUSION AND DISCUSSION

With an ordinary 1/4 wavelength thermoacoustic engine, the pulse tube cooler reached a no-load lowest temperature of 84.4K with a pressure ratio 1.12 at frequency 53.9Hz. This indicates the possibility of obtaining liquid nitrogen temperature with small pressure amplitude, thus leading to a 80K cryocooler with no moving components at all.

Better working conditions still needs to be found, such as better stack meshes, resonance tube diameter and bigger ambient reservoir. Simulation also needs to be improved to help the design.

Meanwhile, pulse tube cooler needs to be carefully designed for thermoacoustically driven condition, which means the pressure ratio available is not high compared with mechanical compressor.

Further experiments will be done to eliminate the inconvenient resonance tube.

ACKNOWLEDGMENTS

The research work is financially supported by the Chinese Academy of Sciences under the Contract of KJCX-SW-W012.

REFERENCES

1. Radebaugh R., "A review of pulse tube refrigeration", Adv.Cryo.Eng 1990, Vol 35(B), P.1191
2. Tang K., Chen G.B. and Kong B., "A 115K thermoacoustically driven pulse tube refrigerator with low onset temperature", Cryogenics(44), P.287
3. Matsubara Y., Dai W., Onishi T., Kushino A. and Sugita H., "Thermally actuated pressure wave generators for pulse tube cooler", Proc.of ICCR 2003, P.57

Experimental investigation on the dynamic character of thermoacoustic engine and parameters of regenerator

Chen X. [*]   Li Q.   Li Z. Y. [*]   Yu Z. B.   Xie X.J.

[*] Cryogenic laboratory, Hua zhong University of Science and Technology, Wuhan, China
Technical Institute of Physics and Chemistry, GSCAS, Beijing, China

As a result of temperature difference and self-organization of regenerator, some oscillatory fluid is filtered whose oscillatory frequency is far from thermoacoustic engine intrinsic frequency in regenerator. And someone is maintained which is close to the engine intrinsic frequency, thus part energy is concentrated on the band of close to the engine intrinsic frequency. Effect of heat conduction and fluid viscosity between the oscillatory fluid and solid interface of regenerator, magnification and attenuation of amplitude of oscillatory fluid simultaneously take place. The dynamic character is achieved with the actions of heat conduction, viscosity and oscillation between the oscillatory fluid and the solid interface, so it is significance to find out the transforming rules of the three objects.

FOREWORD

We used four thermoelectric couples which were even distributed along the regenerator's axis to measure the interior temperature of regenerator. Then, we calculated the parameters of regenerator. Furthermore, we analyzed the variation rules, validated the evolving process of thermoacoustic engine.

EXPERIMENT RESULTS INTRODUCTION

In our experiment, inner diameter and length of the regenerator packed with stainless screens are 25.4 mm and 32 mm respectively. The charge pressure of regenerator is 1.6Mpa. Experiment results can be seen from table 1 below:

Table 1   Experiment results

power(W)	mesh	start oscillation frequency (Hz)	extinguish frequency (Hz)	ultra oscillatory state
223	60	523	maitain	steady high frequency
179	200	528	73	intermittent oscillation of high and low frequency
223	200	528	73	steady low frequency oscillation
223	300	75	maitain	steady low frequency oscillation

From the rows 1, 3, 4, we can see that the temperature of regenerator under high frequency steady state is higher than that of under low frequency. It's well known that the energy to overcome fluid viscosity under low frequency mode is more than that of under high frequency. Therefore, the temperature under low frequency mode would descend firstly then turns to steady, but the temperature under high frequency mode would slowly increase before steady. Row 2 shows that when the system is under transition state the temperature of starting oscillation is higher than that of forming reliable low frequency oscillation mode. It means that after system starting oscillation, part of heat energy converted into acoustic energy. Therefore, the temperature would decrease. And the sound wave must consume energy to get over the resistance of fluid during transferring process. If this part of energy can not be gotten, the temperature of regenerator would decrease continuously until the system can not oscillate.

We can see the movement of thermoacoustic engine is closely related to the temperature of regenerator for it reflects the effect of heat contact and viscosity contact between fluid and solid interface of regenerator. Therefore, we can utilize the extrinsic phenomena of the system to study the dynamic character of thermoacoustic self-excitation oscillation.

VISCOSITY DISSIPATION OF SOUND WAVE

As sound wave transfers in viscous fluid, the viscosity of fluid increase the energy dissipation. This part energy is transferred with diffusion. Now, we discuss the dissipation of sound wave in viscous fluid. Wave equation, diffusion equation, linearized Navier-Stokes equation:

$$\frac{\partial \rho'}{\partial t} = \frac{1}{c_0^2} \frac{\partial p'}{\partial t} \tag{1}$$

$$\frac{\partial p'}{\partial t} + \rho_0 c_0^2 \frac{\partial u}{\partial x} = 0 \tag{2}$$

$$\rho_0 \frac{\partial u}{\partial t} + \frac{\partial p'}{\partial x} = \frac{4}{3}\mu'_0 \frac{\partial^2 u}{\partial x^2} \tag{3}$$

Where $c_0^2 = c_p(k-1)/\beta_0^2 T_0$ for the velocity of sound, β_0 for coefficient of heat expansion, $\mu'_0 = \mu_0(1+3/4 \cdot \mu_{v0}/\mu_0), \mu_{v0}$ for coefficient of viscosity expansion. From formula (1), (2), (3) we can get

$$\frac{\partial^2 p'}{\partial t^2} = c_0^2 \left(1 + \frac{4}{3}\frac{v_0}{c_0^2}\frac{\partial}{\partial t}\right)\frac{\partial^2 p'}{\partial x^2} \tag{4}$$

The solution of (4) can be expressed in vector form:

$$p' = \mathrm{Re}\left[\tilde{p}(x)e^{-j\omega t}\right] \tag{5}$$

Then get

$$p''(x) + K^2 \tilde{p}(x) = 0 \tag{6}$$

Where $K^2 = (\omega/c_0)^2 /[1 - j(4\omega v_0/3c_0^2)]$, it can be written as $K = k_1 + jk_2$, thus we get

$$p(x,t) = A \cdot e^{-k_2 x}\cos(k_1 x - \omega t + \theta) \tag{7}$$

Where, K^2 must satisfy $(c_0/\omega)^2 (k_1^2 - k_2^2 + 2jk_1k_2) = [1 - j(4\omega v'_0/3c_0^2)]^{-1}$, it can be expressed as

$$\overline{\beta}^2 - \overline{a}^2 = (1 + \omega^2\tau_v^2)^{-1}, \overline{a} = k_2 c_0/\omega \tag{8}$$

$$2\overline{a}\overline{\beta} = \omega\tau_v/(1 + \omega^2\tau_v^2), \overline{\beta} = k_1 c_0/\omega \tag{9}$$

Where, $\tau_v = 4v'_0/3c_0^2$, $\overline{a}$ for coefficient of oscillatory attenuation, $\overline{\beta}$ for coefficient of phase shift, ω/k_1 for wave velocity with dissipation, so $\overline{\beta}$ represent the ratio of velocity and describe the infection of dissipation to the velocity of transfer.

We can know from above dissipation makes oscillatory attenuation, which is reflected from the imaginary part of complex wave number K. And

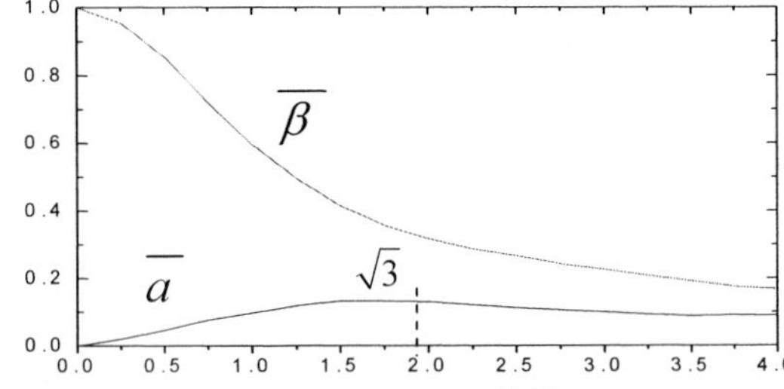

Figure 1   relation of $\overline{a}, \overline{\beta} \sim \omega\tau_v$

variety of wave velocity makes phase shift. Oscillatory attenuation and phase shift rest with $\omega\tau_v$. τ_v depends on the dissipation of sound wave. And ω is the frequency of oscillatory fluid. The product $\omega\tau_v$ reflects the diffusion wave transferring depends not only wave but also diffusion.

Figure 1 shows the relation of $\overline{a}$, $\overline{\beta}$ and $\omega\tau_v$. $\overline{\beta}$ is always less than 1, this means the velocity of sound wave with dissipation is higher than that of without dissipation. The momentum diffusion velocity that results from dissipation is append to c_0.

ANALYSIS OF THE PARAMETERS

The parameters below are close to the temperature of oscillatory fluid in regenerator: depth of heat penetration and viscosity penetration δ_k, δ_v, distribution function of viscosity and heat conduction f_v, f_k, relaxation time of viscosity and heat τ_v, τ_k, coefficient of flow magnify e. From the expression of e, we can get the real part of coefficient of flow magnifying Re[e] and the imaginary part of Im[e]:

$$\mathrm{Re}[e] = \frac{\left(\mathrm{Re}\left[\beta\left(f_\kappa - f_v\right)\right]\mathrm{Re}\left[1 - f_v\right] + \mathrm{Im}\left[\beta\left(f_\kappa - f_v\right)\right]\mathrm{Im}\left[-f_v\right]\right)}{\left|1 - f_v\right|^2\left(1 - \sigma\right)}\frac{dT_m}{dx} \tag{10}$$

$$\mathrm{Im}[e] = \frac{\left(\mathrm{Re}\left[\beta\left(f_\kappa - f_v\right)\right]\mathrm{Im}\left[f_v\right] + \mathrm{Im}\left[\beta\left(f_\kappa - f_v\right)\right]\mathrm{Re}\left[1 - f_v\right]\right)}{\left|1 - f_v\right|^2\left(1 - \sigma\right)}\frac{dT_m}{dx} \tag{11}$$

From the experimental data, we calculated these parameters shown in Figures 2,3,4,5.

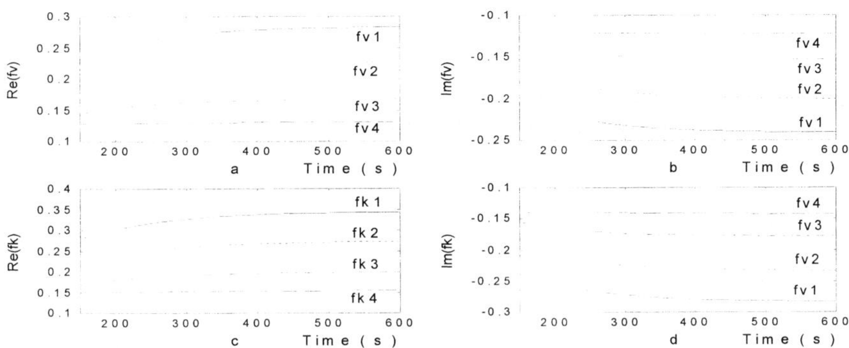

Figure 2   distribution function under high frequency mode

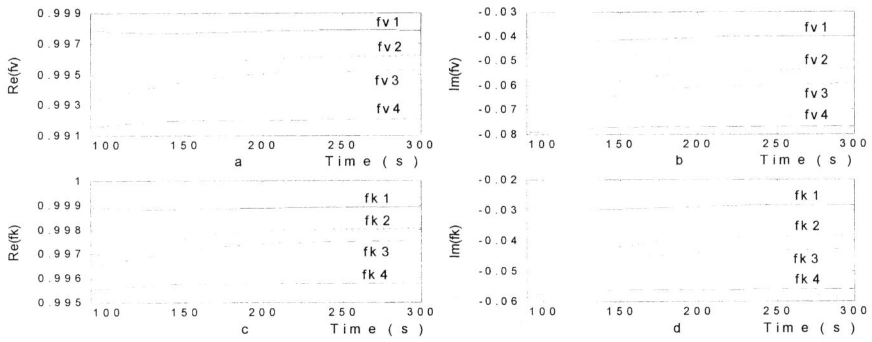

Figure 3   distribution function under low frequency

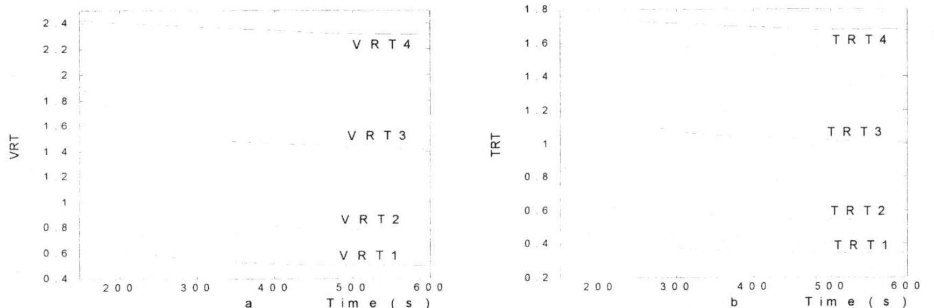

Figure 4   variation of $\omega\tau_v$ under constant high frequency

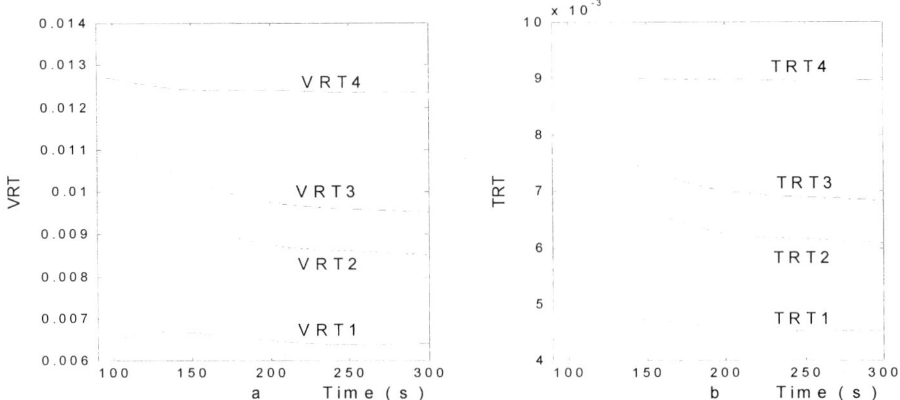

Figure 5　variation of $\omega\tau_v$ under constant low frequency

Where1, 2, 3, 4 for the four temperature sensors, VRT for period of viscosity diffusion, TRT for the period of heat diffusion. From the above we can get some conclusions:

(1) Absolute values of the real part and imaginary part of f_k are larger than those of f_v under high frequency. According to formula (10,11), we can easily get that the real part of coefficient of flow increasing under high frequency mode is less than that under low frequency, but the imaginary part is reverse. Figure 2 shows that sound pressure is thoroughly different under different modes.

(2) The velocity of sound wave is more rapidly in viscous fluid. This relates to microcosmic property of material. And the value of coefficient of phase shift changes small, the phase angle of sound wave varies sharply. For traveling wave, f_v is small, and the phase shift will be small, so the variation of phase angle must small. Therefore, sound wave can maintain a constant speed. Furthermore, the phase angle of sound pressure is different on different position at the same time. But for standing wave, viscosity dissipation may lead to phase angle identical.

(3) $\omega\tau_v$ is correspond to $\bar{a}$ and $\bar{\beta}$. While $\omega\tau_v$ is in the range of $0\sim\sqrt{3}$, $\bar{a}$ increase with it. So, we can understand that the temperature difference of regenerator under high frequency is larger than that under low frequency. This means relaxation time of viscosity determines dissipation of sound wave .The numerical value of $\bar{\beta}$ decrease while $\omega\tau_v$ increasing. Therefore, $\bar{\beta}$ is larger under low frequency. It means that phase shift becomes larger. So, it is possible to keep phase angle stable at the same time.

CONCLUSION

We analyzed and quantified the parameters of regenerator with temperature temporal series of oscillatory fluid in the regenerator. The dynamic character of thermoacoustic engine is performed with these parameters. Mechanism of oscillatory mode transferring is described with no dimension time scale $\omega\tau_v$.

ACKNOWLEDGMENT

The authors gratefully acknowledge the National Natural Science Foundation of China under Grant No. 50276064 and Knowledge Innovation Program of the CAS under Grant No. KJCX2-SW-W08.

REFERENCES

1. Guo Fang Zhong, Thermodynamics Lecture, Hua Zhong University of Science and Technology, 2003
2. G. W. Swift. Thermoacoustics: A unifying perspective for some engines and refrigerators. Fifth draft, available at http://www.lanl.gov/thermoacoustics/,2001
3. Zhang Xiao Qing, Investigation on Modeling and Optimization of Thermoacoustic Engine Systems, 2001, Hua Zhong University of Science and Technology.

Basic operation principle of a novel cascade thermoacoustic prime mover

Luo E., Dai E., Ling H.

Technical Institute of Physics and Chemistry, Chinese Academy of Sciences, Beijing 100080
Graduate School of Chinese Academy of Sciences, Beijing 100039

This paper presents an analysis for basic operation principle of the cascade thermoacoustic engine. Based on the analysis, the paper points out that an efficient cascade thermoacoustic system can be realized by using some special features of a half-wavelength, standing-wave system in which there is usually a high impedance zone with a traveling-wave component being predominant, and has no closed-loop topology. Thus, placing a regenerator-based engine in this zone as a second-stage engine can realize a more efficient thermoacoustic process than a single-stage standing-wave engine. In addition, there is no DC-flow problem due to without a closed-loop topology in the cascade system. Finally, the in-line system can avoid imbalanced thermal stress existing in the thermoacoustic Stirling engines, thus simplifying mechanical consideration and improving lifetime.

INTRODUCTION

Thermoacoustic machines are new kinds of environment-friendly devices. Thermoacoustic prime movers can be classified into two categories, standing wave and traveling-wave movers. In principle, the standing-wave prime mover relies on an intrinsically irreversible thermodynamic process, the so-called medium heat transfer process, consequently, its highest efficiency is bounded to about 60% of the ideal Carnot cycle [1]. On the other hand, the thermoacoustic Stirling prime mover can theoretically achieve the efficiency of the ideal Carnot cycle if heat transfer process is perfect [2]. However, almost any real thermodynamic process is accompanied by various irreversibilities. In the thermoacoustic prime movers, the irreversibilities resulting from imperfect heat transfer and viscous friction are two main ones that usually affect each other. In a one-wavelength, pure traveling-wave prime mover, the in-phase relation between the pressure and velocity is indeed beneficial to the conversion of heat and acoustical power, so is the good heat transfer between the gas and solid matrix of the regenerator of the prime mover. However, the velocity amplitude is too large in the pure traveling-wave system and thus produces significant viscous loss.

The thermoacoustic Stirling prime mover that was developed by Los Alamos Laboratory and other laboratories, can provide a predominantly traveling-wave acoustical field yet with small velocity by making using of the special feature of standing-wave filed and designing a lumped-parameter traveling-wave loop [3]. The demonstrated prime mover achieved the efficiency of about 30% from heat to acoustical power, which can be competitive to modern internal combustion engine. However, the thermoacoustic Stirling engine may exist a significant DC-flow loss and also has imbalanced thermal stresses. The DC-flow loss greatly decreases the efficiency of the looped system and the imbalanced

The research is financially supported by Chinese Academy of Sciences under Contract of KJCX2-SW-W012-01.

thermal stress decreases reliability of the system. Recently, a concept of cascaded thermoacoustic systems was proposed [4], but there is little work reported on its basic operation principle. Thus, the paper tried to present an explanation for fundamental mechanism of the cascaded thermoacoustic systems. Firstly, special features of a half-wavelength, standing-wave system were analyzed. Then, by analyzing the production rate of local acoustical power, the operation principle of the cascaded thermoacoustic engine was described.

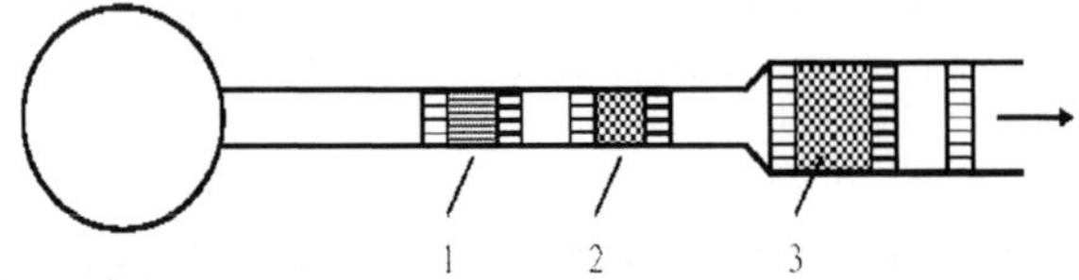

Figure 1. Schematic for one kind of cascade thermoacoustic prime mover [4]
(1. Standing-wave engine 2. First traveling-wave engine 3. Second traveling-wave engine)

MODEL AND ANALYSIS

To understand basic operating principle of the cascade thermoacoustic engine, let us first examine the standing-wave systems shown in Figure 1. Under linear acoustical assumption, the expressions for pressure and velocity distributions $\tilde{p}(x)$ and $\tilde{u}(x)$ can be given in Eq.(1). In addition, Eq.(2) gives all eigenvalue frequencies of the acoustical system.

$$\tilde{p}(x) = i2C_1 \sin(kx); \quad \tilde{u}(x) = -\frac{1}{\rho_0 a_0} 2C_1 \cos(kx) \qquad (k = a_0/\omega, \text{ wave number}) \qquad (1)$$

$$f = \frac{na_0}{2l}, (n = 0,1,2,...) \qquad (a_0, \text{ sound velocity}) \qquad (2)$$

For fundamental oscillation mode, the expressions of the pressure and velocity distributions are given as follows,

$$\tilde{p}(x) = i2C_1 \sin(\pi x / l); \quad \tilde{u}(x) = -\frac{1}{\rho_0 a_0} 2C_1 \cos(\pi x / l) \qquad (3)$$

Furthermore, the local acoustical impedance can be obtained as follows,

$$\frac{\tilde{p}(x)}{\tilde{u}(x)} = -i\rho_0 a_0 \sin(\pi x / l) / \cos(\pi x / l) \qquad (4)$$

Examining the local acoustical impedance can know the phase angle between the oscillating pressure and velocity. In terms of thermoacoustics, the larger the amplitude of the acoustical impedance is, the less viscous loss the thermoacoustic engine has. Actually, the local production rate of acoustical power is a key parameter to evaluate if the thermoacoustic engine can deliver net acoustical power or not. The local production rate of acoustical power is given by the following equation,

$$\frac{dW_x}{dx} = \frac{1}{2}\text{Re}[\tilde{U}(x)\tilde{p}^*(x)f_{WT}]\beta_0 \frac{dT_0}{dx} - \frac{1}{2}\frac{(\gamma-1)A}{\rho_0 a_0^2}|\tilde{p}(x)|^2 g_{wK} - \frac{1}{2}\frac{\rho_0}{A}\omega|\tilde{U}(x)|^2 g_{w\mu} \qquad (5)$$

It should be noticed that only the first item in Eq.(5) can be positive because the other two items denote power losses by imperfect heat transfer and viscous losses. Obviously, the first item is also strongly determined by various factors such as the temperature gradient and direction, the amplitudes and relative phases of pressure and velocity, and the power-production factor f_{WT}, etc.

Obviously, the acoustical field of the thermoacoustic engine is extremely important for thermoacoustic processes. For a standing-wave acoustical field, it is more appropriate to use a so-called "stack" as solid medium; contrariwise, a traveling-wave acoustical field requires a so-called "regenerator". In figure 1, for the fundamental oscillation mode of the half-wavelength standing-wave system, the left and right zones are dominated by standing-wave components whereas in the middle zone a traveling-wave component predominantly exists for the real system. Thus, if designing a first-stage standing-wave engine in the left zone and a second-stage traveling-wave engine in the middle zone, such a cascade thermoacoustic system should be more efficient than a single-stage standing-wave engine. In addition, adjacent zone to the middle zone has larger pressure and smaller velocity, so the first-stage standing-wave should not be far from the middle zone where the second-stage traveling-wave is located.

Now let us consider what temperature-gradient direction should be imposed on the two-stage engine. Assuming the propagation direction of the wave is from left to right. From Eq.(4), it can be shown that the pressure lags the velocity in the left of the middle location while leading in the right of the middle location. Roughly speaking, in the far left side, the pressure lags the velocity by about 90°; in the middle zone, the lagging phase varies from about -90° to about 90°; in the far right side, the pressure leads the velocity by 90°. Thus, there are the following approximate expressions for the pressure distributions in the different zones.

In the far left-side, the pressure distributions of the real cascade thermoacoustic system can be expressed approximately by Eq.(6) to Eq.(8), respectively

$$\widetilde{p}(x) = (a - bi)\widetilde{U}(x), (a, b > 0, and\ b >> a) \tag{6}$$

$$\widetilde{p}(x) = (a - bi)\widetilde{U}(x), (a, b > 0, and\ a \approx b) \tag{7}$$

$$\widetilde{p}(x) = a\widetilde{U}(x), (a > 0, and\ b \approx 0). \tag{8}$$

In addition, the power-production factor f_{WT} is given by the following equation,

$$f_{WT} = c - id, (c, d > 0) \tag{9}$$

For a stack-based solid medium, $c \approx d \approx 0.5$; for a regenerator-based solid medium, $c >> d,\ c \approx 1,\ d \approx 0$.

As mentioned above, it is better to design a stack-based, standing-wave engine in the far left-side zone. Thus, when neglecting the two dissipation items of acoustical power, the production rate of acoustical power can be simplified as follows

$$\sum w'_{s\,\tan d} = \frac{1}{2} \mathrm{Re}[(a + bi)(c - di)]\,|\,\widetilde{U}(x)\,|^2\,\beta_0\frac{dT_0}{dx} \approx \frac{1}{2}bd\,|\,\widetilde{U}(x)\,|^2\,\beta_0\frac{dT_0}{dx}\ . \tag{10}$$

To make the power-production rate be positive, it is necessary to impose a positive temperature gradient dT_0/dx. Moreover, substituting the pressure and velocity distributions into Eq.(10) yields the following analytical expression for the power-production rate

$$\sum w = \frac{C_1^2 A \sin(2\pi x/l)}{\rho_0 a_0}\beta_0\frac{dT_0}{dx}\ . \tag{11}$$

It can be found from Eq.(11) that the local power-production rate reaches its maximum value if the stack is placed one-fourth distance from the middle point.

By making a similar analysis, we can obtain the following expression for the local power-production rate if a regenerator is placed in the middle zone

$$\sum \dot{w}_{travel} = \frac{1}{2}\mathrm{Re}[a(c-di)]\,|U(x)|^2\,\beta_0\frac{dT_0}{dx} \approx \frac{1}{2}ac\,|U(x)|^2\,\beta_0\frac{dT_0}{dx}. \tag{12}$$

To make the power-production rate be positive, it is also necessary to impose a positive temperature gradient dT_0/dx. Because $b < a$ (in the middle point, $\tilde{p}(x)/\tilde{u}(x)$ reaches its maximum) and $c > d$, $\sum \dot{w}_{travel} > \sum \dot{w}_{s\tan d}$, which implies the second-stage traveling-wave engine is more efficient than the first-stage standing-wave engine. Moreover, the second-stage traveling-wave engine has small viscous loss and no DC-flow loss. Thus, the cascade thermoacoustic system would be more efficient than a single-stage standing-wave thermoacoustic engine. This is just the basic operation principle of the cascade thermoacoustic system.

According to the above analysis, it is also possible to design other topology of efficient cascade thermoacoustic system, which was described in the literature [5].

CONCLUSION

The basic operation principle was presented in this paper. The analysis shows that any efficient thermoacoustic system has these requirements: high impedance, traveling-wave acoustical field, and non-looped flow passage. The high impedance can minimize the viscous loss by velocity. The traveling-wave acoustical field can maximize the conversion between heat and acoustic power. And, the non-looped flow passage can avoid the DC-flow loss. The proposed, cascaded thermoacoustic systems can satisfy these requirements, therefore having the potential for high efficiency. In addition, the cascaded thermoacoustic systems do not exist imbalanced thermal stress; therefore, they are more reliable.

REFENRENCE

1. J. Wheatley, T. Hofler, G. Swift and A. Migliori, An intrinsically irreversible thermoacoustic heat engine, J.Acoust.Soc.Am., vol.74(1),1983

2. P.H.Ceperley, A pistonless Stirling engine-Thermoacoustic traveling wave heat engine, J.Acoustic.Soc.Am., vol.66,1979

3. S. N. Backhaus and G. W. Swift,A thermoacoustic-Stirling heat engine: Detailed study, J.Acoustic.Soc.Am.,vol 107, 3148-3166 (2000)

4. S. N. Backhaus and G. W. Swift, New varieties of thermoacoustic engines, Paper number 502, Proceedings of the Ninth International Congress on Sound and Vibration, Orlando FL, July 8-11 2002

5. Luo Ercang, Dai Wei, et al., Operating principle, thermodynamic analysis and optimization of cascade thermoacoustic prime movers or refrigerators, Cryogenics (in Chinese), No.6 (2003), p.10

Study on a linear compressor with metal-bellow replacing conventional cylinder-piston unit

Wu Z., Luo E., Dai W., Hu J.

Technical Institute of Physics and Chemistry of Chinese Academy of Sciences, Beijing 10080, China
Graduate School of Chinese Academy of Sciences, Beijing 100029, China

Linear compressor, which acts as a pressure wave generator in regenerative refrigerator system, is an important element converting electric power to acoustic power. And its high efficiency, flexibility and survivability exhibit multifold advantages over conventional compressors. This paper presents the study of a linear compressor of moving-coil type with metal-bellow unit, which is made up of a wavelike thin wall pipe and is used to produce the pressure oscillation. The influence of mechanical parameters and acoustic loads is investigated when it works resonantly. It provides some guidelines for designing an efficient linear compressor with metal-bellow assembly.

INTRODUCTION

The linear compressor, which is developed to meet the increasing demand of high efficiency and high reliability, has advantages over traditional reciprocating compressor on several aspects including high efficiency, simple mechanism and flexible capacity by controlling the stroke, etc. This paper presents the study of a linear compressor with metal-bellow unit instead of cylinder-piston unit for thermoacoustic refrigerator applications, which have a ring-shaped moving coil and a stationary magnet. Due to non-leaking interface between the front and back sides of bellow unit, the working gas is protected from being contaminated by the gas emitted from the coil and lubrication oil inside. Further more, it can avoid the friction, precision mechanical technology. We use linear bearings other than flexure bearings, because of unlimited displacement and availability; however, it may bring a little friction loss. In the computation, it can be found that the linear compressor has a highest efficiency for optimized design. And the influence of moving mass, acoustic load and working frequency of the linear compressor are investigated.

MATHEMATIC MODEL OF LINEAR COMPRESSOR

The linear compressor is an electro-mechanical system, which transforms electrical power to mechanical energy. Consequently, from Kirchhoff law and Newton second law, the following equations can be given:

$$R_{elec}i + L_{elec}\frac{di}{dt} + B_g lv = V \; , \quad B_g li - \tilde{p} \cdot A - R_{mech}v - Kx = M\frac{dv}{dt}$$

where, R_{elec} and L_{elec} are the resistance and the inductance of the coil, B_g is the magnetic flux density of

---

This work is financially supported by Chinese Academy of Sciences under contract   KJCX2-SW-W012-01.

the gap, where the coil moves, v is velocity, V is the voltage, K is the suspension spring stiffness, R_{mech} is damping coefficient of mechanical system, M is moving mass.

Using complex notations, $i = \tilde{I} \cdot e^{j\omega t}, v = \tilde{v} \cdot e^{j\omega t}, V = \tilde{V} \cdot e^{j\omega t}$. Superscript '~' represents the oscillating value. Substitute these variables into the first two equations, and they can be reduced to:

$$\tilde{V} = (R_{elec} + jX_{elec})\tilde{I} + \frac{\tau}{A}\tilde{U}, \quad \tilde{P} = \frac{\tau}{A}\tilde{I} - \frac{R_{mech} + jX_{mech}}{A^2}\tilde{U}$$

where $X_{mech} = \omega M - K/\omega$ is the imagery of the mechanical impedance. And the acoustic load is:

$$R_{acoust} + jX_{acoust} = \frac{\tilde{p}}{\tilde{U}}$$

Solving these three equations leads to the following expressions:

$$\tilde{I} = \frac{\tilde{V}(R_{am} + jX_{am})}{(R_{elec}R_{am} - X_{elec}X_{am} + \tau^2) + j(R_{elec}X_{am} + X_{elec}R_{am})}$$

$$\tilde{p} = \frac{\tilde{V}A\tau(R_{acoust} + jX_{acoust})}{(R_{elec}R_{am} - X_{elec}X_{am} + \tau^2) + j(R_{elec}X_{am} + X_{elec}R_{am})}$$

$$\tilde{U} = \frac{\tilde{V}A\tau}{(R_{elec}R_{am} - X_{elec}X_{am} + \tau^2) + j(R_{elec}X_{am} + X_{elec}R_{am})}$$

where $R_{am}=R_{mech}+A^2R_{acoust}$ and, $X_{am}=X_{mech}+A^2X_{acoust}$, which can be discovered that the convert coefficient of acoustic impedance to mechanical impedance is square of the swept cross-section area. From the impedance network [1], this relation is much more clear. The electric power into the compressor and the acoustic power out of the compressor are given as follows:

$$W_{elec} = \frac{1}{2}\text{Re}\left[V\tilde{I}^*\right] = \frac{V}{2}\text{Re}\left[\tilde{I}^*\right], \quad \Delta\dot{E}_2 = -\frac{1}{2}\text{Re}\left[\tilde{p}\tilde{U}^*\right]$$

the superscript '*' denotes complex conjunction. The electro-acoustic efficiency is given as follows [2]:

$$\frac{1}{\eta_{trans}} = \frac{W_{elec}}{\Delta\dot{E}_2} = 1 + \frac{R_{mech}}{A^2R_{acoust}} + \frac{R_{elec}}{A^2\tau^2R_{acoust}}\left[R_{am}^2 + X_{am}^2\right]$$

Obviously, setting $X_{am}=0$ leads to the resonance state of the system. The X_{am} depends strongly on not only mechanical system and acoustic load themselves, but also the acoustic-mechanical convert coefficient of impedance.

METAL-BELLOW ASSEMBLY AND PROTOTYPE OF THE LINEAR COMPRESSOR

Metal-bellow assembly is used to replace the conditional cylinder-piston Unit, so as to avoid high accuracy mechanical fabrication and installation and oil lubrication. The metal-bellow unit can be made up by dozens of dishing metal slice welded together at the edges, or formed by a thin-wall wavelike pipe (see Figure 1). The welding metal-bellow has lower spring stiffness than the molding one, and has more potential ability of deformation. Basically, the smaller the deformation of each wave, the longer longevity the bellow has. Figure 2 is our prototype of the linear compressor. Inside the upmost part is the metal-bellow unit with a 75 mm diameter and 100 mm high.

Figure 1   The welded (left) and molder (right) metal-bellows

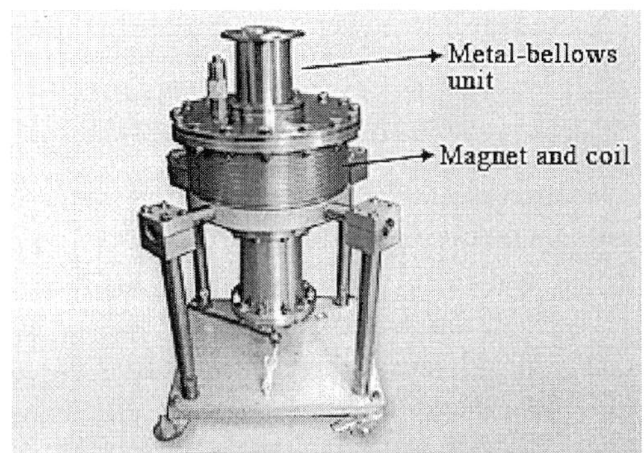

Figure 2   Picture of the linear compressor

COMPUTATIONAL RESULTS AND ANALYSIS

In our design, the magnetic flux density of the gap is 1.1T. The copper wire of the coil is about 180 meters long, which has a diameter of about 0.63 mm, and the electrical resistance is 10.6 ohms. And the maximum swept volume is about 60cc. Under these conditions, this paper investigates the influence of the moving-mass, the acoustic load and frequency response of the system.

Moving mass

The moving mass of the linear compressor is about 2 kg, which includes the mass of shaft, the coil and 1/3 mass of the suspension springs and metal-bellow. As it mentioned above, the efficiency is independent of the moving mass when the system is working resonantly, however, it affects the piston stroke and the phase difference between the voltage and the current.

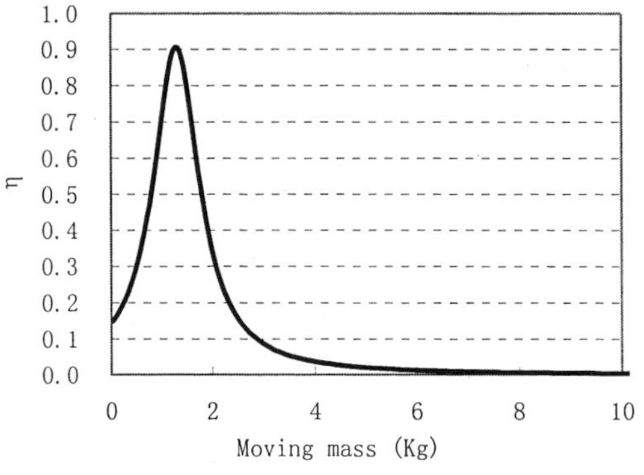

Figure 3   Efficiency varying moving mass

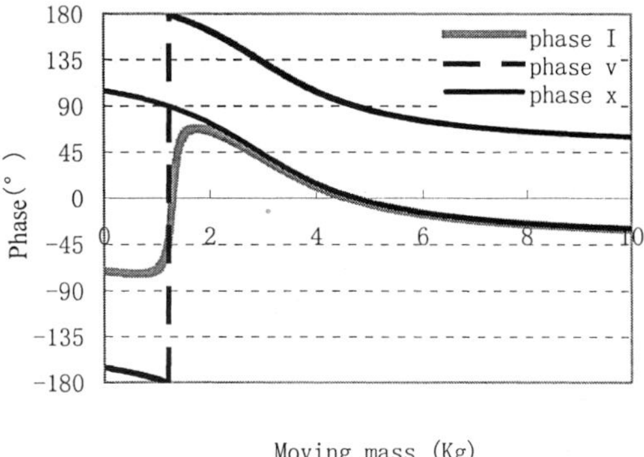

Figure 4 Relative phase of I, v and x vs. mass

324

Keeping the other variable unchanged, we obtain the relationship between the efficiency and the moving mass (See Figure 3). And in Figure 4, it shows the relationship between the relative phase of current, velocity and stroke versus the moving mass respectively. Obviously, with the increase of the moving mass, the efficient increases quickly and then reduces sharply. It can be said that at the peak efficiency, the compressor reaches the "resonance state". At the same time, the voltage and current are in phase and the phase between the current and stroke is near 90° [3], that is, the magnetic force is lag the stroke by 90° . It can be interpreted that when the system works resonantly, the compressor consumes only a little electric power to overcome the power dissipations in the electrical, mechanical and acoustic resistance.

<u>Frequency response and acoustic load</u>

The linear compressor can be treated as single-degree-of-freedom spring-mass-damped system. And the linear compressor is sensitive to the acoustic load. If the acoustic load and the compressor match each other, the system reaches a highest efficiency (see Figure 5, "A.L" means the acoustic load). However, the relative phase diagram of the current, stroke and velocity is similar to the Figure 4 (see Figure 6). In Figure 5, $\eta_{\_max}$ represents the maximum efficiency for different acoustic load. If the acoustic load smaller than the appropriate load, the efficiency drops sharply. Therefore, it is important to match the compressor to the acoustic load.

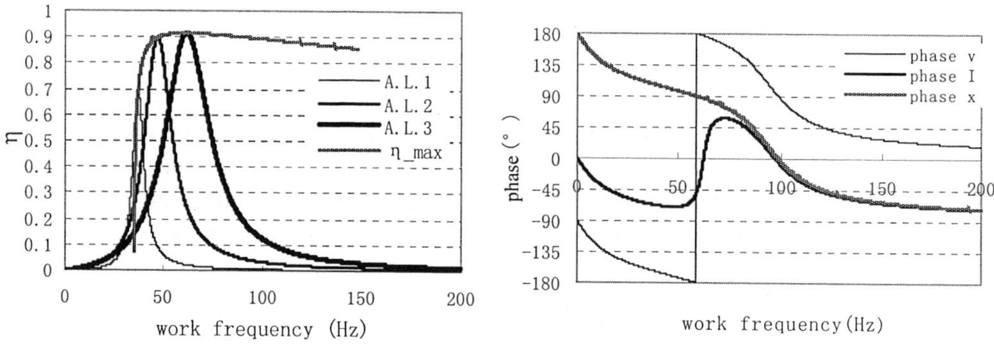

Figure 5 Efficiency Vs. frequency and difference load

Figure 6 Relative phase diagram of I, v, x

CONCLUSION

Theoretically, it is important for the linear compressor to work resonantly to achieve high efficiency and low electrical power consumption. And it's also important to match the acoustic load to the compressor to obtain high efficiency. Furthermore, the phase between current and stroke must be 90° when it works resonantly.

REFERENCE

1. Wakeland, R.S, Use of electrodynamic drivers in thermoacoustic refrigerators, <u>J. Acoust. Soc Am.</u> <u>107(2),</u> 2000
2. Swift, G, Thermoacoustics: <u>A unifying perspective for some engines and refrigerators (4th draft),</u> 184-189, 1999
3. Deuk-Yong Koh, Yong-Ju Hong, Seong-Je Park ect., <u>A study on the linear compressor characteristics</u> <u>of the Stirling cryocooler,</u> Cryogenics(2002) <u>42</u> 427-432

Proceedings of the Twentieth International Cryogenic Engineering Conference
(ICEC 20), Beijing, China. © 2005 Elsevier Ltd. All rights reserved.

Analytical model for onset temperature difference of thermoacoustic Stirling prime movers

Hu J., Luo E., Ling H.

Technical Institute of Physics and Chemistry of Chinese Academy of Sciences, Beijing, 100080
Graduate School of Chinese Academy of Sciences, Beijing, 100039

The onset temperature difference of thermoacoustic Stirling prime movers is a very important parameter for utilizing low quality heat energy. This paper establishes an analytical model to calculate it and finds its dependence on the length and porosity of the regenerator, the working gas, the oscillating frequency and the working pressure.

INTRODUCTION

Thermoacoustic engine is a kind of new machine that can convert heat into mechanical work in the form of an oscillating pressure. They, especially thermoacoustic Stirling prime movers with higher achievable efficiency, have received lots of attention in recent years due to their outstanding advantages of structure simplicity, heat-driven mechanism and environment-friendliness, etc. The onset temperature difference of thermoacoustic Stirling prime movers has crucial influence on the whole device's performance and is a very important parameter for utilizing low quality heat energy, so it is very significant to study the onset temperature difference under different parameters. In this paper we get an analytical expression for it and find out its dependence on various parameters such as the length and porosity of the regenerator, the working gas, the oscillating frequency, the working pressure, and so on.

MODEL

The schematic of the system that we will study in this paper is shown in Figure 1 and the deductions are based on the following assumptions: the temperature distribution along the regenerator is linear; the equation of state of perfect gas is valid; the viscous and thermal dissipations in all components are neglected except in the regenerator; the amplitude of the oscillation is small enough that linear thermoacoustic theory can be used.

According to the theory of literature [1, 2], the onset temperature gradient along the regenerator filled with mesh in thermoacoustic Stirling prime movers is

$$\left(\frac{dT}{dx}\right)_{cr} = \frac{\gamma-1}{a\beta_0}\,\omega\,\frac{g_{wk}+\dfrac{1}{\gamma-1}\left|Y_a\right|^2 g_{w\mu}}{R_e\left[Y_a f_{WT}\right]} \tag{1}$$

The research is financially supported by Chinese Academy of Sciences under Contract No. KJCX2-SW-W12-01

Where, $f_{WT} = (1-\Phi)\,\Omega/[\,j\omega\Phi(1-\Phi)+\Omega]$ denotes the factor of power production; $g_{wk} = Re(\,j*f_{WT}\,)$ denotes the factor of power dissipation by imperfect heat transfer; $g_{w\mu} = a\mu/\rho\omega$ denotes the factor of power dissipation by flow friction; $Y_a = \rho aU/(\varphi AP)$ denotes the local specific acoustic admittance. All the nomenclature for these physical parameters and mathematical operations used in this paper are listed in Table 1.

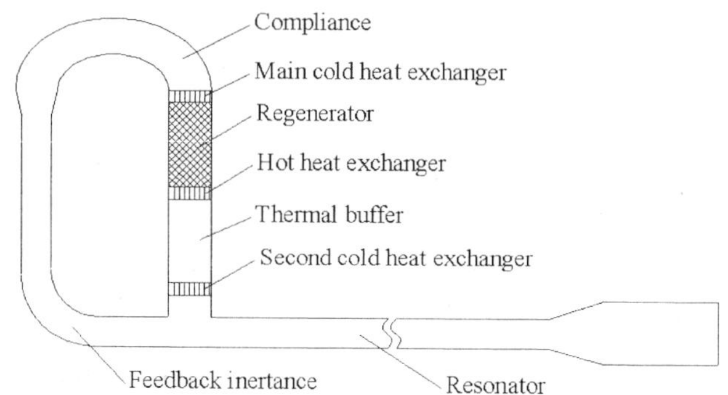

Figure 1 Schematic of a thermoacoustic Stirling prime mover

Obviously, when $dT/dx > (dT/dx)_{cr}$, the local acoustic power flow will be magnified; conversely, when $dT/dx < (dT/dx)_{cr}$, the local acoustic power will be attenuated.

Table 1 Nomenclature

α	Adiabatic sound velocity	A	Cross-sectional area of regenerator
C_0	Compliance of regenerator	C_1	Compliance of feedback tube
j	Imaginary unit	l	length of regenerator
L_1	Inertance of feedback tube	L_2	Inertance of thermal buffer tube
P	Amplitude of pressure	R	Flow resistance of regenerator
R_e	Real part of	T	Temperature
U	Amplitude of volume velocity		
a	Flow viscous coefficient	β_0	Thermal expansion coefficient
γ	Ratio of isobaric to isochoric specific heat	μ	Dynamic viscosity of working gas
ρ	Density of working gas	τ	Ratio of hot to cold end temperature
φ	Volume porosity of regenerator	Φ	Heat capacity ratio
ω	Angular frequency	Ω	Characteristic heat exchanger angular frequency

In the expression for Y_a, the ratio of U to P is unknown, which is the main impediment for us to calculate the onset temperature gradient. Fortunately, literature [3] presents a formula for U/P in the cold end of a regenerator:

$$\frac{U_{cold}}{P_{cold}} = \frac{\omega^2 C_1 L_1 + \dfrac{j\omega C_0 R}{2} g(\tau,b)}{R\dfrac{\tau+1}{2} f(\tau,b) + j\tau\omega L_2 + j\omega L_1} \tag{2}$$

Where, $f(\tau,b)=2*(\tau^{b+2}-1)/[(b+2)*(\tau^2-1)]$; $g(\tau,b)=2*[\tau^{b+2}\ln\tau-(\tau^{b+2}-1)/(b+2)]/[(b+2)*(\tau-1)^2]$, for helium, $b=0.68014$; τ denotes the ratio of hot-end to cold-end temperatures of the regenerator.

The onset temperature difference can be obtained by integrating the local temperature gradient along the regenerator. However, we already assumed that the temperature along the regenerator is linearly distributed, so we only need to calculate the onset temperature gradient in the cold end of the regenerator. Then we can obtain the onset temperature difference in a simpler way as given by equation (3).

$$\Delta T_{cr} = l \times \left(dT/dx\right)_{cr,cold} \tag{3}$$

NUMERICAL RESULT AND DISCUSSION

The main structure parameters used in this calculation are: the diameters of the regenerator and the thermal buffer tube are 8.75cm, 9cm, respectively, and their lengths are 8cm, 24cm, respectively; the feedback tube consists of two sections, one is 7.8cm in diameter and 56.7cm in length, the other is 10.34cm in diameter, 33.7cm in length.

Figure 2 is the distribution of onset temperature difference versus the length of the regenerator. The triangle, square and diamond lines correspond to different working pressures of 1MPa, 2MPa, 3MPa respectively. The working frequency is 80 Hz; the stainless-steel screen filled in the regenerator is 20-mesh. From this figure one can see that: the onset temperature difference always decreases first and then goes up as the regenerator length increases. It means that there is an optimum length of regenerator. For shorter regenerator, a lower working pressure has a lower onset temperature difference; conversely, for longer regenerator, the higher working pressure has a lower onset temperature difference.

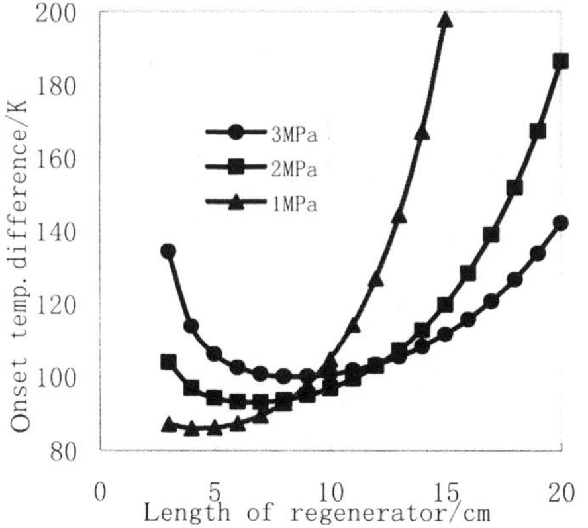

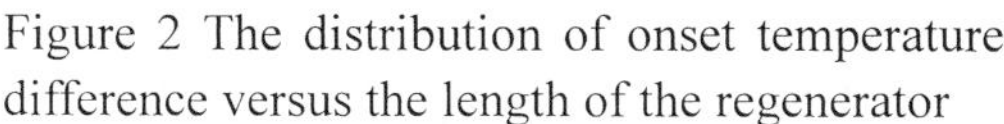

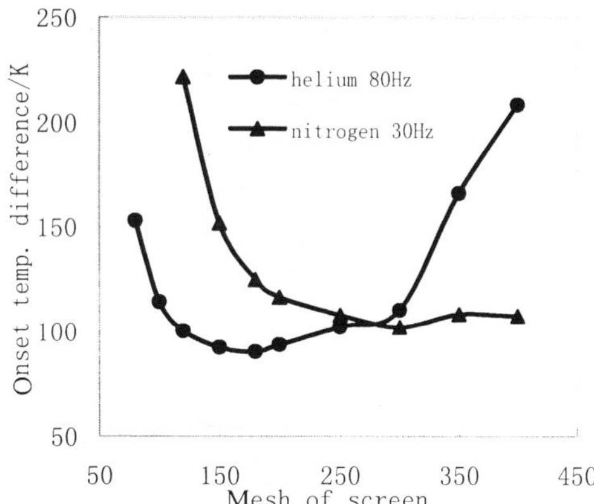

Figure 2 The distribution of onset temperature difference versus the length of the regenerator

Figure 3 The distributions of onset temperature difference versus the mesh of the screen

Figure 3 is the distribution of onset temperature difference versus the mesh of the screen, and the triangle and circle lines correspond to different working gases, helium and nitrogen. The working pressure is 3MPa and the frequency for helium is 80Hz, nitrogen, 50Hz. In general, with the increase of the mesh, the heat exchange between the screen and the working gas is strengthened, which makes the onset temperature difference lower. Conversely, the viscous dissipation increases and makes the onset temperature difference higher. Thus one can see in the figure the onset temperature difference decreases first and then goes up.

Figure 4 shows the onset temperature difference distribution of helium and nitrogen versus different frequencies. Obviously, the onset temperature difference of nitrogen is higher than that of helium, they all increase monotonically and the nitrogen is more sensitive to the frequency.

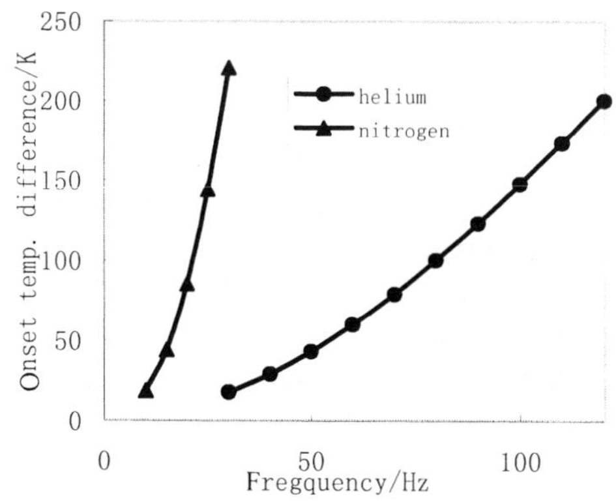

Figure 4 the onset temperature difference distributions versus working frequencies

CONCLUSIONS

From the above calculated results and analyses, some conclusions are drawn as follows:

(1) The onset temperature difference of a thermoacoustic Stirling prime mover deeply depends on the length and porosity of the regenerator, the working gas, the frequency and the pressure (notice: the size of feedback tube, the compliance and the thermal buffer tube is fixed in this study), this analytical model can provide some guidance for designing a low onset-temperature difference engine.

(2) The conclusions are based on some simplified assumptions, in which some factors that may influence the onset temperature difference are not considered in this model, a more accurate model will be developed in future.

REFERENCES

[1] J.H.Xiao, Thermoacoustic theory for cyclic flow regenerators, Cryogenics (1992) 32 895-901

[2] H.Ling, E.C.Luo, et al, Study on the Onset Temperature Gradient of Regenerators Used for Thermoacoustic Prime movers, 12th International Cryocooler Conference, USA (2002) 421-424

[3] M.Yang, E.C.Luo, et al, Thermodynamic Analysis on Traveling -Wave Thermoacoustic Devices with Distributed-parameter Network Method, 12th International Cryocooler Conference, USA (2002) 431-438

Two-dimensional numerical simulation of the inertance tube

Zhang Y[1]., Dai W[1]., Luo E[1]., Radebaugh R[2]., Lewis M[2].

1. Technical Institute of Physics and Chemistry, Beijing 100080, China
2. National Institute of Physics and Chemistry, CO80305, USA

Phase shifting is an important issue in the development of inertance tubes. Computational fluid dynamics (CFD) has the potential to assist in predicting operating conditions and designs that simulate the oscillating behavior inside inertance tubes. This paper reports the construction of a long inertance tube with a gas reservoir and a CFD model of this system using the commercial code, Fluent 6.1. These calculation results are compared with experimental results obtained by our group in cooperation with NIST of USA.

INTRODUCTION

In a pulse tube cooler, the acoustic power flow is proportional to the component of the mass flow in phase with the pressure. In most cases the optimum phase relationship is that phase between mass flow rate and pressure is zero at approximately the midpoint of the regenerator. Many methods were used to get the appropriate phase. Up to now, the introduction of double-inlet phase shifter has greatly improved the efficiency of pulse tube coolers. The multiple bypass inlet pulse tube coolers have also occurred. The Stirling coolers use two pistons to adjust the phase at the regenerator, but the pulse tube coolers only can use other passive methods that are discussed above to adjust the phase.

The inertance tube can serve as a phase shifter[1], which adjusts the phase relation between mass flow rate and pressure. It is important for oscillating flow systems such as high frequency pulse tube coolers and thermoacoustic machines to achieve high efficiency. Up to now, the experimental data on inertance tube is very limited and it is difficult to obtain a simple formula to analyze the experiments due to complex oscillating flow behavior. On the other hand, numerical research reports are also scarce, mostly based on linear model or one-dimensional model.

Most importantly, we want to find a good and relatively simple method to predict inertance tube performance and use it as a guideline for applications. And the first objective of the project is to check out if this design of structure will lead to optimal performance and efficient phase adjustment via an analysis of the power flow, volume flow rate and the pressure in the inertance tube. Computer modeling using commercial Computational Fluid Dynamics (CFD) packages, such as Fluent can generate pressure and mass flow rate data which are then tested experimentally in an inertance tube system. And this paper describes a test system which has been modeled using Fluent 6.1.

CFD MODELS

<u>Inertance tube system geometry and computational grid generation</u>

The research is financially supported by Chinese Academy of Sciences under Contract KJCX2-SW-W012-01.

The physical model is a long inertance tube connected with a reservoir (Figure 1). The geometrical model in this system consists of two subregions. The first is the inertance tube. Due to the long and thin shape of the long tube, a two-dimensional, axis-symmetry model was created. It consisted of a 1689 mm length, 1.016 mm diameter copper tube that allowed for the contained gas to vibrate in. This part was divided into almost 135000 quadrilateral cells, and they are all of the same size. The second subregion consists of the gas reservoir. With a dimension of 40 mm, the gas reservoir is 59mm length. This part was divided into almost 150000 cells. And the standard $k-\varepsilon$ model was used to simulate the system [2].

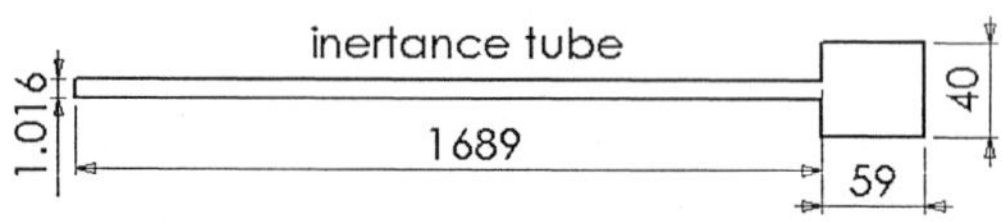

Figure 1   Diagram of the inertance tube system

<u>Initial and boundary conditions</u>
For solving the governing equations, the relevant initial and boundary conditions are as follows. Initially the system is at rest and at uniform temperature T=293K. The velocity is zero throughout the fluid field. The system is filled with He at a mean pressure of 2.5MPa, and the pressure ratio is set from 1.05 to 1.45. This oscillating pressure is set as the pressure boundary condition at the pressure inlet surface, which is at the leftmost of the system.

The thermal conductivity of all the walls was specified. Furthermore, a uniform convective heat transfer boundary condition was defined at the outer side of solid boundary. A uniform convective surface heat transfer coefficient $h = 40\ W/(m^2 \cdot K)$ was assumed at the boundary.

RESULTS AND DISCUSSION

The details of experiments for the verification of the CFD calculation can be found elsewhere[3]. The oscillation in the reservoir is not sinusoidal wave. In order to get the phase, we firstly analyze the data into sinusoidal wave via a fast-Fourier-transform algorithm. Figure 2 illustrates the phase angle difference of pressure between inlet and gas reservoir respectively obtained from experiment and CFD simulation. Figure 3 shows the measured and simulated phase angle of pressure leading mass flow rate at the pressure inlet surface.

Figure 4 shows the contrast of measured and simulated mass flow from the inertance tube to the reservoir. With the increase of the pressure ratio, the amplitude of the oscillating mass flow rate rises gradually. In the experiment, the gas reservoir was treated as an adiabatic gas capacitance. And the mass flow rate and volume flow rate passing in and out the gas reservoir can be calculated out. Just as what the figure illustrates, the experimental data and the simulation data agree well with each other.

Figure 5 shows the measured and simulated mass flow at the pressure inlet surface. The experimental data was measured by hot-wire anemometer. The measuring accuracy is not high due to the difficulty in measuring oscillating flow. So the discrepancy could be attributed to both experimental and calculation errors.

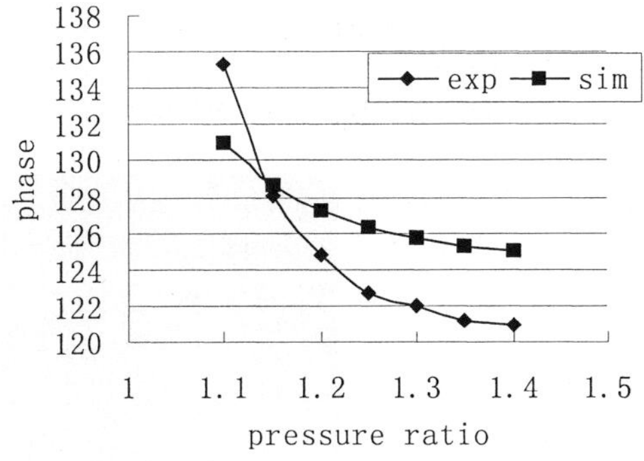

Figure 2　Measured and simulated phase difference of pressure between inlet and the reservoir

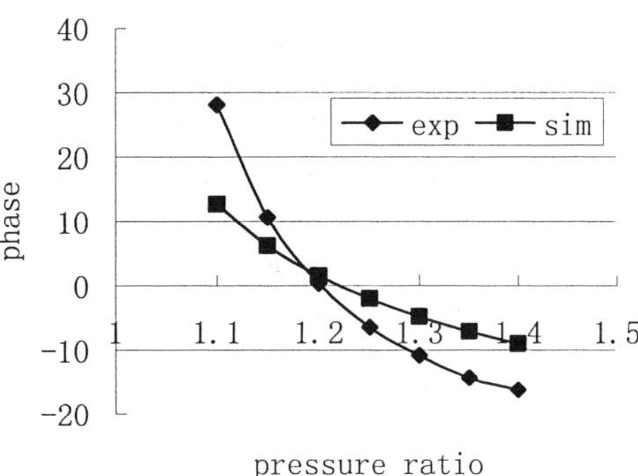

Figure 3　Measured and simulated phase difference between pressure and mass flow rate flow at the inlet

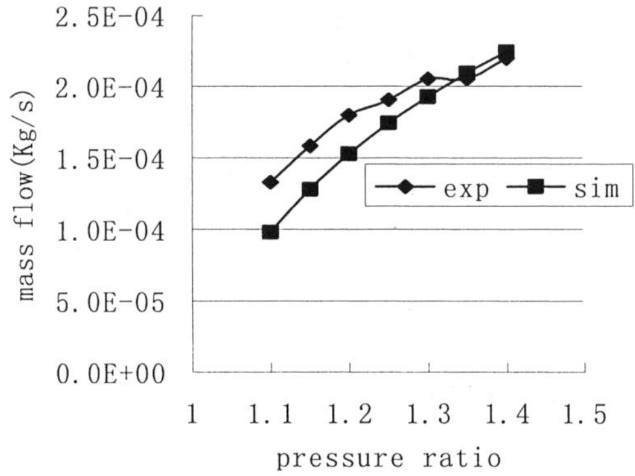

Figure 4　Measured and simulated mass flow rate from the inertance tube to the reservoir

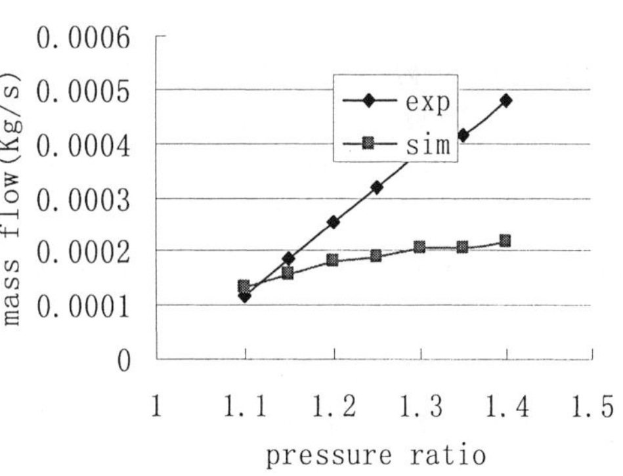

Figure 5　Measured and simulated mass flow rate at the pressure inlet

Figure 6 shows the acoustic power at the inlet. The acoustic intensity across the cross-sectional area of the channel:

$$W = \frac{1}{2}\left|\widetilde{p}_1 \widetilde{U}_1\right|\cos\theta_{Pu} = \frac{1}{2}\left|\frac{\widetilde{p}_1}{p_0}\right|\dot{m}R_M T_0 \cos\theta_{pu} \tag{1}$$

Where θ_{pU} is the phase angle between $\widetilde{p}_1$ and $\widetilde{U}_1$ and the tilde denotes complex conjugation. This is the acoustic power flowing in the x direction. Both the experimental data and the simulation data are all calculated from equation (1). The power is basically in direct proportion to the product of corresponding mass flow rate.

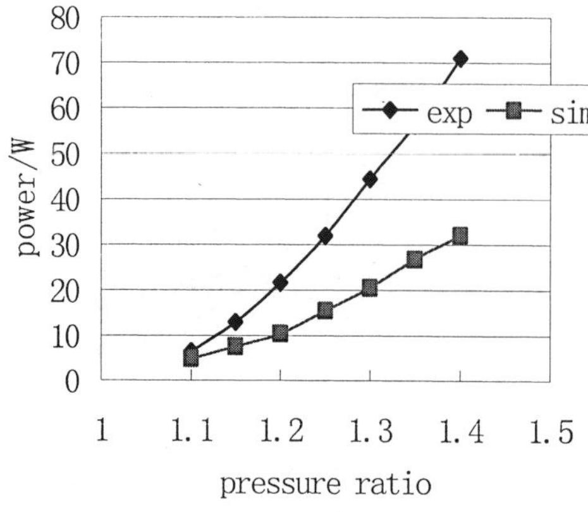

<table>
<tr><td>

Figure 6   Measured and simulated power at the pressure inlet surface

</td><td>

Figure 7   Simulated contour of velocity magnitude at the joint with the reservoir

</td></tr>
</table>

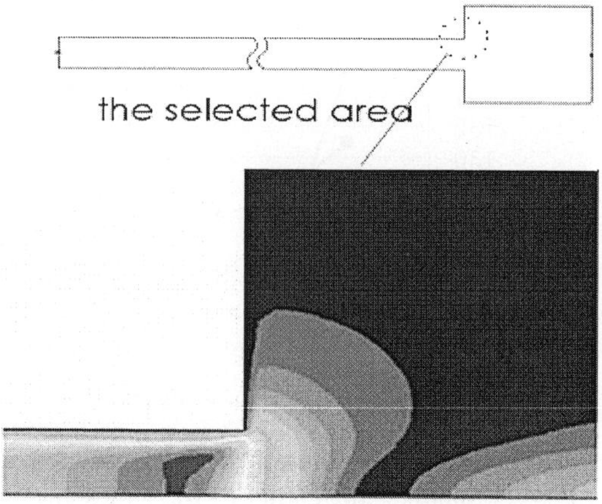

Figure 7 shows the contour of velocity magnitude at the joint of inertance tube and the gas reservoir. When the gas flows out of the inertance tube into the reservoir, the velocity quickly decreases. This will greatly influence the static pressure at this small inlet region in both phase and amplitude.

CONCLUSIONS

We use a two-dimensional, turbulence model to simulate the performance of an inertance tube. The sinusoidal pressure wave causes an unsteady oscillating flow in the tube. We contrast the simulation data, including phase between pressure at the inertance inlet and the inertance outlet, and that between pressure and mass flow rate at the inlet, also mass flow rate at inlet and outlet with the experimental data. Through these data, we can compute out the power at the inlet. The agreement between the values calculated by the previous model and experimental data is quite encouraging. Based on the present model, guidance for the design, operation, and control of inertance tube can be prepared for achieving optimal design, as well as safe and economical operation.

REFERENCES

1.   D. L. Garden, G. W. Swift, Use of inertance in orifice pulse tube refrigerators, Cryogenics (1997) 37 117-121

2.   Jacek Smolka, Andrzej J. Nowak , Luiz C. Wrobel, Numerical modeling of thermal processes in an electrical transformer dipped into polymerized resin by using commercial CFD package fluent, Computers & Fluids (2004) 33 859–868

3.   E. Luo, Ray Radebaugh, M. Lewis, Inertance tube models and their experimental verification, Adv. Cryo. Eng., vol 49 (to be published), American Institute of Physics (2004)

Numerical simulation on a novel cascade thermoacoustic prime mover

Ling H., Luo E., Dai W., Li X.[*]

Technical Institute of Physics and Chemistry, Chinese Academy of Sciences, Beijing 100080, China
Graduate School of Chinese Academy of Sciences, Beijing 100039, China
[*]Cryogenic Lab, Zhejiang University, Hangzhou, 310027, P.R.China

A novel cascade thermoacoustic prime mover is numerically studied based on linear thermoacoustic theory. With the structure of system and the external conditions such as heat input and room temperature given, the fundamental resonant frequency and the distributions of all concerned variables are gained. From the simulation, it is proved that the key components of the machine—the regenerator and stack can work under the traveling-wave mode and standing-wave mode respectively when there is an imaginary load, so the efficiency is improved. Some particular characteristics of this new engine are described.

INTRODUCTION

It is well known that heat and acoustic energy can be converted each other by thermoacoustic process, and a thermoacoustic engine is a device to achieve the conversion. Different modes of machine do not have the same conversion efficiency, for example, a standing-wave engine (SWE) has lower efficiency because of an intrinsically irreversible thermodynamic cycle; and a traveling-wave engine (TWE) has 50% higher efficiency due to excellent thermal contact in regenerator. But it is inevitable that the Gedeon streaming would come into being in the traveling-wave loop, which results in a decreased efficiency. Another point of disadvantage is that there exits a huge thermal stress in the loop. Recently a cascade thermoacoustic engine was invented, whose geometry, only straight-line topology, is simple to build like the standing-wave engine, and achieved efficiency up to 20%[1]. In this paper we will make a numerical analysis for this type of machine. The calculation is based on linear thermoacoustic theory [2], which is a powerful tool to understand the thermoacoustic effect under low amplitude.

MODEL AND NUMERICAL METHOD

In an ideal half-wavelength standing-wave tube, the phase differences between oscillating pressure and volume velocity are always 90°. However, for a lossy one, there is a usual transition zone, where the traveling-wave component becomes dominant. If a regenerator is placed at this special location of a standing-wave engine, the acoustic power coming from a standing-wave stage will be amplified. The total efficiency of the cascade machine could be likely increased, since the traveling-wave Stirling engine is more efficient than the primary stage. However, not any in-phase zone adapt to be inserted a regenerator,

This work is financially supported by Chinese Academy of Sciences under contract KJCX2-SW-W12-01.

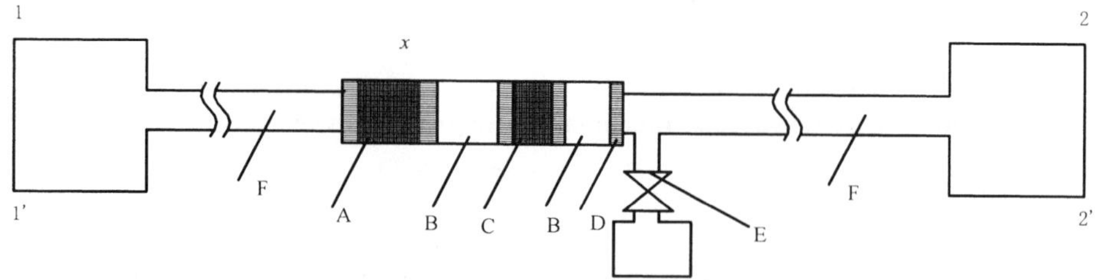

Figure 1   Schematics of a cascade thermoacoustic engine with 2-open ends

or large viscous losses could be produced. In fact, only the one with high acoustic impedance would be a suitable choice. So a standing-wave pipe attached to two infinitely large reservoirs is selected as a model. For this pipe, its fundamental frequency is expected. Figure1 is the schematic of our model cascade engine, which comprises of (A)standing-wave stage(first engine), (B)thermal buffer tube*(TBT)* (C)traveling-wave stage(second engine), (D)ambient heat exchanger, (E)dissipation load and (F)resonator. The two reservoirs simulate open boundary condition.

Using complex notation [3], we have the following universal expressions for any components,

$$\frac{d}{dx}\begin{bmatrix} \tilde{p} \\ \tilde{U} \end{bmatrix} = \overrightarrow{A_F} \cdot \begin{bmatrix} \tilde{p} \\ \tilde{U} \end{bmatrix} \tag{1}$$

$$\frac{d}{dx}\begin{bmatrix} T_0 \\ E_x \end{bmatrix} = \overrightarrow{A_T} \cdot \begin{bmatrix} T_0 \\ E_x \end{bmatrix} + \vec{B} \ , \tag{2}$$

where $\tilde{p}, \tilde{U}$ are the oscillating pressure and volume velocity, T_0, E_x are the mean temperature and second-order total energy flux, which is caused by the hydrodynamic transportation of enthalpy and heat conduction of working medium. We impose the boundary condition of （i）volumetric velocity $\tilde{U} = 0$ (ii) energy flux $E_x = 0$ at the two ends (namely Section11' and 22' in Figure 1). We can obtain the following solution by simple integral

$$\begin{bmatrix} \tilde{p}(x) \\ \tilde{U}(x) \end{bmatrix} = \vec{F}(x,x_0)\begin{bmatrix} \tilde{p}(x_0) \\ \tilde{U}(x_0) \end{bmatrix} \tag{3}$$

$$\begin{bmatrix} T_0(x) \\ E_x(x) \end{bmatrix} = \vec{G}(x,x_0)\begin{bmatrix} T_0(x_0) \\ E_x(x_0) \end{bmatrix} + \vec{S}(x,x_0), \tag{4}$$

where $\vec{F}(x, x_0)$ is a 2×2 flow transmission matrix(FTM), $\vec{G}(x,x_0)$ is a 2×2 thermal transfer matrix(TTM) and $\vec{S}(x, x_0)$ is a 2×1 thermal response vector (TRV). According to Eqs(3),(4), the acoustical variables $\tilde{p}, \tilde{U}$ and thermal variables T_0, E_x could be respectively regarded as a 4-port network element with 2-input and 2-output. Once the FTM, TTM and TRV of all components are obtained, the whole system is decided, including frequency, oscillating variables and temperature distribution. However, there are couplings between the two matrixes and the response vector, the concerned parameters can only be obtained analytically by iteration. Before starting a simulation, we must provide initial values for oscillating frequency and the distribution of temperature. Then the physical property and the frequency-dependent FTM can be calculated. By solving the acoustical equation, a new frequency is obtained. Then, the distribution of pressure and velocity could be obtained quickly according to Eq.(3). The next step is to solve the temperature distribution by calculating the TTM, TRV with boundary condition of energy equation. Generally, the gained temperature is not same as one given before, so iteration has to be used to repeat above steps until temperature discrepancy between two iterations is fallen within the required accuracy.

COMPUTATION RESULTS AND ANALYSIS

According to the aforementioned analysis, we design a cascade apparatus to match the tested traveling-wave thermoacoustic refrigerator, which works appropriately at a frequency of about 50 Hz. The main parameters are given in Table 1. The heat exchangers have the structure of many parallel channels formed by copper fins with a thickness of 1.0 mm and 1.0 mm gap. The stack and regenerator are made from stainless steel screens of 20 mesh and 120 mesh respectively. The other components are straight stainless tubes. The working fluid is helium gas and the flow is laminar.

Table 1   Main geometry parameters of model

	Reservoir (2)	Resonator (2)	Ambient Heat Exchanger (3)	Stack	Regenerator	Hot heat Exchanger (2)	TBT(2)
Length (m)	0.50/0.50	3.85 / 4.70	0.04/0.04/0.04	0.21	0.08	0.05/0.05	0.30/0.10
Diameter (m)	0.35	0.031	0.05	0.05	0.05	0.05	0.05

In order to examine the performance of cascade engine, we add an imaginary dissipation load R_{load} in the output of the ambient heat exchanger. By adjusting a velocity factor α, $U_{load} = |\bar{U}_{load}| = \alpha U_{Resonator}$, we can gradually shift the load, where U the magnitude of $\bar{u}$. $\alpha = 0$ means there is no load and the load resistance is infinity. As the factor α rising, the load resistance will decrease monotonically to zero. In this case, the produced mechanical energy is completely consumed in load.

Figure 2 is the calculated distribution of several interested variables for three loads($\alpha = 0, 0.7, 1.9$) at mean pressure 2.0 MPa and the total heating power 800 W (the standing-wave stage and traveling-wave stage share the quantity averagely). The pressure antinode is roughly located at the centre of the apparatus

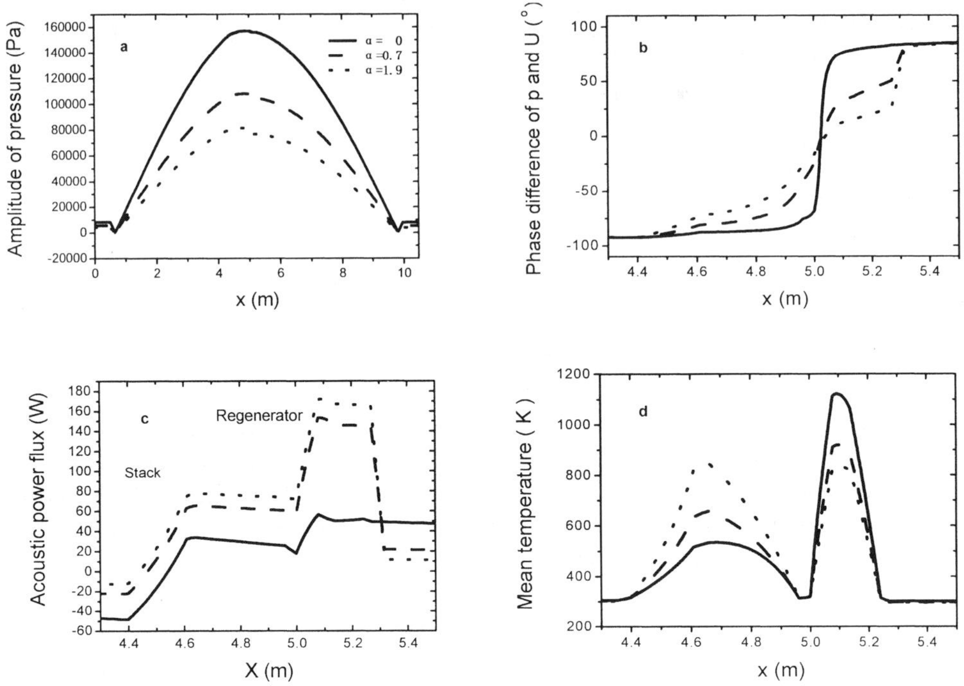

Figue 2   Distributions of variables of (a) Oscillating pressure (b) Phase difference between pressure and velocity (c) Acoustic power flux (d) Mean temperature for different α

(see Figure 2a), where simultaneously is the node of velocity. At the entrances of the two reservoirs, pressure amplitude decrease almost to zero (node) and then rebound to a small value which is nearly stable in reservoir. The oscillation will be weaker while improving the factor α, because more power is used up by the load. Figure 2b is the distribution of phase difference between pressure($\bar{p}$) and velocity($\bar{U}$), which expressly indicates the influence of load to the cascade system. When $\alpha = 0$, the traveling-wave region becomes short abruptly vertical and a majority part of the regenerator is already out of in-phase zone, where the phase difference changes from -71.1° to 70.8°. It can be imagined the working status of second engine would be bad. As α rising, the curve is more evenly and the in-phase zone becomes wider. For example $\alpha = 1.9$, the pressure lags velocity by 18.2° at the cold end of regenerator, and leads velocity by 10.3° at hot end, which means the regenerator is working under the traveling-wave mode mainly. It will be very beneficial for the traveling-wave stage to work normally with gas parcels experiencing Stirling cycle and to produce more acoustic work with high efficiency (see Figure 2c). Figure 2d shows that the fist and second engine have different demand for temperature of heat source despite the same heat power. When no acoustic power is delivered, the standing-wave stage needs a lower one and the traveling-wave stage needs a high one. With more acoustic power consumed by the load, the temperature gradient of the stack will increase and that of regenerator will decrease, which means that there should be a limit to the load, or the hot heat exchanger would be burned out. Figure 3 shows the relations of the thermal efficiency and the factor α, where η_1, η_2, η_3 are respectively that of the first engine stage, the second stage and the entire engine. We find η_1 varies slightly; η_2 increases quickly first and transcends η_1, but keeps stable in succession; η_3 is similar to that of η_2. It indicates that, the final delivered power is limited and controlled by the inherent performance of the system.

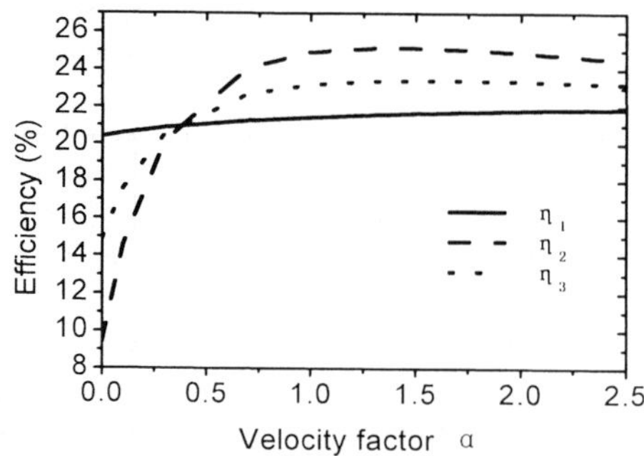

Figure 3   Efficiency versus velocity factor α

CONCLUSIONS

The performance and characteristics of a cascade thermoacoustic engine have been simulated successfully using linear thermoacoustic theory. The calculated results indicate that the cascade engine can operate with a higher thermal efficiency than a single standing-wave stage. It the meantime, because there are different development trend of heat source temperature for the two engine stages, the load has a great influence on the stability of the cascade system. Much efforts on the engine needs to be made in future.

REFERENCES

1. Gardner,D.L. and Swift,G.W., A cascade thermoacoustic engine, J.Acoust.Soc.Am.(2003)114(4),Pt.1, 1905-1919
2. Swift, G.W., Thermoacoustic engine. J.Acoust.Soc.Am.(1988) 84(4),1145-1181
3. Xiao, J.H., Thermoacoustic heat transportation and energy transformation Part 1, Cryogenics (1995) 35(1), 15-19.

New perspective on thermodynamic cycles of oscillating flow regenerators

Luo E.[1], Dai W.[1], Radebaugh R.[2]

1. Refrigeration and Cryogenic Division, Technical Institute of Physics and Chemistry, Chinese Academy of Sciences, P.O.Box 2711, Beijing 100080, CHINA
2. Cryogenic Technology Group, Physical and Chemical Properties Division, National Institute of Standards and Technology, Mail Stop 305, Boulder, CO80305, USA

Regenerative machines usually operate in oscillating flow and the regenerators for the machines are the most important thermodynamic elements. However, the real function of the cyclic regenerator has not been understood thoroughly. This paper tries to reveal a unique working mechanism for a regenerator; that is, the regenerator functions as a recuperator but also finishes a complete thermodynamic cycle.

INTRODUCTION

The operations of regenerative machines rely on the compression and expansion of working gas. Classical thermodynamics thinks that a regenerator used in the regenerative machines has been taken just as a recuperative heat exchanger used in a recuperative cycle. This is not true, however, even for an ideal thermodynamic cycle, i.e., without any loss mechanism but having some 'dead' volume for storing working gas.

The thermoacoustic theory has been recently developed to explain working mechanisms of various thermoacoustic machines including traditional regenerative machines like Stirling machines. From the viewpoint of the thermoacoustic theory, the regenerator is an active thermodynamic element, which can reversibly consume some mechanical power and then pump the heat from the cold side to hot side of the regenerator. This viewpoint distinctly noticed that the regenerator consists of the compression, expansion, heat rejection and heat absorption processes. However, perhaps due to the sophisticated thermoacoustic phenomena and due to the Eulerian viewpoint of the thermoacoustic theory, the intuitional explanation has not been well developed yet so that some aspect of the regenerator is ignored such as the recuperative function. This paper tried to reveal unique function of a cyclic regenerator in terms of Lagrangian viewpoint.

PHYSICAL MODEL AND THERMODYNAMIC ANALYSIS

A regenerator is usually a kind of porous solid matrix filled with working gas. Fig.1 schematically shows a typical regenerator in which presents three typical gas parcels. For simplicity, the following assumptions are made: (1) a perfect heat transfer between working gas and solid medium; (2)an inviscous, ideal gas. We first consider a general case that the phase difference between pressure and velocity oscillations is θ $(-\pi/2 < \theta < \pi/2)$. Having the velocity phase as a base, there are the following expressions for the velocity and pressure.

$$u(t) = u_d \sin \omega t \tag{1}$$

$$p(t) = p_0 + p_d \sin(\omega t + \theta) \tag{2}$$

The research is financially supported by Chinese Academy of Sciences under Contract KJCX-SW-W012-01.

338

Moreover, the displacement of the gas parcel as the function of time can be expressed by Eq.(3), which is schematically shown in Figure 2, and the time-dependent temperature of the gas parcel can be given by Eq.(4).

$$X(t) = \int u\,dt = -\frac{u}{\omega}\cos\omega t = -X_d \cos\omega t \tag{3}$$

$$T(t) = T(X(t)) = T_0 + \frac{dT_0}{dx}X(t) = T_0 + \frac{dT_0}{dx}(-\frac{u_d}{\omega}\cos\omega t) = T_0 + T_d \cos\omega t \tag{4}$$

Once the time-dependent temperature and pressure of the gas parcel are known, it is also readily to have the time-dependent volume and entropy for the gas parcel.

$$v(t) = \frac{RT_0}{p_0}[1 + \frac{T_d}{T_0}\cos\omega t - \frac{p_d}{p_0}\sin(\omega t + \theta)] \tag{5}$$

$$s(t) = C_p \ln\frac{T}{T_{ref}} - R\ln\frac{p}{p_{ref}} = C_p \ln\frac{(T_0 + T_d \cos\omega t)}{T_{ref}} - R\ln\frac{[p_0 + p_d \sin(\omega t + \theta)]}{p_{ref}} \tag{6}$$

According to Eq.(2) and Eq.(5), we can give the p-v diagram of the gas parcel. Similarly, the T-s diagram of the gas parcel can be given by Eq.(4) and Eq.(6).

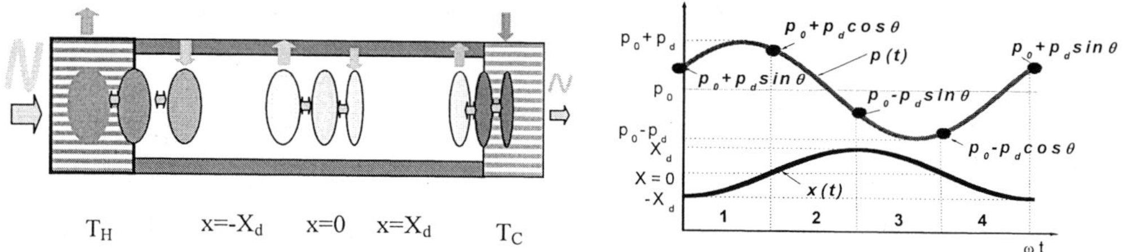

Figure 1. A cyclic regenerator   Figure 2. Pressure and displacement

Here we evenly divide a complete thermodynamic cycle into four processes by time: (1) $0 < \omega t < \pi/2$, when the gas parcel moves from its left peak position to its counterbalance position; (2) $\pi/2 < \omega t < \pi$, when the gas parcel moves from its counterbalance position to its right peak; (3) $\pi < \omega t < 3\pi/2$, when the gas parcel returns from its right peak position back to its counterbalance position; (4) $3\pi/2 < \omega t < 2\pi$, when the gas parcel continues back to its left peak position. So far, the gas parcel finishes a complete thermodynamic cycle. The following section gives the conversion and transportation of thermal energy and acoustical power happening in the four processes.

In terms of thermodynamics, the gas parcel does the p-v power within the time range of t_1 to t_2 can be calculated by the following equation.

$$w(t) = \int_{t_1}^{t_2} p\,dv = \omega RT_0\{(-t\frac{1}{2}\frac{p_d}{p_0}\frac{T_d}{T_0}\cos\theta) + [\frac{1}{\omega}\frac{T_d}{T_0}\cos\omega t - \frac{1}{\omega}\frac{p_d}{p_0}\sin(\omega t + \theta)]$$

$$+ [\frac{1}{4\omega}\frac{p_d}{p_0}\frac{T_d}{T_0}\sin(2\omega t + \theta) + \frac{1}{4\omega}(\frac{p_d}{p_0})^2\cos(2\omega t + 2\theta)]\}\,|_{t_1}^{t_2} \tag{7}$$

Four these four thermodynamic processes（process 1: $0 < \omega t < \dfrac{\pi}{2}$, process 2: $\pi/2 < \omega t < \pi$, process 3: $\pi < \omega t < 3\pi/2$ and process 4: $3\pi/2 < \omega t < 2\pi$）, the p-v power, the change of internal energy and the absorption heat for the gas parcel are given by Eq.(8) to Eq.(19), respectively.

$$\Delta w_1 = RT_0\{(-\frac{\pi}{4}\frac{p_d}{p_0}\frac{T_d}{T_0}\cos\theta) + [-\frac{T_d}{T_0} - \frac{p_d}{p_0}(\cos\theta - \sin\theta)] - [\frac{1}{2}(\frac{p_d}{p_0})^2\cos 2\theta + \frac{1}{2}\frac{p_d}{p_0}\frac{T_d}{T_0}\sin\theta]\} \tag{8}$$

$$\Delta u_1 = -C_v T_d \tag{9}$$

$$\Delta q_1 = \Delta u_1 + \Delta w_1$$

$$= -C_p T_d + RT_0 [(-\frac{\pi}{4}\frac{p_d}{p_0}\frac{T_d}{T_0}\cos\theta) + [-\frac{p_d}{p_0}(\cos\theta - \sin\theta)] - [\frac{1}{2}(\frac{p_d}{p_0})^2\cos 2\theta + \frac{1}{2}\frac{p_d}{p_0}\frac{T_d}{T_0}\sin\theta]\} \tag{10}$$

$$\Delta w_2 = RT_0\{(-\frac{\pi}{4}\frac{p_d}{p_0}\frac{T_d}{T_0}\cos\theta) + [-\frac{T_d}{T_0} + \frac{p_d}{p_0}(\sin\theta + \cos\theta)] + [\frac{1}{2}(\frac{p_d}{p_0})^2\cos 2\theta + \frac{1}{2}\frac{p_d}{p_0}\frac{T_d}{T_0}\sin\theta]\} \tag{11}$$

$$\Delta u_2 = -C_v T_d \tag{12}$$

$$\Delta q_2 = -C_p T_d + RT_0\{(-\frac{\pi}{4}\frac{p_d}{p_0}\frac{T_d}{T_0}\cos\theta) + [\frac{p_d}{p_0}(\sin\theta + \cos\theta)] + [\frac{1}{2}(\frac{p_d}{p_0})^2\cos 2\theta + \frac{1}{2}\frac{p_d}{p_0}\frac{T_d}{T_0}\sin\theta]\} \tag{13}$$

$$\Delta w_3 = RT_0\{(-\frac{\pi}{4}\frac{p_d}{p_0}\frac{T_d}{T_0}\cos\theta) + [(\frac{T_d}{T_0} + \frac{p_d}{p_0}(\cos\theta - \sin\theta)] - [\frac{1}{2}(\frac{p_d}{p_0})^2\cos 2\theta + \frac{1}{2}\frac{p_d}{p_0}\frac{T_d}{T_0}\sin\theta]\} \tag{14}$$

$$\Delta u_3 = C_v T_d \tag{15}$$

$$\Delta q_3 = C_p T_d + RT_0\{(-\frac{\pi}{4}\frac{p_d}{p_0}\frac{T_d}{T_0}\cos\theta) + [\frac{p_d}{p_0}(\cos\theta - \sin\theta)] - [\frac{1}{2}(\frac{p_d}{p_0})^2\cos 2\theta + \frac{1}{2}\frac{p_d}{p_0}\frac{T_d}{T_0}\sin\theta]\} \tag{16}$$

$$\Delta w_4 = RT_0\{(-\frac{\pi}{4}\frac{p_d}{p_0}\frac{T_d}{T_0}\cos\theta) + [(\frac{T_d}{T_0} - \frac{p_d}{p_0}(\cos\theta + \sin\theta)] + [\frac{1}{2}(\frac{p_d}{p_0})^2\cos 2\theta + \frac{1}{2}\frac{p_d}{p_0}\frac{T_d}{T_0}\sin\theta]\} \tag{17}$$

$$\Delta u_4 = C_v T_d \tag{18}$$

$$\Delta q_4 = C_p T_d + RT_0\{(-\frac{\pi}{4}\frac{p_d}{p_0}\frac{T_d}{T_0}\cos\theta) + [-\frac{p_d}{p_0}(\cos\theta + \sin\theta)] + [\frac{1}{2}(\frac{p_d}{p_0})^2\cos 2\theta + \frac{1}{2}\frac{p_d}{p_0}\frac{T_d}{T_0}\sin\theta]\} \tag{19}$$

During the complete thermodynamic cycle, the net p-v power of the gas parcel is given by Eq.(20) and the net heat exchange in the right-half and left-half zones can be given by Eqs.(21) and (22).

$$\sum \Delta w = \Delta w_1 + \Delta w_2 + \Delta w_3 + \Delta w_4 = -\pi RT_0 \frac{p_d}{p_0}\frac{T_d}{T_0}\cos\theta \tag{20}$$

$$\Delta q_c = q_2 + q_3 = RT_0[-\frac{\pi}{2}\frac{p_d}{p_0}\frac{T_d}{T_0}\cos\theta + 2\frac{p_d}{p_0}\cos\theta)] = 2RT_0\frac{p_d}{p_0}\cos\theta(1 - \frac{\pi}{4}\frac{T_d}{T_0}) \tag{21}$$

$$\Delta q_h = q_1 + q_4 = RT_0(-\frac{\pi}{2}\frac{p_d}{p_0}\frac{T_d}{T_0}\cos\theta - 2\frac{p_d}{p_0}\cos\theta) = -2RT_0\frac{p_d}{p_0}\cos\theta(1 + \frac{\pi}{4}\frac{T_d}{T_0}) \tag{22}$$

Furthermore, there is the following correlation for the three parameters.

$$|\Delta q_h| = |\Delta q_c| + |\Delta w| \tag{23}$$

Comprehensively analyzing Eqs.(20) to(23), it is straightforward to achieve such a conclusion that the gas parcel undergoes a complete thermodynamic cycle in which includes compression, heat rejection, expansion and heat absorption; in addition, the gas parcel reversibly consumes a net p-v power $|\Delta w|$, reject a net heat of $|\Delta q_h|$ in hot zone and absorbs a net heat of $|\Delta q_c|$. Actually, Figure 3 and Figure 4 show the p-v and T-s diagrams of the gas parcel, respectively. In a similar way, the p-v and T-s diagrams of all other gas parcels can be given, too. If combining all the diagrams, we can give the p-v and T-s diagram locus of all gas parcels, which is shown in Figure 5 and Figure 6. Two isobaric lines for the locus imply that the thermodynamic cycle is not Stirling cycle. This point is quite different from traditional thermodynamics.

Now let us look at the recuperative process. The net reject heat from the gas to the regenerator solid is given by Eq.(24) when the gas parcel moves from the left-peak position (hot end) to the right-peak position (cold end), and the net absorption heat from the regenerator is given by Eq.(25).

$$\Delta q_{hot-cold} = -2C_p T_d + RT_0 \frac{\pi}{2}\frac{p_d}{p_0}\frac{T_d}{T_0}\cos\theta + 2\frac{p_d}{p_0}\sin\theta \tag{24}$$

$$\Delta q_{cold-hot} = 2C_p T_d - RT_0 \frac{\pi}{2}\frac{p_d}{p_0}\frac{T_d}{T_0}\cos\theta - 2\frac{p_d}{p_0}\sin\theta \tag{25}$$

340

$$\Delta q_{hot-cold} + \Delta q_{cold-hot} = 0 \qquad (26)$$

Moreover, the net recuperative heat is $\Delta q_{hot-cold} + \Delta q_{cold-hot}$, exactly equal to 0! This is just the traditional function of the regenerator.

The analysis above is for a general case. Here we discuss some extreme cases: one is pure a standing-wave mode and the other is a pure traveling-wave mode in terms of thermoacoustics. For a pure standing-wave mode, one can readily see the net p-v power, the net absorption heat in cold side and the net rejection heat in hot side, all three, are zero, when setting $\theta = 90°$. Thus, pure standing-wave mode does not work. For a pure traveling-wave mode it is readily to see the three parameters are maximized, when setting $\theta = 0°$.

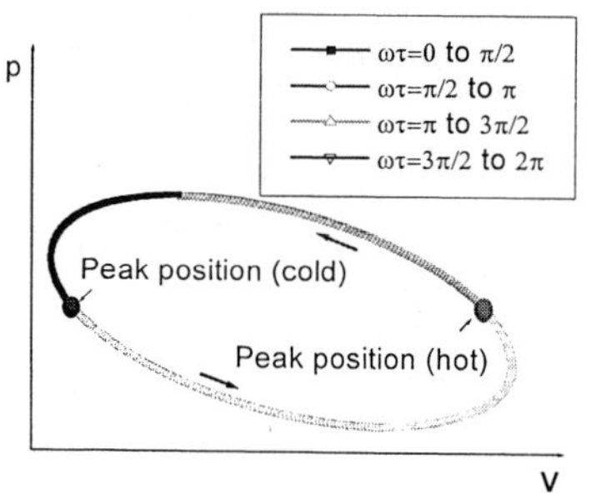

Figure 3. p-v diagram

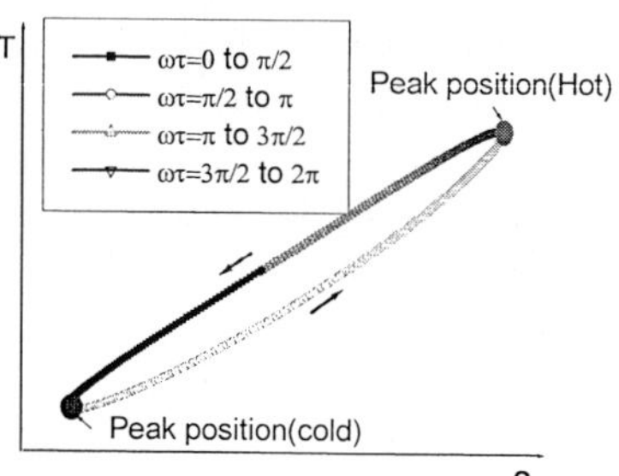

Figure 4. T-s diagram

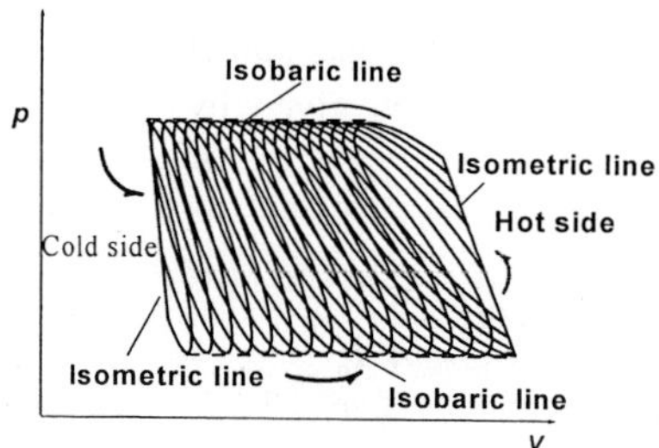

Figure 5. p-v Locus of all gas parcels

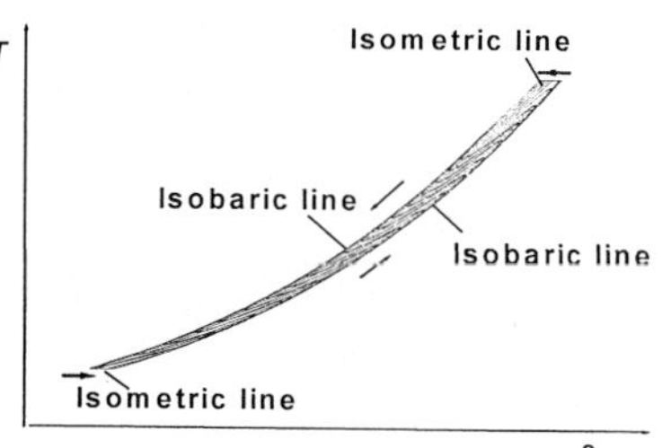

Figure 6. T-s locus of all gas parcels

CONCLUSION

Based on Lagrangian viewpoint, we analyzed the fundamental operation of an ideal regenerator. The analysis shows that the regenerator is active element in which the gas parcel undergoes a complete thermodynamic cycle. The gas parcel does consume a net mechanical power and then does pump heat from cold place to hot place. In the meantime, the regenerator accompanies by a recuperative process. Thus, the regenerator involves compression, heat rejection, expansion, heat absorption and traditional recuperative processes. These processes are parasite together. The regenerator with oscillating flow is definitely not a pure recuperative heat exchanger and it is more like a complete heap pump, which is accompanied by the recuperative function. One of these two functions cannot be separated independently; otherwise the regenerator does not work.

REFERENCES

1. Liang J., Thermodynamic cycles in oscillating flow regenerators, J.Appl.Phys., Vol.82,No.9, pp.4159-4165,1997
2. Xiao J.H., Thermoacoustic theory for cyclic flow regenerator, Cryogenics 32,10(1992), pp.895-901

Experimental investigation on a room-temperature traveling-wave thermoacoustic refrigerator

Huang Y.[1,2], Luo E.[1], Dai W.[1], Wu Z.[1,2]

[1]Technical Institute of Physics and Chemistry of CAS, Beijing, 100080, China
[2]Graduate School of Chinese Academy of Sciences, Beijing, 100039, China

Thermoacoustic refrigerators are considered to be an alternative for food refrigeration, air conditioner and other commercial usages. The design and the performance test of a room-temperature traveling-wave thermoacoustic refrigerator were conducted. The lowest temperature of about -28°C and a cooling power of 108W at 0°C were achieved with helium when driven by a standing wave thermoacoustic engine at a pressure ratio of 1.05. Gedeon streaming was detected in the experiment. The performance was highly improved by suppressing this streaming with a membrane. The measured performances agreed well with our model results.

INTRODUCTION

Thermoacoustic refrigeration is achieved through utilization of energy in acoustic wave forms. They have been considered for civil, military and other industrial applications. This technology could be alternative for conventional technologies, offering the promise of competitive performance combined with an environmentally friendly system with few or no moving parts, and no need for lubrication or sliding seals. Many efforts have been focused on this new technology and some thermoacoustic refrigerator prototypes have been built [1]. Early efforts were focused on standing-wave prototypes. Recent development of "traveling wave" shows a prospect of higher efficiencies and made this technology more viable. However, relatively fewer literatures on traveling-wave thermoacoustic refrigerators can be found.

Thermoacoustic refrigerators are usually driven by two kinds of drivers, electro-dynamic loudspeakers and thermoacoustic engines. The former type of refrigerator usually operates at high frequency and has small size [2]. The latter one needs a good thermoacoustic engine. It has an aspect to use solar energy and other heat of low level, which makes it especially valuable to where electricity is not available.

The object of this study was to develop a thermoacoustic refrigerator in room-temperature range. The investigation was involved with the design, construction, and measurement of a prototype. The operation variables such as the working gas, the frequency and the mean pressure are specially considered.

EXPERIMENTAL PROTOTYPE

The schematic of the prototype is shown in Figure 1. It is made of two subsystems, the refrigerator unit and the standing-wave thermoacoustic driver section. The driver and the refrigerator were designed separately [3]. Here only the refrigerator will be studied. The refrigerator consists of two heat exchangers, a regenerator, a thermal buffer tube (TBT), a compliance reservoir and an inertance tube. Key dimensions of the segments are listed in Table 1. The whole system was entirely constructed from stainless steel, except for the heat exchangers which were made of copper. Several access ports were made for pressure transducers and valves. The hot end of the drive is heated by electrical heaters. The cooling load is measured by four electrical heaters in the cold heat exchanger of the refrigerator.

When the actual temperature gradient of the stack goes beyond the critical temperature gradient of the stack, the driver begins to work. It absorbs heat from the hot heat exchanger, converting it to acoustic power, which is sent to the refrigerator via the resonator. The directions of the acoustic power flows are illustrated by the arrows in the Figure 1. The regenerator of the refrigerator pumps heat from the cold end, while rejecting heat to the water-cooled heat exchanger. The pressure of the inlet of the refrigerator and water-cold heat exchanger were measured. The temperature of several points along the regenerator, the cold heat exchanger and the thermal buffer tube were measured.

The refrigerator is designed according to linear thermoacoustic theory [4]. The predicted performance of the initial design is a cooling power of 80W at -20℃ and a COP of 2.86. It was designed to operate with 2.0MPa helium at a pressure ratio of 1.10, and it was designed to work under a frequency of 50Hz. In practice these working conditions were not realized yet. Thus it led to a difference between the design goals and the experiment.

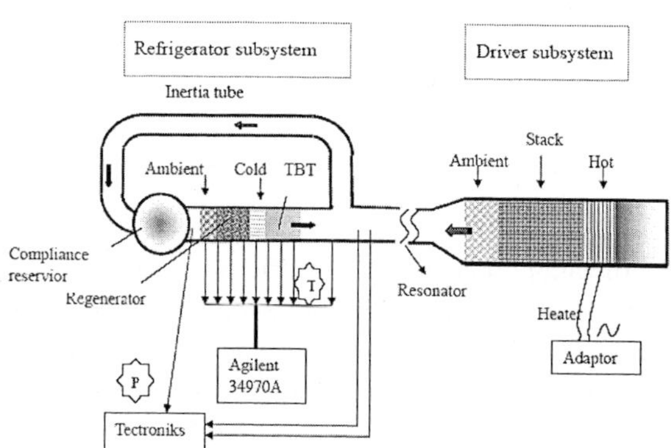

Figure 1: Experimental System

Table 1. Structure Parameters

Section	Inner diameter (mm)	Length (mm)
Regenerator	50	35
Ambient heat exchanger	50	35
Cold heat exchanger	50	30
Thermal buffer tube	50	50
Inertia tube	32	700
Compliance reservoir	0.5 Liter	

EXPERIMENT RESULTS

At the beginning of the experiment, the refrigerator did not work well. Only a small temperature drop was achieved. The cold end temperature of the refrigerator (Tc) went down slowly. In some cases, it worked quite unstably. The cold end temperature came up again after going down for a while. We considered it was caused by the Gedeon Streaming which is a common loss in traveling-wave thermoacoustic systems [5]. In fact we have reserved spaces to take action against possible losses, when the refrigerator was designed. We put a membrane in the refrigerator. Experimental results showed that this membrane worked quite well. It can prevent dc flow while transferring acoustic power. The performance of the refrigerator is greatly improved as illustrated in Figure 2. For 2.1MPa helium under the frequency of 100Hz, the no-load Tc dropped by 24℃ or more.

A series of experiments have been done to study the performance of the system. Different working gases, helium, argon, nitrogen and the mixture of them were used. The mean pressure and the frequency were also changed. Helium achieved the best performance. The most encouraging result is that it got the cooling capacity of about 108W at 0℃ which confirms the viability of this technology for house hold refrigerators. The measured heat pumping capacity (Qc) is shown as a function of the cold end temperature in Figure 3. The data are grouped according to the absolute mean pressure (P_m). The oscillating frequency was 85Hz with a 4.5m long resonator. The no-load Tc decreased and the cooling capacity increased as the mean pressure increased. There are two possible reasons for this trend. Firstly, Figure 4 showed that the drive ratio increased with Pm. This leads to higher amplitude of the pressure wave and, secondly, the property of the different sections, such as the inertance of the inertia tube, the compliance of the reservoir and the resistance of the regenerator, changed with Pm, too. They will change the phase between the pressure and the velocity of the cross sections of the refrigerator. This may also account for the difference.

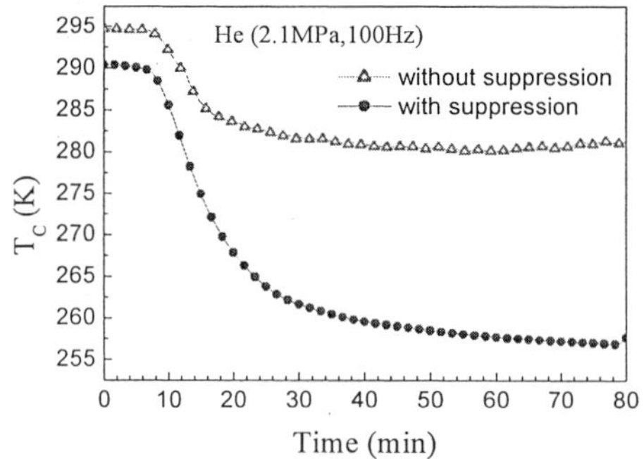

Figure 2 Comparison between with and without the suppression of Gedeon Streaming.

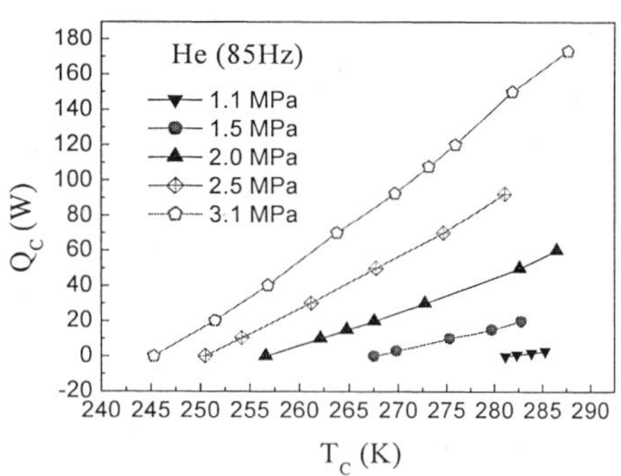

Figure 3 Cooling capacities vs. cold end temperature.

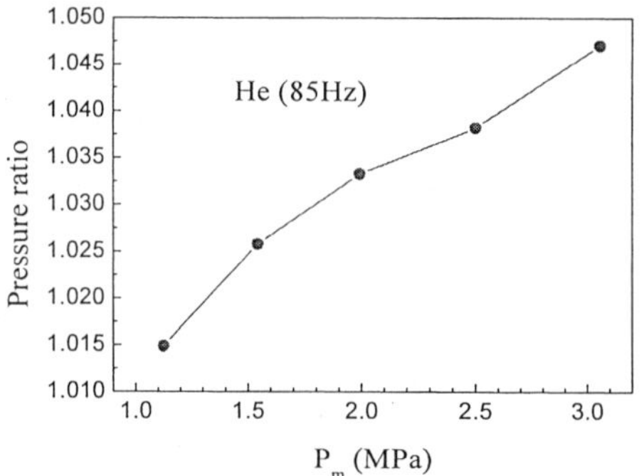

Figure 4 Pressure ratios vs. mean pressure.

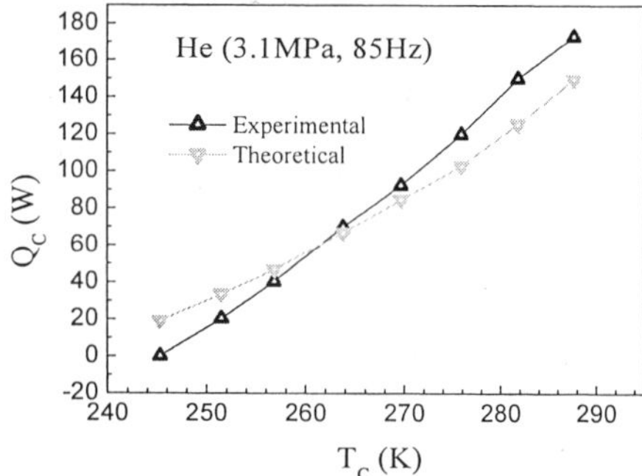

Figure 5 Comparison between the experiment results and the calculation.

To check the numeric model, we first did the experiment, measuring the temperatures and pressures of the refrigerator. Then we did the theoretical calculation. With those given parameters, the cooling power and the acoustic power needed to do it are given out by the model. Figure 5 shows a comparison of the measured and calculated Qc. We found that the experimental result is always smaller than the model result when the Tc below a certain temperature, and larger when Tc goes beyond the temperature. However, the model and the experiment agreed with each other reasonably in general. For the work condition (helium, Tc 0°C, P_m 3.1MPa, pressure ratio 1.05, frequency 85Hz, measured Qc 108W), the calculated Qc is 95.9W and the calculated consumed acoustic power is 46.7W. If the calculation is good enough, then we have the approximate COP of the refrigerator 2.06 (95.9/46.7) at this working condition. In the future we will measure the real work flow to get more precise COP.

CONCLUSIONS

A traveling-wave thermoacoustic refrigerator is designed and tested. Gedeon streaming exists in the traveling-wave thermoacoustic refrigerator and it deteriorated the performance of the refrigerator. Suppression of this streaming greatly improved the performance of the refrigerator. The refrigerator has achieved the no-load Tc of -28℃ and the cooling power of 108W at 0℃, driven by a standing-wave thermoacoustic engine, with helium at a pressure ratio of 1.05.   The design target is not achieved due to the operation frequency and the amplitude of the pressure. But the measured performances agreed reasonably with our model results. Further work will be done to measure the acoustic power in the system directly, to study the pressure and velocity field, and to obtain a better driver.

ACKNOWLEDGEMENTS

This work is financially supported by Chinese Academy of Sciences under the contract number: KJCX2-SW-W12-01.

REFERENCES

1.   Swift, G.W., Thermoacoustics: A unifying perspective for some engines and refrigerators, http:/www. lanl. Gov(2001)
2.   Steven L. Garrett, Thermoacoustic Refrigerator for Space Applications, Journal of Thermophysics and Heat Transfer,(1993),4 595-599
3.   Dai, W., Luo, E., et al., Building a High-efficiency and Compact-sized Thermoacoustically-driven Pulse Tube Cooler, ICC13, (2004)
4.   Huang Y., Luo E. et al., Study on efficient thermoacoustic refrigerator within room temperature range , Cryogenics and Superconductivity (2004), 34(1) 1-5 (in Chinese)
5.   Swift, G., Gardner, D. and Backhaus, S., Acoustic recovery of lost power in pulse tube refrigerators, J.Acoust.Soc.Am, (1999), Vol.105, 711-724

Study on thermoacoustic resonance pipe driven by loudspeaker using two-microphone method

Fan L., Wang B. R., Jin T., Shen Y., Zhang S. Y.

Lab of Modern Acoustics, Institute of Acoustics, Nanjing University, Nanjing, P.R.China, 210093

Generally, the characteristic frequencies of a resonance pipe is the determinant of the working frequency of a thermoacoustic system, however, in a thermoacoustic refrigerator driven by a loudspeaker the match between the loudspeaker and resonance pipe is also significant. The two-microphone method is introduced in this paper to measure the working frequency of such a thermoacoustic system, moreover, it is demonstrated that the lowest match frequency of the loudspeaker and resonance pipe changes little with the length variation of the pipe.

INTRODUCTION

Research efforts have been made to develop thermoacoustic refrigerators for more than twenty years and the performance of the refrigerators has been improved greatly. However, the efficiency of a thermoacoustic refrigerator is still lower than traditional refrigerators, which becomes a major disadvantage of thermoacoustic refrigerators. Therefore, many researchers focus their study on the optimized working frequency of the thermoacoustic system [1,2,3].

In this paper a system is set up to measure the acoustic properties in the thermoacoustic resonance pipe using the two-microphone method, which is similar to the technique presented by Seybert, et al [4]. Then the distributions of the sound pressure, particle velocity and acoustic impedance in the pipe are calculated utilizing the decomposition theory [4]. On the other hand, the acoustic impedance of the loudspeaker is measured using the clio software. Therefore the match frequency between the loudspeaker and resonance pipe can be found, which is just the optimized working frequency of the system. It is shown that the most efficient working frequency of a thermoacoustic refrigerator driven by a loudspeaker is not the resonance frequency of the pipe itself. The optimized working frequency mainly depends on the impedance of the loudspeaker when the lowest antiresonance frequency of the pipe is much higher than the mechanical resonance frequency of the loudspeaker. The lowest match frequency, which is considered to be the best working frequency of the system, changes little with the length variation of the pipe. Besides, the stack position for high efficiency and power of refrigeration can be achieved by analyzing the distributions of the sound pressure and particle velocity in the resonance pipe.

EXPERIMENT METHOD AND SYSTEM

The measurement system of the two-microphone method, which has been used to measure the acoustic impedance distribution in the pipe, is shown in Figure 1. A loudspeaker located at one termination of the pipe is driven by a broadband stationary random signal and two microphones separated by a distance of 3.5cm with each other are mounted in the wall of the resonance pipe to measure the sound pressure of two positions in the pipe. The amplified output signal of both microphones are collected by the sound card and transmitted into a computer. By power spectrum analysis and decomposition theory, the sound pressure, particle velocity and acoustic impedance distributions in the pipe and their variations with the frequency are calculated and consequently the optimized working frequency and appropriate position to set the stack are obtained.

In our experiments, two pipes with the same inner radius of 2.4cm and different lengths of 16.2cm and 32.4cm are used to inspect the influence of the pipe length to the match frequencies. A frequency range of 50Hz-4kHz is covered, in which both microphones have good frequency responses and consistency, meanwhile, the low-frequency parameters of the loudspeaker are valid and the characteristic wave number problem [4] of the two-microphone method, which can cause serious

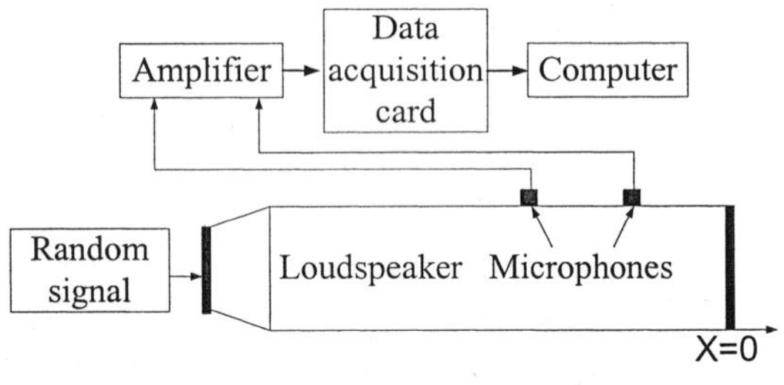

Figure 1   Experiment system

error in the measurement, is avoided. According to the coordinates in Figure 1, the termination of the resonance pipe, which is a rigid iron plate, is indicated as $x = 0$, and the pipe is towards the left, i.e., the negative direction.

On the other hand, the low-frequency parameters of the loudspeaker are measured with the clio software, and it is obtained that the equivalent quality and mechanical compliance of the loudspeaker vibration system are 5.13×10^{-4} kg and 5.97×10^{-4} m/N, respectively, and its mechanical resonance frequency is 288Hz. Therefore, the low-frequency impedance curve of the loudspeaker can be obtained.

EXPERIMENT RESULTS AND ANALYSIS

Match frequencies

In Figure 2 and Figure 3 the dotted lines show the variation of the minus value of the acoustic reactance of the pipe at the pipe mouth where the loudspeaker is equipped and the solid lines show the variation of the equivalent acoustic reactance of the loudspeaker. Therefore, the crossing points of both curves are just the match frequencies since the total input acoustic reactance is zero. It can also be seen that all the match frequencies except the lowest one are near the antiresonance frequencies of the pipe, where the acoustic reactance curve is incontinuous. Meanwhile, the reactance curve of the pipe below the first antiresonance frequency nearly coincides with the transverse axis and the mechanical resonance frequency of the loudspeaker is lower than the first antiresonance frequency of the pipe. Therefore, the lowest match frequency is near the mechanical resonance frequency of the loudspeaker itself and changes little with the length variation of the pipe. In addition, the frequency spectra of the sound pressure level at the termination of the pipes are also measured by the similar system, but which is with one microphone equipped at the termination of the pipe, the results are shown in Figure 5 and Figure 6.

Comparing Figure 2 with Figure 4 and Figure 3 with Figure 5, it can be seen that the sound pressure resonance peaks can be obtained at the match frequencies, moreover, the sound pressure resonance peak at the lowest match frequency is the highest, which is in agreement with the theory presented by Kinsler, et al [5], i.e., the resonance frequency most closed to the mechanical resonance frequency of the loudspeaker can produce stronger resonance. Since the working frequency between 200Hz and 600Hz is typical in a thermoacoustic refrigerator [3], a low-frequency loudspeaker is recommended in a thermoacoustic system. Meanwhile, a shorter resonance pipe is proposed because higher sound pressure can be obtained in it when the output power of the sound source is invariable, which also makes the first antiresonance frequency of the pipe much higher than the mechanical resonance frequency of the loudspeaker. Therefore, the lowest match frequency of the loudspeaker and the resonance pipe, which changes little with the length variation of the pipe, is the best for the working frequency.

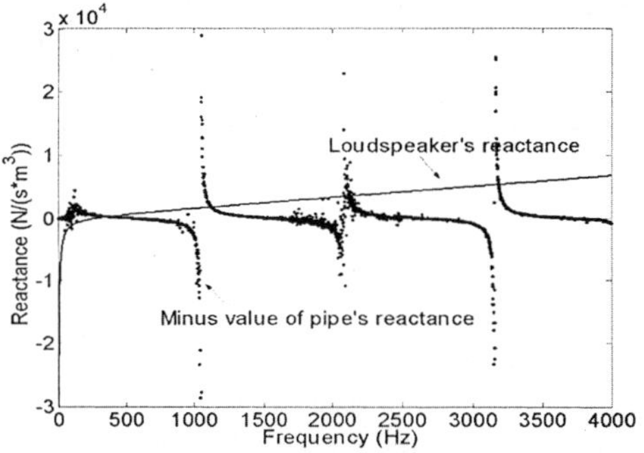

Figure 2   Reactance of the pipe with the length of 16.2cm

Figure 3   Reactance of the pipe with the length of 32.4cm

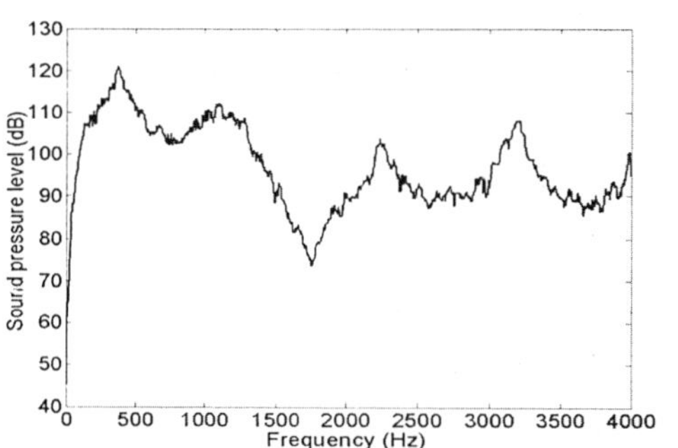

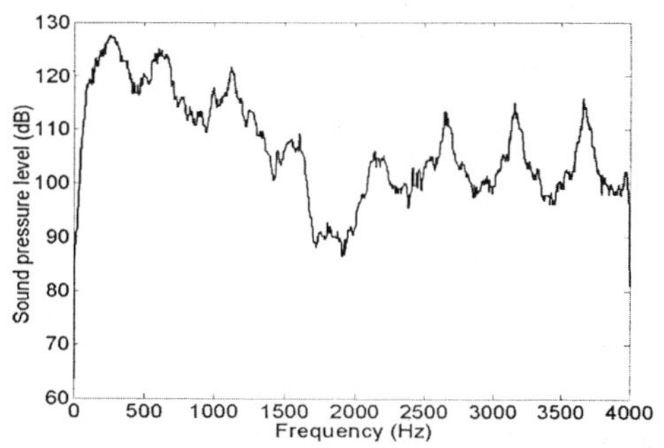

Figure 4   Sound pressure in the pipe with the length 16.2cm

Figure 5   Sound pressure in the pipe with the length 32.4cm

Position of stack

According to the simplest single plate model [1] of the thermoacoustic system, the total heat flux Q along the plate and the efficiency of the refrigerator COP can be expressed as:

$$Q = -\Pi \delta_\kappa T_m \beta p_1^s u_1^s (\Gamma - 1)/4 \tag{1}$$

$$COP = \rho_m c_p u_1^s / \Delta x \beta \varpi p_1^s \tag{2}$$

where T_m, ρ_m, c_p are the mean temperature, density and specific heat of the fluid in the pipe, Γ is the ratio of actual temperature gradient ∇T_m to the critical temperature gradient defined as $\nabla T_{crit} = T_m \beta \varpi p_1^s / \rho_m c_p u_1^s$, β is the ordinary thermal expansion coefficient, δ_κ is the thermal penetration depth, Π is the perimeter of the plate, Δx is the length of the plate, p_1^s and u_1^s are amplitude of the sound pressure and particle velocity in the pipe, respectively. ϖ is the angular frequency. It can be seen from these equations that under a certain condition COP is proportional to u_1^s / p_1^s, i.e., the reciprocal of the acoustic impedance's module and Q is proportional to $p_1^s u_1^s$. Therefore, the distributions of the sound pressure and particle velocity in the pipe can be used to determine a proper position to set the stack. The distributions are obtained from the experiment data utilizing the decomposition theory [4], as is shown in Figure 6 and Figure 7. Figure 6 shows the distribution of u_1^s / p_1^s with both frequency and position in the pipe of length 16.2cm and the distribution of $p_1^s u_1^s$ in the pipe is shown in Figure 7.

Since the working frequency should be the lowest match frequency, only the frequency range around this match frequency is covered in Figure 6 and Figure 7. From the figures, it can be seen that if the stack is set at the position far from the pipe's termination, higher COP can be achieved. In fact, at the sound pressure nodes the highest COP can be achieved since it is proportional to the reciprocal of the module of acoustic impedance. But it is not proper to set the stack at the sound pressure nodes for Q is nearly zero here. Consequently, the proper position for the stack should be a little away from sound pressure nodes, which is recommended be chosen according to the practical refrigerating power required.

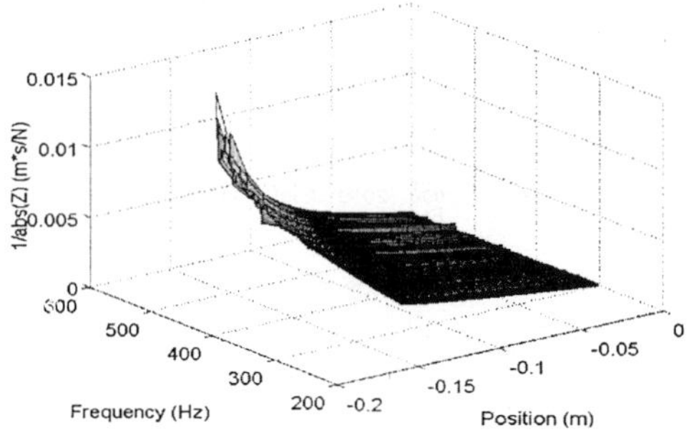

Figure 6   Distribution of $1/|Z|$

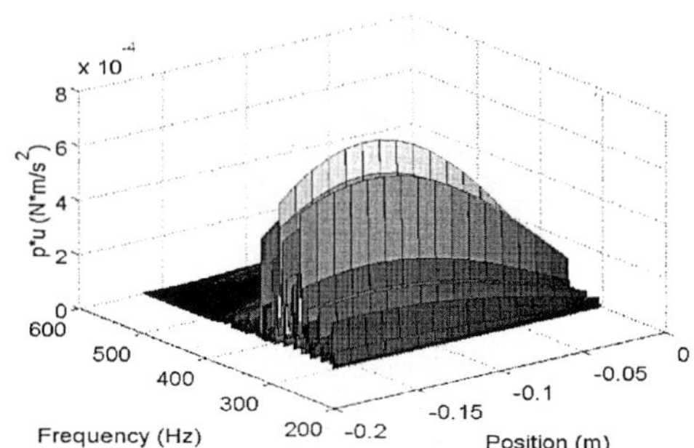

Figure 7   Distribution of $p_1^s u_1^s$

CONCLUSION

The two-microphone method is used to measure the match frequency of a thermoacoustic system and the sound pressure and particle velocity distributions in the pipe. The results demonstrate that the two-microphone method is appropriate to study the characteristics of a thermoacoustic pipe. By this method the optimized working frequency and the proper position of the stack of a thermoacoustic system can be obtained easily. Besides, the study of a quarter-wavelength resonance pipe system terminated by a sphere cavity has been set up and the experiments utilizing the same method on the new system are in progress.

ACKNOWLEDGEMENT

This work is supported by National Natural Sciences Foundation of China (No.50206008).

REFERENCES

1. Swift, G. W., Thermoacoustic engines, J. Acoust. Soc. Am. (1988) 84 1145-1180

2. Chen, G. B., Jiang, J. P., Shi, J. L., Jin, T., Tang, K., Jiang, Y. L., Jiang, N. and Huang, Y. H., Influence of buffer on resonance frequency of thermoacoustic engine, Cryogenics (2002) 42 223-227

3. Tijani, M. E. H., Zeegers, J. C. H. and deWaele, A. T. A. M., A gas-spring system for optimizing loudspeakers in thermoacoustic refrigerators, Journal of Applied Physics (2002) 92 2159-2165.

4. Seybert, A. F., Two-sensor methods for the measurement of sound intensity and acoustic properties in ducts, J. Acoust. Soc. Am. (1988) 83 2233-2239

5. Kinsler, L. E., Frey, A. R., Coppens, A. B. and Sanders, J. V., Fundamentals of Acoustics, John Wiley & Sons, NewYork, USA (1982) 210-214

Stack and frequency's coupling characteristic analysis in thermoacoustic systems

Zhang F. Z., Jiang R.Q., Li Q. [*], Chen X. [#]

College of Power and Nuclear Engineering, Harbin Engineering University, Harbin, China
[*]Technical Institute of Physics and Chemistry, Chinese Academy of Science, Beijing, China
[#]Cryogenics Laboratory, Huazhong University of Science and Technology, Wuhan, China

The performance of a standing-wave thermoacoustic refrigerator system depends on the thermoacoustic stack and frequency modulating phase between the pressure and gas velocity. The quality factor differs for different parameter of stack, so only an effective frequency bandwidth can couple phase to realize heat pumping and excite the resonance. On the contrary, the stack parameter which can realize this goal is an effective bandwidth for a certain frequency too. When frequency or stack parameter is not appropriate to each other, the pumping heat maybe inefficient or vanish.

INTRODUCTION

In the past decades, much attention has been paid on the application of the thermoacoustic technology. Thermoacoustic engines are devices which convert heat energy to acoustic work, and vice versa [1-4]. The phase between oscillating pressure and gas velocity determines voluminous degree of the $p-V$ diagram and the orientation of the cycle for a periodic thermodynamic cycle. Two fundamental conditions must be satisfied in order to pumping heat up the temperature gradient in a standing wave thermoacoustic refrigerator. One is the phase between pressure and velocity can shift from 90 ; the other is the resonance can be excited. The phase between pressure and volume is modulated by the crank in a traditional engine. But for a standing wave refrigerator driven by loudspeaker, there is has no crank. The stack and frequency modulate phase between the pressure and gas velocity. Because the particular dynamic resource and particular manner of phase modulated, it is important to determine the frequency and stack for the goal. On the other hand, the stack is the core to realize the conversion between heat energy and acoustic power. The coupling characteristics between stack and frequency are important for the standing wave thermoacoustic refrigerator. When the frequency of the loudspeaker is not appropriate for a certain system, or the stack parameters is not appropriate for a certain frequency, pumping heat up the temperature gradient maybe inefficient.

THE PHASE CONDITION

From the $p-V$ chart, we can find that the phase between oscillating pressure and gas velocity should be $(-3\pi/2, -\pi/2)$ in order to realize the refrigeration cycle. Acoustics power $\dot{E}_2(x)$ can be expressed as following [5]

350

$$E_2(x) = \frac{\omega}{2\pi} \oint \text{Re}\left[p_1(x)e^{i\omega t} \right] \text{Re}\left[U_1(x)e^{i\omega t} \right] dt = \frac{1}{2}\left| p_1 \tilde{U}_1 \right| \cos\phi_{pU}, \tag{1}$$

with Re: real part, ω: the angular frequency, p_1: oscillating pressure, U_1: oscillating volume flow rate, ϕ_{PU}: phase (p_1, U_1). Using equation (1) for acoustic power $\dot{E}_2$, it is apparent that $\dot{E}_2 = 0$ when the phase between pressure and volume flow rate is 90, i.e., for standing wave phasing. When the phase is departure from standing wave phase, the conversion between heat energy and acoustic power can be caused. This departure is realized by the coupling between stack and acoustic wave.

The momentum equation (including viscosity and arbitrary shape and size of channel) is [6]

$$i\omega\rho_m u_1 = -\frac{dp_1}{dx} + \mu\left(\frac{\partial^2 u_1}{\partial y^2} + \frac{\partial^2 u_1}{\partial z^2} \right) \tag{2}$$

With boundary conditions, $u_1 = 0$ at the solid surface, the solution is

$$u_1 = \frac{i}{\omega\rho_n}\left[1 - h_v(y,z) \right]\frac{dp_1}{dx} = \frac{i}{\omega\rho_m}H_v\frac{dp_1}{dx} \tag{3}$$

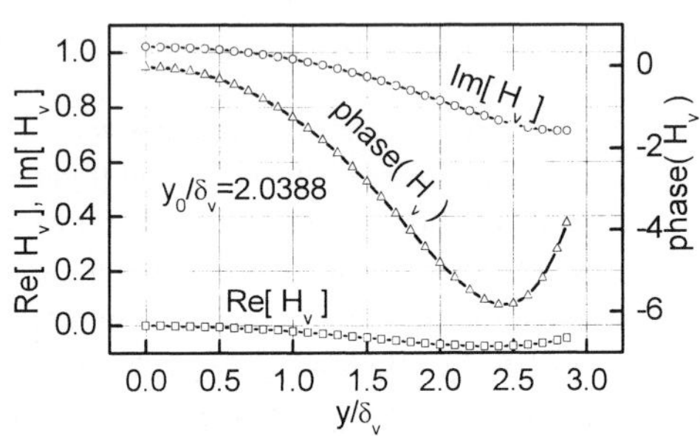

Figure 1 Function H_v and its phase for different layer in Hofler's thermoacoustic refrigerator

The complex function h_v depends on the special channel geometry. There is detailed expression for function h_v in reference [6] for different geometrical channels. Equation (2) indicates that function h_v depends on the hydraulic radius and viscous penetration depth of the system for a given fluid under given thermodynamic conditions. Figure 1 is a plot of H_v and its phase versus the ratio of hydraulic radius r_h and viscous penetration depth δ_v for different fluid layer. According to the calculation model stated in reference [7] (Hofler's thermoacoustic refrigerator, $y_0/\delta_v = 2.0388$). The calculated results indicate that the magnitude and phase of the gas velocity was influenced by the hydraulic radius and frequency. Some traveling wave components were added and the capability of energy conversion was possessed.

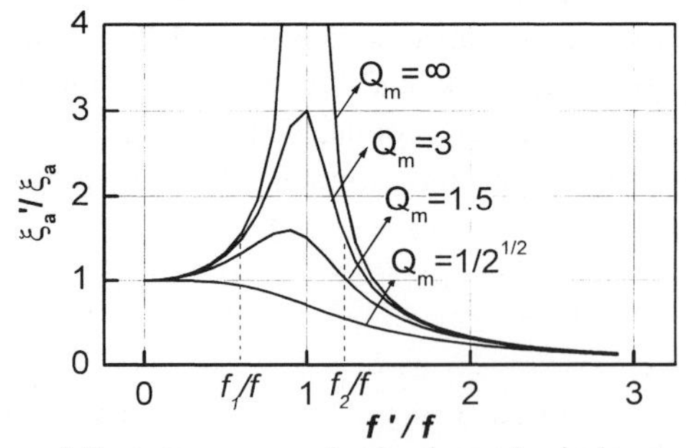

Figure 2 Resonance curve of a simple mechanical system

THE REALIZATION OF RESONANCE

It is a self-sustained oscillation caused by the interaction of the solid with the working fluid at the condition of a gradient temperature existing for thermoacoustic prime mover. The wave equation is

$$M_a \frac{d^2\xi}{dt^2} + R_a \frac{d\xi}{dt} + \frac{1}{C_a}\xi = 0, \tag{4}$$

with M_a, R_a, C_a : acoustic mass, impedance, and compliance respectively. For a thermoacoustic refrigerator driven by a louder-speaker, the louder-speaker provides a periodic driving force to the system. So it is a forced oscillation. The differential equation becomes

$$M_a \frac{d^2\xi}{dt^2} + R_a \frac{d\xi}{dt} + \frac{1}{C_a}\xi = p_a e^{i\omega't} \tag{5}$$

Where $p = p_a e^{i\omega't}$ is acoustic pressure externally driving force, ω' is angular frequency of the driving force. The solution of equation (5) is the sum of two parts (a transient term and a steady-state term) which depends on p_a and ω'. After a sufficient time interval the damping term decreases and vanish in the end. There is only the steady-state term is whose angular frequency ω' is that of the driving force in the end.

The displacement amplitude ξ_a' depends not only on magnitude and frequency of the driving force, but also on some inherent parameter such as thermal properties of thermodynamic working substance, viscous and thermal penetration depth. As stated in reference [8], quality factor Q_m was applied to describe the sharpness of resonance of the forced oscillation system. A plot of ξ_a'/ξ_a versus f'/f is shown in Figure 2, where ξ_a' and f' are the amplitude and frequency of forced oscillation respectively, ξ_a and f are inherent amplitude and frequency of free oscillation respectively. Another definition of the sharpness of resonance can be given in term of the frequency bandwidth $f_2 - f_1$, where f_1 and f_2 are the two frequencies above and below resonance f' for which displacement amplitude has drooped to one-half its resonance value. Figure 2 indicates that frequency bandwidth is corresponding to quality factor.

AN EXAMPLE

The standing wave thermoacoustic refrigerator should follow the principle as stated above. When a certain parameter of the system is changed, its quality factor would change, and then the frequency bandwidth would change. We should search frequency bandwidth of the system when we design a standing wave thermoacoustic refrigerator at first; the effective system parameter bandwidth should be found for a certain frequency, otherwise the pumping heat would not realize.

The stack is the core of conversion between heat energy and acoustic power, so we have more interest in its characteristics. Different length or

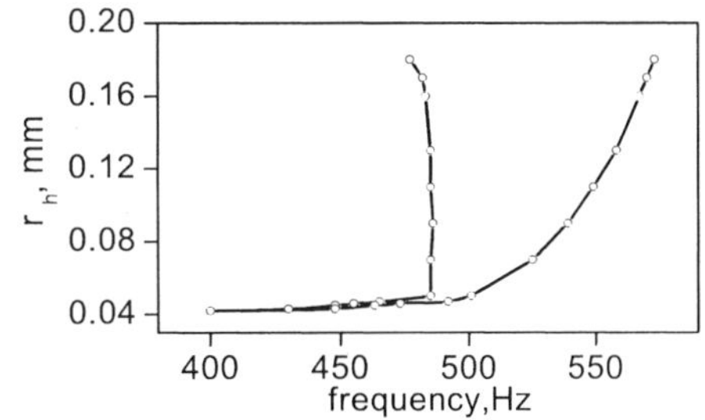

Figure 3 Effective frequency and hydraulic radius bandwidth for Hofler's thermoacoustic refrigerator

hydraulic radius of the stack carries with them corresponding impedance, so their quality factor is different. According to the above principle, the frequency bandwidth of amplitude resonance is different. In the same way, there is an effective stack parameter can realize the goal for a certain frequency.

When we study this kind of problem in a concrete standing wave thermoacoustic refrigerator, same conclusion was obtained. The relation between effective frequency and effective hydraulic radius was shown in Figure 3. The calculation model was Hofler's thermoacoustic refrigerator [8]. The results indicate that the hydraulic radius and frequency is corresponding. i.e. there is an effective frequency bandwidth for a given r_h; there is an effective hydraulic radius bandwidth for a certain frequency. We should pay attention to this problem when we design a thermoacoustic system.

CONCLUSION

For standing wave thermoacoustic refrigerator, the refrigeration performance depends on the phase departing the phase of standing wave and the resonance being excited which as a result of coupling between stack and frequency. Enough attention should be paid to frequency and stack hydraulic radius bandwidth. Resonance frequency and stack hydraulic radius can couple an appropriate phase to realize refrigeration goal.

REFERENCES

1. Ceperley, P.H., A pistonless Stirling engine-the traveling wave heat engine, Journal of Acoustic Soc Am (1979) 66 1508-1513.
2. Rott, N., Thermoacoustics, Adv. Appl. Mech. (1980) 20 135-175
3. H.Ishikawa and P.A. Hobson, Optimization of heat exchanger design in a thermoacoustic engine using a second law analysis, Int. Comm. Heat Mass Transfer (1996) 23 325-334
4. Ashok, G., Donald R. H., An experimental study of heat transfer from a cylinder in low-amplitude zero-mean oscillatory flows, International Journal of Heat and Mass Transfer (2000) 43 505-520
5. Du, G. H., Zhu, Zh.M., Fundamentals of Acoustic, Nanjing University (2001) Nan Jing (in Chinese)
6. Swift, G.W., Thermoacoustics: A unifying perspective for some engines and refrigerators. Fifth draft, available at http://www.lanl.gov/thermoacoustics/(2001)
7. Ward, W. C., and Swift, G.W, Tutorial and User's Guide: Design Environment for Low-Amplitude ThermoAcoustic Engines. Version 5.1, available at http://www.lanl.gov/thermoacoustics/(2001)
8. Lawrence E. Kinsler, Austin R. Frey, Fundamentals of Acoustics, John Wiley & Sons, Inc（1982）

Numerical simulation of loudspeaker-driven thermoacoustic refrigerator

Tu Q., Chen Z.J., Liu J.X., Zhang X.Q., Li Q. [*], Guo F.Z.

Cryogenics Laboratory, Huazhong University of Science and Technology, Wuhan 430074, China
*Technical Institute of Physics and Chemistry, Chinese Academy of Science, Beijing 100080, China

The work formulates a numerical simulation model for loudspeaker-driven thermoacoustic refrigerator (TAR) and uses the network method to calculate temperature difference generated across the heat-pump stack. It provides a theoretical basis for improving the TAR performance.

INTRODUCTION

Numerical simulation of loudspeaker-driven thermoacoustic refrigerator (TAR) provides theoretical basis for evaluating and improving TAR performance. In recent years it has received significant attention. However little work has been carried out for accurately quantitative calculation on TAR. So it is very difficult to determine influence factors and propose improvement measures on TAR performance.

Piccolo [1] adopted Swift`s short-stack approximation theory [2] to obtain the expression of temperature difference, and used standard formulas [3,4] and Kirchoff-Rayleigh`s formula [5] for sound absorption in wide tubes to express pressure and velocity of the stationary wave, then to calculate temperature difference. The approximate method is not accurate and cannot be used to compare influence of different stack geometries on temperature difference.

The purpose of this paper is to use the network model to calculate the temperature difference generated across the heat-pump stack and determine the influence factors on temperature difference.

CALCULATION MODEL

A schematic diagram of loudspeaker-driven TAR is shown in Figure 1.

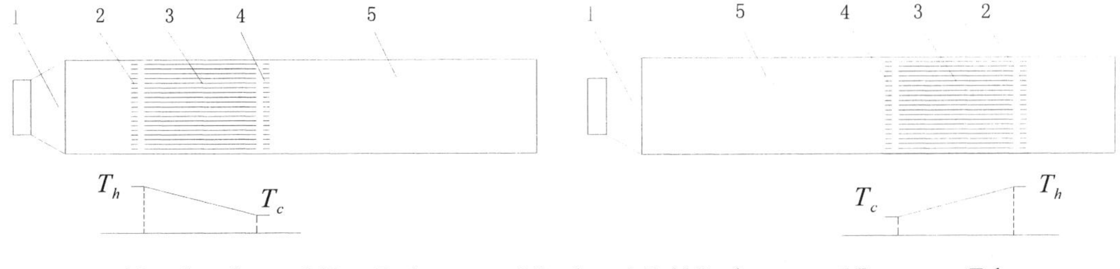

1 Loudspeaker   2 Heat Exchanger   3 Stack   4 Cold Exchanger   5 Resonance Tube

Figure 1   a schematic diagram of loudspeaker-driven

354

Averaged heat flux generated across heat-pump stack due to hydrodynamics transport is

$$Q_2 = \int T_m \rho_m \overline{s_1 u_1} dA \tag{1}$$

with $\quad s_1 = \dfrac{c_p}{T_m} T_1 - \dfrac{\beta}{\rho_m} P_1 \tag{2}$

and $\quad T_1 = \dfrac{1}{\rho_m c_p}(1-f_\kappa)P_1 - \dfrac{1}{i\omega}\dfrac{dT_m}{dx}\dfrac{(1-f_\kappa)-\sigma(1-f_\nu)}{(1-f_\nu)(1-\sigma)}u_1 \tag{3}$

where $T_m, \rho_m, c_p, \beta, \sigma, s_1$ are mean temperature, mean density, isobaric specific heat, thermal expansion coefficient, Prandtl number and specific entropy; P_1, u_1, T_1 are first-order acoustic pressure, oscillating velocity and oscillating temperature; ω is angular frequency; i is the imaginary unit; And f_ν, f_κ are respectively dimensionless spatial-average functions of viscous distribution and heat distribution.

Substituting Eq.(2) and Eq.(3) into Eq.(1) and making use of the relation between acoustic pressure and velocity $z_1 = P_1 / u_1$, we can yield the expression of heat flux

$$Q_2 = \frac{A_f}{2}\frac{|P_1|^2}{|z|^2}\left\{ \mathrm{Re}\left[(1-f_\kappa)z\right] - \frac{\rho_m c_p}{\omega}\frac{dT_m}{dx}\mathrm{Im}\left[\frac{(1-f_\kappa)-\sigma(1-f_\nu)}{(1-f_\nu)(1-\sigma)}\right] - \mathrm{Re}(z) \right\} \tag{4}$$

where A_f is fluid channel area of stack; p_1, z_1 are calculated at the mean position of the heat-pump stack in TAR; and dT_m / dx is temperature gradient expected across the stack. The temperature gradient dT_m / dx can be written as

$$dT_m / dx = (T_R - T_L) / \Delta x \tag{5}$$

where T_R, T_L are the temperatures on the right end and left end of the stack. Δx is the stack length.

Heat conduction flux due to temperature gradient can be written as

$$Q_2 = (A_s K_s + A_f K_s)\frac{dT_m}{dx} \tag{6}$$

where A_s is the cross-sectional area of stack material; K_s, K_f are thermal conductivity of stack material and gas.

According to Eq.(4) and Eq.(6), temperature difference generated across the heat-pump stack can be obtained once oscillating pressure, specific acoustic impedance are known.

NUMERICAL SIMULATION METHOD

Network model of TAR is adopted to calculate the temperature difference generated across the stack.
The network transfer equation of TAR can be written as

$$\begin{bmatrix} P_x \\ J_x \end{bmatrix} = R \begin{bmatrix} P_{end} \\ J_{end} \end{bmatrix} \tag{7}$$

with $\quad R = \displaystyle\prod_{i=1}^{i=3} R_i \tag{8}$

where $R_1 \sim R_3$ are the network transfer matrix of resonance tube, exchangers(hot heat exchanger or cold heat exchanger) and stack respectively. The expressions of $R_1 \sim R_3$ are given in literature [6].

For TAR with rigid end, the boundary condition is $Z_{end} = \infty$ and $P_{end} = DR \cdot P_m$.

where Z_{end} is acoustic impedance on the rigid end of the tube, $Z_{end} = A_f z_{end}$; DR is oscillating pressure ratio.

So the network model of TAR can be adopted to accurately calculate the acoustic pressure, volumetric flow rate and acoustic impedance in arbitrary position of TAR, then to evaluate the temperature difference generated across heat-pump stack.

NUMERICAL SIMULATION RESULTS

Influence of stack position on temperature difference

The system length is 437.5mm. The oscillating frequency is 400Hz. The parallel plate is used to make simulation to compare influence of stack position on temperature difference. The spacing between two plates is 0.8mm and the stack length is 50mm.The calculation results are shown in Figure 2.

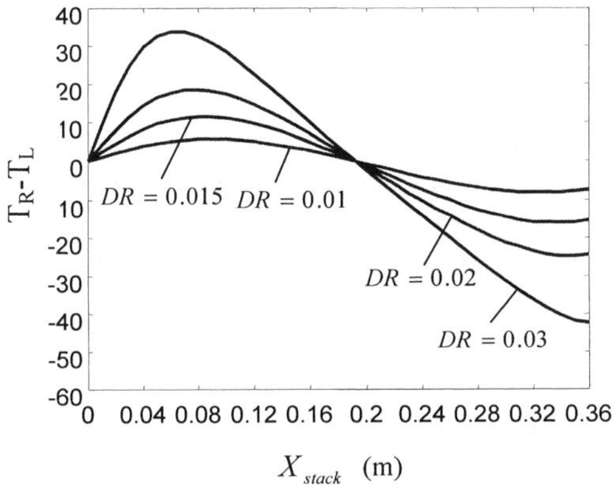

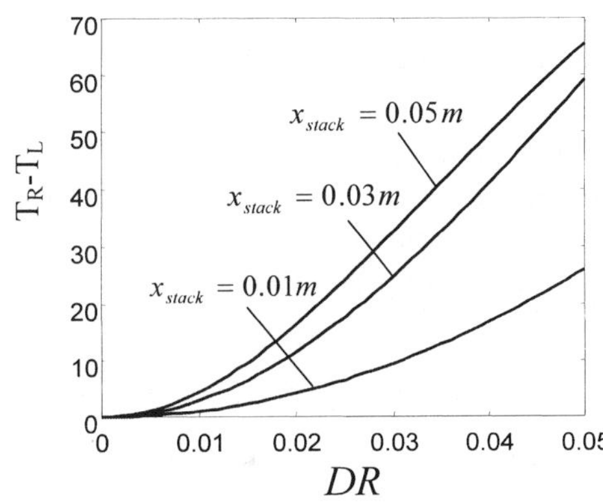

Figure 2 Influence of X_{stack} on temperature difference   Figure 3 Influence of DR on temperature difference

In Figure 2 X_{stack} is the distance from the rigid end of TAR to the right end of stack. Figure 2 shows that the stack position affects the temperature difference greatly. The hot end of heat-pump stack is on its left end and right end respectively when it is closer to the loudspeaker and the rigid end. And no temperature difference is generated when it is closer to the pressure node.

The temperature difference is the highest when the distance is 50mm and 350mm. The theoretically calculated results are in good agreement with the optimization design of loudspeaker-driven TAR in literature [7].

Influence of DR on temperature difference

Oscillating pressure ratio (DR) is also an important factor to affect the temperature difference. The theoretically calculated results are shown in Figure 3.   It can be seen from Figure 3 that the temperature difference is greatly affected by DR and rises with the increase of DR. It also shows that increasing DR is the most effective measure to improve the TAR performance.

Influence of heat-pump stack geometry on temperature difference

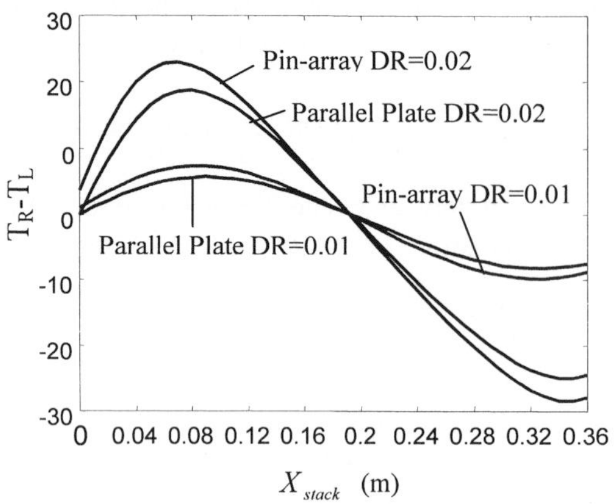

Figure 4 Influence of stack geometry on temperature difference

Parallel plate and pin-array stack are used to make numerical simulation to compare the influence of heat-pump stack geometries on temperature difference.

The steel pins of pin-array stack are made to be rectangular in order to manufacture the special stack feasible. The spacing between pin edges and equivalent radius of pins are optimized to be and 0.6mm and 0.2 mm [8]. The theoretically calculated temperature differences across the heat-pump pin-array stack and heat pump plate stack at DR=0.01,0.02 are shown in Figure 4.

Figure 4 shows that the stack makes some influence on temperature difference and the performance of pin-array stack is superior to that of parallel plate.

RESULTS

The network model can be feasibly used to calculate the temperature differences generated across the heat-pump stack. The theoretically calculated results show that the temperature differences are significantly affected by heat-pump stack position in TAR, different oscillating pressure ratios and different stack geometries. The former two are main factors to affect the temperature difference and the performance of the pin-array stack is superior to that of the parallel plate. The calculated results provide theoretical guidance for designing loudspeaker-driven TAR.

REFERENCES

1   Piccolo.A, Cannistraro.G, Convective heat transport along a thermoacoustic couple in the transient regime, International Journal of Thermal Sciences (2002), 411067-1075

2   Swift. G.W, Thermoacoustic engines, J.Acoust.Soc.Am(1988),84 1145-1180

3   Meyer.E, Neumann.E.G, Physical and Applied Acoustics, Academic Press, New York(1972)

4   Romer.I.C, Gaggioli.R.A, and EI-Hakeem.A.S, Analysis of quadripole methods for the velocity of sound, J. Acoust. Soc. Am.(1966) 86-98

5   Rayleigh.L, The theory of Sound, Dover Press, New York(1945)

6   Qiu Tu, Chih Wu, Qing Li, Feng Wu   and Fangzhong Guo, Influence of Temperature Gradient on Acoustic Characteristic Parameters of Stack in TAE.   Inter. J. Eng. Sci.(2003),411338-1349

7   Qiu Tu, Optimization of thermoacoustic stack and its matching with thermoacoustic engine. Ph.D thesis, Huazhong University of Science and Technology(2003)

8   Qiu Tu, Qing Li, Junxia Liu, Zhengyu Li and Fangzhong Guo, Optimization design of pin-array thermoacoustic stack. International Journal of Heat Exchangers.   Accepted in press

Proceedings of the Twentieth International Cryogenic Engineering Conference
(ICEC 20), Beijing, China.

Study on the optimal characteristic dimension of regenerator in a thermoacoustic engine

Yu Z. B., Li Q., Chen X. [*], Guo F. Z. [*], and Xie X. J.

Technical Institute of Physics and Chemistry, GSCAS, CAS, Beijing, China
[*] Cryogenics Laboratory, Huazhong University of Science and Technology, Wuhan, China

The purpose of this paper is to study the effect of the hydraulic radius of stacked-screen regenerators on the performance of a thermoacoustic engine. Experimental results indicate an optimal ratio of the hydraulic radius versus thermal penetration depth r_h/δ_k which leads to the lowest onset temperature difference ΔT_{onset} in the tested engine. Further more, a simple approximate method is developed to predict this optimal ratio in the regenerator in the case of a traveling-wave phase. The calculated results agree with the experimental results well.

INTRODUCTION

Regenerators are the hearts of traveling-wave thermoacoustic systems. In order to maintain good heat contact between the working gas and solid heat capacity across cross-sectional area, regenerator hydraulic radius should be smaller than thermal penetration depth. However, the further quantitative research on the regenerators in traveling-wave systems is still lacking.

In this paper, according to the network method demonstrated by Backhaus and Swift [1], a thermoacoutic Stirling engine is constructed and tested. A series of experiments have been performed to study the impact of the regenerator hydraulic radius on the onset temperature difference. The relationships between the regenerator hydraulic radius and onset temperature differences are measured and discussed. For each regenerator, the experimental results show an optimal r_h/δ_k, which leads to the lowest onset temperature difference in the tested engine, although the working gas, mean pressure, or regenerator hydraulic radius varies. On the basis of a simple approximate method developed in this paper, the calculated results agree with the experimental results well.

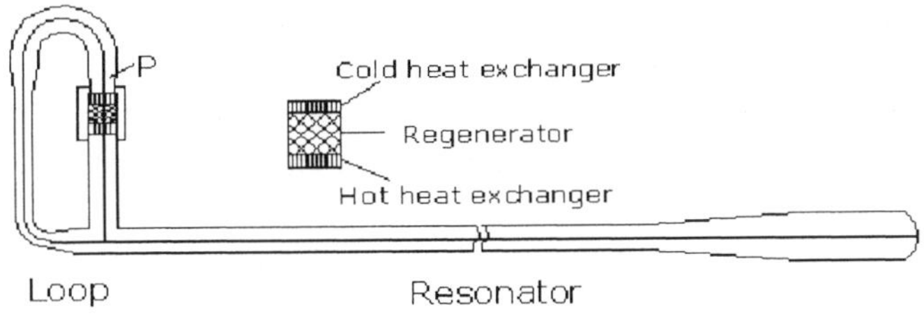

Figure 1 Schematic illustration of the tested engine.

EXPERIMENTAL APPARATUS

The tested engine is shown in Figure 1 schematically. The mean lengths of the looped tube and resonator are about 0.708 and 1.138 m respectively. One piezoelectric pressure sensor is placed above the cold heat exchanger (shown as P in Figure 1) to measure the acoustic pressure. Four thermo-couples are placed along the regenerator to measure the temperatures. The matrix of the regenerator is a pile of stainless-steel screen. The details of the tested engine can be found in our previous paper [2].

EXPERIMENTAL RESULTS

In the following experiments, the input electric power of the heater is fixed at 208 W. The onset points are measured when the mean pressure is changed from 0.1 to 2.6 MPa (absolute pressure) with a step of 0.1 MPa. For the example of 120-mesh regenerator, when the working gas is nitrogen (or helium), the operating frequency is 76 Hz (or 211 Hz). The measured relationships between ΔT_{onset} and p_m are shown in Figure 2 (a). In this figure, there is an optimal mean pressure about 0.4 MPa (or 1.1 MPa) leading to the lowest ΔT_{onset} when the working gas is nitrogen (or helium). Each measured onset point in Figure 2 (a) gives one dimensionless value of r_h/δ_k. So the curves in Figure 2 (a) can be transformed into the ones shown in Figure 2 (b). As shown in this figure, the measured optimal r_h/δ_k is about 0.24 (or 0.20) for nitrogen (or helium).

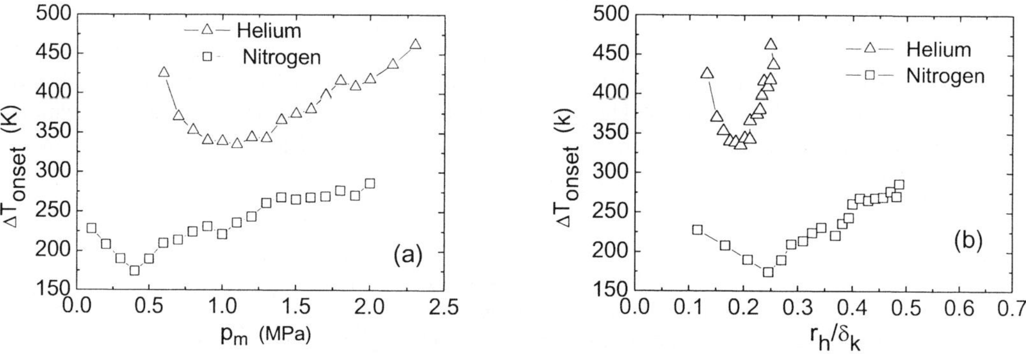

Figure 2   For 120-mesh regenerator, the relationships between ΔT_{onset} and p_m (a). The relationships between ΔT_{onset} and r_h/δ_k (b).   The lines are guide for eyes.

By the same means, for nitrogen, the optimal p_m and r_h/δ_k of the tested engine are measured and shown in Table 1 when the mesh number of the stainless-steel screen in the regenerator varies. From Table 1, it can be realized that there is a measured optimal r_h/δ_k about 0.17 - 0.25 for the tested engine.

On the other hand, the mean pressure is fixed at 0.6 MPa. But, the hydraulic radius of the regenerator is varied by changing the mesh number of the stacked screen. The measured relationship between onset temperature difference and hydraulic radius is shown in Figure 3. It is indicated that 200-mesh regenerator is optimal at this mean pressure. The corresponding r_h/δ_k is about 0.20. Therefore, in this engine, there is an optimal $r_h/\delta_k \approx (0.17 \sim 0.25)$ which leads to the lowest ΔT_{onset}. Note that, this optimal r_h/δ_k is obtained by changing the mean pressure when the hydraulic radius is fixed, and vice versa.

Table 1 The measured optimal p_m and r_h/δ_k when r_h varies

Mesh	80	120	150	200	250	300
r_h (µm)	71.1	47.2	37.7	27.2	22.3	18.2
$p_{m,opt}$ (MPa)	0.26	0.4	0.6	0.6	0.8	1.6
$\left(r_h/\delta_k\right)_{opt}$	0.25	0.24	0.22	0.20	0.17	0.21

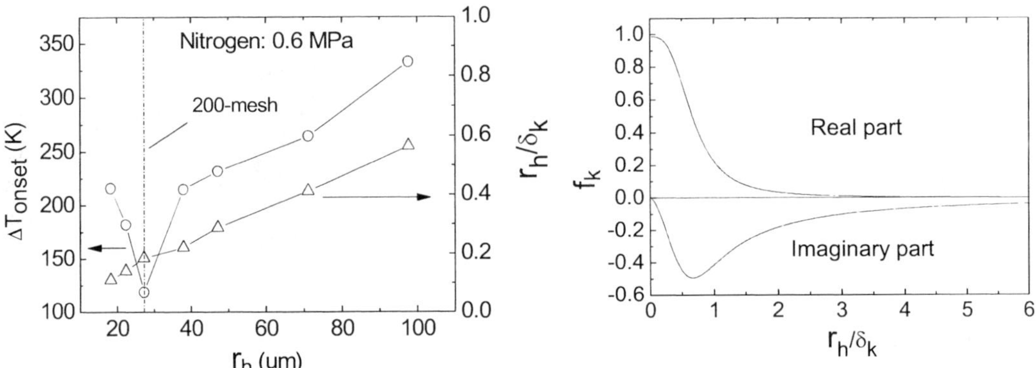

Figure 3 ΔT_{onset} (circles) and r_h/δ_k (triangles) changes as the regenerator hydraulic radius increases.

Figure 4 Spatial-average function f_k for a stacked-screen generator.

CALCULATION AND DISCUSSION

According to the harmonic analysis of regenerators by Backhaus and Swift [3], we obtain the equivalent spatial-average function f_k for stacked-screen regenerators (shown in Figure 4). Note that, this equivalent function is only valid for small r_h/δ_k. It helps to understand the above optimal r_h/δ_k, which is a compromise between the viscous and thermal-relaxation effect in the regenerator. On the other hand, in regenerator, the time-averaged acoustic power $d\dot{E}$ produced in a length dx is defined by

$$\frac{d\dot{E}_2}{dx} = \frac{1}{2}\mathrm{Re}\left[\tilde{U}_1\frac{dp_1}{dx} + \tilde{p}_1\frac{dU_1}{dx}\right], \tag{1}$$

with p_1: oscillating pressure, U_1: volume velocity. Letting the right hand side of equation (1) to zero, one can obtain an approximate relation for critical temperature difference [3]. For stacked-screen regenerators, the details of dp_1/dx and du_1/dx can be obtained in Ref. [3]. Substituting them into equation (1) and letting its right hand side to zero yield the following relation

$$\beta_m \frac{dT_m}{dx} \mathrm{Re}\left[\left(1 - \frac{\varepsilon_s + \left(g_c - g_v\right)\varepsilon_h}{1 + \varepsilon_s + \left(g_c + e^{2i\theta_T} g_v\right)\varepsilon_h}\right)\tilde{p}_1 U_1\right] - \frac{\mu_m}{\phi A r_h^2}\left[\frac{c_1(\phi)}{8} + \frac{c_2(\phi)Re_1}{3\pi}\right]|U_1|^2$$

$$-\frac{\omega\phi A}{p_m}\mathrm{Im}\left[1 - \beta_m^2 T_m^2 \frac{\gamma_m - 1}{\gamma_m}\frac{\varepsilon_s + \left(g_c + e^{2i\theta_p} g_v\right)\varepsilon_h}{1 + \varepsilon_s + \left(g_c + e^{2i\theta_T} g_v\right)\varepsilon_h}\right]|p_1|^2 = 0 \tag{2}$$

According to the research by Swift [1], it can be assumed that $|p_1/U_1| \approx (15 \sim 30)\rho_m c/\phi A$ in the regenerator. Furthermore, a traveling-wave phase ($\mathrm{Im}[\tilde{p}_1 U_1] = 0$) is presumed. Then, we obtain the optimal $r_h/\delta_k \approx (0.18 \sim 0.26)$ for the tested system. The calculated results agree with the measured ones well.

CONCLUSION

In the tested engine, there is an optimal r_h/δ_k leading to the lowest ΔT_{onset}, which is a compromise between the viscous and thermal-relaxation effect in the regenerator. The experimental results show that the optimal r_h/δ_k is about 0.17 to 0.25, although the working gas, mean pressure, or regenerator hydraulic radius varies. Furthermore, according to an approximate method developed in this paper, the calculated result is about 0.18 to 0.26, which agrees with the experimental ones well.

ACKNOWLEDGEMENT

This work was supported in part by the National Natural Science Foundation of China under Grant No. 50276064 and Knowledge Innovation Program of the CAS under Grant No. KJCX2-SW-W08

REFERENCES

1. Backhaus S. and Swift G. W., A thermoacoustic-Stirling heat engine: detailed study, J. Acoust. Soc. Am. (2000), 107 (6) 3148-3166
2. Yu Z. B., Li Q., Chen X., Guo F. Z., Xie X. J., Wu J. H., Investigation on the oscillation modes in a thermoacoustic stirling prime mover: mode stability and mode transition, Cryogenics (2003), 43 (12) 687-691
3. Swift G. W. and Ward W. C., Simple harmonic analysis of regenerators, Journal of Thermophysics and Heat Transfer (1996), 10 (4) 652-662

Investigation on acoustic characteristics of regenerator in the thermoacoustic apparatus

Xie X. J., Li Q., Chen X[*]., Yu Z. B.

Technical Institute of Physics and Chemistry, GSCAS, Beijing, China
*Cryogenic laboratory, Huazhong University of Science and Technology, Wuhan, China

An acoustic-driven thermoacoustic device, which is used to investigate on acoustic characteristics of regenerator, has been built. The transfer function method is used to measure the transfer function H of regenerator over a broadband frequency range. Reflection coefficient R , transmission loss TL , characteristic impedance Z_c and propagation constant γ are calculated from measured H . Furthermore, comparison about acoustic characteristics versus frequency between case a (only regenerator) and case b (regenerator and heat exchanger) is investigated. Different heat power influence on acoustic characteristics of regenerator is also discussed.

INTRODUCTION

Acoustic parameters, i.e. reflection coefficient, transmission loss, characteristic impedance and propagation constant of regenerator, are used to describe acoustic characteristics of regenerator in the thermoacoustic apparatus [1, 2]. Acoustic characteristics of regenerator are important factors influencing on the heat-power transfer efficiency. However, more quantitative investigation on acoustic characteristics of regenerator is still lacking.

In this paper, an acoustic-driven thermoacoustic device for the acquisition on acoustic characteristics of regenerator has been built. A series of experiments has been performed. The transfer function method [3] was applied, which uses a broadband random signal as a sound source to measure the transfer function of regenerator over a broadband frequency range. Based on the model given in this paper, reflection coefficient, transmission loss, characteristic impedance and propagation constant are calculated from the measured transfer function to enrich network parameters of regenerator. Furthermore, heat exchanger and temperature difference influence on acoustic characteristics is investigated in this paper.

EXPERIMENTAL APPARATUS AND RESULTS

The thermoacoustic apparatus, which is filled with nitrogen of 1.6MPa, is shown in Fig. 1. Three pressure sensors were used to measure transient pressures in the tube. The regenerator, 38mm long, is made of stainless steel screen with mesh 300. Case a (only regenerator) and case b (regenerator and heat exchanger) are investigated. Different heat power to the hot heat exchanger is also measured. The experimental results of transient pressure are shown in Fig. 2. Based on the transfer function method, the transfer function H between P1 and P2 was measured using a two-channel fast Fourier transform.

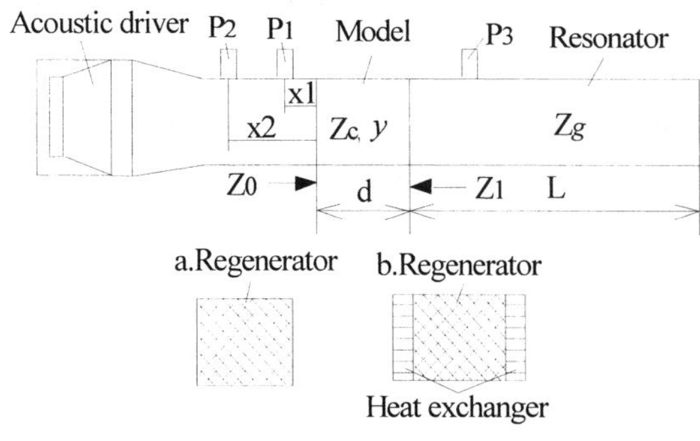

Figure1 Thermoacoustic apparatus for the transfer function method of measuring acoustic impedance

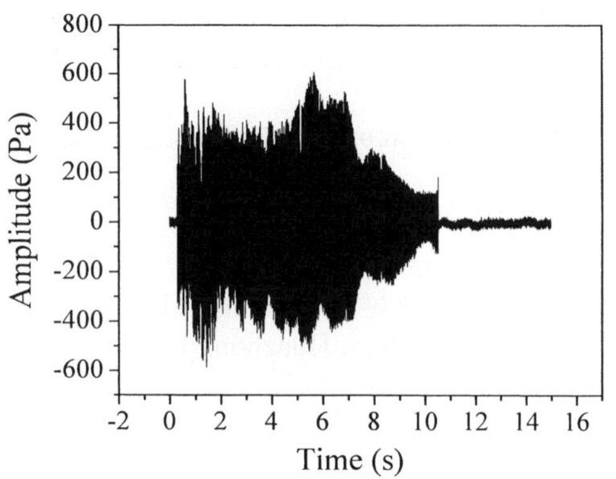

Figure 2 transient pressures in the P1 sensor

DISCUSSION

In order to compare the differences of acoustic characteristics between case a and case b, and investigate temperature gradient influence on acoustic characteristics, reflection coefficient, transmission loss, characteristic impedance and propagation constant are analyzed from the above transfer function.

Comparisons of reflection coefficient and transmission loss at different conditions are shown in Fig. 3. According to Fig. 3 (a), reflection constant R decreases as frequency increases at case $a.Q = 0$ W.

When $f = 400$ Hz, R has minimum, and then increases versus frequency. Comparing between case a. Q=0W and case b. Q=0W, R in case b is larger than that in case a, i.e. increment of R will be caused by additional heat exchanger. At the same time, different heat power to the hot heat exchanger has significant influence on R. Heat power is higher; the curve R is lower at the same frequency range. Furthermore, the minimum of R will shift rightward with the increase of temperature gradients.

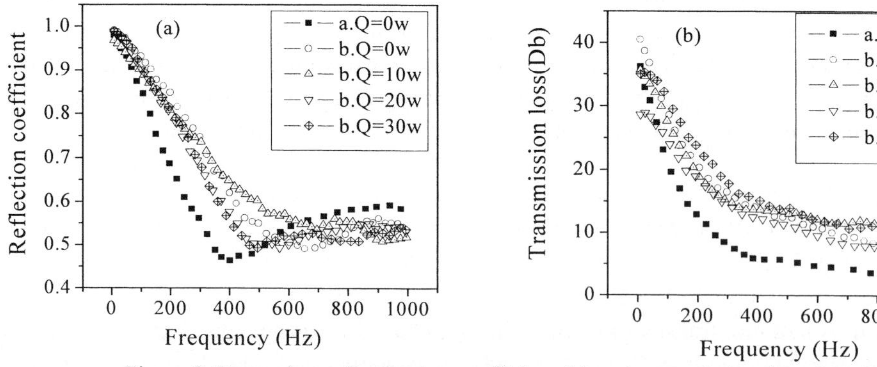

Figure 3 Comparison of reflection coefficient (a) and transmission loss (b) of tested model at different model and heat power.

Transmission loss decreases with frequency at all conditions (see Figure 3 (b)). Additional heat exchanger is equal to the additional impedance to the regenerator, and enlarges transmission loss in the regenerator. In addition, transmission loss increases, when heat power to the hot heat exchanger gradually increases, which is caused by the increase of the thermal resistance.

<u>Investigation on characteristic impedance and propagation constant</u>

Z_c, γ in the regenerator can be related to the transfer function measured from transient pressure, which can be inferred based on Ref. [4]:

$$Z_c = \pm\left(\frac{Z_0 Z_0^*\left(Z_1 - Z_1^*\right) - Z_1 Z_1^*\left(Z_0 - Z_0^*\right)}{\left(Z_1 - Z_1^*\right) - \left(Z_0 - Z_0^*\right)}\right)^{1/2} \tag{1}$$

$$\gamma = \frac{1}{2id}\ln\left(\frac{Z_0 + Z_c}{Z_0 - Z_c}\frac{Z_1 - Z_c}{Z_1 + Z_c}\right) \tag{2}$$

$$Z_0 = iZ_g\frac{-H\sin(kx_1) + \sin(kx_2)}{H\cos(kx_1) - \cos(kx_2)} \tag{3a}$$

$$Z_1 = -iZ_g\cot(kL) \tag{3b}$$

With Z_g : characteristic impedance of nitrogen, k : wave number of nitrogen. The superscript (*) is used when L is changed to another length L*. Using Equations (1) - (3), Z_c and γ can be obtained from the measured transfer function (see Figure 4).

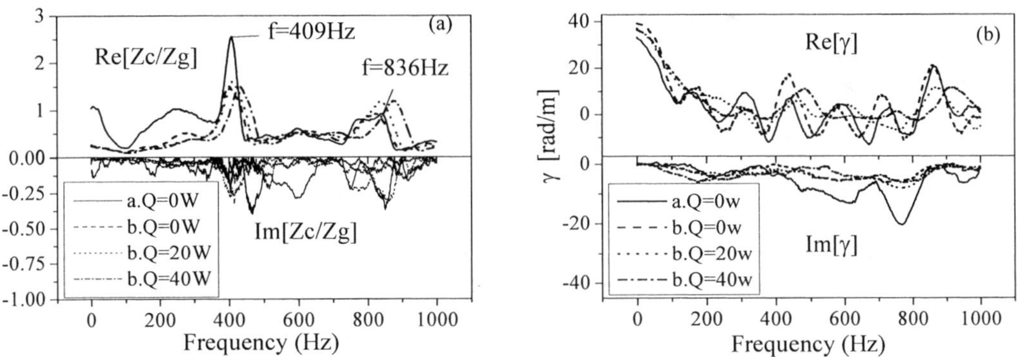

Figure 4 Comparison of characteristic impedance normalized by nitrogen (a) and propagation constant (b) of tested models, where both properties were obtained at case a. Q=0W (solid line), b. Q=0W (dash line), b. Q=20W (dot line) and b. Q=40W (dash dot line)

Figures 4 (a) and (b) show comparisons of characteristic impedance and propagation constant at different conditions. With the reference to Fig. 4 (a), characteristic impedances of regenerator were maximums at $f_1 = 409$ Hz and $f_2 = 836$ Hz. If Z_1 approaches Z_1^* in equation (1), then Z_c approaches Z_1. This incorrect conclusion is caused by elimination of the terms (Z_0-Z_0^*) in equation (1), and that situation arises when the frequency f satisfies $f(L - L^*) = nc/2$, with c: speed of sound in the nitrogen, n: the positive integral. The frequencies of the distinctive peaks f_1 and f_2, shown in Fig. 4 (a), are in complete agreement with the frequencies calculated from the above equation, so the appropriate set of the resonator must be selected so as to not satisfy the above equation. Furthermore, characteristic impedance in case b is smaller than that in case a, i.e., additional heat exchanger will decrease characteristic impedance of tested model. The curve of characteristic impedance will shift rightward with the increase of temperature difference.

Propagation constant of regenerator is shown in Fig. 4 (b) over the same frequency range as characteristic impedance. Propagation constant $\gamma = \alpha + i\beta$, with α: attenuation constant, β: phase constant, which stand for the amplitude and the phase variation of the pressure per length along the tube respectively. When $f < 400\,\text{Hz}$, α falls down quickly; and $f > 400\,\text{Hz}$, α fluctuates diminutively as the frequency increases. When the frequency is about 800Hz, α and β increases sharply, and then level off. Propagation constant of case b is larger than that of case a. It is observed that when the frequency is about larger than 400Hz, heat exchanger has a significant influence on the phase constant β (see Figure 4 (b)). α and β decrease with the increase of heat power to the hot heat exchanger..

CONCLUSION

Heat exchanger has significant influence on acoustic characteristics, which results in acoustic characteristics larger than the case without heat exchanger. Reflection coefficient, characteristic impedance and propagation constant decrease with the increase of heat power to the hot heat exchanger, whereas transmission loss increases with the increase of heat power.

ACKNOWLEDGMENT

The authors gratefully acknowledge the National Natural Science Foundation of China under Grant No. 50276064 and Knowledge Innovation Program of the CAS under Grant No. KJCX2-SW-W08.

REFERANCES

1.   Guo Fang Zhong, Flow characteristics of a cyclic flow regenerator, <u>Cryogenics</u> (1990), <u>30</u> 199-205

2.   Qiu Tu, et al. Influence of temperature gradient on acoustic characteristic parameters of stack in TAE. <u>Int. J. Eng. Sci</u> (2003), <u>41</u> 1337-1349

3.   J. Y. Chung and D. A. Blaser, Transfer function method of measuring in-duct acoustic properties. I. Theory. <u>J. Acoust. Soc. Am.</u> (1980), <u>68</u> 907-913

4.   C. Zwikker and C. Kosten, <u>Sound Absorption Materials</u>. Elsevier, New York, USA (1949)

Discussion on resonant frequency in thermoacoustic systems

Jin T., Fan L.[*], Zhang B.S., Wang B.R.[*], Chen G.B.

Cryogenics Laboratory, Zhejiang University, Hangzhou 310027, P. R. China
*Institute of Acoustics, Nanjing University, Nanjing 210093, P. R. China

Besides two traditional methods of calculating resonant frequency, a new method referred to as Standing Wave Minimal Antinode Method is proposed, which makes it possible to consider the influence of resistance, since zero antinode is no long a necessity as in an ideal standing wave system. Numerical simulation will be made on three types of tubes for their resonant frequencies with three methods, focusing on the influences of impedance composition on the resonant frequency. Analysis and comparison will also be made to these three methods.

INTRODUCTION

Analysis and accurate measurement of resonant frequency is vital for improving the performance of a thermoacoustic system, especially a thermoacoustic refrigerator, for the impedance mismatching will lead to a remarkable reduction of the loudspeaker's electroacoustic efficiency. Impedance Matching Method (IMM) and Acoustic Pressure Maximum Method (APMM) are two traditional calculation methods. In this paper, we will propose a new method referred to as Standing Wave Minimal Antinode Method(SWMAM). With this new method, it is possible to consider the influence of resistance, since zero antinode is no long a necessity as in an ideal standing wave system.

Based on three different methods, numerical simulation will be made on three types of tubes for their resonant frequencies, focusing on the influences of impedance composition on the resonant frequency. Results under different conditions will be used to discuss the equality and difference among these methods. We will also observe the influence of impedance at the tube end on the frequency response, with which the importance of radiation impedance may be evaluated.

TYPICAL STRUCTURES AND CALCULATION METHODS

Before describing the calculation methods, we first introduce three typical tubes in thermoacoustic machines, as shown in Figure 1, where (a) represents for the general case with arbitrary ends(with end impedances Z_{A0} and Z_{Al}), (b) for the case with one rigid end and one compliance volume at the other end(1/4 wave length structure, such as Hofler-type thermoacoustic refrigerator), (c) for the case with rigid ends(1/2 wave length structure, such as standing wave thermoacoustic prime mover). In the following analysis, the sound wave propagating inside the tube will be supposed to be plane wave.

Two traditional methods are applied to calculate the resonant frequency. APMM defines the resonant frequency with which the system's acoustic pressure reaches as high as possible. As shown in Figure 1(a),

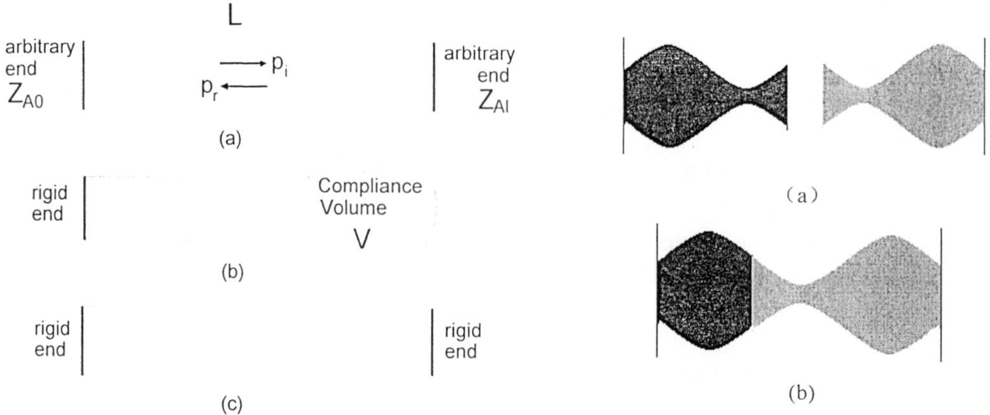

Figure 1   Typical thermoacoustic resonant tubes. (a) arbitrary ends, (b) rigid end + compliance, (c) rigid ends

Figure 2   Schematic of standing wave minimal antinode method(SWMAM)

p_{in} is the sound wave entering the tube from left to right, n means the wave after $(n-1)$ times reflection, e.g., p_{i1} represents for the original input, p_{i2} represents for the wave after one reflection at the right end and then also at the left end to re-enter the tube. Similarly, p_{rn} represents for the sound wave propagating from right to left. Except for the ideal cases, the reflection attenuation α and phase shifts at both ends σ_l and σ_r ($\sigma = 0$ for rigid end) should be considered, based on the acoustic impedance at both ends. The sound pressure can be written as,

$$p_{in} = \alpha^{2(n-1)} p e^{j\{\omega t - k[2(n-1)L+x]+(n-1)\sigma_r \pi + (n-1)\sigma_l \pi\}}$$

$$p_{rn} = \alpha^{2n-1} p e^{j\{\omega t + x - k[2(n-1)L+n\sigma_r \pi + (n-1)\sigma_l \pi\}}$$

(1)

After some reflections, the system obtains a balance, and the sound pressure profile versus frequency may be simulated by numerical means, and then the resonant frequency can be determined according to the maximal pressure principle.

We can also use IMM to calculate the resonant frequency of the tube structure shown in Figure 1(a). The tube length is L, and the acoustic impedance at both ends are Z_{A0} and Z_{Al}. According to the fundamental of acoustics, if the impedance transfer equation as follows is satisfied,

$$Z_{A0} = \frac{\rho_0 c_0}{S} \frac{Z_{Al} + j\rho_0 c_0 \tan kL}{\frac{\rho_0 c_0}{S} + jZ_{Al} \tan kL}$$

(2)

the corresponding frequency can be considered as resonant one, where the reactance equals zero while the resistance is maximal. With the acoustic pressure obtained from Equation (1), we can calculate the acoustic impedance along the tube, and then analyze the profile of the impedance versus frequency.

Besides the above methods, we propose a novel method, Standing Wave Minimal Antinode Method. With the reflection and superposition of sound wave, matching between the frequency and tube length are necessary to form a stable resonant sound field (either standing wave or traveling wave). At any point inside the tube, the sound wave can be regarded as the superposition of two reverse waves, $p = p_i + p_r$, where $p_i = p_{Ai} e^{j(\omega t - kx)}$ is incident wave, $p_r = p_{Ar} e^{j(\omega t + kx)}$ is reflection wave. We then have

$$p_{Ar}/p_{Ai} = \left|r_p\right|e^{j\sigma\pi} \tag{3}$$

$$\zeta = Z_s/\rho_0 c_0 = Z_A S/\rho_0 c_0 = x_s + jy_s \tag{4}$$

$$\left|r_p\right|^2 = \frac{(x_s-1)^2 + y_s^2}{(x_s+1)^2 + y_s^2}, \quad \tan(\sigma\pi) = \frac{2y_s}{x_s^2 + y_s^2 - 1} \tag{5}$$

where r_p is reflection factor, whose module $\left|r_p\right|$ and phase angle $\sigma\pi$ depends on the impedances, ζ is impedance ratio, Z_s is impedance of tube end, $\rho_0 c_0$ is air's impedance, S is cross-sectional area.

The position x, with minimal standing wave, can be determined by σ,

$$x = (1+\sigma)\lambda/4 \tag{6}$$

where λ is wave length of sound wave. If the positions x_1 and x_2 (calculated from both ends) coincide each other, i.e., the sound wave from both ends match inside the tube, we may consider that the system achieves resonant state, when the frequency satisfies $x_1 + x_2 = L + n\lambda/2$, as shown in Figure 2.

SIMULATION RESULTS

In the following simulation, we set the tube length L of 1m, sound velocity c_0 of 340m/s, acoustic impedance of air $\rho_0 c_0$ of 400kg/m²s. In Figure 1(b), the tube length does not include the compliance buffer section. Only the resonant frequency of fundamental is considered as the calculation result.

Figure 3 presents the result of the structure shown in Figure 1(b), with three different methods. Abscissa x/y is the ratio between resistance and reactance. The radii of the buffer and the resonant tube are 0.3m and 0.1m, respectively. We can find that the calculation results by various methods agree quite well, which demonstrates the validity of our new proposal.

Figure 4 is for the resonant frequency of a tube with a buffer at one end. The resistance at tube end has obvious effect on resonant frequency, however, it gets less evident when the ratio of real and virtual part of the normalized impedance is larger than 3. This is attributable to the fact that the rise of reactance leads to a decrease of $2y/(x^2 + y^2 - 1)$, which is dominant for σ. However, this effect can not be found in the void buffer for the resistance is not large enough, unless some extra resistance is put inside the buffer.

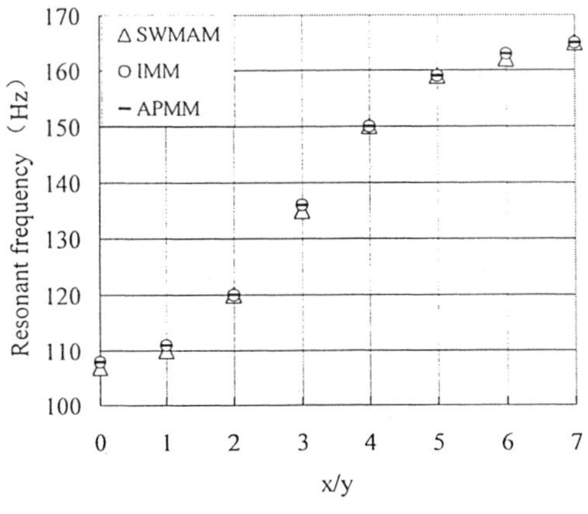

Figure 3　Calculated resonant frequency of the tube with a buffer by three methods

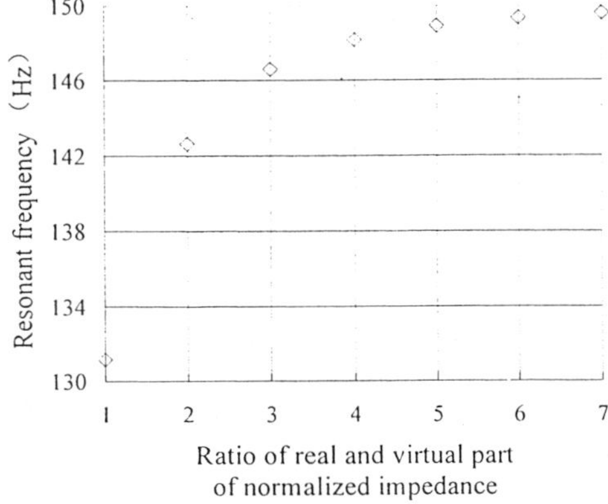

Figure 4　Effect of end impedance on resonant frequency

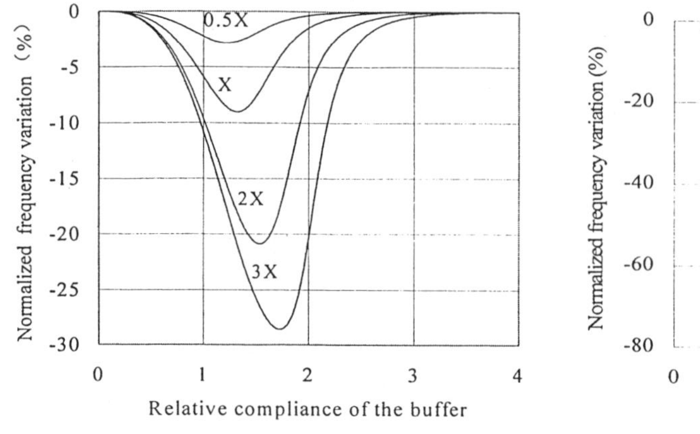
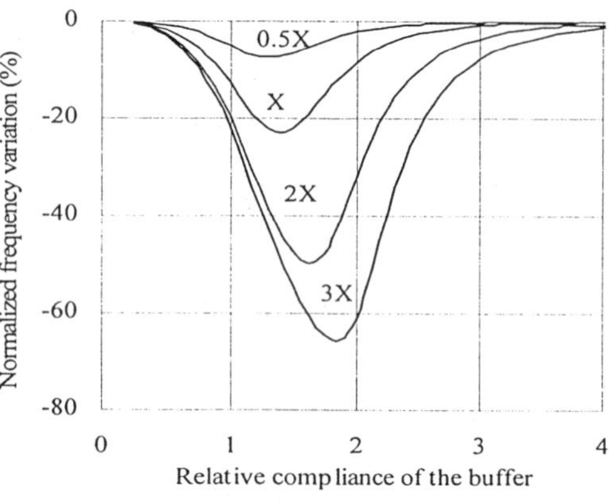

(a) one rigid end and buffer at the other end   (b) two buffers at both end

Figure 5   Normalized resonant frequency variation with relative compliance of the buffer

The influence of the buffer volume on resonant frequency, through relative compliance (the ratio of buffer volume to tube volume), is also analyzed for two cases as shown in Figure 5(a,b), where "X" is the reactance component in the impedance, and "mX" means that the resistance is m times of reactance. We find that the buffer volume may greatly affect the resonant frequency at some certain range (say 0.5-3 times of the tube volume for the present calculated case). Also, the resistance component is another obvious factor influencing the resonant frequency.

CONCLUSION

1. The SWMAM is proved valid for calculating resonant frequency, after comparison with two traditional methods. It is convenient for analyzing the influence of impedance on the resonant frequency.
2. The impedance of tube ends has obvious effect on resonant frequency when the resistance is not so large (say the ratio between the resistance and reactance is lower than 3).
3. When the end buffer volume is 0.5-3 times that of tube, the resistance fraction greatly affects the resonant frequency.

ACKNOWLEDGEMENTS

The financial supports from National Natural Sciences Foundation of China (No.50206008), SRF for ROCS of State Education Ministry, the Third World Academy of Sciences (No.01-284 RG/PHYS/AS) and "985 Program" of Zhejiang University to this work are gratefully acknowledged.

REFERENCES

1. Swift, G. W., Thermoacoustic engines, J. Acoust. Soc. Am. (1988) 84 1145-1180

2. Chen, G. B., Jiang, J. P., Shi, J. L., et al., Influence of buffer on resonance frequency of thermoacoustic engine, Cryogenics (2002) 42 223-227

3. Kinsler, L. E., Frey, A. R., Coppens, A. B. et al., Fundamentals of acoustics, John Wiley & Sons, New York, USA (1982), 210-214

Figure of merit for regenerator working in acoustic field

Guo Fangzhong

Cryogenics Laboratory, Huazhong University of Science and Technology, Wuhan 430074, China

To specify the performance of a TA device, the quantitative description of regenerator is a significant problem in determining the performance of the whole TA system. More attention should be paid on this problem since thermoacoustic system tends to work in high frequency and high amplitude acoustic field, and complexity caused by the last two factors has not been revealed. Aiming to the mentioned problem, a series of experimental and theoretical investigation works have been carried on by the reporter's group.

BACKGROUND

Regenerator is not merely a heat exchanger in the system with pressure oscillation, now this view point has been recognized commonly. To specify the performance of a TA device (including regenerator and stack, we'll call it regenerator later on in this article), quantitative description of regenerator in determining the performance of the whole TA system is a significant problem. Most of the information about regenerator has been obtained only by analysis or computation on simulated models, few of them has been verified or supported by experiments.   More attention should be paid on this problem since thermoacoustic system tends to work in high frequency and high amplitude acoustic field, and the complexity caused by the last two factors has not been revealed..

Aiming to the mentioned problem, a series of experimental and theoretical investigation works has been carried on by the reporter's group [3-8].

EXPERIMENTAL INVESTIGATION ON THE RESONANT MODE OF THE REGENERATE IN A STIRLING CRYOCOOLER SYSTEM

After successful modeling of pressure wave propagation in an isothermal regenerator by a passive network of fluid flow reactance [3], recently the thermoacoutic conversion in a regenerator has been characterized by an active network made up of acoustic impedance and a source element reflecting the contribution of gas parcel velocity variation during its reciprocating displacement along the longitudinal temperature gradient (see Figure 1) [6].

An experimental split Stirling cryocooler with a controllable displacer (driven by a linear motor)

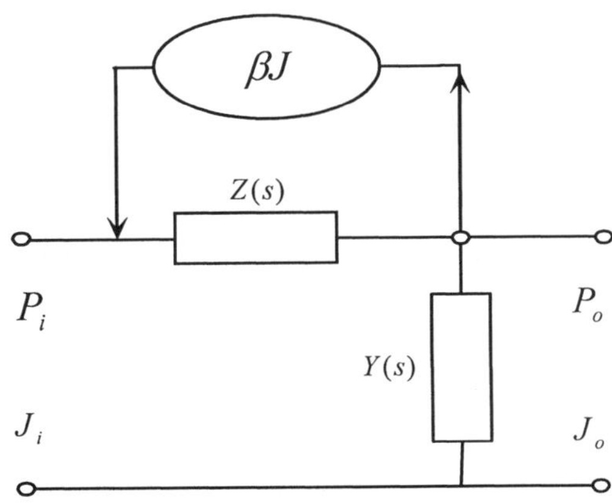

Figure 1 Network modeling for regenerator active network for regenerator with imposed temperature gradient

and an adjustable longitudinal temperature difference ΔT_r exerted between the ends of the regenerator has been built to verify this modeling [7].

The longitudinal temperature difference ΔT_r imposes a significant influence on the pressure wave propagation induced by the motion of the displacer in the regenerator. The experimental result revealed that at zero ΔT_r, P_e has a phase lead over P_w of $\approx 180°$ (shown in Figure 2), and the amplitude of P_e is very small compared to P_w. With the appearance of ΔT_r, the phase leading to about 90° even if the ΔT_r is small (shown in Figure 2). When the temperature difference ΔT_r attains the level of 120K, the phase of P_w shifts abruptly and comes into phase with P_e. This abrupt phase change reminded the appearance of some resonance effect in the system.

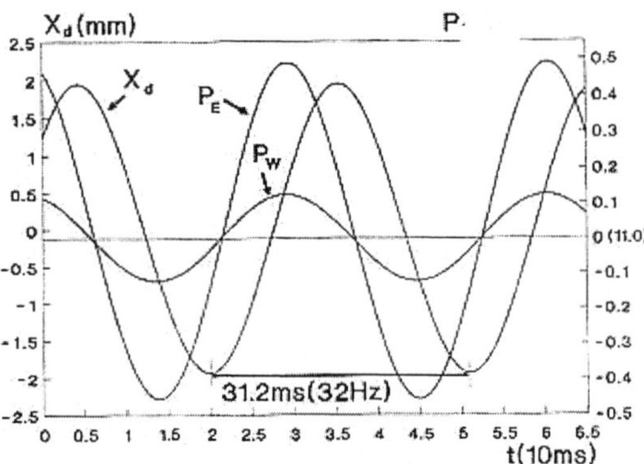

Figure 2   Results of X_d, P_e and P_w oscillation measured at ΔT_r =163K ($T_e = 122K$, $T_w = 255K$)

To explain the revealed resonance effect, a parity simulation circuit of regenerator basing on the network modeling approach, published firstly in [6], has been built. The simulation circuit modeled the regenerator by a coupled LC resonator and the imposed temperature difference to the regenerator by a parallel connected potentiometer which acts as the direct current power source of this circuit (seen in Figure 3).

The imposed ΔT_r cause the periodic variation of density (i.e. impedance L) as shown in Figure4(a) and 4(b),there is an increment of inductance ΔL for each cycle, which converts the heat energy transferred by gas-solid interface into kinetic energy carried by ΔL,thus a parameter called modulation factor can be used to quantify $m = \Delta L/L$,which can be identified directly by dynamical data treatment shown in figure.5. This circuit simulates the modulation and amplification by an analogy between the thermoacoustic and the electro-magnetic wave modulation and amplification process. The simulation circuit has successfully demonstrated that, a resonance mode actually happened at some critical temperature difference. Along with the establishment of a temperature gradient the inductance impedance of regenerator raises because of density increasing, this causes the flow impedance of regenerator varying gradually from the capacitance dominated state to resonant state, which is a resistance dominated state.

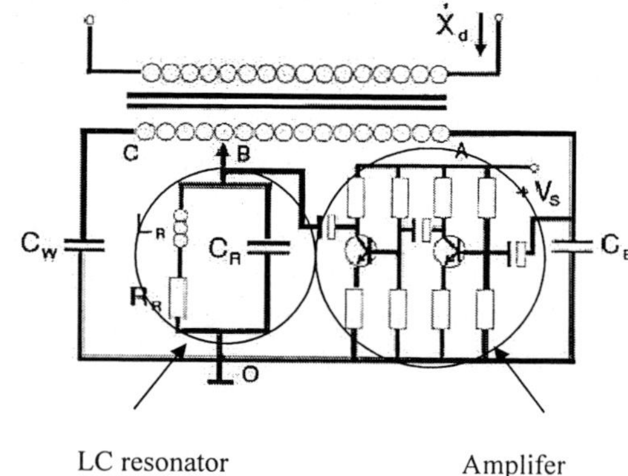

Figure 3 Parity simulation network for regenerator with imposedtemperature gradient

REGENERATOR-RESONANCE CONVERTER WITH HIGH FREQUENCY SELECTIVITY

The network simulating in the circuit shown in Figure3 reminds a widely usable electronic device in electronic communication technique, i.e. the mean frequency active filter. The figure shows the circuit

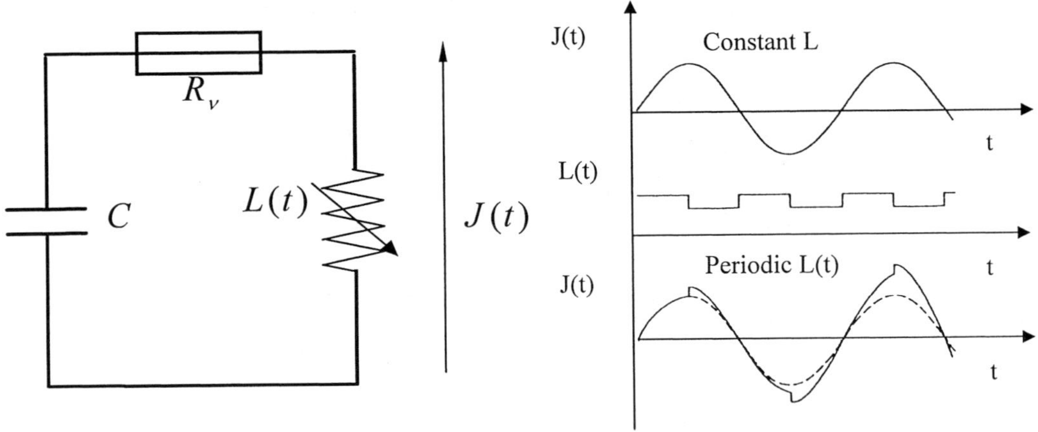

Figure 4 (a) Regenerator—LC coupled converter; (b) Periodic L (t)

coupling a transistor amplifier with a resonance LC circuit, the resonance frequency of it is ω_0 with quality factor of Q.

The regenerator in a TA system plays the same role as the active filter in the above mentioned electronic system. This means that the response of regenerator would have a strong frequency dependence, which has been demonstrated and verified by our experiments in [8] (shown in Figure 6). As shown in these figures, evident peaks of response altitude confirm the existence of optimal frequency for each type of matrix. By network modeling consideration, in an active narrow band amplifier there must be an optimal matching between the frequency spectrum of response given by the transistor amplifier and the connected LC circuit chain. In the regenerator of a thermoacoustic system, the material properties of solid matrix determines the broad band frequency spectrum of transversal entropic wave, while the gas parcels

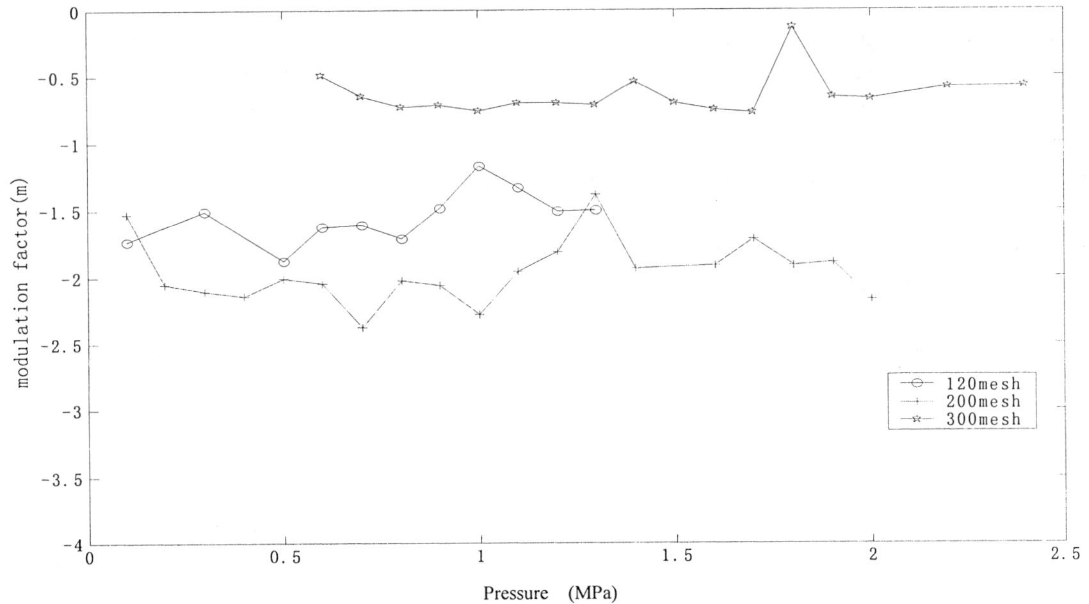

Figure 5 Identification result of modulation factor of different regenerator matrix

carried by the axially propagating acoustic wave   play the role of modulator oscillating in a narrow fixed frequency band.

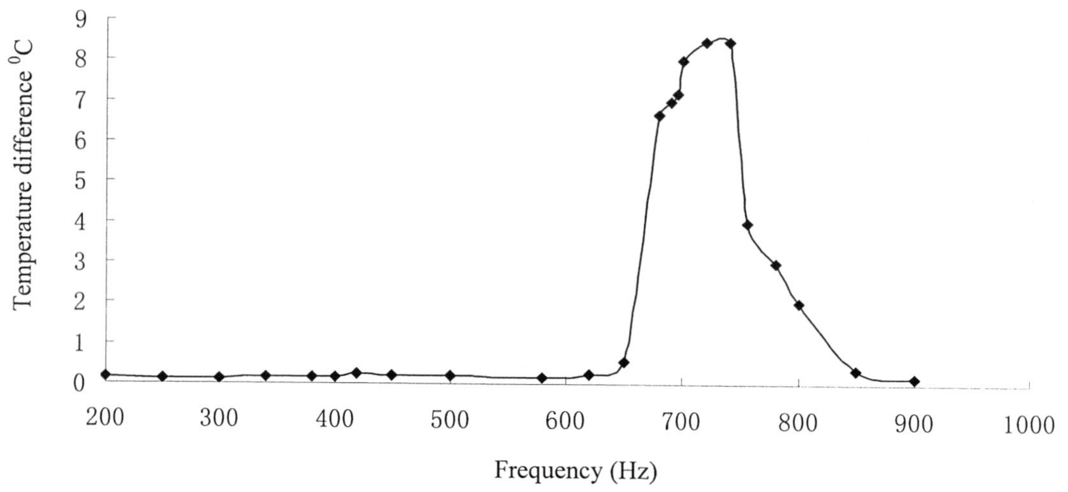

Figure 6 Experiment on the optimal frequency of matrix

THE RESONANCE MODE OF REGENEARTOR IN HEAT DRIVEN THERMOACOUSTIC RESONANCE TUBE

Thus, the recent widely-used active network model of regenerator is only an over simplified description of this nonlinear device, working in a state far from thermodynamic equilibrium especially in the case of heat driven high frequency acoustic field. The heat dynamics of a real high frequency regenerator should take the frequency dependence of its response as one of most important problems. In high frequency application in order to match the entropic wave frequency spectrum of matrix with its LC resonance characteristic, a three frequency parameter network is proposed here to describe our approach to the problem, see figure 7. Initially, the input (left) side in Figure7 is in the state of thermodynamic equilibrium, which is in fact a

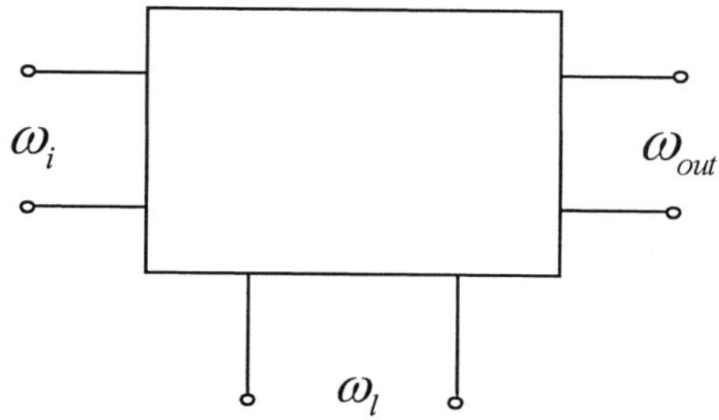

Figure 7 3-frequency parameter network modeling for regenerator in a heat-driven resonance tube

wide broad acoustic field cause by thermodynamic fluctuation of state. The field can be described by mode density [s/cm^3], which is the acoustic energy ρ_m emitted by unit volume of working substance in unit interval of frequency band, i.e. the number of freedom in unit volume. The input mode ω_i that lies most closely to the intrinsic frequency ω_0 of the TA system with regenerator is the most favorite one to be amplified by the gas parcel displacement modulation in regenerator, while the modulated mode by the imposed temperature gradient is also a small part of gas in the system and is shown in the right side of Figure 7, which includes the oscillating mode with frequency ω_p.

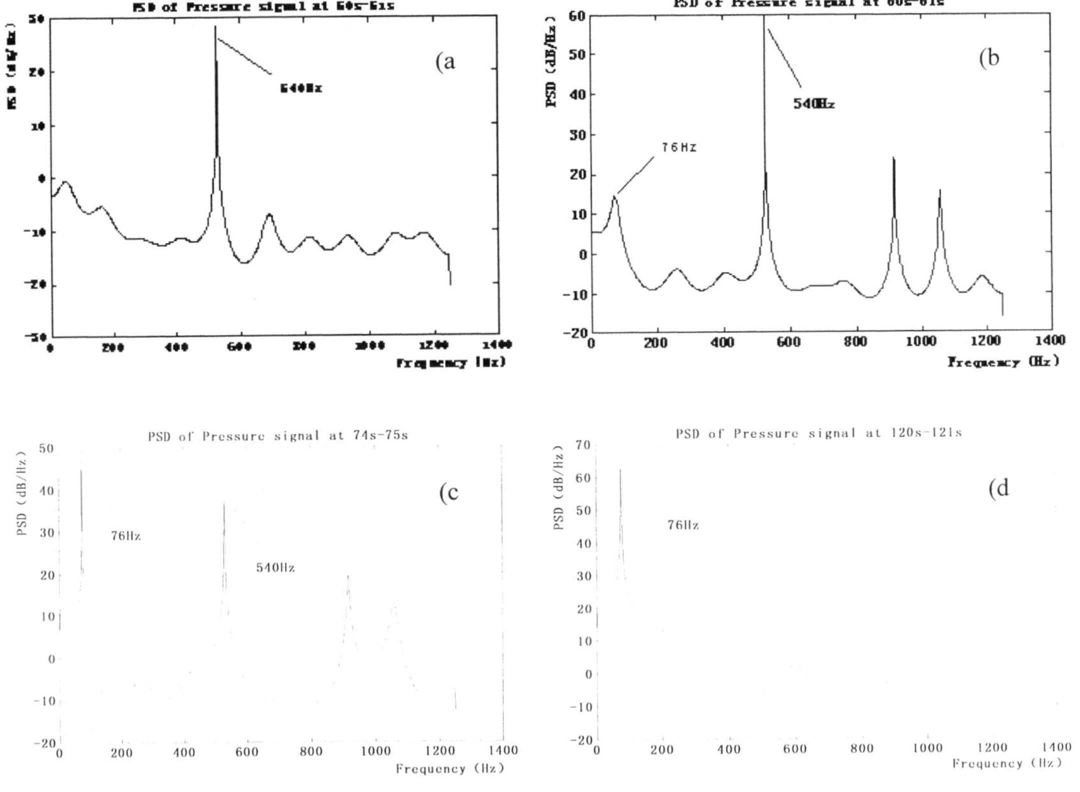

Figure 8 PSD of pressure signal

The connection of regenerator with the LC resonance device shown in figure 3 uses the frequency selectivity feature of the device to provide a narrow band high quality (i.e. high Q) amplifier. The right side of Figure7 shows that among the entropic waves modulated and emitted by the matrix of regenerator only mode with ω_p can be resonated by ω_i according to the rule of parametric resonance

$$\omega_p = n\omega_i , \quad n = 1, 2 \cdots\cdots \tag{1}$$

The resonance between ω_p and ω_i results in an oscillation mode of ω_i in the so-called idler loop of the system. To verify the above mentioned model, a heat driven resonance tube has been tested with an intrinsic frequency of 1100Hz. The time series shown in Figure8 demonstrates the filtering process of the resonance tube, resulting in the appearance of resonance only after 170 seconds of heating. This reveals that the matching between the frequency response of regenerator matrix and resonance tube system may be a way for higher thermoacoustic conversion efficiency.

CONCLUDING WORDS

By following the approach of network modeling, works on the better matching between regenerator and its thermoacoustic heat engine system has been carried on. The view point of considering regenerator as an active amplifier with high frequency selectivity and the resonance mode of regenerator may perhaps be an approach to improve the conversion efficiency of a thermoacoustic heat engine system.

ACKNOWLEDGE

This material is based upon work supported by the National Natural Science Fund of China under the contract No. E060107-50276064.

REFERENCE

1. Xiao, J. H.   Thermoacoustic theory for cyclic flow regenerators: 1. Fundamentals <u>Cryogenics</u> (1992), <u>32</u> 895

2. Xiao, J. H.   Analytical network model on the flow and thermal characteristic of cyclic flow cryogenic regenerator   <u>Cryogenics</u> (1988) <u>28</u> 762

3. Guo F. Z., Chou, Y.M., Lee, S.Z., Wang, Z.S. et al.   Flow characteristic of a cyclic flow regenerator <u>Cryogenics</u> (1987) <u>27</u> 152

4. Xiang Y., Li Q. and Guo F. Z.   Identification of the negative feedback relationship in split cycle free piston Stirling cryocooder system   <u>Proc ICEC 13 Butterworths</u>, Guildford, UK (1990) 216

5. Xiang, Y., Guo F.Z., Kuang, B. and Deng, X.H.   The temperature feedback relation in regenerators <u>Proc JSJS-4 Chinese Association of refrigeration</u>, Beijing, China (1993)

6. Deng, X. H.   <u>Thermoacoustic essence of regenerator and fundamental theory for desilgn of thermoacoutic engine</u>,   PHD Thesis Huazhong University of Science and Technology,   Wuhan, China (1994)

7. Xiang, Y., Kuang, B and Guo F.Z.   Parity simulation of thermoacoustic effect in regeneration of Stirling cryocooler <u>Cryogenics</u> (1995) <u>35(8)</u> 489

8. Tu, Q.   <u>Optimization of thermoacoustic stack and its matching with thermoacoustic engine</u>, PHD Thesis Huazhong University of Science and Technology,   Wuhan, China (2003)

Investigation of Acoustic Streaming in a Thermoacoustic-Stirling Heat Engine

Duthil P., Bretagne E., Wu J., François M.-X., Li Q.*.

LIMSI-CNRS, BP 133, 91403 Orsay Cedex, France
*Technical Institute of Physics and Chemistry, CAS, Beijing 100080, China

Experimental characterization of acoustic streaming in a thermoacoustic-Stirling heat engine is presented. Investigation is indeed carried out with and without Gedeon streaming suppression by means of a rubber balloon positioned inside the loop at a given place. First, results show that whereas the temperature distribution in the regenerator is weakly affected by the streaming flow, heat fluxes measured in the heat exchangers determines the direction of this acoustic streaming. Moreover, they reveal that additional thermal effects are coupled with the streaming. Finally, they show that thermoacoustic conversion efficiency is improved by inserting the rubber balloon.

INTRODUCTION

Gedeon [1] has been concerned with how nonzero second order steady mass flow can arise in Stirling and pulse tube crycoolers whenever a closed-loop path exists. Later on, the term Gedeon streaming has been used [2] to describe such DC flow in the case of a close-loop flow path existing within the device. Theoretical studies of acoustic streaming were extended to annular thermoacoustic prime-mover [3] as well as to channels of arbitrary width in a standing wave [4]. Recent experiments [5-7] confirm the existence of the closed-loop streaming. Moreover, it was demonstrated that suppression of the streaming significantly increases the efficiency of the device. It is interesting to note that some research groups [8-11] study indirectly acoustic streaming through the analysis of jet pumps.

In the current paper, we focus on the Gedeon streaming existence and direction in thermoacoustic-Stirling heat engine through the analyses of temperature profile in the regenerator and heat fluxes in the heat exchangers.

APPARATUS

A thermoacoustic-Stirling heat engine has been designed. A scale drawing of the loop section of the experimental apparatus is shown in Figure 1, without the resonator section - similar to that of [6]. The loop section contains the heat exchangers and the regenerator. We describe briefly the core of the apparatus. Near the top of the loop is the cold heat exchanger. It is made of shell-and-tube construction consisting of 95 brass tubes of 3 mm inside diameter and 35 mm long, welded into a copper flange. The total gas-solid contact area is of 0.0313 m^2. Below the cold heat exchanger is the 75 mm long regenerator which is made of a stainless-steel screens stack [12] machined to a diameter of 56.3mm. The lower end of the regenerator abuts the hot heat exchanger. It is a copper screens stack; the heater is made of six 1-mm-diameter electrically insulated wire [13] with a total resistance of $10\,\Omega$. The thermal buffer tube (TBT) is an upper straight cylinder flaring along its lower part with a 1.35° half-angle taper. This taper is used to minimize boundary layer driven streaming (Rayleigh streaming) [14]. At the lower end of the TBT is another cold heat exchanger of same design as the cold heat exchanger previously described, but of 20-mm-length and 0.0179m^2 total gas-solid contact area.

376

9 type-K thermocouples allow the measurement of the axial temperature profile within the regenerator and the water temperature flowing in/out of the two cold heat exchangers. Absolute accuracy of the thermocouples is of 0.1K but preliminary tests in purely conductive state (mainly linear temperature profile) revealed an accuracy of 2% relatively to their location within the regenerator. In order to determine heat fluxes extracted through the cold heat exchangers, temperature measurements were coupled with mass flow rates measurements (double weighting method) leading to a relative accuracy of 10% on the measured heat power. The input heat power is known from the voltage and intensity of the resistive wires with an accuracy of 1W.

A piezoresistive pressure sensor located at 80mm above the cold heat exchanger allows the measurement of oscillatory pressure - labeled with p_1.

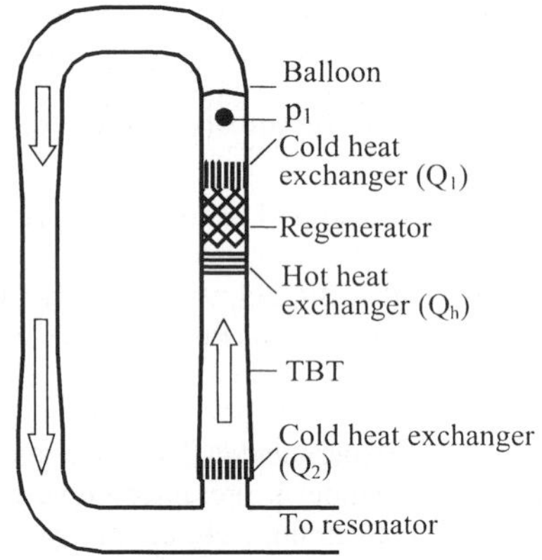

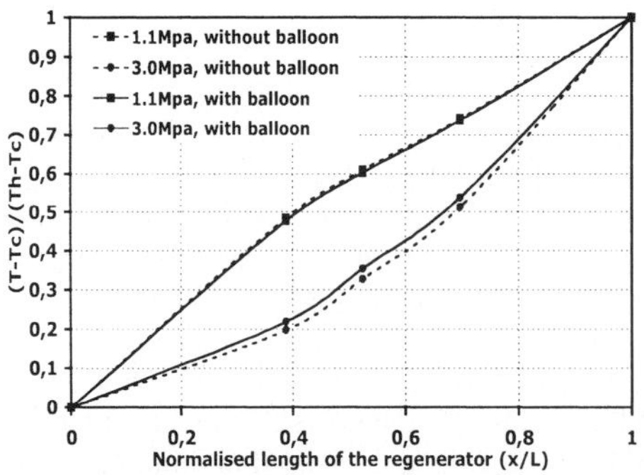

Figure 1. Scale drawing of the loop section.

To quickly confirm Gedeon streaming, we compared experimental results with and without the placement of a rubber balloon at the position shown in Figure 1: we expected the balloon to be acoustically transparent, but to completely block Gedeon streaming. Preliminary experimental work showed good reproducibility of the results for those two cases: with and without balloon.

EXPERIMENTAL RESULTS

It was assumed that the temperature spatial profile along the axis of the regenerator was related to Gedeon streaming in the loop [5, 15]. Hence, in the cases of suppressing or not the Gedeon streaming, we measured the temperature profiles in the regenerator for three different mean pressures (1.1MPa, 2.0MPa and 3.0MPa) and for different input heat powers. With working gas being nitrogen, the normalized temperature $(T(x/L) - T_c)/(T_h - T_c)$ is plotted in Figure 2 and Figure 3 as a function of the normalized length x/L. Here, T_c and T_h are temperatures at the cold and hot ends of the regenerator, respectively. L is the length of the regenerator and x is axial position measured from the cold end of the regenerator. In the case of a free closed-loop path (i.e. without balloon), the normalized temperature profiles have a trend to incurve upwards for a mean pressure of 1.1MPa (input heat power ranging between 172W and 330W; pressure ratio varying from 0.48% to 1.1%) and have an opposite trend at 3.0MPa (input heat power ranging between 810W and 1150W; pressure ratio varying from 1.75% to 2.24%). As for different heat power temperature, profiles are similar for each mean pressure, only two

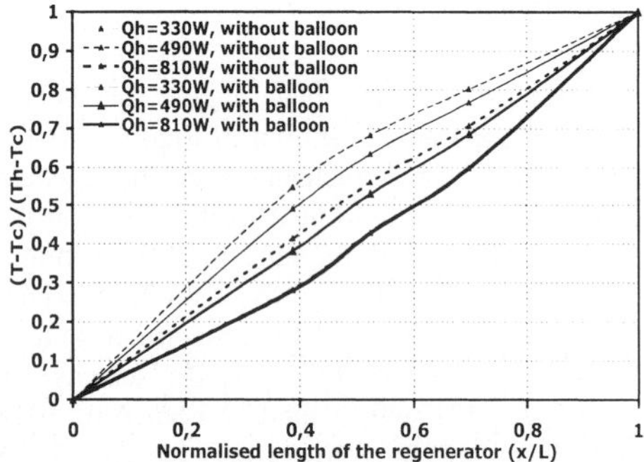

Figure 2. Normalized temperature $(T\text{-}T_c)/(T_h\text{-}T_c)$ vs normalized length x/L at 1.1MPa and 3.0MPa with (solid lines) and without (dashed lines) balloon. For each pressure, representative profile is plotted only.

Figure 3. Normalized temperature $(T\text{-}T_c)/(T_h\text{-}T_c)$ vs normalized length x/L at 2.0MPa with (solid lines) and without (dashed lines) balloon.

representative profiles corresponding to the respective maximum powers 330W (filled squares) and 1150W (filled circles) have been plotted with dash lines in Figure 2. Similarly, in the case of balloon insertion, we plotted with solid lines in Figure 2 the temperature profiles corresponding to the same input power at the same mean pressure (330W at 1.1Mpa - filled squares and 1150W at 3.0Mpa -filled circles). Choosing an intermediate mean pressure of 2.0MPa (shown in Figure 3), the normalized temperature profiles evolve with the increase of input heat power (from 330W to 890W; drive ratio varying from 1.18% to 2.1%) from an upward curve through a nearly linear profile to a downward curve. This is true for the cases with (solid lines) and without (dash lines) balloon.

Considering the comparison between the temperature T_{mid}, at the axial midpoint of the regenerator, and the average temperature $T_{ave}=(T_c+T_h)$ as an indication for determining direction of Gedeon streaming [5,7], it should come at i) 1.1MPa, $T_{mid}>T_{ave}$, $M_2>0$ (anticlockwise through the loop), implying that the Gedeon streaming flows up in the regenerator; ii) 3.0MPa, $T_{mid}<T_{ave}$, $M_2<0$ (clockwise through the loop), implying that the Gedeon streaming flows down in the regenerator; iii) 2.0MPa, the direction of Gedeon streaming changes from anticlockwise to clockwise with the heat power. However, as illustrated in Figure 2 and Figure 3, the balloon has weak effect on the normalized temperature profiles so that Gedeon streaming does not seem to influence it. On the contrary, it seems that both the input power Q_h and the mean pressure have an obvious effect.

For those reasons, we decided to measure lost heat carried out by water flow through the two cold heat exchangers in order to determine the direction of Gedeon streaming. To obtain obvious water temperature difference between entrance and exit of cold heat exchanger, we adjust the water massflow rate to 5g/s. Lost heat powers from the two cold heat exchangers are shown in Figure 4 (a) and Figure 4 (b) where the lost heat is plotted versus the input heat power for three different mean pressure.

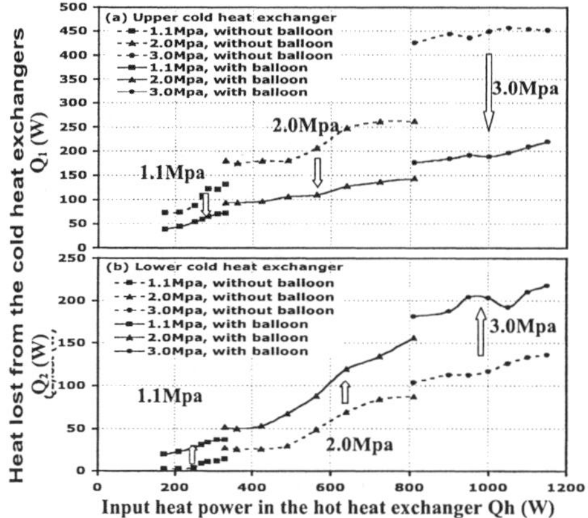

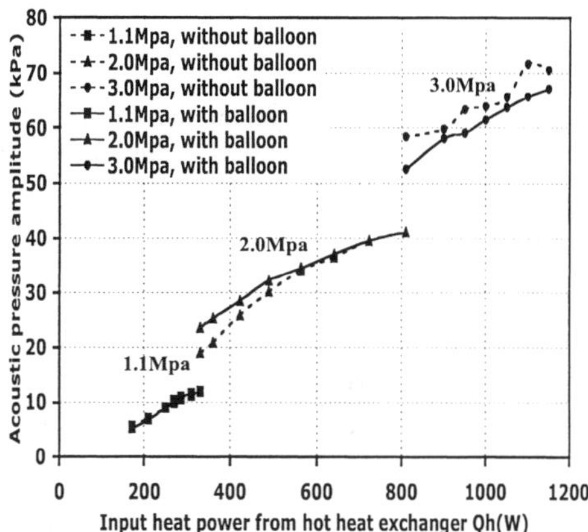

Figure 4. Lost heat power from the upper (a) and lower (b) cold heat exchangers vs input heat power for different mean pressures with (solid lines) and without (dashed lines) balloon.

Figure 5 Dynamic pressure amplitude vs input heat power for different mean pressures with (solid lines) and without (dashed lines) balloon.

Here, subscripts 1 and 2 denote for the upper and lower cold heat exchangers respectively (cf. Figure 1). Solid and dashed lines stand for the cases with and without balloon respectively. For all measurements, adding the balloon in the loop leads to (i) a drop of lost heat power in the upper cold heat exchanger, (ii) a rise of lost heat power in the lower heat exchanger (shown as wide arrow in Figure 4) and (iii) a reduction of total heat lost from the cold heat exchangers. It can be concluded that Gedeon streaming flows in the anticlockwise direction (wide arrows shown in Figure 1) up through the regenerator from hot to cold ends transporting a large amount of heat to the upper cold heat exchanger.

As illustrated in Figure 5, inserting a balloon has no influence on the acoustic pressure amplitudes which could let us suppose that there is no change in the acoustic power at the upper entrance of the regenerator. By use of the first thermodynamic principle over the control volume delimited by he two cold heat exchangers, we calculated an efficiency of the engine (cf. Figure 6 (a)): $\eta=(Q_h-Q_1-Q_2)/Q_h$. Note that heat losses outside the engine (convective and radial), which only depends on temperature levels, can be assumed as identical whenever the balloon is inserted or not (the temperatures levels being not affected by the presence of the balloon – cf. Figure 1 and 2). Thus, we did not take those losses into account in the balance yielding an overestimated

378

efficiency. It hence can be seen in Figure 6 (b) that the efficiency may be increased by placing the rubber balloon. This let us think that Gedeon streaming indeed exists in our thermoacoustic-Stirling heat engine and that inserting a rubber balloon allows to characterize this phenomenon. Moreover, from the temperatures profiles plotted in Figure 2 and 3 we showed that Gedeon streaming does not affect the temperature levels. Hence, we can imagine that other phenomena are responsible for the profiles curvatures in the regenerator, such as regenerator streaming or acoustic power dissipation in the balloon that could alter the acoustic power at the lower entrance of regenerator.

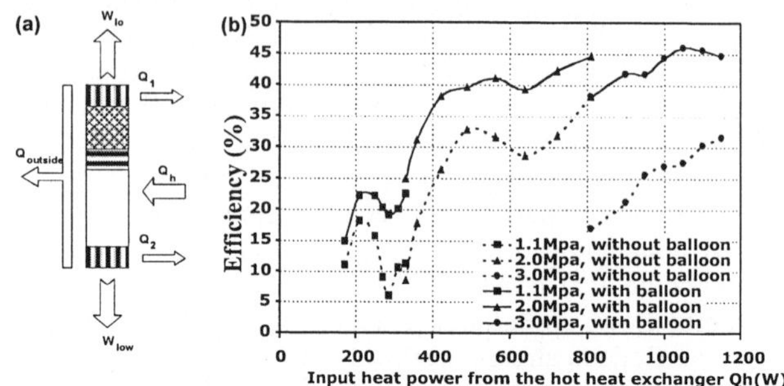

Figure 6. (a) Energy balance for the control volume delimited by the heat exchangers. (b) Efficiency of the thermoacoustic-Stirling heat engine with (solid lines) and without (dashed lines) balloon - radial and convective thermal losses are not taken into account.

CONCLUSION

We have measured temperature profiles in the regenerator and lost heat power carried by water flow in the cold heat exchangers for different working conditions in the presence or not of a rubber balloon inserted in the close-loop resonator. Temperature profiles in the regenerator are not affected by the existence or not of Gedeon streaming leading us to think that another thermal effect predominates in the regenerator. However measurement of heat fluxes through the cold heat exchangers seems to indicate that Gedeon streaming does exist in the engine and that its suppression increases the efficiency. Further quantitative research is required - including radial temperature profile investigation within both the regenerator and the TBT- to place the effect of other streaming losses, e.g., regenerator and Rayleigh streamings.

REFERENCES

1. Gedeon, D.,DC gas flows in Stirling and pulse tube cryocoolers, Cryocoolers (1997) 9 385-390
2. Swift, G.W. , Thermoacoustics: A unifying perspective for some engines and refrigerators, Acoustical Society of America Publications, (2002).
3. Gusev, V., Job, S., Baillet, H., Lotton, P. and Bruneau, M., Acoustic streaming in annular thermoacoustic prime-movers, J. Acoust. Soc. Am (2000) 108(3) 934-945.
4. Hamilton, M.F., Llinskii, Y.A. and Zabolotskaya, E.A., Acoustic streaming generated by standing waves in two-dimensional channels of arbitrary width, JASA (2002) 113(1) 153-160.
5. Swift, G.W., Gardner, D.L. and Backhaus, S., Acoustic recovery of lost power in pulse tube refrigerators, J. Acoust. Soc. Am (1999) 105 711–724.
6. Backhaus, S. and Swift, G.W., A thermoacoustic-Stirling heat engine: Detailed study, J. Acoust. Soc. Am (2000) 107 3148–3166.
7. Backhaus, S. and Swift, G.W., An acoustic streaming instability in thermoacoustic devices utilizing jet pumps, J. Acoust. Soc. Am (2003) 113 1317-1324.
8. Wakeland, R.S. and Keolian, R.M., Influence of velocity profile nonuniformity on minor losses for flow exiting thermoacoustic heat exchangers, J. Acoust. Soc. Am (2002) 1249-1252 .
9. Morris, P. J., Boluriaan, S. and Shieh, C.M., Computational thermoacoustic simulation of minor losses through a sudden contraction and expansion, 7th AIAA/CEAS Aeroacoustics Conference(2001) 2271-2272.
10. Petculescu, A. and Wilen, L., Oscillatory flow in jet pumps: Nonlinear effects and minor losses, J. Acoust. Soc. Am (2003) 113(3) 282-1292
11. Smith, B.L. and Swift, G.W., Power dissipation and time-averaged pressure in oscillating flow through a sudden area change. J. Acoust. Soc. Am (2003) 113 2455-2463.
12. 128 mesh/ (Fr-In) from Gantois.
13. Fabricated by THERMOCOAX
14. Olson, J.R. and Swift, G.W., Acoustic streaming in pulse tube refrigerators: Tapered pulse tube, Cryogenics (1996) 37 769-776.
15. Swift, G.W., Backhaus , S.N., Gardner, D.L., Traveling wave device with mass flux suppression. US Patent (2000) No.6032464.

Investigation on A Thermoacoustically Driven Pulse Tube Cooler Working at 82.5 K

Limin Qiu, Daming Sun, Weilin Yan, Ping Chen, Zhihua Gan, Guobang Chen

Cryogenics Lab, Zhejiang University, Hangzhou 310027, P.R. China

It is an efficient way to increase the performance of the thermoacoustic refrigeration system by improving the frequency matching between thermoacoustic engines and pulse tube coolers (PTCs). Investigation on the performance of the refrigeration system under different operating frequency was carried out by changing the length of the resonance straight tube of a thermoacoustic Stirling engine. With working pressure and heating power of 2.8 MPa and 2500 W respectively, the single stage double-inlet PTC reaches the refrigeration temperature of 82.5K at 45 Hz.

INTRODUCTION

A thermoacoustic engine converts thermal energy into acoustic power, which can be used to drive a pulse tube cooler (PTC) or other kinds of thermoacoustic refrigerator. Research and development of thermoacoustic field is booming in recent years. Thermoacoustic heat engine develops from the standing wave to the traveling wave mode, and its efficiency has increased substantially. To commercialize thermoacoustic heat engines, their output powers need to be increased as well as the efficiencies, which is of importance for large scale applications. Recently, some researchers proposed cascade thermoacoustic heat engine [1], two-end driving traveling wave thermoacoustic heat engine [2] to increase the output power. Concerning thermoacoustic refrigeration, orifice PTC, coaxial single-stage PTC, thermoacosutic refrigerator directly coupled into the torus of a thermoacoustic Stirling engine, and acoustic recovery PTC were developed one after the other. Up to now, there are two main development directions for thermoacoutic refrigeration. One aims to obtain lower refrigeration temperature below 120K, in which common PTCs, such as orifice and double-inlet PTCs, are adopted as refrigeration unit. The other is to obtain large refrigerating capacity above 120K, in which orifice and traveling-wave PTCs are adopted. This is promising for natural gas liquefaction and even for daily life refrigeration.

This paper follows up our former study [3-4], and focuses on the frequency matching between a thermoacoustic heat engine and a single stage double-inlet PTC. The operating frequency was adjusted by lengthening the resonance tube of the thermoacoutic engine. Experimental results show that the operating frequency can greatly influence the performance of the PTC. With working pressure and heating power of 2.8 MPa and 2500 W, respectively, the PTC obtained the minimum refrigeration temperature of 82.5K, which is a new record for the PTC driven by a thermoacoustic engine.

THEORETICAL ANALYSIS

In the thermoacoustic engine there is plane wave acoustic field, where the energy of gas micelle consists of two parts: one is kinetic energy due to motion of gas micelle, the other is potential energy associated with compression and expansion of gas micelle. And the total energy of acoustic field is thus the sum of these two kinds of energy of all gas micelles in acoustic field. The total acoustic energy can be qualitatively written as:

$$E = \eta Q \propto \bar{p}_a^{\,2} \times L \times S$$

Where, η is thermoacoustic conversion efficiency, Q is net heating power, and $\bar{p}_a$ denotes average pressure amplitude in the whole sound wave duct. L and S are the total length and cross-sectional area of the sound propagation pipe, respectively. L influences the performance of thermoacoustic engine in two aspects. One is to determine the operating frequency of the thermoacoustic engine. The other is to change the volume of acoustic field. So there is a tradeoff between the pressure oscillation intensity and the operating frequency by varying L.

EXPERIMENTAL CONFIGURATION

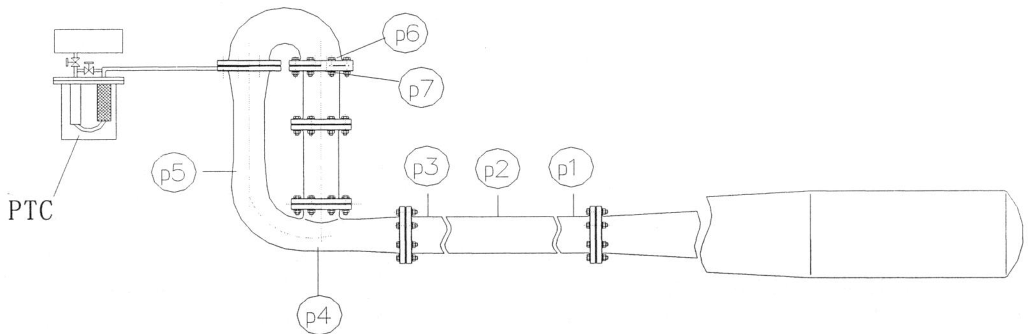

Figure1. Schematic of the thermoacoustically driven PTC system

The experimental apparatus includes a thermoacoustic Stirling engine, a single-stage double-inlet pulse tube cooler, vacuum system, and measurement system. Fig.1 shows the thermoacoustically driven PTC system and the pressure measurement locations. The configuration of each part of the system was introduced in details in reference [3-4].

EXPERIMENTAL RESULTS

<u>Effect of total resonance tube length on working frequency</u>
According to boundary conditions of acoustic field and analysis of pressure amplitude along the length of the engine, the pressure amplitude in acoustic field distributes along the direction shown in Fig.2 as 1/4 standing wave. The total length of the resonance tubes is a sum of the resonance straight tube and tapered tube. The start point (pressure antinode) of the 1/4 standing wave distribution of pressure amplitude is at P7, and terminal point is between P1 and P2, about 100 mm from the entrance of the tapered tube (see Fig.1).We regulated the frequency of the heat engine by lengthening the resonance straight tube. To predict the matching between the PTC and thermoacoustic engine, we calculate the working frequency of the heat engine with different resonance tube lengths as shown in Fig.3. The tendency of predicted frequencies is in good agreement with experimental results.

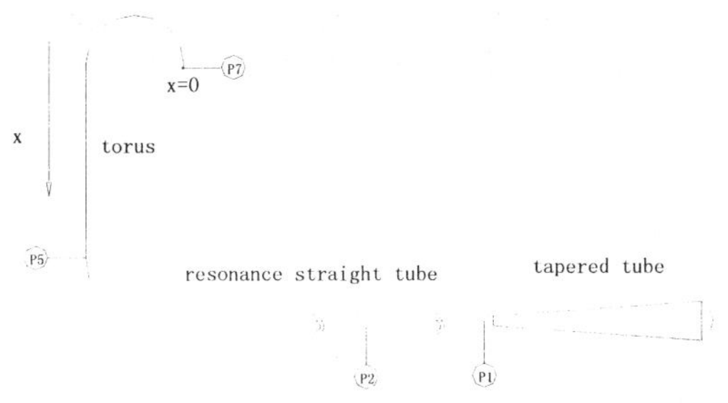

Figure 2 Distribution of pressure amplitude along the heat engine

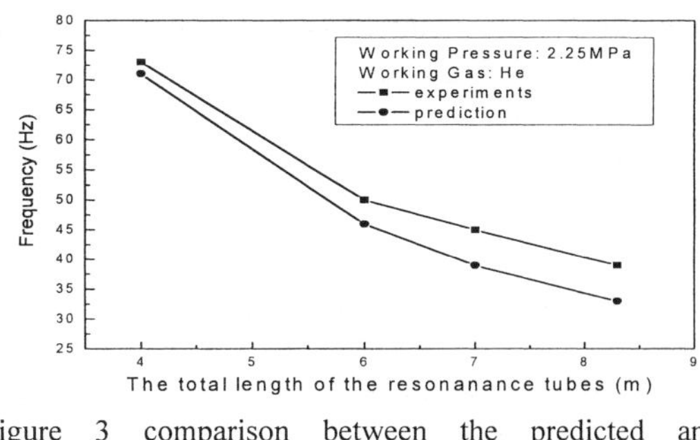

Figure 3 comparison between the predicted and experiments frequency of the engine

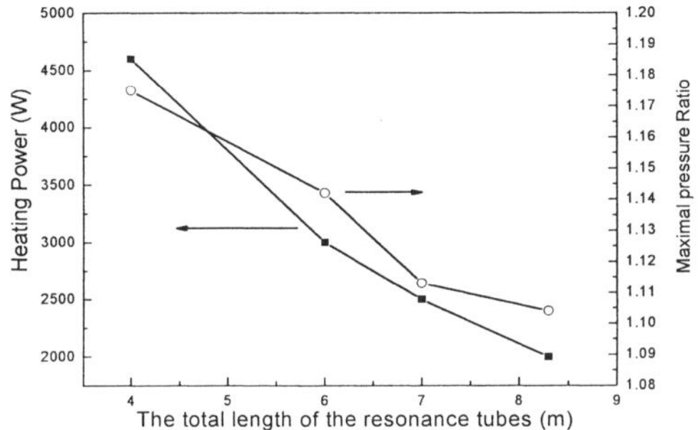

Figure 4 Effect of total resonance tube lengths on heating power, maximal pressure ratio

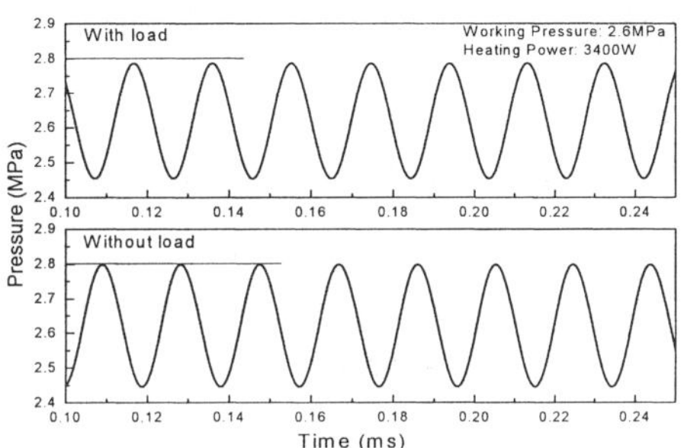

Figure 5 Variations of pressure wave with and without PTC connected (the resonance tubes length, 6 m)

Heating power and maximum pressure ratio

Fig. 4 shows the effect of resonance tube length on the heating power and maximal pressure ratio. The maximum heating temperature is fixed at $675 \pm 5\,°C$. With the length of the resonance straight tube increasing, the heating power absorbed by the thermoacoustic engine decreases, theoretically so does the acoustic power. Consequently, the pressure amplitude declines to some extent. It is consistent with our former theoretical prediction. Therefore, there is a coupling problem among the net heat absorbed, pressure oscillation intensity and operating frequency.

Variations of pressure wave with and without PTC connected

In order to study the effect of the load (PTC) on the thermoacoustic engine, pressure wave with and without load was acquired and analyzed under the same heating power. During the experiment, when the PTC came to a stable state, i.e. when the refrigeration temperature of PTC and the pressure oscillation of the engine were stable, we closed the connection between the engine and the pulse tube cooler. Fig. 5 shows variations of pressure wave with and without PTC at the joint point. We can find that with the PTC connected the pressure ratio falls about 7%. The operating frequencies with and without PTC connected are the same, which means that there is no effect of the PTC on the operating frequency.

Cool-down of pulse tube cooler

Fig. 6 shows the cool-down process of PTC driven by the thermoacoutic Stirling engine with total resonance tube length of 4 m. The refrigeration process of PTC starts once the engine onsets. After about 3 hours, the refrigeration temperature reaches 120 K, and finally reaches a steady state at 110.5 K. The heating power is increased gradually from 3000 to 4900 W. The operating frequency of thermoacoustic engine is 72.5 Hz. Obviously, the operating frequency is too high to get lower temperature although the pressure ratio is big enough. As a tradeoff between the operating frequency and the pressure ratio (see Fig. 4), we selected the engine with the resonance tubes of 7 m to drive the same PTC. Fig.7 shows the cool-down process. The ambient temperature is 12 °C higher than that of Fig. 6. With heating power and working pressure of 2500 W and 2.8 MPa, respectively, the refrigeration temperature reaches a stable state at 83 K after 2 hours. By further valve setting optimization, the lowest refrigeration temperature obtained is 82.5 K, which is the new record of PTC driven by a thermoacoutic engine. It indicates proper frequency matching between the engine and the PTC greatly benefits the refrigeration temperature of the PTC.

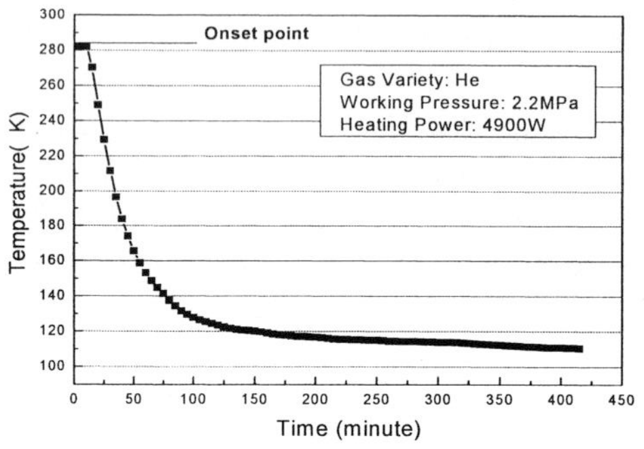

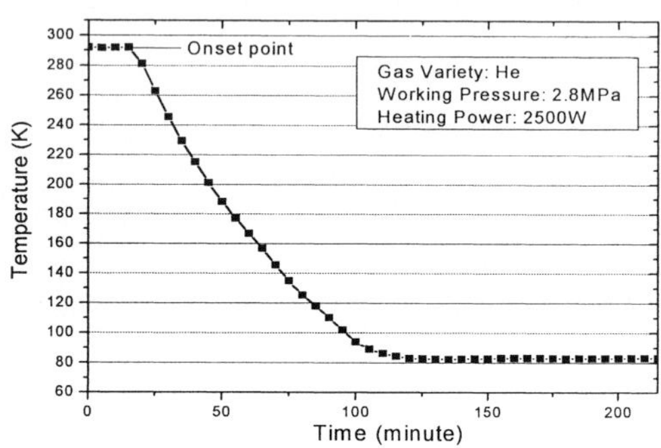

Figure 6 Cool-down process with resonance tubes of 4 m   Figure 7 Cool-down process with resonance tubes of 7 m

CONCLUSIONS

The frequency matching between thermoacoustic engine and pulse tube cooler is of greatest importance to obtain better cooling performance for a pulse tube cooler. It is efficient to adjust the operating frequency by changing the length of the resonance straight tube. A pulse tube cooler driven by a thermoacoustic Stirling engine with the resonance tubes of 7 m obtains a minimum temperature of 82.5 K at 45 Hz.

ACKNOWLEDGEMENT

The research is financially supported by the Foundation of the Author of National Excellent Doctoral Dissertation of P. R. China under contract No. 200033, Natural Science Foundation of Zhejiang Province, and Huo Yingdong Foundation under contract No. 94013.

REFERENCE

1. Gardner. D. L. and Swift. G. W., A cascade thermoacoustic engine, J. Acoust. Soc. Am.(2003) Vol. 114 1905-1919

2. L.M Qiu, D.M Sun, et al, Tow-end driving traveling wave thermoacoustic heat engine, China patent, Application number: 200420020570.9

3. L.M Qiu, D.M Sun, W Zhang, et al, Experimental study of a large-scale multifunction thermoacoustic engine, Cryogenics and Refrigeration-Proceedings of ICCR'2003 157-160

4. D.M Sun, L.M Qiu, W Zhang, et al, Investigation on A Traveling-wave Thermoacoustic Heat Engine with High Pressure Amplitude, Accepted by Energy conversion and management (2004-2)

Thermoacoustic turbulent-flow model for inertance tubes used for pulse tube refrigerators

Luo E.[1], Radebaugh R.[2], Dai W.[1], Lewis M.[2], Wu Z.[1], Zhang Y.[1]

1. Technical Institute of Physics and Chemistry, Chinese Academy of Sciences, Beijing 100080
2. Cryogenic Technology Group, Physical and Chemical Properties Division, National Institute of Standards and Technology, Mail Stop 305, Boulder, CO80305, USA

Inertance-tube phase shifter can avoid DC-flow of the double-inlet phase shifter besides. The flow inside the inertance tube is usually turbulent, which makes linear network mode not accurately workable for describing performance of the inertance tube. Thus, a thermoacoustic turbulent-flow model is proposed to solve the problems, which can consider both turbulent-flow effect and heat transfer effect.

INTRODUCTION

Efficient regenerative cryocoolers operate on proper phase relations between pressure and velocity. A Stirling refrigerator reaches this proper phase relation by complicated mechanical connection of the cranks of both the compressor and the displacer. Pulse tube refrigerators have various devices to achieve these phase relations. A double-inlet phase shifter is a very powerful phase shifter, which is widely used in the pulse tube refrigerators. However, this shifter can bring about some negative effects and sometimes lead to instabilities of operating temperatures due to a so-called DC-flow occurring in this looped pipe system. An inertance tube is supposed to be another powerful shifter, which has the additional function of eliminating DC flow. For typical operating conditions, the flowing of the inertance tube is turbulent. Thus, this paper tried to develop a thermoacoustic turbulent-flow model for describing the inertance tube, by which can both incorporate turbulent-flow effect and heat transfer effect between gas and wall.

PHYSICAL AND MATHMATICAL MODEL

Figure 1 shows the schematic of the inertance tube phase shifter. It consists of an inertance tube and a reservoir. The reservoir is a short cylinder, but with a large cross-sectional area. Because of this special structure, the reservoir can be modeled as a lumped-parameter compliance.   The inertance tube is a long tube with a considerable length, so it has an obvious acoustic behavior. Thus, it is to be modeled here with distributed-parameter models. The operating conditions and the structures for an inertance tube phase shifter are given as follows: (1) the inertance tube has a length of L, a diameter of D and a cross-sectional area of A, and the reservoir has a volume of Vr; (2) the oscillating flow inside the inertance is laminar in an angular frequency of ω; (3) the working gas is helium, having a sound velocity of a_0 and an adiabatic index, γ. In addition, the time-averaged temperature and pressure in the inertance tube are T_0 (~300K) and p_0, respectively. The dynamic viscosity of helium, corresponding to the pressure and temperature, is μ. The wavelength for helium at a frequency of ω is λ.

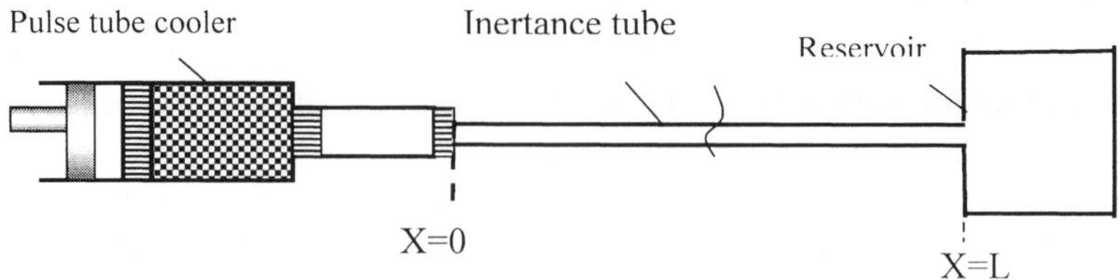

Figure 1   Schematic of Inertance Tube Phase Shifter

In fact, most practical inertance-tube shifters operate in a state of turbulent flow. Therefore, a complete description for such inertance tube shifters may need to be based on some complicated simulations of computational fluid dynamics (CFD) modeling, which is beyond the contents of this paper. In fact, the phase shifting and acoustic power transmission of an inertance tube is still a result of resistive, inertance and compliance whether the flow inside the inertance tube is laminar or turbulent. Thus, Eq.(1) and Eq.(2) are proposed to describe the two cases. The most obvious difference between laminar and turbulent flows is that the turbulent flow makes a significant contribution to viscous flow impedance of an inertance tube. Especially for small-diameter inertance tubes, the viscous component is generally predominant so that the capability of the phase shifter is greatly limited.

$$\frac{d\tilde{p}}{dx} = -(\xi_\mu r_\mu + i\xi_l \omega\, l)\tilde{U} \tag{1}$$

$$\frac{d\tilde{U}}{dx} = -[i\xi_c \omega\, c + \frac{1}{\xi_k r_k}]\tilde{p} \tag{2}$$

where $\tilde{p}$ and $\tilde{U}$ are the acoustical pressure and volumetric velocity in complex number; ξ_μ, ξ_l, ξ_c and ξ_k are the four correction factors for turbulent flow. r_μ, l, c and r_k are the viscous flow resistance, inertia, compliance and thermal-relaxation resistance per unit length for laminar flow, respectively, are given by Eq.(3).

$$r_{\mu,L} = \frac{\omega\rho_0}{A}\frac{\mathrm{Im}[-f_\mu]}{|1-f_\mu|^2}, \quad l = \frac{\rho_0}{A}\frac{1-\mathrm{Re}[f_\mu]}{|1-f_\mu|^2}, \quad c = \frac{A}{\gamma p_0}[1+(\gamma-1)f_k], \quad \frac{1}{r_k} = \frac{\gamma}{\gamma-1}\frac{\omega A\,\mathrm{Im}(-f_k)}{p_0} \tag{3}$$

where f_μ and f_k are the viscous and thermal factors which can be found elsewhere[1].

In principle, the empirical expressions for ξ_μ, ξ_l, ξ_c and ξ_k may be approximately obtained based on numerous experimental data or CFD simulations. However, some simplified expressions for the four correction factors need to be obtained based on some reasonable deductions and qualitative understanding. For turbulent flow, the viscous flow resistance $r_{\mu,T}$ can be described as follows [2].

$$r_{\mu,T} = \xi_\mu r_\mu = \frac{64\rho_0|\tilde{U}|}{3\pi^3 D^5}[f_M - (1-\frac{9\pi}{32})R_e\frac{df_M}{dR_e}] \tag{4}$$

where f_M is the friction factor for steady turbulent flow; R_e is the peak Reynolds number which is defined as $R_e = \dfrac{\rho_0|\tilde{U}|D}{A\mu}$.

$$\tag{5}$$

Qualitatively speaking, under the same operating conditions and the same geometric parameter, the inertia of fluid in turbulent flow may decrease due to the increase of "effective" flowing area. Thus the correction factor for inertia, ξ_l, may approach 1, while the correction factors, ξ_c and ξ_k, may increase due to enhanced heat transfer from turbulent flow. For the case of perfect heat transfer, $f_k=1$ and $1/f_k=0$. As a simplified modeling, a hypothetical, thermal conductivity resulting from this enhanced heat transfer is introduced to make a correction for ξ_c and ξ_k; that is, enhancement of the heat transfer leads to a decreased, effective dimensionless diameter of the inertance tube. The effective dimensionless diameter $\overline{D}_T$ may be approximately evaluated by

$$\overline{D}_T = \overline{D}\sqrt{\frac{Nu_L}{Nu_T}} \tag{6}$$

where $\overline{Nu}_L$ and $\overline{Nu}_T$ are the Nusselt numbers of laminar and turbulent flows, respectively. Therefore, ξ_{ck} and ξ_k can be estimated by calculating $f_{k,T}$ with $\overline{D}_T$. Moreover, Eq.(1) and Eq.(2) can be rewritten in the following forms,

$$\frac{d\widetilde{p}}{dx} = -(r_{\mu,T} + i\omega l_T)]\widetilde{U} \tag{7}$$

$$\frac{d\widetilde{U}}{dx} = -(i\omega c_T + \frac{1}{r_{k,T}})\widetilde{p} \tag{8}$$

where $r_{\mu,T}$, l_T, c_T and $r_{k,T}$ are the flow resistance, inertia, compliance and thermal-relaxation components of the impedance for the inertance tube with turbulent flow. The above-described analysis shows how to evaluate these components in a simplified way. It should be noted here that Eq.(7) and Eq.(8) are highly nonlinear, coupled equations. Moreover, this is usually a two-boundary-value problem, thus, a numerical iteration for their solutions is required. To solve Eq.(7) and Eq.(8), we used the shooting target method for them and adopt the fourth-order Runge-Kutta method for integration. The above turbulent model was compared with the experimental results and is described in the following section.

EXPERIMENTAL VERIFICATION

An experimental set-up was constructed to measure the performance of the phase shift and acoustical power transmission of various inertance tubes and to verify the applicability of the turbulent models previously developed. The experimental system consists of a linear compressor, an inertance tube and a reservoir. In the experiments, the operating frequency for the compressor is adjustable, ranging from 30 Hz to 90Hz. In addition, the pressure ratio can be controlled and varies from 1.05 to 1.40. Several inertance tubes with different diameters and lengths were tested.

A comparison between the turbulent model and the experimental result obtained from the hot-wire anemometer was made, which is shown in Figures 2 to 5. Figures 2 and 3 show the phase angles for different operating frequencies and pressure ratios, and Figures 4 and 5 show the acoustical powers for different operating frequencies and different pressure ratios. There are still some obvious errors between the turbulent flow model and the experimental data. With regard to the errors for the phase angle, one reason is that the phase angle is only for the fundamental frequency of the oscillating turbulent flow; another reason is that the turbulent model developed here is actually semi-theoretical and semi-empirical model that certainly needs to be improved with further study. The reasons are also applicable to errors for the acoustical powers.

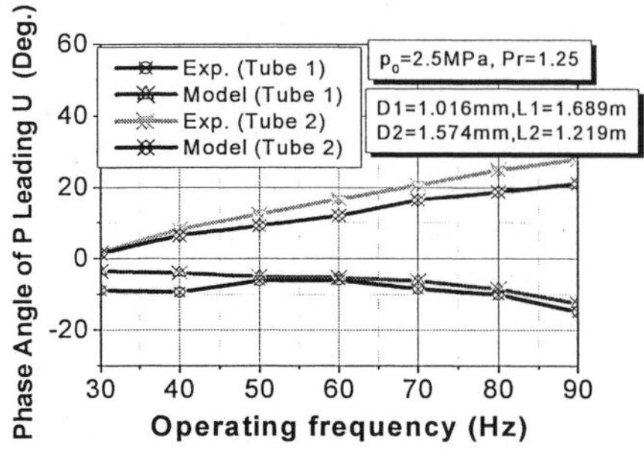

Figure 2   Phase angle

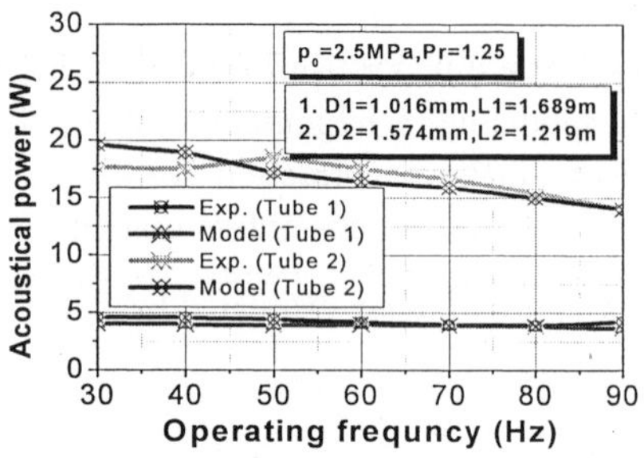

Figure 3   Acoustical power

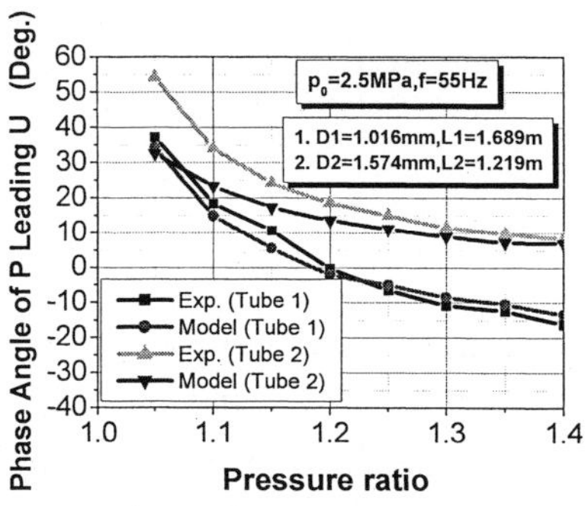

Figure 4   Phase angle

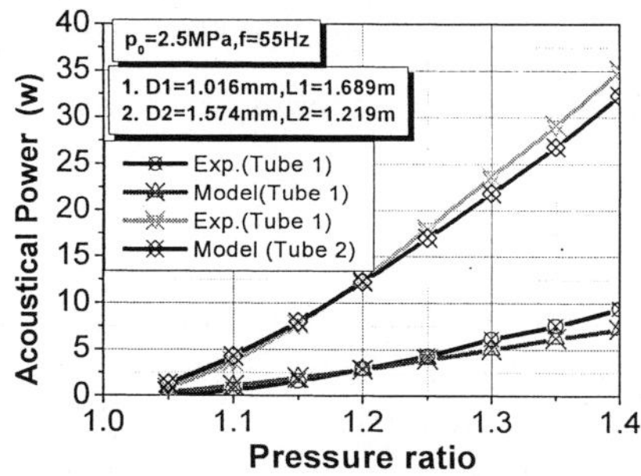

Figure 5   Acoustical power

CONCLUSIONS

The turbulent-flow model in this paper is developed to describe practical inertance tube shifters with acceptable accuracy. In addition, relatively extensive experiments on various inertance tubes for low-capacity pulse tube refrigerators were made and only some of the experiment data are shown in this paper. Comparisons between the simplified turbulent-flow model and the experimental data by hot-wire anemometry were made, which showed a good agreement qualitatively, and also provided an acceptable accuracy quantitatively. Certainly, much more effort is still needed on this topic.

ACKNOWDGEMENT

This work is financially supported Chinese Academy of Sciences under the Contract of KJCX-SW-W12 and the National Institute of Standard and Technology, Department of Commerce, USA.

REFERENCES

1. Xiao J.H., Thermoacoustic hat transportation and energy transformation, Part 1:Formulatiobn of the problem, Cryogenics(1995), vol.35(1), p.15
2. Swift G.W., *Thermoacoustics*, Acoustical society of America, Sewickley, Pennsylvania, 2002.

Thermoacoustically driven pulse tube refrigeration below 90K

Tang K., Chen G.B., Jin T., Bao R.

Cryogenics Laboratory, Zhejiang University, Hangzhou 310027, P.R.China

As a part of our consecutive research efforts on thermoacoustically driven pulse tube refrigeration, recent modification has been made to improve its refrigeration performance, and a refrigeration temperature as low as 88.6K, with helium filling of 2.1MPa as the working fluids, was achieved. The onset temperature was reduced about 200°C (from 550°C to 340°C) by the simple operation of the double inlet valve which would broaden the utilization of low-grade heat energy.

INTRODUCTION

Swift and Radebaugh et al.(1990) invented a thermoacoustically driven pulse tube refrigerator (TADPTR), in which a thermoacoustic engine is used instead of the conventional mechanical compressor [1]. Since it has no moving component in the whole system, this new cryocooler occupies advantages. The achievement of TADPTR attracted wide interest from industry immediately. In 1994, Cryenco started a project of developing thermoacoustic natural gas liquefier, powered by heat energy through burning part of natural gas [2]. A practical thermoacoustic natural gas liquefier prototype has been built up in 1998 [3]. The prototype burns 60% of natural gas to liquefy the rest 40% gas, which is regarded as a milestone for the practical application of TADPTR. Recent development of a higher-efficiency traveling wave thermo-acoustic prime mover [4] provides a possibility of burning 30% natural gas to liquefy the rest 70%.

Our consecutive efforts have been contributed to both experimental and theoretical studies on TADPTR since 1996 [5, 6, 7]. The work reported here will focus on the matching between thermo-acoustic prime mover and pulse tube refrigerator, especially the frequency matching. Recent modification has made an important progress that a refrigeration temperature as low as 88.6K was obtained.

EXPERIMENTAL APPARATUS

The framework of our present experimental apparatus is a symmetrically heated standing wave thermoacoustic engine originally built in 1996 [5]. However, a number of modifications have been realized to improve its performance, such as stack material and its packing density, heater, water cooler and also measuring system. As a result, the output pressure ratio has risen from 1.06 to 1.128 with helium as working fluid. The experimental system includes an orifice type pulse tube refrigerator, as shown in Figure 1. The stack of the prime mover is composed of the brass matrix of 6 mesh and 10 mesh alterna-tively with the ratio of 1:2, while the matrix inside the water cooler is 30 pieces of brass matrix of 6 mesh.

The measuring system is a PC-based digital acquisition system developed with LabVIEW, consisting of temperature, pressure and refrigeration capacity measuring modules, as show in Figure 2. FFT function is also included to analyze frequency spectrum of the pressure wave, and then nonlinear character of the thermoacoustic effect at finite amplitudes.

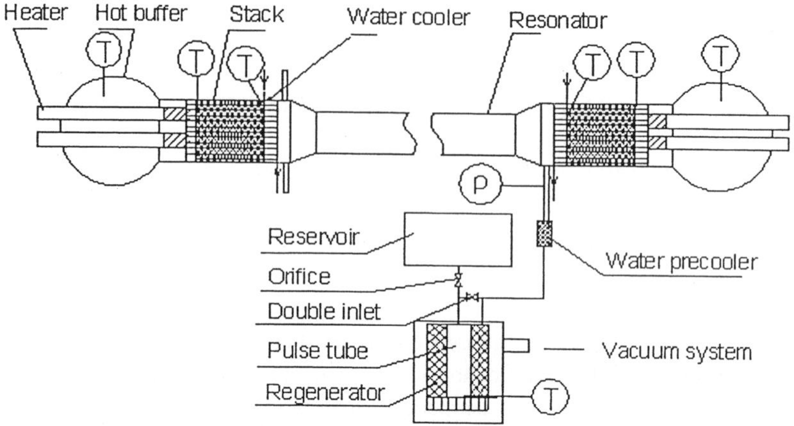

Figure 1. Outline of the thermoacoustically driven pulse tube refrigerator

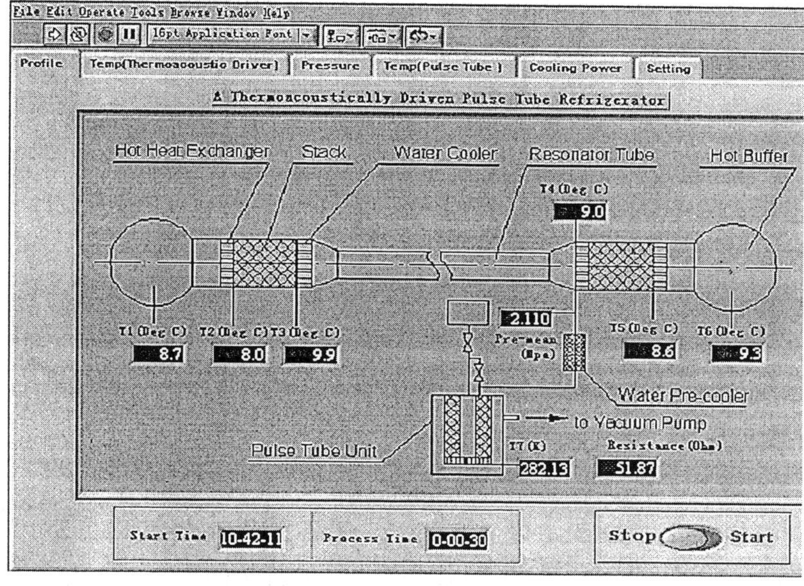

Figure 2. Interface of PC-based measuring system

EXPERIMENTS

In our previous experiments with 4m resonant tube, the operation frequency of the thermoacoustic prime mover is about 70 Hz with helium as working fluid, which brings difficulty for the matching to a pulse tube refrigerator. In order to obtain a better performance of thermoacoustically driven pulse tube refrigeration, the operation frequency was reduced via changing the length of the resonant tube.

With helium (filling pressure of 2.1MPa) as the working fluid and input heating power of 2000 Watts, the operation frequency and refrigeration temperature with different resonant tube lengths are both shown below. From Figure 3, we can see that the operation frequency falls from 70 Hz to 41 Hz, when the resonant tube was extended from 4 m to 9 m, respectively, and the refrigeration temperature drops from 112.8 K to 93.2 K (see Figure 4). Then we made further adjustment on the opening of the orifice and the

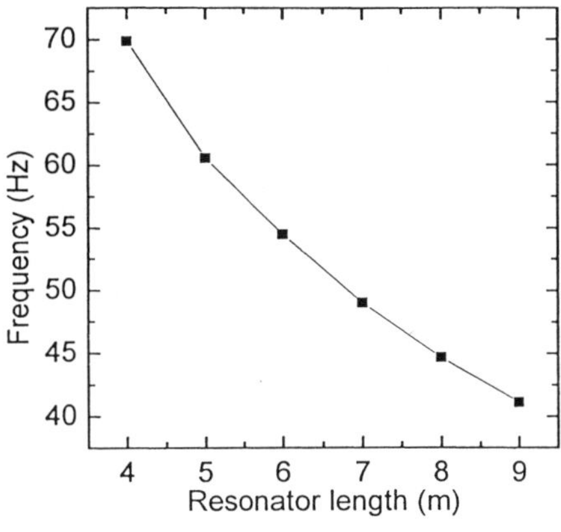

Figure 3. Operating frequency with different length

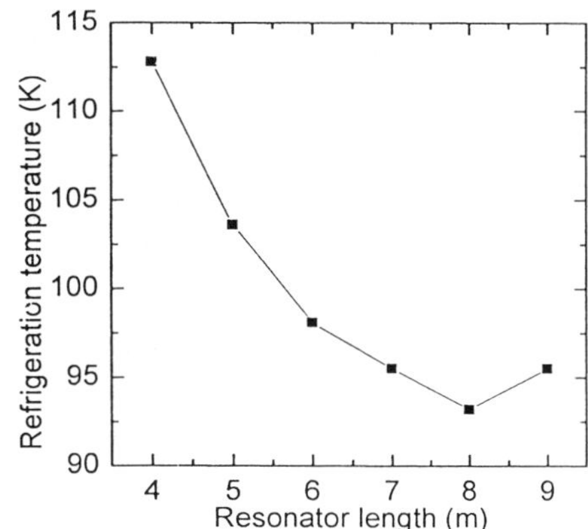

Figure 4. Refrigeration temperature with different length

double inlet, and the input power was increased from 2000W to 2200W, we obtained a lowest refrigeration temperature of 88.6K with 8 m resonant tube (see Figure 5).

Figure 5 shows the typical cooling-down curve of the pulse tube refrigerator. The cooling temperature reaches at 120K half an hour later, and we obtained a lowest refrigeration temperature of 88.6K in 2 hours. The mean operating pressure and the corresponding pressure ratio at this point are 2.64MPa and 1.128, respectively.

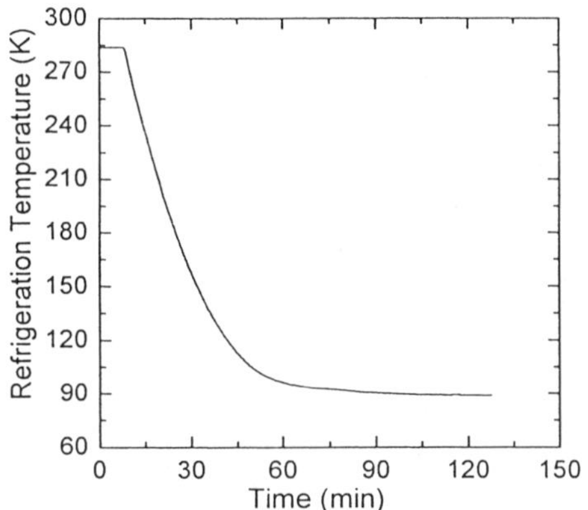

Figure 5. Typical cooling-down curve of the pulse tube

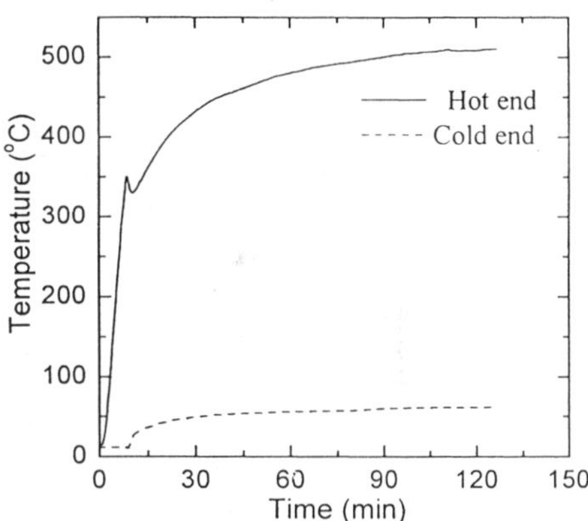

Figure 6. Temperatures vs. time at the hot end and the cold end of the stack

From Figure 6, the temperature profile at the hot end and the cold end of the stack, we can find that 8 minutes after turning on the input power, the temperature at the hot end of the stack reached 340°C, and the system starts to oscillate. Meanwhile, the pulse tube refrigerator begins to work and cools down quickly. When the refrigeration temperature is stable at 88.6K, the temperature of the hot end and the cold end of the stack are 497°C and 64°C, respectively, and the temperature gradient of the stack is 1045.9K/m.

It is worth being mentioned that at the beginning of the experiment, the double inlet valve of the pulse tube refrigerator was closed, and then turned it to the optimal opening value right after the onset of acoustic oscillation. We benefit from this simple operation on the double inlet valve, i.e., the onset temperature of the thermoacoustic system decreases from 550°C to 340°C. The reduction of the onset temperature of the system benefits doubtlessly to the adoption of low grade heat source in thermoacoustic machines.

DISCUSSION

Although a refrigeration temperature as low as 88.6K has been obtained on the self-made thermoacoustically driven pulse tube refrigerator system, there are still many things to do on our system. The refrigeration power is a problem to be solved. The limitation of present heater structure makes it very difficult for us to input more heat and to reach a high heating temperature, which has been one of the bottlenecks to achieve a lower temperature and a higher refrigeration capacity. On the other hand, a new pulse tube should be designed and fabricated to match the relatively high oscillation frequency thermoacoustic prime mover to improve the overall refrigeration performance. These leave us an interesting thing to investigate in the future.

CONCLUSION

1) With helium as working fluid (filling pressure of 2.1MPa), an 8 m resonant tube results in a resonance frequency of 44 Hz (the input power is 2200 Watts), and a refrigeration temperature of 88.6K was obtained in our self-made thermoacoustically driven pulse tube refrigerator system while the mean pressure and the pressure ratio are 2.64MPa and 1.128, respectively.

2) The simple operation on the double inlet valve, which was closed before the onset of acoustic oscillation and turned to the optimal value as soon as the oscillation started, could decrease the onset temperature of the system greatly. It would broaden the access of utilizing the low-grade heat energy.

ACKNOWLEDGEMENTS

The financial support from the National Natural Science Foundation (No. 50376055) and the Special Fund for Doctoral Training in Universities of China (No. 20010335010) are acknowledged.

REFERENCES

1. Radebaugh, R., McDermott K.M., Swift G.W. et al., Development of a thermoacoustically driven orifice pulse tube refrigerator, Proceedings of 4th Interagency Meeting on Cryocoolers, Plymouth, MA, David Taylor Research Center, (1990) Navy Report DTRC91/003 205-220

2. Swift, G.W., Thermoacoustic Natural Gas Liquefier, Proc. of DOE Natural Gas Conference, Houston (1997) 1-5

3. Arman B., Wollan J. J., Swift G. W., et al., Thermoacoustic natural gas liquefiers and recent developments. Cryogenics and Refrigeration-Proceedings of ICCR'2003, International Academic Publishers, (2003) 123-127

4. Backhaus S., Swift G. W., A thermoacoustic-Stirling heat engine, Nature (1999) 399 335-338

5. Jin T, Chen G B and Shen Y. A thermoacoustically driven pulse tube refrigerator capable of working below 120K, Cryogenics (2001) 41 595-601

6. Tang K., Chen G. B., Kong B., A 115K thermoacoustically driven pulse tube refrigerator with low onset temperature, Cryogenics (2004) 44 287-291

7. Chen G. B., Jin T., Experimental investigation on the onset and damping behavior in the thermoacoustic oscillation, Cryogenics (1999) 39 843-846

New generation of cooled infrared space telescopes

Lemke D.

Max-Planck-Institut für Astronomie, Königstuhl 17, 69117 Heidelberg, Germany

Infrared astronomy explores the cold and dusty as well as the young and highly redshifted universe. Observations at mid and far infrared wavelengths require cold space telescopes, the detectors have to be cooled to temperatures $0.1\ K < T < 7\ K$. The first generation of small infrared space telescopes was launched into geocentrical orbits and completely cooled with liquid helium. The next generation has much larger telescope mirrors launched warm. Locating these satellites on an earth trailing orbit or in the Lagrange Point 2 (L2), allows stable thermal conditions and radiation cooling of the telescope mirrors down to 40 K. The new infrared missions will address fundamental questions, like the search for the first light in the universe as well as for earth-like planets at other stars.

WHY INFRARED?

Astronomy is in a golden age: each decade we learn more about the universe than in all centuries before. Astronomy is a technologically driven science. The revolution started about 30 years ago. Solid state physics allowed to develop new sensitive detectors, material science to build much larger and lighter telescopes, the innovation of computer technologies to control complex instrumentation and to handle gigabytes of data. The routine access to space opened all spectral regions blocked by the earth atmosphere from gamma rays via the x-rays, the ultraviolet, the infrared to the millimetre wavelength range. The progress in cryotechniques made it possible to reduce noise and dark current in detectors at all wavelengths and to gain sensitivity by many orders of magnitude. Infrared telescopes in space can now be cooled, too, otherwise their thermal emission would wipe out the faint signals from celestial objects.

The infrared is a wide spectral range covering wavelengths from 0.8 to 1000 μm (for comparison, the visible range comprises 0.3…0.8 μm). Several facts explain the high scientific interest in this range: (i) all "cold" objects (T<3000 K) in the universe emit dominantly here, (ii) the many regions of the sky hidden in the visible by interstellar dust become observable, because the longer wavelengths penetrate the clouds of submicron particles without attenuation, (iii) the large energy output of hot objects (stars, nuclei of galaxies) is often absorbed by surrounding dust clouds and therefore transformed to the far infrared, (iv) molecular lines can be predominantly observed in the infrared and finally (v) the very young universe is highly redshifted due to its expansion.

INFRARED SPACE MISSIONS – PAST AND PRESENT

Table 1 gives an overview of the missions completed, operational, under development and in the planning phase. The pioneering satellite was IRAS, a 60 cm telescope with a camera in its focal plane contained in a liquid helium cryostat. On its 900 km polar earth orbit the whole sky was mapped during 300 days LHe holding time at four wavelength bands between 12 and 100 μm. The spectacular result of this mission was a first inventory of the cold universe, a catalogue of more than 300000 sources including many discoveries. This data base is still a frequently used tool for astronomers and a valuable part of the "Virtual Observatory", containing sky surveys at all wavelength regions.

Launch	Mission	Telescope Diam. [m]	Telescope Temp. [K]	Detector Temp. [K]	Science	Funding
1983	IRAS	0.6	3	3	FIR-Survey	NASA, NL, UK
1989	COBE				Cosmic Background	NASA
1995	ISO	0.6	3	1.6	MIR-FIR Observatory	ESA
2002	WMAP				Cosmic Background	NASA
2003	SPITZER	0.8	5	1.6	MIR-FIR-Observatory	NASA
2005	ASTRO-F	0.7	6	1.6	FIR-Survey/Observ.	ISAS (J)
2007	HERSCHEL	3.5	70	0.3	FIR-Observatory	ESA
2007	PLANCK	~1.9	40	0.1	Cosmic Background	ESA
2011	JWST	6.0	40	6	NIR-, MIR-Observ.	NASA, ESA, CND
2015	SAFIR	10.0	5	0.1	FIR-, Submm-Observ.	NASA
>2010	SPICA	3.5	5		FIR-Observatory	ISAS (J)
>2015	DARWIN	6x1.5		6	MIR-Interferometer	ESA
>2015	TPF	5x3.0			MIR-Interferometer ?	NASA
>2009	SPIRIT	2x3.0				NASA
>2015	SPECS	3x4.0				NASA

Table 1 Cooled space telescopes for infrared astronomy. The three blocks contain (i) completed and in orbit missions, (ii) missions under development or (iii) in the planning phase. The list of future projects is incomplete.
NIR: 1...5 µm, MIR: 5...30 µm, FIR: 30...350 µm, submm: 350...1000 µm.

The first real infrared space observatory was ISO, launched by the European Space Organization ESA [1] in 1995 (see Fig. 1). The 60 cm telescope at T~3 K was surrounded by He-gas cooled radiation shields and contained in a cryostat containing 2300 l of superfluid He at T~1.6 K. The major progress as compared to IRAS and its rigid focal plane camera for mapping were the four complex scientific instruments located in ISO's focal plane. In combination with the 1 arc sec pointing accuracy of the ISO-satellite they allowed by spectroscopy, imaging, and polarimetry a full analysis of the radiation emitted by celestial objects. As an example of the new instrumentation see the ISOPHOT-instrument [2] in Fig. 2. The largest scientific impact was achieved by the first spectroscopic studies (see Fig. 3) and imaging at a wavelength of 200 µm emitted by matter as cold as T~12 K (see Fig. 4). ISO's highly eccentric 24-hour-earth-orbit with an apogee of 70000 km allowed for 17 hours of daily observations, 7 h were contaminated by ionizing particles trapped in the earth radiation belts and stray light near the perigee of only 1000 km above the earth. As a consequence of the less satisfactory lifetime prediction for IRAS using a flow meter, ISO engineers invented the "Direct Liquid Content Measurement": a heat pulse into the superfluid and the measurement of the resulting temperature increase allows a precise determination of the remaining coolant. This type of measurement was applied twice during the mission, and the evaporation of the "last drop" could be predicted with an accuracy of two days for the 29 month mission (see Fig. 5).

Operational at present is the SPITZER observatory [3] launched by NASA in 2003. It is a 85 cm telescope He-gas cooled to T~5.5 K, equipped with three high performance scientific instruments for spectroscopy and imaging in the range 3...180 µm. For first scientific results see Fig. 6. Although the development of SPITZER started prior to ISO, it was delayed for many years because of its high costs. With two innovations the mission became feasible: (i) the selection of a heliocentric orbit (SPITZER trails behind the earth on its orbit) and (ii) the warm launch of the telescope and its passive cool down during the first ~40 days in orbit (see Fig. 7). This concept should enable a five year life-time because the small cryostat containing only 360 l of superfluid helium is well protected by gas cooled radiation shields and suffers only from the small heat dissipation of the scientific instruments.

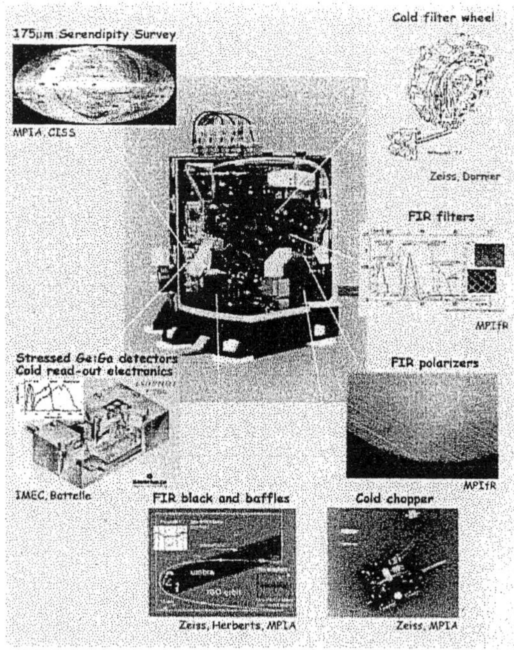

Figure 1
The Infrared Space Observatory during tests at ESA/ESTEC and its launch with an ARIANE 4 rocket on 17 Nov 1995. (ESA)

Figure 2 (right)
The ISOPHOT-instrument of ISO and several technological innovations. The instrument covered the wavelength range 2.5…240 µm. (Lemke et al 1996)

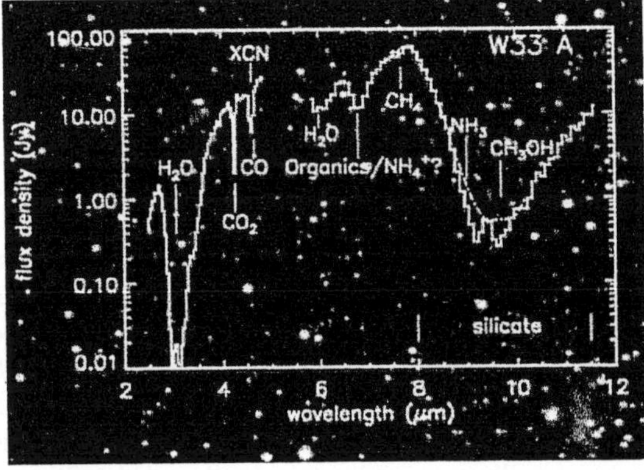

Figure 3 (left)
By spectroscopic studies with ISO, interstellar ices could be identified in star formation regions. (Klaas et al 2004)

Figure 4 (below)
The Andromeda spiral galaxy exhibits rings of cold dust in the far infrared indicating star formation regions. This image is quite different from the visible appearance (upper right) where the star light dominates. (Haas et al 1998)

Figure 5 (below)
The Direct Liquid Content Measurement (DLCM) carried out on ISO twice during the mission allowed for a precise prediction of the " helium boil-off". (ESA)

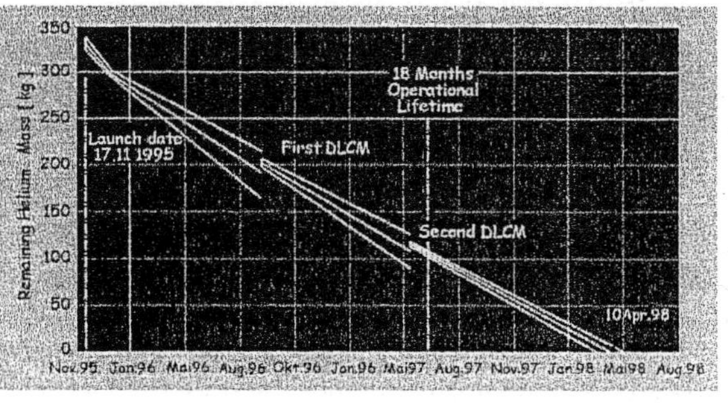

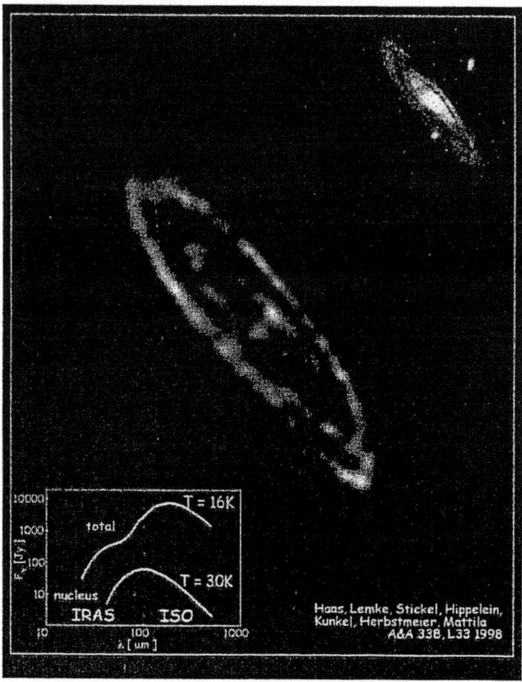

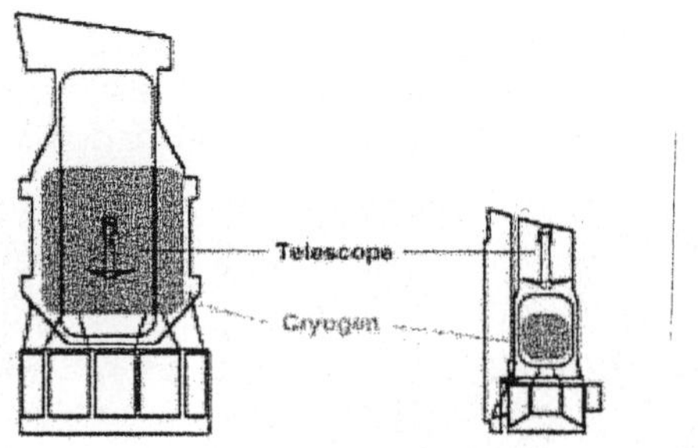

Figure 6
SPITZER on its earth trailing orbit. One side of the telescope tubus is black for efficient radiation cooling. Right: First released images show a galaxy and a comet in the infrared with unprecedented spatial resolution. (NASA JPL)

Figure 7
The gain by the warm launch/heliocentric orbit configuratio clearly demonstrated by SPITZER's (former SIRTF) old an new design. By maintaining the telescope size and the 5 yea lifetime, the cryogenic volume could be reduced from 3800 360 l and the total mass from 5 t to only 1 t. The developme costs were reduced by a factor ~4. (NASA)

A clever operational sequence of the instruments helps to save helium: the most dissipating mid-infrared instruments produce more cold He-gas and prepare a colder telescope for the following long wavelength instrument. An ambitious scientific programme will deliver new insights into the formation of stars and planets and the evolution of galaxies. Details of the thermal behaviour of SPITZER during the early in-orbit phase are given in this volume by Hopkins et al. [4]

MISSIONS UNDER DEVELOPMENT

The ASTRO-F-satellite, dedicated mostly to an all-sky survey, but with higher spatial resolution, sensitivity and larger wavelength coverage than achieved with IRAS, is being prepared in Japan. It will be launched to a 745 km polar orbit with an ISAS M-V rocket in late 2005. This is still a "cold" launch with the scientific instruments and the He-gas cooled 69 cm SiC telescope (T~6 K) within the cryostat. In addition to the 170 l LHe lasting for 550 days, mechanical coolers are onboard a scientific mission for the first time. Even after evaporation of the LHe the telescope will stay moderately cold and the Stirling coolers will allow to extend the mission for several observing modes. Beyond the survey, discoveries of new comets and brown dwarfs (cool stars with masses too low to maintain nuclear fusion in their cores) can be expected from the ASTRO-F-mission.

In the hot phase, with the qualification models presently being tested are two European missions: HERSCHEL [5] and PLANCK, foreseen for a dual launch with an ARIANE 5 rocket in 2007. With its 3.5 m light-weight telescope made of silicon carbide HERSCHEL will become a milestone in high spatial resolution imaging in the far infrared (see Fig. 8). Its focal plane contains three scientific instruments of unprecedented high performance and complexity. Two instruments are equipped with large bolometer arrays which have to be cooled to T~0.3 K by He3-coolers. While the instruments are contained in an ISO-like superfluid helium cryostat, the large telescope is outside and will be launched warm. During the three month travel to the Lagrange Point 2 (L2) it will cool passively to T≤70 K. L2 is located 1.5 million

Figure 8
The HERSCHEL-observatory has a 3.5 m telescope passively cooled to T~70 K in the L2 orbit. (ESA)

km "behind" the earth in antisolar direction. At this position a satellite orbits the sun with the same angular velocity as the earth, i.e. sun and earth are always "seen" along the same line and distance, and therefore L2 offers very stable thermal conditions.

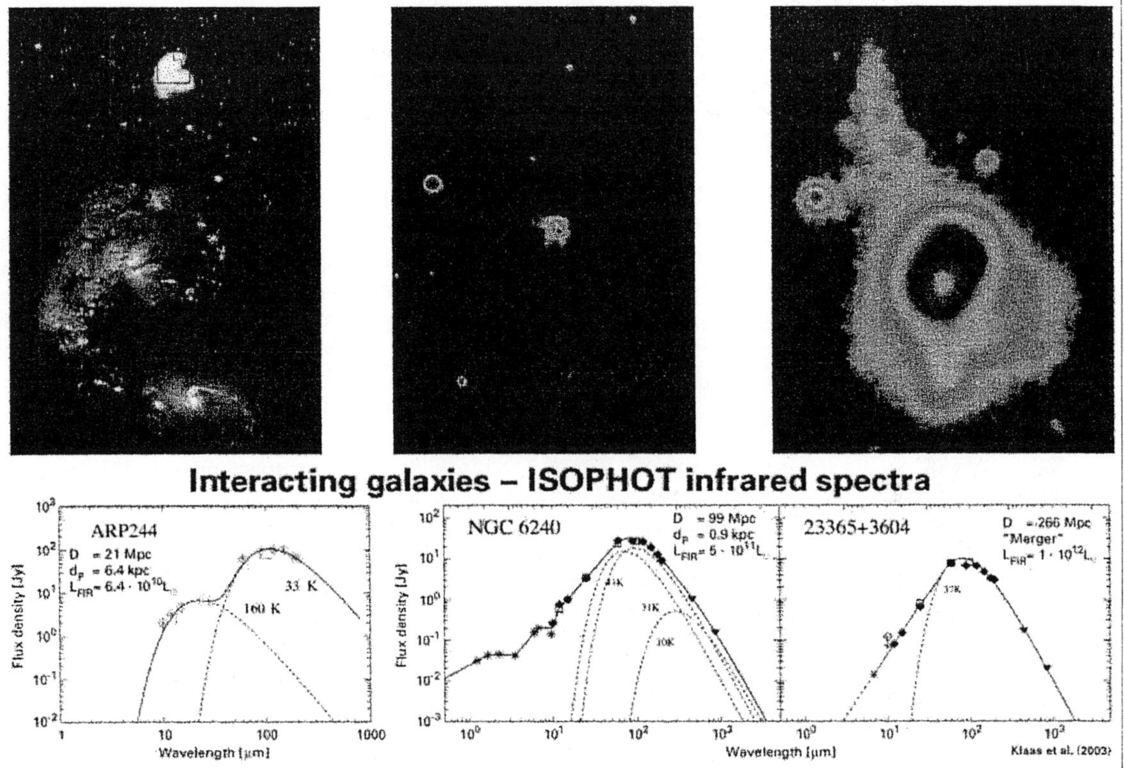

Figure 9
Galaxy collisions trigger bursts of star formation. From the left to the right, three interacting systems can be seen in different stages of approach: the two nuclei of ARP244 are still at a distance d_p=6.4 kpc (~20000 light years), while in the system 23365+3604 the nuclei have almost merged. This approaching is related to an increasing energy output L_{FIR} detected in the far infrared and measured in units of solar luminosity L_o. The transformation of the energy released by the young stars in the ultraviolet to the detected infrared radiation is via heating of the surrounding dust clouds [7]

Dozens of ambitious science projects are being developed for HERSCHEL, making use of the high spatial resolution of the large cooled telescope and its long wavelength (<650 µm) coverage. Examples are the deep surveys dedicated to the evolution of galaxies in the early universe and the detailed study of collisions/merging of galaxies (see Fig. 9).

BACKGROUND MEASUREMENT MISSIONS

Quite different from the dark night sky in the visible, the infrared sky is bright. In the mid infrared the T~250 K warm dust cloud in the solar system (zodiacal light) brightens the sky, at far infrared wavelengths the emission of the T~15...30 K interstellar dust in the Milky Way results in a patchy bright night sky (cirrus). At millimetre wavelength the universe is filled by the 2.7 K black body radiation of the highly redshifted remnant of the Big Bang, the starting point of the expanding universe. With NASA's COBE-satellite all-sky maps were obtained for the first time in 1989. The infrared components were mapped with a LHe-cooled 19 cm telescope (DIRBE instrument) and coarse angular resolution. The mm-measurements obtained with the radiation cooled FIRAS-Instrument demonstrated the ideal black body spectrum, a strong proof of the Big Bang theory. The full sky map also indicated patches on the sky

396

slightly warmer or colder than the average by a few μK – an impressively precise measurement. These patches are fingerprints of the early matter distribution in the fire ball, before it cooled to T~3000 K at an age of ~350000 years after the Big Bang. At that time the original plasma recombined, matter and radiation decoupled and the light started to travel freely.

Presently, the WMAP-satellite of NASA maps the cosmic background radiation with 12 arc min spatial resolution as compared to 7° of COBE. The analysis of the ΔT-maps results in the determination of important cosmological parameters such as the matter content of the universe (only 4% is the usual baryonic matter), the expansion rate etc.

The "3rd generation" cosmic background explorer is the European PLANCK-mission, to be launched jointly with HERSCHEL in 2007. Placed into an L2 orbit, the ~1.9 m telescope (see Fig. 10a) is radiation cooled to ~40 K. Two instruments are aboard: the low frequency instrument (LFI) housing 46 radiometers/transistor amplifiers cooled to 20 K by a hydrogen sorption cooler. Very ambitious is the high frequency instrument (HFI) with 48 bolometers to be cooled to the millikelvin range. Table 2 lists the chain of different coolers to obtain these extremely low temperatures on a long lasting space mission [6]. In order to achieve the $\Delta T/T$-resolution of ~$2 \cdot 10^{-6}$ on the cosmic background, high temperature stability is also required. It can be obtained by damping material of high heat capacity at low temperatures. With an unprecedented spatial resolution of 5 arc min (see Fig. 10b) and the ability to measure also the polarization status of the background radiation, PLANCK will finally mark the change of cosmology from speculations (as a few decades ago) to a high precision observation based field of science.

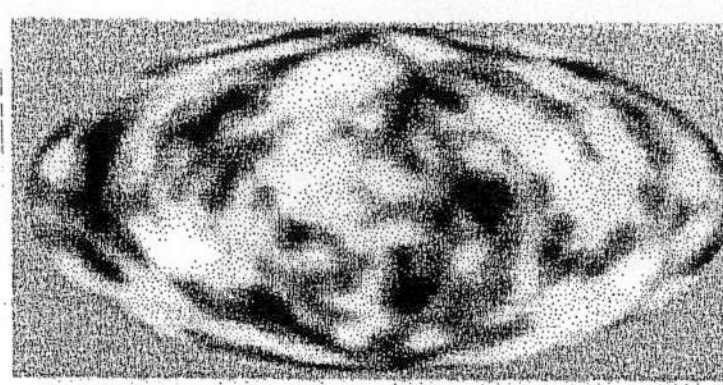

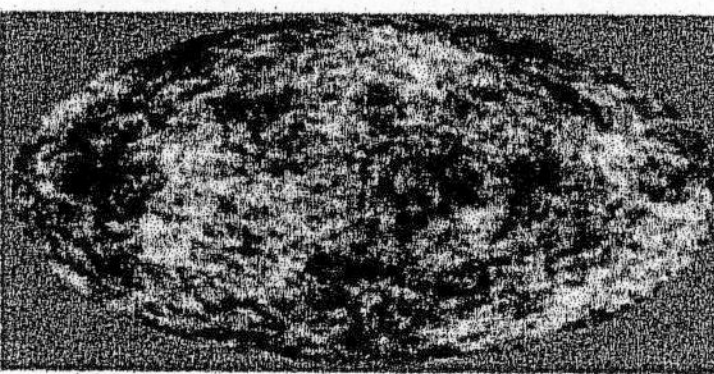

Figure 10
(a) The PLANCK-satellite will map the cosmic background radiation by spinning slowly. It is launched warm, the telescope will be passively cooled to T~40 K (b). The two all-sky maps demonstrate the progress from COBE (top) to PLANCK (bottom, simulation). (ESA)

Optical element	Temperature (K)	Cooler type	Cooling power (W)
Telescope	40	Passive	2
N/A	18	Sorption H2 J-T	1
Entrance horn	4.5	Mechanical He J-T	$2 \cdot 10^{-2}$
Lens and filter#1	1.6	3He/4He J-T	$5 \cdot 10^{-4}$
Horn and Bolometer	0.1	3He/4He Dilution	$2 \cdot 10^{-7}$

Stage temperature	18 K	4.5 K	1.6 K	0.1 K
Required temperature stability	100 mK p-p	$10\,\mu K\,Hz^{-1/2}$	$30\,\mu K\,Hz^{-1/2}$	$20\,nK\,Hz^{-1/2}$

Table 2
The chain of coolers for PLANCK's scientific instruments. The lower table lists the stability requirements. (Lamarre et al 2003)

FUTURE MISSIONS

The big questions (1) where do we come from?, (2) are we alone in the universe?... will be approached by the James Webb Space Telescope (JWST), the successor of the legendary Hubble Space Telescope (HST). JWST has a 6.5 m primary mirror composed of 18 segments which are folded during the "warm" launch and unfolded during the 3-month travel to the L2-orbit (see Fig. 11). Protected by a huge sunshield the mirror will passively cool down to T~40 K. The 450 m^2 sunshield consists of five

Figure 11
The 6.5 m-James Webb Space Telescope in its heliocentric L2-orbit, in the "back" the sun and the earth. A large sunshield reduces the heat input from the sun by a factor of 10^7. (NASA)

Kapton layers coated with Al and Si with 15 cm gaps between the layers to allow the heat to escape. Its attenuation factor is $1.3 \cdot 10^7$, i.e. it reduces the incoming 300 kW solar radiation to only 23 mW on the telescope side. For the first time also the scientific instruments will be launched warm. This is challenging, because they have to be operational/testable under laboratory conditions during the launch campaign as well as in the cryovac of space. Two of the instruments operate at T~35 K achieved by radiation cooling, the mid-infrared instrument requires T≤7 K for its detectors. This will be achieved by a cryostat containing solid hydrogen (pumped by the space vacuum) to be connected via a mechanical heat switch once in orbit.

All instruments contain "cryomechanisms" to change filters, gratings, the focus position etc. with high precision and very low heat dissipation [8]. They have to be extremely reliable, redundant whereever possible and robust against the high launch vibration. Our institute is involved in the development of grating and filter wheels and a flip-mirror for the MIR- and NIRSPEC-instruments of JWST. All wheels are based on a mechanism successfully flown on ISO: a short electrical pulse to a central brushless torque motor moves the wheel by a few degrees until the next position is fixed by a ratchet fitting between the next pair of small ball bearings arranged along the outer circumference of the wheel (see Fig. 12 for details). Each step requires only a few mWs, i.e. the electrical power in this open loop circuit is "on" only for a second per step. Given the long integration time on celestial objects (~10000s), the average heat dissipation of this high precision device (± 2 arc sec) is almost negligible. A flip-mirror for dark measurements and feed-in of the radiation of an onboard calibration source is based on the focal plane chopper design qualified (650 million cold cycles) for the PACS-instrument of HERSCHEL.

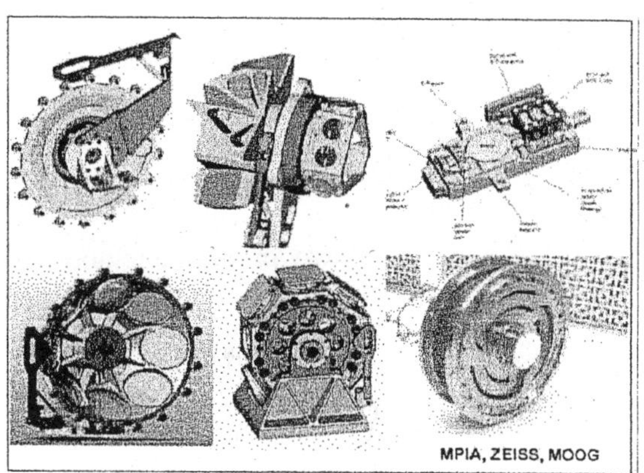

Figure 12
Cryomechanism development for the MIRI- and NIRSPEC-instruments. All filter and grating wheels are based on the ratchet mechanisms developed by ZEISS/MPIA for the ISO-mission. The flip-mirror is identical to a focal plane chopper for HERSCHEL. The MOOG linear actuator is being considered for refocusing. (MPIA, ZEISS, MOOG)

JWST, the flagship of space astronomy of the next decade, will be launched in 2011 by an ARIANE 5 rocket. It will investigate the formation of the first stars and galaxies in the universe a few hundred million years after the Big Bang and the evolution of the cosmos up to its present age of about 13 billion years. The optical spectra of the first hot stars are highly redshifted and have now to be studied in the mid infrared, see Fig. 13. In our Milky Way, JWST is capable to resolve planets around nearby stars and to discover pre-biological molecules in star forming regions.

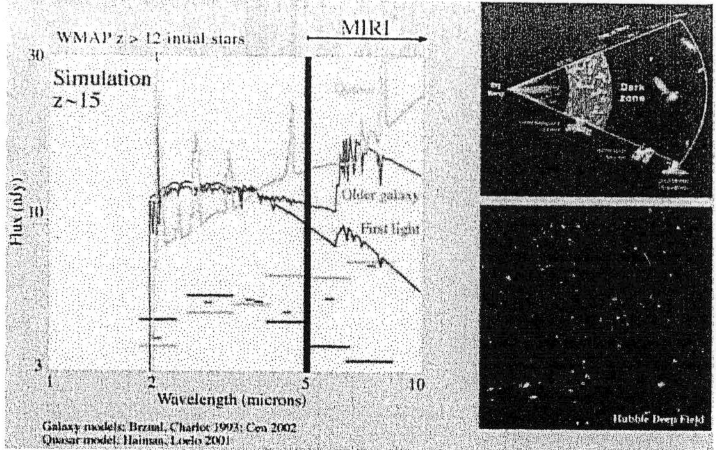

Figure 13

The MIRI-instrument of JWST will enable to identify the first light galaxies in the universe. Originally emitted at ultraviolet and visible wavelengths, the radiation is highly redshifted to the near and mid infrared. The horizontal bars indicate the filter bands and sensitivity of MIRI. In the upper right corner the evolution of the universe from Big Bang (left) to today (right) is sketched. JWST will investigate the dark ages, between the recombination in the fireball and the early galaxies detected with the Hubble Space Telescope (lower right). (NASA, MIRI-Science-Consortium)

Even larger space telescopes will follow JWST after the year 2020. SAFIR, the single aperture far-infrared observatory presently being studied by NASA, has a cooled 10 m primary mirror. It will be launched into L2 or to a heliocentric orbit at 3 AU (three times the distance earth-sun), this latter orbit would reduce the sky brightening caused by the interplanetary dust cloud (zodiacal light) and allow much deeper observations. Active cooling of the huge light weighted primary mirror to T~6 K by refrigerators is a challenge, in particular, if it becomes possible to make the optics of membranes. A concept using two cylindrical membrane surfaces arranged perpendicularly to each other has been proposed by Lockheed.

Since the complexity of enfolding and figure controlling of single dishes dramatically increases with size, the next logical step is the space interferometer. The European DARWIN-mission [9] consists of six free flying medium size telescopes spanning interferometric baselines of ~500 m and enabling spatial resolution of milliarcsec to be achieved in the mid infrared. The precise formation flying technique is under testing. DARWIN will be able to detect earth-like planets around other stars and, by spectroscopy of H_2O, O_3, CO_2 lines, to investigate the chances of life on these planets.

SUMMARY AND OUTLOOK

This journey through past, present and future infrared space telescope development is a story about technological and scientific progress. Each new generation of satellites builds on technology and experience derived from prior missions and adds some novel technology. The scientific discoveries made with each mission raise new scientific questions. In order to address these, larger telescopes, larger and more sensitive detector arrays, more sophisticated instruments and lower temperatures are required. While the first generation of infrared space telescopes was completely enclosed in large LHe cryostats, nowadays telescopes are launched warm and cool down passively. Still, their instruments are cooled by smaller amounts of onboard LHe. The next generation of space telescopes will be launched with warm instruments, too. After passive cool down, they are further cooled by heat-switch connection to an onboard coolant or by mechanical coolers. Temperatures in the mK range will be achieved by sorption, dilution and demagnetization coolers. The progress in cooling and increased lifetime became possible by leaving the earth orbit and launching the satellites into heliocentric orbits, allowing for very large distances also from the earth. The increase in size of the telescopes' size from 0.6 m of the first generation to several meters of the new generation was based on light-weighting the mirrors and actively controlled surfaces of the largest ones. Interferometers consisting of a fleet of passively cooled free flyers are presently studied. These developments are summarized in Fig. 14.

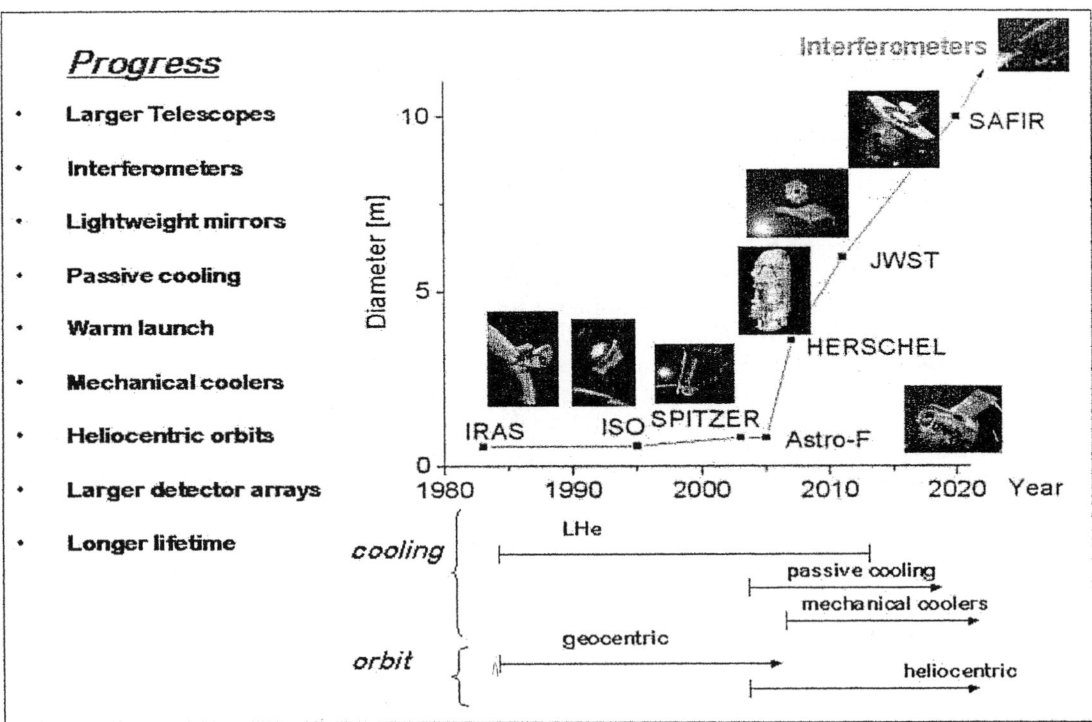

Figure 14
Progress in infrared space technology is a combination of larger telescopes, heliocentric orbits, radiation cooling and the application of more complex cooling devices.

I would like to finish with a recommendation to our Chinese hosts. We were all impressed by the first space flight of a taiconaut last year, and by the announcement that China will go to the moon within the next 20 years or so. The moon would be an ideal site for a cooled infrared telescope [10]. It could be a large "liquid mirror": the surface of a few litres of mercury filled into a slowly rotating bowl forms a "rotation-paraboloid", an excellent shape of a primary mirror. Locating this in a crater in the polar region of the moon enables radiation cooling. The necessary power for the observatory could be gained by solar arrays placed on the rim of the crater. Also, a combination with a second telescope within the crater to form a large interferometer is possible. Although the accessible sky for such a zenith telescope is limited, the discovery space for such a novel and powerful telescope would be much larger than that of any other space telescope under planning.

ACKNOWLEDGEMENTS

This review is based on papers and ideas of numerous colleagues around the world which I had the privilege to summarize here. More information can be found on the web-pages of the space agencies ESA, ISAS and NASA. The author is particularly indebted to the simulating atmosphere created by colleagues in the projects ISO, HERSCHEL and JWST in which he enjoyed to participate. Thanks go to Gustav Klipping in Berlin for more than 25 years of valuable advice on cryoengineering issues.

REFERENCES

1. Kessler, M.F., Steinz, J.A., Anderegg, M.E. et al., The Infrared Space Observatory (ISO) mission, Astronomy and Astrophysics (1996) 315 L27-L31

2. Lemke, D., Klaas, U., Abolins, J. et al., ISOPHOT – capabilities and performance, <u>Astronomy and Astrophysics</u> (1996) <u>315</u> L64-L70
3. Gallagher, D.B., Irace, W. and Michael W. Werner, Development of the Space Infrared Telescope Facility (SIRTF), <u>IR Space Telescopes and Instruments – 4850</u> SPIE, Bellingham, WA (2003) 17-29
4. Hopkins, R.A. et al., <u>ICEC20</u> (2004)
5. Pilbratt, G.L., Cernicharo, J., Heras, A.M. et al. (eds.), <u>The Promise of the Herschel Space Observatory – SP-460</u>, ESA Toledo (2001)
6. Lamarre, J.M., Puget, J.L., Piat, M. et al., The Planck High Frequency Instrument, a third generation CMB probe and the first submillimeter surveyor, <u>IR Space Telescopes and Instruments – 4850</u> SPIE, Bellingham, WA (2003) 730-749
7. Klaas, U., Haas, M., Müller, S.A.H. et al., Infrared to millimetre photometry of ultra-luminous IR galaxies: new evidence favouring a 3-stage dust model, <u>Astronomy and Astrophysics</u> (2001), <u>379</u> 823-844
8. Lemke, D., Grözinger, U., Henning, Th. et al., Cryomechanisms for positioning the optical components of the mid-infrared instrument (MIRI) for NGST <u>IR Space Telescopes and Instruments – 4850</u> SPIE, Bellingham, WA (2003) 544-555
9. ESA, <u>Towards Other Earths – DARWIN/TPF and the Search for Extrasolar Terrestrial Planets – 539</u>, Noordwijk (2003)
10. Angel, R., SPIE Conference "Astronomical Telescopes and Instrumentation", Glasgow June 2004 (Plenary Presentation)

WEB-SITES

1. ESA: <u>http://www.esa.int/esaSC/index.html</u>
2. NASA: <u>http://www.nasa.gov/home</u>,<u>http://www.jpl.nasa.gov/missions/</u>, <u>http://www.gsfc.nasa.gov/indepth/space_ssm.html</u>,
3. ISAS: <u>http://www.isas.ac.jp/e/index.shtml</u>, <u>http://www.isas.ac.jp/e/enterp/missions/astro-f/index.shtml</u>

Study on electricity generation performance of thermoelectric device in extreme environmental conditions

Wu Z-L., Zhou Y-X., Li T., Liu J.

Technical Institute of Physics and Chemistry, Chinese Academy of Sciences,
Beijing 100080, P. R. China

In this paper, studies were performed on the thermoelectric device to generate electricity in extreme outer-space environmental conditions. Mathematical models were established to relate the power generation and temperature difference across the cold and hot surfaces of thermoelectric device. Preliminary experiments were performed to test the practical output of a single stage TED. It was observed when the temperature difference was above one hundred centigrade, the voltaic output of the TED could reach about 5 V, which was capable of driving a small fan. However, a large temperature difference over the TED is hard to retain due to heat conduction, which would lower its electricity generation performance. As a remedy, transient thermal management may help to partially solve this problem.

INTRODUCTION

In a thermoelectric generator, heat is used to generate electricity based on the Seebeck effect. Since such devices are generally simple in structure and fabrication, they have been widely used in a series of practical situations [1]. The major applications of thermoelectric elements are to measure temperatures by thermocouples and thermopiles as heat sensors, or power generation and refrigeration. For the last two cases, efficiency of thermoelectric elements appears rather critical [2].

As is well known, TED plays a significant role in waste heat recovery, although it is still not prepared well for such practices due to insufficient studies [3]. In fact, it has long been a history for the thermoelectric generator to be used in space investigations. Due to the extreme environmental conditions in the outer space, the surface of the spacecraft exposed to the sun could have a temperature higher than 300 K, while the temperature on its other surface exposed to a vacuum space with extremely low temperature around 4 K may be rather low. Making full use of this totally free and huge temperature difference to generate electricity has been a great desire for space vehicle, satellite communication system, space station and other outer space activities. But up to now, little information was reported for such efforts. To better understand the electricity generation performance of the TED in space application, theoretical models were established to estimate the thermoelectric generator's efficiency and simulation experiments were conducted to test the output of a practical device.

THEORETICAL ANALYSIS

Theoretical model for single-layer thermoelectric generator
The thermoelectric generator consists of three major parts: a heat source, a heat sink and a thermopile. It

is from the temperature gradient over the thermopile that electric current is generated and provided to the whole system. Here, the thermopile serves to convert the heat to electricity.

When the heat from the heat source is converted to electricity, it also generates Joule heating inside the thermoelectric materials. At the heat sink, some of the electricity changes back to heat. As shown in Figure 1, the energy circle consists of heat absorption, heat release, heat transfer and heat-to-electricity conversion [3]. Depending on the Peltier and Seebeck effect of thermoelectric element [4,5], one can obtain the transferred heat at the two surfaces of the generator, respectively as

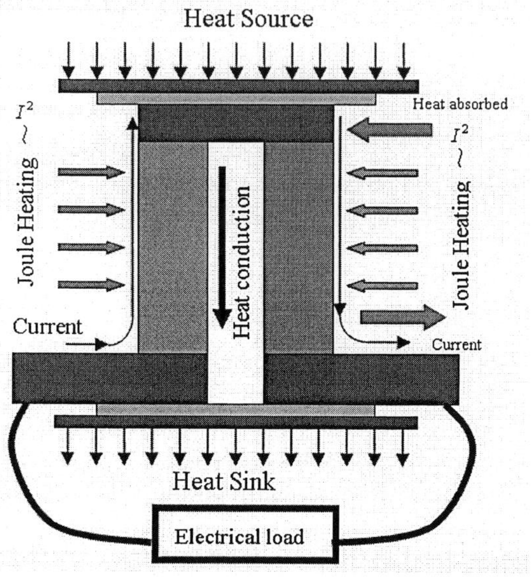

$$Q_H = \alpha I T_H - \frac{1}{2} I^2 R + K \Delta T_{HC} \qquad (1)$$

$$Q_C = \alpha I T_C + \frac{1}{2} I^2 R + K \Delta T_{HC} \qquad (2)$$

where, subscript H stands for the upper surface exposed to the sun, C stands for the rear surface exposed to the space; I is electric current, α the Seebeck coefficient, R the resistance, K the thermal conductivity and T the temperature. Further, the temperature difference writes as $\Delta T_{HC} = T_H - T_C$. In the outer space vacuum, the convective heat can be neglected.

If defining $\quad j = \dfrac{\alpha I}{K}, q_h = \dfrac{Q_H}{KT_H}, q_c = \dfrac{Q_C}{KT_C}, \theta_H = \dfrac{T_H}{T_C}$,

Eqs. (1) and (2) are transformed as a dimensionless form [5], i.e.

$$q_H = j - \frac{1}{2ZT_H} j^2 + 1 - \frac{1}{\theta_H} \qquad (3)$$

$$q_C = j + \frac{1}{2ZT_C} j^2 - 1 + \theta_H \qquad (4)$$

where, $Z = \alpha^2 / RK$ is the Figure of Merit.

Figure 1 Energy circle in a thermoelectric generator [3]

When the thermoelectric element works at its normal temperature, ZT_c is generally close to 1. Therefore, one can derive the electricity generation efficiency as

$$\eta_{Max} = j \cdot \frac{(\theta_H - 1) - 2ZT_C \dfrac{\theta_H - 1}{\theta_H + 1} \cdot (-1 + \sqrt{\dfrac{\theta_H + 1}{2ZT_C}} - 1) \cdot j}{2\theta_H \dfrac{\theta_H - 1}{\theta_H + 1} \cdot (-1 + \sqrt{\dfrac{\theta_H + 1}{2ZT_C}} - 1) \cdot j - 2ZT_C (\dfrac{\theta_H - 1}{\theta_H + 1})^2 \cdot (-1 + \sqrt{\dfrac{\theta_H + 1}{2ZT_C}} - 1)^2 \cdot j^2 + (\theta_H - 1)} \qquad (5)$$

If using Bi_2Te_3 as the arms of the thermoelectric element, the typical parameters are applied as $\alpha = 170 \mu V \cdot K^{-1}$, $K = 2 W \cdot m^{-1} K^{-1}$, and $I = 2 \times 10^2 A \cdot m^{-1}$. For the two cases of $\theta_H = 4$ and $\theta_H = 2$, η_{Max} is predicted as 1.16% and 0.85%, respectively. Both results indicate that the electric generation efficiency for the thermoelectric generator is around 1%.

It should be pointed out that the above calculation is still a rough estimation, because the parameter ZT_C may not be exactly 1, and the arms of thermoelectric material could have many other choices. However, it looks from the calculation that the efficiency of a thermoelectric generator is low. The advantage for the application of thermoelectric generator in outer space is that the solar energy there is totally free and durable, and the temperature difference induced can be high sometimes. To improve the efficiency, a multiple-layer thermoelectric generator can possibly be adopted.

<u>Theoretical model for double-layer thermoelectric generator</u>

When the temperature difference is very large, the single-layer thermoelectric generator only has a low efficiency and will hardly meet the need in power generation. So the double-layer or even multiple-layer thermoelectric generators are often proposed to improve the efficiency [6].

Here a double-layer thermoelectric generator is particularly analyzed for illustration purpose. This device has an upper layer and a bottom layer. Both of them have n pairs of thermoelectric elements. Similar to the analysis of the single-layer device, one gets:

$$q_H = nj - \frac{n}{2ZT_H} j^2 + n - \frac{1}{2}(\frac{1}{ZT_H} j^2 + 1 + \frac{1}{\theta_H}) \tag{6}$$

$$q_C = nj + \frac{n}{2ZT_C} j^2 - n + \frac{1}{2}(\frac{1}{ZT_C} j^2 + 1 + \theta_H) \tag{7}$$

$$\eta = 1 - \frac{Q_C}{Q_H} = 1 - \frac{q_C}{\theta_H q_H} = 1 - \frac{nj + \frac{n}{2ZT_C} j^2 - n + \frac{1}{2}(\frac{1}{ZT_C} j^2 + 1 + \theta_H)}{\theta_H (nj - \frac{n}{2ZT_C} j^2 + n - \frac{1}{2}(\frac{1}{ZT_C} j^2 + 1 + \theta_H))} \tag{8}$$

In the spacecraft or shuttle, the temperature inside is generally 300K, while the outside surface temperature is below 100K. The actual temperature difference depends on whether the surface is exposed to the sun or not. It also relies on the distance between the spacecraft and the earth. With the theoretical model, it is possible to estimate the efficiency under various environmental conditions.

EXPERIMENTS AND DISCUSSIONS

To simulate the extreme environmental conditions and to test the out-put of a practical generator, a copper block pre-cooled by liquid nitrogen or ice-water mixture was adopted as the cooling medium. Meanwhile, heating was applied by an electric heater with power of 80W. The experimental setup is schematically shown in Figure 2. Agilent 34970A (USA) was used as the data acquisition system. Three essential parameters were recorded: the temperatures at both the cold and the hot sides and the voltaic output of the thermoelectric element. The applied exterior load was about 20 ohm. Further, a switch on-off resistor with 1 ohm was introduced to measure the resistance of thermoelectric element. When switching on or off this resistor, a change on the voltaic output is observed. From its variation the inner resistance of the thermoelectric element could be calculated. In this study, two cases were considered: (1) The cold side of the generator was cooled by a pre-cooled copper plate, while the hot side was placed at the room temperature. The test result is shown in Figure 3; (2) The cold side was cooled by the ice-water mixture, while the hot side was heated by an electric heater. The test result is given in Figure 4.

From the data in Figure 3, it was noticed that cooling of the thermoelectric generator was initiated at the time of 100s and a voltaic output was induced. Clearly, the voltaic output was proportional to the temperature difference between the hot and cold sides of the generator. The temperature at the cold side dropped quickly due to contact cooling and the voltaic output then increased gradually due to growth of the temperature difference. At 160s, the temperature difference approached its maximum value of 75°C and the voltaic output becomes 2.8 V. However, such huge temperature difference will not be retained due to heat conduction inside the generator. After 160s, it was noticed that the temperature difference slowly decreased, while the voltaic output also dropped at the same time. At time of 300s, the thermoelectric generator was turned off and a quickly disappearing voltaic output was observed.

From the data in Figure 4, it was seen that there were several voltaic output jumps because of the switch-on or off of the 1 ohm resistor. From the voltaic output variation, the inner resistance of the thermoelectric element was calculated as 2.94 ohm.

CONCLUSIONS

The aim of the present experiment is to simulate the extreme environmental conditions of outer space where the thermoelectric generator is especially useful. The efficiency measured in the present experiment is corresponding to the theoretical prediction. Further, the theoretical model established in this paper can be used to predict the working efficiency of TED in extreme environment. Clearly, liquid helium with much lower temperature can be used to produce a much huge temperature difference and thus test the electricity generation behaviors. The present study warrants further investigations along this direction.

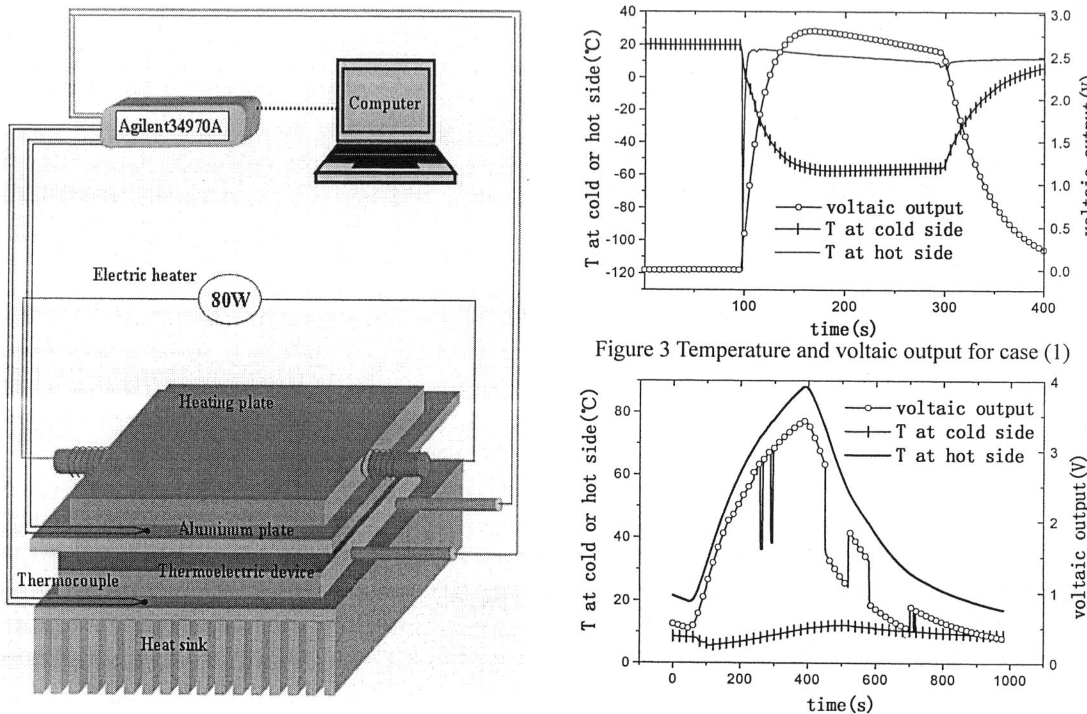

<table>
<tr><td>Figure 2 Schematics of the experimental setup</td><td>Figure 3 Temperature and voltaic output for case (1)</td></tr>
<tr><td></td><td>Figure 4 Temperature and voltaic output for case (2)</td></tr>
</table>

ACKNOWLEDGMENT

This study was supported by the NSFC Grant 50325622 and the Chinese Academy of Sciences.

REFERENCES

1. Snyder G.J. and Tristan S.U., Thermoelectric efficiency and comparatibility, <u>Physical Review Letters</u> (2003) <u>91</u> 148301 1-4
2. Luciana W.S., Massoud K., Micro-thermoelectric cooler: interfacial effects on thermal and electrical transport, <u>International Journal of Heat and Mass Transfer</u> (2004) <u>47</u> 2417-2435
3. Samuel B.S., Aleksander J.F., Klavs F.J. and Martin A.S., A combustion-based mems thermoelectric power generator, <u>The 11th International Conference on Solid-State Sensors and Actuators</u>, Munich, Germany (2001)
4. Jeffrey M.G., Chuac H.T., Chakrabortyc A., The electro-adsorption chiller: a miniaturized cooling cycle with applications to micro-electronics, <u>International Journal of Refrigeration</u> (2002) <u>25</u> 1025-1033
5. Yi X.C., Wang W.Y., The academic analysis of the working parameters of semiconductor refrigerator, <u>Cryogenics</u> (1998) <u>101</u> 26-30 (in Chinese)
6. Bhattacharyya A., Lagoudas D.C., Wang Y. and Kinra V.K., On the role of thermoelectric heat transfer in the design of sma actuators: theoretical modeling and experiment, <u>Smart Materials and Structures</u> (1995) <u>4</u> 252-263

Experimental investigation on He II porous plug liquid-vapor phase separator

Yu X., Li Q., Li Z., Li Qiang

Technical Institute of Physics and Chemistry, GSCAS, Beijing, China

In this paper, the tests of a series of He II porous plug liquid-vapor phase separators for bath temperature at 1.5-1.9K are presented. For one separator, two regions (initial region and hysteresis loop) are observed in the relationship between mass flow rate and temperature (or pressure) difference. For two separators with similar parameters, the repeatability of experimental data could be observed. Furthermore, in order to prepare theoretical analysis of the separator, the data of cold vapor flow rate is obtained when the He II liquid level was below the porous plug.

INTRODUCTION

At zero gravity, superfluid helium (He II) liquid-vapor phase separator (LVPS) is a key device that keeps the liquid phase separated from the vapor for He II cryostat, which is used to cool space superconducting component and far infrared detectors of astronomical observation instruments. Based on a fundamental of thermo mechanical (fountain) effect in restricted geometries, helium II is retained in the tank by LVPS. Boil-off helium gas flows through LVPS and cools the He II tank.

Porous plug type LVPS have been investigated by a number research groups [1-5] and have been successfully operated in some space flight missions. This paper summarizes data taken from tests of several different porous plugs, which are made of sintered stainless steel and fabricated in disc form. The diameter, d, is 25.4 mm and the thickness, l, is 6.0 mm. Table 1 is summarizes the characteristics of these plugs. The radius, r, is calculated from the Blake-Kozeny equation [5] and pore size is measured by bubble method.

Table 1 Characteristics of porous plugs

Plug	Porosity (%)	r (μm)	Permeability (m^2)	Pore size(μm)		
				R_{max}	R_{ave}	R_{min}
1	39.6	1.37	5.01×10^{-14}	7.03	2.99	2.80
2	39.2	1.40	5.05×10^{-14}	8.25	3.05	2.90
3	40.7	1.85	1.03×10^{-13}	8.59	3.10	2.95
4	40.3	1.86	1.00×10^{-13}	8.64	3.08	2.99

The porous plugs are attached to their mount by electric capacity percussion seam welding, the effective cross-sectional area, A, is $5.07 \times 10^{-4} \, m^2$.

EXPERIMENTAL SET-UP

The schematic diagram of experimental apparatus is shown in Figure 1. The apparatus comprises the porous plug, helium dewar and two evacuation systems. The helium dewar is made of stainless steel. A vacuum jacket thermally insulates the He II bath from the helium dewar. The porous plug is fixed at the bottom of the evacuation tube. The inner-vented helium vapor is also thermally insulated from the He II bath by a vacuum jacket. A heater is installed in the He II bath to simulated heat input from a space instrument. The evacuation system for the He II bath and the porous plug is composed of three parts: the mechanical vacuum pump, the gate valve and the regulating valve. The vacuum pump (Pump 1) in the porous plug evacuation system is a lubrication free pump. The temperatures of the He II bath and of the downstream side of porous plug are measured by two germanium resistance thermometers (GRTs1-2). A

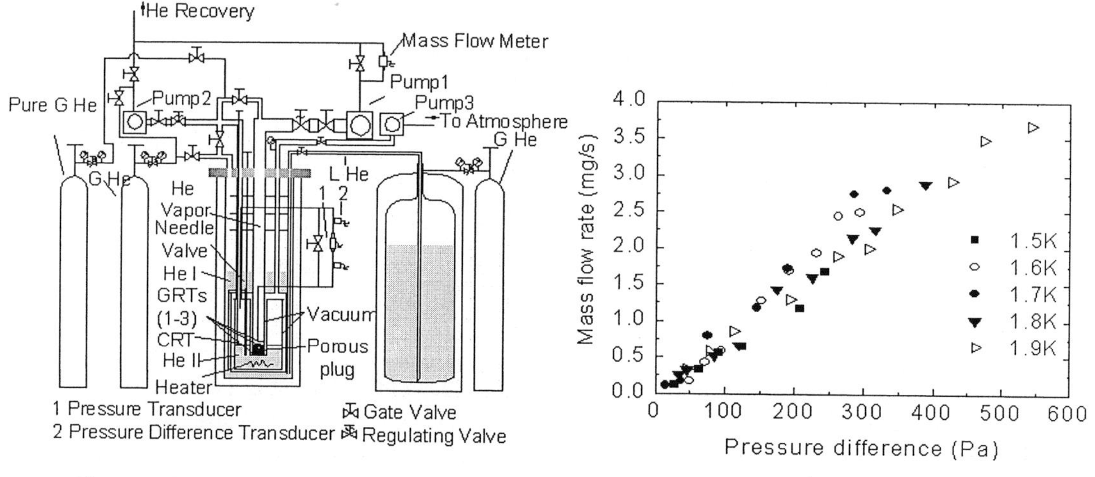

<table>
<tr><td>Figure 1 Schematic illustration of experimental apparatus</td><td>Figure 2 Variation of the mass flow rate of cold vapor with pressure difference</td></tr>
</table>

further GRT 3 is located 4 mm downstream from the upper surface of the porous plug to obtain the temperature of helium vapor. The temperature difference between the GRT 2 and the GRT 3 is used to detect liquid break through phenomena. Each GRT is excited by a $0.5\,\mu$A DC current. The bath pressure and the downstream pressure are measured by resistance pressure transducers. The pressure difference is measured by a precision pressure difference transducer. The mass flow rate of helium vapor was measured by a mass flow rate meter located at the exhaust side of the evacuation pump. The liquid level in the helium dewar was measured by a liquid level sensor. A carbon resistance thermometer (CRT) is fixed at the bottom of thin pipe to monitor the He II liquid level, which is adjustable in the vertical direction. All data were recorded by a Keithley multi meter/data acquisition system and then stored on the hard disk of a personal computer.

EXPERIMENTAL RESULTS AND DISCUSSION

Cold vapor flow rate test

This experiment was performed when the He II liquid level in the inner bath was below the porous plug. The gas flow rate was controlled by liquid boil-off using the heater and the regulating valve. The experimental data of the gas flow rate $\dot{m}_v$ are plotted against the pressure difference between the upstream and the downstream side of the porous plug Δp for several gas temperatures in Figure 2. By fitting to data of figure 2, the linear equation below has been obtained.

$$\dot{m}_v = c_0 \cdot \kappa_p \, \Delta p / l_v \tag{1}$$

The coefficient c_0 of cold vapor was calculated from the fitting data. For the temperature of vapor at 1.5K, 1.6K, 1.7K, 1.8K and 1.9K, c_0 is 739.2, 1023.6, 1068.0, 880.8 and 840.0 mg/(s·m·Pa), respectively, which can be used in the further numerical analysis of the separators.

<u>General flow characteristics</u>
Figure 3 presents the variation of the mass flow rate with pressure difference and the temperature difference for bath temperatures 1.5-1.9 K. Figure 3(a) and (b) show the data obtained from the plug 1. On the other hand, Figure 3c and d show the data obtained from the plug 3. It is found that the mass flow

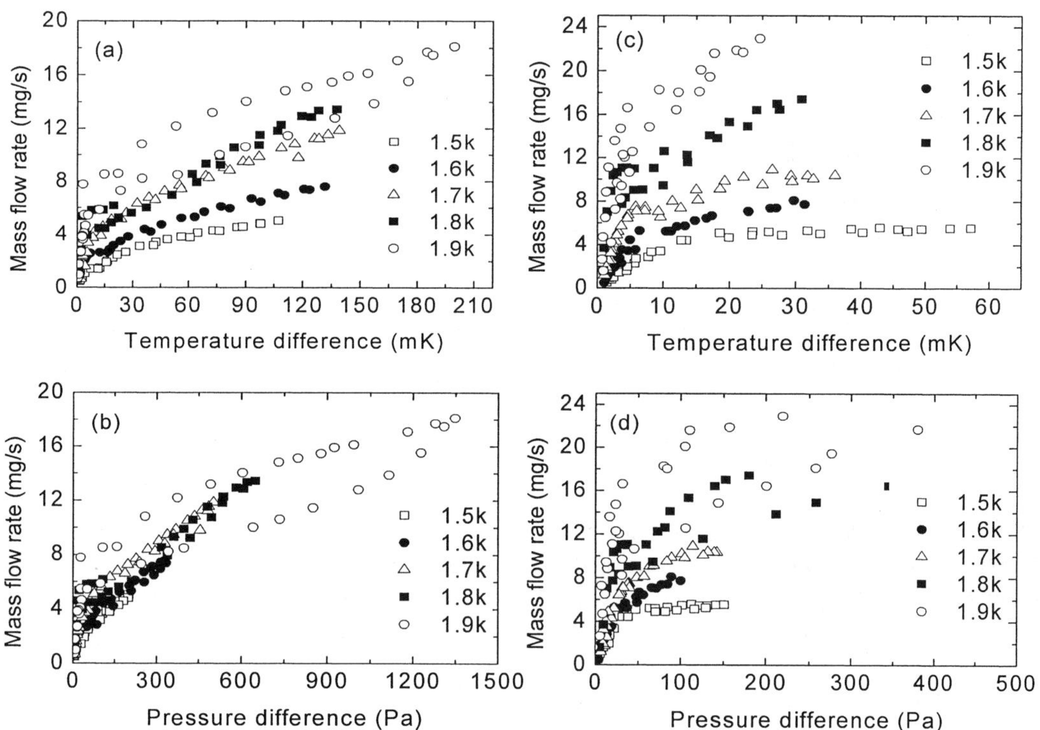

Figure 3 General flow characteristics through porous plug liquid-vapor phase separator
Obtained from Plug 1: (a); (b). Obtained from Plug 3: (c); (d)

rate rises as the bath temperature increases or the permeability of plug increases. Two different regions (initial region and hysteresis) exist in the increases and decreases of mass flow rate. This phenomenon is evident in the variation of the mass flow rate with pressure difference and the temperature difference for each bath temperature. The initial region corresponds to the small pressure and temperature differences. In this region, the mass flow rate dramatically increases as the temperature and pressure differences increase until the plug achieves a choked condition. The flow rate at this point is referred to as the critical flow rate. Above this point, a hysteresis loop is observed in the relationship between mass flow rate and pressure (or temperature) difference. In this loop, the upper branch corresponds to the increasing process and the lower branch corresponds to the decreasing process. For the upper branch, the data points seem unstable because the pressure and the temperature difference increase slightly with the change of time. Furthermore, it is found that the difference between the upper and lower branch becomes small as the bath temperature decreases.

<u>Comparison of two plugs</u>
A number of tests were performed with plug 1 and plug 2 to characterize the repeatability of the flow characteristic through similar porous plugs. Figure 4 shows the data taken for plug 1 and 2 at 1.7 K. The mass flow rate against the temperature difference is plotted in Figure 4a. The relationship between the

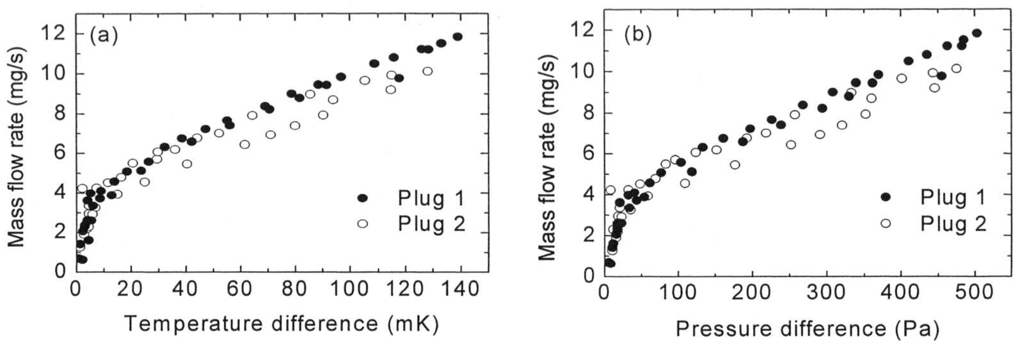

Figure 4 Comparison of the mass flow rates obtained from Plug 1 and Plug 2

mass flow rate and the pressure difference is plotted in Figure 4b. The data obtained from the plug 1 does agree with the data obtained from plug 2. The repeatability of the flow characteristic through the similar porous plugs could therefore be proven.

ACKNOWLEDGMENT

We wish to give special thanks to the colleague in our research group for useful suggestions and technical assistance.

REFERENCES

1. Urbach A. R., Mason P.V., IRAS cryogenic system flight performance report, <u>Advances in Cryogenic Engineering</u> (1984), <u>29</u> 651-655

2. S. W. K. Yuan and T. H. K. Frederking., Non-linear vapor-liquid phase separation including microgravity effects, <u>Cryogenics</u> (1987) <u>27</u> 27-33

3. G. Fujii, S. Tomoya, M. Kyoya, M. Hirabayashi, M. Murakami, T. Matsumoto, T. Hirao, H. Murakami, H. Okuda and T. Kanari., On-orbit thermal behaviour of the IRTS cryogenic system, <u>Cryogenics</u> (1996) <u>36</u> 731-739

4. A. Nakano, D. Petrac and C. Paine., He II liquid/vapour phase separator for large dynamic range operation, <u>Cryogenics</u> (1996) <u>36</u> 823-828

5. A. Nakano, D. Petrac, C. Paine and M. Murakami., Investigation of large dynamic range helium II liquid/vapor phase separator for SIRTF, <u>Cryogenics</u> (1999) <u>39</u> 471-479

Proceedings of the Twentieth International Cryogenic Engineering Conference
(ICEC 20), Beijing, China. © 2005 Elsevier Ltd. All rights reserved.

Novel cryogenic motion control for aerospace

Zhitong Cao, Hongping Chen, Guoguang He, Jiongjiong Cai

Department of Physics, Zhejiang University, China

In this paper two types of novel cryogenic mechanisms are exhibited, which has been designed using high-temperature superconducting (HTS) and magnetostrictive materials. Magnetostrictive materials have lead to a variety of large-stroke, high-force actuators at cryogenic temperature, which was previously difficult to make and is now much easier. Using HTS tapes in magnet takes advantage of the cryogenic temperature environment for added motive efficiency. Innovative actuator or motor designs using HTS magnet with the application of specific drive electronics bring out efficient, compact, and lightweight actuator systems for aerospace.

INTRODUCTION

A single advance in material science can lead to rapid progress in the related technologies of application. The invention of giant magnetostrictive material, called smart materials, is such an advance. A rare earth Giant Magnetostrictive Materials (GMM) including terbium (Tb) and dysprosium (Dy) demonstrate very high performance, magntostrain of 0.1-0.2%. Since having excellent magnetostriction over a broad range of temperatures from near absolute zero to above 400K, they also are ideal for cryogenic device application. These materials have lead to a variety of large-stroke, high-force actuators at cryogenic temperature, which was previously difficult to make and is now much easier. Some of these actuators meet the requirements of applications in special field such as aerospace. Actuators in the aerospace are used for restraint and release of deployable components of the mechanisms such as antennas, booms, and other appendages. Drive mechanisms supply the energy needed to move the components for precision position motion. Solenoids, voice coils and electric motors can be used to provide the desired motion [1, 2].

Another advance in materials science that impacts the application of magnetostriction is a radical reduction in the size and an increase in the current-carrying capabilities of HTS magnets. For cryogenic temperature applications, such as deep aerospace, there is the search program for the origin of the universe. For searching red-shifted stars, which correlate to astronomical time, the more red-shifted a star's spectrum is, the older the star is. In order to observe best the red-shifted stars in the infrared spectrum, the temperature of the telescope as Next Generation Space Telescope (NGST) should be as cold as possible, to keep the background interference as small as possible, because reduced thermal noise is highly desired for improving the instrument resolution. Mission concept studies for NGST suggest that the optical surface should be at 30K. This also requires the mechanisms in the telescope to operate at that temperature. For operation at higher than 10K and below 77K, the HTS offer an attractive opportunity to incorporate the HTS into mechanisms while taking advantage of their persistent mode of operation, i.e., the magnetostrictive material will maintain a train field without power[3].

These two factors work synergistically in generating mechanical movement and producing force at cryogenic temperatures. Consequently, the potential exists for the quick advancement of related techniques for producing cryogenic temperature actuators and mechanisms.

The object of this paper is to review the present situation from the point of view of the working principle of the mechanisms and their applications at cryogenic temperature. Novel constructs and motion control capabilities of two types of the rare earth magnetostrictive devices incorporated with HTS, an actuator and a linear stepping motor, were exhibited at the cryogenic condition.

MAGNETOSTRICTIVE LOW-TEMPERATURE ACTUATOR

The magnetostrictive materials have the coupling characteristic between their magnetic and mechanical states, which means a change in one of these states will produce a change in the other. A designed actuator, comprising a rod of the rare earth TbDyFe magnetostrictive alloy and a HTS solenoid surrounding the TbDyFe rod for magnetic excitation, is shown in Figure 1. One end of the rod is anchored to the backup structure and the other end is extended out. The application of small triggering magnetic field, for instance, will cause the device to elongate and a high force, either push or pull, is activated. The superconductivity of the solenoid minimizes electric power dissipation contributing to energy efficiency and to reduction of waste heat, which must be removed to maintain a cryogenic environment.

The drive electronics consists of a precision current regulator, an overcurrent protection circuit and an efficient DC/DC converter. The combination control of the drive accurately maintains the current through the actuator to hold a position.

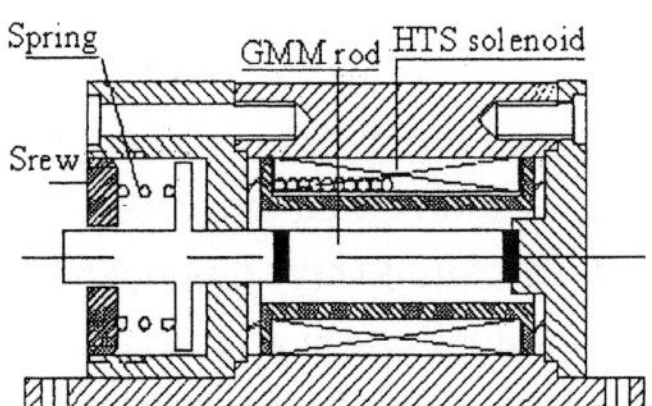

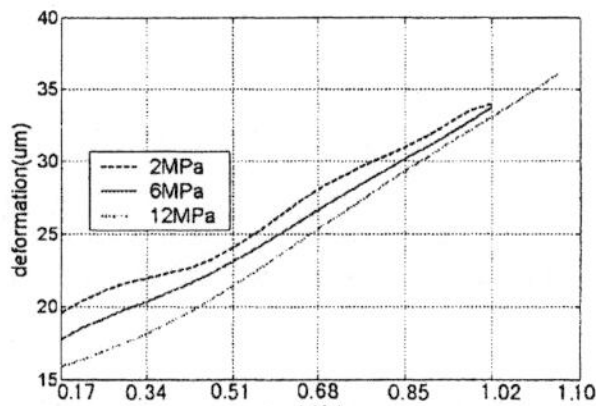

Figure1 an actuator   Figure 2 current- deformation curve under different pressure

This lightweight actuator could be utilized for performing shape and position control functions for the telescope to operate at cryogenic temperature in the deep space. Precision positioning can be achieved by precisely controlling the current energizing the HTS solenoid of the actuator. Such a simple actuator can also be used for precise mechanical positioning, vibration control, or switch and valve operation having strokes up to more than 35 μm (see Figure 2) and force capabilities up to 1200N.

LINEAR STEPPING MOTOR

A magnetostrictive linear translation mechanism has been designed to function as a micropositioning device at cryogenic temperatures, in which there are requirements for high stiffness, increments of motion <1 μm, and long travel of several millimeters. A stepping motor is developed by a traveling active element which is a rare earth magnetostrictive rod in a tight-fitting tube which is exited by the traveling magnetic field from superconducting pancake coils (see Figure 3). These pancake coils are supplied by polyphase current to bring the stepping motion. The stepping motor includes high force motion and the ability to hold position when powered off. These capabilities are completed with a single set of electronics (see Figure 4).

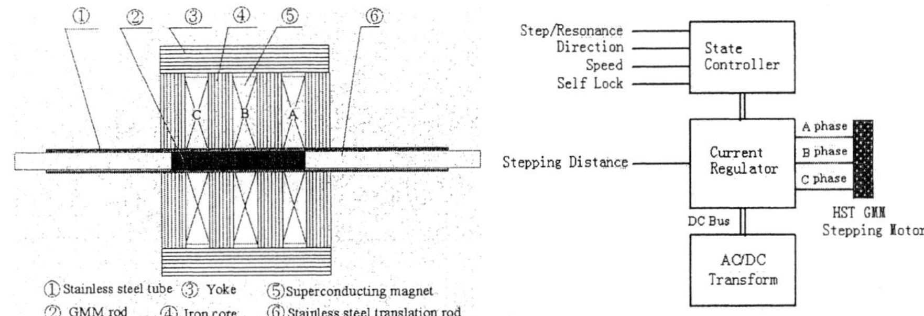

Figure 3 linear stepping motor   Figure 4 electronics of the stepping motor

The peristaltic principle of the linear stepping motor (see Figure 5) is described as below. The powered off condition is depicted in Figure 3, where a length of magnetostrictive element with circular section is confined to the tube. When no magnetic field is applied to the element, there exists a tight fit between the tube and the element. This squeeze preloading produces the normal force necessary for fail-safe joint locking.

When the coils of the phase A are powered, the magnetic field begins to interact with the magnetostrictive element under the phase A ranges. Because the magnetostrictive element under the phase B and C range is embedded, the magnetostrictive element under the phase A range is expanded along the magnetic field, extending to the right to push the right staff (see Figure 5 (a)).

In the second step, the phase A is powered off and the phase B are powered, the magnetostrictive element under the phase A and C range is embedded and the magnetic field begins to interact with the magnetostrictive element under the phase B range. Consequently the magnetostrictive element under the phase B range is expanded along the magnetic field too(see Figure 5 (b)), while keeping the same length as Figure5(a).

In the third step, the magnetostrictive element under the phase C range is expanded along the magnetic field (see Figure 5 (c)), keeping the same length as Figure5(a).

In the next step, phase A again is powered, extending to the right to push the right staff, while phase C is powered off, pulling the left staff from the left to the right. The magnetostrictive element has effectively moved to a right step. If the coils of the stepping motor are powered in the pattern sequence as -A-B-C-A-..., the magnetostrictive element will effectively move to the right to push the loads step by step.

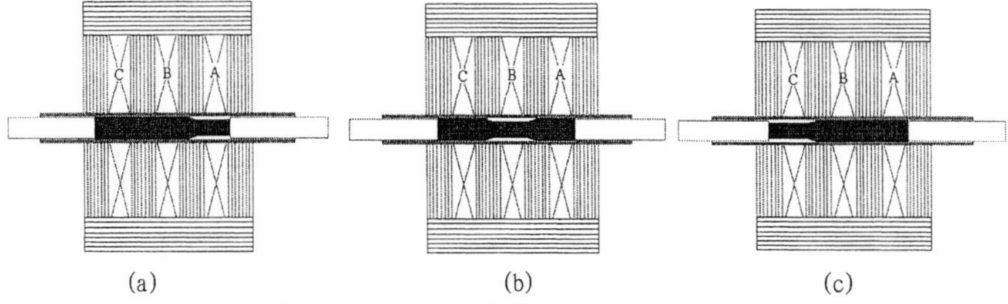

Figure 5 peristaltic principle of the stepping motor

The performances of the high temperature superconducting and giant magnetostrictive linear stepping motor are:

1. Superconducting magnet with the iron core
2. Large stroke

3. Large force

4. Adjustive micro-step

5. Self lock function when powered off

Such micropositioners could be used to make fine position adjustment in diverse scientific and industrial instruments. For example, they could be used to drive translation stages in scanning tunneling microscopes or to move optical elements that must be located at long but precise distances from each other, including telescopes and interferometers for cryogenic deep space application.

OTHER APPLICATIONS

Other applications of these magnetic actuators take advantage of their high force capability, large stroke and desirable repeatability.

1. Resonant frequency control of RF cavities for particle accelerators

High temperature superconducting particle accelerators use a bellows shaped resonant cavity to impart energy to the particle with radio frequency waves. To achieve the particle energies for experiments, hundreds of cavities must work in tandem at the exact same frequency. Frequency matching is accomplished by squeezing the cavities axially. The high force capability combined with the sub-micron positioning capability of the actuators is well suited to this application at the cryogenic environment.

2. Active vibration control

These actuators can be applied to active control of vibration. With state-of-the art accelerometers and control electronics, active vibration control systems are being developed for both cryogenic and room temperature application. Magnetic vibration control is more efficient than the currently available piezoelectric systems for controlling low frequency high amplitude vibrations.

CONCLUSION

Novel mechanisms designed using HTS magnet and magnetostrictive materials, with the application of the specific drive electronics, results in efficient, compact, and lightweight actuator or motor systems. The potential exists for the quick advancement of related techniques for producing actuators and mechanisms at the cryogenic temperature environment.

ACKNOWLEDGMENT

This work is supported by the National Natural Science Foundation of China.(50077019)

REFERENCE

1. Claeyssen, F. and Lhermet, N., Actuator based on giant magnetostrictive materials, 8th international conference on new actuators, Bremen, Germany (2002) 148-153

2. Joshi, C.H., Compact magnetostrictive actuators and linear motors, Actuator 2000 conference, Bremen, Germany (2000) 1-6

3. Joshi, C.H. and Bent, B. R., Application of magnetic smart materials to aerospace motion control, The aerospace mechanisms symposium, Greenbelt, MD.(1999)

Spitzer Space Telescope Thermal/Cryogenic System Flight Performance

Paul T. Finley, Richard A. Hopkins, and Russell B. Schweickart

Ball Aerospace & Technologies Corp., 1600 Commerce Ave., Boulder, CO USA 80301

The Spitzer Space Telescope (formerly called SIRTF) was launched into an Earth-trailing, solar orbit on August 25, 2003. The Cryogenic Telescope Assembly is Spitzer's instrument payload. The design operational lifetime is 5 years, limited by the loss rate from the superfluid helium cryostat that cools the instruments to 1.3 K and the telescope to 5.5 K by vapor cooling. The thermal/cryogenic system flight performance to date is meeting expectations.

INTRODUCTION

The Spitzer Space Telescope, comprised of the Cryogenic Telescope Assembly (CTA) and the Spacecraft, is operating in an Earth-trailing, solar orbit where the influences of the Earth and Moon on the thermal system performance are negligible. This allows for a very efficient thermal system, but creating a test environment to demonstrate the expected performance was very difficult and uncertain[1].

The CTA (Figure 1) consists of four subsystems: the 360-liter superfluid helium cryostat; the multiple instrument chamber that is mounted on the helium tank; the beryllium telescope that is mounted and heat sunk to the cryostat vacuum shell; and the outer shell group[2,3]. The CTA is attached to the Spacecraft with composite supports and miniature electrical cables to control the conducted heat to the telescope and cryostat to a very low level. Shields block radiation from the warm Spacecraft bus and solar panel, which prevents sunlight on any CTA surface at all times. To reject heat the outer shell anti-sun side is coated with black paint having high emittance at low temperature. Inside the outer shell the outer vapor-cooled shield (VCS) surrounds the telescope and cryostat. Supports and electrical cables are vapor cooled between the cryostat vacuum shell and outer shell. This internal thermal system limits the heat flow to the telescope and cryostat vacuum shell to about 4 mW, which allows them to be cooled to the required 5.5 K temperature with helium vapor. This entire system can be thought of as a complex cryostat.

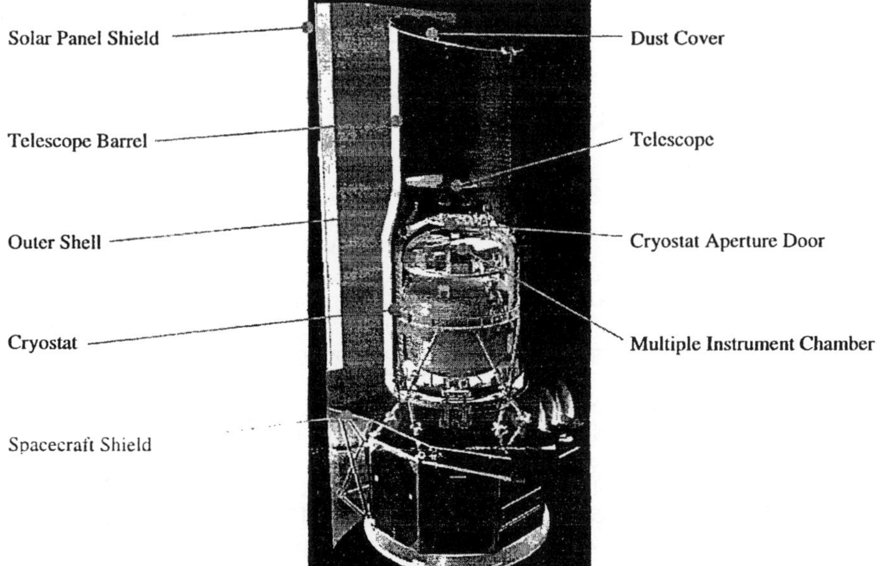

Figure 1 The Spitzer Space Telescope with cutaway view of the CTA. The Sun is always to the left.

About 5 mW heat input to the helium bath is needed to produce the flow rate required to cool the telescope and vacuum shell to 5.5 K. The instruments, which operate one at a time, dissipate between 1 and 3 mW. Parasitic heat inputs to the helium tank are negligible. Therefore, a make-up heater mounted on the tank is used to maintain a bath pressure that will produce the flow needed to cool the telescope. To maintain constant pressure the heater power level is adjusted along with changes in instrument power dissipation. The CTA helium usage rate is nearly 10 times smaller than that of previously flown, helium-cooled telescope systems.

FLIGHT OPERATIONS AND PERFORMANCE

The redundant cryostat vent valves were opened during ascent to prevent liquid breakthrough in the porous plug phase separator. Four days after launch, the telescope dust cover was ejected, and a day later the cryostat aperture door was opened. About 7 weeks after launch, the telescope focus was checked and slightly adjusted. The only CTA operation after that, not including the instruments, is use of the make-up heater to control telescope temperature as discussed below.

<u>Cooldown performance</u>
Figure 2 shows helium bath, porous plug external surface, and inner VCS temperatures during the first 9 days of flight. Temperature response of the inner VCS, which surrounds the helium tank, to aperture door opening can be clearly seen and was used to verify door opening. Bath temperature exceeding the lambda point (2.18 K) might allow catastrophic porous plug breakthrough[4]. Therefore, to account for uncertainty in porous plug performance in the high-flow regime, conservatively high plug impedance was used in the model. This is why the predicted peak bath temperature is higher than the actual value. Change in temperature drop across the plug, not shown in the figure, was used to verify vent valves were open.

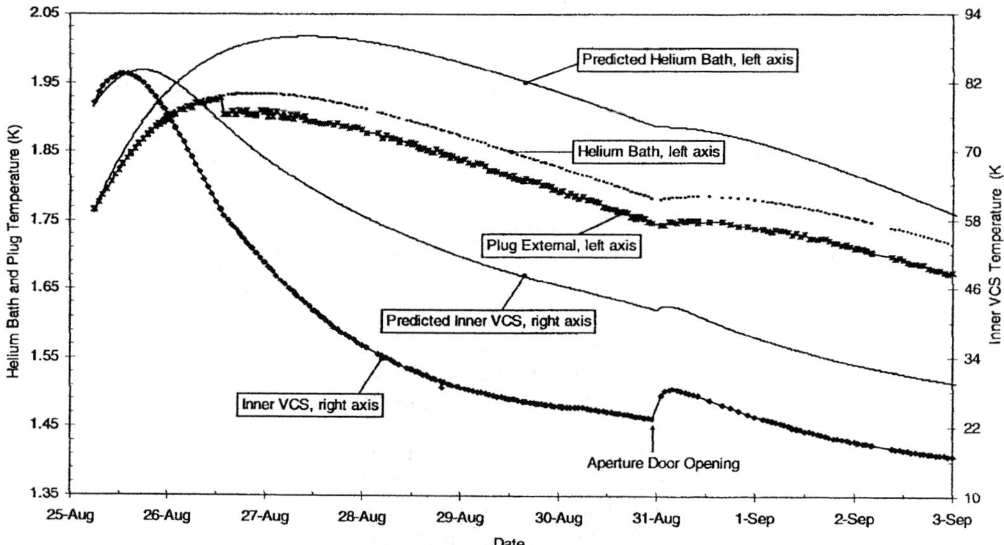

Figure 2 Critical temperatures during the early stages of cool down are shown, along with predictions.

The Spitzer CTA is a unique cryogenic telescope in that it was launched warm. The telescope cooled from 290 K to 5 K in 41 days. This warm launch architecture, made possible by the flight thermal environment, reduced the mass of the cryostat. However, the large changes in material properties over this wide temperature range created optical alignment risks and transient thermal model uncertainties. And the cooldown performance could not be realistically test verified under flight-like conditions.

Two processes drive the cooldown: radiation to space and vapor cooling using the helium effluent. Soon after launch, when the outer shell and telescope were warm, the cooling rate was dominated by radiation. Three weeks after launch, when the outer shell reached its stable temperature, vented helium vapor controlled the remaining telescope cooldown. Once in this regime, flow and telescope cooling rates were controlled with the make-up heater. The outer shell radiates 86% of its incident heat load to space at its 34 K operating temperature; the remainder is transmitted to the outer VCS. Regions between the outer shell and telescope also radiate significant levels of heat to space, but at 5.5 K the telescope does not. Raising or lowering the net power dissipation into the helium tank slowly changes the helium temperature and pressure, thus increasing or decreasing the vapor flow.

The cooldown process was modeled with an integrated thermal math model and fluid flow model using SINDA/FLUINT software from C & R Technologies. This integrated modeling approach provided predictions of helium usage, vapor cooling, and component temperatures as functions of time with instruments and make-up heater power as input. As mentioned, this model could not be strictly validated by test. However, steady-state models were test verified both at room temperature and near flight temperatures. The transient model matched the steady-state models at each end of the temperature range.

Figure 3 compares the flight cooldown data to the pre-launch model predictions. Although there are deviations, the general agreement is very good. After the dust cover was ejected, the outer VCS cooled faster than predicted, indicating the heat rejection to space was underestimated in the model. The predicted telescope and outer shell temperatures match the flight data quite well. Over the cooldown temperature range, thermal conductivities, bolted joint conductances, specific heats, and infrared emittances change substantially. For example, the specific heat of aluminum decreases by a factor of 600 going from 300 K to 10 K. There is remarkable agreement between the flight data and transient predictions considering that the model was not test verified. Note in Figure 3 the impact of dust cover ejection on telescope temperature; this response was used to verify the cover had been ejected.

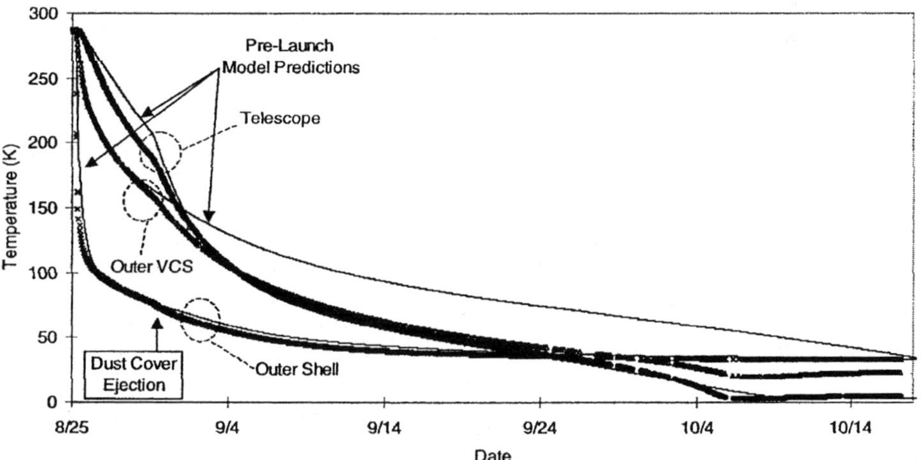

Figure 3 Cooling profiles of telescope, outer VCS, and outer shell after launch are compared with predictions. Thin lines are model predictions. Symbols are flight data. The response of the telescope temperature to dust cover ejection is evident.

Prior to launch, an aggressive schedule of instrument and subsystem flight tests was planned based on pre-launch cooldown temperature predictions using a nominal prescription for make-up heater use. Many of the instrument tests required specific telescope temperatures, and scheduled events were inter-related, leaving little flexibility. We began powering the make-up heater to accelerate cooldown 21 days into flight. The transient model did not have the accuracy to predict telescope temperatures to within a degree, as was needed at times to hold schedule. Therefore, we tracked deviations from the model and used the model to predict the sensitivity of the cooling rate to heater power. Thus, even an inexact model proved useful in making slight modifications to the heater prescription so that all temperature goals were met. Figure 4 shows the instrument temperature requirements and telescope cooldown profile as it cooled below 100 K. Make-up heater power was adjusted along with instrument power dissipation, and levels up to 10 mW were used. The telescope cooled below the 5.5 K requirement 41 days after launch.

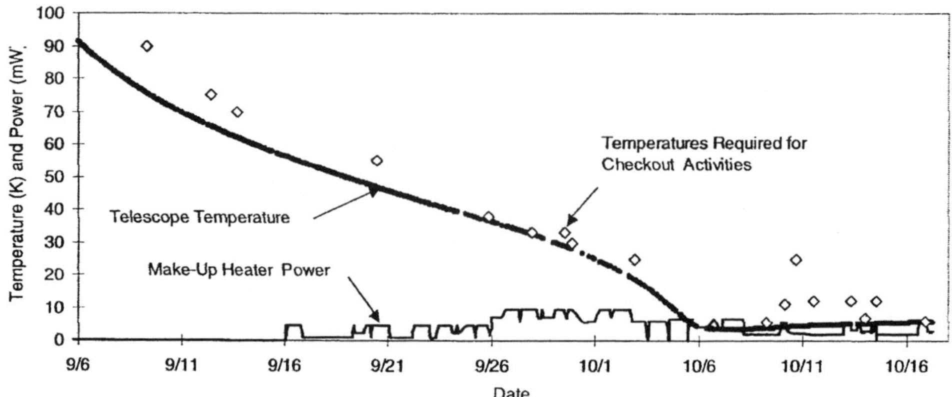

Figure 4 The telescope cooldown profile is shown along with temperatures required for scheduled instrument test activities. Also shown is the heater power used to control the telescope cooling rate to meet requirements.

Helium mass measurement, flight thermal balance test, and lifetime prediction

The cryostat includes a calorimetric helium mass gauge to provide a reliable determination of helium mass at the end of the cool down phase and then occasionally throughout the mission. The operation consists of applying 480 mW of heater power to the tank for 7.5 minutes while measuring the temperature rise. Helium mass was determined to be 43.4 ± 1.8 kg at the end of the 2-month in-orbit checkout phase. This is a serendipitous 6 kg more than the model predicted. This is due to the conservative porous plug characteristics assumed in the model, which resulted in overpredicted flow rate during early stages of cooldown. The measured mass uncertainty comes primarily from the 5 mK absolute accuracy limitation of the temperature measurement.

We performed a thermal balance test to determine the helium flow rate required to hold the telescope at 5.5 K. This 3-day test consisted of stabilizing the telescope temperature at ~5.5 K while putting an accurately known amount of heater power into the tank. The test showed a flow rate of 22 - 28 mg/day is needed, which compares well to the pre-launch model prediction of 21 - 33 mg/day. The test uncertainty comes primarily from the uncertain knowledge of the film flow through the porous plug, which essentially results in flow that is not caused by the heat input and therefore must be estimated by analysis.

Results of the mass gauge measurement and thermal balance test indicate a mission lifetime of 4.0 - 5.3 years (if the telescope is held at 5.5 K). The specified requirement was 2.5 years minimum with a goal of 5 years. Although the helium usage rate is not quite as good as the nominal prediction, the lower-than-expected helium loss during cooldown compensated for it. It is anticipated that another helium mass measurement will be made about a year from launch, and the results will allow a more accurate determination of remaining helium lifetime.

Throughout design and ground testing, we made nominal, worst-case, and operational predictions of lifetime (Figure 5). Because an accurate ground test was not possible, we carried a large uncertainty between nominal and worst-case predictions. However, Figure 5 shows that flight performance is close to nominal predictions. The operational lifetime prediction is based on the nominal model, but with the telescope temperature allowed to float with the instrument needs.[5] Only the longest wavelength channel requires a 5.5 K telescope; most channels require no less than 20 K. This fluctuating telescope temperature operation is achieved through careful use of the make-up heater. Based on preliminary results, it is expected to reduce helium usage rate ~9%, extending the expected mission lifetime to 4.3 - 5.8 years.

Steady-state temperature data

The temperature data and pre-launch predictions are shown in Table 1. The spacecraft shield, solar panel shield, and outer VCS are warmer than predicted. We believe this is due to overestimates of the heat rejection from the shields to space and/or the performance of the insulation on the warm sides of the shields. Since radiation is a small part of the heat flow from the Spacecraft bus to the CTA, the slightly warmer spacecraft shield has little effect on outer shell temperature. Since only 25% of the outer shell

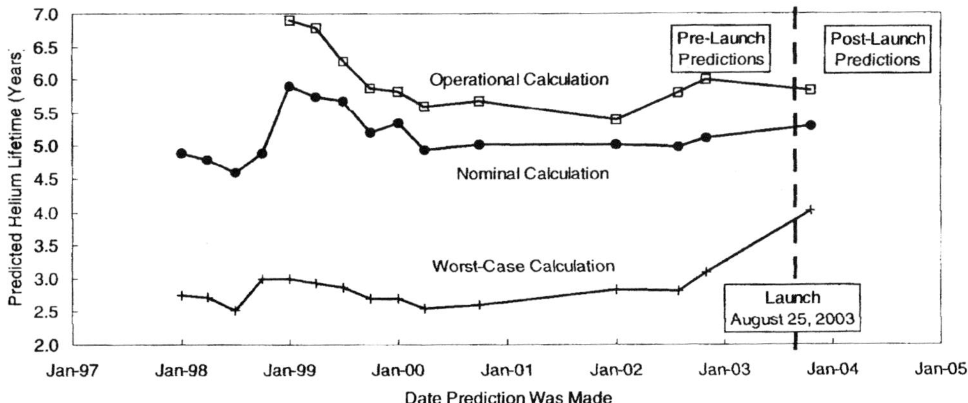

Figure 5 Helium lifetime predictions made over the last six years through design, testing, and flight.

heat load comes from the solar panel shield, the effect of the warm solar panel shield is also small. The heat load on the telescope and cryostat is directly dependent on the outer VCS temperature, explaining why the helium usage during observations is somewhat greater than the nominal prediction. Although the porous plug temperature drop is significantly less than predicted, the plug appears to be functioning properly.

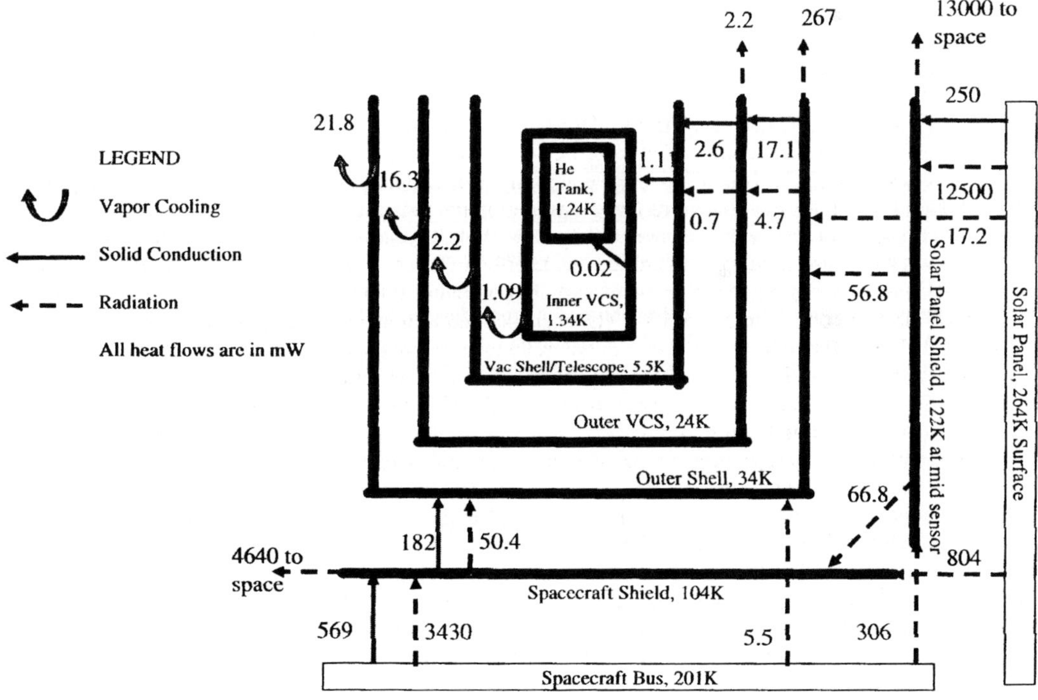

Figure 6. Predicted heat flow diagram for operational steady-state conditions. Combined instrument and make-up heater heat load is 5.3 mW to achieve 5.5 K telescope temperature.

The heat flow diagram, Figure 6, shows that radiation greatly dominates solid conduction in the warmer region of the system, and solid conduction dominates in the colder region. This is of course as expected. The diagram also shows how important vapor cooling is to achieving the telescope 5.5 K temperature.

5.3 mW heat is required to produce enough helium flow to cool the telescope; about half comes from the instrument and half from the make-up heater.

Table 1 Comparison of flight steady-state temperatures and pre-launch predictions (worst-case and nominal).

	Parameter	Value at Launch	Steady State Value	Predicted Value
Temperatures (K)	S/C Shield	285	104	99 - 110
	Solar Panel Shield	285	125	91 - 101
	Outer Shell	285	34	32 - 36
	Outer VCS	285	24	17
	Vacuum Shell & Telescope	285	5.5	5.5
	Inner VCS	79	1.3	1.3 - 1.4
	Helium Bath	1.76	1.24	1.21 - 1.26
Plug Temperature Drop (mK)		0	4	12

CONCLUSION AND ACKNOWLEDGEMENT

The Spitzer Space Telescope cryogenic/thermal system is performing close to pre-launch expectations.

This work was performed for the Jet Propulsion Laboratory, California Institute of Technology, sponsored by the National Aeronautics and Space Administration.

REFERENCES

1. P.T. Finley, R.L. Oonk, R.B. Schweickart, Thermal Performance Verification of the SIRTF Cryogenic Telescope Assembly, Proceedings of SPIE (2003) **4850**, 72-82

2. J.H. Lee, W. Blalock, R.J. Brown, S. Volz, T. Yarnell, and R.A. Hopkins, Design and Development of the SIRTF Cryogenic Telescope Assembly (CTA), Proceedings of SPIE (1998) **3435**, 172-184

3. R.A. Hopkins, P.T. Finley, R.B. Schweickart, and S.M. Volz, Cryogenic/Thermal System for the SIRTF Cryogenic Telescope Assembly, Proceedings of SPIE (2003) **4850** 42-49

4. P.T. Finley, R.A. Hopkins, and R.B. Schweickart, Flight Cooling Performance of the Spitzer Space Telescope Cryogenic Telescope Assembly, to be published in Proceedings of SPIE (2004) *Space Telescope Systems*

5. C.R. Lawrence *et al.*, Operating SIRTF for Maximum Lifetime, Proceedings of SPIE (2003) **4850**, 153-161. A follow-up paper by C.R. Lawrence *et al.* to be published in Proceedings of SPIE (2004) *Space Telescope Systems.*

Thermal validation of the design of the CFRP support members to be used in the spatial framework of the Herschel Space Observatory.

McDonald P.C., and Jaramillo E.*

Institute of Cryogenics, School of Engineering Sciences, University of Southampton, Southampton SO17 1BJ, UK.
*HTS AG, Widenholzstrasse 1, CH-8304 Wallisellen, Switzerland.

FE thermal models of the support members for the low temperature components of the Herschel Space Observatory have been validated by measurements at 1.8K. The Herschel Space Observatory structure is briefly introduced. The thermal modelling of two designs of support member is discussed and the final designs presented. The design of the validation test rig is described together with details of the experimental method. Observed results are presented and discussed with reference to the thermal model. The model was found to be valid to within $1.5 \cdot 10^{-5}$ W/K in the lateral struts and $3.29 \cdot 10^{-5}$ W/K in the interface struts, at a mean temperature of 6K.

HERSCHEL PROJECT BACKGROUND

The Herschel Space Observatory is due for launch by ESA in 2007 and will capture images of the far infrared universe through three instruments:- a camera (PACS), a high resolution spectrometer (HIFI) and a photometer (SPIRE) all of which sit on an optical bench and are cooled to less than 3K. The cooling of these instruments is to be achieved by placing the optical bench inside a large cryostat which contains a superfluid helium tank at 1.6 K and uses a circulation loop to deliver the superfluid helium to the bench. The configuration of the cryostat is shown in the schematic diagram, Figure 1, where it may be seen that the optical bench and helium tank are supported from a spatial framework, consisting of two aluminium frames which straddle the helium tank.

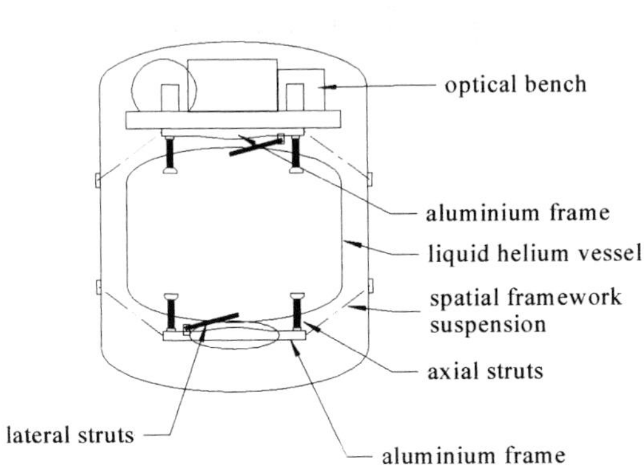

Figure 1 Schematic diagram of Herschel cryogenic assembly

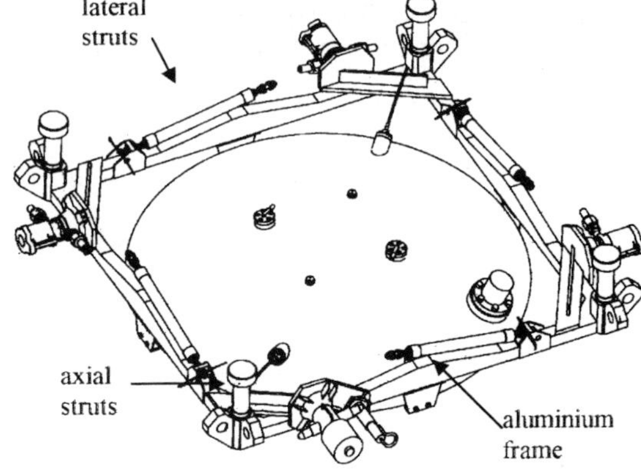

Figure 2 Lower spatial frame showing CFRP support struts

The spatial frames are illustrated in Figure 2.The helium tank and optical bench are attached to the spatial framework by a system of axial and lateral struts and the frames are supported from the wall of the cryostat by an additional assembly of struts. The cryostat wall will be at an estimated temperature of 10K.

This paper is concerned with the interface between the helium tank and the spatial framework. The interface has been designed to minimise heat transfer to the helium by using high strength low conductivity struts of reduced cross section and high aspect ratio. The strut configuration is shown in Figures 3 and 4. They are constructed from carbon fibre reinforced plastic (CFRP) tube with crimped and glued aluminium alloy end fittings.

In addition to providing support with thermal isolation the interface structure must compensate for the thermal contraction of the tank without inducing stresses on the optical bench and to this end the interface struts use coated ball and socket joints as end fittings as shown in Figure 4.

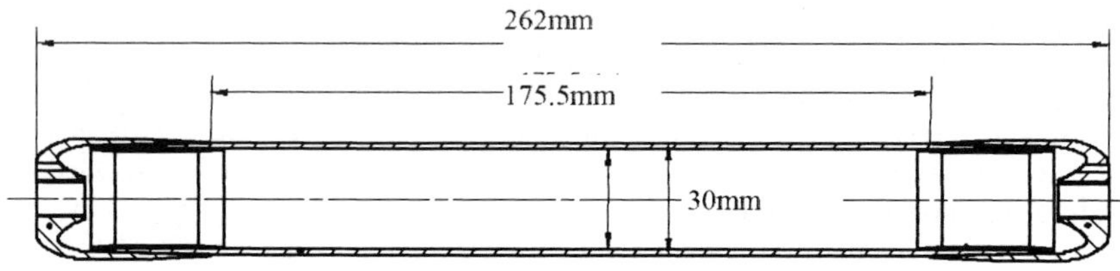

Figure 3 Lateral strut

STRUT DESIGN AND THERMAL MODELING

The designs of the lateral and axial struts are shown in Figures 3 and 4. A Finite Element Model (FEM) was developed to predict the thermal behaviour of the struts under varying temperature conditions and included radiation heat transfer from the cryostat wall.

Thermal modelling was done using thermal simulation software from MAYA [1] and thermo-physical property data obtained by:- [2] Radcliffe and Rosenberg, and [3] Bansemir and Haider.

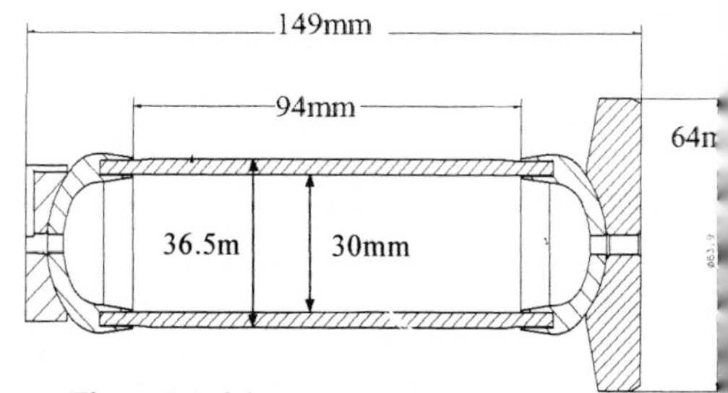

Figure 4 Axial strut

Conductances calculated from the FE model for two different sets of boundary conditions are shown in Table 1. These are for:- cold end temperature: 1.6K and cryostat wall temperatures of 6K and 20K. The predicted operating conditions are for a cold end temperature of 1.6 to 1.8K and high end temperature of 10K, giving a mean temperature in the struts of approximately 6K.

Table 1 Predicted conductance data for axial and lateral struts

High end temperature - K	Conductance – W.K^{-1}	
	Axial struts	**Lateral struts**
6	8.4×10^{-5}	2.1×10^{-5}
20	2.7×10^{-4}	6.7×10^{-5}

VALIDATION MEASUREMENTS

Method

The method of measurement followed, to determine the thermal conductances of the struts, was to:- mount one end of selected samples of the struts on a surface cooled to 1.8K, then apply heat to the free ends, observing their rise in temperature as a function of the heat supplied. The measurements were of very low conductances, measured at low temperatures and care was taken in the design of the measurement system, to ensure that parasitic heat leaks were not significant in the measurements.

Measurements were made in a vacuum insulated cryostat, constructed as illustrated in Figure 5. Two samples of each strut design, were screwed to the base of a copper pot which could be filled with liquid helium at 4.2K and then pumped; using a rotary vane vacuum pump, to reduce the vapour pressure of the helium to 16mbar, thus obtaining a pot temperature of 1.8K. The pot was suspended in the cryostat by two stainless steel fill/vent tubes which were used for the initial filling with liquid helium and subsequent pumping. A valve in the line to the pump was used to control the pumping speed and thus the helium vapour pressure. By this arrangement it was possible to control the helium temperature to 1.8 ±0.1K

An aluminium radiation shield was placed around the pot and was cooled by the first stage of a two stage G-M cryocooler, to approximately 50K. The first stage of the cryocooler was also connected by a thermal link to the fill/vent tubes, to intercept conducted heat from ambient room temperature, which otherwise would have reached the pot. The second stage of the cryocooler was used to cool a charcoal sorption pump to approximately 10K, to maintain a high vacuum in the cryostat and around the test samples. To minimise radiant heat transfer to the test sample, a second, aluminium radiation shield was

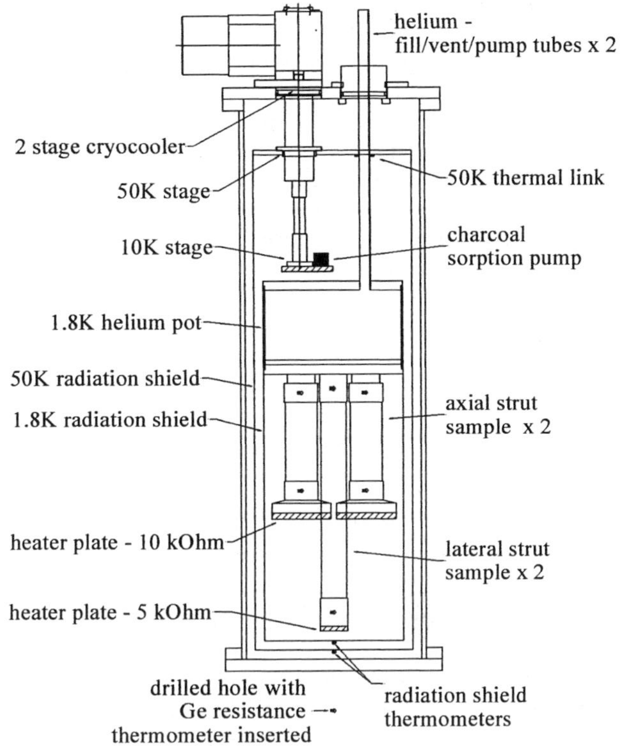

Figure 5. The test cryostat

screwed to the bottom of the helium pot and surrounded the samples. Ge resistance thermometers were attached to the bottom plates of the radiation shields and the shields were wrapped with multilayer superinsulation.

Heaters

Heat was applied to the samples by means of electrical heater plates attached to the free ends of the struts. The heater plates used on the lateral struts were ~10mm thick aluminium disks, to which were glued 5 x 1kOhm ceramic heater chips. The heater plates on the axial struts were constructed using ~5mm thick copper flanges to which were glued 10 x 1kOhm ceramic heater chips. High heater resistances were selected so that the required operating current could be small and could be supplied using 0.127mm diameter insulated phosphor bronze wires within the cryostat, without significant I^2R heating of the wires and without significant thermal conduction through them.

The heater plates were wired in series and therefore received a common current. Their relative resistance was chosen so that the warm end temperatures of the struts would be roughly the same during the tests and there would therefore be no heat transfer between them through the connecting wires.

The voltage across each heater was measured using pairs of non-current carrying voltage taps. These were made using the phosphor bronze wire and were thermally connected to the 1.8K pot at a wire length of ~400mm from the heaters. The thermal conductance of these, based on a material (Cu + 5%Sn) conductivity of < 4W/m.K [4] was calculated to be 1.3×10^{-7} W/K and conduction through the wires, from the heater at a temperature of 20K, was conservatively estimated to be 2.4×10^{-6} W per wire. This represented <0.1% of the heater powers.

Thermometers

The aluminium ends of the struts were drilled and thermometers inserted. The thermometers used were Ge film on GaAs resistance thermometers supplied by the Institute of Semiconductor Physics, Kiev [5]. They were individually calibrated over the range 1.8 to 300K and supplied with individual interpolation polynomials and tables. Their R/T response was exponential with typical sensitivity at 1.8K of ~-6.5kΩ/K reducing to ~-23 Ω/K at 30K. The thermometers were measured using a four-terminal arrangement of independent current (500nA) and voltage wires. As for the heaters, connection was made using phosphor bronze wires thermally anchored at 1.8K and the conduction to the 1.8K pot from each thermometer, for a warm end temperature of 20K was therefore estimated to be 4x 2.4x10^{-6}W.

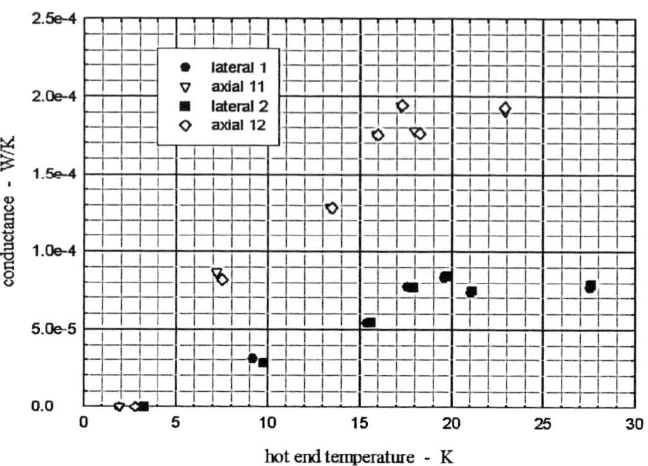

Figure 6 Measured conductance data

Data acquisition

During measurement, the thermometer resistance and heater powers, were logged at three minute intervals using an Agilent 34970A data acquisition unit. The increases in temperature of the heated ends of the samples were monitored until they reached a steady state, at which point the high and low end temperatures were recorded together with the heater power. The heater power was then increased and the measurement process repeated to produce a data table of the steady state temperatures versus heater power.

RESULTS AND CONCLUSIONS

Conductance was calculated from the heater power divided by the temperature difference observed across the samples, in units of W/K. The results are summarised in Figure 6 for varying hot end temperatures. The derived values for the expected operating condition of the struts are shown in Table 2. The model was found to be valid to within $1.5{\cdot}10^{-5}$ W/K in the lateral struts and $3.29{\cdot}10^{-5}$ W/K in the axial struts, at a mean temperature of 6K.

Table 2 Modelled and measured values of conductance for 6K mean strut temperature

Description	Model Conductance at mean T = 6K W/K	Measured at W/K
Axial Strut	$7.21{\cdot}10^{-5}$	$1.25{\cdot}10^{-4}$
Lateral Strut	$1.81{\cdot}10^{-5}$	$3.3{\cdot}10^{-5}$

REFERENCES

1. MAYA Heat Transfer Technologies Ltd., 4999 St. Catherine St. West, Suite 400, Montreal, Quebec, Canada   H3Z 1T3
Thermal modelling was done using the TMG software package from MAYA

2. Radcliffe, D.J: and Rosenberg H.M, The thermal Conductivity of Glass fibre and carbon fibre epoxy composites from 2 to 80 K in Cryogenics pg 245-249 May, 1982

3. Bansemir, H and Haider,O Basic material data and structural analysis of fibre components for space application, Cryogenics (31), pg 298-306, Apr 1991

4. Robert L. Powell and William A. Blanpied Thermal Conductivity of Metals and Alloys at Low Temperatures   National Bureau of Standards Circular 556, September 1. 1954

5. V.F. Mitin, Microsensor Ltd. http://www.microsensor.com.ua

Development and testing of a novel thermal switch

You J.G., Dong D.P., Wang W.Y., Li Z.W.[*]

No.500, Yu Tian Rd, Shanghai, China, 200083
[*]No.1954, Hua Shan Rd ,Shanghai, China, 200030

ABSTRACT

A novel thermal switch, named as Double Driving Devices-based Thermal Switch (DDDTSW) will be described. Some details will be focus on the structure and the work principle of this device, as well as something about experiment for the device's work performance and the test result from it will be introduced. In ground testing of using LN2 as cold resource, the DDDTSW demonstrated an "Off" resistance of 2450K/W and an "On" resistance of 6.0K/W. Finally, it is proposed to advance the DDDTSW design to increase the heat conduction, and it's also pointed out this kind of thermal switch has the potential feasibility used in the space cooling system.

INTRODUCTION

With the development of infrared remote sensing system, there are more requirements in space cryogenic technology, such as better reliability and longer lifetime of   the cryogenic cooling systems using mechanical cryocoolers. One of the feasible approach is to incorporate redundant cryocoolers in the cooling system to protect against individual cryocooler failures.

(see Figure 1)

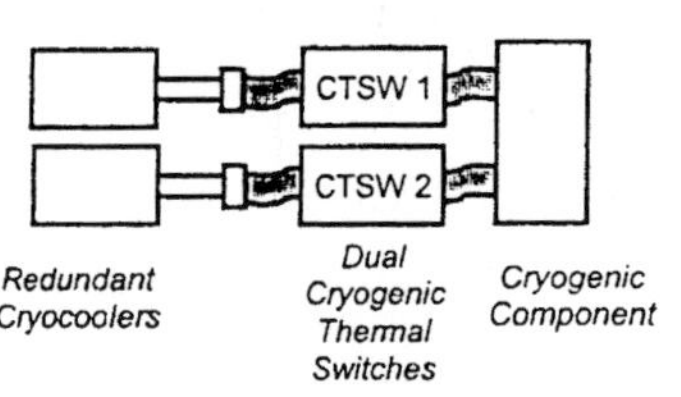

Figure 1. Dual CTSW System.

In applications, the non-operating standby cooler presents an added parasitic load to the operating cooler. So the non-operating cooler needs to be thermally isolated from the focal plane array, while the active cooler is thermally connected to the system. For typical space cryocoolers operating without a thermal switch device, the parasitic load stem primarily from conduction through the non-operating cryocooler expander. The thermal resistance of this conductive path is generally 400-500 K/W for space cryocoolers. As a result, the parasitic load due to the non-operating cryocooler is approximately 0.5W at 60K.[1]

In order to minimize cooling and input power requirements, a reliable cryogenic thermal switch is desirable. A properly designed heat switch increases the thermal isolation between the instrument and expander body of the non-operating crycooler and reduces the parasitic load from the non-operating cryocooler by 65-80% [1].As a result, cryocooler cooling and input power requirements are reduced. [2]

CRYOGENIC THERMAL SWITCH PERFORMANCE REQUIREMENTS

The development of the cryogenic thermal switches presented in this paper was funded by National Innovation Project to improve the reliability and lifetime of space cooling system, which requires consecutive work for 3 years. The design of the DDDTSW is driven by the following performance requirements, outlined in Table 1.

Table 1. DDDTSW Performance Requirements

Operating Temperature	<150K
"Off" Thermal Resistance	>1000K/W
"On" Thermal Resistance	<4K/W
Displacement	>1mm

DOUBLE DRIVING DEVICES-BASED THERMAL SWITCH(DDDTSW) DESIGN

In order to meet the above requirement, a high reliable thermal switch with good performance is needed. So it's very important to invent a novel and promising concept for this purpose. DDDTSW is mainly made up of five parts, which are Shape Memory Alloy Spring(SMAS), small assistant spring, copper strap, link for connecting DDDTSW to cold head, link for connecting DDDTSW to focal plane. (see Figure 2)

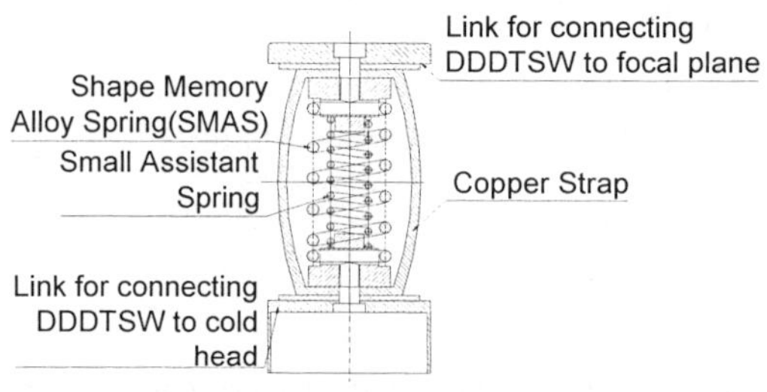

Figure 2. DDDTSW Design

The key part of DDDTSW is SMAS, which is based on SMA effect, that means it can move when certain temperature reaches, and stay stable condition when the working temperature is lower or higher than the certain one. There are four special temperature points for SMA: Ms, the starting temperature of martensite transformation; M_f, the finishing temperature of martensite transformation ; As, the starting temperature of austenite transformation; A_f, the finishing temperature of austenite transformation. If these four special temperature points can be adjusted in line with our requirements, this SMAS will move at certain range of temperature that we need.

In term of DDDTSW, according to our requirements, these four temperature points should be respectively as following: Ms<220K;M_f>150K;As>200K;A_f<250K. These four temperature points can guarantee DDDTSW work effectively and reliably.

The other important part is the small assistant spring, just as its name implies, is to assistant to the move of SMAS. Because SMA we used only has the one-way shape memory effect, which means SMAS only can move towards single one direction, the force need for moving towards another direction should be provided by the small assistant spring. The heat conduction will be considered to resolve by using copper strap. The whole advice is fixed between the cooler's cold head and the focal plane with the link.

The process of DDDTSW is described as following: The temperature will begin to drop after the cryocooler start to work, and DDDTSW is expanded with the force of SMA. When the temperature drops to below Ms, DDDTSW starts to expand with the force provided by small assistant spring until the temperature is lower than M_f. At this moment, DDDTSW completely connects to the focal plane, which is "On" of DDDTSW. Correspondingly, when the cryocooler stop working, the temperature will rise. When the temperature rises to above As, DDDTSW starts to contract with the force of SMA until the temperature is higher than A_f. At this moment, DDDTSW completely disconnects to the focal plane, which is "Off" of DDDTSW. The maximal displacement for DDDTSW is designed to about 2mm.

DDDTSW TEST RESULTS

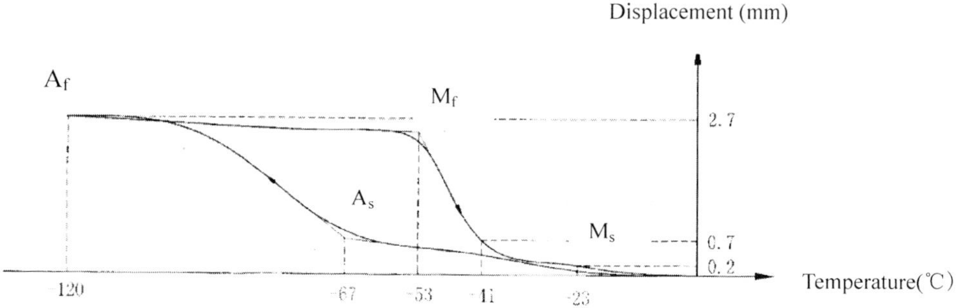

Graph 1   Relation Between the Displacement of SMAS and Temperature

temperature and to see what's the maximal displacement. After experiment, we get a graph showing the relation between the temperature and displacement (Graph 1).

From this graph, we can see clearly the maximal displacement is about 2.5mm. The four special temperature points, respectively, M_s=206K; M_f=153K; A_s=220K; A_f=250K, are nearly in the range the we require.

Another important purpose of the test is to get "on" and "off" performance of DDDTSW. In this experiment, LN_2 is used as the cold source, and the heater is to balance the cold of LN_2 to guarantee keep the temperatures in steady state. So the temperature difference ($\triangle T$) across the switch is measured ($\triangle T=T_o-T_c$, where T_o is the temperature of the warm end and T_c is the temperature of the cold end). And a known heater power Q is obtained. The "on" and "off" conductance is the heater power divided by the change in temperature difference. Eq. (1) provides the analytical relationship.[3]

$$R_{on,off} = Q \big/ \Delta T \tag{1}$$

After some calculation, we get the "On" performance is about 6.0K/W and the "Off" performance is about 2450K/W.

The result is not so satisfactory compared to the other heat switch. Specially, because of poor heat conduction of DDDTSW, the "On" performance is too big. Some measures can be taken to improve it, such as adding the copper thickness and improving the connect condition between DDDTSW and focal plane, and so on.

SUMMARY

The primary objective of this paper is to describe the design, operation, and test results of the Double Driving Devices-based Thermal Switch (DDDTSW). This type of thermal switch features simple construction and reliable work performance, repeatable actuating mechanism. After test, the "Off" performance is about 2450K/W and the "On" resistance is 6 K/W. With further development, this kind of thermal switch will meet the requirements of the space cooling system.

REFERENCES

1. Bugby,D.,Stouffer,C.,Hagood,B.,et.al, " Development and Testing of the CRYOTSU Flight Experiment," Space Technology and Applications International Forum(STAIF-99), M.El-Genk editor,AIP Conference Proceedings No.458,Albuquerque,NM,1999,pp.2-3.
2. B.Marland, D.Bugby,and C.Stouffer,B.Tomlinson and T.Davis, "Development and Testing of a High Performance Cryogenic Thermal Switch " , Crycoolers 11,Plenum Publishers,2001,pp,729-737.
3. Lanping Zhao, "Research on the Heat Transfer between the Solid Interfaces at Low Temperatures" ,Doctoral Dissertation of Jiao Tong University,2000,pp,67-71

CRYOCOOLERS DEVELOPMENT AND INTEGRATION FOR SPACE APPLICATIONS AT AIR LIQUIDE

A. Ravex, T. Trollier, L. Sentis, F. Durand, P.Crespi

Air Liquide, Division Techniques Avancées
2,rue de Clémencière – B.P. 15
38360 Sassenage, France

ABSTRACT

Air Liquide Division Techniques Avancées (AL/DTA) has developed cryorefrigerators for space applications and is integrating those in different payloads to be launched in the next coming years. The covered cooling technologies include Brayton, Joule Thomson, Stirling, Pulse Tube and Dilution cycles. The main characteristics and performances of these coolers are presented in this paper as well as some specific features related to their integration.

INTRODUCTION

AL/DTA has been involved in space cryogenics since several decades in the framework of the European ARIANE program mainly for the launcher liquid cryogens (LH2 and LOx) tanks development, qualification and manufacturing but also for the design and implementation of the liquefaction, storage and distribution infrastructures on the launching pad at Kourou, Guyana.

More recently AL/DTA has extended its space related business by developing and integrating cryocoolers for use onboard the International Space Station (ISS) or to be flown on scientific or commercial satellites payloads.

AL/DTA offer covers a very large panel of technologies (Brayton, Joule Thomson, Stirling, Pulse Tube, Dilution) and a very large range of temperature and cooling power (from a few microwatt at 100 mK till hundred watt at 190K). This new line of products has been developed taking advantage of both the heritage of AL/DTA experience in commercial cryorefrigeration and a strong partnership with National Research Laboratories (CNRS, CEA) and industrial partners (THALES Cryogenics). The main efforts have been dedicated to performance and reliability improvements which are the most critical parameters for space applications.

A special attention has also been paid to the thermal and mechanical integration of these coolers in payloads which is a major concern for systems overall efficiency and reliability.

CRYOCOOLERS DEVELOPMENTS

Brayton Cycle Coolers

At the end of the 80's, AL/DTA has initiated, under an ESA Technological Research Program (TRP), the development of a Brayton Cycle Cooler. A major feature was the use of high isentropic efficiency centrifugal compressor and expender implemented on a common shaft suspended with hydrodynamic gas bearings and driven by a high speed permanent magnets central motor integrated on the shaft between compression and expansion wheels. An other key point to be addressed was the miniature counter flow heat exchanger design which strongly influence the cooler overall efficiency. A cooler prototype has been successfully designed and tested and consequently this technology has been later selected by ESA for the development of a flight system: the Minus Eighty degrees Laboratory Freezer for International space station (MELFI).

The MELFI cooler is based on a reverse Brayton cycle using a very high speed turbo-machine (shown on Fig.1) located inside a toroïdal heat exchanger. It provides up to 90 W of net cooling power at 190 K. The radial compressor and expander wheels of the machine are mounted at both ends of the shaft running up to 100.000 rpm on herringbone groove bearings made in Tungsten Carbide (WC). The DC brushless and sensorless synchronous motor is located in the center between the two journal bearings. The compressor uses a smooth inlet and diffuser while the expander uses a bladed diffuser. The heat from the motor and the bearings is removed by a water heat exchanger wounded around the motor coil. The radial gap in the journal is 10 µm, radius. Herringbones are grooved in the WC using an Argon Ion beam bombardment techniques, using masks. The results is a 14 µm depth groove with an accuracy of ± 1 µm. A two pole permanent magnet is shrunk fitted into the hollow shaft. The thrust bearing is a WC spiral groove runner rotating between two static WC plates with an axial gap of 20µm on both sides. Our shafts are balanced on a dedicated bench using hydrostatic pressurized bearings with large gaps. This provides a very low stiffness allowing a very precise balancing. Displacements of the shaft are measured with capacitance probes with an accuracy better than one micron. Thus, the design of hydrodynamic bearings, the use of a brushless DC synchronous motor, the very low level of induced vibration generated thanks to the high level of balancing, allow to provide high reliable coolers (10 years lifetime without maintenance) and able to support several thousands of start/stop cycles. Presently, more than 17000 hours of operation have been achieved with 2700 start/stop cycles.

The first Flight Unit of the MELFI has been delivered to NASA. Initially planned on STS-114, this payload will be launched in November 2004 or early in 2005.

Figure 1: AL/DTA high speed Turbo-Machine for ISS.

Interesting characteristics of Brayton coolers using centrifugal compression and expansion at high speed is their low level of exported vibrations. Future Astrophysics mission will require cooling in the temperature range of 2 K to 6 K with extremely strong specifications on exported vibrations level. A multistage Brayton cooler could be a solution to fulfill these requirements. AL/DTA has been awarded a TRP by ESA to perform a cooler optimization and preliminary design, to identify critical items and to initiate feasibly demonstrations with the aim of 50 mW net cooling power at 5 K. To fulfill the vibration specification, a high rotational speed (1000 Hz) is mandatory, leading to very small wheel diameters for the compressor and expenders which is a challenging issue. In Fig.2 is shown the selected cycle architecture implementing a two staged centrifugal compressor, radiative precooling at 150K and three stages of expansion (60K, 15K and 5K) with centrifugal turbines. The compressor and the expenders are rotating on hydrodynamic gas bearings. The calculated compression work is 95 Watt for a 305 mg/sec 4He mass flow rate, with estimated isentropic efficiencies for the compressor (67 %) and the expansion turbines (50 %). The identified critical components are obviously the compressor and its high speed motor, the compact heat exchangers and the turbines. In a first step of the TRP it has been decided to demonstrate the feasibility of the compressor. Figure 3 shows a compressor wheel before brazing. The compressor is presently under assembly to be tested by the end of this year.

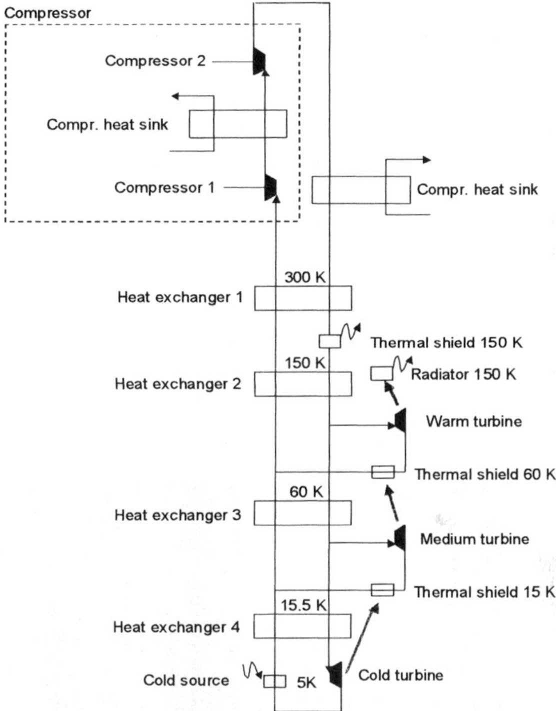

Figure 2: ^{4}He Brayton cycle process flow diagram.

Figure 3: Compressor wheel before brazing ($\varnothing$ 50 mm)

Stirling Cycle Cooler

To achieve the reliability required for space applications, it is now commonly agreed that the flexure bearing technology developed by Oxford University is a unique feature to ensure frictionless operation and clearance sealing in linear Stirling coolers. Several companies in the US (Ball Aerospace, Lockheed Martin, NGST formerly TRW) have developed, space qualified and flown specific coolers generally associating a two back to back pistons compressor with an actively driven cold finger, implementing flexure bearings, linear motors with moving magnets and phase shift control electronics.

In partnership with THALES, AL/DTA has undertaken the development and qualification of a Stirling cooler introducing a new manufacturing approach combining high reliability and reduced development costs by using as much as possible commercial coolers subassemblies. The basic characteristics of commercial THALES coolers are preserved: high reliability flexure bearings twin pistons linear compressor with moving magnet motor (reducing risks of cycle gas pollution by motor coil out-gazing and of flying wires and feed-through failures), pneumatically driven cold finger (reduced complexity: no motor, no active phase shift control electronics). Major improvements have been to adapt the external shelves architecture and materials to the specific mechanical and thermal constraints of space applications and to implement also flexure bearings for the cold finger displacer.

The performances and characteristics of the developed cooler [1] are summarized in Table 1 and a picture of a prototype is shown in Figure 4. This cooler has been selected by EADS for integration in Cryosystem, a deep-freezer to be flown by ESA/NASA on board the ISS (see further information in Integration section of this paper). Several Engineering Models (EM) have been characterized: thermal performances, launch and landing vibration and thermal environment tests. 2 EM coolers are currently run under lifetime test, with 5200 operating hours already achieved without any performance degradation.

Performance	9.3 W @ 80 K
Input Power	150 W
Frequency	50 Hz
Mass	7,3 kg
Compressor OD	90 mm
Compressor length	189 mm
Cold finger OD	110 mm
Cold finger length	162 mm

Table 1: Cooler characteristics.

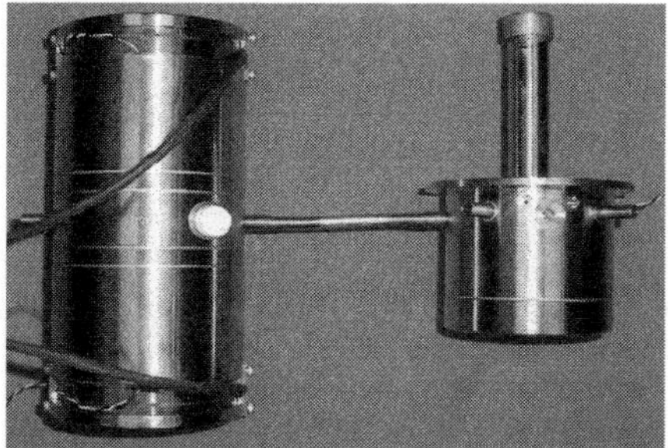

Figure 4: Cryosystem EM Stirling cooler.

Pulse Tube Coolers

The main drawback of the Stirling coolers in terms of reliability and integration easiness is their cold finger: the moving displacer technology and the associated linear motor are a potential source of failure and exported vibrations, the reduced clearance between the moving regenerator/displacer and its shelve induces strict mechanical load restrictions on the cold finger cold tip to avoid any internal friction which generally results in very complicated thermal link design, the active phase shift control loop complicates the drive and control electronics.

The Pulse Tube cooler (PTC) technology with no moving part in the cold finger and passive pneumatic phase shift control is recognized as an attractive alternative to Stirling coolers. Recent developments, such as phase shift control by inertance, have greatly improved the PTCs temperature stability and overall efficiency which is now comparable with Stirling coolers.

AL/DTA, in partnership with CEA/SBT and THALES has undertaken since several years the development of PTC technology either on its own funding or through ESA TRP.

A TRP for a Miniature Pulse Tube Cooler (MPTC) has been recently successfully completed [2, 3]. A U-shaped Engineering Model, with 21 cm transfer line between compressor and cold finger has been designed, manufactured and tested (thermal performances in different orientations and thermal environments) and preliminary qualified for space application (thermal and mechanical environment). Typical performances of this MPTC are reported in Table 2 and a picture is shown in Figure 5. The integration of this MPTC in future space missions is under consideration.

Performance	1 W @ 80 K
Input Power	35 W
Frequency	50 Hz
Mass	2.8 kg
Compressor OD	63 mm
Compressor length	157 mm
Cold finger OD	100 mm
Cold finger length	179 mm

Table 2: MPTC characteristics.

Figure 5: MPTC EM cooler.

The need for a larger cooling capacity PTC has been clearly identified, first for ground based commercial applications (i.e. superconducting filters for telecom) but also for future earth observation or other space applications. AL/DTA and its partners have already completed the development of such a large heat lift capacity PTC demonstrator in the frame work of a PhD work and with the support of French program [4]. The performances achieved with an in-line and a U-shape configuration are reported in Table 3 and a prototype picture of the U-shape is shown on Figure 6. A coaxial version is under industrialization at THALES for ground applications.

In –line Pulse Tube:
• 7,7 W @ 80 K
• 200 W input power
• 50 Hz
U-shape Pulse Tube:
• 5,5 W @ 80 K
• 189 W input power
• 50 Hz

Table 3: PTC characteristics.

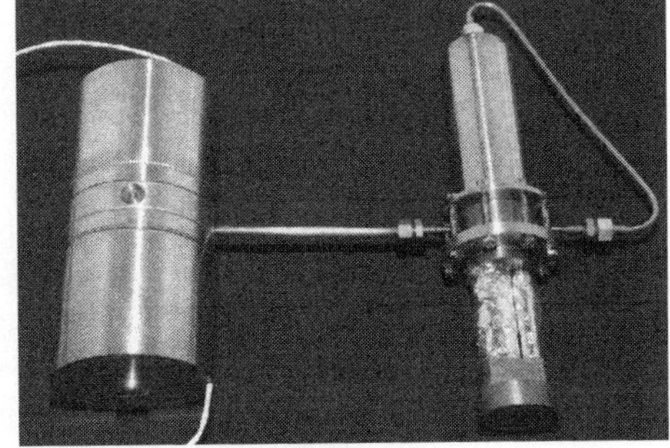

Figure 6: U-shape pulse tube cooler.

In the framework of the European SuperADC program, a two staged PTC is also under development with the aim to produce simultaneous cooling power at 80K and 20K-30K.

Preliminary results have been presented recently elsewhere [5]. It is also foreseen to develop a space version of this PTC in the near future.

3He / 4He Dilution Refrigerator

Planck is a European infrared space observatory dedicated to the mapping of the temperature anisotropies of the cosmic background radiation. AL/DTA and the Centre de Recherches à Trés Basses Températures du Centre National de Recherche Scientifique (CNRS/CRTBT) are involved since 1994 in the design and realization of a dilution refrigerator for the cooling from 4 K down to 0,1 K in microgravity environment of the bolometers array of HFI (one of the Planck mission instruments). Several teams in the world have tried to develop closed cycle dilution refrigerators for space operation. They have generally not been able to solve satisfactorily the phase separation in microgravity and the compressor operation. The CNRS/CRTBT has patented a unique design using 3He and 4He isotopes stored in high pressure vessels to feed a continuous mixing/separation process in a capillary pipe and rejecting the mixture (after a Joule Thomson expansion providing extra cooling at 1.6K) in space. This original process has been already demonstrated on ground against gravity (up side down cryostat) and in the Archeops stratospheric balloon mission. This proven technology is presently implemented by AL/DTA for Planck mission. Some specific features have been especially added for this mission: passive mechanical locking of the cooler during launch with specific memory shape alloy releasing the cooler locking during cooling in orbit, large heat capacity magnetic alloys for focal plane thermal stabilization. The qualification model performances [6] are summarized in Table 4 and a dilution fridge picture is shown in Figure 7. This qualification model is presently going through thermal and mechanical environment tests. The flight model is under assembly to be flow in 2007. A nominal continuous operation in orbit of 18 months is foreseen with a margin for 6 extra months.

Helium 3 flow	Helium 4 flow	Cold end temperature	Cooling power at the cold end	Temperature stability of the focal plane
6.3 µmol/s	21.8 µmol/s	79.9mK	No power applied	15 µK/Hz$^{1/2}$ at 0.02 Hz
6.3 µmol/s	17 µmol/s	101.9 mK	100nW	without large heat
7 µmol/s	19.5 µmol/s	92.55 mK	100nW	capacity magnetic alloy

Table 4: Dilution plate temperature versus ^{3}He and ^{4}He flows and cooling power

Figure 7: The Focal Plane Unit with the bolometer plate

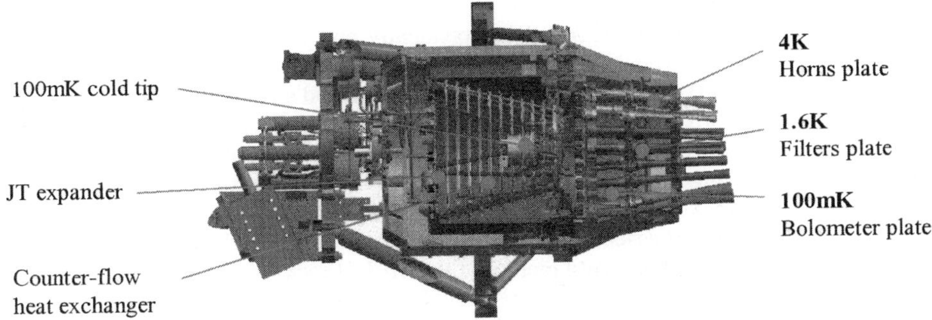

Figure 8. The Focal Plane Unit.

CRYOCOOLERS INTEGRATION

Thermal and mechanical integration of cryocoolers in payloads for space applications is often a complicated and cost effective problem. The best way to solve efficiently this problem is to have a strong collaboration from the initial phases of design between the cryocooler manufacturer and the payload architect. Aware of this important aspect, AL/DTA is not only involved in cryocoolers design and manufacturing but also intents to be involved in integration activities and to develop innovative and performing solutions. The following are some example of payload integration for the above described coolers.

MELFI freezer

As already mentioned, MELFI is a freezer to be operated in a laboratory on board the ISS for biological samples preservation (at –80 °C). The cooling is provided by a Brayton cycle refrigerator previously described. The Brayton cycle is a continuous flow cycle with dedicated components: compressor, compressor aftercooler, counter flow heat exchanger, expander, cold heat exchanger.

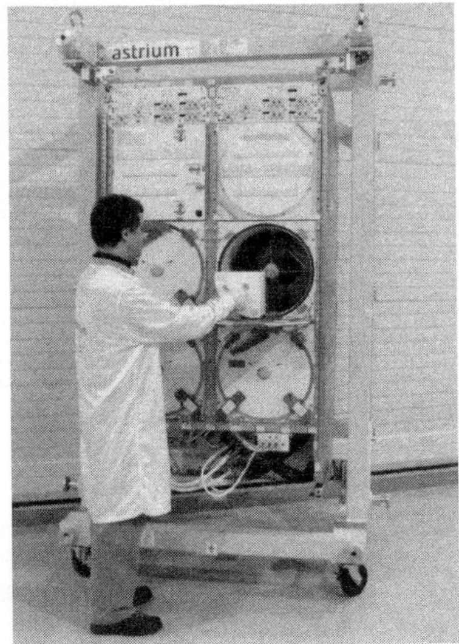

Figure 9. MELFI Rack (courtesy of Astrium).

434

These components can be integrated independently and connected with appropriate piping. This has allowed a particular integration architecture and manufacturing tasks sharing in MELFI project. AL/DTA has developed and manufactured a compact subsystem including all the cycle active parts (turbo compressor, motor and turbo expander on a common shaft) including the compressor aftercooler. The passive parts, the counter flow heat exchanger and the cold heat exchanger, have been directly integrated by the partner in charge of the MELFI dewars and their rack integration: a single compression-expansion unit is able to cool up to three dewars. Both sub assemblies are linked through quick-tight connectors, allowing for easy integration. Figure 9 shows a picture of the MELFI rack.

CRYOSYSTEM freezer

CRYOSYSTEM is also a freezer to be operated in the ISS for biological samples quick/snap freezing and preservation. A Stirling cooler previously described is used to cool a dewar including a rotating massive vials storage vessel. Strict specifications are required by the scientists such has minimal vials freezing time, storage vessel thermal inertia for minimal storage temperature increase during vials loading, capability to survive (minimal temperature) ISS current shut down,... AL/DTA has designed, manufactured and tested a Development Model of this specific dewar and developed simulation tools to reproduce all the potential scientist operations and ISS operation conditions and predict the storage temperature evolution. Thermal performance tests have been successfully performed and reported elsewhere [7]. Due to the challenging cooling performances and electrical consumption specifications and to mechanical integration issues, the thermal link between the Stirling cooler cold finger and the storage vessel has been identified as a key technological point of this project. To combine high thermal conductance with minimal mechanical load on the cold finger, a specific thermal link assembly (TLA) has been designed and tested [8]. Basically a heat pipe like operation is established in a small gap (few millimeters) between a thin walled well mechanically and thermally anchored to the storage vessel and the cooler cold finger: condensation occurs at cold tip surface and evaporation on storage vessel side. The gap is filled with a soft porous media and a heat transfer fluid (nitrogen or argon): the porous media allows for micro gravity operation and liquid draining of the fluid by capillary effect. Figure 10 shows a schematic of the TLA and of CRYOSYSTEM overall assembly.

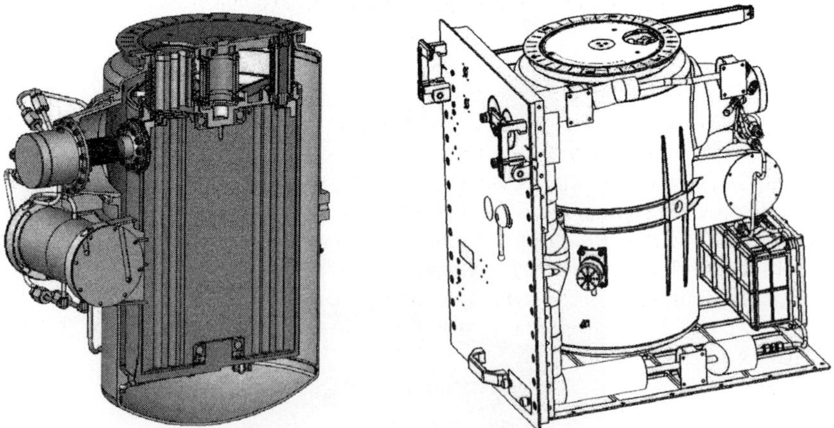

Figure 10: Views of the CRYOSYSTEM Dewar under design (PDR level).

PLANCK HFI Dilution refrigerator integration

The HFI instrument of the Planck mission cooling architecture is of a "cascade" type with four stages. From ambient temperature down to 100 mK, the following different cooling techniques are associated, each one being used as a precooling stage for the next one: a three stages groove type radiator (140 K, 80 K, 50 K), a 20 K Hydrogen sorption compression Joule Thomson cooler (developed by JPL), a 4 K mechanical compression 4He Joule Thomson cooler (developed by RAL) and finally the AL/DTA dilution refrigerator down to 100 mK. A part from the mechanical and thermal interfacing with the upper precooling stages, AL/DTA has faced a specific integration problem with the 1.6 K – 100 mK heat exchanger. This heat exchanger, the efficiency of which is crucial for the dilution refrigerator efficiency, is generally made for ground applications of a tube in tube technology with stainless steel low diameter pipes. To survive launch vibrations, a stiff structure with low thermal conductivity should be used to support this counter flow heat exchanger. NbTi pipes are used as stiffening rods due to their favorable mechanical resistance to thermal conductance ratio in this temperature range. Another point is the minimization of the parasitic conductive heat load transferred down to 100 mK by the bolometers reading electrical wires (64 twisted shielded pair wires) which has been obtained by a continuous thermal anchoring of the wires braid along the counter flow heat exchanger. The resulting hardware, shown on Figure 11, has been mechanically and thermally qualified, fulfilling all the expected specifications and allowing for a smart integration in the instrument with preserved cooling performances.

Figure 11: 0.1K-1.6K dilution heat-exchanger.

CONCLUSION

AL/DTA has put in concrete form its objective to develop an offer for cryocoolers design, manufacturing and integration for space applications. Several technologies, including Brayton, Joule Thomson, Stirling, Pulse Tube and Dilution refrigeration cycles, have been demonstrated. Some of these coolers have been specifically designed and developed to be integrated within ESA or NASA scientific missions (MELFI, CRYOSYSTEM, PLANCK/HFI) to be flown during the next coming years. Space qualification of these coolers has been already achieved or is going on. Development of miniature Brayton and large cooling capacity Pulse Tube coolers have also been initiated to prepare the next generation of AL/DTA space coolers.

REFERENCES

1   T. Trollier, A. Ravex, P. Crespi, J. Mullié, P. Bruins, T. Benschop "High capacity flexure bearing Stirling cryocooler on-board the ISS". Proceedings of the nineteenth International Cryogenic Engineering Conference, ICEC19, p381-384.

2   T. Trollier, A. Ravex, L. Duband, I. Charles, T. Benschop, J. Mullié, P. Bruins and M. Linder, "Miniature 50 to 80 K Pulse Tube Cooler for Space Applications". Cryocoolers 12, Monterey, Kluwer Academic Publisher/ Plenum Publishers, 2003, p 165-171.

3   T. Trollier, A. Ravex, L. Duband, I. Charles, T. Benschop, J. Mullié, P. Bruins and M. Linder, "Performance Test Results of a Miniature 50 to 80K Pulse Tube Cooler". Proceedings of the thirteenth International Cryocooler Conference, ICC13, to be published.

4   I. Charles, J. M. Duval, L. Duband, T. Trollier, A. Ravex and J.Y. Martin, "High Frequency large cooling power pulse tube". Proceedings of the nineteenth International Cryogenic Engineering Conference, ICEC19, p395-398.

5   J. M. Poncet, I. Charles, A. Gauthier, T. Trollier, "Low temperature high frequency pulse tube using precooling". Proceedings of the thirteenth International Cryocooler Conference, ICC13, to be published.

6   L. Sentis, J. Delmas, P. Camus, G. Guyot, Y. Blanc "Cryogenic tests of a 0.1K Dilution Cooler for PLANCK-HFI". Proceedings of the thirteenth International Cryocooler Conference, ICC13, to be published.

7   T. Trollier, C. Aubry, J. Butterworth, S. Martha, A. Seidel and L. De Parolis "Cryogenic tests of a Development Model for the 90 K freezer for the International Space Station". Proceedings of the thirteenth International Cryocooler Conference, ICC13, to be published.

8   T. Trollier, A. Ravex, C. Aubry, A. Seidel, H. Stephan, L. De Parolis, A. Sirbi, R. Kujala "Trade-Off between thermal link solutions for the CRYOSYSTEM cooler on-board the ISS". Proceedings of the Cryogenic Engineering – CEC, Anchorage 2003, to be published.

OVERVIEW OF THE AIR LIQUIDE CRYOGENIC GROUND SPACE ACTIVITIES FOR LAUNCHERS AND SATELLITES

S. Duval, P. Bravais, JC Courty, S. Crispel, G. Marot

Advanced Technology Division, Air Liquide, 38360 Sassenage, France

Air Liquide has been involved since the end of the seventies in Space activities for the development of cryogenic equipment for rocket launching pads (Kourou for Ariane 4 and 5) and ground test benches (for thrusters, satellites,..).

Recently, Air Liquide has been in charge of the upgrade of the cryogenic installations of the ELA3 launching pads in Kourou for the latest version of ARIANE 5 launcher. An overview of theses activities will be briefly presented.

The use of subcooled (densification) propellant allows to increase the capacity of a launcher. Air Liquide has provided a specific liquid oxygen subcooling for a Vinci thruster bench at SNECMA. Process design and test results will be detailed.

The use of Xenon electrical propulsion thruster is a recent development in satellites technologies. Air Liquide has developed an original mobile high pressure (up to 150 bar) cryogenic thermal compressor for the filling of Xenon tanks on board satellites. Air Liquide also developed large pumping capacity cryopumps for Xenon trapping in thrusters test chambers. This specific development will be presented.

ON-SITE INSTALLATIONS

Launching pads for the Ariane rockets (Kourou)

The main functions of cryogenic installations for launching sites are production, storage, distribution and treatment of fluids used on launchers and thruster for the propelling (LH2 and LOX), pressurisation (Ghe), purge, inerting and cooling down (LN2, Lhe). The following equipment was supplied and installed by Air Liquide :

- An LH2 cold box for production on site production (33000 l/day)
- Fixed or mobile storage tanks from $10m^3$ to $360m^3$ for LH2, LOX, LN2 and LH2
- 250 bar compression station of GN2
- vacuum insulated flexible or rigid lines and valves boxes for liquid or gaseous fluids (O_2, H_2, N_2, He)

For the latest upgrade of the pad complex n°3 to be integrated in the Cryo-technical Upper (CSS) of ARIANE 5, study, manufacturing and assembly of the following equipment were carried out over the last 2 years :

- Fixed or mobile storage tanks from $10m^3$ to $360m^3$ for LH2, LOX, LN2 and LH2
- 1200 m of vacuum insulated lines for liquid and gaseous O_2, H_2, N_2 and He (Internal diameter from ½" to 4")
- 22 skids and panels with some 120 valves (among which 110 are cryogenic ones)
- 30 m^3 storage tank for LN2
- Vacuum insulated flexible elements and supports dedicated to the so-called "cryogenic arms".

These Super Insulated flexible lines required some very specific development to comply with customer requirements presented above :

The main requirements are :

- high flexibility to limit the forces on the launcher interface (F<800 N)

- compatibility with kinematic of the arms
- compatibility with high temperature and aggressive chemical environment due to the booster (T>450°C)

Cryogenic arms

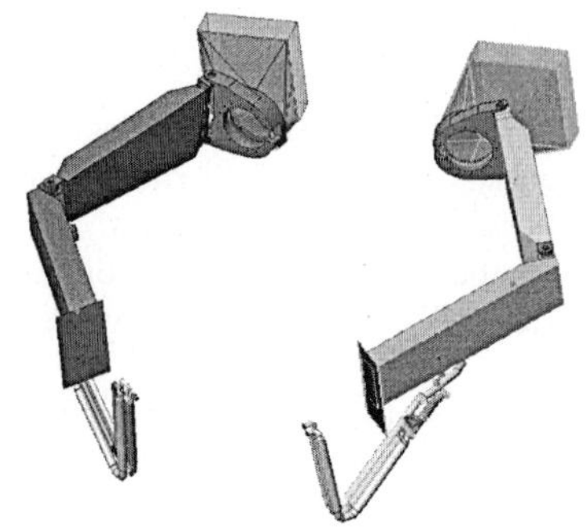

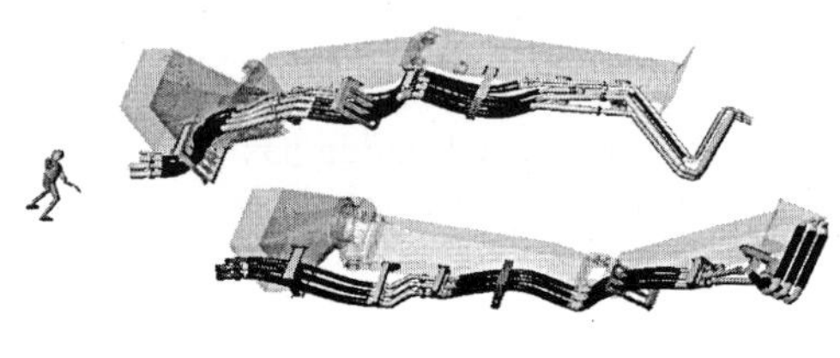

FIGURE 1 LO2 and LH2 arms

FIGURE 2 Cryogenic arm of Ariane 5

To achieve all the technical requirements, a first specific test campaign was carried out during the project in order to :
- measure the stiffness of double flexible envelop
- use these values for customed software modelisation to evaluate the forces on the launcher interface
- characterize the stiffness of manufactured vacuum insulated flexible pipes
- measure the resulting forces on the launcher equipment with the integrated overall system in a qualification step

A second test campaign test was carried out for validating the kinematic of the arms on a scale 1:1 dummy assembly.

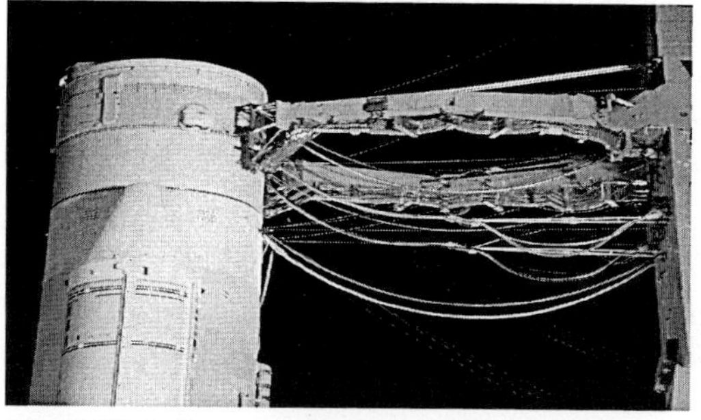

FIGURE 3 Cryotechnical Upper Stage arms

Specific covers were developed for protecting flexible elements against high temperature and aggressive chemical environment due to the booster. Some samples were tested during the launch of the Ariane 5 rocket.

All the requirements were fulfilled and validated during a final validation test campaign carried out by the customer for the first Ariane 5 Cryotechnical Upper stage.

TEST INSTALLATIONS

Cryogenic LH2/O$_2$ thruster test benches

Several Test benches are in operation in Europe for the qualification and the validation of Ariane motors. The main functions of cryogenic installations for test benches are similar to those for launch pads.

The Air Liquide last activities concern conception, fabrication and assembly of cryogenic sections to adjust test benches to the CSS requirements including also the limitation of resultant forces on thruster interface and specific equipments for filling tank storage with subcooled propellant.

The use of subcooled (densification) propellant allows to increase the capacity of a launcher. Air Liquide has designed and installed a specific liquid oxygen subcooling for Vinci motor test bench at SNECMA. A 20 m^3 storage tank is filled (1 to 2 l/s) at 88K with LO2 initially stored at 100K. The process design and test results are presented here after.

Development of a subcooler device for spatial application

In order to perform ignition tests of rocket motors in subcooled conditions, SNECMA needed a device which enables to fill O$_2$ tanks with a well controlled temperature fluid (T = 89 +0/-0.5K). As the liquid O$_2$ is delivered from mobile tankers at a temperature between 90 and 105 K, a process based on a by-pass principle was developed to produce the good conditions at the entry of the tank.

The figure 4 presents the process : the "hot O$_2$" (~2 kg/s) is separated in two flows. The main flow crosses a subcooler filled with liquid nitrogen at ambient pressure (T=77K) and the second one is by-passed. The two fluids are collected, cross a static mixer and are supplied to the tanker. The valve which controls the flow in the subcooler is regulated from the temperature measured just downstream the mixer. During the operation, the subcooler is fed continuously from a LN$_2$ mobile tank through a valve controlled by a level measurement.

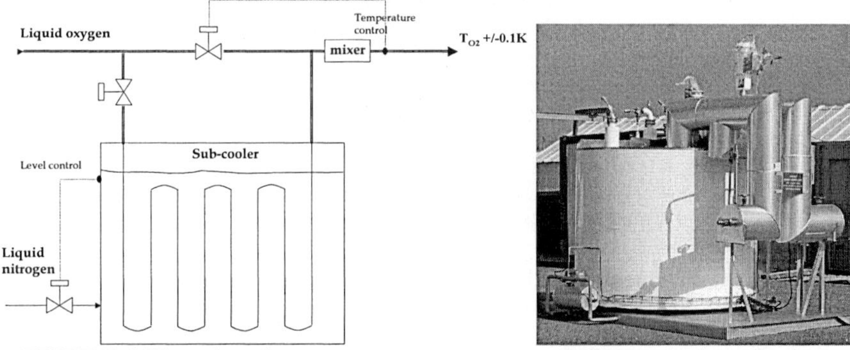

FIGURE 4 The subcooler process   FIGURE 5 The subooler

During the filling phase, the temperature of the O$_2$ flow was controlled with a 0.1 K precision. The device enables to produce subcooled O$_2$ liquid in the range [83-90K].
As the process is installed on a mobile skid, and the connection with the tank is flexible, it could be used in different locations on the customer's site.

Simulation chambers

For Satellites tests, it is required to create space environment in simulation chambers. The Air Liquide is concerned about the supply of turn key cryogenic installations including equipments as cryopumps (from DN250 to DN1250), thermal panels (30KW at80 K) and thermal generators (4 kW at 100 K GN2), all compatible with 100 cleanliness class.

FIGURE 6 Thermal panels

<u>Satellite instruments test bench</u>
For the purpose of simulating the low temperature environment of IASI (meteorological and environmental satellite instrument) during ground tests, Air Liquide delivered to Alcatel Space two cryogenic loops:
- the first one, under operation at Alcatel Space Cannes, has a cooling capacity of 20 W @ 20 K,
- the second one, under operation at ESTEC, has a cooling capacity of 20 W @ 30 K
Each loop is based on the following principle:
- One cold box providing a gaseous helium flow of 4 g/s, and the refrigeration power (1 or 2 cold heads)
- A test panel, cooled by the helium flow, compatible with the cleanliness and vacuum requirements of space test chambers. It provides unique radiation capabilities at low temperature (absorbtivity better than 0,95 @ 20 K), thanks to the pyramid like shapes jointly developed by Alcatel Space Cannes and Air Liquide.

FIGURE 7 View of the test panel under mounting, courtesy of Alcatel Space Cannes

- Low temperature flexible transfer lines between the cold box and the test panel, with heat inputs lower than 0.1 W/m on the helium flow.
Operational advantages of this solution are: no liquid helium consumption, storage and waste, easy management of the test ("on-off" operation, limited training) and limited staff.

<u>Xenon Cryopumping of the satellite motors testing benches</u>
The xenon motors which are mainly designed to correct the position of satellites must be subjected to endurance testing before integration. These tests are performed in a vacuum chamber after cryopumping of the xenon gas ejected by the thruster. Air Liquide has studied and delivered two testing chambers with xenon cryupumps (dia 2,20 m) for 1300 W motors:at at CNRS Orléans and SNECMA Villaroche.

FIGURE 8 The Xenon motor

The following results were obtained
- Xe flow: $m(Xe) = 5{,}4$ mg / s
- Voltage supply: $U = 300$ V
- Output current : $Id = 4{,}2$ A
- Average electrical power :$Pel = 1300$W
- Total Xe pressure with motor ON:
  $P(Xe) = 2{,}5.10\text{-}5$ mb

In 2003, Air Liquide has studied an extension for these casings to be produced in 2005 for 6000 W motors.
The following performances must be obtained :
- Xe flow: $m(Xe) = 20$ mg / s
- Average electrical power: $Pel = 6000$ W
- Total Xe pressure with motor ON: $P(Xe) = 2{,}5.10\text{-}5$ mbar

Thermal Analysis of Cryogenic Rocket Engine with One, Two and Three Dimensional Approaches

Kuzhiveli B.T.[*], Ghosh S.C., Kuruvila G.K. and Gandhi V.G.

Liquid Propulsion Systems Centre, Indian Space Research Organisation,
Valiamala, Trivandrum, India - 695 547
[*]Presently at DCI, 112 Nila, Techno Park, Trivandrum, India - 695 581, Email: btkuzhiveli@yahoo.com

Prediction of heat transfer characteristics in a regenerative cooled cryogenic rocket engine, which uses liquid oxygen (LOX) and liquid hydrogen (LH2) as propellants provides one of the major inputs while designing a rocket engine. The objective of this paper is to present thermal analysis with one, two and three-dimensional approaches. Details of heat transfer analysis program that can predict temperature distribution on the thrust chamber, hot side and coolant side heat transfer coefficients, coolant temperature rise and heat flux have been presented. The thermal analysis program thus developed can be well suited for parametric studies to recommend stable operation during actual hot test conditions.

INTRODUCTION

Modern cryogenic rocket engine thrust chambers are exposed to high pressure and high heat flux environments due to high energy combustion. This has presented some of the most challenging engineering problems due to the presence of high heat flux and high pressure. These extreme conditions are further complicated by the fact that engine should achieve a bare minimum weight and deliver a high specific impulse. A commonly employed method to overcome the problem due to high temperature arising out of high energy combustion is by providing regenerative cooling by passing LH2 through channel passages in the thrust chamber wall.

The design methodology of rocket engine heavily relies on an accurate prediction of temperature and pressure profile on the engine. The physical phenomenon involved in the operation of cryo engine is complicated. The flow in the engine is turbulent, reactive and undergoes both subsonic and supersonic flow regimes. As the hot test of engineering hardware can be prohibitively expensive, numerical simulations of the conditions are necessary. Hence it was essential to develop a program, which could accurately predict the thermal characteristics of the thrust chamber.

COMPUTATIONAL MODEL

To make the analysis computationally traceable it is assumed that the flow through the channel is circumferentially symmetric. As a result, 360°/Number of channel becomes the area of attention. This simplification implies that only half rib and half channel have to be analyzed. The approach holds good if the flow in different channel are the same. If not, it represents the performance of an average channel and assumes that the departure from the performance of an average channel is not significant.

With this in hand, there are three possible ways of having thermal analysis. i.e., one dimensional, two dimensional and three dimensional. In the one dimensional approach, the two possibilities could be, to consider the fin effect or not to consider the fin effect. In both the cases, it is assumed that the heat from hot gas is passed to the coolant channel directly. The major draw back of this method is that conduction factor through the ribs and the outer shell is neglected and it is further assumed that heat passes directly into the cooling channel through a single point entry. It also does not take into account the radiation effects from the hot gas as well as the ambient in an iterative manner. In addition to this, the axial conductivity between the marching sections is also not considered.

In the two dimensional approach, the independent cross sectional domain (x-y direction) is divided into many grids and the grid meeting points are considered as nodes. Each node is connected to the adjacent node by a connecting line. An energy balance is applied to each node, which results in an algebraic equation for the temperature at the node. Separate equations are derived for each node in the control volume. The drawback here is that the axial conduction (in z direction) is neglected.

For the three dimensional approach, the control volume becomes a solid space with thrust chamber cross sectional area as the x-y co-ordinate and the marching distance as the z co-ordinate. The heat transfer model takes into consideration the following five physical aspects (i) heat transfer from combustion gas to the inner wall of the thrust chamber (including convection and radiation effects) (ii) heat transfer through the inner wall, ribs and outer wall through grids (incorporating varying thermal conductivity effects) (iii) heat transfer from the walls and ribs to the coolant through the grids(incorporating varying heat transfer coefficient effects) (iv) heat transfer from the ambient to the outer wall (or vice versa, includes convection and radiation effects) (v) heat transfer between the sections (in the axial direction). Electrical analog of three-dimensional computational model is depicted in Figure1.

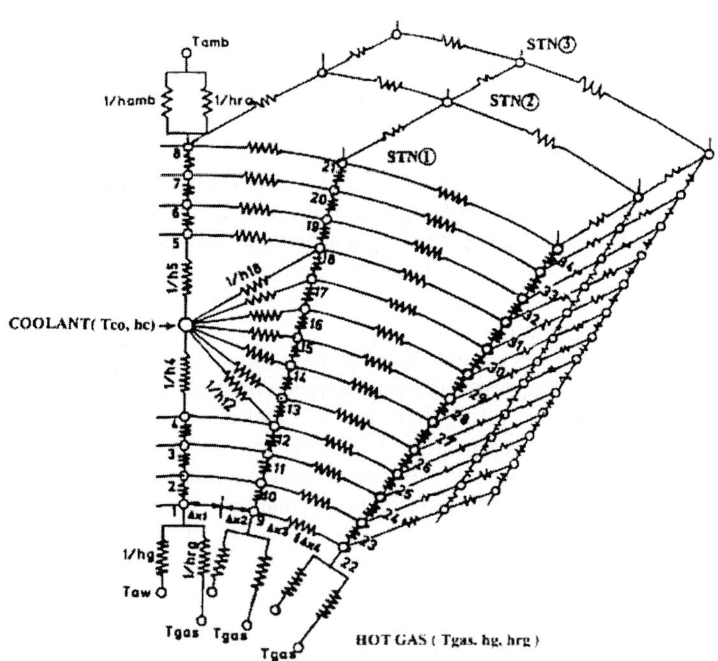

Figure 1 Electrical analog of three dimensional computational model

CALCULATION PROCEDURE

In order to determine the performance of a rocket engine for a given propellant combination and chamber pressure, the composition of the combustion products is assumed to be in chemical equilibrium. The theory presented by Gordan et al [1] is utilised to determine the combustion gas properties.

Throughout the axial length of the engine from the injector end to divergent end, all the combustion gas properties are obtained for a given chamber pressure and mixture ratio. A forward marching method is adopted for finding out the temperature distribution, coolant temperature, heat transfer coefficients, heat flux, and the bulk coolant temperature rise.

Within the thrust chamber, both convection and radiation contribute to the total heat transferred. Before the gas can transfer heat to the wall of the thrust chamber, the heat energy passes through a layer of stagnant gas along the boundary layer. Hydrogen enters the channels in liquid/vapor condition and is in contact with the inner part of the walls and ribs. Hydrogen absorbs the heat energy and thereby raising the bulk coolant temperature.

Heat transfer coefficient hg is predicted with the help of simplified Bartz equation [2]. The coolant heat transfer co-efficient can be estimated by Sieder Tates equation [3].

One dimensional analysis (1D)

In one dimensional analysis, the heat flux has been equated as shown in the equation below.

$$q1 = hg1 \, (Taw1 - Twg1) = (Twg1-Twc1)/(L/kA) = hc1(Twc1- Tco1)$$

Where q - heat flux, hg - heat transfer coefficient of hot gas, Taw - adiabatic wall temperature, Twg - temperature on the hot gas side wall, Twc - temperature on the coolant side wall, L - heat passage length, k - conductivity of the material, A - area.

To find out the wall temperature (Twg), an initial value for Twg is assumed and hg, Taw, Twc, hc are found out. Value of Twg is increased step by step till the equation for q1 satisfies the above condition. The subscript 1 denotes for one-dimensional analysis.

The increase in coolant temperature at each axial location is calculated by Tco1_next = Tco1 + q1 / (mc.Cpc) where mc and Cpc are mass of coolant (hydrogen) and specific heat of coolant respectively.

Two dimensional analysis (2D)

Heat transfer analysis of regeneratively cooled thrust chamber has been carried out using two dimensional finite difference method. The computational domain consists of half the channel width and half the rib width because of axi-symmetry and its repetitive nature. Each subdivided area has a length Δxn in the 'x' direction and length Δyn in 'y' direction. Separate equations are derived for each node in the control volume.

On the hot gas side wall, convective and radiative boundary conditions are imposed. Convective boundary conditions are also imposed on the coolant side. Similarly on the outer skin of the chamber, convective and radiative boundary conditions are imposed. For steady state conditions, an energy balance is applied, as an example, at node 2 (refer Figure 1)

$$\sum_{i=1}^{i=4} qi = 0$$

i.e.; q1-2 + q3-2 + q10'-2 + q10-2 = 0

Applying Fourier's law to each of these terms to express the equation in terms nodal temperatures,

$$q_{1-2} = k_{1-2} \frac{L(\Delta xn + \Delta x^{(-)}n) \, (T1-T2)}{2.\Delta yn} \qquad q_{3-2} = k_{3-2} \frac{L(\Delta xn + \Delta x^{(-)}n) \, (T3-T2)}{2.\Delta yn}$$

$$q_{10'-2} = k_{10'-2} \frac{L(\Delta yn + \Delta yn) \, (T10'-T2)}{2.\Delta xn} \qquad q_{10-2} = k_{10-2} \frac{L(\Delta yn + \Delta yn) \, (T10-T2)}{2.\Delta xn}$$

Three dimensional analysis (3D)

For steady state conditions, an energy balance is applied, as an example, at node 2 (refer Figure 1);

444

$$\sum_{i=1}^{i=6} qi = 0$$

i.e.; q2'-2 + q2''-2+ q1-2 + q3-2 + q10-2 + q10'-2 = 0

$$q_{2'\text{-}2} = k_{2'\text{-}2}.L_z \frac{(\underline{\Delta xn} + \Delta x^{(-)}n)(\Delta yn + \Delta yn)(T2'\text{-} T2)}{2*2*(\Delta z^{(-)}n) + \Delta zn)/2} \qquad q_{2''\text{-}2} = k_{2''\text{-}2}.L_z \frac{(\underline{\Delta xn} + \Delta x^{(-)}n)(\Delta yn + \Delta yn)(T2''\text{-} T2)}{2*2*(\Delta z^{(-)}n) + \Delta zn)/2}$$

$$q_{1\text{-}2} = k_{1\text{-}2}.L_z \frac{(\underline{\Delta xn} + \Delta x^{(-)}n)(\Delta z^{(-)}n) + \Delta zn)(T1\text{-}T2)}{2*2*(\Delta z^{(-)}n) + \Delta zn)/2} \qquad q_{3\text{-}2} = k_{3\text{-}2}.L_z \frac{(\underline{\Delta xn} + \Delta x^{(-)}n)(\Delta z^{(-)}n) + \Delta zn)(T3\text{-} T2)}{2*2*(\Delta x^{(-)}n/2)}$$

$$q_{10'\text{-}2} = k_{10'\text{-}2}.L_z \frac{(\underline{\Delta yn} + \Delta yn)\ (\Delta z^{(-)}n) + \Delta zn)(T10'\text{-} T2)}{2*2*\Delta yn/2} \qquad q_{10\text{-}2} = k_{10\text{-}2}.L_z \frac{(\underline{\Delta yn} + \Delta yn)\ (\Delta z^{(-)}n) + \Delta zn)(T10\text{-} T2)}{2*2*(\Delta xn/2)}$$

The result of the finite difference method is 'N' algebraic equations for N nodes and these equations replace the single partial differential equations and applicable boundary conditions. For each node, the temperature is calculated with changing properties of the material used, in an iterative manner. A computer program has been written down to include computational models of 1D, 2D and 3D analysis.

RESULTS

The results obtained from the three dimensional analysis has been compared with the results of the two dimensional as well as the one dimensional analysis. Temperature profiles for all marching distances (1mm in critical area) over the axial length are found out using the above methods. Inner wall temperatures predicted with 1D, 2D, 3D methods were as follows; at the beginning of the thrust chamber, the predicted inner wall temperatures were 616 K, 519 K, 514 K respectively. Highest temperature observed near the throat was 692 K, 589 K, and 583 K respectively. At the hydrogen inlet (divergent end) these were 606 K, 501 K, and 496 K respectively. The temperatures at 34 points on the cross section (refer Figure 1), T1 to T34 has been found out. As an example, at throat section, the temperatures at points 1 to 34 were 583K, 553 K, 525K, 501K, 335K, 358K, 360 K, 360 K, 582K, 552K, 521K, 489K, 451K, 421K, 397K, 378K, 364K, 353 K, 359 K, 360 K, 360 K, 583 K, 553 K, 523 K, 493 K, 464 K, 438 K, 417 K, 400 K, 388 K, 383 K, 362 K, 360 K, 360 K.

The temperatures predicted by 3D analysis were much lower than that predicted by 1D analysis and slightly lower than the 2D analysis. This difference existed because 3D model takes into account all heat transfer aspects. The experimental measurement of bulk coolant temperature was made to evaluate the suitability of 3D analysis. The bulk coolant temperature measured at exit of the channel was 247.0 K against the predicted value of 258 K (by 3D analysis). A small variation (about 4%) in the prediction could be attributed to the film cooling effect, which was not incorporated in the analysis. Thus the suitability of 3D program has been validated. The program thus developed is well suited for design of regeneratively cooled engines as well as a tool to check the safe operation of the engine during hot test conditions.

ACKNOWLEDGEMENT

Authors greatly acknowledge Dr. G. Madhavan Nair, the chairman of ISRO and former director of Liquid Propulsion Systems Centre for having given constant encouragement for this work The help of Dr. P. Balachandran, Mathew George, S. Sreekumar, A. P. Baiju is also appreciated.

REFERENCES

1. Gordan S., Mcbride B.J., Computer Program for Calculation of Complex Chemical Equilibrium Compositions, Rocket Performance, Incident and Reflected Shocks and Chapman - Jouguet Detonations, NASA SP-273, Washington (1973)
2. Bartz, D.R., A Simple Equation for Rapid Estimation of Heat Transfer Coefficients in Nozzles, Jet Propulsion (1957) 27
3. Philip G. Hill, Carl R. Peterson, Mechanics and Thermodynamics of Propulsion, AIAA Progress in Astronautics and Aeronautics - 147, Addison-Wesley, New York (1992)

Liquid hydrogen droplet generation from a vibrating orifice

Jinquan Xu [a], Dogan Celik [b], M. Yousuff Hussaini [a], Steven W. Van Sciver [b]

[a] School of Computational Science and Information Technology, Florida State University, Tallahassee,
   Florida 32306, U.S.A.
[b] National High Magnetic Field Laboratory, Florida State University, Tallahassee, Florida 32310, U.S.A.

Numerical studies of liquid hydrogen droplet generation in vapor helium from a vibrating orifice are conducted within the framework of incompressible Navier-Stokes equations for multiphase flow. The goal of the study is to unravel the basic mechanisms of droplet formation, so that shape and size can be controlled. The numerical model is validated by experiments, and the numerically predicted droplet shapes show satisfactory agreement with the experimental photographs of droplets. Parametric studies on the influence of injection velocity, nozzle vibration frequency and amplitude on the droplet shape are conducted. The model enables one to obtain a definitive qualitative picture of droplet shape evolution associated with the operating parameters, and may be used for quantitative analysis. In fact, the model could be used to aid the design of future experiments.

INTRODUCTION

One of the advanced propulsion systems for the next generation of space flights is a densified fuel consisting of atomic propellants, such as boron, carbon, or hydrogen [1]. Solid hydrogen particles are preferred for storing atomic propellants in these systems because they significantly reduce the fuel volume and weight in a launch vehicle, and, most of all, have the ability to stabilize and prevent the atoms from recombining. Atomic hydrogen propellant feed systems may require the production of hydrogen particles in cryogenic helium. Experimental and theoretical efforts are underway to investigate the formation of solid hydrogen particles in cryogenic helium. To create solid hydrogen particles in helium, droplets of liquid hydrogen at temperatures around 19 K are injected from a vibrating orifice into a dewar filled with vapor helium at 4 K. Apart from the physics of hydrogen droplet solidification, the shape and size of the ejected droplets and the way to control and optimize them are of particular interest. They pose a formidable task due to very small time scales and other experimental design challenges [2]. However, numerical studies offer a relatively easy, but powerful alternative to study their behavior.

The goal of the numerical simulations (within the framework of the incompressible Navier-Stokes equations for multiphase flow) is to unravel the basic mechanisms of droplet formation with a view towards controlling them. The numerical model is validated by comparing numerical results with experimental ones. Numerically predicted droplet shapes show satisfactory agreement with photographs of experimental droplets. Parametric studies on the influence of injection velocity, nozzle vibration frequency, and nozzle vibration amplitude on droplet shapes are conducted using the model developed.

PROBLEM FORMULATION AND SOLUTION TECHNIQUE

Formation of hydrogen particles in a helium bath involves a complex energy exchange process between hydrogen and the ambient helium, resulting in a reduction of hydrogen temperature and its consequent solidification [3, 4]. As a first step in modeling the process, however, we assume the system is isothermal, eliminating energy exchange between the phases as well as excluding hydrogen or helium phase changes. Also, for the liquid-hydrogen/vapor-helium system, we assume that the two phases are immiscible and non-interpenetrating. A volume-of-fluid two-phase fluid model [5], which solves for the volume fraction of each phase within each control volume, along with a piecewise-linear interface calculation approach [6], is implemented to track interfaces evolving in time between the two phases. A continuum surface force (CSF) formulation [7] is implemented to account for the surface tension along the interface between the two phases. A single momentum equation, with the surface tension as a source term, is solved throughout the computational domain. Dynamic meshing techniques [8, 9] are employed to simulate vibration of the generator orifice.

The computational domain in the study consists of two regions: a vibrating liquid hydrogen nozzle and a bath of vapor helium. Initially, the nozzle is filled with liquid hydrogen, and the helium bath is filled with helium vapor. Both fluids are initially at rest. At computation time zero, liquid hydrogen in the orifice undergoes a velocity jump from zero to a specified velocity. Simultaneously, the orifice vibrates with a velocity governed by a sinusoidal function. No-slip boundary conditions are applied on the walls of the computational domain. For computational purposes, (i.e., easy management of the dynamic mesh), the governing equation systems are solved in a general integral form using a finite volume method [10].

EXPERIMENTAL VALIDATION

An experimental setup, which consists of a vibrating orifice type generator in a cryogenic environment (shown in Fig. 1) and high-speed photography, is established to validate the numerical model. In the experiment, a vacuum jacketed liquid hydrogen reservoir is separated from the helium vapor filled environment by an orifice plate. The helium bath is a dewar with the inner diameter of 0.15 m and the height of 1.22 m. The length of the vibrating orifice is 0.15 m, and its inner diameter d is 1.0 mm. The orifice is attached to a mechanical oscillator vibrating sinusoidally with frequency 13 Hz and the amplitude 2.0 mm. The hydrogen injection rate through the orifice is regulated by use of a pressure gauge that controls the pressure difference at the liquid/vapor interface. The rate was estimated to be about 1.0 m/s based on measurements on the experimentally taken photographs.

For the sake of model validation, parameters in the numerical computation are matched with the experimental operating conditions to our best knowledge, except that a small computational domain is used. The size of the computational domain is smaller than that of the helium bath for reasons of computational expediency. For computational purpose, the resolution involved an unstructured grid with initial 12,225 triangular elements. The dimensions of the computational domain and the relevant physical properties of the fluids are listed in Table 1. Also, instead of simulating a pendulously vibrated orifice that is equipped in the experiment, the entire nozzle is numerically vibrated transversely following a sinusoidal velocity.

Table 1 Simulation parameters

Vapor helium bath	Length (m)	0.04
	Width (m)	0.04
Liquid hydrogen	Length (m)	0.02
nozzle	Width (m)	0.001
Vapor helium [11]	Density (kg/m^3)	18.17
	Viscosity (kg/m·s)	1.276×10^{-6}
Liquid hydrogen [12]	Density (kg/m^3)	70.84
	Viscosity (kg/m·s)	1.392×10^{-5}
	Surface tension (J/m^2)	2.00×10^{-3}

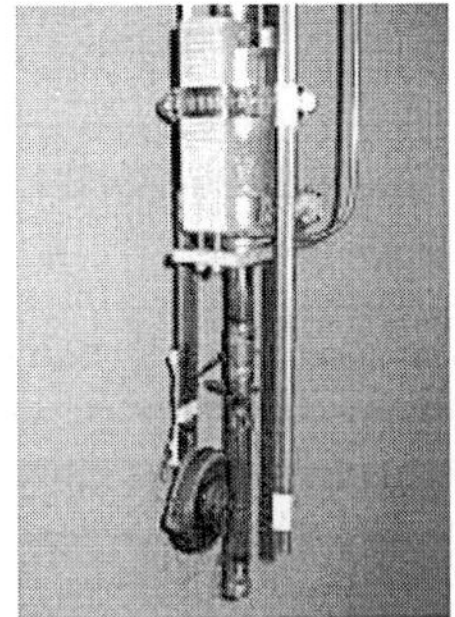

Fig. 1 Vibrating orifice

Figures 2(a)-2(d) show the evolution of a droplet in the experiment. They each are double exposure images, where two individual frames separated by 10 ms are superimposed. In Figure 2(d), there is a second droplet ejected from the orifice following the first ones. Double exposure renders four objects in the figure. Diameters of the droplets in the figures are estimated to be between 1.5 and 2.0 mm.

Correspondingly, Figures 3(a)-3(d) show simulated droplet shapes at various elapsed times, which correspond to the individual experimentally taken photograph in Figure 2. The droplets inside the big circles in Figures 3(a)-3(d) are close-ups of the numerically predicted droplets for clarity of presentation.

The diameter of the simulated droplet in Figure 3(d) is about 1.5-2.0 mm, which roughly agrees with the experimental result shown in Figure 2(d).

Notice that when operating the experiments, unlike running the simulations, it is very difficult to precisely control the orifice location or monitor its velocity under current experimental settings. Thus, the numerical simulation may not duplicate the experimental operating conditions as well as we would like. Nevertheless, droplet shapes that the model predicts catch the characteristic transients of droplet formation, and they are in relatively good agreement with experimental photographs.

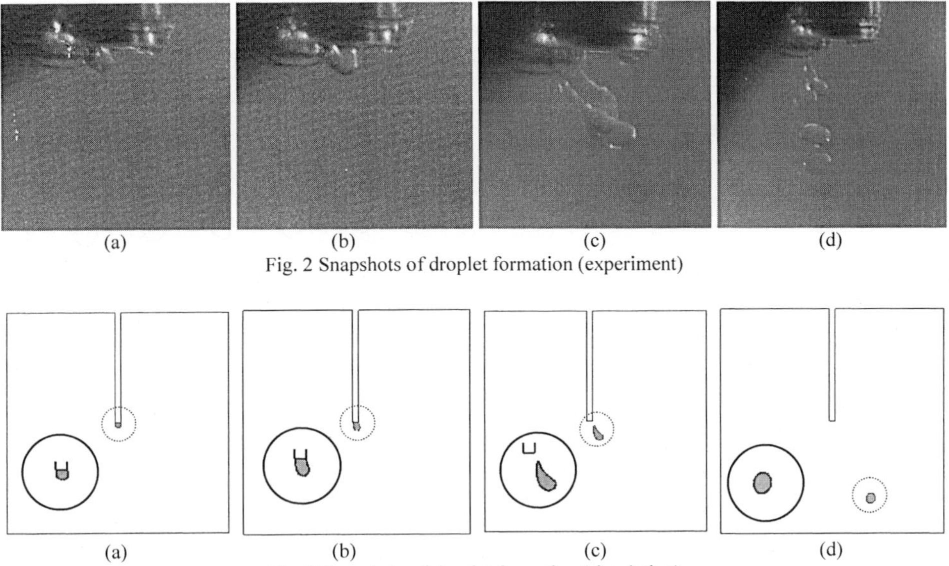

(a)   (b)   (c)   (d)

Fig. 2 Snapshots of droplet formation (experiment)

(a)   (b)   (c)   (d)

Fig. 3 Snapshots of droplet formation (simulation)

PARAMETRIC STUDIES

When liquid hydrogen is injected into vapor helium from a vibrating orifice, the liquid deforms, and, if the deformation is sufficiently large, the liquid breaks into droplets. Parametric studies on the influence of nozzle vibration frequency ω and amplitude A, and hydrogen injection velocity $\bar{v}$ on the droplet deformation are conducted. The studied cases, along with their operating parameters, and the corresponding droplet shape evolutions at different elapsed times are listed in Table 2. Only droplets at their early stages, before detaching from the orifice, are presented in the studies, due to the fact that initial droplet shape is important for completely characterizing drop dynamics in the flow, and it also plays an influential role in determining the final size and shape of a droplet.

Cases 1, 2 and 3 in Table 2 study the influence of orifice vibrating frequencies on the droplet shape; Cases 1, 4 and 5 study the influence of orifice amplitudes; and Cases 1, 6 and 7 study the influence of hydrogen injection velocities. Systematic analysis of the droplet shapes in Table 2 through Weber numbers and orifice-vibrating velocities show that the degree of droplet deformation is largely governed by surface tension stresses, inertial forces and the gravitational force. The orifice vibrating amplitude and frequency influence droplet shape by way of instant vibrating velocities, which determine the inertial force in the vibrating direction and the deformation of the droplet in that direction. The hydrogen injection velocity governs the droplet size considerably and determines the inertial force in the vertical direction. The frequency and amplitude of the vibrations and the injection velocity have to be carefully tuned to obtain a droplet of desired shape and size.

SUMMARY AND CONCLUSIONS

Numerical modeling of liquid hydrogen droplet generation from a vibrating orifice is developed and validated. Parametric studies on the influences of operational conditions on the droplet shape and size are conducted. It has been demonstrated that the computational model enables one to obtain a definitive qualitative picture of droplet shape evolution associated with the operating parameters. The computational model may also be used for quantitative analysis or parametric studies, and, in fact, the model could help design future experiments.

Table 2 Droplet shape evolution

Case	ω (s^{-1})	A (mm)	$\bar{v}$ (m/s)	10 snapshots of initial stages of droplet formation and detachment
1	50	1.6	1.0	
2	5	1.6	1.0	
3	500	1.6	1.0	
4	50	0.32	1.0	
5	50	3.2	1.0	
6	50	1.6	2.0	
7	50	1.6	0.5	

ACKNOWLEDGEMENT

This work was supported by the NASA Hydrogen Research program for Florida Universities. (Contract No. NAG3-2751) and the Office of the Provost at the Florida State University.

REFERENCE

1. Palaszewski, B, Launch vehicle performance with solid particle feed systems for atomic propellants. AIAA-1998-3736
2. Palaszewski, B, Solid hydrogen experiments for atomic propellants: Image analyses. AIAA-2001-3233
3. Xu, J., Rouelle, A., Smith, K.M. Celik, D., Hussaini, M.Y. and Van Sciver, S.W. Two phase flow of solid hydrogen particles and liquid helium. Cryogenics. (2004) 44
4. Xu, J., Smith, K.M. Celik, D., Hussaini, M.Y. and Van Sciver, S.W. Hydrogen particles in liquid helium. Cryogenics. (2004) 44
5. Gueyffier, D., Li, J., Nadim, A. Scardovelli, R. and Zaleski S., Volume-of-fluid interface tracking with smoothed surface stress methods for three-dimensional flows. J. Comput. Phys. (1999) 152 423-456
6. Young, D.L., Time-dependent multi-material flow with large fluid distortion. in: Numerical Methods for Fluid Dynamics. Edited by K.W. Morton and M. J. Baines. Academic Press Inc., New York (1982)
7. Brackbill, J.U., Kothe, D.B., and Zemach, C., A continuum method for modeling surface tension. J. Comput. Phys. (1992) 100 335-354
8. Zhao, Y. and Forhad, A, A general method for simulation of fluid flows with moving and compliant boundaries on unstructured grids. Comput. Methods Appl. Mech. Engrg. (2003) 192 4439-4466
9. Degand, C. and Farhat, C, A three-dimensional torsional spring analogy method for unstructured dynamic meshes. Computers and Structures (2002) 80 305-316
10. Patankar, S.V., Numerical Heat Transfer and Fluid Flow. Hemisphere Publishing Corporation, (1980)
11. Van Sciver, S.W., Helium Cryogenics, Plenum Press, New York, (1986)
12. Souers, P.C., Hydrogen Properties for Fusion Energy. University of California Press (1986)

The Working Performance of Radiant Cooler in Thermal Vacuum Test

Xu Hongyan, Dong Deping, Wang Weiyang, Li Zhong

Shanghai Institute of Technical Physics, Shanghai 200083, China

In this paper a mathematical module is achieved to predict the working performance of radiant cooler in thermal vacuum test. The temperature and the emissivity of the cold shield, two influencing factors are discussed. The working performance of a parabolic cylinder radiant cooler on different thermal vacuum test conditions is given in picture.

INTRODUCTION

It is well known that the radiant cooler is a kind of mini-cooler used in space, with the advantages of light weight, no moving parts, long life and negligible power consumption etc. The radiant coolers have been developed for application to the cooling of infrared detector, optical systems, components, and other devices in spacecraft. Cryogenic radiator temperature can, of course, only be obtained in orbit, which provides sufficiently low sink temperature. The low temperature heat sink provided by a space environment is about 4K and the emissivity is about 1. The test set-up does not totally simulate the on-orbit configuration or environment. To predict the real working performance of radiant cooler in space by test result, the analytical thermal model is established to match the test configuration [1].

The working performance of radiant cooler can be predicated in thermal vacuum test on earth. The thermal vacuum test chamber is a cylinder in certain diameter. A pressure of 5×10^{-5} Pa is maintained inside the thermal vacuum chamber in order to make the convective transfer effects insignificant. The temperature of inter-well of cylinder is as high as that of liquid nitrogen. There is a big piece of black honeycomb board, called cold shield, covering the mouse of radiant cooler. The cold shield could be cooled to lower than 20K by a refrigeration system based on gaseous helium. The side gap between cold shield and cooler is covered with black pained aluminum plate, which connected with cold shield.

THERMAL ANALYSIS

A thermal model is established to analysis heat transfer between the radiant cooler and the cold shield during the thermal vacuum test. The thermal model included a detailed representation of the radiator components as well as appropriate surrounding chamber surface. The radiant cooler discussed here has two stages. The radiant cooler mainly consists of housing, earth shield, first stage radiator, second stage patch and support system. The sketch of the radiant cooler during thermal vacuum test is given in Figure 1.

A closed cavity is made up with the inner surface of the earth shield, the first stage radiator, the second stage patch, the first reflecting screen, the front radiator surface of housing, the cold shield and side covers. In the thermal model, the following assumption is made: The earth shield and the first reflecting screen are low emissivity highly specular surfaces. The other surfaces are diffuse radiation. All surfaces are grey-body radiation.

The thermal balancing temperature can be reached by sending and receiving heat flows in thermal vacuum test. For instance, the heat loading to the patch includes radiant heat from parabolic cylinder reflector and the inner surface of first stage, the conductive heat from support and electric cable, and simulating heat of optical loading. To balance the heat loading, the patch gets cold from the cold shield by

450

radiation. The temperature of the patch and the first radiator can be determined by the energy balance equations [2]. Two equations define the heat flows to the locations of interest:

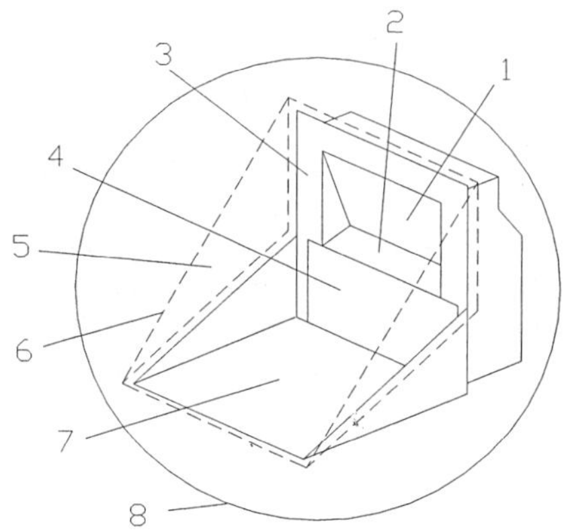

1.First reflecting screen
2.Patch
3.Front white radiator
4.First radiator
5.Side cold shield
6.Cold shield
7.Earth shield
8.Thermal vacuum test chamber

Figure.1 Schematic plan of radiant cooler in thermal vacuum chamber

The first radiator:

$$\sigma A_{one} E_{one-cold}(T_{one}^4 - T_{cold}^4) + \sigma A_{one} E_{one-side}(T_{one}^4 - T_{side}^4) + \sigma A_{one} E_{one-two}(T_{one}^4 - T_{two}^4)$$
$$= \sigma A_{one} E_{one-shield}(T_{shield}^4 - T_{one}^4) + \sigma A_{one} E_{one-screen}(T_{screen}^4 - T_{one}^4) + \sigma A_{one} E_{one-front}(T_{front}^4 - T_{one}^4) \tag{1}$$
$$- F_{m2}(T_{one}^4 - T_{two}^4) + F_{m1}(T_{out}^4 - T_{one}^4) - F_{c2}(T_{one} - T_{two}) + F_{c1}(T_{out} - T_{one}) + Q_{one}$$

The patch:

$$\sigma A_{two} E_{two-cold}(T_{two}^4 - T_{cold}^4) + \sigma A_{two} E_{two-side}(T_{two}^4 - T_{side}^4)$$
$$= \sigma A_{two} E_{two-shield}(T_{shield}^4 - T_{two}^4) + \sigma A_{two} E_{two-screen}(T_{screen}^4 - T_{two}^4) + \sigma A_{two} E_{two-front}(T_{front}^4 - T_{two}^4) \tag{2}$$
$$+ \sigma A_{two} E_{two-one}(T_{one}^4 - T_{two}^4) + F_{m2}(T_{one}^4 - T_{two}^4) + F_{c2}(T_{one} - T_{two}) + Q_{two}$$

The terminology for equations is given below.
Variables：

σ: Stefan-Boltzmann constant, A: Area, E_{x-y}: The radiation transfer coupling coefficient between surface of X and Y, T: Temperature, F_{m1}: Coefficient of radiation transfer between the inner first radiator and inner housing with multiplayer insulation. F_{m2}: Coefficient of radiation transfer between the inner first radiator and inner patch with multiplayer insulation. F_{c1}: Coefficient of conduction transfer between the first radiator and housing through support and wire. F_{m2}: Coefficient of radiation transfer between the first radiator and the patch through support and wire. Q: Constant heat leak

Subscripts：
One: the first radiator, Two: the patch, Cold: the cold shield, Shield: the earth shield, Side: the cover between the cold shield and radiator cooler, Screen: the first reflecting screen, Front: the front radiator surface of housing, Out: the housing.

The radiation exchange factors are calculated with the program utilizing a Monte-Carlo ray trace technique in this paper. A simulating program is compiled to calculate the equilibrium temperature of the radiator cooler by inputting environment parameters of test.

RESULTS

In general, there are two methods to indicate the working performance of radiator cooler in thermal vacuum test. One is the useful cooling capacity, that is, the thermal load when the patch works in the scheduled temperature. The other is the lowest temperature of the patch can be reached while the thermal load is scheduled. In this paper, the second method is used.

In the simulating program, it is supposed that the temperature of the housing is constant. The lowest temperature of the patch depends on many factors, such as the temperature of cold shield, the emissivity of related surfaces and the position of the cold shield. The temperature and the emissivity of cold shield are different according to various test conditions. The equilibrium temperature of the patch is regarded as baseline on the thermal test condition, which the temperature of the cold shield is 4K and the emissivity is 1. On other test conditions, the working performances of radiator cooler are indicated by the temperature difference to baseline.

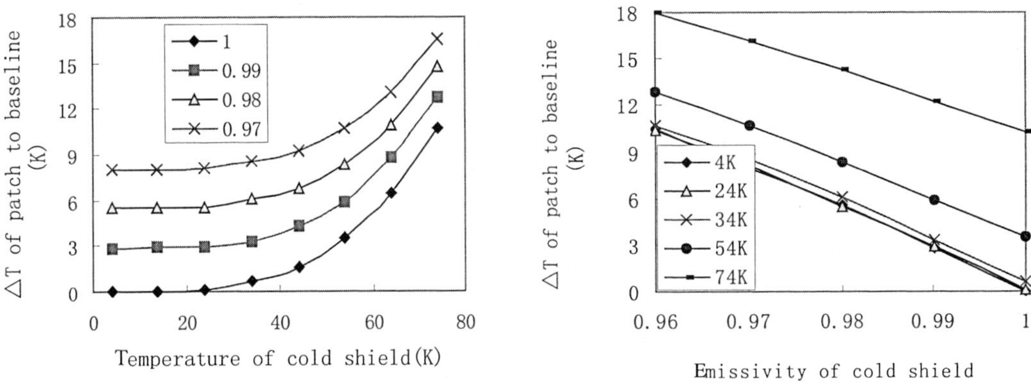

Figure2 Working performance of radiator cooler

The working performance of a kind of parabolic cylinder radiant cooler has been calculated. It is supposed that the housing is 240K in the test process. The cooling capacity is 0.04W. The cold shield temperature varies from 4K to 84K with 10K step length, at the same time its emissivity varies from 1 to 0.96 with 0.01 step length. The effects of these changes in temperature and emissivity of the cold shield on the patch temperature are shown in Figure2.

A smooth curve is drawn through the data points with the same emissivity of the cold shield. There is a little difference in temperature when the cold shield is lower than 24K. The temperature difference increases gradually and greatly with the increasing of the cold shield temperature. For example, as the emissivity is 0.99, the calculated values are 2.78K, 2.89K, 2.92K and 4.26K while the temperature of the cold shield is 4K, 14K, 24K, and 44K respectively. The temperature difference trend of the patch on the temperature of the cold shield is obvious.

A smooth curve is drawn through the data points with the same temperature of cold shield. It can be seen that the working performance of radiator cooler reduces with lowering the cold shield emissivity. The working performance changes linearly with emissivity. A 0.01 change in the emissivity of cold shield results in a change of approximately 2K in patch temperature. The rate of slope is different at different testing condition. The effection of emissivity on the working performance is reducing with raising the cold shield temperature.

To correlating temperature and emissivity of the cold shield, the map of the working performance is obtained, which shown in Figure3. Given the temperature and emissivity of the cold shield in test, the temperature of the patch is determined. For instance, the thermal vacuum test of this kind of parabolic cylinder radiant cooler is conducted. The emissivity of the cold shield is 0.98. The thermal vacuum test is conducted in two stages. In first stage: the cold shield is cooled by a refrigeration system based on liquid Nitrogen to 78K. The equilibrium temperature of patch is 106.1K. In second stage: the cold shield is cooled by a refrigeration system based on gaseous helium to 17K. The equilibrium temperature of patch is

93.7K. The observed temperature difference between those two test conditions is 12.4K. The predicted temperature difference between these two test conditions by the simulating program is 11.9K. The error percentage is within 5%. There is a good agreement between the predictions and the observed test temperature differences. This confirms the validity of the mathematical model.

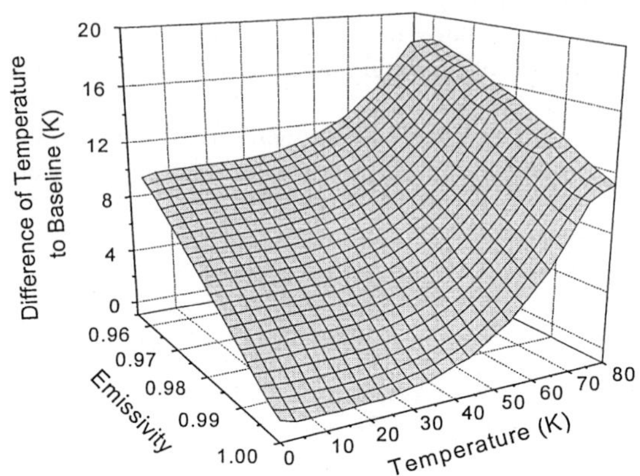

Figure3 Map of working performance

CONCLUSION

The influences of temperature and emissivity of the cold shield to the working performance of radiant cooler in thermal vacuum test have been discussed. It is important to improve the emissivity of the cold shield as it has great influence on the test result. There is little difference in test result when the temperature of the cold shield is lower than 24K. It is well enough that the cold shield is cooled by a refrigeration system based on gaseous helium to about 17K in the thermal vacuum test.

REFERENCE

1. Wang, G. And Han, J., Optimization Design of the Cold-Black Target for Thermal Simulation Test of Space Radiant Cooler, Cryogenics in China (1993) 1 7-13
2. Robert, S. and John, R.H. Thermal Radiation Heat Transfer Hemisphere and McGraw-Hill (1981) 564

A Study on the Conic Support in the Radiant Cooler

Fu Liying, Tu Fenghua, Wang Weiyang

Shanghai Institute of Technical Physics, Chinese Academy of Sciences, Shanghai 200083, P.R, China

We have investigated the conic support in the radiant cooler by using finite element method. Results show that the stiffness of the support increases with an increasing of the coning angle in case of launched state, while the situation is reversed in the orbited state. On the other hand, it is also found that the heat loss of the support caused by the thermal conduction decreases with the reduction of the coning angle. Moreover, the optimized coning angle is provided as a function of the length.

INTRUDUCTION

The cryogenic cooler plays an important role in the infrared remote sensing technology. Up to now, the cryogenic cooler widely used in space mainly includes the radiant cooler and the Stirling cooler. Note that the radiant cooler, emitting heat to space to reach the low temperature, has attracted considerable interests, because it exhibits higher stability, long life, no consuming power, no electromagnetic radiation and no moving parts etc.

In the past forty years, several structures of the radiant cooler have been developed, which include W-shaped, L-shaped, conic style and parabolic style [1]. No matter what structure of the radiant cooler is, it is usually organized as the cooler house, the first stage radiator, the second stage radiator and the structural support [2]. Among these the structural support is the crucial mechanical element, which connects the cooler house with the first stage radiator and brings the first stage radiator and the second radiator together. As a result, it should has the excellent rigidity to meet the demands of mechanics. Meanwhile the heat loss of the structural support, results from the temperature difference of the different side, must be as little as possible for insuring the refrigerating capacity of the radiant cooler.

With the development of the cryogenic technology, the radiant cooler should provide lower temperature and more refrigerating output, which certainly will lead to the size of the radiant cooler become larger. Additionally, the stability of the support can be determined by several factors, like its size and shape etc. In order to better develop the radiant cooler along with the infrared remote sensing technology, a detailed consideration for the support is required. In this contribution, we have investigated a conic support that widely used in the radiant cooler by using finite element method. We present a path to improve the performance of the conic support by optimizing its coning angle and height, which can amend the tradeoff between the mechanical stiffness and the heat loss caused by the thermal conduction.

METHOD

All considerations presented below were performed by finite element method (FEM). The FEM is a

454

numerical technique for solving problems, which are described by partial differential equations or can be formulated as functional minimization. A domain of interest is repented as an assembly of finite elements. Approximating functions in finite elements are determined in terms of nodal values of physical field, which is sought. A continuous physical problem is transformed into a discretized finite element problem with certain nodal values [3]. It is one of the most important methods to solve the continuous medium problem and it can adapt to the complex terminal condition. FEM has been widely used in many kinds of fields, like mechanics, thermodynamics and electromagnetic etc.

Here, a short summary of the calculational conditions is given. Figure 1 displays a sketch of the prototypical conic support considered, which is made from the titanium alloy. In Figure 1, α is the coning angle; T is the thickness of the support, given as 0.0005m; ϕ_1 is the upper diameter of the support, given as 0.02m; L is the length of the support varied from 0.05m to 0.08m; ϕ_3 is the bottom inner diameter of the support ranged from 0.025m to 0.05m, which aims to change the coning angle of the support. The magnitude of ϕ_2 is expressed as $\phi_3 + 0.01$m. The thickness of the upper and the bottom are fixed at a constant (0.001m).

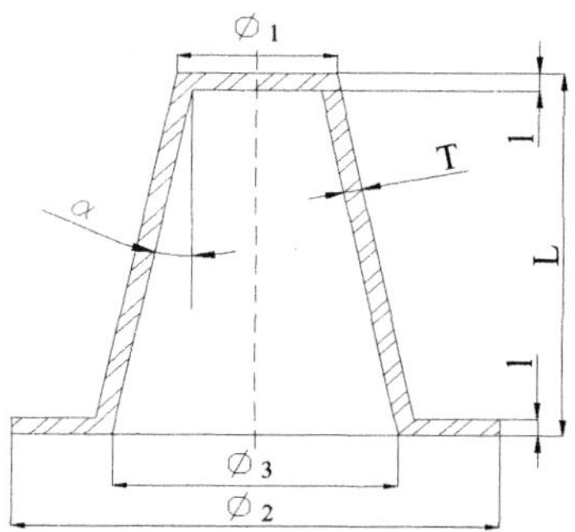

Figure 1 Scheme of the Conic Support

Other necessary parameters are also given. The coefficient of the thermal conductivity is 4.5W/mK, Young's modulus is 100GPa, the thermal expansion coefficient is 1.5e-5K^{-1} and Poisson's ratio is 0.3. Since the temperature-dependence (from 100K to 150K) of the parameters can be neglected, they are adapted as constants for simplifying the computation.

The heat loss of the conic support has two sources. One is the radiant heat loss, which is very little and can be neglected for the conic support locates within the radiant cooler; the other is the heat loss caused by the thermal conduction, which can be written as the following,

$$q = \frac{(\phi_2 - \phi_1)\lambda T \pi}{L \ln\left(\phi_2 / \phi_1\right)}(T_2 - T_1) \tag{1}$$

where λ is the coefficient of thermal conductivity, T_1 is the temperature of the upper surface of the support, T_2 is the temperature of the bottom surface of the support, T, L, ϕ_1 and ϕ_2 are defined above.

In calculations, two kinds of load are considered in order to simulate the support under different load. One is the load in launched state (replaced by LL) with the pressure 1.25e6Pa and the shear force 3.56N applied to the upper surface of the support; the other is the load in orbited state (replaced by LO) with the temperature 100K and 150K applied to upper and bottom surface of the support, respectively. The bottom surface of the support is restrained under both LL and LO.

Based on the information mentioned above, the solid model is established firstly, and then is meshed to get the discretized equations, which are solved by a computer.

RESULTS AND DISCUSSION

As a starting point in the analysis of the conic support, we study the interrelation between its coning angle and its corresponding stiffness. It is well known that the stiffness can be reduced by its maximal deformation (MD). In generally, the smaller the MD is, the better its corresponding stiffness is. Figure 2 shows the MD of the conic support under the LL as a function of the coning angle. It is clear that the MD for all cases decreases with the increasing of the coning angle, whereas the condition is reversed for the MD under the LO, as shown in Figure 3. For a certain coning angle, the larger the length (L) is, the greater the MD is, which is similar to Figure 3. It's suggested that a rather short conic support with a larger coning angle is required under the LL, while a shorter conic support with a smaller coning angle is more suitable to the LO.

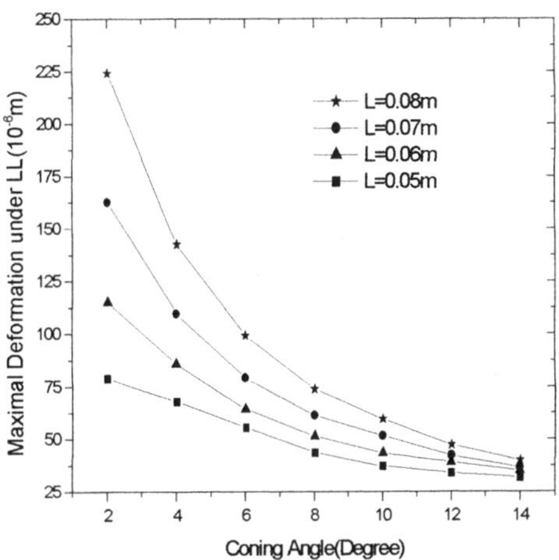

Figure 2 The MD under LL as a function
of the coning angle with various length

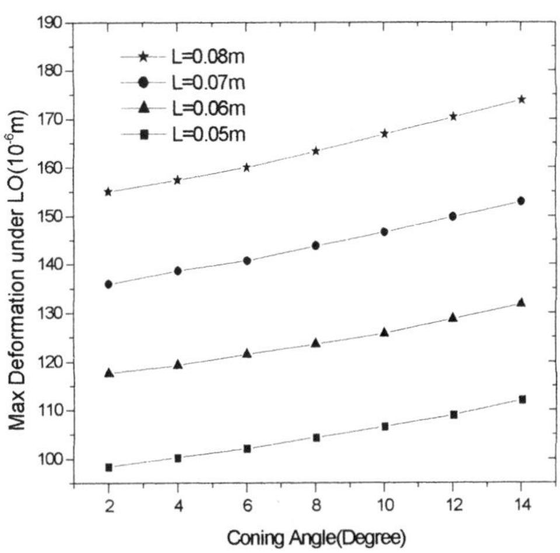

Figure 3 The MD under LO as a function
of the coning angle with various length

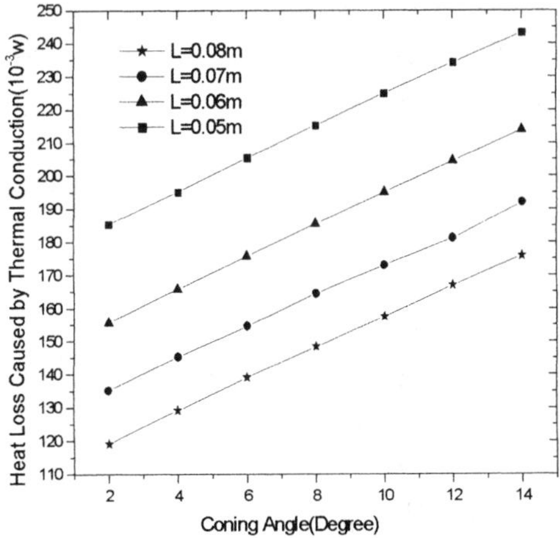

Figure 4 The heat loss caused by the thermal
conduction as a function of the coning angle
with various length

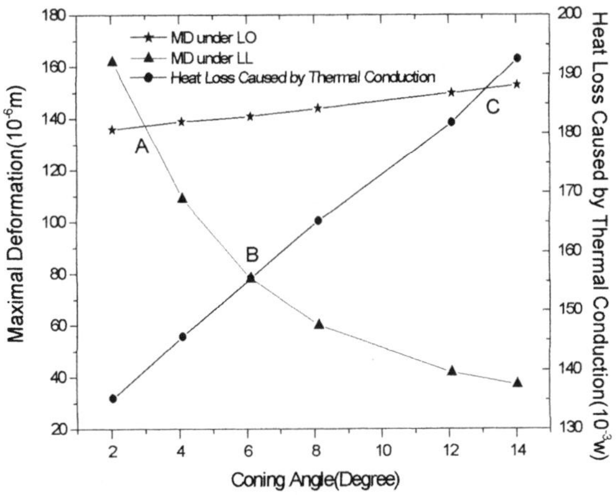

Figure 5 Comparison of MD under LL and LO and the heat
loss caused by the thermal conduction as a function of the
coning angle with the length being0.07m

Next we focus on the heat loss caused by the thermal conduction. Figure 4 predicts the heat loss caused by the thermal conduction as a function of the coning angle. We found that the heat loss caused by the thermal conduction rises mono-linearly with the increasing of the coning angle. The heat loss caused by the thermal conduction decreases with the length increasing at the same coning degree. Hence, to depress the heat loss caused by the thermal conduction, it is necessary to adopt a way that lengthens the conic support and reduces the coning angle.

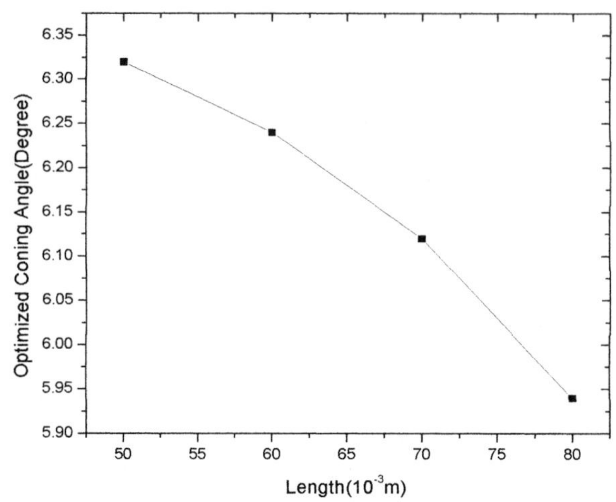

Figure 6 The optimized coning angle as a function of length

However, there is a tradeoff among above obtained results. Taking the length of the conic support being 0.07m for instance, as shown in Figure 5. In Figure 5, it seems that there is no a certain coning angle that can fulfill the requirements of both mechanical and thermal performance at the same time. But note that the MD difference between point A and point C is very small, we can conclude that point B is an approximatively optimized value for the coning degree. According to this, taking varied length into account, an optimized coning angle as a function of the length can be obtained, as shown in Figure 6.

CONCLUSION

In summary, based on finite element method, we have studied the conic support in the radiant cooler. By optimizing the coning angle and the length of the support, we can amend the tradeoff between its mechanical and thermal performance, which presents a possible pass to better develop the radiant cooler.

REFERENCES

1. Zhu H.M, Xu L., Sun H. and Xiao Y.M., Application of thermal stress Analysis in the design of radiant cooler support system, ICCR'2003, 335-338
2. Li Z.P, Wang W.Y. and Quan H.Y., Design and Calculation of Suspension Support System of Radiant Cooler for Space Applications,ICEC17, refrigeration, section6, 327-330
3. Reddy J.N., In: An introduction to the finite element method, McGraw-Hin, New York, USA (1933) 1-10

Analysis of the influences of view angle on paraboloid W style radiant cooler's working performance

Wang X.J., Pan Y.P.[*], He Y.L., Chen C.Z.

State Key Laboratory of Multiphase Flow in Power Engineering, School of Energy & Power Engineering, Xi'an Jiaotong University, 28 Xianning West Road, Xi'an 710049, P.R.China

* National Key Lab. of Vacuum & Cryogenics Technology and Physics, Lanzhou Institute of Physics, 105 Weiyun Road, Lanzhou 730000, P.R.China

In this paper, the performance of a paraboloid W style radiant cooler was numerically simulated under different view angles. By analyzing the data of calculating results, it is found that the temperature of cold stage declined linearly when the view angle decreased, while the fabrication and the inner heat-exchange of radiant cooler are very complicated. Furthermore, the negative affects can be reduced by changing the temperature-controlling mode of ambient housing.

INTRODUCTION

Radiant cooler is one of the cooling methods for the infrared remote sensing systems on spacecrafts. It has advantages of no vibration, no noise, only a very little power consumption, and high reliability, etc. As a passive cooling device, the cooling capacity of radiant cooler is small and the requirements to spacecraft orbit and mounted position are strict [1]. This limited the applications of the radiant cooler. So the study of factors influencing working performance of radiant cooler has become hot spot in radiant cooler research fields.

When a radiant cooler was designed, following considerations should be taken into account: 1) enlarging the view angle of patch to cold space, 2) shielding most input of the external heat flux, 3) decreasing the leakage of thermal flow between different stages [2]. However, in most cases the view angle of radiant coolers applied in spacecrafts can't reach the perfect maximum because of the limitation of spacecraft structure, satellite orbit, and the mounted position. In this paper, the view angle influencing on the cooler performance has been numerically studied.

MODE AND METHODS OF CALCULATION

The paraboloid W style radiant cooler is selected as the calculation mode in this paper. Figure 1 is the physical mode of this radiant cooler [3]. Figure 2 shows two schemes of cold patch and first stage radiator when the horizontal view angle was decreased. In the scheme that the first stage radiator is folded the angle of γ, the areas of first stage radiator and cold patch are bigger. For the reason that the area is very important to the working performance of radiant cooler, this scheme is selected to satisfy the varied view angle in paraboloid W style radiant cooler.

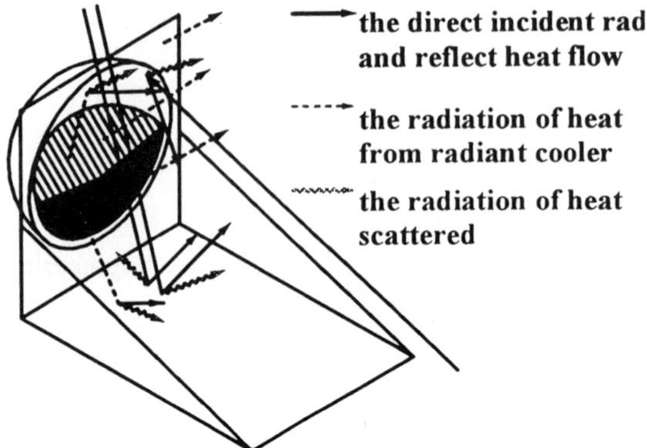

Figure 1　Thermal physical mode of paraboloid W style radiant cooler

Figure 2　Scheme of radiator and cold patch

Based on the thermal physical mode and analysis of the input of external thermal flow, the thermal balance equations for the ambient housing, the first stage, and the second stage were respectively formulated.

By analyzing these thermal balance equations, the fact that only $E_{d\text{-}sho}$, $E_{p\text{-}co}$, A_d, A_p, $E_{d\text{-}sh}$, and $E_{p\text{-}c}$ change with the view angle was found, $E_{d\text{-}sho}$: exchange factor between first stage and open mouth of shield, $E_{p\text{-}co}$: exchange factor between patch and open mouth of cone, A_d: area of first stage, A_p: area of patch, $E_{d\text{-}sh}$: exchange factor between first stage and shield, $F_{p\text{-}c}$: exchange factor between patch and cone.

The calculated results of A_p and A_d are listed in Table 1.

Table 1　Value of A_p and A_d to different horizontal view angle

Horizontal view angle γ	180°	170°	160°	150°	140°
A_p (cm^2)	476.5	437.2	400.1	365.9	335.9
A_d (cm^2)	722.8	691.8	665.4	641.9	621.7

Because the inner heat-exchange system of radiant cooler includes paraboloid revolution surface, the exchange factor can not calculated by a simple function. The theoretic calculation is carried out by the Monte Carlo method. The Monte Carlo method is a numerical method based on the statistical principle, and has been widely used in many research fields.

ANALYSIS OF CALCULATED RESULTS

The working performance of radiant cooler was calculated in two different conditions. One status is the temperature of ambient housing is controlled together with the infrared detecting system. In this case, the temperature of ambient housing is 292 K. Another status is the temperature of ambient housing is controlled separately and the temperature of ambient housing is 270 K. There are three kinds of heat load loaded on cold patch. They are no loads, 60 mW and 100 mW.

The numerical calculation of working performance of paraboloid W style radiant cooler was carried out and several curves were plotted according to the calculated results.

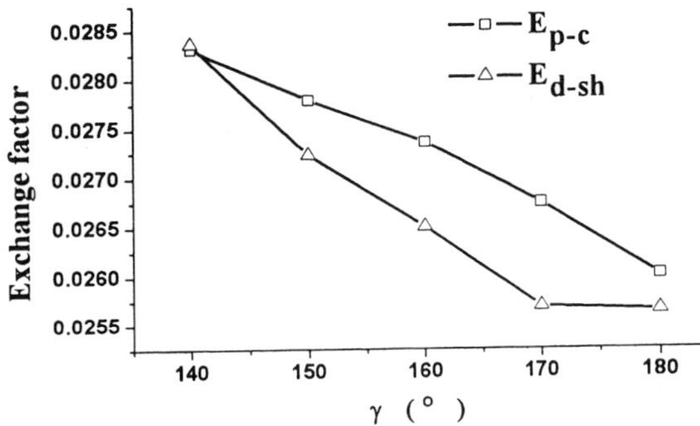

Figure 3　Exchange factors among surfaces dependence on view angle

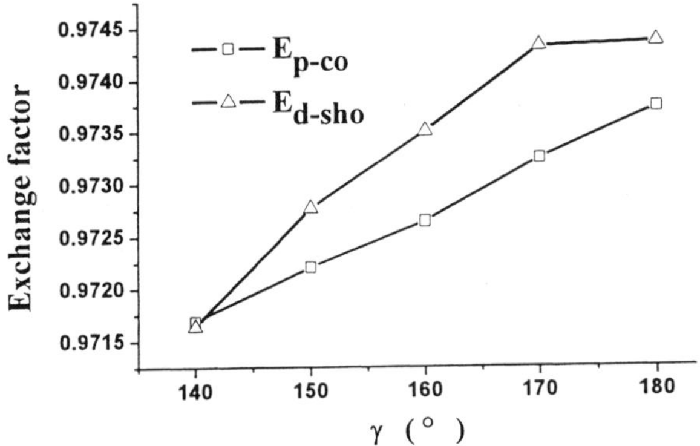

Figure 4　Exchange factors to cold space dependence on view angle

The curves of exchange factors changing with view angle are showed in Fig 3 and Fig 4. E_{d-sh} and E_{d-sho} change remarkably because not only the area of first stage radiator is decreased, but also the position to the shield of earth changes simultaneously, when the horizontal view angle γ varies from 180° to 140°.

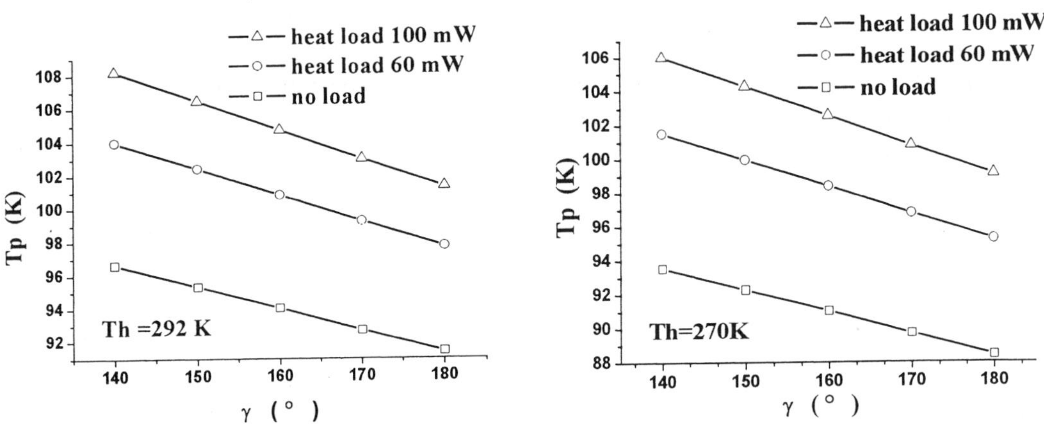

Figure 5　Temperature of second stage dependence on view angle

Figures 5 show that the temperature of second stage T_p change linearly with the horizontal view angle γ. For every 10° γ decrease, T_p increased by 1.27 K with no loads, by 1.57 K with 60 mW heat load, and by 1.72 K with 100 mW heat load. The heat load of cold patch also influences the sensitivity of second stage temperature changing with the view angle. According to the calculated results, when the horizontal view angle γ varies from 180° to 140° and the temperature of ambient housing is controlled at 270 K, the temperature of second stage T_p increases by 5.13 K with no load. But in the same conditions, T_p increases by 6.88 K with 100 mW heat load. The condition of controlled temperature of ambient housing has little effect on the sensitivity of second stage temperature changing.

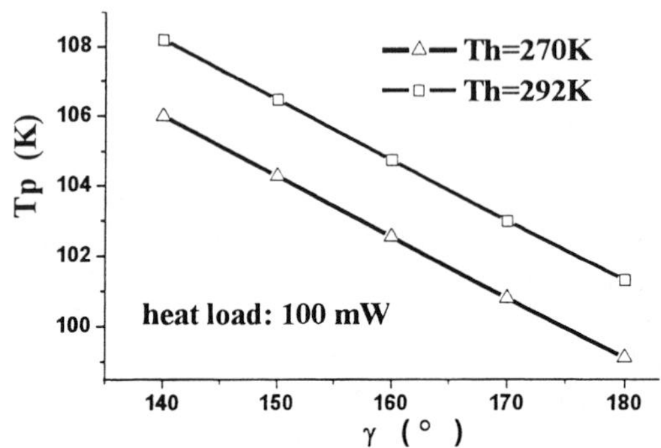

Figure 7   Temperature of second stage under different controlled temperature of ambient housing dependence on view angle

The curve in Fig 7 shows the relation between the temperature of second stage T_p and the view angle in different conditions of controlled temperature of ambient housing. By analyzing this curve, it is found that the radiant cooler can reach the working performance needed by selecting different method to control the temperature of ambient housing when the view angle decreases.

CONCLUSION

When radiant cooler is applied on a spacecraft, in most cases the view angle can not reach the maximum because of some limited conditions. The change of view angle influences the working performance of radiant remarkably. Though the structure and inner heat-exchange of paraboloid W style radiant cooler are very complicated, the fact that the temperature of cold patch changed linearly with the variety of the horizontal view angle was found. This conclusion is very valuable in the thermal and structural design of radiant cooler. Furthermore, changing the temperature-controlling method of ambient housing can reduce this kind of negative affects caused by the decreased view angle.

ACKNOWLEDGMENT

The project was financially supported by the National Key of Fundamental R&D of China (No.G2000026303) and National Natural Science Foundation of China (No. 50276046).

REFERENCE

1.   Da Daoan, In: Space Cryogenic Technology, Space Navigation Press, Beijing, PRC (1991) 29-44

2.   Han Jun, The theory and application of space radiant cooling, Vacuum and Cryogenics (1982), 2 44-49

3.   Lu Yan, The optimal design of radiant cooler in orbit, Vacuum and Cryogenics (1998), 3 165-169

Design and manufacturing of cryosorption pumps for testbeds of ITER relevant neutral beam injectors

Dremel M., Day C., Mack A., Haas H., Hauer V., Speth E.[*], Falter H.D.[*], Riedl R.[*], Gravil B.[**]

Forschungszentrum Karlsruhe, FZK, ITP, PO Box 3640, 76021 Karlsruhe, Germany
[*]Max-Planck-Institut für Plasmaphysik, IPP, 85748 Garching, Germany
[**]CEA / Cadarache, DRFC / STEP, 13108 Saint-Paul lès Durance, France

Special cryosorption pumps based on the adsorption with activated charcoal, coated onto cryopanels in quilted design have been developed and manufactured at Forschungszentrum Karlsruhe (FZK). The cryopanels have been integrated in the 1:2 scaled ITER torus cryopump installed at the FZK testbed TIMO (Test Facility for ITER Model Pump) to determine the pumping properties. This paper describes the design of large cryosorption pumps that are used to pump high gasloads of hydrogen and deuterium needed for Neutral Beam Injectors (NBI). The calculations of the heatloads are pointed out and based on the heatloads the thermo hydraulic behaviours have been examined.

DESIGN OF THE CRYOSORPTION PUMPS FOR THE IPP TESTBED MANITU

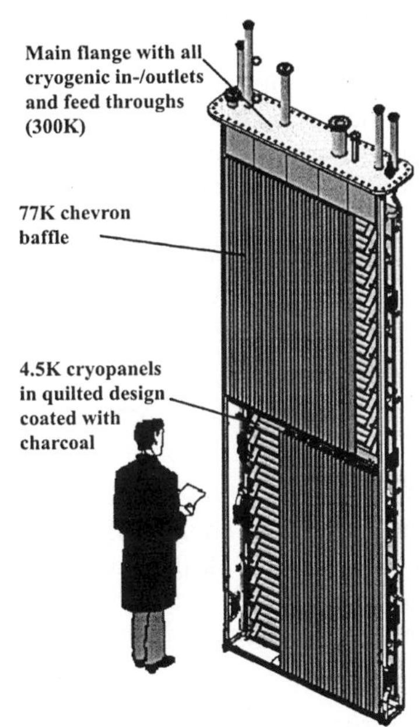

Figure 1 IPP NBI cryopump

The Multi Ampere Negative Ion Test Unit (MANITU) is the Neutral Beam Injector test facility of the Max Planck Institute for Plasma Physics in Garching/Germany. The same NBI systems are used for the IPP Tokamak (ASDEX Upgrade, AUG) with varying beam line components. The test facility is presently equipped with titanium sublimation pumps. To improve the beam line performance closer to ITER relevant requirements FZK designs and leads the manufacturing and installation of two identical cryosorption pumps.

The demand on the cryopumps is to pump a gasload of 3 Pam³/s of hydrogen during 4 hours without a regeneration of the cryopanels. One pump has to establish a mean pressure in the NBI vessel of several 10^{-5} mbar what needs a pumping speed of 350 m³/s per pump. Heating the cryopanels up to about 90K, using electrical heaters will regenerate them to release the pumped hydrogen gas. During the regeneration the radiation shieldings stay at 77K.

The 27 cryopanels at 4.5K are serial connected. Each panel has a length of 1.1m and is 0.2m broad. Both sides are coated with charcoal resulting in a coated surface of 11.88m². Compared to condensation pumps the adsorption with charcoal has the advantage to pump hydrogen at temperatures up to 15K [1]. This reduces the cryogenic demand and makes the pumping efficiency of the cryopanels less sensitive on increases of the temperature [2,3].

A centrifugal blower pumps liquid helium at saturation pressure through the cryopanel system. The main cooling of the cryopanels is than realized by the vaporisation of the forced liquid helium flow. The two- phase flow affects strongly the pressure drop over the 27 serial connected cryopanels and the influence of the heatloads during the different operation scenarios has been carefully studied.

To protect the cryopanels from the 300K radiation of the NBI components a radiation shield at 77K surrounds them. The cryopump is situated next to the NBI vessel so the backside that is orientated to the vessel wall is a closed shield of stainless steel in quilted design actively cooled by liquid nitrogen at saturation pressure. The total front side of the cryopump is designed as chevron baffle resulting in an baffle area of 5.8m². Both radiation shields are supplied from the LN2 bath at the top of the cryopump. The bath is level controlled and separates in addition the vapour- and the liquid- nitrogen phases (see Figure 1 and 2).

The cryopumps will displace the present titanium pumps and must there fore have the same geometry. Within this limitations the cryosorption pumps are designed with a height of about 4.5m, 1.5m in width and 0.25m in depth. The weight of the 77K shielding is about 550kg the coated cryopanels have an overall weight of about 200kg to be cooled down to 4.5K.

CALCULATED HEATLOADS

The temperatures of the surfaces (T) and their emission- (e) and accommodation coefficients (a) define the resulting heatloads onto the cryopump components during the different operation modes (parameters given in figure 2). During the operation of the beam line an additional heatload occurs, due to the pumped gases and the increased surface temperatures of the beam line components.

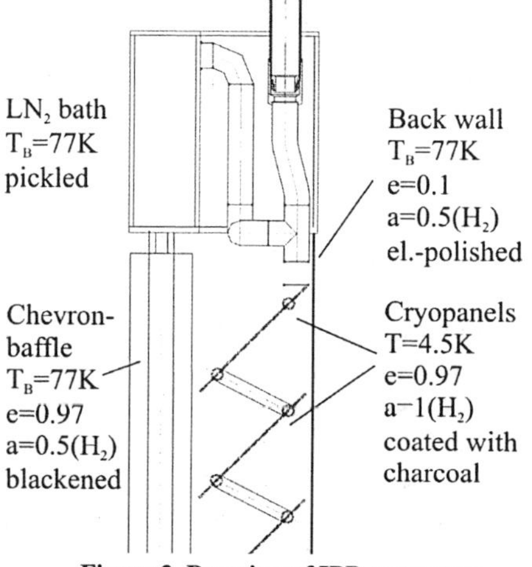

Figure 2 Drawing of IPP cryopump

Heatload onto 77K shielding

During steady state operation the radiation heatloads dominate the heat inleak on the 77K shielding. The chevron baffle is blackened with an Al_2O_3/TiO_2 ceramic leading, to an optical emission coefficient of e=0.97. This results in a low optical transmission through the baffle for the 300K radiation but leads also to a higher radiation heatload from the blackened 77K baffle. The Stefan-Boltzmann law describes the high dependence of the radiation heatloads on the temperature and faces that the advantage of reducing the transmission of the 300K radiation is much higher than the heatload from the blackened chevron baffle at 77K. Equation 1 determines basically the radiation heatloads, where f_S is the geometry coefficient of the radiating surfaces, A_x the area of the radiating surface and T_x the temperatures of the surfaces.

To assess the total heatload during steady state operation of the NBI we took into account a heatload by solid heat conduction of about 100W, resulting in an overall heatload of 2126W.

$$Q_{baffle} := f_B \cdot e_{baffle} \cdot A_{baffle} \cdot \sigma_B \cdot \left(T_{Vessel}^4 - T_{baffle}^4\right) \tag{1}$$

During beam line operation the heat conduction of the residual hydrogen gas and the cooling of the gas throughput increase the heatload. With the accommodation coefficients for hydrogen on the surfaces in the NBI the gas temperature is determined. The operation pressure ($p_{operate}$) is about several 10^{-5} mbar what is in the molecular flow regime. The heat conduction in the molecular flow regime is defined by equation 2.

$$q_{gascond.80K} := \frac{\Theta_{NBI}}{2} \cdot \frac{\gamma + 1}{\gamma - 1} \cdot \sqrt{\frac{R_0}{2 \cdot \pi \cdot M_{H2}}} \cdot \sqrt{\frac{1}{T_{gas.NBI.pulse}}} \cdot \left(T_{gas.NBI.pulse} - T_B\right) \cdot p_{operate} \tag{2}$$

The bimolecular character of the gas is described by γ, Θ_{NBI} is the system accommodation coefficient and T_B is the temperature of the shielding. The additional heatloads by the hydrogen gas are, because of the low pressure and small gasload of 3 Pam³/s, negligible small if they are compared to the heatloads due to radiation. The gaseous heat conduction leads to heatload of 10.4W and the cooling of the hydrogen gas from about 300K to 77K gives additional 5.6W.

Heating up the cryopanels releases the accumulated gas. During this regeneration the desorbed gas leads to a pressure increase in the vessel up to 1300Pa. To assess the highest heatload during regeneration no fore pumping of the NBI vessel is assumed. Therefore the gaseous heat conduction in viscous regime leads to an additional heatload of 760W. Free convection is still to small at this pressure to have an important influence. With these assumptions the overall heatload during the regeneration rises up to 2900W (see table 2).

Heatloads onto 4.5K cryopanels
The heatload onto the 4.5K cryopanels is mainly given by radiation from the 77K shielding and the transmission of the 300K radiation through the chevron baffle. An optimum between the best gas conduction through the baffle and the lowest optical transmission must be designed [4]. With an overlap of the chevrons of about 8% we receive an optical transmission of about 0.1% and a baffle transmission for the cryopump of 21.6% determined by Monte Carlo simulations.
To assess the thermo hydraulic properties, the heat inleak from the flexible supply lines to the cryopumps are taken into account with additional 12W of heatload.

operation	4.5 K cryopanels	77K rad. shielding
steady state	30W	2126W
pulse op.	35.4W	2142W
regeneration	-	2900W

Table 1. Heatloads of IPP NBI cryopump

The heatload during steady state operation is for this design determined with 30W.

During pumping operation the additional heatload is given by heat conduction of the residual gas and the heatloads resulting from the pumped gas. The latter is split into the heatload by the cooling and the adsorption and sublimation enthalpies of the adsorbed gas. The additional amount of heatload due to the gasload is about 5.4W leading to an overall heatload during pulse operation of 35.4W (table 1).

THERMO HYDRAULICS OF THE CRYOPANEL SYSTEM

The cryopanels are cooled by a forced flow of liquid helium at saturation pressure leading to a vaporisation and a two-phase flow. Thus, the different heatloads can affect strongly the pressure drop in the cryopanel system (see Figure 3). The calculation is based on a two-phase flow description using the Reynolds numbers and the helium properties in gaseous and liquid phase to calculate the Froude- and Weber numbers of the two-phase flow [5]. The pressure drop increases from about 80 mbar for a pure liquid helium flow to about 600 mbar for a total flow of vapour crossing a maximum value of 650 mbar at 95% of vaporized helium.

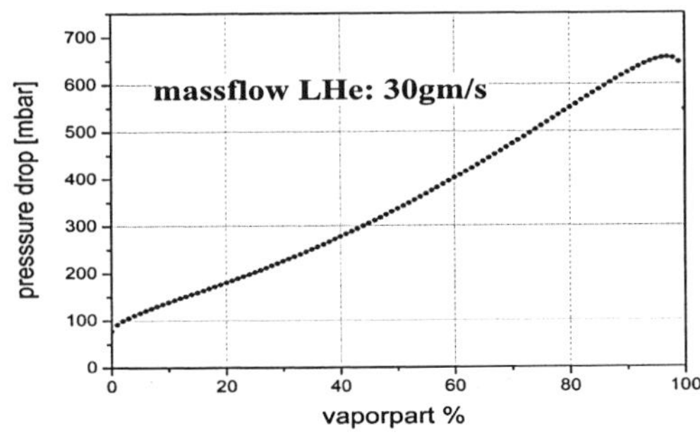

Figure 3 Pressure drop in the 4.5K cryopanels

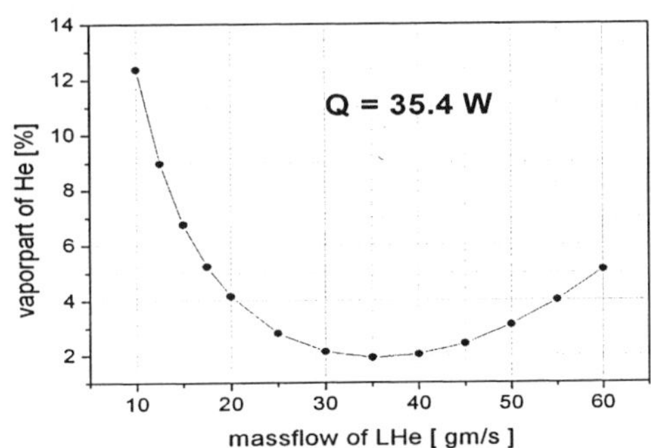

Figure 4 Vaporpart in the 4.5K cryopanels

To reduce the vaporisation one could increase the mass flow of helium but the higher pressure-drop leads again to an additional vaporisation. The resulting thermo hydraulic behaviour of the cryopanel system was determined by an recursive calculation to assess the cryogenic needs for the helium supply (see Figure 4).

The calculations are based on the system parameters, like the pumping head of the pump, the total heatload on the cryopanels, the pressure drop calculations for the quilted design cryopanels and the helium properties of the system. As it would be expected the pressure drop decreases first with increasing mass flow, reaching a minimum of vaporisation at about 35 gm/s and rises up again, due to the vaporisation by the increased pressure drop in the cryopanel system (figure 4).

DESIGN INVESTIGATIONS ABOUT THE CRYOPUMP FOR THE NBI OF THE INTERNATIONAL TOKAMAK EXPERIMENTAL REACTOR (ITER)

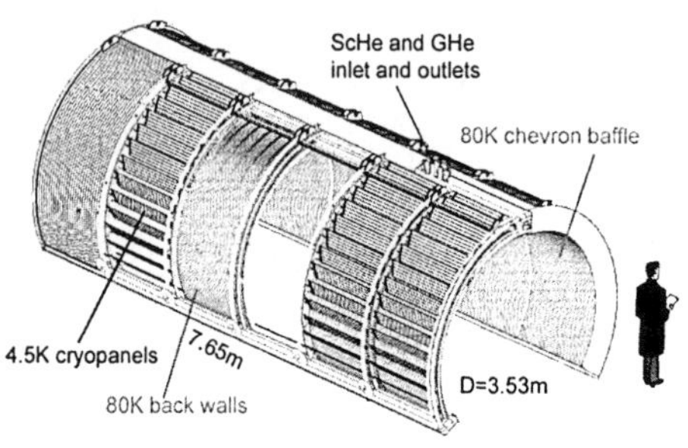

Figure 5 ITER NBI cryopump

operation	4.5 K cryopanels	80K rad. shielding
steady state	35.4W	23.9kW
pulse op.	148.6W	24.5kW
regeneration	-	31.2kW

Table 2 Heatloads of ITER NBI cryopump

For the international nuclear fusion project ITER the NBI system makes great demands on the pumping requirements. The gasload for the hydrogen operation of the NBI is given with 33.83Pam^3/s and during deuterium operation it is about 18Pam^3/s. To handle these gasloads and establish the necessary pressure profile during the operation of the beam line a pumping speed of 3800m^3/s for hydrogen and 2600m^3/s for deuterium is needed. The pressure profile along the beam line is realised with a diaphragm placed in the middle of the cryopump. Thus, the pressure in the second chamber should be thirty times lower than in the first chamber to avoid the reionisation of the neutral gas beam. The NBI components have a temperature of about 373K resulting in a very high radiation heatload onto the blackened baffle of 23.9kW. The maximum heatload occurs again during the regeneration of the cryopanels and reaches up to 31kW (see table 2).

The cylindrical shaped cryopump (see Figure 5) is compared to the IPP cryopumps supplied with ScHe at 4.5K and 4bar and GHe at 80K and 18bar for the radiation shielding. The 80K shielding has a weight of 4.5t and the cryopanels system a mass of 1t. Cooldown requirements are accordingly high depending on ITER requirements for cooldown- and regeneration times.

ACKNOWLEDGEMENT

This work is carried out under the framework of the Nuclear Fusion Project of FZK. The ITER design work is supported by the European Communities under the European Fusion Technology Programme.

REFERENCES

[1] Haas H.: Performance tests of the ITER model pump, Fusion Eng. Des. 69 (2003) 91-95
[2] Hauer V, Day C.: Cryosorbant Characterisation of Activated Charcoal in the COOLSORP Facility, ISSN 0947-8620
[3] Day C.: Pumping Performance of cryopanels coated with activated carbon, Adv. Cryogenic Engineering 43 (1998) 1327 - 1334
[4]   Oka, Y.: Development of cryosorption pump for neutral beam injector, Fusion Eng. Des. 31 (1996) 89-97
[5] Friedel, L.: Improved friction pressure drop correlations for horizontal and vertical two phase pipe flow, 3 R Int. 18, 7, 1979

SIMPLE METHOD FOR CALCULATION OF CRYOGENIC TRANSFER LINE COOL DOWN

A. Alexeev, Th. Pelle

Messer Cryotherm GmbH,
57548 Kirchen, Germany

The cryogenic transfer lines operate often in intermittent mode. A software program for calculation of transient processes, especially cool down, can help to find the most economical operation parameters in some cases. A simple program based on the Mathcad-Software was developed for these purposes at Messer Cryotherm. The thermodynamic model for this program is presented and discussed.

INTRODUCTION

The objective of this work is the development of a software for estimation of cryogenic transfer line cool down time. The main requirements on this software is the simplicity and flexibility. The simplicity means clear and understandable software-syntax, so that each cryogenic engineer could be able to work with this program without special knowledge of programming language. The flexibility means that is possible to modify this software, especially for calculation of some processes or hardware configurations differing from the "standard" transfer line. Consequently the thermodynamic model & mathematics should be simple and understandable.

THERMODYNAMIC MODEL

Table 1 shows assumptions made for the development of the thermodynamic model. Figure 1 shows the calculation idea: the i- wall element has the temperature $T\_wall_i^{old}$ before the calculation step. During the calculation step $\Delta time$ a portion of cryogen $dM\_fluid = \dot{M}\_fluid \cdot \Delta time$ flows through the i-wall element, the temperature of this flow portion changes from $T\_fluid_i^{old}$ to $T\_fluid_i^{new}$, the temperature of the i-wall elements changes from $T\_wall_i^{old}$ to $T\_wall_i^{new}$.

TABLE 1. Assumptions

#	Assumption	Comment
1	The transfer line is long.	This assumption usually means that relationship length / diameter >> 10.
2	Radial heat conductivity of cryogen $\rightarrow \infty$.	This assumption is valid, if the intensive mixing in every cross-section takes place. As a rule this condition is accurate for turbulent flow and areas with intensive boiling of cryogen for lines with small and medium diameter. For large transfer lines this assumption is not absolute correct and should be taken into consideration.
3	Longitudinal heat conductivity of cryogen $\rightarrow 0$ Longitudinal heat conductivity of the wall $\rightarrow 0$	This assumption can be accepted for long transfer line.
4	Radial heat conductivity of the wall $\rightarrow \infty$.	This assumption means that the cooling process in the wall of inner pipe is irrelevant for present calculations method. It is because the order of

		magnitude for thermal diffusivity of stainless steel is 10^{-6} m²/sec. It means the mean transient time of less than 2 sec for 1 mm - wall, and less than 5 sec for 2 mm-wall. The typical total cool down time for a long transfer line is higher than one minute, therefore the assumption 4 can be accepted.
5	Temperature difference between cryogen and the wall is essentially larger than alteration of cryogen temperature during a calculation step for one finite element.	This assumption is valid if the finite elements are appropriate small.
6	Temperature difference between cryogen and the wall is essentially larger than alteration of wall temperature during a calculation step for one finite element.	See # 5
7	Pressure losses → 0.	
8	Cryogen is incompressible.	

The temperature change can be calculated with help of energy conservation equations for fluid (cryogen) and wall-element.

$$T\_fluid_i^{new} = T\_fluid_i^{old} + \frac{\alpha \cdot dF}{\dot{M}\_fluid \cdot Cp\_fluid}\left(T\_wall_i^{old} - T\_fluid_i^{old}\right) \tag{1}$$

$$T\_wall_i^{new} = T\_wall_i^{old} - \frac{\alpha \cdot dF \cdot \Delta time}{dM\_wall \cdot Cp\_wall}\left(T\_wall_i^{old} - T\_fluid_i^{old}\right) \tag{2}$$

These equations can be completed by mass conservation law:

$$\dot{M}\_fluid = \rho \cdot u \cdot A \tag{3}$$

For the basic version of the software the condition at the outlet was used:

$$\dot{M}\_fluid = \rho\left(T^{outlet}\right) \cdot u^{outlet} \cdot A^{outlet}$$

$$u = a\left(T^{outlet}\right) \tag{4}$$

T_fluid	- temperature of cryogen	T_wall	- temperature of the wall
M_fluid	- cryogen mass flow	DM_wall	- masse of a wall finite element (FE)
Cp_fluid	- cryogen heat capacity	Cp_wall	- heat capacity of the wall material
$Alfa$	- heat transfer coefficient	dF	- heat transfer surface of a FE
old	- state before calculation step	new	- state after calculation step
u	- velocity	ρ	- cryogen density
a	- sound velocity	i	- index of the FE
A	- cross section	$outlet$	- acc to the outlet

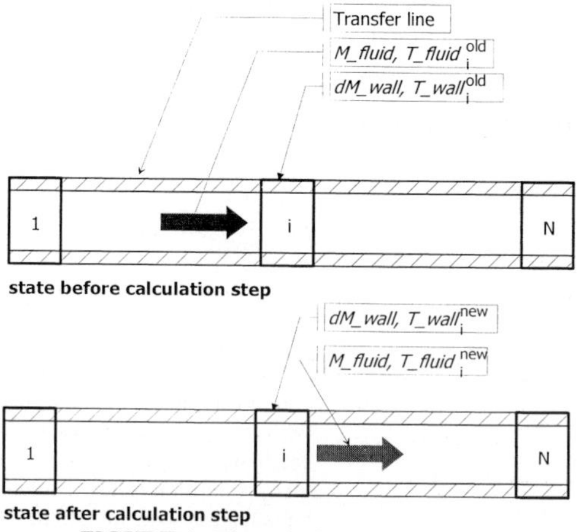

FIGURE 1 The principle of calculation

CRYOGEN PROPERTIES

The equations 1-4 are very simple - this set of equations can be solved directly without any iterations, if the heat capacity of cryogen and the materials is available as a function of temperature. This simplicity means advantages because of little calculation expense and decreased probability of a calculation error. These equations are correct for one-phase area only, because the heat capacity in two-phase area $Cp\_fluid \rightarrow \infty$. In two-phase area these equations cannot be applied. But the cool down process happens mainly in two phase area. The usual approach (to overcome the two phase- problem) is that the equations 1-4 will be completed by an additional set of equations describing the processes in two-phase area. The mathematics becomes difficult and complex. In this paper another approach will be discussed. It was assumed, that the temperature of cryogen in two phase area is not constant, but changes negligible small from boiling temperature Tb to the dew

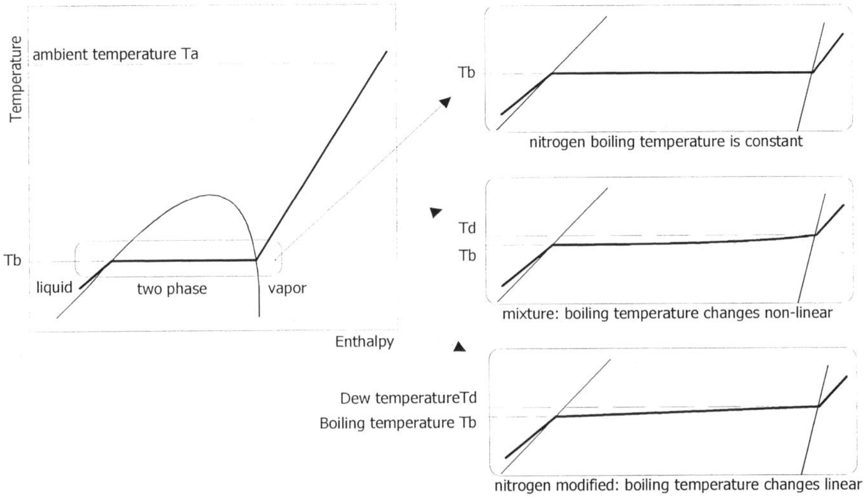

FIGURE 2 Cryogen properties modification

temperature Td[1], so that difference Td - Tb is negligible small (see Figure 2). This modification of cryogen properties has a consequence, that the heat capacity of cryogen are not more unlimited high and equations 1-4 are applicable in two phase area further. This clever solution helps to keep the mathematics simple and understandable.

DISCUSSION

Heat transfer coefficient and cryogen flow

The accuracy in the prediction of the heat transfer coefficient and cryogen flow in the transfer line plays decisive role in calculation of cool down time. It is because of
1. interaction of these both parameters:
    - the heat transfer coefficient depends on the flow velocity and temperature

---

[1] On the first view, this assumption seems to be absurdly. On other side, we know some substances boiling at changing temperature, for example air. It is because air is not a pure substance, but a mixture of nitrogen, oxygen and other atmospheric gases. The mixture of 99% nitrogen und 1% oxygen boils at changing temperature too. But the properties of this mixture differs minimal on properties of pure nitrogen. And the cool down of a transfer line cooled by liquid nitrogen and the cool down of the same line cooled by mixture of 99% nitrogen und 1% oxygen are very similar. This fact is used in the present model.
Moreover to simplify the description of cryogen properties further, it was assumed, that the boiling temperature changes linear in two phase area between Tb and Td. All other properties are equivalent to nitrogen properties.

- the flow depends on the temperature of the cryogen at the outlet, and this temperature depends on the heat transfer inside the transfer line directly.
2. two phase heat transfer, both the heat transfer coefficient and cryogen flow can vary considerably in two phase area:
   - heat transfer coefficient from 100 to some 1000 W/(m^2-K),
   - the change of the flow velocity can amount to three orders of magnitude, because of huge density alteration in two phase area.

Therefore it is necessary to include the calculation methods for forced convection heat transfer as well as for film- and forced convection boiling in the model. Some of these methods are described in [1] and can be recommended to use.

Software Check

A very simple method to check the software for write errors is the calculation of a very long transfer line (l/d >> 10^4). For such a long lines the cryogen outlet temperature is near to ambient for relative long period of time. During this time
-     the outlet conditions are constant, therefore cryogen flow is constant
-     the whole enthalpy of cryogen is used for cooling purposes independent on heat transfer coefficient.

Therefore the total cool down time is close to the

$$cool\ down\ time = \frac{M\_w \cdot Cp\_w \cdot (Tambient - Tb)}{\left(\rho \cdot a \cdot A^{outlet}\right)_{Ta} \cdot \left(h^{liquid} - h^{Ta}\right)} \tag{5}$$

Tb - boiling temperature of cryogen	Ta - ambient temperature
h^{liquid} - enthalpy of liquid cryogen	M_w - masse of the wall
h^{Ta} - enthalpy cryogen at ambient temperature	Cp_w - heat capacity of the wall material
A^{outlet} - cross section	ρ - cryogen density
a - sound velocity	

SUMMARY

A simple method calculation of transfer line cool down was developed. A new approach for description of cryogen properties is presented. It was assumed, that the temperature of cryogen in two phase area is not constant, but changes negligible small from boiling temperature Tb to the dew temperature Tb. This solution helps to keep the mathematics simple and understandable.

LITERATURE

1. A.R.Hasan, K.A.Haque, A.S.M.Rokanuzzaman and M.M.Hasan, "Modeling of cryogenic transfer line cool down", in *Advances in Cryogenic Engineering, Volume 45A*, Plenum, New York, 2000

Study of new aerodynamic foil thrust gas bearing with elastic support for cryogenic turbo-expander

Chen C.Z., Lin S.N., Hou Y., Zhu Z.H.

Institute of Refrigeration and Cryogenic Engineering, Xi'an Jiaotong University, Xi'an 710049, China

In order to meet requirements of the gas bearing cryogenic turbo-expanders development, a new type of aerodynamic foil gas thrust bearing with elastic support was proposed and developed. The experimental study has been conducted on the gas bearing test rig. The factors affected on the main performances of the gas bearing have been analyzed and the ranges of optimum bearing parameters have been obtained.

INTRODUCTION

With the development of small and miniature cryogenic turbo-expander in some fields, the high speed gas bearing has become the key part of evaluating the turbo-expander performance. Compliant foil bearing is one of the innovative bearing technologies for supporting cryogenic turbo-expander. The bearing with elastic support can adjust itself according to the changes of speed and load. It has strong self-adaptability. While the bearing is operating, it can absorb some excess energy by the function of the foil distortion and coulomb friction. It can keep high steady condition at high rotation speed even if it suffers vibration and whirling motion. The Cryogenic turbo-expander, using the foil gas bearings with elastic support, has better development potential and wide application foreground. In recent years, two types of new self-acting foil thrust bearings have been proposed and developed. The foil bearing is supported with double layers copper wires and elastic material respectively. And the domestic experiments have been conducted on the gas bearing test rig. The features of the new aerodynamic foil thrust gas bearing with elastic support are given briefly in this paper.

NEW AERODYNAMIC FOIL THRUST GAS BEARING WITH ELASTIC SUPPORT

The new aerodynamic foil thrust gas bearing with elastic support developed by Institute of Refrigeration and Cryogenic Engineering of Xi'an Jiaotong University is composed with some sectorial bearing housings and elastic foils. The surface of sectorial bearing housings is formed from incline and plane. The import gas film clearance between the surface and elastic foils is defined h_1; the export gas film clearance is defined h_2. When the bearing moves relatively with thrust surface, the supporting load is created by the aerodynamic effect. The aerodynamic foil thrust gas bearings with four bearing housing pads are shown in Figure 1(a). The other geometric structural parameters are defined as follows: r_1: inside radius of pad, r_2: outer radius of pad, β: field angle of pad, β_b: field angle of sectorial incline, S: thickness of the elastic material, t: thickness of the top foil, $b = \beta_b/\beta$: pitch rate.

The top foil is made of Beryllium-bronze, which is a kind of better bearing material. Its shape is sector and the size is suitable to bearing housing pad. Its dimensions are follows: field angle of pad 90°, inside radius 8mm, outer radius19mm and thickness 0.05∼0.10mm. To improve the quality and lubricating property, the surface of the top foil is treated to form the composite coat or wiped by the lubricating material such as MoS_2.

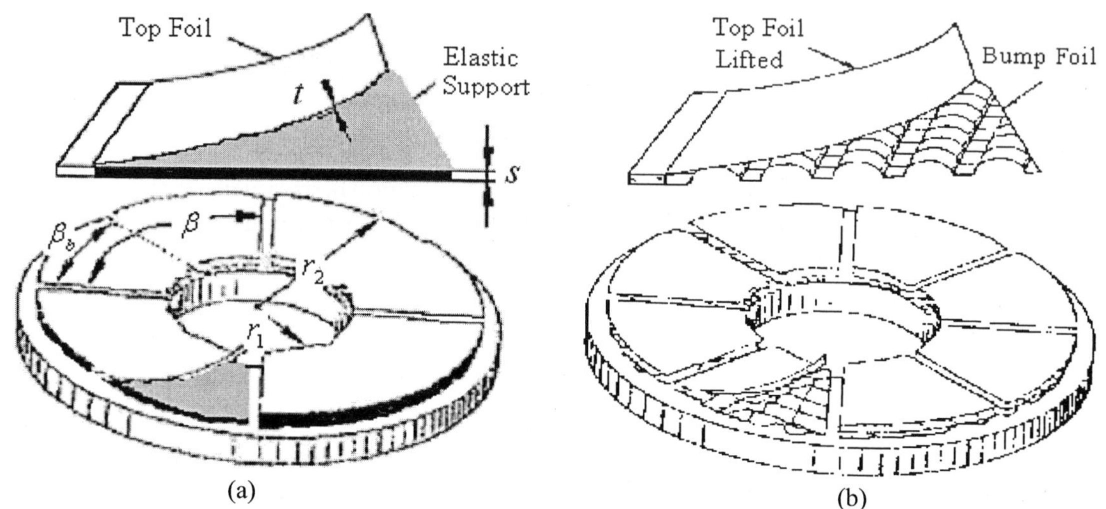

Figure 1   (a) The foil thrust gas bearing with elastic support
(b) The foil thrust gas bearing with bump foil

The support is made of the elastic material that has better elasticity and damp property and can endure high temperature. The shape of it is also sector and the dimensions are follows: field angle of pad 90°, inside radius 9mm, outer radius 18mm and thickness 0.2∼0.5mm. The abridged general view of the new foil thrust gas bearing with elastic support is shown in Figure 1(a). Compared to the bump foil gas thrust bearing shown in Figure 1(b), which is widely used in high speed turbo machines, the support in our scheme is replaced by a kind of new elastic material. Its construction is simple, and it is easy to manufacture. And its surface structural stiffness is more uniform. So it has good practicability and extensibility.

EXPERIMENTAL APPARATUS

The experiments of the new aerodynamic foil thrust gas bearing with elastic support have been conducted on the multifunctional gas bearing test rig shown in Figure 2. The thrust gas bearing test rig is made up of the bearing-rotor system and the load system of the axial piston. The rotor system of the static gas bearing is driven by a reaction wheel. And its rotor speed and axial force can be adjusted by changing the mass flow or pressure. The load system of the axial piston includes piston shaft, cylinder, connecting piece, static journal bearing and eddy current displacement sensor. The experimental foil thrust bearing is mounted on the piston shaft by the connecting piece. The piston shaft is supported by the gas film, which is supplied by a pair of gas bearing. There is no friction in the axial motion. And the axial load is adjusted easily through changing the loading gas pressure of the cylinder.

The GYG01 high precision pressure sensor is adopted in the experiment to measure the axial loaded-gas pressure. In order to measure the motion and amplitude of vibration, the SJ4-1 eddy current sensor and the fore magnifier are used. And using FFT method, the rotor speed can be obtained by analyzing the vibration signal which can be measured by two eddy current type displacement probe mounted on expander housing in X-Y direction respectively.

Figure 2   The multifunctional gas bearing test rig

EXPERIMENTAL RESULTS AND ANALYSES

Many thrust gas bearings of this new elastic support type with different structural parameters have been tested. The better results of the dynamic performance in the experiment are presented. The orbits of rotor center, the time-base and frequency-domain records of the rotor vibrations in the thrust force loading course are shown in Figure 3.The orbits of rotor center are clear and the amplitudes of vibration are small even when the sub-synchronous resonance occurs. The main portion of the rotor response is synchronous. The sub-synchronous resonance is not significant and its frequency is very low. The changes of the shaft vibration amplitudes are rather smooth in the operation. The results indicate that the new aerodynamic foil thrust gas bearing with elastic support is quite stable, and it seems to be suitable for small cryogenic turbo-expander.

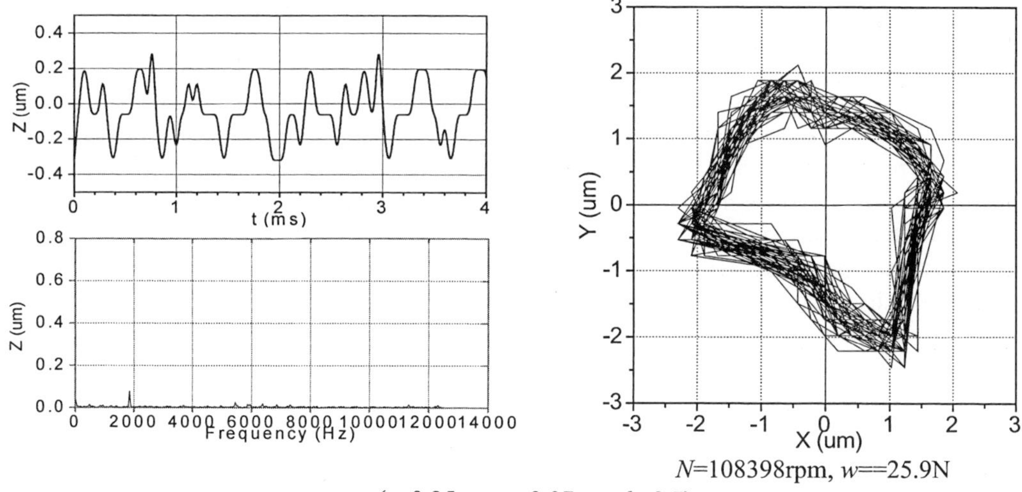

(s=0.25mm, t=0.07mm, b=0.7)

Figure 3   The vibration characteristics of the foil thrust gas bearing

Figure 4 illustrates the axial clearance of the foil thrust bearing as a function of bearing load at the speed 100,000rpm.As shown in the figure, the axial clearance reduces correspondingly with increasing the bearing load when the speed fixed. And the amplitudes reduce slowly and gradually at the greater load.

The results indicate the good damping characteristics.

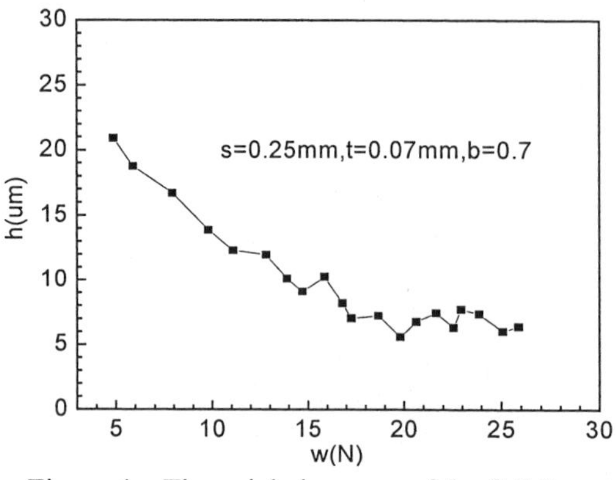

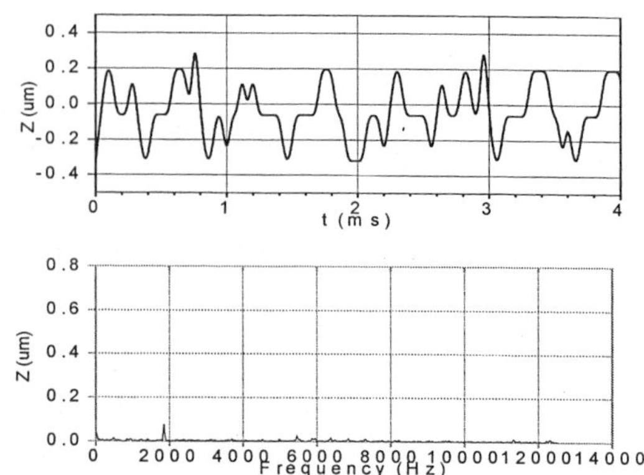

<table>
<tr><td>Figure 4   The axial clearance of the foil thrust bearing with bearing load</td><td>Figure 5   The time-base and frequency-domain records at the highest load</td></tr>
</table>

The relation between the maximum thrust bearing capacity and the structural parameters of foil bearing is obtained in the experiments. It is shown that the thickness of elastic material affects mostly the thrust bearing load. Generally speaking, the thinner the elastic material is, the higher the thrust bearing load is. But the thickness of the material can't be unrestrictedly reduced. On one hand, the restriction of material processing technique in existence must be considered; on the other hand the stability of the bearing is affected with reducing the material damping characteristics because of reducing the elastic support thickness. The highest thrust bearing load 25.9N is obtained (with s=0.25mm, t=0.07mm and b=0.7).The records of the time-base and frequency-domain of the foil bearing at the highest thrust bearing load are shown in Figure 5.Moreover, the material of elastic support is studied preliminarily and some factors affecting the performance of the new foil thrust bearing are obtained.

CONCLUSION

The experimental results indicate that the new aerodynamic foil thrust gas bearing with elastic support possesses good stability and operation performance. The good damping characteristics of the elastic support material restrain the rotor's vibrations and the growth of whirling amplitudes. And the flexible surface of the bearing decreases the damage to the rotor and the bearing surface effectively. The good damping characteristics of the elastic support and the good performance of the bearing are presented sufficiently. With the proper choice of the elastic support material and the parameters, the foil thrust bearing presented here can achieve high stability and high load capacity. The experiments of the cryogenic turbo-expander with new aerodynamic foil thrust gas bearing have also been conducted and some remarkable achievements are obtained. Due to its simplicity and good performance, the new aerodynamic foil thrust gas bearing with elastic support is hopeful to be applied to a small high speed cryogenic turbo-expander successfully.

REFERENCES

1. Chen C. Z., Self-acting Foil Journal Bearing with Elastic Support, 1998, Chinese Patent No. 98233296.3.

2. Xiong L.Y., Wu G, Hou Y. Development of aerodynamic foil journal bearing for a high speed cryogenic turboexpander, Cryogenics (1997) 37(4) 221-30.

Design of small turbo brayton cycle air refrigerator test rig

Hou Y., Sun Y., Wang J., Chen C.Z.

Institute of Cryogenic Engineering, Xi'an Jiaotong University, Xi'an 710049, P.R.China

In order to promote the development of cryocooler in China, the study of reverse turbo brayton cycle cryocooler was firstly begun in 1994 by Xi'an Jiaotong University. This paper introduces the reverse turbo brayton cycle refrigerator test rig established in 2002. The refrigerator adopts air as refrigerant medium and works in single stage, open cycle. The mechanical and thermodynamic performance of the refrigerator was tested and analyzed in this paper. The rotation speed of turbo expander has reached 230,000 rpm. The temperature in cold box can reach below 200 K and the temperature drop in the refrigerator has exceeded 80 K.

INTRODUCTION

In many fields of space application, cryocoolers are used to achieve the temperatures below liquefied nitrogen (77 K) in order to improve the performance of sensors and electronics. Mechanical cryocooler is one of the important choices to provide an environment in low temperature. In the last thirty years, for the increasing demand of infrared remote sensing and multi-object spectrometer technology, more and more efforts were put into the research of small cryocoolers used in space applications. These cryocoolers have been used in wide range of load and temperature, and continuously promote the development of space technology. Without the development of cryogenic technology, especially the development of small cryocooler, it will be difficult to use infrared equipment for detection, tracking and homing. Therefore, the research of small cryocooler used in space applications is significant to military utility and civil space technology.

Many kinds of mechanical cryocoolers used in space, such as stirling cooler, pulse tube cooler and turbo brayton cooler, have been developed. They were used in detectors of different temperature and load. To achieve the long life (10-15 years) of mechanical cryocooler, new type of cryocooler must been developed. Compared with other types, some institute such as NASA considered that the turbo brayton cryocooler matches the demands most superiority. Reverse turbo brayton cycle cryocooler, which includes high speed turbine using gas bearing system and compact heat exchanger, has many advantages such as excellent reliability, vibration free, high efficiency and long life.

There were a lot of researches about stirling cooler and pulse tube cooler in china for many years, whereas not adequate study on turbo brayton cryocooler until now. The Institute of Cryogenic Engineering, Xi'an Jiaotong University, is one of the organizations that take the lead in the field of the reverse turbo brayton system and cryocooler in China and has got remarkable achievements. Xi'an Jiaotong University have been studied the turbo expander with gas bearing since 1970s, which was the most significant component in the reverse brayton cycle cryocooler. The research on turbo brayton cycle cryocooler for space application in China was firstly began in 1994, and the first cryocooler was built in 1995. The Institute of Cryogenic Engineering has carried on elaborate investigation in the cryogenic technology for space application in 1998. The theoretical and experimental researches have been studied thoroughly and some key problems have been solved. In 2002, for the demand of development of prototype of the small reverse brayton cycle cryocooler used in space application, the experimental test rig of the small reverse turbo brayton cycle air refrigerator was built up.

REVERSE BRAYTON CYCLE REFRIGERATOR

This test rig of small reverse turbo brayton cycle refrigerator employs open regenerative cycle and uses the air in atmosphere temperature as its refrigerant, as shown in Fig.1. The turbo expander is the most important component of the reverse brayton cycle refrigerator, and reflects the technological level of this system. The turbo expander provides the refrigeration capability in low temperature for system, so its thermodynamic and mechanical performance are extraordinary important to the economy and reliability of this equipment. At the same time, it is the most difficult component to develop. In a word, the turbo expander is the key component in the development of small reverse brayton cycle refrigerator.

The appearance of this refrigerator is a cylinder box, which dimension is $\phi 500 \times 600$ mm, and its compact structure makes it convenient to be moved, shown in Fig2. The perlite is filled acts as the insulate material. A piston compressor system with filter, sorption and water cooler provides the clean pressure air of 0.8 MPa. A 30 Nm^3/hr air turbo expander is adopted in this refrigerator, shown in Fig.3. A radial impeller with the diameter of 20 mm drives the rotor of the turbine, and a brake blower with a diameter of 22 mm is adopted to absorb the output power of the expander. The pressure of expander inlet is 0.35 MPa, and the outlet of turbine is atmosphere. In order to control the speed of the rotor, the flow rate is controlled through a valve. The rotor weighs about 90 g with a shaft diameter of 12 mm and a total length of 90.0 mm. The normal speed of this turbo expander is about 200,000 rpm. The turbo expander is fixed vertically on the cover of refrigerator by insulated bakelite plate to avoid cold loss, so it can be assembled easily. The compact plate fin heat exchanger is used as the regenerator to recuperate the load of the loop gas in the air refrigerator because of its high effectiveness. The cold box of the refrigerator is a cylinder which dimension is $\phi 100 \times 180$ mm. The temperature in cold box can reach 190 K and it can create 0-1 kW heat by an electric heater. The stainless steel tube in 20mm diameter and copper tube in 19mm diameter are selected as pipelines to connect each component. All the tubes are covered by foaming material to avoid cold loss.

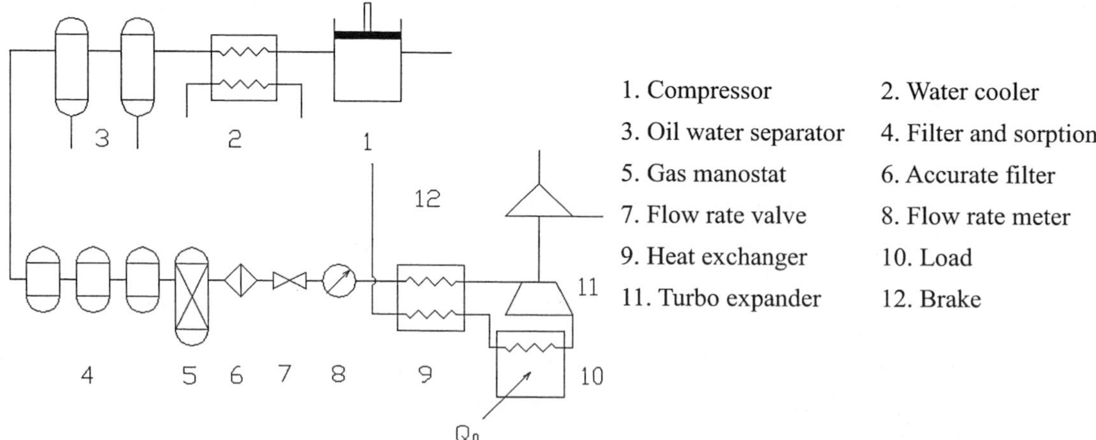

Figure 1   The system schematic diagram of the small air refrigerator system

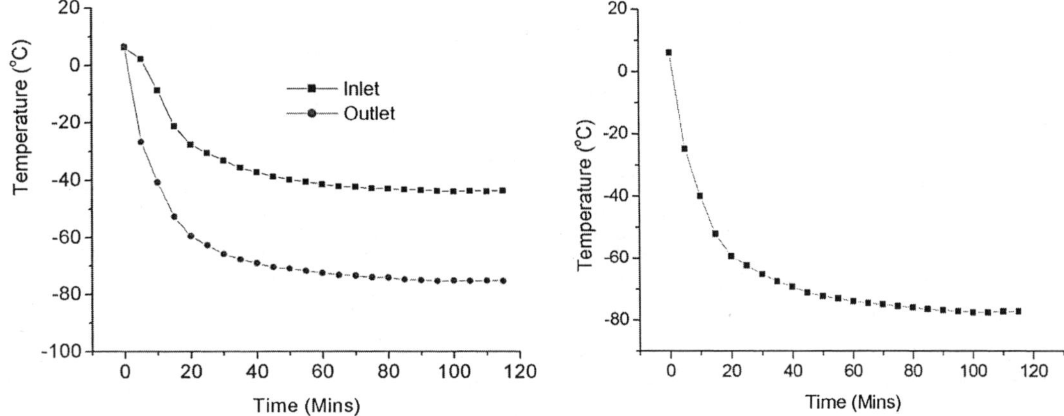

Figure 5   The inlet and outlet temperature of turbine   Figure 6   The temperature of the cryogenic box

CONCLUSION

It can be concluded that this air refrigerator has higher overall performance. The design of this test rig was reasonable and the refrigerator could reach -75℃ in 100 minutes, all of those achieved the design temperature and load capacity. Through the experiments on this test rig, many investigations can be done to resolve the design of compact heat exchanger, increase the efficiency of the turbo expander and the refrigerator system, improve the reliability of rotor-bearing system as well as develop the small reverse turbo brayton cryocooler. With little changes in the insulation and structure of this refrigerator and adopting He or Ne as its refrigerant medium, the test rig can also be used in the study of reverse brayton cycle cryocooler worked on the temperature of 65-80K for space application.

ACKNOWLEDGMENT

This project is supported by the National Nature Science Foundation of China (50206015). The authors also would like to thank Mr. G.J.Liu of Shuzhou Oxygen Plant Manufactory for supplying the experiment parts.

REFERENCES

1. Swift W.L., et al, Developments in turbo brayton technology for low temperature applications. Cryogenics (1999), 39 989-995

2. Collaudin B., et al, Cryogenics in space: a review of the missions and of the technologies. Cryogenics (2000), 40 797-819

3. Hou Y., et al, Comparative test on two kinds of new compliant foil bearing for small cryogenic turbo-expander, Cryogenics (2004), 44 69-72

4. Chen C.Z., et al, Self-acting foil journal bearing with elastic support, Chinese Patent (1998), 98233296.3

5. Xiong L.Y., et al, Development of aerodynamic foil journal bearings for high speed cryogenic turbo-expander, Cryogenics (1997), 37 221-230

6. Henatsch A. and Zeller P., Cold air refrigeration machine with mechanical, thermal and material regeneration, International Journal of Refrigeration (1992), 15 16-30

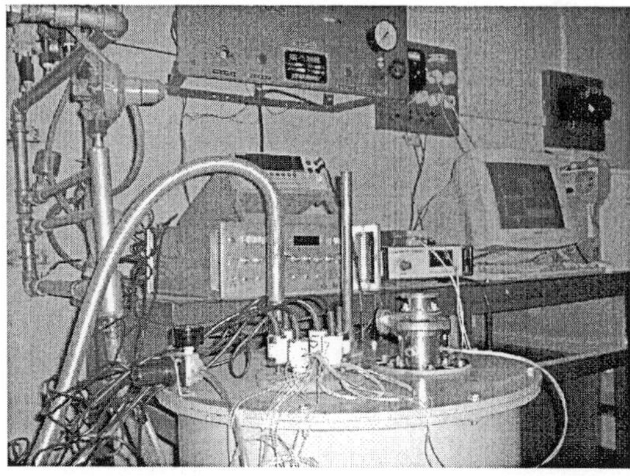

Figure 2   Reverse brayton cycle refrigerator test rig

Figure 3   Gas bearing turbo expander

THE EXPERIMENTAL STUDY

Gas bearings are adopted in this situation because the normal speed of the turbine is about 200,000 rpm. A new type of foil journal bearing, compliant foil bearing with elastic support, is used to support the rotor and the spiral grooved thrust bearing to uphold the axial load for high speed cryogenic turbo expander (Fig.3). The record of rotor vibration at a steady speed and the time base record of the rotor vibration at this speed (Fig.4) indicate that the main portion of the rotor response is synchronous (the fundamental frequency), and the amplitudes of rotor vibration are small in the running period. The rotor bearing system has best dynamic and preferable stability performance and it is clear that the operation reliability of this air refrigerator can be ensured.

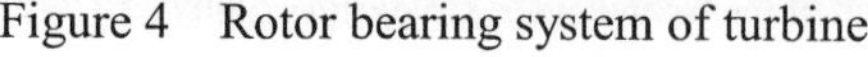

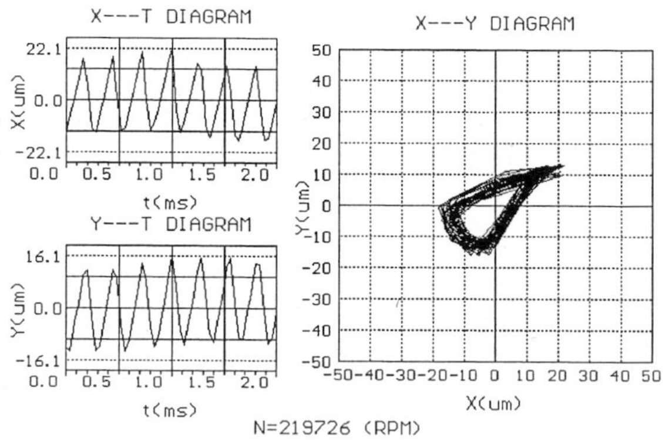

Figure 4   Rotor bearing system of turbine

Time base record of rotor vibration

The thermodynamic performance of this reverse brayton cycle air refrigerator has been tested. The inlet temperature and outlet temperature falling curves of the turbo expander with the increasing of time are shown in Fig.5. The relationship between the temperature in the cryogenic box and the time was shown in the Fig.6. When the environmental temperature was 6℃ and the rotational speed of turbine maintained at 220,000 rpm, the temperature of the cryogenic box could reach -75℃ after 100 minutes and the temperature drop in turbo expander was 32℃ at this time.

Analysis of Centrifugal Stresses in the Rotor of a Cryogenic Turboexpander

Parthasarathi Ghosh, Sunil Kr Sarangi

Cryogenic Engineering Centre, IIT Kharagpur 721 302, India

Cryogenic turbines, because of moderate to high-pressure ratio and low flow rate, rotate at high speeds with peripheral speeds approaching the velocity of sound at prevailing temperatures. At such high rotational speeds, significant centrifugal stresses are generated in the large diameter components – turbine wheel, compressor impeller and shaft collar. The blades of the wheels, which are thin and curved, also expand and deflect, reducing clearances and distorting flow passages. This paper presents an FEM analysis of the stress field and the deflection pattern in the rotor of a high-speed cryogenic turbine, and a recommendation for an appropriate design strategy.

INTRODUCTION

Because the strength of most materials improves at low temperature, the general perception among engineers and scientists is that stress considerations are unimportant in case of cryogenic turbines. On the other hand, cryogenic turbines, because of the moderate-to-high pressure ratio and low flow rates, rotate at high speeds leading to significant centrifugal stresses in the rotor.

A finite element analysis of a complete cryogenic turbine rotor has been carried out using the ALGOR FEM package. The analysis provides a map of the stress field and the deflection pattern in the critical components under design conditions. The effects of geometric features such as blade thickness and fillet radii on the maximum stress have been brought out. The paper presents the formulation of the problem, boundary conditions, analysis and results. A clear recommendation is made on the choice of materials and dimensions of relevant structural features to keep the stresses and deflections within acceptable limits.

DESCRIPTION OF THE ROTOR

The specifications of the turbine system and the dimensions of the components have been obtained from the design of a prototype turboexpander developed at IIT, Kharagpur [1].The schematic of the rotor is shown in the Fig 1. It consists of the shaft, the turbine wheel and the brake compressor impeller. The compressor is used to load the turbine, which produces power by expanding high pressure gas to a lower pressure. The wheel and the impeller are attached to the shaft at both ends "socket head button" screws. The expected rotational speed of the rotor at the designed condition is 140,000 r/min. The shaft is supported by two aerostatic journal bearings and two aerostatic thrust bearings. Orientation of the rotor is vertical with the brake impeller located at the upper side of the assembly.

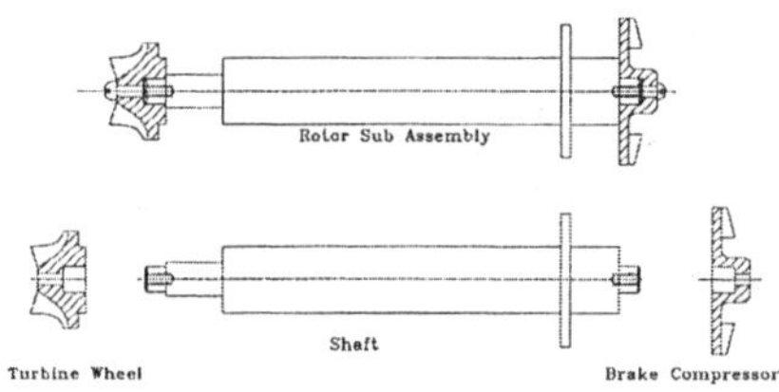

Figure 1: Schematic of the rotor to be analysed

OBJECTIVES OF THE ANALYSIS

A static stress analysis has been performed on the turbine rotor with the following objectives:
1. to map the stress field in the rotating components and to identify points of high stress for possible modification of design,
2. to obtain the deflection pattern under centrifugal load at the nominal rotational speed (The displacement pattern reveals the radial growth of the rotor under centrifugal stress. The deflected shape affects the clearances between the rotating components and the corresponding static components), and
3. to analyse the effect of geometric features such as thickness distribution of the turbine blades and radius of the fillets on the shaft collar on the computed stress pattern.

SOLUTION STRATEGY

The stress analysis has been carried out using a solid model of the rotor, followed by finite element meshing, application of loads and boundary conditions and finally by solution of the resulting finite element equations. The FEM results have been compared with those obtained from simple formulas given in standard textbooks. The detailed steps in the FEM analysis process are :

1. Creation of a solid model using a parametric solid modeling software. [We have used AutoCAD Mechanical Desktop Release 3 and 4 for this step.]
2. Meshing of the solid model (*.dwg file) so generated using the meshing function of the ALGOR finite element analysis package. [An extender provided by ALGOR has established a seamless relation between the AutoCAD MDT4 solid modeller and the ALGOR FEM software. The coordinates of the elements, generated after meshing, are saved in a *.ami file which is exported to the main ALGOR FEM software.]
3. Solution of the finite element equations to yield the complete stress and deflection field.

STRESS ANALYSIS OF THE TURBINE WHEEL

Figure 2 describes the geometry of the turbine wheel used for analysis. The other features of the turbine geometry are:
1. Radial blading;
2. A constant tangential thickness of 1 mm;
3. No taper along the hub-to-shroud direction;
4. Constant hub fillet radius of 0.5 mm.
5. Material of the wheel is Al 6061 T6.

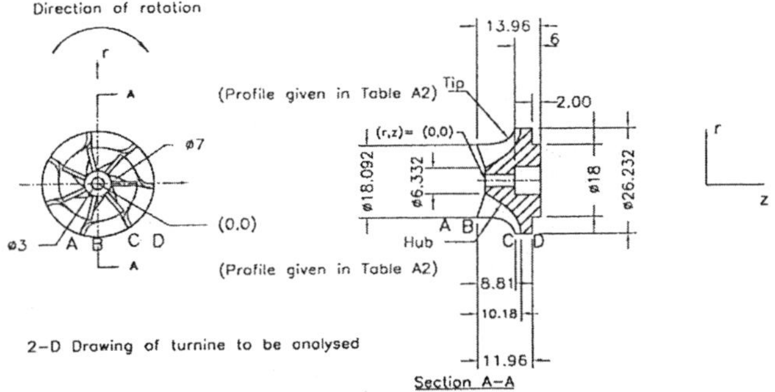

Figure 2 Geometry of the turbine wheel used in the analysis

<u>Geometric modelling of the wheel</u>
Due to an inherent mismatch between the MDT and ALGOR, we followed a somewhat complex route to create the solid model in MDT. A wire frame was created in MDT 3.0 and transferred to ALGOR Superdraw-III using IGES translation. A closed surface was created using the ALGOR surface modeller Supersurf, and was transported to Superdraw-III. The surface mesh was further enhanced using the Merlin Meshing Technology (MMT), a feature provided by the ALGOR package. The turbine wheel has a 3 mm diameter hole with overall external diameter of 26.3 mm. Discontinuities exist at the bore and at the blade-hub intersections. To take into account these features, the surface mesh was refined using the "refinement near short" and 'refinement near gaps" commands of MMT. Quadrilateral mesh type was chosen.

The surface mesh was then transformed into solid mesh using the routine HEXAGEN. For the solid mesh, the maximum aspect ratio has been chosen as 100 and the wrap angle limit has been set at 30^0 (maximum). Figure 3 shows the meshed geometry (pie-slice model) of the wheel, projected on the x-y plane, as seen from the reverse side. The origin is at O. The lines AB and CD are the cut surfaces projected on the x-y plane, making angles of 150° and 70° with the x-axis respectively.

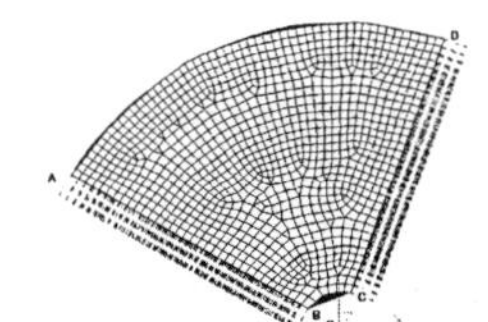

Figure 3 The meshed geometry of the turbine wheel

<u>Displacement boundary conditions</u>
Elastic translation boundary conditions have been chosen on symmetry surfaces. The stiffness vectors on the cut surface AB is inclined to the X axis by 240°, and are give as

$$\hat{n} = -0.5\,\hat{i} - 0.866\,\hat{j}$$

while that on the surface CD are given as (inclination to x axis = -20°)

$$\hat{n} = 0.84\,\hat{i} - 0.529\,\hat{j}$$

<u>Application of loads</u>
Centrifugal load ($m\varpi^2 r$) corresponding to a rotational speed of 1,40,000 r/min about the z-axis is applied to all elements. We have neglected the bending load generated due to the pressure difference at the suction and pressure surfaces.

The turbine was modeled using 3-D solid non-conformal brick elements. Typical solution time for the analysis was three hours on a P-II based PC and the process consumes 34 MB of disk space.

<u>Analysis of the results</u>
Figure 4 shows the displacement profile of the model. Table 1 gives the displacements of a few important points on the wheel surface. These points have been identified in Fig. 3. From the displacement plots the following may be observed:

1. As expected, there is no tangential displacement on the surfaces of symmetry (cut surfaces). At the outer diameter the displacement is fully radial.
2. Displacement is the highest at the exit tip of the blade. There is a radial growth of the wheel, which decreases the clearance between the wheel and the shroud. The radial displacement is 14 μm   at 1,40,000 r/min. The clearance between the turbine and the shroud has been kept as 200 μm as a first approximation. From the vibration study of the machine, the maximum vibration amplitude has been found to be about 5 μm at the rotational speed of 100,000 r/min [1]. The geometric run-out of the rotor has been measured to be 10μm. An increase in the clearance is expected at low temperature due to differential shrinkage of aluminium wheel and the stainless steel shroud. Considering the above observations, it can be concluded that the shroud clearance at the warm conditions can be reduced to 100 μm for a gain in efficiency at the design conditions.
3. It can be seen, from Fig. 4 and Table 1 that there is considerable tangential displacement at the tip of the blade. This is due to straightening of the blade due to centrifugal forces. Thus, we find that the fluid passage between the blades changes under running condition and this may be taken into consideration in design stage.

Figure 5 shows the stress field on the turbine wheel. The critical regions from a stress point of view are the hub-to-blade intersection and the bore region.   In addition, the following conclusions may made from the stress plots.

1. The maximum stress as seen from the back plate occurs at the rim of the counter bore, and has a value of 74 MPa. If the turbine base plate is considered as a uniform disk of diameter 13.16 mm, with a central hole of diameter 3.5 mm, the maximum stress is predicted by formula

$$\sigma_{\theta \max} = \left(\frac{\rho \omega^2}{4} \right)\left[(1 - v)R_1^{\ 2} + (3 + v)R_2^{\ 2} \right] = 84.0 \ MPa$$

    The difference between the two results can be traced to the assumption of plane stress conditions in the latter analysis.
2. In the bladed region, the maximum stress occurs on the blade-hub interface over the convex surface of the blade. High values of stress can be observed up to a radius of $0.4D_2$. The highest stress, however, remains well below the yield stress of the material (270 MPa).

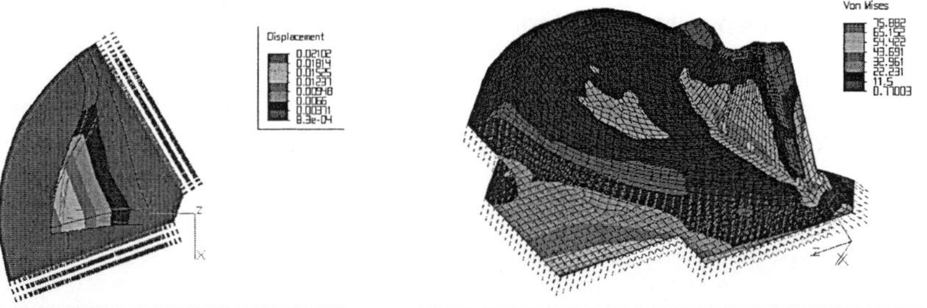

Figure 4 Displacement plot of turbine wheel   Figure 5 Stress plot of turbine wheel

STRESS ANALYSIS OF SHAFT AND BRAKE COMPRESSOR

A major part of the stress analysis of the shaft and the brake compressor has been reported in a previous paper [2]. Some of the major findings which have not been reported are:

1. The highest stress in the shaft occurs at the shaft-collar intersection. The maximum stress is 225.5 MPa with a fillet of radius 0.2 mm. We have changed the fillet radius to 1.0 mm and analysed the stress pattern. The maximum stress has been reduced to 197 MPa (material yield stress 215 MPa), which makes the design reasonably safe.

Table 1 Displacement at important points on the turbine wheel.

Point	Coordinates					Displacements				
	Cartesian			Polar		Cartesian			Polar	
	x(mm)	y(mm)	z(mm)	r(mm)	θ (deg)	dx(mm)	dy(mm)	dz(mm)	dr(mm)	θ (deg)
A	0.23	9.18	-1.21	9.19	88.56	0.011	0.005	0.017	0.012	-0.068
B	-0.23	8.91	-2.06	8.91	-88.51	0.012	0.009	0.015	0.015	-0.079
C	-11.05	7.07	7.18	13.13	-32.61	-0.002	0.002	-0.001	0.003	-0.003
D	-10.49	7.90	7.18	13.13	-36.98	-0.002	0.002	-0.001	0.003	-0.002

2. As the possibility of yielding of the material cannot be confidently ruled out, it is recommended that a stronger material e.g. K-Monel be used for the shaft instead of SS-304. Attempt should also be made to reduce the diameter of the collar and reducing the mass of the collar by cutting a circumferential slit while providing adequate bearing area to the thrust bearings. A more liberal fillet radius is also expected to make a difference.

3. The blade to hub section of the brake compressor requires appropriate fillet distribution to avoid high stress.

REFERENCES

1. Ghosh, P., Analytical and Experimental Studies on a Cryogenic Expansion Turbine, <u>Ph. D Thesis</u> IIT Kharagpur (2002)
2. Ghosh, P. and Sarangi, S., Static Stress Analysis of a Cryogenic Turboexpander Using Finite Element Method <u>International Cryogenic Engineering Conference-18</u> Narosa Publishers, Mumbai, India (2000).

Study of temperature distribution of flat steel ribbon wound cryogenic high-pressure vessel

Cui X. L, Chen G. M

Institute of Refrigeration and Cryogenics of Zhejiang University, Hangzhou 310027, China

By analyzing the heat transfer process of a flat steel ribbon wound vessel (FSRWV) wall, a temperature distribution model of the whole wall is established. Based on the model, the temperature distribution and the length change of vessel walls and flat steel ribbons at low temperature are calculated and analyzed by numerical method. The results show that the flat steel ribbon wound cryogenic high-pressure vessel is simple in structure, safe and easy to manufacture compared with the conventional cryogenic high-pressure vessel.

INTRODUCTION

Cryogenic high-pressure vessels that are used to store liquid hydrogen and liquid oxygen are important equipments in the fields of chemical engineering, astronautic engineering and nuclear power station etc. For the experimental research of astronautic engineering, the inner pressure of the vessel can reach 40 MPa and temperature 20K, so it requires much higher performance of the vessel wall material and it must satisfy some basic requirements such as high mechanical strength, well plastic property, high impact toughness value, high fracture toughness value, good fatigue strength, good forge ability and good harden ability etc. [1-3]. It is difficult for the conventional cryogenic high-pressure vessels to manufacture for that they are mostly forge-welded or multi-layer reel-welded. New type of FSRWV that initiated in China avoids the traditional technology shortcomings and has those merits.

There have been many researches on the flat steel ribbon wound vessel to date, but most of them have only aimed at optimum design [4-5] or safe monitor at ambient temperature [6-7]. Concerning the research of heat transfer, some researchers have calculated and analyzed the temperature difference of interfaces and the rate of evaporation under steady heat flow rate [8]. Up to now, research on temperature distribution, changes of the whole vessel wall and the dimension, stress change of the flat steel ribbons in very low temperatures has little presentation in literature. In this paper, a physical model for heat transfer of the vessel wall in very low temperatures is established under some reasonable hypotheses. Based on the model, the temperature distribution and the length change of vessel walls and flat steel ribbons at low temperature are calculated and analyzed. The results provide a basis for engineering design.

PHYSICAL MODEL AND NUMERICAL SIMULATION OF THE TEMPERATURE DISTRIBUTION

FSRWV model structure that is used in this research is showed in Fig.1. Its design pressure is 35MPa and the temperature is 20K.

This paper mainly discusses the heat transfer of the vessel wall. It is difficult to precisely calculate the heat transfer of the vessel wall, so it is necessary to hypothesize as follows to simplify the calculation: 1) the vessel is an infinite cylinder and the heat transfer is a one-dimensional radial form; 2) the liquid nitrogen isn't filtered into the seams of the flat steel ribbons when cooling the vessel; 3) all of the materials are isotropy.

Outside insulating material of the vessel is polyurethane foam, in which heat transfer is only by conduction and the coefficient of heat conductivity is 0.027[9]. Heat transfer between the insulating layer and outside air is natural-convection.

When liquid nitrogen is charged into liquid nitrogen zone, heat transfer happens between liquid nitrogen and flat steel ribbons and the jacket outside cylinder respectively. The heat transfer process is

484

changed according to the temperature difference ΔT between the liquid nitrogen and the wall: when $\Delta T > 100K$, for example, the film boiling heat transfer will happen; $5K < \Delta T < 100K$, the nucleate boiling heat transfer will happen; $\Delta T < 5K$, the natural-convection heat transfer will happen, in which the nucleate boiling heat transfer is the most intense process [10]. The heat transfer between inner wall and air is the same as the liquid nitrogen with flat steel ribbons and jacket outside cylinder.

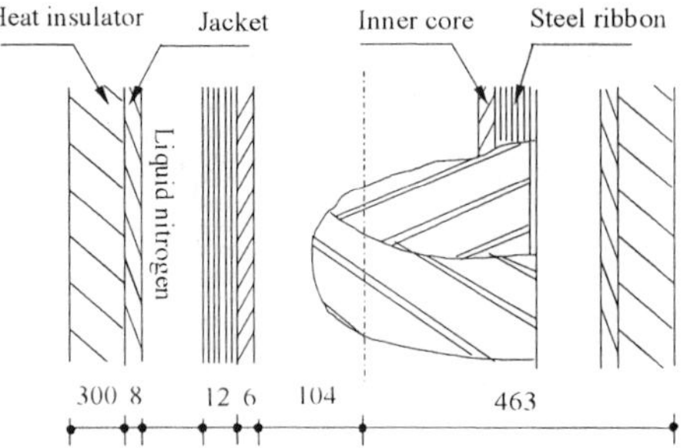

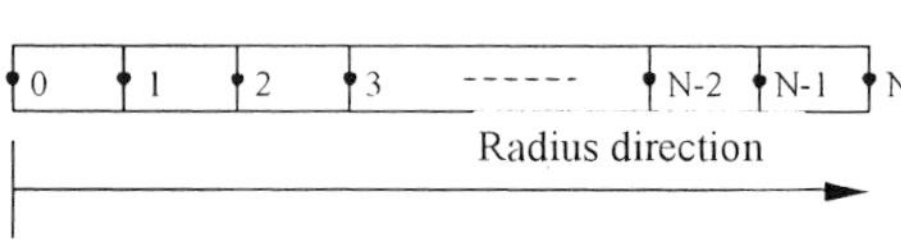

Fig.1. Sketch map of the FSRWV model structure

Fig.2. Diagrammatic sketch of vessel wall

When the liquid nitrogen is pumped out and the liquid nitrogen zone is vacuum-pumped, the zone is regarded as insulating-zone, in which the main heat transfer is radiation and the conduction of the rarefied air can be neglected [10].

According to the physical model, numerical simulation is calculated by using finite difference method. Difference equations of all nodes are established in terms of central difference method. The diagrammatic sketch of the nodes is showed in Fig.2.

RESULTS AND DISCUSSION

<u>Analysis of temperature change of the vessel wall when cooling</u>
To precool the vessel, liquid nitrogen is charged into the nitrogen zone, the temperature distribution of the wall is changed. Based on the results of simulating calculation, we get Fig.3. In the Fig., curves that are arranged from the top down represent temperature change of inner core and six flat steel ribbons from interior to exterior respectively. With the lapse of time, temperature of all nodes goes uniform, the temperature of outside nodes is more prone to stabilizing than the inside nodes and at last the temperature of flat steel and inner vessel reach 77K because the convection heat transfer coefficient of liquid nitrogen h_2 is much greater than that of the inner air h_1.

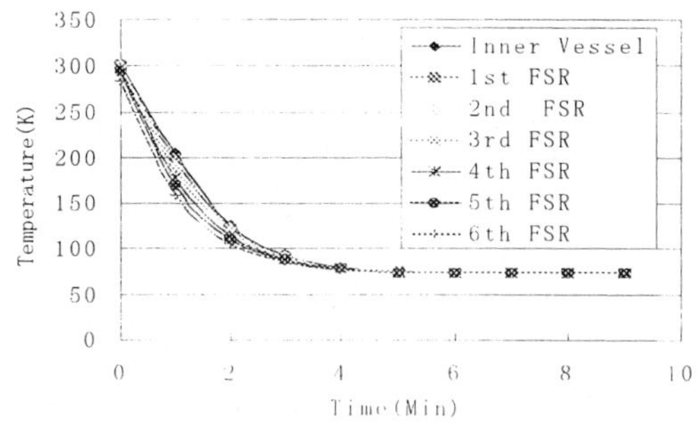

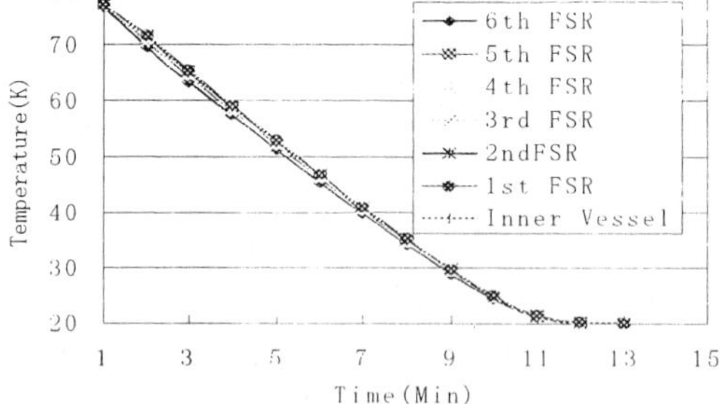

Fig.3. Temperature change of difference nodes when cooling

Fig. 4. Temperature change of different nodes when charged into liquid hydrogen

When the liquid hydrogen is charged into the inner core, the tendency chart of temperature change is showed in Fig.4 that is analogous with Fig.3, but the temperature of inner vessel wall and flat steel ribbons are 22K at last.

<u>Temperature change tendency of all nodes at different time</u>
The temperature distribution is different at different moment when the temperature field of whole vessel wall goes steady and the temperature change of all nodes is showed in Fig.5. Because here h_2 is greater than that of air h_1, the temperature difference between liquid nitrogen and flat steel ribbons is greater than that between air and inner core at initial time, heat transfer between liquid nitrogen and flat steel ribbons is more intense than that between air and inner vessel. But heat transfer goes gentleness with the temperature difference turn small with the lapse of time. It is known that the temperature change of outside flat steel ribbons is greater than that of inside flat steel ribbons at the process of going to steady (of right side slope of the curve is greater than that of left). The curve of temperature change turned into straight line that is also showed in Fig.5 bottom, which illustrates the fact that the heat transfer is changed from intensity to gentleness.

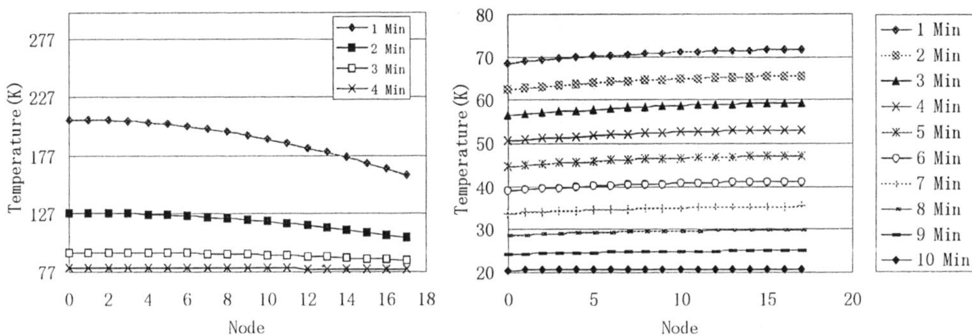

Fig.5. Temperature change of different nodes when cooling

Fig. 6. Temperature change of different nodes when charged into liquid hydrogen

When the liquid hydrogen is charged into the vessel, temperature change of the vessel wall is showed in Fig.6. Since heat transfer between inner core wall and liquid hydrogen is more intense than that of inside of the vessel wall by conduction, the slop of curves at the left part is greater than that of the right part. With the heat transfer going on, temperature changes go to steady at last.

<u>Length change of flat steel ribbons</u>
The length of flat steel ribbons changes when temperature changes. When cooling the vessel, the length change of six flat steel ribbons and inner core is showed in Fig.7, from which we know that the length change of flat steel ribbons in outer layers is greater than that of flat steel ribbons in inner layers. But at last the relative length change of each flat steel ribbon is almost the same. According to the density of the curves we know that the length change of each flat steel ribbon is different. At the beginning of cooling, relative length change of the exterior flat steel ribbon is greater than that of the interior, which tightens the flat steel ribbons and enhances the heat transfer. But it also intensifies the stress of flat steel ribbons and it can be avoided if the vessel is cooled down slowly. At last the relative length change of each ribbon goes uniform.

When the liquid hydrogen is charged into inner core, the relative length change of inner vessel and flat steel ribbons showed in Fig.8 keeps uniform because the cryo-coefficient of thermal expansion of the flat steel ribbons is an approximate constant when the temperature is lower than 80K. At the mean time, the layers of flat steel ribbons have the loosening trend because the temperature difference between inside and outside vessel wall can reach 60K. The trend can be weakened for the initial stress and can be compensated rapidly when the internal pressure is created.

In brief, the flat steel ribbons are safe and they cannot get looseness or fractures when the inner core is charged into liquid nitrogen or liquid hydrogen. they keeps their designed state when the temperature distribution of vessel wall reach steady.

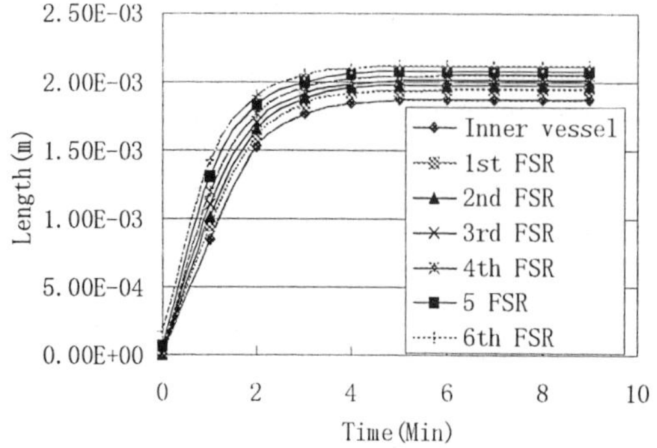

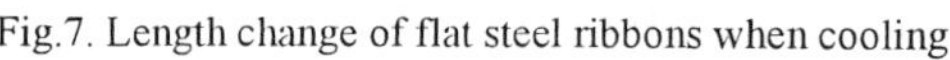

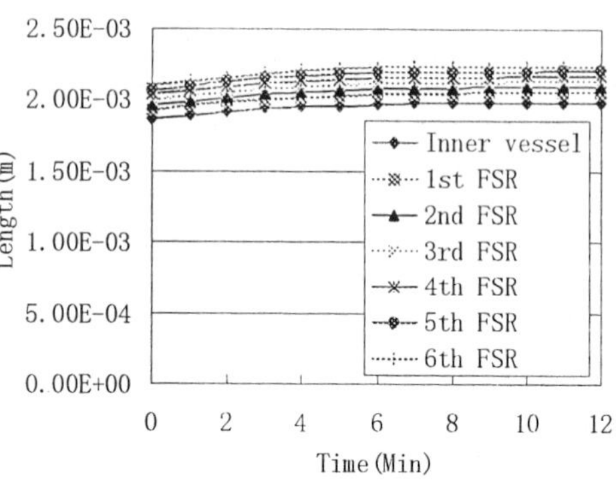

Fig.7. Length change of flat steel ribbons when cooling

Fig.8. Length change of flat steel ribbons when charging liquid hydrogen into inner vessel

CONCLUSIONS

We get the following conclusions by calculating and analyzing the heat transfer process of the new-type flat steel ribbon wound vessel wall.

① Temperature change process that heat transfer undergoes intense to gentle is analogous when the inner core is charged into liquid nitrogen or liquid hydrogen, and the temperature of whole vessel is liquid nitrogen temperature or liquid hydrogen temperature.

② The process of cooling tightens the flat steel ribbons and is propitious to heat transfer. The inner pressure has some change when charged liquid nitrogen for cooling or charged liquid hydrogen into inner core, but it does not have influence upon the vessel and the change can be compensated when the temperature of vessel wall goes steady.

③ The flat steel ribbons do not have the risk of looseness or fracture, their relative length change is the same and they keep their designed state at the steady state.

REFERENCE

1. Zhou Y. Research and development of cryogenic vessel. Cryogenics (1998) 4 15-21(in Chinese)

2. Nichols R W. Developments in pressure technology, Materials and Fabrication, London, Applied Science Publishers Ltd, (1980)

3. Lloyd E B, Edwin H Y. Process Equipment Design. New York, John Wiley 8L Sons, Inc. (1959)

4. Jiang J L. Research on design of newly low temperature and high-pressure vessel wound by steel strip. Cryogenics (2001) 3 44-47(in Chinese)

5. Zheng C X. Optimal design method of flat ribbon wound pressure vessel. Petro-Chemical Equipment (1997) 1 14-19(in Chinese)

6. Zheng J Y, Xu P, Chen C, Investigation on bursting pressure of flat steel ribbon wound pressure vessels, International Journal of Pressure Vessels and Piping (1998) 75 581-587.

7. Zhu R L, Potential developments of pressure vessel technology by using thin inner core and flat steel ribbon winding techniques, International Journal of Pressure Vessels and Piping (1996) 65 7-11

8. Chen Z P, Zheng J Y. Analysis of heat transfer and calculation of evaporation rate of flat steel ribbon wound liquid hydrogen high-pressure vessel. Cryogenics (2001) 6 56-61

9. Xu L. Cryogenic Insulation Heat and Storage and Transportation technology, Beijing, China Machine Press (1995)

10. Chen G B. Cryogenic Insulation and Heat Transfer, Hangzhou, Zhejiang University Press (1989)

The design and study of a reciprocating helium compressor

Lei G.[1,2], Gong L-H.[1], Lu W-H.[1], Zhang L.[1]

[1]Technical Institute of Physics and Chemistry, Chinese Academy of Sciences, Beijing 100080, China
[2]Graduate School of Chinese Academy of Sciences, Beijing 100039, China

This paper discusses the possibility and the performance of a helium compressor modified from a commercial air conditioning, reciprocating compressor. Modifications to the compressor include an oil injection system and new type lubricant. The exhaust helium temperature will be much higher than that allowed if we just operate this compressor without any modifications. Hence a new cooling method has been developed to intensify the heat exchange of the helium gas in the cylinder with the lubricant oil so an acceptable lower discharge temperature can be obtained. A temperature reduction of more than 70°C was achieved in the experiments.

INTRODUCTION

This paper focuses on the modification of a commercial air conditioning compressor with the input power of 7.5 kW. The discharge temperature of the compressed helium gas will exceed 180°C if we do not improve its original cooling system . The new type lubricant oil is processed to remove the water content before we operate the machine. A by-pass valve is set in the oil injection circuit to control the atomizing pressure and mass flow rate of the oil through the atomizers so we can get an optimum value of the injection pressure in various operating conditions.

A new cooling system was developed to solve this problem. An oil pump was adopted to make an oil injection circuit. The lubricant oil injected into the low-pressure chamber of the compressor is atomized when forced through three swirl-type pressure atomizers. Some of the oil particles will be sucked into the cylinders with the helium flow and compressed together with the helium gas, which is called inner cooling. And a high performance filter was also used to remove the contaminant oil from helium gas so pure helium gas should be obtained. The results of our experiments show that this method is feasible.

SYSTEM DESCRIPTION

This experiment system is composed of the compressor module, an oil pump, two plate-type heat exchanger, a multi-stage oil separator (2 or 3 stages), an adsorber and a reservoir. Figure 1 illustrates the setup configuration of the modified compressor system.

The oil injection circuit

A gear pump with the volume flow rate of 0.1 L/s was chosen to circulate the lubricant oil. The power consumption of the motor to drive this pump is 800 W.

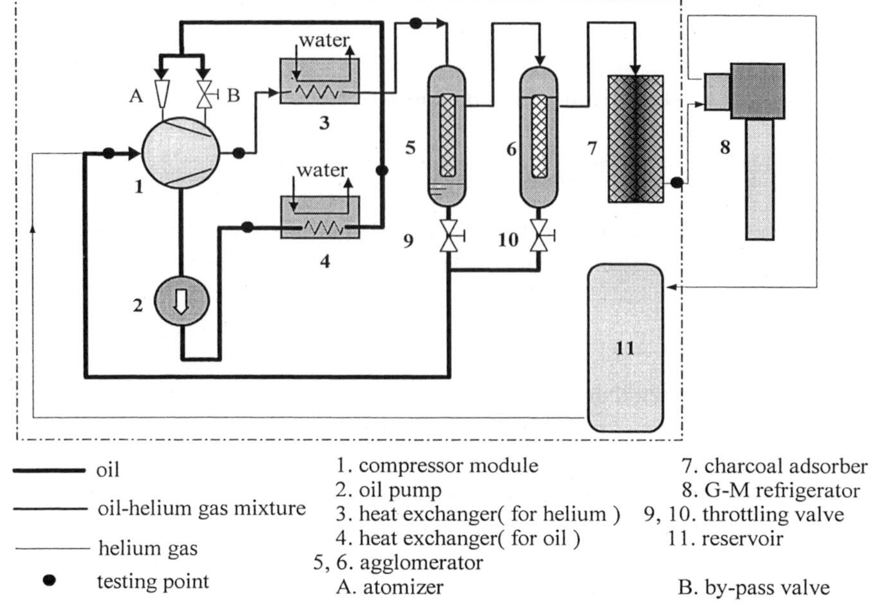

———— oil
———— oil-helium gas mixture
———— helium gas
● testing point

1. compressor module
2. oil pump
3. heat exchanger(for helium)
4. heat exchanger(for oil)
5, 6. agglomerator
A. atomizer

7. charcoal adsorber
8. G-M refrigerator
9, 10. throttling valve
11. reservoir

B. by-pass valve

Figure 1   Modified helium compressor system (inside the dashed)

The atomizers

The atomizers play an important role in the inner-cooling circuit. The lubricant oil forced through the atomizers should be fully atomized so that part of the tiny oil particles can be sucked into the cylinders to remove the compression heat of helium gas. And the pressure drop of the atomizers should not be too high thereby relative low power consumption and lower noise level of the oil pump can be obtained. We select the centrifugal pressure type atomizer that has been wildly used in many industrial facilities as a burner. The cross section of the atomizer is shown in Figure 2.

First the lubricant oil is forced through three tangential inlets symmetrically arranged surround the top end of the swirl chamber and then injected through the outlet by the effect of the centrifugal force and the differential pressure into the low-pressure chamber of the compressor.

The degree of atomization is mainly affected by the tangential and axial velocity components and the ratio of the orifice length to its diameter. A series of nozzles with different size (the diameter of the spin chamber, the radius of the inlet and the length and the radius of the orifice) were manufactured so we can find an optimum combination among these sizes by testing them respectively.

The spraying angle is theoretically given by:

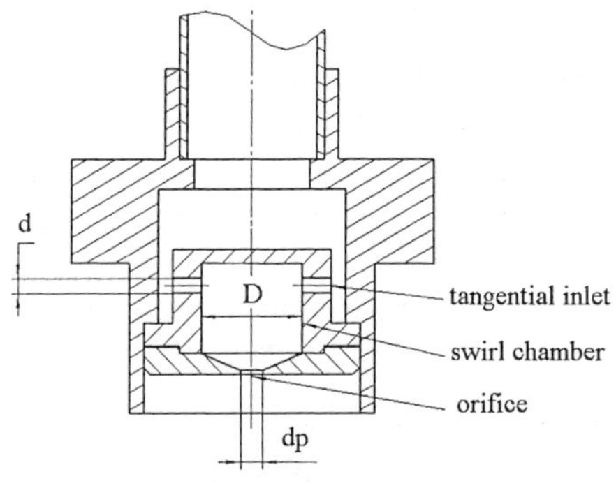

Figure 2   Cross section of the centrifugal
pressure type atomizer

$$\tan(\theta/2) = \frac{(V_t)_0}{(V_v)_0}$$

(1)

with θ : plane cone angel including all the spray, usually measured close to the orifice where it is well defined, $(V_t)_0$: tangential velocity component at orifice wall, $(V_v)_0$: average axial velocity component computed as though the orifice were running full.

Equation (1) is an oversimplification of complex velocity relationships and does not hold for small orifices or a large L/d_p ratio.

The spaying angle from the swirl-type pressure atomizers can be used as a criterion or factor of comparison for fineness of atomization for different atomizers operating under identical conditions of pressure and flow rate. Thus, the atomizer exhibiting the largest cone angle will produce the smallest drop size [1]. This makes it quite easy to identify which nozzle has the better performance of atomization by measuring the angel of the corn.

<u>The oil-separator and adsorber</u>

During the compression process the oil particles entrained the helium gas. Depending on the study of the two-phase flow through porous media [2,3], a three-stage oil separator was designed to remove thoroughly the oil contaminants. The helium-oil mixture first enters a cyclone through a tangential inlet, and the testing results show that more than 98% of the oil was separated from the helium gas by the centrifugal force in this stage. Then the helium gas flowed through a foam-catchers made by stainless steel fiber with fiber diameters of 5 to 10 micrometer, where the relative large size oil particles were trapped. Finally the helium gas entered the agglomerator filled with superfine glass-fiber. The tiny oil particles will accumulate on the fiber surfaces and drip to the bottom of the agglomerator. All the trapped oil was returned to the compressor module. The remaining vaporous contaminants will be adsorbed onto the activated charcoal, allowing only pure helium gas to travel onto the cold head.

TEST RESULTS

Figure 3 shows the variation of the spray angle when the orifice length is changed. All the experiments done are with the same testing pressure of 9.8 bar. The spray angle decreased when the length of the orifice was changed from 0.3mm to 0.2mm. Later investigation indicates that this abnormality results from the irregularity of the orifice caused by the insufficient machining precision of the drilling machine. So a 0.3 mm was chosen as the diameter of the orifice.

The temperature of the exhausting helium gas from the compressor module was measured by a

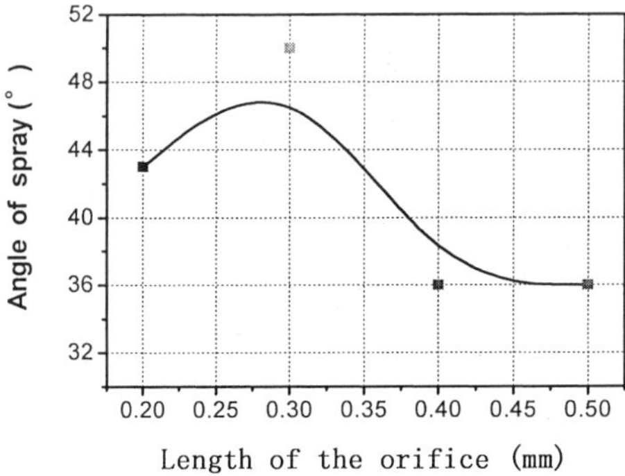

Figure 3   Comparison of spray angle with different orifice length (the other parameters of the atomizer are the same).

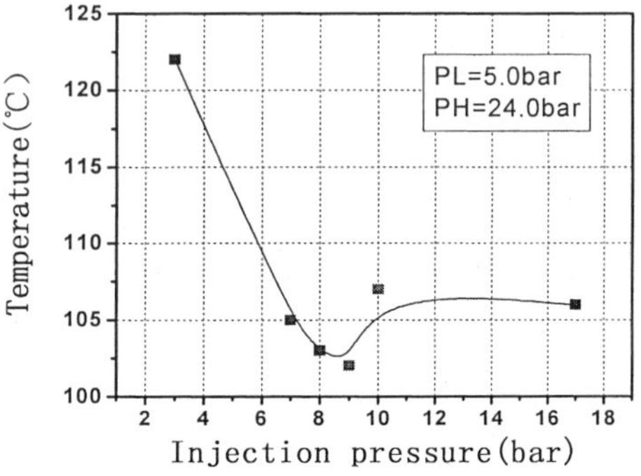

Figure 4   Measured discharge helium temperatures as a function of injection pressure of lubricant oil.

copper-constantan thermocouple located on the helium line. The atomizing pressure was controlled through changing the opening of the bypass valve set parallel with the atomizers in the oil injection circuit. So a pressure range from the minimum pressure (a little more than the pressure in the low pressure chamber of the compressor) to the maximum pressure of 17 bar. The testing results are shown in figure 4. The lowest exhausting helium temperature of 102 degrees centigrade was obtained when the atomizing pressure was set at 9bar. And this optimum atomizing pressure varies a little as the compressor was operated in a series of different conditions. Table 1 shows the specifications and typical operating parameters of the prototype compressor compared with the compressor package of the AL330 cryorefrigerator manufactured by Cryomech Inc. [4].

Table 1   Specifications and typical operating parameters

Unit	Prototype	Compressor package of AL330
Compressor Type	Reciprocating (hermetic)	Scroll (hermetic)
Oil Type	PAG (UCON LB300X)	PAG (UCON LB300X)
Power Supply	380 @ 50 Hz	380 @ 50 Hz
Operating Pressure (bar)	Discharge: 10.5-26.0 Suction: 2.0-5.2	Discharge: 6.5-24.0 Suction: 2.0-9.0
Helium Flow (Nm^3/hr)	102	88.3
Oil Injection Flow (L/s)	0.1	0.1
Nominal Motor Output (kW)	7.2	5.17
Max. Discharge Line Temp. ($^\circ$C)	102	85
Temp. of Helium into Heat Exchanger ($^\circ$C)	86-102	66-85
Temp. of Helium at Heat Exchanger outlet ($^\circ$C)	22	29.4

During an extended service of more than 1,000 hours, the modified compressor ran stably without any problem. With the ongoing improvement of the agglomerator, a lifetime of more than 8,000 hours should be obtained without a replacement of the adsorber.

REFERENCES

1.   Marshall, Jr. W. R., Atomization and Spray Drying, The Science Press, USA (1954)

2.   Scheidegger, A. E., The Physics of Flow Through Porous Media, Toronto ; Buffalo [N.Y.] : University of Toronto Press, (1974)

3.   T. Frising, V. Gujisaite, D. Thomas, S. Callé, D. Bémer, P. Contal & D. Leclerc, Spraying Angle and Spray Polydispersity in Modeling of the Aerodynamic Characteristics of FGD Towers. Filtration and Separation (2004) 41(2) 37-40

4.   Cryomech Inc. The Manual of AL330 Cryorefrigerator.

Strain effects on the critical current of Bi-2223/Ag tape

Wu Y. L., Wang Z. D.[*], Xiong Y. F.

Technical Institute of Physics and Chemistry, Chinese Academy of Sciences, China
[*]Institute of Mechanics, Beijing Jiaotong University, China

Mechanical properties of Bi-2223/Ag tapes have been studies. The conductor's critical current (I_c) keeps nearly constant when the tensile strain is low than a critical value. As tensile strain increase, I_c decreases slowly at the beginning, and then degrades sharply near 0.3% of strain. Degradations of I_c caused by the bending strain are in good agreement with the tensile testing results. Synchrotron radiation shows that the degradation of I_c is related directly with propagation of micro-crack in the oxide layer under tensile tests.

INTRODUCTION

High temperature superconducting (HTS) tapes can be regarded as a typical two component composites consisting of a brittle oxide core, which is composed of ceramic material, covered by a ductile layer. The I_c reduction in strained HTS tapes is irreversible. The explanation for this is that the stress or strain will lead to the appearance and propagation of the micro-cracks, and degrade the I_c of HTS tapes [1-2]. The influence on the I_c during the tensile and bending strain, which decrease the tape cross-section and increase the pinning center concentration, is investigated in some paper [3-5]. In this paper, tensile and bending tests on Bi-2223/Ag tapes were taken to study the effect of the strain on the I_c of HTS tapes. We discuss that the propagation of the cracks plays an important role in the impeding of the superconducting current by using the high-energy ions irradiations photos test.

EXPERIMENTAL RESULTS AND DISCUSSION

The mean thickness and width of Bi-2223/Ag tapes are 0.2 and 4.0 mm, respectively. The tensile and bending tests were taken at liquid nitrogen temperature (77 K). The signals from transducers were measured using a KEITHLEY 2000 multimeter equipped with a scanning card. The entire experiment is software/computer controlled. High-energy ions irradiated by the synchrotron radiation source at Beijing Synchrotron Radiation Laboratory have been used in taking pictures of damaging process of Bi-2223/Ag tapes under tensile test at room temperature.

Typical V-I curves under tensile stress for Bi-2223/Ag tape are shown in figure 1. When the tensile stress below 118 MPa, the critical current has little degradation. However, when the tensile stress is higher than 123 MPa, apparent decrease on critical current was observed, which is thought to be caused by the cracking of the oxide and called multiple cracking [6]. The results for normalized critical current $I_c / I_c(0)$ as a function of applied axial stress under self-field are shown in figure 2. As seen in figure 2, for all the five specimens, the Bi 2223/Ag tapes keep the initial critical current density until the tensile stress up to a critical value, which causes significant damage to the superconductor ceramic core so that the measured critical current decreases greatly. I_c degradation started from 100 MPa tensile stress to 150 MPa, which shows in spite of the overall similarity in structure of the five specimens, there are subtle differences among tapes.

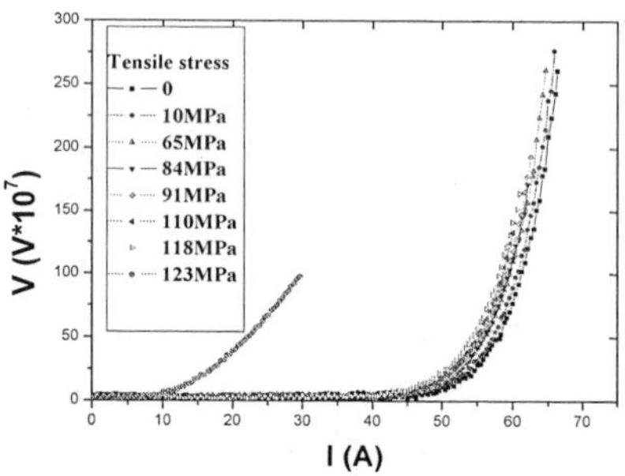

Figure 1 V-I curves of Bi-2223/Ag tape under different tensile stress.

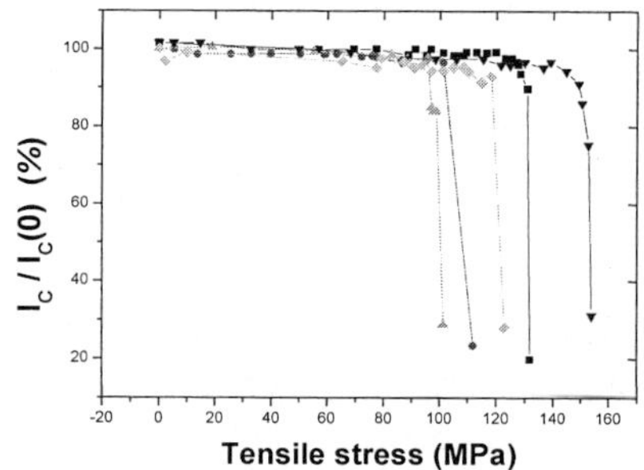

Figure 2 Normalized I_c at 77K under self-field for as a function of axial tensile stress for Bi-2223/Ag tapes.

Figure 3 shows typical results of the strain dependence of the normalized I_c for tensile tests. The I_c of the Bi2223/Ag tape decreases drastically at tensile strain of ε_{irr}=0.3. Therefore, the point of irreversibility in the I_c is about ε_{irr}=0.3%. When the tensile strain is lower than the critical value, the ceramic core and sheath only have elastic deformation, and no fracture happened. When the tensile strain is larger than the critical value, because ceramic materials can undertake lower strain than the sheath, cracks will appear in the core and propagate quickly. During this process, the existing cracks will become larger and new cracks will appear. The existence of these cracks will decrease the capability of the pass of the current and lead to the fracture of the whole superconducting core.

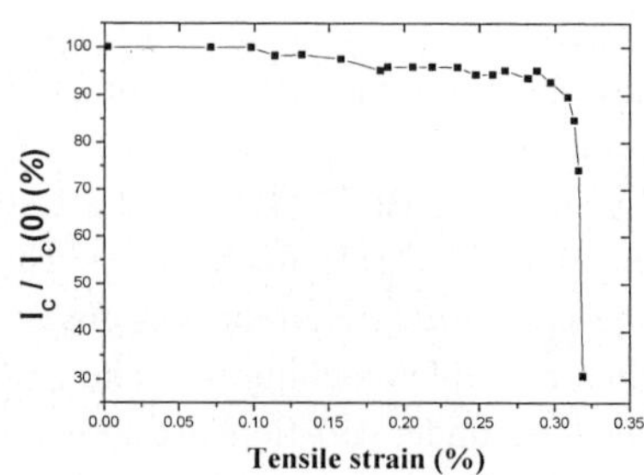

Figure 3 Normalized I_c densities at 77k under self-field as a function of tensile strain (ε).

The bending properties of the Bi2223/Ag tape are given in figure 4. The results of rapid degradation of I_c at bending measurements for the specimens indicate the likely formation of cracks. During the process of bending, the inner side of tape would undergo compressive stress, and the out side would undergo tensile stress, which cause the strain in the oxide core and leads to the decrease I_c.

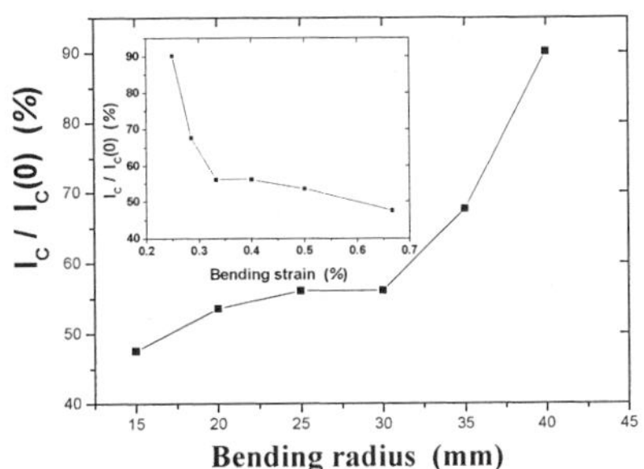

Figure 4 Normalized I_c with bending at 77K under self-field as a function of bending radius for the sample Bi-2223/Ag tape. The insert shows the relationship of the bending strain and normalized I_c.

The bending strain (ε) was determined from the relationship of $\varepsilon = t/2R$, where t is the total thickness of the tape and R is the radius of curvature. In order to study the influence of the bending strain on I_c, the tapes were bent around six different rollers, the diameters change form 30 to 80 mm, corresponding to tensile strains (ε) in the oxide layer from 0.67% to 0.25%. The beginning crack formation and current degradation is observed for a bending radius =35 mm corresponding to 0.29% strain on the convex side of the tape, which is good agreement with the tensile testing results.

Figure 5 Damaging process of Bi-2223/Ag tapes under tensile stress at room temperature.

The tensile and bending strain results can be explained by the expansion of the micro-cracks existed in the tape. Figure 5 gives the inner-structure damaging photo of Bi-2223/Ag tapes under tensile stress at room temperature by synchrotron radiation. The recorded photos show that with increasing of tensile stress, thin and short cracks appear at beginning, propagates along cross-section of oxide core and many new cracks appear with the increase of the stress. The cracks are not totally parallel, which means it might appear and propagate along the poorest connection among ceramic grains. When a certain strain is exerted on samples, there appear micro-cracks in oxide core at beginning, and then the extension of the cracks is the main reason for the decrease of the I_c density.

CONCLUSION

In order to study the influence of the strain on properties of HTS tapes, especially the critical current, tensile, bending tests are taken. It shows when the strain is higher than a critical value (about 0.3% for our specimens), the critical current begin to decrease, slowly at first, then quickly. According to the results got by synchrotron radiation, the existence and propagation of cracks in superconducting core might be the main reason to cause the decrease of the critical current. Therefore, the critical strain is the critical parameter to instruct the rolling and application of HTS tapes.

ACKNOWLEDGEMENTS

This project is funded by NJTU (TJJ03001)

REFERENCES

1. Amm, B.C., Hascicek, Y.S., Schwartz, J. et al., Mechanical properties and strain effects in $Bi_2Sr_2CaCu_2O_x$/Ag composite conductors, <u>Advances in crogenic Engineering (materials)</u> (1998) <u>44</u> 671-678

2. Kiss, T., Van Eck, H., Ten Haken, B., et al., Transport properties of multifilamentary Ag-sheathed Bi-2223 tapes under the influence of strain, <u>IEEE Transactions on Applied Superconductivity</u> (2001) <u>11</u> 3888-3891

3. Nishimura, S., Kiss, T., Inoue, M., et al., Influence of bending strain on Bi-2223 tape, <u>Physica C</u> (2002) <u>372-276</u> 1001-1004

4. Sosnowski J., The influence of the bending process on the critical current of high-T_c superconducting tapes, <u>Physica C</u> (2003) <u>387</u> 239-241

5. Ahoranta, M. Lehtonen, J., Kováč, P., et al. Effect of bending and tension on the voltage–current relation of Bi-2223/Ag, <u>Physica C</u> (2004) <u>401</u> 241-245

6. Ochiai, S., Hayashi, K. and Osamura, K., Improvement of strain endurance of critical current of silver-sheathed superconducting tapes by reducing volume fraction of Bi(Pb)-Sr-Ca-Cu-O oxide core, <u>Cryogenics</u> (1993) <u>33</u> 976-979

Thermal, electrical properties and microstructure of high purity niobium for high gradient superconducting cavities

H.M. Wen*, X.Singer, W.Singer, D. Proch, E. Vogel, D. Hui*

Deutsches Elektronen-Synchrotron(DESY), Notkestrasse 85, 22607 Hamburg, Germany
*Visiting Scientist, Institute of Electrical Engineering, Chinese Academy of Sciences, Beijing 100080, China

For controlling niobium properties for the fabrication of superconducting cavity, the systems for quick measurement of Nb thermal conductivity and residual resistivity ratio RRR are created. Thermal conductivity test can be done in a 100 l transport liquid helium tank. A DC RRR measurement for Nb samples and a specially developed AC method for nondestructive RRR measurement of the cavities in situ are presented. Thermal conductivity, RRR distribution and the microstructure of electron-beam welding seam of high purity niobium are investigated.

INTRODUCTION

Niobium is the favorite metal for the fabrication of superconducting accelerating cavities[1]. The project TESLA XFEL plans to use 1000 niobium superconducting cavities[2]. The proposed TESLA Linac needs over 21,000 niobium cavities[3].

Thermal break down is one of the main limitations on the way to high gradient superconducting cavity. To avoid thermal break down, high thermal conductivity of niobium is needed[4]. For niobium quality control, a cryogenic thermal conductivity measurement system, a DC and an AC RRR measurement techniques are developed. The thermal conductivity, RRR distribution and microstructure of electron beam welding niobium were investigated, and their correlation was obeserved by experiments.

DEPENDENCE OF CAVITY BREAKDOWN FIELD ON THERMAL CONDUCTIVITY

A normal conducting spot triggers quench when it heats the Nb above T_c. Breakdown field is given by [5]

$$H_{tb} = \sqrt{\frac{4k_T(T_c - T_b)}{R_d}} \qquad (1)$$

Where K_T - Thermal conductivity of Nb. R_d - Defect surface resistance. T_c -Critical temperature of Nb. T_b - Bath Temperature.

THERMAL CONDUCTIVITY DEPENDENCE ON RRR

The empirical formula (2) can be used to estimate the thermal conductivity of superconducting niobium in a wide temperature range [6] :

$$\lambda(T, RRR, G) = R(y) \cdot \left[\frac{\rho_{295K}}{L \cdot RRR \cdot T} + a \cdot T^2 \right]^{-1} + \left[\frac{1}{D \cdot \exp(y) \cdot T^2} + \frac{1}{B \cdot G \cdot T^3} \right]^{-1} \quad (3)$$

Where $y = \alpha \bullet T_c / T$, α=1.76, L=2.45×10⁻⁸ WK⁻², a = 2.3 × 10⁻⁵mW⁻¹K⁻¹, 1/D=300 mK⁻³W⁻¹, B=7.0 ×10³Wm⁻²K⁻⁴, G - grain size. From the curve in [6], $R(y)$ can be roughly fitted as

$$R(y) = -0.28401y^3 + 4.6281y^2 + 0.7787y - 0.0131 \quad (4)$$

With formula (3), a theoretical value of thermal conductivity dependence on RRR is calculated and compared to the experimental data. In the range between 4K and 10K, the experimental data are roughly the same as the theoretical values, as shown in Fig.2.

A CRYOGENIC THERMAL CONDUCTIVITY MEASUREMENT SYSTEM

A cryogenic thermal conductivity measurement system is created, as shown in Fig. 1. To avoid the trouble of transferring liquid helium, a 100 litre liquid helium storegy transfer tank is applied as the cryostat. Its advantage is that a simple and quick test can be realized in industry. Due to the limited inlet diameter of the cryostat, the high vacuum sample chamber has a limited space for sample and sensors. It is a challenge to minimize the losses produced by radiation and wires. To block the radiation from the top, a copper flange is fixed on the top of the sample chamber. It has a good contact with liquid helium. To minimize the losses from wires and provides a good tenacity, 0.1 mm diameter of NbTi wires are applied. The steady state method is used to determine the thermal conductivity k. $k = Q / A \bullet dx / dT$, where Q is the heat source power, A the section area of sample, dT / dx the temperature gradient along the sample. To check the system accuracy, a series of niobium samples with known RRR are measured. The results are shown in Fig.2. Comparing the results with the theoretical value and the results measured by the former different system in the range between 4K and 10 K, the results are quite convincing.

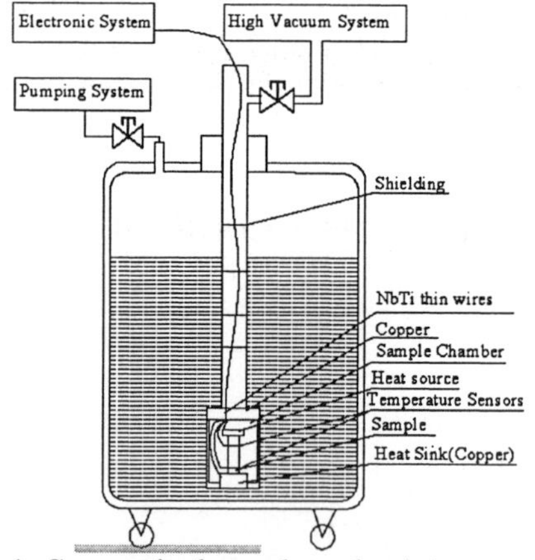

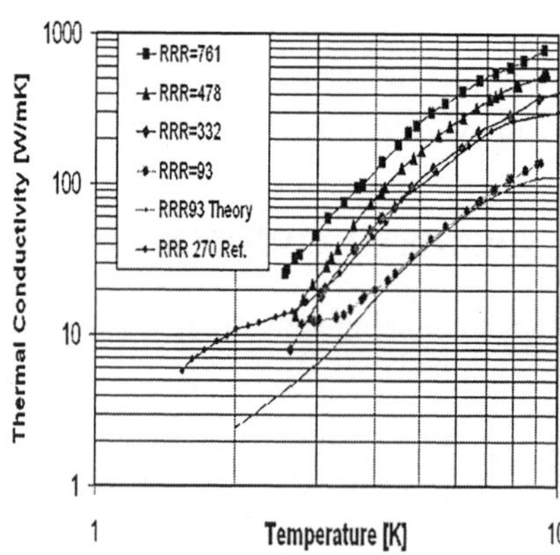

Fig. 1 Cryogenic thermal conductivity measurement apparatus

Fig. 2 Comparison with theoretical value and the results done by conductivity

THERMAL CONDUCTIVITY AND MICROSTRUCTURE OF EB WELDING HIGH PURITY NIOBIUM

Electron beam welding(EBW) is applied in cavity production for welding all parts together with high quality. Its influences on high purity niobium is investigated by measuring its RRR and

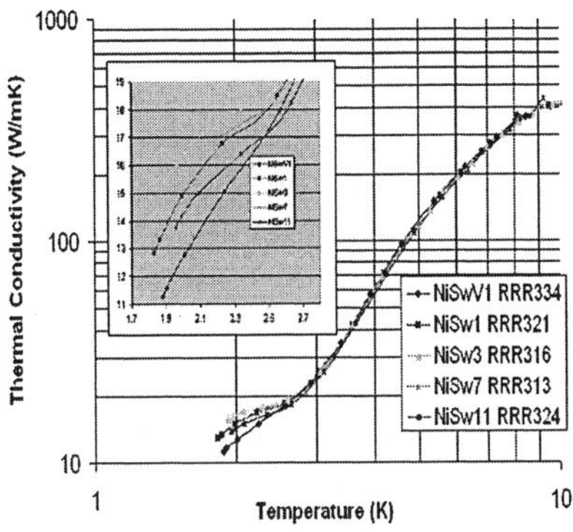

Fig.3 EB welding niobium thermal conductivity

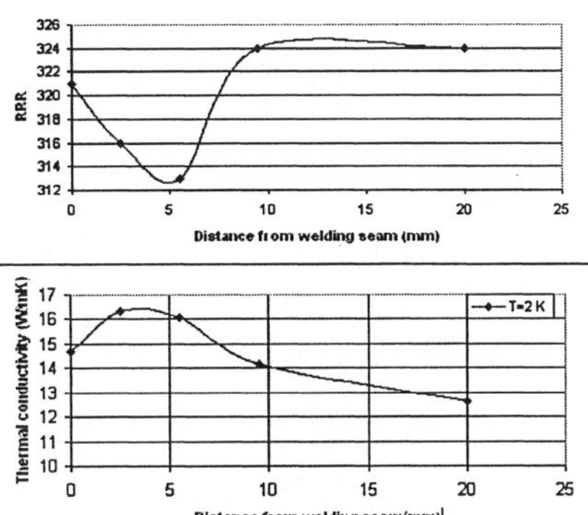

Fig.4 RRR in the welding seam versus distance from the welding seam

thermal conductivity distribution. The influences of EB welding on high purity niobium RRR distribution have already been reported in Ref. [7]. Fig.3 shows the thermal conductivity distribution on an EB welding niobium sample, which was done earlier in a specially designed device. In the range below 3 K, the thermal conductivities are quite different. Fig.4 shows the thermal conductivity distribution in EB welding at 2K. The thermal conductivities in this region do not depend on RRR, but on grain sizes. Except the first point, other points have the same relationship as Fig.5, which gives the theoretical thermal conductivity dependence on grain size at 2K. Fig.6 gives a typical microstructure of EB welding sample. The grain size in the welding seam is much bigger.

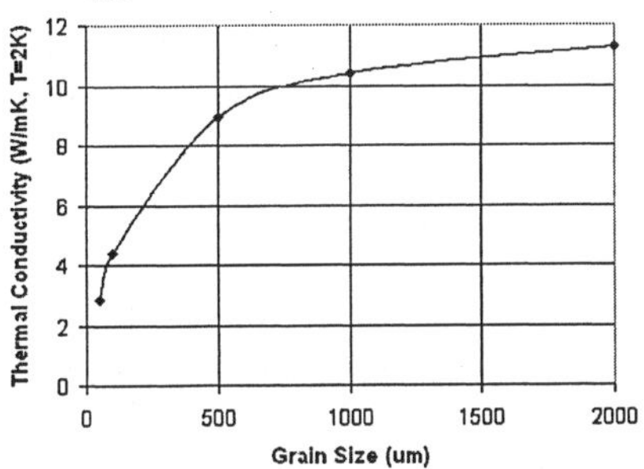

Fig.5 Theoretical thermal conductivity dependence on grain size at 2 K, for RRR=320

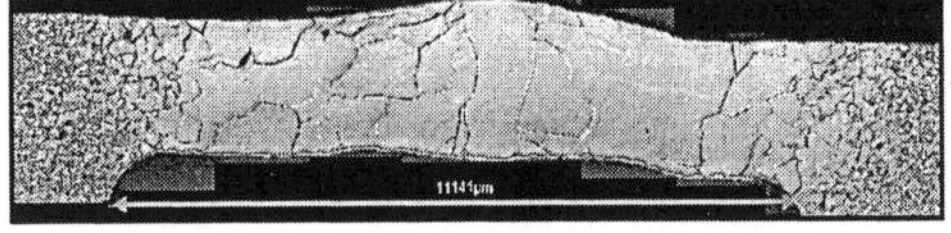

Fig.6 Microstructure of the EB welding area. The grain size G=50-2000 μm

RESIDUAL RESISTIVITY RATIO RRR CONTROL

Since that thermal conductivity of niobium can be estimated by RRR and grain size, RRR can be a tool for estimating the thermal conductivity of niobium. Therefore, a DC RRR system for niobium quality control and an ac RRR test system with eddy current method for cavity RRR control are created, as shown in Fig.7. AC RRR test system is described in Ref. [8].

A DC RRR measurement system with a 1.1 Tesla superconconducting magnet, was created recently. With the superconducting magnet, niobium will be kept on normal conducting state at 4.2K, so that we can measure the sample resistance at 4.2K directly. By measuring the sample's resistance at room temperature and 4.2K, we get the RRR=R(295K)/R(4.2K). In practice, we found that the surface layer of the sample has a big influence on the penetration field, as shown in Fig.8. For samples without etching, the penetration field is much higher than the etched samples. The reason may be that the surface contamination makes more pinning centers and cause a bigger pinning magnetic field. Therefore, for keeping the magnetic field completely penetrate the sample

Fig.7 Cavity RRR measurement system

U4.2K dependence on Magnetic Field

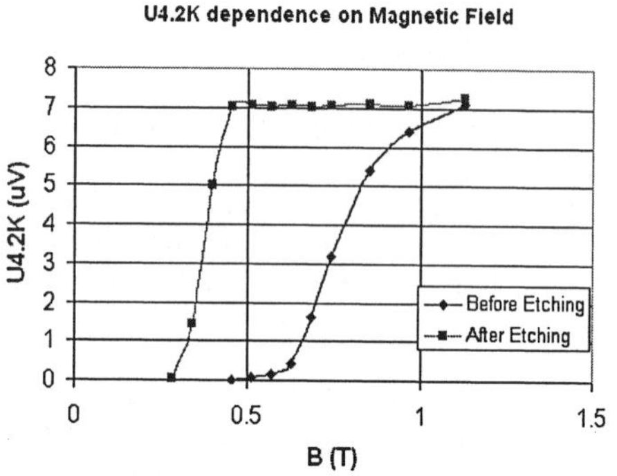

Fig.8 Magnetic field penetration in Niobium Sample

and keep the sample in normal conducting state, the magnetic field is fixed at 1.1 T, much higher than the critical field 0.23 T.

SUMMARY

- An innovative cryogenic thermal conductivity measurement system based on transport liquid helium cryostat was described.
- Thermal conductivity distribution of electron beam welding sample was investigated. Below 3K, the thermal conductivities dependence on grain sizes was observed.
- A RRR measurement system with a 1.1 Tesla superconducting magnet was described. The experiences show that the backgound field up to 1.1 T for niobium is necessary.

REFERENCES

1. X.Singer, High purity niobium for TESLA Test Facility, Matériaux & Techniques, Numéro 7-8-9 2003, pp28-32.
2. TESLA XFEL Technical Design Report, TESLA FEL 2002-9, Oct. 2002, http://tesla.desy.de/new_pages/tdr_update/supplement.html
3. H.Padamsee, Niobium Cavities for TESLA, Workshop on Niobium for RF Cavities, Fermilab June 10-11, 2002, WWW: http://www-bd.fnal.gov/niobium/talks/Padamsee/ fermilab_nb_workshop.ppt.
4. W.Singer, A.Brinkmann, D.Proch, X.Singer, Quality requirements and control of high purity niobium for superconducting RF cavities, Physica C 386(2003) 379-384.
5. H. Padamsee, J. Knobloch, and T. Hays, *RF Superconductivity for Accelerators,* Wiley, New York, 1998.
6. F. Koechlin, B. Bonin, Parametrisation of the Niobium Thermal Conductivity in the Superconducting State, Supercond. Sci. Technol. 9(1996)453-460.
7. W. Singer, X. Singer, J. Tiessen, H.M. Wen, RRR Degradation and Gas Absorption in the Electron Beam Welding Area of High Purity Niobium, *International Workshop on Hydrogen in Material & Vacuum System*, November 11-13, 2002, Jefferson Lab, Newport News, Virginia, USA
8. H.M.Wen, W.Singer, D.Proch, L.Y.Xiao, L.Z.Lin, "Cavity RRR Test with Eddy Current Method for TESLA Test Facility", *Proceedings of the 18th International Cryogenic Engineering conference(ICEC18)*, Mumbai, India 2000

AC loss characteristics of Bi-2212 multifilamentary round wires and their cables

Funaki K., Iwakuma M., Kajikawa K., Hirano N.*, Nagaya S.*, Ohtani N.[#], Hasegawa T.[#]

RISS, Kyushu University, 6-10-1 Hakozaki, Higashi-ku, Fukuoka 812-8581, Japan
*Chubu Electric Power Co., Inc., 20-1 Kitasekiyama, Ohdaka-cho, Midori-ku, Nagoya 459-8522, Japan
[#]Showa Electric Wire & Cable Co., Ltd., 4-1-1 Minami-Hashimoto, Sagamihara 229-1133, Japan

The AC losses of Bi-2212 multifilamentary round wires and the Rutherford-type cables were measured in transverse AC magnetic fields at liquid helium temperature by means of a standard pickup-coil method. The AC losses obtained for the strands were scarcely dependent upon the frequency of the external field in the range from 0.02 Hz to 10 Hz. This suggested that the hysteresis loss in a tightly coupled filament bundle be dominant. A remarkable improvement in the AC loss characteristic by reducing an effective diameter of the filament bundle are observed, while twisting the filaments itself does not contribute to the reduction in the AC loss. It was also found from the comparison between the AC losses in the strands and the cables that the AC losses in the cables mainly came from the individual strands.

INTRODUCTION

Oxide-superconductors such as Bi- and Y-systems have feasibilities of the applications in a wide range of temperature between liquid helium temperature and liquid nitrogen one, because the critical temperatures are higher than liquid-nitrogen temperature. Especially for the Bi-system superconductors whose long wires have been fabricated are expected as advanced materials of windings for developing electrical devices with high performances in near future and the electromagnetic properties including AC losses have been measured in the wide range of temperature from fundamental viewpoints. In the design of the individual wires and the cables, however, it is not always considered to optimize the electromagnetic properties, in particular AC losses, in the environments to be applied to, while the wires and cables are designed and fabricated for constructing R&D models of superconducting electrical devices such as power cables and transformers under the restriction of developing HTS wires.

The present paper describes AC loss properties of Bi-2212 multifilamentary round wires and their Rutherford-type cables, which may be one of candidates for large-capacity windings in the electrical devices such as future superconducting magnetic energy storages (SMES) and so on, from a viewpoint of feasibility study. The strands were not twisted. Their AC loss properties in a transverse magnetic field were to be like those of monofilamentary wires in spite of the multifilamentary structure. The experimental results partially support the prediction that the AC loss is hysteretic. The experiments also give unexpected results that the effective diameters of the filamentary region evaluated from the measured AC losses are much smaller than the actual ones. It is also shown that the electromagnetic coupling among the strands is not dominant in the cables in the transverse field parallel to the flat surface.

SAMPLE PREPARATIONS AND AC LOSS MEASUEMENT

We designed and fabricated Ag-sheathed Bi-2212 multifilamentary round wires and their Rutherford-type cables to evaluate the AC loss properties, which are developed as large-capacity conductors for future HTS coils. They were heat-treated by a wind-and-react method. The characteristics of the wires and the cables are listed in Tables 1 and 2, respectively. Table 1 includes the critical currents measured with a criterion of 10^{-6} V/cm both in the terminal parts of each coiled specimen. The set of wires were prepared for estimating the effects of material of the first-stack sheath and twist of filaments on the AC loss properties. Each pair of twisted and untwisted wires has the common fabrication process except for

Table 1　Characteristics of Bi-2212 multifilamentary wires

No.	Diameter	Filament structure	1st-stack sheath	Silver ratio	Twist pitch	Ic(upper) at 4.2K,0T	Ic(lower) at 4.2K,0T	d_{eff}/d_{fr}
1	0.81 mm	61×7	pure Ag	4.1	--	163 A	179 A	1/3
2	0.71	61×7	pure Ag	4.1	9.5	141	150	1/3
3	0.81	61×7	pure Ag	3.0	--	333	315	1/8
4	0.81	61×7	pure Ag	3.0	9.5	327	165	1/8
5	0.81	61×7	Ag alloy	3.0	--	378	267	1/25
6	0.81	61×7	Ag alloy	3.0	9.5	381	321	1/25
5'	0.81	61×7	Ag alloy	3.0	--	393	372	1/8

Table 2　Specifications of Bi-2212 Rutherford-type cables

No.	Strand Diameter	Filament structure	1st-stack sheath	Silver ratio	No. of strands	Cabling pitch
A	1.02 mm	127×7	pure Ag	2.8	20	73 mm
B	0.81	61×7	pure Ag	3.0	30	87

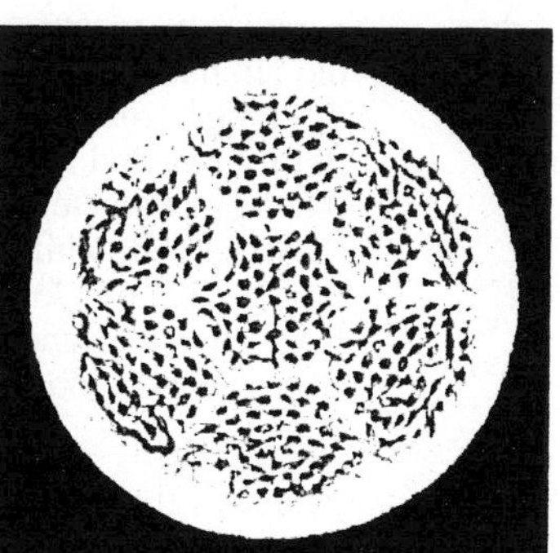

twisting process. Two types of cables listed in Table 2 were also prepared to evaluate some additional losses in which the untwisted strands have no insulation layer and only an oxide layer of the Ag outermost layer. A photograph of the cross section of the wire #5 is given in Fig. 1.

Figure 1　Photograph of cross section of wire #5 after drawing and heat treatment.

We measured the AC losses of the coiled specimens of Bi-2212 the strands and their cables in liquid helium by a standard pickup coil method [1]. AC magnetic field of 0.02 Hz to 10 Hz was applied in the direction parallel to the coil axis. The concentric arrangement of the pickup coils and the coiled specimen are based on the standard method.

RESULTS AND DISCUSSION

Effects of twisting strand on AC loss properties

As seen in Table 1, we prepared a set of strands, twisted and untwisted wires, to estimate the effect of twisting on the AC losses in the strands. Figure 2 shows the comparison between the AC loss properties for the amplitude of the external field in the twisted and untwisted wires, No. 5 and 6 at the frequencies of 0.1 Hz and 10 Hz. It is suggested from the comparison that the effects of twisting can be almost neglected in the internal loss of the strands. This means that the strands behave as monofilamentary ones due to electromagnetic couplings and/or direct touch among the strands in spite of the multifilamentary structure. We obtained the same results for other two sets of specimens, No. 1 and 2, and No. 3 and 4. The frequency dependence of AC loss property, on the other hand, may come from relatively low n value of the specimens. In these cases, we can only estimate the AC loss properties approximately by means of the critical state model and more quantitative evaluation of AC

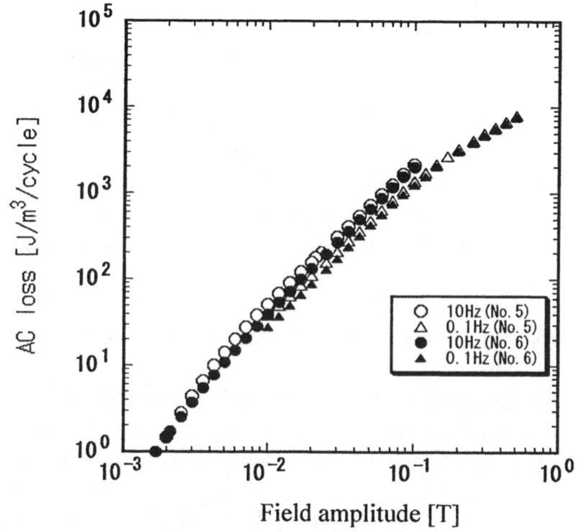

Figure 2　Effects of twisting on AC loss in strands

losses could result from some numerical simulation on the basis of E-J characteristics. For simplicity, however, we use the critical state model to discuss AC loss properties of the strands in the present paper.

AC loss estimation using effective diameter of filamentary region

Under the condition that the Bean model can be applicable to the AC loss estimation for columnar superconductors with diameter d and critical current density J_c exposed to a transverse magnetic field, the hysteresis loss density per cycle [J/m^3] is expressed as [2]

$$W_\perp = \begin{cases} \dfrac{4}{3}\left(2h_{m\perp}{}^3 - h_{m\perp}{}^4\right)\mu_0 H_{p\perp}{}^2 & : h_{m\perp} \leq 1 \qquad (1) \\[3mm] \dfrac{4}{3}\left(2h_{m\perp} - 1\right)\mu_0 H_{p\perp}{}^2 & : h_{m\perp} \geq 1 \qquad (2) \end{cases}$$

$$h_{m\perp} = H_m / H_{p\perp} \qquad (3)$$

where H_m is the amplitude of external magnetic field and $H_{p\perp} = J_c d/\pi$ is a penetration field. The above theoretical prediction shows that the hysteresis loss is almost proportional to the penetration field, namely the diameter, in a region of $h_{m\perp} \gg 1$. If the filaments are electromagnetically independent of each other in the multifilamentary wires, d in $H_{p\perp}$ corresponds to the filament diameter d_f. For the tight coupling among the filaments, on the other hand, d is equal to the diameter d_{fr} of the filamentary region. In this way, if we substitute an effective diameter d_{eff} for d in $H_{p\perp}$, we can estimate the degree of equivalent filament coupling by d_{eff}, whose value is obtained in the region of $d_f \leq d_{eff} \leq d_{fr}$ by fitting the above theoretical prediction to experimental AC loss properties of the wires.

The dependences of AC losses on the field amplitude are shown for the strands #1, #3 and #5 in Figs. 3, 4 and 5, respectively. In each figure, we can find the loss properties are a little dependent upon frequency of applied field in the region between 0.02 Hz and 10 Hz as mentioned in the previous section. Since the loss property is scarcely influenced by the frequency in the region of $h_{m\perp} > 1$, we evaluated the effective diameter d_{eff} by fitting in the higher amplitude region. The effective diameter normalized by d_{fr} is listed for all pairs of wires in Table 1. Each pair of wires fabricated in the common process has the same value. The theoretical prediction with each effective diameter is indicated by a solid line for the strand #1, #3 and #5 in Figs, 3, 4 and5, respectively. In these figures, the theoretical results for the case of $d_{eff} = d_{fr}$ are also indicated by dashed lines for comparison. We can find a remarkable reduction in the effective diameter, which leads to the suppression of the hysyeresis loss in the region of higher amplitude. Since the reduction in the hysteresis loss is not influenced by twisting, this effect is not equivalent to the so-called 'twist effect' observed in twisted in-situ Nb$_3$Sn multifilamentary wires [3].

From the comparisons of the normalized d_{eff} in Table 1, we first find a tendency that the 1st-stack sheath with higher resistivity results in more effective reduction in the normalized d_{eff}, namely hysteresis

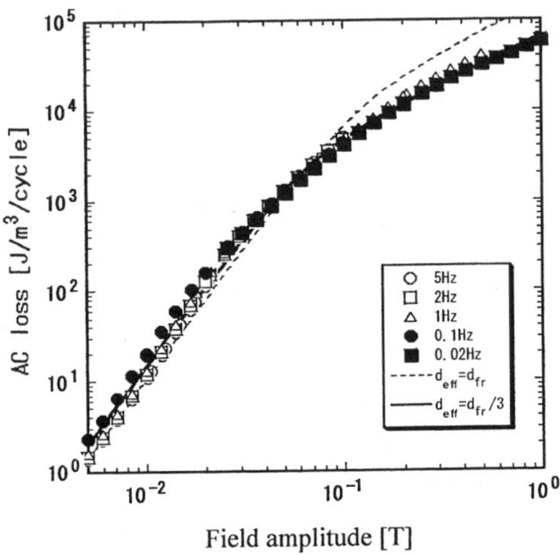

Figure 3   AC loss properties of strand #1

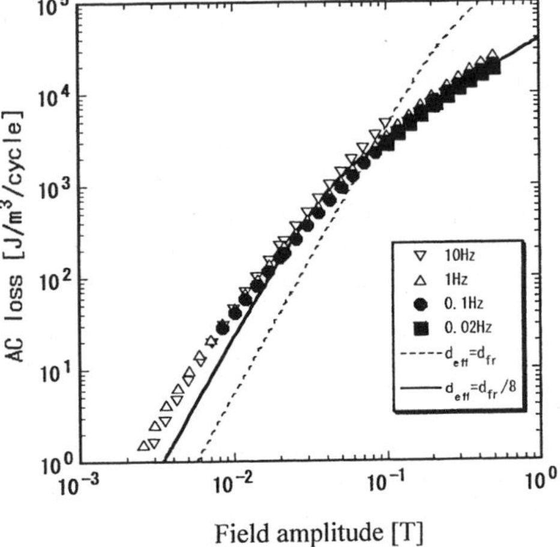

Figure 4   AC loss properties of strand #3

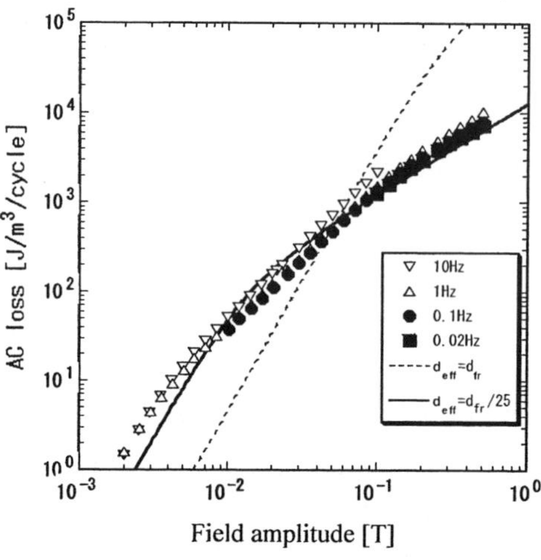

Figure 5   AC loss properties of strand #5

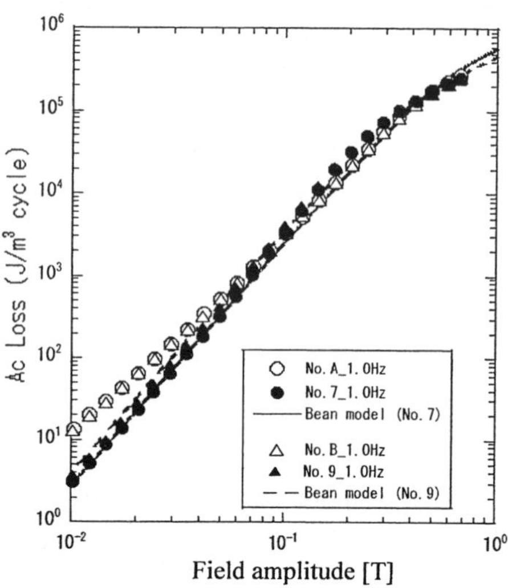

Figure 6   AC losses in strands and cables

loss density. As shown from the further comparisons of the normalized d_{eff} between the wire #1 and #3, or #5 and #5', we cannot control the reduction in the present fabrication processes. We have to need consideration on electromagnetic origin of the reduction in the normalized d_{eff} to design it quantitatively.

<u>Additional AC loss by cabling</u>
According to two types of Rutherford cables, in order to evaluate additional AC losses due to cabling, we also measured the AC losses of the cables and the individual strands by the usual pickup coil method. The coiled specimens of Rutherford-type cables were wound in a flat-wise manner. The results are shown both for the cables #A and #B in Fig. 6, where the AC losses measured for the strands are also plotted. It can be seen from the comparison in Fig. 6 that the difference between the AC losses of the strands and their cable is not dominant in the region of higher amplitude of applied field, where AC losses generated give a major contribution to the total loss in the practical windings.   In this way, the additional loss due to cabling can be almost neglected for the present Rutherford-type cables from the practical viewpoint. We understand this result comes from an oxide layer on the surface of strands.

CONCLUDING REMARKS

The AC losses of Bi-2212 multifilamentary round wires and the Rutherford-type cables were measured in transverse AC magnetic fields at liquid helium temperature by means of a standard pickup-coil method. The AC loss of strands was scarcely dependent upon the frequency. This implies that the AC loss of the strands is almost hysteretic and the filaments are electromagnetically coupled with each other in the filamentary region. A remarkable improvement in the AC loss characteristic by reducing an effective diameter of the filament bundle are observed, while twisting the filaments itself does not contribute to the reduction in the AC loss. Its mechanism is not clear at present. The additional loss due to cabling can be almost neglected for the present Rutherford-type cables from the practical viewpoint, because of an oxide layer formed on the surface of each strand

REFERENCES

1. AC loss measurement--Total AC loss measurement of Cu/Nb-Ti composite superconducting wires exposed to a transverse alternating magnetic field by a pickup coil method, <u>IEC International Standard 61788-8</u> (2003) <u>First Edition</u> 4-43
2. Noda, M., Funaki, K., Yamafuji, K, Hysteresis loss in a superconducting rod with a round cross section exposed to a transverse AC magnetic field, <u>Mem. Faculty Eng. Kyushu Univ.</u> (1986) <u>46</u> 63-76
3. Carr, Jr., W.J., Effect of twist on wires made from *in situ* superconductors, <u>J. Appl. Phys.</u> (1983) <u>54</u>, 6549-6552

Property of Joint Resistance of Bi2223 Multi-filamentary Tape by Using Sn-Pb Solder

Gu C., Zhuang C., Zhang J.Y., Qu T.M., Han Z.

Applied Superconductivity Research Center, Tsinghua University, Beijing, China, 100084

In this paper, the joint resistance between two Bi2223 multi-filamentary tapes was measured by field decay method. In order to achieve this goal, a closed Bi2223 coil was fabricated. The joint was welded with ordinary Sn-Pb solder. The field decay measurement at the center of the coil was carried out at 77K. Most of results followed the curve predicted by the classical R-L circuit. The result is compared with data from commonly used 4-probe method. An equivalent circuit model was also proposed by which the joint resistance could be quickly approximated.

INTRODUCTION

The availability of high performance Bi-2223/Ag multi-filamentary tape makes it a good candidate to substitute for conventional conductors in electrical power applications. Much contrary to the conventional conductor wire, the Bi2223/Ag multi-filamentary tape was usually fabricated with single tape length less than 1km, therefore, the jointing technique is required for long tape applications.

Though some sophisticated superconducting jointing techniques have been developed to meet the need for persistent current applications such as MRI and NMR device[1,2], for economical considerations, soldering the multi-filamentary tape with ordinary Sn-Pb solder directly was still widely accepted in large magnet applications[3]. The preliminary unit of such joint can be simplified as Figure 1 where the total resistance was mostly contributed from Sn-Pb solder as well as the Ag sheath material. Evaluating the resistance of this joint properly is one of basic issues for HTS electrical devices especially operated under low temperature where the refrigeration capacity and the heat generated by this joint as well as other power dissipation have to be considered systematically.

In this study, a closed Bi2223 coil with one joint was first fabricated and its joint resistance was accurately determined by a field decay method. The results were later compared with the data measured by the commonly used 4-probe method. To improve the understanding, an equivalent circuit model was also proposed where the joint resistance was simply regarded as three Ohmic type resistors connected in serial with each other, through which the joint resistance could be quickly approximated.

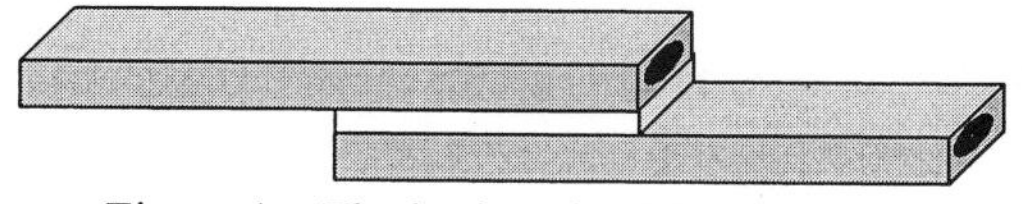

Figure 1   The basic unit of the joint area

EXPERIMENT PREPARATION

504

Sample preparation

For experiment, a HTS coil made of Bi2223/Ag multi-filamentary tape was first fabricated by using single-wound technique. The specifications of the coil are listed in Table 1. The tape was coated with epoxy paint with thickness about $10\,\mu$ m to provide electrical insulation. It should be noted here that a well turn to turn insulation is of critical importance to the accuracy of the measurement, where any current leakage between the turns should be avoided. The end and beginning of tape was later extracted and soldered together with ordinary Sn(60)-Pb(40) solder. A so called "shape hand" structure was applied to the joint area and thus form a closed coil as shown in Figure 2. The overlapped length was about 2.6cm.

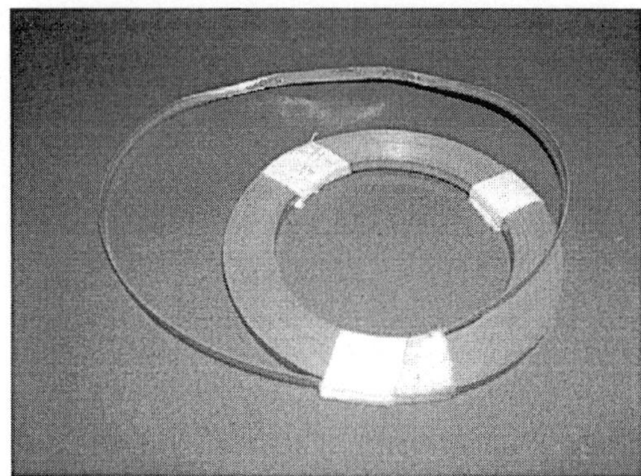

Figure 2   The closed HTS coil

Table 1 Specifications of the HTS coil

Parameters[unit]	Value
Inner diameter [mm]	55.34
Outer diameter [mm]	78.92
Height [mm]	3.99
Turns	56
Critical current [A]	20.03
Magnetic constant [T/A]	0.00105
Inductance [mH]	0.3118

Measuring system

There are three main equipments for the field decay measurement: a copper coil with 480 turns was equipped to provide the background field; a hall probe was utilized to detect the field variation at the center of HTS coil. A Keithley 2400 multi-meter together with a computer was included for data acquisition through the GPIB interface. This system is prove to be capable of distinguish a field variation in a minimum 0.025s.

RESULT AND DISSCUSSING

Field decay measurement

The exciting current of the copper coil was first applied to about 10A, which is able to produce a 400G flux density at the center of the HTS coil. Next, cool the closed HTS coil down to the superconducting state with liquid nitrogen and then shut down the exciting current instantaneously. The field decay curve as a function of time is given in Figure 3. The total measuring time lasts about 10 hours. The measured field was divided by the magnetic constant of the HTS coil and therefore the real current flowing through the coil could be scaled as Y axis.

In a typical R-L circuit whose resistance(R) and inductance(L) are independent to the current, the decay behavior can be described by solving the equation below

$$L\frac{di}{dt} + Ri = 0 \Rightarrow i(t) = i(0)\exp(-\frac{t}{\tau}) \tag{1}$$

where τ is the time constant defined by L/R. It is clear that if the decay line obeys Eq.(1), the curve in Figure 3 should be a straight line. However, an obvious divergence from the linear rule was observed at the beginning of the curve. Furthermore, a slight divergence was also observed at the end of the curve, the reason for which is not discussed here as it involves the consideration of the HTS behavior. For calculating the joint resistance, only middle part of curve from 5483s to 22691s was considered. Referring to the self inductance 0.3118mH of the HTS coil, the joint resistance was calculated to be 45.8 $n\Omega$.

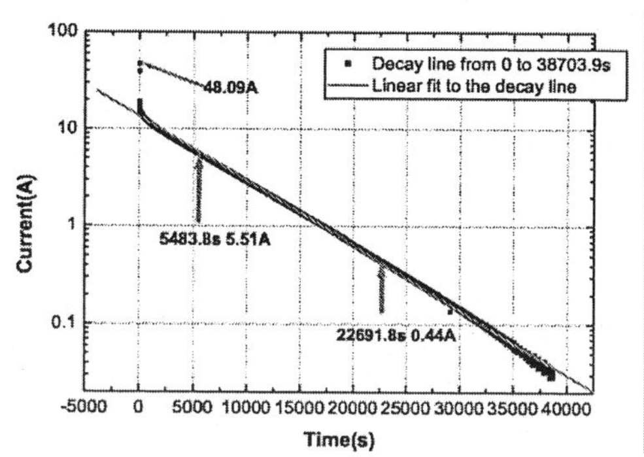

Figure 3   The field decay behavior of HTS coil

<u>Equivalent circuit and 4-probe method</u>
Referring to the joint structure given in Figure.1, in general, the voltage cross over two terminals could be determined as

$$V = R_c I + E_c l_s (\frac{I}{I_c})^n \tag{2}$$

where I is the operation current and the R_c is the joint resistance; E_c is the critical current criterion according to 1 μ V/cm; l_s is the length of the Bi-2223 tape between two terminals. Assume that the current run far below the critical current 40.2A and n index is also large enough, then two parallel superconductor filament as well its outside Ag sheath could be regarded as two equipotent layers. Among this two layers are two inner Ag sheath layers in series with the Sn-Pb layer as depicted in Figure 1. Abased on this assumption, naturally, the applied current flowing through the joint region was uniformly distributed along the contact area. Consequently, the equivalent circuit for joint region was essentially nothing but the application of the Ohm's law. This assumption neglects the real contact resistance between the solder and Ag sheath interface. For a typical Ohmic resistor, the resistance could be determined by $R = \rho l / A$. For this purpose, the resistivity of the Ag alloy was experimentally found to be $1.7093 \times 10^{-8} \Omega \cdot m$ in 77K and the resistivity of the Sn-Pb solder was adopted to be $1.45 \times 10^{-7} \Omega \cdot m$. The length of the joint area was previously mentioned to be 2.6cm. The width of the contact region was assumed to be equal to the tape width 4mm approximately. The thickness of each region was determined by optical micrographs of the joint area as shown in Figure 5, where the averaged thickness of the Ag sheath and Sn-Pb layer are $35.6 \mu m$ and $21.7 \mu m$ respectively. Combination all of these factors give the total resistance of joint area $41.9 n\Omega$, of which Sn-Pb solder layer contributes almost 74.6% of its total resistance. This value is comparable to the measured result $45.8 n\Omega$, but a little smaller than the experimental data.

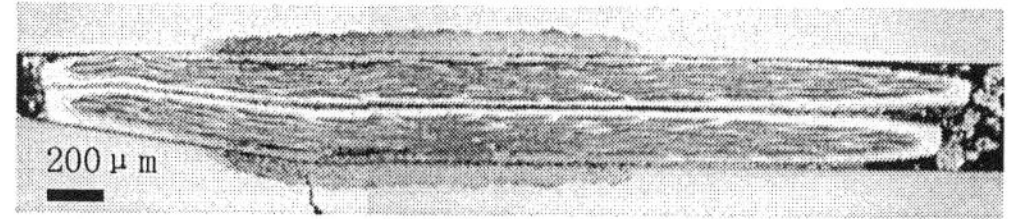

Figure 5   Optical micrographs of cross section of HTS tape

It must be born in mind that the resistivity of the Sn-Pb solder doesn't decrease with the decreasing in temperature as do in conventional pure metal, but it is magnetic field related [4]. Therefore, as long as it was used in high field applications, the joint resistance was suggested to determine experimentally.

In this study, the joint region together with 3cm extension area in each side was cut off from the closed coil and then the commonly used 4-probe method was introduced to measure the $I\text{-}V$ characteristic of the joint area. Figure 6 show the $I\text{-}V$ curve of the joint area measured from A-A'. The A-A' corresponding to the $l_s=0$ in Eq.(2) is exactly contributed from joint region. Therefore, linearizing the curve obtained from A-A' gets the joint resistance $46.6n\Omega$. This value shows a good agreement with the result obtained from the field decay method.

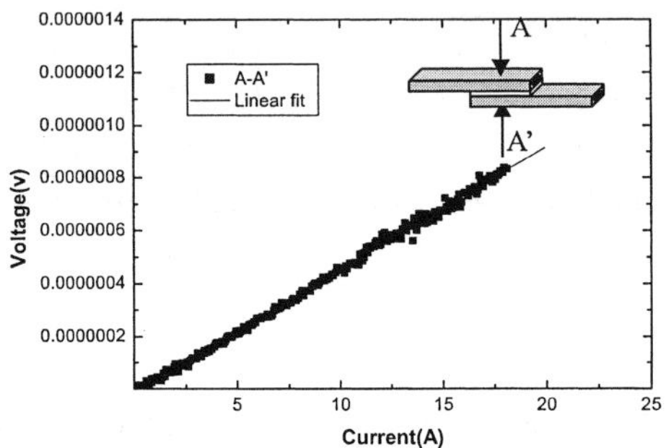

Figure 6   The I-V characteristic of the joint area

CONCLUSION

The joint resistance of the Bi2223/Ag multi-filamentary tape connected with ordinary Sn-Pb solder was studied. The joint resistance with 2.6cm overlapped area was accurately measured to be $45.8n\Omega$ by a field decay method. This result shows a good agreement with data derived from 4-probe method. An equivalent circuit model was proposed in which the joint resistance was regarded as three Ohmic type resistors connected in serial with each other. This model was prove to be valid for estimating the joint resistance in small current region with error less than 10%. Calculation results based on this model also indicate that the Sn-Pb layer dominates the joint resistance mostly.

REFERENCE

1. Horng, L. and Tai, C. H., Critical Persistent Current for a loop Formed by a Bi-2223 Ag-sheathed Superconducting Tape, IEEE Trans. Appl. Supercond. (2001) 11 3006-3009
2. Kim, J. H. and Joo, J., Fabrication and characterization of the joining of BCCO superconductor tape, Supercond. Sci. Technol. (2000) 13 237-243
3. Sneary, A. B, Friend, C. M. and Hampshire, D. P., Design, fabrication and performance of a 1.29 T Bi-2223 magnet,   Supercond. Sci. Technol. (2001) 14 433-443
4. Iwasa, Y., Case Studies in Superconducting Magnets, Plenum Press, New York, USA(1994) 266-270

Proceedings of the Twentieth International Cryogenic Engineering Conference
(ICEC 20), Beijing, China. © 2005 Elsevier Ltd. All rights reserved.

Quench characteristics of high-T_C superconducting tape for alternating over-current using DFT

Sung-Hun Lim[*a], Seong-Woo Yim[b], Jong-Hwa Lee[c], SeokCheol Ko[c], Si-Dole Hwang[b], Byoung-Sung Han[a]

[*a]Research Center of Industrial Technology, Engineering Research Institute, Chonbuk National Uni., 664-14, Duckjin-dong 1Ga, Jeonju 561-756, South Korea. Tel.: [+]82-63-270-2396, Fax.: [+]82-63-270-2394. E-mail address: superhun@mail.chonbuk.ac.kr
[b]Advanced Technology Center, Korea Electric Power Research Institute, 103-16, Munji-dong, Taejon 305-380, South Korea
[c]Division of Electronics and Information Engineering, Chonbuk National Uni., 664-14, Duckjin-dong 1 Ga, Jeonju, South Korea

The quench characteristics of high-T_C superconducting (HTSC) tape for alternating over-current were investigated and its quench developments using discrete fourier transform (DFT) were analyzed. Generally, the voltage-current characteristics of HTSC tape for the alternating over-current are complicated because the quench and the recovery between the superconducting state and the normal state are repeated. In this paper, the numerical formulation for HTSC tape's resistance was obtained by applying DFT for the measured data. By applying its numerical formulation into circuit equation, the quench developments of HTSC tape dependent on alternating over-current could be estimated and well agreed with experimental results.

INTRODUCTION

With the efforts to overcome mechanical weakness and critical current of high-T_C superconducting (HTSC) tape for power application, its critical and mechanical characteristics have been improved [1]. However, quench characteristics of HTSC tape, which occurs when the transporting current exceeds its critical current, are complicated because of the repeated transition between the quench to the normal state and the recovery to the superconducting state [2]. Especially, the analysis for the quench development of HTSC tape is needed to protect it from thermal runaway, and to keep safety operation in the power machine using HTSC tape. In this paper, we investigated the resistance development of HTSC tape after alternating over-currents were applied into it and induced the numerical expression by applying discrete fourier transform (DFT) for the measured voltages and currents. It was confirmed that the resistance development of HTSC tape dependent on the amplitude of alternating over-current could be estimated by introducing its expression into the circuit equation and the similar results to the experimental ones could be obtained.

OVER-CURRENT CHARACTERISTIC TEST AND NUMERICAL FORMULATION

HTSC tape used in this experiment, which was fabricated by power in tube (PIT), had the critical current of 57 A and the critical temperature of 106 K. Its size was 3.81 mm wide and 0.193 mm in thickness. The length of HTSC tape prepared for the over-current test was 110 cm. Two voltage taps were attached with 100 cm distance on the surface of HTSC tape after twisting to minimize the area between the voltage taps and the surface of the tape. Voltage and current signals, which were measured with the voltage and current probes, were obtained by data acquisition system with multi-channels. The over-current whose amplitude was regulated by a power supply with a transformer was applied into HTSC tape during 6 periods. To investigate the resistance development of HTSC tape for alternating over-current, the voltages between two voltage taps and the transport currents were measured after over-currents from 100 A_{peak} to 600 A_{peak} with difference of 50 A_{peak} were applied into HTSC tape. The resistance component of HTSC tape with the fundamental power source frequency could be obtained from the impedance (equation (2)), which was equal to the ratio of the voltage data for the current data (as expressed in equation (1)) extracted using discrete fourier transform (DFT) for the measured voltage and current signal data.

$$X(jkf)=\sum_{n=1}^{N} x(e^{-knfT})=[\,x(T) \quad x(2T) \quad \cdots \quad x(NT)\,]\cdot \begin{bmatrix} e^{-jkfT} \\ e^{-2jkfT} \\ \vdots \\ e^{-jknfT} \end{bmatrix} \tag{1}$$

$$Z=V(jkf)/I(jkf) \tag{2}$$

Where T, N and f represent sampling length, sampling number per period and fundamental frequency, respectively.

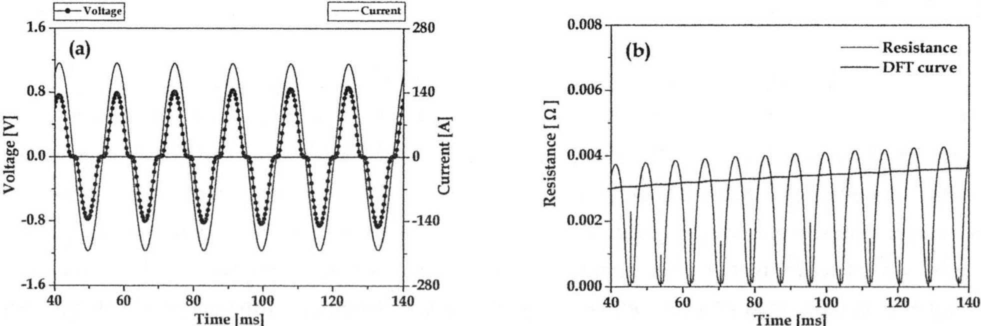

Figure 1 Quench characteristics in case that source current with 204 A_{peak} was applied into HTSC tape (a) Waveforms of alternating over-current and voltage across HTSC tape (b) Resistance and DFT curve of HTSC tape

Figure 1(a) shows the source current waveform of 204 A_{peak} and the voltage waveform across HTSC tape. The resistance waveform, which can be calculated by dividing the voltage waveform of HTSC tape with the source current waveform as seen in Figure 1(a), was shown in Figure 1(b). As seen in Figure 1(b), the HTSC tape at the point that the sinusoidal source current approached to zero recovered to superconducting state, namely, its resistance approached to zero value temporarily. DFT curve for the resistance of HTSC tape obtained by applying the equation (1) for the measured current and voltage waveforms was also added into Figure 1(b). The DFT curve for the resistance of HTSC tape smoothly increased along with its peak resistance value. Another DFT curve for the resistance of HTSC tape in case that the source current of 306 A_{peak} was applied was shown in Figure 2(b), which was obtained from the HTSC tape's voltage and the source current waveforms (Figure 2(a)). As seen in Figure 2(b), DFT curve for the resistance of HTSC tape, which increased as the peak value of HTSC tape's resistance increased, increased with larger value than in case of the source current of 204 A_{peak}.

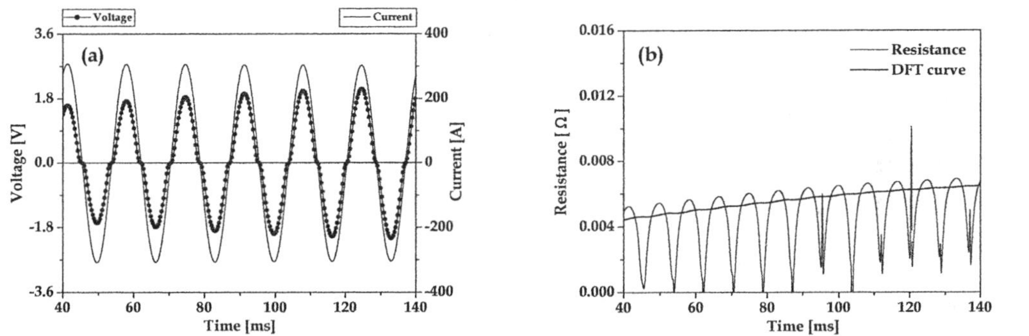

Figure 2 Quench characteristics in case that source current with 306 A_{peak} was applied into HTSC tape (a) Waveforms of alternating over-current and voltage across HTSC tape (b) Resistance and DFT curve of HTSC tape

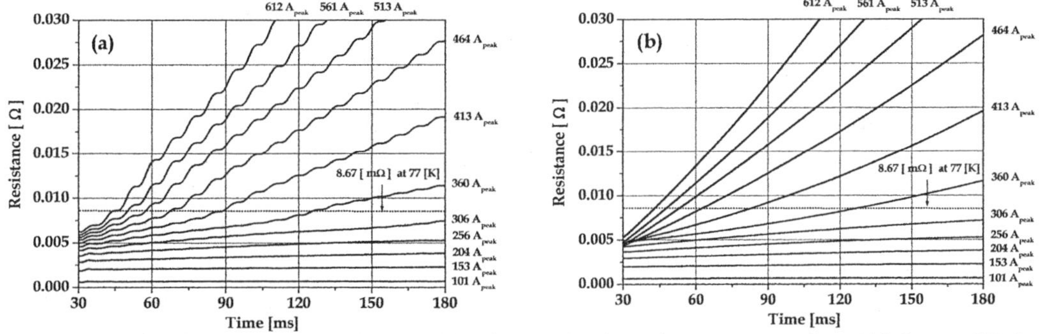

Figure 3 DFT curves for HTSC tape's resistances dependent on the alternating over-current from 101 A_{peak} to 612 A_{peak} (a) DFT curves for HTSC tape's resistances obtained by using equation (1) (b) DFT curves for HTSC tape's resistances obtained by applying the least square method for DFT curves of Figure 3(a)

DFT curves for HTSC tape's resistances dependent on the alternating over-current from 101 A_{peak} to 612 A_{peak}, which could be extracted as explained above, were shown in Figure 3(a). Figure 3(b) shows the DFT curves for HTSC tape's resistances obtained by applying the least square method for DFT curves of Figure 3(a). Each DFT curve in Figure 3(b) was fitted to a quadratic function of time as expressed in equation (3).

$$R(t,I) = C_1(I) \cdot t^2 + C_2(I) \cdot t + C_3(I) \tag{3}$$

510

where C_1, C_2 and C_3 are supposed to be presented as the function of over-current (I). The values of C_1, C_2 and C_3 for different over-currents could be obtained by approximating DFT curves of Fig. 3(b) to polynomial equations for over-current using least square method. By applying the equation (2) for the circuit equation including power source and source resistance, we could estimate the quench development of HTSC tape dependent on alternating over-current. Figure 4(a) and Figure 4(b) show the simulated waveforms for the application of alternating over-currents with 204 A_{peak} and 306 A_{peak} to HTSC tape, whose calculation conditions correspond to Figure 1(b) and Figure 2(b). As compared Figure 4(a) and Figure 4(b) with Figure 1(b) and Figure 2(b), we confirmed that the numerical expression for HTSC tape's resistance derived from the experiments agreed well with the measured results.

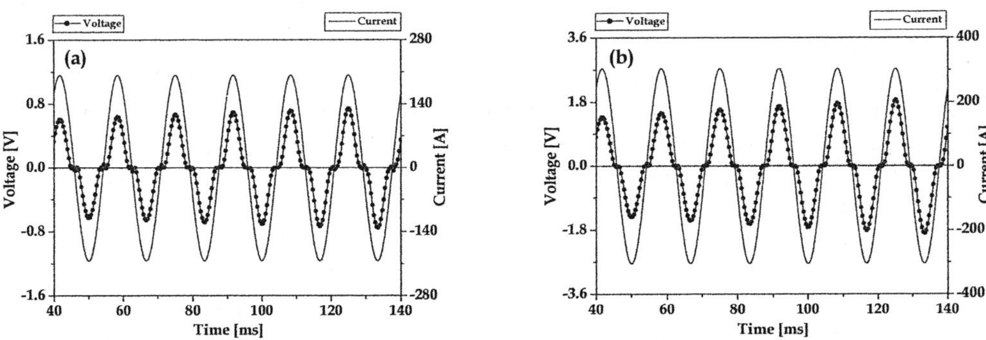

Figure 4 Simulated waveforms of alternating over-current, voltage across HTSC tape (a) In case that source current with 204 A_{peak} was applied into HTSC tape (b) In case that source current with 306 A_{peak} was applied into HTSC tape

CONCLUSIONS

In this paper, we investigated the resistance development of HTSC tape after alternating over-currents were applied into it and induced the numerical expression of it using discrete fourier transform (DFT) from the measured voltages and currents. It was confirmed that the resistance development of HTSC tape dependent on the amplitude of the over-current could be estimated by applying its expression into the circuit equation and the similar results for the resistance development to the experimental ones could be obtained. We will research the current distribution between the superconducting part and the Ag sheath of HTSC tape for alternating over-current application.

REFERENCES

1. A. Shikov, I. Akimov, A. Nikulin, V. Pantscyrnyi and A. Vorobieva, HTS materials for electrical power application, IEEE Trans. Appl. Supercond. (2000) 10 1126-1129

2. Seong-Woo Yim, Hyo-Sang Choi, Ok-Bae Hyun, Si-Dole Hwang and Byoung-Sung Han, Quench Characteristics of HTS Tapes With Alternating Currents Above Their Critical Currents, IEEE Trans. Appl. Supercond. (2003) 13 2968-2971

Dip-coated YBa$_2$Cu$_3$O$_{7-x}$ films on Ag substrate by MOD method

J. Dong, M. Liu, D.M. Liu, Y. Zhao, J.X. Liang, J.L. Sheng, H.L. Suo, M.L. Zhou

The Key Laboratory of Advanced Functional Materials, Ministry of Education, China;
College of Material Sciences and Engineering, Beijing University of Technology, Beijing 100022

The metal organic deposition (MOD) process of YBa$_2$Cu$_3$O$_{7-x}$ (YBCO) using metal trifluoroacetates (TFA) precursors is considered to be a strong candidate as a low cost fabrication process in coated conductors since the TFA-MOD process is a non vacuum process and can provide high J$_c$ films. Metal organic deposition using trifluoroacetates (TFA-MOD) with dip coating was applied for the preparation of YBCO films. The thickness of films increases with the withdrawal speed of substrates. In this work, a triple coated film was fabricated on Ag {110}<110> textured polycrystalline substrate by optimizing the condition of heat treatments in the multi-coating method. The YBCO films have J$_c$ value of 15000A/cm^2 (77K, 0T) measured by the four-probe-method.

INTRODUCTION

The metal organic deposition (MOD) process of YBCO using metal trifluoroacetates precursors is considered to be a strong candidate as a low cost process of coated conductors, since the TFA-MOD process is basically a non vacuum method. Additionally, it has been well confirmed that this process has an advantage to provide a high J$_c$ film of the MA/cm^2 [1] class on the single crystal substrates such as LaAlO$_3$ and SrTiO$_3$ [2] and on the Ni tapes with multi buffer-layer [3]. But, they are difficult to make long single crystal substrates and to make the Ni tapes with multi buffer-layer. We deposited directly YBCO film on Ag {110}<110> textured polycrystalline substrates to solve the above problem, because it is easy to make long Ag substrate with bi-axially texture. On the one hand, in order to develop long tape conductors, it is important to investigate the influence of dip-coating process on YBCO films. On the other hand, in order to obtain high I$_c$, processing for thicker YBCO films was investigated using the multi-coating method.

EXPERIMENT

A solution was prepared by dissolving the acetates of Y, Ba, and Cu into de-ionized water in a 1:2:3 cation ratio with stochiometric quantity of TFA, and then water and acetic acid were removed by an

evaporator to yield a blue glassy residue. The above residue was put into oven at 100 ℃ for 10 hours to

remove water and acetic acid as absolutely as possible. Dissolving the gel into methanol made the coating solutions that had total metallic concentration of 2.0mol/l. The gel films were coated onto the polished Ag substrates by dip coating. A withdrawal speed of 1.2-5.7mm/s was used for coating gel films. The heat treatment was conducted by two stages, which were the low temperature treatment and the successive high temperature treatment. The temperature profiles of the heat treatment are showed in figure 1. In the low temperature treatment, the TFA solution precursor films were decomposed to the mixture consisted of amorphous, oxide, fluoride and oxy-fluoride etc. during slowly heating to 400 ℃ in a moist O_2 atmosphere. In the multi-coating case, after the first low temperature treatment, TFA precursor solution coating was carried out again on the first precursor films. In the high temperature treatment, the films were first heated up to 400 ℃ in a dry mixed gas of Ar/O_2 and heated to 900 ℃ in a wet mixed gas of Ar/O_2 and held for about 90 minutes at 900 ℃ and cooled in a dry gas to the room temperature.

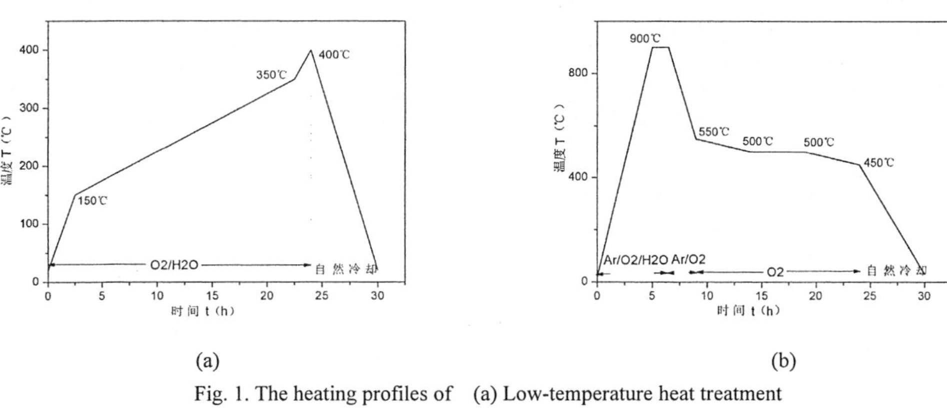

(a)   (b)

Fig. 1. The heating profiles of   (a) Low-temperature heat treatment
(b) High-temperature heat treatment

Crystalline phases in the films were detected by X-ray diffraction. SEM was used to evaluate the surface morphology of films. T_c and J_c were measured using a standard four-probe method at 77K in self-field.

RESULTS AND DISCUSSION

<u>Influence of the polished and unpolished Ag substrates on YBCO films</u>

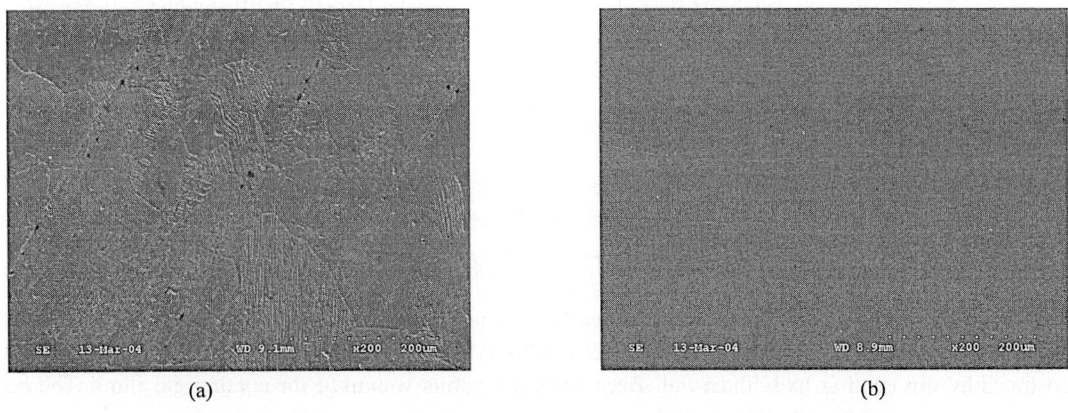

(a)   (b)

Fig.2. SEM images of Ag substrates: (a) unpolished and (b) mechanically polished

Fig.2 shows SEM images of Ag substrates. As shown in these figures, the surfaces are different between (a) and (b). The surface of unpolished Ag is rough. Rolling stripes and crystal boundaries could be seen. On the other hand, the mechanically polished Ag is rather smooth. But the epitaxial growth of YBCO films on substrates is affected strongly by the substrate surface [4].

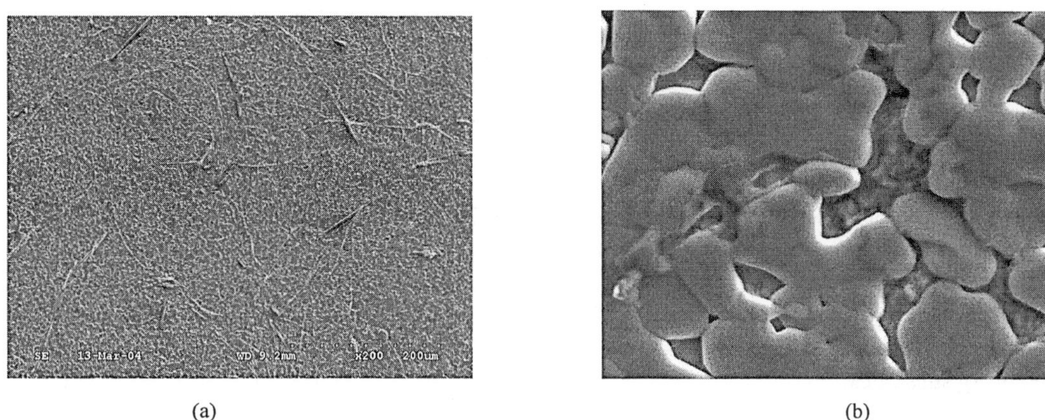

(a)   (b)

Fig.3. SEM images of YBCO films on unpolished Ag substrates: (a) 200× and   (b) 3000×

The influences are shown in Fig.3 and Fig.4 clearly. YBCO grain size, secondary phase particles and pores on the surface of each sample are different. As shown in Fig.3 (a), the surface is not smooth and has many stripes like tree for the rolling stripes and crystal boundaries on unpolished Ag substrates. Comparing Fig.3 (b) with Fig.4 (b), there are many large grains and pores in Fig.3 (b). It means that the connections among YBCO grains are bad and there are fewer chances to form the access of electric current. So polished Ag substrates are more available to grow YBCO films.

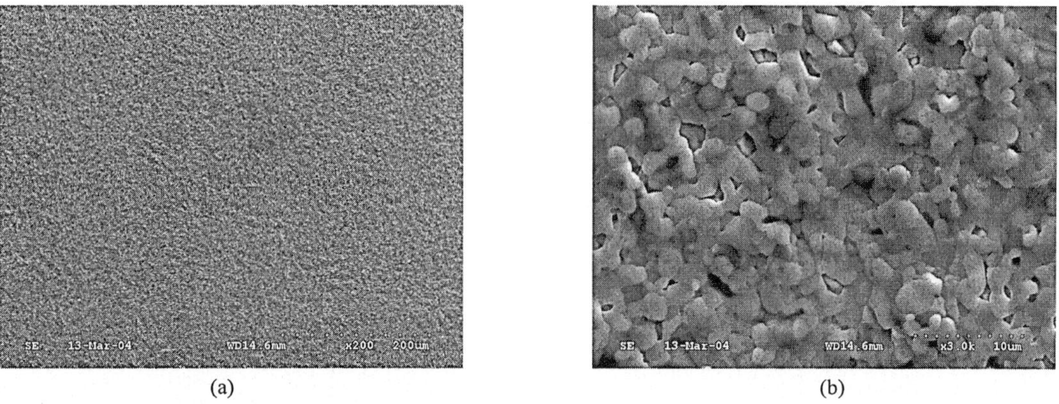

(a)   (b)

Fig.4. SEM images of YBCO films on polished Ag substrates: (a) 200× and   (b) 3000×

The ageing time of the sol solution

In this work, the sol solution, which will be coated, should be lay aside for at least 3-10 days. If the sol solution has not pass through this process, the gel film is easy to crack. It is mainly because the new-made sol is not stable, sol and gel has not achieved equilibrium. On the other hand, by dip coating method, the sol solution should have a certain extent viscosity [5]. The viscosity of new-made sol is too low to make films what we expect. But, when the sol solution is lay aside for long time, the viscosity of it is too high and the fluidness of it is bad. Then, gel films coated by it are inhomogeneous and easy to have small

514

cracks. So, this process is very necessary.

<u>Influence of the withdrawal speed of Ag substrates on YBCO films</u>
According to the theory of dip coating [5], the thickness of YBCO films (t) is proportional to the square root of the withdrawal speed (v). When the thickness exceeds the critical value, YBCO films will crack. Controlling the withdrawal speed is equal to control the thickness of films. On the one hand, when the withdrawal speed is lower, coating film is thinner and homogeneous at a time. It is necessary that TFA precursor solution is coated again on the first precursor films to make the thicker films. On the other hand, when the withdrawal speed is too high, single dip-coating film is so thick as to crack during successive drying and calcinations process. So, it is important to search and confirm the better withdrawal speed to make the better films.

In this work, four different speeds from 1.2mm/s to 5.7mm/s were adopted. The former three films were dark and homogeneous but the last film had cracks for the high withdrawal speed. The YBCO films prepared on polished Ag substrates by TFA-MOD with dip coating show highly c-axis orientation. In Fig.5, XRD patterns show that the intensity of the (00l) peaks is strong and sharp and pure. There are no BaF_2 and the other impurities. The intensity of the YBCO peaks increases with increasing withdrawal speed of Ag substrates.

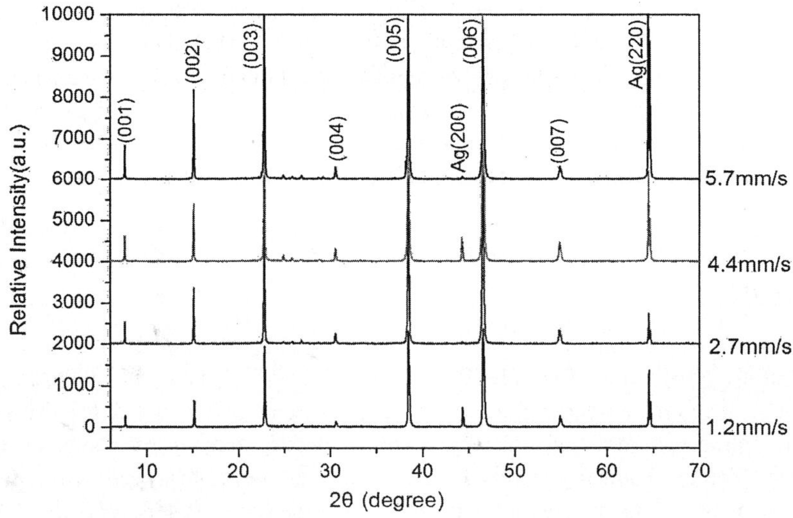

Fig.5. XRD patterns of YBCO thin films on polished Ag substrates

In Fig.6, SEM images show, from a top view, that the YBCO films appear to have a multi-grain structure. With increasing withdrawal speed of Ag substrates, there are more and more joints, fewer and fewer pores among grains. The film with 4.4mm/s withdrawal speed is very dense and homogeneous. So, the withdrawal speed of 4.4mm/s is what we want.

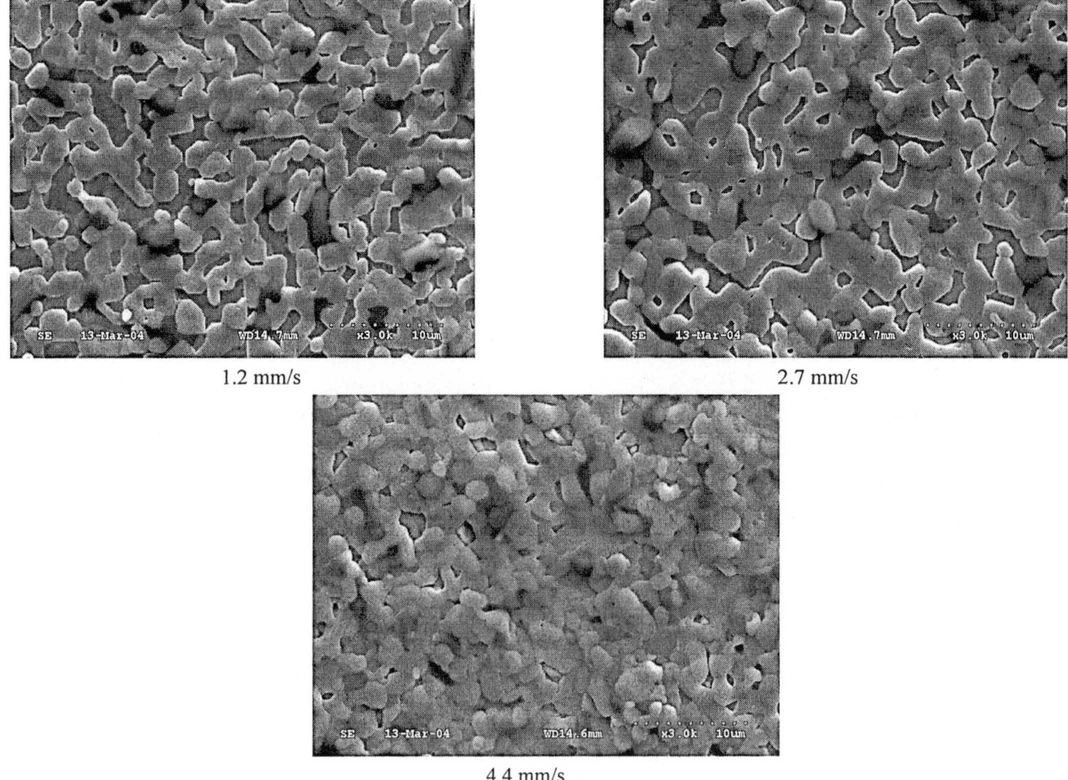

1.2 mm/s   2.7 mm/s

4.4 mm/s

Fig.6. the top surface microstructures for the YBCO samples prepared at different withdrawal speed of polished Ag substrates.

CONCLUSION

We can successfully obtain crack-free YBCO films, which have a J_c value of 15000A/cm^2 (77K, 0T), on Ag {110}<110> textured polycrystalline substrate, by optimizations of the TFA-MOD process: preparation of coating solutions and other processes, including coating, calcining, firing processes and humidity control. The mechanically polished Ag substrates are better for the growth of YBCO films than unpolished Ag substrates. The sol solution should be lay aside for several days to coat YBCO films well. However, considering the future engineering application, it is further required to optimize TFA-MOD method to improve J_c of YBCO films.

REFERENCES

1. Hiroshi Fuji, Tetsuji Honjo, Ryo Teranishi, et al, Physical C (2003)392-396 905-908

2. Tomoaki Ono, Kaname Matsumoto, Kozo Osamura, et al, Physical C (2003) 392-396 917-921

3. Y. Takahashi, Y. Aoki, T. Hasegawa, et al, Physical C (2003) 392-396 887-894

4. A. Takechi, K. Matsumoto, K. Osamura, Physical C (2003) 392-396 895-899

5. S. X. Gu, J. Zhou, J. Liu, et al, BULLETIN OF THE CHINESE CERAMIC SOCIETY（2001）4 18-21

Influence of Ag substrate surface condition on surface morphology of YBCO film

M. Liu, D. M. Liu, J. Dong, Y. Zhao, J. Sheng, J. X. Liang, H. L. Suo, M. L. Zhou

The Key Laboratory of Advanced Functional Materials, Ministry of Education, China;
College of Material Sciences and Engineering, Beijing University of Technology, Beijing 100022

YBCO films were grown on polycrystalline Ag substrates by a metal-organic decomposition (MOD) method using Trifluoroacetate Salt (TFA). The surface defects of the Ag substrates, such as rolling stripe and crystal boundary, were found to have a detrimental effect on the crystal orientation and surface morphology of YBCO films. The surface of YBCO films deposited on Ag substrate annealed in vacuum has many holes and stripes, which are parallel to the rolling stripe on Ag substrates. At the crystal boundary grooves on Ag substrates annealed in Ar environment, the film compositions are not superconducting phases but oxide phases of Cu-O and Ba-Cu-O. To eliminate the rolling stripe and grooves on the Ag surface, we used cold rolling polished Ag as substrates. The film grown on cold rolling polished Ag substrates has a smooth surface and good connectivity of grains without parallel stripes and grooves. The films composition is also uniformity.

INTRODUCE

MOD has a variety of advantages, such as precise controllability of composition, wide flexibility to coating objects and a low cost non-vacuum approach. So many studies about this method have been done in recent years. The long YBCO tapes with high Jc values of over $1MA/cm^2$ were also reported on Ni tapes by this method, with in-plane aligned buffer layers, which were multilayered structures [1-2]. A stabilizing layer of Ag is also required for protection against over current when an insulating buffer layer is used. So, the manufacturing process of buffer layers and over-coated Ag layer make this method time-consuming and complicated. A way to solve this problem is such that the YBCO film is directly deposited on a textured Ag tape without any buffer layers by a MOD method. Meanwhile whole Ag tape can be used as a stabilizing layer for protection against over current. In view of its simplicity, it will be a promising practical preparation method. However, up to now no such a report about this aspect has been seen.

Many factors influence the quality of the YBCO films grown on Ag substrates [3]. One important factor is the substrate-surface roughness. It is known that a strict treatment to reduce the substrate-surface roughness is usually required before the YBCO film deposition even for the oxide single crystal substrate such as STO etc [4]. However, for the soft metal such as Ag, it is much more difficult to reduce the surface roughness and get a smooth surface, especially for polycrystalline silver. Therefore the surface roughness influence on the YBCO film growth will be much more serious, which might be the main reason for the high inhomegeneity and the low J_c of the YBCO films grown on Ag substrates.

In this paper we have prepared YBCO films on polycrystalline Ag substrates by a metal-organic decomposition (MOD) method using Trifluoroacetate Salt (TFA). From experiment we have observed that the surface defects of the Ag substrates annealed in vacuum and Ar environment, such as rolling stripe and crystal boundary grooves, have a detrimental effect on the composition and surface morphology of YBCO films. In order to remove the bad effect, we deposited YBCO films on cold rolled Ag substrates polished by mechanical way.

EXPERIMENT

A TFA precursor solution was prepared by dissolving the acetate of Y, Ba and Cu in distilled water in a 1:2:3 cation ratio with stoichiometric quantity of TFA, and then water and acetic were removed by an evaporator to yield a blue glassy residue. The coating solution with total metallic concentration of 1.5mol/l was made by dissolving the residue into methanal. The gel films were coated onto Ag {110} <110>substrates by spin-coating method or dip-coating method.

The heat treatment of the coating film was applied in two stages with the heating profiles that are showed in our previous paper [5]. In the first calcination stage, the film coated the TFA solution was decomposed to an amorphous precursor film by slowly heating up to 400℃ in a humid oxygen atmosphere. In the second calcination stage, the amorphous precursor film was heated up to 900℃ in humid argon and held for 30 min in dry argon. After growth, films were slowly cooled to 500℃. Post-oxygenation was carried out at 500℃ for 90 minutes followed by naturally cooling to room temperature.

RESULTS AND DISCUSSIONS

Figure 1a presents the optical image of the Ag substrates annealed in vacuum. Clearly, his figure shows parallel striations that may have been formed during the rolling process. Many stripes, which are parallel to the rolling striations on Ag substrates, were also observed on the surface of YBCO film (figure 1b). Additionally, the YBCO films deposited on the Ag substrates are quite rough and connective poorly with many holes (figure 1c).

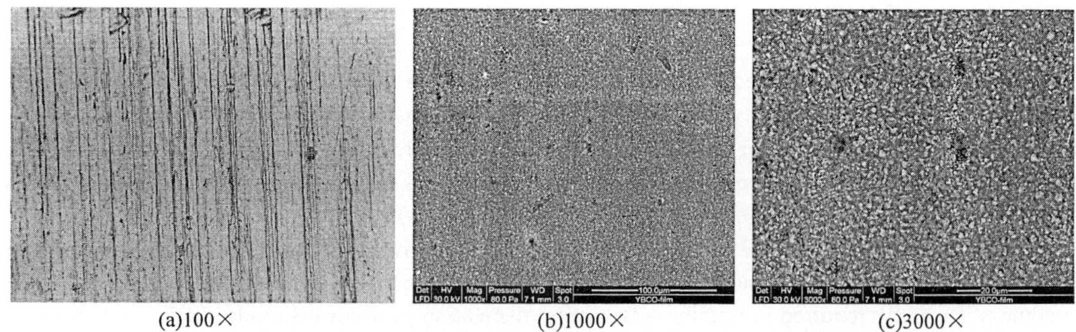

(a)100×   (b)1000×   (c)3000×

Figure 1 The surface morphologies of Ag substrate in vacuum and YBCO film

To eliminate the rolling stripe on the Ag surface, we annealed it in flowing Ar atmosphere. As can be seem from figure 2a, after annealing, grooving of the grain boundaries of the Ag became more pronounced. These too many grooving may contribute to the roughness of YBCO films [6]. Irregular contour of Ag grains boundaries is clearly present on the surface of YBCO films (figure 2b). The EDS results (figure 3a) show that the atom simplified ratio of Y: Ba: Cu: O of point a and b in Fig 2c is 0: 1: 1: 2 (Fig. 3a) and 1: 2: 3: 7(Fig. 3b), respectively, which indicates that the composition of point a is not a superconducting phase, but the phase of Ba-Cu-O. When the grooves are small, the lateral usually can cover it over growth of the YBCO grains surrounding it, with no influence on the good YBCO film growth. For large size grooves, they will strongly influence the quality of the film on them. For the detailed film microstructure and the related forming mechanism, film growth kinetic theory can give a reasonable interpretability.

Due to the anisotropic surface free energy of YBCO crystal nucleus, its shape is usually a thin disc and its surface usually keeps (001) face: the face with lowest free energy [7]. The nature of the thin disc shape of YBCO nucleus decides that YBCO film nucleation will be very sensitive to the substrate surface morphology. For the thin disc nucleus, it has a very large substrate nucleus interface. Therefore, the substrate-nucleus interface energy becomes specifically important to decide the total free energy of the nucleus. The YBCO nucleuses on the flat substrate surface area can remains its lower energy face contacted with the substrate surface and interface energy will be much lower. Therefore, the YBCO nucleus can easily grow to be larger and larger. However, when the YBCO crystal nucleates on the rough substrate grooves will introduce much higher-interfacial energy and YBCO nucleus will be unstable and difficult to grow up. In this case, the other oxide phases such as Cu-O and Ba-Cu-O can nucleate in the surface grooves since these crystal have not so strong anisotropy of the free energy as YBCO. Their substrate/nucleus interfacial energy will not be influenced by the surface roughness so much. In other words, the rough surface will not cause the obvious increase of the free energy for the nucleation of other oxide phases. Therefore, on the rough substrate grooves, it is possible that the free energy for the nucleation of other oxide phases will be lower than that for the nucleation of YBCO crystal. As a result, instead of YBCO crystal, Cu-O and Ba-Cu-O can preferentially nucleate and grow in the surface grooves.

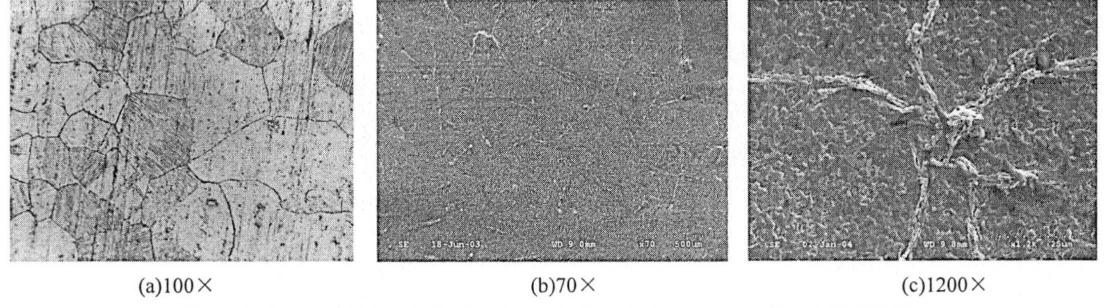

(a)100×   (b)70×   (c)1200×

Figure 2 The surface morphologies of Ag substrate in Ar environment and YBCO film

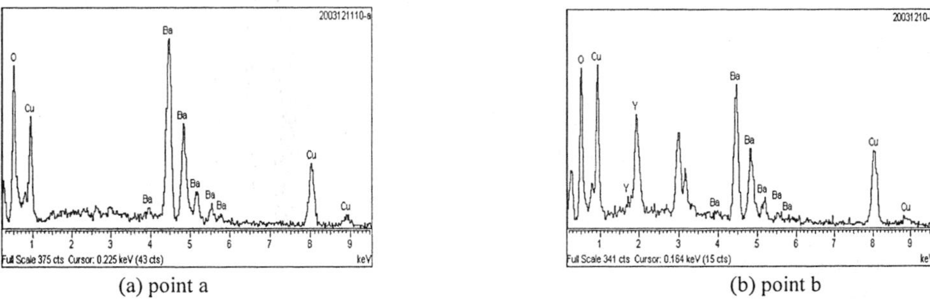

(a) point a   (b) point b

Figure 3 The EDS results of point a and b in figure 2 (c)

Ag substrates annealed in vacuum and Ar environment are so soft that they don't get a smooth surface polished by mechanical way. By controlling rolling conditions, we can gain the cold-rolling Ag substrates whose {110} <110> texture can form during head treatment of the precursor film [8]. The cold-rolling Ag substrate is hard enough to be polished by mechanical way. After being polished, Ag surface becomes very flat and smooth without rolling stripe and grooves. Therefore, the film grown on cold rolling polished Ag substrates has a smooth surface and good connectivity of grains without parallel stripes and grooves. The films composition is also uniformity. Figure 4a, 4b and 4c show the SEM photographs of the surface morphologies of polished Ag substrate and the YBCO film on polished Ag substrates, respectively. From the matter, Polishing of Ag substrates prior to films deposition were found to be useful in improving surface quality of YBCO films.

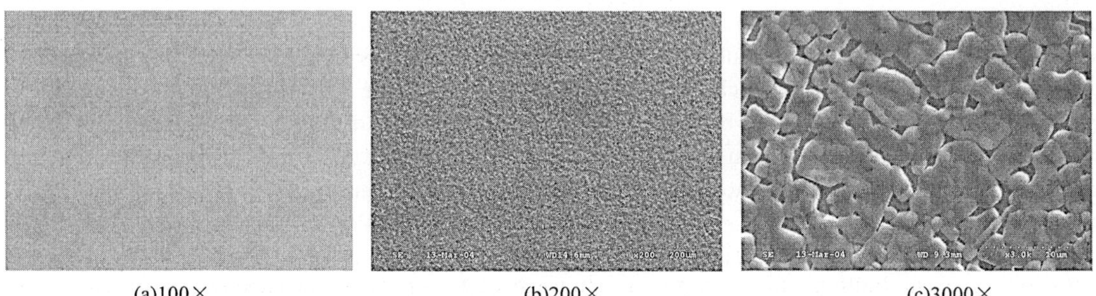

(a)100×   (b)200×   (c)3000×

Figure 4 The surface morphologies of the cold-rolling Ag substrate poloshed by mechanical way and YBCO film

CONCLUSIONS

We have prepared YBCO films on all kinds of Ag substrates by a metal-organic decomposition (MOD) method using Trifluoroacetate Salt (TFA). From experienment we have observed that the surface defects of the Ag substrates annealed in vacuum and Ar environment, such as rolling stripe and crystal boundary grooves, have a detrimental effect on the composition, crystal orientation and surface morphology of YBCO films. Finally, we get YBCO films with a smooth surface and even composition on cold rolled Ag substrates polished by mechanical way.

ACKNOWLEDGMENT

The authors would like to thank the 973 and 863 project of China for funding the work.

REFERRENCES

1. Erdal Celik, Yusuf. S. Hascicek, <u>Materials Science and Engineering B</u> (2004), <u>106</u> 1–5
2. Yasuhiro Iijima1, KazuomiKakimoto, Yasunori Sutoh, <u>Supercond. Sci. Technol.</u> (2004), <u>17</u> S264–S268
3. L. Chen, T. W. Piazza, B. E. Schmidt, <u>J. Appl. Phys.</u> (1993), <u>73</u> 7563.
4. J. P. Contour, D. Ravelosona, C. Fretigny, <u>J. Cryst. Growth</u> (1994), <u>141</u> 141.
5. M. Liu, <u>Transactions of Nonferrous Metals Society China</u> has been accepted.
6. Yanwei Ma, Kazuo Watanabe, Satoshi Awaji, et al, <u>Cryogenics</u> (2002), <u>42</u> 383-386
7. B. Dam, J. H. Rector, J. M. Huijbregtse, et al, <u>Physica C</u> (1998), <u>396</u> 179-187
8. Liu D.M. Li E.D. Hu Y.C. Liu M. Xiao W.Q. and Zhou M.L., <u>Physic C</u> (2004 in publishing)

Boiling Heat Transfer To Liquid Nitrogen Pool From Ag Sheathed PbBi2223 Tapes Carrying Over-Current

Bailey W., Young E.A., Yang Y., Beduz C.

Institute of Cryogenics, School of Engineering Sciences, University of Southampton, UK

Measurements were carried out on boiling heat transfer to a liquid nitrogen pool due to self-heating of Ag sheathed PbBi2223 tapes with increasing over-currents up to 300A, focusing on its difference from increasing heat flux in conventional heated surfaces. Large spontaneous oscillations of the surface temperature were observed and attributed to the interplay between the activation/deactivation of nucleate boiling and the highly non-linear heat generation of the superconductor as a function of temperature. Two distinct steady-states of 8K and 3K superheating were also found for samples carrying 300A depending on the current cycle history.

INTRODUCTION

The significant enhancement to the performance of high temperature superconductor (HTS) composites, has lead intensified research, development and implementation of their applications in power devices, applications such as motors, transformers, cables, fault current limiters and current leads. Although cooling by cryocoolers can be of great benefit in certain cases, some of these devices are designed to operating in a liquid nitrogen (LN_2) pool. There is however little data available [1, 2] on heat transfer to LN_2 to HTS under direct heating by an over-current above the critical current. The distinction between direct heating of superconductors by transport over-current and heating from conventional heater is the focus of the present work. Above the critical current Ic, the heat generation in a superconducting composite can vary sharply by several orders of magnitude with both current and temperature, due to its highly nonlinearly resistivity $\propto (I/Ic)^n$ ($n \gg 1$), the strong temperature dependence of Ic in the vicinity of T_C, and the gradual current sharing with the normal matrix with increasing temperature. In contrast, resistivity of a conventional heater is independent of current and can be considered constant for pool boiling of up to 15K superheating. The unique nonlinear heat generation of superconductors is expected to result in unconventional heat transfer characteristics, as has been shown in [2], where a large fluctuation of heat generation was observed.

It should be noted however that previous works have been limited to a small heat flux due to low critical current of the superconductor used and lack of accurate temperature measurement. In this paper high current BiPb2223 tapes (AMSC 115A) were used. Simultaneous measurements of voltage and temperature were made to determine local value to heat generation and surface superheating with increasing over-current up to 300A. The time dependence of power fluctuations was recorded at 100Hz to investigate the dynamics of the heat transfer process.

EXPERIMENTAL

Four BiPb2223 Ag sheathed tapes (Ic = 115A) were mounted side by side on a base of glass fibre composite to form a meander. Copper current pads recessed into the base and soldered to the tape ends

522

make the meander continuous, and several pairs of voltage taps were soldered at a separation of 10mm along the tapes. Copper-constantan thermocouples were soldered directly to the superconductor between the voltage taps and connected to thermocouples immersed in LN_2 to form differential thermocouples. The tapes were potted with the voltage taps and thermocouples face-down into a recess in the base, in order to ensure a smooth boiling surface and thermal isolation of the thermocouples from the cryogen. The heat leak through the base and wires is negligible. The assembly was mounted vertically in a LN_2 bath, simultaneous temperature and voltage data was taken using voltmeters whilst the current, (constant current mode), was incremented from zero to the maximum value then decreased back to zero. The heat

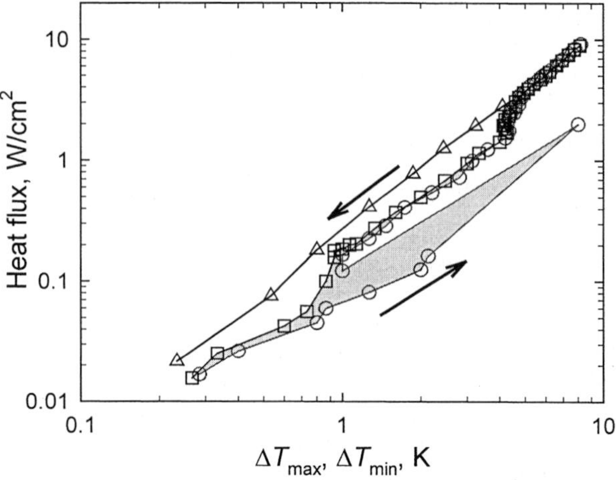

Figure 1 Heat flux vs. surface superheat of a superconducting tape carrying over-current in a LN_2 pool

generation per unit area by the tape is calculated from the voltage current product. Heat transfer stability was recorded at 100Hz using a high resolution Analogue to Digital data logger. A parallel shunt was used to allow stable over-current up to 300A and prevent sample burn-out upon transition to film boiling.

RESULTS

<u>Steady-state characteristics of heat transfer to LN_2 from superconductor directly heated by over-current</u>

The surface superheat $\Delta T = T_S - T_{LN}$ and the corresponding dissipative voltage V were measured with increasing transport current I. The result shown as a conventional plot of heat flux ($\propto V \cdot I$) vs. ΔT in Figure 1. With increasing current up to 200A there is a large fluctuation of surface temperature, and both the maximum (○) and minimum (□) ΔT are shown in Figure 1, where the enclosed grey area indicate the range of fluctuation. Further increase of current above 200A leads to a consistent superheat above 4K with the onset of nucleate boiling, and diminishing temperature fluctuations. Upon reducing current from 300A, surface temperature (△) became well defined as the surface remains activated, and an enhanced heat transfer was found.

As shown clearly in Figure 1, there is no unique correspondence between the heat flux and surface superheat over a large range of current above I_c. As mentioned previously, this is most likely related to the highly nonlinear heat generation of superconductors as a function of both current and temperature. In this case the heat transfer is better understood with the transport current as the primary variable. To this end, superheat ΔT and dissipative voltage V are shown as functions of current in Figure 2a-b respectively. It becomes clear that the large range of temperature

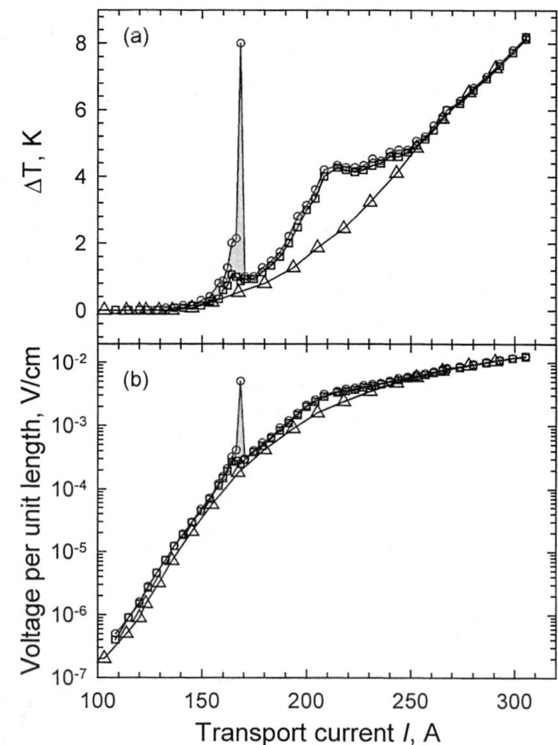

Figure 2 Surface superheat (a) and dissipative voltage (b) of a superconducting tape as a function of over-current in LN_2

fluctuation occurs above 150A when the dissipative voltage is above 0.2mV/cm, corresponding to about 0.1W/cm^2. In the particular case shown in Figure 2, there was a large temperature increase to 8K above the saturation temperature with a transient heat generation of 2W/cm^2. Such a large superheat led to a partial activation of nucleate boiling, which offers much enhanced heat transfer and consequent sharp reduction of superheat. The superheat is further reduced by large reduction in the heat generation. Such an interplay between the superheat and heat flux is unique to superconductors and results in a quick reduction of activated sites. The temperature fluctuation shown in Figure 1 is thus attributed to the cycle of activation and deactivation in the normally non-activated regime for conventional heated surfaces.

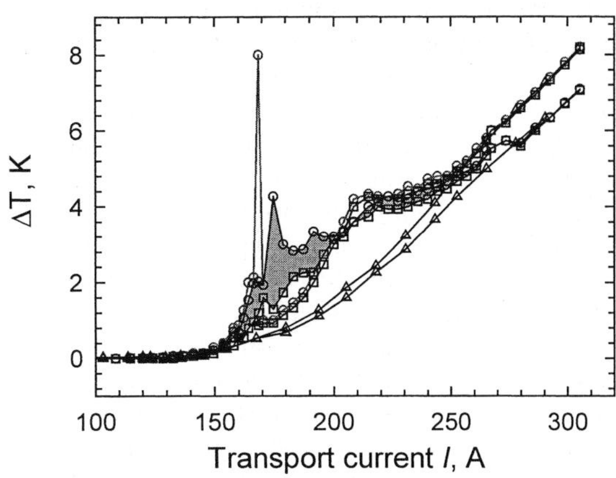

Figure 3 Surface superheat as a function of over-current from two separate locations as an indication of the randomness of the temperature oscillation

It should be noted that the temperature measurement here was very localised, hence the results shown in Figure 2 only indicate the existence of large temperature excursions and does not represent the random variation with location. In Figure 3, ΔT from a different thermocouple located at a different spot showed a lesser degree of fluctuation but a more extended range (dark grey area) to higher current, possibly due to lack of partial activation seen in the other measurement.

According to Figure 2-3, the transition to nucleate boiling occurs between 200A and 250A, where there is little increase of ΔT with increasing current. This is because the surface is still overheated compared to activated nucleate boiling ($\triangle$) upon reducing current from 300A.

<u>Spontaneous behaviour of heat transfer to LN$_2$ from superconductor directly heated by over-current</u>
Further investigation on the dynamics of temperature fluctuation was carried out by time-resolved simultaneous measurement of superheat $\Delta T(t)$ and dissipative voltage $V(t)$. Figure 4(a) shows the time history of ΔT and V for a typical run with ramping current from 180A to 209A with 1A increment. With increasing current to 195A, the superheat increase steadily to 5K with fluctuation in the range of 0.5-1.0K, while the dissipative voltage crept up from 1mV/cm to 3mV/cm. At about 195A, an event of activation occurred with a sharp drop of ΔT by about 1.5K. The activation reduces gradually, even with an increasing heat flux at further increase of current, and the superheat reach 5K again at 209A. With a constant current of 209A, the superheat suddenly dropped by 3K after 500s following another spontaneous incident of activation. The heat generation is reduced from 3W/cm^2 to about 1W/cm^2. Consequently a cycle of activation-deactivation-activation is completed. In addition to the large scale activation events, the fluctuations below 195A can be understood in a similar manner. Figure 4(b) shows the detail of a 2 minutes time history of such smaller oscillations at 190A. A strong correlation between the superheat and the voltage is clearly evident.

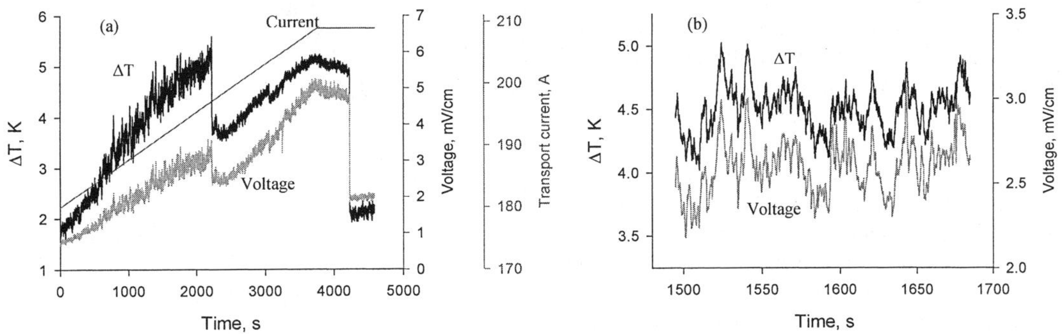

Figure 4 Time history of ΔT and V with increasing current from 180A to 210A (a) and at constant current of 190A (b)

With a larger current increment of 5A, an event of rapid superheating to 8K followed by local activation is captured, as shown in Figure 5(a). The detailed history of the event shown in Figure 5(b) indicates that the large superheat is established with 12s following a 5A current increment. The cooling by activation of boiling is very steep with an initial rate of about 3K/s. While maintaining a constant current of 190A, the superheat in the bulk liquid was removed by introducing vigorous boiling in the pool. As a result the surface of the superconductor became fully activated due to an increase in the effective surface superheat. It is noted that ΔT remained virtually constant at 3K with further increase of current to 215A. The voltage measured during this period (Figure 5(a)) corresponds to isothermal heat generation of the superconductor in the over-current regime and can be used for constructing the heat generation as a non-linear function of temperature and current.

It is clear from Figure 5(a) that with full activation of the boiling surface, the superheating and heat generation of the superconducting tape can be much lower than the sae of partially activated surface (Figure 3). To further highlight such a difference, Figure 6 shows the heat transfer characteristics upon reducing current from 300A with and without full activation. It is clear that a full activation led to a reduction by 4K in the superheat ΔT which remains lower than not fully activated surfaces for current down to 200A. Such a cooler condition is much desirable during recovering from over-current. It remains a challenge to maintain the surface activated at such small superheat.

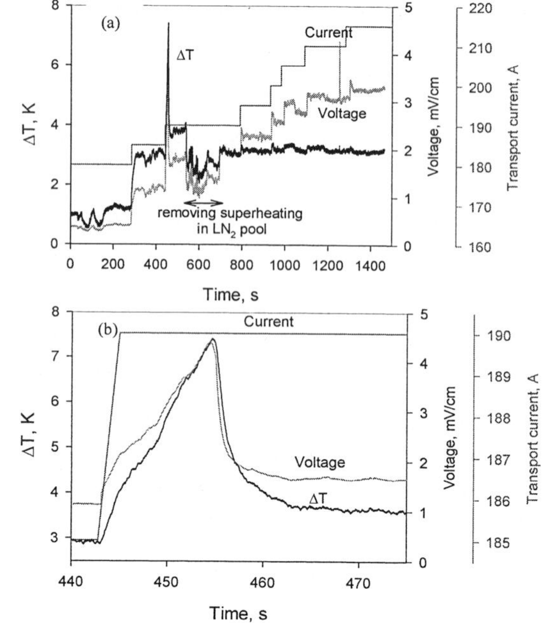

Figure 5 Time history of ΔT and V with large step increment of current (a) and the details of an activation event

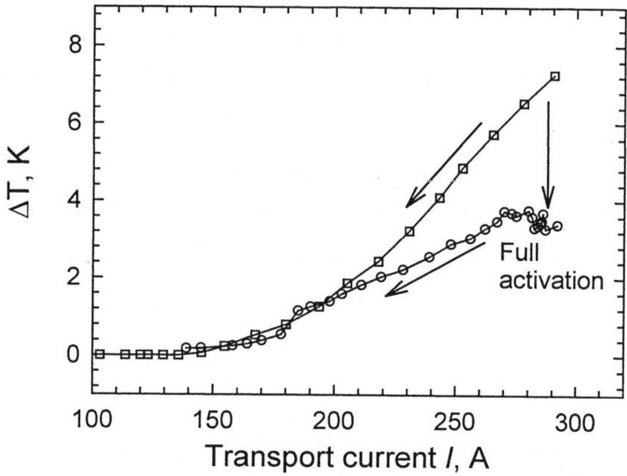

Figure 6 Surface superheat ΔT as a function of over-current for partially activated and fully activated surfaces

CONCLUSION

Unique characteristics were found in the heat transfer from HTS Superconductors carrying over-current due to the interplay between the nonlinear heat generation in the superconductor and the activation-deactivation of nucleate boiling at small heat flux. In addition full activation of surfaces results in not only the expected reduction in the superheat, but also an unconventional lower heat flux. Although such an "over" cooled state is desirable for quench recovery, its stability maybe limited due to a larger tendency to deactivate.

REFERENCES

1. Chovanec, F and Usak, P, Cryogenics **42**, 543 (2002).
2. Skokov V, N, and Koverda, V, P, Cryogenics **33**, 1072 (1993).

Equivalent circuit model for high-T_c superconducting flux flow transistor with a dual-gate structure

Seokcheol Ko[a], Hyeong-Gon Kang*[b], Sung-Hun Lim[c], Jong-Hwa Lee[a], Byoung-Sung Han[a]

[a]Division of Electronics and Information Engineering, Chonbuk National University, 664-14, Duckjin-Dong 1Ga, Jeonju Chonbuk, 561-756, South Korea
*[b]Basic Science Research Institute Chonbuk National University, 664-14, Duckjin-Dong 1Ga, Jeonju Chonbuk, 561-756, South Korea. Tel.: +82-63-270-2396, Fax.: +82-63-270-2394. E-mail address: Joshmoses@hanmail.net (Hyeong-Gon Kang)
[c]Research Center of Industrial Technology, Engineering Research Institute, Chonbuk National University, Jeonju Chonbuk, 561-756, South Korea

We have fabricated dual-gate superconducting flux flow transistor (DGSFFT) with the micro-scale structure from epitaxial superconducting thin films by photolithography. The simulation was performed modulation of the output voltage by varying the penetration depth as a function of the dual-gate currents. This model showed the dependence of the critical current density on the spatial distribution of an applied magnetic field induced by the dual-gate currents. The I-V curves of DGSFFT with a micro-scale channel from the simulation were alike in agreement with the measured curves in the flux creep regime.

The channel of superconducting flux flow transistor (SFFT) is classified into the Josephson junctions [1] and the weak-links [2] with the fabrication method. The Josephson junction flux flow transistors are more complex and more difficult to fabricate than weak-links flux flow transistors. Following the initial work on high-T_c thin film by Martens *et al.* [2], many groups have made the SFFT and studied its properties by the simulation [3,4]. In this paper the current-voltage curves in the DGSFFT (dual-gate superconducting flux flow transistor) with the micro-scale channel were simulated varying of the penetration depth as a function of the dual-gate currents. This paper proposed I-V characteristic simulation using a computational method to analyze the DGSFFT. We applied the Biot-Savart's laws and the Kim-Anderson's model, derived the mean-field description by Bernstein *et al.* from this simulation. The induced output voltage as a function of gate current transporting the dual-gate line could be computed from the derived numerical model. The calculated values were compared in relation to the serial flow of the vortices in the micro-scale channel.

MODEL OF THE DGSFFT

We fabricated the DGSFFT with the micro-scale channel and performed the analytical model. The DGSFFT having a micro-scale channel is composed of dual-gate structure and two serial channels biased with a drain current I_d as shown in Figure 1 (a). Figure 1 (b) shows the equivalent model of a serial DGSFFT. The nucleation and the motion of Abrikosov vortices in the channel can be generated and controlled by the magnetic field induced by the dual-gate currents transporting which flow near the channel. Figure 1 (c) shows the schematic diagram of the SFFT with the micro-scale channel. Figure (d) shows the optical microscope image of the SFFT with the micro-bridge.

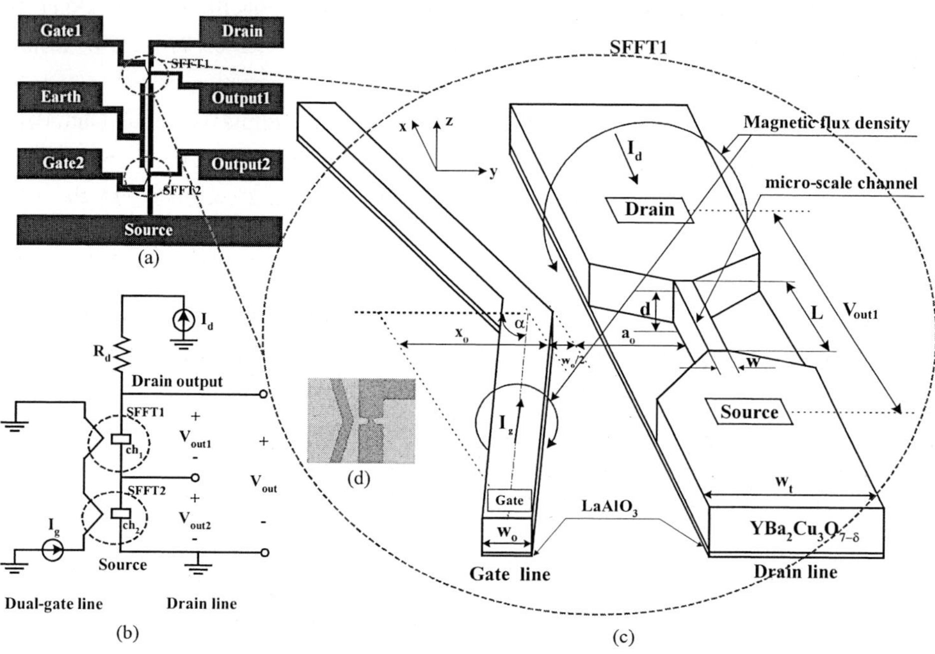

Figure 1 Structure of the DGSFFT with a micro-scale channel. (a) Mask pattern of the DGSFFT. (b) The model of a serial DGSFFT. (c) Schematic of the SFFT with the micro-scale channel. (d) Optical microscope image of the micro-bridge SFFT

The amplitude of the voltage V_{out} induced by the vortex and anti-vortex motion at the micro-bridge terminals can be given as

$$V_{out} = \frac{\Delta \Phi}{\Delta t} = B \frac{\Delta S}{\Delta t} = B \, v(w/2) \, L = \frac{N_{vortex}}{S} \phi_0 \, v(w/2) \, L = n(w/2) \, v(w/2) \, \phi_0 \, L \;, \tag{1}$$

where Φ, t, B, S and N_{vortex} are the flux in the channel, time, the magnetic field, the channel area and the number of vortices, respectively. $n(w/2)$ is the mean surface density of the moving vortices or anti-vortices and L is the length of the micro-scale channel. $v(w/2)$ and ϕ_0 are the average vortex velocity and the absolute value of the flux quantum carried by each vortex and anti-vortex, respectively.

As a conclusion, if the induced voltage from the ch$_1$ is equal to voltage of the ch$_2$ in the equilibrium state, the output voltage induced at the output drain terminals can be written as follows:

$$V_{out} = V_{out1} + V_{out2} = 2\left\{\frac{\mu_0 L k_B T \delta \exp^{-(E_P/k_B T)}}{d\hbar}\right\} \times \sinh\left(\frac{I_d}{w k_B T/(\delta \phi_0)}\right)$$

$$\times \left\{I_d - \left[I_{cr} + \frac{\sin\alpha\, I_g}{\pi}\left\langle\frac{1}{w_g}\sin\left(\tan^{-1}\frac{x_0\tan\alpha}{w_g}\right) - \frac{1}{w_g + w}\right.\right.\right.$$

$$\left.\left.\left.\times \sin\left(\tan^{-1}\frac{x_0\tan\alpha}{w_g + w}\right)\right\rangle \times \left(\frac{\sqrt{2}\lambda\,\sinh\left(d/\sqrt{2}\lambda\right)}{1 + \cosh\left(d/\sqrt{2}\lambda\right)} + \frac{(k-2)d}{2}\right)\right]\right\} \qquad , \quad \text{for} \quad I_d \geq I_{cr}, \qquad (2a)$$

$$V_{out} = V_{out1} + V_{out2} = 0, \qquad\qquad\qquad\qquad \text{for} \quad I_d < I_{cr}, \qquad (2b)$$

where x_0 is a distance of a straight line from the central line of the gate line to both side edge of gate line and α is an angle from the center of gate line. w and λ are the width of the channel and the penetration depth, respectively. w_0 is the width of the gate line. a_0 is the spacing between the gate line and the channel edge of the drain line. d and I_{cr} are the thickness of the micro-scale channel formed by wet etching method and the critical current, respectively. δ and T are pinning potential range and the temperature, respectively. k and I_{cro} are the fitting constant and the initial critical current of the case without applied field, respectively. $w_g = (x_0 + w_0/2 + a_0)$, E_p and k_B are the total distance between the gate line and the drain line, the individual vortex activation energy and the Boltzman's constant, respectively. Then, the penetration depth λ can be expressed as follows [3]:

$$\left\{\mu_0 I_{cro}\right\}\bigg/\left\{4d\left[\frac{1}{2} + \frac{\lambda}{d}\left(\frac{\sqrt{2}\sinh(d/\sqrt{2}\lambda)}{1 + \cosh(d/\sqrt{2}\lambda)}\right)\right]\right\} = \left\{\left[\phi_0/(4\pi\lambda^2)\right]\bigg/\sqrt{\frac{\pi w}{4d}}\right\} \times \ln\left(\frac{\lambda}{\xi}\right), \qquad (3)$$

where the penetration depth λ can be expressed as a function of I_{cro}, d, ξ, and w. If the micro-bridge geometrical dimensions, critical current, and coherence length are given, it is possible to compute λ from Eq. (3).

RESULTS AND DISCUSSION

We have fabricated the DGSFFT with the thickness 350 nm and the channel width 10 μm, and the channel length 5 μm using photolithography process and measured the *I-V* characteristics. The YBa$_2$Cu$_3$O$_{7-\delta}$ films were wet etched by 0.67% phosphoric acid (H$_3$PO$_4$). The most important steps of the patterning process are the etching time for aqueous solution of phosphoric acid. Figure 2 shows the measured and the calculated *I-V* characteristics of DGSFFT with the micro-bridge that plotted from Eqs. (2-3). From the *I-V* characteristics curves, three regimes such as Thermally Assisted Flux Flow (TAFF), the flux creep mode and the flux flow mode were reflected to explain the measured and calculated current-voltage curves. As a simulation result, it showed that the total voltages induced from the ch$_1$ and ch$_2$ increased more than the voltages induced from the ch$_1$ in the equilibrium state. The simulation results of the current-voltage characteristics are similar to the measured results.

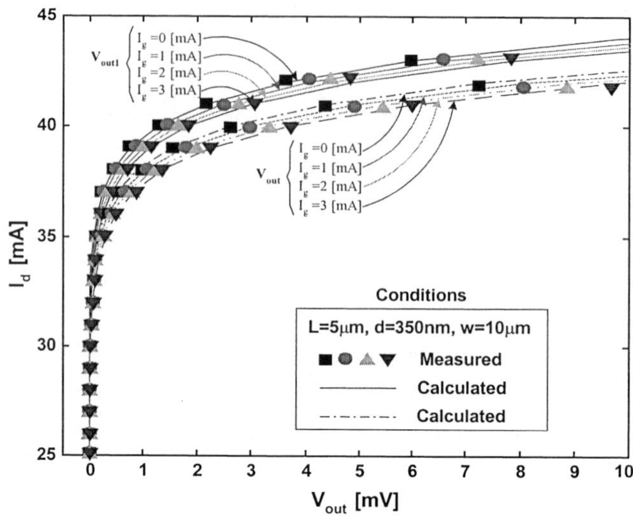

Figure 2 Measured and calculated *I-V* characteristics curves of the output voltage as a function of the dual-gate current of DGSFFT with the micro-bridge at 77K

CONCLUSIONS

We have tried to lead the *I-V* characteristic equation that was modified in proportion to the length, the width and thickness of the micro-scale channel, and a distance between the drain line and the gate line. We proposed analytical model that was composed of the DGSFFT with the micro-scale channel. We worked the simulation to analyze *I-V* characteristics of DGSFFT having a micro-bridge using the computational method. When compared to the calculated and measured values, the induced voltage agreed alike in the DGSFFT with a micro-bridge in the flux creep regime. This model is suitable for prediction of the *I-V* characteristics of the DGSFFT with the micro-scale channel in the flux creep regime.

REFERENCES

1. P. A. C. Tavares, E. J. Romans, and C. M. Pegrum, High current gain HTS Josephson vortex flow transistors, IEEE Trans. Appl. Supercon. (1999) 9 3941-3944

2. J.S. Martens, G.K.G. Hohenwarter, J.B. Beyer, J.E. Nordman, and D.S. Ginley, S parameter measurements on single superconducting thin-film three terminal devices made on high-T_c and low-T_c materials, J. Appl. Phys. (1989) 65 4057-4060

3. P. Bernstein, J. F. Hamet, B. Blanc-Guilhon, S. Flament, C. Dubuc, J. Bok, X. Q. Zhang, J. P. Contour, and F. R. Ladan, A mean field description of the transition to the mixed state of superconducting microbridges with low vortex pinning forces, J. Appl. Phys. (1994) 76 2929-2936

4. Hyeong-Gon Kang, Y.-H. Im, Seokcheol Ko, Sung-Hun Lim, B.-S. Han, and Y.B. Hahn, Superconducting flux flow transistor fabricated by an inductively coupled plasma etching technique, Physica C (2004) 400 111-116

Direct observation of percolation and phase separation in $Pr_{5/8}Ca_{3/8}MnO_3$ system

Cao G. X., Zhang J. C.[*], Li P. L., Yu J., Wang S. P., Jing C., Cao S. X., and Shen X.C.

Department of Physics, Shanghai University, Shanghai 200436, China
*Corresponding author, Email: jczhang@mail.shu.edu.cn

Jump-like magnetization has been study in $Pr_{5/8}Ca_{3/8}MnO_3$ system. The results prove that, above T=5K about, the jump breadth in *M-H* curve suggests a typical AFM-FM transition drove by the magnetic field in different parts of the materials. This can be understood in the picture of phase separation with percolation properties. For lower temperature below 5K, the jumps become ultrasharp with a very narrow width of 1×10^{-4}T. These properties can be explained by magnetostriction and the orbital occupancy of the e_g electron of Mn^{3+}, which reflects the importance of interplay among spin, charge and orbital freedom in present manganites.

INTRODUCTION

In perovskite-type manganese oxides, the strong correlations among spin, charge, orbital, and lattice degrees of freedom play important roles. The charge-order (CO) state can be melted by the application of magnetic/electric fields, x-ray and so on [1-3], which meanwhile is accompanied by some Jahn-Teller distorted MnO_6 octahedra. On the other hand, accumulating theoretical and experimental evidence indicates that the important feature of the manganites is phase separation (PS), which shows the coexistence of the submicrometer CO phase and ferromagnetic (FM) metallic phase [4]. Application of the magnetic field can lead to a collapse of the charge-ordering gap and there is accompanied with a structural PS as well as with the magnetic phase transition [5]. However, this COAFM state is particularly robust in the compounds $Pr_{1-x}Ca_xMnO_3$ systems, owing to its small average A-site cationic radius. In general, very high magnetic fields larger than 20T are needed to overcome this state via a metamagnetic transition. Interestingly, several unexpected and intriguing magnetization steps were recently observed for Mn-site substituted manganese oxides [6-7], where the Mn-substituted weakens the COAFM state and favors the development of PS, then make the required magnetic fields decrease. And it has been proposed that this magnetization transition is related to the martensitic character associated with strain between the PS regions [8]. One should pay attention to the COAFM-FM transition at very low temperature to be ultrasharp and even there appears multiple steps as a function of magnetic field. This intrinsic magnetization steps is different from the traditional metamagnetic transitions. But the satisfying microscopic origins of the observed jumps or steps are not clear at this stage. This set of features has prompted us to undertake a systemic study about this peculiar COAFM-FM transition. Compared to the weakened Mn substitution systems, the $Pr_{5/8}Ca_{3/8}MnO_3$ system was chosen in present work. Based on belonging to phase-separated compound, it should be reasonable to predict that the unusual *M(H)* jumps might be observed in $Pr_{5/8}Ca_{3/8}MnO_3$ as the magnetic field and temperature vary.

EXPERIMENTAL DETAILS

The sample with a $Pr_{5/8}Ca_{3/8}MnO_3$ formula was prepared by conventional solid-state reaction in air. The sample crystallization is good enough and shows good single-phase orthorhombic structure by using XRD analysis (Ragaka 18kWD/max-2500 diffractometer, $Cu-K_\alpha$ radiation). Magnetic and electric measurements were carried out using Physical Property Measurement System (PPMS-9, Q/D Inc.) with the precision 20nV for voltage, and 0.02mT for the magnetic field. The resistivity vs temperature *R-T* curves measured under applied DC magnetic fields from 0-6 T were recorded in a temperature range of

530

1.9-300K with a precision of 0.01K. The experiment can be well repeated.

RESULTS AND DISCUSSION

The temperature dependence of the resistivity is shown in figure 1 under 0-6 T with both cooling and heating process for $Pr_{5/8}Ca_{3/8}MnO_3$ system. Both the resistivity under zero-field and 3 T are insulating, and the curve shows a discernible upturn around 230K, which is corresponding to the CO onset. A superlattice structure has been confirmed below 230K for $Pr_{5/8}Ca_{3/8}MnO_3$ as a further evidence of the $d_{3x^2-r^2}/d_{3y^2-r^2}$ type of charge/orbital ordering by low-temperature transmission electron microscopy [9]. Application of a magnetic field under 6 T drastically modifies such an insulating CO state, the ρ-T curve is metallic and the Curie temperature is $T_C \approx 84K$ (cooling curve). Meanwhile, the transition becomes strongly hysteretic which display the first-order-type transition in the system.

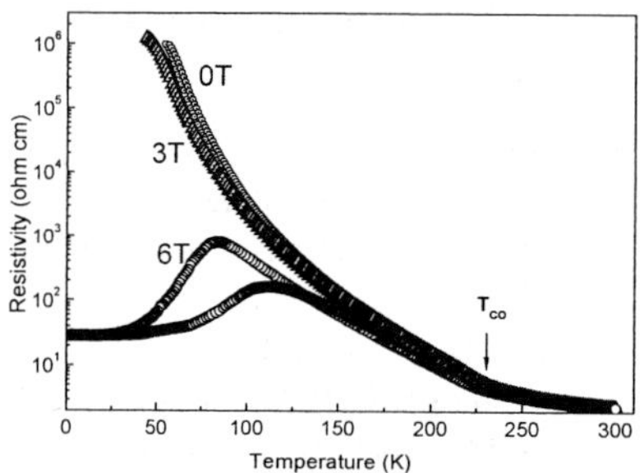

Figure 1. The temperature dependence of resistivity under 0T, 3T, and 6T for $Pr_{5/8}Ca_{3/8}MnO_3$. The resistivity was measured in a FC mode.

Figure 2. DC magnetization M vs temperature in a field of 6T for $Pr_{5/8}Ca_{3/8}MnO_3$ Inset displays resistivity curves recorded in the same conditions.

Figure 2 shows the temperature dependence of the magnetization M measured on heating after the sample was either zero field cooled (ZFC) or field cooled (FC, H=6T). The inset displays resistivity curve recorded in the same conditions. Here, two kinks correspond to the T_{CO} and T_N, around 240 and 160 K, respectively, and at low temperature, an abrupt increasing M is associated with the onset of FM ordering. Below about 100K, ZFC and FC curves separate from each when approaching PS region. This property points to the first order character for FM transition. Another prominent feature is appearance of a steep rise for the ZFC curve with increasing temperature between 4 and 12K. Such behaviors could be connected to the development of the FM phase in the AFM matrix, which shows the temperature dependence of the COAFM/FM coexistent phase. It must be noted that the difference between the values of the M for the ZFC and FC curves around 100K may be related to the large magnetostriction present in this materials. The ρ-T curves recorded in the ZFC and FC modes split at near T<100K, whose behavior is similar to that of magnetization. In the low-temperature region (T<100K), the ρ and M curves vary exactly in opposite ways, which is qualitatively consistent with the simple DE model. This behavior is more difficult to understand for $Pr_{1-x}Ca_xMnO_3$ with a smaller one-electron bandwidth W. According to the DE model, if the localized t_{2g} spins are considered classical and with an angle θ between nearest-neighbor ones, the transfer integral t of e_g electrons between $Mn^{3+}(t_{2g}^3 e_g^1)$ and $Mn^{4+}(t_{2g}^3 e_g^0)$ ions can be expressed as $t = t_0\cos(\theta/2)$, where the t_0 is the transfer integral in a fully spin-polarized state. The value of the t is typically 0.2eV in narrow band manganese oxides such as PCMO [10]. Then the electron-phonon coupling constant λ can be evaluated according to $\lambda = \sqrt{2E_{JT}/t}$, where the static JT energy E_{JT} is estimated as 0.25eV at low temperature. Though the E_{JT} is temperature dependent, in present paper we can conjectured that the application of the H align the Mn t_{2g} spins via the Zeeman coupling, which leads

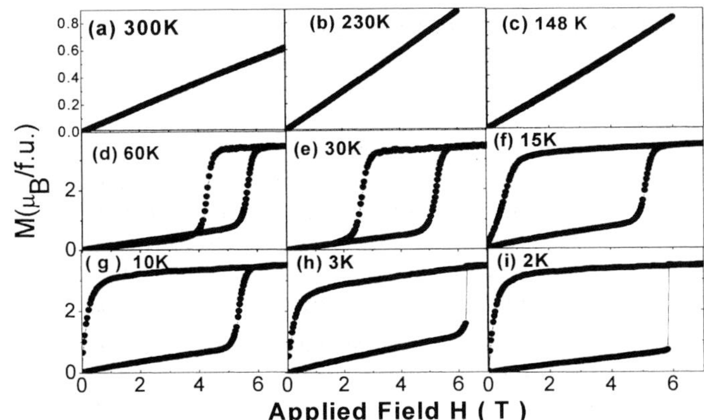

Figure 3. *M(H)* curve at various temperatures under a $0\rightarrow 7T\rightarrow 0$ circular sweeping model for $Pr_{5/8}Ca_{3/8}MnO_3$. For each run, the sample was cooled in zero field and after stabilization, the measurement was done.

to a decrease in θ and an increase in t, and then induces the behavior of ρ and M for $Pr_{5/8}Ca_{3/8}MnO_3$ to become more in line with the DE model. Actually, applied field H in the COAFM state can melt the CO and cause the change of the electronic and lattice structure, and then favor the stabilization of FM phase.

The magnetization isotherms at various temperatures are given in figure 3. It can be seen that the M is strongly dependent on the thermal and magnetic history. The data at 300K, 230K and 148K are all increase nearly linearly in lower field than 6T. However, for the same magnetic field, the values of the M increase as the temperature decreases. Which probably displays the alternation of the magnetic phase as the temperature reduces. The two-phase nature at the low temperature (<60K) region becomes more evident. In the other hand, an initial AFM state increases with H, which probably is resulted from a reduced canting angle due to canted AFM state or of the field-induced melting of CO [1]. At a sufficiently high H, a field-induced kink occurs associating with the conversion of COAFM to FM. The similarity has been observed in other manganites such as $Pr_{0.5}Ca_{0.5}Mn_{0.95}O_3$ and $La_{0.250}Pr_{0.375}Ca_{0.375}MnO_3$ [7]. And a saturate magnetization appears with a saturation moment of 3.45 μ_B/formula unit at 7T, which is consistent with the result excepted for $Pr_{5/8}Ca_{3/8}MnO_3$ and we can regard this highly magnetization as a fully spin-polarization. When the field is reduced from 7T, the *M(H)* for temperature above ~30K remains in the FM state initially and then appears a AFM-FM transition. However, below ~10K, the *M(H)* behaviors similar to a long range ferromagnet only with a rapid decrease H to 0. At 15K, the *M(H)* seem to show the transition process due to its a sign of AFM as magnetic field is down to 1T. As in figure 3, the breadth of the field-induced kink from AFM-FM suggests the process that magnetic field drive this transition in different parts of the sample. This is consistent with the PS picture with percolation properties. This transition may be explained according to the martensitic effects associated with strain between the PS regions [8]. This behavior changes dramatically at below ~3K, which shows an abrupt step near H_c=6.3T and 5.8T for T=3K and 2K, respectively. Whose width are smaller than 1×10^{-4}T and there wasn't seems to display the percolation process for this transition as it did when temperature above ~3K. After the step, *M(H)* increases nearly linearly with a relatively small slop until it reaches to the FM saturation moment. In fact, the COAFM-FM transition is due to the melting of CO in the system because of the application of magnetic field. The CO is accompanied by an ordering of e_g orbital of the $d_{3x^2-r^2}/d_{3y^2-r^2}$ type and the cooperative Jahn-Teller distortion, then the field-induced alter of the superstructure and magnetic structure may determine the change of the magnetization. The charge and /or orbital occupied scenario and the unique magnetic structure have been determined by neutron diffraction [9]. Just contrary to the "x=1/2 plus defects" scenario due to x away from 0.5, there are without defects to be constructed into the specific charge arrangements. For $Pr_{5/8}Ca_{3/8}MnO_3$, the ratio of Mn^{3+} and Mn^{4+} in the sublattice is 5:3 according to the chemical composition, except that the alternation of $(3x^2$-$r^2)$ and $(3y^2$-$r^2)$ orbitals appears in the a-b planes, the extra e_g electrons occupy $3d_{3z^2-r^2}$ orbitals on the Mn^{4+} sublattice which is along the c direction instead of parallel to the ab layers [11]. It is noted that the

Coulomb interaction in this charge lattice state is lowest and play an important role to the experimentally observed CO phase. The magnetic structure at low temperature is commonly refereed to as the pseudo-CE type for the $Pr_{5/8}Ca_{3/8}MnO_3$ [10]. Based on above results, we can conjecture a scenario in which steps may occur. Under normal circumstances, the FM coupling along the c axis would behave like a 1D conductor, because electrons can hop freely in terms of DE mechanism. However, in the a-b plane electron hoping along the FM zigzag chains is arrested by the Coulomb repulsion from the neighboring electrons out of the chain. This is why the solid is an insulator at even very low temperature. If the magnetic field is applied, the ab plane may expand while the c axis is compressed which will improve the orbital overlap within the plane and enhance the probability of the charge transfer. This phenomena will means the possible presence of $d_{x^2-y^2}$ orbital ordering and thus increase the transfer integral t of the e_g electrons and increase the FM interaction. The character of the preferred e_g states under magnetic field are similar to that of $La_{1-x}Sr_xMnO_3$ [12]. Furthermore, when the applied magnetic field is enough large, the ab plane will behave a FM characteristic and make the system a truly 3D conductor. Just in this critical magnetic field, the CO melts completely and the magnetization jumps occur. According to Tokunaga et al. argued [5], the destruction of the orbital ordering by an external field gives rise to the structural phase transition and the recovering of the DE, which promotes the polarization of the t_{2g} spins. In the present case, the field-induced magnetostriction occurs which indicates orbital occupancy of the e_g electron of the Mn^{3+} change as the magnetic filed, and then caused the occurrence of the $M(H)$ jumps. In a word, the central conclusion raised by our data is that all the Magnetic-field jumps/steps should be related to magnetostriction, indicating the presence of a field-induced change in the orbital occupancy of the e_g electron of the Mn^{3+}, which show the possible appearance of $d_{x^2-y^2}$ orbital ordering. Meanwhile, the data also reflect the unusual importance of interplay among spin, charge, and orbital degrees of freedom.

ACKNOWLEGMENTS

This work is supported by the National Foundation of National Science of P. R. China (No.10274049), Developing Foundation of National Science of Shanghai Educational Committee （No.02AK42） and the Key Subject of Shanghai Educational Committee.

REFERENCES

1. Y. Tomioka et al.,-Magnetic-field-induced metal-insulator phenomena in $Pr_{1-x}Ca_xMnO_3$ with controlled charge-ordering instability, Phys. Rev. (1996) B 53 R1689-1692.

2. K. Ogawa et al., Stability of a photoinduced insulator-metal transition in $Pr_{1-x}Ca_xMnO_3$,Phys. Rev. B, (1998) 57 R15 033 −15 036

3. M. Hervieu et al., Charge disordering induced by electron irradiation in colossal magnetoresistant manganites ,Phys. Rev. B (1999) 60 726 -729.

4. M. Uehara, S. Mori, C.H. Chen, and S.-W. Cheong, Percolative phase separation underlies colossal magnetoresistance in mixed-valent manganites.Nature (London) (1999) 399 560-563.

5. M. Tokunaga, N. Miura, Y. Tomioka, Y. Tokura, High-magnetic-field study of the phase transitions of $R_{1-x}Ca_xMnO_3$ (R=Pr, Nd) Phys. Rev. B (1998) 57 5259-5264.

6. A. Maignan,S. Hébert, V. Hardy, C. Martin, M. Hervieu, and B. Raveau,-Magnetization jumps and thermal cycling effect induced by impurities in $Pr_{0.6}Ca_{0.4}MnO_3$,J. Phys.: Condens. Matter (2002) 14 11 809-11819.

7. R.Mahendiran, A. Maignan, S. Hébert, C. Martin, M. Hervieu, B. Raveau, J.F. Mitchell, and P. Schiffer, Ultrasharp Magnetization Steps in Perovskite Manganites, Phys. Rev. Lett. (2002) 89 286602-1-286602-4.

8. V. Hardy, S. Majumdar, S. J. Crowe, M. R. Lees, D. McK. Paul, L. Hervé, A. Maignan, S. Hébert, C. Martin, C. Yaicle, M. Hervieu, and B. Raveau,-Field-induced magnetization steps in intermetallic compounds and manganese oxides: The martensitic scenario,Phys.Rev.B(2004)69 R020407-R020411.

9. T. Asaka, S. Yamada, S. Tsutsumi, C. Tsuruta, K. Kimoto, T. Arima, and Y. Matsui, Phys. Rev. Lett. (2002) 88 097201.

10. Takashi Hotta and Elbio Dagotto, Magnetic ordering effects in the Raman spectra of $La_{1-x}Mn_{1-x}O_3$,Phys. Rev. B (2000) 61 R11879-11882.

11. D. E. Cox, P. G. Radaelli, M. Marezio, S-W. Cheong, Structural changes, clustering, and photoinduced phase segregation in $Pr_{0.7}Ca_{0.3}MnO_3$,Phys. Rev. B (1998) 57 3305-3314.

12. H. Kawano, R. Kajimoto, H. Yoshizawa, Y. Tomioka, H. Kuwahara, Y. Tokura,-Magnetic Ordering and Relation to the Metal-Insulator Transition in $Pr_{1-x}Sr_xMnO_3$ and $Nd_{1-x}Sr_xMnO_3$ with x ~ 1/2 ,Phys. Rev. Lett. (1997) 78 4253-4256.

The reentrant behaviour of spin-glass phase in $(La,Pr)_{0.5}Ca_{0.5}MnO_3$ manganites

Zhang J. C.*, Cao G. X., Sha Y. N., Wang S. P., Yu J., Jia G. Q., Jing C., Cao S. X., and Shen X. C.

Department of Physics, Shanghai University, Shanghai 200436, PR China
* Corresponding author; Email: jczhang@mail.shu.edu.cn

Reentrant spin-glass phase behaviour was reported for $La_{0.5}Ca_{0.5}MnO_3$ and $Pr_{0.5}Ca_{0.5}MnO_3$. ac susceptibility curve for both compounds is very similar and a cusp appears at ~40K after T_C. The $\rho(T)$ measured in a field of 8 T appears M-I transition, but at lower temperature, the resistivity becomes an insulator again. The $M(H)$ gives the presence of FM clusters in the $La_{0.5}Ca_{0.5}MnO_3$. Our data indicate that both $Pr_{0.5}Ca_{0.5}MnO_3$ and $La_{0.5}Ca_{0.5}MnO_3$ compounds have the same ground state, which shows the existence of some FM clusters in the background of remaining region in a spin frozen state

INTRODUCTION

In mixed-valent manganese oxides, the strong correlations among spin, orbital, and lattice degrees of freedom play important roles [1-2]. Especially for the intermediate-hole doped $R_{0.5}Ca_{0.5}MnO_3$ compounds, being close to AFM CE-type charge ordered (CO) phase, is very interesting. Remarkably, CO in the system can be melted by the application of magnetic fields, x-ray, etc [3]. Recent results show that $<r_A>$ can determine the one-electron bandwidth and CO is observed in materials with small $<r_A>$[4]. Both the $La_{0.5}Ca_{0.5}MnO_3$ and $Pr_{0.5}Ca_{0.5}MnO_3$ have small $<r_A>$, and even smaller for $Pr_{0.5}Ca_{0.5}MnO_3$ ($<r_A>$~1.18Å). For $Pr_{1-x}Ca_xMnO_3$ system, though the FMM state is never realized in zero magnetic field, the competition between the FMM and charge ordered antiferromagnetic (COAFM) ground state leads to a an increase of FM tendencies as x decreases below 0.5. While for $La_{0.5}Ca_{0.5}MnO_3$ compounds [5], it has been shown that CO occurs not only in AFM regions but can also occur in FM regions (COFM). On the other hand, accumulating theoretical and experimental evidence indicate that the importance of the phase separation (PS), which shows the coexistence of the submicrometer CO and FMM phase [1-2]. The competition between the coexisting phases opens the possibility for the appearance of locally metastable states. Meanwhile, there are some evidence in the literature tends to the framework of the electronic phase separation, such as spin-glass state et al.[6]. Therefore, studying the magnetic state would be very important for determining the physical properties of mangenties. In this paper, we present resistivity, magnetic susceptibility, and magnetization data for $La_{0.5}Ca_{0.5}MnO_3$ and $Pr_{0.5}Ca_{0.5}MnO_3$ compounds. We have observed the reentrant spin-glass-like phase in both compounds and the relevant result indicates that the two compounds have very similar magnetic structure at low temperature.

EXPERIMENTAL DETAILS

Polycrystalline $La_{0.5}Ca_{0.5}MnO_3$ and $Pr_{0.5}Ca_{0.5}MnO_3$ were prepared by conventional solid-state reaction method. Powder XRD measurement revealed that the samples crystallization is good enough, and both samples show good single-phase orthorhombic structure. Magnetic measurements were carried out using physical property measurement system (PPMS) with the precision 20nV for voltage. All the $M(H)$ curves were recorded after the sample were ZFC from room temperature. Transport measurements were carried out using the conventional four-probe technique. The resistivity vs temperature ($\rho(T)$) curves measured under applied DC magnetic fields ranges from 0-8 T were recorded by PPMS in the range of 1.9-300K.

RESULTS AND DISCUSSION

Figure 1 shows the result of ac susceptibility for $La_{0.5}Ca_{0.5}MnO_3$ vs ac field at 2 Oe, 6 Oe, 8 Oe and 10Oe, respectively. For our measurement, the low ac magnetic field is not supposed to affect the magnetic state of the sample. It must be noted that the curves are obtained by the continuous transform measurement between four frequencies (77, 333, 4577 and 1000Hz) and four fields as the temperature change. The transition of T_C is strongly hysteretic which display the first-order-type transition in the system. From the result of the ac magnetic susceptibility as shown in figure 1, we can also observe that the existence of two kinks for the warming curve which is at around 215 and 40 K, respectively. However, for the cooling

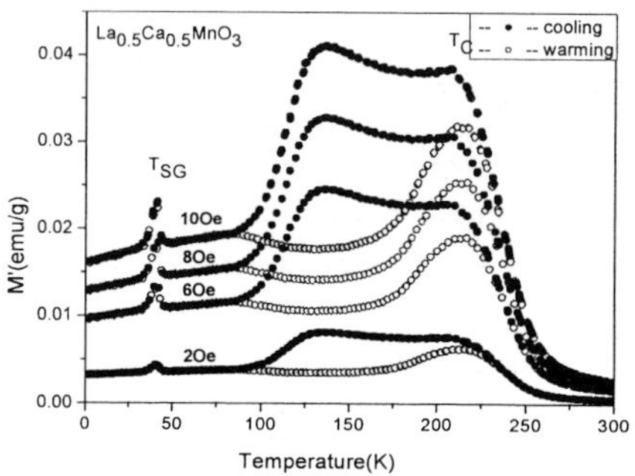

Figure1, Temperature dependence of susceptibility for $La_{0.5}Ca_{0.5}MnO_3$ as a function of ac field at 2 Oe, 6 Oe, 8 Oe and 10 Oe, respectively.

Figure 2, Temperature dependence of susceptibility for $Pr_{0.5}Ca_{0.5}MnO_3$ as a function of ac field at 2 Oe, 6 Oe, 8 Oe and 10 Oe, respectively.

curve, the kink at 215K，which is corresponding to the Curie temperature T_C, becomes more splinted and even appear a plateau between the range of temperature 130~210K. This phenomena may be related to the memory effect which has been detected in bilayer manganite $(La_{1-z}Pr_z)_{1.2}Sr_{1.8}Mn_2O_7$ (z=0.6) single crystal [8] and double layered thin film of LSMO[9]. For the cooling curve, as temperature decreases, the χ_{ac} increases and reaches to its maximum value at around 210K, and then it retain the value until 130K due to the memory effect. While the warming curve retain its AFM state as temperature increases until ~210K. Both behaviors show the FM fraction in the system after the T_C and which is highly temperature sensitive and history-dependent [8]. It should be remarkable that the cusp appeared at ~40K with the temperature decrease as shown in figure 1, it may be related to the low temperature spin-glass behavior which has been reported by De Terai et al [6]. Which presumably is due to the frustration caused by the competition between FM (double exchange) and AFM (super-exchange) interactions. Interestingly, such

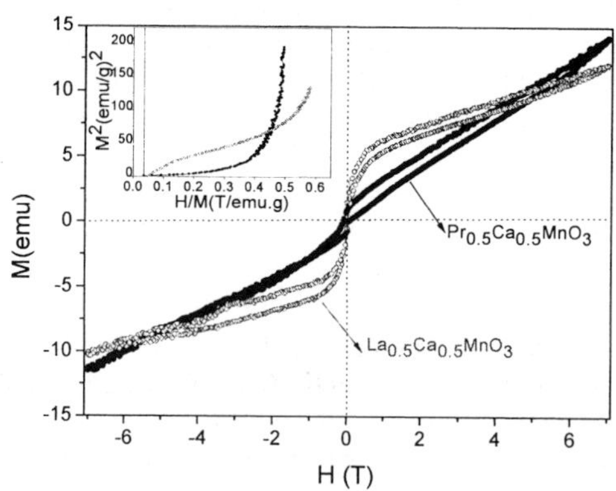

behavior occurs after the T_C, so it should belong to reentrant spin-glass behavior. That is to say, the compound undergoes two phases transition as the temperature decreases. The first is from magnetic disordering to long-range magnetic ordering at T_C, and second is from magnetic ordering to spin freezing disordering at the temperature of spin-glass transition (T_{SG}). Which is typically a feature of reentrant spin-glass. At T_{SG}, the Mn ion spin changes from the ferromagnetic long-range magnetic order to the ferromagnetic short-range order of the spin-glass state at enough low temperature. Similar temperature dependences of M/H due to the spin-glass state have been observed also for other compounds [10] and neutron diffraction and ac susceptibility were done to

Figure 3, Magnetic field (H) dependence of M for $La_{0.5}Ca_{0.5}MnO_3$ (the red curve) and $Pr_{0.5}Ca_{0.5}MnO_3$ (the blue curve) at 3K, respectively.

confirm it. Generally, the T_{SG} is mostly a constant and this is consistent with present property of the cusp at ~40K. Figure 2 shows the ac susceptibility curve for $Pr_{0.5}Ca_{0.5}MnO_3$, which is recorded in the same conditions as that of $La_{0.5}Ca_{0.5}MnO_3$. Interestingly, the behavior of ac susceptibility for $Pr_{0.5}Ca_{0.5}MnO_3$ is very similar to that of $La_{0.5}Ca_{0.5}MnO_3$ as shown in figure 2, except that there didn't appear a plateau between certain ranges of temperature for cooling curve. And the corresponding two kink appear at ~245K and ~40K, respectively. Though the COAFM is the ground state, its small energy difference with the FM state can indicate a tendency towards phase coexistence. This probably can explain the appearance of the cusp at ~40K. There are some trends in the literature pointing to an explanation of the feature in the framework of the electronic phase separation scenario [2], but we suggest it is probably that this electronic phase separation is spin-glass-like state.

In figure 3, the field dependence of magnetization is shown for $La_{0.5}Ca_{0.5}MnO_3$ and $Pr_{0.5}Ca_{0.5}MnO_3$ at 3K, respectively. It can be

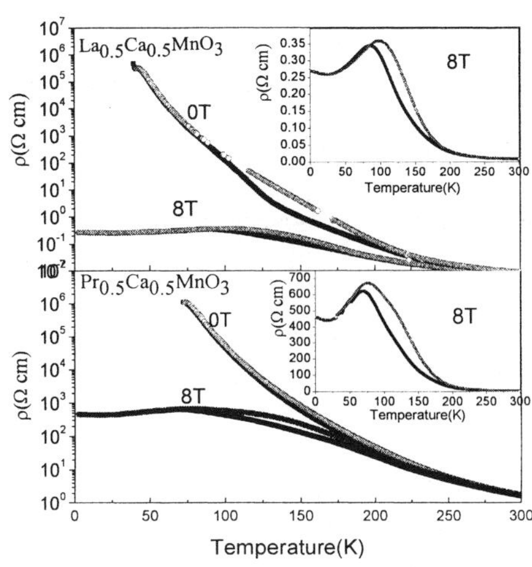

Figure 4, The temperature dependence of the resistivity for $La_{0.5}Ca_{0.5}MnO_3$ and $Pr_{0.5}Ca_{0.5}MnO_3$ in the absence of a magnetic field and under magnetic fields of 8 T. The black and red curves stand for cooling and warming process, respectively

seen that the curve slop increases as the magnetic field increases and there shows a noticeable irreversibility for measurement of increasing and decreasing field. However, there is not remanent magnetization when field is zero, which indicates that present irreversibility is due to spin-glass behavior [11]. Both the M vs H curve didn't appear the sign of saturation when the applied magnetic field is 7T. The absence of saturation at 7T correlate well with the high resistivity value as shown in figure 4 which gives the temperature dependence of the resistivity for $La_{0.5}Ca_{0.5}MnO_3$ and $Pr_{0.5}Ca_{0.5}MnO_3$ in the absence of a magnetic field and under magnetic fields of 8 T. The negative value of the magnetization confirms the existence of AFM interaction. Especially for $La_{0.5}Ca_{0.5}MnO_3$, though behaves a noticeable FM behavior at low temperature, the magnetization continuously increase as the magnetic field, which show the presence of some FM clusters in the background of remaining region in a frozen state. The inset of figure 3 gives the Arrott plots obtained from the relative magnetization isotherms at 3K. From which, we can see that there are no sign of spontaneous magnetization and it is not belong to long-range FM ordering state. According to these results, we may also explain the COFM occurred in $La_{0.5}Ca_{0.5}MnO_3$ at 90K. This character of COFM is very difficult to understand according to the conventional models. The tight relationship between COFM and spin-glass phase in the compound may be explained with the temperature decreasing as follows. At high temperature the compound is paramagnetic, and as the temperature is lowered there is a gradual CO which increases the resistivity. If the enough magnetic field is applied, the compound becomes ferromagnetic and a first-order I-M transition takes place at T_C. However, if the applied magnetic field is zero, the sample is composed of an inhomogeneous mixture of ferromagnetic regions and regions with no net magnetization at 90K[5]. On the other hand, at lower temperature, the spin-glass state rather than the ferromagnetic phase sets in which shows the development of competition between the AFM and FM as the temperature decreases. Applying magnetic field in the spin-glass state brings about a reduction in the AFM and the FM clusters appear in the background of remaining region in a frozen state. From figure 4, it can be seen that both the resistivity of $La_{0.5}Ca_{0.5}MnO_3$ and $Pr_{0.5}Ca_{0.5}MnO_3$ under zero-field is insulating. Application of a magnetic field drastically modifies such an insulating state and a M-I transition take places. Meanwhile, the transition becomes strongly hysteretic which display the first-order-type transition in the system. In the inset of figure 4, we give the magnified curves of $\rho(T)$ measured in a field of 8 T. The present transport results indicate that $La_{0.5}Ca_{0.5}MnO_3$ exhibits M-I transition at about 84K (for cooling curve) and the compound shows a metallic behavior in the temperature range of 84~28K. Whereas at lower temperature, $La_{0.5}Ca_{0.5}MnO_3$

becomes an insulator and show its dramatic increase in resistivity. This may be due to the localization of the charge carriers below temperature of spin-glass transition. We can conjecture a scenario in which localization may occur as follows: According to the DE model, the e_g electron can hop from Mn^{3+} to Mn^{4+} if the two moments are parallel. However, the reentrant spin-glass behavior, as mentioned above, can cause the random freezing of Mn^{3+} and Mn^{4+} moments and thus favors the localization of the e_g electron. In connection with the results of M(H) above, though there are FM cluster formation in the frozen region domain , the system possibly can not find a percolation path through the frozen region and so the resistivity appears second upturn below T_{SG}.

To sum up, using the measurement of resistivity, ac susceptibility and *M(H)*, we have shown that $La_{0.5}Ca_{0.5}MnO_3$ and $Pr_{0.5}Ca_{0.5}MnO_3$ at ~40K, show the existence of reentrant spin-glass behavior. Meanwhile, the result of *M(H)* for $La_{0.5}Ca_{0.5}MnO_3$ at 3K gives the presence of FM clusters. Based on the similar behavior in ac susceptibility, we suggest that there should also exist the competition between AFM and FM in $Pr_{0.5}Ca_{0.5}MnO_3$. All the rich variety of phases can coexist in the manganites is because the bandgap can be locally modified by strain within each grain [12] and other factors. Our results indicate that both compounds have the same ground state which shows the presence of some FM clusters in the background of remaining region in a spin frozen state.

ACKNOWLEDGMENTS

This work is supported by the National Foundation of National Science of P. R. China (No.10274049), Down Project of the Education Committee of Shanghai Municipality（No. 03SG35）and Developing Foundation of National Science of Shanghai Educational Committee（No.02AK42）and the Key Subject of Shanghai Educational Committee.

REFERENCES

1. M. Uehara, S. Mori, C.H. Chen, and S.-W. Cheong, Percolative phase separation underlies colossal magnctorcsistance in mixed-valent manganites, <u>Nature (London)</u> (1999) <u>399</u> 560-563

2. A. Moreo, S. Yunoki, and E. Dagotto, Phase Separation Scenario for Manganese Oxides and Related materials, <u>Science,</u> (1999) <u>283</u> 2034-2040.

3. M. Hervieu *et al.*,Charge disordering induced by electron irradiation in colossal magnetoresistant manganites, <u>Phys. Rev. B,</u> (1999) <u>60</u> 726-729

4. H. Y. Hwang, S.W. Cheong, P. G. Radaelli, M. Marezio and B. Batlogg, Lattice Effects on the Magnetoresistance in Doped $LaMnO_3$, <u>Phys .Rev . Lett .,</u> (1995)<u>75</u> 914-917

5. James C. Loudon, Neil D. Mathur &Paul A. Midgley, Charge-ordered ferromagnetic phase in $La_{0.5}Ca_{0.5}MnO_3$, <u>Nature</u> (2002) <u>420</u> 797-800.

6. J.A. Mydosh, <u>Spin Glasses: An Experimental Introduction</u>, Taylor and Francis Press, London, UK(1993).

7. Ph. Vanderbemden, B. Vertruyen, A. Rulmont, R. Cloots, G. Dhalenne, and M. Ausloos, ac magnetic behavior of large-grain magnetoresistive $La_{0.78}Ca_{0.22}Mn_{0.90}O_x$ materials,<u>Phys. Rev. B</u> (2003) 68 224418-224424.

8. I. Gordon, P. Wagner, V. V. Moshchalkov, Y. Bruynseraede, M. Apostu, R. Suryanarayanan, and A. Revcolevschi, Temperature dependent memory effects in the bilayer manganite $(La_{0.4}Pr_{0.6})_{1.2}Sr_{1.8}Mn_2O_7$,<u>Phys. Rev. B</u> (2001) <u>64</u> 092408-092411.

9. Y. Konishi, T. Kimura, M. Izumi, M. Kawasaki, and Y. Tokura, Fabrication and physical properties of *c*-axis oriented thin films of layered perovskite $La_{2-2x}Sr_{1+2x}Mn_2O_7$,<u>Appl. Phys. Lett.,</u> (1998) <u>73</u> 3004-3006.

10. T. Terai, T. Kakeshita, T. Fukuda and T. Saburi, Electronic and magnetic properties of $(La-Dy)_{0.7}Ca_{0.3}MnO_3$, <u>Phys . Rev . B</u> (1998) <u>58</u> 14908-14912.

11. C. Mitra, P. Raychaudhuri, S. K. Dhar, A. K.Nigam, R. Pinto, S. M. Pattalwar, Evolution of transport and magnetic properties with dysprosium doping in $La_{0.7-x}Dy_xSr_{0.3}MnO_3$ (*x*=0–0.4) ,<u>Journal of Magnetism and Magnetic Materials</u> (1999) <u>192</u> 130-136.

12. P. Levy, F. Parisi, G. Polla, D. Vega, G. Leyva, H. Lanza, R. S. Freitas and L. Ghivelder, Controlled phase separation in $La_{0.5}Ca_{0.5}MnO_3$,<u>Phys. Rev. B</u> (2000) <u>62</u> 6437-6441.

Anisotropic transport and magnetic properties of $Pr_{1/2}Sr_{1/2}MnO_3$ single crystal sample

Cao S.X., Wang X.Y., Zhang Y.F., Yu L.M., Jing C., Zhang J.C., and Shen X. C.

Department of Physics, Shanghai University, Shanghai 200436, China

Resistivity and magnetization for $Pr_{1/2}Sr_{1/2}MnO_3$ single crystal samples were measured in the directions of parallel ($//$) and perpendicular ($\perp$) to the ferromagnetic (FM) plane. The observed anisotropic transport and magnetic properties in antiferromagnetic (AFM) and FM states can be explained by the double exchange model with d_{x2-y2} orbital structure: with the anisotropic orbital structure, carriers in the FM state or AFM state move only in the FM layers and show two-dimensional metallic behavior. As the A-type AFM state undergoes a transition to a canted spin state below T_N, it also shows spin-flop transition at $T<T_N$.

INTRODUCTION

Perovskite manganites $R_{1-x}A_xMnO_3$ (R and A are trivalent and divalent ions, respectively) have attracted much attention since the rediscovery of the colossal magnetoresistance effect. Beyond the classic double-exchange (DE) mechanism, it has been recognized that the interplay of spin, charge, lattice and orbital is important in these compounds. Among the perovskite manganites that showing colossal magnetoresistance (CMR) properties, $Pr_{1/2}Sr_{1/2}MnO_3$ is of great interest because of its particular magnetic and transport properties. Previous studies showed that this compound exhibits two first-order transitions versus temperature, from paramagnetic (PM) state to ferromagnetic (FM) state at $T_C{\sim}265$ K and from FM to antiferromagnetic (AFM) state at about 140 K [1,2]. The FM state in manganite perovskites is attributed to the double exchange mechanism; localized t_{2g} spins are ferromagnetically coupled to itinerant e_g electrons. An interesting feature of $Pr_{1/2}Sr_{1/2}MnO_3$ compound is that the ground state in low temperature studied by neutron scattering is a layered A-type AFM magnetic structure [3,4], instead of the well known CE-type spin and orbital ordered state. Within the ferromagnetic layers in the A-type AFM structure, one can expect that the DE mechanism is in effect, and that it enhances the conductivity within the FM layers. In addition, the crystal structure of the A-type system favors the d_{x2-y2} orbits, giving rise to two-dimensional band for the e_g electrons. Thus, it can be expected the two-dimensional anisotropic transport and magnetic properties for the A-type AFM structure. In this paper, we reported evident anisotropy of transport and magnetic properties for $Pr_{1/2}Sr_{1/2}MnO_3$ single crystal.

EXPERIMENTAL PROCEDURES

For growing the single crystal samples, the precursor is $Pr_{1/2}Sr_{1/2}MnO_3$ polycrystalline prepared using the conventional solid-state reaction technique. High purity Pr_6O_{11}, $SrCO_3$ and MnO_2 were used as starting materials. A prescribed amount of these powers were thoroughly grinded and mixed homogeneously, then precalcined in air at $1100°C$, $1250°C$ for 12 hours respectively, including midst grinding and pressing into pellets. At last, it was sintered at $1300°C$ for 24 hours. All the processes were performed in air and cooled with the furnace. The purity and structure of the polycrystalline $Pr_{1/2}Sr_{1/2}MnO_3$ were checked by powder X-ray diffraction method. Then the obtained samples were crushed, grinded to powder and pressed into rod shape under high pressure of about 150 MPa. Single crystal samples of $Pr_{1/2}Sr_{1/2}MnO_3$ were melt-grown using an four-mirror optical floating zone furnace in flowing air with travelling speed of 1~3 mm/h. The feed and seed rods were well-formed $Pr_{1/2}Sr_{1/2}MnO_3$ polycrystalline, rotating in opposite directions at 30

538

rpm. To release the thermal strain, the as-grown crystal was annealed in flowing oxygen gas at 1200°C for 24 hrs and was slowly cooled to room temperature. The preferential growth direction of the single crystal sample is along the **b** axis, and the **b** axis is known in the FM plane from the studies of structure and orbital schematic of $Pr_{1/2}Sr_{1/2}MnO_3$ compound [3,4]. Thus the boule can be easily divided into pieces along the growth direction and the section has good metallic luster. For all experiments, we use the small sample cut with the edges parallel (//) or perpendicular ($\perp$) to the growth direction (i.e., FM plane) to investigate the anisotropy. The quality of the single crystal samples was characterized by XRD method. Using the Quantum design Physical Property Measurement system (PPMS-9T), Magnetization was measured by means of an induction method using a couple of coaxial pickup coils and resistivity was measured by the standard four-probe method with a dc current.

RESULTS AND DISCUSSION

The crystal structures of the polycrystalline $Pr_{1/2}Sr_{1/2}MnO_3$ at room temperature belong to the monoclinic $P21/n$ structure, consisting of alternating bucking of MnO_6 octahedral along the b-axis. The result of the bulk single crystal sample's XRD diffraction in figure 1 shows only one diffraction peak in the two directions indicates the good quality of the single crystal sample.

In figure 2, temperature dependence of the resistivity at $H=0$ for a current in plane ($\rho_{//}$) and for a current perpendicular to FM plane ($\rho_\perp$) are shown. It shows very different transport properties in the two directions, that is, $\rho_{//}$ shows metallic-type transport behavior and insulating-type behavior for $\rho_\perp$ in the measured temperature region (4.2~350K), even without any insulator-metal transition like the results reported in $Pr_{1/2}Sr_{1/2}MnO_3$ single crystal [1,2,5].

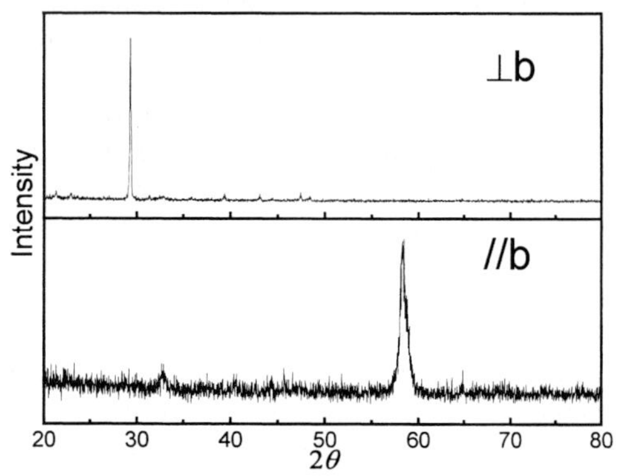

Figure 1 XRD diffraction patterns of $Pr_{0.5}Sr_{0.5}MnO_3$ single crystal sample in different directions measured at room temperature.

Corresponding to the ferromagnetic transition near $T_C\sim275$ K, $\rho_{//}$ decreases dramatically while no visible change for $\rho_\perp$. Around antiferromagnetic transition, $\rho_\perp$ is distinctly enhanced while $\rho_{//}$ continuously decreases; as a result, the transport anisotropy enhanced by the AFM ordering. It shows that in

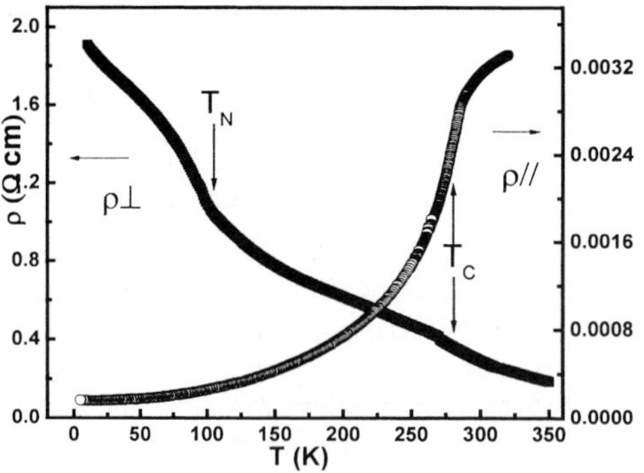

Figure 2 Temperature-dependence of resistivity at $H=0$ field. The arrows denote transition temperature.

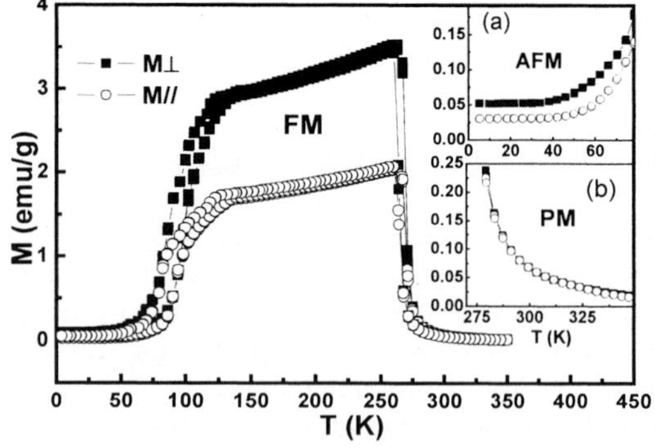

Figure 3 M-T curves of $Pr_{1/2}Sr_{1/2}MnO_3$ at $H=100$ Oe. The inset (a), (b) shows magnetic properties in AFM state and PM state respectively.

$Pr_{1/2}Sr_{1/2}MnO_3$ system two-dimensional anisotropy of the transport properties exists in the measurement temperature region 4.2~350K.

Temperature dependences of magnetization of the $Pr_{1/2}Sr_{1/2}MnO_3$ single crystal sample under an applied field parallel ($M_{//}$) or perpendicular ($M_\perp$) to the FM plane are shown in figure 3. It is clearly shown that there are two transitions, from paramagnetic (PM) state to ferromagnetic (FM) one at $T\sim275$ K and from FM to antiferromagnetic (AFM) one at about 135 K. The thermal hysteresis between heating and cooling process indicate the first order magnetic transition. The PM–FM transition is followed by a slowly decrease of the magnetization indicating an FM canted spin state at low temperature [6]. In figure 3, noticeable different values of magnetization $M_{//}$ and $M_\perp$ indicate the strong anisotropic magnetic properties in the FM state; there is also anisotropy exist in the AFM state (inset of Fig. 3(a)), while there is not anisotropy in the PM state (inset Fig. 3(b)).

For the two-dimensional anisotropy of the transport and magnetic properties at low temperatures, it can explained based on the A-type AFM ground state with the d_{x2-y2} orbital ordering: DE mechanism within FM layers enhances the hopping probability of the itinerant e_g electrons and decrease the resistivity. Due to the couple of orbital and spin, magnetization shows different values in the two directions.

From a viewpoint of the DE mechanism for the CMR effects, one may expect that the system in the pure ferromagnetic metallic state has no orbital ordering and shows an isotropic behavior by gaining a maximum kinetic energy. If the systems show the d_{x2-y2}-type orbital order, their physical properties should exhibit different directional dependences. Indeed, isotropic spin wave excitations were observed in the FM state of $La_{1-x}Sr_xMnO_3$ with $x\leqslant 0.1$[7]. These results seem to indicate that a transfer integral in these systems is isotropic, and there is no orbital ordering. However, Maezono and Nagaosa [8] suggested theoretically that the orbital ordering could be even a dynamical resonant ("orbital liquid") state in the FM metallic state. By using the Monte Carlo simulation, Yunoki et al.[9] also claimed the existence of an orbital ordered state in the FM phase. Quasielastic and dynamical diffuse scattering study by Kawano-Furukawa et al. [4] showed the spins between the planes exhibit AFM correlations even though the system is still in the FM phase at 220 K in the $Pr_{1/2}Sr_{1/2}MnO_3$ compound. The two-dimensional FM spin correlations develop within the FM planes and A-type AFM correlations develop in the perpendicular direction. These are consistent with the idea that the d_{x2-y2}-type orbital order is formed within FM planes in the $Pr_{1/2}Sr_{1/2}MnO_3$ compound. The strong anisotropic magnetic properties in the FM state shown in Fig. 3 provide a further advantage to identify the effects of the orbital ordering experimentally.

Kawano-Furukawa et al. also reported even in the PM phase, the dynamical spin fluctuations are

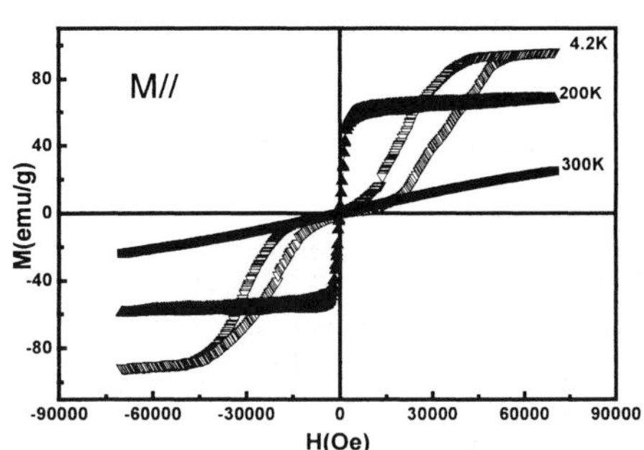

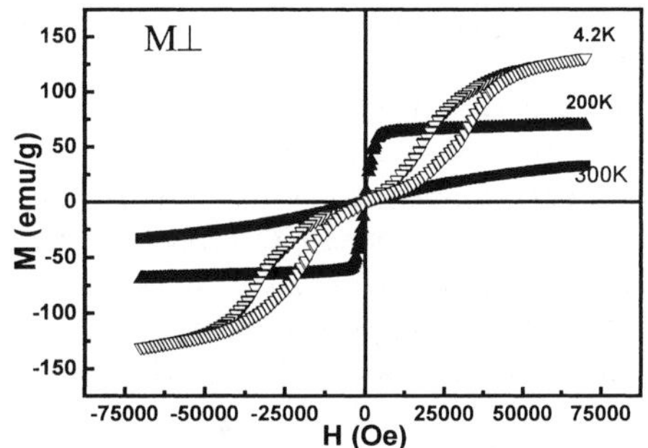

Figure 4 *M-H* Curves of $Pr_{1/2}Sr_{1/2}MnO_3$ at different temperatures.

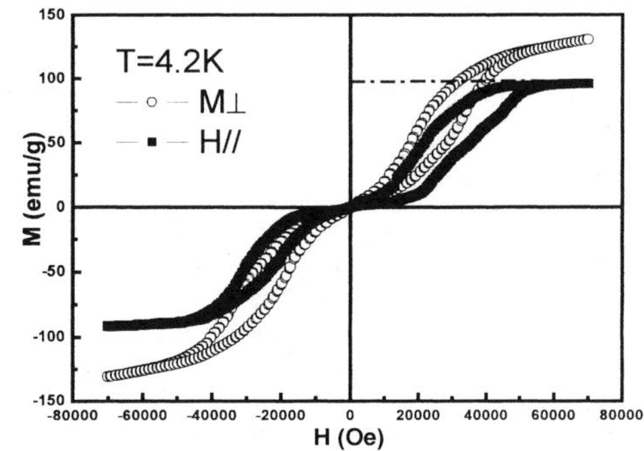

Figure 5 The anisotropy magnetic properties at T=4.2 K

anisotropy due to the polarization of the d_{x2-y2}–type orbital, but the isotropy in PM state of the present experiment result (as can be seen in Fig. 3(b)) is contrary to their explanation.

In Figure 4, field-dependence of magnetization at fixed temperatures for $M_{//}$ and $M_\perp$ are shown. Compare to the saturation magnetization of $M_{//}$ and $M_\perp$ at 4.2 K, an interesting phenomenon is that $M_\perp$ is larger than $M_{//}$, instead of gradually changing to equality at enough large field as usual. This phenomenon is not found at 200 K and 300 K. The magnetization energy ($E = \int_0^{Ms} HdM$) in the FM plane is larger than that of perpendicular direction (see Fig. 5), i.e., the spins perpendicular to the FM plane can align along the applied field easier than those parallel to the FM plane. As the A-type AFM state (AFM1) undergoes a transition to a canted spin state (AFM2) below T_N [6], it also shows spin–flop transition in M-H curve at $T<T_N$.

SUMMARY

We have measured the resistivity and magnetization for $Pr_{1/2}Sr_{1/2}MnO_3$ single crystal samples in the directions of parallel ($//$) or perpendicular ($\perp$) to the FM plane. The observed anisotropy of transport and magnetic properties in AFM and FM states can be explained by the DE model with d_{x2-y2} orbital structure: with the nearly anisotropy orbital structure, carriers in the FM state or AFM state can move only in the FM layers and form two-dimension metallic behavior. In such a classical DE model the presence of spin-canted phase is predicted, which is consistent with the magnetization results. As the A-type AFM state undergoes a transition to a canted spin state below T_N, it also shows a spin –flop transition at $T<T_N$.

ACKNOWLEDGMENT

This work was supported by the National Foundation of the Natural Science of PR China (No.10274049), the Shanghai Nano-technology Promotion Center and the Science & Technology Committee of Shanghai Municipality (No.0352nm036), the Dawn Project (No.03SG35) and the Science & Technology Development Foundation (No.02AK42) of the Education Committee of Shanghai Municipality and the Shanghai Leading Academic Discipline Program.

REFERENCES

1. Pollert E., Jirak Z., Hejtmanek J., Strejc A., Kuzel R., and Hardy V., Detailed study of the structural and magnetic transitions in $Pr_{1-x}Sr_xMnO_3$ single crystals ($0.48 \leq x \leq 0.57$), Journal of Magnetism and Magnetic Materials (2002) 246 290–296
2. Tomioka Y., Asamitsu A., Moritomo Y., Kuwahara H., and Tokura Y., Collapse of a Charge-Ordered State under a Magnetic Field in $Pr_{1/2}Sr_{1/2}MnO_3$, Physical Review Letters (1995) 74 5108-5111
3. Kawano H., Kajimoto R., Yoshizawa., Fernandez-Baca J. A., Tomioka Y., KuwaharaH., and Tokura Y., Two-dimensional anisotropy in a layered metallic antiferromagnet $RE_{1-x}Sr_xMnO_3$ with $x\sim1/2$, Physica B (1998), 241-243 289-294
4. Kawano-Furukara H., Kajimoto R., Yoshizawa H., Tomioka Y., Kuwahara H., and Yokura Y., Orbital order and a canted phase in the paramagnetic and ferromagnetic states of 50% hole-doped colossal magnetoresistance manganites, Physical Review B (2003) 67 174422
5. Argyriou D. N., Hinks D. G., Mitchell J. F., Potter C. D., Schultz A. J., Young D. M., Jorgensen J. D., and Bader S. D., The Room Temperature Crystal Structure of the Perovskite $Pr_{0.5}Sr_{0.5}MnO_3$, Journal of Solid State Chemistry (1996) 124 381-384
6. Hayashi T., Miura N., Tomioka Y., and Tokura Y., Metamagnetic transition and recovery of isotropic transport phenomena in A-type antiferromagnetic $Pr_{0.45}Sr_{0.55}MnO_3$, Journal of Physics and Chemistry of Solids (2002) 63 925-928
7. Martin M. C., Shirane G., Endoh Y., Hirota K., Moritomo Y., and Tokura Y., Magnetism and structural distortion in the $La_{0.7}Sr_{0.3}MnO_3$ metallic ferromagnet, Phys. Rev. B (1996) 53 14285
8. Maezono R., and Nagaosa N., Theory of spin-wave excitation in manganites, Phys. Rev. B (2000) 61 1189-1192
9. Yunoki S., Moreo A., and Dagotto E., Phase Separation Induced by Orbital Degrees of Freedom in Models for Manganites with Jahn-Teller Phonons, Physical Review Letters (1998) 81 5612-5615

Dependence of structure and magnetic properties on magnetic field and temperature for Bi-Mn alloy

Liu Y.S., Zhang J.C*., Jia G.Q., Zhang X.Y., Ren Z.M., Jing C., Li X., Cao S.X., Deng K.

Department of Physics & Department of Ferrous Metallurgy, Shanghai University, Shanghai 200436, China
* Corresponding author; Email: jczhang@mail.shu.edu.cn

The effects of magnetic field and temperature on structure and magnetic properties of *BiMn* compound were investigated. The microstructure showed that *MnBi* (low temperature phase, LTP) in *Bi*-wt.6%*Mn* alloys are all aligned, with the c-axis of the *MnBi* (LTP) crystal along the fabrication magnetic field H_f. H_f explicitly improves the magnetic anisotropy. Magnetic measurements indicated that the *MnBi* (LTP) saturation magnetization M_s and its coercive field H_c change with the temperature. Most of all, the temperature of spin-reorientation increased with the increasing H_f but decreased with the increase of the measuring magnetic field H_m.

INTRODUCTION

Owing to the high uniaxial magnetic anisotropy of its low-temperature phase (LTP) and the good magneto-optical properties of its quenched high-temperature phase (QHTP), the physical properties of the binary compound *MnBi* have been investigated extensively [1~6]. With the development of and the advance of superconducting magnet technologies, a high magnetic field has being operated in various academic fields such as material science and physics [7,8]. Currently, the physics of high field-induced transition is among the most interesting problems in condensed matter physics [9~11]. In order to investigate the effect of magnetic field and temperature on structure and magnetic properties of *MnBi* compound, we fabricated, in different fabrication magnetic field H_f, *Bi*-wt.6%*Mn* alloys. Effects of H_f on structure of *Bi*-wt.6%*Mn* alloys have been reported elsewhere [12]. Here, we mainly report magnetization study of the alloy *Bi*-wt.6%*Mn*, of which *MnBi*(LTP) reveal a spin-reorientation transition (SRT) at about 90K. The research results show that the magnetic field H_f explicitly improves the specimen's magnetic anisotropy. Specifically, we find that dc magnetic fields, H_f and the measuring magnetic field H_m, result in different temperatures of SRT.

EXPERIMENTAL PROCEDURE

Details of experimental techniques are in previous papers [12], and thus only a brief outline of this procedure is presented here. The sample was sealed in a graphite tube and inserted into a resistance furnace, which was placed between poles of the electromagnet. The intensity of the magnetic field between poles of the electromagnet can be adjusted and the temperature in the furnace chamber can be controlled automatically during the experiment. Since the liquidus temperature of *Bi*-6wt%*Mn* alloy is above 630K, the alloy is in a semi-solid state at 548K. The alloy was heated up to the temperature in the mushy zone without the magnetic field, held for 30min and then cooled to the temperature below 535K under various fabrication magnetic fields H_f of 0.0T, 0.3T and 0.5T, respectively. For characterizing the morphology of the *MnBi* phase, the samples obtained in the experiments were mechanically polished parallel and perpendicular to the H_f direction. The following *No.1*, *No.2* and *No.3* samples denote the samples solidified in the fabrication magnetic fields H_f of 0.0T, 0.3T and 0.5T, respectively. The samples were characterized for their structures by optical microscopy, SEM and XRD. Magnetic measurements were performed with H_m up to 9T in a temperature range from 1.9 K to 300K using PPMS.

RESULTS AND DISCUSSION

In the case of solidification with the field, the elongated *MnBi* crystals were oriented and grew preferentially along the applied field in the *Bi* matrix. The hexagonal sections of the crystals only appeared in the section perpendicular to the field. Furthermore, the XRD patterns showed that the *c*-axis of hexagonal *MnBi* crystal (easy magnetization axis) was aligned parallel to the H_f direction, and peaks for *Bi* and *MnBi*(LTP) could only be observed [12]. The XRD results and EDX analysis indicated the formation of *MnBi*(LTP) (approximately 28.656wt.%) and *Bi* phase. By comparison with magnetic powders aligned in a magnetic field in an epoxy resin to form a bonded magnet [5], our method has a prominent feature that *MnBi*(LTP) not only is aligned along the *c*-axis, but also grew preferentially and congregated along the H_f direction. Because *Bi* is diamagnetic, its effect has been ignored and only *MnBi*(LTP) has been considered in the magnetization measurement. The *MnBi* crystals observed in the microstructure micrographs were randomly oriented in the *Bi* matrix in the sample crystallized without the magnetic field (for the *No.*1 sample).

Figure 1 presents the magnetizations of the *No.*3 specimen parallel to and normal to *c*-axis direction at 150K and 300K for the *No.*3 specimen solidified in H_f=0.5T. When the field is parallel to the *c*-axis, the saturation occurs much more easily compared with the field perpendicular to the *c*-axis. The anisotropy fields in *MnBi*(LTP) at 150K and 300K are about 2.5T and 5T, respectively. It can be assumed that there is a very strong anisotropy in the *MnBi*(LTP) compound and the anisotropy field of *MnBi*(LTP) increases with the increasing temperature. Shown in Figure 2 is saturation magnetization M_s and the coercive field H_c of *BiMn*(LTP) compound along the *c*-axis for the *No.*3 specimen at various temperatures. It is apparent from this graph that M_s decreases with the increase of temperature, and H_c decreases with the temperature less than 150K but increases with the temperature above 150K. M_s of *BiMn*(LTP) compound in the *No.*3

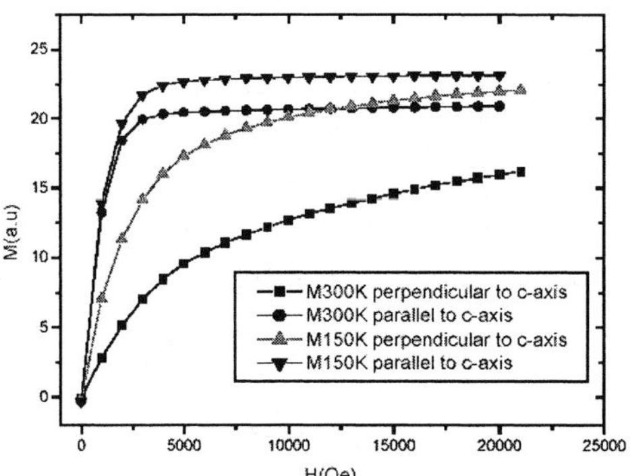

Figure 1 Magnetizations of *MnBi* (LTP) compound parallel to and perpendicular to *c*-axis direction at 150K and 300K for the *No.*3 sample.

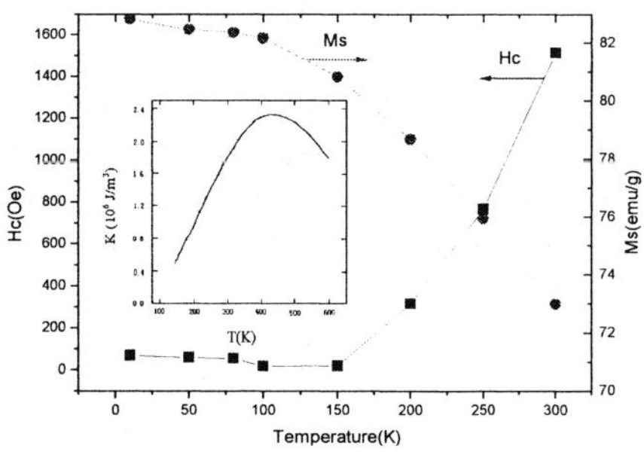

Figure 2 M_s and H_c of *MnBi* (LTP) along the *c*-axis at various temperatures for the *No.*3 sample. Inset: Anisotropy constant *K* vs. temperature in *MnBi*(LTP) [16].

specimen shows spin-wave behavior and holds fairly well the $T^{3/2}$ law especially at low temperature. The increase in coercivity with the increase in temperature is because of the increase in anisotropy energy in *MnBi*(LTP) (see inset in Figure 2.). Chen and Stutius also found that the uniaxial anisotropy energy increases with increasing temperature and reaches a maximum of 2.2×10^7erg/cm^3 at temperature 490K[3].

Figure 3 displays the thermomagnetic curves $M(T)$ of *MnBi* (LTP) compounds in the *No.*1 *No.*2 and *No.*3 specimens at H_m=0.1T. Under an applied dc field H_m, the values of magnetization slowly increase and the shape of the anomaly in magnetization becomes non-peaklike (a peaklike profile indicates a spin reorientation of *MnBi* (LTP), see inset in Figure 5.) with H_m parallel to the *c*-axis. For each sample, the magnetization M decreases with the temperature less than the temperature of SRT, but increases with the temperature above the temperature of SRT. In the meanwhile, M increases with the H_f increasing when the temperature is above about 75K and decreases with the H_f increasing when the temperature is less than about 75K at a certain temperature. As a rule, the higher the H_f is, the better the alignment is and the less the obstacles that impede the spin-reorientation are. So the temperature of SRT increased with the increasing H_f. The effect of the magnetic field H_m on the temperature of SRT can be seen from the

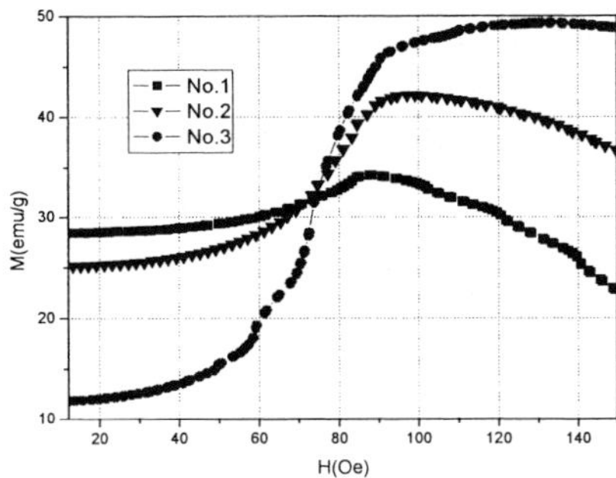

Figure 3   *M* vs. temperature measured on *MnBi* (LTP) in the different samples.

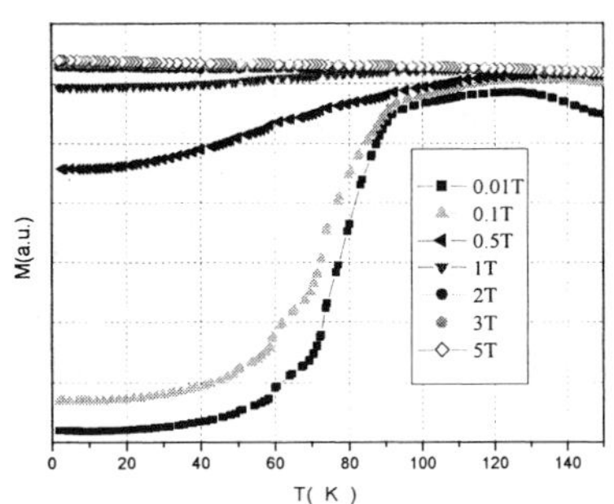

Figure 4   *M* vs. temperature measured on *MnBi* (LTP) for the *No.3* sample when the different H_m is applied along the *c*-axis.

thermomagnetic curves $M(T)$ of *MnBi* (LTP) compound for the *No.3* specimen shown in Figure 4. The $M(T)$ curves of *BiMn*(LTP) compound were obtained under different magnetic fields along the *c*-axis direction of *MnBi*(LTP) for the *No.3* specimen. In Figure 4, the position of the bump in magnetization shifts to lower temperatures when H_m is applied along the *c*-axis direction. But with the increasing H_m, the bump was not more visible, and the spin-reorientation was hardly observed even at H_m=5T. The thermomagnetic curves

$M(T)$ of *MnBi* (LTP) compound in the *No.2* specimen at the magnetic field H_m=9T applied along *c*-axis is plotted in Figure 5. It is easy to see that the bump does not exist, and with the decreasing temperature, the magnetization increases and reaches saturation at about 10K. Those effects mean that although the Zeeman energy ($-M \cdot H_m$) produced by the applied external dc field keeps the magnetization parallel to the field direction, the application of an external dc field H_m modifies the temperature of SRT to some extent. This implies that the anisotropy energy is comparable to the Zeeman energy produced by a certain applied magnetic field.

Generally, the magnetic moments rotated from along the *c*-axis into, or nearly into, the basal plane for *MnBi*(LTP) at about 90K, which is studied by

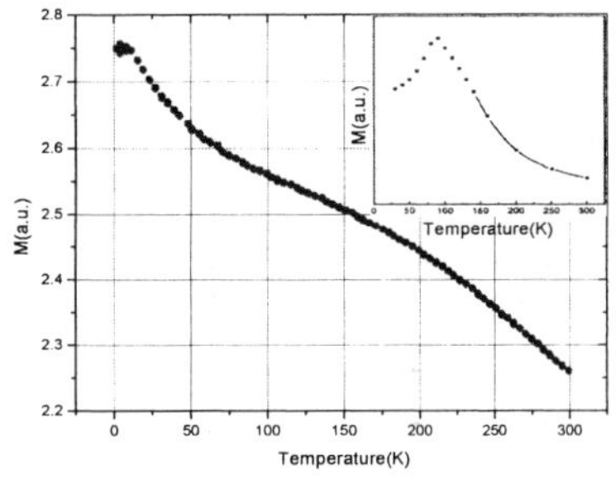

Figure 5 Thermomagnetic curve measured on *MnBi* (LTP) compound for the *No.2* sample. Inset: *M* vs.Temperature for *MnBi* along the *c*-axis [4].

many experiments[3,12,13]. By linear extrapolation of anisotropy constant $K=K_{u1}+2K_{u2}+3K_{u3}$ as a function of temperature (shown in inset in Figure 2.) [16] to temperature below 90K, $K<0$ will probably appear, then K_{u1} and K_{u2} (K_{u1} and K_{u2} are the first and second anisotropy constants) may meet the conditions of the plane anisotropy ($-K_{u2}>K_{u1}\geq0$) or the easy cone ($2K_{u2}>-K_{u1}>0$). The Zeeman energy favors arrangement of the spins along *c*-axis. This competes with the plane anisotropy, causing a continuous rotation of the spins to the *c*-axis with increasing field strength. Finally, with field applied along the *c*-axis, a spin-reorientation transition occurs due to competition between Zeeman energy and the plane anisotropy, causing a slow rotation of the spins into the basal plane. The spin-orbit interaction causes the low-temperature magnetic anisotropy of *MnBi*(LTP) [15], so the effect of magnetic field on the temperature of SRT probably originates from the interaction between the magnetic field and spin-orbit coupling.

CONCLUSIONS

When Bi-wt.6%Mn alloys are solidified in the fabrication magnetic field H_f, the primary phase *MnBi* (LTP) is aligned in the magnetic field, with the *c*-axis of the *MnBi* (LTP) crystal along the direction of H_f. M_s of *MnBi* (LTP) decreases with the increase of temperature, and H_c of *MnBi* (LTP) increases with temperature above about 150K. H_f explicitly improves the magnetic anisotropy. The magnetic moment of *MnBi* gradually deviates from the *c*-axis at approximately 90K. It is interesting that, under a dc field applied H_m parallel to *c*-axis, the temperature of SRT decreases with the increase of magnetic field, and increases with the increasing H_f.

ACKNOWLEDGEMENTS

This work is supported by the National Foundation of the Natural Science of PR China(No.50234020, No.10274049), the Shanghai Nano-technology Promotion Center and the Science & Technology Committee of Shanghai Municipality(No. 0352nm036). This work is also partly supported by the Science & Technology Development Foundation of the Education Committee of Shanghai Municipality（No.02AK42）and the Shanghai Leading Academic Discipline Program.

REFERENCES

1.   Saha, S., Obermyer, R.T., Zande, B.J., Chandhok, V.K., Simizu, S., Sankar, S.G., and Horton, J.A., Magnetic properties of the low-temperature phase of MnBi, J. Appl. Phys (2002) 91 8525-8527
2.   Harder, K.U., Menzel D., Widmer T., and Schoenes, J., Structure, magnetic, and magneto-optical properties of MnBi films grown on quartz and (001)GaAs substrates, J. Appl. Phys (1998) 84 3625-3629
3.   Chen, T. and Stutius, W., The phase transformation and physical properties of the MnBi and Mn1.08Bi compounds, IEEE Trans. Magn. (1974) 10 581-586
4.   Yang, J.B., Kamaraju, K., Yelon, W.B., and James, W.J., Magnetic properties of the MnBi intermetallic compound, Appl. Phys. Lett (2001) 79 1846-1848
5.   Yang, J.B., Yelon, W.B., James, W.J., Cai Q., Kornecki M., Roy S., Ali N., and l'Heritier Ph., Crystal structure, magnetic properties and electronic structure of the MnBi intermetallic compound, J. Phys: Condens. Matter (2002) 14 6509-6519
6.   Xu, Y., Liu, B., and Pettifor, D.G., Half-metallic ferromagnetism of MnBi in the zinc-blende structure., Phys. Rev. B (2002) 66 184435-184439
7.   Lu, X.Y., Nagata, A., Watanabe, K., Nojima, T., Sugawara, K. and Kamada, S., Texture and formation of (Bi,Pb)-2223 phase after partial-melting and solidification in high magnetic fields., IEEE Transactions on Applied Superconductivity (2001) 11 3553-3556
8.   DeRango, P., Lee, M., and Lejayetal, P., Texturing of magnetic materials at high temperature by solidification in a magnetic field, Nature (1991) 349 770-772
9.   Shimamoto, Y., Miura, N. and Nojiri, H., Magnetic-field-induced electronic phase transitions in semimetals in high magnetic fields, J. Phys.: Condens.Matter (1998) 10 11289-11300
10. Lake, B., M.Ronnow, H., Christensen, N.B., Aeppli, G., Lefmann, K., Mcmorrow, D.F., Vorderwisch, P., Smeibidl, P., Mangkorntong, N., Sasagawa, T., Nohara, M., Takagi, H. and Mason, T.E., Antiferromagnetic order induced by an applied magnetic field in a high-temperature superconductor, Nature (2002) 415 299-302
11. Uji, S., Shinagawa, H., Terashima, T., Yakabe, T., Terai, Y., Tokumoto, M., Kobayashi, A., Tanaka, H. and Kobayashi, H., Magnetic-field-induced superconductivity in a two-dimensional organic conductor, Nature (2001) 410 908-910
12. Wang, H., Ren, Z.M., Kang, K., and Xu, K.D., Effects of static magnetic field on alignment structure of MnBi phase in semi-solidified Bi-Mn alloy, The Chinese Journal of Nonferrous Metals, (2002) 12 556-560
13. Roberts, B.W., Neutron Diffraction Study of the Structures and Magnetic Properties of Manganese Bismuthide, Phys. Rev. (1956) 104 607-616
14. Albert, P.A., Carr, Jr W.J., Temperature Dependence of Magnetostriction and Anisotropy in MnBi, J. Appl. Phys. (1961) 32 201s-202s
15. Coehoorn, R. and De Groot, R. A., The electronic structure of MnBi, J. Phys. F: Met. Phys. (1985) 15 2135-2144
16. Guo, X., Chen, X., Altounian, Z., and StrÖm-Olsen, J. O., Magnetic properties of MnBi prepared by rapid solidification, Phys.Rev.B (1992) 46 14578-14582

Physical properties in layered transition-metal oxide crystals and anisotropic transport measurement

K. Q. Ruan, Y. Yu, S. L. Huang, H. L. Li, S. Qian and L. Z. Cao

Structure Research Laboratory, Department of Physics, University of Science and Technology of China, Hefei, Anhui 230026, P. R. China

Two kinds of measurement methods are applied to study the systematic transport property of both $Bi_{2-x}Pb_xSr_2Co_2O_y$ single crystals with various Pb content and $Bi_2Sr_2CaCu_2O_y$ single crystals with various oxygen content. The completely different anisotropic transport properties are observed in both $Bi_{2-x}Pb_xSr_2Co_2O_y$ and $Bi_2Sr_2CaCu_2O_y$ crystals. The result is discussed.

INTRODUCTION

Since the discovery of high-T_c cuprate oxides layered structure quasi-two dimensional transition-metal oxides have attracted extensive attention due to the strong electrical anisotropy manifested in their transport properties. The strongly contrasting anisotropic transport behaviors have been observed in different layered structure systems. For high-T_c cuprates, in-plane resistivity ρ_{ab} keeps metallic in underdoped region, whereas out-of–plane resistivity ρ_c remains non-metallic down to T_c. However, Sr_2RuO_4 has different transport properties from high-T_c cuprates [1]: ρ_{ab} is always metallic, ρ_c is non-metallic ($d\rho_c/dT < 0$) above $T_M \approx 130$ K, and becomes metallic ($d\rho_c/dT > 0$) below T_M. This should indicate that the layers appear as "isolated" at high temperatures, but connected at low temperature to give a 3D system. Many experimental and theoretical studies have investigated the peculiar transport properties. However, anisotropic transport property measurement in strongly layered structure material is a challenge task. A key point is to overcome problems involving non-uniform current flow in the measuring process. In this paper, we employed two kinds of measurement methods to study the systematic transport property of both $Bi_{2-x}Pb_xSr_2Co_2O_y$ single crystals with various Pb content and $Bi_2Sr_2CaCu_2O_y$ single crystals with various oxygen content. $Bi_{2-x}Pb_xSr_2Co_2O_y$ is a misfit-layered compound [2], and Co ions in the conducting CoO_2 layer form a triangular lattice instead of a rectangular one in the conducting CuO_2 layer of high-T_c cuprate oxides. The completely different anisotropic transport properties are observed in both $Bi_{2-x}Pb_xSr_2Co_2O_y$ and $Bi_2Sr_2CaCu_2O_y$ crystals. The result is discussed.

EXPERIMENTAL

Pb-doped $Bi_{2-x}Pb_xSr_2Co_2O_y$ and $Bi_2Sr_2CaCu_2O_y$ single crystals were grown by a self-flux method. Bi_2O_3 was included in excess to act as a flux for crystal growth. The as-grown $Bi_2Sr_2CaCu_2O_y$ crystals were annealed at 450°C and various oxygen partial pressures in the range $10^{-8} - 10$ MPa so as to obtain crystals covering from underdoped to overdoped levels. Analyses of the cation stoichiometry of the single crystals were performed by energy-dispersive X-ray analysis (EDX) using a scanning electron microscopy, with uncertainty ±0.01. All crystals for use in the measurements were verified to be single phase by means of X-ray diffraction (XRD). Two kinds of methods are employed to accomplish anisotropic transport measurement. In-plane and out-of-plane resistivity measurements of Pb-doped $Bi_{2-x}Pb_xSr_2Co_2O_y$ single crystals were carried out by a DC four probe method with current electrodes configurations shown in Fig. 1. A generalization of the Montgomery method shown in Fig. 2 was applied to measure anisotropic resistivities of $Bi_2Sr_2CaCu_2O_y$ single crystals. Good electric contacts were achieved by soldering copper wire onto the surface of the samples on which the pure silver evaporated with the electrical contact of no more than 2Ω.

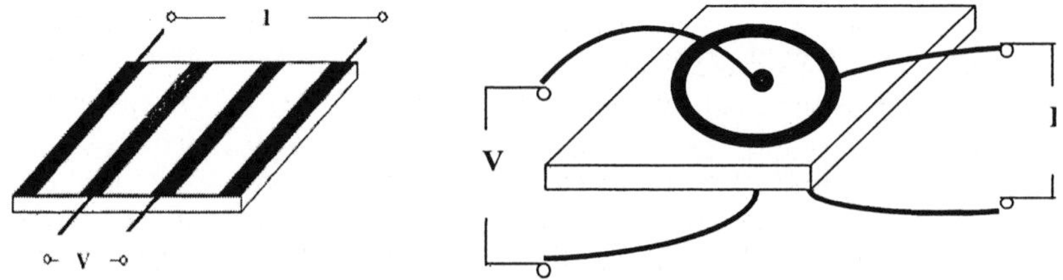

Figure 1   The electrode configurations for ρ_{ab} and ρ_c measurements by a DC four probe method.

Figure 2   A generalization of the Montgomery method for anisotropic resistivity measurement.

RESULTS AND DISCUSSION

Figure 3 and Figure 4 show the temperature dependence behaviors of the in-plane and out-of-plane resistivity of crystals with x = 0.32, 0.41, 0.48 and 0.52, respectively. As Pb content increases, $\rho_{ab}(T)$ become more metallic. $\rho_c(T)$ show rather complex behavior, and the sign change of $d\rho_c/dT$ occurs with the increase of Pb content. For x = 0.32, 0.41 and 0.48 samples, $\rho_c(T)$ changes from non-metallic to metallic at $T_{max} \approx$ 160 K, 180 K and 210 K respectively. For x = 0.52 sample, ρ_c shows metallic behavior up to room temperature.

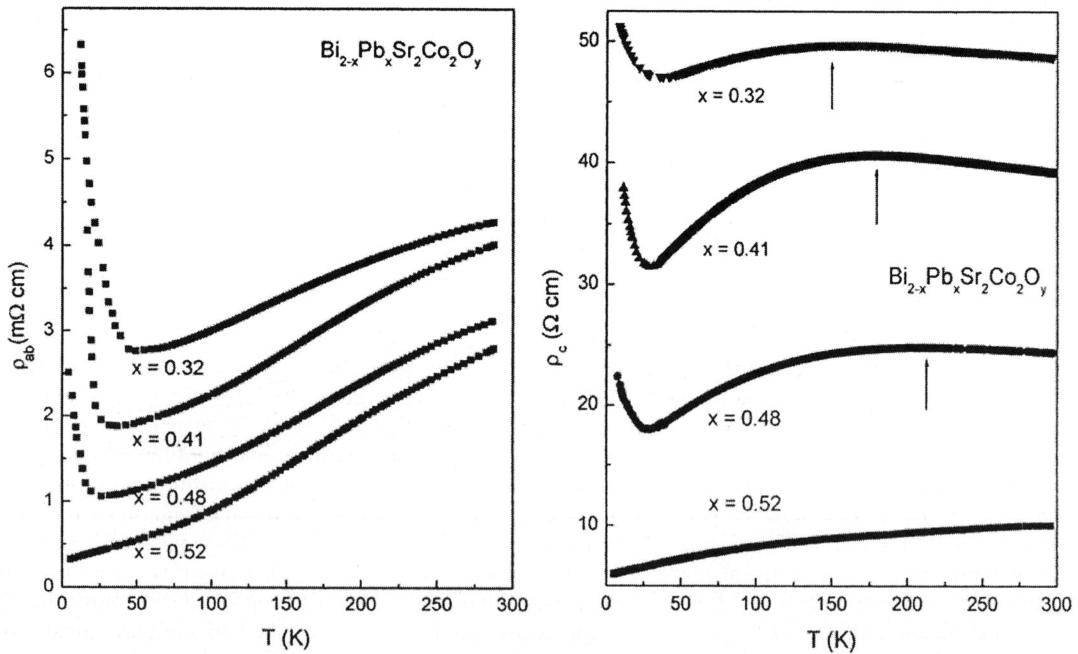

Figure 3   Temperature dependence of in-plane resistivity of $Bi_{2-x}Pb_xSr_2Co_2O_y$ crystals.

Figure 4   Temperature dependence of out-of-plane resistivity of $Bi_{2-x}Pb_xSr_2Co_2O_y$ crystals.

What induces the complex behavior of ρ_c with the increase of Pb content? The magnetic measurements show that there is no magnetic transition in $Bi_{2-x}Pb_xSr_2Co_2O_y$ compounds down to very low temperature about 4 K. This indicates that the crossover behavior of ρ_c is not due to the magnetic origin. The valence-band photoemission as well as the O_{1s} and Co_{2p} X-ray absorption spectra indicate that Co^{3+} and Co^{4+} have the low-spin t_{2g}^{6} and t_{2g}^{5} configurations in $Bi_{2-x}Pb_xSr_2Co_2O_y$ compounds [3]. Angle-resolved photoemission spectra further shows that the dispersion of the t_{2g} feature is very small compared to its width at each angle, and suggest that the electron-lattice coupling energy is much larger than the kinetic energy of the t_{2g} electrons and the carriers in the Co-O triangular lattice are essentially polarons formed by Co^{4+} in the non-magnetic Co^{3+} background. In fact, with the doping of Pb in $Bi_{2-x}Pb_xSr_2Co_2O_y$, there is strong interaction between the carrier and crystal lattice when the carrier concentration reaches some critical value, which forms the small polaron. At low temperature small polaron states overlap sufficiently to allow the formation of a polaron band, in which ordinary band conduction can take place. In the vicinity of T_{max} the bandwidth becomes less than the uncertainty in energy due to the finite lifetime of polaron states. Above this temperature the small polaron can be thought of as localized, and the only way in which the polaron can contribute to conduction is by "hopping" from one lattice to another. For $x = 0.32, 0.41$ and 0.48 samples, the crossover behavior of ρ_c is observed around T_{max} located below the room temperature. For ρ_{ab}, there are two kinds of possibilities. One possibility is that there is no small polaron in the ab plane due to $t_{ab} \gg T_{phonon}$, the small polaron only can be formed in the c-direction. The transport behavior of in-plane resistivity can be interpreted using Boltzmann transport theory. The other is that small polaron can be formed both in ab- and c-direction, but $T_{max}^{ab} \gg T_{max}^{c}$, i. e., T_{max}^{ab} is located far above the room temperature. Both possibilities indicate that ρ_{ab} behaves metallic.

Figure 5 and Figure 6 show the temperature dependences of the in-plane resistivities ρ_{ab} and out-of-plane resistivities ρ_c for $Bi_2Sr_2CaCu_2O_y$ single crystals, respectively. Crystals A, B, C, D, E and F were respectively annealed under oxygen pressures of 10^{-8}, 10^{-6}, 10^{-4}, 10^{-1}, 1 and 10 MPa. T_c and the resistivities for six crystals alter systematically. A typical T-linear behavior of ρ_{ab} and the highest T_c (90K) are observed for the optimally doped C crystal. Although the oxygen contents of the crystals have not been directly determined, it can certainly be deduced that crystals A and B are both in the underdoped regime, while crystals D, E, and F are in the overdoped regime. For the overdoped D, E and F crystals, a slightly upward curve of ρ_{ab} is observed and $\rho_{ab} = \rho_0 + \alpha T^n$ ($n = 1.5$-1.8) is satisfactorily obeyed for the above crystals [4].

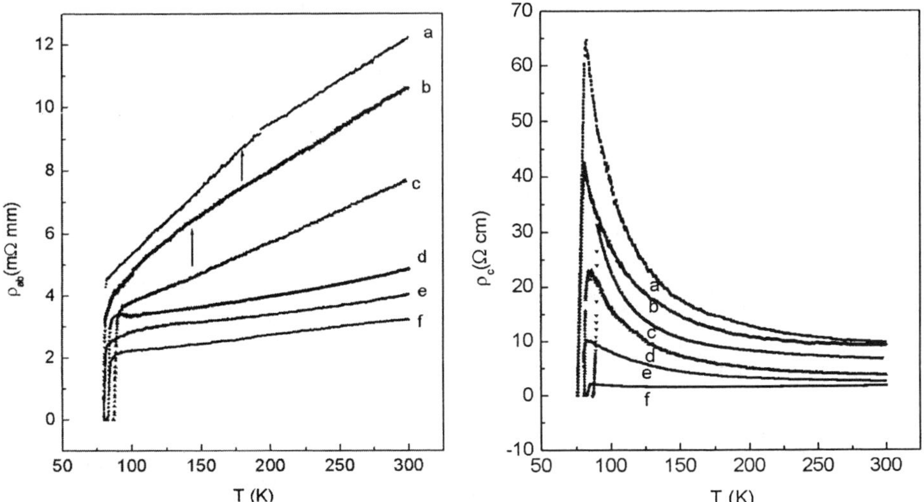

Figure 5   Temperature dependence of in-plane resistivity of $Bi_2Sr_2CaCu_2O_y$ crystals.

Figure 6   Temperature dependence of out-of-plane resistivity of $Bi_2Sr_2CaCu_2O_y$ crystals.

For the underdoped crystals A and B, ρ_{ab} deviates from the high temperature T-linear behavior at a characteristic temperature T^* (shown by the arrows), far above T_c. From Fig. 4 it is observed that, with increasing level of underdoping, the overall magnitude of ρ_c as well as its semiconductive temperature dependence increases. It is worth pointing out that there is always an activation-type component included in the semiconductor behavior of ρ_c (T) even for the slightly overdoped Bi-2212 crystal. In order to interpret a semiconductor behavior along c-axis coexisting with a metallic in-plane resistivity over a wide temperature and carrier concentration range, the origin of a pseudogap was suggested [5]. However, the pseudogap is only observed in the underdoped regime, and the semiconductive behavior of the slightly overdoped crystal casts doubt on the origin of the semiconductive behavior being a pseudogap.

CONCLUSION

The transport properties of $Bi_{2-x}Pb_xSr_2Co_2O_y$ single crystals with various Pb content and $Bi_2Sr_2CaCu_2O_y$ single crystals with various oxygen content were measured. For $Bi_{2-x}Pb_xSr_2Co_2O_y$ crystals, the evolution of ρ_c behavior from semiconductive through crossover to metallic behavior is interpreted by a small-polaron model. As for high-T_c cuprate, there is no consensus about the origin of semiconductor behavior of ρ_c (T) so far. Therefore, more systematic investigation on related layered strongly correlated metals will help to clarify this problem.

ACKNOWLEDGEMENT

This work is supported by the Natural Science Foundation of China (No. 10104013, No. 10174070) and by the Ministry of Science and Technology of China (No. G19990646).

REFERENCES

1. Maeno, Y. et al, superconductivity in a layered perovskite without copper, <u>Nature</u> (1994) <u>372</u> 532-534
2. Yamamoto T. et al, structural phase transition and metallic behavior in misfit layered (Bi,Pb)-Sr-Co-O system, <u>Jpn. J. Appl. Phys.</u> (2000) <u>39</u> L747-L750
3. Mizokawa T. et al, photoemission and x-ray-absorption study of misfit-layered (Bi,Pb)-Sr-Co-O compounds: electronic structure of a hole-doped Co-O triangular lattice, <u>Phys. Rev. B</u> (2001) <u>64</u> 115104 (1-7)
4. Ruan K. Q. et al, the systematic study of the normal-state transport properties of Bi-2212 crystals, <u>J. Phys.: Condens. Matter</u> (1999) <u>11</u> 3743-3749
5. Yan Y. F at al, negative magnetoresistance in the c-axis resistivity of $Bi_2Sr_2CaCu_2O_{8+x}$ and $YBa_2Cu_3O_{6+y}$, <u>Phys. Rev. B</u> (1995) <u>52</u> R751-R754

Design of a cryogenic giant magnetostrictive actuator using HTS

Jiongjiong Cai, Zhitong Cao, Hongping Chen, Guoguang He

Institute of Applied Physics, Zhejiang University, China, 310027

In this paper a Cryogenic Giant Magnetostrictive Materials (CGMM) actuator using HTS is designed, taking into account both the coupled field characteristics of the CGMM and the anisotropy of the investigated Bi2223/Ag HTS tapes. Then an optimal structure, which costs the least HTS tapes while still make the CGMM to the state of saturation, is realized by combining the genetic algorithm (GA) with the coupled field iteration of FEM.

INTRODUCTION

Cryogenic Giant Magnetostrictive Materials (CGMM), especially the family of $Tb_x Dy_y Zn_z$ with magnetostrain of 0.5 percent or more, has Curie temperatures 200K. These materials can be machined to many shapes, afford large transverse load, which provide the great flexibility for application design. They show great promise for cryogenic actuator applications; especially they can be combined with high temperature superconducting (HTS) tapes to create kinds of transducers, actuators, and motors that are characterized by high efficiency and high power density [1].

The properties of the Bi2223/Ag multifilamentry tapes have shown anisotropy in the critical current versus applied external magnetic field [2]. Numerous studies were performed in order to determine the main factors limiting the critical current of the magnets made of anisotropic Bi-2223/Ag tapes [3]. It is shown that special attention has to be paid to the study of the magnetic field distribution in the magnet and in particular to the radial component of the magnetic field when the magnet is wound in the form of a cylindrical solenoid [4].

In this paper the radial component of the magnetic field on the HTS magnet is reduced greatly by a magnetic circuit using laminated silicon iron. So the parallel component become an important factor limiting the critical current too owing to much larger magnitude, which is more than 10 times the radial component. Since both the two components have to be considered, a HTS solenoid of simple cylindrical configuration is designed, which is used to provide the CGMM a magnetic field of good uniformity making it to the state of saturation. According to the previous researches [5], we have developed the valid coupled field iteration of Finite Element Method (FEM) for the coupled field calculation of the smart material like the GMM. In this paper the coupled field iteration of FEM will be combined with an optimal design method, the genetic algorithm (GA) concerning the anisotropy of the HTS tapes, to find the specific size and place of the HTS solenoid costing least materials.

CGMM ESSENTIAL LAW AND THE COUPLED FIELD ITERATION OF FEM

550

The CGMM essential law can be described as follows.

$$B = \lambda T + \mu^T H \tag{1}$$

$$S = s^H T + \lambda H \tag{2}$$

Where s^H, μ^T and λ are compliance at fixed value of H, permeability at a fixed value of T, magnetostrictive coefficient, respectively. The first item of right side in equation (1) is the magnetic flux density resulting from the stress T. The second item of the right side in equation (2) is the strain resulting from the magnetic field H. The nonlinear coupled characteristics of above parameters of CGMM is described through the coupled field iteration of FEM [5], the flowchart is shown as Fig1.

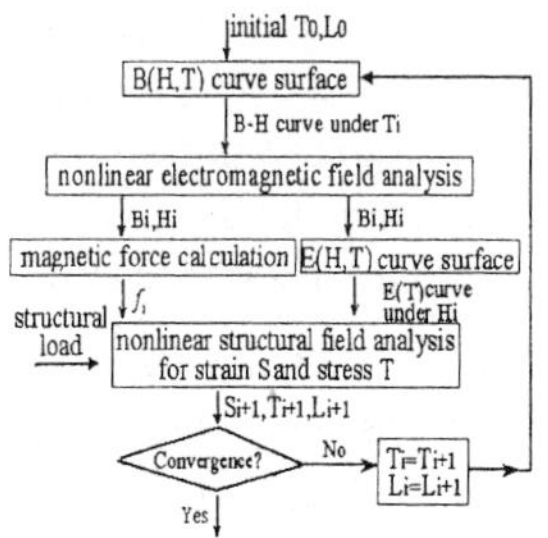

Figure 1 FEM analysis model for coupled field calculation

Figure 2 anisotropic characteristics of the Bi-2223/Ag($Jc(77k,0) = 7000 A/cm^2$)

ANISOTROPY OF THE HTS TAPE

As shown in Fig.2 [6], the approximation function of the B-J characteristic of Bi-2223/Ag at 77k with the magnetic field B is set as follows:

$$J_{c\parallel} = -0.97 B_\parallel^3 + 1.9194 B_\parallel^2 \, 1.5956 B_\parallel + 0.9751 \qquad \text{(for parallel)} \tag{3}$$

$$J_{c\perp} = 2.6021*10^5 B_\perp^6 - 1.8576*10^5 B_\perp^5 + 5.2756*10^4 B_\perp^4 - 7.6118*10^3 B_\perp^3 + 5.9553* B_\perp^2 - 27.1754 B_\perp + 1.0107$$

$$\text{(for perpendicular)} \tag{4}$$

STRUCTURE DESIGN OF AN ACTUATOR

The preliminary design of an actuator is shown in Fig.3. The radial component of the magnetic field, which is perpendicular to the HTS tapes, is exhibited in Fig.4 after electromagnetic field analyzing by coupled field iteration of FEM. Fig.5 is the map of the radical component of HTS when there is no magnetic circuit of laminated iron, which is used for comparison.

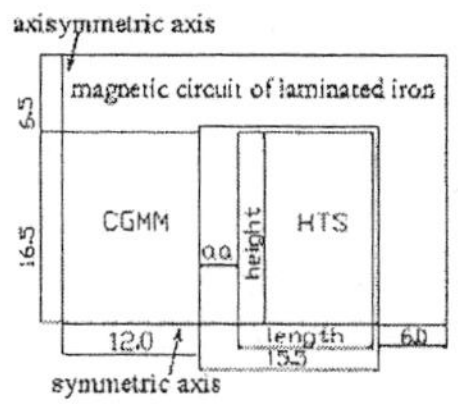

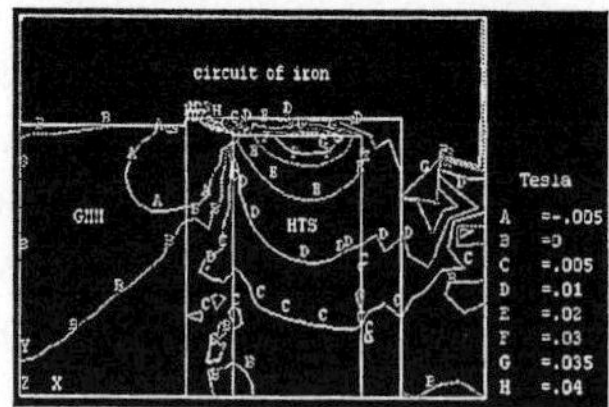

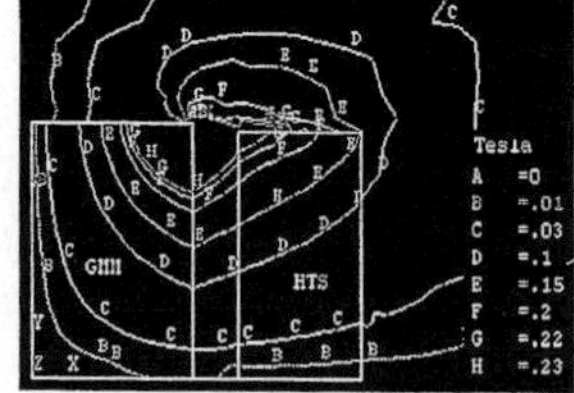

Figure 3 The preliminary design of an actuator

Figure 4 The contour lines of perpendicular component when with the iron made circuit

Figure5 The contour lines of perpendicular component when without the iron made circuit

The radial component of the magnetic field is reduced greatly by a magnetic circuit using laminated silicon iron. The parallel component become an important factor limiting the critical current too owing to much larger magnitude, which is more than 10 times the radial component. So the effects from both the two components have to be considered for the design of an actuator.

OPTIMAL DESIGN USING GENETIC ALGORITHM FOR AN ACTUATOR

In the electromagnetic design of an actuator, the optimization process is realized by genetic algorithm combined with coupled field iteration of FEM. The purpose of optimization is to find out the specific size and place of the HTS solenoid, which can make the CGMM to the state of saturation while costing least HTS tapes.

Mathematical description of the optimization of the investigated actuator

The structure of CGMM is shown in Fig.3, the mathematical description of the above optimization is:

$$\text{min: } S_{hts} = length \times height \quad \text{st:} \quad 0 < length < 0.01550\text{m}, \ 0 < height < 0.01700\text{m}, \ 0 < aa < 0.01550\text{m},$$

$$0 < length + aa < 0.01550\text{m}, \ \text{By}|_{\text{CGMM}} \geq 2.00000\text{T}, \quad J_{permissive} < \min(J_{c\perp}, J_{c//})$$

GA process and flowchart

The optimization of the CGMM actuator is realized using the GA [7], combined with coupled field iteration of FEM. The flowchart is shown as Fig6.

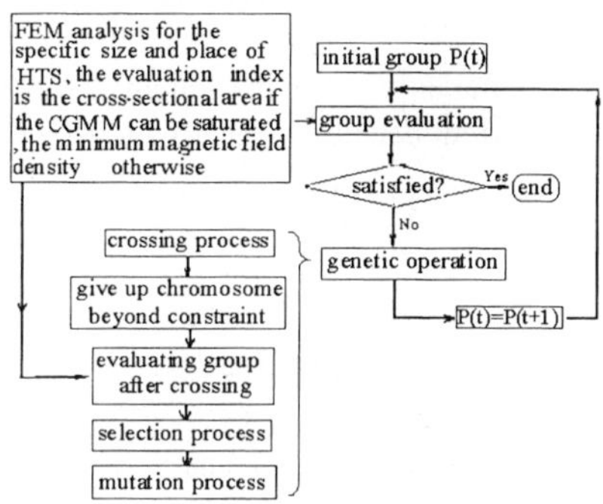

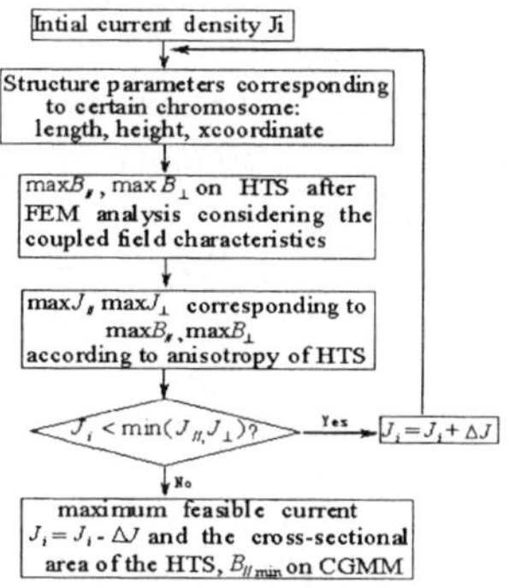

Figure 6 the GA process flowchart   Figure 7 evaluation process flowchart

Evaluation process is shown in Fig.7, the evaluation function is the cross-sectional area S_{HTS} if the CGMM is saturated, the magnetic field density in CGMM if unsaturated.

Optimization result

The parameters of the optimal structure are: $aa = 0.00339$m, $height = 0.01594$m, $length = 0.00920$m.

Contour lines of the parallel component, by (see Fig.8), and the perpendicular component, bx (see Fig.9), are mapped below respectively.

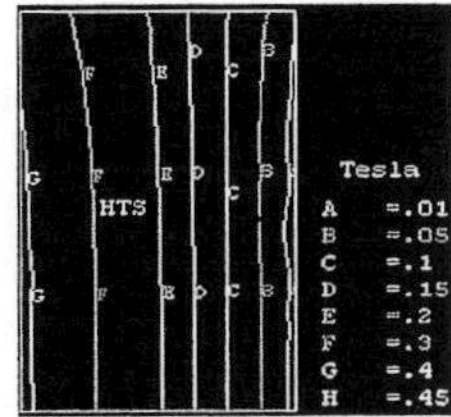

Figure 8 contour lines of the parallel component (by)

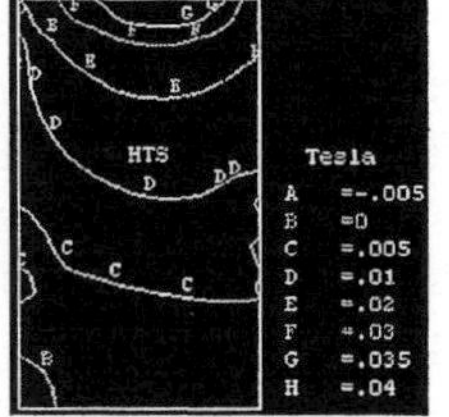

Figure 9 contour lines of perpendicular component (bx)

The maximum values of the parallel component in HTS area, $bx_{max} = 0.04057$T, and the perpendicular component, $by_{max} = 0.40871$T, are gotten and the permissive input current density $J_{permissive} = 3700 A/cm^2$ is found out. Then the critical currents corresponding to the two components of the magnetic field, $J_{c\perp} = J_{bx max} = 3708.439 A/cm^2$ $J_{c//} = J_{by max} = 4041.548 A/cm^2$, are calculated respectively. It is obvious that $J_{permissive} < \min(J_{c\perp}, J_{c//})$.

CONCLUSION

In this paper the radial component of the magnetic field is reduced greatly by a magnetic circuit using laminated silicon iron. So the parallel component become an important factor limiting the critical current too owing to much larger magnitude, which is more than 10 times the radial component.

The genetic arithmetic concerning HTS anisotropy characteristic combined with the coupled field iteration of FEM is used for optimization. The analysis result shows CGMM can be saturated with field of good uniformity when the HTS has the parameters of aa=0.00339m, height=0.01594m, length=0.00920m, which cost the least HTS tapes. It is also found that the maximum current density is permissive $3700 A/cm^2$ with these optimized parameters.

ACKNOWLEDGMENT

This work is supported by the National Natural Science Foundation of China(50077019)

REFERENCES

1. Voccio, J.P., Joshi, C.H. and Lindberg, J.F., application of high-temperature superconducting wires to magnetostrictive transducers for underwater sonar, IEEE trans. On Mag. (1994) ,30 1693-1698

2. Pitel, J., Kovac, P., Melisek,T., et al. Influence of the Winding Geometry on the Critical Current and Magnetic Fields of Cylindrical coils made of Bi(2223)Ag anisotropic tapes, IEEE Trans. On Applied superconductivity (2000), 10 478-481

3. Fabbricatorc, P., Priano, C., Testa, M. P., Musebich, R., et al. Field distribution effect on the performances of coils wound with Ag/Bi-2223 tape, Supercond. Sci. Technol. (1998), 11 304-310

4.So Noguchi, Makoto Yamashita, Hideo Yamashita, et al., An Optimal Design Method for Superconducting Magnets Using HTS Tape, IEEE Trans. on Appl. Superconductivity(2001),11 2308-2311.

5. Zhitong Cao, Jiongjiong Cai, Youtong Fang, FEM Analysis and Design Optimization of an Actuator Made of GMM, record of the ASAEM'2003, oct 22-25,2003, Seoul, Korea, p53

6. http://www.amsuper.com/html/products/htsWire/103419095991.html

7. Ling Wang, Intelligent Optimization Algorithm with Application, Tsinghua University Press, Beijing, China (2003) 36-37

A new method of zooming in the cool power of the room temperature magnetic refrigeration

D.W.Lu, H.B. Wu, G.Q.Yuan, Y.S.Han, X.N.Xu, X.Jin

National Laboratory of Solid Microstructures, Department of Physics, Nanjing University, Nanjing 210093, P.R.China

The cycle speed of a room temperature magnetic refrigerator determines the cooling power. All the materials of giant magnetocaloric effect have much lower heat conductivity compared with copper, so the refrigerator with such a material has a less cooling power. A new method of using the high heat conductivity materials to improve the heat conductivity of the magnetic refrigerants is proposed in the present paper. The new method to be used in magnetic metal Gd with Ag may zoom in about 10 times of the cooling power

INTRODUCTION

In recent years, more attentions have been focused on the room temperature magnetic refrigeration (RTMR) for the breakthrough in the materials and machines [1,2,3,4]. Since the magnetic refrigeration is a reversible cycle and has an advantage of short of pollution, it holds a promising practical usage and extends a wider cooling temperature range from the room temperature to the temperature of liquid hydrogen. So far, there are many milestones in the history of the RTMR in the decade years.

The permanent magnet applied in a demonstrating refrigerator can only leave a quite limited space for the magnetic working materials, so a typical RTMR demonstration machine will only have a little cooling power about 10~100 Watts if it wants to keep a span of 10K. It actually asks us a question: how to increase the cooling power for a real usage? We must notice such a fact that the cooling power of the RTMR depends on the cycle speed, and the quick cycle requires a high heat transfer of magnetic refrigerant. The characters of both high magnetocaloric effect (ME) and high thermal conductivity seldom co-exist in same magneticre refrigerant. So it is necessary to enhance the heat conductivity of the magnetic refrigerant. We may recompose these two kinds of materials to settle this problem, thereby enhancing the cooling power. In the present paper, we study the possibility of increasing the heat transfer and keeping the high ME simultaneously by recomposing magnetic refrigeration materials with the high thermal conductivity materials, and the result of calculation is consistent with this expectation.

CALCULATION MODEL AND RESULTS

Considering a set of gadolinium slices with 0.2mm thickness between them water flowing to transfer the heat from the gadolinium, there is an heat transfer between gadolinium slices and water flow while gadolinium slices move in and out the magnetic field. Generally, given an infinite surface heat exchange coefficient between the water flow and the surfaces of gadolinium slices in such a system, the heat conductivity from the inner of the gadolinium slices to their surfaces plays an important role in the process of refrigeration. According to a simple estimate to the heat conductivity, the typical process time

554

is about 2 seconds, one cycle period is about 5 seconds. In other words, such a heat transfer process determines one cycle period. Because the refrigeration power is a product of cycle times and the cooling capacity in unit time, the shorter the period, the larger the cooling power if the cooling capacity keeps roughly the same each period. Thus, it is necessary to shorten the period of the heat transfer to increase the cooling power. The thermal conductivity time and the size of gadolinium slices is a basic factor to the period, which the high thermal conductivity and small size would carry out a shorter period, and these are the positive influences to refrigerators. Nonetheless, it is not to say the high thermal conductivity and small size have not any other negative influences to refrigeration. In fact, the small size shortens the period at a cost of the flow resistance rising of water, and as a typical magnetic refrigeration material, the gadolinium possesses low thermal conductivity, improving the heat conductivity of the gadolinium slices by recomposing other materials of high heat conductivity also lower the ME. But if the positive influences exceed the negative influences, it may be expected that a proper matching of the size of gadolinium slices with a compound materials could increase the refrigeration power. In following, we will elucidate the design of such compound gadolinium slices. Figure.1 is a sketch map of the compound materials.

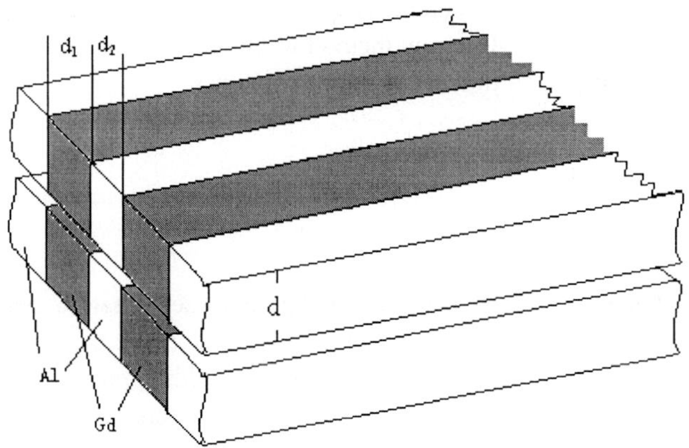

Figure 1 The structure of the compound material

The recomposed Gd-Al magnetic refrigeration material (we call it as compound in the paper) is something like a sandwich in its section. The inner thin slices are thin enough compared with the outside size of 0.2mm. Since the heat conductivity of Al is much larger than that of Gd, we may think that in the Gd-Al compound the heat of magnetocaloric effect of Gd is mainly carried away by Al.

According to the above analysis, the question may be simplified as thermal conductivity in one dimension. By using the heat conductivity equation below,

$$\frac{\partial^2 t}{\partial x^2} = \frac{1}{\alpha}\frac{\partial t}{\partial \tau}$$

where $\alpha = \dfrac{\lambda}{\rho c}$, t is the temperature, τ is the time, x is the position, λ is the heat conductivity, ρ is the density, and c is the thermal capacity, we may calculate the heat transfer capacity of the Gd-Al compound system.

We introduce a character time Γ to denote the typical time of heat transfer from the beginning to the end in a period, then

$$\Gamma \propto \frac{\rho c}{\lambda}$$

The proportion coefficient is a constant independent of materials.

After we define the parameter Γ, We could engage Γ to evaluate the capacity of heat transfer in different systems by comparing the value of Γ in the corresponding system: The small value means the quick heat transfer. The configuration of the compound is illustrated in Figure 1. The value of Γ_{GdAl} (the foot notation $GdAl$ is applied for the compound $GdAl$ and Gd for the pure metal Gd) is quite different from the value of Γ_{Gd}, if the two kinds of materials are of the same figuration.

If the inner sizes of the material are much smaller than the thickness of it ($d_1, d_2 \langle\langle d$), the temperature along the horizontal direction could be thought as uniformity, and the equivalent heat conductivity along the vertical direction could be described as

$$\lambda_{GdAl} = \frac{\lambda_{Gd} d_1 + \lambda_{Al} d_2}{d_1 + d_2}, \quad (\rho c)_{GdAl} = \frac{\rho_{Gd} c_{Gd} d_1 + \rho_{Al} c_{Al} d_2}{d_1 + d_2} \tag{1}$$

Then, $$\frac{\Gamma_{GdAl}}{\Gamma_{Gd}} = \frac{\rho_{Gd} c_{Gd} d_1 + \rho_{Al} c_{Al} d_2}{\lambda_{Gd} d_1 + \lambda_{Al} d_2} \bullet \frac{\lambda_{Gd}}{\rho_{Gd} c_{Gd}}. \tag{2}$$

On the other hand, the change of the configuration also brings a change of difference of the magnetic entropy during a cycle period.

$$\frac{S_{GdAl}}{S_{Gd}} = \frac{d_1}{d_1 + d_2} \tag{3}$$

According to the Eq.(3), the value of S_{GdAl} is smaller than S_{Gd}, but according to the Eq.(2), the heat transfer of the two systems does not work with the change of the entropies in step, if a proper set of d_1 and d_2 is selected, the Γ_{GdAl} can be much smaller than Γ_{Gd}, thus in the time of Γ_{Gd} the compound system could undergo many cycles. Providing the magnetocaloric effect is equal in every cycle for a certain system, the relation of the heat exchange with water flow in the two systems can be written in a simple form

$$\eta = \frac{Q_{GdAl}}{Q_{Gd}} = \frac{d_1}{d_1 + d_2} \bullet \frac{\lambda_{Gd} d_1 + \lambda_{Al} d_2}{\rho_{Gd} c_{Gd} d_1 + \rho_{Al} c_{Al} d_2} \bullet \frac{\rho_{Gd} c_{Gd}}{\lambda_{Gd}} \tag{4}$$

η denotes the amplifying times of cooling power compared with using Gd. It is easy to replace Al as other high thermal conductivity materials such as Ag and Cu. Based on Eq.(4), some values of η in different cases are listed in Table.1

Table.1   The amplifying times of the cooling power of compounds

	Al	Cu	Ag
$d_1 = d_2$	5.41	8.14	10.0
$d_1 = 2d_2$	5.13	8.04	9.52
$2d_1 = d_2$	4.76	6.67	8.33

It is obvious that the value of η varies with the different d_1 and d_2, so there exists an optimal selection of d_1 and d_2. Assuming $d_1 = \mu d_2$, then,

$$\eta_{max} = \frac{\mu^2 + \lambda\mu}{\mu^2 + (1 + \rho c)\mu + \rho c}, \quad (\lambda = \frac{\lambda_{Al}}{\lambda_{Gd}}, \quad \rho c = \frac{\rho_{Al} c_{Al}}{\rho_{Gd} c_{Gd}}) \tag{5}$$

RESULT AND DISCUSSION

Substituting the data of the common high heat conductivity materials into Eq(5), the optimal values of η for these materials are listed in Table.2

Table.2   The maximum amplifying times of the cooling power of compounds

	Al	Cu	Ag
μ	1.166	1.357	1.127
η_{max}	5.44	8.31	10.7

It is obvious that, as to different kinds of high heat conductivity materials, the perfect match of d_1 and d_2 is different (for the three compound materials, the optimal match are about 6：7(GaAl), 3：4(GdCu), and 8：9(GdAg) respectively), and η_{max} is also different correspondingly. The enhancement using GdAg compound materials is highest, about ten times compared with using the pure Gd.

In addition, the method may be also used to improve the performance of the cryogenic magnetic regenerative materials.

REFERENCES

1.Brown, G. V., J. Appl. Phys. (1976), 47, 3673

2.Barclay, J. A., and Steyert W. A. Cryogenics (1982b), 22, 73

3.Zimm C., Jastrab A., Sternberg A., Pecharsky V., Gschneidner K. Jr., Osborne M., and Anderson I., Adv. Cryog. Eng. (1998), 43, 1759

4.Pecharsky V. K. and Gschneidner K. Jr., Phys. Rev. Lett. (1997), 78, 4494

Extended power law of nonlinear transport properties of superconducting materials

Ning Z.H., Hu X., Yin D.L., Qi Z., Wang F.R., Guo J.D., Li C.Y.

Department of Physics, Peking University, Beijing 100871, People's Republic of China

The nonlinear transport properties of superconductors near the transition are usually described by the so called power law $E/E_c = (J/J_c)^n$. We report a wide-range resistive transition equation with the form of an "extended power law". This equation fits the experimental data of MgB_2 and high T_c cuprates.

INTRODUCTION

For designing superconducting magnets, fault current limiters, cables and many other devices thorough knowledge of the electromagnetic response near the critical state is necessary. In principle this should be determined by the Maxwell equations combined with a proper materials equation $J(E, T, B)$. At present, power law $E(J)$ characteristics of the form $E = E_c[J/J_{c0}(T, B)]^n$ are often used (see e.g.[1], [2], [3], [4] and references there-in). In this study we show that the $E(J)$ isothermal characteristics for wider range including the crossover to Ohmic-like regimes have the general form of extended power law which well fits experimental data.

EXTENDED POWER LAW

This form can be derived from the Ginzburg-Landau (GL) free energy density[5]

$$f = f_{n0} + \alpha |\Psi|^2 + \frac{\beta}{2} |\Psi|^4 + \frac{1}{2m^*} \left| \left(\frac{\hbar}{i} \nabla - \frac{e^*}{c} \mathbf{A} \right) \Psi \right|^2 + \frac{h^2}{8\pi} \tag{1}$$

where f_{n0} is the free energy density of normal state, $\Psi(r)$ is the complex order parameter, $\mathbf{A}$ is the vector potential. Gorkov found that the GL theory based on Eq.(1) is derivable as a rigorous limiting case of the BCS microscopic theory with $\Psi(r)$ proportional to the local value of the gap parameter $\Delta(r)$ and the effective charge e^* in Eq.(1) equal to $2e$ [6].

Working in the London limit, Nelson and co-workers showed that according to the GL free energy Eq.(1) a system of N flux lines with a field H along the z direction in a sample length L can be described with the free energy represented by the trajectories $\{\overrightarrow{r_j}(z)\}$ of these flux lines[7] with consideration of the pinning potential $V_P(\overrightarrow{r})$ arising from inhomogeneities and defects in sample[8, 9].

Thermally activated flux motion in the sample can be considered as the sequence of thermally activated jumps of the vortex segments or vortex bundles between the metastable states generated by disorder. Every elementary jump is viewed as the nuclearation of a vortex loop, and the mean velocity of the vortex system is determined by the nuclearation rate[8, 9] $v \propto \exp(-\delta F/kT)$, here δF is the free energy for the formation of the critical size loop or nucleus which can be found by means of the standard variational procedure from the free energy functional due to the in-plane displacement $\overrightarrow{u}(z)$ of the moving vortex during loop formation

$$F_{loop}[\overrightarrow{u}] = \int dz \left[\frac{1}{2} \varepsilon_l \left| \frac{d\overrightarrow{u}(z)}{dz} \right|^2 + V_P(\overrightarrow{u}(z)) - \overrightarrow{f}_s \cdot \overrightarrow{u} \right] \tag{2}$$

558

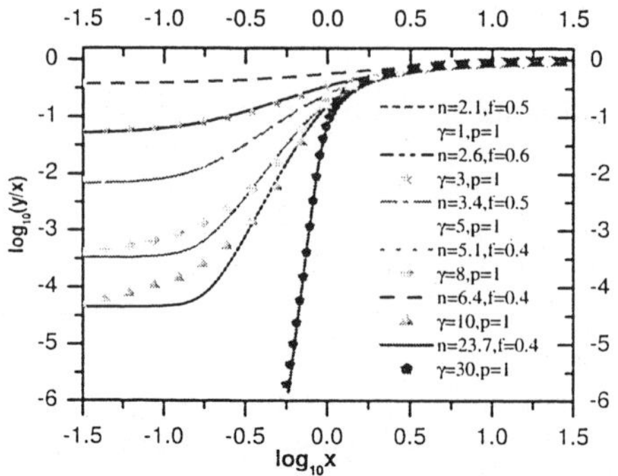

Figure 1: The numerical solutions of Eq.(6)(solid symbols) and Eq.(7)(lines).

with

$$\overrightarrow{f}_s = \overrightarrow{f}_L + \overrightarrow{f}_\eta = \frac{J_P \Phi_0}{c} \times \overrightarrow{e_z} \qquad\qquad J_P = J - \frac{E}{\rho_f} \qquad (3)$$

where $f_L = \overrightarrow{J} \times \overrightarrow{e_z}/c$ is the Lorentz force due to applied current J and f_η is the viscous drag force on vortex, $f_\eta = -\eta v_{vortex}$, with $v_{vortex} = d\overrightarrow{u}/dt$ and viscous drag coefficient $\eta \approx (\Phi_0 B_{c2})/(\rho_n c^2)$ as estimated by Bardeen and Stephen[10].

By a derivation similar to that used Ref.9, one finds the barrier energy

$$\delta F = U(J_p) \approx U_c (\frac{J_c}{J_p})^\mu \qquad (4)$$

which implies a current-voltage characteristic of the form

$$E(J) = \rho_f J exp[-\frac{U_c}{kT}(\frac{J_c}{J_p})^\mu] \qquad (5)$$

where U_c is a temperature- and field-dependent characteristic pinning energy related to the stiffness coefficient and J_c is a characteristic current density related to U_c , μ is an numerical exponent.

Considering the real-size effect Eq.(5) leads to a general normalized form of the current-voltage characteristic[11] in the form

$$y = x \exp\left[-\gamma (1 + y - x)^p\right] \text{ , where } \gamma = \frac{U_c}{kT}\left(\frac{J_c}{J_L}\right) \text{ , } x = \frac{J}{J_L} \text{ , } y = \frac{E(J)}{\rho_f J_L} \text{ , } p = \mu \qquad (6)$$

with J_L the transport current density corresponding to the case where the critical size of loop formation is equal to the real sample size L. The ρ_f in Eq.(5) is the flux flow resistance of a pinning free mixed state as derived by Bardeen and Stephen.

In an earlier work[12], this response equation has also been shown in connection with the Anderson-Kim model.Compared with the widely used power law nonlinear response $E \propto J^n$ (see Refs.[1],[2],[3],[4]), equation(6) can also be asymptotically expressed in an extended power law form

$$\frac{y}{x} = a_0 + f(x - y)^{n-1} \qquad (7)$$

with n the maximal slope at the inflection point of $lny \sim lnx$ curve so

$$\frac{\partial^2 lny}{\partial^2 lnx}|_{x=x_i} = 0 \quad \text{and} \quad n = \frac{\partial lny}{\partial lnx}|_{x=x_i} \qquad (8)$$

and $a_0 = e^{-\gamma}$, f is a numerical factor depending on the parameters γ and p in Eq.(6).

In Fig.1 we show the numerical solutions of Eq.(6) and Eq.(7) for comparison.

COMPARISON WITH EXPERIMENTS

Accounting further the resistive transition between pinning free flux-flow superconducting phase and the normal phase one finds a wide range resistive transition equation of the form

$$\frac{R}{R_n} = exp[-\sum_{i=1}^{2} \gamma_i(1 + y_i - x_i)^{p_i}\theta(T_i - T)] \tag{9}$$

with x_i, y_i, γ_i, and T_i the normalized current, voltage, symmetry-breaking factor, and critical temperature, respectively, defined as

$$x_1 \equiv \frac{I}{I_d(T,B)} \,,\, y_1 \equiv \frac{R}{R_n}[\frac{I}{I_d(T,B)}] \,,\, \gamma_1 \equiv \ln\frac{R_n(T,B)}{R_f(T,B)} \,,\, T_1 \equiv T_c(B)$$

$$x_2 \equiv \frac{I}{I_{c0}(T,B)} \,,\, y_2 \equiv \frac{R}{R_f}[\frac{I}{I_{c0}(T,B)}] \,,\, \gamma_2 \equiv \frac{U_c(T,B)}{kT} \,,\, T_2 \equiv T_m(B), \tag{10}$$

where I_d is the depairing current and I_{c0} is the critical current of vortex solid for overcoming the activation energy barrier $U_c(T,B)$. $\theta(x)$ is the Heaviside function and p_i are exponents[11]. R_f is the unpinned flux-flow resistance of the mixed state in type-II superconductors. In Fig.2 we show

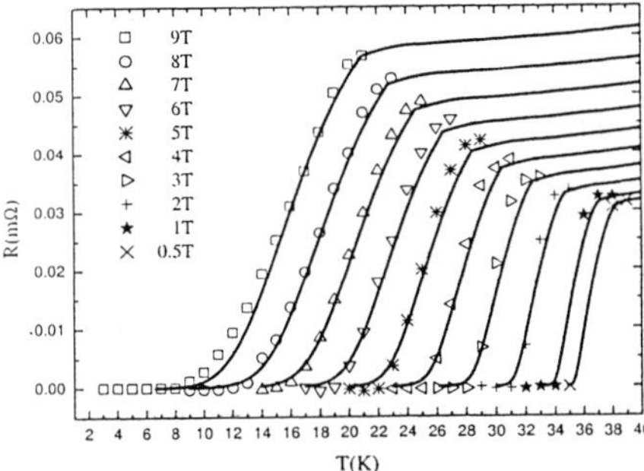

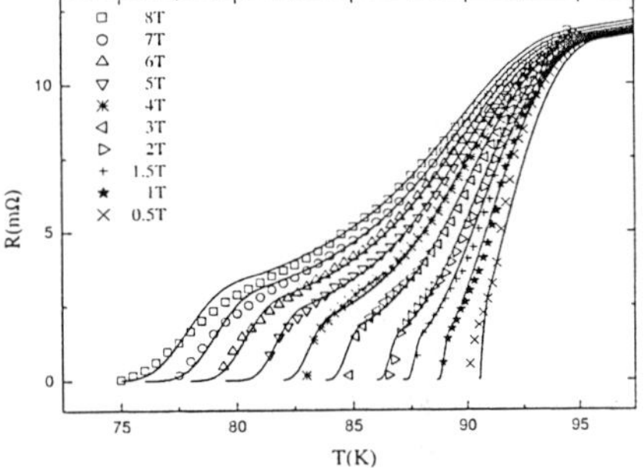

Figure 2: Comparison of Eq.(9) with the experimental resistance data of MgB_2 samples measured by Finnemore et al.(Ref.14). Symbols denote experimental data of resistive transitions in different applied fields and lines denote the theoretical curves in Eq.(9) with corresponding applied fields. The parameters in Eq.(9) are $(1)U_c(T,B) \propto [T_m(B) - T]^{0.8}(B + 5.46)^{-4.3}$; $(2)I_{c0} \propto T_m(B) - T, I_d \propto [T_c(B) - T]^{1.5}\gamma_1^{0.5}$; $(3)T_m(B) = T_c(0)[1 - (B/21.7)^{0.84}], T_c(B) = T_c(0)(1 - B/22.4), T_c(0) = 40.2K$; $(4)\gamma_1 \equiv \ln[R_n(T,B)/R_f(T,B,J \to 0)] = \ln[B_{c2}(T)/B]$, where $B_{c2}(T) = 0.6[T_c(0) - T]^{0.98}$.

Figure 3: Comparison of Eq.(9) with the experimental resistance data of an untwinned YBCO crystal measured in different applied fields for $H_{\parallel c}$ by Kwok et al.(Ref.15). Symbols denote experimental data and lines denote the theoretical curves in Eq.(9) with corresponding applied fields. The parameters in Eq.(9) are $(1)U_c(T,B) \propto [T_m(B) - T]^3 B^{-4.22}$; $(2)I_{c0} \propto T_m(B) - T, I_d \propto T_c(B)[1 - T/T_c(B)]^{2.3}\gamma_1^{0.5}$; $(3)T_m(B) = T_c(0)[1 - (B/1200)^{0.36}], T_c(B) = T_c(0)[1 - (B/3.6 \times 10^6)^{0.3}], T_c(0) = 97K$; $(4)\gamma_1 \equiv \ln[R_n(T,B)/R_f(T,B,J \to 0)] = \ln[B_{c2}(T)/B]^m$, where $m = 0.84B^{-0.59} - 0.07$ and $B_{c2}(T) = 0.6[T_c(0) - T]^{3.3}$.

the comparison of the resistive transition Equation(9) with the experimental data of the temperature dependent resistance of MgB_2 in different applied fields measured by Finnemore et al.[14]. The current density used in their standard four-probe technique is sufficiently low with the value of $0.1 - 0.3A/cm^2$. Kwok et al. studied the width and shape of the resistive transition fo untwined and twinned single crystals of YBCO in fields up to 8T[15]. We compare our resistive Equation(9) with their experimental data of untwinned YBCO crystal in Fig.3.

SUMMARY

We show an extended power law form of current-voltage characteristics which well fits the wide range experimental data of high-Tc MgB_2 and YBCO materials. This equation combined with the Maxwell equation may provide a useful tool for applied superconductivity design.

ACKNOWLEDGEMENT

This work is supported by the Ministry of Science & Technology of China (NKBRSG-G 1999064602) and the National Natural Science Foundation of China under Grant No. 10174003 and No. 50377040.

REFERENCES

1. Cha Y.S., An emperical correlation for E(J,T) of melt-cast-processed BSCCO-2212 superconductor and self field, IEEE Transactions on Applied Superconductivity (2003), 13 2028-2031

2. Inoue K., Sakai N. and Murakami M., E-J characteristics and n value of melt-textured $REBa_2Cu_3O_x$ (RE: Nd, Y), IEEE Transactions on Applied Superconductivity (2003), 13 3109-3112

3. Grilli F., Stavrev S., Dutoit B. and Spreafico S., Numerical modeling of HTS cable, IEEE Transactions on Applied Superconductivity (2003), 13 1886-1889

4. Okamoto H., Kiss T., Nishimura S., Inoue M., Imamura K., Takeo M. and Kanazawa M., Prediction of E-J characteristics in Bi-2223/Ag tapes of low temperature and high magnetic field, IEEE Transactions on Applied Superconductivity (2003), 13 3683-3686

5. Ginzburg V.L. and Landau L.D., On the theory of superconductivity, Ekcperim. Zh. Teor. Fiz. (1950) 20 1064; See also, Tinkham M., In: Introduction to Superconductivity, Second Edition, McGraw-Hill Inc. (1996) Chapter 4

6. Gorkov L.P., Sov. Phys. JEPT (1959)9 1364; see also Ketterson J.B. and Song S.N., Superconductivity, Combridge Univ. Press, (1999) Part III, Chapter 45

7. Nelson D.R., Vortex entanglement in high-Tc superconductors, Phys.Rev.Lett. (1988)60 1973; Nelson D.R. and Seung H.S., Theory of melted flux liquids, Phys.Rev.B(1989) 39 9153

8. Nelson D.R. and Vinokur V.M., Boson localization and correlated pinning of superconducting vortex arrays ,Phys. Rev. B (1993) 48 13060; Vinokur V.M., Physica A (1993) 200 384; Täuber Uwe C., and Nelson D.R., Interactions and pinning energies in the Bose glass phase of vortices in superconductors, Phys. Rev.B (1995) 52 16106

9. Fisher D.S., Fisher M.P.A. and Huse D.A., Thermal fluctuation , phase transition and transport in type-II superconductors, Phys.Rev.B (1991), underline43 , 130-159

10. Bardeen J., Stephen M.J., Viscosity of type-II superconductors, Phys. Rev. Lett. (1965) 14 112-113

11. Chen H.D., Wang Y., Yin D.L., Zhu Y.L. , Chen K.X., Li C.Y. and Lu G., Vortex-boson analogy and the nonlinear response function in high-Tc superconductors near the phase transition, Phys.Rev.B (2000) 61 1468

12. Yin D., Schauer W., Windte V., Kpfer H., Zhang S. and Chen J., Z.Phys.B (1994)94 249

13. Yin D.L., Qi Z., Xu H.Y. and Wang F.R., Resistive transition equation of the mixed state of superconductors, Phys.Rev.B (2003)67 092503

14. Finnemore D.K., Ostenson J.E., Bud'ko S.L., Lapertot G. and Canfield P.C., Thermodynamic and transport properties of superconducting $Mg^{10}B_2$, Phys.Rev. Lett. (2001) 86 2420

15. Kwok W.K., Fleshler S., Welp U., Vinokur V.M., Downey J., Crabtree G.W., and Miller M.M., Vortex lattice melting in untwinned and twinned single crystals of $YBa_2Cu_3O_{7-delta}$, Phys.Rev.Lett. (1992) 69 3370

AC Loss characteristics of HTS tapes subject to bending strains

Tsukamoto O., Ogawa J.[*], Suzuki H.

Yokohama National University, 79-5 Tokiwadai, Hodogaya-ku, Yokohama, 240-8501 JAPAN
*Niigata University, Ikarashi 8050, Niigata, 950-2181 JAPAN

Dependence of AC loss characteristics of Bi2223/Ag-sheatled tapes on bending strains was experimentally investigated. AC transport current losses in the tapes subject to various bending strains were measured in the zero external magnetic field. These losses were compared with the losses of straight tapes. The measurement results show that values of the AC losses are affected by the degradations of the critical currents caused by the bending strain. Based on these results it can be concluded that the AC transport current loss characteristics of Bi2223/Ag-sheathed tapes subject to bending strain can be estimated by knowing values of critical currents of bended tapes and the loss characteristics of straight tapes.

INTRODUCTION

AC superconducting power apparatuses such as power cables and transformers are most promising applications of high temperature superconductor (HTS) because costs to cool the AC losses in the HTS conductors in those apparatuses are much lower than those in low temperature superconductor AC power apparatuses. However, AC losses even in HTS are the major losses in the apparatuses and dominate their efficiency and economic feasibility. Therefore, it is important to understand the AC loss characteristics correctly.

HTS conductors are used in forms of coils and cables in those apparatuses and subject to bending strains. The bending strains affect critical currents and may affect AC loss characteristics. We investigated the AC transport current losses in Bi2223/Ag-sheathed tapes subjected to the bending strains in the zero external field at the first step. We measured the AC losses in a straight Bi/Ag-sheathed tape also and compared with those in the bended tapes. That is because, if we find any relation between the loss data of straight and bended tapes, it is possible to estimate the AC loss characteristics of the tapes subject to the strains in the cables and windings form the loss data of a straight tape which are much easier to measure than those of bended ones.

EXPERIMENT

Measurement arrangement

The transport current losses in Bi/Ag-sheathed tapes subject to bending strains were measured by a four-terminal electric method in the zero magnetic field. An arrangement for the measurement is illustrated in Figure 1 (a) and (b). Figure 1 (a) illustrates the sample arrangement and Figure 1 (b) the measurement circuit. Bi/Ag-sheathed tapes are placed on surfaces of half-cylinders made of GFRP of radii $R = 15, 20, 25$ and 30mm and leads from potential taps attached on the tape edge are arranged 1cm apart from the tape as shown in Figure 1 (a). As shown in Figure 1 (b), AC transport currents were supplied to the sample by a variable frequency power supply, and the transport current losses are measured by measuring the resistive voltage component of the voltage signal from the potential taps using a lock-in amplifier. The frequency of the transport current was controlled by the frequency signal generated in the lock-in amplifier.

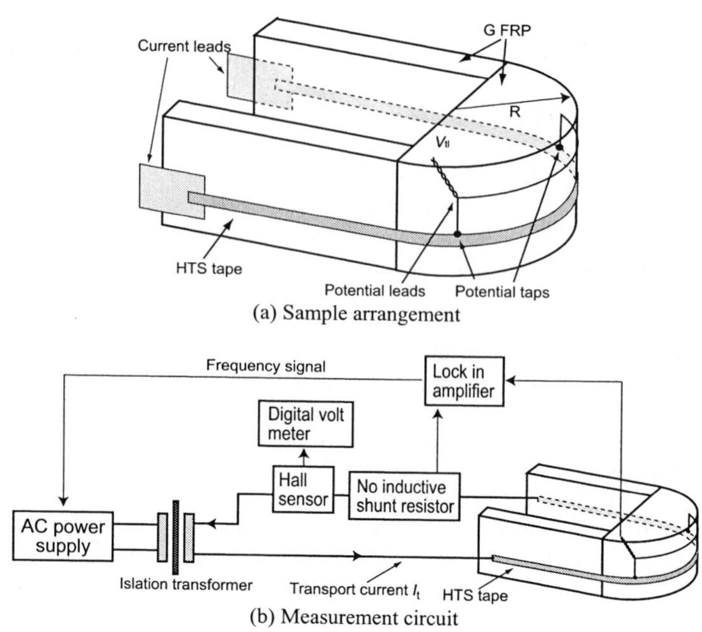

(a) Sample arrangement

(b) Measurement circuit

Figure 1 Schematic illustration of arrangement for electric measurement of AC transport current losses

A sample piece of Bi/Ag-sheathed tape was changed when the radius of the sample holder was changed to avoid any deterioration of performance of the tape caused by heat cycles. The sample pieces were taken from the same long tape. The specifications of the tape used in the experiment are listed in Table 1.

Table 1 Specifications of Bi2223/Ag-sheathed tape used in experiment

Width × Thickness	4.0×0.22 mm^2
Number of filaments	70
Ag / SC	1.5

<u>Measurement results</u>

Critical currents I_c of the samples (defined at 1μV/cm) of different bending radii are plotted against bending strain σ together with I_c of a straight sample in Figure 2. Degradation of I_c is remarkable for

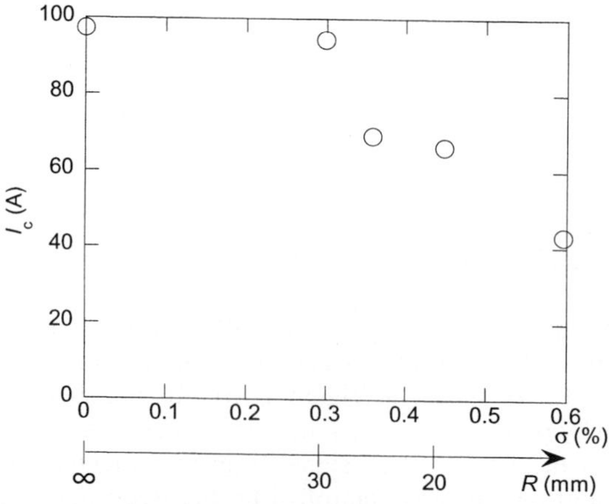

Figure 2 Critical current vs. bending strain σand radius R

$R=15$mm (bending strain 0.6%). AC transport current losses per cycle per unit length of the tape Q_t are shown in Figures 3~5. Figure 3 shows dependence of Q_t vs. the amplitude of the transport current I_t curves on R at 60Hz. Figure 4 shows normalized losses $q_t = Q_t/I_c^2$ vs. normalize transport current $i= I_t/I_c$ for various R at $f = 60$Hz. As seen in Figure 4, curves of q_t vs. i for the samples of different values of R including the straight sample almost fall to one curve which is close to the Norris curve of the elliptical model [1]. Figure 5 shows frequency dependence of q_t for various i and R. As seen in Figure 5, dependence of Q_t on frequency is not remarkable in the range of $f = 30$~720Hz and $I_t < 0.7I_c$, which means that the losses were hysteretic in this range. Also as seen in Figure 5 q_t is dependent on f for $I_t \sim I_c$, which is because the tapes become resistive and dissipate joule heats.

Obviously from these results, the AC transport current losses in Bi/Ag-sheathed tapes subject to bending strain can be estimated by knowing the AC transport current loss characteristics of a straight tape and I_c of a bended tape.

CONCLUDING REMARKS

Dependence of the AC transport current losses on the bending strain was experimentally investigated. The experimental results show that the AC transport current losses in Bi2223/Ag-sheated tapes are hysteretic and that their dependence on the bending strain can be explained by the dependence of the critical currents on the bending strain. This result suggests that the AC losses of Bi2223/ Ag-sheathed subject to bending strains can be estimated by knowing the dependence of the critical currents of tapes on

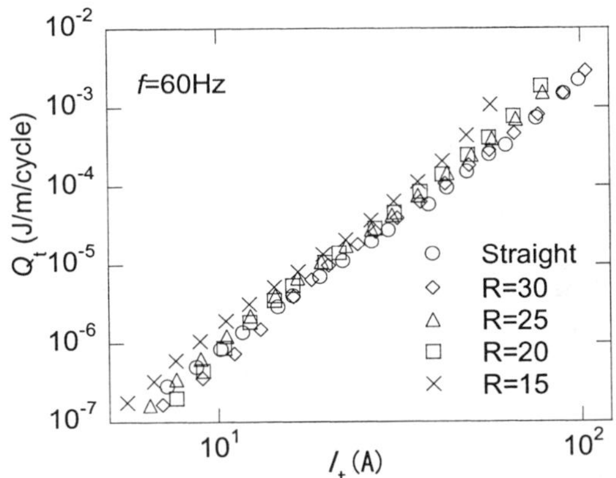

Figure 3 Transport current loss Q_t vs. transport current I_t for various R at 60Hz

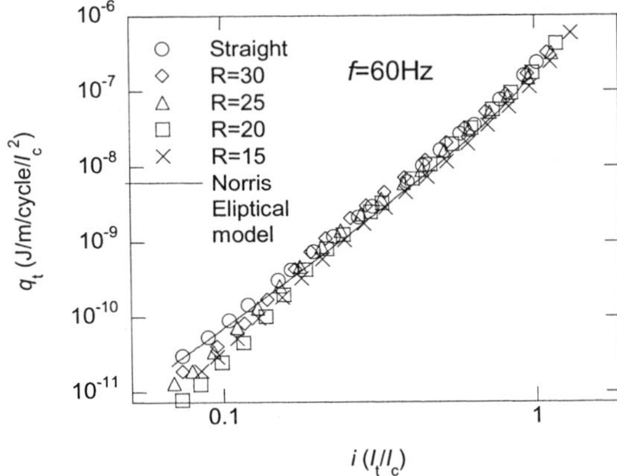

Figure 4 Normalized transport current loss q_t vs. normalized transport current i at 60Hz for various R

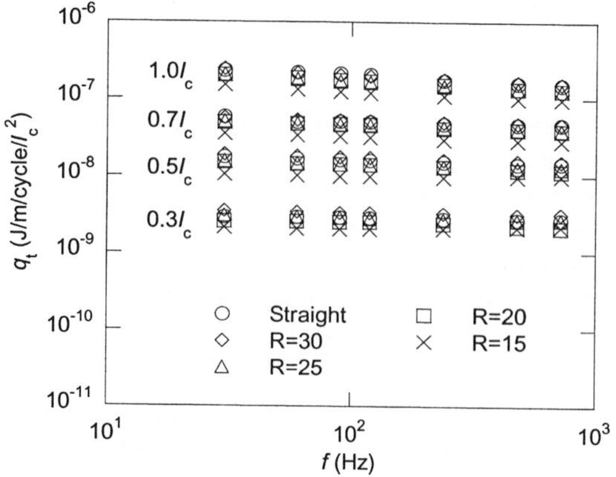

Figure 5 Frequency dependence of q_t for various values of i and R

the bending strain and the AC loss characteristics of straight tapes.

Generally, the HTS conductors in the AC apparatuses are carrying AC transport currents in AC external magnetic fields. Therefore, as the next step, we are investigating AC loss characteristics of HTS tapes subject to bending strains and carrying transport currents in AC external magnetic fields. In the case, it is difficult to measure the AC losses by an electric method but possible by a calorimetric method [2].

ACKNOWLEGEMENT

This work was supported by Grant-in Aid for Basic Research (S) (14102019), The Ministry of Education, Culture, Sports, Science and Technology. The Bi2223/Ag-sheathed tape used in the experiment was supplied by Sumitomo Electric Industries Ltd..

REFERENCES

1. Norris, W.T., Calculation of hysteresis losses in hard superconductors carrying ac : isolated conductors and edges of thin sheets, Journal of Physics D(1970), 3, 489-507.
2. Ogawa,J., Yanagihara, Y., Yamato, Y., and Tsukamoto, O., Measurements of Total AC Losses in HTS Short Sample Wires by Electric and Calorimetric Methods, Advances in Cryogenic Engineering (Materials) to be published.

EAST Superconducting Tokamak Device

P. D. Weng and EAST team

Institute of Plasma Physics, Chinese Academy of Sciences,
P. O. Box 1126, Hefei, Anhui, 230031, P. R. China

Abstract

EAST is a Chinese National Project designed to develop scientific and technological basis on the steady state operation of advanced tokamak [1]. It is a full superconducting tokamak with Superconducting TF and PF magnets, has a long pulse operation capability and appropriate auxiliary heating and non-inductive current drive system, that offer a possibility for steady state advanced performance experiment. The design feature of the EAST and its kW/4K cryogenic system are described in this paper.

Keywords :Tokamak, , Magnet, Cryoplant, helium

1. TOKMAK DEVICE

1.1 General Description

The EAST superconducting tokamak is a full superconducting device, which consists of superconducting toroidal filed (TF) and poloidal Field (PF) magnets, vacuum vessel and in vessel components, thermal shields and cryostat. The cooled components are TF magnets, PF magnets, support structure and thermal shields. Dimensions of the EAST device are 10 m (with the main support) in height and 7.6m in diameter. Its total weight is 360 tons. The main parameters of the EAST device are listed in Table 1

Table 1   Main Parameters

Toroidal Field, B_t	3.5 T
Plasma Current, I_p	1 MA
Major Radius, R_O	1.75 m
Minor Radius, a	0.4 m
Elongation, K_x	$1.6 - 2$
Triangularity, δ_x	0.6 - 0.8
Pulse length	10- 1000 s
Configuration	Double-null divertor
	Single null divertor

1.2 Conductor development

NbTi cable-in-conduit conductor (CICC) is used for both TF and PF magnets. [2] The configuration of the TF cable is (2SC+2Cu)×3×4×5. Two segregated pure copper wires are added in the first stage sub-cable to increase the copper ration in the cable and the last stage cable is formed by five bundles of the third stage sub-cables surround a 21 strands copper core. All of the wires in the cable are coated by Tin alloy. Stainless steel 316LN tube is used as conduit material, the tubes are welded together form a 600 meters long jacket, and the cable is pulled into the jacket.

The conductors for PF1-PF10 have same configuration with the TF cables, instead of Tin alloy, Ni coating on the surface of all strands is adopted to reduce AC losses. The conductor for other two pairs of outer PF coils has less SC strands and higher copper fraction. The sizes of conductors are 20.4mm×20.4mm for TF and PF1-PF10 coils and 18.6mm×18.6mm for PF11- PF 14 coils. The cabling configurations of two types of CICC are shown in Figure 1.

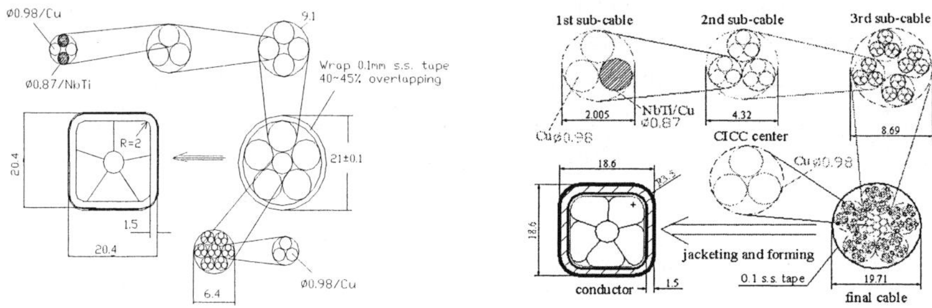

Figure 1   Two types of CICC configuration

1.3 TF Magnet

The TF magnet [3] is consists of 16 D-shaped coils, providing a field of 3.5T at the plasma radius of 1.75 m with a peak field of 5.8T at the TF coils. Table1 listed the parameters of the TF magnet.

Each TF coil contains 130 turns in the forms of 2 of 6 pancakes winding. The entire winding pack is vacuum pressure impregnated (VPI). There are 6 cooling channel in one TF coil and 96 cooling channel in total. All of the channel are connected in parallel and cooled by 3.8 K supercritical helium.

The TF coil cases, which enclose the TF coil winding pack, are welding structure. Stainless steel tubes are embedded and soldered all along the side surfaces of the cases. 4.5 K Supercritical helium will be used for the case cooling.

Table 1   Specifications of the TF system

Magnetic field at the plasma center	3.5T
Maximum field at the coil	5.8T
Number of TF coils	16
Total turns of the TF system	2080 (16×130)
Operating current	14.31kA
Total stored energy	298.39MJ
Conductor length of each coil	1187m
Length of cooling channel	200 m
Operating temperature	3.8 K

1.4 PF Magnet

The EAST PF system [4] consists of a central solenoid (CS) and four pairs of PF coils located symmetrically about the vertical axis of the equatorial plane of the device.

All PF coils are circular. The six inner PF coils form the CS assembly, each coil is vacuum pressure impregnated with epoxy resin. The inlets and outlets for liquid helium are put in the inner side for CS and divertor coils and in the outer side for two pairs of the outer coils.

All PF coils are attached to the TF coil cases to support the electro-magnetic and gravity loads. The main parameters of the PF coils are listed in Table 1.

Table 1   Specifications of the PF system

	PF1- 6	PF 7- 8	PF 9-10	PF 11-12	PF13-14
Out/inner diameter/ Height mm	1418/1085/476	2401/1889/103	2670/188/289	6054/5779/221	6650/6468/179
Turn/cooling channel	140 / 5	44 / 2	204 / 12	60 / 10	32 / 8
Conductor mm	20.8×20.8			18.6×18.6	
B_{max} T	4.3			1.5	
dB/dt_{max} T/s	6.8			0.7	
I_{max} kA	14.5				
Temperature K	3.8				
Total flux swing VS	10				

1.5 Thermal Shields

The thermal shields comprise of the vacuum vessel thermal shield (VVTS), the cryostat thermal shield (CTS) and the ports thermal shield (PTS) [5]. The VVTS is consisting of 16 sectors and the CTS is divided into three parts: upper cap, middle cylinder and bottom platform, each part consists of 8 sectors. The PTS connecting VVTS and CTS are bolted on them during assembling. The Total surface area of thermal shield is 310 m2. All of the thermal shields are double-wall structures. 19× 19mm2 cooling tube is sandwiched between two 3mm thick stainless steel panels. There are 32 parallel cooling circuits for VVTS and PTS and 24 parallel cooling circuits for CTS. The VVTS and PTS are cooled by 58K helium gas at first and then, the return gas will be cooled down to 80 K in a liquid nitrogen sub-cooler and used for CTS cooling.

2. CRYOPLANT AND CRYOGENIC DISTRIBUTION SYSTEM

The cryogenic system includes the cryoplant and the distribution system [6]. The cryoplant provides supercritical helium to cool TF and PF coils, their superconducting bus-lines, magnet support structures and current leads. It also provides cold helium flow to thermal shields.

According to the estimation of the heat loads, the helium refrigerator is designed at the capacity of 1050W/3.5K+200W/4.5K+13g/s LHe + 13KW/80K. An oil ring pump is used to reduce the pressure and obtain the temperature of 3.5K Figure 2 shows a schematic flow sheet of the refrigerator.

The compressor station consists of five screw compressors arranged in two stages; four stages oil removal system and 9000Nm^{3}helium recovery system. A 10000 L Dewar is used as the storage of liquid helium and regulates cooling capacity of the plant when the heat load increases or decreases.

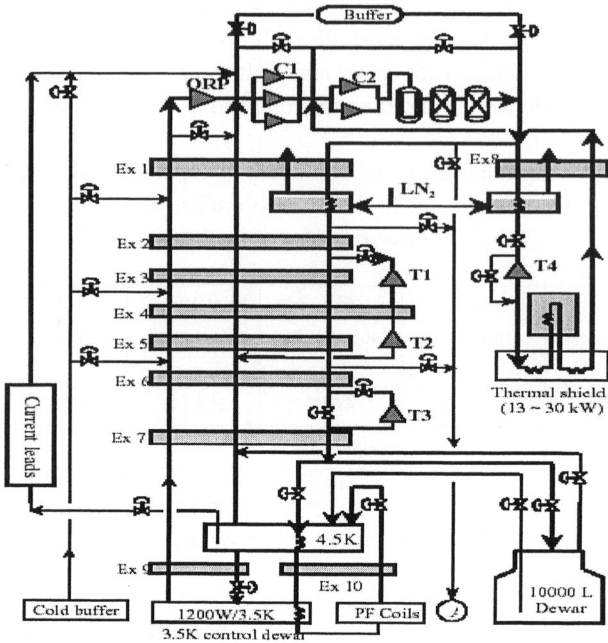

Figure 2   Simplified flow sheet of the helium refrigerator

The cryogenic distribution system comprises the auxiliary control Dewars with a sub-cooler, two circulating pumps and the cryogenic transfer lines which distribute SHe at 4K and 80K helium to the different components of the tokamak.

110g/s-3.8K supercritical Helium flow from refrigerator will be used for the PF coils cooling; 260 g/s-3.8K/ 320 g/s-4.5K supercritical Helium flow circulated by pumps will be used for TF windings and TF coil cases cooling separately; 110g/s-60K Helium flow will be used for cooling of thermal shield.

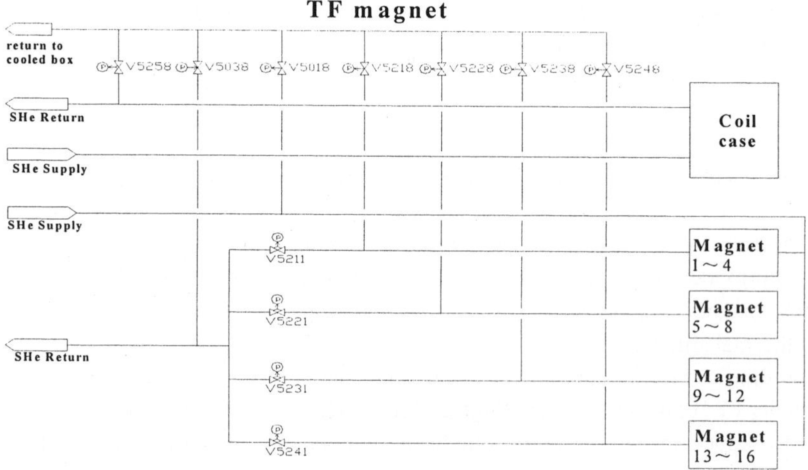

The flow diagram of the cooling cryogenic components is shown in Figure 3,4,5.

Figure 3　　TF cooling

P F　M a g n e t s

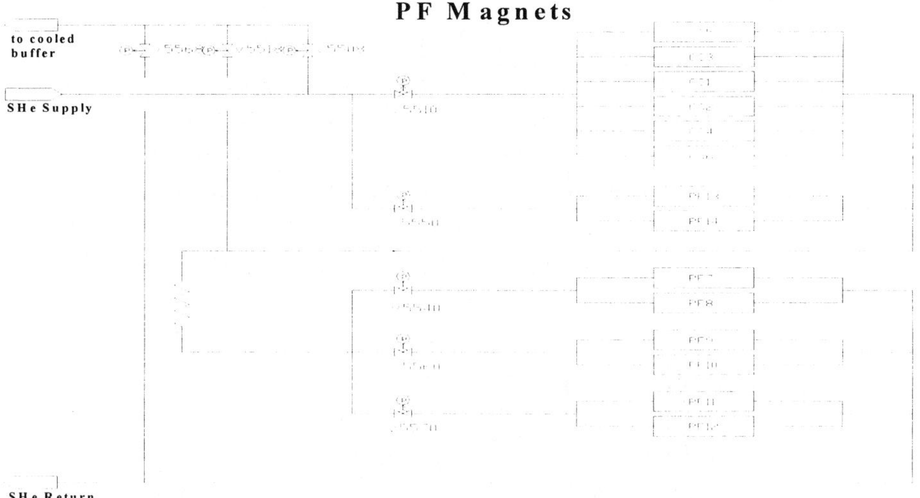

Figure 4　　PF cooling

Thermal shield

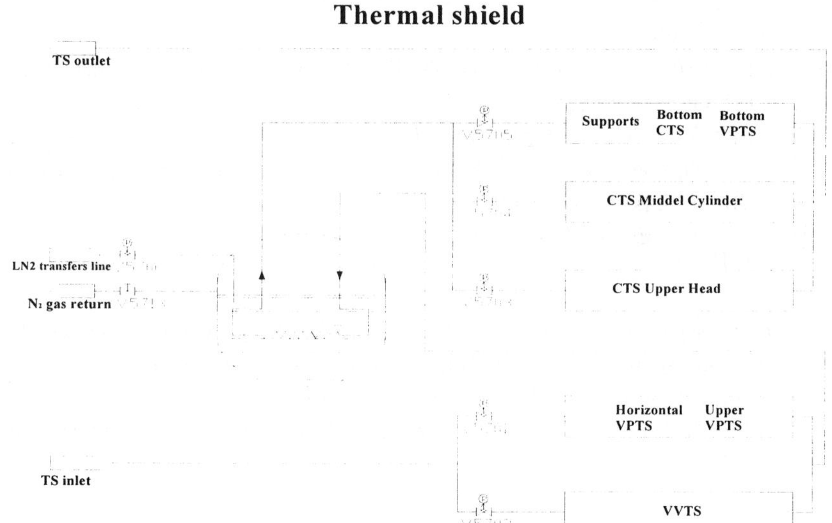

Figure 5　　TS cooling

SUMMARY

The EAST project provides a very good opportunity to develop large-scale supperconducting magnet and cryogenic technology in China. Up to now the device fabrication is going on smoothly, all part of SC Magnet, Cryostat, Vacuum Vessel and Thermal Shield will be completed this year. One CS coil and 10 of TF coil has been tested, the results show that all of the magnets can meet the design requirement. The fabrication and installation of cryogenic system is in progress, the test run of cryogenic system will be made this year. It is hopeful to finish machine assembly and begin startup test of the device in 2005.

ACKNOWLEDGEMENTS

Work presented herein is the work of all of the EAST Project Team, the author many thanks for all of colleagues who made contribution for the project.

REFERENCES

(1) Y.X.Wan and HT-7U team "HT-7U Superconducting Tokamak: Physics design, engineering progress and schedule" 19th IAEA Fusion Energy Conference Proceeding 368

(2) P. D. Weng, et al "IIT-7U TF and PF Conductor Design" Cryogenics, Vol. 40 2000 Iss 8-10, pp 531-538

(3) Y.N.Pan et al "Preliminary Engineering Design of Toroidal Field Magnet system for Superconducting Tokamak HT-7U" IEEE Transactions on Applied Superconductivity 10 (2000) 628-631

(4) W.Y.Wu, Li Baozeng "Design of the poloidal field system and plasma equilibrium of HT-7U Tokamak" Fusion Engineering and Design Volume 58-59 (2001) 277-280

(5) P. D. Weng and the HT-7U team "The Engineering Design of the HT-7U Tokamak" Fusion Engineering and Design Volume 58-59 (2001) 827-831

(6) Hongyu Bai et al "Design of the Cryogenic System for HT-7U Tokamak" 3rd IAEA/TCM on Steady –state Operation of Magnetic Fusion Devices, May 02, 2002, Greifswald.

Challenges of large scale superconducting accelerators

Proch D.

Deutsches Electronen-Synchrotron DESY, Notkestrasse 85, D-22607 Hamburg,
Germany

ABSTRACT

The technology of superconducting RF accelerators has reached a mature state. After
pioneering developments at Cornell several major SRF (Superconducting Radio
Frequency) systems operate/operated reliable at electron storage rings (Cornell,
CERN, KEK, DESY), heavy ion linacs (Legnaro, Argonne) as well as electron linacs
for nuclear physics or FEL application (Jlab, TTF, DALINAC). The first large scale
superconducting proton accelerator is under construction at SNS. Furthermore a
variety of synchrotron light sources using SRF technology are under construction or
planning.
The largest SRF installation is proposed for a linear collider (TESLA). A new X-FEL
user facility is under preparation at DESY. In this talk the SRF accelerator technology
is reviewed. Cryogenic and accelerator challenges are discussed in detail for TESLA
and X-FEL.

Cryogen-free superconducting magnet fabricated by a react-and-wind method employing Nb_3Sn wires with CuNbTi reinforcement

K. Watanabe[1], G. Nishijima[1], S Awaji[1], K. Takahashi[1], K. Miyoshi[2], S. Megro[2],
M. Ishizuka[3], T Hasebe[3]

[1] High Field Laboratory for Superconducting Materials, Institute for Materials
Research, Tohoku University, Sendai 980-8577, Japan

[2] Furukawa Electric Co., Ltd., Nikko 321-1493, Japan

[3] Sumitomo Heavy Industries, Yokosuka 237-8555, Japan

Practical multifilamentary Nb_3Sn wires with CuNbTi reinforcement ($CuNbTi/Nb_3Sn$) exhibit the splendid mechanical tolerance of 340 MPa proof stress.   We intended to develop a Nb_3Sn coil by a react-and-wind method for a wide bore cryogen-free superconducting magnet, and made a react & wind coil employing $CuNbTi/Nb_3Sn$ wires. The performance test for the $CuNbTi/Nb_3Sn$ coil was carried out, and the coil generated 2.2 T at 180 A as the individual test. In the combination test in a background field of 5.6 T, the react & wind processed $CuNbTi/Nb_3Sn$ coil generated 7.5 T in a 220 mm room temperature bore.   The wide bore cryogen-free superconducting magnet wound with highly strengthened $CuNbTi/Nb_3Sn$ wires is now being used for a high field heat-treatment equipment at 7 T. This wide bore furnace provides special heat-treatment conditions of high temperature 1500 ' C in fields up to 7.0 T in a 40 mm sample bore.

INTRODUCTION

A coil fabrication process for high field Nb_3Sn superconducting magnets usually adopts a wind-and-react (W&R) method, in which a coil wound with unreacted Nb-Sn wires is heat-treated for the Nb_3Sn compound reaction.   This is because the reacted Nb_3Sn wire is very sensitive for stress and strain, and is apt to degrade the critical current due to the applied mechanical stress through a coil winding.   However, it is impossible to construct a large Nb_3Sn coil like a nuclear fusion reactor coil by a W&R method.   If we make possible to fabricate a react-and-wind (R&W) processed coil employing the pre-reacted Nb_3Sn wire, a heat-treatment process in a large furnace and an epoxy impregnation process in a large vacuum equipment can be eliminated.   It is expected that the cost reduction due to the simplification of the coil fabrication process is very large.   In particular, the time-related saving cost of the heat-treatment for the Nb_3Sn coil should be concentrated on.

On the other hand, we have developed a highly strengthened Nb_3Sn wire with an internal reinforcement of CuNbTi ($CuNbTi/Nb_3Sn$).   A $CuNbTi/Nb_3Sn$ wire reveals the strong proof stress of

340 MPa, even after the heat-treatment at 650 'C for 240 h [1]. This means that a CuNbTi/Nb₃Sn wire is 2-3 times as strong as a conventional Nb₃Sn wire with an ordinary copper stabilizer (Cu/Nb₃Sn). Therefore, a R&W method is now applicable for fabricating a Nb₃Sn coil employing highly strengthened Nb₃Sn wires.

This paper describes a cryogen-free superconducting magnet wound with pre-reacted CuNbTi/Nb₃Sn wires. A newly developed high temperature heat-treatment furnace combined with the wide bore cryogen-free superconducting magnet is introduced for an in-field process application.

CHARACTERISTICS OF HIGH-STRENGTH CuNbTi/Nb₃Sn WIRE

Figure 1 shows the fabrication process of high-strength CuNbTi/Nb₃Sn wires. The starting CuNbTi reinforcement was incorporated into the central area of a typical bronze-route Nb₃Sn wire, which is a so-called internal reinforcement process. After the heat-treatment at 670 'C for 200 h, the CuNbTi/Nb₃Sn wire has wire parameters such as the outer diameter of 1.0 mm, filament diameter of 3.5 μm, number of filaments of 9600, copper ratio of 26.3 %, and twist pitches of 32 mm, and reveals the critical current properties of 247 A at 12 T and 133 A at 15 T at 4.2 K. Although the CuNbTi/Nb₃Sn wire surely exhibits a strong proof stress of 340 MPa, it unfortunately has poor elongation properties in comparison with an another kind of high-strength Nb₃Sn wire with CuNb reinforcing composite [2]. This results in small shearing elasticity for CuNbTi/Nb₃Sn. In the R&W process, we may have to developed a new high-strength Nb₃Sn wire with a strong proof stress over 300 MPa and a good elongation property. However, the CuNbTi/Nb₃Sn wire does not reveal any degradation of the critical current for the repeated pre-bending treatment in bending strains up to 0.8 % and at 5 repeated times [3].

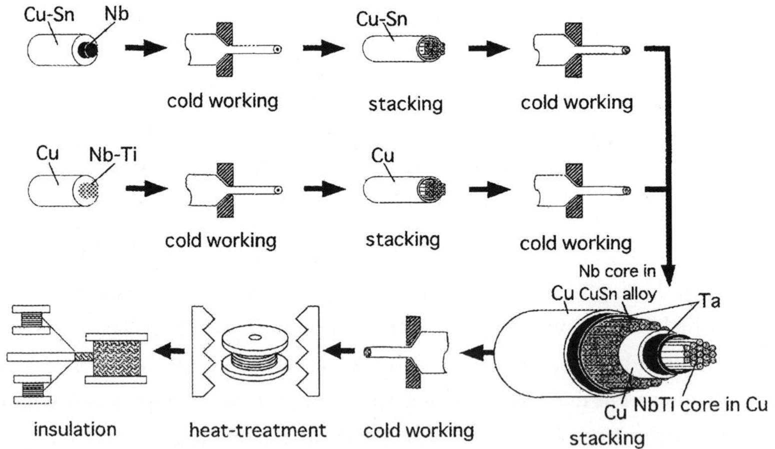

Figure 1   Fabrication process of a high-strength multifilamentary Nb₃Sn wire with internal reinforcement of CuNbTi composite.

CuNbTi/Nb₃Sn COIL FABRICATED BY A R&W METHOD

In the R&W coil fabrication process, the pulleys with 300 mm in diameter and the coil bobbin with 260 mm in diameter, which correspond to bending strains below 0.4 %, were utilized. During the coil winding, the pre-reacted CuNbTi/Nb₃Sn wire was wrapped in polyimide insulation tape and the coil between windings was impregnated by epoxy resin. This proves that the R&W processed Nb₃Sn superconducting magnet no longer needs a coil heat-treatment furnace and a vacuum impregnation furnace for a reacted coil.

The R&W processed CuNbTi/Nb₃Sn coil, which has coil parameters such as 260 mm in inner diameter, 289 mm in outer diameter and 319 mm in height, was set into a background NbTi coil. The performance test was carried out as a cryogen-free superconducting magnet cooled conductively by GM-cryocoolers. Figure 2 shows the load line of the cryogen-free CuNbTi/Nb₃Sn superconducting magnet energized by using dual current supply system. The cryogen-free CuNbTi/Nb₃Sn insert coil quenched at a central field of 7.5 T at an operation current of 158 A in a 220 mm room temperature bore under the background field of 5.6 T. When the CuNbTi/Nb₃Sn coil was energized independently at currents up to 180 A as an individual test, the coil temperature rise due to ac loss was about 2 K. However, the temperature rise of the CuNbTi/Nb₃Sn coil was about 8 K due to the ac loss of the NbTi coil, when the outer NbTi coil was energized first. This means that there unfortunately exists a certain conduction cooling path with a poor connection in the CuNbTi/Nb₃Sn coil. The coil quench at 7.5 T may be related with Joule's heating at such a poor conduction cooling part in the background field of 5.6 T.

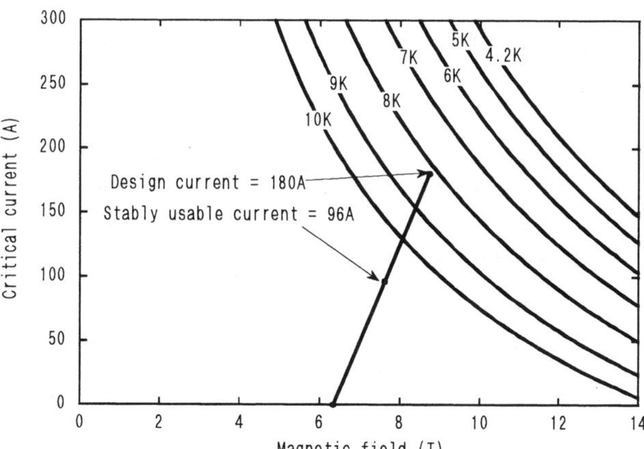

Figure 2   Temperature dependence of the critical current for CuNbTi/Nb₃Sn wire, and the load line of a CuNbTi/Nb₃Sn insert coil for the maximum field at the coil windings.

IN-FIELD HEAT-TREATMENT FURNACE AT 1500 ' C AND AT 7 T

Figure 3 shows the high temperature heat-treatment furnace combined with the cryogen-free CuNbTi/Nb$_3$Sn superconducting magnet.   The R&W processed CuNbTi/Nb$_3$Sn superconducting magnet is now being used for the in-field long-term heat-treatment at 7.0 T.   A cylindrical heater was used to reduce an electromagnetic force applied for a wide bore heater in high fields.   This furnace produces high temperatures up to 1500 ' C in a flowing oxygen gas in fields up to 7.0 T.   A heat-treated sample space with 40 mm in diameter and 50 mm in height is available within temperature homogeneity of 5 ' C.

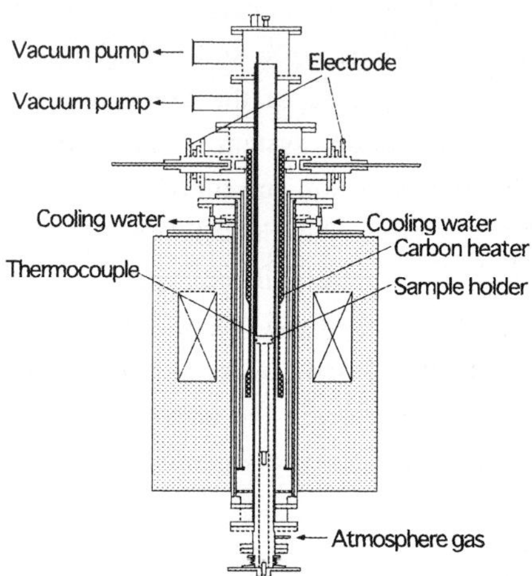

Figure 3   1500 ' C high temperature heat-treatment furnace equipped with a 220 mm wide bore cryogen-free CuNbTi/Nb$_3$Sn superconducting magnet.

CONCLUSIONS

A react-and-wind method to fabricate a wide bore cryogen-free superconducting magnet was intended employing pre-reacted high-strength Nb$_3$Sn wires reinforced with CuNbTi composite. The react-and-wind processed CuNbTi/Nb$_3$Sn coil revealed a good performance and generated the total central field of 7.5 T in a background field of 5.6 T in a 220 mm room temperature bore.

A high temperature heat-treatment furnace combined with the newly developed CuNbTi/Nb$_3$Sn superconducting magnet was successfully demonstrated at 1500 ' C in fields up to 7.0 T.

REFERENCES

1. Sakamoto, H., Higuchi, M., Endoh, S., Nagasu, Y., Kimura, A., Wada, K., Meguro, S. and Ikeda, M., IEEE Trans. Appl. Supercond. (2000) 10, 1008-1011.

2. Watanabe, K., Hoshi, A., Awaji, S., Katagiri, K., Noto, K., Goto, K., Saito, T. and Kohno., IEEE Trans. Appl Supercond. (1993) 3, 1006-1009.

3. Awaji, S., Watanabe, K., Nishijima, G., Katagiri, K., Miyoshi, K. and Meguro, S., Jpn. J. Appl. Phys. (2003) 42, L1142-L1144.

Design and Manufacture of a NbTi Insert Module Relevant for the ITER PF Magnets Conductor

A. della Corte, L. Affinito, S. Chiarelli, P. Gislon, G. Messina, L. Muzzi, G. Pasotti, S. Turtù, M. Mariani*, A. Matrone*, S. Rossi**

Superconducting Division, ENEA, Via E. Fermi 45, 00044 Frascati (Rome), Italy
*Ansaldo CRIS, Via Nuova delle Brecce 260, 80147 Napoli, Italy
**Europa Metalli Superconductor, Via della Repubblica 257, 55052 Fornaci di Barga, Italy

A NbTi insert coil has been manufactured to give a contribution, through its characterisation, to the discussion about the better layout of the full-size conductor for the Poloidal Field Coils for the ITER machine. The coil has been designed and constructed by ENEA and Ansaldo CRIS, with a conductor provided by Europa Metalli Superconductor and consisting of 36 Ni-coated strands twisted in a three-stage cable, jacketed and compacted in a SS round tube. A self consistent unit has been manufactured assembling the insert module and its mechanical supporting structure, to allow easy installation inside the background magnet of the ENEA test facility where the coil will be tested under representative conditions for the ITER PF magnets. The design and the description of the manufacturing process is presented in this paper, together with an overview of the experimental facility.

INTRODUCTION

A magnet wound with a sub-size NbTi conductor has been realized to study its characteristics in view of the realisation of the full-size conductor for the ITER Poloidal Field Coils.

For this reason ENEA has built up an *ad hoc* experiment, called ASTEX (Advanced Stability Experiment) whose experimental campaign shall start within 2004.

The conductor has been realized by Europa Metalli Superconductor (EMS) with 36 Ni-coated strands twisted in a three-stage cable, while ENEA and Ansaldo CRIS designed and manufactured the module.

The main goal of ASTEX is to study the influence of the current distribution on the conductor properties such as critical current, AC losses and stability. To reach the scope, during conductor cabling, one strand of a triplet has been marked in order to make the sub-stages recognizable at coil ends after winding. Then, when mounting the module in the experimental set-up, both of its terminations have been opened and subdivided into different groups of strands, to feed the magnet with a controlled non-uniformly distributed current by using a system of external resistors.

This paper illustrates the design and the manufacturing of the insert coil and gives also a description of the test facility layout.

CONDUCTOR

Strand

A cross section of the EMS NbTi strand is shown in Fig 1. It is a 0.81mm diameter strand, composed by 54 bundles with 1μm thickness individual barrier and a total number of 6534 filaments, 8mm twist pitch, 6μm diameter and a Cu/non-Cu ratio of 1.9. The strand is Ni coated with 1μm thickness.

A critical current of 398A at 4.2K, 6T background magnetic field, corresponding to a $2.2 \cdot 10^9 A/m^2$ current density and hysteretic losses of 163 kJ/m^3 at 4.2K on a ±3T cycle, have been measured at ENEA [1]. Coupling losses on the strands have been also measured, and results indicate a 6.8μm effective filament diameter and 6.7ms coupling time constant.

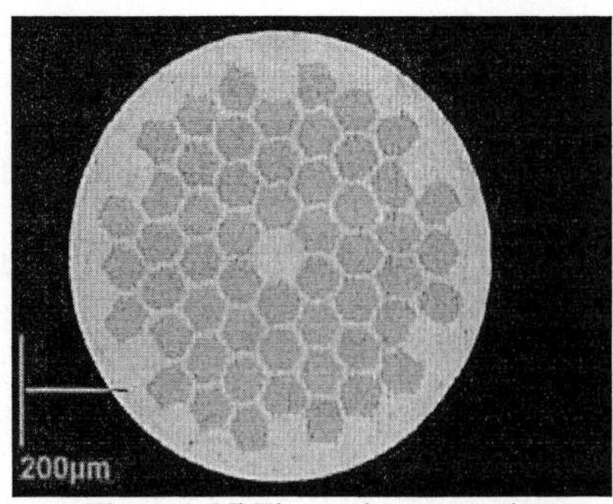

Figure 1 NbTi strand cross-section

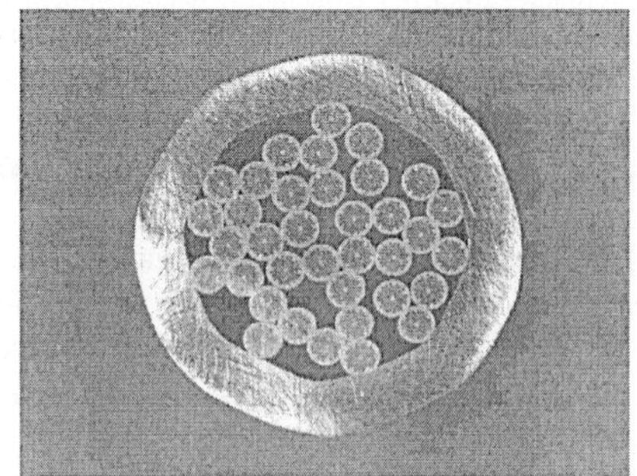

Figure 2 Conductor cross-section

<u>Cabling and Jacketing</u>
The cable layout is a three stages configuration, 36 strands, with the following twist pitches (TP):

stage	specified TP(mm)	measured TP(mm)
3	45±5	42±1
3x3	85±5	83±1
3x3x4	125±5	126±1

All stages have been cabled with a right hand pitch. One strand of one of the twelve triplet has been painted in order to make the sub-stages recognizable at both coil ends after winding. The final cable has been compacted to a circular cross section diameter of about 6.5mm (die diameter).

The jacketing has been carried out by pulling the cable through a straight tube, 120m length, supplied by the FINE TUBE Co UK in an unit coiled length. The jacket was an AISI 304 stainless steel tube 10x8mm. The cable was inserted into the tube by hand and compacted by Turk's head to an average outer diameter of 8.55mm (Fig.2). The void fraction obtained was about 36.1%. An overall 22.5ms coupling time constant of the conductor has been measured by Twente University [2].

MAGNET MANUFACTURING

In Table 1 the main coil and conductor characteristics are reported.

Table 1 Coil and conductor characteristics

Strand type	NbTi
Strand diameter	0.81mm
Cu/noCu ratio	1.9
Ic (6T,4.2K)	398A
RRR	100
Number of strands	36
Conductor diameter (not insulated)	8.55mm
Jacket thickness	1.0mm
Conductor insulation thickness	0.18mm (glass tape, half overlapped)
Layer insulation thickness	0.20mm (glass cloth)
Conductor length	106,60m
Coil axial length	230mm
Coil inner diameter	115mm
Coil outer diameter	249mm
Number of layers	8
Number of turn per layer	23
Total number of turns	184
Outer insulation thickness	4 layers, glass tape, 0.18 mm thick
Voltage taps	n°13 (AISI 304 strip with 0.05mm thickness and 5mm width)
Compensation coil	co-wound insulated copper wire ϕ=0.5mm

The conductor has been insulated by glass tape, 0.18mm thick, hand wrapped half overlapped and wound on a steel mandrel without any pre-bending tool (Fig.3). The interlayer insulation has been obtained by glass cloth with a nominal thickness of 0.20mm. Thirteen voltage taps made by AISI 304 strip (0.05x5mm) have been soft soldered on the conductor jacket and brought out of the winding (Fig. 4) while an insulated copper wire has been co-wound with the conductor to be used for compensating the inductive voltage during field variations.

The impregnation process has been carried out by full immersion of the coil in the epoxy resin under vacuum-pressure cycling. The winding has been previously coated by detaching material then inserted in a mould and filled with resin at 80°C under vacuum, for 24 hours, to guarantee its full penetration.

After that the whole system has been pressurized at 3bar, the temperature has been increased in a first step to 100°C to obtain the gelling of the epoxy and in the second one at 130°C to allow its solidification.

It took one week for the full impregnation cycle.

Figure 3 Module winding

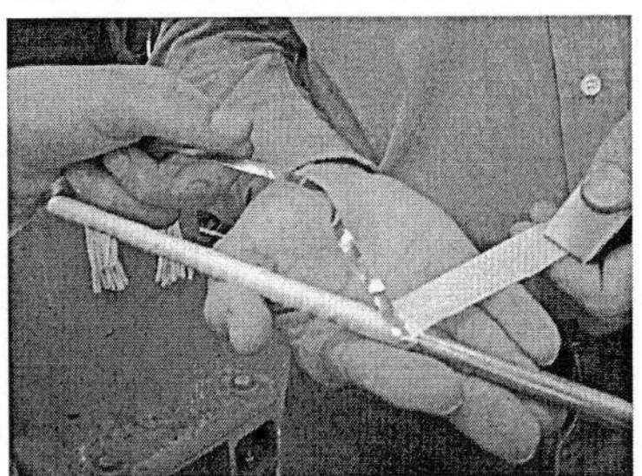

Figure 4 The voltge tap soldered on the jacket and insulation tape

TEST FACILITY

Termination and mechanical supporting structure

Two special boxes have been manufactured to host the opened terminations. The 36 strands have been divided in five groups to feed the magnet with a controlled non-uniformly distributed current: three with 9, one with 6 and the last with 3 strands (Fig.5, 6).

To allow easy installation inside the background magnet of the test facility a mechanical structure supporting the module has been manufactured. The magnet hangs to the top flange by four stainless steel bars and it is thermally insulated from irradiation by four copper screens (Fig.7). Two holes on the top flange host the two groups of five current leads each.

Figure 5 The five termination at one module end

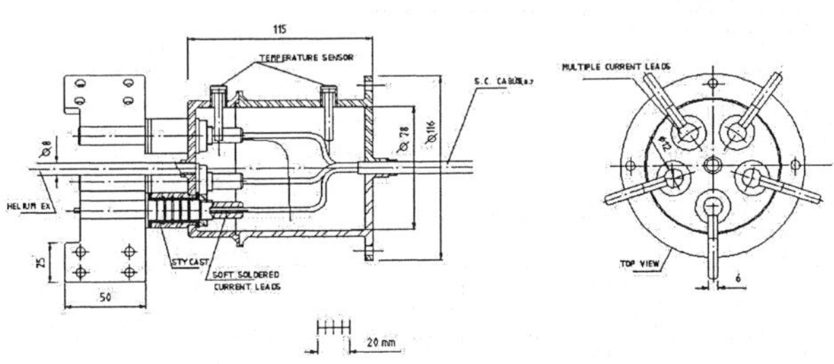

Figure 6 Schematic drawing of termination

Instrumentation

The module has been instrumented as follow:

- 14 fast-response Cernox-type thermometers have been glued on the conductor jacket, after removal of the insulation. Two additional thermometers of the same kind have been installed on each termination box, sensing the inlet and outlet temperature through cold fingers.
- 13 voltage taps have been soldered on the conductor jacket along its length during the winding process. At the 10 module terminations the superconducting strands/sub-bundles have been

brought out of the boxes and voltage taps have been directly soldered on them to measure the inter-bundle transverse resistances, as well as to follow the resistive voltage developing along each of the sub-bundles during DC tests.

- 2 concentric solenoidal pick-up coils have been installed for the magnetic determination of AC losses.
- 2 Micro-Motion Coriolis type flow-meters have been installed at the module inlet and outlet, allowing helium flow measurements over a wide range. The declared response time is 20ms.
- 2 pressure sensors have been mounted at the module inlet and outlet.
- 1 resistive heater will be used to vary and control the helium inlet temperature.

<u>Hydraulic and Electrical Set-up</u>

A Linde 500W@4.2K refrigerator is used to cool down the coil (P_{inlet}=10bar, T_{inlet}=5K) and the background magnet. Two flow-meters located at the module inlet and outlet can work with the helium flowing in both directions, so that counter-flow effects can be observed. Due to this feature, the hydraulic circuit has been designed in such a way that helium flow direction can be reversed in the testing magnet through a 4-valve system. In this way the module stability will be tested in different cooling conditions, in particular varying the quench initiation zone.

Two DC power converters, 6kA and 5kA, will be used to feed the module and the Nb_3Sn background magnet respectively reaching a peak field of about 6T. The 6kA is connected to the module ends by a panel (Fig. 8) manufactured to allow several different resistance configurations in order to obtain the desired unevenly distributed current among the conductor sub-bundles. Ten DCCTs have been mounted to measure the current flowing inside each of the inlet and outlet current leads, as one can see in Fig. 8.

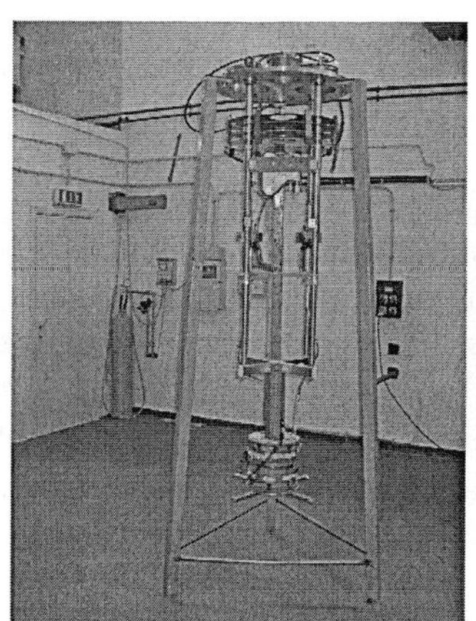
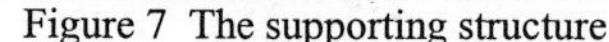

Figure 7 The supporting structure

Figure 8 The resistor panel and the 10 DCCTs

CONCLUSIONS

This NbTi module has been manufactured to qualify the strand for its use in the ITER Poloidal Field Coils also in case of uneven current distribution inside the conductor.

In fact, after having determined stability conditions with uniform transport current, the same runs will be repeated parametrically varying non-uniform current distribution inside the cable. The same kind of measurements will then be repeated after reversing the helium flow. This will cause the transition region to move along the conductor length, owing to the interplay between the self-field profile and the steep temperature profile along the conductor length.

The experimental campaign is foreseen to start in autumn 2004.

REFERENCES

1 P. Gislon, L. Muzzi, S. Chiarelli, A. Di Zenobio, M.V. Ricci, M. Spadoni <u>IEEE Transactions on Applied Superconductivity</u> (2003),vol 13 n°2, June, 1429
2 A. Nijuis, Y. Ilin, W. Abbas <u>Report N°UT-EFDA</u> (2000-1)

Multi-Tube Power Leads Tower for BEPCII IR Magnets

Jia L.X.[1], Zhang X.B.[2], Wang L.[2], Wang T.H.[2], Yao Z.L.[2]

[1]Brookhaven National Laboratory, Upton, New York 11973, USA
[2]Institute of Cryogenics and Superconductivity Technology, Harbin Institute of Technology, Harbin 150001, CHINA

A power lead tower containing the multi-tube power leads is designed and under fabrication for the superconducting IR quadrupole magnets in the Beijing Electron Position Collider Upgrade (BEPCII). The lead tower consists of six pairs of gas-cooled leads for seven superconducting coils at various operating currents. The power lead is designed in a modular fashion, which can be easily applied to suit different operating current. The end copper block of the tube lead has a large cold mass that provide a large time constant in case of cooling flow interruption. A novel cryogenic electrical isolator is used for the leads.

INTRODUCTION

The gas cooled single tube lead was first introduced to the superconducting magnet at the Brookhaven National Laboratory in early 1960s [1]. The gas cooled multi-tube lead was developed at BNL in 1994. Comparing with the single tube leads with a single flow passage, the multi-tube leads consisting of nesting tubes have the advantages of large wetted perimeter and then can carry more current flux with the same cross section area. A pair of multi-tube leads of 6300A for the Muon storage ring magnets and a pair

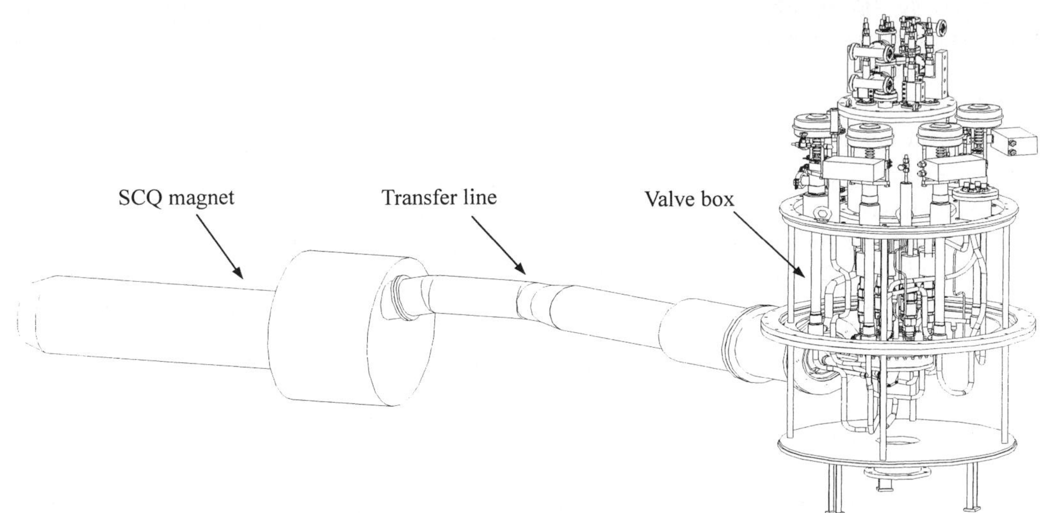

Figure 1   SCQ magnet, transfer line and distribution valve box

of 3300A multi-tube leads for Muon Inflector magnets were installed and operated successfully at BNL in 1994 [2]. The similar design was also applied to the EVA detector solenoid magnet at BNL in 1996 [3]. These magnets were all indirectly cooled by forced flow two-phase helium. For the BEPCII project, because of the similarity between BESIII detector magnet and the EVA magnet, the same design was applied. In the case of the BEPCII quadrupole magnet (SCQ) in the interaction region, which is cooled by supercritical helium, the multi-tube lead is modified. The SCQ magnet has seven coils requiring six pairs of leads. Table 1 gives the parameters of these coils and corresponding tube leads. These 6 pairs of leads are integrated into a lead tower. The superconducting cables run from the magnet end can through the transfer line and valve box to the lead tower (see Figure 1).

Table 1   Parameters of the coils and corresponding leads for the SCQ magnet in BEPCII

Magnet Circuit	SCB (HDC)	SQC	SKQ	VDC	AS1,AS2,AS3
Nominal Current (A)	550	550	65	65	1150,65,65
Design Current (A)	630	630	150	150	1600,150,150
Nominal Lead (#×A)	2×630	2×630	2×150	2×150	2×1600,2×150

Table 2   Dimensions of each copper tube for the BEPCII SCQ magnet leads
(OD: outer diameter; T: thickness; S: total area)

Leads	OD1 (mm)	T1 (mm)	OD2 (mm)	T2 (mm)	OD3 (mm)	T3 (mm)	OD4 (mm)	T4 (mm)	S (mm^2)
1600A	28.6	0.9	25.4	0.9	22.22	0.9	19.05	0.5	207.9
630A	19.05	0.65	15.88	0.65	12.7	0.7	9.52	0.5	95
150A	7.94	0.65	4.76	0.6					22.7

The multi-tube lead composes a set of concentric thin wall copper tubes. Copper tubes carry electrical current and the annular spaces between adjacent tubes provide helium flow channels. For the 1600A leads the four tubes are used, three are current carriers and the very inner one is used to only constitute the flow channel and filled with glass wool to dump the possible acoustic oscillation. The optimized wall thickness of copper tube and the annular space are configured according to the numerical computational simulation, which provides the guidance in selecting the commercial copper tubes [3]. Table 2 gives the dimensions of each copper tube for the BEPCII SCQ magnet leads.

MULTI-TUBE LEAD TOWER

The lead tower is constructed as a single unit that is installed at the center of the valve box (see Figure 2). The lead tower can be disassembled from the vacuum chamber of the valve box. On the top warm

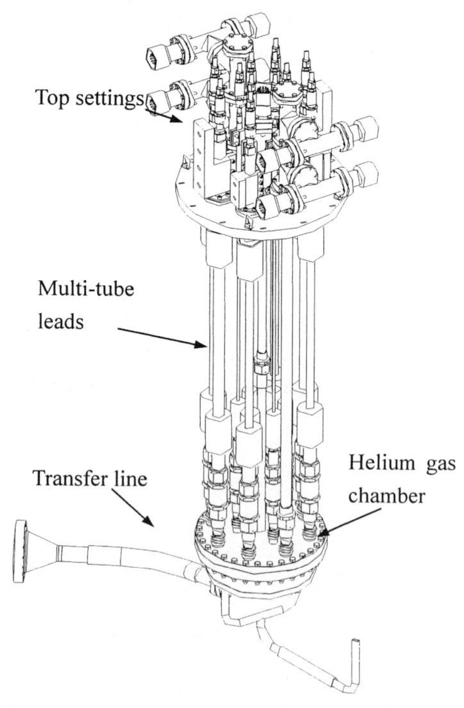

Figure 2   Lead tower of SCQ magnet

flange of the lead tower, the six pairs of leads are mounted with the G-10 electrical insulators. Its top setting includes the copper bus bars in different sizes and currents, the helium cooling tubes electrically insulated through the G-10 tube insulators to the refrigeration system, and the electrical feedthroughs for the voltage tap wires from each coil and bus work in the cable transition lines. At the cold end, each tube leads are connected to the helium gas chamber that provides the helium cooing fluid. Each tube lead is electrically insulated to the helium chamber through the G-10 electrical insulators. The solder joint of the superconducting cable and tube lead is built in the helium chamber. All the tube leads are designed in the same length and connected to the common helium chamber regardless their operating currents. To compensate the differences of possible thermal stresses in different tube sets, the flexible sections are used in each tube lead.

CONFIGURATION OF TUBE LEADS

To provide the electrical carrier and the cooling flow passage through the multi-tubes is not a simple task. At lead ends, both at the top warm side and the bottom cold side, each tube must have large enough weld surface with a copper bock to reduce the electrical resistance. The copper block services as a tube holder. The stack of the copper blocks must be arranged as to provide a large enough internal flow cross section area for each annular cooling flow. And these copper blocks must be welded as a single unit to hold the tube set. Therefore, the welding procedure becomes critical in fabrication of the leads. A copper insert with a long tail at the cold end is used for attaching the superconducting cable. The smallest tube made of stainless steel provides an annular space at its outside. There is no cooling flow inside of this stainless steel tube. Figure 3 shows the configuration of the 1600A power lead for the SCQ magnet.

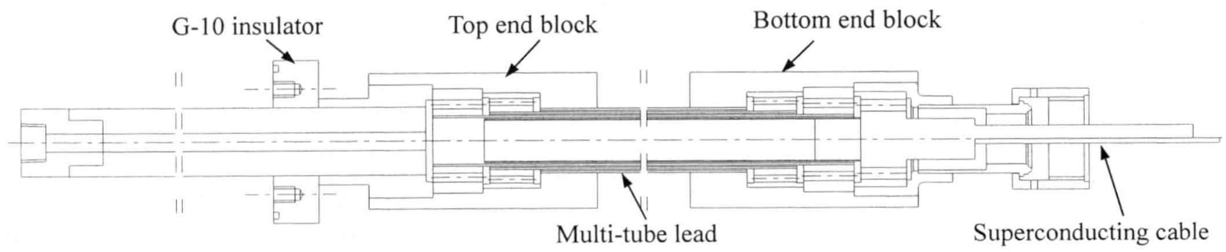

Figure 3   Configuration of 1600A power lead for SCQ magnet

The function of the bottom copper block is also to provide a large mass cold sink, which can stabilize the temperature in case of thermal disturbance. It can also delay the thermal run-away for a long enough time in case of cooling flow interruption. The whole copper block comprises four small cylindrical blocks which splice one by one and the most upper block envelopes the below ones with the hatch directing down. A group of holes is drilled on each small block, and a concaved room is provided to constitute the flow passages.

For electrical insulation to the cooling pipes connecting to the refrigeration system, the G-10 cryogenic tube insulators, instead of the ceramic ones, are used to prevent from crack failure due to thermal stress at low temperature. A cryogenic tube fitting is also used at the cold end for the convenience of reassembling the lead tower or replacing the lead set if a failure of a single lead occurs. Another feature of the design is to use a flexible section in each lead to compensate the differential stress among the tube sets.

The difficulties of fabricating the power lead tower are how to solder the copper block, especially the bottom block. Special attention should be pay to the soldering quality in the following aspects: 1) large contacting surface to reduce the electrical contact resistance; 2) the effective soldering strengths to withstand the thermal stress when being cooled down from the room temperature to the operating temperature; 3) good tightness for the outer soldering area to prevent helium leak.

CONCLUSION

The multi-tube power leads are used for the BEPCII superconducting magnets, including the IR quadrupole magnets and the detector solenoid magnet. The lead tower for quadrupole magnets consists of 6 pair of leads made in multi-tube sets. The leads was designed and under constructed at the Institute of Cryogenics and Superconductive Technology of HIT.

REFERENCES

1.   Smits, R.G. et al, Gas Cooled Electrical Leads for Use on Forced Cooled Superconducting Magnets, <u>Advances in Cryogenic Engineering</u> (1981), <u>27</u> 169

2.   Jia, L.X., Addessi, L.J. et al, Design Parameters for Gas-cooled Electrical Leads of the g-2 Magnets, <u>Cryogenics</u> (1994) <u>34</u> 631-634

3.   Zhang, X.B., Wang, L., Jia, L.X., Numerical analyses on transient thermal processes of gas-cooled current leads in BEPC II, to be published in <u>Advances in Cryogenic Engineering</u> (2004), <u>49</u>

THE MICE FOCUSING SOLENOIDS AND THEIR COOLING SYSTEM

M. A. Green, G. Barr, W. Lau, R. S. Senanayake, and S. Q. Yang

University of Oxford Department of Physics, Oxford OX1 3RH, UK

This report describes the focusing solenoid for the proposed Muon Ionization Cooling Experiment (MICE) [1]. The focusing solenoid consists of a pair of superconducting solenoids that are on a common bobbin. The two coils, which have separate leads, may be operated in the same polarity or at opposite polarity. This report discusses the superconducting magnet design and the cryostat design for the MICE focusing module. Also discussed is how this superconducting magnet can be integrated with a pair of small 4.2 K coolers.

INTRODUCTION

MICE consists of a muon cooling channel and two detectors [1]. Three types of modules will be used in MICE. These are the focusing and absorber module, the coupling coil and RF module, and the detector module [2]. The focusing module for MICE consists of a pair of superconducting solenoids that are around a liquid hydrogen absorber, which provides ionization cooling for the muons within the cooling channel [3]. The warm bore tube and one of the focusing solenoid cryostat vacuum vessel form the vacuum vessel that this liquid hydrogen absorber.

The nominal design configuration for MICE calls for the focusing solenoids to be operated in the gradient mode, with two coils each carrying 2.01 MA while operating at opposite polarities. When the MICE focus coils are operated in the gradient mode, a large inter-coil force is developed in the direction to push the two coils apart along the magnet axis. This force must be carried by a cold aluminum structure between the two coils. There are configurations of the experiment that call for the two focusing coils to operate at the same polarity. When the magnet operates in the solenoid mode, the inter-coil force pushes the two coils together along the magnet axis. This report presents a number of design parameters for the MICE focusing solenoid system and its cryogenic cooling system.

The high cost of a central refrigeration system suggests that the MICE magnets can economically cooled using small 4.2 K coolers. As a result, the magnets for MICE have been designed so that they can be cooled using one or more small 4.2 coolers. The attachment of these coolers to the magnet can have a strong negative effect on the magnet temperature at the high field point in the magnet, if this attachment is not done correctly. A section of this report will deal with the attachment of the coolers to the magnet.

DESIGN PARAMETERS FOR THE MICE FOCUSING MAGNETS

Figure 1 shows a cross-section of the MICE focusing magnet in its warm bore cryostat. The liquid hydrogen absorber fits into the warm bore of the magnet with its liquid hydrogen and liquid cryogen cool down pipes passing around the end of the magnet. The magnet coil and the liquid hydrogen absorber will be cooled with small coolers. Since the cool down time of the system is of importance, the cool down of both the magnet and the hydrogen absorber will be done using liquid nitrogen and liquid helium.

Table 1, presents the basic physical and electrical parameters for the MICE focusing magnet during its normal operating mode (when the muon average momentum = 200 MeV/c). When the two coils are powered at opposite polarity, a 2350 kN force that is pushing the coils apart is created in the longitudinal direction [4]. When the coils operate at the same polarity, the force is about 1000 kN pushing the coils together. The cold mass support peak force is about 200 kN, for any of the MICE operating modes.

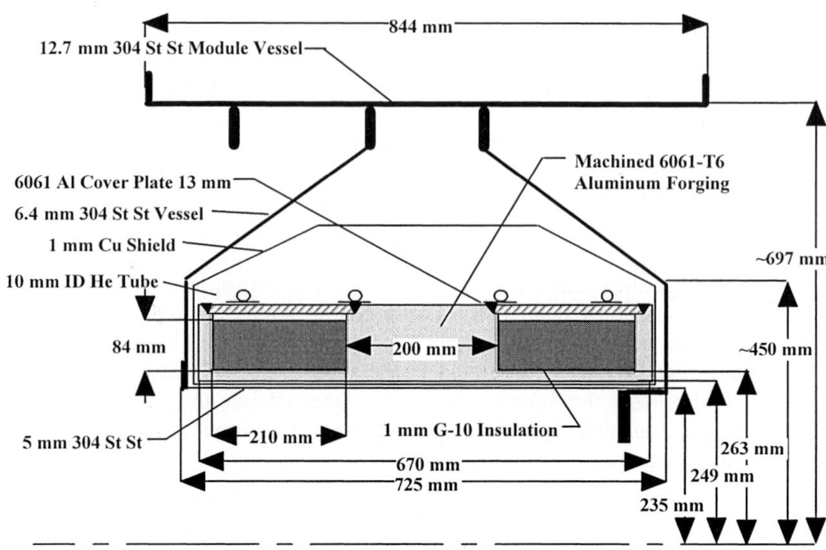

Figure 1. A cross-section of the focusing magnet, showing the key dimension of the coil and cryostat

Table 1. The basic design parameters for the MICE focusing magnet

Magnet Physical Parameters		Magnet Electrical Parameters	
Coil Inner Radius (mm)	263	Magnet Self Inductance (H)	~138
Coil Length (mm)	210	Magnet Design Current* (A)	208.3
Coil Thickness (mm)	84	Matrix Current Density J* (A m^{-2})	156.6
Distance between the Coils (mm)	200	Magnet Stored Energy* (MJ)	~3.0
Number of Layers per Coil	76	EJ^2 Limit for Protection* (J A^2 m^{-4})	7.3x10^{22}
Number of Turns per Coil Layer	127	Peak Induction in the Winding* (T)	~6.39

* The magnet design current is for the solenoid in the focus mode a β = 420 mm and p = 200 MeV/c

Figure 2 shows the load line and the temperature margin for the superconductor chosen for focus coils. The design points represent the standard operating case and a case the takes the conductor nearly to its critical current. Figure 3 shows the temperature distribution in the magnet when it is cooled by a small cooler at one place on the focus coil package [4]. Large temperature differences can result from this.

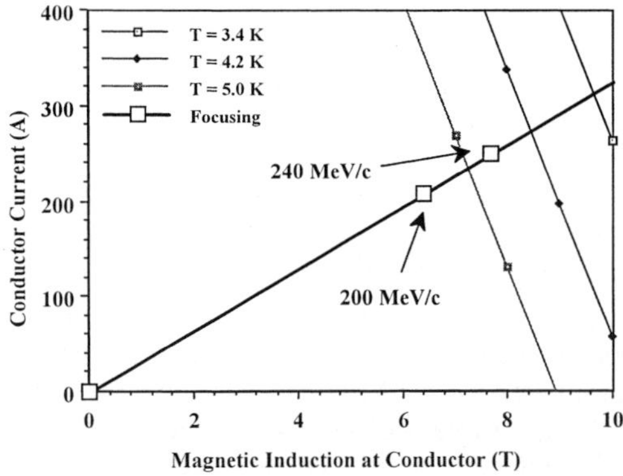

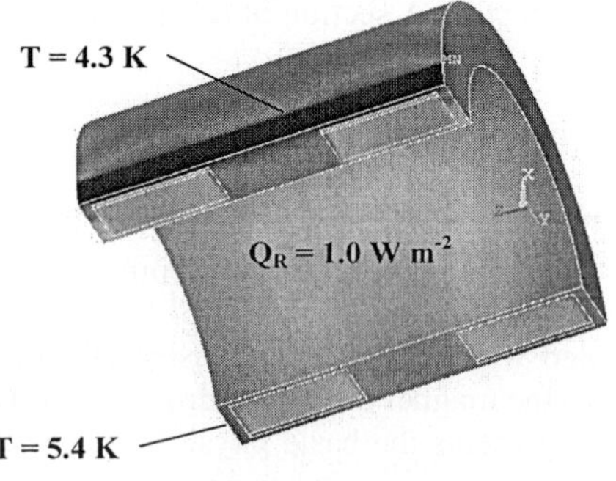

Figure 2. Magnet load line and the Ic of the magnet superconductor (The magnet design points are shown.)

Figure 3. The temperature distribution in the MICE focus coil package when cooled in one place by a cooler

The focusing solenoid Nb-Ti conductor has the following properties; 1) The insulated dimensions are 1.00 mm by 1.65 mm with round ends. 2) The copper to superconductor ratio is four. 3) The copper has a minimum RRR = 75. 4) The twist pitch is 12.7 mm. 4) There are 55 filaments that are 78 μm in diameter. 6) The conductor critical current is 760 A at 5.0 T and 4.2 K (J_c(5.0T, 4.2K) = 2940 A mm^{-2}). The proposed conductor was designed for use in MRI magnets that operate in persistent mode. Since the MICE magnet is a DC magnet with a long charge time AC losses in the conductor are not an issue.

The MICE focusing solenoid has two pairs of 300 A current leads. This allows one to operate the magnet in either the solenoid mode (both coils at the same polarity) or the gradient mode (both coils at opposite polarity). The highest current that the magnet is designed for is about 250 A when the magnet is pushed for the high momentum cooling tests. There may be other experimental modes that require that both coils be powered separately. As the experiment is better defined, the lead current may be reduced.

COOLING THE FOCUSING MAGNET WITH SMALL COOLERS

The MICE focusing solenoid is designed to be cooled using two 1.2 to 1.5 W (at 4.2 K) two-stage coolers. The 1st stage of one of these coolers will provide 30 to 40 W of cooling at 50 K, when the compressor runs on a 50 Hz motor. The 1st stage cooling goes up 20 to 25 percent when the compressor is run on a 60 Hz motor. The 1st stage absorbs the heat leak down the cold mass supports [5] and the radiation heat loads to the shield. The largest load on the first stage of the cooler is the heat down the conduction-cooled current leads. The current lead heat load is about 70 W per kA lead pair at 50 K.

In order to cool the focusing magnet with small coolers, the magnet cryostat must have heat leaks that are less than 1 watt at 4 K. The 2nd stage heat load is primarily from cold mass supports [5], thermal radiation, pipes, and the leads. The current leads between the 50 K intercept and the 4.2 K region must be made from HTS material. Even with HTS lead, the lead heat leak into the 4 K region will be the largest single load at 4 K. The heat loads into the second stage of the cooler are a function of the first stage temperature. If one operates the focusing solenoid on one cooler the first stage temperature would be greater than 60 K and the second stage temperature will be from 4.6 to 4.8 K. By using two coolers, one can reduce the first stage temperature to 40 K and reduce the second stage temperature to 3.7 to 3.9 K.

Figure 3 illustrates the effect of heat leaks into the cold mass on the temperature distribution within the magnet [4]. The figure shows that a radiation heat load of 1.0 W m^{-2} will result in a ΔT of 1.1 K within the magnet. This ΔT is added to any ΔT that occurs between the magnet cold mass and the cooler second stage cold head. For example, if one connects the magnet to the 2nd stage cold head with a copper strap (RRR = 100) and one wants a ΔT of 0.1 K down the copper strap, a 0.15 m long strap must have a cross-section area of about 0.0025 m^2. The ΔT along strap is proportional to its length and inversely proportional its cross-section area [6]. A magnet can be cooled down through the strap.

It is desirable that the cooling from the second stage cold head be spread around the circumference of the focusing magnet and it is desirable to locate the cooler is a position on the magnet cryostat where cooler maintenance can occur without taking the whole magnet apart. Figure 4 shows a method for distributing the cooling using a thermal siphon heat pipe between the cold head and the load.

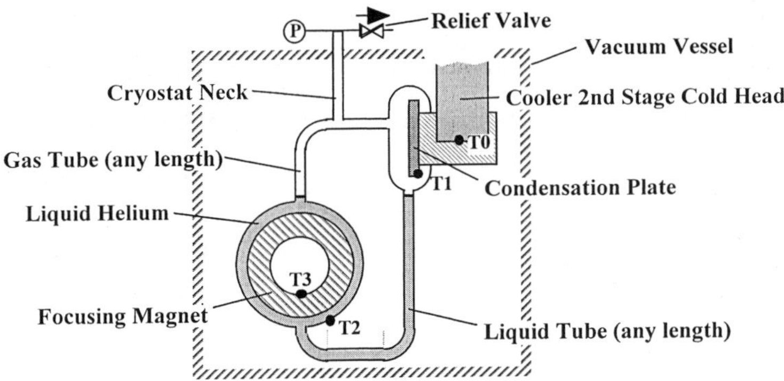

Figure 4. A schematic of a gravity heat pipe for distributing 4 K cooling from the cooler 2nd stage cold head to the magnet (Note: the liquid helium can be in channels around the cold mass. The channels must be attached to the focus coil spacer.)

588

The thermal siphon heat pipe has the following advantages over a copper strap for distributing cold to the focusing magnet coils [7]: 1) Liquid helium can be all around the coil so that the ΔT within the magnet is minimized. 2) If the heat pipe is correctly designed, the ΔT between the 2nd stage cold head and the magnet can be as low as 0.05 K. 3) The heat pipe filters out the 0.25 K temperature oscillations in the cooler 2nd stage cold head. 4) The coolers can be located some distance from the coil package. As a result the cooler position can be optimized. 5) The space occupied by the helium pipes is small compared to a copper strap with less desirable performance. 6) A heat pipe provides good vibration isolation of the magnet from the cooler cold head. 7) The coolers can be connected separately, each with its own condenser. This permits one to keep the magnet cold (at 5 K) on a single cooler while the second cooler is turned off. 8) Liquid helium in the magnet permits the magnet to recover from a short power outage. The disadvantage of using a heat pipe to connect the cooler to the magnet is the magnet must be cooled down to 4 K using liquid cryogens. Because the focusing magnet cold mass is 650 kg, liquid cryogen cooling will be used to cool the focus coils anyway. The cooler first stage can be used to cool down the 50 K shields and intercepts.

CONCLUDING COMMENTS

The design of the MICE focusing magnet is well under way. The magnet is designed to withstand the magnetic stress within the coil package. The temperature distribution within the focusing magnet cold mass has been studied. If one cools the spacer between the coils, the ΔT between the cooled point on the spacer and the hot spot on the aluminum support structure will be less than 0.1 K. A magnet cold mass support system that carries 200 kN in the longitudinal direction has been designed so that its heat leak at 4.2 K is less than 0.1 W [5].

With proper design and execution of that design, the focusing magnet can be kept cold using a pair of small 4.2 K coolers (rated at 1.2 to 1.5 W at 4.2 K). The focusing magnet must be cooled down using liquid cryogens. With proper design, the ΔT between the cooler second stage cold head and the hottest point in the magnet can be less than 0.2 K [7].

ACKNOWLEDGEMENT

This work was supported by the Oxford University Physics Department and the Particle Physics and Astronomy Research Council of the United Kingdom.

REFERENCES

1. "A Proposal to the Rutherford Appleton Laboratory, an International Muon Ionization Cooling Experiment (MICE)," proposed by the MICE Collaboration, 15 December 2002
2. Green, M. A. and Rey, J. M., "Superconducting Solenoids for an International Muon Cooling Experiment," IEEE Transactions on Applied Superconductivity 13, No. 2, p 1373, (2003).
3. Green, M. A., Baynham, E., Barr, G., Lau, W., Rochford, J. H., and Yang, S., "Focusing Solenoids for the MICE Cooling Channel," Advances in Cryogenic Engineering 49, AIP Press, New York (2003)
4. Green, M. A. and Yang S. Q. "Heat Transfer into and within the 4.4 K Region and the 40 K Shields of the MICE Focusing and Coupling Magnets" an Oxford University Report for the MICE collaboration (2004)
5. Green, M. A. and Senanayake, R. S., "The Cold Mass Support System for the MICE Focusing and Coupling Magnets," an Oxford University Report for the MICE collaboration (2004)
6. Green, M. A., "The Integration of Liquid Cryogen Cooling and Cryocoolers with Superconducting Electronics Systems," Superconducting Science and Technology, Volume 16, No. 12, p 1349, (Dec. 2003)
7. Green, M. A., Dietderich, D. R., Marks, S., Prestemon, S. O., and Schlueter, R. D., "Design Issues for Cryogenic Cooling of Short Period Superconducting Undulators," Advances in Cryogenic Engineering 49, AIP Press, New York (2003)

Evaluation of the ITER Cable In Conduit Conductor heat transfer

Nicollet S., Ciazynski, D., Duchateau J.L., Lacroix, B., Renard, B.

Association EURATOM-CEA, CEA/DSM/DRFC, CEA-Cadarache, F-13108 Saint Paul-lez-Durance,
France

Convective heat transfer correlations in dual channel Cable In Conduit Conductor
(CICC) are presented as functions of friction factor, and based on the Reynolds-
Colburn analogy using the Stanton number. The developed thermohydraulic
model determines helium temperatures in both channels, with the real geometrical
spiral perforation. It is applied with pertinence to the Poloidal Field Full Size Joint
Sample (PF-FSJS) and shows good agreement between the experimental
measurements and the calculated temperatures, characterised by a 0.43 m space
constant. The heat load applied in the bundle region induces density imbalance;
the gravity effect in this vertical sample is evaluated and discussed.

INTRODUCTION

The ITER Cable In Conduit Conductors, cooled by forced flow supercritical helium, are characterised by
a dual channel : the bundle region where the superconducting strands are located and the central hole
region delimited by the central spiral. During normal operation of the coils or transient safety discharge,
heat loads are induced in the strands (AC losses) or transferred from the stainless steel case and plates
(nuclear heating, radiation and eddy currents) to the conductor. An overheating of the bundle region can
decrease the temperature margin, and the heat transfer from bundle to central hole is then a key factor.

HEAT TRANSFER COEFFICIENT BETWEEN BUNDLE AND CENTRAL HOLE REGIONS

The main purpose is to determine the convective heat transfer between the bundle and central hole regions.
Since little experimental data are available, the idea is to use the Reynolds-Colburn analogy (1) between
fluid friction and heat transfer [1], associated with the friction factor data base. Expressed with the Stanton
number St (2), the friction factor f_{EU} (3) and indirectly the Nusselt number Nu (4), this analogy is valid for
fully developed turbulent flow in central hole region (Re_h near 10^5) as well as for laminar flow in bundle
region ($Re_b < 5000$), where Re (5) is the Reynolds number. The well known Colburn equation (6) for the
smooth tube is a particular case of these analogy. Combining equation (1), (3) and (4), leads to equation
(7), expressing the convective heat exchange coefficient h_{conv} (W/m²K) from the experimental bundle
region friction factor $f_{EU,b}$ on one hand [2] the central spiral $f_{EU,h}$ on the other hand [3], [4]. Superimposed
on forced-convection, natural-convection effect influences the laminar heat-transfer coefficient [1]: this
point is further developed in the gravity effect section.
The global heat exchange coefficient between the two region h_{perfor} (8) can then be expressed in
function of the spiral perforation ratio (perfor) of the gap length over the twist pitch length, and the heat
exchange coefficients h_{open} (9) and h_{close} (10) corresponding to the open and closed zone of the spiral
respectively. In the latter, the conduction through the spiral is taken into account with the ratio of the
stainless steel conductivity λ_{SS} (W/mK) over the spiral thickness e (m).

$$St . \mathrm{Pr}^{2/3} = f_{EU} / 8 \qquad (1)$$

$$St = Nu / \mathrm{Re} . \mathrm{Pr} \qquad (2)$$

$$\Delta P_f = (f_{EU}.m^2.U.L)/(8.\rho.A^3)\tag{3}$$

$$Nu = h_{conv}.Dh/\lambda\tag{4}$$

$$\mathrm{Re} = \rho \cdot v \cdot Dh/\mu = 4 \cdot m/\mu \cdot U\tag{5}$$

$$Nu = 0.023.\mathrm{Re}^{0.8}.\mathrm{Pr}^{1/3}\tag{6}$$

$$h_{conv} = (f_{EU}.\lambda.\mathrm{Re}.\mathrm{Pr}^{1/3})/(8.Dh)\tag{7}$$

$$h_{perfor} = h_{open}.perfor + h_{close}.(1 - perfor)\tag{8}$$

$$1/h_{open} = 1/h_{conv,b} + 1/h_{conv,h}\tag{9}$$

$$1/h_{close} = 1/h_{open} + e/\lambda_{SS}\tag{10}$$

Where : Pr (-): Prandtl number; ΔP (MPa): pressure drop; m (kg/s): mass flow, ρ (kg/m^3): helium density; U (m): wetted perimeter; L (m): Length; A (m²): cross section; Dh (m): hydraulic diameter of the channel; λ (W/mK): helium thermal conductivity; μ (Pa.s): helium dynamic viscosity.

A STEADY STATE CHARACTERSITIC SPACE CONSTANT MODEL

Using a steady state model of heat transfer [5], and under a few simplifications (helium specific heat independent of the temperature), the energy balance (11) is expressed in both channels of the CICC: the bundle and central hole regions, with mass flows m_b and m_h and temperatures T_b and T_h respectively. That leads to the differential equation (12) and introducing the characteristic space constant Λ (13), the temperature gradient (T_b-T_h) and the bundle temperature are determined analytically. In the heated region ($0<x<L_h$), equation (14) and (15) are valid, with the maximal temperature gradient at the end of the heated region (16). In the non-heated zone (Q_x=0 W/m for $L_h<x<L_h+L_{nh}$), the temperature gradient (17) and bundle region temperature (18) are also expressed.

$$m_b.cp.dT_b = \left(Q_x - hU_{perfor}.(T_b - T_h)\right)dx \quad\text{and}\quad m_h.cp.dT_h = hU_{perfor}.(T_b - T_h).dx\tag{11a) and (11b}$$

$$d(T_b - T_h) = \left(\frac{Q_x}{m_b.cp} - \frac{hU_{perfor}.(T_b - T_h).(m_b + m_h)}{m_b.m_h.cp}\right).dx = (\beta - \gamma.(T_b - T_h)).dx\tag{12}$$

$$\Lambda = \frac{1}{\gamma} = \frac{\alpha.(1-\alpha).m_{tot}.cp}{hU_{perfor}}\tag{13}$$

$$(T_b - T_h) = (Q_x.\Lambda/m_b.cp).(1 - e^{\frac{-x}{\Lambda}})\tag{14}$$

$$T_b = T_0 + \frac{Q_x.x}{m_{tot}.cp} + \frac{\Lambda.Q_x}{cp}.\frac{m_h}{m_b.m_{tot}}.(1 - e^{\frac{-x}{\Lambda}})\tag{15}$$

$$\Delta T_{max} = (Q_x.\Lambda/m_b.cp).(1 - e^{\frac{-L_h}{\Lambda}})\tag{16}$$

$$(T_b - T_h) = \Delta T_{max}.e^{-\frac{(x-L_h)}{\Lambda}}\tag{17}$$

$$T_b = T_0 + \frac{Q_x.L_h}{m_{tot}.cp} + \frac{\Lambda.Q_x}{cp}.\frac{m_h}{m_b.m_{tot}}.(1 - e^{\frac{-L_h}{\Lambda}}).e^{\frac{-(x-L_h)}{\Lambda}}\tag{18}$$

Where : x (m): abscissa along the conductor axis (x=0 at the beginning of the heated zone); L_h and L_{nh}(m): length of the heated and non-heated zone, respectively; cp (J/kgK): helium specific heat at constant pressure; T_0 (K): inlet temperature (x=0); Q_x (W/m): linear heat power deposition in the bundle region; hU_{perfor} (W/mK): heat transfer coefficient per length at the perimeter U_{perfor} (associated with diameter Dh_h-e) , ΔT_{max} (K): maximal temperature gradient at the end of the heated zone (x=L_h), α (-): ratio (equation (15) of [4]) of bundle mass flow m_b to total mass flow rate m_{tot} (m_{tot}=m_h+m_b).

APPLICATION TO THE PF-FSJS

The PF-FSJS prototype using NbTi conductors [6] (with two legs: left (L) and right (R)) was tested at the Sultan Facility. On each leg, were installed one heater (length L_h=0.4 m) at the sample top feeder entrance (x=0 at the heated zone beginning) and four sensors : T2 (x=0.747 m), T3 (x=1.017 m), T4 (x=1.564 m) measuring the bundle region temperature and T5 (x=2.190 m) measuring the mixing temperature T_{inf}, at the end of the joint. Figure 1 presents the temperature measurements for the calibration tests with a power Q of 12 W (Q_x=30 W/m) and a mass flow rate m_{tot} of 8 g/s in each leg (inlet pressure P_{in}=1.02 MPa and temperature T_0=4.5 K). The calculation performed with the Gandalf code (with the associated subroutine concerning friction factor, heat exchange coefficient, heat loads) are also presented. The characteristic space constant (13) Λ=0.437 m is calculated at the average temperature $T_{ave}=T_0+(Q/m_{tot}.cp_0)$=4.96 K ($cp_0$=3270 J/kgK), with a heat transfer coefficient per unit length hU_{perfor}=13.9 W/mK (α=0.297). The bundle and central hole temperatures are deduced from analytical relation (14), (15), (17) and (18). The good agreement between experimental measurements and calculation confirms the pertinence of the heat transfer model developed with the characteristic space constant.

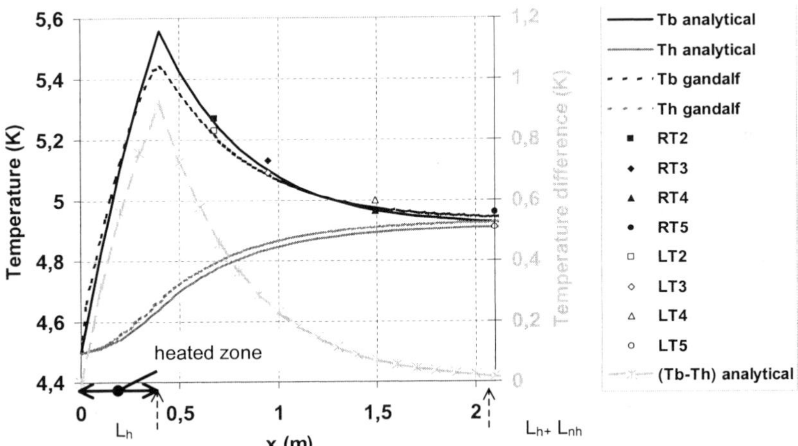

Figure 1 : PF-FSJS calculated and measured bundle and central hole temperatures (Q_x=30W/m, m_{tot}=8g/s)

DISCUSSION AND EVALUATION OF GRAVITY EFFECT

When heating the PF-FSJS vertical sample with AC losses, some abnormal increases of the upstream temperature (T2) was observed. This was interpreted [7] as a signature of a blocked (zero) or reverse (thermosiphon) flow in the bundle region of the heated zone, which is schematised as in Figure 2. The pressure difference ΔP_{tot} equilibrium for both channels is expressed (19) over the corresponding hydraulic length (L=min (L_h,Λ)) as a function of the gravity g (9.81 m²/s) with θ angle between the CICC axis and horizontal direction; for the bundle region, $\Delta P_{f,b}$ (3) should be taken with a minus sign for direct (regular) flow pattern and a plus sign for reverse flow. The resolution of equation (19) is complex; a simple ratio r (20) can nevertheless be introduced to evaluate the importance of the gravity effect compared to the friction factor. A global bundle mass flow ratio α is determined as well to verify (19) with the assumption, that there is no mass flow redistribution over the hydraulic length. Table 1 sums up the experimental observations on the PF-FSJS (choke when overheating of T2) and the corresponding calculated values of α and r in function of Q and m_{tot}; the heat transfer coefficient hU_{perfor} (near 10 W/mK at m_{tot}=4 g/s and 13 W/mK at m_{tot}=8 g/s) and the characteristic space constant Λ (0.24 m at m_{tot}=4 g/s and 0.43 m at m_{tot}=8 g/s) are calculated with correlation (8) and (13) of first and second section respectively.

$$\Delta P_{tot} = P_{out} - P_{in} = \rho(T_h).g.x_{out}.\sin\theta - \Delta P_{f,h} = \rho(T_b).g.x_{out}.\sin\theta -/+\Delta P_{f,b} \qquad (19)$$

$$r = (\rho(T_h) - \rho(T_b)).g.\sin\theta.x_{out}/\Delta P_{f,h} \qquad (20)$$

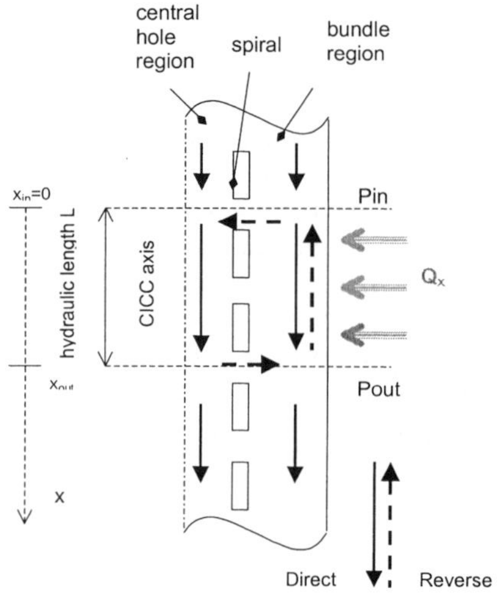

Table 1 : Influence of gravity effect on the PF-FSJS

experimental observation		$m_{tot}=4g/s$ $10<hU_{perfor}<10.9$ W/mK $0.233<\Lambda<0.244$ m	$m_{tot}=8g/s$ $12.9<hU_{perfor}<14$ W/mK $0.421<\Lambda<0.437$ m
direct	no choke	Q=0 W, $\alpha=0.239$, r=0	Q=0 W, $\alpha=0.297$, r=0
flow	no choke	Q=1 W, $\alpha=0.194$, r=0.32	Q=6 W, $\alpha=0.254$, r=0.28
pattern	no choke	Q=2 W, $\alpha=0.106$, r=0.74	Q=12 W, $\alpha=0.171$, r=0.65
reverse	choke	Q=2.2 W, $\alpha=0.079$, r=0.83	

Figure 2 : Direct or reverse CICC flow pattern

For the ITER Project and the vertical part of the Toroidal Field Coils, such a phenomenon should be estimated, especially during the 400 s power deposition (mainly nuclear heating) of plasma operation [8]. Considering the 18 coils with each 14 CICC turns on the Inner Leg Plasma Facing Wall first layer (with length near 8m), the maximal CICC power is $Q_x=2W/m$. With the interpolated correlation for the new spiral friction factor [9] ($Dh_h=9mm$, perfor=0.5 and $f_{EU,h}=0.45.Re_h^{-0.034}$), $m_{tot}=8$ g/s, Pin=5 MPa, and Tin=5 K, the bundle mass flow ratio, heat transfer coefficient and space constant are respectively $\alpha=0.656$, $hU_{perfor}=20.2$ W/mK and $\Lambda=0.41$ m. The resulting ratio r=0.023 allows to conclude that gravity effect is negligible compared to friction, and since α remains largely positive, no local overheating and reverse flow should occur in the bundle region.

CONCLUSION

New correlations for the convective heat exchange coefficient in CICC were applied with pertinence to express the global heat transfer. Depending on this last parameter and on the bundle mass flow ratio, a characteristic space constant was introduced in a steady state model, and analytical expressions of the bundle and central hole temperatures as well as the maximal temperature gradient were deduced. The gravity influence on the mass flow distribution was estimated and compared with experimental results for the PF-FSJS, especially in the case of a reverse-thermosiphon phenomenon at low mass flow. On the ITER TF Coils however, this phenomenon is expected to only slightly reduce the bundle mass flow.

REFERENCES

1. Holmann, J.P., Heat Transfer, ninth Edition, International Edition, Mc GrawHill, pp.230 and pp.272.
2. Nicollet S., Cloez H., Duchateau J.L., Serries J.P., Hydraulics of the ITER Toroidal Field Model Coil Cable-in-Conduit Condutors, in proceedings of the 20th Symposium on Fusion Technology, Marseille, France (1998) 771-774.
3. Nicollet S., Duchateau J.L., Fillunger H., Martinez A., Parodi S., Dual Channel Cable in Conduit Thermohydraulics : Influence of some Design Parameters, in IEEE Transactions on Applied Superconductivity (2000) 10 1102-1105.
4. Nicollet S., Duchateau, J.L., Fillunger, H., Martinez, A., Calculations of pressure drop and mass flow distribution in the toroidal field model coil of the ITER project, Cryogenics (2000) 40 569-575.
5. Park, S.H. Duchateau, J.L., The effect of Perforation between Hole and Bundle in CICC, Internal Note AIM/NTT-2003.035.
6. Decool, P., Design and manufacture of a prototype NbTi full-size joint sample for the ITER poloidal field coils, in proceedings of the 22nd Symposium on Fusion Technology, Helsinki, Finland (2002) 1165-1169
7. Ciazynski, D., Test of PF-FSJS in Sultan, Themohydraulics and Calibration, CRPP Workshop, Gstaad, January 2003.
8. Shatil, N., Thermo-hydraulic Analysis of TF Magnets, ITER FDR DDD, 11 Magnet, 2.1 CDA, Annex 10. Table 7 and 8
9. Nicollet, S., Cloez, H. Duchateau, J.L. Serries, J.P., Task CODES : Results of ITER type central spirals friction factor measurements in the OTHELLO facility and application to ITER Coils, Internal Note AIM/NTT-2003.018.

Proceedings of the Twentieth International Cryogenic Engineering Conference
(ICEC 20), Beijing, China. © 2005 Elsevier Ltd. All rights reserved.

The BESIII detector magnet

Zhu Z.[a], Zhao L.[a], Wang L.[a], Hou Z.[a], Huang S.[a], Yang H.[a], Hu J.[a], Zhou J.[a], Han S.[a], Yi C.[a], Chen H.[a],
Xu Q.[b,e], Liu L.[b], Makida Y.[c], Yamaoka H.[c], Tsuchiya K.[c], Wang B.[d], Wahrer B.[d], Taylor C.[d], Chen C.[d]

[a] Institute of High Energy Physics(IHEP), Chinese Academy of Sciences, Beijing 100039, China
[b] Technical Institute of Physics and Chemistry, Chinese Academy of Sciences, Beijing 100080, China
[c] High Energy Accelerator Research Organization, KEK, 1-1 Oho, Tsukuba, Ibaraki,305-0801,Japan
[d] Wang NMR Inc., 550 North Canyons Parkway, Livermore, CA 94551, USA
[e] Graduate School of the Chinese Academy of Sciences, Beijing 100039, China

BESIII (Beijing Spectrometer III) is a detector designed to run in the autumn 2007 at a 1×10^{33} cm^{-2}s^{-1} luminosity @1.89 GeV at BEPCII (Beijing Electron-Positron Collider II) at IHEP Beijing. It has a 1 T superconducting solenoid magnet with the inner winding diameter of 2962 mm, winding length of 3532 mm and a 600 tonne flux return yoke. It will be assembled by the end of 2004 and tested in the middle of 2005. The indirectly cooled, pure aluminum stabilized single layer coil is internally wound with a 4 kA superconductor. The magnet design is described.

INTRODUCTION

BESIII detector magnet is one of the largest superconducting magnets in China. It generates 1.0 T magnetic field with a uniformity of 5% within the drift chamber. Rectangular aluminum stabilized NbTi/Cu superconductor, made by Hitachi Cable Ltd., is adopted to wind the one-layer coil inside a support cylinder. Furthermore, the winding is indirectly cooled by forced flow of two-phase helium. The main structure disign of BESIII magnet is shown in Figure 1. The important parameters of the magnet are listed in Table 1.

Table 1.   The important parameters of BESIII magnet

Items	Value
Central field	1.0 T
Uniformity in the tracking region	5%
Operating current	3250 A
Inductance	2.1 H
Stored energy	9.5 MJ
Winding structure	single layer
Winding length	3532 mm
Winding mean radius	1490 mm
Total turns	905

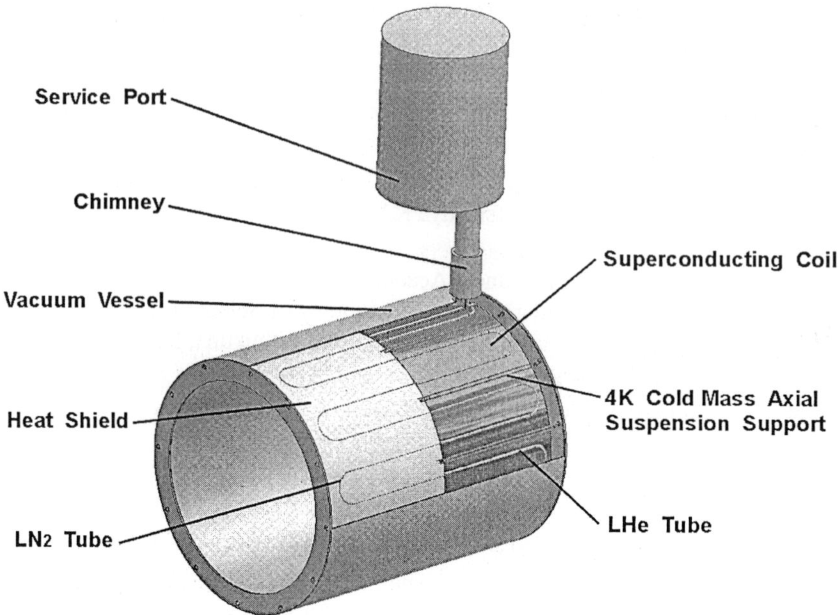

Figure 1.　The design of BESIII magnet

CONDUCTOR

Total length of the superconductor is about 8500 m , it is composed of three long superconductors because of the limit of the manufacturing technique. The operating current is designed to be 3250 A and the critical current test of a short sample of the superconductor has been performed, whose result is about 7300 A at 4 T, so it has a large margin of safety. The NbTi/Cu superconducting cable is embedded in the center of the aluminum stabilizer. The coil temperature will be kept below 70 K after quench.

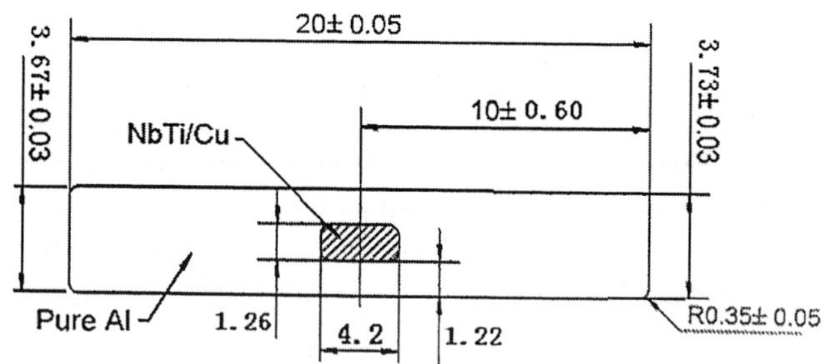

Figure 2.　The cross section of the conductor(in mm)

COIL SUPPORT SYSTEM

The loads to be supported are the self weight of the cold mass and the magnetic forces due to the decentring and misalignment of the coil with respect to the return yoke, the contraction of the coil during cooling must also be taken into account. The 4.5 K cold mass is about 3900 kg. There are 12 radial supports at each end and the total radial load is 10,065 kg. The capacity of each radial support is rated at 2516 kg with a safety factor. There are 12 axial supports in each direction to support 53291 kg. Each axial support is rated at 4540 kg with a safety factor. The race-track shape GFRP (Glass Fiber Reinforced Plastic) supports are adopted in order to decrease the heat leak through the supports.

Table 2. The main characters of the conductor

	Items	Value
SC cable	Total length	8.5 km
	SC core wire material	NbTi/Cu
	Aluminum stabilizer material	Al purity 99.998%
	RRR of copper in core wire	70
	RRR of Aluminum in cable	500
	Interface shearing stress between core wire and aluminum	20 MPa
	Cross section area ratio	NbTi/Cu/Al~1:0.9:28
SC core wire	No of strands	12
	Strand diameter	0.7 mm
	Pitch of twisted wire	50 mm
	Critical current	6800 A @ 4 T, 4.2 K
Single strand	Filament diameter	15~20 μm
	Pitch of twisted filaments	20 mm

WINDING

An inner winding machine is under construction. The superconductor will be wound onto the inner surface of the support cylinder. Before the winding, the helium cooling tubes will be welded onto the outer surface of the support cylinder, and the ground plane insulation will be mounted on the inner surface of the support cylinder.

One layer of 0.075 mm thick UG (Upilex-Glassfiber) film is designed for turn-to-turn insulation of the conductor with 100% coverage. The 0.025 mm thick upilex layer has dielectric strength of 7.4 kV at 25°C. Taking into consideration the reality that the dielectric strength of polyimide is higher at cryogenic temperature than that at the room temperature, it is estimated that the mentioned UG turn-to-turn insulation design will fully satisfy the required insulation strength above 100 V. Two layers of GUG (Glassfiber-Upilex-Glassfiber) are employed as ground plane insulation., with 0.07 mm epoxy resin and non flatness, the total thickness of the ground plane insulation is 0.4 mm. The GUG sheets will be attached onto the inner surface of the support cylinder warmed to 80°C. The conductor joints will be made with TIG (Tungsten Inert Gases) welding technique, and the joint resistance is designed to be less than $1*10^{-9}$ Ω at room temperature. Some pure aluminum strips will be mounted onto the inner surface of the coil along the axial direction, it will serve as a quench propagator.

CRYOSTAT

Thermal shield cooled by liquid nitrogen is designed to reduce the radiation heat load to the 4.5 K cold mass. According to the estimation on the heat loads and using a margin factor of 1.5, the mass flow rates of nitrogen and helium have been determined, which is 1.89 g/s and 10 g/s respectively. There are 50 layers of super insulation between the heat shield and the vacuum vessel, and 15 layers between the heat shield and the coil.

SUMMARY

The BESIII magnet will be fabricated and assembled in Beijing by the end of 2004 and will be tested in the middle of 2005. The NbTi/Cu superconductor has been delivered and the critical current test of a short sample of the superconductor has been performed. An inner winding machine is under construction.

ACKNOWLEDGMENT

The authors would like to thank particularly Kurokawa S., Yamamoto A., Xu S., Zhang L., Lin L. and Wang Q. for their continuous supports and advices on various technical aspects during design.

REFERENCES

1. Yamamoto, A. et al., A thin superconducting solenoid wound with the internal winding method for colliding beam experiments, Journal de Physics – C1 (1984) 337-340
2. Yamamoto, A. et al., Performance of the TOPAZ Thin Superconducting Solenoid Wound with Internal Winding Methods, Japanese Journal of Applied Physics (1986)
3. Lucio Rossi, Superconducting magnets for accelerators and detectors, Cryogenics (2003) 43 281-301
4. Goldacker, W. et al., Development of superconducting and cryogenic technology in the ITP of Research Center Karlsruhe, Cryogenics (2002) 42 735-770
5. Paola Miele, et al., The superconducting magnet system for the ATLAS detector at CERN, Fusion Engineering and Design (2001) 58-59 195-203
6. Herve, A. et al., CMS-The Magnet Project Technical Design Report, CERN/LHCC 97-10, (1997)
7. Wilson M., Superconducting Magnets, Oxford University Press, Oxford, UK (1983) 200-232

Stability Analysis for Cryocooled Aluminum-Stabilized Superconducting Magnet

Hongwei Liu, Qiuliang Wang, Yunjia Yu and Ye Bai

Institute of Electrical Engineering, Chinese Academy of Sciences, Beijing, 100080, China

Aluminum-stabilized superconductors have been employed in superconducting magnet, such as SMES and detector magnets etc. The stability and homogeneity of temperature in superconducting magnet fabricated by these kinds of conductors are considerably improved due to the advantage of high RRR and high thermal conductivity at low temperature. Because of current redistribution, there is finite-length normal zone propagation in the conductors during quench. The transient recovery process is affected by the slow current diffusion in the bulky aluminum stabilizer.

INTRODUCTION

For the development of high current density and compact superconducting coils, superconducting wires with low copper superconductor ratio are usually wound in close-packed and epoxy-impregnated formats. It is a simplest way to increase the volume of the copper stabilizer to enhance the coil stability margin. However, it brings some drawbacks such as increase of coil volume and weight, and reduction of coil current density [1]. The great concern here is to increase the coil stability margin while keeping the coil weight as light as possible and the current density as high as possible. To overcome these contradictions, using high purity aluminum stabilizer is a good solution [2]. The development of conductors with high pure aluminum as the super-stabilizer was proposed for such applications of detector magnets for high-energy physics and superconducting magnet storage energy devices. It adopted cryogenics stabilization with coolant. The stability and homogeneity of temperature in superconducting magnet fabricated by these kinds of conductors are considerably improved due to the advantage of high RRR, and high thermal conductivity at low temperature [3]. With the widely use of cryocooler for superconducting magnets, our interest is focused on cryocooled aluminum-stabilized superconducting magnet. A numerical method has been developed to investigate the thermal and electromagnetic behaviors of cryocooled aluminum-stabilized superconducting magnet.

SIMULATION METHOD AND MODEL

The dynamics process in composite superconductors is determined by both temperature and current density distributions. A complete treatment of the problem requires the solution of heat diffusion equation which defines the dynamics of the temperature field, and a set of Maxwell equations which define the dynamics of the current density distribution. The equations are formed by a set of three-dimensional and

598

time-dependent nonlinear equations. We developed a numerical method to investigate the thermal and electromagnetic behaviors of cryocooled aluminum-stabilized superconducting magnet.

The parameters of conductors and superconducting magnet are presented in Table 1. The conductor model used in this analysis is shown in Figure 1.

Table 1　Parameters of conductor and superconducting magnet

Conductor	NbTi/Cu/Al	Magnet	Solenoid
Cable	Φ3. 0 mm	Inner dia.	1.0m
Conductor	Φ15. 0 mm	Outer dia.	1.015m
Cu/SC ratio	1.2	High	0.3m
Al/ NbTi/Cu ratio	23.8	Total turns	20
Current density (in NbTi/Cu)	6×10^8A/m^2	layer	1
External field	6.25T	Current	7.1KA

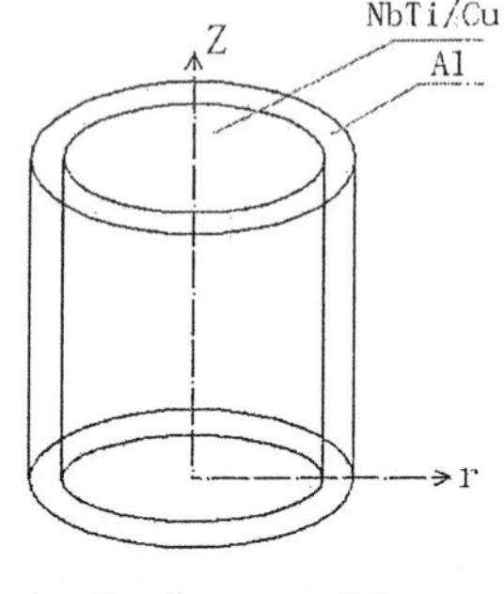

Figure 1　Conductor model

A simulation of 3D quench propagation was done by using a logical coordinate system in which each magnet is transformed into a single long conductor and it is divided into finite-length elements. Due to high thermal conductivity of ultra-purity aluminum and its volumetric ratio is larger in the conductor, uniform temperature distribution over the cross section is assumed. The governing equation of heat conduction is as following:

$$\gamma C(T)\frac{\partial T}{\partial t} = \frac{\partial}{\partial z}\left(k(T)\frac{\partial T}{\partial z}\right) + Q_J + Q_D \tag{1}$$

Where T is temperature, t is time, γ is density, C is specific heat, k is thermal conductivity, Q_J is joule heating density, Q_D is energy density of disturbance.

Basic equation for the current diffusion in stabilizer of aluminum matrix is:

$$\mu_0\frac{\partial J}{\partial t} = \nabla \bullet (\rho(T)\nabla J) \tag{2}$$

Where J is current density, ρ is electrical resistance, μ_0 is permeability of free space.

The initial condition is:

$$J(r,z,0) = J_{op}, \quad (r,z)\in\Omega_{sc} \tag{3}$$
$$J(r,z,0) = 0, \quad (r,z)\notin\Omega_{sc}$$

Though current density may be changing, the current through the conduct is constant. The magnetic field derivative produced by the current is zero at the conductor boundary.

$$\frac{\partial J_{SC}}{\partial r}\Big|_{r=0} = \frac{\partial J_{ST}}{\partial r}\Big|_{r=b} = 0 \tag{4}$$

Where b is radius of conductor, ST represents the whole conductor, SC represents the superconducting section.

The conductor in this paper is considered as 'quasi-adiabatic' during the normal zone propagation process. The assumption holds true since a cryocooler delivers cooling only through thermal conduction, therefore the cryocooler's response to a rapid temperature change within the magnet is attenuated [4]. Equation (1) and (2) can be solved by alternating direction implicit (ADI) with globally convergent methods. The nonlinearity of electrical resistance, heat capacity and thermal conductivity are also taken into account.

NUMERICAL SIMULATION RESULTS

The stability of a cryocooled superconducting magnet will be determined by two following factors.

Current Diffusion into the Stabilizer

If any disturbance causes the temperature to rise above the current sharing temperature, current will begin to transfer into the copper and then to the aluminum stabilizer. Because the time constant for the magnetic field diffusion in aluminum stabilizer is relatively long due to the low resistivity of aluminum at low temperature, the slow diffusion of the current will generate joule heat pulse. And the most important, it is much larger than uniform current distribution over the stabilizer cross-section (see Figure 2). The front of normal zone is then propagating while the back is recovering as the current diffuses deeper into the stabilizer and joule heat drops below the helium heat removal rate when the conductor is immersed in the coolant. Thus there is a finite length normal zone propagating in the conductor during the transient process shown in Figure 3.

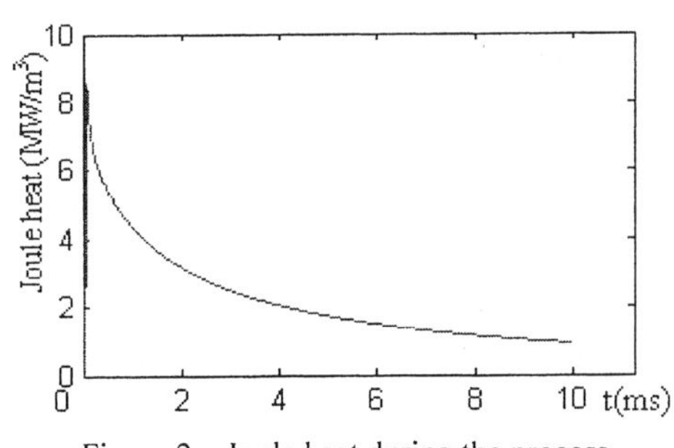

Figure 2   Joule heat during the process

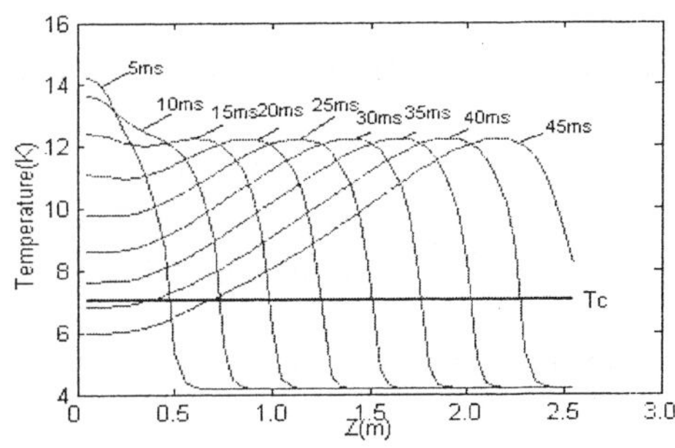

Figure 3   Temperature distributions

But when the conductor is cooled by crycooler, any losses or heat load will be conducted only along the conductor. In this circumstance, there also exists the current diffusion phenomenon (see Figure 4). The relationship between heat generation and thermal conduction determine the conductor stability margin.

Heat Transfer to Neighboring Turns and Layers

In our 'quasi-adiabatic' model, under the disturbance with the length of 0.25 m and duration of 0.25 ms, it was found that any disturbance caused the temperature to reach current sharing temperature will quench in the end (see Figure 5).

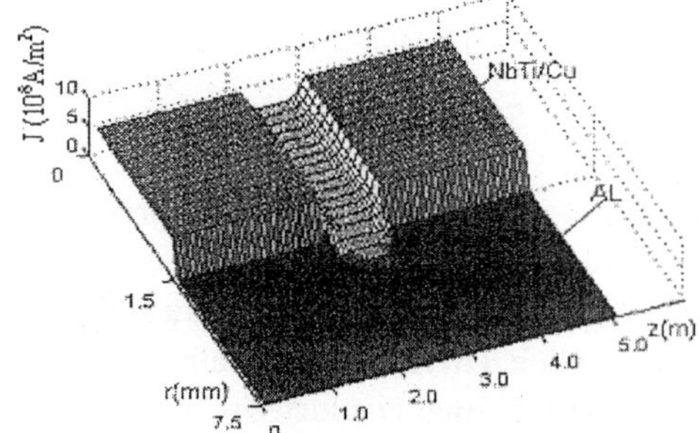

Figure 4   Current distributions with respect to space and time

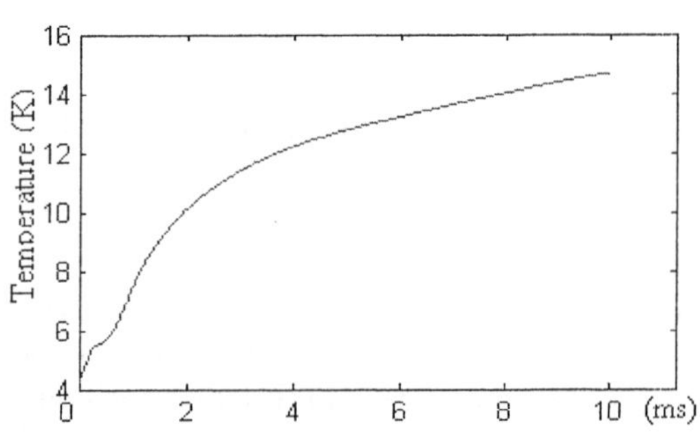

Figure 5   Hot spot temperature in 'quasi-adiabatic' model

As for magnet, the thermal conductivity between turns or layers play an important role in magnet stability. The following graphs (see Figure 6 and Figure 7) are the temperature distributions of the magnet with our logical coordinate, in which the whole magnet is transformed into a single long conductor. We can see the heat transfer between turns is much more quickly than that along azimuthal direction.

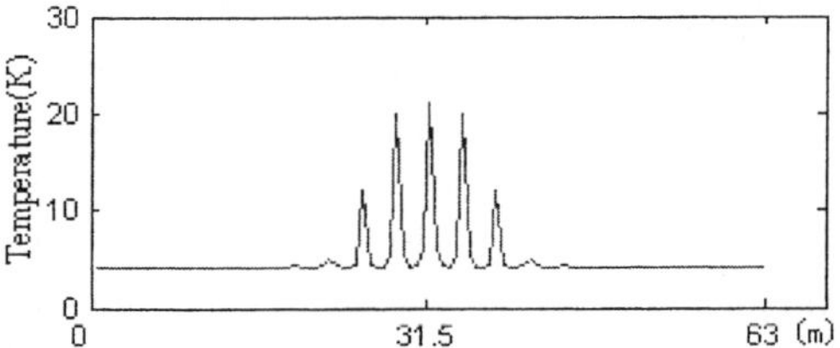

Figure 6   Temperature distribution in the magnet at 15ms during transient process

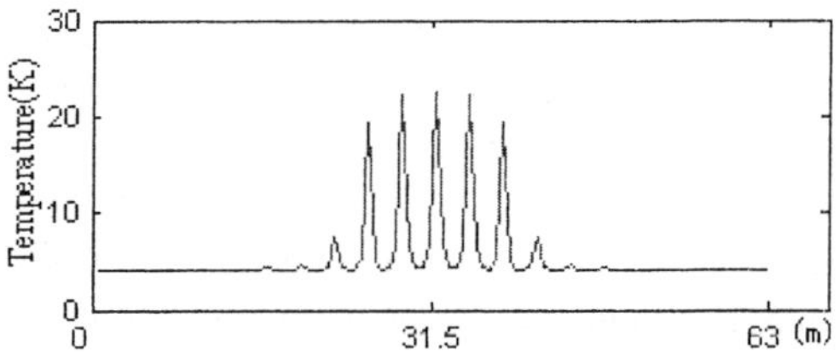

Figure 7   Temperature distribution in the magnet at 20ms during transient process

CONCLUSION

The stability of cryocooled aluminum-stabilized superconductor was investigated by numerical simulation method. It is the most important for cryocooled superconducting magnet to improve the thermal conductivity structure, especially for aluminum-stabilized magnet. We will investigate several kinds of ways that can improve the stability margin and increase the quench propagation velocity as the following step.

REFERENCES

1、 Wilson, M.N., Superconducting Magnets, Clarendon Press, Oxford (1983)
2、 Takahashi, R., Iida, F. and Hotta, Y., Development of aluminum-stabilized superconducting coil, IEEE Trans.Magn(1994) 30 1915-1918
3、 Wang, Q.L., Influence of current diffusion in superconducting magnet fabricated by high stabilizer aluminum on quench propagation, IEEE Trans.Magn.(2002) 38 1197-1200
4、 Lim, H., Iwasa,Y. and Smith,J.L., Normal zone propagation in cryocooler-cooled Nb3Sn tape-wound magnet, Cryogenics (1995) 35 367-373

Thermal design of BESIII magnet

Xu Q.[1,4], Zhu Z.[2], Liu L.[1], Wang S.[3], Yi C.[2], Chen H.[2] and Zhang L.[1]

[1]Technical Institute of Physics and Chemistry, Chinese Academy of Sciences, Beijing 100080, China
[2]Institute of High Energy Physics, Chinese Academy of Sciences, Beijing 100039, China
[3]Wang NMR Inc., 550 North Canyons Parkway, Livermore, CA 94551, USA
[4]Graduate School of the Chinese Academy of Sciences, Beijing 100039, China

This paper presents the thermal analysis of Beijing Spectrometer Magnet (BESIII
magnet) which will be used as a superconducting detector magnet in BEPC-II.
BEPC-II is an upgrade project to the Beijing Electron Positron Collider (BEPC).
BESIII Magnet is cooled by forced flow of two phase helium and the inner winding
technique is adopted in the coil winding process. A 3D model of calculation for the
magnet was built. The heat load and the temperature distribution of the magnet were
analyzed by a finite element method, and the upper limits of several heat leak sources
were calculated.

INTRODUCTION

In order to provide an axial magnetic field of 1.0 T over the tracking volume of the third upgraded
Beijing Spectrometer (BESIII) in BEPC-II, a superconducting magnet scheme is adopted due to its
superior performance and stability during operation. The superconducting magnet is designed to run at
4.5 K and be cooled by forced flow of two phase helium. High purity (>99.99%, RRR>500)
aluminum-stabilized NbTi/Cu (1:1) superconductor is adopted, and the superconductor is encapsulated
inside a restraining cylinder made of aluminum alloy 5083 for reason of radiation transparency and for
compatibility of thermal contraction. During coil winding the inner winding technique will be used. A
liquid nitrogen thermal shield is designed to minimize the radiation heat load to the superconducting
coil. The main structure of BESIII magnet is shown in Figure 1.

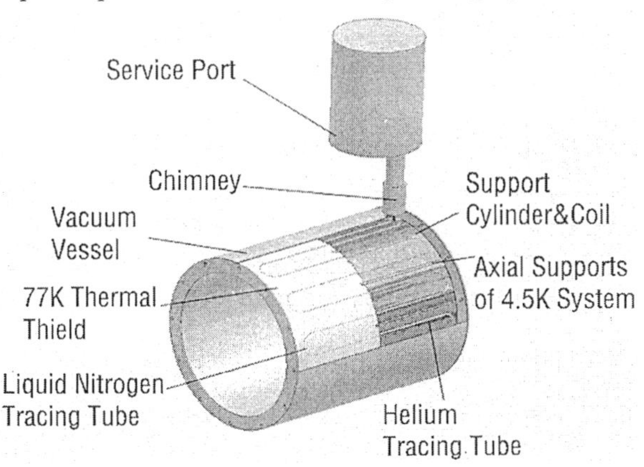

Figure 1   Overall view of BESIII magnet

STATIC HEAT LOAD

There are mainly five kinds of heat loads when BESIII Magnet is in static operating condition: Joule
heating of the joint turns, radiation heat leak, heat leak through axial and radial supports, current leads

heat leak and measuring wires heat leak. Detail is listed in Table 2 (Heat load in the interface of magnet system and cryogenic system: chimney and service port is also listed here). BESIII superconductor is composed of three long conductors, design data of electric resistance of the joint turn is 10^{-9} Ω. On the outside of the support cylinder and the nitrogen thermal shield, the multilayer superinsulation (S.I) system is used to decrease the radiation heat load. Double side aluminized mylar or polyester film is used. The aluminized mylar is insulated by polyproparene. The design wrapping density is about 50 to 60 layers per inch, and mylar is cut off for every 10 layers. Thus, the design of S.I. system has the effective thermal conductivity and Q/A value as Table 1.

Table 1 Experience data of superinsulation system adopted in BESIII Magnet

Range (K)	Layer (#)	Vacuum Gap (mm)	K (W/m.k)	Q/A (W/ m^2)
300 – 80	50	25.4	8.3 x 10^{-5}	0.6
80 – 4.5	30	19	1.447 x 10^{-5}	0.043

There are totally 24 axial supports and 24 radial supports for the 4.5 K cold mass system and 8 radial supports for nitrogen thermal shield, all these supports are connected to 300 K system directly. Material of these supports is unidirectional fiberglass. Heat load is calculated using following equation:

$$Q = A/L \int KdT \tag{a}$$

Where Q: heat load through supports, A: cross section area of supports, L: length of supports, K: thermal conductivity of suppports and T: temperature of supports. Designed operating current of BESIII Magnet is 3250 A. According to Martin N. Wilson's book <Supercondcting magnets>, heat leak of optimum current leads is about 1.04 mW/A, using the safe factor of 1.1, we have the current leads heat leak calculated. There are totally about 370 measuring wires (thermometers, strain sensors, heaters and so on) in BESIII Magnet system. Heat load of these wires is also calculated using equation (a).

Table 2   Calculated data of heat loads of BESIII magnet (W)

Items of heat load	77 K system	4.5 K system
Joule heat of the joint turn	/	0.02
Radiation heat leak	44.75	3.50
Heat leak through supports	0.3	0.97
Current leads heat leak	/	7.44
Measuring wires heat leak	5.31	0.83
Heat load in chimney & SP	60	13.43
Total	110.36	26.66
Adopted heat load（×1.5）	165.54	40

MASS FLOW RATE OF HELIUM AND NITROGEN

The 4.5 K cold mass system is cooled by two phase helium forced flow. Considering that too much liquid helium boiling away causes many bubbles in the tracing tube, which may greatly decrease the convective heat transfer coefficient between liquid helium and tracing tube, also according to the operating experience of BELLE Magnet in KEK Japan, vaporization rate of helium is determined to be 20 percent in the magnet. And liquid helium at outlet of magnet should not less than 50 percent. Then the mass flow rate of helium is designed to be 10 g/s, and the pressure should be about 1.25 bar. The radiation thermal shield is cooled by liquid nitrogen and the mass flow rate is designed to be 1.89 g/s

TEMPERATURE DISTRIBUTION

The static temperature distribution of BESIII magnet was calculated with a commercial FE method software-Ansys. A 3D model of the magnet was built. Following assumptions were made in the calculation: There are no contact heat resistances between 4.5 K system and axial&radial supports, superconducting coil is well bond to the support cylinder, and the temperature of the inner surface of helium tracing tube is 4.5 K. Although the first assumption is very different to the fact, but it is a safe consideration. The second assumption is not a safe consideration, but it is a basic request for the operation of the magnet. If the coil is not well bond to the support cylinder, much trouble may occur: temperature of the joint turn will rise, coil temperature will rise, and during quenching, coil temperature will increase to a very high level. Calculated heat transfer temperature difference between two phase helium and tracing tube is about 0.01 K, so the third assumption is also acceptable. Calculating result is shown in Table 3. We still find that with the assumption that the coil is well bond to the support cylinder, the temperature of the joint turn hot point is not high even the joule heating of the joint turn increases 100 times larger than before. But if the coil is not well bond, things will become troubled. Figure 3 presents the detail. In the calculation of radiation heat load, experience data of 0.043 W/ m² (77 K to 4.5 K) is used, but as everyone is painfully aware, it is the way in which superinsulation system is applied that really makes the difference in heat transfer, the above heat flux data may differ greatly with the real condition. Figure 4 presents the influence of radiation heat load on the main hot points.

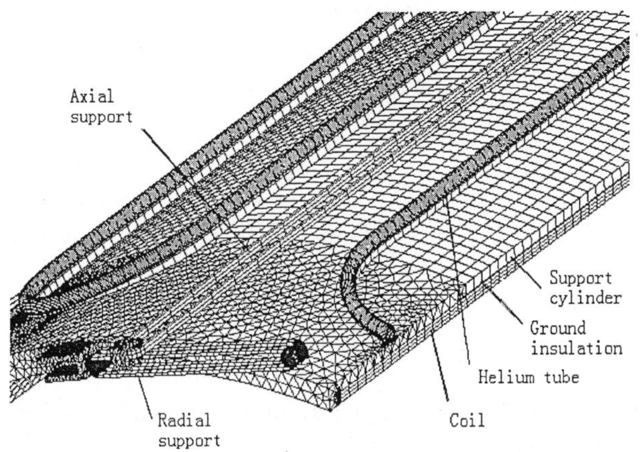

Figure 2   1/12 part 3D model of BESIII magnet

Table 3   The static temperature distribution of BESIII magnet

Items of hot point	Temperature (K)
Hot point due to joint turn joule heat	4.51
Hot point due to radiation heat leak	4.51
Hot point due to axial support heat leak	4.58
Hot point due to radial support heat leak	4.78
Temperature difference due to the welding of the helium tracing tube	0.01
Temperature difference due to the ground plane insulation and turn to turn insulation	0.02
Temperature difference due to the side wall insulation	0.22

DYNAMIC HEAT LOAD

Eddy current loss in support cylinder during the magnet charging/discharging period of 30 minutes was

604

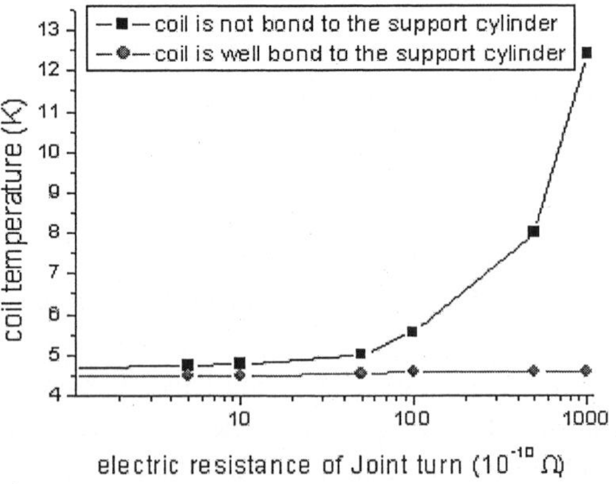

Figure 3   The influence of electric resistance of
joint turn on coil temperature

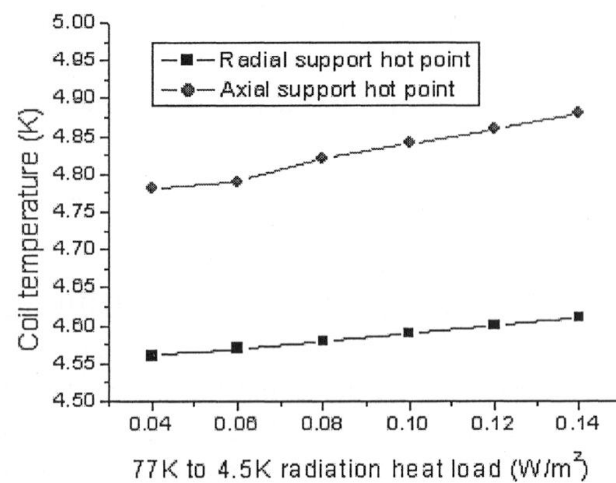

Figure 4   The influence of radiation heat load
on the main hot points

calculated by the following equations:

Induced voltage $\qquad E = - \, d\Phi / dt$ (b)

magnet flux $\qquad \Phi = SB = S\mu H = S\mu nI / L$ (c)

Dynamic heat load $\qquad Q\mathrm{dy} = E^2 / R$ (d)

Where t: Charging/discharging time, S: Cross section area of the support cylinder, n: net turn of the coil, L: Axial length of the coil, R: Electricity resistance of the support cylinder. The calculating result of the dynamic heat load is about 4.95 W. In case of emergency fast ramping, the dynamic heat load would be increased

SUMMARY

Heat load and temperature distribution of BESIII magnet were analyzed by FE method. The main heat loads are current leads heat load and heat load in chimney and service port, radiation heat load is also an important one. The main hot points are hot points due to radial supports and hot points due to axial supports. Dynamic heat load is about 4.95 W during the magnet charging/discharging period of 30 minutes.

ACKNOWLEDGMENT

This work was carried out under BESIII superconducting magnet group of IHEP and active support of Wang NMR Inc. in USA.

REFERENCES

1.   Lucio Rossi, Superconducting magnets for accelerators and detectors, Cryogenics (2003) 43 281-301

2.   Goldacker,W. et al., Development of superconducting and cryogenic technology in the ITP of Research Center Karlsruhe, Cryogenics (2002) 42 735-770

3.   Wilson M., Superconducting Magnets, Oxford University Press, Oxford, UK (1983)

A Numerical Code to Simulate Quenching in Conduction Cooled Superconducting Magnet

Yinming Dai, Bertrand Blau[*], Hans Hofer[*], Yunjia Yu, Qiuliang Wang

Institute of Electrical Engineering, CAS, P.O.Box 2703, Beijing, China
* ETHZ, Zurich, Switzerland

A numerical code (1D+2D) has been developed to simulate quenching in conduction cooled superconducting magnet. The code is formed based on the numerical thermal-physical model to describe an initial hotspot evolution in a practical superconduting coil. The code is in an integrated frame that couples a 1D model and a 2D model, describing the normal zone propagations along the superconductor and among winding turns respectively. Material characteristics of the magnet winding constituents are involved as the basic data to simulate normal zone propagation process. Information such as maximum temperature rise and voltage evolution can be output as the result of code execution. This code is used to perform a quench simulation on the Alpha Magnetic Spectrometer (AMS) superconducting magnet and the simulation results are presented in this paper.

INTRODUCTION

Conduction-cooled superconducting magnet system can be found in many applications[1,2,3] due to their simple operation. In contrast to traditional bath cooling, conduction-cooled superconducting magnets are cooled by cryocooler directly via thermal conduction circuit. In the vacuum environment, the superconducting magnet is operated with no liquid cryogen at all inside the coil windings. During a quench incident, all the energies stored in the magnet have to be dumped into the superconducting windings, resulting in a rapid temperature rise after quench. Severe damages to the magnet should be prevented by appropriate protection measures based on a detailed quench analysis.

A complete description on the physics of normal zone propagation during a quench incident has been summarized by Wilson[4]. Some quench simulation methods to simulate quenching in specified superconducting magnet have been presented by authors[3,5], in which finite difference and finite element method are used. Rough and fast estimation on the hotspot temperature can be obtained by quench integral value and 2D method. Complicated 3D finite element method is preferred but a huge simulation work is needed. Here, we proposed a 1D+2D quench simulation model to evaluate the normal zone propagation in a conduction cooled multi-coil superconducting magnet. As an example, simulation results on a dipole coil in the Alpha Magnetic Spectrometer (AMS) superconducting magnet are presented.

1D+2D SIMULATION MODEL

Superconducting coils are usually wound with a continual superconducting wire. Once a big disturbance happens in the superconducting coil, the initial normal zone will enlarge its size and propagate longitudinally along the superconducting wire. This process can be fairly described by a one dimensional thermal balance model. Meanwhile, the heat will be conducted from the hotspot of the initial normal zone to the adjacent turns/layers. A 2D thermal balance model can be used to define this thermal conduction process. Figure 1 shows the coil circuit and 2D element discretization used in the 1D+2D model.

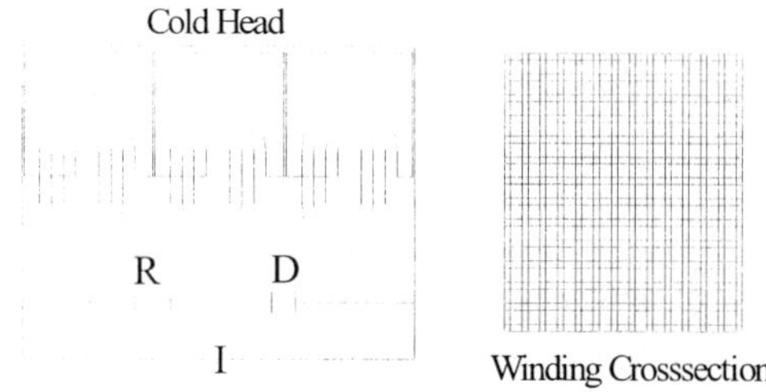

In the 1D+2D quench simulation model, the coils or winding sections are connected in series and powered by an external power supply. Protection resistor R or diode are connected in parallel with the magnet. Terminals of each coil are anchored to the cold head of cryocooler. Winding cross section of the coil is discretized into rectangular elements. Each conductor is regarded as an element and surrounded by 8 insulation elements.

For the 1D and 2D thermal balance model, the following two differential equations can be used to describe the thermal diffusion process in each discretized element, with boundary conditions of $T_b=T_0$ and adiabatic boundary for equation (1) and (2) respectively.

$$\rho C_p \frac{\partial T}{\partial t} = \frac{\partial}{\partial x}(k_x \frac{\partial T}{\partial x}) + S \tag{1}$$

$$\rho C_p \frac{\partial T}{\partial t} = \frac{\partial}{\partial x}(k_x \frac{\partial T}{\partial x}) + \frac{\partial}{\partial y}(k_y \frac{\partial T}{\partial y}) \tag{2}$$

where the temperature and time are denoted by $T(x,y,z,t)$ and t respectively. ρ is material density, C_p is specific heat, k is thermal conductivity and S is source term which is a function of operating current $i(t)$ and normal zone resistance $R(t)$.

When a quench happens, the power supply shuts off immediately and the current will decay through the protection resistor R or diode D. The quenched coil in the magnet has a self inductance of L_q and a mutual inductance M with the rest coils. The total inductance of the magnet is denoted by L_m. The current i(t) and voltage V(t) across the quenched coil can be described by equation (3) and (4) respectively.

$$L_m \frac{di(t)}{dt} + i(t)(R(t) + R) + V_D = 0 \tag{3}$$

$$V(t) = i(t)R(t) + (L_q + M)\frac{di(t)}{dt} \tag{4}$$

1D+2D QUENCH CODE PROGRAMMING

The numerical 1D+2D quench code programming is based on the finite differential method. The 1D and 2D are programmed into two separate subroutines. Each subroutine transfers their results via the main program. Thermal physical properties such as thermal conductivity, specific heat and resistivity, which are functions of temperature, are compiled into data input subroutines.

For the 1D numerical model, the superconducting wire is discretized into N equal elements, with each elements length dx. In a time step dt, the temperature of an arbitrary element i can be expressed as following set of equations, where T_i^0 is the temperature of element i at last dt, j is the current density. ρ_i is the average resistivity of element i at T_i^0. Thermal physical properties such as k, Cp and the density ρ_0 are also averaged on the conductor component ratio.

$$a_i T_{i-1} + b_i T_i + c_i T_{i+1} = R_i \qquad (5)$$

$$a_i = -\frac{2k_{i-1}k_i}{dx(k_{i-1}+k_i)} \qquad (6)$$

$$c_i = -\frac{2k_{i+1}k_i}{dx(k_{i+1}+k_i)} \qquad (7)$$

$$b_i = -(a_i+c_i) + \frac{C_p \rho_0 dx}{dt} \qquad (8)$$

$$R_i = \frac{C_p \rho_0 dx}{dt} T_i^0 + j^2 \rho_i dx \qquad (9)$$

$$i = 1,2,\ldots\ldots N$$

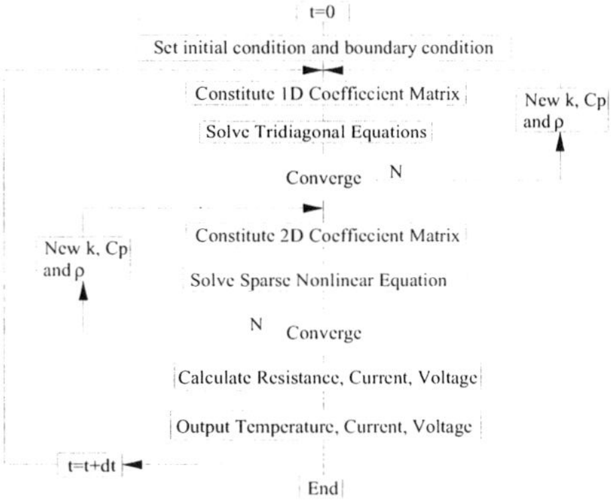

Figure.2 1D+2D Quench Simulation Program Structure

Finally, the 1D numerical model is transformed into a set of nonlinear equations in the tridiagonal system and can be solved by Thomas algorithm for a given time step. And similarly, the 2D numerical model can be also expressed as a set of nonlinear equations and be solved by conjugated gradient method.

The structure of the 1D+2D quench simulation program is presented in Fig.2.

SIMULATION EXAMPLE

The 1D+2D quench simulation program has been performed to analyze the quench process in a dipole coil of AMS02 superconducting magnet. AMS02 superconducting magnet [6] is an assembly of 2 big dipole coils and 12 racetrack coils that are distributed circumferentially to provide a magnetic dipole field. The dipole coil is wound with rectangular aluminum stabilized NbTi/Cu superconducting wire. The cryogenic scheme on the AMS02 superconducting magnet system has been described in detail elsewhere [7]. Parameters of the AMS02 superconducting magnet and the dipole coil are collected in Table 1.

Table 1 Parameters of AMS02 Superconducting Magnet and Dipole Coil

AMS02 Magnet		Dipole Coil	
Total Coils in Assembly	14	Length of Straight Part (cm)	40
Operating Current Io (A)	459	Inner Radius of End Part (cm)	19.5
Peak Magnetic Field (T)	6.59	Winding Cross-section (cm x cm)	8.76 x 14.55
Total Inductance (H)	48.9	Number of Turns	3360
Energy Storage (MJ)	5.15	Conductor Length (m)	8342

The aluminum stabilized NbTi/Cu superconducting wire has a cross section dimension of 2.0×1.546 mm^2. Matrix components are aluminum and copper that each accounts for 84.51%, 7.57% respectively. The RRR of high purity aluminum and copper stabilizer are greater than 1000 and 100 respectively. Thermal physical properties such as thermal conductivity, specific heat and resistivity are estimated based on their purity level.

The quench simulation on the dipole coil is executed based on assumptions that the magnet is closed by a superconducting switch and mutual inductance between the dipole coil and the rest coils in the magnet is excluded in the voltage evolution equation. Simulation results for the temperature rise, current decay and voltage evolution are presented in Fig.3 and Fig.4 respectively.

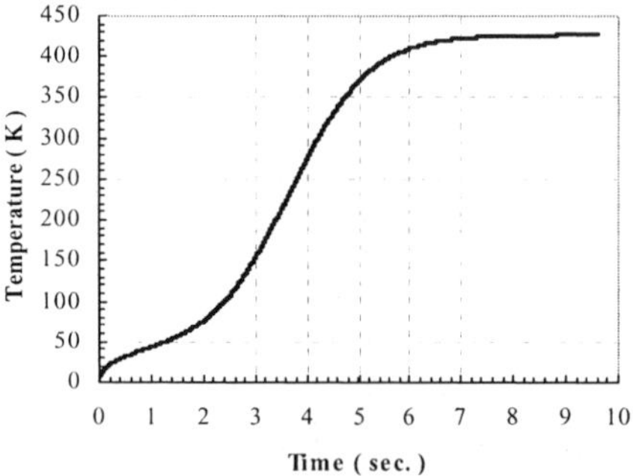

Figure.3 Temperature Rise after Quench

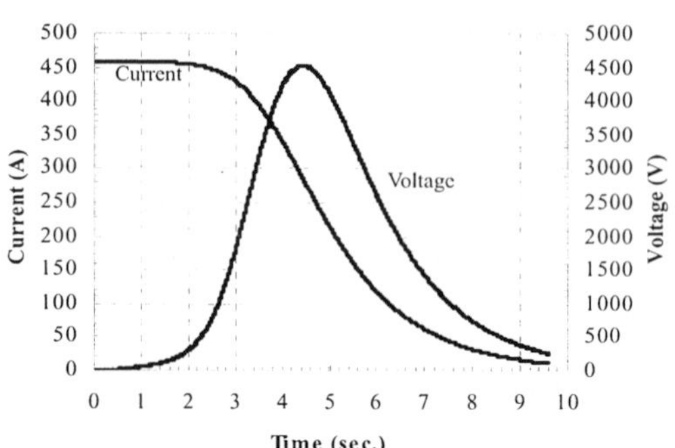

Figure.4 Current Decay and Voltage Evolution

CONCLUSIONS

A numerical quench simulation code 1D+2D is built based on the differential equations describing the one dimensional normal zone propagation along the superconductor and two dimensional thermal diffusion among turns/layers in the coil windings. The nonlinear differential equations are solved iteratively by Thomas algorithm and conjugated gradient method for the 1D and 2D subroutines respectively. During the iteration solving process, the nonlinear thermal physical properties such as thermal conductivity, specific heat and resistivity are updated simultaneously. By coupling the 1D and 2D subroutines through the main program, element temperatures are updated interactively for each time step. Involving the protection circuit, execution of the code can output the maximum element temperature, normal zone resistance, current and voltage at every time step, that are very important data for evaluating the quench protection design.

As an example, the 1D+2D quench simulation code is applied to the safety evaluation of quenching in a dipole coil in AMS02 Superconducting magnet.

REFERENCES

1.Giebeler, F., Thummes, G. and Best, K-J, A 5 T persistent current niobium–titanium magnet with a 4 K pulse tube cryocooler, Supercond. Sci. Technol. (2004) 17 135-139

2. Watanabe, K., Awaji, S., Takahashi, K., Nishijima, G., Motokawa, M., Sasaki, Y., Ishikawa, Y., Jikihara, K., Sakuraba, J., Construction of the cryogen-free 23 T hybrid magnet, IEEE Trans. Appl. Supercond. (2002), 12 678 - 681

3. Korpela, A., Kalliohaka, T., Lehtonen, J., Mikkonen, R., Protection of conduction cooled Nb$_3$Sn SMES coil, IEEE Trans. Appl. Supercond. (2001), 11 2591 – 2594

4. M. N. Wilson, Superconducting Magnet, Oxford University Press, New York (1983)

5. Kim, S. W., Quench Simulation Program for Superconducting Accelerator Magnets, Proceedings of the 2001 Particle Accelerator Conference, 3457-3459

6. Blau, B., Harrison, S. M., Hofer, H., Horvath, I. L., Milward, S. R., Ross, J. S. H., Ting, S. C. C., Ulbricht, J., and Viertel, G., The Superconducting Magnet System of AMS-02—A Particle Physics Detector to be Operated on the International Space Station, IEEE Trans. Appl. Supercond. (2002), 12 349 – 352

7. Harrison, S.M., Ettlinger, E., Kaiser, G., Blau, B., Hofer, H., Horvath, I. L.,Ting, S.C. C., Ulbricht, J., and Viertel, G., Cryogenic System for a Large Superconducting Magnet in Space, IEEE Trans. Appl. Supercond. (2003), 13 1381 – 1384

Numerical Simulation on Thermal Stress and Magnetic Forces of Current Leads in BEPC II

Zhang X.B.[1], Jia L.X.[2]

[1]Institute of Cryogenics and Superconductivity Technology, Harbin Institute of Technology, Harbin 150001, CHINA
[2]Brookhaven National Laboratory, Upton, New York 11973, USA

Six pairs of low current leads are used for the superconducting interaction quadruple magnets (SCQ) in the Beijing Electron Positron Collider Upgrade (BEPCII). This paper presents analyses on the magnetic field induced by the current lead bundle as well as the magnetic forces on the leads. The thermal stress of the single lead is investigated. The stress of interference fitting for a novel insulator used for the current leads at various operating temperatures is analyzed.

INTRODUCTION

One pair of current leads with nominal current 1600A, two pairs with nominal current 630A and three pairs with nominal current 150A have been designed for the SCQ magnet in BEPCII. Figure 1 shows the schematic diagram of the current lead bundle. A novel cryogenic electrical isolator has also been designed for the leads. With the FEM software package ANSYS, this paper presents numerical simulation on the magnetic field of the current lead bundle induced by the operating current. The thermal stress of the 1600A current lead is investigated in detail. The stress of interference fitting for the insulator at various operating temperatures is analyzed to avoid the failure of the involved material.

MAGNETIC FIELD OF CURRENT LEAD BUNDLE

The numerical model of the current lead bundle for the magnetic field simulation is shown in Figure 2. The relative location of the current leads carrying different current flux is indicated. The symbols "+" and "-" indicate the current direction. The equivalent inner diameter d_e is introduced to simplify the

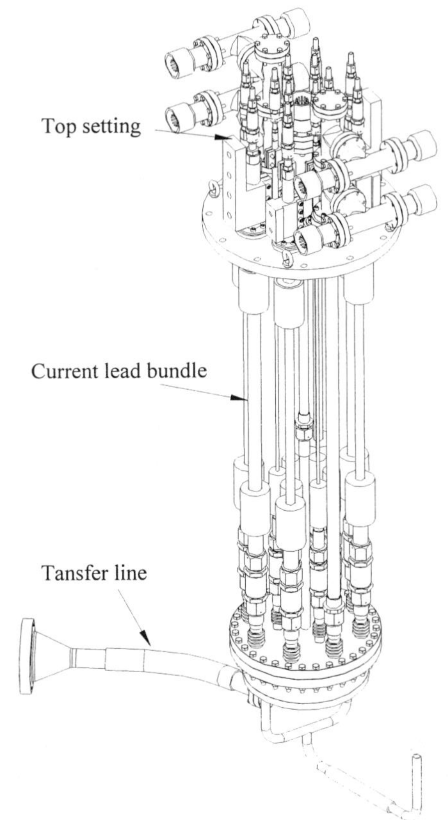

Figure 1   Current lead bundle for the SCQ magnet in BEPC II.

multi-pipe structure [1] of the current leads based on the same cross section area. Both the copper and vacuum have the same relative magnetic permeability of 1, and no external magnetic field is applied. The vacuum shield and other transfer tubes are neglected for their little effects on the magnetic field. The far field is used to explain the infinite field beyond the interesting zone of the current lead bundle. Figure 3 shows the magnetic intensity and the isochronous magnetic force applied to the left

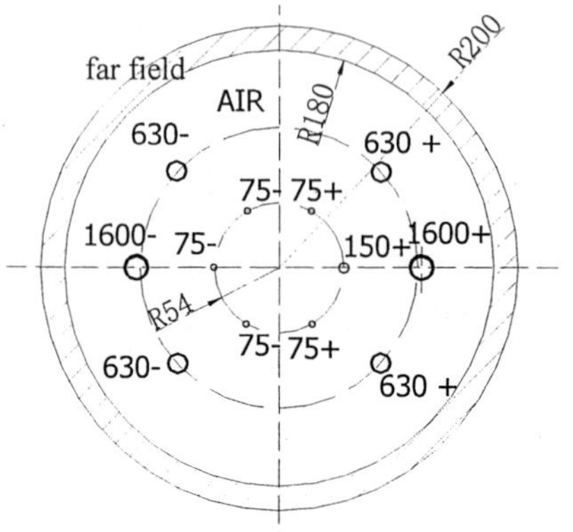

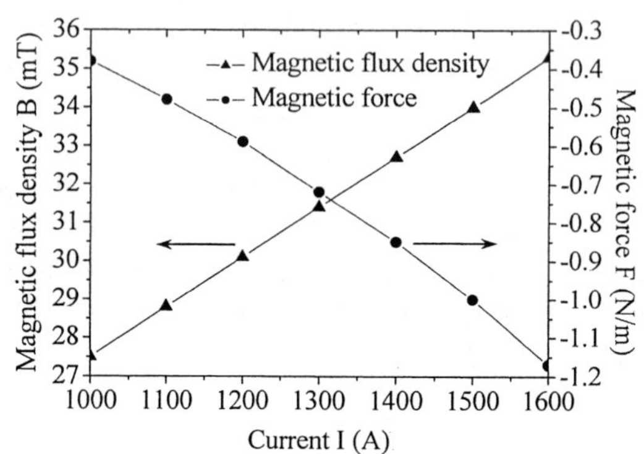

Figure 2   Model of magnetic field simulation of the current lead bundle.

Figure 3   Variation of magnetic force F and magnetic flux density B as function of maximum current I.

1600A current lead at different current I. The maximum magnetic flux density of 0.035T is found in the vicinity of the 1600A current leads for the normal operation. This value reveals that the effect on the accuracy of the silicon diode thermometers is inappreciable [2]. The symbol minus for the magnetic force means that the lead is pushed back from the right 1600A lead. When $I = 1600A$, the maximum magnetic force -1.17N/m is applied to the 1600A current leads. This corresponds to the stress of 52.7pa with its outer diameter of 22.22mm.

THERMAL STRESS OF 1600A CURRENT LEAD

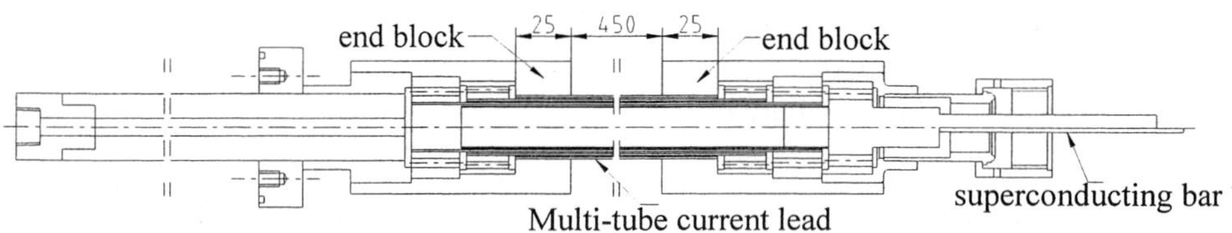

Figure 4   Schematic diagram of the 1600A current lead for the SCQ magnet

The thermal stress of the leads cooled down from room temperature to the operating temperature has great effluence on the lead performance. The temperature distribution along the gas-cooled current lead by the numerical analysis on thermal process has been studied [3]. For the multi-tube current lead of 1600A (see Figure 4), axisymmetric condition is adopted for simplifying the simulation model. Considering their similar temperature distribution and connecting fashion, only outer tube of the multiple tubes is modeled. The length of the lead is 0.45m excluding the 25mm soldering length with the end copper block (see Figure 4). The tube has the outer diameter of 22.22mm with the wall thickness of 1.5mm. The temperature of the hot end block is set at 300K and is considered stable along the soldering length for all degrees of freedom. A ripple tube is added below the cold end block to allow thermally shrinking and for convenient assembling of the leads, thus no constraint of degree of freedom is set at cold end block. The Young's modulus and coefficient of thermal expansion of copper are the function of temperature and the Poison ratio of copper is set to 0.3 here [4]. Figure 5 shows the Von Mises stress profile along the outer wall of the tube. It reveals clearly that the stress has almost no effect on the majority of the length of the lead. However, the stress goes up to about 10MPa just at the beginning of the soldering surface at hot end. Fortunately, this value of thermal stress is very small comparing with the soldering strength we usually acquired.

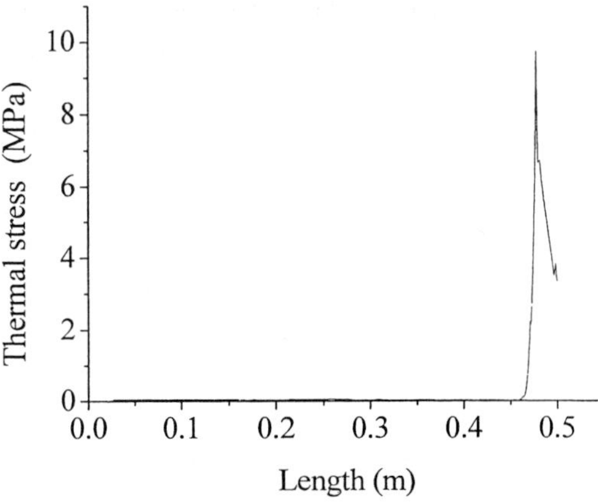

Figure 5 Von Mises stress profile of outer wall along the 1600A current lead for the SCQ magnet.

TRESS OF INSULATOR

The leads must be electrically insulated at both ends while the cooling gas can go through them. One type of insulator has been designed to meet this requirement. The insulator is consisted of two coaxial tubes. The inner tube is made of g-10 with its outer diameter greater than the inner diameter of the outer aluminum casing at room temperature. The function of aluminum casing is to strengthen the insulator. The helium transfer pipes are connected to the inner tube through the fittings. A quarter of model is appropriate to simulate the contact because of the symmetry of the insulator. Axis-direction displacement of one end of the inner tube is set at zero and no other constraint of degree of freedom is applied.

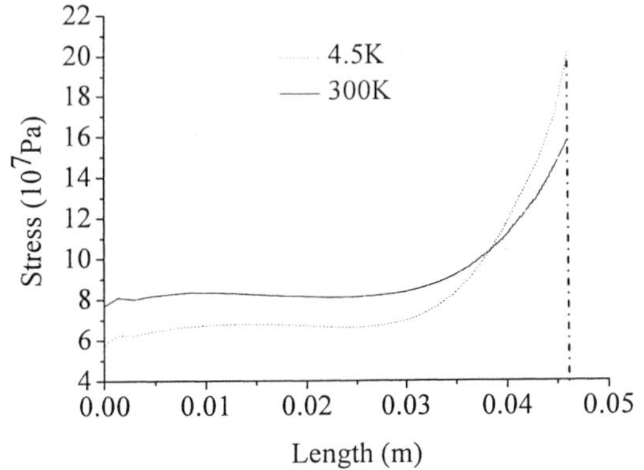

Figure 6 Von Mises stress along the length of the contact surface of the inner tube at 300K and 4.5K.

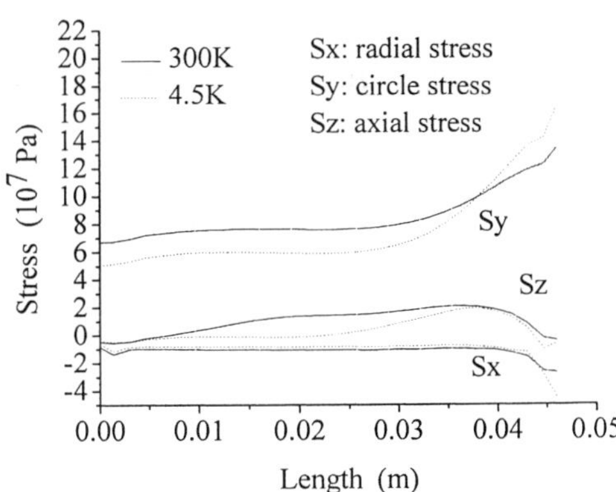

Figure 7 Stress components along the length of the contact surface of the inner tube at 300K and 4.5K.

During simulation two different load steps are defined: The objective of the first load step is to observe the interference fitting stress of the insulator at room temperature. The second load step is to observe the contact stress when the insulator is cooled down to 4.5K. Figure 6 shows the Von Mises stress profile along the length of the contact surface of the inner tube at 300K and 4.5K. The fixed end of the inner tube has the maximum stress value of 0.135GPa and 0.2Gpa at 300K and 4.5K, respectively. A strange phenomenon is found that the stress of the contact surface at 300K is larger than that at 4.5K when the length is less than about 40mm, but reversed at the rest part. The reason is that aluminum shrinks theoretically more than g-10 when they are simultaneously cooled down to 4.5K and the two will contact more tightly than that at 300K, while the Young's module of g-10 is decreased from 73.5GPa at 300K to 68.6GPa at 4.5K [4]. Besides the radial displacement, there also exits the axial frictional stress and the circle stress (see Figure 7). The circle stress component holds dominant position among these stress components. The constraint on the zero axial displacement at the right end of the inner tube also resists the radial displacements during cooling process, so the circle stress component increases greatly as it approaching to the fixed end, which offsets the reducing of the Young's module of g-10 and finally exceeds the stress at 300K.

CONCLUSION

The numerical simulation for the magnetic field induced by the current lead bundle of the SCQ magnet in BEPCII shows that the effect of the magnetic field intensity on the accuracy of silicon diode thermometers is inappreciable, and the magnetic force on each of current leads is negligible by comparing with the thermal stress. When the 1600A current lead is cooled down from room temperature to the normal operating temperature, the maximum stress of 10MPa is found at the soldering place of the warm end from the results of thermal stress simulation. For copper, this stress value is acceptable. The results of the numerical simulation for the novel cryogenic electrical insulator reveal that the maximum stress of the inner tube is 0.135GPa and 0.2GPa at 300K and 4.5K respectively. However, the maximum stress value may be overestimated for the boundary condition of no axial displacement at one end.

REFERENCE

1. Jia, L.X. et al, Design Parameters for Gas-cooled Electrical Leads of the g-2 Magnets, Cryogenics (1994) 34 631-634

2. Jia, L.X. and Wang, L., Cryogenic Engineering, Harbin Institute of Technology, 20 (interior material)

3. Zhang, X.B., Wang, L. and Jia, L.X., Numerical analyses on transient thermal processes of gas-cooled current leads in BEPC II, to be published in Advances in Cryogenic Engineering (2004), 49

4. Ying X., Handbook of Mechanical design, Mechanical Industrial Press, Beijing, China (1991) 2-47

Design and manufacture of a low-current ADR magnet for a space application

Brockley-Blatt C., Harrison S.*, Hepburn I., McMahon R.*, Milward S.*, Stafford Allen R.*

Mullard Space Science Laboratory, Dorking, Surrey RH5 6NT, UK
*Space Cryomagnetics Ltd, E1 Culham Science Centre, Abingdon OX14 3DB, UK

Adiabatic demagnetisation refrigerator (ADR) systems offer a convenient method for achieving temperatures below 50 mK. The magnet system described in this paper is a key component of the ADR engineering and qualification model for the proposed European Space Agency (ESA) XEUS mission, in which the ADR will be used to cool x-ray detectors. The magnet has been designed as a collaborative project between Mullard Space Science Laboratory (University College London) and Space Cryomagnetics Ltd. This paper describes the design, manufacture, and testing of the coils.

INTRODUCTION

The number of applications - such as x-ray detectors - requiring temperatures less than 1 K is increasing [1]. Currently the most widely-used systems for achieving these low temperatures are dilution refrigerators and ADRs. But dilution refrigerators are generally complicated, while ADRs are often heavy and cannot provide continuous cooling. However, if the cold stage temperature of an ADR is used in a feedback loop to control the rate of magnetic field discharge, relatively long-term operation at low temperatures is possible. Moreover, with careful design of the superconducting magnets in this particular ADR system, a light, strong assembly suitable for a satellite application has been manufactured.

XEUS requires a two-stage ADR (Figure 1), equipped with salt pills of chromium potassium alum (CPA) for the cold stage and dysprosium gallium garnet (DGG) for the intermediate stage. The salt pills are supported in the clear bore of the magnet on Kevlar strings [2]. The paramagnetic refrigerants were selected to provide an ultimate temperature below 50 mK (target 30 mK) for 24 hours with an ADR cycle time of 4 hours [3]. The heat sink is maintained at a temperature of approximately 4.5 K by a space-qualified cryocooler. A thermal budget of 5 mW is allowed for the ADR and magnets.

Ultimately, the ADR will be part of the proposed ESA XEUS x-ray observatory, and will be launched on an Ariane 5 rocket. The observatory will consist of two satellites: one for the mirrors and one for the focal plane assembly. Together they will orbit the Earth to work as an x-ray telescope. A cryogenic spectrometer, with detectors operating between 30 and 100 mK, is the current baseline system for the ESA XEUS study.

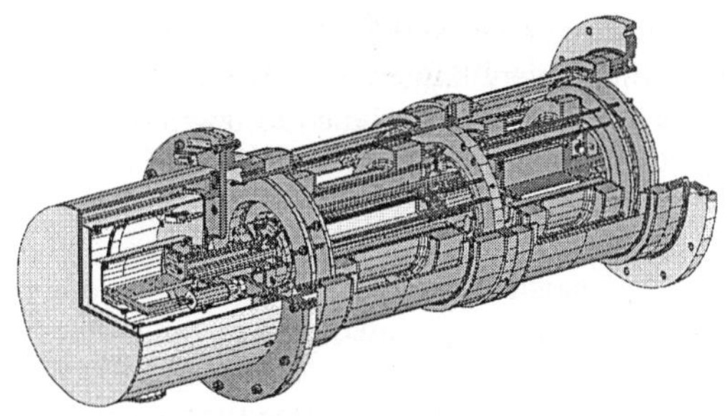

Figure 1 Cut-away view of the ADR assembly

MAGNET SYSTEM DESIGN

Magnetic design and shielding

The components of the magnet system were designed and developed with emphasis on the satellite requirements, and on constraints on the overall length and the interactions between the two ADR stages. A number of design iterations were required before the optimum configuration for the solenoids and coils could be identified (Table 1).

Table 1 Parameter list for the superconducting magnet system

Clear bore	65.5 mm
Overall length	400 mm
Outer diameter	215 mm
Maximum field on the salt pill	3.0 T
Maximum operating current	2.5 A
Superconductor	0.1 mm diameter NbTi (235 km total length)
Total magnet system mass	24.08 kg
(of which coil mass)	11.19 kg
Configuration	Solenoids with compensating and active shielding coils
Magnetic shielding	Active with additional passive material
Fringe field	$< 67\ \mu T$ 500 mm from the centre
Paramagnetic pill size	0.5 moles (CPA) 1.0 moles (DGG)

The CPA and DGG magnets both consist of a major solenoid with compensating coils at each end, surrounded by the active-shielding coil pairs. All the coils are connected in series, which minimises the number of current leads required. Additional shielding outside the coils is provided by high-permeability material used in the construction of the cryostat. The combination of the shielding materials and Helmholtz biasing coils results in a fringe field of less than $5\ \mu T$ at the detector focal plane, and less than $67\ \mu T$ at a distance of 500 mm from the centre of the ADR.

Current leads

The leads - which supply current to the magnet coils – operate in vacuum, and therefore have to be thermally anchored to the 4.5 K stage of the cryocooler. However, the cooling power available from the cryocooler is very limited, so it is vital to minimise the heat load from the current leads. For this reason, the magnet has been designed to operate at very low current, using very small (0.1 mm diameter) NbTi superconducting wire, running at 60% of its short sample performance.

Structural design

The magnet structure consists of a one-piece outer former to maximise rigidity and optimise alignment, with segmented middle and inner formers mounted from it. Sensitivity analyses were performed to determine the effect of relative displacements of each coil in the magnet system, to minimise body forces and define the tolerances required in the alignment of the magnet formers. Stress analyses at three stages of fabrication and operation (winding, cool down and full field operation) were undertaken to assist in the choice of former materials.   The CPA and DGG magnets were analysed at field separately, and again with all coils operating simultaneously. Finite Element Analysis was used in conjunction with 3-D CAD to produce the final design of the magnet system. The formers are manufactured from electrically-conductive materials, so ground plane insulation has been incorporated in the construction: insulation components are made from fibreglass and Kapton.

The two solenoids were removed from their mandrels after winding, and inserted in aluminium alloy cylinders for support. The other coils were conventionally wound on formers made from metal matrix composite. The composite material has a number of advantages, including high strength and stiffness for relatively low weight. The thermal contraction of the formers was also tailored to match the contraction of the coils, to minimise thermal stresses.

<u>Quench protection</u>

For quench protection, the solenoids and coils are subdivided into sections with resistor and diode protection circuits.

<u>Thermal design</u>

The magnet system consists of two major solenoid coils and eight secondary coils. All are cooled by conduction through heat shunts to the 4.5 K stage cryocooler: there are no liquid cryogens. Thermal modelling and eddy current analysis have been used to determine the optimum geometry for the heat shunts.

MAGNET TESTING

The completed magnet system is about to be tested at Space Cryomagnetics Ltd (Figure 2). The test cryostat has a central, room-temperature bore tube through which a Hall probe will be used to plot the field along the axis of the magnet. The DGG magnet has already been tested in isolation, and reached its design field after a single training quench.

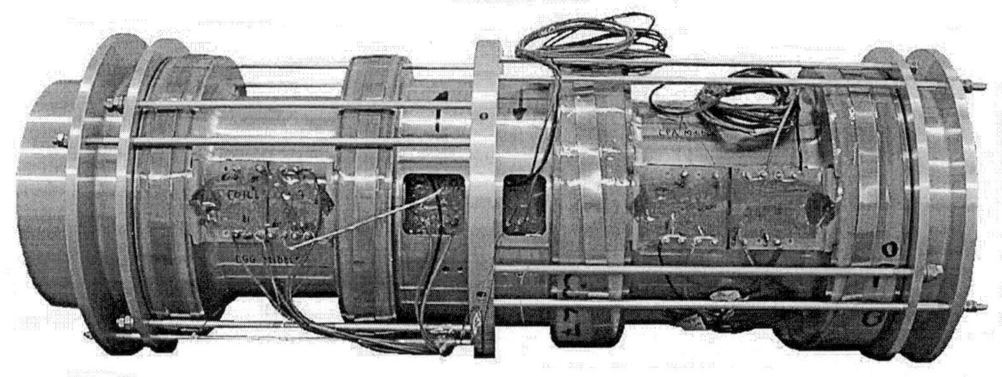

Figure 2.   The ADR magnet partly assembled

SUMMARY

A low-operating current ADR magnet has been constructed as part of the development of the XEUS x ray observatory satellite. The requirements for a cryogen-free, low current, lightweight, actively shielded ADR system are met by this design. Use of metal matrix composites in the formers, and the unusually small size of the superconducting wire, are novel aspects of the construction of this system.

REFERENCES

1.   Proceedings of the Ninth International Workshop on low temperature detectors, <u>AIP conference proceedings</u>, **605**, 2001
2.   Hepburn, I.D., et al "Space Engineering Model Cryogen Free ADR For Future ESA Space Missions" <u>Proceeding of CEC 2003</u>
3.   Bromiley, P.A., Hepburn, I.D. and Smith, A. "Ultra Low Temperature Cryogen Free Refrigerators" in <u>6th European Symposium on Space Environmental Control Systems</u>, ESA SP-400 Netherlands 1997 pp507-513

Proceedings of the Twentieth International Cryogenic Engineering Conference
(ICEC 20), Beijing, China. © 2005 Elsevier Ltd. All rights reserved.

Study of stress, strain in super-conducting magnet by Fiber-Bragg Grating

Feng Z.A.[a] ,   Wang Q.L. [a] ,   Dai F. [b], Huang G.J.[b]

[a]Institute of Electrical Engineering，CAS，Beijing 100080
[b]Institute of mechanics，CAS，Beijing 100080

Stress and strain are one of the sophisticated problems in superconducting magnet system. Although some simulation can be done to predict its performance, great concerns are taken about the real action. In this paper the performances of the Fiber Bragg Grating are investigated in liquid nitrogen temperature and are applied to measure the strain in the superconducting magnet. The experiment result agrees well with simulation result.

INTRODUCTION

Stress and strain are one of the sophisticated problems in superconducting magnet system. The prediction of stress and strain is very important for the magnet designers. Although some simulation can be done to predict its performance, great concerns are taken about the real action. Fiber optic sensors have been used extensively in measurement of strain, temperature, pressure, magnetic field and electrical current over the past years [1]. It has many advantages over conventional electrical strain gauges, such as the absence of electrical interference, high resolution and small size, especially the sensor signal is not influenced by disturbances on the signal lines from the location of the probe in the cryostat to the detector outside [2]. In this paper the performances of the Fiber Bragg Grating strain are investigated in liquid nitrogen temperature and are applied to measure the strain of superconducting magnet wound by Bi2223 tapes. The experiment result agrees well with simulation result.

THEORY OF FIBER BRAGG GRATING AND DEMARCATION EXPERIMENT

Fiber Bragg Grating (FBG) is a longitudinal periodic variation of the index of refraction in the core of an optical fiber. The spacing of the grating determined the wavelength of the reflected light. The Bragg Condition is the result of two requirements:

1.Energy Conservation: Frequency of incident radiation and reflected radiation is the same.

2.Momentum Conservation: Sum of incident wave vector and grating wave vector are equal to the wave vector of the scattered radiation (Figure 1).

$$K + k_i = k_f \tag{1}$$

The resulting Bragg Condition is: $\quad \lambda_B = 2\Lambda n_{eff}$

(2)

The grating reflects the light at the Bragg wavelength (λ_B). λ_B is a function of the grating periodicity

(Λ) and effective index (n_{eff}).Typically, $\lambda_B = 1.5$ mm, $\Lambda = 0.5$mm

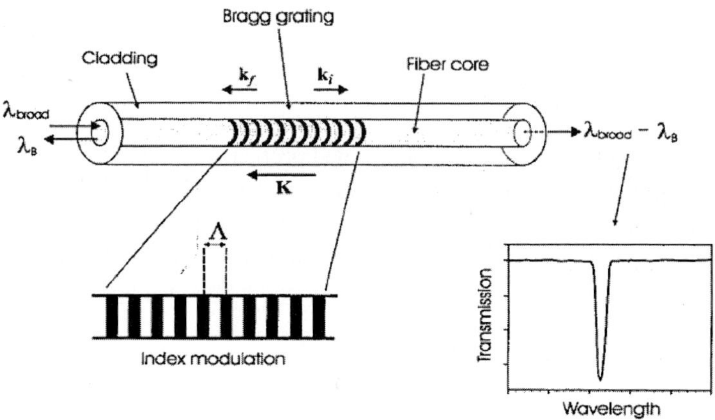

Figure 1 Fiber Bragg Grating

The shift of Bragg Wavelength can be expressed as following:

$$\Delta \lambda_B = 2nL(\{1-(n^2/2)[P12 - n(P11 + P12)]\} \varepsilon + [a + (dn/dT)/n] \Delta T \tag{3}$$

Where:

ε : applied strain,

$P_{i,j}$:Pockel's coefficient of the stress-optic tensor,

n: Pisson's ratio,

a :coefficient of thermal expansion ,

ΔT :temperature change,

$\Delta [P12 - n(P11 + P12)] \sim 0.22$,

If we keep the temperature constant, the displacement in Bragg Wavelength is approximately linear with respect to strain and temperature.During the experiment we change the strain of the beam by increasing the poise tied on the beam. Every time we increase a poise a $\Delta\lambda$ is got, at the same time the strain of the beam can been calculated by the follow function:

$$\varepsilon = \frac{M}{EI} \times \frac{Y}{2}$$
$$M = F \cdot L \tag{4}$$
$$I = \frac{bY^3}{12}$$

Where, Y: the depth of the beam, b: the width of the beam, ε :strain, $E=20 \times 10^9 Pa$

Figure 2 shows the result of the experiment. The abscissa presents the $\mu\varepsilon$ of the beam, and the y-axis presents the wavelength that is reflected. Based the experiment we can get the $\mu\varepsilon$ - $\Delta\lambda$ function:

$$\mu\varepsilon = 0.48\Delta\lambda \tag{5}$$

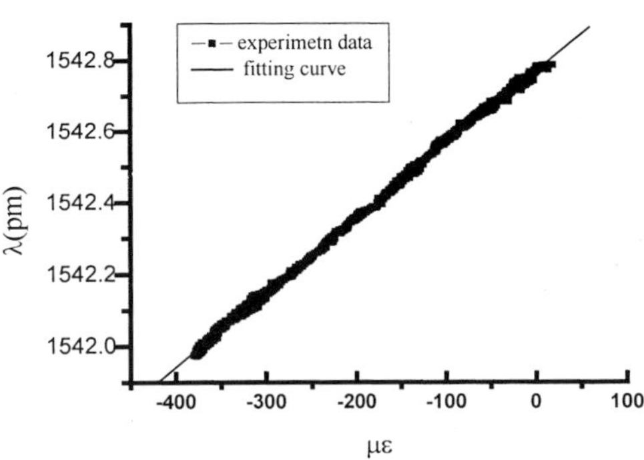

Figure 2: the relationship between wavelength and strain

Table 1: the simulation parameter

Parameter	Value	Error
A	1542.769	3.317E-4
B	0..00207	1.732E-6
R	0.9939	
SD	0.00934	
N	1744	
P	<0.0001	

MEASUREMENT OF THE REAL MAGNET STRAIN

After the demarcation experiment was done, we use it in a superconducting magnet wound by Bi2223 tapes. The Fiber Bragg Gating sensor is stuck on the flank of the magnet (Figure 3). When the current is below 20 ampere, due to the small magnet and low magnetic field, the stress and strain are too small to be measured. We regard the strain of the magnet as the strain zero point when its operating current is 20 ampere. We measure the strain of the magnet at different operating current that is more than 20A ampere by FBG strain sensor. The result of the experiment is shown in Figure 5.

Figure 3: The FBG strain sensor stuck on the magnet

SIMULATION

The finite element method (FEM) is used to simulate the magnet strain. A model that is averaged all the mechanical property according to the proportion of every material of the magnet was conducted widely. This time we build a detail model for every material of the magnet. Because our magnet is axisymmetric, a planar model is build. The analysis coupled the magnetic field and structure. The couple analysis is that the input of one physics analysis depends on the results from another analysis. We chose sequentially coupled physics analyses method using the concept of a physics environment. The term physics environment applies to both a file we create which contains all operating parameters and characteristics for a particular physics analysis and to the file's contents.

The simulation parameter is presented in Table 1. The tapes and the resin are created respectively. The detail model is created (Figure 4). The average model regard the hole magnet as one material, and the mechanical property of it depends on the proportion of each material of the magnet. The mechanical

620

property of each material of the detail model is attributed separately. So the detail model can simulate the magnet more accurately. The material property is nonlinear. The property of the tapes and the resin is plastic and elastic. The total strain components $\{\varepsilon_n\}$ are used to compute an equivalent total strain measure:

$$\varepsilon_e^t = \frac{1}{\sqrt{2}(1+v)}[\,\varepsilon_x - \varepsilon_y)^2 + (\varepsilon_y - \varepsilon_z)^2 + (\varepsilon_z - \varepsilon_x)^2 + \frac{3}{2}(\varepsilon_{xy})^2 + \frac{3}{2}(\varepsilon_{yz})^2 + \frac{3}{2}(\varepsilon_{xz})^2\,]^{\frac{1}{2}} \tag{6}$$

ε_e^t is used with the input stress-strain curve to get an equivalent value of stress σ_e.

The elastic (linear) component of strain can then be computed:

$$\{\varepsilon_n^{el}\} = \frac{\sigma_e}{E\varepsilon_e^t}\{\varepsilon_n\} \tag{7}$$

And the "plastic" or nonlinear portion is therefore:

$$\{\varepsilon_n^{pl}\} = \{\varepsilon_n\} - \{\varepsilon_n^{el}\}) \tag{8}$$

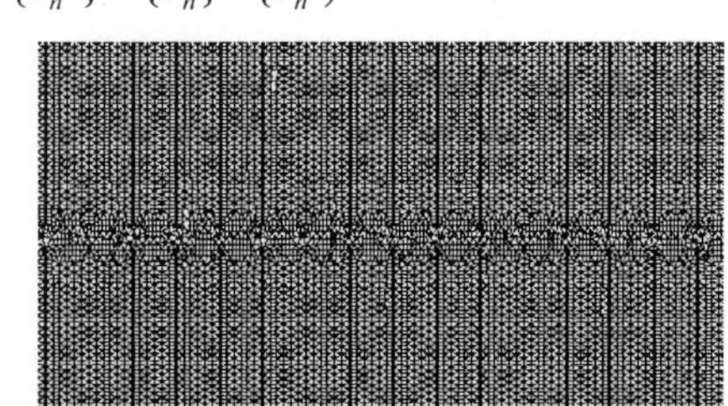

Figure 4: A part of the detail FEM model

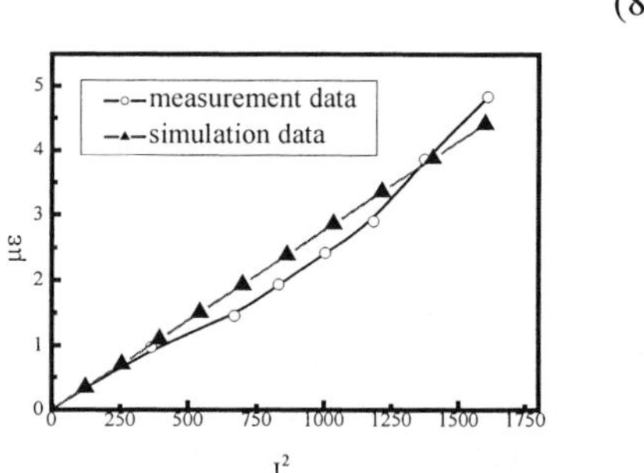

Figure 5: The experiment and simulation data

CONCLUSION

The FBG strain sensor is applied in measuring the superconducting magnet stain firstly. The simulation result well agrees with the measurement result.

ACKNOWLEDGMENT

This work is supported by the laboratory of the applied superconduct Institute of Electrical Engineering, CAS , and Institute of mechanics，CAS .

REFERENCES

[1]. R. O. Claus, K. D. Bennet , A.M. Vengsarkar and K. A. Murphy，Embedded Optical Fiber Sensors for Materials Evaluation ，Journal of Nondestructive Evaluation , (1989) 8
[2]. Johannes M. van Oort , Ronald M. Scanlan , Herman H. j. ten Kate , A Fiber-Optic Strain Measurement and Quench Location System for Use in Super-conducting Accelerator Dipole Magnets, IEEE TRANSACTIONS ON APPLOED SUPERCONDUCTIVITY, (1995) 5
[3]. K Osamura , M Sugano , A Nyilas , H S Shin, The tensile property of commercial Bi2223 tapes: a report on the international round-robin test , Supercond. Sci .Technol . (2002)15,888-893.

Design study of conduction-cooled high temperature superconducting magnet

Qiuliang Wang, Hui Huang, Shouseng Song, Yingming Dai, Baozhi Zhao, Yunjia Yun, Luguang Yan

Institute of Electrical Engineering, CAS, P.O.Box 2703, Beijing 100080, China

We are designing and fabricating a separator using the conduction-cooled high temperature superconducting (HTS) magnet for a National high-technology program in China. The magnet is made of Bi-2223 double pancake coils, and has inner and outer coil diameters of 120 mm and 211.2 mm, respectively and coil height of 202.8 mm. The superconducting magnet includes 20 double pancake coils. The operating current is about 87.8 A. This magnet is cryocooler-cooled with no liquid helium and should generate a magnetic filed of 3 T at a temperature of 20 K. We developed and tested a model Bi-2223 double pancake coil. The design and technique of the magnet is reported in the paper.

INTRODUCTION

With advent of Bi-based high temperature superconducting (HTS) tape, the critical current density for Bi-based superconducting tape is $J_C=10^4$-10^5 A/cm^2 in the operating temperature of 20-30 K and self-field. The applications of HTS tape for the superconducting magnet have been employed in fabrication of magnetic separator system. An HTS magnetic separator has no moving parts, consumes no power beyond refrigeration requirements, has no liquefying system to maintain, takes up less space, and costs less to purchase and operate than a conventional magnetic separator[1]. The HTS magnetic separators have a variety of industrial applications, most notably in the pharmaceutical, environmental, chemical fields and treatment for waste water in steel industry[2]. A high gradient magnetic separator has been conducted using a HTS magnet in China. It is supported by high technology project of China. The conceptual design and fabrication details of a conduction-cooled HTS Reciprocating Magnetic Separator are reported. Reciprocating magnetic separators are used in waste water treatment for special steel factory. The salient features and results of the electromagnetic, thermal, and quench protection analyses are discussed. The model coil has been fabricated and tested.

DESIGN OF HTS MAGNET FOR MAGNETIC SEPARATOR

The HTS magnet has 202.8 mm in length and 120 mm in inner diameter. The central operating magnetic field is a nominal 3.2 T with a design operating current of 87.78 A. The HTS magnet has a stored energy of 10 kJ. The specific parameters of the HTS magnet are listed in Table I. The configuration for the magnet is illustrated in Fig.1. The system operates in a vacuum and is conduction cooled via two-stage GM cryocooler with a nominal operating temperature of 20 K. The HTS conductor uses Bi-2223 tape and coil will be reinforced by a stainless steel tape. The HTS magnet is made of 3960 meters of Bi-2223 superconducting tape. The main parameters for the high temperature superconducting tape are listed in the Table II. The HTS tape was tested in liquid nitrogen. The profile of the critical current with respect to the magnetic field in various angles is illustrated in Fig.2 under the external field and 77 K. The HTS current leads to be used in the magnetic separator were designed to improve shock resistance. The current leads were designed to operate with the warm end at about 40-50 K and the cold end at 20 K. The HTS magnet, current leads, and thermal shield are supported by G-10 tubes and hung from the lid of the vacuum vessel. The cryocooler is mounted on the vessel lid and connected to the thermal shield top plate and the magnet cooling plate with flexible links to provide vibration isolation. The upper stage of the cryocooler cools the thermal shield and the heat pipe thermal intercepts. The heat pipe thermal intercepts combine high thermal conductivity with good electrical insulation and ensure that the upper end of the HTS portion of the current leads is adequately cooled. The lower stage of the cryocooler cools the HTS magnet and the bottom of the HTS current leads. The bottom of the leads is connected to the cooling plate by copper braid. The magnet is bolted to the cooling plate. The bottom plate of the HTS magnet and the magnet bore tube are made of copper and provide the conductive path into the windings for cooling. The magnet consists of individual double pancake windings stacked on the copper bore tube. Current lead mounting pads are provided to bolt the current leads to the magnet.

TABLE I MAGNET DESIGN PARAMETERS

Design Parameter	Value
peak radius Br-field	1.52 T
central B-field (B_{op})	3.29 T
operating current (I_{op})	87.74 A
operating temperature(T_{op})	20 K
HTS conductor	Bi-2223 PIT
Self Inductance (H)	4.355
stored energy (kJ)	10
no. of double pancakes	20
total conductor length (km)	~3.9
coil height (m)	0.2028
coil inner diameter (m)	0.12
coil outer diameter (m)	0.2112

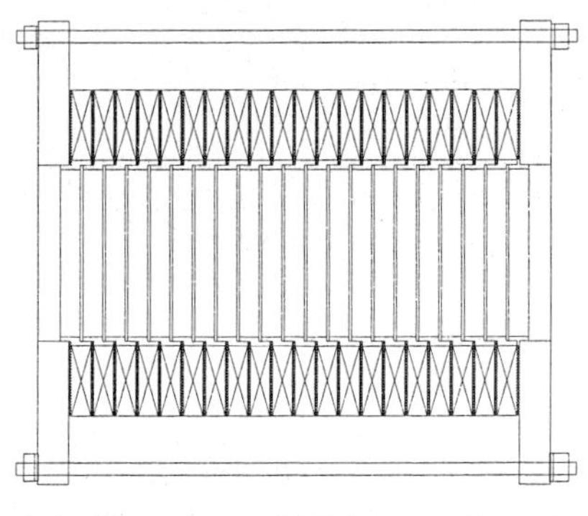

Fig.1 Configuration of HTSC magnet for separator

TABLE II Main parameters for Bi-2223 tape

Wideness	4.2 ± 0.2mm
Thickness	0.24 ± 0.02mm(with insulation thickness in 0.01mm)
Length	200m
Filamentary number	61
Filling factor	0.3 ~ 0.35
Density	4.5g/cm^2
Engineering current density	$\geqslant$ 60A(77K, self field)
Max. tensile stress	100Mpa(5%Ic degradation)
Max. tensile strain	0.15%(5%Ic degradation)
Min. bending radius	30mm(5%Ic degradation)
Critical temperature	110 K
Insulator	Maylar
Breakout voltage	300V(10μm, 300 K)
Thickness of insulator	$\leqslant$ 10μm

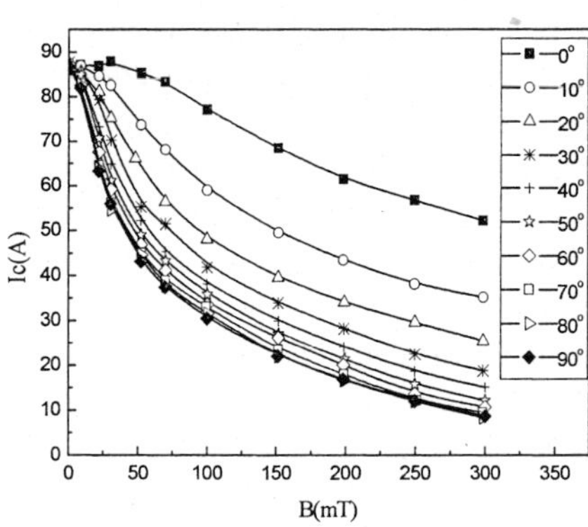

Fig.2 Critical current versus magnetic field

For the turn-to-turn coil insulation, a coating of 0.01mm in thickness was applied to the Bi-2223 tape as the tape was being wound into its double pancake coil. The double pancake coil was separated from its winding mandrel and spiral wrapped with glass insulation. The wrapped pancake coil was then placed on top of a copper sheet. The purpose of the copper sheet was to improve axial thermal conduction. The double pancake coil and its copper insert were then epoxy impregnated. After epoxy resin DW-3 impregnation, the double pancake coils were stacked and spliced together on the inner diameter to form a continuous double pancake coil. The double pancakes were then stacked in a precision fixture and subsequently spliced at the outer diameter using a copper transition piece.

The magnetic field analysis was performed on the HTS magnet. There were the primary areas of interest in the analysis: axial B-field strength and homogeneity to establish the waste water processing parameters, radial B-field strength to establish the critical current margins and performance of the HTS tape. The distribution for magnetic field in the HTS magnet is plotted in Fig.3. Where the center field in the magnet is about 3.23 T. The maximum radius magnetic field in the magnet is shown in the Fig.4, where the maximum radius field is located at the center of top in the magnet.

The analysis in HTS magnet on mechanical characteristics is shown in the Fig.5 for strain and Fig.6 for the hoop stress, the maximum hoop strain in the HTS magnet is about 2.0×10^{-6} . The maximum hoop stress is located at the inside middle-plane in the magnet. Its value is about 4.747 MPa.

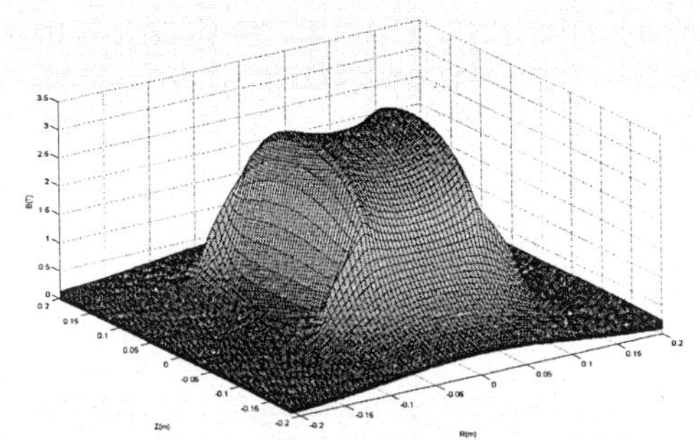

Fig.3 Magnetic field distribution in magnet (unit: T).

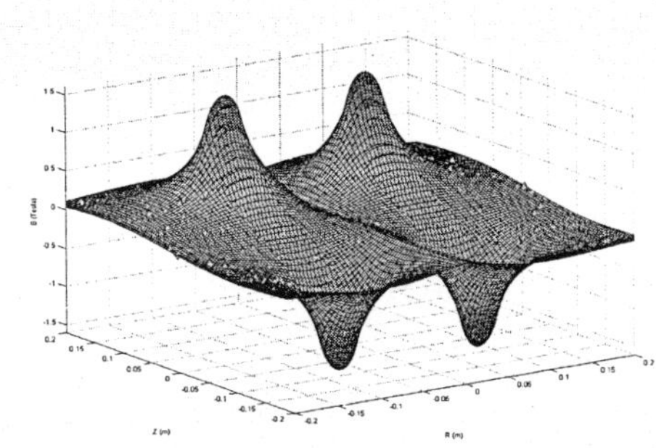

Fig.4 Radius component of magnetic field distribution(unit: T).

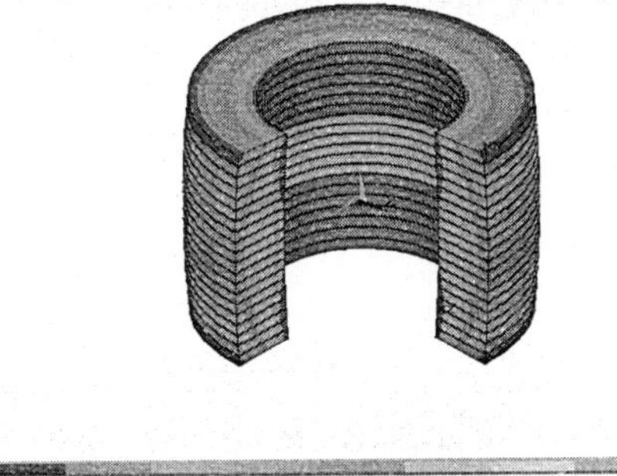

Fig 5 Hoop strain in HTS magnet.

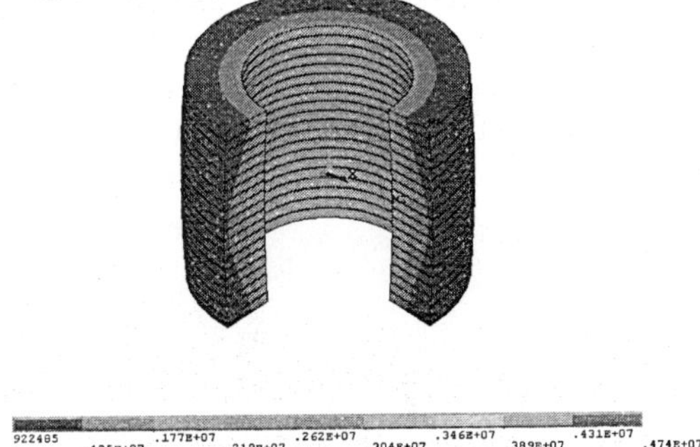

Fig 6 Hoop stress in HTS magnet (Unit: Pa).

CRYOSTAT, THERMAL LOAD AND MODEL HTS COIL

The cryostat for the high temperature superconducting magnet has the outer diameter of 650 mm and height of 805mm and weight of 160 kg. The cryostat had a penetrating room-temperature bore of 61 mm in diameter. The two-stage GM cryocooler was adjacently mounted on one side of the cryostat. In order to avoid a decrease of cooling capacity caused by the influence of the magnetic field, the magnet was cooled down using high purity copper braid for a flexible thermal link which connected magnet with second stage. Super-insulation of 15 layers was provided around the superconducting coil to minimize thermal losses. Coil terminals was cooled through pieces of ALN chips. The magnet was protected by using 4 diodes. A pair of copper current leads was anchored between the room temperature and the first stage of cryocooler. A thermal radiation shield was mounted on the flange of first-stage to reduce heat radiation to the coils and HTS current leads. A slit was installed in the wall of the thermal radiation shield to cut the eddy current, and two reinforcements with stainless ring were mounted to the two ends of the thermal radiation shield. Operational heat loads for the thermal analysis were calculated. The heat load at the first cool head of cryocooler is about 11.08 Watt and the heat load at second-stage cool head is about 2.075

Watt. The cooling capacity of the cryocooler as a function of temperature was obtained from the commercial manufacturer, and at 20 K the cooling capacity is about 20 Watt. Fig.7 shows the temperature different (ΔT) between the magnet and second-stage cool head with respect to the heat load during thermal contact length (L) and cross-sectional area (A) given. It shows that the highly pure copper as the thermal contact can reduce the temperature different. Fig.8 shows ΔT with respect to cross-sectional area of the thermal contact during heat load given.

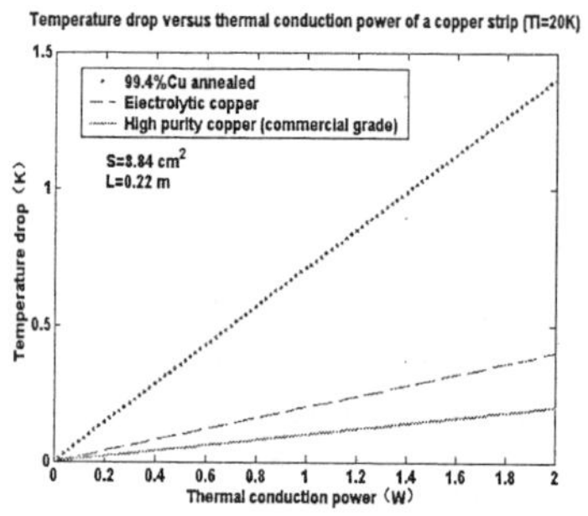
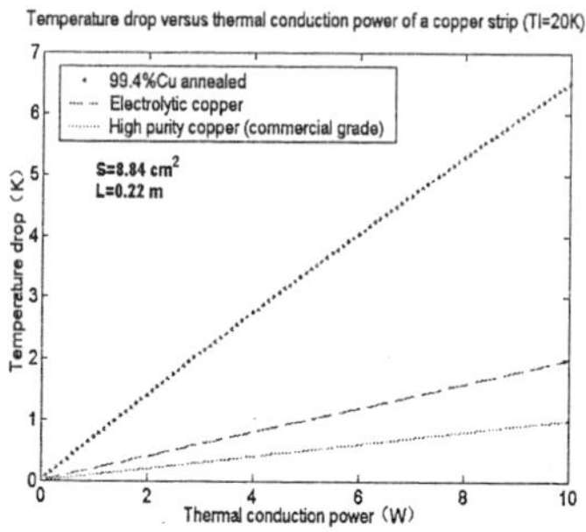
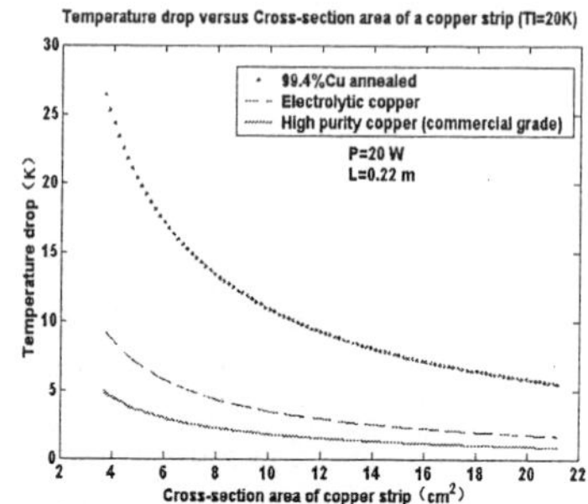

Fig.7 Profiles of temperature different with respect to heat load.　　Fig.8 Profile of ΔT versus cross-sectional area.

In order to study the fabricating technology, a double pancake coil has been fabricated and tested. The photograph of the double-pancake coil is shown in Fig.9. The fabrication technology is based on the wet-winding technology, and it is impregnated by DW-3. The first double pancake coil is with an inner diameter of 80 mm, outer diameter of 199.3 mm, thickness of 8.75 mm and total turn of 222. It was tested in liquid nitrogen with voltage criteria 1 μ V/cm. The critical current is 36.5A. The calculation shows the maximum radius field 0.15T, and axis magnetic field is 0.3T. The E-J curve is shown in Fig.10.

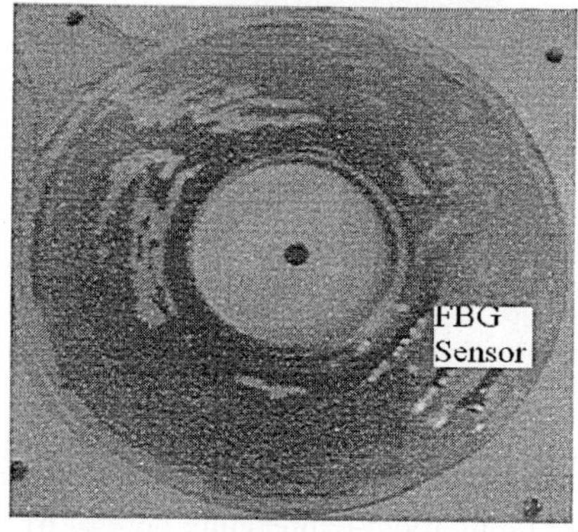

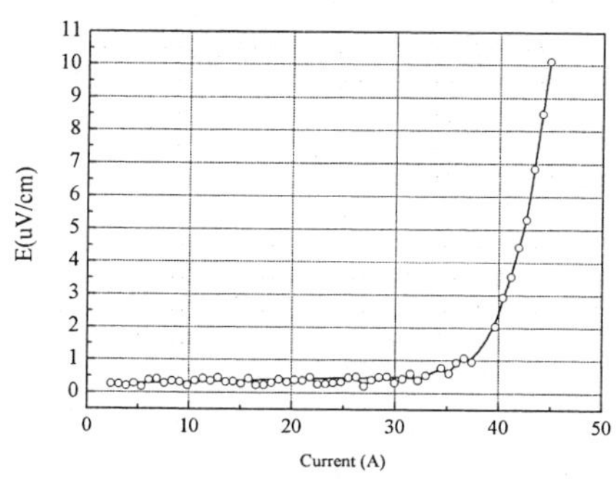

Fig.9 Photograph of fabricated HTS coil.　　Fig.10 Test results for HTS coils

CONCLUSION

The design of HTS magnet for the magnetic separator has been completed, The fabrication of the full size HTS magnet will be finished by the end of the July, 2004.

REFERENCES

[1] H. Kumakura etc. PhysicaC 350(2001)76-82.
[2]K.Ohmatsu etc. IEEE Transactions on Applied Superconductivity 9(1999)924-928.

Testing the coils of the superconducting magnet for the Alpha Magnetic Spectrometer

McMahon R., Harrison S., Milward S., Stafford Allen R., Hofer H.*, Ulbricht J.*, Viertel G.*,
Ting S.C.C.†

Space Cryomagnetics Ltd, E1 Culham Science Centre, Abingdon OX14 3DB, UK
*Eidgenössische Technische Hochschule, CH-8093 Zürich, Switzerland
†Massachusetts Institute of Technology, 51 Vassar Street, Cambridge, MA 02139-4307, USA

The Alpha Magnetic Spectrometer (AMS) is a particle physics experiment for use
on the International Space Station (ISS). At the heart of the detector is a large
superconducting magnet system cooled to a temperature of 1.8 K by 2500 litres of
superfluid helium: both the magnet and cryogenic system are currently under
construction by Space Cryomagnetics Ltd of Culham, England. The magnet
consists of 14 superconducting coils arranged around the 1 m diameter warm
bore. Each coil is tested under flight-like cryogenic conditions before assembly
into the final configuration. This paper describes the design of the testing facility,
and the results from the coil tests.

INTRODUCTION

The AMS experiment is designed to examine the fundamental physics of the universe, in particular
through the search for antimatter and dark matter. Following a successful precursor mission on the US
Space Shuttle (STS-91) the AMS collaboration decided to increase the sensitivity of the detector by
upgrading the original permanent magnet arrangement to a superconducting system.

THE AMS MAGNET

The magnet consists of 14 coils arranged around
the 1 m diameter ambient temperature bore
(Figure 1). The two larger "dipole" coils generate
magnetic field perpendicular to the axis of the bore
which is useful for resolving incoming charged
particles. The remaining twelve "racetrack" coils
constrain the return flux to reduce the stray field
from the magnet. This is vital for operation in
space; otherwise the field from AMS would
interact with the Earth's magnetic field to put a
significant torque on the ISS [1]. A list of the
principle magnet parameters is given in Table 1.

CRYOGENIC SYSTEM

The experiment will be launched fully cold with
an inventory of 2500 litres of superfluid helium.
The helium gradually boils away throughout the
mission, and the experiment will end when it finally runs out.

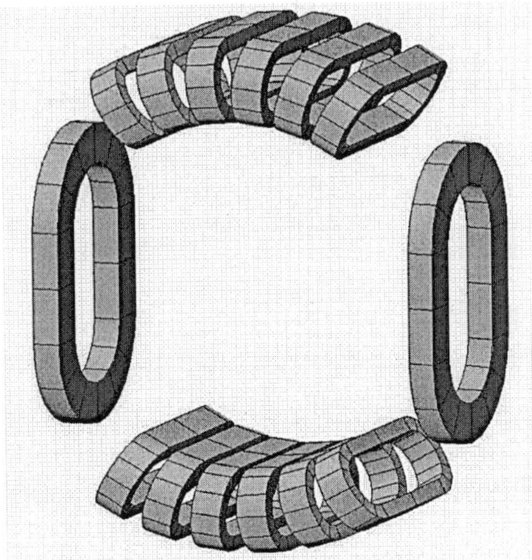

Figure 1 Layout of the AMS Magnet Coils

One of the most important features of the cryogenic system is that the coils are not located inside the helium tank. This feature means that the magnet is able to quench in space without losing all of its helium inventory, and so can be re-cooled and operated again without requiring another mission to refill the helium vessel. The costs of a launch are so high that, no matter how unlikely a magnet quench may be, it is necessary to provide this fallback position.

For this reason, each coil is suspended inside the vacuum space, with copper heat shunts for removing heat from the winding. The heat shunts are connected in a number of positions to a thermal bus bar, which simply consists of a copper pipe filled with superfluid helium. Internal convection inside the pipe is sufficient to transfer the heat away from the coils to the helium vessel.

Details of the design of the magnet cryogenic system have been published previously [2].

Table 1 Key magnet parameters

Parameter	Value
Magnet bore	1.115 m
Outer diameter of vacuum case	2.771 m
Length of vacuum case	1.566 m
Central magnetic flux density	0.87 T
Maximum magnetic flux density	6.59 T
Maximum operating current	459.5 A
Stored magnetic energy	5.15 MJ
Inductance	48.4 H
Operating temperature	1.8 K
Target endurance on orbit	3 years
Total mass (excluding vacuum case)	2300 kg

COIL TESTING REQUIREMENTS

The AMS magnet coils will all be tested individually before assembly into the flight configuration. It is advantageous to be aware of any potential problem at the earliest possible stage, and the cost of rectifying a problem after the magnet has been assembled could be prohibitive.

To obtain the most benefit from the experiments, the coils need to be tested in conditions similar to flight. This means that the mechanical and cryogenic environments should be modelled as closely as possible (although, unfortunately, zero-gravity is not feasible).

The mechanical loadings on the coils are very different when they are charged individually, compared with the forces generated when the adjacent coils are present. In particular, the forces on the racetrack coils are much lower when tested singly than they are when the fully-assembled magnet is charged to the same current, but the forces on the dipole coils are higher during the single coil test. It is most important not to overload any parts of the coils during testing, so all are charged until some part of the winding experiences the same load that it will see in the full magnet assembly. During single coil testing, the racetrack

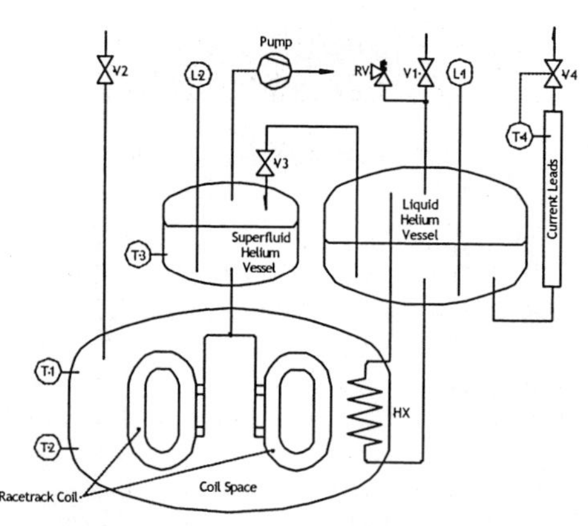

Figure 2 Schematic diagram of the coil test facility

coils are charged to currents higher than the nominal magnet current (four to 562 A, and eight to 600 A), but the dipole coils can only be operated at 335 A.

To make the test cryogenically representative of the flight system, the coil is mounted in vacuum, with its heat shunts connected to a pipe filled with superfluid helium. Figure 2 is a schematic diagram of the arrangement. (The liquid nitrogen shield and outer vacuum case have been omitted for clarity.)

TEST PROCEDURE AND RESULTS

The test facility is large enough to accept either two racetrack coils simultaneously (as in Figure 2) or a single dipole coil. Although this allows two racetrack coils to be cooled down together – saving a considerable amount of time – the coils have to be energised separately because of the large forces which would otherwise exist between them.

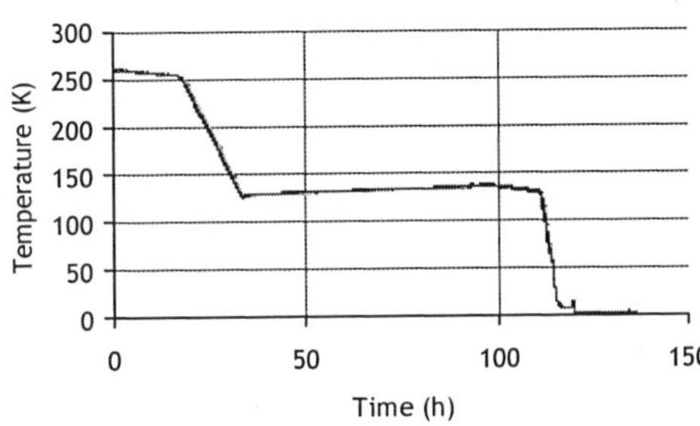

Figure 3 Racetrack coil cool down

Figure 4 The last two racetrack coils with the test facility

Cool down

Cooling the coils to their operating temperature of 1.8 K in vacuum is a necessarily complicated process. The objectives are to cool the coils as quickly as possible, while minimising the consumption of liquid helium, and avoiding large thermal gradients at temperatures above 100 K (which could damage the coils by differential thermal contraction).

First the coil space vessel is filled, through valve V2, with helium at ambient temperature as an exchange gas. The liquid helium vessel is then filled with liquid nitrogen, filling heat exchanger HX and setting up convection in the coil space which cools the coils. Once below about 120 K, the liquid nitrogen is removed and the coils are cooled directly by transferring liquid helium into the coil space until the temperature drops to about 10 K. Then the coil space is evacuated, and the liquid helium vessel filled with liquid helium at about 4.5 K. Once this vessel is full, some of the helium is transferred through valve V3 to fill the superfluid helium vessel. Finally, the helium in this vessel is pumped down to 16 mbar or less to reduce the temperature below 1.8 K. This helium fills the pipe connected to the heat shunts on the coils, and thus the coils are cooled to 1.8 K in the vacuum of the coil space vessel.

Figure 3 shows a typical cool down profile for one of the racetrack coils. The gentle cooling for the first 20 hours was simply a result of filling the radiation shield of the test rig with liquid nitrogen. The more rapid cooling between 20 and 30 hours was achieved by cooling the heat exchanger with liquid nitrogen transferred into the liquid helium vessel, as described above. About 70 hours later (the delay was due to a holiday period) helium was applied directly to the coil space: this corresponds to the rapid cool down from 130 to 10 K. Finally, liquid helium was transferred into the helium vessel and pumped down. Note that Figure 3 actually consists of the traces of 7 temperature sensors located at various positions on the surface of the coil: cooling was so evenly distributed that the coil was virtually isothermal.

<u>Energisation</u>
The coil is charged from an external power supply through vapour-cooled current leads. The changing current causes eddy currents to flow in the aluminium coil former (see Figure 4), and the warming effect of these can be seen from the thermometry to be a strong function of the current sweep rate. This warming effect will not be a problem for charging the magnet in orbit since the current sweep rate is much lower for the full magnet than for a single coil.

<u>Training behaviour</u>
Some of the racetrack coils have experienced training quenches at currents between the nominal magnet current (459.5 A) and the test current (either 562 or 600 A). Where such quenches have occurred, the coil has been re-cooled and re-charged. All twelve racetrack coils have been successfully charged to the maximum test current: in the worst case four training quenches were needed, but six of the coils required no training at all. None of the coils has quenched below the magnet operating current of 459.5 A.

<u>Quench testing</u>
The AMS magnet uses an electronic protection system to ensure that, in the event of a quench of the fully-assembled magnet, the stored energy is distributed evenly among the 14 coils. Each coil is therefore equipped with two heaters which can be powered externally to initiate a quench. One of the objectives of the tests was to check the operation of the heaters, and to measure the energy required to force a quench. Each coil – regardless of its training history - has therefore been deliberately quenched from the magnet operating current of 459.5 A, then re-cooled and re-energised.

Quench heater pulses with a power of 200 W were applied to the coils for periods increasing in increments of 10 ms. Although there was some scatter in the results from different coils, most of the racetracks required a pulse lasting 90 ms to initiate a quench. This means that the energy to quench is 18 J, so the coils are very stable.

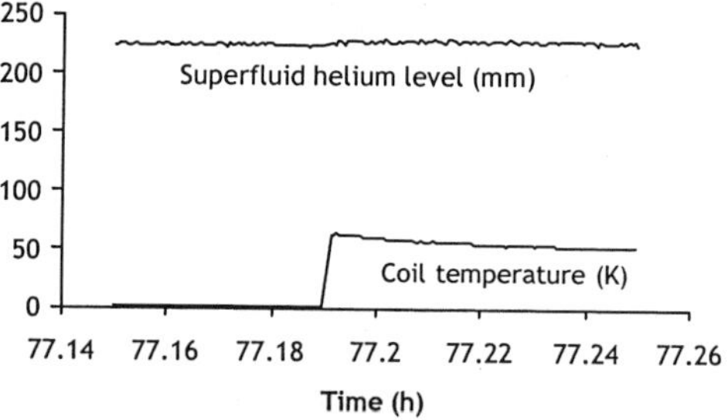

Figure 5 Temperature and level in a quench

It is of particular importance to AMS that a magnet quench does not lead to rapid loss of the helium inventory, because it is a requirement of the system that it should be possible to re-cool and energise it in space without the option of re-filling with liquid helium. Figure 5 shows how this has been achieved with the present design of indirectly-cooled coils: the level of the superfluid helium in the vessel remains almost constant, even though the temperature of the coil itself approaches 60 K.

CONCLUSIONS

A special test facility has been designed and constructed which allows the AMS superconducting magnet coils to be tested individually in a cryogenic environment which approximates to conditions on the International Space Station. The twelve racetrack (flux return) coils have been tested at currents exceeding the assembled magnet current, to simulate the mechanical loadings in service. Quench testing has also been performed on each coil to check the operation of the quench heaters and the stability of the coils. All tests have been successful, and the coils are now being assembled into the flight configuration.

REFERENCES

1.   Blau, B., Harrison, S.M., et al., The superconducting magnet system of AMS-02 – a particle physics detector to be operated on the International Space Station, <u>IEEE Transactions on Applied Superconductivity</u> (2002) <u>12</u> 349-352
2.   Harrison, S.M., Ettlinger, E., et al., Cryogenic system for a large superconducting magnet in space, <u>IEEE Transactions on Applied Superconductivity</u> (2003) <u>13</u> 1381-1384

Characterization of the quenching process and the protection of an ac high temperature superconductor coil

Li X. H., Wang Y. S., and Xiao L. Y.

Applied Superconductivity Laboratory, Institute of Electrical Engineering, Chinese Academy of Sciences, Beijing 100080, China

Experimental characterization and numerical simulation of the quenching process in a test high temperature superconductor (HTS) coil were made. The results show that heat accumulation is the key point that causes quenching. The effective critical current of the coil decreases with the temperature increasing. This enhances both the ac loss and the current sharing effect. Consequently, the first step of quenching will occur within a second. Detecting this and making a quick response are important for protecting the ac coil. Based on our results, some criteria were concluded for detecting this first step of quenching.

INTRODUCTION

Devices based on the HTS are nowadays more and more widely utilized in a variety of areas, especially in electrical power applications, such as transformers, power cables, motors and SFCLs [1]. Many of these applications consist of a HTS coil. Therefore, the stability of the HTS coil working at high ac currents becomes more and more important. The modeling and simulating work done recently based on a test HTS coil confirms the important role of heat accumulation in the quench processes [2]. The heat sources can be very different, a typical one is the Joule heat generated by the point at which the working current $I_w >$ I_c. Here I_c is the critical current of the superconductor at the point. Other sources such as vortex motion and mechanical disturbance must be also considered in the evaluation of the coil stability. However, in dc cases, the heat accumulation normally starts only after $I_w > I_c$. The heat generated in the "hot" point can be transferred to the environment, and the point is then cooled down. A thermal runaway quench only occurs after the cooling rate is exceeded by the heating rate. In ac cases, the competition of cooling and heating is in principle the same except that the coil is heated up continuously by the ac loss as long as I_w is applied. Since the ac loss is proportional to I^α, where α is decided by the frequency [3], with increasing of I_w, an "unexpected" quench can occur without any obvious disturbance. This is very harmful for the coils used in power applications and magnets. To evaluate the effects of ac losses on the quench process, simulation based on the model of the test coil [2] and experimental investigation on the same coil were made. Combining the simulation and experiment results, some criteria of quench detection can be concluded for the quench protection system. All ac currents and voltages used here are RMS values.

SIMULATION AND EXPERIMENT

The simulation model based on the test coil was developed as described in a previous work [2]. A working current slightly beyond the maximum safe working current [4] was applied to make sure that the coil would quench after working for several minutes. As shown in Fig.1, the temperature of the hottest point in the coil reaches the zero-field T_c at about 400 second. However, as shown in Fig.2, the resistance and the voltage across the coil started to rise much earlier. Therefore, it is rather difficult to define when the quench starts. Actually, according to the simulation result, we argue there are two steps of quench in the coil. The first step occurs at part of the coil becomes non-superconductive, either due to heat accumulation or an external disturbance. This step is reversible as the part can be cooled down and become superconductive again. However, the second step is non-reversible, since it occurs after the heating rate exceeds the cooling rate. To protect a coil from quenching, it is necessary to cut off I_w before the second step. Hence, detecting of the first step of quenching is very important.

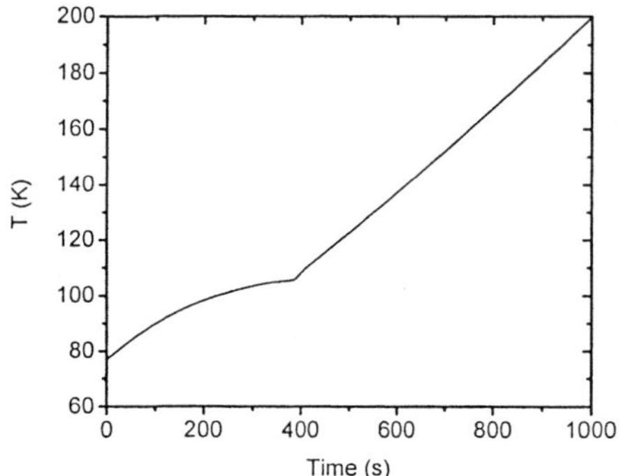

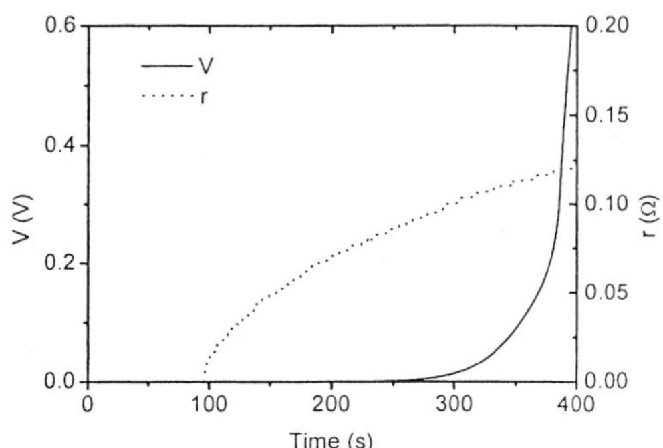

Figure 1 The simulated temperature T of the hottest point in the test coil quenching by over-current

Figure 2 The simulated voltage V and resistance r across the coil during the quench process

The heating rate of a certain section in the coil is not a constant. The critical current I_c at a certain section is a function of the magnetic field B and the temperature T at that section. It will drop to nearly zero at a temperature close to T_c. When $I_c < I_w$, current sharing will occur and causes Joule heat. Moreover, as I_c dropping, the ac loss is also increasing. This has to be added to the heat accumulation, too. Therefore, finding out when the first step of quenching occurs by simulation demands a lot of calculation.

On the other hand, to investigate the quench process experimentally is also difficult. A simple test circuit was setup in our work. It consisted of the coil, which was soaked in liquid N_2, a programmable ac power source and an ac voltage meter measuring the voltage across the coil. Variant I_w was applied to the coil by the power source. As shown in Fig. 3, the voltage across the coil V increases with I_w, nearly linearly. No clear quenching effect of V as shown in Fig. 2 can be detected, even if I_w already exceeds the dc I_c of the coil, which is 2.8 A. Besides this, as shown in Fig. 3, the V/I value is also much higher than the commonly expected resistance of the coil. This is the effect of the inductance. As shown in Fig. 4, a series of total resistance r can be measured by changing frequencies f of I_w. The inductance of the coil can then be calculated from the relation of r and f. However, the calculated inductance is not the same for different frequencies. This indicates the voltage across the coil was generated from more complicated sources besides the inductance and the dc resistance of the coil. As shown in Fig. 2, the dc resistance of the coil is about $0.12\ \Omega$, if it is in the normal state. One of these extra voltages across the coil comes from the ac loss. As the current and frequency increase, the voltage generated by the ac loss becomes more and more significant. A lock-in technique can be used to divide this voltage and the effect of the inductance, because the ac loss voltage has the same phase of I_w, while the effect of the inductance is 90° later. But it is not possible to distinguish the ac loss voltage and voltage rise of quenching. Although it was thought that the quenching process will happen in a sudden, as shown in Fig. 2, the voltage across the coil V increases slowly and mach later after the first step of the quenching. On the other hand, as shown in the simulation, the increase of V caused by quenching, which is about 0.5 V, is also not very obvious compared to the effect of the inductance and the ac loss voltage. Actually, in our experiment, the sharp increasing of V due to quenching was not clearly detected at all.

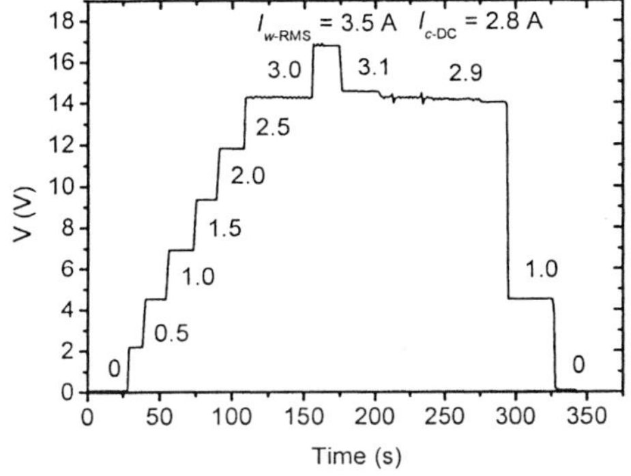

Figure 3 The experimental voltage V across the test coil while applying variant ac currents

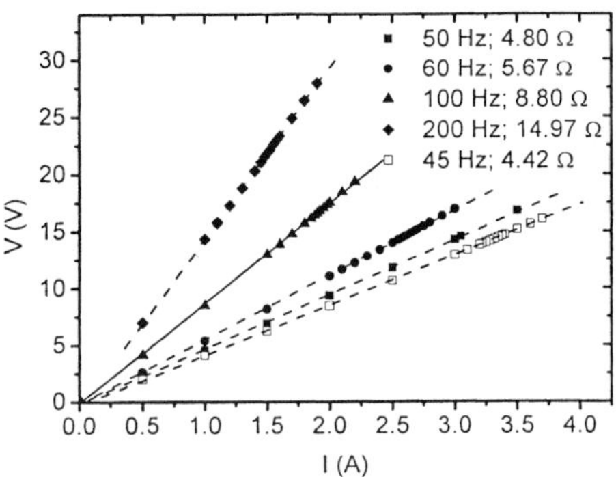

Figure 4 The I-V correlation of the coil at different frequency f of the working current.

By analyzing the experimental data more carefully, we found that, at a certain I_w, the voltage across the coil V is not as stable as at other I_ws. In Fig. 5 this phenomenon is illustrated. The voltage oscillation is not very significant, but it is beyond the error level of the ac power source. Tentatively, we take this as an evidence of the ac critical current I_c' of the coil. At I_c', the coil starts to quench. The quenching process can be described as below: At first, because of heat accumulation or disturbance, some section of the coil reaches a temperature where $I_c = I_w$. Current sharing then starts at this section, which leads to an increase of the resistance and voltage. After this, the section is heated up to a higher temperature. With higher temperature, the cooling rate of the section can be a little faster, too. If the cooling rate exceeds the heating rate, the section can be cooled down again. Finally, at the cooled section, the heat will accumulate again, and starts another cycle of the oscillation. As shown in the figure, the period of the oscillation suggests that this reversible first step of quenching can occur within one second.

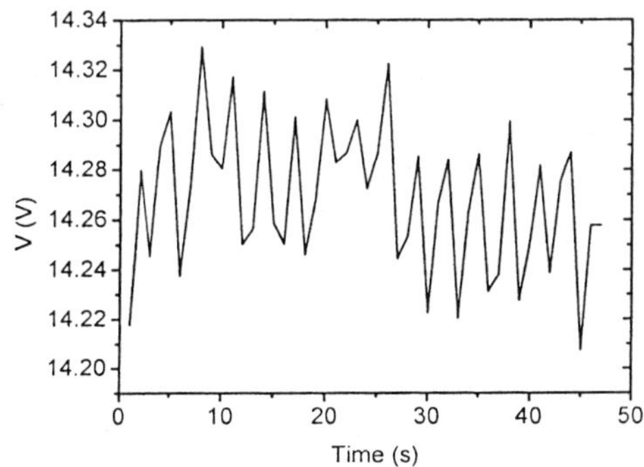

Figure 5 A typical oscillation of the voltage across the coil at the ac critical current.

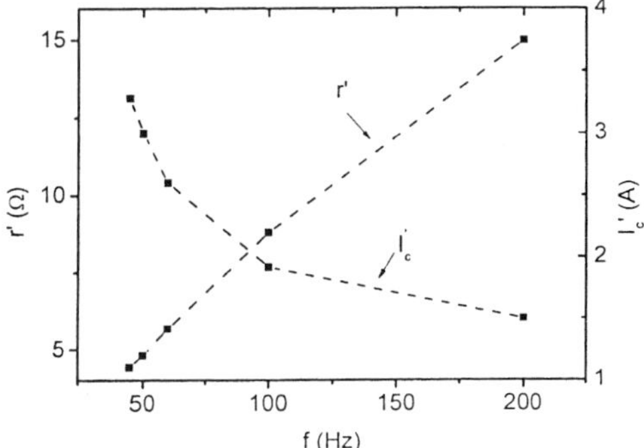

Figure 6 The frequency effect on the impedance r and the ac critical current I_c'

According to the phenomenon described above, the ac critical current I_c' of the test coil was measured at different frequencies. The result is shown in Fig. 6, along with the total impedance r' calculated by linear fitting of the I-V correlation curves shown in Fig. 4. From this figure, it is clear that, with increasing frequency, I_c' nonlinearly decreases, while r' increases nearly linearly. This demonstrates that at the higher frequency part, the ac loss heating, which is proportional to f, is more pronounced. For the test coil used here, an extra part of the ac losses, the vortex loss, which is not included in the simulation may also be significant at the higher frequency part [3].

DISCUSSION

From the simulation and experimental results, it is demonstrated that in an HTS coil, the quenching process can have two steps. The coil maybe heated up by disturbance and ac losses. As the temperature rises, the current sharing starts at the section where $I_c <= I_w$. Then the first step of quenching begins. Since the resistivity of the sheath metal is low at low temperatures, the increase of the resistance due to current sharing is not obvious. Despite of this, the sections where the current sharing happens and the adjacent area will be further heated and this will finally lead to a dangerous heating up if the cooling conditions are not very good. This is the second step of quenching. To avoid it, the protecting system must be started before the coil loss its thermal stability. However, it is difficult to detect by common methods if there are only a few sections in the coil start to lose their superconductivity. From our results, a key point which can help in such detection is a pronounced oscillation of the voltage across the coil at I_w close to I_c', especially when the coil is connected to a power source running in current constant functions. Besides this, the calculated total resistance is also an evidence for the first step of quenching because that in most metals, the resistivity is proportional to the temperature, if the coil is heated up by current sharing, its resistance will rise continuously, although maybe very slowly. Based on these criteria, the I_c' of the test coil used here was detected manually as described above.

CONCLUSION

In this paper, a two-step model of the quenching in the HTS coil was suggested. The heat accumulation comes from ac losses and current sharing was attributed for the quenching development in the coil. According to the modeling and the experimental results, a possible method to detect the first step of quenching was developed, which can be helpful in the power usage of the ac HTS coils.

ACKNOWLEDGEMENT

This work was supported by Chinese National Fund of Sciences, grant No. 50137020.

REFERENCES

1. Wang, Y. S., Zhao, X., Xu, X., Xiao, L. Y., Lin, L. Z., Lu, G. H., Hui, D., Dai, S. T., Angular and Magnetic Field Dependence of Critical Current Density of Multifilamentary Bi-2223 Tapes, Supercond. Sci. Technol. (2004) 17 705-709
2. Li, Y., Li, X. H., Song, N. H., Xiao, L. Y., Numerical study on stability of HTS coil under high pulse currents, Chinese J. of Low Temp. Phys. (2004) 26 76-80
3. Zhang, G. M., Lin, L. Z., Xiao, L. Y., Qiu, M., and Yu, Y. J., The angular dependence of ac transport losses for a BSCCO/Ag tape in dc applied field, IEEE Trans. on Appl. Superdond. (2003) 13 2972-2975
4. Li, X. H., Chen, G., Li, Y., and Xiao, L. Y., Evaluation of the stability profile for a HTS coil by numerical simulation, to be published

Proceedings of the Twentieth International Cryogenic Engineering Conference
(ICEC 20), Beijing, China. © 2005 Elsevier Ltd. All rights reserved.

Technical analyses of HTS bulk used in a rotating machine

Qiu M., Huo H.K., Xu Z., Yao Z.H., Xia D., Lin L.Z.

Institute of Electrical Engineering, Chinese Academy of Sciences, Beijing 100080, P.R. China

HTS bulks are used for developing novel rotating machines. In these systems, the qualities of HTS bulk are closely related to the structure, start-up and operation of motors. Based on the experimental results and theoretical simulations, some technical considerations, such as pre-magnetization, dynamic performance and structural strengthening of HTS bulk motor, were discussed in details.

INTRODUCTION

There is a continuing need for new motors with reduced size and weight, higher efficiency, and increased reliability. Commercial HTS bulks provide excellent potential to design high-efficiency motors with improved performance to respond to these new requirements. Based on peritectic solidification processes and different philosophies on improving the performance, various techniques have been adopted to produce large-domain HTS bulks with high Jc. It was reported that MTG-YBCO bulk can trap a strong magnetic field over 1 T at 77 K, which may be dramatically enhanced by irradiation and lowering the temperature. A large light rare-earth (LRE) BCO bulk is believed to trap very high magnetic fields of more than 5 T at 77 K. Such bulks are tried to construct novel rotating machines, such as hysteresis, reluctance, permanent-magnet (PM) and linear motors. At present, a maximum output power of HTS bulk motor has reached 38 kW at 77 K. In these motors, the characteristics of HTS bulk should be fully taken into account in the design, construction and operation. Based on our experiments and theoretical analyses, some technical considerations for HTS bulk motor were proposed and discussed in the paper.

PRE-MAGNETIZATION

As we know, the generated torque of a hysteresis machine is equivalent to the loss in the HTS bulk rotor. It is important to allow adequate fluxes to permeate the rotor. For situations where the penetration is less than optimal, the torque can be increased by partly-magnetization of HTS bulk rotor beforehand. As an alternative to field windings on the rotor, HTS bulks in the form of cylinders carrying currents which circulate an iron member can be used as PM in a synchronous machine. This necessarily requires HTS bulks in the assembly to have trapped a field by means of pre-magnetization. For these kinds of motor, we have no alternative but to perform impulse magnetization by using stator / extra auxiliary coils.

Figure 1 shows an equivalent circuit of pulse magnetizing system. R and L are the external circuit resistance and leakage inductance of the magnetizing coil, The impulse field is generated by discharging a large capacitor bank into the coil. The circuit equation is written as:

$$\oint_c \frac{\partial \vec{A}}{\partial t} \cdot d\vec{s} + (R + R_c)\frac{dQ(t)}{dt} + L\frac{d^2Q(t)}{dt^2} + \frac{Q(t) - Q_0}{c} = 0 \tag{1}$$

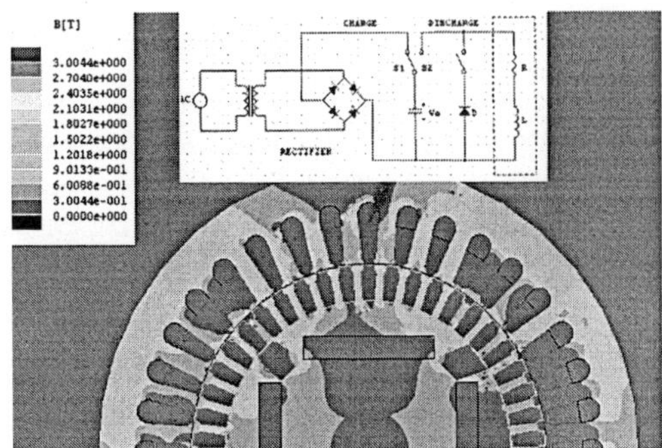

Figure 1   Equivalent circuit of pulse magnetizing system
and Instantaneous field distribution inside HTS motor

Table 1   Specification of the analysis model

Section	Item	Value (Unit)
Stator	Number of phases	3
	Number of slots	36
	Outer diameter	245 (mm)
	Inner diameter	162 (mm)
	Number of turn / phase / pole	28 (turn)
Rotor (PM)	Number of pole	4
	Outer diameter	161 (mm)
	Radial thickness	10 (mm)
	Axial Length	58 (mm)
	Residual Flux density	1.0 (T)
Air gap	Mechanical air gap	0.5 (mm)

where A is the magnetic vector potential, R_c is the resistance of the magnetizing yoke, $Q(t)$ and Q_0 are the electric charge of the capacitor at the time t and its initial value. Considering $M(H)$ curve of HTS bulk obtained from Kim model, instantaneous field distributions in the regions of the magnetizing coil and HTS bulks were simulated by FEA method (see Figure 1) using fundamental equations as follows:

$$rot(vrot\vec{A}) = \vec{J}_0 + \vec{J}_e + v_0 rot\vec{M} \qquad \vec{J}_e = -\sigma(\frac{\partial \vec{A}}{\partial t} + grad\phi) \qquad div\vec{J}_e = 0 \tag{2}$$

where J_0, J_e are the exciting and eddy current density, v, σ, ϕ, M and v_0 are the reluctivity, conductivity, electric scalar potential, magnetization of HTS bulk and reluctivity of vacuum respectively. Our analytical model is shown in Table 1. HTS bulks in the rotor were magnetized successively pole by pole using each neighboring stator coils. It was observed that the required magnetizing mmf is generally higher since the stator coils are further from HTS bulk in the rotor, whilst the problem may be aggravated by the presence of iron components which may shunt the magnetic field and / or cause eddy-current screening of iron rotor. The field penetration was initially inhibited by eddy currents induced in the rotor core, and gradually began from HTS bulks' flanks near flux barriers containing no magnetic material. The pole transition region is the most difficult to be magnetized since the radial alignment of the magnetizing field is very poor. It is desirable to make the pole width similar to the HTS bulk width in order that magnetizing fluxes flow HTS bulk regions uniformly.

Theoretically, HTS bulks could be magnetized to saturation by increasing the capacitance. The magnetized volume can be calculated from $M(H)$ curve. However, our experiments indicated that the thermo-magnetic instability occurs during impulse magnetization, and results in a drastic reduction of magnetization due to a strong heating of HTS bulks, which may be well described by the flux-creep model. Prominent fringing effects would be exhibited as the iron core becomes saturated, causing an uneven magnetization of HTS bulks. One impulse field configuration exists to make the remnant magnetization maximal. Additional 2-3 pulses would be helpful to improve the magnetization by further flux penetration. So an optimum pre-magnetization strategy, in terms of the level of saturation achieved, should be selected subject to constraints such as the subsequent temperature rise in HTS bulks due to flux movements, and the peak current withstand of stator / auxiliary coils. Larger number of stator coil-turns is better to reduce eddy current in the rotor. Some other issues may require consideration, such as the possible insulation degradation due to large induced emf, mechanical shocking to the bearings, and effects of the impulse on Hall devices and any other drive electronics which are housed within the motor case.

DYNAMIC PERFORMANCE

It was proven that the shielding-currents redistribute with time inside HTS due to dynamic disorder. The law of distinct regions has obvious differences, which is attributed to local distributions of effective pinning energies. It is possible to improve the time stability of the magnetization profile by consciously introducing ordered-array pinning centers, but the energy dissipation cannot be completely avoided by the quenched disorder. Once the motor runs, HTS bulk in the rotor will also be exposed to a complicated rotating field produced by the stator currents, which isn't always sinusoidal and sometimes contains harmonics. The onset of flux penetration into HTS bulk happens if only the angle of the rotating field exceeds some threshold value. The dynamic magnetic behavior and the losses strongly rely on the flux-cutting effect and anisotropy of HTS as well as the instantaneous direction of rotating-field. Rotating fluxes would penetrate $\sqrt{2} \sim \sqrt{3}$ times further than an alternating field with the same magnitude. The magnetic instability may extend to inner regions quickly by correlated flux motion. The induced unsymmetrical magnetization / magnetic shield levels of HTS bulk in the rotor have an effect on dynamic performances such as torque ripples and unbalanced magnetic force acting on the rotor. As for the HTS PM motor, torque pulsation and the unbalanced magnetic force can be calculated by [1]

$$T = r \sum_{i=1}^{u} \frac{1}{\mu} \left[B_n^{(i)} B_s^{(i)} \right] (Dl_i) \ [\text{Nm}] \qquad\qquad f_b = \sum_{i=1}^{u} \frac{1}{2\mu} \left[\left(B_n^{(i)} \right)^2 - \left(B_s^{(i)} \right)^2 \right] (Dl_i) \ [\text{N}] \qquad (3)$$

where u is the number of incremental integration path l on the rotor surface, μ stands for permeability, r is the rotor radius, D is the stator core length, B_n and B_s are the normal and the shear component of average flux density on an incremental integration path. The simulation result of torque ripple was shown in Figure 2 for our motor model. It can be seen that the harmonics of torque pulsation change due to the asymmetric magnetization distribution. Furthermore, it was observed that the difference of force in the opposite direction at HTS inner surfaces is greater than that of symmetric magnetization. The increase of the unbalanced force will affect the bearing life cycle and the vibration in the motor drive.

It is general preferable to eliminate pulsating torque by improving the machine design. The screening current in HTS bulk keeps a value larger than a critical threshold J_0 for all times, i.e. $J_s(t)/J_0 \leq 1 - \phi_{max}$, where ϕ_{max} is the maximum value of the instability parameter. As a precaution for the safety of motor, a lost of synchronization caused by a fault in the control must be considered, especially the most critical condition when the stator and rotor fluxes are in phase opposition. The stator current is limited so that the power losses do not exceed the cooling power. The increase in heat transfer coefficient of HTS would result in smaller ϕ_{max}, improving stability. In order to reduce HTS magnetization degradation, non-linear magnetic bypass by magnetically soft material can be used at the opposite of the edge of HTS bulk. The configuration of slotless motor seems to be more adaptable for HTS bulk motor. Meanwhile, active cancellation control-based techniques may be adopted for minimizing pulsating torque, which depend on accurate knowledge of machine parameters provided by either careful tuning or adaptive control [2].

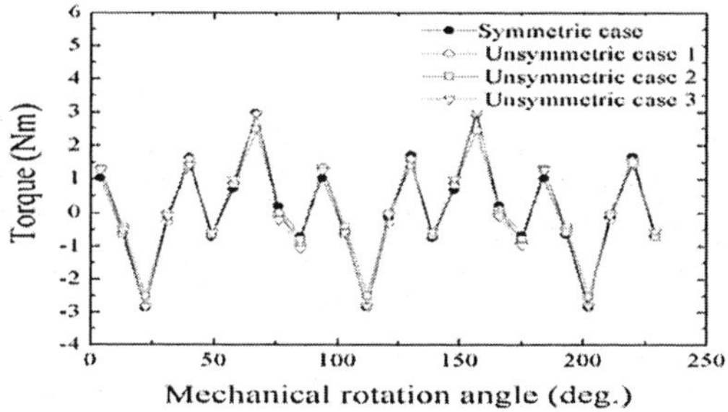

Figure 2   Torque ripple at 1500 rpm. Unsymmetric cases are that 10% degradation of trapped fields happened in one pole (1) and adjacent/opposite two poles (2)/(3), respectively

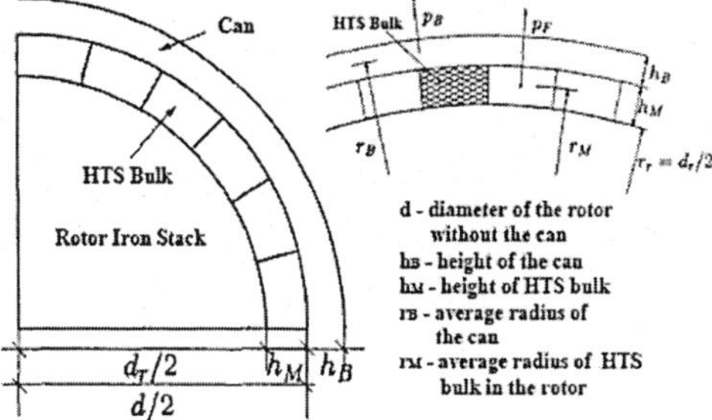

Figure 3   Cutout of the rotor with HTS bulk and the can

636

STRUCTURAL STRENGTHEN

In high speed applications, the rotor must have sufficient mechanical strength to withstand large magnetic pressures and centrifugal forces during pre-magnetization and rotation. However, HTS bulk exhibits poor mechanical characteristics: a low value of Young's module and brittleness. The fracture strength for MTG YBCO is just about 40 MPa. All the solutions using shaft notches and keys lead to unfavorable stress concentration. Any discontinuity in magnetic circuit is more vulnerable to the magnetization / magnetic shield degradation of HTS bulk. Our experiments showed epoxy-impregnation and joining is one of the feasible and effective methods in toughening, reinforcing and combining HTS bulks as well as locking HTS bulks and shaft. As an index of structural feasibility, the static safety coefficient can be evaluated by the maximum Von Mises stress amount in HTS bulk over the material crack stress.

Magnetic and nonmagnetic thin cans can be fitted around the HTS bulk to provide the rotor with the required mechanical strength (see Figure 3). The residual pressure p_{res} for fixing HTS bulks to the rotor surface must be larger than zero, and the tangential strength must be lower than the material limit, i.e.

$$p_{res} = 2\sigma h_B d^{-1} - 4\pi^2 n_{os}^2 \left(r_M \rho_M h_M + r_B \rho_B h_B \right) > 0 \qquad d\left(2h_B\right)^{-1}\left[2\sigma h_B d^{-1} + 4\pi^2 n_{os}^2 \left(r_M \rho_M h_M + r_B \rho_B h_B \right)\right] < \sigma_{limit} \qquad (4)$$

where σ, ρ_M, ρ_B and n_{os} are the initial tangential tension within the can, the mass densities of HTS bulk and the can as well as an over-speed of 120% of the nominal speed for the motor, respectively. An increase in the magnetic can thickness increases flux leakage between the poles, with the result that the airgap flux density and the synchronous torque are reduced. The pulsations of flux density in the rotor would cause eddy currents in the HTS and in the can. Rotor losses are particularly undesirable in HTS bulk motor since the can is heated and in direct contact with HTS bulks. Comparatively, the use of a nonmagnetic can avoids short-circuit of magnetic fluxes, but reduces the overall output of motor. So an optimal design requires careful consideration of the conflicting requirements of high output and acceptable losses.

CONCLUSION

The optimal impulse pre-magnetization strategy should be selected for theses constraints such as the temperature rise of HTS bulk, peak current withstand of magnetizing coils and mechanical strength of the motor. The asymmetrical magnetization / magnetic shield levels of HTS bulk have serious effects on dynamic performances of the motor. It is necessary to improve the motor design pertinently or apply active control schemes to eliminate induced pulsating torques. Epoxy-impregnation and the application of can may be the feasible and effective methods to strengthen the HTS bulk rotor.

ACKNOWLEDGMENT

The work was supported by the National Nature Science Foundation of China under Grant 50107010.

REFERENCES

1. Chen, S.X., Low, T.S., Mah, Y.A., and Jabbr, M.A., Super convergence theory and its application to precision force calculation, IEEE Trans. Magn. (1996) 32 4275-4277.

2. Roux, W.L., Harley, R.G. and Habetler, T.G., Detecting rotor faults in permanent magnet synchronous machines, Symp. on Diagnostics for electric Machines, Power Electronics and Drives, Atlanta, USA (2003).

Investigation of the subcooled LN2 characteristics in the prototype cryostat for the resistive superconducting fault current limiter

Cho S., Yang H.S., Kim D.L., Jung W.M., Kim D.H., Kim H.R.*, Hyun O.B.*

Korea Basic Science Institute, 52 Yeoeun-dong, Yuseong-gu, 305-333, Daejeon, Korea
* Korea Electric Power Research Institute, 103-16 Moonji-dong, Yuseong-gu, 305-380, Daejeon, Korea

This paper presents the experimental results of the subcooled liquid nitrogen characteristics in the prototype cryostat. A subcooled liquid nitrogen of 68 K and 1 bar were easily obtained by using cryocooler and helium gas. The temperature and pressure increase during cool down and after quench were measured. Here the quench for the SFCL was simulated by a pulsed power input to a heater in the experiment. The pressure increase due to bubbles generated after quench is measured and compared with the calculated values. They agreed each other reasonably within the range of 35 % on the average.

INTRODUCTION

Liquid nitrogen is an excellent coolant for the cooling of HTS superconducting devices, such as superconducting fault current limiter (SFCL) and power cable system. As subcooled liquid nitrogen has been frequently used for better cooling performance, a subcooled liquid nitrogen cryostat is adopted for the cooling of a 6.6 kV, 200 A resistive type SFCL, which has been developed under a project of the 21st Century Frontier R&D Program started since 2001 in Korea.

As a method of obtaining subcooled liquid nitrogen of 68 K and 1 bar, a direct cooling using cryocooler is selected. Since bubbles are generated due to quench under a fault condition, the phase change of nitrogen from liquid to gas occurs. So it is important to investigate the effect of the pressure increase due to volumetric expansion of the nitrogen on the cryostat robustness. Also, since the transient and steady state temperature variation of the subcooled liquid nitrogen affects the performance of SFCL, the cryostat needs to be fabricated to provide a uniform temperature and to endure abrupt pressure increase.

To help understand and design a practicable cryostat for investigating the characteristics related to SFCL, a prototype cryostat was fabricated [1]. The capacity of LN_2 Vessel is about 0.02 m^3. The thermal conduction cylinder made of copper connected to the cooling stage of the cryocooler is immersed in the subcooled LN_2 bath to obtain the uniform temperature. The cooling capacity of the cryocooler is controlled by a heater.

In the previous work [1] some similar experiments had been performed. This work, however, expands the scope of the experiments by providing higher input power using different SFCL simulating heater and by measuring the pressure using more accurate sensors. This paper presents the newly obtained experimental results of the subcooled liquid nitrogen characteristics in the prototype cryostat.

EXPERIMENTAL APPRATUS

The cryostat is mainly composed of vacuum vessel, LN_2 vessel, conduction cylinder and a GM cryocooler. The overall configuration of the cryostat is shown in Figure 1. As a method of obtaining subcooled liquid nitrogen, a conductive cooling using GM cryocooler was selected. The cooling load of the cryocooler is controlled by a heater installed at the first stage cold head. Detail specifications are mentioned in Ref. [1] except a ceramic heater used as a substitute for SFCL with the resistance of 20.8 ohm at 77 K, which are shown in Figure 1(b). The dimension of heater is 6 mm x 3 mm x 85 mm. In order to verify the creation of subcooled LN_2 and to investigate the influence of bubbles generated due to quench in the SFCL cryostat, it has to measure the temperature of the liquid nitrogen and the pressure inside LN_2 vessel. The temperature and pressure sensors used in this experiment are also shown in Figure 1(a).

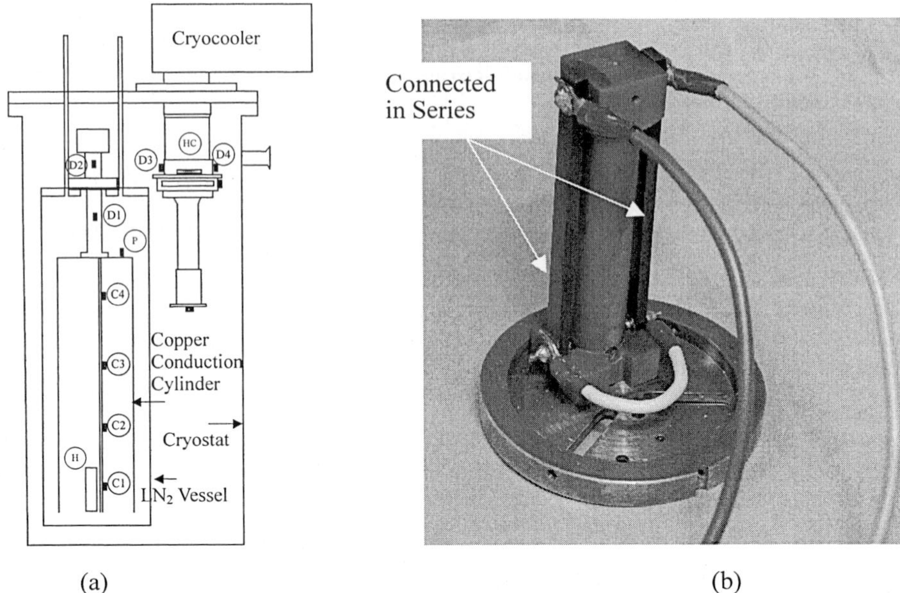

(a)   (b)

Figure 1   SFCL prototype cryostat. (a) Location and notation of the temperature and pressure sensors along with heaters, C1 – C4, D1 – D4: temperature sensors; P: pressure sensor; H: SFCL simulating heater; HC: control heater; (b) Ceramic heater simulating SFCL.

CALCULATION OF PRESSURE INSIDE OF LN$_2$ VESSEL

One of the major SFCL prototype cryostat design parameters is the pressure increase inside of LN$_2$ vessel after quench. The bubble creation during SFCL quench under fault condition induces a volumetric expansion. LN$_2$ vessel thickness has to be determined to endure this pressure increase. This pressure increase can be estimated from the following thermodynamic correlation [2]:

$$\frac{P_2}{P_1}\left(1 + \frac{V_{LN2}}{V_1}\right) = \left(1 + \frac{m_{VN2}}{m_{He}}\frac{M_{He}}{M_{VN2}}\right)\frac{T_2}{T_1} \tag{1}$$

with P_1 and P_2: pressure inside LN$_2$ vessel before and after quench, V_1: initial gas volume inside LN$_2$ vessel, V_{LN2}: LN$_2$ volume equivalent to the evaporated bubble mass due to quench, m_{VN2} and m_{He}: mass of vaporized nitrogen and helium, M_{He} and M_{VN2}: mass molecular weight of helium and nitrogen, T_1 and T_2: temperature of gas region of LN$_2$ vessel before and after quench.

In order to obtain the pressure required for obtaining subcooled LN$_2$ during steady state, helium gas was used. The mass of helium gas can be estimated from the following ideal gas equation [2]:

$$m_{He} = \frac{P_1 V_1 M_{He}}{R T_1} \tag{2}$$

with R: universal gas constant. Also the temperature increase, ΔT, after quench can be obtained from the heat transfer equation [3]:

$$\Delta T = \frac{Q}{m_{LN2}\cdot c_p} \tag{3}$$

with Q: input energy, c_p: nitrogen heat capacity, m_{LN2}: mass of LN$_2$. Using equations (1) through (3), the pressure inside LN$_2$ vessel after quench, P_2, can be calculated.

GENERATION OF SUBCOOLED LIQUID NITROGEN

A temperature variation representing the process of obtaining subcooled liquid nitrogen during cool down is shown in Figure 2. The temperature increases at each location (see Figure 1) of the cryostat are plotted in this figure. Before LN_2 was injected into the cryostat, the cryocooler was operated first in order to obtain cold atmosphere inside LN_2 vessel. After about 70 min, LN_2 was flowed into the cryostat. Thereafter the temperature (C1-C4) inside LN_2 vessel dropped quickly. When LN_2 was filled about 90% of the vessel, valves were closed keeping the cryocooler running until the temperature of the LN_2 dropped down to 68 K. Then helium gas was injected to obtain the pressure of 1 bar generating subcooled LN_2.

The temperature fluctuation at two sensors, D3 and D4, represents that the contacting surface between cryocooler head and copper block fits tight during cool down. Therefore the thermal contact resistance decreases and the heat is transferred efficiently since then.

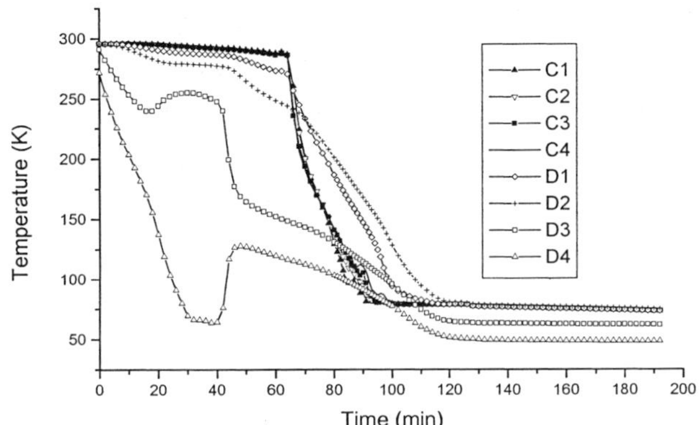

Figure 2   Temperature variation of the prototype cryostat during cool down

SFCL QUENCH TEST RESULTS

When the fault current flows in SFCL the superconductivity is broken up, and then a quench occurs. To investigate the quench phenomenon in the subcooled liquid nitrogen three SFCL quench test runs have been performed using a SFCL simulating heater, as shown in Figure 1(b). Four heat pulse shots were applied to the heater in series with the time period of 0.25 s through 2 s per each test run. The test results for several heat pulse conditions including input energy are summarized in Table 1.

Table 1   Temperature increase of the liquid nitrogen at each temperature sensor and pressure increase inside LN_2 vessel for several heat pulse conditions simulating SFCL quench

Run Number	Pulse Shot Number	Power (W)	Heating Time (ms)	Input Energy (J)	Temperature Increase (K)				Pressure Increase (measured/calculated)	
					C1	C2	C3	C4	(Pa)	|difference| (%)
1	1-a	469	250	117	0	0.028	0.033	0.047	75/61	23
	1-b	470	500	235	0.01	0.001	0.08	0.066	295/122	142
	1-c	466	1000	466	0.005	0.001	0.08	0.120	280/242	16
	1-d	467	2000	935	0.013	0.015	0.174	0.246	431/485	11
2	2-a	1,241	250	310	0.005	0.009	0.081	0.054	185/161	15
	2-b	1,219	500	609	0.007	0.008	0.063	0.167	421/316	33
	2-c	1,252	1000	1,252	0.058	0.086	0.232	0.202	803/650	24
	2-d	1,189	2000	2,378	0.878	0.457	0.6	0.689	1079/1235	13
3	3-a	2,135	250	534	0.006	0.001	0.078	0.144	526/277	90
	3-b	2,029	500	1,014	0.04	0.069	0.207	0.297	557/526	6
	3-c	2,061	1000	2,061	0.527	0.296	0.487	0.517	982/1070	8
	3-d	2,130	2000	4,260	1.239	0.706	0.57	0.544	1124/2217	49

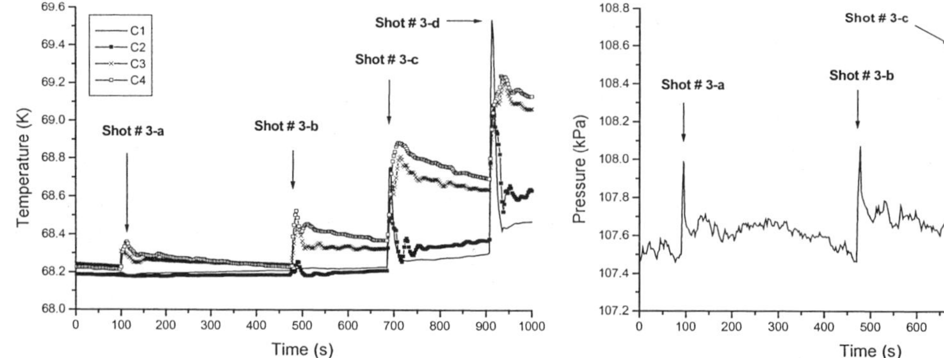

Figure 3   Temperature variation of the liquid nitrogen at each location for the test run #3.

Figure 4   Pressure variation inside liquid nitrogen vessel for the test run #3.

A representative transient temperature variation of the liquid nitrogen at each sensor mentioned in Figure 1(a) for the test run #3 is shown in Figure 3. A temperature peak was found at each heat pulse shot and the values of temperature increase are summarized in Table 1. A higher peak was obtained for high input energy. It is also found that the temperatures at C3 and C4 located relatively in upper position represent the peak phenomena well because the bubble generated during quench moves upwards. However, at very high heat input of 4260 J the temperature at C1 shows the highest peak, because a huge amount of bubbles generated during quench touch the nearest sensor directly.

A representative transient pressure variation inside LN_2 vessel for the test run #3 is shown in Figure 4. The measured peak values of the pressure at each shot and the values calculated from the equation (1) – (3) are summarized in Table 1. Their differences are expressed by percentage. The average of the differences was estimated to be about 35% from the results of 3 sets of test runs. Based on the pressure measurement results, it can be found that the calculated pressure increase during quench agrees with the experimental results reasonably. Therefore the pressure calculation correlation could be used for the design of cryostat for SFCL.

CONCLUSION

To design a practicable cryostat for SFCL, a prototype cryostat was fabricated. The experimental results for the prototype cryostat have been presented. Especially the characteristics of the temperature and pressure inside the cryostat have been investigated. The subcooled liquid nitrogen was obtained by using cryocooler and helium gas. The transient temperature variations before and after pulse heating were obtained. The temperature sensors near SFCL simulating heater showed temperature peaks representing the influence of bubble generated during heat pulse. The pressure increase due to bubbles generated after pulse heating was measured and compared with the calculated values. The calculated pressure increase during quench agrees with the experimental results reasonably. It was found that the pressure calculation correlation could be applied to the design of subcooled liquid nitrogen cryostat for SFCL.

ACKNOWLEDGMENT

This work is carried out under a grant from Centre for Applied Superconductivity Technology of the 21st Century Frontier R&D Program funded by the Ministry of Science and Technology, Republic of Korea.

REFERENCES

1. Cho S., Experimental investigation of design parameters of the cryostat for resistive superconducting fault current limiter, 6th European Conference on Applied Superconductivity (2003) 136.

2. Van Wylen G. and Sonntag R., Fundamentals of Classical Thermodynamics, John Wiley and Sons, New York (1978) 43-44.

3. Mills A., Heat and Mass Transfer, IRWIN, Chicago (1995) 29-37.

LN2 forced flow cooling on HTS power cables

Deuk-Yong Koh, Han-Kil Yeom, Kwan-Soo Lee[*]

HVAC & Cryogenic Engineering Group, Korea Institute of Machinery & Materials, 305-600, P. O. Box 101, Yu-Sung, Taejeon, Korea
[*]Mechanical Engineering Department of Hanyang University, 17 Haengdang-dong, Sungdong-gu, Seoul, 133-791, KOREA

A high temperature superconducting power cable requires forced flow cooling. Liquid nitrogen is circulated by a pump and cooled back by cooling system. Typical operating temperature range is expected to be between 65K and 77K. A subcooler heat exchanger uses saturated liquid nitrogen as a coolant to cooling the circulating liquid nitrogen stream that cools the HTS power cable. The HTS power cable needs sufficient cooling to overcome its low temperature heat load. To achieve successful cooling, it is required to investigate the hydraulic characteristics to design the cables. Especially, the pressure drop in the cable is an important design parameter, because the pressure drop decides the length of the cable, capacity of the coolant circulation pump and circulation pressure, etc. This paper describes measurement and investigation of the pressure drop of the cooling system.

INTRODUCTION

The discovery of high temperature superconductors in 1986 by Bednorz and Muller has refreshed the research in the field of superconducting cable. In the superconducting state, the maximum current density for which current flows with no electrical resistance can be made extremely large.

The HTS power cable is cooled with forced flow of subcooled liquid nitrogen. The heat leak into the cable or dissipated heat in the cable itself is absorbed by liquid nitrogen. The amount of heat, which can be absorbed by liquid nitrogen at a given mass flow rate, is limited by the freezing temperature of nitrogen at the cold end and the critical temperature of the conductor at the warm end. Thermal load of the HTS cable is determined on the basis of a unit length. In other words, the total refrigeration load depends on the length of the cable. The thermal load from surroundings and electrical heating load depend on the thermal performance of the cryostat installed in the system. Therefore the hydraulic characteristics of the HTS power cable must be well investigated to design the cables. Especially, the pressure drop in the cable is an important design parameter, because the pressure drop decides the length of the cable. Corrugated pipes may be adopted as the guide pipes of refrigerant due to flexibility of those pipes. The current cable cooling system design is based upon corrugated flexible cryostat.

Unfortunately the friction factor in corrugated pipes has not been studied sufficiently[1]. Weisend II and Van Sciver reported the dependence of the friction factor on Reynolds number, however their results are for smaller diameter bellows[2]. The modified Blasius equation which is proposed by S. Fusino et al. shows a good agreement with the experimental data in above 1.0×10^4 Reynolds number region[3]. However, geometric shape of S. Fusino et al. is not suitable for the current cable cooling system. Consequently, we measured the pressure drop through the cable cooling passage, control valve, and mass flowmeter for various mass flow rates.

DESCRIPTION OF THE COOLING SYSTEM

642

The cooling system of the HTS power cable is shown in Figure 1. The length of cryostat is 30m, the maximum flow rate of liquid nitrogen is 0.5kg/s. Sub-cooled liquid nitrogen is divided into cable passage and termination cooling passage in the distribution box. Liquid nitrogen of the cable cooling passage flows into a inner tube(former) and returns through the space between the former outer surface and cable cryostat inner surface. The termination cooling passage is composed of two supply passages for two terminations and one return passage. The liquid nitrogen supply and return lines are connected using bayonets. The pressure tank serves to maintain the circulating loop pressure. The internal circulation loop consists of the circulation pump, followed by the subcooler. The subcooler heat exchanger uses saturated liquid nitrogen boiling on the shell side to subcool the circulating liquid nitrogen stream that cools the cable system. The nitrogen gas boiled off in the subcooler can be directly vented to atmosphere or can be discharged through a vacuum pump system. The vacuum pump system is used to produce sub-atmospheric pressure on the shell side of the subcooler heat exchanger to keep temperature below 77 K. The auto filling system is used to replenish the liquid nitrogen boiled off in the subcooler. The heat loads were determined by measurements of the temperature drop across the subcooler and measured flow rates of liquid nitrogen. The pressure drops in the cooling system are calculated from pressure gauge measurements(see Figure 1).

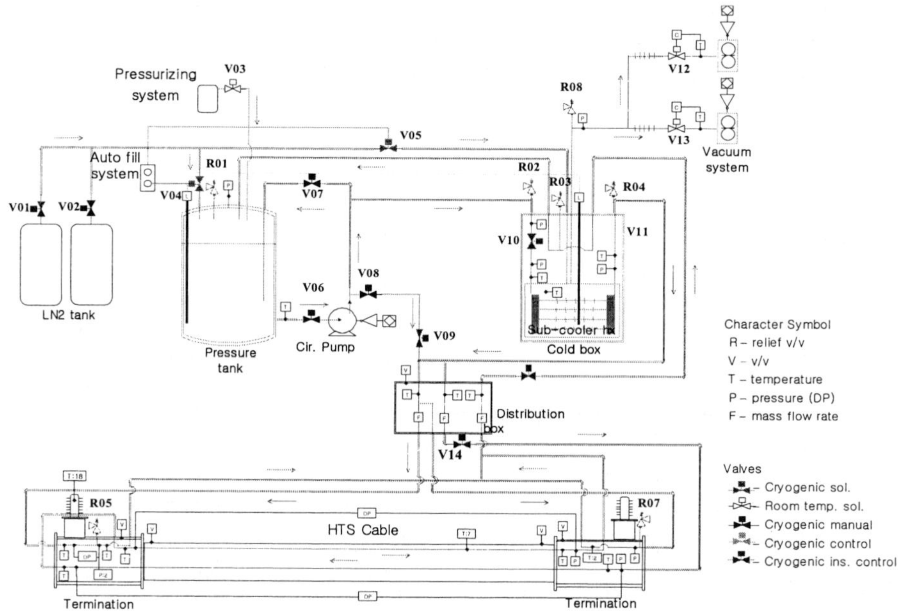

Figure 1   The schematic diagram of the HTS power cable cooling system

EXPERIMENTS

The pressure drop along the cooling passage is main design parameter of long cable cryostat, because structural strength, length, circulation pressure and pump power are decided by the pressure drop. In general, a pressure drop of a flow passage can be expressed by mass flow rate, hydraulic diameter, length of flow passage and friction factor[4]. The steady state pressure drop of liquid nitrogen dP(MPa) in the cooling passage is given by the following equation,

$$\Delta P = 2 f \rho v^2 L / D \tag{1}$$

where f is the friction factor of the cooling passage, ρ (kg/m^3) is the density of subcooled liquid nitrogen, v(m/s) is the velocity of liquid nitrogen in the cooling passage, L/D is ratio of the cooling passage length and diameter. The pressure drop of the corrugated cooling passage is larger than the smooth passage, because generally the complex flow patterns (such as the re-circulation, flow reattachment, and etc.) enlarge the friction. The friction factor f is expressed by the following experimental correlation[5],

$$f = 2.596(e/d_e)^{1.08}(P/d_e)^{-0.57} \tag{2}$$

where e(mm) is the rib height of corrugated tube(cooling passage) , p(mm) is the pitch of the rib, d_e is the average diameter of the corrugated tube.

The pressure drop through the supply and return cooling passages of the cable is shown in Figure 2. The pressure drop increases as the mass flow rate increase. Especially, the pressure drop of the supply passage is bigger than that of the return passage because the area of the supply flow passage is bigger than that of the return flow passage. Figure 3 shows the variation of the pressure drop of the termination cooling passage when the mass flow rate increases. Since the flow passage of high pressure region is designed more complex, the pressure drop of the high pressure region is bigger than that of the low pressure region.

Figure 4 shows pressure drop through the flowmeters(coldbox, cable, termination) when mass flow rate increases. The pressure drop is proportional to the square of the mass flow rate and the increases of the mass flow rate yields the huge increases of the pressure drop. The whole liquid nitrogen flows through the cold box flowmeter and is divided into cable passage and termination flow passage in distribution box. Therefore the mass flow rate passing through each flowmeter is different.

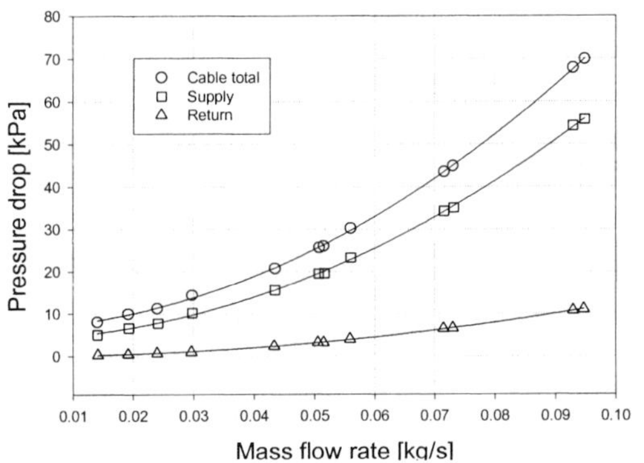

Figure 2   Pressure drop of the cable cooling passage versus mass flow rate.

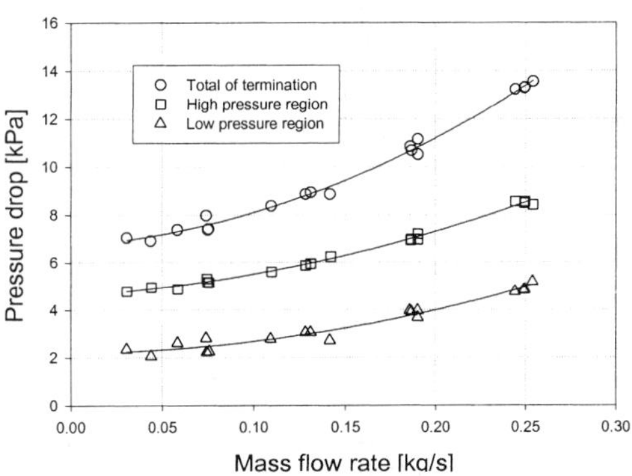

Figure 3   Pressure drop of the termination cooling passage versus mass flow rate.

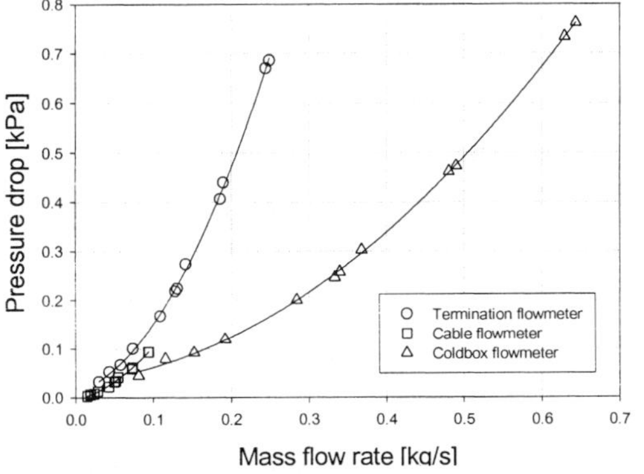

Figure 4   Pressure drop through the each flowmeter versus mass flow rate.

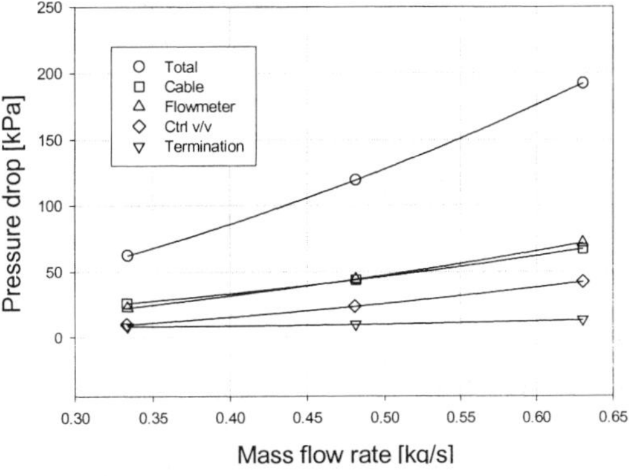

Figure 5   Pressure drop of the each compoment of the cooling system versus mass flow rate.

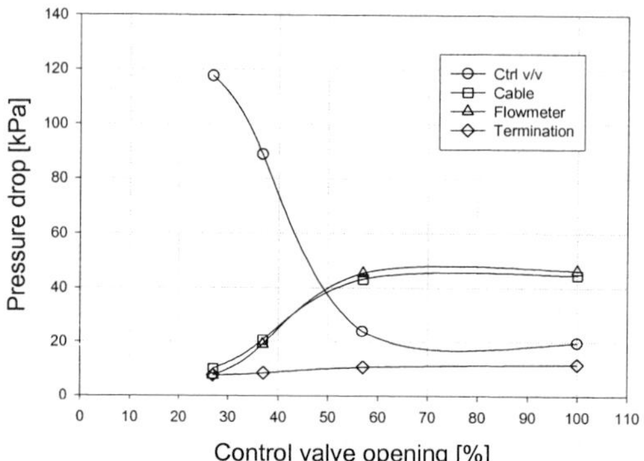

Figure 6   Pressure drop of the each component for the control valve opening.

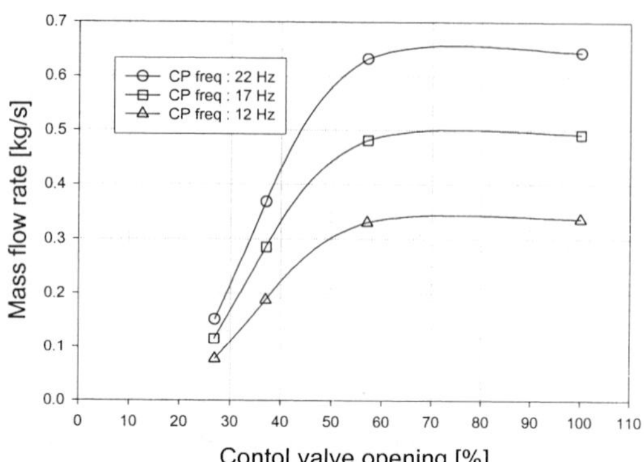

Figure 7   Mass flow rate versus control valve opening rate with various pump frequency.

Figure 5 shows pressure drops of the components(cable, termination, flowmeter, control valve) when flow control valve opens 57%. Pressure drops through the cable and flowmeter have almost same value and it is bigger than the other components. Because the cable flow passage is longer than the termination flow passage and the passage three flowmeters installed in the system. The pressure drop through the flow control valve shows small when valve opens 57%. On the other hand, the pressure drop through the valve is biggest when the flow control valve opens below 45%.

It shows that pressure drop of the components for the control valve opening in Figure 6. The pressure drop through the flow control valve is very high as valve opens below 45%. In order to reduce the total pressure drop of the cooling system, liquid nitrogen flow rate is controlled by rotational speed of the circulation pump as the flow control valve has been kept opening over 45%. Variation of the mass flow rate with the flow control valve opening rate is shown in Figure 7. When the flow control valve opens over 60%, the mass flow rate becomes nearly constant and it means that pressure drop is nearly constant.

CONCLUSION

Pressure drop characteristic of the cooling system for the HTS power cable is analyzed by experiments with the following results.

(1) The pressure drop through the flowmeters is proportional to the square of the mass flow rate.
(2) The pressure drop through the flow control valve is very high as valve opens below 45%.
(3) In order to reduce the total pressure drop of the cooling system, the flow rate of liquid nitrogen must be controlled by rotatonal speed of the circulation pump as the flow control valve opens over 45%.

ACKNOWLEDGEMENT

This research was supported by a grant from Center for Applied Superconductivity Technology of the 21st Century Frontier R&D Program funded by the Ministry of Science and Technology, Republic of Korea.

REFERENCE

1. R.C. Hawthorne, H.C. von Helms, Prod. Engineering 34, 475.
2. J.G. Weisend, S.W. Van Sciver, Cryogenics (1990) 30 935.
3. S.Fuchino, N. Tamada, I. Ishii, N. Higuchi, Hydraulic characteristics in superconducting power transmission cables, Physica C(2001) 345 125-128
4. G. F. Hewitt, G. L. Shires and T. R. Bott, Process Heat Transfer, CRC Press (1994).
5. KERI, Development of Distribution Level HTS Power Cable, 2003 DAPAS Program Workshop, CAST, 2003

Shape Optimization of Field Windings in a High Temperature Superconducting Synchronous Generator with Finite Element Method

Song F.C., Zhang G.Q., Chen B., Yu S.Z., Gu G.B.

Institute of Electrical Engineering, Chinese Academy of Sciences, Beijing 2703, China

Based on magnetic field analysis by finite element method, the optimal design of high temperature superconducting (HTS) field windings in a demonstrative HTS generator model is presented. The studies are focused on pursuing the minimal volume of HTS magnet, as well as the minimum vertical component ($B_\perp$) of magnetic flux density passing through the surface of HTS tapes. Two types of HTS rotors are analyzed. It is proved that the $B_\perp$ can be decreased effectively by adopting flux diverters for the configuration with iron core and by adjusting cross-sectional shape of HTS magnet for the configuration without iron core respectively.

INTRODUCTION

High temperature superconducting (HTS) technologies enable generators and motors more compact and lighter than their conventional counterparts. Substituting HTS conductors for conventional conductors will eliminate energy losses caused by resistance, and the efficiency of HTS generators will increase 0.5%-1.0% approximately [1]. However, the vertical field $B_\perp$ has a strong impact on the critical current I_c of HTS tapes. How to reduce $B_\perp$ becomes an important task. In this paper, two types of HTS rotors are analyzed, which are the HTS rotors with or without iron cores.

There are two steps in optimization procedures. In the first step, the magnetic flux distribution excited by the initial HTS conductors in which two rotors are analyzed accurately on the basis of Biot-Savart's law and two-dimensional finite element method, and then the maximum $B_\perp$ in the initial magnets are obtained. In the second step, structures and configurations of the HTS magnets are adjusted by changing design variables. The optimized models are compared with previous ones till a qualified scheme is found out.

OPTIMAL DESIGN OF THE HTS MAGNET

For HTS generator, there are two types of rotors with HTS field windings, one is traditional magnetic iron core rotor and the other is nonmagnetic rotor.

Initial HTS magnets

The cross section of the initial HTS magnets consisting of six rectangle shape single-layer pancake coils is shown in Figure 1 and Figure 2, and the magnet is wound on the iron core rotor or on the nonmagnetic module (nonmagnetic rotor) respectively. Each pancake coil has 50 turns. The cross section of HTS tapes is assumed as 4.5*0.5 mm, in which a DC current of 54 A runs through.

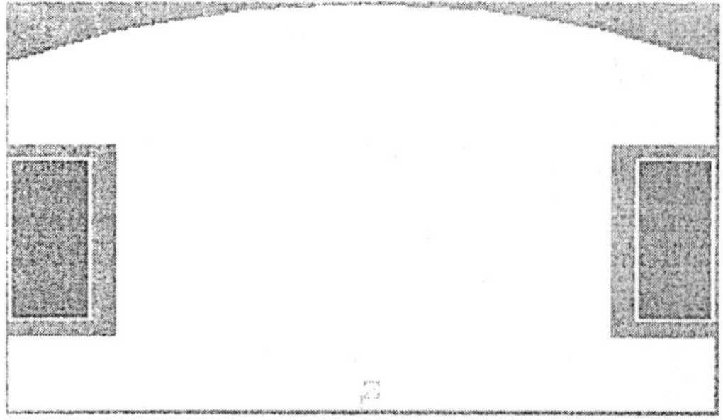

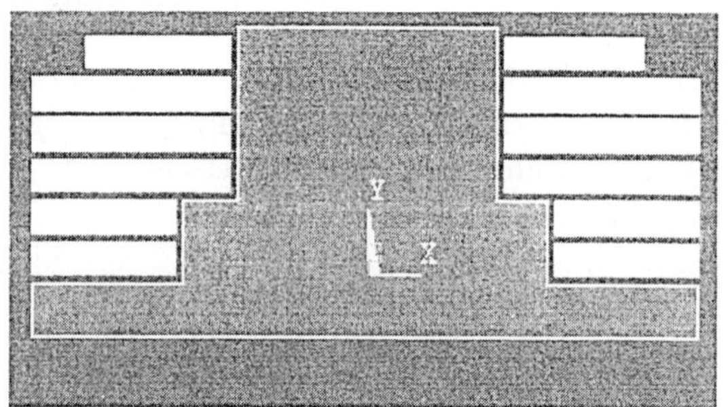

Figure 1 Initial HTS magnet wound on traditional rotor

Figure 2 Initial HTS magnet wound on nonmagnetic rotor

With above two configurations, the maximum $B_\perp$ of about 0.52 T and 0.42 T can be achieved respectively in the iron core rotor type HTS magnets and nonmagnetic rotor type ones. The magnetic field distribution of initial HTS magnet wound on traditional rotor and nonmagnetic rotor is shown in Figure 3 and Figure 4, and the value of $B_\perp$ according to the location of points on the HTS magnet edge is shown in Figure 5 and Figure 6.

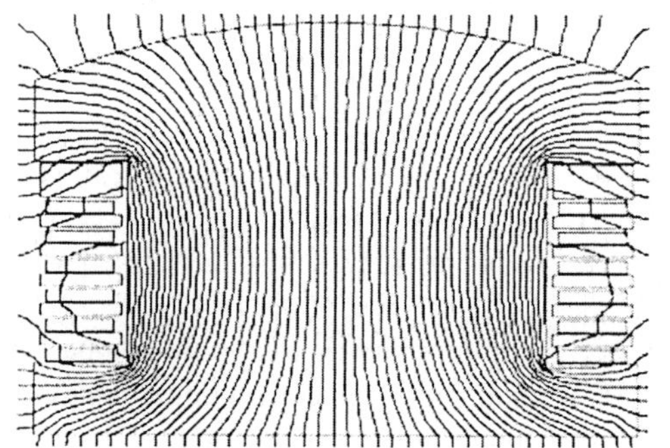

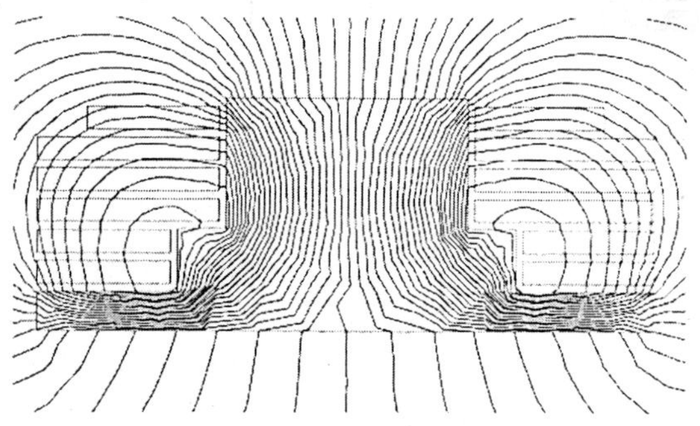

Figure 3 Magnetic field of traditional rotor

Figure 4 Magnetic field of nonmagnetic rotor

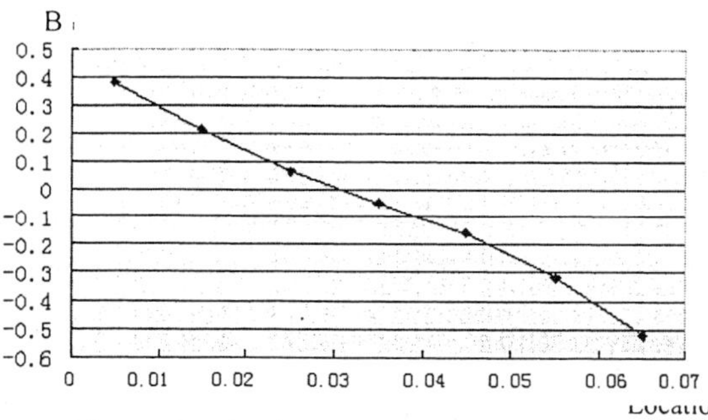

Figure 5 $B_\perp$ through HTS magnet of traditional rotor

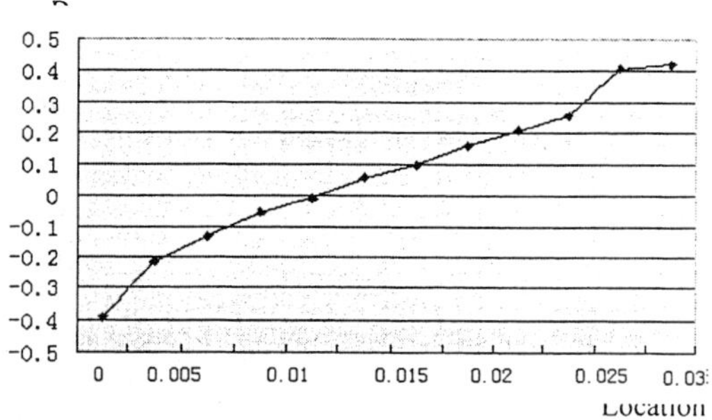

Figure 6 $B_\perp$ through HTS magnet of nonmagnetic rotor

From Figure 3 and Figure 4, it can be concluded that a lot of flux lines go through the HTS magnets, and that proves the value of $B_\perp$ is too great and has effects on the operation stability of initial HTS magnets. From Figure 5 and Figure 6, the value of maximum $B_\perp$ can be obtained.

Optimized HTS magnets

Based on initial design, flux diverters are employed for iron core rotor configuration, which are made of laminated steel sheets. Ring-shaped flux diverters are inserted between the HTS coils to reduce $B_\perp$ of the magnetic field in the coils. For this configuration, the maximum value of $B_\perp$ in the coils, with the same exciting current of DC 54 A, is reduced to 0.02 T. The shape of optimized HTS magnet wound on iron core rotor and its magnetic field distribution are shown in Figure 7 and Figure 8 respectively, and Figure 9 shows the value of $B_\perp$ according to the location of the spots on optimized HTS magnet.

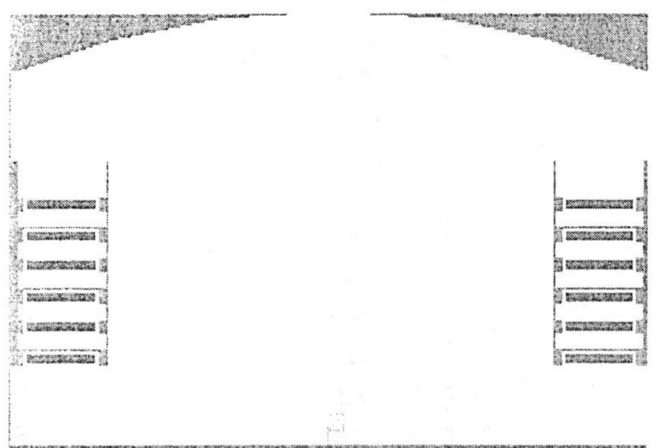

Figure 7 Optimized magnet wound on iron core rotor

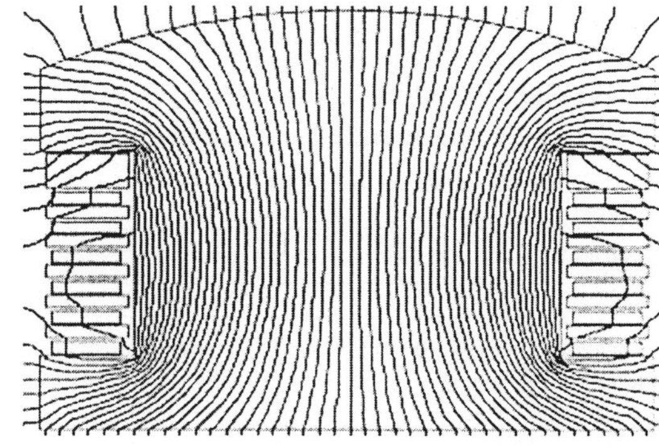

Figure 8 Magnetic field of optimized magnet of iron core rotor

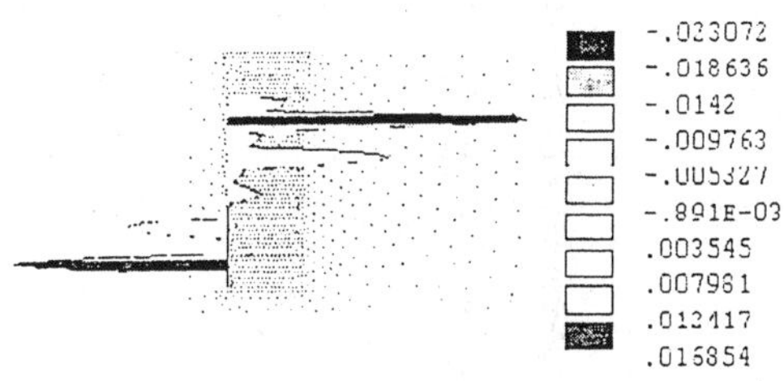

Figure 9 $B_\perp$ of optimized magnet in iron core rotor

For nonmagnetic core rotor, a step-shaped magnet is introduced and the shape of the initial magnet is optimized by finite element analysis software ANSOFT. The objective function is the HTS magnet volume, and the constraint is the value of the outer width of each coil, therefore we can get the step-shaped model of the optimized magnet. The objective function and constraints are shown as formulas (1), (2) and (3), in which V is volume of magnets, r_i is bend radius at winding end of each pancake, l is axial length of the rotor, and w_i and h are width and height of each pancake. After optimization, the volume of optimized step-shaped magnet is 93% of the initial magnet wound on nonmagnetic rotor. The maximum value of $B_\perp$ in the coils is 0.53 T. For further study, two iron plates, which are used to reduce the value of $B_\perp$, are installed on above optimized HTS magnets, and the value of $B_\perp$ is reduced to 0.3 T. However, magnetic saturation of iron plates must be considered by selecting proper materials and suitable shape and dimensions.

Figure 10 illustrates the step shape of optimized HTS magnet, and Figure 11 and Figure 12 are magnetic field distribution of step shape magnet and the value variation of $B_\perp$.

$$V = \min F(x) \tag{1}$$

$$F(x) = 2\sum_{i=1}^{n} (\pi r_i + 2l) w_i h \tag{2}$$

$$0 \le w_i \le 16.5 mm \tag{3}$$

And the optimized step-shaped magnet with iron plates is shown in Figure 13, Figure 14 and Figure 15 illustrate its magnetic field distribution and the value of $B_\perp$ according to location of the spot on HTS magnet edge.

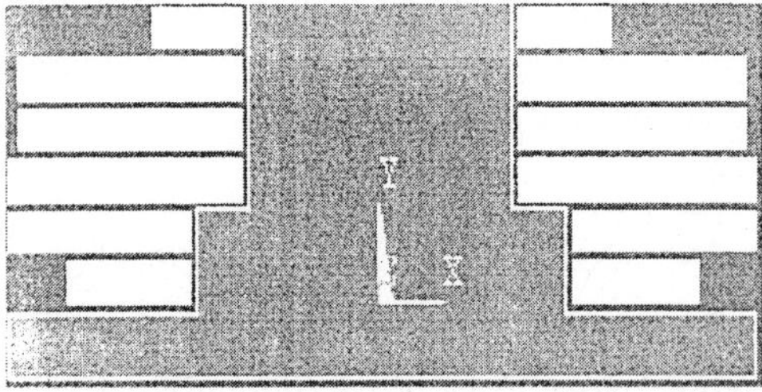

Figure 10 Step shape magnet

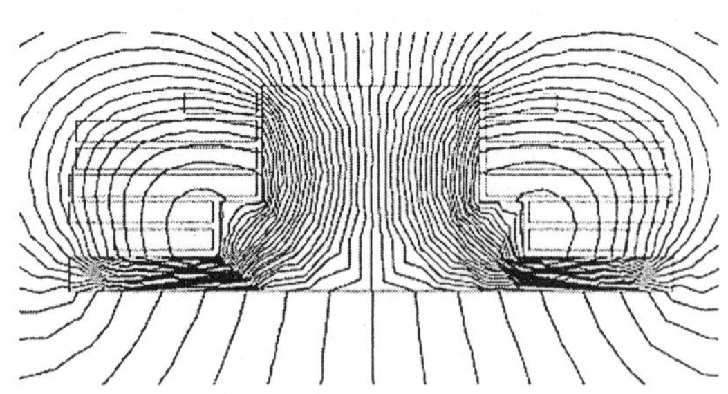

Figure 11 Magnetic field of step shape magnet

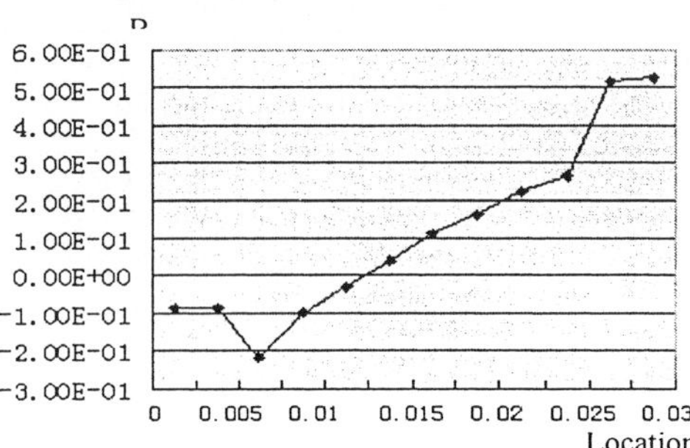

Figure 12 The variation of $B_\perp$ of step shape magnet

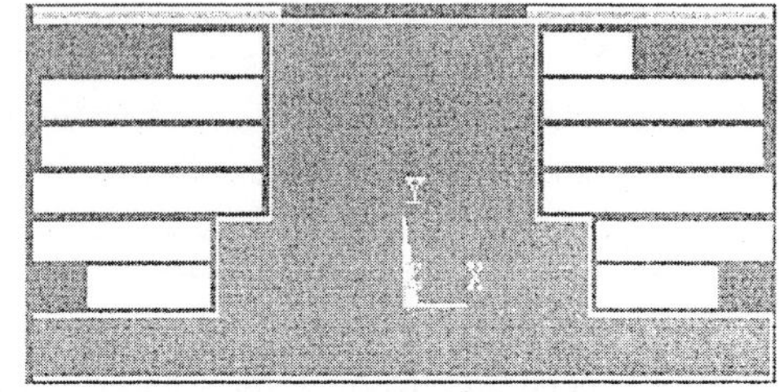

Figure 13 Step shape magnet with iron plates

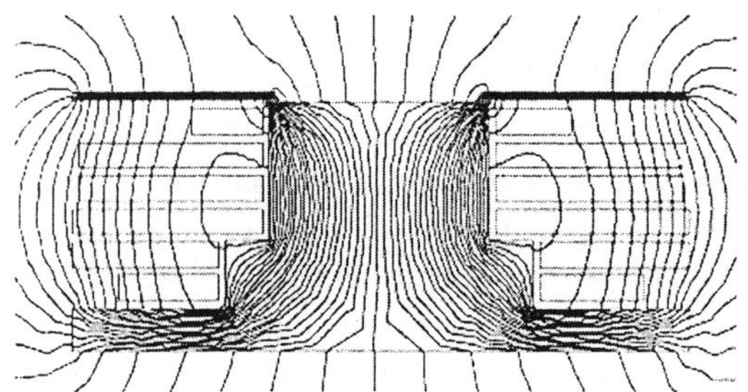

Figure 14 Magnetic field of step shape magnet with iron plates

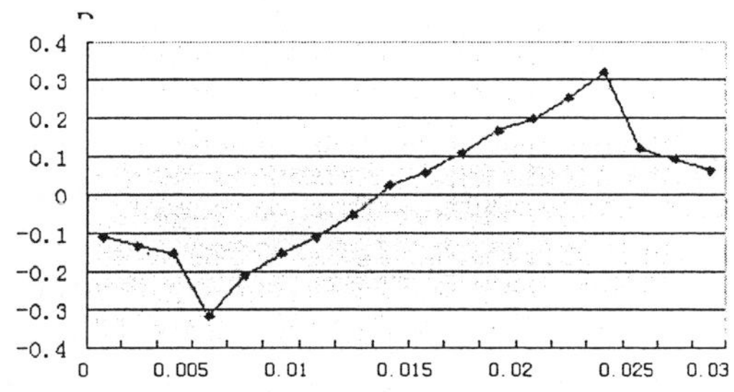

Figure 15 Variation of $B_\perp$ of magnet with iron plates

CONCLUSIONS

The optimal shapes of HTS magnet with iron core and without iron core are studied, and the value of $B_\perp$ in each kind of HTS magnet is achieved. It can be conclude that the iron core type HTS magnet with diverters and the nonmagnetic core type HTS magnet with step-shaped HTS magnet and iron plates are two effective ways to reduce $B_\perp$.

REFERENCES

1. Al-Mosawi, M.K., Beduz, C., Goddard, K. etc, Design of a 100kVA high temperature superconducting demonstration synchronous generator, Physica C (2002) 372-376 1539-1542
2. Young-Sik, Jo., Young-Kil, K., Myung-Hwan, S. etc, High temperature superconducting synchronous motor, IEEE TRANSACTIONS ON APPLIED SUPERCONDUCTIVITY (2002) 12 833-836

Development and testing of 10.5kV/1.5kA HTS power cable

Lin Y. B., Gao Z.Y., Ning Z., Dai S. T., Teng Y. P.*, Xu L., Lin L. Z., Xiao L. Y.

Institute of Electrical Engineering, Chinese Academy of Sciences, Beijing 100080, China
*Gansu Changtong Cable Science & Technology Co., Ltd., Gansu, China

A 10-m long, 3-phase, 10.5 kV/1.5 kA HTS power cable system has been built
and tested. The warm dielectric cable includes 3 single-phase HTS cables using
different winding schemes. The critical current of the cable is more than 2800 A
and the total joint resistance of the conductor is less than 0.12 $\mu\Omega$ at 74 K. The
AC loss measurements showed the loss is less than 0.85 W/m at 74 K and 1.5
kA $_{rms}$ / 50 Hz. The cable has been operated stably and reliably at rated current for
more than 5 hours continuously many times.

INTRODUCTION

Along with the rapid growth of the national economy in China, the electric utility is facing an
ever-growing demand for electricity and the necessity of large capacity power transmission. HTS power
cables, with 3 to 5 times more transmitted power than conventional cable of the same size, can meet
increasing power demand in urban areas without renewing the existing cable ducts. The power
transmission loss in China accounts for 8.5 % of the total electricity output each year, and is roughly
equivalent to the future annual output of the Three Gorges Hydropower Station. Using HTS power cables,
the power loss in the transmission line can be reduced at a percentage of 50 %, and this will save a large
amount of electricity.

Under the support of the High Technology & Development Program of China and Gansu Changtong
Cable Science & Technology Company Limited, R&D project of a 75-m long, 3-phase, 10.5 kV/1.5 kA
HTS power cable system was initiated in April, 2002. Taking the lead, the Institute of Electrical
Engineering, Chinese Academy of Sciences is developing the cable system in close collaborations with
the Technical Institute of Physics and Chemistry, Chinese Academy of Sciences, and Gansu Changtong
Cable Science & Technology Co., Ltd. As an important stage of the project, a 10-m long, 3-phase, 10.5
kV/1.5 kA HTS power cable system was developed and tested successfully in August, 2003. This paper
presents the development and testing results of the 10-m long HTS power cable.

DESIGN OF HTS CABLE

Structure of the cable
The cable is warm dielectric. Compared with cold dielectric cable, about half amount of HTS tapes can be
saved, and the conventional dielectric technique can be adopted, thus the cost of HTS cable is reduced.
Another advantage of the HTS cable with the warm dielectric is that its impedance matches that of a
conventional cable of similar diameter, so it can easily replace a conventional cable in cable retrofit.

10-m long, 10.5 kV/1.5 kA HTS power cable includes 3 single-phase HTS cables with the warm dielectric. Figure1 shows sketch of each single-phase HTS cable. The conductor is consisted of 4 layers of HTS tapes spirally wound on a flexible stainless steel former. Thin Kapton tape was used for insulation between conductor layers to reduce ac losses due to the electromagnetic coupling of each layer. The cable core was housed in a cryogenic envelope. Sub-cooled liquid nitrogen flowed through the inside of former and the space between the cable core and the inner pipe of cryogenic envelope. The warm dielectric construction, a copper shield and a PVC cover were applied over the cryogenic envelope.

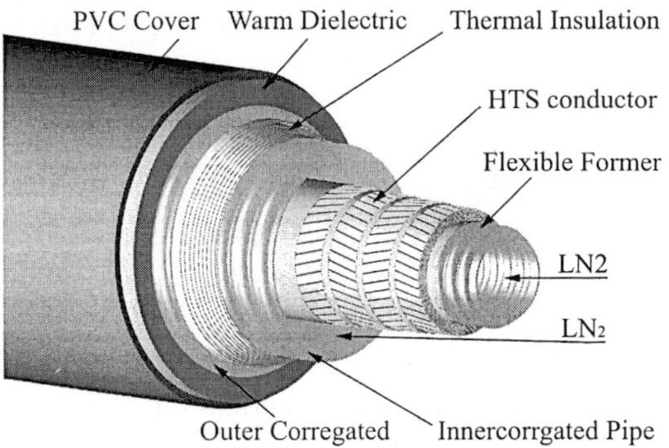

Figure 1   Sketch of a single-phase HTS cable

Table 1   Specifications of the VAC tapes

Parameter		Value
Width and thickness	(mm)	4.0 × 0.23
Number of filaments		121
Twist of filaments	(mm)	8
Length of single tapes	(m)	100 to 200
Dc critical current at 77 K	(A)	65 to 75
Critical tensile stress	(MPa)	100
Critical tensile strain	(%)	0.21
Minimum bend diameter	(mm)	70

Design of cable conductors

HTS tapes for our cable were multifilamentary Bi-2223/Ag/AgMg tapes supplied from Vacuum schmeltze, Germany. Specifications of the tapes are listed in Table 1. Three single-phase cable conductors rated current of 1.5 kA $_{rms}$ needed 4 layers of the HTS tapes for each one.

Based on our previous works [1, 2], the optimum design method of the cable conductor has been developed. The essentials of optimum design can be generalized as follows:

• Both I_c and n values should be used as criterion of the selected HTS tapes to ensure cable nice performance.

• The magnetic field acting on cable tapes should be minimized to reduce I_c degradation and AC losses.

• The current distribution among conductor layers should be as uniform as to gain most efficient use of HTS tapes and to reduce AC losses.

• The strains experienced by HTS tapes due to thermal contraction and bending must be limited in order to minimize the I_c degradation of HTS cable.

• The HTS tapes or insulation tapes in cable conductor should be wound at adequate butt gaps to avoid the damage to HTS tapes and crinkling of the insulation due to cool-down and bending of the cable.

Design of conductors with very low AC losses

Various methods to reduce AC losses of cable were developed and used in three single-phase cable conductors. A special method with the equal winding pitch was used in phase-A conductor. In this method, the critical currents in outer conductor layers are higher than that in inner layers to gain most efficient use of the tapes, and the winding direction of adjacent layers is opposite to eliminate the axial magnetic field. Schemes with uniform current distribution were used

Table 2   The design parameters of cable conductors

Parameter	Phase-A	Phase-B	Phase-C
first layer pitch(mm)	306	305	190
2nd layer pitch(mm)	306	282	463
3rd layer pitch(mm)	306	239	445
4th layer pitch(mm)	306	193	134
Number of tapes	74	72	70

in other phases. After the radius of conductor layers are given, the uniform current distribution can be realized by adjusting the winding pitches and directions of conductor layers. For phase-B, all conductor layers were wound in the same direction. For phase-C, two inner layers and two outer layers were wound in the opposite direction. The design parameters of cable conductors are shown in Table 2.

FABLICATION OF HTS CABLE

The cable conductors were wound according to the designed parameters. The HTS tapes were soldered to the copper terminals using the low-melting-point alloy solder. The skid wires were wound on the conductors for protection. The cryogenic envelopes comprised double stainless steel corrugated tubes that provide high vacuum and super-insulation without maintenance. The cross-linked polyethylene insulation triple-extruded, the copper shield and PVC cover were applied over the cryogenic envelope. There were the semiconductors on each side of the dielectric to smooth electrical field.

The general view of the HTS cable with its terminations is shown in Figure 2. The main parameters of the HTS cable are given in Table 3.

Figure 2   The general view of the HTS cable system

Table 3   The main parameters of the HTS cable

Parameter		Value
Outer diameter of former	(mm)	25.7
Inner/Outer of conductor	(mm)	25.8/28.1
Outer diameter of skid wires	(mm)	33.0
Inner/Outer of Cryogenic envelope	(mm)	39/83
Thickness of warm dielectric	(mm)	5.0
Thickness of copper shield	(mm)	0.4
Thickness of PVC cover	(mm)	6.0
Outer diameter of HTS cable	(mm)	113.8

TESTING OF HTS CABLE

Integrated with a cryogenic system, two terminations, the power supplies and monitoring system, the 3-phase cable was tested. The cryogenic system for cable cooling can supply circulating subcooled LN_2 at temperatures of 67-77K and pressures up to 4 bars. LN_2 from cryogenic system went into phase-B and phase-C in parallel, returned in phase-A. The monitoring system includes a real-time monitoring system for temperature and pressure, V-I measurement system and AC operation monitoring system for cable.

Dc V-I characteristic and joint resistances

E-I characteristics of 3 single-phase HTS cables were measured. All their electric fields are equal to zero at the currents up to 2710A, which is the limit of our dc power supply. E-I curve of phase-C cable is shown typically in Figure 3.

The I_c of more than 2800 A at 1μV/cm criterion can be estimated. Total joint resistances of cable conductors were measured also. The resistances of phase-A, phase-B and phase-C are 0.119μΩ, 0.066 μΩ and 0.072μΩ respectively.

AC losses of conductors

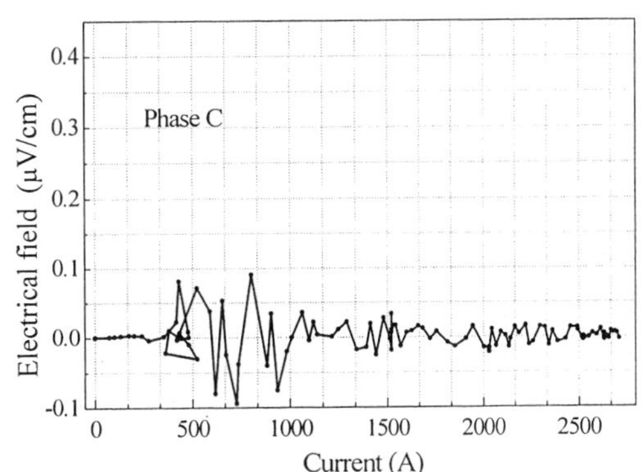

Figure 3 E-I curve of phase-C cable at 74K

AC losses of conductors were measured electrically. The AC loss curves are shown in Figure 4 and compared with theoretical curve based on monoblock model [3] at 74 K and 50 Hz. At 1.5 kA $_{rms}$ the ac losses of the phase-A, phase-B and phase-C conductors are respectively 0.42W/m, 0.5W/m and 0.85W/m, which are less than the theoretical value.

<u>Long-time operation and thermal cycle tests</u>
The cable has been operated stably and reliably at rated current for more than 5 hours continuously 5 times. Typical Long-time test curves of cable are shown in Figure 5. The cable has experienced 6 thermal cycle tests up to present and not any I_c degradation of the cable is observed.

<u>AC withstand tests of dielectric</u>
The dielectrics of HTS cable were subjected to the ac withstand tests to 35 kV for 4 hours without any breakdown before and after bending of 1.8 m in bending diameter.

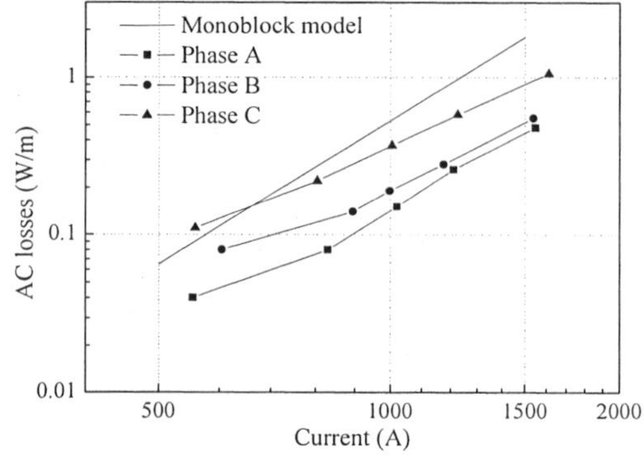

Figure 4　AC loss curves of cable conductors

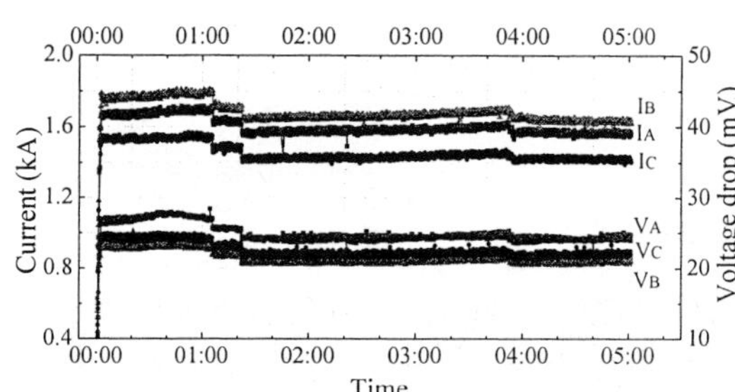

Figure 5　Long-time test curves of cable

CONCLUSIONS

Critical current of the HTS cable is more than 2800 A which exceeds the designed target. Long-time operation and thermal cycle tests indicated that the performance of the cable is stable.

The joint resistances and ac losses of cable conductors are very low. This proved that joint weld technique and the methods to reduce AC losses of cable are very useful.

The tests of HTS cable indicated that the design and development are successful. The reliable data and practice got from the development and tests of the cable can be used for development and operation of the 75m, 10.5kV/1,5kA HTS power cable.

ACKNOWLEDGEMENT

The authors appreciate Dr. D. Hui, Dr. F.Y. Zhang, Mr. H.D. Li and Q. Bao for their support and assistance during the testing of the cable.

REFERENCES

1. Lin, Y.B., Lin, L.Z., Li, S.P., Zhang, F.Y., Chen, S.F., Wang, Y.S., Xu, L., Song, N.H., Wen, H.M., and Li, J., Development of A 1000A Class Bismuth-Based HTS Model Cable, <u>Advances in Cryogenic Engineering</u> (2000) <u>45</u> 1589-1595
2. Lin, Y.B., Lin, L.Z., Gao, Z.Y., Wen, H.M., Xu, L., Shu, L., Li, J., Xiao, L.Y., Zhou, L., and Yuan, G.S., Development of HTS Transmission Power Cable, <u>IEEE Trans. on Applied Superconductivity</u> (2001) <u>11</u> 2371-2374
3. Vellego, G. and Metra, P., An analysis of the transport losses on HTSC single-phase conductor prototypes, <u>Supercond. Sci. Technol.</u> (1995) <u>8</u> 476-483

Normal transition in YBCO thin film and its effect on current limiting characteristics

Onishi T., Sasaki K., Kayoda T.

Graduate School of Engineering, Hokkaido University, Sapporo 060-8628, Japan

An active method for quenching the whole area of superconducting thin film homogeneously by a pulsed field was investigated. From measurement of the dissipative power triggering off a spontaneous SN (super-to-normal) transition, it is shown that the substrate will affect considerably the SN transition. In the active quenching method, it is revealed that the film carrying 50 Hz transport current over the critical current will be quickly quenched by a pulsed field. A method for producing pulsed field automatically when a transport current exceeds around $1.3\,I_c$ is also proposed and proved.

INTRODUCTION

The various types of fault current limiter (SFCL) have been investigated for application to a reliable and flexible power line system [1]-[4]. The resistive type fault current limiter with YBCO thin film is thought to be one of the promising candidates for the high performance SFCL among them. When an ac transport current increases, however, a sudden normal transition will occur spontaneously at around $2.5\,J_c$ of YBCO film [1]. At such a high current density, a part of the film of lower critical current density may turn into normal state faster than in other part, resulting in damage of the film by a hot spot. In order to avoid such damage, a whole area of YBCO film has to quickly turn into normal state at very early stage of a fault, that is, at a current as close as the critical current density.

In the present study, the dissipative power triggering off spontaneous SN transition of the film was firstly studied as function of frequency of ac current in order to clarify the mechanism of SN transition. Secondary, the experiments on the active quenching of the whole area of superconducting thin film by pulsed fields were conducted. Finally, a method producing a pulsed field automatically at the current over I_c is proposed and proved.

EXPERIMENTAL

The characteristics of the conductors used in the experiments are as follows; the superconductor is YBCO thin film in which the thickness is 300 nm and the width is 2×10^{-3} m, the Au coating thickness is 100 nm and the substrate is sapphire whose thickness is 300μm. The resistivity of Au is 9.21×10^{-9} Ωm at a temperature just above T_c of 87 K. The strip conductor length is 70 mm and the electrodes were soldered over 10 mm at both ends of the conductor by Indium. In the experiments on active quenching, a pulsed field was applied perpendicularly to the film surface. The conductor was set between a split coil in which the size of each coil is 50 mm in diameter, 55 mm high and 52 turns. Each coil was connected in series with separation of 19 mm and a coil current was supplied from 1,940 μF condenser which was charged up to a maximum voltage of 250 V. The field strength is 6.89×10^{-4} T/A. A free-wheeling diode was connected in parallel with the condenser and the split coil. Therefore, the rate of increase of field is as high as 700 T/s and the decay is slow and the pulse width at a half-value is around 2.6 ms.

RESULTS AND DISCUSSIONS

Spontaneous SN transition

The spontaneous SN transition is defined as a transition in which a quench occurs first when a current in the film increases from 0. In this study, the spontaneous one occurred at around $2.2\,I_c \sim 2.5\,I_c$ when 50 Hz ac voltage is applied to the film. Typical waveforms of current and voltage are shown in Fig.1 together with the frequency dependency of the flow voltage just before the SN transition. The voltage was measured between the voltage taps of 10 mm separation. J_c of the film is 2.67×10^{10} A/m^2. The SN transition occurred at ~ 0.019 s and the film returned to the flux flow state just after the transition due to the rapid decrease of current, as shown in Fig. 1(a). It is also shown in Fig.1(b) that the flow voltage increases with increasing frequency and hence depends strongly on dI/dt.

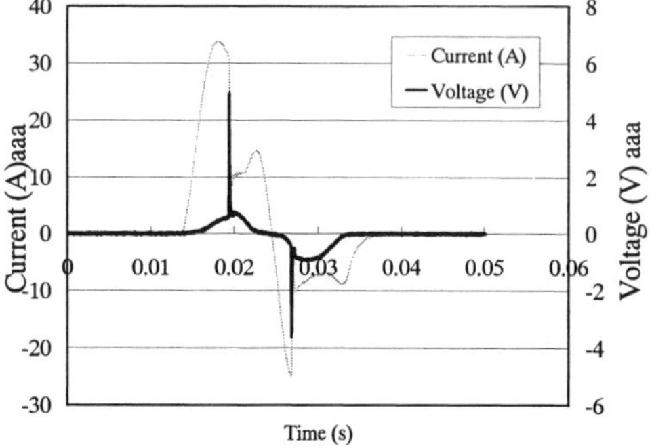

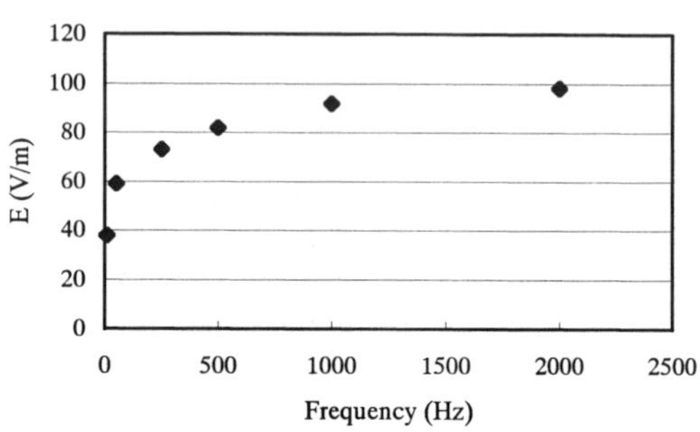

Fig.1. (a) Waveform of flow voltage and current.   Fig, 1 (b) Electric field vs. frequency.

The dissipative instantaneous powers just before the SN transition occurs and the total amount of heat generation until the SN transition occurs were measured as a function of frequency of ac current. In those experiments, the SN transitions during first half cycle of current were taken into account. The result is shown in Fig. 2. The instantaneous power g increases with increasing frequency, that is, dI/dt, but the total amount of heat q decreases

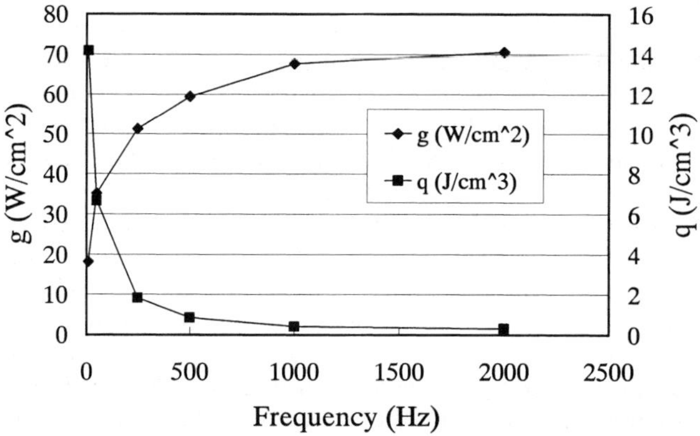

Fig. 2.   Dissipative instantaneous power and total amount of heat generation.

rapidly with increasing frequency because the period of time when the heat is generated becomes shorter for higher frequency. The magnitude of heat q_{sc} required to heat the unit volume of YBCO conductor including its substrate over T_c is around 2.4 J/cm^3. From Fig. 2, the heat generation q is less than q_{sc} in frequency range over 250 Hz. This means that only a portion of the substrate near YBCO film will be heated up over T_c in a higher frequency due to the effect of thermal diffusion. Accordingly, it shows that when a disturbance of high frequency is applied to YBCO conductor in order to quench it, the energy may be much less than the one heating up the whole conductor including the substrate over T_c. It is concluded that the mechanism of SN transition will be mainly due to the heating of the conductor including its substrate over T_c.

Active SN transition

AC current of 15 A (peak) which is approximately equal to I_c was supplied to the film in a constant current mode and then a pulsed field was applied perpendicularly to the face when the current attained to a specified value in the range of 0.5 I_c to 1.5 I_c. Quenches were triggered by the pulsed field whose rate of increase is as high as 700 T/s and pulse width at the half value of the amplitude is around 2.6 ms. Typical results on the active SN transitions are shown in Fig. 3 for the cases of B = 220, 254, 338 and 424 mT which are applied at $I = 0.75\ I_c$. The quenches occurred at around the peak of the current and those SN transition points are shown by the arrow in the figures. The voltage was measured between the voltage taps of 15 mm separation. J_c is 2.54×10^{10} A/m.

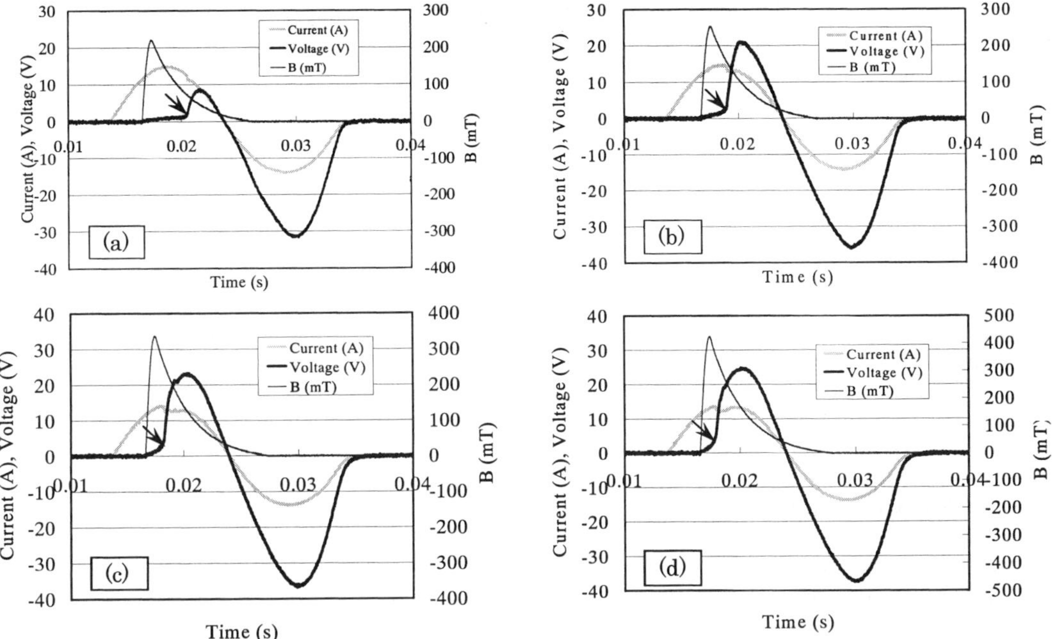

Fig. 3. Active SN transition triggered by pulsed field. (a) B=220 mT, (b) B=254 mT, (c) B=338 mT, (d) B=424 mT.

The result indicates that the whole area of the YBCO film used for, for example, a fault current limiter can be quenched quickly and homogeneously by the pulsed field at a current slightly larger than the rating current of FCL, which is usually designed to be $0.9\ I_c \sim 0.95\ I_c$. When the field of 220 mT is applied, the electric field of 50 V/m is generated at a transport current density J which is equal to J_c, as shown in E - J curves in Fig. 4. Therefore, the calculated heat generation by the pulsed field is estimated to be over 3 J/cm^3 per unit conductor volume. Since this value is almost the same as q_{sc}, the conductor can be heated over T_c by the field above 220 mT. The total amount of heat and time before SN transition starts were obtained as a function of field strength by using the data in Fig.3 and is plotted in Fig.5 in case of I_{peak} = 15 A. It was found that the heat and time become longer with decreasing field and there occurred no SN transition below B = 220 mT. The reason is mainly due to the cooling effect. It occurred when I_{peak} was increased to 20 A (~ 1.3 I_c) even in case of B = 120 mT because the heat generation increases with increasing current. Therefore, the field strength of less than 220 mT will be sufficient for quenching the film for the

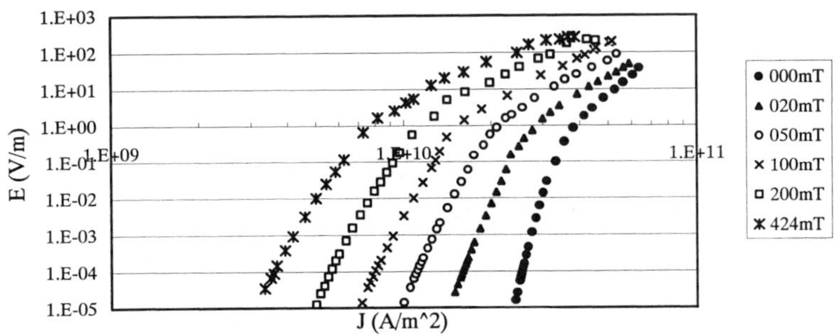

Fig. 4. E - J characteristics of YBCO thin film.

case of $I_{peak} > 1.3\ I_c$. The amount of heat for quenching is $3 \sim 4$ J/cm^3, which is almost equal q_{sc}, for the fields except 220 mT. For the case of 220 mT, the heat generation is small and hence more energy will be required for quenching because a fairly large part of energy generated is cooled by liquid nitrogen. These considerations point out that the film can be easily quenched by the heat disturbance of the order of q_{sc} when the peak current of 1.3 I_c or more flows in the film.

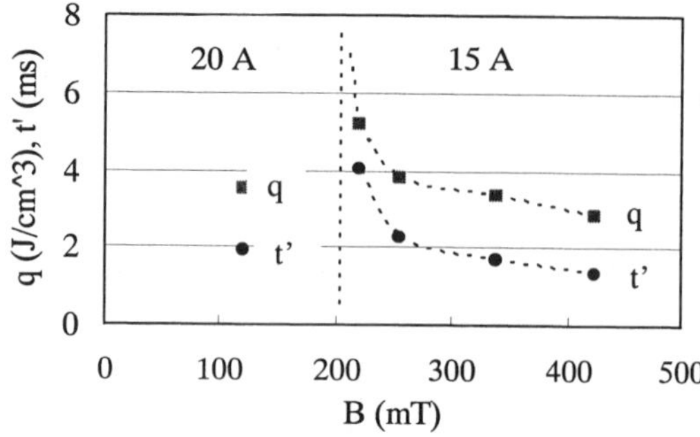

Fig. 5. Field dependency of q and t'.

Automatic generation of pulsed field

A method in which the energy stored in a condenser is discharged automatically to a pulsed coil and produces a pulsed field was investigated. A SCR is automatically fired when the current flowing in the conductor attains to a prescribed value in the range of $(1.3\sim1.5)\ I_c$ or more. The circuit used is shown in Fig. 6. A 1,940 μF condenser was first charged to a specified value of voltage. Then, ac current was supplied to the conductor. As an example, when it attained to $1.5\ I_c$, the flux flow voltage of ~ 0.9 V/m will appear in the conductor, as shown in Fig. 4. In the present

experiment where the current of $1.5\ I_c$ is supplied to the film, the voltage of 0.015 V appeared across the voltage taps of 15 mm separation. It was amplified to 1 V and then the signal was supplied to the gate of the SCR. The condenser was successfully discharged and hence a pulsed coil was produced automatically. The input to the gate shown in Fig. 6 is effective only for the positive half cycle of ac current. For the negative half cycle, another amplifier will be required but is not shown there.

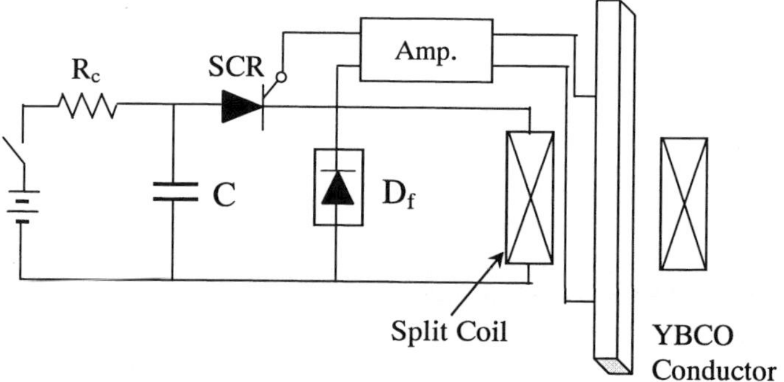

Fig. 6. A circuit for automatic generation of pulsed field.

CONCLUSIONS

The spontaneous and active quench phenomena were investigated. In the spontaneous one, it is shown that 300 nm-YBCO film coated with 100 nm Au suddenly turns into normal state at $(2.2 \sim 2.5)\ I_c$. The active quench was also studied and it is revealed that the film carrying 50 Hz transport current larger than around I_c can be quickly quenched by pulsed field. In these studies, it is pointed out that the mechanism of SN transition will be mainly due to the heating of the YBCO film above T_c and its substrate will affect considerably the heating characteristics. A method producing a pulsed field automatically when a transport current exceeds around $1.3\ I_c$ is proposed and proved. This method will be effective for quenching quickly and homogeneously the whole area of thin film which will be used for, for example, the fault current limiter and preventing the hot spot problem.

REFERENCES

[1] Decroux, M., Antognazza, N., Musolino, N., de Chambrier, E., Reymond, S., Triscone, J.-M., Fischer, Φ., Paul, W. and Chen, M., Properties of YBCO films at high current densities: implications for a fault current limiter, IEEE Trans. on Applied Superconductivity (2001) 11(1) 2046-2049
[2] Onishi, T., Sasaki, K. and Akimoto, R., A proposal of fast self-acting and recovering magnetic shield type superconducting fault current limiter and the analyses of their characteristics, Cryogenics (2001) 41 239-243
[3] Boenig H. J., Paice D. A., Fault current Limiter Using a Superconducting Coil, IEEE Trans. on Magnetics (1983) MAG-19 1051-1053
[4] Sasaki, K., Yamagata, A., Nii, A. and Onishi, T., Thermal Design and Performance Tests of a Current Limiter with a Conduction Cooled Nb$_3$Sn Screen, IEEE Trans. on Applied Superconductivity (2001) 11(1) 2114-2117

Study on a Hybrid Superconducting Magnetic Bearing System

Fang J., Lin L., and Yan L.

Institute of Electrical Engineering, Chinese Academy of Sciences Beijing, 100080, China

High temperature superconducting magnetic bearings have several significant advantages over conventional magnetic bearings. However, the low stiffness, low damping, and the uncertainty of working displacement make it impracticable to realize a superconducting magnetic bearing. In order to deal with these problems, a hybrid superconducting magnetic bearing system which consists of superconducting magnetic bearings, permanent magnetic bearings, and active magnetic bearings has been developed. In this paper, the five-degree motion equation of the rotor has been built, and the distributed control method is used to control active magnetic bearings. The experimental results of the small hybrid superconducting magnetic bearing system are presented.

INTRODUCTION

One of the practical applications of high temperature superconductors (HTS) is the superconducting magnetic bearing (SMB) [1]. In comparison with the active magnetic bearings (AMB), however, the SMB has two main problems: one is low stiffness and damping because of no feedback control system, and another is the initial loading equipment due to magnetic flux creep or flux flow, and the differences of levitation forces between zero-field cooling (ZFC) and field cooling (FC).

In order to resolve the above problems, we have designed and constructed a small experimental flywheel system which consists of three kinds of magnetic bearings: SMB, AMB, and permanent magnetic bearings (PMB).

In this paper, the five-degree motion equation of the rotor is discussed by using force analyses and rotor dynamics. The distributed control method is used to control active magnetic bearings and some experimental results of the small hybrid superconducting magnetic bearing system are presented.

STRUCTURE

We designed and built a small hybrid SMB experimental device. The schematic illustration of hybrid SMB is shown in Figure 1. In this system, an axial PMB is located at the top of the rotor; two radial AMBs are added in the radial directions; a SMB is located at the bottom of the rotor. The design specifications of three kinds of magnetic bearings were reported in [2], and the numerical and experimental analyses of SMB and PMB were presented in [3]. In this design, the axial PMB is applied to levitate the weight of the rotor, and initially position the rotor in the axial direction. In addition, two radial AMBs are applied to improve the magnitude of radial stiffness and damping of the rotor and initially position the rotor in the radial direction.

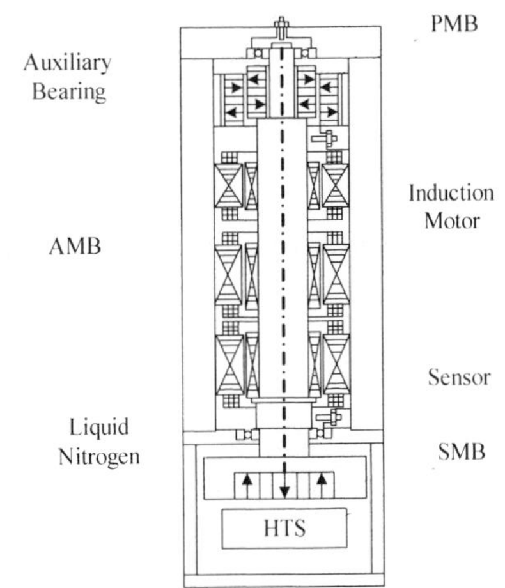

Figure 1   Schematic illustration of hybrid SMB

ANALYSES OF FORCES

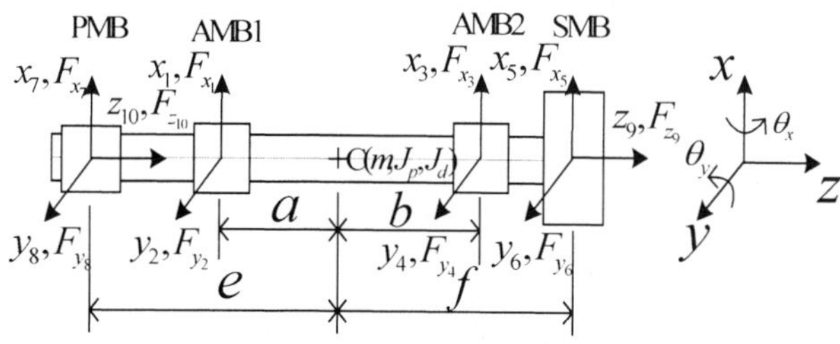

Figure 2   The rotor model

As shown in Figure 2, the rotor receives totally ten forces from magnetic bearings with exception of gravity along the axial degree z. F_{x1}, F_{y2}, F_{x3}, and F_{y4} are the forces acting on the rotor along the four radial degrees x_1, y_2, x_3, and y_4 by two AMBs: AMB1 and AMB2. F_{x5}, F_{y6}, and F_{z9} are the forces acting on the rotor along the two radial degrees x_5, y_6, and the axial degree z_9 by SMB. F_{x7}, F_{y8}, and F_{z10} are the forces along the two radial degrees x_7, y_8, and the axial degree z_{10} by PMB.

SMB can obtain intrinsic stability in all three directions due to the Meissner effect and flux pinning center. With the axial PMB, the hybrid system can suspend the rotor in FC condition. For SMB and PMB, the interaction forces between superconductors and permanent magnets along the three degrees x_5, y_6, and z_9, and the interaction forces between the stator and rotor permanent magnets along the three degrees x_7, y_8, and z_{10} can be linearized around their nominal equilibrium operation points by

$$\begin{cases} F_{x5} = -k_{x5} \cdot x_5 \\ F_{y6} = -k_{y6} \cdot y_6 \\ F_{z9} = -k_{z9} \cdot z_9 \\ F_{x7} = -k_{x7} \cdot x_7 \\ F_{y8} = -k_{y8} \cdot y_8 \\ F_{z10} = -k_{z10} \cdot z_{10} - mg \end{cases} \tag{1}$$

where m is the mass of the rotor; k_{x5}, k_{y6}, and k_{z9} are SMB stiffness constants; and k_{x7}, k_{y8}, and k_{z10} are PMB stiffness constants.

For AMB, the nonlinear current-force dependence is linearized by supplying a bias current into both coils at opposite sides of the rotor. Neglecting the coupling of sensor locations and magnetic circuits, the forces of two radial AMBs along the four radial degrees x_1, y_2, x_3, and y_4 can be linearized by

$$\begin{cases} F_{xj} = k_{cj} \cdot i_{cj} + k_{dj} \cdot x_j \\ F_{yj} = k_{cj} \cdot i_{cj} + k_{dj} \cdot y_j \end{cases} \tag{2}$$

where $j = 1,2,3,4$ refer to the four radial degrees x_1, y_2, x_3, y_4, k_d is the force-displacement coefficient, and k_c is the force-current coefficient.

SYSTEM MODEL

Assuming that the rotor is rigid and the coordinate for the rotor is set as shown in Figure 2, the dynamic five-degree motion equations for the rotor of the hybrid superconducting magnetic bearing system are

$$\begin{cases} m\ddot{x} = F_{x1} + F_{x3} + F_{x5} + F_{x7} \\ m\ddot{y} = F_{y2} + F_{y4} + F_{y6} + F_{y8} \\ J_d\ddot{\theta}_x + J_p\dot{\theta}_y\omega = aF_{y2} + eF_{y8} - bF_{y4} - fF_{y6} \\ J_d\ddot{\theta}_y - J_p\dot{\theta}_x\omega = -aF_{x1} - eF_{x7} + bF_{x3} + fF_{x5} \\ m\ddot{z} = F_{z9} + F_{z10} + mg \end{cases} \tag{3}$$

where J_d is the transversal inertia moment, J_p is the polar inertia moment, ω is the angular velocity of the mass center, θ_x and θ_y are rotation angles for the rotor, and a, b, e, and f are distances between the center of upper AMB1, lower AMB2, PMB, SMB and the mass center, respectively.

After linearizations of all three kinds of magnetic bearings, substituting (1-2) and the transfer equations from the coordinate $(x, y, \theta_x, \theta_y)$ and (x_5, y_6, x_7, y_8) to (x_1, y_2, x_3, y_4) into (3), we can obtain the motion equations of the rotor based on the coordinate (x_1, y_2, x_3, y_4, z), and design the controller of AMB to get the desirable higher stiffness than that of SMB and PMB by nearly two orders of magnitude [2-3]. Therefore, the stiffness of the hybrid system will be mainly dependent on the stiffness of AMB in the four radial degrees x_1, y_2, x_3, and y_4. According to the principle of the rotor dynamics, the motion along the axial degree z is independent of the motions along the radial degrees. In (3), neglecting F_{x5} and F_{y6} acting on SMB, and F_{x7} and F_{y8} acting on PMB, with the transfer equations from the coordinate $(x, y, \theta_x, \theta_y)$ to (x_1, y_2, x_3, y_4), we can obtain the motion equations of the rotor in the four radial degrees x_1, y_2, x_3, y_4. Assuming that the state vector, the control vector, and output vector are

$$X(t) = [x_1, x_3, y_2, y_4, \dot{x}_1, \dot{x}_3, \dot{y}_2, \dot{y}_4]^T, \quad U(t) = [i_1, i_3, i_2, i_4]^T, \quad \text{and} \quad Y(t) = [x_1, x_3, y_2, y_4]^T,$$

respectively, the state space model is presented as

$$\begin{cases} \dot{X}(t) = A(t)X(t) + B(t)U(t) \\ Y(t) = CX(t) \end{cases} \tag{4}$$

where

$$A(t) = \begin{bmatrix} 0 & 0 & I_2 & 0 \\ 0 & 0 & 0 & I_2 \\ K_{13} & 0 & 0 & -G \\ 0 & K_{24} & G & 0 \end{bmatrix}, \quad B(t) = \begin{bmatrix} 0 & 0 \\ 0 & 0 \\ H_{13} & 0 \\ 0 & H_{24} \end{bmatrix}, \quad C = [I_4 \quad 0],$$

where

$$G = \frac{\omega \cdot J_p}{1 \cdot J_d}\begin{bmatrix} a & -a \\ -b & b \end{bmatrix}, \quad K_{ij} = \frac{1}{m}\begin{bmatrix} k_{di}\left(1+\dfrac{a^2}{\rho^2}\right) & k_{dj}\left(1-\dfrac{ab}{\rho^2}\right) \\ k_{di}\left(1-\dfrac{ab}{\rho^2}\right) & k_{dj}\left(1+\dfrac{b^2}{\rho^2}\right) \end{bmatrix}, \quad H_{ij} = \frac{1}{m}\begin{bmatrix} k_{ci}\left(1+\dfrac{a^2}{\rho^2}\right) & k_{cj}\left(1-\dfrac{ab}{\rho^2}\right) \\ k_{ci}\left(1-\dfrac{ab}{\rho^2}\right) & k_{cj}\left(1+\dfrac{b^2}{\rho^2}\right) \end{bmatrix},$$

where $i = 1,2$, $j = 3,4$ and $\rho^2 = \dfrac{J_d}{m}$.

CONTROL DESIGN

Through the structure design, some mechanical couplings in (4) can be reduced to a large extent to be negligible, and motions of the rotor along the four radial degrees can be decoupled. At low speeds, neglecting the gyroscopic effect G, the transfer function of AMB $W_o(s)$ in every degree relating the state output to control can be simplified as

$$W_o(s) = \frac{X(s)}{U(s)} = \frac{k_c}{m_i s^2 - k_d} \tag{5}$$

where $m_i = m\dfrac{\rho^2}{\rho^2 + i^2}$, and $i = a, b$.

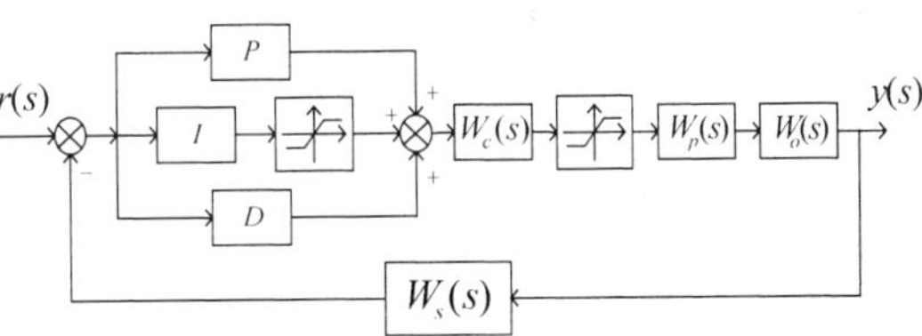

Figure 3   Block diagram of nonlinear control system

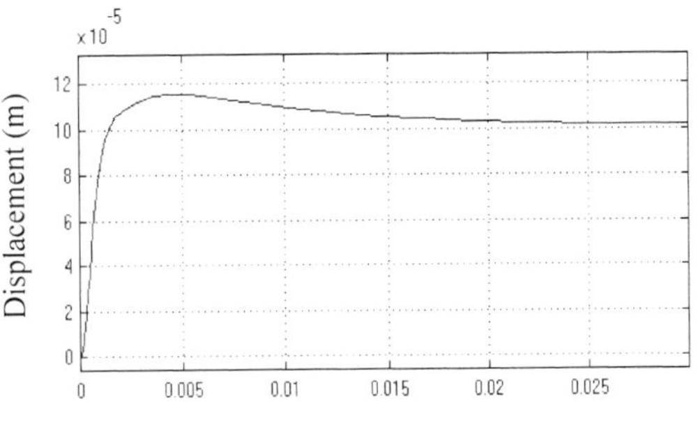

Figure 4   The step response of nonlinear control system

660
The characteristics of the stiffness and damping of AMB mainly depend on the electrical control system, and the controller can be designed as the desired stiffness and damping. Figure 3 shows the block diagram of the applied nonlinear control system, where $W_c(s)$, $W_p(s)$, $W_o(s)$, and $W_s(s)$ are transfer functions of the phase-lead compensator, power amplifier, plant, and sensor, respectively. The integrator (I) is added into the PD controller in order to reduce the steady errors. Two nonlinear saturation components are joined to limit the effect of the integrator and simulate the saturation of electromagnets. The phase-lead compensator is applied to improve the stability of control system. The unit step response simulation of the nonlinear control system is shown in Figure 4. The overshoot of the step response is about 15%, and the settling time is about 20ms.

EXPERIMENTAL RESULTS

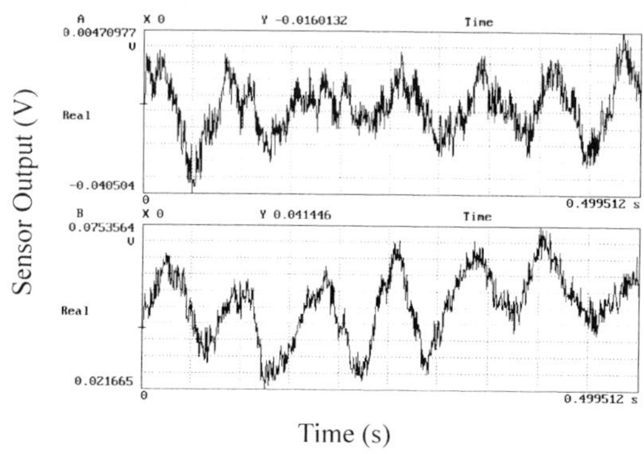

Figure 5   Vibration characteristics of the rotor along x_3 and y_4 at steady state

Figure 6   Vibration characteristics of the rotor along x_3 and y_4 at speeds of 9,600 rpm

Using above nonlinear controller, the 3.5 kg rotor of the hybrid superconducting magnetic system has been levitated by PMB and SMB in axial direction, and can be accelerated with the assistance of AMB in radial directions. The vibration characteristics of the rotor along the degree x_3 and y_4 at steady state is shown in Figure 5. The transfer function of the sensor is 20mV/μm, so the rotor vibration is within 5 μm at steady state. Figure 6 shows the vibration of the rotor along the two degrees x_3 and y_4 at speeds of 9,600 rpm with vibration amplitude of about 30 μm.

CONCLUSION

In this paper, the five-degree motion equation of the rotor has been built by means of the linearization of the forces of three kinds of magnetic bearings, and rotor dynamics. The stiffness of the hybrid system is mainly dependent on the stiffness of active magnetic bearings in the four radial degrees, and the distributed control method is used to control active magnetic bearings. Using a nonlinear controller with phase-lead compensator, the rotor vibration along the degree x_3 and y_4 is within a 5 μm at steady state, and about 30 μm vibration amplitude at speeds of 9,600 rpm. The experimental results show that the hybrid superconducting magnetic bearing system is effective in the view of the distributed control system.

REFERENCES

1. Moon, F., Superconducting Levitation, 1st edition, John Willey & Sons, Inc., New York, USA (1993)
2. Fang, J., Lin, L., Yan, L. and Xiao, L., A new flywheel energy storage system using hybrid superconducting magnetic bearings, *IEEE Trans. on Appl. Supercon* (2001) Vol. 11, No. 1, 1657-1660
3. Fang, J., Lin, L. and Yan, L., Numerical analysis of a new hybrid superconducting magnetic bearing flywheel system, Advances in Cryogenic Engineering, Proceeding of CEC (2002) 47, 465-472

Electromagnetic design and analysis of a PM motor with HTS bulk

Huo H. K., Qiu M., Xia D., Xu Z., Lin L.Z.

Institute of Electrical Engineering, Chinese Academy of Sciences, Beijing 100080, P.R. China

Melt-processed HTS bulk with approximately single-domain structure can trap a magnetic field exceeding 2 T at 77 K. So the flux-trapped superconductor is regarded as a magnet superior to conventional permanent magnet, which can be used for electrical motor. A PM electrical motor with HTS bulk was designed in theory and analyzed by FEA. The magnetization method for HTS bulk in the motor was discussed in detail and simulated using the flux flow-creep model. The experiment was also carried out to test the validity of the simulation.

INTRODUCTION

Melt-Processed YBCO bulk superconductor with approximately single domain structure can trap a magnetic field exceeding 2 T at 77 K [1]. Hence, the flux-trapped superconductor has been regarded as a magnet superior to conventional permanent magnet such as Nd-Fe-B. We designed a theoretic model of permanent synchronous electrical motor, in which conventional permanent magnets are replaced by YBCO HTS bulks and simulated it with ANSOFT. The magnetization method for HTS bulk in the motor was discussed, and simulated by flux flow-creep model.

STRUCTURE DESIGN OF THE HTS MOTOR

The stator of the HTS PM motor is the same as that of the conventional synchronous motor. The stator coils are wound from a bundle of 528 Cu wires, each 0.8 mm in diameter and electrically insulated. As we know, conventional 3 phases winding in the slots of the stator can produce rotating field, if there are symmetrical 3 phases electrical current flowing in them. The rotor of the motor is made up of laminated irons, axle, squirrel cage, magnetizing coil and YBCO bulks serving as the permanent magnet. The rotor and the stator are immersed in liquid nitrogen. Before staring the motor, the YBCO bulk is magnetized first by the magnetizing coil. The rated capacity and rated voltage of the motor in theory are 0.8 kW and 220 V, respectively. Inner diameter and outer diameter of the stator are 162 mm and 245 mm, respectively. Table 1 shows the parameters of the HTS synchronous motor that we designed in theory.

Table 1　Parameters of the HTS PM synchronous motor designed in theory

Rated power P_N=800 W	Rated voltage U_N=220 V	Rotating speed n_N=1000 r/min
Frequency f=50 Hz	Inner diameter D_1=0.162 m	Outer diameter D_2=0.245 m
Pole distance τ=0.0848 m	No of phase m=3	Height of bulks h=8 mm
Length of axis c=58 mm	Width of bulks a=35 mm	Width of air gap g=0.8mm
Bundle of Cu wires in stator W=528	Diameter of Cu coil D_3=0.8 mm	Number of the stator slots Z_1=36
Number of the rotor slots Z_2=46	Connection mode Y	Trapped field of bulks B=1 T

The rotor speed is synchronous to the rotating stator speed, which is 1000 rpm at 50 Hz for a 6-pole motor. The interaction between rotating field and rotor field results in a force along the surface and produce the output torque.

PULSE FIELD MAGNETIZATION

The YBCO bulks used in this motor are magnetized by the magnetizing coil with pulsed current, which is produced by voltage source through discharging a series of capacitors. Remnant Magnetic fields are measured by three Hall sensors placed at the center, periphery, and the midpoint on the top surface of the bulk at 77 K. We use flux flow-creep model to simulate the magnetization process, and compare the results with our experiments.

The governing equations for both superconductors and normal conductors are derived form Maxwell's equations. In cylindrical coordinates, the axisymmetric scalar equation is written as

$$\left(\frac{1}{r}\frac{\partial}{\partial r}\left(r\frac{\partial}{\partial r}\right) + \frac{\partial^2}{\partial z^2} - \frac{1}{r^2}\right)A = -\mu_0 J \tag{1a}$$

$$E = -\frac{\partial A}{\partial t} \tag{1b}$$

where μ_0 is the magnetic permeability in vacuum, E is the electric field, B is the magnetic flux density, and A is the magnetic vector potential. Nonlinear relationship between shielding current density and electric field is described by the flux flow-creep model and power law model.

$$E = 2\rho_c J_c \sinh\left(\frac{U_0}{kT}\frac{J}{J_c}\right)\exp\left(-\frac{U_0}{kT}\right) \quad \text{for flux creep region} \tag{2a}$$

$$E = \rho_c J_c + \rho_f J_c\left(\frac{J}{J_c} - 1\right) \quad \text{for flux flow region} \tag{2b}$$

where ρ_c is the creep resistivity, ρ_f is the flow resistivity, U_0 is the pinning potential, T is the temperature, and k is the Boltzman constant. The nonlinear dependence of critical shielding current on magnetic field is considered by using the Kim model and on the temperature is described also. The temperature distribution in the bulk is obtained through solving the heat diffusion equation:

$$J_C(|B|) = \frac{J_{C0}B_0}{(|B| + B_0)} \tag{3}$$

$$J_{C0}(T) = \alpha\left\{1 - \left(\frac{T}{T_{C0}}\right)^2\right\}^2 \tag{4}$$

$$m_0 C \frac{dT}{dt} = \kappa_{ab}\frac{\partial^2 T}{\partial r^2} + \kappa_{ab}\frac{1}{r}\frac{\partial T}{\partial r} + \kappa_c\frac{\partial^2 T}{\partial z^2} + Q \tag{5}$$

where $m_0, C, \kappa_{ab}, \kappa_c$ are the density, specific heat, heat conductivities along the a-, b-axes, and heat conductivities along the c-axes, respectively. An iterative method is applied for the calculation of the shielding current and the magnetic field, ie.,

1. The artificial electrical conductivity σ is set to a special value.
2. The shielding current density is obtained by equations (1a), (1b)
3. The electric field is obtained by solving equations (2a), (2b), (3), (4) and (5)
4. The conductivity is modified by equations (1a) and (1b).
5. Steps 2-4 are repeated until σ is converged.

Table 2 shows parameters in the present numerical analysis.

Table 2　Parameters used in numerical simulation

$J_c = 10^7\ A/m^2$	$U_0 = 0.1eV$	$\rho_f = 7.0\times10^{-10}\ \Omega m$	$C = 1.32\times10^2\ J/kg\cdot K$
$m_0 = 6.31\times10^3\ kg/m^3$	$\kappa_{ab} = 14.5W/m\cdot K$	$\kappa_c = 3.0W/m\cdot K$	$B_0 = 0.4$

The pulsed current plotted in Figure 1 is produced by voltage source through discharging a series of capacitors, which has been charged in advance. The maximum of the magnetic field in the coil during the pulsed field magnetization is 1 T. Magnetic fields are measured by Hall sensors placed at center, periphery, and the midpoint on the top surface of the bulk at 77 K. Figure 2-4 show the experimental results which are recorded, and used to compare with the simulation results.

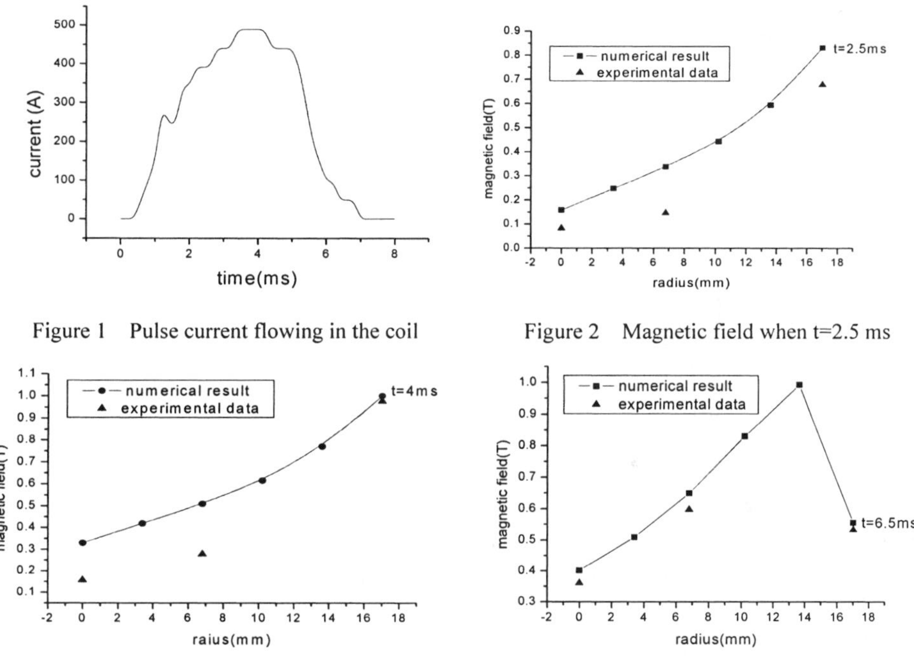

Figure 1　Pulse current flowing in the coil

Figure 2　Magnetic field when t=2.5 ms

Figure 3　Magnetic field when t=4.0 ms

Figure 4　Magnetic field when t=6.5 ms

When the pulsed field is applied to the bulk, the field changes from zero to a small value and the shielding current comes forth, which prevents the magnetic field from penetrating into the bulk. So the magnetic flux cannot enter the center of bulk at the beginning. With the increase of pulsed field, the magnetic field begins to penetrate into the center of the bulk step by step. We can see in Figure 4 that the magnetic field penetrates into the center of the bulk when time is 2.5 ms. On the decrease of the pulse field, shielding current is produced from outside of the bulk. The inner field of the bulk is still increasing after the peak of the external field, which was caused by the flux flow. From these figures we can see that the FFC model is better than the power law model in describing the PFM.

SIMULATION RESULTS OF THE MOTOR

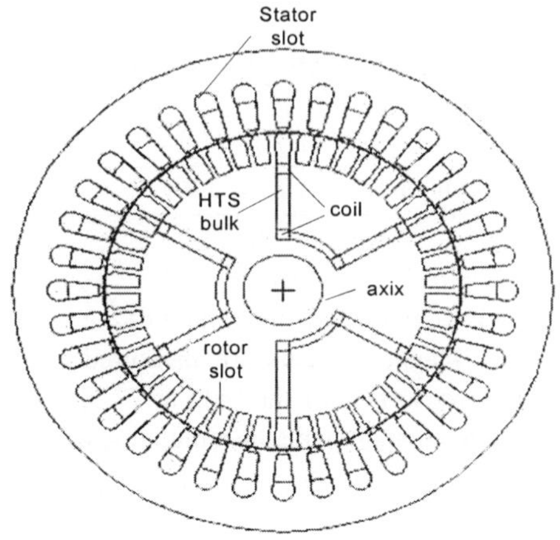

Figure 5   Cross section of the HTS PM motor

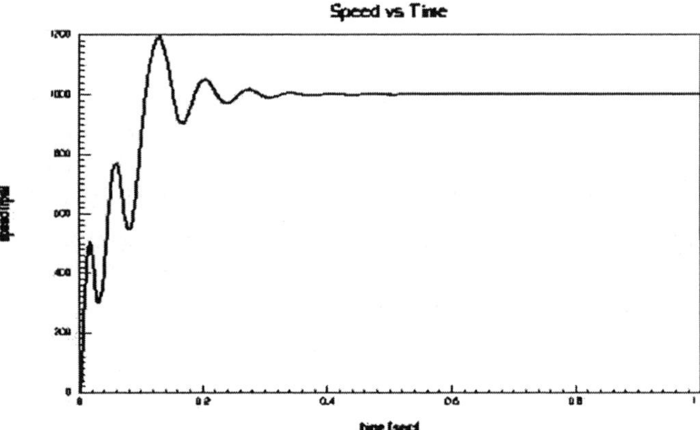

Figure 6   Speed curve of the HTS PM motor with time

Figure 5 shows the cross section of the HTS asynchronous motor. In Figure 6, the speed serving as a function of the time is nearly constant and equal to the synchronous speed after 0.5 s. It can be seen that this HTS PM synchronous motor will be stable after 0.7s.

ACKNOWLEDGMENT

This work was supported by the National Nature Science Foundation of China under Grant 50107010.

REFERENCE

[1] Itoh Y., and Mizutani U., Pulsed Field Magnetization of Melt-processed Y-Ba-Cu-O Superconducting Bulk Magnets, Jpn. J. Appl. Phys. (1996) 35 2114-2125.
[2] Tsuchimoto M., Morikawa K., Macroscopic Numerical Evaluation of Heat Generation in a Bulk High Tc Superconductor during Pulsed Field Magnetization, IEEE Trans. Appl. Superconductivity (1999) 9 66-70.

Finite-element simulation of HTS bulk reluctance motor

Xu Z., Qiu M., Yao Z.H., Xia D.

Institute of Electrical Engineering, Chinese Academy of Sciences, Beijing 100080, P.R. China

A Demo HTS reluctance motor with the rotor containing bulk YBCO elements is presented. Based on YBCO bulk magnetic characteristics, the field distribution in the motor and the resulting torque are evaluated by finite-element analysis. The feasibility of our method is proved by comparisons with the simulation and the following experiment.

INTRODUCTION

The development of HTS materials, both tape and bulk, has allowed the design of systems in several areas of electro-mechanical engineering. In particular, YBCO bulks have been used directly in electrical motors, such as hysteresis motors. It is expected that these novel motors possess higher output power, efficiency and power factor than the conventional ones [1]. For simplicity, HTS bulks are usually regarded as a material with constant permeability in the design. However, HTS bulks show strongly anisotropic and non-linear behaviors in magnetic fields. Based on the critical model, the finite-element method is presented to analyze a small Demo HTS reluctance motor in the paper. Its feasibility is proved by the experiments.

SPECIFICATIONS OF DEMO RELUCTANCE MOTOR

A two-pole, three-phase HTS demo reluctance motor has been developed in IEE/CAS. The technical parameters and basic structure are shown in Figure 1. Here P_2 is the output power, $U_{N\varphi}$ is the phase voltage (Y-connection), I_N is the phase current, f_1 is the frequency, D_{i1} is the inner diameter of the stator, l_{ef} is the effective length of the rotor, σ is the gap length. The difference from the conventional one is that the motor uses YBCO bulks instead of aluminum or air filled in the slots of the rotor.

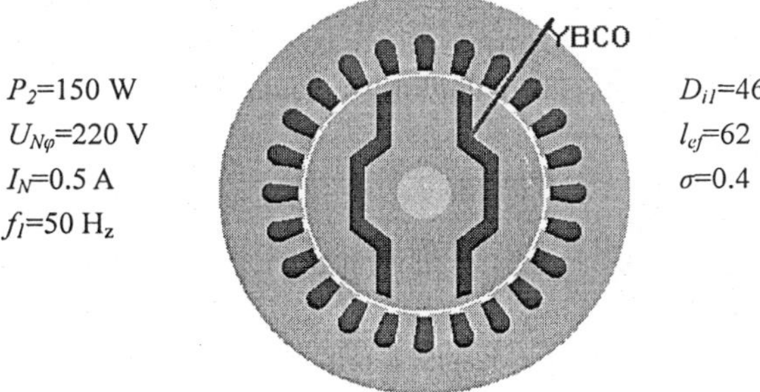

Figure 1   Cross section of the HTS reluctance motor

FINITE-ELEMENT ANALYTICAL METHOD

The operation of HTS bulk reluctance motor depends on the direct-axis (d-axis) synchronous reactance X_d, the quadrature-axis (q-axis) synchronous reactance X_q and the phase resistance of the stator winding R. In order to evaluate its operating characteristics, it is necessary to obtain values of X_d and X_q, by solving the field distribution in the motor firstly. For the demo motor, its length is much greater than its radius. The analytical model can be two-dimensional by neglecting end-effects. It is assumed that the motor operates in synchronous regime and no fundamental eddy current would be induced in the rotor. So the inner magnetic field can be approximately treated as stationary. HTS bulks are treated as nonlinear magnetic mediums rather than idealized conducting materials. Magnetic qualities of YBCO bulk is defined by the magnetization curve [4].

The field distribution in the cross section of the motor is calculated in a rectangular coordinate system, with the x-axis parallel to the q-axis of the rotor and the y-axis parallel to the d-axis. Considering the non-linearity of silicon-steel and the anisotropy of YBCO, the two-dimensional quasi-Poisson's equation is adopted to describe the mathematical model of the magnetic field in the cross-section:

$$\frac{\partial}{\partial x}(v_y \frac{\partial A}{\partial x}) + \frac{\partial}{\partial y}(v_x \frac{\partial A}{\partial y}) = -J \qquad \text{in the section area of the motor} \qquad (1a)$$

$$A = 0 \qquad \text{on the outside border of the section area} \qquad (1b)$$

where A is the magnetic vector potential; v_x, v_y are the x-axis and y-axis reluctivity at arbitrary point in the cross section; and J is the source current density. When solving Equation (1) by finite-element method, local B-H curves of each finite element is calculated by expression (2) and then assigned as a property to that finite element.

$$B(H) = \mu_0[H + M(H)] \qquad (2)$$

where B is magnetic flux density, H is magnetic field intensity, M is magnetization vector and μ_0 is free space permeability. local $M(H)$ data of each YBCO finite element is achieved from the expressions given in [5], which is based on the Kim model of the Type-II superconductors. Some parameters are corrected by our experimental measurements. After the field distribution is worked out, X_d and X_q can be computed exactly. Subsequently, the operating characteristics of the motor can be evaluated easily.

RESULTS AND DISSCUSSIONS

Figure 2 shows the distribution of the magnetic field under $\theta=0°$ and $\theta=90°$ after rated current is injected into the stator winding, where θ is the angle between the rotary magneto-motive potential vector and the d-axis of the rotor. Thereout, the radial air-gap induction B_r is computed by the following expression:

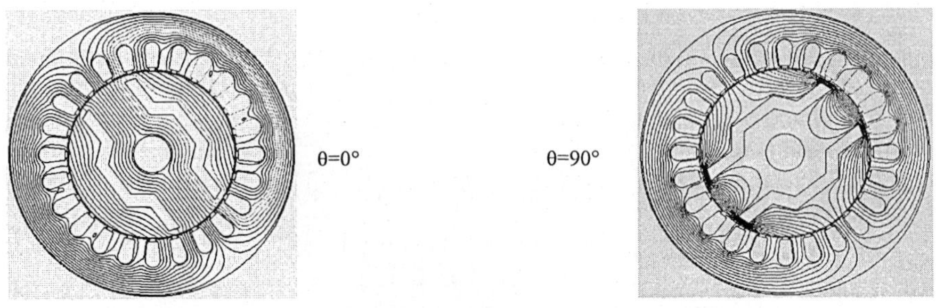

Figure 2   Magnetic flux distribution within the motor

$$B_r = \boldsymbol{B} \bullet \boldsymbol{R}^0 = (\frac{\partial A}{\partial y}\boldsymbol{i} - \frac{\partial A}{\partial x}\boldsymbol{j}) \bullet (x_r\boldsymbol{i} + y_r\boldsymbol{j})/\sqrt{x_r^2 + y_r^2} \qquad (3)$$

where (x_r, y_r) is the coordinate of arbitrary point at the circle line in the gap. The fundamental component of B_r can be obtained further by Fourier decomposition. The B_r after ABS transform is shown in Figure 3. It can be seen that YBCO bulks have obviously blocked the fluxes in the direction of q-axis.

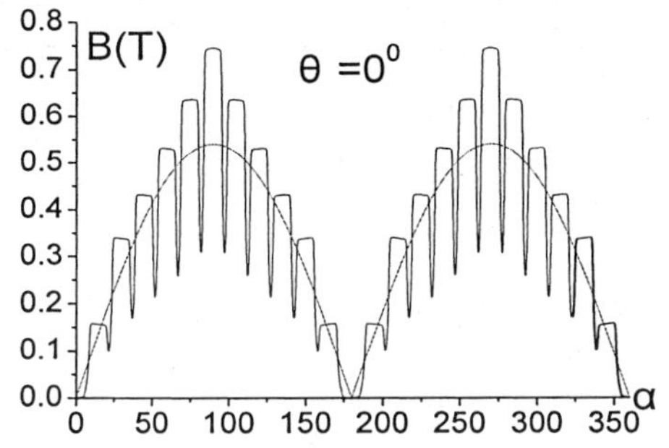

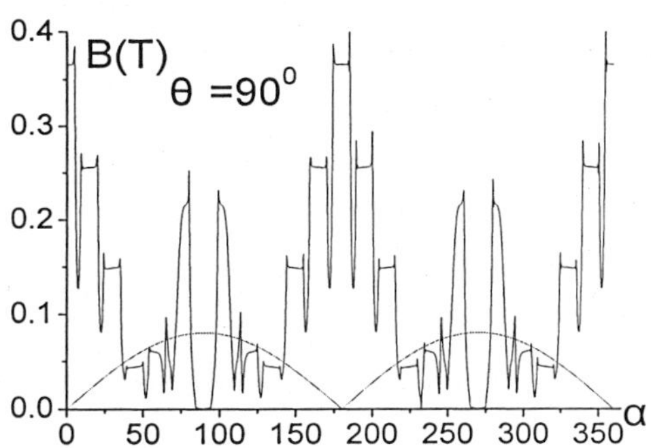

Figure 3   Radial air-gap induction within the motor

It's supposed that $i_A=I_m$, $i_B=-0.5I_m$, $i_C=-0.5I_m$ and $\theta_0=0$ at a particular time. Here i_A, i_B, i_C are stator phase currents with the amplitude of I_m, θ_0 is the angle between the d-axis of the rotor and the axis of A phase winding. In terms of the definition of the direct-axis synchronous reactance, X_d can be expressed as:

$$X_d = 2\pi\pi\frac{\sum_{k=1}^{N}\iint_S \boldsymbol{B} \bullet d\boldsymbol{s}}{aI_m/\sqrt{2}} \qquad (4)$$

where S is the area of the single turn coil, a is the number of parallel branches of stator winding, f is the frequency of stator fundamental current, and N is the number of total turns of the single-phase winding. Likewise, X_q may also be determined as shown in Figure 4. It can be seen that both X_d and X_q decline with the increase of their respective exciting currents. Also, we find the X_q-I_q curve drops rapidly when I_q is lower than 0.45 A. All these observations from Figure 4 are in accordance with the experimental data, which proves the validity of the analytical method we presented above.

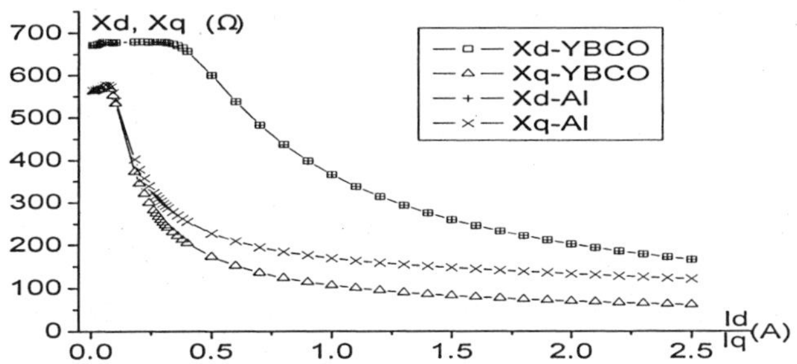

Figure 4 X_d-I_d and X_q-I_q curves of the motor

Based on curves of X_d-I_d and X_q-I_q, the working curves of the HTS reluctance motor can be completely determined by the electrical machine theory, as shown in Figure 5. In the figure, P_2 is the output power, T is the output torque, I is the stator current, $\cos\varphi$ is the power factor, η is the efficiency and θ is the power angle. The corresponding experimental results are shown in Table 1. The comparison reveals the feasibility of our finite-element analytical method.

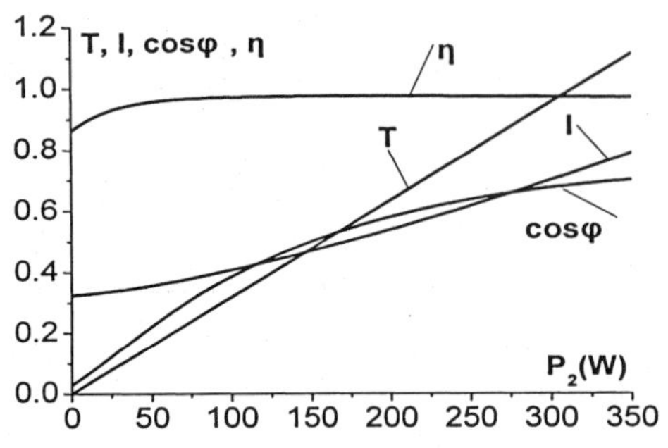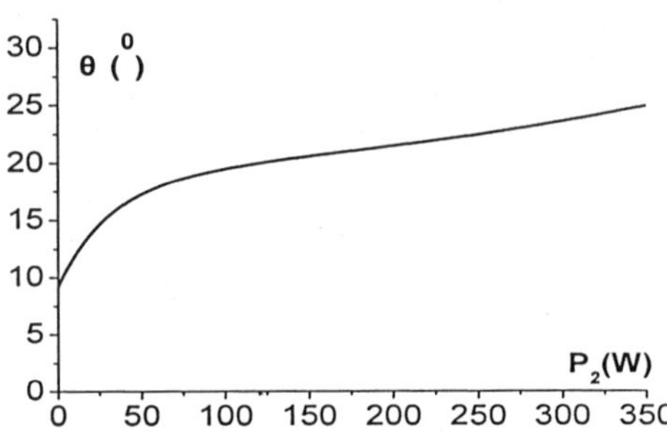

Figure 5   Working curves of HTS bulk reluctance motor

Table 1   The comparisons of the computational and experimental data (220V)

	P_2 (W)	I (A)	$\cos\phi$	η
Experimental value	150	0.50	0.49	0.948
computation value	150	0.47	0.50	0.975
Experimental value	300	0.72	0.70	0.907
computation value	300	0.70	0.68	0.972

CONCLUSION

Considering that the strongly anisotropic and non-linear magnetic behaviors of HTS bulk, a finite-element method is presented to analyze a small Demo HTS reluctance motor. Its feasibility is proved by the comparisons between the experimental and simulated results of X_d-I_d, X_q-I_q curves and the operating characteristics.

ACKNOWLEDGMENT

The work was supported by the National Nature Science Foundation of China under Grant No. 50107010.

REFERENCES

1.   Kovalev L. K., Oswald B., Gawalek W., Superconducting Reluctance Motor with YBCO Bulk Materials, IEEE Transactions on Applied Superconductivity (1999) 9 1201-1204

2.   Kovalev L. K., Gawalek W., Oswald B., Hysteresis and Reluctance Electrical Machines with the Bulk HTS Rotor Elements, IEEE Transactions on Applied Superconductivity (1999) 9 1261-1263

3.   Kovalev L. K., Gawalek W., Oswald B., New types of electric machines on the basis of the bulk HTS elements, Proc. of ICEC-18, Bornbay, India, (2000)

4.   Vajda I., Mohacsi L., Advanced hysteresis model for levitating applications of HTSC materials, IEEE Transactions on Applied Superconductivity, (1997) 7 916-919

5.   Chen D. X., Goldfarb R. B., Kim model for magnetization of type-II superconductors, Journal of Applied Physics (1989), 66 2489-2500

Proceedings of the Twentieth International Cryogenic Engineering Conference
(ICEC 20), Beijing, China. © 2005 Elsevier Ltd. All rights reserved.

A Current Compensation Type Superconducting Fault Current Limiter

Danfei Chen, Caihong Zhao, Liye Xiao

Institute of Electrical Engineering, Chinese Academy of Sciences, Beijing, 100080, P.R.China

Current compensation type superconducting fault current limiter is a novel topology of FCL, which consists of an equivalent AC current source with a limiting resistor in parallel. It connects power systems in series with a transformer. At normal state, the current of equivalent AC current source is consistent with the system current and the limiter has no effect on the system. But under fault condition, the system current is greater than the current of equivalent AC current source and the limiting resistor takes effect immediately. The application circuit and its control system are presented. Simulation results show that the SFCL can reduce both transient and steady-state fault currents significantly.

INTRODUCTION

Fast growing networks with increasing fault levels require the use of FCL techniques. Many investigations to develop the FCL have been carried out. Recently, superconducting FCL and solid state FCL attract more attentions. In this paper, a new topology of FCL named current compensation type superconducting fault current limiter [1] that combines superconducting technology and power electronic technology is proposed. It is a promising device because of following several features. 1) It is possible to limit the current at once. 2) It can reduce both transient and steady-state fault currents. 3) Low losses and high current reduction rate.

PRINCIPLE

Model

Figure 1 shows principle diagram of a current compensation type SFCL. It consists of an equivalent AC current source with a limiting resistor in parallel and then it connects power systems in series with a transformer.

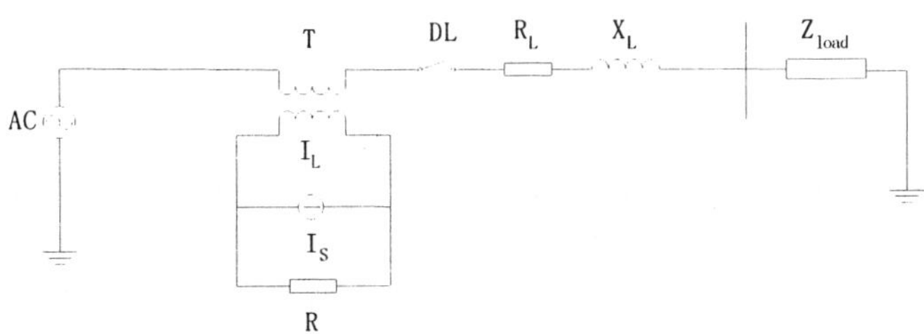

Figure 1 Principle diagram of a current compensation type SFCL

Figure 2 shows circuit diagram of a current compensation type SFCL. A superconducting reactor HTS-L is used because of its advantages such as with a larger inductance, smaller volume and lower losses. The inductance of HTS-L is so large that we can consider the current through the coil to be invariable, which makes it work as an equivalent constant current source. The superconducting reactor and a fully controlled bridge constitute a current source inverter (CSI). Then the CSI is in parallel with a limiting resistor and a transformer.

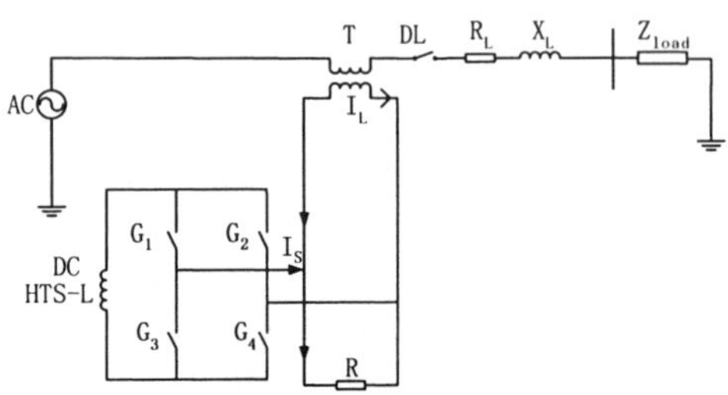

Figure 2 Circuit diagram of a current compensation type SFCL

Operating principle

Figure 3 shows the operating principle scheme of a current compensation type SFCL.

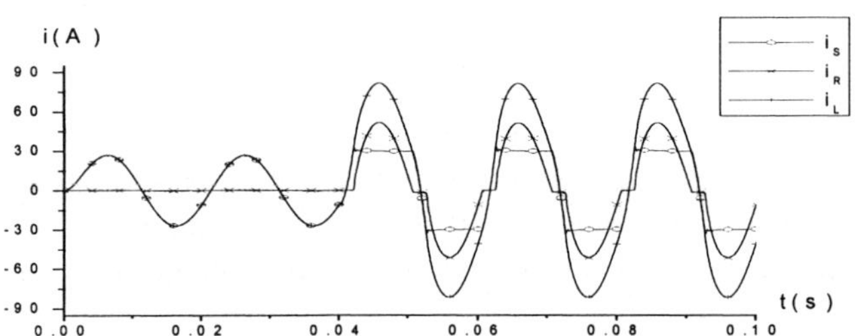

Figure 3 Operating principle scheme of a current compensation type SFCL

The value of superconducting reactor current is set to I_d. I_d is kept larger than the value of normal system current but much smaller than fault system current.

At normal state, the CSI generates equivalent AC current source i_s which is consistent with the system current i_L and there is no current through the limiting resistor ($i_R=0$). The transformer's secondary winding becomes shorted and the limiter has no effect on the system.

During fault, the peak value of the output current of the CSI is I_d, so I_L becomes greater than I_s and the difference current passes through the limiting resistor. Correspondingly, there is a voltage drop on the transformer and the voltage on R_L and X_L becomes lower, which effectively limits the short current.

Control strategy

Current source inverter has many advantages such as inherent short circuit protection and ruggedness, but received less attentions because of the difficulties associated with gating the switches. In this paper, we use on-line generation of gating signals for the CSI [2]. The control system is shown in Figure 4.

The power switches in the CSI must be operated so as to avoid an open circuit on the dc link or a sudden short circuit on the output capacitor. An alternate way is to add the required shorting pulse to obtain the gating signals. These pulses create a dc bus short through one leg of the inverter whenever either all top or all bottom switches are open.

SIMULATION

<u>Analysis conditions</u>
Simulation studies were carried out with the circuit shown in Figure 4. Each IGBT was in series with a diode. The short circuit was supposed to occur at the end of the transmission line. Table 1 shows parameters for the simulation.

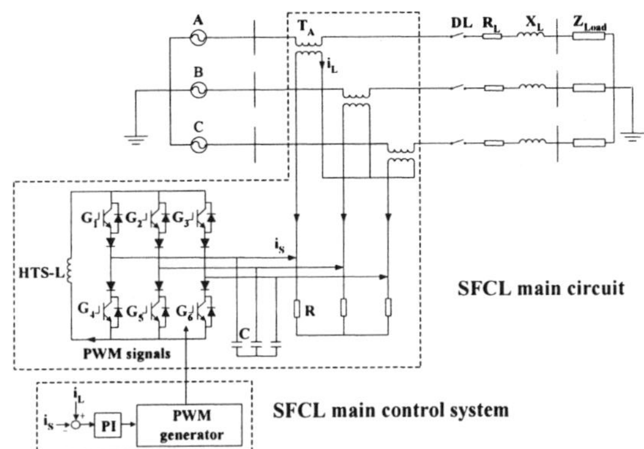

Figure 4 Application circuit and its control system of a current compensation type SFCL

Table 1 System simulation parameters settings

Power supply	220V	System frequency	50Hz
Line resistance	0.47Ω	Line inductance	4.49mH
Load resistance	10Ω	Load inductance	10.468mH
Limiting resistance	5Ω	Filter capacitance	50μF
Initial current of HST-L	30A	HST-L inductance	0.5H

<u>Simulation results</u>
Figure 5 shows the simulation results. Figure 5(a) shows a single-phase-to-ground short circuit without SFCL. Figure 5(b) shows a single-phase-to-ground short circuit with SFCL. Figure 5(c) shows a phase-to-phase short circuit without SFCL. Figure 5(d) shows a phase-to-phase short circuit with SFCL. Figure 5(e) shows a three-phase short circuit without SFCL. Figure 5(f) shows a three-phase short circuit with SFCL.

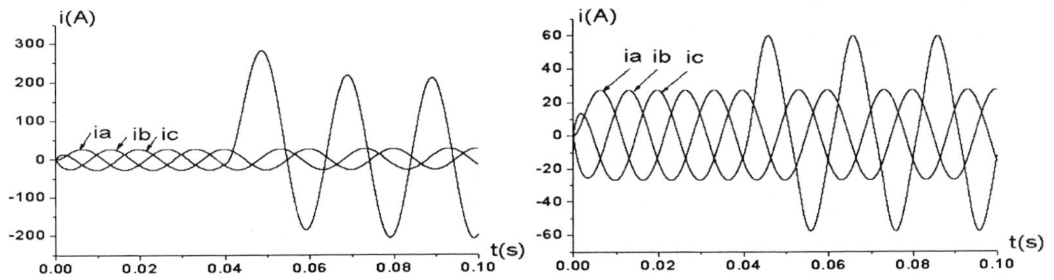

(a) Single phase-to-ground short circuit without SFCL   (b) Single phase-to-ground short circuit with SFCL

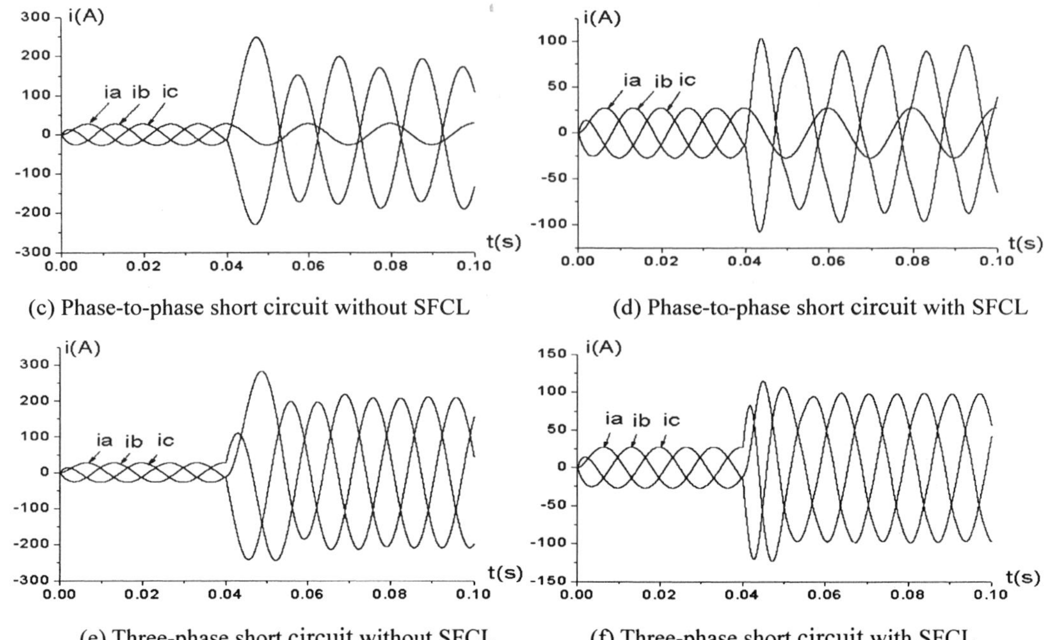

(c) Phase-to-phase short circuit without SFCL   (d) Phase-to-phase short circuit with SFCL

(e) Three-phase short circuit without SFCL   (f) Three-phase short circuit with SFCL

Figure 5 Simulation results of different faults in the three-phase system

We easily noticed that there was a great contrast between the short circuit currents with and without SFCL. The current compensation type SFCL could limit the faulted current to a desired level. The fault current reduction rate of the first peak short circuit current was between 50% and 80%.

For example, three-phase system worked without SFCL. At normal state, the peak current of A phase was 27.1A, THD was 1.8%. When a single line-to-ground short circuit happened, the first peak fault current of A phase was 281A and the steady-state fault current was 209A. By contrast, three-phase system worked with SFCL. The normal peak current of A phase is 27A, THD was 1.95%, The first peak fault current of A phase was limited to 59.5A and the steady-state fault current was limited to 57A. The current loss was only 0.37% and the harmonics that came from the CSI was very small when the system operated normally. The first peak fault current reduction was 79% and the steady-state fault current reduction was 73%. All above showed that the SFCL had a good current limiting characteristic.

CONCLUSION

It is verified that the current compensation type SFCL is an efficient device for reducing fault current under different fault conditions by the analysis and simulation in the paper. The superconducting characteristics have been used in the current compensation type SFCL, which produce low energy losses at normal state. With the development of superconducting material and power electronics, it will become a promising application device in the future.

REFERENCES

1. Caihong Zhao, Liye Xiao, Liangzhen Lin, A Kind of Circuit of FCL, Patent Application Number, 021568251
2. Espinoza, J. and Joos, G., On-line Generation of Gating Signals for Current Source Converter Topologies, IEEE International Symposium (1993), 674-678

Evaluation on power system transient stability with resistor type SFCL and parameter design

Ye Ying, Xiao Liye

Institute of Electrical Engineering, Chinese Academy of Sciences, China

With the development of high temperature superconducting fault current limiter (SFCL), it is necessary to investigate its effects on the power system with SFCL installed. In this paper a theoretical analysis of power system transient stability with resistor type SFCL using Lyapnov energy function is presented, and a guideline for SFCL's parameter design is given. Digital simulation was also made. Both theoretical analysis and simulation results demonstrate that resistor type SFCL could improve the transient stability of power system.

INTRODUCTION

SFCLs as the devices for limiting fault currents have been progressing thanks to the development of superconducting technology [1]. Power system transient stability with resistor type SFCL was already studied [2,3], but it still needs deep investigation. In this paper, we put forward a detailed theoretical analysis of transient stability with resistor type SFCL using Lyapnov energy function. In order to investigate the effects of SFCLs on the power system, we propose a simple single-machine infinite-bus power system with SFCL, a generator, and transformers being mathematically modeled. Digital simulation is made to observe the transient performance of the circuit. In view of improving transient stability, the parameter of the resistor type SFCL is also designed.

MODEL SYSTEM

OMIB system
The one machine and infinite bus system (see Figure 1) consists of one generator supplying power to an infinite network via a transformer and double-circuit transmission lines. The generator is a salient pole synchronous machine, and its mechanical behavior is described by the classical swing equation.

SFCL model
Here we use the resistor type SFCL with YBCO thin film, which is one of the most promising current limiting devices for its effective operation in the power system. In case of short circuit fault, when excess current flows in the YBCO film, the superconducting film changes to the resistive state and generates the joule heating. In this condition, the weakest zone becomes resistive and heats firstly. Then the heat generation expands the zone by thermal propagation, and the expansion is limited at the millisecond time scale because the conductivity is low. It is possible that the voltage-current characteristics of superconductor rise exponentially to a stable value within some dozens of milliseconds during the fault [2]. The SFCL has an extremely fast current transition in comparison to electro-mechanic time constants. When the transport current exceeds the critical current under the condition that the temperature of the superconductor is above the critical temperature, the resistance of the superconductor reaches its normal state resistance. When the transport current is below the critical current, the resistance of the superconductor keeps zero. Considering that the S/N transition of the superconductor can be completed in one or two steps during the transient stability simulation, the resistor type SFCL could be modeled as a

resistor with a shunt switch (see Figure 2). If the current of the circuit exceeds the critical current of the superconductor, the shunt switch is open. Otherwise the shunt switch is closed. The SFCL is installed at the Y-side of transformers T-1.

ANALYSIS OF TRANSIENT STABILITY WITH SFCL

Direct method and energy function

Lyapnov direct method solves the problem of nonlinear systems stability based on the structure of differential equations and initial values. For autonomous differential equations:

$$X = F(X), X(0) = X_0, X \in R^N, F \in R^N$$

if we can find a positive scalar v(x) and a region D, which satisfy:

$$V(X) > 0, V(0) = 0, X \in D \tag{1a}$$

$$\dot{V}(X) \le 0, X \in D \tag{1b}$$

Then, for any $x_0 \in D$, solution of differential equations (1) is stable. We could get the Lyapnov energy function of one machine infinite bus system by means of first integrals [4]. Considering that the power system is conservative, we get:

$$V(\delta,\omega) = \frac{1}{2}M\omega^2 + \frac{1}{M}\int_0^\delta (p_e \sin(u+\delta^s) - p_e \sin\delta^s)du = V_k(\omega) + V_p(\delta) = constant \tag{2}$$

Assessment

When the power system is in stable state, the kinetic energy V_k is zero because the warp of synchronous rotor speed $\omega = 0$. When the fault occurs, energy is injected into the power system. Assuming that potential energy reference point is set at $\delta_c(t_c)$ (t_c is the time when the fault is cleared), the total transient energy V after disturbance is:

$$V_c = V_k\big|_c + V_p\big|_c = \int_{\delta_0}^{\delta_c} M\frac{d\omega}{dt}d\delta = \int_{\delta_0}^{\delta_c}(p_m - p_{11}\sin\delta)d\delta \tag{3}$$

The maximal energy that the system can endure is $V(0,\delta'')$, where δ'' is the saddle point given by $\delta'' = \pi - \delta^s$. $V(0,\delta'')$ is the critical transient energy:

$$V_{cr} = \int_{\delta^c}^{\delta''}\{P_{111}\sin\delta - p_e \sin\delta^s\}d\delta = \frac{E_q U}{X'_{d\Sigma}}(\cos\delta^c - \cos\delta'') - P_m(\delta'' - \delta^c) \tag{4}$$

if $V_c \le V_{cr}(\delta'',\delta_c)$, the system keeps stable; if $V_c > V_{cr}(\delta'',\delta_c)$, the system loses synchronization. If the resistor type SFCL works at the fault beginning without time delay and we neglect the quench time of the superconductor, the transient energy when the fault is eliminated is:

$$V_c = \frac{1}{2}M\omega_c^2 = \int_{\delta_0}^{\delta_c} M\frac{d\omega}{dt}d\delta = \int_{\delta_0}^{\delta_c}(p_m - p'_{11}\sin\delta)d\delta \tag{5}$$

It could be obviously seen that $V_c - V'_c = Area(b'bcc') > 0$, which means that the SFCL decreases the initial energy injected into the power system. The recovery time of resistor type SFCL is relatively long (about a few seconds). Taking the recovery characteristics into consideration, the critical transient energy with SFCL applied is:

$$V_{cr}' = \frac{E_q^2}{|Z|}\sin\alpha(\delta^{b-o} - \delta^s) + \frac{E_q U}{|Z|}\{\cos(\delta^s - \alpha) - \cos(\delta^u - \alpha)\} + \frac{E_q U}{|Z|}(\cos\delta^{b-o} - \cos\delta^u) - P_m(\delta^u - \delta^s) \tag{6}$$

Compare the critical transient energy with SFCL installed and without SFCL installed, we could see that: $V_{cr}' - V_{cr} = Area(mefh) > 0$. With SFCL installed, the initial transient energy injected into power system is reduced. Meanwhile, the maximal energy that the power system can endure is enlarged taking recovery characteristics of the SFCL into account.

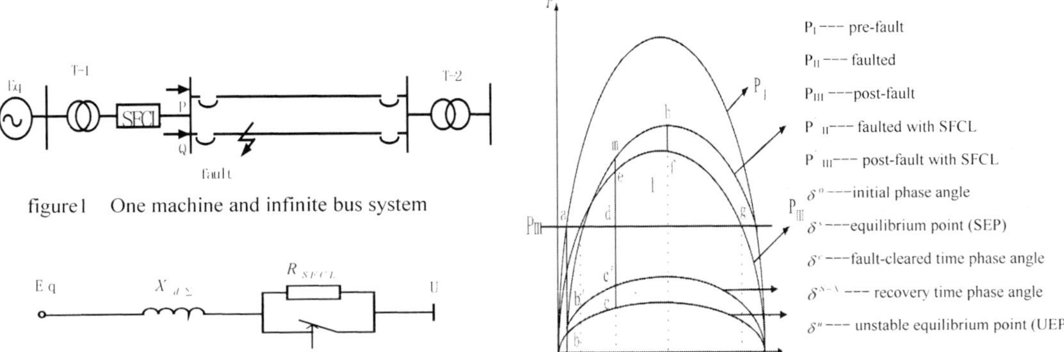

figure1   One machine and infinite bus system

figure2   Equivalent circuit

figure3   Power-angle curves with SFCL

Parameter design

The analysis above shows that it is good to the power system transient stability if the SFCL do not quit immediately after the fault clear. The SFCL is put at the side of transformer in order to avoid its influence on the transmission line auto-reclosed brake. Thus the resistor type SFCL functions like breaking resistors except that the recovery time of the SFCL could not be controlled. But if the resistance of SFCL is too large, the power system will lose synchronism after the second swing. When energy is concerned, the resistance value of SFCL should be restricted by the equation: $V_{k2} = V'_{p1} < V_{cr}$.

$$\int_{\delta^c}^{\delta^u} \{P_{III}'(\delta) - p_e \sin\delta^s\}d\delta \le \int_{\delta^c}^{\delta^u} \{P_{III}(\delta) - p_e \sin\delta^s\}d\delta \tag{7}$$

We can approximately get:

$$R_{SFCL} \le X_{d\Sigma}' \tan(\frac{\pi}{2} - \arcsin\frac{U(\cos\delta^s - \cos\delta^c) - P_m(\delta^u - \delta^s)X_{d\Sigma}'}{E_q(\delta^u - \delta^c)}) \tag{8}$$

Where $X_{d\Sigma}' = X_d + X_{T-1} + X_L + X_{T-2}$. Thus, there are three factors we should take into consideration when we design the parameter of SFCL: affects on protection relay, switching off ability of circuit breaker and enhancement of power system stability.

SIMULATION RESULTS AND DISCUSSION

System simulation parameters are shown in Table 1. A short circuit fault occurred on one of the transmission lines. After 100ms the circuit breaker operated, and cut off the faulted single line. Two cases, SFCL being applied or not, were simulated using Matlab simulink. Fig.4 and Fig.5 show that the swing curve with SFCL applied is lower than the curve without SFCL, and the same for the rotor angular velocity. This demonstrates that the resistor type SFCL can protect the synchronism of the generator.

When the resistance value of SFCL is too large, that is the value exceeds the restriction of equation, the power system lost synchronism. The simulation results are shown in Fig.6 and Fig.7. The resistance value of SFCL is 2 per unit.

Table 1   Model system parameters

Generator					Transformer	Transmission line
S_N	P_0	δ_0	M	X_d	X_T	X_L
500MVA	385.2MW	49.34deg	3.7.p.u	0.238.p.u	0.13.p.u	2.93.p.u

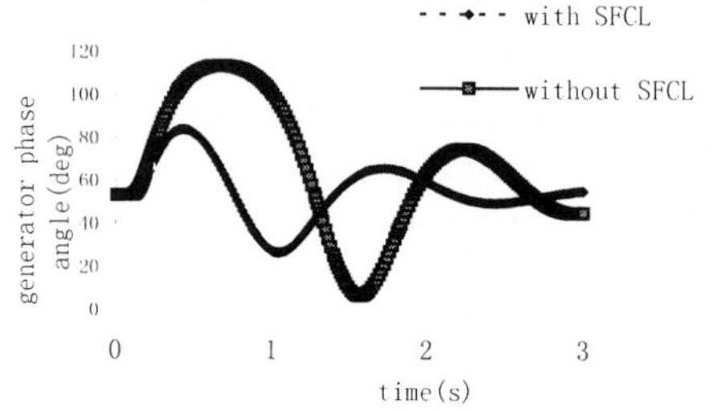

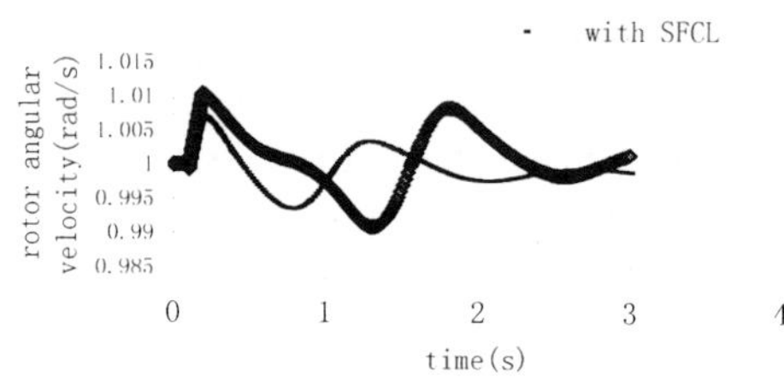

Figure 4   Generator swing curve with and without SFCL

Figure 5   Gnerator rotor angular velocity curve with and without SFCL

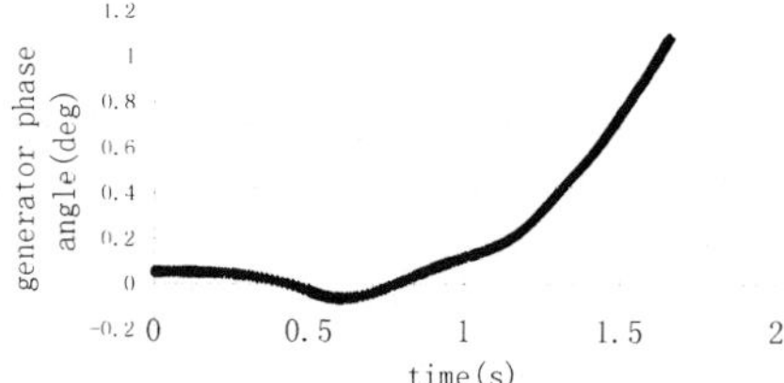

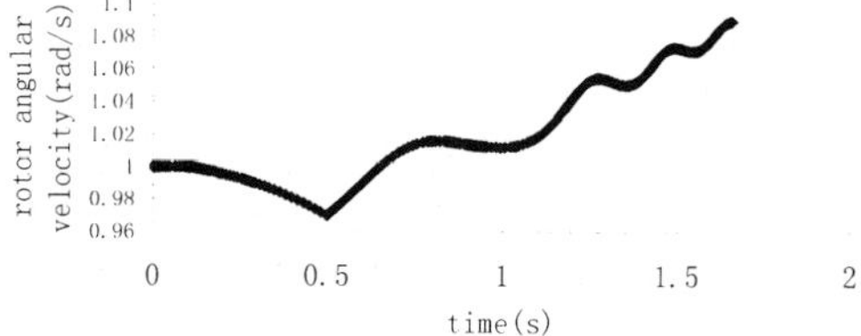

Figure 6   Swing curve with $R_{SFCL} = 2$ .p.u

Figure 7   Angular velocity curve with $R_{SFCL} = 2$ .p.u

CONCLUSION

Through theoretical analysis and digital simulation, it is found that the resistor type superconducting fault current limiters not only limit fault current but also protect synchronization of generators. The SFCL can decrease the initial transient energy V_c injected into the power system and increase the critical transient energy V_{cr}, which enhances the transient stability. Considering the second swing stability, the resistance value of SFCL cannot be too large, and should be restricted.

REFERENCES

1. R.F.Giese and M.Runde, Assessment study of superconducting fault current limiters operating at 77k, IEEE Trans on Power Delivery (1993), 18 1138-1147

2. Mårten Sjöström, Rachid Cherkaoui, Bertrand Dutoit, Enhancement of power system transient stability using superconducting fault current limiters, IEEE Trans on Applied Superconductivity (1999), 9 1328-1330

3. Lin Ye, LiangZhen Lin, and Klaus-Peter Juengst, Application studies of superconducting fault current limiters in electric power systems, IEEE Trans on Applied Superconductivity (2002), 12 900-1330

4. M.A.PAI, In: Power System Stability-Analysis by the Direct Method of Lyapnov, North-Holland Publishing Company, New York, USA (1981).

Effects of bias power supply on bridge type superconducting FCL

Zhang zhifeng, Xiao liye

Institute of Electrical Engineering, Chinese Academy of Science 100080, Beijing, China

The effects of bias power supply (BPS) on bridge type of SFCL are studied in this paper. The main research method is a superconducting reactor charging analysis. With shift of charging path and change of charging power source in a process of short fault, on the condition of the SFCL with BPS in two connection modes, load current distortion is eliminated and the SFCL limiting capability is enhanced, And the parameters of the BPS is also optimized. Simulation and experimental results validate the necessary of the BPS to enhance the limiting capability of the SFCL.

INTRODUCTION

For enhancing reliability of power system, fault current limiter is becoming an essential part in the modern power grid. More and more researchers and scholars are focused on the R&D of bridge type of high temperature superconducting fault current limiter. There are different viewpoints of introducing bias power supply (BPS) into the bridge. Some researchers are interested in designing such SFCL without using BPS. Taking no account of the BPS, the simpler makeup of SFCL and the lower investment cost of the SFCL is available [1]. Meanwhile, others are paying more attention to designing such SFCL with using BPS. In order to enhance quality (e.g. mitigating line voltage sags, decreasing line phase-angle jumps, reducing line current distortion, and etc.) of distribution networks, improve the limiting capability of the SFCL, and conceal the effects of load changing, it is necessary to analyze the effects of BPS on the

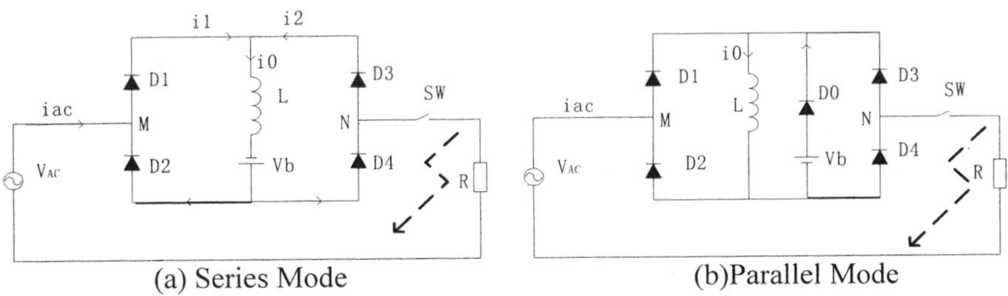

(a) Series Mode   (b)Parallel Mode

Fig.1 Circuit diagram of FCL with BPS

SFCL and on the networks [2,3,4]. Two kinds of typical connection modes are described [4], in which the connection modes are determined by the connection relationship between the BPS and a superconducting reactor L. One of them is series mode (Fig.1 (a)), and another is parallel mode (Fig1 (b)). In order to design a large scale SFCL used in three phases 10.5kV, 400A system, a diodes bridge is chosen and further study of the function of the BPS in the SFCL is present in this paper. The two-mode SFCL are discussed. The reactor charging law is analyzed in theory. Based on the law in the two circuits, the interrelationship of enhancing limiting capability of the SFCL in fault state and removing steady-state current distortion in steady-state is discussed.

CHARGING PRINCIPLE OF THE REACTOR IN THE SFCL

For a FCL to have a good performance, the two requirements must be matched: in steady state, load current distortion must be removed so that the effects of the SFCL to the load current must be minimized. In fault state, the limiting capability of the SFCL must be big enough so that the fault current is restricted availably. Though, both the limitation of fault current and the elimination of load current distortion are carried out by the reactor, the contradiction of them is still existing. If the voltage amplitude of the BPS is smaller, the load current distortion can't be eliminated in steady state, and the reactor current can't keep constant. If the amplitude of the BPS is larger, the fault current limiting capability of the SFCL can be weakened in fault state because the reactor is constantly charged by the BPS. The contradiction can be solved by investigating a special charging law of the reactor in the SFCL.

It is noted that a current increment of a single reactor is determined by charging voltage, charging time and the inductance. For a FCL, there are two operating states. It is the current relationship of the load and the reactor that determining the change of the operating states and selection of charging power

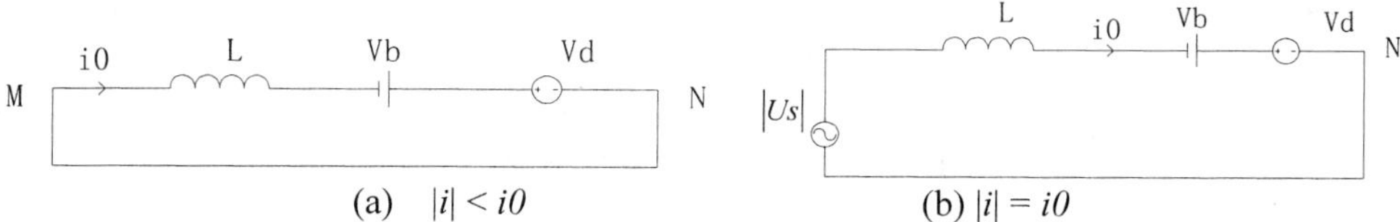

(a) $|i| < i0$ (b) $|i| = i0$

Fig.2 Reactor charging equivalent circuit diagram of FCL in series mode

sources. Both the BPS and the power source act as the reactor charging power source and they work at different state and different time. When Vd and $|Us|$ are defined as the forward voltage of two of four diodes and the equivalent voltage of the power source, respectively, the charging law can be described by equations. For the series mode, the reactor charging equivalent circuits are shown in Fig.2 (a) and (b) and the equivalent circuits are presented as:

$$Vb - Vd = L\frac{di0}{dt} \qquad (|i| < i0)$$
$$|Us| - Vd + Vb = L\frac{di0}{dt} \qquad (|i| = i0)$$
\qquad (1)

In steady state or when the reactor current is larger than the load current, in order to keep the reactor

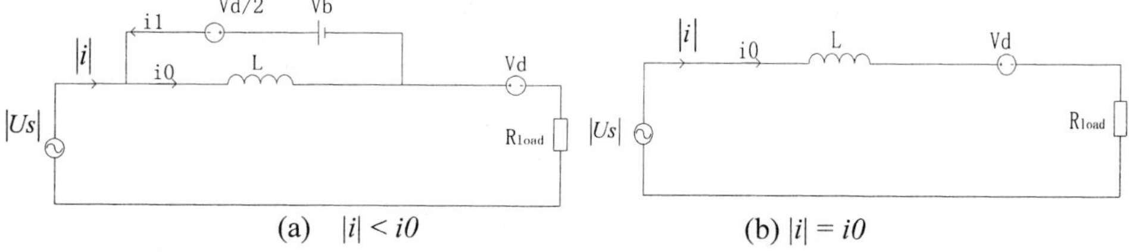

(a) $|i| < i0$ (b) $|i| = i0$

Fig.3 Reactor charging equivalent circuit of SFCL in parallel mode

current constant and also avoid distortion occurrence near the peak of the load current, the BSP voltage Vb must match the following relationship: $Vb \geq Vd$. In fault state, in order to restrict the rate of the increment of load current, the voltage of BSP should be small or even a negative voltage so as to restrict the charging speed of the power source and limit the fault current of the grid. So the constant value of the voltage of the BPS is of benefit to the steady process and goes against the fault process. For the parallel mode, the charging equivalent circuits are shown in Fig.3 (a) and (b). The BPS act as a fly-wheel path and the fly-wheel current through the BPS reduces to zero in limiting process. In steady state or when the load current $|i|$ is smaller than the reactor current $i0$, the charging sources of the reactor take on by both the power source of the grid and the BPS in parallel. So the current relationship can be described:

$$Vb - Vd/2 = L\frac{di1}{dt}$$

$$|Us| - Vd = L\frac{d|i|}{dt} + |i|Rload \qquad (|i| < i0) \qquad\qquad (2)$$

$$i0 = |i| + i1$$

The BPS is necessary to hold on the reactor current $i0$ constantly and the load current waveform distortion can also be removed by the BPS. In fault state, when the load current $|i|$ arrives to the value of the reactor current $i0$, the fly-wheel current $i1$ will decreases to zero accordingly and the charging function is only taken on by the power source. The charging equation is obtained:

$$|Us| - Vd = L\frac{d|i|}{dt} \qquad (|i| = i0) \qquad\qquad (3)$$

From the above discussion, the key of implementing the SFCL performance is how to regulate the reactor current to control the current paths change, and restrict fault current increasing, and minimize load current waveform distortion.

SIMULATION AND EXPERIMENT

Using SIMULINK of MATLAB, the SFCL performance analysis and the load current distortion discussion both in steady state and in fault state are carried out. We have considered a 10.5kV, 400A and 50Hz system with 10% source impedance. The inductance value of the reactor and the forward voltage of the diodes are assumed 80 mH and 16V respectively. Per unit of current and voltage are 400A and 16V respectively. If the fault takes place at a zero crossing time of the current waveform, the value of the BPS is chosen in order to remove load current waveform distortion in steady state. The peaks of the load

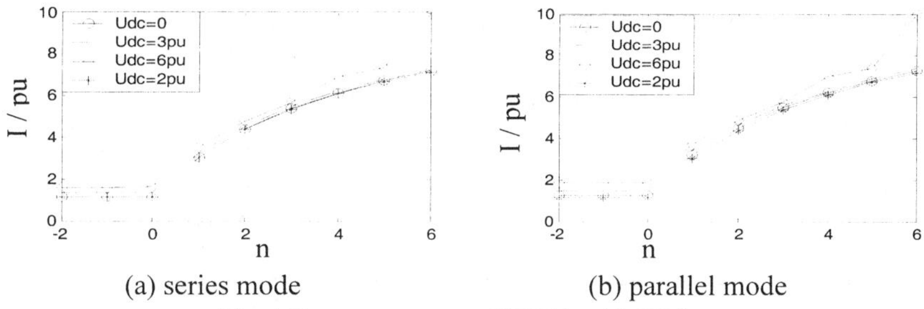

(a) series mode   (b) parallel mode

Fig.4 Reactor current of SFCL with BPS

current of two modes of the SFCL are showed in Fig.4 (a) and (b) respectively. The parameter n represents the times of the peaks. When the BPS voltage increases, whether the SFCL in series mode or in parallel one, the distortion of the steady state load current can be removed satisfactorily, but the limiting capability of the SFCL is decreased. If the voltage of the BPS is higher, the steady current of the reactor is no longer constant and tends to increasing, especially in parallel mode.

Considering the influence of the BPS to the limiting capability, we make the voltage of the BPS to be zero when fault occurs. The difference of the limiting results of the change is negligible because the power source is very bigger than the BPS. It is noted that the small change of the BPS voltage makes few effects to the limiting capability. So we can optimize the parameters of the BPS to remove the distortion of the steady state load current and at same time enhance the limiting capability. The optimized results are shown in table 1. The current through the BPS in series mode is far great than in parallel mode because, in parallel mode, the BPS acts as a flywheel path and no fault current through the path. The voltage of the

BPS is to remedy the loss of the diodes forward voltage drop. In conclusion, the BPS in series mode is more advantageous to enhance the limiting capability, but its volume is far larger than the BPS in parallel mode.

The experiment of the SFCL with the BPS in different connection modes were made in the single-phase 220V, 50Arms system. The inductance of the superconducting reactor is 23mH. The voltage of the BPS is two times of the forward voltage of the diodes. Data acquisition

TABLE 1. Parameters of BPS

	Series mode	Parallel mode
Current /A	3900	$400\sqrt{2}$
Voltage / V	32	48

and monitoring process is carried out on a LabVIEW platform. Fig. 5 (a) and (b) show the reactor current and the load current waveform when the SFCL in different connection with the BPS. The load current distortion is removed in steady state in two circuits. The influence of the BPS to the limiting capability is not obvious if the voltage of the BPS is nearly equal to the forward voltage drops of the diodes.

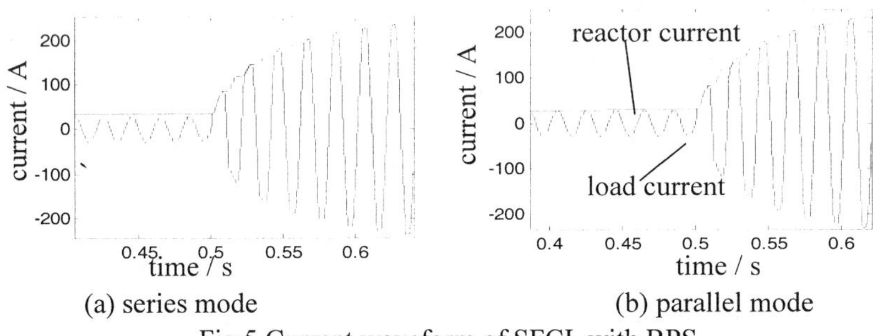

(a) series mode　　　　　　(b) parallel mode

Fig.5 Current waveform of SFCL with BPS

CONCLUSION

By analysis of superconducting reactor charging, the function of BPS to the SFCL is studied. It is clear that regulation of the reactor charging can eliminate load current distortion and enhance the SFCL limiting capability. So the negative effects of the SFCL to power grid can be obviously eliminated.

ACKNOWLEDGMENT

This work is supported by National Science Foundation of China under grant number of 50225723, 50137020.

REFERENCE

1. Leung E, Burley B, Chitwood N, *et al*. Design and development of a 15 kV, 20 kA HTS fault current limiter. Applied Superconductivity, IEEE Transactions on (2000), 10 832 -835
2. Hoshino T, Mohammad Salim K. DC reactor effect on bridge type superconducting fault current limiter during load increasing. Applied Superconductivity, IEEE Transactions on (2001), 11 1944 -1947
3. Eung Ro Lee, Seungje Lee, *et al*. Test of DC reactor type fault current limiter using SMES magnet for optimal design. Applied Superconductivity, IEEE Transactions on (2002), 12 850 -853
4. T. Nomura, M. Yamaguchi, *et al*. Study of single dc device type FCL for three-phase power system, Cryogenics (2001), 41 125-130

20kA HTS current leads for EAST tokamak project

BI Yanfang, CHEN Xingqian, MA Dengkui, LIU Xiaolong, WU Songtao and LI Jiangang

Institute of Plasma Physics, Chinese Academy of Sciences, P.O. Box 1126 Hefei 230031, China

Two 20kA current leads are made of different materials. Module 1 consists of melt cast processed Bi-2212 tube and two layers of Bi-2223/Ag-Au alloy tapes as the shunt and also to enhance current-carrying capacity. The critical current of the Bi-2212 tube alone is 9.052 kA at 77 K. After adding 104 Bi-2223 tapes the critical currents increase up to 10.2 kA at 78 K and >20 kA at 70.7K. Module 2 consists of 48 stacks of CryoBlock tapes. The critical currents of 192 tapes reach over 14 kA at 78 K and over 20 kA at 70.7 K.

INTRODUCTION

EAST tokamak (original name HT-7U) under construction in our institute is a full superconducting device for fusion experiments. The tokamak has a major radius of 1.7 m and a minor radius of 0.35 m. The rated currents of 16 D-shaped toroidal field coils made of NbTi/Cu superconductors are 14.3 kA for phase I and 16.5 kA for phase II. There is one pair of current leads for 16 TF coils. In order to obtain a single-hull divertor field configuration the 6 solenoids for Ohmic heating, and 6 round-shaped coils for the divertor field and equilibrium field must be fed power supplies independently. These coils have the same maximum current of 14.5 kA. So 12 pairs of current leads will be equipped for these coils.

EAST tokamak has a total current lead capacity of 33 kA in steady mode and 348 kA in pulsed mode. According to the optimum design of a nominal current lead to cool the leads needs the helium vapor consumption of 1.7 g/s for the TF coils and ~11.6 g/s for all the PF coils (assuming the pulse operation mode to require 2/3 of steady-mode mass flow). It equivalents a refrigeration power about 1.2 kW/4.5K.

According to investigation [1,2] the heat load of HTS current leads (HTSCL) at 4.5K can reduce to 0.1-0.18W/kA of the nominal leads. Using HTSCL for EAST tokamak can reduce refrigeration power of 900 W/4.5K at least and save operating cost greatly. Institute of Plasma Physics, Chinese Academy of Sciences starts developing HTSCL from the 2003 March. Now we design and manufacture two test leads of 20 kA aimed at a cryogenic test facility for EAST coils.

Both Bi-2212 tube and Bi-2223 tape sheathed with Ag-Au alloy are suitable for 15-20 kA leads. The melt cast processed (MCP) Bi-2212 tube has a lower thermal conductivity, lower critical temperature (~92 K) and poor strength. The Bi-2223 tapes have a higher critical temperature (110 K), and Ag-Au alloy sheath possesses much higher thermal conductivity and better stability. In order to compare their heat load at the cold end, contact resistance, stability and performance degradation due to manufacture process and cooldown-warmup cycles. We try to obtain experiences on more HTSCL manufacture and knowledge of making low contact resistance and higher stability, so both HTS materials are selected. In this paper the design consideration, manufacture technology and some test results of 20kA HTSCL are described.

LEAD COOLING MODE AND COPPER SECTION DESIGN

A 500W/4.5K refrigerator is equipped for our test facility for large-scale SC magnets. One pair of 20kA copper current leads spend more than 2g/s of liquid helium, which almost is half of the refrigeration capacity. In order to solve the 4.5K cooling power shortage, the up-ends of the 20kA HTSCL and copper sections will be cooled with liquid nitrogen (LN_2) and vapor, and the lower ends of HTS sections are cooled by 4.5K supercritical helium.

Evaporating latent heat of liquid nitrogen is much greater comparison with liquid helium, and enthalpy difference of nitrogen vapor from 70 K to 280 K is much smaller than helium from 4 K to 280 K. So design and manufacture of the heat exchanger for this kind of copper current leads should be as simple as possible. Helical fins heat exchanger is selected. To decrease vapor flow resistance the surface of heat exchanger is machined three helical parallel cooling channels (see Fig. 1).

The copper lead with lower current density has better stability when over current and coolant stoppage. A low rated current density of 10.9 A/mm^2 for the copper leads is chosen. The ratio of length to cross-section is optimized according to a rated current of 16 kA in order to minimize LN_2 consumption while leads standby. The copper for the section contains total impurity of less than 0.1% and annealed.

Between the heat exchanger and joint with the HTS section there is 27cm-long copper tube immerged in LN_2 as a heat sink to keep temperature of the HTS up-end stable. If the leads carrying current continuously, the tube section should be solder with Bi-2223/Ag tapes to reduce the Joule-heating. Through a buffer in the cryostat LN_2 is supplied to the current leads. To keep LN_2 level in the buffer in a suitable range is important for the normal operation.

The copper section has a vacuum jacket, and also is electrically insulated from coolant transfer pipes and the cryostat.

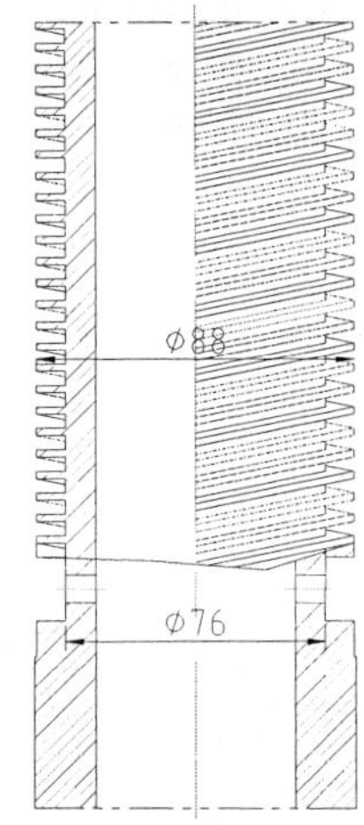

Fig.1 Heat exchanger with tree helical cooling channels

HTS SECTION DESIGN

One of the HTS modules is based on MCP Bi-2212 tube with 80mm outer diameter made by Nexans SuperConductors. According to the data provided by the vendor the critical current (Ic) of the Bi-2212 tube is 9.051 kA at 77K, and increases twice at 70 K. For a lead rated current higher than 5 kA adding a shunt is necessary. The vendor told us that Cu-Ni or Ag-Au alloy sheet could be the shunt material. So we try to use Bi-2223/AgAu alloy tapes for the shunt. Two layers of this kind tapes made by InnoST in China were added. Dimension of the tapes is 4.1 mm wide and 0.21 mm thick, and its Ic is ~ 71 A. So the 106 tapes can enhance current-carrying capacity of ~4kA at 77 K and 6kA at 70 K at least. The sheath of Ag-Au alloy contains 5 wt. % Au. These tapes also can be a shunt when Bi-2212 tube quench. The tapes in two layers are arranged on Bi-2212 tube surface in helical lines (see Fig. 2). The ends of HTS section are directly soldered to the lead copper section and to a copper clamp joint, respectively.

Another HTS module (Fig. 3) is based on Bi-2223 tapes sheathed with Ag-Au 5.3 wt.% alloy (CryoBlock) provided by AmSC. A support cylinder with an outer diameter of 83 mm and 320 mm long consists of two copper terminals and a brass cylinder, which has 48 grooves for tape stacks. The intervals between the stacks should be as small as possible in order to minimize perpendicular field on the HTS tapes and to obtain higher Ic. Each stack consists of 7 CryoBlock tapes. Four of them have full length, and others have shorter length. The tapered configuration can reduce heat leak at 4.5 K and required amount of HTS tape. The tape dimensions are 4mm wide and 0.2mm thick. The Ic of the tapes are not lower than 106 A. The module Ic of more than 20 kA at 70 K is expected. Table 1 lists the major parameters of the two modules.

According to the heat leak data provided by Nexans SuperConductors, the tube with 200 mm length and 8 mm thick has a conduction heat of ~0.5 W at the lower end when up-end at 77 K. It is quite low comparison to the full Bi-2223 wire module.

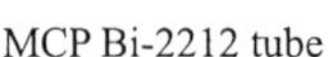

MCP Bi-2212 tube

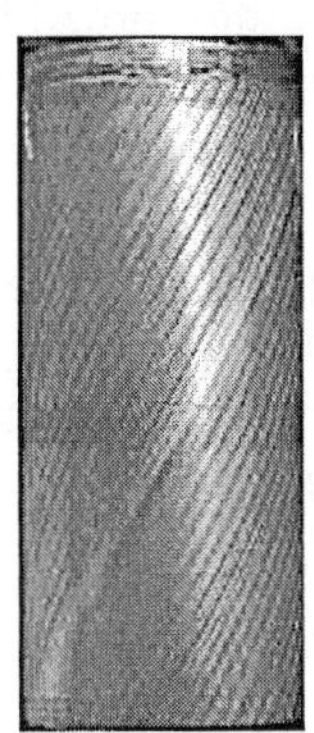

Covering with 2 layers
of Bi-2223/AgAu tapes

Fig. 2 HTS module 1

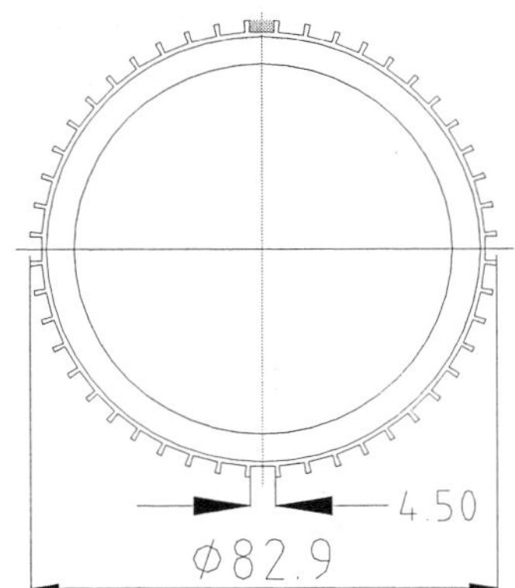

Cross-section of support cylinder

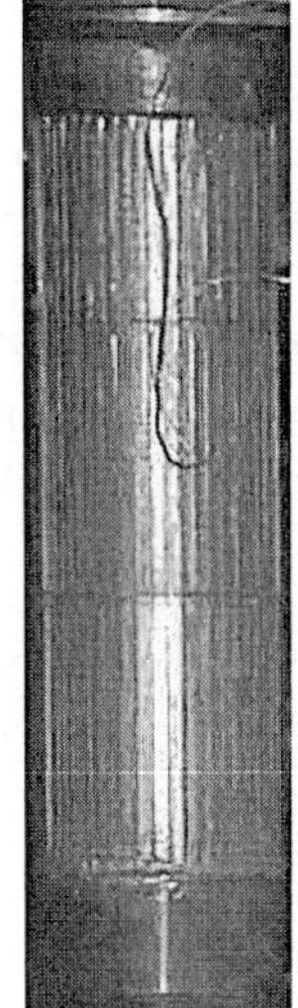

Fig. 3 HTS module 2

Table 1. Major parameters of 20 kA HTSCL

	Copper part	HTS module 1	HTS module 2
Effective length (mm)	700	200	320
Effective diameter (mm)	88	81.2	83
Conductor material	Copper	MCP Bi-2212+ Bi-2223/Ag-Au wt.5%	CryoBlock
Configuration	Tube with 3 helical fins	Cylinder	Tapered cylinder
Cooling mode	68K Sub-cooled LN$_2$ and vapor	4.5K supercritical He	4.5K supercritical He

MANUFACTURE

Soldering is most important technique for HTSCL manufacture and assembling because incorrect soldering results in degradation of HTS Ic and too high contact resistance. To heat and cool HTS tube or tapes must be slow enough, and avoid heating above the temperature limit (300 ℃) for both Bi-2223 wire and MCP Bi-2212. According to the vendor suggestion In-Bi solder was used for module 1. The joints between the HTS module and copper cap and the copper lead part are soldered in atmosphere, and a few of citric acid are applied as flux. The two layers of Bi-2223 tapes are soldered on the Bi-2212 tube in a vacuum vessel. For module 2 62Sn-36Pb-2Ag-paste solder was used between Bi-2223 tape stacks and the support cylinder. This HTS module was soldered into the copper caps also with In-Bi alloy due to its low melt point.

TEST RESULTS OF THE HTS MODULES COOLED WITH LN$_2$

We concern performance of the HTS modules, the contact resistance of ends and the optimum operation current of the heat exchanger. The two modules were cooled down to 78 K and immerged in LN$_2$ bath. A computer acquired the potential difference data when current increasing steps by step. For HTS module 2 the critical current was higher than 14kA at 78 K. But for module 1 linear-increasing potential difference appeared when the current higher than 10 kA, it meant more and more current transferred from Bi-2212 tube into Bi-2223 tapes, and Ic of the module 2 was 10.2 kA at 78 K according to 1 μV/cm criterion (see Fig. 4). When LN_2 temperature decreased from 73 K to 70.7 K, the current increased to 20 kA in 2 kA steps. Module 2 showed a stable performance, and had an Ic higher 20 kA at 70.7 K. Module 1 also had an I_C higher than 20 kA at 70.7 K too, but part of current carried by Bi-2223 tapes changed while the temperature decrease (see Fig, 5).

The contact resistances were 130 nΩ for module 1 and 44 nΩ for module 2 at 78 K. The two current leads stayed at 13 kA for 20 minutes. The voltage spanning the heat exchanger was 55 mV. It meant the optimum current near 13 kA for the copper lead, which was smaller than expected. Other measurements require LHe cooling the lower ends of HTSCL, and will be done in future.

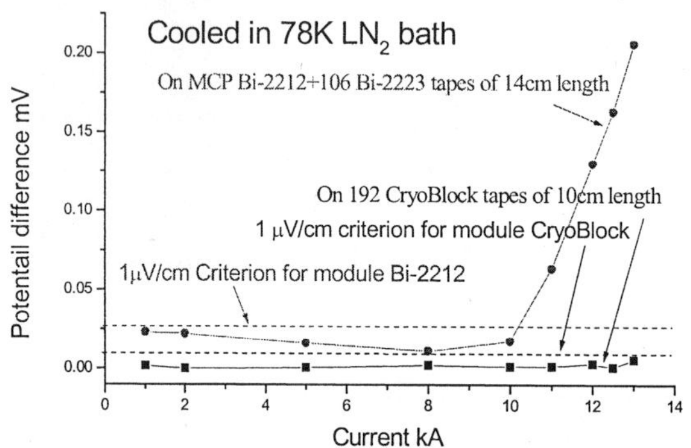

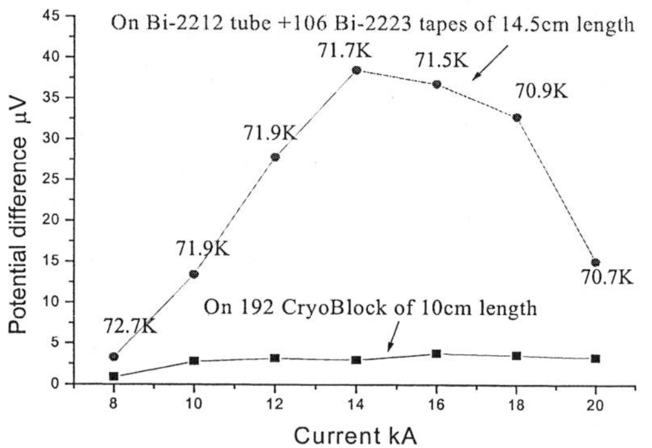

Fig. 4 Ic measurements of HTS modules at 78K LN_2 Fig. 5 Ic measurements of HTS modules below 73K LN_2

SUMMARY

The preliminary test results show that module 2 made of CryoBlock has better performance on the higher critical current and lower contact resistance, and the Bi-2223/AgAu alloy tapes as the shunt can enhance current carrying capacity of MCP Bi-2212 tube module to reach 20kA at 70.7K.

Correct solder technique and intervals between the stacks as small as possible are most important for HTSCL to reach a high performance without degradation of Bi-2223/AgAu tapes.

ACKNOWLEDGMENT

This work has been supported under the framework of Knowledge Innovation Program of Chinese Academy of Sciences.

REFERENCES

1. Heller R. et al., Development of a 20 kA high temperature superconductor current lead, Cryogenics (2001) 41 539-547
2. Jeremy A. Good et al., 13000 A current lead with 1.5 W heat load to 4.2 K for the Large Hadron Collider at CERN, IEEE Trans. on Applied of Superconductivity (2000) 10 1474-1476

Design of Cooling System and Temperature Characteristics Analysis of High Temperature Superconducting Demonstrative Synchronous Generator

Chen B., Zhang G.Q., Song F.C., Yu S. Z., Gu G.B.

Institute of Electrical Engineering, Chinese Academy of Sciences, GSCAS, Beijing, China

The cooling system of a demonstrative high temperature superconducting (HTS) generator, in which solid nitrogen (SN_2) was used as cryogen to keep the operating temperature of HTS synchronous machine, is presented in this paper. Operation process of cooling system is illuminated and main circulation parameters of cooling system are also investigated. Furthermore, temperature characteristics of rotor are calculated by finite element method (FEM) software ANSOFT. The simulated results show that temperature field was uniform and the maximum temperature of HTS coils is lower than 30 K during normal operation.

INTRODUCTION

Applying HTS conductors in the rotor of synchronous machines allows the design of future motors or generators that are lighter, more compact and features an improved coefficient of performance [1]. There are two types of rotor configurations in existing conceptual designs and actual prototypes of HTS machines: iron core rotor [2] and nonmagnetic rotor [1, 2]. Almost all HTS rotating machines were fabricated as racetrack coils of Bi-2223/Ag HTS tapes operating at temperature range between 20 K and 30 K in order that the HTS machines could possess sufficient over current capacity and endure bigger vertical component ($B_\perp$) of magnetic flux density passing through the surface of HTS tapes.

A 100KVA HTS demonstrative synchronous generator was designed in this paper. Its tasks were to pursue the minimal volume of HTS magnets and utilize the configuration characteristic of conventional generator. Its hybrid rotor has three novel aspects: the HTS magnets are installed on a warm iron core; four Dewar flasks having racetrack outline were distributed symmetrically and SN_2 was used as cryogen. Furthermore, evaporative cooling technology was applied to solve the cooling problem of the stator, which is more effective than air-cooling technology and more safety than water inner cooling technology. Magnetic invar rings were used between adjacent HTS coils of the field windings to divert the normal component of the magnetic field away from the Bi2223 superconducting tapes. Design parameters of the generator model are shown in Table 1. Figure 1 is the 3-D sketch of the 100KVA HTS demonstrative generator. Consumed HTS conductors in this model were one seventh of the generator with nonmagnetic core rotor [4]. Comparing with generators, which had conventional iron core, the values of magnetic flux density at air-gap were identically approximate. However, AC losses were decreased sharply, and the loads of cooling system were consequently reduced. It is easy to exclude AC magnetic field from the stator by placing a cold copper shielding cylinder around the rotor. The main difficulty of this configuration is to design and manufacture Dewar flasks of racetrack type because of their special outline.

Table 1 Parameters of the HTS generator model

Rating	100 kVA
Speed	1500 rpm
Armature	380V/152A Evaporative cooling
Housing dimensions	Diameter 660 mm Length 520 mm
Frequency	50Hz
Number of pole	4
Field winding	57A/30K SN_2+GHe
Magnetic flux density at air-gap	0.74 T

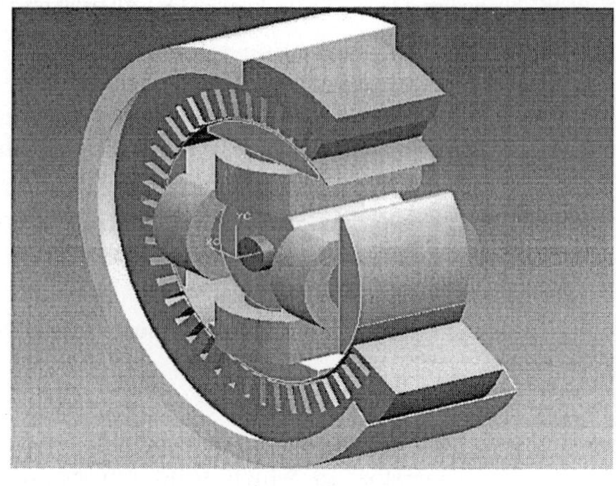

Figure 1 3-D sketch of HTS demonstrative generator

DESIGN OF ROTOR COOLING SYSTEM

The cooling system is one of the important components of HTS machine, of which function is to take heat away rapidly and maintain the superconducting property. Therefore, the robust and reliable cooling system is required for HTS generators to operate for a long time. In the design, SN_2 as cryogen and GHe as circulation medium were used after investigating the cooling system of HTS rotating machines [1, 2, 3, and 4]. The sketch is shown in Figure 2.

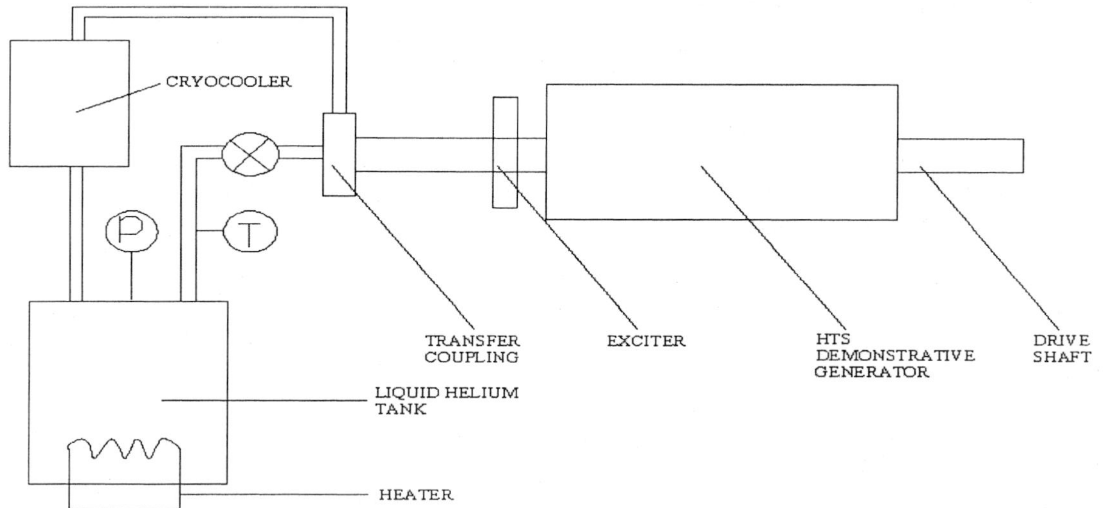

Figure 2 Sketch of the cooling system

<u>Operation process of cooling system</u>
In HTS generator model, liquid nitrogen (LN_2) was transported into the space of magnets, and then liquid helium (LHe) was forced into the heat exchanger of racetrack Dewar flasks to solidify LN_2 before generator working. During normal operation, LHe in liquid helium tank was boiled to gas by the action of heater. By controlling the status of heater, GHe (10-15 K, 0.13-0.20 MPa) was flowed through transfer

coupling into heat exchanger to absorb heat from HTS windings. As a result, GHe was heated up and forced into cryocooler through transfer coupling, and then became LHe again returning to liquid helium tank. GHe (15 K) was forced into the heat exchanger continuously and cooled SN_2 down to maintain the temperature of magnets. The rotor of HTS synchronous rotating machine was cooled by thermal capacity of SN_2 and its temperature was kept below 30 K. The delivered cooling power could be determined from measured mass flow, inlet and outlet temperatures. By varying the valve, the cooling power could be regulated directly.

Calculation of main parameters of cooling system

The load on the cooling system is composed of two parts. One part is the leakage of Dewar flasks, which is mainly caused by the conduction of supporting and cervix configuration of Dewar flasks. The leakage of one Dewar flasks was estimated about 0.8 W according to the level of fabricative technology. Another part is AC losses of HTS windings and magnetic invar rings. There are three loss mechanisms in AC losses of HTS windings: hysteresis loss, eddy current loss and coupling loss; however, it is difficult to obtain the accurate results by theoretical analysis. According to the references [1, 3, 4], it was estimated as 2 W in each pole.

The pressure loss of GHe through rotor is an important parameter for the design of GHe circulation of cooling system. Fluid flowing in rotating pipe was influenced by pressure, centrifugal force and Coriolis force. The pressure loss is 0.028 MPa because of centrifugal force, which can be calculated from formula 1. In fact, the effect of Coriolis force ($F_c = 2mv \times \omega$) is to change the direction of the velocity vector and increase the friction coefficient as pressure does. However, Comparing with centrifugal force, the effects of pressure and Coriolis force can be ignored because the circulation medium is gas. Therefore, the GHe pressure loss through rotor is about 0.028 MPa.

$$\Delta P_c = 5600(\frac{n}{1000})^2 \rho r^2 \tag{1}$$

with ΔP_c: pressure difference, n: speed of rotor, ρ: density of GHe , r: maximal distance to shaft center line of heat exchanger pipe.

Convective heat transfer coefficient h (turbulent flow) between GHe and SN_2 can be obtained from formula 2. This formula is simplified from Dittus-Boelter equation [5].

$$h = \left(\dot{m} \right)^{0.8} \Big/ d^{1.6} \tag{2}$$

with h: convection heat transfer coefficient, $\dot{m}$: mass flow of GHe, d: diameter of heat exchanger.

INVESTIGATION ON TEMPERATURE CHARACTERISTICS OF ROTOR

The temperature characteristics were simulated at steady state by ANSOFT. The calculated results indicate that temperature range is between 24.14 K and 24.38 K while the GHe (15 K) velocity is 0.2 m/s. Thermal gradient of field is shown in Figure 3 with vectors and contour respectively. Thermal flux of field is illustrated in Figure 4 with vectors and contour. The field dimensions are 80 mm $\times$ 20 mm. Each pancake is assumed to occupy the 5 mm$\times$7 mm space (including insulation layer), and the space of magnetic invar ring is 4 mm×7 mm. The diameter of heat exchanger is 10 mm.

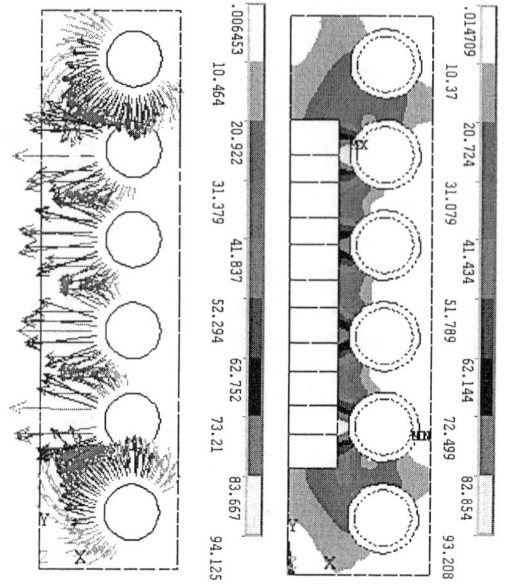

Figure 3 Thermal gradient vectors (left), contour (right)

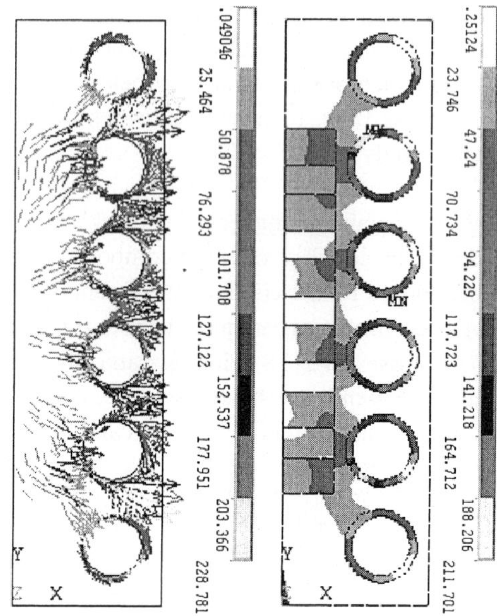

Figure 4 Thermal flux vectors (left), contour (right)

CONCLUSIONS

The cooling system of 100kVA HTS demonstrative synchronous generator has been designed. The pressure losses of GHe through rotor were analyzed. It can be inferred that centrifugal force has significant impact on GHe flow, and the effects of pressure and Coriolis force can be ignored. The steady temperature field was also simulated by ANSOFT, and the results indicate that the maximal temperature of HTS coils was lower than 30 K and distributed uniformly, while the GHe (15 K) velocity was 0.2 m/s.

REFERENCES

1. Nick, W., Nerowski, G., Neumüller, H.-W., 380kW synchronous machine with HTS rotor windings-development at Siemens and first test results, Physical C (2002) 372-376 1506-1512

2. Al-Mosawi, M.K., Beduz, C., Goddard, K., Design of a 100kVA high temperature superconducting demonstration synchronous generator, Physical C (2002) 372-376 1539-1542

3. Swarn, S.K., Development status of superconducting rotating machines, IEEE PES Meeting, New York, USA (2002) 27-31

4. Young-Sik, J., Young-Kil, K., Myung-Hwan, S., High temperature superconducting synchronous motor, IEEE Transactions on Applied Superconductivity vol.12 No.1 (2002) 833-836

5. White, G. K., In: Experimental techniques in low temperature physical, 2nd Ed, Oxford, UK (1968)

A prototype 4m, 2kA, AC HTS power cable system

Xi H.X.[1],Hou B.[1],Bi Y.F.[2],Yang X.C.[1]and Ding H.K.[3],Xin Y.[1]

[1]Innopower Superconductor Cable Co., Ltd., Longsheng Industrial Park, Beijing Economic &
  Technological Development Area, Beijing 100176, China
[2]Institute of Plasma Physics, Chinese Academy of Science, Hefei 230031, China
[3]Hefei Research Institute of Cryogenics and Electronics, Heifei 230043, China

A project to develop a 30m. 35kV/2kV, 3 phase, warm dielectric HTS power cable
system has been installed in a power grid at Puji substation of Kunming, Yunan
province at April of 2004. To enhance the understanding of the basics of a HTS
cable system and to prepare the fabrication techniques and skills, a 4m system
with terminations and cryogenics was built and tested. In this paper, we will
present the detailed parameters of the cable system, update the progress of the
work. Some experimental results, and analytical discussions are included.

INTRODUCTION

As compared to conventional power cables, High Temperature Superconductor (HTS) cables have larger
capacity, about 3-5 times to a conventional cable at a same cross section, and smaller transmission line
ioss, about 50% of that of a conventional cable[1]. Therefore, HTS cables are especially attractive for
metropolitan areas, where increasing load dictates that the existing conventional cables have to be
replaced. Till now, 2 sets of HTS cable system have been successful installed in a power grid in the world
[1]. In the spring of 2004, a 3 phase 30m warm dielectric (WD) HTS cable with 35kV/2kArms capacity
has also been installed at Puji substation of Kunming, Yunan Province of China.

To enhance the understanding of the basics of a HTS cable system and to prepare the fabrication
techniques and skills, a 4m system with terminations and cooling system was built and tested. In this
paper, the detailed parameters and structure of the prototype system are introduced, and some test results
are also given.

DESCRIPTION OF THE 4m HTS POWER CABLE SYSTEM

A 4m HTS cable system, which include the cooling system, terminations, and cable was built in April of
2003, the detailed parameters are listed in Table 1 and the flow diagram of the cooling system and the
schematic diagram of the HTS cable and terminations are shown in Figure 1 and Figure 2 respectively.

The cooling system is a closed cycle system as shown in Figure 1. The coolant LN_2 is pumped into
the sub-cooling tank, then into the HTS cable and brings heat out, and discharging to the LN_2 pump tank.
The LN_2 pump is a partial emission centrifugal type with a hermetically sealed motor. The motor is
controlled by a variable frequency drive (VFD). This allows adjustment of the pump speed to produce any
desired head and flow within the available power range. The pump is so designed that the pump housing
and impeller are in LN_2 and the motor is at atmospheric conditions. The pump and motor are separated by
a 31-inch, thin-walled shaft and housing, which minimize heat input from the motor into the LN_2 [2]. The
capacity of the GM refrigerator is 250W@70K-50Hz (300W@70K-60Hz). As shown in Figure 1,
Refrigerator cools LN_2 in sub-cooling tank, then the sub-cooled LN_2 cools the cycling LN_2 of the system.
The sub-cooling tank is at negative gauge pressure during operating, so a vacuum pump can be prepared
to compensate unexpected cooling power.   This is very economical and reliable.

Table 1　Parameters of prototype 4m HTS cable system

Components	Item	Parameter
Conductor of cable	Material	Bi-2223
	Dimension	4.2×0.22mm
	Ic (77K,0T)	>55A
	Bending Strain	0.5%
	Minimum bending radii	25mm
	Tension stress	100Mpa
Cable	Former	SUS304,18/26.0mm,flexible
	Numbers of conductor layers	6
	Winding angle of conductor (inside to outside)	35°/30°/19°/-20°/-34°/-44°
	Dielectric between conductor layers	Kaputon,25×0.1mm
	Spacer	Teflon,2/4mm
	Cryostat	SUS304,40/76.6mm,flexible
	Cooling structure	Counter-flow cooling
	Rated current	2000A
Termination	Current lead	Cu
	Epoxy pipe	Epoxy+ Silica aerogel
	Contact resistance	<0.4μΩ
	Cryostat	SUS304, rigid
Cooling system	Cooling mode	Closed LN_2 Cycle
	Cooling capacity	250W@70K (only refrigerator) 750W@70K(with the aid of a vacuum pump)
	Refrigerator	GM (AL300)
	LN_2 pump	Bncp-30c-000

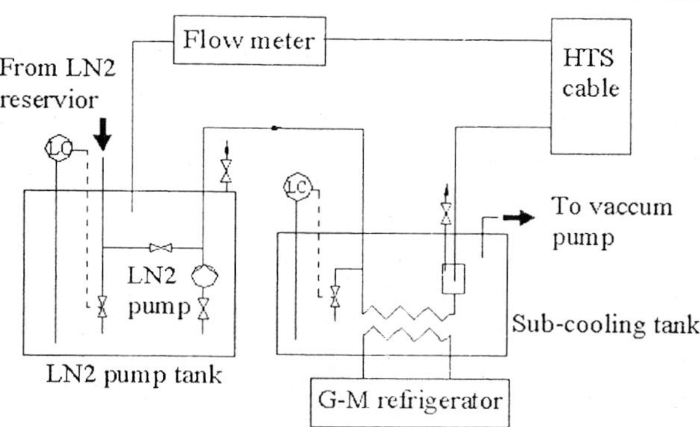

Figure 1　Flow diagram of cooling system

　　The 4m HTS cable system is composed of termination 1, termination 2 and the cable as shown in Figure 2. The epoxy pipe, cryostat and current lead constitute termination 1. Current lead connects the HTS conductor to the electric source, its upper end is in atmosphere and lower end is immerged in LN2. The current lead will breed some heat during operation of the system and should be designed carefully. In our system, the current lead was made of Cu braid and has a length of 0.8m and a cross section of 400mm^2. At the joint between current lead and the cable, silver was plated to lower the contact resistance, and the joint was also designed skillfully so that the termination and the cable can be easily assembled at the installation field. The cryostat of termination isolates the heat from the system, and was made of SUS304 stainless steel. It was in high electric potential during operating, so a pipe made of fiber-epoxy and silicon aerogel is used to isolate the heat from system and also isolate the high electric potential of termination from the zero electric potential of the cooling system. On top of the termination, a heat exchanger was installed which helps release the heat from the upper of the current lead [3-4]. The cable is a HTS cable with warm dielectric and has a counter-flow cooling structure, which means LN2 getting into the cable through the former and returning from the annular room between the inner wall of cryostat and the outer wall of the HTS tapes[5]. Termination 2 is almost the same as termination 1, but without the epoxy pipe. In the 4m HTS cable system, high voltage toroid or ground potential toroid and outdoor porcelain are also included to satisfy the demands of electric parameters.

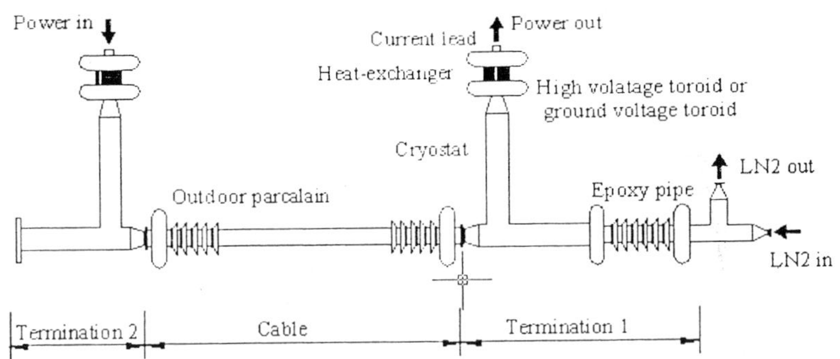

Figure 2   Schematic diagram of HTS cable and terminations

EXPERIMENT OF THE 4m HTS CABLE

In order to investigate the properties of a HTS cable system, some experiments are designed and executed before and after building the 4m HTS cable system. In this section, we give some test data on the heat loss.

Table 2   Test results of the components.

Current(A)	0	500	1000	1500	2000
Heat from conductivity of current leads （W/unit）	41.03	43.56	44.4	45.26	47.26
Heat from joule loss of current lead （W/unit）	0	4.87	19.56	44.17	79.12
Heat-leak of cryostat of termination 1(W/unit)	8.7				
Heat-leak of cryostat of termination 2(W/unit)	9.9				
Heat-leak of epoxy pipe(W/unit)	34.6				
Heat-leak of cryostat of the cable(W/m)	1				
Ac loss of the cable （W/m）	0	0.13	0.62	1.33	2.3
Total heat loss of the cable (W)	139.26	154.59	187.61	241.27	319.12
Heat loss of termination 2 (W)	49.73	57.1	72.62	98.1	135

Test results of the components of the 4m HTS cable system are listed in Table 2. The tests were done before the assembling of the system. At the last two rows of Table 2, we roughly predicted the total heat loss of the HTS cable and termination 2. Among these results, the heat-leak of the cryostats and the epoxy pipe were tested by calorie method which means by the rate of evaporation of LN_2 [3-4]. The ac loss of the cable was obtained by testing the current, voltage and the phase angle between them [6]. The heat from conductivity and joule loss of the current leads are gained through the verified mathematical models and some test results [3-4].

After building the 4m HTS cable system, we did some test to get the total heat loss of the HTS cable and termination 2. The results are shown in Table 3.

Table 3　Test results of the 4m HTS cable system

Current(A)	0	500	1000	1500	2000
Heat loss of the HTS cable (W)	307	323	348	410	498
Heat loss of termination 2(W)	108	127	138	165	195

Comparison between the predicted results and the measured results is shown in Figure 4. Measured results have the same rising trend as the predicted results with the increasing current, but are higher in value. For termination 2, the measured results are about 60~70W higher than the predicted results at different operating current. For the cable, the measured results are about 160~180W higher than the predicted results at different operating current. The difference of heat-loss between measured and predicted results for HTS cable and termination 2 vary little from one current value to another, so they are nearly irrelative to the current value. What caused this almost constant discrepancy? We quickly identified the problem that there was a bad thermal insulation on the terminations and the LN2 transmission line. The evidence for this conclusion was that the surface of termination 2 and the LN_2 transmission pipe at inlet and outlet of HTS cable are covered with frost during operation. This was the result of bad workmanship in assembling the termination and the LN_2 transmission pipeline. Taking out the discrepancy, we can see that the results by modeling and calculation are in good agreement with the experimental data, which indicates the designs of the cable conductor and the current leads meet our specifications. Now, the work to improve the thermal insulation of the terminations and the LN_2 transmission lines was done.

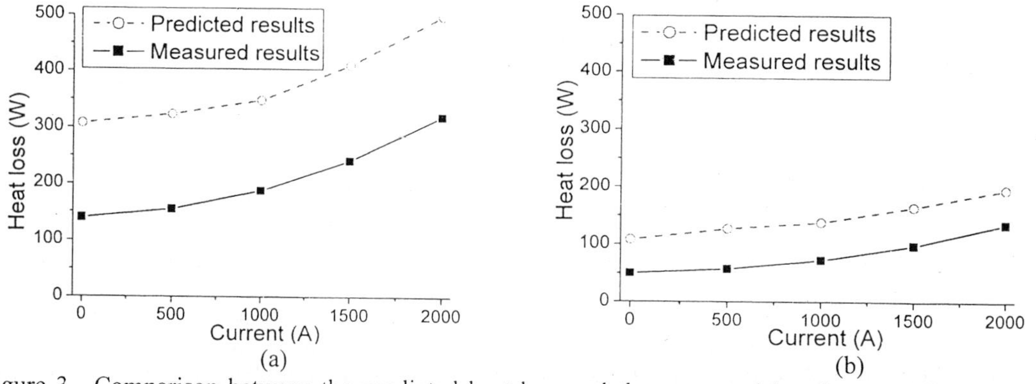

Figure 3　Comparison between the predicted heat-loss and the measured heat-loss at different current, and the y-axis represents the heat-loss of HTS cable in (a) and the heat-loss of termination 2 in (b)

SUMMARY

By building and testing a prototype 4m HTS power cable system, we have learnt the techniques of fabricating a HTS cable system. On the other side, we should pay a great attention to the workmanship in the manufacturing, transportation and installation processes. Till now, 30m,35kV/2kA HTS cable system have been installed and some tests are also finished, the results will be released in the near future.

REFERENCES

1. Mujibar M. Rahman, High-capacity cable's role in once and future grids, IEEE spectrum(1997), 31-35
2. Barber-Nichols, Liquid nitrogen pump BNCP-30C-000 Installation, Operation, and Maintenance Manual.
3. Institute of Plasma Physics, CAS. Design of 35kV/2kA HTS power cable prototype termination(2003)
4. Institute of Plasma Physics, CAS. Report of performance testing of 35kV/2kA HTS power cable prototype termination(2003)
5. Xi H.X , Xin Y., LN_2 cooling system for a 3 phase 35kV/2kA HTS cable, cec-icmc'03
6. Zhang Y., Liu Y., Investigation of the transmission loss in HTS cable, eucas 2003

Development of a three-phase HTS power transformer

Yinshun Wang[1,2], Xiang Zhao[3], Huidong Li[1], Junjie Han, , Liye Xiao[1], Liangzhen Lin[1], Ying Guan[1], Qing Bao[1], Guanghui Lu[3], Zhiqin Zhu[1], Xi Xu[1], Yan Lu[1], Shaotao Dai[1], Dong Hui[1]

[1] College of Physics Science and Technology, Hebei University, Baoding 071002, Hebei Priovince, P.R.China
[2] Applied Superconductivity Lab., Inst. of Electrical Engineering, CAS, Beijing 100080, P.R.China
[3] Xinjiang Tebian Electric Apparatus Stock CO., LTD., Changji 831100, Xinjiang , P. R. China

We have developed a 26kVA (400V/16V) three-phase HTS transformer cooled by liquid nitrogen. The primary and secondary windings were wound by transposed conductors, which are made from the stainless steel-reinforced multifilamentary Bi2223/Ag tapes. The structures of primary and secondary windings are solenoid and double-pancake respectively. Fundamental characteristics of the transformer are obtained through no-load and short circuit tests. The gas-cooled current leads of secondary windings were optimally designed. The cryostats are made from electrical insulating materials Fiberglass Reinforced Plastics (FRP), which have excellent heat-preservation. The iron cores are made of common scratched Fe-Si steel and operate at room temperature. Based on results of this transformer, we will develop 630 kVA HTS transformer with same rated currents in near future.

INTRODUCTION

The development of high-temperature superconducting (HTS) devices has made remarkable progress in recent years [1-3]. The superconducting transformer is one of the most promising applications in electric power systems. And it offers several merits such as reduced size and weight, high efficiency, oil-free, nonflammable and less environmental hazards. The next step is to investigate and design the apparatus which is near commercialization. For HTS materials, technique to make practical long and stainless steel-reinforced tapes with critical current density J_c more than 10^4 A/cm^2 in liquid nitrogen and self-field have been achieved, which supplies possibility for large capacity HTS transformers.

We designed and fabricated a three-phase 26kVA(400 V/16 V) HTS transformer for the analysis of its fundamental characteristics. The primary and secondary windings were wound by transposed conductors, which are made from the stainless steel-reinforced multifilamentary Bi2223/Ag tapes fabricated by American superconductor (AmSC). The main aims are improvements in basic technology for application of HTS transformers by analyzing the characteristics of electro-magnetism, insulation, thermodynamics and stability for the transformer. Our near goal is to design and build a 630 kVA (10.5 kV/0.4 kV) distribution HTS transformer with amorphous cores which would be tested in field in Xinjiang Tebian Electric Apparatus Stock CO., LTD., China.

The transformer is tested under AC and DC rated current, no-load and short circuit conditions for four hours at power frequency 50Hz in liquid nitrogen. Also we measure the current distributions of double pancakes in secondary windings with and without iron cores. In the paper, we report on developing the basic technology for the HTS transformer, and present the results of individual tests of the HTS transformer.

DESIGN AND CONSTRUCTION OF WINDINGS

Specifications of windings

The primary and secondary windings of the transformer are composed of 2-strand parallel stainless reinforced-steel Bi2223/Ag multifilamentary tapes that have 55 untwisted filaments. The cross-section of the tape is 4.1×0.31mm^2.

The HTS tape must be insulated because it is not insulated by AmSC. We developed a wrapping machine which can be used to wrap the tape automatically with single, double and triple wrapping by polyimide films. The insulation technique can successfully wrap tape about 100m/hr without any degration of the critical current. The thickness of polyimide film is about 25 micrometer. The polyimide film was tested at nitrogen temperature 77 K. The breakdown voltage is 208.4kV/mm and the AC withstand voltage is 147.2kV at power frequency 50Hz for 1min., which is enough for this transformer. Proposed conductors of windings are composed of 2-strand parallel conductors with proper transposition. In the parallel conductors, the strands are electrically insulated with each other.

We designed and fabricated 6 coils, three solenoid coils are for primary windings and the other three are double pancake coils for secondary winding in a three-phase 26 kVA HTS transformer respectively. The secondary winding is made from 24 double pancakes connected in parallel, and the primary coil winding is solenoid. The strand in two kinds of windings is consisted of two parallel transposed multifilamentary tapes in order to prevent unbalanced current flowing because it may cause instability of the HTS coils as well as much of AC loss. Solenoid and double pancake coils are concentric cylindrical. The helical coil is wound with 4 layers and the double pancake one is wound with 3 layers, and the double pancake coil is located coaxially outside the helical coil, Fig. 1 shows the overview of three phase windings.

Fig.1 Overview of one phase windings. Inner solenoid coil: Primary winding. Outer double pancake coil: Secondary winding

Table I

Parameters of the transformer

Parameters	Design values	Unit
Capacity	26	kVA
Voltage (Primary/Secondary)	400/16	V
Current (Primary/Secondary)	37.5/938	A
Average Diameter (Primary/Secondary)	152.5/178.5	mm
Core Diameter	90	mm
Height (H_w)	650	mm
Width (M_0)	360	mm
Magnetic flux density	1.27	T
Cryostat		
Height	550	mm
Diameter (outermost/innermost)	300/92	mm
Vector Group	Yyn0	
Impedance x_{cc}	2.83%	
No-load current	1.26%	
Operation Temperature	77	K
Operation Frequency	50	Hz

Specifications of transformer

We designed and fabricated a three-phase HTS transformer with capacity of 26kVA operated in liquid nitrogen of 77K for the primary/secondary voltage of 400V/16V and the primary/secondary current of 37.5A/938A. The main parameters are summarized in Table I. The primary winding is located coaxially outside the secondary one in a FRP cryostat around an iron core of room temperature with diameter of 92mm. The turn ratio is 25. Design of electrical insulation for present transformer is attributed to electrical properties of nitrogen gas in low temperature 77K. The current leads on low voltage side are gas-cooled and those on high voltage side are not cooled. Fig. 2 shows the overview of the three phase 26kVA HTS transformer before installed in box. The level of liquid nitrogen in three cryostats are monitored by sensors which act on the switchgear in case of liquid nitrogen level problems and protect the windings from being destroyed.

CRYOSTAT AND CURRENT LEADS

Since metal cryostat locating around the magnetic path will form a closed loop, same as one turn short circuit winding, it cannot be used in AC field. The Fiber Glass Plastics (FRP) should be used to fabricate cryostat which is electrical insulated and strong enough even it requires permanent use of vacuum pump to maintain thermal insulation between inner and outer walls. We designed three FRP cryostats with room temperature cores for iron cores.

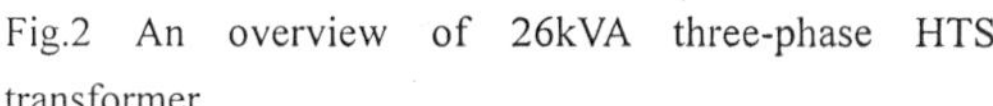

Fig.2 An overview of 26kVA three-phase HTS transformer

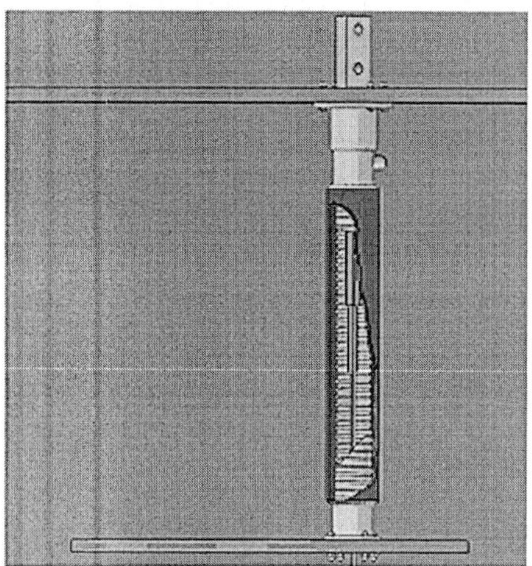

Fig. 3 Schematic of gas-cooled current leads

Current leads made up from copper span from room temperature to 77 K temperature, and heat leakage is important loss considering consumption of liquid nitrogen. We optimized the gas-cooled current leads for secondary windings [4]. In primary winding, the current leads are non cooled since the cross sections of secondary current leads are about 25 times larger than those of primary current leads, heat-conduction in primary leads is very small compared to secondary case. So the current leads of secondary windings are only chosen as traditional boiling gas-cooled current leads in order to reduce consumption of liquid nitrogen. Fig.3 shows schematic diagram of the gas-cooled current lead. The cross section of current lead is different, the top cross section in room temperature is designed as that of traditional transformer, the end cross section near liquid nitrogen is optimized with different loss of transformer, optimal cross section is 64mm^2. Test was done by caloric method, calculated and experimental results are shown as Fig.4.

TEST RESULTS OF THE TRANSFORMER

Firstly, critical currents of windings were measured by standard four probes method with criterion 1μV/cm at liquid nitrogen temperature 77K. Critical currents of windings are about 180A, which means that critical currents of six windings in the transformer are almost uniforms. In AC operation, AC current in primary winding is same; but since the secondary windings is consisted of 24 double pancakes connected in parallel, the current distribution between double pancakes is different due to different inductance. The currents of double pancakes are very different in both cases. The maximum current is about three times of the average value without iron core, but the currents of the secondary with iron core are almost uniform [5].

In no-load test, we had usual tests of the transformer cooled in saturated liquid nitrogen at 77K in order to get steady characteristics in the rated operation. The transformer was excited from primary side at rated voltage 400V. The exciting current was 0.48A (1.264%) and the total no-load loss was 320W in usual procedure with three power Wattmeters, which can be attributed to a core loss. At the rated condition the designed value for the magnetic induction of the core is 1.27T, we obtained the transformation ratio of 25.01, as shown in Fig.5.

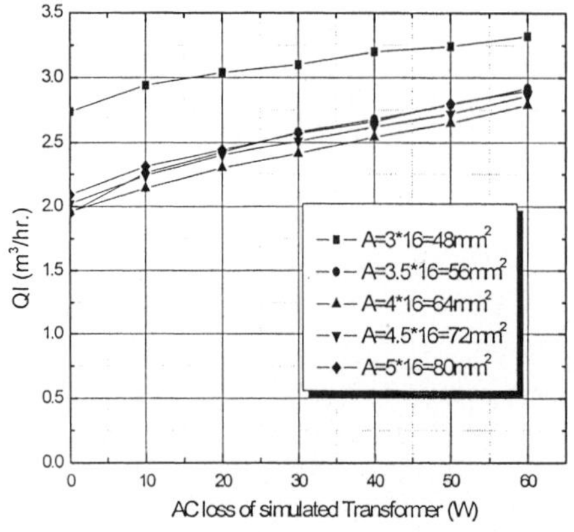

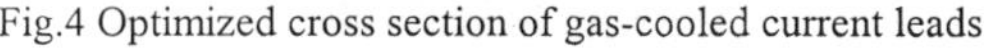

Fig.4 Optimized cross section of gas-cooled current leads

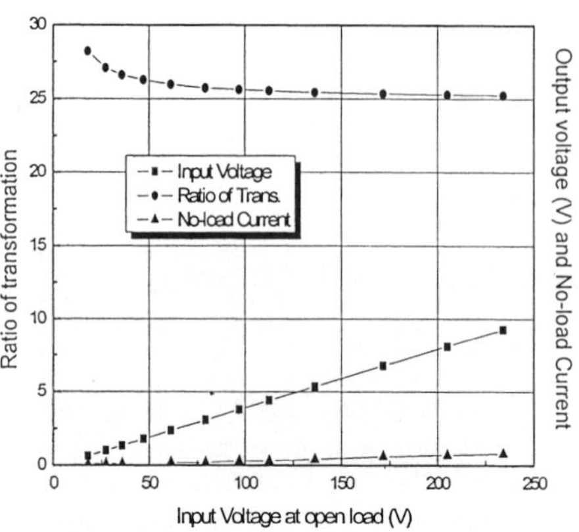

Fig.5 Experimental results of single transformer

Short-circuit test was performed from secondary side at the rated current 938A for 4 hours. The impedance voltage was estimated as 2.8% of the rated level. All of the parameters are in good agreement with designed ones. The ratio at lower input voltage is larger than designed value 25 since there is an error in output voltage measurement, but the ratio is almost exactly 25 at high input voltage. Finally, we performed an overload test of the transformer at 77K, where the current of the windings was increased up to 1200A for 2 hours, about 128% of the rated current. In the overload test, we had a stable operation of the transformer with a 26kVA secondary resistive load.

SUMMARY AND CONCLUSION

A three-phase 26kVA HTS transformer with room temperature iron core was developed and tested at 77K. The primary winding was solenoid coil and the secondary coil was consisted of 24 double pancakes connected in parallel, and the strand of windings was consisted of two parallel transposed stainless steel-enforced multifilamentary Bi2223/Ag tapes which were insulated by wrapping with polyimide films. The rated primary and secondary voltages are 400V and 16V, and rated currents are 37.5A and 938A respectively. The transformation ratio is 25, short circuit impedance is 2.8% and the excited current is 1.26%.

ACKNOWLEDGEMENT

This work was supported in part by the Chinese Ministry of Science & Technology under Grant No.2002AA306381, Xinjiang Tebian Elec. Co. Ltd. and Doctoral Foundation of Hebei University. The authors appreciate Mr. Liang Lin for his help in the transformer test.

REFERENCES

1. Funaki, S., Iwakuma, M., Kajikawa, K. et al., Development of a 500kVA-class oxide-superconducting power transformer operated at liquid-nitrogen temperature, Cryogenics (1998), 38 211–220
2. Zueger, H, 630kVA high temperature superconducting transformer, Cryogenics (1998), 38 1169–1172
3. Funaki, F, Iwakuma, M, Kajikawa, K et al., Development of a 22kV/6.9kV single-phase model for a 3MVA HTS power transformer, IEEE Trans. Appl. Supercond.(2001), 11 1578-1581
4. Xiangchun, X, On the Optimal Design of Gas-Cooled Peltier Current Leads, IEEE Trans. Appl. Supercond.(2003), 13 48-53
5. Yinshun, W, Xiang, Z, Huidong, L, Guanghui, L, Liye, X, Liangzhen, L et al., Development of Solenoid and Double Pancake windings for a Three-phase 26KVA HTS Transformer, IEEE Trans. Appl. Supercond.(2004), 14

Proceedings of the Twentieth International Cryogenic Engineering Conference
(ICEC 20), Beijing, China. © 2005 Elsevier Ltd. All rights reserved.

THERMAL STABILITY AND QUENCH PROPAGATION IN A CRYOCOOLER-COOLED PANCAKE COIL OF BI-2223 TAPE OPERATING BETWEEN 35K AND 65K

Johnstone A P, Yang Y, Beduz C

Institute of Cryogenics, Dept. of Mechanical Engineering, University of Southampton, Southampton, SO17 1BJ, UK

A cryostat for thermal stability and quench propagation studies of HTS pancake coils has been constructed with an operating temperature range between 35K and 65K. The transient thermal response of a test coil to an applied energy pulse has been measured and the minimum energy needed to initiate a quench is presented as a function of temperature and heater length. It has been observed that a normal zone can be formed with a maximum temperature below T_C. The radial normal zone propagation (NZP) velocity is also presented as a function of temperature.

INTRODUCTION

The thermal stability and quench propagation characteristics of HTS coils are important considerations when designing and operating HTS magnets and machines. The minimum energy needed to initiate a quench for HTS coils has been shown to be in the order of joules [1,2]. Despite the high stability of HTS coils their response to transient thermal disturbances must be better understood so that suitable design can be made for quench protection [3].

The concept of a minimum propagating zone (MPZ) was first proposed by Wipf and Martenelli [4], it is a measure of how sensitive a superconductor is to applied energy pulses [5] and has been successfully used to describe the thermal stability of LTS magnets. If a disturbance causes a normal zone to form in a coil with a size greater than the MPZ the normal zone will propagate and the coil will quench. The analysis of the formation and propagation of a normal zone in a HTS coil is complex because a quench occurs over a wide temperature range where the properties of the coil composite are highly temperature dependent. This work investigates the formation and propagation of a normal zone in a 38 turn Bi-2223 pancake coil operating between 35K and 60K.

EXPERIMENTAL

A cryostat for thermal stability and quench propagation studies of HTS pancake coils has been constructed, as shown schematically in Figure 1 and detailed in Table 1 respectively. The pancake coil was cooled at its inner and outer boundaries by conduction via copper thermal links which were thermally anchored to the cold-head of a single stage GM cryocooler. Two small liquid nitrogen vessels located in the vacuum space were used to provide cooling for the radiation shield and AuAg sheathed HTS current leads, and acted as current leads between room temperature and 77K. The current leads were tested up to 300A with the temperature of the cold-head remaining constant at 30K over this current range.

The pancake coil was wound on a thin cylindrical copper former fixed to a Tufnol inner ring using NST PbBi-2223 tape and the copper was used as one of the current terminals. Copper tabs to be used for temperature and voltage measurements were soldered to the tape at different positions. A constantan ribbon to be used as a heater was co-wound with the middle turn of the coil. After winding copper sheets were soldered to the outer boundary and the coil was vacuum impregnated with epoxy. Copper-constantan thermocouples and voltage taps were soldered to the copper tabs. The characteristics of the test coil are shown in Table 1.

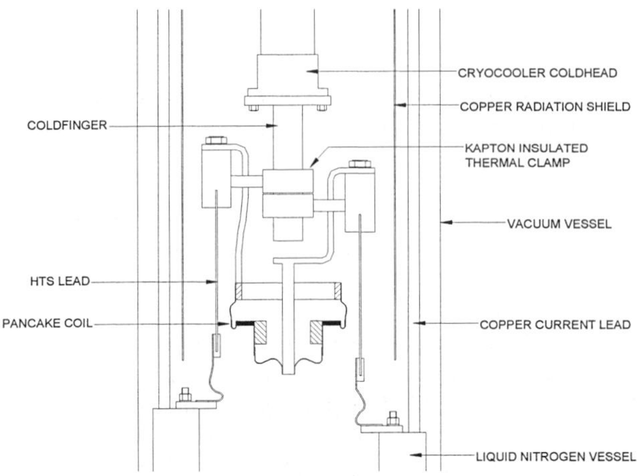

Figure 1, Schematic of part of the cryostat.

Table1. Test coil characteristics

Items	Specifications
Insulation	Fibreglass cloth ~100μm
Former	Tufnol
Number of turns	38
Coil inner radius	30mm
Coil outer radius	45mm
Coil fill factor	~50%
Conductor type	NST 37-multifilamentary Bi2223 tape
Conductor sheath	Outer 0.1MgNi-Ag alloy Inter-filaments: Ag
SC filling factor	27%
Tape dimensions	4.1mm x 0.25mm
I_C	60A at 77K (1μV/cm)

For each test coil boundaries were set at a given temperature between 35K and 60K, and an appropriate transport current was applied that produced an average electric field of 0.7μV/cm across the ends of the coil. For the test results at 60K, 45K and 35K shown in the following, the corresponding transport currents were 87A, 121A and 150A. Upon reaching the steady-state for a given temperature and transport current, transient heating pulses were applied to the different length segments (2.5cm, 8.0cm and 12.5cm) of constantan heater to simulate local and distributed heating. The heat pulses generated by an audio amplifier were a 50Hz sinusoidal wave modulated by a square wave pulse lasting about a second. The transient thermal and electrical response of the coil was recorded using a 32 channel amplifier and data acquisition system. The energy of the heating pulses was increased gradually until a runaway quench was induced. The current supply was programmed to cut out once a set voltage across the quenched coil was reached.

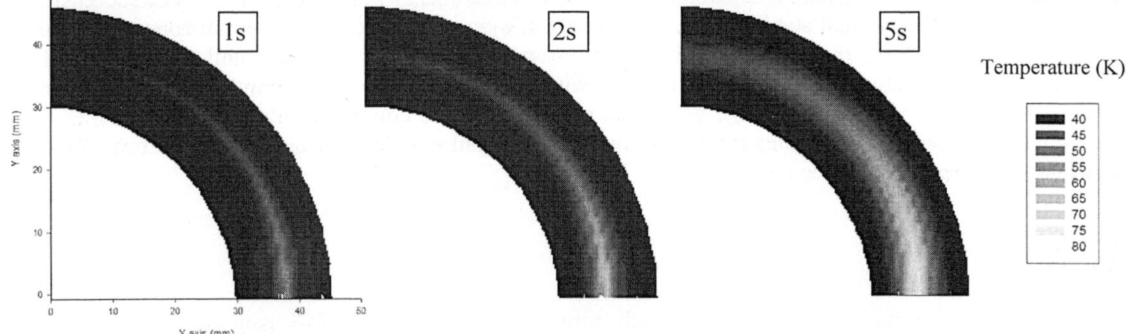

Figure 2 Evolution of the temperature distribution in ¼ of the coil at 35K

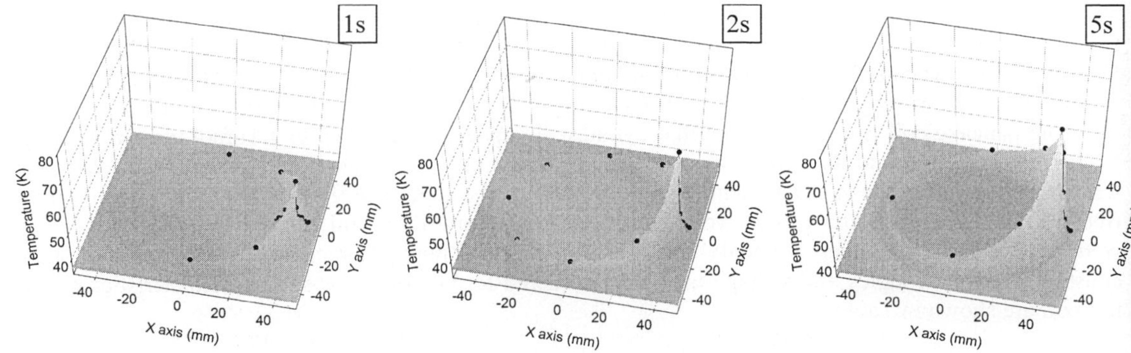

Figure 3. Evolution of the temperature distribution in the entire the coil at 35K (●=Measured data points).

RESULTS AND DISCUSSION

Normal zone growth and propagation

Contour maps of the coil temperature distribution after each heat pulse were obtained using the 18 differential thermocouples distributed across the coil. Figure 2 shows a set of examples at 1s, 2s and 5s after a heat pulse was applied to the coil operating at 35K. In Figure 3 the same maps are presented as a surface distribution with the positions and measured temperatures of thermocouples shown at the same time. Such maps should be used only as visual aids for the qualitative behaviour of propagating normal zone. For the case shown in Figure 2-3, a well defined normal zone can be seen to grow and propagate in the radial and tangential directions. As expected, the normal zone is much narrower in the radial direction than in the tangential direction due to the anisotropic thermal conductivity of the coil.

Figure 4 shows the temperature responses at 60K of thermocouple T6 located at the centre of the heater, and thermocouple T5 located 3 turns away along the same radial direction. It can be seen that for a heating pulse of 13.7J T6 reached a maximum temperature of 80K followed by a full recovery. For a heat pulse of 14.2J, T6 shows the temperature rising within 3s to 81K and keeping almost constant for 8s, during which the temperature at T5 increased steadily to a same temperature of 82K. Sharp temperature increase followed after this point, indicating a classic quench event with the normal zone reaching the size of minimum propagation zone (MPZ).

A constant maximum temperature T_Q (Figure 4) signifies a well defined uniform normal zone temperature which increases very little during its growth to MPZ. Consequently T_Q can be regarded as the quench temperature. It is significant that T_Q is significantly below the critical temperature $T_C = 110K$, even for coils operating at 60K. The reduction of quench temperature from T_C to T_Q leads to a smaller current sharing regime with the quench of coil before full current sharing by the normal matrix, hence reduces stability of such HTS coils. The quench temperature T_Q for 35K and 45K were found to be 68K and 76K respectively.

Minimum energy

The minimum energy at 35K needed to initiate a quench in the coil was obtained as a function of heater length at various temperatures (Figure 5). It can be seen that the minimum energy to quench the coil increases with the length of the heated section, suggesting a smaller MPZ. The conventional minimum energy for a point disturbance was estimated using the value extrapolated to zero heater length shown as a function of temperature in the inset of Figure 5. The increase of minimum energy with temperature is expected due to the temperature dependent heat capacity of the coil composite. The measured values for the minimum energy are consistent with previous work [1,2] and indicate a high level of stability to transient disturbances.

Propagation velocities

The measured temperature profiles at 35K in the radial direction shown in Figure 6 highlights the

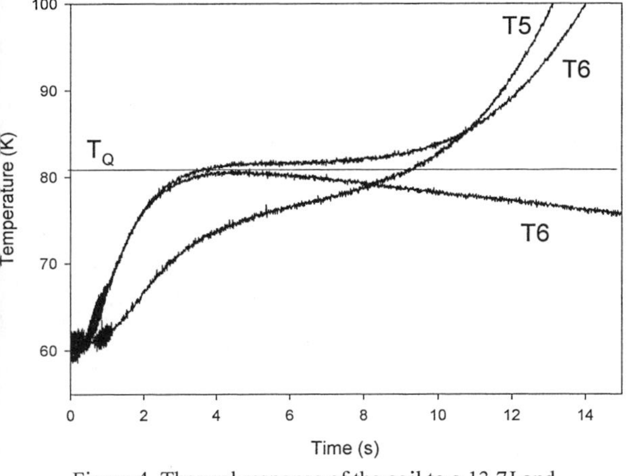

Figure 4. Thermal response of the coil to a 13.7J and 14.2J energy pulse over a 2.5cm heater length at 60K

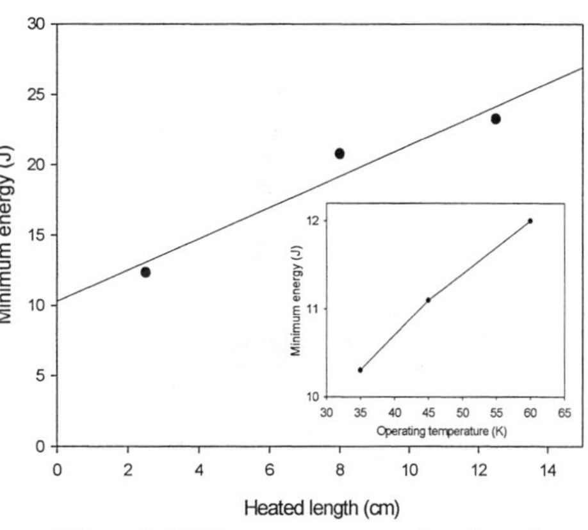

Figure 5. Minimum energy as a function of heater length and temperature

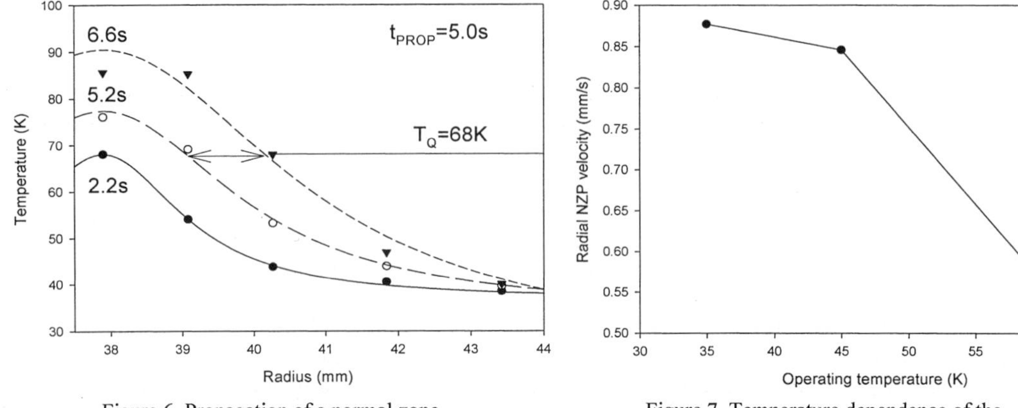

Figure 6. Propagation of a normal zone created by a 2.5cm heater length at 35K

Figure 7. Temperature dependence of the radial NZP velocity

evolution of the propagating form of a normal zone. The radial normal zone propagation (NZP) velocity is obtained at T_Q for each operating temperature and shown in Figure 7. A radial NZP velocity about 0.5-0.9cm/s between 35K and 60K is consistent with previous measurement [2,6]. The observed increase of the NZP velocity with decreasing temperature is due to the increase in the current density of the superconductor and decrease of the heat capacity of the coil composite at lower temperatures. Due to the low radial NZP velocity the normal zone did not propagate far towards the coil boundaries before the temperature of the normal zone began to rise rapidly and the current supply switched off. Therefore the results presented here are consistent with the adiabatic assumption of the classic quench theory.

CONCLUSION

The formation and propagation of a well defined normal zone has been observed in a HTS coil. The maximum temperature of the initial normal zone (T_Q) has been shown to be well below the critical temperature T_C of the superconductor. The lower normal zone temperature $T_Q<T_C$ means that the temperature range for current sharing during normal zone formation is reduced. Less energy is therefore needed to raise the coil temperature to T_Q instead of to T_C. The minimum energy was measured and found to increase as the length of the heated section and operating temperature increased. The propagation of the normal zone was analysed by calculating the radial NZP velocity, which decreases with increasing operating temperature. Both the minimum energy and radial NZP velocity were found to be consistent with previous work.

REFERENCES

1. Lue J.W., Lubell M.S., Aized D., Campbell J.M. and Schwall R.E., Quenches in a high-temperature superconducting tape and pancake coil, Cryogenics (1996) 36 379-389
2. Penny M., Beduz C., Yang Y., Manton S. and Wroe R., Normal zone propagation studies on a single pancake coil of multifilamentary BSCCO-2223 tape operating at 65K, Proc. 3rd Applied Superconductivity Conference, Netherlands, IOP publishing (1997) 1551-1554
3. Iwasa Y., HTS magnets: stability; protection; cryogenics; economics; current stability/protection activities at FBML, Cryogenics (2003) 43 303-316
4. Wipf S. and Martenelli A., Proc. 1972 Applied Superconductivity Conference, Anapolis, IEE, New York (1972) 331-340
5. Wilson M.N. and Iwasa Y., Stability of superconductors against localized disturbances of limited magnitude, Cryogenics (1978) 18 17-25
6. Kim S.B., Ishiyama A., Okada H. and Nomura S., Normal-Zone propagation properties in Bi-2223/Ag superconducting multifilament tapes, Cryogenics (1998) 38 823-831

A new concept of bridge fault current limiter-SMES for interline application to improve power quality

Zhao Caihong[1,2], Xiao Liye[1], Lin Liangzhen[1], Yu Yunjia[1]

[1]Applied Superconductivity Lab, Inst. of Electrical Engineering, CAS, Beijing 100080, P.R.China
[2]Graduate School of CAS, Beijing 100039, P.R.China

By replacing the bias power source for bridge SFCL with the chopper for SMES, SFCL and SMES are integrated into a new concept: fault-current-limiting SMES (FCL-SMES). It can be used for interline application, when the fault appears on the feeders for common loads, SMES can absorb the incoming energy into the superconducting coil to compensate the sag to improve the power quality of the critical load. Since the current of the coil can not be charged all along, FCL-SMES can not only limit the peak current, but also the steady current. The analysis of FCL-SMES is given and the simulation validates it.

INTRODUCTION

Power quality problems are becoming more and more important these years, it is reported that they cost US manufacturers between $12 billion and $26 billion annually. Among all the problems, voltage sag and momentary outages are the most serious ones faced by industrial and commercial customers. SMES offers a solution to this problem with its many significant advantages, especially the capability to charge and discharge with very high power as compared with other power quality equipments like battery, fly-wheels and so on[1]. Furthermore, voltage sags are mainly caused by fault in distribution network, so limiting fault current also provides a solution to this problem. Some kinds of superconducting fault current limiter(SFCL) have been developed. Combined SMES and SFCL may be a new way, and various investigations and feasibility study were reported in [2]. In this paper, a new concept of fault-current-limiting SMES (FCL-SMES) for interline application is proposed, and it is analyzed and validated by the computer simulation.

BASIC PRINCIPLE OF FCL-SMES

Figure1 illustrates the configuration of FCL-SMES. For the common distribution substation, most loads are common ones, but some are critical ones which are sensitive to power quality problems caused mainly by faults in the feeders for common load. As shown in Figure1, the proposed FCL-SMES is interlined between two buses for these two kinds of loads. By replacing the bias power source for bridge SFCL with the chopper for SMES, SFCL and SMES are integrated into fault-current-limiting SMES (FCL-SMES). When fault appears on one of the feeders of common loads, the bridge SFCL can work automatically to limit the current. However, since its inherent disadvantage is that the longer the fault current limiting time, the larger the superconducting coil current. Some cycles after the fault the limiting function of SFCL is very little, so the voltage of bus decreases very much, and the severe voltage sag for the neighbor critical load is produced. Therefore, improving the fault limiting function is the first way to increase the bus voltage, reducing the compensated voltage of the sag for critical load at the same time.

During the process of fault limiting of SFCL, the superconducting coil is charged all along with rectified voltage by diode bridge. In [3] a resistor has been put to consume the energy in a three phases system to improve the function of FCL, and FCL-SMES is also based on this principle, since its chopper can work like a controlled resister r(t), absorbing the energy of the superconducting coil to compensate the sag of the bus for critical loads, as shown in Figure 2. FCL-SMES overcomes the shortcoming of common bridge SFCL. It not only can limit the fault current all the time, but at the same time compensates the sag with reduced energy needed. Thus, FCL-SMES is integrated into a promising equipment to provide a complete solution to power quality problems.

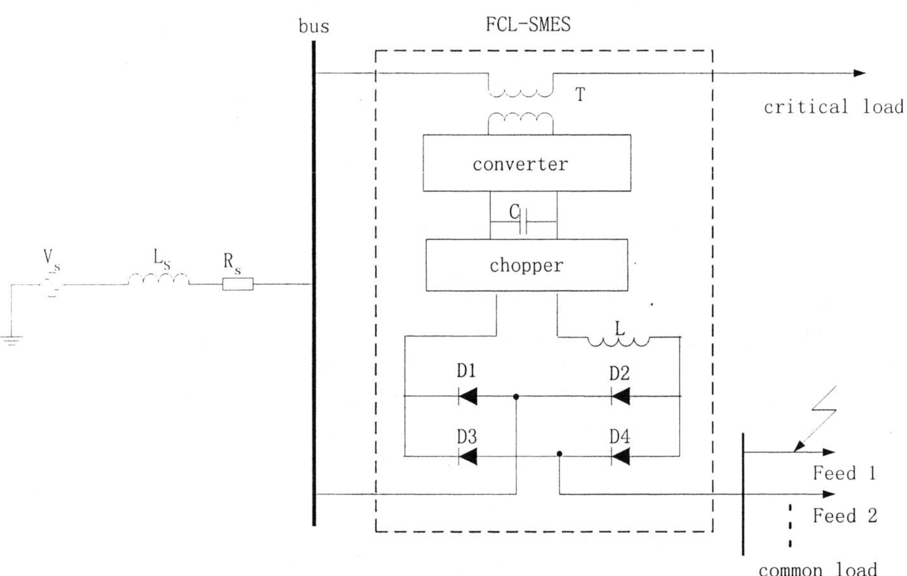

Fig. 1: Proposed configuration of bridge FCL-SMES

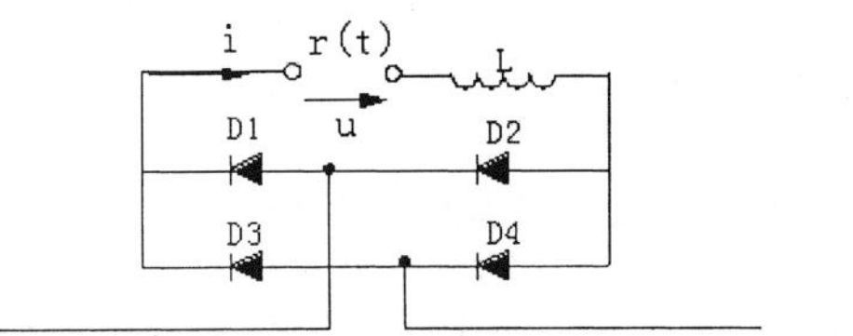

Fig.2: Principle circuit of FCL-SMES for its limiting function

Table1   Parameters of simulation for FCL-SMES

V s: 6.06kV, 50Hz

Ls: 1.11mH, Rs: 0.01 Ω ,common load: 198A(RMS)

Short circuit: 0 Ω

FCL-SMES: L: 0.002H, r(t)=1 Ω (0.001s after fault), ideal diode.

SIMULATION OF FCL-SMES

The main circuit of proposed FCL-SMES is shown in Figure 1, and Table 1 lists simulation parameters. The simulation is directed to the principle of FCL-SMES, and some assumptions are made: the compensating function, that is the discharging power, is represented by r(t) which works after 0.001s after fault; the diodes bridge is assumed to be ideal; and the critical load is neglected as compared with the fault current capacity. In Figure 3 and Figure 4, the simulation results show the principle of FCL-SMES by comparing the current and voltage between FCL-SMES and common bridge FCL. The line current of the system and the maximum current of magnet with SFCL are 25kA, but those of FCL-SMES are just 6.5kA with r(t) which equals 1 ohm and represents 16 MW discharging power for critical loads. The needed maximum coil capacity of FCL-SMES is reduced to 25% of the common FCL, and the fault current is also limited to only 26% of that. Moreover, the voltage bus of the system with FCL-SMES increases to about 0.8pu, leaving only 20% of the voltage to be compensated. That is, the energy needed for FCL-SMES is reduced to just 20% of common SMES.

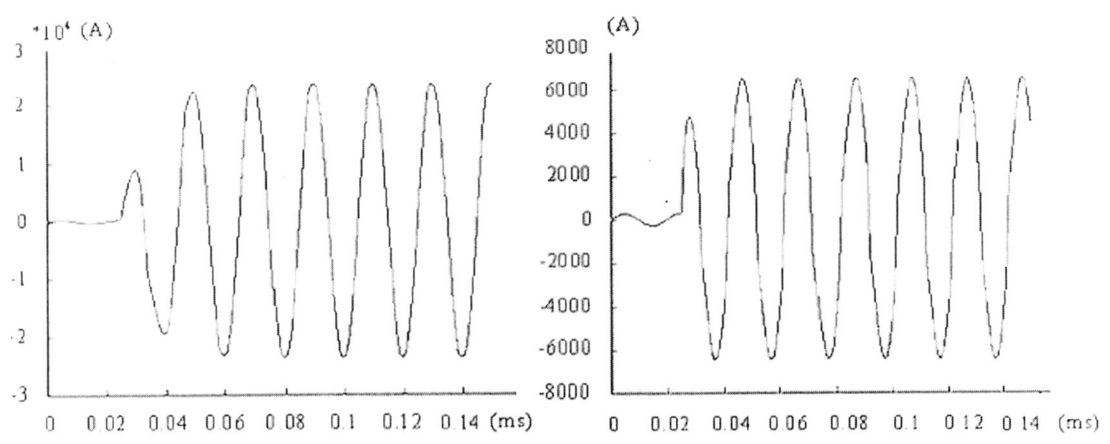

(a) Line current of the system with bridge SFCL   (a) Line current of the system with FCL-SMES

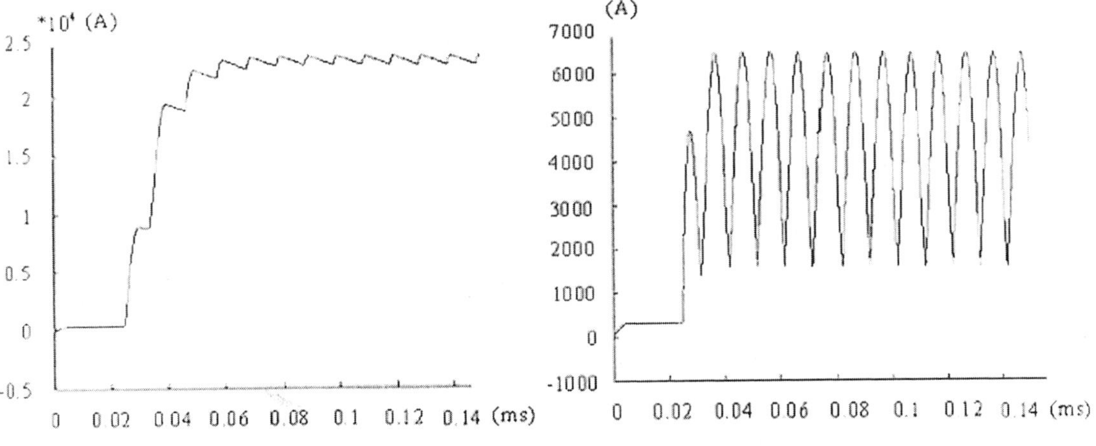

(b) Magnet current of bridge SFCL   (b) Magnet current of FCL-SMES

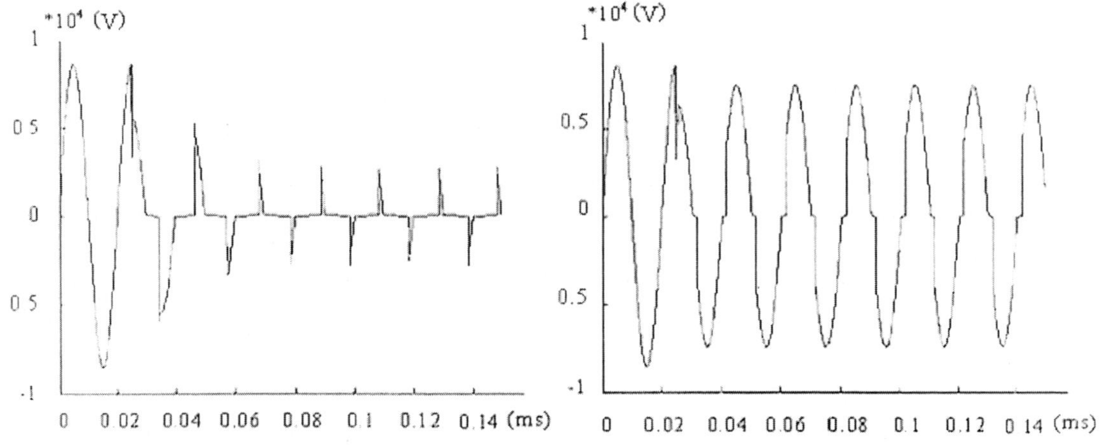

(c) Bus voltage of the system with bridge SFCL

Fig.3: Voltage and current of the system with bridge SFCL

(c) Bus voltage of the system with FCL-SMES

Fig.4: Voltage and current of the system with SFCL-SMES

CONCLUSIONS

A new concept of FCL-SMES for interline application is proposed, analyzed, and validated by simulation. It can be a promising equipment for its performance and economical design. The contents about the mathematical model analysis, the superconducting magnet optimization, the prototype development and the experiment results of FCL-SMES, will be described in other papers.

ACKMOWLEDGEMENTS

This work was supported by National Natural Science foundation of China (Grant#: 50137020), National outstanding young scientist foundation of China (Grant#: 50225723), and "863 "High Technology Project (Grant#: 2002AA306351).

REFERENCES

1. Parizh, M., et al, Superconducting magnetic energy storage for substation applications, IEEE Trans. Appl. Supercond.(1997) 7 849 –852
2. Zanji Wang, et al, The feasibility study on the combined equipment between micro-SMES and inductive/electronic type fault current limiter, IEEE Trans. Appl. Supercond.(2003) 13 2116 –2119
3. Toshifumi Ise, et al, Reduction of inductance and current rating of the coil and enhancement of fault current limiting capacity of a rectifier type SFCL, IEEE Trans. Appl. Supercond.(2001) 11 1932 –1935

Design and primary experiment for a single phase 220V/100A/6kW bridge Fault Current Limiter-SMES

Zhao Caihong[1,2], Zhang Yigong[3], Li Shun[1,2], Huang Xiaohua[1,2], Chen Danfei[1,2], Xiao Liye[1], Lin Liangzhen[1], Yu Yunjia[1]

[1]Applied Superconductivity Lab, Inst. of Electrical Engineering, CAS, Beijing 100080, P.R.China
[2]Graduate School of CAS, Beijing 100039, P.R.China
[3]Department of Electrical Engineering, North China Electric Power University, Beijing102206, P.R.China

A single phase 220V/100A/6kW bridge Fault Current Limiter-SMES (FCL-SMES) demonstrator was developed. The system consists of a series linking transformer, a 6kW IGBT voltage converter with 20kHz PWM control method, a 6 kW IGBT current regulator with 20kHz phase-shifted control method, a 26mH/25A Bi-2223 coil, a rectifying diode bridge and a DSP-based controller. The current regulator can not only charge the HTS coil, but also absorb the energy from the system and the coil to compensate the voltage sags through the converter. In the controller of the system two very fast DSPs are used to implement the control algorithms for FCL-SMES.

INTRODUCTION

Bridge Fault-current-limiting SMES (FCL-SMES) has been studied in [1]. As a new equipment, FCL-SMES integrates the common diode bridge superconducting fault current limiter (SFCL) and SMES together by replacing the bias voltage source for SFCL with a new current regulator. It improves the limiting function of bridge SFCL and compensates the sags caused by the fault, reducing significantly superconducting coil capacity at the same time. A single phase 220V/100A/6kW bridge Fault Current Limiter-SMES (FCL-SMES) demonstrator was developed, and a brief description of its various subsystems, including the converter, current regulator, Bi-2223 coil, rectifying diode bridge and the DSPs-controller, is given. Some experimental results with the FCL-SMES are shown, too.

PRINCIPLE OF THE SINGLE PHASE FCL-SMES

Figure 1 shows the configuration of the single phase FCL-SMES. It is interlined between two buses for the common and critical loads respectively, and it is composed of a series linking transformer, a converter unit, a current regulator, a DC capacitor between the converter and regulator, a HTS coil, a diode bridge and a DSPs-based controller. When fault appears on one of the feeder of common loads, FCL-SMES can work automatically to limit the current all along. At the same time, the current regulator is introduced and starts to work as a controlled resister, absorbing the energy from the system and the superconducting coil itself to compensate voltage of the sag for critical load. The energy is transferred through the regulator to be controlled DC voltage for the DC capacitor, then the DC voltage of the capacitor is converted into 50Hz AC voltage, which is injected into the system through the series linking transformer to implement the voltage compensating function. So far, the fault current is limited and the problem of voltage sags is solved. The following is a brief description of the main components of FCL-SMES.

CONVERTER and ITS CONTROL SYSTEM

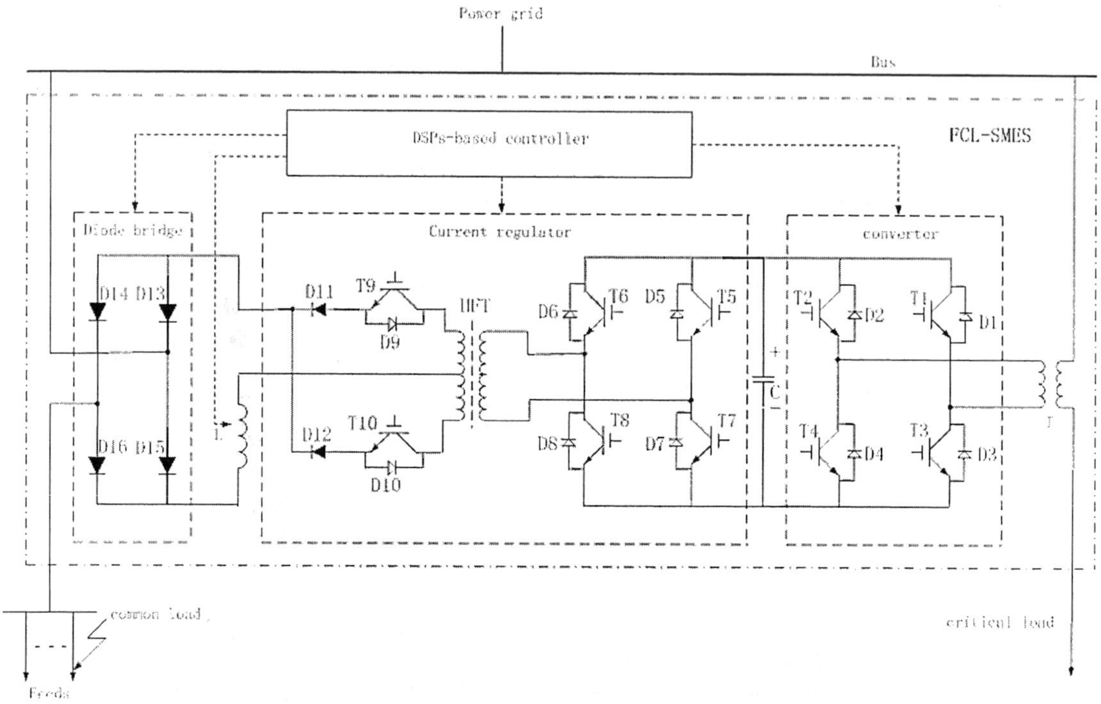

Fig.1: Main topology of the proposed single phase FCL-SMES

Table 1   Parameters of the converter and transformer for FCL-SMES

AC voltage: 0~110V (RMS)
AC Current: 60A
THD: <2%
DC voltage: 600V
Rated Power: 6kW
Transformer: 6kVA, 110/220(n=2),50Hz

Fig.2: An overview of the single phase 220V/100A/6kW bridge Fault Current Limiter-SMES

A 6kW IGBT voltage source converter (VSC) was developed to serve as the interface between the AC power network and the DC current regulator. Fig.1 shows its topology, and the single phase test is just our Phase 1 program. To improve the Total Harmonics Distortion (THD) of the AC voltage, an 20kHz SPWM (Sinusoidal Pulse Width Modulation) switching strategy and 3-Dimensional Voltage Space Vector PWM

Algorithm based on the DSP TMS320F240 are achieved [2]. The parameters of the converter and transformer are listed in Table 1.

CURRENT REGULATOR AND ITS CONTROL SYSTEM [3]

A 6kW IGBT current regulator was developed to work as the interface between the DC voltage of the converter and the superconducting coil. In fact, two equal units work in parallel. Fig.1 shows its topology. A 20kHz phase shifted control method based on DSP TMS320F2812 is adopted, and the zero current switched (ZCS) principle for the bi-directional power of the regulator is realized. In this experiment, the regulator works as a controlled resister r(t) to perform its function for FCL-SMES. R(t) is controlled to be 2.5 Ω or the average power of regulator is 3.5kW. The parameters are shown in Table 2.

Table 2   Parameters of the current regulator for FCL-SMES

Input DC voltage : 600V
Input DC: 10A
Output DC Voltage: 0~100V
Output DC current: 60A
Rated Power: 6kW
High frequency transformer: 6 kW, 20kHz, 600/60 (n=10)

DIODE BRIDEGE and HTS COIL

The diode bridge is a common rectifying one in Figure 1. The HTS coil was wound with Bi-2223 Ag-sheathed type from AMSC. It consists of 2 double pancakes and the total number of turns is 436, the inductance is 26mH, and the central field of the coil is $3.644 \times 10\text{-}3T/A$. During the experiment, the coil was immersed in the liquid nitrogen. The parameters in detail can be seen from [4].

EXPERIMENTS

Fig.6 compares the line currents between the system with common bridge SFCL and with FCL-SMES. The peak line current with SFCL is 75A about 2 cycles after the fault, and it is the same as the one without SFCL. It is verified that the function of limiting steady current of bridge SFCL is very little, but the peak line current with FCL-SMES is still 45A after the fault. Fig.7 shows the current of the coil of bridge FCL and FCL-SMES, and they are 67A and 43A respectively. Thus, the coil capacity for FCL-SMES is reduced. Fig.8 is the voltage of the coil. It can be seen that FCL-SMES not only can limit the peak line current, but limit the steady current, especially in the case of the reduced coil capacity.

CONCLUSIONS

A single phase 220V/100A/6kW bridge Fault Current Limiter-SMES (FCL-SMES) demonstrator was developed, and the primary experiment was done. Its principle of the function to limit the current was verified. The components of FCL-SMES, such as converter, current regulator, the HTS coil and so on were preliminarily tested. The further experiment is underway and the stage phase II will proceed with the construction and test of a three phase prototype.

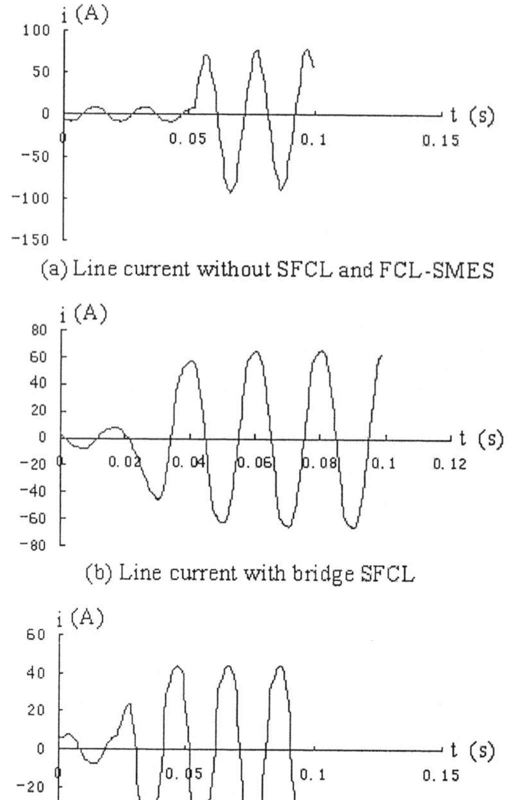

(a) Line current without SFCL and FCL-SMES

(b) Line current with bridge SFCL

(c)   Line current with FCL-SMES

Fig.6 Experiment results of line current

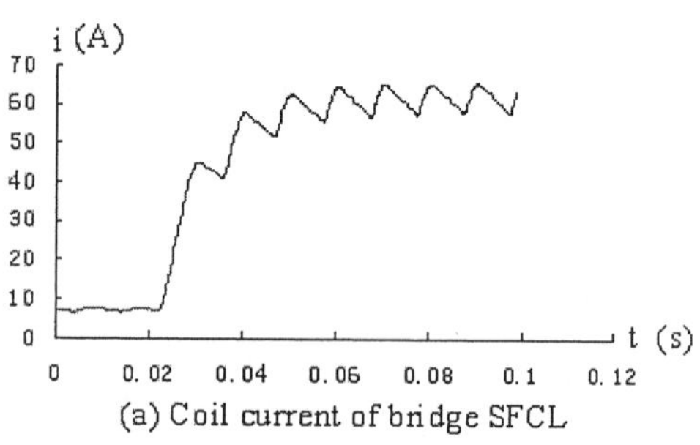

(a) Coil current of bridge SFCL

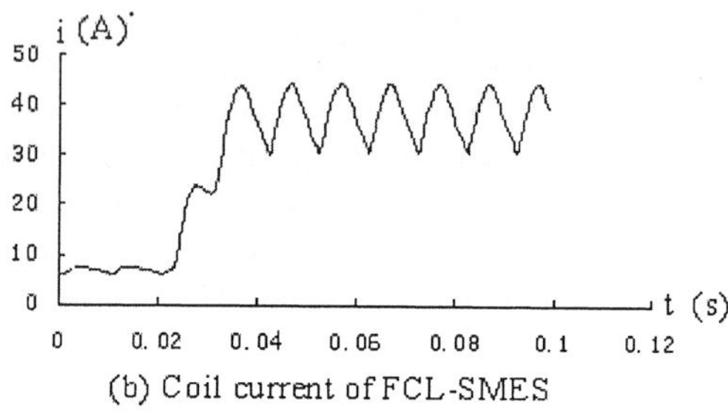

(b) Coil current of FCL-SMES

Fig.7 Experiment results of coil current

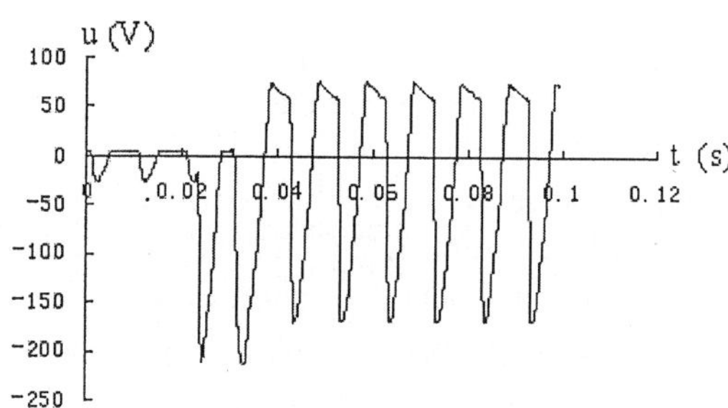

Fig.8: Coil voltage of FCL-SMES

ACKMOWLEDGEMENTS

This work was supported by National Natural Science Foundation of China (Grant#: 50137020), National outstanding young scientist foundation of China (Grant#: 50225723), and "863 "High Technology Project (Grant#: 2002AA306351).

REFERENCES

1. Caihong Zhao, Liye Xiao, Liangzhen Lin, Yunjia Yu, A new concept of bridge Fault Current Limiter-SMES for interline application to improve power quality, Submitted to ICEC 20.
2. Xiaogang Wang, Study of a voltage converter for FCL-SMES, Master Thesis, Institute of Electrical Engineering, Chinese Academy of Sciences.
3. Caihong Zhao, Xiaogang Wang, Liye Xiao, Liangzhen Lin, A new current regulator for superconducting magnet, Chinese patent, No. 03137460.3.
4. Jiacheng Zhang; Zhiyuan Gao; Naihao Song, et al. Dynamic simulation and tests of a three-phase high Tc superconducting fault current limiter, IEEE Trans. Appl. Supercond.(2002) 12 896 -899

Fabrication and characterization of Bi-2223 coils for generator applications

Xu B, Al-Mosawi M K, Yang Y, Beduz C

School of Engineering Sciences, University of Southampton,
Southampton, SO17 1BJ, UK

A set of race-track pancake coils for a 100 kVA superconducting generator was constructed using high strength reinforced Bi-2223 tapes with co-wound fibre glass cloth and epoxy impregnation. Mechanical and thermal properties of the coil composites were conducted. *I-V* characteristics and the critical current at 77 K of the superconducting coils were measured. The onset and progress of thermal runaway due to over-current were investigated by measuring the temperature rise and overall voltage increase. The outcome of this work contributed greatly to the design stage and quality control of the superconducting generator.

INTRODUCTION

The development of High Temperature Superconducting (HTS) tapes is making significant progress in terms of availability, high Ic, long length and cost. Tapes are often being wound as pancake-shaped coils for use in small or medium scale HTS machines. The design and construction of these coils should minimize degradation of the critical current during construction and improve robustness during operation. Large mechanical and electromagnetic forces during operation as well as forces induced by the manufacturing process and thermal cycling should be taken into consideration.

Comprehensive characterization of the mechanical, thermal and superconducting properties of the coils is conducted as part of a project at Southampton University to design, construct and test a 100 kVA HTS superconducting synchronous generator [1]. The results of this work are crucial for modelling and predicting performances during operation of the generator and also represent a mean of quality control to qualify the coils for use in the machine.

COIL WINDING TECHNIQUE

Ten 40-turn single layer race-track coils are constructed to form the rotor winding of the generator, see Figure 1. The radial and axial lengths of the coils are 188 mm, 364 mm, respectively, whereas the minimum bending radius at the corners is 39 mm. These coils are produced by the react and wind method using a purpose built apparatus. American Superconductor Corporation (AMSC) Bi-2223 stainless steel reinforced tapes with nominal critical current of 115 A (77 K, self field) are used. The tape is co-wound with a fibre glass sheet of thickness 100 μm to provide turn-to-turn insulation as shown in Figure 2.

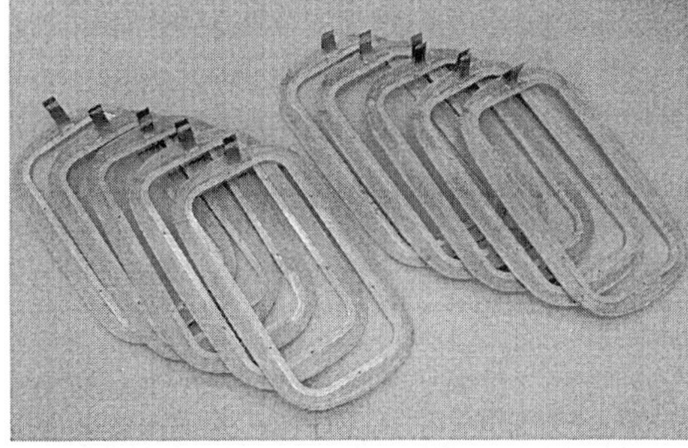

Figure 1 Photo of ten race-track pancake coils.

710

The winding apparatus, driven by a DC motor at variable speeds between 4 and 8 rpm, is designed and constructed to handle the race-track shaped coils by resting the feeding spool on a reference race-track former to trace the contour of the coil. A winding tension of 10N is maintained by the use of an adjustable soft brake, which also moves with the reference coil former. Such arrangements help to maintain a constant winding speed and reduce the potential damage caused by the changing diameter of the race-track coil.

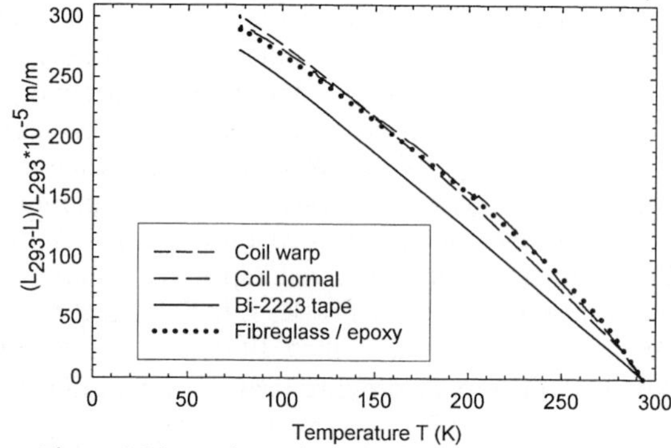

Figure 2, Winding of race-track superconducting coils

Copper current leads are soldered to the ends of coils, with voltage taps and thermocouples soldered at various locations within the coil. An extra twelve layers of fibre glass sheet is wound to provide further support to the turns. Finally, epoxy resin is vacuumed impregnated to enhance structural integrity of the coils.

COIL CHARACTERIZATION

Mechanical properties

HTS coils are essentially a composite comprising of HTS tape, insulation material of and impregnation epoxy. The configuration and results for the axial tensile, transverse tensile, and axial shear strengths are given in Table 1. For the axial tensile test, the specimens are prepared via vacuumed impregnation of 4 lengths of tapes interleaved by fibre glass sheets. The final specimen is about 1.6 mm thick and 110 mm long. For the transverse tensile test, two superconducting tapes were soldered to brass holders and glued together with a sheet of fibre glass in between. In the shear test, the specimen is formed by overlapping two tapes over a 20 mm long (see Table 1). All the specimens are tested at 77K with a tensile machine and loaded using a purpose built grip chucks.

Table 1 Mechanical test of coil composite

Test Configurations		Test Results
axial tensile		$\sigma_{0.2} = 290\text{MPa}$ $\sigma_u = 380\text{MPa}$
transverse tensile		$\sigma_u = 3.8\text{MPa}$
shear		$\sigma_u = 3.9\text{MPa}$

The axial proof tensile strength ($\sigma_{0.2}$) and ultimate tensile strength (σ_u) were obtained using axial load-elongation curve, and are calculated to be 290 MPa (0.2% offset) and 380 MPa, respectively. The ultimate (fracture) transverse tensile strength is 3.8 MPa and the ultimate sheer strength is found to be 3.9MPa. These results, as listed in Table 1, provide crucial data in the design of superconducting machines.

Thermal contraction

Thermal contraction of the coil composite and its constituents (superconducting tape, fibre glass/epoxy) from 77 K to 293 K are measured using a quartz tube dilatometer with a linear variable differential transformer (LVDT) [2]. Due to inhomogeneous nature of the coil composite, the thermal contraction measurements are conducted for both warp and normal directions of the coils, using similar specimens to those for the axial tensile test.

Figure 3 Linear thermal contractions of superconducting coil and its constituents.

The results of these measurements are shown in Figure 3. It can be seen that there are rather small but noticeable difference between the various constituents and the coil composite itself, with the sole superconducting tape having the lowest contraction for the same range of temperature. The total linear thermal contractions from room temperature to 77 K of coil in warp and normal directions are 0.28% and 0.29%, respectively. The total thermal contractions for the sole Bi-2223 tape and fibreglass/epoxy composite are found to be 0.28% and 0.30, respectively.

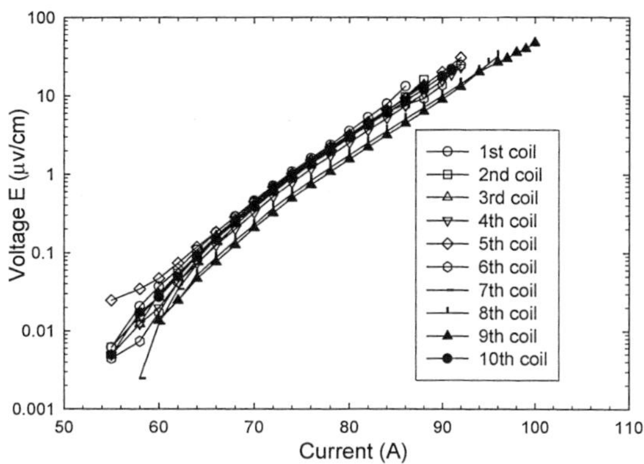

Figure 4 *I-V* characteristic of 10 Bi-2223 pancake coils.

Superconducting properties

The *I–V* characteristics of the 10 coils are determined at 77 K and coil self-field (see Figure 4). The critical currents of the coils are in the range of 75 A ± 3 A, which is about 60% of the critical current of a short sample and broadly consistent with the predicted values using the field dependence of the critical current. The power exponent n of the *I–V* curves for the 10 coils found to about 16. It can be concluded that the construction process and the use of coil composite materials prove to be adequate for our performance criteria. The coils also show consistency over several cooling cycles.

In a superconducting winding tape-to-tape joints sometimes become unavoidable. In the generator winding, a lap-splice tape joint in one of the coils is deliberately introduced using Sn–Pb solder. The total joint resistance is measured and found to be $0.15\mu\Omega$, which gives contact resistance $0.58\ \mu\Omega cm^2$.

Instability above critical current

As indications to coil stability, the excursions of coil temperature and voltage with time of one of the coils subjected to currents above the critical value were monitored using differential thermocouples in the middle of the coil and end-to-end voltage taps. Such over-current measurements were conducted by submerging the coils in a liquid nitrogen bath at 77K, before and after coil impregnation to examine the effect of epoxy resin to overall heat transfer. The activation of nucleation sites has been varied by cycling the current through the coil for several times.

Figure 5 & 6 shows the temperature and voltage traces of the pre-impregnated and post-impregnated coils with various applied over-currents. It can be seen that the temperature and voltage runaway occurs at 106 A for the pre-impregnated whereas the runaway occurs at lower current of 94 A for the post-impregnated coil. As expected, the temperature and voltage runaway for pre-impregnated coils showed strong dependence on the boiling conditions, as activation of the surfaces inhibits the thermal runaway at 106A (thick lines in Fig. 5a-b). In contrast, post-impregnated coils exhibited only a weak improvement at 93 A (thick lines in Fig. 6a-b), as the impregnation epoxy becomes the dominant thermal resistance. The heat flux of thermal runaway for the post-impregnated coil is $0.2\ W/cm^2$, well below the onset nucleate boiling. This suggests that thermal runaway is initiated in the natural convection regime, where heat transfer in generally poor. It remains unclear whether full recovery is possible with established nucleate boiling at higher heat flux below the critical heat flux CHF. However the runaway tests were terminated to prevent burnout of coils, which have to be used as part of the superconducting rotor. To verify whether this runaway will stabilize or it is a true runaway, further tests are required to clarify the true nature of this behaviour.

CONCLUSIONS

Ten single layer race-track vacuum impregnated coils are successfully manufactured, and the mechanical, thermal and superconducting properties of the coils are measured. The choices of impregnation and insulation materials and matching thermal contractions of the coil components are crucial for reliable performances of the coils. The coils show a constant performance after several cooling cycles with critical currents in the range of 75 ± 3 A. The proof and ultimate axial tensile strength are 290 MPa and 360~390 MPa, which indicate satisfactory strength for safe operation of the 100 kVA superconducting generator.

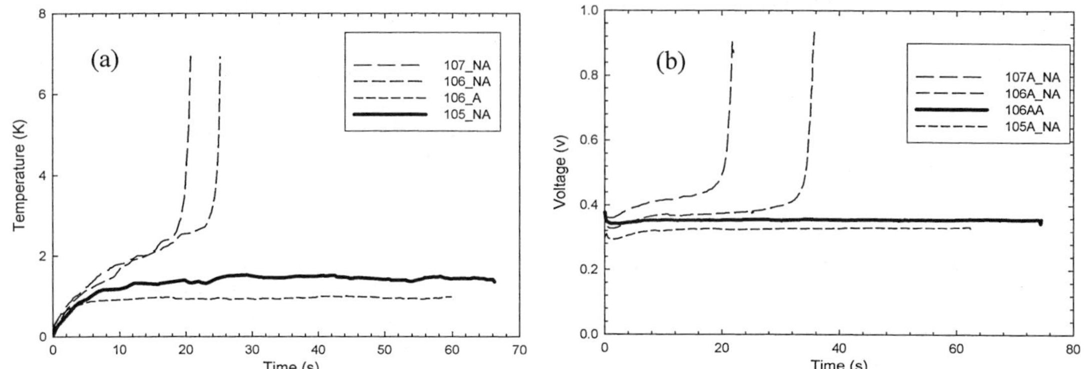

Figure 5 Temperature (a) and voltage (b) excursions at various applied currents for the pre-impregnated coil. (NA: not activated, A: activated.)

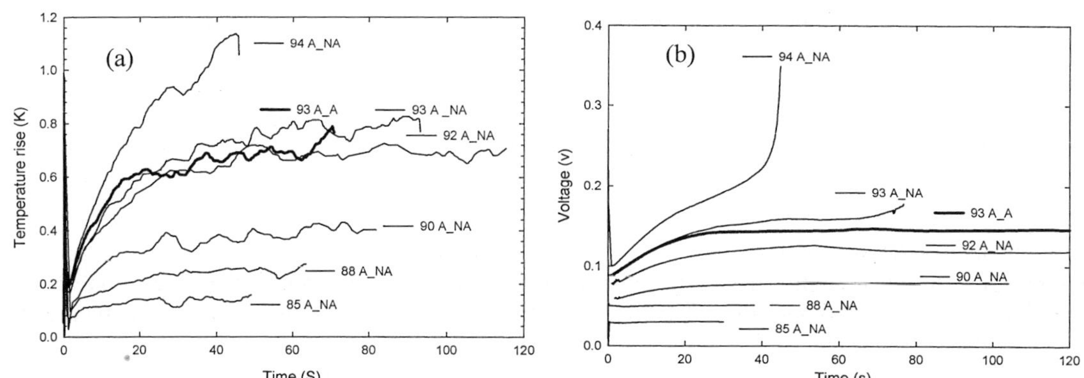

Figure 6 Temperature (a) and voltage (b) excursions at various applied currents for the post-impregnated coil. (NA: not activated, A: activated.)

ACKNOWLEDGEMENTS

The authors would like to express their appreciation to Dr. P. MacDonald for the use of the test machine and helpful discussions. This work was supported by EPSRC research grant GR/N21253/01.

REFERENCES

1. Al-Mosawi, M.K., Xu, B., Beduz C., Goddard, K., Sykulski, J. K., Yang Y. and Stephen N.G., Webb, M., Ship, K. S. and Stoll, R., 100 kVA high temperature superconducting generator. *In*: G.G. BAGUER, ed. <u>Proceedings of 19th International Cryogenic Engineering Conference</u> (2002), Grenoble, Narosa. 237-240.
2. Macdonald, P. C., <u>Cryogenically cooled amplifiers for deep space communication.</u> PhD Thesis (2001), University of Southampton.

Shape optimization of HTS magnets using hybrid genetic algorithms

Wang C.[1,2], Wang Q.L.[1]

[1]Institute of Electrical Engineering, Chinese Academy of Sciences, Beijing, 100080,China
[2]Graduate School of the Chinese Academy of Sciences, Beijing, 100039,China

Two kinds of optimal methods and test of their abilities for searching optimal solutions are presented in the paper. A 12 T high temperature superconducting (HTS) magnet by Bi-2223/Ag tape is designed according to the methods. A new configuration of the HTS magnet which can reduce the winding volume and improve the efficiency of superconductor utilization effectively is suggested with consideration of the constraints, such as central magnetic filed, field homogeneity, critical current characteristic and so on.

INTRODUCTION

Bi-2223/Ag tapes are extensively used in HTS engineering field such as HTS magnets. The J_c-B characteristics of Bi-2223/Ag tapes hardly decrease at all in 20~30 K under high magnetic field and HTS magnets can set the operating temperature within a wide range. However, the properties of the Bi-2223/Ag tapes have an anisotropy in the critical current versus applied external magnetic field characteristic.

In Figure 1, the critical currents versus the magnetic flux density which was measured against the flux angle of 0-90^0 are shown for a Bi-2223/Ag tape at liquid nitrogen temperature. A critical current I_c was specified by the measured current-voltage characteristics of the tape, where I_c was defined by 1 μv/cm. The critical current is the lowest at 90 degree from Figure 1. It is necessary to consider the anisotropic B-I characteristic of Bi-2223/Ag tapes at the design stage. As the perpendicular component of the external magnetic field increases, the critical current is reduced largely. In the case of the HTS magnet with rectangular cross sectional shape, due to the higher radial magnetic field (the perpendicular component of magnetic flux density on HTS tapes) near the end of the coil, a large amount of superconductor is underused and only the terminal part of the coil works near the tape critical conditions. This lack of efficiency calls for a more thorough optimization of the magnet design, able to reduce the amount (volume) of superconductor while achieving the same performance. In this paper, we propose two kinds of optimal design methods for HTS magnets with consideration of the constraints, such as central magnetic filed, field homogeneity, B-I characteristic and so on.

OPTIMIZTION METHODS

The Genetic Algorithm (GA) is to mimic some of the processes observed in natural evolution. It can escape from local minima and deal with constrained nonlinear optimization problems. However, the solution obtained by GA is generally near the global minimum in the whole solution region. GA can easily appear premature at early iterations and evolution stalling at late iterations. It blocks seriously the GA to find the global optimum, so we adopt a fitness scaling technique which is from the simulated annealing (SA) method. The modified GA is named of GASA. Sequential Quadratic Programming (SQP) is an efficient method in finding local optima for constrained nonlinear optimization problems, but it can not guarantee that the solution is the global optimum for the problem. We combined GASA with SQP and it is called hybrid GA.

Figure 2 shows the flowchart of the first hybrid GA. The SQP is inserted in the GASA method as a local search operator to improve the local search ability. The method is called GASASQP.

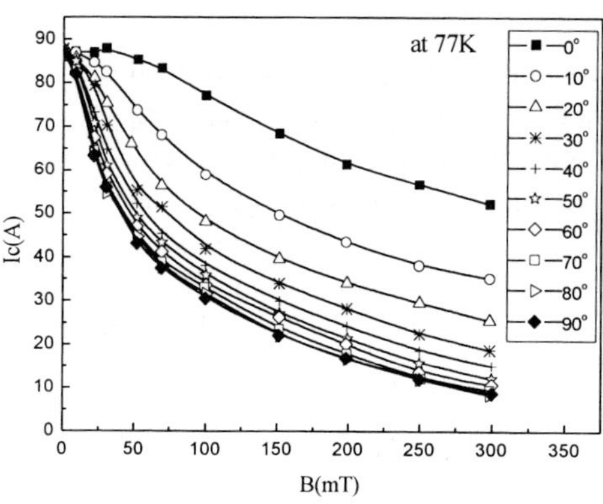

Figure 1   Critical current characteristics of the Bi-2223/Ag tape

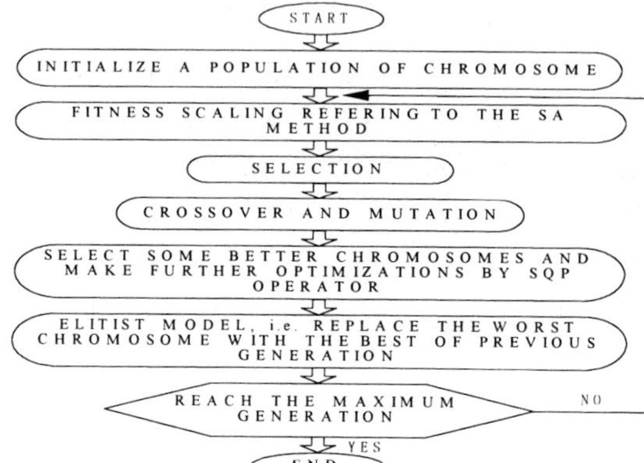

Figure 2   Flowchart of the first hybrid genetic algorithm (GASASQP)

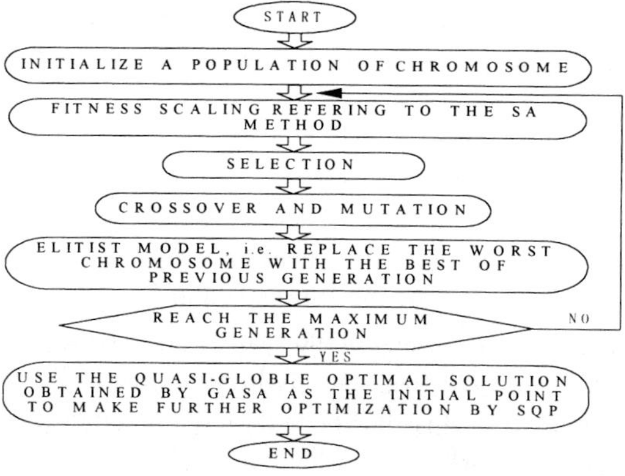

Figure 3   Flowchart of the second hybrid genetic algorithm (GASA+SQP)

Talble 1   Comparision of the solutions

optimal method	GA evolution generation / solution	100	200
GASA		147. 1185	0. 4036
GASA+SQP		118. 4385	1. 2730e-4
GASASQP		1. 2728e-4	———

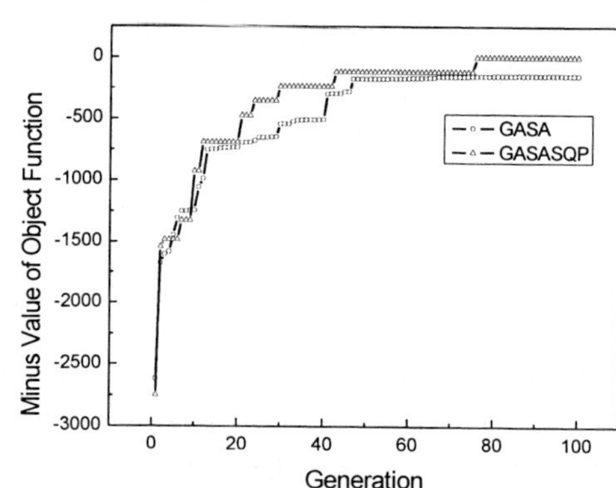

Figure 4   Comparison of optimum search abilities (Generation vs. Current Minus Value of Objective Function)

Figure 3 shows the flowchart of the second hybrid GA. At the first, GASA searches the global optimum in the whole solution region to obtain a quasi-optimal solution, and then, the global optimal solution can be obtained by SQP. The method is named as GASA+SQP.

To contrast two kinds of methods, a function which is often used in optimization test is selected

[1]:

$$f(x) = 4189.829 - \sum_{i=1}^{10} (x_i \sin \sqrt{|x_i|}) \qquad x_i \in [-500,500] \tag{1}$$

The function has many local minima so that it is very difficult to obtain the global minimum value of zero. The solutions through the methods are listed in Table 1. It shows that the GASASQP has stronger searching ability from Table 1 and Figure 4, but it takes long time to obtain the optimal solutions.

EXAMPLE OF OPTIMAL DESIGN

A 12 T HTS magnet operating at 4.2 K with the field homogeneity of 1% in the radius of 20mm region, bore size of 100mm and Bi-2223/Ag tape with the size of 0.23 mm×3.5 mm is designed. Figure 5 shows the configurations of HTS magnets [2,3], where X_i ($i = 1,....,6$) are the length and the thickness of each coil, B_0 is the center magnetic flux density, and B_i ($i =1,2$) is the magnetic fields located at the radius of 20 mm.

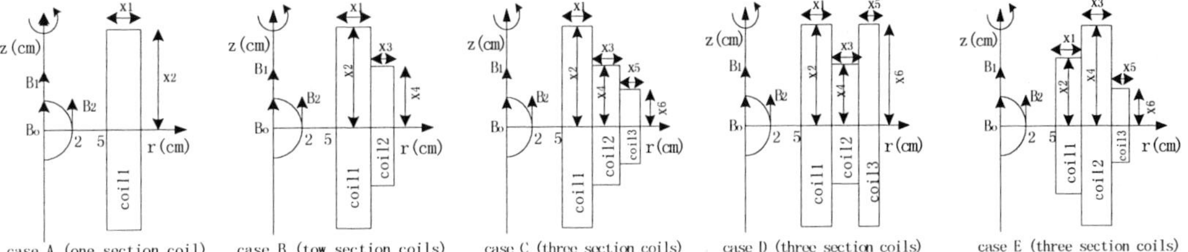

Figure 5   Models of superconducting magnets to be optimized

Minimize : V (the winding volumes of HTS magnets)

Subject to:

$$B_0 = 12 \text{ T} \tag{2}$$

$$I_{op} \leq I_c(B) \tag{3}$$

$$I_c(B) \approx I_c(B_r) = 118 - 107 \log_{10}(B_r) \tag{4}$$

$$\frac{|B_i - B_0|}{B_0} \times 100 \leq 1.0 \quad (i=1,2) \tag{5}$$

The constraints of B_0 and the field homogeneity are represented by equations (2) and (5), respectively. The function of $B-I$ characteristic for the HTS tape is approximately expressed as equation (4), where B_r is the radial magnetic field [4]. Equation (3) represents the constraint about the $B-I$ characteristic of the tape, where I_{op} is the operating current.

Table 2 lists the optimal results and θ represents the angle between the z-axis and the magnetic flux density at the critical point. It should be noted that the length and thickness of each coil are discrete parameters given by the number of tape layers and turns, respectively. However, the discrete character of these parameters is neglected in our optimizations, so we select three better configurations (case A, case B, case E) from Table 2 based on the number of section coils and do some work on the length and thickness of each coil to determine the number of turns and layers. Table 3 shows the final results of optimizations of HTS magnets. And the values of winding volume in each design case are shown in Figure 6, the operating currents and maximum radial magnetic fields are shown in Figure 7.

The optimal results show that the winding volume of the HTS magnet can be reduced effectively and the maximum radial magnetic field decreases while the operating current increases by using the configuration in case E. The winding volume in case E is about 64.49% in case A, and the operating current in case E is about 30% higher than in case A.

Table 2   Optimal results of HTS magnets

	case A	case B	case C	case D	case E
$V(cm^3)$	49516.67	37913.36	33091.73	33566.94	31854.39
$I_{op}(A)$	57.70	68.11	74.99	75.42	74.73
$B_{r\max}(T)$	3.66	2.93	2.52	2.50	2.54
$\theta(\deg)$	50.78	69.84	64.54	46.10	21.30

Table 3   Final optimal results of HTS magnets

	case A	case B	case E
$V(cm^3)$	49663.70	38019.57	32028.77
Thickness/turns per layer of coil1 (cm)	16.63/723	8.05/350	4.65/202
Length/layers of coil1 (cm)	35.70/102	50.40/144	30.80/88
Thickness/turns per layer of coil2 (cm)	——	6.37/277	4.32/188
Length/layers of coil2 (cm)	——	23.10/66	54.60/156
Thickness/turns per layer of coil3 (cm)	——	——	4.30/187
Length/layers of coil3 (cm)	——	——	18.20/52
$I_{op}(A)$	57.66	68.22	74.80
$B_{r\max}(T)$	3.66	2.91	2.53
$\theta(\deg)$	50.77	48.13	55.17
$B_{r\max}$ position (r, z) (cm)	(14.04, 17.85)	(9.42, 25.20)	(11.67, 27.30)
Calculating Time (minute)	3.50	5.20	24.80

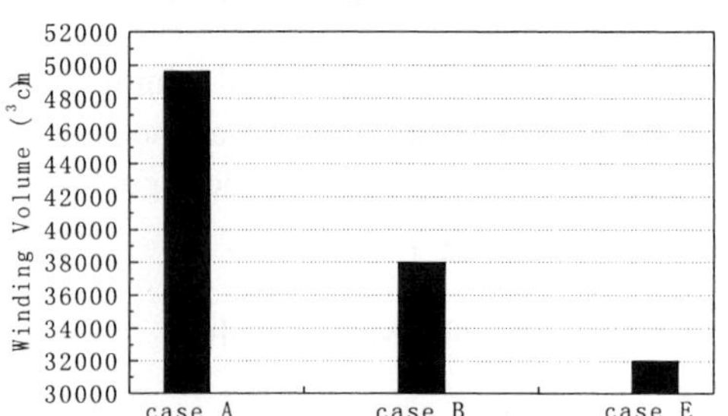

Figure 6   Winding volumes of the optimized HTS magnets

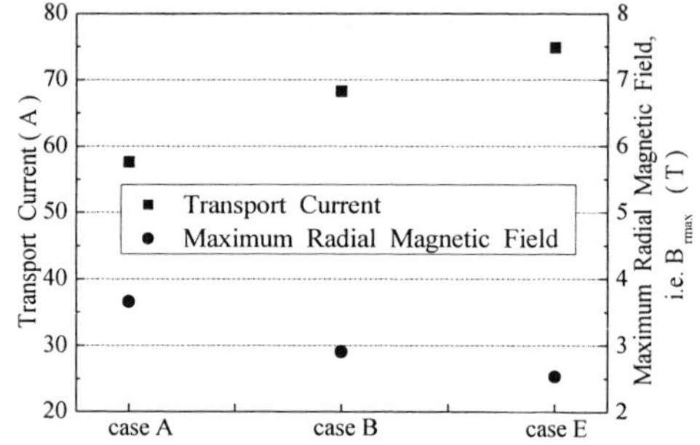

Figure 7   Operating currents and maximum radial magnetic fields of the optimized HTS magnets

CONCLUSIONS

Two kinds of optimal design methods for HTS magnets is proposed. As an example, a 12 T HTS magnet wound with Bi-2223/Ag tapes was designed. A new configuration of HTS magnet (case E) that can reduce the winding volume effectively is found. By using optimal methods, the efficiency of superconductor utilization can be improved, the amount of superconductor needed in the magnet and the electrical power wasted in refrigeration can be saved.

REFERENCES

1. Schwefel H P. Numerical optimization of computer models[M], Chichester: Wiley & Sons, (1981).

2. So Noguchi, Makoto Yamashita, Hideo Yamashita, and Atsushi Ishiyama, An Optimal Design Method for Superconducting Magnets Using HTS Tape, IEEE Trans. Appl. Superconduct. (2001) 11 2308-2311.

3. Mitsugi Yamaguchi, Atsushi Honma, Shinichi Ishiguri, Satoshi Fukui, Itsuya Muta, and Takatsune Nakamura, A Study on Performance Improvements fo HTS Coil, IEEE Trans. Appl. Superconduct. (2003) 13 1848-1851.

4. J. Pitel, P. Kovac, T. Melisek, A. Kasztler, and R. Kirchmayr, Influence of the Winding Geometry on the Critical Currents and Magnetic Fields of Cylindrical Coils Made of Bi(2223)Ag Anisotopic Tapes, IEEE Trans. Appl. Superconduct. (2000) 13 478-481

An adaptive neuro-fuzzy control strategy for the bridge type superconducting fault current controller

Wenyong Guo[1,2], Caihong Zhao[1,2], Liye Xiao[1]

[1]Applied Superconductivity Lab, Inst. of Electrical Engineering, CAS, Beijing 100080, P.R.China
[2]Graduate School of CAS, Beijing 100039, P.R.China

This paper presents an adaptive neuro-fuzzy control strategy for the bridge type superconducting fault current controller (SFCC), which can automatically decide the phase-delay angle according to the fault current amplitude, limit the fault current amplitude to the degree that the system can endure, and recover to the normal situation immediately after the fault current ceases. Simulation results show that this design's performance is good.

INTRODUCTION

The bridge-type fault current controller (FCC), which was previously called fault current limiter (FCL), consists of a full-wave bridge, an inductance, and an optional bias power supply. The FCC can make the inductor switched automatically into the ac circuit and limit the amount of fault current, when values are higher than a preset current value; and present no impedance to the ac current flow, when load current values are smaller than the preset value [1]. By using a superconducting core, no loss will occur, here this is called SFCC. By changing the phase-delay angle, it can limit the fault current to any degree. It may be appropriate to preset the phase-delay angle if only short-circuit will occur, yet in some situation, for example, sudden overload may occur, in this situation the preset phase-delay angle is no longer appropriate.

This paper proposes an advanced adaptive neuro-fuzzy control strategy for SFCC, which can automatically determine the proper phase-delay angle. Simulations of single-phase mode will also be presented. Simulation results demonstrate that this design's performance is good.

THE BRIDGE-TYPE SUPERCONDUCTING FAULT CURRENT CONTROLLER AND THE PROPOSED ADAPTIVE NEURO- FUZZY CONTROL STRATEGY

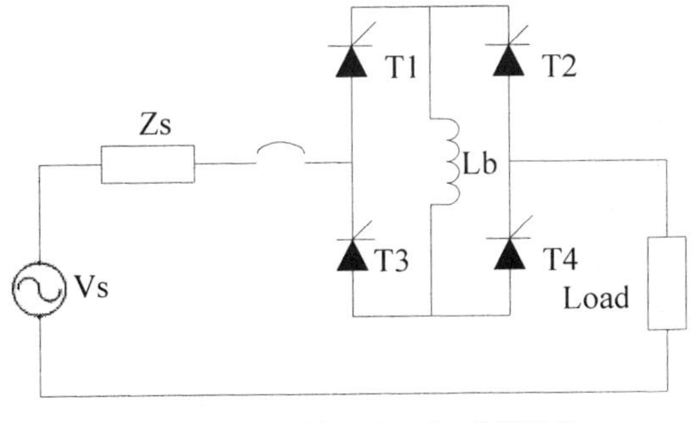

Figure 1 The circuit of SFCC

The bridge-type superconducting fault current controller without a bias power supply is shown in Figure 1. Consider the most common 10KV/400A subsystem in China, the parameters are set as follows:

$$V_S = 10KV / \sqrt{3} \quad , \quad L_S = 6.6mH \quad , \quad R_S = 264m\Omega \quad ,$$

$$L_b = 20mH \quad , \quad Load = 25\Omega \quad .$$ And the system frequency is 50Hz. By setting the phase-delay angle, the fault current amplitude can be adjusted. Changing

the phase-delay angle α from 0 to 90 degrees results in a gradual decrease of the fault current. The ac current has a sinusoidal wave shape for phase angles up to $90°$. The short-circuit current will decrease further for angles greater than $90°$. At $α = 90°$ the effect of the SFCC is the same as if it were replaced by a series connected inductor Lb. For angles $α > 90°$ the SFCC produces ac currents that are discontinuous and adjustable in amplitude[1].

What is interested here is how to set the phase-delay angle automatically according to the overload or short-circuit level. As it's known, fuzzy reasoning is capable of handling imprecise and uncertain information whilst neuro networks are capable of being identified using plant data, and the neuro-fuzzy networks combine the advantages of both fuzzy reasoning and neuro networks[2], so the neuro-fuzzy networks can be used to model SFCC and obtain the proper phase-delay angle.

In simulation, in large scale (during the first two cycles of the fault current), the neuro-fuzzy inference structure is used. In the neuro-fuzzy inference structure, three input parameters are used. They are the current phase angle minus the voltage source phase angle θ, the fault current amplitude A, and the fault current amplitude minus nominal current amplitude ΔA. When the inductance is switched into the circuit, the parameter θ is changed according to overload level. For example, if the load resistance is changed to 2.5Ω in the above-mentioned systems, and the other parameters is not changed, θ will be $-71.7°$, and 5Ω will be $-57.8°$. So the parameter θ can indicate the overload level. To measure θ quickly and precisely, a very fast phase angle estimation algorithm for a single phase system having sudden angle jumps [3] is applied. In small scale (the fault current amplitude is near the nominal one), PI control strategy is better suited, so a PI regulator is used to modify the phase-delay angle. The input of the PI regulator is the parameter ΔA.

SIMULATION PROCESSES

To obtain the inference structure of the neuro-fuzzy networks, data for training must first be collected. And then these data are trained to generate the neuro-fuzzy inference structure.

<u>data obtaining and training</u>

To obtain the data described before, simulations are done on different overload level. In project, experiments should be done in similar condition to obtain these data. Here the overload resistance is set to: 20.83Ω, 10Ω, 5Ω, 3.33Ω, 2.5Ω, 0Ω, corresponding to the fault current of 1.2 times, 2.5 times, 5 times, 7.5 times, 10 times of the nominal one and the short-circuit current . Table1 is part of the data acquired when the overload resistance is set to 5Ω.

Table1   data when overload resistance is 5Ω

θ (rad)	-0.9825	-0.9825	-0.9825	-0.9825	-0.9825	-0.9825	-0.9825	-0.9825
A	1359.6	1258.7	1171.8	942.3	888.9	795.6	577.1	460.0
ΔA	971.2	870.3	783.4	553.9	500.5	407.2	188.7	71.6
α (deg)	50	60	70	90	100	110	130	140

By using the data obtained before, membership functions of the inputs and the output can be obtained. To obtain the neuro-fuzzy inference structure precisely, up to 60 epochs of training is applied. After training, the neuro-fuzzy inference structure is obtained.

<u>control strategy</u>

The strategy used here is that when the current detector detects that the load current is 20% larger than the nominal level, the fault current control program starts to work. The phase-delay angle is preset to 90°, thus before the fault current falls to zero, the inductance is completely switched into the circuit. At the peek amplitude of the fault current, the parameters A, ΔA , θ are measured and these parameters and the preset phase-delay angle $\alpha =90^0$ are used as the checking data to modify the membership function parameters of the neuro-fuzzy inference structure, so as to make the neuro-fuzzy inference structure to adapt the real-time situation. Then the nominal value of the load current , here A=400, $\Delta A =0$,and the measured θ are used as the input of the adaptive neuro-fuzzy inference structure， so as to produce the phase-delay angle α and it is applied to the next half-cycle of fault current. During the following three half-cycles, the same method is used except that the previously obtained α is used as the checking data. After that, a PI regulator is used to produce the phase-delay angle instead.

SIMULATION RESULT

Using the strategy mentioned above, the simulation results are obtained. Here, two cases are shown. In the first case, at 0.06s , the overload resistance is set to 5Ω (five times overload)，at 0.16s, the fault current ceases(the load resistance is reset to 25Ω), at 0.24s, the overload resistance is set to 2.5Ω (ten times overload)，and at 0.34s, the fault current ceases again. In the second case, at 0.1s, the overload resistance is set to 12.5Ω (two times overload), at 0.2s, short circuit occurs (the overload resistance is 0Ω), and at 0.3s, fault current ceases. These two cases are shown in Figure 2 and Figure 3.

From the figures, it is can be found that whatever the fault current amplitude is, after the first half-cycle of the fault current (when the phase-delay angle is preset to 90°,and the inductance is completely switched into the circuit), the fault current amplitude can be always limited to the amplitude near the nominal level, and after the fault current ceases, it will recover to the normal situation immediately (within 1~2 cycles).

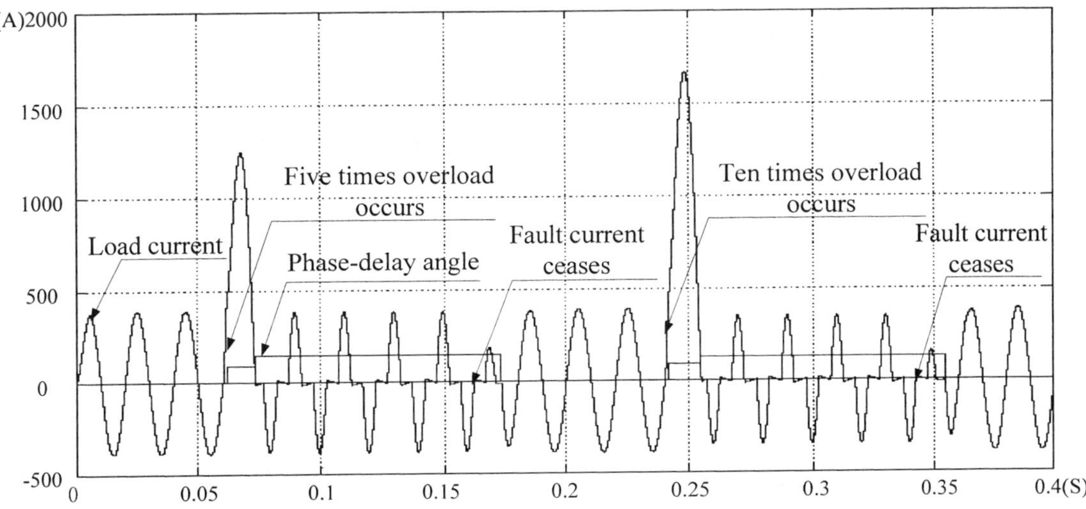

Figure 2　The first case

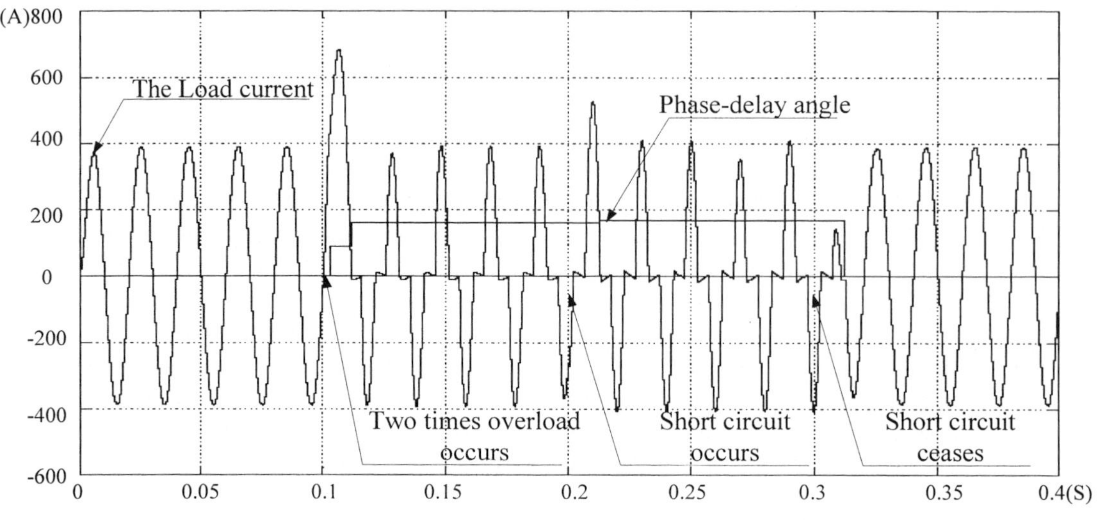

Figure 3   The second case

CONCLUSION

The simulations show that the adaptive neuro-fuzzy control strategy for SFCC can suit with any overload and short-circuit situation, and can limit the fault current amplitude near the nominal level. After fault current ceases, it can also recover to the normal situation immediately .The transient fault current limiting ability can be increased by selecting a larger inductance L_b, and the stable fault current limiting ability is always good by using this control strategy. This control strategy will then greatly enhance the bridge-type superconducting fault current controller's performance, and provide a potential application to the electrical system.

REFERENCES

1. Boenig, H.J.; Mielke, C.H.; Burley, B.L.; Chen, H.; Waynert, J.A.; Willis, J.O.; The bridge-type fault current controller - a new FACTS controller, Power Engineering Society Summer Meeting (2002) 1 455-460
2. Jie Zhang, Morris, J, Neuro-fuzzy networks for process modeling and model-based control, IEE Colloquium on Neural and Fuzzy Systems: Design, Hardware and Applications (1997) 9 1-6
3. Hong-Seok Song, Kwanghee Nam, Mutschler P, Very fast phase angle estimation algorithm for a single-phase system having sudden phase angle jumps, Industry Applications Conference (2002) 2 925-931

Cryoelectronics – An experimental examination of the behaviour of power electronic devices at low temperatures

Hawley C.J., Gower S.A.

SECTE, University of Wollongong, Northfields Ave, Wollongong, NSW 2522, Australia

ABSTRACT

The operating characteristics of power electronic devices vary greatly with the operating temperature at which the device is used. The characteristics for devices operating at room temperature or above are well documented. However, only limited data is available for operation at cryogenic temperatures, especially for recently released devices. Operation of power electronic devices at cryogenic temperatures is known as cryoelectronics and operation in this temperature region can be beneficial as the operating characteristics of some devices can improve markedly. The potential use of cryoelectronics is quite broad, but it is particularly relevant in applications where a cryogenic environment exists such as spacecraft design and superconducting technologies.

INTRODUCTION

This study examines the two main semiconductor-switching technologies, Metal Oxide Silicon Field Effect Transistors (MOSFETs) and Insulated Gate Bipolar Transistors (IGBTs). It has already been demonstrated that when operating at 77K, MOSFETs show a large drop in on-resistance ($R_{ds(on)}$) [1] but also a proportional drop in breakdown voltage (V_{BD}) [2]. This might suggest that at cryogenic temperatures MOSFETs are only suitable for low voltage applications, however devices that exhibit very high V_{BD} and current carrying (I_{ce}) ratings have recently become available. If the reduction in breakdown voltage is within an acceptable level then paralleling such devices may be a viable solution for high power applications. IGBTs are specifically designed for high power applications, and whereas MOSFETs would need to be paralleled to satisfy the current rating, a single IGBT device would be required.

Past work on cooling IGBTs to cryogenic temperatures showed that some devices did exhibit reduced losses whilst others would not work below certain temperatures – typically 120K [3]. Development of the Non Punch Through (NPT) technology in IGBTs has produced devices that are capable of operating at cryogenic temperatures down to 5K [3]. Whether or not the cooling of these devices results in a substantial drop in power loss is yet to be determined.

In this paper, operating characteristics including V_{BD}, $R_{ds(on)}$ and forward collector-emitter voltage (V_{ce}) of various MOSFETs and IGBTs operating at cryogenic temperatures will be presented and discussed.

EXPERIMENTAL SET-UP

Devices Under Testing (DUTs)

The DUTs used in the experiments were selected with the following three objectives in mind. 1) To confirm the decline in $R_{ds(on)}$ in MOSFETs and investigate the resulting reduction in V_{BD} in highly rated devices. 2) To confirm that Punch Through (PT) IGBTs are unable to operate below certain temperatures and quantify the operating characteristics within their temperature limitation. 3) To investigate the claims of consistent positive temperature coefficients of NPT IGBTs by some manufacturers and confirm that these devices do not cease to conduct at any cryogenic temperature.

Using the three objectives as a guide, six DUTs were selected for experimentation; the specifications for these devices are shown in Table 1.

Table 1 Device Characteristics as Specified by Manufacturer (Ambient Temperature Operation)

Brand	Type	Device No.	V_{BD}	I_{CEmax}	Package
STM	MOSFET	STY15NA100	1000	10	TO-247
IXYS	MOSFET	IXTH13N80	800	13	TO-247
Fairchild	PT	SGL40N150DTU	1500	40	TO-264
IRF	PT	IRG4PH50S	1200	57	TO-247
Infineon	NPT	BUP314	1200	52	TO-218
Infineon	NPT	SKW25N120	1200	46	TO-247

It should be noted that these devices were selected due to the package type. This was for two reasons; firstly the physical size of the cold head used for the V_{ce} experiments was quite small and secondly, previous experiments with large package devices revealed that the devices were destroyed due to the thermal shock when immersed in liquid nitrogen (LN2).

V_{ce} and $R_{ds(on)}$ Test Rig

To determine the increase or decrease in power losses by a device operating at cryogenic temperatures, accurate measurements of the V_{ce} of the IGBTs and the $R_{ds(on)}$ of the MOSFETs are needed. The experimental rig was the same for both measurements and a schematic of the set-up is shown in Figure 1.

The DUTs were placed on a cold head cooled by a gaseous helium CTI Cryogenics Model 8300 cryocompressor. The chamber around the DUT was evacuated to 2 Pa, enabling the device to be operated in an environment ranging from 300K to 30K. The gate of the devices remained open as a single 20A pulse of current was applied across the collector-emitter and the resulting V_{ce} measured by a Lecroy 9304AM 4 channel CRO. For the $R_{ds(on)}$ measurements, a 10A pulse was used and a current transformer monitored the current waveform enabling resistance of the device can be accurately determined. For each of the DUTs, the value was then confirmed by re-measuring it in LN2.

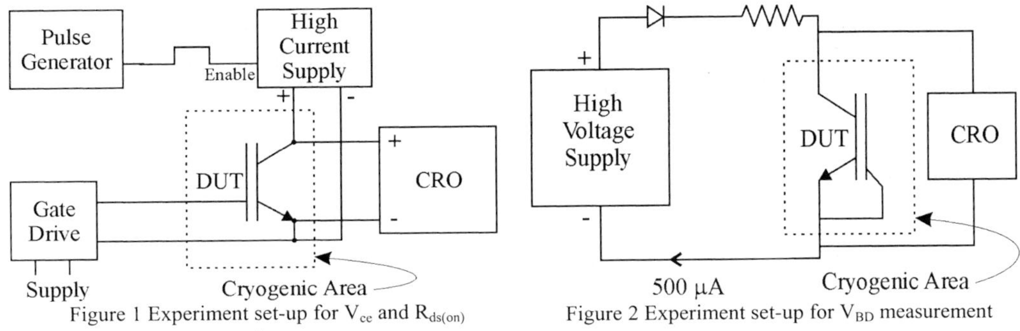

Figure 1 Experiment set-up for V_{ce} and $R_{ds(on)}$ measurement

Figure 2 Experiment set-up for V_{BD} measurement

V_{BD} Test Rig

Operating semiconductor devices at cryogenic temperatures causes an increase in the carrier mobility, leading to an increase in the ionization impact rate and hence a decreased V_{BD} rating [4, 5]. To effectively use these semiconductors at cryogenic temperatures it is important to determine how much each device is de-rated.

To determine the V_{BD} for each device, the experimental set-up shown in Figure 2 was used. Due to the destructive nature of the tests, the V_{BD} value for each DUT was measured at two temperatures. LN2 provides an easily available and stable cryogenic environment at 77K and hence this was the temperature chosen as a comparison to room temperature.

The high voltage supply was applied across each of the DUTs and the voltage increased until avalanche occurred. At this point, the V_{BD} was measured.

RESULTS

V_{ce} and $R_{ds(on)}$ Test Results

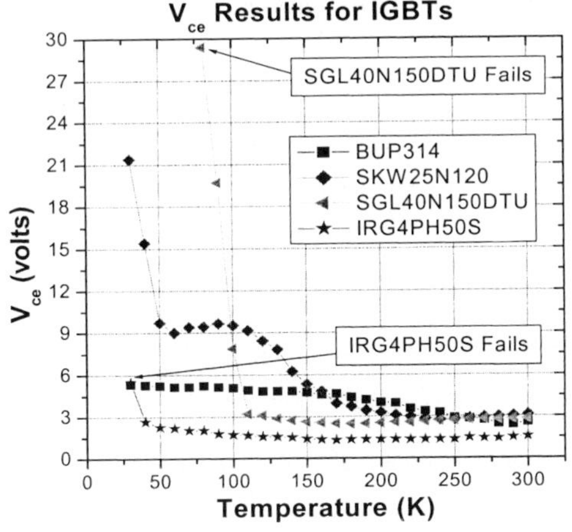

Figure 3 Graph of V_{ce} results for the IGBTs

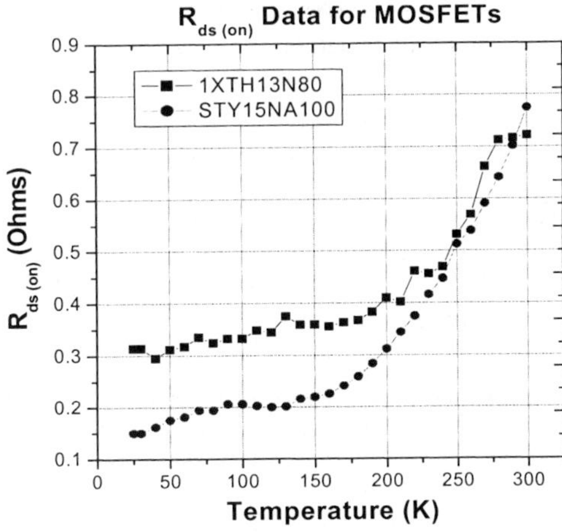

Figure 4 Graph of $R_{ds(on)}$ results for MOSFETs

Each of the devices were tested from 300K to 30K in increments of 10K. To ensure repeatability of the data, measurements on the samples were repeated as the devices warmed to ambient temperature. The V_{ce} results for the NPT and PT IGBTs are given in Figure 3 and the $R_{ds(on)}$ measurements for the MOSFETs in Figure 4.

V_{BD} Test Results

The measured V_{BD} at ambient temperature and at LN2 temperatures is recorded in Table 2 for each device.

Table 2 V_{BD} Results at 300K and 77K

Device	V_{BD} 77K (V)	V_{BD} 300K (V)	V_{BD77}/V_{BD300} (%)
STY15NA100	-	1182	-
IXTH13N80	723	960	75.3
SGL40N150D	1286	1612	79.7
IRGPH50S	1067	1332	80.1
BUP314	1105	1433	77.1
SKW25N120	1098	1307	74.9

Note there was no value recorded for STY15NA100 as the package design was not able to withstand the thermal shock associated with immersion in an LN2 bath. It was also observed that the V_{BD} measured at ambient room temperature was consistently higher than that specified by the manufacturers.

CONCLUSIONS

From the results presented, the following conclusions can be drawn; 1) NPT IGBTs are able to operate at cryogenic temperatures down to 30K, however do so at an increase in continuos power losses and a reduction in V_{BD}. 2) PT IGBTs show stable operation in regard to temperature and show a slight reduction in power loss at some temperatures; however they have an intrinsic cut-off temperature below which they will not function. 3) MOSFETs are the most suitable device for cryogenic operation. In this sample, some showed a reduction in $R_{ds(on)}$ of greater than five times. A reduction in V_{BD} was apparent but was no more than other types of semiconductors. The advent of high V_{BD} MOSFETs means this is not as significant a problem.

REFERENCES

1. Mueller, O., Properties of High-Power CryoMOSFETs, IEEE Conference Record of the 31st Industry Aplications Conference (1996) 3, 1443-1448.
2. Jackson, W.D., Mazzoni, O.S., Schempp, E., Characteristics of semiconductor devices at cryogenic temperatures, Proceedings of 31st Energy Conversion Engineering Conference (1996) 2, 676-681.
3. Rosenbauer, F., Lorenzen, H.W., Behaviour of IGBT Modules in the Temperature Range 5 to 300K, Advances in Cryogenic Engineering (1996) 41, 1865-1872.
4. Gutierrez-D, E.A., Jamal Deen, M., Claeys, C., Low Temperature Electronics: Physics, Devices and Applications, Academic Press, USA (2001) 105-259.
5. Caiafa, A., Wang, X., Hudgins, J.L., Santi, E., Palmer, P.R., Cryogenic Study and Modeling of IGBTs, IEEE Proceedings of 34th Power Electronics Specialist Conference (2003) 4, 1897-1903.

Proceedings of the Twentieth International Cryogenic Engineering Conference
(ICEC 20), Beijing, China. © 2005 Elsevier Ltd. All rights reserved.

Operational Amplifiers in the Temperature Range from 300 to 4 K

Nawrocki W., and Pajączkowski J.

Institute of Electronics and Telecommunications, Poznan University of Technology
ul. Piotrowo 3A, PL-60695 Poznan, Poland; E–mail: nawrocki@et.put.poznan.pl.

In this paper the amplifying properties of operational amplifiers (bipolar, Bi-FET and CMOS) in the function of temperature are presented and discussed. We measured operational amplifiers between 300 K and 4 K. The Bi-FET opamps work well in the temperature range down to 60-70 K. The CMOS operational amplifier (type of AD8594) operate correctly in the whole temperature range from 300 K to 4 K.

INTRODUCTION

Operational amplifiers are structural elements of many analog circuits. The ability of the above circuits and devices to work properly in a large range of environment temperature depends on the thermal properties of opamps. The low temperature range is divided into three sub-ranges: from 0 °C to −50 °C (from 273 K to about 220 K), from 220 K to 77 K and below 77 K. In this sub-range bipolar devices do not operate.

According to catalogue data, the sub-range below 220 K is no longer the nominal work range for operational amplifiers. The operating temperature range is limited because operational amplifiers consist bipolar transistors. Literature studies show that bipolar transistors are not able to operate at 77 K and below [1]. The authors of various papers state that at 77 K the current amplification factor β of commercial bipolar transistors is close to 1 or at most not greater than 10-20. There exist some exceptions. There were presented bipolar transistors manufactured by polyemitter bipolar processing technology, for which β even increases with the decrease of temperature from 300 K to 77 K [2]. The second exception is a bipolar transistor with a tunnel MOS emitter and an induced base [3].

However, many technical instruments should work in the low temperature range. Control systems and communications systems of aircraft and spacecraft must work at temperatures below 220 K, as well as meteorological instruments and a number of instruments in low temperature physics. In our research we considered commercial operational amplifiers. We determined an open-loop gain and an offset of input voltage in the function of temperature in the range from 300 K to 4 K. The aim of the research was to determine the lower limit of work temperature for all types of operational amplifiers.

Bi-FET AMPLIFIERS

Operational amplifiers always include bipolar transistors in their structure, because of the large value of amplification of the latter. Bi-FET opamps consist of an input stage with field effect transistors and of the remaining stages with bipolar transistors. Figure 1 shows amplification properties of a bipolar transistor (BC108C) and a FET transistor (BF245B) in the function of temperature. We measured a transconductance g_m for the FET because a value of g_m describes amplification of a JFET transistor.

$$g_m = \mathrm{d}I_D/\mathrm{d}V_{GS}$$

After cooling down the JFET from 300 K to 77 K its transconductance g_m increases twice (Fig. 1, right). We measured a β-coefficient of current amplification of a bipolar transistor BC108C (Fig. 1, left). The current amplification decreases from 600 at 300 K to 7 at 77 K.

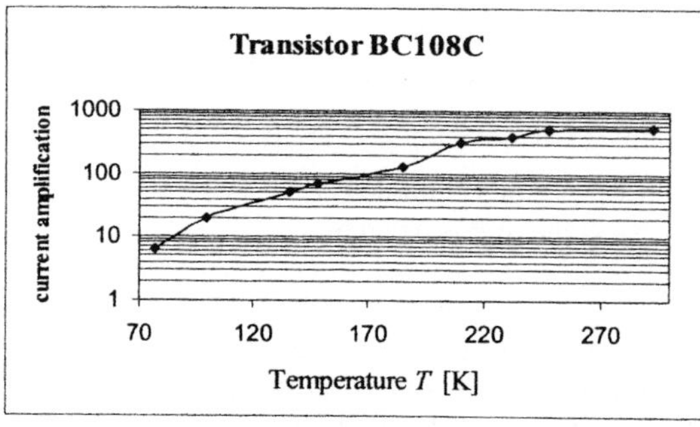

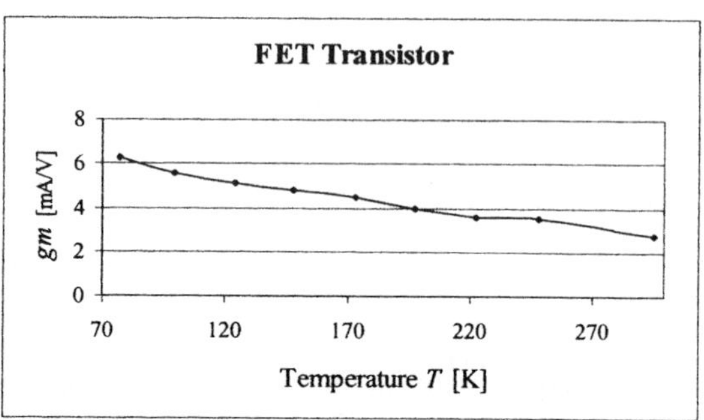

Figure 1 Amplification properties of transistors in the temperature range 77-300 K. At left: a transconductance g_m of FET transistor BF245 (on the right); a current amplification β of a bipolar transistor.BC108C (on the left)

We constructed and tested a 2-stage Bi-FET amplifier with the transistors described above. The amplifier circuit is presented in Fig. 2 (left). A bias of a base circuit of a bipolar transistor is realised using two diodes. Such method of a bias can stabilised a collector current in a bipolar transistor. A voltage gain of the amplifier is shown in Fig. 2 (right). The gain of the amplifier varied from 56 V/V to 80 V/V in the temperature range 77-300 K. The gain is satisfactorily constant for many applications. An input impedance of the Bi-FET amplifier is rather high because of a FET transistor in the input stage of the amplifier.

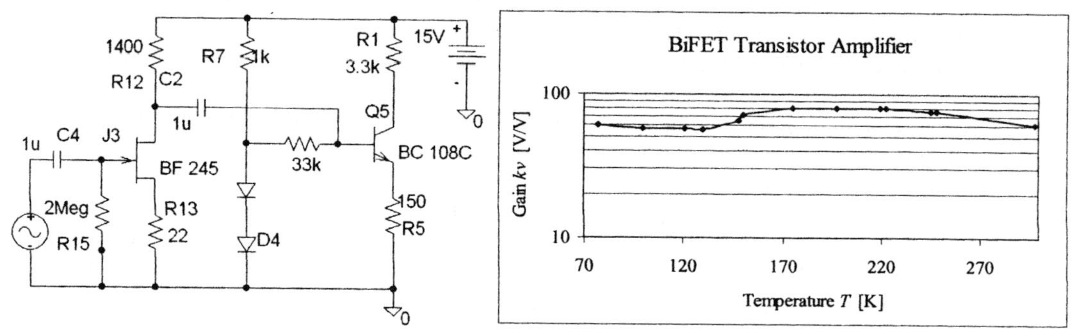

Figure 2 A circuit of the Bi-FET amplifier (on the left) and its voltage gain in the range 77-300 K (on the right)

BIPOLAR AND BI-FET AMPLIFIERS

Bipolar operational amplifiers

We took a general purpose operational amplifier μA741 and a ultralow noise operational amplifier AD797 (manufactured by Analog Devices) for investigation. We studied the amplification factor and the offset voltage of opamps in the function of temperature in the range from 293 K to 77 K. We assumed 100% amplification at room temperature. For all opamps examined, the amplification factor decreases steadily with the decrease of temperature below 150 K. Below 100 K the amplification factor decreases sharply and reaches a zero value. The offset of input voltage slightly changes in the temperature range from 293 K to about 100 K and increased sharply below 100 K.

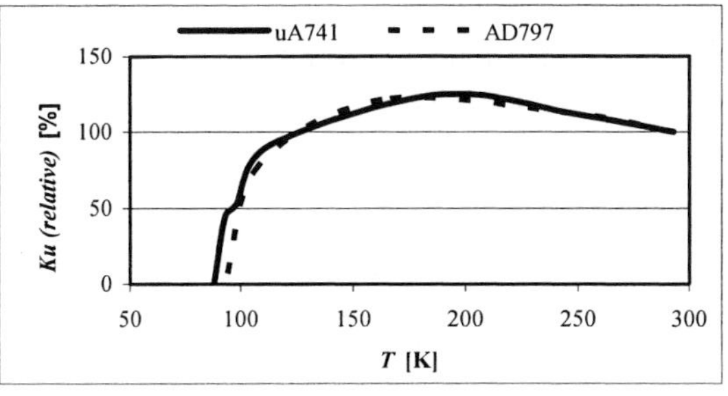
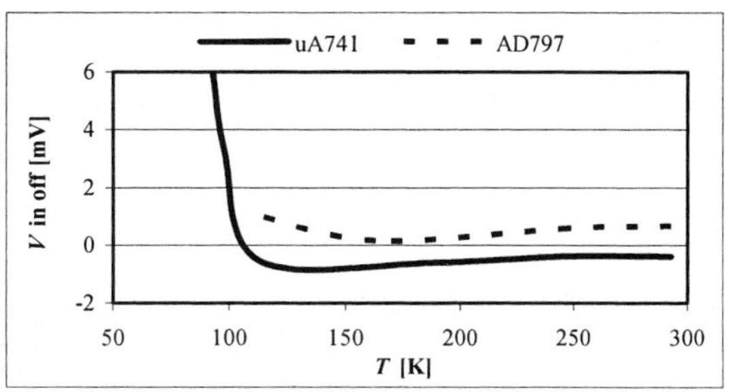

Figure 3 A relative amplification factor (on the left) and an offset of input voltage (on the right) in the function of temperature for two types of bipolar opamps (μA741 and AD797)

BiFET operational amplifiers

Bi-FET operational amplifiers show similar temperature properties to bipolar amplifiers. We investigated two types of Bi-FET amplifiers: TL082 (made by Texas Instruments) and LF411 (made by National Semiconductor). The same parameters, i.e. the amplification factor of an open-loop amplifier and the offset of the input voltage, were measured for Bi-FET opamps in the function of temperature. In the whole temperature range the opamps examined preserve their amplification properties. The relative amplification factor decreases from 100% at 293 K to about 40% at 77 K – Fig. 4. The maximum of the function $K_u = f(T)$ at about 100-150 K results from the highest mobility of electrons in silicon in this temperature range. The actual temperature at which the amplification is equal to 0 falls in the range 50 K – 70 K and differs for different samples of opamps. The offset of input voltage slightly changes in the temperature range from 293 K to about 100 K and decreased dramatically to −300 mV below 100 K.

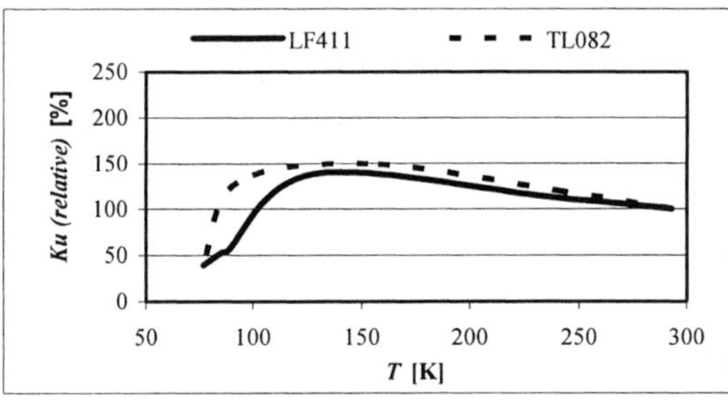
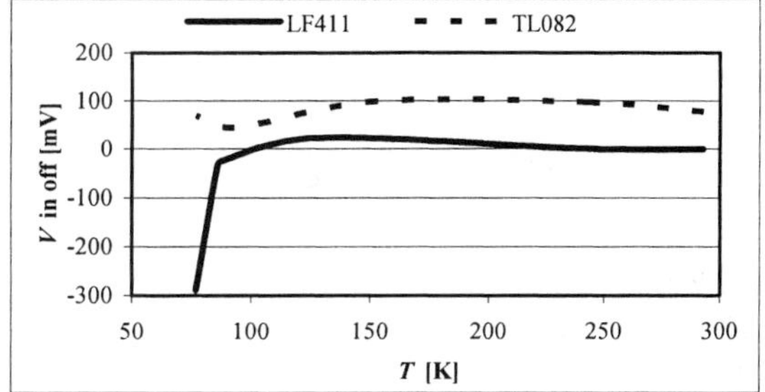

Figure 4 A relative amplification factor (on the left) and an offset of input voltage (on the right) in the function of temperature for two types of Bi-FET opamps (TL082 and LF411)

We also noticed some fluctuations of the offset voltage of the order of 1 mV. These fluctuations might be due to temperature fluctuations of the device environment. We must note that the introduction of a negative feedback component in the opamp circuit reduces dramatically the influence of the offset voltage on the output voltage.

CMOS OPERATIONAL AMPLIFIERS

We investigated an analog CMOS IC at low temperatures. It is known that CMOS digital integrated circuits are able for an operation in liquid He. We investigated an operational amplifier of a type of AD8594 (manufactured by Analog Devices) in the temperature range from 300 K to 4 K. The AD8594 device consists 4 operation amplifiers in one chip. According to catalogue data of AD8594 opamps the operating temperature range is −40 °C to +85°C, the junction temperature range is −60 °C to +150 °C, an offset of input voltage is 25 mV (maximal), a gain bandwidth product is 3 MHz (typical) – [4]. We measured two parameters: an open-loop gain and an offset of input voltage in the temperature range down to 4 K. We determined the open-loop gain indirectly, from measurements of a gain bandwidth product at several temperature points. The operational amplifier operated with a negative feedback (the gain with a feedback is of 500 V/V. The results of the measurements are presented in Fig. 5.

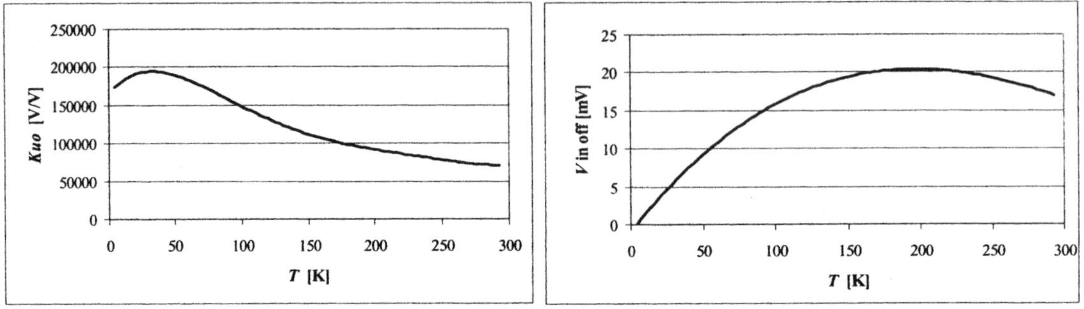

Figure 5 An open-loop gain (on the left) and an offset of input voltage (on the right) in the function of temperature for CMOS opamp of AD8594 type

CMOS opamp AD8594 not only keep its amplifying ability down to 4 K but its amplification factor is even greater at 4 K (open-loop gain is of 19×10^4 V/V) than at 300 K (open-loop gain is of 7×10^4 V/V). The offset of input voltage decreases below 200 K. Its value approaches zero at 4 K.

CONCLUSIONS

For bipolar operational amplifiers the amplification drops to zero at the environment temperature of about 90 K. Bi-FET operational amplifiers keep good amplifying properties in the range down to about 70 K. The CMOS operational amplifiers AD8594 not only keep its amplifying ability down to 4 K but its amplification factor is even greater at 4 K than at 300 K. It is possible to use CMOS operational amplifiers for instruments operated in a liquid He bath.

ACKNOWLEDGEMENT

Measurements on the CMOS operational amplifier below 77 K were carried out in Cryothermometry Laboratory of Institute of Low Temperature and Structure Research Polish Academy of Sciences in Wrocław, Poland. Authors thank for this cooperation.

REFERENCES

1. Lengeler, B. Semiconductor devices suitable for use in cryogenic environments, Cryogenics (1974) 14 439-447
2. Jayadev, T.S., Ichiki, S. and Woo, J.C.S. Bipolar transistors for low noise, low temperature electronics, Cryogenics (1990) 30 137-140.
3. Grekhov I.V., Schmalz K., Shulekin A.F., Tittelbach-Helmricht and Vexler M.I. Operation of a bipolar transistor with a tunnel MOS emitter and an induced base from 4.2 to 300 K, Cryogenics (1998) 38 613-618.
4. Analog Devices, National Semiconductor, Maxim, Dallas Semiconductor – Catalogues 2004

Development and test of a cryopulverizer for reprocessing plastic wastes

Jacob S., Kasthurirengan S., Karunanithi R., Upendra Behera, Nadig D. S.

Centre for Cryogenic Technology, Indian Institute of Science, Bangalore, India

The paper describes the development and test of an indigenous cryopulverizer for cryogrinding of PVC/polypropylene scraps and other waste plastics. This cryopulverizer has been designed for PVC scraps for a throughput of $\approx$60 kg/hr with about 85-90% of the ground product to be less than 212 μm. With this cryopulverizer, polypropylene balls of around 5 mm diameter have been ground successfully with a throughput of $\approx$35 kg/hr. Apart from PVC and polypropylene scraps, the above system will be useful for size reduction of many materials, which are difficult to grind at room temperature.

INTRODUCTION

The modern technology in milling process is cryomilling/cryogrinding. Cryogrinding is a useful technology for recycling of waste plastics for industrial applications. Cryogrinding of PVC scraps/polypropylene/waste plastics has lots of advantages over conventional grinding methods because of brittle fracture of these materials at low temperatures. Some of the significant advantages of cryogrinding over conventional grinding are: no chemical degradation of the product, finer particle size, higher throughput, lower energy consumption, no clogging and gumming of the mill, possibility to grind variety of materials which are difficult to grind by conventional techniques, inert atmosphere to prevent explosion in the grinding chamber, etc [1].

Cryogrinding is a process in which the plastic material to be ground is cooled to a desired temperature especially below the glass transition temperature by means of a cryogenic fluid such as liquid nitrogen (LN2) and ground in a low temperature compatible mill.

Three cryogenic cooling methods can be adopted for cryogrinding. They are: (i) the feed is cooled in a LN2 bath and then transported from the bath to the mill by means of a screw conveyor. The mill is cooled by the feed itself and also by the nitrogen vapour released during the cooling process. The process is grossly inefficient and the grinding temperature cannot be controlled optimally, (ii) LN2 is injected to cool-down the feedstock while it is transported through a screw conveyer. Here again, the chilled feeds as well as evaporated nitrogen cool the mill. The available enthalpy of the evaporated nitrogen is not fully utilized due to the absence of pre-cooling of the feedstock. In spite of the process inefficiency, this process can achieve all the advantages of cryogrinding and the plant is not complex, (iii) the stock is initially pre-cooled by the enthalpy of the evaporated nitrogen and the operating temperature of the mill can be controlled suitably for a specific application. Utilization of the cold gas circuit makes the process efficient. However the system becomes more complex.

The consumption of liquid nitrogen is an important factor to optimize the cost of cryogrinding. For continuous operation of the mill, the consumption of LN2 comprises of the following components:

(i)   Cooling of the raw material and mill (initially) to operating temperatures.
(ii) Compensating the heat liberated in the grinding process.
(iii) Thermal insulation loss of the grinding installation and cryogen systems.
(iv) Residual coldness of the ground product, which leaves the mill.
(v)   Residual refrigeration in the gaseous nitrogen being vented out.

Optimization of the above five parameters can result in the economical use of liquid nitrogen for cryogrinding.

OVERALL DESCRIPTION OF THE CRYOPULVERIZER

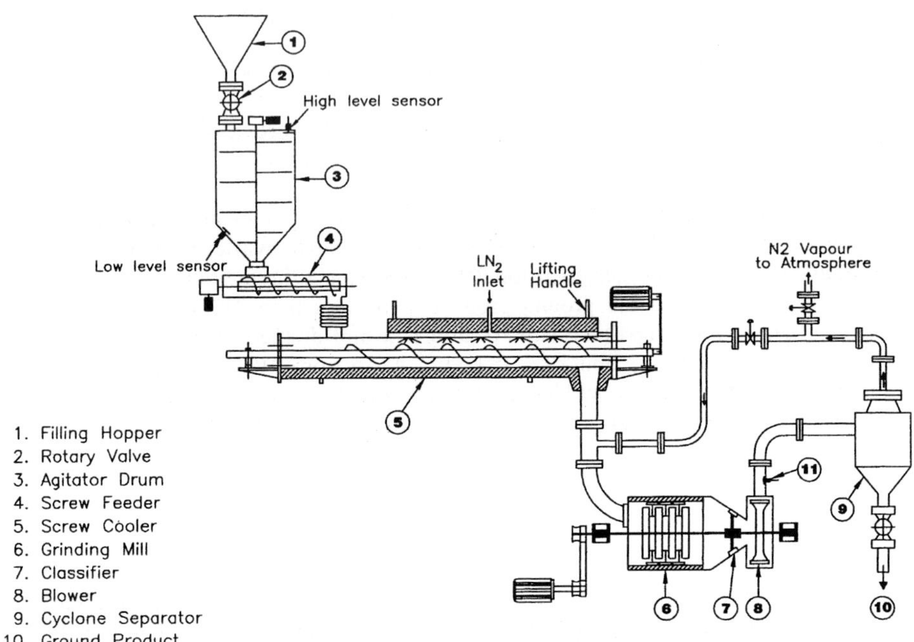

Figure 1   Schematic diagram of cryopulverizer

A cryopulverizer (Figure 1) has been designed and developed for the specific need of grinding ≈60 kg/hr for PVC scraps of less than 4 mm in size to obtain 85-90% of the ground product less than 212 μm which is required for manufacturing of battery liners.

The grinding mill is provided with 12 numbers of SS-304 hammers out of which 8 numbers are L-shaped hammers and 4 are T-shaped hammers. The cryopulverizer is totally designed and fabricated using stainless steel and materials compatible to liquid nitrogen temperature. The bearings of the cryopulverizer operate at room temperature. The gland seals of the cryopulverizer are provided with special arrangement such that ambient nitrogen vapour is supplied continuously through the gland seals so that no ambient moisture is sucked into the grinding chamber during grinding as well as warm up period [2]. A temperature sensor mounted on the blower exit is connected to the PID controller for regulating the liquid nitrogen supply by switching on and off of the solenoid valves in the liquid nitrogen supply line.

The working of the cryopulverizer can be described as follows. From the filling hopper (1), the material to be ground enters to the agitator drum (3) via the rotary valve (2) electric power of which is controlled by the low and high level sensors mounted in the agitator drum. A controlled rate of material is transferred from the agitator drum to the screw cooler (5) by the screw feeder (4), which has a provision to control the feed rate. While the material to be ground is transported in the screw cooler before entering the grinding zone of the mill (6), liquid nitrogen from the storage container is supplied through insulated transfer line and is sprayed on it to cool it to the desired low temperature preset by the PID controller. The material is ground to the desired particle size as per the adjustment of the classifier blades (7) in the mill housing. The ground product is separated from the cold nitrogen vapour in the cyclone separator (9) and is collected in the collecting bin. Part of the cold nitrogen vapour from the cyclone separator is recycled to the mill for utilization of enthalpy and the rest is vented to atmosphere.

In the present design, part of the cold nitrogen from the cryopulverizer is vented to the atmosphere for maintaining the system pressure. The enthalpy of this cold nitrogen vapour can be utilized for pre-cooling of the room temperature raw material by suitable heat exchanger to achieve optimal consumption of LN2.

EXPERIMENTAL STUDIES WITH THE CRYOPULVERIZER

Throughput and fineness
Experimental studies have been conducted with PVC scraps of less than 4 mm size and polypropylene balls of 5 mm diameter to determine the optimum speed of the mill for maximum throughput and particle size distribution. It has been observed that at around 4000 rpm mill speed and cooling the PVC scraps to approximately to -90°C, the desired throughput ($\approx$60 kg/hr) and fineness (85-90% of the ground product less than 212 μm) are achieved.

Experimental studies on cryogrinding of polypropylene balls of 5 mm diameter show that this material has to be cooled to -130°C and ground at a mill speed of 3400 rpm to achieve a throughput of 35kg/hr. The particle size distribution of the cryoground PVC and polypropylene are shown in table 1.

Table 1 Particle size analysis of cryoground products

Partice size (μm)	Residual weight (%) (PVC powder)	Residual weight (%) (Polypropylene powder)
>1000	0.0	0.5
850 –1000	0.4	7.3
500 – 850	3.3	31.5
300 – 500	2.8	23.7
212 – 300	5.9	15.1
150 – 212	10.5	7.0
106 – 150	34.8	10.1
<106	42.3	4.8

LN2 consumption
From the experimental studies it has been estimated that approximately 2.5 kg of LN2 is required per kg of PVC ground product and 3.5 kg of LN2 is required per kg of polypropylene ground product. However, there is a great scope to reduce the LN2 consumption by incorporating pre-cooling section using waste cold nitrogen vapour from the mill [2].

Electrical power consumption
Typical power consumption of the cryopulverizer for achieving 60 kg/hr PVC grinding is approximately 15 kW-hr.

CONCLUSION

A cryopulverizer that could be operated down to liquid nitrogen temperature (-196°C) has been designed, developed and tested for cryogrinding of PVC scraps and polypropylene balls. The throughput of this cryopulverizer is $\approx$60 kg/hr for PVC scraps with fineness of 85-90% of the ground product less than 212 μm and a throughput of 35 kg/hr for popypropylene balls. The cryopulverizer has been used successfully to grind 4 tons of PVC scrap so far. The technology has been transferred to a leading industry in India.

A remarkable advantage of the developed cryopulverizer is the absence of fiber in the ground product in comparison to conventional grinding wherein considerable fibers are generated by the high grinding zone temperature causing chemical degradation of the plastics. It is very difficult to grind polypropylene to fine powder in conventional grinding mills. Apart from PVC and polypropylene materials, the cryopulverizer will be quite useful for size reduction of many materials, which are difficult to grind by conventional techniques.

ACKNOWLEDGEMENT

The authors sincerely thank the staff of Centre for Cryogenic Technology, Indian Institute of Science, Bangalore for their help in setting up, operation and experimental studies of the cryopulverizer for reprocessing of plastic wastes.

REFERENCES

1. Jacob S., Kasthurirengan S., Karunanithi R. and Upendra Behera, Development of pilot plant for cryogrinding of spices: A method for quality improvement, Advances in Cryogenic Engineering (2000), 45 1731-1738
2. Jacob S., Kasthurirengan S., Karunanithi R., Upendra Behera and Nadig D. S., Cryopulverizer and a method of cryogrinding, Indian Patent Application No. 64/CHE/2004

A cryostat with optics windows for study of ICF cryogenics target

Liu Y., Zhu Y. Q., Zhou G., Jiang C.

Cryogenic Laboratory, Huazhong University of Science and Technology, Wuhan 430074, China

An optical cryostat used for fabricating inertial confinement fusion (ICF) cryogenic target is described. The cryostat holds a target cell. With the temperature control system, a desired thermal gradient vertically crossing the cell can be formed. The specifications of the optical cryostat are presented.

INTRODUCTION

Gravity can affect the distribution of the liquid fuel layer inside the target. This result the layer on the target's bottom is thicker than that on the top.   In the ICF technology, one key is to produce a cryogenic target with an evenly distributed liquid fuel layer inside it. Utilizing thermal gradient vertically crossing the target to overcome the gravity is an effective method [1]. The manufactures of this kind of cryogenic target call for a cryostat of a special design, which must satisfy some requirements. First, it can provide a low-temperature environment within 10~40 K [2]. Second, it must hold a target cell, in which an ICF target can be loaded. Third, it must possess a temperature control system that can be utilized to create a desired thermal gradient vertically crossing the target cell. Fourth, it must have optical windows. Through these windows, holographic interferometer can be used to measure the distribution of the liquid fuel layer inside the target.

THE STRUCTURE OF THE CRYOSTAT

The cryostat diagram is shown in Figure 1. A liquid helium tank (4) is suspended from the upper flange of the cryostat by two thin-wall stainless steel tubes ($\Phi15\times0.3$ mm). One of these tubes is acted as liquid helium input pipe (1); the other is gaseous helium output pipe (14). The level detector (16) can be inserted into the tank through the output pipe. The liquid helium tank is enclosed by liquid nitrogen circular bath (5), which is hanged from the mid flange by four stainless steel tubes ($\Phi6\times0.8$ mm). By this way, the bath can decrease greatly the incoming radioactive heat from inner wall of the dewar flask to the liquid helium tank. The target cell can be set under the liquid helium tank. There is a device between the tank and the cell, which can be utilized to promptly adjust the position of the cell by three directions. With the refrigeration supplied by the helium tank, the cell can be frozen to a desired temperature. A copper screen (9) used to shield the target cell is connected with liquid nitrogen bath, and it is able to intercept radiation between the cell and dewar flask's inner wall. Of course, there are holes in the screen in order the light beam can go through. Three metal (1Cr18Ni9Ti) cylinders (12,16,10) comprise the dewar flask. Several windows are set in the lower part cylinder of the dewar. Some are used as optical windows (8) and another can be used as adjustment windows (7). The lower cylinder is connected with middle one by screws, so it is easy to be detached. This brings convenience to the replacement of the target. The dewar is a evacuated sealed body [3] Using high-vacuum, the dewar can seal a low-temperature environment in it.

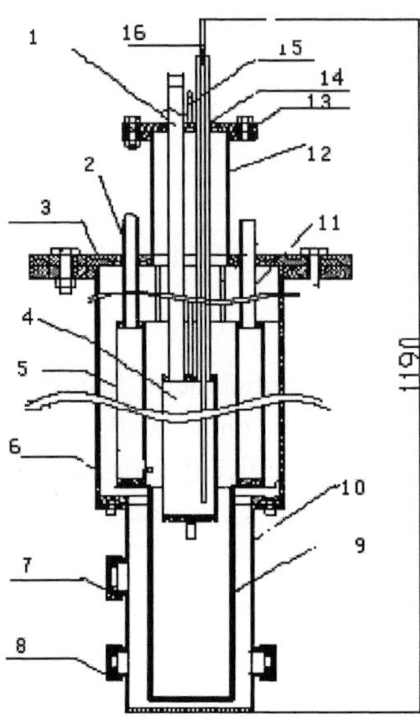

1. LHe input pipe
2. LN input pipe
3. mid flange
4. LHe tank
5. LN circular bath
6. mid cylinder
7. direction-adjust window
8. optical window
9. copper screen
10. lower cylinder
11. GN output pipe
12. upper cylinder
13. upper flange
14. GHe outlet pipe
15. GHe pipe to target cell
16. level detector

Figure 1 the cross-sectional view of
the cryostat

TARGET CELL WITH TEMPERATURE CONTROL SYSTEM

The target cell is shown in Figure 2. The cell is an evacuated sealed body too. The walls of the cell are made of cryogenic glue. The target is placed near the center of the cell and horizontally supported by a thin glass fiber glued to the end of a nylon screw (3). The cell has two optical windows, so the measure light beam can go through. The top and the bottom of the cell are made of copper plates (5). Two heat conduct rods respectively connect these plates with the liquid helium tank. Each plate has a heater (6) and a temperature sensor (1) fitted into. The temperature sensor is Cryogenic Liner Temperature Sensor (CLTS). [4] Thus through controlling the electrical power of each heater, the refrigeration, which is conducted by heat conduct rod to each plate, can be adjusted respectively. If the sizes of these two plates are chosen properly,

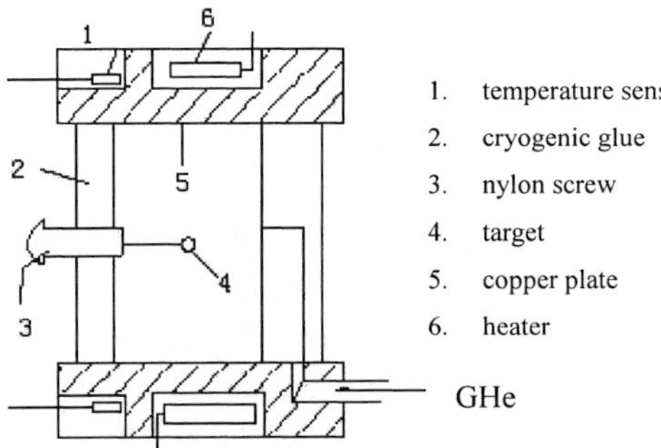

1. temperature sensor
2. cryogenic glue
3. nylon screw
4. target
5. copper plate
6. heater

GHe

Figure 2   Cross-sectional view
of the target-freezing cell

each plate can reach at a stable temperature desired.

In the process of the experiment, gaseous helium ($10^{-4} \sim 10^{-5}$ Pa) as the heat exchange gas will be filled into the cell. Driven by the top plates temperature and the bottom's, a stable thermal flux going through from the top to the bottom can be formed, thus a desired thermal gradient, which vertically cross the inner space of the cell, can be built.

SPECIFICATIONS

Capacity of liquid helium	0.9 L
Capacity of liquid nitrogen	4.3 L
Operating temperature	10K~40 K
The temperature gradient achieved	0~20 K/cm
Stability of operating temperature	±0.1 K
Power of the heater	0~10 W
Operating time	10 h

ACKOWNLEGEMENT

This work is supported by National Natural Science Foundation of China.

REFERENCES:

1. Miller .J. R., A new method for producing cryogenic laser fusion targets, <u>Advanced in Cryogenics Engineering</u> (1987), <u>23</u> 669~674

2. P. Clark Souers. <u>Hydrogen properties for fusion energy</u>. Los Angeles, USA: University of California Press.1986

3. Kim. K., Mok. L., and Bernat. T. P. et al, Non-contact thermal gradient method of fabrication of uniform cryogenic inertial fusion target, <u>Journal of Vacuum Science and Technology</u> (1985), A 3(3) 1196~1200

4. J.J. Winter. Variable temperature multimode magnetometer. <u>Rev.Sci. Instr.</u> (1978), <u>49</u> 845~849

Proceedings of the Twentieth International Cryogenic Engineering Conference
(ICEC 20), Beijing, China. © 2005 Elsevier Ltd. All rights reserved.

Temperature effects on formation of a uniform layer of isotopes inside a cryogenic ICF target

Zhou G., Zhu Y. Q., Liu Y., Jiang C., Li Q. [*]

Cryogenic Laboratory, Huazhong University of Science and Technology, Wuhan 430074, China
*Technical Institute of Physics and Chemistry, Chinese Academy of Science, Beijing 100080, China

The steady-state motion of a gas bubble inside a non-isothermal, spherical, liquid filled cryogenic target is investigated by taking into account the effects of gravity, the thermally induced gradient of the gas-liquid interfacial tension and the finite size of the liquid container. The net result is an expression for the temperature gradient at the target exterior, which will sustain a uniform liquid layer of hydrogen isotopes inside an ICF target. A simple model was established on the basis of the calculation and analysis above.

INTRODUCTION

An optimal configuration for ICF targets is a spherical shell containing a uniform layer of fusion fuel condensate on the interior surface. Such targets, in particular those containing a thick fuel layer, are difficult to fabricate because the fuel sags due to gravity, thus making the condensate layer thicker at the bottom of the target than at the top [1].

The non-contact thermal gradient method [2] is one such technique that is designed to counteract the gravity induced fuel sagging. This technique employs a vertically imposed temperature gradient across the target. In particular, the magnitude and direction of this gradient are chosen such that the thermally induced migration of the liquid fuel may precisely counterbalance its slumping due to gravity, bringing about a uniform liquid fuel layer on the inner surface of the target.

The purpose of this work is to study the steady-state motion of a gas bubble inside a spherical micro-shell under the influence of gravity, a thermally induced gradient of gas-liquid interfacial tension, and the finite size of the micro-shell--a situation frequently encountered in fabricating high-compression inertial confinement fusion (ICF) targets.

THEORY

To conveniently describe the thermally induced behavior of hydrogen isotopes inside a spherical ICF target, the system under investigation is divided into four regions. Starting from the innermost region they are the fuel vapor, liquid, spherical glass shell (SGS), and helium gas regions, and all considered concentric spheres. The helium exchange gas envelope responsible for cooling the target is assumed to be quiescent.

Both the vapor and liquid of the hydrogen isotopes are assumed to be Newtonian and incompressible. The flow fields are assumed to be laminar, and azimuthally symmetric. The thermodynamic and transport properties, namely, density, viscosity and thermal conductivity are defined at average target temperature, which is nearly the temperature at the target equator since the temperature gradient across the target is

738

linear and in the vertical direction. Also we assume that there is no mass transfer between the liquid and the gas. Another assumption is that the bubble has already attained its terminal velocity when it becomes concentric with the micro-shell.

Calculation about the fuel thickness of ICF target

According to the mass conservation in cryogenics target, the total mass in the cryogenic target is constant whether the status of the fuel is liquid or gas. When the infilling processing is completed, the whole mass in cryogenic target is

$$\rho_f V_0 = \rho_l(T)V_c + \rho_g(T)V_g \tag{1}$$

where ρ_f is the density of the initial gas. ρ_l (T) and ρ_g (T) are respectively densities of condensed and saturated vapor fuel. V_0, V_c, V_g are total volume, the condensed and the saturated vapor volume in the target, respectively.

If δ =W/R, here R and W is respectively the inner radius and the fuel thickness of the target, then

$$\rho_f = \rho_l(T)\left[1-(1-\delta)^3\right]+\rho_g(T)(1-\delta)^3 \tag{2}$$

It is obviously that when the initial gas density and the temperature T are constant, we can get the δ, then the thickness of fuel in target. To demonstrate this relation more clearly, we figure out the relation curve between the relative volume of vapor bubble and temperature of normal hydrogen (see Figure 1).

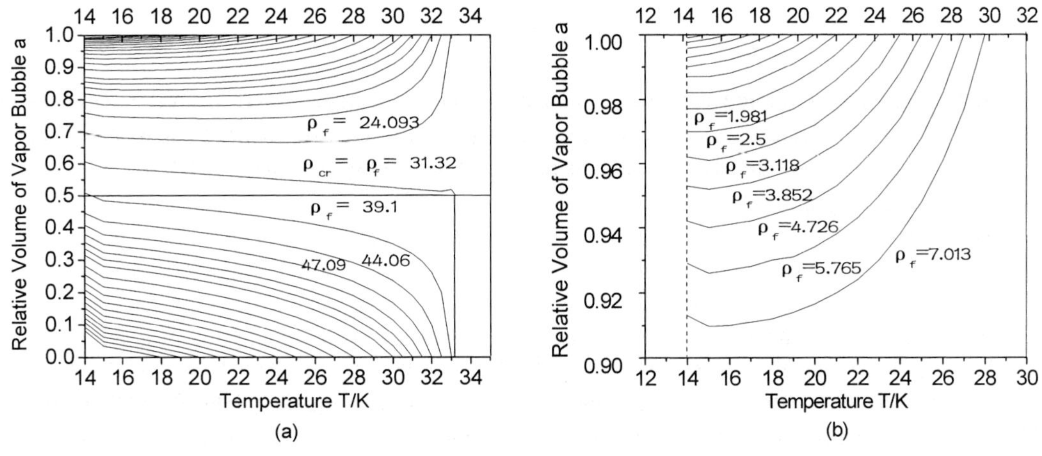

Figure 1.   The relation curve between relative volume of vapor bubble and temperature of normal hydrogen(ρ_f g/cm^3)

Governing equations of fluid field

Since the fluid flows are azimuthally symmetries, in terms of the Stokes stream function ψ defined as $\upsilon_r = (-1/r^2 \sin\theta)\partial\psi/\partial\theta$, and $\upsilon_\theta = (1/r\sin\theta)\partial\psi/\partial r$, the low Reynolds number approximation of the fluid equation in the spherical coordinate system is

$$E^2\ (E^2\ \psi_1)\ =0 \quad \text{and} \quad E^2\ (E^2\ \psi_g)\ =0 \tag{3}$$

where

$$E^2 = \frac{\partial^2}{\partial r^2} + \frac{\sin\theta}{r^2}\frac{\partial}{\partial\theta}\left(\frac{1}{\sin\theta}\frac{\partial}{\partial\theta}\right), \quad r=\tilde{r}/R_B, \quad \psi=\tilde{\psi}/(R_B V_{ref}),$$

V_{ref} is the reference velocity, and R_B is the radius of the gas bubble. The title denotes that the variables are expressed in real units, υ_r and υ_θ are the r component and the θ component of the flow respectively. The general solution to Eq.(3) also can be obtained in terms of the Happel & Brenner [3].

At $r=b^{-1}$: $\upsilon_{r,l}=-U\cos\theta$, $\upsilon_{\theta,l}=U\sin\theta$,

where U is the steady-state velocity of the bubble, $b=R_B/R$.

at $r=1$: $\upsilon_{r,l}=\upsilon_{r,g}=0$, $\upsilon_{\theta,l}=\upsilon_{\theta,\,g}$ and $\mu_l\left[\dfrac{1}{r}\dfrac{\partial V_{r,l}}{\partial\theta}+\dfrac{\partial V_{\theta,l}}{\partial r}-\dfrac{V_{\theta,l}}{r}\right]-\mu_g\left[\dfrac{1}{r}\dfrac{\partial V_{r,g}}{\partial\theta}+\dfrac{\partial V_{\theta,g}}{\partial r}-\dfrac{V_{\theta,g}}{r}\right]=-\dfrac{\partial\gamma_{gl}}{\partial\theta}$

where μ is the viscosity of the fluids, and the γ_{gl} is the gas-liquid interfacial tension.

at r$=0$: $\psi_g(0,\theta)$ is finite.

Following Haberman & Sayre (1950) [4], the drag force on the bubble is given by

$F_d=-4\pi\mu_l R_B D_n\big|_{n=2}$, then

$$G_g + F_d + F_b=0 \tag{4}$$

where G_g, F_b respectively stand for the weight and buoyancy forces.

After a straightforward, yet tedious, algebra, the expression for the steady-state velocity U of the bubble is obtained as

$$U=\left[\frac{2}{3\mu_l}\frac{R_B^{\,2}(\rho_c-\rho_g)g}{K_2}-\frac{1}{6}I_2(2-5b^3+3b^5)\right]\left[3+2\sigma-3(1-\sigma)b^5\right]^{-1} \tag{5}$$

where $\sigma=\mu_l/\mu_g$.

The function I_2 in (5) represents the effect of the interfacial tension gradient on the velocity of the bubble. Because the interfacial tension gradient results from the non-isothermal temperature field, if the temperature gradient across the target is conformed, the I_2 then can be estimated as follow:

$$I_2=-\frac{3}{\mu_g}\int_0^\pi\frac{\partial\gamma_{gl}}{\partial\theta}\sin^2\theta d\theta$$

also the velocity of the bubble can be gotten.

Governing equations of temperature field
Due to the fact that Reynolds numbers of the liquid and gas flows are low and that the Prandtl number for liquid hydrogen is approximately unity, which allows one to ignore the convection term in the heat equation and therefore, reduces the heat equation for the fluids to

$$\frac{1}{r^2}\frac{\partial}{\partial r}\left(r^2\frac{\partial}{\partial r}\right)+\frac{1}{r^2\sin\theta}\frac{\partial}{\partial\theta}\left(\sin\theta\frac{\partial}{\partial\theta}\right)T=0 \tag{6}$$

This is an exactly the Laplacian equation.

According to the continuities of the temperature fields:

At $r=d^{-1}$ $T_W(\frac{1}{d},\theta)=T_{ext}(\theta)$; at $r=b^{-1}$ $T_W(\frac{1}{b},\theta)=T_l(\frac{1}{b},\theta)$; at r$=1$ $T_l(1,\theta)=T_g(1,\theta)$; at r$=0$,

$T_g(0,\theta)$ is limited, where $d=R_B/R_0$, R_0 is the outer radium of the target.

Also according to the condition of the heat fluxes:

At r$=1$, $k_1\dfrac{\partial T}{\partial r}\Big|_{r=1}=k_g\dfrac{\partial T}{\partial r}\Big|_{r=1}$; at $r=b^{-1}$, $k_w\dfrac{\partial T}{\partial r}\Big|_{r=\frac{1}{b}}=k_1\dfrac{\partial T}{\partial r}\Big|_{r=\frac{1}{b}}$

Then we can get the temperature field of the target.

740
DISCUSSIONS

Let the temperature field on the outer surface of the spherical shell be specified as

$$T_{ext}(\theta)=T_0+R_0\sum_{n=1}^{\infty}t_n P_n(\cos\theta) \tag{7}$$

We note that for a small temperature change across the target, the interfacial tension can be expressed as a linear function of the temperature T at the liquid-gas interface:

$$\gamma(T)=\gamma_0+\gamma_1 T \tag{8}$$

where γ_0 is the interfacial tension at some reference temperature and γ_1 is a constant. The constant γ_1 represents the magnitude of the rate of change in surface tension with respect to temperature.

Therefore I_2 can be evaluated analytically, and the temperature gradient needed to hold a bubble stationary is

$$(t_1)_s=-\frac{2}{3}\frac{N_2-d^3}{N_2-b^3}\frac{\left[1+2\varepsilon_1-(1-\varepsilon_1)b^3\right]}{\varepsilon_1\sigma K_2(2-5b^3+3b^5)}\frac{R_B(\rho_l-\rho_g)g}{\gamma_1} \tag{9}$$

CONCLUSIONS

The steady-state migration velocity of a gas bubble located at the center of a spherical shell has been calculated by considering the combined effect of gravity, interfacial tension gradient, and the finite size of the shell. An analytical expression for the temperature gradient that will sustain a stationary bubble at the center of the spherical shell has been derived. Also a calculating function relation about liquid H_2 thickness of cryogenic target is proposed using the mass conservation in cryogenic target. The analytic method for the model is also fit for the D_2, T_2 or the mixture of the D-T.

It must be pointed out that the present work does not include the corresponding release and absorption of heat in the processes of condensation and evaporation, also not consider the isotope effects such as the differences in the thermodynamic properties and transport coefficients of the isotopic species. A theory intended to include factors above is currently being formulated and will be reported in a future publication.

ACKOWNLEGEMENT

This work is supported by National Natural Science Foundation of China.

REFERENCES

1. Miller.J.R, A new method for producing cryogenic laser fusion targets, <u>Advanced in Cryogenics Engineering</u> (1987), <u>23</u> 669~674

2. Kim. K., Mok. L., and Bernat. T. P. et al, Non-contact thermal gradient method of fabrication of uniform cryogenic inertial fusion target, <u>Journal of Vacuum Science and Technology</u> (1985), <u>A 3(3)</u> 1196~1200

3. Happel.J, Brenner.H. <u>Low Reynolds Number Hydrodynamics.</u>2nd ed. Martinus Nijhoff Publishers, Boston (1983)

4. Haberman. W.L., Sayre. R.M., Motion of rigid and fluid sphere in stationary and moving liquid inside cylindrical tubes. <u>David W. Taylor Model Basin Rep No.1143</u>, Washington DC, 1950

Optimization Analysis of Peakshaving Cycle to Liquefy the Natural Gas

Shi Y., Gu A., Wang R., Zhu G.

School of Mechanical Engineering, Shanghai Jiao Tong University, Shanghai 210003, PR China

In this paper, three kinds Peakshaving LNG cycles are optimized, namely nitrogen-methane expansion liquefaction cycle, propane pre-cooled mixed refrigerant liquefaction cycle, two stages mixed refrigerant liquefaction cycle. The three kind cycles are optimized with specific power consumption of unit LNG product as objective function. The optimum results are the optimum value of the objective function and the corresponding process parameters.

INTRODUCTION

Peakshaving liquefied natural gas (LNG) plays an important role in peak loads and increasing the reliability of gas supply.

In this paper, three kinds Peakshaving LNG cycles are optimized. The three LNG cycles are nitrogen-methane expansion cycle, propane pre-cooled mixed refrigerant cycle (MRC), two stages mixed refrigerant cycle. The specific power consumption of unit LNG production is used as the objective function [1].

LNG CYCLE

N_2-CH_4 Expansion Liquefaction Cycle

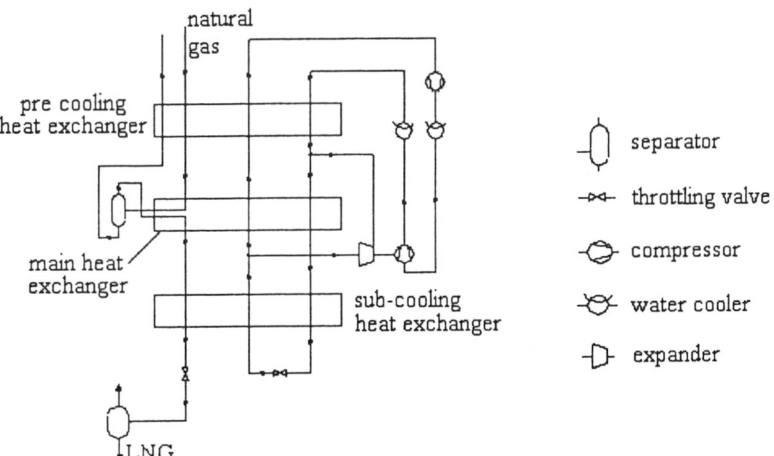

Figure 1 N_2-CH_4 expansion liquefaction cycle

Figure 1 is the N_2-CH_4 expansion liquefaction cycle. The purified natural gas is cooled in A1, then the liquid is separated in A2. The gas is cooled and liquefied in A3, sub-cooled in A4, then flows through the throttle valve. The produced Liquefied natural flows into storage tank.

Propane Pre-cooling Mixed Refrigerant Liquefaction Cycle

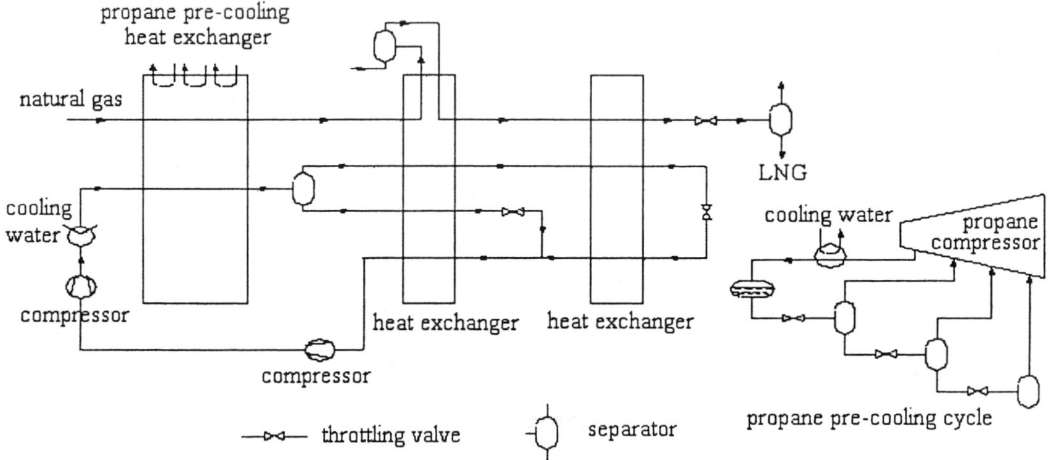

Figure 2 C_3H_8 pre-cooling mixed refrigerant liquefaction cycle

As shown in Figure 2, C_3H_8 pre-cooling mixed refrigerant liquefaction cycle mainly consists of 3 loops: two closed refrigerating cycles, that is, C_3H_8 pre-cooling cycle which is used to pre-cool natural gas and refrigerant mixture and mixed refrigerant refrigerating cycle which is used to condense and sub-cool natural gas; one open cycle-natural gas liquefaction cycle.

Two Stages Mixed Refrigerant Liquefaction Cycle

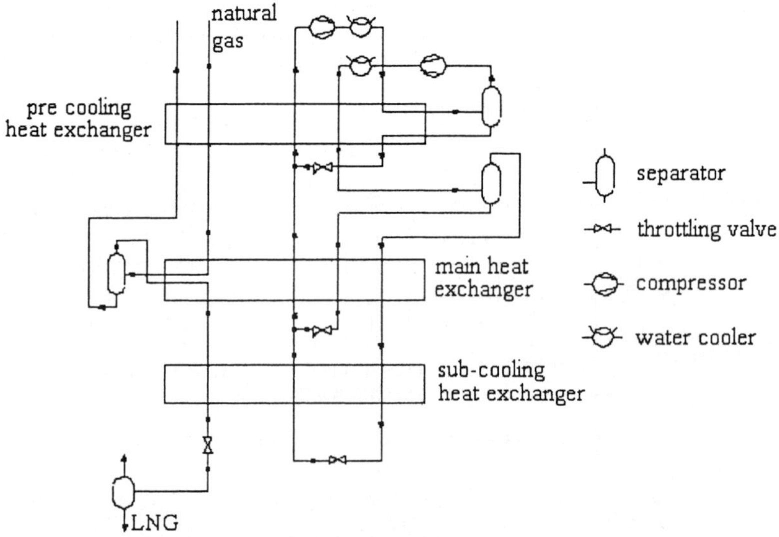

Figure 3   Two-stages mixed refrigerant liquefaction cycle

Figure 3 is the two-stages mixed refrigerant liquefaction cycle. This cycle includes refrigerating cycle and natural gas cycle.

OPTIMIZATION ANALYSIS OF CYCLES

The sequential modular is used to calculated the thermodynamic parameter of the three LNG liquefaction cycle showed as figure 1~3. The optimization analysis is based on the thermodynamic simulation.

Objective Function
During optimization analysis of three kinds of cycles shown in Figure 1-3, the power consumption per unit LNG product is chosen as the objective function of optimization.

Known Parameters
Inflow rate of natural gas: 1kmol/s
    Pressure of natural gas: 4.8MPa
    Components and mole percentage of natural gas: N_2 (0.7%), CH_4 (82%), C_2H_6 (11.2%), C_3H_8 (4%), i-C_4H_{10} (11.2%), n-C_4H_{10} (0.9%);
    LNG storage pressure: 0.12MPa

Iterative Step
In the optimization calculation, the pressure iterative step is 100kPa, the iterative step of mol percentage is 0.02, and the temperature iterative step is 2K.

Restricted Conditions
In the optimization calculation, the restricted conditions are as follows:
    (1) The sum of mol percentage is equal to 1.
    (2) The mixed refrigerant flowing into the compressor is vapor phase only.
    (3) The temperature of refrigerant gas after expansion is higher than the dew point of vapor in low pressure circuit.
    (4) The liquid and vapor must be produced when mixed refrigerant flows through every separator.
    (5) The temperature difference between hot and cold fluid in every heat exchanger should not be minus.
    (6) The energy of every heat exchanger must be balance.

Optimization Results
In N_2-CH_4 expansion liquefaction cycle, the designed variables and gained optimization values are as follows: the high pressure of refrigerant is 4400kPa; the low pressure is 600kPa, the mol percentage of N_2 and CH_4 is 0.56 and 0.44 respectively, the temperature before expansion is 244.2K, the temperature of natural gas at cold side of main heat exchanger is 157.15K.

In C_3H_8 pre-cooling mixed refrigerant liquefaction cycle, the designed variables and gained optimization values are as follows: the high pressure of mixed refrigerant is 2550kPa, the low pressure is 310kPa, the mole percentage of N_2, CH_4, C_2H_6, C_3H_8 is 0.02, 0.42, 0.41, and 0.15 respectively.

In the two stags mixed refrigerant liquefaction cycle, the designed variables and gained optimization values are as follows: the high pressure of mixed refrigerant is 4260kPa, the low pressure is 360kPa, the mole percentage of N_2, CH_4, C_2H_6, C_3H_8, n-C_4H_{10}, i-C_4H_{10}, n-C_5H_{12}, i-C_5H_{12} is 0.052, 0.246, 0.295, 0.204, 0.048, 0.0554, 0.0486 and 0.051 respectively, the temperature of separator S1 is 269.75K, the temperature of S2 is 243.35K.

<u>Comparison of Peakshaving Liquefaction Cycles</u>
Based on the optimization results of liquefaction Cycles, the parameters standing for the performance of the three kinds of peakshaving liquefaction natural gas cycles could be calculated, shown as Table 1.

Table 1   Comparison of three liquefaction cycles

Liquefaction Cycles	Natural gas flow rate (kmol/s)	Refrigerant flow rate (kmol/s)	Cooling load of natural gas (kW)	Compressor power consumption (kW)	Liquefaction rate	Unit power consumption (J/mol)
C_3H_8 pre-cooling MRC	1.0	1.463	13945.3	14260.5	0.904	20069.9
N_2-CH_4 expansion LC	1.0	4.749	14080.0	28635.5	0.923	39434.4
Two stages MRC	1.0	3.138	14201.6	21934.4	0.941	29642.0

CONCLUSION

Three kinds of peakshaving natural gas liquefaction cycles are optimized in this paper, the specific power consumption of unit LNG product is used as the objective function.

The results of optimization analysis indicate the power consumption of C_3H_8 pre-cooling mixed refrigerant liquefaction cycle is the least and power consumption of N_2-CH_4 expander liquefaction cycle is the largest.

Based on the optimization results, we can get conclusions as follows:

(1) N_2-CH_4 expansion liquefaction cycle is compact, flexible and with good applicability but its power consumption is large, so it's suitable for small and medium-sized LNG system, which is used to recycle natural gas resources of marginal gas field.

(2) C_3H_8 pre-cooling mixed refrigerant liquefaction cycle has complicated equipments, its power consumption is low so it is suitable for large-sized, base load LNG system.

Two stages mixed refrigerant liquefaction cycle is simple and flexible, its power consumption is low and the liquefaction rate is high, so it's suitable for small and medium-sized peakshaving LNG system.

REFERENCE

1. Zhu G., The Optimization Research of Peakshaving liquefaction cycle and the transfer properties of natural gas, Shanghai Jiaotong University, Shanghai, (2000)

The experimental study on the performance of a small-scale oxygen concentrator by psa

Zhang Y.W., Wu Y.Y., Gong J.Y., Zhang J.L.

Institute of Cryogenic Engineering, Xi'an Jiaotong University, Xi'an 710049, P.R.China

A small-scale adjustable rich oxygen concentrator by pressure swing Adsorption (PSA) was designed and manufactured to study influences of nozzle size, purge quantity on characteristics of the production oxygen, such as its pressure, purity and volume flux. The purity of oxygen produced by this device could reach above 90% and up to 96% in volume. The flux of rich oxygen product could be adjusted intelligently in the range of 0.5l/min~3.5l/min. The device is quite usable in hospital, domestic application, oxygen bar and other places in shortage of oxygen. It would have great economic benefit in the future.

INTRODUCTION

The basic principle of producing oxygen from air by PSA is based on the fact that the adsorbed nitrogen quantity of Q_n by adsorbents at higher pressure is larger than it at lower pressure [1,2,3]. The operation can be performed as the following processes [4,5]. Firstly, the pressurized, dry and filtered feeding air flows into the adsorption column which is filled with 5A zeolite molecular sieves (ZMS). Then most of nitrogen will be adsorbed by 5A ZMS and most of oxygen flows through the column. In this paper, a small-scale oxygen concentrator is constructed to investigate the effects of the operating parameters on the performance of the device so as to optimize them [4,5].

Nomenclature

Q_O The flow of production oxygen, L/min
Q_P The flow of purge quantity, L/min
Q_e The flow of oxygen- enriched air exiting out of adsorption column, L/min
Q_a The flow of air feeding into adsorption column, L/min
Q_n The flow of adsorbed nitrogen in adsorption column, L/min
P The high plateau adsorption pressure (HPAP) , MPa.

EXPERIMENTAL ARRANGEMENTS

The experiments are performed to investigate how to get either large flow of production oxygen or high purity of it by adjusting HPAP and purge quantity of Qp. The schematic drawing of the experimental system is shown in Figure 1.Firstly the feed air is passed through the air filter 1 to remove dust and other solid substances, then enters the compressor 2 to be pressed to the required HPAP. Finally, it flows into the adsorption columns 5 to produce oxygen-richened air. Under the controlled by the control system 7, Two columns of A and B, operate alternately to be controlled by the controlling system 7. While A is adsorbing, B is desorbing at the same time, then the valve V1 will be opened and valve V2 be closed with V3 closed and V4 opened .The high- pressure air flows into column A via V1 and the component of nitrogen in the air is adsorbed in column A. Then one part of the oxygen-richened air is supplied to the consumer as product oxygen of Qo. Another part of the oxygen-richened air will act as the purge gas of Qp to enter the column B. At last the purge gas is discharged into atmosphere via valve V4 and via muffler 6.

In next step, the column B will be in adsorbing and the column A will be in desorbing with the similar process .The flow and purity of the production oxygen are measured by 8 and 9 respectively.

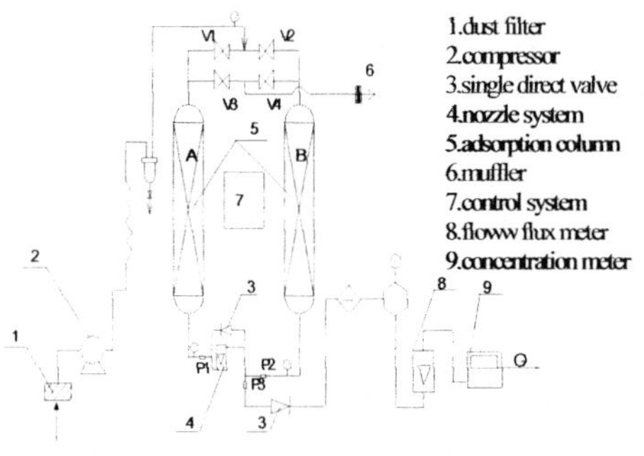

Figure 1 Experimental system

The testing condition are as follows:
The dimension of the adsorption column (two): 82 mm (diameter); 650 mm (height);
Adsorption period: 10 ~ 30 seconds (adjustably);
Void fraction: 0.32 ~ 0.47;
Temperature: ambient temperature;
Adsorbent: 5A Molecular sieve (d=1.6~2.5mm);
Experimental pressure: 0.11Mpa ~ 0.28 MPa.
New packing arrangement for molecular sieves.
An important improvement has been made in packing arrangement for molecular sieve in the two adsorption columns.
The traditional packing arrangements are uniform packing by using only one kind of molecular sieve.
Namely, the same size of diameter of one kind of molecular sieve is uniformly distributed in the adsorption column. The disadvantage of this uniform packing is insufficient usage of molecular sieve for whole column. It is caused by the mal-distribution of flow velocity in the column. The maximum velocity lies in the central core of the column. While the velocity outside the central core, it is much smaller. Therefore the molecular sieve in central core is firstly saturated by adsorbing enough nitrogen and thus firstly penetrated by feed-air. However the molecular sieve outside the central core is still not saturated, and is not sufficiently used. In order to use the molecular sieve sufficiently, the flow velocity in the central core must be reduced in comparison with the traditional packing.

The new packing arrangement is called multi-layer and multi-section packing method. The small diameter ofmolecular sieves is packed in the central core while the larger diameter of sieve lies in the outer layers outside of the central core. According to previous experimental results the arrangement of layers will be adjusted so as to obtain

Figure 2 New packing arrangements in radial and axial direction

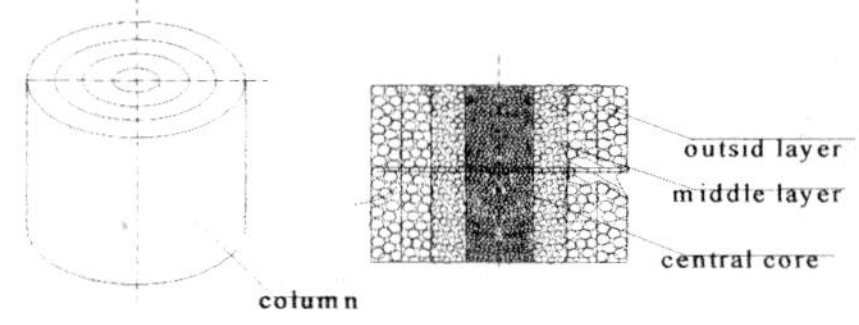

Figure 3 Adsorption quantity of Qe related to P

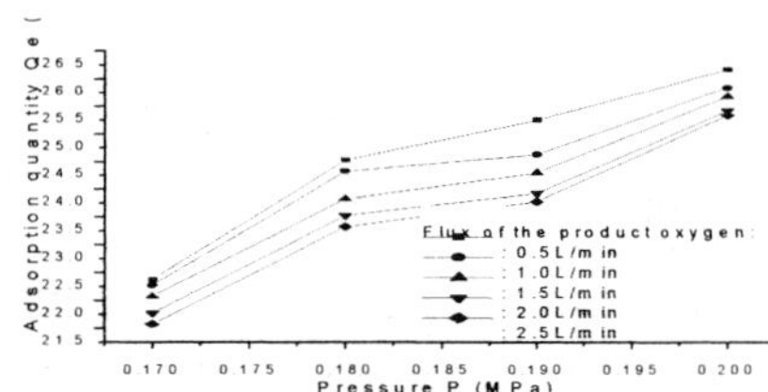

appropriate flow resistance and flow velocity distribution at last. In our experiment, there are three layers in radial direction and two sections in axial direction of the column, which is shown in Figure 2. The size of molecular sieve in the central core, in outside layer and in middle layer are of 1.6 mm, 2.5mm, and 2mm , respectively. By this new packing arrangement the flow velocity distribution has been improved greatly.

EXPERIMENTAL RESULTS AND DISCUSSIONS

The influence of HPAP on Q_e

The discharge flux out of adsorption column of Q_e, that is total flux of high purity oxygen, is equal to the flux of production oxygen of Q_O plus the flux of purge quantity of Q_P. The Q_e is related to the HPAP as shown in Figure3.The relationship between the discharge flux of Qe and HPAP shows the same tendency as the typical

Langmuir curve. The results also indicate that the higher the volume flux of the production oxygen of Qo, the lower the Q_e in the condition that the HPAP was kept constant, and the flow of feed air of Qa was also kept constant. When the flux of the production oxygen is higher the purge quantity of Qp will drop. The quantity of the purge gas of Qp has decisive effect on the process of regeneration of adsorbent. Thus the regeneration of the adsorbent in the adsorption column is not sufficient due to the insufficient quantity of the purge gas of Qp. The adsorption ability will be decreased. It will directly lead to the decrease of the effective adsorption, that is, the decrease of Qe. Therefore, the Q_e decreases with the increase of the volume flow of the product oxygen when the purity of production oxygen and the HPAP were kept constant.

<u>The effect of HPAP on the flux of production oxygen Qo</u>
Figure 4 also shows the relationship between the flow of the production oxygen and the purity of the production oxygen with different adsorption pressure. Only in case the pressures were kept in medium range both the purity of production oxygen and the flux of production oxygen can obtain optimum performance. Neither the higher nor the lower HPAP would make the performance of the device better according to the Figure 4.The main reason is as follows. When the HPAP is below a minimum value, the adsorbent will have a little adsorption ability. It will directly lead to the drastic drop of the purity of the production oxygen according to basic theory of PSA. Inversely, when the HPAP is too high, the purge quantity will decrease based on above analysis. When the purge quantity is so small to be less than the quantity required by the regeneration of the adsorbent, the adsorption ability will be decreased. It will cause the decrease of the effective adsorption so as to decrease the purity of the production oxygen. Therefore the HPAP should be in an appropriate range to ensure both the high flow flux and purity of the product oxygen at the same time. The appropriate HPAP is found to be 0.17Pa to 0.23Pa according to the Figure 4. The corresponding nozzle diameter is from 0.7mm to 1.2mm.

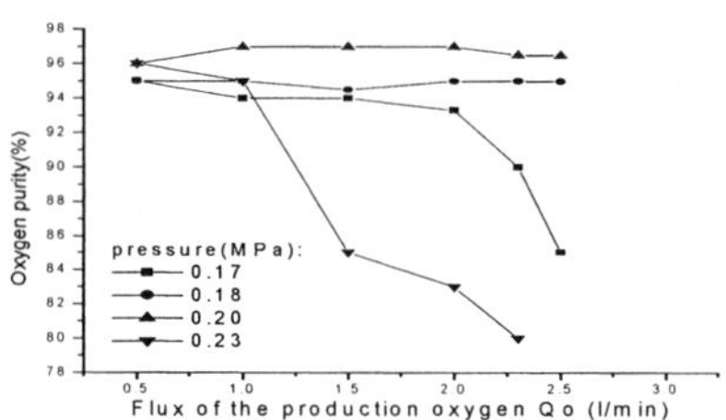

Figure 4 The purity of oxygen related to Qo

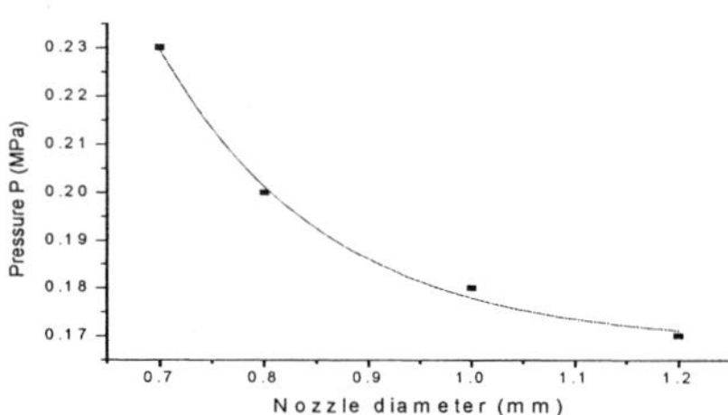

Figure 5 Pressure related to nozzle size

<u>The effect of nozzle size on HPAP</u>
A series of nozzles are designed and manufactured to ensure the stable and effective operation of the small device. The nozzle size is related to the adsorption pressure as shown in the Figure 5 .The smaller the nozzle diameter is, the higher the HPAP is. The higher HPAP can be attained in short adsorption time by the small diameter nozzle so that the energy consumption will be reduced. According to the experimental results in Figure 5 the right size of nozzle can be selected in accordance with the required HPAP.

<u>The effect of the purge quantity Qp on the purity of oxygen</u>
With the nozzle size unchanged the purity of the production oxygen is related to the purge quantity as shown in the Figure 6. When the purge quantity Qp is moderate, the high purity of the product oxygen Qo can be obtained. Too small purge quantity leads to the decrease of the regeneration of the adsorbent in the adsorption column and cause the remarkable drop of the purity of the production oxygen Qo. On the other hand, when the purge quantity Qp is too high to exceed the quantity required by the adsorbent, the excessive quantity of pressurized oxygen- richened air is filled into the adsorption column. It will cause a higher pressure in the adsorption column to reduce the adsorbing ability of the adsorbent. Therefore the purity of the production oxygen is also reduced.

<u>The effect of the purge quantity Qp on the flux of production oxygen Qo</u>

Figure 7 shows the relationship between the flow of the purge quantity Qp and the flow of the product oxygen Qo for different nozzle sizes when the purity of the production oxygen is kept above 93% in volume. The higher the flow of the production oxygen Q_O, the lower the purge quantity Q_P, when the nozzle size is kept constant. As is known, the flow of the product oxygen Q_O subtracted from Q_e gives the purge quantity Q_P. Therefore the purge quantity Q_P decreases with the increase of the flow of the production oxygen Q_O when the HPAP is kept constant. Only when the ratio of the purge quantity Qp to the flow of the product oxygen Qo is maintained above 2.5 shown in the curve of the nozzle 0.8mm in Figure 7, the device can be ensured to operate properly.

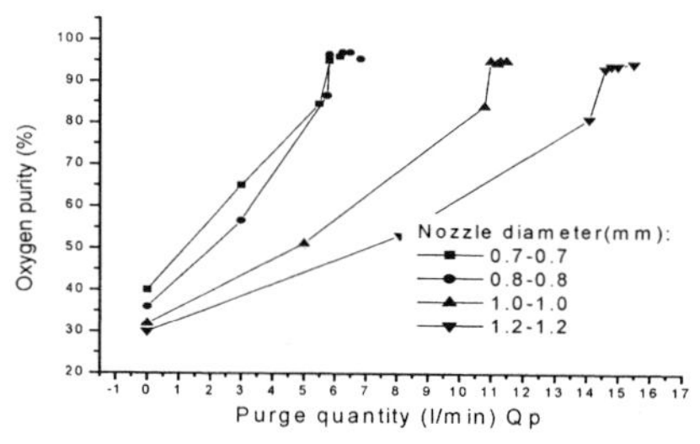

Figure 6 Oxygen purity related to Qp

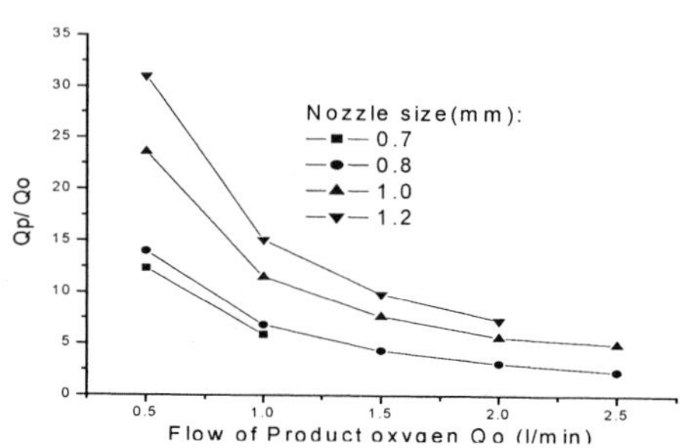

Figure 7 Qp/Qo related to Qo

CONCLUSIONS

(1) It is found that the small-scale concentrator could operate in optimum performance, with the purity of oxygen being above 90%. The relationship between the HPAP and the Q_e is a typical curve of Langmuir. (2) The optimal operating pressure of the experiment device is 0.18Mpa to 0.2MPa. (3) The ratio of the purge quantity Qp to the flow of the product oxygen Qo is maintained above 2.5. (4) The smaller the nozzle diameter, the higher the HPAP. The optimal nozzle diameter is from 0.8mm to 1.2mm while the adsorption time is 10 seconds (5) The highest purity of the production oxygen is about 96% in volume by using this small-scale concentrator while the flow of the production oxygen is of 2.0 l/min. The small-scale concentrator is quite suitable for hospital, domestic usage and oxygen bar.

REFERENCES

1. Han Hui, and others he study of pressure swing adsorption, Physical experiment (2000), 4 16~18, in Chinese
2. Zhu Xuejun, Guo Tong. The mathematic simulation to medical oxygen-enriched by pressure swing adsorption, Iatrical device (1998), 616~18,in Chinese
3. Yang Chunyu. The mathematic simulation to the pressurization and depressurization step during separating air by PSA, Petroleum and chemical engineering (1998), 5 335~340, in Chinese
4. Feng Jinzhe and others. The semi-experiential formulation of the relationship between the pressure and oxygen concentration in the rich oxygen device by using PSA method, Cryogenics engineering (1997), 5 12~17, in Chinese
5.Critten Guan j,W.N.Ng,W.J.Thomas. Dynamics of pressurization and depressurization during pressure swing adsorption. Chem. Eng.Sci (1994), 49 2657

The thermodynamic analysis and adjusting methods for reliquefaction plant on board LEG carriers

Zaili Zhao, Ailin Jiang, Linglong Xiong

School of Energy and Power Engineering, Wuhan University of Technology, Hubei 430063, China

The real capacity of the reliquefaction plant on board the ethylene carrier depends on the knowledge and skills of the ship cargo engineer to a great extent. Aimed at optimum operation, thermodynamic analyses are made on refrigerating cycles of the plant. The maximum cooling rate mode and the optimum energy-conservation mode are brought forward. Ethylene tank temperature is forecasted; optimum-operating mode is chosen and the division point is determined. Meanwhile, various influencing factors and working parameters of the plant are discussed in the paper.

INTRODUCTION

LEG (liquefied ethylene gas) carrier is a semi-pressurized ship for loading and transporting the cargo of ethylene. Normally, the storing temperature of ethylene is within 169.5-175 K on board. The daily ethylene evaporation rate is about 0.2% - 0.3% of their weight in sailing. The vapor not only causes the tank pressure rise, but also discharge problems because most of the terminals demand the ethylene discharge temperature (T_D) to be kept below 170.5 K. So the reliquefaction plant onboard should run at good performance and in correct operation to maintain the T_D as required.

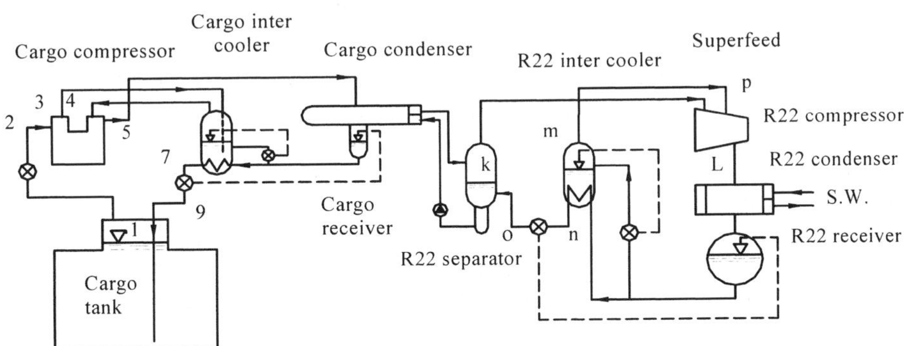

Figure 1   LEG reliquefaction plant system

The system on board MV Norgas Energy is shown in Figure 1. A LEG carrier has three sets of the plant (3 units), whose total power is about 1000 kW. There are two working cycles in each plant, their cooling processes are plotted into a Mollier diagram (see Figure 2).

If ethylene temperature is not low enough when carrier arrives at a terminal, the carrier has to anchor for cargo cooling down; as a result, the ship owner will lose about 10,000-14,000 dollars a day. In order to avoid this, the plant should run at the maximum cooling rate mode (MCRM). But if there is a long voyage or a very low cargo-loading temperature (T_c), the only needs are to keep the tank temperature constant and make the plant run at minimum energy consumption, here termed as optimum energy-conservation mode (OECM). Until now, no LEG carrier has distinguished the above two different operating modes, and a lot of energy has been wasted.

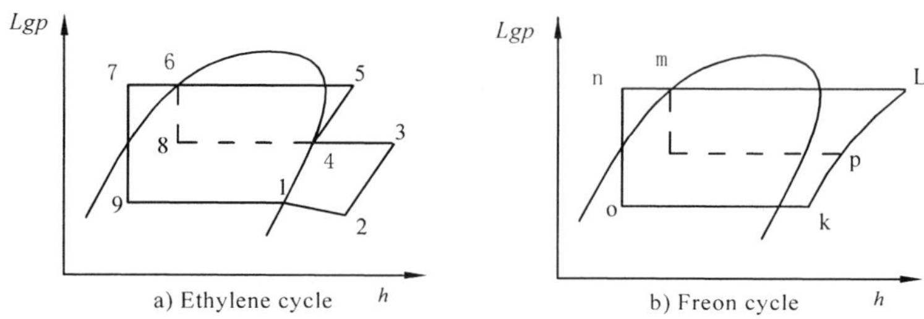

Figure 2 Refrigeration cycles

THERMODYNAMIC ANALYSIS OF THE PLANT

The working parameters of the plant

The key thermodynamic parameters of liquid & vapor include saturation pressure and temperature, specific volume, enthalpy, entropy, exergy, available energy, etc.. Taking the system in Fig.1 for an example, and from the running data of the plant onboard during its 158th voyage, ethylene property table and the data calculated, the parameters of each point in Fig.2 have been derived, and some of them are shown in Table 1.

Table 1 Working parameters of the plant

Running time	T_1 (K)	P_1 (Mpa)	T_2 (K)	P_2 (Mpa)	Enthapy$_2$ (KJ/kg)	Entropy$_2$ (KJ/kg·K)	T_7 (K)	P_7 (Mpa)	Enthalpy$_7$ (KJ/kg)	Entropy$_7$ (KJ/kg·K)
1st day	172.1	0.1185	172.186	0.118	507.922	3.349	205.495	20.195	109.963	0.628
2nd day	171.4	0.115	171.594	0.114	507.423	3.361	203.147	20.02	104.251	0.613
3rd day	170.8	0.112	171.019	0.111	506.937	3.347	201.101	19.51	99.235	0.598

Note: 1. The data are based on seawater temperature 301 K and ambient temperature 306 K.

2. Subscripts are consistent with points in Fig. 1 & 2.

Thermodynamic analysis

Tab.1 shows that tank pressure (P_1) and tank temperature (T_1) obviously change with the plant running. That is to say, the plant working condition varies all the time. According to Mollier diagram and regarding refrigeration coefficient (ε) as goal function, the thermodynamic calculation has been carried out (computational process omitted). When three sets of plant cool down three tanks filled with a total of 3000 tons of ethylene, the ethylene cooling curve in tank and instantaneous ε of the plant are shown in

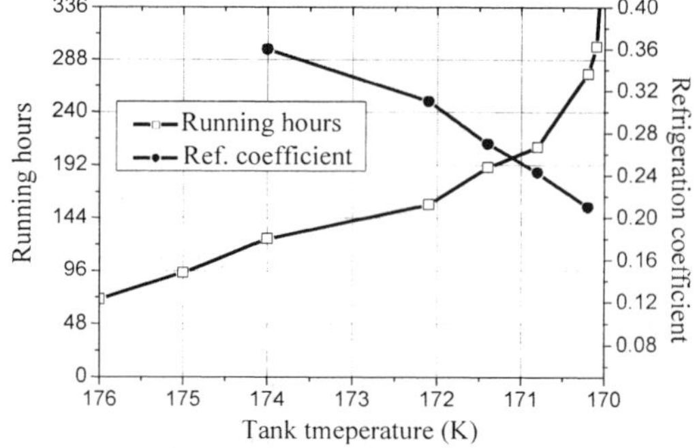

Figure 3 Cooling curve & refrigeration coefficient

Figure 3. In the Fig., the ε value is relatively low when T_1 is below 172 K. It shows OECM should be used after the running hours corresponding to 172 K.

In Fig.3, both the ε and temperature drop per day in tank descend with T_1 decreasing. That means from the beginning, MCRM should be used and then followed with OECM. For a whole voyage, one or both of the operating modes may be adopted according to different sailing conditions. So there are three different operating modes for the plant, i.e. MCRM, OECM, and the combination of MCRM and OECM.

TEMPERATURE FORECAST AND DIVISION POINT DETERMINATION

Operating mode and T_1 forecast

In order to choose the best one of the three operating modes, T_1 should be forecasted. The forecast model in the grey system theory can be used here. A simple program for T_1 forecast has been developed [1]. The basic equation of GM (1,1) used is as follows:

$$\hat{x}^{(1)}(k+1) = \left(x^{(1)}(0) - \frac{u}{a} \right) e^{-ak} + \frac{u}{a} \qquad (1)$$

Table 2   Forecast temperature & real temperature

Running hours	3 units T'_1 (K)	3 units T_1 (K)	2 units T''_1 (K)	1 unit T^o_1 (K)
0	174.25	174.25	174.25	174.25
24	173.18	173.01	174.16	174.51
48	172.04	172.10	174.08	174.80
72	171.36	171.40	173.96	175.11
96	170.72	170.80	173.89	175.43
120	170.12	170.14	173.83	175.80

Where $\hat{x}$ means T_1 sequence and a & u are endogenous variable.

The computed results of GM (1,1) for above system are shown in Table 2 (computational process omitted). According to the table, the forecasted tank temperature (T'_1) is very close to real T_1. So the varying tendency of T_1 can be determined after a short period of the plant running.

If only two sets of plant (2 units) are used, the tank temperature (T''_1) will descend slowly, which may rise when ambient temperature (T_a) and cargo amount increase or no boost compressor is used. If only one set of plant (1unit) is used, the temperature (T^o_1) will rise obviously (the heat leakage from ambience into the tank is more than the capacity of a plant). When one or two sets of plant are used, it is usually run at OECM.

After forecasting T_1, the enthalpy, entropy, exergy and system power consumption in each point can be pre-calculated. As a result, the best one among the 3 operating modes can be chosen.

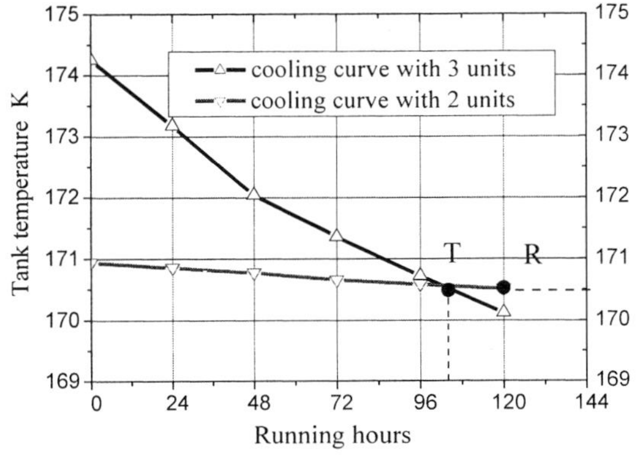

Figure 4   Division point determine

Determination of the division point

The number of the plant to be used depends on T_c, T_a, capacity of cooling plant, length of voyage and insulation of cargo tanks. Normally, three sets of plant at MCRM should be given to the highest priority.

If the combination of MCRM with 3 units and OECM with 2 units is used for a voyage, the transfer time should be at a properly moment (termed as division point). The division point can be determined as follows (see figure 4): By T_c, the cooling curve in tank with 3 units is first forecasted. By considering both the T_D and sailing days of a voyage (point R), the curve of T_1 with 2 units can also be forecasted. Their intersection point (T) will be an optimum division point.

REGULATING METHODS AT DIFFERENT OPERATING MODES

Regulating methods of condenser

There are two condensers in one set of plant (Fig.1), i.e. cargo condenser and R22 condenser. According to Lu Xuesheng's calculation [2], the available energy loss in a condenser is relatively large. For reducing the energy loss, the T_6, T_m and ΔT (difference in temperature) between hot and cold liquids should be reduced. The basic equation for condenser heat transference is as follows:

$$Q = KA(t_k - t_w) = KA[t_k - (t_{w1} + \frac{1}{2}\Delta t_w)]\tag{2}$$

In Eq. 2, increasing of condensation temperature (t_k) and decreasing of average seawater temperature (t_w) will make Q greater and MCRM applicable; decreasing of t_k and increasing of t_w will save energy and make OECM applicable. The t_w in R-22 condenser can be controlled by adjusting seawater (SW) current; the t_w in cargo condenser can be controlled by adjusting freon superfeed current.

For R-22 condenser, decreasing of P_m will reduce ΔT between R-22 and SW. For cargo condenser, increasing T_o (corresponding to P_o) will slightly change the value of ΔT between ethylene and R-22 since T_6 will automatically match with T_o. The ΔT between ethylene and R-22 normally is 2~4 K. If ΔT exceeds 6 K, it means some troubles have occurred in the condenser.

Regulating methods on OECM and MCRM

Besides the regulating methods of condenser, the following should be noted.

If MCRM is used, the plant should be adjusted as follows: (1) With P_1 falling, increase the opening of the cargo compressor suction valve step by step. If the suction valve is not opened properly in time, the compressor will be impossible to run at full capacity. (2) Keep the superfeed current full open. (3) Keep P_S, P_L, T_S & T_L as high as possible. (4) Use boost compressor for making P_1 steady, if any.

If OECM is used, the plant should be adjusted as follows: (1) Properly choose the number of plant to be used. (2) Properly determine the division point, if necessary. (3) Close the superfeed current gradually, while the cargo plant running under partial load. (4) Adjust SW current smaller than that of MCRM.

During the operation of all the modes, the following should be adjusted: (1) Keep the compressor satisfactorily cool. (2) Keep the condenser at good performance and clean the SW side of R-22 condenser in time. (3) Keep R-22 compressor at 100% of capacity all the time. (4) Purge void space with dry air at appropriate intervals to protect the insulation from water or humidity.

ACKNOWLEDGMENTS

This work was carried out under Engine Department of Norgas Energy and active support of Norwegian Gas Carriers AS.

REFERENCES

1. Zhao Zaili, The Temperature Forecast Based on Grey System Theory in Cargo Tank of LEG Vessel, Journal of Wuhan University of Technology (2003) 27 655-658
2. Lu Xuesheng, Study on Reliquefaction Plant in LPG Carrier, Low Temperature Engineering, China (1999) 4 310-314

A study of the optimum temperature for hydrogen storage on Carbon Nanostructures

Zheng Q.R., Gu A.Z., Lu X.S., Lin W.S.

Institute of Refrigeration and Cryogenics, School of Mechanical and Power Engineering, Shanghai Jiao Tong University, 1954 Huashan Rd., Shanghai, 200030, PRC

The energy of intermolecular interaction is used to probe into the optimum temperature for hydrogen storage by adsorption on carbon nanostructures. Thermodynamic analysis is undertaken based on the lattice theory to the adsorption data of hydrogen on Multi-walled Carbon Nanotubes (MWCNTs) over a temperature range of 123-310 K and pressure up to 12.5 MPa. Results show that the hydrogen-hydrogen interaction energy captures characteristics of physical adsorptions of supercritical gases; almost linearly increases with increases of adsorption temperatures and surface loadings. However, the result cannot reveal much information about the optimum temperature for hydrogen storage in the MWCNTs.

INTRODUCTION

Recently a great number of reports of molecular hydrogen on a novel class of graphitic materials, carbon nanostructures, which include Carbon Nanofibers, Single-walled Carbon Nanotubes (SWCNTs) and Multi-walled Carbon Nanotubes (MWCNTs), have attracted a lot of attention by media and the automotive industry [1]. However, from the presented literatures, progresses achieved seem not so great as supposed to be, there still have such unclear points as the maximum adsorption capacity and the interaction mechanism of hydrogen molecules on these carbon nanostructures [2].

The work reported here is to study the temperature dependent state of hydrogen molecules on MWCNTs at a temperature range from 123-310 K. It was undertaken for several reasons. Firstly, as mentioned above [1-2], most of the hydrogen adsorption data in recent literatures needs reasonably theoretical explanations [3]. The second stimulus for us to do this work is still lack of hydrogen adsorption data on carbon nanostructures in continuous variations of temperature and pressure over a wide range. Thirdly, systematic analyses, which include the hydrogen-hydrogen interaction energy and the hydrogen-carbon interaction potential as well as the isosteric heat of adsorption, may much cogently evaluate the adsorption capacity of the adsorbent and clarify the state of adsorbed hydrogen molecules. Besides these, the research is also beneficial to locating the optimum storage temperature for hydrogen on the carbon nanostructure [3].

EXPERIMENT

Isotherms of excess adsorption amount of hydrogen on the MWCNTs were volumetrically measured in our lab, resulted isotherms are shown in Fig. 1; the pore size distribution (PSD) function $f(r)$ of the MWCNTs determined by the adsorption isotherm of N_2 on the MWCNTs at 77 K is shown in Fig. 2. The detailed information about the experiment can be referred to [3].

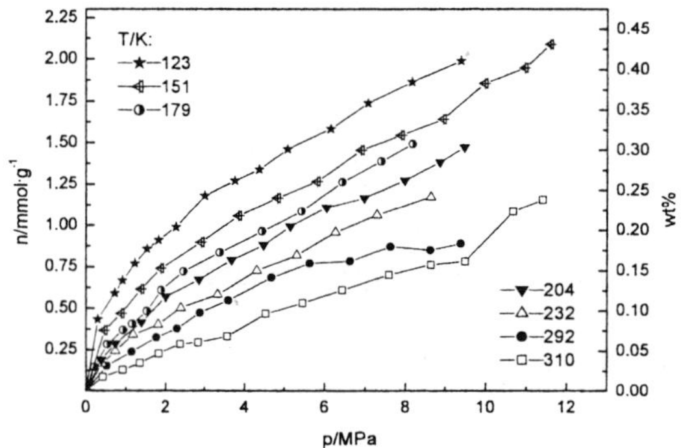

Figure 1   Isotherms of excess amount of hydrogen
adsorption on the MWCNTs

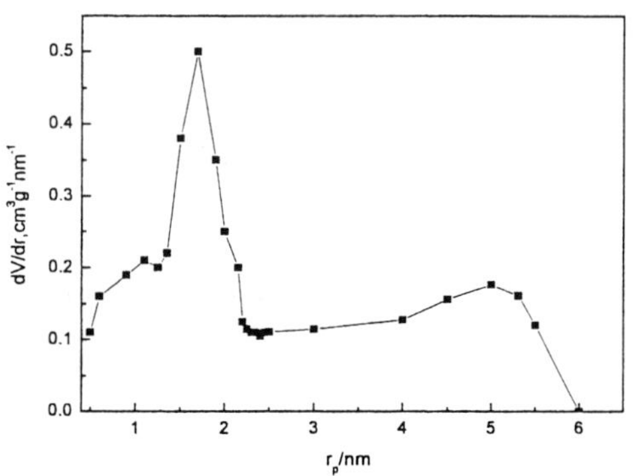

Figure 2   PSD of the MWCNTs determined by the
analysis of nitrogen adsorption isotherm at 77 K

ADSORPTION EQUILIBRIUM OF HYDROGEN ON MWCNTS

Determination of the monolayer coverage

Based on the lattice theory [4], as discussed in the paper [5], if the adsorbate-adsorbate interaction energy and the adsorbate-adsorbent interaction potential are included and calculated by the Lennard-Jones potential function, one can get an approximate adsorption model as

$$kT \ln \frac{(\alpha_i/\alpha_m)(1-x_b)}{(1-\alpha_i/\alpha_m)x_b} + \Delta u(i) = 0, \quad \Delta u(i) = \varepsilon_{sf}(i) + \phi(i), \quad x_b = \alpha/\alpha_m, \quad x_i = \alpha_i/\alpha_m. \tag{1}$$

Here k is Boltzmann's constant; T the absolute temperature; α, a_i, a_m respectively the number of adsorbate molecules per unit surface area in the bulk gas phase, *ith* adsorption layer and the maximum adsorption capacity state; $\varepsilon_{sf}(i)$ and $\phi(i)$ respectively the interaction potential from the adsorption wall and the molecular interaction to an adsorbed molecule in *ith* adsorption layer. From Eq. (1), one can obtain the excess amount of adsorbtate molecules per unit surface area Γ_1 in the first adsorption as [3]

$$\Gamma_1 = C \frac{x_b(1-x_b)\{1-\exp[\Delta u(1)/kT]\}}{x_b + (1-x_b)\exp[\Delta u(1)/kT]}. \tag{2}$$

Here $\Delta u(1)$ is the change of the internal energy of adsorbate molecules in the first adsorption layer; C the correlating parameter. We can further get the following expression from Eq. (2)

$$\frac{\alpha}{\Gamma_1} = \frac{\alpha_m}{C\{1 - \exp[\Delta u(1)/kT]\}} \times \frac{\alpha}{\alpha_m - \alpha} + \frac{\alpha_m \times \exp[\Delta u(1)/kT]}{C\{1 - \exp[\Delta u(1)/kT]\}} .$$ (3)

Due to the prominent monolayer adsorption of supercritical hydrogen, under our experimental conditions, α_m can be considered temperature dependent constants, a series of points $\left(\dfrac{\alpha_{(j)}}{\alpha_m - \alpha_{(j)}}, \dfrac{\alpha_{(j)}}{\Gamma_{(j)}} \right)$ from the same measured isotherm will be in linear relationship [3]. Thereby, by the optimization, we obtain α_m and the correlating parameter C (see Table 1). In Table 1, it shows that α_m decreases with the temperature increasing and the value is smaller than that of liquid hydrogen upon the surface of the MWCNTs (26.864 nm^{-2}) [3], adsorbed hydrogen molecules are therefore likely in a compressed gas state.

Table 1   Determined values of hydrogen adsorption on MWCNTs

T [K]	B_{2S} [m³/g]	q_{st}^{0} [J/mol]	α_m [/nm²]	C [/nm²]
123	8.245E-7	2135.932	14.233	9.944
151	6.223E-7	2368.724	13.269	9.262
179	6.559E-7	2601.516	12.429	8.853
204	5.587E-7	2809.366	11.851	8.276
232	5.987E-7	3042.158	11.338	7.872
292	4.789E-7	3540.998	10.655	6.434
310	3.486E-7	3690.650	10.457	5.661

Determination of the interaction potential of hydrogen-MWCNTs

We firstly used adsorption data in very low surface concentration region and got the second virial adsorption coefficient B_{2S} (Table 1). Then, by plotting the $\ln(B_{2S}/\sigma_{ff}S_{BET})$ versus $1/T$ (σ_{ff} and S_{BET} are respectively the collision diameter of a hydrogen molecule and the specific surface area of the MWCNTs) [3], we got the interaction potential $\varepsilon_{sf}(r)= -505.86\ K\cdot k$ and the isosteric heat of adsorption in low limit surface concentration q_{st}^{0} (Table 1). From Table 1, we can find that q_{st}^{0} are smaller than 5596.68 J/mol of hydrogen on the graphitized carbon black [6].

Determination of the interaction energy of hydrogen-hydrogen

By algebraic manipulations and introducing the common expression presented by Do [7], we obtained following expressions from Eq. (3) for calculating the hydrogen-hydrogen interaction energy u in the first adsorption layer

$$\phi(1) = -\Delta u(1) - \varepsilon_{sf}(1) = kT \ln \frac{C \times \alpha \times (\alpha_m - \alpha) - \Gamma \times \alpha \times \alpha_m}{(\Gamma \times \alpha_m + C \times \alpha)(\alpha_m - \alpha)} - \varepsilon_{sf}(r),$$ (4)

$$u = -\phi(1) \cdot \frac{M}{n^2 N}, \quad \alpha_m = M, \quad n = \frac{\alpha_m}{\alpha_1} .$$ (5)

Here M is the number of independent active sites upon the adsorbent surface; n the number of active sites occupied by an adsorbed molecule [7]. Results calculated by Eqs. (4)-(5) are shown in Figs. 3-4.

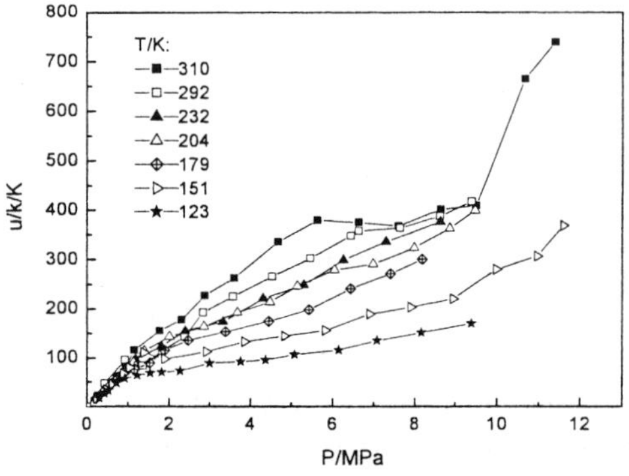

Figure 3　Variation of the hydrogen-hydrogen
interaction energy with adsorption pressures at different
temperatures on the MWCNTs

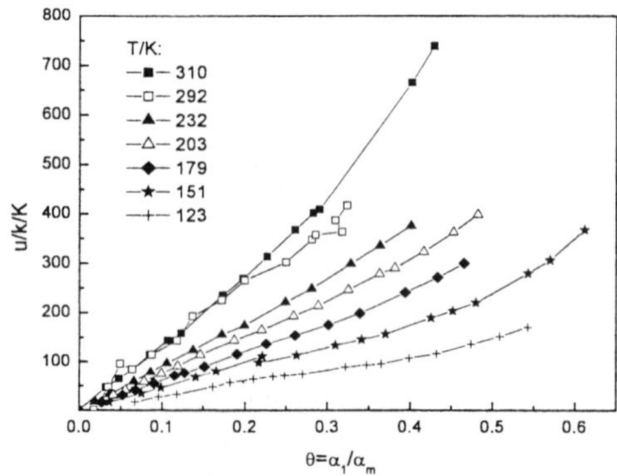

Figure 4　Temperature dependence of the
hydrogen-hydrogen interaction energy at different
surface loading on the MWCNTs

CONCLUSIONS

Adsorbed hydrogen molecules on the MWCNTs are likely in a compressed gas state. The isosteric heat of hydrogen adsorption in low limit of the surface concentration on the MWCNTs is smaller than that on the graphitized carbon black. The hydrogen-hydrogen interaction energy shows characteristics of physical adsorptions of supercritical gases, the optimum adsorption temperature has not been revealed by the determined results and should still be in researching.

ACKNOWLEDGEMENTS

This work was financially supported by China Postdoctoral Science Foundation (No. 2003034260).

REFERENCES

1. Züttela, A., Sudana, P., Maurona, P., Kiyobayashib, T., Emmeneggera, C., Schlapbacha, L., Hydrogen storage in carbon nanostructures, International Journal of Hydrogen Energy (2002) 27 203-212

2. Shiraishi, M., Takenobu, T., Ata, M., Gas-solid interactions in the hydrogen/single-walled carbon nanotube system, Chemical Physics Letters (2003) 367 633-636

3. Qingrong, Z., A study of hydrogen storage by adsorption on multi-walled carbon nanotubes: [Dissertation], Shanghai Jiao Tong University, PRC (2002) 22-40

4. Ono, S., Kondo, S., Molecular theory of surface tension in liquids, Springer Press, Berlin-Göttingen-Heidelberg, Germany (1960) 23-41

5. Aranovich, G.L., Donohue, M.D., Adsorption compression: an important new aspect of adsorption behavior and Capillarity, Langmuir (2003) 19 2722-2735

6. Constabaris, G., Sams, Jr., J.R., Halsey, Jr., G.D., The interaction of H_2, D_2, CH_4 and CD_4 with graphitized carbon black, Journal of Physical Chemistry (1961) 65 367-369

7. Do, D., Adsorption analysis: equilibria and kinetics, Imperial College Press, London, UK (1998) 35-38

Design of vortex tubes and experimental program on LOX separation using cryogenic vortex tubes

Jacob S., Upendra Behera*, Paul P. J.*, Kasthurirengan S., Karunanithi R., Ram S. N., Dinesh K.

Centre for Cryogenic Technology, Indian Institute of Science, Bangalore-560 012, India
*Aerospace Engineering, Indian Institute of Science, Bangalore-560 012, India

The use of CFD techniques to arrive at optimum design parameters of vortex tubes to fabricate them is described. Experimental studies have shown that for a 12 mm diameter straight vortex tube with six conical nozzles, maximum temperature difference of $\approx$109 K between hot and cold end flows was obtained for length to diameter ratio (L/D) > 25 and with optimum cold end diameter (d_c) of 7 mm. Studies of LOX separation from pre-cooled air flow show that conical vortex tube gives highest LOX purity of $\approx$96% and the higher separation efficiency of $\approx$61% compared to straight vortex tubes.

INTRODUCTION

Ranque-Hilsch vortex tubes are well known devices having no moving mechanical parts in which, compressed gas injected through tangential nozzles into a vortex chamber results in the separation of inlet flow into two streams, one of which is warmer than the inlet gas while the other is colder. The strong circular flow field in the inlet area causes pressure distribution of the flow in radial direction. As a result, a free vortex is produced as the peripheral warm stream and a forced vortex as the inner cold stream. The schematic diagram of the flow pattern is shown in Figure 1.

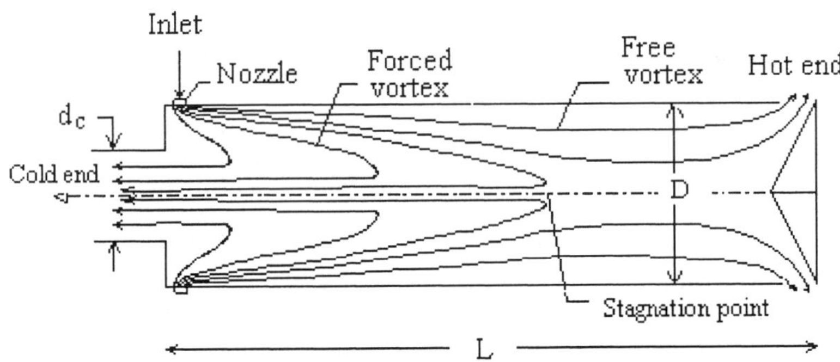

Figure 1 Schematic flow pattern of Ranque-Hilsch tube

There are several applications of vortex tubes for moderate heating and cooling requirements provided there is ready availability of compressed air. They include cooling applications in machining sensitive materials, environment for explosive chemicals, temperature control of diver's air supplies, cooling of electronic components, controlling the temperature of personnel suits in hostile environment, drying of food grains, etc. Apart from these, vortex tubes for mixture separation are finding increased use for drying and purification of gas mixtures as well as separation of liquid oxygen (LOX) from pre-cooled air stream [1].

Cockerill [2] has provided a simple theory for separation of binary mixtures of gases. It has been shown that the swirl velocity and the molar mass difference of the constituents primarily influence any change in the mole fraction composition across the vortex radius. Therefore centrifugation is the key for the mixture separation in vortex tubes. The axial flow pattern in the vortex tube influences the resident time period of the flow, increasing the higher molar mass concentration distribution at the warm end of the vortex tube.

Many investigators, Kurosaka [3], Gutsol [4] have suggested various theories to explain the Ranque effect. However till today no exact theory has come up to explain the phenomenon satisfactorily. Thus much of the design and development of vortex tubes have been based on empirical correlations leaving much scope for optimization of critical parameters.

In the present studies, Computational Fluid Dynamics (CFD) techniques have been used to analyze the flow behavior in the vortex tube and to arrive at optimized design parameters. Experiments have been conducted to validate the design parameters evolved through CFD. The study also describes the experiments on LOX purity and separation efficiency when pre-cooled air stream is injected to straight and conical vortex tubes at controlled conditions.

TEMPERATURE SEPARATION: CFD ANALYSIS AND EXPERIMENTAL PROGRAM

CFD studies have been conducted by modeling a 12 mm diameter vortex tube using 'Renormalization Group' version of k-ϵ turbulence model by Star-CD code [5]. The analysis is carried out to arrive at the optimum number of nozzles, nozzle profile, cold end diameter (d_c) and length to diameter ratio (L/D). The diameter of vortex tube and ratio of nozzle inlet area to vortex tube area (=0.07) are kept nearly constant in the analysis. The L/D ratios for the studies ranged from 10 to 35.

Experimental studies have been conducted with the experimental setup, which has been described in reference [1].

Nozzle profile and number of nozzles
Earlier, investigators [6] had to carry out laborious fabrications and experimental program to arrive at the optimum nozzle profiles and evaluate their performances. However, no clear guidelines could be evolved due to large number of parameters affecting the results. CFD analysis can minimize these difficulties.

CFD analyses were conducted for different types of nozzle profiles and numbers such as convergent type (two and six nozzles), straight (six nozzles) and helical type (circular and rectangular single nozzle) to arrive at maximum temperature separation between hot and cold end discharges by optimizing the swirl velocity magnitude and its profile. The studies showed that six numbers of convergent nozzles provide good radial symmetry of flow along with optimum swirl velocity resulting in maximum temperature separation compared to other nozzle profiles. Hence vortex tubes having six numbers of convergent nozzles have been selected for fabrication of vortex tubes for experimental programs.

Cold end orifice diameter (d_c) and length to diameter ratio (L/D)
Vortex tubes can be used in such a way that they can produce maximum hot gas temperature and minimum cold gas temperature by selecting suitable d_c. Experiments have been conducted for straight vortex tubes with different d_c of 5, 6, 7 and 7.5 mm and L/D ratio ranging from 10 to 35.

A maximum hot gas temperature of 391 K is obtained for a 12 mm vortex tube injected with dry compressed air at 7 bar (absolute) and 300 K, for d_c of 7 mm and L/D of 30 (Figure 2a). To achieve minimum cold gas temperature of 268 K the optimum d_c is 6 mm and L/D ratio is 25 (Figure 2b).

The temperature difference between hot and cold gases for different L/D ratios with d_c of 7 mm obtained by experiments and CFD analysis is shown in Figure 3. The maximum temperature difference is obtained for L/D more than 25 as the stagnation point of the forced vortex lies with in the vortex tube (visualized in CFD studies) giving rise to higher thermal interaction between forced and free vortex regimes. The present CFD studies predict the temperature difference to $\approx$87% accuracy compared with experimental results.

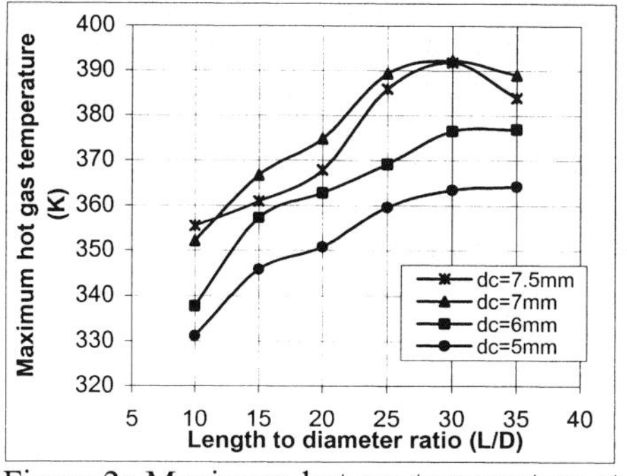

Figure 2a Maximum hot gas temperature at different L/D ratios and cold end diameters

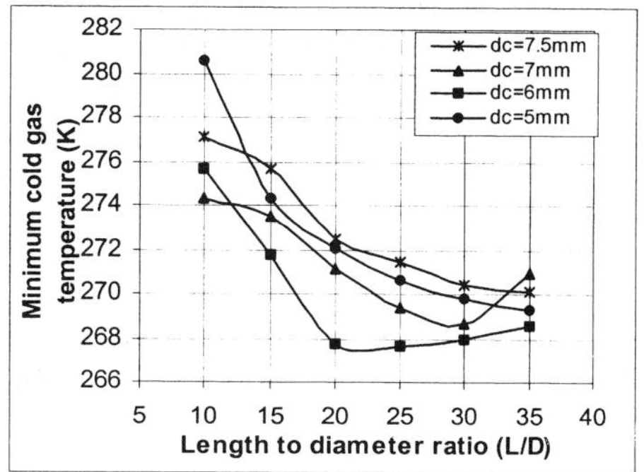

Figure 2b Minimum cold gas temperature at different L/D ratios and cold end diameters

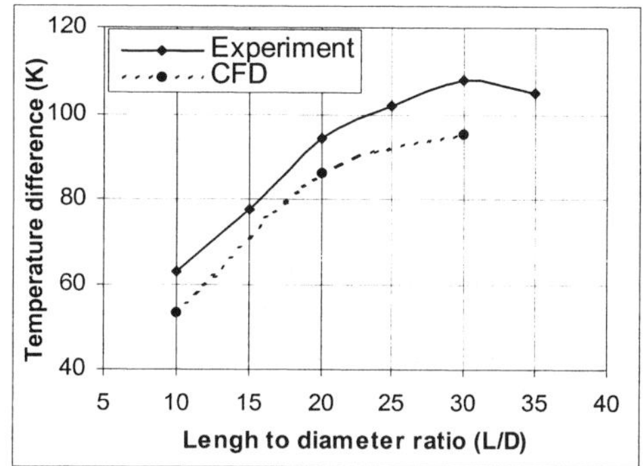

Figure 3 Temperature difference between hot and cold gas for different L/D ratios

MIXTURE SEPARATION IN VORTEX TUBE

Vortex tubes are potential candidates for separation and condensation of LOX from pre-cooled air stream. This process may have many ground and space applications due to non-moving components in vortex tube and its performance remaining unaffected by orientation and gravity. Experiments have been conducted to optimize the parameters for achieving high purity and separation efficiency of LOX from pre-cooled air stream in vortex tubes.

The separation efficiency is defined as,

$$H_{sep} = f\,(C_o/C_a)$$

Where, f is the ratio of oxygen mass flow rate of enriched air to that of inlet air.

C_o is the mass percentage of oxygen in the enriched airflow.

C_a is the mass percentage of oxygen in ambient air = 23.15%.

In the present studies, pre-cooled air at 95-100 K and about 4 bar (absolute) pressure is injected into the vortex tube experimental system [1]. Under this condition of two-phase flow, oxygen enriched liquid is thrown to periphery due to the centrifugation and large liquid to vapour specific gravity ratio and flows to its conventional hot end discharge. More volatile nitrogen boils from the liquid film into the vapour core around the vortex tube axis flowing towards the cold end. In turn oxygen from the vapour flow condenses into the liquid film increasing its concentration.

LOX separation experiments have been conducted for straight vortex tubes with optimum d_c of 7 mm for different L/D ratio ranging from 10-35, as well as for conical vortex tube with a divergence angle of 2.5° towards the hot end and having L/D of 10. The highest LOX purity of 96% is obtained for conical vortex tube (Figure 4).

To optimize the separation efficiency as well as LOX purity, a series of experiments have been conducted using straight and conical vortex tubes. From the results, (Figure 5) it is observed that conical vortex tube of L/D=10 provide the highest separation efficiency as well as better purity compared to straight vortex tubes. Also, in straight vortex tubes as L/D ratio increases, the separation efficiency and oxygen purity increase.

The better performance of conical vortex tube as compared to straight vortex tube may be due to the increased surface for condensation-evaporation phenomenon of oxygen and nitrogen molecules at the interface between the free and forced vortices. The fact that in straight vortex tube LOX separation efficiency increases with increase in L/D ratio could be attributed to longer residual time of flow in the tube giving higher LOX concentration. Further studies are in progress to optimize the conical vortex tubes for different values of d_c and L/D ratios.

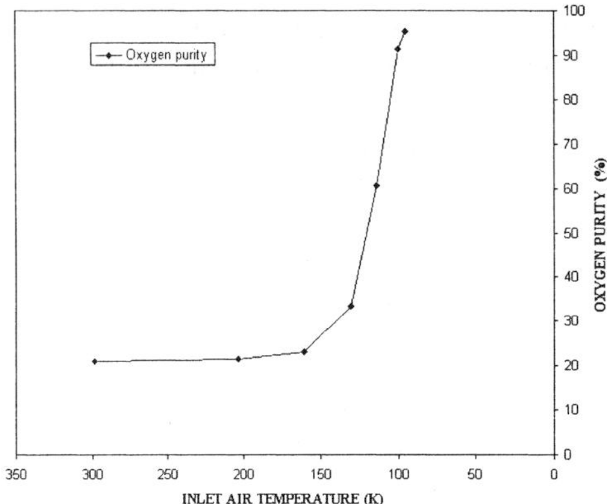

Figure 4 Oxygen purity vs. inlet temperature

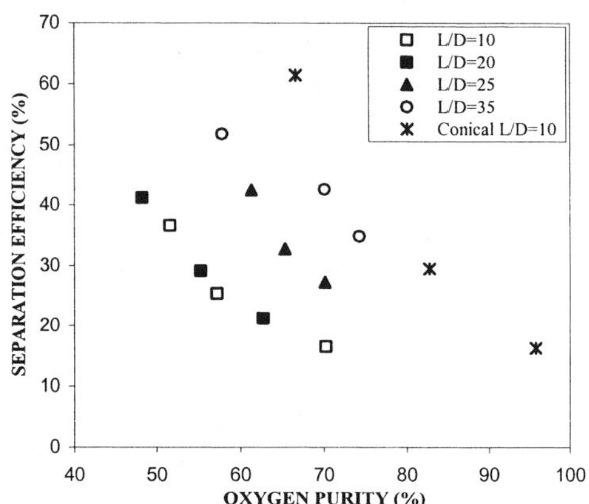

Figure 5 Separation efficiency vs. oxygen purity

CONCLUSIONS

The studies provide a new approach in designing vortex tubes using CFD technique rather than using empirical correlations of critical parameters. CFD studies conducted on 12 mm diameter straight vortex tube showed that six numbers of conical nozzles give better thermal performance compared to other profiles. Experiments conducted yielded a maximum temperature difference $\approx$109 K between the hot gas and cold gas flow for vortex tube with L/D=30 and d_c=7 mm, the results could be predicted to 87% accuracy by CFD analysis.

Experiments were conducted to study LOX purity and separation efficiency of straight and conical vortex tubes. The maximum LOX purity of 96% was obtained for conical vortex tube along with 14% separation efficiency. A maximum of 61% separation efficiency was obtained for conical vortex tube when the LOX purity was controlled at $\approx$66% Straight vortex tubes gives inferior performance in LOX separation compared to conical tubes. For straight tubes, the studies show that the purity and separation efficiency are dependent on L/D ratios.

REFERENCES

1. Jacob S., Kasthurirengan S., Karunanithi R. and Jagadish T., Oxygen separation using cryogenic vortex tube, <u>Advances in Cryogenic Engineering</u> (2000), <u>45</u>, 1771-1777
2. Cockerill T. T., Thermodynamics and fluid mechanics of a Ranque Hilsch vortex tube, <u>Masters Thesis</u> (1995), University of Cambridge, England, 1-44
3. Kurosaka M., Acoustic streaming in swirl flow and the Ranque-Hilsch (vortex-tube) effect, <u>J. Fluid Mech</u> (1982), <u>124</u>, 139-172
4. Gutsol A. F., The Ranque effect, <u>Physics-Uspekhi</u> (1997), <u>40</u> (6), 639-658
5. Star-CD methodology and user guide, <u>Computational Dynamics Limited</u> (1999), Version 3.10A
6. Metenin V. I., Investigation of vortex temperature type compressed gas separators, <u>Translated from Zhurnal Tekhnicheskoi Fiziki</u> (1960), <u>30</u> (9), 1095-1103

Heat Transfer Enhancement of He II Co-current Two-phase Flow in The Presence of Atomisation

B. Rousset*, P. Thibault**, S. Perraud*/**, L. Puech**, P. E. Wolf**, R. van Weelderen***

* CEA-Grenoble/DRFMC/SBT, 17 rue des Martyrs, 38054 Grenoble Cedex 09, France
** CNRS /CRTBT, 17 rue des Martyrs, 38054 Grenoble Cedex 09, France
*** Accelerator Technology Department, CERN, 1211, Geneva, Switzerland

Previous experiments performed on HeII co-current two-phase flow at CEA-Grenoble have shown the existence of a transition from stratified two-phase flow to droplet mist flow at high vapour velocities. The realisation of a new refrigerator/liquefier able to produce up to 20 g/s of single phase superfluid helium at 1.8 K (instead of 7 g/s previously) was achieved. Benefit was taken of the necessary junction between the existent test line and the refrigerator to introduce some new instrumentation [1]. Results of preliminary experiments performed on this new configuration are given. The response of liquid level and vapour density on droplet flow is presented. First results on pressure drop obtained for a total mass flow rate of 15 g/s are also presented. Finally, the use of various capacitive level gauges glued at different azimutal positions along the inner pipe give access to the perimeter wetted by a continuous thin liquid film.

INTRODUCTION

Large size superconductive magnets are generally made of NbTi or of Nb3Sn and require to be operated at liquid helium temperature. Moreover, in the case of high magnetic fields, it may be necessary to decrease the temperature below 2.2 K and thus to enter the field of superfluid helium (e.g. the future LHC at CERN or the tokamak Tore Supra at CEA). To avoid the presence of vapour in contact with the magnets, He II around the magnets will have to be subcooled (i.e. maintained with a pressure higher than its saturated vapour pressure). This subcooling will be done through a heat exchanger separating the pressurized He II and the saturated He II (i.e. the cold source). The safe operation of the magnet depends on the effectiveness of this heat exchanger. In the case of LHC, the system studied by CEA/SBT is made of a heat exchanger pipe traversing right through the magnets[2]. The magnets will be immersed in a static bath at 1 Bar of pressurized superfluid helium. The heat exchanger pipe is cooled by a He II two-phase flow, the tube being almost entirely filled with liquid at a magnet string inlet and being dry at exit (heat losses collected along the flow evaporating the liquid). Previous experiments consisted of a two phase co-current stratified flow through a 40 mm inner diameter, 10 m long tube, with a descending slope ranging between 0 and 1.4%. Studies highlighted an improvement of heat exchange at high vapour velocities, while at the same time only a small fraction of liquid remains present in the tube. This excess of heat exchange appears simultaneously with the presence of droplets within the vapour phase. One can think that the spray generated above the level of the free surface will deposit on the wall and will strongly increase heat exchange. After deposition, the drops can flow and contribute to a liquid film.

Due to the limitation in available mass flow (7 g/s), the range in vapour velocities was limited, in particular for temperatures higher than 2 K, were velocity values higher than 8 m/s could not be achieved. For the same reason, it was not possible to have both high vapour velocities and sufficient residual liquid mass flow, preventing the investigation of the influence of liquid level on droplet flow.

In order to overcome these limitations, the use of a new refrigerator/liquefier able to produce up to 20 g/s of single phase superfluid helium at 1.8 K was scheduled. Other improvements in this new configuration include the possibility to obtain steady state two-phase flow regimes from very low vapour quality up to pure vapour. Preliminary results are presented hereafter.

EXPERIMENTAL FACILITY AND INSTRUMENTATION

The previous cryoloop experiment (figure 1) has been connected to a new refrigerator able to remove up to 400 Watt at 1.8 K. From inlet to outlet, the test line mainly consists of series of thermometers, a heat exchange box (from which the inner wetted perimeter of the tube is calculated), a first capacitive level gauge device (see figure 2a), and two optical sectors.

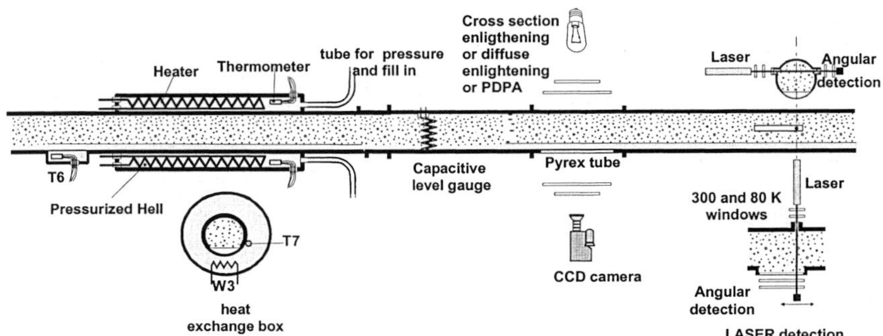

Figure 1. Scheme of the cryoloop instrumentation

A new instrumental sector including a second capacitive level gauge of different gaps and capacitive flags was also introduced in the test line (figure 2b). Description and fabrication of the capacitive sensors were given elsewhere[3]. The capacitive response of this sensor depends on the wet surface and on the thickness of the liquid film covering the wet surface. Figure 2c shows this dependence for various gaps between electrodes. Calculations were performed using the ANSYS code. In first approximation, once the liquid thickness of the wet part is higher than the gap between the two electrodes, the measurement directly indicates the wet area.

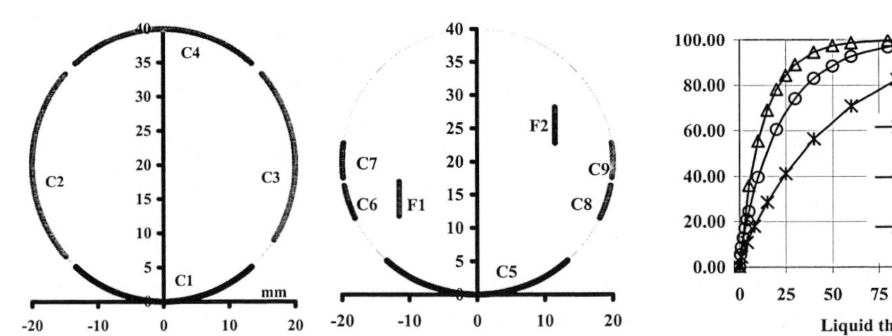

Figure 2a Position of the capacitive gauges of 100 µm gap inside the first sector.

Figure 2b C5 is identical to C1, C6 and C7 have a 30 µm gap and C8, C9, F1 and F2 have a 50 µm gap. F1 and F2 are placed in a separated sector.

Figure 2c Relative variation (in %) of the capacitance with the thickness of a uniform liquid helium film

The response of these sensors will be presented in percentage, 0 % corresponding to the dry situation and 100 % to the situation where the sensor is wetted over its whole surface with a film thicker than the gap between electrodes.

RESULTS AND ANALYSIS

All measurements presented here are acquired at the end of a 11 m long tube (40 mm I.D.) for a 0.6 % slope. Heat losses along this straight tube are estimated to 10 Watt.

<u>Heat transfer results</u>

Typical accuracy of capacitive sensors can be appreciated from the comparison between C1 and C5 (figure 3); these two sensors being identical and located at the same altitude inside the pipe.

As already mentioned and explained[4], wall heat transfer is directly proportional to the inner wetted perimeter of the tube. Heat transfer is improved at high vapour velocities (figure 4). However, this tends to be reduced when heat flux increases, which can be explained by a dry out of the upper part of the pipe.

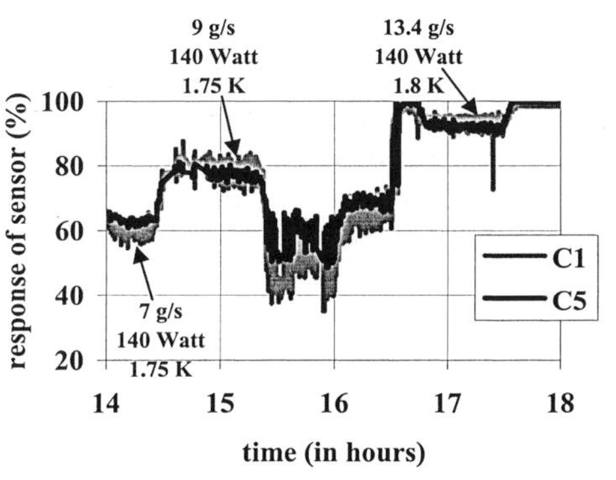

figure 3 Perimeter wetted by the liquid at the bottom of the tube

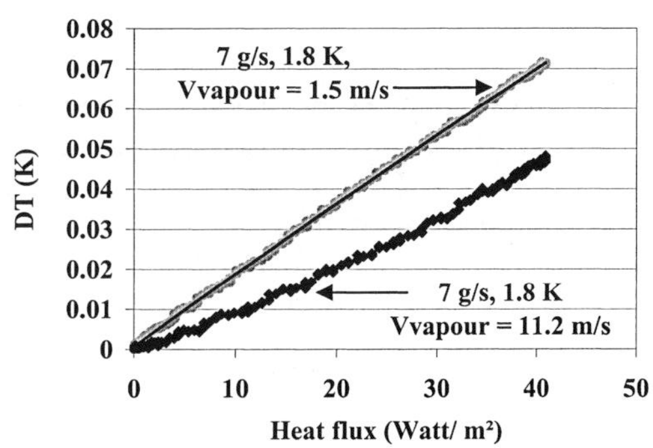

figure 4 Increase in temperature in the pressurized bath

Table 1 Summary of experimental results. m is the total mass flow (in 10^{-3} kg/s), W the inlet power injected (in Watt), T the temperature (in Kelvin) and H the heat transfer measured in the heat exchange box (measured for a heat flux of 40 Watt/m^2) converted in wetted perimeter (in %).

line	m	W	T	H	C1	C2	C3	C4	C6	C7	C8	C9	F1	F2
1	7	140	1.75	57.3	60	3.6	3	0.3	20.5	15.5	14.7	11.7	6.8	5.1
2	9	140	1.75	64.3	79	5.1	4.8	0.6	24.6	18.9	17.7	14.3	8.6	7.4
3	13.4	140	1.80	66.5	92	10	5.6	0.9	28	21.1	20.0	15.4	9.4	8.0
4	15.3	80	1.84	37	100	11.4	2.1	1.6	1	3	X	X	2.4	1
5	15.3	174	1.84	73	95	13.4	8	3	29	24	X	X	10.0	8.2
6	15.3	200	1.84	71	90	11.6	7.8	3.4	32	24	X	X	10.2	9.0
7	15.3	280	1.84	71	45	5.3	5	3.2	22.5	17	X	X	10.5	10.5

Comparison between lines 1 to 3 of table 1 gives access to the influence of free surface level (given by C1). For approximately the same vapour velocity, values of capacitive sensors increases as the liquid level increases. Comparison between C6 and C8 (located at the same altitude), indicates that sensors are covered by a layer of liquid thinner than 30 μm (C6>C8 and C6<100 %). Comparison between F1 and C8 indicates that liquid is probably flowing down to the free surface, C8 being also partly covered with liquid at its upper part.

Looking at the values of F1 and F2 for all the lines, one can have an estimate of the mist stratification. Mist was always found stratified except for line 7, this latter corresponding to the maximum velocity and the minimum free surface level obtained.

<u>Pressure drop results</u>

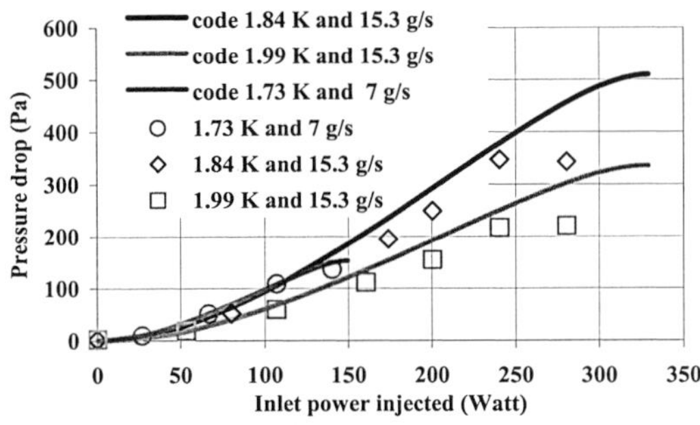

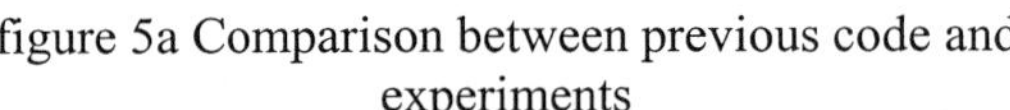

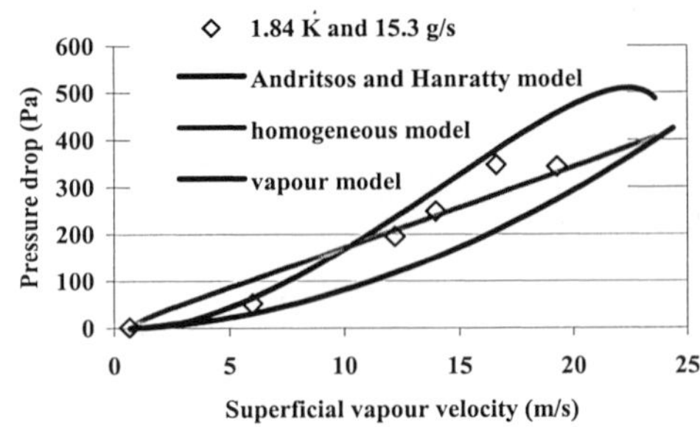

figure 5a Comparison between previous code and experiments

figure 5b Comparison between various models and experiments performed at high velocities

Previous experiments were used to validate a model using the Andritsos Hanratty[5] correlation in the range of 1 to 7 g/s and 1.8 to 2 K. We check here its validity for a wider range of mass flow and vapour velocities (figure 5a). Furthermore, our new test station enables experiments with a larger void fraction. It was found that for the largest void fraction (i.e. the largest power injected), discrepancies with the model appear. Comparison with various models (figure 5b) suggests the use of a homogeneous model in this last case, which is compatible with the low free surface level observed.

CONCLUSION

He II two-phase flow measurements were performed at a mass flow ranging from 7 to 15 g/s and a temperature ranging from 1.8 to 2 K. Improvement of wall heat transfer at high vapour velocities was confirmed, with a corresponding wetted inner tube value as high as 70 % for the highest velocities. Droplet deposition on wall forms a thin liquid film. Film thickness depends on altitude and velocities. Above the free surface, the mist of droplets is stratified, with almost no droplets at the top of the tube. Furthermore liquid film is falling along the wall and tends to increase its thickness until it reaches the free surface. For the highest velocities, the mist was found more homogeneous.

Pressure losses were found in good agreement with our previous code derived from the Andritsos and Hanratty model[5] except for the highest vapour velocity corresponding to the largest void fraction. In that case, the major fraction of liquid phase may flows as a spray and a homogeneous model would be more convenient.

ACKNOLEDGMENTS

Authors would take here the opportunity to thanks Roser Vallcorba for its works on ANSYS and the SBT/GRTH technician team for the assembly and test of all the components of the test loop.

REFERENCES

1. Thibault P. et al., Description of a new experiment to explore HeII two phase flow behavior at high mass flow rate International Cryogenics Engineering Conference 19 Grenoble, France(2002), 825-828

2. Lebrun P. et al., Cooling String of Superconducting Devices below 2 K : the Helium II Bayonet Heat Exchanger Advances in Cryogenics Engineering (1998) 43 419-426

3. Thibault P. et al., Probing the Wetted Perimeter in a Pipe-Flow Experiment Using a Capacitive Sensor, Advances in Cryogenics Engineering (2002) 47 1683-1690

4. Rousset B. et al., HeII two phase flow in an inclinable 22 m long line, Advances in Cryogenics Engineering (2000), 45 1009-1016

5. Andritsos, N. and Hanratty, T.J., "Influence of interfacial waves in stratified gaz-liquid flows", AIChE Journal (1987) 33 444-454

Design consideration validation of cryo-components for Current Feeder System of SST-1 using flexibility analysis

Gupta, N.C., Sarkar, B., Gupta.G. and Saxena Y.C.

Institute for Plasma Research, Bhat, Gandhinagar 382 428, India

The current feeder system (CFS) of steady state superconducting tokamak (SST-1) has been designed for its normal, cool down/ warm up as well as for emergency operations. The system will transit in a temperature range of 300K to 4.5K during the operation. Components with different materials of construction will experience the differential thermal contraction and thermal stress. Therefore an optimized route for the superconducting (SC) bus bars and layout for piping has been designed to feed cryogen to respective current leads and SC bus bars. The design consideration, salient features and analysis of cold piping network are discussed .

INTRODUCTION

The Current Feeder System (CFS) [1,2] of SST-1 consists of 10 pairs of 10 kA vapour cooled current leads and 20 no.s of SC busbars. Forced flow supercritical helium (SHe) at 4 bar and 4.5K will be used as a coolant for bus bars and liquid helium (LHe) at 1.3 bar and 4.5K will used as a coolant for current leads. Under such low temperature condition, the pipings are subjected to induce thermal stress due to the prevention of free contraction as CFS being an integrated system. Therefore, an optimized hydraulic network for LHe and SHe with in the current lead assembly chamber (CLAC) has been designed and analyzed. The analysis includes hydraulic and thermal stress considering different loops for the cryogen carrying hydraulic network as well as interconnections details. Design of such a complex system needs many iteration and ANSYS [3] has been used for this purpose. The input for this analysis is the modeling of piping layout, material properties and the temperature loading.

SYSTEM DESCRIPTION

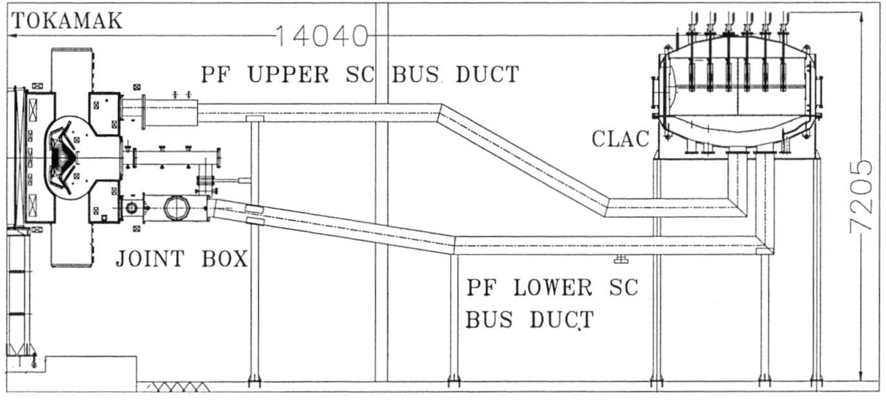

Figure 1 Elevation view of CLAC and SC bus ducts with SST-1

The CLAC of diameter 3 meter and of height ~2.2 meter houses 20 Nos of vapour cooled current leads, joints between current leads and SC bus bars, helium feed lines and isolators along with their supply headers. 20 nos of bus bars are divided in 3 segments and assembled inside 3 individual vacuum ducts. All bus bars are assembled inside three vacuum ducts. The bus ducts are designated as TF, PF upper and PF lower as shown in figure 1. The flow direction of coolant in the bus line will be from bottom end of current lead to SCMS. Therefore, at lower end of current leads SHe will be fed. The SHe & LHe are supplied by individual transfer lines to their respective headers in CLAC. TF and PF bus bars have their separate hydraulic headers for SHe to avoid any instability caused by PF in TF but all the current leads have a common header for LHe. Hydraulic analysis shows that TF bus bars require approximately 1.2 g/s flow and PF bus bars require approximately a total of 11 g/s of SHe. The consumption of LHe for per pair current lead varies from 15 l/h for I=0 to 35 l/h for I = 10 kA. Thus a hydraulic network of cryogen distribution pipes and tubes, with bends is required to distribute equal mass flow rate of SHe and LHe in each SC bus bar and current lead respectively without inducing higher thermal stresses in the piping.

DESIGN DRIVERS

Cryogen carrying process tubes/piping shall be capable to withstand high-pressure (~40bar) and should have enough flexibility (low thermal contraction coefficient, 3-mm/m upto 4.5 K) to withstand transient events. The designed process line should have acceptable level of pressure drop (1 mbar/m) and piping material shall be high vacuum (10^{-5} mbar) compatible.

THE THERMAL ANALYSIS

The stress analysis is often made synonym to flexibility analysis, since the thermal stress can be reduced by making the system as flexible as possible. This is done by providing a number of U, O and L bends, which allow more deflections in the lines and make it less rigid. In order to design piping layout with loops, for SHe and LHe network we have used ANSYS software. In ANSYS software we employ the 'pipe 16' element for bends and 'pipe 18' for straight part. Both elements were used for modeling. The boundary conditions are fixed end of a pipe with U bend and load of temperature difference is 295.5 K. The stresses calculated from empirical formula [4] were compared with ANSYS result. The empirical relation for maximum stress is given as

for $\alpha > 1/2$

$$\frac{\sigma_{max} L}{E\varepsilon_t D_0} = \frac{1.5\beta^2(1+\beta)}{\alpha^2(1-3\beta^2)+\alpha\beta(2+3\beta)}$$

and for $\alpha < 1/2$

$$\frac{\sigma_{max} L}{E\varepsilon_t D_0} = \frac{1.5\beta^2[(\beta/\alpha)(1-\alpha)+1)]}{\alpha^2(1-3\beta^2)+\alpha\beta(2+3\beta)}$$

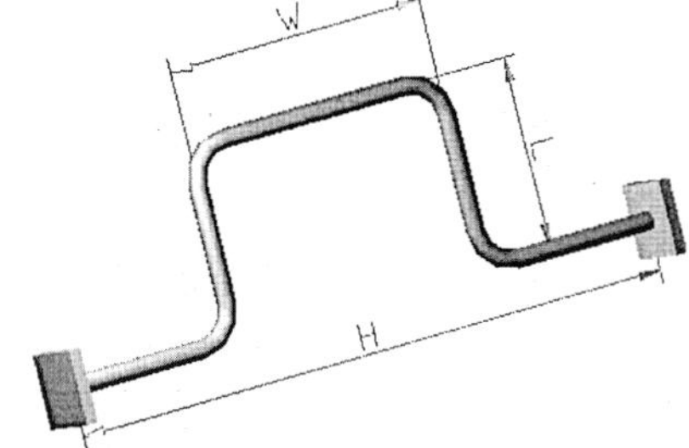

Figure 2 3D view of a U bend

ε_t :unit thermal strain, Do: diameter, α: W/L and β: W/H. In our calculation we have taken α=1/2 for minimum thermal stress. For W = 200 mm, L = 400mm and considering maximum allowable stress as 60Mpa, calculated H is 153.8mm. The same geometry was modeled in ANSYS with identical boundary conditions and loads by using pipe16 and pipe 18 elements. Maximum stress σ is estimated as 68 MPa, which is in agreement with emperical estimation validating the modelling.

HYDRAULIC ANALYSIS

Hydraulic analysis decides the flow diameter, velocity, and pressure drop. When a single phase fluid flows through a tube, then there will be pressure drop along the length of the line. The mjor contributions are from pressure drop at entry & exit, friction and bends along the line. Pressure drop in turbulence flow due to friction can be estimated by

$$\Delta P_{fric+bend} = \left[\frac{fl}{D} + nK\right]\left[\frac{8m^2}{\pi^2 \rho D^4}\right] \tag{1}$$

$K = 13f + 1.85(r/R)^{0.35}(\alpha/180)^{0.5}$, $f = 1.325/[\ln\{(e/3.7D) +(5.74/R_e^{0.9})\}]^2$ for turbulent flow in rough tubes/ pipes, with l: total length of pipe, D: hydraulic diameter, n: total number of bends, K: loss coefficient in the bend, Re: Reynold number, e: surface roughness of pipe, α: bend angle and R: bend radius.

DESIGN OF SUPER-CRITICAL HELIUM HEADER

In CLAC, SHe will be fed to PF and TF bus bars from individual transfer lines. The TF supply transfer line is connected to 2 nos of bus bars for TF magnet and PF supply transfer line is connected to supply header of 18nos of bus bars for PF magnet. Header connected with PF bus bars has been designed for equal mass distribution of SHe to each current lead, acceptable pressure drop and minimum thermal stress. After each tapping points from SHe header the size of reduced diameter can be estimated by

$$D_2 = D_1 \times (m_2/m_1)^{0.5} \tag{2}$$

The estimated diameters of the header changes slightly after each distribution and we have divided the header in to four parts. Table 1 shows the change in diameter of header after each tapping of SHe for SC bus bars.

Table 1 Diameter variation of SHe header

	D1	D2	D3	D4
Flow Diameter (mm)	23.37	18.04	15.8	10.42
Velocity (m/s)	0.210	0.235	0.210	0.235

DESIGN OF LOOPS FOR SHe HEADER USING FLEXIBILITY ANALYSIS (FINITE ELEMENT ANALYSIS AND RESULTS)

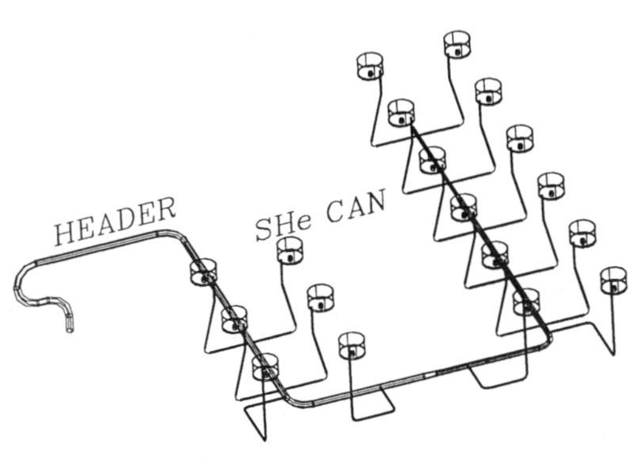

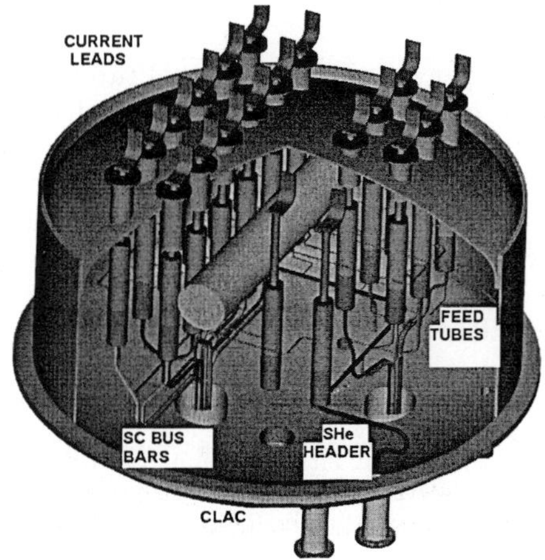

Figure 3 3D view of SHe header with feeding tubes in CLAC

Figure 4. 3D cut view of CLAC

The flexibility of a system is nothing but the ability to resist the loads experienced by it with the help of its meticulous way of construction. Adding more number of bends increases the flexibility of a hydraulic network and allows the thermal contractions. The SHe header with bends is located in between the current leads as shown in figure 3. The weight of the header is ~ 4.5 Kg and can be supported from its taping points connected to 18 current leads, therefore total 19 points, including supply end can be considered as fixed end to design the feed tubes. The main supply end is considered as a fix end, and a loop has been provided for the required flexibility. With this configuration header is free to contract and will move up due to thermal contraction of feeding tubes. In order to feed the SHe to PF bus bar system with one main supply of maximum 12 g/s from the main SHe header, total 18 numbers of tapping are required to feed in lower end of current leads. Hydraulic analysis shows that the required flow diameter of each feeding tube is ~6mm. An optimized layout was modeled in ANSYS with identical boundary conditions and loads by using pipe16 and pipe 18 elements. The result with maximum stress σ is 100 MPa, and is acceptable.

DESIGN OF LOOPS FOR CURRENT LEADS

The LHe header which supply liquid helium to all the current leads, supported from the top dish end inside the CLAC and all the LHe feeding tubes with their loops (as shown in figure 5) are connected to their respective current leads. The bottom edge of liquid helium header is ~175mm above the supply

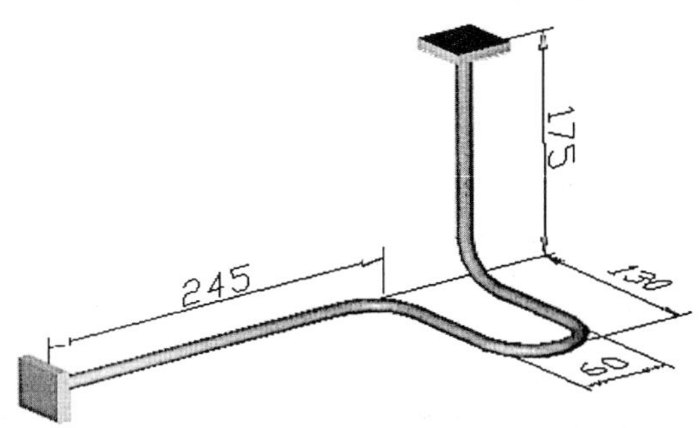

tapping of current lead can and inter-space in between two loops is more then 30 mm from each other to avoid any electrical break down of insulation. LHe feeding loop has been designed keeping a considering simple design, identical, easy to form, easy to assemble and induced thermal stress in the loop is less then 120 MPa. Moreover the axial loading on fixed point connected to current lead shall be less than 40N to avoid bending of current lead. Therefore an optimized loop was modeled in ANSYS. The result with maximum stress σ is 42 MPa, which is acceptable.

Figure 5 3D view of LHe feeding tube.

SUMMARY

This paper describes the thermal stress, flexibility analysis and layout design of cryogen carrying piping/ tubing and their distribution network inside CLAC for CFS of SST-1. At low temperature conditions, the pipelines contracted and thermal stress induces due to prevention limit of thermal contraction. The layout, which is an optimized and aesthetic layout, has been designed by considering the space limitation and restriction, which may rises during the fabrication & assembly. The whole distribution network has been designed using ANSYS software.

REFERENCES

1 V. L. Tanna, et al Super conducting Current Feeder System with associated test results for SST- 1 Tokamak Proc. 18th Magnet Technology (2003)
2 Saxena, Y.C. et al. Present Status of SST-1, Nucl . Fus. (2000), 40 (6), 1069.
3 ANSYS, Revision 5.3, Swanson Analysis System Inc., Houston, PA, USA.
4 Cryogenic Heat Transfer, Barron & Randall F.

Contractible thermosyphon for conduction cooled superconducting magnets

Noh C.[*], Kim S.[*], Jeong S.[*], Jin H.[**]

[*]Dep. of Mechanical Engineering, KAIST, 373-1, Kusung-Dong, Yusung-Ku, Taejon, Korea (S)
[**]Duksung Co. Ltd, 272-1, Sucksudong, Manangu, Anyangsi, Kyungido, Korea (S)

A thermosyphon, as a thermal switch of a conduction cooled superconducting magnet, was designed, fabricated and tested. When the temperature decreases below the triple point temperature, the working fluid inside the thermosyphon freezes and the thermosyphon does not work as a thermal shunt anymore. Although a usual thermosyphon is at OFF-state below the triple point temperature, there may be heat leak through the thermosyphon wall. The contractible thermosyphon, which has a metal bellows as a contraction component, is proposed to completely eliminate such a heat leak through the wall by mechanical detachment of the thermosyphon at low pressure.

INTRODUCTION

A Conduction cooled superconducting magnet uses a cryocooler for its operation without using any cryogenic fluid such as liquid nitrogen or liquid helium. However, it takes a long cool-down time to operate the magnet in spite of its conveniences. Especially, using the cooling capacity of the second stage only in the two-stage cryocooler is not an efficient way of cool-down because of its small cooling capacity. Thermosyphon is a very useful and effective heat transfer device due to its simple structure and high heat transfer characteristics. The thermal diode characteristic of thermosyphon can be used to connect the first and second stages of cryocooler to control heat transfer in superconducting magnet system.

There have been several researches [1]-[3] on reducing the cool down time by using thermosyphons as thermal shunts between the first and second stages of the cryocooler. In general, a thermosyphon installed between the first and second stages of cryocooler can decrease the cool-down time of the magnet by utilizing the large first stage cooling capacity of the first stage. It is also noticeable that the pressure of a thermosyphon is reduced to be very small below the triple point value due to the solidification of the working fluid. The thermosyphon does not work as a heat transfer device in that condition.

Although the thermosyphon does not work at low temperature as an OFF-state, there still exists heat leak through the thermosyphon wall. Therefore, a novel concept of contractible thermosyphon is suggested to eliminate such a heat leak. This paper describes the experimental procedure and the cool-down characteristic during non-isothermal operation of the thermosyphon.

EXPERIMENTAL APPARATUS

Liquid nitrogen container and mock-up magnet

A small liquid nitrogen container was used instead of an actual cryocooler in the experiment. To simulate an actual cryocooler, which will be used for a real conduction cooled superconducting magnet system at the later stage of this research, the liquid nitrogen boiling surface area was designed to give a similar cooling power of cryocooler. Instead of an actual superconducting magnet, a mock-up magnet was made by a 1.8 kg cylindrical copper block to test the cool down performance of the thermosyphon. Fig. 1 shows the system configuration of nitrogen container, the thermosyphon and the mock-up magnet.

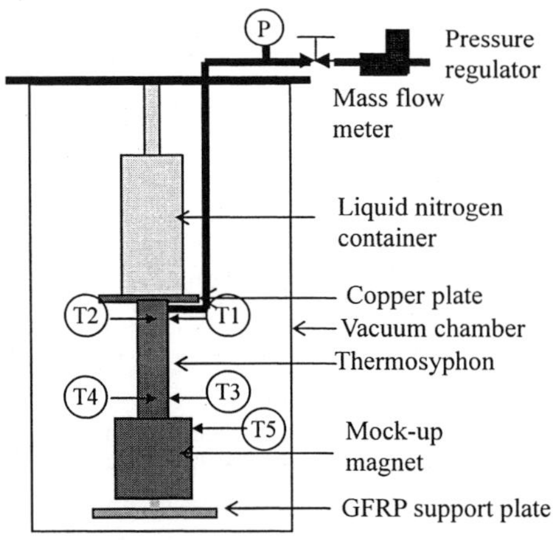

Figure 1 Schematic diagram of experimental apparatus
and temperature measurement

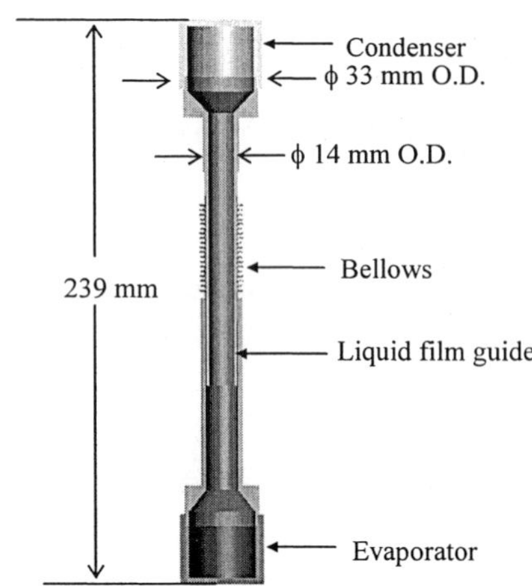

Figure 2 Cross section of the thermosyphon

Thermosyphon

Since a thermosyphon operates at two-phase state, the maximum operating temperature range is limited by the fluid property. In case of nitrogen and argon, which would be used as the working fluid of the thermosyphon, the operating temperature and the pressure lie between the critical and the triple points as listed in Table 1. In addition to the critical temperature and pressure limit, the maximum operating pressure is also limited by the strength of a thermosyphon wall.

Although a thermosyphon cannot operate below the triple point temperature, there might be some conduction heat leak through the wall of thermosyphon. Therefore, the thermosyphon was designed to be contractible and detachable from the mock-up magnet at reduced pressure and temperature. This novel feature eliminates conduction heat leak at the OFF-state. For the contraction capability, the adiabatic region of the thermosyphon was composed of a copper tube and the formed metal bellows (MDC, Part No. 470003) as shown in Fig. 2. In principle, the thermosyphon should be designed to withstand the high pressure more than 3 or 4 MPa considering critical temperature to use the full operating range of the working fluid. This means that much higher charging pressure is required at room temperature. However, since the allowable maximum pressure of the thin-walled bellows was only 2.5 MPa, the operating temperature of the thermosyphon was practically limited by this pressure. In the experiment, the thermosyphon was pressurized at 1.6 MPa and a continuous charging method was applied.

Experimental set up

As shown in Fig. 1, the thermosyphon and the mock-up magnet were installed in a vacuum chamber. Soft metal such as indium wire was used to reduce the thermal contact resistance between the evaporator and the mock-up magnet. During the cool-down process, the surface temperatures (T1, T3), the inside fluid temperatures (T2, T4) and the magnet temperature (T5) were measured by E-type thermocouples as shown in Fig. 1. The thermosyphon pressure and the mass flow rate were also measured continuously.

Nitrogen or argon gas was continuously charged from a high pressure gas cylinder at constant pressure of 1.6 MPa until the thermosyphon's filling ratio reached 25 % to avoid the dry out limit. For precise control of the filling ratio, the charged mass flow rate of the working fluid was instantaneously measured and integrated.

Table 1 Critical States and Triple Points of Nitrogen and Argon

	Nitrogen	Argon
Critical state	126.1 K,3.39 MPa	150.7 K,4.89 MPa
Triple point	63.2 K,12.85 kPa	83.8K, 68.8 kPa

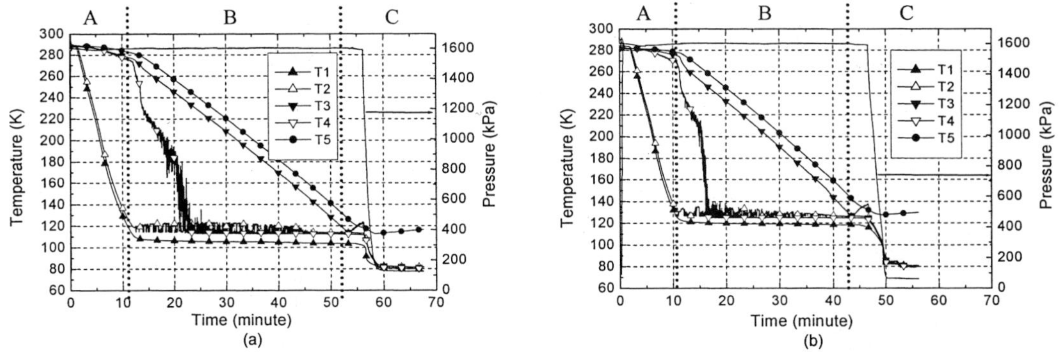

Figure 3 Cool down history of thermosyphon (a) with nitrogen and (b) with argon, at charging pressure of 1.6 MPa

EXPERIMENTAL RESULTS

The cool-down process of the thermosyphon is divided into 3 periods according to heat transfer mode of the thermosyphon (Fig. 3).

Conduction period; A

When the liquid nitrogen container was filled with liquid nitrogen, the temperature of the condenser started to be cooled by thermal conduction to the bottom copper plate of the liquid nitrogen container. Since the temperature of the condenser was higher than the critical temperature, there was no liquid phase at all during this period. The dominant heat transfer mechanisms in the thermosyphon were natural convection and wall conduction. Therefore, the temperature decrease of the evaporator and the copper block was very small and most of the cooling capacity was used to cool down the copper plate and the condenser.

Non-isothermal period; B

When the condenser temperature reached the saturation state, the condensation of the working fluid started at the condenser surface and the incoming mass flow to the thermosyphon was increased slightly around 10 minute as shown in Fig. 4. Nevertheless, the evaporator temperature was still higher than the saturation temperature of the operating pressure and the condensed liquid did not accumulate at the evaporator. Therefore, there was little liquid in the evaporator. The condensed liquid film might immediately evaporate by hot liquid film guide or the evaporator when it came down. This period can be called as the non-isothermal operation of the thermosyphon. Although the operation of the thermosyphon was non-isothermal, most of the cool-down time occupied this period. This period continued until the evaporator was cooled down to the saturation temperature.

Isothermal period and detachment; C

The evaporator temperature also decreased to the saturation state and the liquid started to accumulate from the bottom of the evaporator. During this period, the gas flow rate increased very rapidly as shown in Fig. 4 to compensate for the increase of liquid phase and to keep the constant pressure. The valve of gas flow into the thermosyphon was closed when the total mass of the working fluid reached to 25 % of the filling ratio.

During the large gas flow, the evaporator surface temperature was not decreased due to continuous condensation at the condenser. Moreover, the gas temperature at the condenser was a little bit increased by the large inflow of the working fluid. After the end of the gas charging process, the evaporator and the condenser temperatures started to decrease simultaneously. The pressure of the thermosyphon also decreased rapidly as the temperature decreased. From this moment, the thermosyphon was at isothermal state. It means that there is enough liquid phase at the evaporator and the condenser.

When the pressure decreased below 0.26 MPa, the bellows was contracted and the evaporator was detached from the mock-up magnet. As a result, the temperature of mock-up magnet was slowly increased by a parasitic heat leak from environment.

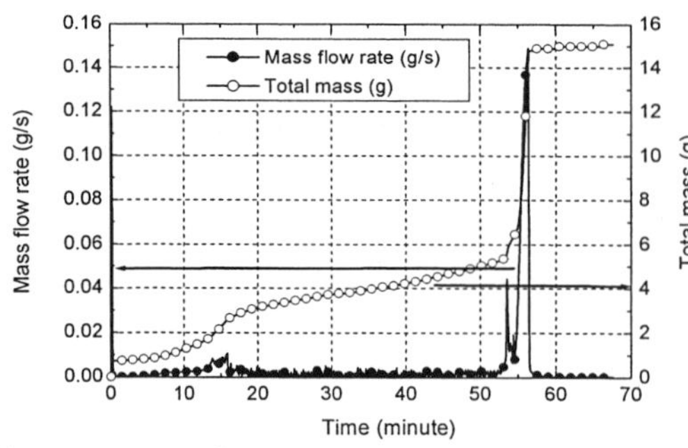

Figure 4 Mass flow rate and total mass into the thermosyphon in 1.6 MPa nitrogen thermosyphon

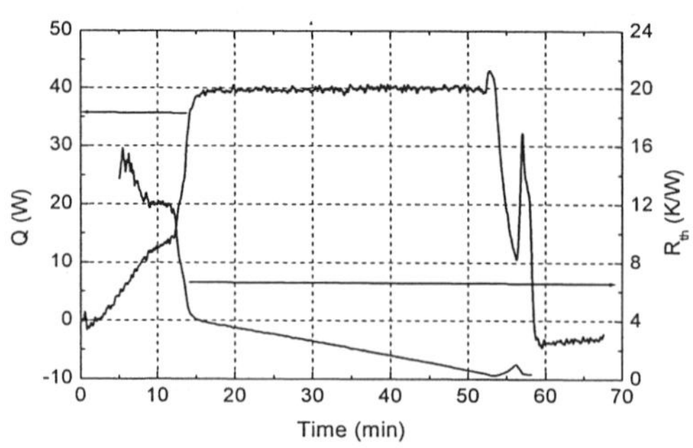

Figure 5 Heat transfer rate and thermal resistance through thermosyphon calculated from Fig. 3 (a)

<u>Heat transfer rate and thermal resistance</u>
The heat transfer rate through the thermosyphon can be calculated by equation (1).

$$\dot{Q} = mC\frac{dT}{dt} \tag{1}$$

where m and C is the mass and the heat capacity of the mock-up magnet. dT/dt is its temperature rate. Fig. 5 shows the calculated heat transfer rate through the thermosyphon and the overall thermal resistance between the condenser and the evaporator. In the beginning of the non-isothermal period, heat transfer rate was rapidly increased to 40 W and kept constant during this period, which was limited by the stable film boiling at the copper plate of liquid nitrogen container. The thermal resistance was rapidly decreased around 12 minute by the transition to the non-isothermal period. It was also slowly decreased during the non-isothermal period and it reached 0.28 K/W at isothermal thermosyphon.

CONCLUSION

Two types of contractible thermosyphons, which used nitrogen and argon as working fluid, were tested to cool-down a mock-up superconducting magnet. About 40 W was the cooling rate through the thermosyphon during the non-isothermal period. After the end of the cool down proccess, the thermosyphon was successfully detached from the mock-up magnet. Although more effective cooling can be achieved by high working pressure, the entire cooling range is not covered by a single fluid thermosyphon due to limitation of the critical temperature and the triple point temperature. Therefore, thermosyphons of various working fluids such as nitrogen, argon, oxygen and ethane should be used together to cover wide operating temperature range of a cryocooler. Moreover, the optimized design is necessary for effective operation of thermosyphon in consideration of its transient response.

ACKNOWLEDGEMENTS

This work was supported by Duksung Co., the Combustion Engineering Research Center of the KAIST and the Brain Korea 21 project.

REFERENCES

1. Prenger, F.C., D. D. Hill, D. D., Daney, D. E., Daugherty, M. A., Green, G. F., and Roth, E. W., Nitrogen thermosyphons for cryocooler thermal shunt, <u>Adv. Cryogenic Engineering</u> (1996), <u>Vol. 41</u>, 147-154
2. Prenger, F. C., Hill, D. D., Daugherty, M. A., Green, F. G., Chafe, J. and et al, Performance of cryocooler shunt thermosyphons, <u>Adv. Cryogenic Engineering</u> (1998), <u>Vol. 43</u>, 1521-1528
3. Prenger, F. C., Hill, D. D., Daney, D. E., Daugherty, M. A., Green, G. F. and et al, Thermosyphons for enhanced cool down of cryogenic systems, <u>Cryocoolers</u> (1997) <u>9</u>, 831-839

Experimental Characterization Of The Hydraulic Behavior Of Cooling For Toroidal Field Magnet Of SST-1

Sarkar B., Gupta N. C., Sahu A. K., Bansal G., Pradhan S., Dhard C. P., Misra Ruchi, Panchal Pradip, Tank Jignesh, Gupta G. and Y. C. Saxena

Institute for Plasma Research, Bhat, Gandhinagar 382 428, India.

In order to validate the cooling configuration of the Steady State Superconducting Tokamak (SST-1) Toroidal Field (TF) magnet system, an experiment has been proposed on the actual TF magnet without having cooling channels for the casing, which houses the winding pack. A special cryostat has been designed for this purpose. The test magnet has been cooled with the Helium Refrigerator/Liquefier (HRL) in automatic mode. Appropriate temperature, pressure and flow sensors have been implemented at strategic locations. The paper will describe the test arrangements as well as results emphasizing the understanding on mechanism of SST-1 TF magnet cool down.

INTRODUCTION

The SST-1 superconducting magnet system (SCMS) [1] comprises of sixteen Toroidal Field (TF) coils and nine Poloidal Field (PF) coils. Cooling of large scale forced flow magnet system has been a point of debate and limited experience is available as far as superconducting magnet in Tokamak configuration is concerned. In order to build confidence and gather experience before cooling the SST-1 SCMS, it is planned to perform a test on integrated system of actual TF magnet with electrical isolators/helium feed-through, joints without actively cooling the casing of the magnet. The aim of the experiment is to study the temperature profile of the casing of the TF magnet without actively cooling the same while maintaining flow in the hydraulic path, identical to the actual flow requirement. Each of the TF coil is modified D shaped and made up of six double pancakes. Therefore, 12 nos. of flow paths, each with 48 m path length, are available for passing the coolant. An NbTi based Cable-in-Conduit-Conductor with 40 % void fraction for flowing the coolant, has been used for winding both the TF and PF coils of SST-1 [2]. The TF coil casing consists of two side plates, an inner ring and outer ring. The thickness of casing is 10 mm everywhere except at the inner leg, where it is 25 mm. The TF winding pack (vacuum pressure impregnated) is shrink fitted inside the casing, keeping a gap of 5 mm for the winding pack to move under thermal/electromagnetic stress. The bore dimensions are 1190 mm (radial) and 1746 mm (vertical), whereas the outer dimensions are 1560 mm (radial) and 2120 mm (vertical).

SYSTEM DESCRIPTION

Figure 1 shows the schematic diagram of the test set up. The single TF magnet is assembled inside a specially designed cryostat (3 m diameter, 3.2 m height) and hydraulically connected to the helium refrigerator/Liquefier (HRL) through vacuum insulated transfer line mounted with appropriate cryogenic valves for the process flow control. The cryostat has been specially designed in 3 parts, namely the top dished end, bottom dished end, and the shell in order to facilitate ease in assembly of the TF single magnet with all sensors and diagnostics. Four support columns with heat intercepts are raised from the bottom-dished end and the TF magnet is assembled from the support bars with specially designed hangers for minimum contact. Load on per support hanger is approximately 500 kg. The TF magnet assembly is

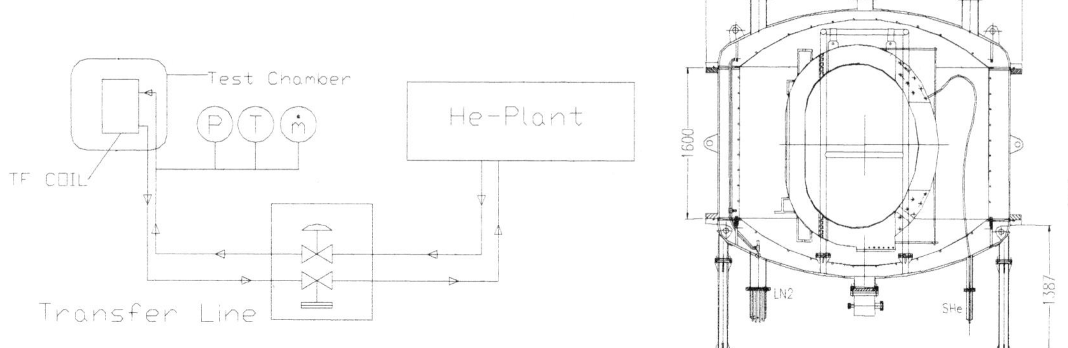

Figure 1 Schematic diagram of Test Setup

Figure 2 Assembly of TF magnet in cryostat

surrounded by an actively cooled liquid nitrogen (LN$_2$) shield, which is also made in 3 parts (top & bottom dished end and shell). LN$_2$ cooling paths are adjusted in such a way so that all the 3 parts of the LN$_2$ shield are able to cool simultaneously. Figure 2 shows the assembly of TF magnet in the cryostat on the support bars with hangers. Figures 3 (a) and 3 (b) show the location of the temperature sensors mounted on the hydraulic paths and on the casing of the TF magnet. All the temperature sensors are calibrated CERNOX sensors suitable to measure temperature in the range of 300 K to 4.2 K. Temperature Sensors T1 to T8 have been mounted on the hydraulic paths and T9 -T10 have been mounted on inter - pancake joints. Temperature sensors T11 to T 27 have been mounted at distributed locations on the TF magnet casing. Two numbers of KELLER make absolute pressure sensors at the inlet and outlet paths have been mounted to monitor the entry and exit pressures of the TF magnet. A differential pressure sensor across the magnet has been mounted to monitor the pressure drop. The total flow across the magnet is measured with a calibrated orifice flow meter. Four nos. of LAKESHORE make PT-102 (one no. each at top and bottom dished end and two nos. at the opposite sides of the shell) have been mounted on the LN$_2$ shield to monitor the temperature. A supervisory control and data acquisition system has been developed with programmable logic controller (PLC) to acquire all the measured parameters for analysis. All the necessary software has been developed indigenously on a base platform "WONDERWARE". In order to acquire the data, all the process parameters are first converted into the engineering range of 4-20 mA and fed to the PLC for processing. MIMIC pages display the on-line data as well as trend plots obtained continuously. The cryostat along with a 12 m long transfer line is pumped to a base vacuum of 2 X10^{-5} mbar with the help of 1000 l/s turbo-molecular pump. The background for helium gas inside the cryostat has been continuously monitored with a quadrupole mass analyzer.

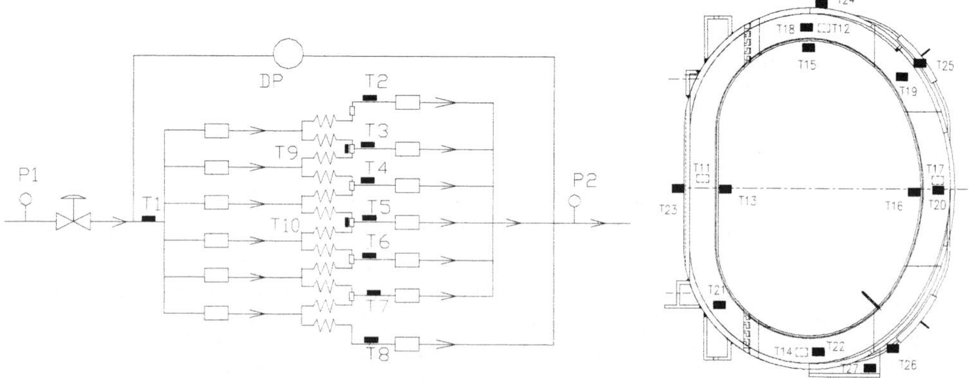

3(a)   3(b)

Figure 3 Location of the temperature sensors mounted on the hydraulic paths (3a) and on the casing of the TF magnet (3b)

COOLING SCHEME

Figure 4 shows the cooling scheme of the TF magnet along with the HRL. A 1 kW class HRL has been commissioned [3,4,] to cater the cryogen requirement at 4.5 K. On the down stream of the cold end of the HRL, an integrated flow distribution and control system (IFDC) [5,6] has been commissioned to distribute the SHe/LHe for the SST-1 SCMS. A branch line has been taken from the main supply and return line within the IFDC system and suitably connected to the test TF magnet with flow control valve. It is proposed to cool the single TF magnet in the same way as in the SST-1 SCMS, that is, by maintaining 50 K temperature difference between the inlet and outlet. Moreover, the pre-commissioning mode of the SCMS along with the HRL and on-line purifier also used for preparation of cleanliness inside of CICC.

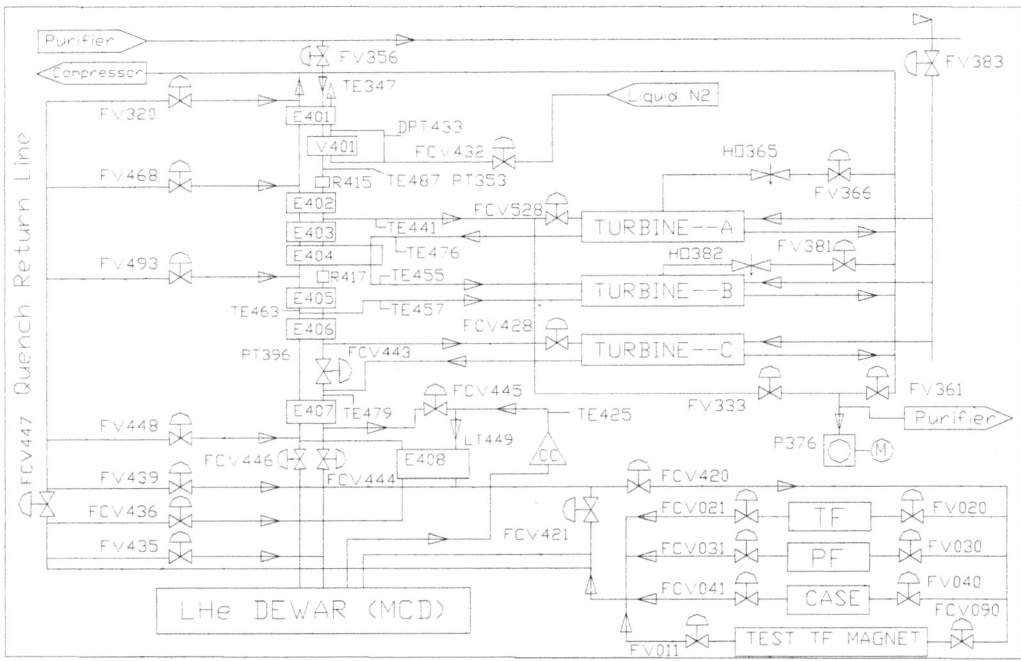

Figure 4: The cooling scheme of the TF magnet along with the HRL

The HRL consists of 7 nos. of heat exchangers to take care of the whole process. E 401/V401 are the first stage heat exchangers in the cold-box of the HRL. These have been designed with large margin to take care of the cool-down of SCMS of SST-1, for a cool-down flow rate adjusted at a maximum value of helium gas flow 50 g/s. The refrigeration capacity of the HRL, considering only E 401 is in the range of 15 kW and it is possible to cool-down the SCMS of SST-1 up to 100 K using the LN_2 exchanger. Turbines are, therefore, not operated. The outlet temperature is automatically controlled using valve FCV 432 which looks at the control loop for regulation. The control loop sequentially looks at the controlling parameters and adjusts the valve opening. The same procedure has been followed for cooling down the test magnet. The maximum of all the 27 temperature sensors is fed to the control loop as one of the controlling parameter for opening of valve FCV 432. The temperature-controlled flow of helium gas is diverted through FCV 443 – FCV 445 – FCV 420 and the return flow is back to the process circuit via FCV 447. The SCMS cool-down circuit of the HRL is adjusted in such a way that the return flow from the load joins the process circuit at appropriate location depending upon the temperature. The only problem, which was anticipated and was also observed, that for a single coil, the flow requirement at 300 K is very low and remains low throughout the whole temperature range of cooling in comparison to the SST-1 SCMS. Therefore, if the flow at the outlet of the HRL is regulated as per the requirement of the test TF coil, the coldbox circuits are warmed up. Therefore, in order to cool the cold-box circuits as well as to maintain the required flow in the test coil, sufficient flow has been taken from FCV 445 - FCV 420 and part of the excess flow is by-passed through FCV 031 – FV 030 of the IFDC system. The orifice flow meter at the outlet of the test coil ensures the flow rate requirement for cool down.

RESULTS AND DISCUSSIONS

Vacuum of the order of 5×10^{-5} mbar was successfully achieved in the cryostat. The TF coil inside circuit was pumped and purged 3 times with 10 ppm, pure helium gas before connecting the same with the HRL. The inlet temperature to the test TF magnet has been first set at 250 K keeping 50 K temperature difference. It has been observed that the HRL has automatically regulated the inlet temperature keeping 50 K temperature difference. The inlet temperature has been successfully reduced to 163 K while reporting the results. Thermo-hydraulic calculation shows that the flow requirement is 2.5 g/s at 300 K and increases to 4.5 g/s at 80 K and the actual flow was found to be in agreement. Figure 5 shows the typical cooling data obtained from the experiment. The results show that the overall average temperature is following the inlet to the hydraulics. The cooling time is, however, restricted due to the lagging in **maximum temperature**.

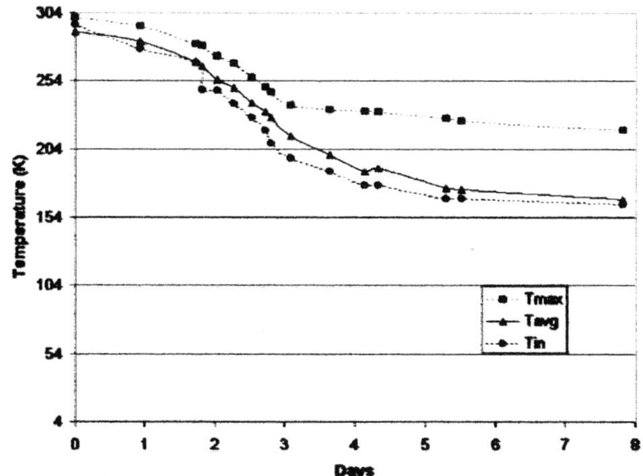

Figure 5 Typical cooling data obtained from the experiment

CONCLUSION

In order to develop confidence in operation of large volume superconducting magnets, such as in SST-1, it is necessary to perform test on at least one coil in actual configuration. In the present experiment, it is observed that the hydraulic paths are well cooled and follow the inlet temperature, whereas the casing lags. However, it is not possible to conclude the significant contribution of radiation cooling from the LN_2 shield to the casing as the inlet temperature is still above 90 K. In case any such contribution is observed during the experiment, active case cooling might probably reduce the lagging in temperature.

REFERENCES

1. Saxena Y. C. et al., Present Status of SST-1, Nucl. Fus. (2000), 40 (6), 1069
2. Saxena Y. C. et al, Magnet system of SST-1, Proc. of CRYOWORK, I.P.R., Gandhinagar, India
3. Sarkar B. et al., 1 kW class Helium refrigerator/liquefier for SST-1, Proc. 18th International Cryogenic Engineering Conference, Narosa Publishining House, New Delhi, India (2000) 123-126
4. Dhard C. P. et al., Commissioning and operational experience with 1 kW class helium refrigerator/liquefier for SST –1, Presented at CEC/ICMC-2003, Alaska (To be published in Advances in Cryogenics)
5. Das, S et. al., Integrated flow distribution and control system of super-critical and liquid helium for SST-1, Proc. 19th International Cryogenic Engineering Conference, Narosa Publishining House, New Delhi, India (2002) 165-1168
6. Sarkar B. et. al., Integrated cryogenic fluid flow distribution and cooling scheme with helium liquefier/refrigerator for SST-1 magnet system, Presented at MT-18, Morioka (to be published in IEEE trans. Mag.)

Performance analysis of counter flow plate-fin heat exchangers considering the effect of longitudinal conduction

Mi T.C., Li Y.Z., Wang J., Zhu Y.H.

School of Energy and Power Engineering, Xi'an Jiaotong University, Xi'an 710049, P.R. China

The performance of cryogenic plate-fin heat exchangers will degrade due to longitudinal heat conduction. A group of governing equations for a counter-flow plate-fin heat exchanger with double cold channels were put forward on the basis of theoretical analysis and eventually solved by using differential method to confirm the heat loss caused by longitudinal heat conduction. A dimensionless longitudinal heat diffusion coefficient is proposed to indicate the effect of longitudinal heat conduction in heat exchangers. A degradation factor τ is introduced to consider the deterioration in the performance of the heat exchanger due to longitudinal heat conduction.

INTRODUCTION

Plate-fin heat exchangers are used more and more widely because of their high performance and compactness. The longitudinal heat conduction through the fluids is in general negligible except in liquid metals at normal temperature. However, heat exchangers in cryogenic system have a large temperature gradient between the hot and cold ends. Longitudinal heat conduction through the wall, on the other hand, is an important irreversibility in the case of high effectiveness heat exchangers and those having high temperature gradient along the flow direction [1,2]. In cryogenic systems the desired heat load, i.e. the refrigeration load, is just a small fraction of the heat transfer within heat exchanger. In this case, the effect of axial heat conduction and parasitic heat transfer can lead to ineffectiveness of the heat exchanger and may dominate the performance of the device [3]. The attainable effectiveness is largely controlled by the specific heat capacity of fluids as well as the longitudinal thermal resistance [4]. In this paper, the influence of longitudinal heat conduction through counter flow plate-fin heat exchangers with double cold channel is discussed.

MATHEMATICAL MODEL

The following assumptions are made in this analysis [5]:

    (1) Constant fluid heat capacity, $mc_{p,h}$, $mc_{p,c}$;

    (2) No leak heat loss;

    (3) The entry length effects are not considered in the present analysis;

    (4) The thickness of the fin wall and parting sheet are small, so that the temperature in the fin thickness direction is uniform, the same to the parting sheet temperature in the direction normal to the fluid flows.

(5) Temperature profile along fins is determined according the equation:

$$t_w(y) - t_\infty = (t_{w,0} - t_\infty)\frac{ch[m(y-H)]}{ch(mH)} \tag{1}$$

Figure 1 illustrates a control volume consisting of the solid wall of the heat exchanger. Assuming steady state operation, there are four streams of energy flows that must be taken in account: the heat flowing into and out of the solid wall, the heat transferring from the hot fluid to the solid wall and the heat transferring from the cold fluid to the solid wall. The four streams of energy flowing into control volume compose the governing energy equation as follows:

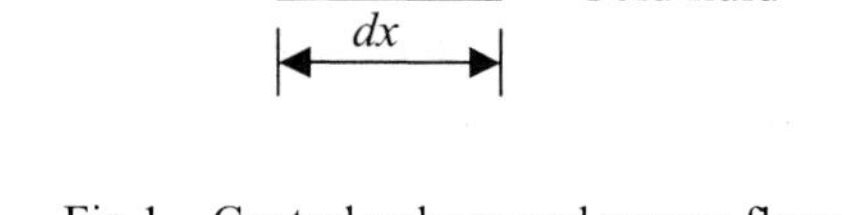

$$Q_h + Q_c + Q_{f,in} + Q_{f,out} = 0 \tag{2}$$

So the dimensionless temperature equation can be derived on the basis of equation 2.

Fig.1 Control volume and energy flows for solid wall

$$\frac{d^2\theta_w}{dX^2} + \frac{A_{ce}}{A_{we}}\frac{d^2\theta_c}{dX^2} + \frac{A_{he}}{A_{we}}\frac{d^2\theta_h}{dX^2} - \frac{[(KA)_c + (KA)_h]L^2}{\lambda_w A_{we}}\theta_w + \frac{(KA)_c L^2}{\lambda_w A_{we}}\theta_c + \frac{(KA)_h L^2}{\lambda_w A_{we}}\theta_h = 0 \tag{3}$$

The governing equations of fluids can be obtained from energy conservation as follows:

Cold fluid: $$\frac{d\theta_c}{dX} = \frac{(KA)_c L}{(\dot{m}c_p)_c}\theta_c - \frac{(KA)_c L}{(\dot{m}c_p)_c}\theta_w \tag{4}$$

Hot fluid: $$\frac{d\theta_h}{dX} = \frac{(KA)_h L}{(\dot{m}c_p)_h}\theta_h - \frac{(KA)_h L}{(\dot{m}c_p)_h}\theta_w \tag{5}$$

The parameters in above equations are defined as follows:

$$\theta = \frac{t - t_{c,in}}{t_{h,in} - t_{c,in}}, \quad X = \frac{x}{L}, \quad \Delta X = \frac{\Delta x}{L}. \tag{6}$$

$$A_{we} = A_w[1 + \frac{1}{ch(mH)_c}] + A_{2c}\frac{th(mH)_c}{(mH)_c} + A_{2h}\frac{th(mH)_h}{(mH)_h} \tag{7}$$

$$A_{ce} = A_{2c}[1 - \frac{th(mH)_c}{(mH)_c}] + A_w[1 - \frac{1}{ch(mH)_c}], \quad A_{he} = A_{2h}[1 - \frac{th(mH)_h}{(mH)_h}] \tag{8}$$

The boundary conditions can also be derived as follows:

$$X = 0; \quad \frac{d\theta_w}{dX} = 0 \quad \theta_h = 1 \qquad X = 1; \quad \frac{d\theta_w}{dX} = 0 \quad \theta_c = 0 \tag{9}$$

RESULTS AND DISSCUSSION

It is found from equation 3, 4 and 5 that $\frac{\lambda_w A_{we}}{L^2}$ is denoted as the capability of absolute longitudinal heat conduction of a heat exchanger. On the basis of dimensional analysis, a dimensionless longitudinal heat diffusion coefficient φ is defined as:

$$\varphi = \frac{\lambda_w A_{we}}{L(\dot{m}c_p)_{min}} \qquad (10)$$

which presents the relative longitudinal heat conduction ability, and is influenced by the wall thermal conductivity λ_w, the wall equivalent area A_{we}, the length of heat exchangers L and the minimum fluid heat capacity $(\dot{m}c_p)_{min}$. Furthermore, the heat transfer effectiveness of secondary surface is not as high as that of first surface, so the value of A_{we} is less than the practical cross-sectional area of plate-fin heat exchangers. Smaller A_{we} and longer L make φ smaller and benefit to reduce longitudinal heat conduction if heat exchanger material and flow condition are certain. Ideal condition is $\varphi = 0$ in which the longitudinal heat conduction won't exist. Therefore the dimensionless longitudinal heat diffusion coefficient φ possesses clear physic meaning, and indicates longitudinal heat conduction ability of plate-fin heat exchangers with secondary surface.

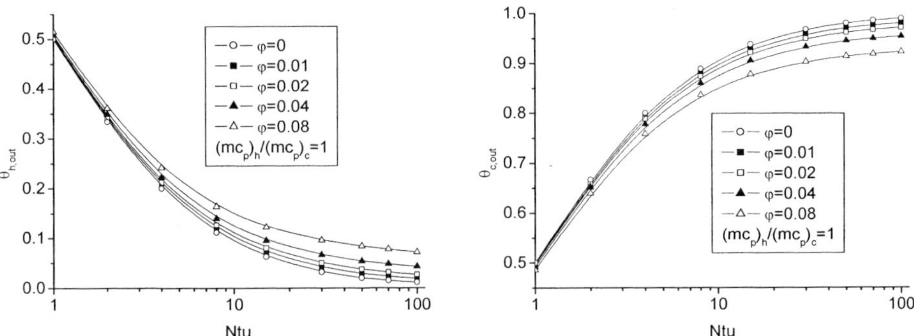

Fig.2　Relation of hot fluid outlet temperature and *Ntu*　Fig.3　Relation of cold fluid outlet temperature and *Ntu*

Figs. 2 and 3 show the relation of the hot and cold fluid outlet temperatures to the number of heat transfer units under different dimensional longitudinal heat conduction parameter φ and certain heat capacity rate $\dot{m}_h c_{p,h} / \dot{m}_c c_{p,c} = 1$. The results show that the outlet temperature of hot fluid will decrease along with the increase of NTU at a certain φ, while the cold fluid outlet temperature displays the reverse tendency. It is found that the increase of φ at a certain NTU will enhance the outlet temperature away from its desired point for both hot and cold fluid. It is evident as shown in those figures that longitudinal heat conduction results in the outlet temperature deviation from ideal outlet temperature because of the internal thermal dissipation, which degrades the performance of plate-fin heat exchangers. It is also derived that the longitudinal heat conduction has more influence on the effectiveness of a balanced flow $\dot{m}_h c_{p,h} / \dot{m}_c c_{p,c} = 1$ than that of an unbalanced flow. The reason is that the temperature gradient across the length of the wall is the maximum, which results in the maximum thermal dissipation.

In order to demonstrate the effect of longitudinal heat conduction, a degradation factor τ is defined to consider the deterioration in the performance of the heat exchanger due to longitudinal heat conduction through wall. It is mathematically expressed in terms of effectiveness as follows:

$$\tau = \frac{\varepsilon_0 - \varepsilon}{\varepsilon_0} \qquad (11)$$

Figs. 4 and 5 show the relation of the degradation factor τ and NTU, the dimensionless longitudinal heat conduction parameter and the ratio of heat capacity rates are denoted as φ and $\dot{m}_h c_{p,h} / \dot{m}_c c_{p,c}$, respectively. It is found that the degradation of heat exchangers effectiveness is the maximum for the balanced flow heat exchangers. Less effect happens for unbalanced flow heat exchangers. This is due to the fact that the temperature gradient across the length of the wall is a maximum for this value. It is also shown in figs. 4 and 5 that the degradation factor has the peak value at a certain NTU for balanced flow or unbalanced flow, but at different NTU position.

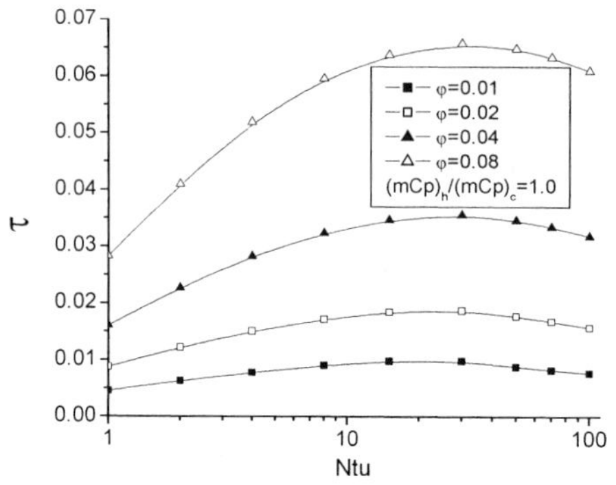

Fig.4   Relation of degradation factor τ and Ntu for a balanced flow

Fig.5   Relation of degradation factor τ and Ntu for an unbalanced flow

CONCLUSIONS

(1) A group of governing equations for a counter-flow plate-fin heat exchanger with double cold channel were put forward on the basis of theoretical analysis and eventually solved by using differential method.

(2) A dimensionless longitudinal heat diffusion coefficient is proposed to indicate the effect of longitudinal heat conductivity in heat exchangers. The effect of the longitudinal heat conduction on heat exchangers increases along with the increase of φ and Ntu, which will degrade the performance of Heat exchangers.

(3) The degradation of the heat exchanger effectiveness will get a maximum at $(\dot{m}c_p)_h / (\dot{m}c_p)_c = 1$, it is because the temperature gradient across the length of wall reaches the maximum value.

REFERENCES

1. S.Pradeep Narayanan, G. Venkatarathnam, Performance degradation due to longitudinal heat conduction in very high NTU counterflow heat exchangers, Cryogenics (1998) 38 927-930

2. Mi Tingcan, Li Yanzhong, Influence of Longitudinal Heat Conduction on a Counter-flow Plate-Fin Heat Exchanger, J. of Xi'an Jiaotong University, (2003) 37 1142-1145

3. Prabhat Gupta, M.D. Atrey, Performance evaluation of counter flow heat exchangers considering the effect of heat in leak and longitudinal conduction for low-temperature applications, Cryogenics (2000) 40 469-474

4. S.Pradeep Narayanan, G.Venkatarathnam, Performance of a counterflow heat exchanger with heat loss through the wall at the cold end, Cryogenics (1999) 39 43-52

5. G.F. Nellis, A heat exchanger model that includes axial conduction, Cryogenics (2003) 43 523-538

Investigation of header configuration and its effect on flow maldistribution in plate-fin heat exchanger

Wen J., Li Y.Z., Zhang K.

School of Energy and Power Engineering, Xi'an Jiaotong University, Xi'an 710049, P.R. China

In order to enhance the uniformity of flow distribution, a baffle with small holes of different diameters is recommended to install in the header. The flow maldistribution parameter S is obtained under different header configuration. When the baffle is properly installed with an optimum length, with stagger arranged and suitably distributed holes from axial line to baffle boundary, the ratio of the maximum velocity to the minimum drops from 3.04~3.44 to 1.57~1.68 for various Reynolds numbers. The improved configuration is of great significance in the improvement of plate-fin heat exchanger.

INTRODUCTION

Plate-fin heat exchangers are widely used in process industries because of their higher efficiency, more compact structure and lower costs than two-stream heat exchanger networks [1,2]. In the design of plate-fin heat exchanger, it is usually presumed that the inlet flow and temperature distribution across the exchanger core are uniform and steady. However, the assumption is generally not realistic under actual operating conditions due to various reasons. The design of the header significantly affects the velocity distribution approaching the face of exchanger core. The flow maldistribution effects have been well recognized and presented for heat exchangers. While the literature of improved configuration to enhance the flow uniformity in plate-fin heat exchanger is little in recent years. Zhang[3] proposed a structure of two-stage-distribution and the numerical investigation shows the flow distribution in plate-fin heat exchanger is more uniform if the ratios of outlet and inlet equivalent diameters for both headers are equal. In this paper, a simple way is put forward to homogenize the flow distribution. A baffle with small-size holes is installed in the traditional header to optimize the header configuration. The investigation on the effect of the configuration of the baffle on the flow distribution is presented.

BASIC CONFIGURATION AND ITS IMPROVEMENT

A schematic view of conventional header (denoted as configuration A) presented in this study is shown in Fig.1. There are 43 micro-passages in the outlet of the header. Composite constructive grids are used in the analog computation and the finest implemented grid involved about 245,817 cells. There are selective refined grids in some local place where parametric variation is severe. In this work, CFD software FLUENT was employed to simulate the fluid flow distribution and pressure drops in the header of plate-fin heat exchanger. Continuity equation and momentum equation are discretized using finite volume method and two-equation K-ε flow turbulent model is used in the calculation [4]. Semi-implicit SIMPLER Algorithm is used in the velocity and pressure conjugated problem and second order upwind difference scheme is used in convective terms [5]. Boundary conditions and convergent condition are as follows: Inlet fluid Reynolds numbers and pressure are given. The wall condition is adiabatic and no slip occurs on the wall. Convergence criterion is specified to residuals $\leqslant 1.0 \times 10^{-6}$.

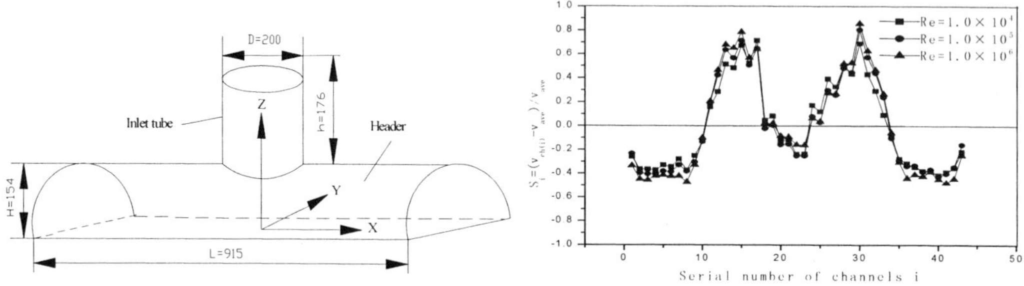

Figure 1   Model of header construction

Figure 2 S_i at different Re

Two parameters are introduced in this paper to evaluate the flow maldistribution, namely, relative flow maldistribution parameter S_i and absolute flow maldistribution S, which are defined as follows:

$$S_i = \frac{v_{ch(i)} - v_{ave}}{v_{ave}} \qquad (1)$$

$$S = \sqrt{\frac{1}{N-1}\sum_{i=1}^{N}\left(v_{ch(i)} - v_{ave}\right)^2} \qquad (2)$$

Where N stands for the passage number (here is 43), $V_{ch(i)}$ stands for the velocity of each passage and V_{ave} stands for the average velocity of all the passages.

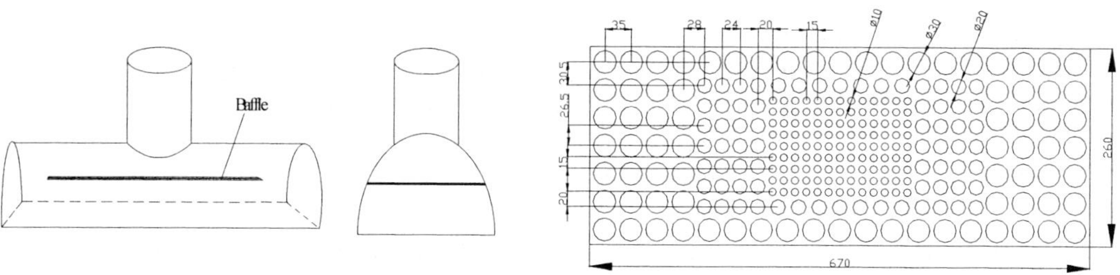

Figure 3   Definition of the baffle position in the header

Figure 4   Baffle construction of configuration B

Figure 5   Baffle construction of configuration C

Figure 6   Outlet velocity of different headers

Fig.2 presents the numerical results of conventional configuration A. The flow maldistribution of the outlet along the x direction is very serious and the absolute flow maldistribution S is equal to 0.95, where the average value for y direction is adopted. Because the flow header has larger dimension comparing to the inlet tube diameter, the fluid tends to go preferentially into the channels in the center. And it has been found that the best position for the perforated grid is midway between the inlet tube and the core of the header [6]. So a baffle with small holes is put forward to install at the 1/2 height of the header symmetrically, which is demonstrated in Fig.3. The small holes are arranged in the baffle according to the velocity distribution, and the punched ratio is gradually increasing in symmetry from the axial line to the boundary. It is presumed that $\int vdA$ is equal to a constant value under ideal condition. Thus the fluid flow is distributed uniformly before it reaches the header outlet and the expected object of uniform distribution is achieved.

The baffle configurations are demonstrated in Fig.4 and Fig.5, in which the baffle with in line arranged holes is denoted as configuration B (Fig.4) while the one with in stagger arranged holes is denoted as configuration C (Fig.5). For the improved configurations, the velocities increase in the zone of two ends of header and decrease in the zone near the axial line. Thus, the fluid flow is distributed more uniformly. Unfortunately the pressure drop may increase and result in the decrease of mean velocities to some extent, which is inevitable but not anticipated. So it is obliged to get the suitable baffle configuration for getting the optimum point of uniform flow distribution and pressure drops.

OPTIMIZATION OF BAFFLE CONFIGURATION

The velocity distribution of three header configurations is shown in Fig.6 in order to compare the effects of different hole distributions. The curves in Fig.6 illustrate the distribution characteristics of flow velocity and their differences for three configurations at similar working conditions. The inlet conditions are the same at Re=1.0×10^5 and p = 27kPa. It is indicated that the average velocity is 1.94m/s and the absolute maldistribution parameter is 0.36 for configuration B, while they are 2.67m/s and 0.32 for configuration C, respectively. It shows that the average velocity of configuration C is much larger than that of configuration B. It is easily understood that when the hole distribution in the baffle is changed from in-line arrangement to staggered arrangement, the punched ratio on the baffle will increase from 47% to 53%, and the flow resistance brought about by baffle necessarily decreases. Moreover, the increase of punched ratio leads to the increase of flow area on the baffle and further results in the decrease of S. The improvement of header configuration with a stagger arranged baffle should be selected firstly.

The location where the baffle is installed has been determined to the 1/2 height of the header, and the thickness of the baffle is determined to 5mm, so the baffle is 260mm in width. The diameters of the three kinds of holes are the same as described above. Fig.7 shows the flow distribution performance along with the change of relative length of baffle to header at Re=1.0×10^5. The average velocity decreases when the baffle length increases, and also the flow resistance increases and brings about the increase of pressure drops. The influence of the baffle on flow distribution is significant along with the increase of baffle length since the absolute maldistribution parameter S decreases. Combinative consideration of the relationship between the average velocity and the absolute maldistribution parameter S, leads to selection of the baffle length as 670mm, which is just 3/4 of the length of header.

For the baffle configuration as above mentioned, the average velocity of configuration C is larger than that of configuration A due to the decrease of the highest fluid velocity. For configuration C, the velocity distribution is mostly concentrated in the range of 2.5 and 4.0 m/s. The numbers of passages with the flow velocity between 2.5 to 4.0 m/s takes about 72% of the whole passages at Re=1.0×10^5. While for configuration A, the velocity distribution is concentrated between 1.0~2.5 m/s, which takes about 65% of the numbers of whole passages. The flow velocity ratio of the maximum to the minimum drops from 3.04~3.44 of configuration A to 1.57~1.68 of configuration C, which reflects a more uniform flow for the improved header. From the above discussion, the effect of header configurations on flow maldistribution is prominent. The flow velocity of the passages near the boundary can be enhanced effectively by changing the header configuration from A to C. The flow velocity distribution of configuration C gives the most uniform result among the cases considered in this paper.

From above discussion, it can be concluded that the determination of baffle configuration has the relationship with the diameter of inlet tube, the length and diameter of header, the diameters and distribution of holes when the baffle installation location has been defined. Fig.8 shows the distribution of punched ratio, in which the staircase curve is drawn according to the realistic condition and the smooth curve is simulated from the former. The correlation of punched ratio along with the position in the direction of baffle length under the ideal situation should be established as follows:

$$\delta = 47.75 + \left(\frac{X}{100}\right)^2 \tag{3}$$

where δ stands for the punched ratio (%) and X stands for the X position along with the baffle length (mm). In practical condition, if the holes on baffle can be punched according to the simulated smooth curve, we can get $\int v dA$=const and the flow can be distributed more uniformly than the configuration C.

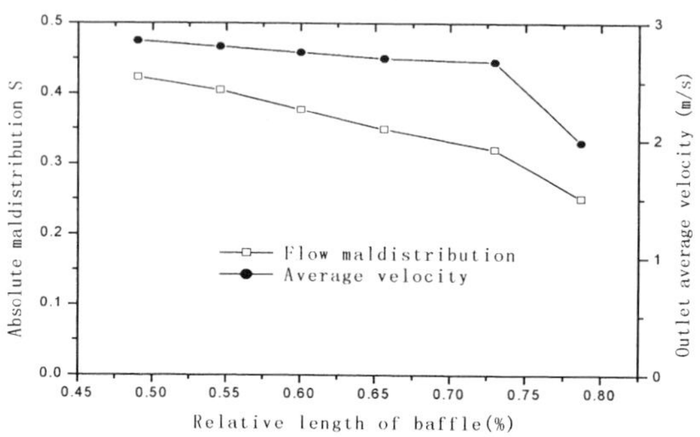

Figure 7 S and average velocity versus baffle length

Figure 8   Punched ratio versus x position

CONCLUSION

The results of calculation indicate that the improved header configuration can effectively enhance the fluid flow uniformity. The flow absolute maldistribution parameter S in plate-fin heat exchanger has been reduced from 0.95 to 0.32 by installing the baffle. When the baffle is proper in length, the holes are distributed in staggered arrangement, and the punched ratio gradually increases from axes along with the dam board length, the ratio of the maximum flow velocity to the minimum flow velocity may drop from 3.04~3.44 to 1.57~1.68 for various Reynolds numbers. The fluid flow distribution in plate-fin heat exchanger is more uniform by the optimum design of the header configuration. The baffle is lower in cost and convenient in assembly, while the effect of the fluid flow distribution uniformly by the improved configuration is obvious. The conclusion of this paper is of great significance in the improvement of plate-fin heat exchanger.

REFERENCES

1. T.F. Yee, I.E. Grossmann, Z. Kravanja, Simultaneous optimization models for heat integration I, Area and energy targeting and modeling of multi-stream exchangers, Comput. Chem. Eng (1990) 14 1151 – 1164.

2. B.-D. Chen, Z.-W. Lu, B.-Y. Liu, A study on heat exchanger network with multi-fluid heat exchanger, in: K. Chen, G. Sarlos, T.W. Tong (Eds.), Energy and Environment, China Machine Press, Beijing, China (1998)313 – 319.

3. Zh. Zhang, Y.Z. Li, CFD simulation on inlet configuration of plate-fin heat exchangers, Cryogenics (2003) 43 673-678.

4. John D. Anderson, Computational Fluid Dynamics, The Basic With Applications, McGraw-Hill Companies (1995)

5. Tao Wenquan, Numerical Heat Transfer, Xi'an: Xi'an Jiaotong University Press (2002)

6. Lalot S, Florent P, Flow maldistribution in heat exchangers, Applied Thermal Engineering (1999) 26 847-63

Experimental Investigation on High Performance Cryogenic Heat Transfer Based upon Natural Circulation Cooling

Limin Qiu, Weizheng Li, Zhihua Gan, Xuejun Zhang, Bo Jiao, Guobang Chen, Canjun Yang

Cryogenics Lab，Zhejiang University，Hangzhou，310027, China

A novel cryogenic heat transfer unit is proposed, which combines NCC and CHP together. The unit takes advantages of NCC and CHP, and eliminates the disadvantage of NCC for its low thermodynamic efficiency. Furthermore, it has also the merit of a self-feedback process, which makes the object come back to the objective temperature quickly. In order to verify the principle, a compound system of NCC and CHP was set up and tested. And experimental results are compared with that simply cooled by a NCC or a CHP, which show the promising potential for further applications. An application to the Power MEMS cooling is also presented.

INTRODUCTION

With the rapid developments of cryogenic engineering, it is of the most importance to build an efficient "bridge" between the cryogenic source and applications. It is always a problem to transfer heat within some distance efficiently and quickly, especially for some special occasions such as space and superconducting applications. Up to now, there are several means to solve the problem, such as thermal conduction by pure metal (e.g. copper), cryogenic heat pipe (CHP), natural circulation cooling (NCC) and so on. Recently, we have studied the principles of natural circulation cooling [1-2]. Theoretical and experimental results show that the NCC can cool down an object fleetly. Essentially, the direct driving force of NCC comes from the temperature difference between the object to be cooled and the cold source, which generates the pressure difference through the cryogenic fluid evaporation. Obviously, the driving force of NCC will gradually weaken and vanish at last during the cool-down process. So the NCC only serves for fast cooling with low efficiency instead of continuous operation. Moreover, there are some problems of temperature instability and mechanical vibration due to the exquisite evaporation of cryogenic fluid. The cryogenic heat pipe (CHP) is an acknowledged alternative to realize continuous cooling with high efficiency, because of its small axial temperature difference[3-4]. However, the cool-down process of the CHP is slow and it will not work if over-loaded.

Therefore, we proposed to combine the NCC and the CHP together and make them into a cryogenic heat transfer unit[5], which synthesizes the advantages of the NCC and the CHP. We make use of NCC to shorten the cool-down process and also simultaneously cool down the CHP. Once the NCC nearly stops, the CHP starts to work. It serves for continuous operation with high efficiency and keeps the temperature stable. Especially, once the temperature of the object sharply increases (e.g. quench), the NCC may automatically start again and cool down the object to the objective temperature fleetly. So the new cryogenic heat transfer unit has a self-feedback function, which improves the security of the whole systems. In order to verify the principle, an experimental setup was designed and tested. The preliminary results accords with those we expected. The heat transfer unit has been successfully applied to Power

MEMS as a cold source.

EXPERIMENTAL SETUP

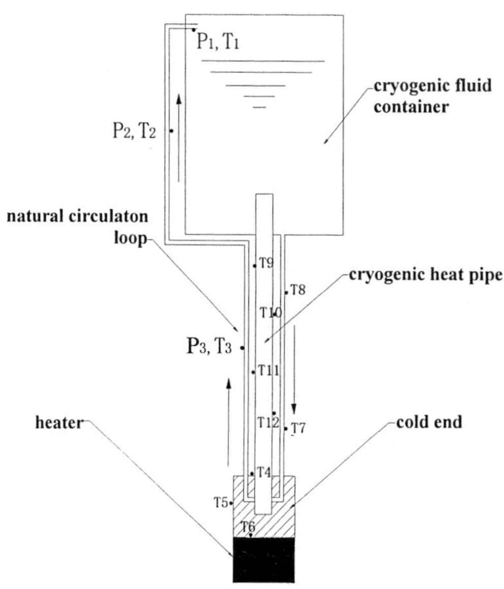

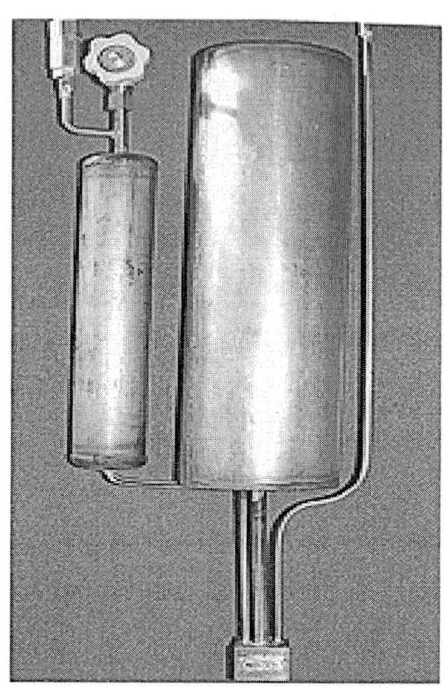

Figure 1 Schematic of the experiment set-up and photograph

The NCC+CHP unit consists of cold source, cryogenic heat pipe, natural circulation loop and heater. For convenience, liquid nitrogen is used for the cold source. The liquid nitrogen is stored in a stainless steel (1Cr18Ni9Ti) container with volume 5×10^{-4} m^3. The insulation material surrounded the container is polyurethane foam whose thermal conductivity is around 0.0026 to 0.028 W/m•K. The thickness of insulation material is 3×10^{-2} m. Two copper thermosyphons with a length of 0.2 m and inner diameter of 6×10^{-3} m are used. The condensation section of the thermosyphon is immersed in the bottom of liquid nitrogen container, while the evaporation section shares the same copper body with the cold end as shown in Figure 1. The working fluid in the thermosyphon is nitrogen with high purity. The natural circulation loop (inner diameter of 4×10^{-3} m, length of 0.8 m) is made of stainless steel to decrease the axial thermal conduction. The inlet of the natural circulation loop is located at the bottom of the container, while the outlet is above the liquid nitrogen level. Themosyphons, the natural circulation loop and the cold end are also insulated. A heater with maximum input power of 50 W is installed at the bottom of the cold end. The input power to heaters is measured by a digital power-meter.

Twelve calibrated copper-constantan thermocouples are arranged along the natural circulation loop (T1, T2, T3, T7, T8), the thermosyphon (T9, T10, T11, T12), and the cold end (T4, T5, T6), as show in Figure 1. A Keithley 2700 digital multi-meter is used to acquire the temperature signal. Three piezoelectic pressure sensors (Siemens, KPY-45R) are installed along the natural circulation loop to observe the fluid flow (see Figure.1). Pressure and temperature data are processed by a computer with the software of Labview 6.1 and a DAQ card (NI 6023E).

EXPERIMENTAL RESULTS AND DISCUSSION

Figures 2 shows the cool-down process of the cold end with $Q_0=0$ W. Q_0 is the initial heating power applied to the cold end. It takes about 22, 30 and 47 minutes to reach stable states at 81, 83 and 87 K for the NCC+CHP, the NCC and the CHP, respectively. In respect of the cool-down time and the final objective temperature, the NCC+CHP is the best. For NCC+CHP, the NCC cools down not only the cold end but also the CHP, which enhances the heat transfer performance of CHP. Simultaneously, the CHP directly transfers heat from the cold end to the cold source, which improves the flow of cryogenic liquid in natural circulation loop. Because the CHP is filled with gas at the beginning and needs more time to pre-cool itself as well as the cold end, the cool-down time is much longer and the final objective temperature is higher than those of NCC+CHP.

Figures 3 shows the cool-down process of the cold end with $Q_0=10$ W. The tendency of cool-down processes is similar with those in Figure 2. The effect of initial heating power on the heat transfer performance of the NCC+CHP and the NCC is not significant. The final objective temperatures at the cold end are almost the same for the NCC+CHP and the NCC. The cool-down time is a little longer than those in Figure 1 only because of the initial heating power. The effect of initial heating power Q_0 on the heat transfer performance of CHP is remarkable. Because there is a reservoir connected with thermosyphons (Figure 1), the CHP needs more time to cool down the gas besides the cold end. It is estimated that the cool-down time will be shortened and the objective temperature will be lower if we cut the connection between the reservoir and thermosyphons.

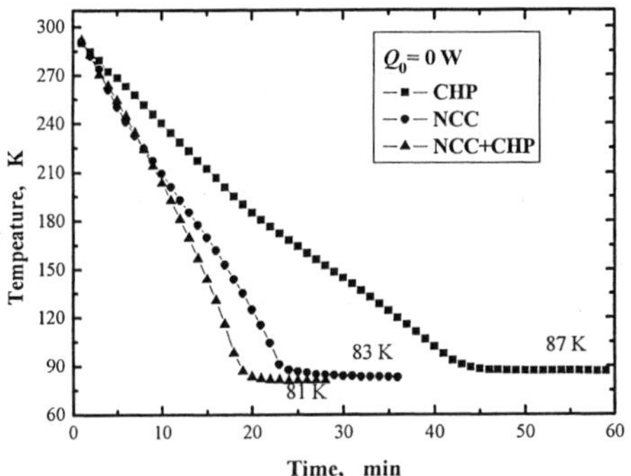

Figure 2   Cool-down process of cold end with $Q_0=0$ W

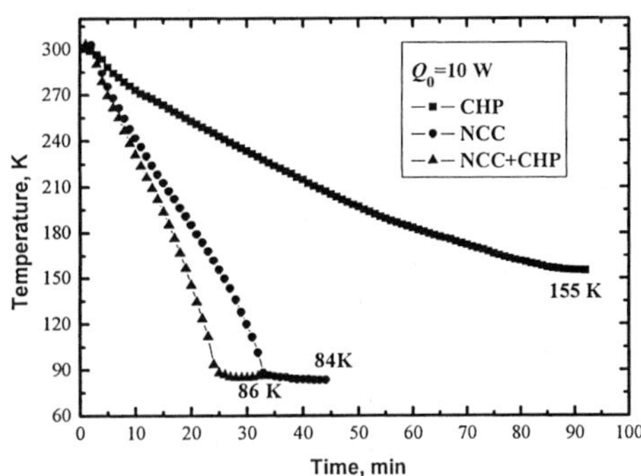

Figure 3   Cool-down process of cold end with $Q_0=10$ W

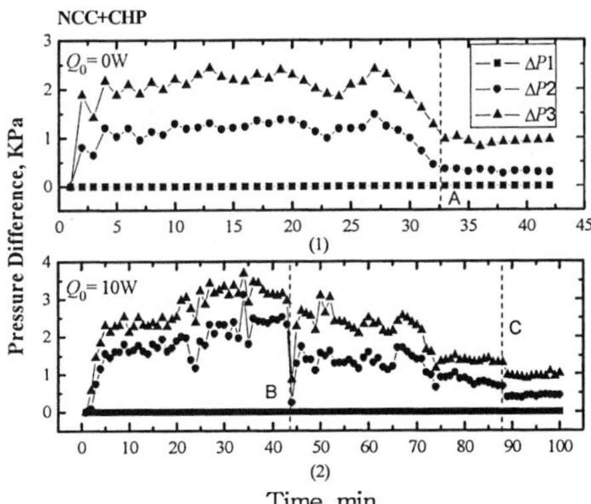

Figure 4   Pressure difference variations of NCC+CHP during cool-down process

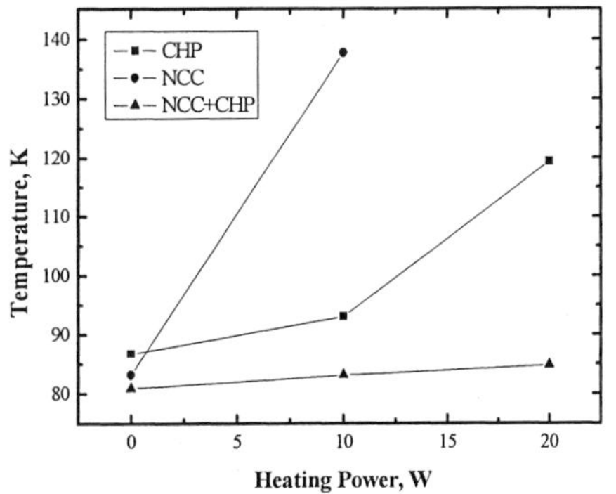

Figure 5   Cooling performance of CHP, NCC and NCC+CHP

Figure 4 shows the pressure difference variation along the NCC during the cool-down process of NCC+CHP. The benchmark of the pressure difference is based on P1 as shown in Figure 1. The pressure difference gradually attenuates and reaches the minimum, such as the point A, B and C (Figure 4) which means the NCC can hardly circulate. In Figure 4(2), the NCC may circulate again from point B when some nitrogen gas at room temperature is charged into the circulation pipe. And the circulation still terminates finally.

Figure 5 shows the comparison of the cooling performance among CHP, NCC and NCC+CHP. The heating power was applied to the cold end after the cold end cooled down without heat load. With the same heat load, the NCC+CHP reaches the lowest temperature, where the NCC is the highest (except Q_0 $=0$ W). The driving force of NCC greatly reduces after the cool-down process. Once heated, the temperature of cold end increases rapidly, which means that the NCC is not good for steady operation. On the contrary, the CHP works well to transfer heat from the cold end to the cold source and keeps the temperature of cold end lower after the cool-down process for the NCC+CHP.

Furthermore, the new cryogenic heat transfer unit was successfully applied to the research of power MEMS, which makes use of Seebeck effect to generate electricity by semiconductor chips. The cryogenic heat transfer unit serves as a cold source and ensures the temperature difference between the two sides of the semiconductor chip. With a temperature difference of 400 K, 4.9 V is produced by two pieces of semiconductors with the size of 25×25mm.

CONCLUSIONS

1. A new type of cryogenic heat transfer unit is proposed, which makes full use of the advantages of natural circulation cooling and cryogenic heat pipe. It can not only cool down an object fleetly but also keep the object temperature stable with high efficiency. The NCC may automatically circulate again when the object temperature increases sharply.

2.   The NCC plays important role during the cool-down process, while the CHP is primary during the steady operation to maintain the cold end at a lower temperature.

ACKNOWLEDGEMENT

The project is financially supported by the Foundation for the Author of National Excellent Doctoral Dissertation of China (200033) and Excellent Young Teacher Project of Department of Education, China.

REFERENCES

1.   G.B. Chen, Y.K. Zhong, X.L. Zheng, Q.F. Li, X.M. Xie, Z.H. Gan, Y.H. Huang, K. Tang, B. Kong, L.M. Qiu, Experimental study on natural circulation precooling of cryogenic pump system with gas phase inlet reflux configuration, Cryogenics (2003). 43 693-698.

2.   G.B. Chen, Y.K. Zhong, Y.L. Jiang, Z.H. Gan, Chinese patent: The means of natural circulation cooling, Patent Number: 01100968.3.

3.   Nakano, A., Shiraishi, M., Nishio, M., Murakami, M., An experimental study of heat transfer characteristics of a two-phase nitrogen thermosyphon over a large dynamic range operation, Cryogenics (1998) 38 1259-1266.

4.   W.Z. Li, L.M. Qiu, X.J. Zhang, P. Chen, Y.L. He, Experimental Investigation on Heat Transfer Performance of A Cryogenic Thermosyphon, ICEC20 (2004).

5.   L.M. Qiu, W.Z. Li, Z.H. Gan, G.B. Chen, Chinese Patent : The Apparatus Of High Performance Cryogenic Heat Transfer Based On Natural Circulation Cooling, Application Number: 200310109516.1.

Film Boiling Modes in Weakly Subcooled He II around Lambda Pressure

Nozawa M., Kimura N.[*], Murakami M., Zhang P.[**]

Institute of Engineering Mechanics and Systems, Univ. of Tsukuba, Tsukuba 305-8573 Japan
[*]Cryogenics Science Center, Applied Research Lab., High Energy Accelerator Research Organization, Tsukuba 305-0801 Japan
[**]Institute of Refrigeration and Cryogenics, Shanghai Jiao Tong Univ., Shanghai 200030, P.R.China

Film boiling modes in both subcooled and saturated He II were experimentally investigated. It was found that there were two film boiling modes clearly bounded by a line just above the lambda pressure (p_λ). The one is a typical subcooled film boiling mode that normally appears at still higher pressure than p_λ. The other film boiling mode resembles noisy film boiling in saturated He II. It is found that in the vicinity of the region of the lambda point or the lower heat flux, noisy film boiling does not occur and subcooled region directly connects with silent film boiling one.

1. INTRODUCTION

For cooling of large superconducting magnets that are required for generating a high magnetic field, He II has been frequently utilized as a coolant in its subcooled state that is referred to as He II_p. When a superconducting magnet is quenched, boiling may occur in coolant as a result of large amount of heat generation due to Joule heating. It is a natural understanding that for He II in a pressurized state at a pressure higher than the lambda pressure p_λ (=5.04kPa) the subcooled film boiling occurs. On the other hand, when He II is in nearly saturated state when the pressure is lower than p_λ, the noisy or silent film boiling occurs. In the past, studies on saturated film boiling in He II have been conducted frequently [1,2,3,4]. But little studies have been attempted to reveal the physical nature of subcooled film boiling [5,6].
In the present study the characteristics of film boiling in both subcooled and saturated He II have been experimentally investigated. It is a characteristic feature of the present study that a single cryostat was used for the measurement of the change of boiling modes in both subcooled and saturated He II. A pressurized He II cryostat was utilized, where the pressure (p_{bath}) can be set arbitrarily in a range from the atmospheric pressure (101kPa) down to the saturated vapor pressure of He II. In our previous study, it was reported that the noisy film boiling occurred even at the pressure just above p_λ [7]. This result suggested that the boundary between the weakly subcooled film boiling and the noisy film boiling should be investigated in more detail. In the present study, the boiling mode map was drawn by taking the temperature (T) and the heat flux (q) as parameters. The condition for the appearance of the subcooled, the noisy and the silent film boilings in He II was investigated, and the influence of He I layer on boiling modes, which appeared in the case of subcooled film boiling was examined.

2. EXPERIMENTAL APPARATUS

The cryostat shown in Figure 1. is of a Claudet type designed for pressurized superfluid helium experiments. The detail of the cryostat system was described in Ref.[7]. The He II_p vessel is equipped with optical windows for visual observation. The visualization optical system, a compact Schlieren system, is also shown in Figure 1. The visualization image is taken by the high-speed video camera.
In the test section shown in Figure 2., the planar heater (25 mm x 25 mm) is located horizontally. The pressure and the superconductive temperature sensors are fixed at 10 mm and 5 mm above the center of the heater, respectively. For an experimental run, the planar heater is heated in the form of a square wave in time for 0.20 second.

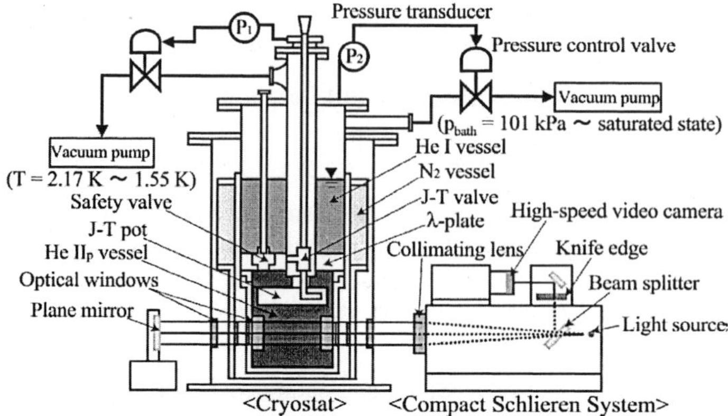

Figure 1. Schematic illustration of the pressurized superfluid cryostat system and the optics for visual study.

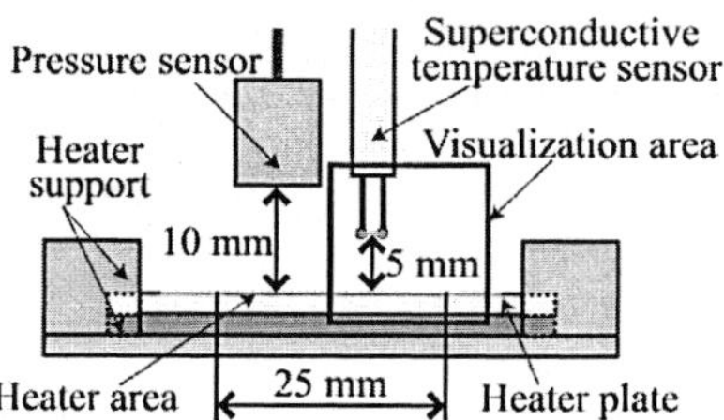

Figure 2. Schematic illustration of the test section in the He II$_p$ vessel and the locations of the heater, and the pressure and superconductive temperature sensors.

3. RESULTS AND DISCUSSION

3.1 Two boiling modes just above p$_\lambda$

Two boiling modes appeared at the pressure just above p$_\lambda$.   One mode is observed even at relatively higher pressure than p$_\lambda$.   We understand this boiling mode is classified into the weakly subcooled film boiling in He II.   A very thin vapor layer weakly oscillates only at the vapor-liquid interface.   And the pressure oscillation is also very weak with a frequency around one thousand Hz, but the temperature variation is not detected.   This film boiling mode is referred to as A-mode for the present.   The other boiling mode appears when the pressure is slightly decreased below the A-mode state but still higher than p$_\lambda$.   This mode change between the two is rather drastic.   This mode is now identified to the noisy film boiling in saturated He II, which is characterized by a large vapor bubble to repeat formation and crush, and accompanied with loud noise and hard vibration with a frequency of several hundred Hz.   As for the pressure and the temperature variation data compared with the A-mode boiling, the amplitude of the pressure oscillation is increased tremendously and the frequency is rather decreased.   The temperature oscillation is also very large in magnitude and the frequency is just the same as the pressure oscillation. It is confirmed from the visualization study that the frequency just corresponds to the vapor bubble cycle to repeat formation and crush.   It is reported that in the noisy film boiling in saturated He II the condition of the vapor layer is quite unstable and the vapor layer does not always exist on the heater [2]. Nevertheless, this boiling mode is referred to as B-mode boiling for the present.

3.2 Boiling mode map

The boundary dividing the two boiling modes is examined in detail.   Figure 3. shows the boiling mode map plotted on a pressure-temperature (p-T) diagram, which is experimentally obtained.   It is found that the boundary between A-mode and B-mode always exists at the pressure a little higher than p$_\lambda$, and the pressure value of the boundary increases as the drop of He II temperature.   The boundary line adjacent to the saturated vapor pressure line (S.V.P) is the boundary between the noisy film boiling and the silent film boiling [8], which is designated as the lower boundary.   The silent film boiling is characterized by the fact that significant pressure and temperature variations are not detected and by quite a weak thin vapor layer oscillation [3]. In this sense, the A-mode boiling is regarded as identical to the silent film boiling. On the other hand, when the upper and lower boundary lines are extrapolated toward high temperature, the two boundary lines seem to meet at a little lower temperature than λ-point. In fact, in the experiment in the vicinity of the λ-point, the noisy film boiling never arose, and the regions of the weakly subcooled and the silent film boiling modes are regarded as directly connected. Therefore, it is indicated that A-mode boiling and the silent film boiling are identical.

Figure 4. shows the boiling mode map plotted on a pressure-heat flux (p-q) diagram at He II temperature of 1.9 K. It is seen that the pressure value of the upper boundary increases with the heat flux. On the contrary, if the boundary line is extrapolated to the lower heat flux, it is, however, seen that the pressure value of the upper boundary will reach p_λ at the heat flux of about 5 W/cm^2 at 1.9 K. It is experimentally confirmed that the film boiling occurs when the heat flux is larger than about 5 W/cm^2 at 1.9 K. Therefore, it is thought that the pressure value of the upper boundary is certainly p_λ, for the case of small heat flux. However, the noisy film boiling, which is a kind of noise-induced instability, occurs even at higher pressure than p_λ due to a non-ideal effect of large heat flux. Shown in Figure 5. is the enlarged diagram around the lower boundary between the noisy and the silent film boilings as shown in Figure 4. It is indicated that the pressure of the boundary is the lowest for the heat flux about 10 W/cm^2, and it increases as the heat flux decreases. So, it is natural to consider that around the heat flux at which the film boiling begins to occur, 5 W/cm^2 at 1.9 K, the upper and the lower boundaries are connected.

From the discussion mentioned above, some suggestion for the nature of each boiling mode is given. The condition that fixes the upper boundary is whether the influential He I layer exists to the degree being decisive of the boiling mode adjacent to the heater. The noisy film boiling is a strongly unstable film boiling mode of which thermo-fluid dynamic state is determined exclusively by He II and helium vapor. However, when the heat flux is large, the noisy film boiling occur even at the pressures just above p_λ. In this state, He I certainly exists between helium vapor and surrounding He II, but the influence of He I on

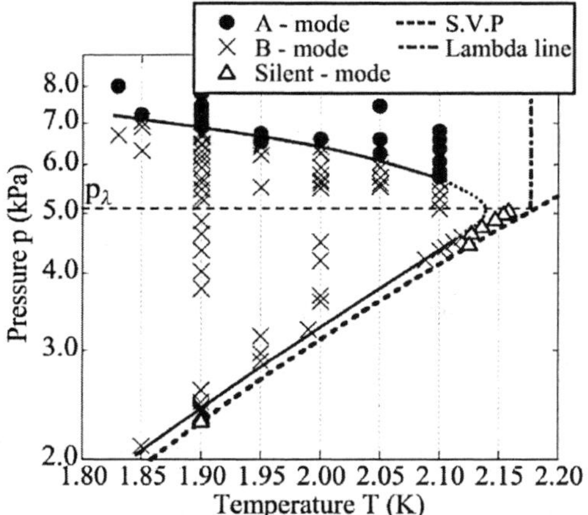

Figure 3. Boiling mode map on p-T diagram. q = 15 W/cm^2, planar heater of 25 mm x 25 mm.

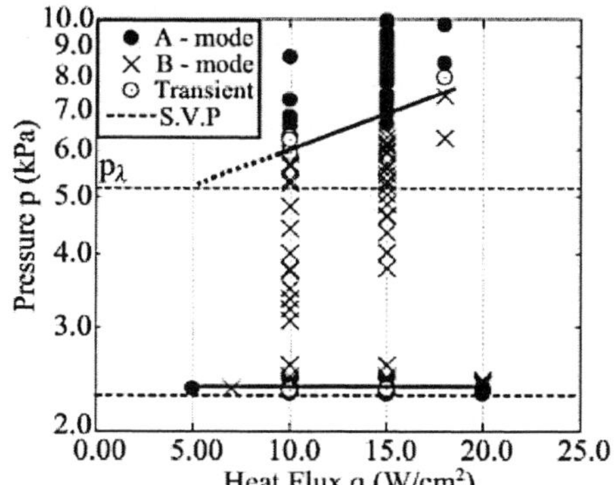

Figure 4. Boiling mode map on p-q diagram near critical q for the appearance of film boiling. T = 1.9 K.

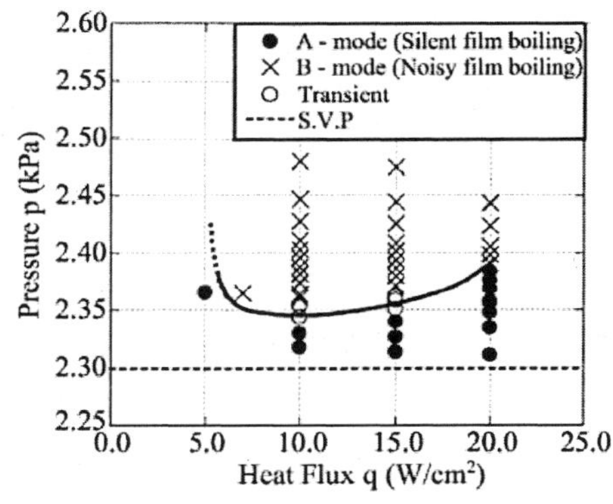

Figure 5. Detailed boiling mode map on p-q diagram near S.V.P. curve for the data presented in Fig.6. T = 1.9 K.

the boiling condition is quite a little.　It is assumed that in the regions adjacent to the λ-point and of the small heat flux which gives the onset condition for film boiling where the A-mode boiling and the silent film boiling are connected, the boiling is in a normal film boiling mode that occurs in common fluids. In the A-mode boiling including the silent film boiling, unstable behavior is recognized only on the surface of vapor film.　On the other hand, in the B-mode boiling, that is to say the noisy film boiling, the vapor violently repeats the formation, growth and crush.　In this sense, the noisy film boiling is a kind of large-scale unstable phenomenon that generates the large oscillation of the heater temperature and the heat transfer coefficient.

4. CONCLUSIONS

The variation of the film boiling mode at the pressure around the p_λ or lower was investigated in detail. The following conclusions are drawn.
1. The subcooled and the silent film boiling are the same boiling mode and these film boiling modes are identical to the normal film boiling that occurs in common fluids.　And the noisy film boiling is a kind of large-scale unstable phenomenon that generates the large oscillation in the heater temperature and the heat transfer coefficient, and it is the characteristic boiling mode in He II
2. In the regions very close to the λ-point and of the small heat flux which gives the onset condition for film boiling, the noisy film boiling does not occur.
3. The boundary between the weakly subcooled and the noisy film boilings in He II exists just above p_λ. This boundary increases as the He II temperature decreases or the heat flux increases.

ACKNOWLEDGEMENT

The authors would like to thank Cryogenics Science Center and Mechanical Engineering Center at KEK for their professional support and assistance during this work.
This study was partially supported by the Japan Society for the Promotion Science, Grant-in Aid for Science Research and the auspices of the NIFS Collaborative Research Program.

REFERENCE

1) Zhang P., Murakami M., Wang R.Z. and Takashima Y., Investigation of noisy film boiling under various thermal conditions in He II, Advances in Cryogenic Engineering (2000), 45 1017-1024
2) Yamaguchi M. and Murakami M., Study of pressure oscillation during noisy film boiling in He II, Cryogenics (1997) 37 523-527
3) Katsuki Y., Murakami M., Iida T. and Shimazaki T., Visualization study of film boiling onset and transition to noisy film boiling in He II, Cryogenics (1995) 35 631-635
4) Kobayashi H. and Yasukochi K., Maximum and minimum heat flux and temperature fluctuation in film boiling state in superfluid helium, Advances in Cryogenic Engineering (1980), 25 372-377
5) Takashima Y., Murakami M. and Toyoshima T., Experimental study on the transition of He II boiling modes with the hydrostatic pressure around the lambda temperature, Advances in Cryogenic Engineering (2000), 45 1057-1064
6) Murakami M., Yamaguchi M., Yanase N. and Inaba H., Various film boiling states in He II at hydrostatic pressure from saturated vapor pressure to 1 atm, Advances in Cryogenic Engineering (1998), 43 1425-1432
7) Nozawa M., Kimura N., Murakami M. and Yamamoto I., Variation of Subcooled Film Boiling State in He II with the Pressure, Advances in Cryogenic Engineering (2004), 49 (to be published)
8) Zhang P., Murakami M., Wang R.Z. and Inaba H., Study of film boiling in He II by pressure and temperature oscillation measurements, Cryogenics (1999) 39 609-615

Modeling of multilayer vacuum insulation – complexity versus accuracy

Chorowski M., Polinski J.

Wroclaw University of Technology, Wybrzeze Wyspianskiego 27, 50-370 Wroclaw, Poland

A thermodynamic analysis of different insulation systems used in low temperature installations is presented and the desired features of cryogenic insulation are derived. The most efficient and best matched to cryogenic conditions is a multilayer vacuum insulation (MLI). A simple mathematical model of heat transfer through MLI is proposed. Available experimental data and are compared with the model output. The limitations of the model applicability are defined and the possibilities of its upgrade discussed. A sensitivity analysis of heat flux variation with the selected physical properties of materials used to MLI production is done.

INTRODUCTION

Even the most efficient thermal insulation can not prevent parasitic heat flows to low temperature parts of any cryogenic system. To maintain a constant temperature of the cryostated object, a heat flux through thermal insulation must be compensated by cooling power generated by a refrigerator. The minimal specific input power w of the reference Carnot refrigerator, which compensates parasitic heat flux from ambient temperature T_H to cryostatic temperature T_C through the insulation with thermal conductivity k, can be described by equation 1:

$$w = k \frac{(T_H - T_C)^2}{T_C} \tag{1}$$

It can be derived from engineering experience, that independently on a low temperature level, the refrigerator input power to compensate heat flows through $1 m^2$ of any insulation is of the order of $1 W$. Figure 1 presents typical thermal conductivity values and operation temperature ranges of cryogenic and refrigeration insulations. They lay along the line representing the value of thermal conductivity k calculated from equation (1) on the assumption that $w = 1$ W. A vacuum insulation is characterized by the best thermal performance, however this insulation is very sensitive to the vacuum level and hence it is rather rarely used, except for special cases like very low temperature research systems. In most cases instead of vacuum insulation a multilayer vacuum insulation (MLI) is implemented. There is also a tendency observed, to replace powder vacuum insulation by MLI due to the cost reduction and industrial scale availability of the multilayer insulation.

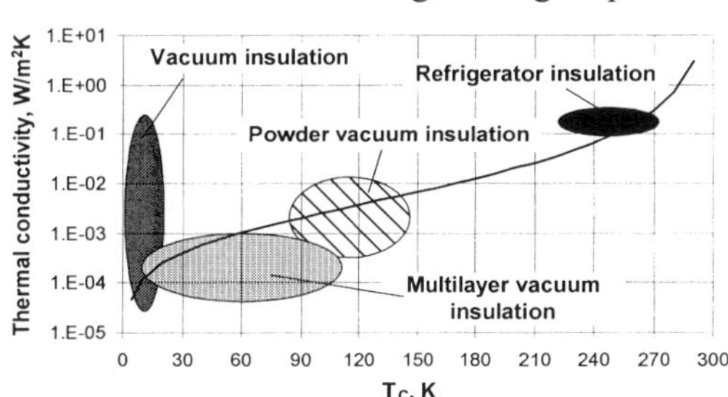

Figure 1. Typical thermal conductivity values and work temperature ranges of cryogenic and typical refrigerator insulation, line shows k calculated for $w = 1$ W.

MODELLING OF HEAT TRANSFER THROUGH MULTILAYER INSULATION

Multilayer insulation consists of a number of thin aluminum or metalized (aluminized or goldized) plastic foil radiation shields, alternated with a low-conductivity spacer material and placed in a vacuum space between the insulated boundary walls (see Figure 2a). A heat flux through multilayer insulation involves of three heat transfer modes: residual gas conduction, thermal radiation and solid conduction via the spacer material. Total heat flux between two adjacent layers is a sum of all three transfer modes, except for the outermost shield not thermally bridged with the boundary wall by the spacer. (see Figure 2b). To estimate numerically the heat flux through MLI, a mathematical model was developed and a dedicated computer code was created. The mathematical model is based on fundamental equations of heat transfer between two boundaries of T_1 and T_2 temperature, namely: residual gas conduction in both molecular and transient conditions (2), heat radiation (3) and solid conduction (4).

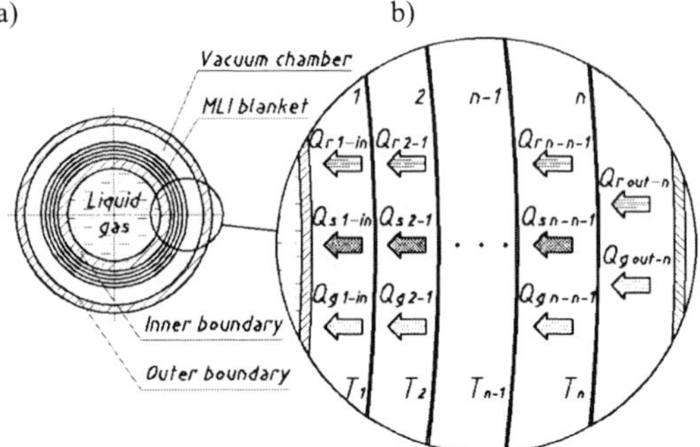

Figure 2. Structure and heat flux through multilayer insulation scheme

$$\dot{Q}_g = \frac{1}{N_a \cdot d_0^2}\left(\frac{R_u}{\pi}\right)^{1,5}\frac{9\kappa - 5}{4(\kappa - 1)}\left(\frac{T_0}{M_0}\right)^{0,5}\frac{F}{L + \dfrac{9\kappa - 5}{\kappa + 1}\dfrac{2 - \alpha}{\alpha}\dfrac{k_B}{\pi}\dfrac{T_0}{2d_0^2\, p}}(T_1 - T_2) \tag{2}$$

$$\dot{Q}_r = F\frac{\sigma}{\dfrac{1}{\varepsilon_1} + \dfrac{1}{\varepsilon_2} - 1}(T_1^4 - T_2^4) \tag{3}$$

$$\dot{Q}_s = F\frac{\lambda}{g}(T_1 - T_2) \tag{4}$$

where: d_0 - average gas molecule diameter, κ - adiabatic exponent, M_0 - gas molecular weight , α - accommodation factor, L - distance between boundaries, F - boundaries area, p - gas pressure, N_a - Avogadro number, R_u - universal gas constant, k_B - Boltzmann's constant, T_0 - average gas temperature, g - spacer thickness, λ - spacer thermal conductivity coefficient, σ - Stefan – Bolzmann constant, $\varepsilon_1/\varepsilon_2$ emissivity of hot/cold boundary.

The set of equations (2)-(4) was solved iteratively on the assumption that the sum of all heat transfer modes is constant in a steady state condition. It was also assumed, that the gas concentration in the vacuum space and between MLI layers (interstitial gas concentration) is constant, as well as the contact resistance between individual radiation shields and spacer for all the layers is the same.

Modeling results

Figures 3 and 4 show a comparison between the computed results and literature experimental data. In the calculations the inner and outer wall emissivities were equal to 0.16 for 300 K, 0.12 for 77.3 K and 0.074 for 4.2 K, the spacer material thermal conductivity coefficient was given by equation (5) and the foil shields emissivities were calculated according to (6) [1, 2].

$$\lambda = 8.823 \cdot 10^{-6} + 1{,}04 \cdot 10^{-7} \cdot T \tag{5}$$

$$\varepsilon = 0.0035 \cdot T^{0.5} \tag{6}$$

Figure 3 presents the calculated and measured [3, 4] heat flux through multilayer insulation in a function of residual gas pressure for two boundary temperature ranges: 300 – 77.3 K and 77.3 – 4.2 K. The number of layers was $N=10$ in temperature range 77.3 K – 4.2 K, and $N=30$ in 300 K – 77.3 K temperature range. Figure 4 shows the measured [5] and computed results of heat transfer through multilayer insulation as a function of the number of layers for the temperature range 300 K – 77.3 K and a high vacuum i.e. for residual gas pressure lower than 10^{-3} Pa. A good agreement between the calculated and measured results is observed.

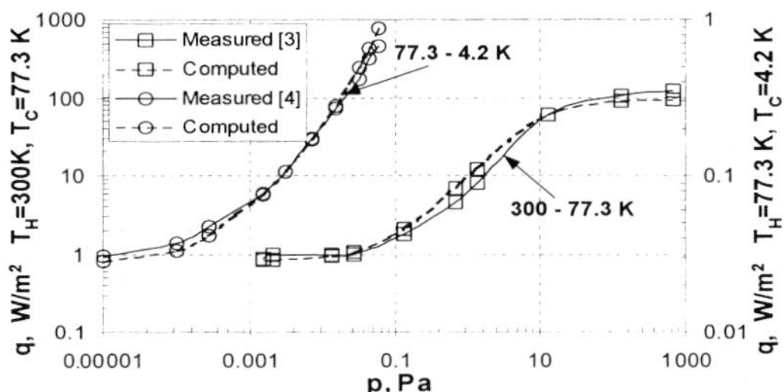

Figure 3. The MLI heat transfer variation with residual gas pressure

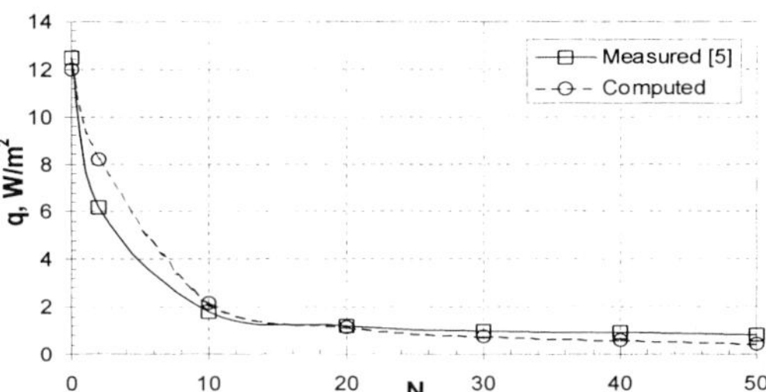

Figure 4. The MLI heat transfer variation with number of layers, 300 – 77.3 K case

Based on the model, the sensitive analysis of heat flux variation with selected physical properties of materials used to MLI production have been done, and the results are presented in Figure 5. The variated parameters were: ε_{hb} – hot boundary emissivity, ε_{cb} – cold boundary emissivity, ε_{sh} – shield surface emissivity, λ – spacer material thermal conductivity. Nominal heat fluxes are q_{nom}=0.9 W/m^2 and q_{nom}=0.03 W/m^2 for 300 – 77.3 K and 77.3 – 4.2 K temperature range adequately. Nominal values of the variated parameters were the same as given above.

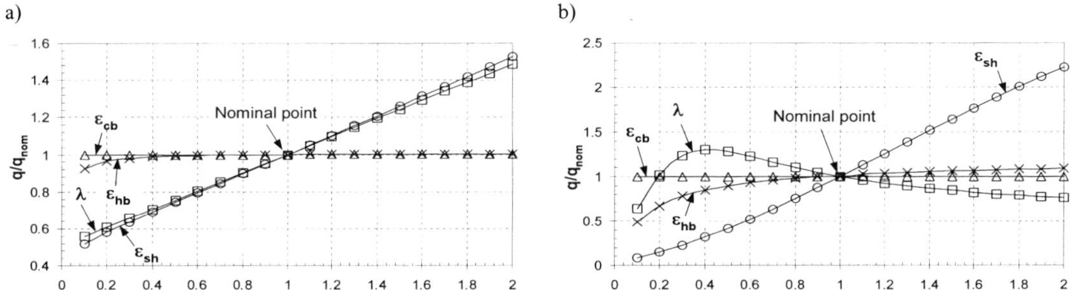

Figure 5. Heat flux variation with selected physical properties of materials used to MLI production. a) N=30 layers MLI, 300 K – 77.3 K case; b) N=10 layers MLI, 77.3 K – 4.2 K; ε_{hb} – hot boundary emissivity, ε_{cb} – cold boundary emissivity, ε_{sh} – shield surface emissivity, λ – spacer material thermal conductivity

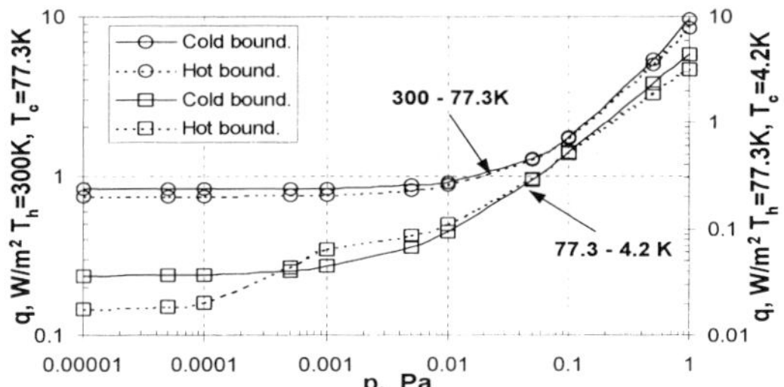

Figure 6. The influence of the location of radiation shields on the heat flux

We have also investigated the influence of the location of radiation shields on the heat flux. Due to technical reasons radiation shields are usually placed beside a cold wall. But in some cases, e.g. cold boxes, the shields can be located beside a warm boundary wall. The results calculated for MLI comprising 30 layers, for 77.3 K – 4.2 K and 300 K – 77.3 K temperature ranges are presented in Figure 6.

LIMITATIONS OF THE MODEL

S.L. Bepat et al. measured the dependence of the heat flux on the layer density expressed in layers/cm [6]. The calculated results are in the agreement with the measurements only if the layers density is below 20 – see Figure 7. It can be concluded that for higher layer densities some physical parameters having an influence on the heat transfer are missing in the model.

The dependence of heat flux on the layer density is shown schematically in Figure 8. If the density is low (left side region on Figure 8), there are only a few shield-spacer contact points and the contact thermal resistivity can be treated as infinite. Moreover, there are good conditions to evacuate residual gas from interlayer space during vacuum pumping process. This situation is corresponding to the presented above model assumptions.

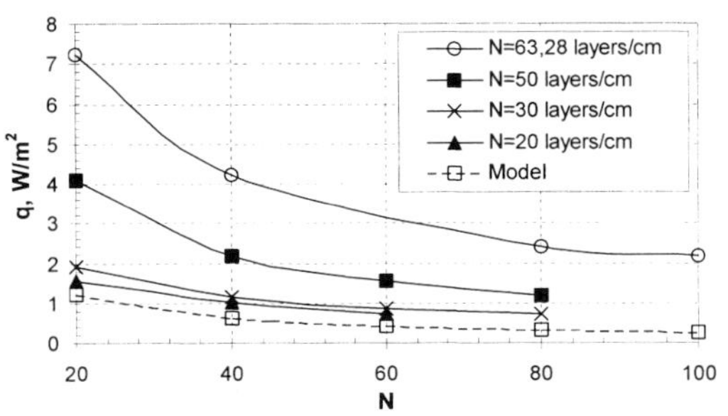

Figure 7. The MLI heat transfer variation with layers density for different layers number, 300 – 77.3 K case

Figure 8. The influence of layer density variation on the heat flux expected character

To provide accurate results for layer density higher than 20 the following phenomena should be taken into account into the model: out gassing rate of radiation shield and spacer materials, the conductivity of radiation shield orifices, interlayer passage conductivity and contact thermal resistance in function of the layer density. Interstitial pressure in the *i-th* interlayer passage p_i can be described by:

$$p_i = 2 j_o \cdot L_p \cdot \left(\frac{1}{C_0} + \frac{L_p}{c} \right) \sum_{i=1}^{n} (n - i + 1) + p_{ch} \tag{7}$$

where: j_0 – specific outgassing rate, C_0 – orifice conductivity, L_p – length of passage, c – interlayer passage conductivity, n – layer amount, i – number of interlayer passage, p_{ch} – vacuum chamber pressure.

CONCLUSIONS

Thermal insulations used in refrigeration and cryogenics are characterized by a similar work demand to compensate unit area heat fluxes to low temperature parts of the installations.

It is sufficient to consider residual gas conduction, radiation and solid conduction when describing heat transfer through MLI when layer density is below 20 layer/cm. For higher layer densities it is necessary to take additional phenomena into account, namely: out gassing rate of radiation shield and spacer materials, the conductivity of radiation shield orifices, interlayer passage conductivity and contact thermal resistance.

REFERENCES

1. Chorowski, M., Grzegory, P., Parente, Cl., Riddone, G., Optimization of multilayer insulation – an engineering approach. Proc. Cryogenics 2000, International Institute of Refrigeration, Praha, October 2000
2. Zhitomirskij, I.S., Kislov, A.M., Romanenko, V.G., A theoretical model of the heat transfer processes in multilayer insulation, Cryogenics (1979) MAY 265-268
3. Benda, V., Lebrun, Ph., Mazzone, L., Sergo, V., Vollierme, B., Qualification of multilayer insulation system between 80K and 4.2K, Proc. Cryogenics 2000, International Institute of Refrigeration, Praha, October 2000
4. Fesmire, J., Augustynowicz, S., Darve, C., Performance characterization of perforated multilayer insulation blankets, International Cryogenic Engineering Conference 19, Narosa Publishing House 2002, 843-846
5. Santhil Kumar, A., Krishna Murthy, M. V., Jacob S., Kasthurirengan S., Thermal performance of multilayer insulation down to 4.2 K
6. Bapat, S. L., Narayankhedkar, K. G., Lukose T.P, Experimental investigation of multilayer insulation, Cryogenics (1990) 30 711 - 719. Butterworth & Co (Publishers) Ltd.

Experimental study of the narrow channel heat transfer in liquid nitrogen

Zhang P., You G. C., Ren X., Wang R. Z.

Institute of Refrigeration and Cryogenics, Shanghai Jiao Tong University, Shanghai, 200030, China

In the present study, the investigation of the narrow channel heat transfer in liquid nitrogen is carried out. The experiments are conducted at the orientation angles from $0°$ to $180°$ with $45°$ intervals with three different gap widths. The experimental results show that both the channel gap width and the orientation angle have the influential effect on the heat transfer. It is suggested that the combined effect of the channel gap width and the orientation angle of the channel contribute together to the heat transfer enhancement in the nucleate boiling region. However, the narrow channel may decrease CHF (the critical heat flux).

INTRODUCTION

With the further development of the science and technology, the materials were found with higher transition temperature up to liquid nitrogen temperature range. Liquid nitrogen has been frequently used for the cooling of the high temperature superconducting devices. For this application, the good heat

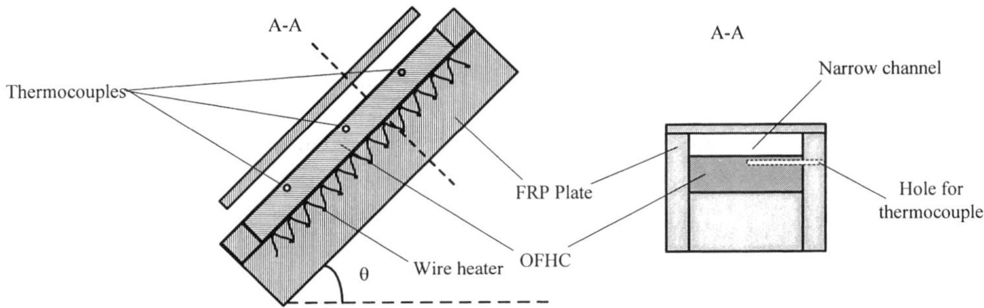

Figure 1   The schematic illustration of the narrow channel (not to scale)

transfer performance should be maintained in order to keep the stable working condition of the devices. Narrow channel is often used to make the design more compact, because it holds the potential for the nucleate boiling heat transfer enhancement, which has been proved by using water and some room-temperature liquids as the test fluids [1]. However, the information of the confinement and orientation effects on the heat transfer of the cryogenic liquids is rarely available. Nguyen [2] reported the confinement effect on CHF (critical heat flux) of a heater in liquid nitrogen. Chen [3] studied heat transfer characteristics of a channel immersed in He I. It is expected that different heat transfer characteristics will be shown under different orientation angles of the narrow channel. However, the combined effect has not been systematically investigated yet. In the present study, an experimental investigation on the boiling heat transfer of liquid nitrogen in narrow rectangular channels is carried out. The experiments are conducted at the orientation angles of the channel from $0°$ to $180°$ with $45°$ intervals and with the gap widths of 0.5 mm, 1 mm and 2 mm.

EXPERIMENTAL SETUP AND PROCEDURE

Shown in Figure 1 is the schematic illustration of the test section. The narrow channel was formed by placing a FRP plate over the heat transfer surface, which was made of the OFHC and was heated by the wire heater immersed below. The longitudinal ends were open to the liquid nitrogen bath, and the transverse ends and the bottom were insulated by the FRP plates. The length of the channel was 90.0 mm and the width was 15.0 mm. The heating applied to the heat transfer surface was generated by the wave generator and the power amplifier, the temperature of the heat transfer surface was measured by the calibrated T–type (Copper vs Constantan) thermocouples which were immersed in the drilled holes 2 mm below the heat transfer surface. Some silicon grease was used to enhance the thermal contact between the thermocouples and the OFHC block. The experiments were conducted under one atmosphere pressure condition, and the liquid nitrogen in the dewar was kept at almost constant level. The orientation of the narrow channel varied from $0°$ (horizontally upward-facing) to $180°$ (horizontally downward-facing) with $45°$ intervals.

RESULTS AND DISCUSSION

Shown in Figure 2 is the result of the heat transfer curve for the channel gap width of 0.5 mm under different orientation angles. It is understood from the figure that all the curves almost coincide with one another when the heat flux is smaller than 0.2 W/cm^2. The heat transfer in this region is dominated by the single phase natural convection, in which the heat transfer surface temperature rise ($\Delta T = T_W - T_{LN_2}$) increases almost linearly with the heat flux. When the heat flux is further increased, it is seen that the curves begin to diverge and the heat transfer mode begins to enter nucleate boiling region. In general, the isolated vapor bubble in the narrow channel can not grow naturally when the Bond number is small (of the order of unity or less) because the channel may be narrower than the vapor bubble diameter. The Bond number in Figure 2 is about 0.5. The Bond number is defined as the ratio of the channel gap width to the departure diameter of the isolated bubbles and is formulated as:

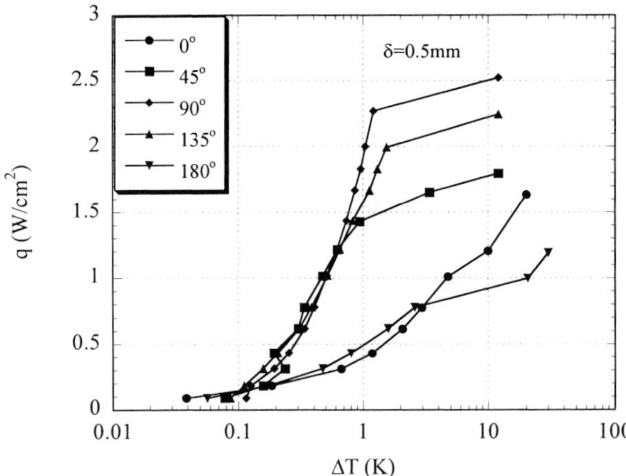

Figure 2   The results of the heat transfer curve of 0.5mm gap width channel

$$Bo = \delta(\sigma / g(\rho_l - \rho_g))^{-1/2},$$ where, δ is the gap width of the channel, σ is the surface tension, ρ is the density, subscription l and g represent the liquid and vapor, respectively. When the vapor bubble

grows in the narrow channel, the shape of the vapor bubble deforms from sphere-like to elliptical sphere which enlarges the microlayer area under the vapor bubble and in turn, the latent heat transfer is increased. As the heat flux is further increased, nucleate boiling fully develops. Comparing to the present results and that of the open pool boiling [2], it is found that the heat transfer is better in the narrow channel for the orientation angle at 45°, 90° and 135°, which is due to the effect of the sliding vapor bubble in those cases [4]. The sliding vapor bubble contributes to the heat transfer enhancement mainly in the following three aspects.

1. the sliding bubble in the inclined channel moves upwards due to the buoyancy and it collides with other bubbles in the channel, which will cause the growing bubbles detach from heat transfer surface in advance. In this way, the detaching frequency of the vapor bubble is increased;

2. the sliding bubble induces a enhancement of the local convection around the spot where it detaches from the heat transfer surface, which brings the cooler liquid from the nearby pool. In turn, it will also enhance the heat transfer;

3. the sliding bubble causes a thinner microlayer under the bubbles [5] as it slides along the heat transfer surface. Obviously, this reduces the thermal resistance to manifested heat transferring from the heat transfer surface.

It is obvious from the figure that the heat transfer curves could be divided into two groups: the curves under the inclined conditions and the curves under the horizontally-set conditions. The heat transfer in the former group is better. It is seen from the figure that the heat transfer coefficient in the case of the orientation angle at 90° is the largest when the heat flux is above 1.5 W/cm^2, which may be attributed to more efficient escape of the sliding vapor bubble from the narrow channel.

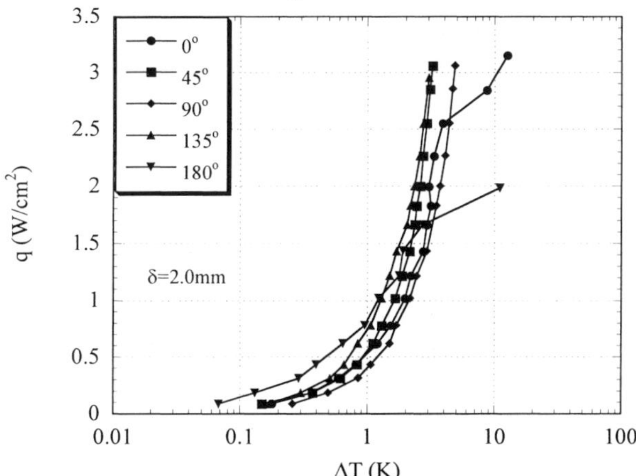

Figure 3   The results of the heat transfer curve of 2.0mm gap width channel

When the orientation angles are 0° and 180°, heat transfer begins to be deteriorated at around 1.0 W/cm^2. The temperature of the heat transfer surface increases drastically as the heat flux increases. In higher heat flux region, the vapor bubble is difficult to detach from the heat transfer surface because of the confinement of the upper plate of the channel, and it is confined inside in narrow channel and grows into flat shape. Finally, many deformed vapor bubbles coalesce into a big one and it sticks in the narrow channel, which makes the cooler liquid nitrogen difficult to be replenished into the narrow channel. Thus, some portion of the heat transfer surface is partially dry-out because the vapor bubble in the narrow channel acts as a thermal insulation layer which increases the thermal resistance drastically. It is implied from this result that the narrow channel may decrease CHF in some cases, as shown in Figure 2, which has also been observed in the experiments [6] by using water as a test fluid.

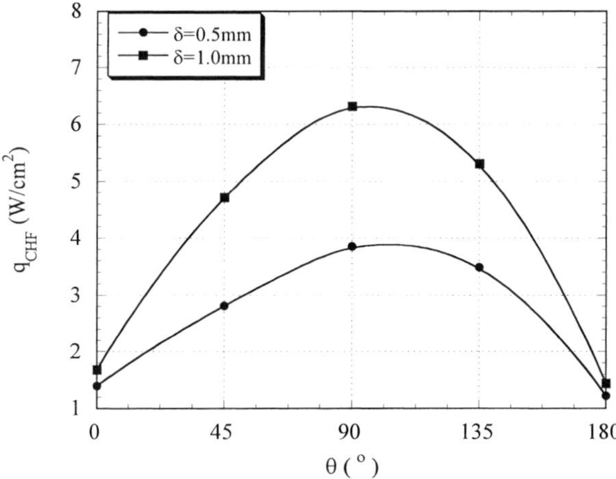

Figure 4   Variation of the critical heat flux with the orientation angle of the channel

The heat transfer curves of 2.0 mm gap width channel are shown in Figure 3. It is seen from the figure that the results display different feature from that of 0.5 mm gap width channel. Bond number in this case is about 1.9, which means the confinement of the narrow channel is not as strong as that in the former case. The results in this case are somewhat similar to that of the pool boiling case. The vapor bubble inside the channel can grow and detach from the heat transfer surface relatively freely and easily. Moreover, the vapor bubble does not deform heavily to take the flat shape, and thus, the area of the microlayer under the vapor bubble is smaller. Sliding vapor bubble detaching from the heat transfer surface is the dominant heat transfer enhancement mechanism in the case of wider channel. It is further confirmed that the heat transfer at 0° and 180° is deteriorated earlier, as shown in Figure 3, resulting in lower CHF compared to that of the open pool boiling.

Shown in Figure 4 is the variation of the critical heat flux with the orientation angle for two gap widths. It is obvious from the figure that the gap width has the influential effect on the critical heat flux. The critical heat flux increases with the increasing of the gap width since wider gap has larger capability of replenishing the cold fluid into the channel. The inclined channel augments the perturbation to the vapor bubble and promotes the vapor bubble sliding along the heat transfer surface, which enables more cold fluid to be replenished into the channel, thus, the critical heat flux reaches the peak at 90°. As the gap width is increased to above 2 mm, the critical heat flux in the narrow channel is already close to that of the unconfined case [2], which implies that the confined effect is weak in the wider channel.

CONCLUSIONS:

The experimental study of the nucleate boiling heat transfer of liquid nitrogen in narrow channel is carried out. It is shown that both the gap width and the orientation angle of the channel are responsible for the heat transfer enhancement. In the narrower channel, the larger area of the microlayer and the sliding bubble contribute together to the heat transfer enhancement, while the sliding bubble effect is predominant in the wider channel. Nevertheless, the narrow channel may decrease CHF.

ACKNOWLEDGEMENTS

This research is jointly supported by Shanghai Sci.&Tech. Young Star Program (contract No. 01QF14027), A Foundation for the Author of National Excellent Doctoral Dissertation of PR China (200236) and National Natural Science Foundation of China (50306014).

REFERENCES

1. Ishibashi, E. and Nishikawa, K., Saturated boiling heat transfer in narrow spaces, Int. J. of Heat Mass Transfer (1969) 12 863-894

2. Nguyen, D. N. T., Chen, R. H., Chow L. C. and Gu C. B., Effects of heater orientation and confinement on liquid nitrogen pool boiling, J. of Thermophysics and Heat Tranfer (2000) 14 109-111

3. Chen, Z. and Van Sciver, S. W., Channel heat transfer in He I-Steady state orientation dependence, Advances in Cryogenic Engineering (1986) 31 431-438

4. Bayazit, B. B., Hollingsworth D.K. and Witte, L. C., Heat transfer enhancement caused by sliding bubbles, ASME J. of Heat Transfer (2003) 125 503-509

5. Koffman, L. D. and Plesset, M. S., Experimental observations of the microlayer in vapor bubble growth on a heated solid, ASME J. of Heat Transfer (1983) 105 625-632

6. Bonjour, J. and M. Lallemand., Flow patterns during boiling in a narrow space between two vertical surfaces, Int. J. of Multiphase Flow (1998) 24 947-960

Observation of various boiling states in nearly saturated He II

Zhang P.[1,2], Murakami M.[1], Nozawa M.[1]

[1]Institute of Engineering Mechanics and Systems, University of Tsukuba, Tsukuba, 305-8573, Japan
[2]Institute of Refrigeration and Cryogenics, Shanghai Jiao Tong University, Shanghai, 200030, China

He II boiling phenomenon is one of the least understood aspects of He II. In the present investigation, boiling phenomenon in He II is studied by the simultaneous measurement of the heater surface temperature and the pressure oscillation in He II bath. Several boiling states can be encountered with the variation of the immersion depth of the heater at small heat flux. The heat transfer coefficients of the various boiling states are measured. It is observed that there exists correlation between the dominant frequencies of the pressure oscillation and the heat transfer coefficient in these boiling states.

INTRODUCTION

Superfluid helium (He II) is widely utilized in many cryogenic cooling application cases for its high cooling capability, such as cooling of superconducting magnets and IR detectors in space cryogenic applications, and so on. However, its cooling performance is sometimes deteriorated by the appearance of the gas phase, and even boiling phenomena, for example, superconducting magnet quenching in He II. Boiling phenomena in He II are of both academic and applied interests, because their occurrence may lead to catastrophic events for the cryogenic systems where good heat transfer performance should be maintained and because it still remains one of the least understood aspects of He II. Obviously, film boiling is the worst situation from the heat transfer performance point of view. Recently, a lot of effort has been made to understand He II film boiling. Zhang [1] conducted film boiling experiments under different thermal conditions, and constructed a three dimensional boundary map to classify noisy film boiling, transition boiling and silent film boiling. Pressure oscillation and temperature oscillation in He II were measured by a pressure sensor and a superconducting temperature sensor, which showed strong correlation between each other. It was found from the visualization video during the experiments that the vapor bubble was oscillating on the heater surface and the size could be comparable to the heater size. It was further confirmed that every contact of the cold He II with the higher temperature heater surface would generate an audible loud noisy sound, after which noisy film boiling was named. There was another non-noisy counterpart of He II film boiling mode, which was so-called silent film boiling. The appearance of He II film boiling modes depended strongly on the thermal conditions. Although a lot of research has been carried out, the information about He II film boiling is still inadequate [2] from the heat transfer point of view. In the present study, experiments are carried out to study the heat transfer performance in different boiling states by employing a stainless steel heater.

EXPERIMENTAL SETUP AND MEASURING METHODS

A glass cryostat of 9 cm inner diameter was used in the experiments. The boiling process could be observed through the unsilvered strips along the body of the cryostat. An evacuation system was used for regulating the bath temperature. In order to measure the heater surface temperature, a stainless steel foil heater of about 10 μm in thickness was employed in the present experiments. The resistance of stainless steel foil (SUS-304) is temperature dependent, and the average temperature can be obtained by measuring the resistance. The stainless steel foil was cut into a zigzag form in order to obtain higher resistance, as shown in Figure 1. The width of the heating element was about 3.2 mm, and the width of the gap between the heating elements was about 0.8 mm. As the width of the gap was much smaller comparing to that of the heating elements, the total heating part could still be regarded as a square planar heater. After the heating part was cut out, it was pasted carefully to a FRP base plate by using a thin layer of Stycast. Total net heat transfer area of the heater in the present measurement was about 8 cm^2. A Kulite-piezoelectric pressure sensor (CCQ-093) was placed right above the heater surface to measure the pressure oscillation during film boiling. The pressure sensor was calibrated against the saturated vapor pressure of liquid helium. The

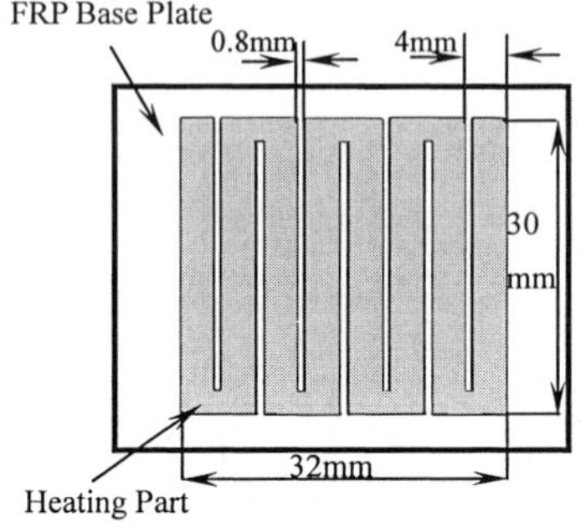

Figure 1 Schematic illustration of the stainless steel foil heater (not to scale)

experiments were conducted under nearly saturated pressure condition. The heater was mounted horizontally in the cryostat. The helium bath temperature was controlled by regulating the vapor pressure in the cryostat with the pressure controlling system. The electric voltage and the electric current applied to the heater were simultaneously measured. The signal measured by the pressure sensor was 100 times amplified by the low noise pre-amplifier. All data were recorded by a storage oscilloscope and transmitted to the computer through GPIB interface and then stored there for a further analysis.

RESULTS AND DISCUSSION

Shown in Figure 2 is one example of the pressure oscillation and the heater surface temperature oscillation in noisy film boiling. It is seen from the figure that the pressure spike and the temperature spike are rather strong, and there exists strong correlation between the pressure and the temperature spikes, which indicates the pressure spike results from the direct contact of He II with the heater surface. However, the hydrodynamic information reflected from the pressure and the temperature oscillations is different. The pressure oscillation displays both global and local hydrodynamic information, while the temperature oscillation displays

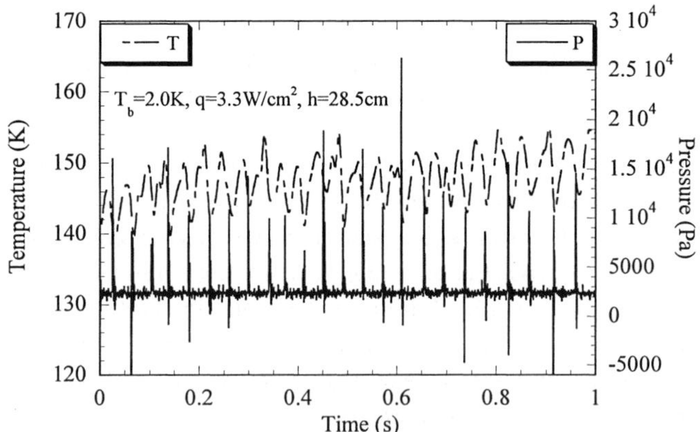

Figure 2 The pressure oscillation and the heater surface temperature oscillation in noisy film boiling

only local hydrodynamic information [3]. The heater surface temperature drops about 10 K at the moment of the He II contacting with the heater, because the evaporation of He II is an effective way for the heat removal.

One of the interesting features of He II film boiling is that the boiling state varies with the immersion

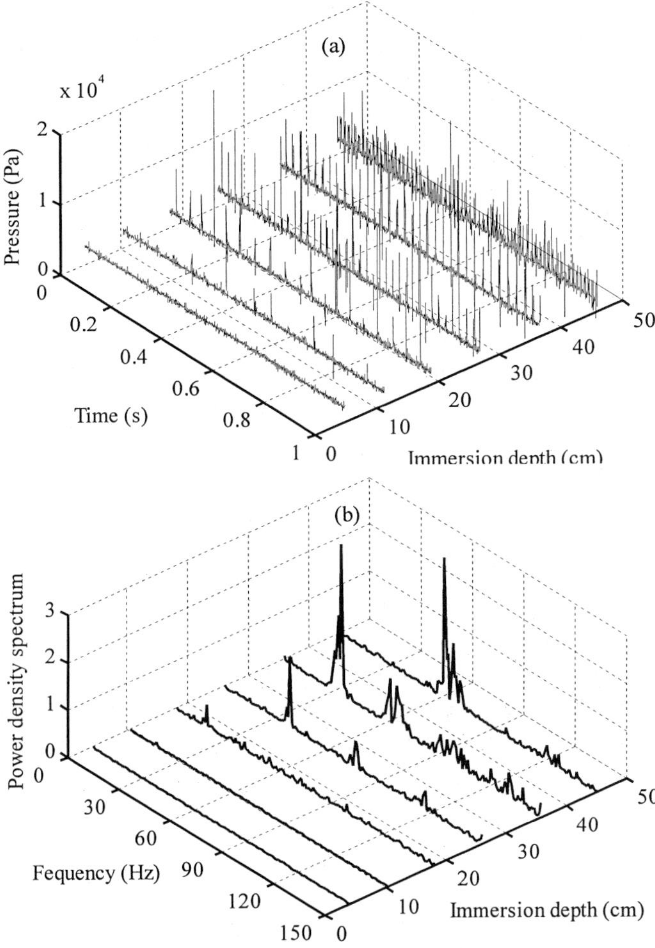

Figure 3   Variations of the pressure oscillations with the immersion depth (a) and their FFT analysis results (b), T_b=2.0 K, q=3.3 W/cm^2

depth of the heater. In general, boiling state is noisy film boiling when the immersion depth is deep; on the contrary, boiling state is silent film boiling when the immersion depth is small. When the heat flux is moderate and the immersion depth of the heater is deep, say above 38 cm, the general outlook of the boiling state in this case is different from that in noisy film boiling or silent film boiling. It is observed that a very small vapor bubble oscillates on the heater surface and the boiling sound is fizzy and in high frequency, as shown in Figure 3 (a). (In the following figures, only the data of the pressure oscillation is shown because the fundamental frequency of the pressure oscillation and the temperature oscillation is essentially the same.) It is also seen from the figure that the amplitude of the pressure oscillation is much smaller than that in noisy film boiling, which occurs at the immersion depth below 38 cm. It could be concluded that the boiling state at deeper immersion depth is a kind of undeveloped film boiling (pre-mature film boiling state). Shown in Figure 3 (b) are the FFT analysis results of the pressure oscillation data. It is obvious

that the fundamental frequency of the pressure oscillation in the undeveloped film boiling is much higher than that in noisy film boiling. When the immersion depth decreases to the range of 15-38 cm, the boiling state is noisy film boiling, in which the fundamental frequency is around 20-30 Hz. As the immersion depth further decreases to below 15 cm, the boiling state turns into an intermittent state: transition boiling, in which noisy film boiling and silent film boiling appear intermittently. And thus, the fundamental frequency is smaller. In general, silent film boiling occurs at lower immersion depth without detectable pressure oscillation. It is seen form Figure 3 (b) that the immersion depth for silent film boiling in the present case is around 4 cm.

Shown in Figure 4 are the variations of the heat transfer coefficient and the fundamental frequency with the immersion depth. The shadowed area indicates the transition boiling region, wherein the lightly shadowed area indicates the transition boiling dominated by noisy film boiling and the heavily shadowed indicates the transition boiling dominated by silent film boiling. It is understood from the figure that there exists some correlation between the heat transfer coefficient and the fundamental frequency. At the deeper immersion depth, the boiling state is the undeveloped film boiling and the corresponding heat transfer coefficient is also larger. The heat transfer coefficient decreases with the decreasing of the immersion depth as well as the fundamental frequency does. An interesting phenomenon is that the heat transfer

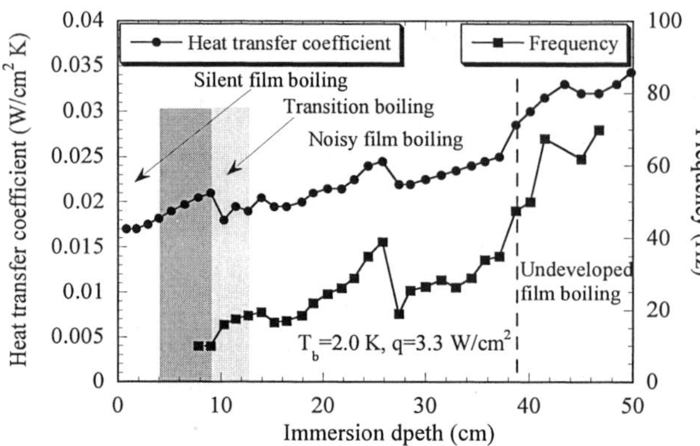

Figure 4   Variations of the heat transfer coefficient and the fundamental frequency with the immersion depth

coefficient suddenly increases when the immersion depth decreases to around 25 cm, and the fundamental frequency counterpart also shows a jump, which may be due to some resonant frequency component of the system being excited by the oscillation of the vapor bubble in noisy film boiling region and then gets sympathetic vibration. It can be further concluded that the frequent contact of He II with the heater surface may enhance the heat transfer performance. With the decreasing of the immersion depth, the boiling state turns into transition boiling in which the heat transfer coefficient shows a small jump and then decreases continuously to silent film boiling in which no obvious fundamental frequency is detected.

CONCLUSIONS:

The film boiling in He II is investigated in the present study. It is found from the experimental results that the boiling states display different features under different thermal conditions. At the immersion depth above 38 cm, the boiling state is undeveloped film boiling and it is characterized by higher fundamental frequency and larger heat transfer coefficient; as the immersion depth decreases, the boiling state is noisy film boiling in which the fundamental frequency and the heat transfer coefficient are weakly dependent on the immersion depth. It is interesting to note that there exists a correlation between the fundamental frequency and the heat transfer coefficient, and frequent contact of He II with the heater surface will enhance the heat transfer performance. The boiling state turns into transition boiling as the immersion depth decreases to the range of 15-4 cm, below which the boiling state is silent film boiling. The corresponding heat transfer coefficient displays a jump in transition boiling and decreases continuously in silent film boiling.

ACKNOWLEDGEMENTS

This research is jointly supported by Shanghai Sci.&Tech. Young Star Program (contract No. 01QF14027), A Foundation for the Author of National Excellent Doctoral Dissertation of PR China (200236) and National Natural Science Foundation of China (50306014) and the Grant-in-Aid for JSPS fellows.

REFERENCES

1.   Zhang, P., Murakami, M., Wang, R. Z. and Takashima, Y., Investigation of noisy film boiling under various thermal conditions in He II, Advances in Cryogenic Engineering (2000) 45 1017-1024

2.   Van Sciver, S. W., Helium Cryogenics, Plenum Press, New York, USA (1986)

3.   Zhang, P. and Murakami, M., On the wavelet analysis of the oscillatory phenomena in He II film boiling, Cryogenics (2003) 43 679-685

Heat transfer during natural convective boiling in a vertical uniformly heated closed tube submerged in saturated helium

B. Baudouy

CEA/Saclay, DSM/DAPNIA/SACM
91191 Gif-sur-Yvette Cedex, France

An experimental study was carried out on heat transfer in natural boiling convection in a vertical tube closed at the bottom and connected to a liquid helium bath at the top. A 14 mm inner diameter copper tube is submerged in liquid helium near atmospheric pressure. The two phases are counter-flowing in which the vapor exits the system upward through the liquid bath while the liquid moves downward replacing the vapor. The wall temperature at different locations, the pressure drop and the vapor mass flow rate was measured, and critical heat fluxes (CHF) are determined. Evidence of flooding is witnessed in the light of the measurements and CHF values are compared with existing flooding correlations.

INTRODUCTION

Many studies related to the heat transfer in convective boiling counter-flow have been reported for critical heat flux (CHF) in a heated vertical closed tube. The closed two-phase thermosiphon, *i.e.*, the heat pipe, is the most studied boiling system (see [1, 2], for example,) and the closed bottom tube that is opened to a saturated liquid bath has been studied by few groups including, Barnard *et al.* on the refrigerant R-113 [3], Nejat on four different fluids [4] and Katto *et al.* on water [5]. In such boiling configuration, the vapor and the liquid flow in opposite directions, that is, the liquid flows downward to replace vapor. A simplistic view of such fluid flow is presented in Figure 1 (a) where the annular counter-flow type boiling configuration is assumed. The vapor flows upward at the center of the tube while a thin liquid layer on the wall flows downward. The operation limit of such a counter-flow is known as "flooding" where the liquid film expelled by the upward vapor flow disappears from the wall. This limit is achieved at a certain vapor velocity unique to a given heat flux value. This study presents experimental data on heat transfer in boiling counter-flow of liquid helium in the closed bottom tube configuration. The experimental results show the evidence of flooding in liquid helium and the critical heat transfer is compared to the open two-phase thermosiphon configuration.

EXPERIMENTAL SET-UP

The experimental apparatus, as illustrated in Figure 1 (b), consists of a liquid helium (He) reservoir and a test tube closed at the bottom with its associated instrumentations. A complete description of the apparatus is given elsewhere [6]. The test section consists of a 1.2 m long copper tube with 14 mm inner diameter. The instrumentation includes heaters, Germanium thermometers and a differential pressure sensor. Five thermometers are inserted in small copper blocks brazed to the tube at 0.07, 0.3, 0.6, 0.9 and 1.2 m, respectively, from the entry of the flow. All wiring is wound around and glued with GE varnish to the copper thermal anchor held at 4.2 K. In the range of ΔT investigated, in the order of 10 K, the heat loss through the temperature sensor wiring is negligible; therefore we consider that the measured temperatures are those of the inner wall. A gas mass flow-meter is capable of measuring flow rates up to 4.2 g/s of helium at room temperature with a precision of ± 0.01 g/s. Pressure differences are obtained with a sensor at room temperature with a precision of 1 Pa.

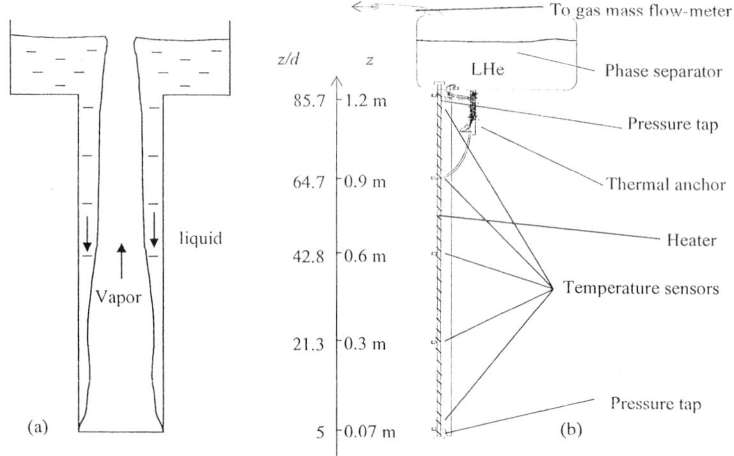

Figure 1 (a) Boiling configuration and (b) Schematic of the experimental set-up.
(z is the height from the entry of the tube bottom and d the inner diameter of the tube)

EXPERIMENTAL RESULTS

For each experiment, the power input, q, to the tube is increased stepwise. At each step, the steady state in the wall temperature, the pressure drop and the vapor mass flow are established for a sufficiently long time. The heater power incrementing is continued until the critical heat flux is reached.

Tests have been performed with steady state final heat fluxes of 125, 130, 140, 150 and 160 W/m². The evolution with time of the various measured physical parameters is similar. The Figure 2(a) presents the evolution of the wall temperatures at five different locations (z). Corresponding to five different values of the z/d ratio (Figure 1 (a)), the pressure drop, Δp, across the tube length) (Figure 2c) and the vapor mass flow rate, m_v, at the exit of the tube (Figure 2d) are measured as functions of time. The time here corresponds to the elapsed time since the initial heat flux was set at 120 W/m². After a period of 330 s, shown by the vertical dashed line on the graphs, the heat flux was set to 130 W/m². During this period (where the heat flux was set constant at 120 W/m²), it can be observed that the wall temperature has a very small increase (Figure 2 (b)), Δp decreases slowly to 910 Pa from 960 Pa (Figure 2 (c)) and m_v increased slightly to 0.2 g/s (Figure 2 (d)). The value of the vapor mass flow rate and the wall temperature rise (between 50 to 100 mK for q=120 W/m²) are similar to the values found in the steady state measurements in an open thermosiphon configuration [6]. On the other hand, due to the counter flow configuration, that creates a larger friction between the vapor and the liquid, Δp is roughly 1.5 times smaller in our configuration than that in the boiling thermosiphon configuration for the same heat flux [7].

Once the heat flux is set to 130 W/m², an increase in temperatures, comparable to the one found in the boiling thermosiphon configuration, is observed for a period of 50 s. Then the wall temperatures gradually decreased (Figure 2 (b)) until finally after 172 s, a significant change of several tens of Kelvin is noted at z/d=21.3, 42.8, and 5. The wall temperature at z/d=42.8 falls towards 4.2 K after the temperature increase at z/d=5. The wall temperature increase is an order of magnitude higher here than in the boiling thermosiphon configuration [6]. Meanwhile, Δp decreases rapidly and reaches 85 Pa indicating that the tube contains very little liquid and superheated vapor. The brutal temperature increase occurs simultaneously with the Δp drop that can be attributed to flooding. The large decrease in Δp combined with the wall temperature drop towards the saturation temperature immediately preceding the large jump is also an evidence of flooding. Flooding is the phenomenon that manifests when the velocity of the vapor is fast enough to generate large waves at the interface between the falling liquid film and the rising vapor core. It has the effect of reducing the liquid film thickness, explaining that the temperature drops towards the saturation temperature, and finally expelling the liquid from the tube. Moreover, the onset of the temperature rise at z/d=5 corresponding to the temperature decrease at z/d=42.8 also suggests the occurrence of flooding: as the liquid is expelled at z/d=5, it may rewet the tube at z/d=42.8.

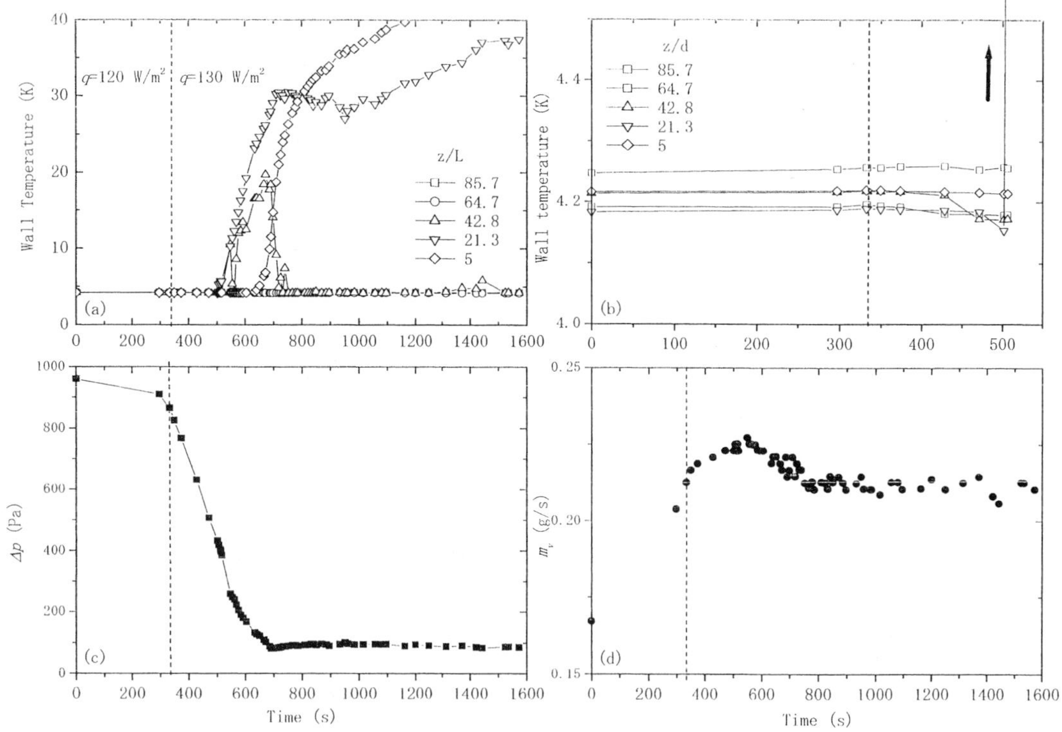

Figure 2 Wall temperature (a) and (b), pressure drop (c) and vapor mass flow rate (d)

CRITICAL HEAT FLUX DETERMINATION

The location of the onset of the CHF is defined as the location where the first temperature rise is seen such as z/d=42.8 in Figure 2 (a) for q=130 W/m^2, as it is done by other authors[2, 5]. The results of CHF onset for different heat flux values (q=125, 130, 140, 150 and 160 W/m^2) are presented in Figure 3 (a). The onset of temperature increase is after 1218 s at z/d=21.3 (q=125 W/m^2), after 172 s at z/d=21.3 (q=130 W/m^2), after 75 s at z/d=21.3 and z/d=42.8 (q=140W/m^2), after 38s and 32s at z/d=42.8 (q=150 W/m^2), and after 28 s at z/d=42.8 (q=160 W/m^2). Separate experiments have been performed at q=150 W/m^2 with two different liquid heights, 150 mm and 210 mm. The onset of the temperature increase, however, remained unchanged. For q=140 W/m^2, we find that the temperature rise appears simultaneously at the locations z/d=21.3 and 42.8. The general trend of the onset of CHF is similar to that in boiling water [5]. The CHF appears between the bottom and the top end of the tube. The smaller the applied heat flux is the lower the CHF onset is situated in the tube and the longer the time is required before the temperature change appears. The decrease of the temperature towards the saturation temperature before the onset is also found in two-phase water [5]. It is interesting to remark that the CHF values for helium in such a configuration are ten times smaller than for the boiling thermosiphon configuration [8].

To compare CHF data with existing flooding correlations, we only need to consider the minimum CHF value for the whole length of tube, L, and for this study, q_c=125 W/m^2. In an annular two-phase counter flow, Wallis gave an empirical criterion describing the velocity limit of the two-phases for flooding [9]. For the closed bottom tube configuration, the magnitude of the velocities at the top end of the tube is usually applied to this criterion. Katto presented a critical heat flux expression constructed from the continuity and energy balance equations [5],

$$q_c = \frac{C_w^2}{4} \rho_v L_v \left(\frac{g(\rho_l - \rho_v)d}{\rho_v} \right)^{1/2} \frac{1}{\left[1 + (\rho_v/\rho_l)^{1/4} \right]^2 (L/d)} \tag{1}$$

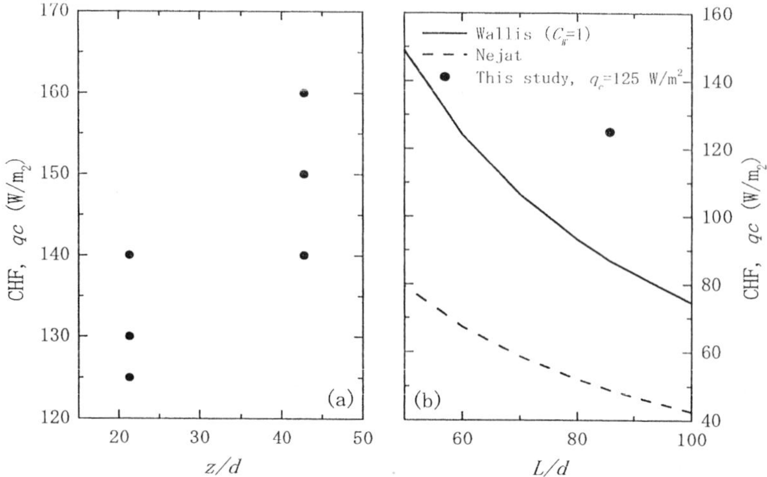

Figure 3 (a) Critical heat flux, q_c as a function of z/d and (b) comparison with (1) as a function of L/d

where ρ_v and ρ_l are the vapor and liquid density, g the gravitational acceleration and L_v the latent heat of vaporization and C_w a constant between 0.725 and 1. Nejat also proposed an empirical correlation based on the original Wallis configuration where he found that $C_w^2=0.36(L/d)^{0.1}$ fits his own data better [4].

The Figure 3 (b) depicts the two correlations with our minimum CHF value, which is 1.5 - 2.5 times higher than the value given by the correlations. This discrepancy can come from the error in the CHF determination. Firstly, the determination of the CHF onset brings uncertainty notably at low heat flux. Lower CHF could be found for much longer time that the ones investigated in this study. Secondly, the Wallis criteria is based on water and air in open counter-flow, which is different from our configuration where the liquid and vapor do not have uniform mass flow rate along the tube. Note that Equation (1) is developed with velocities taken at the top end of the tube. In the closed tube configuration the velocity of the vapor at the top end is higher than at lower location and therefore, it would reduce the CHF value.

CONCLUSION

In a closed bottom tube connected to a saturated helium bath, it is demonstrated that critical heat flux occurs as a consequence of flooding. Temperature increase and the variation of Δp are much higher in this configuration than for the thermosiphon configuration. The onset of CHF is not induced immediately after the application of the heat flux; rather it appears after a time duration that increases inversely with heat flux input. The CHF found in this study is much higher than what is expected from the CHF flooding correlations. To understand the mechanism behind the present findings, more investigation is needed.

REFERENCES

1. Imura, H., *et al.*, Heat transfer in two-phase closed thermosyphons, Heat Transfer - Japanese Research, (1979) 8 41-53
2. Fukano, T., *et al.*, Operating limits of the closed two-phase thermosyphon, ASME-JSME Th. Eng. Conf., (1983) 1 95-101
3. Barnard, D.A., *et al.*, "Dryout at low mass velocities for an upward boiling flow of Refrigerant-113 in vertical tube," UKAEA, AERE-R 7726, 1974
4. Nejat, Z., Effect of density ratio on critical heat flux in closed end vertical tubes, Int. J. Multiphase Flow, (1981) 7 321-327
5. Katto, Y. and Hirao, T., Critical heat flux of counter-flow boiling in a uniformly heated vertical tube with a closed bottom, Int. J. Heat and Mass Transfer, (1991) 34 993-1101
6. Baudouy, B., Heat and mass transfer in two-phase He I thermosiphon flow, Adv. in Cryo. Eng., (2001) 47 B 1514-1521
7. Baudouy, B., Pressure drop in two-phase He I natural circulation loop at low vapor quality, (2002) International Cryogenic Engineering Conference 19 817-820
8. Baudouy, B., Heat transfer near critical condition in two-phase He I thermosiphon flow at low vapor quality, accepted for publication in Adv. in Cryo. Eng., (2003) 49
9. Wallis, G.B., In: One-dimensional Two-phase flow, MacGraw-Hill, New York (1969)

Several Problems in the Thermal Contact Resistance Research of Solid Interfaces at Low Temperature

Xu L., Li Z.C.

Institute of Refrigeration and Cryogenics Engineering, Shanghai Jiao Tong University, Shanghai 200030, People's Republic of China

According to our experiences in the research of the thermal resistance between solid interfaces at low temperature, some modeling problems such as surface topography, the interface deformation and the improvement of thermal contact of the solid surfaces are introduced in this article. A new thermal contact fractal model is developed. The novel measure equipment is developed which is capable to measure the thermal conductivity and the thermal contact resistance between solid surfaces simultaneously and the test error is analyzed. The experimental results were compared with the prediction data.

INTRODUCTION

In many new technologies, problems of thermal contact resistance are often met. For example, cooling a detector in a satellite, cooling large scale integrated circuit chips and exchanging heat with great efficiency of heat flow in a satellite, are all done via contact heat exchangers.   In these fields, the thermal contact resistance determines the work condition of some key equipment. Especially in many spacecraft electronic components, such as focal-plane detector arrays used in satellite imaging systems, operate efficiently only at low temperatures within narrow limits.

Prediction of operating temperatures needs correct modeling of the heat path from source to sink. A primary factor influences heat transfer is the thermal contact resistance across metal contacts. Therefore, proper model and accurately measurement of thermal contact resistance between metal interfaces at low temperatures is necessary.

In this article, the surface roughness of the metal samples was measured by using STAR-1 profile meter. A prediction model was developed by applying a discretization method to obtain the rough height and deformation mode based on the measurement of surface roughness. Then, the thermal contact resistance between a pair of stainless steel samples and a pair of Al samples were measured experimentally.

MECHANISM OF THERMAL CONTACT RESISTANCE

Engineering surfaces are never absolutely smooth and surface irregularities are apparent when observed under a microscope. As a result, when two solid faces are pressed together, contact is made only at a few discrete spots separated by relatively large gaps. Therefore, the heat flow shrinks at the interface and the temperature difference between the two contact faces appears, as shows in Figure 1. It indicates that the thermal contact resistance exists between two solid interfaces.

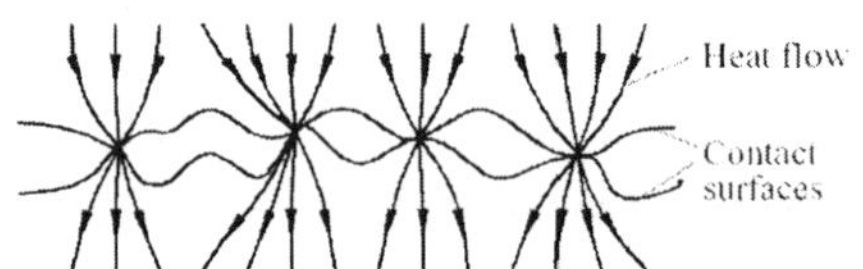

Figure 1 The constriction of heat flux at the interface

PREDICTION MODELS

The thermal contact conductance h_c (the reciprocal of the thermal contact resistance) between two metal interfaces can be predicted theoretically by used the following equation.

$$h_c = \frac{2\bar{a}N\lambda}{\psi} \qquad (1)$$

Here, h_c is the thermal contact conductance; $\bar{a}$ is the average dimension of the contact points; N is the contact point's number; λ is the conductivity of the material; and ψ is contraction coefficient between the two contact surfaces. The profile model and the deformation model about the surfaces determine these parameters.

However, the routine contact deformation models about the surfaces are based on some ideal assumptions. For example, the assumptions in G-W elastic contact deformation model are: the heights of the surface micro-heaves distribute as normal type, the curvature of the surface micro-heaves is a fixed value and the most of the surface micro-heaves deform elastically [1]. The Y (Yovanovich [2]) elastic contact model is based on the assumptions that all the gradients of the micro-heaves distributing random and all the micro-heaves deforming elastically.

Due to the complexity of the surface profiles and the contact conditions, the assumptions are different from the practical ones. The results based on all the former models deviate from the experimental results. Based on the measurement of the practical surfaces profiles, the mathematic model on the thermal contact conductance is established by application profile discretization method to calculate practical deformations at different heights about micro-heaves [3].

The contact between two rough surfaces can be simplified by introducing the concept of a equivalent rough surface. This condition could regard as the contact between an equivalent rough surface and a rigid and smooth surface. The equivalent rough surface may be discretized into many small sections, and then the contact pressure on each section could be calculated. Then, the thermal contact conductance could be calculated.

MEASUREMENT OF THE SAMPLE SURFACE

The STAR-1 surface profile gauge with computer data processing system is used to measure the surface roughness of the samples. The end of detector the STAR-1 surface profile gauge is $4\mu m$ in diameter, and its sampling distance is $2.5\mu m$. In the experiments, the scanning distance is 17.5mm in length; therefore, there are 7000 data on each scanning. These data recorded in computer are enough to analyze the surface topography. Figure.1 and Figure.2 show the measured surface topographies of a stainless steel sample and an Al sample respectively.

Based on the analysis of the profile measuring results, the thermal contact conductance (or resistance) can be predicted by used the models above-mentioned.

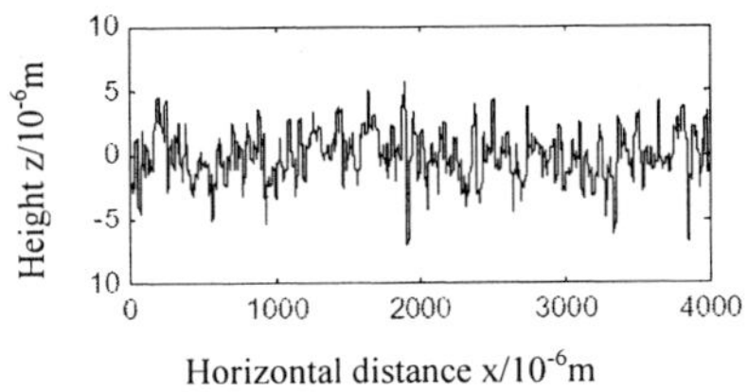

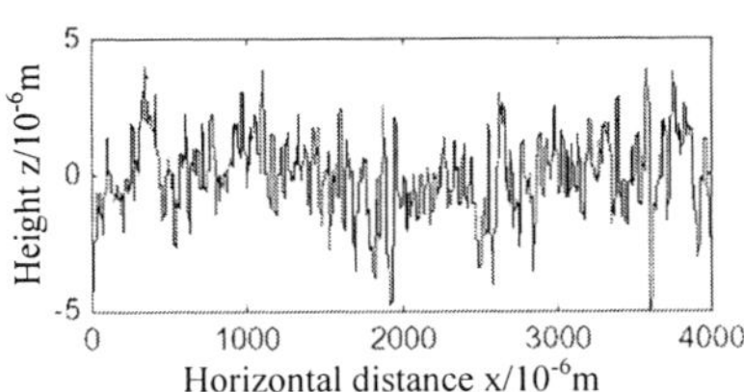

Figure 2 The actual profile of stainless steel sample   Figure 3 The actual profile of Al sample

EXPERIMENTAL MEASUREMENT

Experimental method and apparatus

The contact resistance of the solid material at low temperature and vacuum has been studied and measured widely, and a lot of experimental data has been obtained. But there are some common defects remain. Some of the experimental apparatus can only measure rod shaped samples, and some can only measure thin-disk-shaped samples, but hitherto no apparatus can measure both sample shapes. In some experiments, the input power had been regarded as the sole heat flux to the sample. In fact, many factors cause a heat leakage.

In our research, a new apparatus that overcomes these defects is introduced [4]. In this experimental apparatus, double heat flow meters were used and the heat flow through the sample and the temperature different between the interfaces can be obtained by the outer-expansion method. The apparatus can measure not only rod-shaped samples but also thin-disk-shaped samples. Also, the thermal conductivity of the materials at low temperature can be measured on this apparatus.

The relative error of the temperature difference ΔT between the contact surfaces is less than 9%, and the total relative error is less than 12.21%.

Experimental results and discussion

The experimental results about the two pairs of samples were shown in Figure 4 ~5. In order to compare with the prediction data, the data predicted by the profile discretization method, the G-W model and Y model were shown in the figures also.

It can be seen from Figure 4~5 that predicting values of the G-W model are usually less than the experimental values, and the predicting values of the Y model are far more than the experimental values. From the results of the stainless steel samples in Figure 4, it can be seen that the maximum error of the predicting values of Y model is exceed 100% of experimental values; the maximum error of the predicting values of G-W model is more than 50% of experimental values. In the mean time, the maximum error of the values predicted by discretization method is only about 10% to experimental values. As for the Al samples (see Figure 5), the maximum error of the predicting values of Y model is exceed 200% of the experimental values; the maximum error of the predicting values of G-W model is more than 50% of the experimental values. In the mean time, the maximum error of the values predicted by discretization method is less than 50% of the experimental values. However, the data predicted by Profile discretization method is in good agreement with the experimental results.

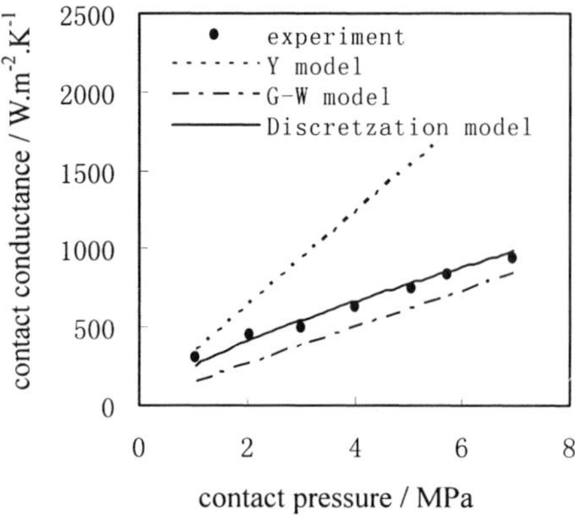
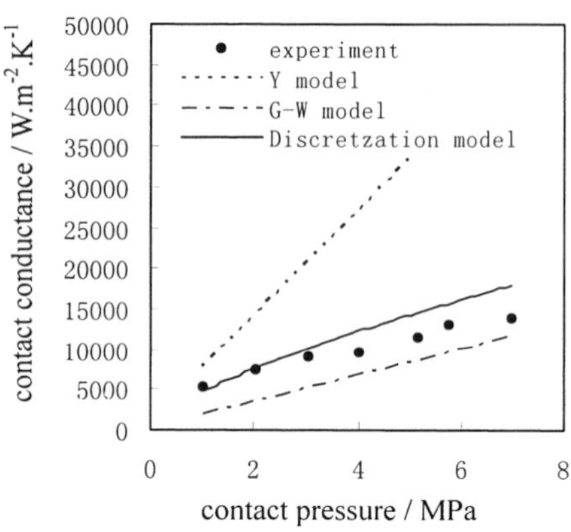

Figure 4 Experimental and theoretical predictions data of the stainless steel samples (T=155K)

Figure 5 Experimental and theoretical predictions data of the Al samples (T=155K)

Compared Figure 5 with Figure 4, it can be found that the error of the values predicted by discretization method for Al samples is more than it for stainless steel samples. For an easily oxidable metal, the oxidation will cover the machining surfaces of the Al samples inevitably. The thick of the oxidation layer is about 0.075~.236 μm and on influence on the profile of the samples, but it harder than the main body material and the conductivity is smaller. So the actual deformation and thermal conductance between the interfaces is different in theoretical. The detail mechanism about this phenomenon is waiting for more research.

CONCLUSIONS

Based on the measurement of the profile of the surfaces, it is discovered that the height distribution of micro-heaves on the rough surfaces is random. Then the profile discretization model without any assumption was put forward to predict the thermal contact conductance (or resistance) between the metal interfaces.

The thermal contact conductances between two pairs of samples were researched experimentally. The experimental results were compared with the predicting values of G-W model, Y model and the discretization model. It shows that simulated results of the discretization model are in best agreement with experiments.

REFERENCES

[1]. J.A. Greenwood and J. B. P. Williamson. Contact of nominally flat surfaces. Proc. Roy. Soc (London), 1966, A295:300-319

[2].M.R. Sridhar and M.M. Jovanovich, Elastoplastic contact conductance model for isotropic confirming rough surfaces and comparison with experiments, Journal of Heat Transfer, (1996) 118 3-9

[3].Xu Lie, Zhao Lan-ping, et al. Application profile discretization method to investigate on thermal contact conductance between rough interfaces. Proceeding of ICEC'19, Grenoble, France. (2002,7)

[4]. Lie X. Using double heat flux meter method to measure resistance of solid material simultaneously at low temperature and vacuum. 20th Proc. Int. Refrigeration Congress, Sydney, Australian (1999)

CFD approach to thermal pulse propagation and heat transfer in forced flow field of He II

Yanase M., Murakami M

Institute of Engineering Mechanics and Systems, Univ. of Tsukuba, Tsukuba 305-8573, Japan

For understanding the effects of forced flow on heat transfer in superfluid helium (He II), a numerical simulation was carried out, where Landau's two-fluid equation supplemented with Vinen's vortex line density evolution equation was numerically solved.   An improvement is introduced with respect to a superfluid turbulence model to oncoming forced flow for more realistic simulation.   The effect of forced flow on the propagation of a thermal pulse and on the heat transfer in superfluid turbulent state were computed by taking the forced flow velocity as a parameter.

INTRODUCTION

He II is an excellent coolant for cryogenic cooling of superconducting magnets and IR detectors.   A forced flow has been considered as a heat transfer enhancing technique even for He II , which in fact, was experimentally confirmed [1].   It is known there are two heat transport modes, that is the second sound wave mode and the diffusive one, in superfluid turbulent state.   The former dominates in early transient phase for unsteady heating, and the latter does in steady phase.   Our numerical simulation for the forced flow heat transfer problem based on Landau's two-fluid equation [2] straightforwardly supplemented with Vinen's vortex line density (VLD) evolution equation [3] resulted in a poor agreement with experimental result for the wave mode heat transfer, though it did in a good agreement for steady heat transfer.   So, an improvement is introduced with respect to a superfluid turbulence model for oncoming forced flow for more realistic numerical simulation.

It is thought the numerical simulation is a useful method to elucidate the detail mechanism of heat transfer enhancement by a forced flow, though the analysis of the transient heat transport mechanism with the existence of a forced flow has not been well performed yet.   In the present study, the one-dimensional numerical simulation was conducted on the basis of Landau's two fluid equation and Vinen's VLD equation in order to investigate heat transfer in He II   forced flow field.

NUMERICAL METHODS

One-dimensional Landau's two fluid equations including the mutual friction term are written as,

$$\frac{\partial \rho}{\partial t} = \frac{\partial \rho v}{\partial x} = 0 \tag{1}$$

$$\frac{\partial \rho s}{\partial t} + \frac{\partial \rho s v_n}{\partial x} = A \rho_n \rho_s v_{ns}{}^2 \left(v_{ns} - v_c\right)^2 / T \tag{2}$$

814

$$\frac{\partial v_s}{\partial t} + \frac{\partial}{\partial x}\left(v_s^2/2 + \mu\right) = A\rho_n v_{ns}\left(v_{ns} - v_c\right)^2 \tag{3}$$

$$\frac{\partial \rho v}{\partial t} + \frac{\partial}{\partial x}\left(\rho_n v_n^2 + \rho_s v_s^2 + p\right) = 0 \tag{4}$$

Here, ρ is the density, v the velocity, p the pressure, T the temperature, μ the chemical potential, v_{ns} the relative velocity between the normal and super components and v_c is the critical velocity for superfluid break-down. The subscripts n and s are referred to as the quantities of normal and super components. The right hand sides of equations (2), (3) are related to the mutual friction. The coefficient A as a function of vortex line density, L, is defined as [3], $A = hBL/(3\rho m(v_{ns} - v_c)^2)$, where h is the Plank constant, m the mass of a ^{4}He atom, and B Vinen's mutual friction constant. Growth of L is governed by Vinen's vortex line density evolution equation [3] written as

$$\frac{\partial L}{\partial t} + \frac{\partial v_L L}{\partial x} = \frac{\chi_1 B\rho_n}{2\rho}|v_{ns}|L^{3/2} - \frac{\chi_2 h}{2\pi m}L^2 \tag{5}$$

Where v_L is the drift velocity of vortex tangle, and χ_1 and χ_2 are functions of temperature. The coefficient A depends on L, which means that A also depends on the time, though the numerical calculation for a steady state, A is treated as a time independent quantity.

Numerical calculation

Equations (1)-(5) are numerically solved by using the finite difference two-step explicit predictor-corrector MacCormack scheme having second-order accuracy in both time and space. In order to avoid numerical oscillation, the FCT (Flux-Corrected Transports) algorithm is combined with the MacCormak scheme [4]. The tabulated data, HEPAK (CRYODATA) are used for the calculation of thermodynamic quantities with the aid of the linear interpolation to the discrete data sets. The quantity B was given by Vinen [3]. We have $\chi_1 = 0.29$ and χ_2 is obtained from the following relationship [3], $A = \pi B^3 \rho_n^2 m (\chi_1/\chi_2)^2 / (6\rho^3 h / 2\pi)$. In the present study, we define $v_L = v_s$ based on a physical consideration, though its expression is still an open question. The critical velocity v_c is neglected because of smallness compared to v_{ns}.

The calculation field is a one-dimensional one, and thus there are no side boundaries in it. The inlet forced flow velocity U_∞ is given between 0 and 20 m/s at the left end of the calculation field, $x = -0.2$m. The initial condition of the uniform forced flow is given as follows in the most part of this study; $P_0 = 1.0 \times 10^5$ Pa, $T_0 = 1.7$ K, $L_0 = 1.0 \times 10^9$ m^{-2} and $v_n = v_s = U_\infty$. The effect of L_0 will be discussed later. The heat source is located at $x = 0$ m with a width of 5mm. The stepwise heat flux is imposed at t = 0 with a constant heat flux of $q_{in} = 40 \times 10^4$ W/m^2, and is kept at this constant value.

RESULT AND DISCUSSION

Figure 1 shows the numerical result of the transient variations of the temperature and VLD distributions in the case of $U_\infty = 8$ m/s. The second sound wave portion propagating in the upstream direction seems to be spatially compressed, and consequently the amplitude is augmented. The propagation speed decreases roughly by U_∞. Conversely, that propagating toward the downstream is stretched, and the amplitude is diminished, and the propagation speed increases by U_∞. In the diffusive heat transfer region around the heat source, the temperature rises up to a quasi-steady plateau value and the plateau area, this is the thermal boundary layer, expands toward downstream with a speed roughly equal to U_∞. The plateau value is found to

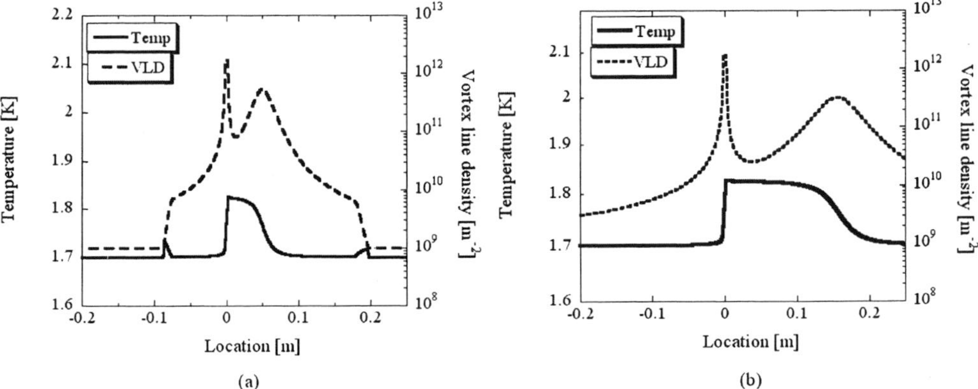

Figure 1 Transient variation of spatial distributions of the temperature and VLD in the case of $U_\infty = 8$ m/s. Figures (a) for $t = 8$ ms, and (b) for 20 ms.

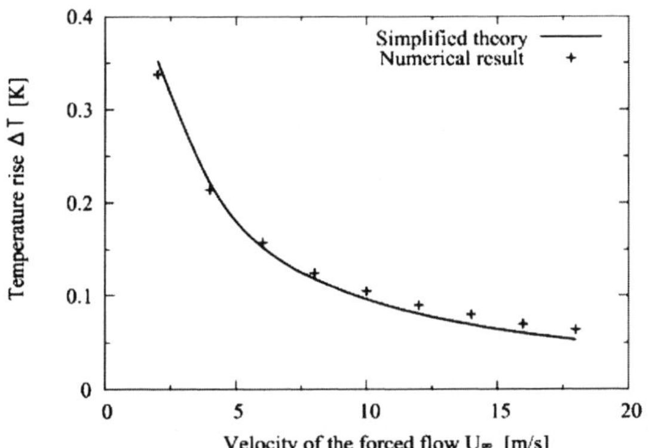

Figure 2 Relationship between U∞ and the quasi-steady value of the temperature rise at the position of heat source, x = 0.

drop with the increase of U_∞, which means the heat transfer in the diffusive region is enhanced due to a forced flow. It seems unreasonable that L singularly drops at the center of the heat source, $x = 0$, in a very early stage and then the trough of L expands between the heater and the leading edge of the thermal boundary layer as a result of the convection. The cause of this can be understood on the basis of the fundamental equations. The relative velocity v_{ns} changes the sign there, and thus $v_{ns} = 0$ at $x = 0$, and is small around the point. It is seen from equation (5) that L cannot develop in the point of $v_{ns} = 0$. This fact may suggest that Vinen's equation must be modified, though the effect on the temperature is not so significant because L value is still very large even in the trough area.

Shown in Figure 2 is the numerical result of the quasi-steady temperature value at the heat source plotted against U_∞. It is understood from the result that the heat transfer is enhanced with the increase of U_∞. The simple theoretical prediction of the quasi-steady temperature rise is given by $\Delta T_d = q_{in} / (\rho\, c_P\, U_\infty)$, where c_P is the specific heat. It is the main reason for the small discrepancy between the theoretical and numerical results that the contribution of the internal counter flow between the two components is ignored in the simple theoretical treatment.

It is seen in the above result the amplitude of the second sound pulse propagating in the upstream direction is augmented by the compression action of the forced flow. It was, however, pointed out that this result disagreed with some experimental results in the point that the second sound wave propagating toward the upstream was not observed. This fact suggests that an oncoming flow brings vortex tangle with rather high density presumably generated by (classical) turbulent motion of He II. Thus, we attempted to modify Vinen's equation so as to include the additional VLD, L_v, generated by forced flow as,

$$\frac{\partial L}{\partial t} + \frac{\partial v_L L}{\partial x} = \frac{\chi_2 h}{2\pi m}\left(\sqrt{L_v + \left(\frac{\chi_1 B \rho_n \pi m}{\chi_2 \rho h} v_{ns}\right)^2} \, L^{3/2} - L^2 \right) \tag{6}$$

816

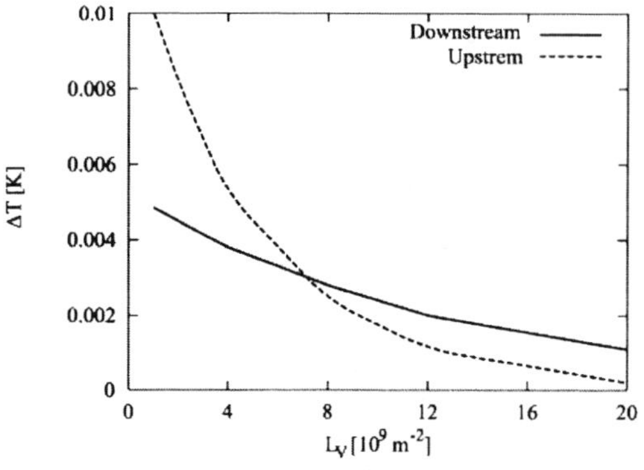

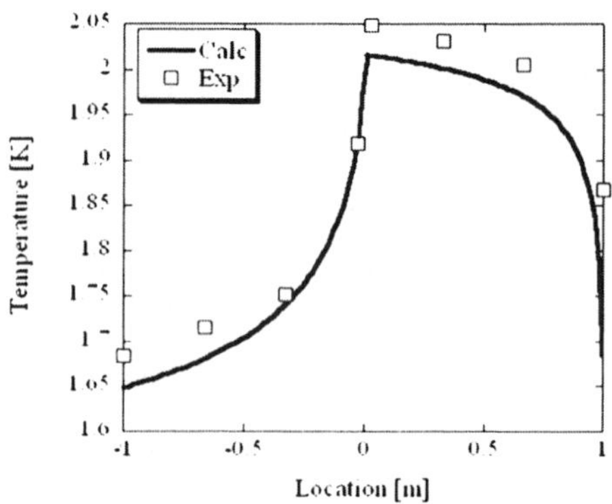

Figure 3 Variation of the amplitudes of second sound thermal pulses propagating in both upstream and downstream directions recorded at x = ±0.1 m with Lv.

Figure 4 Comparison between the present numerical result and the experimental one [6] of the temperature distribution in the diffusive heat transfer zone in a quasi-steady phase (t = 10 s).

The numerical result for the same case as above is shown in Figure 3, where the temperature amplitudes of thermal pulse propagating in both directions recorded at x =±0.1 m are plotted against Lv. The result indicates the thermal pulse propagating toward the upstream is strongly attenuated for large Lv, and will not be detected in experiments. It is now understood the numerical computation became much more realistic as a result of the modification.

Shown in Figure 4 is the comparison between the present result and the experimental one [6] of the temperature distribution in the diffusive heat transfer zone in a quasi-steady phase (t = 10 s). The agreement seems to be fairly good, considering the experimental boundary condition can not fully be reproduced in numerical computation.

CONCLUSION

A fairly good numerical simulation can be conducted based on Landau's two fluid equations supplemented with Vinen's vortex line density evolution equation with some modification to take the effect of high density quantized vortices created by classical turbulence in a forced flow into account for the problem of heat transfer in a forced He II flow field. It can be concluded that the mechanism of heat transfer enhancement by a forced He II flow can be made clear through the present study.

REFERENCES

1. Fuzier, S. and van Sciver, S. W., private communication (2002).

2. Landau, L. and Lifshitz, E., Fluid Mechanics,, 2nd ed. Pergamon Press, Oxford (1990).

3. Vinen, W., Mutual friction in a heat current in liquid helium II III, Proc. R. Soc. London (1957) A243 400-413

4. Fletcher, C., In: Computational Technique for Fluid Dynamics, Spring-Verlag (1988) 143-170

5. Arp, V. and McCarty, R., HEPAK, CRYODATA.

6. Kashani, A. and van Sciver, S. W., Transient forced convection heat transfer in turbulent He II flow, International Cryogenic Engineering Conference-12 Butterworths, Goildford, UK (1988) 299-304.

Investigation of thermal insulation for HTS cable systems

Kim D. H., Kim D. L., Yang H. S., Jung W. M., Hwang S. D.*

Korea Basic Science Institute, 52 Yeoeun-dong, Yuseong-gu, 305-333, Daejeon, Korea
*Korea Electric Power Research Institute, 103-16, Moonji-dong, Yuseong-gu, 305-380, Daejeon, Korea

It is well known that the capability of HTS cable system in electric power transmission is increased as the temperature of the cable conductor is decreased. LN_2 is used to cool down the cable conductor. Vacuum and MLI(Multi-Layer Insulation) is employed to minimize heat leak and keep low temperature. In this study, heat leaks into LN_2 vessel are measured using boil-off calorimetry and performances of MLI related to the number of layers, patterns and layer density are studied.

HEAT LOADS ON THE HTS CABLE SYSTEM

The HTS cable system is a highly feasible technology among the applications of high temperature superconductivity and a number of studies have been undergoing for a practical use. A HTS cable system is divided into two parts : conductor and refrigeration systems. Bi-2223 is widely used as a conductor and is refrigerated by sub-cooled LN_2. Subsequently the cable system is operated around 77K and the heat leak from the environment takes place. Multilayer insulation(MLI) is extensively used as insulation material at low temperature and it is well known that the performance of MLI is depend on the number of layers, material and layer density.[1] In this experiment, the heat leaks into the model cable cryostat are measured using boil-off calorimetry and compared to the result of calculation. Then the relations between heat leak and features of MLI which are the number of layers, patterns and layer density is investigated .

EXPERIMENTAL SETUP

The model cable cryostat is shown in Figure 1. The upper and lower show plane and side view of the cryostat, respectively. The LN_2 and vacuum vessel are made of stainless steel. The length, outer diameter and thickness of LN_2 vessel are 1m, 63.5mm and 1.5mm respectively and are 1.3m, 127mm and 2mm for the vacuum vessel. The LN_2 vessel has a surface area of $0.2m^2$ and has a volume of 2.9 litres. LN_2 vessel is also provided with fill and vent lines which are open to atmosphere. MLI is wound over the surface of LN_2 vessel and the cross section is shown in Figure 2. It is shown from the figure that MLI is overlapped in circumferential direction and the length overlapped is 5cm. In order to fix MLI on LN_2 vessel it is tied at 4 points in longitudinal direction using cotton thread. Two piece of the supports made by Bakelite are employed to support the LN_2 vessel in the vacuum vessel. The pressure in the vacuum vessel is measured using a cold cathode vacuum gauge and maintained less than 8.0×10^{-4} Pa while a measurement was going on. The flow rate of nitrogen gas evaporated in LN_2 vessel is measured using wet type flow meter.(WS-

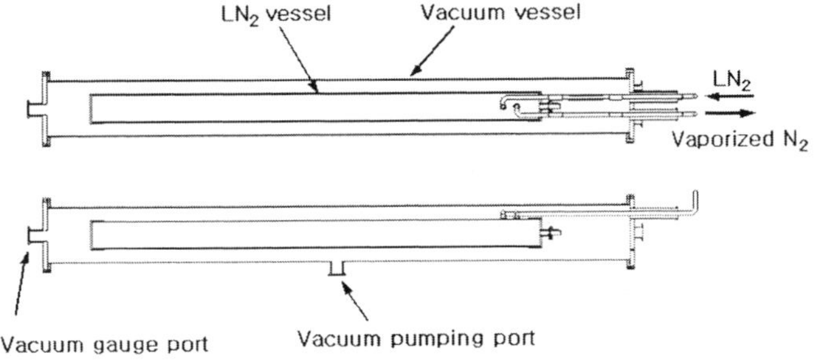

Figure 1   Physical configuration of the model cryostat

1A, Shinagawa) Because the level of liquid nitrogen is lowered as time goes on, 3 units of platinum temperature sensors(PT-111, Lakeshore) are used to examine the temperature of surface of LN_2 vessel. Figure 3 shows the locations where the sensors are set up. Data acquisition system using LabVIEW program is used to record all of the measured data into personal computer. Figure 4 shows schematic of the measuring system used in the experiment.

Figure 2   Cross section of MLI wound on LN_2 vessel

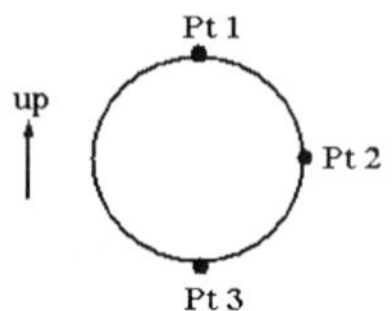

Figure 3   The locations of the temperature sensors

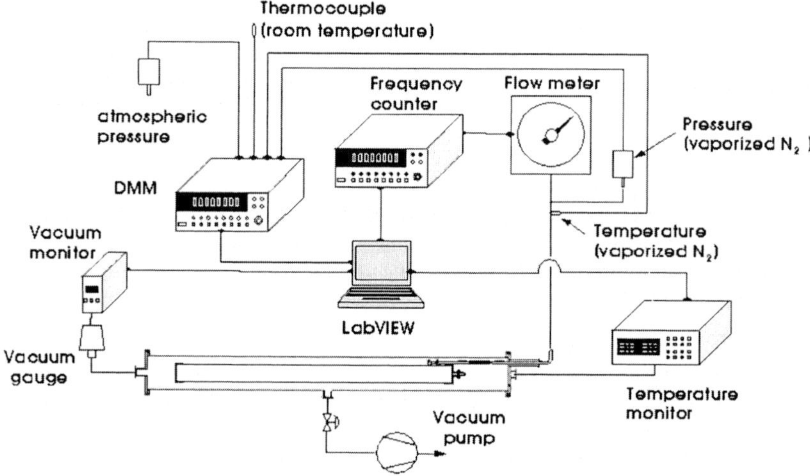

Figure 4   Schematic of the measuring system

Installation of MLI

In the experiment, two type of MLI are tested. The specification of the materials is shown in Table 1. MLI is installed as shown in Figure 2 and heat leaks on 1, 5, 10, 20 and 30 layers of NRC-2 are measured. To examine the effect of spacer to the performance of MLI, another measurements are carried out for 10 and 20 layers of IR-305.

Table 1   Specifications of the MLI

Product name	Manufacturer	Specification
NRC-2	MPI(USA)	Crinkled single side aluminized polyester film Thickness 6μm(thickness of aluminum coating : 250Å)
Insulray IR-305	Jehier(France)	Double aluminized polyester film with polyester tulle Thickness 6μm(thickness of aluminum coating : 400Å)

CALCULATION OF RADIATION THROUGH MLI

If MLI is ideally established without contact between adjacent layers and emissivities of all layers are equal and constant, the radiation heat transfer through MLI of the heat transfer is expressed as follows[2] :

$$\dot{Q} = \frac{\sigma\left(T_H^4 - T_L^4\right)}{\dfrac{1-\varepsilon_H}{\varepsilon_H A_H} + \dfrac{1}{A}\left(\dfrac{1}{\varepsilon_L} + \dfrac{2N}{\varepsilon_s} - N\right)} \tag{1}$$

where, σ : Stefan-Boltzman constant, N : the number of layers, A : area, ε_H, ε_L, ε_s : emissivity of surfaces of room temperature(300K), low temperature(77K) and layer of MLI, respectively. Since equation (1) is derived from the ideal condition, the results calculated from the equation are able to use as a reference to estimate the heat leaks measured from the experiment.

RESULTS AND DISCUSSION

In all the experiment, surfaces of LN_2 and vacuum vessel are the cold and warm boundary, respectively. In the case without MLI, it takes about 7 hours until LN_2 vessel becomes empty. The evaporating times for the cases with MLI are twice or more than without MLI. The flow rates are stabilized after 1~5 hours after filling LN_2 vessel. In the case without MLI, difference in temperatures between Pt1 and Pt3(see Figure 3) exists. It is about 6K and hold on during the experiment. In the cases with MLI, the difference does not exist so that the temperature of surface is considered to be uniform and to have no concern with the level of LN_2. The difference of pressure between the inside of LN_2 vessel and atmosphere is less than 200Pa which is averaged for experiment.

Determination of conduction heat through support
Radiation and conduction due to the supports made of Bakelite are constitute the total heat leak. Because the temperatures of LN_2 and vacuum vessel do not changed during the experiment, the heat leak by means of conduction is equal in all the experiments and measured to classify the radiation heat leak. The heat leak through supports is 1.1W.

Heat leak for the number of MLI(NRC-2)
Figure 5 shows heat flux with respect to the number of layers. It is known from the figure that heat flux is exponentially decreased as the number of layers is increased. Especially, although 1 layer is applied, the heat leak is sharply decreased. It is less than 50% of the results without MLI. On the other hand, difference between the calculation and the measurement is observed and the difference converged as the number of layers is increased. In the experiment, there is a contact between layers so that this causes an additional heat leak through MLI.[3] Therefore the difference between the results of NRC-2 and the calculation exists.

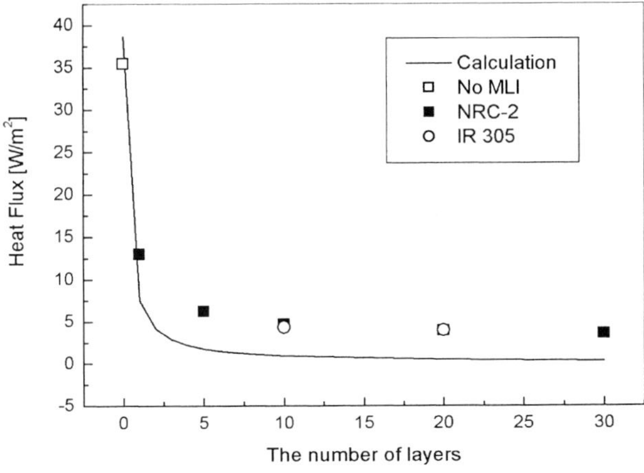

Figure 5   Heat flux with respect to the number of layers

<u>Heat leaks for different patterns of MLI</u>
As noted in Table 1, IR-305 has a polyester tulle(net) as a spacer. In order to examine the effect of spacer, the experiments are performed for 10 and 20 layers of IR-305 and compared to the result of NRC-2. From Figure 5, it is shown that there is no difference between the two for each layers. This result is expected due to the crinkled surface of NRC-2 which reduces contact heat transfer between layers. Consequently, it is considered that the crinkle surface plays the role of spacer.

<u>Influence of layer density on heat leak</u>
Heat leak for 10 layers of NRC-2 is measured at two different layer density. The heat leaks for 33 and 67 layers/cm are 4.7 and 9.8 W/ m^2, respectively. Heat leak is proportional to layer density. Because increased layer density results in increasing a contact area between layers, conduction heat transfer in MLI is increased. It is deduced from the experiment that layer density is more important factor in thermal insulation of LN_2 vessel than patterns of MLI.

REFERENCES

1. Timmerhaus, K. D. and Flynn, T. M., <u>Cryogenic Process Engineering</u>, Plenum press, New York, USA(1989) 387-398
2. Incropera, F. P. and DeWitt, D. P., <u>Introduction to Heat Transfer</u>, 3rd Ed., John Wiley & Sons, New York, USA(1996) 678-714
3. Cunnington, G. R. and Tien, C. L., A study of heat-transfer processes in multilayer insulations, <u>AIAA 4th Thermophysics Conference</u>, San Francisco, Calif., USA (1969), Paper 69-607

Experimental investigation on pool boiling heat transfer of pure refrigerants and binary mixtures

Sun Z.H.[1,2], Gong M.Q.[1], Qi Y.F. [1,2], Luo E.C. [1], Wu J.F. [1]

[1] Technical Institute of Physics and Chemistry, Chinese Academy of Sciences, Beijing, 100080，China
[2] Graduate School of the Chinese Academy of Science, Beijing, 10039, China

Heat transfer coefficients in nucleate boiling on a smooth flat surface were measured for pure fluids of R-134a, propane, isobutane and their binary mixtures at different pressure from 0.1 to 0.6 MPa. A wide range of heat flux and mixture concentrations were covered in the experiment. The influences of pressure and heat flux on the heat transfer coefficient for different pure fluids was studied. Isobutane and propane were used to make up binary mixtures. Compared to the pure components, binary mixtures showed lower heat transfer coefficients. This reduction was more pronounced as heat flux was increased.

INTRODUCTION

Heat transfer performance is one of the important factors for the efficiency of refrigeration applications. Therefore, it is necessary to know the mechanism of heat transfer in the whole refrigeration system. Extensive studies of the boiling heat transfer of pure refrigerant have been made. And many generalized correlations for predicting the coefficients have been proposed, which can be applicable to various substances[1]. Boiling behavior of binary mixtures is more complex than that of pure refrigerants, because it is dependent upon many combinations of substances and their liquid-vapor equilibrium curves. So the study of binary systems is much more tedious due to the large number of experiments required to cover the whole composition range [2]. No complete set of data was published where the Heat Transfer Coefficient (HTC) are presented as a function of mixture concentration and heat flux.

The object of this study is to measure heat transfer coefficient in nucleate boiling of pure fluids and their binary mixtures. Special emphasis is laid on the question, how the influence of the heat flux and pressure on the heat transfer coefficient is predicted at different saturation pressures. Several pool boiling heat transfer correlations for pure refrigerants will be compared, and a new correlation will be proposed based on the experimental data. Mixture concentrations are fully covered and heat flux also varied for a wide range of nucleate boiling. Based on the measured data the mixture effects on boiling heat transfer coefficients will be discussed.

EXPERIMENTAL APPARATUS

Figure 1 shows the schematics of experimental apparatus used in the present measurements. It consists of a boiling vessel, a refrigerant tank and a liquid nitrogen condensing system, an AC regulator and a data acquisition system. The boiling vessel is a hollow vertical tube of stainless steel with an inner diameter of 151 mm and a height of 300 mm. Boiling takes place on the upper end of a copper cylinder, 25 mm in diameter, which is fixed on the bottom of boiling vessel. Heat flux is supplied by a loop heater intertwined at the bottom of copper cylinder. And different power can be obtained by adjusting the AC regulator.

Nine platinum resistance thermometers were installed in the boiling vessel at four different depths from the top to end. The boiling surface temperature and heat flux were determined from measured copper cylinder temperatures assuming one-dimensional heat conduction along the copper cylinder. Axial temperature distributions at the center and at a half radius of the copper cylinder agreed well within measurement errors, conforming one-dimensional heat flow and negligible heat loss from the side of cylinder. One thermometer was set in the liquid pool and another one in the vapor in equilibrium with it. These liquid and vapor temperatures conform the system being maintained at the saturation state during the experiments. And the internal pressure of boiling vessel was measured by a pressure transducer. The boiling vessel and heating unit were well insulated inside the stainless vacuum chamber by a vacuum pump. Electrical signals from the platinum resistance thermometers were processed by a data acquisition system (a 22 bit Model 2700 Keithley Multimeter with 40 channels).

Mixture supplied to the boiling vessel was prepared by mixing pure fluids on a weight base. Sampling liquid and sampling vapor were extracted for measurement of their composition by a gas chromatograph. The measured values agreed within ± 0.01 accuracy with the prepared concentrations. The repeatability of the experiment was always within 5% of the measurement error.

EXPERIMENTAL RESULTS AND DISCUSSIONS

Pure refrigerants

In this paper, pure refrigerant R134a was used to verify the experimental apparatus. Measured heat transfer coefficients of R-134a were compared with correlations by Nishikawa et al. [2] and by Fujita [3], which run as:

$$\alpha = \frac{31.4 P_c^{1/5}}{M^{1/10} T_c^{9/10}} (8R_p)^{0.2(1-P/P_c)} \frac{(P/P_c)^{0.23}}{[1-0.99(P/P_c)]^{0.9}} q^{4/5} \tag{1}$$

where M is average molecular weight, P_c and T_c is critical pressure and density, and $R_p = 0.4 \mu m$.

$$\alpha = 1.21 q^{0.83} \tag{2}$$

Figure 2 shows the comparison of experimental results and the correlations from references for R-134a. The present measured heat transfer coefficients agree well with the predication of Nishikawa et al. and Fujita correlations. Therefore the experimental apparatus and experimental method were found appropriate in performing other pure refrigerants and their binary mixtures experiment.

Figure 3 shows the heat transfer coefficient of pure refrigerant propane at 0.5 MPa. The heat transfer coefficients agreed well with Nishikawa et al. correlation. And the experimental data are very close to the heat transfer correlation of literature [4]. A correlation with a simple expression was also obtained from this experimental measurement, which is expressed as:

$$\alpha = 4.84 \cdot q^{0.73} \tag{3}$$

Figure 4 is the comparison of heat transfer coefficients between R-134a and isobutane at 0.5MPa. It

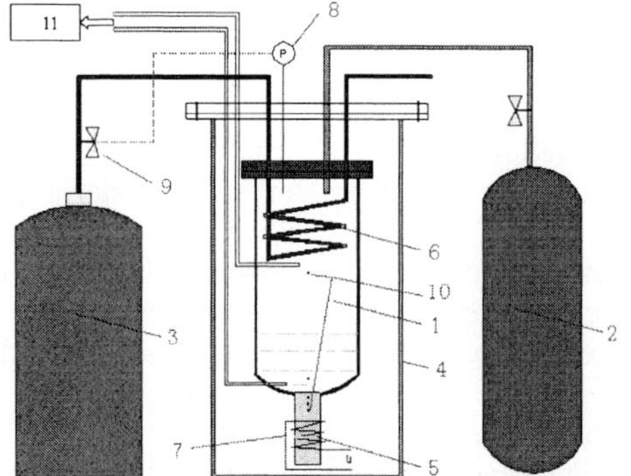

1. boiling vessel   2. refrigerants tank
3. liquid nitrogen cryostat   4. vacuum chamber
5. copper cylinder   6. condenser
7. electric heater   8. pressure transducer
9. electric-magnetic valve
10. platinum resistance thermometers
11. gas chromatograph instrument

Figure1   Experimental Apparatus

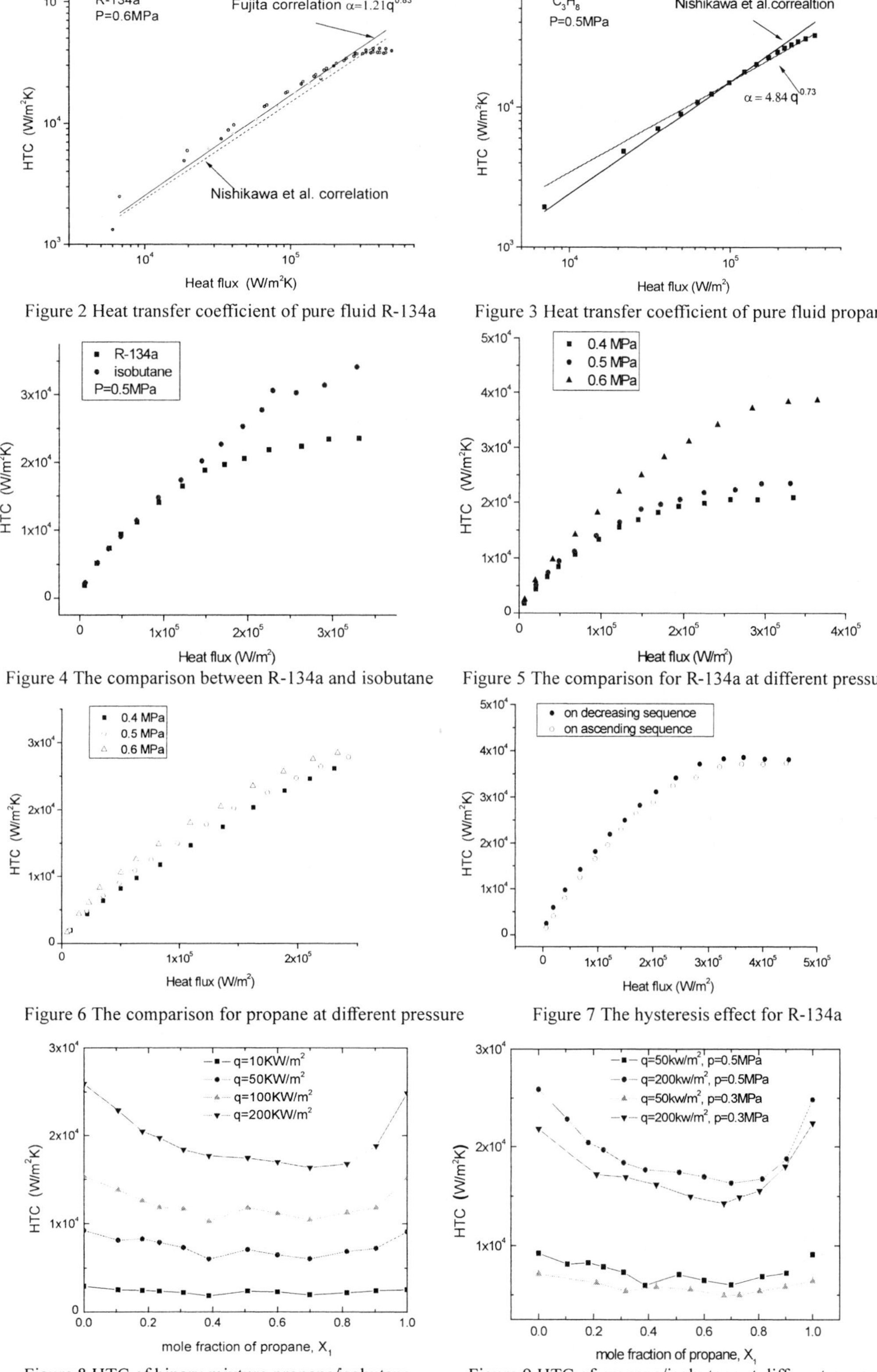

Figure 2 Heat transfer coefficient of pure fluid R-134a

Figure 3 Heat transfer coefficient of pure fluid propane

Figure 4 The comparison between R-134a and isobutane

Figure 5 The comparison for R-134a at different pressures

Figure 6 The comparison for propane at different pressure

Figure 7 The hysteresis effect for R-134a

Figure 8 HTC of binary mixture propane/isobutane

Figure 9 HTC of propane/isobutane at different pressures

is shown that the heat transfer coefficient of isobutane is higher than that of R-134a. And the trend become bigger as the heat flux increases.

Figure 5 shows the heat transfer coefficient curve of R-134a at different pressures from 0.4 MPa to 0.6 MPa, and Figure 6 shows the results of pure fluid propane at the same conditions. From the HTC curve, when the pressure increases, heat transfer coefficient of R-134a will enhance accordingly. And it is more distinct at higher pressure. But for pure fluid propane, it becomes less sensitive to the change of pressure.

Figure 7 shows the experimental data on ascending sequence and decreasing sequence of heat flux. It can be noted that heat transfer coefficients on the decreasing sequence heating are higher than that on ascending sequence heating. In order to obtain the proper pool nucleate boiling data, decreasing sequence heating method should be adopted to avoid the effect of boiling hysteresis.

<u>Binary mixtures</u>

Figure 8 shows the measured heat transfer coefficients against the mixture concentration with heat flux as parameter for propane (X_1). As shown in Figure 8, heat transfer coefficient of binary mixture is reduced in an intermediate range of the mixture concentration. As the heat flux increases, the reduction becomes larger and the minima are more pronounced at a higher heat flux. In other words, the heat transfer coefficient of mixtures are significantly lower than those of single component substances and dramatically deteriorate in the vicinity of single component substances, reaching their lowest values in the range of $0.3<X_1<0.7$ in which the coefficients are independent of the concentration. There are two reasons. One is the change of physical and transport property for binary mixture. The other is the mass diffusion effect. And the latter one is mostly the main reason on pool nucleate boiling. One important parameter required to characterize the mixture is the boiling range, that is dew-point temperature minus the bubble-point temperature for the bulk liquid mixture. Boiling range has a large influence on pool boiling heat transfer, so many heat transfer correlations include it to simulate pool nucleate boiling.

Figure 9 shows the boiling curve of binary mixture of propane and isobutane at different pressure. It is explained that the effect of pressure on heat transfer coefficient tends to be smaller in the mixtures than in the single component substances.

CONCLUSIONS

An experimental apparatus is designed and built to investigate pool boiling heat transfer characteristics for pure refrigerants and their binary mixtures. Experimental data are obtained for pure fluid R-134a, propane, isobutane and binary mixture of propane and isobutane. Based on the present measured data, pressure, heat flux and different refrigerants are studied as influencing factors of pool boiling heat transfer. Heat transfer reduction is found in binary mixtures due to the mixture effects. More binary mixtures and ternary mixtures will be investigated and new heat transfer correlations will be developed according to the experimental data in future.

ACKNOWLEDGEMENTS

The authors acknowledge with gratitude the continued support received from the National Natural Sciences Foundation of China under the contract number of 50206024.

REFERENCE

1.Dongsoo Jung, Youngil Kim,Younghwan Ko, Kilhong Song. Nucleate boiling heat transfer coefficients of pure halogenated refrigerants, <u>International Journal of Refrigeration</u> (2003) <u>26</u> 240 -248

2. K. Nishikawa, Y. Fujita, H. Ohta, S. Hidaka. Effect of the surface roughness on the nucleate boiling heat transfer over the wide range of pressure, <u>Proc. 7th Int. Heat Transfer Conf</u> (1982) <u>4</u> 61- 66

3. Yasunobu Fujita. Experimental investigation in pool boiling heat transfer of ternary mixture and heat transfer correlation. <u>Experimental Thermal and Fluid Science</u> (2001) <u>26</u> 237-244

4. J.Shen,K.Spindler,E.Halne. Pool boiling heat transfer of propane from a horizontal wire. <u>Int. Comm. Heat Mass Transfer</u> (1997) <u>24</u> 633 - 641

MODELING THE THERMAL MECHANICAL BEHAVIOR OF A 300 K VACUUM VESSEL THAT IS COOLED BY LIQUID HYDROGEN IN FILM BOILING

S. Q. Yang, M. A. Green, and W. Lau

Oxford University Department of Physics, Oxford OX1 3RH, United Kingdom

This report discusses the results from the rupture of a thin window that is part of a 20-liter liquid hydrogen vessel. This rupture will spill liquid hydrogen onto the walls and bottom of a 300 K cylindrical vacuum vessel. The spilled hydrogen goes into film boiling, which removes the thermal energy from the vacuum vessel wall. This report analyzes the transient heat transfer in the vessel and calculates the thermal deflection and stress that will result from the boiling liquid in contact with the vessel walls. This analysis was applied to aluminum and stainless steel vessels.

INTRODUCTION

The proposed Muon Ionization Cooling Experiment (MICE) [1] has three liquid hydrogen absorbers that remove both the transverse and longitudinal momentum from a muons beam with an average momentum of about 200 MeV/c. Eight 201-MHz RF-cavities accelerate the muons back to their original longitudinal momentum without adding their back the transverse momentum lost in the cooling process. Liquid hydrogen has been selected as an absorber material because it has twice the dE/dx (energy loss per unit mass per square meter) of any other material. Hydrogen has only single proton in its nucleus. Thus, the coulomb scattering, which reintroduces transverse momentum into the muons, is minimized. Hydrogen is the best material for muon ionization cooling by a factor of two.

For safety reasons, the LH_2 vessel must be surrounded by a vacuum that is blanketed from the outside air by an inert gas shield. This vacuum is separated from any of the other vacuum systems in MICE. This means that the muons must pass through four windows, two on the absorber vessel and two on the absorber vacuum vessel. These windows, which are made from a high z material (compared to hydrogen), will contribute to the coulomb scattering that produces of transverse momentum in the beam. Very thin windows must be made from a low z material. Of the low z materials available, only beryllium, magnesium and aluminum can be used for vacuum leak tight thin windows. MICE will use 180-μm thick 6061-T6 aluminum windows on both vessels. These windows have a design burst pressure of 0.68 MPa.

This report discusses the safety implications that result from the rupture of a liquid hydrogen thin window. A rupture will result in 20-liters liquid hydrogen ending up in the absorber vacuum vessel [2]. The liquid hydrogen spills onto the walls and bottom of the 300 K absorber vacuum vessel with dome shaped heads. Some of the spilled hydrogen will be splashed onto the 300 K aluminum vacuum windows. The spilled hydrogen boils removing thermal energy from the vessel walls and the windows.

Figure 1 shows a cross-section of the MICE focusing magnet with a liquid hydrogen absorber installed within the magnet warm bore. Figure 1 shows where the liquid hydrogen ends up when a liquid hydrogen window is ruptured. A window rupture at either end of the absorber will result in a liquid hydrogen pool on the bottom of the absorber vacuum vessel at the lowest point. A hydrogen window rupture can also result in liquid hydrogen flowing along the magnet warm bore. A rupture at either end will result in splashing the hydrogen on a vacuum window. The boiling hydrogen will pressurize the absorber vacuum vessel to until the pressure can be relieved through a relief device. This report shows a FEA analysis of the transient heat transfer into the 300 K vessel walls and the thin windows. The thermal deflections and stress that result from the contact with boiling liquid hydrogen are calculated for the time that the liquid hydrogen is in contact with the wall or window. The thermal and stress analysis was applied to vacuum vessels fabricated from 6061-aluminum and 304-stainless steel.

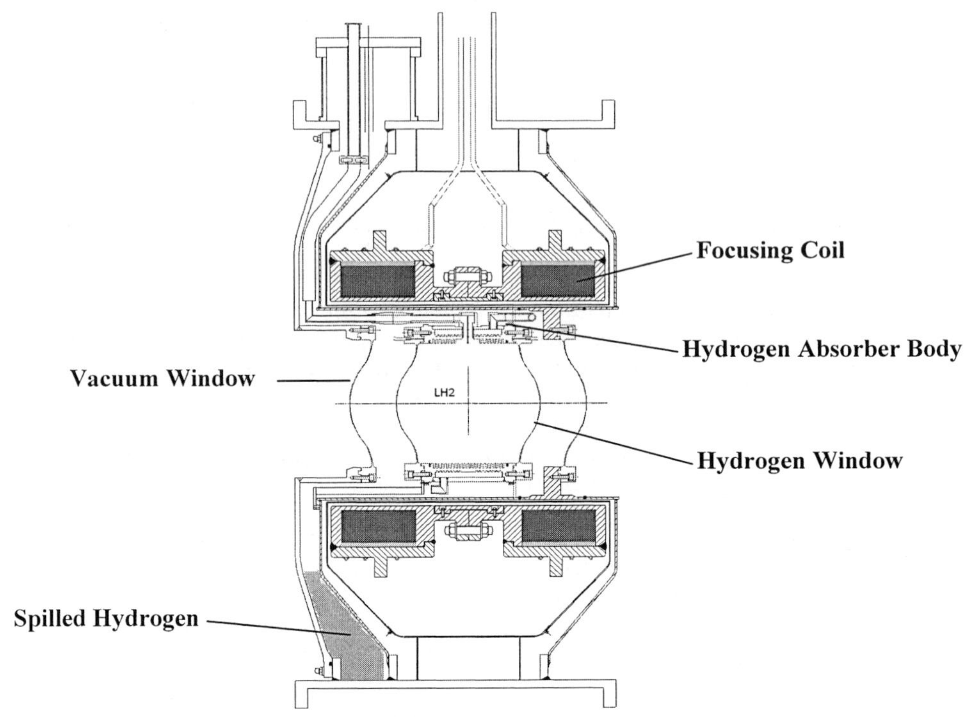

Figure 1. Cross-section of the MICE absorber focus module showing the liquid hydrogen after a window rupture.

WALL THERMAL TIME CONSTANTS AND BOILING HEAT TRANSFER

In order to determine the rate that cold is transferred through the metal in contact with the hydrogen one should look at the thermal diffusivity and the thermal penetration depth for the cryostat material. Since the temperature change for the vessel is relatively small, the room temperature values for the thermal diffusivity and thermal penetration time constants can be used to judge the problem.

The thermal penetration depth λ as a function of time t is directly related to the thermal diffusivity α. An expression for the thermal penetration depth in a solid λ (t) is given as follows [3]:

$$\lambda(t) = (\alpha t)^{0.5} = \left(\frac{kt}{\rho c}\right)^{0.5} \tag{1}$$

where k is the material thermal conductivity; ρ is the material density; and c is the material specific heat. For 6061 Aluminum, k = 180 W m^{-1} K^{-1}; ρ = 2713 kg m^{-3}; c = 963 kJ kg^{-1}; and α = 6.89 x 10^{-5} m^2 s^{-1}. For 304 stainless steel, k = 14.9 W m^{-1} K^{-1}; ρ = 7900 kg m^{-3}; c = 477 kJ kg^{-1}; and α = 0.395 x 10^{-5} m^2 s^{-1}. Since the time constant for the boiling of 20 liters of liquid hydrogen spilled into the absorber vacuum vessel is about 20 s, it is useful to look at the thermal penetration depth for key materials at a time of 20 s. For 6061 aluminum λ(20) = 37.1 mm. For 304L stainless steel λ(20) = 8.9 mm. The thermal penetration depth for aluminum is a factor of four larger than for stainless steel, which suggests that the depth of a rib or a plate is less easily penetrated for the stainless steel.

Thermal deflection is a function of the contraction coefficient β as well as the temperature difference across a plate. The thermal stress involves the modulus of elasticity E as well the contraction coefficient and temperature difference across the plate. For 6061 aluminum, β= 2.15 x 10^{-5} K^{-1} and E = 69 GPa. For 304 stainless steel, β= 1.43 x 10^{-5} K^{-1} and E = 200 GPa.

Since the hydrogen enters the 300 K vacuum chamber as a liquid, the starting ΔT is 280 K. Since the temperature differences are large, a film boiling equation such as the Breen Westwater equation [4] should be used. For hydrogen, heat flux per unit area for film boiling $[Q/A]_f$ can be estimated using the following simplified expression [5];

$$\left[\frac{Q}{A}\right]_f \approx 333\Delta T \tag{2}$$

The linear relationship between film boiling heat transfer per unit area the ΔT between the wall and the bulk hydrogen means that film boiling can be treated just like convection heat transfer. This heat transfer coefficient was applied to the surfaces that are in contact with liquid hydrogen in order to calculate the thermal stress and deflection. When there is 20 l of liquid hydrogen in the vacuum vessel, about 0.55 m^2 of 300 K surface is exposed to the liquid. The peak heat flux to the hydrogen is about 51 kW. The peak hydrogen boil off rate is about 0.116 kg s^{-1}. All of the hydrogen boils away in just over 20 seconds.

THE FINITE ELEMENT CALULATIONS OF TEMPERATURE, STESS AND DEFLECTION

A worst-case analysis of the effect of thermal shock was studied for the absorber vacuum vessel end shell. The edges of the end shell are fixed to the end plate flange. Figure 1 shows the depth of the liquid hydrogen as it pools at the end of the absorber vacuum vessel. For the worst-case simulation, the liquid hydrogen was assumed to be in contact with the vacuum vessel end plate at its maximum depth for 20 s. The maximum liquid hydrogen depth is 455 mm. The heat transfer coefficient for the part of the plate in contact with the liquid was 333 W m^{-2} K^{-1}. The heat transfer coefficient used for the rest of the plate was zero. The study was done for 10-mm thick 304 stainless steel and a 6061-aluminum end plates. Table 1 shows the calculated maximum temperature change, temperature differences through the shell, the maximum thermal stress, and deformation within the vacuum vessel end plates.

Table 1. The maximum ΔT in the vacuum vessel end plate, the ΔT through the plate, the maximum plate deflection and the maximum thermal stress in the MICE vacuum vessel end plate for hydrogen in contact with the plate for 20 s.

Parameter	304 St Steel	6061 Aluminum
Maximum Temperature Change (K)	58.8	62.4
Temperature Difference Through Shell (K)	25	3.4
Maximum End Shell Displacement (mm)	0.79	1.60
Maximum Stress in the End Shell (MPa)	176	115

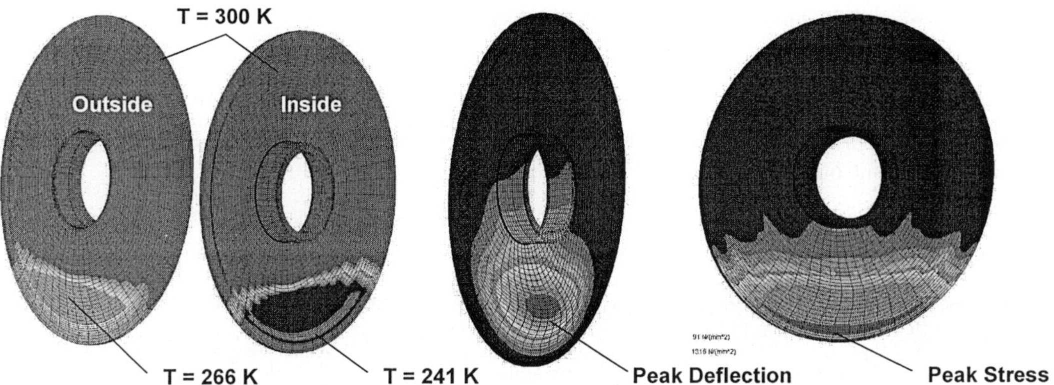

Figure 2. The minimum temperature on the inside and the outside and the peak deflection and stress points due to LH$_2$ spill

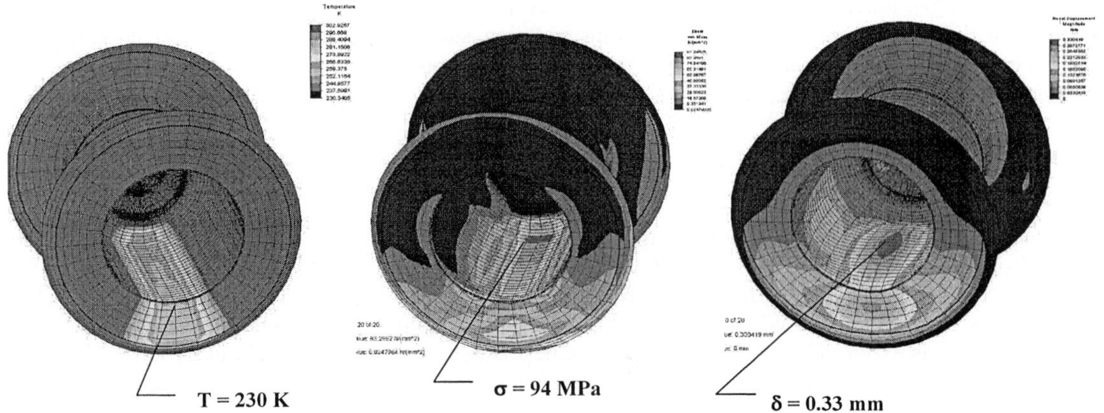

Figure 3. Minimum temperature, peak stress and peak deflection on the magnet warm bore tube due to a LH$_2$ spill in the bore

Figure 2 shows the temperature on the inside and the outside of a 304 stainless steel end shell. Also shown in Figure 2 are the locations of the peak stress and peak deflection points for the stainless steel end shell. The values of the temperature peak stress and peak deflection are given in Table 1.

Figure 3 shows the warm bore and one end of the focusing magnet 304 stainless steel cryostat. For the case shown in Figure 3, the cryostat vacuum vessel was set at 5 mm and the time of contact for the liquid hydrogen is 10 seconds. As in Figure 2, one can see that the vessel deforms in the direction of the cold hydrogen. The peak stress point is where the inner bore cylinder connects to a stiff ring, upon which one of the vacuum windows is mounted. The minimum temperature point is the lip where the liquid hydrogen runs out of the magnet warm bore.

Splashing liquid hydrogen on the thin windows causes them to deform toward the source of the cold. The levels of stress and deflection in the windows due to thermal gradients are over an order of magnitude lower than the stress and deflection due to pressure.

CONCLUDING COMMENTS

If the contents of the absorber suddenly spills into the absorber vacuum vessel about 0.55 m^2 of the absorber vacuum vessel will be in direct contact with the liquid hydrogen. The maximum heat transfer rate to the liquid hydrogen is about 51.2 kW, which in turn will cause the hydrogen to boil off at the rate of 0.116 kg s^{-1}. As long as the inside diameter of the pipe carrying hydrogen gas from the absorber vacuum to the buffer vacuum is larger than 60 mm, the peak pressure drop from the absorber vacuum to the buffer vacuum will be less than 0.3-bar. As a result, the peak pressure in the absorber vacuum should be less than 2-bars during a rupture of an absorber hydrogen window.

ACKNOWLEDGEMENTS

This work was supported by the Oxford University Physics Department and the Particle Physics and Astronomy Research Council of the United Kingdom.

REFERENCES

1. "A Proposal to the Rutherford Appleton Laboratory, an International Muon Ionization Cooling Experiment (MICE)," proposed by the MICE Collaboration, 15 December 2002
2. Green, M. A. and Yang S. Q. "The Effect of a Hydrogen Spill inside of a 300 K Vacuum Vessel on MICE Absorber Hydrogen Safety," an Oxford University Report for the MICE collaboration (2004)
3. Heat Transfer with Applications, Hagen, K. D. Prentice Hall, Upper Saddle River NJ (1999)
4. Breen, B. P. and Westwater, J. W. Chem. Eng. Progress 38, No. 7, p 67 (1962)
5. E. G. Brentari and R. V. Smith, "Nucleate and Film Pool Boiling Design Correlations for O$_2$, N$_2$, H$_2$, and He," Advances in Cryogenic Engineering 10, p 325, Plenum Press, New York, USA (1964)

Fractal description of thermal contact resistance between rough surfaces at low temperature

Xu R.P., Feng H.D., Jin T.X., Xu L., Zhao L.P. *

Institute of Refrigeration and Cryogenics, Shanghai Jiao Tong University, Shanghai, 200030, P.R. China
*Thermal Engineering Department, Tongji University, Shanghai, 200092, P.R. China

In order to model the heat transfer of cryogenic contact refrigeration system, the fractal recursive thermal contact resistance model is established. In this model real contact surfaces are described based on developed Cantor set fractal theory and the volume conservation of plastically deformed asperities is considered. It is concluded from comparison with the experimental results at low temperature that this model can predict thermal contact resistance well.

INTRODUCTION

Complex scientific instruments such as the space infrared detective facility or miniature thermal contact switch apparatus in satellite are often cooled through bolted or pressed links to their refrigeration systems. When two rough nominally flat surfaces are brought together under load, the discrete real contact points impede the heat flow through contact surfaces and result in the temperature drop and thermal contact resistance (TCR) at the interface. In the past many TCR models based on statistics theory had been proposed to predict this resistance. However, the principal statistical roughness parameters such as roughness height, slope, and curvature in those models are always dependent of length scale and resolution of the instrument. And at the same time, roughness measurements on a variety of surfaces have demonstrated that their structure follows fractal geometry where similar rough images of surfaces appear under repeated magnification [1]. This implies that Fractal geometry may be an effective method to study TCR phenomenon. In this paper the Cantor set fractal theory is used to describe the surface morphology of the interface and the fractal TCR network model is obtained based on the elastic-plastic theory. This model considers the volume conservation of plastically deformed materials and the constriction resistance of asperities. The calculating results agree well with the experimental ones at low temperature.

CANTOR SET FRACTAL THERMAL CONTACT RESISTANCE MODEL

Topographic description of contact surfaces

The self-affined Cantor set fractal method of Warren and Krajcinovic [2,3] is further developed to represent isotropic contact surface. At each step of Cantor set construction of the surface (see Figure 1), the middle section of the initial segments are removed so that the remaining horizontal length of segments at the (i)th generation is $1/fr$ of the length constructed at (i-1)th generation ($fr>1$). Similarly, the recess depth at the (i)th generation of the Cantor set surface is $1/fz$ of the depth at (i-1)th generation ($fz>1$). It is also shown from Figure 1 that the horizontal length L_i in x or y direction and recess depth h_i in z direction of the (i)th generation are $L_0(1/fr)^i$ and $h_0(1/fz)^i$ respectively.

At (i)th generation, the Cantor set profile contains the $N=s^i$ asperities and the Cantor set surface contains $N=s^{2i}$ segments where s is the number of asperities on a repeating segments. It is noted that Figure 1 is just an example of s=3. And the length of each asperity at the (i)th generation is $d_i=L_0(1/sfr)^i$. The height is $z_i=h_0(fz-1)(1/fz)^i$. Further, the gap between adjacent asperities at (i)th generation which is generated on the single asperity at (i-1)th generation is $g_i=L_0(fr-1)/[(s-1)fr^i s^{i-1}]$. Here L_0 corresponds to the profile length, and h_0 is equal to twice of the R.M.S. roughness height.

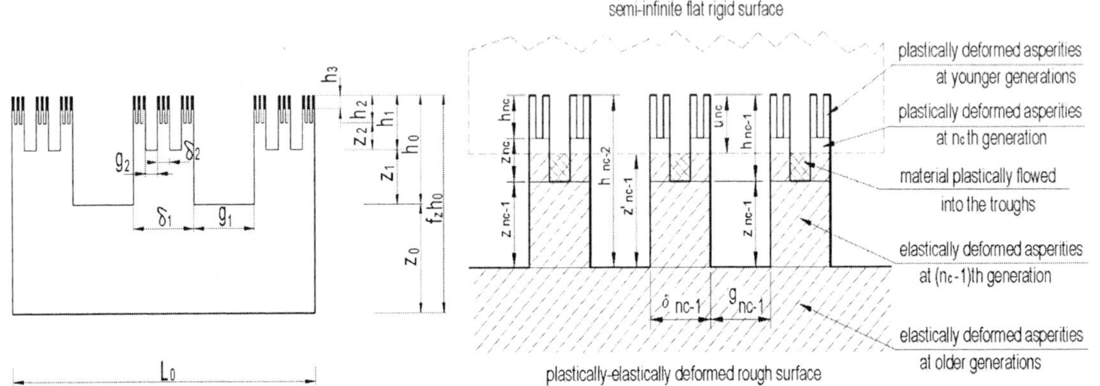

Figure 1   Fractal surface profile of Cantor set (s=3)   Figure 2   Elastic-plastic contact model of asperities

Deformation of the cantor set surfaces

Here the real contact surfaces are equivalent to that an elastic-plastic solid surface with Cantor set fractal structure is in contact with an idealized smooth rigid semi-infinite surface. Since the Cantor set fractal surface is composed of the small asperities stacking upon the large asperities self-similarly, each asperity behaves as an elastic-plastic axial loaded column.

When the normal load P is given, the smooth rigid surface will first be in contact with the youngest generation ($i\rightarrow\infty$). If $P>P_{c,i}$ (critical load of plastic deformation), the ith generation asperities will flow plastically into troughs surrounding the asperities and then the smooth surface will be in contact with (i-1)th generation asperities. Such process will not stop until the (n_c-1)th generation of asperities whose critical load is large than the exerted load. Thus the oldest plastically deformed asperities is the (n_c)th generation (see Figure 2). And the corresponding contact area is

$$A_c=\left(\frac{1}{f_r^2}\right)^{n_c}L_0^2=\frac{P}{\sigma_y} \tag{1}$$

$$n_c=\begin{cases}\zeta,\zeta\in Z^+\\1+trun(\zeta),\zeta\notin Z^+\end{cases} ; \quad \zeta=\frac{\ln\xi}{\ln(1/f_r^2)},\xi=\frac{P}{L_0^2\sigma_y},(0\leq\xi\leq1) \tag{2}$$

where Z^+ represents the positive integers, σ_y is the yield stress, $trun(\zeta)$ truncates ζ to an integer value. Therefore, for the given normal load, the deformation of asperities at the whole generations can be classified into two parts: (a) asperities at $[n_c,\infty]$ generations deform plastically and (b) asperities at $[0, n_c-1]$ generations deform elastically. The contact interface is deforming from plastically to elastically. This is absolutely different from the traditional theory of Greenwood and Williamson [4] in which deformation of the identical asperities transits from elasticity to plasticity.

Figure 3 shows the change of the ratio between real and apparent contact area under increasing load. From it, you can see that how small the real contact area is compared with apparent ones. Since Equation (2) includes the truncation function, the calculating results of real contact area are discrete and increase with load like stairs. In addition, because aluminum is softer than stainless steel, the ratio of aluminum is much larger than stainless steel.

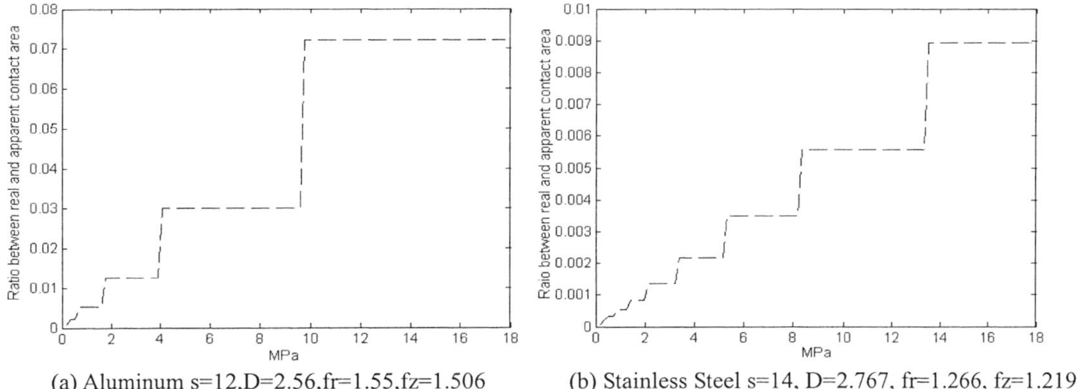

(a) Aluminum s=12,D=2.56,fr=1.55,fz=1.506   (b) Stainless Steel s=14, D=2.767, fr=1.266, fz=1.219

Figure 3   Ratio between numerical results of real and apparent contact area of aluminum and stainless steel

<u>Thermal contact resistance model</u>

It is assumed that the plastically deformed asperities at $[n_c,\infty]$ generations will fill in adjacent troughs and increase the height of the (n_c-1)th generation by $h_{nc-1}-u_{nc}$ (see Figure 2). Hence, the increased volumes of asperities at (n_c-1)th generation is equal to the plastically flowed volume of ones at $[n_c,\infty]$ generations:

$$\delta_{nc-1}^2 \cdot s^{2(nc-1)} \cdot (h_{nc-1} - u_{nc}) = \sum_{i=nc}^{\infty} V_i \tag{3}$$

$$l_i = \begin{cases} h_0 f_z^{(1-nc)}(f_z - \alpha\beta)(1-\xi\varepsilon f_r^{2(nc-1)}), i = n_c-1 \\ h_0 f_z^{-i}(f_z-1)(1-\xi\varepsilon f_r^{2i}), i = 0,1,2,...,n_c-2 \end{cases} ; \alpha = 1-\frac{1}{f_r^2}; \beta = \sum_{i=0}^{\infty}\left(\frac{1}{f_r^2 f_z}\right)^i; \varepsilon = \frac{\sigma_y}{E} \tag{4}$$

Considering the volume conservation of plastic deformation and elastic deformation of asperities, the actual height of asperities at $[0,n_c-1]$ generation under the given load is derived to equation (4).

According to Fourier law and CMY model [5], heat conduction resistance $R_{a,i}$ and constriction resistance $R_{c,i}$ of single asperity at (i)th generation in $[0,n_c-1]$ can be obtained. Because of recursive construction of Cantor set fractal contact surfaces, the total thermal contact resistance $R_{t,0}$ is the recursive serial and parallel resistance of the constriction resistance and heat conduction resistance.

$$R_{a,i} = \frac{l_i}{k \cdot \delta_i^2}; \quad R_{c,i} = \frac{\sqrt{\pi}(\delta_i + g_i)^2}{4k\delta_i} \cdot f\left(\frac{\delta_i}{\delta_i + g_i}\right); \quad R_{t,0} = \frac{\sum_{i=0}^{nc-1} s^{2i}(R_{a,(nc-1-i)} + R_{c,(nc-1-i)})}{s^{2 \cdot nc}} \tag{5}$$

COMPARISON AND DISCUSSION

In order to check the Cantor set fractal TCR model, the experiment was conducted at the interface of Al5052 and stainless steel 304 respectively. The temperature of experiments was 155K and the vacuity was less than 1.5 Pa. The normal contact pressure is from 1.0307 to 6.9675 MPa. The surface morphology of the specimens was measured by STRA-1 stylus profilometer. The physical properties of materials were measured in the experiment, which is listed in Table1.

The experimental results are illustrated in Figure 4 in the form of Thermal contact conductance (reciprocal of TCR). Because of the discrete real contact area, the calculating results are discrete, too. The fitting curves of staircase numerical results are used to compare and predict the real conductance. The equations of fitting curves are $y=5470x^{0.5121}$ for aluminum and $y=318.5x^{0.5562}$ for stainless steel respectively. From Figure 4, it is shown that the fitting curve of Cantor set fractal TCR model can predict the experiment results well.

Table 1   Experimental parameters of Aluminum and Stainless steel samples

Material	Thermal conductivity (W/m^{2}K)	Elastic modulus (GPa)	Yielding strength (MPa)	R.M.S height (μm)	h_0 (μm)	Fractal dimension D
Al-5052	107	73.3	450	3.2692	6.5384	1.71
SS-304	10.7	201	2400	2.1328	4.2656	1.85

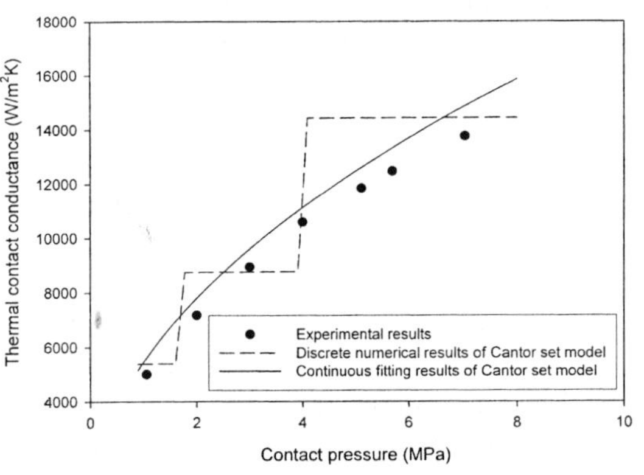

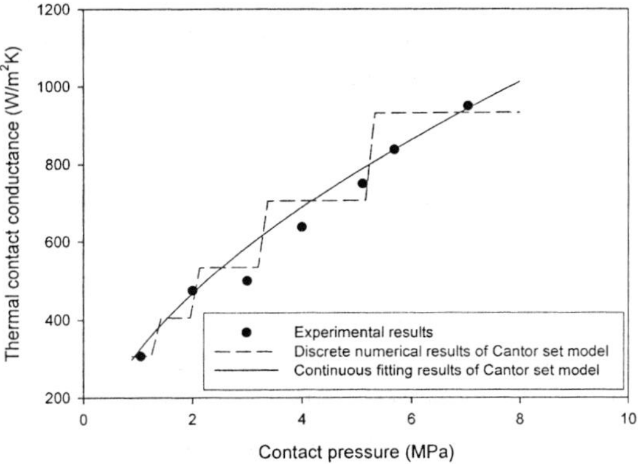

(a) Aluminum s=12, D=2.56, fr=1.55, fz=1.506   (b) Stainless Steel s=14, D=2.767, fr=1.266, fz=1.219

Figure 4   Comparison between experiment and numerical results

CONCLUSION

In this paper Cantor set fractal theory is used to describe rough surface topography. Considering the volume conservation of plastically deformed asperities, the fractal recursive TCR model is established. It is concluded from the model that the asperities of the contact interface are deforming from plastically to elastically under normal load. The simulated results are in good agreement with the experimental results at low temperature. This model provides a new way to study the phenomenon of TCR.

REFERENCES

1. Xu, L., Zhao, L.P., Application of a profile discretization method to investigate on thermal contact conductance between rough interfaces, Proceedings of International Cryogenic Engineering Coference-19 Grenoble, France (2002) 835-838
2. Warren, T.L. and Krajcinovic, D., Random Cantor set models for the elastic-perfectly plastic contact of rough surface, Wear (1996) 196 1-15
3. Warren, T.L. and Krajcinovic, D., Fractal models of elastic-perfectly plastic contact of rough surfaces based on the Cantor set, Int. J. of Solids Struct. (1995) 32 2907-2922
4. Greenwood, J.A. and Williamson, J.B.P., Contact of nominally flat surfaces. Proc. Roy. Soc. London, UK (1966) A295 300-319
5. Cooper, M.G., Mikic, B.B. and Yovanovich, M.M., Thermal contact conductance, Int. J. of Heat Transf. (1969) 12 279-300

Study of Two Phase Flow Distribution and its Influence in a Plate-fin Heat Exchanger

Xu Q., Li Y.Z., Zhang Z.

School of Energy and Power Engineering, Xi'an Jiaotong University, Xi'an 710049, China

In this paper, the experimental investigation on the effects of inlet flow rate and dryness on the two-phase flow distribution near entrance of plate-fin heat exchangers is presented. The results indicate that the liquid phase non-uniformity is more serious than that of gas and the flow non-uniformity in crosswise direction is more serious than that in ordinate direction. The effects of both gas Reynolds number and inlet flow dryness are also presented.

INTRODUCTION

The maldistribution in the core of plate-fin heat exchanger caused by improper entrance configuration, such as poor design of header and distributor conformation, manufacturing tolerances, fouling, frosting of condensable impurities, especially in multi-phase flow, will lead to the performance deterioration of heat exchanger. For two-phase flow heat exchangers, especially for cryogenics heat exchangers working under little temperature difference, fluid flow distributing non-uniformly happens. A. C. Mueller[1] indicated that the types of two-phase maldistribution including (1) vapor quality difference in parallel circuits; (2) density-wave instability, which is a more complex phenomenon of oscillating flows and pressure drops. Wu Jianghong [2] theoretically analyzed the effect of two-phase maldistribution in plate-fin heat exchanger by setting up a model to numerically calculate the performance of distribution. P. Vlasogiannis [3] recorded construction of a flow regime map of air-water two-phase flow in a plate heat exchanger by a high-speed video camera. The experiment studies of realistic factors influencing two-phase maldistribution are fewer than that of single-phase [4,5]. In this paper, the experimental investigation on the effects of gas flow Re (i.e. inlet gas flow rate) and flow dryness in both crosswise direction (perpendicular to the inlet flow) and ordinate direction (parallel to the inlet flow) are presented.

EXPERIMENTAL SYSTEM

The experimental system is shown in Figure 1, which includes the fluid flow system and the data acquisition system. The fluid flow system is composed of air compressor, water tank, pump, stabilization tank, filter, test section, passage-switching device and separator. The data acquisition is composed of a series of pressure and pressure difference sensors (silicon crystal film), air turbine flow meter and liquid turbine flow meter, which are connected to a computer through a Keithley plug-in data acquisition board.

Single-phase and two-phase fluid flow distribution research can be taken in the experimental system. Water constitutes the liquid phase of the gas/liquid mixture and is provided by pumping from a water tank. Air is supplied by a screw air compressor and is mixed with the water in a joint, located on the pipe leading to the heat exchanger (see Figure 1).

Air and water flow rates are measured by a series of turbine meters and the accuracy of the measurement is less than 1%. The airflow rate recorded is corrected for deviations from standard conditions according the actual pressure and temperature. The test unit (see Figure 2), which is manufactured by Kaifeng Air Separation Group Company Limited, has the volume of $200 \times 250 \times 178$ mm^3. The flow at cross-section is divided into 30 zones, and the flow distribution in each zone is assumed to be uniform. The arrangement of channels (zones) is shown in Figure 3.

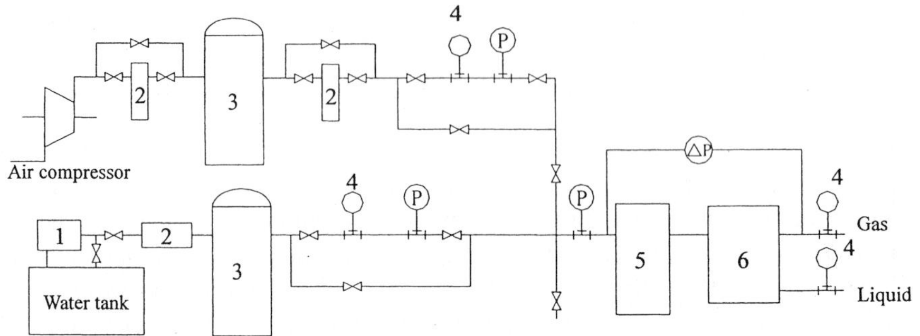

1-Pump; 2-Filter; 3-Stabilization tank; 4-Flow meter; 5-Test section; 6-Separator

Figure 1   Schematic diagram of experimental system

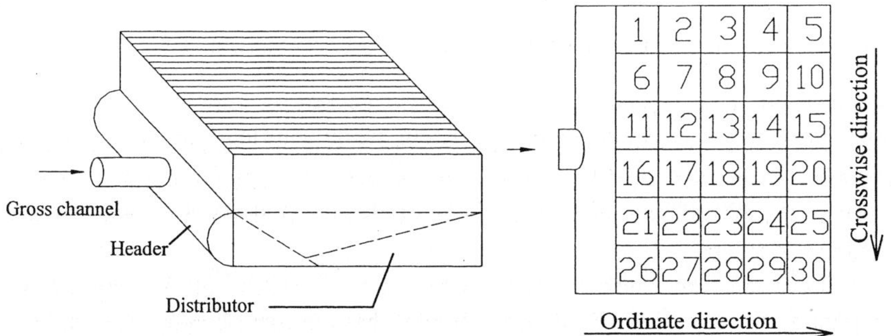

Figure 2   Schematic diagram of test section   Figure 3   Channels arrangement and direction

DATA ANALYSIS AND DISCUSSION

All two-phase flow data are reported in terms of superficial air and water velocities V, which are defined from the respective volumetric flow rate, Q_G and Q_L, and the area S of zone shown in the Figure 3. In particular, the airflow rate is converted to standard state (1 atm and 25°C). Thus

$$V_G = \frac{Q_G}{S}, V_L = \frac{Q_L}{S} \tag{1}$$

For convenient investigation of the flow distribution in whole cross-section, the fluid flow is discomposed into those in crosswise and in ordinate direction (see Figure 3) as follows,

$$U_{j,i}' = (V_{j,5i-4} + V_{j,5i-3} + V_{j,5i-2} + V_{j,5i-1} + V_{j,5i})/5 \qquad (i=1,2,3,4,5,6; j=G \text{ or } L) \tag{2}$$

$$U_{j,i}'' = (V_{j,i} + V_{j,i+5} + V_{j,i+10} + V_{j,i+15} + V_{j,i+20} + V_{j,i+25})/6 \quad (i=1,2,3,4,5; j=G \text{ or } L) \tag{3}$$

where U' and U'' are the average flow velocities in crosswise and in ordinate direction, respectively.

The dimensionless velocity deviation, θ' and θ'' indicates the flow uniformity in cross-section of the test unit as shown in eq. (4). The more the value is, the worse the flow distribution is.

$$\theta_i' = (U_i' - U_{ave})/U_{ave}, \theta_i'' = (U_i'' - U_{ave})/U_{ave} \tag{4}$$

where U_{ave} stands for average flow velocity of cross-section.

<u>Single-phase flow distribution experiment</u>

In this set of experiments the test section has been tested with air at different flow rates from $1.73 \sim 2.34$ m^3/min. Figure 4 shows the relationship of velocity deviation vs. airflow Re. According to the figure, the single-phase flow maldistribution in both directions is an increasing function of inlet flow Re. The figure indicates that in both directions the velocity deviation is below ±0.3 and the deviation in ordinate direction is little greater than that in crosswise direction, which means that the velocity profile in ordinate direction is more non-uniform.

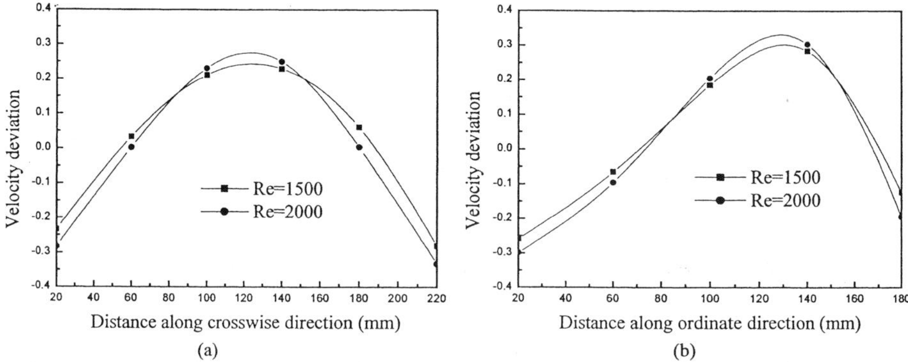

Figure 4   Velocity deviation in both crosswise and ordinate direction with different Re

<u>Two-phase flow distribution experiment</u>

In two-phase flow operation, we maintain the mass dryness of the flow at 18% by varying air and water flux synchronously to get the flow profiles in the cross-section with different gas Re. Then the Re of gas flow is kept constant at 2600, and we obtain the flow profiles in both directions with different fluid mass dryness by changing water flux.

Distribution of gas and liquid flow for different gas Re and for different fluid dryness is displayed in Figure 5 and Figure 6. According to the Figures, liquid phase non-uniformity, especially in crosswise direction (normal to gross channel), mainly represent the two-phase flow non-uniformity in the entrance of heat exchanger. Both gas and liquid components of the flow maldistribution increase with the increases of gas flow Re in Figure 5. And in Figure 6, the gas flow maldistribution decreases, especially in crosswise direction, with the inlet dryness increasing. However, the liquid flow maldistribution increases with the inlet dryness increasing, and it's more obvious in ordinate direction.

Distribution of single-phase flow shown in Figure 4 with 100% dryness is much distinguishing with those of gas phase in two-phase flow shown in Figures 5 and 6. The gas phase velocity distributions in both crosswise and ordinate directions have more than one peaks, and moreover, the distributions in ordinate direction at different Re and those in crosswise with different dryness show various patterns, which are different with single-phase distribution. The figures also illustrate that single-phase gas flow distribution is more serious than gas phase distribution in mixture. The reason of this behavior can be explained as follows: the complex interaction between gas and liquid causes the change of flow pattern and liquid phase play major role in distribution of two-phase flow.

Figures 5 and 6 also illustrate that the variation of gas flow Re mostly affects the liquid flow profile in the crosswise direction and the dryness mostly affects that in ordinate direction.

CONCLUSIONS

The two-phase flow distribution at entrance of plate-fin heat exchangers is non-uniform for its configuration. The experimental results show that the liquid phase non-uniformity is more serious than that of gas and the flow maldistribution in the crosswise direction (perpendicular to the inlet flow) is more serious than that in ordinate direction. The velocity profile in crosswise direction plays a significant role in distribution of two-phase flow.

The flow maldistribution of both gas and liquid components will increase with the increase of gas flow Re. With the increase of inlet dryness, the gas flow maldistribution decreases but the liquid flow maldistribution increases simultaneously.

The variation of gas flow Re mostly affects the liquid flow profile in crosswise direction and the dryness mostly affects that in ordinate direction, which is significant for the improvement of two-phase flow non-uniformity at the entrance of plate-fin heat exchangers. It is known that header configuration and distributor conformation affect distribution in crosswise direction and ordinate direction, respectively.

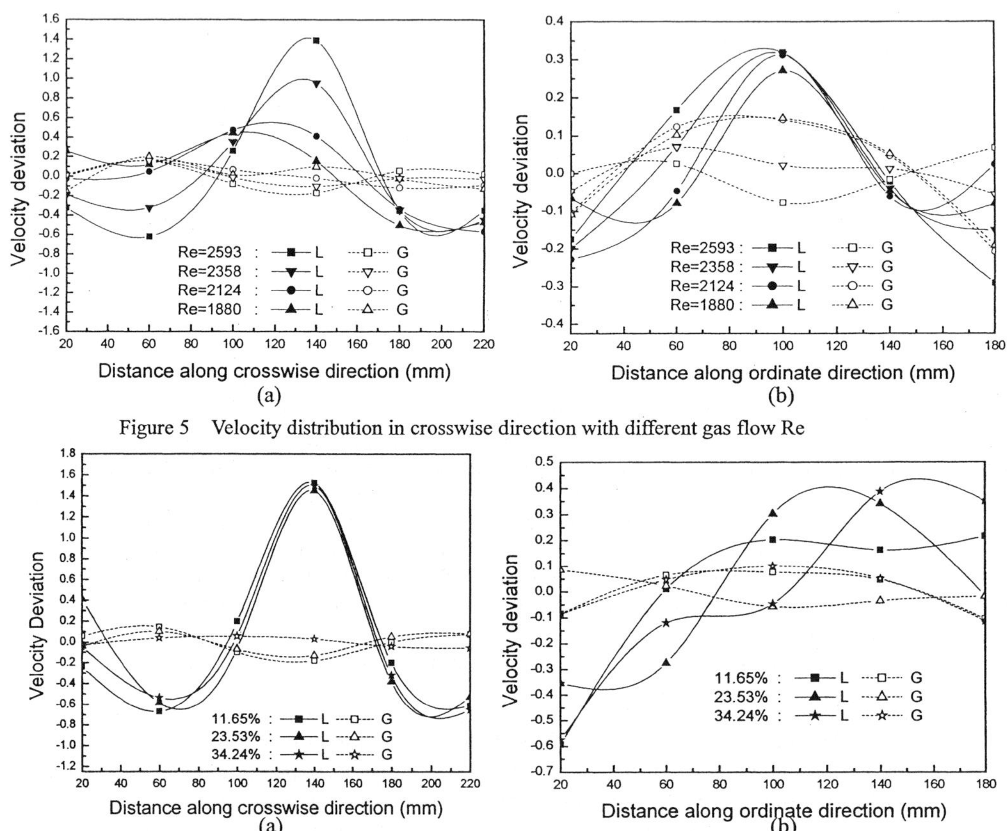

Figure 5　Velocity distribution in crosswise direction with different gas flow Re

Figure 6　Velocity distribution in crosswise direction with different inlet dryness

REFERENCES

1. Mueller A. C., Chiou J. P., Review of Various Types of Flow Maldistribution in Heat Exchangers, Heat Transfer Engineering (1988) 9 36-50

2. Wu Jianghong, Chen Changqing, Wu Yezheng, Analysis of Two-Phase Flow Distribution in Plate Fin Heat Exchanger and Experimental Research, Journal of Xi'an Jiaotong University (1995) 29 10-14

3. Vlasoginnis P., Karagiannis G., Air-water two-phase flow and heat transfer in a plate heat exchanger, International Journal of Multiphase Flow (2002) 28 757-772

4. Zhang Z., Li Y.Z., Xu Q., Experimental Research on Flow Maldistribution in Plate-Fin Heat Exchangers, Chinese Journal of Chemical Engineering (2004) 12 1-7

5. Zhang Z., Li Y.Z., Xu Q., Experimental Study of the Impact of Deflector Plate Configuration on the Performance of Material Flow Distribution, Journal of Engineering for Thermal Energy & Power (2003) 18 612-614 (in Chinese)

A thermodynamic optimization of counterflow recuperative-type heat exchangers

Wang Q., Zhang H. F., Chen G. M.

Institute of Refrigeration and Cryogenics, Zhejiang University, Hangzhou, China

The counterflow recuperative-type heat exchangers have been used widely in cryogenic systems. Thermodynamic optimization of a steady heat transfer process with constant specific-heat fluids in a counterflow recuperative-type heat exchanger is conducted in this paper. The analytical results show that the irreversibility of a heat transfer process is minimized when the log-mean heat transfer temperature difference equals to the difference of mean thermodynamic temperatures of two fluids. These optimum relations can be conveniently used for optimization design of heat exchangers in cryogenic systems, such as natural gas separation systems, gas liquefaction systems, gas refrigerators, and so on.

INTRODUCTION

The counterflow recuperative-type heat exchangers have been used widely in cryogenic systems, such as natural gas separation systems, gas liquefaction systems, gas refrigerators, and so on. The efficiency of these heat exchangers often directly affects the performance of the cryogenic systems. The irreversibility of these heat exchangers can be minimized by thermodynamic optimization of them. The thermodynamic optimization of a heat exchanger could be defined as the minimization of irreversibility of heat transfer processes through modifying the thermodynamic properties of fluids when specifying heat exchanger structure and total heat transfer rate. From the results of researches [1-3], we can see that the COP of refrigeration systems could be improved at specified size of heat exchangers. But the condition of perfect configuration of heat transfer has been generally accepted as that the heat transfer in the heat exchanger should be carried out with a constant heat transfer driving force, namely, the local temperature difference between heat transfer fluids along the heat transfer surface should be constant [1-4].

In fact, the constant local temperature difference between heat transfer fluids along the heat transfer surface does not definitely lead to minimizing total entropy generation rate in the heat transfer process. The total entropy generation rate and the heat transfer temperature difference in a heat transfer process are independent of each other. Therefore, it leaves room for thermodynamic optimization.

The aim of this paper is to conduct a thermodynamic optimization of a steady heat transfer process which often happens in counterflow recuperative-type heat exchanger with constant specific-heat fluids. The results of this research will be helpful for further research on the heat transfer process with variable specific-heat fluids.

HEAT EXCHANGER MODEL

Consider a counterflow recuperative heat exchanger shown schematically in Figure 1. Following assumptions are employed:
(a) Both of the fluids flow steadily through the heat exchanger at a constant pressure. Their heat capacity rates are kept constant along the heat transfer surface.
(b) The overall heat transfer coefficient is constant throughout the heat exchanger.
(c) Heat losses to the surroundings and longitudinal heat conduction are all neglected.
(d) All frictional pressure drops of both fluids are negligible.

838

(e) The changes in kinetic energy and potential energy of fluids within the heat exchanger are negligible.

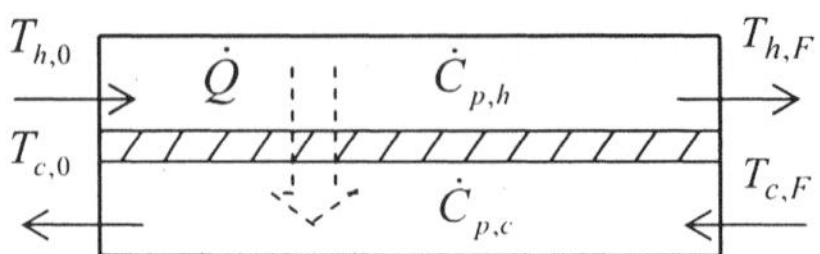

Figure 1   Sketch of heat exchanger model

Thus the heat exchanger model could be written as:

<u>Heat transfer equation</u>

$$\dot{Q} = KF\Delta t_{lm} \tag{1}$$

where $\dot{Q}$ is the total heat transfer rate in the exchanger, K is the overall heat transfer coefficient, F is the total heat transfer area and Δt_{lm} is the log-mean temperature difference across the heat exchanger, defined as:

$$\Delta t_{lm} = \frac{T_{h,0} - T_{c,0} - T_{h,F} + T_{c,F}}{\ln \dfrac{T_{h,0} - T_{c,0}}{T_{h,F} - T_{c,F}}} \tag{2}$$

where $T_{h,0}$ and $T_{h,F}$ are the absolute temperatures of hot fluid at the hot and cold ends of the exchanger, $T_{c,0}$ and $T_{c,F}$ are the absolute temperatures of cold fluid at the hot and cold ends of the exchanger.

<u>Heat balance equation</u>

$$\dot{Q} = \dot{C}_{p,h}(T_{h,0} - T_{h,F}) = \dot{C}_{p,c}(T_{c,0} - T_{c,F}) \tag{3}$$

where $\dot{C}_{p,h}$ and $\dot{C}_{p,c}$ are the heat capacity rates of hot and cold fluids at constant pressure.

<u>Entropy balance equation</u>

$$\dot{S}_g = \left(\frac{1}{T_{c,m}} - \frac{1}{T_{h,m}} \right)\dot{Q} \tag{4}$$

where $\dot{S}_g$ is the total entropy generation rate in the exchanger. $T_{h,m}$ and $T_{c,m}$ are mean thermodynamic temperatures of hot and cold fluids in the exchanger. Since $\dot{C}_{p,h}$ and $\dot{C}_{p,c}$ are uniform, we have:

$$T_{h,m} = \frac{T_{h,0} - T_{h,F}}{\ln \dfrac{T_{h,0}}{T_{h,F}}} \quad \text{and} \quad T_{c,m} = \frac{T_{c,0} - T_{c,F}}{\ln \dfrac{T_{c,0}}{T_{c,F}}} \tag{5}$$

OPTIMIZATION

Based on the above heat exchanger model, following Lagrange function is constructed:

$$L = \dot{S}_g + \lambda\left(\dot{Q} - KF\Delta t_{lm}\right) = \left(\frac{\ln\alpha_c}{T_{c,0} - T_{c,F}} - \frac{\ln\alpha_h}{T_{h,0} - T_{h,F}}\right)\dot{Q} + \lambda\left(\dot{Q} - KF\frac{T_{h,0} - T_{c,0} - T_{h,F} + T_{c,F}}{\ln\beta}\right) \qquad (6)$$

where λ is the Lagrange multiplier, and

$$\alpha_h = \frac{T_{h,0}}{T_{h,F}}, \quad \alpha_c = \frac{T_{c,0}}{T_{c,F}}, \quad \beta = \frac{T_{h,0} - T_{c,0}}{T_{h,F} - T_{c,F}} \qquad (7)$$

When $\dot{Q}$, F, K are specified, using the necessary conditions for this multi-objective optimization problem, we can get the optimum solution as:

$$\frac{T_{c,0}}{T_{c,F}} = \frac{T_{h,0}}{T_{h,F}} \qquad (8)$$

DISCUSSION

Other important optimum relations in the heat transfer process could be derived from the above optimum solution of equation (8).

Mean thermodynamic temperature difference ΔT_m
Combining equation (8) with equations (5), we can obtain the optimum mean thermodynamic temperature difference between the hot and cold fluids in the heat exchanger:

$$\Delta T_m = T_{h,m} - T_{c,m} = \Delta t_{lm} \qquad (9)$$

Equation (9) shows that $\dot{S}_g$ is minimized when ΔT_m equals to Δt_{lm}. In addition, from heat transfer equation (1), we can see that Δt_{lm} is always constant under the assumptions discussed. Therefore, ΔT_m is minimized while Δt_{lm} keeps constant in the optimization processes. This indicates that ΔT_m is independent of Δt_{lm}. It can be concluded that ΔT_m is an important factor which indicates the value of $\dot{S}_g$ during heat transfer prcesses.

Ratio of heat capacity rate γ
Combining equation (8) with heat balance equation (3), we can express the optimum ratio of heat capacity rate of cold fluid to hot fluid as:

$$\gamma = \frac{\dot{C}_{p,c}}{\dot{C}_{p,h}} = \frac{T_{h,0} - T_{h,F}}{T_{c,0} - T_{c,F}} = \frac{T_{h,m}}{T_{c,m}} \qquad (10)$$

Substituting equation (9) into (10) results in:

$$\gamma = 1 + \frac{\Delta t_{lm}}{T_{c,m}} = \frac{1}{1 - \dfrac{\Delta t_{lm}}{T_{h,m}}} \qquad (11)$$

Equation (11) shows that the optimum γ is a function of the ratio of Δt_{lm} to the absolute temperature of hot or cold fluid. This general relation could be conveniently used for thermodynamic optimization of a heat transfer process.

Assuming that $T_{h,m}$ and $T_{c,m}$ are finite values, more conclusions can be derived:

(a) If $\Delta t_{lm} > 0$, from equation (11), $\gamma > 1$, i.e., the optimum heat capacity rate of cold fluid is greater than that of hot fluid in heat transfer process with finite temperature difference.

(b) If $\Delta t_{lm} = 0$, from equation (11), $\gamma = 1$, i.e., when two fluids undergo a zero temperature difference heat transfer process, the optimum heat capacity rates of both fluids are equal. From these results, we can draw that $\gamma = 1$ is only the optimum condition for a reversible heat transfer process, but not for an irreversible one.

(c) From equation (10), $\dot{C}_{p,c} = \dfrac{T_{h,m}}{T_{c,m}} \dot{C}_{p,h}$. When $\dot{C}_{p,h} \to +\infty$, $\dot{C}_{p,c} \to +\infty$, and vice versa. This accounts that the optimum heat capacity rate of a fluid should be infinite when it exchanges heat with another fluid whose heat capacity rate is infinite.

CONCLUSIONS

For a heat exchanger, the heat transfer area is always finite, which means the mean heat transfer temperature difference is impossible to be zero for a specified total heat transfer rate. Consequently, there must exist a definite difference between the mean thermodynamic temperatures of heat transfer fluids, which results in an existence of certain exergy loss in heat transfer process. When the ratio of heat capacity rate γ conforms to the relation (10), the optimum heat transfer is achieved, so does the minimum of ΔT_m, and ΔT_m equals to Δt_{lm}.

The analytical results indicate that the ultimate reason of power saving contributed from thermodynamic optimization is the decrease of the difference of mean thermodynamic temperature between two fluids, even though there is no reduction of the log-mean heat transfer temperature difference between two fluids. In other words, the optimum configuration of heat transfer temperature difference does not need to keep the local temperature difference constant throughout the heat exchanger.

Therefore, it should be emphasized that only the total entropy generation rate should be the fundamental criterion for the thermodynamic optimization. The concepts proposed in this paper can also be used to the optimization of a steady heat transfer process with variable specific-heat fluids.

REFERENCES

1. McLinden M. O., Radermacher R., Methods for comparing the performance of pure and mixed refrigerants in the vapor compression cycle, International Journal of Refrigeration (1987) 10 318-325

2. Miyara A., Koyama S., Fujii T., Consideration of the performance of a vapor-compression heat-pump cycle using nonazeotropic refrigerant mixtures, International Journal of Refrigeration (1992) 15 35-40

3. Mulroy W. J., Domanski P. A., Didion D. A., Glide matching with binary and ternary zeotropic refrigerant mixtures: part 1. an experimental study, International Journal of Refrigeration (1994) 17 200-225

4. Venkatarathnam G., Mokarish G., Srinivasa Murthy S., Occurrence of pinch points in condensers and evaporators for zeotropic refrigerant mixtures, International Journal of Refrigeration (1996) 19 361-368

A new type of condenser-evaporator safely operated in large air separation plant

Wu Y.Y., Chen L.F., Wu T.H.

College of Energy and Power Engineering, Xi'an Jiaotong University, 710049, China

A new condenser–evaporator for 30000 m^3/hr air separation plant is developed. Its heat transfer coefficient is increased by 50%. Both its volume and weight are reduced by 1/3 compared with the traditional one. Its total temperature difference for heat transfer is only 0.57K. According to the theoretical analysis, the flow velocity of oxygen gas is increased to 6.74m/s. Thus the oxygen gas could carry the heaviest particles of N_2O of 0.4mm in diameter out of the flow channel in comparison with the traditional products in which the particles of greater than 0.2mm in diameter could not be carried away from the channel.

INTRODUCTION

The condenser- evaporator is also called as the double phase-change heat exchanger. Its importance lies in the fact that all the high purity products of oxygen and nitrogen in oxygen plant are produced in this equipment. In oxygen production plant, acetylene is most likely accumulated in the condenser –evaporator. The behavior of fluid flow in the equipment has a direct influence on the accumulation of acetylene. Therefore, many engineers and experts in the area of air separation are making their efforts now in improving the performance of heat transfer of the condenser–evaporator and in preventing it from explosive possibility. Due to their backward mechanism of heat transfer, the temperature difference of heat transfer of the old type of condenser-evaporator is so large and the heat transfer coefficient is so small that the diameter of the condenser –evaporator is larger than the allowed standard for railway transportation. The distillation column can only be installed in situ, so that the quality of the installation cannot be assured. Additionally the backward mechanism of heat transfer brings about an increase in pressure of the lower distillation column, so that the work consumption of the air compressor becomes larger. In order to meet the requirements for the developments of the iron-steel industry, the modern oxygen plants are expanding to a very large scale during the 1990's. However, the large-scale plants present a demand to increase the size of condenser –evaporator greatly.

But there is a limitation to increase the size of condenser –evaporator due to the difficulties existed in transportation and in construction of very large distillation column. The backward technique of traditional condenser –evaporator could not resolve this contradiction to develop large-scale plants.

A new type of condenser –evaporator was invented, manufactured and installed in an air separation plant with oxygen output of 6500m^3/hr in Hangzhow Iron-Steel Company in 1997', which has been successfully operated for 4 years already. Another large condenser –evaporator was manufactured and installed in Baoshan Iron-Steel Company in 2002', which has been successfully operated for one year already. It has a compact structure and thus meets the urgent requirements for developing large scale oxygen production plants, large scale ethylene production plants and so on.

THE PRINCIPLE

The invention makes a breakthrough in traditional mechanism of nucleation boiling heat transfer in which the liquid flow occupies the most part of the boiling channel. It creates a new mechanism of boiling heat transfer with quasi-annular flow and results in a phenomenon of stimulation heat transfer enhancement, and accordingly it innovates a new structure. Thus it enables the boiling channel to be operated at an optimum state with quasi-annular flow in boiling channel, which is called as stimulation state of boiling heat transfer. Therefore the invented condenser-evaporator is called quasi-annular flow condenser-evaporator. It not only increases the critical heat flux of the boiling channel greatly up to $10000W/m^2$, but also increases the heat transfer coefficient up to the maximum and decreases the temperature difference to the minimum of 0.57K compared with the current products. The invention also makes a breakthrough in the traditional mechanism of condensing heat transfer with laminar condensing liquid film, and creates a new mechanism of the enhanced condensing heat transfer with turbulent condensing liquid film. Afterwards, it also innovates a new structure of condensing channel. As a result of above modifications the condensing heat transfer coefficient is remarkably increased and the overall heat transfer performance of the condenser–evaporator has a marked improvement.

As is known, the temperature difference of Dt for the new condenser–evaporator is decreased due to the modification of mechanism of boiling heat transfer, the modification of condensation mechanism and the invention of new coupling method of condensing channel and evaporating channel. Though it benefits to the decrease of energy consumption of the plant, it would bring about decreasing in heat flux of q. However the heat flux of q must be increased greatly in. order to develop large-scale oxygen production plant. Fortunately the heat flux of q for the new condenser–evaporator is able to increase greatly due to the increase of product of (h · Dt).

THE MAIN TECHNICAL PARAMETERS

In the December of 2002', a new condenser-evaporator of $30000m^3/hr$ air separation plant was manufactured and installed in Baoshan iron steel company to replace the original one which was imported outside of China. The volume of the old one was too large to be installed between the upper and lower distillation columns. Due to mismatch of the distillation column with the condenser-evaporator, the oxygen output of the old plant was only 20000 m^3/hr and never attained 30000 m^3/hr. The diameter of the new condenser-evaporator is reduced by 1200 mm (the relative ratio of 25%). Thus it could be installed into the distillation column so as to simplify the piping and the instillation process.

All the main technical parameters of the new condenser–evaporator are superior to the current products. In comparison with the designing standard set by the most advanced companies in the world, the heat transfer coefficient is increased by 50%, the output of the oxygen production is increased by 15~20%. The total temperature difference for heat transfer is only 0.57K, which is the minimum one among the same products currently operated in the world. The most outstanding characteristic of the new product is its recovery ability of oxygen purity after an undesired shut down of electric power system. It takes only 20 minutes for recovery in comparison with 3 to 4 hours of the recovery time for old products. The oxygen output of the new product attains 30000 m^3/hr more. The power consumption of the new product is reduced by 6.18%.In following tables, the specific volume is defined as the volume of the condenser-evaporator required for 1 m^3 of oxygen gas output when the total temperature difference of heat transfer of the condenser-evaporator is 1 K. So is the definition of the specific weight or the specific surface area. According to the theoretical analysis, the flow velocity of oxygen gas in the narrow channel is increased to 6.74m/s. Thus the oxygen gas could carry the heaviest particles of N_2O of 0.4mm in diameter out of the flow channel to prevent it from accumulating inside the flow passage. In comparison with the traditional products the particles of greater than 0.2mm in diameter could not be carried away from the channel.

Table 1 The main technical performance indices of new condenser –evaporator in comparison with the traditional products for 6000m³/hr oxygen plant

Technical parameters	New condenser Evaporator	traditional Main condenser
Total heat transfer coefficient(W/m².K)	930	620
Total heat transfer temperature difference(K)	0.57	1.2—1.4
Pressure of lower column(MPa)	0.42	0.5
Purity of oxygen gas(O_2%) above	99.8	99.7
Purity of nitrogen gas(N_2%) above	99.99	99.99
Output of oxygen gas(m³/h)	6600	6000

Table 2 The operating performance of new condenser–evaporator in comparison with the traditional products

Technical parameters	New condenser evaporator	Traditional products
Starting period	14~16 hours	20~24 hours
Operability	Convenient	Difficult
Safety	Good	General
Contamination of heat transfer surface	Not easy	Easy
Blockage in tube	No	Yes for sintered tube type
Energy consumption	Reduced by4.2%	Large
Operation state	Stable	Not stable
Adjust- ability	Good	Bad
Recovery period after shutdown	20 minutes	At least 3 hours
Continuous running time	More than2 years	Less than 2 years

Table 3 Feasibility of manufacture of new condenser –evaporator in comparison with the traditional products for 6000m³/hr plant

Technical parameters	new condenser- Evaporator	Traditional products	Reduced by (%)
Manufacturing procedure	Simple	Complicated	
Specific* volume （L/m³O_2K）	0.531	1.0296	48.4%
Specific* weight （kg/m³O_2K）	0.6435	0.9724	33.8%
Specific* Surface area （m²/m³O_2K）	0.3645	0.559	44.8%
Cost r (RMB1000 Yuan)	514	754	31.8%

Table 1 and Table 2 show the main technical performance indice and the operation performance of the new condenser-evaqporator which was installed in a 6500 m^3/hr oxygen plant in Hangzhow Iron-steel Company, respectively. The data were collected from the original records in the factory.Table 3 shows the advantages of the new condenser- evaporator in manufacturing of the 6500 m^3/hr oxygen plant.Table 4 shows the fluid flow parameters and heat transfer coefficient of both 6500 m^3/hr oxygen plant in Hangzhow Iron-steel Company and the 30000 m^3/hr oxygen plant in Baoshan Iron-Steel Company .In the 30000 m^3/hr oxygen plant in Baoshan Iron-Steel Company ,the total amount of carbon-hydrogen mixtures is only 80 ppm, which is lower than allowed value.

The invention has been successfully applied in large-scale oxygen production plants, whose outstanding performances are superior to all the current products in the world.

Table 4   The fluid flow parameters and heat transfer coefficient for two products

	Boiling channel (Oxygen)			Condensing channel (Nitrogen)		
	Heat transfer coefficient ($W/m^2.K$)	Oxygen vapor flow velocity (m/s)	Reynolds number	Heat transfer coefficient ($W/m^2.K$)	Nitrogen vapor flow velocity (m/s)	Reynolds number
New products	2200	6.74	4400	3600	1.94	11150
Old products	1500	1.88	2200	1960	0.92	6700

CONCLUSION

By changing the boiling and condensing mechanism of heat transfer and accordingly modifying the structure of boiling and condensing channels the new condenser –evaporator has become more efficient to operate, more compact in structure, advantageous to environmental protection, beneficial to energy saving and economical in manufacturing costs. Its working state is stable, easier to achieve normal state from start and convenient to be adjusted, and could be rapidly recovered after an undesired shut down. The authors are grateful to the State Natural Science Foundation for the financial support to the projects of 50176036 and 502766048.

REFERENCES

1 Chen L. F., Wu Y. Y., Sun S. F., Zhang R.and Liu Y. Z., The study on performance of heat transfer in quasi- annular flow condenser –evaporator, Cryogenics (1999) 39 209-216

Heat transfer in gas filled pipes with closed warm end under different orientations

Hieke M., Haberstroh Ch., Quack H.

Technische Universitaet Dresden, Lehrstuhl fuer Kaelte- und Kryotechnik, D 01062 Dresden, Germany

In cryogenics it is a standard arrangement, that pipes filled with a stagnant gas, which lead from a high temperature to a low temperature, are installed with the warm end up. But sometimes the overall configuration does not allow this and pipes have to be arranged e.g. horizontally or even with the warm end down. How large is the additional heat transport by internal natural convection in such lines? To find this out, CFD-simulations in straight pipes have been done. The heat transport was evaluated from the simulations as function of orientation, diameter, length and wall thickness.

SETUP AND BOUNDARY CONDITIONS

Based on a straight pipe with a closed warm end and an open cold end, a model (Fig. 1) for a CFD-simulation is created. The model contains the fluid region as well as the solid walls. The warm and cold temperatures are applied to the pipe ends. The other outer walls are specified as adiabatic walls. The inner walls are set as domain interfaces between fluid and solid. This means there can be a heat flux but no flow. To simulate the cold open end, the length of the tube is extended by a section with fixed cold wall temperature.

A structured grid is applied to the model and the basic parameters are set. The warm end is connected to the surrounding temperature at 298 K, the cold end corresponds to a liquid hydrogen tank at 28 K and 6 bar absolute pressure. The pipe is filled with gaseous para-hydrogen and the wall consists of stainless steel.

The geometrical parameters of the pipe are varied as follow:

inner diameter
  D = 8, 12 mm
wall thickness
  s = 0.4, 0.8 mm
pipe length
  L = 500, 1000 mm
orientation
  $\alpha = -90° \ldots +90°$

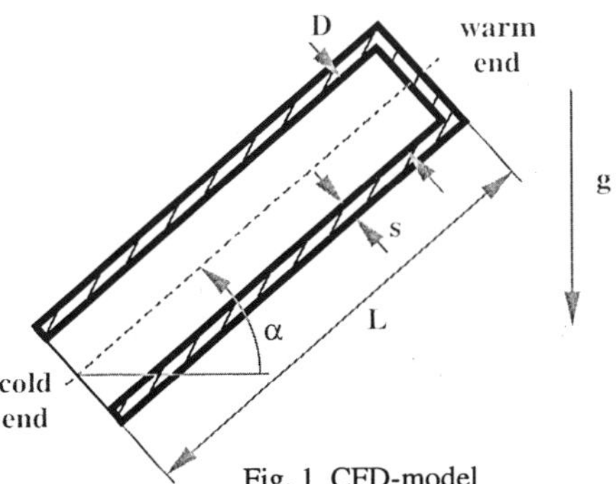

Fig. 1 CFD-model

A negative orientation means that the warm end is below the cold end. An orientation of 0° stands for a horizontal pipe. The inclination is varied in 15° steps for the short pipe and in 45° steps for the long pipe. The simulations are made with the commercially available software ANSYS-CFX v5.6. All calculations are steady state, 3-dimensional and use a combined turbulence model of k-ε and k-ω.

The physical properties of para-hydrogen are calculated with GASPAK v3.20. Because of the small pressure range within the pipe the properties are specified at constant pressure as a function of temperature. For the solid wall the properties are taken from CRYOCOMP v3.01 as a function of temperature.

MODEL VERIFICATION

To verify the abilities of the CFD-software for solid and fluid heat transfer a simple one-dimensional problem is modeled. A vertical pipe with the warm end up is simulated by CFD as well as calculated analytically. For the pipe aligned vertically a stable temperature stratification is expected. The heat is transported exclusively by axial heat conduction in gas and solid. Radial temperature gradients will be neglected. The axial temperature profile can be evaluated in a one-dimensional model assuming a constant axial heat flux $\dot{Q}$.

The pipe is discretized in axial direction and the local temperatures are calculated from

$$T_{x+dx} = T_x + \frac{\dot{Q} \cdot dx}{\left(A_{solid} + A_{fluid}\right) \cdot \lambda_{eff}}$$

with an effective heat conduction coefficient of

$$\lambda_{eff} = \frac{\lambda_{solid} \cdot A_{solid} + \lambda_{fluid} \cdot A_{fluid}}{A_{solid} + A_{fluid}}.$$

The local temperatures are then calculated iteratively by varying the axial heat flux until the warm end temperature has reached its given value.

In comparison with the CFD-simulation the axial temperature distributions are almost identical. Near the cold end there is a slight temperature difference as an effect from different temperature dependencies of the thermal conductivities of wall and fluid. There is a small radial temperature gradient in the simulation giving the divergence between simulation and analytical solution. For this reason the heat fluxes are taken from the warm side. The comparison will show a difference smaller than 1 % between analytical and CFD-solution:

	analytical	CFD	
heat flux (warm end)	0.7582	0.7512	W
difference		- 0.93	%

CFD-SIMULATIONS

A general converging problem is the relatively small cross area of the wall compared to the pipe length. For the axial heat transport it takes a long real time until a stable temperature profile is formed in the wall and later in the fluid. This has a strong influence on the overall computing time. On one hand a very small timestep is necessary for an accurate solution of the flow field, on the other hand a long timestep is important for the temperature field. To reduce the computing time a good start condition is very important. At negative orientations the solution from - 45°, once calculated with appreciable processing time, is used to start. The analytical solution for the vertical pipe is used for positive orientations instead.

The extrema + 90° and - 90° converge very badly. At + 90° a nearly stable temperature stratification exists. The only stimulation for a flow comes from different thermal conductivities of wall and fluid respectively. The flow velocities are nearly zero. For - 90° we find a fully convective flow. But there is no stable solution because of the symmetries in the pipes. The software can not solve the problem exactly. In the outer gas region a heated flow heading upwards and in the center a downward straight flow would be expected. The stable solution for - 90° could possibly being forced by an appropriate starting condition.

For all other orientations a stable converged solution could be found.

RESULTS

To post-process the results, an averaged heat flux is calculated from the heat fluxes through the warm and the cold end. In the chart (Fig. 2) the results are shown for different geometric parameters. Beside the heat fluxes through the standard pipe the heat fluxes through deviating geometries are plotted in extra curves.

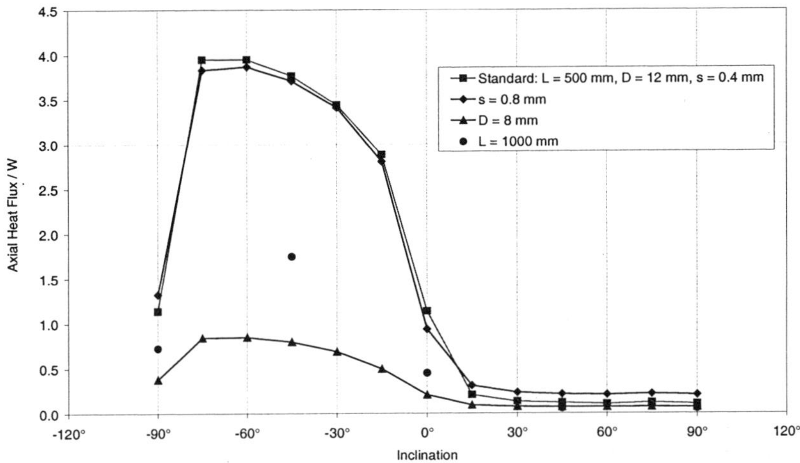

Fig. 2 Heat fluxes through pipes at different orientations

As an overall view all variants have a similar character with different absolute values. At positive inclinations the heat flux is small. Only heat conduction through solid and gas occur. With decreasing inclinations lower than + 30° the heat flux increases rapidly. A convective flow starts within the pipe and increases the heat flux through the gas region. The maximum heat flux occurs around - 75°. At lower inclinations no exact values are known because no stable solution could be found.

As a rough estimation three regions can be defined:

I.	- 90° ... 0°	free convection
II.	0° ... + 30°	free convection / heat conduction
III.	+ 30° ... + 90°	heat conduction

Some detailed local flow and temperature behavior are given in the following pictures. It can be shown (Fig. 3), that at lower inclination angles after start of the convection the temperature differences across the pipe will increase.

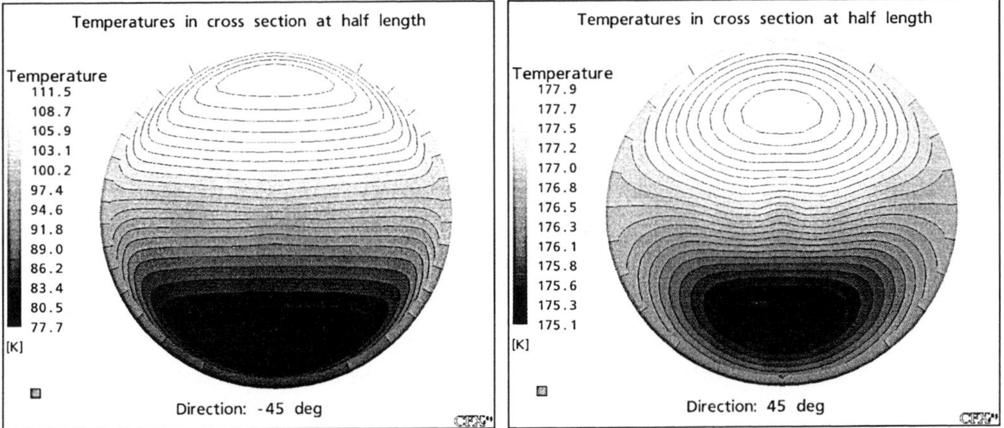

Fig. 3 Temperatures in cross section at half length (note: different scales)

In the upper half of the pipe a flow will occur from the hot to the cold end (Fig. 4). In the pictures the warm end is on the right side. A second flow from cold to hot end will be in the lower half of the pipe. These flows are responsible for the increase in heat transport. At positive orientations they are not fully developed. There are mainly local eddies negligible to the overall heat transport.

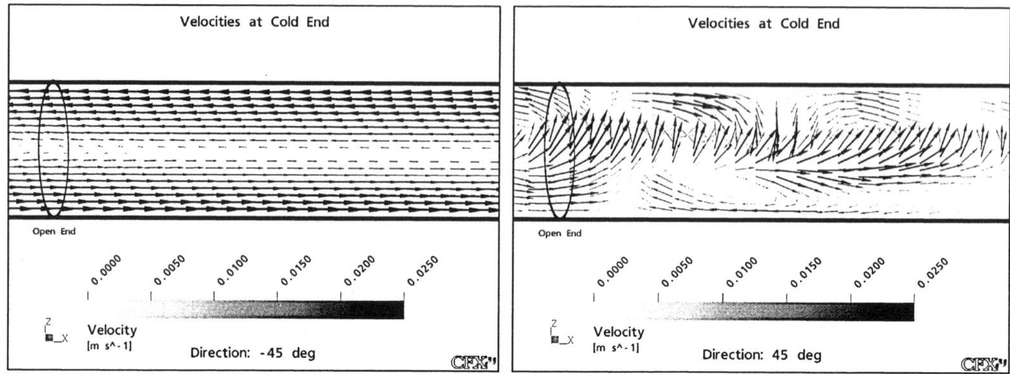

Fig. 4 Velocity vectors at cold end

Close to the warm end the gas is heated and ascends (Fig. 5). For positive inclinations the velocities are one order of magnitude smaller than at negative inclinations.

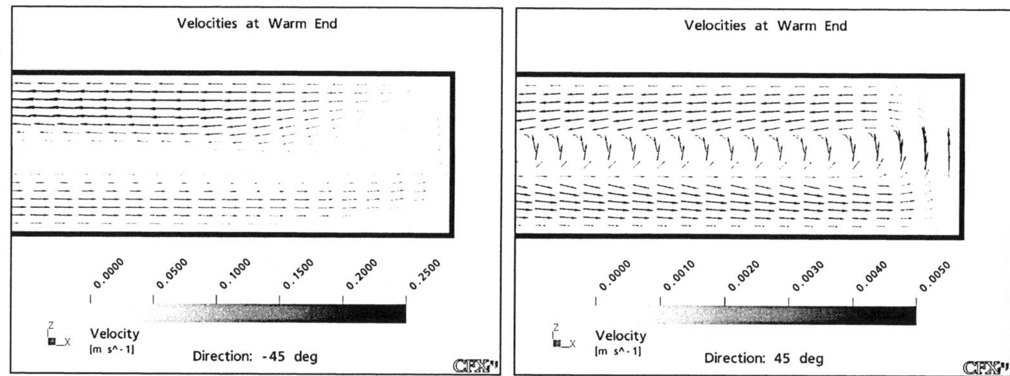

Fig. 5 Velocity vectors at warm end

CONCLUSIONS

The following dependencies are found by the CFD-simulations:

- The heat flux through pipes by convection compared to conduction can be up to 30 times higher.
- The doubling of the pipe length will reduce the heat flux by 50 %.
- A doubling of the wall thickness increases the heat flux for positive inclinations. The cross section for heat conduction becomes higher.
- A doubling of the gas cross section gives a more than 4 times higher heat flux if convection occurs.

As a result from these CFD-simulations more detailed hints for construction can be deduced. Usual design rules (as long as possible, minimum wall thickness, low heat conductivity...) can be optimized more precisely. Moreover the influence of the inclination can be taken into account.

As an option for future work the results could be cumulated into simple equations. Then the heat transport through complex pipe systems can be easily estimated with approximations of bent pipes by straight pipes.

Normalized representation for steady state heat transport in a channel containing He II covering pressure range up to 1.5 MPa

Sato A., Maeda M., Yuyama M. and Kamioka Y. *

TML, NIMS, 3-13 Tsukuba, Ibaraki, 305-0003, Japan
*Taiyo Toyo Sanso Co. Ltd., 3-1-1 Kyobashi, Chuo-ku, Tokyo, 104-0031, Japan

Steady state heat transport along a channel containing He II was measured up to 1.5 MPa. The thermal conductivity functions (TCF) were determined letting exponent m in the Gorter-Mellink equation equal to 3.4. The result was represented uniquely by using a normalized TCF. It has a good agreement with the Bon Mardion's heat conductivity function at 0.1 MPa.

INTRODUCTION

The turbulent heat transport in a channel containing He II is described by

$$dT/dx = f(T, P) \, q^{m},$$ (1)

where $f(T, P)$ is a function of He II properties, temperature and pressure, and m is a numerical coefficient. The quantity $f^{-1}(T)$ is called the thermal conductivity function (TCF) . Regarding the m-value, theory indicates that m should be equal to 3 [1]. Van Sciver analyzed the TCF letting $m = 3$ on the basis of the experiments up to 0.23 MPa [2, 3, 4]. The commercially based data base "HEPAK" also uses exponent m of 3 [5]. On the other hand, Bon Mardion represented his data by using $m = 3.4$ [6, 7].

It is of interest to know which exponent m is proper for the representation of the heat transport in He II. Therefore, we undertook determining the TCF in a wide range of pressure from saturated pressure to 1.5 MPa. Temperature gradients along the channel containing He II were measured in the temperature range from 1.4 K to 2.1 K [8]. Furthermore, we found that the experimental data deviation from the TCF had a minimum when assuming m was equal to 3.4 [9]. In the process of measurement error estimation, we recognized that the largest error came from the estimation of the inner diameter of the tube. Therefore, we measured the diameter precisely and checked the data again in order to get a universal representation of the thermal conductivity functions.

MEASUREMENT APPARATUS

A test channel was formed in a stainless steel tube with 6 mm or 10 mm inner diameters and 0.5 mm wall thickness. The length of the test section was 19 cm or 10 cm. The inner diameters were measured precisely at several locations along the tube using a three point internal micrometer. The effective cross-sectional areas were estimated considering the size of thermometers and the lead wires, and the thermal shrinkage when cooled at 2 K. The estimated results are shown in Table 1 with the sample tube dimensions.

A test vessel including the channel was immersed in a saturated superfluid helium bath for cooling and was pressurized through a filling tube up to 1.5 MPa [8]. Pressure in the test vessel was measured *in situ* using a piezo-resistive pressure sensor, FPS51B by Fujikura Ltd. [10].

A heater producing heat flux is at the bottom of the channel. The upper end of the channel is open to the subcooled helium bath. Thermometers are located along the channel to measure temperature profiles in the channel. The distance between the thermometers was 20 mm in channels No.1 and

No.2 and 10 mm in the channel No.3. Ruthenium oxide resistors were used as thermometers. The bath temperature is measured using a germanium resistor thermometer. Accuracy of the temperature measurement is within 3 mK. Concerning the detail of the measurement, refer the previous report [8, 9].

Table 1 Estimated cross-sectional area of the channel at 2 K

	Cross-sectional area	Length	Diameter
No. 1	0.225 ± 0.002 cm^2	19 cm	6 mm
No. 2	0.736 ± 0.004 cm^2	19 cm	10 mm
No. 3	0.225 ± 0.002 cm^2	10 cm	6 mm

DATA ANALYSIS

Temperature profiles along the channel were measured in the presence of heat flux at temperatures from 1.4 to 2.1 K. The series of temperature profile data is the same as reported in the previous paper [9]. Heat flux data were checked according to the precise measurement of the cross-sectional area of the channel. The error in determining the temperature gradient mainly depends on the location of the channel where the gradient is determined. Note that the error at the end of the channel is large. Figure 1 shows the measured temperature gradient versus heat flux density for pressurized He II at 0.1 MPa and 1.71 K, compared with the predicted lines by Van Sciver ($m = 3$). The probable errors are indicated as error bars for the channel No.2. The errors vary from 1.4 to 2 % in the heat flux range from 10^4 to 2×10^4 W/m^2. They are 1.6 - 3.5 % in the higher heat flux region from 2×10^4 to 3×10^4 W/m^2. The straight line indicates the fitted result using the least squares method assuming m is equal to 3.4. The slope of our results suggests that the exponent m should be 3.4.

We checked the exponent m with another method used in the previous work [9]. The data of temperature gradient calculated from the newly estimated heat flux densities were fitted with a TCF by using the least squares method for a certain value of m. The experimental data deviation from the fitted TCF changes depending on m. The exponent m giving the minimum deviation is considered to be the most suitable exponent m. Actually, a series of temperature profile in the same pressure is expressed smoothly in the TCF when it is determined with exponent m giving the minimum deviation. The result at 0.1 MPa is shown in Figure 2, which indicates the exponent m is approximately 3.4 considering the estimation error. Almost the same results were found at the other pressures.

Based on the above analyses, the TCF was determined by using the exponent m of 3.4.

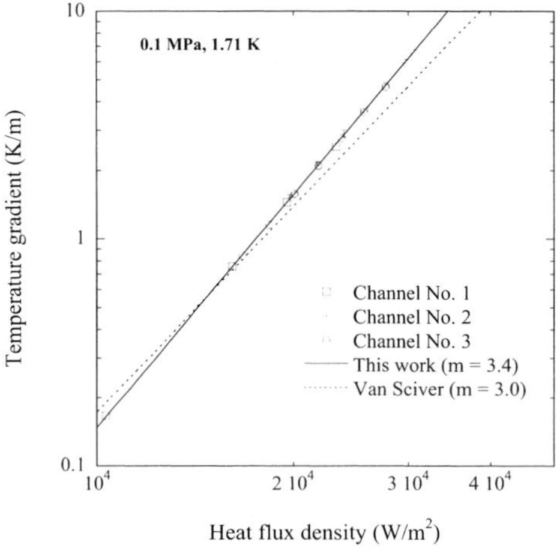

Figure 1 Temperature gradient versus heat flux in pressurized He II at 0.1 MPa and 1.71 K, compared with the predicted line by Van Sciver ($m = 3$).

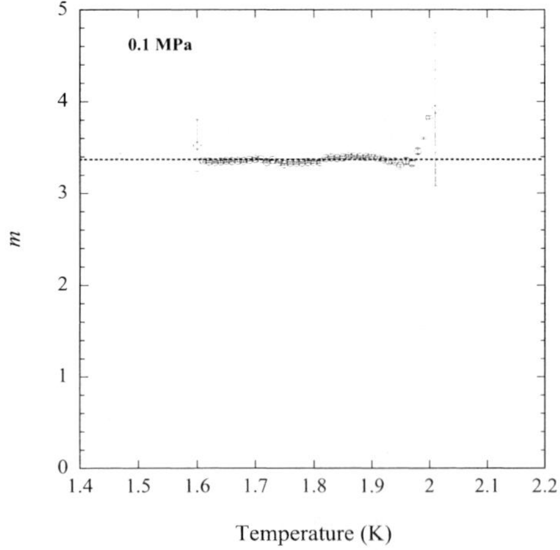

Figure 2 Temperature dependence of the exponent m with the minimum deviation of the experimental data at 0.1 MPa.

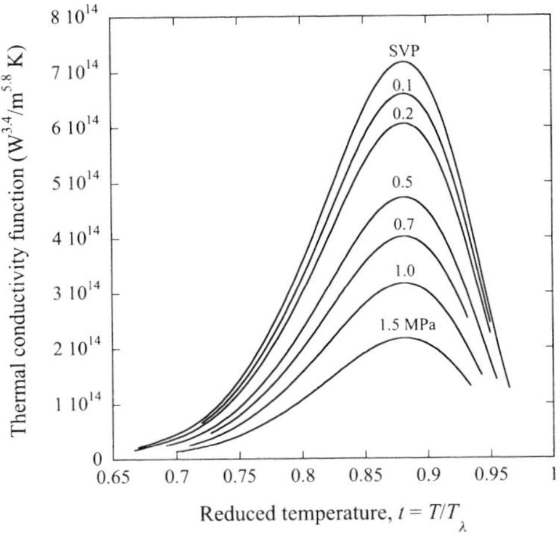

Figure 3 Reduced Temperature dependence of the thermal conductivity function at various pressures.

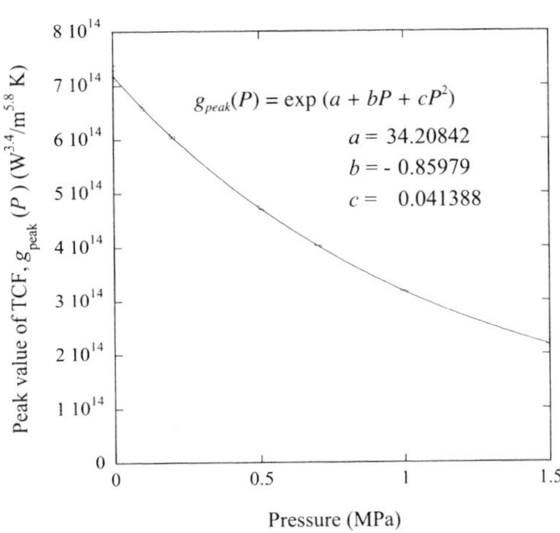

Figure 4 Pressure dependence of the peak value of the thermal conductivity function.

RENORMALIZATION OF HEAT CONDUCTIVITY FUNCTION

Plotted as smooth lines in Figure 3 is the thermal conductivity functions as they depend on reduced temperature, $t = T/T_\lambda$. The function $f^{-1}(T, P)$ is rewritten as a function of reduced temperature t as

$$f^{-1}(T, P) = g(t, P) . \tag{2}$$

Peak values occur at $t = 0.882$ which does not depend on pressure. Figure 4 shows the pressure dependence of the peak values. This function is fitted to the equation

$$g_{peak}(P) = \exp (a + bP + cP^2) . \tag{3}$$

The functions $g(t, P)$ in equation (2) are normalized at various pressures by the peak value g_{peak} as

$$h(t, P) = g(t, P) / g_{peak}(P) . \tag{4}$$

The normalized thermal conductivity functon, $h(t)$, is shown in Figure 5. All data appear to be expressed as a unique function which does not depend on pressure. Therefore, $h(t, P)$ is expressed as $h(t)$. This function is fitted to the equation

$$h(t) = 1 + (t - 0.882)^2 \sum_{n=0}^{9} \{a_n(t - 1)^n\} , \tag{5}$$

where

$$a_0 = - (0.118)^{-2}, \qquad a_1 = 1.2172617 \times 10^3,$$
$$a_2 = - 1.4992321 \times 10^4, \qquad a_3 = - 3.9491398 \times 10^5,$$
$$a_4 = - 2.9716249 \times 10^6, \qquad a_5 = - 1.2716045 \times 10^7,$$
$$a_6 = - 3.8519949 \times 10^7, \qquad a_7 = - 8.6644230 \times 10^7,$$
$$a_8 = - 1.2501488 \times 10^8, \qquad a_9 = - 8.1273591 \times 10^7 .$$

As a result, the heat transport characteristics are represented uniquely by using the normalized TCF as

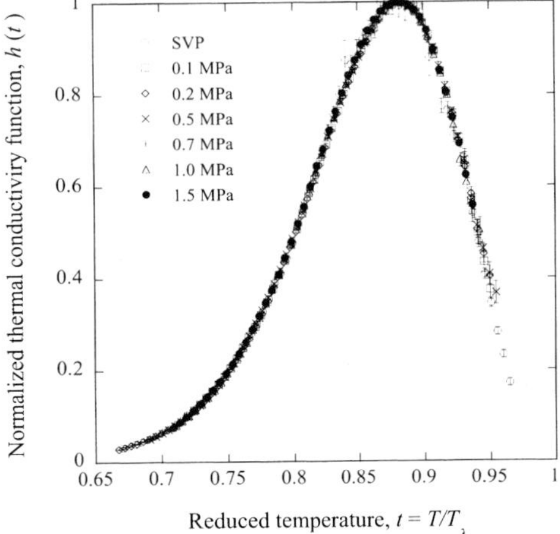

Figure 5 Normalized thermal conductivity functions at various pressures. Peak value appears at $t = 0.882$.

$$dT/dx = q^{3.4}/g_{peak}(P) \cdot h(t) \qquad (6)$$

Figure 6 shows the thermal conductivity function at 0.1 MPa compared with the Bon Mardion's result . The line calculated from the normalized TCF, $h(t)$, explains the Bon Mardion's data very well.

SUMMARY

Steady state heat transport through He II in wide channels was investigated up to 1.5 MPa, The exponent in the Gorter-Mellink equation is determined to be 3.4 from the data analysis. The thermal conductivity functions were determined by using $m = 3.4$. The results in a wide range of pressure from saturated pressure to 1.5 MPa were represented uniquely by using the normalized TCF.

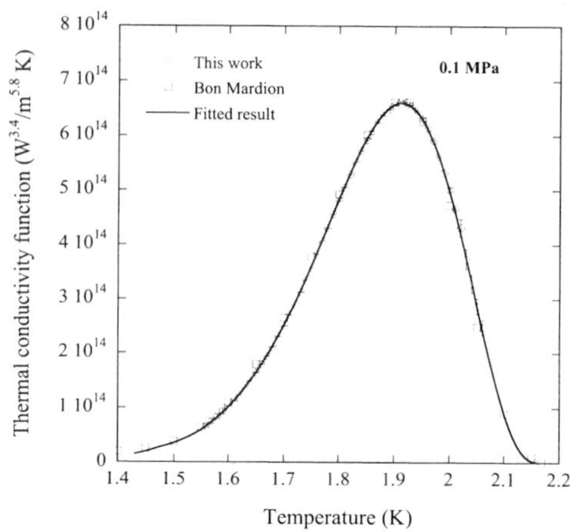

Figure 6   Comparison of the thermal conductivity function at 0.1 MPa with Bon Maridon's results. The straight line is calculated from the normalized TCF, $h(t)$.

Agreement with the Bon Maridon's result is quite good. This normalized equation will be used practically for the design of superfluid equipments. The physical meaning of $m = 3.4$ needs to be investigated in a future.

REFERENCES

1. Gorter, C. J. and Mellink, J. H., On the irreversible processes in liquid helium II, <u>Physica</u> (1949) <u>15</u> 285-304

2. Van Sciver, S. W., <u>Handbook of Cryogenic Engineering</u>, Taylor & Francis, Philadelphia, USA (1998) 455-456

3. Van Sciver, S. W., Kapitza Conductance of Aluminum and Heat Transport from a Flat Surface through a Large-diameter Tube to Saturated Helium II, <u>Advances in Cryogenic Engineering</u> (1978), <u>23</u> 340-348

4. Van Sciver, S. W., Kapitza conductance of aluminium and heat transport through subcooled He II, <u>Cryogenics</u> (1978) <u>18</u>, 521 - 527

5. HEPAK is licensed by Cryodata, Niwot, Colorado.

6. Bon Mardion, G., Claudet, G. and Seyfert, P., Steady State Heat Transport in Superfl uid Helium at 1 bar, <u>Proceedings of 7th International Cryogenic Engineering Conference</u>, IPC Technology Press, London, UK (1978) 214-221

7. Bon Mardion, G., Claudet, G. and Seyfert, P., <u>Cryogenics</u> (1979) <u>19</u> 45-47

8. Sato, A., Maeda, M. and Kamioka, Y., Steady State Heat Transport in a Channel containing He II at High Pressures up to 1.5 MPa, <u>Proceedings of 19th International Cryogenic Engineering Conference</u>, Narosa Publishing House, New Delhi, India (2003) 751-754.

9. Sato, A, Maeda, M. and Kamioka, Y., Steady State Heat Transport in A Channel Containing He II at High Pressures up to 1.5MPa, CEC2003, C1-H-03

10. Maeda, M., Sato, A., Yuyama, M., Kosuge, M., Matsumoto, F. and Nagai, H., Characteristics of a silicon pressure sensor in superfluid helium pressurized up to 1.5 MPa, <u>Cryogenics</u> (2004) <u>44</u> 217-222

Thermal insulation of the Wendelstein 7-X superconducting magnet system

Schauer F., Nagel M., Bozhko Y., Pietsch M., Shim S.Y., Kufner F*, Leher F.*, Binni A*., Posselt H.**,

Max-Planck-Institut fuer Plasmaphysik, Greifswald Branch, Euratom Association, Wendelsteinstrasse 1, D-17491 Greifswald, Germany
* MAN DWE GmbH, Dpt. AB – Plant Construction, Werftstrasse 17, D-94469 Deggendorf, Germany
** Linde AG, Engineering Div., Dr.-Carl-von-Linde-Strasse 6-14, D-82049 Hoellriegelskreuth, Germany

The cryostat thermal insulation of the stellarator Wendelstein 7-X (W7-X) is being built by MAN DWE and its main sub-contractor Linde AG. Development and engineering of the plasma vessel (PV) insulation is finished and production is starting. The complex geometry and narrow gaps to the superconducting coils require innovative solutions concerning the multi-layer insulation (MLI) as well as the shield and its supports. Crinkled Al-coated Kapton® foils with glass paper spacers were chosen as MLI, and the shield consists of epoxy-glass resin containing copper meshes for temperature equalisation. Shield supports are made of Torlon® which exhibits extremely low thermal conductivity and good mechanical strength.

INTRODUCTION

The main assembly activities of Wendelstein 7-X at the Max-Planck-Institut fuer Plasmaphysik, Greifswald Branch, start this autumn with stringing the first superconducting coil over a plasma vessel (PV) sector which by then must be thermally insulated with multi-layer insulation (MLI) and an actively cooled thermal shield [1]. Subsequent insulation of the other cryostat components has to be performed step by step in accordance with the machine assembly until the torus is completed in 2009.

Basically, three regions with different conditions have to be discerned: Firstly, the PV and adjacent "inner" port sections (ports allow access to the plasma vessel from the outside world) are exposed to strong magnetic field changes in case of coil emergency switch-offs. Space is extremely restricted there, and the insulation has to endure occasional PV baking at 150°C. Secondly, the "outer" port sections adjacent to the outer vessel are exposed to relatively small stray fields but they are also baked. Thirdly, the outer vessel always remains at room temperature and encounters small stray fields only.

In order to keep insulation effort and costs within reasonable limits, the heat flow from the MLI-covered cryostat walls being at 300 K to the shield at around 60 K is specified to relatively undemanding 6 W/m². This limit increases to 9 W/m² at the PV and port walls when they have an average temperature of 60°C during plasma operation. Since the shields of the inner port sections are cooled indirectly only, some hot spots there can reach up to 150 K. Some layers of MLI are thus applied at those shields also to their cold sides. Generally, heat influx from the shields to the coils and their support structure is ≤1.5 W/m².

Following the assembly sequence of the W7-X basic device, the PV insulation and corresponding port interfaces have to be manufactured first. Therefore, development work has been concentrated onto these components. Now the critical issues are solved and production is starting. This summer, the thermal insulation components will be ready to be assembled onto the first PV segment before stringing over the first coil. The insulation of the outer vessel, the ports, and other components like feedthroughs, pipes, instrumentation wires, etc., are considered to be less demanding and less time critical. The insulation concepts for these constituents basically exist; the detail engineering is straightforward and will be based on experience from current work with the plasma vessel insulation, from the Demo-Cryostat [2], and from general know-how in the field.

THERMAL INSULATION OF THE PV

W7-X assembly starts with mounting the first coil in the middle of the 1st PV half-module (HM) which is delivered in two sectors. After insulating segment 1 (Fig. 1) on top of the larger sector and stringing the middle coil over, both PV parts are welded together. Then insulation segment 2 is mounted and the next coil strung over, etc. Every plasma vessel HM insulation is built up from four segments. Each segment in turn is sub-divided into shield and corresponding MLI panels, adding up to 4 x 5 panels in poloidal and toroidal directions, respectively, per HM. Each panel is cooled by two tubes as indicated in Fig. 1.

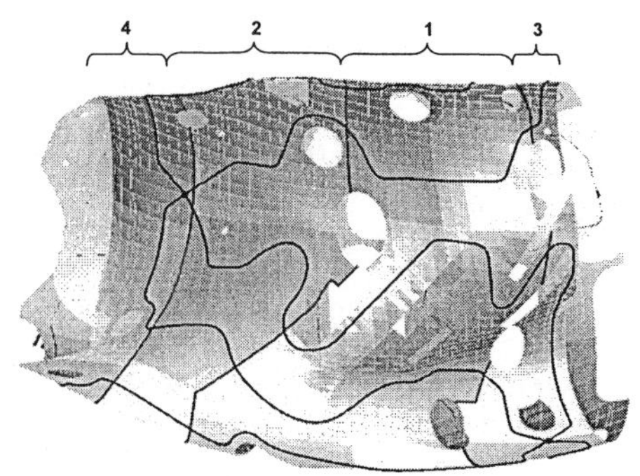

Figure 1 Insulation of a plasma vessel HM. The segment assembly sequence is indicated

MLI of the PV

The most challenging requirements to the MLI of the plasma vessel are adaptability to the complex PV surface and heat resistance. This lead to the choice of aluminized crinkled polyimid (Kapton®) foils with glass paper layers (DBW Type 50111) in between. The low thermal conductivity within the foils allows simple overlap design between panels, and undemanding interfaces to the port tube insulation. This MLI is also flame retardant, and eddy current forces caused by magnet emergency discharges are negligible.

The MLI of the PV is built up from panels consisting of two shifted mats. Each mat contains ten crinkled Al-coated Kapton®-foils, alternating with glass paper. This MLI is well adaptable to the PV surface which is not developable into a plane (Fig. 2).

The MLI panels are laid into corresponding shield panels and are fixed by plastic pins. The MLI mats are then cut to right size, and these assemblies are mounted onto the PV surface with optically tight overlaps.

Loss tests of a 20-layer MLI system were performed by Linde Co. on a 0.32 m diameter, 3.4 m long cylinder at 80 K. In one case the MLI was laid perfectly loose with a thickness of about 15 mm, and with one overlap (two narrow gaps) along the cylinder length. In the second case, the insulation was compressed to a

Figure 2   Crinkled Kapton® MLI prototype mat adapts easily to the PV surface

thickness of 10 mm, and the overlap gaps were opened to 10 mm width. At the cylinder ends, a PV to port insulation interface was simulated. In the first experiment, the losses were 0.62 W/m², and in the second, realistic one, the losses were 0.93 W/m². These excellent results were better than expected.

Thermal shield of the PV

Following the experience from the Demo-Cryostat, the first idea was to use panels made up by canted stainless steel stripes approximating the PV surface. Attached copper stripes were planned for temperature equalization. This steel – copper combination avoids excessive eddy currents and associated forces in case of a magnet emergency shut-off. However, during the detail design phase it turned out to be quite difficult to fulfill the stringent tolerance requirements, and this concept was dropped.

One promising alternative appeared to be "dieless forming", a process where a spherically tipped forming tool incrementally presses securely clamped metal sheets over a stationary swage into the complex shapes. The final panel contour has to be cut out of the formed sheet. Experiments were performed with stainless steel and brass. However, the required tolerances could not be achieved without elaborate measures for compensation of spring-back. In both cases, the sheet was out of tolerance already after forming, and subsequent annealing was required to avoid spring-back during cutting the panel out of the formed sheet. As expected, brass was easier workable. This material has the additional advantage that its thermal and electrical conductivities are such that copper stripes for thermal equalization can be omitted and eddy currents can be kept within acceptable limits at the same time.

In parallel, the possibility to use epoxy-fiber panels containing copper meshes was investigated. After optimization of small samples, a real size glass-fiber epoxy panel was built (Fig. 3) with excellent surface tolerances of ±1.5 mm. The copper net was tightly integrated within the homogeneous laminate. No cracks or any signs of disintegration appeared after many thermal shocks in liquid nitrogen.

Technical and economical evaluation led to the decision to use epoxy-glass resin laminates for the PV shield to be produced by the sub-contractor Wethje Co. Each panel contains three copper mesh layers for sufficient thermal conductivity. Every copper layer will

Figure 3 Epoxy-glass resin PV shield panel not yet cut to final shape, without Al coating

be cut into two or three electrically insulated parts in order to reduce eddy currents. FEM analysis confirmed the thermodynamical, electrodynamical and mechanical behaviour of this shield [3]. The emissivity of the panel surface facing the cold coils will be reduced by applying adhesive aluminium tape.

Shield supports

Stringent space requirements call for short shield supports (Fig. 4). They must be rugged in order to withstand eddy current forces as well as the weight of assembly personnel stepping on it, and must exhibit at the same time low thermal conductivity.

One type considered uses thin polyimid ropes (Kevlar®) as supporting elements (Fig. 5, left). Two U-shaped steel stripes are arranged coaxially in opposite.

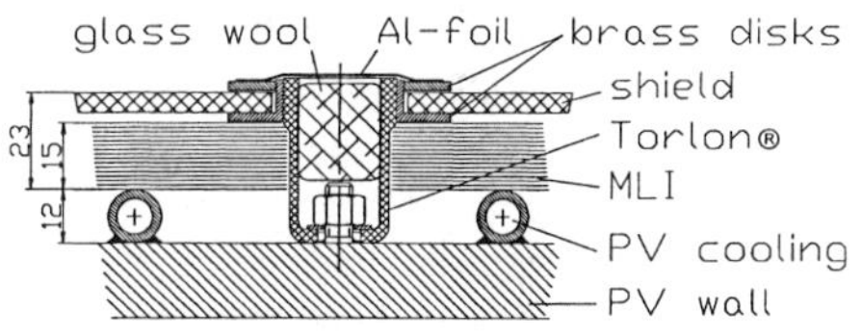

Figure 4 PV thermal insulation with shield support

The bottoms and tips of the U's are connected by ropes to each other, respectively, to prevent any in-axis movements. Forces perpendicularly to the axis are supported by the skewed arrangement of the ropes. These tensile-loaded support elements can be made very thin and thus with low thermal losses.

The second - and finally selected - spacer consists of the polyamid-imid Torlon 4203® (Fig. 5, right) whose thermal conductivity is extremely low and varies almost linearly from $0.08\ \mathrm{Wm^{-1}K^{-1}}$ at 30 K to $0.25\ \mathrm{Wm^{-1}K^{-1}}$ at 300 K [4]. Also the mechanical properties, measured by FZK, are excellent: tensile strength ≥155 MPa at 7 K and 300 K, and fracture toughness K_{IC} ranging from $4.3\ \mathrm{MPa \cdot m^{1/2}}$ at 7 K to $5.7\ \mathrm{MPa \cdot m^{1/2}}$ at 300 K. The design of the support can be gathered from Figs. 4 and 5: A thin-walled Torlon® tube is bolted at its warm bottom to the PV, and at the cold end the shield is being inserted between two circular brass disks which are screwed onto the tube.

Each shield panel has a firm support near its center of gravity, and a few additional "loose" supports which

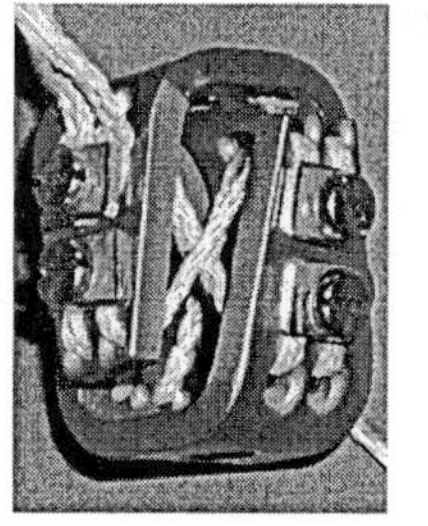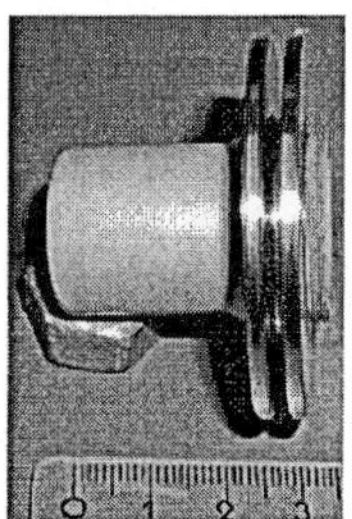

Fig. 5 Kevlar® (left) and Torlon® shield supports

allow sliding of the shield between the brass disks for thermal contraction compensation. The tube OD is 22 mm, and its wall thickness is 2 mm and 1 mm for the fixed and loose supports, respectively. Both support types were tested under real thermal and mechanical conditions, inter alia with forces perpendicular to the tube axes. The resulting breaking strengths were ≥4.9 kN and ≥2.5 kN which by far exceed the specified fixed point load of 1.0 kN.

Cooling circuit

The four toroidal shield panel rows of a half-module are cooled in series via two parallel He-tubes each which merge at the HM ends (Fig. 1). The meandering shape of the tubes results from evasion of ports and constrictions between PV and coils. Flexible copper braids are soldered to the tubes, and at the other braid ends, attached cable shoes are riveted elastically to the shield. Sufficient heat transfer is provided by thermally conductive paste between the cable shoes and the glass-fiber epoxy panels.

THERMAL INSULATION OF THE PORTS

The port tubes welded to the PV are insulated using the same MLI system as at the PV. The interface between the MLIs of the PV and a port is basically performed by cutting staggered strips into the port MLI, and by overlapping these strips with the PV-MLI for an optically tight cover of the intersection.

The shield of this "inner" port section is made of brass which is loosely bolted to the PV shield panel(s) such as to allow sliding for thermal contraction compensation. On the other end, the inner port shield is fixed by spacers. It will be cooled indirectly from the PV and the OV shields.

The "outer" port section adjacent to the outer cryostat vessel contains bellows for compensation of PV movements and thermal expansion. The insulation of this assembly adjoins to the OV. Since the magnetic field is small in this region, copper is allowed as shield material. MLI and shields of the inner and outer port parts are overlapping such as to allow relative movements of almost 30 mm in any direction.

THERMAL INSULATION OF THE OUTER VESSEL

The MLI of the outer vessel is not yet finally decided, it will be either the same as on the PV or consist of pure aluminum foils having also glass paper spacer layers in between. The low magnetic field in this area allows copper as shield material. The shield will be divided into 5 x 4 panels per module toroidally and poloidally, respectively. The same Torlon® supports will be used as described above. Also the cooling loops will be in principal the same as with the PV shield. Each panel will be cooled by two tubes in parallel, and the tubes of a panel-row merge at each end of a half-module. The four poloidal HM panel-rows are cooled in series. The cooling tubes are soldered via copper adapters to the copper shields. Corresponding PV and OV half-module cooling loops are connected in series. Each such HM shield cooling loop can be controlled by a valve.

SUMMARY

Development and detail engineering of the main components of the plasma vessel thermal insulation and the interfaces to adjacent port tubes is finished and production is starting. The chosen MLI system is crinkled Kapton® foils with vapor deposited aluminium on both sides, and with glass paper as spacers in between. The actively cooled thermal shield consists of epoxy-glass resin panels which contain three copper mesh layers each for sufficient thermal conductivity. Eddy current induction is controlled by one or two cuts per copper mesh. The surface facing the coils is covered by ahesive Al-foil in order to reduce emissivity. Each shield panel is supported by one firm support near the center and several "loose" supports which allow in-plane gliding for thermal contraction compensation. The shield supports consist of Torlon® tubes. They are bolted to the PV, and at the cold ends the panels are clamped between two disks. Each panel is cooled by two cooling pipes which are thermally contacted via flexible copper braids.

Concepts were worked out for the thermal insulation of the ports and outer vessel. Detail design and assembly tests of critical components will be performed in a sequence corresponding to the overall W7-X mounting activities.

REFERENCES

1. Schauer, F., Bau, H., Bozhko, Y., Brockmann, R., Nagel, M., Pietsch, M., Raatz, S., Cryotechnology for Wendelstein 7-X, Fusion Eng. and Design 66-68 (2003) 1045-1048
2. Schauer, F., Bau, H., Bojko, I., Brockmann, R., Feist, J.-H., Hein, B., Pieger-Frey, M., Pirsch, H., Sapper, J., Sombach, B., Stadlbauer, J., Volzke, O., Wald, I., and Wanner, M., Assembly and test of the W7-X demo cryostat, Fusion Eng. and Design 56-57 (2001) 861-866
3. Nagel, M., Shim, S.Y. and Schauer, F., Thermal and structural analysis of the W7-X magnet heat radiation shield, to be presented at the 23rd Symposium on Fusion Technology, 20 – 24 September 2004, Venice, Italy
4. Barucci, M., Olivieri, E., Pasca, E., Risegari, L., and Ventura, G., Thermal conductivity of Torlon between 4.2 and 300 K, to be presented at the TEMPMEKO 2004, 9th International Symposium on Temperature and Thermal Measurements in Industry and Science, 22 - 25 June 2004, Cavtat - Dubrovnik Croatia

MODELING FREE CONVECTION FLOW OF LIQUID HYDROGEN WITHIN A CYLINDRICAL HEAT EXCHANGER COOLED TO 14 K

S. Q Yang, M. A. Green, and W. Lau

Oxford University Department of Physics, Oxford OX1 3RH, United Kingdom

A liquid hydrogen in a absorber for muon cooling requires that up to 300 W be removed from 20 liters of liquid hydrogen. The wall of the container is a heat exchanger between the hydrogen and 14 K helium gas in channels within the wall. The warm liquid hydrogen is circulated down the cylindrical walls of the absorber by free convection. The flow of the hydrogen is studied using FEA methods for two cases and the heat transfer coefficient to the wall is calculated. The first case is when the wall is bare. The second case is when there is a duct some distance inside the cooled wall.

INTRODUCTION

Neutrino factories require that muons be accelerated to high energy before they are allowed to decay into neutrinos and electrons. The muon must be cooled in an ionization-cooling channel before they can be accelerated to high energies [1,]. This cooling must occur in a time that is substantially less than the decay time for a muon (~2.1 μs). Ionization cooling of the muons requires a large number of liquid hydrogen absorbers that remove both the transverse and longitudinal momentum from a muons beam. RF-cavities reaccelerate the muons back to their original longitudinal momentum without adding back the transverse momentum lost in the cooling process. Liquid hydrogen has been selected as an absorber material because it has twice the dE/dx (energy loss per unit mass per square meter) of any other material. Hydrogen has only single proton in its nucleus. Thus, the coulomb scattering, which reintroduces transverse momentum into the muons, is minimized. Hydrogen is the best material for muon ionization cooling by at least a factor of two over any other material.

A liquid hydrogen absorber for muon cooling requires that up to 300 W be removed from liquid hydrogen that has been heated by an intense muon beam. The absorber is a 350-mm long cylindrical hydrogen tank with the cylinder axis in the direction of the muon beam. The cylindrical wall of the absorber is cooled with helium gas that enters in channels on the outside of the wall at 14 K (just above the freezing temperature for para hydrogen). The temperature of the helium gas leaving the absorber is determined by the heat put into the liquid hydrogen and the mass flow of the helium gas. The helium gas will enter the tubes on the outside of the wall at the bottom and it will leave at the top forming a counter flow heat exchanger with the hydrogen in the absorber. It has been proposed that hydrogen heated by the muon beam rises and then it is circulated down the cylinder walls of the absorber by free convection.

Free convection hydrogen absorber cooling has been studied using conventional heat transfer methods [2, 3]. Conventional heat transfer studies suggest that the inner surface of the liquid hydrogen absorber has to be extended in order to remove anywhere near the 200 to 300 W of beam energy that is required. The conventional heat transfer studies were done using a duct 10 to 20 mm from the inner wall of the absorber to improve the heat transfer to the wall. Conventional wisdom suggests that the duct can be used to direct the flow of hydrogen down and along the cylindrical wall surface to improve the heat transfer coefficient between the hydrogen flowing down the wall and the wall.

The free-convection flow of the hydrogen was studied using finite element methods is presented in this report. The temperature drop and heat transfer coefficient at the wall was calculated for various beam powers. The free convection calculations were done using a bare cylindrical wall with no duct to direct the hydrogen flow. A duct 20 mm from the cylindrical wall surface was added to see if the heat transfer coefficient between the liquid hydrogen and the cylindrical wall could be improved.

DESCRIPTION OF THE MODEL

The liquid hydrogen absorber was modeled in three dimensions. The Computational Fluid Dynamic (CFD) model [4] consists of a cylindrical body that is 300 mm in diameter. The body length is 235 mm. The heat from the hydrogen is transferred out of the absorber only through the body. The area of the heat transfer surface on the inner wall of the model is 0.221-m^{-2}. At the ends of the absorber are two thin windows with a diameter of 300 mm. The thickness of the windows at their center is 180 μm. The length of the curved window section is 57.5 mm. The distance between the window centers is 350 mm. The windows are adiabatic. The muon beam is modeled as a 10-mm diameter cylindrical heat source that goes down the axis of the absorber. The length of the cylindrical heat source is 340 mm.

The absorber body is modeled as a 300-mm ID aluminum cylinder with a thickness of 5 mm and a thermal conductivity of 235-W m^{-1} K^{-1}. The length of this wall is 235-mm. The outside temperature of the cylindrical wall was set at a constant value of 14 K. The hydrogen in the CFD model starts out at an initial temperature of 17 K. In each case, the model was run until steady state equilibrium is reached. The hydrogen density used for the model is 71 kg m^{-3}. The hydrogen specific heat used in the model is 9680-J kg^{-1} K^{-1}.

In the cases with the duct, the duct is 3 mm thick and 20 mm from the inner wall of the cylindrical part of the absorber. The duct length is the same as the aluminum cylindrical body length. The duct is open at the top and the bottom so that hydrogen can flow into the space between the duct and the wall. The duct is located in a region that is ± 70 degrees on either side of the horizontal centerline of the absorber body. The duct material has the properties of glass at 300 K; density = 2500 kg m^{-3}, specific heat = 750 J kg^{-1} K^{-1} and thermal conductivity = 1.4 W m^{-1} K^{-1}. The thermal diffusivity of the duct material is nearly the same at 20 K as it is at 300 K. Figure 1 shows the CFD model for the absorber.

THE RESULTS OF THE SIMULATION

The CFD model for a beam power of 30 W is shown in Figure 1 below. The temperatures shown in Figure 1 are global and include the temperature of the wall as well as the temperature of the liquid hydrogen that is within the absorber. The hydrogen peak temperature where the beam strikes is about 16.8 K. At the point where the hydrogen enters the duct, the temperature is reduced to about 16.3 K. The hydrogen temperature at the bottom that leaves the duct is about 15.6 K.

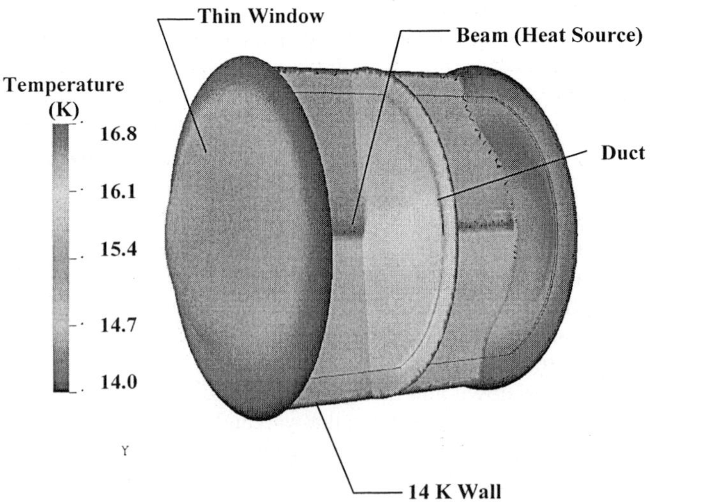

Figure 1. The CFD model of the Absorber showing the window, the duct, and heat exchanger wall. Beam power = 30 W

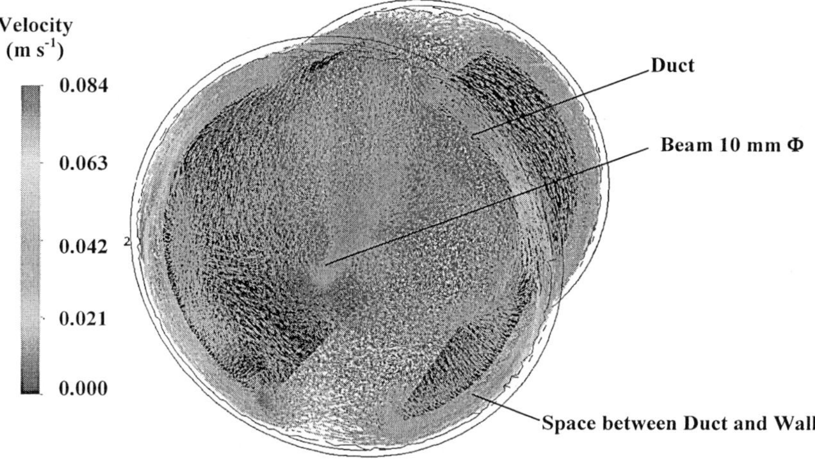

Figure 2. The velocity vector for the same case shown in Figure 1. This case has a duct and the beam power = 30 W

Figure 2 shows the velocity vectors for the case shown in Figure 1. One can see that the peak velocity in the system occurs within the duct, where the hydrogen flows down the wall. When there is no duct, the velocity of the flow next to the wall is between 0.01 and 0.015 m s⁻¹. This shows that the ducting really does direct the hydrogen flow along the heat transfer surface. The down side of ducting is that the flow can be choked within the duct, because most of the driving pressure is created by the buoyancy force when the hydrogen is heated is taken up by the increased velocity between the duct and the wall.

The CFD model was run with beam powers at 30, 60, 90, 120, 150, and 200 W. The model was run with and without a duct to direct the hydrogen flow along the wall. The model generates a temperature distribution in the liquid hydrogen and in the solid portions of the absorber (the cylindrical heat transfer surface and the windows). Velocity vectors were calculated for the liquid in the absorber. From the average hydrogen temperature along the wall one can estimate the log mean delta T for the heat exchanger. Using the calculated log mean delta T, the average heat exchanger U factor was estimated.

For both the ducted and un-ducted absorbers, the average temperature at the top T2 and the bottom T1 of the absorber goes up with beam power (see Figure 3 left). For the duct case, T2 and T1, one can estimate the heat exchanger average U factor directly. The U factor calculated for the case with no duct is probably not valid for Q > 120 W. In both cases, the U factor goes up as Q increases (see Figure 3 right). The temperature rise from the bottom of the absorber to the absorber hot spot ΔT also increases with heat load (see Figure 4 left). This temperature rise is not linear with heat load. Studies of free convection flow driven (ducted) heat exchangers suggests that ΔT and hydrogen mass-flow both increase as $Q^{0.5}$[2]. A plot of ΔT versus $Q^{0.5}$ fit to the value ΔT at Q = 90 W is shown in Figure 4 left for comparison.

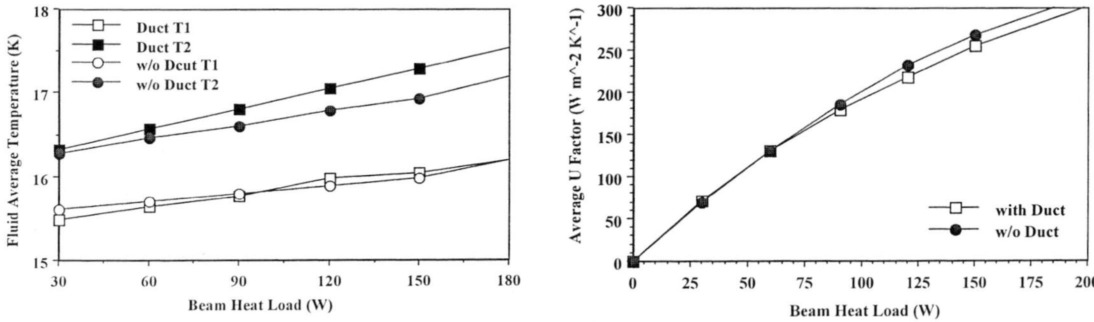

Figure 3. On the left, the fluid average temperatures at the top of the absorber T2 and bottom of the absorber T1
On the right, the average heat exchanger U factor versus the beam heat load Q using log-mean ΔT calculated from T1 and T2

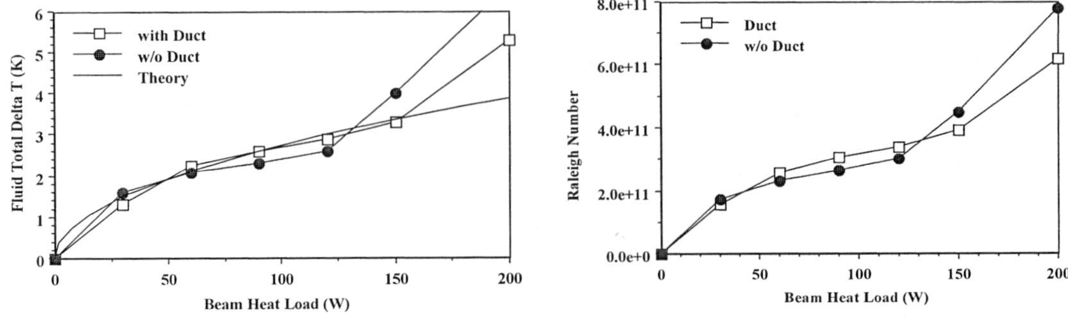

Figure 4. The total hydrogen temperature rise ΔT and the Raleigh number of the flow as a function of the beam heat load Q

From Figure 4 right, one can see that the calculated value of the Raleigh number has the same shape the curves for the hydrogen total ΔT versus the beam heat load Q. At high heat loads (>135 W), the peak temperature of the hydrogen stream (where the beam interacts with the hydrogen) is higher for the case with no duct (see Figure 4 left).   This corresponds to the Q where the Raleigh number is also higher (see Figure 4 right). Despite having a higher Raleigh number at Q > 135 W, the heat exchanger average heat transfer coefficient on the inner surface for the case without the duct must go down (relative to the case with the duct) for Q >135 W. Increased mixing means that T2-T1 is lower (see Figure 3 left) for the case with no duct, but the added turbulence is not on the heat transfer surface.

CONCLUDING COMMENTS

One can model a three-dimensional liquid hydrogen absorber using a CFD model. The results from the CFD model appears to be well behaved for beam heat loads Q less than 150 W. At Q > 150 W, the heat build up in the hydrogen will cause it to boil, even with a duct. At high heat loads, ducting does not appear to improve the calculated heat exchanger U factor, but on the other hand, the duct appears to retard the temperature build up in the hydrogen due to the beam heating. The results show that the cooling performance is driven by the velocity of the flow along the wall. The ability to create a turbulent flow on the wall is the key to improving cooling. The duct appears to allow this to happen.

Further work is needed to see how the distance between the wall and the duct affect the heat transfer. The model must also be run for Gaussian beams that are up to 250 mm in diameter (with the same total heating in a larger beam). Finally the cooling of the wall must be properly modeled (with cold helium gas going in at the bottom at 14 K and warm helium gas at 17 K coming out of the top). It is clear that the transfer surface must be extended a factor of 3 or more, if one is going to remove 300 W from a helium gas cooled liquid hydrogen absorber.   One good way to extend the heat transfer surface is to put part of the heat exchange surface inside the absorber. The extended heat exchanger acts as a duct on the cooled wall as well as added heat exchange surface. This study suggests that the duct can cause no harm.

ACKNOWLEDGEMENTS

This work was supported by the Oxford University Physics Department and the Particle Physics and Astronomy Research Council of the United Kingdom.

REFERENCES

1.   Ozaki, S, Palmer, R. B, Zisman, M. S. and J. Gallardo, J. Eds. "Feasibility Study II of a Muon Based Neutrino Source," BNL-52623, June 2001
2,   Green, M. A., "Comments on Liquid Hydrogen Absorbers for MICE," LBNL Report LBNL-52082, Jan. 2003
3.   Green, M. A., Ishimoto, S., Lau, W, and Yang, S., "A Heat Exchanger between Forced Flow Helium Gas at 14 to 18 K and Liquid Hydrogen at 20 K circulated by Natural Convection," <u>Advances in Cryogenic Engineering</u> **49**, (2003)
4.   <u>CFX-4.2 Solver</u>, AEA Technology, CFX International, Didcot OX11 0RA, United Kingdom

Experiment on natural convection of subcooled liquid nitrogen

Yeon Suk Choi [1,2], Ho-Myung Chang [1,3], and Steven W. Van Sciver [1,2]

[1] National High Magnetic Field Laboratory, Tallahassee, FL 32310, USA
[2] ME Department, FAMU-FSU College of Engineering, Tallahassee, FL 32310, USA
[3] ME Department, Hong Ik University, Seoul, 121-791, Korea

The natural convection of subcooled liquid nitrogen between two vertical plates is investigated by experiment. A vertical plate is thermally connected to a cryocooler and uniform heat flux is supplied to the other plate so that subcooled liquid in the gap may generate cellular flow. The wall-to-wall heat transfer coefficients are measured and compared with the existing correlations. Good agreement is observed between two data sets, when the heat flux is small. As the heat flux increases, the measured heat transfer becomes greater because of the vertical temperature gradient with the cooling from the top, which generates a multi-cellular flow.

INTRODUCTION

Liquid nitrogen is an excellent cooling media for HTS electric application because of its thermodynamic and dielectric properties as well as low price. Most commercial applications of HTS systems cooled by liquid nitrogen also require refrigeration by a cryocooler in order to eliminate the boil-off and maintain the liquid at the subcooled state. We have proposed a new cryogenic concept for HTS cooling, operating in the range of 63~66 K by natural convection of subcooled liquid nitrogen utilizing a cryocooler [1]. In order to confirm the feasibility of the proposed concept, we designed and constructed a natural convection cooling experiment [2].

In this experiment, liquid nitrogen is cooled to nearly the freezing temperature (63 K) by a vertical copper plate thermally anchored to the coldhead of a cryocooler. As the cryocooler is located at the top, the temperatures of solid bodies including copper plate and HTS magnet decrease upwards which are obviously different thermal boundary conditions from the most studies on heat transfer in a cavity [3-5]. In a previous paper [2], we reported preliminary values for the heat transfer coefficient in a vertical cavity whose surface temperatures decrease upwards. We have now completed measuring the temperature distributions along the vertical surfaces in the cavity for various heat fluxes. In the present paper, the detailed heat transfer characteristics are investigated and the multi-cellular flow patterns with augmentation of wall-to-wall heat transfer are discussed.

EXPERIMENTAL APPARATUS

A schematic overview of the experiment is shown in Figure 1. A single-stage GM cryocooler is mounted directly at the top plate of cryostat and a rectangular shaped heating plate is vertically located at the center of the cryostat. Two parallel cooling plates are positioned at a given distance symmetrically on both sides of the heating plate and thermally anchored to the coldhead of the cryocooler through a horizontal plate.

The two vertical cooling plates are bolt-jointed at the top with a horizontal copper plate. The flexible tinned copper braids are used for connection between the coldhead and the horizontal plate as well as to protect the coldhead from thermal contraction during cool-down. A Thermofoil™ heater is sandwiched between two identical stainless steel plates with 1 mm thickness, and cryogenic epoxy is applied to ensure good contact between the heater and the plates. The heater covers the entire surface and supplies a

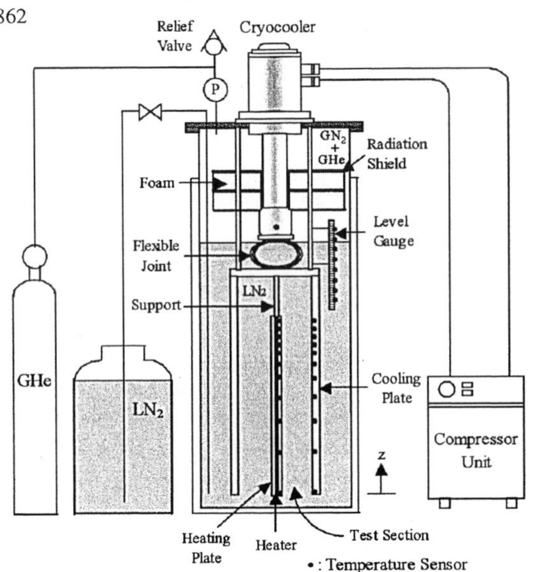

Figure 1 Schematic of experimental apparatus

constant heat flux to the plates. The heating power is regulated with a DC power supply. The cooling plates with 10 mm thickness are suspended at the top plate of cryostat, and the heating plate is suspended with gravitational and lateral supports made of threaded GFRP rod.

The surface temperatures of cooling and heating plates are measured with E-type thermocouples at a number of vertical locations as shown in Figure 1. The lead wires of the temperature sensors are connected through the holes from the opposites sides of the test section. Holes housing for thermocouple beads are located on the vertical centerline on each plate. The sensing beads of thermocouple are dipped in cryogenic thermal grease in order to prevent reaction with the fluid.

At the initial phase of the experiment, the cryostat is filled with liquid nitrogen and cooled down to near its freezing temperature (63 K) using the cryocooler and the extended cooling surfaces. Once the cryostat is cooled down, a uniform heat flux is supplied so that liquid nitrogen between the heating and cooling plates experiences natural convection. The vertical temperature distribution on both plates is measured in steady state, from which the heat transfer coefficients are calculated. The variable in this experiment is the magnitude of heat flux (8~160 W/m^2) for a given gap distance between the heating and cooling plates, L = 60 mm.

RESULTS AND DISCUSSIONS

The measured temperature distributions along the vertical surface of heating and cooling plate are presented with error bar and compared with analysis [1] in Figure 2(a) when the heat flux is 8 W/m^2. The largest value of temperature error is ± 0.025 K at z = 320 mm on heating plate, which is one-order of magnitude smaller than the wall-to-wall temperature difference, approximately 0.2 K. As the heat flux increases, the temperature error maintains within ± 0.025 K while the wall-to-wall temperature difference increases. Therefore, the sensitivity of temperature sensors is quite sufficient for our experimental purpose. In Figure 2(a), the top temperature of cooling plate is matched with the measured value as one of the boundary condition in the analysis. The heat transfer coefficient in the analysis was evaluated from the existing correlations for rectangular cavity [6] where each vertical surface has a uniform temperature. Good agreement is observed between two data sets, because the temperatures of heating and cooling plates are relatively uniform when the heat flux is small.

Shown in Figure 2(b) are the vertical temperature distributions of each plate when the heat flux is 80 W/m^2. A noticeable discrepancy is observed between two data sets, mainly because the heat transfer coefficient evaluated from the existing correlation is assumed constant vertically in the analysis. In the experiment, on the other hand, the heat transfer coefficient is not vertically uniform because of wavy temperature distribution that may result from multi-cellular flow and the details of which are discussed below. Also, the heat transfer is more active in thermal boundary condition of our experiment where the surface temperature decreases upwards, so the heat transfer coefficient is greater than that of the existing correlation. The discrepancy gets larger with increasing heat flux since the natural convection in a vertical cavity is accelerated and the wall-to-wall heat transfer is augmented.

Figure 3 displays the averaged Nusselt numbers (Nu) against Rayleigh numbers (Ra) for all heat fluxes (8~160 W/m^2) and compared with the existing correlations for a rectangular cavity [6] where each vertical surface has a uniform temperature. The height-to-gap ratio for the experimental conditions (H/L) is 8.3. When the heat flux is smaller than 40 W/m^2 or the corresponding Ra is smaller than 1.6×10^8, good agreement is observed between the experiment and correlation because the plate temperatures are

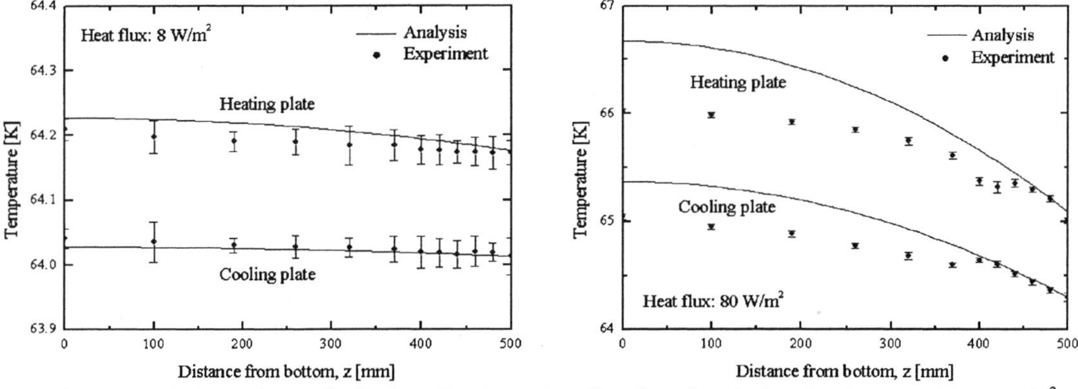

Figure 2 Vertical temperature distribution of heating and cooling plate when the heat flux is (a) 8 and (b) 80 W/m^2

relatively uniform in vertical direction. As the heat flux increases over 40 W/m^2 or Ra exceeds 1.6×10^8, however, the measured heat transfer becomes noticeably greater than the existing correlations. This discrepancy may be explained by the thermal boundary conditions in the experiment where the surface temperatures decrease upwards. Such boundary conditions can cause vertically segregated cellular flows and enhanced heat transfer in the vertical cavity. Once multi-cellular flow occurs the effective H/L ratio is smaller than the actual H/L ratio. Plotted in Figure 3 are correlations for H/L approximately 5 and 3 with the corresponding flow patterns of two and three cellular flow, respectively. These correlations compare more favorably with the experimental data for the high Ra number flows.

The temperature along the streamlines of natural convection is approximately sketched in Figure 4 for the heat fluxes at 8 and 160 W/m^2. When the heat flux is small as shown in Figure 4(a), the surface temperature of each vertical plate is relatively uniform and single cellular flow is formed in the cavity, as the fluid simply ascends along the heating plate and descends along the cooling plate. The heat transfer on the heating plate is active at the bottom and becomes less active along the ascending flow, and the temperature of fluid is getting closer to that of heating plate along the vertical direction. The opposites are true on the cooling plate.

When the heat flux is large as shown in Figure 4(b), however, the temperatures of vertical plates decrease upwards since the cryocooler is located above the vertical plates as shown in Figure 1. The ascending fluid near the heating plate becomes warmer than the surface nearly z = 250 mm, and the heat is then transferred from the fluid to the surface so the density of fluid increases, which makes the cellular flow vertically segregated with the fluid moving downward. The descending fluid near the cooling plate becomes cooler than the surface nearly z = 100 mm, and heat is then transferred from the surface to fluid so the density of fluid decreases, which make the cellular flow vertically segregated with the flow moving upwards.

The number of cells in the multi-cellular flow should be related to the vertical temperature gradient and the wall-to-wall temperature difference. A necessary condition for the cellular flow to be vertically segregated is that the minimum temperature on heating plate must be lower than the maximum temperature on cooling plate. As a result, the segregated positions of fluid are near at the local minimum temperature on heating plate or the local maximum temperature on cooling plate. In Figure 4(b) the second cellular flow is segregated nearly z = 420 mm and the third cellular flow did nearly z = 500 mm on the heating plate. Since the cellular flows are overlapping each other, it is hard to tell the exact height of each cellular flow. However, as

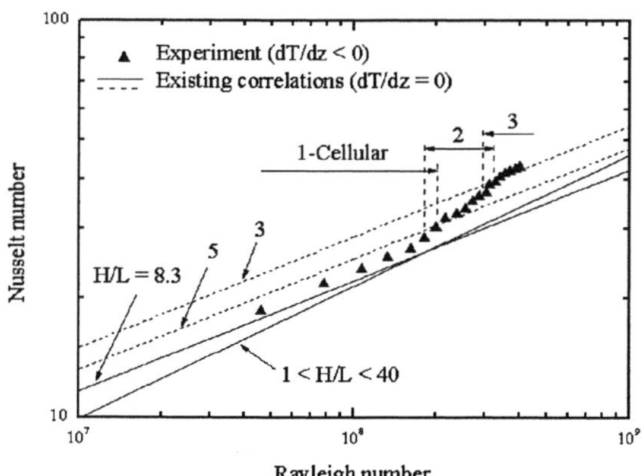

Figure 3 Average Nusselt number versus Rayleigh number

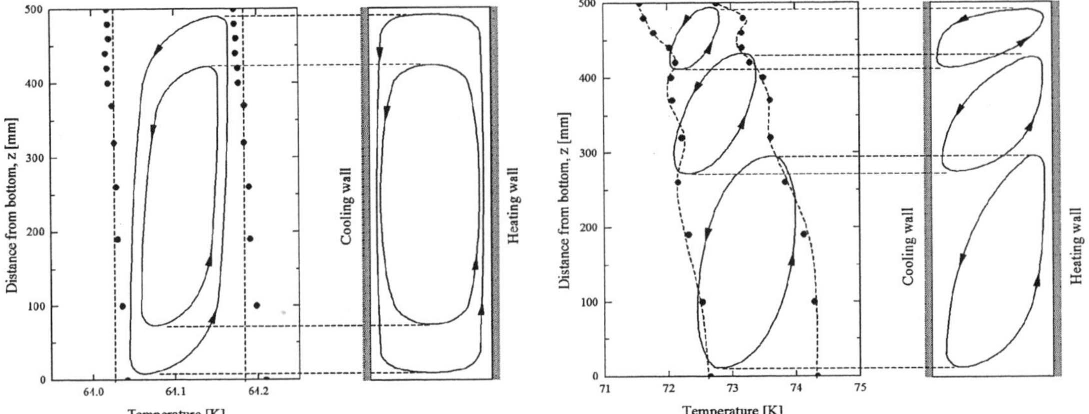

Figure 4 Temperature along streamlines of natural convection when heat flux is (a) 8 and (b) 160 W/m^2

discussed earlier, three cellular flow occurs and the effective H/L ratio is approximately 3 when heat flux is 160 W/m^2.

The multi-cellular flow pattern leads to a wavy temperature distribution along the vertical surface and enhanced heat transfer between top to bottom and plate to plate. As described above the heat transfer is active at the bottom and top when a convection cell is formed in vertical cavity as Figure 4(a). In the multi-cellular flows as Figure 4(b), the heat transfer is active not only at top and bottom but also near at z = 250 and 420 mm, which results in a smaller temperature difference between top to bottom and plate to plate.

CONCLUSION

Experiments were successfully performed to investigate the heat transfer characteristics in a vertical cavity filled with subcooled liquid nitrogen. At high Ra number, the heat transfer coefficients are approximately 20~30 % greater than the existing correlations. The thermal boundary conditions in experiment with surface temperature decreasing upwards may cause vertically segregated cellular flows in the cavity. These multi-cellular flow patterns can lead to the augmentation of wall-to-wall heat transfer by reducing the effective height of the cavity.

ACKNOWLEDGMENT

This research was supported by a joint grant from the CAST in Korea and CAPS with NHMFL in USA.

REFERENCES

1.   Chang, H.-M., Choi, Y.S., and Van Sciver, S.W., Cryogenic cooling system of HTS transformers by natural convection of subcooled liquid nitrogen, Cryogenics (2003) 43 589-596
2.   Choi, Y.S., Chang, H.-M., and Van Sciver, S.W., Natural convection of subcooled liquid nitrogen in a vertical cavity, Advances in Cryogenic Engineering, (accepted for publication vol.49 2004)
3.   Ostrach, S., Natural convection heat transfer in cavities and cells, Heat Transfer-1982 (1982) 365-379
4.   Kimura, S., and Bejan A., The boundary layer natural convection regime in a rectangular cavity with uniform heat flux from the side, Journal of Heat Transfer (1984) 106 98-103
5.   Turkoglu H., and Yucel N., Natural convection heat transfer in enclosures with conducting multiple partitions and side walls, Heat and Mass Transfer (1996) 32 1-8
6.   Barron, R.F., Cryogenic Heat Transfer, Taylor & Francis, Philadelphia, USA (1999) 121-123

Numerical simulation of flow boiling of cryogenic liquids in tubes

Li X.D., Wang R.S.*, Gu A.Z.

Institute of Refrigeration and Cryogenics, Shanghai Jiaotong University, Shanghai, 200030, China

The two-fluid model of two-phase flow, which is included in a commercial CFD code, was used to numerically predict the boiling flows of cryogenic liquid in a vertical tube. Important information, including heat transfer coefficient, void fraction and pressure loss was obtained. The numerical results were in satisfactory agreement with the experimental data available in the literature, which indicated that CFD offers an economical and effective method for cryogenic engineering designing and studies.

INTRODUCTION

Flow boiling is widely encountered in cryogenic systems. Accurate knowledge about cryogenic two-phase flows is very important to the design and optimization of these systems. Experimental studies are very costly and, furthermore, are subject to many limitations. On the contrary, CFD (Computational Fluid Dynamics) simulations may complement plenty of useful information without the considerable expense of experimental facilities. In recent years, CFD is increasingly utilized in studies of cryogenic systems, such as the jobs by Ishimoto [1] and Boukeffa [2]. However, there is no any report on the successful application of CFD in the flow boiling of cryogens. To overcome this, a two-fluid model of two-phase flows, which is included in a commercial CFD code CFX-4 (AEA Technology), was utilized to numerically predict the boiling flows of liquid nitrogen in a vertical tube. The numerical results were in an encouraging agreement with the experimental data available in the literature.

INTERPHASE TRANSFER MODELS

In a two-fluid model, two sets of conservation equations governing the balance of mass, momentum and energy of each phase are solved. Sine the macroscopic fields of one phase are not independent of the other phase, the interaction terms which couple the transport of mass, momentum and energy across the interfaces appear in the field equations.

Inter-phase momentum transfer

Previous studies [3] have indicated that unless inter-phase momentum transfer terms are accurately modeled, the advantage of the two-fluid model over other two-phase flow models disappears and, numerical instabilities result. The inter-phase momentum transfer is usually modeled with the interfacial forces, which include the drag force and the "non-drag" forces. All of these forces are modeled using the correlations recommended by CFX-4 [4].

Inter-phase heat transfer

Heat transfer across a phase boundary is usually described in terms of an inter-phase heat transfer coefficient, and the heat transfer coefficient is usually expressed in terms of a dimensionless Nusselt number, which can be calculated using the Ranz-Mashall [5] correlation.

Inter-phase mass transfer

The mass transfer between phases could be modeled using the RPI boiling model. In the model, the total wall heat flux is split into three parts: heat transfer rates due to convection, due to quenching and to evaporation, among which, the last one determines the mass transfer rate to the vapor phase at the walls. In the interior of the flow, the mass transfer rate depends on the liquid temperature. When the liquid is subcooled, there is a bulk condensation, otherwise, evaporation occurs. In addition, in the issue investigated in the present job, the pressure loss is negligibly small relative to the system pressure, therefore, the saturated temperature and the latent heat of the liquid can be assumed as equal along the tube.

In order to describe the change of the bubble diameter, it is assumed that the bubble diameter varies linearly depending on the liquid temperature between two reference diameters calculated using the equation proposed by Zeitoun and Shoukri [6], as recommended by Tu [7].

NUMERICAL SIMULATION

Experiments in reference [8] were numerically investigated. In the experiments, subcooled liquid nitrogen flows upwards into an electrically heated vertical tube with a diameter of 10mm and a length of 1850mm. According to the geometry of the tube, a full-scale three-dimensional computational domain was built to perform the simulations, and the domain was dispersed into about 150000 meshes in the body-fit coordinate, as shown in Figure 1.

The inlet pressure was 0.7MPa and the inlet liquid subcooling was 7.8K. Other boundary conditions are shown in Table 1.

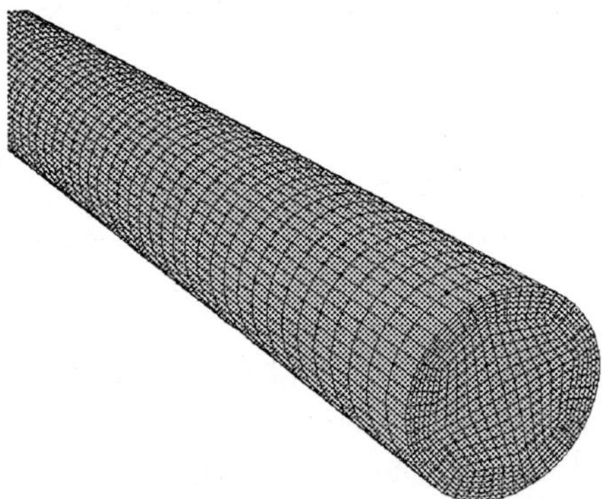

Figure 1   the computational domain and the meshes

Table 1   Boundary Conditions of the Simulations

Case No.	Mass flow rate [kg/(m^{2}s)]	Heat flux [W/m^2]
1	310	17530
2	330	20980
3	310	24710
4	300	29060
5	320	33120
6	320	37510

The two-fluid model was solved using the Inter-Phase Slip Algorithm (IPSA). The turbulence was modeled using the k-ε model and the bubble-induced turbulence in the liquid was modeled using the model proposed by Sato [9].

RESULTS AND DISCUSSION

<u>Heat transfer coefficient</u>

The circumference-averaged heat transfer coefficients along the tube length are given in Figure 2. Where, the solid lines are the simulated heat transfer coefficients and the marks stand for the experimental values. It can be seen that the simulated heat transfer coefficients at all heat fluxes investigated are in satisfactory agreement with the experimental data.

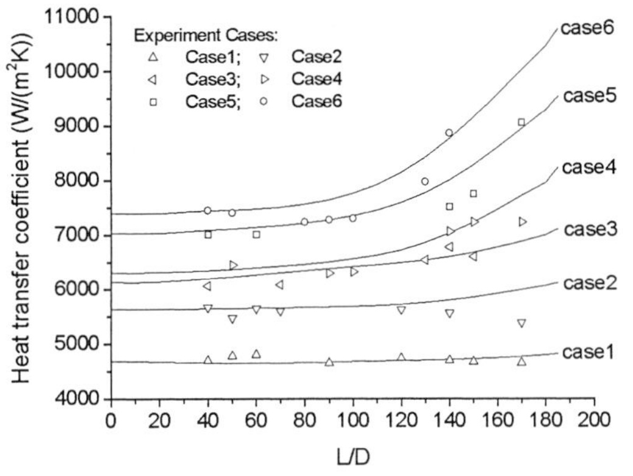

Figure 2   circumference-averaged heat transfer coefficient along the tube length

Figure 3   void fraction along the tube length

<u>Void fraction</u>

Area-averaged void fractions along the tube are shown in Figure3. Since in the experiments, the authors did not measure the void fraction and the pressure loss that will be discussed in the next section, the correctness of these parameters could be evaluated according to the heat transfer coefficients.

Take the related heat transfer coefficients in Figure 1 into consideration, it could be safely concluded that only when the void fraction is higher than a certain value could the heat transfer coefficient of a two-phase flow be obviously higher than that of its single-phase counterpart.

<u>Pressure loss</u>

The pressure losses of the two-phase flows are shown in Figure 4, in which, the maximum pressure loss is 3.2% of the inlet pressure. Under this value of pressure loss, the changes of the latent heat and the saturated temperature are negligible. Therefore, the hypothesis of constant physical properties is acceptable. It also can be found out that the pressure loss over the length decreases as the heat flux increases. This is due to the fact that the gravitational one is dominant among the three components of the pressure loss in the present issue. Since the density of the vapor is much smaller than that of the liquid, as the volume of vapor increases, the gravitational pressure loss gets smaller, which makes the total pressure loss decreases.

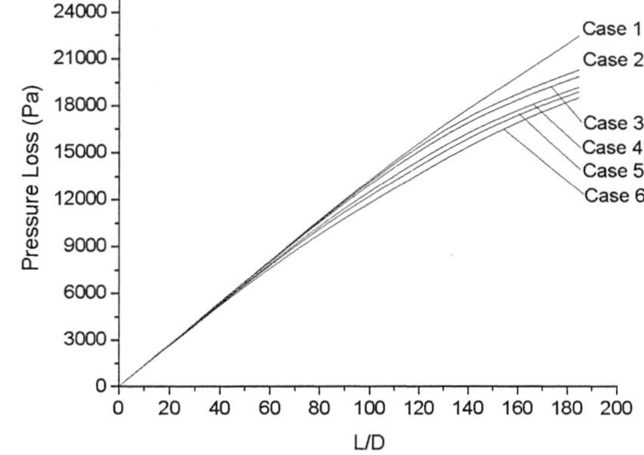

Figure 4   pressure loss along the tube length

CONCLUSION

Flow boiling of subcooled liquid nitrogen in a vertical tube was numerically simulated using a two-fluid model of two-phase flows, which is included in a commercial CFD package CFX-4. Comparison between the simulated heat transfer coefficients and the experimental values available in the literature showed an encouraging agreement. Since other parameters such as the void fraction and the pressure loss, which are not available in the related literature, are essentially related to the heat transfer coefficients, the validity of them therefore could be rationally explained according to that of the heat transfer coefficients.

It also can be concluded from the present job that CFD prediction can offer plenty of important information which is difficult to be measured in experiments, therefore it offers an effective and powerful method for cryogenic engineering designing and studies.

REFERENCES

1 Ishimoto, J. and Kamijo, K., Numerical simulation of cavitating flow of liquid helium in venturi channel, Cryogenics(2003), 43　9-17

2 Boukeffa, D. and Boumaza, M., Experimental and numerical analysis of heat loss in a liquid nitrogen cryostat, Applied Thermal Engineering(2001) 21 967-975

3 Ishii, M. and Mishima, K., Two-fluid model and hydrodynamic constitute relations. Nuclear Engineering and Design (1984) 82 107-126

4 AEA Technology, CFX-4.3 User Guide Volume 3. AEAT, U.K., (1997)

5 Ranz,W.E. and Marshall, W.R., Chemical Engineering Progress(1952) 48 141-148

6 Zeitoun, O. and Shoukri, M., Bubble behavior and mean diameter in subcooled flow boiling, ASME Journal of Heat Transfer, (1996), 118 110-116

7 Tu, J.Y. and Yeoh, G.H., On numerical modelling of low-pressure subcooled boiling flows, International Journal of Heat and Mass Transfer(2002), 45 1197-1209

8 Klimenko, V.V. and Sudarchikov, A.M., Investigation of forced flow boiling of nitrogen in a long vertical tube, Cryogenics, (1983), 23 379-385

9 Sato, Y. and Sekoguchi, K., Liquid velocity distribution in two-phase bubble flow, International Journal of Multiphase Flow(1975), 2 79-95

* Corresponding author. Tel: +86-21-6293 2602

E-mail address: rswang@sjtu.edu.cn (R.S. Wang); leexiangdong@sjtu.edu.cn (X. D. Li)

A density equation for saturated helium-3[*]

Li X.Y., Huang Y.H., Chen G.B., Arp V.[†]

Cryogenics Laboratory, Zhejiang University, Hangzhou 310027, P. R. China
[†]Cryodata Inc. / NIST, Colorado 80027, U.S.A.

An accuracy satisfying density equation of saturated vapor and liquid ^{3}He is obtained by nonlinear regression based on experimental data collected after a thorough survey of literatures. This equation not only can be used to calculate saturated density of ^{3}He independently, but also is to be of great significance in building the equation of state for ^{3}He in both gas and liquid regions. For a better understanding of the equation, the saturated density curves of ^{3}He and ^{4}He are compared.

INTRODUCTION

^{3}He is one of the two stable isotopes of helium in nature. Because of its unique physical properties, ^{3}He has many important applications in various fields. Almost all the applications of ^{3}He involves its thermodynamic properties, therefore scientists and engineers need a convenient, comprehensive, and accurate database for ^{3}He properties in their design and research. From the mid 20th century, studies on properties of ^{3}He have been carried out with intense interests, however, the problem is that the experimental data were scattered in literatures from 1950s until now, and there is no reported work of systematic collection of experimental data and equations covering full regions of gas, liquid and solid. A project on establishing a database for thermodynamic and transport properties of gaseous and normal liquid ^{3}He, is being carried out in Cryogenics Laboratory, Zhejiang University. Although saturated vapor pressure equation was published in 1964, which was used to define the 1962 temperature scale, no accurate saturated density equation in a wide range has been presented. This paper is to introduce the work of collecting experimental data and building the density equation of saturated vapor and liquid ^{3}He, an important step of the whole project. This equation not only can be used to calculate saturated density of ^{3}He and act as a boundary to identify the gas and liquid phase of ^{3}He, but also will be a significant reference for ^{3}He equation in both gas and liquid regions and provide the initials for properties calculation.

^{3}He PHASE DIAGRAM

The atomic weight of ^{3}He is 3.016; the atomic nucleus is made up of 2 protons and 1 neutron; ^{3}He is a colorless, odorless, innoxious, and nonflammable inert gas under normal conditions. Under normal pressures, ^{3}He gas can only be liquefied until the temperature reaches below 3.191K, and it is impossible

[*] Funded by the National Natural Science Foundation (Grant No. 50376055) and the Special Research Fund for Doctoral Training in Universities by the National ministry of Education of China (Grant No. 20010335010)

to obtain solid ^{3}He at atmospheric pressure at any temperature including the absolute zero. Like liquid ^{4}He, liquid ^{3}He doesn't have a triple point and can exist in either of 2 very different states --- the normal liquid and the superfluid liquid; the λ-transition temperature is 2.6mK. The phase diagram of ^{3}He is shown in Figure1 while the characteristic points on the diagram are shown in Table 1. (The dot line on the diagram separates the regions with positive and negative thermal expansion coefficient α.)

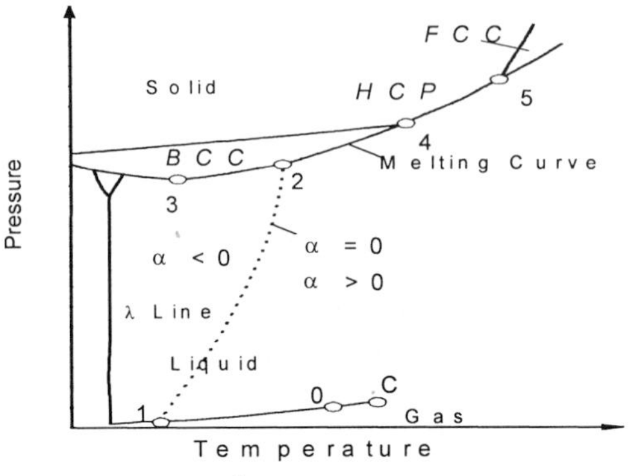

Figure 1 ^{3}He P-T Phase Diagram

Table 1   Characteristic Points Data on ^{3}He Phase Diagram [1]

Characteristic Points	T (K)	P (Mpa)
C (Critical point)	3.324	0.1165
λ Line	0.0026	
0 (Normal boiling point)	3.191	0.101325
1 ($α=0$)	0.502	$0.2736×10^{-4}$
2 ($α=0$)	1.26	4.7623
3 ($P=P_{min}$)	0.32	2.9303
4 (^{3}He Liquid-BCC-HCP equilibrium)	3.138	13.7234
5 (^{3}He Liquid-FCC-HCP equilibrium)	17.78	162.93

Table 2   Density or Specific Volume of Saturated Vapor and Liquid ^{3}He

Source of Data	Temp. Range	Phase and Number
Grilly et al. (1949) [2]	1K-3.34K	Liquid & Vapor (14)
Kerr (1954) [3]	1K-3.3K	Liquid & Vapor (28)
Пешков (1957) [4]	1.4K-3.29K	Liquid & Vapor (12)
Sherman & Edeskuty (1960)[5]	0.8K-3K	Liquid (23)
Kerr & Taylor (1962) [6]	0.2K-3.2K	Liquid (57)
Sherman (1965) [7]	2.4K-3.324K	Liquid & Vapor (28)
Wallace & Meyer (1970) [8]	3.0K-3.310K	Liquid & Vapor (34)
Chase & Zimmerman (1973) [9]	3.2K-3.309K	Liquid & Vapor (43)

DENSITY DATA OF SATURATED ^{3}He

Experimental density data of ^{3}He on the saturated curve are shown in Table 2. Before nonlinear regression we made a preliminary filter to the collected 239 points of experimental data. Considering that the data by Grilly et al. [2] were made relatively earlier when the measuring technique was relatively unreliable and the understanding of ^{3}He properties was just on the horizon, and their data seem to deviate away from other data, it is decided not to use their data. In addition, some data near the critical point with great deviation were also eliminated. After filtering, 205 points of experimental data were left.

DENSITY EQUATION OF SATURATED ^{3}He

After comparison, the nondimensional form of density equation was determined and can be expressed as (the only difference between liquid and vapor branches is the sign before the middle terms):

$$\rho / \rho_c = 1 \pm (c_1\tau^\beta + c_2\tau^{1+\beta} + c_3\tau^{2+\beta} + c_4\tau^{3+\beta}) + c_5\tau + c_6\tau^2 \tag{1}$$

where $\tau = (T_c - T)/T_c$, T_c is the critical temperature, ρ_c is the critical density, $c_1 \sim c_6$ are the constant coefficients obtained by least squares regression, and β is a critical index to describe the density dependence near critical point. As the fitting result is very sensitive to the critical parameters, it is then very important to determine the values of these parameters. After careful comparison, we decide to use critical parameters by Chase and Zimmerman [9]: $T_c = 3.3093K$, $\rho_c = 41.191Kg/m^3$, and $\beta = 0.3653$. The results of regression are shown in Table 3 and Table 4.

Table 3   Technical Results of Regression

Number of observations	205
Convergence tolerance factor	1.00000000E-10
Number of iterations performed	7
Final sum of squared deviations	4.1179188E-3
Final sum of deviations	2.2542236E-1
Standard error of estimate	0.00454896
Average deviation	0.1727264
Maximum deviation for any observation	1.040135
Average relative deviation	0.360%
Maximum relative deviation	1.995%
Number of points with relative deviation larger than 2%	0
Number of points with relative deviation larger than 1%	16
Proportion of variance explained (R^2)	1.0000 (100.00%)
Adjusted coefficient of multiple determination (Ra^2)	1.0000 (100.00%)
Durbin-Watson test for auto-correlation	0.535

Table 4   Calculated Coefficient Values

Coefficient	Value	Standard dev.
c_1	1.34598418	0.002580188
c_2	-0.0738744808	0.02311163
c_3	-0.506778465	0.05545185
c_4	0.225069671	0.03864127
c_5	0.00700440506	0.004916735
c_6	-0.0104634733	0.007278005

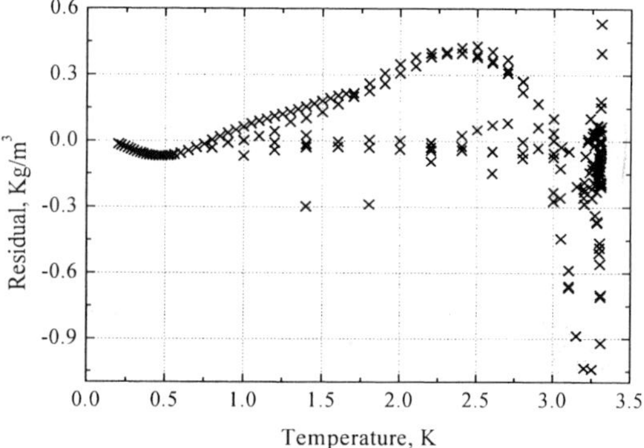

Figure 2   Residual ($\rho_{cal} - \rho_{exp}$) Distribution

By comparing the calculated values obtained by Equation (1) with the experimental data, it is found that the relative error of any observation is within 2%, and the average relative error of all data is 0.360%. There are only 16 points with relative error larger than 1%, in which the maximum is 1.995%. Most of these 16 points are in the critical region close to critical point, because the properties of ^{3}He in this region are very complicated so that the measurement becomes difficult to be accurate. The distribution of absolute residual is shown in Figure 2.

Figure 3 shows all the 239 experimental data of saturated vapor and liquid ^{3}He and the curve generated by Equation (1) of this work. It indicates that this equation can represent the property of ^{3}He density on saturated curve accurately.

SATURATED DENSITY COMPARISON BETWEEN ^{3}He AND ^{4}He

Figure 4 shows the comparison of saturated density curves of ^{3}He and ^{4}He. From the diagram, it is noted that the most significant differences of the two curves are the different critical points which lead to different normal boiling points, different temperature and density ranges. It is also noted that there is an interesting similarity between the two saturated curves so it is then possible to make a scaling from ^{4}He which is relatively known to ^{3}He. This is also a possible work for us in the future.

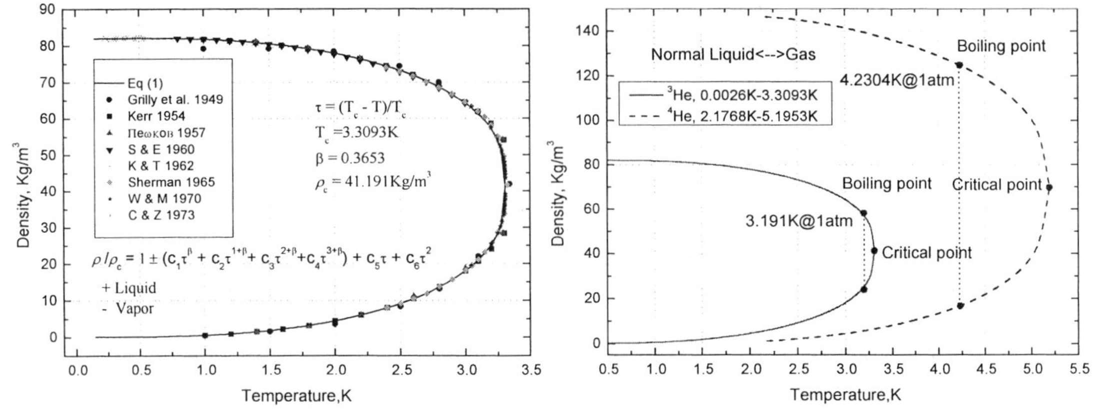

Figure 3 ^{3}He Saturated Density Curve   Figure 4   Saturated Density Curves of ^{3}He and ^{4}He

CONCLUSION

The present work made a collection of density data of saturated vapor and liquid ^{3}He, and obtained an accuracy satisfying density equation. The average relative error of calculated values from experimental values is 0.360%, while the maximum relative error is 1.995%. There are only 16 points out of 205 with the relative error above 1%. This equation not only can be used to calculate saturated density of ^{3}He independently, but also is to be of great significance in building the equation of state for ^{3}He in both gas and liquid regions. The similarity of saturated density curves of ^{3}He and ^{4}He suggests a possibility of scaling.

REFERENCES

1. Timmerhaus, K. D. and Flynn, T. M., <u>Cryogenic Process Engineering</u>, Plenum Press, New York, USA (1989) (The critical parameters in this table are from Ref. 7, which we think are not accurate enough compared to Ref. 9.)

2. Grilly, E. R., Hammel, E. F. and Sydoriak, S. G., Approximate Densities of Liquid He^3 between 1.27° and 2.79°K, <u>Phys. Rev.</u> (1949) <u>75</u> 1103-1104

3. Kerr, E. C., Orthobaric Densities of He^3 1.3°K to 3.2°K, <u>Phys. Rev.</u> (1954) <u>96</u> 551-554

4. Пешков, В. П., Определение плотности He^3 оптическим методом, <u>Ж. Экспер. и теро. физики</u> (1957) <u>33</u>(4) 833-838

5. Sherman, R. H. and Edeskuty, F. J., Pressure-Volume-Temperature Relations of Liquid ^{3}He from 1.00 to 3.30K, <u>Ann. Phys.</u> (1960) <u>9</u> 522-547

6. Kerr, E. C. and Taylor, R. D., Molar Volume and Expansion Coefficient of Liquid He^3, <u>An. Phys.</u> (1962) <u>20</u> 450-463

7. Sherman, R. H., Behavior of He^3 in the Critical Region, <u>Phys. Rev. Letters</u> (1965) <u>15</u> 141-142

8. Wallace, Jr., B. and Meyer, H., [0]Measurement of P-V-T Relations and Critical Indices of ^{3}He, <u>Phys. Rev. A</u> (1970) <u>2</u> 1563-1575

9. Chase, C. E. and Zimmerman, G. O., [0]Equation of State of He^3 Close to the Critical Point, <u>J. Low Temp. Phys.</u> (1973) <u>11</u> 551-579

The characterization of carbon nanofibres based on N_2 adsorption isotherms at 77K

Chao Zhang   Xuesheng Lu   Anzhong Gu

Institute of Refrigeration and Cryogenics, Shanghai JiaoTong University, shanghai, China, 200030

The BJH method is applied to determine the pore size distribution of several carbon nanofibres based on N_2 adsorption isotherms at 77K. The study results show that the carbon nanofibres used in the present study include abundant mespores of 20nm. The adsorption data between 0.01 and 0.99 of relative pressure show features of gas adsorption in mespores adsorbents, and only the mesopores size distribution could be determined with the BJH method. The adsorption data at lower relative pressure (<0.01) must be collected if the micropores size distribution is determined.

INTRODUCTION

The determination of the structural proprieties of porous adsorbents is important in the field of gas and liquid adsorption. Indeed, the pore shape, the pore dimensions and the chemical structure of the walls are the main characteristics to be considered when studying the potentiality of a given adsorbent to fit a given application. The problem of relating such basic structural properties to the thermodynamic and kinetic behavior of a system involving the considered adsorbent and a complex liquid or gas of industrial interest is far from being solved. Carbon nanofibres are porous adsorbents specially developed for hydrogen adsorption storage [1]. The microstructure characterization of carbon nanofibres is critical for its application in hydrogen storage and understanding of supercritical hydrogen adsorption mechanism. In this paper gas adsorption method [2-6] is used to the determination of pore size distribution of carbon nanofibres.

EXPERIMENTS

The materials used in the present study are carbon nanofibres made from acetylene with flowing catalytic method [7]. The FESEM images are shown in Figure 1. The sample was respectively purified with nitric acid and sulfuric acid oxidation method (qb5), kali permanganate, water and sulfuric acid oxidation method (qb7), carbon dioxide oxidation method (qb2), water vapor oxidation for 4 hours (qb8) and water vapor oxidation for 5 hours (qb9). The N_2 adsorption isotherms at 77K have been measured with Micromeritics ASAP2000 auto-analytic instrument. The adsorption isotherms are typical type-II isotherms, as showed in Figure 2.

874

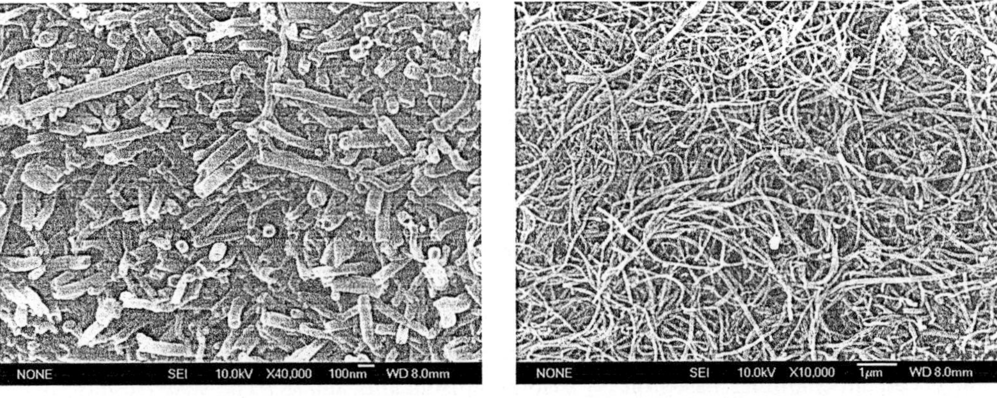

Figure1. The FESEM images of the carbon nanofibres made from acetylene

DETERMINATION OF MESOPORE SIZE DISTRIBUTION

The BJH method is used to determine the mesopore size distribution of carbon nanofibres. The BJH method is based on Kelvin equation and Halsey equation, and it has been proved that the BJH method could acculately determine the mesopores size distribution [3]. The detailed calculation equation is shown in reference [8]. The calculated results are shown in Figure 3. As shown in Figure 3, all the carbon nanofibres include abundant mesopores of 20nm, and the difference is just the volume of 20nm mesopores. The calculated results are consistent with the FESEM images of samples, and prove that different purification methods could not change the basic structure of carbon nanofibres, and only get rid off impurifies in different degrees. The volume of 20nm mesopores of carbon nanofibres purified with water vapor oxidation for 5 hours is maximal, then it indirectly justify the validity of water vapor oxidation.

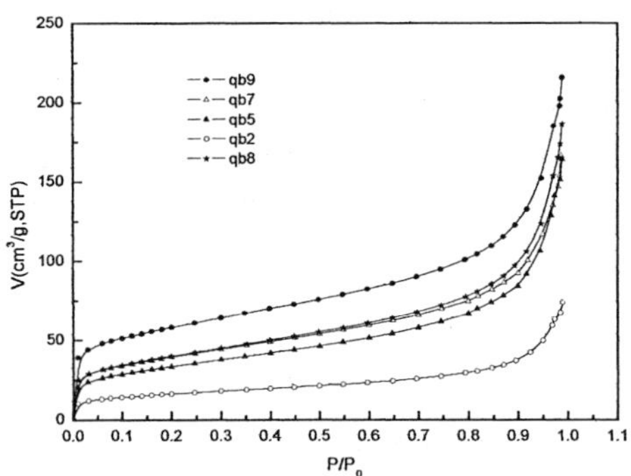

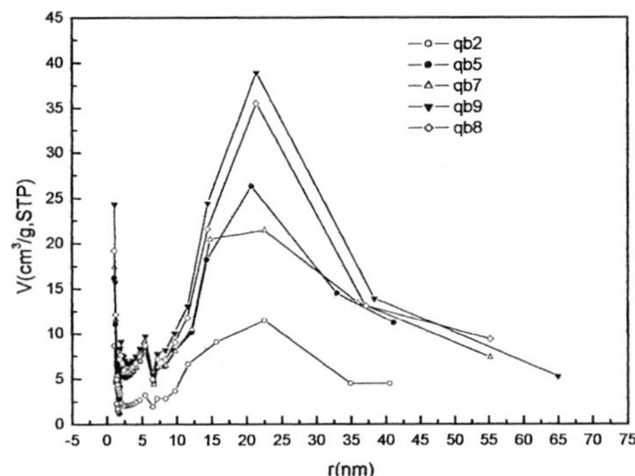

Figure2. The N_2 adsorption isotherms at 77K   Figure3. The pore size distribution of carbon nanofibres

DETERMINATION OF MICROPORE SIZE DISTRIBUTION

For porous adsorbents including micropores and mesopores, at lower adsorption pressure, gas adsorption may occur both in micropores and mesopores at the same time. So the adsorption isotherms in micropores must be determined before characterization of micropores size distribution. At some adsorption pressure, the adsorption amount can be described as,

$$V = V_{mic}{}^{0}\theta_{mic} + \frac{V_{mes}{}^{0}V_{s}(0.4)}{V_{s}^{0}} \frac{V_{s}}{V_{s}(0.4)}$$

(1)

Which $V_{mic}{}^{0}$ is the max adsorption in micropores, θ_{mic} are defined as micropore filling degree, and $V_{mes}{}^{0}$ is the adsorption of single molecular layer in mesopores. $V_{s}{}^{0}$ and V_{s} is respectively the saturation adsorption of single layer and the adsorption in nonporous adsorbents, and $V_{s}(0.4)$ is the adsorption in nonporous adsorbents when relative pressure is 0.4. The data of N_2 adsorption in nonporous adsorbents is shown in references [9-11]. When the relative pressure is enough, the micropores are filled and the adsorption V and the $V_{s}/V_{s}(0.4)$ is in line relationship. Then the relative pressure while the micropores are filled, the $V_{mes}{}^{0}V_{s}(0.4)/V_{s}{}^{0}$ and the micropores volume $V_{mic}{}^{0}$ could be determined with the line slope and intercept. So the adsorption isotherms may be corrected according to equation (1). The relationship between the adsorption V and the $V_{s}/V_{s}(0.4)$ of the N_2 adsorption in carbon nanofibres is shown in Figure4.

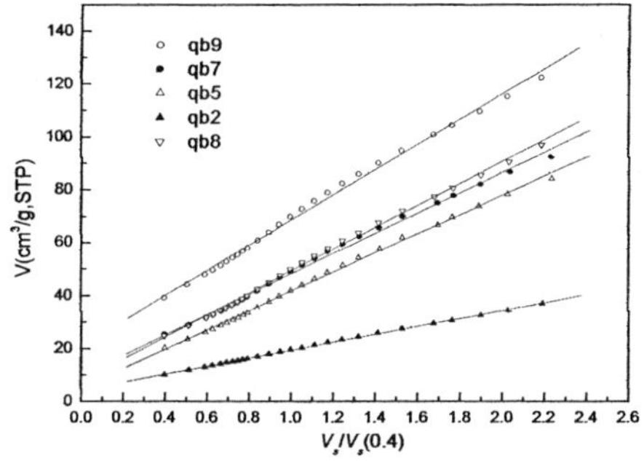

Figure4. The relation curves of the V and the $V_{s}/V_{s}(0.4)$ of N_2 adsorption in carbon nanofibres

As shown in Figure 4, the adsorption data are all on a line among measured pressure, which is the feature of gas adsorption in mesopores. But the lines do not go through the origin of coordinate, which proves that the adsorption of N_2 in micropores occurs. But at the point of the initial measure pressure the micropores are already filled. So the pressure when the micropores are filled could not be determined. But we can get the micropores volume according to the intercept, as shown in Table1. The maximal micropores volume is only 20.94cm³/g (STP, Standard Temperature and Pressure). Because the adsorption data at the relative pressure less than 0.01 could not be obtained with equipments, so the micropores size distribution can't be determined. But it is unnecessary to determine the micropores size distribution because of little micropores volume.

Table1 The micropore volume of carbon nanofibres

Materials	Micropores Volume (cm³/g, STP)	Materials	Micropores Volume (cm³/g, STP)
Qb9	20. 94	Qb7	9. 77
Qb5	5. 22	Qb2	4. 47
Qb8	7. 81		

CONCLUSION

The mesopores size distributions are determined with BJH method based on N_2 adsorption isotherms at 77K. The results show that the carbon nanofibres includes abundant mesopores of 20nm, which is consistent to the FESEM images of samples. By correcting the original adsorption isotherms, it is found that among relative pressure of 0.01~0.99, the adsorption data show features of gas adsorption in mesopores. And the micropores volume are determined with the intercept of the line between V and $V_s/V_s(0.4)$. The micropores volume of carbon nanofibres purified by water vapor oxidation for 5 hours is maximal, about $20.94cm^3/g$. And in order to determine the micropore size distribution, the adsorption data at relative pressure less than 0.01 must be gotten. But it seems unnecessary to determine the micropores size distribution because of little micropores volume.

ACKNOWLEDGEMENT

This work was supported by the Special Funds of the Science and Technology Development of Shanghai City in China. (#0116nm044)

REFERENCES

1. Kaylene A, Siegmar R, Michael H, etc., Carbon nanostructures: An efficient hydrogen storage medium for fuel cells? Fuel Cells Bulletin (2001) 4(38) 9-12

2. Carrott P.J.M., Ribeiro Carrott M.M.L., Evaluation of the Stoeckli method for the estimation of micropore size distributions of activated charcoal cloths, Carbon (1999) 37 647-656

3. Valladares D.L., Rodriguez Reinoso F. and Grablich G.Z., Characterization of active carbons: The influence of the method in the determination of the pore size distribution, Carbon (1998) 36(10) 1491~1499

4. Stoeckli F., Guillot A., Hugi-Cleary D., Slasli A.M., Pore size distributions of active carbons assessed by different techniques, Letters to the editor/Carbon (2000) 38 929~941

5. Bereznitski Y., Gangoda M., Jaroniec M., and Gilpin R.K., Adsorption Characterization of Active Carbons Modified by Deposition of Silica, Langmuir (1998) 14: 2485~2489

6. Carrot P.J.M., Ribeiro Carrott M.M.L., Mays T.J., Comparison of methods for estimating micropore sizes in active carbons from adsorption isotherms, Fundamentals of adsorption (1998) 6 677-682

7. Huiming Chen. Carbon nanotubes-synthesis, Microstructure, Properties and Applications, Chemistry industry Press. Beijing, China (2002)

8. Jiming Yan, Qiyuan Yan. Adsorption and condensation- surface and porous of solid, Science publisher, Beijing, China (1979)

9. Carrott P.J.M., Roberts R.A. and Sing K.S.W., Standard nitrogen adsorption data for nonporous carbons, Carbon (1987) 25(6) 769~77

10. Mietek Jaroniec, Michal Kruk, Standard Nitrogen Adsorption Data for Characterization of Nanoporous Silicas, Langmuir (1999) 15 5410~5413

11. Carrott P.J.M., Roberts R.A. and Sing K.S.W., Adsorption of nitrogen by porous and non-porous carbons, Carbon (1987) 25(1) 59~68

Experimental investigation into storage of confined cryogenic liquids without evaporation venting

Huang Z., Wang R., Shi Y., Gu A.

Institute of Refrigeration and Cryogenics, Shanghai Jiao Tong University, SH 210003, PR China

Abstracts: In this study, a two-phase thermo-dynamical model was presented to evaluate the heat transfer and pressurization of cryogenic liquids in a closed container, and programs were formulated to predict the effects of various factors on the cryogenic storage performance. Experiments were carried out to verify the simulation. The experimental results agreed well with the theoretical prediction. The research helps to design long-term cryogenic vessels with minimum evaporation loss while abiding safety standards.

INTRODUCTION

Cryogenic liquids storage vessels are widely used in aerospace, transportation, and energy industries. Boil-off vapor is a great concern on cryogenic storage. It will concentrate in the vapor part of the inner tank, and will be eventually drained before causing tank overpressure. Substantial mass and cost savings could be achieved if evaporation venting can be avoided. Furthermore, inflammable, explosive or toxic liquids must be stored in a closed container without venting. Therefore, it is very important to study the storage feathers of confined liquids. Many researches dealt with the problem of thermal stratification and interface instability of propellants [1-2]. This study handles general cases of cryogenic liquid pressurization for engineering practice.

EXPERIMENTAL SETUP AND PROCEDURES

The experimental setup was schematically shown in Fig. 1. The setup consisted of a cryogenic vessel, a vacuuming system, data acquisition system and auxiliary facilities. The vessel was a 2-cubic meter, hi-vacuum multi-layer insulated (MLI) pressure vessel. The vacuuming system mainly included an oil diffusion pump and vacuum meter. Data acquisition system included computer, computer-based data logger, sensors and connection cables. All the sensors, viz. thermocouples, pressure transmitter, level meter, and flow meter, were linked to a Keithley data logger, whose voltage outputs were sent to PC computer via a GPIB bus.

Temperature profile of cryogenic liquids was recorded with copper-constantan thermocouples. In all the tests, the tank was initially filled with liquid nitrogen saturated at pressures close to local atmospheric pressure. The wall heat flux was then obtained from a boil-off calibration and the test begun immediately afterward. The experimental data was considered after the thermal equilibrium of system. During the test, a continuous record of the various sensors reading was made.

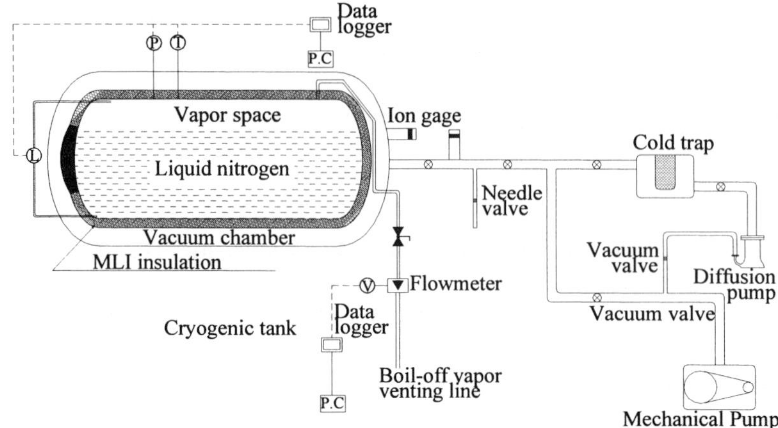

Figure 1   Experimental setup (L-level meter, P-pressure transmitter, T-thermocouples, V-flow meter)

RESULTS AND ANALYSES

Analytical models

The heating rate of MLI tank is usually less than 10 W/m², which is much too small for nucleate boiling to occur [3]. Confined cryogenic liquid is heated by inner shell. The liquid next to the container surface is warmed and move upward via a boundary layer causing slight circulation in the bulk liquid. A warmer stratified layer collecting around the vapor-liquid interface and continually thickness increasing result in a higher temperature at the surface than in bulk liquid. Some parts of the layer evaporate into vapor chamber; the other parts lose heat and move inward [4].

Confined cryogenic liquid storage can be divided into 2 successive phases: the steady expansion of cryogenic liquids, and the liquid pressurization at a constant volume. The initial cryogenic liquid in the tank is assumed as saturate and isothermal. Heat input to the liquid through the inner wall gives rise to the increase of temperature, pressure and enthalpy of the cryogenic contents. The liquid temperature rise results in its expansion. Taking the cryogenic contents as a control volume, we have energy equation 1.

$$Q = q\tau = M(u_f - u_i)$$

(1)

Where, u_f, u_i is the final and initial enthalpy of cryogenic contents in the tank, respectively; Q is the heat leakage to the inner tank. Fluid density and inner energy can be derived from the pressure of cryogenic vessels. Hence, the final inner energy and density takes the forms of equation 2 and equation 3, respectively. Thus, enthalpy increase will be equation 4.

$$u_f = \rho_{lf}\phi_f u_{lf} + \rho_{vf}(1-\phi_f)u_{vf},$$

(2)

$$\phi_f = \left.(\rho - \rho_{vf})\middle/(\rho_{lf} - \rho_{vf})\right.,$$

(3)

$$Q = V\left[\rho_{lf}u_{lf}\left.(\rho - \rho_{vf})\middle/(\rho_{lf} - \rho_{vf})\right. + \rho_{vf}u_{vf}\left.(\rho_{lf} - \rho)\middle/(\rho_{lf} - \rho_{vf})\right.\right] - V\left[\rho_{li}\phi_i u_{li} + \rho_{vi}(1-\phi_i)u_{vi}\right],$$

(4)

where, ρ and m is the density and mass of cryogenic contents, respectively; V is tankage, ρ is constant.

The confined liquid was heated in a constant volume when the tank was 100% full. The tank is assumed as a rigid body without deformation, the liquid enthalpy increase will be

$$\int_{\tau_0}^{\tau} q F_w \, d\tau = \rho \int_0^{V_L} c_v (T - T_0) \, dV$$

(5)

where, q is the heat flux to the cryogenic liquids; τ is the storage duration; F_w is the surface area of inner tank shell; C_v is the specific heat at constant volume; ρ is the liquid density; V_l is the tank volume and T is the liquid temperature. The above equations are numerically solved, and results are as follows.

Results and discussion

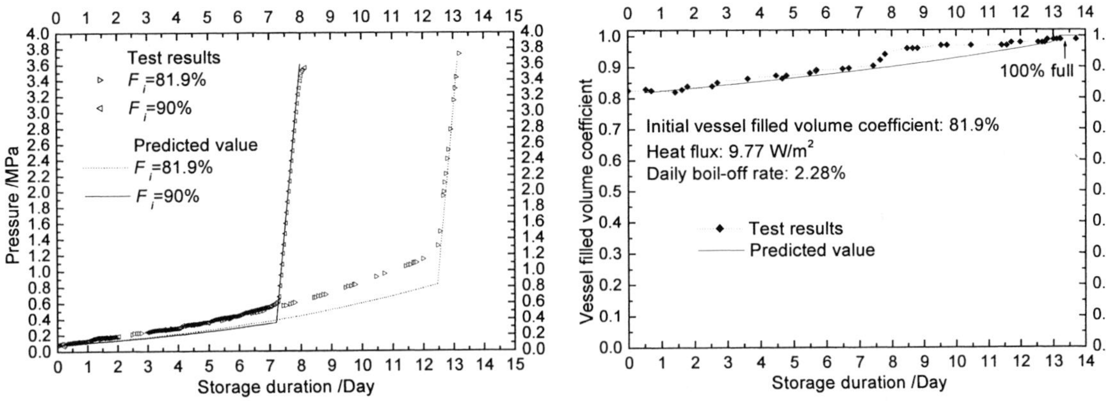

Figure 2   Pressure-time correlation

Figure 3   Variation of vessel filled volume coefficient

The pressure, the vessel filled volume coefficient, and the temperature of confined LN$_2$ is recorded as in Fig. 2, Fig. 3 and Fig. 4, respectively. In conformity with the former expectation, 100% full is the turning point of pressurization. Confined LN$_2$ expands steadily and tank pressure jumps after zero ullage. Furthermore, there is a clear temperature gradient in the LN$_2$, the maximum temperature difference is about 6 ℃. The interface temperature is higher than that of liquid core. The LN$_2$ temperature is lower than its saturate temperature corresponding to the tank pressure. The cryogenic contents are consumed as saturate and homogenous, so the predicted pressure is lower than the test pressure. However, the test vessel filled volume coefficient is higher than the predicted due to the surface flashing and instability.

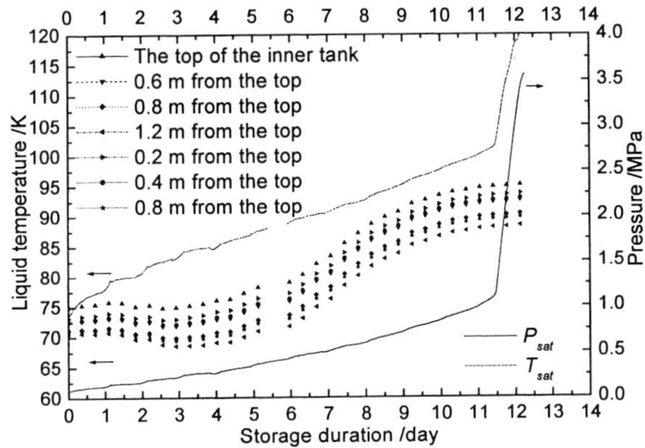

Figure 4   Temperature profiles of liquid nitrogen

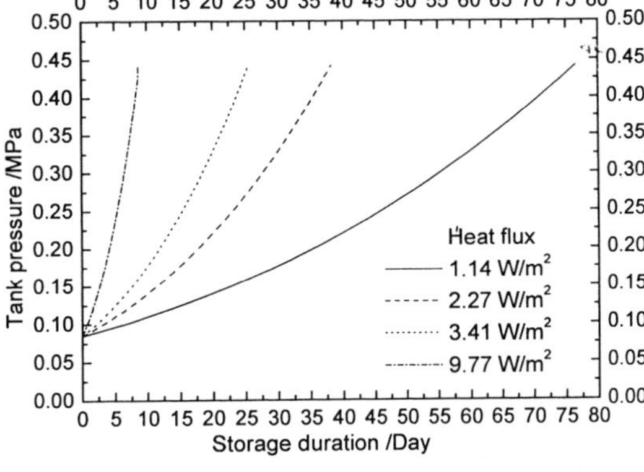

Figure 5   Heat flux effects on the pressurization

The confined cryogenic liquids pressurization is affected by various factors, viz. heating rate, initial vessel filled volume coefficient, and thermodynamic properties of cryogenic liquids. As shown in Fig. 5, the larger heating rate will lead to further pressure increase. The storage cycle will double if the heating rate halves. Therefore, insulation quality advance will effectively prolong storage cycle. As shown in Fig. 6, the cryogenic liquid temperature increase will cause its volume expansion. Higher Initial vessel filled volume coefficient will be quicker to reach the 100% full of cryogenic tank. However, if the initial vessel filled volume coefficient is low enough, the liquid content will decrease till the 100% full of vapor. Fig. 7 shows initial vessel filled volume coefficient and properties effects on storage cycle of cryogenic liquids. Given the heating rate is 9.77 W/m², the storage cycle fluctuates with initial vessel filled volume coefficient and liquid properties. The storage cycle will be the largest if the initial coefficient is 60%. Compared with liquid nitrogen, liquid oxygen, argon, and methane have longer storage cycle, while liquid hydrogen has much shorter storage cycle.

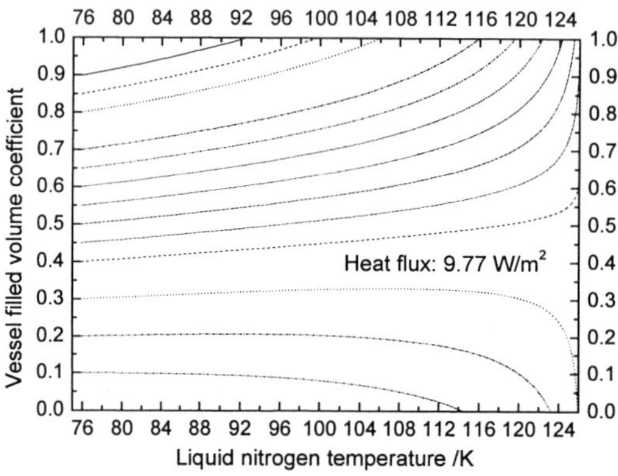

Figure 6   Vessel filled volume coefficient vs. T_l

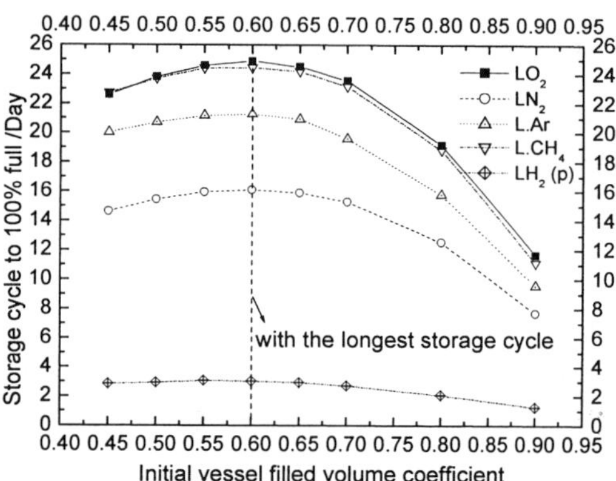

Figure 7   Property effects on liquid storage cycle

CONCLUSION

The 2-phases thermodynamic model can be used to predict the storage duration of cryogenic liquids with a fair agreement. Test tank pressure and ullage is slight higher than predicted value. Confined liquid temperature is lower than its saturate temperature, and the liquid-vapor interface temperature is higher than that of other parts.

Various factors affect the vent-less storage duration. The higher heating rate will cause shorter storage cycle. Vessel filled volume coefficient affects the storage duration of cryogenic tanks. For the experimental tank, its optimum Vessel filling volume coefficient is about 60%. Storage performance also varies with the properties of cryogenic liquids. Oxygen and methane have much longer storage cycle than nitrogen, while hydrogen storage cycle is quite shorter than that of nitrogen.

REFERENCES

1.   Anderson John E., Fester Dale A. and Czysz Paul M., Evaluation of long-term cryogenic storage system requirements, Proceedings of the 1989 Cryogenic Engineering Conference UCLA, Los Angeles, California, USA (1989) 795-803
2.   Hochstein John I., Hyun-Chul Ji and Aydelott John C, Prediction of self-pressurization rate of cryogenic propellant tankage, J PROPUL POWER (1990) 6 11-17
3.   Beduz C., Rebai R. and Scurlock R.G., Thermal overfill and the surface vaporization of cryogenic liquids under storage conditions, Advances in Cryogenic Engineering 29A, Clark A.F. and Reed R.P., Plenum Press, New York (1984) 795-803
4.   Rongshun Wang and Anzhong Gu, A study of drastic shock-resistant cryogenic tank, Chinese Ph.D. dissertation, Institute of Refrigeration and Cryogenics, SJTU (2001) 95-100

Thermal Stress Analysis of Cryogenic Adhesive Joints for Non-Magnetic Dewar

Zhu H. M., Jin T. X., Xu L., Sun H., Xiao Y. M.

Institute of refrigeration and cryogenics, Shanghai JiaoTong University, ShangHai 200030, China

The thermal stress analysis of screwed-adhesive joints of the non-magnetic Dewar was carried out in this paper. The serious stress concentrations occurred at the adhesive free ends and the external thread roots of the lower adhesive joint. The effects of LN_2 volume fraction and screw pitch on the stress distribution of the adhesive joints were discussed. The results provided herein some important references for the design of the non-magnetic Dewar.

INTRODUCTION

SQUID is the most sensitive magnetometer, which can detect 10^{-16}T magnetic field. It has been widely applied in nondestructive examination, scanning SQUID microscope and satellite, etc. The non-magnetic Dewar can provide the cryogenic and non-magnetic environment, which is quite necessary for SQUID. The different parts of the non-magnetic Dewar mainly made of glass fiber reinforced plastics (GFRP) are adhesively bonded. The brittleness of adhesive will be increased at cryogenic temperatures. In addition, the differences of the thermal expansion coefficients and Young's moduli of adhesive and adherends can produce high thermal stress when temperature changes. These may cause adhesive crack and destroy the vacuum tightness. Therefore, it's necessary to perform the thermal stress analysis of the adhesive joints.

Although many studies on thermal stress analysis of adhesive joints have been reported already, the thermal stress analysis at cryogenic temperatures is very rare. Takao and Qiang [1] tested the strength of the adhesive joint between metal and FRP and carried out thermal stress analysis in 1991. Gorbatkina and sulyaeva [2] investigated the effect of cyclic cooling from ambient temperature down to liquid nitrogen (LN_2) temperature on the shear adhesive strength of fiber/polymer joints.

Since the adhesive joints in the non-magnetic Dewar require high strength and high vacuum tightness, the screwed-adhesive joint, rectangular thread-adhesive joint and bellows-adhesive joint are applied in the non-magnetic Dewar. However, the thermal stress analysis of these adhesive joints has not been reported yet. Thus, the present work is to carry out thermal stress analysis for the screwed-adhesive joint used extensively in the non-magnetic Dewar at cryogenic temperatures, and discuss the effects of LN_2 volume fraction and screw pitch on the stress distributions of the adhesive joints.

JOINT CONFIGURATION AND FINITE ELEMENT MODEL

The main dimensions of the non-magnetic Dewar and screwed-adhesive joints are shown in Figure 1. An adhesive thickness t=0.2mm was applied in the study. The screw pitch was 1.5mm for the first analysis. In production of adhesively bonded joints, the adhesive layer is squeezed out and accumulated around the free ends of the adhesive layer, called adhesive fillets. It has a considerable effect on the peak adhesive stresses arising at the adhesive free ends. Here, the shape of adhesive fillet was idealized to a triangle with

882

a height and a width twice the adhesive thickness (f_t=0.4mm). In order to avoid the stress singularities, the corners of adherends were rounded with a radius r of 0.2t [3] as shown in Figure 1. The pitch angle of the screwed-adhesive joints in the non-magnetic Dewar usually is less than 2°. In this case, the effect of the pitch angle on the load distribution along threads can be neglected [4]. Therefore, the thermal stress analysis can be simplified to a two-dimensional axi-symmetrical problem.

The finite element software ANASYS was used for the thermal stress analysis. An eight-noded quadrilateral axi-symmetric plane element was used to model the Dewar. Since the stress distribution in the adhesive region changes greatly, mesh areas of the adhesive was refined as shown in Figure 2. The sequential coupling method was used to solve the thermal stress problem. The main properties of adhesive and GFRP was from literatures.

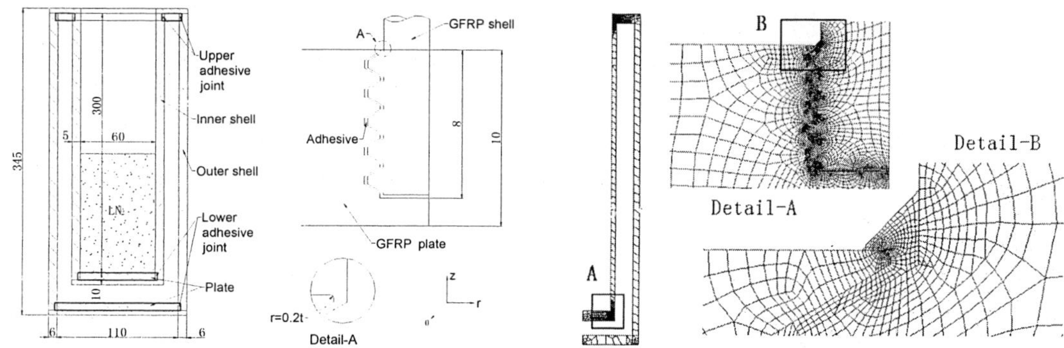

<table>
<tr><td>

Figure 1   Dimensions of the Dewar and Joints

</td><td>

Figure 2   Mesh details of the FEM model of the Dewar

</td></tr>
</table>

THERMAL ANALYSIS AND STRESS ANALYSIS

Thermal analysis

The screwed-adhesive joints experience the thermal load caused by the LN_2 in the Dewar. The LN_2 volume fraction was assumed to be 80% in the first analysis. The initial temperature through the Dewar was at 300K. The thermal boundary conditions are as follows: the lower part of the inner surface subjected to LN_2 was at 77 K; the upper part of the inner surface exposes to nitrogen whose temperature distribution is linear from 77K to 250K which was pre-measured by thermocouple; the outer surface of the outer shell exposes to air at 300K. The natural convection takes place at the later two boundaries and the heat transfer coefficient was calculated according to [5]. The steady-state thermal analysis was carried out and the temperature distribution calculated is shown in Figure 3. It is found that great temperature gradient occurs at the upper part of the inner shell. However, the temperature difference between the final temperature and the initial temperature at the lower joint is much larger than that of the upper.

Stress analysis

The Dewar experience three loads, the thermal load obtained from thermal analysis, the weight of LN_2 and the atmospheric pressure. The applied structural boundary condition is that the outer surface of the bottom of the Dewar was fixed in the z-direction. In the analysis, the joint members, i.e. GFRP shells and plates and adhesive, were assumed to have linear elastic properties at cryogenic temperatures..

The radial stress distributions of the upper and lower screwed-adhesive joints are compared in Figure 4 (The unit of all stress is Pa in this paper). Figure 4 (a) and (b) are for the upper and lower adhesive regions, respectively. It is found that the stress in the lower adhesive region is greater than that in the upper, because the temperature difference of the lower adhesive joint is much larger than that of the upper. The serious stress concentrations occur around the adhesive free ends of both the upper and lower adhesive joints. In addition, the high stress is observed at the external thread roots of the lower screwed-adhesive joint and is more evident through thread close to the bottom of the joint. Similarly, the GFRP also experiences stress concentrations in the corresponding regions. However, the stress distribution of the upper adhesive joints is relatively even.

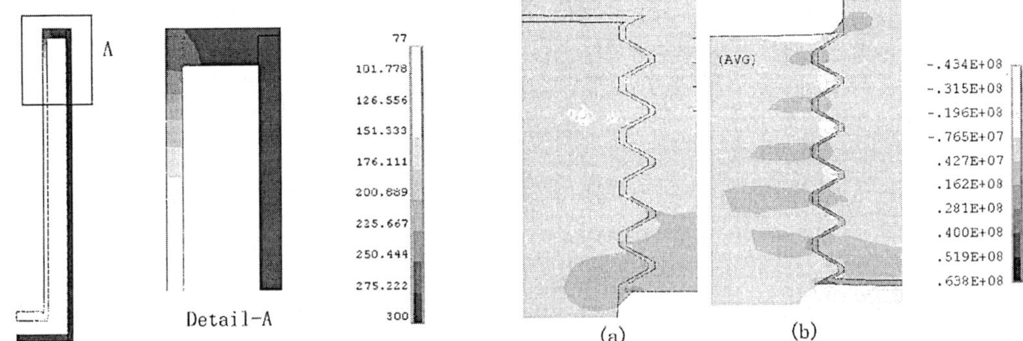

Figure 3   Temperature distribution of the Dewar   Figure 4   Comparison of σ_{rr} distribution for upper and lower joints

The stress distributions of different directions of the lower adhesive joints were investigated. Figure 5 (a), (b) and (c) are the radial σ_{xx}, hoop $\sigma_{\theta\theta}$ and shear σ_{rz} stress distributions of the lower adhesive region, respectively. From Figure 5 and Figure 4 (b), it is found that the σ_{zz} stress is dominant while the σ_{xx}, $\sigma_{\theta\theta}$ and σ_{rz} stresses are similar, but still serious level. The stress concentrations are observed at the adhesive free ends for all the directions. The distributions of the σ_{zz} and $\sigma_{\theta\theta}$ stresses along the thread are relatively even while those of the σ_{rr} and σ_{rz} stresses change greatly. The σ_{rz} stress reaches higher level through the thread close to the top of the joint. Since the high stress concentrations occur around the adhesive free ends and external thread roots, the distributions of the σ_{rr} and σ_{zz} stresses of free ends and the first thread of the lower screwed-adhesive joint are plotted in Figs. 6 (a) and (b), respectively.

As a result, it is concluded that the free end of the lower adhesive joint is the most critical region of the whole Dewar. The high stress occurs at the external thread roots of the lower screwed-adhesive joint and is more evident close to the bottom of the joint.

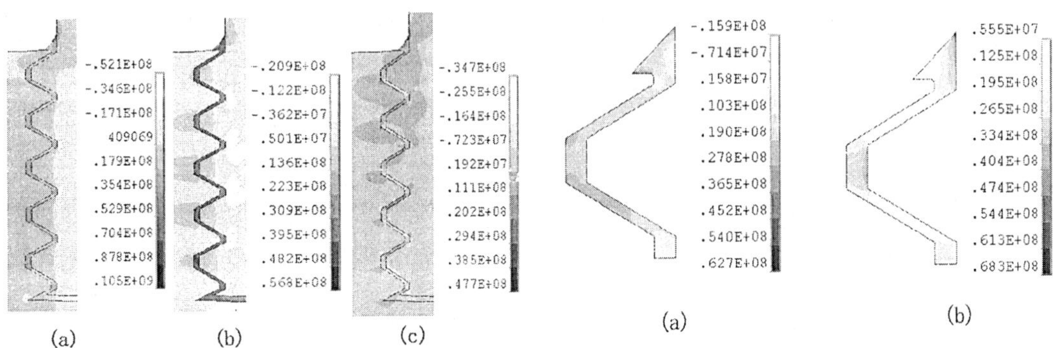

Figure 5 σ_{zz}, $\sigma_{\theta\theta}$ and σ_{rz} distribution of lower region   Figure 6 σ_{rr} and σ_{zz} distribution near the fillet for lower joints

EFFECTS OF LN$_2$ VOLUME FRACTION AND SCREW PITCH ON STRESS DISTRIBUTIONS

Effect of LN$_2$ volume fraction on stress distributions

The LN$_2$ volume decreases continuously during the usage process. Hence, the effect of fraction of LN$_2$

884

volume on stress distribution was studied. Comparison of the Von-Mises stress distributions of the lower adhesive joint for variable volume fractions is shown in Figure 7. It is found that the stress distribution of the lower adhesive joint hardly changes. This is because the temperature difference which causes thermal stress changes very little when LN_2 volume fraction decreases. Therefore, it can be drawn that the effect of LN_2 volume fraction on stress distribution in the adhesive region is very minor.

<u>Effect of screw pitch on stress distributions</u>
The screw pitch is an important parameter of the screwed-adhesive joint. Comparison of the Von-Mises stress distributions for variable screw pitches under same thermal and mechanical boundary conditions is plotted in Figure 8. In case of the 3mm screw pitch, the stress distribution is very similar to that with 1.5 mm screw pitch, and is relatively lower. The main stress concentration occurs at the free end of the adhesive and the stress at external thread root is larger than that of other places of thread.

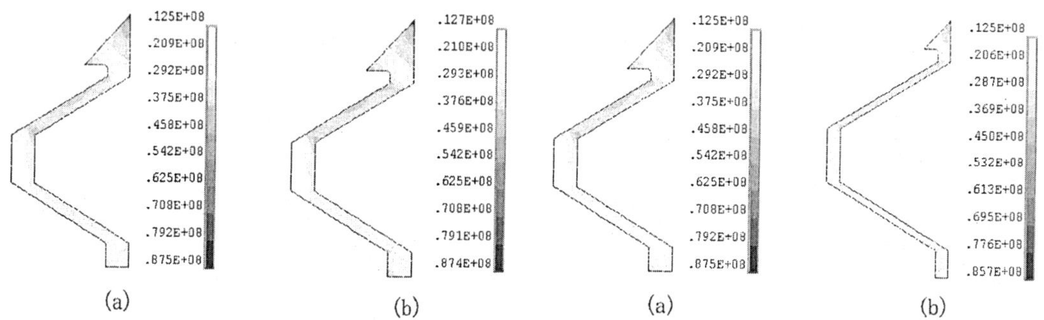

Figure 7   Von-Mises stress distributions for (a) 80% LN_2 and (b) 40% LN_2 volume fraction

Figure 8   Von-Mises stress distributions for (a) 1.5mm and (b) 3mm screw pitch

CONCLUSION

In this study, the thermal stress analysis of screwed-adhesive joints of the non-magnetic Dewar was conducted. The serious stress concentrations occur in the adhesive free ends, and the stress of the lower adhesive joint is higher than that of the upper. Thus, the most critical region is the adhesive free end of the lower adhesive joint of Dewar. Besides, the stress at the external thread roots of the lower adhesive joint is larger than that of other places of the thread. The maximum Von-Mises for 1.5mm screw pitch will be larger than for 3mm screw pitch whilst the stress distributions keep similar. The LN_2 volume fraction has negligible effect on the stress distribution of the adhesive joints. The results reported in this study can provide some important references to the design of the non-magnetic Dewar.

REFERENCES:

1. Mori, T., Yu, Q., Takahana S., Shiratori M., Low temperature strength of metal-FRP bonded joints, <u>JSME Int. J. Ser I: Solid Mech.</u> (1991), <u>34</u> 257-263
2. Gorbatkina, Y .A., Sulyaeva, Z. P., Strength of the fiber/thermoplastic-matrix interface under cyclic cooling to low temperatures, <u>Compos. Sci. Technol.</u> (1997), <u>57</u> 995-1000
3. Apalak, M. K., Günes, R., Fidanci, L., Geometrically non-linear thermal stress analysis of an adhesively bonded tubular single lap joint, <u>Finite Elem. Anal. Des.</u> (2003), <u>39</u> 155-174
4. Zhao H., Analysis of the load distribution in a bolt –nut connector. <u>Comput. Struct.</u> (1954), <u>53</u> 1465-147
5. Zhao Z. N. In: <u>Heat Transfer.</u> Higher Education Press, Beijing, China(2003)

Development of cryogenic structural materials

Li L.F.[1], Zhang Y.T.[2], Li Y.Y. [2]

1 Technical Institute of Physics and Chemistry, Chinese Academy of Sciences, Beijing, China
2 Institute of Metal Research, Chinese Academy of Sciences, Shenyang, China

During the last decades, cryogenic structural materials have experienced a sustaining development by institutes, universities as well as industries. Besides cryogenic metals and alloys, significant progress has been made in cryogenic composites and cryogenic ceramic materials. The world development in cryogenic material research and applications is overviewed. Firstly, an overview of the applications of cryogenic materials will be presented. Cryogenic materials are being used in a wide variety of fields, such as space technology, high energy particle accelerator, fusion reactor development, cryocoolers and gas industry. Secondly, the microstructures and properties of cryogenic structural materials will be discussed. Among the cryogenic structural materials, metallic base materials take on a predominant role. The chemical composition and the mechanical properties of different alloys at low temperatures as well as their advantages and disadvantages are compared with each other. Composites and ceramic materials are the prospective structural materials for cryogenic application and their microstructures and properties as well as potential applications at cryogenic temperatures are also discussed in this paper.

INTRODUCTION

Cryogenic structural materials are used at temperatures below 120K in order to meet the development of cryogenic engineering.

Cryogenic engineering started near the end of the 19th century. The main task during the first half of the 20th century was in connection with the building up of the modern gas industry through the application of cryogenic separation of gaseous mixtures. Once started, this technology of air separation played crucial roles in the development of the steel industry and the chemical industries[1]. As refrigeration techniques improved and liquid gas became readily available, suitable materials were needed for storage dewars, industrial processing equipment, and refrigerators[2].

In 1911 Onnes discovered the superconductivity, which stimulated the fundamental studies of condensed matter physics and superconductors. Both low temperature superconductivity (LTS) and high Tc superconductivity (HTS) attracted great interests for researchers. LTS superconductors are used extensively today in many applications where high magnetic fields are required, including superconductive magnets for controlled thermonuclear fusion, for nuclear magnetic resonance, for high energy physics research, for magnetic refrigeration, for Maglev system as well as for magnetic ore separation. For HTS[1], studies today concentrate on the material side, which is certainly the fundamental issue. Claims about future applications abound, some are being tested, and it takes time, at least, to check on their practicality. In applied superconductivity, high strength/thermal conductivity ratio, adequate toughness, and weldability alloys are required, and also for insulating purpose, nonmetallics, structural composites were also developed.

Another chief application field of cryogenic structural materials is the space technology. Cryogenic materials specific for the rockets and the spacecrafts has been largely developed in some countries in the world. Use of liquid oxygen and liquid hydrogen fuels, led to the development of high strength, lightweight materials and heat insulating composites, Chinese efforts in space technology started in late 1950s. Manned spacecraft with LH_2/LO_2-engine was launched in 2003..

Although many materials have been successfully used for a wide range of structures and devices that operate at low temperatures, in recent years, many large cryogenic engineering projects bring new requirements on materials. This is a new challenge for cryogenic materials R&D, the practical way is how to design and how to tailor present materials, upgrading them to a higher level by combining the considerations of good performance, environmental-friendly, and less-cost. With the developments of the advanced material synthesizing method, the new structure characterization and computer simulation technology, it will be possible to realize this task.

In this paper, we will give an overview on applications and developments of cryogenic structural materials. Focusing on some specific applications, metal, composites and ceramics materials will be discussed by emphasizing mechanical and thermal properties as well as microstructures.

APPLICATIONS OF CRYOGENIC STRUCTURAL MATERIALS

Table 1 is a brief summary of cryogenic structural materials used in different fields. It indicates the order of preference with asterisk marks.

Table 1 Various of cryogenic structural materials used for different fields

Fields	Structural steels	Al alloys	Other alloys	Composites	Ceramics
Space ● Ground and rockets	★ ★ ★ ★	★ ★ ★			
● Spacecraft	★	★ ★ ★	★ ★	★ ★	★
Superconducting engineering	★ ★ ★ ★	★ ★ ★		★ ★	
Gas industry	★ ★ ★ ★	★ ★ ★ ★			
Cryocoolers	★ ★		★ ★	★	★

<u>Materials used for space technology</u>
Materials used for space technology mainly involve in large power launch vehicles for liquid oxygen and liquid hydrogen storage tanks and rocket engine components such as valves, turbopump, conduits, bearings and seals. Space cryogenic materials include metal and composites materials. For metal materials, the high strength/weight ratio, high toughness, weldability, high resistance to hydrogen

brittleness at cryogenic temperature are required. Typical materials are Inconel 718, Ti-5Al-2.5Sn, Ti-6Al-4V, Al-Mg alloys, Cu-Zr alloys. There were lots of Ti-alloys used in USA Apollo project for liquid hydrogen tank, conduit and high pressure vessels. Inconel 718 was also widely used for space shuttle of USA, H-2 rocket of Japan as key materials for LH_2/LO_2 engine in recent year.[3] For polymer composite materials, high specific strength, high heat insulation value, high modulus and low density is required. Typical materials are GFRP, CFRP, polyimide, epoxy and CFCs (CFC-11 has high ozone depletion potential and thus, many companies such as Solvay Germany, Elf Atochem France and Allied Signal USA are searching new substitutes, for example, HFC-245fa、HFC-365mfc and liquid CO_2).

Space cryogenic materials are highly related to the special application environments, including temperature, pressure, load, medium and medium state, so that the research activity on mechanical, physical properties under practical using environment is very important. Since 1960, researchers from US, USSR, China, France, Japan, et al. have made a lot of investigations on mechanical properties as strength, modulus, fatigue, toughness, creep, impact, and on thermal properties as expansion, thermal conductivity, specific heat as well as on electromagnetic properties, cryotribology and hydrogen embrittlement. Current research mostly focuses on how to further decrease the weight and how to decease environmental pollution. Attempts to use composites instead of metallic alloy for LH_2/LO_2 tank and use low ozone depletion potential value materials instead of CFCs are carried up.

Structural materials in applied superconductivity

Structural materials are used for structural support and electrical insulation in large superconducting devices. In addition to HTS power cable, most of applications focus on superconducting magnets. Structural materials related to superconducting magnet technology, are mainly austenitic stainless steels and composites.

For fusion application, nowadays, many projects are on establishing in the world, such as ITER (multi-country joined project), LHD-Japan, EAST-China and KSTAR-Korea. Austenitic stainless steels and epoxy base resin composites as candidates will be used for supports and insulating materials. For instances, both ITER and EAST have decided to use 316LN as CICC conduit and winding case materials, 316L as upper and lower supports, and resins developed in Rutherford Appleton Laboratory as vacuum pressure impregnation (VPI) insulation. It is estimated that hundreds of tons of austenitic stainless steels and epoxy resins will be used for EAST device.

For particle accelerator, there are also many big projects on establishing, such as LHC (CERN), TESLA(DESY), CEBAF (USA), VLHC(USA), KEKB(Japan), BEPCII (Beijing), and SSRF (Shanghai). Similar to fusion applications, structural materials used for high field magnets are also austenitic stainless steels and composites.

Structural materials for gas industry

The need to transport and store large quantities of liquefied natural gas (LNG) led to the development and use of tough 9Ni steel and Al alloys that are less expensive than the austenitic stainless steels . The safety and long service lifetime demanded of these large containers required an increased understanding of materials properties at low temperatures.

In addition to traditional materials, modern gas industry brings materials to some new requirements because modern gas industry has been extended to many fields. For example, Japan's WE-NET(the world energy network system) projects are aiming at constructing clean energy systems using hydrogen as fuel, focusing on the development of hydrogen production, transport, storage and utilization since 1993 They have established goals for Japan with respect to fuel cell vehicles and stationary fuel cell systems of 50,000 vehicles by 2010, and 5 million vehicles by 2020. The R & D work on various of properties of austenitic stainless steels and Al alloys under liquid hydrogen environment has been carried up these

888
years for WE-NET projects[4].

Structural materials for cryocoolers

Cryocoolers, especially small scale coolers become much more important in novel refrigeration technology since they will be deeply involved in information technology.

Basically, cryogenic environment is essential to high quality information handling. As we keep finding new information carriers, bearing the signature of quantum physics, turning up in physics research, we should expect the field of 'cryoelectronics' to become widened with 'electronics' in a broad sense. We might also expect that more subtle cryogenic backup will be required besides the cooling function, and cryogenic structural materials will be part of the 'electronics' system[1].

We have to accept the facts that there really exist some barriers between person engaged in cryocooler investigations and person engaged in materials and /or physics studies. The former does not pay more attention on the development of materials and selections on proper materials. The latter, does not know much about refrigeration and its appropriate requirements on materials. In recent years, ICMC and CEC are regularly held together, and sometimes they were also held together with ICEC. These connections are helpful to reduce the obstacles between the two subjects. We are glad to see some progresses on pulse tube cooler in which some components are instead by ceramic materials and composites[5-6]. An attempt to use toughening ceramics to fabricate the rotary valves in G-M cooler is undergoing[7].

MICROSTRUCTURES AND PROPERTIES OF CRYOGENIC STRUCTURAL MATERIALS

Cryogenic structural alloys

Commonly used cryogenic structural alloys are austenitic stainless steels and Al-alloys. Ti-alloys is under further developing and the application is very limited.

a) Austenitic stainless steels

The austenitic stainless steels are Fe-Cr (16-26%)-Ni (8-24%) alloys sometimes with Mn and N in order to stabilize the austenitic phase. They have higher ultimate strength, toughness at cryogenic temperatures coupled with reliability, ease of fabrication, and good service. Their disadvantages are that they are more expensive and their machinnability is poorer than that of Al-alloys. According to the strengthening mechanism, austenitic alloys can be classified as basic type (AISI 300-series), nitrogen-strengthened type (21-6-9,22-23-5), and precipitation-hardened type (A286, Inconel 718). Inconel series are nickel based alloys, however, their cryogenic mechanical behaviors are very similar to austenitic stainless steels.

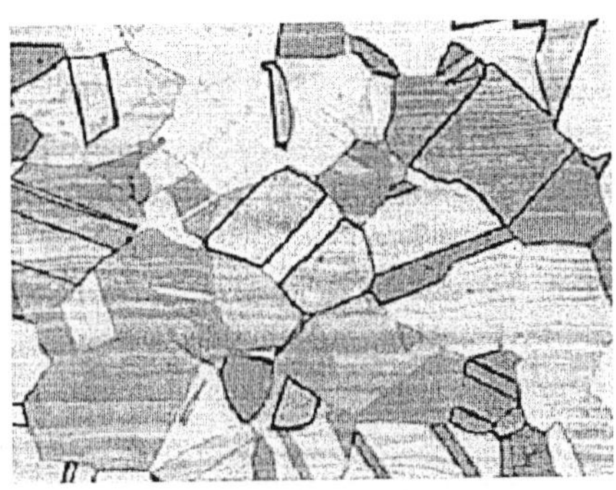

Figure 1 Microstructure of austenitic phase[8]

Table 2 Typical compositions of austenitic stainless steel

Type	Cr (%)	Ni (%)	Mn (%)	N (%)	C (%)	Others (%)
304	18-20	8-12			<0.08	
304LN	18-20	8-12	2.0	0.10-0.16	< 0.03	
316	16-18	10-14			< 0.08	2-3Mo
316L	16-18	10-14			< 0.03	2-3Mo
316N	16-18	10-14	2.0	0.10-0.20	<0.08	
316LN	16-18	10-14	2.0	0.10-0.20	< 0.03	
21-6-9	19-21.5	5.5-7.5	8-10	0.15-0.40	< 0.08	
A286	14-16	23-27			< 0.08	1-2Mo
JBK-75A	15.0	29.0			0.015	Al0.3,Ti 2.4
Inconel 718	19.0	bal	0.35	Mo3.0 Nb5.0 Fe18.0	0.08	
Incoloy 907	-	37.4		-		Co 13.0

Figure 2　Microstructures of Alloy 21-6-9 [9]: (a) After solution at 293K, (b) Lath martensites after deformation at 4K

Table 3 Tensile properties of 316LN and 21-6-9

Alloy	Temp. (K)	Yield Strength (MPa)	Tensile Strength (MPa)	Charpy Impact (J/cm^2)	Elongation (%)	Reduction of Area (%)
316LN	298	318	626	477	67	84
	77	802	1364	277	59	71
	20	964	1624	285	51	61
	4.2	1067	1592	283	51	58
21-6-9	298	384	701	484	68	79
	77	881	1360	289	66	70
	20	1076	1596	300	52	61
	4.2	1163	1576	293	51	59

Typical compositions of austenitic stainless steels are summarized in Table 2. Basic microstructure of austenitic phase is shown in Figure 1. Meta-stability austenite steels deformed or cooling at low temperature, will experience stress induced martensitic transformation, so as to enhance the strength and decrease toughness. Figure 2(a) is the austenite steel 21-6-9 after solution treatment. Figure 2(b) shows lath martensites after deformation at 4K for the same steel..

Mechanical properties of austenitic steels are excellent at cryogenic temperatures, one can easily get relevant the data from many handbooks. Here listed some materials' data obtained recently from IMR and TIPC (see Table 3).

b) Al-alloys

Commonly used aluminum alloys are Al-Cu (2000-series, Cu: 4-6.5%), Al-Mg (5000-series, Mg:2.5-6%), Al-Mg-Si (6000-series, Mg: 0.6-1%, Si: 0.6-1%) as well as Al-Li alloys developed in 90s. They have been used for many structural applications such as rocket propulsion tanks (LH_2, LO_2), LNG tanks, and superconducting structural supports.

Aluminum alloys have moderate strength, high toughness and ductility at cryogenic temperatures. They have low density, nonmagnetic behavior, stable microstructure and good machinability. The disadvantages of aluminum alloys are low strength in weldments, low elastic modulus, and high thermal expansion, which limited their use in large structure, such as land-based storage tanks. Typical microstructure of Al-Li alloy is shown in Figure 3.

Generally, Aluminum alloys strengthening mechanisms are precipitation-hardened. Their ultimate strength is in a level of 300-700MPa, yield strength 150-520MPa, no big change at cryogenic temperatures.

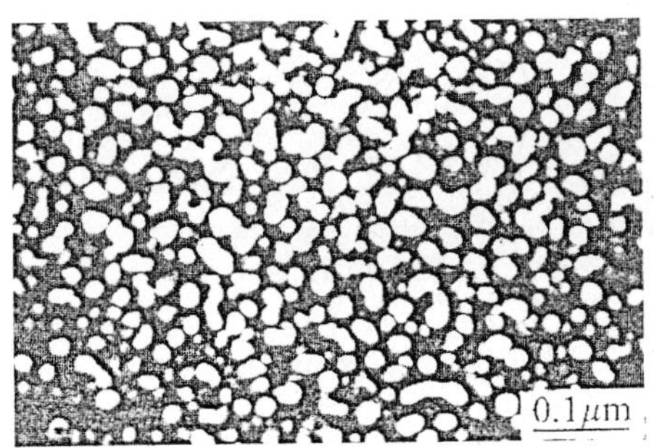

Figure 3   Microstructure of Al-Li alloys[10]

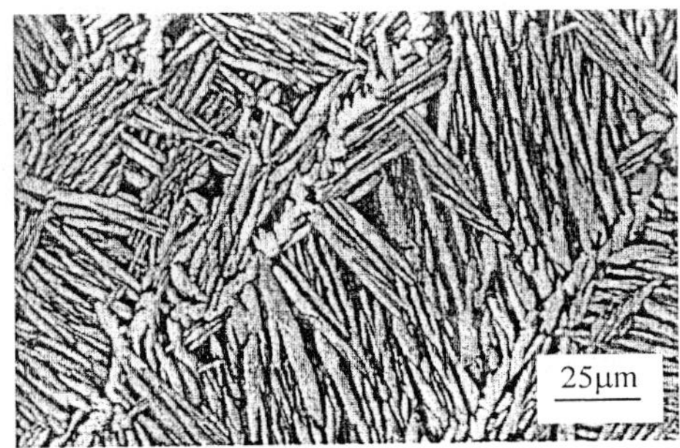

Figure 4. Micrograph picture of Ti-6Al-4V alloy showing α+β phase[11]

c) Titanium alloys

Titanium alloys are classified as three types according to their compositions. They are α type (Ti-5Al-2.5Sn), β type (Ti-13V-11Cr-3Al), and α+β type (Ti-6Al-4V). Among them, Ti-6Al-4V alloys have excellent strength, ultimate strength in a level of 1200MPa (RT), 1600MPa (77K), 1900MPa (4K), and yield strength in a level of 1000MPa (RT), 1300MPa (77K), 1800MPa (4K). Typical micrograph of Ti-6Al-4V alloy is shown in Figure 4. Although Ti-alloys have high strength both at room and cryogenic temperatures, the use at cryogenic temperatures has been limited to aerospace applications where strength to weight ratio is primary concern. This is due to their high costs and difficult to fabricate. In addition, Ti-alloys have low ductility, low fracture toughness. So it is limited to use widely.

Studies on metallic base materials in the future will be concentrated on enhancing their mechanical properties to another order of magnitude through grain-refined, homogenized and purified, so as to satisfy the requirements of large cryogenic engineering.

Cryogenic Composites

Composites always act as both structural and functional materials used for cryogenic technology. Superconducting technology, space technology, and fusion technology push the development of cryogenic composites since 1970s. It is mainly due to that composites can satisfy the requirements of reduced weight and increased electrical/thermal insulations. The advantages of composites can be described by using strength to thermal conductivity and density ratio, i.e., $\sigma/(\lambda.\rho)$, the higher the ratio value, the less refrigeration cost. The ratio of composites is about 2 order of magnitude over stainless steels. However, composites also have many disadvantages[12-13]: low toughness, heterogeneous strength and high thermal contraction, neutron radiation damage and low bonding strength between fiber and matrix.

Composites are classified as fiber reinforced and particle reinforced plastics. Commonly used fibers are E-glass, S-glass, Kevlar, boron, carbon, alumia; particles are SiC, SiO_2, Al_2O_3; matrices are epoxy, polyester, Teflon, polyimide. Glass fiber reinforced plastics (GFRPs) are well-developed, commercial products are G-10, G-10CR, G-11CR. Carbon fiber reinforced plastics still need to be improved. It is noted that some materialists are trying to use new materials as reinforced agents (such as carbon nanotube) and new technology (such as nanotechnology). It has been shown in laboratory scale tests that the physical properties and performance of composite materials can be tremendously improved by the addition of small percentages ~2% of carbon nanotubes. However, there have not been many successful for large applications that show the advantage of using nanotubes as fillers over traditional carbon fibers. The main problem is in dispersing the nanotubes. The other problem is not match well between the nanotube and the polymer matrix [14-15]. Plasma polymer coating on fibers or particles might be a good method to solve the dispersing problem (see Figure 5).

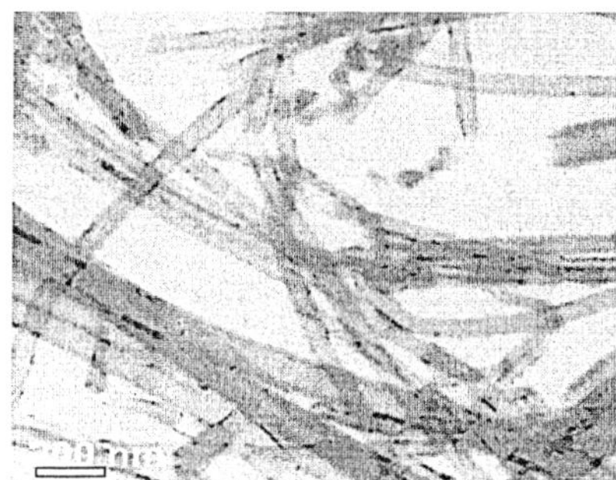

Figure 5 Thin polymer film plasma coating of the carbon nano-tubers (left). The magnification picture (right) shows the thickness of coating[15].

Recently, under the application backgrounds of EAST project and HTS power cable project in China, researchers in TIPC (Technical Institute of Physics & Chemistry, CAS) are carrying on the modification of epoxy resins and polyimide films by the addition of various shapes and sizes of nanometer SiO_2 particles and fibers (see Figure 6). Mechanical property tests on these synthesized composites show that there are more than 30% increases both in tensile strength and modulus. For HTS power cable, polyimide insulating films shall satisfy the matching of the thermal expansion to that of Ag-shield BSCCO conductor at ~77K. This work is undergoing at TIPC. A novel type of toughening epoxy resin for low temperature application was studied.

Another significant progress developed in TIPC is about active toughening epoxy resin[16]. An active toughening component was used to improve the properties of bisphenol-A-epoxy resin and form two-phase structure in the reaction. A complex amino cure agent was investigated to control the inhomogeneity of molecular cross-links in the chemical structure, in which the combination of stiffness and flexibility decreased the sensitivity to temperature, so that the system can be used at low and middle high temperature. The excellent bonding at interface between the dispersed phase and matrix phase is performed by adding the ether-chain coupling in the curing reaction. The applied temperature region is from 4.2 K to 373 K. The mechanical property such as lap shear strength is 20 MPa at 4.2 K. The glass transition temperature (Tg) is at 393 K. The coefficient of liner thermal contraction is 3.88×10^{-5} (1/K) in the region of 300K to 77K. It has the excellent vacuum hermetic property and the leakage rate is less than 1×10^{-8} Pa M^3/S. It has been successfully applied in EAST project of IPP, CAS.

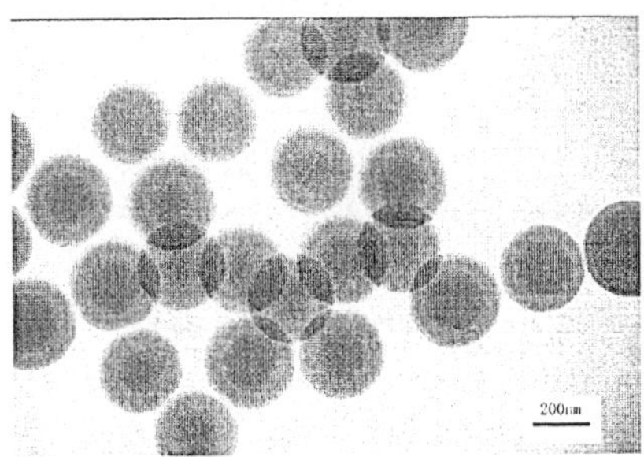
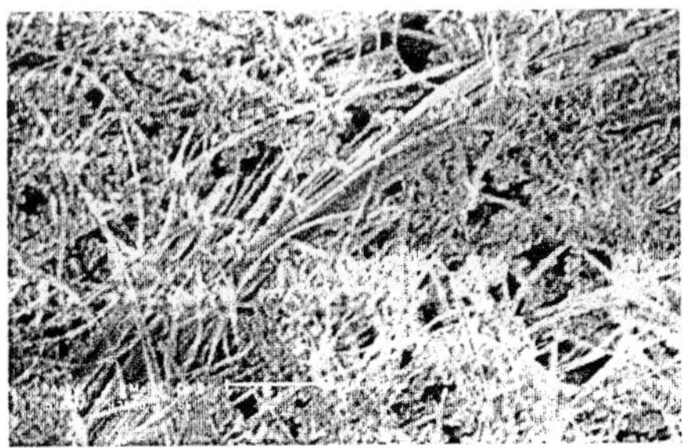

Figure 6 Fillers to be added to epoxy resin and polyimide, the left picture is the uniform spherical nano silica particles by sol-gel technology; the right picture is the silica tubes by self assembly.

Cryogenic ceramic materials

Ceramic material is comparatively a new approach. Comparing ceramics to FRPs and metals, ceramics appear to be feasible as cryogenic structural materials because they have high strength, elastic moduli and low thermal expansion coefficients, as well as low thermal conductivity and high electrical resistivity. However, the brittleness of most ceramics seems to be a significant disadvantage in engineering application. The ZrO_2 based system is suitable for gains in toughness and strength because the character is stress-induced toughening. Since the phase transformation temperature can be held significantly below room temperature by carefully controlling the chemical composition and grain size, the toughening and strengthening are able to occur at cryogenic temperatures.

Table 4 Property dada of some materials (the average values of 4-77K)

Materials	Density ρ (g/cm^3)	Strength σ (MPa)	Thermal conductivity λ(W/m.K)	$\sigma/(\rho.\lambda)$	K_{1c} (MPa.m$^{1/2}$)
14.5Ce-Zirconia	5.77	730	0.6	211	12
16.5Ce-Zirconia	5.69	720	0.6	211	13
50CeZTA	4.62	710	0.6	256	6.5
CFRP-T60	1.50	1050	0.15	4667	3-5
GFRP-P	2.00	978	0.29	1686	3-5
SUS 304	8.03	490	5	12	300

Experimental results obtained by researchers at Cryogenic Laboratory of TIPC, CAS, show a great increase of fracture toughness and strength from 298 to 4.2K for ceria stabilized tetragonal zirconia polycrystals (Ce-TZPs)[17], and then researchers at Japan also found similar results in yttria stabilized tetragonal zirconia polycrystals (Y-TZPs)[18]. Mechanical properties of some ceramics, steels and composites are compared in Table 4. It is clear that the ceramic materials have moderate values both on ratio of strength to thermal conductivity and fracture toughness. In fact, they have already been examined in small scale cryocoolers[5-7]. In final, we give two micrographs of ceramic materials in Figure 7.

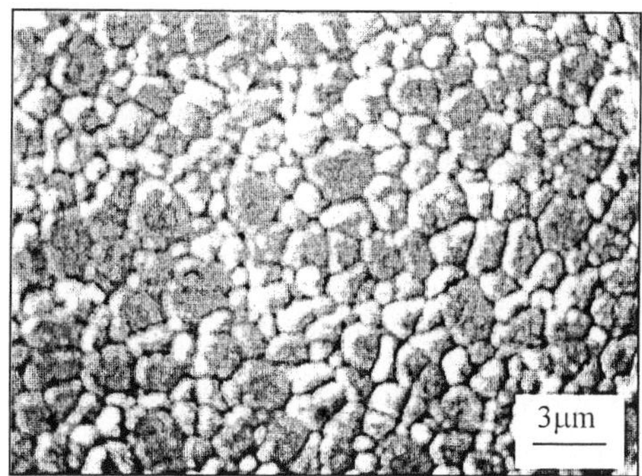

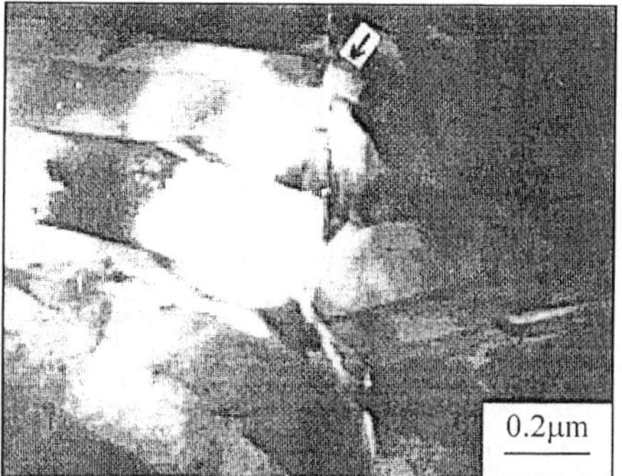

Figure 7 Micrographs of ceramics. The left one is SEM picture of the as sintered Ce-TZP ceramics. It shows high sintering density with tetragonal phase and the average grain size is about 1 micrometer. The right one is TEM picture on the fracture surface of Ce-TZP ceramics fractured at 4.2K, which shows that the crack propagation was blocked by t to m phase transformation.

SUMMARY

Among the cryogenic structural materials, metal cryogenic materials are still in a predominant position. However, how to decrease the grain size and increase the strength without change of the composition is the aim of metallic scientists. So, the developments of metal materials in the future are purification, homogenization and grain-refinement. Composites and ceramics will be prospective structural materials for cryogenic applications in the future. Anyhow, the cryogenic structure materials have to continue into a higher level.

ACKNOWLEDGMENT

We would like to thank Prof. C.S. Hong for valuable discussions.

REFERENCES

1. Hong, C.S. , "Cryogenics for China Tomorrow? --a reverie--", Proc. of ICEC 18, Mumbai, India, edited by K.G. Narayankhedkar, (2000), 59-62.
2. "Materials at low temperatures", edited by Reed, R.P. and Clark, A.F. , 1983.
3. Liu, C.L., et al, "Study on Physical and Mechanical Properties of Materials Under Hydrogen Circumstance Conditions in USSR and USA", Chin. Cryogenics, (2000), 113, 1-6
4. Ohria, K., "Research and Development Work on Liquid Hydrogen Technologies in Japan's WE-NET projects", Proc. of ICEC 19, Grenoble, France(2002), 557-562.
5. Ju, Y.L., Zhou, Y., "On the Development of Non-Metallic Non Magnetic Minature Pulse Tube Cooler", Proc. of 12th Inter. Cryocooler Conf. , Cambridge, Massachusetts, USA(2002).

6. Dang, H.Z., Ju, Y.L., Zhou, Y., "Design and Construction of a Non-metallic, Non-magnetic Miniature Pulse Tube Cooler", Proc. of ICEC 19, Grenoble, France (2002), 423-426

7. Li, L.F., Su, X.T., Gong, L.H., et al, "A Kind of Ceramics Ued at Cryogenic Temperatures", Chin. Patent, No. 03160054.9, 2003.

8. From:http://www.outokumpu.com/

9. Li, Y.Y., Chai S.S., Xu, Y.K., Qian, B.N. et al. The microstructure and properties of a cryogenic steel Fe-21Cr-6Ni-9Mn-N. Advances in Cryogenic Engineering Materials, 30, Edited by A.F.Clark and R.P.Reed, Plenum Press, New York, (1994), 237-244

10. Jiang, X.J. et al. The effect of Zn, Ag, Sc on microstructure and fracture mechanism of Al-Li alloy. Ph.D dissertation of IMR, 1993

11. In: Colour metallographic technology. Defence industry press, 1991

12. Nishijima, S. "Cryogenic Properties of Advanced Composites and their Applications", Proc. of ICEC16/ICMC, Kitakyushu, Japan, (1996), 19-26

13. Evans, D., Reed, R.P. ,"Low temperature adhesive bond strength of electrical insulation films", Adv. Cryo. Eng. (Mat.), (2000), 46A , 243-250.

14. Thostenson, E.T., Ren, Z.F., Chou, T.W., "Advances in the science and technology of carbon nanotubes and their composites: a review," Composites Science and Technology, (2001), 61, 1899-1912.

15. Shi, D., Ooij, W.J., "Uniform Deposition of Ultrathin Polymer Films on the Surface of Aluminum Nanoparticles by a Plasma Treatment," Appl. Phys. Lett, (2001) , 78, 1234 -9.

16. Zhao, L.Z., A novel toughening epoxy adhesive for cryogenic application, In: Interface of Composites, Harbin Univ. Press, Harbin, China (1997)72-77.

17. Li, L.F., Hong, C.S., Li, Y.Y. et al., "Strength and Toughness of ZrO_2-based Ceramics at Cryogenic Temperatures", Cryogenics,(1994), 34, Suppl., 469-472, and "Martensitic Transformation in ZrO_2-based Ceramics at Cryogenic Temperatures", Cryogenics, (1996), 36, 7-11.

18. Ueno, S., Nishijima, S., Nakahira, A., Sekino, T, Okada, T., Niihara, K., "Mechanical and Thermal Properties of Zirconia at Cryogenic Temperature", Proc. of ICEC16/ICMC, Kitakyushu, Japan, (1996) 2041-4.

Thermal conductivity of materials and measuring system at cryogenic temperature

B. Li, B. M. Wu, D. S. Yang, W. H. Zheng, L. Z. Cao

Structural Research Lab., Department of Physics,
University of Science and Technology of China, Hefei, 230026, P.R. China

A measuring system, including the measuring setup and corresponding software, is developed to perform the measurement of the thermal conductivity of rod samples with millimeter scale in temperature range 4.2-300K. The precisions and applicability are proved by experiments on a number of materials including superconductors, SDW alloys, CMR materials, as well as engineering materials.

INTRODUCTION

The thermal conductivity of materials is of high importance in both science and engineering fields. There are several mechanisms by which heat can be transmitted through materials and many processes that limit the effectiveness of each mechanism. In a non-metal heat is conducted by means of the thermal vibrations of the atoms. In a simple metal this mode of heat transport makes some contribution, but the observed thermal conductivity is almost entirely due to the electrons. It does not follow that thermal conductivity of the two different materials must be very different in magnitude, as they are in the case of electrical conductivity, but their dependences on temperature and on the imperfections in individual specimens are rather different. In many materials, such as alloys and semiconductors, both transport mechanisms can make comparable contributions to the observed conductivity, and the relative proportions vary with composition and temperature. In superconductors the proportions are different in the moral and superconducting states, so that below the transition temperature they can be changed by an appropriate magnetic field.

In engineering, automotive, aerospace and related industries, thermal conductivity is important design parameters. These industries require access to reliable standards of thermal conductivity to facilitate material development, evaluation and improvement of component design as well as to ensure good quality control of manufactured products. There is also a demand for Certified Reference Materials and Transfer Standards for the calibration of comparative thermal conductivity apparatus that are mainly supplied by US companies. At present the manufacturers supply these apparatus with standards or reference materials that are uncertified and primarily refer to book values of thermal conductivity typical for that material. In the current climate of increasing accreditation of measurement testing facilities auditors are requiring test laboratories to have reference materials whose calibrations are traceable to national standards.

At any rate, it is necessary to measure thermal conductivity exactly. We have developed a thermal conductivity measuring system, including the measuring setup and corresponding control software, to measure the thermal conductivity of rod samples with millimeter scale in temperature range 4.2-300K. The precision and applicability of the system are proved by experiments on kinds of materials including superconductors, SDW alloys, CMR materials, as well as some engineering materials.

EXPERIMENTAL CONSIDERATIONS

Method
The method of longitudinal steady thermal flow [1] is used in the measurement. In this simplest steady-state experimental arrangement, heat is supplied at one end of a rod of uniform cross-sectional

area A at a known rate $\dot{Q}$ and is removed at the other end. The temperature difference ΔT is established with a distance L. The thermal conductivity is then derived from the relation

$$K = \frac{\dot{Q}}{\Delta T} \cdot \frac{L}{A} \qquad (1)$$

<u>Setup</u>

A schematic view of the experimental chamber setup and sample holder is shown in Figure 1 [2]. Vacuum of the sample chamber was kept 10^{-4} Pa during the measurement. The temperature difference on the sample generated by the gradient heater is monitored directly with a differential thermocouple. The temperature difference on the sample is controlled about 1 K.

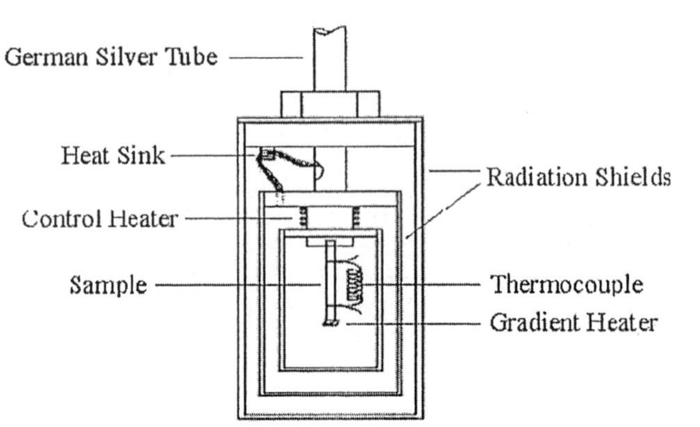

Figure 1   Schematic diagram of the experimental setup

<u>Automation</u>

The real-time read and write are realized by a GPIB interface board and insolated analog output boards respectively, as shown in Figure 2. The temperature instability of the sample is controlled within $\pm 2mK$ by a PID temperature controller [3]. Corresponding software is developed to collect and process the experimental data.

The Borland C++ Builder 6.0 environment is used to develop the real-time direct digital control software that practices the automation of data collection and process. We pay most attention to the convenience, security and stability of the software as BCB 6.0 provides perfect visualization component libraries (VCL). The measuring system can work properly in temperature range 4.2-300K. The experimental results of different materials are probed continuous and smooth as temperature varies.

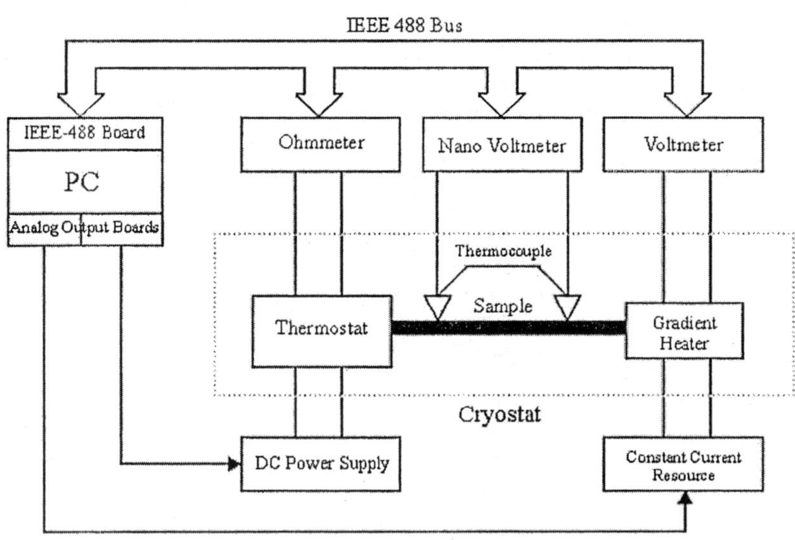

Figure 2   Schematic diagram of the real-time data collection and process

EXPERIMENTAL RESULTS

On this measuring system, we have obtained the thermal conductivity of a number of materials including superconductors [4-10], SDW alloys [11,12], and CMR materials [13, 14], as well as some engineering materials. The results meet the requirement of numerical analyses and are repeatable. Some typical thermal conductivity temperature dependences are given below.

Thermal conductivity of superconductors

The thermal conductivity peak in high T_C superconductors (HTS) has been validated [4 and its references] under T_C. Below T_C superconducting electrons condensing into Cooper pairs carry no heat, the thermal conductivity is thus ensured by normal electrons. Since the normal electron scattering rate in HTSs has been observed to decrease rapidly below T_C, thermal conductivity should increase under T_C. As temperature kept decreasing, the number of carriers decreases, that leads to the decreasing of thermal conductivity. The thermal conductivity of high T_C superconductor $Hg_{0.9}Tl_{0.2}Ba_2Ca_2Cu_3O_y$ [4] is shown in Figure 3. The thermal conductivity peak under T_C is remarkable. The thermal conductivity of new superconductor MgB_2 [6] is shown in Fig. 4. No obvious variation is observed near T_C.

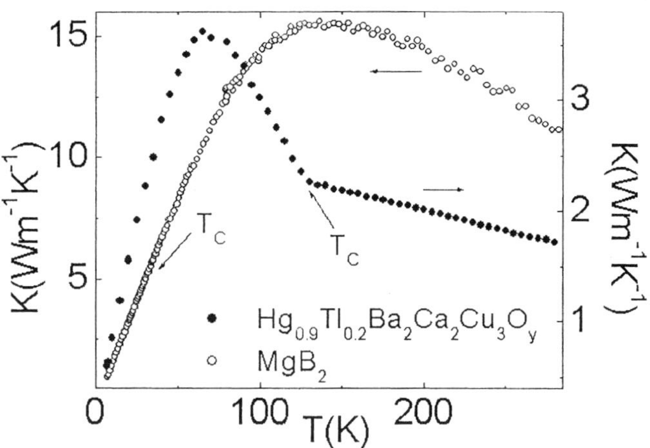

Figure 3   The thermal conductivity of high T_C superconductor Hg1223 and new superconductor MgB_2

Thermal conductivity of CMR materials

The thermal conductivity of CMR material $La_{0.7}Ca_{0.3}Mn_{0.95}Cr_{0.05}O_3$ [12] is shown in Figure 4. A minimum is observed at the Curie temperature (T_C), and a phonon-induced peak [1] is observed at about 20K. It is remarkable that the thermal conductivity and electrical conductivity show similar temperature dependence about T_C. The similarity is more interesting knowing the fact that electron contribution educed from Wiedemann-Franz law [1] is less 1%. Because of the strong electron- phonon interaction in perovskite structures, electron contributes to the thermal conductivity indirectly as well. The unusual similarity between thermal conductivity and electrical conductivity temperature dependence results from the indirect electron contribution. Above T_C, electrons are localized around phonons forming polarons. Thus modes of phonon are limited by these localized electrons. Under T_C, electrons delocalize. As temperature decreasing, more electrons become mobile. Phonons are no more bound to electrons. Thus the thermal conductivity increases with a decreasing temperature.

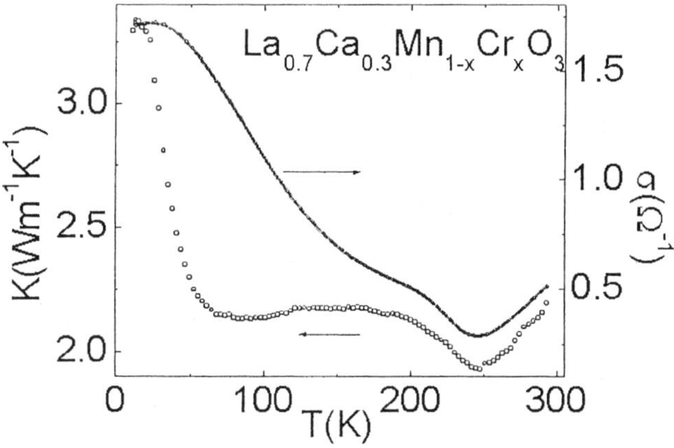

Figure4   The thermal conductivity of CMR material

Thermal conductivity of engineering materials

Both good and poor thermal conductors are demanded in different engineering fields. The thermal conductivity of Ag sheathed Bi-2223 superconducting cable and insulator in superconducting loop are shown in Figure 5. The difference in thermal conductivity is about two orders of magnitudes.

CONCLUSION

A measuring system, including measuring setup and corresponding software, is developed to measure the thermal conductivity of rod samples with millimeter scale in temperature range 4.2-300K. The precisions and applicability of the system are proved by the measurement performed on kinds of materials.

ACKNOWLEDGEMENT

This work is supported by the National Nature Science Foundation of China (No.10174070), the Ministry of Science and Technology of China (NKBRSF-G19990646).

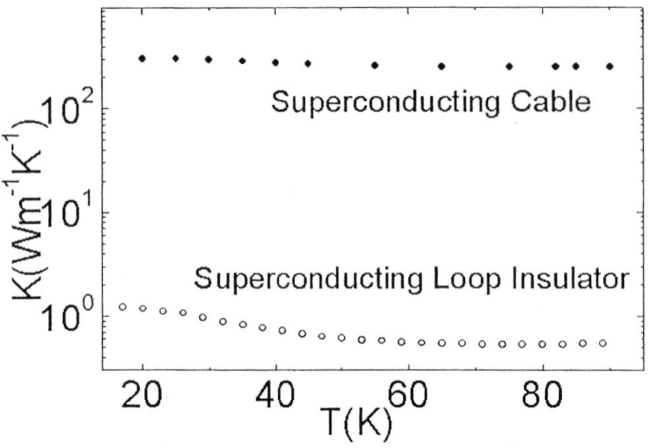

Figure 5   The thermal conductivity of some engineering materials

REFERENCES

1 R. Berman, Thermal conduction in Solids, Clarendon, Oxford, (1976).

2 Bai-Mei Wu, Dong-Sheng Yang, Bo Li, Wei-Hua Zheng, Song Sheng, The construct and automatization of thermal conductivity measurement system, Chinese Journal of Low Temperature   (2003) 25 supp. 248.

3 Dong-Sheng Yang, Bai-Mei Wu, Computer based temperature controller in low temperature, Chinese Journal of Low Temperature, (1999) 21 156.

4 Dong-Sheng Yang, Bai-Mei Wu, Song Sheng, Ping Zheng, Lie-Zhao Cao, Qin-Lun Xu, Thermal conductivity of superconducting $Hg_{0.9}Tl_{0.2}Ba_2Ca_2Cu_3O_{8+d}$, Chinese Journal of Low Temperature, (1999) 21 405.

5 Bai-Mei Wu, Dong-Sheng Yang, Ping Zheng, Song Sheng, Zhao-Jia Chen, Qin-Lun Xu, Thermal conductivity of $Hg_{0.9}Tl_{0.2}Ba_2Ca_2Cu_3O_{8+d}$ in magnetic field, Acta Physica Sinica, (2000) 49 267.

6 Dong-Sheng Yang, Bai-Mei Wu, Bo Li, Wei-Hua Zheng, Shi-Yan Li, Rong Fan, Xian-Hui Chen, Lie-Zhao Cao, Thermal Conductity of Two-Energy-Gap Superconductor MgB_2, Acta Physica Sinica, (2003) 52 683.

7 Dong-Sheng Yang, Bai-Mei Wu, Bo Li, Wei-Hua Zheng, Shi-Yan Li, Xian-Hui Chen, Lie-Zhao Cao, Anomalous Thermal Conductivity Enhancement in the Mixed State of MgB_2, Acta Physica Sinica, (2003) 52 2015.

8 Bai-Mei Wu, Bo Li, Dong-Sheng Yang, Wei-Hua Zheng, Shi-Yan Li, Lie-Zhao Cao, Xian-Hui Chen, The Thermal Conductivity in New Superconductor MgB_2 and $MgCNi_3$, Acta Physica Sinica, (2003) 52 3150.

9 Bo Li, Bai-Mei Wu, Dong-Sheng Yang, Wei-Hua Zheng, Hua Jin, Hui-Ping Gao, Thermal Conductivity of Ce Doped Bi-2212, Chinese Journal of Low Temperature (2003) 25 supp. 253.

10 Dong-Sheng Yang, Bai-Mei Wu, Wei-Hua Zheng, Hong-Shun Yang, Bo Li, Lie-Zhao Cao, Thermal conductivity of excess-oxygen-doped La_2CuO_4, Chinese Journal of Low Temperature (2001) 23 45.

11 Bai-Mei Wu, Dong-Sheng Yang, Song Sheng, Wei Liu, Ying-Lei Du, Wei-Ming Xu, The transport proerties and internal friction related to the antiferromagnetic transition in $Cr_{75}(Fe_xMn_{1-x})_{25}$ alloys, J. magn. Magn. Mater, (1999) 202 426.

12 Bai-Mei Wu, Dong-Sheng Yang, Song Sheng, Wei-Ming Xu, Zhao-Jia Chen, Thermal Conductivity of Cr75FexMn25-x(x=16,18) Alloys, Acta Physica Sinica, (1999) 48 1147.

13 Wei-Hua Zheng, Bai-Mei Wu, Dong-Sheng Yang, Jian-Yang, Li, Xue-Rong, Liu, Bo Li, Ya-Qin Zhou, thermal conductivity and electrical conductivity in $La_{1-x}Ca_xMnO_3$, Chinese Journal of Low Temperature, (2001) 23 216.

14 Wei-Hua Zheng, Bai-Mei Wu, Bo Li, Dong-Sheng, Yang, Lie-Zhao Cao, Thermal conductivity in $La_{0.7}Ca0.3Mn_{1-x}Cr_xO_3$ at low temperature, Chinese Journal of Low Temperature, (2002) 24 230.

Cryogenic properties of polymer composite materials – a review

Fu S. Y., Li Y., Zhang Y. H., Pan Q. Y., Huang C. J.

Cryogenic Materials Division, Technical Institute of Physics and Chemistry, Chinese Academy of Sciences, Beijing 100080, China

Polymer composite materials are used in a wide variety of cryogenic applications to replace metals because of their unique and highly tailorable properties. In order to develop high performance polymer composites for cryogenic applications, it is necessary to understand how polymer composites behave at cryogenic temperatures and how their cryogenic properties are affected by factors such as filler content and matrix type etc. This review presents detailed discussions on the effects of various factors on the cryogenic mechanical and thermal properties at the unique operating environments.

INTRODUCTION

Polymer composite materials are being increasingly employed to manufacture structural components that are exposed to low temperature environments in space and superconducting applications [1-18]. The involved composites include fiber and particle reinforced thermoplastic and thermosetting polymer composites. The cryogenic applications of polymer composites can be classified as support structures, vessels and electrical insulation. Fundamental mechanical, thermal and electrical requirements should be met by polymer composites for specific cryogenic applications. Composites are subjected to uncommon synergetic conditions in cryogenic environments: high mechanical stress by electromagnetic force, thermo-mechanical stress caused by the cryogenic environment, phase transition of coolant and high energy radiation etc. The cryogenic mechanical and thermal properties of polymer composites are important in cryogenic applications. In order to meet the requirements of mechanical and thermal properties for cryogenic applications, various polymer composites have been studied extensively. This review presents detailed discussions on the effects of various factors on the cryogenic mechanical and thermal properties at the unique operating environments.

MECHANICAL PROPERTIES

Fiber reinforced plastic composites

Fiber reinforced plastics (FRP) were widely used in cryogenic applications because of their high strength, high stiffness, low weight and good thermal properties. Fiber type, fiber direction and interfacial properties play very important roles in determining the mechanical properties of FRP.

Hartwig et al [19] discussed the influence of the fiber type on the fatigue behavior of unidirectional cross-plied composites. The fatigue behavior was studied on composites with the same epoxy matrix but different types of fibers. The fatigue behavior at 77 K of epoxy composites with different fiber types (carbon AS4, ceramic Al_2O_3 and Kevlar 49 fibers etc) is shown in Figure1 (so-called S-N curves, namely stress-load cycles diagram). The highest strength is achieved for carbon fiber composites.

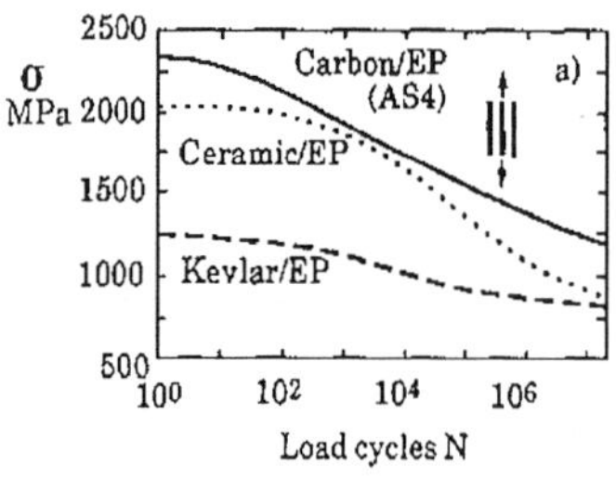

Figure1. S-N curves of unidirectional fiber reinforced epoxy composites at 77 K [19]

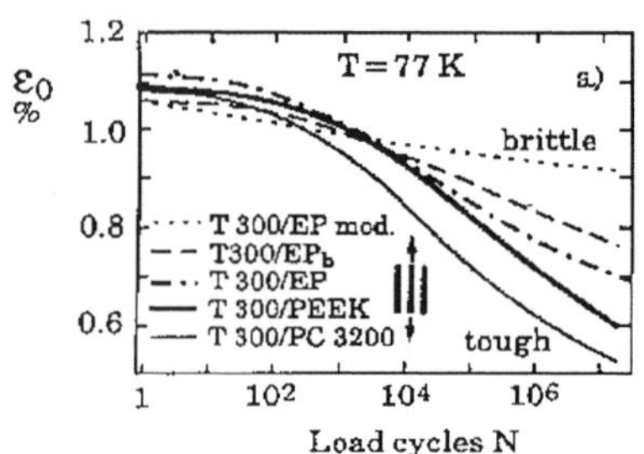

Figure 2. Strain-life curves of UD carbon fiber composites under tensile loading [19]

Composites having a poor transverse strength are sensitive to microcracks induced in the matrix. The formation of microcracks at fatigue loading is different for different matrix types, and so is the fatigue endurance limit. In Figure 2 the influence of the different matrix was shown on the fatigue behavior of unidirectional carbon fiber reinforced plastic (CFRP) composites by strain-life diagrams [19].

The influence of fiber directions on the mechanical properties of FRP wasstudied by Hussain et al [20] and Baynham et al [21].The mechanical properties of FRP in parallel to fiber direction is higher as the properties are mainly controlled by the fibers. However, the mechanical properties of FRP in transverse to fiber direction is inferior because the properties are mainly governed by the properties of the matrix and the fiber-matrix interface. Hussain et al suggested that Young's modulus in the transverse to fiber direction can be improved by incorporating Al_2O_3 particles into the matrix (Figure 3) [19]. The Al_2O_3 filler dispersions act as secondary reinforcement.

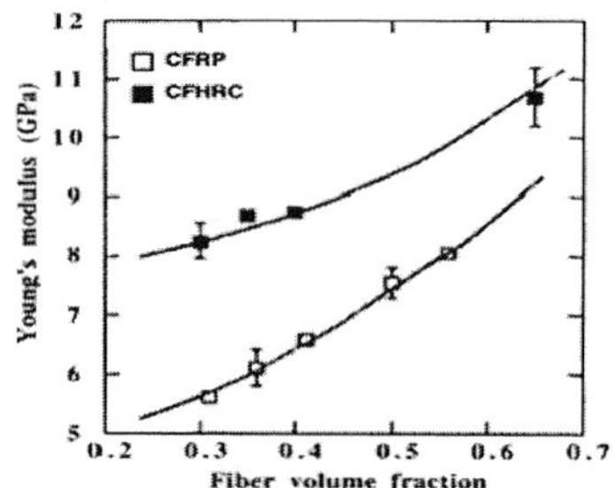

Figure 3. Young's modulus in the transverse to fiber direction as a function of fiber content □ CFRP [19]
■ CFHRC (CFRP contained 10 vol % Al_2O_3 particles)

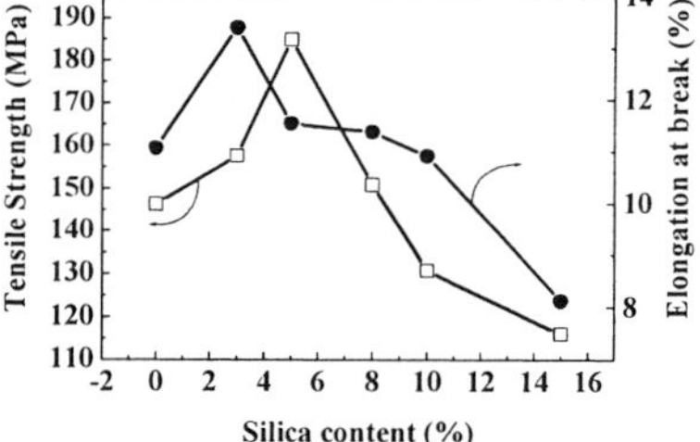

Figure 4. The tensile properties at 77K of polyimide/silica hybrid films

Particulate reinforced polymer composites

The mechanical properties at liquid nitrogen temperature of various particle reinforced polymer nano-composites have been studied recently by our research group. These nanocomposites include epoxy/SiO_2, polyimide/SiO_2 and polyimide/clay nanocomposites. It has been shown that the cryogenic mechanical properties can be effectively enhanced by incorporating proper quantities of nano-fillers into the matrix (as an example, see Figure 4) [22].

THERMAL PROPERTIES

Thermal properties of polymer composites are important design parameters in cryogenic applications. The factors influencing the thermal expansion and conductivity of composites at low temperatures will be reviewed below.

Thermal expansion

Nadeau and Baschek et al [23,24] studied the factors influencing the thermal expansion. The influencing factors include thermal cycling, mechanical creep loading and mechanical geometrical shape (plates, half-tubes and tubes). The results showed that the expansion was influenced in different manners by thermal cycling and mechanical creep loading. The influence of thermal cycling on expansion was shown in Figure 5. [23]. The result indicated that the integral thermal expansion was lowered by more than 20%. In addition, the influence of thermal cycling on the coefficient of thermal expansion was significant as well. The influence of mechanical creep loading on expansion of carbon reinforced plastics with different fiber angles was also reported. The results showed that in the range of $\omega = \pm 30°$ the thermal expansion was not sensitive to the change of the fiber angle.

Thermal conductivity

The thermal conductivity of fiber reinforced plastic is much lower than that of metals and shows anisotropic. Hence, in general, it is much more difficult to dissipate heat in fiber reinforced plastic than in metals. This is an important consideration in some situation [25].

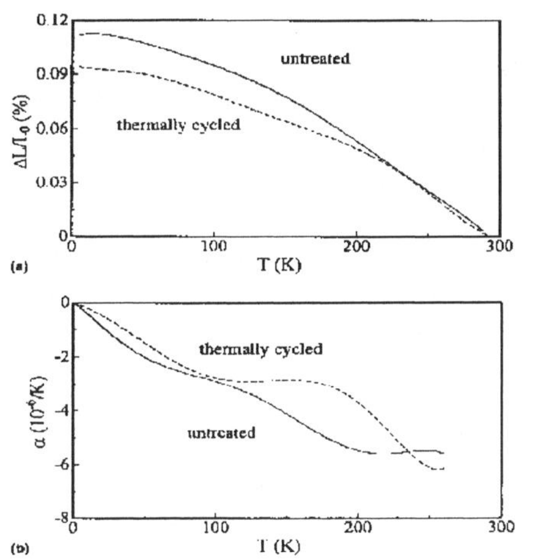

Figure 5. Thermal expansion of carbon fiber reinforced plastics (±30°) before and after thermal cycling (100 cycling,77-293 K). (a)Integral thermal expansion; (b) coefficient of thermal expansion [23]

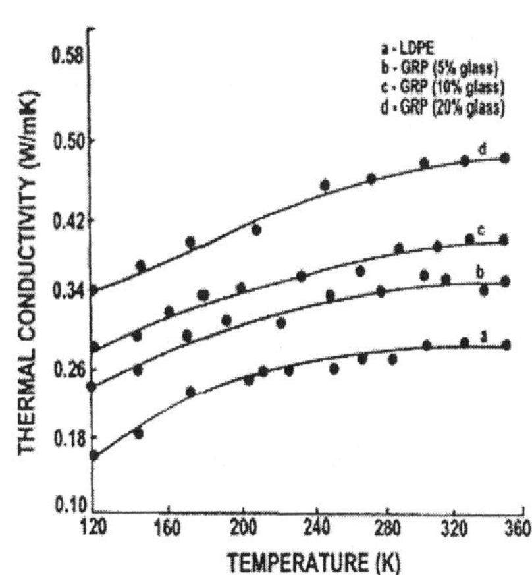

Figure 6 The thermal conductivity of LDPE and GRP composites [25]

SUMMARY

Polymer composite materials are being increasingly employed for cryogenic applications. Cryogenic properties of the polymer composites are influenced by a number of factors. This review has given detailed discussions on the effects of various factors on the cryogenic mechanical and thermal properties.

REFERECES

1. R. P. Reed and M. Goldat, Cryogenic composite supports: a review of strap and strut properties, Cryogenics (1997) 37 233-250.
2. R. P. Reed and M. Golda, Cryogenic properties of unidirectional composites, Cryogenics (1994) 34, 909-928.
3. D. Evans and J. T. Morgan, Low temperature mechanical and thermal properties of liquid crystal polymers, Cryogenics (1991) 31 220-222.
4. Yasuhide Shindo, Hitoshi Tokairin, and Kazuaki Sanada et.al, Compression behavior of glass-cloth/epoxy laminates at cryogenic temperatures, Cryogenics (1999)39 821-827.
5. H. Yamaoka, K. Miyata and O. Yano, Cryogenic properties of engineering plastic films, Cryogenics (1995)35,787-789.
6. Vernon T. Bechela, Mark B. Fredina, Steven L. Effect of stacking sequence on micro-cracking in a cryogenically cycled carbon/bismaleimide composite, Composites :Part A (2003)663-672.
7. John F. Timmerman, Brian S. Hayes and James C. Seferis, Nanoclay reinforcement effects on the cryogenic microcracking of carbon fiber/epoxy composites, Composites Science and Technology (2002) 62 1249-1258.
8. Y. Shindo, K. Horiguchi and R. Wang, Double cantilever beam measurement and finite element analysis of cryogenic mode 1 interlaminar fracture toughness of glass-cloth/epoxy laminates, Journal of Engineering Materials and Technology, (2001)123 191-197.
9. K. Bittner-Rohrhofer, K. Humer and H.W. Weber, Low-temperature tensile strength of the ITER-TF model coil insulation system after reactor irradiation, Cryogenics (2002) 42 265-272.
10. Takefumi Horiuchi and Tsutomu Ooi, Cryogenic properties of composite materials, Cryogenics (1995)35 677-679.
11. N. Albritton and W. Young Babcock, Cryogenic evaluation of epoxy bond strength, Cryogenics (1996) 36 713-716.
12. T. Nishiura, S. Nishijima and T. Okada, Synergistic effects of radiation and stress on mechanical properties of organic and composite materials, Cryogenics (1995) 35 747-749.
13. Hirokazu Yokoyama, Thermal conductivity of polyimide film at cryogenic temperature, Cryogenics (1995) 35 799-800.
14. Masakatsu Takeo, Seiki Sato and Masaaki Matsuo, Dependence on winding tensions for stability of a superconducting coil, Cryogenics (2003) 43 649-658.
15. K. Pannkoke and H. -J. Wagner, Fatigue properties of unidirectional carbon fibre composites at cryogenic temperatures, Cryogenics (1991)31,248-251.
16. K. Ahlborn, Cryogenic mechanical response of carbon fibre reinforced plastics with thermoplastic matrices to quasi-static loads, Cryogenics (1991)31 252-256.
17. K. Ahlborn, Durability of carbon fibre reinforced plastics with thermoplastic matrices under cyclic mechanical and cyclic thermal loads at cryogenic temperatures, Cryogenics (1991)31 257-259.
18. Y. Iwasaki , J. Yasuda and T. Hirokawa ,Three-dimensional fabric reinforced plastics for cryogenic use, Cryogenics (1991) 31 261-264.
19. G. Hartwig, R. Hübner, S.Knaak et.al, Fitigue behaviour of composites, Cryogenics (1998) 38 75-78.
20. M. Hussain, A. Nakahira and S. Nakahira et.al, Evalution of mechanical behavior of CFRC transverse to the fiber direction at room and cryogenic temperature, Composites: Part A (2003) 31 173-179.
21. D. E. Baynham, D. Evans and S. J. Gamage, Transverse mechanical properties of glass reinforced composite materials at 4 K, Cryogenics (1998) 38 61-67.
22. Y. Li, S. Y. Fu, Y. H. Zhang and Q. Y. Pan, A new process for the preparation of PI/silica hybrid films. ICEC 20, 10-14 May 2004, Beijing, China.
23. J. C. Nadeau and M. Ferrari, Effective thermal expansion of heterogeneous materials with application to low temperature environments, Mechanics of Materials (2004) 36 201-214.
24. G. Baschek and G. Hartwig, Parameters influencing the thermal expansion of polymers and fibre composites, Cryogenics (1998) 38 99-103.
25. G. Kalaprasad,   P. Pradeep and George Mathew, Thermal conductivity and thermal diffusivity analyses of low-density polyethylene composites reinforced with sisal, glass and intimately mixed sisal/glass fibres, Composites Science and Technology (2000) 60 2967-2977.

CRYOFLUIDS – A Software for Physical Property Data of Cryogenic Fluids

P Lalkoshti, Raja Banerjee, Abhijit Tarafder and Sunil Kr Sarangi

Cryogenic Engineering Centre, IIT Kharagpur 721 302, India

Thermodynamic and transport properties of process fluids are essential for design of cryogenic equipment. There are several authentic sources of data for thermophysical properties of cryogenic fluids; but many of them do not cover the entire useful range of pressure and temperature. To address to these difficulties, we have developed a user-friendly data bank in our laboratory. The data bank takes data from well-known data sources available in open literature and computes the rest by intelligent extrapolation from available data using standard thermodynamic relations. The paper presents the features of the software and the structure of the program. The program is available in the intranet of the Institute and will be shortly available over Internet.

INTRODUCTION

Thermodynamic and transport properties of process fluids are essential for design of cryogenic equipment. A computerized database can be very helpful to the designer, particularly in a computer aided design environment. There are several authentic sources of data for cryogenic fluids, the well known among them being (a)ALLPROPS from the University of Idaho, USA, (b)MIPPROPS and HELIUM from NIST, and (c)GASPAK and HEPAK from Cryodata, USA. The programs often cover different domains of pressure and temperature and different properties. No single program covers the full range or temperature and pressure that a cryogenic engineer is ordinarily interested in. Further, the cost of the programs also becomes a factor for occasional users, students and researchers in developing countries

To address to these difficulties, we have developed a computerized databank named CRYOFLUIDS, which is currently available over our Institute's intranet and will be served to the cryogenic community over the Internet. The program is not based on any new computational approach or empirical correlation; instead, we have computed and stored property data at selected points on the p-T plane in computer memory and calculate the values at desired points by intelligent interpolation or extrapolation. The software contains a user-friendly interface and facility for addition of new fluids and revision of source data.

THE STRUCTURE OF CRYOFLUIDS

CRYOFLUIDS covers seven thermodynamic and two transport properties of fourteen cryogenic fluids. The thermodynamic properties are pressure, temperature, density, enthalpy, entropy, isobaric specific heat C_p and isochoric specific heat C_v. The two transport properties are viscosity and thermal conductivity. The chemical species included in the software are Nitrogen, Oxygen, Argon, Normal Hydrogen, ParaHydrogen, Helium, Methane, Ethane, Propane, Iso-Butane, Normal Butane, Water and Carbon-Dioxide. Water and Carbon-Dioxide are not cryogenic fluids, but are used frequently in many cryogenic processes. Some of the spcial features of CRYOFLUIDS are the following.

(1)   It predicts properties in both tabular and graphical forms. The input conditions may be given in one of the four combinations: (1)pressure and temperature (ii)pressure and vapour fraction (iii)pressure and enthalpy or (iv)pressure and entropy.

(2)   The user can examine the source data and make changes as he thinks necessary. The program has the provision of updating fluid source data and of adding new fluids if more reliable data are available.

INPUT OF BASIC DATA

Basic thermodynamic and transport property data has been provided to the software at selected points in the p-T plane. The points have been expressed in terms of reduced temperature and reduced pressure. As the non-ideality of gas behaviour is highest near the critical point and close to the phase boundaries, we have provided a larger concentration of points in this region, and a somewhat sparse distribution away from these points. Reduced pressures have been chosen in geometric progression and reduced temperatures in arithmetic progression. Tables 1 and 2 give the distribution of source data points in the reduced p-T plane.

Table1: Number of source data points over the specified range of reduced pressure

Range of Reduced Pressure	Number of Points
0.0001 to 0.8	15
0.8 to 1.2	15
1.2 to 3	10
3 to 25	5

Table 2: Number of source data points over the specified range of reduced temperature

Range of Reduced Temperature	Increment in Reduced Temperature
Melting Point to 0.5	0.025
0.5 to 0.8	0.030
0.8 to 1.3	0.010
1.3 to 2.0	0.014
>2.0	0.240

Thermodynamic Properties

Basic thermodynamic properties have been taken from the software ALLPROPS 4.2 [1] over the range of temperature and pressure covered by the software. Outside the range, we have used empirical relations and standard thermodynamic relations to compute the values and provide to CRYOFLUIDS as basic data. The isobaric specific heat C_p is a fundamental quantity from which other properties such as enthalpy and entropy can be calculated. We have computed C_p at these points by using the empirical relation (Eq. 1) suggested by Scott and Sontagg [4].

$$Cp = 1/M [A + B\,\theta + C\theta^2 + D\,\theta^3 + E/\theta^2]\tag{1}$$

where C_p is given in kJ/ kg K, $\theta = T/100$, the temperature T being expressed in Kelvin. M is the molecular weight (kg/kmol) and A, B, C, D, E are constants taken from Reference [4].

Other thermodynamic properties are computed by using equations (2), (3), (4) and (5).

Density:
$$\rho = (\rho^* T^*) / T \tag{2}$$
Enthalpy:
$$h = h^* + \int C_p \, dT \tag{3}$$
Entropy:
$$s = s^* + \int C_p \, dT / T \tag{4}$$
Isochoric Specific Heat:
$$C_v = C_p - R/M \tag{5}$$

The variables with an asterisk refer to known points, e.g. the nearest data point provided by ALLPROPS 4.2 [1]. In equations (3) and (4) the integration is carried over the limits T* to T.

Transport Property Data Source: -

Transport property data such as viscosity and thermal conductivity are not provided by ALLPROPS. We have taken them from the programs MIPROPS and HELIUM by the NIST [2]. Outside the range of MIPROPS, we have used the following empirical relations from Ref [4] to extend the data.

Gas Viscosity: -

Several methods are available to predict gas phase viscosity, the well known among them being the empirical relations suggested by Lucas [3,5,6], Chung et.al. [3], Reichenbug [3]and Brute and Startling [3]. The Lucas and Chung et al's methods employ the same set of relations for both low and high-pressure gas. Lucas method requires T_c, P_c, M, Z_c and μ as input data. However Chung et al's method requires V_c and the acentric factor in addition to those variables. V_c is not readily available for all the fluids and Reichenbug's method cannot be used for inorganic gases. Therefore, we have adopted the Lucas method to compute gas phase viscosity

Liquid Viscosity: -

To predict liquid phase viscosity, we have adopted the method suggested by Van Velzen etal [3,7] and Yaws et.al. [3,8]. This method predicts the saturated liquid viscosity which is modified to incorporate the effect of pressure by the Lucas method [3,6]. This scheme has been employed for all fluids except helium and hydrogen, for which saturated liquid viscosity was determined using the relation from DIPPR databank [9]. The effect of pressure has not been included becasue the negative value of the Pitzer acentric factor for these fluids resulted in negative values of viscosity at many points.

Gas Thermal Conductivity: -

There are several method are available to predict pure gas thermal conductivity and all of them use a dimensionless factor known as the Eucken Factor defined as:
$$\text{Eucken Factor} = (\lambda M^1)/(\eta \, C_v)$$
where

λ = Thermal Conductivity (W/mK)
$M^1 = M/100$
M = Molecular weight (kg/ kmol)
η = Viscosity (Pa.s)
Cv = Isochoric specific heat (kJ/ kg K)

We have adopted the method of Ely and Hanley [10, 11] because it was found to predict the correct trend in variation of Eucken factor with temperature.

Liquid Thermal Conductivity: -
We have employed different methods for computing thermal conductivity of different liquids in saturated state at a given temperature. Variation due to pressure is incorporated by the Missenard Method [12]. We have used the method of Miller, McGinely and Yaws [13] for nitrogen, oxygen, argon, water, carbon dioxide, methane, ethane, and propane, Latini's method [14,15,16] for isobutene and normal butane, and the empirical relation from the DIPPR databank [9] for helium and hydrogen.

Rules for interpolation and extrapolation
For computing values of the properties from the stored data, we need to use interpolation and extrapolation routines. Best results are obtained if appropriate physical relationships are employed to linearise the interpolation or extrapolation process. Table 3 gives a summary of those relationships.

Table 3: Interpolation relationships employed to interpolate or extrapolate thermophysical property data

Property	Liquid Phase (temperature)	Gas Phase	
		Temperature	Pressure
Density	$\rho=\alpha+\beta T$	$\rho=\alpha+\beta/T$	$\rho=\rho+\beta P$
Enthalpy	$h=\alpha+\beta T$	$h=\alpha+\beta T$	$h=\alpha+\beta P$
Entropy	$s=\alpha+\beta T$	$s=\alpha+\beta\ln(T)$	$s=\alpha+\beta\ln(P)$
Isochoric Specific Heat	$Cv=\alpha+\beta T$	$Cv=\alpha+\beta T$	$C\ Cv=\alpha+\beta P$
Isobaric Specific Heat	$Cp=\alpha+\beta T$	$Cp=\alpha+\beta T$	$Cp=\alpha+\beta P$
Viscosity	$\eta=\alpha+\beta\sqrt{T}$	$\eta=\alpha+\beta\sqrt{T}$	$\eta=\alpha+\beta\sqrt{P}$
Thermal Conductivity	$\lambda=\alpha+\beta T$	$\lambda=\alpha+B\sqrt{T}$	$\lambda=\alpha+\beta P$

ORGANIZATION OF THE PROGRAM

The software CRYOFLUIDS consists of two important parts: data retrieval and data updating. Seven C++ classes, namely: *DataRetrieval, Interpolation, TempSec, EnthalpySec, EntropySec, QualitySec* and *Enter* have been written and linked together to form the data retrieval part. The basic architecture is based on hybrid inheritance. A schematic representation of class hierarchy is shown in Fig 1. *DataRetrieval* is the main base class from which the class *Interpolation* has been derived. Four classes *TempSec, EnthalpySec, EntropySec* and *QualitySec* have been derived from *Interpolation* by declaring the later as a virtual base class. Finally, *Enter* is derived from the last four classes. These classes have been developed by using fairly standard procedures Object Oriented Programming.

Updating of existing fluid data tables and adding new fluids to the data bank is the second most important feature of CRYOFLUIDS. *Intialisation* is the main basic class from which the class store is derived. Five classes: *GasViscosity, GasThend, LiqViscosity, LiqThend* and *PropGen* are derived from this class. Finally, *Mgenet* is derived from the last five classes. A Schematic representation of the class hierarchy is shown in Fig 2.

ERROR IN ESTIMATION OF PROPERTY VAUES

Table 2 shows the error in thermodynamic and transport properties computed by CRYOFLUIDS for Nitrogen at different pressure and temperature combinations. It can be taken as an example of performance of the software. Errors in density, C_p, C_v are calculated on percent basis by using following formula.

$$\% \text{ Error in Property}=(P_{CRYOFLUIDS}-P_{ALLPROPS4.2})*100/\ P_{ALLPROPS4.2}$$

The error in Viscosity and Thermal Conductivity is calculated by using following formula.

$$\% \text{ Error in Property} = = (P_{\text{CRYOFLUIDS}} - P_{\text{MIPPROPS}}) * 100 / P_{\text{MIPPROPS}}$$

The error in Enthalpy and Entropy is not computed on percent basis but is expressed as a difference in temperature given by two sources for the same value of enthalpy or entropy. It is computed by the relation:

$$\text{Difference in Temperature } \Delta T \text{ (K)} = T_{\text{CRYOFLUIDS}} - T_{\text{ALLPROPS4.2}}$$

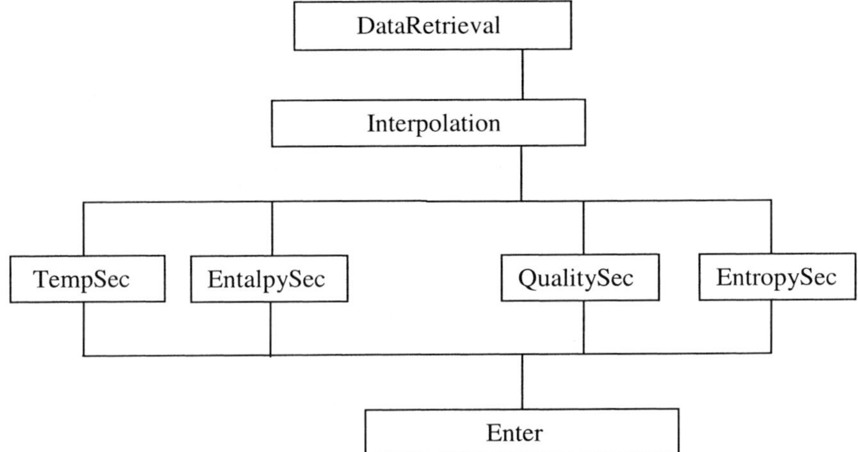

Figure 1:   Class Hierarchy for Data Retrieval routines

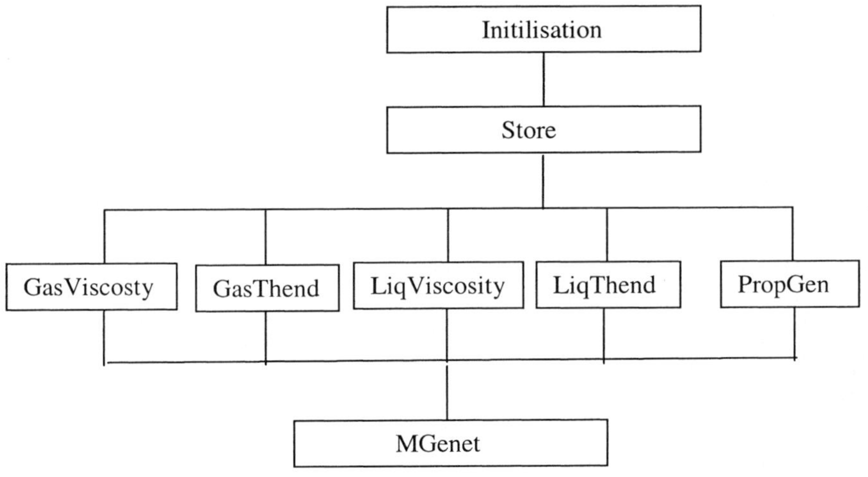

Fig 2: Class Hierarchy of routines for Updating data or Adding New Fluid

CONCLUSION

CRYOFLUIDS is a user-friendly software which can be employed for design and analysis of cryogenic equipment and processes. It should serve as a useful to the cryogenic community, particularly to the students and researchers without access to expensive data banks.

Table 4:   Error in prediction of thermophysical data of Nitrogen

P (bar)	T (K)	Density (%)	Enthalpy ΔT (K)	Entropy ΔT (K)	Cp (%)	Cv (%)	Viscosity (%)	Thermal Conductivity (%)
1	70	-0.00345	-0.001	0.016	0.02978	0	0.302109	-0.0181185
1	77.2	-0.0009	-0.003	0.001	0	0	0.030164	0.18596
1	77.4	0.026352	0.008	0.01	0.233662	0.282799	0.377358	-1.21053
1	126.2	0.022252	0	0	0	0.001342	0.011765	0.35
1	300	-0.02671	0	0.02	0.009603	0.00942	-0.5	0.05814
15	100	-0.02799	-0.02	0	0.127004	0.021174	0.175202	-0.08342
15	110.3	-0.00969	0	0	0.07595	0.017286	0.08365	1.03102
15	110.4	0.043637	0.01	0.01	0.213134	0.068688	-0.64706	-1.39189
15	126.2	0.381274	0.03	0.09	0.426591	0.075608	0.391304	0.32
15	300	0.02369	0.02	0.04	0.037594	0.013394	0.076923	0.31203
30	110	-0.00031	0	0	0.023964	0.004133	-0.04895	-0.02817
30	123.5	-0.26083	-0.03	-0.03	4.450659	0.429261	-0.46512	1.75537
30	123.7	0.373689	0.02	0.02	3.84973	0.311833	0.118644	-9.84416
30	126.2	0.247355	0.03	0.03	1.489074	0.146628	-0.07143	0.121457
30	300	0.041409	0.04	0.04	0.06432	0.018663	-0.25946	-0.1449275
34	120	-0.01687	0	0	0.137907	0.015334	-0.05432	-0.027818
34	126.2	0.98518	0	0	14.69403	1.134175	N/P	N/P
34	300	0.04697	0.05	0.05	0.063993	0.014645	0.281081	0.1474820
100	120	-0.0087	-0.01	-0.01	0.040486	0.003189	-0.01066	-0.039953
100	126.2	-0.01545	-0.01	0	0.068599	0.007548	0.593429	0.5019455
100	300	0.111897	0.13	0.15	0.192469	0.041848	-0.06965	0.2163
500	120	-0.03999	-0.02	0.04	0.068142	-0.01974	-0.04794	-0.117331
500	126.2	-0.04591	-0.03	0.02	0.075166	-0.01	-0.06042	0.14091
500	300	-0.23435	0	0.12	0.021882	0.039424	0.255814	0.277487

REFERENCES

1.   **Lemmon, E.W., Jacobsen, R.T., Penoncello, S.G. and Beyerlein, S.W.:** *ALLPROPS4.2: Computer Programs for Calculating Thermodynamic Properties of Fluids of Engineering Interest,* Center for Applied Thermodynamic Studies, University of Idaho, 1995.

2.   **McCarty, R.D.: MIPPROPS:** *Interactive FORTRAN Programs for Micro Computers to Calculate the Thermophysical Properties of Twelve Fluids,* NBS Technical Note 1097, May 1996.

3.   **Reid, R.C., Prausnitz, J.M. and Poling, B.E.:** *The Properties of Gases and Liquids,* McGraw-Hill International Editions, 1988.

4.   **Van Wylen, G and Sontagg, R.E:** *Fundamentals of Classical Thermodynamics: Table A.9,* Wiley Eastern Ltd, pp. 683 Sept. 1990.

5.   **Lucas, K.:** *Phase Equilibrium and Fluid Properties in Chemical Industry,* Dechema, Frankfurt, pp. 573, and 1980.

6.   **Lucas, K.:** *Chemical Ing. Tech.* 53: 595, 1981.

7.   **Van Velnez, D., Cardozo, R.L., and Langenkamp, H.:** *Liquid Viscosity and Chemical Constitution of Organic Compounds: A New Correlation and a Compilation of Literature Data,* Euratom, 4735e Joint Nuclear Research Center, Ispra Establishment, Italy, 1972

8.   **Yaws, C.L., Miller. J. W. Shah. P.N., Schorr, G.R. and Patel, P.M.:** *Chem. Eng.,* 83 (25): 153 (176).

9.   **B-JAC Design System for Windows Version 5.42:** *B-JAC International Inc.,* 1973.

10.   **Ely, J.F. and Hanley, H.J.M.:** *Ind. Engg. Chem. Fundam.* 22:90, 1983.

11.   **Hanley, H.J.M.:** *Cryogenics* 16 (11) 643, 9176.

12.   **Missennard, A.:** *Rev. Gen. Thermodyn.* 101 (5): 649, 1970.

13.   **Baroncini, C., Di Filippo, P. and Latini, G.:** *Comparison between Predicted and Experimental Thermal Conductivity Values for Liquid Substances and the Liquid Mixtures at Different Temperatures and Pressures,* paper presented at the Workshop on Thermal Conductivity Measurement IMEKO, Budapest, March 14-16, 1983.

14.   **Baroncini, C., Di Fillippo, P. and Latini, G.:** *Intern. J. Refrig.* 6 (1):60, 1983.

15.   **Baroncini, C., Di Fillippo, P. Latini, G. and Pacetti, M.:** *High temp-High press.* 11:581, 1979.

Discussion on Gibbons' Equation of State for Helium-3[*]

Huang Y.H., Chen G.B., Li X.Y.

Cryogenics Laboratory, Zhejiang University, Hangzhou 310027, P.R. China

A thorough literature survey was made to obtain all useful information relating to helium-3. Based on this fundamental work, the Gibbons' equation of state for ^{3}He in the gaseous region from 4K to 20K at pressures 0.1MPa to 10MPa was tested and refitted, and a modified form was presented.

INTRODUCTION

Helium-3, as the most expensive rare gas, is being gradually applied in the realm of high-tech and fundamental science. Many scientists have showed their interests in the thermodynamic and transport properties of ^{3}He.Although most of the experimental and theoretical research work on ^{3}He had already been done since the middle of the last century, many valuable experimental reference data and theoretical models were scattered among various literatures. Due to lack of a comprehensive, united database for the thermodynamic and transport properties of ^{3}He, a project on these properties of normal liquid and gaseous ^{3}He has been carried out in Cryogenics Laboratory, Zhejiang University. A thorough literature survey was made to obtain all existing data relating to ^{3}He for the following properties: PVT data, equilibrium data, melting point, critical point, latent heat, specific heat, entropy, expansion coefficients, compression coefficients, velocity of sound, thermal conductivity and viscosity. Based on this fundamental work, the Gibbons' equation of state (EOS) [1] for ^{3}He in the region from 4K to 20K at pressures 0.1MPa to 10MPa was tested and refitted, and a modified form was presented.

COLLECTION OF EXPERIMENTAL DATA

Firstly, all available literatures on ^{3}He since 1949 which contain useful information were classified and indexed by the authors for convenience. Then valuable data, figures and correlations or EOSs were inputted into computer as computer-readable documents. This work concerns the information of the saturated vapor and liquid, the normal liquid region, the gas region, the melting curve and the critical region etc of ^{3}He. Here for examples, Figures 1 and 2 show the 3D distribution of PVT in 0.2K-3K and specific heat, respectively.

GIBBONS' EQUATION OF STATE FOR GASEOUS ^{3}He

In 1967, Gibbons and Nathan brought forward an EOS of ^{3}He (See Equation (1)) for the gas region from

---

[*] Funded by the National Natural Science Foundation (Grant No. 50376055) and the Special Research Fund for Doctoral Training in Universities by the National ministry of Education of China (Grant No. 20010335010)

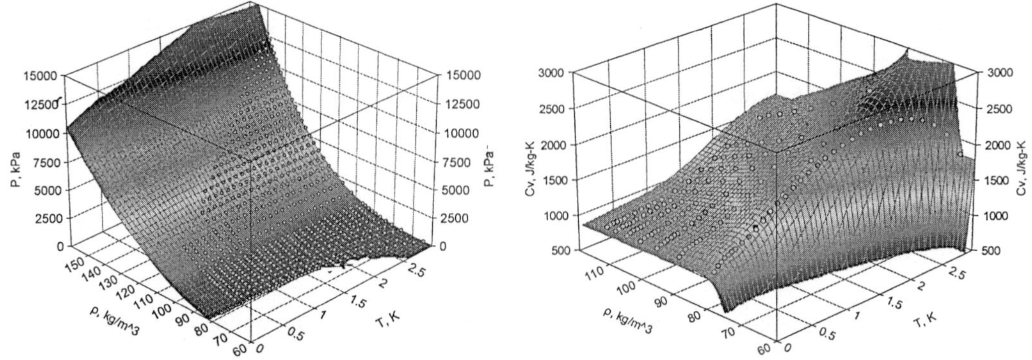

Figure 1 PVT distribution of ^{3}He
Figure 2 Specific heat distribution of ^{3}He

4K to 20K at pressures 0.1 to 10MPa. The corresponding coefficients are listed in Table 1 after a least square fitting to 336 experimental data points. Gibbons said that the maximum error in the pressure was 2.5% and the average error was 0.7% [1].

$$T(Z-1) = (N_1 + \frac{N_2}{X} + \frac{N_3}{X^2} + \frac{N_4}{X^3})\rho + (N_5T + N_6 + \frac{N_7}{X})\rho^2 + N_8\rho^3 + N_9\rho^4 + N_{10}\rho^5$$
$$+ \left[(\frac{N_{11}}{X^2} + \frac{N_{12}}{X^3} + \frac{N_{13}}{X^4})\rho^2 + (\frac{N_{14}}{X^2} + \frac{N_{15}}{X^3})\rho^4\right]\exp(-\frac{N_{16}}{T}) \tag{1}$$

where $X=T+5.6906$, ρ is density in mol/l and $Z=P/\rho RT$ is the compressibility factor. P is pressure in atm. If we use SI unit MPa for pressure, $Z=9.8692 \cdot P/\rho RT$.

Table 1 The initial coefficients of Gibbons' equation

$N_1 = 0.711661079\times10^{-1}$	$N_5 = -0.103027537\times10^{-4}$	$N_9 = 0.111831781\times10^{-8}$	$N_{13} = -0.467828934\times10^2$
$N_2 = -0.264145152\times10^1$	$N_6 = 0.123947764\times10^{-2}$	$N_{10} = 0.63176889\times10^{-8}$	$N_{14} = -0.89772816\times10^{-4}$
$N_3 = 0.205080136\times10^2$	$N_7 = -0.641439718\times10^{-2}$	$N_{11} = -0.203977574$	$N_{15} = 0.470376428\times10^{-3}$
$N_4 = -0.32981763\times10^3$	$N_8 = 0.158411494\times10^{-5}$	$N_{12} = 0.668353927\times10^1$	$N_{16} = 0.73596\times10^{-2}$

We compare the pressure predicted by Equation (1) with the coefficients given in Table 1 and the collected experimental data, which partly used by Gibbons et al themselves. Unfortunately, the results indicated that the equation can not express the experimental data distribution accurately. It has a maximum error 182.03% and an average error 27.27%. There are 498 points out of total 600 data records with an error greater than 5%. The authors firstly thought that some literal errors might be occurred, but we found that the equation and coefficients in other literatures [2, 3] turn out to be the same as Equation (1) and Table (1).

MODIFICATIONS TO EQUATION (1)

Refitting Equation (1)

First, the basic form of Gibbons' equation (1) is still used to make a nonlinear regression with those experimental data we collected. Then we refitted the coefficients, given in Table 2, and compared the predicted values and the experimental observations. It is found that only 16 points out of total 600 experimental records have a relative error greater than 5% and 45 points greater than 3%. The maximum and average errors are 14.18% and 1.194%, respectively. Obviously, the statistical results are much better than those generated by Equation (1) with the original coefficients.

Table 2　Refitted coefficients for equation (1)

$N_1 = 6.17194320552383E\text{-}1$	$N_5 = 4.32113547815425E\text{-}4$	$N_9 = \text{-}2.91237878499204E\text{-}5$	$N_{13} = 9.72104073531524E\text{+}0$
$N_2 = \text{-}1.91531493414952E\text{+}1$	$N_6 = \text{-}2.22439026331829E\text{-}2$	$N_{10} = 3.56069188211502E\text{-}7$	$N_{14} = \text{-}1.94416104080400E\text{-}4$
$N_3 = 1.23487003861953E\text{+}2$	$N_7 = 1.84867792880005E\text{-}1$	$N_{11} = \text{-}4.22210226133859E\text{-}1$	$N_{15} = 1.97960175514069E\text{-}3$
$N_4 = \text{-}1.35913310975649E\text{+}3$	$N_8 = 1.09489564515292E\text{-}3$	$N_{12} = 1.14520483939277E\text{+}0$	$N_{16} = \text{-}5.67114133250345E\text{+}0$

Modifying Equation (1)

On the basis of the above work and plenty of empirical trial, we obtained a modified form of Gibbons' equation, which can express the PVT surface of gaseous ^{3}He in the temperature region 4K-20K very well, see Equation (2). The coefficients are listed in Table 3.

$$T(Z-1) = \{N_1 + N_2 T + \frac{N_3}{T+N_0} + \frac{N_4}{(T+N_0)^2} + \frac{N_5}{(T+N_0)^3} + \frac{N_6}{(T+N_0)^4}\}\rho + (N_7 T + N_8 + \frac{N_9}{T+N_0})\rho^2$$

$$+\{N_{10}T + N_{11} + \frac{N_{12}}{T+N_0} + \frac{N_{13}}{(T+N_0)^2} + \frac{N_{14}}{(T+N_0)^3} + \frac{N_{15}}{(T+N_0)^4}\}\rho^3 + (N_{16}T + N_{17} + \frac{N_{18}}{T+N_0})\rho^4 + N_{19}\rho^5 \tag{2}$$

$$+\{[\frac{N_{20}}{(T+N_0)^2} + \frac{N_{21}}{(T+N_0)^3} + \frac{N_{22}}{(T+N_0)^4}]\rho^2 + [\frac{N_{23}}{(T+N_0)^2} + \frac{N_{24}}{(T+N_0)^3} + \frac{N_{25}}{(T+N_0)^4}]\rho^4\}\exp(-\frac{N_{26}}{T})$$

Table 3　Coefficients for Equation (2)

$N_0 = 1.37913054616371E\text{+}01$	$N_9 = \text{-}1.78125335499170E\text{+}01$	$N_{18} = \text{-}5.66286520059774E\text{-}03$
$N_1 = \text{-}8.47271117750707E\text{+}00$	$N_{10} = 2.18996380714379E\text{-}04$	$N_{19} = 4.32113723739915E\text{-}07$
$N_2 = 1.68593794065112E\text{-}01$	$N_{11} = 8.01717914362068E\text{-}04$	$N_{20} = \text{-}9.18876483522023E\text{+}02$
$N_3 = 5.62665344888040E\text{+}00$	$N_{12} = \text{-}1.38065195665170E\text{+}00$	$N_{21} = 1.90660089948584E\text{+}04$
$N_4 = 1.35205917030388E\text{+}04$	$N_{13} = 7.82901309576907E\text{+}01$	$N_{22} = \text{-}1.24336066467970E\text{+}05$
$N_5 = \text{-}3.40048252460412E\text{+}05$	$N_{14} = \text{-}1.53157150932023E\text{+}03$	$N_{23} = \text{-}3.19768785811710E\text{-}01$
$N_6 = 2.51846515212280E\text{+}06$	$N_{15} = 1.02688572589399E\text{+}04$	$N_{24} = 6.89110717561145E\text{+}00$
$N_7 = \text{-}1.63437319145173E\text{-}02$	$N_{16} = \text{-}4.93385536008283E\text{-}06$	$N_{25} = \text{-}4.85698353122027E\text{+}01$
$N_8 = 1.07275069012551E\text{+}00$	$N_{17} = 3.08888230842748E\text{-}04$	$N_{26} = 1.20249749298560E\text{+}01$

The maximum and average error of pressures predicted by Equation (2) with the coefficients listed in Table 3 is 9.94% and 1.11676%, respectively. There are only 10 points out of total 600 data records with an error greater than 5% and 40 points greater than 3%. So, this model is much better than Equation (1).

RESULTS AND ANALYSIS

To compare the performance of the equations mentioned above clearly, Table 4 and Figure 5 give their parameters and surface distribution. We can see that the refitted Gibbons' equation improved the agreement with the experimental observations obviously. And the accuracy can be further improved by using Equation (2) but with more coefficients. Further study showed that although the refitted Equation (1) and Equation (2) can express the PVT surface very well, their first and second derivative properties such as the specific heat could not be derived from both of them accurately. Hence, a new structure EOS is needed to describe all the thermodynamic parameters uniformly.

Table 4   Contrast of the three equations (Total 600 experimental data records)

Statistical Parameters	Original Gibbons' equation	Refitted Gibbons' Equation	Equation (2)
Average error	27.27%	1.194%	1.117%
Max error	182.03%	14.18%	9.94%
Number of points >5%	498	16	10
Number of points >3%	531	45	40

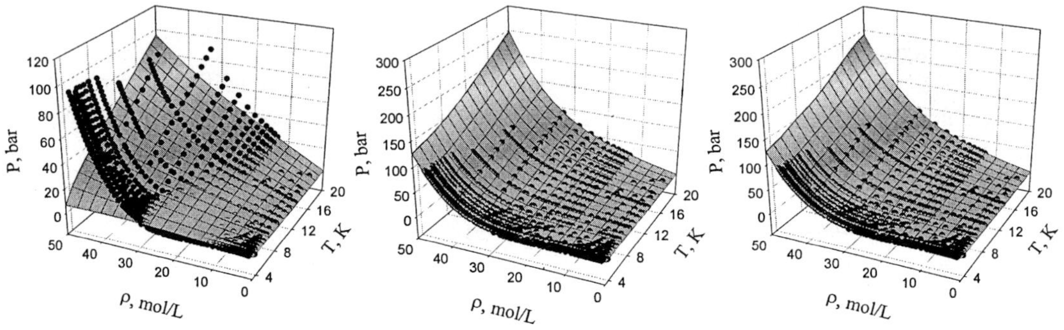

(a) Original Gibbons' equation   (b) Refitted Gibbons' Equation   (c) Equation (2)

Figure 5 Contrast of the surface generated by the three equations

CONCLUSIONS

The refitted Gibbons' equation can express the PVT properties of ^{3}He much better than the initial one in the gaseous region from 4K to 20K at pressures up to 10MPa. Then a modified form of the equation of state was brought forward with the average and maximum relative error 1.117% and 9.94% respectively. This equation can be used to predict the PVT properties of low temperature gaseous ^{3}He for technical application and to be a good reference to future study.

ACKNOWLEDGEMENT

Thanks for the enthusiastic help from Dr. Vincent Arp, a former senior researcher of NIST in USA, Prof. de Waele of Eindhoven University of Technology in Holland, Dr. Radebaugh of NIST in USA and Prof. Thummes of Gissen University in Germany who provided partial literatures and information on ^{3}He.

REFERENCES

1.   Gibbons, R.M. and Nathan, D.I., Thermodynamic Data of Helium-3, In: Technical Report AFML-TR-67-175, Air Products and Chemicals Inc., USA (1967) 29-30

2.   Gibbons, R.M. and McKinley, C., Preliminary Thermodynamic Properties of Helium-3 between 1 and 100K, Advances in Cryogenic Engineering (1968) 13 375-383

3.   Maytal, B. Z., ^{3}He Joule-Thomson Inversion Curve, Cryogenics (1996)36 271-274

An internet-based working fluid properties database

Huang J., He Y.L., Zhang Y.W.

State Key Laboratory of Multiphase Flow in Power Engineering
School of Energy & Power Engineering, Xi'an Jiaotong University, Xi'an 710049, China

In this paper, an internet-based working fluid property calculation system is proposed. An internet database is set up and a series of programs using database method to calculate the properties of working fluids are developed. To validate the system, property databases for Nitrogen and Oxygen were set up. In the programs, special attention was paid to some special conditions. These conditions might appear when the points to be calculated are near two-phase area. A set of correction methods to deal with these points are developed. Lastly, the calculation error is assessed by two different methods.

INTRODUCTION

The methods to calculate thermal properties of working fluids can be divided into two categories: database method and equation method. Equation method uses equations of working fluid state to get the properties of unknown state points. The programs developed on basis of this method are compact and flexible. But to develop this kind of program needs a lot of works and large numbers of iterations are used in the calculation which makes the calculation relatively slow and unstable. Another kind of method, database method, has no such shortcomings. This method depends on interpolation to get the unknown properties on specific state point. This is feasible because the property of working fluid in a single phase is continuous. But this kind of method has the disadvantage that a property database consumes relatively large space in computer. With the developing of internet technology, we can solve this problem by using internet database. The database could be stored in an internet server. Users get necessary information from the server through internet and complete the calculation at either the server side or the user side. This calculation system avoids complex processing of the properties data got by experiments. At the same time, no iteration is needed and no special software is needed on the user's computer but internet browsers only.

DATABASE METHOD FOR PROPERTY CALCULATION

In the database, properties data on discrete state points are stored in a series of data tables which are arranged by pressure and temperature. To calculate the thermophysical properties of a state point, the phase of the point is confirmed firstly by comparing the known properties such as pressure and enthalpy of the point with the data in the saturation curve table. Then the nearest points to this specific point in the database are found out whose properties are already known. The number of needed surrounding points is determined by the method used for interpolation, typically linear or least square interpolation. Then the properties of these points are interpolated to get the unknown properties of the specific point. For example, if we want to

916

know enthalpy of nitrogen at p=2.5MPa, T=273K, and bilinear interpolation is adopted, the calculation process is shown in Figure 1. In the figure, points 1, 2, 3 and 4 stand for the nearest known points to the point to be calculated. Points 1 and 4 have the same pressure while points 2 and 3 have the same pressure. This convention is followed for all the subsequent methods. The equation is:

$$h = \left[\frac{T-T_1}{T_4-T_1}(h_4-h_1) - \frac{T-T_3}{T_3-T_2}(h_3-h_2) \right] \frac{p-p_2}{p_1-p_2} \tag{1}$$

If the known properties are pressure and enthalpy, the state point may be located in two-phase area. In this case, dryness number x is calculated firstly and is used to calculate other properties. The calculation process is shown in Figure 2.

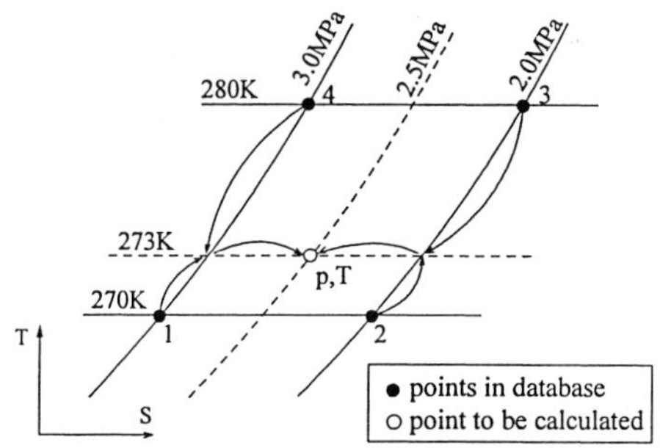

Figure 1   Basic calculation process

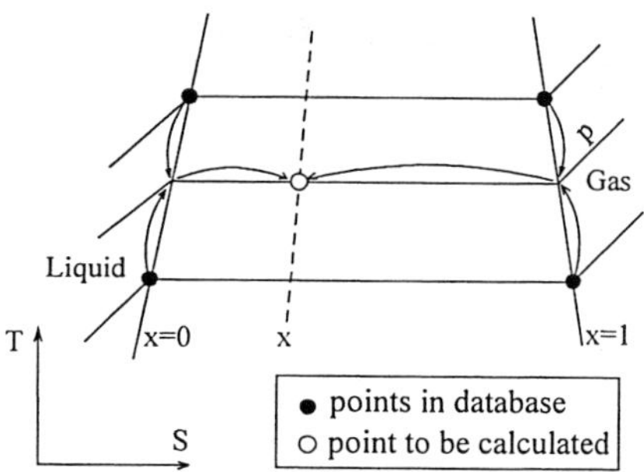

Figure 2   Case of two-phase area

SYSTEM STRUCTURE

In the system, web browsers such as Internet Explorer are used as user-interface to input calculation condition. The web browser transfers these data to the web server. Programs at the server-side then carry out calculation and transfer readable results back to the web browser. The structure of the system is shown in Figure 3.

Apache is used as web server in the current system. The database is developed on the basis of Mysql which has good compatibility with different operation system. Calculation programs and user interface are developed with PHP and JavaScript. The entire developing environment is free of charge which decreases the cost of the system.

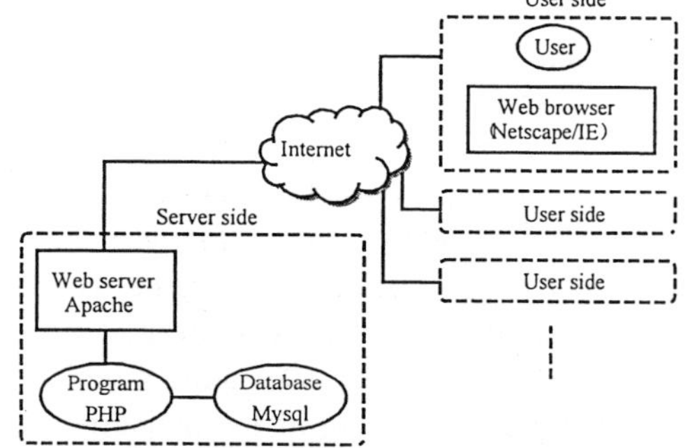

Figure 3   Structure of the system

STRUCTURE OF THE DATABASE

In the current system, property database for oxygen and nitrogen has been set up. The tables store the values of density, compressibility, enthalpy, entropy, isochoric and isobaric specific heat for the liquid and gaseous phases and on the solidification and saturation curves. An additional field "con" is added to the table to indicate the phase state of the point. The value of "con" may be 0, 1 or 2, which respectively stand for liquid area, gas area and the area whose temperature or pressure is higher than the critical point. We use the data from [1] and [2] which is supposed to be very reliable. The data are arranged by temperature and pressure. Table 1 is part of the data tables. The points are in the temperature interval from the triple point to 1500K and pressures between 0.1 and 100 MPa. An independent table is used to store the data of points on saturation curve. All the values are in SI units.

Table 1   Structure of the database (Oxygen)

T (K)	p (MPa)	ρ (kg/m^3)	z	h (kJ/kg)	s (kJ/(kg·K))	c_v (kJ/(kg·K))	c_p (kJ/(kg·K))	con
				...				
135	20.0	976.37	0.5840	226.9	3.537	0.831	1.710	2
135	21.0	980.40	0.6106	227.2	3.532	0.831	1.710	2
				...				
140	20.0	950.78	0.5783	235.5	3.600	0.824	1.735	2
140	21.0	955.29	0.6043	235.8	3.594	0.824	1.721	2

SPECIAL POINTS

Because the data tables are arranged by temperature and pressure, points of different phase are successive in the table. So the points to be interpolated may be located in different phase, which may lead to large error. Special steps must be taken to treat with these conditions. The following methods are based on two-dimensional linear fit.

The case can be divided into 2 categories according to the phase of the point to be calculated. To be concise, only the case of gas phase will be illustrated here. The case of liquid phase is similar to the case of gas phase.

The rule of the methods to treat with these special conditions is to discard the points whose phase is not the same as the point to be calculated and replace them with appropriate point in the saturation curve table. For example, in Figure 4, although the point 1, 2 and 4 is the nearest points in the data table to the point to be calculated, they are in different phase. So we replace them with the points on saturation curve that have the same temperature and pressure with the point to be calculated and use them in interpolation.

Totally there are 12 special conditions to be considered. In fact, we find that these conditions can be processed with only six sub programs which can make the program compact and the calculation more efficient.

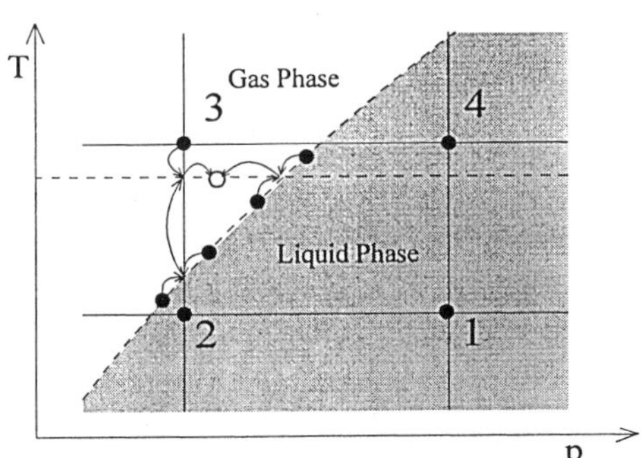

Figure 4   Processing of special points

ERROR ASSESSMENT

Two methods of error assessment are used to check the accuracy of the system. The first could be called self-comparison method. A known point in the database is calculated by the system and then compared with itself. Figure 5a shows error of enthalpy assessed by this method when temperature equals to 140K. It can be found that in area far from double phase area, error of enthalpy calculated by the system is less than 0.8kJ/kg. In the area that is near to double-phase area, the error may be too high to be acceptable. To solve this problem, the density of points in this area should be increased in the database and higher order interpolation such as bicubic interpolation should be adopted. But higher order interpolations need more points in calculation. For example, bicubic interpolation uses an average of 16 surrounding points. At the same time, conditions will be much more complex near double-phase area. The second method is to compare the calculation result with the result of the calculation program recommended by [3]. Figure 5b shows the deviation of enthalpy assessed by this method. The conclusion is similar to that of method 1.

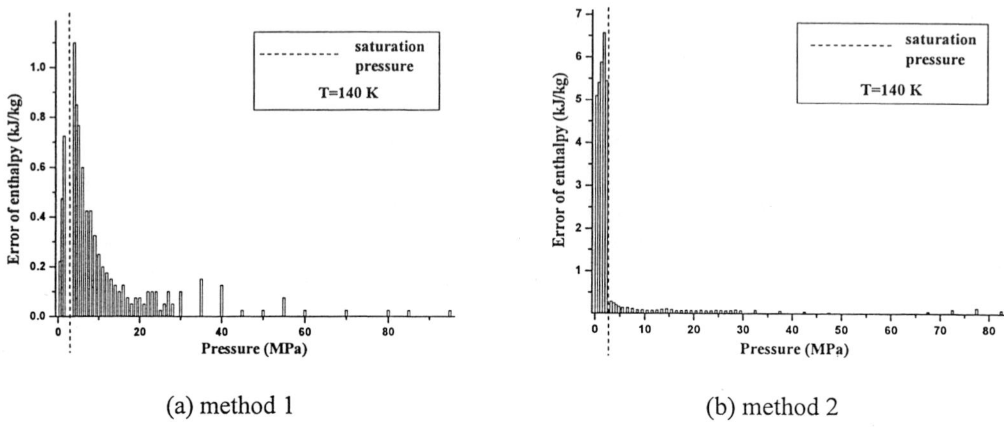

(a) method 1   (b) method 2

Figure 5   Error assessment of enthalpy

To check the accuracy for changes in properties over smaller temperature and pressure steps than the step size of database, we calculated the enthalpy change while temperature changes from 137K to 139K and pressure change is half of the database step size. The result is compared with [3] and shows a highest relative error of 4.7% while in most cases the error is less than 1.0%.

CONCLUSION

An internet-based working fluid property calculation system has been set up and the error of the calculation is assessed. The result shows that to calculate the thermophysical properties of working fluids through internet is feasible and the error is acceptable if the density of points in database is high enough.

ACKNOWLEGEMENT

This work was supported by the National Key Project of Fundamental Research of R & D of China (G2000026303) and the National Natural Science Foundation of China (50276046).

REFERENCE

1. V. V. Sychev, A. A. Vasserman, A. D. Kozolov, G. A. Spiridonov, V. A. Tsymarny, <u>Thermodynamic Properties of Oxygen</u>, Hemisphere Publishing Corporation, Washington, USA, (1987) 98-210

2. V. V. Sychev, A. A. Vasserman, A. D. Kozolov, G. A. Spiridonov, V. A. Tsymarny, <u>Thermodynamic Properties of Nitrogen</u>, Hemisphere Publishing Corporation, Washington, USA, (1987) 130-216

3. Richard T Jacobsen, Steven G. Penoncello, Eric W. Lemmon, <u>Thermodynamic Properties of Cryogenic Fluids</u>, Plenum Press, New York, USA, (1997) 293-300

Proceedings of the Twentieth International Cryogenic Engineering Conference
(ICEC 20), Beijing, China.

A new process for the preparation of polyimide/silica hybrid films

Li Y., Fu S. Y.[*], Pan Q. Y., Zhang Y. H., and Lin D. J.

Cryogenic Materials Division, Technical Institute of Physics and Chemistry,
Chinese Academy of Science, Beijing 100080, China

In this article a new process was introduced to prepare a series of transparent PI/silica hybrid films. The mechanical properties at 77 K were studied and compared with the films prepared by the traditional process. It was shown that the tensile strength of the films prepared by the new process were superior to the films prepared by the traditional process. The morphology of the films was characterized by scanning electron microscopy (SEM) observation.

INTRODUCTION

In recent years, the cryogenic properties of polymers and polymer composites have drawn much attention with rapid developments in space, superconducting magnet and electronic technologies[1-3]. Polyimide (PI) films are used as insulating materials in superconducting magnet systems because of their high mechanical properties and excellent electrical properties etc. Electrical insulation in superconducting magnet systems is subjected to uncommon synergetic conditions: high electrical stress, high magnetic stress, high mechanical stress by electromagnetic force, thermo-mechanical stress caused by the cryogenic environment, phase transition of coolant and high energy radiation etc. So study on the mechanical properties of the insulating materials at cryogenic temperature is of great importance. Recently, the properties of PI/SiO_2 hybrid films prepared by sol-gel process has been studied extensively[4-10]. The PI/silica hybrid films possess lower coefficient of thermal expansion, higher thermal stability and better mechanical properties at room temperature compared with neat PI films. However, the mechanical properties of PI/SiO_2 hybrid films by the traditional process decreased fast by the addition of SiO_2 particles at higher silica content because of the fast increase of particle size; moreover, few studies have been carried out on cryogenic properties of the PI/ SiO_2 hybrid films.

In this article, a new process, which would lead to a smaller particle size than the traditional process, was developed for the preparation of PI/SiO_2 hybrid films. The mechanical properties at 77 K of the films prepared by the new process were studied. The morphology of the films was characterized by scanning electron microscopy (SEM) observation.

EXPERIMENTAL

Materials

Pyromellitic dianhydride (PMDA,$C_{10}H_2O_6$) and diphenylene diamine (ODA,$C_{12}H_{12}N_2O$) were provided by Tecnidd Enterprise co. Ltd. N,N-dimethylacetamide (DMAc, $[(CH_3)_2NCOCH_3)]$) was desiccated by molecular sieve before use. Tetraethoxysilane (TEOS), ethanol (EtOH) and catalyst were purchased from Beijing Chemical Co. LTD., and used without further purification.

920

Preparation of PI/SiO$_2$ hybrids

A: Traditional process

ODA was first dissolved in DMAc. PMDA was then added to the reaction mixture under a nitrogen atmosphere and stirred at room temperature for 6 h. Then, the TEOS, water and catalyst were added to the solution. After the addition of TEOS, water and catalyst, further stirring was needed to recover a homogeneous solution. The transparent solution was spun onto a glass plate and subsequently dried respectively at 80, 100, 120, 150, 180, 240 and 270 °C for 1 h. The formulations for the PS1 hybrid films prepared using the traditional process are shown in Table 1.

Table 1.   Preparation of PI/SiO2 hybrid films

Sample name	PAA[a] (g)	TEOS (g)	H$_2$O (ml)	Catalyst[b]	EtOH (ml)	Silica content (wt%)	Remarks[c]
PI	15	0	—	—	—	0	T
PS1-1	15	0.035	0.012	5.0	—	1	T
PS1-3	15	0.106	0.037	5.0	—	3	T
PS1-5	15	0.182	0.063	5.0	—	5	T
PS1-8	15	0.301	0.103	5.0	—	8	S
PS1-10	15	0.385	0.132	5.0	—	10	O
PS1-15	15	0.612	0.210	5.0	—	15	O
PS2-3	15	0.106	—	37	0.301	3	T
PS2-5	15	0.182	—	37	0.494	5	T
PS2-8	15	0.301	—	37	0.791	8	T
PS2-10	15	0.385	—	37	0.987	10	T
PS2-15	15	0.612	—	37	1.571	15	O

a: 15wt% DMAC solution.

b: Weight percentage of hydrochloric acid based on the amount of water added.

c: The appearance of hybrid films, T: transparent, S: translucent, O: Opaque.

B: New process

ODA was first dissolved in DMAc under a nitrogen atmosphere. TEOS, ethanol (EtOH) and catalyst were then added. PMDA were added after stirring for about 30 min. Subsequently stirring of about 6h was needed in order to gain a homogeneous solution. The hybrid films can be obtained through the same thermal treatment as above. The formulations for the PS2 hybrid films prepared using the new process are also given in Table 1.

Measurement and characterization

Mechanical properties were measured with a RGT-20A testing machine at a rate of 2 mm/min. The morphology of the fracture surfaces of samples was investigated by scanning electron microscopy (SEM) with a Hatachi S-4300 microscope (Japan), operating at 10.0 kV.

RESULTS AND DISCUSSION

Mechanical properties

The influence of the silica content on the tensile strength at 77 K is shown in Figure 3 (a). The tensile strength at 77 K increased initially with the increase of silica content for both systems. The tensile strength at 77 K of the PS1 films was increased by about 8% when the 3 wt% silica content was added. A further increase was observed for the PS2 system. When 5 wt% silica was added, the tensile strength of the hybrid films reached 184 MPa (an increase of about 34%). This effect might result from the stronger physical interaction between organic and inorganic phase in the PS2 system.

The influence of the silica content on the elongation at break at 77K is shown in Figure 3 (b). The elongation at break at 77 K of the PS1 system decreased with the increase of the silica content. While the elongation at break of the PS2 system increased initially and decreased with the increase of SiO_2 content. The elongation at break was increased by about 21% when 3 wt % silica was introduced.

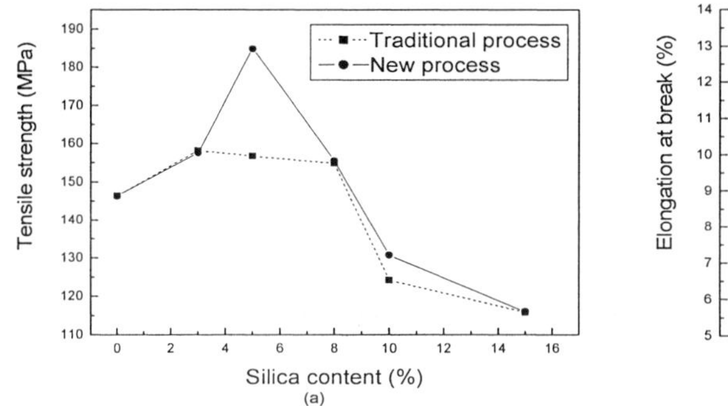

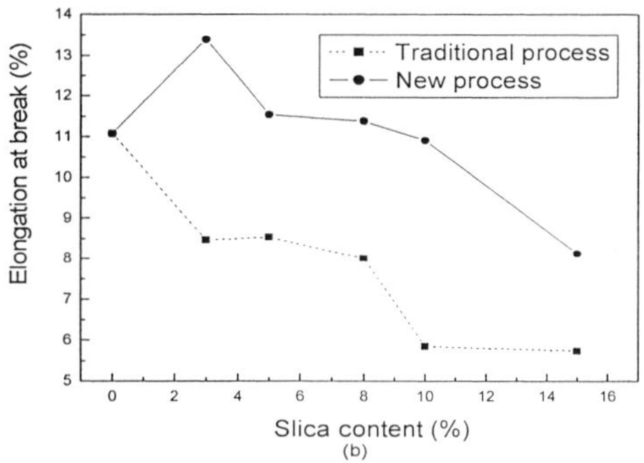

Figure 1.   The mechanical properties at 77K of hybrid films for two systems.
(a) Tensile strength, (b) elongation at break.

SEM analysis

Figure 2 shows the SEM photographs of the tensile fracture surfaces of PI/SiO_2 hybrid films. The dispersed silica particles could be seen as white beads with an average diameter of about 40 nm, 1 μ m, 30 nm and 600 nm respectively for PS1-3, PS1-15, PS2-3 and PS2-15 films. The increase in silica particle size clearly resulted from the increase in the aggregation tendency as the increase of silica content. A fine interconnected phase morphology of PS1-3 and PS2-3 could be seen in Figure 2 (a) and (b), The morphology of PS1-15 and PS2-15 reveals that the silica particles are very smooth and completely debonded from the surrounding polyimide matrix, indicating a very poor interfacial adhesion between particles and matrix.

On the basis of the morphological observations, it is evident that the reduction in ultimate properties observed for the films with high SiO_2 contents can be attributed to weak interfacial adhesion between particles and matrix, which allows the silica particles to act as stress-concentration defects, rather than as effective reinforcing filler. On the contrary, for the films with low silica contents, the increased tensile

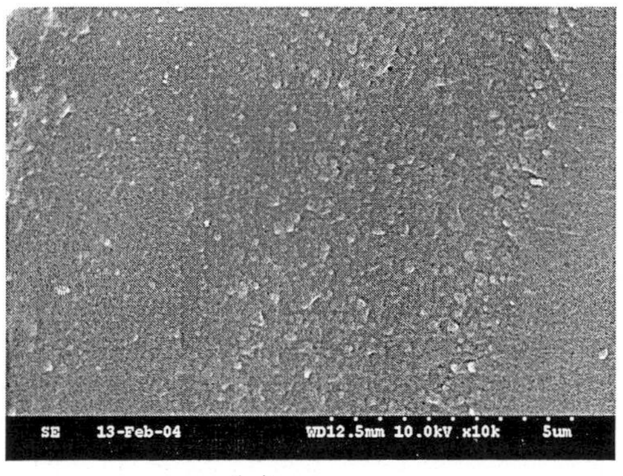

(a)

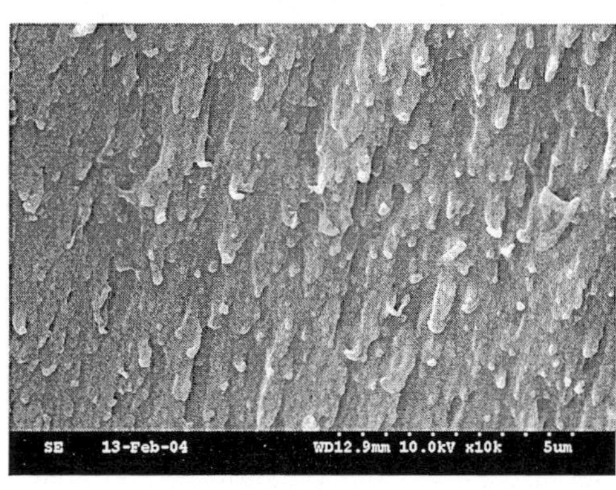

(b)

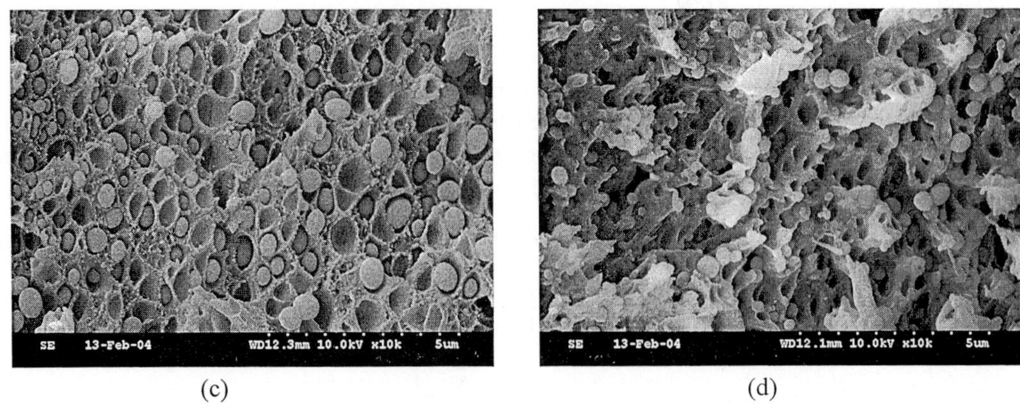

(c)   (d)

Figure 2.   The morphology of the tensile section of the films by SEM.
(a) PS1-3, (b) PS2-3, (c) PS1-15 and (d) PS2-15 films

strength is the result of both a better interfacial adhesion and the formation of co-continuous morphologies, which improve the efficiency of stress transfer mechanisms between the two components. At a given silica content, the size of the SiO_2 particles is smaller and then the interfacial area is larger in the PS2 films than in the PS1 films, so the silica particles would have better reinforcing and toughing effects for the PS2 films than for the PS1 films.

CONCLUSIONS

PI/silica hybrid films have been prepared with the traditional and newly developed sol-gel process. It was shown that the newly developed sol-gel process led to smaller silica particle sizes than the traditional sol-gel process did. The PI/SiO_2 hybrid films prepared by the new process showed better cryogenic mechanical properties than the hybrid films prepared by the traditional sol-gel process.

REFERENCES

1. H. Yamaoka, K. Miyata and O,Yano, Cryogenic properties of engineering plastic films, <u>Cryogenics</u> (1995) <u>35</u> 787-789.

2. Takefumi Horiuchi and Tsutomu Ooi, Cryogenic properties of composite materials ,<u>Cryogenics</u> (1995) <u>35</u> 677-679.

3. Okimichi Yano and Hitoshi Yamaoka ,Cryogenic properties of polymers, <u>Prog.Polym.Sci.</u>(1995)<u>20</u> 585-613.

4. Yi Huang, Yi Gu, New Polyimide-silica Organic-Inorganic Hybrids, <u>J.Polym. Sci.</u>(2003) <u>88</u> 2210-2214.

5. Ging-Ho Hsiue，Jem- Kun Chen，Ying-Ling Liu, Synthesis and Characterization of Nanocomposite of Polyimide-Silica Hybrid from Nonaqueous Sol-Gel Process, <u>J.Appl.Polym.Sci.</u>(2000) <u>76</u> 1609-1618.

6. Jing Liu, Yan Gao, FuDong Wang, Ming Wu, Preparation and Characteristics of Nonflammable Polyimide Materials <u>J.Appl.Polym.Sci.</u>(2000) <u>75</u> 384-389.

7. Zi-Kang Zhu, Yong Yang, Jie Yin, Zong-Neng Qi, Preparation and Properties of Organosoluble Polyimide/Silica Hybrid Materials by Sol-Gel Process, <u>J. Appl. Polym. Sci.</u>(1999)<u>73</u> 2977-2984.

8. Atsushi Morikawa, Yoshitake Iyoku, and Masa-aki Kakimoto etc.Preparation of a new Class of Polyimide-silica Hybrid Films by sol-Gel process   <u>Polym. Jour.</u> (1992)<u>24</u> 107-113.

9. L.Mascia and A.Kioul, Influence of siloxane composition and morphology on properties of polyimide-silica hydrids, <u>Polymer </u>(1995) <u>36</u> 3649-3659.

10. T.A.Shantalii, I.L.Karpova, K.S.Dragan,etc,Synthesis and thermomechanical characterization of polyimides reinforced with the sol-gel derived nanoparticles , <u>Sci. and Tech. of Adv. Mater.</u> (2003) <u>4</u> 115-119.

Design of a fibre reinforced plastic anticryostat for magnetorelaxometric measurements

Koettig T., Weber P., Prass S., Seidel P.

Institut für Festkörperphysik, Friedrich-Schiller-Universität Jena, Helmholtzweg 5, 07743 Jena, Germany

A new method for the characterization of magnetic nanoparticles is based on the analysis of the temperature dependence of the Néel relaxation sample signal. The presented cryostat extends the investigated temperature range from 300 K to 77 K down to the boiling point of liquid helium at 4.2 K. We designed an anticryostat to use only one liquid helium cryostat for the sample as well as for the Low-T_C SQUID. Therefore it is necessary to study the permeation process through the used fibre reinforced plastic (FRP) material and the adhesive joints between the components.

INTRODUCTION

The magnetic properties of ferrofluids are strongly influenced by the distribution density of energy barriers in these many particle systems [1]. The low signal level of the magnetisation relaxation demands the application of non-magnetic and electrical insulating materials due to prevent self-induced eddy currents within the cryostat walls. The high permeation constant of helium through any kind of synthetic material inhibits long measurement periods. It is necessary to characterise the permeation conductivity through the used FRP materials. Because of the competing influence of the heat conductance along the wall of the cryostat small wall thicknesses are useful. This results in a high pressure gradient and a high helium permeation through the wall material. This permeation process restricts the lifetime of the insulating vacuum.

EXPERIMENTAL

We created a permeation measurement set-up to characterise the gas transport due to the permeation process through composites used as wall materials. The basic arrangement of the whole measurement set-

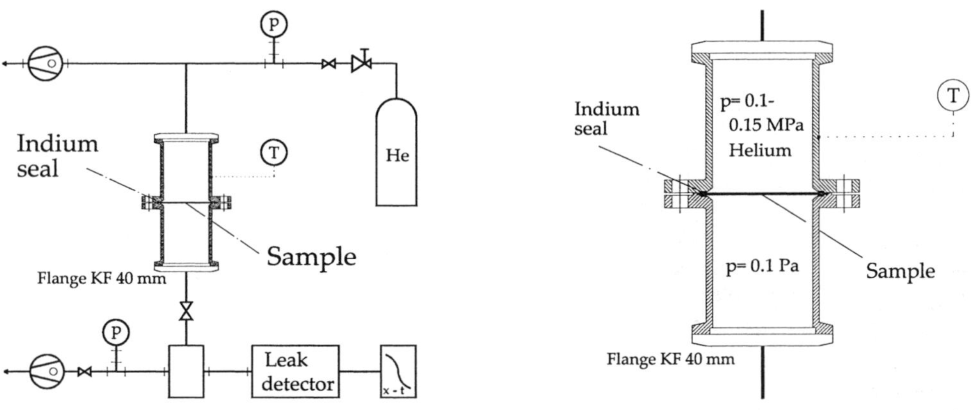

Figure 1 Permeation measurement assembly (a) and the permeation cell in detail (b)

up and the permeation cell are shown in Figure 1. We tested thin plates of various FRP materials with a diameter of 41 mm and a thickness between 1 and 1.5 mm depending on the different distributors. All permeation measurements have been made with helium. The permeation process is strongly influenced by the ratio of glass and epoxy resin, the glass composition and the reinforcement structure. In addition the effect of machining and the adhesive joint material must be taken into consideration. In order to compare these measurements we conditioned the same FRP-material with various adhesive and lacquer layers. We have used a commercial leak detector with a detection limit of $5 \cdot 10^{-11}$ mbar $\cdot$ dm^3/s leak rate. The leak rate q_L corresponds to a p-V-current regarding to standard temperature and standard pressure (STP). The measured leak rate has to be standardised to an area of A = 1 m^2 and a pressure gradient across the wall thickness of $\Delta p/\Delta x = 1$bar/ 1mm [2]. The temperature dependence of the permeation can be described by an Arrhenius equation [3] (see Equation 2):

$$q_{Perm} = q_L \cdot \frac{\Delta x}{A \cdot \Delta p} \tag{1}$$

$$q_{Perm} = q_{Perm,0} \cdot exp(-E_P/RT) \tag{2}$$

With q_{Perm}: permeation conductivity, E_P: activation energy of permeation, R: gas constant, T: absolute temperature. Hence it is sufficient to study the permeation process from 290 K to 310 K (see Figure 2). The cold parts of the cryostat at a wall temperature of about 4.2 K can be neglected because the permeation values approximately decrease by a factor of 10^{10} already at the liquid nitrogen temperature of 77 K [4,5].

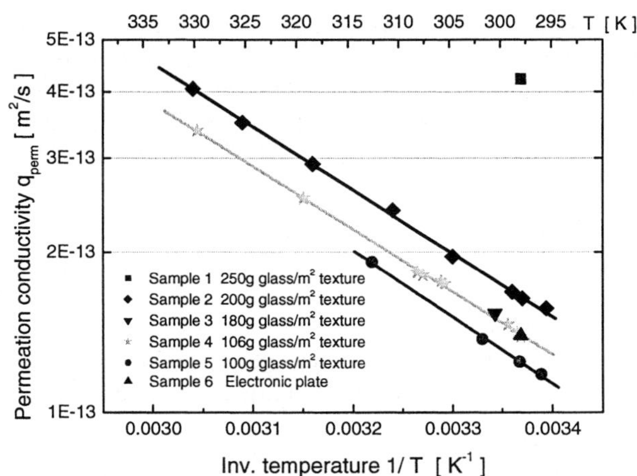

Figure 2 Temperature dependence of the helium permeation for different FRP materials

The legend in Figure 2 explains the amount of glass at every inserted texture inside the epoxy resin. The higher the amount of glass per square meter the coarse is the glass texture. Every FRP sample has consequently the same glass fibre volume inside the epoxy resin (55 %) except sample 1. This sample additionally consists of a higher glass partition of about 69 % fibre volume.

In a further step we studied the influence of the adhesive joints and of one first lacquer film coating (see Figure 3). In order to investigate the gas permeation through adhesive joints we conditioned one 200 g glass/m^2 sample with 100 holes having a diameter of 1 mm and filled them with adhesive. The calculated value of the adhesive material is shown in Figure 3. The permeation value is four times higher than the untreated FRP material. The same 200g glass/m^2 sample was used as a substrate of the deposited lacquer films. Film thicknesses below 0.05 mm have no influence on the permeation value. The result is an identical permeation amount compared to the untreated sample.

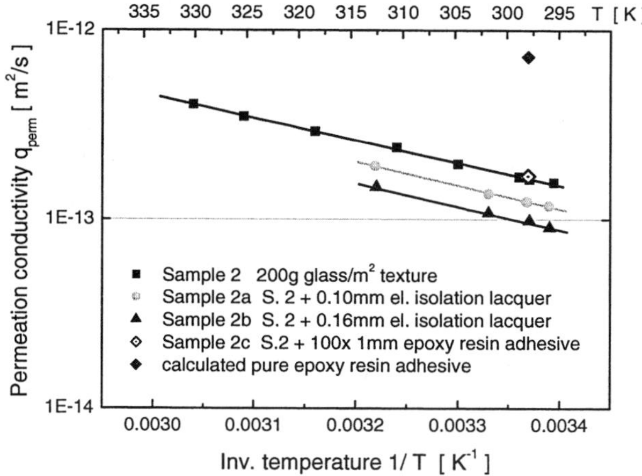

Figure 3 Modification of the helium permeation by coating methods and adhesive joints

RESULTS AND DISCUSSION

By means of the designed permeation gauging set-up we are able to investigate the influence of the surface coating and adhesive joints at different FRP materials. The thinner the glass texture the lower is the permeation value q_{Perm}. Adhesive joints have to be dimensioned five times wider along the pressure difference direction to compensate the higher permeation value. Machined surfaces and adhesive joints can be additionally coated with thin lacquer films to decrease gas permeation. That requires a deposited film thickness greater than 0.1 mm. In consideration of all these results we designed an anticryostat that extends the measuring range below 77 K down to the boiling temperature of liquid helium (see Figure 4).

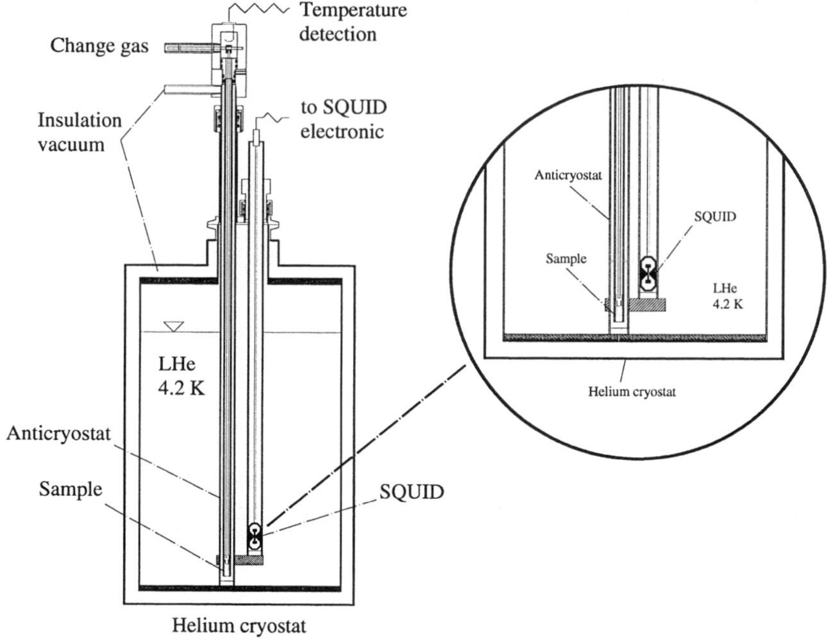

Figure 4 Relaxation measurement set-up including the anticryostat

First measurements were accomplished with a ferrofluid consisting of Maghemit (γ-Fe$_2$O$_3$)-particles whereby a temperature of 77 K confines the measurement range to a particle size above 15 nm. The maximum of these distributions moves to lower temperatures with decreasing particle size. Figure 5 shows the energy barrier density distribution of γ-Fe$_2$O$_3$ versus the temperature. The maximum at 13 K corresponds to a particle size of about 7 nm. That's why measurements below 77 K are of physical note.

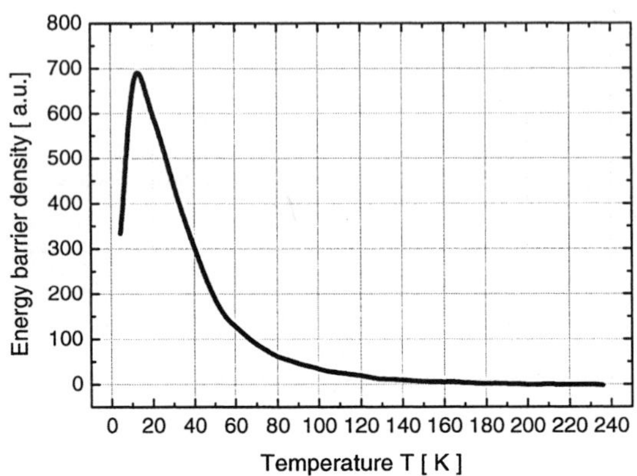

Figure 5 Energy barrier density distribution of γ-Fe$_2$O$_3$ versus temperature

CONCLUSIONS

We designed an FRP anticryostat for magnetorelaxometric measurements with a minimum distance between the SQUID-gradiometer and the sample position. The experimental set-up has been tested successfully and a first particle size distribution at about 7 nm is presented.

To improve the lifetime of FRP helium cryostats other coating materials and methods have to be tested. The mechanical properties especially the adhesive strength down to 4.2 K limits the material variety. There are eligible candidates among these, like lead borate glasses and lead borosilicate glasses. The films will be sputtered as surface layers with a thickness of more than 5 µm.

REFERENCES

1. Romanus, E., Berkov, D.V., Prass, S., Groß, C., Weitschies, W. and Weber, P., Determination of energy barrier distributions of nanoparticles by temperature dependent magnetorelaxometry, Nanotechnology (2003), 14 1251-1254
2. Norton, F.J., Permeation of gases through solids, Journal of Applied Physics (1957), 28 34-39
3. Humpenöder, J., Gas permeation of fibre reinforced plastics, Cryogenics (1998), 38 143-147
4. Okada, T. and Nishijima, S., Gas permeation and performance of an FRP cryostat, Advances in Cryogenic Engineering (1988), 34 17-24
5. Evans, D. and Morgan, J.T., The permeability of composite materials to hydrogen and helium gas, Advances in Cryogenic Engineering (1988), 34 11-16
6. Ianno, N.J. and Makovicka, T.J., Measurement of the permeability of thin films, Review of Scientific Instruments (1999), 70 2072-2073
7. Mende, F.F., Gorbunov, V.M., Bondarenko, N.N., Logvinov, V.N. and Zhuravel, T.I., Broad-neck liquid helium cryostat with long lifetime, Cryogenics (1989), 29 998-1001
8. Shelby, J.E., Helium migration in alkali borate glasses, Journal of Applied Physics (1973), 44 3880-3888
9. Abbot, P.J. and Tison, S.A., Commercial helium permeation leak standards: Their properties and reliability, Journal of Vacuum Science Technology A (1996), 14(3) 1242-1246

MEASUREMENT OF HEAT LOAD ON THE CURRENT LEAD USING BOIL-OFF CALORIMETRY

Yang H. S., Kim D. L., Kim D. H., Jung W. M., Cho S.

Device Operation Division, Korea Basic Science Institute
52 Yeoeun, Yuseong, Daejeon 305-333 Korea

In order to measure heat load through the current lead for HTS power cable system, we developed a new type of cryostat system applying boil-off calorimetry method. One of the main characteristics of the system is to avoid gas convection cooling effect. A prototype current lead inserted in the system was made of copper. Heat conduction loads on the current lead were measured at different heat load conditions. The experimental results were in good agreement with the numerical ones within 3% error.

INTRODUCTION

The development of HTS (High Temperature Superconducting) power cable system is in progress as one of the project of the 21st Century Frontier R&D program in Korea since 2001. The power cable system consists of termination cryostats, HTS power cable and their cooling system. The current lead located in the termination cryostat is used to transmit the power from the room temperature power cable to the low temperature HTS power cable. Because the current lead is the major heat source of cooling load, the heat load of the current lead has influence of the liquid nitrogen consumption and the refrigeration capacity. For this reason, the design of current lead with smaller heat load is very crucial for the economic refrigeration system. Generally, the measurement of heat load through current lead by boil-off calorimetry method is affected by gas convectional cooling effect in the cryostat. This is one of the factors hindering the precise measurement of the heat load through the current lead.

In the present study, in order to measure heat load through the current lead, we developed a new type of cryostat system applying boil-off calorimetry method [1, 2], which was designed to avoid gas convection cooling effect. As the first step, we verified the usefulness of the system. And we measured the conduction heat loads through the current lead under different heat load conditions.

EXPERIMENTAL APPARATUS

The schematic overview of the experimental apparatus and the photograph of the cryostat are illustrated in figures 1-(a) and (b) respectively. The experimental apparatus consists of two major parts, a cryostat, where a current lead with a diameter of 26 mm and a length of 1500 mm is inserted in liquid nitrogen vessel and a measurement system for the data acquisition and storage. The size of the current lead was determined by optimum calculation [3,4] for 1.26 kA of HTS power cable.   The temperature of the current lead is monitored by several temperature sensors attached at the top and bottom end. The top end temperature of the current lead can be controlled by a temperature controller for the purpose of changing the heat load through the current lead. The measurement system consists of a gas flow meter for measuring boil-off gas volume flow rate, a pressure transducer and a thermometer for measuring the pressure and temperature of the boil-off gas, respectively, a barometric pressure transducer and a thermometer for measuring atmosphere conditions. The heat load is computed by the boil-off gas mass obtained by relations as following.

$$m = \rho \cdot V \tag{1}$$

$$Q = m \cdot h_{fg} \tag{2}$$

Where ρ: density, m: boil-off gas mass, V: boil-off gas volume, Q: heat load, h_{fg}: latent heat

Major characteristic of this apparatus is that the heat load of the current lead is maintained constant by contacting the liquid nitrogen at the lower end only. It removes the vapor cooling effect from convection heat transfer and conduction heat load variation appeared by changing free surface location of liquid nitrogen. It is very important point to measure the heat load through current lead correctly.

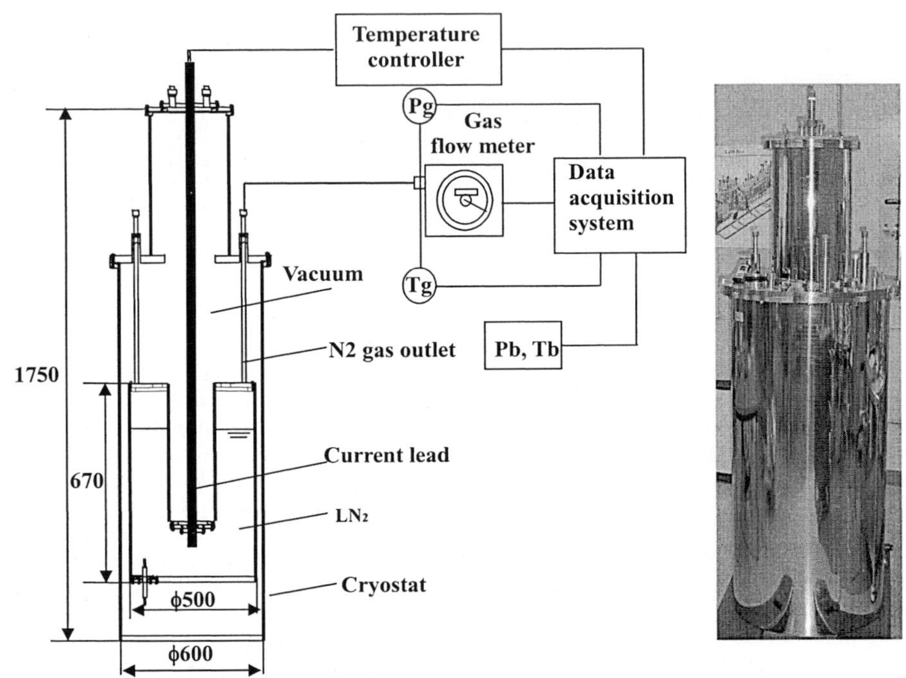

(a) Schematic overview of the experimental

(b) Photograph of the cryostat

Figure 1 Experimental apparatus

Pg: Pressure transducer, Pb: Barometric pressure transducer, Tg and Tb: Temperature sensor for N_2 gas and barometric temperature measurement, respectively.

RESULTS AND DISCUSSION

<u>Verification of boil-off calorimetry accuracy</u>

Before measuring the heat load of the current lead, we verified the accuracy of boil-off calorimetry method. In order to verify the accuracy, we measured the background heat load of the cryostat without inserting the current lead. The heat loads are presented in Table 1.

Table 1 Verification of Boil-off Calorimetry Method Accuracy

Background Heat Load (W)	Input Heat Load (W)	Estimation Heat Load (W)	Measured Heat Load (W)	Error (%)
8.20	9.20	17.40	17.31	0.5%
8.08	23.01	31.09	30.97	0.4%

At first, the background heat load of the cryostat reaches about 8 W, which are mostly from the heat conduction load through support loads of the liquid nitrogen vessel and the radiation heat load. And then we installed a heater and changed the heat load in the liquid nitrogen vessel. For two cases of input heat loads 9.20 W and 23.01 W, the values measured by boil-off calorimetry method are in good agreements within 0.5 % comparing to the estimated value obtained from adding input heat loads to the background heat loads.

The Heat Load of the Current Lead

Figure 2 shows the heat loads of the current lead measured from different four top temperatures of the current lead under vacuum pressure of 6.6×10^{-5} Pa for changing conduction heat load. The results are compared to the values calculated using by well-known heat conduction equation :

$$Q = \frac{A}{L} \int_{T_1}^{T_2} \kappa(T) dT \tag{3}$$

Here, A is the cross-sectional area of the copper current lead, L is the length of the current lead and the thermal conductivity, $\kappa(T)$, is a function of temperature. T_1 is bottom temperature and T_2 is top temperature of the current lead. The measured heat loads coincide with the calculated ones within 3% error. From these results, it is known that the heat load through current lead can be accurately measured and that the heat load can also be exactly estimated by the heat conduction equation (3).

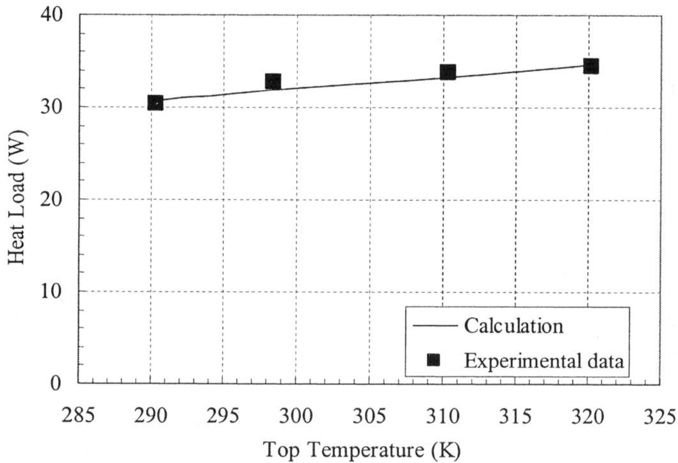

Figure 2 Measured Heat Loads in Comparison with the Calculation Values

CONCLUSION

We developed a new type of cryostat system applying boil-off calorimetry method to measure accurately heat load through the current lead for HTS power cable system. The main characteristic of the system is to avoid gas convection cooling effect. Heat loads through the copper current lead were measured for four different heat load conditions. The experimental results coincide with the numerical ones within 3% error.

ACKNOWLEDGMENT

This research was supported by a grant from Center for Applied Superconductivity Technology of the 21st Century Frontier R&D Program funded by the Ministry of Science and Technology, Republic of Korea.

REFERENCES

1. Hanzelka, P. and Musilova V., Influence of changes in atmospheric pressure on evaporation rates of low-loss helium cryostat, Cryogenics (1995) 35 215-218
2. Ohi, T. Calorimetry by Boil-Off Measurement, Cryogenic Engineering in Cryogenic Association of Japan (1992) 27 315-324
3. Change, H.-M. and Van Sciver, S. W., Optimal Integration of Binary Current Lead and Cryocooler, Internation Cryocooler Conference-10 (1999) 707-716
4. Cho,S. et al, Calculation of Heat Loads and Temperature Distribution for HTS Temperature Current Lead, presented at CEC/ICMC 2003 Anchorage, USA (to be published)

Alignment of the ISAC-II Medium Beta Cryomodule with a Wire Monitoring System

Rawnsley W.R., Bylinski Y., Fong K., Laxdal R.E., Dutto G., Giove D. *

TRIUMF, 4004 Wesbrook Mall, Vancouver, BC, Canada, V6T 2A3
* LASA - INFN, Via Fratelli Cervi, 201 - 20090 Segrate, Milano, Italy

TRIUMF is developing ISAC-II, a superconducting (SC) linac. It will comprise 9 cryomodules with a total of 48 niobium cavities and 12 SC solenoids. They must remain aligned at liquid He temperatures: cavities to ±400 μm and solenoids to ±200 μm after a vertical contraction of ~4 mm. A wire position monitor (WPM) system based on a TESLA design measures the signals induced in stripline pickups by a 215 MHz signal carried by a position reference wire. The sensors, one per cavity and two per solenoid, monitor their motion during pre-alignment, pumping and cool down. System accuracy is ~7 μm.

INTRODUCTION

TRIUMF is now constructing an extension to the ISAC facility, ISAC II, to permit acceleration of radioactive ion beams up to energies of at least 6.5 MeV/u for masses up to 150. The proposed acceleration scheme will use the existing ISAC RFQ (E = 150 keV/u) with the addition of an ECR charge state booster to achieve the required mass to charge ratio ($A/q \leq$ 30) for masses up to 150. A new room temperature IH-DTL will accelerate the beam from the RFQ to 400 keV/u followed by a post-stripper heavy ion superconducting linac designed to accelerate ions of $A/q \leq$ 7 to the final energy. The superconducting linac is composed of two-gap, bulk niobium, quarter wave rf cavities, for acceleration, and superconducting solenoids, for periodic transverse focussing, housed in several cryomodules. A total of 48 cavities and 12 solenoids will be used. The center line of each cavity must be aligned to within ±400 μm of the true beamline centre while that of the solenoid must be within ±200 μm .

We will discuss the system that has been designed to monitor changes in the alignment of the cavities and solenoids during pump out and cool down. The system has been tested in the first of five medium beta cryomodules, each containing four cavities and a single solenoid, see Figure 1 [1].

METHOD

Wire Position Monitors (WPM's)

A stretched wire alignment system based on a TESLA Test Facility system has been developed at TRIUMF [2,3]. Six WPM's, one per cavity and two on the solenoid, are positioned along a wire displaced 30.48 mm horizontally from the beam axis to measure lateral displacements. The wire, stretched between the tank walls, provides a position reference and carries a 215 MHz rf signal. The WPM's are supported from the cold masses by stainless steel brackets.

Each WPM is similar to a beam position monitor, Figure 2. It contains four Cu plated Al antennas supported by SMA jacks. The upstream ports are connected to the readout electronics which

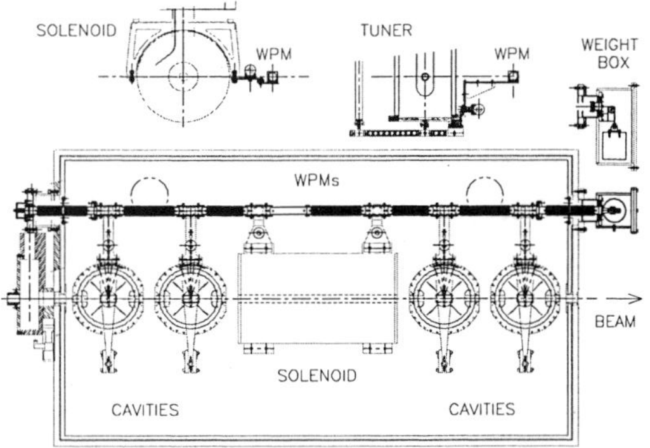

Figure 1 A plan view of a medium beta cryomodule and cross-sections of the wire weight box and WPM mounting brackets.

measure the signal amplitudes while the downstream ports are terminated by a 50 Ω loads. The striplines are 0.254 mm wide, and 61 mm long. Their heights are set to give 50 Ω impedance using a network analyzer in time domain reflectometry (TDR) mode. The Au plated brass Johnson SMA jacks use Teflon insulation. The jacks were found to contract by 50 μm overall when cooled with LN_2 but this can be accounted for by the brass alone. The thermal expansion coefficient of Teflon is six times that of brass for this temperature change, however, the internal geometry of the jack appears to prevent this from being a problem. The electronics measures differentially, cancelling common mode contraction effects to first order. Bench tests were performed on a single WPM using rf applied to a rod rigidly supported 3 mm off centre by end caps. The

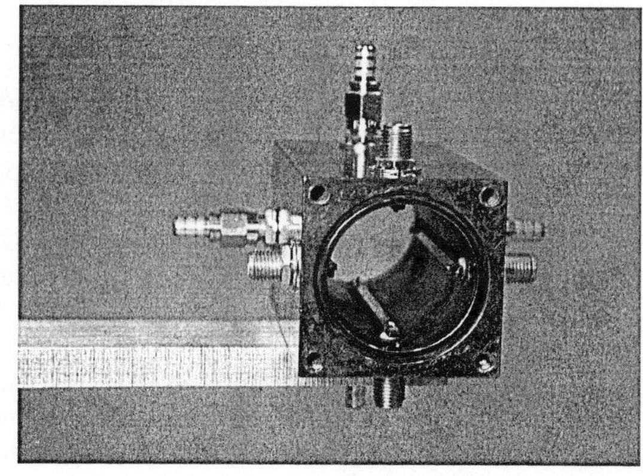

Figure 2 The stripline pickups of a WPM are spaced at 90° in the 28 mm bore and supported by SMA jacks.

apparatus was cooled using LN_2 in a small cryostat in air and an average change in readings of 16 μm was measured. Some of the change may have been due to condensation or temperature differentials.

Position Reference Wire

The 0.5 mm diameter bronze-Cu wire has a sag of 0.162 mm over a length of 2 m with a stretching load of 4.55 Kg provided by a pulley and weight in the vacuum. The wire passes through pin holes in dielectric disks at each end to define its path. It runs inside thin walled stainless steel bellows between monitors in order to form a coaxial transmission line. The corrugations of the bellows create a slow wave structure. The ends of each bellows are welded to square plates which are screwed to the monitors. Pump out holes are placed symmetrically around the circumferences. Two special bellows pieces carry the rf signal through the tank ports. They pass through the magnetic shield and have a floating flange in the middle for thermal contact with the LN_2 shield.

A TDR measurement indicated a wire impedance of 251 Ω. The sum of the signal strength from each WPM varies by about ±7% along the wire indicating little loss. The rf signal source is not matched to the wire impedance. Instead the rf passes through 10 dB of attenuators and a vacuum feedthru on the weight box and is connected to the wire by a jumper. The far end of the wire passes through a vacuum feedthru and is terminated by a 220 Ω resistor. This provides a directivity of 7 dB at the end WPM and is a measure of the quality of the termination. We wish to minimize the reflection as it provides a contribution to the WPM signal strength which varies with the WPM load resistance. The DC resistance of the Pasternack loads varied by 0.07 Ω on average when cooled with LN_2.

Cables

The signals are carried to the tank lid SMA feedthrus by 2.28 m long RG-303 cables with 0.5 dB loss. The cables use FEP and Teflon insulation and are vacuum compatible. Their impedance was found to decrease by 1.7 Ω when cooled with LN_2. 18.3 m long RG-223 double shielded cables with an attenuation of 3.5 dB carry the signals to the electronics.

Data Acquisition

The response of the stripline signals for a centred wire was measured to be about -40.5 dB using a network analyzer. An accidental short of the wire to a stripline would decrease the loss to only -3.6 dB. Attenuators at the amplifier output insure that fault

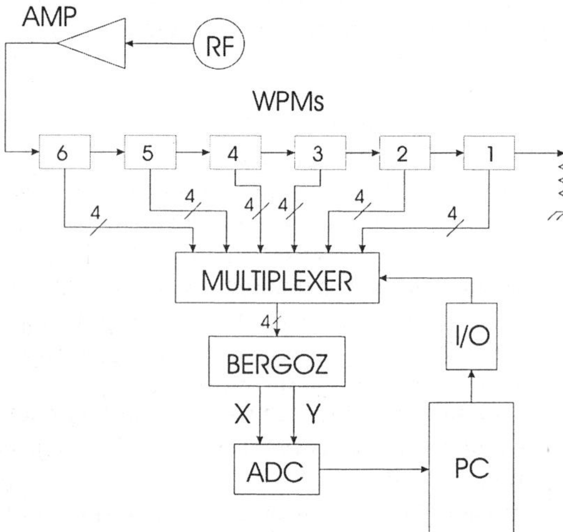

Figure 3 A block diagram of the data acquisition electronics.

levels cannot exceed the maximum input to the Bergoz card of 24 dBm and that the SWR remains less than 2:1 to protect the amplifier.

The TRIUMF built rf multiplexer uses M/A-Com SW-221 GaAs switches yielding greater than 60 dB isolation and 1.2 dB insertion loss, see Figure 3. Each of the four channels (l,r,d,u) is housed in a single width NIM module and selects one of the WPM's. The other inputs are switched to 50 Ω loads. The Bergoz Instrumentation card provides a single detector front end and is based on a four channel rf multiplexer, a down converter, a single AGC IF amplifier and a homodyne amplitude detector [4]. The card takes care of the rf to dc conversion and is insensitive to rf phase. The electrical noise specification is given as 8.9×10^{-6} diameter/$\sqrt{Hz}$ for signal levels above -50 dBm. Given our 28 mm

Figure 4 The calibration stand with the weight box on the left, a WPM, bellows and the Oriel interface behind.

diameter and a bandwidth of 10 Hz the electronics would contribute a noise to the position measurement of only 0.8 µm RMS. A National Instruments PCI bus 16 bit ADC card reads the dc signals and a digital I/O card controls the multiplexer. The PC uses Windows XP and runs a LabVIEW program written by the SIDeA Corporation of Milan.

Calibration

The WPM's are non-linear and must be calibrated. A pair of Oriel translator stages mounted a right angles are used to move a WPM about a stretched wire, see Figure 4. The servo motor units contain optical encoders with a resolution of 0.1 mm. The weight box and flanges from the cryomodule are used and the scan is computer controlled. A raster scan with 0.2 mm steps over a range of ±6 mm requires about 2 hours. A 2D, third order polynomial curve fit is used to reduce the data to a set of 20 polynomial coefficients. Beyond ±4.8 mm the electronics sharply compresses the response and though still useable, these points were not included in the curve fits. A fitting error of better than 20 µm was achieved over most of the ±4.8 mm range, exceeding it only near the ends of the range.

RESULTS

Preparation for Alignment Tests

The WPM system was installed in the cryomodule for alignment measurements during March and April of this year, Figure 5. All accelerator elements in the cryomodule are suspended from a support frame that in turn is suspended from the cryomodule lid with stainless steel struts. The cryomodule lid is installed in an assembly frame that replicates the cryomodule vacuum tank. The assembly frame doubles as an alignment stand. The top flange is machined flat and dowel pins are used to precisely position the top plate on the stand. An alignment jig is used to transfer the lid reference to a line of sight replicating the reference beam line and the WPM wire line. End plates on the assembly frame are fastened to replicate the vacuum tank beam port and WPM flanges. Both cryomass elements and WPM striplines are pre-aligned at room temperature to the theoretical line of sights with telescopes and alignment targets fitted in the beam tubes of the cavities and solenoid.

The cryomodule tank is also outfitted with a pair

Figure 5 The cryomodule components in the assembly stand. The WPM system is in the foreground.

of optical windows and alignment targets to set up and monitor an external optical reference line with a telescope. A pair of optical targets are installed in the upstream and downstream cavities. After the warm alignment in the assembly frame the top assembly is transferred to the vacuum tank, the wire is attached to the end flanges and tensioning device and the cryomodule is prepared for pump down. Optical measurements are taken periodically. They serve to check for unexpected differences between the WPM position and the position of the cold mass. During cool down they provide a calibration of the thermal contraction of the WPM brackets.

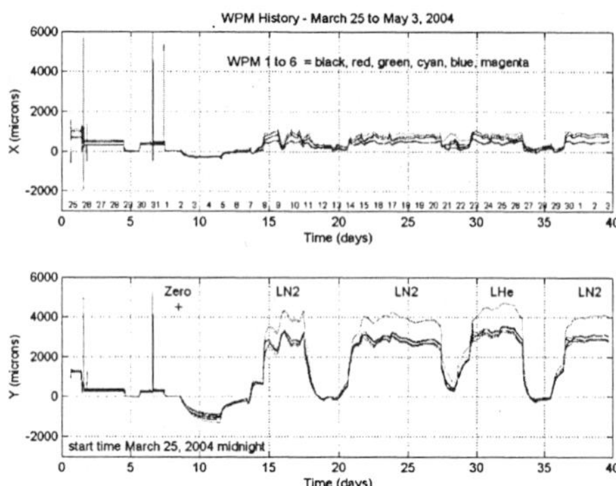

Figure 6 A history of the WPM readings over four temperature cycles.

Measurements

The WPM is useful for both warm and cold studies. Initially the repeatability of the lid bolt up procedure was investigated. The lid was bolted down then unbolted and lifted by a few mm before being bolted down again. This was repeated several times to get an idea of the repeatability of the lid positioning. Warm studies also include the effect on alignment of the tank pump down and the repeatability of the pump down.

The cold tests included three temperature cycles from room temperature to LN_2 temperature and one cool down to He temperature. The main goal is to look at repeatability of the cool down process and to establish cold offset values for each cavity and the solenoid to facilitate warm alignment at initial positions compatible with alignment at cold temperatures. A record of the cold cycle data is given for horizontal and vertical data over several weeks in Figure 6. The cavities contracted about 3 mm while the solenoid position contracted about 4 mm during the test. The position of the cold mass during the three LN_2 cycles proved to be repeatable to within 80 μm vertically and 120 μm horizontally. The graph of the relative position of the WPM's during the cold cycle is shown in Figure 7. A comparison between the optical targets and the visual targets gives the calibration for WPM position relative to the beam ports.

CONCLUSIONS

A WPM six monitor system has been developed at TRIUMF and is now operational on the first ISAC-II medium beta cryomodule. The device is giving a wealth of information over and above the data collected with the installed optical targets. The use of optical targets involves personnel and the readings can be taken only periodically. Conversely the WPM data is monitored continuously providing detailed data that is extremely valuable to help characterize a new structure.

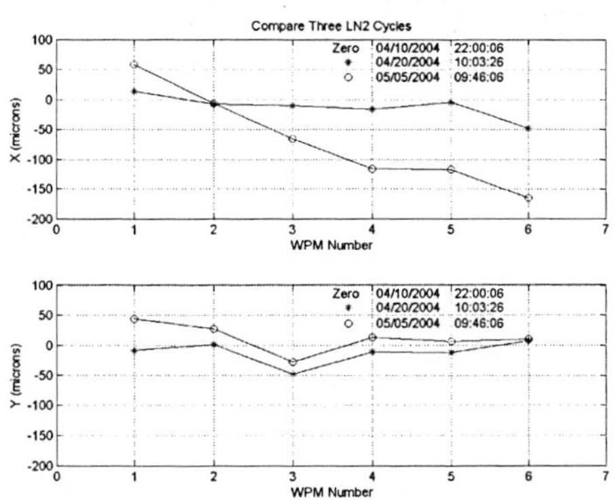

Figure 7 A comparison of the positions at the lowest temperatures achieved during three LN_2 cold cycles.

REFERENCES

1. Rawnsley, W.R., Giove, D., ISAC II Cryomodule Alignment Monitor, TRIUMF Design Note TRI-DN-03-02 (2003)
2. Giove, D., Bosotti, A., Pagani, C. and Varisco, G, A Wire Position Monitor (WPM) System to Control the Cold Mass Movements Inside the TRF Cryomodule, Particle Accelerator Conference (1997)
3. Bosotti, A., Pagani, C., Varisco, G., On Line Monitoring of the TTF Cryostats Cold Mass with Wire Position Monitors, INFN/TC–00/02 (Mar 17, 2000)
4. Unser, K. B., New Generation Electronics Applied to Beam Position Monitors, Beam Instrumentation Workshop (1996)

Proceedings of the Twentieth International Cryogenic Engineering Conference
(ICEC 20), Beijing, China. © 2005 Elsevier Ltd. All rights reserved.

Digital Temperature Sensors in the Range of 77-300 K

Nawrocki W., Jurkowski A., and Pająkowski J.

Institute of Electronics and Telecommunications, Poznan University of Technology
ul. Piotrowo 3A, PL-60695 Poznan, Poland; E–mail: nawrocki@et.put.poznan.pl.

We investigated the digital sensors of temperature (integrated circuits)
manufactured by Analog Devices (AD7416) and National Semiconductor (LM74).
We have investigated both lower limits of the operate temperature of digital
sensors and measurement accuracy. AD7416 sensors operate as low temperature as
−128 °C (145 K). Their measurement error was below ±2 °C. Lower limit for
LM74 sensors was different for two investigated sensors: −140 °C and −70 °C.
The accuracy is of 5 °C at temperature below −100°C. We have not found any
sensor able for a work at the temperature as low as 77 K.

SEMICONDUCTOR TEMPERATURE SENSORS

The operation of integrated temperature sensors is based on the dependence of voltage V_{BE} in the p-n junction on temperature, for a constant value of junction current. For the p-n junction of the diode or the transistor biased for conductance, the current of the junction is a function of voltage bias V_{BE} and temperature T (formula 1).

$$V_{BE} = \frac{k_B T}{e} \ln \frac{I_E}{I_0(T)} + V_G , \tag{1}$$

where: I_E – the current flowing through the junction, e.g. the emitter current,
$\quad$ $I_0(T)$ – the inverse current, k_B – Boltzmann constant,
$\quad$ T – the temperature in the absolute scale,
$\quad$ V_{BE} – the voltage on the p-n junction, e.g. the base-emitter voltage,
$\quad$ V_G – the material constant, a potential difference resulting from the energy gap width of the solid
$\quad$ (for Si: $V_G = 1.205$ V at $T = 300$ K).
$\quad$ The voltage U_{BE} on the p-n junction of the silicon transistor increases together with its temperature drop from 0.6 V at a temperature of 350 K to 1.3 V at a temperature of 50 K. The sensitivity of the device is therefore equal to −2.3 mV/K. Below the temperature of 50 K the voltage-temperature characteristic of the junction becomes strongly non-linear and in this temperature range the p-n junction is not useful for thermometry.
$\quad$ The semiconductor temperature sensor contains two p-n junctions, usually base-emitter junctions of two transistors, made of one block of semiconductor. Voltage difference V_{BE} for two integrated transistors conducting the emitter current $I_{E1} \neq I_{E2}$, is expressed by formula (2):

$$\Delta V_{BE} = V_{BE1} - V_{BE2} = \frac{k_B T}{e} \ln \frac{I_{E1}}{I_0(T)} - \frac{k_B T}{e} \ln \frac{I_{E2}}{I_0(T)} = \left(\frac{k_B}{e} \ln \frac{I_{E1}}{I_{E2}} \right) T . \tag{2}$$

Integrated thermosensor contains in one structure a couple of transistor-sensor as well as circuits of amplifiers converting the signal ΔV_{BE} to the required level of the voltage $V_{out} = f(T)$ or the output current $I_{out} = f(T)$. Due to a very considerable roll-off in the amplification of internal amplifiers within a temperature range below 200 K, the measuring range of integrated sensors has the lower temperature limit of −50°C ($\approx$ 220 K).

DIGITAL TEMPERATURE SENSORS

Apart from integrated temperature sensors with output analog signal (V_{out} or I_{out}), sensors with output digital signal are also manufactured. An integrated circuit of such sensor is formed of the following elements: a couple of junctions (transistors or diodes), the voltage of which (ΔV_{BE}) is the measuring signal, the signal conditioner (amplifier), the analog-to-digital converter, memory, and the digital interface – Fig. 1.

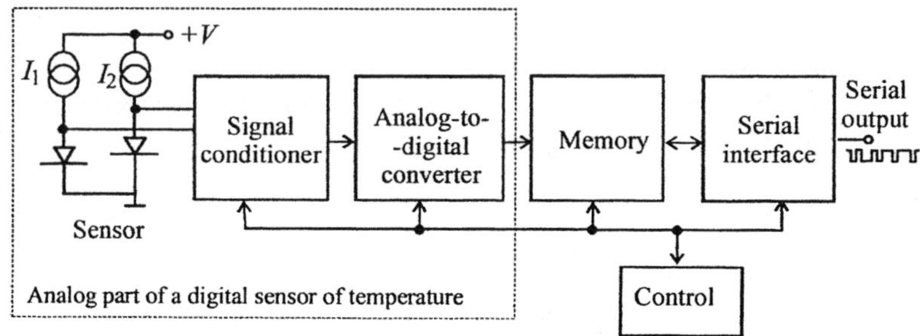

Figure 1. Block diagram of a digital sensor of temperature

The analog part of the sensor is sensitive to temperature change. The digital part should be insensitive to the influence of temperature in a wide range, especially when digital circuits are manufactured in the CMOS technology. Usually integrated digital sensors are allowed to be connected to various types of digital interface systems. Integrated digital sensors are offered by such manufactures as Analog Devices, National Semiconductor or Maxim. A review of digital temperature sensors is tabulated in Table 1.

Table 1. Digital integrated sensors of temperature (AD – Analog Devices, LM – National Semiconductor, MAX – Maxim, DS – Dallas Semiconductor)

Type	Measuring range	Accuracy (narrow range)	Accuracy (wide range)	Resolution	Output signal	Interface type
AD7416	−40°C to +125°C	±2°C	±3°C	0.25°C	10-bit	I²C
AD7818	−40°C to +125°C	±2°C	±3°C	0.25°C	10-bit	I²C
LM74	−55°C to +150°C	±1.25°C	±5°C	0.0625°C	12-bit + sign	SPI
LM92	−55°C to +150°C	±0.5°C	±1.5°C	0.0625°C	12-bit + sign	SPI
MAX6635	−55°C to +150°C	±1°C	±2.5°C	0.0625°C	12-bit + sign	2-Wire, SMB
MAX6662	−55°C to +150°C	±1°C	±2.5°C	0.0625°C	12-bit + sign	3-Wire, SPI
DS1624	−55°C to +125°C	±0.5°C	+3°C/−2°C	0.0625°C	12-bit + sign	2-Wire, SMB
DS18B20	−55°C to +125°C	±0.5°C	±2°C	0.0625°C	8 to 12-bit	1-Wire, SPI

Semiconductor temperature sensors – both analog and digital – are characterised with big conversion error, usually not less than ±1°C (without calibration), and with relatively narrow measuring range, at best included between −55 °C and +150 °C. For this regard, the area of their applications comprises watchdogs and monitoring systems as well as temperature indicators; they are quite rarely applied to industrial control systems. Though the manufacturers determine the lower limit of the measuring range as −40 °C or −55 °C, for most types of digital sensors the measuring range may be much wider in lower values. Our testing of digital sensors was done for AD7416 integrated sensors (Analog Devices) and LM74 digital sensors (National Semiconductor).

The AD7416 digital sensor has its measuring range from −40°C to +125°C, the measuring resolution 0.25°C and the output signal in the form of a 10-bit word. Analog-digital converter of SAR type (successive approximation register) was applied in the circuit. In a temperature range of −25°C to +100°C the measuring error is equal to ±2°C and in the range of −40°C to +125°C the error is equal to ±3°C (catalogue data). The digital 10-bit output word enables to record 2^{10} combinations. For the AD7416 sensor, for the measuring resolution of 0.25°C it gives the measuring range from −128°C (to this temperature corresponds the output word Y_{min} = 10 0000 0000) to +127°C (the output word Y_{max} = 01 1111 1100). The sensor may be connected to a microprocessor system with a 2-wire line of the I^{2}C serial interface.

The digital thermosensor LM74 has its measuring range from−55°C to +150°C, the measuring resolution 0.0625°C and the output signal in the form of a 12-bit word plus bit of a sign (+ or −), 13 bits altogether. Analog-digital converter of Δ-Σ type was applied in the circuit. The maximum measuring error for the LM74 sensor depends on the measuring range:
  − range: −10°C to +65°C, error ±1,25°C
  − range: −25°C to +110°C, error ±2.1°C
  − range: −40°C to +110°C, error +2.65°C/−2.0°C
  − range: −55°C to +125°C, error ±3°C.

The format of the digital output word allows to send the information on temperature of even −255°C, if the sensor could work in such low temperature.

MEASUREMENTS OF DIGITAL TEMPERATURE SENSORS

We conducted research of selected types of digital temperature sensors in the range of +10°C to −196°C (77 K). The lower limit of temperature range, in which the sensor was still operating, was verified. The measurement error of the ΔT digital sensor was also measured in the function of temperature of the T_{Dig} digital sensor. For reference temperature measurement the Pt100 thermometer was used.

$$\Delta T = T_{Dig} - T_{Pt100} \tag{3}$$

The measurements were carried out on a group of 4 sensors of AD7416 type (Analog Devices) and 2 sensors of LM74 type (National Semicon.). The Pt100 sensor and each digital temperature sensor tested (AD7416 or LM74) were mechanically connected to ensure good thermal contact between them.

All the AD7416 sensors tested preserved their working ability in a temperature reduced to −128°C. Further temperature reduction did not change the defined output word Y_{min} = 10 0000 0000, but the heating of the sensor to a temperature higher than −128°C restored its proper working. The sensors tested were exposed to repeated cooling-heating cycles (up to 10 cycles), which did not change the proper working of the sensors.

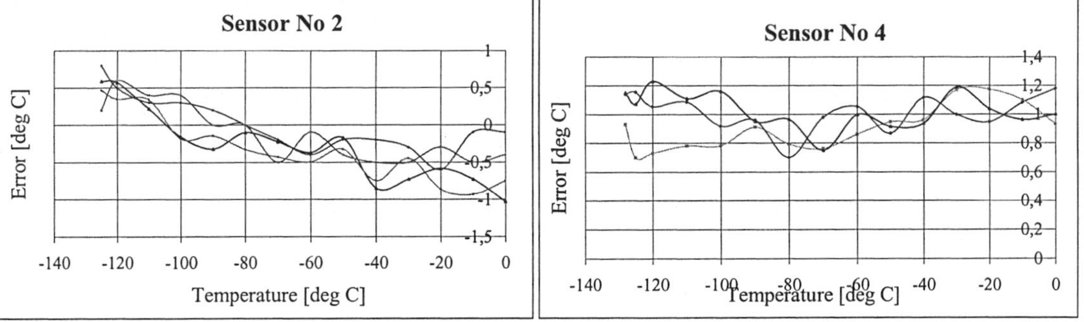

Figure 2. The error ΔT for temperature measurement with AD7416 digital sensors: No 2 (4 cycles) and No 4 (3 cycles)

Figure 2 shows an exemplary diagram of the difference between the temperature measurement using AD7416 sensors (No 2 and No 4) and the Pt100 thermometer in the temperature function T_{Dig} for 4 measuring cycles. The testing revealed a surprisingly good coincidence of temperature measuring with a digital sensor and the Pt100 thermometer. The error ΔT in measurement realised by means of a digital sensor was less than ±1.5°C in the whole range from −125°C to +10°C. The measurement error was not greater than ±2°C for the remaining AD7416 sensors tested. In the measuring range given by the manufacturer up to −40°C the measurement error was less than the limit value ±3°C for all 4 sensors tested.

For the LM74 sensors their working ability in low temperatures was confirmed as well. According to the manufacturer, the lower limit of the measuring range was different. For the first investigated sensor LM74 the limit was −140 °C instead of −55 °C. The lower limit was only −70 °C for the second sensor LM74. In the temperature range given by the manufacturer (up to −55°C) the measurement error amounted to +2°C and was less than the limit error ±3°C. A diagram of the measurement error in temperature function for two LM74 sensors tested was shown in Fig. 3. In the range of low temperatures the measurement error amounted to +5°C. The sensor placed in a temperature below −140°C generated a constant output word corresponding to temperature −140°C. It means that the digital part of the sensor does not work in this temperature range.

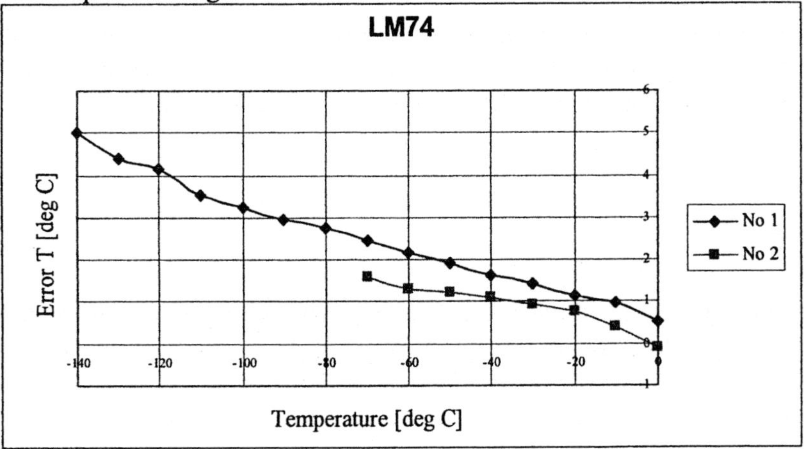

Figure 3. The measuring error ΔT for temperature measurement with the LM74 digital sensors No 1 and No 2
(sensor No 2 does not operate below −70 °C)

CONCLUSIONS

The testing of sensors realised on a sample of AD7416 and LM74 integrated circuits showed that digital temperature sensors preserve their working ability in temperature much lower than that defined by the manufacturers. For the group of AD7416 sensors the lower limit of working temperature was equal to −128°C. For one LM74 sensor the limit was still lower: −140°C. The reduction of lower limit of the measuring range is impossible for the AD7416 sensors. The reduction of this lower limit for the LM74 sensors would be useful because of changes in parameters of the analog part of the sensor in temperature function. For some devices of the sensors tested the measuring error did not exceed ±1.5 °C, for others it amounted to +5 °C. Single devices of digital temperature sensors may be therefore used to construct cryogenic thermometers of less accuracy. It is useful especially in such measurements in which a signal from analog sensor is notably distorted, for example by a strong electromagnetic field.

REFERENCES

1. Michalski L. et al., Temperature Measurement, John Wiley, Chichester, UK (2001) 116-122
2. Analog Devices, National Semiconductor, Maxim, Dallas Semiconductor – Catalogues 2004

Proceedings of the Twentieth International Cryogenic Engineering Conference
(ICEC 20), Beijing, China. © 2005 Elsevier Ltd. All rights reserved.

Measurement of characteristic feature of cavitation flows of He I and He II

Harada K., Tsukahara R., and Murakami M.

Institute of Engineering Mechanics and Systems, University of Tsukuba, Tsukuba, 305-8573, Japan

In the present experimental study cavitation phenomena in both He I and He II flows were investigated through the pressure and temperature measurements and the application of optical visualization. The cavitation flow is generated in the downstream region of a Venturi channel driven by contracting the metal bellows. The cavitating flows can be observed through the optical windows of the cryostat. Tests were carried out for both He II and He I. The pressure loss and the temperature drop caused by cavitation are measured in a wide range of the flow velocity. The spatial distribution of the cavitation bubble velocity is also measured by the application of PIV (Particle Image Velocimetry) technique.

INTRODUCTION

Use of cryogenic fluids such as liquid helium has been expanded in large scale cryogenic plants relating to aerospace and superconductivity applications. There occurred a serious accident of the H-II rocket in 1999 of which cause was cavitation in the liquid hydrogen turbo-pump. However, there have been a little cavitation researches of cryogenic fluids [Reference 1, 2, 3]. It is, thus, the purpose of this research to experimentally study cavitation flow of cryogenic fluids, in particular of liquid helium. We measured the temperature drop induced by cavitation and the pressure loss as a pressure difference between the upstream and downstream points of the channel in a wide range of the flow velocity. Visualization pictures were also taken for both cavitating He I and He II flows for the purpose of visualization and PIV measurement.

EXPERIMENTAL SET-UP

A schematic illustration of the key area of the experimental set-up is shown in Fig. 1. This section composed of a Venturi channel, a metal bellows, a pressure transducer and a temperature sensor is immersed in liquid helium. The flow is generated by the metal bellows pump. The detail of the two-dimensional Venturi channel is represented in Fig. 2. Cavitation is generated in the downstream region of the Venturi channel, which consists of two pieces of thin stainless steel plate shaped into a Venturi profile placed between two parallel plates of quartz glass. It is of a rectangular cross section with a thickness of 3 mm between the glass plates, a maximum width of 15 mm at the downstream exit, and a minimum width of 5 mm at the throat. The velocity of the flow was kept constant during each experiment. The maximum velocity was V_t=30.70 m/s at the throat. The cavitating flow can be observed through the cryostat windows of 60 mm-diameter and recorded with a high-speed video camera or a digital still camera. The absolute pressure transducer was used to measure the pressure loss between the upstream chamber and the outside of the channel. A thin film resistance temperature sensor (Cernox) fixed at the tip of the plastic probe was inserted into the downstream diverging section at the center of channel below the throat by 25 mm.

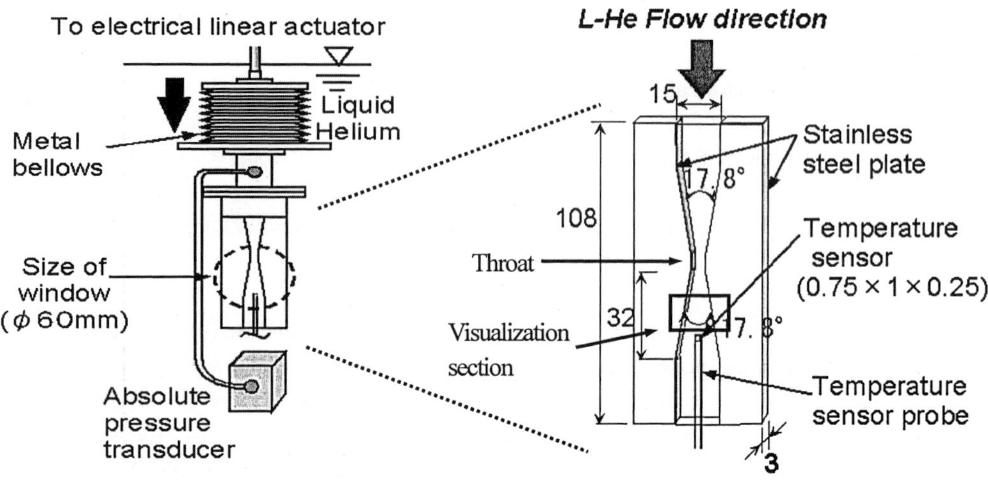

Fig. 1   The schematic illustration of the key area of experimental set-up

Fig. 2   The detail of two-dimensional Venturi channel

RESULTS AND DISCUSSION

Pressure loss data

The relation between the pressure loss, P_{loss}, and the velocity at the throat, V_t, is presented in Fig. 3 for the case of fully developed cavitating flows.   It is seen that there is a difference in the magnitude of the pressure loss between He I and He II.   In the case of He II, the data are almost independent of the temperature, but in the case of He I, temperature dependency is clearly recognized.   It is considered that the difference in the void fraction causes the difference in P_{loss} between He I and He II.   According to the visualization study by a high speed camera, it was clear that cavitation flows of He I were more intermittent than those of He II.   It is indicated, consequently, that the void fraction of He I flow is smaller than that of He II in an averaged sense.   The pressure loss becomes large as the bath temperature drops toward the λ-temperature, T_λ in the case of He I.   The fact that the pressure loss curve for 2.2 K coincides with the He II curve can be explained by the λ-transition from He I to He II as a result of cavitation bubble formation.   It is necessary to measure the void fraction in order to quantitatively understand the difference in the pressure loss between He I and He II.

Cavitation number plot

The P_{loss} data include the effect of the saturated vapor pressure as a strong function of temperature.   Elimination of the effect of the saturated vapor pressure is more appropriate for more discussion.   So the pressure loss data are rearranged in terms of the cavitation number σ defined by $2[p_\infty - p_v(T_\infty)] / \rho_l V_t^2$.   The cavitation inception and the magnitude of cavitation can be also quantitatively investigated in terms of σ.   Here p_∞ and $p_v(T_\infty)$ are the pressures in the upstream chamber and of the saturated vapor at the liquid temperature T_∞, and ρ_l is the liquid density.   The relation between σ and V_t is presented in Fig. 4 for a wide range of the velocity, V_t.   It is distinguishable there are two branches in the data plot, non-cavitating and cavitating branches.   Along the cavitating branch of He II σ is nearly constant independently of V_t though it is weakly dependent on T_∞.   It was also seen that the magnitude of σ is clearly different between He I and He II in the cavitation inception region with small V_t, and both σ values for He I and He II close to each other in the fully developed case.   In the case of He II flows, the cavitation inception causes sudden jump in cavitation number.   However, in He I fows, cavitation number rises continuously irrespectively of cavitation inception.   It seems the critical value V_t for the inception is not definitely decided between 5 and 10 m/s, but it is rather sensitive to disturbances.

Temperature drop data

The temperature drop, $-\Delta T$, from the initial temperature caused by cavitation is plotted against V_t in Fig. 7. It is seen that the He II branches can be clearly distinguished from the He I branch. In the case of He I, the temperature drop rapidly increases with the increase of V_t. However, in the case of He II, the temperature drop increases more moderately. The occurrence of the cavitation-induced λ-transition is clearly recognized by the sudden change of the gradient of the data in the cases of T, 2.2 K and 2.3 K. This was fully reasoned in Ishii's articles [Reference 2,3].

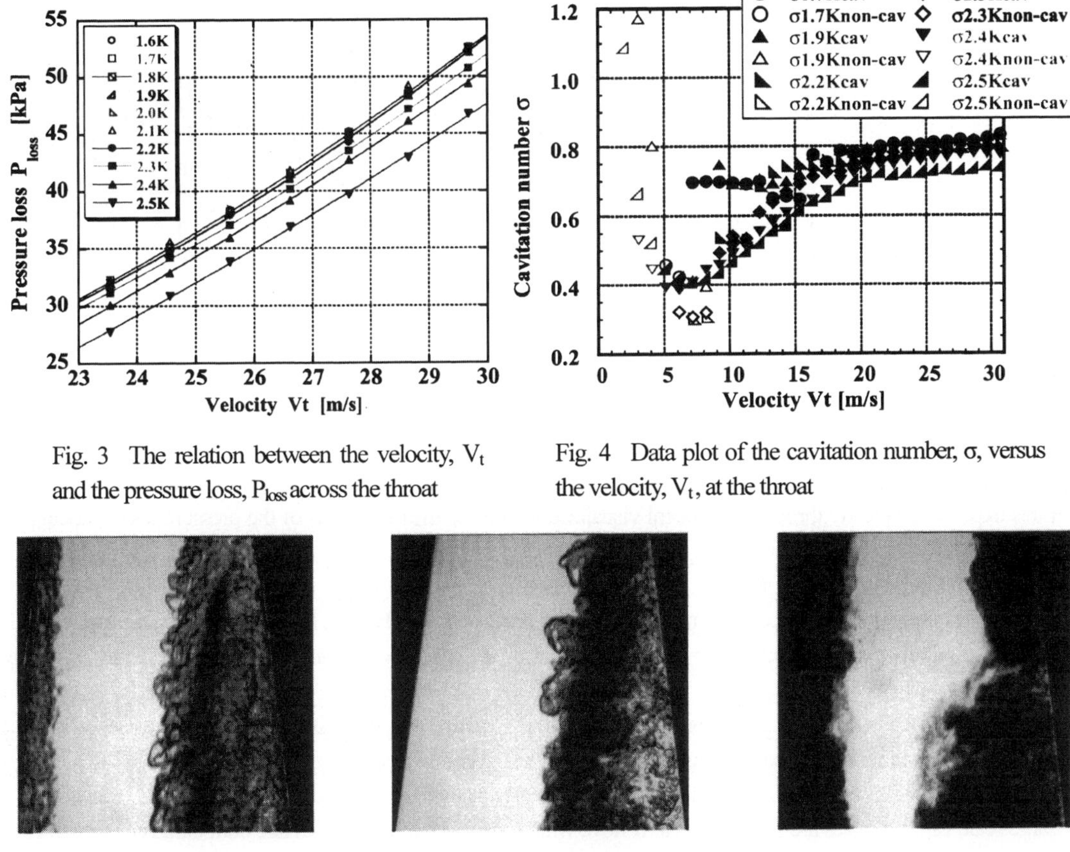

Fig. 3   The relation between the velocity, V_t and the pressure loss, P_{loss} across the throat

Fig. 4   Data plot of the cavitation number, σ, versus the velocity, V_t, at the throat

T=1.90K (He II)   T=2.20K (He I)   T=2.50K (He I)

Fig. 5   Pictures of the cavitation flow of He I and He II for V_t =12.28 m/s. The visualization area is a part of the Venturi channel(10×10 mm) at 11 mm downstream of the throat indicated by a square in Fig. 2.

Visualization and PIV results

The comparison of the visualization pictures of cavitation flows of He I and He II is presented in Fig. 5. Large scale vapor bubbles are seen in the He II flow at 1.9 K, while massive minute bubbles like dark cloud are seen in He I flow at 2.50 K. The image at 2.20 K seems to be a mixture of them. It is because bubbly flow was partly converted into He II as a result of temperature drop. It was seen from the photographs taken by the high speed camera that cavitation flows of He I were highly intermittent, on the other hand He II flows were rather steady and exhibit definite formation of separation shear layer. The PIV technique is successfully applied to cavitating flows of He I and He II. Cavitation bubble velocity field is shown in Fig. 6, which is the PIV result for the picture at 2.20 K shown in the middle of Fig. 5.

942

It is found that the PIV result for He II cavitating flow, which is not shown in this paper due to lack of space, is also different from that for He I as recognized by the visualization pictures. It is suggested by the picture of He I cavitating flow shown in Fig. 5 that large scale vortex shedding causes flow intermittency. The PIV result also evidently shows the flow structure with large scale vortex shedding.

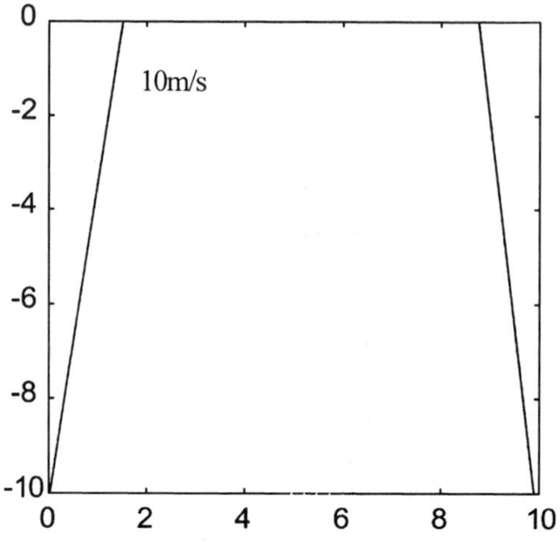

Fig. 6   PIV result of He II cavitatng flow at 2.50K measured in the downstream flow field (10×10mm)

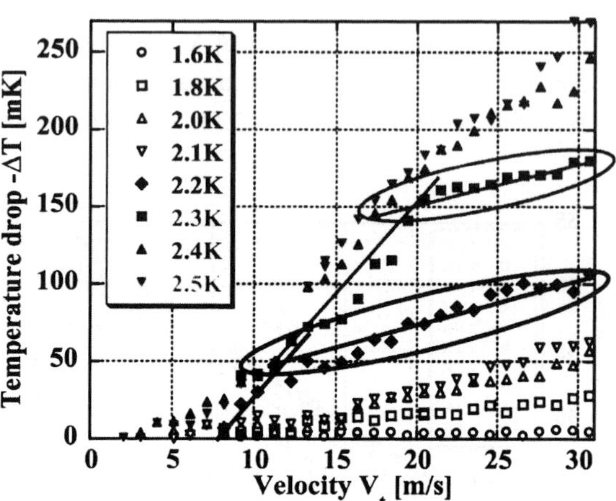

Fig. 7   The temperature drop plotted as a function of V_t

CONCLUSIONS

From this experimental study through the optical visualization and the measurements of the pressure and temperature, the following conclusions are drawn.

1. It is seen that there is a difference in the magnitude of the pressure loss between He I and He II. In the case of He II, the data are almost independent of the temperature, but in the case of He I, temperature dependency is clearly recognized.
2. In the case of He II flows, the cavitation inception causes sudden jump in cavitation number. However, in He I flows, cavitation number rises continuously irrespectively of cavitation inception.
3. The occurrence of the cavitation induced λ-transition is clearly recognized by sudden transfer to He II branch.
4. Large scale vapor bubbles are seen in the He II flow, while massive minute bubbles like dark cloud are seen in He I flow.
5. It was confirmed that bubbly flow in He I was partly converted into He II as a result of temperature drop.
6. It was seen from the photographs taken by the high speed camera that cavitation flows of He I were highly intermittent, while He II flows were rather steady and exhibit a definite formation of separation shear layer.
7. The PIV technique is successfully applied to cavitating flows of He I and He II. It was found that the PIV result for He II cavitating flow is different from that for He I as recognized by the visualization pictures.

REFERENCES

1.   T. Ishii and M. Murakami, Comparison of cavitation flows in He I and He II, Cryogenics (2003) 43 507-514

2.   Jun Ishimoto and Kenjiro Kamijo, Numerical simulation of cavitating flow of liquid helium in venturi channel, Cryogenics (2003) 43 9-17

3.   Takashi ISHI, Masahide MURAKAMI, Temperature Measurement and Visualization Study of Liquid Helium Cavitation Flow Through Channel, Advances in Cryogenic Engineering (2002) 47 1421-1428

Proceedings of the Twentieth International Cryogenic Engineering Conference
(ICEC 20), Beijing, China. © 2005 Elsevier Ltd. All rights reserved.

Application of PIV technique to cavitating flows of liquid helium

Tsukahara R., Murakami M. and Harada K.

Institute of Engineering Mechanics and Systems, University of Tsukuba, Tsukuba, 305-8573, Japan

PIV technique was applied to the measurement of bubbly flow of cavitating liquid helium driven by a bellows pump. In the present PIV application the velocity of cavitation bubbles is measured without adding any seeding particles. The flow field, the downstream of the Venturi throat, is separately illuminated by two stroboscopes through an optical window of the cryostat, and subsequent two successive pictures of cavitaion bubbles are taken by a digital CCD camera. The image information of the two pictures is analyzed with the aid of a PIV algorithm to yield a 2-D velocity distribution of cavitation bubbles. It may be concluded that the application of PIV technique is of great use in the quantitative study of cavitaion flows of He I and He II.

INTRODUCTION

Liquid Helium shows some unique physical characteristics, such as lamda phase transition, thermal counter flow, and quite a low dynamic viscosity. Only a few researches based on velocity measurement have been conducted. They are Laser Doppler Velocimetry (LDV) application with seeding particles made from mixture of hydrogen and deuterium [1], and Particle Image Velocimetry (PIV) application with hollow glass spheres [2] or polymer micro-spheres [3] as seeding particles, both of which were applied to the thermal counter flow measurement of He II. Though LDV enables a point measurement, two dimensional axial velocity can be measured with much difficulty. On the other hand, PIV enables essentially a two dimensional measurement. In both researches, some discrepancies have been still pointed out between the measurement results and theoretical ones. One of the reasons for the discrepancy is attributed to low dynamic viscosity. It seems rather hard to select suitable tracer particles in He II that must be as close as neutrally buoyant and capable of flowing with sufficiently small slip with respect to He II flow compared with ordinary fluids like water. Consequently, we first applied the PIV technique to cavitating flow through the Venturi channel without adding tracer particles. In this PIV technique we don't use laser light of which scattered light from tracer particles is used in conventional PIV technique, but, instead, we used bubbly flow patterns illuminated by the back light from two stroboscopes. By doing this, we could avoid the risk arising from the difficulty concerning tracer particle, and examine the validity of the PIV technique for the application in liquid helium flows measurement.

EXPERIMENTAL APPRUTUS

The cavitating flow in the Venturi channel is driven by the bellows pump. In the present PIV application no seeding particles are added because cavitation bubbles play a role of particles [4]. The optical system is composed of four parts, that is the digital CCD camera, the light source, the timing synclonyzer, and the image memory system. The Digital camera, MegaPlus Model ES 1.0 (REDLAKE MASD Inc.) is capable of taking successive two pictures with an interval at the least $1\,\mu$ s at the maximum repetition frequency of 12 Hz. The image resolution is 1008x1018 pixels. In this experiment, we took two pictures with 10 μs-interval at 10 Hz. The light source, a fiber video flash Model FB-05J30 (Nissin electronic Co. Ltd.) is composed of two independent channels and the output power is 0.5 J/flash with a half value width of 3 μs. The setting of the light source is easy due to independent two units of fiber light guides. The timing synclonyzer and the image memory system are controlled by a personal computer. A number of sets of pictures are temporarily accumulated on an interface board, and then are stored in a hard-disk memory.

PIV BASIC THEORY AND APPLICATION

For the application of the PIV technique, synchronized timing for each device is quite important. The timing chart is shown in Fig.1. The timing of two pictures is decided according to the formula as

$$U_{throat} = \frac{L}{k}\frac{\Delta X}{\Delta t}, \tag{1}$$

where U_{throat} is the throat velocity of liquid helium, L the characteristic length of the field of view, ΔX the displacement of an image measured in pixels between two successive pictures, k the pixel resolution of a camera, Δt the time interval between two pictures. Here, ΔX is decided depending on the particular method of PIV analysis. We adopted the cross-correlation method software developed by Sakakibara Laboratory of University of Tsukuba, in which appropriate ΔX is recommended to be about 10 pixels. The two pictures are taken at the instances of the stroboscope emissions.
In this experiment, we have U_{throat} about 20 m/s, L about 10 mm, k about 1000 pixels, and ΔX about 10 pixels, and consequently Δt is about 5 μs. We select the value of Δt to be 10 μs because we estimate that the actual Δt becomes larger due to a little slow down of the flow velocity in the high bubble area compared with the throat velocity.

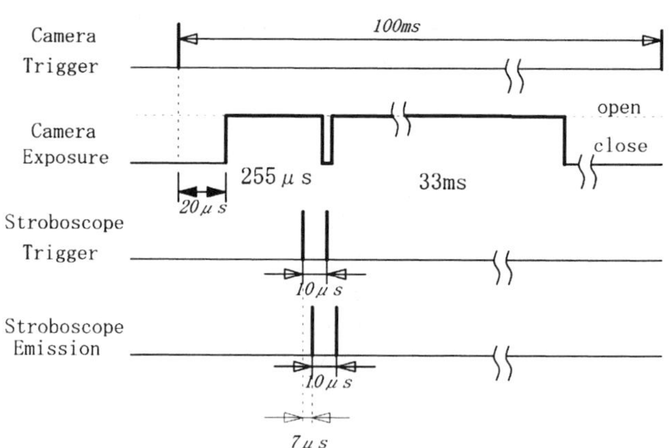

Figure 1   PIV timing chart

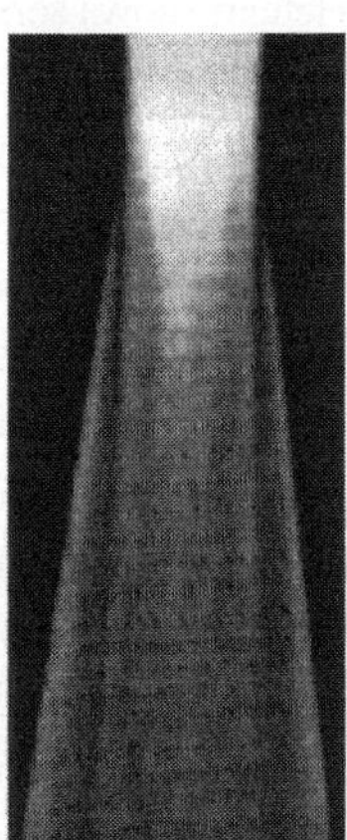

Figure 2 Whole cavitating flow area of He I (left, temperature=2.5 K), and He II (right, temperature=2.1 K), both U_{throat}=21.5 m/s

RESULTS AND DISCUSSION

The still pictures of the cavitating flows of He I and II in the Venturi channel are respectively shown in Fig. 2. The PIV analysis area is the middle square region of the downstream of the Venturi channel, 11 mm below the throat with a size of 10 mm x 10 mm.

<u>He I cavitating flow</u>

The picture of a typical He I cavitating flow and its PIV analysis result that is an ensemble average of 6 data are shown in Fig. 3. It is seen that the PIV result has a good enough spatial resolution to distinguish the thin shear layers indicating the boundary between the central and the separated layers on both sides of the flow-field. In both separated flow regions, reverse flows can be recognized. This can be also seen from the velocity distribution shown in Fig. 5 which is drawn on the basis of the PIV result. The PIV result indicates the average velocity of cavitating flow in the potential core region is about 16.1 m/s.

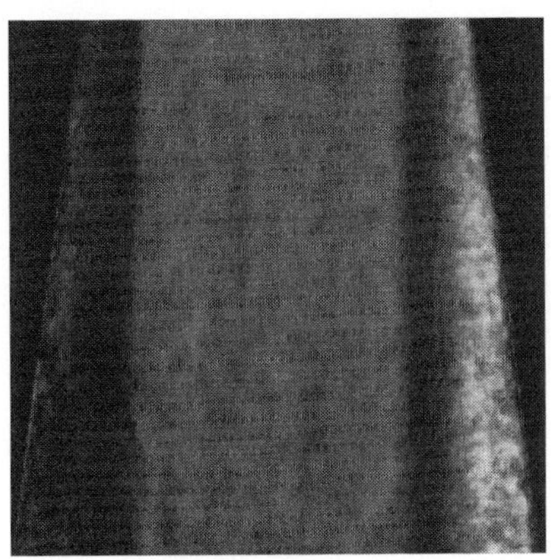
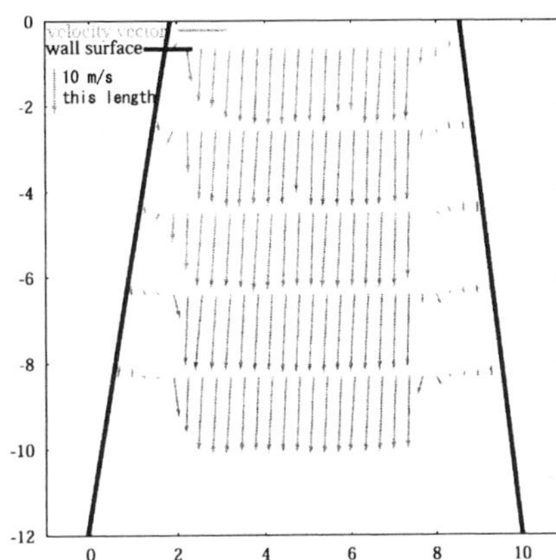

Figure 3   Typical He I cavitating flow picture and the PIV result shown in a velocity vector form (ensemble average of 6 data), T=2.5 K, U_{throat}=21.5 m/s

<u>He II cavitating flow</u>

The similar result for He II flow is presented in Fig. 4. Clear difference from He I cavitating flow is seen with respect to less clear shear layer and apparently no reverse flow in the separated flow region. The latter result of no reverse flow is also seen in Fig. 5. The trough in the middle of the potential core region is an erroneous result presumably caused by lack of bubbles there. The average velocity of cavitating flow in the potential core region is about 15.8 m/s.

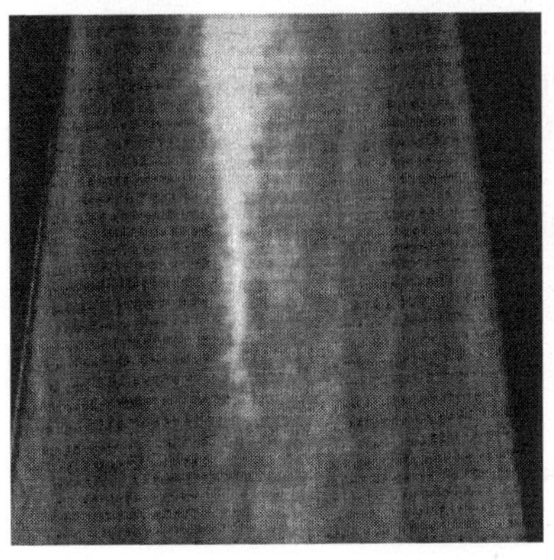
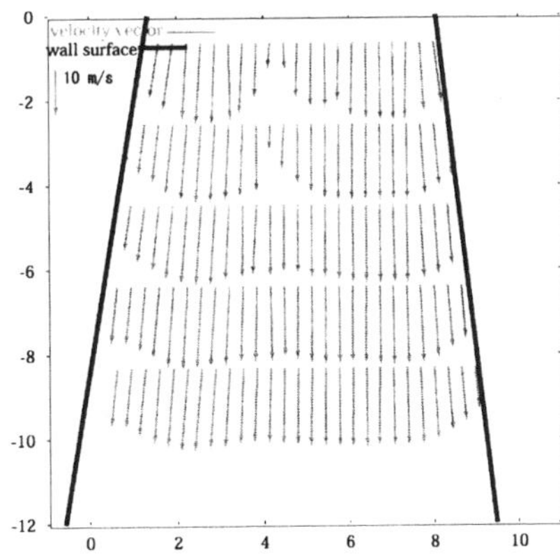

Figure 4   Typical He II cavitating flow picture and the PIV result shown in a velocity vector form (ensemble average of 6 data), T=2.1 K, U_{throat}=21.5 m/s.

<u>Discussion</u>

It may be concluded that the present PIV application yielded rigorous result as presented in Figs. 3 and 4. In fact, the average correlation coefficient in the procedure determining the spatial displacement of an image to calculate the flow velocity was all larger than 0.7 and a single strongest peak value could uniquely resulted in every interrogation region throughout the flow field. And apparent error vectors were not found in both Figs.3 and 4. Figure 6 shows relationship between a measurement area and a number of interrogation regions. A velocity vector in each interrogation regions is calculated from the average bubble movement during the interval Δt for two successive pictures, that is decided by the maximum correlation principle. In the present PIV analysis, an interrogation region of 64 x 64 pixels (one pixel is about 10 μ m) is used with an overlap of 50% of each interrogation region. The spatial and time resolutions may be around 320 μ m and about 0.1 sec, respectively.

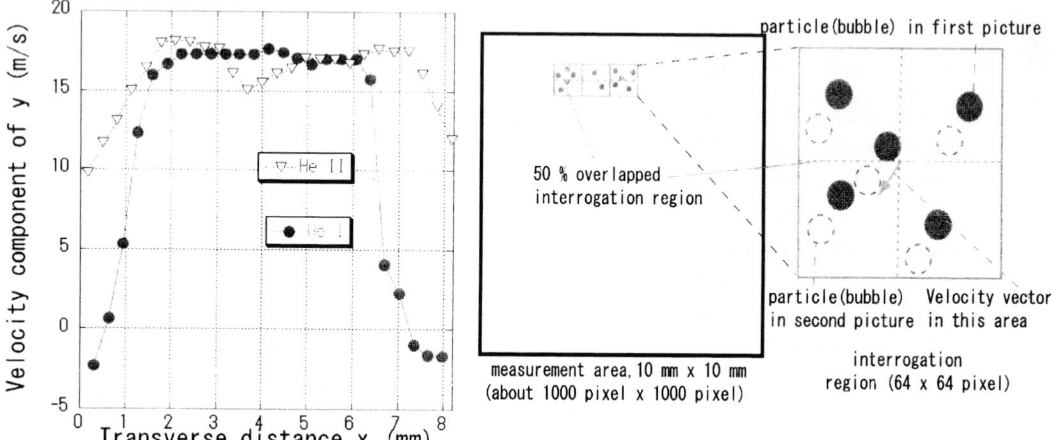

Figure 5 Spatial distributions of the y-component of bubble velocity across the flow field at y=-5.44.

Figure 6   PIV image analysis

CONCLUSIONS

1. PIV technique can be successfully applied to cavitating flows of liquid helium, both He I and He II.
2. Using PIV technique, we can measure the bubble velocity with a 320 μ m spatial resolution and 0.1 s time resolution.
3. Difference between He I and He II cavitating flows can also be clearly indicated by the PIV result.

REFERENCES

1. Murakami, M., Nakano, A. and Iida, T., Applications of a Laser Doppler Velocimetry and Some Visualization Methods to the Measurement of He II Thermo-Fluid Dynamic Phenomena, In: Flow at Ultra-High Reynolds and Rayleigh Numbers, Springer-Verlag, New York, USA (1998), pp.159-178

2. Celik, D., Zhang, T. and Van Sciver S. W., Application of PIV to Counterflow in He II, Advances in Cryogenic Engineering, (2002), 47, pp.1372-1379

3. Zhang, T. and Van Sciver S. W., Measurements of He II thermal counterflow using PIV technique, presented at Cryogenic Engineering Conference Anchorage, Alaska, USA (2003)

4. Harada, K., Tsukahara, R. and Murakami, M. Measurement of characteristic feature of cavitation flows of He I and He II, presented at International Cryogenic Engineering Conference – 20 Beijing, China (2004)

Measurement of the cryogenic tensile properties of polymer composites

Pan Q.Y., Fu S.Y. , Zhang Y.H., Li Y.

Cryogenic Materials Division Technical Institute of Physics and Chemistry, Chinese Academy of Sciences, Beijing 100080, China

In this paper, measurement of tensile strength, modulus and strain of polymer composites was conducted on a universal testing machine using our self-designed tensile testing jigs at cryogenic temperature. The cryogenic tensile properties of nanofiller reinforced polymer composites have been studied as a function of filler contents and compared with that at room temperature. The results showed that there exists an optimal filler content corresponding to the maximum strength and modulus. And it was shown that the tensile strength and Young's modulus of the polymer composites were generally higher at 77 K than that at room temperature except at the 20 wt % of clay for clay/polyimide nanocomposites. Moreover, the elongation to failure has much lower values at the cryogenic temperature (77 K) compared with that at room temperature.

INTRODUCTION

The research and development of polymeric materials for cryogenic applications have been intensified in recent years, especially in the fields of space science, superconducting magnet technology and also the other advanced cryosurgery and cryobiology in the medical field technologies [1-5]. However, at the present time, a judgment on the utility of a specific polymer material under cryogenic conditions usually requires a test demonstration because of the limited condition available and the uncertain influence of many factors such as sample geometries, test methods and environmental variables. There is, therefore, much to be done to develop the test technology for measurement of the cryogenic properties of polymers. In this paper, tensile testing jigs have been designed for measurement of cryogenic tensile properties of polymer composites. The tensile properties of nanofiller reinforced polymer composites have been studied then at room and cryogenic temperatures (77 K) taking into account the effects of filler contents.

EXPERIMENTAL DETAILS

The polymer composites were synthesized in our laboratory. The PI/MMT composite films were prepared via in-situ polymerization. The sizes of film specimens were respectively 10 mm×90 mm (room temperature) and 10 mm×120 mm (cryogenic temperature). The gauge length was 50 mm. The thickness of films was 25-39 um, and the specimens were cut from free films. Another, the silica nano-particles were incorporated into epoxy resin by sol-gel process and SiO_2/epoxy nanocomposites with different silica contents were then prepared. Cryogenic tensile properties of polymer composites were measured at a loading rate of 2 mm/min with a RGT 20A Testing Machine using our cryogenic system.

In general it's very hard to measure the cryogenic mechanical performance of polymer composites with high accuracy and credibility because the polymer composite samples are not easy to be clamped using commercial tensile jigs. Moreover, the commercial jigs have other disadvantages such as complex structure and big volume etc. In order to study the cryogenic properties of polymer composites, the tensile jigs suitable for measurement of the cryogenic tensile properties of plastic materials are designed in terms of the standard of ASTM D638-01, see Figure 1. The self-designed jigs have some advantages over commercial jigs. Its volume is about 1/10 and its weight is about 1/16 of the commercial jigs. The self-designed jigs are simple and convenient to be fixed for tensile testing at low temperature. Moreover, the entire measurement system can be conveniently connected to the cryostat which allows the low temperature to be reached and maintained. Cryogenic strain gauge extensometers were used to measure the specimen deformation. Then, the results for cryogenic tensile strength, modulus and strain of polymer composites can be obtained using a tensile testing machine. Another set of self-designed jigs have also been designed for measurement of the cryogenic tensile properties of polymer films, which are not given because of the limit of the paper length.

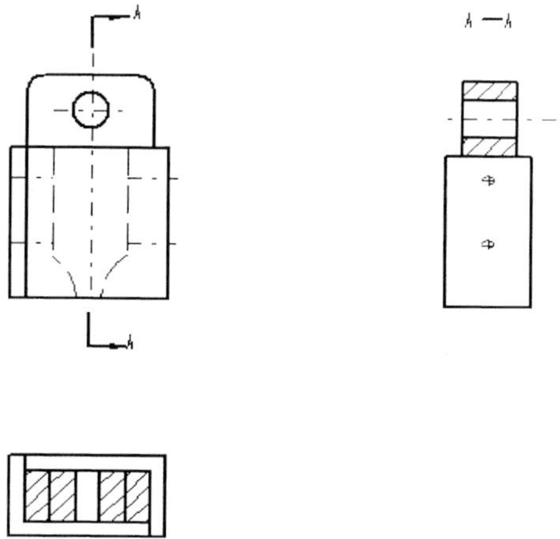

Figure 1. Drawing of the self-designed tensile jig for measurement of cryogenic tensile properties of polymer composites.

RESULTS AND DISCUSSION

In this study, the results for the tensile properties of the PI/MMT composite films and the SiO_2/epoxy nanocomposites measured using our self-designed system at room and cryogenic (77 K) temperatures are shown in Figure 2 and 3 taking into account the effects of filler contents. Similar results for the tensile properties at room temperature have also been obtained using normal commercial jigs that are not suitable for measurement of cryogenic tensile properties of polymer composite films. Figure 2 and Figure 3 showed that the effects of the clay content on the tensile strength and the elongation at break of PI/clay hybrid films at room and cryogenic (77 K) temperatures. It is exhibited that the tensile strength of the PI/clay hybrid films was generally higher at 77 K than that at room temperature except at the 20 wt % of clay. This observation is reasonable because the polymer molecules are tightly arranged at cryogenic temperature so that the strength at cryogenic temperature is higher than that at room temperature. The exception at the 20 wt % of clay for clay/polyimide nanocomposites can also be reasonably explained. Since at such a high clay content the clay aggregates would be very severe, the cracks formed easily during cryogenic tensile testing would then extremely readily propagate along the aggregate/polyimide interfaces at cryogenic temperature, leading to a lower strength. Moreover, the elongation to failure has much lower values at the cryogenic temperature (77 K) when compared with that at room temperature. This is due to the fact that the polymer molecules become frozen and are much more brittle at cryogenic temperature, bringing about lower ductility at cryogenic temperature when compared with room temperature.

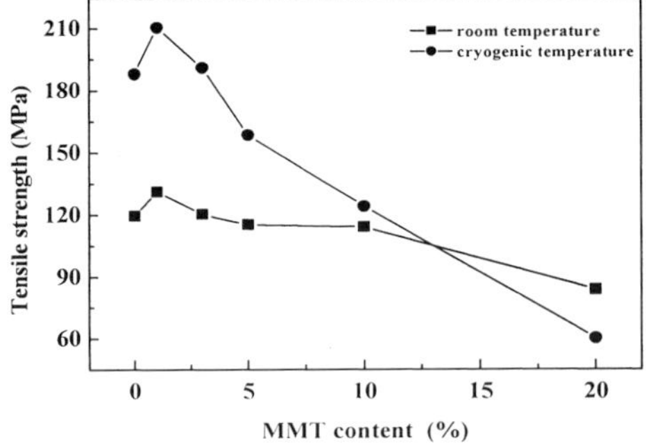

Figure 2 Effects of the clay content on the tensile strength of PI/clay hybrid films at room and cryogenic (77 K) temperatures.

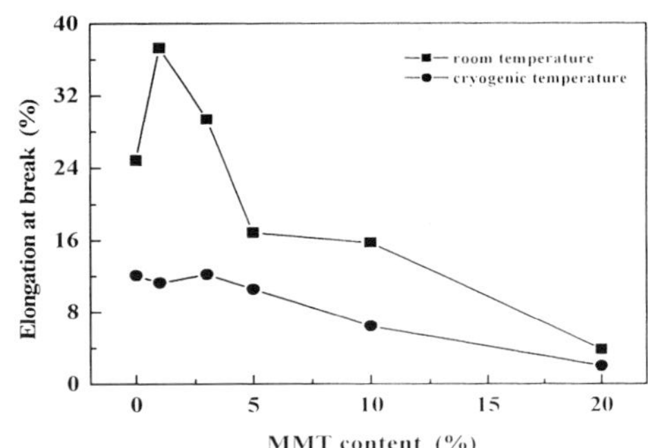

Figure 3 Effects of the clay content on the elongation at break of PI/clay hybrid films at room and cryogenic (77 K) temperatures.

Table 1 Tensile properties of SiO_2/epoxy nanocomposites at room and cryogenic temperature (77 K).

The specimen composition	Tensile strength（MPa）		Tensile modulus（GPa）		Elongation at the break（%）	
	300 K	77 K	300 K	77 K	300 K	77 K
RAL-230（epoxy）	47.5	72.7	1.50	4.85	6.5	1.6
2 wt % SiO_2/RAL-230	50.2	84.6	1.54	5.44	16.6	1.8
4 wt % SiO_2/RAL-230	46.3	102.5	1.51	6.30	15.5	2.3

Table 2 Comparison of the tensile properties of SiO_2/epoxy nanocomposites at room temperature (300 K) using the self-designed and commercial jigs

The specimen composition	Tensile strength（MPa）		Elongation at the break（%）	
	Commercial	Self-designed	Commercial	Self-designed
RAL-230（epoxy）	45.5	47.5	6.0	6.5
2 wt % SiO_2/RAL-230	51.2	50.2	16.2	16.6
4 wt % SiO_2/RAL-230	48.1	46.3	14.8	15.5

The cryogenic tensile properties of SiO_2/epoxy nanocomposites at room temperature (300 K) and cryogenic temperature (77 K) were investigated (see Table 1). Similar results for the tensile properties at room temperature have also been obtained using normal commercial jigs that are not suitable for measurement of cryogenic tensile properties of polymer composites. Table 2 showed the tensile properties of SiO_2/epoxy nanocomposites at room temperature measured with the two different jigs. It can be seen that the self-designed jigs have little effect on measured results for the tensile properties of SiO_2/epoxy nano-composites, indicating that our self-designed jigs can be employed to precisely measure the tensile properties of polymer composites. Moreover, it can be seen that the tensile strength and Young's modulus of the SiO_2/epoxy nanocomposites were generally higher at 77 K than that at room temperature. The results showed that the addition of 2wt% SiO_2 leads to an increase of 16% in tensile strength and 12% in Young's modulus, the addition of 4wt% SiO_2 brings about an increase of 36% in tensile strength and 30% in Young's modulus compared with room temperature, while the elongation-to-break nanocomposites have a dramatic reduction with the addition of SiO_2 because the materials become brittle. The improvement of the strength and modulus at cryogenic temperature by the addition of silica nanoparticles can be easily understood because the filler-matrix interface adhesion becomes stronger at cryogenic temperature because of clamping stress while the stiffness of inorganic nano-particles have a much higher modulus than that of polymer matrix. The above results exhibited that self-designed tensile testing jigs suited well for measurement of tensile properties of composite materials at low temperature.

CONCLUSIONS

In summary, as shown above, all the measured results can be reasonably explained. It is thus believed that our self-designed mechanical testing system can be credibly used to measure the cryogenic tensile properties of polymer composites.

REFERENCES

1. S. Nishijima, T. Okada, Y. Honda. Evaluation of epoxy resin by annihilation for cryogenic use. Advances in cryogenic engineering. Vol 40:1137.
2. D. Evans. Turn, layer and ground insulation for superconducting magnets. Physica C. 354 (2001):136-142.
3. S. Usami, H. Ejima, T. Suzuki, et al. Cryogenic small-flaw strength and creep deformation of epoxy resins. Cryogenics. Vol 39(1999):729.
4. H. Yamaoka, K. Miyata and O. Yano. Cryogenic properties of engineering plastic films. Cryogenics (1995) 35:787-789.
5. Takefumi Horiuchi and Tsutomu Ooi Cryogenic properties of composite materials. Cryogenics (1995) 35:677-679.

High precise thermopower measurement system and its applying to organic conductors $(TMTSF)_2ClO_4$ in a direction

Y. S. Chai, H. S. Yang, J. Liu, L. Zhu and L. Z. Cao

Structure Research Laboratory, Department of Physics, University of Science and Technology of China, Hefei, Anhui 230026, P. R. China

We present a carefully designed apparatus for high precise measurement of the thermopower. The resolution of the system can reach to $0.01\mu V/K$. The thermopower of organic conductors $(TMTSF)_2ClO_4$ single crystals were measured. A linear behavior at high temperature and a deviation at 140 K attributed to 1D—2D crossover were observed. Different cooling rates are realized to study the anion order-disorder transition in thermopower at $T_{ao}=24K$. There is an obvious upturn just below T_{ao} when the cooling rate reaches 0.3 K/s. A clear trace of SDW transition at 4.5 K can be spotted by the fastest cooling process.

INTRODUCTION

The thermoelectric power S (thermopower or TEP) is the voltage generated across two points on a material divided by the temperature difference between the two points [1]. Besides being an important transport property, the thermoelectric power is very sensitive to the sign of charge, the density of states at Fermi level, composition, structure, pressure and external fields. It is essential for understanding the physics of materials both theoretically and experimentally. However, the measurement of thermopower on the $(TMTCF)_2X$ (C=Se for TMTSF or S for TMTTF, and X= PF_6, AsF_6, ReO_4, ClO_4, Br etc)family of organic quasi-one-dimensional (1D) conductors is very hard to perform, due to their needle-like shapes and fragile body condition.

Due to the strong anisotropy and subtle balance between interchain and intrachain electronic coupling, these quasi-1D conductors exhibit a variety of electronic ground states, including SDWs, superconductivity, magnetic-field-induced spin-density waves (FISDWs), anion order-disorder transition and so on [2]. In the case of $(TMTSF)_2ClO_4$, which exists the anion order-disorder transition ($T_{ao}= 24$ K), if quenching above T_{ao}, its ground state is SDW state ($T_{SDW}=6$ K). In relaxed states (very slow cooling procedure), its ground state is metallic/superconductivity state ($T_C \approx 1$ K) [3].

The object of this paper is to present a carefully designed apparatus for the high precise studying of the thermopower in the temperature range from 4-300 K. The thermopower data of $(TMTSF)_2ClO_4$, at different cooling rates through T_{ao}, are reported.

DESIGN OF THE THERMOPOWER MEASUREMENT SYSTEM

The schematic diagram of the computer-interfaced system is shown in Figure 1. A sample is connected between two copper blocks. A thermal gradient across the sample is applied by heating the two blocks to different temperatures. Thermometers were used to monitor the temperatures of the blocks. In order to control the temperature of either of the copper blocks, first, by using 7150 Digital Multimeters, computer

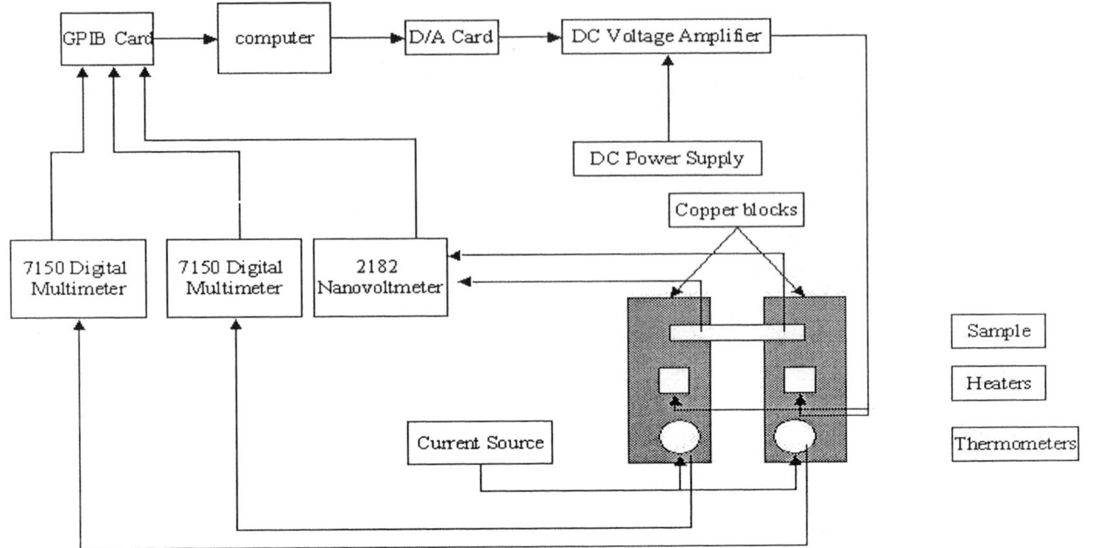

Figure 1   The schematic diagram of the computer-interfaced measuring system

can read the voltages from GPIB port and calculate their temperatures, then, the computer calculate the heating power through PID arithmetic and control the D/A card to output certain heating current to the heaters. After reaching thermal equilibrium, ΔT can be calculated or directly measured between the blocks. The electric potential difference ΔV is measured by the Keithley 2182 nanovoltmeter with contact leads on the sample. In general, the thermopower is:

$$S=\Delta V/\Delta T$$

The thermoelectric effect of the Cu lead wires was calibrated using a type of metal as a reference. In fact, each thermopower data point is obtained by averaging over 100 ΔV and ΔT points to depress the electric random error, and then by inversing temperature gradient to eliminate contacting thermopower.

The high vacuum and two thermal shields were performed in suppressing the temperature fluctuation of sample and increasing signal-to-noise performance. The temperature of either of copper blocks, which was controlled by computer, can be stabilized to less than 0.003 K or even less depending on the time. A temperature gradient $\Delta T/T\sim0.6$ % across the samples was carefully controlled between   6 K and 300 K. Every leads coming from the computer were added filters to screen high frequent electric noise. So the resolution of thermopower is proven to be $0.01\mu V/K$ at best condition. When combined with a soft-supporting platform, our procedure leads to effectiveness and accuracy for the thermopower of small samples, with direct application to organic conductor $(TMTSF)_2ClO_4$.

EXPERIMENT

To demonstrate the techniques described here, we consider the thermopower of organic conductor $(TMTSF)_2ClO_4$ single crystal. The typical size of the sample is about $3\times0.1\times0.05$ mm^3. The sample is mounted along a direction. A small soft-supporting copper block, as shown in Figure 2, is used to prevent the needle like single crystal from cracking in a fast cooling rate. Even for the fastest cooling rate in our experiment of 1K/s, the data around the anion order-disorder transition (~24 K) of $(TMTSF)_2ClO_4$ is credible without cracks.

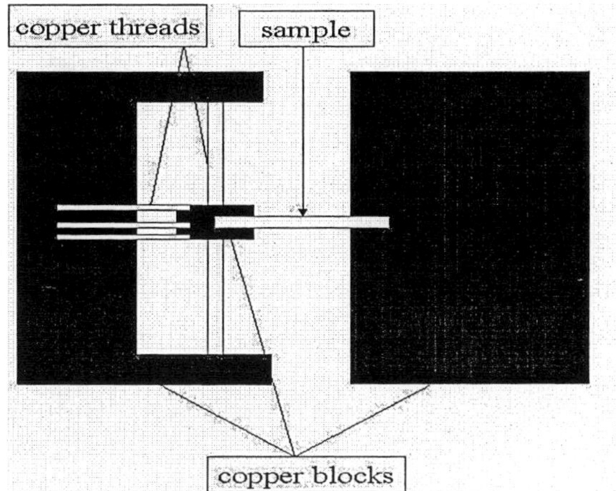

Figure 2   Soft-supporting platform

The small soft-supporting platform mounted over two parallel copper threads to support one end of the sample softly. 30 threads were attached between both a big copper block and the end of the small platform so much as to maintain a same temperature between the big and small blocks.

Figure 3 is temperature dependent thermopower of $(TMTSF)_2ClO_4$ in a direction, from 6 K to 280 K. The cooling rate through the T_{ao} is 0.0005 K/s, which make it getting into the relaxed state.

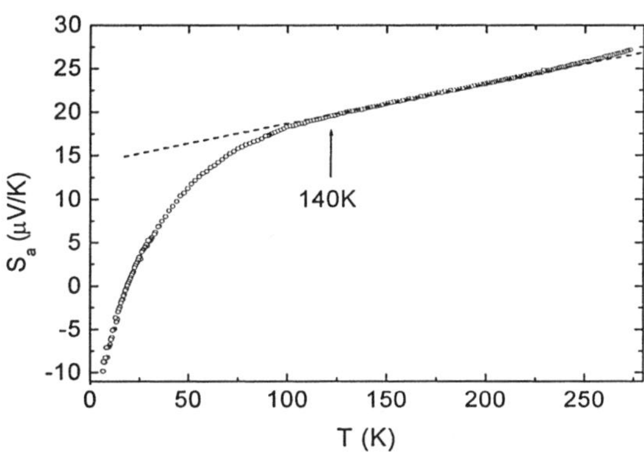

Figure 3   The temperature dependent thermopower of $(TMTSF)_2ClO_4$ metallic state along the a direction

There is a linearly temperature dependent thermopower at high temperature. Below 140 K, the curve deviates from linear behavior and goes down all the way through. Such a deviation behavior can be attributed to a 1D—2D crossover, which is observed in EPR (electron paramagnetic resonance)[4]. The overall behaviors agree well with the previous results of Choi's [5]. The sign of thermopower S_a is positive at high temperature, implying the carriers are holes.

Figure 4 shows the temperature dependence of the thermopower below 30 K with different cooling rate through T_{ao}. Each cooling process below 35 K was reached by filling of He exchanging gas and the rate was controlled by the computer. The thermopower was measured by warming after bumping the gas for 12 hrs. For the fastest cooling rate of 1 K/s, it yields the SDW state at 4.5 K [3]. Comparing with the result of the slowest cooling process (relaxed state), there is an obvious upturn just below T_{ao} when the cooling rate reaches 0.3 K/s. It is evident that random potential of the frozen anion disordering introduces

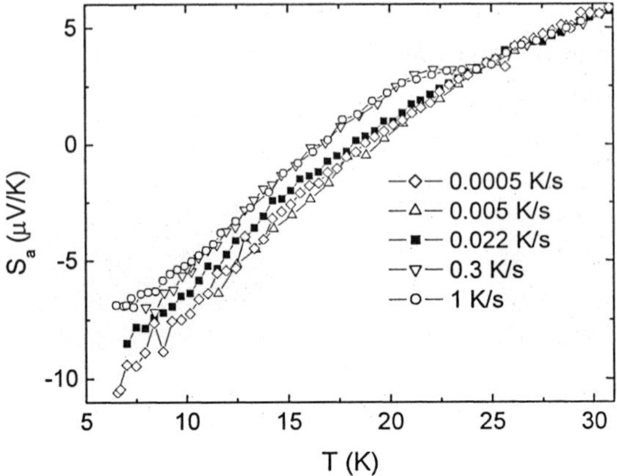

Figure 4   Temperature dependence of the thermopower with different cooling rate through the T_{ao}

the scattering. A clear trace of prelude to the SDW transition at 4.5 K can be spotted by the fastest cooling process, for there is an obvious increase in the thermopower below 10 K.

CONCLUSION

The design of a high precise measurement system of the thermopower is reported. The thermopower of organic conductors $(TMTSF)_2ClO_4$ needle single crystals were measured successfully. A linear behavior at high temperature and a deviation at 140K attributed to 1D—2D crossover were observed. There is an obvious upturn just below T_{ao} when the cooling rate reaches 0.3 K/s. A clear trace of prelude to the SDW transition at 4.5 K can be spotted by the fastest cooling process.

ACKNOWLEDGEMENT

Project is supported by the National Natural Science Foundation of China (Grant No. 10374082) and by the Ministry of Science and Technology of China (No.G19990646)

REFERENCES

1.   MacDonald D. K. C., In: Thermoelectricity. Wiley & Sons, New York, USA (1962) 1-36

2.   Yang H. S., Lasjaunias J.C. and Monceau P., Specific heat measurement of the lattice contribution and spin-density -wave transition in $(TMTSF)_2X$ ($X=PF_6$ and AsF_6) and $(TMTTF)_2Br$ salts, J.Phys: Condens. Matter (1999), 11 5083-5098

3.   Yang H. S., Lasjaunias J.C. and Monceau P., Specific heat measurements of the lattice contribution and the spin-density -wave and anion-ordering transitions in the $(TMTSF)_2ClO_4$ salt (TMTSF =tetramethyltetraselenafulvalene), J.Phys: Condens. Matter (2000), 12 7183-7198

4.   Lee C.E. et al, EPR Observation of the dimensional crossover in $(TMTSF)_2ClO_4$, Solid State Commun. (1999), 109 69-72

5.   Choi E. S. et al, Thermoelectric power of anisotropic electron system in quasi-one-dimensional organic conductors, Synth. Met. (2001), 120 1069-1070

Multichannel Temperature Monitor Compatible with PC

Filippov Yu.P., Alpatov S.V. and Sveshnikov B.N.

Joint Institute for Nuclear Research - JINR, Dubna 141980, Russia

A temperature monitor to measure cryogenic temperatures by means of resistive temperature sensors is presented. This is a box of $172\times76\times24$ mm^3 connected to COM-port (RS232) of a personal computer. A four-lead technique is used to measure the resistance of the temperature sensor with respect to the precision reference resistor. The main characteristics are as follows: the number of measured channels/sensors can be from 1 up to 15 or 16; the range of measuring direct current is from 0.5 μA to 5 mA; the range of measured resistances, R, is from 1 Ohm to 100 kOhm; the accuracy of measurements is $\Delta R/R < 0.01$ %. The temperature monitor was successfully used during the 18 months operation with the rhodium-iron, carbon- glass and TVO temperature sensors.

INTRODUCTION

Modern superconducting installations require the systems to control operation characteristics of cooled devices - magnets, cavities, detectors etc. and to monitor the thermodynamic state of cryogens with rather high accuracy. The value of temperature is one of the main parameters to determine the state of a cryogen or a cooled device. The aim of this work was to create a measuring system to operate with a big number of temperature sensors to test the quality of a superconducting cavity like [1] for steady state conditions. The total number of sensors at the tested cavity should be 256. In principle, there are some ways to solve this problem. We have chosen a schematic of multichannel system shown in Figure 1. This variant supposes the usage of several remote devices (with standard 12/15 V d.c. power supply) which are connected with a usual PC via standard interface RS232. To realize this variant, one needs several electronic modules, each of them allows to measure signal of 15 or 16 sensors. In principle, 16 devices can be connected to the COM-port of the PC: in this case the maximum number of sensors can be 256. A 15/16 channel temperature monitor (TM) is a module of multichannel system to measure temperatures by means of resistive temperature sensors whose resistance can be of a wide range in principle. This module is described below.

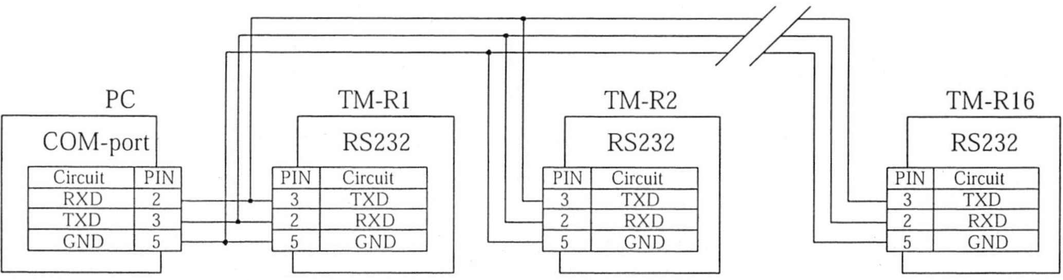

Figure 1 Schematic of multichannel system (16×16=256 temperature sensors) with remote temperature monitors connected to PC via RS232

15/16 CHANNEL TEMPERATURE MONITOR

Principle of measuring

A four-lead technique is used to measure the resistance of the temperature sensor with respect to the precision reference resistor. To avoid the influence of parasitic voltages, measurements are performed with a direct current source whose polarity can be changed. In this case the measured resistance R_m can be found as $R_m = [(U_m/U_r)^+ + (U_m/U_r)^-]\times R_r/2$, where U_m is a voltage drop at the measured resistor, U_r – a voltage drop at the reference resistor R_r, symbols «+» and «-» correspond to the positive and negative polarities. The advantage of this method of relative measurements is that it does not require any calibration of the measuring device or verification of its zero offset. In order to check the accuracy of measurements from time to time, it is necessary to connect the certified precision reference resistor of 0.001 %, for example, instead of the measured resistor R_m.

Structural schematic

The structural schematic of the temperature monitor is shown in Figure 2. The sixteen measured resistive sensors (or only one sensor) can be connected to the measuring device via 44-pin connector. These sensors are connected in series. The measuring device has a temperature sensor/chip whose readings allow to correct the value of the reference resistances depending on temperature. The electronic board includes, in particular, a direct current source with reversible polarity, multiplexers for 15 or 16 resistive temperature sensors, reference resistors of 0.005 % and ADC. The value of the current can be regulated from 0.5 µA to 5 mA. This device can be supplied with reference resistors of 10 Ohm, 100 Ohm, 1 kOhm and 10 kOhm which allows to measure the signals of the sensors whose resistance can be from 1 Ohm to 100 kOhm. In principle, there is no problem to support ISA-interface if one needs to manufacture an electronic board which should be inserted into the industrial PC.

Design

The described variant of the electronic board is mounted within a plastic box of $172\times76\times24$ mm^3 shown in Figure 3. Its weight is about 0.19 kg. A RS232-connector and power supply input are located in the front side of the box, and the 44-pin connector for temperature sensors is mounted at its reverse side. A standard 220/110 V 50/60 Hz adapter with 12 V d.c. and 500 mA output is used for power supply. Figure 3 also shows a connecting cable for the sensors with the cable connectors and hermetic wire connector which should be mounted at the cap of the cryostat. If a standard 32-pin wire connector is used, one can realize a 15 channel variant of the temperature monitor: 2 pins – for current leads and 15×2 =30 pins – for potential leads. The next Russian standard for connectors is 50-pin variant which allows to realize the 16 channel version. If it is necessary, one can manufacture one more 16 channel version - with two independent 19-pin wire connectors when each group of 8 sensors, connected in series, is mounted at a separate connector at the cap of the cryostat.

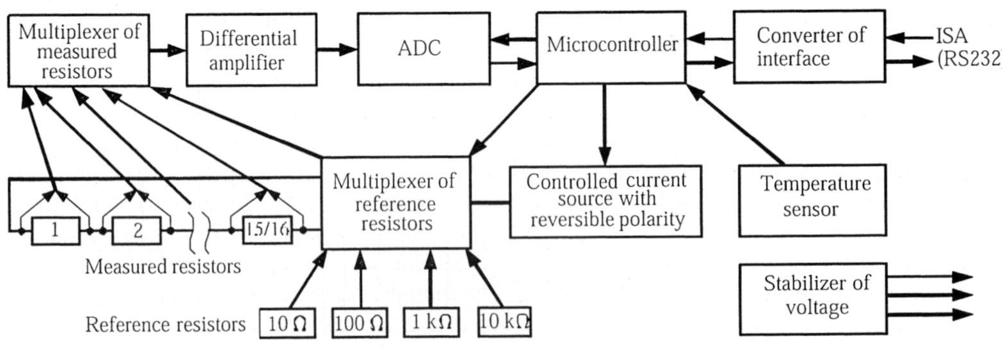

Figure 2 The structural schematic of the temperature monitor

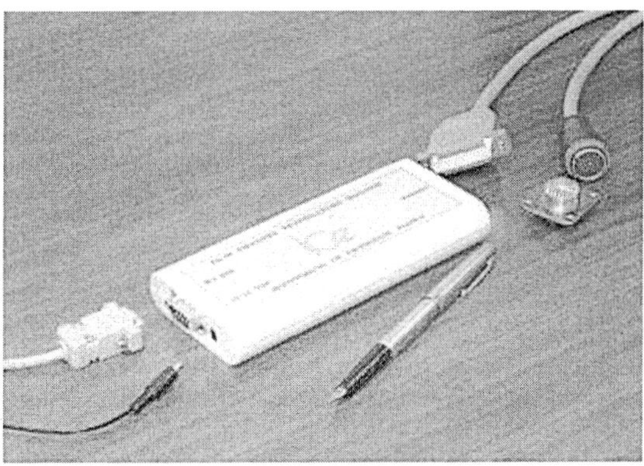

Figure 3 Temperature monitor with interface RS232 and connecting cables

Software

A DOS-program allows one to find: values of the resistances (Ohm-meter regime), temperatures T(R-resistance), temperatures depending on time T(R, t-time), to compare a found value of the temperature with a reference value T_r (in this case any connected sensor can be used as the reference one or the reference temperature T_r can be entered as the reading of the chosen thermometer), to enter a new calibration curve T(R) for a replaced sensor if necessary. If the system operates with TVO temperature sensors under magnetic field, one can find the corresponding temperature shift due to magnetic field ΔT(R, B-magnetic field) for each TVO-sensor in accordance with [2]. While operating in the regimes "Measurements of resistances" or "Measurements of temperatures", one can change the measuring current according to the certificate for the connected sensor. There are automatic regimes to choose the current for the TVO and carbon-glass temperature sensors [3].

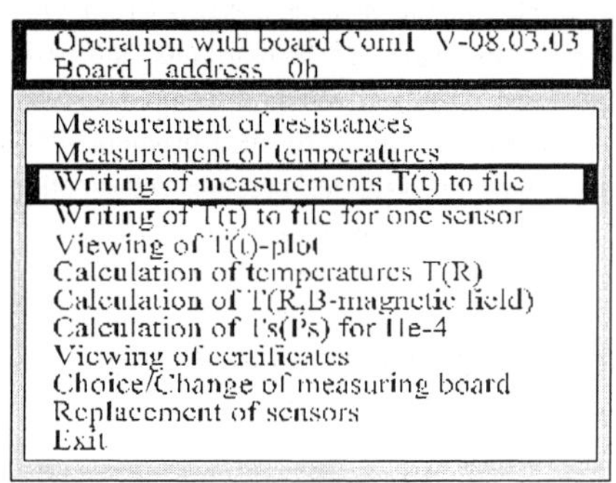

Operation with board Com1 V-08.03.03
Board 1 address   0h

Measurement of resistances
Measurement of temperatures
Writing of measurements T(t) to file
Writing of T(t) to file for one sensor
Viewing of T(t)-plot
Calculation of temperatures T(R)
Calculation of T(R,B-magnetic field)
Calculation of I's(P's) for He-4
Viewing of certificates
Choice/Change of measuring board
Replacement of sensors
Exit

Figure 4 The main menu

The software consists of several files: T_meter.exe, T_meter.ini, T_board.ini, alarm.ini, HIEW.EXE. T_meter.exe is the main program, its possibilities are briefly described above. The main menu is shown in Figure 4. The T_meter.exe-program operates in accordance with all items of the menu if there are files containing the information about calibration points and polynomial coefficients. They can look as "jinr_001.pas", for example, where 001…128 is the identification number of the sensor. Such information is written into the file "T_meter.ini". This file defines the correspondence between the really connected sensors and their certificates/calibration data. In principle, this is the text file whose lines look as follows: *"nkk file_name"* where "n" is the number of the board (1..16), "kk" is the number of the measuring channel/connected sensor (1..15/16), "file_name" is a certificate of the sensor which can look as "JINR\jinr_120.pas". The ancillary file "T_board.ini" provides an operation with several boards - up to 16 pieces. In principle, this is also the text file whose lines look as *"board n = xxx"*, where "n" is the number of the board and "xxx" is the hexadecimal base address. A range of base addresses is from 0h to 15h which can be changed by four jumpers at the electronic board. The file "alarm.ini" activates the low and high alarm capability. The program "HIEW.EXE" reviews the certificates of the used sensors.

The measurement process is arranged as follows. The software transfers commands of tuning (values of current, reference resistor and so on) to the electronic board and sends the command to start measurements. Then it receives and represents the results after the readiness signal. The data swapping with the board is done by using the interruption mode since the measurement process is relatively slow and the board is connected with COM-port which has a standard interruption. This enables one to use a processor for another applications while starting the operation under Windows-system.

This software and hardware can be used for operation with sensors calibrated not only by us but for sensors of another firms. In this case one needs to convert the calibration data to a compatible format.

RESULTS OF TESTING

The temperature monitor was tested with rhodium-iron, carbon-glass and TVO resistive temperature sensors [3]. It provides two modes of measurements: the continuous mode when measurements are done during all the time and the single one when only one measurement for the sensors could be performed. A kind of measurements can be carried out with one chosen sensor and with all 15/16 sensors. The time to measure all the channels is about 1.6 sec. and 0.3 sec. for one chosen sensor. The tests have shown that the effective resolution of the analog to digital converter is 20 bit. The biasing current can be regulated as follows: 0.5, 1, 2, 5, 10, 20, 50, 100, 200, 500 µA and 1, 2, 5 mA. All the measurements have been carried out with the connecting cable (between the temperature monitor and the cryostat) of 10 m. In principle, the results with a shorter cable could be better. However we have not seen any difference for the 3 m and 10 m cable at our laboratory. The accuracy of measurement is estimated as $\Delta R/R < 0.01$ % for all the channels. This accuracy allows to use this temperature monitor not only for routine measurements but also for calibration of cryogenic temperature sensors with moderate accuracy – not worse than ± 10 mK at 4.2 K. The time to reach the operation regime is up to 15 minutes, and the time of continuous operation is not restricted. The service life is estimated as 10 years and longer. One can note that there was no error revealed during the 18 months period of operation with the tested temperature monitor.

CONCLUSIONS

The designed, manufactured and tested temperature monitor, TM, of the compact size ($172 \times 76 \times 24$ mm^3) meets the technical requirements for measurements with any cryogenic resistive sensors whose resistance can be from 1 Ohm to 100 kOhm or wider. The TM is connected with a personal computer via COM-port (RS232). It provides an operation with 15 or 16 sensors connected in series. The accuracy of measurements of the resistance, R, is $\Delta R/R < 0.01$ %. The software allows one to carry out writing of the temperature dependence on time (t) to file and observing of the corresponding T(t)-plot on the screen, to glance at the certificates of the sensors with calibration data, to calculate deviations due to the magnetic field for TVO sensors and to compare the measured temperatures with the reference value.

In principle, 16 temperature monitors can be connected to COM-port of the PC to provide measurements with 256 resistive temperature sensors.

ACKNOWLEDGEMENTS

Many thanks to our colleague V.M. Miklayev for the suggestions to improve the software based on his long-term experience to operate with the tested temperature monitor.

REFERENCES

1. Filippov, Yu.P. and Kovrizhnykh, A.M., Cryostat to Test Superconducting RF Gun for the Drossel Project, Advances in Cryogenic Engineering, Kluwer Academic/Plenum Publisher, New York, (2000), 45A, 919-924
2. Filippov, Yu.P. and Shabratov, V.G., Measurement of Helium Temperatures by TVO-Sensors under Magnetic Fields, Cryogenics (2002) 42 127-131
3. Dedikov, Yu.A. and Filippov, Yu.P., Characteristics of Russian Cryogenic Temperature Sensors, Proc. of the ICEC18, Mumbai, India (2000) 627-630.

INTERCHANGEABILITY OF DIODE TEMPERATURE SENSORS FOR VARIOUS NON-STANDARD EXCITATION CURRENTS

Courts S., Yeager C.

Lake Shore Cryotronics, Inc., 575 McCorkle Blvd., Westerville, OH 43082, USA

Diode thermometers interchangeable to an "average" curve are commonly used in cryogenic thermometry. Typically an excitation of 10 µA is chosen to minimize self-heating at low temperatures while maintaining an acceptable signal-to-noise ratio. For some applications, it is desirable to use a different excitation current but still use the interchangeability feature. Eleven sensors from each of two diode thermometer models, the DT-470 and the DT-670, were calibrated at currents ranging from 0.05 µA to 10 mA at 4.2 K, 30 K, 77 K, and 300 K. This paper examines the "average" curve and interchangeability of diode temperature sensors for non-standard excitations.

INTRODUCTION

Silicon diodes are widely used as general-purpose cryogenic temperature sensors due to the temperature dependence of the forward voltage drop across a p-n junction. Silicon diodes have many useful features: they are usable over a wide range (1 K to 500 K), have high signal output, have simpler and more rugged packaging, and diodes from each manufacture will all have the same response curve. The last result allows diodes to be binned into tolerance bands with accuracies between ±0.25 K to ±3 K depending on tolerance band and temperature range.

The "average" curve to which the thermometers are interchangeable is developed from the calibration of hundreds or thousands of samples from the particular diode wafer lot. This is an expensive task, and the "average" curve is generally defined only for a single excitation current. For most cryogenic applications the excitation current used to define the voltage response curve is pre-set at 10 µA. This is done to minimize self-heating at the lowest temperature and still maintain a good signal to noise ratio. However, a larger excitation current is desired for some applications (e.g., higher temperatures, instrumentation constraints). While it is impractical to develop a new "average" curve for a new excitation current, it is necessary to estimate the curve and the level to which the sensors remain interchangeable.

DESCRIPTION AND TESTING

For the testing in this study, eleven DT-470-SD diodes and eleven DT-670-SD diodes from Lake Shore Cryotronics, Inc. were used.

Temperature calibrations from 2 K to 325 K were performed in the Lake Shore commercial calibration facility. Temperature was measured using standards grade platinum and germanium thermometers in conjunction with a Keithley Model 224 current source, a Hewlett Packard Model 3458 DVM, and Guildine Model 9330 standard resistors. The excitation current was varied from a minimum of 0.05 µA to a maximum of 11 mA. In total, 22 discrete excitation currents were tested.

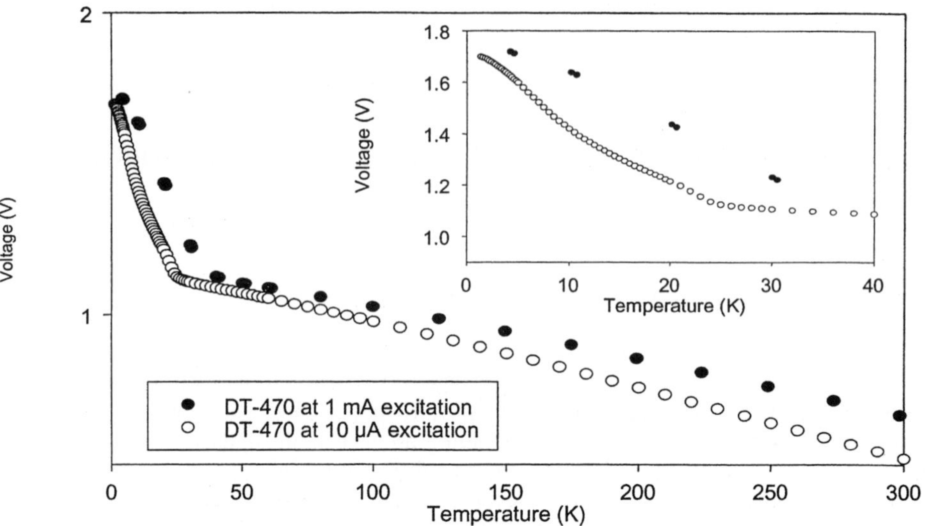

Figure 1. Voltage as a function of temperature for the DT-470 at two excitations: 10 µA and 1 mA. Inset shows detail over a narrow temperature range.

RESULTS AND DISCUSSION

At the lowest excitation of 0.05 µA, the results showed erratic behavior. It is believed that this excitation is too low to turn on the diode. At the higher excitations—greater than 1 mA—there was obvious evidence of self-heating. The 10 mA excitation showed self-heating effects at 20 K to 30 K. For this paper, discussion is limited to comparison between 10 µA and 1 mA excitation at 4.2 K, 30 K, 79 K, and 305 K.

Voltage response curves are shown in Figure 1 and Figure 2 for the DT-470-SD and DT-670-SD, respectively. Shown are results for 10 µA and 1 mA. What is noticed is that the 'knee' region occurs at higher temperatures. The crossover region is approximately 25 K for 10 µA while it is about 35 K for 1 mA. As the excitation increases, the 'knee' region increases in temperature. Figure 3 shows the voltage verses current for a DT-670-SD at various temperatures. When examining the results at 30 K, a crossover region is observed.

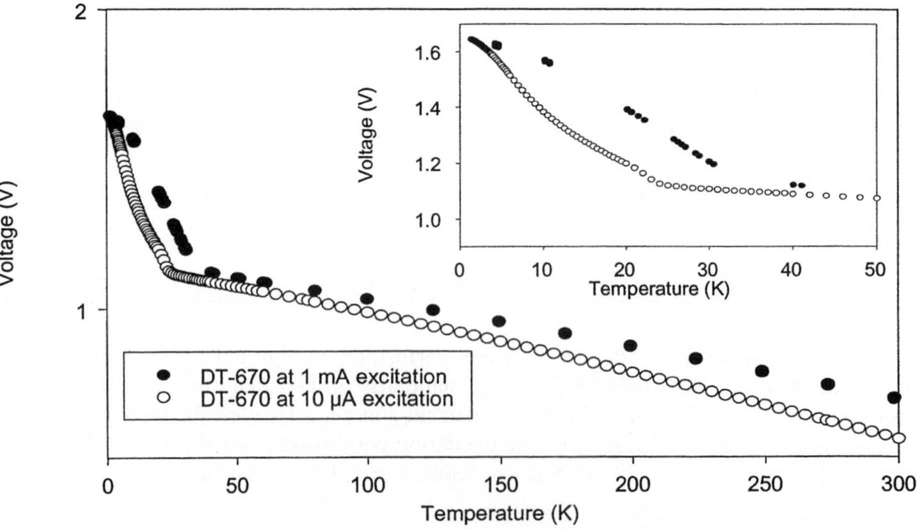

Figure 2. Voltage as a function of temperature for the DT-670 at two excitations: 10 µA and 1 mA. Inset shows detail over a narrow temperature range

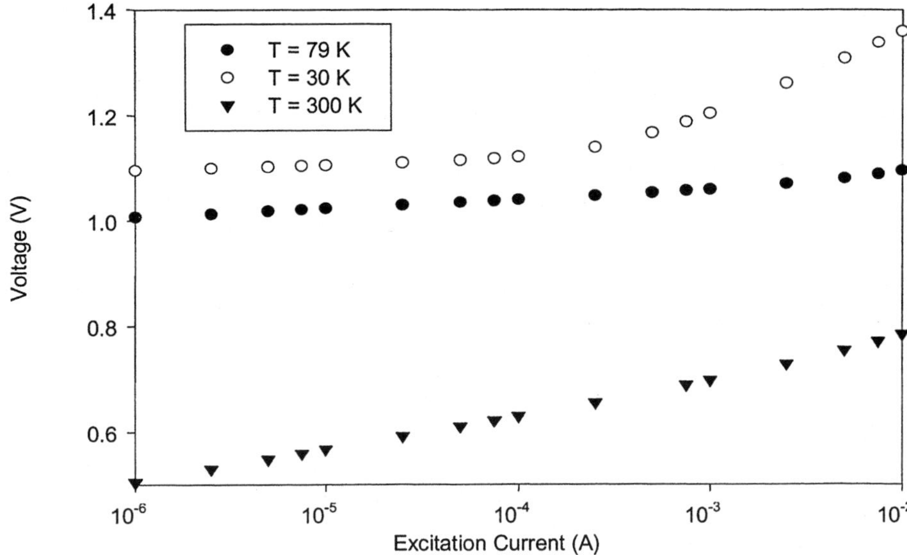

Figure 3. Voltage as function of excitation current for various temperatures. Shown is data for the DT-670.

For the lowest temperatures, there is evidence of self-heating as the sensitivity is significantly decreased at 4.2 K. For a 1 mA excitation the joule heating is 1.4 mW. Using past data for thermal resistance for the SD package at 4 K, this would correspond to a self-heating bias of more than 1 K. However, thermal resistance decreases as T^{-3}, and by 12 K the self-heating error is less than 0.05 K. This indicates a 1 mA excitation could be used to 10 K to 15 K without significant self-heating.

With a 1 mA excitation the DT-670-SD sensitivity from 40 K to 300 K is approximately 1.6 mV/K. This is compared to 2.03 mV/K over the same range for a 10 µA excitation. Specifically, at 79 K the sensitivity at 10 µA is 1.75 mV/K while at 1 mA it is 1.4 mV/K. Below 30 K at 1 mA the sensitivity is approximately 18 mV/K. This is slightly greater than the sensitivity with 10 µA below 20 K. This trend appears to continue to 10 K before the self-heating begins to dominate the results.

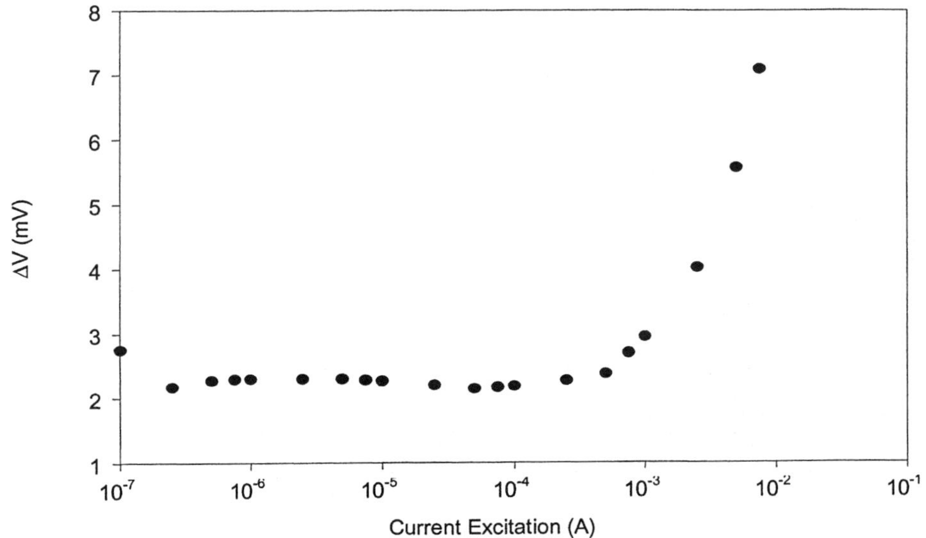

Figure 4. Difference between maximum and minimum voltage for the 11 DT-670s
tested at 79 K as function of excitation.

At 10 µA and 79 K, the 11 DT-670-SDs under test had an average value of 1.0234 V at 79 K. The difference between maximum and minimum values was a spread of 0.95 mV. At 1 mA the average value was 1.0603 V with a spread of 0.768 mV. The tolerance in temperature units is about 0.5 K for both 1 mA and 1 µA. In fact, the spread between the maximum value and minimum value of the 11 DT-670-SDs is fairly uniform from the lowest excitation up to 2.5 mA. This is shown in Figure 4. Above 2.5 mA the increase can be attributed to self-heating effects biasing the results.

Similar effects are also seen in the DT-470-SD. The spread at 79 K with 10 µA is 2.26 mV while the spread at 1 mA is 2.94 mV. While the excitation changed by a factor of 100, the tolerance band of the diode only changed by 15%.

The voltage tolerance at 30 K did vary between 10 µA and 1 mA. At 10 µA the tolerance was less than 0.1 mV while at 1 mA it was 1 mV. However, the sensitivity changed from 2 mV/K at 10 µA to 18 mV/K at 1 mA. The order of magnitude change in voltage tolerance is offset by an order of magnitude increase in sensitivity. The result is that the temperature tolerance is unchanged. This is the same for the DT-470-SD at 30 K. The sensitivity increased by a factor of 10 while the voltage tolerance band increased by a nearly equal amount.

For completeness, we looked at the changes at 4.2 K. The voltage tolerance band remained relatively constant from 10 µA to 1 mA. This was seen in both DT-470-SDs and DT-670-SDs.

CONCLUSIONS

The results of this study show that the tolerance band at 10 µA is predictive of the same tolerance at 1 mA. As expected the I-V curve and V-T curve do change with excitation but the spread, or tolerance bands, of the diodes do not change significantly. The implication is that a Lake Shore standard curve diode with a ±1 K tolerance that is defined at 10 µA will also have a tolerance of about ±1 K at 1 mA excitation. Because the 1 mA "average" curve is only defined by a few sensors, this will introduce some variation in the final temperature accuracy.

The voltage temperature response curve at 1 mA has a slightly lower sensitivity down to 40 K. However, the knee region is seen at a higher temperature than in 10 µA. This allows for the interesting option of increasing the sensitivity in the 20 K to 40 K region by using a larger excitation.

Further studies will examine the dI/dV and the voltage vs. temperature curve of other excitations. Work will continue to develop and provide a basic "average" curve at 1 mA and other excitations.

REFERENCES

1.   Lake Shore Cryotronics Inc. Westerville, OH. 43082

The Laser Megajoule CryoTarget Thermal Regulation

Lamaison V., Brisset D., Cathala B., Paquignon G., Bonnay P.*, Chatain D.*, Communal D.*, Périn J-P.*

CEA-CESTA/DLP/SELP/Laboratoire Assemblage et Intégration des Cibles - Route des Gargails, BP2, 33114 Le Barp France ; lamaison@drfmc.ceng.cea.fr
*CEA-Grenoble/DSM/DRFMC/Service des Basses Températures - 17, rue des Martyrs, 38054 Grenoble Cedex9 France

The Laser Megajoule requires a high resolution temperature regulation to obtain the cryotarget temperature conditions: a temperature slope of 1 mK/min with ±1 mK to reach the triple point 19,79 K and a regulation at constant temperature with ±1 mK. This regulation required a thermal model elaboration for the target and the process, a regulation module development with a ±50 μK resolution and a specific algorithm of regulation. It runs on the prototype "Echelle 1", the Cryotarget Positioner mock-up built at the Service des Basses Températures - CEA Grenoble. The temperature stability obtained on the cryostat "Echelle 1" is presented.

INTRODUCTION

The Laser Megajoule facility is the French equipment intended for studies of inertial fusion. Thermonuclear fusion is obtained by focusing 240 laser beams with an energy of 1.8 MJ on a cryotarget (see Figure 1). This target, made up of a two millimeter diameter microballoon, filled with a Deuterium Tritium mixture freezed at 19.79 K. The microballoon implosion, following laser beams impact, generates in its centre the ignition conditions of temperature and pressure. The success of implosion strongly depends on the geometrical characteristics of the Deuterium Tritium ice layer: margin 1% for thickness, 1μm for roughness. The procedure to obtain this ice layer requires a temperature control of the target base with a margin of ±1 mK. This regulation includes 3 phases: a temperature slope of 1 mK/min with ±1mK to reach the triple point 19.79 K, a regulation at constant temperature with ±1mK, and finally a fast decrease (< 10 s) of temperature from 19.79 K to 18.2 K.

These specifications must be maintained during the target transfer to the centre of the experimental vacuum chamber by the cryotarget positioner (see Figure 2).

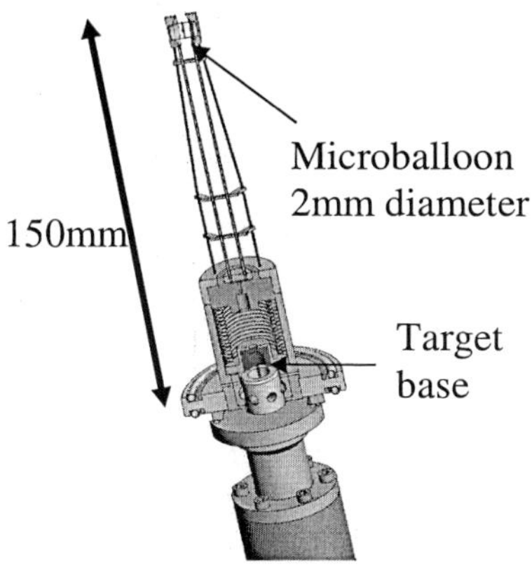

Figure 1 Cryogenic Target

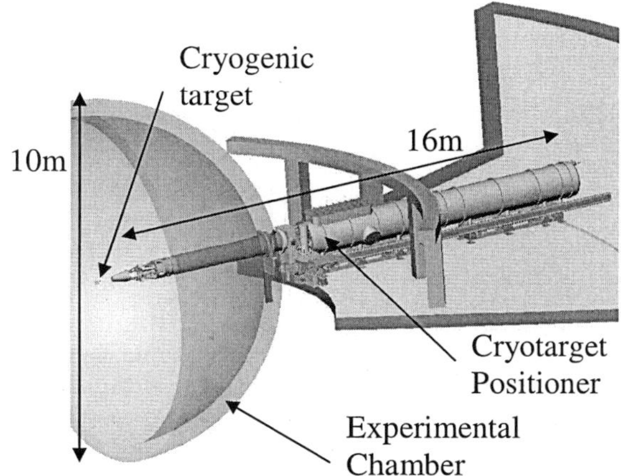

Figure 2 Cryotarget Positioner and Experimental Chamber

THERMAL OPERATION OF THE « ECHELLE 1 » PROTOTYPE

The cryogenic arm of the "Echelle 1" prototype consists of a target base, a gripper and a heat exchanger (see Figure 3). Thermal sensors mounted on each element follow temperatures evolution and heaters on target and heat exchanger are used for regulation.

Target base is cooled by conduction from a cold source made up of heat exchanger cool down by a helium gas flow. A pressure of 300 mbar above the liquid helium in the reservoir (100 liters) generates the helium flow [1]. Figure 4 presents the moving part of the cryostat bringing the target to the centre of the vacuum chamber.

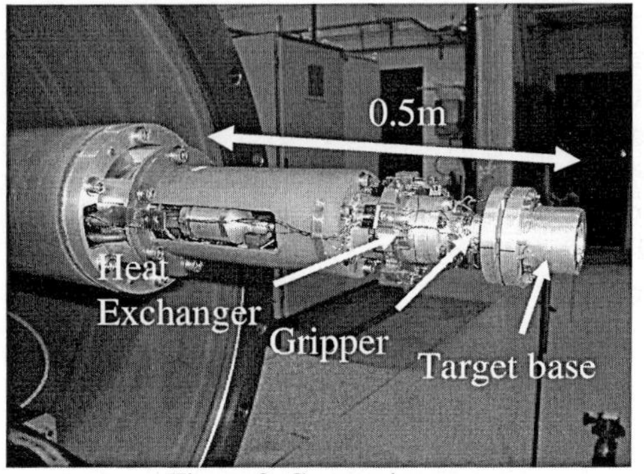

Figure 3 Cryogenic arm

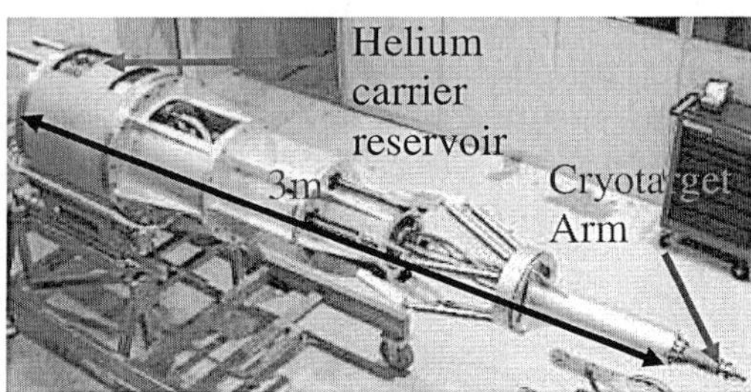

Figure 4 Moving part of the cryostat "Echelle 1"

THERMAL MODEL OF THE CRYOGENIC ARM

The thermal model includes the target base, the gripper and the heat exchanger. An electrical analogy is done (see Figure 5).

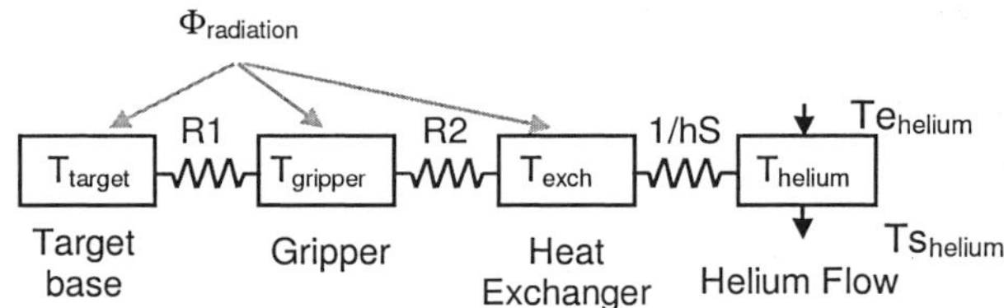

Figure 5 Thermal model scheme

A heat balance is carried out on each element by neglecting the heat diffusion. For example, target base balance is given by:

$$M_{t\,arg\,et}Cp_{t\,arg\,et}\frac{dT_{t\,arg\,et}}{dt} = P + \phi_{radiation} - \frac{T_{t\,arg\,et} - T_{gripper}}{R_1} \qquad (1a)$$

with R1: thermal resistance target base/gripper, P: power injecting on target base, R2: thermal resistance gripper/heat exchanger, h: convective coefficient between helium flow and heat exchanger.
The system of equations is solved by Laplace transform under the Matlab/Simulink software.

REGULATION MODULE

The regulation module was developed at the Service des Basses Températures - CEA/Grenoble. The needed high sensitivity for temperature measurement ($\Delta T < 100$ µK) required the implementation of a synchronous detection and the use of low noise special cables with double shielding. Thanks to these precautions, the regulation module resolution on the "Echelle 1" cryostat is ±50 µK.

The regulation algorithm of target base uses simultaneously two temperatures measurement and acts on two heaters (target base and heat exchanger).

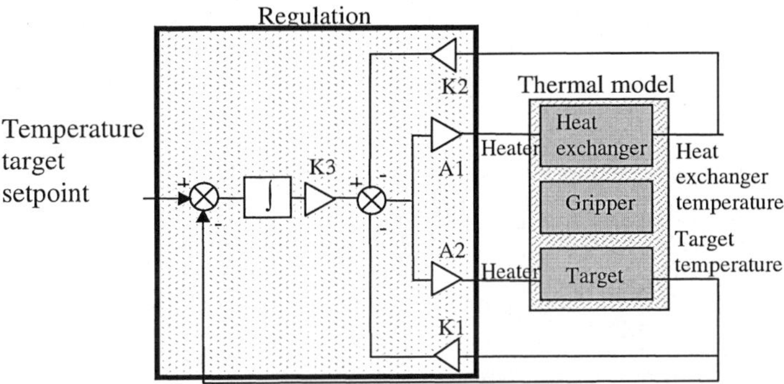

Figure 6 Regulation scheme

EXPERIMENTAL RESULTS ON « Echelle 1 » PROTOTYPE

The validation of the cryogenic arm model was carried out in open loop with power step on the target base and by comparing the target base, gripper and heat exchanger temperature evolution with calculations. Moreover, the response to an instruction level (closed loop) shows the good prediction of the system reaction associated with the regulation module (see Figure 7). The good agreement between measurement and thermal model made possible the optimization of the regulation module parameters.

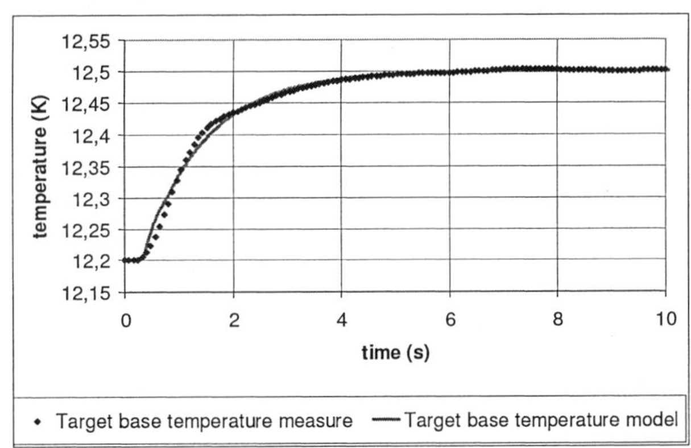

Figure 7 Regulation temperature step from 12.2 K to 12.5 K

The regulation performance obtained at constant temperature is a 55% reduction on temperature fluctuations. The experimental conditions are the following: Helium flow = 1.9 l_{liq}/h, T_{target} = 12.5 K. Without regulation, the total amplitude of temperature disturbances is ±2.9 mK (95% confidence variation with a Gaussian distribution) and is reduced to ±1.3mK with regulation. The spectrum analysis of target temperature signal ranges from 0 to 0.5 Hz (see Figure 8). The regulator reduces the low frequency disturbances in the interval of [0;0.2 Hz]. The fluctuations beyond 0.2 Hz are not modified. The performance obtained in temperature slope of 1 mK/min is almost the same with a 44% reduction on temperature fluctuations.

Moreover, the regulation robustness has been tested under displacement of the liquid helium reservoir and the cryogenic arm. This test simulates the transfer of the target towards the centre of the vacuum chamber. The speed of the carriage is 50 mm/s. Without regulation, the movement generates fluctuations of ±6 mK on the heat exchanger temperature due to helium flow fluctuation. With the regulation, the temperature fluctuation decrease to ±3 mK on heat exchanger (see Figure 9). The target base temperature is maintained within the interval ±2 mK (see Figure 10).

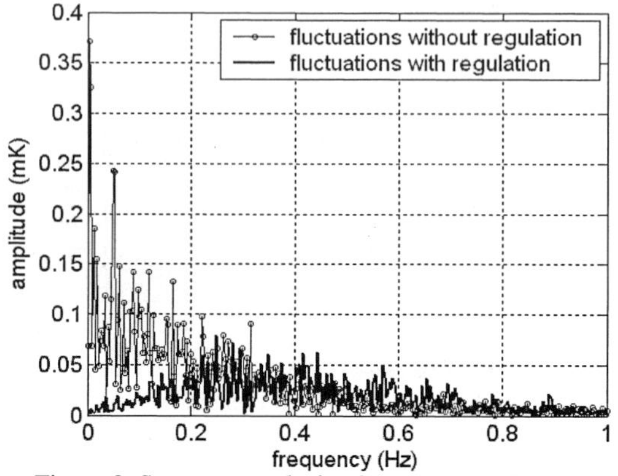

Figure 8 Spectrum analysis of temperature fluctuations

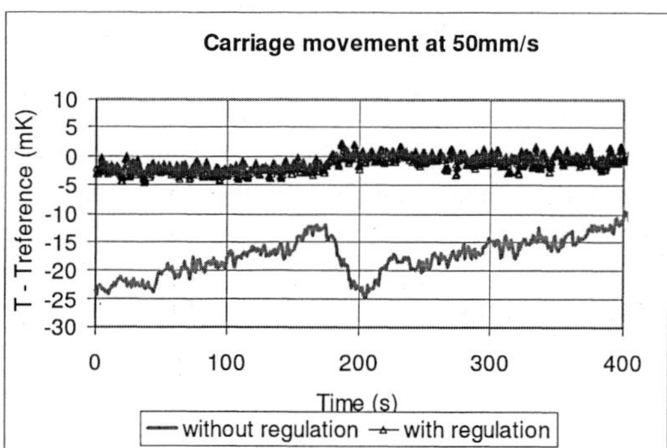

Figure 9 Heat exchanger temperature

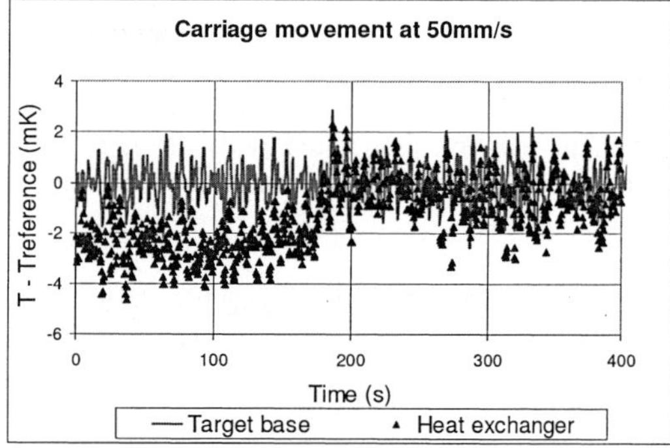

Figure 10 Target base and heat exchanger temperature with regulation

CONCLUSION

The regulation module, developed at the Service des Basses Températures, has a sensitivity of ±50 μK with a low noise system of measurement. The regulation algorithm allows an accurate, fast and robust temperature control. We reached a temperature stability of ±1.3 mK with a 55% reduction of the fluctuations. We still study the origin of perturbations to improve temperature stability.

ACKNOWLEDGEMENT

We thank Marc CHICHOUX and Pierre NIVELON for their reliable technical support. This work is supported by the CEA/LMJ Program.

REFERENCES

1. Paquignon, G., Brisset, D., Cathala, B., Lamaison, V., Chatain, D., Bonnay, P., Bouleau, E., Périn, J-P., First results on the prototype of the Laser-Megajoule Cryotarget Positioner, Fusion Science Technology 15 (2004), 45 282-285

Discussion of the protection of pressure vessels by using safety valves-rupture disc-combinations

Süßer M.

Forschungszentrum Karlsruhe, Institut für Technische Physik, D76021-Karlsruhe, Germany

Due to safety requirements cryogenic facilities are equipped with safety valve-rupture disc-combinations against overpressure. Spring loaded safety valves in combination with rupture discs are designed as the "last resort" in the safety hierarchy to protect life and property. Under certain conditions while discharging, so called fluttering or chattering may occur. This behavior is defined as an extremely opening and closing of the valve. The associated mechanical loads can damage the piping and the equipment or cause pressure oscillations with high amplitudes in the inlet piping to the safety devices. This paper presents safety considerations in cryogenic test facilities for superconducting magnets and shows the reasons, the risks and furthermore the prevention of the oscillations.

CONCEPTUAL DESIGN OF PRESSURE RELIEF SYSTEMS

The consideration of the pressure relief system is an important step in the design of a safe and reliable facility. The conceptual design consists of different steps: At first, a step wise interaction against pressure increase has to be defined (see figure1). Then a decision concerning the location and capacity of pressure relief device has to be made. The selection of the general type of pressure relief devices for each identified location, i.e. safety valve and rupture disc is followed. A further step is the choice of the special features for the chosen devices. At last, in case of releasing medium to the atmosphere, a consideration with respect to economic and environment has to be included.

SPECIFIC CRYOGENIC CONDITIONS

Compared with other facilities, cryogenic plants have specific characteristics. The most noticeable one is the long inlet piping to the safety devices, because the devices are mounted on the warm side of the facility. Another one is the oversize of the safety valve system; usually it is a worst case design with extra charge for not exactly determined conditions. The probable mass flow during discharging is much lower as during the worst case and can be very wide. In addition, at the worst

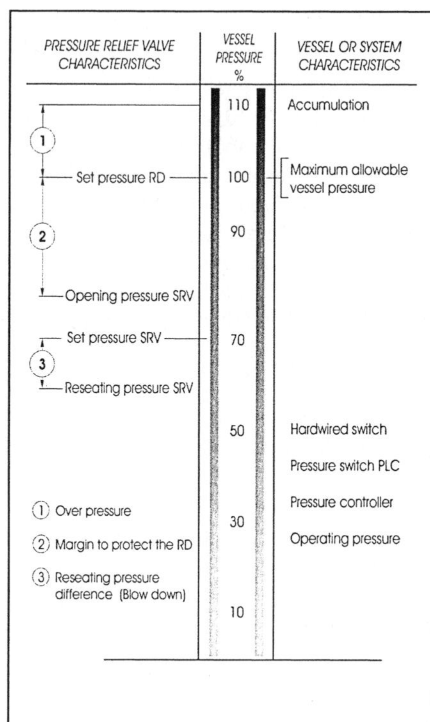

Figure1: Pressure terminology

case, pressure rates up to10 bar/s may occur. A further specific peculiarity is the cool down of the inlet piping at the beginning of discharging with very high pressure drop and pressure fluctuations. A validation of the safety system, to ensure that the safety system has the required functionality, is only possible under limited conditions. Therefore, there is less change of a retrofit of the system. The safety devices have to work on the highest safety level with very different real case conditions, however without reliable pretests of the safety system in any case. A risk of loss of coolant caused by an opening of the rupture disc should be absolutely minimized, because it produces high costs, needs also a lot of time for bringing the system aga in operation and there is also a high risk for contaminations of the cryogenic system.

SAFETY DEVICES

The types of pressure relief devices considered in this paper are limited to a combination of spring loaded safety valves and rupture discs (RD) mounted in parallel on a common inlet piping. In this combination the RD is provided as: an additional safeguard if there are some doubts concerning the efficacy of the safety valve, or to provide additional discharge capacity, or where a larger safety valve may be impractical, the safety valve is designed for a more likely contingency and the RD is designed for the rare contingency, or the pressure rise is too rapid for the safety valve alone. In order to get a safety margin for the protection of the rupture disc, the set pressure of the safety valves is 70% of the burst pressure of the RD. There exists two different types of safety valves, the so called full lift valves (SRV) and the relief devices (RV).

SRVs are characterized by rapid opening or pop action. The valves opens rapidly within 5 % pressure rise, the amount of lift up to the rapid opening should not be more than 20 % of the total lift. The opening times of SRVs are on the same order of magnitude as the burst disc. These are results of new specific investigations on different types of SRVs from different suppliers. The consequences of these astonishing results are discussed in the next chapter.

RVs are characterized by a more or less steady opening in relation to the pressure increase. A sudden opening within a 10 % lift range will not occur without pressure increase. A further distinctive feature is the small coefficient of discharge (about factor 3) compared to a SRV.

RDs are non-closing devices. The thin diaphragm is designed to rupture (or burst) at a determined pressure. They are fast acting safety devices; the rupture time is related to size and burst pressure and in the time range of about 10 msec [7]. There are different discs in use, simple metallic dome discs manufactured on ductile materials, subjected to tension and reverse buckling types, with the dome against the direction of flow. Reverse types are subjected to compression so that the discs inverts fully and than separates from the disc holder. The disc can fully open or remains partly open. Graphite discs are flat, made of brittle material and exposed to tension and shear stress. They shatter almost as soon as the burst pressure is passed and the fragments swept into the downstream piping or in a cage. The subject to premature failures if operating pressure exceeds 70% of set pressure is very important and needs great care as well as the sensitivity of the discs against pressure oscillations and temperature.

SAFETY VALVE OSCILLATIONS

During discharging undesirable instabilities of the valve discs may occur. They are distinguished between fluttering and chattering.

Fluttering frequencies are in the range of approx. 1 Hz. The principal causes are an over sizing of the SRVs or a high pressure drop in the inlet piping. A further reason for flattering is a high back pressure in the outlet piping. In general, this behavior can also be considered as the normal behavior caused by wide flow rates.

Chattering is the more dangerous oscillation, caused by shock waves due to the rapid opening or closing of the valve. Resulting vibrations may cause misalignment, valve seat damage and failure of valve internals and associated piping. Furthermore under certain conditions very strong pressure oscillations in the inlet piping may occur, with chattering frequency in the range of 100 Hz.

There are some design rules to avoid chatter: The 3 % loss rule is the most commonly applied rule in practice [1, 2, 3]. On one hand, the pressure loss of 3 % of the set pressure is allowable in the inlet piping, on the other hand, the difference between the pressure loss in the inlet piping and the reseating pressure difference must amount to at least 2 % of the set pressure for a safe and proper function. This states the International standard ISO4126 as well the American ASME code. This rule is very conservative but independent of safety valve types and behavior [4]. Due to the lack of conclusive experimental evidence, industry has generally accepted this rule as the standard; but to fulfill this rule is very difficult. The consequences are very large diameters of the inlet piping, because the 3 % rule is rated to the maximum flow capacity of the SRVs at the beginning of discharge.

Another criterion is the so called pressure surge criterion [5, 6]. Due to the rapid opening of the safety valves, in form like a pop action, pressure waves in the inlet line are generated

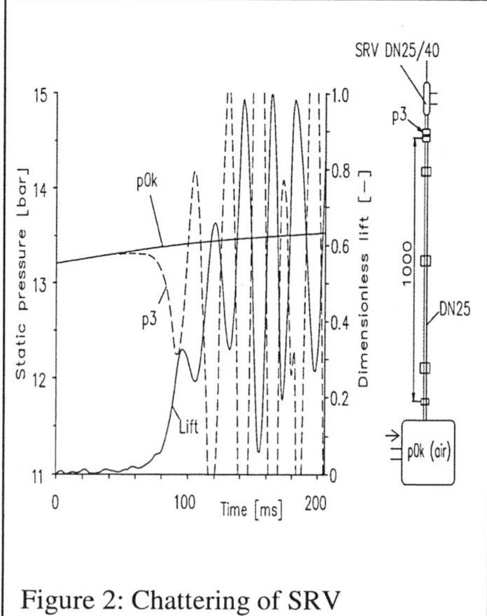

Figure 2: Chattering of SRV

and cause chatter under certain conditions. The important parameters are: value of discharge coefficient, flow diameter, length and diameter of the inlet piping, opening time of the safety valve and fluid parameters like sound velocity and density. The opening time of a SRV is an important parameter to avoid chattering. The longer the opening times and the reseating difference, the smaller the probability of chattering. Manufacturer's data and research paper suggest about 20-100 msec for SRVs in normal industrial context. The measured data [7] are contrary to these published statements which makes the probability of chattering much higher. Unfortunately in cryogenic services there are also big changes of the fluid parameters during blow up and there is no experience with this criterion in the cryogenic community.

Figure 2 shows a very impressive example for chattering of a SRV mounted on a test rig [8]. The SRV is mounted on a vertical inlet piping with the same diameter as the nominal diameter of the SRV. The medium is air at room temperature and the pressure drop is about 9 % of the set pressure. As shown in the figure, during the chattering the pressure in the vessel increases and there are large pressure amplitudes in the inlet piping. Such behavior must be prevented in any case, because such pressure amplitudes may be the reasons for unexpected response or injury of the RD.

SUPPLEMENT OF THE SYSTEM FOR AN IMPROVED SAFETY DEVICE ARRANGMENT

For a reliable safety system, flattering or chattering of the valve disc has to be absolutely prevented and therefore special actions have to be taken. A pressure rise must be handled in sequential steps. The first step is on the PLC-level, the second with hard wired pressure switches and at least using safety devices (see figure 1). The installation of the safety devices must be considered as an integral part of the safety system. Important is the reduction of the pressure drop in the inlet piping by careful attention of all the other cryogenic border conditions, like heat input, stiffness of piping or leak tightness. To equip the SRV with an oscillation damper and also with a holdup cylinder is reasonable, because these are two important devices to prevent chattering. The damper is the most effective, reliable and economic method to achieve valve stability. The pneumatic holdup cylinder, triggered by the system pressure, keeps the valve disc for a chosen time in an upper position after a first lifting. Furthermore, an extension of the opening time and an increasing of the reseating pressure difference (blow down) can be useful. A bellow is necessary to compensate the backward pressure and also for the protection of the spring. The over design of the valve capacity must be avoided. This action reduces the inlet piping diameter, because the 3 % rule is capacity related. A separate piping for the RD is recommended, because pressure oscillation caused by chattering is only in the inlet piping of the SRV. There is a reflection of the pressure wave on the nozzle of the inlet piping. A large margin between the set pressure of SRV and RD is necessary. An important action to reduce the loss of coolant after an ignition of the RD is the use of a change-over valve in the inlet piping of the RD. The switch over to the second disc reduces the loss of coolant and prevents the deposit of humidity. After an ignition of the RD, a switch over to the second disc can be done, if the arrangement is equipped with the necessary actuators and sensors (see figure 3). Because of dangerous fragments, a strong cage for the RD is necessary and the mounting position of the RD must be in such a way, that there is no possibility for a deposit of fragments in the system.

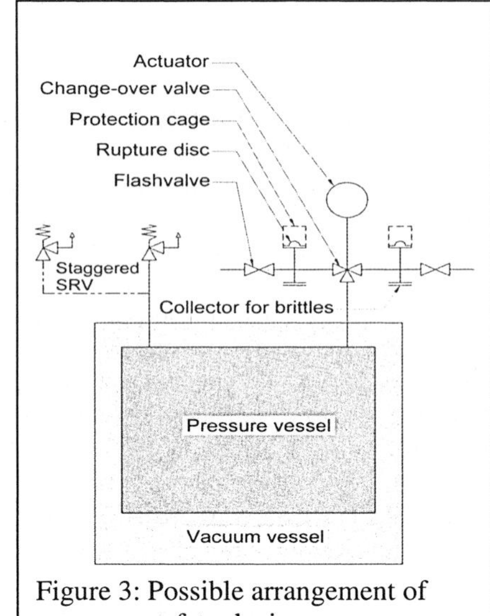

Figure 3: Possible arrangement of safety devices

CONCLUSION

The protection of cryogenic facilities against overpressure with safety valve-rupture disc-combinations includes certain risks. Safety valves manufactured from castings may look not very sophisticated, but in their design, accuracy and function they are delicate instruments and perform an essential role. The installation of safety devices is an integral part of the safety system and must be done very carefully. Special care must be taken in order to satisfy the stability criterion and for the prevention of chattering of the safety valves discs. A broad discussion about safety devices installations is needed in the cryogenic community, because there are not enough information's published about this topic, even in specific journals.

REFERENCES

[1] API RP520: Sizing, selection and installation of pressure-relieving systems in refineries 1993

[2] AD2000-Merkblatt A1 (2000): Sicherheitseinrichtungen gegen Drucküberschreitungen –
Berstsicherungen

[3] AD2000-Merkblatt A2 (2001): Sicherheitseinrichtungen gegen Drucküberschreitungen –
Sicherheitsventile

[4] Schmidt J., Giesbrecht H., Grenzen der sicheren Funktion von Vollhub-Sicherheitsventilen –
Bewertung des 3 %-Druckverlust-Kriteriums, Chem.-Ing.-Techn, (1997) 69 1281

[5] Cremers J., Friedel L., Pallaks B., Validated sizing rule against chatter of relief valves during
gas service, J. Loss. Prev. Proc. Ind., (2001) 14 261-267

[6] Frommann O., Friedel L, Analysis of safety valve chatter induced by pressure waves in gas flow,
J. Loss. Prev. Proc. Ind., (1998) 11 279-290

[7] Testing and analysis of relief device opening time, Offshore Technology report 2002/2003,
www.hsebooks.co.uk

[8] Frommann O., Experimentelle und theoretische Untersuchungen des dynamischen Verhaltens
federbelasteter Vollhubsicherheitsventile bei anlaufender Gasströmung, Shaker Verlag Aachen 1997

Proceedings of the Twentieth International Cryogenic Engineering Conference
(ICEC 20), Beijing, China. © 2005 Elsevier Ltd. All rights reserved.

New temperature and magnetic field sensors for cryogenic applications developed under a European Project[*]

Mitin V.F.[1,2], McDonald P.C.[3], Pavese F.[4], Boltovets N.S.[5], Kholevchuk V.V.[1,2], I.Yu. Nemish[1], Basanets V.V.[5], Dugaev V.K.[6], Sorokin P.V.[7], Venger E.F.[1], Mitin E.V.[1,2]

[1]V.Lashkarev Institute of Semiconductor Physics, National Academy of Sciences, Kiev, Ukraine
[2]Research & Production Company "MicroSensor", Kiev, Ukraine
[3]Institute of Cryogenics, University of Southampton, Southampton, UK
[4]CNR - Istituto di Metrologia "G.Colonnetti", Torino, Italy
[5]State Research Institute "Orion", Kiev, Ukraine
[6]Institute of Materials Science Problems, Ukrainian Academy of Sciences, Chernovtsy, Ukraine
[7]National Science Centre, Kharkov Institute of Physics &Technology, Kharkov, Ukraine

Progress is reviewed, in the development of various semiconductor sensors for measurement of temperature and magnetic field that are intended for use in cryogenic engineering and low-temperature physics and have been developed in an international collaboration supported by the European Union. A range of resistance thermometers based on both bulk and film Ge and SiC has been developed, to provide high sensitivity over complementary temperature ranges, within the overall range 0,03 to 500 K. New types of Si and GaAs diode temperature sensor have also been produced. A novel multisensor for concurrent measurements of temperature and magnetic fields has been designed, which consists of a Ge-film resistance thermometer and a InSb-film Hall generator.

INTRODUCTION

In July 2001 an EU project (European Commission INTAS Contract n° 2000-0476) was established to develop, characterize and establish production of, novel temperature and multifunctional sensors, for use in various cryogenic applications. This project connects seven groups of researchers from West- and East-European research centers, universities, institutes and industrial companies with complementary activities in the field of semiconductor technology, microsystems and sensors, low-temperature physics, cryogenic thermometry and metrology. The primary objective of this collaborative project was the development of new technological approaches to the solution of some urgent problems in cryogenic thermometry, namely, measurement in the presence of high magnetic fields and stability in the presence of ionizing radiation. The project aims to provide a number of types of new temperature and multifunctional sensors that cover the temperature range from 0,03 to 500 K. The first results for this international collaboration and development have been reported by N.S. Boltovets et. al. [1].

This paper presents a review of sensors that are being produced and characterized within the European Project. They are: (i) resistance thermometers based on both film and bulk Ge and SiC; (ii) diode temperature sensors based on Si and GaAs; (iii) dual element resistance thermometers (DERTs), in which the overlapping T/R characteristics of two temperature sensitive elements are combined to provide a measurement range from 0,1 K to 400 K; and (iv) dual function sensors (DFS) for concurrent measurements of temperature (1,5 K to 400 K) and magnetic field.

---

[*] Work partially funded under European Commission INTAS Contract n° 2000-0476.

DESIGN OF SENSORS AND THEIR OPERATING CHARACTERISTICS

Ge-film resistance thermometers

The basis of the resistance thermometer is a Ge-film resistor deposited on a semi-insulating GaAs substrate. The principles of designing such sensors and the fabrication technology involved have been reported previously by V.F. Mitin et. al. [2] and the characteristics of fabricated thermometers have also been reported [3-5]. The fabrication technology of such Ge-film thermometers is now well developed and new types of thermometers and multisensors have been produced, using micro-electronic design and production methods, to meet the requirements of the modern cryogenic sensor market.

Ge-film resistance thermometers have been produced with cylindrical canister packages, made from gold plated copper. The dimensions of this package are 3 mm in diameter and 5 mm long. The thermometers have four copper or phosphor bronze contact leads. A micro-package has also been designed for temperature measurements with high spatial resolution and a fast thermal response time. These micro-thermometers measure 1.2 mm in diameter by 1,0 mm long. Typical dependencies of resistance and sensitivity $S=dR/dT$ on temperature for the different types of thermometer are shown in Figures 1.

Ge and SiC bulk resistance thermometers

The Ge thermometers are made from heavilly doped and compensated bulk Ge. Doping is carried out using a set of impurities with various activation energies by both the metallurgical method and neutron transmutation method. These thermometers cover the temperature range of operation from 0,3 to 300 K. The temperature dependencies of resistance and sensitivity $S=dR/dT$ for the different types of the Ge-bulk thermometers are shown in Figures 2

A new high sensitive thermometer for the 50 to 400 K temperature range has also been developed and produced on using bulk SiC. The typical resistance-temperature dependence for this thermometer is shown in Figure 1 (curve 7).

Si and GaAs diode temperature sensors

Shown in Figures 3 are the typical temperature dependencies of forward voltage, U, and sensitivity, $S=dU/dT$, for the Si and GaAs diodes. These diodes show quite different U-T dependencies at temperatures below 80 K. Their behaviours in magnetic fields are now under characterization.

Dual element thermometers

New, dual element thermometers (DERTs) have been designed and produced. These provide temperature measurements over wide range, from ultralow to high temperatures, with high sensitivity and resolution over the whole range. The DERT contains two Ge-film resistor elements which have high sensitivity over complimentary ranges within the extended temperature range. The elements are incorporated in the same parallelepiped package, made from gold plated copper. The dimensions of this package are 3,5 mm wide, 2,2 mm high and 10,1 mm long. The dual element thermometer has eight copper contact leads:- four leads for each element. Using one element, for example, in the 0,1 K to 4 K range and the another for the 4 to 400 K range one can obtain high sensitivity and resolution both at ultralow and high temperatures. Typical dependencies of resistance and sensitivity $S=dR/dT$ on temperature for the elements of the dual element thermometer are shown in Figures 1.

Dual function sensors

The dual function sensors (DFSs) for temperature and magnetic field measurements consist of a Ge-film resistance thermometer and an InSb-film Hall-effect magnetic field sensor. The package for the DFS is the same as that for the DERT (3,5 mm wide, 2,2 mm high and 10,1 mm long). The dual function sensor also has eight copper contact leads:- four leads for the thermometer and four leads for the Hall "generator". At constant current the Hall generator provides an output voltage proportional to magnetic field induction. The main aim for development of the dual function sensor is to provide accurate temperature measurements in high magnetic fields. Typical thermometric characteristics for the DFS are included in Figures 1. Operating characteristics for magnetic field measurements are listed in Table 1.

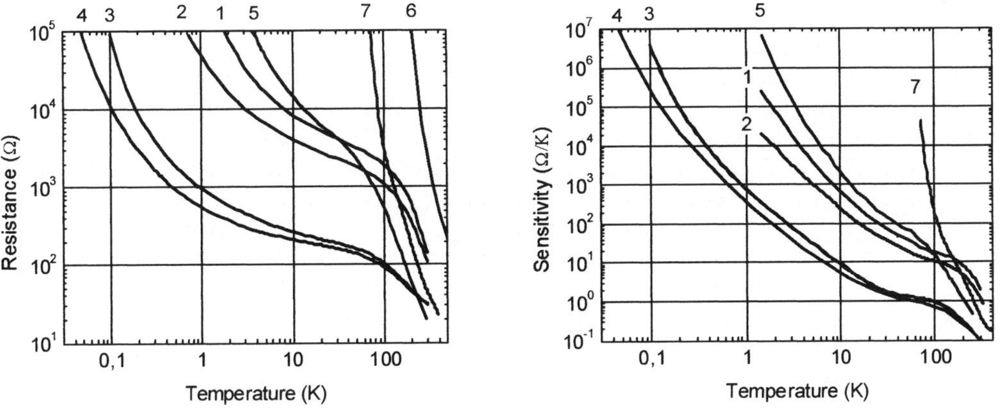

Figure 1 Resistance and sensitivity vs. temperature curves for DFS (1), DERT (1, 3), and Ge-film thermometers of different models: TTR-G (2), TTR-D (4), TTR-M (5), TTR-3 (6) and SiC-bulk sensor (7)

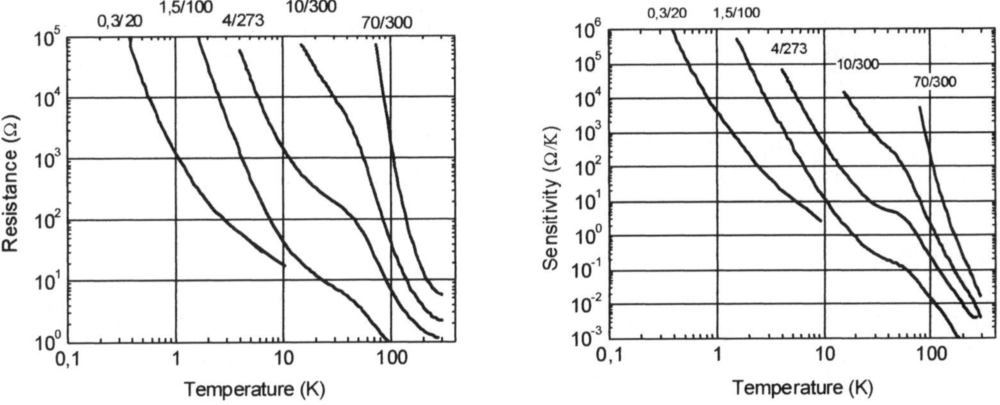

Figure 2: Resistance and sensitivity vs. temperature curves for Ge-bulk thermometers of different models

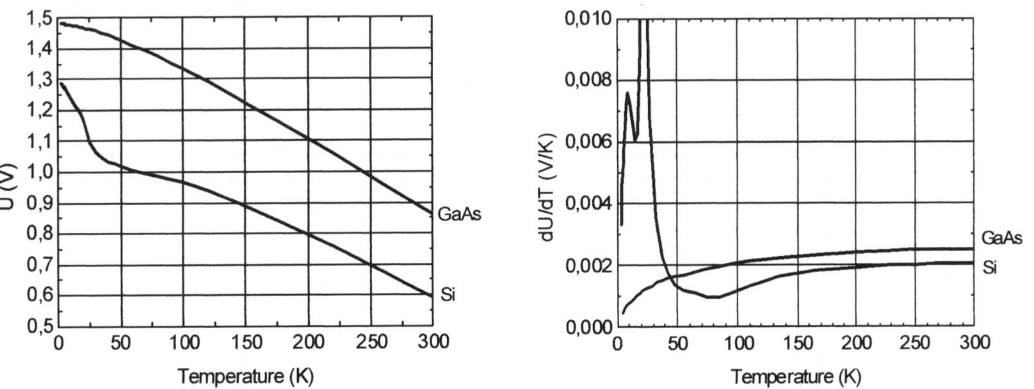

Figure 3: Forward voltage and sensitivity vs. temperature curves for the Si and GaAs diodes at dc. current of 10 µA

Table 1   Hall generator characteristics

Operating temperature range	1 to 400 K
Input resistance	20 to 70 Ω
Output resistance	50 to 150 Ω
Rated control current	0,5 mA
Magnetic sensitivity (at rated control current)	5 mV/Tesla
Zero field offset voltage (at rated control current)	0,2 mV (max.)
Temperature coefficient of magnetic sensitivity	0,03 %/K (max.)

EFFECT OF MAGNETIC FIELDS AND GAMMA IRRADIATION

The magnetic field effect varies for different models of Ge thermometers and depends on the nature of the conduction mechanisms in the sensor materials. It may be negative or positive and is dependent on temperature and magnetic field ranges. The behaviour of some models of Ge film thermometers in magnetic fields can be found in ref. [1, 3-5] and for Ge-bulk in ref [6]. New models of thermometers based on both bulk and film Ge have been developed with low magneto-resistance for use in high magnetic fields. The temperature measurement error is not more than 1 % in magnetic fields up to 7 T at 4,2 K; 9 T at 3,0 K and 14 T at 2,0 K for the Ge-bulk thermometers, and for the Ge-film thermometers this error (1 %) is in the fields of 6 T at 4,2 K; 7 T at 1,0 K and 6 T at 0,3 K.

The gamma irradiation has very little effect on the Ge-film resistance thermometers up to a dose of 1.5×10^8 rad. Their sensor resistance increases after a dose of 1.5×10^8 rad and at the very large dose of 7.6×10^8 rad, the error in the thermometer reading is ~2 %, i.e. 100 mK at 4.22 K and 2.0 K at 77.4 K.

CONCLUSIONS

A number of types of new temperature sensor have been developed, produced and characterized. They are: (i) resistance thermometers based on both film and bulk Ge and SiC; (ii) diode temperature sensors based on Si and GaAs; (iii) dual element resistance thermometers (DERTs), in which the overlapping T/R characteristics of two temperature sensitive elements are combined to provide a measurement range from 0,1 K to 400 K; and (iv) dual function sensors (DFS) for concurrent measurements of temperature (1,5 K to 400 K) and magnetic field. Further studies are planned for the comprehensive evaluation of these temperature sensors under various environmental conditions.

REFERENCES

1   Boltovets N.S., Dugaev V.K., Kholechuk V.V., McDonald P.C., Mitin V.F., Nemish I.Yu., Pavese F., Peroni I,. Sorokin P.V., Soloviev E.A., and Venger E.F, New generation of resistance thermometers based on Ge film on GaAs substrates, Thermperature: Its Measurement and Control in Science and Industry, **7,** edited by Dean C. Ripple, AIP, Chicago, 2003, pp.399-404

2   Mitin V.F., Tkhorik Yu.A. and Venger E.F., All-purpose technology of physical sensors on the base of Ge/GaAs heterostructures, Microelectronics Journal (1997) 28 617-625

3   Mitin V.F., Miniature resistance thermometers based on Ge films on GaAs, Advances in Cryogenic Engineering (1998) 43 749-756

4   Mitin V.F., Resistance thermometers based on the germanium films, Semiconductor Physics, Quantum Electronics & Optoelectronics (1999) 2 No.1 115-123

5   Boltovets N.S., Kholevchuk V.V., Konakova R.V., Mitin V.F. and Venger E.F., Ge-film resistance and Si-based diode temperature microsensors for cryogenic applications, Sensors and Actuators A (2001) 92 191-196

6   Zarubin L.I., Nemish I.Y., Szmyrka-Grzebyk A., Germanium resistance thermometers with low magnetoresistance, Cryogenics (1990) 30 533-537

CHARACTERIZATION STUDIES OF SPUTTER DEPOSITED CERMET CRYOGENIC HEATERS

Yeager C., Courts S., Chapin L.

Lake Shore Cryotronics, Inc., 575 McCorkle Blvd., Westerville, OH 43082, USA

Characterization studies of a new cryogenic heater will be presented. The material is a sputter-deposited ceramic-metal composition. It has a constant resistance to within 0.1% between 77 K and 50 mK. At liquid helium temperature, the reduced sensitivity is 2×10^{-3}. The nominal die thickness is 837 nm. The sheet resistance is 7.27 Ω/square and the room temperature resistivity is 615 $\mu\Omega\cdot$cm. The die resistance is 1000 Ω. The die size is 1.02 mm by 0.76 mm. We will show results for maximum operating power, operational life, and reproducibility due to thermal cycling. Critical current densities will be presented.

INTRODUCTION

At present, cryogenic heaters are made from winding wire alloys that have a low temperature coefficient of resistivity (Evanohm, phosphor-bronze, and nichrome)[1,2]. Heaters are wound from long lengths of small gauge wire, making for a tedious process and a relatively large heater bobbin.

Techniques used to develop the Cernox [3] sensor can be applied to develop a sputter-deposited thin-film material with an extremely low temperature coefficient of resistivity. This method allows for controlling the resistivity through die design and meander patterns. Additionally, dies can be mounted into packages used for cryogenic RTDs. These devices could be used for applications requiring a fixed resistance. This includes cryogenic heaters and fixed resistance standards.

DESCRIPTION

The new material is a proprietary ceramic-metal composition (cermet). It is reactively sputter-deposited onto a sapphire wafer. The nominal thickness is 837 nm. The sheet resistance is 7.27 Ω/square and the room temperature resistivity is 615 $\mu\Omega\cdot$cm. The electrodes were patterned to make an individual die with a target resistance of 1000 Ω, and the line width was approximately 0.0063 cm [4].

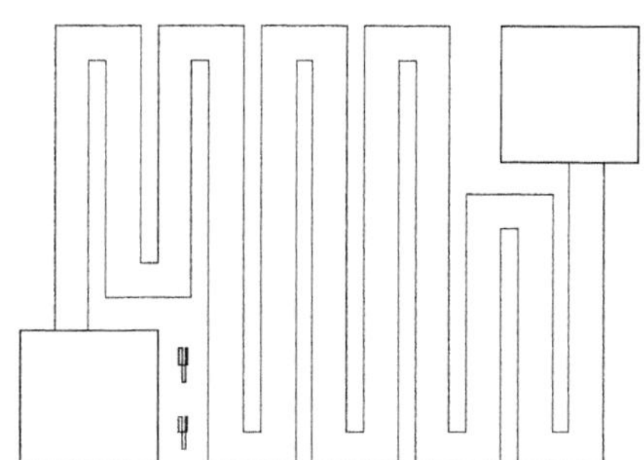

Figure 1 . Serpentine Meander Pattern for Novel Resistors.

Test resistors were fabricated from each wafer using standard photolithography techniques to define the active sensor area and subsequent contacts. The sensors were patterned in a serpentine meander. Figure 1 shows a finished sensor chip layout. The die size was 1.02 mm by 0.76 mm.

A total of nine devices were tested in a package. Three were from an original test and six additional new devices were built. The individual chips were packaged into Lake Shore standard canister packages, which are gold-plated copper canisters 3.048 mm in diameter and 8.5 mm long. The device is mounted inside the canister using a sapphire header. The device is connected by gold leads and is sealed into the canister with Stycast 2850 epoxy.

After packaging, the devices are thermally cycled from room temperature to liquid nitrogen temperature 200 times. In addition to the packaged devices, two bare dies were tested for maximum power load.

TESTING

Temperature calibrations from 0.05 K to 325 K were performed in the commercial calibration facility of Lake Shore Cryotronics. Temperature and resistance was measured using standards grade platinum and germanium thermometers in conjunction with a Keithley Model 224 current source, Hewlett Packard Model 3458A DVM, and Guildline Model 9330 standard resistors (resistance values from 10 Ω to 1 MΩ in decade steps). The device under test was placed in series with a standard resistor of comparable value. The current was varied to a minimum of 0.05 μA to maintain a nominal 2 mV signal level across each sensor during calibration. The voltmeter was used in a ratiometric form with readings taken across both the standard resistor and device under test. Current reversal was performed to eliminate thermal EMFs. Calibrations from 0.05 K to 1.4 K were performed in a dilution refrigerator using a Lake Shore Model 370 AC Resistance Bridge. The nominal excitation was 100 μV.

To determine their reproducibility and stability due to thermal stress, the packaged devices were thermally cycled and measured at three temperatures: 4.2 K, 77 K, and 305 K. The 4.2 K and 77 K readings were taken in an open dewar of liquid helium and liquid nitrogen. The 305 K reading was taken in a temperature-controlled oven. Readings at liquid helium were repeated 10 times for each of the nine packaged sensors.

RESULTS

Figure 2 shows the resistance as a function of temperature for three select devices. At room temperature, the devices range from 905 Ω to 925 Ω. From 305 K to 4.2 K their resistance changes by 2.5%, with most of this change appearing from 77 K to 305 K. From 77 K to 4.2 K the resistance changes less than 0.1%.

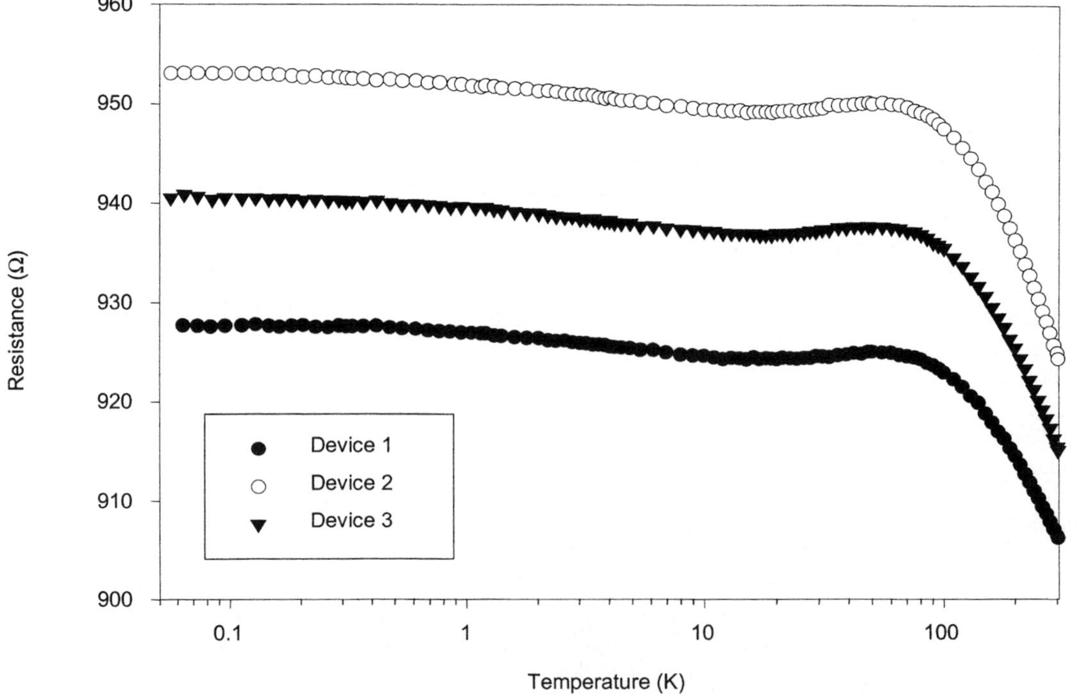

Figure 2. Resistance as a function of temperature for three selected heater devices.

The temperature coefficient of resistivity, (1/R)(dR/dT), at 4.2 K is approximately 2×10^{-3} K^{-1}.

Results from the thermal reproducibility tests are summarized in Table 1. Over the course of 10 thermal cycles into an open dewar of liquid helium, the standard deviation for all the devices was ± 0.0195 Ω. Using an average resistance of 936.8 Ω, this is a reproducibility at 4.2 K of approximately 20 ppm in resistance. It should be noted that results were taken in an open dewar and the temperature was assumed to be constant during the duration of testing. For these tests, there was no attempt to correct for temperature drifts.

Table 1. Reproducibility results at 4.2 K. Results at 4.2 K based on 10 thermal cycles. All devices come from the same wafer. Devices A – C were packaged 9 months prior to Devices 1 – 6.

	Resistance at 305 K	Resistance at 77 K	Average Resistance at 4.2 K (Ω)	Standard Deviation at 4.2 K (Ω)
DEVICE A	905.8	924.244	925.318	± 0.0575
Device B	914.819	936.839	937.955	± 0.0181
Device C	924.339	949.381	950.483	± 0.0487
Device 1	924.377	946.268	947.132	± 0.0133
Device 2	932.653	954.435	955.366	± 0.0079
Device 3	926.201	948.925	949.880	± 0.0083
Device 4	885.541	905.081	905.818	± 0.008
Device 5	898.665	921.018	921.010	± 0.0065
Device 6	914.762	938.346	938.339	± 0.0068

Also seen in Table 1 is that the change in resistance for 77 K to 4.2 K is, on average, better than 0.1% in resistance. This confirms earlier results.

To test maximum power we used both the packaged devices and bare die devices. Two packaged devices were tested with a Keithley 220 current source and HP voltmeter. Power was incremented from 1 mA up to 100 mA to determine the maximum current. All four devices tested could sustain a constant current of 70 mA. This corresponded to approximately 4.5 W. This same device was tested in LN2 to provide a thermal shock while at maximum power. There were no damaging effects due to the thermal shock. When the current was increased to 100 mA (9 W) the device burned out and failed. This was duplicated on both packaged devices.

To confirm the failure was due to the thin film resistor and not some other component of the package, we tested bare dies to maximum power. A strip of 10 dies was cut from the original wafer. The dies were left on the cutting tape and placed in a 4-point probe station. Using a Keithley 220 current source and HP voltmeter, current was increased from 1 mA to 100 mA. Again, the devices could sustain a constant current of 70 mA without damage. At 100 mA the die under test flashed as arcing developed across the traces.

To estimate the critical current density we will use the value of 70 mA, based on a cross-sectional area of 5.3×10^{-11} m^2. This gives an average current density of 1.3×10^9 A/ m^2. The actual current density of the cermet film may be higher. The geometry of the meander pattern had sharp corners where there could be excessive current build up. In this case, the effective cross-sectional area would be smaller. To explore this further we will study different patterns.

APPLICATIONS

The results indicate these devices can be packaged as cryogenic heaters with power ratings up to 5 W. Increasing the film thickness will result in larger current densities and a larger power rating. Long term rating at maximum power still needs to be tested and alternate meander patterns will be explored.

The new cermet thin films offer many advantages over conventional techniques for cryogenic heaters. Because of the fabrication techniques, the new cermet thin-film resistors can be packaged similar to cryogenic RTDs and also as very small bare die heaters. Standard packaging allows greater flexibility in attaching and mounting the devices. A bare chip would allow the heater to be used for applications that would be impossible for wire-wrapped heaters.

Additionally, different resistances could be designed by changing the electrode mask, using a laser trimmed meander pattern, or changing the film thickness. Changing the film thickness can also change the effective power rating. The canister configuration is not ideal for a cryogenic heater and a simpler package will need to be used to best transfer heat out from the resistive element.

Another application would include a small moderately powered heater used by cryostat developers for temperature control over a wider range. The resistance is nearly constant over the whole range for easier power calculation based on room T values. Small bare die device heaters can be used in precision thermal properties studies.

CONCLUSIONS

A novel cermet material has been presented that shows a very low temperature dependence of resistivity. The temperature coefficient of resistance at 4.2 K is 2×10^{-3} K^{-1}. The material is very stable under multiple thermal cycles. The reproducibility at 4.2 K is 20 ppm in resistance or better.

This material can be packaged into standard cryogenic sensor packages and is useful as a small cryogenic heater for power rating up to 5 W.

REFERENCES

1. Cieloszyk, G.S., Cote, P.J., Salinger, G.L., and Williams, J.C., <u>Rev. Sci. Instrum.</u>, 46, 1182 (1975)
2. Cimberle, M.R., Michi, U., Mori, F., Rizzuto, C., Siri, A., and Vaccarone, R. <u>6th International CEC, Grenoble</u>, IPC Science and Technology Press, (1976)
3. Lake Shore Cryotronics Inc. Westerville, OH. 43082
4. Yeager, C.J., Courts, S.S., Chapin, L., Cermet Materials with Low Temperature Coefficients of Resistivity, to be published in <u>Advances in Cryogenic Engineering</u> 49, Plenum, New York (2000), pp. 1699–1706

Study of skin burns due to contact with cold mediums at extremely low temperature

Deng Z-S., Liu J.

Technical Institute of Physics and Chemistry, Chinese Academy of Sciences,
Beijing 100080, P. R. China

Although very important, the freezing burns caused by extremely low temperature
have received few attentions up to now. In this study, a three-dimensional
multi-layer bioheat transfer model is developed to predict the tissue damage
following a freezing burn. This mathematical model is numerically solved using
finite-difference method. The results indicate that the production of freezing burns
is dependent on both the surface area exposed to cold medium and the duration of
exposure to freezing. Further, a preliminary experiment is also conducted to
validate the theoretical results. The experimental results qualitatively agree with
the theoretical ones.

INTRODUCTION

The burn injury is one of the most commonly encountered types of trauma in both civilian and military
communities. It is typically caused due to interaction with various high temperature sources, including
exposure to flames, contact with a hot solid or liquid, inhalation of a hot vapor, exposure to harmful
radiations, or electrical energy dissipation. With the widespread use of cold liquids and gases at extremely
low temperature in many industries and scientific researches, the freezing injuries caused by these
coolants become more and more common [1]. The pathophysiological mechanism of such injury suggests
the treatment analogous to that of burns; customarily, it is also named as "burn". Although the thermal
burns caused by high temperature have been intensively studied over the last half-century [2-4], the
freezing burns caused by extremely low temperature receive few attentions besides several case reports
described in the surgical literatures [1, 5, 6].

To better understand and accurately predict the tissue damage following a freezing burn, a
three-dimensional multi-layer bioheat transfer model is developed in this study. In the model, multiple
factors to respectively present the properties of the different layers of the biological tissues and
solidification features due to freezing are considered. The model is solved by a finite-difference algorithm
based on the effective heat capacity method. The extent of freezing burns is then determined from the
numerical results. In order to validate the theoretical results, a preliminary experiment is also performed
to illustrate the effect of freezing duration on the extent of burn damage.

THEORETICAL ANALYSIS

Mathematical model and numerical algorithm
It is reasonable to partition the biological tissue into three-layers (including the skin, fat and flesh layer
respectively). In each layer, the thermal parameters are treated as constant however different from each

980

other. The simplified three-layer geometry used for the analysis is shown as Fig. 1.

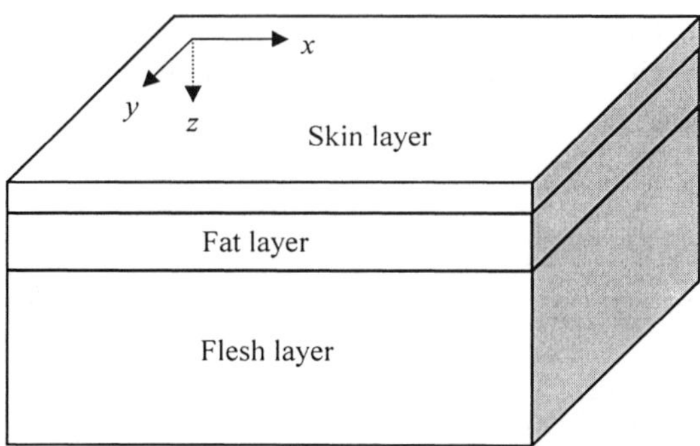

Figure 1 Simplified three-layer model of tissue (z denotes the tissue depth from the skin surface, and three layers are skin, fat and flesh layer respectively)

For brevity, the governing equations describing the phase change problem of biological tissue are not presented here. Readers can refer to [7] for more details. In order to avoid complex iteration at the front of phase change, the effective heat capacity method is applied in this study. A detailed development of the numerical algorithm for such phase change problem is similar to that for the freezing problem involved in cryosurgery (which has been developed in our previous work [7]). The derivation is not repeated here.

<u>Theoretical results and discussion</u>

In calculations, the typical thermal physical properties of tissue are applied as given in reference [7]. The thicknesses of skin and fat layers are respectively taken as 2 mm and 4 mm [8]. The tissue dimension parameters are taken as 0.1 m in x and y directions and 0.05 m in z directions. Considering that at the positions far from the center of the domain, the temperature field there is almost not affected by the center domain, the adiabatic conditions are assumed at the boundaries along x and y directions. The other boundary conditions are prescribed as follows:

$$-k\frac{\partial T}{\partial z} = h_f\left[T_f - T\right] \text{ at } z = 0; \qquad T = 37\,°\text{C at } z = 0.05\,\text{m} \tag{1}$$

For the area of skin surface contacted with the cold medium (liquid nitrogen assumed in this study), the convective parameters are taken as h_f=200 W/m$^2\cdot$°C and T_f=-196 °C, respectively; and for the other area of skin surface, h_f=10 W/m$^2\cdot$°C, T_f=20 °C. The area contacted with the cold medium is prescribed as a rectangular region located at the center of skin surface.

Figure 2 shows the temperature distributions at cross-section x=0.05 m, in which the size of skin surface contacted with cold medium is 2 cm×2 cm. Here, as is expected, the temperature distributions of tissues around the area contacted with cold medium are much lower than that of tissues far from this area. After the temperature distribution inside the tissue is given, the possible extent of the freezing burn can thus be determined by the freezing burn threshold of tissues. It had been proved that the value of the freezing burn threshold depends on many factors, and ranges from -2 °C to -70 °C [9]. The determination of practical value of the freezing burn threshold for a given tissue needs tremendous experimental works. In this study, the freezing burn threshold is taken as the solidification point of tissues for simplicity. Figure 3 gives the fronts of freezing burn threshold at cross-section x=0.05 m, which are determined by the isotherms at solidification point of tissues. The results indicate that the extent of freezing burns is dependent on both the surface area exposed to cold medium and the duration of exposure to freezing. The extent and severity of the freezing burn is important in medical treatment, and its determination will guide the clinician to apply a specific skin grafting. Unfortunately, due to the complex mechanisms for freezing damage, it is hard to establish a quantitative model similar to that for high-temperature thermal burns to evaluate the severity of freezing burn (namely the degree of freezing damage). Such important issue needs tremendous researches in future.

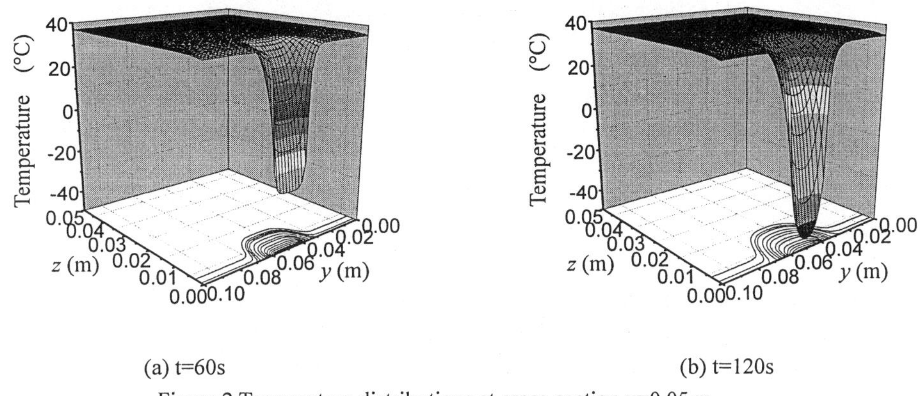

(a) t=60s   (b) t=120s

Figure 2 Temperature distributions at cross-section x=0.05 m

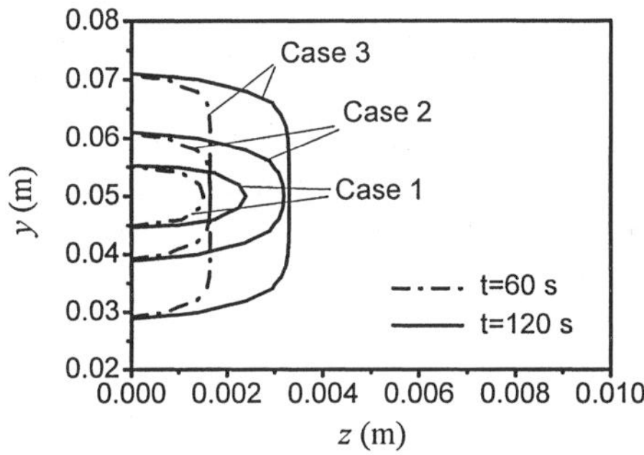

Figure 3 The fronts of freezing burn threshold at cross-section x=0.05 m. Cases 1, 2 and 3 respectively present 1cm×1cm, 2cm×2cm and 4cm×4cm skin surface contacted with cold medium

PRELIMINARY EXPERIMENTS ON FREEZING BURNS

Considering that the physiological response to freezing burns involves complicated and coupled reactions which are not well understood up to now, a preliminary experiment is also conducted. In the experiment, dehaired rabbit under anesthesia is burned by liquid nitrogen under different conditions, and the corresponding freezing damage degrees are respectively examined after 72 hours postburn.

Figure 4 shows the photograph of rabbit's ear and thigh 72 hours after the corresponding freezing burn. During dealing with the photographs, the scaling for the four photos is taken as identical as possible. Comparing these photos, it is clear that the extents of the freezing burns for the cases presented in Figs. 4(a) and 4(c) are much less than that presented in Figs. 4(b) and 4(d). So it can be concluded that the experimental results qualitatively agree well with the theoretical ones presented above. It needs to be pointed out that the present experiment only demonstrated qualitatively the effect of freezing duration on the extent of burn damage. Further researches along this direction are needed in the near future.

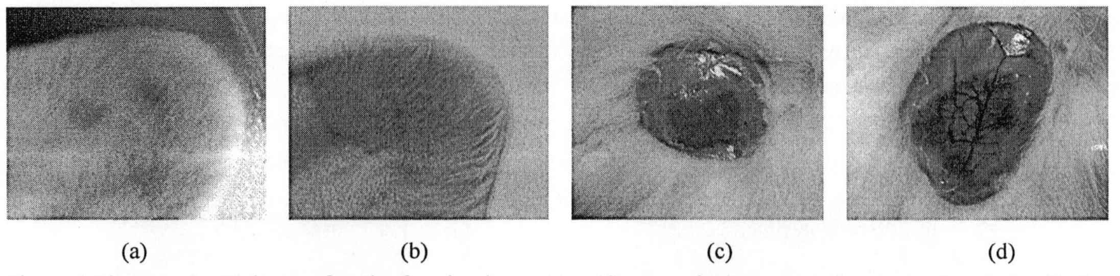

(a)   (b)   (c)   (d)

Figure 4 Photographs 72 hours after the freezing burn: (a) - (d) respectively present the cases of contact with liquid nitrogen 5, 15, 30, 60 seconds, in which (a) and (b) are photographs for rabbit ear, (c) and (d) for rabbit thigh

CONCLUSIONS

This study develops a three-dimensional multi-layer bioheat transfer model for predicting the tissue damage following a freezing burn. The theoretical results indicate that the extent of freezing burns is dependent on both the surface area exposed to cold medium and the duration of exposure to freezing. A preliminary experiment qualitatively agrees with the theoretical prediction. It concludes that the theoretical analysis presented in this article can be applied to quantitatively predict the extent of skin burns resulted by cold mediums at extremely low temperature.

ACKNOWLEDGMENT

This work was partially supported by the National Natural Science Foundation of China under grant No.50306027 and 50325622.

REFERENCES

1. Roblin, P., Richards, A., and Cole, R., Liquid nitrogen injury: a case report, <u>Burns</u> (1997) <u>23</u> 638-640

2. Henriques Jr, F. C. and Moritz, A. R., Studies of thermal injury: I. The conduction of heat to and through skin and the temperature attained therein – A theoretical and an experimental investigation, <u>American Journal of Pathology</u> (1947) <u>23</u> 531-549

3. Diller, K. R. and Hayes, L. J., A finite element model of burn injury in blood-perfused skin, <u>ASME Journal of Biomechanical Engineering</u> (1983) <u>105</u> 300-307

4. Ng, E. Y. K. and Chua, L. T., Comparison of one- and two- dimensional programs for predicting the state of skin burns, Burns (2002) <u>28</u> 27-34

5. Kuyven, C. R., Gomes, D. R., Serra, M. C. F., Macieira, L., and Pitanguy, I., Major burn injury caused by helium vapour, <u>Burns</u> (2003) <u>29</u> 179-181

6. Kumar, P. and Chirayil, P. T., Helium vapour injury: a case report, <u>Burns</u> (1999) <u>25</u> 265-268

7. Deng, Z.S. and Liu, J., Numerical simulation on 3-D freezing and heating problems for the combined cryosurgery and hyperthermia therapy, <u>Numerical Heat Transfer, Part A: Applications</u> (2004) in press

8. Deng, Z.S. and Liu, J., Mathematical modeling of temperature mapping over skin surface and its implementation in thermal disease diagnostics, <u>Computers in Biology and Medicine</u> (2004) in press

9. Gage, A. A. and Baust, J., Mechanisms of tissue injury in cryosurgery, <u>Cryobiology</u> (1998) <u>37</u> 171–186

A new method for uniformly cooling engineered tissues

Yu L-N., Deng Z-S., Liu J., Zhou Y-X.

Technical Institute of Physics and Chemistry,
Chinese Academy of Sciences, Beijing 100080, P. R. China

Aiming at the uniform cooling of the 3-D cultured tissues, we proposed in this study a new cooling method by perfusing low temperature fluid through the micro-pipes scaffold pre-configured inside the tissues. To illustrate the benefit of the new method and to test its advantages over the conventional surface cooling, theoretical modeling on the heat transfer process of the tissues subject to uniform spatial cooling was presented. Further, preliminary experiments were performed to test the practical output of the uniform cooling.

INTRODUCTION

In tissue engineering, cultured tissue must be preserved properly to ensure its viability and availability when needed in clinics. Cryopreservation has been one of the best choices to meet the request for long-term preservation. But still this approach is facing technological challenge of the unavoidable damage of the tissue when subject to freezing and thawing process [1]. During the process of decreasing temperature, the method of surface cooling is often adopted [2]. Therefore the cultured tissue is cooled from outside to inside, which will induce a non-uniform temperature distribution due to the large heat capacity of the cultured tissue and low thermal conductivity. And a relatively large temperature gradient will be produced, which will result in strong thermal stress and thus damage the tissues. To alleviate such adverse effect, we proposed in this study a new cooling method by perfusing low temperature fluid (gas or liquid, ethanol for example) through the micro-pipes pre-configured inside the tissues to realize uniform cooling. To illustrate the advantages of the new method over the commonly applied surface cooling, theoretical modeling on the heat transfer process was presented. Preliminary experiments were also carried out, and promising results were reported and discussed.

METHOD DESCRIPTION

This uniform cooling method is borrowing the idea from the concept of high heat transfer efficiency of blood vessel networks. As is well known in natural biological tissues, the spatially distributed vascular structure plays an important role in nutrition transport as well as heat transfer. This feature is of great importance and has been applied for resolving the problem of nutrition transport in tissue engineering [3,4]. There, to transport efficient nutrition substances and oxygen to the internal of the cultured 3-D tissues, special designed micro fibers are deliberately buried through the tissue scaffold, which serves as the flow pipes and bioreactors during the process of culture. The method was proved to be promising in solving the problem of nutrition supply in tissue culture. In analogy to this strategy, the spatially distributed pipes with running cooling fluid inside, similar to the blood vessel network, may also provide a new method of uniformly cooling the cultured tissue and thus realize its successful cryopreservation.

For the process of implementing this uniform cooling strategy, two main steps should be contained as follows. First, the material, amount and dimension of pipes should be strictly selected according to the specific needs of the tissues to be cultured and cryopreserved, then the pipes are spatially constructed and buried inside the tissue scaffold before culture. After days of successful culture, the pipes are spatially buried into the tissue. Note that the pipes will serve as the flow channels in the cooling stage, thus the material of the pipes to be selected and the spatial construction should be optimized to fulfill the request of freezing tolerance and efficient heat transfer respectively. For the second step, the cultured tissue to be cryopreserved is cooled by immersing it into low temperature fluid (gas or liquid), at the same time, the fluid with the same temperature flows though the pipes to achieve a uniform cooling of the whole tissue. During the process of cooling down, the tissue temperature decreasing rate should be precisely controlled by changing the fluid temperature so that the minimum freezing injury can be obtained. Furthermore, different kinds of fluids can be used one after another to ensure the efficient heat transfer through the pipes at different temperatures. When the pre-designed temperature is reached, the fluid is removed from the pipes and then the tissue is preserved at the pre-designed temperature. Similar to the freezing process, the tissue can be thawed by perfusing warm fluids through the pipes.

COMPUTATIONAL SIMULATION

In order to compare the uniform cooling method with the commonly used surface cooling method, the mathematical model of heat transfer of the tissue is made and computational simulations based on it were carried out. Because the mathematical model is very similar to the bioheat transfer model in our previous work, the detailed description is not repeated here for brevity (see [5, 6] for reference). The tissue domain is prescribed in a rectangular geometry with $20\times50\times100\text{mm}^3$ in the x, y and z directions respectively. Four pipes is fixed in a proportional spacing manner at the horizontal plane throughout the tissue and the vessel diameter is set as 1 mm. Ethanol is adopted as the experimental fluid whose flow velocity is set as 10 cm/s and temperature -100 ℃. The results of the computational simulation are presented as follows.

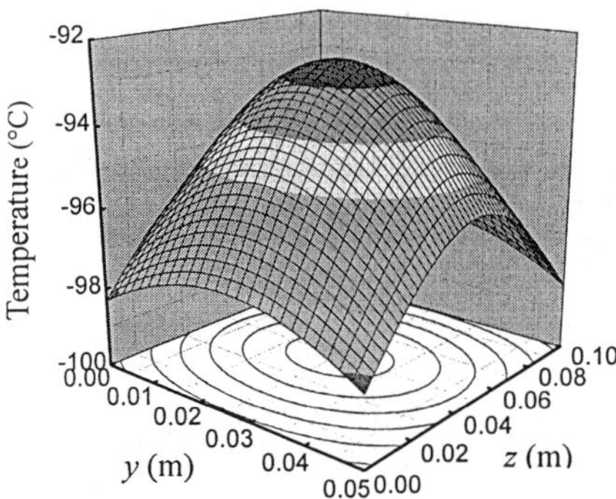

Figure 1 Temperature distribution at cross-section x=0.01m at t=960s for surface cooling method

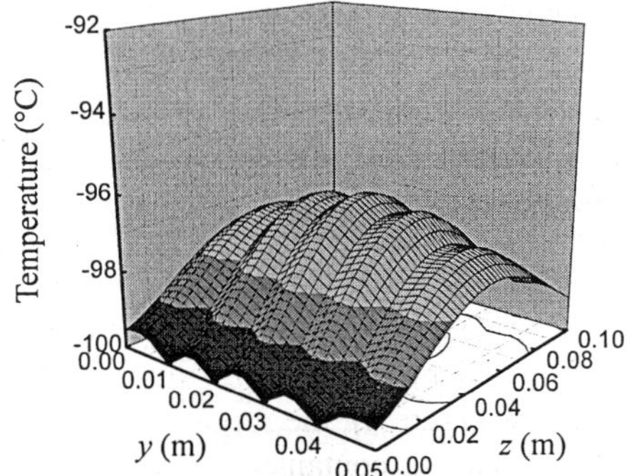

Figure 2 Temperature distribution at cross-section x=0.01m at t=960s for uniform cooling method

Figure 1 demonstrates the temperature distribution at the horizontal plane of the tissue by using the surface cooling method. At the surface of the tissue temperature drops dramatically due to large heat transfer coefficient of convection cooling. But inside the tissue, as a consequence of the large heat capacity of the tissue and relatively low thermal conductivity, large temperature gradient exists during the process of cooling. As shown in Figure 1, even after 960 s of surface cooling, the temperature difference between the center point and the edge point is 5 ℃. By contrast, the tissue subject to uniform cooling method has a rather uniform temperature distribution (see Figure 2) inside which the maximum temperature difference is about 2.5 ℃. The results indicate that by introducing the strategy of perfusing low temperature fluid through the tissue, heat transfer inside the tissue could be significantly enhanced due to the large heat transfer coefficient of convection. Thus smaller temperature gradient can be achieved, which is very beneficial for the successful tissue cryopreservation.

Figure 3 gives the temperature history of the center point of the tissue by using different cooling method. It can be seen from Figure 3 that compared to that of surface cooling method, a rather rapid cooling rate can be obtained by using this uniform cooling method, and the phase change region is much diminished in the curve. That is to say, the tissue with uniform cooling strategy experienced a shortened stage of phase transition region, which may reduce the strong freezing injury of the tissue during the phase change process. It also can be concluded from the figure that to reach a pre-designed temperature, less time is needed by using uniform cooling method. In other words, the uniform cooling strategy has a shorter respond time.

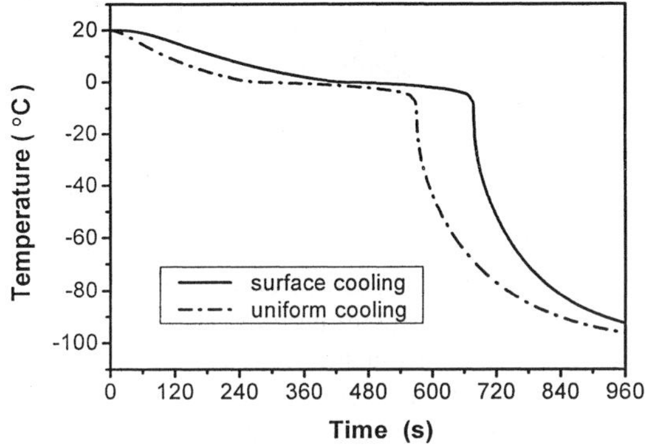

Figure 3 Temperature history at the center
point x=0.01m, y=0.025m, z=0.05m

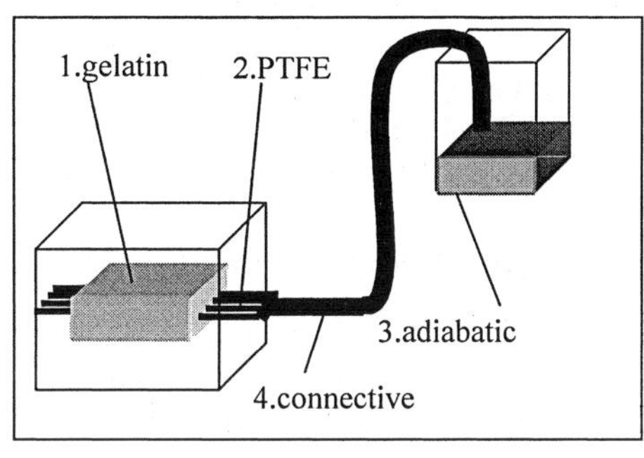

Figure 4 Schematic illustration of the setup for
uniform cooling

QUALITATIVE EXPERIMENT

To compare the practical effect of the new method and the commonly used surface cooling method for cooling cultured tissues, two qualitative experiments were designed and carried out. The setup is composed of three main parts (see Figure 4): a gelatin block 1 (14mm×40mm×80mm) with four parallelly aligned polytetrafluoroethylene (PTFE) pipes 2 (1mm for diameter) buried in, an adiabatic container 3, and a plastic pipe 4 to connect these two parts. The block A (see Figure 5) was placed in cold ethanol which is pre-cooled to -65 ℃ for 15 minutes. Its upper surface was not immersed in ethanol for the purpose of convenient observation. Note that there was no ethanol flowing in the pipes in Block A. For the block B (see Figure 6), the former step was repeated, at the same time, the cold ethanol (-65 ℃) was added to the adiabatic container and flew through the plastic pipe and PTFE pipes. The ethanol flux through each PTFE pipe was controlled at the velocity of about 4.5 ml/s.

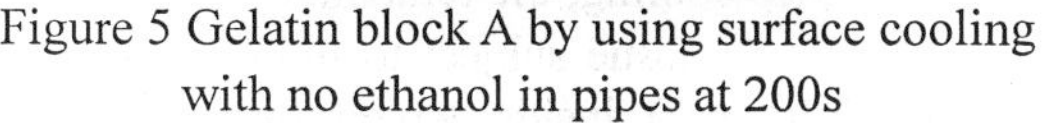

Figure 5 Gelatin block A by using surface cooling
with no ethanol in pipes at 200s

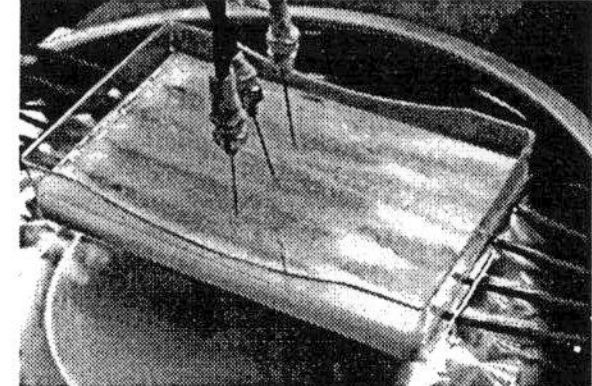

Figure 6 Gelatin block B by using uniform cooling
with ethanol flowing through pipes at 200s

As shown in Figure 5 and Figure 6, the difference of cooling effect by these two methods is evident. For block A, ice is only formed around the gelatin, and the center of the gelatin remains unfrozen. But for block B, ice is not only formed around the gelatin, but also surrounding the PTFE pipes. Further, three thermal couples were placed in the gelatin to examine the temperature history of the two method. Take the center point for example, as shown in Figure 5, the differences in temperature transient resulted by two cooling strategy were also large. Compared to that of surface cooling method, a rather rapid cooling rate

986

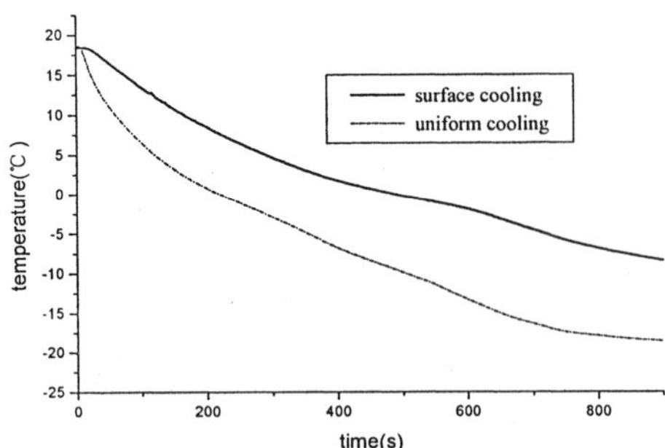

Figure 7 Temperature history for the center point of block A and block B

is obtained in the center similar to other regions of the tissue by using this uniform cooling method, and no evident phase change region exists in the curve. That implies that the tissue by using uniform cooling method may experience a shorter phase change stage, which will reduce the freezing injury of the tissue. Note that in both blocks, the phase change stage was relatively short compared to the computational results. This may be caused by the relatively small size of the block, which dramatically affects the consequence of the cooling process. All the results above imply that uniform cooling method can achieve a rather uniform temperature distribution inside the tissue, and freezing injury caused by large thermal stress due to large temperature gradient will thus be significantly decreased.

DISCUSSIONS AND CONCLUSIONS

This study proposed a uniform cooling method by perfusing cold fluid into micro-pipes specifically constructed over the tissues during culture. Its validity and advantages over the commonly used surface cooling method and limitations are discussed. It should be pointed out that the model of the four parallel pipes in this paper is just for illustration. Actually, various structures (especially the special net works structure) of the pipes can be adopted to meet the heat transfer request. But on the other hand, introducing of such pipes into tissues will increase the complexity of culturing tissues. So the selection of the pipes must be cautious and the biological effect caused by the pipes should also be further discussed. The possibility of using such pipes as channels to load CPA is another aspect worthy of consideration. Besides, combining the nutrition transport in culture as well as flow cooling in later preservation by using the similar pipes is also worth of trying in later study. Clearly, with the development of the tissue engineering technology, three-dimensional tissue culture in large scale would come true in the near future. This uniform cooling method may open a new strategy for successful cryopreservation and thus promise the tissue engineering in culturing large-scale biological objects.

ACKNOWLEDGEMENT

This work was partially supported by the National Natural Science Foundation of China under Grants: 50306027 and 50325622.

REFERENCES

1. Ishine N., Rubinsky B. and Charles Y. L., A histological analysis of liver injury in freezing storage. Cryobiology (1999) 39 271-277

2. Xu X. and Cui Z. F., Modeling of the co-transport of cryoprotective agents in a porous medium as a model tissue, Biotechnology Progress (2003) 19 972-981

3. Tabata Y., Recent progress in tissue engineering, Drug Discovery Today (2001) 6 482-487

4. Xu C.Y., Inai R., Kotaki M., Ramakrishna S., Aligned biodegradable nanofibrous structure: a potential scaffold for blood vessel engineering, Biomaterials (2004) 25 877-886

5. Deng Z.S. and Liu J., Numerical Simulation on 3-D Freezing and Heating Problems for the Combined Cryosurgery and Hyperthermia Therapy, Numerical Heat Transfer, Part A, in press, 2004

6. Deng Z. S. and Liu J., Effects of large blood vessels on 3-d temperature distributions during simulated cryosurgery. Submitted to ASME Journal of Biomechanical Engineering

Comparative analysis of the cryogens used in cryomedical applications

M. Chorowski, A. Piotrowska

Wroclaw University of Technology, Wybrzeze Wyspianskiego 27, 50–370 Wroclaw, Poland

Low temperature medicine or cryomedicine is becoming a wide appreciated therapy method in rheumatology, dermatology, gynecology, surgery and sport medicine. Cryomedicine can be divided into cryotherapy and cryosurgery. The paper gives the overview of the methods, equipment and cryogens used in cryotherapy and cryosurgery. The cryosurgical apparatuses are usually supplied with liquid nitrogen LN_2 or compressed nitrous oxide N_2O. We have experimentally investigated the thermal interactions of both cryogens with a tissue during a simulated cryosurgical intervention. The conclusions concerning the applicability of both cryogens and related equipment used in specific cryosurgical interventions have been formulated.

INTRODUCTION

Low temperatures are used in many different branches of science, technology and medicine. It has been known for a long time that low temperatures alleviate pain, decrease swelling and bleeding. The effect of low temperatures does not encumber the blood circulatory system and does not cause side-effects. Low temperatures can be applied with ice, cold compress or sprays both in the first aid and in an ambulatory treatment. The use of low temperatures in medicine can be in treatment (cryosurgery), rehabilitation (cryotherapy) and diagnostics (magnetoresonance tomography – cooling of superconducting magnets by liquid helium). Biological specimens can be kept for a long time in liquid nitrogen or in solid carbon dioxide.

To provide sufficient heat transfer intensity for medical treatment of the tissue, the temperatures lower than 200 K should be applied. The cryoliquids have boiling temperatures lower than 200 K but not all o f t hem a re c onvenient f or m edical a pplications. T he t emperatures o f l iquid h elium, h ydrogen a nd neon are too low for precise control of heat exchange during cryo-treatments. Besides helium, neon, argon and xenon are very expensive. Liquid hydrogen and methane can generate explosive mixture with air. Oxygen is chemically active and it is a powerful oxidizer. Therefore for medical application only nitrogen, nitrous oxide and carbon dioxide are commonly used in practice.

CRYOTHERAPY

Cryotherapy is a stimulating therapy (cryostimulation), where a patient body is subjected to an effect of low temperature (as a rule below 150 K) within less than 3 – 4 minutes, in order to activate defensive reactions. These reactions are therapeutically beneficial and very effective in restoring the natural balance of the organism. There are two main ways of cryotherapy specified in medical nomenclature: whole body cryotherapy and local cryotherapy.

Local cryotherapy

During local treatment (Figure 1) only a part of the body, like a joint or a muscle, is affected by low temperature. Cryotherapy is an effective way to support physiotherapy of moving organs.
The most important therapeutic effects are:
1. Raise of pain level threshold,

2. Muscle's tension decrease,
3. Muscles strength and joints mobility increase,
4. Shortening of convalescence time after contusions.

Besides nitrogen, carbon dioxide and nitrous oxide, the cryotherapy apparatuses can be supplied with cold air (cooled down in a compressor refrigerating system). A small group of cryotherapy equipment presents thermoelectric modules based on the Peltiere effect.

Whole body cryotherapy

Whole body cryotherapy treatments are carried out in cryochambers (figure 2a). A cryochamber consists of two rooms, vestibule and main cabin (figure 2b). The vestibule is a transitional room where temperature level is of about 210 K (- 60°C). It is a place where the patients can get used to much more extreme thermal conditions. After about 30 seconds spent in the vestibule the patients proceed into the main cabin. In cryochamber main cabin the temperature is maintained from 150 K (-120°C) to 110 K (-160°C) [4]. One session of the whole body cryotherapy can last no more than 3 minutes. The heat exchangers in cryochambers are supplied with liquid nitrogen. One working hour of a cryochamber requires of about 90-100 dm^3 of LN$_2$. The air vented into both cabins is purified, dried and cooled down in a dedicated installation located outside the cryochamber.

Figure 1. Local cryotherapy
(courtesy Kriosystem Ltd.)

Figure 2a. Cryochamber view
(courtesy Creator Ltd.)

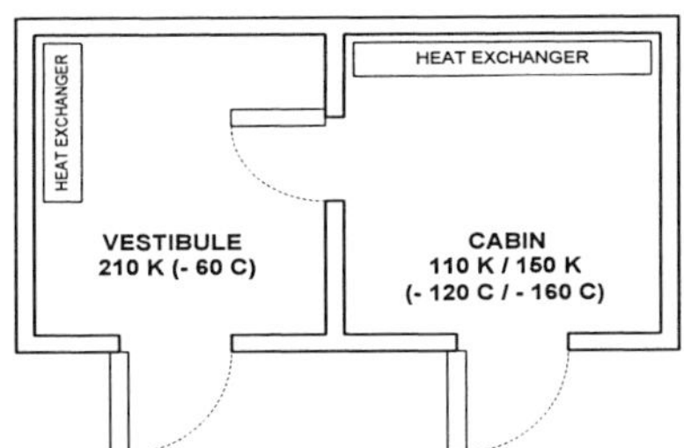

Figure 2b. Cryochamber scheme

CRYOSURGERY

The necrosis temperature of the majority of human cells is 250 K, on the condition that the fall of the temperature is fast, about 1 K/s. When the decrease of the temperature is significantly slower, the cells die in the temperature below 235 K. Therefore this effect has found the application in cryosurgery, which is the therapeutic use of cold to induce tissue necrosis with ablative intent. The first report of the use of local freezing as a treatment modality is attributed to Dr James Arnott [1], who described in 1850 the direct application of a salt-ice mixture to various skin lesions. For almost a century, cryosurgery was practiced by a handful of surgeons. Thanks to the development of cryogenics, easily available liquid gases have replaced the salt-ice mixture. Cryosurgery is now used in many medical fields e.g. dermatology, neurosurgery, gynecology, urology, ophthalmology and oncology. There are three methods of cryosurgery treatment: direct evaporating, spray and contact method [3].

Direct evaporating method

In this method a small amount of cryoliquid is applied directly on the surface of a pathology tissue.
Usually a sterile wooden rod with a cotton ball on its end is used as cryogen applicator. When the cotton ball contacts with a surface of the tissue the cryoliquid will evaporate and it will cool down the tissue. This method is easy to practice, but the heat capacity of the cotton ball is small and it can be used only for local, shallow and small pathologic changes. Due to its availability and low price, liquid nitrogen is used in this method.

Spray method

In this method liquid refrigerant is sprayed directly on the surface of the tissue. Cryogen evaporates very fast a nd c ools d own r apidly t he t issue. T he c ryoliquid i n s pray apparatus s hould b e u nder p ressure t o create a jet when relieved from the vessel. Cryosurgery sprays are fed with liquid nitrogen, nitrous oxide or carbon dioxide.

Contact method

The method is similar to spray way of cryo-treatment. It can be applied by use of the same apparatus, but with different, closed applicator. The heat exchange proceeds through the wall of the applicator contact surface. In this method the liquid does not come into direct contact with the tissue.

CRYOSURGERY EQUIPMENT

The cryosurgical apparatuses are usually supplied with liquid nitrogen LN_2 or compressed nitrous oxide N_2O. The liquid nitrogen is stored in a dewar and transferred to cryosurgical tip, where it is evaporated providing a cooling power at 78 K. The compressed nitrous oxide is stored in a gas cylinder at ambient temperature. To obtain a low temperature source it is throttled at the Joule-Thomson valve, then partly liquefied and vaporized at 185 K. In spite of much higher phase transition temperature in comparison with LN_2, nitrous oxide N_2O is often used due to its non-limited storage time in a compressed gas cylinder.

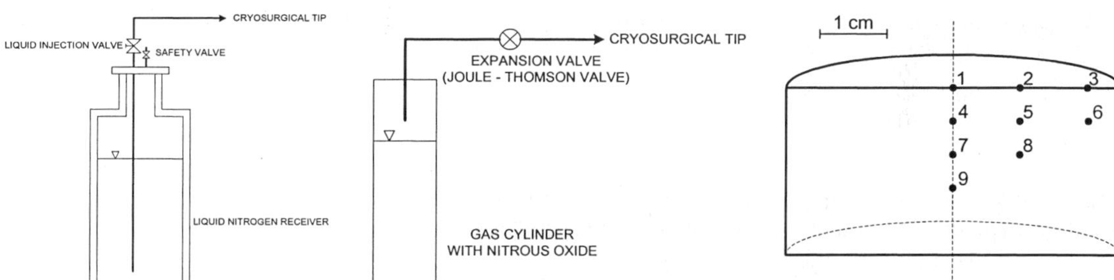

Figure 3. LN_2 cryosurgical apparatus   Figure 4. N_2O cryosurgical apparatus   Figure 5. Location of measured points

In spite of the fact that boiling LN_2 and N_2O differ in the temperature by about 100 K, both cryogens are commonly used in cryosurgical treatment. We have experimentally investigated the thermal interactions of both cryogens with a tissue during a simulated cryosurgical intervention.

The tissue was modeled with a water-gelatin and the heat transfer between the cryosurgical tip and the solution was measured. We have measured simultaneously the temperature at 9 points (figure 5).

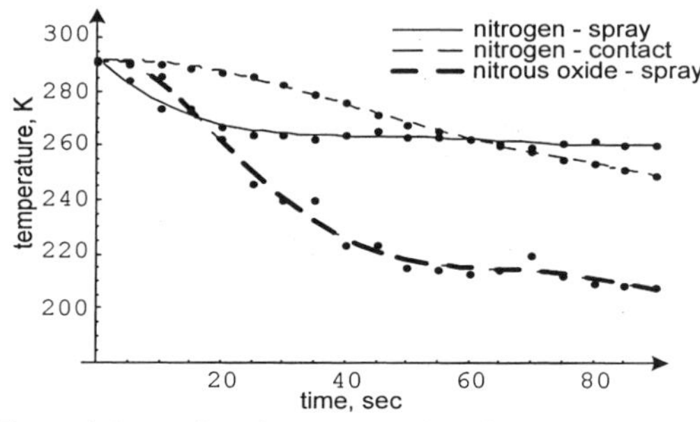

Figure 6. Dynamics of temperature in point 2

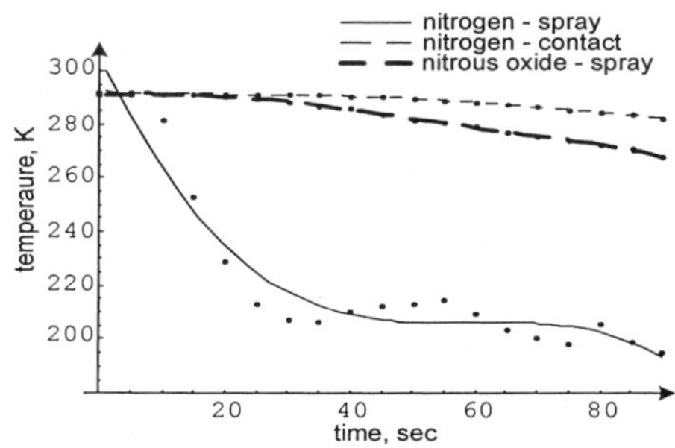

Figure 7. Dynamics of temperature in point 4

Every treatment simulation has lasted for 90 seconds. We have assumed that the cells are dying in 235 K and have tried to find out the dynamic geometry of the tissue frozen to this temperature. In the point 1 we have always measured the temperature lower than 238 K (spray N_2 – 158.9 K, contact N_2 – 222.4 K and spray N_2O – 190.8 K). The temperature below 238 K has been measured in point 2 (figure 6) only with N_2O spray – 207.8 K (spray N_2 – 260.1 K, contact N_2 – 248.9 K). The similar result was observed for point 3. It means that the wide pathogenic change should be destroyed with a spray method using a nitrous oxide. The development of temperature measured in points 4 (figure 7) and 7 indicated that deep changes can be destroyed only by a spray method using liquid nitrogen. If the pathologic cells are located below 1.5 cm, the time of the treatment has to be extended.

A dynamic creation of an ice ball was observed (figure 8), giving the basis for further mathematical modeling of cryosurgical treatment of different morbid changes.

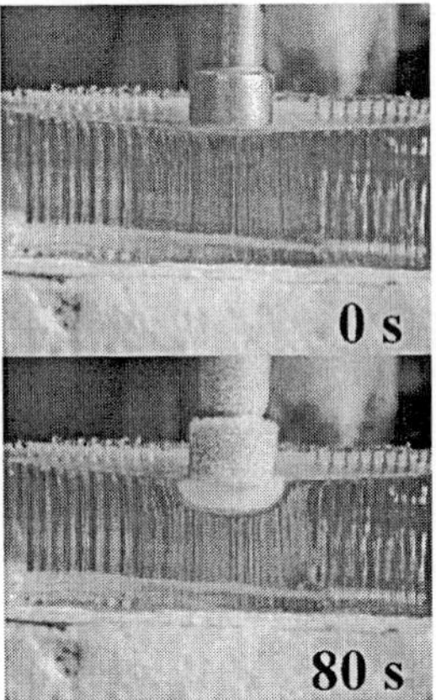
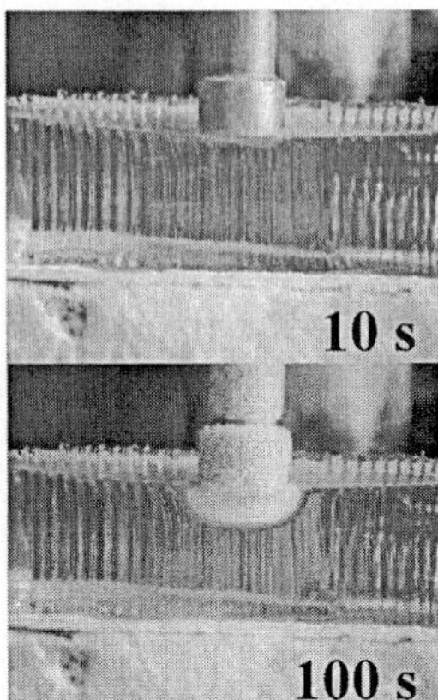
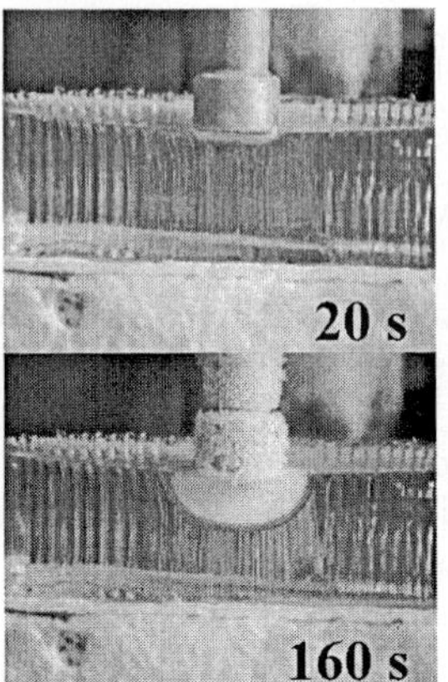
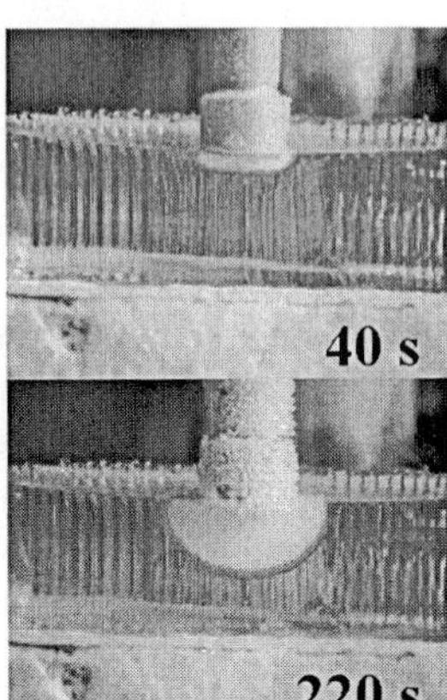

Figure 8. Dynamics of "ice ball creation"

CONCLUSIONS

Cryosurgical treatments have been simulated using liquid nitrogen and nitrous oxide as the cooling agents. In liquid nitrogen case both spray and contact treatment methods were investigated. On the basics of the obtained results our experiments recommendations of cryosurgical methods and refrigerants suitable for the type of pathogenic changes were formulated – table 1.

Table 1. Summary of cryosurgical treatment recommendations.

	NITROGEN		NITROUS OXIDE
	SPRAY	CONTACT	SPRAY
Local surface change (radius up to 0.5 cm)	***	*	***
Surface change (radius over 0.5 cm)	*	*	***
Changes up to 0.5 cm deep into tissue	***	*	*
Changes up to 1.0 cm deep into tissue	**	-	-
Changes up to 1.5 cm deep into tissue	*	-	-

Symbols: *** – perfect, ** – good, * – not recommended, - – not applicable

ACKNOWLEDGEMENTS

We thank Dr J. Gawlik and Mr. B. Adamowicz for their kind advices and help in the measurements.

REFERENCES

1. Arnott JM., Practical illustrations of the remedial efficiency of a very low or anesthetic temperature in cancer, Lancet (1850), 2:257-316
2. Holden HB, History and development of cryosurgery, In: Holden HB, editor, Practical cryosurgery, Chicago, Pitman Medical Publication (1975), 1-9
3. Kaźmierowski M., Kriochirurgia w chorobach skóry, Wydawnictwo Czelej, Lublin 1997
4. Zagrobelny Z., Krioterapia miejscowa i ogólnoustrojowa, Wydawnictwo Medyczne Urban & Partner, Wrocław
5. Gabryś M. S., Popiel A., Krioterapia w medycynie, Wydawnictwo Medyczne Urban & Partner, Wrocław

Measured Performance of Four New 18 kW@4.5 K Helium Refrigerators for the LHC Cryogenic System

Gruehagen, H., Wagner, U.

CERN, CH-1211 Geneva 23, Switzerland.

The cryogenic system for the Large Hadron Collider (LHC) under construction at CERN will include four new 4.5 K-helium refrigerators, to cover part of the cooling needs of the LHC at the 4.5 – 20 K and 50 -75 K levels. Two refrigerators are delivered by Air Liquide, France, and two by Linde Kryotechnik, Switzerland. During the last three years, all four refrigerators have been installed and commissioned at four different points along the LHC. The specified requirements of the refrigerators are presented, with special focus on the capacities at the various temperature levels. The capacities of the refrigerators were measured using a dedicated test cryostat, and the measured performance for all four installations is presented, and compared to the guaranteed performance in the original proposal of the suppliers. Finally, the process design of the two supplies is compared, and their differences and similarities briefly analysed.

INTRODUCTION

The total cryogenic capacity necessary for the operation of the Large Hadron Collider (LHC) [1] will be delivered by eight refrigerators supplying cooling capacity down to a temperature of 4.5 K and eight refrigeration units that are coupled to each of these refrigerators supplying cooling capacity down to a temperature of 1.8 K [2]. Of the eight refrigerators four will be recovered from the now decommissioned LEP accelerator and upgraded for the LHC needs. Four new refrigerators supplying an equivalent capacity of about 18 kW at 4.5 K were specified by CERN in 1997 [3] and purchased from European Industry in 1998. Two of these refrigerators have been supplied by Air Liquide, France [4] and two have been supplied by Linde Kryotechnik, Switzerland [5]. The first of these four new refrigerators was commissioned in 2001, the last at the end of 2003.

SPECIFIED COOLING NEEDS OF THE NEW LHC REFRIGERATORS

During LHC operation, the refrigerators will work for a large portion of the time far from the maximum capacity. Therefore, CERN specified to test the refrigerators not only in its design capacity mode, called "installed" but as well in a turn-down mode called "low intensity". The capacities required for these two operation modes are also listed in Table 1.

Table 1 Specified capacity of the refrigerators

Operation mode	4.5 K isothermal	4.5 – 20 K non isothermal	50 -75 K non isothermal	20 – 280 K non isothermal
	[W]	[W]	[W]	[W]
Installed	4400	20700	33000	55400
Low intensity	1600	7700	22000	36500

The non-isothermal load between 20 and 280 K corresponds to the current leads cooling and, together with the corresponding fraction of the 4.5 to 20 K load, is seen as a liquefaction rate by the refrigerator. This liquefaction rate corresponds to 41 g/s for the installed mode and 27 g/s for the low intensity mode.

PROCESS DESIGN OF THE MANUFACTURERS

Independent of CERN, the two suppliers chose the same compressor manufacturer. Except for the final oil removal system, the compressor station from the two suppliers are therefore identical, operating at three pressure levels, a high pressure (HP) of 20 bar, a medium pressure (MP) of 3.9 bar, and a low pressure (LP) of 1.05 bar. Figure *1* shows the principle arrangement of the compressors.

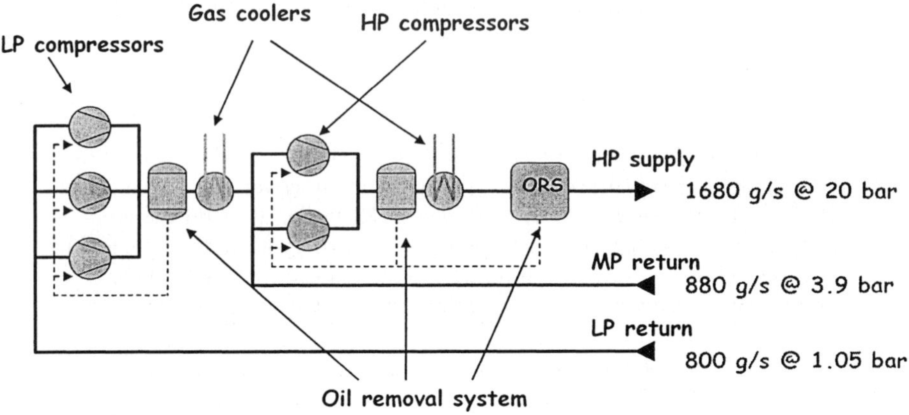

Figure 1 Compressor system of the four refrigerators

In addition to the capacity mentioned above, the typical flow scheme of a refrigerator cold box was specified, including switchable adsorbers at the 80 K level, one adsorber at the 20 K level, liquid nitrogen precooler, and a 4.5 K phase separator including a helium sub-cooler.

From a process point of view the design of the two refrigerator cold boxes are almost identical down to the 20 K level. The main compressor flow is sent through a heat exchanger, and part of it is branched of through a turbine circuit working between the HP and MP levels. Linde uses three turbines in series while Air Liquide has two turbines with heat exchange in between. After the 80 K adsorbers, a second heat exchanger block coupled with two turbine strings brings the temperature down to 20 K.

The principle flow diagram of the process realised by Linde Kryotechnik is shown in Figure 2, the one realised by Air Liquide in Figure 3.

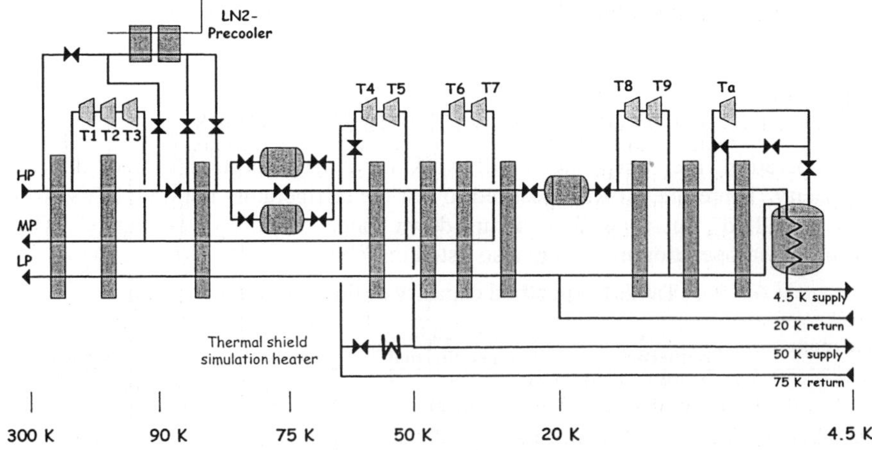

Figure 2 The Linde process with ten cryogenic turbines

As the helium supplied to the LHC (235 g/s) at 4.5 K is returned at 20 K, the lower end heat exchangers will see a considerable unbalanced flow, requiring several turbines to compensate for this.

To produce the required cooling capacity at low temperature, both suppliers use a turbine string (T8 and T9 for Linde, T7 for Air Liquide) working between HP and LP, down to a temperature of 10 K.

For Linde, the main helium flow is then sent through the last turbine, after which part of it is sent through a Joule-Thomson-Valve down to approx. 1.3 bar and fed into the phase separator. There it is used to sub-cool the helium that is supplied to the LHC at 4.6 K and 3 bar.

Air Liquide has chosen a different concept to compensate for the unbalanced heat exchangers. A part of the flow bypasses the lower end heat exchangers through T8, effectively making the coldest heat exchangers balanced, and avoiding the inherit losses of unbalanced heat exchange. The flow from T8 is fed directly into the phase separator to sub-cool the helium supplied to the LHC.

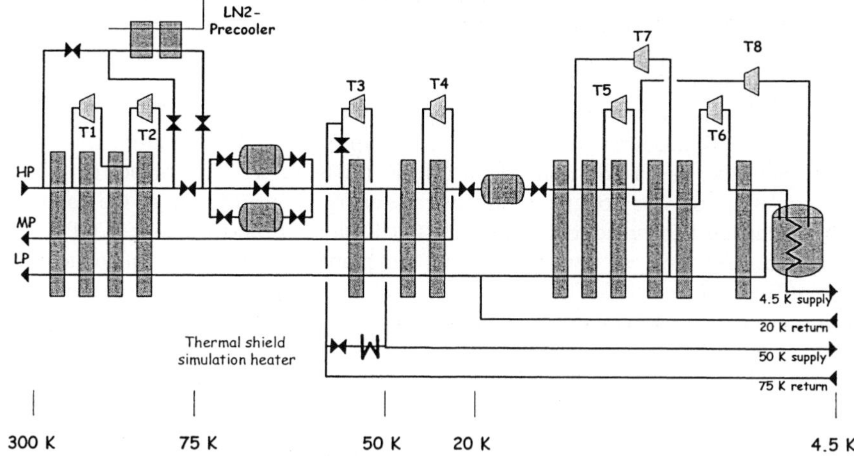

Figure 3 The Air Liquide process including eight turbines

CERN asked for guaranteed values for the power consumption for each installation from the two suppliers. These guaranteed values for the compressor station in the two operation modes described above can be found in Table 2.

Table 2 Guaranteed power consumption for both suppliers

Mode	Air Liquide	Linde
Installed	4204 kW	4275 kW
Low Intensity	2262 kW	2461 kW

Please note that the values in Table 2 only take into account the power consumption of the compressors themselves, and not that of utilities.

TEST CRYOSTAT

To verify the capacity and power consumption in all four installations, a dedicated test cryostat was built and has been used for all four installations. It simulates the thermal loads of the LHC and has been used for the two operational modes described above.

In Figure 4 a schematic lay-out of this test cryostat can be found. Dedicated heaters are installed to simulate the heat loads at the 4.5 K (Q10), 4.5 – 20 K (Q11) and the current lead flow at 20 – 280 K (Q13). Together with the shield simulation heater in the cold box, the conditions represented in Table 1 can be re-produced for the capacity tests. Heaters Q10 and Q11 are slowly ramped up to their nominal

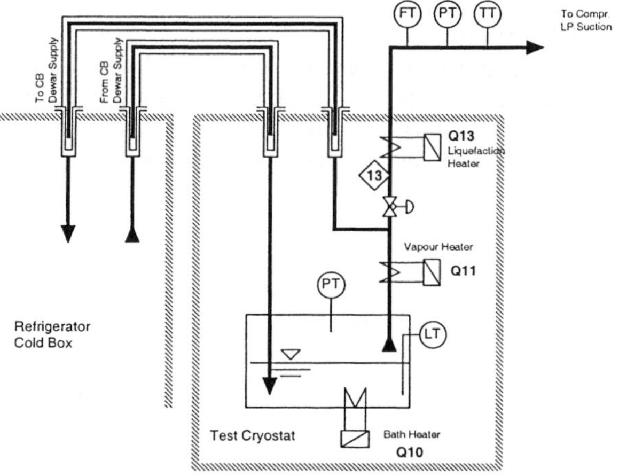

Figure 4 Flow diagram of the Test Cryostat

value and thereafter the flow and temperature of the flow through Q13 are controlled to 41 (27) g/s and 280K respectively.

MEASUREMENTS

During the commissioning, each refrigerator was connected to the test cryostat as shown in Figure 4. The refrigerators were operated in fully automatic mode for 40 to 48 hours for the installed mode and about 24 hours for the low-intensity mode. During this time, the power consumption of the compressor station was monitored, as well as temperatures and pressures in the cold box to assure that stable conditions were maintained.

During the capacity tests in installed mode one regeneration of the 80 K adsorbers and one regeneration of the 20 K adsorbers were performed. During this regeneration, the adsorbers were cooled down, which acts as an additional load on the refrigerator. The impact of this load on the overall energy consumption was noted, and the average energy consumption of the compressor station calculated. This value has been used to calculate the Coefficient Of Performance (COP).

To calculate the efficiency, or performance, of the plant, it is usual to calculate the equivalent capacity of the refrigerator at 4.5 K. The COP can then be given by the energy consumption divided by this equivalent capacity. In other words it is the number of watts consumed at the compressor station needed to produce one watt of cooling power at 4.5 K.

During the capacity test in Point 1.8, due to problems with the CERN electrical supply, it was not possible to power the heaters in the test cryostat to the specified value. In the end, the tests were accepted with about 3% less capacity than specified. Spare capacity was available, and doing an analysis as described above gives about the same COP as for the refrigerator in Point 4. For all other points the refrigerators achieved the specified capacity during the capacity tests. The energy consumption and the COP of all refrigerators are calculated using the capacities measured by the test cryostat and can be found in Table 3. These measurements and the guaranteed values for the power consumption were the basis of a commercial bonus/malus calculation.

Table 3 Measured energy consumption and COP for all refrigerators

	Point	PA18 Air Liquide	PA4 Air Liquide	PA6 Linde	PA8 Linde
Installed mode	Energy consumption (kW)	4297	4474	3964	4095
	% of Guarantee	*102*	*106*	92.7	95.7
	COP	248	247	222	231
Low Intensity mode	Energy consumption (kW)	2491	2560	2179	2203
	% of Guarantee	*110*	*113*	88.5	89.5
	COP	338	339	289	298

CONCLUSIONS

Four new helium refrigerators have been installed and commissioned on CERN during the last three years. Their capacities have been measured using a dedicated test cryostat simulating the future cooling needs of the LHC accelerator. All four refrigerators have the specified capacity. Even though their COP and energy consumption varies considerably, they are all adapted to the future needs of the LHC accelerator.

REFERENCES

1. Schmidt, R., Status Of The LHC, <u>CERN-LHC Project Report 569</u>, Geneva, Switzerland (2002), <u>Eight European Particle Accelerator Conference (EPAC)</u>, Paris, France (2002),
2. Claudet, S; Gayet, P; Jäger, B; Millet, F; Roussel, P; Tavian, L; Wagner, U; Specification of Eight 2400 W @ 1.8 K Refrigeration Units for the LHC, <u>ICEC 18</u> , Mumbai, India , (2000) 207-210
3. Claudet, S; Gayet, P; Wagner, U; Specification of Four New Large 4.5 K Helium Refrigerators for the LHC, <u>ICMC '99</u> , Montreal, Canada , (1999)
4. Dauguet P; Guerin C and Monneret E; Construction and start up of two 18 kW at 4.5 K helium refrigerators for the new CERN accelerator; LHC, <u>ICEC19</u>, Grenoble, France, (2002) 187-190
5. Boesel, J., Chromec, B., Meier, A. Two large 18 kW (equivalent power at 4.5K) refrigerators for CERN's LHC project supplied by Linde Kryotechnik AG <u>CEC 1999,</u> Montreal, Canada, (1999), 1285- 1292

OVERVIEW OF THE AIR LIQUIDE CRYOGENIC SYSTEMS DESIGNED FOR CERN LHC

Dauguet P., Briend P., Delcayre F., Hilbert B., Mantileri C., Marot G., Monneret E., Walter E.

Advanced Technology Division, Air Liquide, BP 15, 38360 Sassenage, France

The 27-km-long Large Hadron Collider (LHC) will make use of superconducting magnets. It requires refrigeration systems at temperatures ranging from 1.8 K to 80 K. Two refrigerators delivering 18 kW equivalent power at 4.5 K were installed and commissioned. Connected to these refrigerators, the pre-series unit of the so-called Cold Compression System was installed and commissioned. These systems are able to absorb up to 2.4 kW at 1.8 K. An overview of these projects is given. Other related cryogenic equipments are also delivered for the LHC project and are listed at the end of the present paper.

LHC CRYOGENIC ARCHITECTURE

The LHC cryogenic architecture is presented in figure 1.

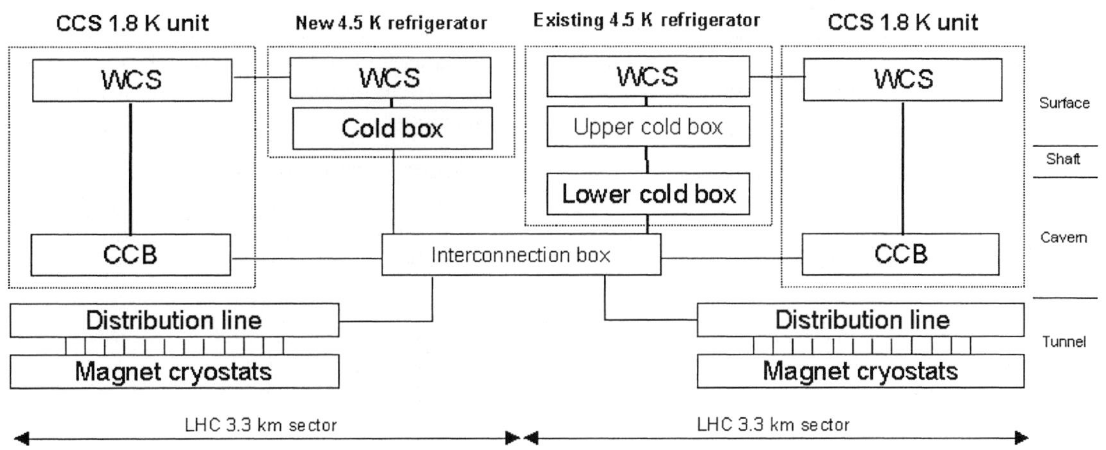

FIGURE 1 LHC cryogenic architecture

THE 18 kW AT 4.5 K REFRIGERATORS

The cryogenic duties to be provided by each 4.5 K refrigerator are presented in Table 1. Each refrigerator has to fit very different operating loads. A large flexibility of operation of the machine is thus requested. As the CERN call for tender was giving a very high importance to the electrical consumption of the plant as adjudication criteria, the choice of a very efficient cycle was mandatory [1]. The third request to be taken into account in the design of the machine is the high availability of the cryogenic system needed by the LHC project. This is insured by the choice of reliable and technically proven components. The cycle design presented hereafter is a compromise between efficiency, flexibility and reliability of operation.

<u>The cycle design</u>
The cycle design is presented in figure 2. The cycle design has already been described in details in [2].

TABLE 1 Cryogenic loads of the LHC refrigerators in main operating modes

Loads / Operation mode	4.5 K – 20 K (W)	20 K – 280 K (W)	50 K – 75 K (W)
Installed	25100	55400	33000
Normal	15000	36500	22000
Low Intensity	9300	36500	22000
Injection Standby	6900	14900	22000
75 K standby	0	0	22000

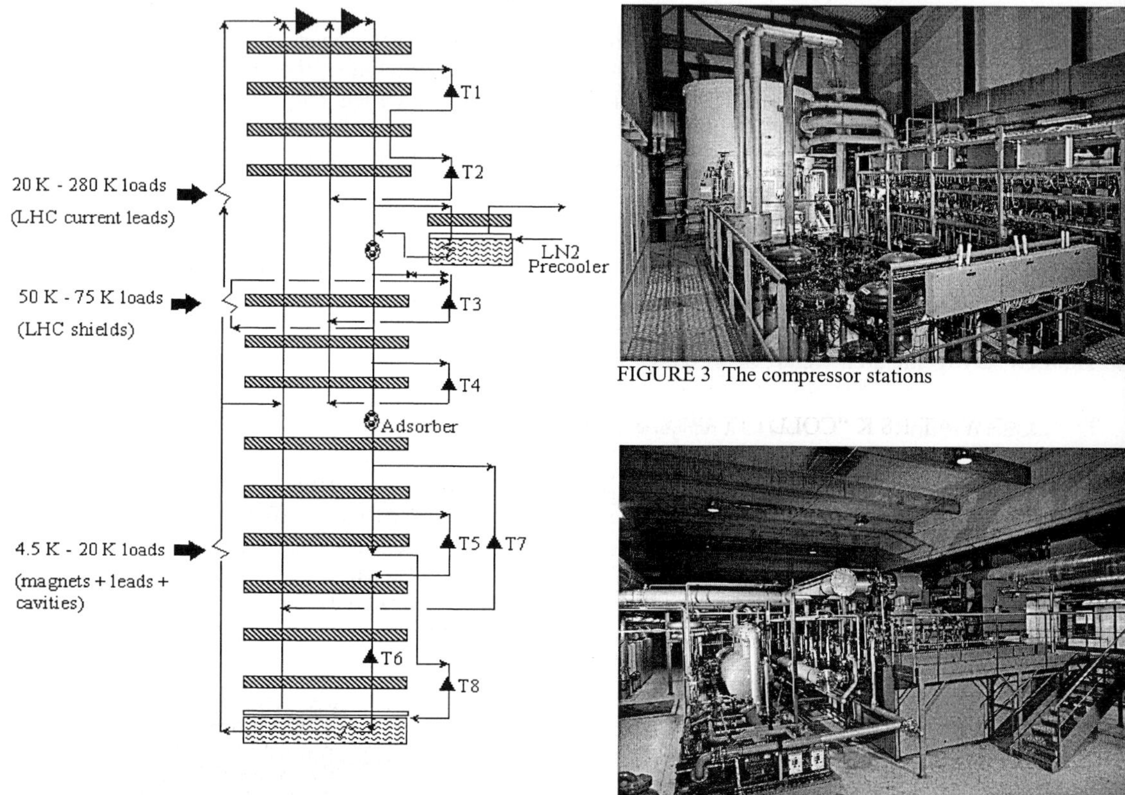

FIGURE 2 LHC 4.5 K refrigerators cycle design

FIGURE 3 The compressor stations

FIGURE 4 The cold boxes

Measured Performances

The refrigerators have been successfully operated in all the operating modes necessary to the LHC operation, as requested in the CERN specifications. Transitory situations between the static operating modes defined in the CERN specification have also been tested.

The cryogenic load that the refrigerators are able to deliver is 105% of the defined "Installed" exergetic power, that is more than 18kW.

Performances obtained in term of cycle efficiency have been precisely measured during contractual reception tests, in "installed" and "stand by mode". The results are presented in Table 3. The measured values of electrical power consumptions are a few percent more than the expected values, and about 1% more than the guaranteed values of the contract. The reason is that the efficiency of the room temperature screw compressors is around 5% less than expected. The efficiency of the cold box however, conforms to expectations.

TABLE 2 Expected electrical power consumptions of each LHC refrigerator versus the cryogenic load.

Operating mode	Cryogenic load (Exergetic equivalent at 4.5 K) kW and (% of max value)	Electrical consumption kW and (% of max value)	Factor of merit (W/W) (electrical consumption divided by cryogenic load)
Installed	17.5 (100)	4200 (100)	240
Normal	10.7 (61)	2800 (67)	260
Low Intensity	7.6 (43)	2300 (55)	300

TABLE 3 Expected, guaranteed and measured performance of the refrigerators

Operating mode	"Expected" power consumption (kW)	"Guaranteed" power consumption (kW)	"Measured" power consumption (kW)
Installed	4200	4275	4297
Low Intensity	2300	2461	2491

Conclusion

A very high efficiency cycle has been custom designed for the LHC accelerator. The factor of merit of this cycle at maximum cryogenic power is 245 W/W, which corresponds to an efficiency of 27 % compared to Carnot ratio. The way the plant is operated for reduced cryogenic duty enables it to keep a high cycle efficiency even at partial loads down to 40 % of the installed power.

THE 2.4 KW AT 1.8 K "COLD COMPRESSOR SYSTEM"

The Large Hadron Collider (LHC) at CERN makes intensive use of superconducting magnets operated below 2 K. It requires high-capacity (i.e. 2.4 kW) 1.8 K refrigeration systems, the so-called "Cold Compressor System" (CCS). Air Liquide's Advanced Technologies Division designed and built a pre-series of these CCS for CERN - to be followed by three additional series. This pre–series unit was manufactured in 2001 and installed at CERN during the first 2002 trimester. It has been commissioned over the spring of 2002 and has been tested during 2002 / 2003. The 1.8 K refrigeration units are described in more details in [3, 4].

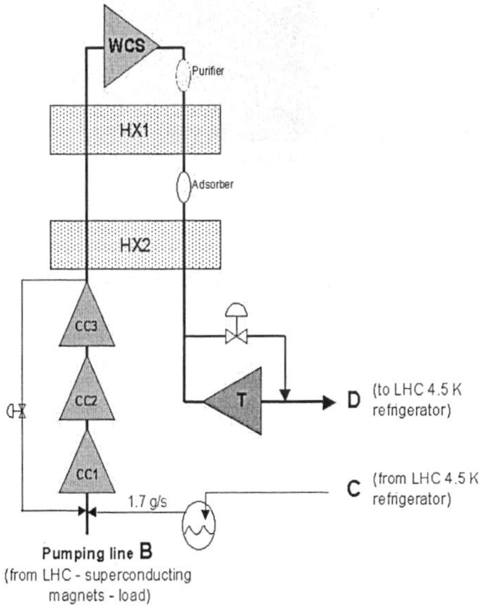

FIGURE 5 LHC/CCS Air Liquide cycle design

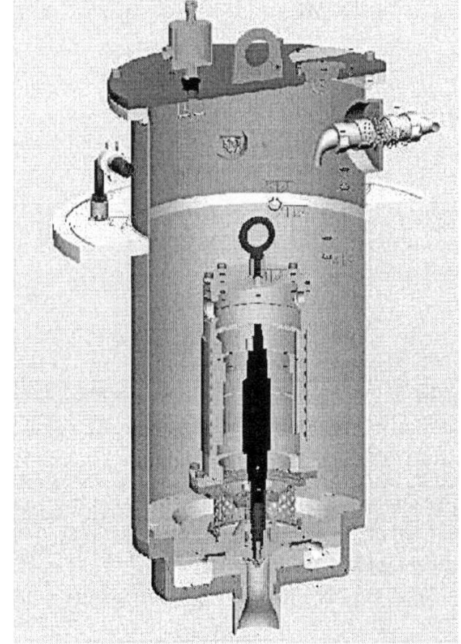

FIGURE 6 Cryogenic centrifugal-compressor simplified cross-section

Cycle design :
Figure 5 shows the Air Liquide CCS cycle. The very low pressure (1.5 kPa) needed on the He II saturated bath at 1.8 K is obtained using three cryogenic centrifugal compressors (CC) coupled with a one-stage ambient temperature (warm) compressor station (WCS) in a series arrangement. This article focuses on experimental results only. More details on the CERN specification and the Air Liquide cycle design are given in [3,4].

Compression at cryogenic temperature using centrifugal compressors :
Cryogenic centrifugal-compressors fitted with magnetic bearings (see Figure 6) -already successfully used in Tore Supra (CEA Cadarache, France) and CEBAF (TJNAF, USA) installations - turn out to be the key technology for achieving temperatures less than 2.17 K on a large scale (these units allow high rotation speeds, up to 600 Hz in the present case). All of these units uses 3D (axial-radial) impeller, designed in collaboration with the Czech company PBS. Their measured and guaranteed efficiencies in the different operating modes are given in Table 4.

TABLE 4 Measured parameters of the WCS for the main operating modes [3]

Operating mode	Installed	Normal	Low Intensity	Injection sb	Cold sb
Mass flow rate [g/s]	126	85,2	63,3	43,2	38.3
Inlet pressure [kPa] (compressor flange)	35	24	18	12,7	11,4
Pressure ratio	12,6	11,5	12,8	17,7	19,6
Measured volumetric efficiency	**0,81**	**0,80**	**0,79**	**0,77**	**0,76**
Expected power consumption [kW]	529	384	363	283	271
Expected isothermal efficiency	0.38	0.33	0.27	0.25	0.24
Measured power consumption [kW]	**446,5**	**316**	**267,5**	**250**	**246**
Measured isothermal efficiency	**0,45**	**0,41**	**0,37**	**0,31**	**0,29**

Conclusion
The CC measured efficiencies are better than expected. The required stability (i.e. +/- 0,05 kPa maximum pressure variation at the CC1 inlet in steady state regimes) and mass flow ramp-ups (i.e. +/- 6 g/s per min) could be demonstrated. The system is now in operation at CERN. Air Liquide is selected for the series which is now under way. Three other CCS are now under construction.

OTHER PROJECTS
Air Liquide also delivers for LHC project : the 27 km transfert lines to distribute the requested cryogenic power around the accelerator, a 600 mm vacuum jacketed transfer lines with a 5 pipes in 1 design, the 5 interconnection valve boxes connecting the refrigerators with the cryogenic lines and including each a 600 kW heater, the CMS particle detector helium refrigeration system, the ATLAS particle detector nitrogen refrigeration system, and also 12 test benches for the LHC magnets.

REFERENCES

1. Claudet S. et al., Specification of four new large 4.5 K refrigerators for the LHC, Advances in Cryogenic Engineering (1999) 45 1269-1276
2. Dauguet P. et al., Design, construction and start up by Air liquide of two 18 kW at 4.5 K helium refrigerators for the new CERN accelerator , Advances in Cryogenic Engineering (2003) 49
3. Claudet S. et al., Specification of eight 2400 W @ 1.8 K refrigeration units for the LHC, ICEC 2000 proceedings, Mumbai, India
4. Hilbert B., et al., Air Liquide 1.8 K refrigeration units for CERN LHC project, Advances in Cryogenic Engineering (2003)49

1.8 K Refrigeration Units for the LHC: Performance Assessment of Pre-series Units

Claudet S., Ferlin G., Millet F.*, Tavian L.

Accelerator Technology Department, CERN, CH-1211 Geneva 23, Switzerland
*SBT, DRFMC, DSM, CEA, 17 avenue des martyrs, 38054 Grenoble, France

The cooling capacity below 2 K for the superconducting magnets of the Large Hadron Collider (LHC), at CERN, will be provided by eight refrigeration units of 2400 W at 1.8 K, each of them coupled to a 4.5 K refrigerator. The two selected vendors have proposed cycles based on centrifugal cold compressors combined with volumetric screw compressors with sub-atmospheric suction, as previously identified by CERN as "reference cycle". The supply of the series units was linked to successful testing and acceptance of the pre-series temporarily installed in a dedicated test station. The global capacity, the performance of cold compressors and some process specificities have been thoroughly tested and will be presented.

INTRODUCTION

The cooling capacity below 2 K for the superconducting magnets of the Large Hadron Collider, at CERN, will be provided by eight 2400 W @ 1.8 K refrigeration units, each of them coupled to a 4.5 K refrigerator. The 1.8 K refrigeration units have been specified in 1998 [1] and ordered in 1999 to IHI-Linde (four units) and Air Liquide (four units) [2,3].

According to CERN technical specification and procurement scenarios, the proposed cycle and machinery had to provide the minimum investment plus operating costs [1]. This principle leads IHI-Linde to propose a process including four cold compressors and two turbines whereas Air Liquide installs three cold compressors and only one turbine. Figure 1 shows the simplified schemes of both 1.8 K refrigeration units as well as the interfaces with the LHC (header B) and a 4.5 K refrigerator (headers C and D). The 1.8 K units are composed of :
– one warm compression station (WCS) including oil lubricated screw compressors (WC) associated with the oil removal system (ORS).
– one cold compressor box (CCB) including mainly a train of cold compressors (CC), heat exchangers (HX) and turbo-expanders (Tu).

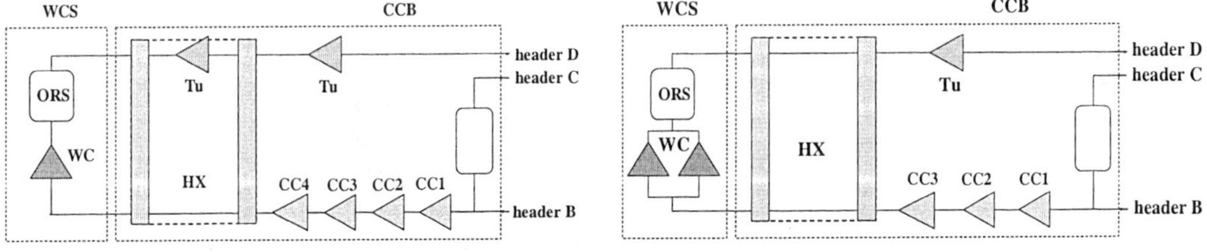

Figure 1 Simplified scheme with the LHC interfaces of a 1.8 K refrigeration unit by IHI-Linde (left) and Air Liquide (right)

In order to validate the process and components, a first unit called "pre-series unit" had to be validated by extensive testing in a dedicated test facility [4] at CERN before launching production of series unit. Acceptance tests were performed in 2002 and 2003 in order to measure and validate the overall and detailed performance of the pre-series units.

WARM COMPRESSOR STATION

IHI-Linde installed one Mycom screw compressor of two-stage compound type using slide valves whereas Air Liquide has installed two Kaeser one-stage screw compressors without any slide valve control operating in parallel. Even though sub-atmospheric conditions are required for the system, isothermal efficiency around 45 % have been successfully achieved for both screw compressors. The corresponding volumetric efficiency of each machine was higher than expected and in the range of 85 to 90 %. Figure 2 shows the measured isothermal efficiency of the screw compressors which have to compress helium gas from few kPa to high pressure (> 0.2 MPa).

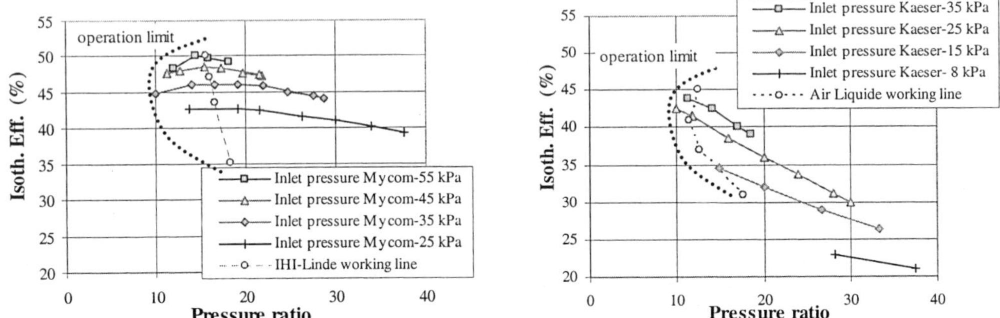

Figure 2 Isothermal efficiency of the warm compression station for IHI-Linde (a) and Air Liquide (b) pre-series units

One can note that the IHI-Linde cycle is defined for the optimum efficiency of the Mycom compound compressor (Pressure Ratio PR $\cong$ 15) and that operating margins are comfortable. The Air Liquide cycle operating at lower warm compressor suction pressures imposes a pressure ratio higher than the usual optimum for single stage machines (PR $\cong$ 7). The operation margins are kept at the minimum acceptable.

COLD COMPRESSORS

Cold compressors fitted with active bearings and equipped with individual frequency drive were already successfully used in Tore Supra [5] and CEBAF [6] installations. The cold compressor design of both suppliers are similar and based on technically proven components such as magnetic bearings, frequency drive and electrical motor cooled with water. All cold compressors are of the 3D (axial-radial) impeller type with fixed diffuser. Special attention was paid to prevent any air inleaks to sub-atmospheric helium circuits. Therefore a helium guard system (associated with high-vacuum feedthroughs) isolating the process from air is installed for IHI-Linde cartridges while Air Liquide cold compressors are mounted in individual vacuum-pumped housings. To reduce the heat inputs to the process flow, IHI-Linde has installed a cold intercept (50 to 85 K) on their short shaft, supplied with cold gas subtracted from the main HP flow. Air Liquide preferred a longer shaft and a housing design with dissociated mechanical and sealing functions.

Each cold compressor has its individual working field indicating the combination of rotating speed, mass flow, pressure ratio and suction conditions as shown in Fig. 3. In multistage configuration, the different cold compressor stages are interacting and have to enable safe operation of each compressor within their respective working field. Special automatic control strategies [3,7] based on the volumetric behaviour of the warm compressor, the variable frequency drive and (for transients) the suction temperature adjustment were developed by each supplier to keep the suction pressure below 1.5 kPa ± 50 Pa in steady-state operation or below 1.7 kPa during transient modes as specified.

The fully automatic pump-down towards 1.5 kPa is started when the cold compressor box is cold, with the screw compressor initially close to the required pressure at the inlet of first CC stage (typically 20 to 25 kPa for 64 g/s) allowing the final pump-down with the cold compressors to start. Such automatic pump down from 20 to 1.5 kPa is achieved in less than 30 min at fastest for the present pumped volume and would automatically adapt to any volume to be pumped down.

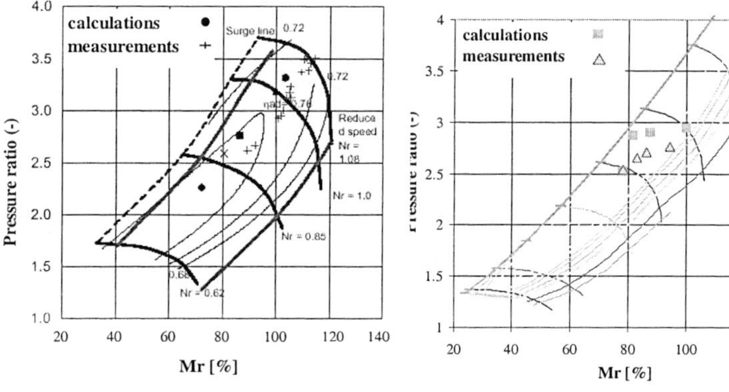

$$Mr = \frac{M}{Mo} \times \frac{Po}{P} \times \sqrt{\frac{T}{To}}$$

$$Nr = \frac{N}{No} \times \sqrt{\frac{To}{T}}$$

Figure 3 Typical cold compressor working field for a cold compressor cartridge from IHI-Linde (left) and Air Liquide (right)

Table 1 summarises the acceptance test results in the Installed Mode for each cold compressor train. The values were measured several times with good reproducibility (± 2 %). In most of the operation modes and for most of the cold compressors, the isentropic efficiency are above 75 %. One can note that larger pressure ratios (around 4) have been achieved without any disturbance of the cold compressors and preserving high efficiency above 70 % for any pressure ratio > 1.5.

Table 1 : Main performances of cold compressors : Pressure ratio / Isentropic efficiency for the Installed mode

Operation Mode		IHI-Linde					Air Liquide			
		CC1	CC2	CC3	CC4*	CC chain	CC1*	CC2*	CC3	CC chain
Installed	Pressure ratio [-]	3.26	2.9	2.55	1.61	38.75	3.13	2.96	2.76	25.56
mode	Isentropic efficiency [%]	77	78	75	70	66	77	74	75	65

** sensor correction required for outlet temperature of CC4 IHI-Linde and inlet temperature of CC2 Air Liquide*

The test cryostat also allows performing transient tests called "LHC daily" simulating the expected heat load change during a nominal working day for LHC physics runs. Fig. 4 shows the typical specified transients and corresponding measured values.

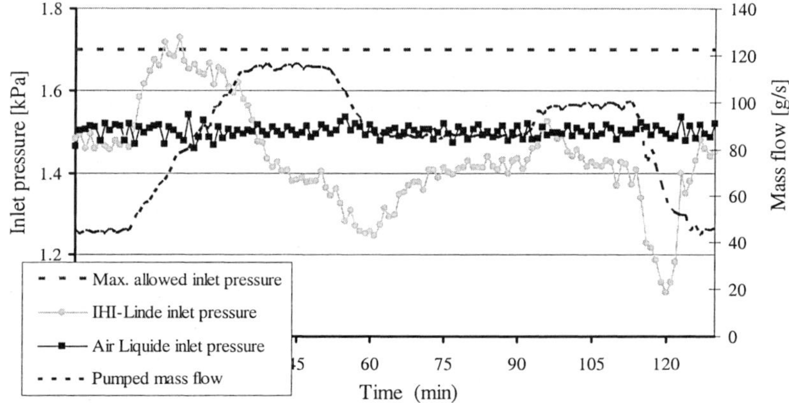

Figure 4 Typical specified transients and corresponding measured values for IHI-Linde and Air Liquide

Both solutions have provided efficient control of the mass flow variation : the suction pressure stays all the time below the specified value of 1.7 kPa. Flow variations above 8 g/s per minute were successfully tested and have shown promising capability of the 1.8 K refrigeration unit to achieve the required turn down range of 3 in less than 15 min.

THERMODYNAMIC EFFICIENCY

The thermodynamic efficiency of the 1.8 K refrigeration units is as expected and guaranteed by suppliers during the conceptual design phase. To illustrate the thermodynamic efficiency, it is usual to calculate the Coefficient Of Performance (COP) which is equal to the overall electrical power consumptions divided by the cooling power at 1.8 K. Table 2 presents, for three steady-state modes, the measured electrical power consumption of the 1.8 K pre-series units and the averaged measured values for the 4.5 K refrigerators [8] absorbing the non isothermal loads from the 1.8 K unit between 4.5 K (header C supply) and 20 K (header D return) as specified. Moreover, by reducing the header D return temperature down to 17.5 K and assuming the same COP for the 4.5 K refrigerator, the best achieved COP for the IHI-Linde 1.8 K unit is as low as 870 W/W in the Installed mode.

Table 2 : Achieved COP (W/W) for the three LHC operation modes

		Installed mode	Normal mode	Low Intensity mode
1.8 K heat loads	1.8 K isothermal heat loads [kW]	2.4	1.6	1.2
4.5 K refrigerator	COP [8] [W/W]	237	277 *estimated*	316
1.8 K refrigeration pre-series units	Non isothermal heat loads [kW]	13.45	11.26	9.33
	Equivalent iso. 4.5 K loads [kW]	7.6	5.5	4.3
	4.5 K ref. input power [kW]	1800	1530	1360
IHI-Linde / AL	1.8 K unit measured input power [kW]	427 / 466	322 / 345	272 / 290
IHI-Linde / AL	**COP [W/W]**	**930 / 940**	**1160 / 1170**	**1365 / 1380**

CONCLUSION

Two 2400 W@1.8 K refrigeration "pre-series units" have been successfully tested at CERN and accepted in 2002 and 2003. Their achieved COP are in the range of 900 W/W with cold compressor isentropic efficiency higher than 70 % for all modes and warm compressor isothermal efficiency around 45 % in sub-atmospheric conditions. Series units delivery at CERN are now finalised for IHI-Linde and are expected in the coming months for Air Liquide allowing then the acceptance tests of the series units.

ACKNOWLEDGEMENTS

The authors would like to thank all colleagues involved in defining, building-up and operating the test station and also the pre-series unit teams from IHI-Linde, Air Liquide and their industrial partners.

REFERENCES

1. Claudet S., Gayet P., Jager B., Millet F., Roussel P., Tavian L., Wagner U., Specification of eight 2400W@1.8K refrigeration units for the LHC, Proceedings of ICEC18 Mumbaï, India (2000)
2. Asakura H., Honda T., Mori M., Yoshinaga S., Bosel J., Kündig A., Kurtcuoglu K., Meier A., Senn A.E., Four 2400 W@1.8 K refrigeration units for CERN-LHC: the IHI/Linde system, Proceedings of ICEC18 Mumbaï, India (2000)
3. Hilbert B., Delcayre F., Maccagnan M., Monneret E., Walter E., Experimental results obtained with Air Liquide cold compression system for pre-series unit for CERN LHC project, paper presented at CEC 2003 Anchorage, USA (2003)
4. Claudet S., Ferlin G., Gully P., Jager B., Millet F., Roussel P., Tavian L., A cryogenic test station for the Pre-series 2400 W@1.8 K Refrigeration units for the LHC, Proceedings of ICEC19 Grenoble, France (2002), p.83-86
5. Claudet G. and Aymar R., Tore Supra and helium II cooling of large high-field magnets, Advances in cryogenic engineering, 35A (1990), p.55-67
6. Rode C.H., Arenius D., Chronis W.C., Kashy D., Kesee M., 2.0 K Cebaf cryogenics, Advances in cryogenic engineering, 35A (1990), p.275-286
7. Kundig A. and Asakura H., Control consideration of multi stage cold compression system in large helium refrigeration plants, Proceedings of ICEC19 Grenoble, France (2002), p.681-686
8. Gruehagen, H. and Wagner, U., Measured performance of four new 18 kW@4.5 K helium refrigerators for the LHC cryogenic system, paper presented at this conference ICEC20 (2004)

The applicability of ultrasonic Oxygen Deficiency Hazard detectors in the LHC accelerator tunnel

Brodzinski K.*, Chorowski M., Gizicki W., Ostropolski W., Riddone G.*

Wroclaw University of Technology, Wybrzeze Wyspianskiego 27, 50- 370 Wroclaw, Poland
* Accelerator Technology Department, CERN, 1211 Geneva 23, Switzerland

The Large Hadron Collider (LHC), will contain about 96 tonnes of high-density helium, mostly located in the underground components of the LHC machine. Some of potential LHC cryogenic system failures might be followed by helium discharge to the tunnel and potential decrease of the oxygen concentration below the safety level of 18 % cannot be excluded [1]. A novel concept for oxygen deficiency detection can be based on measurements of sound velocity in the atmosphere. The paper presents the test results of ultrasonic ODH detection prototype system in radiation environment similar to that predicted for the LHC.

INTRODUCTION

The cryogenic system of the LHC machine is characterized by specific design features, which make it inherently safe. Nevertheless, as it was specified in the preliminary risk analysis [1], some of potential LHC cryogenic system failures might be followed by helium discharge to the tunnel and potential decrease of the oxygen concentration below safety level of 18 % in underground tunnel and caverns cannot be excluded. The LHC oxygen deficiency detection system will comprise of about 200 ODH detectors installed throughout the LHC machine. Standard commercially available ODH detectors usually operate on the galvanic cell principle and they measure directly the oxygen concentration.

A novel concept for oxygen deficiency detection is based on the dependence of sound velocity on the atmosphere composition [2]. The oxygen concentration is derived on the assumption that the gas added to the air is helium. The sound velocity a in a perfect gas mixture is described by equation (1). Figure 1 presents the dependence of sound velocity on the helium concentration in the helium-air mixture, calculated and measured at temperature equal to 300 K. The sound velocity for helium-air mixture depends strongly on the mixture composition and it is about three times higher for pure helium than for the air alone.

$$a = \sqrt{\kappa R T} = \sqrt{\left(\frac{1}{\sum \frac{z_i}{\kappa_i - 1}} + 1\right) \cdot \frac{\bar{R} T}{\sum z_i M_i}} \qquad (1)$$

where:

$\bar{R}$ - universal gas constant
R - gas constant of the mixture
κ - specific heats ratio,
T - temperature,
z_i, M_i, κ_i - molar concentration, weight and specific heats ratio of the i-th component

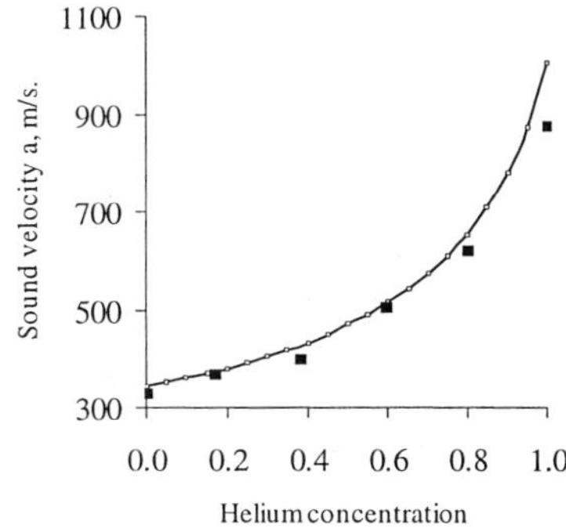

Figure 1. Sound velocity as a function of helium concentration in the air (the line shows eq. (1) calculations and the marks show experimental results)

It results from equation (1) that it is necessary to compensate the temperature influence on the sound velocity. This means that for helium concentration measurement at reasonable accuracy it is necessary to monitor simultaneously the mixture temperature.

EXPERIMENTAL SETUP

In order to qualify the system to be used in the LHC tunnel, the stability of the ultrasonic transducers and electronics has been investigated in the radiation environment, in the conditions similar to those predicted for the LHC. The measurements took place in the TCC2 radiation test zone, situated in the SPS complex at CERN. In the target hall, protons at 400 GeV are dumped on targets to generate secondary particles for the fixed target experiments situated further down stream. The radiation produced is typical of a proton accelerator. A radiation environment in this facility is very similar to that predicted for the arcs of the LHC. For equipment sensitive to particle fluences above E_{cut} = 1 MeV, radiation environments of TCC2 test area and LHC are the same. In case of particle fluences with E_{cut} < 1 MeV, the TCC2 test area has a higher neutron dose ratio than the LHC radiation environment. TCC2 irradiation facility provides good radiation environment for testing electronics, which could by used in the LHC [3].

Two prototypes of the ultrasonic sound velocity measurements system were designed and built at Wroclaw University of Technology. The systems are aimed at detection of helium relieved to the air in a quantity that may cause Oxygen Deficiency Hazard. The systems have been tested both in active (working continuously) and passive (being triggered once a week for half an hour) mode.
The test set-up is shown schematically in Figure 2. Each system consisted of three parts: sound-velocity measuring unit, bunch of supply and signal cables and control unit. The control unit has triggered the measurements, performed the readout of the results and transmitted the data to the PC. The sound velocity measuring unit, placed in the irradiation zone, comprised two piezoelectric ultrasonic transducers (transmitter and receiver) working at 108.5 kHz resonance frequency, signal processing electronics and Ni160 thermocouple. The electronic system was shielded with 12 layers of one-millimeter sheet of lead. The entire system was closed in the aluminium housing. A mirror placed on an extension arm of 800 mm length reflected the signal emitted by the transmitter to the receiver (Figure 3).

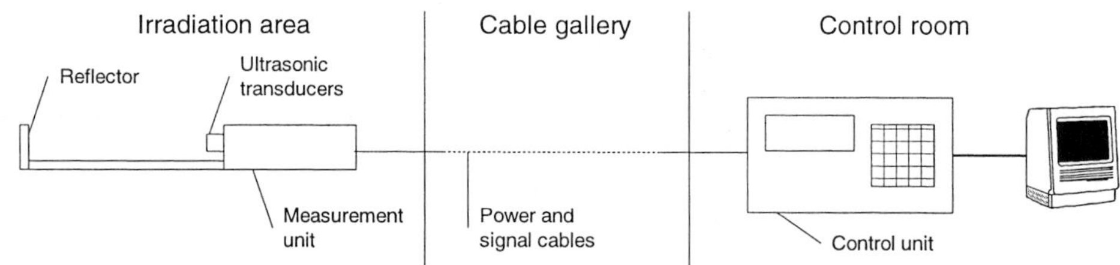

Figure 2. Test set-up schematic view

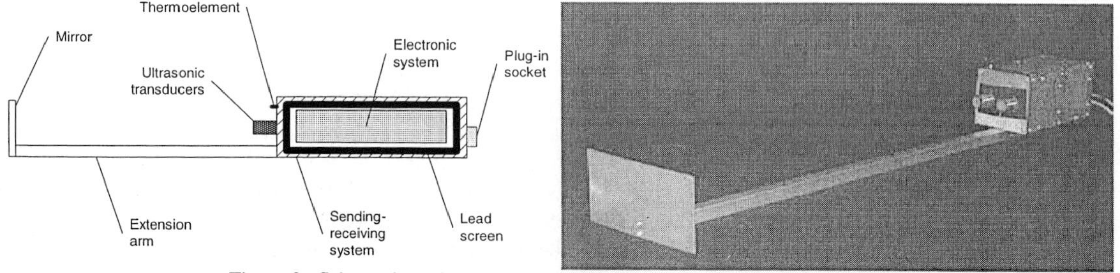

Figure 3. Schematic and general view of the sound-velocity measuring unit

After triggering the measurements by the control unit located in the SPS control room, the electronic system had to emit an ultrasonic burst and receive a signal reflected from the mirror. The received and processed signal was sent back to the control unit to evaluate time of wave propagation along a known distance. Additionally the unit was measuring a temperature of the air in the ultrasonic wave propagation area.

The electronic system of the emitter consisted of quartz stabilized frequency oscillator, modulo-16 counter, logical gate and power amplifier. The receiver path consisted of selective amplifier, amplitude detector and 2-threshold comparator.

The communication between the measuring unit and the control unit was made by a RS485 port. This was a current loop (4 - 20 mA) on electromagnetic interferences.

Figure 4 shows a block diagram of the electronic system of the measuring unit.

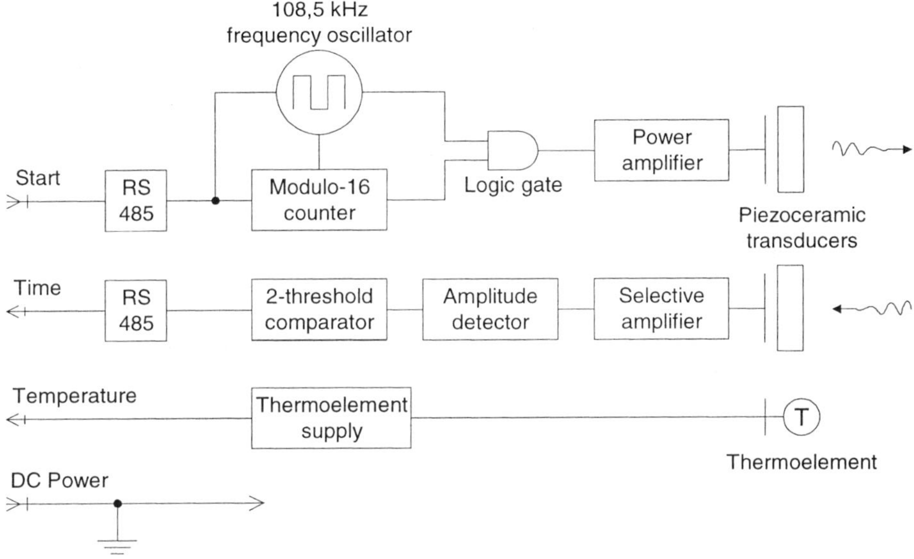

Figure 4. Block diagram of the electronic system of the sound-velocity measuring unit placed in the TCC2 irradiation zone

The electronic system of the control unit was made of a SIEMENS SAB 80537 microprocessor, E PROM chip, LCD, keyboard, communication ports to measuring unit (RS 485) and to computer (RS 232) and a power supply. A dedicated software was installed to start measurements in a specific period of time and to preliminary data handling. The operation of starting the measurement and reading out answers was repeated 100 times and then the microprocessor, using algorithm written in E-PROM chip, was computing average response time.

RESULTS

The results from both systems are given in Figure 5. The measured sound velocities are converted to the base temperature T = 16 °C. The obtained results are very close to each other and both systems gave satisfactory values. The two systems operated reliably for a similar period of time before they failed. Cumulative absorbed doses until failure for each measurement system was of about 70 Gy. The largest measured deviation in the average values of the measured propagation time was 0.08 ms in the case of the passive mode unit. This difference corresponds to helium concentration of about 4 % in the air and the corresponding oxygen concentration of 20.16 % (drop by 0.84 %). In the case of the active mode unit the largest measured deviation in the average value of the measured time of propagation was 0.02 ms (see Figure 6). The difference in this case corresponds to helium concentration of about 1 % in the air and the corresponding oxygen concentration of 20.79 % (drop by 0.21 %). The variations in the measured propagation times correspond to the variations in oxygen concentrations in the air below 0.5 %. Hence the tested systems did not create a danger to trigger false alarms of ODH.

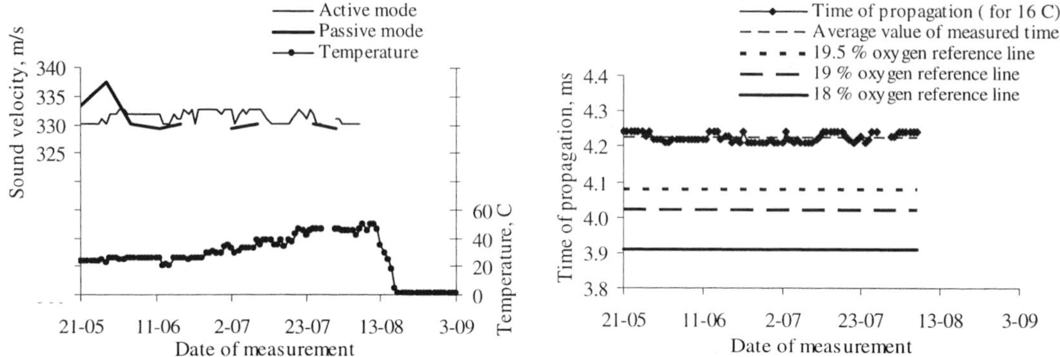

Figure 5. Results of passive and active mode systems, velocity of sound converted to the base temperature $T = 16\,°C$

Figure 6. Comparison between average values of propagation time measured by active system and 18, 19 and 19.5 % oxygen concentration thresholds

CONCLUSIONS

Helium gas presence in the air can be detected directly by measuring sound velocity in the mixture. The performed experiments confirm theoretical dependence of the air/helium mixture sound velocity on the helium content.

Prototypes of the acoustic sensors to monitor helium presence in the air have been successfully tested in the TCC2 radiation test zone in an ionizing radiation environment similar to that predicted for the arcs of the LHC [3]. The accumulative absorbed doses, until the system failures, were of about 70 Gy. As the annual dose in the LHC arcs is of the order of a few Gy/year, this predicts the possibility of working for about 10 years in the LHC tunnel. Preliminary analysis of the failure reasons indicates a defect in piezoceramic crystal, caused by zirconium impurity excitation in the crystal structure. It can be concluded that continuous triggering of the active system did not cause additional danger of the electronics degradation, due to the ionizing radiation.

The oxygen concentration readouts were very stable and there is no danger to provoke false alarms of ODH. The developed technology can be considered to be used in the installations were the oxygen deficiency hazard can be result of helium relief to the air. This non-intrusive and quick way of measurement is an alternative for electrolyte oxygen sensors, which are characterised by a relatively long response time and in presence of helium may manifest erroneous behaviour [4].

ACKNOWLEDGEMENTS

Research performed in framework of collaboration agreement between European Organization for Nuclear Research CERN and Wroclaw University of Technology under the contract K944/AT/LHC. The authors thank Miguel Angel Rodriguez Ruiz for help in the measurements.

REFERENCES

1. Chorowski M., Lebrun Ph., Riddone G., Preliminary Risk Analysis of the LHC Cryogenic System, Advances in Cryogenic Engineering 45B, edited by Quan-Sheng Shu et al., Plenum New York, 2000, pp. 1309-1316.
2. Chorowski M., Gizicki W., Wach J., Application of acoustic tomography in helium-air mixture content monitoring, Proc. ICEC 19, edited by Guy Gistau Baguer, Peter Seyfert, Narosa Publishing House, New Delhi, 2003, pp.671-674.
3. Fynbo C. A., Stevenson G. R., Qualification of the radiation environment in the TCC2 experimental test area, LHC Project Note 235, EST/LEA, CERN, Switzerland (2000).
4. Arenius D., Curry D., Hutton A., Mahoney K., Robertson H., Helium sensitivity in oxygen deficiency measurement equipment, Jefferson Lab, Newport News, USA (2001).

Numerical study of the final cooldown from 4.5 K to 1.9 K of the Large Hadron Collider

Riddone G., Liu L.* and Tavian L.

Accelerator Technology Department, CERN, CH-1211 Geneva 23, Switzerland
*Technical Institute of Physics and Chemistry, Chinese Academy of Sciences, Beijing 100080, China

To simulate and analyze the final cooldown process of a LHC standard cell (106.9 m length) including the filling operation at 4.5 K and the cooldown from 4.5 K to 1.9 K, a mathematical model is proposed and validated by experimental data. In this model, the mass equations for liquid and vapor, the momentum equation for liquid and the general equation for mixed flow are taken into account for the different filling and cooldown phases. As a result of the simulation, the temperature profiles of the cold mass of the cell as well as the evolution of the helium void fractions, filling ratio and HeII position are obtained. The filling and cooldown times have also been estimated, and some characteristics of heat transfer of the HeII bayonet heat exchanger have been presented.

INTRODUCTION

The cooldown operation of the eight LHC sectors from 300 K to 4.5 K has been simulated and analyzed numerically [1]. However, to model and analyze the final cooldown process including the filling operation at 4.5 K and the cooldown from 4.5 K to 1.9 K, a new mathematical model is required due to the two-phase flow and superfluid helium involved. In this paper, as a basis of simulating the whole sector, the mathematical model describing the final cooldown process of a standard cell of LHC is presented.

MATHEMATICAL MODEL

Main assumptions
1) One-dimensional compressible flow for both liquid and vapor phases;
2) One-dimensional wall made of several kinds of materials with zero longitudinal thermal conductivity (due to the laminated structure of the magnet) and infinite transverse thermal conductivity;
3) Stratified flow with a smooth interface between liquid and vapor phases;
4) Saturation state at any position of channel where both liquid and vapor exist;
5) The local liquid pressure increases hydrostatically with the vertical distance from the surface;
6) No jump of pressure across the liquid-vapor interface;
7) Heat input entirely used for the vaporization of liquid helium;
8) Negligible flow between nodes of the cold mass during cooldown.

Flow scheme
The simplified flow scheme for the final cooldown of a LHC standard cell [2] is shown in Figure 1.

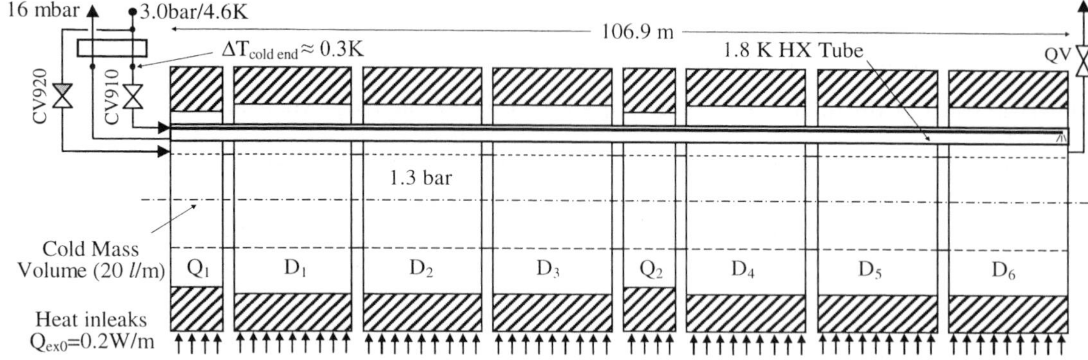

Figure 1 Simplified flow scheme of a standard cell for final cooldown

The final cooldown process is composed of two phases: filling phase at 4.5 K and cooldown phase from 4.5 K to 1.9 K (which also includes the last part of the filling). During the filling phase, the liquid helium is taken from header C and flows into the cold mass volume (the volume of the helium inside the cold mass is 20 l/m) via CV920. The liquid helium receives heat from the cold mass and then the vapor is discharged from QV to the return header. During the cooldown phase from 4.5 K to 1.9 K, two phase helium at about 16 mbar enters the bayonet heat exchanger tube (1.8 K HX tube). The helium inside the 1.8 K HX tube, after vaporizing, is pumped away via the subcooling heat exchanger. During the two phases, distributed heat inleaks Q_{ex0} have also been taken into account.

Models

For the sake of simplicity, only key equations which are applicable to the helium inside the 1.8 K HX tube and to the helium inside the cold mass volume (during the first filling phase only) are listed hereafter.

(1) Mass continuity equation for liquid helium flow:

$$\frac{\partial[(1-\alpha)\cdot\rho_l]}{\partial t}+\frac{\partial[(1-\alpha)\cdot\rho_l V_l]}{\partial x}=-\frac{\dot{m}_{vl}}{A_l+A_v} \tag{1.1}$$

where subscripts l and v represent liquid and vapor phases respectively, A is cross-section area, α is void fraction, V is velocity, ρ is density, $\dot{m}_{vl}$ is net mass transfer rate per unit length between vapor and liquid.

(2) Momentum equation for liquid helium flow:

$$\frac{\partial[(1-\alpha)\cdot\rho_l V_l]}{\partial t}+\frac{\partial[(1-\alpha)\cdot\rho_l V_l^2]}{\partial x}=$$

$$-\frac{\dot{m}_{vl}(V_v-V_l)}{A_{tot}}-\frac{\tau_l S_l}{A_{tot}}+\frac{\tau_i S_i}{A_{tot}}+(1-\alpha)\cdot\rho_l g\sin\beta-\frac{\partial[(1-\alpha)\cdot p_l]}{\partial x}+p_v\frac{\partial(1-\alpha)}{\partial x} \tag{1.2}$$

where τ, S are shear stress and wetted perimeter of the helium with the wall, τ_i, S_i are interface shear stress and interface perimeter between the phases, β is channel inclination (positive or negative), g is acceleration of gravity, p_l is mean pressure of liquid helium.

(3) Mass continuity equation for helium vapor flow:

$$\frac{\partial(\alpha\rho_v)}{\partial t}+\frac{\partial(\alpha\rho_v V_v)}{\partial x}=\frac{\dot{m}_{vl}}{A_l+A_v} \tag{1.3}$$

(4) General energy equation for helium flow:

$$\frac{\partial[\rho_l(1-\alpha)\cdot e_l+\rho_v\alpha e_v]}{\partial t}+\frac{\partial[\rho_l(1-\alpha)\cdot V_l e_l+\rho_v\alpha V_v e_v]}{\partial x}-\frac{\partial[(1-\alpha)\cdot p_l+\alpha p_v]}{\partial t}=$$

$$-\frac{1}{A_l+A_v}\frac{\partial Q_i}{\partial x}-\frac{\tau_l S_l V_l}{A_l+A_v}-\frac{\tau_v S_v V_v}{A_l+A_v}+\frac{Q_h}{A_l+A_v} \tag{1.4}$$

where $e = h + V^2/2 + g\cdot x\cdot\sin\beta$ is the total energy transferred by the convection of unit mass of He, h is specific enthalpy and Q_h is heat flow in liquid He.

SIMULATION RESULTS AND DISCUSSION

Filling phase

During the filling phase, the mass flow rate of 4.5 g/s for normal operation is considered. Figure 2 gives the void fraction in the cold mass volume as a function of time at each cell position. From this figure we can see that, after 14.5 h, about 2/3 of the volume is filled with liquid helium. Due to the position of the inlet and outlet tappings, 2/3 of the volume is the limitation for the filling phase and to complete the filling, the cooling by the 1.8 K HX has to be started. Figure 3 gives the temperature of cold mass as a function of time. This figure shows that before the liquid helium arrives, the temperature of the cold mass increases over time, and that the highest rise is less than 1.6 K. This happens because the helium vapor warms up as it flows past the warm magnets.

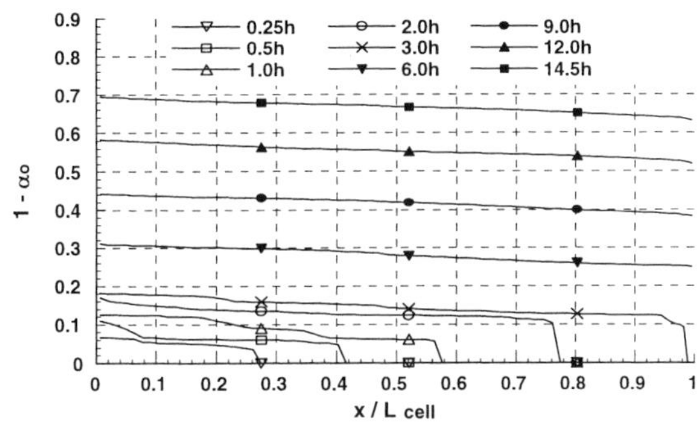

Figure 2 Development of liquid helium during filling of a standard cell (α_o is void fraction of He outside HX)

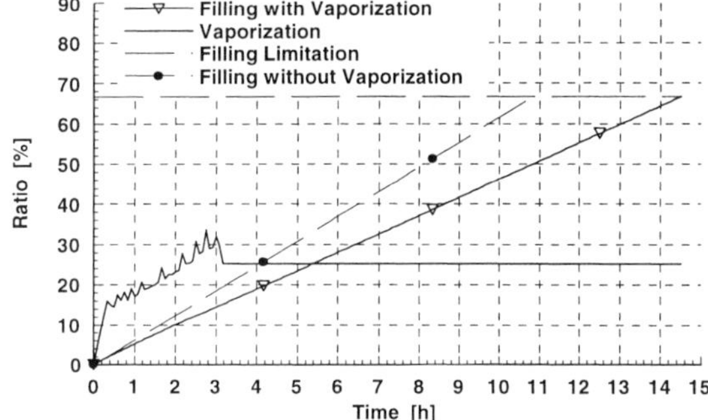

Figure 3 Temperature of each magnet versus time versus filling of a standard cell

Figure 4 shows the overall filling ratio of liquid helium versus time. At the beginning the filling ratio does not rise linearly with time as the vaporization increases gradually (finite number of nodes during discretization explains the small waves of the curve). The vaporization postpones the accomplishment of the filling of the 2/3 of the volume of the cold mass tube from 10.8 h to 14.5 h. As soon as the liquid helium reaches the opposite end of the cold mass (after about 3.1 h), the temperature of cold mass at any cell position reaches 4.5 K. As a consequence, the vaporization remains fixed and the filling curve becomes linear.

Figure 4 Overall filling ratio of liquid helium versus time during filling

Final cooldown phase

In order to compare the calculation results with the test data of String 2 [3], we use the mass flow rate of 7.8 g/s to simulate the final cooldown phase and validate thus the mathematical model. Figures 5 and 6 give the simulation and the test results respectively. The calculation shows that the further cooldown will last about 7 h, which is in good agreement with the test results. Some differences between simulation and test curves can be explained by the assumption No.8. As a consequence of this assumption, the discretization nodes are filled (last part of the filling) and cooled down one by one. This difference may be also explained by the instabilities during tests caused by the large flow. In the simulation only the pumping pressure of 16 mbar was used. A phenomenon can be observed from the computing results is that once the T_λ is reached, the temperature of cold mass reduces to 1.8 − 1.9 K rapidly due to the very high heat conduction in HeII.

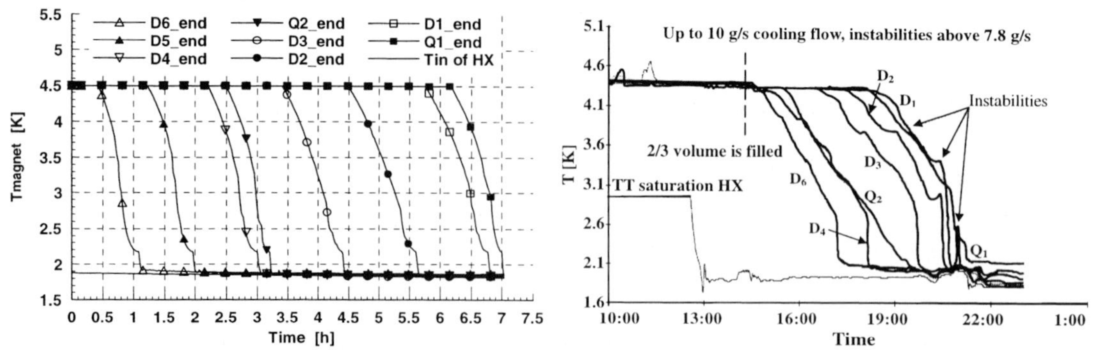

Figure 5 Temperature profiles of magnets during final cooldown of a standard cell

Figure 6 Test results of final cooldown of a standard cell

The normal final cooldown, performed with a mass flow rate of 4.5 g/s, has also been simulated and results are shown in Figure 7, which demonstrates that this phase will take about 13.3 h. It should be recalled that before T_λ is reached, helium in the cold mass tube is cooled mainly by the heat convection from the 1.8 K HX tube. From Figure 7 it can be also noticed that the position of the HeII front inside the 1.8 K HX tube fluctuates with time. This is due to the discretisation used in the mathematical model.

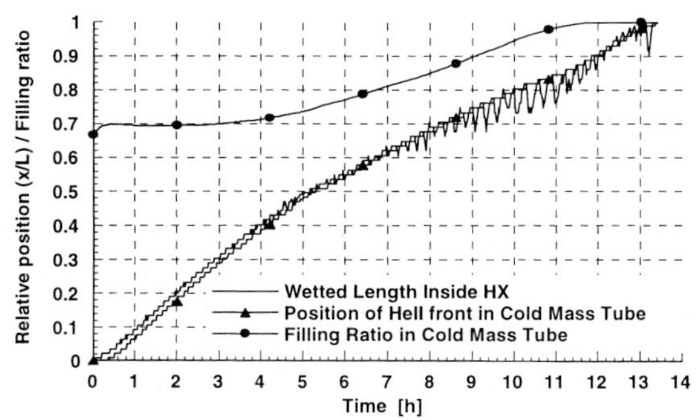

Figure 7 Position of the HeII front and overall filling ratio during final cooldown of a standard cell

CONCLUSIONS

A mathematical model developed for simulating the last phase of cooldown (filling at 4.5 K and cooldown from 4.5 K to 1.8 K) of a LHC standard cell is proposed and the results are in good agreement with experimental data. The filling at 4.5 K up to 2/3 of the cold mass volume takes 14.5 h and the last part of the filling and cooldown from 4.5 to 1.8 K takes 13.3 h. Before T_λ is reached, helium inside the cold mass tube is cooled mainly by the heat convection from the 1.8 K HX tube.

ACKNOWLEDGEMENTS

The authors would like to thank L. Serio for his fruitful technical support.

REFERENCES

1. Liu L., Riddone G. and Tavian L., Numerical Analysis of Coolddown and Warmup for the Large Hadron Collider. Cryogenics, 2003, 43(6): 359~367
2. Lebrun, Ph., Superconductivity and cryogenics for the Large Hadron Collider, in Proc. of Beijing International Conference on Cryogenics, Beijing (2000): 28-34
3. Blanco, E., et al., Experimental validation and operation of the LHC test string 2 cryogenic system, Advances in Cryogenic Engineering 49, American Institute of Physics, 2004 (to be published)

Flow and thermo-mechanical analysis of the LHC sector helium relief system

Chorowski M., Fydrych J., Riddone G.*

Wroclaw University of Technology, Wyb. Wyspianskiego 27, 50-370 Wroclaw, Poland
* Accelerator Technology Department, CERN, 1211, Geneva 23, Switzerland

A simultaneous resistive transition of a LHC magnet sector, which may occasionally occur, will cause rapid helium outflow from magnet cryostats to a special relief system composed of long pipes, buffer volumes and accessories. The paper presents some safety and operational aspects of this system. The results show helium dynamic property distributions along the pipes as well as FEM calculations of thermo-structural stresses in the pipe walls.

INTRODUCTION

The Large Hadron Collider, presently under construction at CERN, will make an extensive use of superconducting magnets located in the underground tunnel of about 26.7 km length and divided into eight sectors. The amount of helium stored in the magnet cold masses located in the sector will be of about 6400 kg. A simultaneous resistive transition of the sector magnets has been defined as an event that may occasionally occur during a life-time of the machine and is called a sector quench. In case of such an event the helium would be vented from the cold masses to cold recovery header D in the cryogenic distribution line (QRL). The capacity of header D will not allow gathering all the helium blown from the sector cold masses. The excess helium would have to be relieved via the pressure valve (PV) and quench line (QL) to two buffer volumes, each composed of four 250 m^3 tanks and located on the surface at both extremities of the sector. To fulfill safety requirements, there is also a possibility to discharge the helium from header D via the safety valve (SV) or two bursting disks (BD) and further through the vent line (VL), which directly opens to the environment. A set of SV and two BD constitute the connection between header D and the VL. The PV, SV and BD are located at the QRL extremities (return module, RM and cryogenic interconnection box, QUI) as shown in Figure 1. Header D, lines QL and VL as well as the accessories compose the helium relief system (see Figure 1).

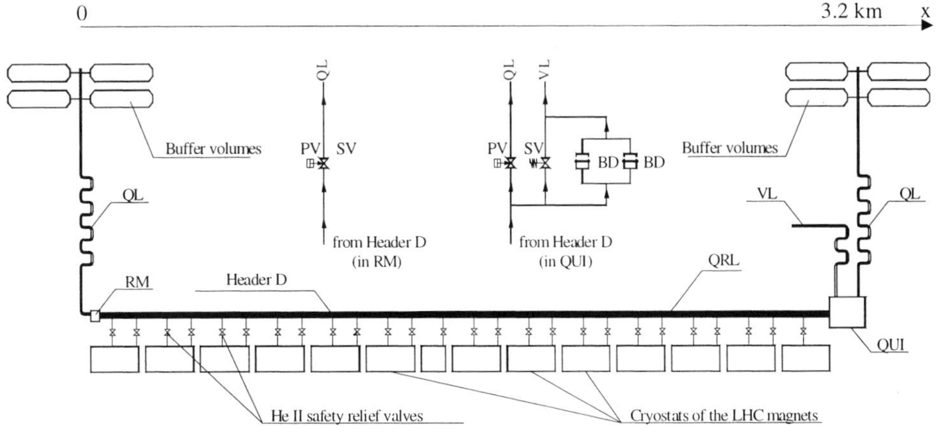

Figure 1 Simplified scheme of helium relief system

The set pressures of PV, SV and BD are equal to 6 bar, 9 bar and 14 bar respectively. If the pressure inside header D exceeds 6 bar, the PV will open and the helium will be evacuated to the buffer volumes. If the pressure still increases and reaches the level of 9 bar, then the SV will open. In case the pressure inside header D exceeds the value of 14 bar, one BD will rupture. If the BD ruptures, most of the helium will be lost and vented into environment.

The main issue of the analysis is to study safety, operational and mechanical aspects of the helium relief system. The QL and VL are not thermally insulated and their nominal stand-by temperature is equal to 300 K. Therefore after cold helium rapid inflow into the lines, their walls will be exposed to impact of thermal and pressure loads.

HELIUM FLOW FROM COLD MASS TO HEADER D AFTER A SECTOR QUENCH

The helium flow into QL (or VL) will be triggered by the pressure growth in header D. After a sector quench a part of the magnetic energy will be dissipated in the cold mass helium and will cause the increase of its temperature and pressure. Based on experimental data gathered from String 1 [1] and String 2 [2] the heat flux to the helium in the cold mass after a sector quench was estimated (Figure 2a). The heat flux will reach the highest value during the first second and it will rapidly go down to 3.7 MW. It will then decrease to about 1 MW during a period of 20 s, and afterwards it will constantly and gradually reduce with a much slower rate.

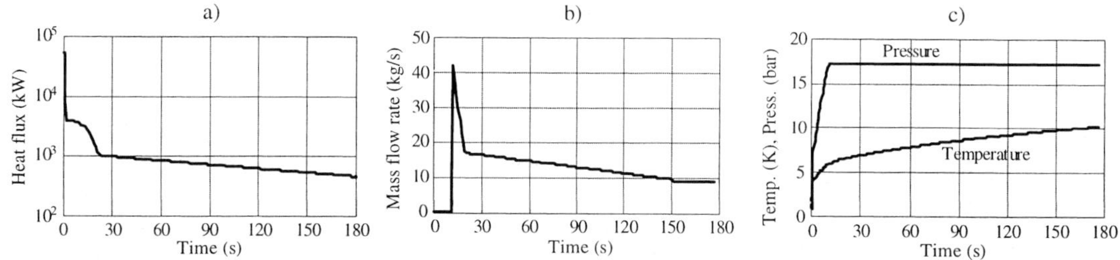

Figure 2 Evolutions of: a) heat flux to the cold mass helium, b) helium mass flow out of the cold mass and c) helium temperature and pressure in the magnet cryostats after a sector quench

The helium internal energy U of the helium in the magnets can be calculated from the equation hereafter:

$$U_2(T_2, p_2) = U_1(T_1, p_1) + \int_{t1}^{t2} P(t)dt - h(T_1, p_1) \cdot \Delta m \; , \tag{1}$$

where P refers to power dissipated in the helium, h is specific enthalpy of helium and Δm denotes the mass of helium that flows out from the magnet cryostats to header D. Equation (1) together with the helium equation of state [3] enables to calculate the helium mass flow rate $q_m = \Delta m / \Delta t$ from the cold mass to header D (Figure 2b) as well as the helium temperature and pressure evolutions (Figure 2c). After 10.5 s following a sector quench, the helium pressure in cold mass will reach the value of 17 bar and HeII safety relief valves will open and will allow helium to flow into header D. The helium mass flow rate will then increase rapidly to 42 kg/s. During the following 9 s the flow rate will sharply decrease to the value of 18 kg/s, and afterwards it will further go down, but much more smoothly and gradually. Based on the calculated helium outflow from cold mass, the helium temperature and pressure evolutions in header D

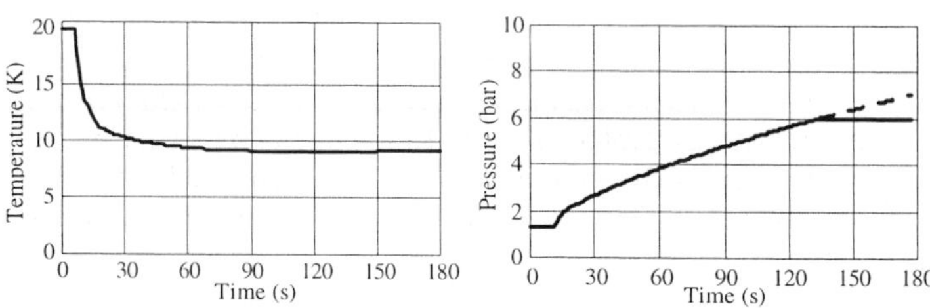

Figure 3 Evolutions of helium temperature and pressure in header D after a sector quench

were calculated taking into account the thermal capacity of header D (Figure 3). The calculation revealed that the lowest value of temperature will be of about 9 K and it will be reached after 85 s following a sector quench. At the same time the pressure in header D will exceed 6 bar and PV will open.

HELIUM OUTFLOW FROM HEADER D

To estimate the values of helium flow rate from header D to QL a complex numerical flow model has been elaborated. The model describes unsteady turbulent compressible and thermal helium flow from header D through QL (about 400 m long) to the buffer volumes. The initial helium temperature and pressure inside header D were equal to 9 K and 6 bar respectively. The initial helium temperature and pressure inside QL and the buffer volumes were equal 300 K and 1.3 bar respectively. The selected model results are shown in Figures 4 and 5 which gives the helium properties along header D and QL for the chosen time periods.

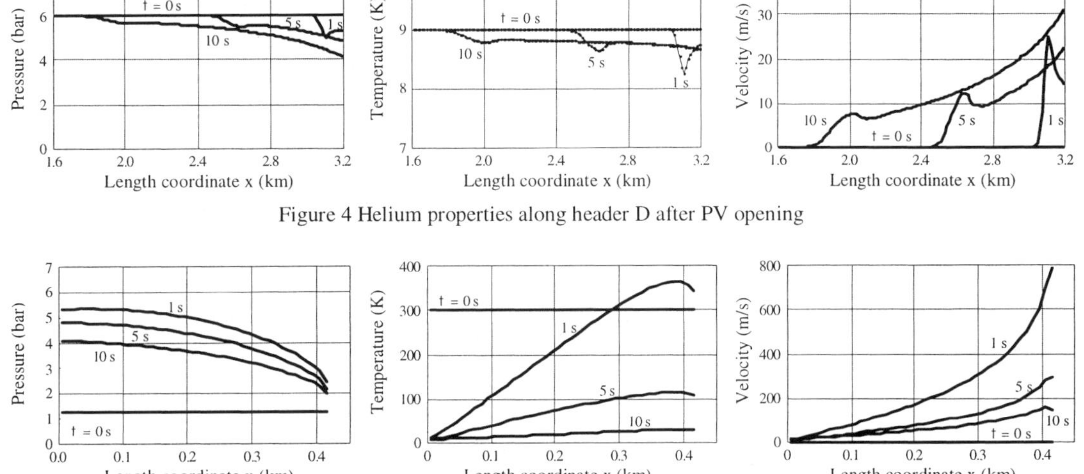

Figure 4 Helium properties along header D after PV opening

Figure 5 Helium properties along QL after PV opening

When the helium flow starts, the pressure inside header D will suddenly decrease by almost 1 bar in the vicinity of the header extremities (Figure 4). It should be emphasized that due to continuous helium flow from cold mass, the pressure in the middle region of header D will slightly increase, and will reach the maximum value of about 2.5 bar above the initial pressure of 6 bar. The local decrease of helium temperature due to the helium expansion will not exceed 1 K.

The evolution of the overall helium mass outflow from header D at the inlet of QL is shown in Figure 6. After the PV opening the flow will rapidly increase to almost 20 kg/s and then drop for a few seconds to about 16 kg/s. This drop is caused by the helium thermal expansion, and when the QL pipe wall is cold enough the flow rate will increase again and reach asymptotically 24 kg/s.

The comparison of Figures 2b and 6 reveals that at about 90 s after a sector quench (5 s after the PV opening) the helium mass inflow to header D starts to be lower by 10% than the potential helium outflow rate. The temperature inside header D results to be stable (see Figure 3), and after the PV opening the mean pressure in header D will be at the level of 6 bar (see the solid line in the graph of Figure 3 giving the pressure evolution in header D). If due to unexpected reasons both PV valves do not open, the pressure in header D would further increase, and it would reach the SV set pressure (9 bar) at about 230 s after a sector quench (value obtained by extrapolating the dashed line of Figure 3).

Table 1 gives the masses of helium that can be kept within the main components of the helium recovery system on the assumption that temperature and pressure reach the maximal values.

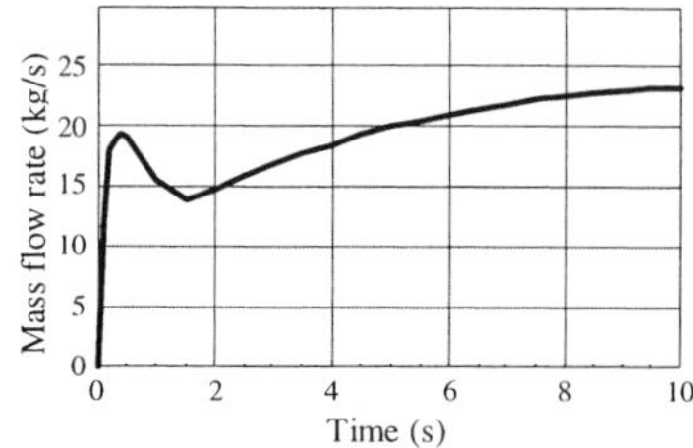

Figure 6 Helium mass outflow
from header D into QL after PV opening

Table 1 Helium masses that can be kept by the main components
of helium recovery system after a sector quench

	Volume m^3	Pressure Bar	Temp. K	Density kg/m^3	Helium mass Kg
Cold masses	43	17	30	26.1	1130
Header D	66	9	13	36.4	2415
QL	8	9	300	1.44	12
Buffer vol.	2000	9	300	1.44	2875
				Total mass:	6432

THERMO-MECHANICAL STRESSES IN PIPE WALL MATERIAL

The results of helium flow model reveal that the worst condition with respect to thermo-mechanical strength of the QL pipe wall will occur at the inlet region of the line just after beginning of the helium flow (see Figure 5). The temperature of the helium blown into the QL will rapidly reach the value of about 10 K, and the pressure will remain slightly below 6 bar. To analyze the thermo-mechanical strength of QL and VL the dynamic FEM calculations of thermo-structural stresses in the walls of the pipes were carried out. In the pipe strength model we assumed forced convection on the inner pipe wall surface, with a convective heat transfer coefficient of about 2400 W/m^{2}K [4], and natural convection on the outer surface, with bulk temperature of 300 K and film heat transfer coefficient of 15 W/m^{2}K [5]. The results of the thermo-mechanical strength analysis are shown in Figure 7.

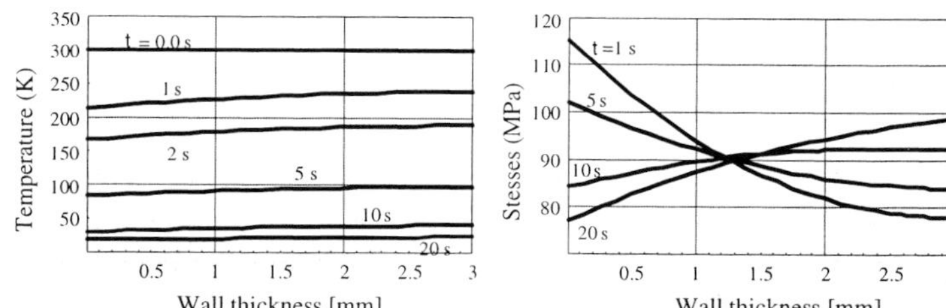

Figure 7 Temperature and stress distributions along the QL wall thickness
after the start of the inflow of cold helium to the pipe

CONCLUSIONS

The LHC sector helium relief system is designed to recover a simultaneous resistive transition of a magnet sector (sector quench) without releasing helium in the atmosphere. The cold buffer constituted by header D can retain up to 46 % of the helium discharged from the magnets.
The stresses in QL and VL pipe wall material have been calculated and they will not exceed 120 MPa. This value is twice lower than the yield strength of the applied material (240 MPa).

REFERENCES

1. Chorowski, M., Lebrun, Ph., Serio, L., van Weelderen, W., Thermohydraulic of quenches and helium recovery in the LHC prototype magnet string, Cryogenics (1998) 38, 533-543
2. Blanco, E., Calzas, C., Casas, J., Gomes, P., Knoops, S., Serio, L., Van Weelderen, R., Experimental validation and operation of the LHC test String2 cryogenic system, Presented at CEC/ICMC (2003)
22-26 September 2003, Anchorage, Alaska
3. HEPAK Version 3.4, code by Cryodata, Inc., P.O. Box 173, Louiaville, CO 80027-0173
4. Barron R.F., In: Cryogenic heat transfer, Ann Arbor, MI, USA (1999) 100 –106
5. Lindon C.T., In: Heat transfer, Professional version, Prentice-Hall, Inc., USA (1993) 563-591

Proceedings of the Twentieth International Cryogenic Engineering Conference
(ICEC 20), Beijing, China. © 2005. Published by Elsevier Ltd

Cryogenic Infrastructure for Testing of LHC Series Superconducting Magnets

J. Axensalva, V. Benda, L. Herblin, JP. Lamboy, A. Tovar-Gonzalez and B. Vullierme

CERN-AT Department, CH-1211 Geneva 23 Switzerland

The ~1800 superconducting magnets for the LHC machine shall be entirely tested at reception before their installation in the tunnel. For this purpose and in order to reach the reliability and efficiency at the nominal load required for an industrial operation for several years, we have gradually upgraded and retrofitted the cryogenic facilities installed in the early nineties for the testing at CERN of prototypes and preseries magnets. The final infrastructure of the test station, dedicated to check industrially the quality of the series magnets, is now nearly complete. We present the general layout and describe the overall performance of the system.

INTRODUCTION

In the 1990's, in order to fulfil the development and preparatory work program for the design of the LHC arc superconducting magnets, hereafter called cryomagnets, CERN has constructed a dedicated cryomagnet test facility [1] in an existing 7200 m^2 floor space hall, so-called SM18. The first cryogenic test bench [2] started to operate in spring 1994, for the test of the first 10-metre long, 18-ton prototype superconducting twin-aperture dipole operating at 1.8 K. In the years that followed, many few prototype-cryomagnets from different versions and suppliers were tested on this prototype test bench.

Early 1997, in order to deal with the updated cryomagnets characteristics and test program (*e.g.* 15-m long, 27.5-t series arc dipoles with new position of hydraulic interfaces), we launched the upgrade of the cryogenic infrastructure in terms of installed test capacity & reliability (redundancy of sub-systems), and the procurement of the modular interfaces between the infrastructure and the cryomagnets to be tested. The final number of test benches was set to twelve, as a maximum number for the available space in the test area forming a reasonably adequate plant for the entire reception-testing in turn and in a 3-year time of the 1722 series high-field, twin-aperture, superconducting arc cryomagnets for the LHC machine.

SYSTEM LAYOUT

The basic objective was to deliver or circulate the required existing or upgraded cryogenic utilities to a modular distribution system of which individual elements (the test benches) offered the widest possible range of independent relative operating conditions, *i.e.* giving the highest possible flexibility for operation. From the experience gained with the intensive operation of the test station with the prototype and the first two preseries test benches [3], we clearly finalized how each type of utility could be shared among the benches, according to criteria such as reliability, built-in redundancy, possible operation errors, controllability, flexibility, efficiency, available space and last but not least, budgetary conditions.

Cryogenic Utilities
The following central utilities or systems are shared, each of them being given, according to the selected criteria, not any, limited or full redundancy: the liquid nitrogen storage, the liquid nitrogen distribution, the cryomagnets Cooldown Warmup System (CWS), the refrigeration plant, the helium storage, the Cryogenic Compound Line (CCL) for helium distribution & return and the 1.9 K pumping facility.

The liquid nitrogen storage consists of two 50'000 litre vessels connected in parallel. The first one is filled up to a daily basis according to the actual consumption of the test station while the second one is maintained full enough in order to ensure an additional 2-day autonomy in case of a delayed liquid nitrogen delivery. A 120-metre long liquid nitrogen transfer network feeds the CWS with up to 500 g/s of liquid nitrogen withdrawn from the selected storage vessel.

The CWS can be schematized as a box including two twin manifold terminals able to circulate to the selected test benches gaseous helium at 80 K for cooldown and at 320 K for warmup of cryomagnets. It consists of 2 twin 100 g/s @ 0.2 MPa helium screw compressors-oil removal units delivering an outlet pressure of 1.2 MPa, connected in parallel to twin cooldown units (CWU1 and CWU2) designed in order to extract a cooling power of up to 120 kW @ 80 K, thanks to the helium circulated through their counter-flow heat exchanger-liquid nitrogen vaporiser. On the other hand, the built-in 25 kW warmup function of both CWU, definitely insufficient for the series test duty, is disabled in order to dedicate both CWU to the cooldown function only, thus doubling the average total cooldown power of these two formerly procured units. The CWS warmup function is now performed by dedicated units consisting of enhanced industrial immersion electrical heaters fitted in vacuum-isolated envelopes. A first 30 kW heater heats up from ambient temperature up to 320 K the gaseous helium delivered by the compressors to be circulated to the cryomagnets while the second 200 kW one heats up from its actual temperature in the range [5 K – 295 K] up to 295 K the circulated helium returned from the cryomagnets. Eventually, a 30 m^3 gaseous helium buffer allows the charge or the discharge of the amount of helium gas in the circulation loop (variable number of connected cryomagnets, at variable pressure and temperature). Most of the above-mentioned subassemblies are integrated by means of a 60-metre long, 4-header, compound vacuum isolated network housed in a 400-mm diameter vacuum envelope and equipped with twelve valve boxes. Each one of the CWS valve boxes is connected *via* a vacuum barrier to one Cryogenic Feed Box (CFB), which represents the complex (cryogenic-mechanic-electrical) interface between the infrastructure and the cryomagnet under test. Thanks to the isolating vacuum, the acoustic vibrations emitted by the inner header passing the warmup flow rate of up to 180 g/s @ 320 K are significantly damped. The very last upgrading of the CWS is scheduled for end-2004, when the two oil removal systems forming the present limitation to the installed circulation capacity will be replaced, in order to handle individual helium flow rates of up to 150 g/s @ 1.0 MPa. This will result in a CWS maximum capacity of 300 g/s @ 0.3 MPa with the two existing compressors only, *i.e.* more than three cryomagnet cooldown and three cryomagnet warmup can be carried out daily. For additional flexibility of operation and redundancy reasons, a third 150 g/s @ 0.3 MPa helium screw compressor-oil removal unit will be installed by the same time.

The CCL consists mainly of three headers housed in a 400-mm diameter vacuum envelope and twelve valve boxes, each of them being connected to one CFB *via* a vacuum barrier. Two of these headers replace the original network of flexible helium transfer lines as well as the former helium phase separator-liquid distribution box (LDB). The main drawback of the former helium distribution system was that it imposed a LDB operating pressure at 0.135 MPa for the withdrawing of the required flow rate of helium from the liquid helium storage. This intermediate pressure stage was creating a loss of efficiency, as the flash gas due to the 150 kPa isenthalpic expansion had to return directly to the refrigerator. Moreover, it would have been impossible to increase the operating pressure of this liquid helium distribution system at a value high enough compatible with the 0.13 MPa cold return pressure imposed by the process of the new 18 kW @ 4.5 K refrigerator cold box [4]. Therefore, the first 30-mm diameter header of the CCL distributes to the twelve CFB up to 60 g/s of saturated liquid helium at 0.14 MPa withdrawn from the 25'000-litre dewar. The dewar operates at 0.16 MPa and is fed through a ~200-metre long, 2-header cryogenic line, by the refrigerator operating mostly in liquefier mode. The second 86-mm diameter header of the CCL collects the gaseous helium returned from the twelve CFB in the [5 K – 90 K] temperature range while ensuring the active cooling of the CCL thermal shield. As a rule, for flow rates of up to 60 g/s at temperatures of up to 50 K, this gaseous helium is sent to the cold return line of the refrigerator, at 0.13 MPa. For higher flow rates or temperatures, the gaseous helium flow rate is partly redirected to the warm return line to the refrigerator, at 0.1 MPa, *via* an electrical heater of a power up to 100 kW. The third 150-mm diameter header of the CCL collects the cold gaseous helium from the pumping ports of the CFB 1.9 K subcooling heat exchangers. Its outlet is connected to the 1.9 K pumping facility.

The 1.9 K pumping facility consists of two parallel branches. The first one, of a pumping capacity of 18 g/s @ 1.2 kPa, 5 K, includes a precompression stage performed by a cold centrifugal compressor unit [5], a combined valve box-32 kW electrical heater [6] and the low-pressure gaseous helium pumping unit installed in 1993 in the neighbouring building. The second branch with a present pumping capacity of 7.2 g/s @ 1.2 kPa includes since 2003 the twin of the former 32 kW electrical heater in series with the twin of the former low pressure pumping unit. This branch is ready to receive a combined valve box-precompression stage unit [7] presently under integration, in order to ensure the full redundancy of the 18 g/s @ 1.2 kPa, 5 K pumping capacity from end-2004 on. Special attention was paid for the design of circuits (use of helium guards, monitoring of oxygen residual content …). The present pumped helium flow rate of up to 25.2 g/s is presently redirected to the warm return line to the refrigerator.

<u>Cryogenic Feed Boxes</u>
From the experience gained with the construction at CERN and operation of the first feeder unit so-called Magnet Feed Box (MFB), and with the turn-key procurement from industry of the two preseries feeder units, so-called the Cryogenic Feeder Units (CFU), we decided to reorganize the production for the procurement of twelve series units, the CFB. We decided to procure the critical components of the CFB separately and to specify their final integration by the contractor of CFB, the control system remaining under CERN responsibility.

As seen above, the CFB is the complex interface between the infrastructure and the cryomagnet under test. Each CFB is precisely positioned and bolted on a metallic inertia structure and fits a formerly allocated space delimited by a platform and other ancillary equipment. A CFB comprises mainly helium pipework and helium-cooled, high-current circuits installed within a thermally insulated vessel. Its subassemblies are the following: a vacuum vessel, an actively-cooled thermal shield covered with multilayer insulation, a helium vessel, a gas-liquid subcooling heat exchanger [8], a set of main current lines, a pair of 15 kA current leads [9], four auxiliary superconducting current lines, four 600 A current leads [10], inner supports, internal and external pipework, a rough pumping set, a turbomolecular pumping set, a helium leak detection set, a purge panel and industrial instrumentation for remote control and monitoring. The innovative features of the CFB are the following:
- a single inner vessel ensures the functions of phase separation, level and flow rate measurement of liquid helium (two measurements based on head of liquid/orifice principle), buffering of liquid helium for feeding the current-lead baths by head of liquid and finally retro-transfer of liquid helium and storage at up to 1.8 MPa of supercritical helium after resistive transition (quench) of the cryomagnet;
- the whole set of 15 kA power circuits can be easily located in two positions over 290 mm in a nearly vertical plane in order to be connected to the main busbars of the two different families of cryomagnets, *i.e.* arc dipoles, short straight sections;
- a dry pump is used for the heavy-duty repetitive rough pumping (~10'000 ppm weigh of humidity in the initially evacuated air) in a self-limiting way (the multilayer insulation tolerates maximum −10 kPa/s) down to 1 Pa of the ~10-m^3 volume of the CFB-cryomagnet assembly vacuum vessel. The dry pump tolerates 5 years of operation without any maintenance;
- the six hydraulic interfaces to the cryomagnet (the only non-welded inner pipes of the CFB-cryomagnet assembly) make use of compact, single-use double metallic sealing. This sealing technique for such circuits operating in vacuum with superfluid helium at 1.9 K allows fast automatic helium leak detection of the flanged connections and even allows the operating at cryogenic conditions with significant −but unusual− leaks, without impairing the mandatory global helium leak measurement of the cryomagnet under test, by means of temporary or continuous pumping of the interspaces;
- the two hydraulic interfaces housing the *in situ*-soldered provisional electrical junctions of the cryomagnet busbars to the CFB main and ancillary superconducting lines, respectively, consist of two especially designed retractable sleeves (diaphragm bellows working under an external pressure of up to 2.5 MPa) for the quickest possible hydraulic connection;
- all the instrumentation of the CFB high voltage or high current electrical circuits, those in warm gaseous helium at 0.13 MPa included, are designed in order to withstand in normal operation and with significant contingency the repetitive electrical insulation tests at up to 3.1 kV of the cryomagnets coils .

CONCLUSION

This test infrastructure is presently operated 24 hours a day, 7 days per week. The achieved peak capacity is two LHC 15-m long, 27.5-t arc dipoles tested daily. Thanks to the control system and to the recently developed dedicated tools for coordinating the work of the involved teams of operators [11], thus taking the maximum advantage of the installed or upgraded systems resulting from this layout, we do believe that the final average capacity will fulfil the requirements for the cryogenic test of all LHC cryomagnets.

Photograph 1 SM18, partial view of infrastructure for testing of LHC superconducting magnets

ACKNOWLEDGEMENTS

The authors would like to express their warmest thanks to Ch. Berthelier, Ch. Gueroult, G. Mouron, B. Poulnais, A. Wiart and particularly to G. Bonfillou and A. Delattre for their participation to this project.

REFERENCES

1. Benda, V., Duraffour, G., Guiard-Marigny, A., Lebrun, P., Momal, F., Saban, R.I., Sergo, V., Tavian, L., Vullierme, B., Cryogenic Infrastructure for Superfluid Helium Testing of LHC Prototype Superconducting Magnets, International Cryogenic Engineering and Materials Conference Albuquerque, NM, USA (1993)
2. Benda, V., Granier, M., Lebrun, P., Novellini, G., Sergo, V., Tavian, L., Vullierme, B., Cryogenic Benches for Superfluid Helium Testing of Full-Scale Prototype Superconducting Magnets for the CERN LHC Project, International Cryogenic Engineering Conference Genova, Italy (1994)
3. Benda, V., Vullierme, B., Schouten, J.A., Experience with a Pre-Series Superfluid Helium Test Bench for LHC Magnets, International Cryogenic Engineering Conference – 18 Mumbai, India (2000)
4. Calzas, C., Chanat, D., Knoops, S., Sanmartí, M., Serio, L., Large Cryogenic Infrastructure for LHC Superconducting Magnet and Cryogenic Component Tests : Layout, Commissioning and Operational Experience, Cryogenic Engineering Conference and International Cryogenic Materials Conference Anchorage, AK, USA (2003)
5. Decker, L., Löhlein, K., Schustr, P., Vins, M., Brunovsky, I., Lebrun, P., Tavian, L., A Cryogenic Axial-Centrifugal Compressor for Superfluid Helium Refrigeration, International Cryogenic Engineering Conference / International Cryogenic Materials Conference – 16 Kitakyushu, Japan (1996)
6. Benda, V., Sergo, V., Vullierme, B., Electrical Heater For Very-Low Pressure Helium Gas, Kryogenika Prague, Czech Republik (1996)
7. Saji, N., Asakura, H., Yoshinaga, S., Itoh, K., Nogaku, T., Bézaguet, A., Casas-Cubillos, J., Lebrun, P., Tavian, L., A One kPa Centrifugal Cold Compressor for the 1.8 K Helium Refrigeration System of LHC, International Cryogenic Engineering Conference – 17 Bournemouth, UK (1998)
8. Roussel, P., Bezaguet, A., Bieri, H., Devidal, R., Jager, B., Moracchioli, R., Seyfert, P., Tavian, L., Performance Tests of Industrial Prototype Subcooling Heat Exchangers for the Large Hadron Collider, Advance in Cryogenic Engineering (2001) Volume 47b 1429-1436
9. Andersen, T P., Benda, V., Vullierme, B., Series Production of 13 kA Current Leads with Dry and Compact Warm Terminals, Paper presented at this conference
10. Andersen, T P., Benda, V., Vullierme, B., 600 A Current Leads with Dry and Compact Warm Terminals, 7th Cryogenics IIR International Conference Prague, Czech Republic (2002)
11. Axensalva, J., Herblin L., Lamboy JP., Tovar-Gonzalez, A. and Vullierme, B., Control System and Operation of the Cryogenic Test Facilities for LHC Series Superconducting Magnets, Paper presented at this conference

Final Testing of the ATLAS Central Solenoid before Installation

Y. Doi[1], T. Haruyama[1], M. Kawai[1], T. Kondo[1], Y. Kondo[1], Y. Makida[1], A. Yamamoto[1]
F. Haug[2], J. Metselaar[2], G. Passardi[2], O. Pavlov[2], M. Pezzetti[2], O. Pirotte[2]
R. Ruber[3], E. Sbrissa[3], H. Ten Kate[3], H. Tyrvainen[3]

[1] KEK, Tsukuba, Ibaraki 305-080, Japan
[2] Accelerator Technology Department, CERN, CH-1211 Geneva 23, Switzerland
[3] Experimental Physics Department, CERN, CH-1211 Geneva 23, Switzerland

The central solenoid is part of the superconducting magnet system of the ATLAS experiment at the CERN LHC collider. It provides a 2 tesla axial magnetic field for the inner 24 m^3 volume centre particle tracker. Design and construction was done in Japan by KEK and Toshiba in collaboration with CERN. Factory tests were made in Japan with the proximity cryogenics in a geometrical arrangement corresponding to the final installation and, a full magnet test. After shipment to CERN the proximity cryogenics has been installed at a surface hall and re-commissioning with load simulations and the instrumentation adapted for radiation hard requirements at the final underground area. The solenoid has recently been integrated in the common cryostat vessel of the liquid argon barrel. Cool down for final surface testing has started. The final control systems architecture and process logics are applied which is tested.

INTRODUCTION

At CERN the 27 km circumference Large Hadron Collider (LHC) is under construction. ATLAS is one of four large particle experiments to exploit the capabilities of colliding beams after commissioning in 2007. It uses a complex array of superconducting toroid magnets and a central solenoid (CS) for momentum analysis of charged particles produced in the 14 TeV proton-proton collisions. The solenoid is housed in the cryostat of the liquid argon barrel calorimeter. This paper summarises the tests of the proximity cryogenics and the solenoid magnet made in Japan and at CERN.

MAGNET AND CRYOGENICS DESIGN

The CS has 5.3 m length with an inner diameter of 2.5 m and provides 2 T at 7.6 kA for the inner tracker. The 5.5 ton magnet is designed for high "transparency" of particles with thin coil and support cylinder (1). In its final arrangement it is placed at short distance in front of the barrel liquid argon detector sharing the same cryostat vessel (Fig. 1). The 44 m^3 volume of liquid argon is cooled to 87 K and its vessel serves as external thermal shield to the solenoid cold mass while at the inner radius an active 40 – 80 K shield is installed. The solenoid cold mass is indirectly cooled with a two-phase flow helium in inclined serpentine shaped cooling pipes welded to the outer support cylinder. The proximity cryogenic system has two major components: the control dewar and the valve unit. In the final underground installation the control dewar will be placed on top of the ATLAS detector on a support structure at a distance of 13 m with respect to the central axis. Cryogenic connections between thesolenoid and the dewar is done with a chimney which houses also the superconducting bus. In normal operation mode the helium refrigerator (2) provides asupercritical helium flow

. After being sub-cooled in the 250 l control dewar it is expanded for two-phase cooling of the cold mass. In case of emergency (refrigerator failure) this stored quantity is directly supplied to the magnet in gravity assisted thermo-syphon cooling mode providing autonomy for slow discharge of the magnet. The valve unit, as second major component, houses the warm control valves, instrumentation and the electronic equipment. As proven by radiation experiments, the expected ionisation and hadron radiation in the detector cavern can severly harm to the electronic equipment. It has, hence, been decided to install this unit at a protected area at 150 m distance in the technical side cavern, thus requiring adaptation of instrumentation.

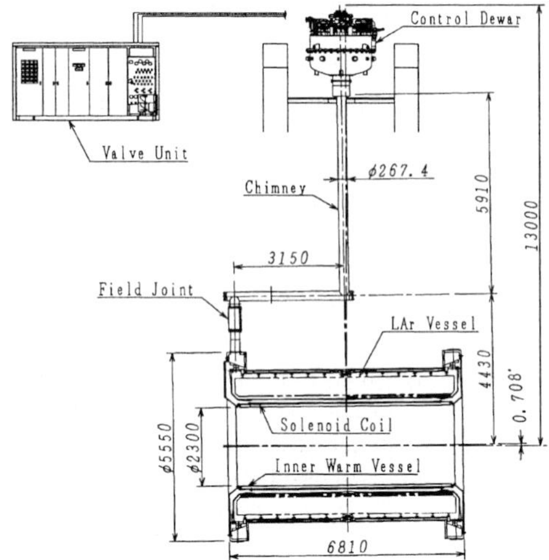

Figure 1 Final arrangement of the proximity cryogenic system for the central solenoid and the common liquid argon barrel cryostat. The valve unit will be installed at 150 m distance in a side cavern.

TESTS IN JAPAN

Proximity Cryogenics

For the first cryogenic test of the proximity cryogenic system at Toshiba works the chimney was lowered into a vertical pit and connected to the control dewar at its top to simulate the vertical arrangement corresponding to the final installation (see fig. 1). At the bottom of the chimney the cooling pipes and the superconducting bus were short-circuited and heaters placed for load simulation. Being the only location permitting such tests prior to final underground installation at the ATLAS cavern, particular attention to the emergency thermo-syphon cooling operation mode was given, which was confirmed.

Magnet Test

A full test was conducted with the magnet in a temporary cryostat and the chimney mounted horizontally. The temporary control system used was based on a Yokogawa Astnex PLC system with programmed operation modes. At 4.5 K operation, pressurised mobile dewars supplied the magnets cooling circuits. The return gas flow was used after phase separation in the control dewar to cool the inner and outer thermal shields connected in series. Performance tests were conducted and quenches induced quenches by heaters. Ultimately 8.4 kA was reached without quench (nominal 7.6 kA). At 4.5 K the magnet static heat load measured was 11 W. Eddy current losses amount to 25 W at nominal ramp rate.

PRELIMINARY TESTS AT CERN

Re-commissioning

After shipment to CERN the proximity cryogenic system was installed in a surface test hall for re-commissioning and a first test series. The horizontally positioned chimney was equipped with a cap containing the shorts for the cooling lines and superconducting bus and, local heaters. Already at an early stage of the new LHC UNICOS standardisation project for industriel control systems, a Schneider Quantum PLC was applied for process control with the adapted functional logics. The current was ramped to 9 kA and quenches simulated for testing the magnet control (MCS) and magnet safety systems (MSS). Thermal load simulations showed an effective refrigeration capacity of 70 W at the terminus of the chimney which is considered to be sufficient for the final cooling of the magnet when compared to the actual thermal budget of 36 W as measured in Japan (magnet static load 11 W, dynamic load 25 W)

<u>Instrumentation adaptation and testing</u>

To protect the sensitive equipment of the valve unit from harmful radiation in the detector cavern its installation at 150 m distance in a non-radiation technical side cavern is needed. In order to limit modifications of the existing instrumentation, an investigation was done to verify if 150 m long capillaries could be used to bridge the measurement pick-up in the control dewar to distant pressure transducers. The test program consisted of signal response experiments in a laboratory set-up and a field test after implementation of the complete proximity cryogenics. To quantify the pressure signal delay, transducers were connected to 150 m long transmission capillaries of different diameter and the signal response compared with a reference. The results indicated that the response time is almost independent from the amplitude of the reference signal, both for absolute and relative pressure transducers, even at very small pressure differences of a few mbars. Figure 2 shows representatively the response time of a relative pressure transducer for three different diameter capillaries at a reference pressure rise of 10 mbar. The fastest response is obtained with the largest tube of 8 mm diameter exhibiting 10 seconds delay. For the application this response time was considered to be sufficient and equivalent capillaries were installed in the proximity cryogenic system for verification of the overall process. The positive results proved the feasibility of this solution which was adopted.

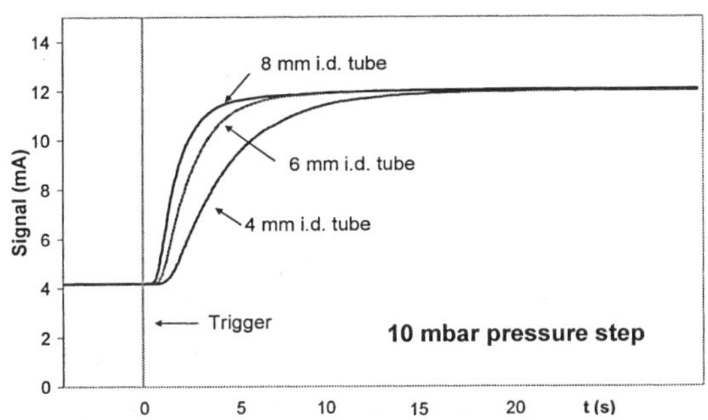

Figure 2 Pressure transmission in 150 long capillaries

Figure 3 Integration of the solenoid in the cryostat

FINAL TEST PREPARATION

After the installation of the two calorimeter detector wheels in the liquid argon barrel cryostat the solenoid was integrated and the chimney connected to the top of the cryostat. Figure 3 shows a photograph of the solenoid and the common cryostat during the integration phase and figure 4 a principle test lay-out. The configuration of the cryogenic controls corresponds to the final UNICOS standard with PVSS supervision. Signal and interlocks are exchanged between the solenoid, the liquid argon system and the MCS, which also use the UNICOS standard. Interlocks from MSS are directly hardwired to the solenoid cryogenic system.

The cooling of the 110 t detector cold mass has started on April 20 at rate of about 0.3 K/h. At detector temperatures close to the final 87 K the solenoid will be cooled down to 4.5 K. All the different operational scenarios foreseen for the final installation will be applied and validated. After filling of the calorimeter with 44 m^3 of liquid argon, the final combined surface testing of the two cryogenic systems will start. Static and dynamic load measurements of the solenoid will be made and the magnet operated with the final MCS and MSS systems. A series of fast discharges and heater initiated quenches will be made. The overall schedule being limited, a fast re-cooling is required after each magnet temperature excursion (up to 70 K). For this purpose an extension to the actual cooling system has been made in using a 10000 liter mobile helium dewar which is connected to the control dewar phase separator of the proximity cryogenics (Fig. 5). For re-cooling liquid helium will be withdrawn from the dewar and injected in the 250 l phase separator to cool the refrigerator J.T. flow going to the magnet and the vaporized helium from the phase separator is taken back by the cold box. By using this facility the overall system is "boosted" and the re-cooling of the magnet to operation temperatures shortened down to a few hours. After the completion of the testing of the calorimeter, the cryostat will be heated up to ambient temperatures and the systems will be ready for final underground installation

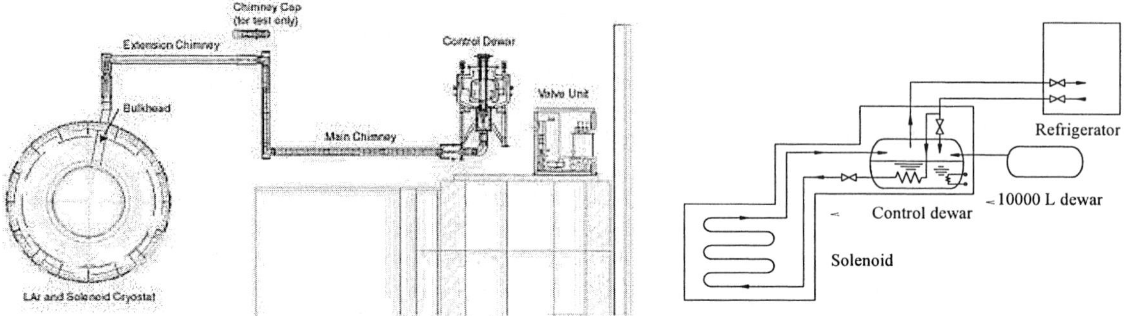

Figure 4 Principle experimental set-up   Figure 5 Simplified flow scheme

ACKNOWLEDGEMENTS

The authors would like to thank all their colleagues at KEK, Toshiba and CERN for their contributions and excellent work to help this project progressing to the advanced current state.

REFERENCES

1) Mizumaki, S., et.all, "Fabrication and mechanical performance of the ATLAS central solenoid", IEEE Trans. on Applied Superconductivity, 2002

2) Delruelle, N., Haug, F., Passardi, G., ten Kate, H., "The helium cryogenic system for the ATLAS experiment", IEEE Trans. on Applied Superconductivity, 2000.

3) Doi, Y., et all. "Performance of the proximity cryogenic system for the ATLAS central solenoid magnets", Proceedings ICEC19, 2002

4) Ruber, R.J.M.Y., et all, "On-surface integration and test of the ATLAS central solenoid and its proximity cryogenics", to be published in IEEE Trans. on Applied Superconductivity, 2003

Control System and Operation of the Cryogenic Test Facilities for LHC Series Superconducting Magnets

J. Axensalva, L. Herblin, JP. Lamboy, A. Tovar-Gonzalez and B. Vullierme

CERN-AT Department, CH-1211 Geneva 23 Switzerland

ABSTRACT

Prior to their final preparation before installation in the tunnel, the ~1800 series superconducting magnets of the LHC machine will be entirely tested at reception on modular test facilities using dedicated control systems. The test facilities are operated by teams of high-skilled and trained operators. This paper describes the architecture of the control & supervision system of the cryogenic test facilities as well as the tools and management systems developed to help in real time all involved operation teams in order to reach the required industrial production level.

INTRODUCTION

The LHC ring consists mainly of ~1800 twin-aperture superconducting magnets made by the industry. The 1248 15-m long arc dipoles and the 474 short straight sections quadrupoles, hereafter called cryomagnets, shall pass a sequence of conformity tests. The final acceptance tests, including the cryogenic ones, are conducted at CERN by the means of a dedicated infrastructure [1], before final preparation and installation of the cryomagnets in the tunnel. The ACR group/Magnet Test section of the LHC Accelerator Technology Department is in charge of providing the cryogenic infrastructure and of running the cryogenic facilities required for such cryogenic tests.

Cryomagnets are designed in line with the LHC machine requirements, not with the test ones and apart from the constraints linked to intensive sharing of cryogenic utilities and budgetary conditions, another very strong constraint is the time allocated for these tests: most of the 1800 LHC cryomagnets have to be tested over a 3-year period. This challenging objective has lead CERN to setup industrial methods and support, the current target being to reach a peak test capacity of up to 3 cryomagnets per day. This paper describes briefly the typical test sequence and how CERN has organized and is running the concurrent operation of 12 cryogenic test benches, with the help of the control and supervision systems.

CONTEXT

For the complex and numerous jobs repeatedly required to perform the complete cryogenic test of LHC cryomagnets, CERN has taken the approach of splitting the tasks by area of expertise, i.e. mainly mechanics, magnetic & electrical measurements and cryogenics, to 3 teams. For the mechanical tasks, mostly those of (dis)connection of cryomagnet electro-mechanics interfaces, the activities are outsourced in the frame of results oriented Work Packages (WP) of an industrial contract. Each WP is described in an engineering specification document in which the details of the tasks as well as those of the quality control are exhaustively set. The operation of magnetic & electrical measurements, which has to comply with strict procedures, is conducted by CERN and supported through a collaboration program with India. Last but not least, the operation of cryogenics, which relies on predefined procedures for acting on modular or shared infrastructure, is outsourced in the frame of a dedicated resource-oriented WP of an industrial contract in the field of cryogenics, the technical management of which being closely followed-up by CERN. CERN insures the training of operators, as well as the on-the-job transfer of know-how to the 3 teams, thus leading to guarantee the accurate execution of the WP's and the correct application of the procedures, respectively. Moreover, we have setup software tools to monitor and improve, and in some extent formalize, the communication between the 3 teams involved in the execution of the interleaved tasks of the whole cryogenics tests sequence, in order to make the operators working together in the most efficient way.

CRYOMAGNET TEST SEQUENCE

The cryomagnet is equipped with a return box and, when required, with two anti-cryostats [3]. It is transported by the means of a special wire-guided transport vehicle to the selected test bench. Then it is positioned, aligned and levelled on the supporting structure of the Cryogenic Feed Box (CFB). The power cryomagnet electrical interfaces are *in-situ*-soldered and mechanically secured. The retractable sleeves housing the electrical joints are stretched out and locked in closed position. The six flanged hydraulic connections are tightly bolted. An automatic helium purge of the just-connected hydraulic circuits and a helium leak check on tightness of their double-sealed interfaces are performed. The dismountable thermal shield and the multilayer insulation (MLI) blanket surrounding the connection area are fitted. Lastly, the vacuum sleeve is pulled out and tightly bolted. At this stage, the vacuum enclosure starts and runs down to ~1 Pa, then the secondary pumping down to ~0.1 Pa and a global helium leak rate measurement is done, while instrumentation cables are plugged and electrical measurement & protection circuits checked on conformity. The cooldown to ~90 K is done by means of helium circulated at 80 K. Subsequent cooldown to ~4.5 K and filling are done through a 2-phase helium distribution, and lastly, down to superfluid helium temperature by the means of the 1.9 K pumping facility. The cryomagnet coils are maintained at 1.9 K for the duration of the electrical measurements and of the magnetic ones, if any. It is re-cooled down to 1.9 K in case of training quenches (resistive transitions) or of a thermal cycle, ramped up back to nominal current, and lastly quenched at lower energy. The so emptied cryomagnet is warmup up to 295 K and the vacuum enclosure is pressurized up to atmospheric pressure with nitrogen gas. The vacuum sleeve is pulled away, the MLI & thermal shield are removed. The helium circuits to be disconnected are automatically purged with nitrogen gas. Hydraulic and electrical interfaces are disconnected. The tested cryomagnet is then transported back for stripping off (test tooling & configuration). A typical summary of 9 concurrent test sequences running in cog wheeling over 9 test benches (TB) is given in Table 1 hereafter.

TB	Period n	n+1	n+2	n+3	n+4	n+5	n+6	n+7	n+8
1	Install	Setup	To 90 K	To 1.9K	Test 1.9 K	Test 1.9 K	Test 1.9 K	Warmup	Remove
2	Test 1.9 K	Test 1.9 K	Test 1.9 K	Warmup	Remove	Install	Setup	To 90 K	To 1.9K
3	Remove	Install	Setup	To 90 K	To 1.9K	Test 1.9 K	Test 1.9 K	Test 1.9 K	Warmup
4	To 1.9K	Test 1.9 K	Test 1.9 K	Test 1.9 K	Warmup	Remove	Install	Setup	To 90 K
5	Warmup	Remove	Install	Setup	To 90 K	To 1.9K	Test 1.9 K	Test 1.9 K	Test 1.9 K
6	To 90 K	To 1.9K	Test 1.9 K	Test 1.9 K	Test 1.9 K	Warmup	Remove	Install	Setup
7	Test 1.9 K	Warmup	Remove	Install	Setup	To 90 K	To 1.9K	Test 1.9 K	Test 1.9 K
8	Setup	To 90 K	To 1.9K	Test 1.9 K	Test 1.9 K	Test 1.9 K	Warmup	Remove	Install
9	Test 1.9 K	Test 1.9 K	Warmup	Remove	Install	Setup	To 90 K	To 1.9K	Test 1.9 K

Table 1 Typical arrangement of 9 concurrent LHC cryomagnet test sequences

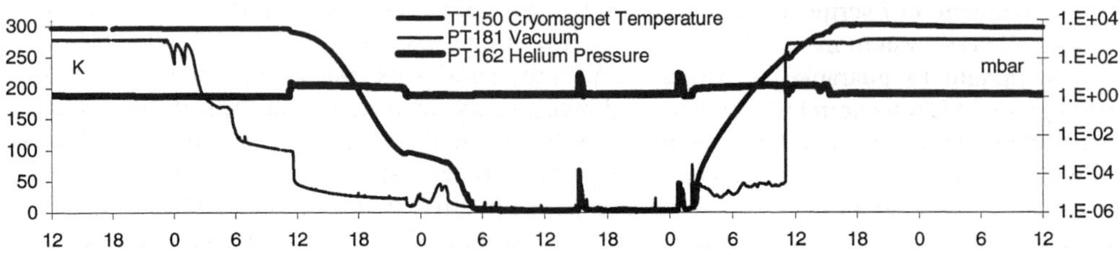

Fig 1 Real chart of a standard cryogenic test sequence

CONTROL & SUPERVISION SYSTEMS OF THE CRYOGENIC TEST FACILITIES

The control system is designed and built like any industrial one: PLC units and Supervisory Control and Data Acquisition (SCADA) applications running on PC. A network of PLC's linked together by TCP/IP and Profibus™ protocols control the processes of the test facilities. This system of PLC's hosts ~850 analog inputs, ~1650 digital inputs, ~350 analog outputs and ~1200 digital outputs. The PLC system is the first logical level of control, it handles all the real-time tasks, runs the ~380 PID's controls of all regulated objects, assure the safety and the reliability of the installation through programs sequences and background processes and transmits, using the TCP/IP protocol, the state and the physical measurements values to the second layer, the supervision system (SCADA) which by turns, displays the animated synoptics, the control, trends & alarms windows and sends back all the control parameters and the operators' commands. An exhaustive set of flags and alarms centralized and summarized by system, helps the operators in diagnosing process deviation or possible equipments malfunctioning. The majority of the program parameters are customizable through the SCADA application to allow a tuning of the process. The SCADA applications are running on 8 PC, 6 for the 6 clusters (pairs of CFB) of neighbouring test benches and 2 for the utilities of the cryogenic test station.

Whereas all the PLC hardware is made with off-the-shelf components, the software is deeply tailored for our applications and was totally designed and written by CERN. The heart of the system is a central PLC, the main tasks of which are to control the processes of the centrally available utilities (liquid helium distribution & return, circulated helium for cooldown & warmup) and to run the master software of the test facilities, the so-called Cryogenic Test Handling (CTH) software. Connected to this central PLC, 4 individual PLC control separate pumping, circulation and compression units of the test facilities, and 12 identical PLC & software control the CFB. The CFB program is structured into phases, themselves divided into steps, resulting from the test sequence of the cryomagnet. Transitions between steps and/or phases arise when predefined values and/or states are reached, some of them requiring the acknowledgement of the operator.

The numbers of benches (12), of process phases (21) and of utilities of limited capacity (4), i.e. liquid helium & return, subcooling, cooldown and warmup, have naturally introduced the concept of "priority" among the test station. It is essential that all cryomagnets are assigned, at any time, an exclusive priority value in the range [1-12] compatible with a logical and efficient production flow. Practically, the magnetic measurement team updates and transmits its priority list accordingly with its tests objectives and with its own shared resources. The cryogenic operation team then acknowledges this request by updating the present input priority list to CTH. The CTH priority module, which discriminates the present phase/step of each of the 12 CFB, e.g. CFB of cryomagnets powered at 1.9 K, so their respective requirements with respect to the 4 shared utilities, reacts instantly by recalculating the weight of the corresponding 4 utilities set points of each of the 12 CFB. As a result, the production flow can make use of the full installed capacity of the 4 above-mentioned utilities in a self-limiting way.

Like the PLC software, the development of the SCADA applications was done by CERN. Both the PLC programs and the SCADA applications were designed and developed jointly, so that we have a complete compliance between the databases, objects, etc... of the PLC and the SCADA. This conformity is very timesaving when upgrading and maintaining the overall control system.

Additional improvements to the supervision systems have been developed through a central database stored on a web server. A set of web pages gives access to the on-line information about the status of the production, anywhere on the CERN's web intranet. Reports on current or previous tests (test duration, statistics, etc...) can be queried. To achieve this, the SCADA system is used as an event generator launching a collection of scripts updating the central database. The refresh time is a few seconds, as the load on the SCADA system mainly limits this updating frequency; nevertheless the data refresh occurs at least every minute to guarantee consistent synchronization with the actual production status. This methodology (SCADA → scripts → database) allows an easy maintenance, any change or upgrade can be implemented and tested independently. A bridge with a CERN's reference database –Manufacturing and Test Folders (MTF)– has also been implemented into the SCADA application of the CFB for downloading the parameters of the anticryostats [3] in an automated way and writing them back to the dedicated PLC. An e-mailing of important events or errors is also implemented through this system of scripts.

OPERATION OF THE CRYOGENIC TEST BENCHES

The cryogenic test station is operated 24 hours per day, 7 days per week, by a team of 2 or 3 operators. In addition to its field activities, mainly to witness the liquid nitrogen deliveries, the watch patrol for installations, the follow-up of helium leak measurements, its mandate is to complete and/or check the correct execution of the concurrent test sequences from the control room by the means of the SCADA with the help of the Task Tracking System, another web database application beside the database, automatically fed by the control system. The Task Tracking System (TTS) has been implemented to standardize the communications between the 3 –mechanics, magnetic & electrical measurements and cryogenics– teams and to delegate to it the repetitive action of broadcasting information. Technically, the TTS is housed in the same web server as mentioned in the above paragraph. All the key production tasks, grouped by area of expertise, are rationally ordered in a to-do list following the test sequence. This to-do list of tasks is the backbone of the TTS. The TTS is field-oriented operator production follow-up application, it means that the data recorded in it is entered by the operators –via password protected web forms– on the field, right after the completion of their current task. Every test bench has its TTS list where the teams have to record the progress of their job.

When the "hand" passes to the next team in the to-do list, this one is automatically and immediately notified by a SMS message sent by the TTS, that its action is requested (the bench and actions are recalled in the SMS message and, of course, on the TTS main web page, accessible elsewhere on the CERN's intranet). The TTS is used from the arrival of a cryomagnet on a bench up to its removal. By this mean, the history of the tasks achieved on a bench –and on a cryomagnet– is recorded into the TTS database so that data can be later queried for statistical purpose. The TTS guarantees that the production information is automatically and always recorded and forwarded on time. The safety and quality functions are also present in the TTS, because the system requires that the tasks are strictly and chronologically executed in the way they were defined in the to-do list. For example, the opening of the cryomagnet's helium circuit by the mechanical team cannot be done while the cryogenic operators have not signed the safety and lock tasks.

CONCLUSION

Significant improvements have been implemented in the control & SCADA systems. We now have stable foundations for reliable and efficient operation of the cryogenic facilities for the test of series cryomagnets. The present production of the test plant demonstrates that a messaging system embedded in a web-based follow-up system can help the coordination of 3 teams from different area of expertise. More experience is still needed for outputting meaningful statistics and to identify hidden bottlenecks in the test production flow. For that purpose additional analysis work has do be done on the recorded data.

ACKNOWLEDGEMENTS

We would express our thanks to A. Raimondo, B. Poulnais, V. Chohan and G-H. Hemelsoet, for their help and advices in this project.

REFERENCES

1. Axensalva, J., Benda, V., Herblin, L., Lamboy, J-P., Tovar-Gonzalez, A. and Vullierme B., Cryogenic Infrastructure for Testing of LHC Series Superconducting Magnets Paper presented at this conference
2. Momal, F., Bienvenu, D., Brahy, Lavielle, D., Saban, R., Vullierme, B., Walckiers, L., A Control System based on industrial Components for Measuring and Testing the Prototype Magnets for LHC, CERN-AT Department, CH-1211 Geneva 23 Switzerland
3. Dunkel, O., Legrand, P., Sievers, P, A Warm Bore Anticryostat for Series Magnetic Measurements of LHC Superconducting Dipole and Short Straight Section Magnets, 2003) Cryogenic Engineering Conference and International Cryogenic Materials Conference, Anchorage, AK, USA (2003)

Studies on cooling of the TOTEM particle detector at the LHC

F. Haug

Accelerator Technology Division, CERN, CH-1211 Geneva 23, Switzerland

TOTEM is an experiment at the CERN LHC collider to measure the total p-p cross section and elastic scattering of protons. For the detection of the elastically scattered particles at very small angles with respect to the beams edgeless silicon detectors will be contained in 36 small vacuum vessels called "roman pots" to be installed inside the LHC beam vacuum pipe. The final -not yet fixed- operating temperature is expected to be between 130 K and 220 K. The preferred cooling method applies a combination of cryogenic heat pipes and pulse tube refrigerators.

INTRODUCTION

The Large Hadron Collider (LHC) is CERN's major project. The commissioning of this 27 km circumference superconducting proton-proton collider with its four large scale experiments ATLAS, CMS, LHC, ALICE is expected for 2007. The fifth experiment TOTEM (1) is particular in its composition with inelastic detectors in the forward region of CMS and with numerous small sub-detectors extending over large distances along the LHC tunnel. Figure 1 gives a principle lay-out of the LHC with its five experiments and figure 2 of the TOTEM experiment. The distant sub-detectors called "roman pot detectors" require cooling. Different cooling methods have been studied and their practicality investigated for the LHC collider environment. The adopted solution is described.

THE TOTEM PARTICLE EXPERIMENT

TOTEM measures the inelastic proton-proton collision rate and the elastic scattered protons up to 220 m downstream of the intersection point 5 (IP5), (Fig. 1). It consists in the very forward of a series of small sub-detectors ("Roman Pots") placed inside the beam pipe and of ambient temperature tracking telescopes T1 and T2 close to CMS. The Roman Pot stations RP1, RP2, RP3 are installed in the collider tunnel symmetrically from IP5 on either side at 147 m, 180 m and 218 m (Fig. 2). Each station is identically composed of two units being 4 m apart. A unit is made of two Roman Pots that move vertically and one that moves horizontally (Fig. 8), (2).

An individual roman pot consists of an evacuated vessel holding a set of 10 silicon strip detectors close to the thin-walled 0.2 mm thick particle window. The ± 45 degrees strip orientation of the detectors form an x-y coordinate system for two-dimensional resolution (Fig. 3). With a high precision step motor they can be retracted from the beam at injection and then approach the beam to 1 mm in stable run conditions.

The Roman Pots are vacuum sealed from the beam pipe. The dissipated power is approximately 20 W. Detailed cooling requirements and temperatures for the system depend on the final design solution of the silicon detectors. In case the radiation hardness of the detector is an issue operating temperatures of 130 K are required to make use of the Lazarus effect (a phenomenon of natural self repairing of the lattice of the silicon chips disrupted by high-energy particles). In addition low temperatures would reduce parasitic edge surface currents of the edgeless silicon strips. A flexible cooling system able to operate at different temperature is needed.

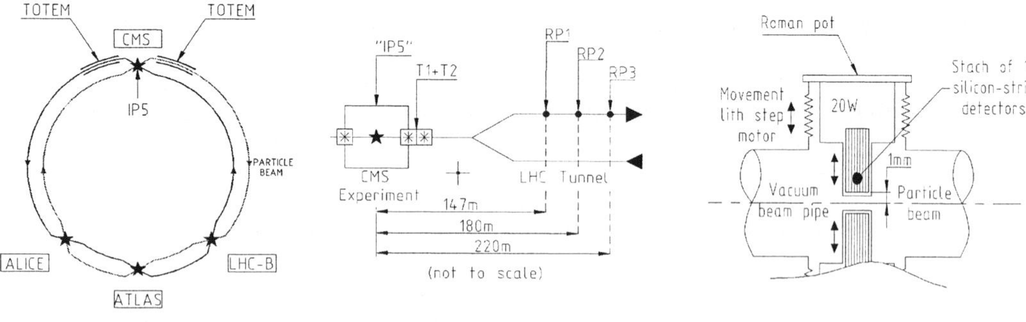

Figure 1 Lay-out of the LHC experiments

Figure 2 Lay-out of TOTEM

Figure 3 Simplified Roman pot assembly

COOLING SYSTEMS PRINCIPLES

Several cooling principles and schemes have been studied as potential solutions which are discussed.

System with a micro-pump for coolant circulation

The RD39 collaboration at CERN has built a system to cool edge-sensitive silicon detectors of similar type as are the ones of TOTEM. A Gifford McMahon refrigerator cools a secondary closed loop with two-phase argon at 130 K circulated with a cryogenic micro-pump (3), (Fig. 4). The flow is split and distributed to micro cooling tubes of 0.3 mm i.d. attached to the detector mother boards for direct cooling. The high heat transfer obtained due to the evaporative direct cooling are balanced by the requirements for permanent sub-cooling of the suction flow to avoid cavitation in the pump.

System with coolant circulation by compressor flow

A proposal was made with a single, non-condensable fluid (3) (Fig. 5). A central compressor supplies the individual pulse tube refrigerators (one per station). A small by-pass flow is pre-cooled at the cold head before extracting the dissipated heat in the detector. Counter-flow heat exchangers are used. With helium as working fluid the system can operate at different temperatures. Some complication is due to the parallel fluid distribution to the sub-systems which may require valves and process control.

System with J.T. refrigerator

The local cooling stations consists of a counter-flow heat exchanger and a J.T. orifice expansion where part of the fluid is condensed. The two-phase mixture is pushed through the cooling pipes in the roman pot by the compressor flow. The systems advantageous of simple circuitry and two-phase cooling is balanced by the proportionally small condensate quantity produced at expansion and the low efficiency of J.T. refrigerators requiring high pressure ratios and the limited operation temperature range. This could partially be overcome by using fluid mixtures.

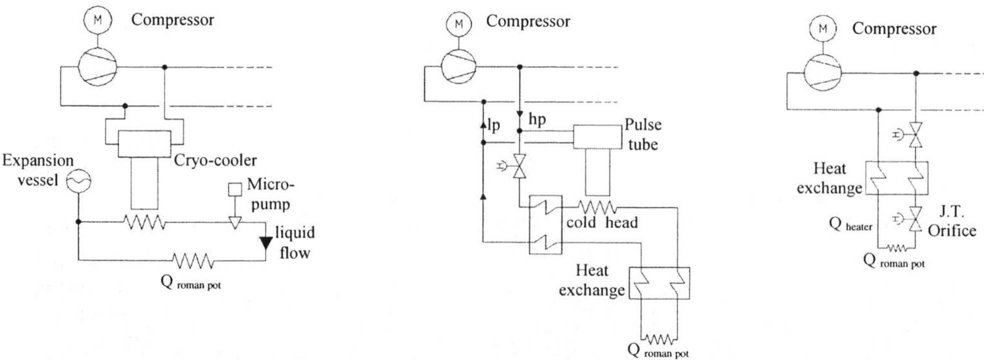

Figure 4 System with micro-pump

Figure 5 System with compressor circulation

Figure 6 System with J.T.-refrigerator

<u>Discussion</u>

The described systems apply active circulation of a coolant fluid to extract the dissipated heat of the Roman Pots and a "network" of pipes are needed for coolant distribution requiring instrumentation and process control. For the hostile LHC radiation environment with difficult access and repair options these comparatively complex systems might be vulnerable. The small diameter pipes may be obstructed by solids. Direct circulation of a fluid in the roman pot bears the potential risk of leaks with subsequent contamination of the beam vacuum pipe in case the thin walled particle window brakes. The system based on heat pipes, as shown in figure 7 and described in the following chapter, tries to overcome these limitations. This preferred solution is driven by reliability aspects.

THE PROPOSED COOLING SYSTEM

The design concept is based on passive heat transfer with a minimum of active elements. Each of the 2 x 3 roman pot stations (RP1, RP2, RP3) are identically equipped with a modular cooling system consisting of a central pulse tube refrigerator and six heat pipes connecting to the six roman pots. Figure 7 shows the principle scheme. The dissipated heat in a roman pot is collected by thermal contacts and transferred to the outside by conduction of bulk metal to the evaporator of the connected heat pipe. The 2.5 m long heat pipe transfers the heat to a central cryostat where it is extracted by re-condensation of the working fluid at a heat sink. The cryostat contains the six heat pipe terminals connected to the cold head. The operating temperature is controlled via a film heater attached to the cold head and the excess cooling capacity is dissipated. The refrigerator compressor will be installed in the vicinity in the tunnel. Figure 8 shows the lay-out of a roman pot station with the mechanical devices and the modular cooling system.

As the heat pipe principle is based on a self-sustaining continuous evaporation and condensation process of a working fluid with a very effective thermal conductivity it is well suited to stabilise the temperature of the roman pot connection at a given value even at distance. Operation and transfer of heat starts as soon as a heat source and sink are applied. The heat pipe inner wall is lined with a wick to enhance capillarity and radial heat transfer and the temperature gradient in the fluid phase is expected to 1-2 K over its entire length of 2.5 m. The pulse tube refrigerator used during the current development phase is based on a design for a 165 K application (6). Its cooling capacity is 140 W at 130 K which can be increased with a larger compressor. Helium is the cycle gas and the operating temperature can be varied from below 100 K to 250 K which is useful during the development phase.

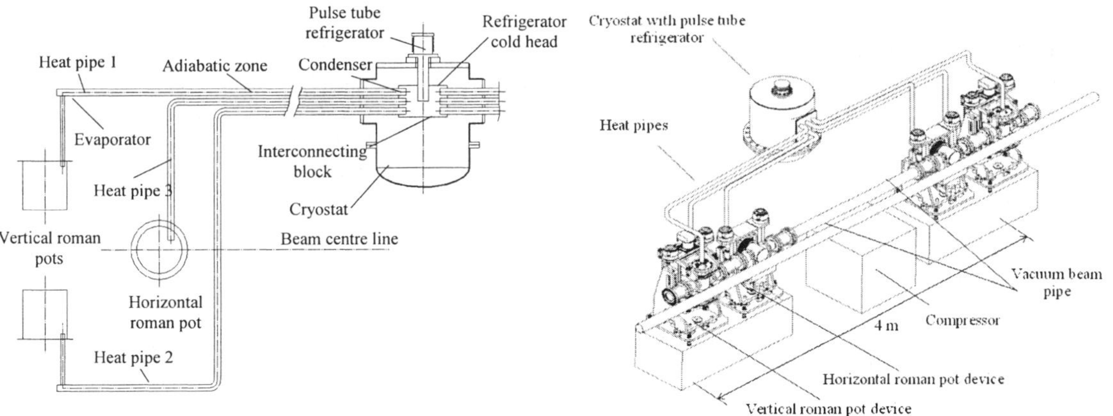

Figure 7 System with heat pipes

Figure 8 Cooling system lay-out with heat pipes and a central pulse tube refrigerator for an individual roman pot station

Also the heat pipes can be adapted to different temperatures by selecting the appropriate working fluid. The theoretical temperature limits are between the triple and the critical point of the respective fluid. However, in practice operation close to the triple point should be avoided as heat transfer performances are reduced with potential failures due viscous and sonic limitations of the low density vapour. Close to the critical point the heat of vaporisation becomes smaller and boiling becomes a limiting factor for the performance. Further limitations are entrainment and capillary pumping. Figure 9 shows the principle performance limitations of a heat pipe as a function of temperature. In Table 1 several low temperature working fluids are listed being potential candidates for this application. Furthermore the overall two-phase cycle pressure drop must be compensated for by the driving capillary and gravitational forces that pump the condensate back to the evaporator. Apart from the fluid properties the geometrical arrangement is of importance. The current design applies longitudinal grooves as wick structure and the outer pipe diameter is 15 mm. The heat pipe has an average inclination of 5% for gravity assist of the condensate return flow.

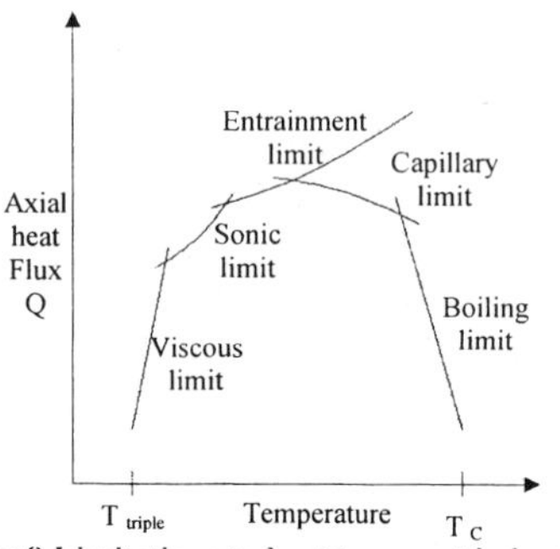

Figure 9 Limitations to heat transport in heat pipes

Table 1 Heat Pipe Working Fluids proposed

Fluid	Triple point Temp. (K)	Critical point Temp. (K)
Argon	83.8	150.86
Krypton	115.76	209.38
Xenon	161.36	289.74
Methane	90.69	190.55
Ethane	90.35	305.53
Propane	85.47	369.85

To allow the operation of the TOTEM cooling system at different temperature ranges a design feature permits to empty and change the working fluid for adaptation. This is done with valves installed at the pipe appendix after the condenser allowing the evacuation of the fluid and the refill with a different one.

The design principle will be validated with the installation of a prototype Roman Pots detector unit consisting of a pair of vertical pots at the SPS collider beam in the fall of 2004. Two heat pipes and the above described pulse tube refrigerator will be used for this purpose and operation and data acquisition will be done at distance. This permits the verification of the cooling requirements in view of the final installations in 2006. If necessary the refrigerator design may be reviewed for increase of cooling capacity.

SUMMARY

Compared to the rather complex forced flow systems the proposed design mainly applies passive elements for heat extraction and transport with heat pipes and a pulse tube refrigerator. It seems to be particularly suited for the LHC tunnel hostile environment. A prototype will be constructed for tests at the CERN SPS accelerator.

ACKNOWLEDGEMENTS

Acknowledged is the cooperative and fruitful collaboration with the TOTEM team headed by K. Eggert. Thanks are due to my colleagues T. Niinikoski, J.M. Rieubland, G. Passardi for discussions and proposals made and to Prof. Groll, Dr. Brost and Dr. Mertz from the "Institut für Kernenergetik und Energiesysteme" (IKE), University of Stuttgart, for consulting in matters of heat pipe technology.

REFERENCES

1) TOTEM LOI, Letter of intend, CERN/LHCC 97-49, 1997
2) "TOTEM" Technical Design Report, TOTEM-TDR-001, CERN-LHCC-2004-002, 7.1.2004
3) Grohmann, S., Niinikoski, T.O., Perea Solano, B., Herzog, R., Wobst, E., Vögele, G., "Cooling power distribution from a small cryocooler", CERN Report, 2002
4) Rieubland, J.M., private communication
5) Dunn, P., Reay, D.A., "Heat Pipes", Pergamon Press, 1994
6) Haruyama, T., et all., "Development of a high-power coaxial pulse tube refrigerator for a liquid xenon calorimeter", Proceedings of the CEC, Anchorage, 2003 (to be published)

Cryogenics Safety Review of the ATLAS Experiment at CERN

Haug F., on behalf of the ATLAS collaboration
Accelerator Technology Division, CERN, CH-1211 Geneva 23, Switzerland

The ATLAS detector at CERN to be installed at 90 m depth in a 50,000 m^3 underground cavern is of unprecedented size and complexity. This is reflected in the helium and nitrogen cryogenic systems required respectively by the magnets (three large superconducting toroids and the central solenoid with 1.6 GJ stored energy) and by the argon calorimeters containing 82 m^3 of liquid which can be drained into two 50 m^3 dewars in case of emergency. Further coolants of 11 m^3 of liquid helium and 15 m^3 of liquid nitrogen are stored underground. The potential hazards of the large quantities of cryogens in underground areas require specific attention. Design, construction and quality assurance strictly follow applicable safety rules and the cryogenic process and controls are conceived to actively cope with a number of faults. In severe cases of accidental coolant loss (helium, nitrogen) or argon, detection systems produce alarms which result in the activation of emergency gas extraction. Reviews with international experts confirmed the good safety standard of the systems.

INTRODUCTION

At CERN the 27 km circumference Large Hadron Collider (LHC) under construction uses large scale cryogenic systems for collider magnets and for two of the detectors (CMS, ATLAS). Basic cryogenic safety issues are addressed in CERN guidelines (1) or in comprehensive handbooks like (2). However, they are not exhaustive for this new endeavour with very large quantities of cryogens used in underground areas. Specific studies were needed and made over the past years at CERN and collaborating institutes to evaluate potential hazards. This permitted the implementation of appropriate safety measures. In particular ATLAS uses large amounts of three different cryogens stored in the proximity of personnel work. This paper summarises the preventive measures to reduce the risk of accidents for the ATLAS experiment.

RISKS

The largest risks to be found are in the underground areas, in particular in the almost hermetic experimental cavern. A quantity of 82 m^3 of liquid argon is permanently stored in the three calorimeter cryostats. A 15 m^3 LN_2 phase separator dewar is needed for its uninterrupted year-round cooling. The magnet system (toroids and solenoid) operate during approximately 9 months per year with the 11 m^3 helium storage dewar at full. In comparison the fluid quantities in the remaining equipment and cooling circuits are small (Table 1).

The three cryogens argon, helium, nitrogen are non toxic fluids. Instead their risks to humans results from their physical properties. At cold even a brief contact with the cryogen can cause severe injuries. The large expansion ratio of the fluids is a potential risk for asphyxiation in confined areas when air oxygen is replaced. An oxygen deficiency hazard (ODH) exists already at oxygen contents of few percent less than the normal 21 % in air. Owing to the different densities at ambient temperatures helium rises, argon accumulates at ground level and nitrogen mixes with air. A liquid spill leads to a mixture of the cryogen with air resulting into the formation of a cold mist with condensed air moisture expanding rapidly with a movement depending on the fluid. Liquid nitrogen and argon fall to ground while a helium air mixture moves rapidly upwards. Experiments simulating large liquid helium leak (3) demonstrated the rapid expansion and upward move with mist formation while the tests with argon confirmed the potential danger in particular at the detector floor level (4). The tests revealed that decreasing temperatures and oxygen contents were of concern but that the major problem arises from the cloud formation with impaired visibility for persons trying to use escape routes. The primary objective is, hence, at utmost to avoid cryogen spills by design and process with inherent safety features. However, in the unlikely event of an accident external systems are installed to reduce risks for personnel.

Table 1

Experimental cavern	quantity
Barrel Toroid (BT)	0.6 m³ LHe
2 End Cap Toroids (ECT's)	0.4 m³ LHe
Storage dewar	11 m³ LHe
Phase seperator cryostat	1 m³ LHe
Central Solenoid (CS)	0.02 m³ LHe
Control dewar CS	0.25 m³ LHe
Barrel Calorimeter (Barrel)	44 m³ LAr
2 End Cap Calorimeter (EC's)	38 m³ LAr
2 x 50 m³ Argon dewars	normally empty
LN₂ phase separator dewar	15 m³ LN₂

CRYOGENIC SYSTEMS WITH BUILT-IN SAFETY FEATURES

Cryogenics for magnets

For the magnets two helium refrigerators are used; the main (MR) and the shield refrigerator (SR) (5). The MR keeps the magnets 600 ton cold mass at 4.5 K. The SR safety function is used in case the MR fails in recovering the vaporised helium from the magnets by one of the two redundant SR compressors. For autonomy of 2 hours during slow discharge the 11000 l dewar supplies the toroids via the phase separator cryostat from which the liquid is withdrawn and circulated with one of two redundant pumps. During the 20 minute slow discharge the solenoid is cooled in gravity-driven thermo-syphon mode by 250 l stored in the control dewar on top of the detector. Figure 1 illustrates this emergency operation with the circuitry of the two refrigerators and the two proximity cryogenic systems. No helium is discharged in the underground cavern. In case of an unlikely accidental magnet energy fast dump the two-phase mixture in the cooling circuits would be expelled without discharging in the underground area (6) to the phase separator cryostat that rises in pressure. All the circuits are designed and tested accordingly.

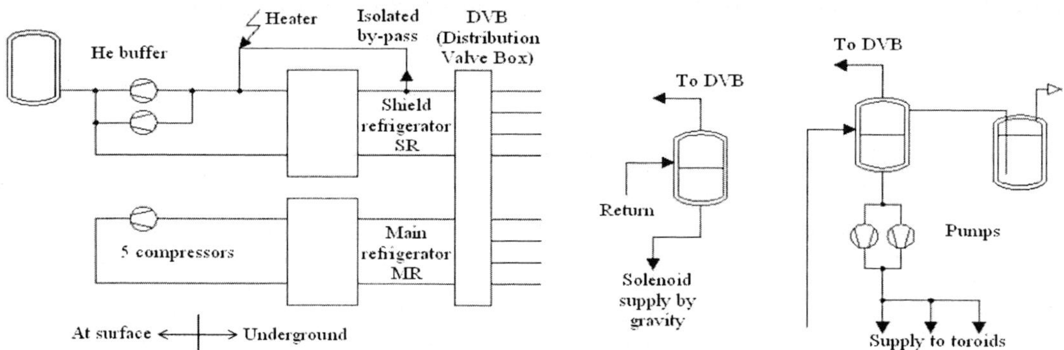

Figure 1 Emergency operation mode (MR refrigerator failure). Supply of toroids (pump) and solenoid (thermo-syphon) with stored helium. He recovery with SR shield compressor.

Cryogenics for Calorimeters

The three liquid argon calorimeter cryostats operating at 89 K dispose each of an expansion vessel at a higher level connected with transfer lines (flexible for the two end caps EC which can be moved and rigid ones for the central barrel) to provide sub-cooling of the liquid by hydrostatic pressure (7). Cooling of the calorimeters is done with a LN₂ refrigerator with surface dewars back-up in case of failure. The LN₂ is circulated in the calorimeter heat exchangers by centrifugal pumps with bypass circuits for allowing cooling directly from surface dewars. These measures have been adopted to keep the calorimeter cryostats cold at any time. In the unlikely event of their total failure or degradation of vacuum the argon pressure in the cryostat will rise and relief valves on the respective expansion vessel activate to blow out to the 500 mm diameter vent line connecting to the surface area. Also the nitrogen pressure relief valves of the respective equipment are connected to this line. In addition, in case of emergency, the liquid argon can be drained into the two adjacent 50 m^3 dewars by gravity. To accelerate the emptying process centrifugal pumps can be used.

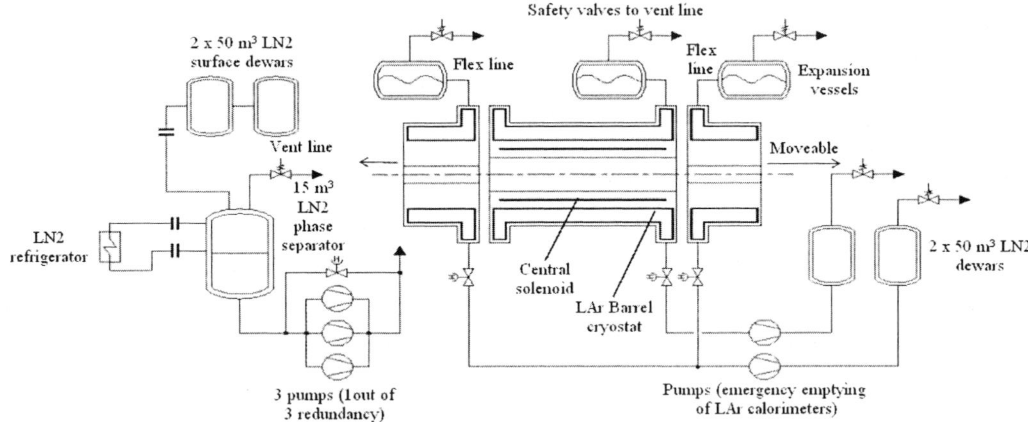

Figure 2 Redundancy in LN$_2$ coolant supply (left) to the liquid argon cryostats (right) showing equipment for emergency drain and safety valves connecting to vent line

Services Redundancy

For vital equipment, services supply of electricity, cooling water, compressed air is equally redundant. CERN benefits from the rapid switch-over capability between the electricity networks of France and Switzerland in case of failure of either. For all of ATLAS, un-interruptable power supplies (UPS, 2 x 600 kVA) and Diesel generators (2 x 900 kVA) are installed. Cryogenics uses this supply for the PLC's and the proximity cryogenics with the centrifugal coolant pumps. Two redundant air compressors with buffer volume and back-up 200 bar nitrogen batteries are installed. Two-out-of-three redundancy is applied for the water supply with 3 cooling towers and 3 water pumps. The water pumps, the air compressors and the two helium shield (SR) compressors benefit from the electricity grid switching capability.

IN CASE OF ACCIDENTAL CRYOGEN LOSS

Retention areas and volumes

The entire ATLAS detector is placed on a retention area formed by trenches in the concrete floor. In case of argon spillage of the calorimeter vessels the fluid is conducted to a low point pit. Also the two 50 m^3 argon and the 15 m^3 nitrogen dewars are placed in pits. Ventilation ducts are installed for extraction. In case of helium loss the air-gas mixture moves upwards towards the vault of the cavern. Estimates have shown that in worst case assumptions of rapid release of the entire 11 m^3 of helium the air-gas mixture expands downwards from the ceiling when warming up. The cloud with air moisture and little oxygen contents could reach the detector upper level endangering personnel working there. In order to provide a passive safety feature in parallel to the ventilation extraction systems for such severe accident the vertical access shaft PX 16 (fig. 3) is kept open at its bottom and closed at the top. The helium can escape to this 7000 m^3 retention volume sufficiently large to accommodate the amount of helium stored in the cavern.

Detection and Alarms

To ensure safety of personnel by detecting abnormal situations two independent systems are installed: the "SNIFFER" and the oxygen deficiency hazard meters "ODH". The SNIFFER is multipurpose to simultaneously detect smoke from fire, flammable gas and oxygen deficiency. A network with 200 air sampling tubes are installed within the detector envelope for all gases with aspiration points for the argon calorimeters in particular at the most vulnerable feed-throughs regions. In case of detection of leaks the respective location is known. This information can be used for warning of personnel working at proximity and for the rescue and localising by the fire brigade.

The ODH meters system is independent and will be installed outside the detector limits, e.g. at the floor level in retention pits and shafts and close to big vessels to detect ODH from leaks or spills. At oxygen levels of 18.5 % both ODH and SNIFFER systems generate a level-3 alarm. A level-3 is the highest alarm level and its significance is a hazardous, serious abnormal situation or accident where people's lives may be in danger and an immediate action by the fire and rescue group is eminent.

<u>Ventilation system</u>
The level-3 alarm is sent with additional information to the safety control room of the rescue and fire group. An activation of the emergency operation mode of the ventilation system may be done. The ventilation system for ATLAS is rather complex and it can be categorised to provide for the following functions: Air Pressurisation, Air Ventilation and Extraction. Pressurisation is done in the underground areas and the access shaft at three different pressure levels with a difference of approximately 50 Pa. The main detector cavern exhibits the lowest, slightly sub-atmospheric pressure. A permanent air extraction of 5000 m^3/h with several intakes at either side of the cavern wall is made for small detector or nitrogen leaks. For the retention pits and trenches of the argon calorimeters and tanks the permanent extraction flow is 7000 m^3/h. In case of emergency with ODH following a leak or spill the extraction rate is increased to 32000 m^3/h. A large ventilation system with laminar injectors at the floor level of the cavern and two extraction vaults at the ceiling provide a permanent flow of 60,000 m^3/h of air which under normal operation is partially re-circulated flow. In case of emergency (fire, helium leaks…) the flow is increased by a factor of two to 120,000 m^3/h in open circuit with a 100 % fresh air intake from the injectors.

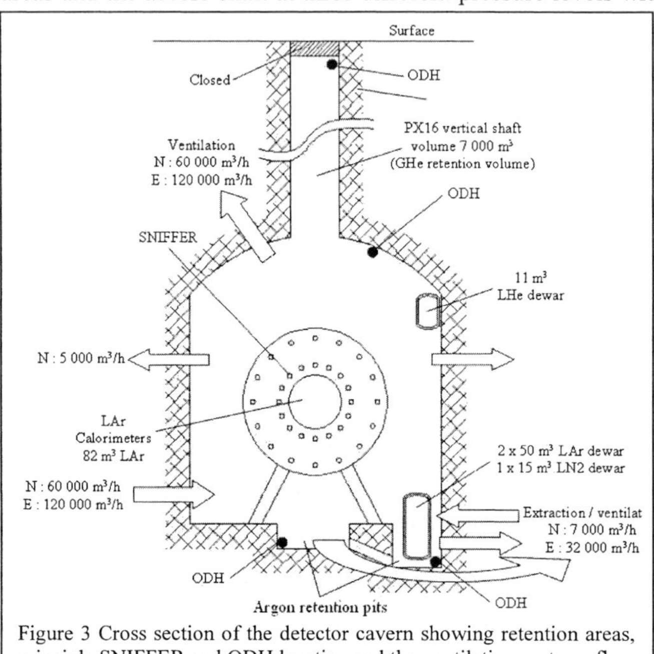

Figure 3 Cross section of the detector cavern showing retention areas, principle SNIFFER and ODH location and the ventilation systems flow rates (N = normal operation, E = emergency operation)

ACKNOWLEDGEMENTS

Particular thanks are due to G.-P. Benincasa who is the ATLAS safety group leader for his engagement to bring the experiment to a high safety standard comprising the cryogenics part. Acknowledged are all the many contributions in matters of safety from colleagues at CERN and collaborating institutes.

REFERENCES

1   Technical Inspection and Safety Service, "The Use of Cryogenic Fluids", Safety Instructions, TIS IS 47, CERN internal report, published 1998, revision 2004.
2   Edeskuty, F.J., Stewart, W.F., "Safety in Handling of Cryogenic Fluids", Plenum Press, New York, 1996
3   Chorowski, M., Konopka, G., Riddone, G., Rybkowski, D., "Experimental Simulation of Helium Discharge into the LHC Tunnel", Proceedings ICEC 19, Grenoble, France, 2002
4   Vadon, M., Perea Solano, B., Balda, F., "Simulations of Liquid Argon Accidents in the ATLAS Cavern", Proceedings CEC, Madison, 2001
5   Delruelle, N., Haug, F., Passardi, G., ten Kate, H., "The helium cryogenic system for the ATLAS experiment", IEEE Trans. on Applied Superconductivity, 2000
6   Haug, F., Bottura, L., Broggi, F., Junker, S., "Quench induced Pressure Rise in the Cooling Pipes of the ATLAS Barrel Toroid Model", Proceedings CEC, Anchorage, 2003, to be published
7   Bremer, J., "The Cryogenic System for the ATLAS Liquid Argon Detector", Proceedings ICEC 18, Bombay, 2000

Series Production of 13 kA Current Leads with Dry and Compact Warm Terminals

Andersen T.P.[1], Benda V.[2], Vullierme B.[2]

[1]Mark & Wedell, Oldenvej 5,DK-3490 Kvistgaard, Denmark
[2]Accelerator Technology Department, CERN, 1211 Geneva 23, Switzerland

For the LHC magnet test benches 13 pairs of conventional helium vapour-cooled 13 kA current leads are required. The current leads have been designed and built by industry. Attention was given to economical and reliable design and to a design of the warm terminal in order to avoid any condensation. Three pairs of them were tested at CERN. The dry warm terminal enables voltage test at 4.1 kV at cold condition. The paper describes construction details and compares calculated and measured values of the main parameters.

INTRODUCTION

For the LHC magnet test benches [1] 13 pairs of conventional helium vapour-cooled 13 kA current leads are required. The current leads are operating at 4.5 K saturated liquid helium. Table 1 summarises the design parameters of leads.

Table 1 Design parameters of the current leads

Nominal current	13 kA DC
Maximum current (Imax) for 10 minutes	15 kA DC
Leakage current at 4.1 kV DC at normal working condition	$<10^{-6}$ A
Design pressure	2 MPa
Working pressure	0.12 MPa
Total pressure drop at maximum current	<10 kPa
Heat load at nominal current at self cooling condition	<1.2 W/kA (0.80 g/s)
Heat load in Stand By (I=0 A) at self cooling condition	<10 W (0.51 g/s)
Insert length	1910 mm
Insert diameter	<100 mm
Time of coolant gas flow interruption at Imax without quench	>60s
Warm terminal at any working condition	compact and dry

For series production the vacuum brazing technology as well as the principle of the warm terminal was similar for the 13 pairs of 13 kA current leads discussed in this paper as for 26 pairs of 600 A current leads described in [2]. Including test leads, one of 600 A and one of 13 kA, 80 in total were built. Principle of the lead optimisation is given in [2]. We will describe the various elements and the performance of the leads.

DESCRIPTION OF THE LEAD

The lead consists of the warm terminal, the main heat exchanger, the cold terminal, electrical insulation and instrumentation.

<u>Warm terminal</u>

A schema of the warm terminal is shown in Figure 1.
A massive silver plated copper (SE-Cu) part (1) equipped with an electrical terminal (3) for water-cooled cables is vacuum brazed to 61 copper pipes housed by the hexagonal tube of the main heat exchanger (2). The copper part of the terminal is electrically insulated from the cryostat by a glass fibre ground isolator (4). The outlet gas tapping (5) as well as temperature transmitters (6) including electrical connectors (7) for high voltage taps, (8) for thermometers and (9) for heaters are insulated from copper part of the warm terminal by a massive glass fibre insulator (10).

Figure 1 Scheme of the warm terminal

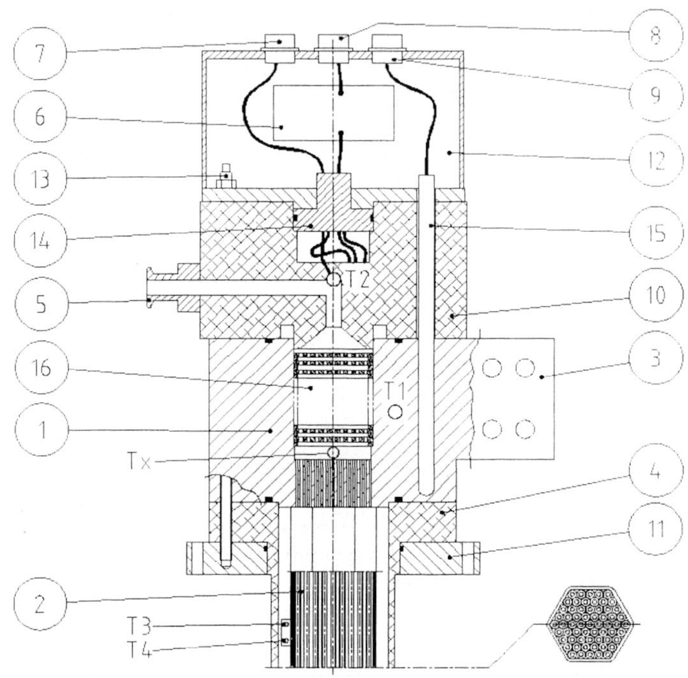

The whole sandwich, including connecting stainless steel flange (11), ground insulator, copper part, massive insulator and electrical box (12) are bolted together with electrically insulated double-end stud bolts (13) and sealed by O-rings in between.

<u>Heating element</u>

In order to keep the warm terminals at ambient temperature at any working condition five heating elements (15) of 250 W each are installed directly into the warm terminal. Ceramic heaters (Al_2O_3) provide excellent electrical insulation while heat transfer is good. The small remaining gap between the warm terminal and the ceramic heater is filled with a heat conductivity compound to ensure a good heat transfer.

<u>Auxiliary heat exchanger (AHX)</u>

As the permitted leakage current at test voltage of 4.1 kV is low, no humidity on the warm terminal is acceptable. To achieve this, an auxiliary heat exchanger (16) is integrated in the terminal. This exchanger is designed as a perforated plate heat exchanger. Forty-four perforated disks interlaced by silver plated spacers are vacuum brazed in a compact cylinder pressed-in into the terminal copper part. The perforation diameter is 1 mm.

Main heat exchanger (HX)

The body of the heat exchanger consists of 61 copper tubes housed in the hexagonal copper tube all together acting as the conductor. The cooling helium vapour passes externally along the 61 tubes. Material for the conductor is copper (ES-Cu) with a residual resistance ratio (RRR) of 27.

Cold terminal

The copper (OF-Cu) cold terminal of cylindrical shape is vacuum brazed to the copper pipes the hexagonal tube of the main heat exchanger included. The terminal is equipped with a groove and silver plated for high quality soft soldering of superconducting cable. As installation required horizontal soldering in situ a dedicated soldering device and procedure was developed (controlled and constant temperature of the whole soldering area). The surface area of the terminal immersed in liquid helium has been dimensioned to avoid film boiling.

Technology and material check

Modern CNC manufacturing equipment was utilised for cutting operations, which made precise manufacturing possible. This precision allows a high degree of interchangeability between parts before brazing. The vacuum brazing technique was chosen to obtain perfect and long term stable mechanical, thermal and electrical contact. The brazing process is precisely controllable and programmable. There is no temperature gradient in brazing parts. As the brazing material was used alloy Ag-Cu eutectic, melting temperature 780 °C. Six 13 kA bodies together with twelve 600 A bodies [2] were brazed in one batch. Design of the rig for assembly and test enabled smooth flow during the assembly and test.
The choice of the material is based on two factors. The copper must be oxygen free and possible impurities must have low vapour pressure at brazing temperature. Impurities which tune RRR of copper, must be stable during the brazing process. In our case impurity in copper was phosphorus (about 0.003%). RRR measured before and after brazing shown are nearly identical (27).

High voltage taps

The current lead is equipped with 4 voltage taps. Tap 1 is connected to the warm terminal. Tap 2 is connected to the cold terminal. Tap 3 is prepared to be connected to superconducting bus bar connecting the lead to the magnet coil. Tap 4 is prepared to be connected to the magnet coil. Voltage drop of the current leads is U_{1-2}, resistance of the contact between cold terminal and superconducting cable is given by U_{2-3} while contact resistance between the superconducting bus bar and magnet coil is given by U_{3-4}.

Thermometers

Four current lead thermometers are located as follows. Thermometer T1 measures temperature of the warm terminal, thermometer T2 measures the temperature of outlet gas and thermometers T3 and T4 (redundancy) measure temperature of main exchanger at 90% of exchanger length from cold side where the burn region is expected. Thermometer T1 is installed on the air side. Other three thermometers are located on helium side. All four are connected to the insulated transmitters and further to the connector.

Instrumentation feedthrough

Wires for voltage taps and 3 thermometers pass from the helium side to the air side through a 13-pin feedthrough (14), designed specifically for this purpose. Sufficient distances between live and ground parts ensure that the leakage current at 4.1 kV does not exceed the permitted value.

PERFORMANCE OF THE LEAD

Test conditions and procedure

In order to measure parameters of the current lead a dedicated test set up was designed and built at CERN. The pair of current leads was electrically short-circuited by a superconducting cable and connected to a power supply. The helium level was maintained constant, just below the gas inlet of the main heat exchanger. Heat inleak at self-cooling condition was measured by boil-off method considering that LHe in the cryostat is replaced with cold gas helium, eliminating heat inleak of the cryostat itself and subtracting inleak of resistive connection of the cold terminal/superconducting bus bar. Heat inleak of resistive connection was measured by electrical method; self-heat inleak of the cryostat was measured separately. Only for test purpose one extra thermometer Tx is installed in a tapping between both heat exchangers. T3 monitored at this condition is the future set point of the temperature control at normal working condition.

Results of measurement

In Table 2 are summarised measurement results and the corresponding calculated values.

Table 2 Calculated and measured results

Parameters		Calculated	Measured
Consumption of liquid helium at I=0A	[g/s]	0.40	0.52
Consumption of liquid helium at I=13 kA	[g/s]	0.75	0.79
Consumption of liquid helium at I=15 kA	[g/s]	0.87	0.86
Pressure drop at I=15 kA, total (mainHX)	[kPa]	5 (1.7)	2 (0.7)
Leakage current at 4.1 kV DC at working condition	[A]	-	16×10^{-9}
Gas outlet temperature Tx at I=0/13/15 kA of the main HX with set point of warm terminal T1=295K	[K]	291	289
Gas outlet temperature T2 at I=0/13/15 kA of the auxiliary AHX with set point of warm terminal T1=295K	[K]	295	294
T3 at I=0/13/15 kA respectively	[K]	140/154/165	134/145/150
Interruption of coolant gas at Imax without any degradation		>60s	>60s

CONCLUSION

Current leads rated to 13 kA continuously and to 15 kA for more than 10 minutes have been constructed by industry and tested. The measurements of the performance indicate that the design is reliable. The glass fibre insulator is robust and cheap and allows optimisation of its geometry. Sophisticated design of the warm terminal avoids any humidity, which ensures the low leakage current. The low leakage current of the leads allows a magnet high voltage test. Elaborated choice of the material and technology was optimised for series production. One pair of the leads installed in one test bench lost signals of T3 and T4 after heavy quench training. Reason will be investigated in near future. All twelve benches using these leads are operational, most of them for more than one year without any problem related to the current leads except in the point mentioned above.

REFERENCES

1. Axansalva J., Benda V., Herblin L., Tovar-Gonzales A., Vullierme B., Cryogenic infrastructure for testing of LHC series superconducting magnets, ICEC 20, Beijing, China
2. Andersen T.P., Benda V., Vullierme B., 600 A current leads with dry and compact warm terminal, Cryogenics 2002, Prague Czech Republic

Measurement of the thermal properties of prototype lambda plates for the LHC

Marie R., Metral L., Perin A., Rieubland J.-M.

CERN Accelerator Technology Department, CH-1211 Geneva 23

In order to power the LHC superconducting magnets, thousands of busbars will be routed from electrical feedboxes containing saturated helium at 4.5 K to magnets operating in pressurized superfluid helium at 1.9 K. Between those two volumes, the busbars will pass through special feedthroughs also called lambda-plates. This article presents heat flow measurements performed on several configurations of vertical prototype lambda-plates feedthroughs. The results show that the heat flow is strongly influenced by the configuration of the busbar insulation in the saturated helium.

INTRODUCTION

The assessment of heat loads and their resulting needs in cooling liquids mass flows are among the important requirements of machines operating in cryogenic conditions such as the Large Hadron Collider (LHC). In the LHC, the electrical currents are supplied to the magnets through electrical feedboxes [1] operating at 4.5 K, 1.3 bar. The lambda-plates ensure the electrical continuity of the circuits while keeping the hydraulic and thermal separation between the normal (4.5 K) and superfluid (1.9 K) helium volumes. The geometrical configuration of the feedboxes does not allow the usage of a stratified helium bath, but requires the usage of vertical lambda plates (and horizontal superconductive busbars). In order to limit the heat flow between these two temperature levels, it is therefore necessary to maximize the thermal impedance of the lambda plates. An obvious way to obtain such result is to use long lambda plates, but this approach is limited. Firstly, the plug length is restricted to be less than 100 mm by the stability requirements of the bus bars' superconductive state when conduction cooled from the ends only. Secondly, the plug is limited to lengths of the order of several tens of mm due to the mechanical difficulty coming from differential thermal contraction of the busbars and insulating material of the lambda plates.

Another possible way is to act on the heat exchange between the helium and the busbars: busbars are usually electrically insulated with several layers of polyimide tapes that also act as a thermal insulation towards the surrounding helium. This insulation increases the thermal impedance between the two temperature levels while still being far from acting as an adiabatic wall. We investigated the influence of this insulation on the thermal properties of the lambda plates; heat flow measurements were performed on simulated bus bars with varying numbers of polyimide tape layers corresponding to existing LHC electrical insulation configurations, and for two temperatures of the normal liquid helium. The results were then confirmed on prototype lambda plates with real busbars.

EXPERIMENTAL

The configuration of the lambda plate used for our study is shown in Figure 1. It is made of:
- An insulating plug that ensures the hydraulical and thermal separation of the two helium volumes.
- One or several busbars crossing the plug simulated with a round section copper rod with a diameter of 10.8 mm.

Heat flows were measured for the conditions listed in Table 1. The plug is made of polyetherimide (PEI) [2], the busbar is simulated by a copper bar. The RRR of the copper rod is 70 and it is insulated by layers of 50 µm-thick polyimide (PI) tape for cases 3 & 4.

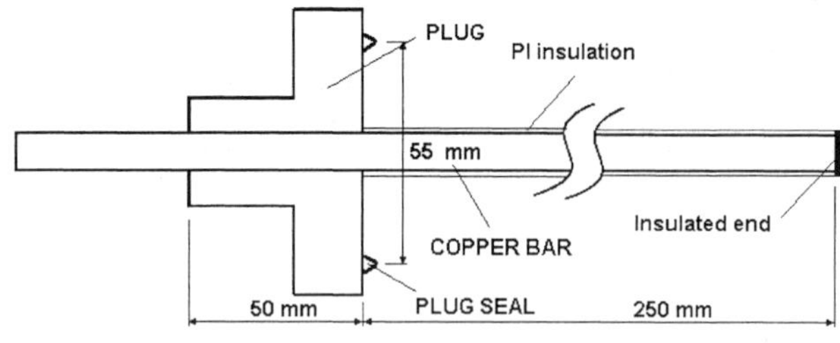

Figure 1 Schematic configuration of the lambda plates

Table 1 List of the test cases

Case	Configuration
C1	Full plug (no cable)
C2	Cu bar not insulated
C3	Cu bar with 1 PI layer
C4	Cu bar with 2 PI layers

The 20 mm-wide PI tape is wrapped around the bar with a 50% overlap to match the configurations of the LHC bus bars. One layer of PI corresponds to a total thickness of 0.1 mm around the busbar. A thermometer T_e is placed at the He I extremity of the rod (see Figure 2), glued to the copper and under the insulation.

The test system, shown in Figure 2, was designed to reproduce the thermohydraulic operating conditions of the lambda plates in the LHC. The whole system is installed in a vacuum vessel insulated by a 20 K thermal screen. It is composed of:

- A volume of saturated helium at 1.9 K which acts as the cold source and cools the pressurized He bath through a copper heat exchanger acting as a Kapitza resistance heatmeter [3]
- A pressurized bath of superfluid helium at 1.9 K
- The lambda-plate, shown in Figures 1 and 2, separating pressurized 1.9 K helium II and helium I
- A volume of helium I at 1 bar. The temperature of this volume is controlled with a heater.

Each volume is equipped with thermometers to ensure an accurate temperature control. Heat flows are determined by measuring, the temperature difference across the Kapitza resistance heatmeter. The system was calibrated by performing a measurement on a lambda plate without busbar (test case 1). This configuration was then used as a reference, in order to discriminate between the heat flow due the plug and the conduction along the copper bars. Measurements were performed for 2 temperatures of the 1 bar helium I, T_{Hel}. The temperature of the pressurized helium II was always 1.9 K.

RESULTS AND DISCUSSION

The heat flow through the lambda-plate $Q_\lambda(T)$ can be separated into:

- The heat flow through the full plug without any copper bar $Q_{plug}(T)$: it includes the heat flow due to helium leaks and the heat flow through the plug material and stainless steel piping.
- The heat flow through the copper bar(s) $Q_{Cu}(T)$

In order to distinguish the two contributions to the total heat flow, a measurement was performed without

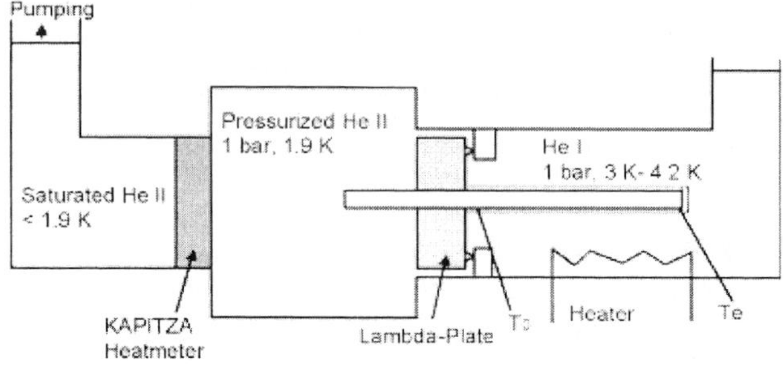

Figure 2 Schematic view of the test system

Table 2 Heat flows in mW for several configurations.

T_{HeI} [K]	$Q_{plug}(T)$	$Q_{Cu}^{C2}(T)$ no PI	$Q_{Cu}^{th}(T)$ no PI	$Q_{Cu}^{C3}(T)$ 1 layer PI	$QC_{Cu}^{C4}(T)$ 2 layers PI
4.1	163	867	1007	327	205
3.1	100	300	403	125	-

any copper rod (test case C1 in Table 1, $Q_{plug}(T)$ in table 2) . For the other measurements the contribution of the copper rod $Q_{Cu}(T)$ was calculated by subtracting $Q_{plug}(T)$ from the total heat flow. The results obtained for 0 ($Q_{Cu}^{C2}(T)$), 1 ($Q_{Cu}^{C3}(T)$) and 2 ($QC_{Cu}^{C4}(T)$) layers of PI insulation are shown in Table 2 for $T_{HeI}=4.1$ K and $T_{HeI}=3.1$ K ($QC_{Cu}^{C4}(3.1\,K)$ is missing due to an experimental problem). The results are also shown graphically on Figure 3. For comparison, the theoretical value $Q_{Cu}^{th}(T)$, estimated assuming perfect heat exchange between the copper rod and the surrounding helium is also shown in Table 2. For this calculation the temperature T_0 of the copper rod at the warmer end of the plug (see Figure 2) is thus assumed to be equal to T_{HeI}.

In the experimental configuration, T_{HeI}-T_e (Figure 2) decreases with increasing distance from the plug and for an infinitely long rod $T_e = T_{HeI}$. For a given T_{HeI}, the distance required to transfer a defined part of the heat flow is proportional to the heat impedance between the helium I and the rod surface. The effect of the experimentally limited rod length (250 mm) is to underestimate the heat flow with respect to an infinitely long rod. A qualitative evaluation of this effect can be obtained by comparing T_e with T_{HeI}.

Without insulation, the heat flow for $T_{HeI} = 4.1K$ is very similar to $Q_{Cu}^{th}(4.1\,K)$, confirming a good convective heat exchange between the helium and the copper at 4.1 K. For $T_{HeI} = 3.1$ K, the heat exchange does not appear to be as good: this is most probably due to a reduced natural convection resulting from the smaller temperature difference.

At 4.1 K, the first layer of PI insulation reduces the heat flow by a factor close to 3 and a second layer divides the heat flow by a factor greater than 4. At 3.1 K the effect of the insulating layers is somewhat smaller. For insulated rods, the heat exchange between the helium and the copper rod depends on essentially two mechanisms: natural convection at the PI / helium interface and conduction across the PI layers. The measurements show that even one PI layer is already very effective in decreasing the total heat flow by insulating the copper rod and therefore also decreasing the convection heat exchange thanks to reduced temperature gradients at the interface between the PI and the helium I. This effect is most apparent for $T_{HeI} = 4.1$ K where the convection heat exchange is stronger than at 3.1 K where the temperature gradients are smaller. We can estimate that even with only one layer of PI the temperature T_0 (figure 2) of the copper rod is significantly lower than 3.1 K ($Q_{Cu}^{C2}(3.1\,K) > QC_{Cu}^{C3}(4.1\,K)$).

However, the results of T_{HeI}-T_e (0.6 K for $T_{HeI} = 4.1$ K, and 0.5 K for $T_{HeI} = 3.1$ K) show also that the rod is not long enough to be representative of very long busbars for which a smaller gain due the insulation is expected.

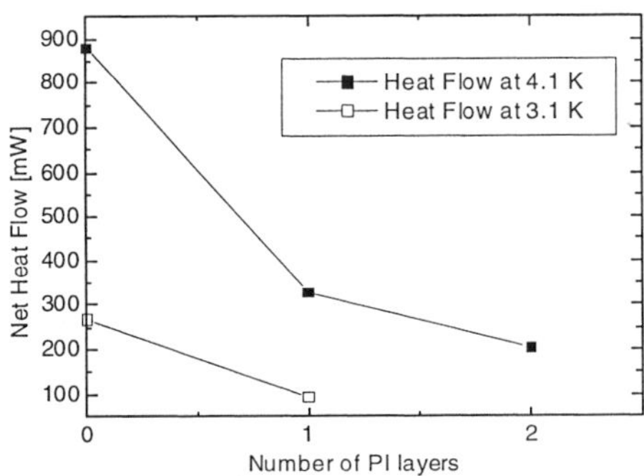

Figure 3 Heat flows through the Cu rod as a function of the number of PI layers

Table 3 Heat flows for the prototype lambda plate in mW

He I temp. [K]	$Q_{Cu}^{th}(T)$ Proto	Proto	Proto corrected
4.1	880	362	215
3.1	366	140	93

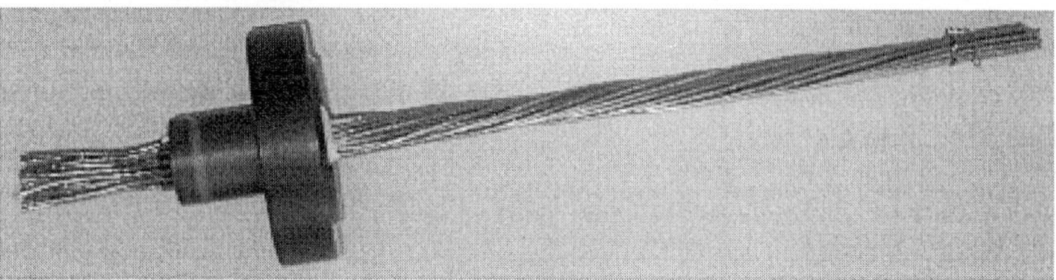

Figure 4 Lambda plate prototype with 22 superconducting wires

In order to confirm the validity of the results for the LHC busbars, we measured the properties of a prototype lambda plate with the same plug configuration (see Figure 1) but with 22 superconducting LHC busbars of a cross section of 2 mm^2 (Cu stabilization RRR=100) insulated with 2 layers of 50 μm PI (Figure 4). This configuration is the original insulation of the busbars, determined by electrical requirements. Table 3 presents the heat flows for T_{HeI} = 4.1 K and T_{HeI} = 3.1 K. For practical reasons the extremities of the busbars were not insulated, but to have a comparison with the test rod, we also present a roughly corrected value for which the estimated contribution of the conduction from the extremities has been subtracted from the measured heat flows (value estimated from heat exchange of the copper rod without insulation). Table 3 also shows the estimated value for a theoretical configuration without insulation and perfect heat exchange between the helium and the busbars $Q_{Cu}^{th}(T)$. As the extremities are not insulated $T_e=T_{HeI}$, the measured values are higher than the expected values for long (>250mm) busbars. The corrected values represent an "insulated extremities" situation and are therefore smaller than the expected values for long busbars. The relative estimated magnitude of the heat flow through the extremities emphasizes the importance of extending the insulation as far as possible from the plug.

CONCLUSIONS

Our measurements show that the electrical insulation around the busbar is very effective in reducing the heat flow between the superfluid helium and the normal helium volumes in vertical lambda plates. These results, confirmed on a multi-busbars lambda plate, demonstrate that it is possible to optimize the thermal properties of a lambda plate by improving the thermal insulation of the busbars on the He I side, without increasing the length of the lambda plate itself.

ACKNOWLEDGEMENTS

The authors gratefully acknowledge the support of the team of the Cryolab at CERN and especially S. Prunet and L. Dufay for their support.

REFERENCES

1.Goiffon, T., Lyngaa, J., Metral, L., Perin, A., Trilhe, P., van Weelderen, R., paper presented at this conference (ICE20)
2.Perin, A., Macias, R., Study of materials and adhesives for superconducting cable feedthroughs, <u>Adv. Cryo. Eng.</u>, vol.47, p. 551
3.Kuchnir, M., Gonczy, J. D. and Teague, J. L., <u>Adv. Cryo. Eng.</u>, vol. 31, Plenum Press, New York (1985), 1285.

The cold-compressor systems for the LHC cryoplants at CERN

A. Kuendig, H. Schoenfeld, Th. Voigt, K.Kurtcuoglu
Linde Kryotechnik AG, Daettlikonerstrasse 5, CH-8422 Pfungen, Switzerland

S.Yoshinaga, N.Saji, T.Shimba, T.Honda
IHI, Ishikawajima-Harima Heavy Industries Co., Ltd. Tokyo / Japan

The IHI-Linde Consortium is reviewing the successfully completed tests of the cold-compressor prototype plant at CERN and looks forward to the commissioning of the serial plants. The paper informs on experiences and conclusions for future cold-compressor applications.

INTRODUCTION

CERN, the European Laboratory for Particle Physics, is presently engaged in the construction of the Large Hadron Collider (LHC). This LHC will make use of high field super-conducting magnets operating in super-fluid helium and thus will require a huge refrigeration capacity at several temperature levels. The refrigeration capacity will be generated and distributed by eight refrigerator stations located along the colliders tunnel.

Table 1: The Specification of the approximate process requirements for the CERN–LHC process vacuum systems. In the transient modes the process can be supported with a make up helium flow of 300 kPa and 4.5 K.

	Flow	Intake condition	Return condition
Steady State Modes: Nominal Load Part load operation down to	 126 g/s 43 g/s	 1.45 ± 5 kPa , 4.0 K 1.45 ± 5 kPa , 4.7 K	 130 kPa, < 20 K 130 kPa, < 25 K
Transient Modes: Connecting to the cold experiment Recovery from a limited quench	 0 >20 g/s	 1.45 – 100 kPa 1.45 – 20 kPa, 4–25 K	 130 kPa, < 30 K

At the lowest temperature level of 1.8 K, the total specified refrigeration capacity is 19.2 kW, thus 2.4 kW have to be generated by each refrigerator station. In order to reach this low temperature, a process vacuum interface is part of each refrigerator. The approximate process data of this interface are listed in table 1.

Four of the cold-compressing units are supplied by the consortium IHI-Linde. They were subject of two previous papers. At the 18th ICEC conference in Mumbai the rough technical solution to the CERN specification was presented[2]. Later, at the 19th conference in Grenoble, just after the first successful tests of the pre-serial unit, the control philosophy was published. Currently the commissioning of the four units is upcoming. It is a good time now to make a review to a successful project and to me-

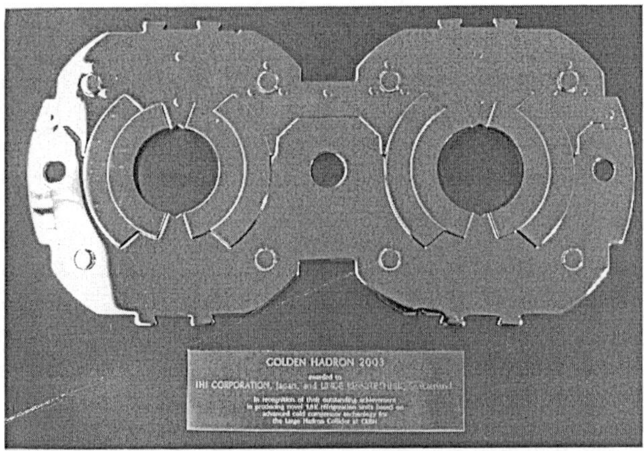

Fig. 1:   Golden Hadron 2003 awarded to
IHI Corporation, Japan and Linde Kryotchnik, Switzerland
In recognition of their outstanding achievement
in producing novel 1.8 K refrigeration units based on
advanced cold compressor technology for
the Large Hadron Collider at CERN

ditate on improvements in future cold compressing applications. These are the milestones in the chronology of the of cold-compressing systems for LHC at CERN:

1994:   Tender for a single stage prototype cold-compressing unit. Different suppliers were requested to develop a cold-compressor cartridge performing the following process data: suction pressure 1.5 kPa, suction temperature 3.5 to 4.4 K, pressure ratio 3, mass-flow 18 g/s. IHI and Linde Kryotechnik, at that time, were independent suppliers of two different units.

1996:   Commissioning and tests of the prototypes. CERN presented a paper on the results of these tests at the 17th ICEC conference in Bournemouth[1].

1998:   Tender for the eight cold-compressing systems for LHC. IHI and Linde Kryotechnik decided to cooperate and quoted as the IHI-Linde consortium.

1999   The consortium received the order for one pre-serial unit with the option of three further units.

2001:   Installation of the pre-serial unit at a test stand at CERN. Warm tests and adjustments at the magnetic bearings were performed.

April 2002:   Start of cold operating tests. Verification of the specified capacity.

Mai 2002:   Provisional acceptance of the system by CERN.

Mai-October: Further tests and a modification of the phase separator.

October 2002:   End of the tests. Confirmation of the order for three further systems
The consortium is honoured by CERN with a bonus for a low power input.

March 2003:   The IHI-Linde consortium is awarded for the excellent commissioning of the pre-serial plant (Fig 1.)

Mai 2004   Start of the installation of all the four units in the tunnel.

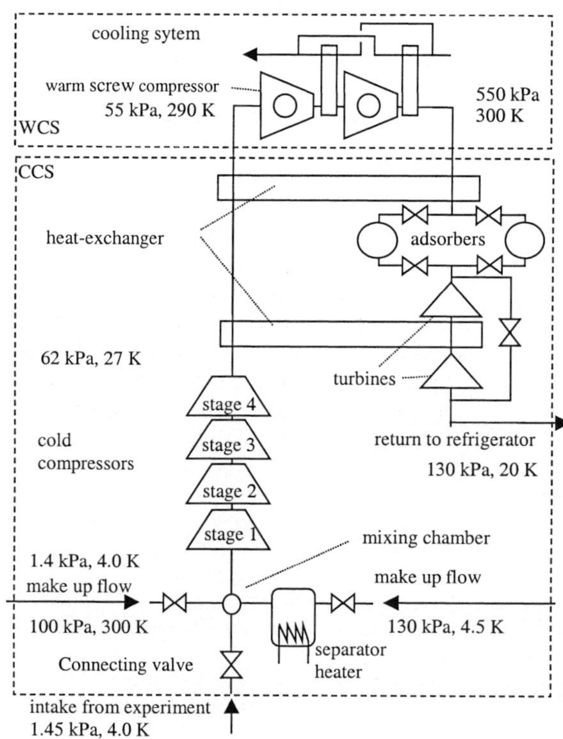

Fig 2: Rough process flow diagram of the LHC cold compressor system supplied by the IHI Linde consortium

THE PROCESS ARRANGEMENT

A simplified flow diagram of the vacuum compression system supplied by the IHI-Linde consortium is shown in figure 1. The approximate gas conditions at nominal load are indicated in this diagram.

The mixing chamber upstream of the intake of the first stage is only used during transient modes.

Before the cold compression system can be connected to the experiment, the suction pressure has to be equalized with the pressure of the experiment. If the pressure in the experiment is lower, the cold compressors have to be started while the connecting valve is still closed. In this case the matching flow conditions for the cold compressors are established by evaporating liquid helium with a heater and mixing the saturated vapor with a very small quantity of warm gas. The separator vessel is continuously refilled up to a constant liquid helium level.

The main components are the four cold turbo compressors. Each machine has its own frequency controlled drive and the shaft of each machine is equipped with active magnetic bearings. The impeller diameters are 250, 165, 115 and 105 mm. The overall pressure ratio on the four cold compressor stages is almost forty five. These machines are flanged from outside to the top-plate. In case of failure they are easily replaced within hours.

A warm screw compressor unit is the last link in the compression chain. It is a two stage compound machine. The volumetric suction flow of the screw compressor performs a pressure at the outlet of the forth cold compressor stage which is almost proportional to the circulated flow. This behavior is well matching with the requirements of the turbo-compressors. It enables to operate the turbo compressors in a wide range of part-load conditions without using a make-up flow.

Two switchable cold adsorber units on the high pressure side remove traces of impurities from the helium gas which may have leaked into the sub-atmospheric parts of the system.

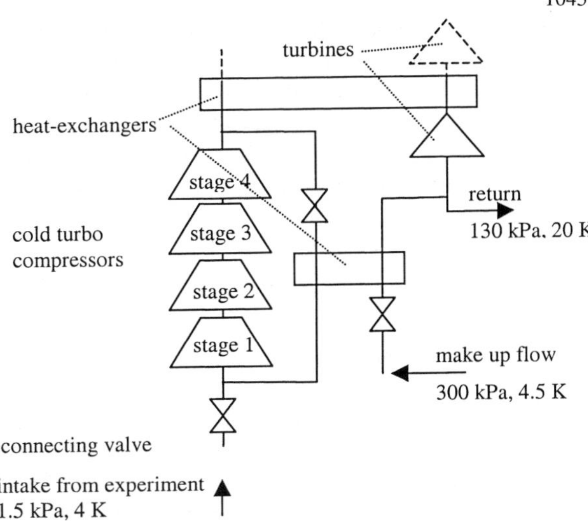

Fig 3: Alternative solution for startup flow, without separator and without mixing chamber

The two turbines compensate exergetic losses of the heat-exchangers and adsorbers and they perform a return temperature which is lower than the discharge temperature of the forth cold compressing stage.

Special attention has been paid to the speed control of the compressors. Unlike piston or screw compressors, turbo compressors are not volumetric type machines. Each turbo compressor has its individual working field indicating the correlation of rotating speed, flow, pressure ratio and suction conditions (Fig 4). Leaving the permitted field across the surge line will separate the flow from the blades resulting in a sudden drop of pressure ratio and attended by considerable rotor vibrations. Leaving the permissible field across the choke limit results in a loss of pressure ratio as well. The task of the speed control is to maintain the operating points of all the four cold turbo compressor stages within their permitted field.

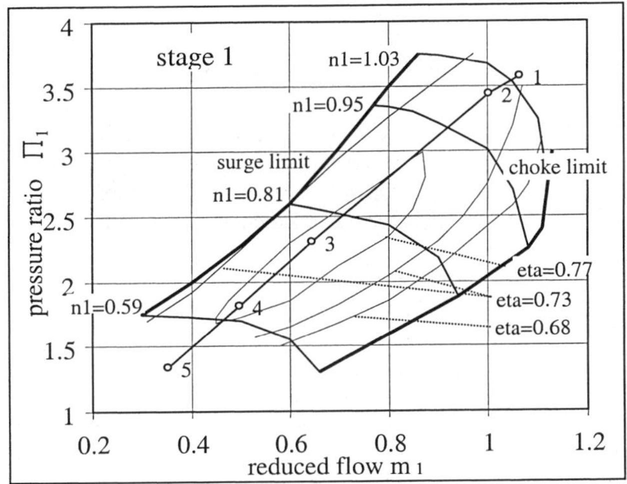

Fig.4: The graph of a typical turbo compressor working field

EXPERIENCES

Strong points of this system are the layout of the turbo-cartridges, their assembling in the coldbox and their sophisticated speed control which provides stable and optimized running conditions for all the four stages in a wide range of helium flow and suction pressure conditions[3].
However, even a very successful system has some weak points and for the design of future systems, we will give here some recommendations.

- Not only the cold compressors, but also the warm screw compressor should be equipped with a variable frequency drive and at design conditions, there should be a margin of 10 to 20 percent for capacity increase. The adjustable suction flow volume combined with sophisticated control operations, would help to run the cold compressors at optimal operating conditions within their working field. The flow control by using the slide valve is not precise enough.

- The helium separator holds the risk that droplets of liquid helium reach the first cold-compressing stage. While passing the phase change to super-fluid helium, the Lambda point, such droplets can de-stabilize the system. It results that the cold compressors drop into search conditions what requires to stop and to restart. A solution performing the necessary startup flow with a re-cooled cold-compressor bypass like Fig 3 would be preferable.

CONCLUSIONS

A ten year development period for cold compressor systems nears the completion. It has been an ambitious project and the risk to fail was some times omnipresent. With the successful commissioning of a four stage cold compressor application for LHC, the state of prototypes is left. A lot of know how is gathered concerning machinery as well as concerning process arrangement and speed control.

The cold compressing systems have reached the state of marketable products. But their market, particularly the market for refrigerators with super-fluid helium in the capacity range of several Kilowatt, is small. Further investigations are required to design smaller cold compressor units which are applicable for a lower capacity range. The target is a flow of 5 g/s at 1.5 kPa.

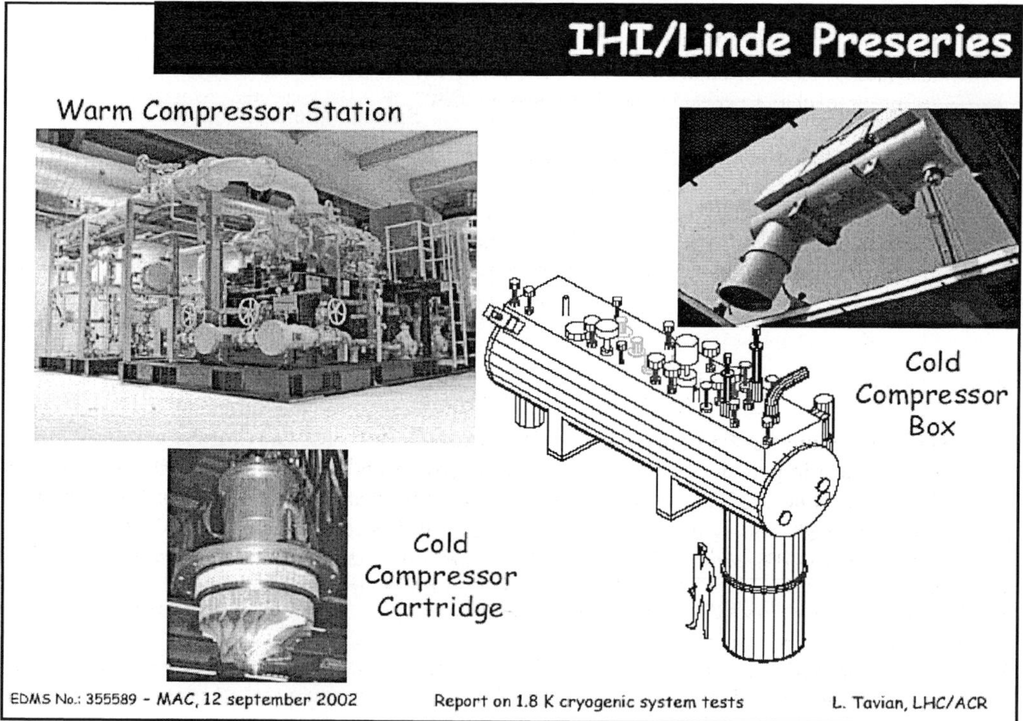

Fig.5: Illustration of the process vacuum system.

REFERENCES

1. A. Bézaguet, P. Lebrun, L.Tavian
   Performance assessment of industriel prototype cryogenic helium compressors for the Large Hadron Collider., ICEC17
2. H. Asakura, J. Boesel, T. Honda, A. Kuendig, K. Kurtcuoglu, A. Meier, M. Mori,, A.-E. Senn, S. Yoshinaga.
   Four 2400 W / 1.8 K Refrigeration Units for CERN-LHC: The IHI-Linde System, ICEC18
3. Hiroshi Asakura, Andres Kuendig
   Control Considerations of Multi Stage Cold Compression Systems in Large Helium Refrigeration Plants, ICEC19

Two 100 m Invar® transfer lines at CERN: design principles and operating experience for helium refrigeration

Claudet S., Ferlin G., Millet* F., Roussel** E., Sengelin** JP.

Accelerator Technology Department, CERN, CH-1211 Geneva 23, Switzerland
* SBT, DRFMC, DSM, CEA, 17 avenue des martyrs, 38054 Grenoble, France
** Air Liquide DTA, BP 15 ,38360 Sassenage, France

The distribution of helium for the Large Hadron Collider (LHC), at CERN, will require a large variety of transfer lines. At the time of qualification of possible technologies, Invar® was investigated as potential material for internal tubes. Intensive developments were made in industry to qualify the use of Invar® M93 and its associated welding parameters. Although all tests showed good perspective, the risk associated with the lack of proven reference turned out to be dissuasive with respect to the possible cost savings for the LHC cryogenic system.

However, since DN100 transfer lines were necessary for the supply and return of a test facility over a distance of 100 m, an Invar® based solution was considered, as repair or exchange would have been less dramatic than in the LHC accelerator tunnel.

After recalling the technical requirements, the required material qualification will be presented as well as the design principles and operating features. This equipment has been first cooled-down to 4.5 K and accepted in 2001. More than sixty complete thermal cycles from 300 K to 4.5 K have been performed since then without change of performance including helium leaktightness.

INTRODUCTION

The distribution of helium for the Large Hadron Collider (LHC), at CERN, will require a large variety of transfer lines. The most important one in terms of size, complexity and low heat inleak demand is the main cryogenic distribution line [1]. It will feed helium at different temperatures and pressures to the LHC elementary cooling loops over more than 25 km, mostly with four independent headers housed in one vacuum enclosure. The critical cost-to-performance ratio imposed thorough engineering from the beginning of the design phase onwards, with clearly identified milestones such as a preliminary design review prior to any detailed engineering or manufacturing activity. This was supposed to validate mechanical and thermal hypothesis with their corresponding assumptions. At this stage, Air Liquide, one of the three selected contractor for the pre-series test cell (~ 112 m), investigated an alternative based on Invar as potential material for internal tubes [2]. Together with IUP-Arcelor, they conducted a program to qualify the welding parameters of Invar® and evaluated the corresponding performance.

Although all tests showed good perspective, the risk associated with the lack of proven reference turned out to be dissuasive with respect to the possible cost savings for the LHC cryogenic system. However, since DN100 transfer lines with design pressure of 1.0 MPa were necessary for the supply and return of a test facility [3] over a distance of 100 m, an Invar® based solution was considered, as repair or exchange would have been less dramatic than in the LHC accelerator tunnel.

BASIC DESCRIPTION OF INVAR®

As shown on Figure 1 around 36% of nickel content, the iron nickel alloy features a very low coefficient of thermal expansion (CTE), which originated the given name of Invar®. With mechanical properties slightly lower than austenitic stainless steel, it is widely used for LNG activities at temperatures around 100 K.

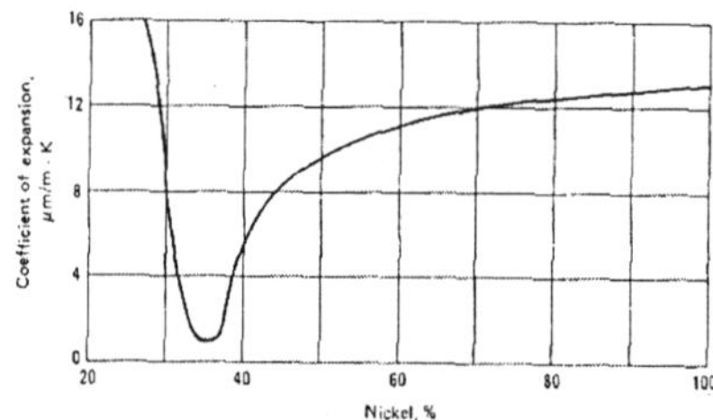

	Invar® M93	Stainless steel
CTE	1.5 10⁻⁶/K	16 10⁻⁶/K
Young Modulus	140'000 MPa	200'000 MPa
Theoretical thermal stress (CTE x E x ΔT) for 200 K	42 MPa	640 MPa

Figure 1 Invar® properties and comparison with austenitic stainless steel

One of the main problems encountered when assembling alloys with acceptable properties comes from its tendency to cracking under thermal or mechanical stress. This comes mostly from accumulation of impurities at the primary austenitic joints which occurs during the solidification phase after welding. This effect has been quantified for austenitic stainless steel where the sulphur and phosphorus contents is compared with the chromium to nickel ratio, as shown on Figure 2. Extrapolating these results to the very specific case of Invar® and its 36% nickel content simply imposes that sulphur and phosphorus content have to be extremely low. As this specific feature was not always reachable in the 1970s or 1980s, some previous experience might have shown problems at cryogenic temperatures.

With the technical progress made in industry, IUP-Arcelor developed a new product called Invar® M93 which fulfil the requirements preventing getting cracks in welds.

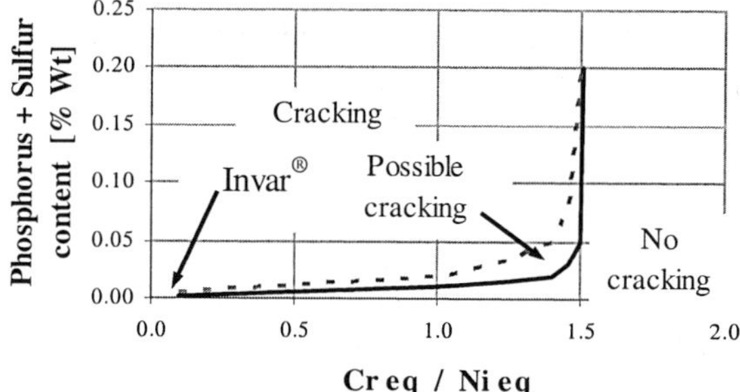

Material	Content
C	0.03 %
Mn	0.36 %
Si	0.20 %
S	< 10 ppm
P	< 25 ppm
Ni	36.0 %
Co	0.04 %
Fe	Balance

Figure 2 General tendency to cracking and composition of Invar® M93

MATERIAL AND WELDING CERTIFICATION

Once the base material qualified, European Directive EN 288-3 imposes at least a simplified qualification of the welding operating mode. This includes micrography, ductility, tension tests of the welded zone at room and working temperature. Using samples from the manufactured batch, Air Liquide and CERN [4] performed such tests on plates and tubes. TIG orbital welding without additional metal shows the greatest performance. When necessary, TIG manual welding with Invar® M93 is acceptable. The main results are presented in table 1 for circular welds. Although some values were lower than the specified ones, the qualification was granted as the line load case is more driven by the working temperature which show good performance, rather than any load case at ambient temperature.

Table 1 Results of mechanical properties for circular welds (TIG manual welding with Invar[®] M93)

	Specified	Measured 20 C	Measured 4.2 K
Maximum load Rupture limit (MPa)	≥ 450	300 ± 25	740 ± 30
Strength 0.2% yield stress in tension (MPa)	≥ 300	180 ± 10	560 ± 10
Ductility Elongation to rupture in tension (%)	> 14	17.5 ± 1.5	19.0 ± 2.0

DESIGN PRINCIPLES

From a thermal point of view, the design of this line was not a real challenge, as existing industrial standard would easily fulfil the requirements. Therefore it was possible to standardise both the supply and return lines (DN100) with separate vacuum enclosure (DN200). End boxes allowed for recombining the two lines towards the single interface port at both ends. Moreover, it was possible to place several internal fixed points linking the inner tube to the vacuum jacket in the bent elements in order to minimise the efforts on the internal spacers. These unusual additional internal fixed points helped handling and placing in position the bended pipe elements. Better positioning was achieved, without the need for additional specific tools to guarantee the position of the internal tubes with respect to the outer jacket.

A specific point was the design of the compensation system for the longest straight part of the vacuum jacket made of stainless steel (AISI 304), as in case of accidental loss of vacuum, it would contract more than the inner tube!

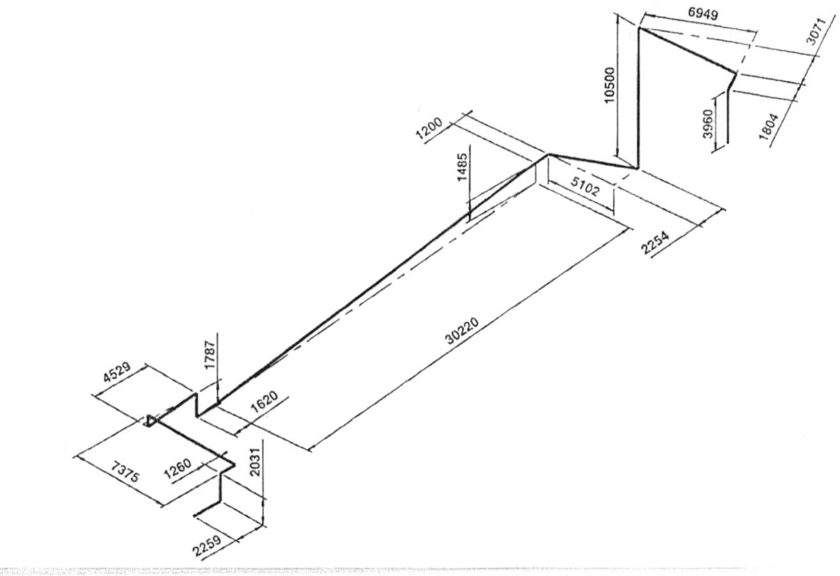

Figure 3 Isometric view of the Invar[®] transfer line

OPERATING SEQUENCE

The weekly-based operation of the test station fed by these lines defines the cool-down and warm-up frequency. The complete cycle down to 4 K and back to room temperature has been performed about 25 to 30 times yearly over the past two years (See Figure 4).

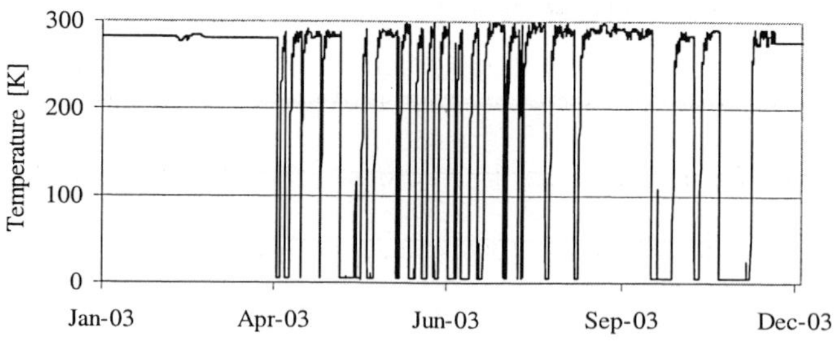

Figure 4 Typical annual operating history

MAIN RESULTS

At the end of the installation phase, a global helium leak test was performed to validate the inner lines as well as the welded joints and vacuum connections on the outer jacket. The residual leak rate was monitored during the cool-down to 4 K and during the heat load measurements. All tests were redone earlier this year, some 30 months later showing no change in performance (Table 2).

Table 2 Evolution of performances of the Invar® transfer line system

		Technical specification	Initial measurements (2001)	Last measurements (2004)
Heat loads	[W]	≤ 140 / supply line	≤ 100	70 ± 30
He leak rate	[mbar.l/s]	$\leq 1.0 \times 10^{-6}$	$\leq 1.0 \times 10^{-7}$	$\leq 1.0 \times 10^{-7}$

CONCLUSION

An Invar® M93 based transfer line system (2 x 100 m, DN100 int. diam.) has been designed and built for a test facility at CERN, after specific material and welding qualifications. This equipment has been first cooled-down to 4.5 K and accepted in 2001. More than sixty complete thermal cycles from 300 K to 4.5 K have been performed since then totalling over 5'000 hours of integrated operation without change of performance including helium leaktightness.

ACKNOWLEDGEMENTS

The authors would like to thank all parties involved, from LHC management for their authorisation and support all along this project to material and welding specialists for the required qualifications. Special recognition is due to Air Liquide and project engineer Ph. Hirel for daring this première with us.

REFERENCES

1. Erdt, W.K., Riddone, G., Trant, R., The cryogenic distribution line for the LHC: functional specification and conceptual design, Advances in Cryogenic Engineering (2000), 45 1387-1393

2. Hirel, Ph et Al, Invar variant for the LHC cryogenic distribution line, Air Liquide preliminary design review documentation (1999)

3. Claudet, S., Ferlin, G., Gully, Ph., Jager, B., Millet, F., Roussel, P., Tavian, L., A cryogenic test station for the pre-series 2400W @ 1.8K refrigeration units for the LHC, Proceedings ICEC19, Grenoble, France, (2002), 83-86

4. Letant D, Couturier K, Sgobba S, Caractérisation de l'Invar® M93 et de ses soudures pour application a la ligne cryogénique QLLC, Rapport de stage IUT Annecy et CERN EST (2001)

Experience with the String2 Cryogenic Instrumentation and Control System

Gomes P., Balle Ch., Blanco E., Casas J., Pelletier S., Rodriguez M.A., Serio L., Suraci A., Vauthier N.

Accelerator Technology Department, CERN, 1211 Geneva 23, Switzerland

String2 was a 120 m full-scale model of a regular cell of the LHC accelerator arc. It was composed of eight superconducting main magnets, fed by a separate cryogenic distribution line (QRL); an electrical feed box (DFB) which supported the superconducting current leads that powered the magnets [1].
Bearing an intensive experimental programme, String2 was heavily instrumented. The cryogenic instrumentation and control system, whose complexity was close to a full 3.3 km LHC sector, have already been described in [2].
String2 was a useful facility to validate design choices and to gain knowledge on installation and commissioning procedures. This paper reports on the experience of four years of designing, installation, commissioning and maintenance, and outlines the lessons learned for the LHC.

THE CRYOGENIC CONTROL SYSTEM ARCHITECTURE

About 700 cryogenic sensors and actuators were distributed along the 120 m length of String2; their signals were conveyed by seven fieldbus segments. Intelligent devices were directly connected to Profibus®-PA, whereas conventional devices were connected to analogue or digital modules on remote I/O stations, which communicated on Profibus®-DP. The cryogenic process automation was handled by two Siemens-S7® controllers (PLCs), running 126 closed control loops.

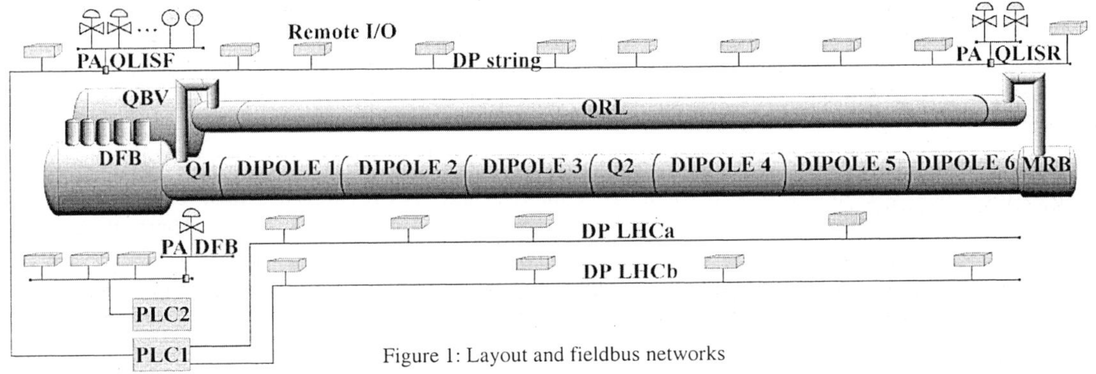

Figure 1: Layout and fieldbus networks

THE PROGRAMME

Design
The LHC cryogenic control system and instrumentation are based on commercially available components whenever possible; specific developments were necessary in some cases, such as for high accuracy electronic signal conditioning for thermometry [3], or to cope with the ionising radiation in the accelerator tunnel [4]. For String2, a particular effort was put to integrate components and systems as close as possible to their final LHC design. It was intended to investigate the accuracy and long-term behaviour of various types of sensors, validate signal conditioning units and their integration with fieldbuses, evaluate advanced control techniques, and the performance of the control system in general.

Installation
The deployment of the front-end electronics and fieldbuses for cryogenics started by the end of 1999, simultaneously with the first phase of the mechanical assembly. During more than one year, most of the String2 Phase1 components were installed (Figure 1): 2 quadrupole and 3 dipole prototype cryomagnets, the QRL, the DFB, and some ancillary systems (QBV, QLISF, QLISR, MRB).

Commissioning
While the last components were still being assembled, the commissioning started in April 2001: every instrument had its signal chain systematically verified; this included cabling, signal processing electronics, communication, synoptic view, and coherence between the databases of the PLC and of the supervision (SCADA).

Before the final closure of the String2 vacuum vessel, an important number of sensors was found to be swapped, damaged or malfunctioning, due to wires inverted, broken, or shorted to ground. They were repaired whenever the sensor or cables were still accessible.

Cool down & powering
In the beginning of August 2001 the cool down could start, as all the required instrumentation was available. At nominal operating conditions, the control loops were fine-tuned, and the main thermometers and pressure sensors were verified for high accuracy.

The process was gradually set into automatic mode, and eventually the facility could be left unattended, except during experiments or powering; the on-call cryogenic operator would be alerted by SMS, doubled by a hardwired general-fault alarm to a continuously manned central control room.

During cool-down, the analysis of the profiles of temperature vs. time and vs. location showed additional inversions and misidentifications in all the 32 current leads of the DFB. An extra effort was therefore deployed in order to power the magnets to nominal current in September.

Phase 2 & Phase 3
During the 2002-03 winter shutdown, the String2 was completed to a full cell by addition of 3 LHC pre-series dipoles, introducing some 30 new analogue channels. Cool down took place in May 2002 and powering started in the following month. For the third and final run in 2003, several upgrades were made on the control system, in order to move closer to the final LHC configuration, hardware and software wise.

THE EXPERIMENTS & THE LESSONS

Instrumentation
The String2 instrumentation was particularly complex and non-regular as most components were prototypes, from multiple origins, and with frequently changing specifications. There were many unexpected problems, which took too much time and resources to identify and repair.

After one month of commissioning, only 64% of the thermometers in the magnets and QRL were acknowledged to perform within specifications; three months later, 24% more became operational. Only after 8 months of commissioning the instrumentation diagnosis could be frozen; for the 215 thermometers in the magnets and QRL: 64% OK since day one, 26% repaired, 7% working with reduced performance, 3% lost; for the currents leads (220 thermometers): 0% OK at day one, 90% identified as permutated in several ways, and 10% lost. Other instruments were in smaller amount and had fewer problems.

This laborious commissioning of the instrumentation emphasized the risk of degrading or losing the control of several magnets temperatures, and stimulated urgent actions within different teams:
• magnet reception test procedures were reviewed in order to avoid thermometer damage;
• thermometer owners or installers were urged to thoroughly follow installation guides and to properly document all phases of assembly;
• emphasis was put on the thermometry databases, in order to finally have working interfaces for inserting and analysing thermometer calibrations and for assembly follow-up, and also to have automatic generation of PLC interpolation tables; otherwise there was a high risk of having temperatures measured in ohm;
• people became aware that commissioning time must not be neglected and that it cannot be entirely squeezed in the shadow of other activities; the goal is to have a machine running within specifications and to minimise the interventions and the down time.

<u>EMC</u>
Unexpectedly, when the main superconducting magnets were powered, their temperature measurement was found to be strongly correlated to the magnet current. Several studies and measurements showed:
- a magneto-resistance phenomenon would have required a magnetic field of about one Tesla at the thermometer location; this was ruled out by the magnet designers;
- when two independent magnet circuits were powered, the individual temperature offsets were adding up quadratically; this hinted at the cumulative effect of two uncorrelated noise sources;
- the power converters voltage to ground and current indeed exhibited high-frequency noise, with amplitude increasing with the current; this noise was also seen at the thermometer leads, superimposed to the DC thermometer excitation; this meant electromagnetic coupling from magnet coils into the thermometer, and was confirmed by direct signal injection on the current leads;
- the temperature offset was proportional to the noise power, indicating Joule effect; this was compatible with previous laboratory measurements (30-100 mK / μW);
- for higher noise power, the signal conditioner's input reaches its saturation threshold.

This temperature dependence on magnet current (Figure 2) was solved by adding capacitive filters to the power converters; they will therefore be also required for the LHC.

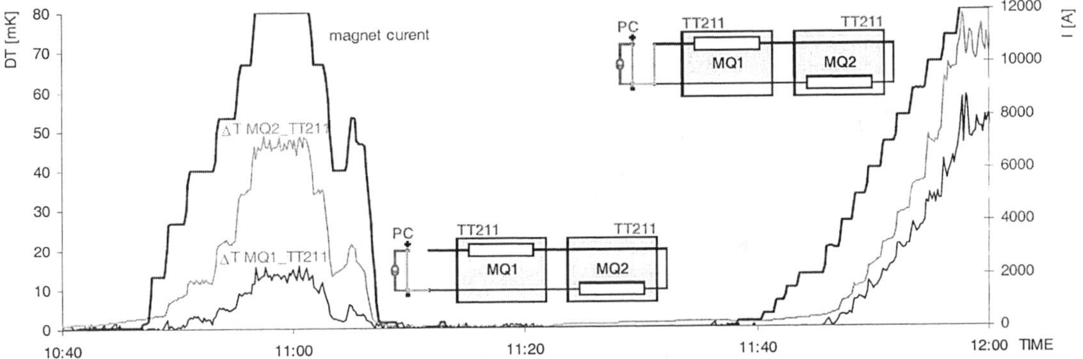

Figure 2: Correlation between the quadrupoles temperatures offset and the DC current in short-circuited Power Converter

Further electromagnetic compatibility were performed, to seek for possible effects of the electrical noise generated by kicker magnets and their pulsed power supplies. These experiments confirmed the stronger interference rejection of double shielded cables, when compared to single shielded ones. Otherwise, a digital communication board was blocked by every kicker pulse, if its auto-restart was not enabled; this warns that the radiated power is still a real concern for electronics near the kickers.

<u>Process Control</u>
The temperature of the superconducting magnets is a control parameter with tight operating constraints, which implied the development of an advanced process regulator based on the Model-Based Predictive Control paradigm (MBPC), in order to reduce the control band and improve stability [5].

The dynamics involved in the magnet temperature regulation presented the following characteristics:
- asymmetric inverse response, where temperature excursion initially reacts opposite as expected;
- variable dead-time depending mostly on the heat load situation;
- non-uniform cold mass temperature across magnets, due to a constrained heat transfer through the cold mass interconnections.

These complex characteristics, together with the severe operational constraints, suggested the need of a more advanced control technique than a simple proportional, derivative and integral controller (PID) or any other linear regulator. String2 operation evidenced the poor performance of PID regulation where re-tuning was absolutely necessary every time the heat load varied. LHC will have more than 200 temperature regulators and tuning becomes a fundamental issue.

Extensive MBPC tests showed a better wide-range performance than the PID. Fig 3 compares the response of both controllers under the same heat load disturbance; MBPC is able to halve the initial temperature excursion and to stabilize the temperature twice faster than the PID regulator [6].

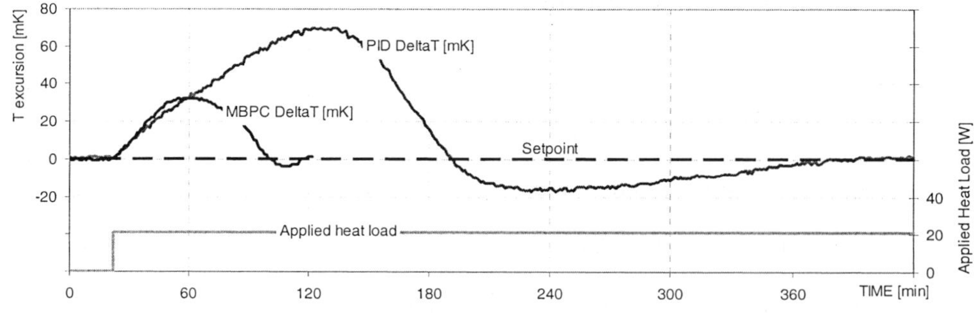

Figure 3: Temperature excursion of MBPC vs. PID, following a step in applied heat load

<u>Closer to the LHC</u>

Due to the high level of ionising radiation in the LHC tunnel, all front-end electronics for cryogenics has been designed to be radiation tolerant. Thus, new WorldFIP® fieldbus interfaces [7] and signal conditioners, for temperature and pressure, were introduced in String2, together with a gateway between WorldFIP® and the PLCs; they performed flawlessly.

Under CERN's request, Siemens developed a version of their intelligent valve positionners that allows the pneumatic actuator to be separated from the electronics, which must be out of the tunnel radiation. In String2, a set of these conditioners was introduced, with 1 km of cable.

Furthermore, a large part of the control system was migrated into the standard LHC control framework (UNICOS [8]), which comprises new programming methodology, new PLC and new SCADA. Specification and reception were performed by our team, and implemented by the UNICOS team.

FOR THE LHC…

The installation and commissioning of String2 cryogenic instrumentation and controls was very educative and will have a direct impact on the commissioning of the LHC sectors, permitting to save precious time.

The String2 facility provided validation and additional knowledge for instruments, front-end electronics, electromagnetic compatibility, fieldbus and PLC architectures, remote electronics positionners, control algorithms, programming methodology, and SCADA.

In some cases weaknesses were highlighted, triggering improvements in issues such as inter-team communication, respect of instrumentation installation procedures, and documentation production or access. It was also clear that wireless access to local area network will be essential in the LHC tunnel.

All in all, the skills of the involved teams and the robustness of components and process allowed keeping an overall cryogenics availability of more than 97% of total operation time [6]. Furthermore, no correction of the thermometers transfer function was ever needed, due to the quality of the original calibrations, the performance of the built-in thermalisation and the accuracy of the signal reading chain.

We wish to express our acknowledgement to the Operators, the UNICOS team and other String2 crews, who gave us valuable help during the commissioning of instruments and control system, contributing for the success of the String2 venture.

REFERENCES

1. Bordry F. et al., The LHC Prototype Full-Cell: Design Study, <u>LHC Project Report 170</u>, CERN, March 1998.
2. Suraci A. et al, Instrumentation, Field Network and Process Automation for the Cryogenic System of the LHC Test String, <u>ICALEPCS01-TUAP065</u>, Nov 2001.
3. Gomes P. et al , Signal Conditioning for Cryogenic Thermometry in the LHC, In: <u>Proc. CEC/ICMC99</u>, Montreal, Canada.
4. Agapito J. A. et al, Instrumentation amplifiers and voltage controlled current sources for LHC cryogenic instrumentation , <u>6th Workshop on Electronics for LHC Experiments LEB 2000</u> , Cracow, Poland , 11 - 15 Sep 2000.
5. Blanco E. et al., Nonlinear Predictive Control in the LHC accelerator, <u>ADCHEM-03</u>, Hong Kong, January 2004.
6. Serio L. et al, Experimental Validation and Operation of the LHC test String 2 Cryogenic System, <u>LHC Project Report 681</u>, CERN, Jan 2004
7. Casas J. et al, SEU tests performed on the digital communication system for LHC cryogenic instrumentation, <u>Nucl. Instrum. Methods Phys. Res., A 485 (2002) 439-43</u>
8. Gayet Ph. et al, Application of Object-Based Industrial Controls for Cryogenics, <u>8th EPAC</u>, Paris, France, Jun 2002.

The management of cryogens at CERN

D. Delikaris, K. Barth, G. Passardi, L. Serio, L. Tavian

CERN, Accelerator Technology Department, 1211 Geneva 23, Switzerland

CERN is a large user of industrially procured cryogens essentially liquid helium and nitrogen. Recent contracts have been placed by the Organization for the delivery of quantities up to 280 tons of liquid helium over four years and up to 50000 tons of liquid nitrogen over three years. Main users are the very large cryogenic system of the LHC accelerator complex, the physics experiments using superconducting magnets and liquefied gases and all the related test facilities whether industrial or laboratory scale. With the commissioning of LHC, the need of cryogens at CERN will considerably increase and the procurement policy must be adapted accordingly. In this paper, we discuss procurement strategy for liquid helium and nitrogen, including delivery rates, distribution methods and adopted safety standards. Global turnover, on site re-liquefaction capacity, operational consumption, accidental losses, purification means and storage capacity will be described. Finally, the short to medium term evolution of the Organization's requirements will be reviewed.

INTRODUCTION

The European Organization for Nuclear Research (CERN) is an international organization, seated in Geneva, Switzerland, with 20 Member States. Its objective is to provide for collaboration among European States in the field of high energy physics research and to this end it designs, constructs and runs the necessary particle accelerators and the associated experiments.

To perform its scientific program, CERN has to operate several helium cryogenic plants and their ancillary equipment. Liquid helium plants are used for superconducting components of accelerators, physics experiments and tests bench facilities. Liquid nitrogen is used for pre cooling of super conducting magnets for accelerator and experiments, boosting the plant liquefaction capacity, purification systems and cooling of liquid argon and krypton calorimeters.

The procurement of these cryogens is achieved via industrial contracts placed by CERN with several companies.

INVENTORY OF CRYOGENIC PLANTS & ANCILLARY EQUIPMENT

With the implementation of the LHC helium refrigeration system, Lebrun [1] for the cooling at 1.9 K of superconducting magnets distributed over the 26.7 km of the underground accelerator, CERN has reached an unprecedented level of overall helium refrigeration capacity in research facilities. The refrigeration system consists of eight plants each providing a capacity of 18 kW @ 4.5 K. In order to achieve operation at 1.9 K, eight additional 2.4 kW @ 1.8 K refrigeration units, Tavian [2] are implemented in the accelerator complex infrastructure.

In the framework of the LHC collider physics program, the two main experiments ATLAS and CMS have decided to use superconducting magnets as particle spectrometers and three independent refrigeration units are in the installation process, Passardi [3]. A new refrigeration system of a capacity of 1.5 kW @ 4.5 K is implemented for the operation of the superconducting solenoid of CMS. Two helium independent refrigeration systems are installed for the operation of the ATLAS toroidal magnetic system including a thin central solenoid: the main unit with a capacity of 6 kW @ 4.5 K and the thermal shields unit with a capacity of 20 kW @ 40-80 K. In addition to fulfill the ATLAS liquid argon calorimeter cooling requirements, a nitrogen refrigeration unit with a capacity of 20 kW @ 84 K will be used.

Several other helium cryogenic systems are in continuous operation supplying refrigeration capacity for the cryogenic test benches of the LHC superconducting wires, cables and magnets, the spectrometers of fixed target physics experiments, the LHC detector components and various cryogenic laboratory test facilities. Table 1 summarizes the number of helium refrigeration units actually installed at CERN either in operation or in installation and commissioning phases. Refrigeration power is provided at several temperature stages but for homogeneity reasons equivalent capacity @ 4.5 K is shown.

Table 1 Number and capacity of helium refrigeration units at CERN

Helium refrigeration capacity @ 4.5 K [kW]	Number of units	Total helium refrigeration capacity@ 4.5 K [kW]
18	8	144
6	2	12
1.5	1	1.5
1.2	1	1.2
0.8	2	1.6
0.4	9	3.6
0.1	1	0.1
Total	**24**	**164**

Helium refrigeration capacity@ 1.8 K [kW]	Number of units	Total helium refrigeration capacity@ 1.8 K [kW]
2.4	8	19.2

With the increase of the installed helium refrigeration capacity and ancillary infrastructure, the needs for helium and nitrogen storage have been upgraded. New 250 Nm^3 (2.1 MPa maximum operating pressure) horizontal gas tanks are installed for the LHC cryogenic project including the experiments. The existing helium storage capacity in using gas tanks of 80 Nm^3 (1.5 & 2.1 MPa maximum operating pressure), recovered for the former LEP project, are maintained in operation. Table 2 summarizes the helium gas storage capacity presently at CERN.

Table 2 Helium gas storage capacity at CERN

Tank capacity [Nm^3]	250 (2.1 MPa)	80 (1.5 & 2.1 MPa)
Number of units	60	65
Total capacity [Nm^3]	393000	

The storage of helium in liquid phase is used for supplying local refrigeration capacity to large and medium test bench areas. Four fixed units are operational, including one large container of 25000 liter capacity and three smaller of respectively 6000, 5000 and 2000 liter capacities. In addition two 11000 liter transportable liquid helium containers allow operational flexibility following specific requirements. At laboratory scale and for small applications, liquid helium is stored and distributed CERN-wide by means of a fleet of transportable containers. The fleet includes containers of a wide range capacity, starting at 100 liter (10 units), 450 liter (3 units), 500 liter (14 units), 1000 liter (2 units) and 2000 liter (1 unit).

For liquid nitrogen the storage capacity has been increased by installing 13 additional vertical containers of 50000 liter capacity, distributed around the LHC accelerator and experiment facilities. Table 3 summarizes the present liquid nitrogen storage capacity at CERN, including all existing infrastructures.

Table 3 Liquid nitrogen storage capacity at CERN

Container capacity [l]	50000	40000	27000	20000	15000	7000	6000
Number of units	16	1	2	2	2	1	9
Total capacity [l]	1025000						

The LHC helium cryogenics, for both accelerator and experiments, following an accidental or scheduled plant shutdown, foresees helium recovery to the 250 Nm^3 (2.1 MPa maximum operating pressure) horizontal gas tanks, thus avoiding investment on capital and maintenance expensive low-pressure recovery with associated purification systems.

For the existing cryogenic test facilities and physics experiments with wide geographical distribution, CERN uses a low and high-pressure helium recovery network interconnecting several sites over long distances. A 5 km long, 20 MPa maximum operating pressure, high-grade helium distribution line has been implemented as well two independent low-grade helium recovery lines (3 kPa and 20 MPa maximum operating relative pressure) each about 3 km long. The actual overall capacity of helium low-pressure recovery and purification at CERN totals: 2020 m^3 storage volume (21 units ranging from 20, 80 and 600 m^3 gas bags, 2 kPa maximum operating pressure), 1350 Nm^3/h recovery flow (ten helium reciprocating compressor units ranging from 50 to 400 Nm^3/h, 20 MPa maximum operating pressure) with five associated purification units ranging from 100 to 200 m^3/h, 20 MPa maximum operating pressure.

CRYOGEN PROCUREMENT AND USE

To procure the cryogens to the above facilities, CERN places renewable contracts with highly qualified industrial partners complying with the Organizations' selection criteria and purchasing rules as well with the International Standards for the management of helium or nitrogen facilities including production, trading, transport and delivery. As a general rule, CERN has always promoted the selection of several contractual suppliers (from two to three at the present time) in order to secure the deliveries.

<u>Procurement of High-Grade Helium</u>

Liquid helium (in 11000 US gallons transportable containers) is delivered to CERN technical sites. The minimum quantity to be delivered, at CERN, at one time is 1250 kg of liquid helium. Helium is stored indoors in liquid phase in the 25000 liters fixed container or in the two 11000 liters transportable units. Transfer of liquid helium directly into the suction side of the cryogenic plant compressors is performed by means of special charging devices provided by CERN. In addition to liquid helium procurement, CERN has the contractual possibility of specific deliveries for high-grade gaseous helium (quality 4.6) or low-grade gaseous helium (2% max. impurities) in high-pressure tube trailers with a minimum delivery at one time of 3000 Nm^3. The contractors assume the full responsibility for helium delivery. The normal minimum delay for contractual delivery time is one working week. CERN also implemented a contractual option allowing short-notice delay (48 hours) for delivery of up to 3000 kg of liquid helium in case of major accidental loss. The normal average unloading time is 24 hours per container (11000 US gallons).

In the last six years CERN has placed six contracts in two independent contractual periods of 3 years. Three contracts were placed for the 1998 to 2000 period with Air Liquide (FR), Carbagas (CH) and Messer Griesheim (DE). After renewal of the tender procedure three new contracts have been placed for the 2001 to 2003 period with BOC (UK), Carbagas (CH) and Messer Griesheim (DE). In total, 157422 kg of liquid helium have been delivered. Table 4 summarizes the yearly helium deliveries to CERN for the 1998-2003 period.

With the commissioning of LHC, the need for helium deliveries at CERN is considerably increasing and the procurement adapted accordingly. New contracts have been placed with Air Products (FR) and Carbagas (CH) covering the operational period 2004 to 2007 for an estimated quantity to be delivered of 280000 kg including an option for an additional quantity of 50000 kg. The first full load of the LHC collider (about 96000 kg of liquid helium) is included in the above mentioned quantities as well the delivery requirements for completing the tests and commissioning of the LHC accelerator and experiment superconducting devices with associated cryogenic plants.

<u>Procurement of liquid nitrogen</u>

Liquid nitrogen in transportable containers is delivered to the various CERN technical sites. The minimum quantity to be delivered, at CERN, at one time is 20 ton of liquid nitrogen. Liquid nitrogen is stored outdoors in several (33 units at present time) vertical containers with a capacity range from to 6000 to 50000 liter (Table 3). CERN requires liquid nitrogen of cryogenic grade quality corresponding to a maximum impurity level of 1 ppm for H_2, CO_2, C_xH_y, 10 ppm for H_2O, O_2 and 150 ppm for Ar. The contractors assume the full responsibility for nitrogen delivery. The normal minimum delay for contractual delivery time is fixed to 24 hours. The normal average unloading time is 3 hours per 20 ton delivery of liquid nitrogen.

Since 1998, CERN has strengthened the safety requirements of its liquid nitrogen storage containers improving both the installations on its premises and the unloading procedures. The implementation of high-pressure liquid nitrogen storage containers (maximum operating pressure 1.8 MPa) equipped with an additional inlet safety filling valve provide protection of the storage container from over pressurization during transfer. The installation of two identical circuits constituted by safety valves and bursting discs working on switchable mode by means of a diverter valve allow periodic inspection and test of the relief system without interruption of operation.

For the period from 1999 to 2001, CERN has placed three contracts with Air Products (FR), Messer Griesheim (FR) and Praxair (FR) with a total delivered quantity of 12617 ton of liquid nitrogen. For the commissioning of the LHC refrigeration systems and cryogenic tests of LHC components, the procurement was adapted accordingly by placing contracts for the period 2002 to 2005 for an estimated quantity to be delivered of up to 50000 ton of liquid nitrogen. Table 4 summarizes the liquid nitrogen delivery distribution at CERN for the 1998 to 2003 operational period.

Table 4 Helium & Nitrogen deliveries at CERN (1998-2003)

Year	1998	1999	2000	2001	2002	2003	Total
Helium deliveries [kg]	21980	26874	29682	21598	21613	35675	**157422**
Nitrogen deliveries [t]	3982	3046	5388	4183	4344	5147	**26090**

1058

<u>In situ helium liquefaction and management</u>
Liquid helium deliveries to small applications in the range of 100 to 2000 liter are performed internally at CERN by means of the transportable containers. For this purpose, CERN is operating a central helium liquefier allowing local liquid helium distribution at the level of about 250000 liter per year (mean value over the last four years) to more than 30 users. Helium gas is recovered via a dedicated network with an on-line purification and quality control allowing a continuous re-liquefaction. Including all CERN cryogenic activities the helium management in terms of total inventory, deliveries, operational losses and purification load as observed for the 2003 year is summarized in Figure 1. Operational losses reached the level of about 24000 kg for a final inventory of 12500 kg while the annual purification load, based on recovery compressors' running, was evaluated to 194000 Kg. No major accidental losses have been observed. However, many uses concern component tests, with frequently opened cryostats.

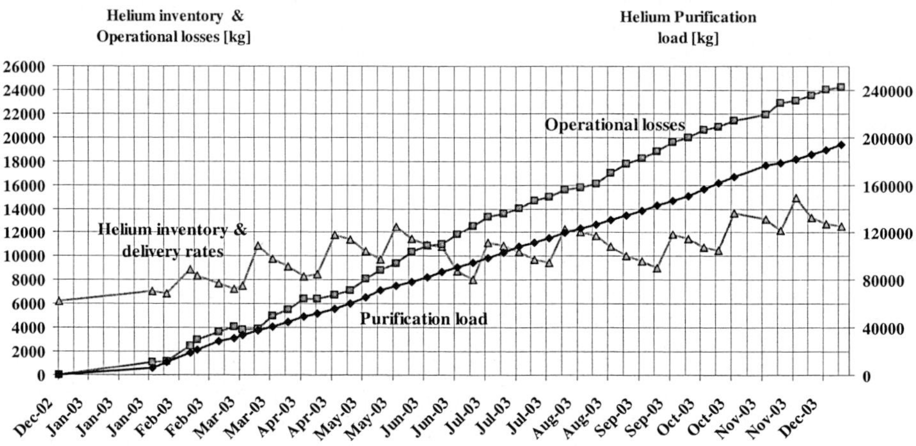

Figure 1 Helium management at CERN in 2003

OUTLOOK

With respect to the test and commissioning program of the LHC cryogenics, CERN implemented new contracts for helium deliveries for an estimated quantity of up to 280000 kg over the 2004-2007 periods, including the first full load of the LHC. For the liquid nitrogen deliveries the present contracts placed for an estimated quantity of up to 50000 ton covers the needs for all commissioning activities over the 2002-2005 periods. New contracts have to be placed at later stage covering also the need of the cool down of the full LHC magnetic system requiring about 10000 ton of liquid nitrogen for each complete thermal cycle.

ACKNOWLEDGEMENTS

The authors wish to thank their colleagues from the ACR(Accelerator Cryogenics) and ECR(Experiments and Test Facilities Cryogenics) groups for updated information as well all industrial support services from the Air Liquide, Linde, Serco consortium performing cryogenic operation & maintenance activities at CERN.

REFERENCES

1. Lebrun, Ph., Large Cryogenic Helium Refrigeration System for the LHC, 3rd International Conference on Cryogenics & Refrigeration (ICCR'2003)
2. Tavian, L., Large Cryogenics systems at 1.8 K, 7th European Particle Accelerator Conference (EPAC'2000)
3. Passardi, G. and Tavian L., Cryogenics at CERN, 19th International Cryogenic Engineering Conference (ICEC'2002)

Conceptual design of the Cryogenic Electrical Feedboxes and the Superconducting Links of LHC

Goiffon T. , Lyngaa J., Metral L., Perin A. , Trilhe P., van Weelderen R.

AT Department, CERN, CH-1211 Geneva 23

Powering the superconducting magnets of the LHC arcs and long straight sections is performed with more than 1000 electrical terminals supplying currents ranging from 120 A to 13'000 A and distributed in 44 cryogenic electrical feedboxes (DFB). Where space in the LHC tunnel is sufficient, the magnets are powered by locally installed cryogenic electrical feedboxes. Where there is no space for a DFB, the current will be supplied to the magnets by superconducting links (DSL) connecting the DFBs to the magnets on distances varying from 76 m to 510 m.

INTRODUCTION

The LHC will include several thousand superconducting magnets operating either at 1.9 K in superfluid helium or at 4.5 K in saturated helium, powered with electrical currents ranging from 60 A to 13 000 A. Most magnets are installed in 8 continuous strings of magnets in the arc regions of LHC, but a certain number of superconducting magnets are also required between the arcs, in the long straight sections and in the final focusing regions of the LHC (inner triplets), close to the main experimental areas. CERN will supply 16 cryogenic electrical feedboxes (DFBAs) for the arcs and 28 feedboxes for the long straight sections (DFBMs and DFBLs) while the 8 feedboxes for the inner triplets (DFBXs) will be supplied in the framework of the US-LHC collaboration and are described by Zbasnik et al. [1].

When the integration of a DFB is not possible close to the superconducting magnets, the magnets are powered through superconducting links (DSL) that connect the DFBs and the superconducting magnets on distances varying from 76 to 510 m.

CRYOGENIC ELECTRICAL FEEDBOXES

Powering the LHC will require 3 families of cryogenic electrical feedboxes:

- The DFBAs, connected to each end of the LHC octants, ensuring also the mechanical and cryogenic functions of arc termination. There are 16 DFBAs in the LHC.
- The DFBMs, powering standalone magnets in the long straight sections. There are 23 DFBMs in the LHC, located in the LHC tunnel, alongside the magnets that they power.
- The DFBLs, powering the superconducting links. They also supply the cryogenic fluids to the DSLs. There are 5 DFBLs in the LHC, located in underground caverns close the LHC tunnel.

The main function of the DFBs is to transfer high currents from room temperature cables to superconducting busbars via current leads. The current leads are gas cooled devices designed to transfer high currents with limited transmission of heat to the 4.5 K liquid helium in which the busbars are immersed [1]. The types of DFBs and the types of current leads and their respective number for each type of DFB are listed in Table 1. Table 2 shows the total number of current leads required for the LHC.

Table 1 List of the DFBs and their current leads

DFB type	Number	Type of leads (nb/DFB)
DFBA	16	13kA (2-6), 6kA (12-15), 600A (44-62), 120A(0-4)
DFBM	23	6kA (3-5), 600A(4-12)
DFBL	5	6kA (0-5), 600A(0-44), 120A(0-12)

Table 2 The current leads for the DFBs

Type of lead	Number
13'000 A	64
6'000 A	258
600 A	692
120A	196

The DFBs also ensure other functions which are not directly related to the electrical powering but are crucial for the operation of LHC, like terminating the LHC arcs for the DFBAs and supplying the cryogenic fluids to the superconducting links.

Design of the DFBs

In addition to providing a number of functionalities for the LHC operation, the DFBs design must take into account the very stringent constraints of the LHC tunnel (geometry, radiation, transport, etc.) and must also allow the in-situ exchange of the current leads in all long straight sections of the LHC.

The DFBs are of modular design: they are assembled from two families of current leads modules, assembled together with interface modules and other specific equipment that depends on the requested configuration (see Figure 2 and Figure 3).

The high-current module integrates 13 kA and 6 kA leads while the low-current module integrates 6 kA, 600 A and 120 A leads. The number of leads and their arrangement is different for each DFB but the same basic design is used for all modules. The current lead modules support the leads and the busbar system and integrate the cryogenic circuits for their operation. To power the magnets, the busbars pass trough specially developed cryogenic superconducting feedthroughs. Big removable doors allow a good access to the cryogenic piping and to the electrical connections.

There are two types of interface modules: DFBAs are connected to the LHC arcs with an interface module called "shuffling module". It ensures the arc termination functions and also allows the rerouting of the busbars to the current lead modules. It shall withstand all vacuum and pressure forces due to its position at the end of the arc, while ensuring a very precise positioning of the beam pipes. Its support system is designed to allow realignment in LHC operating conditions. It also provides the supports of the high-current module. DFBL are connected to the superconducting links and to the cryogenic fluid supply with a specific interface module, whose function is essentially to distribute the busbars and cryogenic piping.

A summary of the basic configurations for the 3 types of DFBs is shown in Figure 1. All DFBs are built by combining the two types of current lead modules and interface equipment specific to each DFB in the combinations shown in Figure 1. The resulting typical design of the DFBAs (1 shuffling module and 1 or 2 current moduels), the DFBMs (1 low-current module) and of the DFBLs (2 low-current modules side by side and 1 interface module to the DSL) are shown in Figure 2 and Figure 3..

A view of the DFBA located at the left side of IR8, therefore powering sector 7-8 of LHC, is shown in Figure 2. This DFBA is a typical example of the above cited modular design and consists essentially of:

- An interface (shuffling) module
- A high-current module, placed besides the beam pipes. connected to the shuffling module and integrating a jumper connection to the cryogenic distribution line of LHC [2]
- A low-current module.

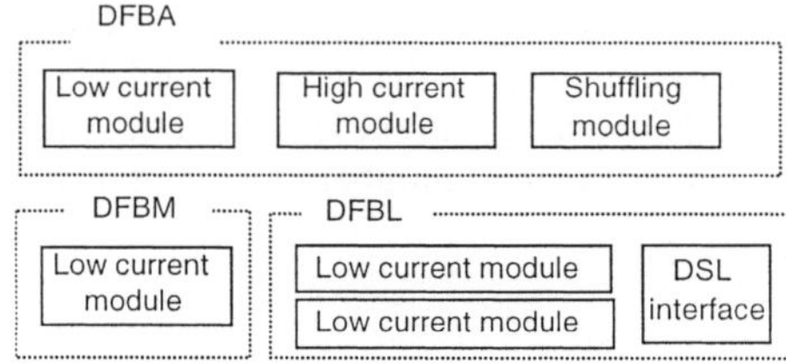

Figure 1: Schematic representation of the DFB configurations

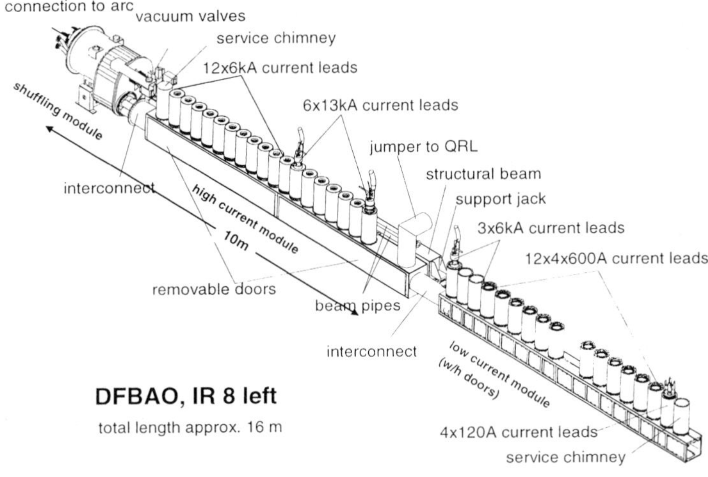

Figure 2: DFBA of IR 8 left, powering sector 7-8 of the LHC.

SUPERCONDUCTING LINKS

When the integration of a DFB close to the magnets is not possible, the electrical current is transferred from the DFBs to the LHC magnets through superconducting links (DSL).

Five DSLs will be needed for the LHC. One of them will be exceptionally long, about 510m in length without any intermediate branches. It will link the 3 km long continuous cryostat of accelerator magnets of Arc 3-4 of the LHC to a current feed box located in UJ33 some 510 m away. Besides its power transmission function, the link will also need to provide the cryogens for this current feed box. An additional four, significantly shorter, links will be used at points 1 and 5 of the LHC machine to bring power from current feed boxes to individual magnet cryostats (Q6, Q5 and Q4D2). Each of those four links will be about 76 m in length with two intermediate branches, roughly 3 m in length, to individual magnet cryostats. A summary of the characteristics of the superconducting links of LHC is shown in Table 3.

Design of the DSLs

The DSL consist essentially of cryogenic, vacuum-insulated, transfer lines housing one or more superconducting cables. Superconducting links are used for several circuits with current ranging from 120 A to 6kA. Nominal operation temperatures will be from 4.5 K to 6 K for the part which houses the cable, and about 76 K for the heat shielding. Cross sections of the 510 m DSLC is shown in Figure 4. In the DSLs, thermal contraction compensation will be performed by elbow-shaped metallic hoses at each

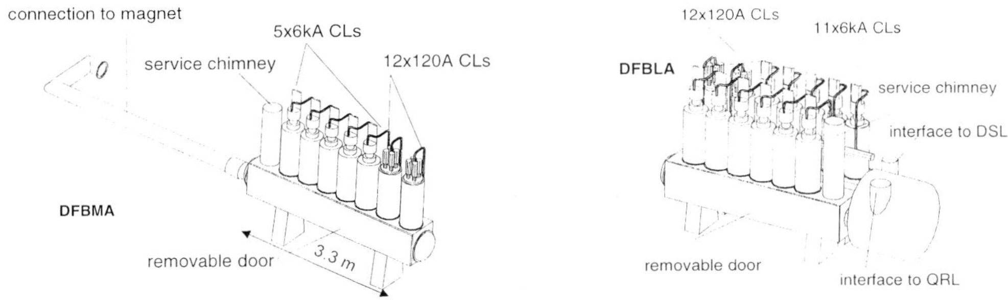

Figure 3 View of the typical configurations of DFBMs and DFBLs

1062

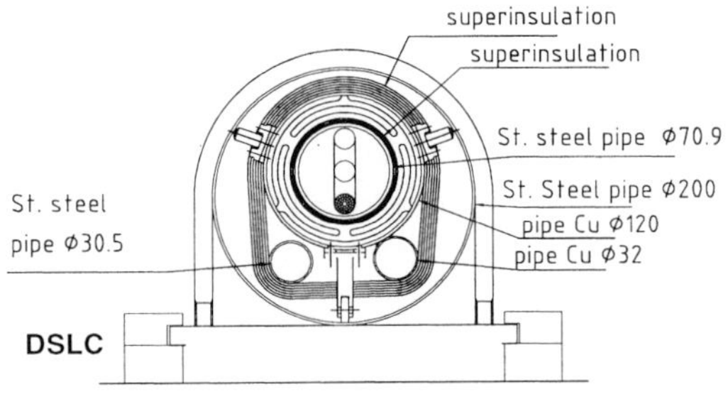

Figure 4: cross sections of the 510 m DSL

Table 3 : List of the DSLs

DSL type	length [m]	SC cables number	rating [A]
DSLA, DSLB, DSLD, DSLE	76	12 12	600 6000
DSLC	510	44	600

branch for the shorter types (76 m) and by a combination of bellows for the cryogenic piping and a distributed "waving" of the superconducting cable for the longer type (510 m).

PROJECT STATUS

An intensive development program has been carried out for validating the design of the DFBs and of their subsystems. Several models of the cryostats, vacuum envelopes, busbar feedthroughs [4], interconnects and other components have been built and successfully tested. The first two DFBs are currently under construction at CERN for a first installation in the LHC during the first half of 2005. The remaining DFBs will be produced under the responsibility of an external contractor. The installations will follow the construction of the LHC, ending at the end of 2006. It is planned to test one of the series DFBAs during the second quarter of 2005.

After a specification, pre-design and tendering phase, an order has been placed with European industry in March 2004 for the production and installation of the superconducting links. It is planned to test a 30 m model of the 510 m DSL during first quarter of 2005. The first 76 m DSL (DSLA) will be installed in the LHC tunnel in the second quarter of 2005, the 510 m DSLC installation being planned during the first quarter of 2006.

CONCLUSION

The complex task of powering the LHC superconducting magnets in the very limited underground available space will be performed with a combination of locally installed cryogenic electrical feedboxes and the use of superconducting links for the locations where the space limitations do not allow the installation of DFBs.

ACKNOWLEDGEMENTS

The author gratefully acknowledge the contributions of the DFB project team at CERN, V. Benda, C. Davison, V. Fontanive, R. Marie, L. Metral of AT-ACR group and of L. Baliev, P. Issiot, V. Kleimenov, of TS-MME group.

REFERENCES

1. Zbasnik J., Corradi C, Gourlay S., Green M, Hafalia A., Kajiyama Y, Knolls M., LaMantia R., Rasson J., Reavill D., Turner W., _IEEE Trans. Appl. Supercond._ 13 (2003), 1906
2. Ballarino A, Mathot S., Milani D, _Proceedings of the 6th European Conference on Applied Superconductivity (EUCAS 2003)_(2003)
3. Riddone G., Trant R., _Proceedings of the 19th International Cryogenic Engineering conference (ICEC19),_ (2002), 59
4. Marie R., Metral L., Perin A., Rieubland J.-M., paper presented at this conference (ICEC20)

AUTHOR INDEX

Abe, I. 103
Adams, A. 55
Affinito, L. 577
Akutsu, T. 281
Alexeev, A. 11, 19, 465
Al-Mosawi, M.K. 709
Alpatov, S.V. 955
Anand Bahuguni 253
Andersen, T.P. 1035
Andersson, W. 111
Antipenkov, A. 95
Arp, V. 869
Awaji, S. 573
Axensalva, J. 1015, 1023

Bahuguni Anand 253
Bai, Y. 597
Bailey, W. 521
Bak, J.S. 99
Balle, Ch. 1051
Banerjee, R. 903
Bansal, G. 773
Bao, Q. 693
Bao, R. 83, 285, 301, 387,
Barr, G. 585
Barth, K. 1055
Basanets, V.V. 971
Baudouy, B. 805
Beduz. C. 521, 697, 709
Behera Upendra 245, 253, 729,
757
Benda, V. 1015, 1035
Bi, Y.F. 681, 689
Binni, A. 853
Blanco, E. 1051
Blau, B. 605
Boltovets, N.S. 971
Bonnay, P. 963
Bonneton, M. 115
Bozhko, Y. 51, 853
Bravais, P. 437
Bretagne, E. 375
Briend, P. 43, 99, 103, 995
Brisset, D. 963
Brockley-Blatt, C. 613
Brodianski, V.M. 11
Brodzinski, K. 1003
Bylinski, Y. 931

Cai, J.J. 409, 549
Cai, J-H. 201, 225, 229, 233, 237
Caillaud, A. 115
Candia, A. 67
Cao, G.X. 529, 533
Cao, L.Z. 545, 895, 951
Cao, S.X. 529, 533, 537, 541

Cao, Y. 165
Cao, Z.T. 409, 549
Casas, J. 1051
Cathala, B. 963
Celik, D. 445
Chai, Y.S. 951
Chang, H-M. 861
Chang, H.S. 43, 99, 103
Chang, S.H. 35
Chapin, L. 975
Chatain, D. 963
Chen, A.B. 75
Chen, B. 645, 685
Chen, C. 593
Chen, C.Z. 249, 457, 469, 473
Chen, D.F. 669, 705
Chen, G. 209, 379
Chen, G.B. 83, 197, 221, 285, 301,
365, 387, 785, 869, 911
Chen, G.M. 483, 837
Chen, H. 593, 601
Chen, H.P. 409, 549
Chen, J.R. 35
Chen, L.F. 841
Chen, N. 241, 261
Chen, P. 269, 379
Chen, X. 313, 349, 357, 361
Chen, X.Q. 681
Chen, Y. 305
Chen, Z.J. 353
Cheng, Z.Z. 221
Chesny, P. 87
Chiarelli, S. 577
Chiou, W.S. 35
Cho, K.W. 43, 99, 103
Cho, S. 637, 927
Choi, Y. 205
Choi, Y.S. 861
Choi, Y-D. 141
Chorowski, M. 793, 987, 1003,
1011
Chromec, B. 47
Ciazynski, D. 589
Claudet, S. 999, 1047
Communal, D. 963
Cordier, JJ. 31
Corte, A. della 577
Courts, S. 959, 975
Courty, JC 437
Crespi, P. 427
Cretegny, D. 119
Crispel, S. 437
Cui, X.L. 483
Cun, Z 91

Dai, E. 317

Dai, F. 617
Dai, S. 693
Dai, S.T. 649
Dai, W. 309, 321, 329, 333, 337,
341, 383
Dai, Y. 605
Dai, Y.M. 621
Dang, H-Z. 237, 277
Dauget, P. 103
Dauguet, P. 995
Day, C. 461
Day, Chr. 95
de Waele, A.T.A.M. 293
Delcayre, F. 103, 115, 995
Delikaris, D. 1055
della Corte, A. 577
Deng, K. 541
Deng, Z-S. 979, 983
Dhard, C.P. 51, 773
Dinesh, K. 757
Ding, H.K. 689
Ding, W.J. 289
Doi, Y. 27, 1019
Dong, D.P. 423, 449
Dong, J. 511, 517
Dremel, M. 31, 95, 461
Du, H.P. 75
Duchateau, J.L. 87, 589
Dugaev, V.K. 971
Durand, F. 427
Durgesh Nadig 245
Duthil, P. 375
Dutto, G. 931
Duval, S. 437

Falter, H.D. 461
Fan, L. 345, 365
Fan, Y.F. 71
Fang, J.R. 657
Fang, Z-C. 153, 213
Fauve, E. 103
Feng, H.D. 829
Feng, S. 91
Feng, Z.A. 617
Ferlin, G. 999, 1047
Fietz, W.H. 87
Filippov, Yu P. 955
Finley, P.T. 413
Fleck, U. 47
Fong, K. 931
Fossen, A. 43, 99, 103
François, M.-X. 375
Fu, L.Y. 453
Fu, S.Y. 899, 919, 947
Fuenzalida, P. 43, 99, 103
Fujisawa, Y. 273

Funaki, K. 499
Fydrych, J. 1011

Gai, X.Y. 75
Gan, Z. 379
Gan, Z.H. 221, 285 , 297, 785
Gandhi, V.G. 441
Gao, Z.Y. 649
Ghosh, P. 477
Ghosh, S.C. 441
Giove, D. 931
Gislon, P. 577
Gistau Baguer, G.M. 43, 99
Gistau, G. 103
Gizicki, W. 1003
Goiffon, T. 1059
Gomes, P. 1051
Gong, J.Y. 745
Gong, L. 39
Gong, L-H. 71, 153, 157, 213,
    487
Gong, M.Q. 161, 165, 169, 173,
    177, 821
Gower, S.A. 721
Grabié, V. 115
Gravil, B. 31, 87, 461
Grechko, A. 63
Green, M.A. 585, 825, 857
Grillot, D. 115
Gruehagen, H. 991
Gu, A. 63, 741, 873, 877
Gu, A.Z. 753, 865
Gu, C. 503
Gu, G.B. 645, 685
Guan, Y. 693
Guo, F.Z. 353, 357, 369
Guo, J.D. 557
Guo, W.Y. 717
Gupta, G. 765, 773
Gupta, N.C. 765, 773

Haas, H. 95, 461
Haberstroh, Ch. 845
Han, B-S. 507, 525
Han, G. 75
Han, J. 693
Han, S. 39, 593
Han, Y.S. 181, 553
Han, Z. 503
Harada, K. 939, 943
Harrison, S. 63, 613, 625
Haruyama, T. 27, 281, 1019
Hasebe, T. 573
Hasegawa, T. 499
Hauer, V. 461
Haug, F. 1019, 1027, 1031
Hawley, C.J. 721
He, G.G. 409, 549
He, Y.L. 249, 269, 289, 457, 915
Heller, R. 87
Hemsworth, R. 31
Henry, D. 31
Hepburn, I. 613
Herblin, L. 1015, 1023
Hieke, M. 845
Hilbert, B. 115, 995
Hirano, N. 499

Hofer, H. 63, 605, 625
Honda, T. 1043
Hong, G. 145
Hong, Y-J. 141, 217
Hopkins, R.A. 413
Hou, B. 689
Hou, X-F. 225
Hou, Y. 469, 473
Hou, Z. 593
Howell, G. 67
Hsiao, F.Z. 35
Hu, J. 321, 325, 593
Hu, Q.G. 177
Hu, X. 557
Huang, C.J. 899
Huang, G.J. 617
Huang, H.H. 621
Huang, J. 249, 915
Huang, M.X. 75
Huang, S. 593
Huang, S.L. 545
Huang, X.H. 705
Huang, Y. 341
Huang, Y.H. 285, 869, 911
Huang, Z. 877
Hui, D. 495, 693
Huo, H.K. 633, 661
Hussaini, M.Y. 445
Hwang, S-D. 507, 817
Hyun, O.B. 637

Ikushima, Y. 281
Ishizuka, M. 573
Iwakuma, M. 499

Jacob, S. 245, 253, 729, 757
Jaramillo, E. 419
Jeong, S. 149, 205, 769
Ji, Z.M. 75
Jia, G.Q. 533, 541
Jia, L.X. 75, 107, 123, 581, 609
Jia, Z.Z. 301
Jiang, A.Z. 749
Jiang, C. 733, 737
Jiang, N. 197, 301
Jiang, R.Q. 349
Jiang, W. 75
Jiang, Y. 209
Jiao, B. 785
Jignesh Tank 773
Jin, H. 769
Jin, T. 345, 365, 387
Jin, T.X. 829, 881
Jin, X. 181, 553
Jing, C. 529, 533, 537, 541
Jing, W. 229
Johnstone, A.P. 697
Ju, Y.L. 277
Jung, J. 205
Jung, W.M. 637, 817, 927
Jung, W-S. 217
Jurkowski, A. 935

Kajikawa, K. 499
Kamioka, Y. 849
Kang, H-G. 525

Karunanithi, R. 245, 253, 729,
    757
Kasthurirengan, S. 245, 253, 729,
    757
Kawai, M. 1019
Kayoda, T. 653
Kholevchuk, V.V. 971
Kim, D.H. 637, 817, 927
Kim, D.L 637, 817, 927
Kim, H.R. 637
Kim, H-B. 141, 217
Kim, S. 769
Kim, S-Y. 217
Kim, Y.S. 43, 99, 103
Kimura, N. 27, 789
Klipping, G. 1, 19
Klipping, I. 1
Ko, S. 525
Ko, S-C. 507
Kobayashi, T. 27
Koettig, T. 923
Koh, D-Y. 141, 217, 641
Kondo, T. 1019
Kondo, Y. 1019
Kuendig, A. 47, 51, 1043
Kufner, F. 853
Kurtcuoglu, K. 1043
Kuruvila, G.K. 441
Kuzhiveli, B.T. 441

Lacroix, B. 589
Lalkoshti, P. 903
Lamaison, V. 963
Lamboy, JP. 1015, 1023
Lau, W. 585, 825, 857
Laxdal, R.E. 79, 111, 931
Lee, G.S. 43, 99
Lee, J-H. 507, 525
Lee, K.-S. 641
Lee, L.A. 103
Leher F. 853
Lei, G. 487
Lemke, D. 391
Lewis, M. 329, 383
Li, B. 895
Li, C.Y. 557
Li, G. 185
Li, H. 693
Li, H.C. 35
Li, H.L. 545
Li, J.G. 681
Li, L.F. 71, 185, 213, 885
Li, P.L. 529
Li, Q. 145, 313, 349, 353, 357,
    361, 375, 405, 737
Li, Q.F. 83
Li, Qiang 405
Li, R. 281
Li, S. 705
Li, T. 401
Li, W.Z. 269, 785
Li, X. 333, 541
Li, X.D. 865
Li, X.H. 629
Li, X.Y. 869, 911
Li, X.Z. 249
Li, Y. 899, 919, 947

Li, Y.Y. 885
Li, Y.Z. 777, 781, 833
Li, Z. 405, 449
Li, Z.C. 257, 809
Li, Z.W. 423
Li, Z.Y. 313
Liang, J.X. 511
Liang, J.X. 517
Liang, J-T. 201, 225, 229, 233, 237, 277
Liang, W. 39
Liang, W.Q. 157, 507
Lim, S-H. 525
Lin, D.J. 919
Lin, L.Z. 633, 649, 657, 661, 693, 701, 705
Lin, S.N. 469
Lin, W. 63
Lin, W.S. 753
Lin, Y.B. 649
Lindemann, U. 197
Ling, H. 317, 325, 333
Liu, C.S. 75
Liu, D. 75
Liu, D.M. 511, 517
Liu, H. 597
Liu, H.Z. 221
Liu, H-J. 185
Liu, J. 401, 951, 979, 983
Liu, J.X. 353
Liu, J.Y. 165
Liu, L. 39, 593, 601, 1007
Liu, L-Q. 153
Liu, M. 511, 517
Liu, X. 305
Liu, X.K. 75
Liu, X.L. 681
Liu, Y. 733, 737
Liu, Y.S. 541
Liu, Y.W. 249
Loehlein, K. 119
Lu, D.W. 181, 553
Lu, G. 693
Lu, G.H. 137
Lu, W-H. 487
Lu, X. 873
Lu, X.S. 753
Lu, Y. 693
Luo, E. 317, 321, 325, 329, 333, 337, 341, 383
Luo, E.C. 161, 165, 309, 821
Luzheng, Y. 91
Lyngaa, J. 1059

Ma, D.K. 681
Mack, A. 95, 461
Maeda, M. 849
Makida, Y. 27, 39, 593, 1019
Mantileri, C. 995
Mariani, M. 577
Marie, R. 1039
Marot, G. 437, 995
Marshall, C. 79, 111
Matrone, A. 577
Matsubara, Y. 189
McDonald, P.C. 419, 971
McMahon, R. 613, 625

Megro, S. 573
Messina, G. 577
Metral, L. 1039, 1059
Metselaar, J. 1019
Mi, T.C. 777
Mikheev, V. 55
Millet, F. 87, 999, 1047
Milward, S. 613, 625
Misra Ruchi 773
Mitin, E.V. 971
Mitin, V.F. 971
Miyoki, S. 281
Miyoshi, K. 573
Monneret, E. 995
Murakami, M. 273, 789, 801, 813, 939, 943
Murdoch, D.K. 95
Muzzi, L. 577

Nadig Durgesh 245
Nadig, D.S. 253, 729
Nagaya, S. 499
Nagel, M. 853
Nakamoto, T. 27
Nakano, A. 273
Nam, K. 149, 205
Nawrocki, W. 725, 935
Nemish, I.Yu. 971
Nicollet, S. 87, 589
Nie, Z.S. 145
Ning, Z. 649
Ning, Z.H. 557
Nishijima, G. 573
Noh, C. 769
Noonan, P. 55
Nozawa, M. 789, 801

Ogawa, J. 561
Ogitsu, T. 27
Ohhata, H. 27
Ohtani, N. 499
Onishi, T. 653
Ostropolski, W. 1003

Pająkowski, J. 725, 935
Pan, Q.Y. 899, 919, 947
Pan, Y.P. 457
Panchal Pradip 773
Paquignon, G. 963
Park, S-J. 141, 217
Pasotti, G. 577
Passardi, G. 1019, 1055
Paul, P.J. 757
Pavese, F. 971
Pavlov, O. 1019
Pelle, Th. 465
Pelletier, S. 1051
Perin, A. 1039, 1059
Périn, J-P. 963
Perraud, S. 761
Pezzetti, M. 1019
Pietsch, M. 853
Piotrowska, A. 987
Pirotte, O. 1019
Polinski, J. 793
Posselt, H. 853
Pradhan, S. 773

Pradip Panchal 773
Prass, S. 923
Praud, A. 115
Proch, D. 495, 571
Puech, L. 761

Qi, Y.F. 161, 821
Qi, Z. 557
Qian, S. 545
Qingfeng, Y. 91
Qiu, L. 379
Qiu, L.M. 221, 269, 785
Qiu, M. 633, 661, 665
Qu, T.M. 503
Quack, H. 845
Quanke, F. 91

Racine, M. 67
Radebaugh, R. 329, 337, 383
Ram, S.N. 757
Ravex, A. 127, 427
Rawnsley, W.R. 931
Ren, X. 797
Ren, Z.M. 541
Renard, B. 589
Riddone, G. 1003, 1007, 1011
Riedl, R. 461
Ries, T. 79
Rieubland, J.-M. 1039
Rodriguez, M.A. 1051
Rossi, S. 577
Roussel, E. 1047
Rousset, B. 761
Ruan, K.Q. 545
Ruber, R. 1019
Ruchi Misra 773

Sahu, A.K. 773
Saji, N. 1043
Sarangi, S. K. 477 , 903
Sarkar, B. 765, 773
Sasaki, K. 653
Sato, A. 849
Sato, N. 281
Saxena, Y.C. 765, 773
Sbrissa, E. 1019
Schauer, F. 51, 853
Schoenfeld, H. 119, 1043
Schweickart, R.B. 413
Seidel, P. 923
Sekachev, I. 79, 111
Senanayake, R.S. 585
Sengelin, JP. 1047
Sentis, L. 427
Serio, L. 1051, 1055
Sha, Y.N. 533
Shen, X.C. 529, 533, 537
Shen, Y. 345
Sheng, J. 517
Sheng, J.L. 511
Shi, L.H. 75
Shi, Y. 63, 741, 877
Shim, S.Y. 853
Shimba, T. 1043
Shintomi, T. 281
Shiraishi, M. 273
Singer, W. 495

Singer, X. 495
Song, F.C. 645, 685
Song, S.S. 621
Sorokin, P.V. 971
Speth, E. 461
Stafford Allen, R. 613, 625
Stanford, G. 79, 111
Su, X-T. 153, 213
Sun, D. 379
Sun, H. 881
Sun, Y. 473
Sun, Z.H. 161, 821
Suo, H.L. 511, 517
Suraci, A. 1051
Süßer, M. 967
Suzuki, H. 561
Suzuki, T. 281
Sveshnikov, B.N. 955

Takahashi, K. 573
Tamizhanban, K. 245
Tang, H.M. 75, 123
Tang, K. 285, 301, 387
Tank Jignesh 773
Tao, W.Q. 289
Tarafder, A. 903
Tavian, L. 999, 1007, 1055
Taylor, C. 593
ten Kate, H. 1019
Teng, Y.P. 649
Thibault, P. 761
Thummes, G. 197, 209, 297
Ting, S.C.C. 63, 625
Tomaru, T. 281
Tovar-Gonzalez, A. 1015, 1023
Trilhe, P. 1059
Trollier, T. 127, 427
Tsuchiya, K. 39, 593
Tsukahara, R. 939, 943
Tsukamoto, O. 561
Tu, F.H. 453
Tu, Q. 353
Turtù, S. 577
Tyrvainen, H. 1019

Uchiyama, T. 281
Ulbricht, J. 625
Upendra Behera 245, 253, 729, 757

Van Houtte, D. 31
Van Sciver, S.W. 445, 861
van Weelderen, R. 761, 1059
Vauthier, N. 1051
Venger, E.F. 971
Viertel, G. 625
Vogel, E. 495
Voigt, Th. 47, 1043
Vostrikov, S. 63
Vullierme, B. 1015, 1023, 1035

Wagner, U. 991
Wahrer, B. 593
Walter, E. 115, 995
Wang, B. 593
Wang, B.R. 345, 365
Wang, C. 265, 713

Wang, F.R. 557
Wang, G.P. 229
Wang, J. 473, 777
Wang, L. 75, 107, 123, 581, 593,
Wang, Q. 597, 605, 837
Wang, Q.L. 617, 621, 713
Wang, R. 63, 741, 877
Wang, R.S. 865
Wang, R.Z. 797
Wang, S. 601
Wang, S.P. 529, 533
Wang, T.H. 75, 581
Wang, W.Y. 423, 449, 453
Wang, X.J. 457
Wang, X.Y. 537
Wang, X-L. 201, 225, 237
Wang, Y.S. 629, 693
Wang, Y-Y. 233
Wang, Z.D. 491
Watanabe, K. 573
Weber, P. 923
Weber, T.B. 67
Weisend II, J.G. 67
Wen, H.M. 495
Wen, J. 781
Weng, P.D. 565
Will, M.E. 293
Wolf, P.E. 761
Wu, B.M. 895
Wu, H.B. 181, 553
Wu, J. 375
Wu, J.F. 161, 165, 169, 173, 177,
821
Wu, S.T. 681
Wu, T.H. 841
Wu, Y. 137
Wu, Y.L. 491
Wu, Y.Y. 745, 841
Wu, Z. 321, 341, 383
Wu, Z-L. 401

Xi, H.X. 689
Xi, Y.M. 145
Xia, D. 633, 661, 665
Xiao, H. 75
Xiao, L.Y. 71, 629, 649, 669, 673,
677, 693, 701, 705, 717,
Xiao, Y.M. 881
Xie, X.J. 313, 357, 361
Xie, X.M. 83
Xin, Y. 689
Xing, Z.H. 75
Xiong, L.L. 749
Xiong, Y.F. 491
Xu, B. 709
Xu, F.Y. 75
Xu, H.Y. 449
Xu, J. 241, 257, 261, 445, 649,
809, 829, 881
Xu, Q. 593, 601, 833
Xu, R.P. 241, 829
Xu, X. 693
Xu, X.D. 71
Xu, X.N. 181, 553
Xu, Z. 633, 661, 665

Yamamoto, A. 27, 281, 1019

Yamaoka, H. 39, 593
Yan, L.G. 621, 657
Yan, W. 379
Yanase, M. 813
Yang, C.G. 241, 261
Yang, C.J. 785
Yang, C.Z. 75
Yang, D.S. 895
Yang, G.D. 75, 123
Yang, H. 593
Yang, H.S. 637, 817, 927, 951
Yang, S.Q. 585, 825, 857
Yang, X.C. 689
Yang, Y. 521, 697, 709
Yao, Z.H. 633, 665
Yao, Z.L. 75, 581
Ye, Y. 673
Yeager, C. 959, 975
Yeom, H.-K. 641
Yi, C. 593, 601
Yim, S-W. 507
Yin, D.L. 557
Yoshinaga, S. 1043
You, G.C. 797
You, J.G. 423
Young, E.A. 521
Yu, J. 529, 533
Yu, L.M. 537
Yu, L-N. 983
Yu, S.Z. 645, 685
Yu, X. 405
Yu, Y. 545, 597, 605
Yu, Y.J. 701, 705
Yu, Z.B. 313, 357, 361
Yuan, G.Q. 181, 553
Yun, Y.J. 621
Yuyama, M. 849

Zahn, G.R. 87
Zhang, B.S. 365
Zhang, C. 873
Zhang, F.Z. 349
Zhang, G.Q. 645, 685
Zhang, H.F. 837
Zhang, J.C. 529, 533, 537, 541
Zhang, J.L. 745
Zhang, J.Y. 503
Zhang, K. 781
Zhang, L. 39, 71, 153, 157, 213,
487, 601, 789
Zhang, P. 789, 797, 801
Zhang, S.Y. 345
Zhang, W. 137
Zhang, X.B. 75, 123, 581, 609
Zhang, X.J. 269, 785
Zhang, X.Q. 305, 353
Zhang, X.Y. 541
Zhang, Y. 329, 383, 705
Zhang, Y.F. 537
Zhang, Y.H. 899, 919, 947
Zhang, Y.T. 885
Zhang, Y.W. 745, 915
Zhang, Z. 833
Zhang, Z.F. 677
Zhao, B.Z. 621
Zhao, C.H. 669, 701, 705, 717
Zhao, L. 593

Zhao, L.P. 829
Zhao, M-G. 201, 225
Zhao, X. 693
Zhao, Y. 511, 517
Zhao, Z.L. 749
Zheng, Q.R. 753
Zheng, W.H. 895

Zhou, G. 733, 737
Zhou, J. 593
Zhou, M.L. 511, 517
Zhou, Y. 201, 225, 237, 277, 309
Zhou, Y-X. 401, 983
Zhu, G. 741
Zhu, H.M. 881

Zhu, L. 951
Zhu, W.X. 309
Zhu, Y.H. 777
Zhu, Y.Q. 733, 737
Zhu, Z. 39, 593, 601, 693
Zhu, Z.H. 469
Zhuang, C. 503